# The MOLECULES of LIFE

## Physical and Chemical Principles

John Kuriyan Boyana Konforti David Wemmer

Garland Science
Taylor & Francis Group
NEW YORK AND LONDON

*Garland Science*
Vice President: Denise Schanck
Editor: Summers Scholl
Senior Editorial Assistant: Kelly O'Connor
Primary Illustrator: Lore Leighton
Additional Illustration: Laurel Muller, Cohographics, and Tiago Barros
Production Editor and Layout: EJ Publishing Services
Cover and Text Design: Matthew McClements, Blink Studio, Ltd.
Developmental Editors: Sherry Granum Lewis, John Murdzek, and Miranda Robertson
Copyeditor: John Murdzek
Proofreader: Sally Huish
Indexer: Merrall-Ross International Ltd.



ISBN 978-0-8153-4188-8

**Library of Congress Cataloging-in-Publication Data**

Kuriyan, John.
The molecules of life : physical and chemical principles / John Kuriyan, Boyana Konforti, David Wemmer.
p. ; cm.
Includes bibliographical references and index.
ISBN 978-0-8153-4188-8 (alk. paper)
I. Konforti, Boyana. II. Wemmer, David. III. Title.
[DNLM: 1. Molecular Biology--methods. 2. Biochemical Processes--physiology. 3. Genomics--methods. QH 506]
572'.33--dc23

2012008865

Published by Garland Science, Taylor & Francis Group, LLC, an informa business, 711 Third Avenue, New York, NY 10017, USA, and 3 Park Square, Milton Park, Abingdon, OX14 4RN, UK.

Printed in the United States of America

15 14 13 12 11 10 9 8 7 6 5 4 3 2 1

Visit our website at http://www.garlandscience.com

The cover illustration shows the bacterial ribosome (*purple* and *gray*) in the act of decoding the sequence of a messenger RNA molecule (*blue*). Three tRNA molecules are bound to the ribosome (*red, green,* and *yellow*). The growing protein chain, which is hidden within the ribosome, is attached to the *green* tRNA. The *red* tRNA is delivering a new amino acid for incorporation into the protein, and the *yellow* tRNA is about to depart. (Based on X-ray crystallographic analysis by V. Ramakrishnan and colleagues at the MRC Laboratory of Molecular Biology, Cambridge, UK. )

JOHN KURIYAN is Professor of Molecular and Cell Biology and of Chemistry at the University of California, Berkeley. His laboratory uses x-ray crystallography to determine the three-dimensional structures of proteins involved in signaling and replication, as well as biochemical, biophysical, and computational analyses to elucidate mechanisms. Kuriyan was elected to the US National Academy of Sciences in 2001.

BOYANA KONFORTI is the launch Editor of Cell Reports, an open-access journal that covers all of biology with a focus on short papers. Over her career, Konforti has researched the mechanisms of DNA recombination and RNA splicing. She has been a professional editor for over 13 years; most recently she was Chief Editor of Nature Structural & Molecular Biology.

DAVID WEMMER is Professor of Chemistry at the University of California, Berkeley and has served as Vice Chair, Assistant Dean, and Executive Associate Dean since joining the faculty in 1985. His research in structural biology uses magnetic resonance methods to investigate the structure of proteins and DNA toward a better understanding of how these molecules function. Wemmer is a Fellow of the AAAS and a member of Phi Kappa Phi and Sigma Xi.

# Preface

The field of biochemistry is entering an exciting new era in which genomic information is being integrated into molecular-level descriptions of the physical processes that make life possible. Our understanding of how biological macromolecules work at the level of atoms and interactions is also enabling great strides to be made in molecular medicine—where the path between the identification of a target and the development of therapeutics that modulate its functions is becoming ever shorter. Key to making future advances in these areas is a new generation of molecular biologists and biochemists who are able to harness the tools and insights of physics and chemistry to exploit the emergence of genomic and systems-level information in biology. This book is the result of a decade-long series of discussions among the three of us, in which we considered how biology students should best prepare themselves to take advantage of the growing depth of information concerning molecular mechanisms in biology.

The central theme of this book is that the ways in which proteins, DNA, and RNA work together in a cell are connected intimately to the structures of these biological macromolecules. These structures, in turn, depend on interactions between the atoms in these molecules, and on the interplay between energy and entropy, which results in the remarkable ability of biological systems to self-assemble and control their own replication. This book is not intended to be a comprehensive reference, nor does it contain the most recent biological breakthroughs and discoveries. Our goal in this textbook is to integrate fundamental concepts in thermodynamics and kinetics with an introduction to biological mechanism at the level of molecular structure. We have done so by choosing biological examples to illustrate the basic physical and chemical principles that underlie how biological molecules function.

We have written this textbook with an undergraduate audience in mind, particularly those students who have chosen biology or the health sciences as their principal area of study. We assume that students have taken introductory courses in physics and chemistry, and have been introduced to differential calculus at a basic level. We anticipate that the book will also be useful for graduate students in biology who have not taken courses in physical chemistry, or who seek to learn more about structural biology. We also hope that the book will be useful for scientists wishing to refresh their knowledge of the elementary principles of biological structure, thermodynamics, and kinetics.

The development of this textbook has been anchored, over the last few years, by the creation of a one-semester undergraduate course at the University of California at Berkeley, offered jointly by faculty in the departments of Chemistry and of Molecular and Cell Biology. This course has merged the first part of a traditional course in biochemistry with a new way of teaching physical chemistry to biology undergraduates. There are two aspects of this course that are a departure from past practice. The first is the integration of structural biology with physical chemistry, as mentioned earlier. The second aspect, and perhaps the more radical one for a course aimed at biology undergraduates, is to develop the laws of thermodynamics and the concept of free energy through statistical analysis of molecular interactions and behavior rather than on the more abstract concepts underlying heat engines. It is our experience that biology students take to the statistical treatment of energy and entropy more readily because this approach allows us to link thermodynamics and structure in an intuitively obvious way. Our initial hesitation concerning the implementation of this approach reflected a concern that the mathematical preparation of typical biology students might leave them ill-prepared to grapple with the statistical approach to thermodynamics. But, to our satisfaction, we have found that students understand these concepts readily, as witnessed by the growing enrollment in this class each year since its inception at Berkeley. The majority of these students are majors in Molecular and Cell Biology, with another large group of them majoring in Bioengineering.

The organization of our textbook follows how a course could be developed over one semester. We begin by introducing the nature of biological macromolecules and the structures that they form, placing these ideas in the broad context of how evolution proceeds while obeying physical laws. The first chapter provides an overview of DNA, RNA, and proteins and also reviews the processes of replication, transcription, and translation. A more detailed discussion of the structures of biological molecules is provided in Chapters 2 through 5, including a discussion of how evolutionary processes have shaped the architecture of proteins. Chapters 6, 7, and 8 provide a quantitative treatment of energy and the statistical basis for the concept of entropy, culminating in the development of the Boltzmann distribution and the idea that the energies of different molecular configurations determine the probabilities of observing them. The concept of free energy is introduced in Chapter 9 and, along with chemical potential, is developed further in Chapter 10, which applies these ideas to acid–base equilibria and to protein folding. Chapter 11 takes the concept of chemical potential one step further, by linking it to voltages through applications in redox chemistry and an analysis of how action potentials are transmitted in nerve cells.

Chapters 12 to 14 are concerned with the principles of molecular recognition, developing the ideas of affinity and specificity, with applications to drug interactions, protein–DNA, protein–RNA and protein–protein interactions, followed by a treatment of allosteric systems. Chapters 15 to 17 introduce kinetic concepts, including an analysis of enzyme mechanisms and transport properties (the material in these chapters could be presented in a course before Chapters 12 to 14 are covered). Finally, Chapters 18 and 19 bring together all of the ideas introduced in the earlier chapters by discussing two particularly interesting aspects of the self-assembly of biological systems: the folding of proteins and RNA, and the fidelity of replication and translation.

We have organized the book in a modular fashion, with each chapter broken into separate parts, some of which could be omitted according to instructor preference. While Chapters 6 to 19 assume that the student is familiar with the structural principles introduced in Chapters 1 to 5, an instructor could begin with Chapter 6, provided that the students have been introduced to proteins, DNA, and RNA in an earlier course (we believe that the earlier chapters could then serve as an excellent refresher). Each chapter has an associated set of problems—as anyone who has taken physical chemistry knows, working through problems is an important aspect of learning the material, and we hope that the problems at the end of each chapter can serve as a nucleus for generating assignments for the students to work through on their own.

There are two topics that might belong in an undergraduate biophysical chemistry course that we have purposely omitted. One is quantum mechanics, and the other concerns methods of instrumental analysis and structure determination in biochemistry. At Berkeley, students are introduced to these topics in a separate course that follows the one based on our book.

## ONLINE RESOURCES

Accessible from www.garlandscience.com/TMOL, the Student and Instructor Resource Websites provide learning and teaching tools created for *The Molecules of Life*. The Student Resource Site is open to everyone, and users have the option to register in order to use book-marking and note-taking tools. The Instructor Resource Site requires registration and access is available to instructors who have assigned the book to their course. To access the Instructor Resource Site, please contact your local sales representative or email science@garland.com. Below is an overview of the resources available for this book. On the Website, the resources may be browsed by individual chapters and there is a search engine. You can also access the resources available for other Garland Science titles.

### FOR STUDENTS

#### Animations and Videos

The animations and videos dynamically illustrate important concepts from the book, and make many of the more difficult topics accessible.

#### Flashcards

Each chapter contains a set of flashcards, built into the Website, that allow students to review key terms from the text.

#### Glossary

The complete glossary is available on the Website and can be searched and browsed as a whole or sorted by chapter.

### FOR INSTRUCTORS

#### Figures

The images from the book are available in two convenient formats: PowerPoint® and JPEG. Figures are searchable by figure number, figure name, or by keywords used in the figure legend from the book. There is one PowerPoint presentation for each chapter.

#### Animations and Videos

The animations and videos that are available to students are also available on the Instructor's Website in two formats. The WMV formatted movies are created for instructors who wish to use the movies in PowerPoint presentations on Windows® computers; the QuickTime formatted movies are for use in PowerPoint for Apple® computers or Keynote® presentations. The movies can easily be downloaded to your computer by using the "download" button on the movie preview page.

#### Solutions Manual

A complete solutions manual is provided for all problems in the text.

# Acknowledgments

This book could not have been developed without essential input from the following people in particular: Stephen K. Burley (with whom John Kuriyan developed the inaugural set of HHMI lectures entitled "Da Vinci and Darwin in the Molecules of Life") and the late Carl Brändén. Both were instrumental in moving very early stages of this project forward; Lore Leighton, who worked in John Kuriyan's lab, developed the illustrations from the earliest stages of writing this book; Tiago Barros helped with figure work and rendered the cover ribosome; James Fraser developed the problem sets; Samuel Leachman checked the solutions manual; Rachelle Gaudet developed a similar course at Harvard University based on early drafts of this book and provided valuable feedback; Krzysztof Kuczera, at the University of Kansas, carefully read and checked all the chapters; Tom Alber, Jamie Cate, and Bryan Krantz (who also teach the Berkeley undergraduate course); Susan Marqusee (who uses parts of this book for a graduate course at Berkeley); Ken Dill (whose masterly introduction to statistical mechanics in a graduate course at the University of California, San Francisco motivated our own simplified treatment of this material); and a large group of undergraduates at Berkeley who provided constant feedback as the book metamorphosed from a collection of notes into its present form. We hope this book will help many other students to come. Sherry Granum Lewis and John Murdzek provided helpful editorial suggestions. We thank the students who participated in focus groups at Berkeley: Bob Bellerose, Aron Kamajaya, Kotaro Kelly, Melinda Mathur, and Jayasree Sundaram; and at Harvard: Meng Xiao He, Koning Shen, Helen Yang, and Angela Zhang.

The following people also provided valuable commentary as readers, reviewers, class testers, and advisors during the development of the project:
Jochen Autschbach (State University of New York, Buffalo); Philip Bevilacqua (Pennsylvania State University); Phil Biggin (University of Oxford); Mark Braiman (Syracuse University); Charles Brenner (Dartmouth College); Angus Cameron (University of Bristol); Wei-Jen Chang (Hamilton College); Yun-Wei Chiang (National Tsing Hua University); King-Lau Chow (Hong Kong University of Science & Technology); Mads Hartvig Clausen (Technical University of Denmark); James Cole (University of Connecticut); EJ Crane (Pomona College); Ivan Dmochowski (University of Pennsylvania); Martha Fedor (Scripps Research Institute); Ruben Gonzalez, (Columbia University); Stephen Harrison (Harvard Medical School); Lars Bo Stegeager Hemmingsen (University of Copenhagen); ChulHee Kang (Washington State University); Katherine Kantardjieff (California State University, Fullerton); Roderick MacKinnon (Rockefeller University); Jeffry Madura (Duquesne University); Dmitrii Makarov (University of Texas, Austin); MK Mathew (National Centre for Biological Sciences, Bangalore); Kimberly Matulef (University of San Diego); Kevin Mayo (University of Minnesota); Ann McDermott (Columbia University); Megan McEvoy (University of Arizona); Stephanie Mel (University of California, San Diego); Daniel Moriarty (Siena College); Donald Nelson (deceased); Hung Kui Ngai (The Chinese University of Hong Kong); Timothy Nilsen (Case Western Reserve University); Patricia Pellicena (Catalyst Biosciences); Jack Preiss (Michigan State University); Margot Quinlan (University of California, Los Angeles); Venkataraman Ramakrishnan (MRC Laboratory of Molecular Biology, Cambridge); Ruth Reed (Juniata College); David Rueda (Wayne State University); Gordon Rule (Carnegie Mellon University); Paul Schettler (Juniata College); Kevin Schug (University of Texas, Arlington); Lawrence Shapiro (Columbia University); Kunchithapadam Swaminathan (National University of Singapore); Martha Teeter (Peace Films); Greg Tucker (University of Nottingham); Hiroshi Ueno (Nara Women's University); Didem Vardar-Ulu (Wellesley College); Kam Bo Wong (The Chinese University of Hong Kong); Sarah Woodson (Johns Hopkins University); Michael Yaffe (Massachusetts Institute of Technology).

JK—I am deeply grateful to my wife, Devaki Chandra, and my mother, Anna Kuriyan, who made it possible for me to write this book by giving me the supported mental space in which to work. I thank Ruth Reed and Paul Schettler, my teachers at Juniata College, and Greg Petsko and Martin Karplus, my graduate school advisors, for introducing me to the connection between biochemistry and statistical thermodynamics. Miranda Robertson's guiding hand was instrumental in allowing me to find my own voice. Denise Schanck and Summers Scholl at Garland displayed the patience of saints, keeping this project alive over many years.

BK—I would like to thank my family for their patience and understanding. For my youngest daughter Niki the book has been a part of her life for as long as she can remember. My oldest daughter Sophie has viewed my working on the book with a mixture of pride and incomprehension as she has veered as far away from the biological sciences as possible in her academic pursuits. And my husband Richard has had to put up with a lot—in particular, many prolonged absences at book retreats and when I have holed myself up for days at a time struggling to meet a deadline. Now that the textbook is done, if it reflects even a small amount of the time and effort that went into it, then we will have accomplished something to be proud of.

DEW—First I need to thank John and Boyana for inviting me to participate in the writing of this book. If I had known the full scope of what was to be done I might have hesitated, but now see it as having been an adventure of a new kind and feel great satisfaction in seeing it completed. I also need to thank my family and lab members for their understanding in times when work on the book had priority. Help and encouragement from the Garland editors was invaluable in getting this done, as were many other kindnesses such as my sister-in-law Teresa's loan of the beach house for writing retreats.

# Contents

# Detailed Contents

# How Do We Understand Life?

The ultimate goal of molecular biology and biochemistry is to understand in molecular terms the processes that make life possible. In his "Lectures on Physics," Richard Feynman famously remarked that in order to understand life "the most powerful assumption of all ... (is) that everything that living things do can be understood in terms of the jigglings and wigglings of atoms."1 Feynman made this statement about 50 years ago, shortly after James Watson and Francis Crick had discovered the double-helical structure of DNA, and Max Perutz and John Kendrew were working out the first structures of proteins.

How do we even begin to make good on Feynman's assertion that life can be understood in terms of the "jigglings and wigglings" of atoms? Our purpose in this book is to connect fundamental principles concerning the structure and energetics of biologically important molecules to their function. The concepts that we shall introduce provide a start towards establishing a complete understanding of the physical basis for life. As we move through these concepts, we shall assume that you are familiar with the essential principles of chemical structure, reactivity, and bonding, as covered in a typical introductory chemistry course. We shall also assume familiarity with basic concepts in molecular biology, again at the level encountered in introductory courses. If you find some of the material presented in the earlier chapters of this book difficult to follow, you may wish to consult elementary textbooks in chemistry and biology, such as those listed at the end of Chapter 1.

Any living cell is, ultimately, a collection of different kinds of molecules. The molecular structure of a particularly well-studied bacterium, *Escherichia coli*, is shown in **Figure 1**. This rendering of the cell, by the scientist and artist David Goodsell, is based on three-dimensional molecular structures that have been determined by many scientists, piece by piece, over the 50 years since Feynman's assertion about life. The particular cell shown in Figure 1 is encapsulated by two lipid membranes that are coated by a layer of glycans (carbohydrates). The interior of the cell is densely packed with many different kinds of macromolecules, which are very large molecules consisting of thousands of atoms each. Prominent among these is DNA, which is depicted as long yellow strands. The macromolecules with irregular shapes are various kinds of proteins and RNAs, as well as glycans.

Like all living cells, *E. coli* takes in nutrients and catalyzes chemical reactions that release energy from the nutrients. The cell is able to harness this energy to grow and to reproduce. By dividing into two cells, the mother cell passes on the blueprint for life, encoded in the DNA, to its two daughter cells, along with all of the other kinds of molecules that are necessary for these cells to live. It is apparent, even from looking at this relatively simple bacterial cell, that it is a formidable challenge to work out how such a molecular system can grow and reproduce, using only information contained within itself. We shall start by understanding the properties and interactions of the biological macromolecules that are the nanoscale machines of the cell.

The four types of macromolecules in the cell (DNA, RNA, proteins, and glycans) are polymers—that is, they are constructed by forming covalent linkages between

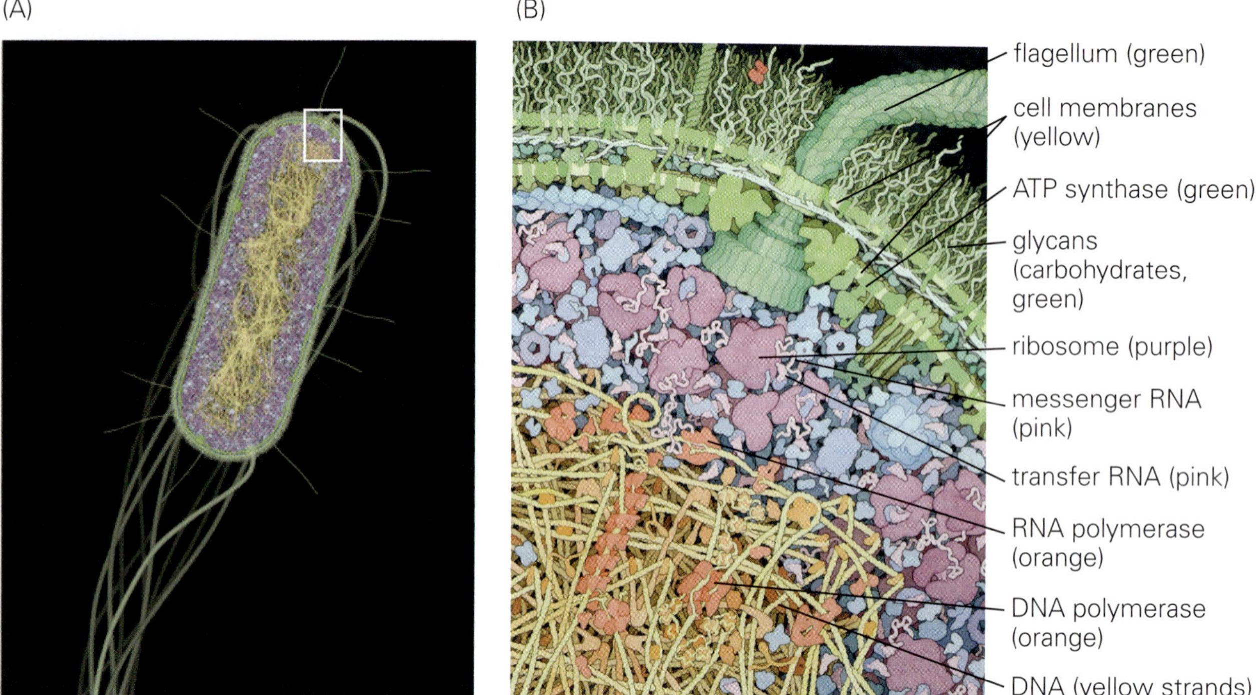


**Figure 1 Molecular structure of a bacterial cell.** Shown here is an *Escherichia coli* cell, illustrated by David Goodsell of The Scripps Research Institute. (A) Cross section of an *E. coli* cell. The main body of the cell is approximately 1 μm wide and has long whip-like flagella, which power the movement of the cell. (B) Expanded view of the region outlined in white in (A). Many of the macromolecules in the cell are shown here, drawn to scale. Some of the many protein machines in the cell are identified: DNA polymerase makes copies of DNA strands, RNA polymerase generates messenger RNA (mRNA) from DNA, and ATP synthase stores energy in the form of adenosine triphosphate (ATP). Transfer RNA (tRNA) is involved in the translation of the sequence of a messenger RNA to the sequence of a protein, by a particularly large machine called the ribosome. You can appreciate the scale of this drawing by considering that each ribosome is ~300 Å in diameter. (From D.S. Goodsell, *Biochem. Mol. Biol. Educ.* 37: 325–332, 2009. With permission from John Wiley & Sons, Inc.)

smaller molecules. RNA and DNA are formed by linking nucleotides together, proteins are formed from amino acids, and glycans are polymers of sugars. Of these, DNA, RNA, and proteins are special because they are the three components of the process by which genetic information is translated into the machinery of the cell.

DNA, RNA, and proteins are linear polymers in which the linkage between the component units extends in only one direction without branching. The order of specific kinds of nucleotides in DNA or RNA, or of specific amino acids in proteins, is called the sequence of the polymer. All living cells store heritable information in the form of DNA sequences, which are copied through the process of DNA replication and transmitted to progeny cells. The sequences of particular segments of DNA are also copied during the process of transcription to make RNAs with different kinds of functions. Messenger RNAs (mRNAs) are used to synthesize proteins. Other kinds of RNAs carry out diverse functions in the cell.

Most of the molecular machines that carry out the various processes essential for life are proteins. These include enzymes that catalyze chemical reactions, motor proteins that move things inside the cell, architectural proteins that give the cell its dynamic shape, and regulatory proteins that switch cellular processes on and off. Two particularly important kinds of protein enzymes are DNA polymerases, which replicate DNA, and RNA polymerases, which make RNAs based on the sequence of DNA. Another important protein enzyme is ATP synthase, which stores energy by synthesizing adenosine triphosphate (ATP). Some enzymes are made of RNA. The ribosome, which synthesizes proteins based on the sequences of messenger RNAs, is made of both proteins and RNA, with RNA being the functionally more important part. These molecular machines are identified in Figure 1, and we shall study some of them in this book.

There are four kinds of nucleotides in DNA and also in RNA, and 20 kinds of amino acids in proteins. Although this basic set of molecular building blocks is limited, they can generate a vast number of possible sequences. The *E. coli* bacterium has ~4.5 million ($4.5 \times 10^6$) nucleotides in its DNA. A DNA molecule of this length corresponds to ~$10^{2,700,000}$ possible sequences ($4 \times 4 \times 4 \times 4 \times \ldots 4.5 \times 10^6$ times), which is an unimaginably large number. A typical protein molecule is made from ~300 amino acids. The total number of different sequences possible for proteins of this length is $20^{300} \approx 10^{390}$, also an enormously large number. It is from this vast diversity of possible sequences that evolution is able to select the much smaller number of sequences of DNAs, RNAs, and proteins that are used in life.

There are two central themes underlying the concepts in this book. The first is that the function of a molecule depends on its structure and that biological macromolecules can assemble spontaneously into functional structures. The second theme is that any biological macromolecule must work together with other molecules to carry out its particular functions in the cell, and this depends on the ability of molecules to recognize each other specifically. Clearly, to understand the molecular mechanism of any biological process, we must understand the energy of the physical and chemical interactions that drive the formation of specific structures and promote molecular recognition.

You may be familiar with the concept of entropy, which is a measure of the likelihood of a particular arrangement of molecules. The flow of energy is governed by a very general principle, which is that the entropy of the universe always increases in any process. This statement is known as the second law of thermodynamics, and you have encountered it in some form in introductory chemistry. Another way of stating the second law is that a system always tends towards increased disorder, unless there is an input of energy. The relevance of the second law to living systems should become apparent if you study Figure 1 again. A living cell is a highly organized entity, with the cell membrane surrounding a specific collection of macromolecules that are where they need to be in order to function efficiently. Cells require a constant supply of energy to carry out the processes associated with living. Without energy, they would quickly go into a quiescent state and eventually disintegrate. The increase in entropy (disorder) upon disintegration overcomes the energetically favorable interactions that enable the cell to function.

In the first part of this book (Part I, Biological Molecules), we introduce the important classes of biological macromolecules and discuss the details of their structures. With the architectural principles of macromolecular structures in hand, we turn our attention to the physical principles that govern the interactions between molecules. As we explain in Part II of this book (Energy and Entropy), considerations of the energetics of interactions must always go hand in hand with consideration of the entropy (taken together, energy and entropy control the "jigglings and wigglings" of the atoms). By combining energy and entropy we arrive at a parameter known as the free energy, which allows us to predict whether a molecular process will occur spontaneously. This concept is developed in Part III of the book (Free Energy), and applied to processes such as the spontaneous adoption of specific structures by proteins and the transmission of electrical signals in nerve cells. In Part IV (Molecular Interactions), we focus on the idea that molecular interactions in living systems have to be highly specific. By drawing on the descriptions of protein and nucleic acid structure that we developed in Part I and the idea of free energy developed in Part III, we explain how molecules that need to interact find each other in the crowded environment of the cell.

Living systems change with time. Another way of saying this is that living systems are never at equilibrium: they would be dead if they were. In Part V (Kinetics and Catalysis), we turn to a study of kinetics, which describes the time dependence of molecular processes such as chemical reactions and diffusion. This part of the book provides us with several essential ideas about how enzymes work. Finally, in Part VI (Assembly and Activity), we focus on two particularly fascinating aspects of cellular processes: how proteins and RNA fold into specific three-dimensional structures, and how the processes of replication and translation achieve very high fidelity.

[1] Feynman RP, Leighton RB, and Sands ML (1963) The Feynman Lectures on Physics. Reading, MA: Addison-Wesley Publishing Co.

# PART I

## BIOLOGICAL MOLECULES

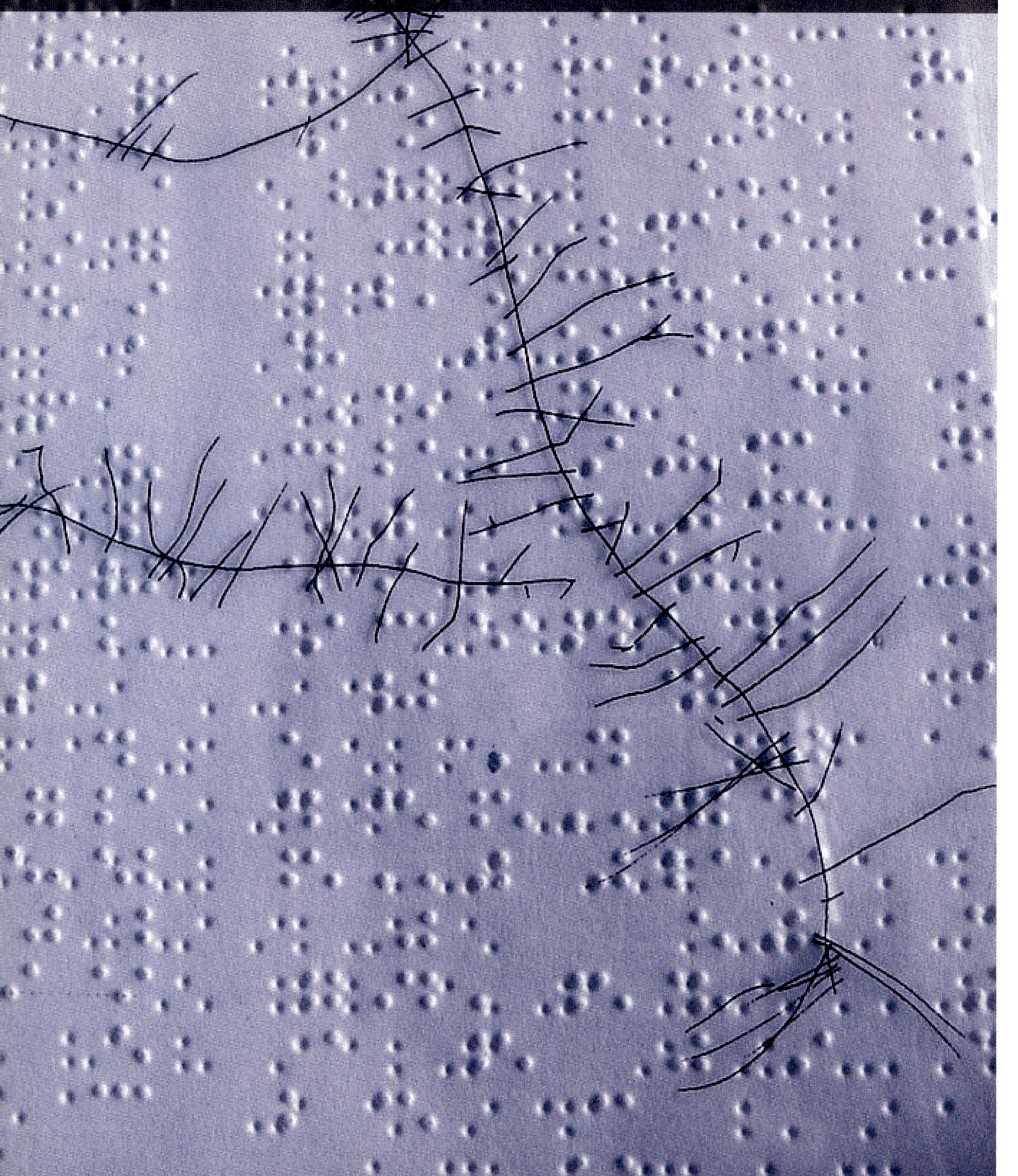

# From Genes to RNA and Proteins

There are four main types of macromolecules in the cell: two kinds of nucleic acids [deoxyribonucleic acid (**DNA**) and ribonucleic acid (**RNA**)], **proteins**, and **glycans** (carbohydrates). All of these macromolecules are polymers that are constructed through the formation of covalent bonds between smaller molecules. DNA and RNA are polymers of nucleotides, proteins are polymers of amino acids, and glycans are polymers of sugars. DNA and RNA are each made of four different kinds of nucleotides, and proteins are made of 20 kinds of amino acids. Examples of DNA, RNA, and proteins are shown in **Figure 1.1**. Glycans are described in Chapter 3.

DNA, RNA and proteins are different from glycans because they are linear polymers—that is, the polymer chain does not branch, as it does in glycans. DNA, RNA, and proteins have defined sequences, and the sequences of DNA are related in a precise way to the sequences of the RNA and protein molecules that are generated from it. A segment of DNA that codes for functional RNA or protein molecules is known as a **gene**. The precise relationship between the sequences of genes and that of RNA and proteins enables the heritable transmission of information from parent to child.

**Gene**

A gene is a segment of DNA that is transcribed into RNA. The RNA could be functional on its own or, if it is a messenger RNA, used for the production of proteins.

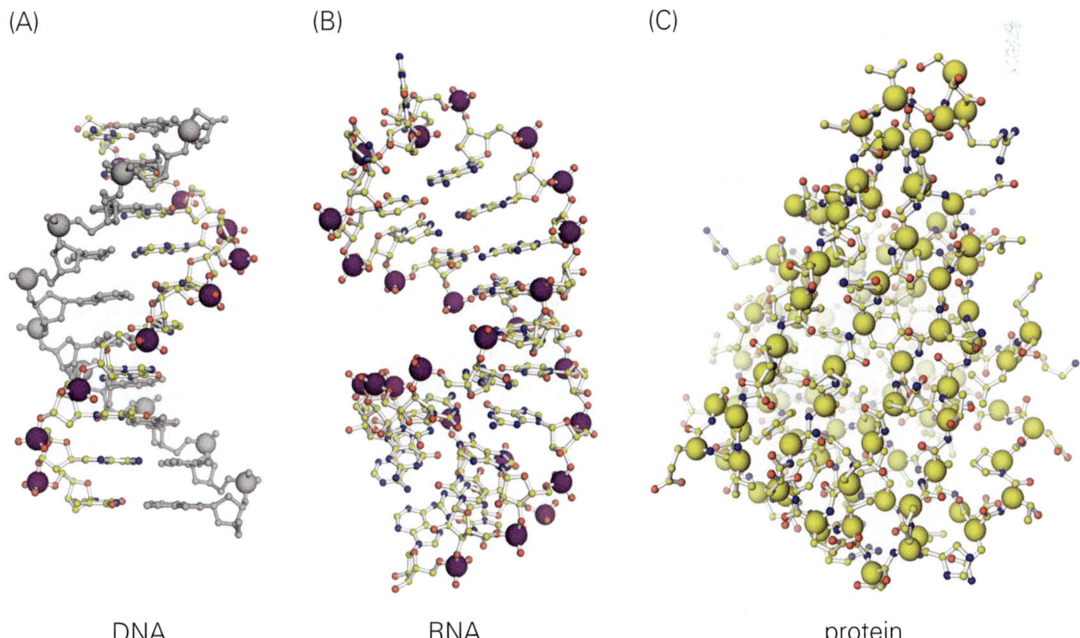


**Figure 1.1 DNA, RNA, and protein**. Covalent bonds are shown as sticks in this diagram and atoms as spheres, with carbon *yellow*, nitrogen *blue*, oxygen *red*, and phosphorus *purple*. Hydrogen atoms are not shown. (A) The structure of a DNA molecule. Two molecules of DNA usually form a double-helical structure (one molecule is colored *gray*). A phosphorus atom in each nucleotide unit is shown as a larger sphere. (B) A portion of an RNA molecule, colored as in (A). (C) A protein molecule, colored as in (A). One carbon atom in each amino acid is shown as a larger sphere. (PDB codes: B, 1HR2, C, 3K4X.)

DNA, RNA, and proteins can be further classified as informational and operational molecules. Informational molecules carry the instructions needed to make other polymers. Informational molecules would be useless, however, if there were not a way for them to be read and interpreted. This is one of the tasks of the operational polymers, which form the molecular machines that carry out the many functions required for life. Within the cell, DNA is an informational polymer, proteins are operational, and RNA can act in either capacity.

In part A of this chapter, we begin a discussion of the structures of DNA, RNA, and proteins that is continued in three subsequent chapters (Chapter 2, Nucleic Acid Structure; Chapter 4, Protein Structure; and Chapter 5; Evolutionary Variation in Proteins). Glycans, along with the lipid molecules that form cell membranes, are discussed in Chapter 3 (Glycans and Lipids). Before we discuss the structures of nucleic acids and proteins, we provide a brief description of the different kinds of interactions between atoms, without which one cannot understand the molecular logic of the various kinds of structures described in this part of the book (Part A of this chapter). A more detailed description of molecular energy is provided in Part II (Chapter 6, Energy and Intermolecular Forces). In part B of this chapter, we introduce the structures of nucleic acids and proteins. In part C, we review how the sequence of DNA specifies the synthesis of RNA and DNA.

## A. INTERACTIONS BETWEEN MOLECULES

In this part of the chapter we provide a qualitative picture of some of the key types of interactions between molecules, so that we can begin to discuss the architectural principles of nucleic acids and proteins. At this point we shall not concern ourselves with a precise and quantitative description of the relative strengths of the interactions, but simply characterize them as relatively "strong" or "weak." A more detailed discussion is provided in Chapter 6.

### 1.1 The energy of interaction between two molecules is determined by noncovalent interactions

All atoms potentially attract or repel each other because of their charges. Consider what happens to the energy of an interaction between two molecules when the distance between them varies, as shown in **Figure 1.2**. When the molecules are far apart, they do not influence each other. As they move closer together, they may

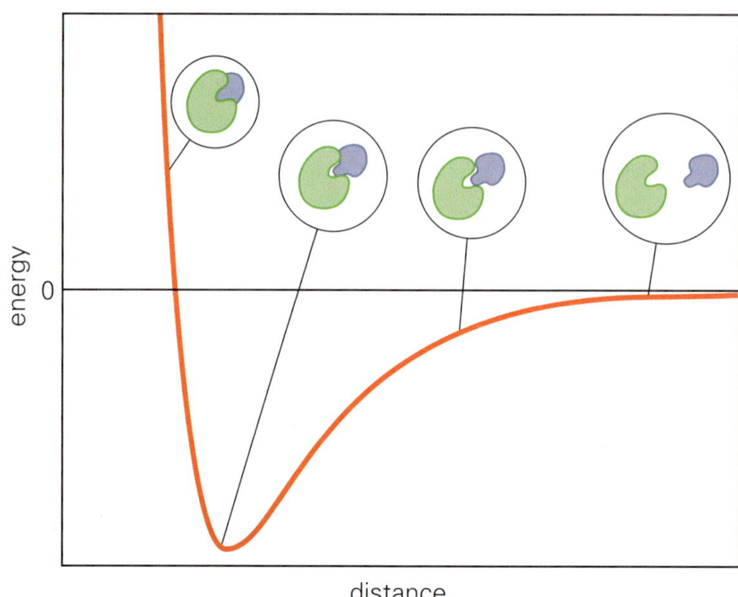


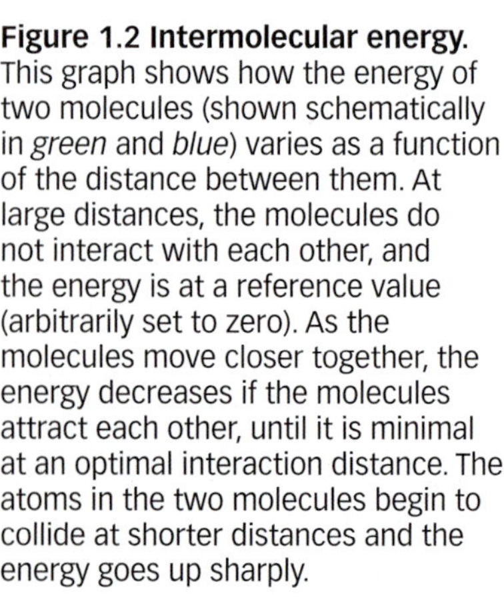

**Figure 1.2 Intermolecular energy.** This graph shows how the energy of two molecules (shown schematically in *green* and *blue*) varies as a function of the distance between them. At large distances, the molecules do not interact with each other, and the energy is at a reference value (arbitrarily set to zero). As the molecules move closer together, the energy decreases if the molecules attract each other, until it is minimal at an optimal interaction distance. The atoms in the two molecules begin to collide at shorter distances and the energy goes up sharply.

attract or repel each other. If they attract each other, the energy will decrease until it reaches a minimum value at an optimal interaction distance. If they approach each other more closely, the atoms in the two molecules will start to collide and the energy will increase sharply.

In order to calculate the energy as a function of the distance between two molecules, we decompose the total energy into a set of pair-wise interactions between the atoms in one molecule and the atoms in the other (**Figure 1.3**). This simplifies the problem of calculating the energy by reducing it to understanding how individual atoms, or small groups of atoms, interact with each other. The total energy of the interacting molecules is obtained by summing up over all the pair-wise interactions.

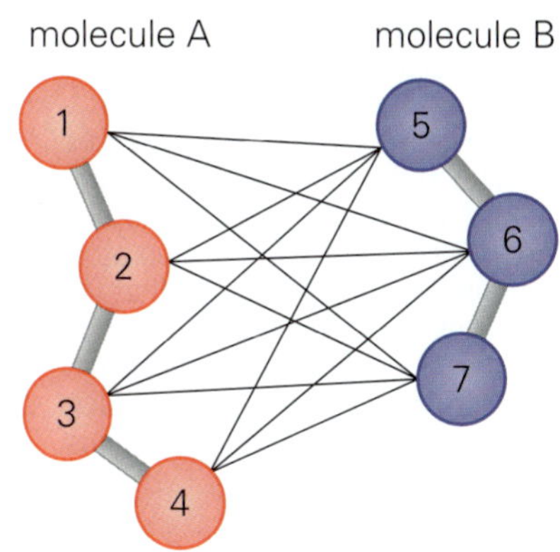


**Figure 1.3 Pairwise interactions between atoms in two molecules.** The energy of interaction between two molecules can be calculated by summing up the interaction energies between pair-wise combinations of atoms in the two molecules.

The interactions between the atoms in molecule A and those in molecule B, as depicted in Figure 1.3, are referred to as **noncovalent interactions** because the atoms in molecule A (labeled 1 to 4) are not covalently bonded to those in molecule B (labeled 5 to 7). Noncovalent interactions can occur between atoms in different molecules or, in the case of polymers, between atoms in different parts of the same molecule that are not covalently bonded to each other.

Compared with covalent interactions, noncovalent interactions are weak and can form or break during molecular collisions and vibrations at room temperature, as shown in **Figure 1.4**. Covalent bonds, in which electrons are shared between the atoms, are so strong that atoms that are held together in this way stay together through all of the fluctuations and collisions that occur when two molecules interact with each other at room temperature.

**Noncovalent interactions**

Noncovalent interactions are interactions between atoms that are not covalently bonded to each other. Noncovalent interactions can be attractive or repulsive, and they arise from interactions between transient or stable charges on atoms.

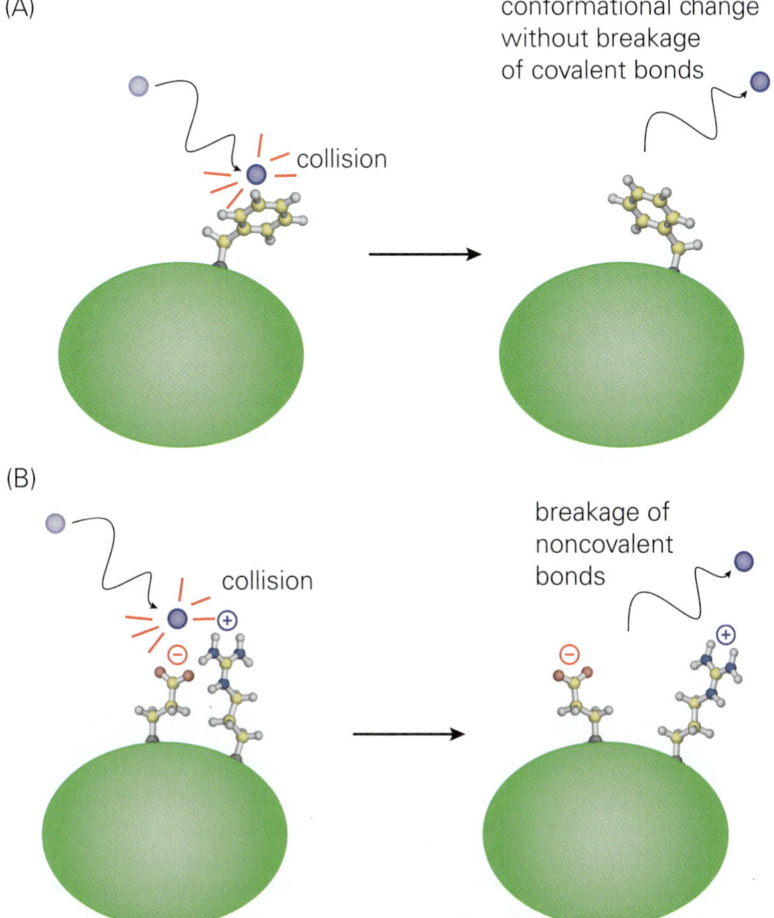


**Figure 1.4 Noncovalent interactions are broken and remade due to thermal fluctuations.** (A) A protein is represented by a *green* oval, and the covalent bonds of part of an amino acid in the protein are shown as sticks. A water molecule (*blue* sphere) collides with the amino acid, resulting in a change in conformation. None of the covalent bonds is broken or rearranged. (B) An example of a noncovalent interaction. Two amino acids in a protein have opposite charge and attract each other electrostatically. Collisions with water molecules can break such a noncovalent interaction at room temperature.

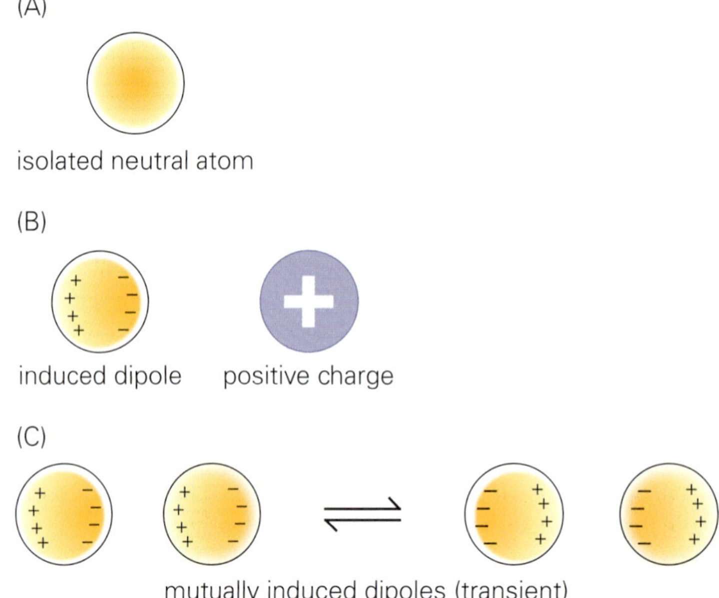


**Figure 1.5 Induced dipoles.** (A) An isolated neutral atom has no net charge. (B) The presence of a charge near a neutral atom induces a dipole in the atom. (C) When two neutral atoms approach each other, the electron clouds of the two atoms are polarized, which means that the electrons become redistributed so as to produce a dipole moment in the atom. Such dipoles attract each other, but the charge polarization is transient.

## 1.2 Neutral atoms attract and repel each other at close distances through van der Waals interactions

When two neutral atoms (that is, atoms with no net electrical charge) approach each other closely, they attract each other. This attraction is due to an **induced dipole** effect. Recall that a dipole is a pair of opposite charges separated by a distance. An induced dipole is the result of a transient separation of charge in an atom that causes the atom to behave as if it were a dipole (**Figure 1.5**). Induced dipoles occur when an atom is subjected to an electrical field that causes the redistribution of electrons in an asymmetric way. For example, when a neutral atom is next to a positive charge, the external charge will attract the electron cloud of the atom towards itself and tend to push the nucleus of the atom away. The result is the formation of an induced dipole in the atom (see Figure 1.5B).

When two neutral atoms approach each other, transient fluctuations in the electron clouds of each atom set up transient dipoles (see Figure 1.5C). The transient dipoles mutually reinforce each other, leading to an attractive force between the atoms. This attractive force is called the **London force**, after the physicist Fritz London, or the **van der Waals attraction**, after Johannes van der Waals. The spatial range of the van der Waals attraction depends on the nature of the interacting atoms. For the atoms commonly found in biological molecules, van der Waals attractions are optimal at distances between 3 and 4 Å. They are of negligible strength beyond 5 Å (**Figure 1.6**).

**van der Waals radius**

The van der Waals radius is a measure of the size of an atom. The energy due to the van der Waals attraction between two atoms is optimal when they are separated from each other by the sum of their van der Waals radii. If they move closer, the energy increases sharply.

Despite the attraction that brings atoms together, electron repulsion prevents atoms from getting much closer to each other than ~3 Å. This strong repulsion at small distances is called the **van der Waals repulsion**, and it causes atoms to behave like hard spheres when they approach each other very closely. Once two atoms are at the distance of minimum energy, further decreases in the interatomic distance cause the energy to rise very sharply, as shown in the graph in Figure 1.6.

Because of this hard repulsive aspect to the van der Waals interaction, it is often useful to describe each atom as if it behaved like an impenetrable sphere. The radius of the hard sphere corresponding to an atom is known as its **van der Waals radius** (the van der Waals radii of several atoms are given in **Table 1.1**). When two atoms approach each other, the energy is at a minimum (that is, the atoms interact optimally) when the distance between the two atoms is equal to the sum of

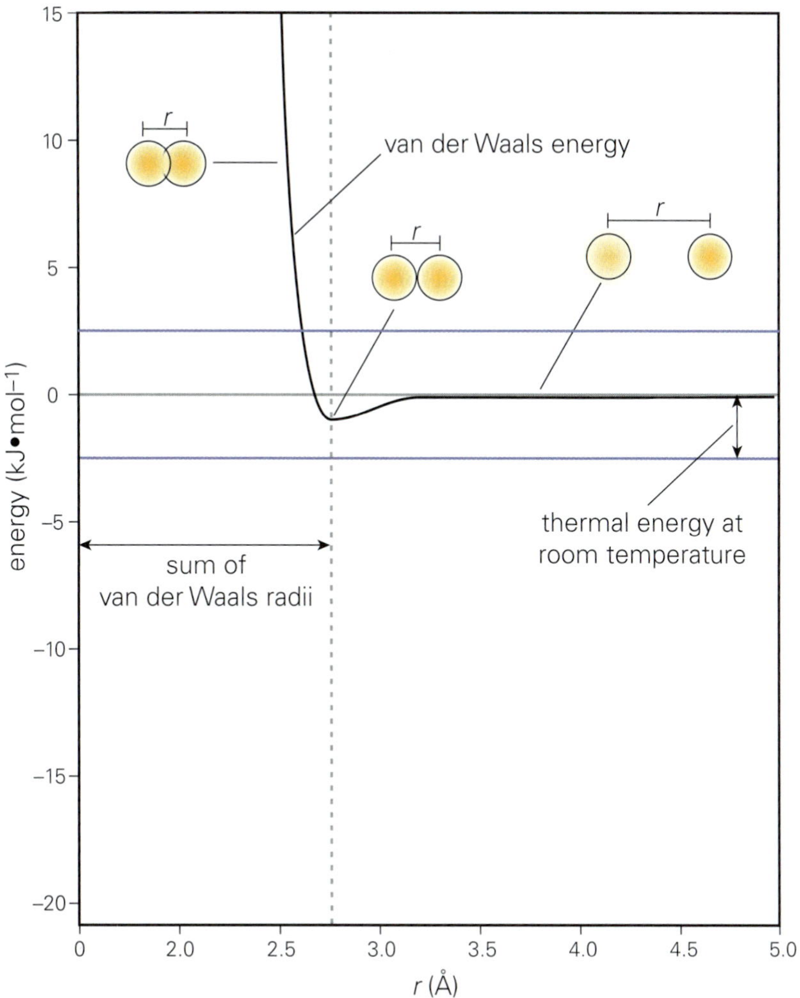


**Figure 1.6 Graph of energy as a function of interatomic distance for the van der Waals interaction.** This graph shows how the energy changes as the distance between two neutral atoms (for example, oxygen and hydrogen) is varied. The energy is lowest when the atoms are separated by the sum of their van der Waals radii (see Table 1.1). The horizontal *blue* lines represent thermal energy at room temperature, which is the energy that is easily transferred between molecules during random collisions. Notice that the magnitude of the thermal energy is greater than the stabilization afforded by the van der Waals interactions, so that the interactions between atoms can be easily disrupted by collisions.

their van der Waals radii (see Figure 1.6). If the atoms approach each other more closely, they begin to repel each other. Two atoms that are separated by the sum of the van der Waals radii are said to be in **van der Waals contact.**

The repulsion between atoms at short distances is responsible for many fundamental aspects of the structures of DNA, RNA, and proteins. These **steric effects** provide important constraints on the three-dimensional structures of biological macromolecules. For example, van der Waals repulsions prevent RNA from adopting the standard double-helical structure adopted by DNA, as noted by Watson and Crick in their original paper describing the DNA double helix (**Box 1.1**; RNA adopts a somewhat different double-helical structure, as explained in Chapter 2). Steric effects also determine the principal architectural elements of proteins, as we will describe in Chapter 4.

The van der Waals attraction between atoms is very weak. We shall defer discussion of energy units until Chapter 7, but for now you should note that we will use joules per mole (J•mol$^{-1}$; 1 J = 0.24 calories) or kilojoules per mole (kJ•mol$^{-1}$, $10^3$ J•mol$^{-1}$), as units of energy per quantity of matter. As you can see from Figure 1.6, when two atoms are in van der Waals contact, the **stabilization energy** is about −1 kJ•mol$^{-1}$. The stabilization energy is the amount by which the energy at the optimal distance is lower than when the atoms are far apart.

To understand why we say that the van der Waals attraction is very weak, consider that thermal motion causes constant collisions between molecules that are in solution. In Section 8.11, we shall explain the concept of **thermal energy**. This is the amount of energy that is readily transferred between molecules by random

**Table 1.1 Van der Waals radii and the electronegativities for atoms commonly found in biological molecules.**

| Atom | van der Waals radius (Å) | Electro-negativity (Pauling scale) |
|---|---|---|
| O | 1.5 | 3.4 |
| Cl | 1.9 | 3.2 |
| N | 1.6 | 3.0 |
| S | 1.8 | 2.6 |
| C | 1.7 | 2.6 |
| P | 1.8 | 2.2 |
| H | 1.2 | 2.1 |

The atoms are listed from largest electronegativity (electron withdrawing ability) to smallest, as determined by Linus Pauling.

collisions, and its value depends on the temperature. At room temperature (which we shall take to be ~300 K), the value of the thermal energy is ~2.5 kJ•mol$^{-1}$. This means that if an interaction between two atoms is stabilized by less than ~2.5 kJ•mol$^{-1}$, then this interaction is very easily disrupted by collisions at room temperature. It is by this criterion that the van der Waals attraction is very weak.

Even though van der Waals interactions are individually very weak, a considerable degree of stabilization can be achieved in the central structural core of proteins or in the stacked nucleotides of DNA by the additive effect of many such interactions between closely (but not too closely) packed atoms. The importance of van der Waals attractions in biological molecules will be illustrated elsewhere in this book, but it is instructive to consider a macroscopic example of how these individually weak forces can add up to a substantial net attraction when many atoms are brought into play. Such an example is provided by small lizards known as geckos, which are able to walk along smooth surfaces in a way that seems to defy gravity (**Figure 1.7**).

Various mechanisms had been invoked to explain the ability of geckos to adhere to surfaces, such as suction, friction, or capillary effects on wet surfaces. It turns out, however, that none of these explanations is correct. Instead, experiments have shown that geckos use van der Waals attractions to form tight junctions between pads on their feet and and the surfaces that they walk along. As shown in Figure 1.7, the foot pads of geckos contain millions of tiny hair-like protrusions with flat tips, called spatulae. The tips of the spatulae stick to surfaces using van der Waals attractive forces. Even though the surface may be rugged or corrugated, as long as each tiny spatula is able to engage a microscopically smooth area of the surface, numerous van der Waals contacts are formed that together enable the gecko to support the entire mass of its body against gravity.

## 1.3 Ionic interactions between charged atoms can be very strong, but are attenuated by water

We have seen, in Section 1.2, that the van der Waals attraction arises from flickering redistributions of electric charge on atoms. Other kinds of noncovalent interactions are also fundamentally electrostatic in nature and are due to the interactions between full or partial charges on atoms (one full charge corresponds to the magnitude of the charge on the electron). The simplest kind of electrostatic

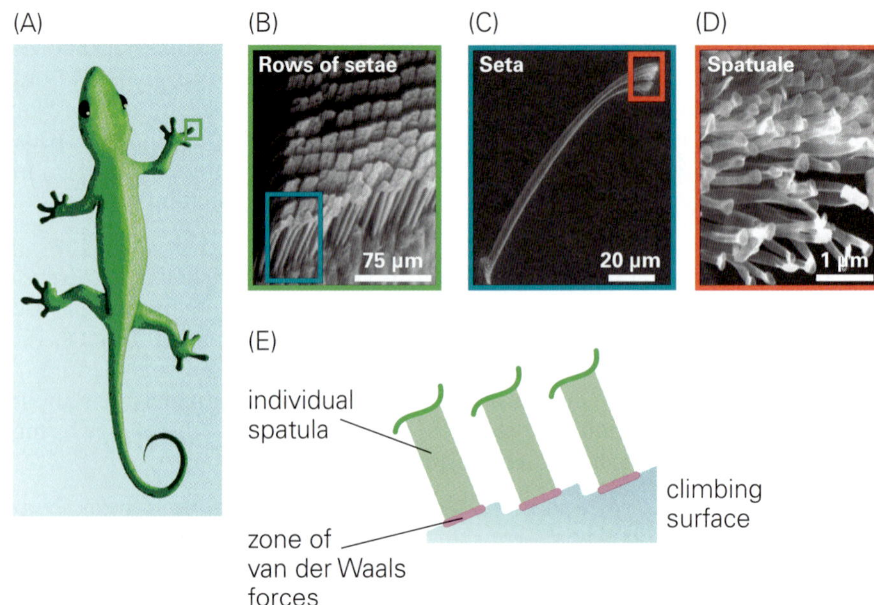


**Figure 1.7 Van der Waals interactions allow geckos to cling to surfaces.** (A) Schematic diagram of a gecko walking along a surface, against the force of gravity. (B) Electron microscopic view of a portion of the toe of a gecko, outlined in *green* in (A). Note the columnar structures, known as setae. (C) Expanded view of a single seta. (D) Expanded view of the tip of one branch of a seta. The tip consists of many tiny hair-like structures with flat tips, known as spatulae. (E) Schematic diagram of the interaction of three spatulae with a surface. Even though the surface is rugged, each spatula is able to make contact with a smooth portion of the surface, with the formation of van der Waals interactions between large numbers of atoms. (A–D, adapted from K. Autumn et al., and R.J. Full, *Nature* 405: 681–685, 2000. With permission from Macmillan Publishers Ltd.)

interaction is between two atoms, or groups of atoms, each of which bears a full positive or full negative charge. When two oppositely charged groups are close to each other, the interaction is called an **ion pair** or a **salt bridge**, and an example involving two amino acids in a protein is shown in **Figure 1.8**.

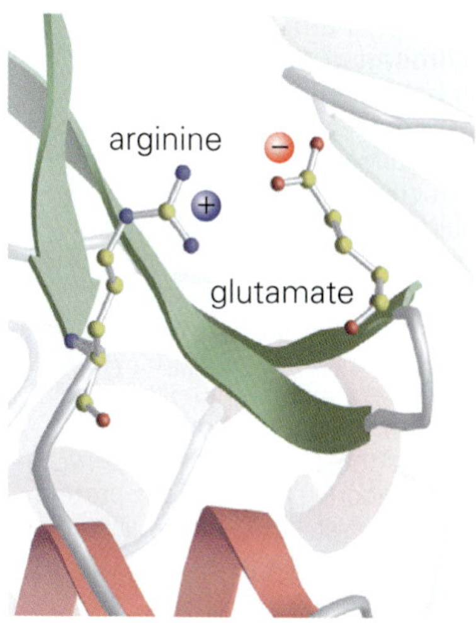


**Figure 1.8 A salt bridge or ion pair in a protein.** An electrostatic interaction between a positively charged amino acid (for example, arginine) and a negatively charged amino acid (for example, glutamate) is called a salt bridge when the charges are close together.

As you may have learned in elementary physics, the energy of interaction between two charges is described by Coulomb's law. The energy becomes increasingly negative (that is, more favorable) as the distance between two opposite charges decreases. This causes two opposite charges to attract each other more strongly as they come closer. But, just as for the van der Waals interaction, when the atoms get very close, the energy starts to go up steeply because of electronic repulsions. As a result, a graph of energy versus distance for the interaction between two oppositely charged atoms looks somewhat similar to that for the van der Waals interaction (**Figure 1.9**). There are some important differences, however. As you can see in Figure 1.9, the stabilization energy is much greater, and the attractive force makes the optimal distance smaller for an ion pair than for a van der Waals interaction (see Figure 1.8). The electrostatic attraction also falls off more slowly with increasing distance for the ion pair.

If we use Coulomb's law, the interaction energy for a negative charge that is separated from a positive charge by 3 Å in vacuum turns out to be about –500 kJ•mol$^{-1}$ (**Figure 1.10**A; we shall explain how such a calculation is done in Chapter 6). This result might make it appear that electrostatic interactions are extremely strong (this value is 200 times larger than the value of thermal energy at room tempera-

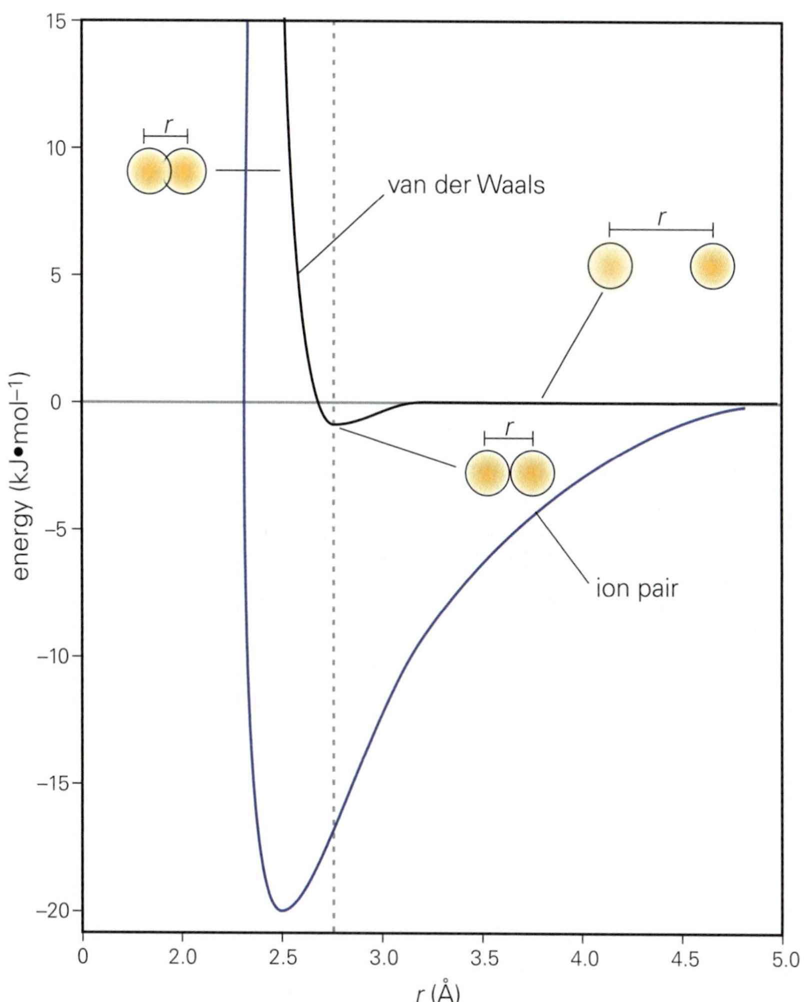


**Figure 1.9 Graph of energy for an ion pair.** This graph shows how the energy changes as the distance between two oppositely charged atoms is varied (*blue*). The energy for the van der Waals interaction between two neutral atoms is also shown (*black*; compare with Figure 1.6).

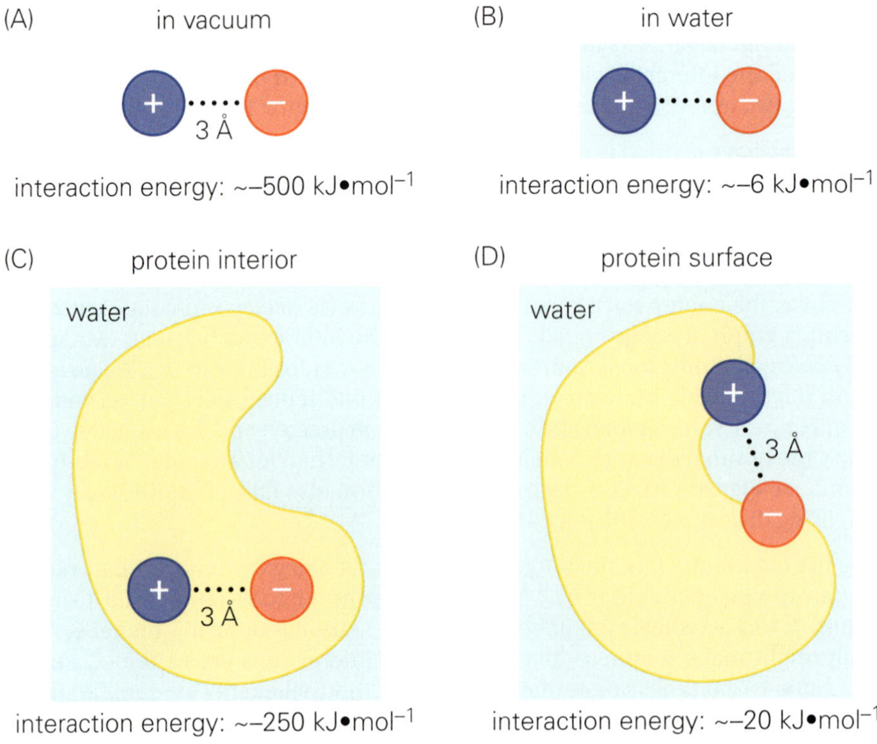


**Figure 1.10 Effect of the environment on the strength of ion pairing interactions.** (A) Two opposite charges separated by 3 Å in vacuum are shown, along with the interaction energy calculated using Coulomb's law. (B) The interaction energy is attenuated by a factor of 80 if the two charges are in water. (C) Two charges buried inside a protein interact strongly. (D) Charged groups are typically found at the surface of a protein. The interaction energy is closer to that in water and depends on the neighboring atoms of the protein and other associated molecules.

ture). But electrostatic effects are strongly modulated by the shielding provided by the medium in which the charged groups are dissolved. Water and ions can weaken electrostatic interactions, reducing both their strength and the distance over which they operate. If the same two ions are separated by 3 Å in water, the interaction energy is reduced by a factor of 80, to about –6 kJ•mol$^{-1}$ (see Figure 1.10B).

The environment within the interior of a protein is closer to that of vacuum than water in terms of its effect on electrostatic interactions. The interaction energy of two charges separated by 3 Å in the interior of a protein can be very large, reduced by only a factor of two compared with the energy in vacuum (see Figure 1.10C). It turns out, however, that fully charged groups are very rarely found in the interior of proteins because there is an energetic penalty associated with separating them from strongly bound water molecules. Instead, charged groups are usually found on the surfaces of proteins. The interaction energy depends on the extent to which they are exposed to water and, for a 3 Å separation, is expected to be in the range of –6 to –20 kJ•mol$^{-1}$. The graph of energy versus distance shown in Figure 1.9 is for a relatively strong ion pair on the surface of a protein in water.

## 1.4 Hydrogen bonds are very common in biological macromolecules

**Electronegativity**

Electronegativity is a parameter that describes the tendency of an atom to attract electrons towards it when it is participating in a covalent bond. The greater the electronegativity of an atom, the greater its tendency to withdraw electrons.

Covalent bonds are often **polarized**, which means that one of the atoms participating in the bond withdraws electrons towards it. This leads to the generation of an electric dipole, with the electron-withdrawing atom having a partial negative charge and the other atom in the bond having a partial positive charge (**Figure 1.11**A). Recall from introductory chemistry that **electronegativity** is a property of an atom that describes its tendency to attract electrons. Linus Pauling developed a useful numerical scale for the electronegativity of atoms (see Table 1.1). The more positive the value of the electronegativity, the greater the tendency of the atom to attract electrons. The most common electronegative atoms in biological systems are oxygen (3.4 on the Pauling electronegativity scale) and nitrogen (3.0),

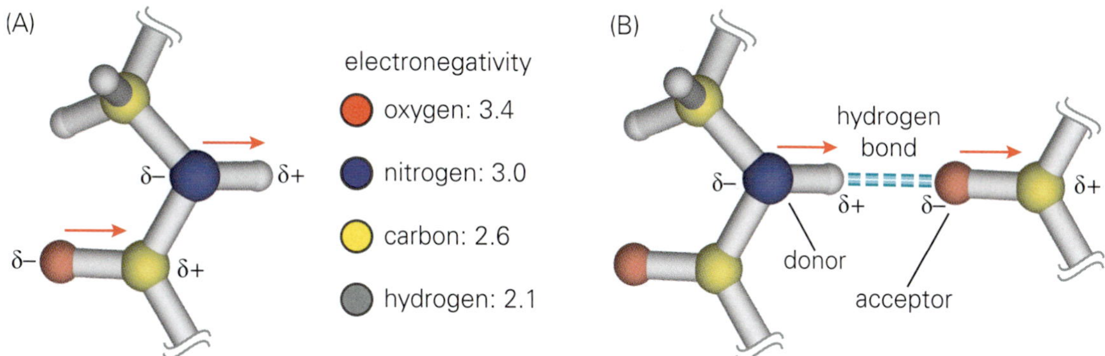


**Figure 1.11 Dipole moments and hydrogen bonds.** (A) A small portion of a protein structure is shown, with polarized covalent bonds. The partial positive and negative charges associated with the atoms are indicated by δ+ and δ–, respectively. The electronegativities of the atoms on the Pauling scale (see Table 1.1) are shown on the right. The dipoles formed by two of the bonds are indicated by *red* arrows that point towards the positive pole.
(B) A hydrogen bond (dashed line) can be thought of as a favorable interaction between two dipoles. One dipole is formed by the hydrogen and the atom that it is bonded to, typically oxygen or nitrogen. The other dipole is formed by a more electronegative atom (for example, oxygen or nitrogen) bonded to a less electronegative atom (for example, carbon). The nitrogen bearing the hydrogen is the donor and the oxygen with the partial negative charge is the acceptor.

while carbon (2.6) and hydrogen (2.1) are relatively electropositive (that is, they have a lesser tendency to attract electrons).

If two atoms that are covalently bonded have different electronegativities, then the bond will be polarized, with the more electronegative atom being the one with the partial negative charge. The extent of polarization is related to the difference in the electronegativities of the atoms. The polarization of the bond results in an electric dipole, as shown in Figure 1.11A. Molecules or groups of atoms containing polarized covalent bonds are called **polar molecules** or **polar groups**. Conversely, molecules or groups that do not contain strongly polarized covalent bonds are called **nonpolar** molecules or nonpolar groups.

Two dipoles interact favorably when they are aligned so that the positive pole of one points towards the negative pole of the other. A particularly important example of dipoles interacting favorably with each other is the formation of **hydrogen bonds**, as shown in Figure 1.11B. When a hydrogen is covalently bonded to a more electronegative atom (for example, nitrogen or oxygen), then the bond is polarized and the hydrogen has a partial positive charge. If the hydrogen is close to an electronegative atom (for example, oxygen or nitrogen) covalently bonded to a less electronegative one (for example, carbon), then a favorable dipole–dipole interaction can result, which is called the hydrogen bond. The atom bearing the hydrogen is called the **hydrogen-bond donor** and the atom that interacts closely with the hydrogen is called the **hydrogen-bond acceptor**.

Figure 1.11B illustrates a hydrogen bond formed between an N-H group (with the nitrogen as the donor) and a carbonyl group (C=O, with the oxygen as the acceptor). Such hydrogen bonds have optimal energy when the dipoles are colinear, with the donor nitrogen and the acceptor oxygen separated by ~2.4–2.7 Å. The dipole–dipole interaction falls off quickly with distance (**Figure 1.12**), so that hydrogen bonds are energetically favorable only when the atoms are very close to one another and are also oriented appropriately (the interaction energy falls off if the angle between the dipoles is too far from linear).

Just as for ion pairs, the presence of water weakens the strength of hydrogen bonds. In addition to electrostatic shielding, which weakens the Coulomb interaction energy between charges, an additional attenuation arises because water is a very polar molecule that forms strong hydrogen bonds with itself and with other polar molecules. When a polar group in a biological molecule forms hydrogen bonds with another polar group, it gives up hydrogen bonds with water (**Figure 1.13**). This leads to a reduction in the effective strength of the hydrogen bond, which is the difference in energy between the actual hydrogen bond and the hydrogen bonds that these groups form with water.

Hydrogen bonds between polar groups that are net neutral are typically 5–20 times stronger than van der Waals attractions, after accounting for attenuation by water (see Figure 1.12). Polar groups bearing full charges can also form hydrogen bonds, and these are stronger than those between groups that do not have an overall charge.

**Hydrogen bonds**

Hydrogen bonds are interactions between polar groups in which a hydrogen atom with a partial positive charge is located close to an atom with a partial negative charge, called the acceptor. The partial positive charge on the hydrogen atom is a consequence of a polarized covalent bond with a more electronegative atom, called the donor. See Figure 1.11.

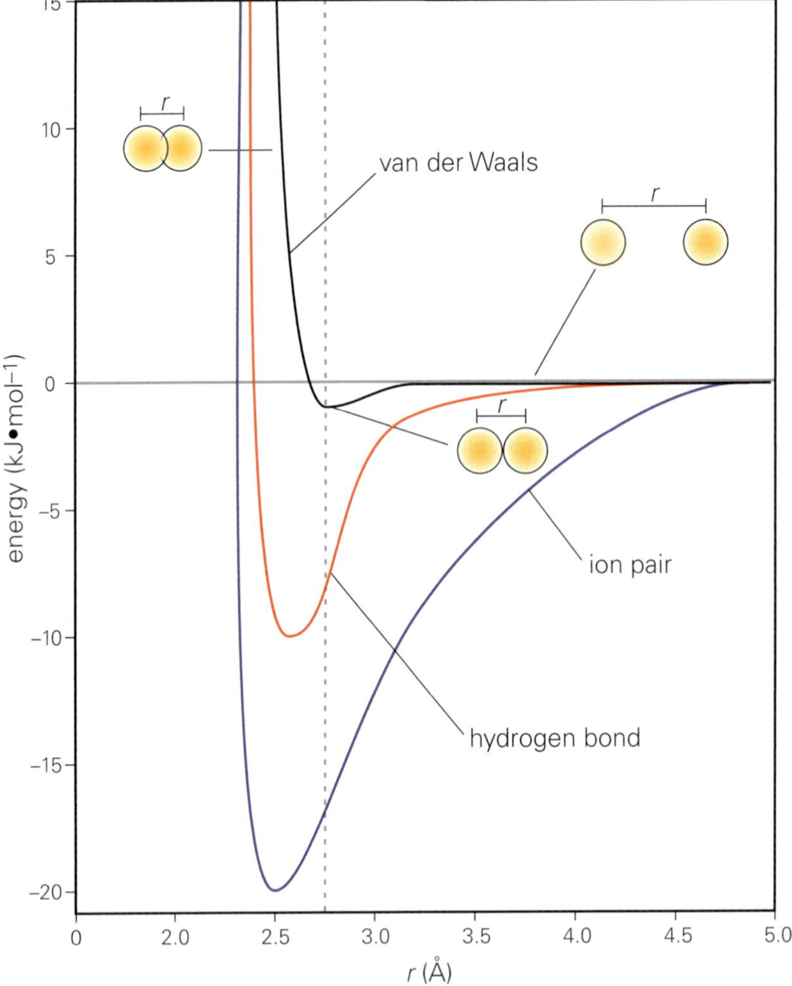


**Figure 1.12 Graph of energy for a hydrogen bond.** This graph shows how the energy of a N–H•••O–C hydrogen bond changes as the distance between the nitrogen and the oxygen atoms is varied (*red*). The energy for van der Waals and ion-pairing interactions between two atoms is also shown (*black* and *blue*; compare with Figures 1.6 and 1.9).

In situations where hydrogen bonds are formed between pairs of atoms (as in the C=O•••H-N hydrogen bond), it is easy to see how the dipole-dipole model can account for the hydrogen-bonding interaction (see Figure 1.11). But what about the situation shown in **Figure 1.14**A? In this case a hydrogen bond is formed between an N-H group and a nitrogen in an aromatic ring system. In such cases

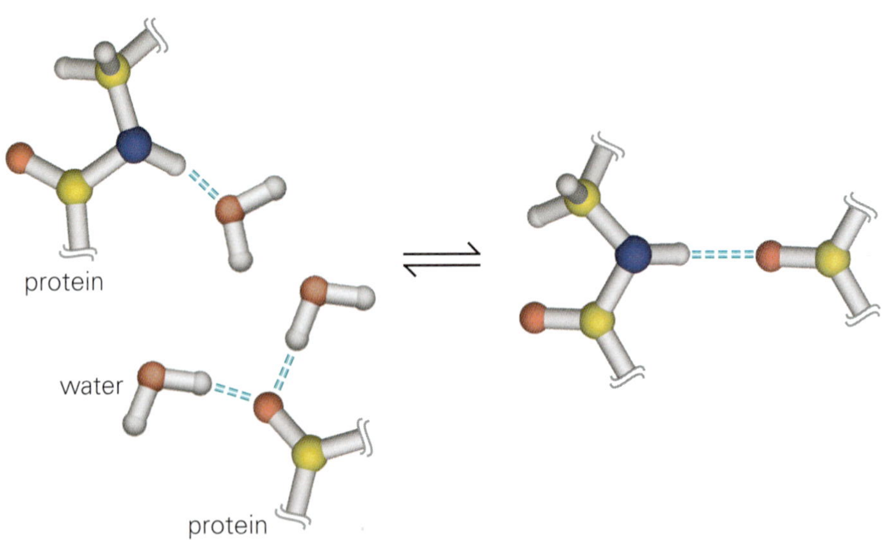


**Figure 1.13 Water molecules weaken the effective strengths of hydrogen bonds.** Water can form strong hydrogen bonds with polar groups in biological molecules, as shown here. Formation of hydrogen bonds between polar groups requires the release of water, which reduces the effective strength of the hydrogen bond.

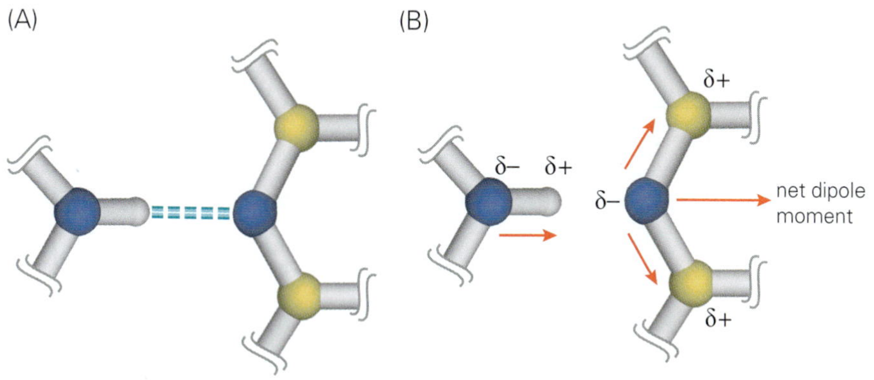


**Figure 1.14 A hydrogen bond involving atoms in a ring system.** (A) The hydrogen bond is indicated by a dotted line. (B) Dipoles generated by the polarization of the covalent bonds are shown.

the hydrogen bond can still be thought of as a dipole–dipole interaction because the distribution of partial charges in the ring generates a net dipole moment that arises from adding up component dipole moments, as shown in Figure 1.14B.

Although we have emphasized partial charges and dipoles in our discussion of hydrogen bonds, we should keep in mind that the dipole–dipole model for hydrogen bonds is a simplification. As we explain in Chapter 6, the hydrogen bond, like all molecular interactions, originates from the quantum mechanical properties of atoms and molecules. The strength and directionality of hydrogen bonds can depend on factors such as molecular orbitals, which we have ignored here.

# B. INTRODUCTION TO NUCLEIC ACIDS AND PROTEINS

## 1.5 Nucleotides have pentose sugars attached to nitrogenous bases and phosphate groups

DNA and RNA are both polymers of nucleotides. As shown in **Figure 1.15**, a **nucleotide** consists of a sugar covalently bonded to a phosphate group and to a heterocyclic aromatic ring system. The sugar contains five carbon atoms and is a

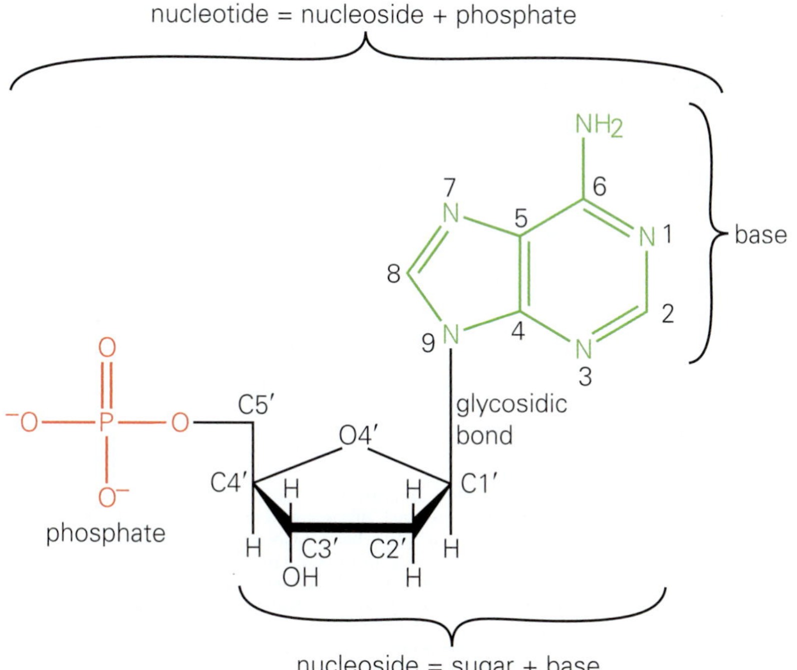


**Figure 1.15 Structure of a nucleotide.** Nucleotides consist of three parts: a five-carbon sugar (pentose, *black*), a nitrogen-containing aromatic ring system called a base (adenine in this case, *green*) attached to the 1′-carbon atom of the pentose and one (as shown, *red*), two, or three phosphate groups attached to the 5′-carbon atom of the pentose. Sugar atom numbers are distinguished from those associated with atoms in the base by a prime.

## Box 1.1 Reprinted from *Nature*

No. 4356 April 25, 1953 NATURE 737

# MOLECULAR STRUCTURE OF NUCLEIC ACIDS

## A Structure for Deoxyribose Nucleic Acid

WE wish to suggest a structure for the salt of deoxyribose nucleic acid (D.N.A.). This structure has novel features which are of considerable biological interest.

A structure for nucleic acid has already been proposed by Pauling and Corey[1]. They kindly made their manuscript available to us in advance of publication. Their model consists of three intertwined chains, with the phosphates near the fibre axis, and the bases on the outside. In our opinion, this structure is unsatisfactory for two reasons: (1) We believe that the material which gives the X-ray diagrams is the salt, not the free acid. Without the acidic hydrogen atoms it is not clear what forces would hold the structure together, especially as the negatively charged phosphates near the axis will repel each other. (2) Some of the van der Waals distances appear to be too small.

Another three-chain structure has also been suggested by Fraser (in the press). In his model the phosphates are on the outside and the bases on the inside, linked together by hydrogen bonds. This structure as described is rather ill-defined, and for this reason we shall not comment on it.

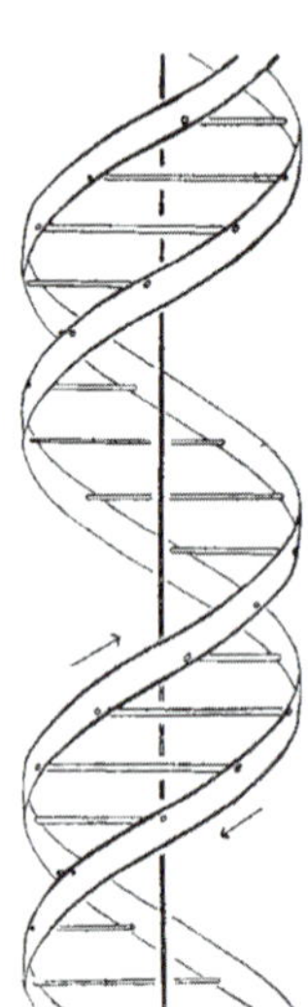

This figure is purely diagrammatic. The two ribbons symbolize the two phosphate—sugar chains, and the horizontal rods the pairs of bases holding the chains together. The vertical line marks the fibre axis

We wish to put forward a radically different structure for the salt of deoxyribose nucleic acid. This structure has two helical chains each coiled round the same axis (see diagram). We have made the usual chemical assumptions, namely, that each chain consists of phosphate diester groups joining β-D-deoxyribofuranose residues with 3′,5′ linkages. The two chains (but not their bases) are related by a dyad perpendicular to the fibre axis. Both chains follow right-handed helices, but owing to the dyad the sequences of the atoms in the two chains run in opposite directions. Each chain loosely resembles Furberg's[2] model No. 1; that is, the bases are on the inside of the helix and the phosphates on the outside. The configuration of the sugar and the atoms near it is close to Furberg's 'standard configuration', the sugar being roughly perpendicular to the attached base. There is a residue on each chain every 3·4 A. in the $z$-direction. We have assumed an angle of 36° between adjacent residues in the same chain, so that the structure repeats after 10 residues on each chain, that is, after 34 A. The distance of a phosphorus atom from the fibre axis is 10 A. As the phosphates are on the outside, cations have easy access to them.

The structure is an open one, and its water content is rather high. At lower water contents we would expect the bases to tilt so that the structure could become more compact.

The novel feature of the structure is the manner in which the two chains are held together by the purine and pyrimidine bases. The planes of the bases are perpendicular to the fibre axis. They are joined together in pairs, a single base from one chain being hydrogen-bonded to a single base from the other chain, so that the two lie side by side with identical $z$-co-ordinates. One of the pair must be a purine and the other a pyrimidine for bonding to occur. The hydrogen bonds are made as follows: purine position 1 to pyrimidine position 1; purine position 6 to pyrimidine position 6.

If it is assumed that the bases only occur in the structure in the most plausible tautomeric forms (that is, with the keto rather than the enol configurations) it is found that only specific pairs of bases can bond together. These pairs are: adenine (purine) with thymine (pyrimidine), and guanine (purine) with cytosine (pyrimidine).

In other words, if an adenine forms one member of a pair, on either chain, then on these assumptions the other member must be thymine; similarly for guanine and cytosine. The sequence of bases on a single chain does not appear to be restricted in any way. However, if only specific pairs of bases can be formed, it follows that if the sequence of bases on one chain is given, then the sequence on the other chain is automatically determined.

It has been found experimentally[3,4] that the ratio of the amounts of adenine to thymine, and the ratio of guanine to cytosine, are always very close to unity for deoxyribose nucleic acid.

It is probably impossible to build this structure with a ribose sugar in place of the deoxyribose, as the extra oxygen atom would make too close a van der Waals contact.

The previously published X-ray data[5,6] on deoxyribose nucleic acid are insufficient for a rigorous test of our structure. So far as we can tell, it is roughly compatible with the experimental data, but it must be regarded as unproved until it has been checked against more exact results. Some of these are given in the following communications. We were not aware of the details of the results presented there when we devised our structure, which rests mainly though not entirely on published experimental data and stereochemical arguments.

It has not escaped our notice that the specific pairing we have postulated immediately suggests a possible copying mechanism for the genetic material.

Full details of the structure, including the conditions assumed in building it, together with a set of co-ordinates for the atoms, will be published elsewhere.

We are much indebted to Dr. Jerry Donohue for constant advice and criticism, especially on interatomic distances. We have also been stimulated by a knowledge of the general nature of the unpublished experimental results and ideas of Dr. M. H. F. Wilkins, Dr. R. E. Franklin and their co-workers at

738 NATURE April 25, 1953 VOL. 171

King's College, London. One of us (J. D. W.) has been aided by a fellowship from the National Foundation for Infantile Paralysis.

J. D. WATSON
F. H. C. CRICK
Medical Research Council Unit for the Study of the Molecular Structure of Biological Systems,
Cavendish Laboratory, Cambridge.
April 2.

[1] Pauling, L., and Corey, R. B., *Nature*, **171**, 346 (1953); *Proc. U.S. Nat. Acad. Sci.*, **39**, 84 (1953).
[2] Furberg, S., *Acta Chem. Scand.*, **6**, 634 (1952).
[3] Chargaff, E., for references see Zamenhof, S., Brawerman, G., and Chargaff, E., *Biochim. et Biophys. Acta*, **9**, 402 (1952).
[4] Wyatt, G. R., *J. Gen. Physiol.*, **36**, 201 (1952).
[5] Astbury, W. T., Symp. Soc. Exp. Biol. 1, Nucleic Acid, 66 (Camb. Univ. Press, 1947).
[6] Wilkins, M. H. F., and Randall, J. T., *Biochim. et Biophys. Acta*, **10**, 192 (1953).

(From J.D. Watson and F.H. Crick, *Nature* 171: 737–738, 1953. With permission from Macmillan Publishers Ltd.)

cyclized form of a pentose (**Figure 1.16**; the structures of sugars are described in Chapter 3). Four of the five carbon atoms in the pentose form a five-membered ring along with an oxygen atom. These carbons are labeled C1′ to C4′, and the oxygen is labeled O4′. The fifth carbon atom (C5′) is attached to C4′ and is not part of the ring.

The nucleotides found in RNA (ribonucleic acid) are derived from ribose, a pentose sugar in which hydroxyl groups are attached to the C1′, C2′, C3′, and C5′ carbon atoms (Figure 1.16A). Thus, the nucleotide building blocks of RNA are called **ribonucleotides**, and these have the C1′-OH group of ribose replaced by a base and the C5′-OH group replaced by a phosphate group. The nucleotides

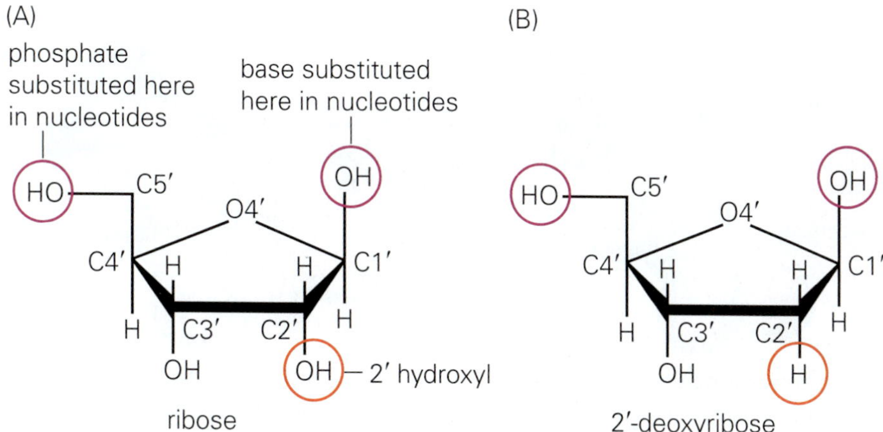


**Figure 1.16 Ribose and 2′-deoxyribose.** (A) Structure of ribose, indicating how the nucleotides found in RNA are derived from it. (B) 2′-Deoxyribose, from which the nucleotides found in DNA are derived, with the C1′ and C5′ sites again used for the addition of phosphate and base.

that make up DNA (deoxyribonucleic acid) are derived from 2′-deoxyribose, in which the 2′-OH group in ribose is replaced by hydrogen, and they are called **2′-deoxyribonucleotides** (Figure 1.16B).

The phosphate group that is attached to C5′ can have one or two additional phosphates attached to it, as shown in **Figure 1.17**. Nucleotides with one, two, or three phosphate groups are referred to as nucleotide mono-, di-, or triphosphates, respectively. The three phosphate groups, starting with the one attached to C5′, are called the α-, β-, and γ-phosphates. The combination of the sugar and a base, without the phosphate, is called a **nucleoside**.

There are five specific kinds of bases in the nucleotides that are the building blocks of DNA and RNA, the structures of which are discussed in the next section. All of these bases are covalently bonded to the C1′ atom of the sugar, as shown for one particular nucleotide (deoxyadenosine) in Figure 1.17. You might wonder why the heterocyclic ring system is called a "base." These ring systems contain nitrogen atoms with lone pairs of electrons that are not part of the aromatic system and extend outward in the plane of the ring. This means that the ring systems can act as electron-pair donors, and so in the language of chemistry, they are referred to as Lewis bases.

## 1.6 The nucleotide bases in RNA and DNA are substituted pyrimidines or purines

The nucleotide bases in RNA and DNA are substituted forms of two heterocyclic molecules known as **pyrimidine** and **purine**. As shown in **Figure 1.18A**, pyrimidine has a single six-membered aromatic ring, with two nitrogens and four carbons. The numbering system used for pyrimidines has the nitrogens at the 1 and 3 positions. Purine has two fused rings, one with five atoms and one with six (Figure

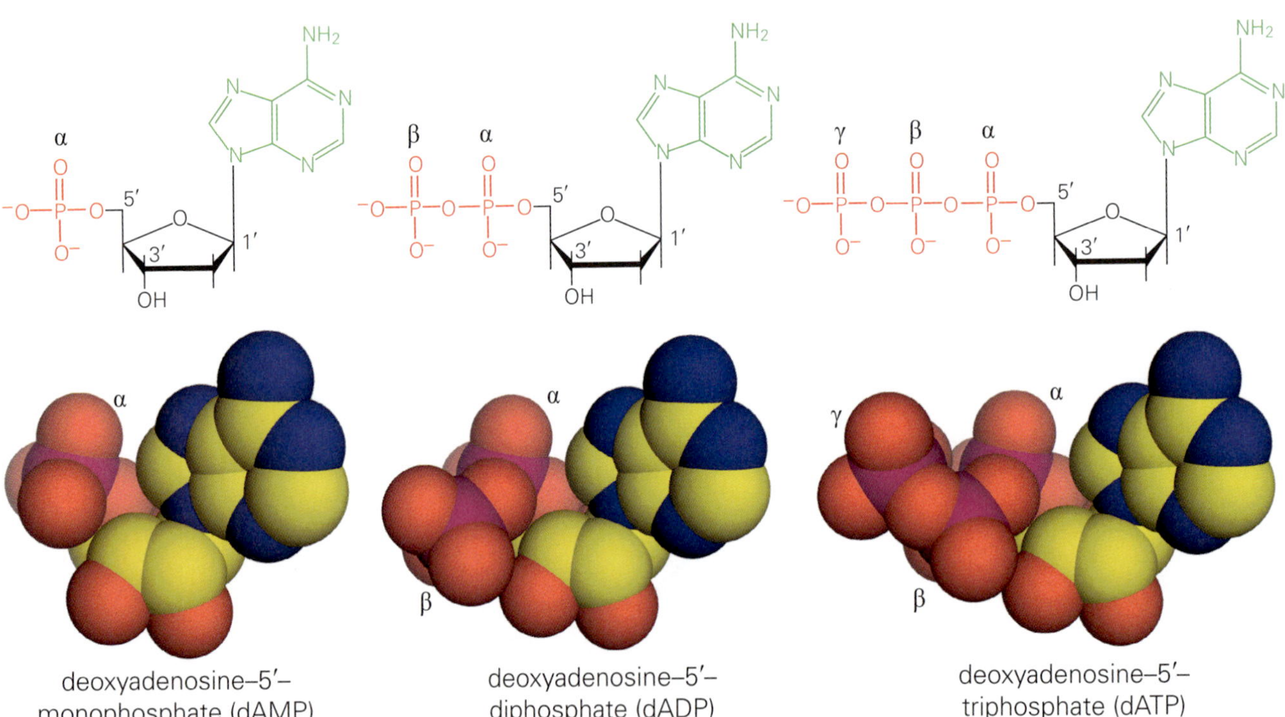


**Figure 1.17 Mono-, di-, and triphosphate forms of nucleotides**. Left, deoxyadenosine 5′-monophosphate (dAMP). Middle, deoxyadenosine 5′-diphosphate (dADP). Right, deoxyadenosine 5′-triphosphate (dATP). The three phosphoryl groups, starting from the one closest to the sugar, are called α, β, and γ phosphates. The prefix "deoxy" indicates that the nucleotide is derived from deoxyribose (see Figure 1.16).

1.18B). Purine has four nitrogens, at positions 1, 3, 7, and 9 in the ring system. In the corresponding purine nucleotides the hydrogen attached to the nitrogen at position 9 is replaced by the sugar. In the pyrimidine nucleotides, the sugar is attached to the nitrogen at position 1 on the base.

DNA contains two substituted purines (**adenine** and **guanine**) and two substituted pyrimidines (**cytosine** and **thymine**). Adenine, guanine, and cytosine also occur in RNA, but in RNA thymine is replaced by **uracil**. The five substituted bases are illustrated in **Figure 1.19**.

Each base has a unique pattern of hydrogen-bond donors and acceptors, which you will see is the key to the ability of DNA and RNA to serve as templates for the transfer of genetic information. Adenine contains an amino group ($-NH_2$) at position 6. The amino group can serve as the donor for two hydrogen bonds (indicated by the *blue* arrows in Figure 1.19). In addition, the lone pairs of electrons on the nitrogens at positions 1, 3, and 7 (see Figure 1.18) give these nitrogens a partial negative charge, enabling them to serve as hydrogen-bond acceptors. Guanine contains a carbonyl group (C=O) at position 6 and an amino group at position 2. The oxygen of the carbonyl group is a hydrogen-bond acceptor, and the amino group is a hydrogen-bond donor. In addition, the nitrogen at position 1 of guanine

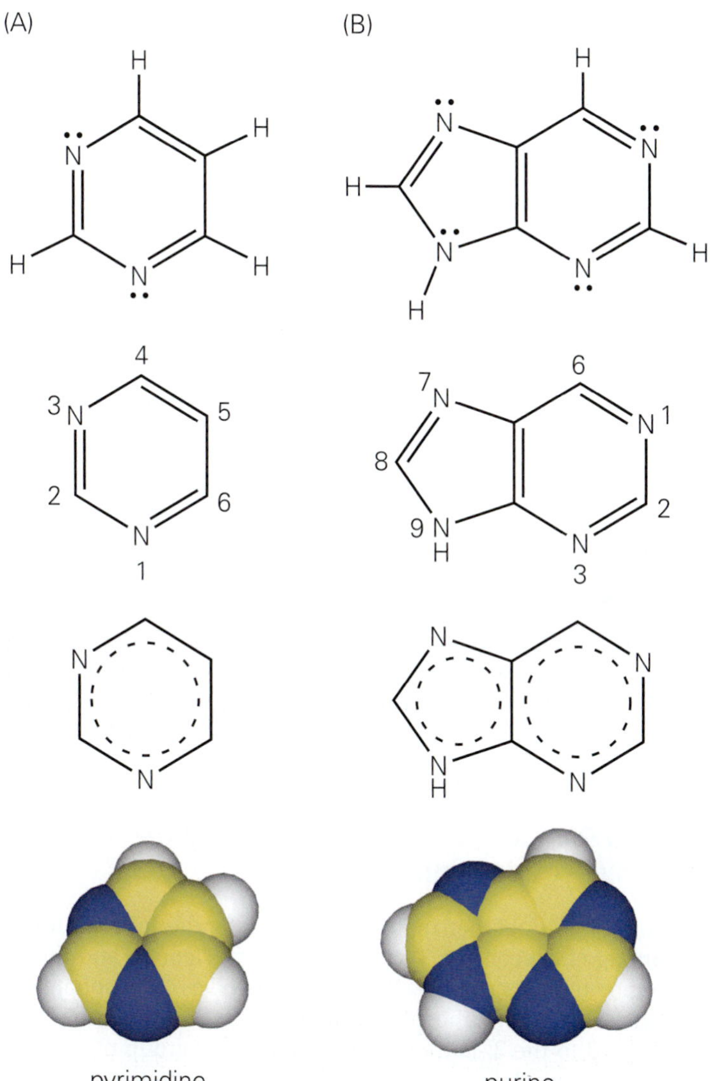


**Figure 1.18 Pyrimidine and purine.** Four representations of the structures of (A) pyrimidine and (B) purine.

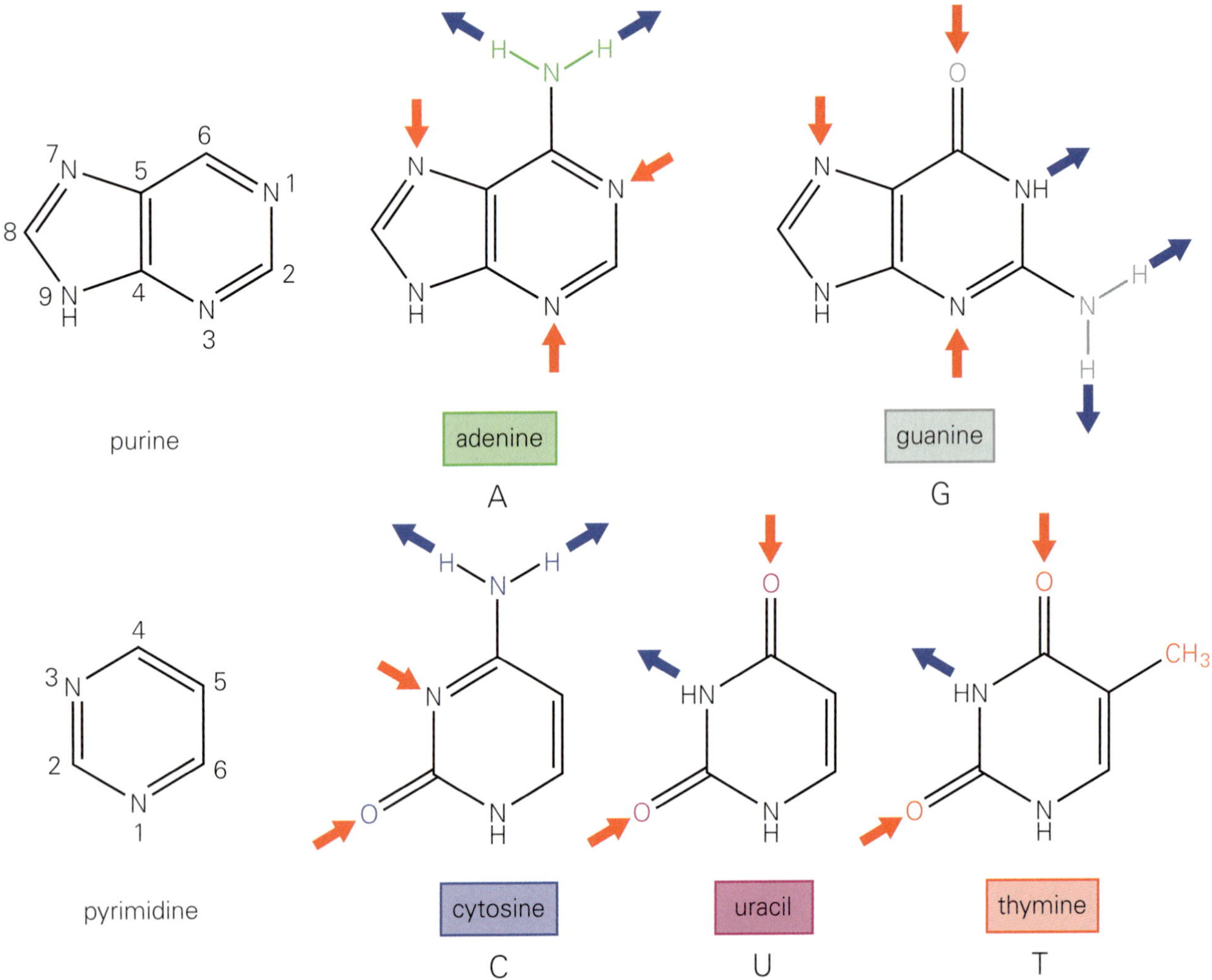


**Figure 1.19 Structures of the substituted purines and pyrimidines found in DNA and RNA.** *Red* arrows point towards hydrogen-bond acceptors and *blue* arrows point away from hydrogen-bond donors (compare with Figure 1.11). The sugars are attached to the nitrogens at the 9 and 1 positions in purine and pyrimidine nucleotides, respectively. The single-letter codes for each base are shown below the full name. The colors of the boxes around the names of the bases are the colors we will use to identify specific nucleotides throughout this book.

is protonated and can serve as a hydrogen-bond donor. The nitrogens at positions 3 and 7 are hydrogen-bond acceptors, as they are in adenine.

Cytosine, uracil, and thymine all have carbonyl groups that are hydrogen-bond acceptors at position 2 of the pyrimidine ring. In addition, cytosine contains an amino group at position 4, uracil contains a carbonyl group at position 4, and thymine contains a carbonyl group at position 4 and a methyl group ($-CH_3$) at position 5.

**Phosphodiester linkage**

The nucleotides in chains of DNA and RNA are connected by phosphodiester linkages. There are two ester bonds in each linkage. These connect the phosphate group to the 3′-oxygen atom of the first nucleotide and the 5′-oxygen atom of the second nucleotide, respectively.

## 1.7 DNA and RNA are formed by sequential reactions that utilize nucleotide triphosphates

Nucleotides are joined together in DNA and RNA by the formation of a **phosphodiester linkage** between the 3′ carbon of one nucleotide and the 5′ carbon of another, as shown in **Figure 1.20**. This linkage involves a phosphate group that bridges two nucleotides. The phosphorus atom forms two ester bonds, one with the oxygen bonded to the 3′-carbon atom of the sugar group of the first nucleotide and the other with the oxygen bonded to the 5′-carbon atom of the second

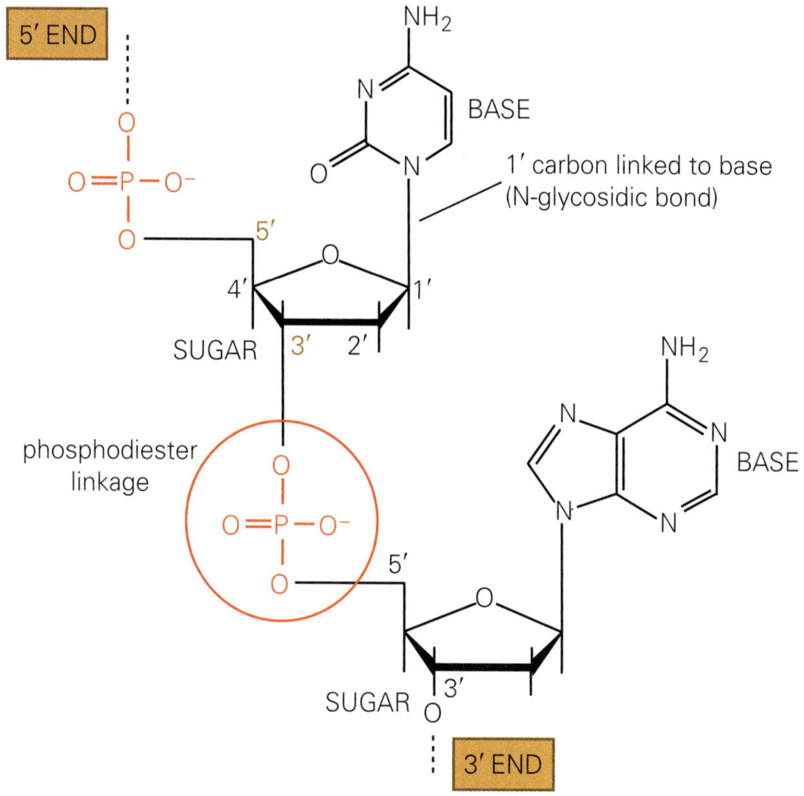

**Figure 1.20 The phosphodiester linkage in DNA.** Two nucleotides in a DNA chain are linked by a phosphate group. The phosphate group and the two ester bonds that it forms with the 5′ and 3′ carbons is called the phosphodiester linkage. A similar phosphodiester linkage is found in RNA.

nucleotide. The linked chain of sugar and phosphate groups is called the **backbone** of the DNA or RNA molecule. The phosphate groups are negatively charged, and this is a critical determinant of the three-dimensional structures formed by DNA and RNA, which are constrained by the need to keep phosphate groups as far away from each other as possible, due to electrostatic repulsion.

The synthesis of new molecules of DNA and RNA involves the stepwise addition of nucleotides to one end of the chain. In each step of the reaction, the 3′-OH group of the nucleotide at the end of chain attacks the α-phosphate group of an incoming nucleotide triphosphate (**Figure 1.21**). The triphosphate group is high in energy,

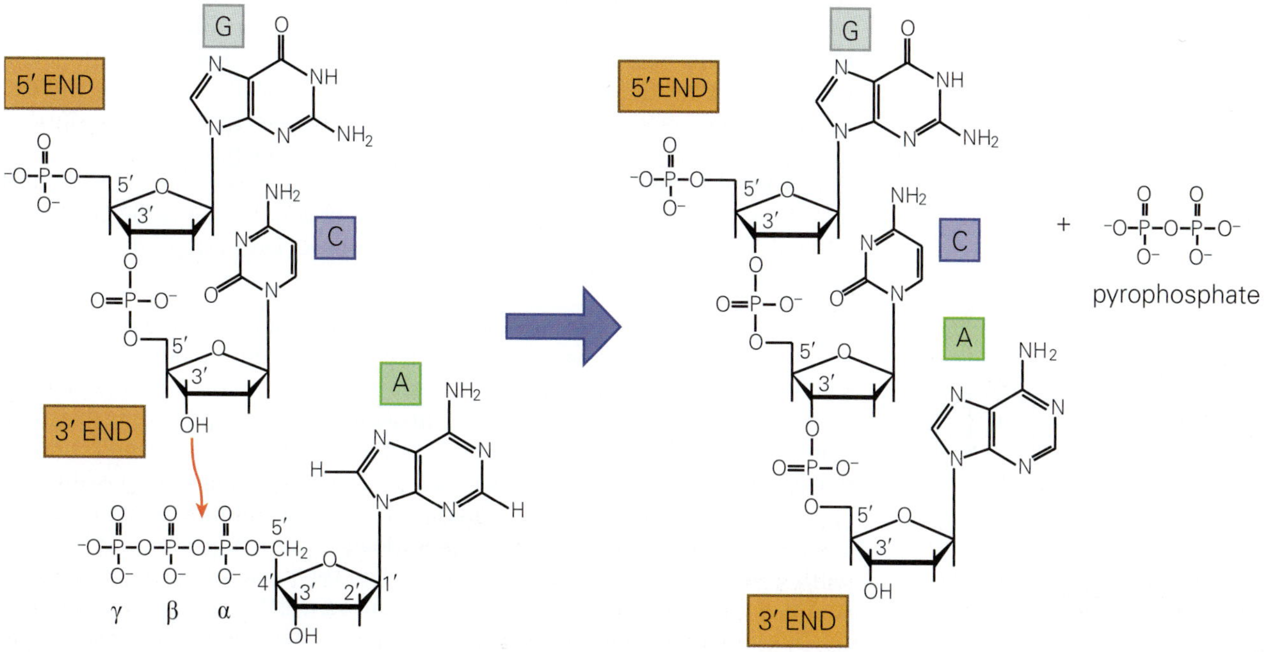

**Figure 1.21 Growth of a DNA chain by addition of a nucleotide.** The hydroxyl group at the 3′ position on the terminal nucleotide attacks the linkage between the α- and β-phosphate groups of an incoming nucleotide triphosphate. The products of the reaction are a DNA chain that is longer by one nucleotide, and a pyrophosphate ion.

**Template strand**

The order of addition of nucleotides to a growing chain of DNA or RNA is dictated by the sequence of another strand of DNA or RNA, known as the template strand.

and its hydrolysis drives the reaction. As the nucleic acid chain elongates, each incoming nucleotide triphosphate is converted into a nucleotide monophosphate that is linked by an ester bond to the previous nucleotide in the chain, and one molecule of pyrophosphate is released.

The unique feature of DNA and RNA synthesis, and the basis for the transmission of genetic information, is that their syntheses are template-directed. The enzymes that synthesize nucleic acids—DNA polymerase and RNA polymerase—select each new nucleotide in the growing chain on the basis of an existing strand, called the **template strand**. The mechanism of template-directed DNA synthesis is discussed in detail in Chapter 19.

The 3′ → 5′ phosphodiester linkage between nucleotides imparts directionality to chains of RNA or DNA. One end of the nucleic acid polymer has a phosphate group attached to the C5′ atom of the sugar, and this end is called the 5′ end of the DNA or RNA chain. The other has a free hydroxyl attached to the C3′ carbon, and is called the 3′ end of the chain (see Figure 1.21). By convention, the sequence of DNA is written (or read) from the 5′ to the 3′ end. There are a number of different ways to represent the primary structure of DNA (and RNA), some examples of which are shown in **Figure 1.22**.

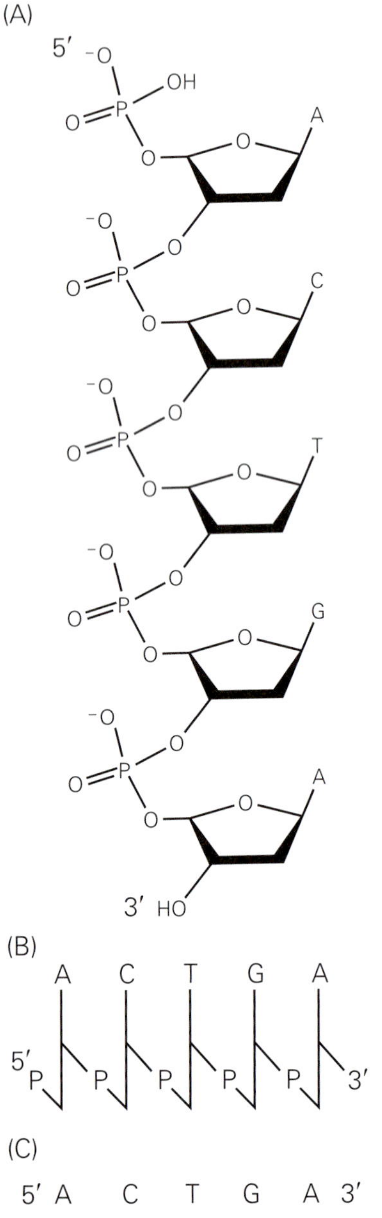

(A)

5′ -O OH O=P O O A -O O O=P O O C -O O O=P O O T -O O O=P O O G -O O O=P O O A 3′ HO

(B)

A C T G A

5′ P P P P P 3′

(C)

5′ A C T G A 3′

**Figure 1.22 Three ways of representing a DNA chain**. Each is written from the 5′ to the 3′ end.

## 1.8 DNA forms a double helix with antiparallel strands

For many years, DNA was thought to be too simple a molecule to be the basis of inheritance: the quantity of genetic information in a living organism, the reasoning went, is simply too great for it all to be encoded in a substance constructed entirely of just four nucleotide building blocks. Even after DNA was shown conclusively to be the genetic material, its detailed three-dimensional structure was still unknown. How was the genetic information stored in this polymer, and how was the information passed on from parent to offspring? A number of scientists attempted to solve the structure of DNA in the hope that understanding how the molecule was constructed would give some insight into how it functioned in the cell. Among them were Francis Crick, James Watson, Rosalind Franklin, and Maurice Wilkins.

It was known by 1953 that DNA was a polymer of just four types of nucleotide units—A, C, G, and T—and Erwin Chargaff had shown that the amount of G in DNA equaled that of C and the amount of A equaled that of T. Scientists were also fairly confident that the basic construction of the DNA molecule included a sugar-phosphate backbone, with 3′ → 5′ phosphodiester linkages, from which the nitrogenous bases protruded. Combining this and other information, along with crucial x-ray diffraction data from Franklin and Wilkins, Watson and Crick proposed a model for DNA in which two DNA chains—oriented in opposite directions and held together by hydrogen bonds between the protruding bases—form a right-handed double helix (**Figure 1.23**), with the bases on the inside of the double helix and the sugar-phosphate backbones running along the outside. The original paper describing the Watson-Crick model of DNA is reproduced from the journal *Nature* in Box 1.1.

In the Watson-Crick model, the hydrogen bonds holding the two chains together always join a purine on one chain and a pyrimidine on the other. Most significantly, a given purine always pairs with the same pyrimidine: A always pairs with T, and C always pairs with G. Near the end of the short paper proposing this model, Watson and Crick make the memorable statement: "It has not escaped our notice that the specific pairing we have postulated immediately suggests a possible copying mechanism for the genetic material." This was a stunning demonstration of insight about function resulting from knowledge of structure: the complementary pairing of the bases (A-T and G-C) suggested that, when DNA replicates, each strand of the parent DNA serves as a template on which a complementary strand can by synthesized, resulting in two exact copies of the original. For this discovery,

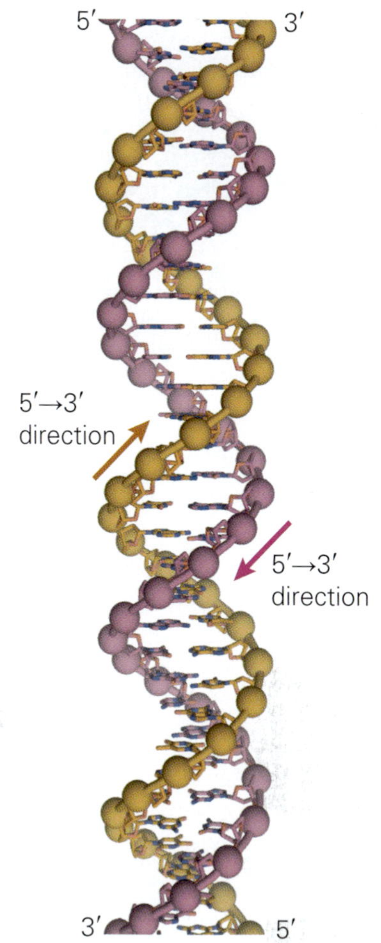


**Figure 1.23 The structure of DNA.** The phosphate groups of the two strands are shown as *purple* and *orange* spheres. The 5′-to-3′ direction of each strand is indicated by the arrows.

Watson and Crick shared the 1962 Nobel Prize for Physiology or Medicine with Maurice Wilkins. Rosalind Franklin had died before the prize was awarded.

There are three essential features of the Watson and Crick model (see Figure 1.23). First, the two polynucleotide chains run in opposite directions (the 5′ end of one strand is adjacent to the 3′ end of the other), and the two strands together wind around a common axis to form a right-handed double helix. Second, the bases are on the inside of the helix and the sugar-phosphate groups are on the outside. This structural feature allows the negatively charged phosphate groups to interact with water and metal ions, which compensates for the electrostatic repulsion between phosphates. Third, base-pairing holds the DNA strands together and is strictly complementary: adenine is paired exclusively with thymine (held together by two hydrogen bonds), and guanine is paired exclusively with cytosine (held together by three hydrogen bonds) (**Figure 1.24**). Complementary base-pairing between the two strands of the DNA helix is an essential feature of replication and transcription. In particular, the high symmetry of the structure of DNA (**Figure 1.25**) is a consequence of close geometrical similarities in the structures of the base pairs. This fact is exploited by mechanisms that ensure fidelity in replication and transcription, which are described in Chapter 19.

As we shall see in Chapter 2, RNA can also form double-helical structures that resemble the DNA double helix in general terms, with the phosphate groups on the outside and with base pairs on the inside. Primordial life forms are thought to have used RNA rather than DNA as the genetic material, and many viruses that

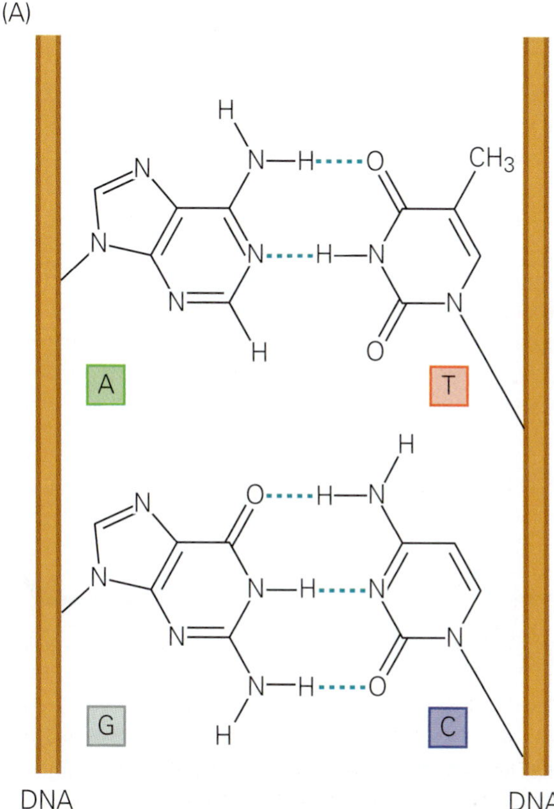


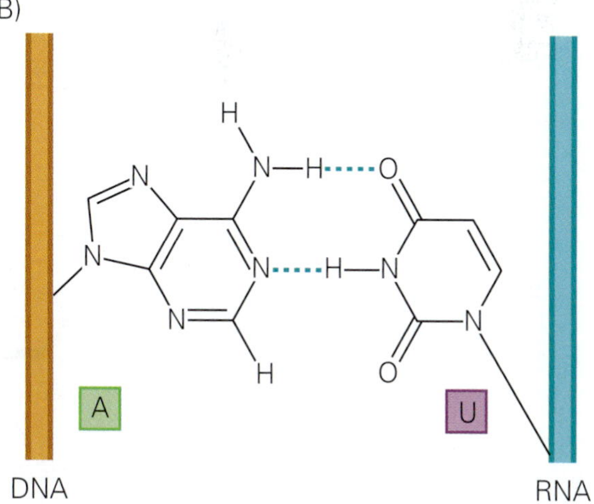


**Figure 1.24 The Watson-Crick base pairs: A-T, G-C, and A-U.** The sugar-phosphate chains are indicated by vertical lines. Hydrogen bonds (two for A-T and three for G-C) are shown by dotted lines. Thymine is absent from RNA; in its place is the pyrimidine uracil (U). Base pairs in double-stranded RNA or in DNA-RNA hybrids are therefore G-C and A-U. DNA-RNA hybrids also have A-T base pairs, where the T is on the DNA strand.

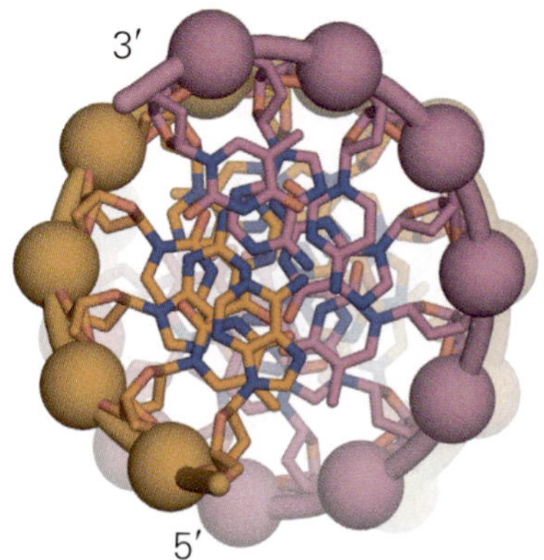


**Figure 1.25 The symmetry of DNA**. This view of the double helix—looking down the central axis of the double helix—is orthogonal to the standard view shown in Figure 1.23 and in Watson and Crick's original paper (see Box 1.1). Note the way in which the base pairs align so that they overlap closely, even though there are four different kinds of base pairs in projection (A-T, T-A, G-C, and C-G). This superposition emphasizes the similarity in the width of the base pairs and the angle made by the base pair and the phosphate backbone.

are extant today have chromosomes that are made of RNA. All known living cells, however, store genetic information in the form of DNA rather than RNA. One reason for this could be that DNA is chemically more stable than RNA. The 2′-OH group in an RNA nucleotide can attack and break the phosphodiester linkage at the 3′ position, as shown in **Figure 1.26**. DNA lacks the 2′-OH group, and so DNA is more chemically stable.

## 1.9 The double helix is stabilized by the stacking of base pairs

Molecules that contain aromatic ring systems, such as the phenylalanine and tyrosine sidechains of proteins and the bases of nucleic acids, are polarized in a special way. The polarization in these molecules is distributed around the ring, and the net electrostatic effects are more complex than those of isolated dipoles. Even though these molecules are net neutral, individual atoms in the molecules have partial charges that are distributed asymmetrically (the distribution of partial charge in the four bases of DNA is illustrated in Figure 19.10).

These partial charges play a crucial part in stabilizing the double helix. The bases in double-helical DNA and RNA stack on top of one another, as shown in **Figure 1.27**. Two adjacent base pairs that are stacked on each other are rotated with respect to each other, because of the helical twist of the double helix (see Figure 2.10 in Chapter 2). This rotation of the stacked base pairs brings the partial charges on the top and bottom base pair into favorable overlap, with positively charged regions of the rings interacting with negatively charged regions. This helps to stabilize the stacked base pairs.

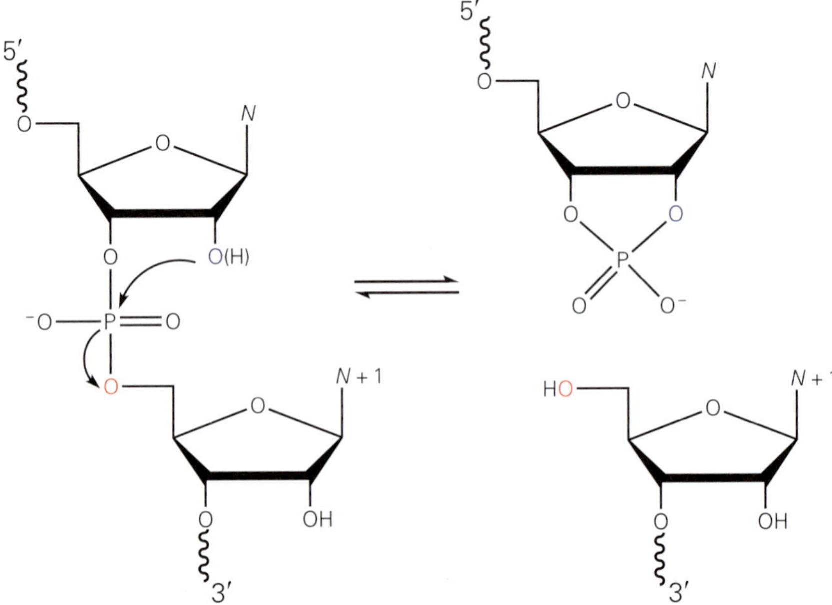


**Figure 1.26 The self cleavage of RNA**. The 2′-OH group attacks the phosphate group as shown and generates products with 5′-OH and 2′-3′-cyclic-phosphate termini.

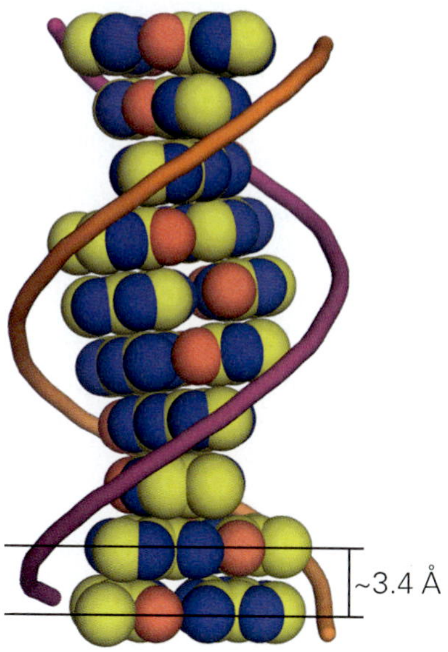

**Figure 1.27 Base-stacking interactions.** The structure of double-helical DNA is shown, with the atoms in the bases shown as spheres and with the radius of each sphere given by the van der Waals radius of the atom. Notice that the atoms of adjacent base pairs are in van der Waals contact. The phosphate backbone is indicated in *orange* and *purple*. The rise of the helix per base pair is ~3.4 Å, as indicated.

Stacking is also favored by van der Waals attractive forces because the atoms in stacked bases are just touching each other, as you can see in Figure 1.27. Recall from Section 1.2 that, although any single van der Waals contact provides only a small amount of stabilization energy, the combined effect of several such contacts can be substantial. Because the base pairs are planar, the stacking of one base pair on another results in ~15 simultaneous van der Waals contacts. One way to appreciate that the base pairs are in optimal van der Waals contact with each other is to consider that the rise per base pair in double-helical DNA is ~3.4 Å (see Figure 1.27). If you refer to Table 1.1, you will see that the van der Waals radii of carbon and nitrogen atoms are 1.7 Å and 1.6 Å, respectively. The rise per base pair of 3.4 Å is just what we would get if the carbons and nitrogen atoms in stacked base pairs were touching each other at close to their optimal van der Waals contact distance. The combination of electrostatic and van der Waals attractions makes base stacking the dominant energetic term favoring the formation of helical structures in nucleic acids, as we discuss further in Chapters 2 and 19.

## 1.10 Proteins are polymers of amino acids

With a few rare exceptions, all of the proteins of all life-forms on Earth are polymers of the same set of just 20 different amino acids (the exceptions include un-usual amino acids in a few specialized enzymes). The general form of the amino acids is simple: an amino group ($-NH_3^+$), and a carboxylate group ($-COO^-$), a hydrogen atom (–H), and an all-important variable **sidechain** (denoted R) are all attached to a carbon atom (**Figure 1.28**). At physiological pH (~pH 7), both the carboxylate group and the amino group are almost completely ionized, thus forming a zwitterion—a dipolar molecule that contains charged groups but is electrically neutral overall.

The 20 amino acids found in proteins are illustrated in **Figure 1.29**. All 20 are α-amino acids, in which the sidechain is attached to the central carbon atom, which is denoted the $C_\alpha$ atom. Nineteen of the 20 amino acids are primary amines (that is, they contain a $-NH_3^+$ group) and differ only in the nature of the sidechain. The exception is proline, which is a secondary amine ($-NH_2^+-$) because its nitrogen and α-carbon atoms are part of a five-membered pyrrolidine ring.

The name of an amino acid depends on its sidechain. The sidechain of alanine, for example, is a methyl group, $-CH_3$. Each amino acid is also denoted by a three-letter code (for example, Ala for alanine) and a one-letter code (for example, A for alanine). The full names, three-letter codes, and one-letter codes for all 20 amino acids are given in Figure 1.29.

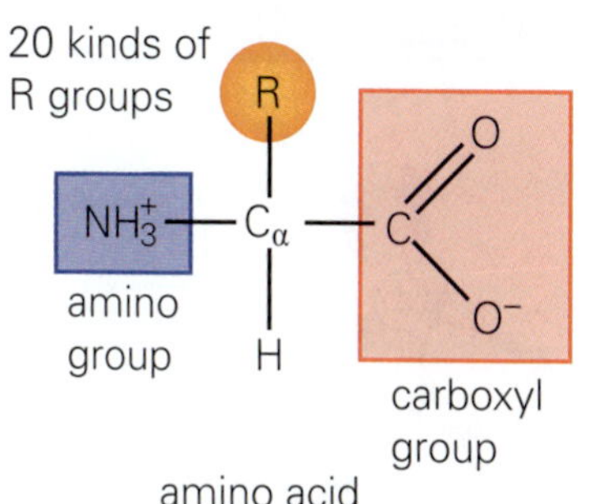


**Figure 1.28 General structure of amino acids.** A central carbon atom ($C_\alpha$) is attached to an ammonium group ($-NH_3^+$), a carboxylate group ($-COO^-$), a hydrogen atom (–H), and a sidechain (–R). Each of the 20 amino acids encoded by DNA has a unique sidechain.

## 1.11 Proteins are formed by connecting amino acids by peptide bonds

The synthesis of proteins involves a condensation reaction in which the amino group of one amino acid combines with the carboxyl group of another, with the formation of a **peptide bond** and the elimination of a water molecule (**Figure 1.30**). Amino acids linked by peptide bonds are called **amino acid residues** (they are no longer amino acids because the amino and carboxyl groups that reacted to form the peptide bond no longer exist). The reaction is repeated over and over again to form a chain. This reaction is catalyzed by a very large assembly of proteins and RNA called the **ribosome**. The ribosome ensures that the sequence of amino

residues with polar groups

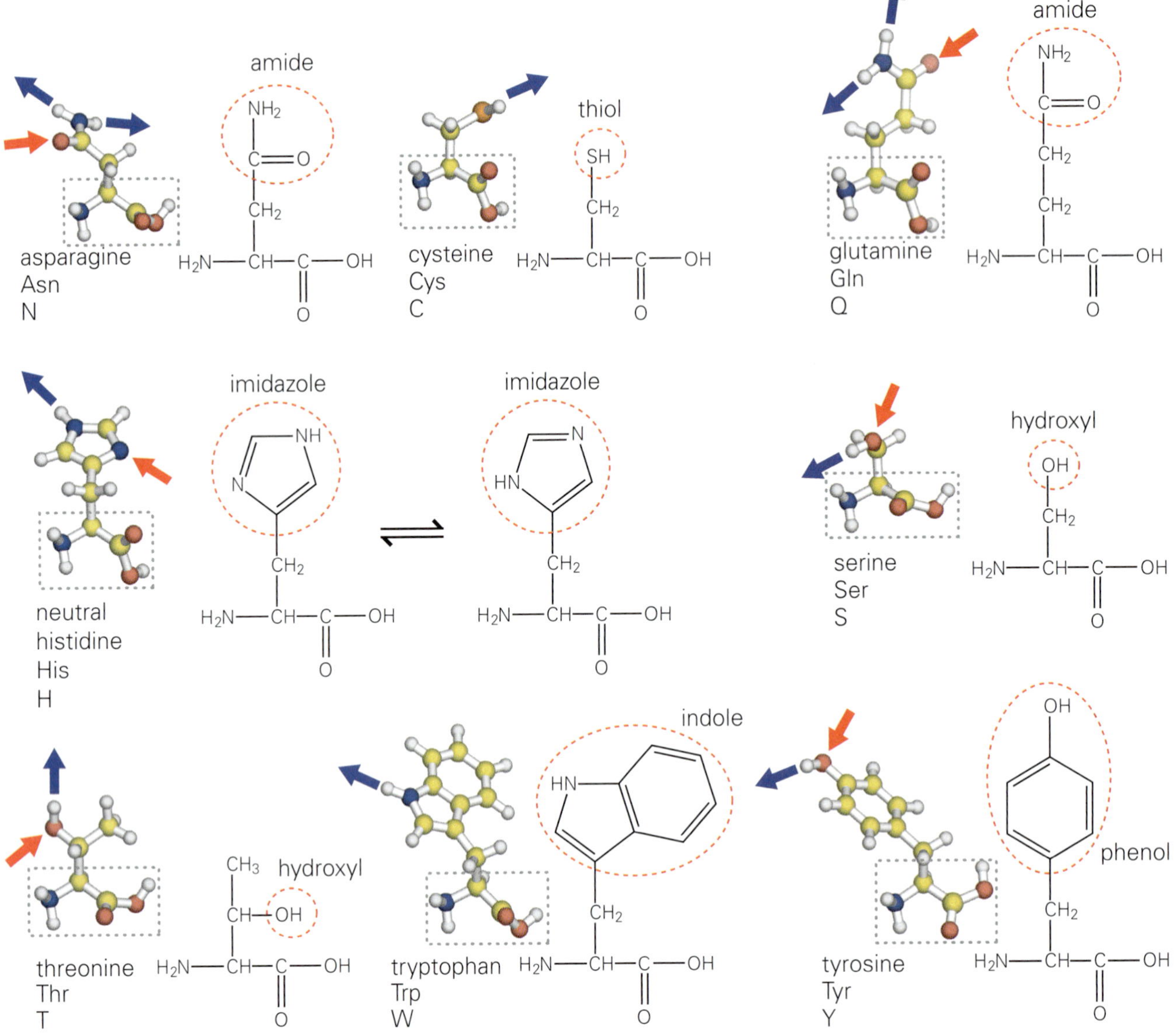

amide
$NH_2$
C=O
$CH_2$
$H_2N$—CH—C—OH
O
asparagine
Asn
N

thiol
SH
$CH_2$
$H_2N$—CH—C—OH
O
cysteine
Cys
C

amide
$NH_2$
C=O
$CH_2$
$CH_2$
$H_2N$—CH—C—OH
O
glutamine
Gln
Q

imidazole
NH
N
$CH_2$
$H_2N$—CH—C—OH
O
⇌
imidazole
N
HN
$CH_2$
$H_2N$—CH—C—OH
O
neutral
histidine
His
H

hydroxyl
OH
$CH_2$
$H_2N$—CH—C—OH
O
serine
Ser
S

$CH_3$
CH—OH
hydroxyl
$H_2N$—CH—C—OH
O
threonine
Thr
T

indole
HN
$CH_2$
$H_2N$—CH—C—OH
O
tryptophan
Trp
W

OH
phenol
$CH_2$
$H_2N$—CH—C—OH
O
tyrosine
Tyr
Y

positively charged, hydrophilic residues

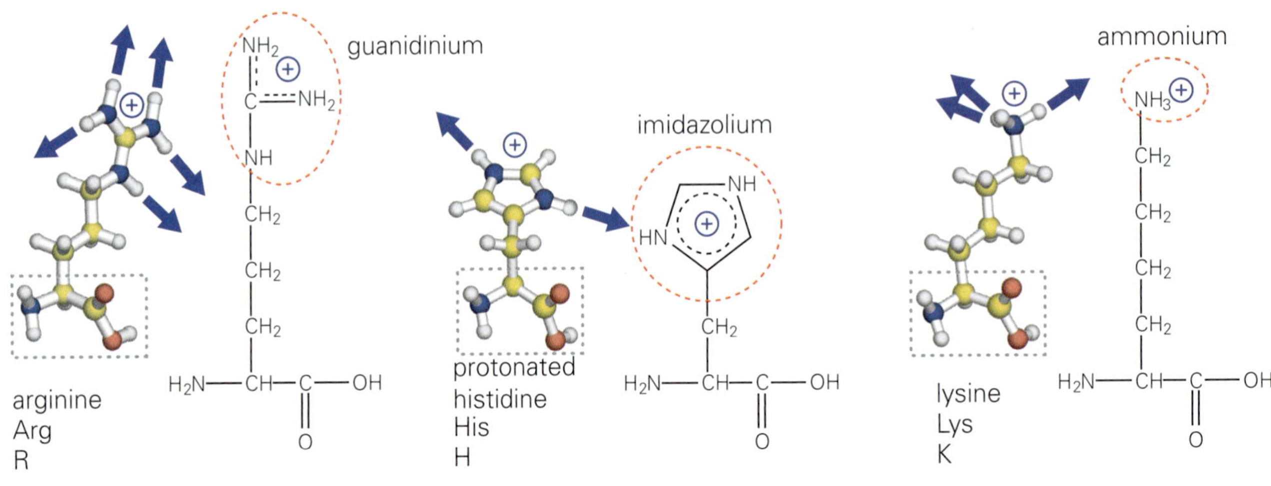

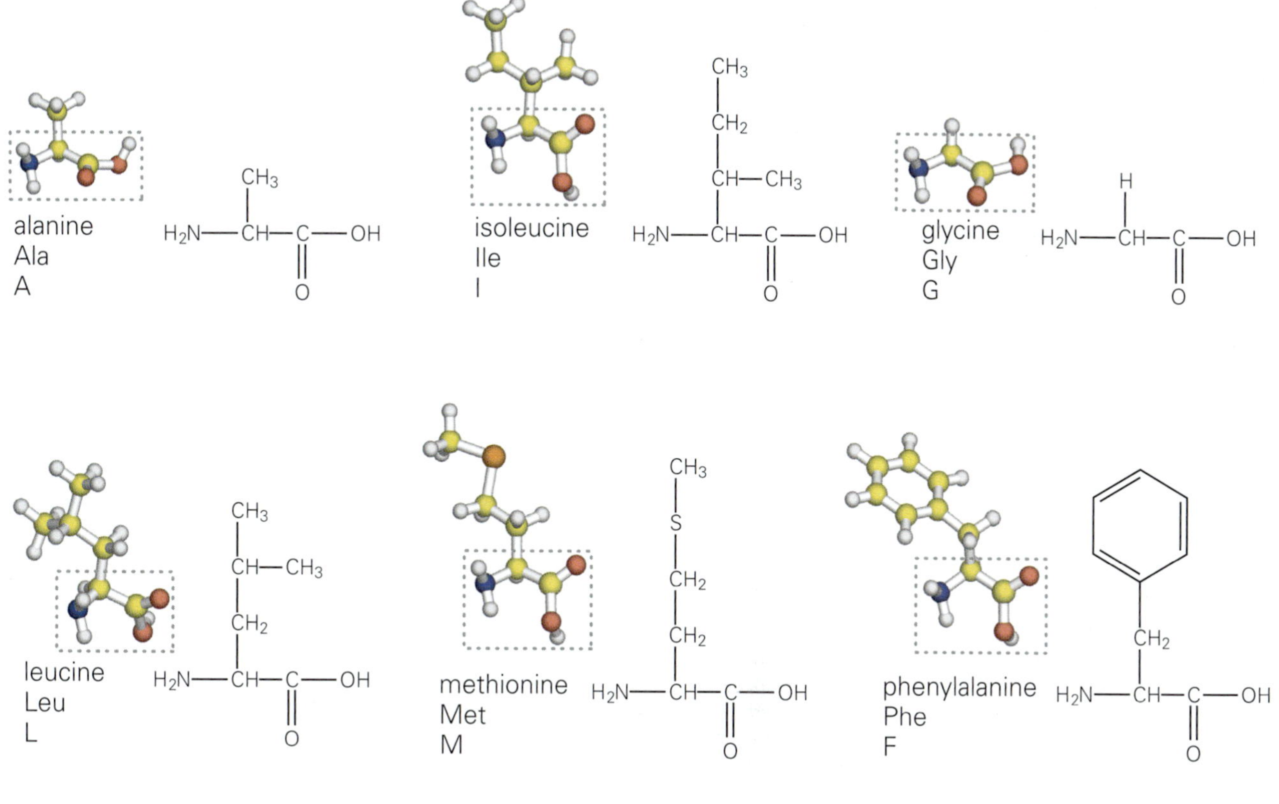


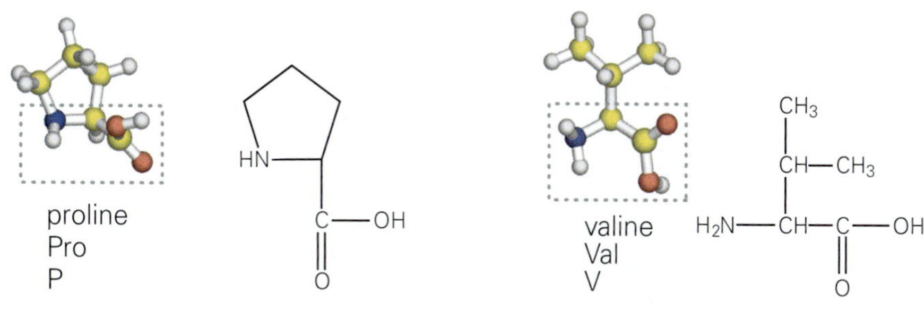


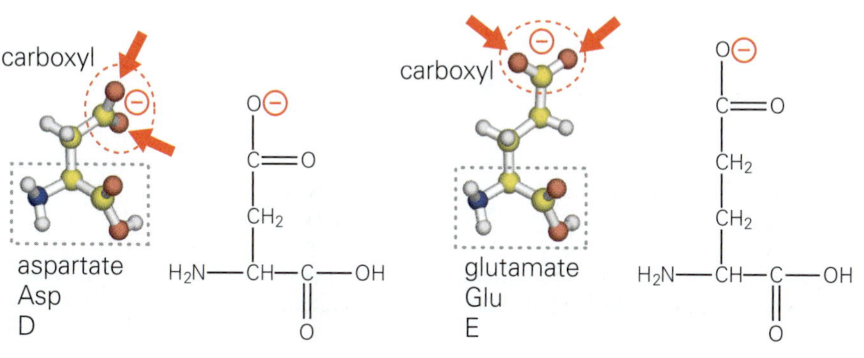


**Figure 1.29 The amino acids.** The 20 amino acids that are encoded by DNA are illustrated. For each amino acid, a three-dimensional representation is shown on the left, and a chemical structure is shown on the right. Each amino acid is identified by its full name, as well as by its three-letter and single-letter codes (for example, arginine, Arg, R). Hydrogen-bond donors are indicated by *blue* arrows and hydrogen-bond acceptors by *red* arrows. Polar chemical groups are circled and named. Although free amino acids form zwitterions in solution at pH 7 (see Figure 1.28), the structures shown here correspond to the uncharged state of the carboxyl and amino groups. Each of the amino acids, with the exception of glycine, is a chiral molecule that has a specific handedness, as explained in Figure 4.13.

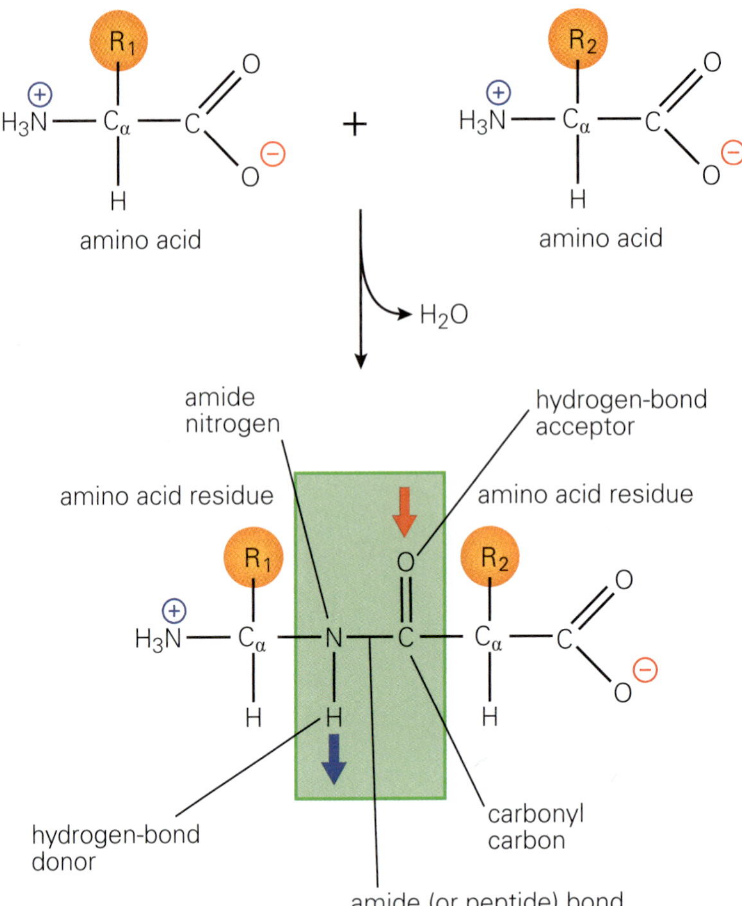


**Figure 1.30 Condensation of two amino acids to form a peptide.** The peptide bond is highlighted in *green*. The oxygen atom of the carbonyl group involved in the peptide bond is a hydrogen-bond acceptor (*red* arrow) and the amide nitrogen is a hydrogen-bond donor (*blue* arrow).

acids in the protein that is synthesized corresponds to the sequence specified by the messenger RNA that carries the genetic instructions from the DNA. We shall see, in Section 1.20, that this involves the engagement of specialized adapters for translating the sequence of nucleotides in the RNA into an amino acid sequence.

When the chain contains only a small number of residues (fewer than ~50) it is referred to as a **peptide** or a **polypeptide**, with the term "protein" usually being reserved for longer chains. The repetitive sequence of -NH-CH-CO- atoms that runs the length of the polymer is called the **peptide backbone**. The sequence of a protein is written with the **N-terminal amino acid** (the one with the free $-NH_3^+$ group) on the left and the **C-terminal amino acid** (the one with the free $-COO^-$ group) on the right, as shown in **Figure 1.31**.

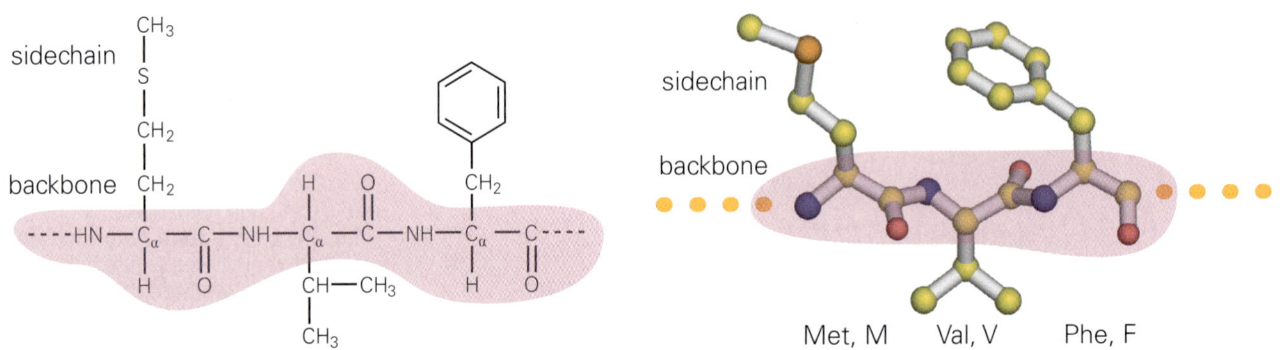


**Figure 1.31 The structure of a peptide.** The chemical and three-dimensional structure of a short peptide sequence (Met-Val-Phe) are shown here. Hydrogen atoms are not shown in the three-dimensional drawing on the right. The peptide backbone is highlighted in *purple*.

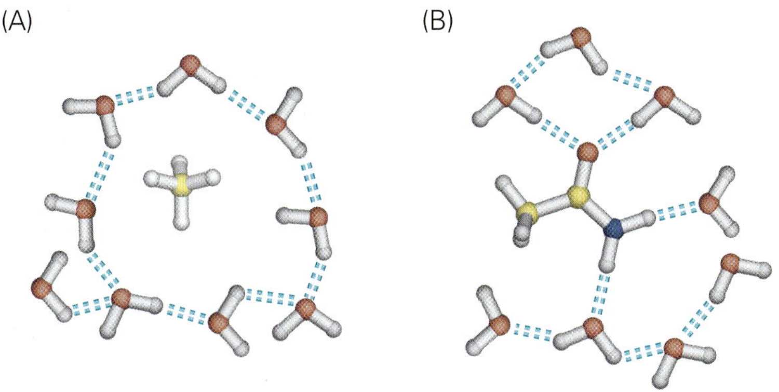


**Figure 1.32 Hydrophobic (nonpolar) and hydrophilic (polar) groups.** (A) Hydrophobic groups or molecules, such as methane, do not form hydrogen bonds with water and prefer to interact with each other. (B) Hydrophilic groups, such as the carbonyl and amino groups in acetamide, can readily form hydrogen bonds with water.

## 1.12 Amino acids are classified based on the properties of their sidechains

The classification of amino acids into functional categories is based on the chemical nature of the sidechains. The most important characteristic of a sidechain is the extent to which its covalent bonds are polarized, because this determines how the sidechain interacts with water. Nonpolar sidechains, such as methyl or other alkyl groups, are **hydrophobic**. Hydrophobic groups do not interact well with water molecules, in part because they cannot form hydrogen bonds with them (**Figure 1.32**). **Hydrophilic** sidechains, on the other hand, contain **polar groups** such as the carboxyl, amino, amide, guanidino, and hydroxyl groups, that can form hydrogen bonds with water and thus readily interact with it.

**Hydrophobic groups**

Hydrophobic groups are nonpolar—that is, they do not form hydrogen bonds. Hydrophobic groups prefer to cluster with other hydrophobic groups rather than interact with water, in part because they cannot form hydrogen bonds with water. Molecules that are hydrophobic do not dissolve readily in water.

The sidechains of the hydrophobic amino acids are hydrocarbons that are poorly soluble in water: alanine (Ala, A), valine (Val, V), leucine (Leu, L), isoleucine (Ile, I), phenylalanine (Phe, F), proline (Pro, P), and methionine (Met, M) all have hydrophobic sidechains. The methionine sidechain contains a sulfur atom (see Figure 1.29), but it has several aliphatic carbon atoms that make it hydrophobic overall.

Hydrophilic amino acids have sidechains that are very water soluble and may be either charged or uncharged (neutral). The charged amino acids have sidechains with a net negative or positive charge. The negatively charged amino acids are aspartate (Asp, D) and glutamate (Glu, E). The positively charged amino acids are arginine (Arg, R) and lysine (Lys, K). Histidine (His, H) is special because it can be either positively charged or neutral depending on whether it is protonated or not. The histidine sidechain readily takes up or loses an extra proton at normal pH (~7), so it is positively charged when it takes up the proton, and neutral when it loses the proton.

**Hydrophilic groups**

Hydrophilic groups are polar and they can form hydrogen bonds with water. Molecules that are hydrophilic dissolve readily in water.

The other amino acids with neutral but polar sidechains are cysteine (Cys, C), serine (Ser, S), threonine (Thr, T), asparagine (Asn, N), glutamine (Gln, Q), histidine (His, H), tyrosine (Tyr, Y), and tryptophan (Trp, W). Although the sidechains of tryptophan and tyrosine contain groups that can form hydrogen bonds with water, other parts of these sidechains are quite hydrophobic.

Glycine (Gly, G), with just a hydrogen atom as its sidechain, is the simplest of the 20 amino acids. It can be grouped with the hydrophobic amino acids or treated as the sole member of a fourth class because of its special properties. These arise from the absence of a bulky sidechain, which allows glycine to adopt three-dimensional conformations that are energetically unfavorable for other amino acids, so that segments of protein chains that contain glycine can be much more flexible than those that do not.

Cysteine is a unique sidechain because the terminal thiol (-SH) group is quite reactive. The sulfur atoms in two cysteine residues can react in the presence of oxygen or other oxidizing agents to form a covalent bond, known as a **disulfide bond** (**Figure 1.33**). Two cysteine residues linked in this way are sometimes referred to as a cystine residue.

**Disulfide bond**

A disulfide bond is a covalent bond between the sulfur atoms of two cysteine residues. Disulfide bonds cross-link different parts of a protein chain and can increase the stability of the three-dimensional structure.

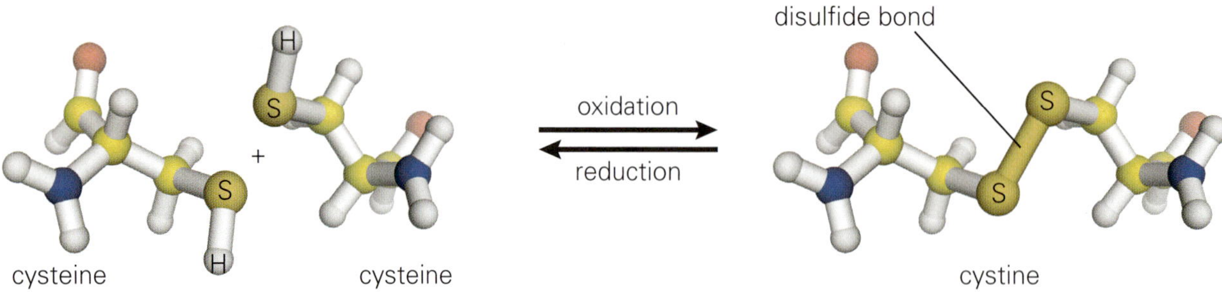


**Figure 1.33 Cysteine residues can form disulfide bonds.** The sulfur atoms in two cysteines can form a covalent bond, called a disulfide bond.

The formation of disulfide bonds can make cross bridges within the protein chain, or between two different protein chains, and is one mechanism for increasing the stability of proteins. This reaction is readily reversed by reducing agents, such as other molecules with thiol (–SH) groups. Disulfide bonds are important structural elements of many proteins, particularly those that are secreted out of the cell: the interior of the cell has high concentrations of molecules with reduced thiol groups, which tend to break disulfide bonds and, consequently, disulfide bonds are rarely found inside the cell.

## 1.13 Proteins appear irregular in shape

In 1958, five years after the discovery of the structure of DNA, the first three-dimensional structure of a protein was determined—that of myoglobin, an oxygen-binding protein. **Figure 1.34** shows a historical set of photographs of the original clay model of the structure. This structure came as a shock to those who had hoped for simple, general principles of protein structure and function analogous to the elegant structure that Watson and Crick had determined for DNA. John Kendrew, who solved the myoglobin structure, expressed his disappointment thus: "Perhaps the most remarkable features of the molecule are its complexity and its lack of symmetry. The arrangement seems to be almost totally lacking in the kind of regularities which one instinctively anticipates, and it is more complicated than has been predicted by any theory of protein structure."

**Globular protein**

A water-soluble protein that folds into a compact three-dimensional shape is referred to as a globular protein.

We now know that such structural irregularity is actually required for proteins to fulfill their diverse functions. Information storage and transfer from DNA is essentially linear, so DNA molecules of very different information content nevertheless have essentially the same gross structure. Proteins, on the other hand, must recognize many thousands of different molecules in the cell by detailed three-dimensional interactions. So, proteins themselves require diverse and irregular structures. In spite of these requirements, there are regular features in protein structures, reflecting chiefly the constraints on the arrangements that polypeptides can adopt in an aqueous environment.

Myoglobin, like most of the proteins discussed in this book, is a **globular protein**—that is, a water-soluble protein with a compact folded state. Proteins can also be embedded in lipid membranes, or have fibrous or irregular structures, but

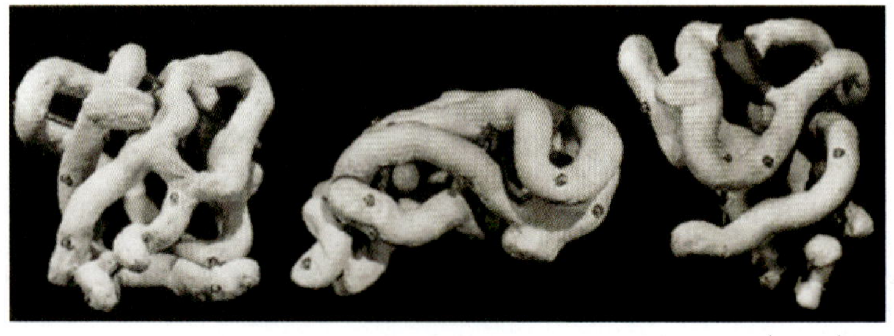


**Figure 1.34 Kendrew's model of myoglobin**. These photographs show three different views of the first protein structure to be determined. The sausage-shaped regions represent α helices (see Figure 1.36), which are arranged in a seemingly irregular manner to form a compact globular molecule. (From J.C. Kendrew et al., *Nature* 181: 662–666, 1958. With permission from Macmillan Publishers Ltd.)

the general principles underlying their structure are the same, and for simplicity in this chapter the discussion of these principles will be confined to the globular proteins.

## 1.14 Protein chains fold up to form hydrophobic cores

When the structure of myoglobin was determined, it was noticed that the amino acids in the interior of the protein had almost exclusively hydrophobic sidechains. This was one of the first important general principles to emerge from studies of protein structure. That is, the main driving force for folding water-soluble proteins is to pack hydrophobic sidechains into the interior of the molecule, thus creating a **hydrophobic core** and a hydrophilic surface.

All water-soluble proteins, not just myoglobin, have a hydrophobic core. Proteins that are in the lipid membranes of cells do not have hydrophilic surfaces within the membrane, as we shall discuss in Chapter 4, but they too have hydrophobic cores. The hydrophobic core of a protein known as triose phosphate isomerase is illustrated in **Figure 1.35**. Notice the partitioning of the sidechains, with most of the residues in the interior of the protein being hydrophobic and closely packed. Given the constraints of the different shapes of the hydrophobic sidechains and, considering that their positions must be compatible with the three-dimensional folding of the protein backbone, fitting these shapes into a densely packed core is like solving a three-dimensional jigsaw puzzle.

**Hydrophobic core**

The three-dimensional structure of most proteins is characterized by an interior hydrophobic core, consisting mainly of hydrophobic sidechains. The exclusion of these sidechains from water stabilizes the protein.

## 1.15 α helices and β sheets are the architectural elements of protein structure

There is a major problem that the protein chain must overcome in order to form a hydrophobic core. To bring the sidechains into the core, the protein backbone must also fold into the interior. The backbone is highly polar and therefore hydrophilic, with one hydrogen-bond donor (NH) and one hydrogen-bond acceptor (C=O) for each peptide unit. These groups form hydrogen bonds with water when the protein is unfolded, and their hydrogen-bonding needs must be satisfied as the protein chain crosses through hydrophobic regions.

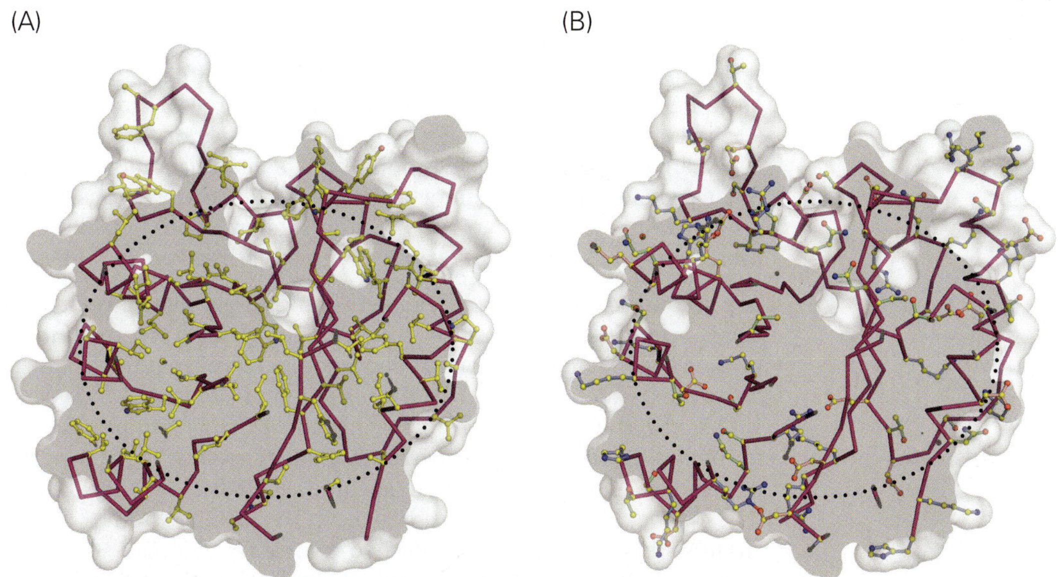


**Figure 1.35 Hydrophobic core of a protein**. The diagrams show a slice through a protein structure. The *light gray* shading represents the surface of the protein, as defined by the van der Waals radii of the atoms. The *darker gray* shading indicates the interior of the protein. The hydrophobic core is outlined with a dotted line. Sidechains are shown in a stick representation, with carbon *yellow*, nitrogen *blue*, and oxygen *red*. The polypeptide backbone is in *purple*. (A) Hydrophobic sidechains, including tyrosine and tryptophan. (B) Polar sidechains. (PDB code: 1TIM.)

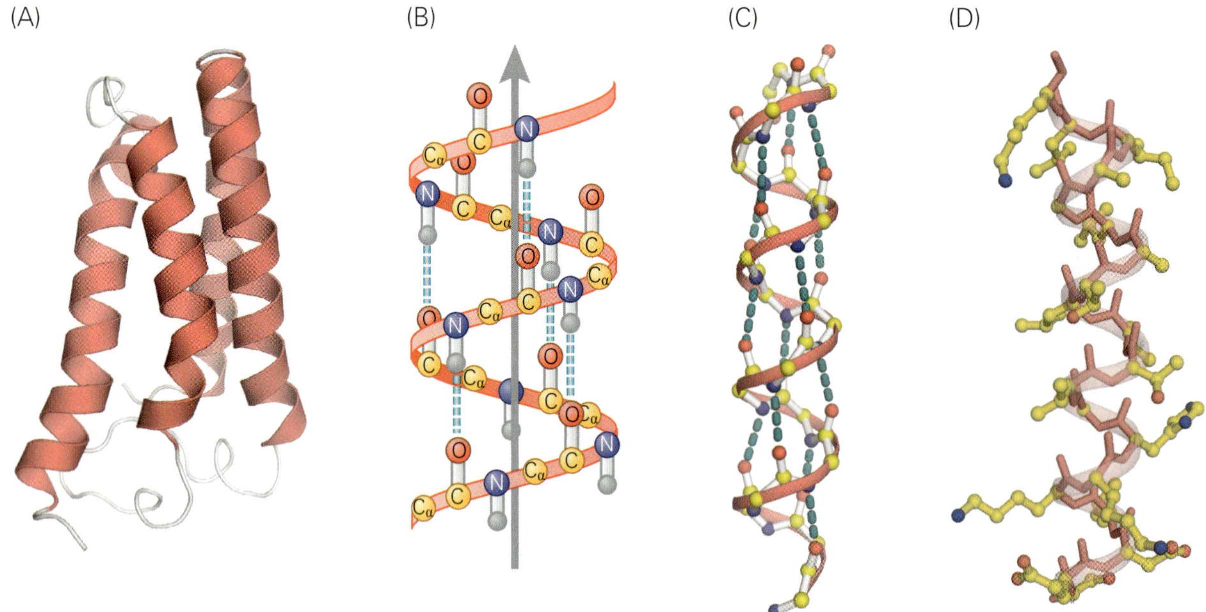


**Figure 1.36 The α helix is one of the major structural elements in proteins.** (A) Ribbon diagram showing the path of the peptide backbone in a protein containing several α helices. (B) Schematic diagram of the backbone of an α helix. Sidechains are not shown, and hydrogen bonds are in *cyan*. The arrow indicates the direction of the helix, from the N-terminus to the C-terminus. (C) Structure of the backbone of an actual α helix. (D) An α helix, showing the sidechains. (Adapted from C. Brändén and J. Tooze, Introduction to Protein Structure, 2nd ed. New York: Garland Science, 1999; PDB codes: A, 2CCY; D, 1MBC.)

This problem is solved in a very elegant way by the formation of two kinds of structural elements that accommodate the hydrogen-bonding requirements of the backbone. These structural elements are **α helices** and **β sheets**. The folded structures of most proteins are formed by specific arrangements of α helices and β sheets, as we discuss in detail in Chapter 4.

The α helix, illustrated in **Figure 1.36**, is a right-handed helical structure adopted by the protein backbone, which is enforced by the formation of hydrogen bonds between the C=O group of one residue and the NH group of another residue four positions ahead of it in the chain (see Figure 1.36). Thus, all NH and C=O groups are joined with hydrogen bonds, except the first three NH groups and the last three C=O groups at the ends of the α helix. Because backbone hydrogen bonds are satisfied, an α helix that bears hydrophobic sidechains can pass through the interior of the protein without an energetic penalty. α helices vary considerably in length, ranging from four or five to over 40 amino acid residues in globular proteins. The average length is around 10 residues, corresponding to three helical turns.

The second major structural element found in globular proteins is the β sheet, which is formed by two or more β strands. The β sheet is formed through interactions between noncontiguous amino acids in the polypeptide chain, in contrast to the α helix, which is formed from one contiguous segment of the chain. β strands are usually from five to 10 residues long and are in an almost fully extended conformation. To satisfy their hydrogen-bonding capacity, they form sheets in which β strands are aligned adjacent to each other (**Figure 1.37** and **Figure 1.38**) such that hydrogen bonds can form between C=O groups and NH groups across the strands. The β sheets that are formed from several such β strands are "pleated", with $C_\alpha$ atoms alternately a little above and below the plane of the β sheet. The sidechains follow this pattern, pointing alternately above and below the β sheet.

The orientations of β strands within a sheet can be antiparallel, as shown in Figure 1.37, or parallel (see Figure 1.38). Antiparallel β sheets have narrowly spaced hydrogen bonds that alternate with widely spaced ones. Parallel sheets have

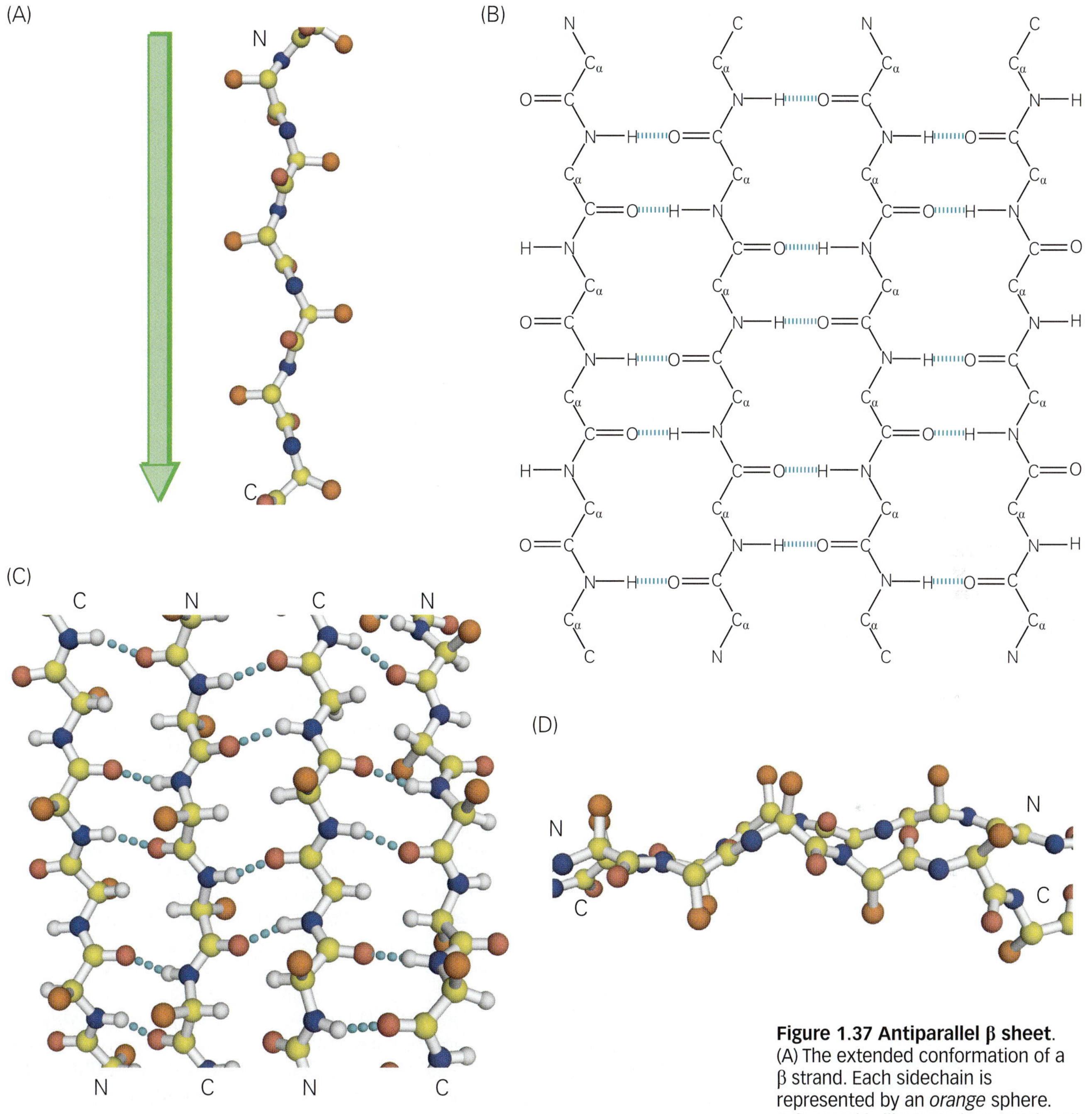

**Figure 1.37 Antiparallel β sheet.** (A) The extended conformation of a β strand. Each sidechain is represented by an *orange* sphere. A β strand is illustrated conventionally as an arrow pointing from the N- to C-terminus. (B) The hydrogen-bond pattern in an antiparallel β sheet. The sidechains are not shown. (C) The structure of an actual β sheet. The orientation of the β strands is orthogonal to that in (A). (D) Side view of the pleat of a β sheet. Two antiparallel β strands are viewed from the plane of the β sheet. Note that the orientations of the sidechains follow the pleat. (Adapted from C. Brändén and J. Tooze, Introduction to Protein Structure, 2nd ed. New York: Garland Science, 1999.)

evenly spaced hydrogen bonds that bridge the β strands at an angle. Within both types of β sheets, all possible backbone hydrogen bonds are formed, except in the two outside strands of β sheet, which have only one neighboring β strand each, but are exposed and so can form hydrogen bonds with water.

β strands can also combine into mixed β sheets with some β-strand pairs parallel and some antiparallel. There appears, however, to be a strong bias in nature against mixed β sheets. Finally, the plane of almost all β sheets is itself twisted. This twist is always right-handed, as shown in **Figure 1.39**.

Most protein structures, such as the one shown in Figure 1.39, contain specific arrangements of α helices and β sheets that are connected by loop regions of various lengths and irregular shapes. The loop regions are typically at the surface of

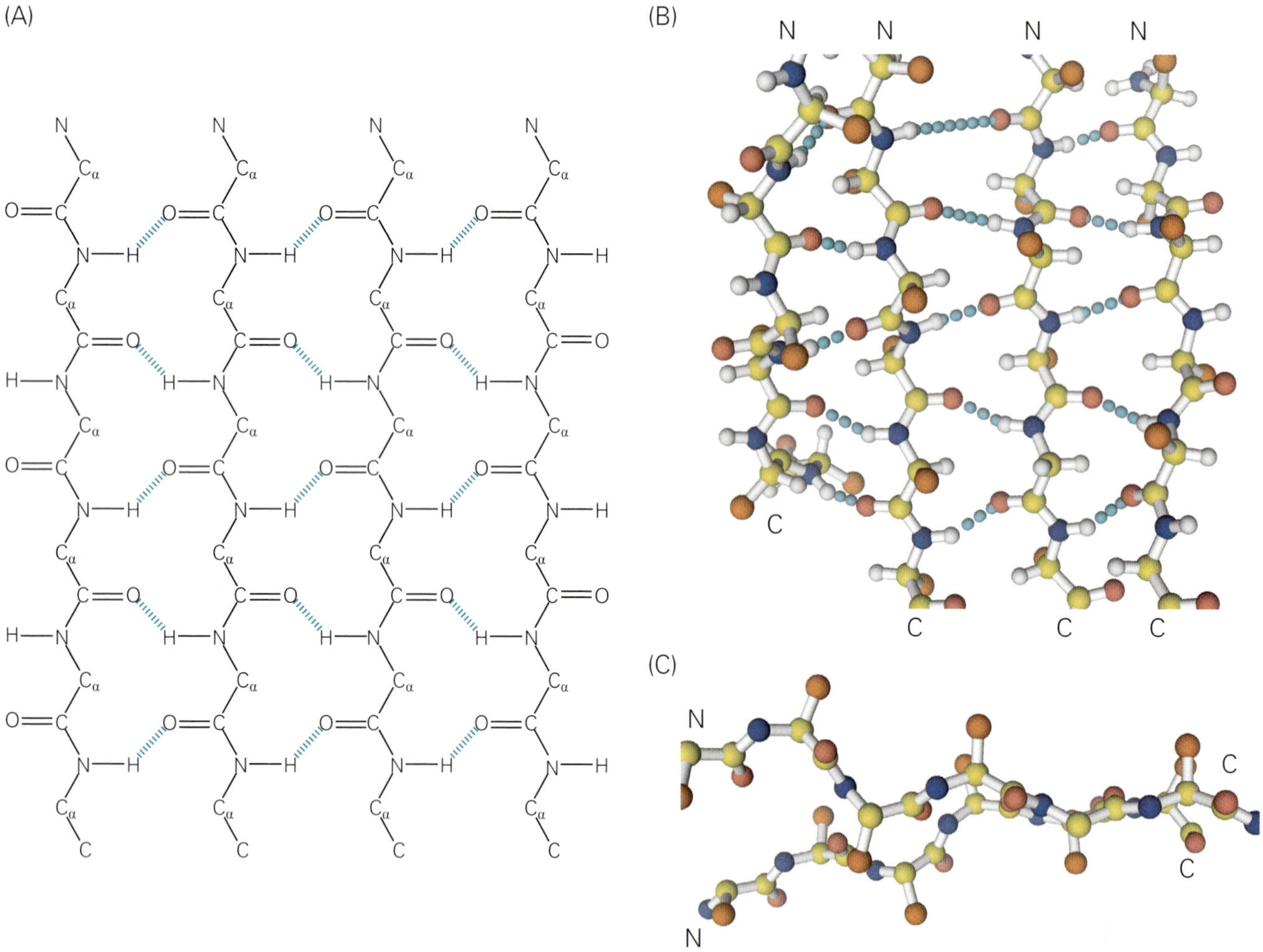

**Figure 1.38 Parallel β sheet.** (A) The hydrogen-bond pattern in a parallel β sheet. (B) Structure of an actual parallel β sheet. Sidechains are represented by *orange* spheres. (C) Side view of the pleat of a parallel β sheet, as seen from within the plane of the sheet. (Adapted from C. Brändén and J. Tooze, Introduction to Protein Structure, 2nd ed. New York: Garland Science, 1999.)

the molecule. The backbone C=O and NH groups of these loop regions, which in general do not form hydrogen bonds to each other, are exposed to the solvent and can form hydrogen bonds to water molecules. As we shall discuss in Chapters 4 and 5, the loop regions provide the diversity of function in proteins.

**Figure 1.39 The right-handed twist of β sheets.** In the illustration on the left, the *E. coli* protein thioredoxin is shown, with the β strands drawn as arrows. The right-handed twist of the β sheet can be appreciated by comparing it to a right hand. If the thumb of the hand is aligned along one of the strands, then the other strands are seen to curl in the directions of the fingers of the right hand. (PDB code: 2TRX.)

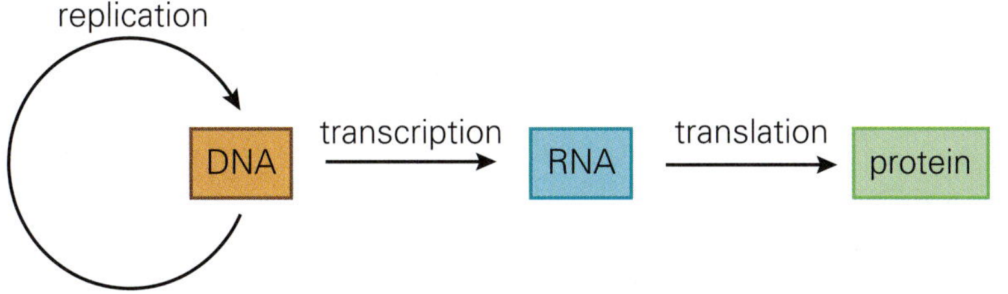


**Figure 1.40 The flow of genetic information.** In all living cells, information is stored and transmitted to progeny in the sequence of nucleotides in DNA. The processes of transcription and translation convert this information into the sequences of nucleotides in RNA and amino acids in proteins.

# C. REPLICATION, TRANSCRIPTION, AND TRANSLATION

We shall complete our survey of DNA, RNA, and proteins by describing, briefly, the processes of replication, transcription, and translation. An organism's genome contains vast stores of information that the cell must access in order to maintain itself and build its operational molecules, the proteins and RNAs that carry out specific functions. The flow of genetic information is described by the **central dogma of molecular biology**, which states that information flows from DNA into RNA and then into protein (**Figure 1.40**). DNA stores the master copy of genetic information, RNAs represent working copies of that information, and the proteins (sometimes along with RNA) make up the machines and many of the structural elements of the cell. While this might seem obvious to us today, it is remarkable that the central dogma is a universal fact of life on Earth: it describes how every known living cell functions.

**Replication, transcription, and translation**

Replication, transcription, and translation are, respectively, the processes by which the nucleotide sequence of DNA is duplicated, copied into RNA sequences, and used to synthesize protein molecules.

The transmission of genetic information and its transfer into RNA and then proteins occurs through the processes of DNA **replication** and **transcription** and RNA **translation**. These processes are illustrated schematically for a eukaryotic cell in **Figure 1.41**. DNA replication is the process by which identical (or nearly identical) copies are made of existing DNA in preparation for cell division. The replication of DNA preceding cell division results in one new copy of the entire genome. Transcription is the process by which information in a strand of DNA is copied into a complementary strand of RNA, and occurs over small discrete regions of DNA comprising specific genes. Several different sorts of RNA are made. The type that serves as a template for synthesizing protein is called messenger RNA or mRNA. Translation is the process by which mRNA directs the assembly of amino acids into protein. This occurs in the ribosome, a huge macromolecular complex composed of RNAs and proteins.

## 1.16 DNA replication is a complex process involving many protein machines

Before a cell divides, it needs to replicate (duplicate) its DNA so that each of its two daughter cells can receive one complete copy of the genome. Because the strands in a DNA helix are complementary, each parent strand can serve as the template on which a new strand is synthesized (**Figure 1.42**). This idea sounds simple enough, and Watson and Crick alluded to it in a single sentence in their famous paper in 1953 (Box 1.1).

It turns out, however, that even for small DNA molecules, replication is a complex, multistep process that involves many different proteins working together. In Chapter 19, we shall discuss one aspect of DNA replication in detail, concerning the mechanisms by which DNA polymerases achieve high fidelity in copying the sequence of DNA. Here we provide a simplified overview of the process.

DNA replication is particularly complicated in eukaryotic cells, which have so much DNA that it is compacted and packaged with proteins into structures called chromatin. In its most open conformation, chromatin resembles beads on a string: helical DNA is wrapped tightly twice around a protein core (there are

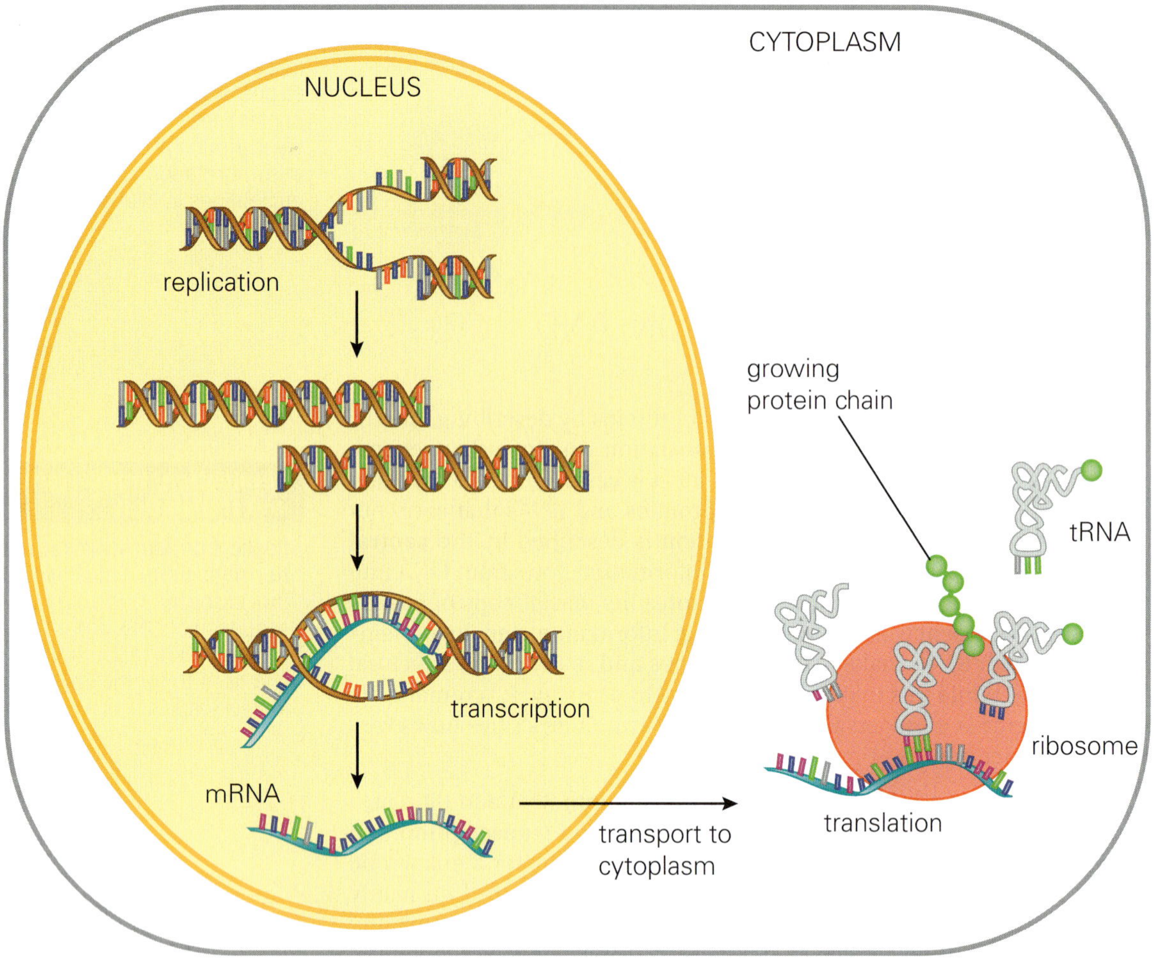

CYTOPLASM

NUCLEUS

replication

growing protein chain

tRNA

transcription

ribosome

mRNA

translation

transport to cytoplasm

**Figure 1.41 Replication, transcription, and translation in a eukaryotic cell**. A double-stranded DNA molecule is shown at the top within the nucleus. Replication of the parent DNA yields two daughter DNA molecules. DNA is transcribed into mRNA, which is then transported out of the nucleus into the cytoplasm, where it is translated into protein (represented by a string of *green* beads) by the ribosome. Adapter molecules known as transfer RNAs (tRNAs) are essential for the conversion of the sequence of nucleotides in mRNA into the sequence of amino acids in proteins.

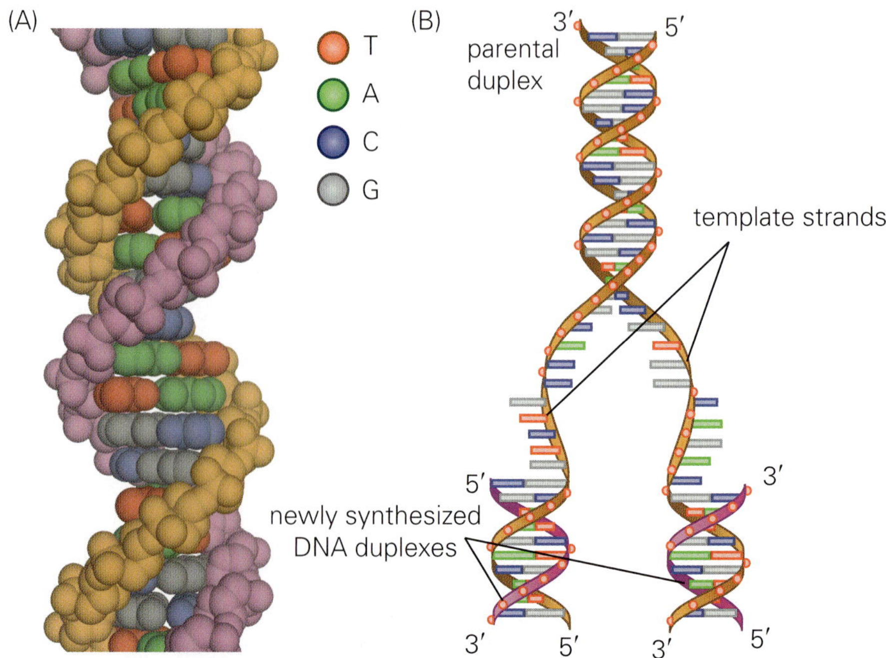


**Figure 1.42 Replication relies on the complementarity of DNA strands**. (A) The four bases are shown in distinct colors to emphasize the complementarity between the two DNA strands. (B) The parental DNA duplex is opened up during replication, and a new strand is synthesized on each parent strand, thus generating two new molecules of DNA.

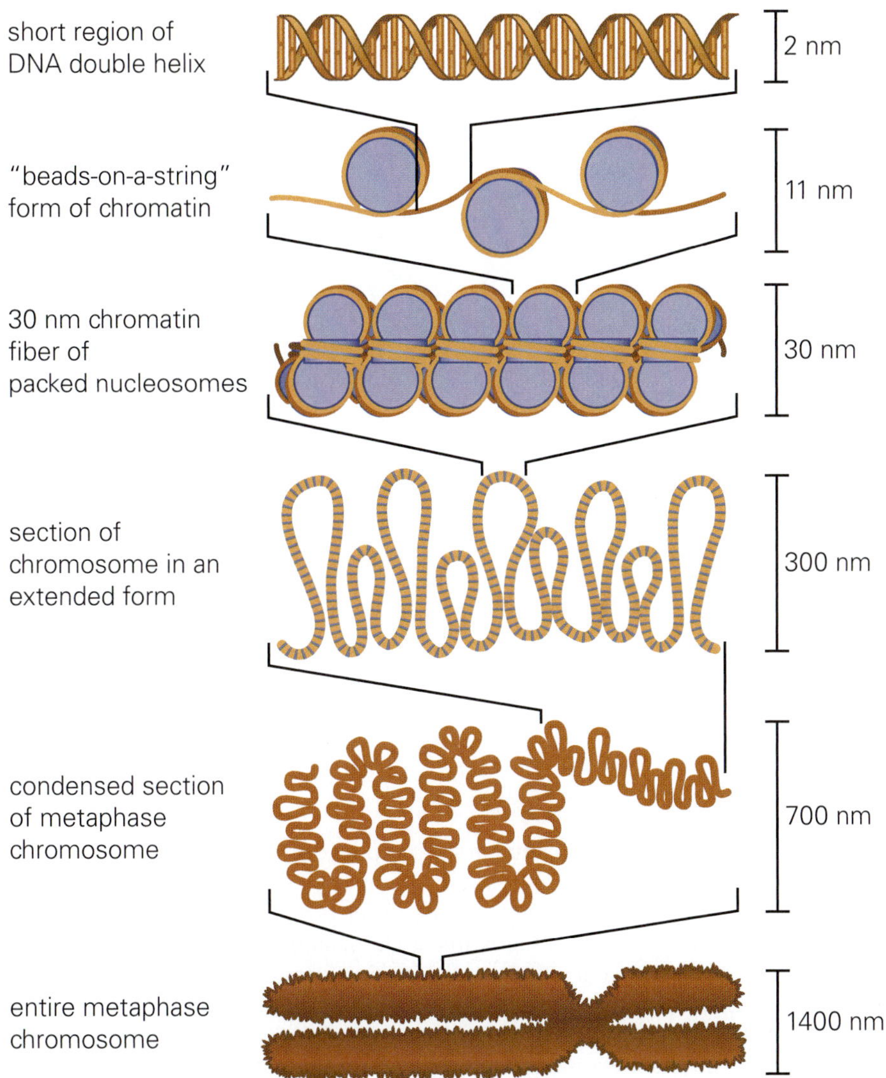


**Figure 1.43 Eukaryotic cells package DNA into chromatin.** The repeating units within chromatin—nucleosomes—consist of DNA wrapped tightly around protein cores, resembling beads on a string. Multiple rounds of further compaction eventually result in the structures of chromosomes that are visible in light microscopy. (Adapted from B. Alberts et al., Molecular Biology of the Cell, 5th ed. New York: Garland Science, 2008.)

about 146 base pairs per core), forming **nucleosomes** (**Figure 1.43**). Nucleosomal DNA is further compacted into increasingly higher-order structures—eventually achieving about a 10,000-fold contraction of the length.

For replication to occur, the DNA must be unpackaged and unwound. Large protein complexes assemble on the DNA at specific sequences known as origins of replication, stimulating local unwinding, and enabling motor proteins known as DNA helicases to prise open the DNA strands, thus forming the hallmark Y-shaped replication forks that are the sites on DNA where replication is taking place (**Figure 1.44**).

It is a quirk of evolution that the DNA polymerases that replicate DNA can only extend an existing DNA or RNA chain, but cannot create a new chain. Because of this limitation, the first step in the synthesis of a new copy of a DNA chain is the generation of a short piece of RNA by an RNA polymerase (see Figure 1.44). This short RNA is called a primer, because it primes the action of DNA polymerase, which elongates the RNA to make a new DNA chain. The primer RNA is later removed and replaced with DNA.

Another complication is that DNA synthesis proceeds only in the 5′-to-3′ direction (see Section 1.7), and this has consequences for replication. The template DNA strands are oriented in opposite directions, but for both new strands (at each replication fork), the overall direction of strand growth must be in the direction

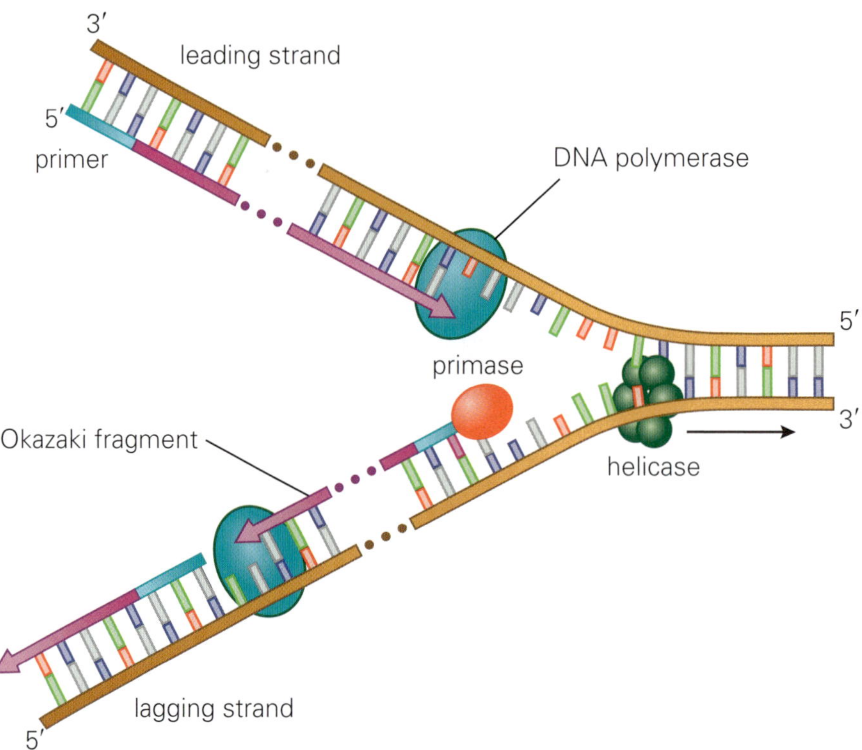


**Figure 1.44 A replication fork.** A DNA helicase opens up the DNA, generating a replication fork. The primase (small *red* sphere) synthesizes short stretches of complementary RNA on a single-stranded DNA template. DNA polymerases (large *blue* spheres) add nucleotides to the 3′ ends of these RNA primers. Synthesis proceeds only in the 5′-to-3′ direction, so the DNA polymerase replicates the leading strand continuously. Lagging-strand replication is discontinuous, with Okazaki fragments being formed that are later linked together. Each Okazaki fragment is hundreds of nucleotides long.

the fork is moving. The result is that at each replication fork one new strand—the leading strand—can be synthesized continuously just behind the moving fork, but synthesis of the lagging strand, on the other template, must be discontinuous. As a result, discontinuous segments of DNA, called **Okazaki fragments**, are synthesized one fragment at a time on the lagging strand template. Each Okazaki fragment originates from a separate RNA primer. Thus, DNA polymerase need only initiate DNA synthesis once on the leading strand, but must do so repeatedly on the lagging strands. Eventually, the Okazaki fragments are linked seamlessly together by enzymes that remove the RNA primers and fill in the missing segments of DNA.

## 1.17 Transcription generates RNAs whose sequences are dictated by the sequence of nucleotides in genes

The basic unit of transcription is the gene, a segment of DNA that codes for RNA or protein. Gene expression is the process of making the RNA or protein product for which a gene codes. The sequence of nucleotides in the DNA specifies the sequence of RNA, produced by RNA polymerase. Many RNAs produced from genes are the end product of transcription by themselves, and are not translated into protein. Genes that encode proteins produce messenger RNA, and the sequence of nucleotides in the mRNA specifies the sequence of the amino acids that are assembled into protein on the ribosome.

In the transcription of DNA into RNA, as in DNA replication, Watson-Crick base pairing is the basis for the selection of nucleotides to add to the growing chain. As in replication, nucleotides are added to the growing strand in a 5′-to-3′ direction, in this case by RNA polymerase. One difference between the two processes is that only one of the two strands of DNA is copied during transcription. This reflects the fundamental difference in the functions of replication and transcription. Whereas DNA replication must produce a single complete copy of each strand of the DNA double helix once in each division cycle of the cell, the function of transcription is to allow specific proteins or RNAs to be produced in appropriate quantities when they are needed. For single-celled organisms, this will vary according to growth conditions; for multicellular organisms, it will depend upon cell type and physiological conditions; and for both, it will vary according to the phase of the cell division cycle. Transcription thus produces many RNA copies of discrete segments of the genome from just one strand of the DNA, and is subject to regulation, determining which genes are transcribed according to conditions and cell type.

In all organisms, the overall transcription process is orchestrated by regulatory proteins (transcription factors) that determine where RNA polymerases bind to DNA, and thus which genes are transcribed. RNA polymerases bind to DNA at a specific DNA sequence called the promoter, adjacent to the coding sequence of each gene (**Figure 1.45**). Transcription factors bind close by and recruit the RNA polymerase to the DNA, where it can begin to transcribe the DNA into RNA. The process of transcription elongation continues until the RNA polymerase encounters a transcription termination signal, which sometimes takes the form of a hairpin structure formed by the newly synthesized RNA chain. This causes the completed RNA transcript, the RNA polymerase, and the DNA to dissociate from each other.

## 1.18 Splicing of RNA in eukaryotic cells can generate a diversity of RNAs from a single gene

The nucleotide sequences of messenger RNAs specify the linear sequence of amino acids in protein. In the mRNAs of bacteria, the nucleotides specifying the sequence of an entire protein form a contiguous block. In eukaryotes, however, protein-coding regions of genes (known as **exons**) are interrupted by noncoding regions (known as **introns**). When these genes are transcribed, the primary RNA transcripts, known as precursor mRNAs or **pre-mRNA**, contain both the exons and introns. The RNAs are then processed in a reaction known as **RNA splicing** in which the introns are cleaved from the primary transcript and the exon sequences are ligated together (**Figure 1.46**).

**Exons and introns**

The regions of eukaryotic genes that code for functional RNA or proteins are known as exons. These are interrupted by noncoding regions known as introns. The splicing process removes introns and joins the exons to produce messenger RNA.

The splicing of precursor mRNAs occurs via a two-step transesterification reaction that is carried out by the spliceosome, a large complex composed of proteins and RNAs. The first step is a nucleophilic attack by the 2′ hydroxyl of an adenosine (A) within the intron on the 5′-terminal phosphodiester linkage of the intron (indicated in *green* in Figure 1.46). In the second step, the 3′ hydroxyl of the upstream exon attacks the phosphodiester linkage at the 3′ end of the intron. This results in the fusion of the two exons.

Alternative splicing allows several different mRNAs to be generated from a single pre-mRNA, as shown in **Figure 1.47**. The processing of the pre-mRNA can result in exons being removed from it, as well as introns, so that a different selection of exons can be included in mRNAs from a single gene. This means that many different proteins can be encoded by a single gene.

Interrupted genes and the potential for alternative splicing provides a basis for generating protein diversity in evolution. As we shall see in Chapter 5, proteins are constructed in a modular fashion from structural units known as domains. These domains have different functions, and are usually encoded in different exons or combinations of exons. Alternative splicing allows these modules to be combined in different ways, allowing natural selection to choose combinations with improved function.

## 1.19 The genetic code relates triplets of nucleotides in a gene sequence to each amino acid in a protein sequence

Translation is the process whereby the nucleotide sequence of the messenger RNA is read into the amino acid sequence of a protein. The linear sequence of nucleotides in the mRNA determines the linear sequence of amino acids in a

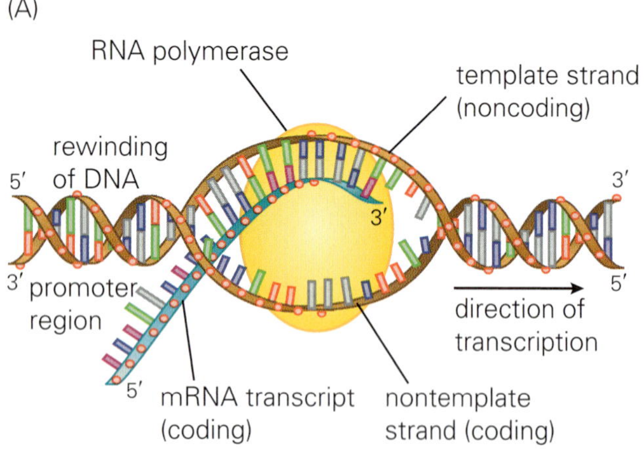


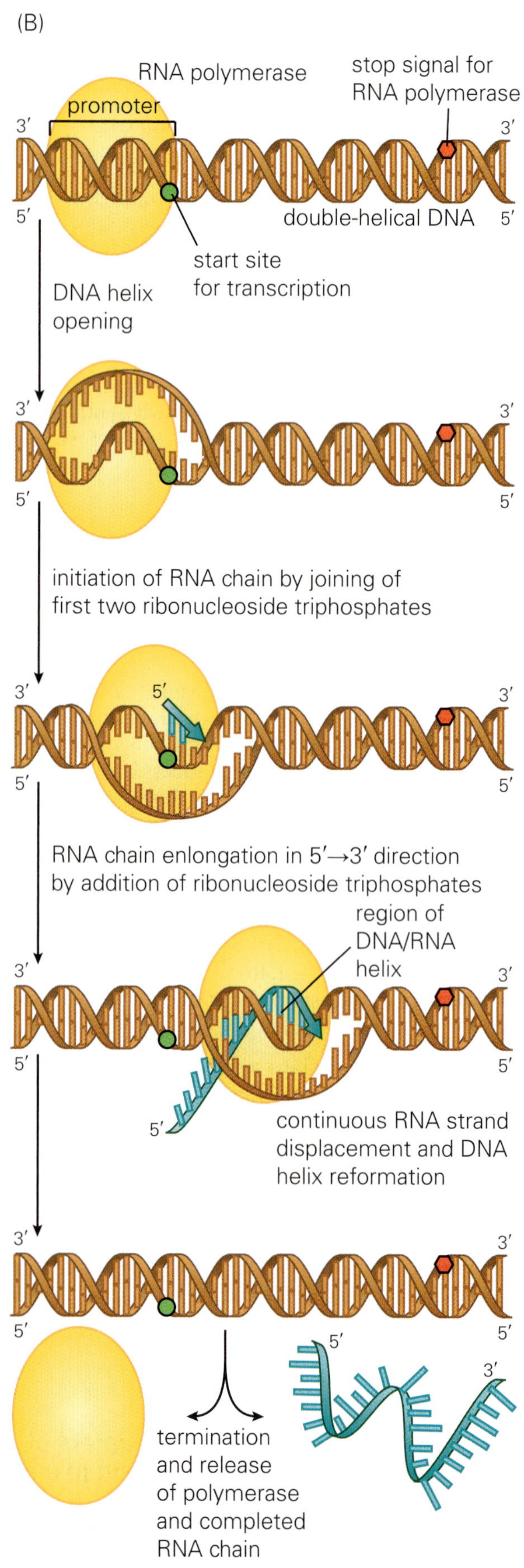


**Figure 1.45 The process of transcription.** (A) Schematic view of RNA polymerase transcribing DNA into RNA. The promoter is the DNA region to which the RNA polymerase enzyme is recruited by transcription factors. (B) Steps in a complete transcription cycle, which is initiated at a specific sequence in the DNA and continues until a sequence that signals termination.

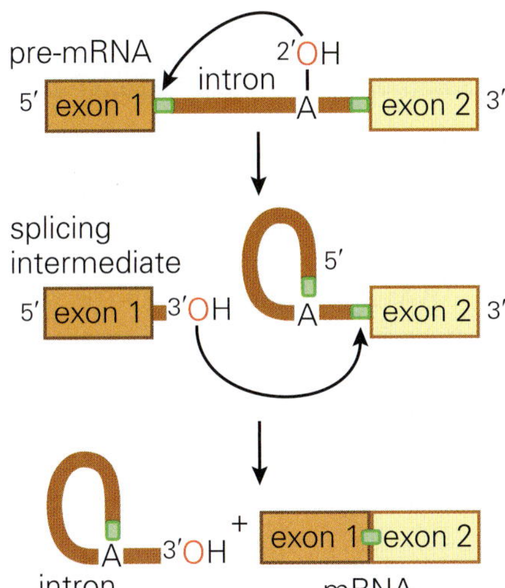


**Figure 1.46 Pre-mRNA splicing.** A schematic diagram of pre-mRNA is shown as two protein-coding exons (boxes) flanking a single noncoding intron (line). A large RNA–protein machine, the spliceosome (not shown here), catalyzes the removal of the introns and splices together the exons.

protein through the **genetic code**—a mapping rule that matches the sequence of three consecutive nucleotides in the mRNA, known as a **codon**, to a single type of amino acid. By the rules of this code, three consecutive ribonucleotides constitute one codon, and each codon is specific for a given amino acid. For example, the mRNA sequence 5′-UUA-3′ codes for the amino acid leucine (**Figure 1.48**). The code is unpunctuated, in the sense that there are no gaps between codons, and it is non-overlapping. The genetic code was elucidated by the combined insights of several scientists, including Robert Holley, Har Gobind Khorana, and Marshall Nirenberg, who were awarded the Nobel Prize in 1968 for their discoveries concerning the relationship between gene sequences and protein synthesis.

Given that there are four bases in RNA (A, C, G, and U), there are $4^3 = 64$ possible codons. Since only 20 amino acids generally occur in proteins, this means that some must be encoded by more than one codon. In fact, as you can see from Figure 1.48, most amino acids are encoded by two to four different codons. Sixty-one of the 64 possible codons correspond to amino acids, and three (UAA, UGA, and UAG) are stop codons: they trigger the termination of translation of the mRNA by the ribosome. One codon (AUG) doubles as an amino acid codon (for methionine, Met) as well as the start codon—that is, translation of an mRNA by the ribosome is initiated at this codon.

**Codon**

The sequence of messenger RNA is read one codon at a time by the ribosome during the synthesis of proteins. Each codon consists of three consecutive nucleotides that are specific for a given amino acid. There are 64 possible codons, three of which encode stop signals. The mapping of codons to amino acids is called the genetic code.

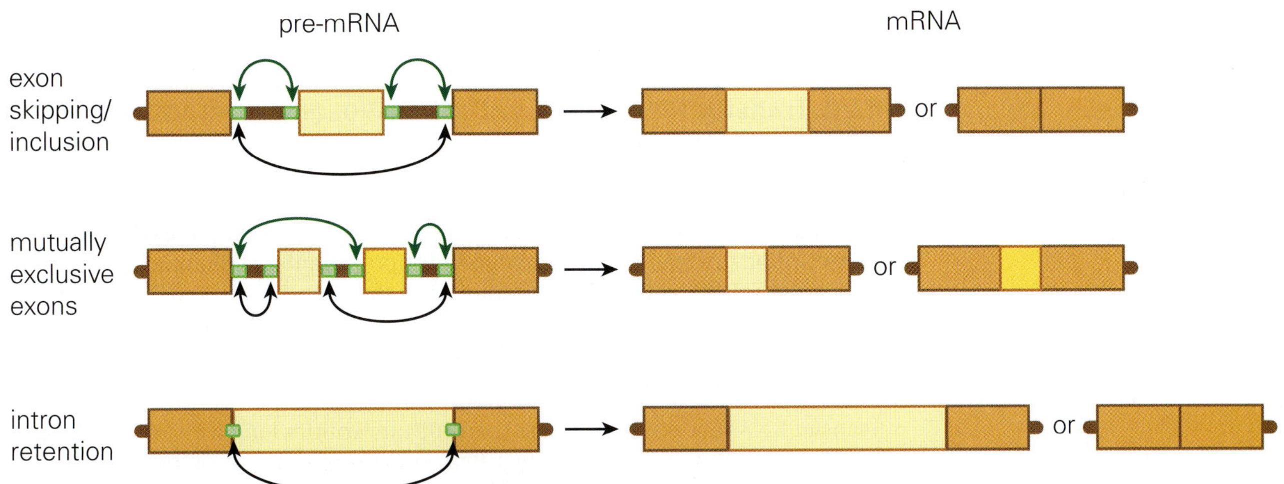


**Figure 1.47 Some modes of alternative RNA splicing.** In each diagram, alternative splicing options are indicated by *green* and *black* arrows. (Adapted from L. Cartegni, S.L. Chew, and A.R. Krainer, *Nat. Rev. Genet.* 3: 285–298, 2002. With permission from Macmillan Publishers Ltd.)

**Figure 1.48 The genetic code.** Each codon corresponds to only one amino acid. Thus, the genetic code is unambiguous. The code, however, is redundant (degenerate) and synonymous codons specify the same amino acid. The first two nucleotides of a codon are sometimes enough to specify a given amino acid. Codons with similar sequences often specify similar amino acids. Only 61 of the 64 possible codons specify amino acids and the other three—UAA, UGA, and UAG—instruct RNA polymerase to stop and are known as stop codons. The methionine codon (AUG) specifies the initiation site for protein synthesis on the ribosome.

second position

| first position (5′ end) | U | C | A | G | third position (3′ end) |
|---|---|---|---|---|---|
| U | UUU Phe F | UCU Ser S | UAU Tyr Y | UGU Cys C | U |
| U | UUC Phe F | UCC Ser S | UAC Tyr Y | UGC Cys C | C |
| U | UUA Leu L | UCA Ser S | UAA Stop | UGA Stop | A |
| U | UUG Leu L | UCG Ser S | UAG Stop | UGG Trp W | G |
| C | CUU Leu L | CCU Pro P | CAU His H | CGU Arg R | U |
| C | CUC Leu L | CCC Pro P | CAC His H | CGC Arg R | C |
| C | CUA Leu L | CCA Pro P | CAA Gln Q | CGA Arg R | A |
| C | CUG Leu L | CCG Pro P | CAG Gln Q | CGG Arg R | G |
| A | AUU Ile I | ACU Thr T | AAU Asn N | AGU Ser S | U |
| A | AUC Ile I | ACC Thr T | AAC Asn N | AGC Ser S | C |
| A | AUA Ile I | ACA Thr T | AAA Lys K | AGA Arg R | A |
| A | AUG Met M | ACG Thr T | AAG Lys K | AGG Arg R | G |
| G | GUU Val V | GCU Ala A | GAU Asp D | GGU Gly G | U |
| G | GUC Val V | GCC Ala A | GAC Asp D | GGC Gly G | C |
| G | GUA Val V | GCA Ala A | GAA Glu E | GGA Gly G | A |
| G | GUG Val V | GCG Ala A | GAG Glu E | GGG Gly G | G |

Redundant codons typically differ in their third positions. For example, AGU and AGC both code for the amino acid serine. Notice that a single base change in a codon often results in a codon specifying the same amino acid as the original codon, or for one with similar biochemical properties, so that the structure and function of the protein that is made may remain relatively unchanged. This feature of the genetic code limits the effect of mutations that are introduced by errors in replication or transcription. For example, the codons CUU and CUC both encode leucine, and so a mutation at the third position of this codon that replaces U by C does not change the sequence of the protein. Such a mutation is called a **silent mutation**.

## 1.20 Transfer RNAs work with the ribosome to translate mRNA sequences into proteins

Because there is no chemical relationship between the mRNA codon and the amino acid specified by it, translation is not a simple template-based process like transcription. Instead, translation depends upon adapters that match each amino acid to its mRNA codon. The adapters are **transfer RNAs** or **tRNAs**, so-called because they transfer amino acids to the growing polypeptide in the ribosome. Insights regarding the adapter mechanism by which mRNA is translated into protein came especially from Francis Crick.

Transfer RNAs are small RNAs less than 100 nucleotides long with an **anticodon**—a sequence complementary to the mRNA codon—at one part of the molecule and an attachment site for an amino acid at the other end (**Figure 1.49**). The tRNAs are recognized specifically by enzymes known as aminoacyl-tRNA synthetases that link the appropriate amino acid to the tRNA that contains the triplet anticodon for that amino acid: the tRNA is then said to be charged.

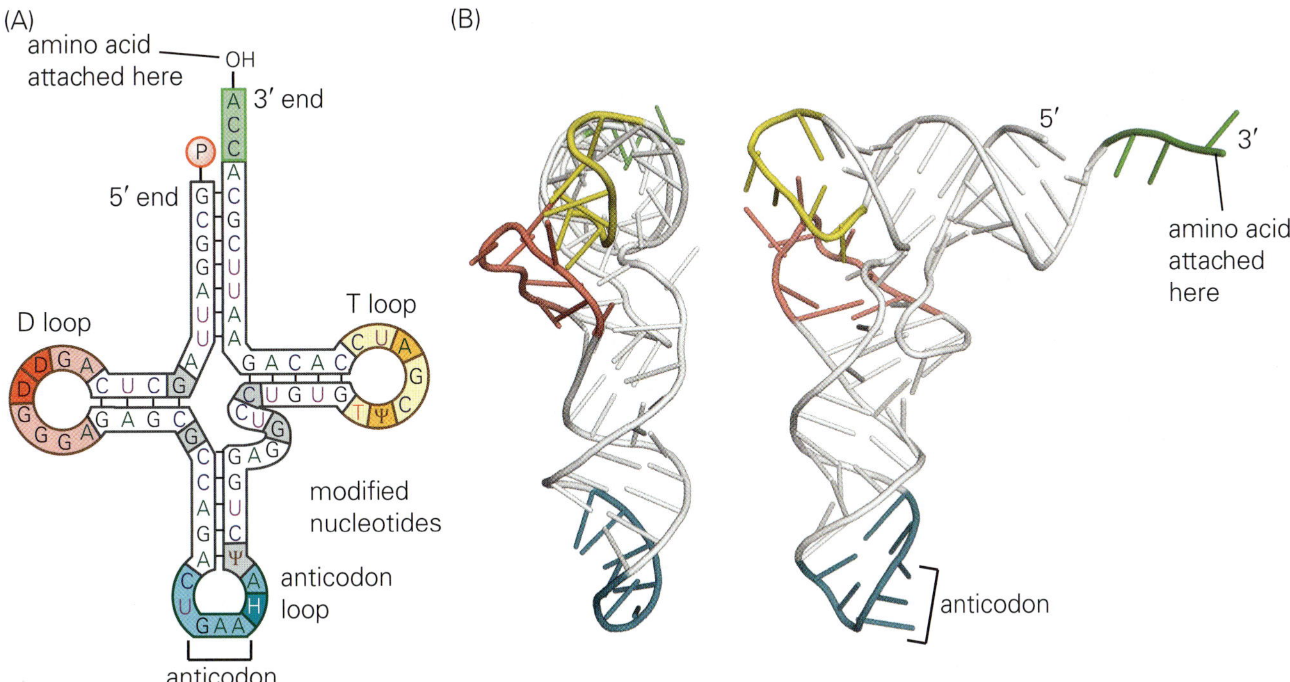


**Figure 1.49 Transfer RNA (tRNA).** A tRNA specific for phenylalanine is shown, but all tRNAs are similarly constructed. (A) Schematic showing the cloverleaf structure, dictated by base-pairing interactions between different parts of the RNA chain. Base pairs in double-helical segments are indicated by lines. The symbol ψ denotes a modified nucleotide, as explained in Chapter 2. (B) Three-dimensional structure of a tRNA. (Adapted from B. Alberts et al., Molecular Biology of the Cell, 5th ed. New York: Garland Science, 2008.)

Once the synthetase has attached an amino acid to its appropriate tRNA molecule (forming aminoacyl-tRNA), the anticodon in tRNA can base pair with the complementary sequence in the mRNA bound to the ribosome. We will discuss ribosomes in some detail in Chapter 19, as part of a discussion on the fidelity of translation.

The ribosome consists of two separate subunits that assemble around mRNA to form the intact ribosome. Protein synthesis is initiated when the start codon (AUG, which also codes for Met) on the mRNA is correctly positioned in a site in the ribosome called the peptidyl-tRNA (P) site and is paired with the tRNA with an anticodon complementary to AUG (the initiator tRNA). Protein synthesis starts with the binding of a tRNA carrying a specific amino acid to the A (for aminoacyl-tRNA) site of the ribosome. The ribosome then catalyzes the formation of a new peptide bond linking the carboxyl group of the methionine to the amino group of the newly arrived amino acid. A series of conformational changes in the ribosome then translocates the ribosome one codon further along the mRNA, thus shifting the most recently added amino acid (still attached to its tRNA) to the P site. Finally, the empty "spent" tRNA from the previously added amino acid is released from the E site.

Successive amino acids are added to the growing peptide chain according to the sequence specified by the mRNA, accompanied by the movement of the ribosome along the mRNA (**Figure 1.50**). Termination of protein synthesis occurs when the A site reaches a stop codon, where the ribosome pauses and the polypeptide chain is released from the ribosome, the mRNA is also released, and finally the ribosomal subunits dissociate from each other. The subunits are then free to reassemble on mRNA at another site and start another cycle of translation initiation, elongation, and termination.

## 1.21 The mechanism for the transfer of genetic information is highly conserved

The ribosome is at the center of the translation machinery in all organisms, and the genes encoding ribosomal RNAs display an extraordinarily high degree of sequence conservation. This evolutionary conservation is reflected, in turn, in the

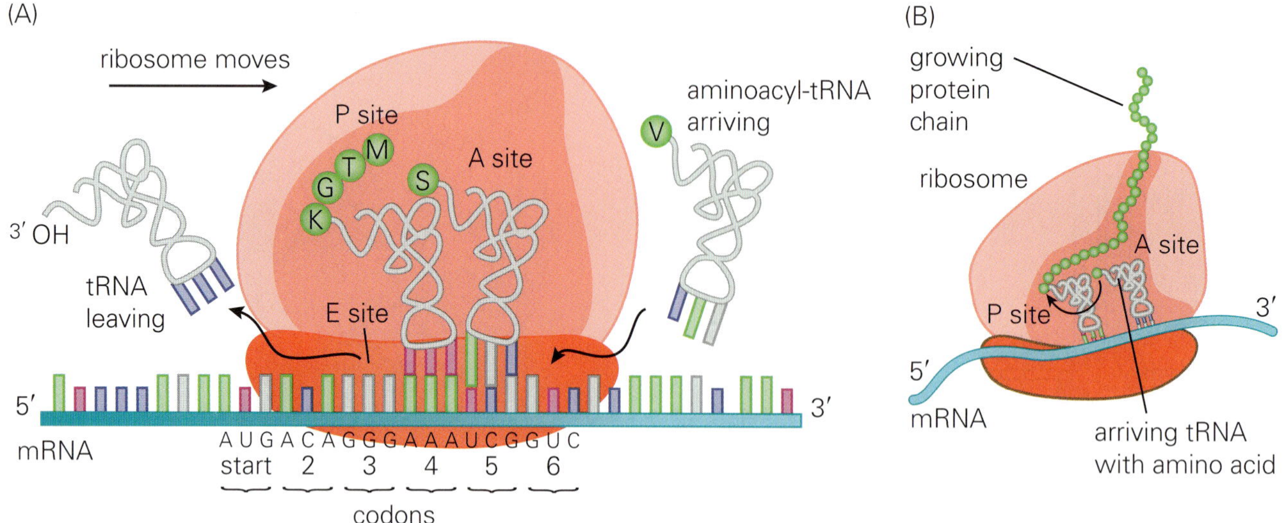


**Figure 1.50 Ribosomes synthesize proteins.** Ribosomes consists of two subunits, shown here in different shades of *red*. (A) Schematic diagram indicating how mRNA and tRNAs move through the ribosome. (B) Simplified mechanism of protein synthesis within the ribosome. The amino group of the amino acid attached to the A-site tRNA acts as a nucleophile and attacks the carbonyl carbon of the C-terminal amino acid P-site-bound tRNA peptide ester. As a result, the peptide chain is attached to the tRNA in the A site and the new amino acid is added to the C-terminal end of the polypeptide chain. This process is described in more detail in Chapter 19.

similarities in the structure of the ribosomes themselves in all organisms. Because of this, the phylogenetic analysis of ribosomal RNAs forms the basis of the division of living organisms into the domains Bacteria, Eukarya, and Archaea.

The fundamental mechanism of protein synthesis is highly conserved throughout the three domains of life. A similar evolutionary stability is seen in the majority of the ribosomal proteins, other proteins involved in assisting or regulating protein synthesis, the tRNAs, and the aminoacyl-tRNA synthetases. There are other genes that have similar phylogenetic patterns to those of ribosomal RNA. Most such genes encode proteins and RNAs that are involved with the transfer of genetic information, such as in DNA replication and in DNA-to-RNA transcription. Products encoded by other highly conserved genes produce proteins that directly interact with the ribosome. Together, these findings demonstrate that the fundamental features of these reactions were established early in evolution.

The genetic code is essentially identical in all organisms, with very slight variations in some lower organisms and in mitochondrial genes. This means that the same codons are used to specify the same amino acids in all organisms. This is a profound observation about the nature of life, and points to a common evolutionary origin for all living things.

## 1.22 The discovery of retroviruses showed that information stored in RNA can be transferred to DNA

The central dogma of molecular biology, which states that information is transferred from DNA to RNA to proteins (see Figure 1.40), has to be extended to include the transfer of information from RNA to DNA. Many viruses demonstrate that information stored in RNA can be converted back to DNA. **Retroviruses**, for example, store genetic information in RNA, which is then copied into DNA when these viruses infect a suitable host cell (**Figure 1.51**).

Retroviruses were originally identified as the causative agents of tumors (the first retrovirus to be discovered was Rous sarcoma virus, which was isolated from tumors in chickens). The human immunodeficiency virus (HIV) is today's best known example. A retrovirus genome is a single-stranded linear RNA molecule, about 8000 to 11,000 nucleotides long, present within each virus particle in two copies. Each virus also contains about 50 copies of reverse transcriptase, an RNA-dependent DNA polymerase that transcribes RNA into DNA. After the viral

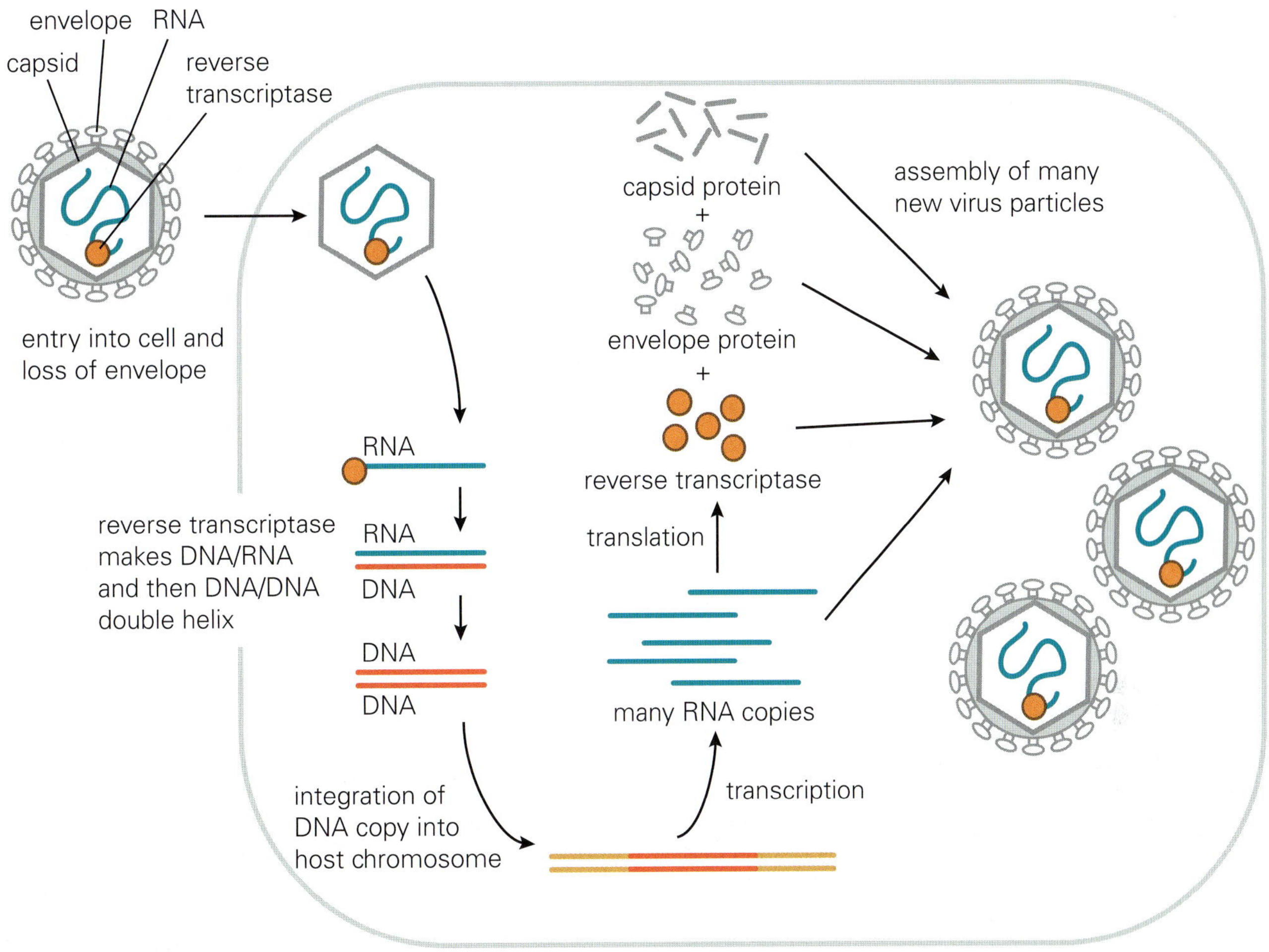

**Figure 1.51 Retrovirus life cycle.** A retrovirus stores its genetic information (typically ~10,000 nucleotides) in the form of RNA. Reverse transcriptase copies the RNA into DNA in a host cell, and an integrase inserts the viral DNA into the host genome. Host transcription and translation machinery then produces many more copies of the viral RNA and proteins. (Adapted from B. Alberts et al., Molecular Biology of the Cell, 5th ed. New York: Garland Science, 2008.)

capsid enters the cell, the reverse transcriptase transcribes the single-stranded RNA genome first into a double-stranded RNA-DNA hybrid, and then into a double-stranded DNA molecule. An integrase enzyme inserts this double-stranded DNA version of the viral genome into the host genome, from where it can direct the synthesis of many more viral RNAs (as well as viral proteins), and the reverse transcription cycle can begin again. Repeated insertions of the DNA version of the viral genome can be made into either another part of the same host genome or the genome of another cell.

In addition to the RNA-dependent DNA polymerase found in single-strand RNA retroviruses, there are also RNA-dependent RNA polymerases observed in a number of double-stranded RNA viruses and in some eukaryotic organisms. These polymerases allow replication of RNA into RNA without the need for a DNA intermediate (**Figure 1.52**).

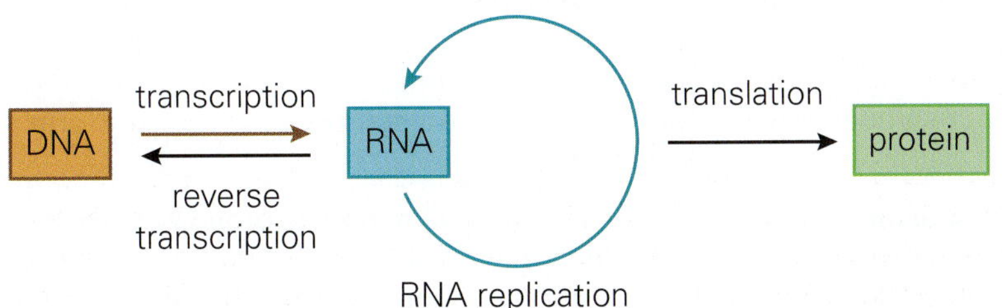


**Figure 1.52 Modifications to the central dogma.** RNA can act as a template for DNA synthesis through the action of a RNA-dependent DNA polymerase or reverse transcriptase (left arrow). RNA can also be directly synthesized using RNA as a template by RNA-dependent RNA polymerases (circular arrow).

A retroviral genome can be transcribed from RNA to its DNA equivalent and inserted into host DNA multiple times in multiple retro-transcription cycles. The extent to which this seems to have happened in evolution, and to have left permanent traces in modern cells, is remarkable. Genome sequencing of a number of different organisms (including humans) has revealed the existence of large numbers of genetic elements that appear to have been integrated into these genomes by a retro-transcription mechanism similar to that described above for retroviruses. These so-called retrotransposons are estimated to account for ~20% of the human genome—vastly more than the mere ~1.5% of the genome that codes for protein.

## Summary

As we noted in the introductory essay before this chapter, our purpose in this book is to explain how the structure and energetics of molecules are connected to biological function. We began this chapter by providing a qualitative description of the nature of interactions between atoms and molecules, as a prelude to a more quantitative analysis of this topic in Chapter 6. All of the molecules in a cell, including DNA, RNA, and proteins, interact with each other through noncovalent interactions. These include van der Waals interactions, ionic interactions, and hydrogen bonds. These interactions are very weak compared with covalent bonds, and they can be broken and made again at room temperature. The dynamic nature of intermolecular interactions is a key to life, without which DNA could not be replicated or specific structures formed by proteins and RNA.

DNA, RNA, and proteins are polymers of nucleotides (DNA and RNA) and amino acids (proteins). The order in which the four nucleotides or 20 amino acids occur in DNA, RNA, and proteins is known as the sequence of the polymer. All three are linear polymers with defined sequences. The sequences of these macromolecules determine their structures, which determine, in turn, their function.

Each nucleotide in DNA is composed of a 2′-deoxyribose sugar, a phosphate group, and a base (A, T, C, or G). DNA is synthesized by DNA polymerase in the 5′-to-3′ direction, with nucleotides in the chain connected by phosphodiester linkages. DNA forms a double-stranded helix and the two strands of DNA are complementary in sequence and antiparallel. The formation of the DNA double helix obeys specific base-pairing rules—namely, A pairs with T and G pairs with C. As a consequence, each strand of DNA can serve as a template for the synthesis of its complementary strand. This results in two new DNA molecules (daughter strands) that are identical in sequence to the two complementary strands of the parent DNA.

The nucleotides in RNA are structurally similar to those in DNA, except that the sugar is ribose rather than deoxyribose. The presence of the 2′-hydroxyl group makes RNA less stable chemically than DNA. A strand of RNA can form double helices with another RNA strand or with DNA. The same base-pairing rules apply, except that A pairs with U (which is found in RNA instead of T).

Proteins are polymers built up from 20 different amino acids linked together by peptide bonds. Proteins are synthesized in the N-terminal to C-terminal direction. The covalent backbone of the protein is formed by atoms that are common to all of the amino acids. In addition, the $C_\alpha$ atom of the backbone is attached to a sidechain that is unique. These sidechains can be hydrophobic (nonpolar), polar, or charged. When proteins fold into specific three-dimensional structures, they typically adopt two types of structural elements, known as $\alpha$ helices and $\beta$ sheets. In water-soluble proteins, these pack to form a hydrophobic core, the formation of which stabilizes the folded structure.

The flow of information in the cell is from DNA to RNA to proteins, with the genetic code specifying how triplet sequences of nucleotides in the DNA, called codons, correspond to amino acids in a protein chain. The genetic code is redundant, which reduces the effects of mutations. The genetic code and the processes of replication, transcription, and translation are conserved in all cells. The universality of these processes is reflected in the central dogma of molecular biology, which states that information in living systems flows from DNA to RNA to proteins. An extension to the central dogma is provided by some viruses that store their genetic information in the form of RNA that is transcribed into DNA by a specialized reverse transcriptase enzyme. These viruses depend, however, on the machinery of the cell to produce the proteins specified in the DNA copies of their RNA genomes. The conservation of these basic processes in all living systems suggests that all known life emerged from a single origin.

## Key Concepts

### A. INTERACTIONS BETWEEN MOLECULES

- The energy of interaction between two molecules is determined by noncovalent interactions, which are electrostatic in origin.
- Neutral atoms attract and repel each other at short distances through van der Waals interactions.
- Ionic interactions between charged atoms can be very strong, but are attenuated by water.
- Hydrogen bonds are very common in biological macromolecules and are a consequence of the polarization of covalent bonds.

### B. INTRODUCTION TO NUCLEIC ACIDS AND PROTEINS

- DNA and RNA are the informational polymers in the cell—that is, they encode genetic information in a way that can be read by macromolecular machines, to direct the synthesis of other molecules.
- RNA and proteins are operational polymers; they carry out the functions of the cell.
- Nucleotides have pentose sugars attached to nitrogenous bases and phosphate groups.
- The nucleotide bases in RNA and DNA are substituted pyrimidines or purines.
- DNA and RNA are polymers of deoxyribonucleotides and ribonucleotides, respectively. There are four deoxyribonucleotides in DNA, denoted A, T, C, and G. There are four ribonucleotides in RNA, denoted A, U, C, and G.
- DNA and RNA are synthesized in the 5′-to-3′ direction by sequential reactions that are driven by the hydrolysis of nucleotide triphosphates.
- DNA forms a double helix with antiparallel strands.
- The structure of DNA is highly symmetrical, and the double helix involves complementary base pairing (A-T and C-G).
- The double helix is stabilized by the stacking of base pairs.
- Proteins are formed from amino acids connected by peptide bonds.
- The 20 genetically encoded amino acids are classed as either hydrophilic (charged, or polar but neutral) or hydrophobic (nonpolar) depending on the properties of their sidechains.
- Proteins fold into specific three-dimensional structures that are highly diverse.
- The hydrophobic effect is the principal driving force underlying protein folding.
- $\alpha$ helices and $\beta$ sheets are the regular structural elements on which the structures of proteins are based.

### C. REPLICATION, TRANSCRIPTION, AND TRANSLATION

- The processing of genetic information involves the replication of DNA, the transcription of DNA into RNA, and the translation of RNA into protein.
- DNA replication is a complex process involving many protein machines.
- Transcription generates RNA copies of the DNA sequence in genes.
- The splicing of RNA in eukaryotic cells generates a diversity of RNAs from a single gene.
- The genetic code relates specific triplets of nucleotides in a gene sequence to specific amino acids in a protein sequence.
- The genetic code is redundant, making it relatively robust to mutation.
- Transfer RNAs are adapters that couple codons to specific amino acids during the synthesis of proteins in the ribosome.
- The transfer of genetic information is a highly conserved process across all of the branches of life.
- The discovery of retroviruses revealed that genetic information stored in RNA can be converted into a DNA form.

## Problems

### True/False and Multiple Choice

1. When two atoms approach each other closely, the energy goes up because the nuclei of the atoms repel each other.
   True/False

2. Ionic interactions are stronger in water than in vacuum because water forms strong hydrogen bonds with polar molecules.
   True/False

3. An N–H•••O=C hydrogen bond has optimal energy when it:
   a. is bent.
   b. is linear.
   c. has a donor–acceptor distance of 4 Å.
   d. has a donor–acceptor distance of 2 Å.

4. A by-product of forming a peptide bond from two amino acids is water.
   True/False

5. Circle all of the polar amino acids in the list below:
   a. Phenylalanine
   b. Valine
   c. Arginine
   d. Proline
   e. Leucine

6. Proteins fold with their hydrophobic amino acids on the surface and their hydrophilic amino acids in the core.
   True/False

7. Which of the following is not a unit of structure found in proteins?
   a. β sheets
   b. Loop regions
   c. α helices
   d. γ arches

8. The central dogma of molecular biology states that RNA is translated from proteins.
   True/False

9. Which of these types of molecules serve as a template for messenger RNA?
   a. Protein
   b. DNA
   c. Transfer RNA
   d. Carbohydrates
   e. Ribosomes

10. DNA primase synthesizes DNA molecules.
    True/False

### Fill in the Blank

11. When two atoms approach closer than _______________ the interaction energy goes up very sharply.

12. The van der Waals attraction arises due to _______ induced dipoles in atoms.

13. At room temperature, the value of _________ is about 2.5 kJ•mol$^{-1}$.

14. ________ is an operational nucleic acid, whereas ________ is strictly an informational nucleic acid.

15. Amino acids are zwitterions: molecules with charged groups but an overall ______ electrical charge.

16. The ______ consists of two subunits that assemble around messenger RNA.

### Quantitative/Essay

17. Order the following elements from lowest electronegativity to highest: P, C, N, H, O, S. Use your ordering of the atoms to rank the following hydrogen bonds from weakest to strongest:
    a. SH----O
    b. NH----O
    c. OH----O

18. The stabilization energy of a bond or interatomic interaction is the change in energy upon breakage of a bond between two atoms (that is, the change in energy when the atoms are moved away from each other). We can classify bonds into the following categories, based on their dissociation energies:

    Strong: > 200 kJ•mol$^{-1}$

    Medium: 20–200 kJ•mol$^{-1}$

    Weak: 5–20 kJ•mol$^{-1}$

    Very weak: 0–5 kJ•mol$^{-1}$

    Consider the bonds highlighted in purple in the diagram below.

    a. First consider the bonds in molecules isolated from all other molecules (in a vacuum). Classify each of them into the four categories given above, based on your estimation of the bond strength.

    b. Which of these bonds could be broken readily by thermal fluctuations?

c. Next, consider what happens when these molecules are immersed in water (fully solvated). For each bond, indicate whether it becomes weaker, stronger, or stays the same in water.

d. Which of these bonds could be broken readily by thermal fluctuations in water?

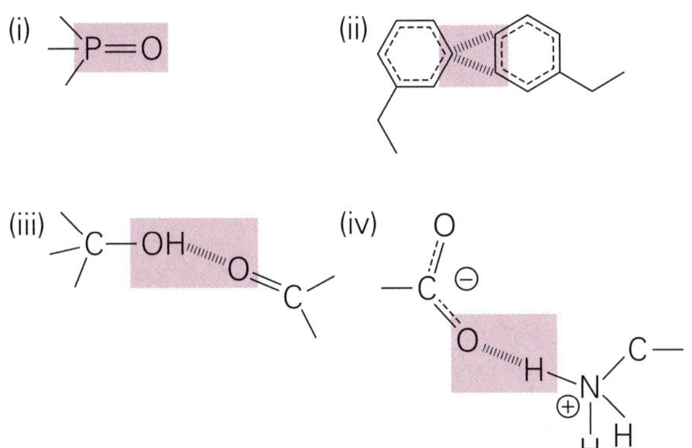


19. The diagram below shows a representation of the structure of a peptide segment (that is, a short portion of a larger protein chain). Hydrogen atoms are not shown. Nitrogen and sulfur atoms are indicated by "N" and "S." Sidechain oxygen atoms are indicated by "*."

a. Identify each of the amino acid residues in the peptides.

b. Draw a linear chemical structure showing the covalent bonding, including the hydrogen atoms, of the entire peptide.

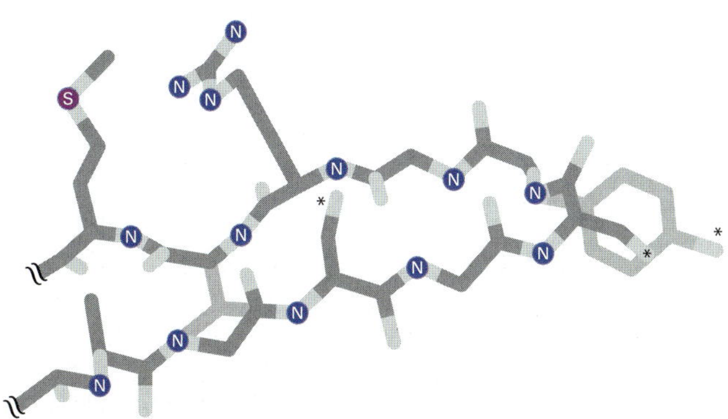

20. With 64 possible codons and 21 options to code (20 amino acids + stop), (a) what is the average number of codons per amino acid/stop codon? (b) Which amino acids occur more often than expected in the actual codon table?

21. There are approximately 3 billion nucleotides in the human genome. If all of the DNA in the entire genome were laid out in a single straight line, how long would the line be? Express your answer in meters.

22. The human immunodeficiency virus (HIV) is a retrovirus with an RNA genome.

a. Assume that each HIV contains two RNA genomes and 50 molecules of the reverse transcriptase enzyme.

b. Assume that each reverse transcriptase molecule acts on each RNA genome 10 times to produce DNA.

c. Assume that an integrase enzyme successfully integrates 1% of the available reverse transcribed HIV genomes into the genome of a human host cell.

d. Assume that each integrated copy of the viral genome is transcribed 500 times/day.

How many HIV RNA genomes are created per day from one infected cell?

23. What is a step in RNA processing that occurs in eukaryotes but not in prokaryotes?

24. What changes to the central dogma were necessary after the discovery of retroviruses?

25. What chemical properties have led to DNA being selected through evolution as the information molecule for complex life forms instead of RNA?

26. How do size considerations forbid G-A base pairs in a double helix?

27. Why are DNA chains synthesized only in the 5′-to-3′ direction in the cell?

28. Chemists are able to synthesize modified oligonucleotides (that is, polymers of nucleotides) in which the phosphate linkage is replaced by a neutral amide linkage, as shown in the diagram below. (Adapted from M. Nina et al., and S. Wendeborn, *J. Am. Chem. Soc.* 127: 6027–6038, 2005. With permission from the American Chemical Society.)

a. Such modified oligonucleotides are able to form double-helical structures similar to those seen for DNA and RNA. Often these double helices are more stable than the natural DNA and RNA double helices

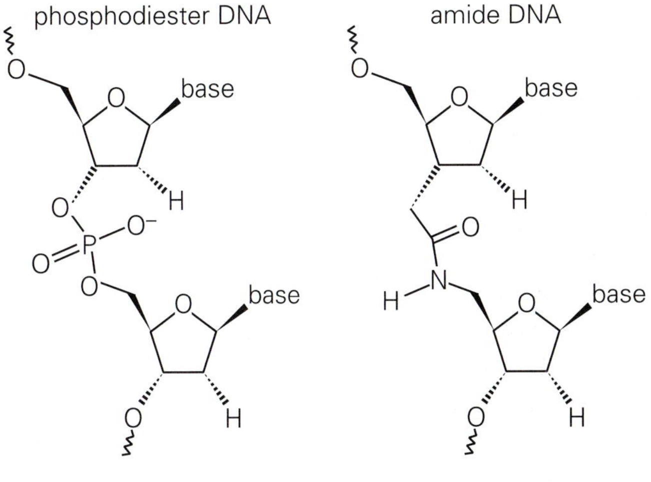

phosphodiester DNA

amide DNA

with the same sequence of bases. Explain why such helices can form, and why they can be more stable.

b. Given the increased stability of such modified nucleotides, why has nature not used them to build the genetic material? Provide two different reasons that could explain why these molecules are not used.

**29**. Suppose you were told that the two strands of DNA could in fact be readily replicated in opposite directions (that is, one in the 5′ → 3′ direction, and one in the 3′ → 5′ direction). Would you have to postulate the existence of a new kind of nucleotide? If so, what would the chemical structure of this compound be?

## Further Reading

### General

Alberts B, Johnson A, Lewis J, Raff M, Roberts K & Walter P (2008) Molecular Biology of the Cell, 5th ed. New York: Garland Science.

Atkins PW & Jones L (2010) Chemical Principles: The Quest for Insight, 5th ed. New York: W.H. Freeman.

Brändén C & Tooze J (1999) Introduction to Protein Structure, 2nd ed. New York: Garland Science.

Cox MM, Doudna J & O'Donnell M (2011) Molecular Biology: Principles and Practice, 1st ed. New York: W. H. Freeman.

Judson HF (1996) The Eighth Day of Creation: Makers of the Revolution in Biology, Expanded ed. Plainview, NY: CSHL Press.

Oxtoby DW, Gillis HP & Campion A (2007) Principles of Modern Chemistry, 6th ed. Belmont, CA: Thomson Brooks/Cole.

Pauling L (1960) The Nature of the Chemical Bond and the Structure of Molecules and Crystals, 3rd ed. Ithaca, NY: Cornell University Press.

Saenger W (1984) Principles of Nucleic Acid Structure. New York: Springer-Verlag.

Voet D & Voet JG (2004) Biochemistry, 3rd ed. New York: John Wiley & Sons.

Vollhardt KPC & Schore NE (2011) Organic Chemistry: Structure and Function, 6th ed. New York: W.H. Freeman.

# Nucleic Acid Structure

A guiding principle that helps us understand mechanism in biology is that *form follows function*. At the molecular level, a fully functioning protein or a noncoding RNA (one that does not code for a protein) must have not only the proper sequence, but also the correct three-dimensional shape. All of the information required to acquire the specific three-dimensional structure of a protein or noncoding RNA molecule is present in the primary structure (that is, its sequence). The largely spontaneous process of assuming this precise three-dimensional structure is called self-assembly, and is a general property of proteins and noncoding RNAs—even some surprisingly complex ones.

Underlying the accurate transfer of information from nucleic acid to nucleic acid is Watson-Crick base-pairing. The strict base-pairing rules for DNA (A pairs only with T, and C pairs only with G) help ensure that a template strand of DNA is copied into a complementary strand of DNA or RNA of precisely defined sequence. When the information is passed from nucleic acid to protein (translation), the genetic code comes into play. Triplets of nucleotides (codons) in messenger RNA (mRNA) specify the sequence of amino acids that are joined together to form a protein (**Figure 2.1**). Each codon specifies one particular amino acid (out of the 20 occurring in cells) and, using the genetic code (see Chapter 1), the nucleotide sequence of the DNA gene is thus translated into the amino acid sequence of a protein via mRNA.

To function correctly, DNA, RNA, and protein must all adopt appropriate three-dimensional structures. The self-assembly of RNA and protein molecules is crucial because the blueprints for the cell that are encoded in the DNA cannot find expression unless the RNA and protein molecules fold up properly. In contrast to the simple and linear relationship between DNA sequences and the sequences of the corresponding RNA and protein products, the rules governing folding are

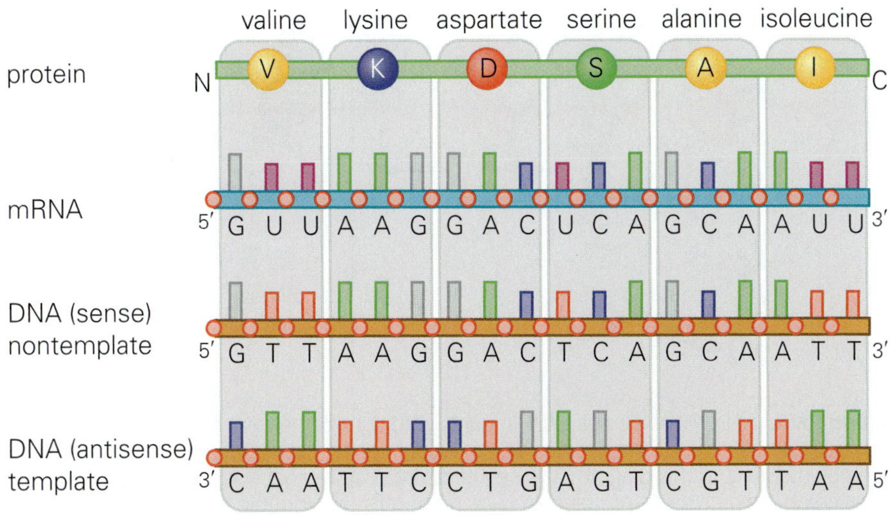


**Figure 2.1 Complementary base-pairing and the genetic code.** Watson-Crick base-pairing (A with T, and C with G) underlies the fidelity of DNA replication. When DNA is transcribed, the template (or antisense) strand is used to make the coding strand of RNA, also called the sense strand (the complement of the template strand). Each codon, shaded in *gray*, is translated into a specific amino acid, and the sequence of codons specifies the sequence of amino acids.

complex. Even though these rules are based on simple physical principles (which we shall study in this book), we cannot as yet predict protein structure from amino acid sequence in a reliable way. Computer modeling of RNA structure from nucleotide sequence is somewhat more successful, but still very imprecise. Deciphering this indirectly encoded genetic information remains a major challenge in modern molecular biology.

In this chapter, we continue the discussion we began in Chapter 1 on the basics of DNA, RNA, and protein structure by focusing on DNA and RNA. We discuss protein structure in Chapters 4 and 5. Protein folding and RNA folding are discussed in more detail in Chapter 18. In this chapter, we will briefly describe the structures of DNA and RNA and how they differ from one another. We will consider the ability of RNA to form intricate tertiary structures that lead ultimately to a diverse array of functions. Most importantly, we will begin to appreciate how the differences in the physical properties of RNA and DNA determine their different functions in the cell.

## A. DOUBLE-HELICAL STRUCTURES OF RNA AND DNA

### 2.1 The double helix is the principal secondary structure of DNA and RNA

Nearly all DNA occurs in long, continuous, right-handed double-helical form, produced in complementary pairs by the replication process. RNA, in contrast, is usually produced as a single chain and, therefore, cannot form long, continuous double helices. Nevertheless, even though RNA molecules consist of unpaired chains, these chains do form short, discontinuous stretches of double helix interspersed with single-chain stretches (**Figure 2.2**). Thus, the double helix is the principal type of secondary structure in both DNA and RNA.

Key to the formation of the DNA double helix is the character of the base pairs. Not only is the pairing strictly complementary (A always pairs with T, and G pairs only with C, as shown in **Figure 2.3**), but each pair also forms a purine-pyrimidine set of the same shape and size as any other correctly matched purine-pyrimidine set. The glycosidic bonds (the bonds that join the bases to the 1′ carbon atom of the sugar) are all in the same orientation with respect to the sugar-phosphate backbone. The base pairs can therefore stack neatly on top of one another in any order.

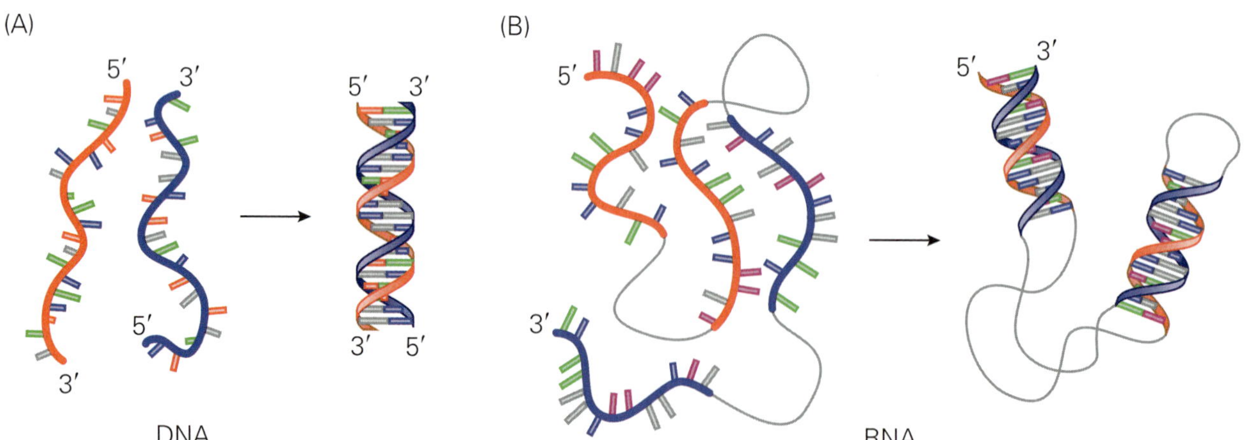


**Figure 2.2 Double-helical structure in DNA and RNA.** (A) DNA molecules exist in pairs, with complementary sequences that enable the two molecules to form a continuous double helix. (B) RNA molecules do not exist in pairs, but structured RNA molecules usually contain internal regions that are self-complementary. In this diagram, the segments colored *red* have sequences that are complementary to sequences within the regions colored *blue*. The RNA molecule folds up to form two double helices.

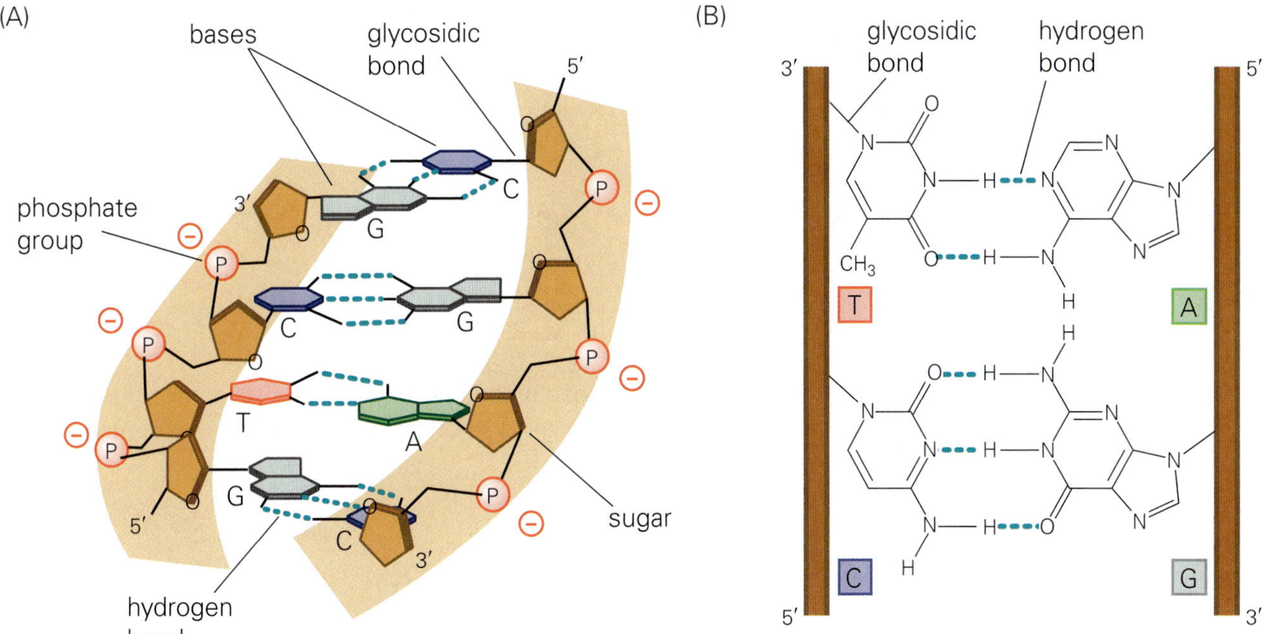


**Figure 2.3 DNA structure.** (A) The sugar-phosphate backbone of DNA is charged due to a single negative charge on each phosphodiester linkage. (B) The Watson-Crick base pairs are shown with the hydrogen bonds formed by them indicated in *blue*. (A, adapted from C. Brändén and J. Tooze, Introduction to Protein Structure, 2nd ed. New York: Garland Science, 1999.)

Because each of the Watson-Crick base pairs has the same shape, the three-dimensional structure of DNA does not depend on its primary sequence. Complex genomes can evolve and be accommodated within the simple double helix. Folded proteins, on the other hand, have complex three-dimensional structures that are strongly influenced by the linear sequence of amino acids in the polypeptide chain (see Chapter 4). RNA lies in a middle ground, with helical secondary structure organized into complicated tertiary structures. RNA can also hold information, but in general it is not self-complementary and thus cannot form a single, continuous double helix.

## 2.2 Hydrogen bonding between bases is important for the formation of double helices, but its effect is weakened due to interactions with water

Hydrogen bonding between bases contributes to the stability of double-helical nucleic acids. These hydrogen bonds are cooperative: establishing one of them favors the formation of others through geometrical factors (**Figure 2.4**A). Despite being such a striking feature of DNA, the hydrogen bonds do not contribute as much to the stabilization of the double helix as does the base stacking. Hydrogen bonds between water and bases in single-stranded DNA are replaced by hydrogen bonds between bases once the double helix forms (Figure 2.4B). As we explain in Section 19.7, where we discuss the energetics of DNA base-pairing, the effective contribution of hydrogen bonds to the stability of DNA is less than the intrinsic strength of the hydrogen bonds because of competition with water.

Water forms reasonably strong hydrogen bonds with the functional groups of DNA or RNA. The net stabilization afforded by the hydrogen bonds between bases is the *difference* in stability between the hydrogen bonds that the bases form with water (for example, in single-stranded DNA) and the hydrogen bonds between bases in duplex DNA. This difference is small, so the net contribution that hydrogen bonds make to the stabilization of the DNA is also small.

Despite this competition with water, the formation of hydrogen bonds between properly paired bases is crucial for the formation of the double helix. This apparent contradiction is resolved if one considers what happens if a base is brought into the double helix without an appropriate hydrogen-bonding partner. The

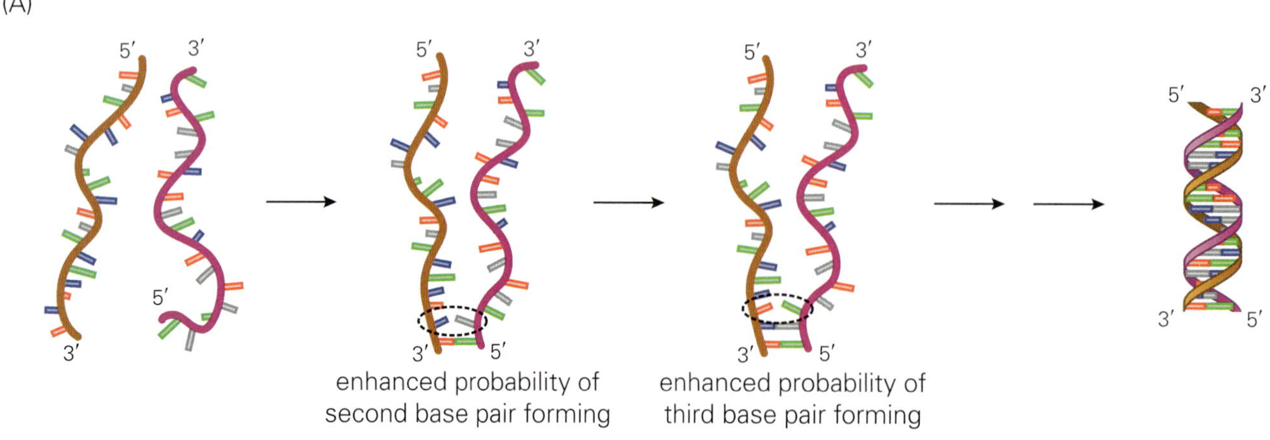


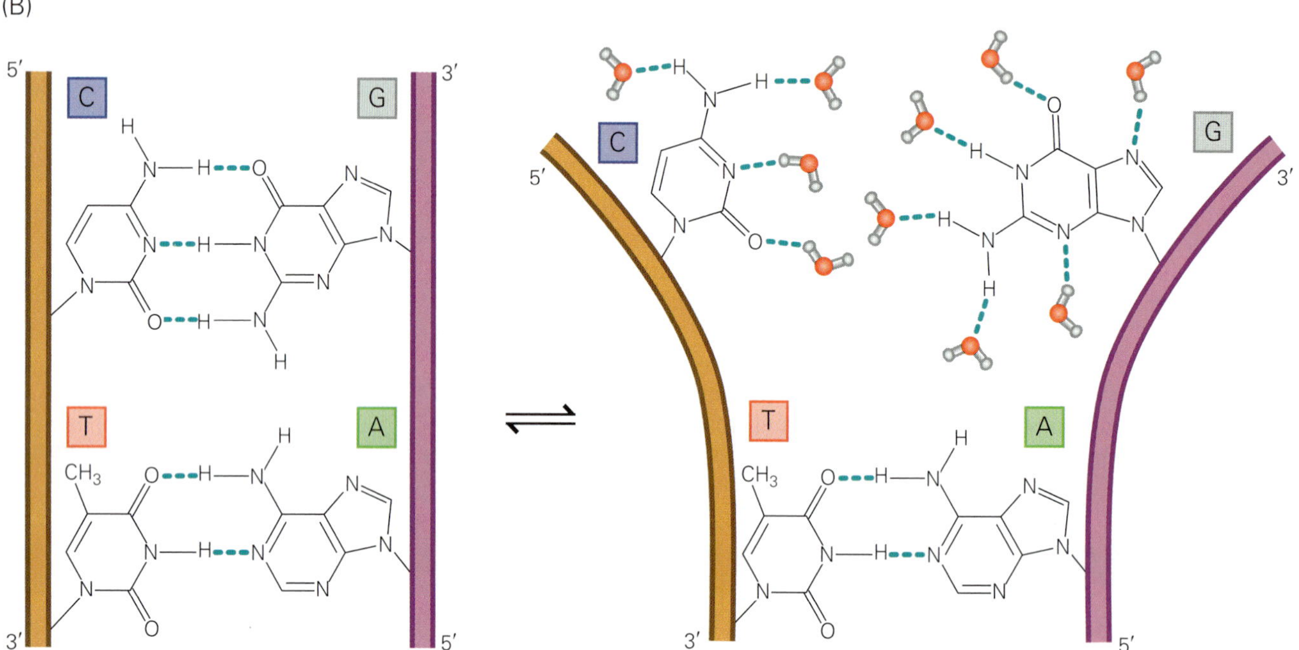


**Figure 2.4 The formation of base pairs in DNA.** (A) One feature that stabilizes RNA and DNA double helices is that the formation of one base pair favors the formation of additional ones; that is, DNA base-pairing is cooperative. As shown in this diagram, the formation of the first base pair between two nucleic acid strands favors the formation of the second one, because the nucleotide bases in the two strands are brought into proximity (dotted oval). Once the second base pair is formed, the third base pair falls into place, and so on, like the teeth of a zipper. (B) Water competes with the hydrogen bonds between bases. The net stability of the hydrogen bonds is reduced by competition with water.

energetic penalty for such a mismatched base pair can be substantial, because the base loses the hydrogen bonds that it makes with water in the single-stranded state without forming compensating hydrogen bonds with the partner chain in the double-stranded state.

## 2.3 The electronic polarization of the bases contributes to strong stacking interactions between bases

The bases, both purines and pyrimidines, are flat and planar, so they can stack on top of one another. In addition, the atoms of the aromatic rings are very polarizable, and many of the atoms have a partial charge. These two factors make the combination of van der Waals and electrostatic interactions between stacked bases particularly strong. These effects are known collectively as base-stacking interactions, and they provide the dominant contribution to stabilizing the double-helical conformation of nucleic acids.

The importance of base stacking can be appreciated by studying **Figure 2.5**, in which two kinds of base stacks are illustrated for a noncoding RNA molecule. Some of the base stacks are similar to those seen in DNA, with the bases stacked and hydrogen bonded to each other in a double-helical structure. Other base

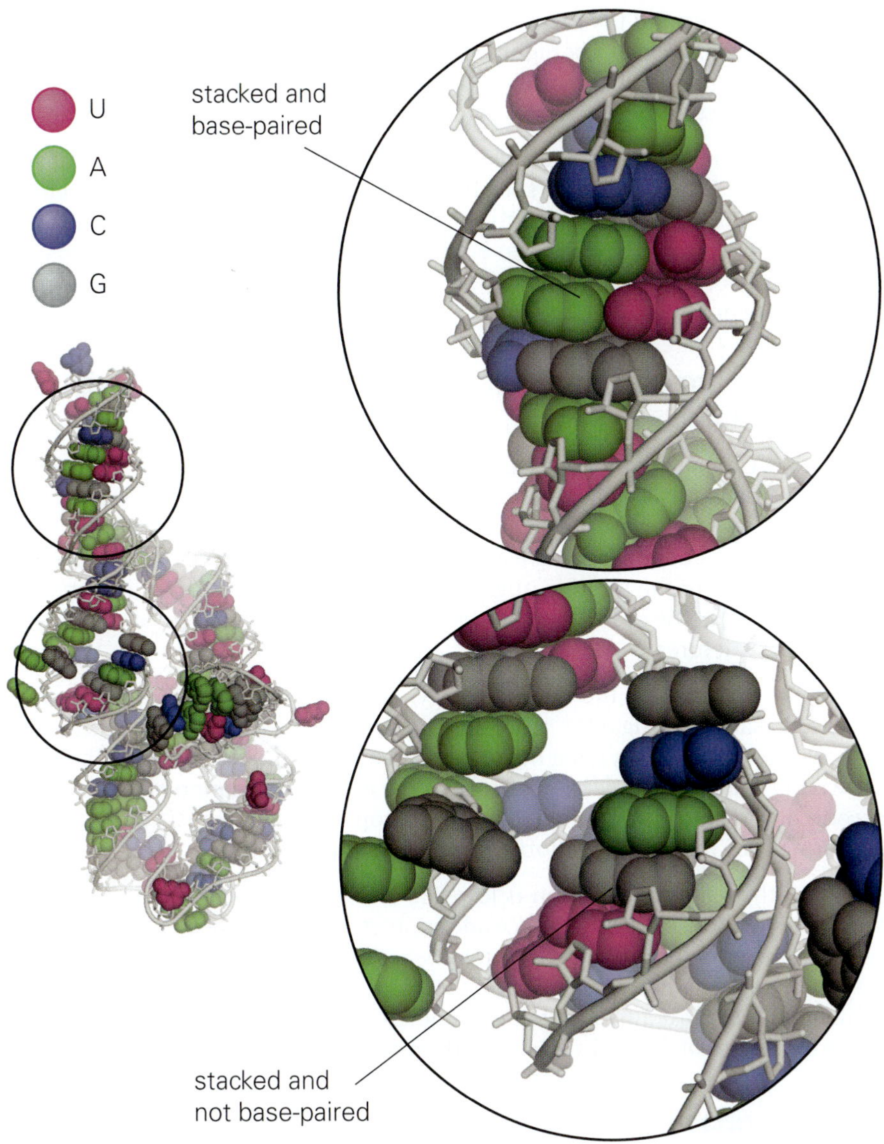


**Figure 2.5 Base stacking in a noncoding RNA molecule.** The favorable energetics of stacking is demonstrated by the fact that even nonbase-paired regions of RNA are often found to be well-stacked, as shown in this example. (PDB code: 1HR2.)

stacks are formed by bases that are not part of a double-helical structure (that is, they do not form Watson-Crick base pairs). Instead of being flexible, which might be anticipated without the complementary bases present, the RNA chain in this region is helical (without being double-helical), with the bases of the strand packing on top of one another. This kind of helix formation by single-stranded (that is, not base-paired) RNA is driven by the stabilizing energetics of base stacking. We compare the strengths of hydrogen bonding and base stacking quantitatively in Section 19.7.

## 2.4 Metal ions help shield electrostatic repulsions between the phosphate groups

The sugar-phosphate backbone of nucleic acids is charged (see Figure 2.3A). The backbone contains one phosphate group per residue and, under physiologically relevant conditions, these phosphates will be negatively charged and will tend to repel each other. As will be described further in Section 2.23, these negative charges can be shielded by metal ions. Both monovalent (sodium and potassium) and divalent (particularly magnesium) ions have been found to interact with various sites on nucleic acids. These interactions may be through direct metal-ion coordination or they may be mediated by water molecules that form a

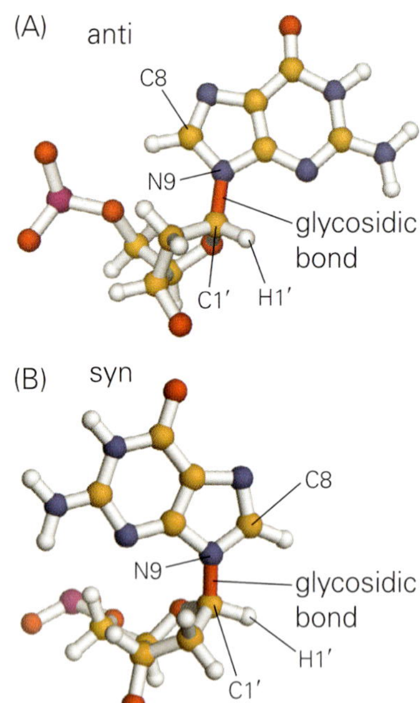


**Figure 2.6 Anti and syn conformations of a nucleotide.** The two orientations of the base relative to the sugars are shown for an adenine. These orientations have different rotations about the glycosidic bond, which is indicated in *red*. (A) The more common anti conformation. (B) The less common syn conformation. See Figure 1.15 for the numbering scheme used to identify the atoms in the nucleotides.

hydration shell around the metal ion. In general, the metal ions around DNA are best described as a cluster of highly mobile cations that hop rapidly among various sites. More localized salt-bridge-like interactions occur quite often in RNA, as discussed in Section 2.23.

In summary, the structure of double-stranded, helical nucleic acids depends on several factors. The backbones (specifically the phosphates) need to be separated as far as possible and be exposed to solvent to minimize electrostatic repulsions. The bases, in turn, need to stack to optimize the van der Waals and electrostatic interactions between them, as well as to form mutually stabilizing hydrogen bonds where possible.

## 2.5 There are two common relative orientations of the base and the sugar

As noted in Chapter 1, nucleosides in DNA and RNA have the base attached covalently to the sugar by a single bond from the C1′ atom of the sugar to the N1 or N9 atom of the pyrimidine or purine base, respectively. The glycosidic bond is a single bond, and so rotations about it are possible. Nevertheless, steric clashes occur for many values of the rotation angle around this bond. These clashes constrain the orientation of the base in double-helical nucleic acids into one of two distinct families of structures.

The conformation with the lowest energy, and hence the one most frequently observed, places the sugar H1′ atom and the base C6 or C8 atoms (for pyrimidines and purines, respectively) in a *trans* conformation about the glycosidic bond, as shown in **Figure 2.6**A. This is referred to as the **anti** conformation of the base. The second observed conformation has the H1′ atom and the C6 or C8 atoms in a *cis* conformation. This conformation is termed **syn** (Figure 2.6B).

Although the anti conformation is lower in energy and occurs more frequently, there are situations that can enforce adoption of the syn conformation. For example, in some structures there is much better stacking of the bases in the syn conformation, as seen in a form of DNA known as Z-DNA (see Section 2.12) and in some RNA loop structures. Chemical modifications of bases may also affect the relative energies of the anti and syn conformations.

## 2.6 The ribose ring has alternate conformations defined by the sugar pucker

Another important conformational parameter for DNA and RNA is known as the **sugar pucker**, which refers to the different out-of-plane distortions in the deoxyribose or ribose rings of nucleosides, as shown in **Figure 2.7**. Ring pucker is discussed further in Chapter 3 for other sugars. The sugar group consists of the five atoms that form the ring (C1′, C2′, C3′, C4′, and O4′) and one carbon atom that extends from the ring (C5′). The base also extends from the plane of the ring on the same side as C5′ (see Figure 2.7).

**Sugar pucker**

The sugar ring in polynucleotides is generally nonplanar and it displays a preferred puckering mode, C3′ endo, found in A-form helices, or C2′ endo, found in B-form helices. The terms endo and exo specify the nature of the out-of-plane atom of the sugar ring, with endo indicating displacement toward the side with the C5′ carbon and exo indicating displacement toward the opposite side.

All of the atoms in the ring have $sp^3$ hybridization, and so the bonds around each atom point toward the corners of a tetrahedron, with bond angles around 109°. A planar five-membered ring has bond angles of about 108°, close to the optimum value. However, with all of the atoms in a plane the substituent atoms of the ring are in high energy eclipsed conformations. The energy is lower when one atom moves out of the plane, reducing steric clashes, and this is the geometry normally observed. Most commonly it is the C2′ atom or the C3′ atom that is out of the plane of the other four atoms. Four such conformations, which are termed "half-boat" conformers, are shown in Figure 2.7.

When the out-of-plane atom is located on the same side of the plane as C5′, the conformation is referred to as **endo** ("inside"). When it is located on the side opposite C5′, the conformation is referred to as **exo** ("outside"). Thus, a sugar

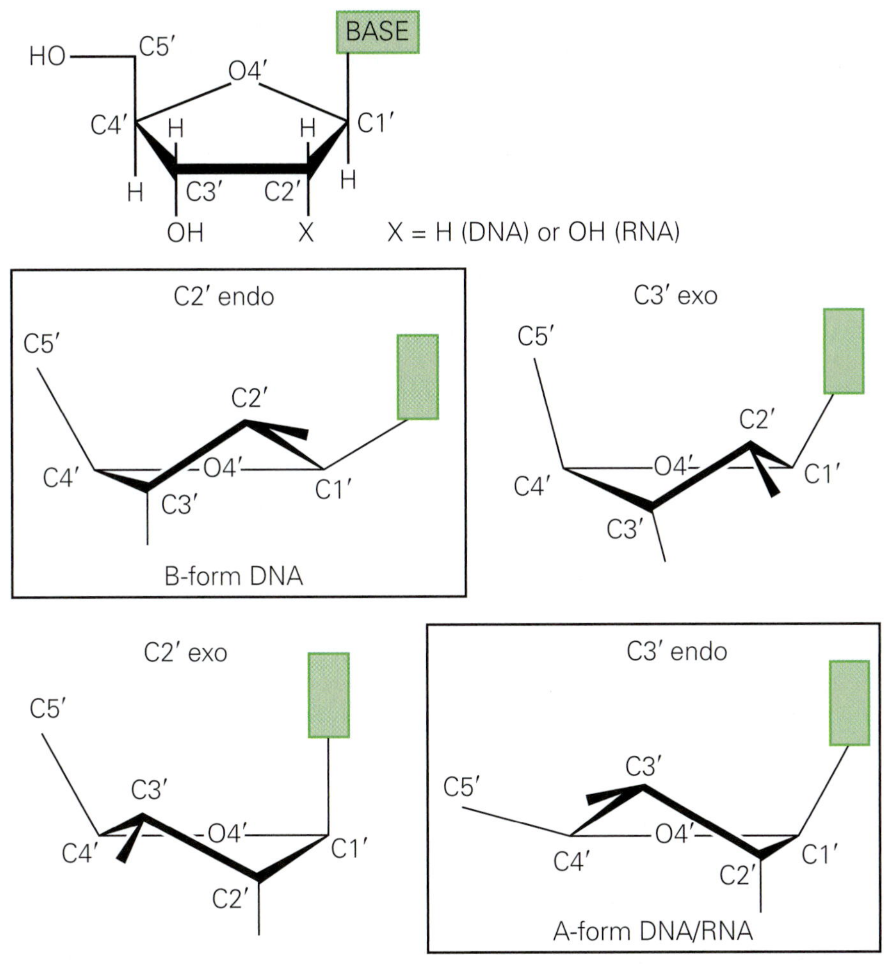


**Figure 2.7 Sugar pucker in DNA and RNA.** Five atoms of the sugar group (C1′, C2′, C3′, C4′, and O4′) form a ring. In energetically favorable conformations, four of these are roughly coplanar and one is out of the plane. If the atom that is out of the plane is on the same side as the C5′ atom and the base, the conformation is referred to as *endo*. For example, in the C2′ endo conformation illustrated here, the C2′ atom is out of the plane and is on the same side as the atom C5′ and the base. If the atom that is not in the plane is located on the side opposite to that of the base, the conformation is referred to as *exo*. In the C2′ exo conformation shown here, it is the C2′ atom that is not in the plane, and it is on the other side of the plane with respect to the base.

pucker in which the C2′ atom is out of the plane and is on the same side of the plane as C5′, is referred to as C2′ endo, whereas a sugar pucker in which C2′ is on the side opposite C5′ is called C2′ exo. Likewise, a C3′ endo sugar pucker has C3′ out of the plane, but on the same side as C5′, whereas a C3′ exo sugar has C3′ on the side opposite C5′.

In nucleic acids, endo sugar puckers are more common than exo. In DNA, the sugar pucker can be either C2′ endo (in B-form DNA) or C3′ endo (in A-form DNA). In RNA, however, the C2′ endo sugar pucker cannot be adopted because of steric hindrance between the OH group on C2′ and the phosphate group on C3′ (**Figure 2.8**).

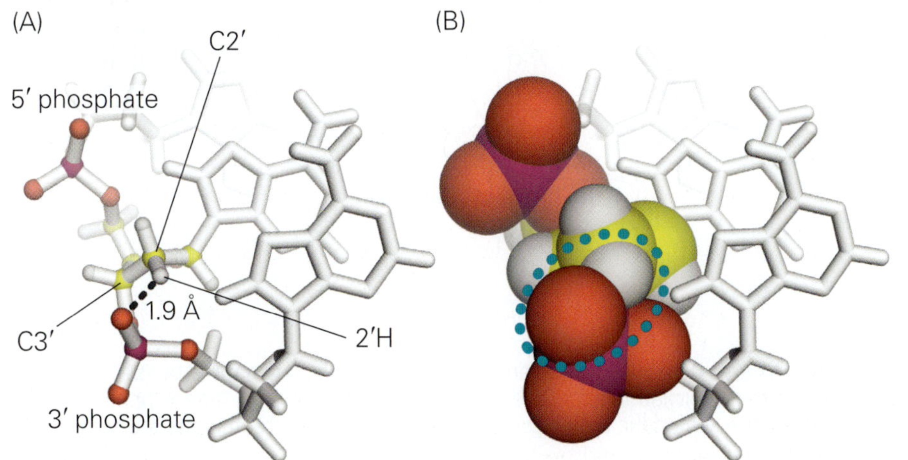


**Figure 2.8 Close contact between the 3′ phosphate and the 2′ hydrogen in standard B-form Watson-Crick DNA**. In (A), a nucleotide and its phosphate groups are shown in a ball-and-stick representation. In (B), the sugar and the phosphates are shown as space-filling models, in which the sizes of the spheres correspond to the van der Waals radii of the atoms. There is no room to accommodate a hydroxyl group at the 2′ position in this conformation, as can be seen from the close contact between the 3′ phosphate group and the hydrogen at the 2′ position (dotted *cyan* oval).

## 2.7 RNA cannot adopt the standard Watson-Crick double-helical structure because of constraints on its sugar pucker

Why is the difference in sugar conformation between the sugar pucker in RNA and DNA important? In the standard Watson-Crick model of DNA, all of the sugar groups have the C2′ endo conformation. This results in close contact between one of the oxygen atoms of the 3′ phosphate group and the hydrogen atom at the 2′ position of the ribose ring in DNA (Figure 2.8). The distance between the phosphate oxygen and the hydrogen atom is 1.9 Å. If the 2′ hydrogen were replaced by a hydroxyl group, as in the ribose ring of RNA, the two oxygen atoms would repel each other strongly. Instead, in RNA, the sugar adopts a C3′ endo conformation (**Figure 2.9**), which causes a change in phosphate separation and a general modification of the double-helical structure (discussed in detail in Section 2.10).

As discussed earlier, noncoding RNA molecules fold so that segments of the same chain pair up to form double helical structures (Figure 2.2B). RNA also forms double-helical structures when paired with DNA. Recall from Section 1.16 that DNA polymerases cannot initiate replication. Instead, a primase enzyme lays down a short strand of RNA on the DNA template, forming a short RNA-DNA hybrid duplex from which DNA synthesis is initiated. Finally, many viruses store their genetic information in double-helical RNA. What is the nature of the duplex in these RNA-DNA and RNA-RNA structures? To answer this question we must first understand the Watson-Crick B-DNA model in a little more detail.

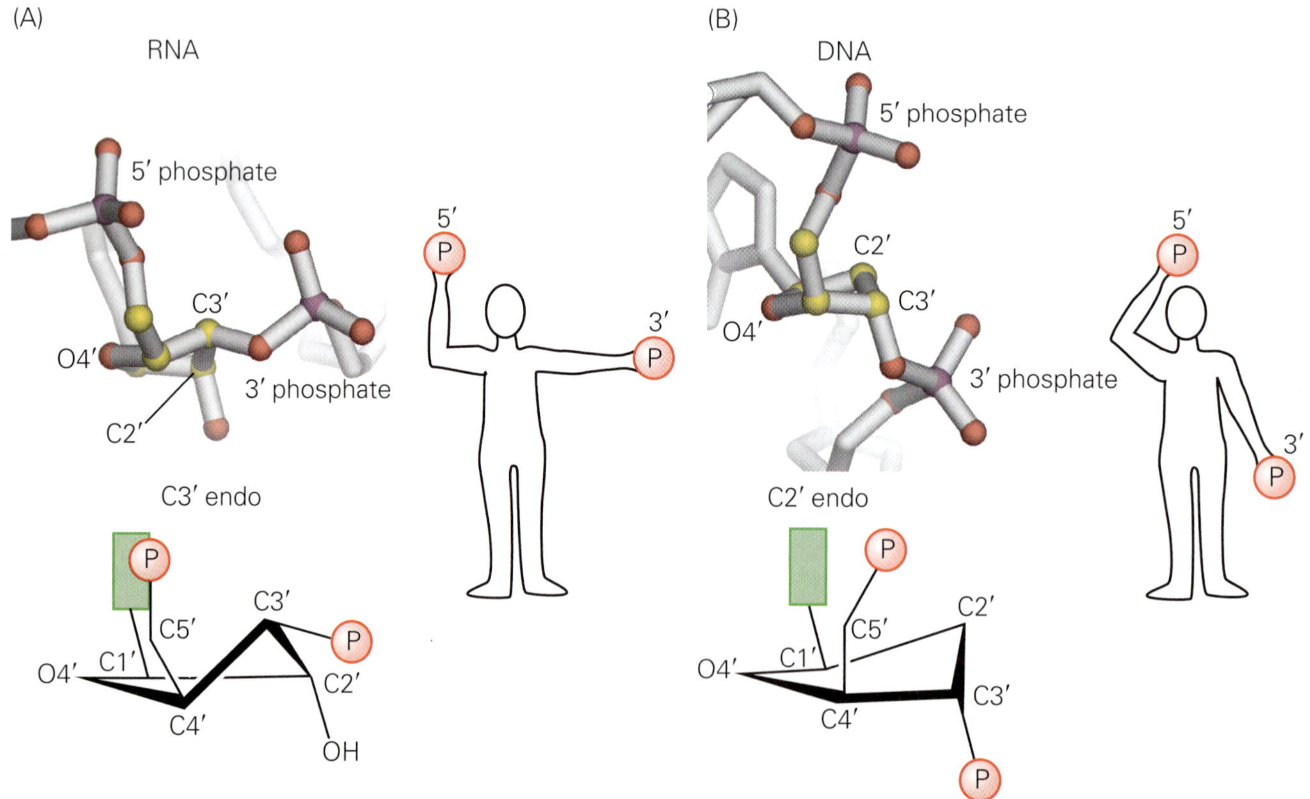


**Figure 2.9 Molecular view of the sugar pucker in RNA and DNA.** (A) The C3′ endo conformation is shown for RNA. (B) The C2′ endo conformation is shown for DNA. The change in sugar conformation alters the relative orientation of the 5′ and 3′ phosphate groups drastically, as shown by the arms of the person in the schematic drawings. The body of the person represents the sugar group, and the arms represent the two phosphodiester linkages.

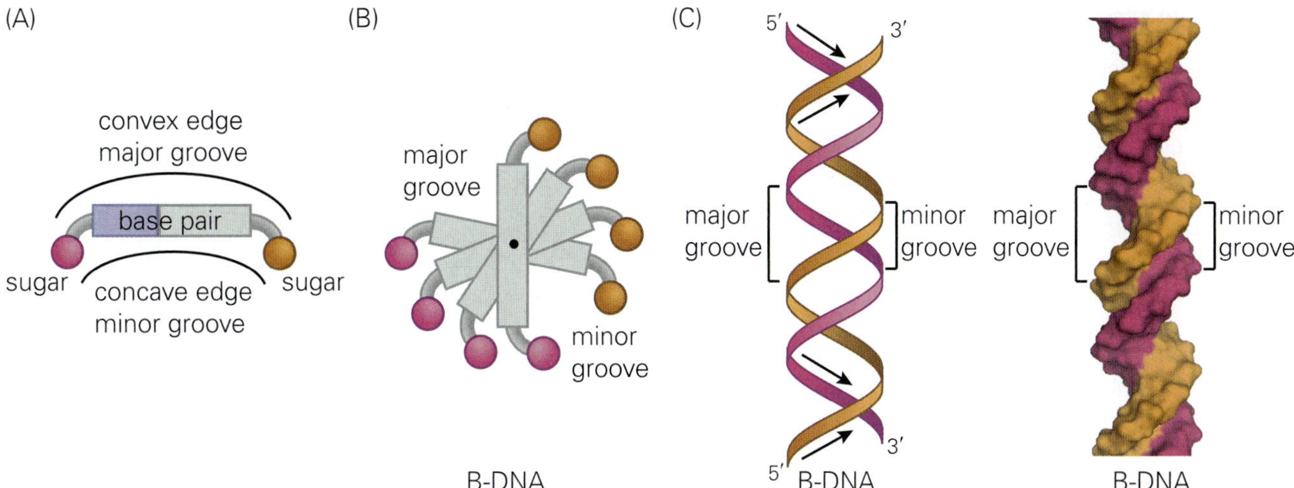


Figure 2.10 The grooves of B-DNA. (A) This diagram shows one base pair. One edge of the base pair has a convex shape, and it faces the major groove in double helical DNA. The other edge has a concave shape, and it faces the minor groove. (B) The ladder of steps formed by the base pairs leads to the formation of the major and minor grooves. Each base pair is indicated schematically as a *gray plank*. (C) The structure of B-form DNA, showing the major and minor grooves. (Adapted from C. Brändén and J. Tooze, Introduction to Protein Structure, 2nd ed. New York: Garland Science, 1999.)

## 2.8 The standard Watson-Crick model of double-helical DNA is the B-form

The structure of double-helical DNA, as originally described by Watson and Crick on the basis of Franklin and Wilkins' x-ray diffraction data, is known as **B-form DNA** or **B-DNA**. There are several characteristic features of this structure that stand out visually. The base pairs, for example, are perpendicular to the axis of the helix. Each base pair is centered on the helix axis, so that when viewed from above, the base pairs all cross each other, in projection, at the central axis (**Figure 2.10**).

The complementary chains are parallel, but run in opposite directions, and together they twist into a right-handed double helix. The helix forms two grooves of unequal size (Figure 2.10). The wider and more accessible **major groove** (**Figure 2.11**) allows regulatory proteins or other molecules to gain access to nucleotide functional groups on the edges of the groove. The narrower, less accessible **minor groove** (**Figure 2.12**) allows much more limited access to the functional groups lying within the groove.

**B-form DNA**

B-form DNA is the standard conformation of double-helical DNA. The sugar pucker is C2′ endo, which is inaccessible to RNA. The base is in the anti conformation with respect to the sugar.

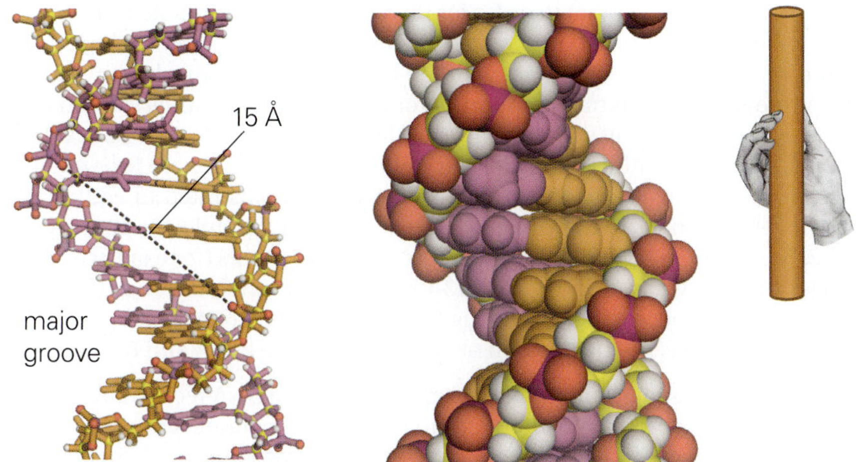


**Figure 2.11 The major groove of B-DNA.** The right-handed nature of the double helix is illustrated by a right hand grasping a cylinder. The phosphate groups are far apart across the major groove, and the edges of the base pairs are accessible. The "width" of the major groove depends on which atoms are used to measure the distance across the groove. Here the distance between two phosphate oxygen atoms across the groove is indicated. Considering the oxygen van der Waals radius of 1.5 Å there is ~12 Å available for interactions. The structure shown here is for ideal B-form DNA. In real DNA structures the width of the groove varies somewhat.

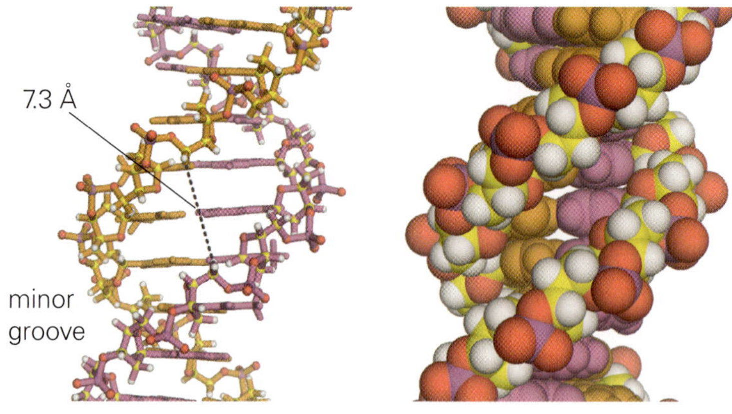


**Figure 2.12 The minor groove of B-DNA.** A rotated view of Figure 2.11 shows that the phosphate groups are closer together, and the edges of the base pairs are less accessible in the minor groove than they are in the major groove. As for the major groove, the width of the minor groove depends on which atoms are used to measure the distance across the groove. Here, the distance between two hydrogen atoms attached to C5′ carbon atoms is indicated, corresponding to an open space of ~6 Å. The DNA structure is that of an ideal B-form helix.

The B-form helix rises ~34 Å per helical turn and there are 10–11 base pairs per turn, so the rise per base pair is ~3.1–3.4 Å. The stacked base pairs form favorable van der Waals interactions and pack tightly together. Some of the structural parameters that characterize B-form DNA are listed in Table 2.1.

## 2.9 B-form DNA allows sequence-specific recognition of the major groove, which has a greater information content than the minor groove

As we emphasized in Chapter 1, sequence information stored in DNA can be directly copied into new DNA (replication) or into RNA (transcription) by forming complementary Watson-Crick base pairs on a template strand of DNA. In both replication and transcription, the helix must be opened for the information within to be read out nucleotide by nucleotide. There is, in addition, a fundamentally different way of reading information stored in DNA that neither opens the helix nor copies the sequence, but instead reads the base sequence from the "outside" (via the grooves), leaving the helix largely intact. This process of reading the surface signposts of the base pairs is particularly important in the mechanisms by which DNA replication and transcription are controlled.

**Major and minor grooves**

DNA and RNA double helices have two characteristic grooves, denoted the major and minor grooves.
In B-form DNA, the major groove is wide and can accommodate α helices, which is important for the sequence-specific recognition of DNA by proteins. The major and minor grooves can be identified by looking at the connections of the base pairs to the sugars. The major groove is at the convex edge of the base pair, while the minor groove is at the concave edge, as shown in Figure 2.10.

Recall from Section 1.17 that in order for RNA polymerase to initiate transcription it has to be recruited to promoter sites on DNA by proteins that bind to specific regions on DNA and then interact with the RNA polymerase machinery. How do these proteins, known as transcription factors, recognize their target sites on DNA? They do so by interacting with the edges of the base pairs in double-helical DNA, which are exposed within the major and minor grooves. There are four potential types of interactions with complementary base pairs, illustrated in **Figure 2.13**. These involve, on the DNA, hydrogen-bond donors (for example, the amino group of adenine in Figure 2.13C and D), hydrogen-bond acceptors (for example, the N7 nitrogen atom of guanine in Figure 2.13A and B), hydrophobic groups (for example, the methyl group of thymine in Figure 2.13C and D), and nonpolar hydrogen atoms (for example, the ring proton of cytosine in Figure 2.13A and B).

The pattern of potential contact sites forms the basis of a DNA recognition mechanism, in which the protein positions complementary functional groups at the recognition sites shown in Figure 2.13. The patterns of recognition elements in the major groove are unique for each of the four base pairs (G-C, C-G, A-T, and T-A). In the minor groove, however, the pattern for G-C is indistinguishable from that for C-G, and the pattern for A-T is the same as that for T-A. Thus, the major groove of DNA allows each of the four kinds of base pairs to be distinguished from each other, whereas the interactions in the minor groove do not.

The major groove of B-DNA readily accommodates several common structural elements of proteins, such as the α helix (**Figure 2.14**), and most proteins that

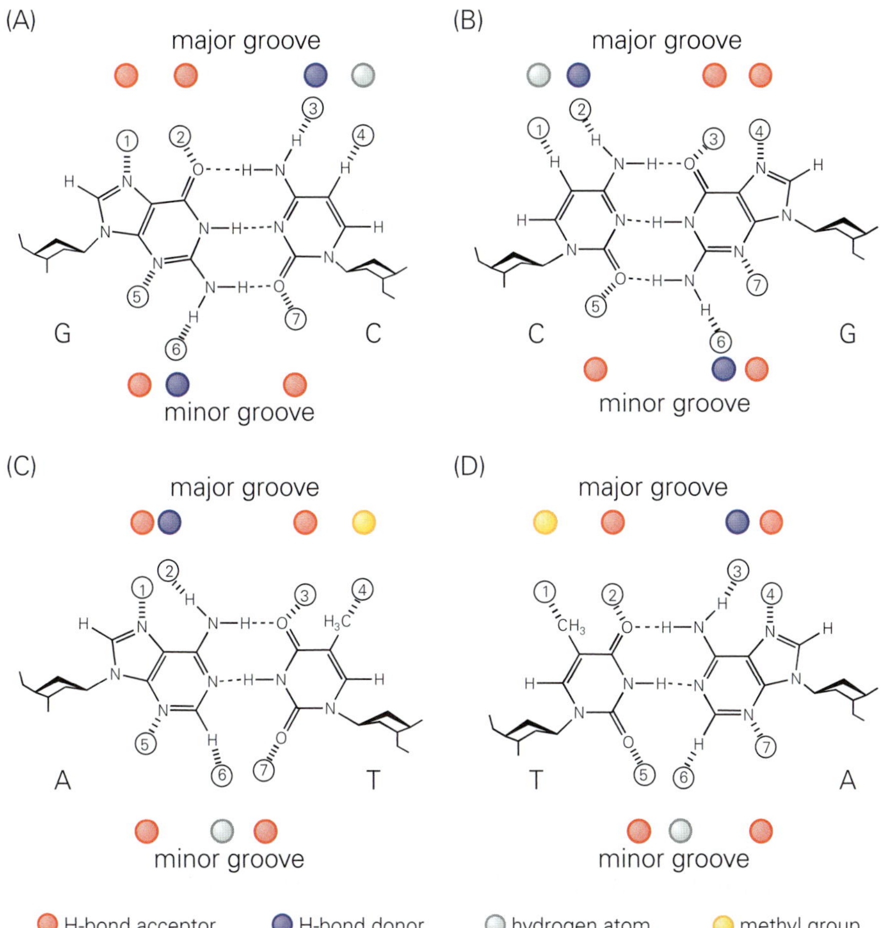


**Figure 2.13 Potential interaction sites at the edges of Watson-Crick base pairs.** There are four interaction sites in the major groove (numbered 1–4) and three in the minor groove (numbered 5–7). These sites include hydrogen-bond acceptors (*red dots*), hydrogen-bond donors (*blue dots*), methyl groups (*yellow dots*), and nonpolar hydrogen atoms (*white dots*). (Adapted from C. Brändén and J. Tooze, Introduction to Protein Structure, 2nd ed. New York: Garland Science, 1999.)

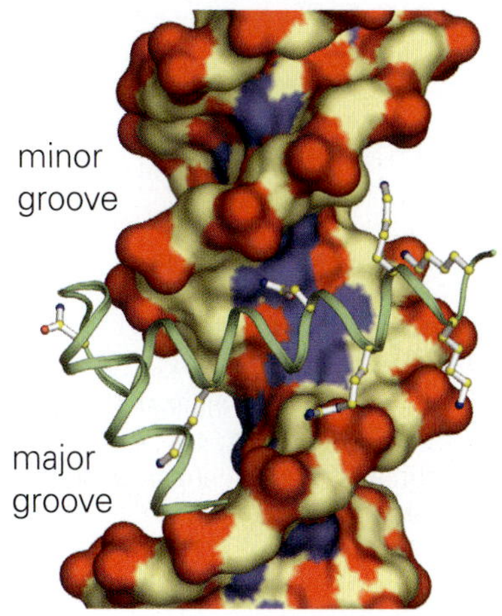


**Figure 2.14 Interaction between a protein and the major groove of DNA.** The molecular surface of DNA is shown. The surface is colored according to the nature of the underlying atoms, with nitrogens *blue*, oxygens *red,* and carbons *yellow*. (PDB code: 1DU0.)

bind to sequence-specific regions of DNA do so in the major groove. How proteins recognize nucleic acids, both DNA and RNA, is discussed in more detail in Part C of Chapter 13.

## 2.10 RNA adopts the A-form double-helical conformation

Recall from Section 2.7 that RNA cannot readily adopt the C2′ endo sugar pucker characteristic of B-DNA. However, RNA does form a double-helical structure known as the **A-form helix**, in which the sugar pucker is C3′ endo. DNA can also adopt the A-form helical structure (**Figure 2.15**), but it is usually only favored when limited water is available to hydrate the DNA. When DNA molecules that are in the B-form are allowed to dry out slowly, they switch spontaneously to A-DNA. The A-form helix is also adopted by RNA-DNA hybrids, as shown in **Figure 2.16.**

A-form DNA or RNA is similar to B-DNA in several aspects. The two polynucleotide strands are antiparallel, and form a right-handed spiral with the phosphate groups on the outside and the bases on the inside. The precise number of base pairs per turn depends on whether the double helix is DNA-DNA, DNA-RNA, or RNA-RNA, but it is generally close to the ~10 base pairs per turn of B-DNA (see Table 2.1).

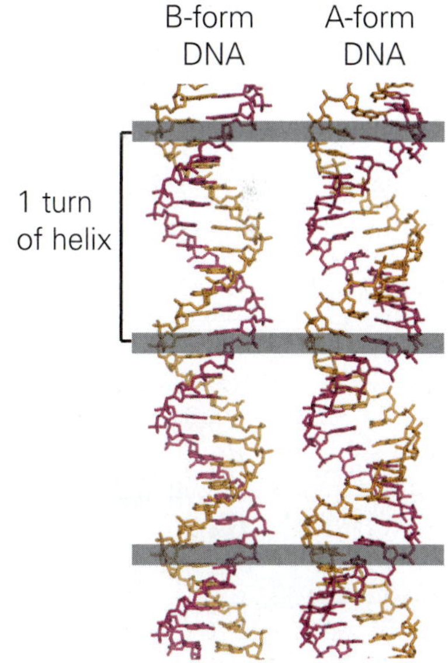


**Figure 2.15 B-form and A-form DNA.** DNA can adopt either the B or A conformation, with the B-form more stable under normal conditions. RNA only adopts the A-form double helix. The general nature of the A and B helices is similar, but there are significant differences in the orientation of the base pairs with respect to the helix axis, and in the size and shape of the major and minor grooves. These features are illustrated in more detail in Figures 2.18 and 2.19.

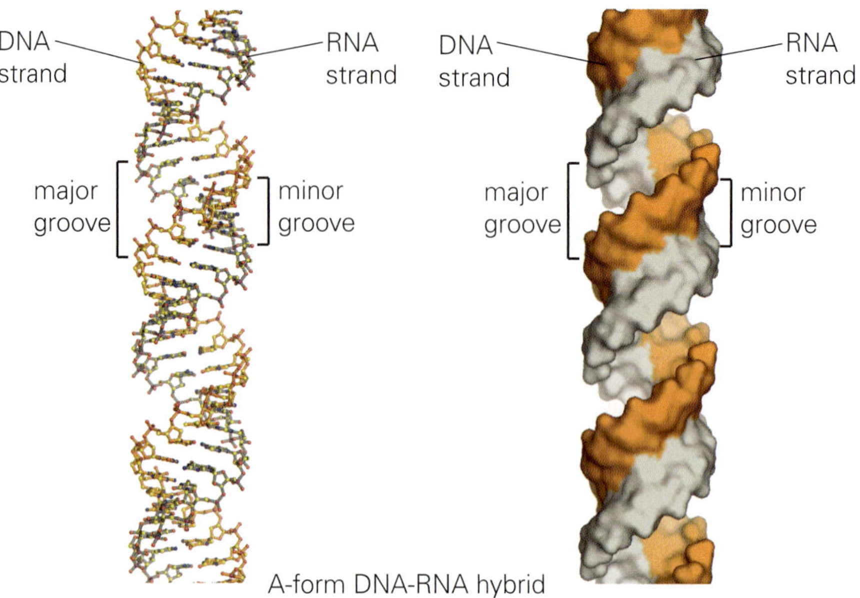


**Figure 2.16 A DNA-RNA hybrid A-form double helix.** As in B-form DNA, the hybrid A-form structure has a very deep major groove, whereas the minor groove is very shallow. The DNA strand is *orange* and the RNA strand is *gray*.

**A-form double helix**

The standard double-helical conformation of RNA, it is also adopted by DNA. The sugar pucker is C3′ endo.

## 2.11 The major groove of A-form double helices is less accessible to proteins than that of B-form DNA

One important difference between B-form DNA and A-form DNA or RNA is the position of the base pairs with respect to the helix axis. In B-DNA, they are all parallel to each other and perpendicular to the axis of the helix. In A-form helices, however, the base pairs are tilted away from perpendicular and are moved away from the center of the helix (**Figure 2.17**). As a result, the major groove in the A-form is deeper and narrower than it is in B-form DNA, while the minor groove is wider and shallower (**Figure 2.18** and **Figure 2.19**).

The narrowing and deepening of the major groove in A-form helices means that it is more difficult for proteins to read out the sequence-specific information at the edges of the bases in A-form helices (that is, α helices cannot readily enter the narrower major groove of the A-form helix). By storing genetic information in DNA, nature takes advantage of the higher chemical stability of polynucleotides that lack a 2′ hydroxyl substituent (see Figure 1.26). At the same time, the presence of a hydrogen atom rather than a hydroxyl group at the 2′ position allows DNA to readily adopt the C2′ endo sugar pucker and the B-DNA structure. This facilitates the control of genetic processes via the readout of sequence-specific information in the major groove of DNA, which is broad and accessible to proteins such as transcription factors. In contrast, the double-helical structure of RNA impedes access to the major groove, so access to sequence-specific information in RNA is usually through its nonpaired regions.

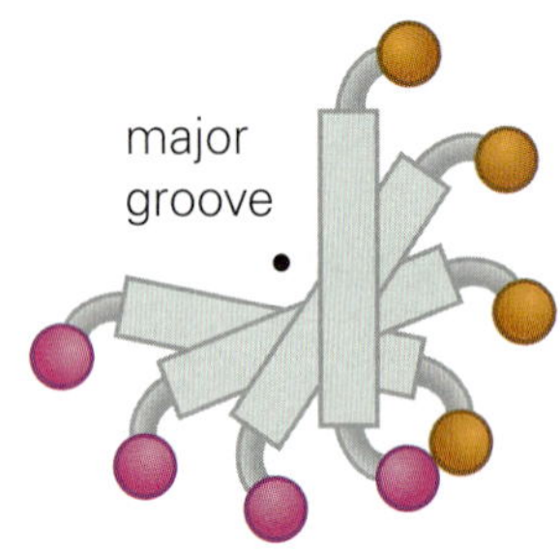


**Figure 2.17 The base pairs and grooves in A-form helices.** The base pairs in A-form helices are displaced from the helix axis, which leads to a deepening of the major groove. Compare this to the positioning of the base pairs in B-form DNA, which is illustrated in Figure 2.10 B. (Adapted from C. Brändén and J. Tooze, Introduction to Protein Structure, 2nd ed. New York: Garland Science, 1999.)

## 2.12 Z-form DNA is a left-handed double-helical structure

A third completely distinct type of double helix is **Z-form DNA** or **Z-DNA**, discovered by Alexander Rich. Unlike the A- and B-forms, Z-DNA is a left-handed helix (**Figure 2.20**). The Z-form is adopted preferentially by segments of DNA that have strictly alternating C and G nucleotides. The base pairs in Z-form DNA obey the Watson-Crick rules, so the alternation of C and G must occur on both strands. The

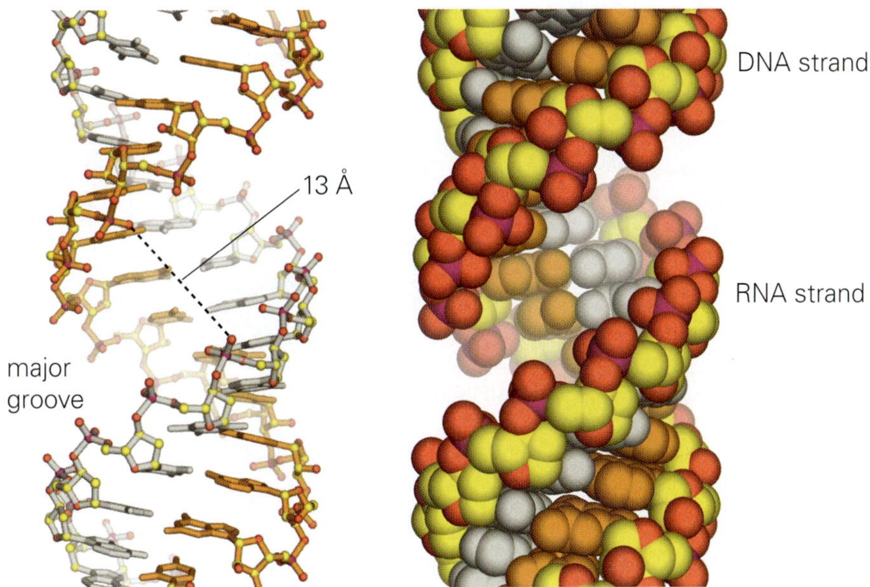


**Figure 2.18 The major groove of an A-form RNA-DNA hybrid helix.** The bases are colored *orange* in the DNA strand and *gray* in the RNA strand. The phosphate-phosphate distance across the major groove in this double helix is ~13 Å (compared to ~15 Å in B-DNA; see Figure 2.11). The base pairs are pushed away from the axis of the helix, away from the major groove (see Figure 2.17). This makes the groove significantly deeper than in DNA, as well as narrower (<10 Å opening for interactions).

requirement for alternation of C and G is not absolute: substitution of A for G (and correspondingly T for C) destabilizes the Z-form, but with a small number of such substitutions, Z-form conformations can still form.

In the Z-form structure, there are 12 base pairs per turn, the sugars alternate between 2′ endo and 3′ endo puckers, and the G (or A) residues are in the syn conformation, while the C (or T) residues are in the normal anti conformation (**Table 2.1**). This combination of features makes the repeating unit *two* nucleotides on

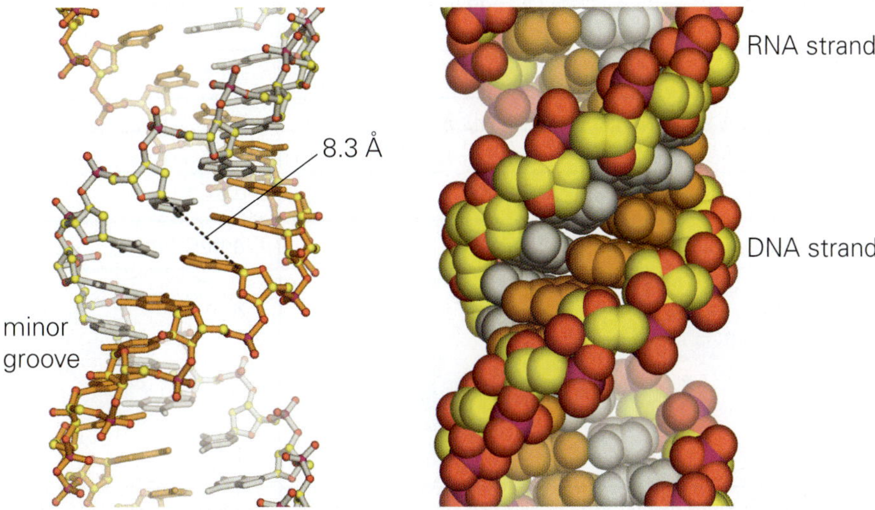


**Figure 2.19 The minor groove of an A-form RNA-DNA hybrid helix.** The base pairs in A-form helices are pushed away from the helix axis and towards the minor groove. The distance between C1′ atoms across the minor groove is 8.3 Å in this RNA-DNA helix and the groove is very shallow.

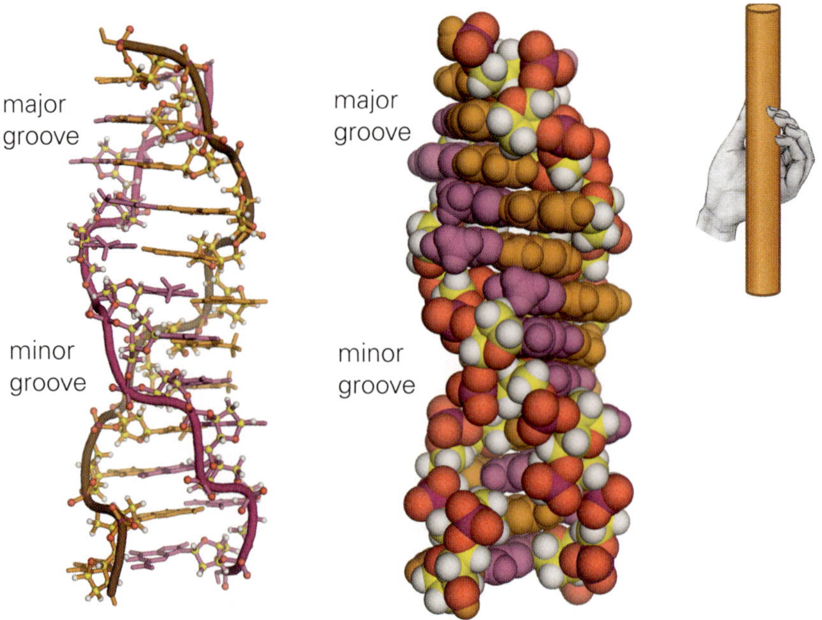


**Figure 2.20 Left-handed Z-form DNA.** Carbon atoms in the bases are *magenta* in one strand and *orange* in the other. The Z-form is a left-handed helix with a repeating structural unit of two base pairs. The major "groove" is so shallow that it is not really a groove and the minor groove is deep and very narrow.

each strand, rather than one nucleotide, as in B- or A-DNA. The alternation of sugar pucker gives the helix backbone a zigzag appearance, which gives Z-form DNA its name. Because RNA cannot adopt a C2′ endo sugar pucker, this conformation is restricted to DNA.

The physiological role of Z-form DNA in the cell is still not completely understood. One possibility is that it triggers the opening of DNA base pairs. If a region of B-DNA switches to Z-DNA, then the change in handedness of the double-helical

**Table 2.1 Structural features of A-, B-, and Z-form helices.**

| Helical form | A | B | Z |
|---|---|---|---|
| Helical sense | Right | Right | Left |
| Diameter | ~ 26 Å | ~ 20 Å | ~ 18 Å |
| Base pairs per turn | ~ 11 | ~ 10 | ~ 12 |
| Helical twist (rotation per base pair for A and B, per two-base repeat for Z) | ~ 34° | ~ 36° | ~ 60° (CpGp) |
| Helix pitch (rise per helical turn) | ~ 25 Å | ~ 33 Å | ~ 46 Å |
| Helix rise (along helix axis; per base pair for A and B, per two-base repeat for Z) | ~ 2.3 Å | ~ 3.3 Å | ~ 7.4 Å (CpGp) |
| Base tilt (with respect to helix axis) | ~ 20° | ~ 0° | ~ – 9° |
| Base orientation (with respect to sugar) | Anti | Anti | C anti/G syn |
| Base pair positions (helix axis indicated by black dot) | minor<br>major | minor major | minor major |
| Features of base pair positions | Base pairs displaced from axis; deep major groove, less accessible | Base pairs on axis; both major and minor grooves accessible | Base pairs stick out into the major groove, the minor groove is deep and narrow |

(Adapted from R.E. Dickerson et al., and M.L. Kopka, *Science* 216: 475–482, 1982. With permission from AAAS.)

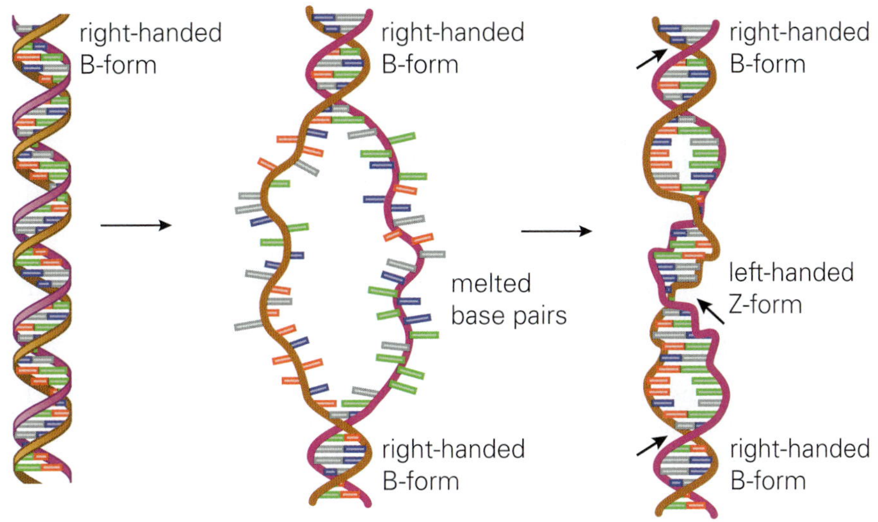


**Figure 2.21 Z-form DNA results in the melting of base pairs.** In this diagram, the central portion of a B-form DNA melts and then anneals into Z-form DNA. The left-handed sense of the Z-form helix requires that the segments of DNA that are at the junctions of the two forms of DNA are melted—otherwise, geometric constraints would prevent the formation of the Z-form structure.

structure would require the base pairs of the helix to open up in segments that intervene between the B-form and Z-form segments (**Figure 2.21**). Z-DNA *in vivo* has been difficult to study since it seems to be a transient structure. Intriguingly, a correlation has been found between transcription and Z-DNA formation at some genes.

## 2.13 The DNA double helix is quite deformable

Comparison of the structures of A-, B-, and Z-form double helices demonstrate the wide range of double-helical conformations that DNA can adopt. Although the B-form double helix is the preferred conformation of DNA, it is an idealization. In reality, local regions of DNA deviate considerably from the B-form while still maintaining a double-helical structure that preserves the Watson-Crick base pairing and has average helical parameters that correspond well to those of B-form DNA.

The ease with which the structure of DNA can be deformed locally (that is, around a few base pairs) is demonstrated by the structures of certain small molecules bound to DNA (**Figure 2.22**). These molecules bind with high affinity to DNA

**Figure 2.22 Small molecules that distort DNA.** (A) An intercalator DNA complex. The intercalator TOTO (bis-thiazole orange) is shown bound to DNA. The aromatic parts of the intercalator stack with DNA bases both above and below it. The increased separation of the base pairs causes local unwinding of the DNA. Some DNA sequences, rich in A-T base pairs, will accommodate either one (B) or two (C) minor groove ligands (such as distamycin in this case). The width of the minor groove must increase substantially for the second ligand to bind. (A, PDB code: 108D; B and C, courtesy of the laboratory of David Wemmer.)

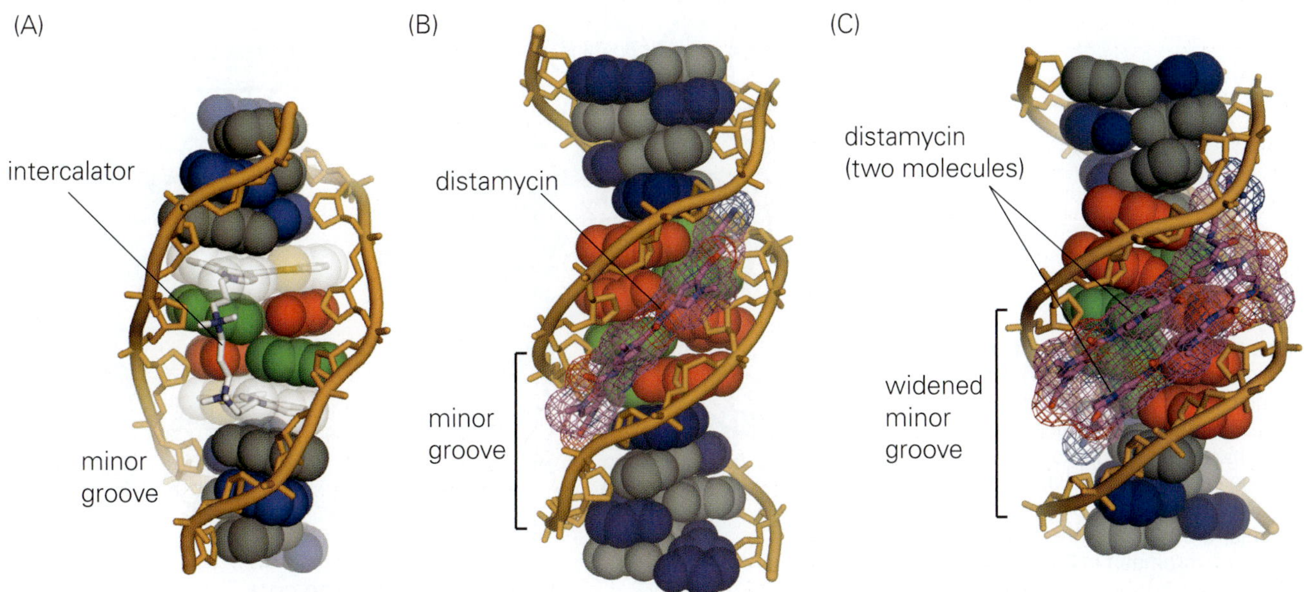

and deform the DNA away from the ideal B-form conformation when they do so. The DNA structure must be quite malleable to accommodate these molecules, because otherwise they would not be able to bind with such high affinity.

Molecules containing heterocyclic aromatic rings (usually with N or O atoms in them) can bind to DNA through intercalation—that is, slipping between base pairs of the DNA and stacking with the bases. An example of an intercalator (bis-thiazole orange, known as TOTO) bound to DNA is shown in Figure 2.22A. The complementarity of the size and shape of the aromatic rings of TOTO to those of the base pairs is apparent in the illustration. As you might imagine, such molecules are often quite toxic, because they can interfere with the normal replication and transcription of DNA.

Other aromatic molecules prefer to bind in the minor groove of DNA, interacting with the sugars and the edges of the bases, rather than stacking between bases. One such molecule is the antibiotic distamycin, which binds in the minor groove of DNA segments that have specific sequences (Figure 2.22B and C). Up to two molecules of distamycin can bind side by side in a single region of the minor groove, but the minor groove has to expand considerably in order to accommodate two distamycin molecules. Since both types of complex can form at the same DNA sequence, with similar binding energetics, the change in groove width must be a fairly low-energy process.

**Persistence length**

The persistence length of DNA corresponds to the maximum length of a segment of DNA that behaves as a rather rigid rod. DNA segments that are shorter than the persistence length behave as relatively rigid rods. DNA segments that are longer than the persistence length can be bent.

The deformability of double-stranded DNA is critical for its packaging in the cell, for its recognition by other molecules, and for the opening up of base pairs for replication, repair, and transcription. One measure of deformability is how easily DNA can be bent. The stiffness of DNA is characterized by its **persistence length**. DNA segments that are longer than the persistence length can bend without a significant energy penalty, whereas DNA segments shorter than the persistence length are relatively rigid. The persistence length of B-form DNA is approximately 500 Å. This corresponds to 14 or 15 helical turns, or roughly 140–150 base pairs.

In nucleosomes, which compact chromosomal DNA (discussed in Chapter 1), the DNA is bent much more sharply than would be expected from its the persistence length; roughly 150 base pairs of DNA wrap *twice* around the nucleosome core (**Figure 2.23**). This deformation costs energy, which is provided by the formation

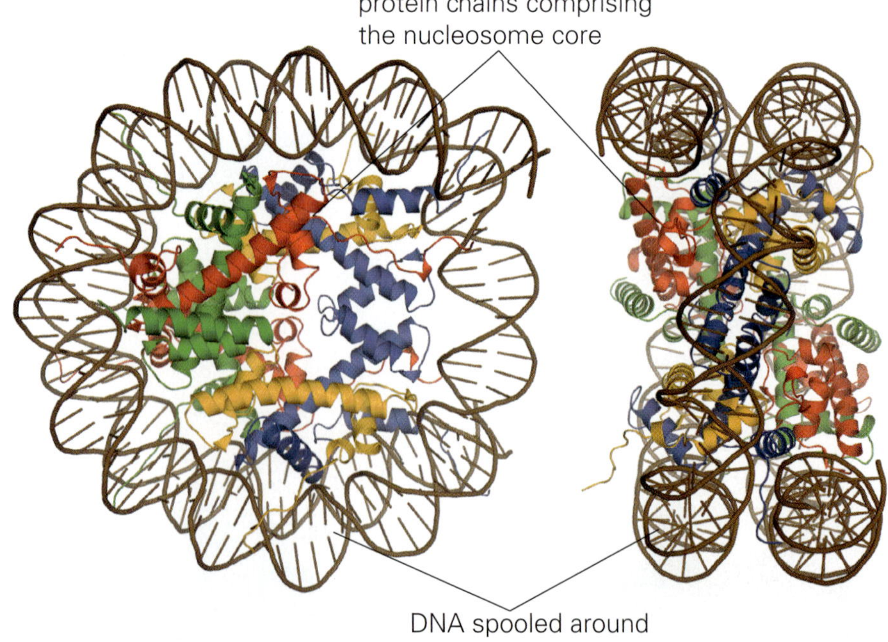


**Figure 2.23 The packaging of DNA in chromatin.** DNA wraps around the nucleosome core, which contains eight different protein chains. The DNA, which contains approximately 150 base pairs, is spooled around the nucleosome core. Compare this molecular drawing with the schematic diagram of chromatin in Figure 1.43. (PDB code: 1AOI.)

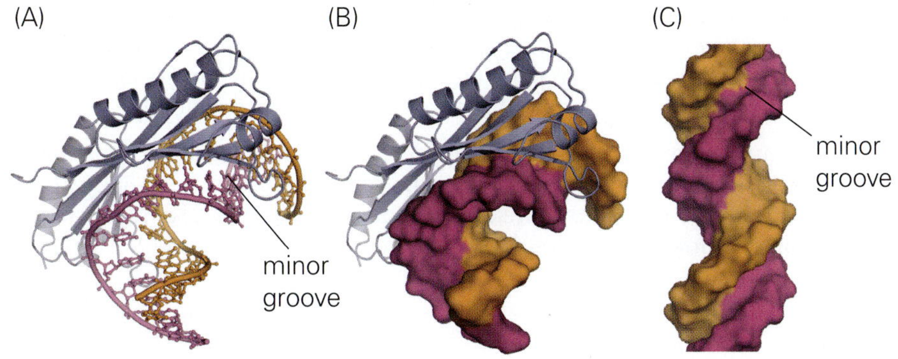


**Figure 2.24 TBP deforming DNA.** (A) and (B) The TATA-box binding protein (TBP) induces a sharp bend in the double helix and drastically widens the minor groove. (C) Standard B-DNA, included here for comparison. (A and B, PDB code: 1CDW.)

of favorable electrostatic interactions between the DNA and the histone proteins that make up the nucleosome core.

B-DNA may also be distorted when regulatory proteins, such as transcription factors, bind to it. A striking example of DNA distortion by one such transcription factor is shown in **Figure 2.24**, which compares the structure of standard B-form DNA to the structure of DNA bound to this transcription factor. The protein shown in Figure 2.24 is the TATA-box binding protein (TBP), which binds to regions of DNA that have a central TATA sequence, inducing a sharp bend in the double helix and drastically widening the minor groove. The DNA bending is linked to recruiting RNA polymerase and the opening of the transcription bubble in the promoter where TBP is bound. The degree to which the DNA can be bent or distorted varies for different sequences, and this sequence-specific deformability can, in turn, affect the affinity of a protein for a particular DNA sequence. These ideas will be explored in detail in Chapter 13.

## 2.14 DNA supercoiling can occur when the ends of double helices are constrained

Supercoiling is a large-scale conformational effect in DNA in which the whole double helix winds into a superhelix, as shown in **Figure 2.25**. This effect is a consequence of the elastic deformation of DNA when the ends of the double helix cannot rotate relative to one another. It is easiest to describe supercoiling for

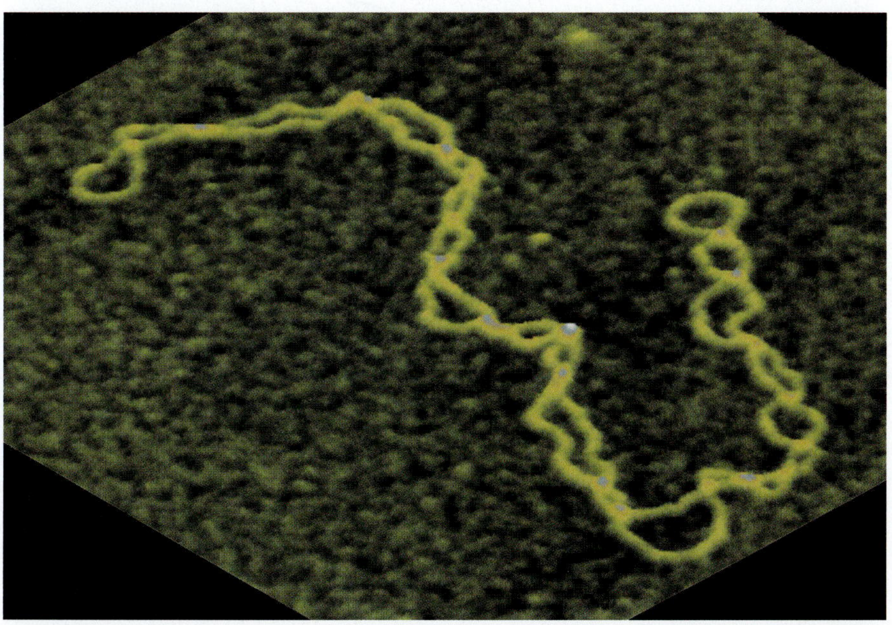

**Figure 2.25 Supercoiled plasmid DNA.** In this image, generated using atomic force microscopy, a circular plasmid containing double-helical DNA appears like a piece of string laid out on a surface. The supercoiling of the double helix is particularly apparent at the right side of the figure. (Adapted from Y. Lyubchenko, *Cell Biochem. Biophys.* 41: 75–98, 2004. With permission from Springer Science+Business Media.)

*circular* DNA molecules, such as plasmids (extrachromosomal DNAs that replicate in cells, often containing genes). These occur frequently in bacteria and are generally a few thousand base pairs in length. In a circular molecule, the ends cannot rotate relative to each other because they are linked covalently.

To begin to understand supercoiling, consider the process of taking a linear piece of double-helical DNA, in which each end has a 5′ phosphate and a 3′ hydroxyl, and bending it smoothly into a circle. The ends are then joined to make normal phosphodiester linkages (**Figure 2.26**A). If we consider a DNA with 1040 base

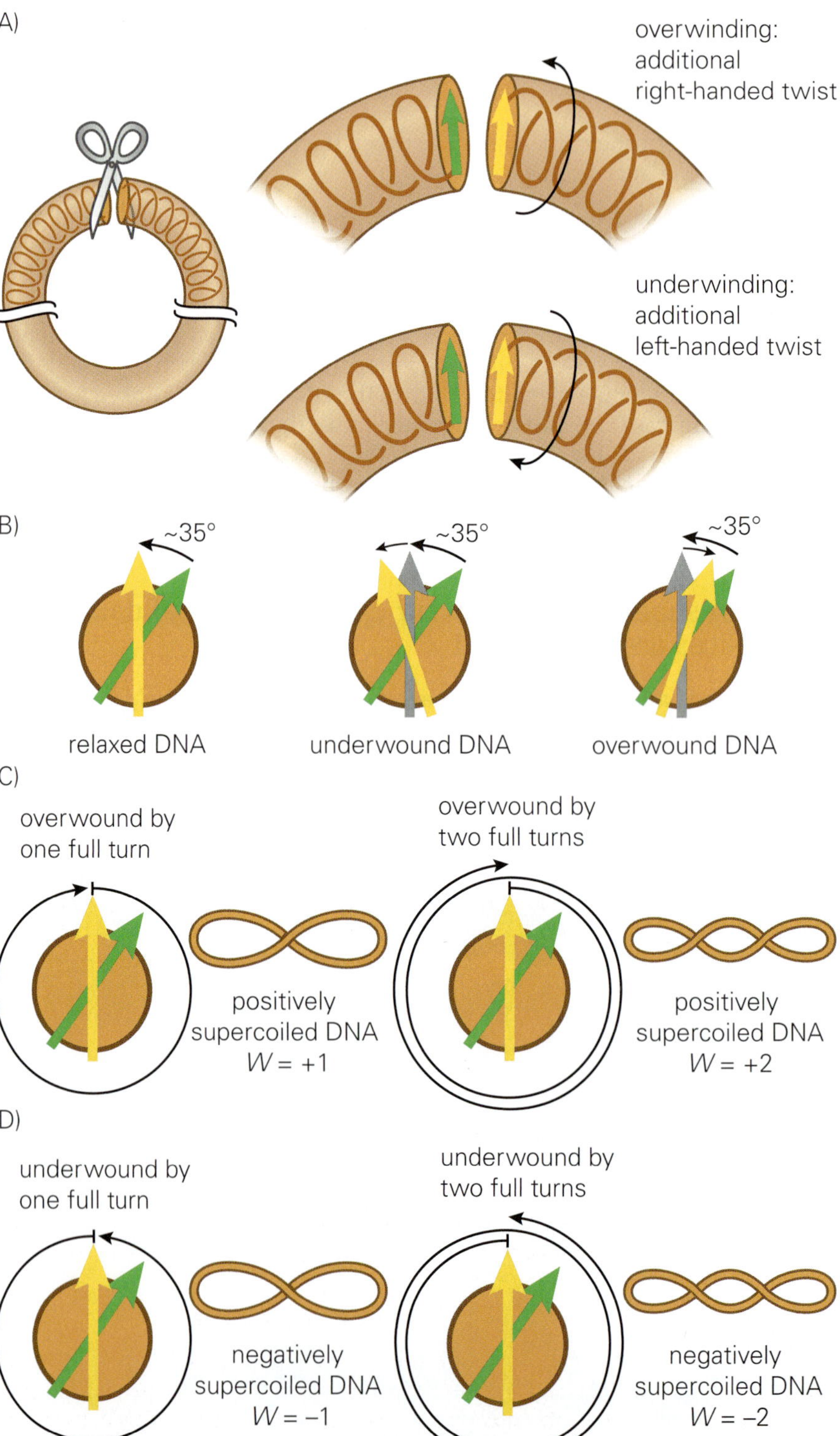


**Figure 2.26 Supercoiling of DNA.** (A) A circular double-helical DNA is illustrated, with only one of the two strands of DNA shown explicitly at the top of the circle. The expanded views illustrate what happens when the DNA is cut, one end is rotated with respect to the other, and then the two ends are rejoined. The base pairs at the junction point are indicated by *green* and *yellow arrows*. If the rotation is right-handed (that is, in the same sense as the rotation of the DNA), then the DNA is overwound. If the rotation is left-handed, then the resulting DNA is underwound. (B) Rotation of the base pairs at the junction point in underwound and overwound DNA, compared to relaxed DNA. (C) The result of overwinding the DNA by one full turn (left) or two full turns (right). The circular DNA relaxes into forms with one superhelical turn with a right-handed sense. This is known as positive supercoiling, and the writhe, $W$, as defined in the main text, is +1. For two full turns before rejoining, the value of $W$ is +2, as shown on the right. (D) Underwinding the DNA by one or two full turns. The DNA now has one or two negative superhelical turns, each with a left-handed sense ($W$ = −1 or −2).

pairs, and assume that there are 10.4 base pairs per turn for B-DNA, then there should be exactly 100 turns in the linear form. As the DNA is bent to bring the ends together, the ends will be positioned perfectly to be joined into one continuous double helix, with the two strands wrapping around each other 100 times. Because there is no twisting of the ends before joining them, this is called a "relaxed" circle of double-helical DNA. Its lowest energy conformation is a flat circle .

Now suppose that, before closing the circle, one end of the DNA is rotated about the helix axis by one full turn and then the ends are joined. Depending on the direction of twisting, the DNA can be **overwound** or **underwound** (corresponding to 101 or 99 wraps of the DNA strands around one another). If the rotation is made in a right-handed sense, then extra turns are introduced into the DNA, which is then overwound (see Figure 2.26A and B). Conversely, if the rotation is made in a left-handed sense, then some of the turns in DNA are removed, and the DNA is underwound.

The extra twist introduces potential energy into the DNA, and the base pairs will seek to return to their optimal stacking geometry (that is, a rotation of ~35° between adjacent base pairs, as shown in Figure 2.26B). If the DNA is constrained to lie in a plane, and if the extra twist is distributed uniformly among all the base pairs, then each base pair would be away from its optimum twist by a small amount—namely, 360°/1040 = 0.35°/base pair. DNA in solution is not constrained to a plane, and changing the path of the double helix axis can release the local strain introduced by this twist by altering the local twist while increasing the overall bending of the DNA.

Based on experimental measurements, it takes less energy to bend DNA than twist it. In the lowest energy structure, the DNA helix axis forms a spiral, which allows the base pairs to return to very close to their optimal twist value at the expense of increased bending. For DNA that is underwound or overwound by one turn, the lowest energy state is then an almost planar figure-eight structure, in which the DNA strands cross over once (see the left-hand diagrams of Figure 2.26C and D).

If one end of the DNA is rotated by two turns (that is, 720°) before joining the ends, then the helix axis twists more. When laid out on a plane, there will be two crossings, as shown in the right-hand diagrams of Figure 2.26C and D. Progressively more deviation in twist from the relaxed value leads to more crossings, as for the real DNA shown in Figure 2.25. Underwound and overwound DNAs undergo helix axis twists in opposite directions. These different forms of DNA are termed *topo*isomers (that is, identical in chemical form but with different *topo*logies).

For this length of DNA (that is, much longer than the persistence length), the bending and twisting behavior is very well described by assuming that the double helix behaves elastically. That is, we can assume that the DNA behaves just like a piece of rubber tubing, which is an excellent model with which to explore the behavior of DNA supercoiling.

## 2.15 Writhe, linking number, and twist are mathematical parameters that describe the supercoiling of DNA

The analysis of how the release of twisting stress along the helix axis is coupled to the spiraling of the helix axis is not straightforward. This type of problem has been analyzed in the mathematical field of topology, which teaches us that there are three variables that describe the problem, but only two of them are independent. These parameters are the **writhe**, $W$, the **linking number**, $L$, and the **twist**, $T$.

The writhe counts how many times the local helix axis crosses itself when the DNA is projected onto a plane. To understand the meaning of writhe, study **Figure 2.27**, in which a circular tube is used to illustrate supercoiling. We start with the tube closed in a relaxed circle. A red line is drawn on the surface of the tube, and

**Writhe, linking number, and twist**

These three parameters characterize the nature of supercoiling in DNA. The writhe describes the rotation of the axis of the DNA helix. The linking number counts how many times the DNA strands cross each other. The twist describes how many times the DNA strands cross the helix axis, and it changes depending on how twisted the helix axis is.

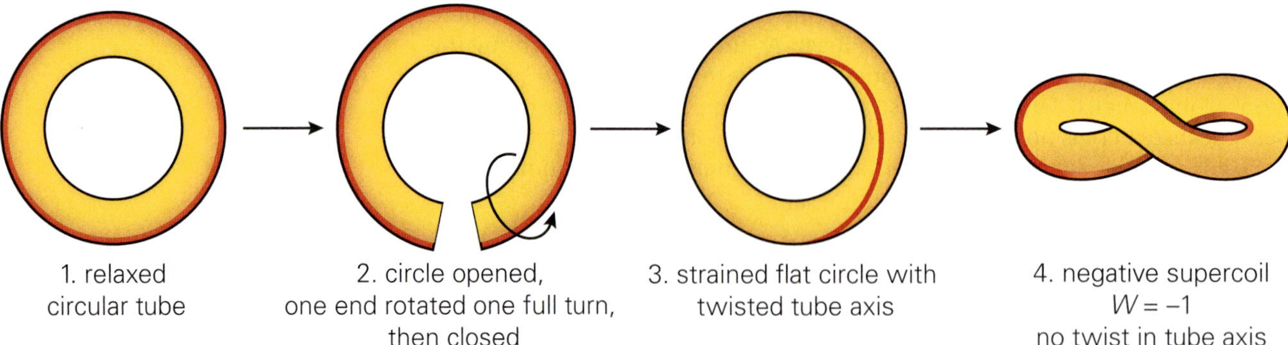


**Figure 2.27 How writhe compensates for twist.** (1) We start with a tube closed in a relaxed circle. The *red* line runs along the outer circumference, and is parallel to the axis of the tube. (2) The tube is cut and, while one end is held fixed, the other end is rotated through a full turn. The ends are then closed. In this diagram, the sense of rotation is in the direction of DNA unwinding (compare Figure 2.26). (3) If the tube is constrained to be in a plane, it will be strained because the axis of the tube is twisted by one full turn. This is shown by the distortion in the *red* line, which moves from the front of the tube to the back and then returns to the front. (4) If the tube is allowed to relax, it adopts a supercoiled structure in which the axis of the tube is no longer twisted. The tube is now negatively supercoiled, with the value of the writhe, *W*, equal to –1. You may want to demonstrate this phenomenon for yourself by using a piece of elastic rubber tubing with a line drawn around the outer circumference. You will need a way to hold the two ends of the tube together. A cylinder that just fits inside the tube works well.

initially the red line does not cross itself (the writhe, *W*, is zero). The tube is then cut in one place and, while one end is held fixed, the other end is rotated through a full turn (the sense of rotation is in the direction of DNA unwinding, which you can verify by comparison with Figure 2.26). The ends are then closed. If the tube is constrained to be in a plane, it will be strained because the axis of the tube is twisted by one full turn (notice how the red line now wraps around the tube in the structure marked "3" in Figure 2.27). If the tube is allowed to relax, it adopts a supercoiled structure in which the axis of the tube is no longer twisted (the red line no longer wraps around the surface of the tube). But now the axis crosses itself once. Because the sense of the supercoiling is left-handed, the value of the writhe, *W*, is equal to –1, which corresponds to negative supercoiling (the left-handed sense of the supercoiling is also illustrated in Figure 2.29).

The linking number, *L*, of DNA strands is the number of times one strand wraps around the other. For the 1040 base-pair circle with 10.4 base pairs per turn, the 100 turns of the double helix correspond to a linking number of 100 when it is closed into a circle. If the DNA is unwound by one rotation before closing, the linking number, *L*, would then be 99, and so on.

The linking number is explained in **Figure 2.28**, which shows a short segment of a double helix, with the two strands colored yellow and magenta. To generate the linking number, we count how many times the yellow and magenta strands cross each other. In order to count in a consistent way, we have to choose a reference line along the helix and count crosses that only occur on this line. For example, in Figure 2.28, we count crosses that occur whenever the yellow strand is in the front of the helix. There are three such crosses, and so the linking number is 3 for this short segment.

The twist parameter, *T*, counts the number of turns that the strands make around the helix axis, projecting onto the local helix axis at each base pair. At first it may seem that the linking number and the twist are the same, and indeed for the relaxed circle they are. However, when writhe is introduced, the twist is not the same as the linking number because the rotation of the helix axis enters into the calculation of the twist.

## 2.16 The writhe, twist, and linking number are related to each other in a simple way

The writhe, twist, and linking number are related in the following way:

$$W = L - T \tag{2.1}$$

We shall not prove this relationship, which involves more advanced considerations of geometry than are useful for this textbook. Nevertheless, we can get a sense of how this relationship arises by considering the situation shown in Figure 2.26D. In this case, the DNA is underwound by one full turn before rejoining the ends. The linking number, $L$, is therefore 99, and this cannot be changed without breaking the DNA circle.

The base pairs are initially underwound ($T = 99$), which introduces strain into the circle. The strain is released by local adjustments in the structure so as to bring the rotation between adjacent base pairs back to the optimum value of ~35°. The value of $T$ is 100 after this process is completed. Thus, according to Equation 2.1, the value of $W$ will be –1, which is what happens when a supercoiled loop is formed. In other words, the change in the writhe from 0 to –1 generates the twist that relaxes the rotation between base pairs back to optimum local geometry.

The figure-eight structures shown in Figure 2.26 correspond to the minimum energy conformations for supercoiled circular DNA without any proteins bound to it. There are, however, other geometries that have the same value of writhe but a different shape. For example, DNA that is packaged in chromatin is wound into a solenoid (a solenoidal form has the DNA wrapping around a virtual cylindrical tube, as shown in **Figure 2.29**). Each turn around the solenoid gives one unit of writhe. For underwound DNA (negative writhe, negatively supercoiled), the solenoid formed is left-handed and, in the interwound (figure-eight) form, the crossover points go from left to right. The higher compaction of the solenoidal form is apparent (compare Figure 1.43).

The solenoidal structures have DNA that is more sharply bent, and hence are intrinsically somewhat higher in energy than the interwound forms. This energy difference can be more than compensated for by interaction with proteins such as histones (see Figure 2.23). The solenoidal geometry leads to a higher compaction of the DNA than interwound structures do, as shown in Figure 2.29B, which is important, particularly during cell replication.

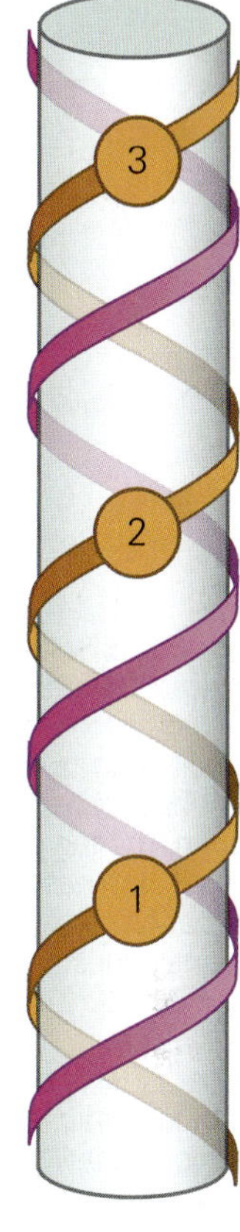


**Figure 2.28 Linking number.** The linking number is the number of times the DNA strands wrap around each other. The linking number for this short segment is 3.

## 2.17 The DNA in cells is supercoiled

The writhe can be determined experimentally by putting interwound DNA onto a surface and counting the number of times the DNA molecule crosses over itself in the electron microscope pictures, with each unit of writhe corresponding to one crossing. The writhe of DNA in cells is usually negative. If $L_0$ is the linking number of the relaxed DNA, and $L$ is the linking number for the supercoiled DNA, then we can define a superhelical density, σ, as follows:

$$\sigma = \frac{L - L_0}{L_0} \qquad (2.2)$$

DNA in cells usually has σ values from about –0.04 to –0.08. Note that in Section 2.16 we argued that writhe compensates for changes in linking number, so that twist is at (or very near) its optimum value, which is $L_0$. In this limit, $(L - L_0) = W$. For an average value of $\sigma = -0.06$, this means that for the DNA with 1040 base pairs that we discussed above, the value of $W$ would be –6.

The discussion of supercoiling has so far focused on circular DNA molecules, such as plasmids. In these molecules, the ends are constrained (that is, not allowed to rotate relative to each other), because they are joined covalently, defining the linking number $L$. For a free linear DNA, rotation of the ends allows the local twist to adjust to the minimum free-energy value. This would make it seem that supercoiling would not be relevant for eukaryotic cells, in which DNA in the chromosomes is intrinsically linear. However, such DNA is not really free in the cell, because there are proteins that bind to the DNA and anchor it to the nuclear membrane (see Figure 17.25). If both ends of a linear segment of DNA are anchored (that is, they *cannot* rotate about the helix axis), then a topological constraint is created that is equivalent to circularization and so all of the features of supercoiling appear.

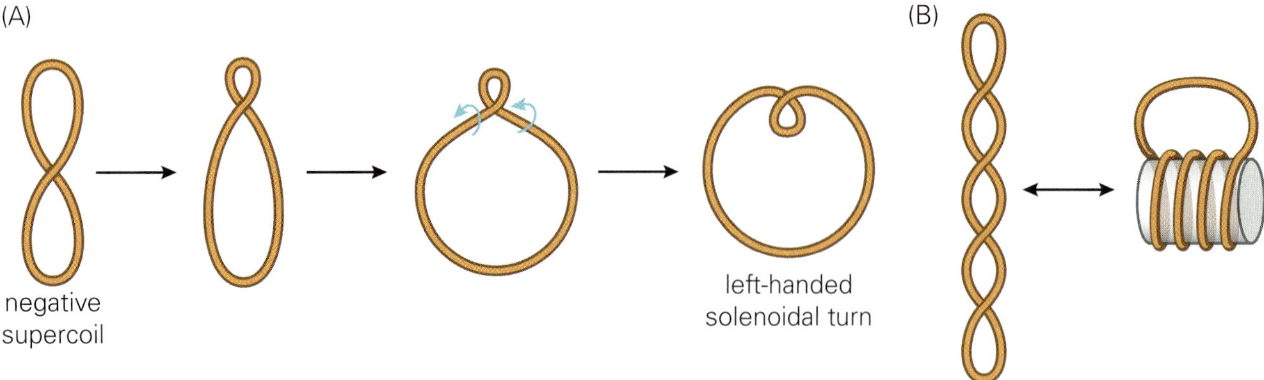


**Figure 2.29 Solenoidal vs. interwound structures of superhelical DNA.** (A) The circular DNA at the left has one negative supercoil. Going across the diagram, it is continuously deformed to make a small loop. The lowest energy state for the small loop has it on the inside, so the rotation to accomplish this corresponds to introducing a twist into the helix axis (small *blue arrows*). (B) In this example, DNA that is underwound by four turns (that is, writhe = –4) is shown in solenoidal and interwound forms that are topologically equivalent.

How is the supercoiling of DNA in cells created and controlled? As with most other processes in cells, supercoiling is both introduced and removed by enzymes. The enzymes that affect supercoiling are called **topoisomerases** (some are also called **gyrases**). There are two types, which act through distinct mechanisms. For type I topoisomerases, one of the strands of the DNA duplex is cut, and the cut strand is allowed to rotate about the intact one, in which the nucleotides are connected by single bonds that allow essentially free rotation. There is no energy source, and so the rotation occurs in the direction that lowers energy, removing any superhelical twist, whether positive or negative. The enzyme then rejoins the ends of the cut strand to return the DNA to its original chemical state.

In the second class of enzymes, type II topoisomerases, both of the DNA strands are cut (but are tightly held by the protein), and another double-stranded segment of DNA is passed through the opening in the cut strand, as shown in **Figure 2.30**. Depending on the directionality of the crossing (that is, going left to right or right to left in Figure 2.30), this can either insert or remove superhelical turns. After strand passage, the cut ends are rejoined to restore the DNA to its original state, with an altered level of supercoiling. Each strand passage changes $W$ by $\pm 2$.

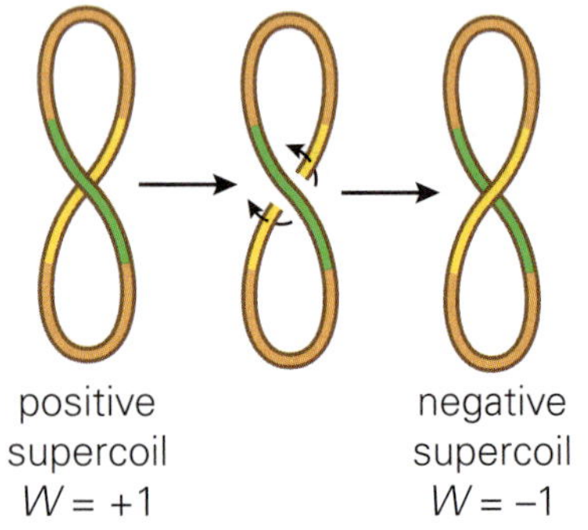


**Figure 2.30 The effect of type II topoisomerase enzymes on DNA topology.** Type II topoisomerases cut a superhelical DNA on both strands (making bonds to the protein so that the strands cannot be released), then allow the strands to pass through one another. This process changes the writhe by 2 (shown here in the direction of increasing negative superhelical density). The directionality for a topoisomerase is determined by whether it prefers to bind a left-handed crossover (for example, $W = -1$) or a right-handed crossover (for example, $W = +1$). The topoisomerase is not shown at all in this picture, which just depicts what happens to the DNA.

## 2.18 Local conformational changes in the DNA also affect supercoiling

Although topoisomerases are the regulators of supercoiling, any protein that binds DNA and alters its local structure will affect supercoiling. A protein that locally unwinds DNA upon binding must increase the winding somewhere else (the linking number, which is the number of times the strands wind around each other, is constant unless the DNA is cut and rejoined as by topoisomerases). An extreme example of this is provided by the actions of enzymes like RNA polymerase, which separates the strands and then moves along the template strand. Polymerase is relatively large and cannot rotate sufficiently quickly to track the DNA, and so it compresses coils in front of it and stretches coils behind it, dramatically affecting local supercoil density. The effect of histones in creating solenoidal forms of DNA was already mentioned above (see Figure 2.29).

Local modifications in the structure of DNA, such as formation of a segment of Z-form DNA or a cruciform structure, will modify the superhelicity of the DNA (**Figure 2.31**). Consider, for example, the DNA discussed above with 1040 base pairs, and assume that it contains a sequence of alternating C's and G's that can form one turn of left-handed Z-form DNA. If this DNA is circularized under high-salt conditions, in which the C-G repeat is in the Z-conformation, then it would have $L = 98$; the one left-handed turn cancels one right-handed turn from the 99 turns of B-form DNA (assuming the transition between B and Z takes no base pairs).

Under high-salt conditions, bending this DNA smoothly into a circle would give $W = 0$, because the twist at each base pair is at its optimal value, $T = 98$. The lower

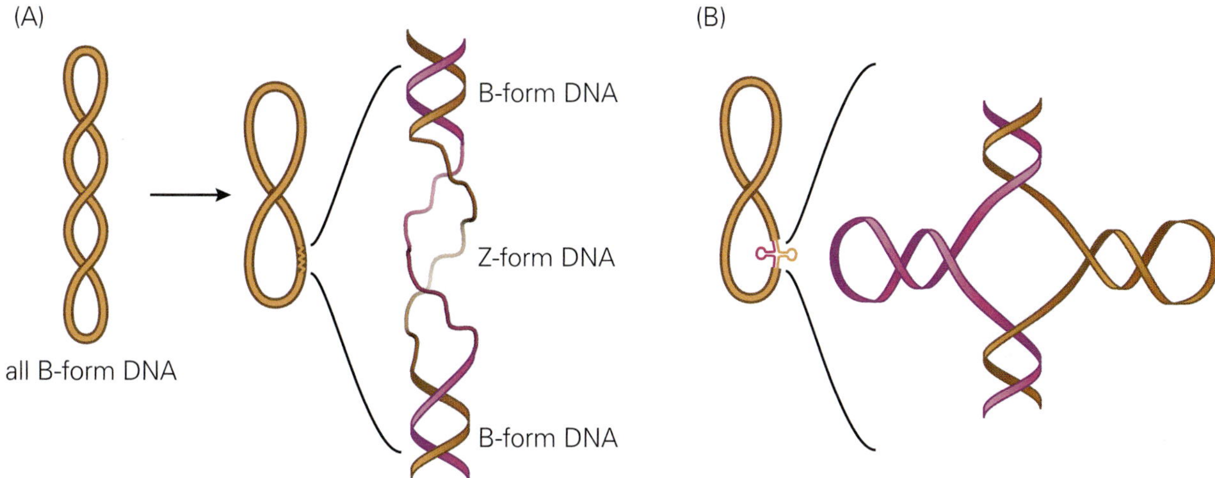


**Figure 2.31 Local structure affects overall supercoiling.** Negatively supercoiled DNA is underwound. A local change that reduces the intrinsic twist reduces the writhe that occurs in the DNA. Two such local changes are shown, changing a turn of DNA from right-handed B-DNA to left-handed Z-DNA (A), or forming a cruciform in which two turns of DNA are extruded into the arms, where they do not contribute to the twist of base pairs going around the circle (B). In this example, both of these changes in structure reduce the writhe by two units. The actual change in writhe depends on the length of the Z-DNA segment, or the length of the cruciform arms. (Note that in a real DNA, the B-to-Z junction is not zero length as idealized here).

optimal value of the twist arises because the left-handed twist of the Z-form DNA cancels two turns of right-handed twist for the B-form DNA. If the salt concentration is lowered after covalent closure, then all of the DNA will convert to the B-form, and the optimum value of $T$ will be 100 again. But note that $L$ is still 98 (because the strands are not broken in the change of salt concentration), and so in this condition $W$ will be −2. The change can also go the other way. If the DNA is initially negatively supercoiled, and if one turn of Z-form DNA is generated, then the change in writhe will be +2. Such a situation is illustrated in Figure 2.31. Note that because supercoiling corresponds to the introduction of torsional strain energy, the ability of Z-form DNA to reduce the strain means that Z-form DNA is stabilized by negative supercoiling—that is, the transition of a segment of DNA from the B-form to the Z-form is energetically more favorable in a negatively supercoiled plasmid than in a linear DNA. At high negative superhelical density, the transition to Z-form can be favorable even at low salt concentrations. The same ideas apply to the formation of a cruciform structure, which also reduces the number of turns in the DNA by two for each 10–11 base pairs in the "arms" (see Figure 2.31).

# B. THE FUNCTIONAL VERSATILITY OF RNA

## 2.19 Wobble base pairs are often seen in RNA

Watson-Crick base pairs, also called **standard** or **canonical base pairs**, have a uniform geometry. Each of the four standard base pairs (A-T/U, T/U-A, C-G, and G-C) has the same width, and each glycosidic bond makes similar angles to the sugar-phosphate backbone (**Figure 2.32**). The result is the beautiful regularity of the DNA double helix—a regularity also seen in A-form RNA-RNA or RNA-DNA helices. As we discuss in Chapter 19, this regularity underlies the fidelity of DNA replication.

**Standard base pairs and nonstandard base pairs**

The standard base pairs, or Watson-Crick base pairs, are A-T and C-G in DNA and A-U and C-G in RNA. Folded RNA structures often contain alternative base pairs (for example, G-U), which are referred to as nonstandard.

RNA structure is by no means constrained, however, to regular double helices. RNA molecules exist in a large variety of sizes and shapes—a variety encouraged by an expanded repertoire of building blocks, including modified bases, modified sugars, and **nonstandard base pairs** that do not occur in DNA. Noncoding RNA molecules do not have to satisfy the demands of fidelity because they do not serve as templates, and so alternative base pairing interactions are a common feature of their structures. These base-pairing interactions do not have the same geometry as the standard Watson-Crick base pairs, as shown in Figure 2.32.

Nonstandard base pairings in RNA often have specific functions as recognition elements for proteins, nucleic acids, and ligands. They can also serve as ion-binding sites. Such noncanonical base pairs are linked by at least one interbase

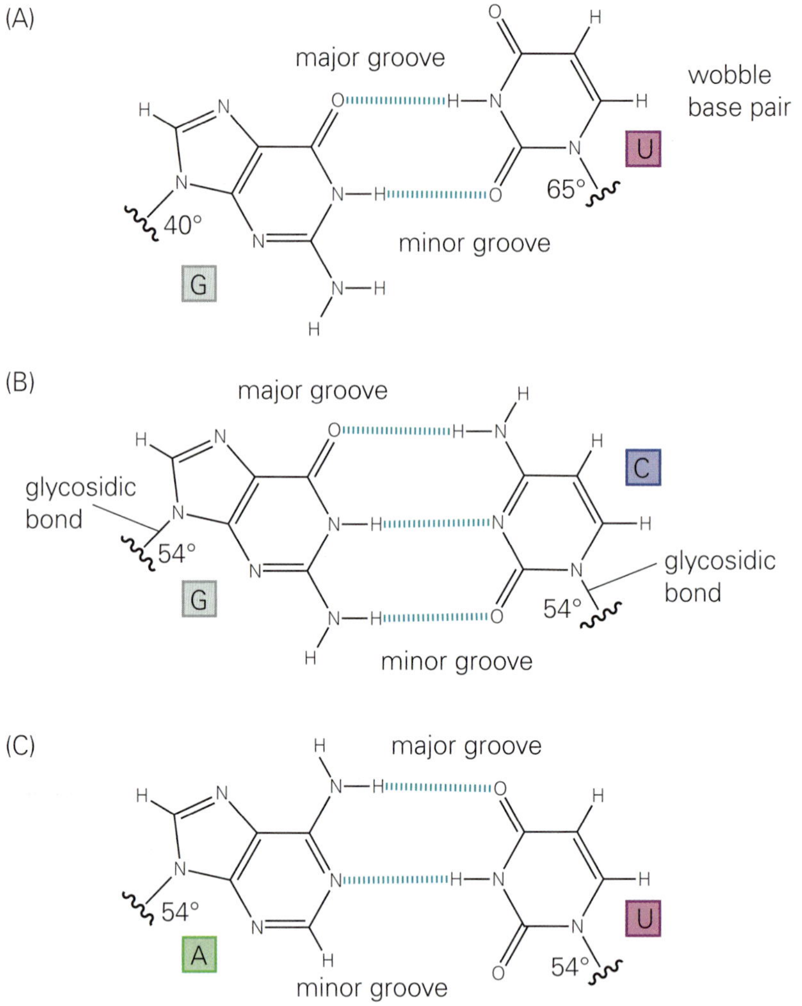

**Figure 2.32 Comparison of wobble and standard base pairs.** (A) The wobble G-U base pair is compared to the standard (B) G-C and (C) A-U base pairs. They all have generally similar dimensions, but the angles at which their glycosidic bonds meet the sugar-phosphate backbone are markedly different (a constant 54° in the standard base pair versus 40° and 65° in the wobble pair). The altered shape of the pyrimidine-purine pair (compare G-U with G-C), causes a regional change in the shape of the double helix. See also Figure 2.33.

**Wobble base pair**

A nonstandard base pair, first identified in codon–anticodon interactions. The base pair involves two hydrogen bonds (between G and U, as illustrated in Figure 2.32), but the geometry of the base pair is different from that of a standard base pair.

hydrogen bond and occasionally involve water-mediated base-base interactions. G-U **wobble base pairs** are the most common and conserved non-Watson-Crick base pairs in RNA (see Figure 2.32 and **Figure 2.33**).

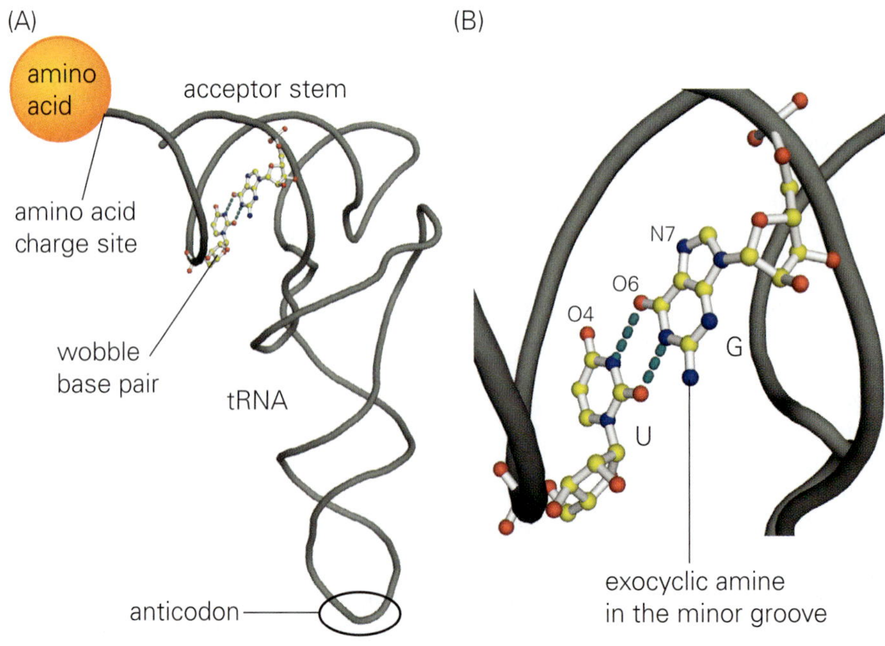


**Figure 2.33 A wobble (G-U) base pair in a tRNA molecule.**
(A) Structure of the tRNA, showing the location of the wobble base pair.
(B) Expanded view of the wobble base pair. (PDB code: 6TNA.)

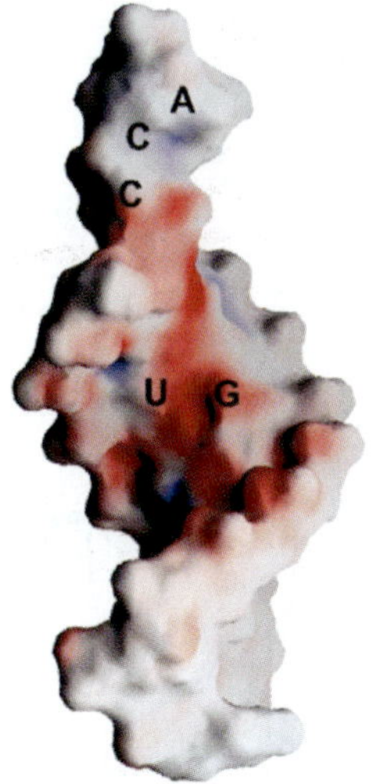

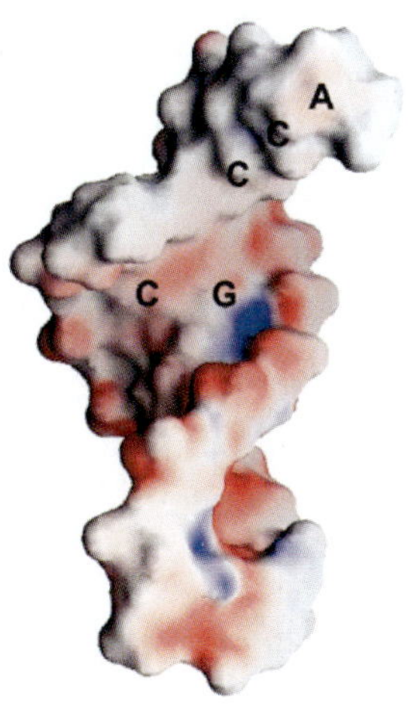


wobble U-G base pair

standard C-G base pair

**Figure 2.34 Changes in the surface of a tRNA molecule upon introducing a wobble base pair.** A wobble base pair influences the surface of a tRNA molecule. The structure of a tRNA acceptor stem that normally contains a U-G wobble base pair is shown on the left. Replacement of this wobble base pair by a canonical Watson–Crick C-G base pair changes both the shape of the tRNA and the properties of its surface. Regions of the molecular surface with negative and positive electrostatic potential are colored *red* and *blue*, respectively (see Section 6.24 for an explanation). (Adapted from G. Varani and W.H. McClain, *EMBO Rep.* 1: 18–23, 2000. With permission from the European Molecular Biology Organization.)

The G-U pairing was first proposed to occur in codon–anticodon interactions when Francis Crick noticed that the G and U bases would be able to form two hydrogen bonds by interacting through the same face of the base involved in Watson-Crick pairing (see Figure 2.33). The G-U wobble base pair occurs not just in codon–anticodon interactions, but also in every class of RNA from all organisms.

Why are G-U base pairs so ubiquitous? Their unique chemical and structural properties make them particularly well suited to binding RNA molecules, proteins, or other ligands. For example, the exocyclic amino group of guanosine projects into the minor groove (see Figure 2.33) and the N7 and O6 groups of guanosine and the O4 of uridine contribute to the negative electrostatic potential in the major groove (**Figure 2.34**). This latter feature increases the RNA molecule's ability to bind divalent metal ions. In fact, tandem G-U base pairs are often sites of metal-ion binding.

The stability of G-U base pairs approaches that of Watson-Crick pairs, and they are more stable than most other mismatched base pairs. This stability allows G-U pairs to substitute for Watson-Crick base pairs in phylogenetically conserved double-stranded regions in RNA. For example, a G-U pair in the acceptor stem of tRNA helps it load the correct amino acid. One such G-U pair within the acceptor stem of $tRNA^{Phe}$ is shown in Figure 2.33. Figure 2.34 shows how the shape of the tRNA and the electrostatic potential of its surface change as a result of incorporating this non-Watson-Crick base pair. One function of a tRNA molecule is to ensure that it is recognized with high fidelity by the appropriate amino acyl tRNA synthetase, so that only the correct amino acid is attached to it; alterations in surface shape and charge caused by the presence of the G-U pair can help the recognition process.

G-U pairs are also found in catalytic RNA molecules that are involved in RNA self-splicing. One such RNA molecule, called the self-splicing group I intron, contains an almost universally conserved G-U pair at one of the sites of cleavage, **Figure 2.35**. The structure of this RNA is shown in **Figure 2.36**, which also serves to illustrate the folding of a complex RNA molecule.

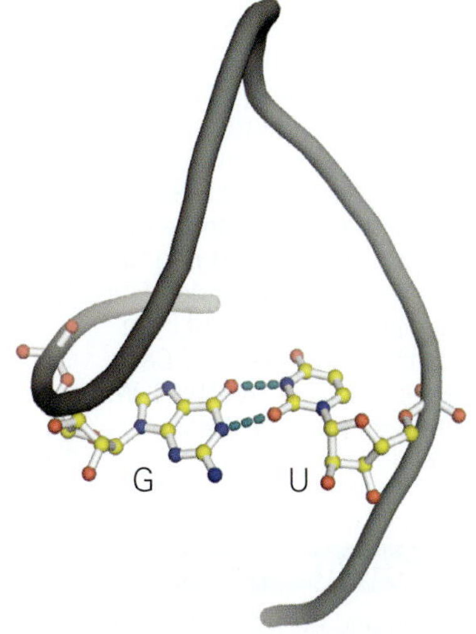


G-U wobble base pair

**Figure 2.35 G-U wobble base pair near the catalytic site of the self-splicing group I intron.** The structure shown is the helical stem denoted "P1" in Figure 2.36. (PDB code: 1C0O.)

## 2.20 Nonstandard base-pairing is common in RNA

There are many different kinds of non-Watson-Crick base pairing interactions, each with its own pattern of hydrogen bonds between the bases. These interactions can help stabilize the sorts of complex RNA conformations that might not

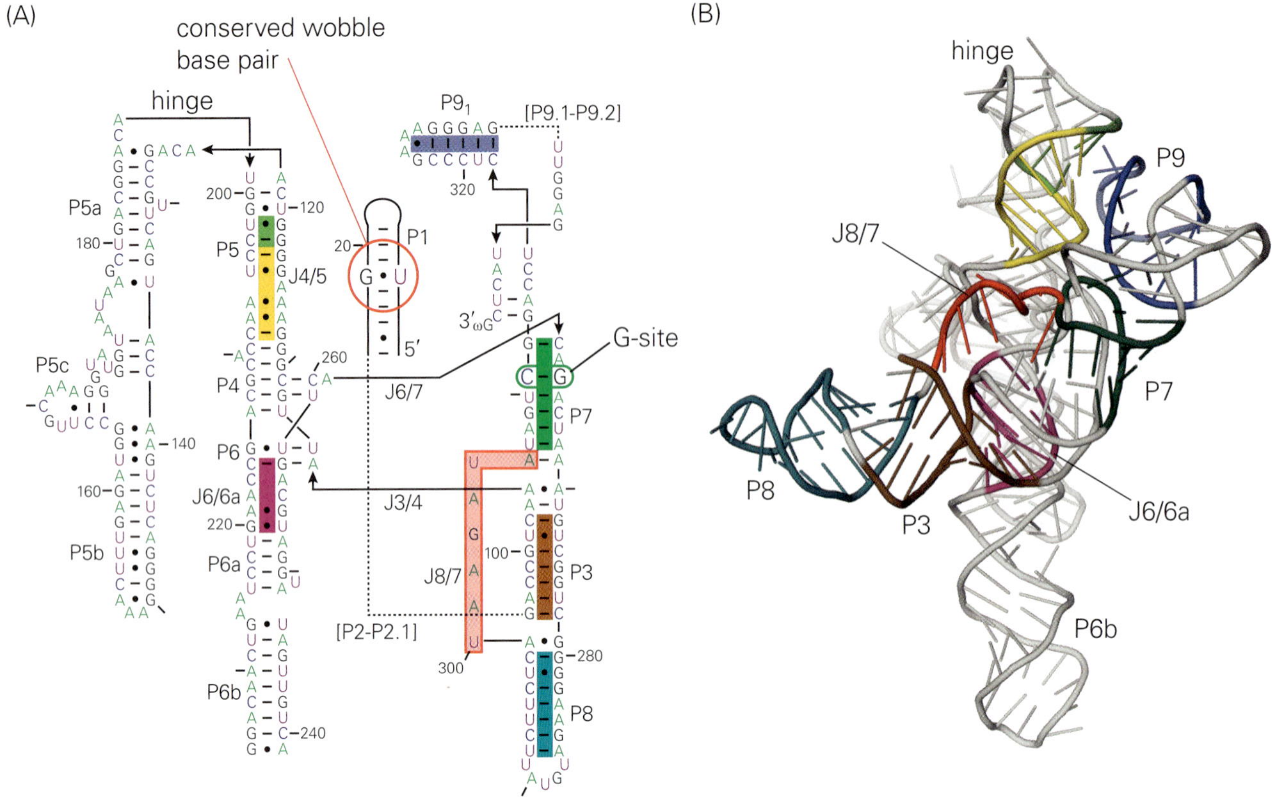


**Figure 2.36 Structure of a self-splicing group I intron.** RNA introns are segments of RNA that are cleaved out of precursor RNA molecules (see Chapter 1). The intron shown here has the unusual ability to catalyze its own excision. One of the cleavage sites is at the 5′ end, near a G-U wobble base pair (see Figure 2.35). (A) The secondary structure diagram shows how a single RNA molecule forms many double-helical elements using internal complementary sequences. (B) The structure of this RNA molecule, showing how the helical elements pack together. Some of the helical segments in (A) are colored to indicate where they are in the folded structure shown in (B). (A, adapted from B.L. Golden et al., and T.R. Cech, *Science* 282: 259–264, 1998. With permission from the AAAS; B, PDB code: 1GRZ.)

**Hoogsteen base pair**

A nonstandard base pair in which the hydrogen-bonding interactions utilize the Watson-Crick base-pairing edge on one base and the edge corresponding to the major groove in the other (see Figure 2.37).

be possible using only standard pairing and unmodified nucleotides (see Section 2.21). We shall not list all types of nonstandard pairing, but simply note one other important class of noncanonical base pairs—the **Hoogsteen base pairs** named after Karst Hoogsteen. These involve hydrogen bonds between the Watson-Crick base-pairing edge of one base, and the major groove edge of another base (**Figure 2.37**).

Hoogsteen base pairs are possible because within the major groove the exposed edges of the purines and pyrimidines contain several atoms that are available for hydrogen bonding. Hoogsteen base pairs are formed when a single-stranded piece of DNA or RNA enters the major groove of double-stranded nucleic acid and forms hydrogen bonds with one of the strands in the double helix. The result is the formation of a triple helix (**Figure 2.38**). These kinds of interactions can be used to detect specific sequences in DNA and, for example, shut down transcription of certain genes.

## 2.21 Some RNA molecules contain modified nucleotides

While the primary structure of RNA is derived from a template DNA sequence, most eukaryotic RNA molecules are post-transcriptionally modified. Some nucleotides are eliminated (and some are inserted), and the remaining nucleotides at certain positions have bases that are chemically distinct from the normal RNA complement of A, U, C, and G (the sugar groups are always ribose). The flourishes added to the RNA product after it is generated mean that the correlation between the nucleotides in the template DNA and the transcribed RNA is not always simple.

Some of the covalent modifications that RNA molecules can undergo are shown in **Figure 2.39**. These include the methylation of nucleotide bases and the 2′-OH groups of ribose sugars and the formation of unusual bases, such as pseudouridine (Ψ) and dihydrouridine (D). Thus, to determine the sequence of a mature RNA molecule accurately, it is not enough just to know the sequence of the gene

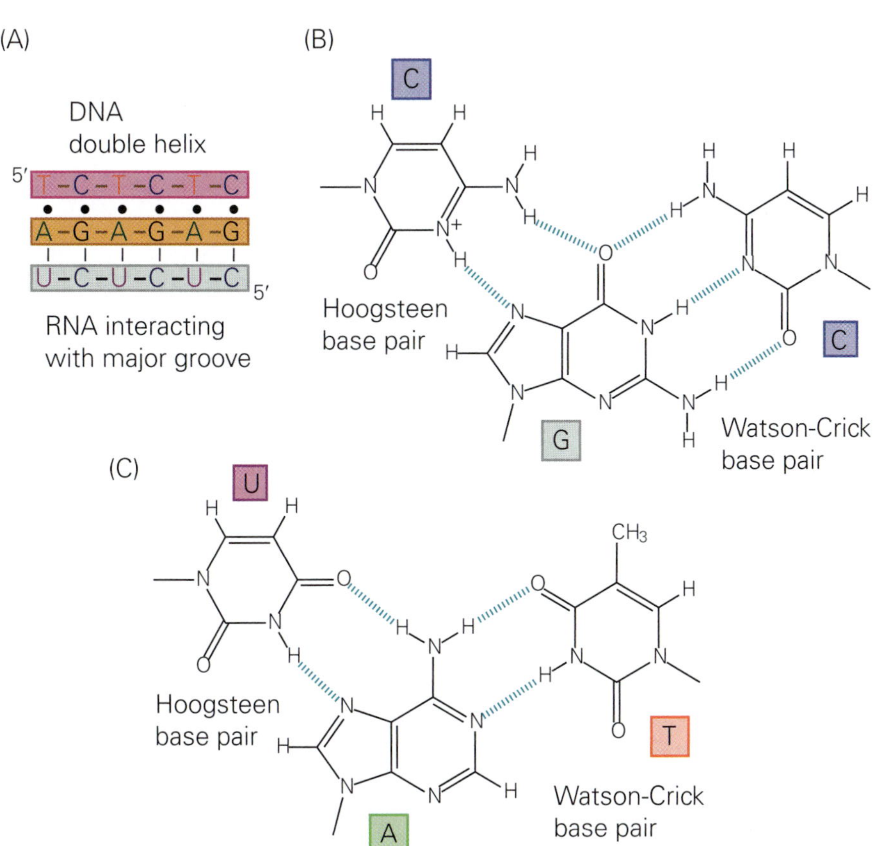


**Figure 2.37 Hoogsteen base-pairing.** (A) The sequence of a double-helical segment of DNA is shown (*magenta* and *orange*), to which is bound a strand of RNA (*gray*). The RNA interacts with bases in the major groove of double-helical DNA, forming C-G (B) and U-A (C) Hoogsteen base pairs. The cytosine has to be protonated (that is, positively charged) to form the Hoogsteen base pair shown here.

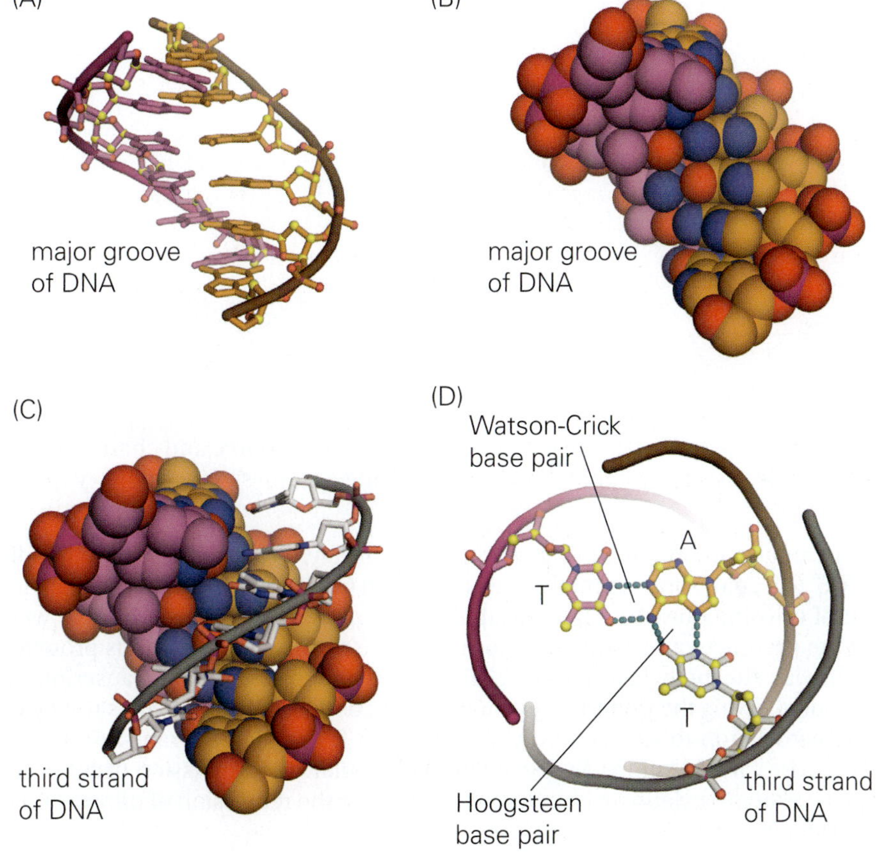


**Figure 2.38 Triple-helix formation by Hoogsteen base-pairing.** A DNA double helix is shown in stick representation (A) and as a space-filling model (B). (C) A third strand of DNA is shown interacting with the edges of the bases in the double helix. (D) Hoogsteen base-pairing between a T on the third strand and an A in the double helix. (Adapted from J.P. Bartley, T. Brown, and A.N. Lane, *Biochemistry* 36: 14502–14511, 1997. With permission from the American Chemical Society; PDB code: 1AT4.)

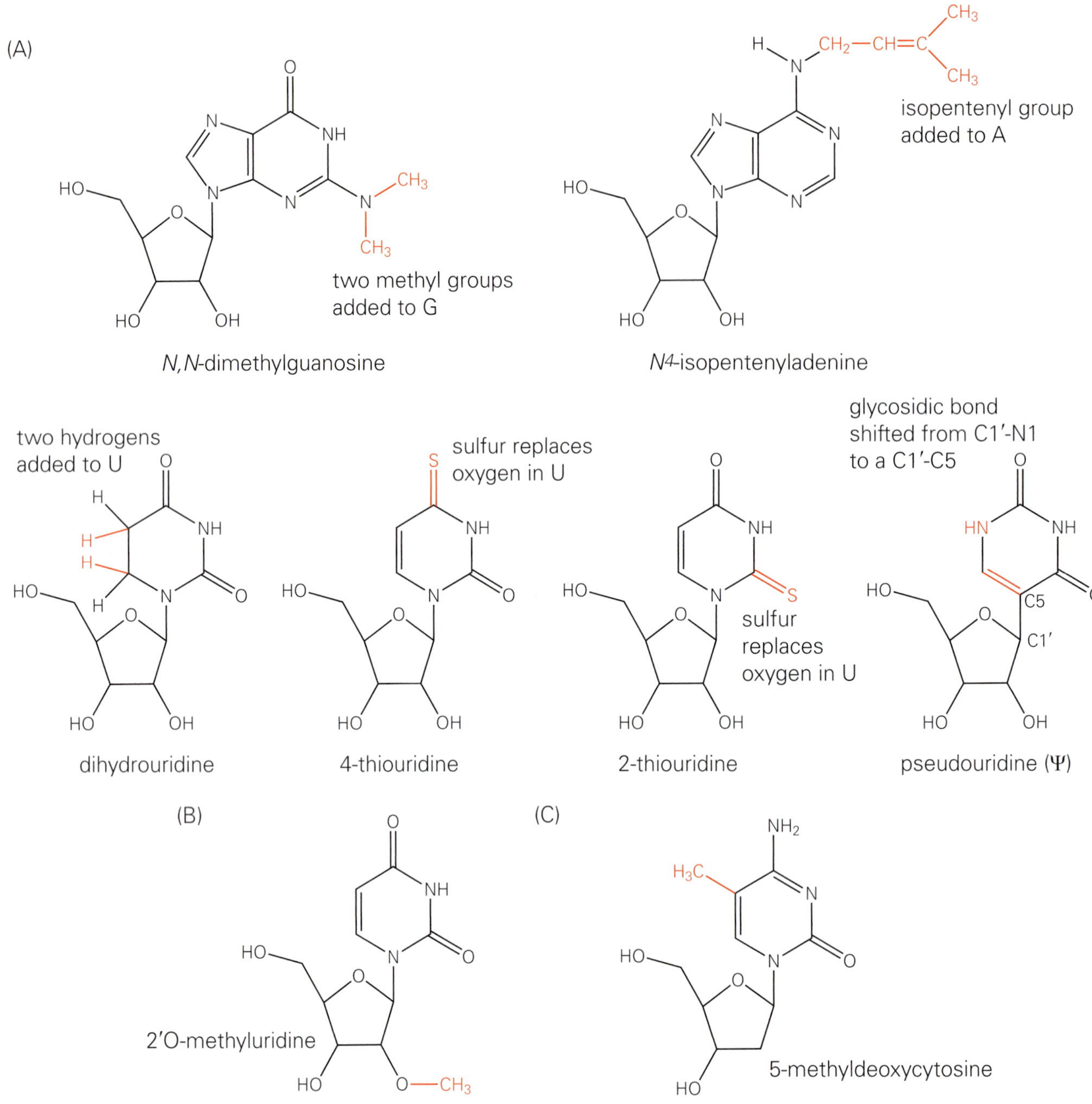

(A)

two methyl groups added to G

*N,N*-dimethylguanosine

isopentenyl group added to A

*N4*-isopentenyladenine

two hydrogens added to U

dihydrouridine

sulfur replaces oxygen in U

4-thiouridine

sulfur replaces oxygen in U

2-thiouridine

glycosidic bond shifted from C1′-N1 to a C1′-C5

pseudouridine (Ψ)

(B)

2′O-methyluridine

(C)

5-methyldeoxycytosine

**Figure 2.39 Post-transcriptional modifications of RNA and DNA.** (A) Various modifications of the bases in RNA. (B) Methylation of the 2′-OH group in the ribose sugar of RNA. (C) Methylation of the cytosine base in DNA.

encoding it: the RNA must be extracted from its native source and characterized, for example, using a combination of sequencing and mass spectrometry methods.

DNA can also be modified (Figure 2.39C). Typically, cytosines that lie immediately 5′ to a guanosine are methylated (so-called CpG methylation). The methylation of cytosine does not affect its ability to form base pairs in the normal way, but the presence of methylated cytosine can be recognized by various proteins. In this way, the methylation of DNA can regulate processes, such as transcription, without affecting the underlying genetic instructions. Many eukaryotes use cytosine methylation to silence foreign elements in DNA, such as transposons and retroviruses. Methylation can also be used to regulate the expression of genes—a high level of methylation within a gene can lead to the repression of transcription of the gene.

Why are RNA molecules modified chemically? The answer presumably has to do with increasing the repertoire of what the RNA is able to do. The available set of nucleotide building blocks in nucleic acids is still rather limited compared to the 20 amino acid building blocks of proteins, and the spectacular functional versatility of proteins is probably one reason that they took over from RNA most of the catalytic and structural functions within the cell. The greater variety of sidechains in proteins allowed proteins to become more highly specialized at carrying out nuanced tasks in molecular recognition and catalysis, or to develop more diverse structural capabilities.

Despite the general usurpation of functional tasks by proteins, there are several key processes in the cell that critically depend on RNA. These include the synthesis of peptide chains in the ribosome, the splicing of precursor mRNA, and the coupling of codons in mRNA to protein synthesis by tRNAs. The chemical modification of nucleotides in RNA has improved the ability of some of these RNA molecules to carry out these tasks.

## 2.22 A tetraloop is a common secondary structural motif that caps RNA hairpins

The structures of some noncoding RNA molecules are complex (for example, see Figure 2.36). A primary sequence of nucleotides folds into one or more helices and intervening single-strand regions, all of which finally fold together into a specific tertiary three-dimensional shape. The helices and the elements that connect them together are called **RNA secondary structural elements** or **motifs**, and include helices, single-stranded regions, hairpins, bulges, internal loops, and junctions. We will consider a few different ways in which some of these RNA secondary structural motifs are assembled to form the more complicated tertiary structures.

Hairpin loops are formed when an RNA strand folds sharply back on itself and the resulting "stem" is stabilized by complementary base pairing (**Figure 2.40**). Internal loops are also formed when one or more bases within a duplex segment remain unpaired. A particularly common RNA motif is one that stabilizes the

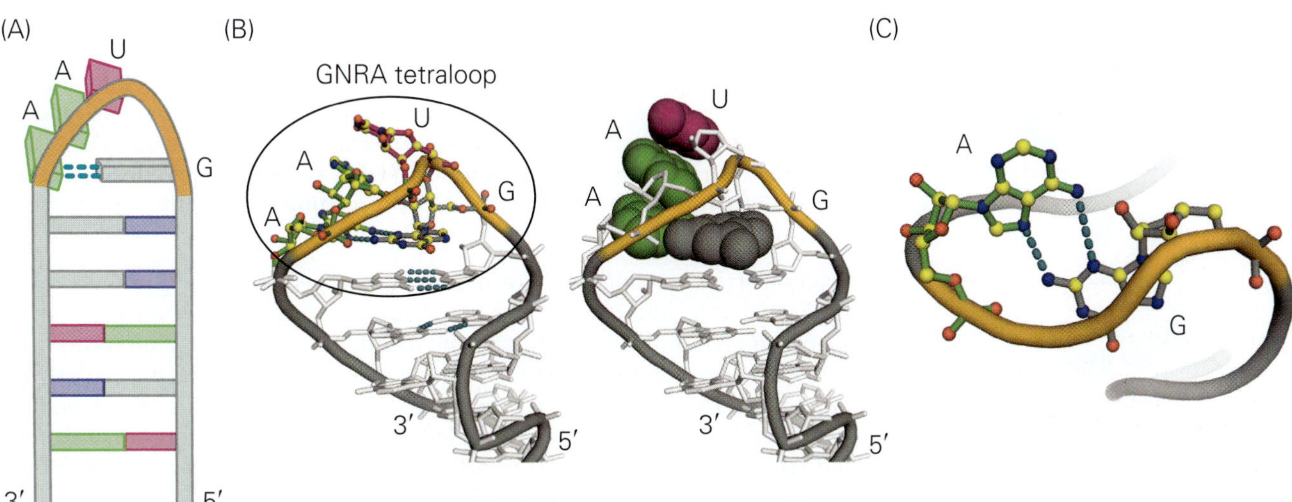


**Figure 2.40 A GNRA tetraloop motif.** A schematic diagram of a stem-loop structure is shown in (A). Three views of the structure are shown in (B) and (C). The particular GNRA tetraloop shown here has the sequence GUAA. The G of the GNRA tetraloop stacks against the top base pair of the double-helical stem, while the U and the two As stack against each other and are swung away from the stem. The turn structure is stabilized by the formation of nonstandard hydrogen bonds between the first G and the fourth A, as indicated by *dotted blue lines* in (A) and (C). (PDB code: 1MSY.)

sharp turn of the phosphate backbone in a stem-loop substructure. Such a loop structure is stabilized by the formation of base-pairing interactions in the stem, but the loop region itself often forms a specific structure that drives the formation of a sharp turn. In the tertiary structures of protein molecules, such turns are stabilized primarily by the backbone of the protein chain (see Section 4.12), but in RNA the negative charge on the phosphate groups tends to keep the sugar-phosphate groups away from each other. Instead, interactions between the bases in loop structures can induce the backbone to undergo a sharp turn.

**GNRA tetraloop**

A GNRA tetraloop is a structural motif in RNA that helps induce a sharp turn in the backbone. The name refers to the sequence of the motif, which has a guanine (G) in the first position, any nucleotide (N) in the second position, a purine (R, either A or G) in the third position, and adenine (A) in the fourth position, as illustrated in Figure 2.40.

One common turn-inducing motif occurring within a stem-loop structure is the **GNRA tetraloop** (see Figure 2.40). The name stands for the four-nucleotide sequence motif that is characteristic of these tetraloops: guanine (G) is followed by any nucleotide (N), a purine (R), and a final adenine (A). There are a number of other tetraloop motifs in RNA, but we do not describe these here.

The turn formed by the GNRA tetraloop is stabilized by hydrogen bonds and by base stacking. The first G stacks with the base pairs of the main stem. The final A is located on the same level of the stem as the G, but is swung out into the minor groove, where it provides a platform onto which the second and third bases of the tetraloop are stacked (see Figure 2.40C). The loop is capped by nonstandard hydrogen bonds between the G and the last A (see Figure 2.40A and C). The hydrogen-bond geometry depicted in Figure 2.40C requires G and A at the first and fourth positions, respectively; and the purine at the third position provides better stacking interactions with the fourth A than would a pyrimidine. These factors explain the preferred sequence motif of GNRA for this tetraloop.

The GNRA structure is seen in many different RNA molecules, where it occurs in a variety of structural contexts. The tetraloop is a common recognition element in RNAs, because the three flipped-out bases provide a number of interaction possibilities for other proteins or RNA molecules. One such tetraloop in the ribosome is the recognition target of the deadly toxin ricin. Produced by the castor bean, ricin is a protein that is toxic to many animals and insects, including humans. It binds to a GNRA tetraloop in the ribosome, catalyzing a depurination (removal of a purine base from its sugar) in the ribosomal RNA that inactivates the ribosome. The resulting cessation of protein synthesis causes the acute toxicity of ricin. The importance of the ricin-sensitive RNA loop for ribosome function is explained in Section 19.30.

(A)

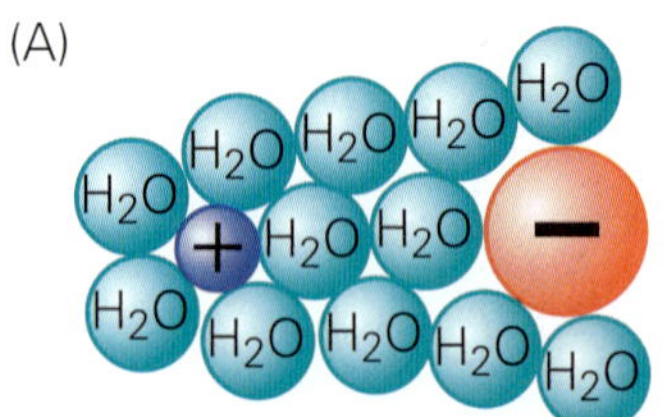


(B)

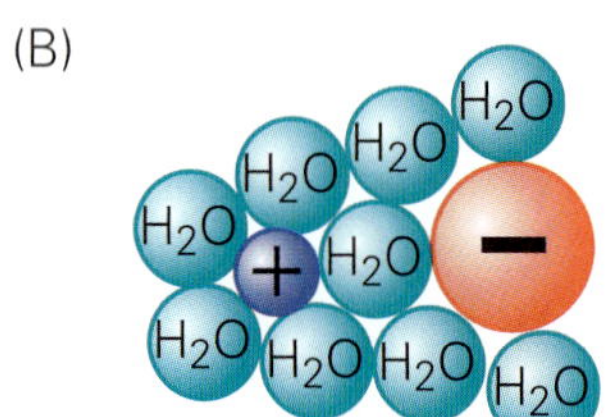


(C)

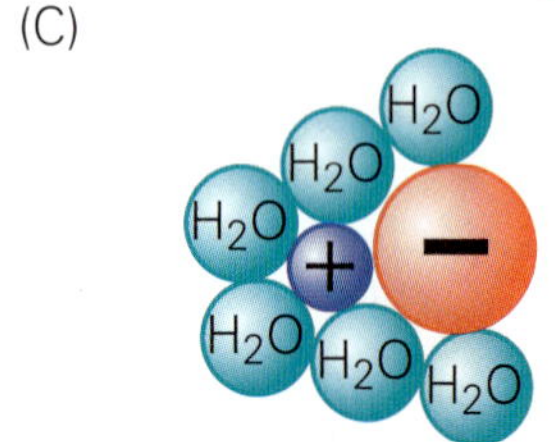


**Figure 2.41 Schematic representations of different types of metal-ion interactions with RNA.** A phosphate oxygen atom of RNA is indicated by the *red sphere* and a metal ion by the *dark blue sphere*. (A) Diffuse ion interaction. (B) Outer-sphere interaction. (C) Inner-sphere interaction. (Adapted from V.K. Misra and D.E. Draper, *Biopolymers* 48: 113–135, 1998.)

## 2.23 Interactions with metal ions help RNAs to fold

The negative charge of the phosphate backbone of RNA hinders RNA from folding into compact structures unless it is at least partially neutralized. In physiological solutions, $Mg^{2+}$, $Na^{+}$, and $K^{+}$ ions are present at significant concentrations, and can interact with the phosphate backbones of DNA and RNA. These positive metal ions (cations) promote the folding of RNA by screening the negative charge of the phosphate groups.

Interaction of these cations with the negatively charged phosphate groups is energetically favorable, and may involve fully hydrated cations. However, if the phosphate of the RNA directly coordinates the metal, then a coordination water must be removed from the metal, which may be energetically costly, particularly for divalent metals such as $Mg^{2+}$. The penalties for dehydration can be comparable to the enhancement in strength of the electrostatic interaction between the metal ion and the RNA, and hence are important to consider when trying to understand which sites in RNA will bind metal ions.

Three types of metal–RNA interactions are illustrated schematically in **Figure 2.41**. The first is the **diffuse ion interaction**, a general charge screening of the negative phosphate groups by a "sea" of monovalent or divalent cations (see Figure 2.41A). These interactions are typically mediated by several water molecules. Cations involved in diffuse interactions move and exchange positions frequently. The

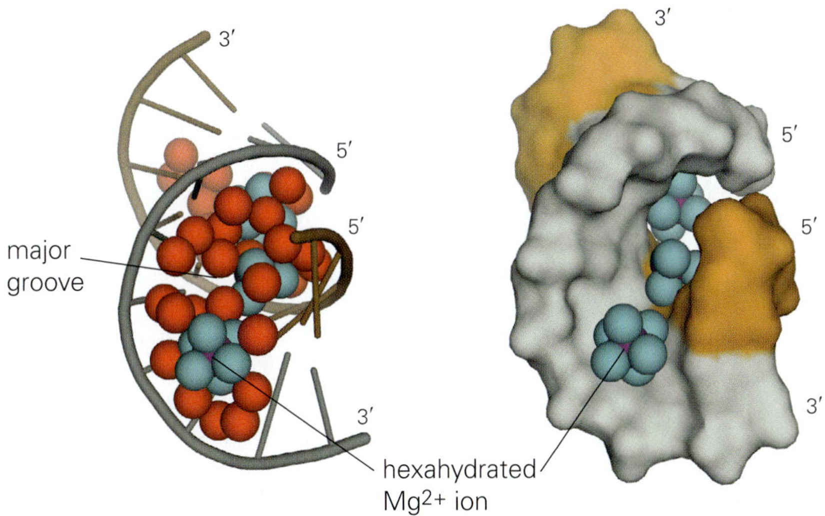


**Figure 2.42 Hydrated magnesium ions in the major groove of an RNA-DNA hybrid.** In the drawing on the left, the octahedrally coordinated $Mg^{2+}$–water clusters are colored *aqua*, and the water molecules surrounding them are colored *red*. The octahedral $Mg^{2+}$–water clusters are shown in isolation on the surface of the major groove in the drawing on the right. (PDB code: 1DNO.)

positions of such ions can be quite widely spread around the RNA, or they can be localized to fairly specific regions, such as within the major groove (**Figure 2.42**).

The second type of metal-RNA interaction is called **outer-sphere coordination** because a ligand on the metal, generally a water molecule, also interacts with a functional group on the RNA (see Figure 2.41B and **Figure 2.43**). The third type of interaction is called **inner-sphere coordination** because the metal ion loses a water ligand and binds directly to the RNA molecule (Figure 2.41C).

**Diffuse, outer-sphere, and inner-sphere ions**

The metal ions that interact with RNA (for example, $Mg^{2+}$) are classified by how directly they interact with the nucleic acid. Diffuse ions have water between the hydrated ion and the nucleic acid. Outer-sphere ions are those where a water that is coordinated to the metal also makes contact to the nucleic acid. Inner-sphere ions make direct contact with the nucleic acid.

The interactions of magnesium ions ($Mg^{2+}$) with RNA and DNA are particularly important from a structural point of view. $Mg^{2+}$ prefers to be octahedrally coordinated, with four ligands in one plane and two aligned axially above and below the plane. $Mg^{2+}$ ions with six water molecules bound to them (hexahydrated $Mg^{2+}$) are commonly observed in RNA structures (see Figure 2.43). The water molecules coordinating the $Mg^{2+}$ can interact with functional groups on the bases or with the phosphate groups (see Figure 2.43). Oxygen atoms from phosphate groups are also seen to directly coordinate the $Mg^{2+}$ ion, as illustrated for one such ion in the ribosome in **Figure 2.44**. The ribosome is constructed from a dense organization of RNA molecules that are tightly packed together and, without this kind of metal-ion coordination, the formation of the structure would be disfavored due to charge repulsion between the phosphate groups.

## 2.24 RNA tertiary structure involves interactions between secondary structural elements

The tertiary structure of RNA results from interactions between distinct secondary structural elements. The development of the tertiary structure may entail the

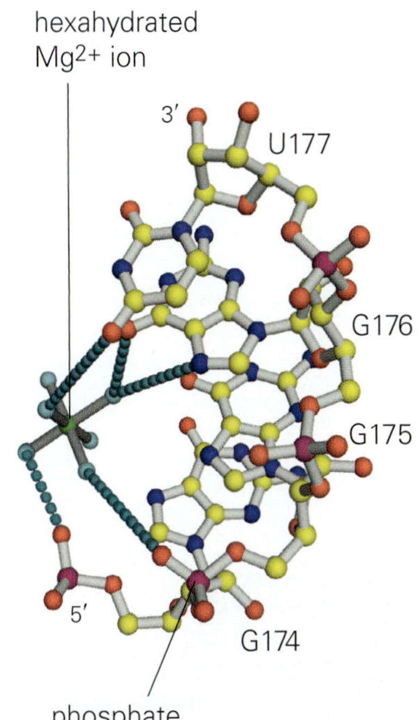


**Figure 2.43 Structural organization provided by an octahedrally coordinated metal ion.** The nucleotides shown here are from the self-splicing group I intron illustrated in Figure 2.36. The magnesium ion is coordinated by six water molecules that form hydrogen bonds with bases and with the phosphate backbone. This illustration is based on a crystal structure of the intron bound to osmium hexamine, a mimic of hexahydrated $Mg^{2+}$. (PDB code: 1HR2.)

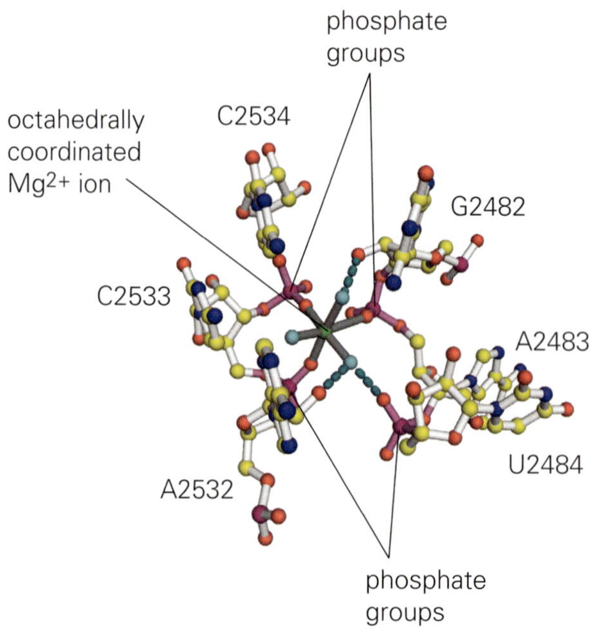


**Figure 2.44 Coordination of $Mg^{2+}$ by phosphate groups in ribosomal RNA.** The $Mg^{2+}$ ion is octahedrally coordinated. Three of the water molecules that normally coordinate the $Mg^{2+}$ have been replaced by oxygen atoms from three phosphate groups. An additional phosphate group is coordinated by a water molecule bound to the $Mg^{2+}$ ion. These interactions allow the tight packing of the phosphate backbone, which would otherwise be impossible due to charge repulsion. (PDB code: 1FFK.)

formation of numerous van der Waals contacts and hydrogen bonds as a consequence of base-pairing interactions between secondary structural elements (which may or may not conform to the Watson-Crick rules). These tertiary interactions may involve hairpin loops or internal bulges.

There are numerous ways in which disparate secondary structural elements of an RNA molecule can come together to form the organized tertiary structure. In broad outline, they can involve interactions between two unpaired regions, between an unpaired region and one helix, or between two helices. Tertiary interactions can be between bases, between a base and a functional group on the backbone, or between functional groups on different segments of the backbone. We will briefly discuss a few examples of such interactions.

## 2.25 Helices in RNA often interact through coaxial base stacking or the formation of pseudoknots

There are basically two types of interactions between helices in RNA. The interacting helices can form what is in effect one continuous helix, in an arrangement known as **coaxial base stacking**. This conformation, in which the base pair at the end of one helix stacks against the base pair at the beginning of the other helix, is favored by base-stacking interactions analogous to those that stabilize duplex RNAs or DNAs. These stacking interactions are an important feature of the overall tertiary structure of RNAs. A pair of coaxially stacked helices found in the self-splicing group I intron RNA is depicted in **Figure 2.45**.

**Coaxial stacking**

When two regular helical elements (A-form for RNA) stack end-to-end, they are said to be coaxially stacked.

Another tertiary structural motif formed by two interacting helices is the **pseudoknot**, in which nucleotides within the loop of a hairpin form base pairs with a segment at the other end of the stem (**Figures 2.46** and **2.47**). In the classic pseudoknot, the hairpin loop pairs with a complementary sequence next to the

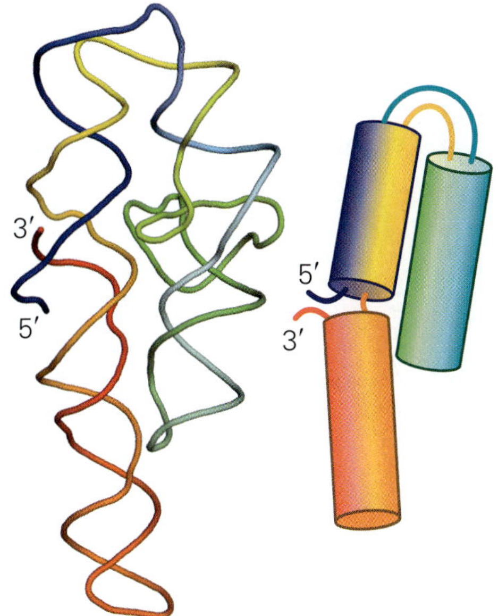


**Figure 2.45 Coaxially stacked helices in RNA.** The structure shown here is that of a segment of the group I self-splicing intron that is illustrated in Figure 2.36. The phosphate backbone in the RNA molecule (left backbone trace) is colored in a spectrum from *blue* to *red* as the chain progresses from the 5′ end to the 3′ end. The coloring makes clear that the two vertical coaxial helices in the center are not one continuous double-helical segment. Schematic diagram (right) of the same coaxial stack. (PDB code: 1GID.)

hairpin stem to form a contiguous coaxially stacked helix. The structure of a small pseudoknot region is shown in Figure 2.47. It can be quite difficult to trace the topology of the RNA backbone in this compact structure, but if you study the drawing carefully you will see that the formation of the pseudoknot leads to the formation of a nearly continuous ladder of stacked base pairs, which results from the two discontinuous helices stacking on top of each other.

**Pseudoknot**

A pseudoknot is an RNA structure containing a hairpin loop, with a segment at one end of the stem folding back to base-pair with residues in the loop.

A peculiar structural motif in RNA is the **adenosine platform** (A-platform). This consists of two consecutive adenosine residues that are arranged side by side rather than stacked on top of each other. The two adenosines form hydrogen bonds to each other to create a "pseudo-base pair" (**Figure 2.48**), resulting in a relatively large, flat surface. The bases in the A-platform stack against base pairs

**A-platform**

Adenosine platforms are formed by two consecutive A residues in a sequence. Rather than stacking in the normal fashion, they form a pseudo-base pair. The two bases are arranged side by side and form a platform onto which other RNA elements can stack.

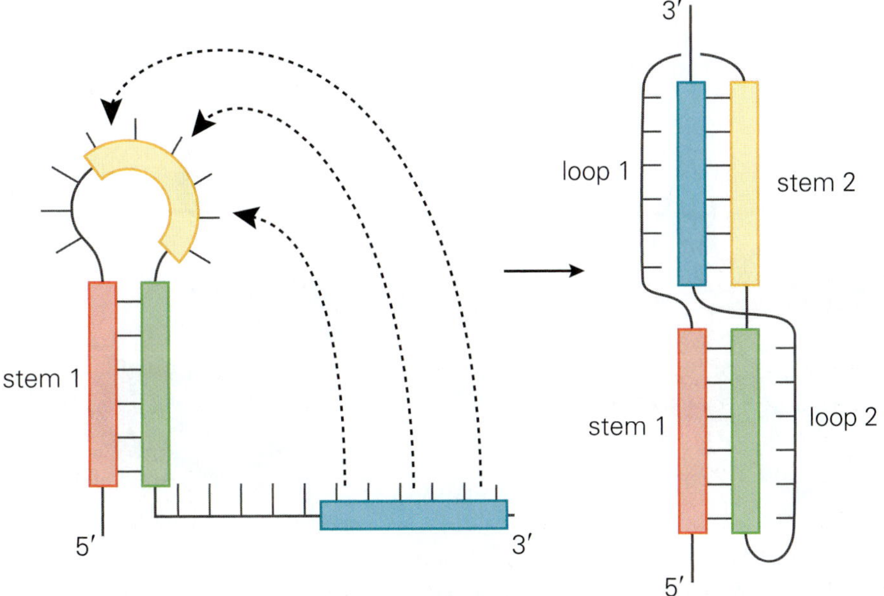


**Figure 2.46 Formation of an RNA pseudoknot.** A pseudoknot arises from sequence complementarity between an otherwise unpaired region (the 3′ segment colored *blue* in the illustration) and the loop of a stem-loop hairpin (colored *yellow*). The pseudoknot results in the formation of two helices that can stack on top of each other.

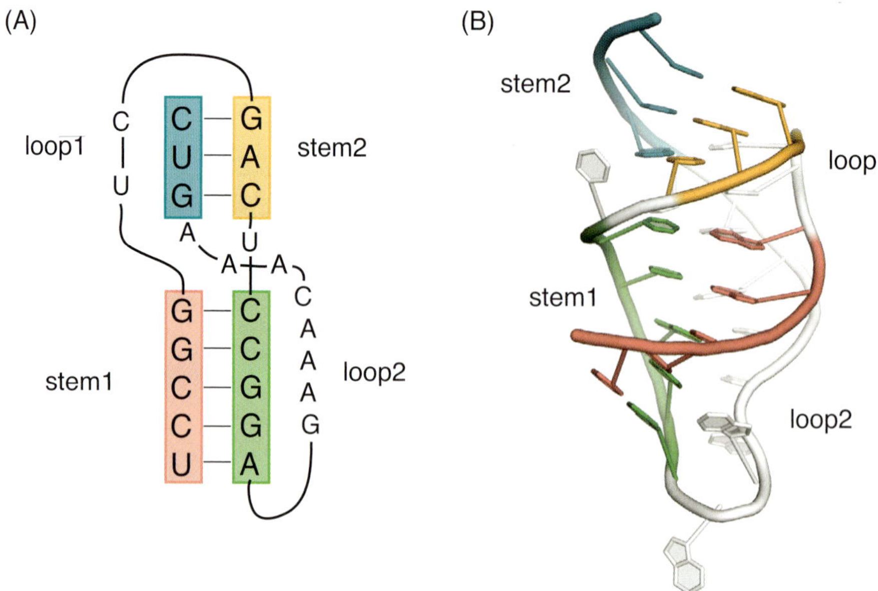


**Figure 2.47 Structure of a pseudoknot.** (A) Secondary structure diagram of a pseudoknot. (B) Three-dimensional structure of the pseudoknot. The color of the backbone corresponds to the colors in (A). (PDB code: 2RPI.)

**Base triples**

Base triples are formed when three bases come together in a near-planar arrangement, with hydrogen bonding. Two of the bases usually form a Watson-Crick base pair, and the third base comes into either the major or the minor groove to form the triple.

from other elements of the structure to form coaxially stacked double helices. However, because these two bases come from immediately adjacent positions on the same chain, they allow the formation of highly interdigitated structures in which distant elements of the RNA molecule are brought together.

## 2.26 Various interactions between nucleotides stabilize RNA tertiary structure

**A-minor motif**

A tertiary interaction that involves the minor groove edge of an A residue.

Interactions between helical and unpaired motifs often come in the form of **base triples**. One common structural motif of this kind uses Hoogsteen base pairs, as discussed in Section 2.20. A second kind is called the **A-minor motif** because it involves the interaction of the minor groove edges of adenine bases with the minor groove of neighboring helices (**Figure 2.49**). The minor groove edge of adenine refers to the edge that adenine reveals to the minor groove when it is in a double-helical structure (see Figure 2.13). In the case of the A-minor motif, the

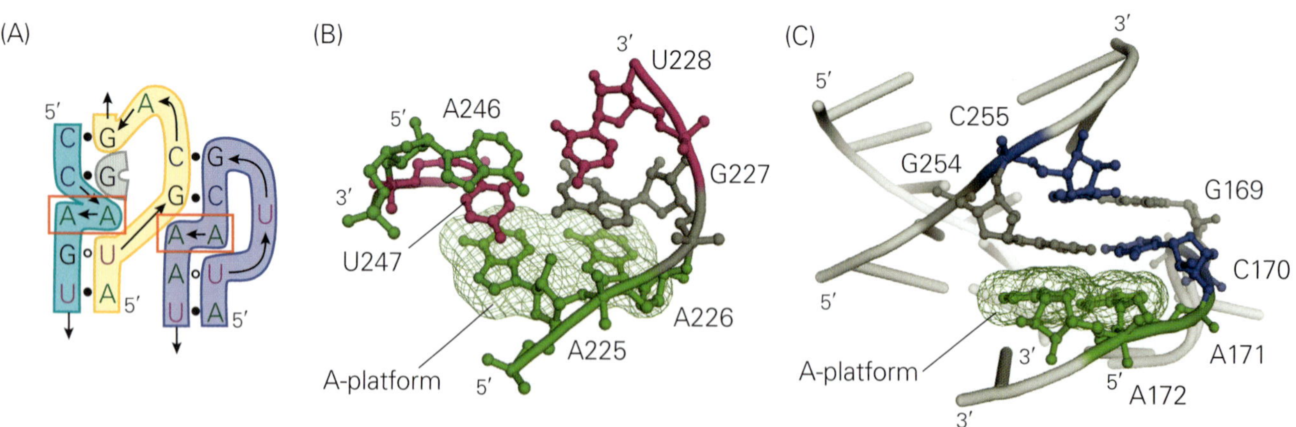


**Figure 2.48 The structure of the adenosine platform (A-platform) motif.** (A) Schematic diagram showing four segments of an RNA chain that come close together to form an interdigitated structure. Each segment of the chain is shown in a different color. There are two A-platforms in this structure (*red boxes*). (B) Structure of the adenosine platform indicated by the *red box* in the left side of the schematic in (A). The two adenosines that form the A-platform are outlined in *green*. A G-U wobble base pair stacks on top of the A-platform (*gray* and *magenta*). (C) Structure of the A-platform in the *red box* on the right in (A). (Adapted from J.H. Cate et al., and J.A. Doudna, *Science* 273: 1696–1699, 1996; PDB code: 1GID.)

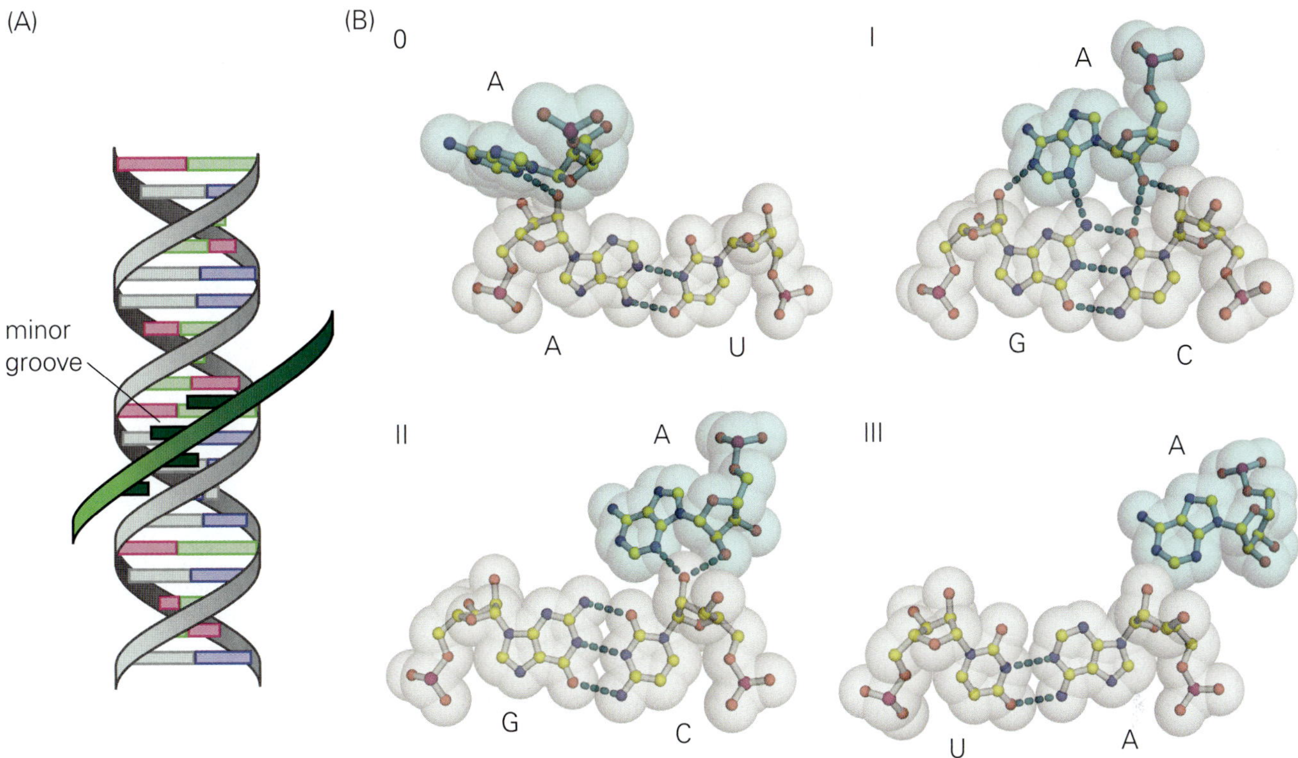


**Figure 2.49 Examples of four major kinds of A-minor interactions.** (A) A schematic diagram of a double-helical segment of RNA. A single strand of RNA (*green*) from another portion of the molecule reaches in and interacts with the edges of the Watson-Crick base pairs in the minor groove. (B) Four types of interactions made by adenine bases presented by the third strand are shown. (Adapted from P. Nissen et al., and T.A. Steitz, *Proc. Natl. Acad. Sci. USA* 98: 4899–4903, 2001. With permission from the National Academy of Sciences.)

strand that is interacting with the helix is not itself in a double helix. Nevertheless, by using the minor groove edges of a string of adenine bases, the incoming strand can cause several of its own bases to stack on top of each other favorably.

Four kinds of A-minor interactions, denoted type 0, I, II, and III, are illustrated in Figure 2.49. These interactions typically involve a third strand interacting with the minor groove of RNA, a variation on the theme of triple-helix formation discussed in Section 2.20 (see Figure 2.49A). Each type of A-minor motif is defined by the position of the 2′-OH group of the interacting adenosine relative to the positions of the two 2′-OH groups of the receptor base pair. The type I and II interactions are A-specific, while type 0 and type III are observed for other bases as well.

As we discussed in Section 2.22, a tetraloop is a common structural motif that forms a cap for helical hairpin structures. The turn region of the tetraloop leaves some nucleotide bases unpaired, and these bases can interact with the minor groove of adjacent double-helical segments. The docking site for the tetraloop is referred to as the tetraloop receptor, and such interactions help stabilize the tertiary structure of RNA molecules, as illustrated in **Figure 2.50**.

The tertiary structure of RNA can also be stabilized by interactions made by the sugar-phosphate backbone. For example, the ribose groups of two RNA strands that are running alongside each other in an antiparallel orientation can form a structural motif known as a ribose zipper, in which the ribose sugars of the two strands are interdigitated. Hydrogen bonds between the 2′-OH group of a ribose from one helix and the 2′-OH group and the N3 of a purine or the O2 of a pyrimidine in the opposite helix create a "zipper" of ribose sugars in the minor grooves of two helices (**Figure 2.51**).

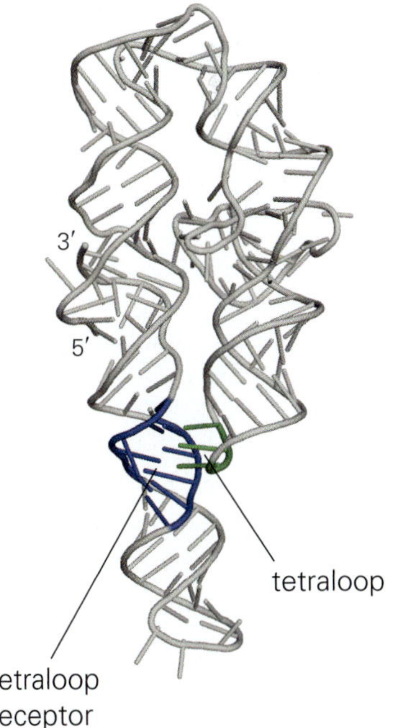


**Figure 2.50 Tertiary structure of an RNA molecule showing the interaction of a tetraloop with its receptor.** The tetraloop and its receptor are colored *green* and *blue* respectively. (PDB code: 1GID.)

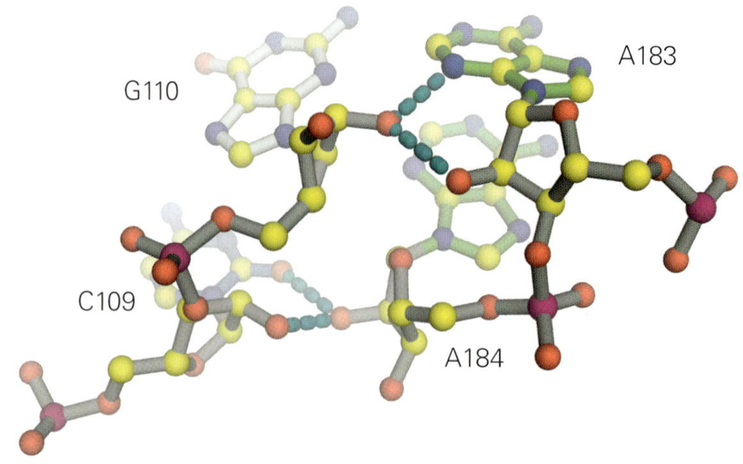


**Figure 2.51 The ribose zipper motif.** This motif involves the formation of a hydrogen-bond network between 2′-OH groups of the ribose sugar and nucleotide bases. (PDB code: 1GID.)

## Summary

Nearly all naturally occurring DNA is found in a double-stranded, right-handed helical form. In this double helix, the bases point towards the inside of the spiral, the backbone is on the outside, and the two complementary strands are antiparallel. Key to the formation of this double-helical structure (and especially for its replication) is specific complementary base pairing—A pairs with T and G pairs with C. RNA contains uracil rather than thymine, so A pairs with U in RNA. The sugar moieties in the two nucleic acids are different, too. In RNA, the sugar is a ribose, while in DNA it is a deoxyribose.

The formation of both RNA and DNA double helices is driven by Watson-Crick base pairings, base-stacking interactions, and electrostatic interactions. The aromatic rings of the bases are stacked against one another, while the functional groups not involved in base pairing protrude into the two grooves between the backbones. Double-helical DNA and RNA have two distinct grooves, called the major and minor grooves. The major and minor grooves differ in their width and depth, and also in the nature of the functional groups on the base-pair edges that are exposed within them. These differences affect how other molecules, including proteins, nucleotides, and small molecules, interact with DNA and RNA.

DNA usually forms a B-form helix in which the major groove is wider and more accessible than in A-form double helices formed by RNA. The patterns of potential hydrogen-bonding sites within the major groove are unique for each of the four base pairs. In contrast, the minor groove is smaller and less accessible, and the hydrogen-bonding sites are more uniform. Thus, molecules that bind to double-stranded DNA (for example, regulatory proteins) are more able to distinguish specific nucleotide sequences in the major groove.

An important conformational parameter that governs the structure of the double helix is the sugar pucker. The sugar pucker in B-form DNA is C2′ endo. The presence of the 2′-OH group on the ribose in RNA makes it difficult for RNA to adopt the C2′ endo conformation. A-form helices have the C3′ endo sugar pucker, and both DNA and RNA can adopt the A-form conformation. DNA can also form a left-handed double-helical structure, known as the Z-form. Z-form DNA is rare and is favored by sequences that have stretches of alternating C and G nucleotides on both strands.

Circular DNA, or DNA that has its ends constrained, undergoes supercoiling to relieve tension from under- or overwinding of the helix. Torsional strain in DNA is a natural consequence of many cellular processes, including replication, and cellular DNA is extensively supercoiled. Supercoiling is also an essential aspect of the packaging of DNA in chromatin. Enzymes known as topoisomerases modify the supercoiling of DNA.

Cellular RNA, initially synthesized in a single-stranded form, undergoes considerable processing. This may include splicing, chemical modification, being complexed with proteins, and folding into compact secondary and tertiary structures. The many short duplex regions formed in an RNA molecule are usually in the A-form, in which the major groove is deep and narrow, and the minor groove is shallow and broad.

The binding of divalent metal cations stabilizes RNA and is therefore essential in forming the global tertiary structure. There are three main types of interactions between metal ions and nucleic acids. The first is a general charge-screening of the negative phosphate groups along the nucleic acid backbone by diffuse metal ions that are hydrated. The second is an interaction between a ligand bound to the metal ion, such as $H_2O$, and a functional group on the RNA. The third is a direct coordination of the metal by one or more groups in RNA, which replace water molecules that normally coordinate the metal. The displacement by the RNA of some of the waters that are bound to the metal ion results in an energetic cost for direct ion coordination.

In addition to A-form duplexes, RNA can readily adopt more complicated secondary structures. Some of these are hairpins, internal loops, and bulges. These structural elements often feature nonstandard (non-Watson-Crick) base pairs, modified nucleotides, and unpaired nucleotides. Secondary structure elements can then interact with other secondary structures or with single-stranded regions to form intricate tertiary structures that give RNA its distinctive shape and function. A few of the many motifs contributing to the tertiary structure of RNA are coaxial stacking between helices, the use of tetraloops to induce bends, and the formation of base triples.

The complex tertiary structures that RNAs form are essential to their function as biological catalysts (as in the case of RNA enzymes like the ribosome), as translators of genetic information, and as structural scaffolds. How these properties of RNA are similar to and different from those of proteins are worth keeping in mind as you study Chapters 4 and 5.

## Key Concepts

### A. DOUBLE-HELICAL STRUCTURES OF RNA AND DNA

- The double helix is the principal secondary structural element of DNA and RNA.
- The structure and stability of double-stranded nucleic acids is determined by hydrogen bonding, base stacking, and electrostatic interactions.
- Hydrogen bonding between bases is important for the formation of double helices, but its effect is weakened due to competing interactions with water.
- The electronic polarization of the bases contributes to strong stacking interactions between bases.
- Metal ions help shield electrostatic repulsion between the phosphate groups.
- The ribose ring can have alternate conformations defined by the sugar pucker, with the C2′ endo and C3′ endo conformations being common in DNA and RNA, respectively.
- The most common conformation for DNA double helices is the B-form helix, which has a wide major groove, with the base pairs arranged nearly

perpendicular to the helix axis. The sugar pucker is C2′ endo, which is energetically disfavored in RNA.

- B-form DNA allows sequence-specific recognition of the major groove by proteins. Each base pair has a unique set of interacting elements in the major groove, but not in the minor groove.
- RNA adopts a double-helical conformation known as the A-form, with a narrower and deeper major groove than the B-form double helix. The sugar pucker is C3′ endo, which is accessible to both DNA and RNA, both of which can form A-form helices.
- The major groove of double-helical RNA is less accessible to proteins, making sequence-specific recognition by proteins more difficult than in DNA.
- Z-form DNA is a left-handed double-helical structure.
- The DNA double helix is deformable.
- DNA can form supercoiled structures when its ends are constrained.
- Writhe, linking number, and twist are mathematical parameters that describe the supercoiling of DNA.
- Local conformational changes in the DNA also affect supercoiling.

**B. THE FUNCTIONAL VERSATILITY OF RNA**

- RNA forms nonstandard base pairs, such as G-U wobble base pairs.
- RNA molecules can contain modified nucleotides, which increases the functional versatility of RNA.
- A tetraloop is a common secondary structural motif that caps RNA hairpins.
- Metal ions, particularly $Mg^{2+}$, help RNA molecules to fold.
- RNA tertiary structure involves interactions between secondary structural elements, such as coaxial helices and pseudoknots.
- Various interactions between nucleotides stabilize RNA tertiary structure, such as adenosine platforms, A-minor motifs, tetraloop–receptor interactions, and ribose zippers.

## Problems

### True/False and Multiple Choice

1. The tertiary structure of functional RNA molecules is easily predicted.

   True/False

2. Which of the following is not a stabilizing force for the structure and stability of double-stranded nucleic acids?
   a. base stacking
   b. hydrogen bonding
   c. disulfide bonds
   d. electrostatic forces

3. H-bond acceptor, H-bond donor, H-bond acceptor, methyl group is the pattern of potential interactions at the edge of which Watson-Crick base pair?
   a. A-T
   b. G-C
   c. A-A
   d. C-G
   e. T-A

4. Genomic DNA can become deformed from its normal B-form by DNA binding proteins, such as the histone proteins and the TATA-box binding protein.

   True/False

5. Which nonstandard base pair is most likely to form a wobble base pair?
   a. U-U
   b. A-G
   c. A-A
   d. G-U
   e. G-G

6. Classify the following RNA structural elements as secondary or tertiary structure:
   a. coaxial helices
   b. pseudoknot
   c. hairpin
   d. junction
   e. adenosine platform
   f. ribose zipper
   g. bulge

### Fill in the Blank

7. The _____________ structure of a nucleic acid is the sequence of nucleotides in the DNA or RNA molecule.

8. The modified RNA base in which two methyl groups are added to guanine is _________.

9. B-form DNA has a C2′ _______ sugar pucker.

10. Hoogsteen base pairs, where the hydrogen-bonding interactions utilize the Watson-Crick base-pairing edge on one base and the major groove edge in the other base, can be utilized to form an RNA ______ helix.

11. Metal ions, such as $K^+$, $Na^+$, and $Mg^{2+}$, typically interact with the _________ group on the backbone of nucleic acids.

## Quantitative/Essay

12. What physical factors force RNA to adopt only the C3′ endo sugar configuration, but allow DNA to adopt either the C2′ endo or the C3′ endo sugar configurations?

13. What type of RNA interaction is shown below?

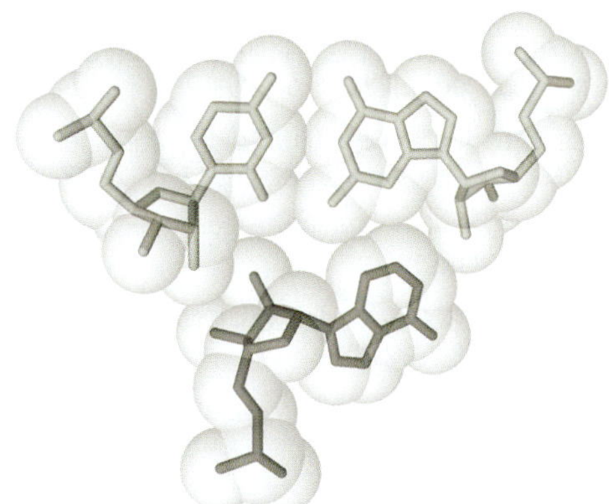

Label the major and minor grooves of the Watson-Crick base pair and the potential hydrogen bonds between all bases.

14. Label the atoms, the bases, the hydrogen bonds across bases, the major and minor grooves, and the interactions along the grooves for the following base pair (oxygen and nitrogen atoms are not identified explicitly):

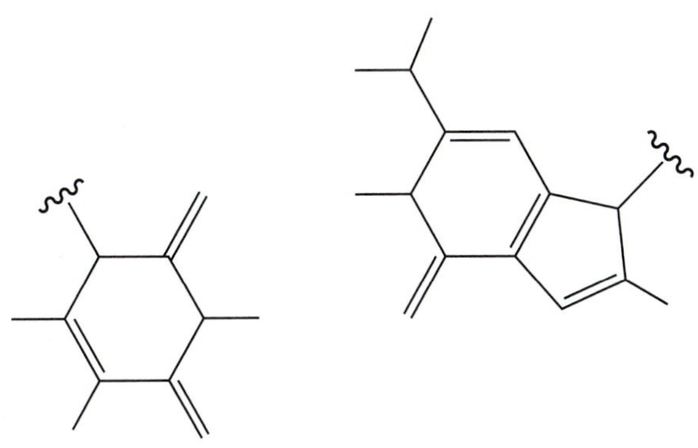

Is the base pair a standard Watson-Crick base pair?

15. List at least three physical features that distinguish A- and Z-form DNA.

16. In water, why is base stacking relatively more important than hydrogen bonding to forming a DNA double helix?

17. How are the interactions between a nucleic acid and $Mg^{2+}$ mediated by water?

18. Why is the R (purine) in GNRA tetraloops required rather than a pyrimidine?

19. Why do most transcription factors interact with the major groove of B-form DNA rather than the minor groove?

20. Which of the following DNA sequences is most likely to adopt the Z-form? Why?

    a. `GCGCGCGCATATGCGCGCGCC`
    b. `AGAGAGCTCTCTCTCTAAAAT`

21. Consider a relaxed, closed-circular DNA plasmid that has 1040 base pairs with writhe = 0. An intercalator is added, such that there is one intercalator per 104 base pairs. The effect of the intercalator is to cause the twist between the base pairs that flank it to be reduced to zero. Will the resulting intercalator-bound DNA be positively or negatively supercoiled?

22. The structure of the large ribosomal subunit from *Haloarcula marismortui* (PDB code: 1FFK) has been solved by x-ray crystallography. The 23S RNA contains 2922 nucleotides (758 A, 889 G, 739 C, and 536 U).

    a. Assuming a random distribution of nucleotides, how many four-mer sequences with the sequence G-N (any base)-R (purine)-A are possible?

    b. There are actually 21 GNRA tetraloops in the structure. What percentage of possible GNRA sequences actually formed tetraloops in the structure?

23. A bacterial DNA polymerase moves at approximately 1000 base pairs per second when replicating DNA. The polymerase holoenzyme is approximately 110 Å long. By what multiple of its length does the polymerase move forward along the axis of the DNA double helix in 3 seconds? Assume the DNA stays fixed and ignore the rotational component of the motion of the polymerase along the DNA.

# Further Reading

## General


Bloomfield VA, Crothers DM & Tinoco I (2000) Nucleic Acids: Structures, Properties, and Functions. Sausalito, CA: University Science Books.

Brändén C & Tooze J (1999) Introduction to Protein Structure, 2nd ed. New York: Garland Science.

Calladine CR & Drew HR (1997) Understanding DNA: The Molecule & How It Works, 2nd ed. San Diego: Academic Press.

Saenger W (1984) Principles of Nucleic Acid Structure. New York: Springer-Verlag.


## References

### A. Double-Helical Structures of RNA and DNA


Cozzarelli NR & Wang JC (1990) DNA Topology and Its Biological Effects. Cold Spring Harbor, NY: Cold Spring Harbor Laboratory Press.

Dickerson RE, Drew, HR, Conner, BN, Wing, RM, Fratini, AV & Kopka, ML (1982) The anatomy of A-, B-, and Z-DNA. *Science* 216, 475–485.

Harrison SC & Aggarwal AK (1990) DNA recognition by proteins with a helix-turn-helix motif. *Annu. Rev. Biochem.* 59, 933–969.

Richmond TJ & Davey CA (2003) The structure of DNA in the nucleosome core. *Nature* 423, 145–150.

Saenger, W, Hunter WN & Kennard O (1986) DNA conformation is determined by economics in the hydration of phosphate groups. *Nature* 324, 385–388.

Seeman NC, Rosenberg JM & Rich A (1976) Sequence-specific recognition of double helical nucleic acids by proteins. *Proc. Natl. Acad. Sci. USA* 73, 804–808.

Wahl MC & Sundaralingam M (1997) Structures of A-DNA duplexes. *Biopolymers* 44, 45–63.

Wang JC (2002) Cellular roles of DNA topoisomerases: a molecular perspective. *Nat. Rev. Mol. Cell. Biol.* 3, 430–440.

Wemmer DE (2000) Designed sequence-specific minor groove ligands. *Annu. Rev. Biophys. Biomol. Struct.* 29, 439–461.

Werner MH & Burley SK (1997) Architectural transcription factors: proteins that remodel DNA. *Cell* 88, 733–736.


### B. The Functional Versatility of RNA


Ban N, Nissen P, Hansen J, Moore PB & Steitz TA (2000) The complete atomic structure of the large ribosomal subunit at 2.4 Å resolution. *Science* 289, 905–920.

Batey RT, Rambo RP & Doudna JA (1999) Tertiary motifs in RNA structure and folding. *Angew. Chem. Int. Ed. Engl.* 38, 2326–2343.

Correll CC & Swinger K (2003) Common and distinctive features of GNRA tetraloops based on a GUAA tetraloop structure at 1.4 Å resolution. *RNA* 9, 355–363.

Gold L, Polisky B, Uhlenbeck O & Yarus M (1995) Diversity of oligonucleotide functions. *Annu. Rev. Biochem.* 64, 763–797.

Mathews DH, Moss WN & Turner DH (2010) Folding and finding RNA secondary structure. *Cold Spring Harb. Perspect. Biol.* 2, a003665.

Misra VK & Draper DE (1998) On the role of magnesium ions in RNA stability. *Biopolymers* 48, 113–135.

Staple DW & Butcher SE (2005) Pseudoknots: RNA structures with diverse functions. *PLoS Biol.* 3, e213.

Varani G & McClain WH (2000) The G x U wobble base pair. A fundamental building block of RNA structure crucial to RNA function in diverse biological systems. *EMBO Rep.* 1, 18–23.

Wyatt JR, Puglisi JD & Tinoco I (1989) RNA folding: pseudoknots, loops and bulges. *BioEssays* 11, 100–106.

Zuker M (2003) Mfold web server for nucleic acid folding and hybridization prediction. *Nuc. Acids Res.* 31, 3406–3415.

# CHAPTER 3

# Glycans and Lipids

Nucleic acids and proteins are the molecules whose synthesis is specified by the genetic code. There are two other broad classes of molecules of central importance to life—namely, **glycans** (carbohydrates) and **lipids**. These molecules, which are the focus of this chapter, are not directly coded for by DNA, but their presence is essential for the proper functioning of all cells.

**Glycans**

Glycans are sugars or sugar derivatives, and their polymers. The simplest individual subunits of glycans have roughly the formula $(HCOH)_n$ with $n$ between 3 and 9, but most commonly 5, 6, or 7. Many glycans also have sugars with additional functionality relative to these simplest subunits. Glycans are also known as carbohydrates.

Glycans are sugars and sugar polymers, in which the predominant chemical unit is hydroxylated carbon (HCOH). The building blocks of typical glycans have three to nine carbons, and are generally very soluble in water because of the presence of hydroxyl groups. Most of the carbons in glycans are chiral, leading to many stereochemical variations in the structures of sugars. The building blocks of glycans are linked to form large, complex, and diverse polymers that serve a wide variety roles in cells, including energy storage, structural support, and cellular signaling. Glycans are among the most abundant organic compounds on Earth.

Lipids are diverse organic compounds that are soluble in nonpolar organic solvents but are insoluble in water. Examples of lipids include fats, waxes, and oils. The lipids that are most relevant for this chapter are **amphipathic** molecules—that is, they have both polar, hydrophilic parts and nonpolar, hydrophobic parts, a feature critical to their functions. A distinctive and critical property of such molecules is their ability to form cellular membranes. In contrast to the **biopolymers** (that is, the proteins, nucleic acids, and polysaccharides, all of which have many covalently linked residues), lipid structures are noncovalent assemblies of large numbers of molecules of moderate size (the molecular weight of a typical lipid molecule is roughly 1 kD). Membrane bilayers assembled from lipid molecules are critical for the formation of cells, and for forming compartments within cells. The development of membrane bilayers was necessarily an early feature in the evolution of life.

**Lipids**

Lipids are amphipathic molecules (also called amphiphilic, that is, part hydrophobic and part hydrophilic). Lipids are the primary components of biological membranes. Although membranes contain embedded proteins, most of the critical properties of membranes arise from the lipid components.

## A. GLYCANS

### 3.1 Simple sugars are comprised primarily of hydroxylated carbons

The term "carbohydrate" comes from hydrated carbon. Addition of water molecules to carbons gives an empirical formula of $(CH_2O)_n$, which reflects the composition of the core elements in the molecules of this class. Among compounds with this approximate formula, there are many isomers (that is, molecules with the same atoms but with differences in bonding and/or stereochemistry) that occur in biological systems. The molecules with small values of $n$ are generally called **sugars** (one of which is sucrose, the common table sugar; see **Figure 3.1**). Many sugars taste sweet, but this term is used whether they taste sweet or not.

The simple sugars, and modified versions of them, also form polymers. Both homopolymers, made of just one kind of sugar building block, and heteropolymers, made from different kinds of building blocks, play important roles in

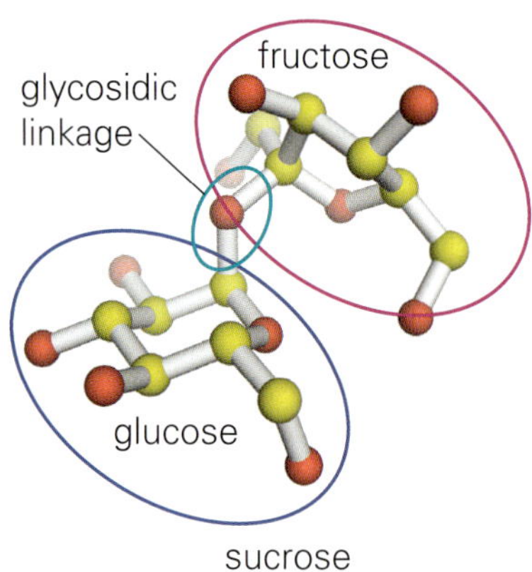


**Figure 3.1 The structure of sucrose, common table sugar.** The chemical formula of sucrose is $C_{12}H_{22}O_{11}$. Sucrose is made from two smaller building blocks, glucose and fructose, joined by a glycosidic linkage. The six-membered ring is glucose, and the five-membered ring is fructose. This depiction, like most molecular illustrations in this book, has *yellow* carbon atoms and *red* oxygen atoms. The hydrogen atoms are not shown.

living systems. Most of the common sugar building blocks contain five, six, or seven "hydrated carbon" units ($CH_2O$), which are best thought of as H-C-O-H to reflect the typical bonding pattern.

Carbon always forms four bonds, so the hydrated carbons can form two additional bonds to other carbons, forming chains (for example, a six-carbon chain of $CH_2O$ units is shown in **Figure 3.2**A). Note that there are unsatisfied bonds at the ends of the chain, so this would not actually be a stable molecule. One modification that satisfies the bonding is to shift one hydrogen from an OH to a terminal carbon, and add a second bond to the oxygen, making a carbonyl group, as shown in Figure 3.2B. When the oxygen double bond is at the end of the chain, the molecules is known as an **aldehyde**, whereas if the oxygen double bond is somewhere in the middle, the molecule is known as a **ketone**. Sugars are classified as **aldoses** or **ketoses**, respectively, depending on whether they contain aldehyde or ketone groups.

**Anomers**

Anomers are forms of sugars that differ only in the equatorial or axial position of the hydroxyl at the position of ring closure.

(A)

```
   H    H    H    H    H    H
   |    |    |    |    |    |
 — C —— C —— C —— C —— C —— C —
   |    |    |    |    |    |
   OH   OH   OH   OH   OH   OH
```

(B)

```
     H    H    H    H    H    H
     |    |    |    |    |    |
 H — C —— C —— C —— C —— C —— C
     |    |    |    |    |    ||
     OH   OH   OH   OH   OH   O
```

aldehyde

(C)

```
     H    H         H    H    H
     |    |         |    |    |
 H — C —— C —— C —— C —— C —— C — H
     |    |    ||   |    |    |
     OH   OH   O    OH   OH   OH
```

ketone

**Figure 3.2 Sugar aldehydes and ketones.** (A) A chain of bonded hydroxylated carbons leaves bonding unsatisfied at the ends. (B) In some sugars, a double bond to a terminal oxygen is formed, making an aldehyde. (C) In other sugars, the double-bonded oxygen is not at the end of the chain, making a ketone.

Two examples of common sugars are **glucose** (an aldohexose that is a major energy source in eukaryotic cells), and **fructose** (a ketohexose, also often called fruit sugar because of its common occurrence in fruit). The chemical structures of glucose and fructose, drawn using several different chemical conventions, are shown in **Figure 3.3**. There are many different ways to represent chemical structures, including their stereochemistry (consult an organic chemistry textbook to review these conventions). The different schematic representations shown in Figure 3.3 are those most commonly used for describing sugars and glycans. Some representations, such as the **Fischer projection**, are very convenient for comparing the stereochemistry of different sugar isomers, but are poor for seeing the three-dimensional relationship of atoms and the real geometry of the molecules. The representation indicating the three-dimensional structure provides a better sense of the conformational features, but may be less suited for seeing relationships between isomers.

## 3.2 Many cyclic sugar molecules can exist in alternative anomeric forms

The presence of the carbonyl carbon allows for the facile cyclization of sugars, as shown in **Figure 3.4**. In this reaction, one of the hydroxyl groups of the sugar carries out an intramolecular nucleophilic addition to the carbonyl group, leading to ring structures. For the commonly occurring sugars in biological systems, the cyclic forms are favored (for glucose, about 99.98% is cyclic in water), but the open and ring structures are in equilibrium.

The closure to the ring form can occur through attack of a hydroxyl with the carbonyl in either of two orientations, leading to a different position of the hydroxyl derived from the oxygen of the carbonyl. The two isomers (that is, the two structures on the right hand side of Figure 3.4) are called **anomers**, and they differ only in the orientation of the hydroxyl group at the position of ring closure.

Alternative anomers are a general feature of the cyclic forms of sugars. The carbon at which this isomerism occurs is called the **anomeric carbon**. The anomers can

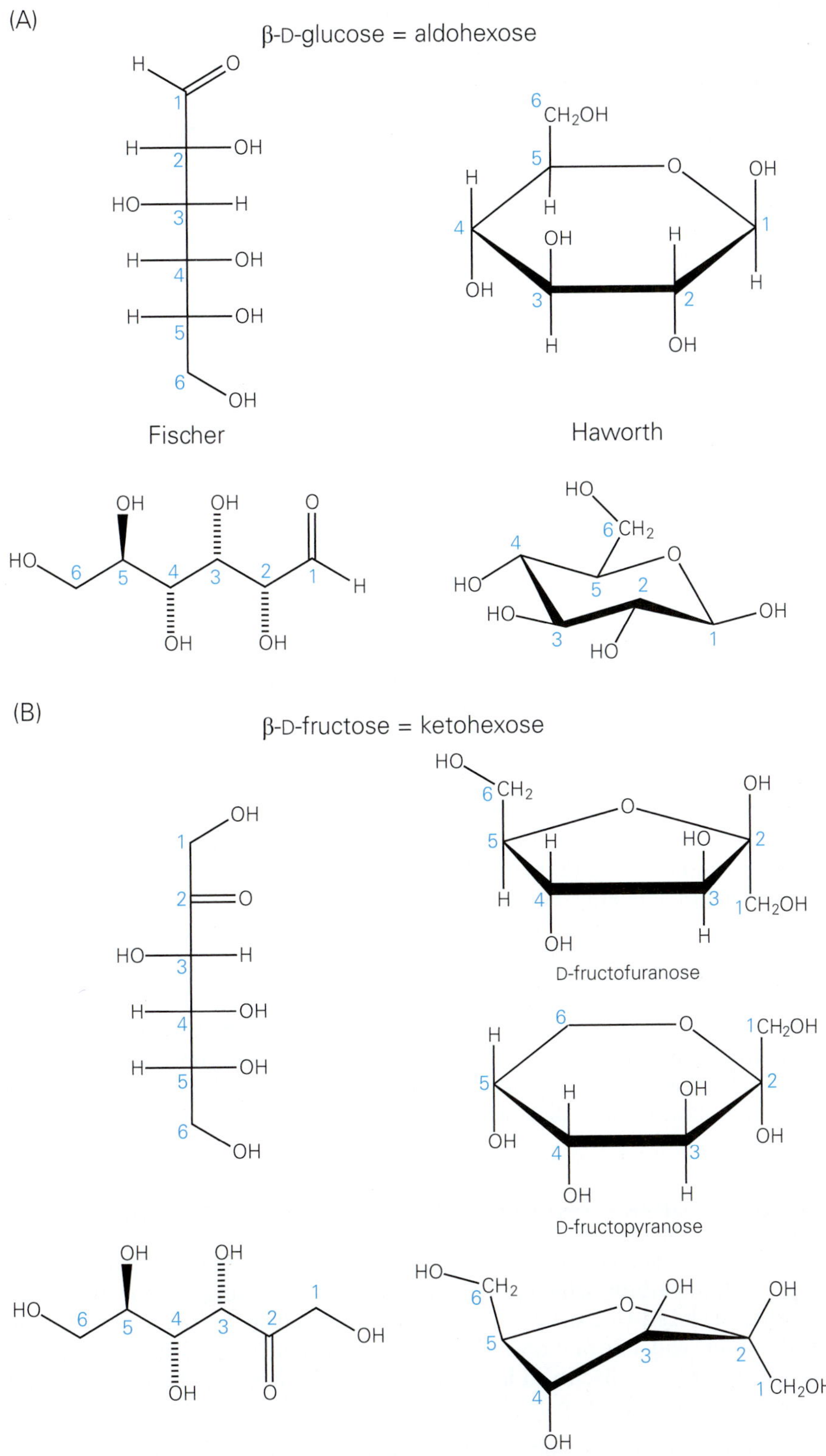


**Figure 3.3 The structures of two sugars are shown.** (A) Glucose (an aldohexose) and (B) fructose (a ketohexose that can cyclize in two different ways) are shown in several representations. Both exist as an equilibrium mixture of the open, linear chain and the closed, cyclic form (the cyclic forms being >99% of the mixture). In addition to molecular structures that specifically indicate stereochemistry, sugars are also commonly drawn in representations known as Fischer and Haworth projections. These are often easier for comparing the relative stereochemistries of different sugars. Carbons are numbered consecutively from the top carbon in the Fischer projection.

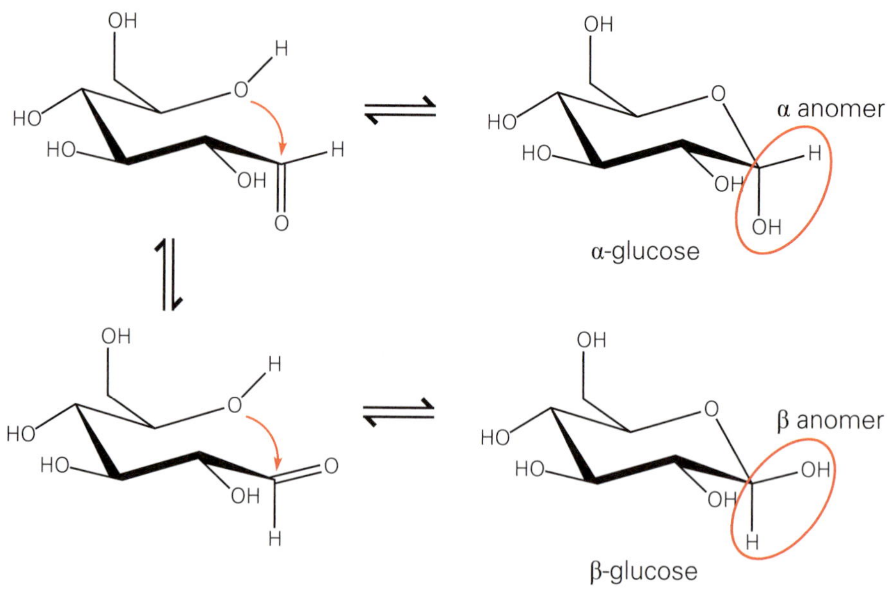

**Figure 3.4 Linear and cyclized forms of sugars.** The linear form of glucose is shown at the left in conformations that lead to the ring closure reaction. Depending on the orientation of the aldehyde relative to the attacking C5-hydroxyl, closure can lead to either the α or β anomer, shown at the right.

**Axial and equatorial**

Axial and equatorial are terms that define the directions of substituents of rings such as sugars, relative to the plane of the ring.

interconvert by opening to the linear form and reclosing, as shown in Figure 3.4. This process is relatively slow, taking minutes to hours in neutral water, but it can be accelerated greatly by enzymes. The different anomers of a sugar have distinct chemical properties, and play particularly important roles when polymers are formed, as discussed in later sections of this chapter.

The relative abundance of the anomers depends on interactions between atoms that affect their energy. For glucose, the two anomers are known as α and β (see Figure 3.4) and the ratio of their abundance is ~1:2, reflecting a small difference in conformational energy. For most sugars both anomers are significantly populated.

The other option for cyclizing sugars is to satisfy the bonding by joining the carbon atoms at the two ends of the sugar. This generates a carbon-to-carbon bond, as shown for the six-carbon sugar called inositol in **Figure 3.5**. Although inositol is synthesized from a linear sugar, the cyclization is not readily reversible and anomeric forms of inositol do not occur. Inositol is often multiply phosphorylated *in vivo*, and is an important signaling molecule both when free and when attached to lipids.

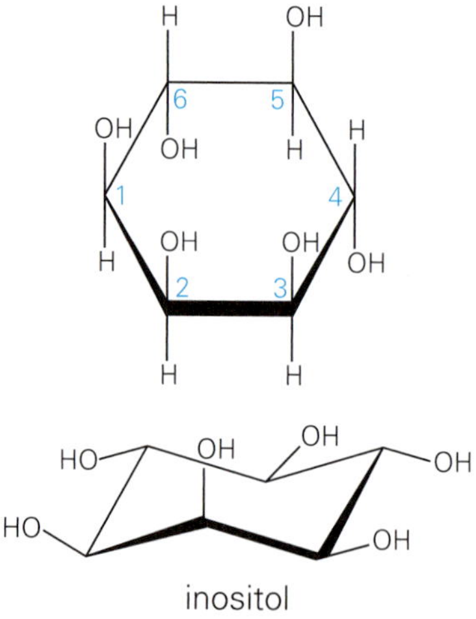

**Figure 3.5 Inositol, a six-carbon sugar, is atypical in not having an oxygen in the ring.** The stereochemistry of the natural isomer is shown.

## 3.3 Sugar rings often have many low energy conformations

Carbon atoms that make four single bonds should have bond angles near the value in a tetrahedron, which is 109°. To maintain bond angles near this value, the atoms in closed-ring structures of sugars cannot all lie in a plane. Instead, the rings "pucker," with one or more atoms moving out of the plane. The phenomenon of sugar pucker was discussed in Section 2.6 for the five-membered ribose and deoxyribose rings that are part of nucleic acids.

The most stable conformation for six-membered rings has the atoms at opposite ends of the ring out of the plane, one atom above the plane and another below the plane of the other four, leading to a **chair conformation**. For a ring of six atoms, there are two distinct chair conformers, which are shown for α-glucose in **Figure 3.6A** and B. In a chair conformation, substituents, such as the hydroxyls, can either point out from the ring (termed **equatorial**) or point up or down (termed **axial**). There is less space for nonhydrogen atoms in the axial positions, so the conformational energy is lower when most of the hydroxyls are in equatorial positions. The chair conformer shown in Figure 3.6A has just one axial hydroxyl, while

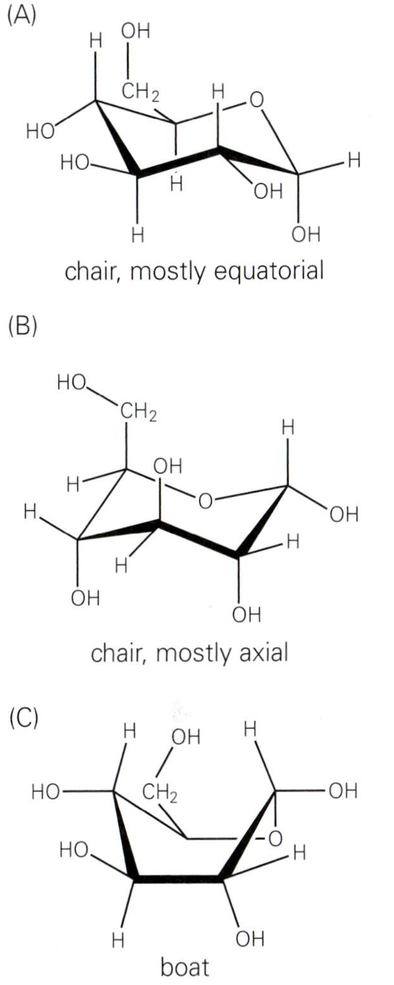

**Figure 3.6 Conformations of α-glucose.** The ring forms of sugars can interconvert between two different chair conformations, which exchange the positions of substituents between mostly equatorial (A) and mostly axial (B) positions. (C) The boat conformation.

the conformer shown in Figure 3.6B has three axial hydroxyls and an axial methylene group. In this case the "mostly equatorial" form has a considerably lower energy, and is therefore the most populated.

Another nonplanar geometry has both ends of the ring out of the plane of the central four carbons, but on the same side of the plane rather than opposite. This is referred to as a **boat conformation** (Figure 3.6C). For a six-atom ring, there are three combinations of atoms that can be moved above or below the plane, giving rise to a set of distinct conformational isomers. The boat conformation is generally higher in energy than the most stable chair conformer due to unfavorable steric interactions.

For five-membered rings, one atom usually moves out of the plane, either up or down, as seen in the structure of fructose (Figures 3.1 and 3.3). Alternatively, two atoms can be out of plane, one moving slightly above the ring and the other slightly below it. Recall from Chapter 2 that the deoxyribose in DNA and the ribose in RNA have different stable ring geometries (puckers), leading to differences in the double helices formed by DNA and RNA (see Figure 2.7).

## 3.4 Many sugars are structural isomers of identical composition, but with different stereochemistry

Another very important aspect of sugar chemistry is that, for any given molecular formula, there are many structural isomers. All simple hexoses have the formula $C_6H_{12}O_6$. In the family of aldohexoses (for example, glucose, as shown in Figure 3.3), there are four carbons that are chiral centers (**Figure 3.7**). Each of these can have either handedness, specified as *R* or *S* stereochemistry (see Figure 3.7A; consult an organic chemistry textbook for the rules governing the designation of chiral centers). For example, the first chiral carbon atom could be in the *R* configuration, the second in the *S* configuration, the third one in the *S* configuration, and the fourth one in the *R* configuration, giving the pattern *RSSR*. There are 16 such configurations, two of which are illustrated in Figure 3.7B.

Within this set of 16 configurations, there are pairs that are mirror images of each other—that is, they are **enantiomers**. For example, the *RSSR* configuration is the mirror image of the *SRRS* configuration. Molecules that are enantiomers have identical physical properties (for example, density, melting temperature, and vapor pressure). Thus, there are eight stereochemical variants of a simple hexose (that is, eight sets of two enantiomers each) that will have distinct physical and chemical properties.

The fact that sugars are chiral was shown more than 100 years ago when it was demonstrated that their solutions rotate polarized light (the renowned scientist Louis Pasteur was the first to connect molecular chirality to the rotation of polarized light). For glyceraldehyde, with just one chiral carbon, the *R* isomer is designated as D (dextrorotatory), because it rotates the polarization of light in a clockwise fashion. Likewise, the *S* isomer is designated L (levorotatory) for rotating the polarization counterclockwise.

The stereochemical designation of sugars is based on the chirality of the carbon furthest from the aldehyde. If it matches the chirality of D-glyceraldehyde, then it is denoted a D sugar, whereas if it matches L-glyceraldehyde, then it is an L sugar. It turns out that only the D forms of the aldohexoses occur in nature. Of the eight possible D aldohexose sugars, three of them—glucose, mannose and galactose—occur quite commonly in biological systems; their structures are shown in **Figure 3.8** in both linear (Fischer) and cyclic forms.

(A)

H OH CH2 H O HO HO H H OH H OH

chair, mostly equatorial

(B)

HO CH2 H H OH O H OH H H OH OH

chair, mostly axial

(C)

H OH H HO CH2 OH O HO H H OH

boat

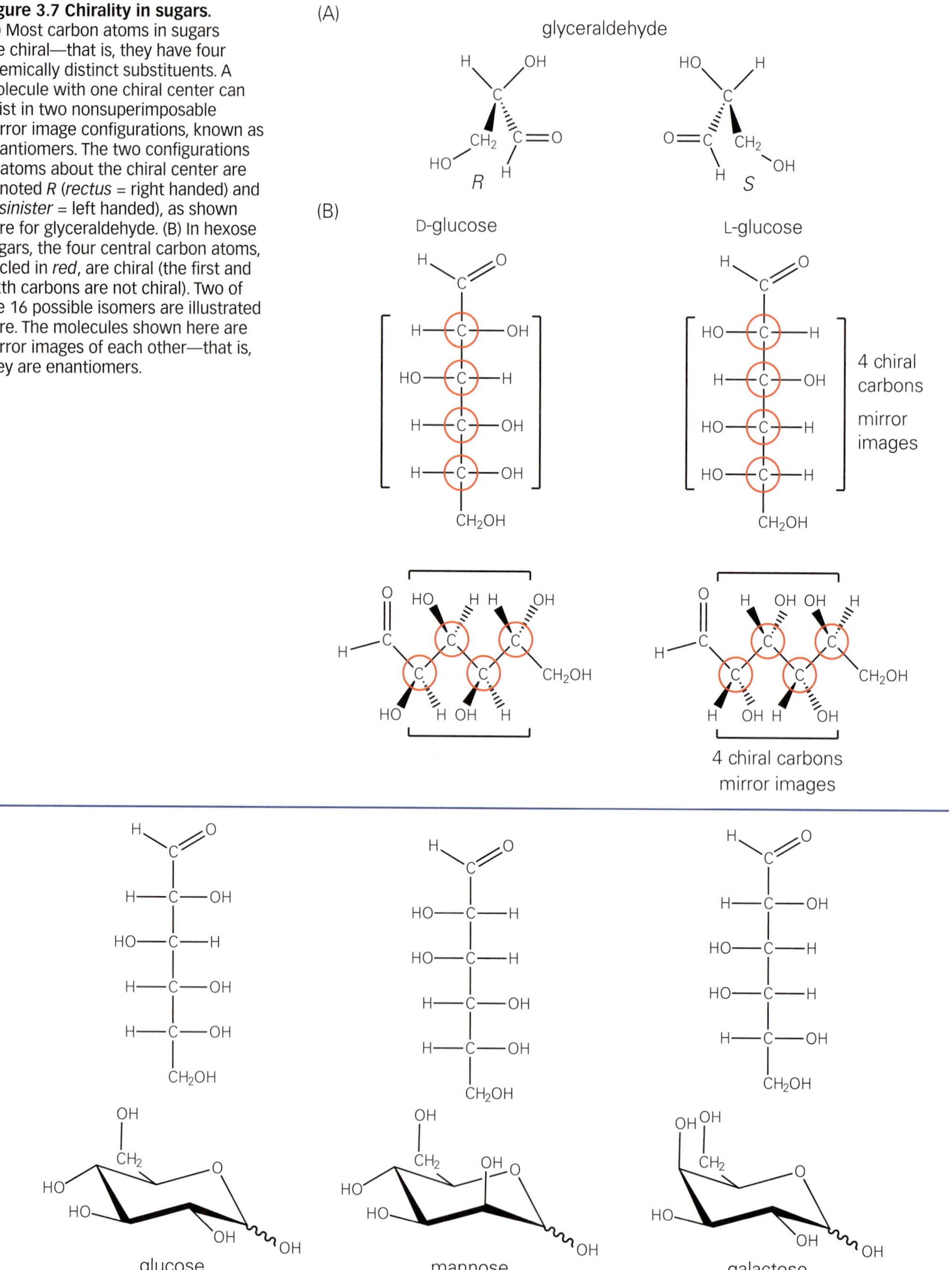


**Figure 3.7 Chirality in sugars.** (A) Most carbon atoms in sugars are chiral—that is, they have four chemically distinct substituents. A molecule with one chiral center can exist in two nonsuperimposable mirror image configurations, known as enantiomers. The two configurations of atoms about the chiral center are denoted *R* (*rectus* = right handed) and *S* (*sinister* = left handed), as shown here for glyceraldehyde. (B) In hexose sugars, the four central carbon atoms, circled in *red*, are chiral (the first and sixth carbons are not chiral). Two of the 16 possible isomers are illustrated here. The molecules shown here are mirror images of each other—that is, they are enantiomers.

**Figure 3.8 The structures of three common simple hexoses—glucose, mannose, and galactose—shown in both Fischer and stereochemical representations.** Each sugar has α and β anomers. A "squiggly" line is used to indicate either possibility (that is, hydroxyl axial or equatorial) at the anomeric position.

## 3.5 Some sugars have other chemical functionalities in addition to alcohol groups

**Oligosaccharides** are polymeric glycans that appear on cell surfaces (the term **polysaccharide** is used if there are more than about 30 sugar units in the polymer). Oligosaccharides are built up of repeating units of the simple sugars mannose and galactose (see Figure 3.8), but also contain other sugars with additional chemical functionality. Several of the sugars that are present in oligosaccharides arise from relatively small modifications of the simple sugars. For example, replacing the commonly occurring $-CH_2OH$ group with a methyl group ($-CH_3$) gives **fucose** (which is also called 6-deoxy-L-galactose), and replacing one hydroxyl with an amino group gives **glucosamine** or **galactosamine**. These are usually acetylated to give *N*-acetylglucosamine, or *N*-acetylgalactosamine (**Figure 3.9**).

**Oligosaccharide and polysaccharide**

Oligosaccharides are glycans that are polymers of sugars. Specific small polymers are usually designated by the number of sugar units, such as monosaccharide, disaccharide, trisaccharide, etc. The term polysaccharide is usually used when there are more than about 30 sugar units in the polymer.

Some sugar residues have more extensive modification. Sialic acid, for example, has additional hydroxylated carbons, a carboxylic acid, and an acetylated amine. The structures of the most commonly occurring sugars in oligosaccharides are shown in Figure 3.9, and are listed in **Table 3.1**. Note that, as for amino acids and nucleotides, there are abbreviations for sugars that are a useful shorthand for specifying sugars and their combinations. It is important to remember that there are many additional modified sugars that occur in different particular contexts, some of which contain other distinct functional groups including carboxylates, phosphates, and/or sulfates.

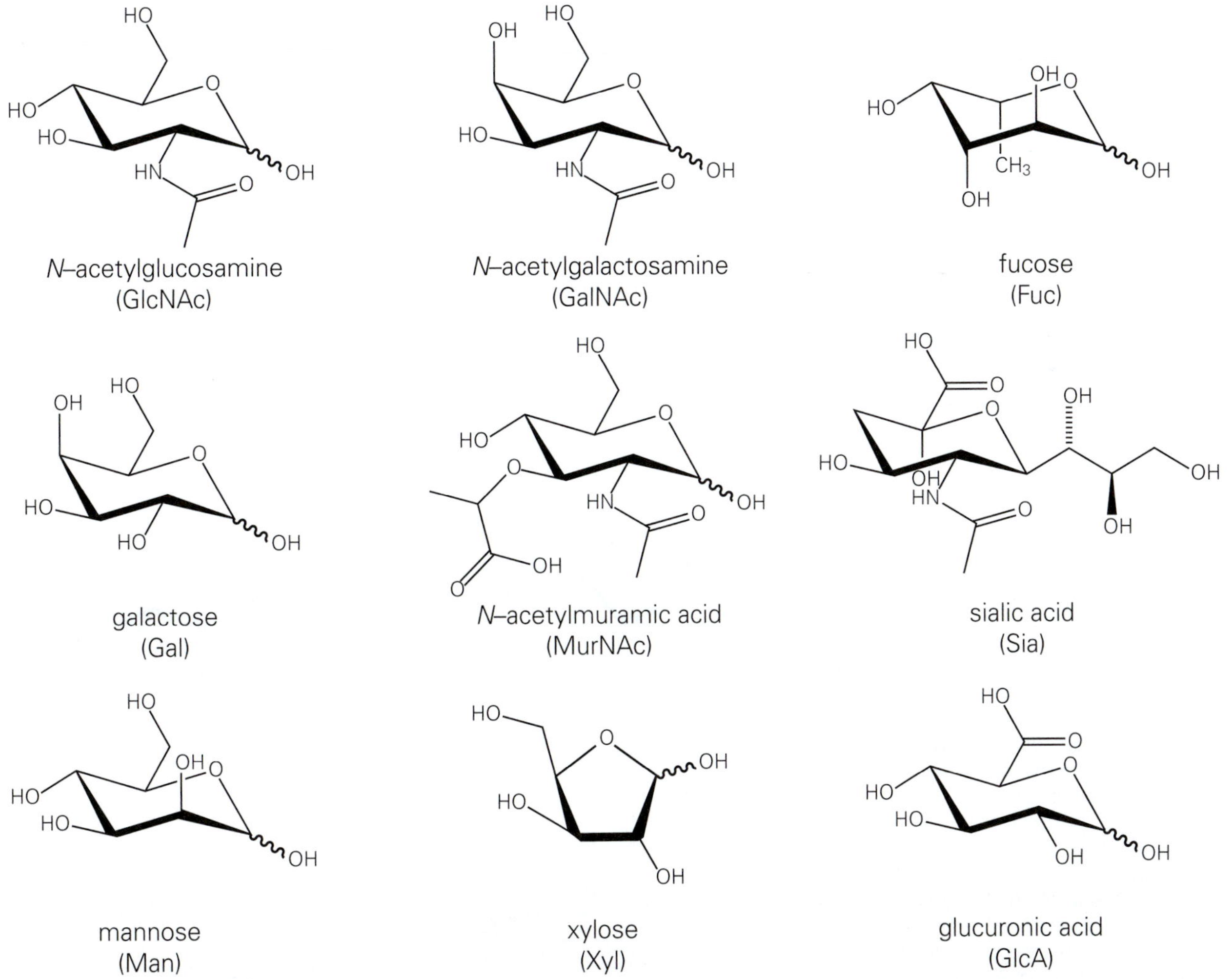

*N*-acetylglucosamine (GlcNAc)

*N*-acetylgalactosamine (GalNAc)

fucose (Fuc)

galactose (Gal)

*N*-acetylmuramic acid (MurNAc)

sialic acid (Sia)

mannose (Man)

xylose (Xyl)

glucuronic acid (GlcA)

**Figure 3.9 Nine sugars that occur commonly in oligosaccharides attached to proteins.** The names of these sugars are shown together with their abbreviations. The squiggly line to oxygen indicates an anomeric position and that the oxygen can be either axial or equatorial at that position.

**Table 3.1 Abbreviations used for the names of the sugars that commonly occur in oligosaccharides.**

| | |
|---|---|
| Glucose | Glc |
| Mannose | Man |
| Galactose | Gal |
| Fucose | Fuc |
| Xylose | Xyl |
| Sialic acid | Sia |
| *N*-Acetylglucosamine | GlcNAc |
| *N*-Acetylgalactosamine | GalNAc |
| *N*-Acetylmuramic acid | MurNAc |
| Glucuronic acid | GlcA |

## 3.6 Glycans form polymeric structures that can have branched linkages

The bonds linking sugar residues into polymers are made by the elimination of a water molecule, resulting in the formation of ether bonds between the sugar residues, often referred to as the **glycosidic bonds** or **glycosidic linkages** (**Figure 3.10**). Although water can react with these bonds to break them, once formed they generally persist for long times in neutral solutions. As is common for most biological reactions, both the joining and the breaking of sugars is done by enzymes, which greatly accelerate these processes and allow the amounts of polymeric forms present to be regulated (see Chapter 16 for more about enzymes and regulation).

In proteins and nucleic acids, essentially all the bonds between residues involve identical chemical groups from linkage to linkage. For example, peptide bonds in genetically encoded proteins are made between backbone amines and backbone carboxylates, never through sidechains. For polysaccharides, however, any of the hydroxyls around a sugar ring may be used to link monomer subunits, giving oligosaccharides a variable covalent "skeleton" (in this case the term "backbone" no longer suffices). There may be more than one linkage to a particular sugar, leading to branching of the structure.

**Glycosidic linkage**

Glycosidic linkages are the attachments between sugars in polysaccharides. They are specified by the atom linked on each sugar and the orientation of the bond to the ring (when needed). These linkages are often called glycosidic bonds.

The linkages in oligosaccharides are specified by indicating the sugars involved and the positions that are linked (see Figure 3.10). Glycosidic bonds often involve the anomeric carbon, and the stereochemistry of the oxygen must then also be specified. For example, the disaccharide **maltose**, shown in Figure 3.10, is glucose-α(1→4)-glucose. In this case the glycosidic bond locks the first glucose into the α anomer (the ring can no longer open to the aldehyde form), but the second glucose residue still occurs in both α and β forms. The linkage to the α anomer of the first sugar creates a bend between the sugars. When this linkage is repeated in a polymer, called **amylose**, the resulting structure is quite highly bent and forms helical spirals, as shown in **Figure 3.11**A. The curvature is high enough that with just six or seven glucose residues the chain can easily be closed into a circle, a form called **cyclodextrin** that is illustrated in Figure 3.11B.

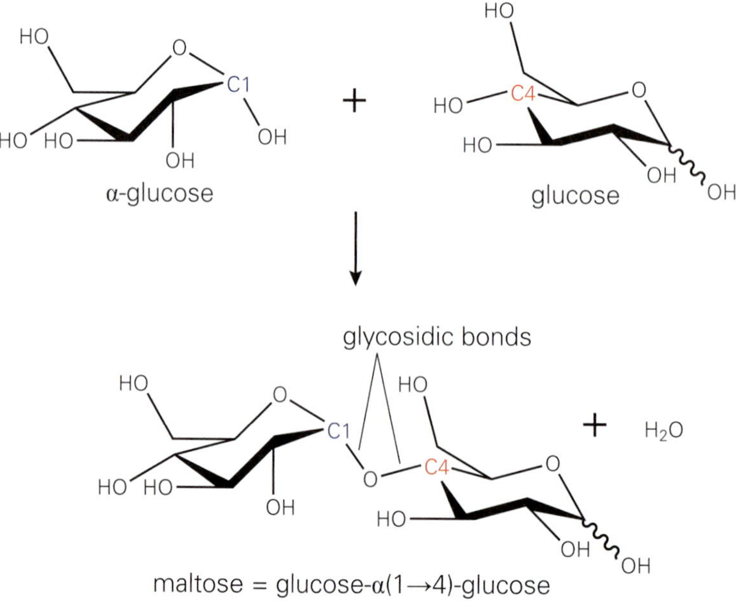


**Figure 3.10 The reaction of two glucose molecules, eliminating water to form the disaccharide maltose.** To specify a polysaccharide structure we must specify the positions linked, the specific sugars involved, and the stereochemistry when an anomeric position is involved. Maltose is thus glucose-α(1→4)-glucose. The terminal glucose can still have both α and β anomers (giving α- and β-maltose), while linked anomeric positions are locked into the anomeric form present when they are linked.

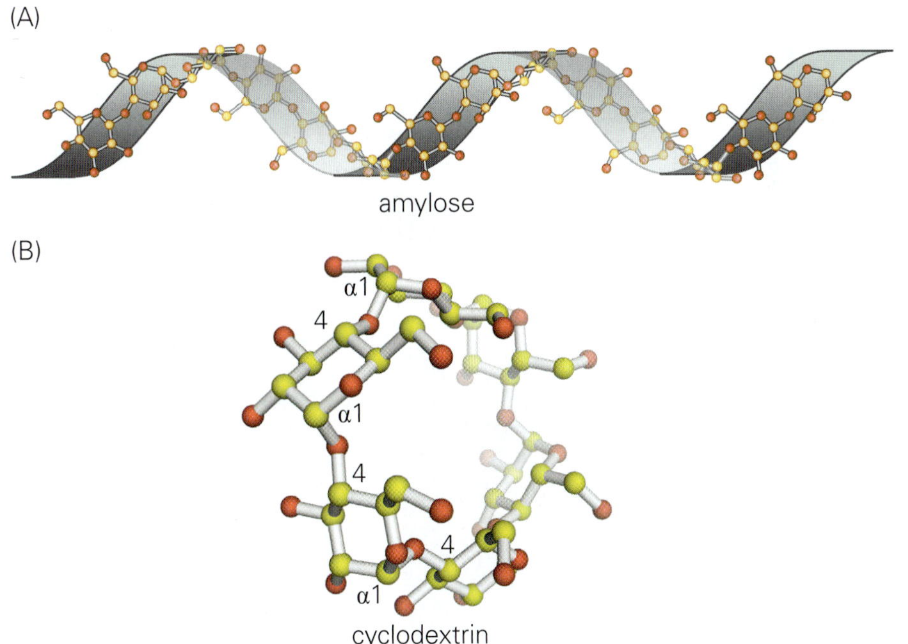


**Figure 3.11 Oligosaccharides with glucose-α(1→4)-glucose linkages.** (A) A helical spiral of amylose. Amylose is polymeric glucose-α(1→4)-glucose and is a major component of starch. The structure has six glucose units per turn. (B) Cyclodextrin is a cyclic oligosaccharide comprised of six glucose units, all α(1→4) linked as occurs in amylose, but cyclized. (For B, PDB code: 1BTC.)

## 3.7 Differences in anomeric linkages lead to dramatic differences in polymeric forms of glucose

Two other examples of biologically important homopolymers of glucose are **glycogen** and **cellulose**. Glycogen is a primary energy storage compound in animal cells that can be rapidly broken down to glucose and then metabolized (that is, "burned") to produce $CO_2$ and water with the release of energy. Glycogen is made up of repeating glucose-α(1→4)-glucose units, but with occasional glucose-α(1→6)-glucose branches, yielding a high-molecular-weight polymer (**Figure 3.12**).

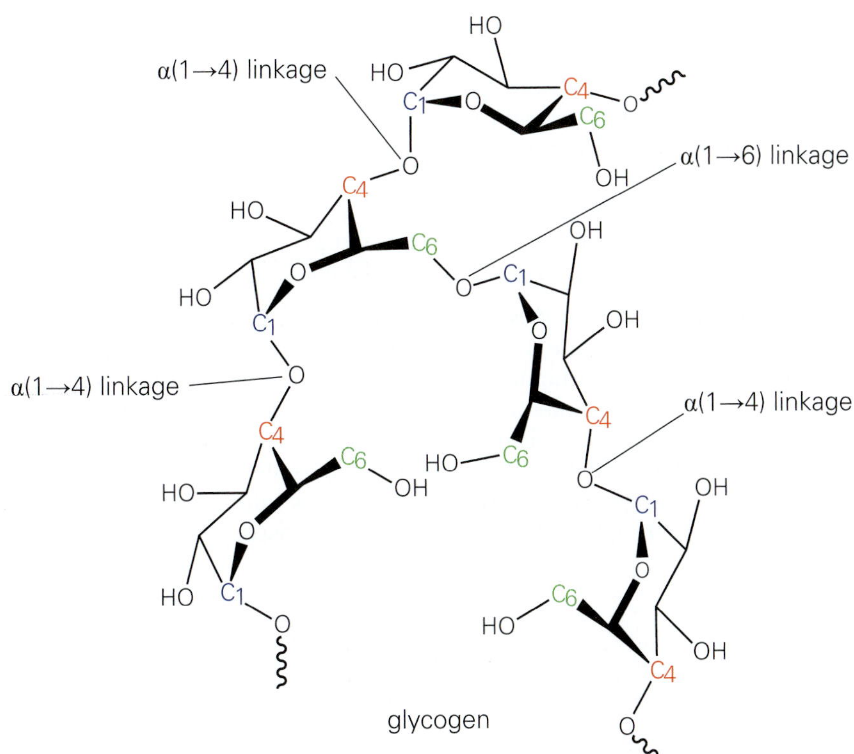


**Figure 3.12 The structure of glycogen.** Glycogen, like amylose and cyclodextrin, is a polymer of glucose with α(1→4) glycosidic linkages. In addition, glycogen has occasional α(1→6) linkages, as occurs in the central sugar in this structure. The C1, C4, and C6 carbons are colored *blue*, *red*, and *green*, respectively.

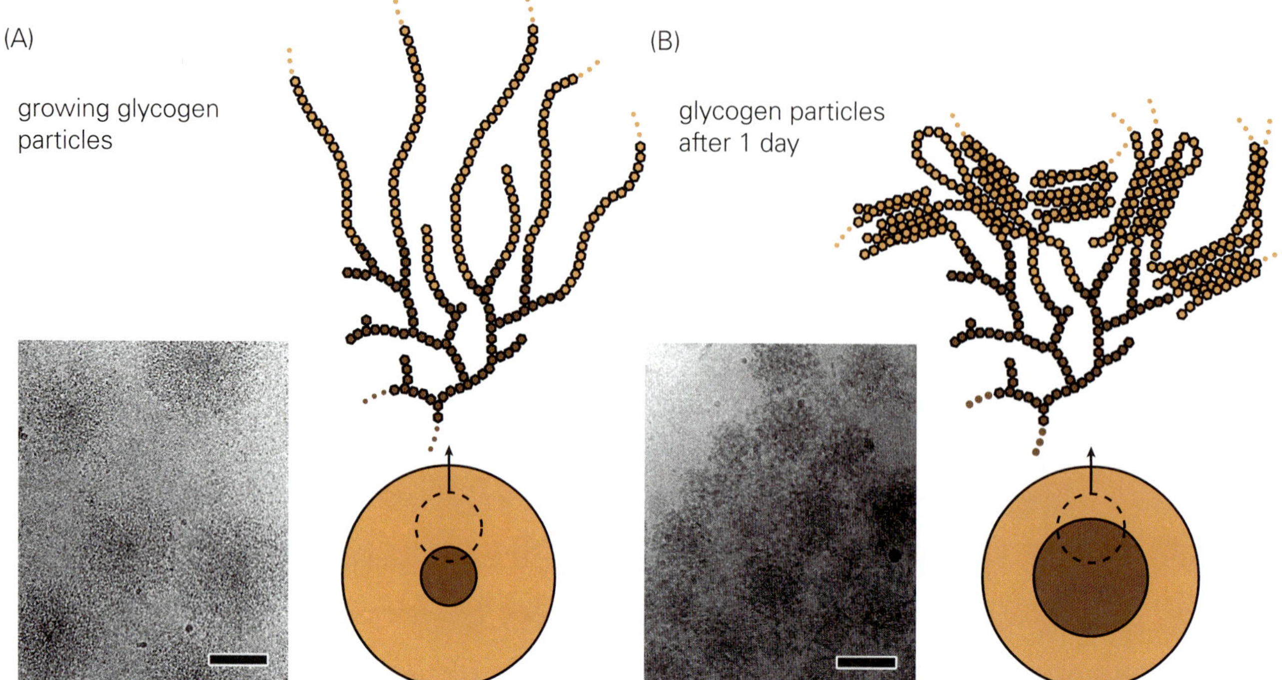


**Figure 3.13 Glycogen particles.** These electron microscope images show glycogen particles growing due to the action of an enzyme. (A) Glycogen particles are visible as dark gray circular objects in the electron micrograph on the left. The scale bar indicates 500 Å. Drawings of the sugar chains are shown, with each dot corresponding to one glucose unit. The inner circle of the schematic (*brown*) shows the initial branched glycogen core, while the outer region (*orange*) contains longer unbranched chains that have been added more recently. (B) The same glycogen particles, one day later. The chains are more highly branched and are denser. (From A. Buleon, G. Veronese and J. Putaux, *Aust. J. Chem.* 60: 706–718, 2007. With permission from CSIRO.)

The primary linkages in glycogen are the same as those in amylose and, like amylose, glycogen also forms spiral structures. This spiral structure leaves space for additional glucose residues to be connected to the ones in the spiral, using the 1→6 branches. These glucose residues are part of separate spirals, generating a highly branched structure. About one glucose residue in 12 has a branch, and glycogen polymers typically have about 50,000 glucose units. Glycogen chains are extended by the action of enzymes, as shown in **Figure 3.13**.

In plants, the cellulose, like glycogen, is a polymer made from 1→4 linked glucose subunits. The key difference between cellulose and glycogen, though, is that the β anomer of glucose is used to form the linkages in cellulose, rather than the α linkages in glycogen (**Figure 3.14**). Cellulose is thus polymeric glucose-β(1→4)-glucose, without the branching that glycogen has. The subtle difference in stereochemistry between glycogen and cellulose leads to dramatic differences in physical and chemical properties. Glycogen is a soft, gel-like substance that forms large storage granules in animal cells, whereas cellulose forms stiff fibers that provide much of the strength in plant cell walls. Glycogen is a readily mobilizable energy store crucial in animal metabolism, whereas cellulose cannot be metabolized at all by humans (or by most other organisms).

The α(1→4) linkages in amylose, cyclodextrin, and glycogen lead to a bend between residues (see Figure 3.11, in which the bending is apparent) that generates a gradual helical structure in the polymer that is not stabilized by any particular intramolecular interactions. Cellulose, in contrast, consists of linear chains, assembled into stacked sheets; intramolecular hydrogen bonds between neighboring residues stabilize individual chains, as shown in Figure 3.14, and intermolecular hydrogen bonds organize the chains into layers stacked on one another, as shown in **Figure 3.15**. The result is an almost crystalline structure that

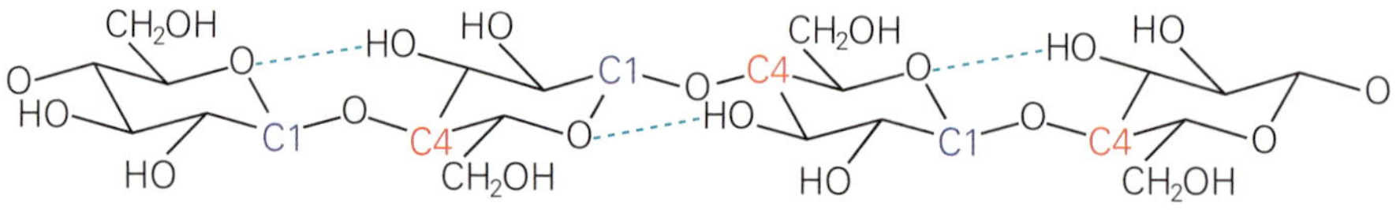

cellulose = poly(glucose-β(1→4)glucose)

**Figure 3.14 A chain of cellulose.** In regions of "crystalline" cellulose, multiple chains hydrogen bond to one another, making it mechanically stiff.

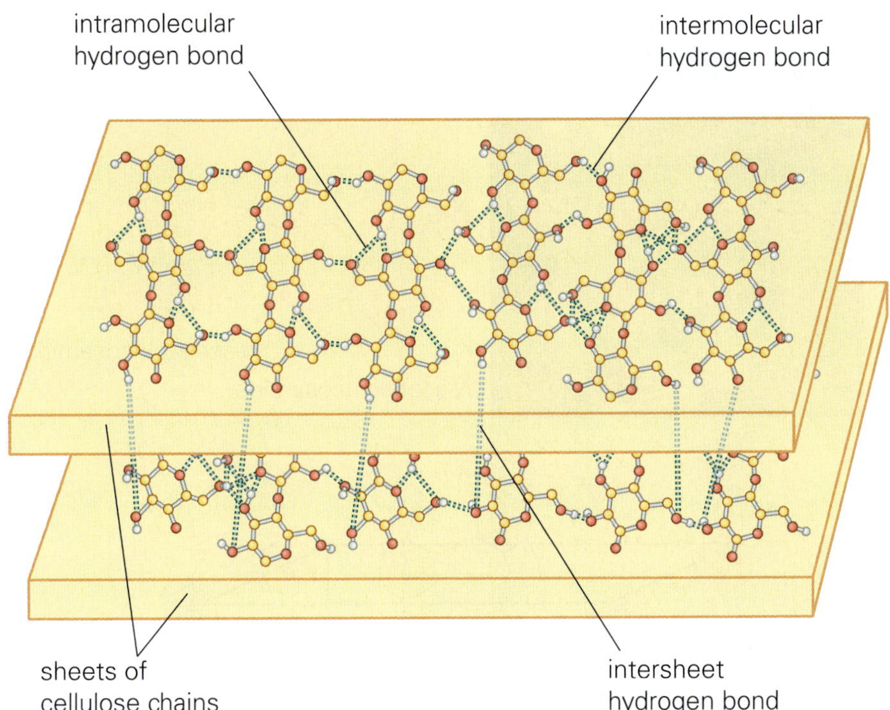


**Figure 3.15 Layers in crystalline cellulose.** Chains of cellulose molecules form sheets in which the chains run side by side and form hydrogen bonds with each other. These sheets are stacked upon one another (drawn artificially separated for clarity), and hydrogen bonds are also formed between molecules in different sheets.

is essentially insoluble in water. These interactions in cellulose are critical to its mechanical stability, and thus to its structural role in plants, where cellulose is responsible for much of the strength of the cell walls. Cellulose is the most abundant biopolymer in nature.

## 3.8 Acetylation or other chemical modification leads to diversity in sugar polymer properties

Another extremely abundant polymer (probably second only to cellulose in the biosphere) is **chitin**: poly-β(1⟶4)-*N*-acetylglucosamine. Chitin is a primary component of the exoskeleton of insects (**Figure 3.16**A) and the cell walls of fungi. Unmodified chitin is leathery, pliable, and tough, serving as an outer "skin" for insect larvae. Modified chitin, embedded in a protein matrix or mineralized through calcium carbonate deposits, is still resilient but much harder, forming the exoskeleton of adult insects.

The linkage between sugar units in chitin is the same as in cellulose, β(1⟶4), and the stereochemistry of the functional groups on the ring is also the same (Figure 3.16B). The difference in chemical and physical properties between chitin and cellulose is due largely to the conversion of the C2-OH group of cellulose to an *N*-acetyl group in chitin. As with cellulose, the individual polymer chains in chitin can align into layered sheets with near crystalline order that contributes to its mechanical properties, as shown in Figure 3.16C.

Another relatively simple sugar polymer is **agarose**, a gelatinous polysaccharide isolated from seaweed. Agarose dissolves in water at high temperatures, but as the solution cools, the agarose forms a gel that is useful for separating macromolecules, such as DNA fragments, by electrophoresis (see Chapter 17). Agarose is a linear disaccharide repeat polymer of galactose, linked 1⟶4 to 3,6-anhydrogalactose (anhydroGal, which is formed from galactose that has been cyclized by eliminating water and linking C3 and C6). The anhydrogalactose, in turn, is linked 1⟶3 to the next galactose, as shown in **Figure 3.17**. Agarose is also a major component of agar, used for culturing bacteria in petri dishes.

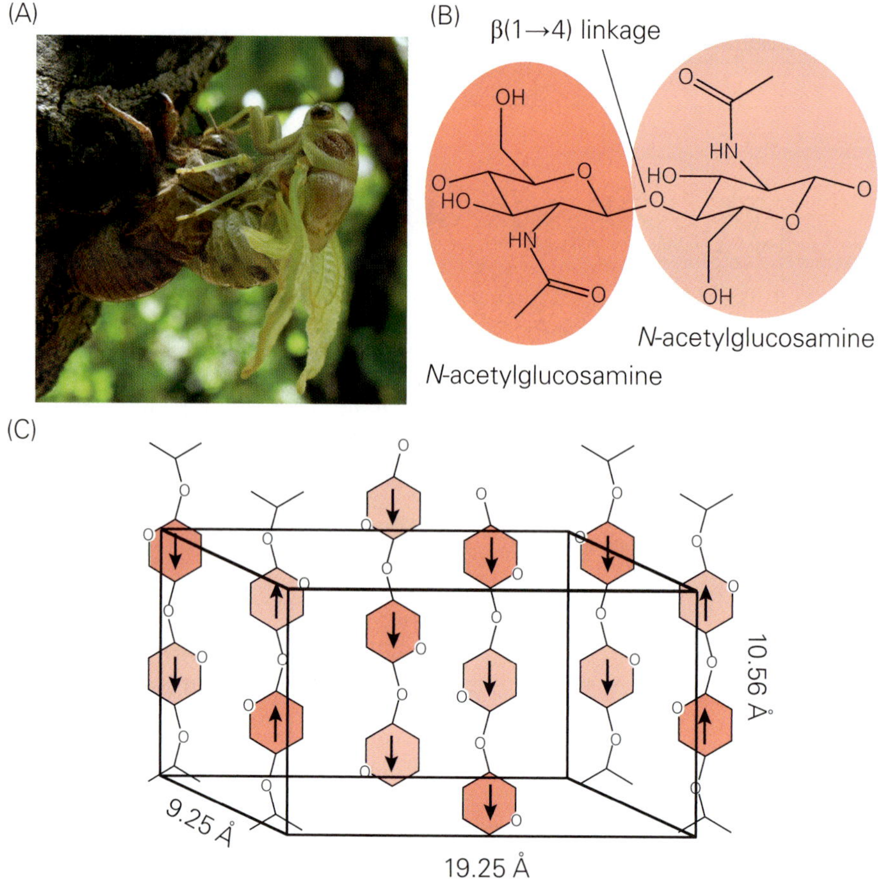


**Figure 3.16 Chitin, a fundamental component of the exoskeletons of insects.** (A) A chitin-rich exoskeleton is shed by a cicada. (B) Chitin consists of homopolymers of *N*-acetylglucosamine with β(1→4) linkages. (C) The basic organization of chains of polymers in chitin. The *N*-acetylglucosamine units are shown as hexagons, with the arrows pointing to the C1 atom. (A, courtesy of Jodelet/Lépinay, Wikipedia; C, from G.L. Clark and A.F. Smith, *J. Phys. Chem*. 40: 863–879, 1936. With permission from the American Chemical Society.)

## 3.9 Glycans may be attached to proteins or lipids

Many proteins in eukaryotic cells are "decorated" with covalently attached glycans. The process of attaching carbohydrate to a protein is called protein glycosylation, and it occurs in two principal forms. The glycan can be linked to the sidechain amide of an asparagine (Asn) residue, which is referred to as **N-linked glycosylation**. Alternatively, in **O-linked glycosylation**, the glycan is linked to the sidechain hydroxyl of a serine (Ser) or threonine (Thr) residue (**Figure 3.18**). Almost all proteins that are destined either to be secreted or to remain membrane bound are glycosylated, but only a small fraction of the Ser, Thr, and Asn residues in any particular protein are modified. Proteins destined to remain in the cytosol are rarely glycosylated, and then often with only a simple monosaccharide.

**O-linked and N-linked glycosylation**

O-linked and N-linked glycosylation refer to the attachment of polysaccharides through the sidechain hydroxyl of serine or threonine (O-linked), or the sidechain amide of asparagine (N-linked).

N-linked glycosylation accounts for about 90% of glycoprotein modifications, and is a co-translational process that starts in the endoplasmic reticulum with a nascent protein chain being synthesized and translocated across the rough endoplasmic reticulum membrane. As the protein emerges on the other side of the membrane, the enzyme oligosaccharyl transferase moves a 14-mer oligosaccharide *en bloc* from a lipid anchor (dolichol—see Figure 3.40) to the target Asn residue within the emerging protein, as shown in **Figure 3.19**. Asparagine residues

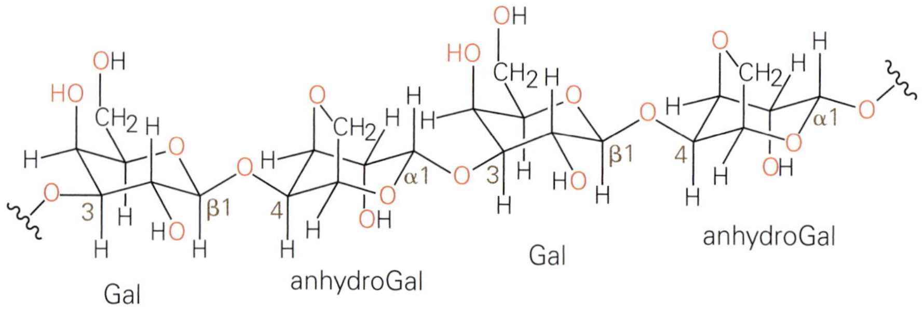

**Figure 3.17 Agarose is composed of alternating galactose and anhydrogalactose units.** Agarose is a polymer used to make gels that are useful in bioanalytical methods.

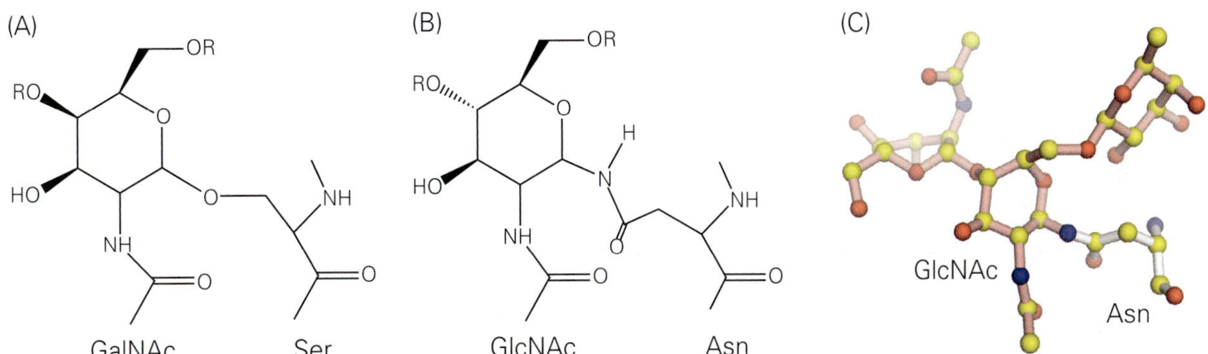


**Figure 3.18 O-linked and N-linked glycosylation.** (A) A Ser residue with O-linked glycosylation. (B) An Asn residue with N-linked glycosylation. (C) A ball-and-stick model of three sugar residues (colored bonds) closest to the Asn (*white* bonds) of an N-linked polysaccharide, the full structure of which is shown in Figure 3.22.

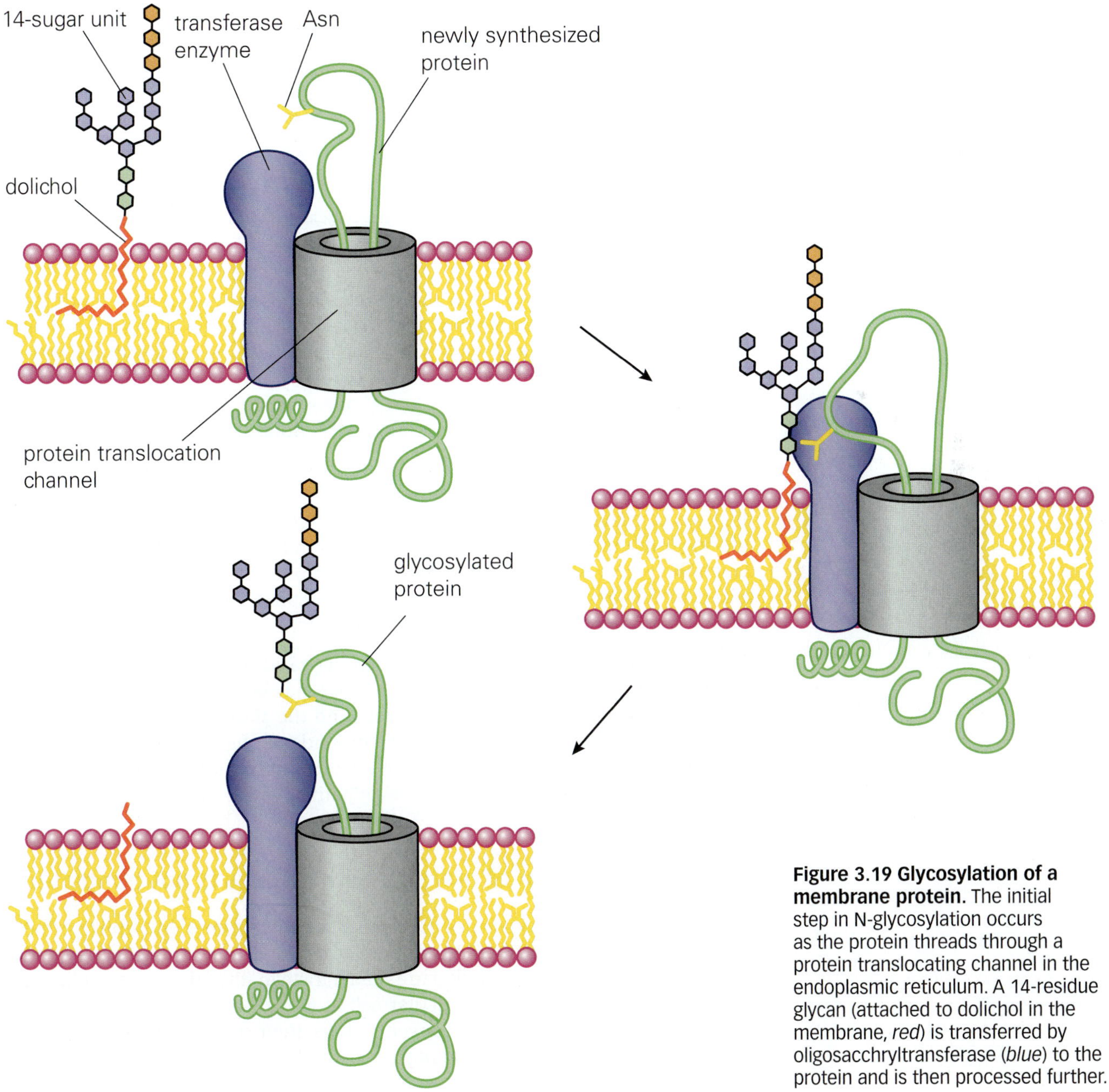

**Figure 3.19 Glycosylation of a membrane protein.** The initial step in N-glycosylation occurs as the protein threads through a protein translocating channel in the endoplasmic reticulum. A 14-residue glycan (attached to dolichol in the membrane, *red*) is transferred by oligosacchryltransferase (*blue*) to the protein and is then processed further.

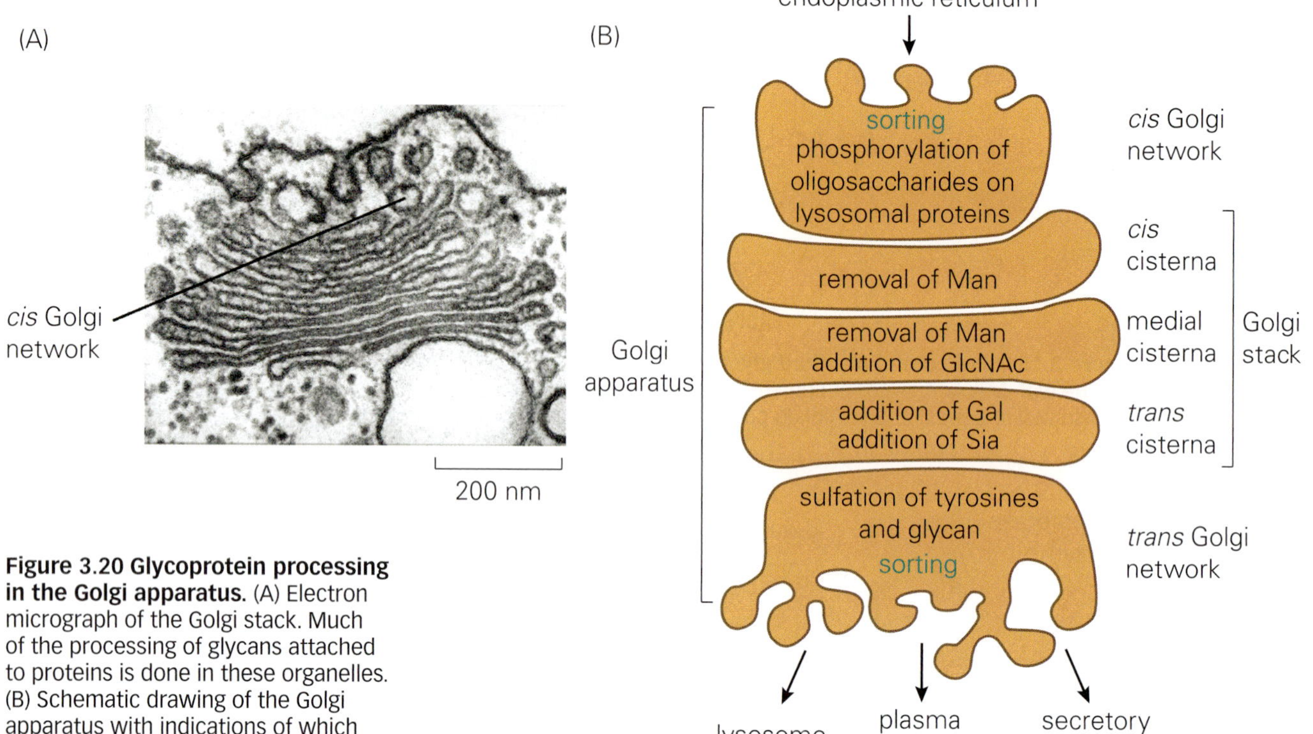


**Figure 3.20 Glycoprotein processing in the Golgi apparatus.** (A) Electron micrograph of the Golgi stack. Much of the processing of glycans attached to proteins is done in these organelles. (B) Schematic drawing of the Golgi apparatus with indications of which processes occur in which regions as proteins go through them. See Figure 3.9 for the abbreviations used for sugars. (A, courtesy of George Palade; B, adapted from B. Alberts et al., Molecular Biology of the Cell, 5th ed. New York: Garland Science, 2008.)

that are glycosylated are found within a specific consensus sequence motif: Asn-X-Ser/Thr, where X is any residue. Modification of the initial 14-sugar structure begins almost immediately, with some sugar units being removed and others being added back as the protein undergoes processing in the endoplasmic reticulum. Further modification of the N-linked oligosaccharide continues later, when the glycoprotein is being processed through the Golgi apparatus (**Figure 3.20**).

The process of O-linked glycosylation differs from that of N-linked glycosylation. O-linked oligosaccharides are formed almost entirely in the Golgi apparatus, and the class of enzymes carrying out the glycosylation—the glycosyl transferases—add sugar monomers (for example, glucose or mannose) to Ser or Thr residues, which then grow by one sugar at a time. The monosaccharides to be added are linked to nucleoside diphosphates. The diphosphate provides a good leaving group to facilitate the sugar transfer reaction.

The two major categories of N-linked oligosaccharides are called high mannose and complex glycans (**Figure 3.21**). Both types may be present on the same protein. Because both the sequence of sugars and the sites and stereochemistry of linkage can vary, vast numbers of different kinds of oligosaccharides are possible. With eight major sugars, 36,000 different tetrasaccharides could be made; the number of variants possible increases exponentially with the number of residues in the oligosaccharide.

## 3.10 The decoration of proteins with glycans is not templated

The synthesis of oligosaccharides in cells is done by enzymes, but unlike protein or nucleic acid synthesis, there is no template (that is, no explicitly stored code) for the carbohydrate polymers that are made. Instead, the structures of oligosaccharides synthesized by any particular cell at any given time is determined by what enzymes are present, their concentrations, their substrate specificities, and their spatial localization (in different parts of the Golgi apparatus, for example, as indicated in Figure 3.20).

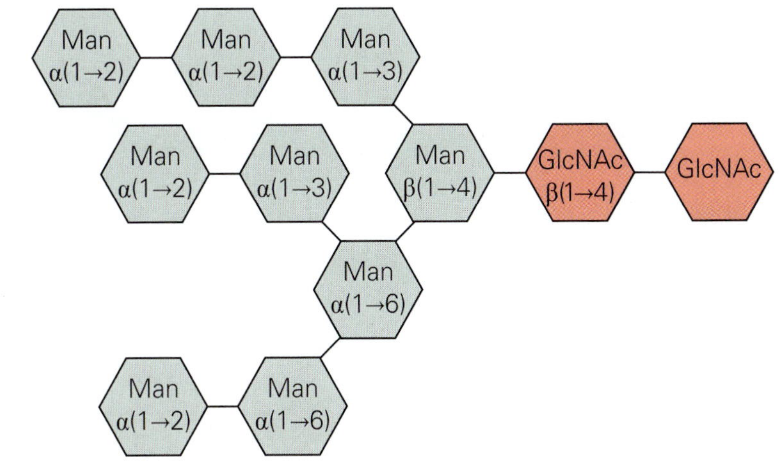


(B) complex

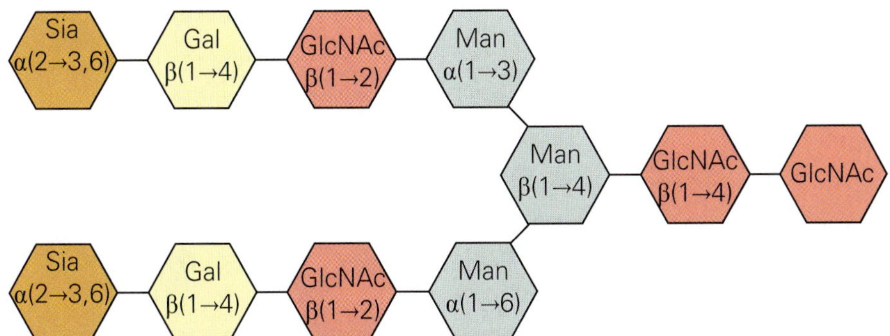


**Figure 3.21 Representative glycans of the (A) high mannose and (B) complex types.** Many variations on these structures occur in natural oligosaccharides. The 14-sugar glycan shown schematically in Figure 3.19 corresponds to the high mannose sugar in (A), with three additional glucose residues added to the upper mannose branch.

The end result of the nontemplated synthesis of glycans is that polymers synthesized in a particular location share the same core structure but vary in the finer details of their structure. For example, individual glycans may contain different numbers of terminal mannose or sialic acid residues. The control of which enzymes are present and where they are localized (both ultimately resulting from gene expression) does lead to fairly well-defined and reproducible products under defined conditions, but fine structure variations do occur. Glycosylated proteins, therefore, are quite heterogeneous.

## 3.11 Glycan modifications alter the properties of proteins

Because polysaccharides do not, in general, fold into compact structures, the addition of even a modest molecular weight oligosaccharide can cover a large surface area on the protein, as shown in **Figure 3.22**. Many of the proteins and lipids that are exposed at the surface of eukaryotic cells are glycosylated (**Figure 3.23**), and so the predominant surface presented to the surroundings may be due to glycan rather than to lipid or protein. Cell-cell interactions are therefore often mediated through glycan-protein interactions.

Glycans on proteins play important roles both inside and outside cells. For example, erythropoietin, the hormone that stimulates red blood cell production in the bone marrow, has multiple glycosylation sites (**Figure 3.24**), and its biological activity increases with the extent of glycosylation. One way in which glycosylation can improve the efficacy of a protein such as erythropoietin is to protect it against degradation by protease enzymes. These enzymes cleave the backbone of proteins, and glycans can make it harder for proteases to access the backbone. The presence of glycan can also stabilize the folded state of a protein, further protecting against degradation.

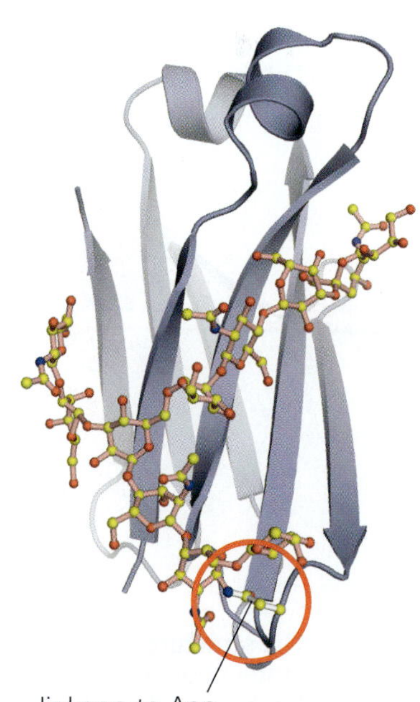


**Figure 3.22 An N-linked oligosaccharide attached to an asparagine residue in a protein.** The oligosaccharide covers one face of the protein. (PDB code: 1MCO.)

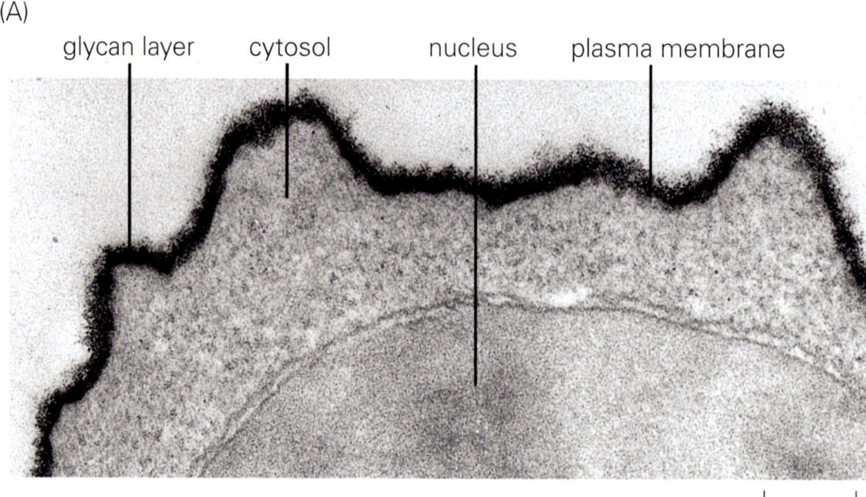


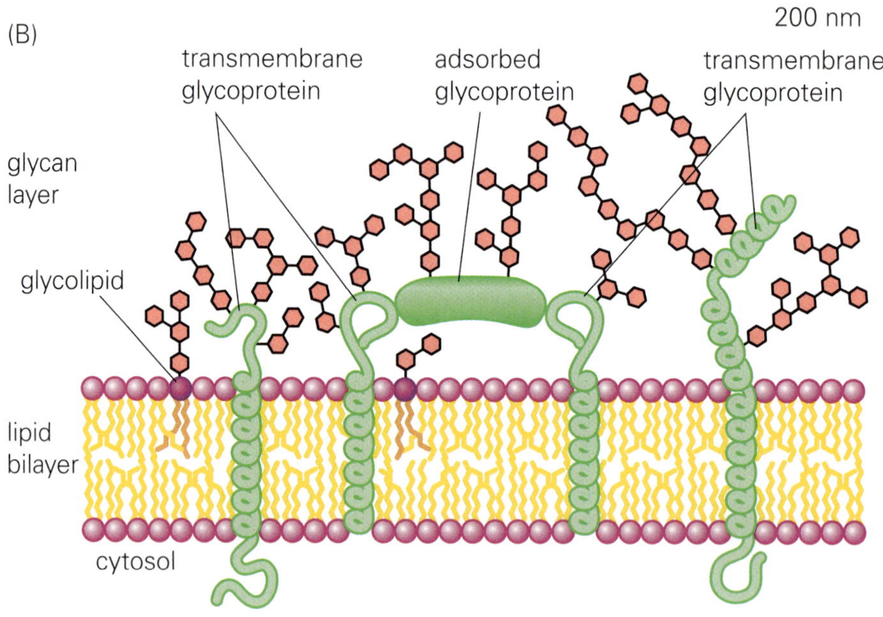


**Figure 3.23 Cell-surface glycans.** (A) An electron micrograph of a fibroblast cell is shown with the outer glycan layer stained (dark layer). (B) Cell-surface glycolipids, transmembrane glycoproteins, transmembrane proteoglycans, and adsorbed glycoproteins all contribute to this thick glycan layer. (A, courtesy of Audrey M. Glauert and G.M.W. Cook; B, adapted from B. Alberts et al., Molecular Biology of the Cell, 5th ed. New York: Garland Science, 2008.)

**Lectins**

Lectins are proteins that recognize specific sugars, usually binding with moderate affinity and high specificity.

Glycosylation is also important for the trafficking of proteins inside cells. Within the endoplasmic reticulum, the attached sugars are indicators of the progress of protein folding, determining when a protein is allowed to progress from the endoplasmic reticulum to its next destination. For organelles in general (particularly those of secretory and endocytic pathways—namely, the endoplasmic reticulum, the Golgi apparatus, endosomes, lysosomes, and secretory vesicles), the oligosaccharides attached to proteins can act as "address labels," determining where the proteins are transported to, including whether they are delivered to the cell surface (glycosylated proteins are enriched on cell surfaces). Glycans are also very important in the formation of cell walls, which is discussed in Section 3.23.

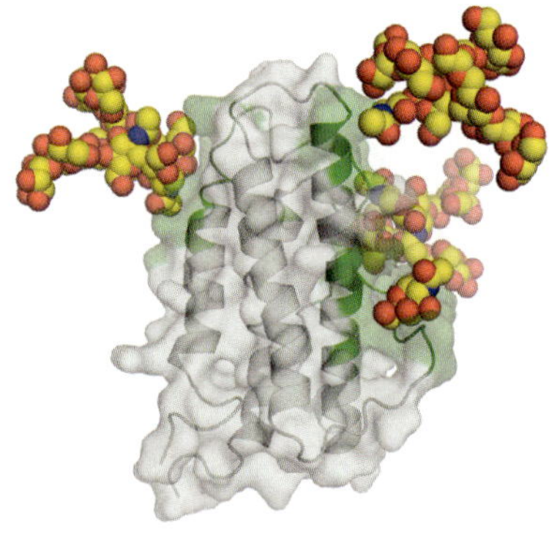

**Figure 3.24 The structure of the human protein hormone erythropoietin.** Each erythropoietin molecule has three N-linked complex carbohydrates and one O-linked GalNAc residue. Regions of the protein surface that interact with the sugars are colored *green*. (PDB code: 1BUY; glycans modeled by R.J. Woods.)

## 3.12 Protein–glycan interactions are important in cellular recognition

Cell surface glycans can interact with proteins on other cells, establishing stable connections between cells. A large group of proteins, known collectively as **lectins**, recognize specific sugars in glycans (**Figure 3.25**). In general, lectins bind only a small portion of a surface carbohydrate, but do so with high specificity. The binding of a single polysaccharide to a single protein is often weak, but because there are many glycans on each cell, and many copies of proteins that bind them, these multivalent interactions lead to both tight and specific cell–cell associations.

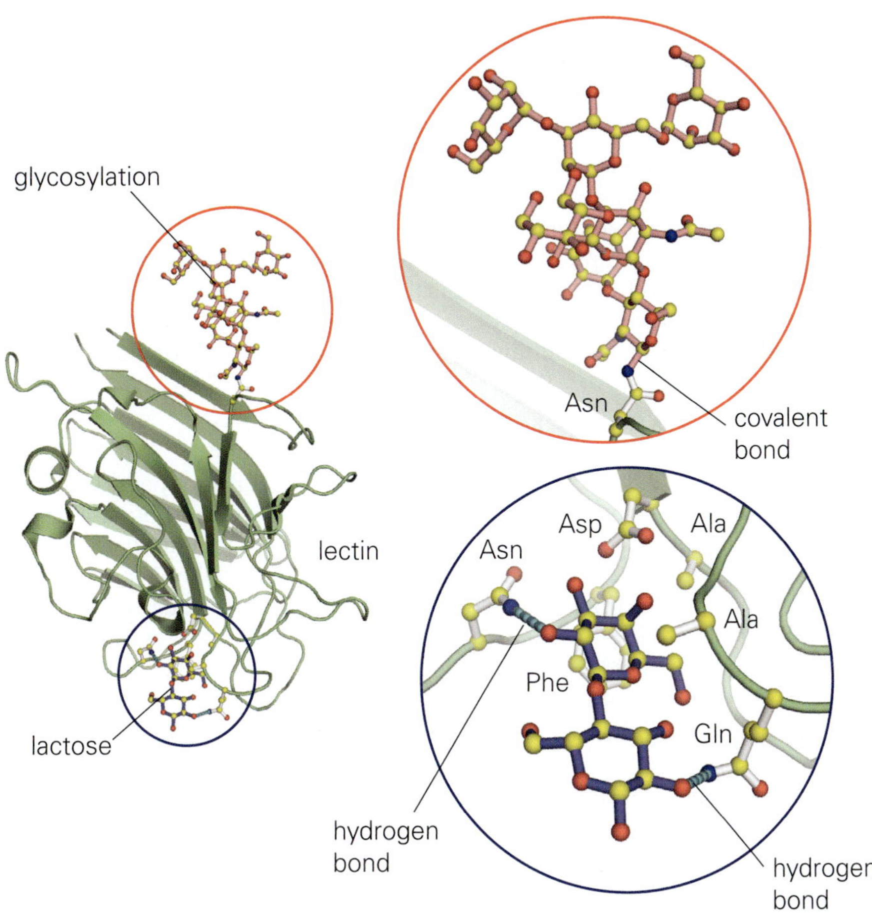


**Figure 3.25 A lectin isolated from a legume.** The lectin is shown in complex with lactose [the disaccharide β-galactose-(1→4)-β-glucose, *blue* circles]. This lectin recognizes primarily the galactose end of the lactose, making both hydrogen bonds and hydrophobic contacts. The lectin itself is glycosylated (see the sugars at the top of the protein, *red* circles). (PDB code: 1LTE.)

Cell-surface glycoproteins can also be recognized by antibodies in the immune system. For example, glycans present on red blood cells define the "blood groups" A, B, AB, and O (**Figure 3.26**). Subtle variations in a polysaccharide on red cells arise from genetic differences in the enzymes that control the glycosylation.

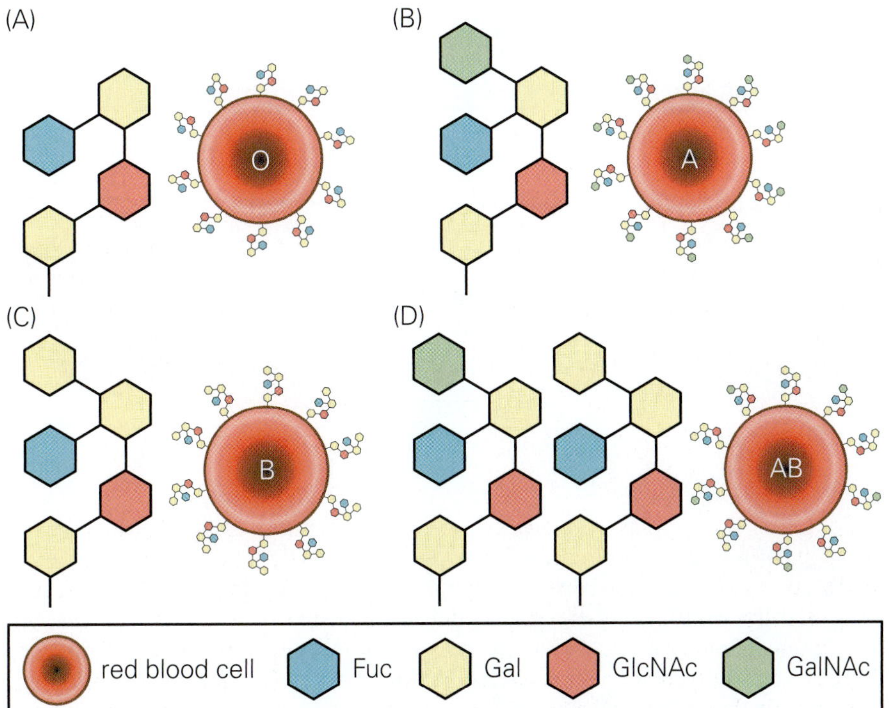


**Figure 3.26 Glycans that are attached to proteins on the surface of red blood cells.** These glycosylation patterns define an individual's blood type. (A) The O blood type has a non-antigenic tetrasaccharide that is common to the surface antigens of the other blood groups. A-type (B) and B-type (C) individuals produce oligosaccharides with an additional Gal or GalNAc residue, respectively. (D) AB-type individuals produce a mixture of the A- and B-type oligosaccharides.

Type O individuals synthesize a cell-surface tetrasaccharide. Type A individuals augment the tetrasaccharide with an extra Gal-NAc residue, whereas type B individuals add an extra Gal residue. AB type individuals produce both A and B pentasaccharides.

As the immune system develops in an infant, the substances that are present and natural to the body (proteins and glycans, for example) are defined as "self," and antibodies that recognize them are eliminated from the immune system. If an individual is exposed to non-self antigens, as might happen if mismatched blood is transfused into the body, the cells displaying the foreign antigens are recognized as foreign by the immune system. The ensuing immune reaction can cause lysis of the red blood cells, resulting in renal failure, shock, and sometimes death.

## B. LIPIDS AND MEMBRANES

**Amphiphiles**

Amphiphiles are molecules that have parts that are hydrophilic (loving water) and other parts that are hydrophobic (hating water).

**Lipids** are a class of biologically important molecules that are grouped by the common property that they are soluble in nonpolar organic solvents, but are insoluble in water as individual molecules. The distinguishing feature in the molecular structures of lipids is that they always have a strongly hydrophobic part (usually hydrocarbon chains or rings, with most of the carbons being saturated) and they also have a hydrophilic part (charged, or very polar, but this part of the molecule may be as small as a single hydroxyl group). This dual character leads to the classification of these molecules as **amphiphiles**.

The very different solvation preferences of these two parts of lipid molecules drives them to form clusters in which the hydrophobic parts are brought together (**Figure 3.27**). This excludes water from the hydrophobic parts of the lipids, while

(A)

single-chain lipids in water

micelles

(B)

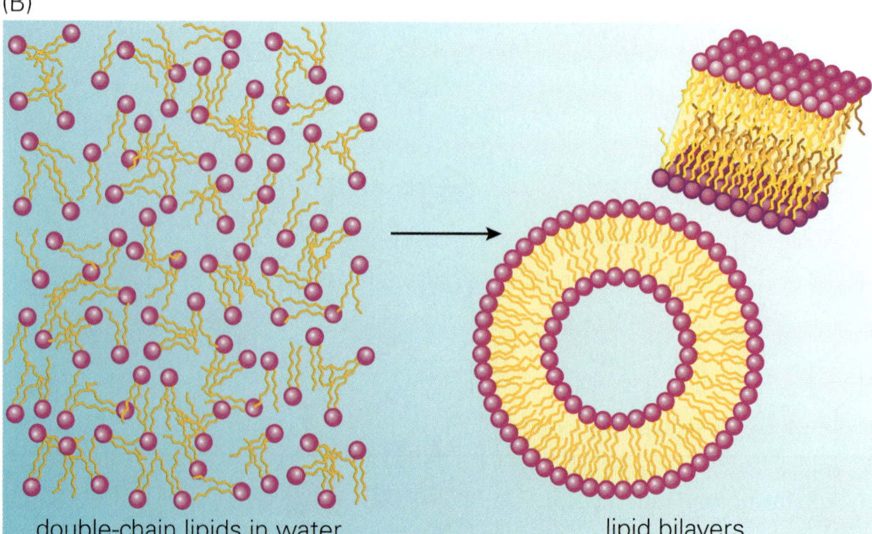


**Figure 3.27 The phase separation of lipids in water.** Lipids are amphiphilic, so they tend to aggregate (form separate phases from water) when they are dissolved in water. (A) Single-chain lipids aggregate to form micelles. (B) Double-chain lipids aggregate to form planar or spherical bilayer structures.

at the same time allowing the hydrophilic parts to interact well with water. This clustering is a phase separation, just as oil forms a separate phase from water. For many kinds of lipids this phase separation leads to arrays of molecules organized into **lipid bilayers**, called membranes, which are critical to the function of cells. Depending on the specific sizes and shapes of both the hydrophobic and the hydrophilic parts, these molecules can form other structures, including droplets known as **micelles** (see Figure 3.27). Although forming membranes is a critical function of lipids, specific lipids serve many other functions in cells, ranging from energy storage to signaling.

**Lipid bilayer**

A lipid bilayer is comprised of lipids packed into two parallel layers with the head groups exposed to water and the alkyl chains packed together away from water. Lipid molecules move relatively freely within the plane, but do not easily flip between the two layers.

## 3.13 The most abundant lipids are glycerophospholipids

**Glycerophospholipids** are the most abundant molecules in biological membranes. As the name suggests, they are built from a glycerol unit, $HOCH_2$–CHOH–$CH_2OH$, as shown in **Figure 3.28**. One of the terminal hydroxyls is linked to a phosphate, which itself carries a polar or charged substituent (an R group). The phosphate group, along with the R group, is hydrophilic and is called the **head group**.

**Head group**

A head group is the hydrophilic part of a glycerophospholipid, which is attached to the glycerol. The head group remains in contact with water when the lipids are in bilayers or micelles.

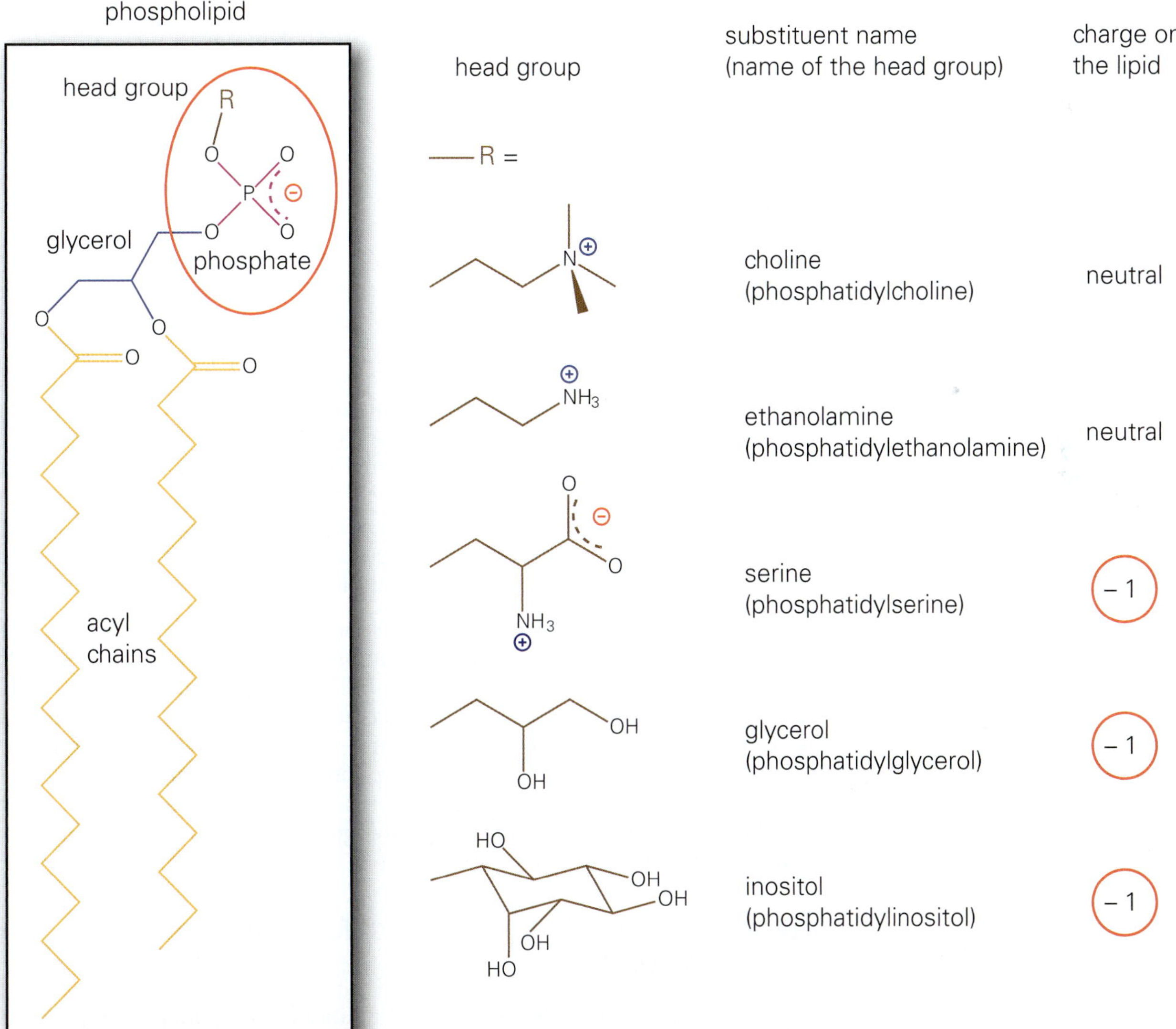


**Figure 3.28 Diacylglycerol phospholipid.** Phospholipids contain a glycerol backbone linking two acyl chains to head groups, which are phosphoesters. The acyl chains vary in length and in the number of double bonds they contain. The substituents that define some of the different head groups are shown, as are the charges on the lipids with these head groups. Note that because the phosphate group bears a negative charge, lipids with R groups that have a single positive charge are net neutral overall.

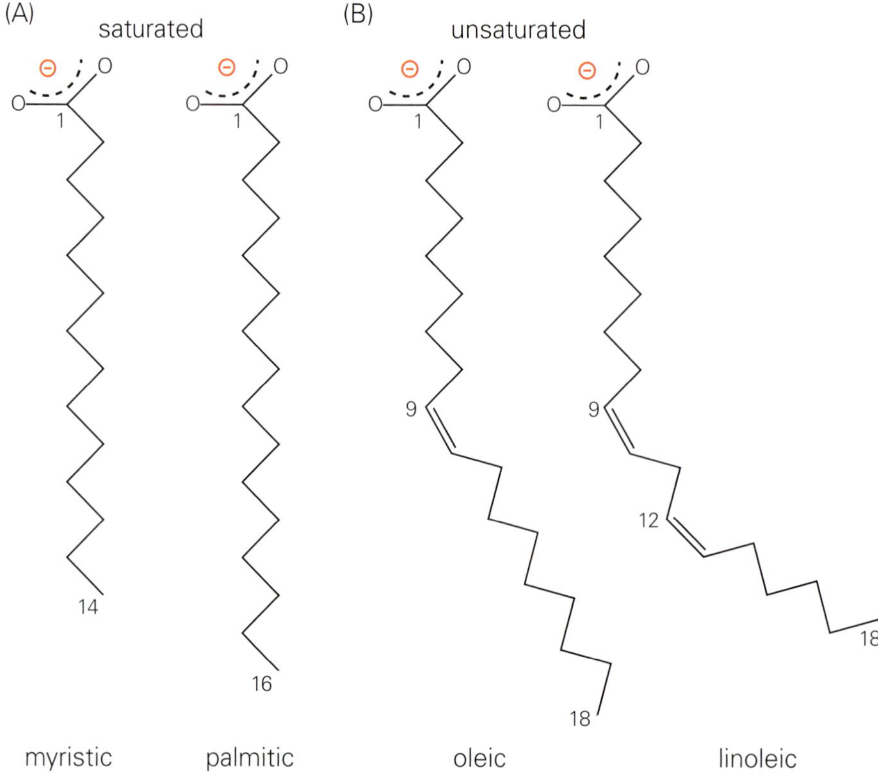


**Figure 3.29 Structures of several fatty acids.** These form glycerol esters to become components of lipids. The length and degree of unsaturation are variable. Lipids in cells contain many different fatty acids.

The R groups on these lipids can be derived from an amino acid (for example, serine), a sugar (for example, inositol), or some other small, polar moiety (for example, choline, ethanolamine, or glycerol). Some examples of head groups are presented in Figure 3.28. The other two glycerol hydroxyls are esterified to fatty acids, which are long (C12 to C24) hydrocarbon chains terminating with a carboxyl group. These fatty acids may be saturated (all C–C single bonds), or singly or multiply unsaturated (with C=C double bonds), as shown in **Figure 3.29.** Lipids are often colloquially referred to as fats—and the terms "saturated fat," "unsaturated fat," and "polyunsaturated fat" may be familiar from nutritional information. Unsaturation in natural fats occurs as *cis* double bonds, as shown in Figure 3.29. The consumption of artificially unsaturated fats with *trans* double bonds (also known as *trans* fats) is thought to be a health risk.

Cells contain a considerable variety of phospholipids, with different head groups and alkyl chains of various lengths and numbers of double bonds (see Table 3.2 in Section 3.22 for the lipid composition of different kinds of membranes). The molecular structures of phospholipids affect the characteristics of the membranes they form, as discussed below. The net charge of the membrane surface is determined by the composition of the head groups present. These can be positive, zwitterionic, negative, or neutral. Having an excess of either positive or negative lipids gives the surface of the membrane a net charge, which can profoundly affect the interactions of proteins with the membrane.

## 3.14 Other classes of lipids have different molecular frameworks

While glycerophospholipids are the most abundant component in most membranes, other kinds of lipids also play important roles in determining the physical properties of the membranes. One set of examples is provided by the **sphingolipids**, which occur in eukaryotes and have structural similarities to phospholipids. Sphingolipids have one "built-in" hydrophobic chain from the molecule sphingosine that in essence replaces the glycerol unit of glycerophospholipids

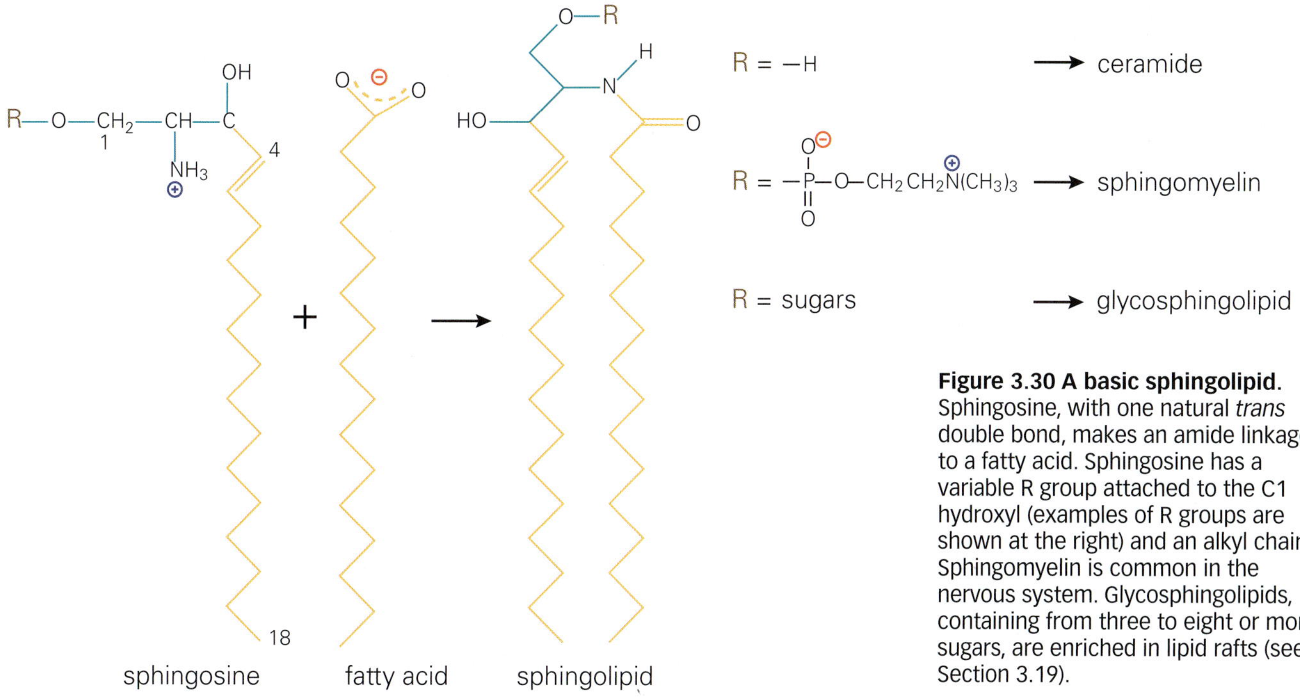

**Figure 3.30 A basic sphingolipid.** Sphingosine, with one natural *trans* double bond, makes an amide linkage to a fatty acid. Sphingosine has a variable R group attached to the C1 hydroxyl (examples of R groups are shown at the right) and an alkyl chain. Sphingomyelin is common in the nervous system. Glycosphingolipids, containing from three to eight or more sugars, are enriched in lipid rafts (see Section 3.19).

(**Figure 3.30**). The sphingosine forms an amide bond with one fatty acid chain that can vary in length. The C1 hydroxyl of the sphingosine has an R group attached to it, analogous to phospholipid head groups, creating a molecule that looks remarkably similar to a glycerophospholipid.

Sugar head groups are more commonly found in sphingolipids than in glycerophospholipids. Sphingolipids with one single sugar are **cerebrosides**, and those with multiple sugars (typically three to eight) are **gangliosides**. A large number of different gangliosides have been identified. Their distribution is cell-type specific. Cerebrosides were first identified in brain tissue, where they occur at a level of ~5% of the membrane lipids.

In eukaryotic cells, another important lipid is **cholesterol** (**Figure 3.31**A). Cholesterol is a **steroid**, containing a conserved core of four fused carbocyclic rings, one with a polar OH group attached and, at the opposite end, one with a short alkyl chain. The linked-ring core is relatively rigid (particularly in comparison to the intrinsically very flexible alkyl chains of other membrane lipids). Cholesterol is mixed with other lipids, sometimes to a substantial fraction of the lipid present, thus modifying the membrane properties. The rigid core of cholesterol packs against alkyl chains of neighboring lipid molecules, reducing their flexibility. The net effect is to stiffen the membrane bilayer against distortion, while retaining fluidity.

A storage form of cholesterol also occurs, made by the esterification of its OH group with a fatty acid. The resulting molecule, having lost the OH, is fully hydrophobic and aggregates into droplets rather than forming bilayers. These cholesterol esters are transported in the bloodstream, packed into low-density lipoprotein (LDL) particles (see Section 3.21).

**Triglycerides**, three fatty acids esterified to a glycerol backbone (Figure 3.31B), are the main storage form for lipids in eukaryotic cells. Triglycerides serve to transport the dietary fats that are important in metabolism. Triglycerides are a major component of very low density lipoproteins (discussed further in Section 3.21). **Cardiolipin** is a double phospholipid with four alkyl chains, essentially a dimer of a phosphoglycerol lipid (Figure 3.31C). It occurs in mitochondrial membranes.

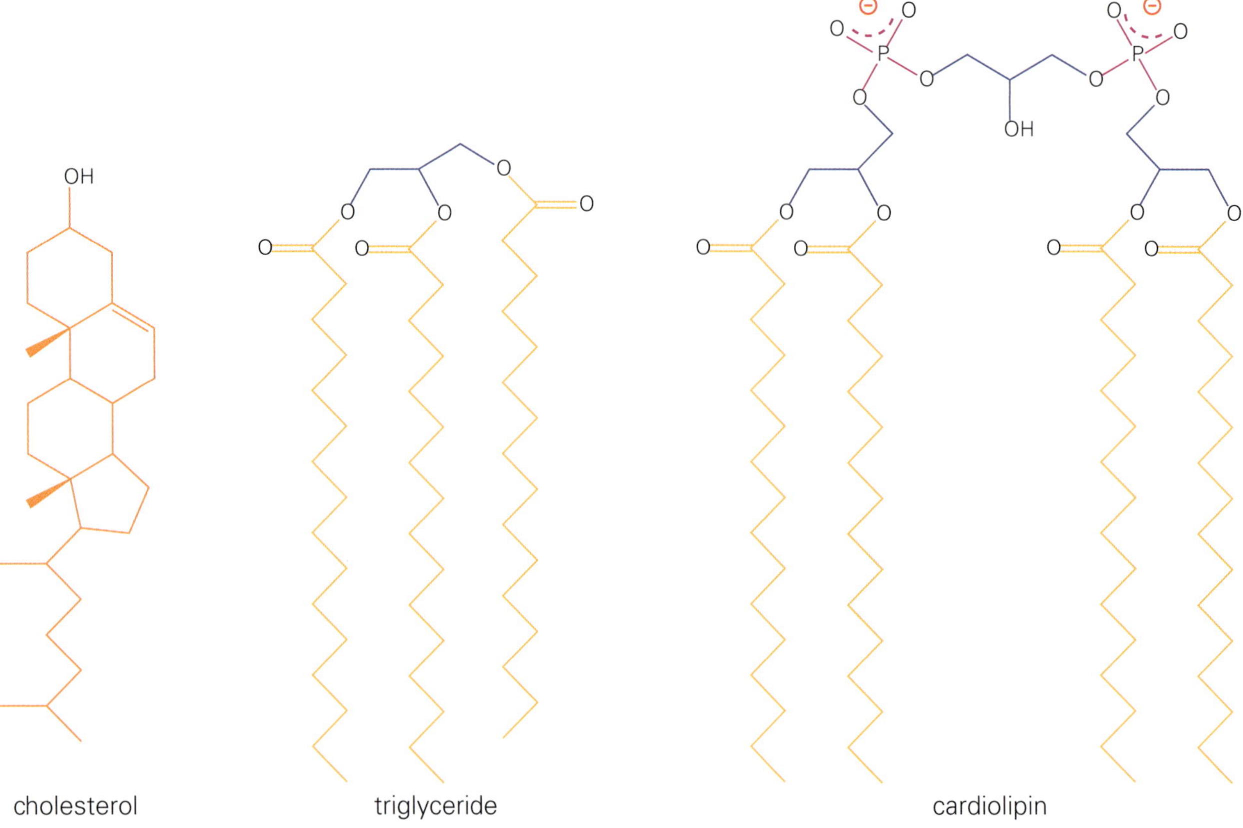


**Figure 3.31 Cholesterol, triglyceride, and cardiolipin.** Cholesterol is an important component of eukaryotic membranes. Triglycerides are important in fat metabolism. Cardiolipin comprises ~20% of the inner mitochondrial membranes.

Some lipids have biological activity on their own—that is, activity not associated with forming membranes or micelles. One example of bioactive glycolipids are the Nod factors, which are lipo-chito-oligosaccharide lipids that are produced by bacteria that live symbiotically in the root nodules of plants. These lipids have one

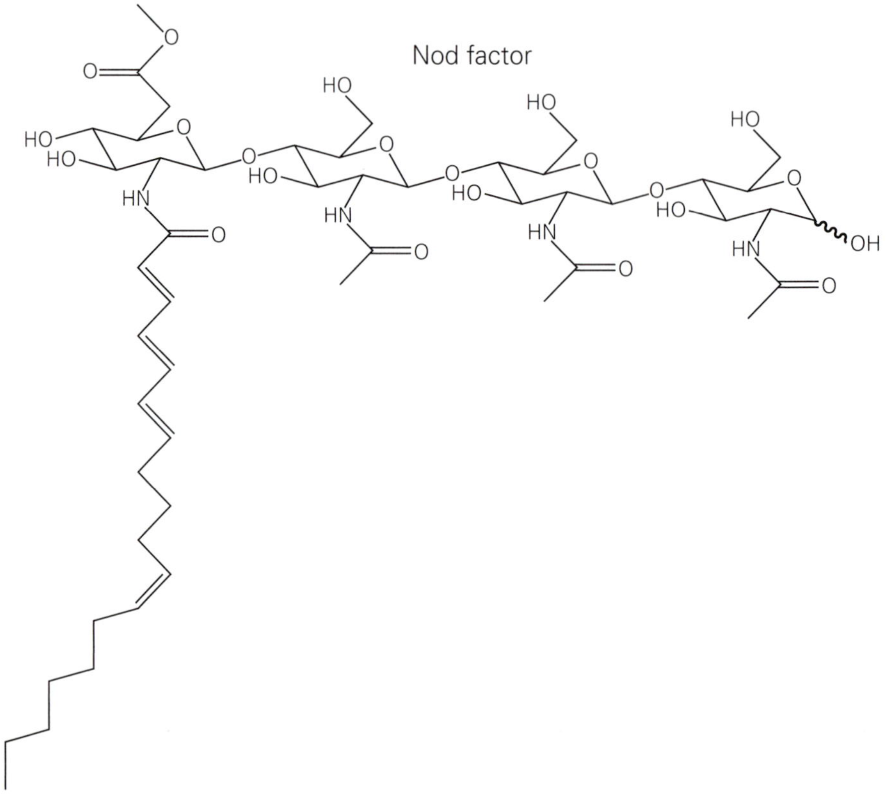

Nod factor

**Figure 3.32 A Nod factor.** In this case, the fatty acid is 18:4 (18 carbons with four sites of unsaturation), attached to a tetrasaccharide, a derivative of GlcNAc–GalNAc–GalNAc–GalNAc.

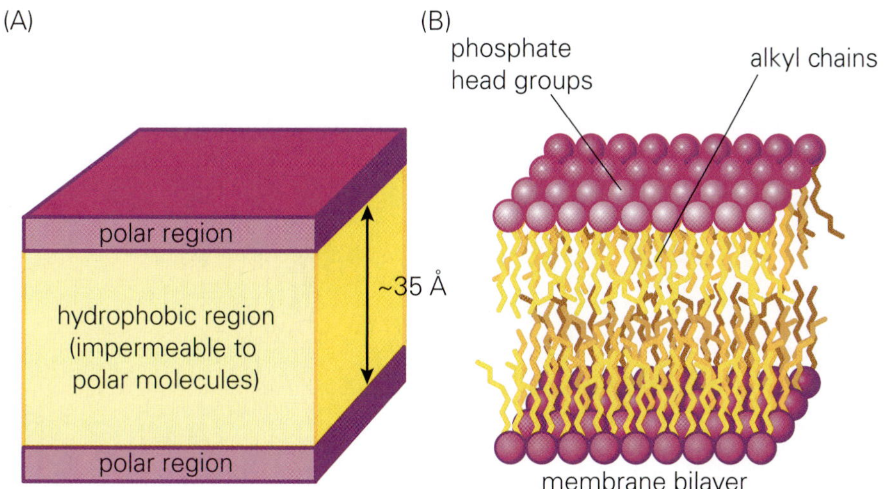


**Figure 3.33 Lipid membranes.** (A) A schematic representation of a lipid bilayer, showing a hydrophobic interior sandwiched by polar regions. (B) This diagram shows the individual lipids in the bilayer. The head groups are represented by spheres, while the alkyl chains extend into the regions between the head group layers. The bilayer is about 35 Å thick.

unsaturated fatty acid, which is attached directly to a small oligosaccharide (with four to five sugar groups), as shown in **Figure 3.32**. These factors stimulate host-specific responses in the plants, inducing the development of the root nodules. The symbiotic arrangement benefits the plant by fixing nitrogen needed for plant growth, and it benefits the bacteria by providing nutrients and a protected place to live.

## 3.15 Lipids form organized structures spontaneously

The key behavior of lipid molecules is their strong tendency to form separated phases spontaneously when in contact with water. The aggregated lipid molecules are *not* covalently linked (as the amino acids in a protein are), but the same hydrophobic effects that drive protein folding also cause the hydrophobic parts of lipids to cluster together.

Oil and water separate spontaneously, in a way that minimizes the area of the contact surface. Oil shaken with water forms small droplets, which then gradually merge to give two completely separate layers. For lipids, however, the polar head groups want to remain in water, while the hydrophobic parts want to minimize their interaction with water. Phospholipids, therefore, form organized structures, the most important of which for biology are lipid bilayers, which are illustrated in **Figure 3.33**. In the lipid bilayer, the head groups point out and form a favorable interface with water. The hydrophobic alkyl chains point in and interact with each other laterally, and also pack against the hydrophobic chains of the other layer. Water is excluded completely from the interior region. This structure satisfies, therefore, the interaction preferences for both segments of the lipid molecules.

Bilayers serve as the membranes that partition cells into defined regions. These membranes are effective barriers, and the transport of many molecules (particularly those that are very polar or charged) through the membrane is slow, unless mediated by specific transmembrane proteins. The two layers of the bilayer (often called the **leaflets** of the bilayer) are in contact with different compartments of the cell and may themselves have a rather different composition. For example, glycolipids are not found in the leaflet in contact with the cytoplasm, but are found in the exterior leaflet.

## 3.16 The shapes of lipid molecules affect the structures they form

Lipids dispersed in water can form different kinds of superstructures that contain membrane bilayers. When placed on a surface that is wetted well by water (indicating that it has hydrophilic groups on its surface), lipids form simple planar

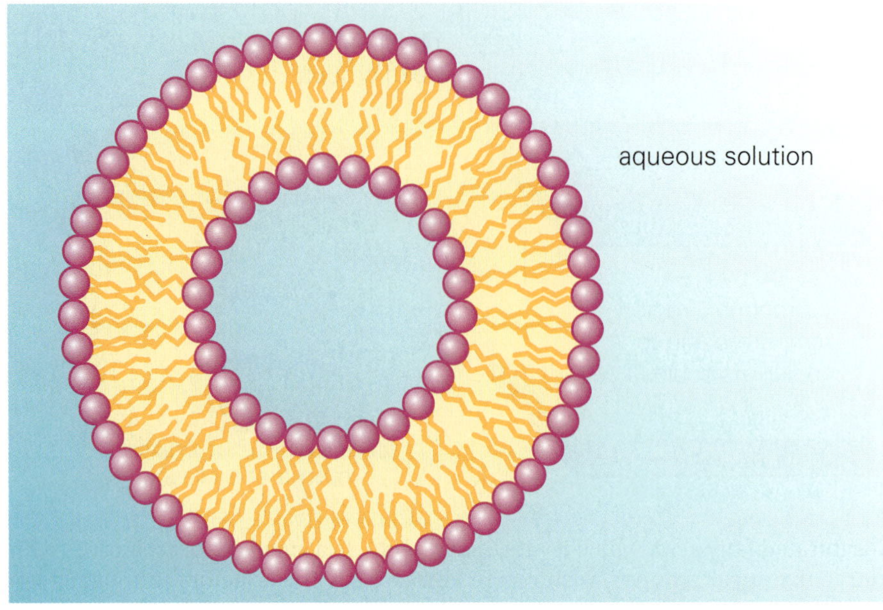


**Figure 3.34 A cross section of a vesicle formed by a phospholipid bilayer.** Water is both inside and outside the vesicle.

bilayers. If dispersions of lipids in water are sonicated (with an ultrasound source analogous to those used for cleaning jewelry), **lipid vesicles** are formed instead of planar bilayers. These are closed spherical bilayers rather like bubbles, but with water inside and outside rather than air (**Figure 3.34**). By varying the lipid concentration and conditions, **multilamellar vesicles** can also be formed. These are concentric spheres of bilayers, rather like the layers of an onion.

**Micelle**

A micelle is a spherical array of lipids with head groups on the outside and chains on the inside (see Figures 3.35 and 3.36). Micelles are typically formed by single-chain lipids, or by detergents.

The curvature of the bilayer (which is highest in small vesicles) makes these structures intrinsically less stable than extended bilayers. However, vesicles are often kinetically trapped, and can therefore be stable for long times. Vesicles can be made with various compositions of lipids and can mimic many aspects of real biological membranes. Lipid vesicles are often used in the laboratory as models for characterizing the physical properties of membranes under well-controlled conditions. Under similar conditions, some lipids will also form long tubes, for which a cross section would also look like that shown in Figure 3.34.

The phase separation of lipids is driven by the hydrophobicity of the alkyl chains and is a common property of strongly amphiphilic molecules. The formation of membrane bilayers is, however, a property of only some lipids. The ability to form a bilayer is determined by the size and shape of both the head groups and the hydrophobic chains. Lipids that have a polar head group but only a single alkyl chain have a higher cross-sectional area in the head group than the hydrophobic section. For these kinds of lipids, packing is optimized by forming a highly curved surface. The radius of curvature is often quite small (the surface is much more curved than for a bilayer vesicle), leading to structures called **micelles**, which are spherical arrays of the molecules with head groups on the outside and chains on the inside, as shown schematically in **Figure 3.35**.

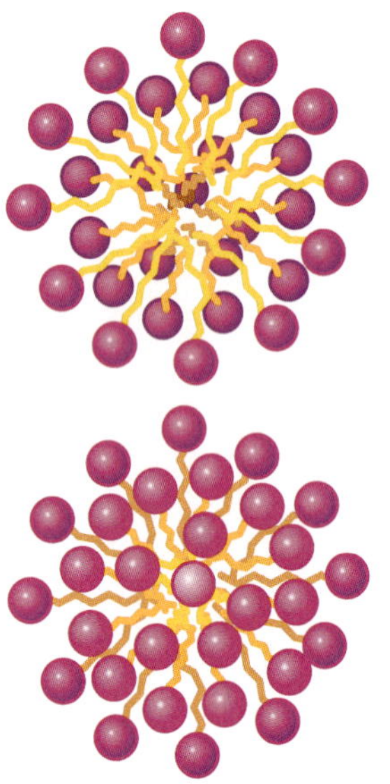

**Figure 3.35 A micelle in cross section (top) and from the outside (bottom).** Circles represent polar head groups and wavy lines represent alkyl chains.

The type of structure formed by micelles removes the hydrophobic alkyl groups from water, but unlike membrane bilayers, the micelles are small and move quite freely. The structures and the interface with the solvent water are quite disordered, as shown by a computational model of a dodecylphosphocholine micelle in **Figure 3.36**.

The tendency of amphiphilic substances to form micelles depends on the size of the hydrophobic portion of the molecule. Amphiphilic compounds are soluble as individual molecules up to a specific concentration, called the **critical micelle concentration**. When that concentration is exceeded, micelles begin to form. The

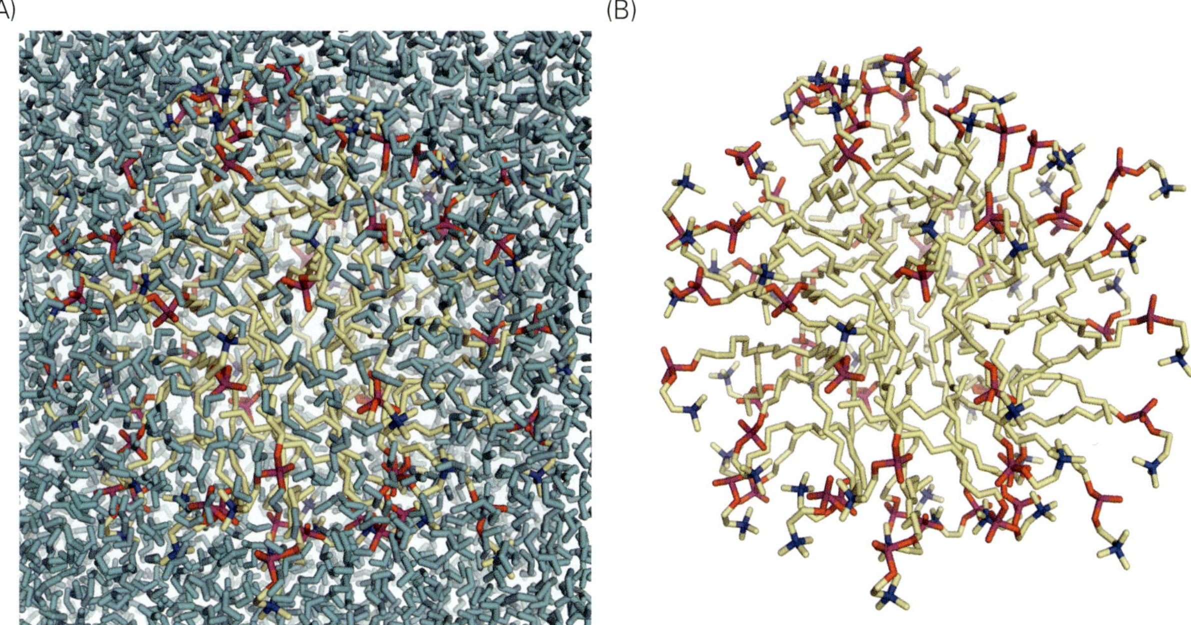


**Figure 3.36 Structure of a micelle.** (A) A solvated micelle made from dodecylphosphocholine. Water molecules are shown in *cyan*. (B) Water surrounding the micelle in (A) has been removed to show the micelle more clearly. As with lipid bilayers, the positions of both the head groups and the alkyl chains are poorly defined, exchanging rapidly among many conformations. (Atomic coordinates: m54.pdb from http://moose.bio.ucalgary.ca)

size of the micelles formed (both in dimensions and number of molecules) is a characteristic of the substance, although it can be affected by solution conditions (that is, by ionic strength and the presence of other compounds).

## 3.17 Detergents are amphiphilic molecules that tend to form micelles rather than bilayers

The molecules that typically form micelles rather than bilayers are generally called **detergents** rather than lipids, even though they are chemically and structurally similar (**Figure 3.37**). The idea of a detergent is familiar from everyday life as a substance that helps remove oil and dirt from clothing and household objects. Dirt particles that are hydrophobic are coated effectively by detergents, producing an outer layer around them that allows the particles to be solubilized in water and removed in the process of washing and rinsing (**Figure 3.38**). Basically, the detergents form a micelle around the dirt particle.

Detergent molecules can also solubilize proteins by surrounding exposed hydrophobic residues, creating hydrophilic surfaces. The ability to bind hydrophobic sidechains can be so effective that some detergents can unfold (denature) water-soluble proteins by clustering around hydrophobic residues in the protein. The tendency to do this varies considerably with the structure of the specific detergent molecule.

Detergents have a wide variety of chemical structures. The simplest of these are the alkyl sulfates. Sodium dodecylsulfate, for example, is a denaturing detergent used commonly in biology labs during electrophoresis, as discussed in Chapter 17. Longer chain alkyl sulfates are also used in products such as shampoo. The head groups of some detergents are the same as those in lipids (for example, DPC; see Figure 3.37). While many detergents are charged, some are not, and these tend to be less strongly denaturing towards proteins than ionic ones. Nondenaturing detergents are particularly useful for extracting membrane proteins from lipid bilayers into micelles for protein purification and structural studies. The hydrophobic surfaces of membrane proteins that would normally be in contact with the alkyl chains of lipids are coated instead with the alkyl chains of the detergent. The hydrophilic parts of the detergent, in turn, make favorable contacts with water, solubilizing the complex.

**Figure 3.37 Micelle-forming detergents.** Dodecylphosphocholine (DPC) has a lipid-like head group. Sodium dodecylsulfate (SDS) is an ionic detergent that unfolds (denatures) proteins. Dodecylmaltoside (DDM) and Triton X-100 are both nonionic and are relatively nondenaturing towards proteins.

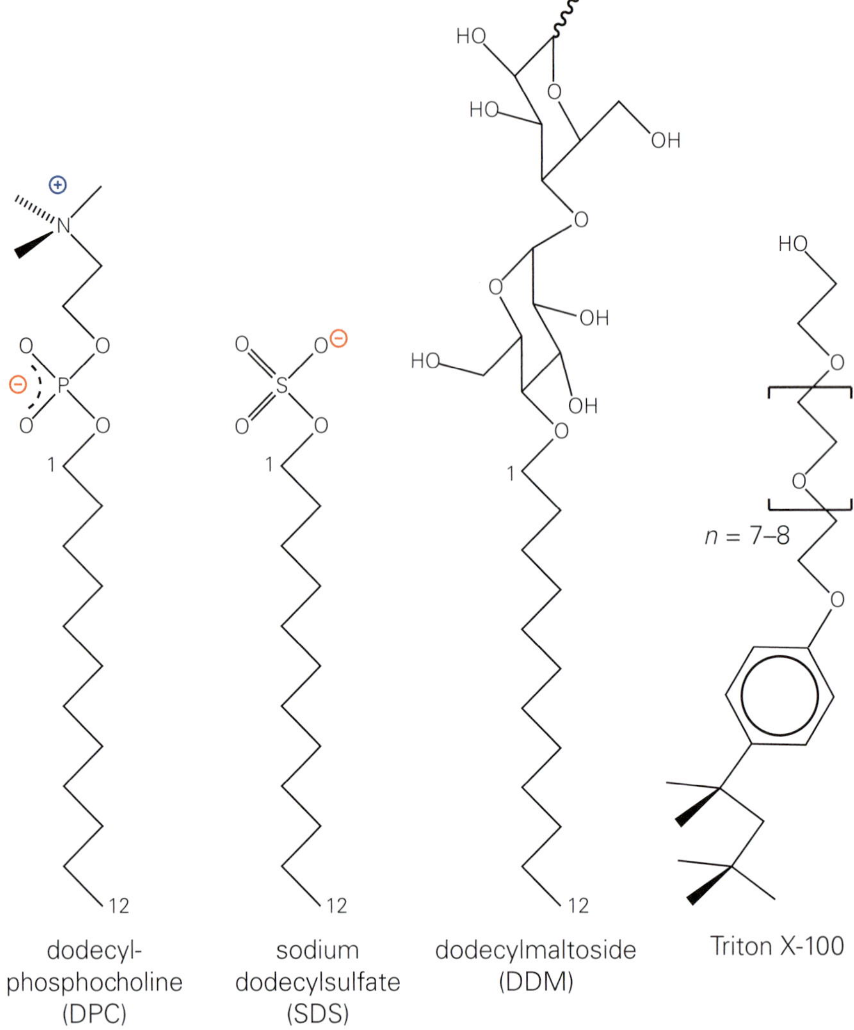

**Figure 3.38 A "dirt" particle covered by a detergent micelle.** The micelle makes the dirt particle soluble in water.

## 3.18 Lipids in bilayers move freely in two dimensions

Membrane bilayers are usually drawn as a pair of uniform planes containing the head groups of the lipids (see Figure 3.33). Real membranes are organized into clear layers, but these are not highly ordered at the atomic level. The individual lipid molecules take on many different conformations, and there are significant deviations in the positions of the head groups with respect to the average plane they occupy. The details of water molecules solvating the region around the head groups are also highly variable, with the amount of water present on average decreasing as the region containing the hydrophobic chains is approached.

The hydrophobic center of the membrane is more fluid than the head-group region, with individual alkyl chains undergoing very rapid rotations about the C–C single bonds. The alkyl chains are less densely packed than the head-group region, leaving room for the alkyl chains to exchange rapidly among various conformations. Molecules in a realistic computational membrane model are shown in **Figure 3.39**, in which many of these features can be seen.

Lipid molecules have no positional order within the planes of the bilayer, although the average spacing between molecules is well defined. The lipid molecules act as a two-dimensional liquid, and individual lipid molecules diffuse relatively rapidly within the bilayer (**Figure 3.40**).

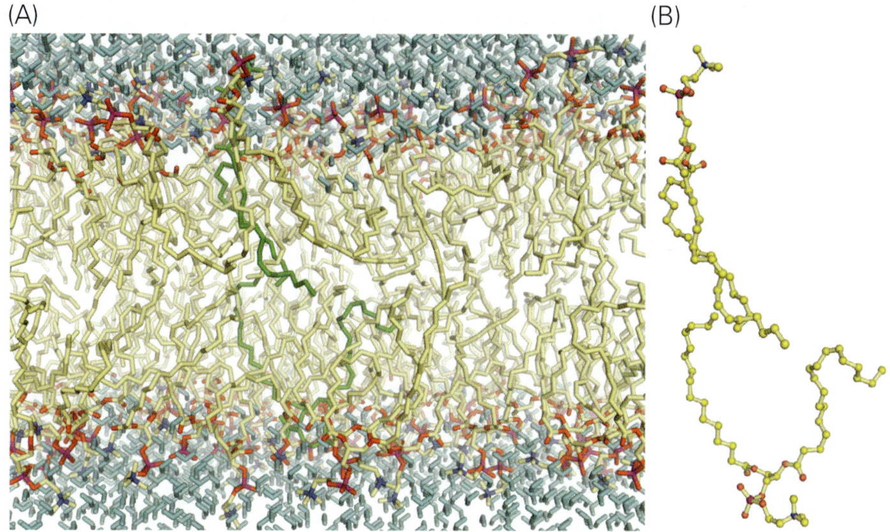


**Figure 3.39 A slice through a bilayer membrane.** (A) A snapshot from a computer simulation of a lipid bilayer. The glycerophospholipid molecules have choline head groups. Water molecules (*cyan*) can be seen at the top and bottom, with some waters interspersed into the head group region. Two of the lipid molecules are colored in *green* (one from each layer of the membrane). These same molecules are shown separately in (B). Note the variability in conformation of the alkyl chains and the head group. (Atomic coordinates: plpc128.pdb from http://moose.bio.ucalgary.ca)

The diffusion of molecules in the lipid bilayer can be observed directly by using fluorescent molecules that are attached to the lipids. By viewing the lipid bilayer through a microscope that detects fluorescence, one can observe where the fluorescent molecules are. If the density of fluorescent molecules is high, it can be technically challenging to resolve the positions of individual molecules. Fortunately, there are experimental methods that allow one to deduce how fast the fluorescent molecules are moving.

One such experiment is described in **Figure 3.41**, and it involves a protein that is fluorescent and is attached to a lipid in the membrane. The microscope objective is focused so as to detect fluorescence within a very small region on the membrane (a few microns across). At a certain point in time, a flash of very intense light is transmitted through the microscope and onto the small region under observation. The bright light bleaches the fluorescent molecules, and they lose the ability to absorb and re-emit light. The bleached molecules take an extremely long time to recover their fluorescence, so they become invisible after the flash of light. This leads to a drop in the fluorescence observed through the microscope, as shown in Figure 3.41 (at the 1 second time point). But, because the fluorescent proteins are mobile in the plane of the membrane, within a short time the bleached proteins move out of the field of view of the microscope. At the same time, unbleached proteins move into the field, and the observed fluorescence increases (this takes a few seconds in the experiment shown in Figure 3.41). This kind of experiment, which is called **Fluorescence Recovery After Photobleaching** (FRAP), confirms that lipids in the membrane are diffusing freely.

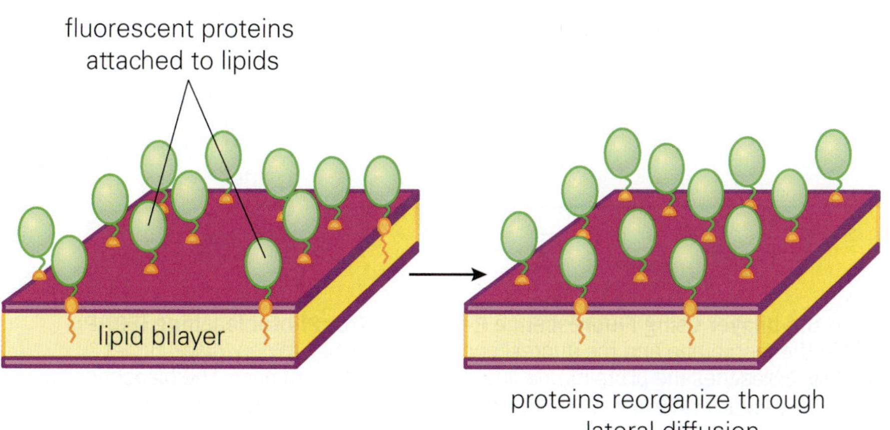


**Figure 3.40 Lateral mobility of lipids in a bilayer.** The movement of lipids in a bilayer can be monitored by attaching fluorescent molecules or proteins to a few lipids in the membrane. In the example shown here, green fluorescent proteins are attached to lipids. These proteins diffuse rapidly in the plane of the membrane.

For pure lipids, there are phase transitions between gel states (almost solid, at lower temperatures) and liquid-like states (fluid or liquid crystalline, at higher temperatures). These are loosely analogous to the melting phase transition of an ordinary solid. In biological systems, the membranes are probably always in the more fluid state. The amount of unsaturated lipid and the degree of unsaturation affect the chain packing and fluidity of the membrane. More double bonds leads to higher fluidity (giving rise to the difference between butter and olive oil, for example).

## 3.19 Lipid composition affects the physical properties of membranes

Just as the structure of amphiphilic molecules determines whether they preferentially form micelles or bilayers, the lipids present in a membrane determine the physical properties of that membrane (its thickness, fluidity, and surface

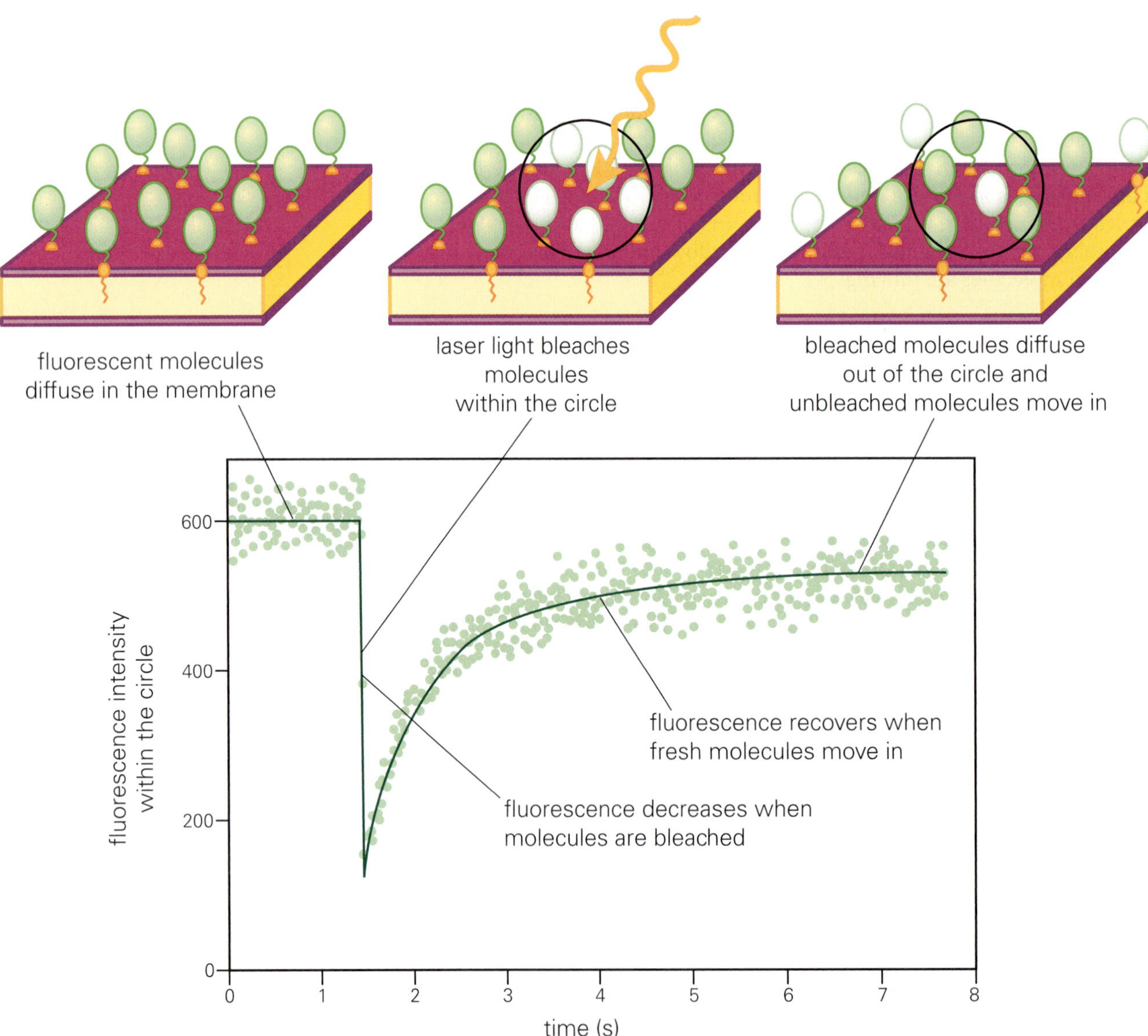


**Figure 3.41 Observing the mobility of lipids in the bilayer using Fluorescence Recovery After Photobleaching (FRAP).** A green fluorescent protein is attached to lipids in the membrane and the fluorescence within a small region (*black* circle) is observed using a microscope. A bright flash of light bleaches the proteins, leading to a drop in fluorescence. The bleached molecules move out of the field of view, and the fluorescence recovers as unbleached molecules move in from elsewhere. (Adapted from Y.I. Henis, B. Rotblat, and Y. Kloog, *Methods* 40: 183–190, 2006. With permission from Elsevier.)

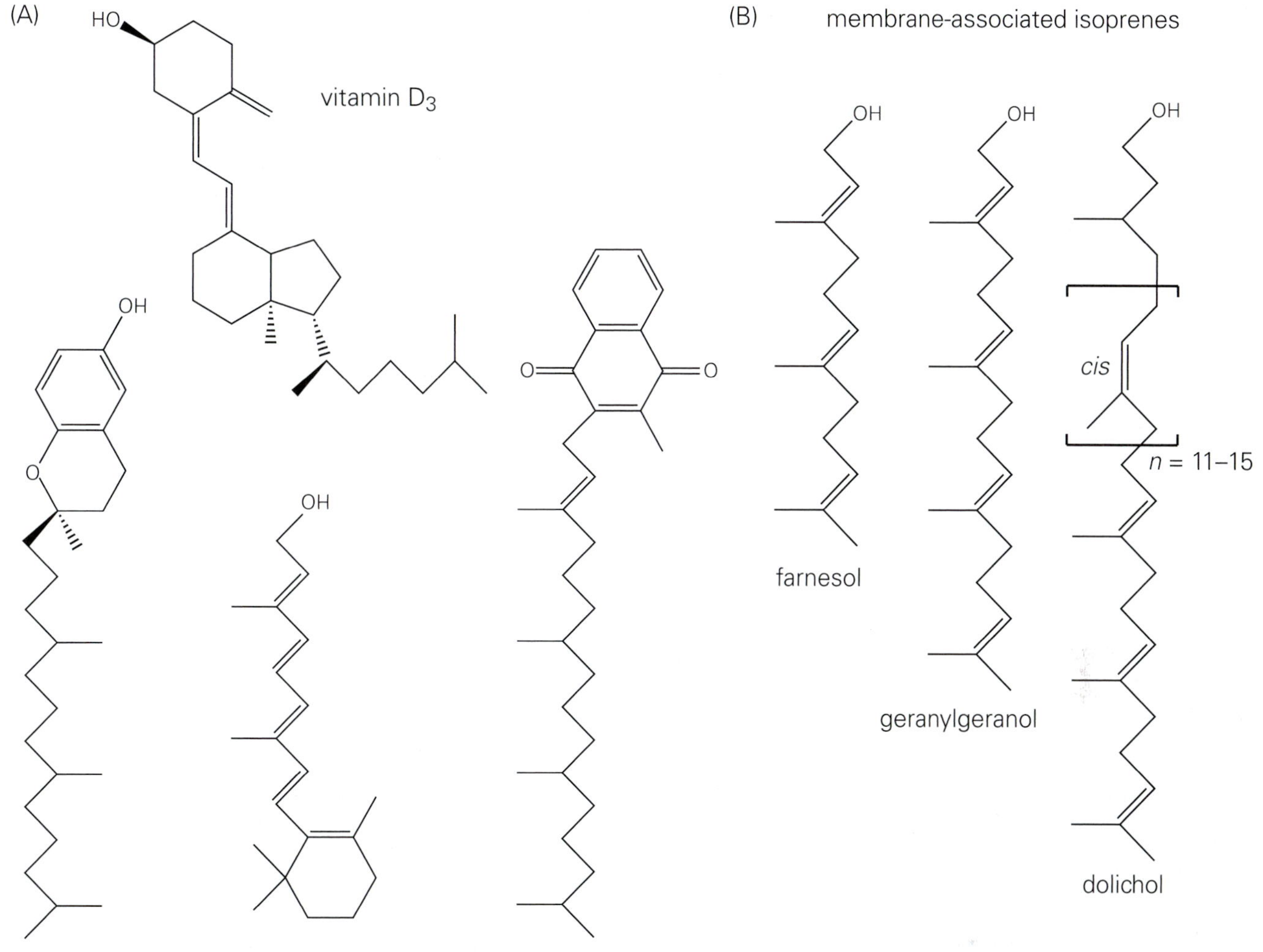

charge, for example). Real membranes in cells are complex mixtures of lipids and, by varying the composition of the lipids, the properties of the membrane can be adjusted by the cell.

Many components of membranes have to move (rotate and translate within the membrane) in order to function. As noted in Section 3.13, many of the fatty acid chains within lipids have carbon–carbon double bonds (that is, they are unsaturated). These double bonds are always *cis* in natural lipids, so each double bond puts a kink in the alkyl chain. Kinked chains pack less well than straight ones, making the membrane more fluid. The cholesterol present in the membranes of animal cells also affects the movement of lipid chains. The structure of cholesterol, which contains fused hydrocarbon rings, makes it relatively rigid. The presence of cholesterol in a membrane restricts the movement of lipid chains.

In addition to the major lipids, such as glycerophospholipids, glycolipids, and cholesterol (in eukaryotes), other hydrophobic molecules are also present in cell membranes. These include the fat-soluble vitamins (A, $D_3$, E, and K; **Figure 3.42**A). Such compounds affect both the chemistry and the physical properties of the membranes in which they occur. Polyisoprenes, such as farnesol, geranylgeranol and dolichol, are also components of some membranes and often serve as covalently attached membrane anchors for proteins or glycans (Figure 3.42B). Dolichol derivatives were mentioned in Section 3.9 (see Figure 3.19), and other polyisoprenes are discussed in the following section.

**Figure 3.42 Fat-soluble vitamins and polyisoprenes.** (A) The structures of vitamins $D_3$, E, A, and K. These are primarily membrane associated, as they have amphiphilic characteristics analogous to lipids (compare vitamin $D_3$ with cholesterol, Figure 3.31). (B) Isoprene corresponds to a branched five-carbon building block with a double bond. Farnesol (three isoprene units) and geranylgeranol (four isoprene units) occur attached to proteins. The much longer dolichol is used as a carrier during carbohydrate synthesis (see Figure 3.19).

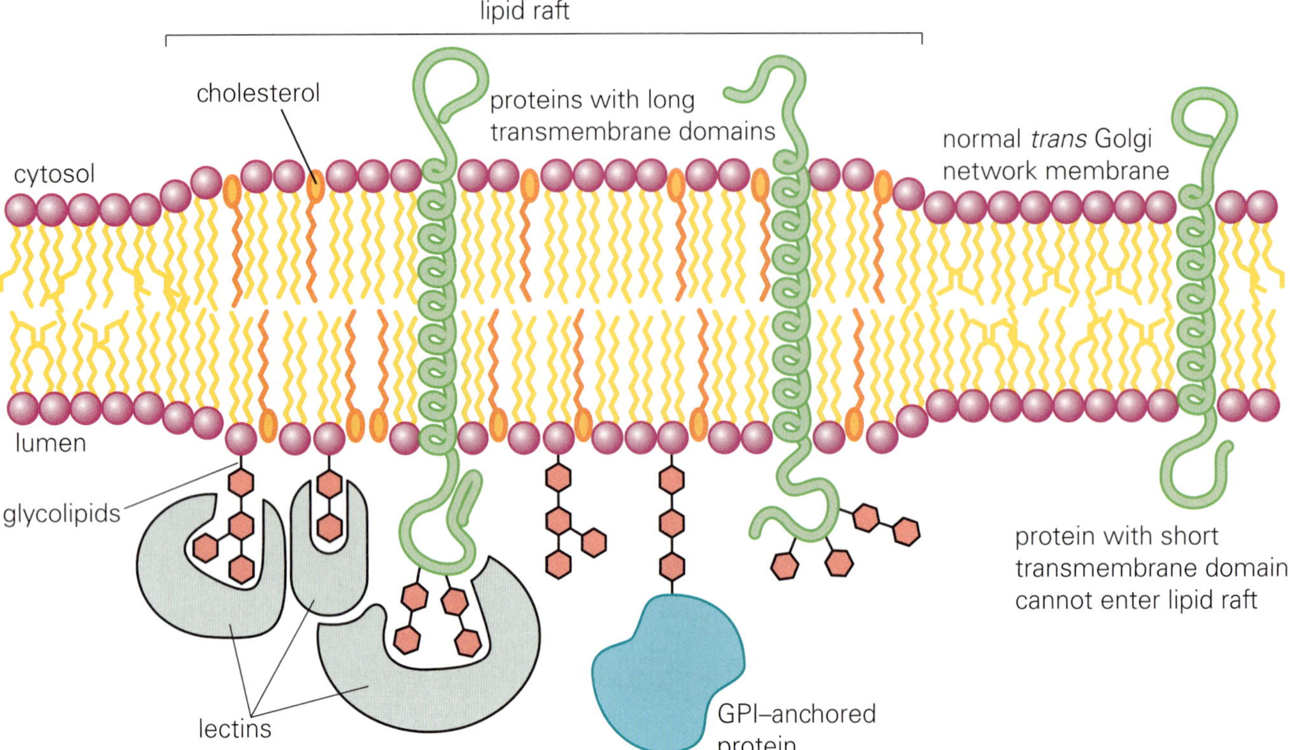


**Figure 3.43 Lipid raft.** The raft is a section of the membrane bilayer that is enriched in specific lipids (particularly glycolipids) and proteins that segregate together. These membrane regions have different physical and chemical properties from the rest of the membrane. (Adapted from B. Alberts et al., Molecular Biology of the Cell, 4th ed. New York: Garland Science, 2002.)

Certain regions of cell membranes, called patches or **rafts**, are enriched in particular lipid components, such as glycolipids, sphingolipids, and cholesterol, and specific proteins. These membrane patches have distinct physical properties, including different density and resistance to solubilization by detergents (**Figure 3.43**).

It is thought that interactions among the components of the membrane lead to a kind of phase separation—that is, regions of membrane with distinct composition that separate spontaneously from other regions. This separation allows optimization of the most favorable interactions among particular combinations of molecules. The distinct physical character of these regions is apparent when an image is made using atomic force microscopy, which detects differences in the position of the membrane surface, reflecting its thickness, as shown in **Figure 3.44**. Nevertheless, these rafts, or patches of distinct composition, are difficult to characterize

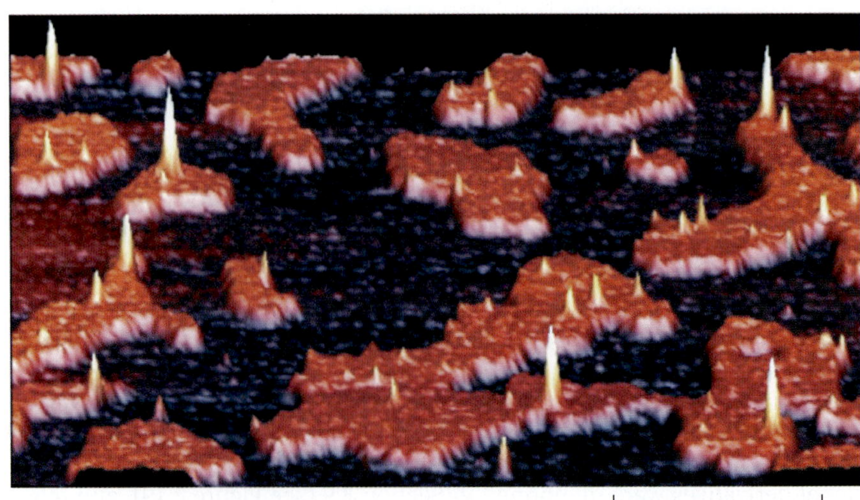


**Figure 3.44 An atomic force microscope image shows "rafts" of distinct lipid composition.** These regions of thicker membrane protrude above the surface of the surrounding regions. (Courtesy of Robert M. Henderson and J. Michael Edwardson; data from D.E. Saslowsky et al., *J. Biol. Chem.* 277: 26966–26970, 2002.)

*in situ* in the cell membrane. This makes the determination of their functions and properties (including size) difficult, and so the importance of rafts in various cellular processes remains poorly understood.

## 3.20 Proteins can be associated with membranes by attachment to lipid anchors

Many proteins carry out their functions while localized to the membrane, which can be accomplished in a number of ways. Part or most of the protein can be embedded into the membrane; these are called intrinsic membrane proteins (discussed in detail in Chapter 4). Other proteins are, instead, covalently attached to lipids or lipid-like molecules, which are strongly hydrophobic and anchor the protein to the membrane. Molecules acting as membrane anchors include myristic acid (a 14-carbon fatty acid), palmitic acid (a 16-carbon fatty acid), farnesol, and geranylgeranol. A complex anchoring molecule, glycosylphosphatidylinositol (GPI), is attached to some proteins. GPI is a glycerophospholipid with a variable glycan containing a core structure: inositol-GlcN-Man-Man-Man. The GPI anchor is attached by an amide bond to the C-terminus of the protein, as shown in **Figure 3.45.**

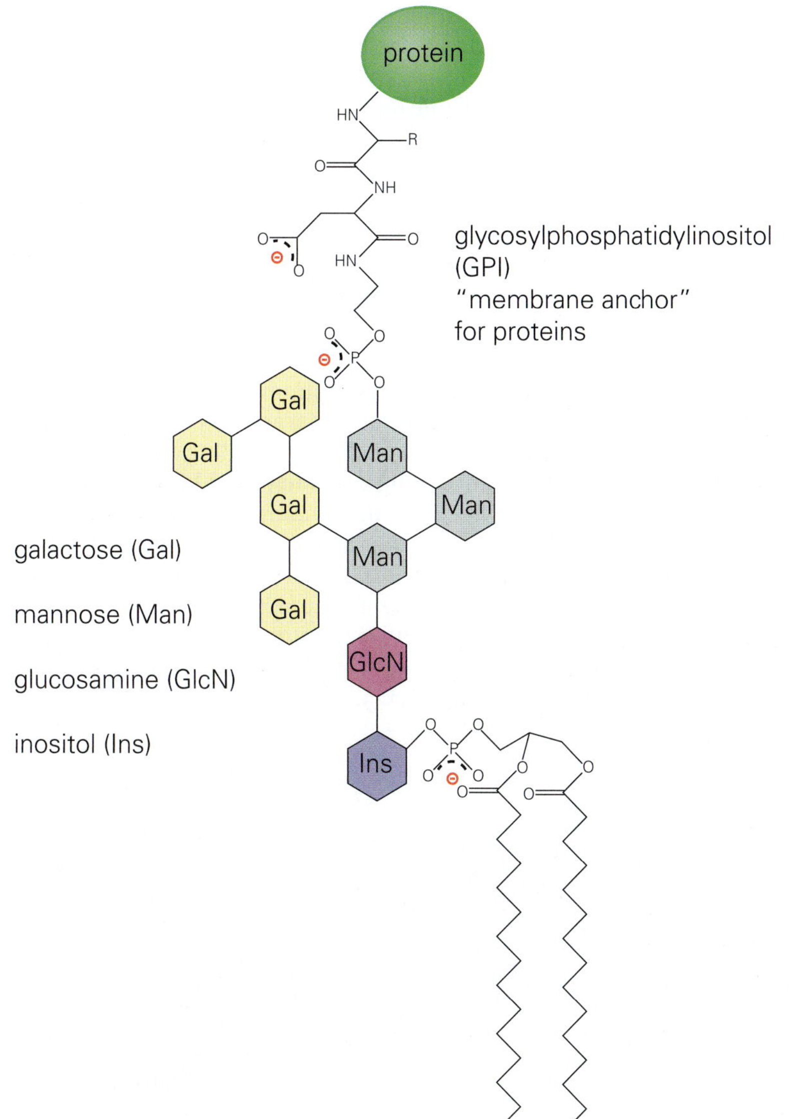


**Figure 3.45 A glycosylphosphatidylinositol (GPI) membrane anchor.** The C-terminus of the protein is attached via an amide bond to a phosphoethanolamine on the last sugar of the core glycan. The modifications of the mannose nearest the lipid (with four Gal residues in the example shown here) and of the last mannose (no modifications are present in this example) are variable.

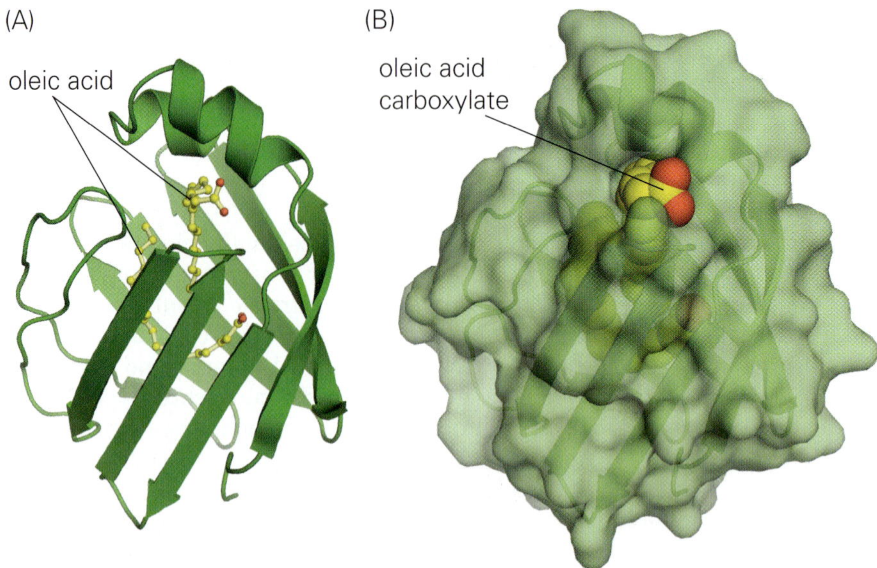


**Figure 3.46 Structure of a fatty acid binding protein.** (A) Backbone structure and (B) molecular surface of the protein. The protein forms a hydrophobic cavity that binds two molecules of oleic acid (*yellow*), sequestering the alkyl chains from water. (PDB code: 1LFO.)

Membrane anchors are attached to proteins post-translationally in the endoplasmic reticulum (ER) in response to signal sequences contained in the protein itself. Specific enzymes bind these sequences and attach the anchoring group. For example, proteins that are to be derivatized with GPI anchors have a hydrophobic sequence of amino acids at the C-terminus that keep the proteins associated with the membrane of the endoplasmic reticulum after they are synthesized. This signal sequence is enzymatically cleaved and replaced by the glycolipid. The GPI anchor can carry the attached protein into membrane regions that are enriched in specific lipids, causing the GPI-attached proteins to cluster with a specific set of other proteins (creating lipid "rafts," as discussed in Section 3.19).

## 3.21 Lipid molecules can be sequestered and transported by proteins

Fatty acids and lipids have low solubility as monomers in water. To allow them to be moved as needed within and between cells, and to control where they go, fatty acids and lipids are generally transported by binding to specific proteins. Individual molecules of fatty acids can bind to **fatty acid binding proteins** that have hydrophobic crevices into which the alkyl chains fit, as shown in **Figure 3.46**. There are many variations of such proteins in different organisms. There are also proteins that interact with cholesterol, or bind and transport specific hydrophobic vitamins. In all of these, the lipid components interact with a hydrophobic surface of the protein, usually in an interior cavity. In the structure of a human cholesterol transport protein, both cholesterol and lipid can be seen bound together near the center of an elongated protein (**Figure 3.47**).

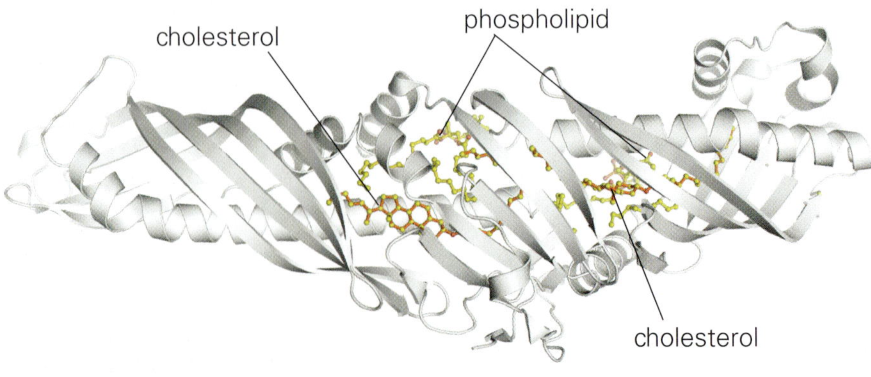


**Figure 3.47 A cholesterol transport protein.** Two molecules of cholesterol (*orange*) and two molecules of phospholipid (*yellow*) are bound to the protein. (PDB code: 2OBD.)

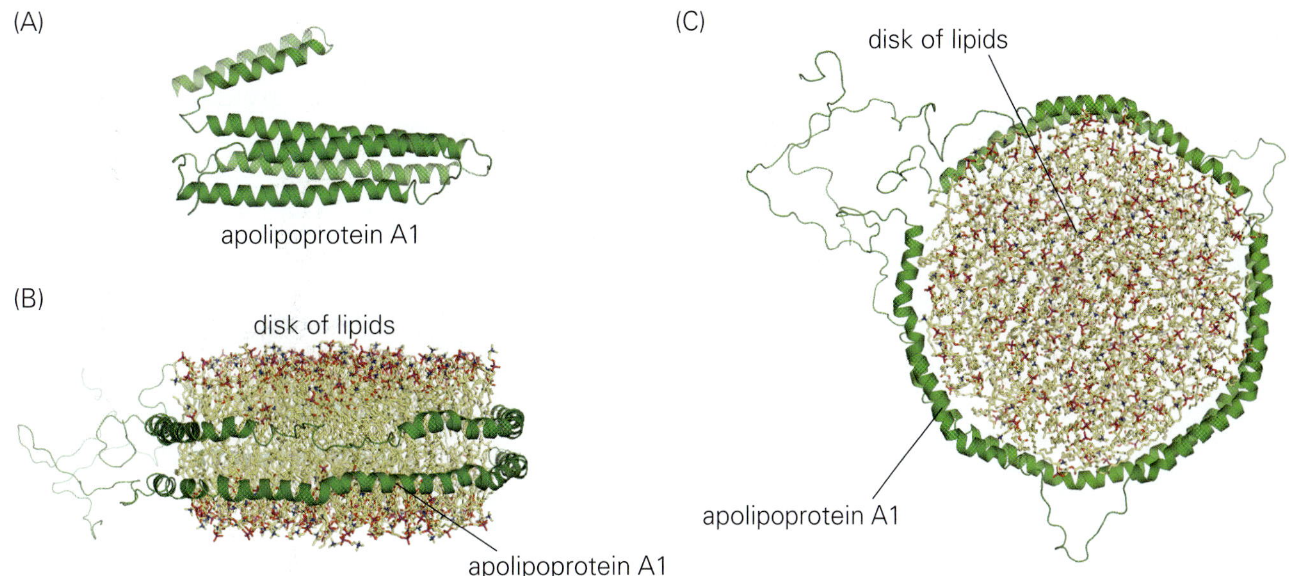


**Figure 3.48 Apolipoprotein A1.** (A) The structure of the protein in the absence of lipid. (B, C) Two views of a model for the structure in the presence of lipid. The helices move apart and expose a surface that interacts with the edge of a small disk of lipid. There are two protein molecules per lipid disk. (See Protein Model Database mi.caspur.it/PMDB entry PM0074956; Z. Wu and S.L. Hazen, *Nat. Struct. Mol. Biol.* 14: 861–868, 2007.)

The lipid carrier proteins described above transport individual lipid molecules primarily. There is another class of transport proteins, called **lipoproteins**, that carry relatively large quantities of lipid per protein. In their apo form (that is, in the absence of lipid), these proteins fold into typical globular domains. The structure of one such protein, apolipoprotein A1, is shown in **Figure 3.48**A. In the presence of lipid, the protein partially unfolds so that hydrophobic sidechains can interact with the hydrophobic alkyl tails of phospholipids that are arranged in a disklike shape (Figure 3.48B and C).

## 3.22 Different kinds of cells and organelles have different membrane compositions

The segregation of space into compartments must have been a critical step in the development of self-replicating systems of molecules. Lipid-like molecules (even as simple as fatty acids) can assemble spontaneously into membrane structures that could contain and concentrate molecules, encouraging selective reactions with other chemicals present. In the simplest independently living organisms, bacteria, there is basically just one membrane defining the volume of the cell (though most have an additional cell wall to reinforce the barrier to the outside). In higher life forms, each cell contains many segregated regions bounded by membranes that make it possible to maintain sophisticated functional specialization. Most **organelles**, including the nucleus, mitochondria, the Golgi apparatus, the endoplasmic reticulum and lysosomes (**Figure 3.49** and **Figure 3.50**), would be unable to function properly without being segregated within asymmetric membranes.

**Organelle**

An organelle is a subcellular structure or chamber, often bounded by a membrane, that is specialized for particular functions. Examples include the nucleus, the endoplasmic reticulum, the mitochondria, the lysosomes, and the Golgi apparatus.

The membranes of different organelles are in general distinct (that is, they have different compositions of both lipids and proteins), and functionally specialized. For example, lysosomes (which degrade proteins, old organelles, ingested bacteria, or viruses, etc.) function at acidic pH (pH $< 7$), while the cytoplasm must stay near neutral pH (pH $= 7$). Lysosomal membranes allow this pH difference to be created and contain proteins that are proton pumps to maintain it. To give a sense of the variability in overall membrane composition, some approximate values of lipid content for different cell types and organelles are listed in **Table 3.2**.

In addition to differences in average compositions between cell types, the inner and outer layers (or leaflets) of membranes are also quite different from one another, accommodating different needs for interactions within the cell and with

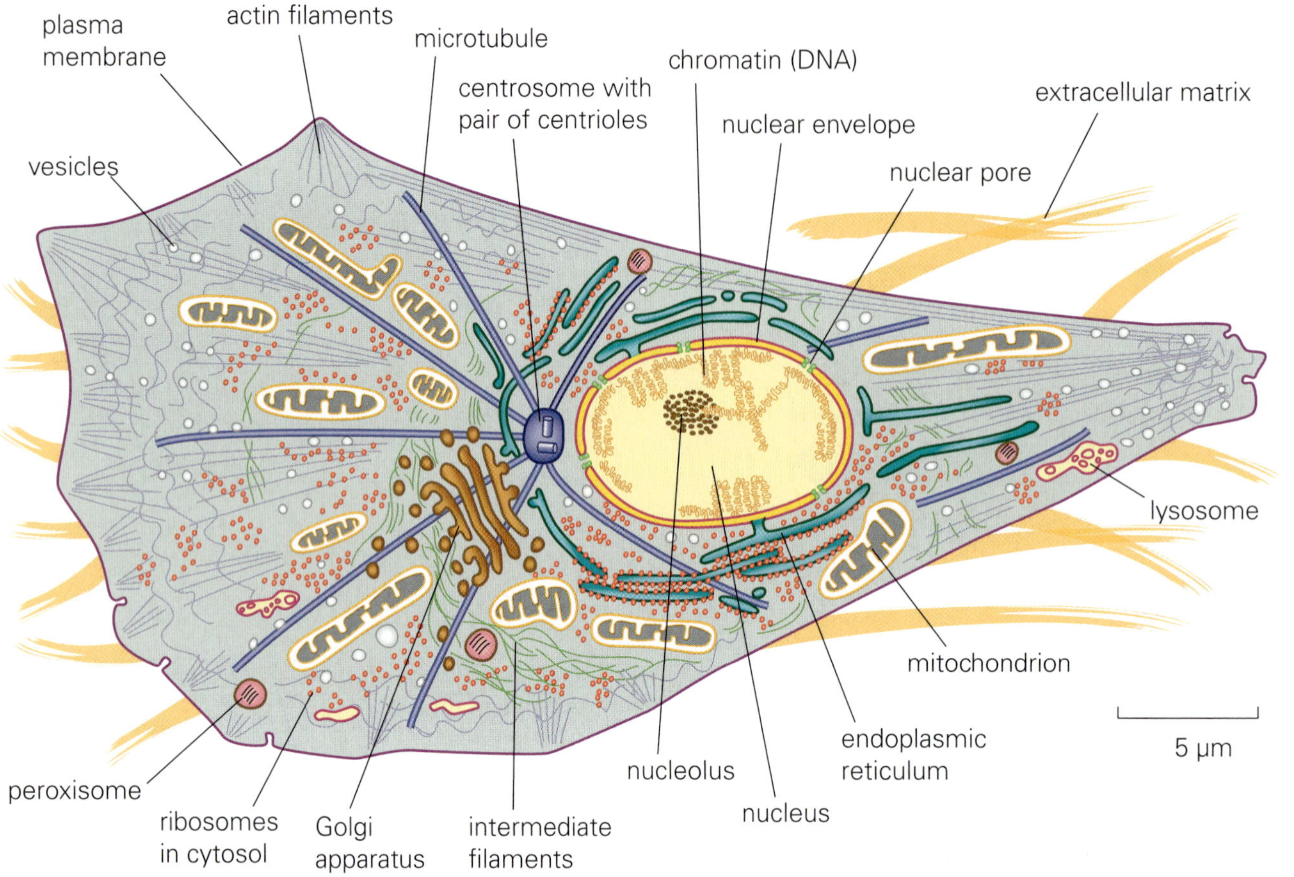

plasma
membrane
actin filaments
microtubule
centrosome with
pair of centrioles
chromatin (DNA)
nuclear envelope
extracellular matrix
nuclear pore
vesicles
lysosome
mitochondrion
5 μm
endoplasmic
reticulum
nucleolus
nucleus
peroxisome
ribosomes
in cytosol
Golgi
apparatus
intermediate
filaments

**Figure 3.49 A eukaryotic cell**. Some of the important structures and organelles are shown. Most organelles are separated from the rest of the cell by membranes, which allows for distinct compositions and, thus, distinct functions in each organelle. (Adapted from B. Alberts et al., Molecular Biology of the Cell, 5th ed. New York: Garland Science, 2008.)

other cells on the outside. This asymmetry is apparent in Figure 3.43, in which the glycolipids are all on one side of the membrane. As noted above, glycolipids tend to be particularly important for cell–cell interactions, and so glycolipids are found in the outer leaflet of the cell membrane.

We noted in Section 3.18 that membranes are fairly fluid, allowing rapid translation of individual lipid molecules within one leaflet of the membrane. However, the transfer of a lipid molecule from one leaflet to another requires moving the polar (and often charged) head group across the hydrophobic region, a process that is energetically unfavorable, and hence is very slow. Enzymes called **flippases** recognize and "flip" specific lipids between the inner and outer leaflets, a process needed for generating and maintaining membrane asymmetry.

nuclear envelope
rough
endoplasmic
reticulum
vesicular tubular
clusters
*cis* Golgi network
1 μm

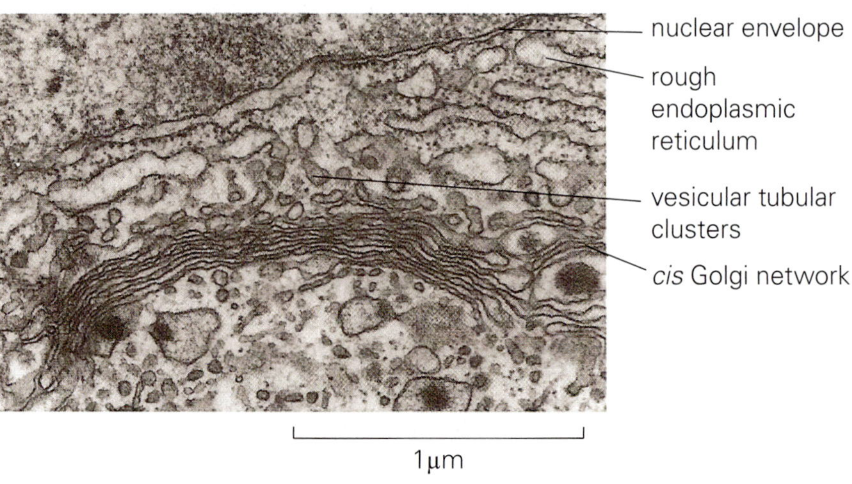

**Figure 3.50 Organelle membranes.** This electron micrograph of part of a cell includes part of the nucleus (upper left), part of the endoplasmic reticulum with associated ribosomes (dark dots), and part of the Golgi apparatus. Each dark line surrounding a region is a bilayer membrane. (Courtesy of Brij Gupta.)

**Table 3.2 The lipid compositions of different membranes.**

| Lipid | Source | | | | | |
|---|---|---|---|---|---|---|
| | *E. coli* | Red cell | ER | Mitochondrion | Myelin | Liver |
| Phosphatidylcholine | 0 | 17 | 40 | 39 | 10 | 24 |
| Phosphatidylserine | 0 | 7 | 5 | 2 | 9 | 4 |
| Phosphatidylethanolamine | 70 | 18 | 17 | 25 | 15 | 7 |
| Sphingomyelin | 0 | 18 | 5 | 0 | 8 | 19 |
| Glycolipids | 0 | 3 | ~0 | ~0 | 28 | 7 |
| Cholesterol | 0 | 23 | 6 | 3 | 22 | 17 |

Many of these membranes have additional minor components, and so the percentages do not add to 100%. (Adapted from B. Alberts et al., Molecular Biology of the Cell, 5th ed. New York: Garland Science, 2008.)

## 3.23 Cell walls are reinforced membranes

Lipid membranes are sufficient to partition spaces within cells, but the lipid membranes are relatively fragile because they are only held together by noncovalent interactions. Cells that are exposed to the surroundings often need additional reinforcement to maintain their integrity. This takes the form of a **cell wall** in bacteria and plants.

All bacterial cells are surrounded by a membrane that is the primary barrier to the outside world. A majority also have a cell wall made from a peptidoglycan. This material has a linear polysaccharide [β(1→4)-linked GlcNAc and MurNAc], with peptides containing three to five amino acids attached to the MurNAc residues. The peptides are unusual in that they have D-isomer versions of alanine and glutamate in addition to L-alanine, and glycine in some. The D amino acids reduce susceptibility to protease enzymes, which digest normal proteins but are ineffective against peptides containing D amino acids.

**Cell wall**

A cell wall is a polymeric outer layer just outside the cell membrane of some cells (especially in plants, bacteria, fungi, algae, and archaea). The cell wall generally has a high polysaccharide content.

Cell walls in bacteria fall into two classes. Some bacteria have thick peptidoglycan layers and others have thinner peptidoglycan layers associated with a second lipid membrane. In either case, the peptidoglycan matrix is relatively porous, so that small molecules can get to the membrane for transport into the cell, but it is also a dense and strong enough web around the membrane to resist mechanical stress (**Figure 3.51**).

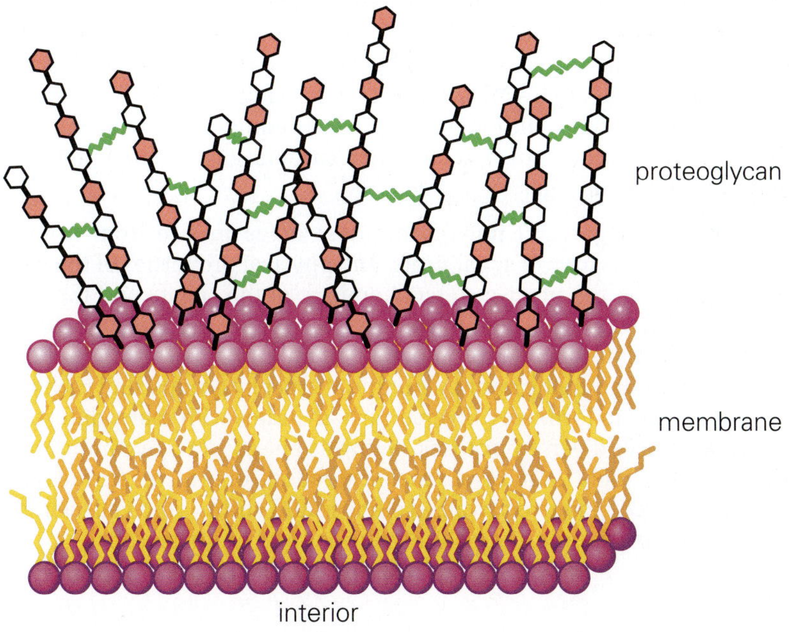

**Figure 3.51 The proteoglycan (also called peptidoglycan) cell wall in bacteria.** Polysaccharide chains extend from the membrane and are cross-linked by short peptides (*green*).

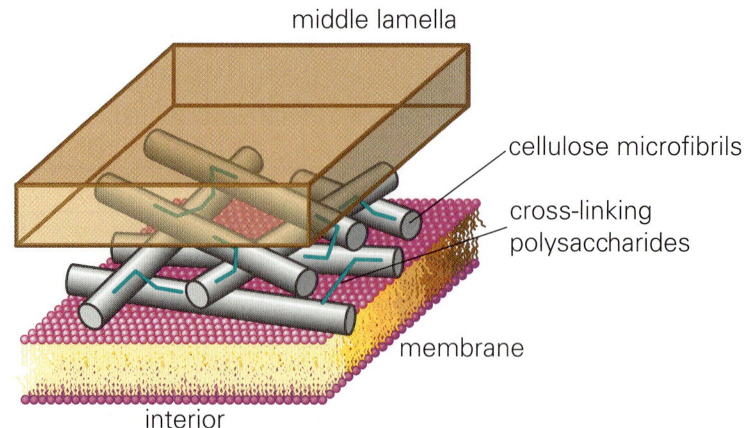


**Figure 3.52 A plant cell wall.** Outside the cell membrane are cellulose microfibrils with cross-linking polysaccharides. Pectin is interspersed throughout, filling and strengthening the wall. The middle lamella is rich in pectin and helps hold neighboring cells together. (Adapted from B. Alberts et al., Molecular Biology of the Cell, 5th ed. New York: Garland Science, 2008.)

These two kinds of bacteria were distinguished originally by their color when treated with the "Gram" stain, developed by the bacteriologist Hans Christian Gram. The bacteria with the thinner peptidoglycan layers produce light pink staining when treated with the Gram stain, and were classified as **Gram negative** bacteria. In contrast, the bacteria with the thicker peptidoglycan layers produce dark purple staining and were classified as **Gram positive**.

Plant cell walls are more complex than those of bacteria. They contain microfibrils of cellulose (described in Section 3.7) and hemicellulose (a soft, slightly branched heteropolymer containing many different monomer sugars), with **pectin** [another polysaccharide, $\alpha(1\rightarrow4)$-galacturonic acid with ~4% branching with neutral sugars, galactose, arabinose, and xylose]. Other polysaccharides crosslink the cellulose fibrils to add mechanical strength, as shown in **Figure 3.52**. There is also a filler of lignin, a complex, cross-linked polymer with a high content of aromatic carbons. As noted previously, the cellulose provides high mechanical strength, critical for supporting the complex morphology of the whole plant.

Fungi also have cell walls to protect them. These are made of chitin (described in Section 3.6) and other polysaccharides, but no cellulose. Chitin is also in insect exoskeletons, but there it is mixed with proteins and inorganic mineral deposits.

## Summary

Glycans (carbohydrates) and lipids are essential components of all cells, but they differ from proteins and nucleic acids in that they are not directly genetically encoded. The simplest glycans are monosaccharide sugars, compounds with the approximate formula $(H\text{–}C\text{–}OH)_n$, with $n$ in the range of 3–9. These sugars primarily form closed rings, although these are in equilibrium with linear forms. Variations in the stereochemistry at different carbons give rise to many structural isomers of each size of sugar.

Sugar polymers (oligo- and polysaccharides) range in size from a few sugar units to tens of thousands of sugar units. The physical properties of the polymers depend strongly on the precise stereochemistry of the sugars and the chemical linkages between the sugar rings. In contrast to proteins and nucleic acids, glycan polymers can be both linear and branched. Many cell types produce chemically modified sugars (containing amines, methyl and acetyl groups, and carboxylates) that are combined to form very diverse carbohydrates with many different functions. Glycans serve diverse purposes, acting as energy storage compounds (glycogen), "address labels" that dictate the destinations of proteins (proteoglycans), structural reinforcement in cell walls (cellulose and chitin), and cell–cell recognition elements (for example, lipopolysaccharides and proteoglycans).

Lipids are a critical component of all cell membranes. Although quite variable in their structures, lipids always have a strongly hydrophobic part, as well as some hydrophilic part. The most common membrane lipids, the glycerophospholipids, have a glycerol backbone with ester bonds attaching two fatty acids (commonly with 16–20 carbons) and a head group with a negatively charged phosphate with a variable R group. Unlike proteins, nucleic acids, and glycans, lipids do not form chemically linked polymers; rather, they assemble noncovalently into organized structures.

Natural lipids form bilayer membranes in water, which are planar structures with the head groups exposed to the water and the alkyl chains of the fatty acids clustered together to exclude water completely. Such bilayers are two-dimensional fluids (molecules move quite freely within the plane of the bilayer). Molecules analogous to lipids, but with just one alkyl chain, tend to form spherical droplets (micelles) with the hydrocarbon tails on the inside, rather than bilayers. Such compounds are called detergents, and they can surround other molecules to solubilize them. Because of their largely hydrophobic character, lipid molecules must be associated with proteins to be transported in and between cells.

Membranes play a critical role in partitioning space into distinct regions that can have different chemical properties. Internal regions in eukaryotic cells are partitioned by membranes into organelles, which carry out specific biochemical functions. Many different kinds of proteins are associated with membranes. Some proteins embed into the membrane, while others are covalently attached to lipid molecules and are thereby tethered to the membrane surface.

## Key Concepts

### A. GLYCANS

- Sugars are primarily "hydrated carbon," $(H–C–OH)_n$, with $n$ = 3–9 (but mostly 5 or 6), and are generally cyclic.
- Individual sugars are covalently linked to form oligomeric or polymeric glycans (also called oligosaccharides or polysaccharides).
- Many sugars are structural isomers with the same chemical formula (for example, $C_6H_{12}O_6$ = glucose, mannose, galactose).
- Some sugars have additional chemical functional groups, including methyl or acetyl groups, amines, acids, phosphates, etc.
- Polymers of different sugars (or different isomers of the same sugar) can have very different physical properties.
- Cell-surface proteins often have covalently attached glycans; such proteins are said to be glycosylated and are called glycoproteins.
- Glycosylation can modify the physical properties and activities of proteins.
- Noncovalent protein–carbohydrate interactions are important for cell–cell recognition and interaction.

### B. LIPIDS AND MEMBRANES

- Lipid molecules are amphiphilic (that is, they have both hydrophobic and hydrophilic parts).
- Lipid molecules form the basis of cell membranes, accounting for over 50% by weight of biological membranes.
- Glycerophospholipids, with a variety of attached polar head groups, are abundant in cell membranes.
- Sphingolipids and cholesterol are also common in the cell membranes of eukaryotes.
- Fatty acids in lipids vary in length and the degree of unsaturation.
- Lipids assemble spontaneously into organized bilayer structures.
- Detergents are similar to lipids, but form micelles rather than bilayers.
- Membranes with different lipids have distinct thicknesses and fluidities; thus, regions of different lipid compositions and rafts can exist within cell membranes.
- Membrane-associated proteins can either embed in the bilayer (integral membrane proteins) or be tethered by attachment to a fatty acid or lipid molecule.
- Lipid molecules are generally transported by proteins.
- Membranes define spatial compartments, thus defining individual cells or organelles within cells.
- Lipids and carbohydrates are key components of cell walls.

# Problems

## True/False and Multiple Choice

1. In most carbohydrate monomers, the ratio of carbon:hydrogen:oxygen is
   a. 1:1:1
   b. 2:1:1
   c. 1:2:1
   d. 1:1:2
   e. 1:1.5:1

2. Which of the following is not a lipid?
   a. cholesterol
   b. sphingolipid
   c. apolipoprotein
   d. vitamin A
   e. triglyceride

3. Many sugars are structural isomers of other sugars.
   True/False

4. Most detergents will organize spontaneously into bilayers when placed in aqueous solutions.
   True/False

5. Which of the following fatty acids are unsaturated?

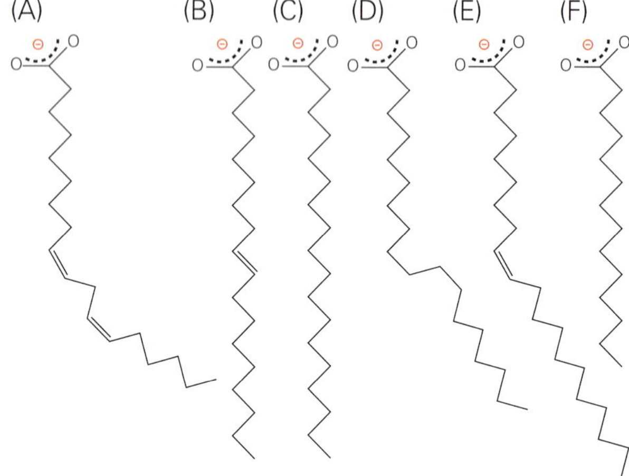


6. An individual lectin-binding domain is likely to bind with high affinity to many diverse sugars.
   True/False

7. In plants, the cell wall is generally higher in percent polysaccharide content and lower in percent lipid content than the cell membrane.
   True/False

## Fill in the Blank

8. Simple sugars can exist in both linear and ______ forms.

9. ________ is an acetylated polymer found in the exoskeletons of insects and the cell walls of fungi.

10. Inositol is an unusual six-carbon sugar because it does not contain an _______ atom in its ring.

11. A lipid _____ is a section of membrane with both composition and physical properties that are distinct from the surrounding membrane.

12. In eukaryotic cells, intracellular compartments bounded by membranes are called ________.

## Quantitative/Essay

13. Why are the most stable conformations of sugars most often nonplanar with hydrogens in the equatorial positions?

14. What information can be gained more readily from the Fischer representation than from the Haworth representation?

15. How do the homopolymers cellulose and glycogen differ at the level of: (a) the stereochemistry of the monomer, (b) physical properties, and (c) the intramolecular interactions that maintain those physical properties?

16. A student performs a biochemical fractionation of a cell and isolates the membrane-containing fraction. An analysis of the glycoproteins in the fraction reveals only N-linked sugars.
    a. What type of amino acid residue are the sugars linked to?
    b. What is the enzyme responsible for transferring the first sugar to that residue?
    c. What organelle has the student likely isolated?

17. The O blood type group is known as the universal donor. What property of the glycosylation pattern of O-type individuals allows for the blood to be transfused into individuals of any blood type without eliciting an immune response?

18. What molecular forces drive the formation of bilayers when phospholipids are immersed in water?

19. How does the packing of cholesterol against other lipids alter the biophysical properties of a membrane?

20. Many proteins need to be in close proximity to the membrane in order to perform their biological functions. What are two strategies employed by the cell to ensure that certain proteins are anchored to the membrane?

21. Without the aid of an enzyme, is a carbohydrate-modified lipid more likely to (a) move in an undirected lateral motion for 500 Å or (b) transfer across the ~40-Å thick cell membrane from one leaflet to the other?

22. Scientists* have proposed a model for how yeast cells determine the length of a class of fatty acids that

are synthesized using a family of fatty acid synthase enzymes. In different variants of the enzyme, a key lysine residue can be positioned at various places along an $\alpha$ helix of the synthase to control the length of the fatty acid chain that is produced.

Imagine a similar enzyme in which moving the position of a catalytic lysine residue along a $\beta$ sheet creates a synthase that catalyzes four additional carbons into the fatty acid chain. The lysine residue can only be moved in two-residue units along the $\beta$ sheet and still face the substrate (that is, it can be moved from position 1 to positions 3 or 7, but never to positions 2 or 4). Thus, movement from position 1 to position 3 results in a four-carbon change in the length of the fatty acid produced.

There are 14 residues in the $\beta$ sheet. The maximum length, a 40-carbon fatty acid, is created when the residue is placed at the end of the $\beta$ sheet (position 1). What is the shortest possible fatty acid that the synthase can create?

*[Denic V & Weissman JS (2007) A molecular caliper mechanism for determining very long-chain fatty acid length. *Cell* 130, 663–677.]

## Further Reading

### General

Taylor ME & Drickamer K (2011) Introduction to Glycobiology, 3rd ed. New York: Oxford University Press.

Vance DE & Vance JE (2008) Biochemistry of Lipids, Lipoproteins and Membranes, 5th ed. Amsterdam: Elsevier.

Varki A, Cummings R, Esko J, Freeze H, Hart G & Marth J (2009) Essentials of Glycobiology, 2nd ed. Cold Spring Harbor, NY: Cold Spring Harbor Laboratory Press.

### References

#### A. Glycans

Drickamer K & Taylor ME (1998) Evolving views of protein glycosylation. *Trends Biochem. Sci.* 23, 321–324.

Helenius A & Aebi M (2004) Roles of *N*-linked glycans in the endoplasmic reticulum. *Annu. Rev. Biochem*. 73, 1019–1049.

Laughlin ST & Bertozzi CR (2009) Imaging the glycome. *Proc. Natl. Acad. Sci. USA* 106, 12–17.

Petrescu A, Wormald MR & Dwek RA (2006) Structural aspects of glycomes with a focus on *N*-glycosylation and glycoprotein folding. *Curr. Op. Struct. Biol.* 16, 600–607.

#### B. Lipids and Membranes

Grecco HE, Schmick M & Bastiaens PIH (2011) Signaling from the living plasma membrane. *Cell* 144, 897–909.

Mayor S & Rao M (2004) Rafts: Scale-dependent, active lipid organization at the cell surface. *Traffic* 5, 231–240.

Maxfield FR & McGraw TE (2004) Endocytic recycling. *Nat. Rev. Mol. Cell Biol.* 5, 121–132.

Mouritsen OG & Zuckermann MJ (2004) What's so special about cholesterol? *Lipids* 39, 1101–1113.

Sanyal S & Menon AK (2009) Flipping lipids: Why an' what's the reason for? *ACS Chem. Biol.* 4, 895–909.

Schleifer KH & Kandler O (1972) Peptidoglycan types of bacterial cell walls and their taxonomic implications. *Bacteriol. Rev.* 36, 407–477.

Simons K & Toomre D (2000) Lipid rafts and signal transduction. *Nat. Rev. Mol. Cell Biol.* 1, 31–41.

van Meer G, Voelker DR & Feigenson GW (2008) Membrane lipids: where they are and how they behave. *Nat. Rev. Mol. Cell Biol.* 9, 112–124.

Zhang FL & Casey PJ (1996) Protein prenylation: molecular mechanisms and functional consequences. *Annu. Rev. Biochem*. 65, 241–269.

# Protein Structure

This chapter is concerned with the architectural principles underlying the three-dimensional structures of protein molecules. We shall first discuss some general principles, including the hierarchical organization of protein structure and the role of the hydrophobic effect and hydrogen bonding in stabilizing it. We then study the nature of conformational restrictions on the peptide backbone. These restrictions arise from collisions between backbone atoms in certain conformations, which are therefore disallowed. We shall see how the restricted set of allowed conformations, when combined with hydrogen bonding, lead naturally to the secondary structural elements known as the α helix and the β strand as the fundamental building blocks of protein structure.

There are two major classes of proteins, called water-soluble proteins and membrane proteins, which are categorized based on whether or not they interact with cell membranes. Water-soluble proteins include globular proteins, which have compact and well-defined structures, and are not associated with membranes. Globular proteins have mostly hydrophobic sidechains on the inside and mostly hydrophilic sidechains on the outside (see Figure 1.35). In this chapter, the only soluble proteins we shall focus on are the globular proteins. We shall not discuss fibrous proteins, such as collagen. Nor do we study intrinsically disordered proteins—a class of proteins that do not have well-defined structure on their own.

Membrane proteins, which are discussed in this chapter, are embedded either partially or completely in the interior of the hydrophobic part of a lipid bilayer, and they have hydrophobic sidechains on the outside as well as on the inside. The principles underlying the structures of membrane proteins are a little different than those of globular proteins, but both consist of α helices and β strands. We shall first survey the principles governing the folded structures of globular proteins and describe some of the commonly occurring kinds of protein structures. These principles will help us understand how the architectures of membrane proteins, discussed at the end of the chapter, differ from those of globular proteins. Membrane proteins are particularly important for the generation of energy and for transporting molecules into and out of the cell, and we discuss two examples of proteins that carry out these functions.

## A. GENERAL PRINCIPLES

### 4.1 Protein structures display a hierarchical organization

Recall from Chapter 1 that the folded structures of protein molecules are built up from the packing together of two kinds of secondary structure elements: α helices and β strands. The terms "primary," "secondary," and "tertiary" structure are used to emphasize the hierarchical levels in the structures of proteins. **Primary structure** is the amino acid sequence—that is, the pattern of amino acids along a linear polypeptide chain (**Figure 4.1**). The local conformation of the polypeptide chain is referred to as the **secondary structure**, which occurs mainly as α helices and β strands, with loops connecting them.

**Primary, secondary, tertiary, and quaternary structure**

The amino acid sequence of a protein is referred to as its primary structure. The local conformation of the protein backbone (α helix, β strand, or loop) is the secondary structure. The three-dimensional fold of a protein is the tertiary structure. Finally, the arrangement of subunits in a multi-subunit protein complex is the quaternary structure.

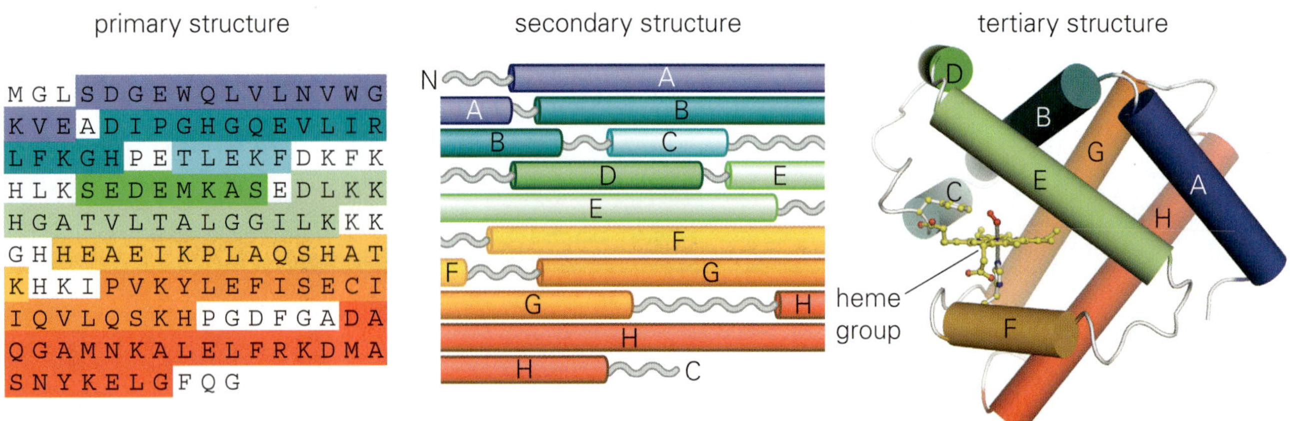


**Figure 4.1 Primary, secondary, and tertiary structure.** The amino acid sequence of human myoglobin, an oxygen-binding protein, is shown on the left. The secondary structure of myoglobin is shown in the middle. The α helices are indicated as cylinders and the loops connecting them as wavy lines (myoglobin contains no β strands). The tertiary structure of myoglobin is shown on the right. The particular three-dimensional arrangement of α helices that is characteristic of myoglobin is known as the globin fold and is also seen in hemoglobin (see Figure 4.2). Oxygen binds to an iron atom that is part of a cofactor known as a heme group (*yellow* bonds), which is discussed in more detail in Chapters 5 and 14. The heme group is not part of the protein chain. The sidechains of two amino acid residues that interact with the heme group are also shown. (PDB code: 1MBC.)

**Structural domains**

Structural domains are the fundamental units of three-dimensional protein structure. A protein domain is typically 50–200 residues long and contains a well-defined hydrophobic core. Domains are mixed and matched in evolution to produce more complex proteins.

As a general rule, the residues that form the interfaces between secondary structural elements are hydrophobic, and the packing of secondary structural elements results in the formation of a protein **structural domain** with a hydrophobic core. Protein molecules contain one or more such domains. **Tertiary structure** refers to the three-dimensional organization of the secondary structure elements in these domains. The particular three-dimensional arrangement of the α helices and β strands in the tertiary structure is known as the **protein fold**.

Many proteins consist of several different polypeptide chains (subunits) that associate into a multimeric complex in a specific way called the **quaternary structure**. These subunits can function either independently of each other or cooperatively so that the function of one subunit depends on the functional states of the other subunits. The quaternary structure of the oxygen transport protein hemoglobin is illustrated in **Figure 4.2**. Hemoglobin consists of four subunits that are structurally similar to each other. In other protein complexes, such as RNA polymerase (**Figure 4.3**), the subunits are structurally distinct.

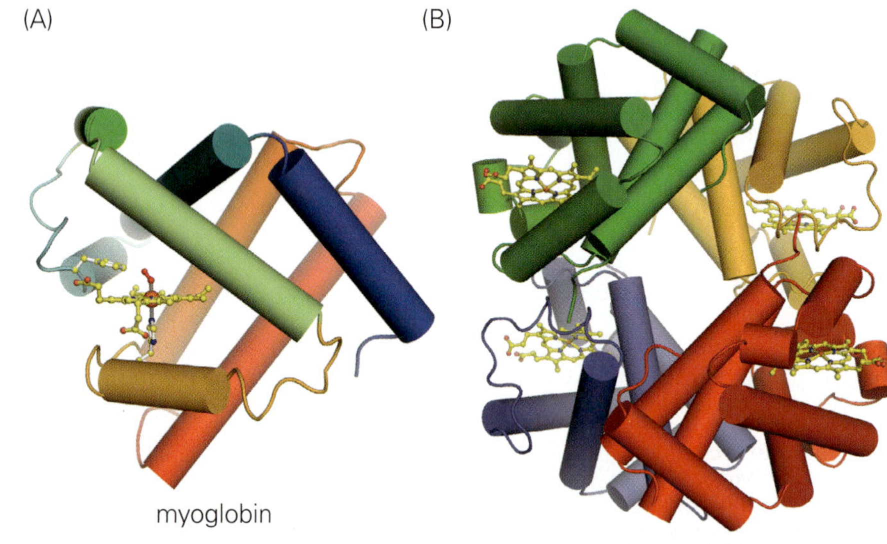


**Figure 4.2 Quaternary structure of human hemoglobin.** Hemoglobin, the oxygen transport protein, is made up of two copies each of two different subunits, both of which look like myoglobin. (A) The tertiary structure of myoglobin, illustrating the globin fold. (B) The quaternary structure of hemoglobin. (PDB codes: 1MBC and 1A00.)

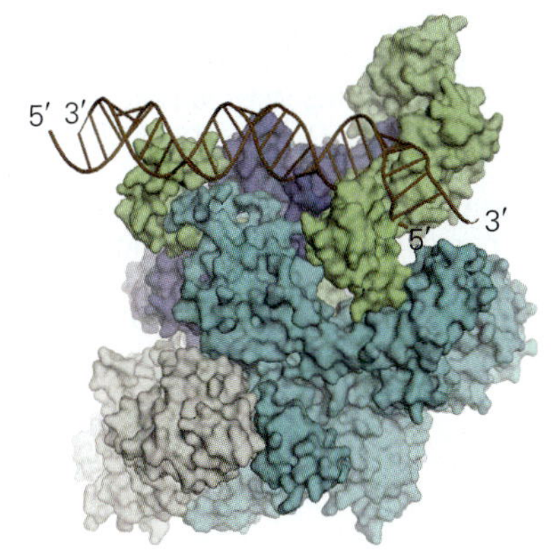


**Figure 4.3 Quaternary structure of bacterial RNA polymerase.** This enzyme, which transcribes genetic information from DNA into RNA, is composed of several different kinds of subunits, each of which is shown in a different color. A DNA molecule bound to RNA polymerase is also shown. (Adapted from K.S. Murakami et al., and S.A. Darst, *Science* 296: 1285–1290, 2002. With permission from the AAAS; PDB code: 1L9Z.)

## 4.2 Protein domains are the fundamental units of tertiary structure

A protein structural domain is a polypeptide chain or a part of a polypeptide chain that forms an independent structural unit, typically with a well-defined hydrophobic core. Domains are often units of function, and proteins may comprise a single domain or several. Protein domains range in size from ~50 to ~200 amino acid residues and represent a great many, perhaps thousands, of distinct polypeptide chain folds.

**Protein fold**

The visually recognizable arrangement of α helices and β strands in the three-dimensional structure of a protein is referred to as the protein fold. Different proteins can have the same fold, as is the case for myoglobin and the subunits of hemoglobin (see Figure 4.2).

Domains are the fundamental building blocks in the evolution of complex protein structure, as discussed in Chapter 5 (Part D). Gene duplication and genetic recombination have resulted in the combinatorial creation of multifunctional proteins by mixing and matching structural domains. We illustrate this idea by using three particular domains as examples. These domains are important in the processes that control signaling between cells, and are known as the Src homology (SH) domains—namely, the SH1, SH2, and SH3 domains (**Figure 4.4**). These names reflect the original identification of these domains in a family of proteins that are related to (that is, are *homologous* to) a protein known as Src (pronounced *sark*). The Src protein is so named because it causes a type of cancer known as a sarcoma (hence the abbreviation, *Src*, for sarcoma).

SH1 domains are enzymes known as tyrosine kinases, and they catalyze the transfer of phosphate groups from ATP to tyrosine residues on other proteins. SH2 domains bind to phosphorylated tyrosine residues in other proteins, and SH3 domains bind to peptide segments containing proline residues at specific positions. There are dozens of distinct SH1, SH2, or SH3 domains in a mammalian genome. Individual SH1, SH2, and SH3 domains are combined with each other or with a variety of other domains to generate a diversity of proteins with specialized

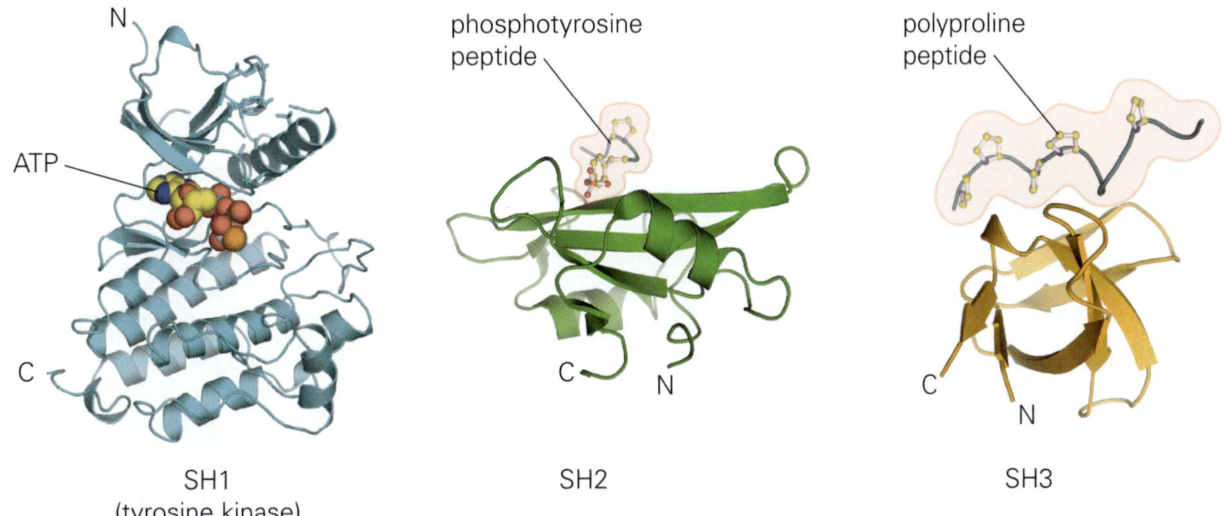


**Figure 4.4 The Src homology domains SH1, SH2, and SH3.** The SH1 domain (*blue*) is a tyrosine kinase enzyme and is shown here bound to ATP. Tyrosine kinases catalyze the transfer of the terminal phosphate of ATP to proteins that they bind to. SH2 domains (*green*) bind to segments of proteins that contain a phosphorylated tyrosine residue (shown shaded in *red*). SH3 domains (*yellow*) bind to protein segments containing proline residues (shaded *red*). [PDB codes: 1M14 (SH1 domain), 1SHA (SH2 domain), and 1SRL (SH3 domain).]

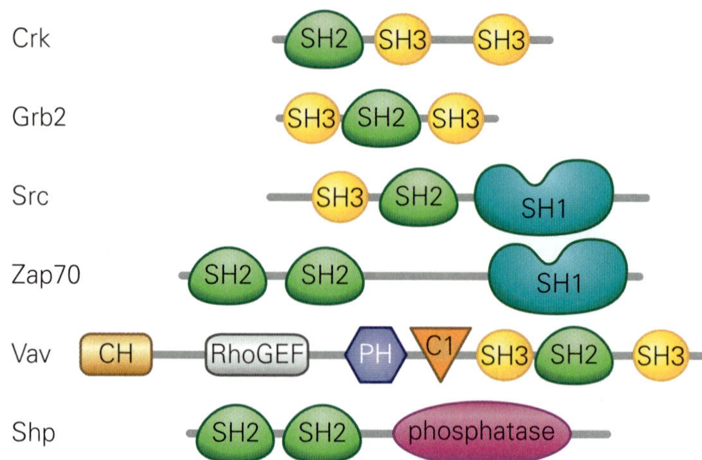


**Figure 4.5 Diversity of proteins generated by combining SH1, SH2, and SH3 domains with various other domains.** Only a very small subset of the large number of proteins containing SH1, SH2, and SH3 domains is shown here. The proteins are illustrated schematically, with the polypeptide chain indicated by a line. Different domains are indicated by different symbols. Phosphatase is an enzymatic domain that removes phosphate groups from phosphorylated tyrosine residues. Rho-GEF is a domain that causes the dissociation of GTP or GDP bound to proteins of the Rho family. CH domains bind to the actin cytoskeleton, and PH and C1 domains bind to specific lipids in the cell membrane. (Adapted from T. Pawson, M. Raina, and P. Nash, *FEBS Lett.* 513: 2–10, 2002. With permission from Elsevier.)

functions (**Figure 4.5**). Notice that while the Src protein has an SH2 domain, an SH3 domain, and an SH1 domain connected in that order, some of the other proteins illustrated in Figure 4.5 lack one or two of these domains, have multiple copies of one kind of domain, or have other kinds of domains included in their structures.

The SH1, SH2, or SH3 domains in the proteins shown in Figure 4.5 have different amino acid sequences. As we discuss in the next chapter, changes in the sequences of the domains, during evolution, result in slightly different structures and in altered specificities for the proteins that they interact with. Although the general fold of each of the domains is conserved, the three-dimensional arrangement of these domains in different proteins can be quite different. This is illustrated in **Figure 4.6**, which compares the structure of a protein that contains an SH2 domain and two SH3 domains, known as Grb2, and the Src protein. Note, as shown in Figure 4.6, that the SH2 and SH3 domains are arranged quite differently in Grb2 and Src.

## 4.3 Protein folding is driven by the formation of a hydrophobic core

When secondary structural elements pack against each other in folded proteins, they bring together the hydrophobic sidechains that form the **hydrophobic core** (see Section 1.14). Protein folding is a process that is similar to the oiling out of greasy molecules in water, except that the greasy molecules (the hydrophobic sidechains) are linked together by the peptide backbone (**Figure 4.7**). As discussed

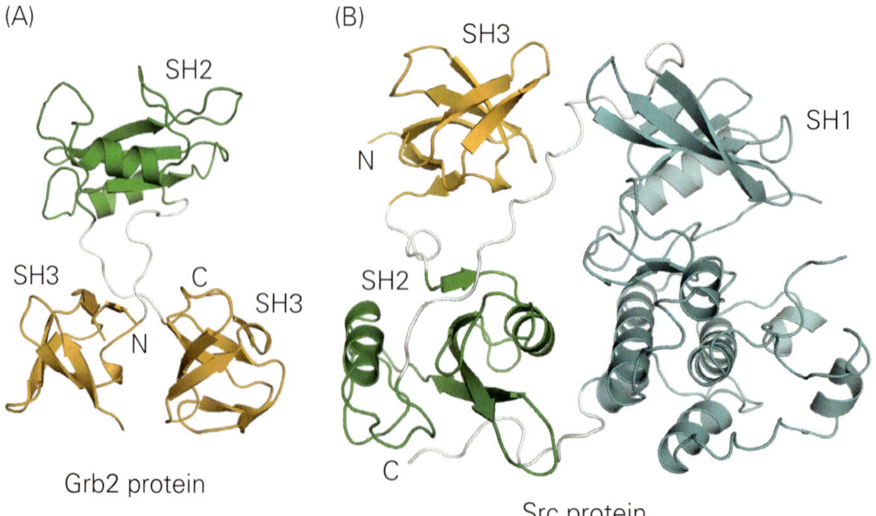


**Figure 4.6 Similar domains in different proteins.** (A) The Grb2 protein has an SH2 domain and two SH3 domains. (B) The Src protein has an SH2 domain and an SH3 domain attached to an SH1 domain. (PDB codes: 1GRI and 2HCK.)

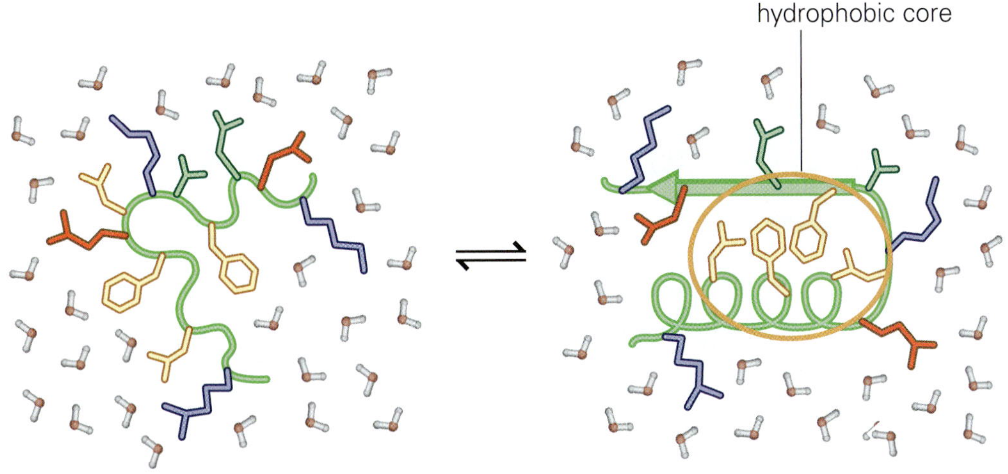


**Figure 4.7 The packing of secondary structural elements results in the formation of the hydrophobic core.** The backbone of a small protein molecule is shown schematically here, along with a few of the sidechains. In the unfolded state, all of the sidechains are exposed to water. When the protein folds up, most of the hydrophobic sidechains (*yellow*) are buried in the interior of the protein and are removed from water. The charged and polar sidechains remain exposed to water.

in Chapters 10 and 18, the stability of the folded structure results primarily from the preference of the hydrophobic sidechains to be clustered together and away from water. Smaller contributions to stability are made by the favorable van der Waals interactions between atoms in the folded state (see Section 6.16), as well as by hydrogen bonds and ionic interactions.

The hydrophobic effect, and not hydrogen-bond formation, is the dominant factor that drives the folding of protein molecules. Because proteins form α helices and β sheets when they fold up, with intricate networks of hydrogen bonds (see Figures 1.36, 1.37, and 1.38), it is natural to think that hydrogen-bond formation might drive protein folding. Nevertheless, hydrogen bonding is not necessarily the dominant feature in the energetics of protein folding. To see why this is so, consider an unfolded protein chain in water, as illustrated in **Figure 4.8.** The backbone –NH and –C=O groups of the unfolded chain form hydrogen bonds with water. When the chain folds up, the formation of α helices and β sheets results in the hydrogen

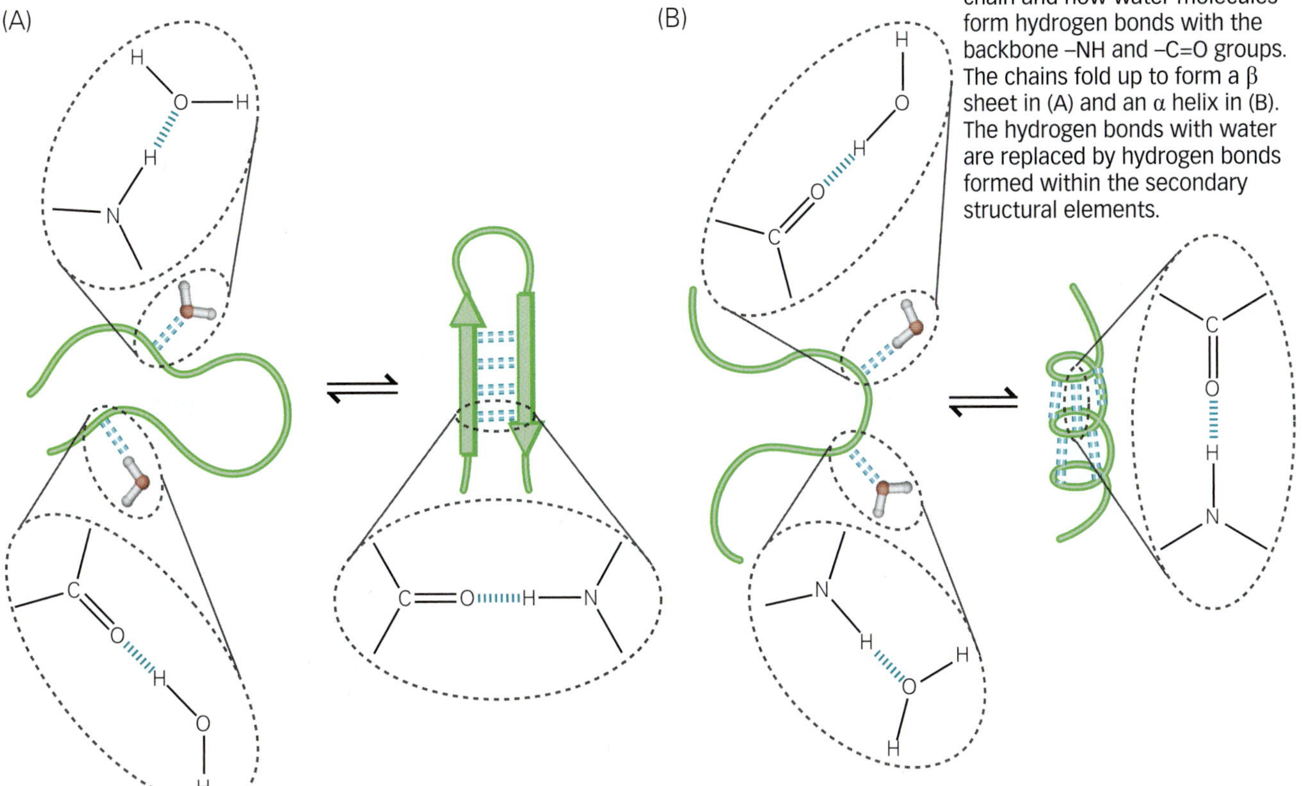


**Figure 4.8 Exchange of hydrogen bonds between backbone groups and water molecules.** The left-hand side of each panel in this figure shows an unfolded protein chain and how water molecules form hydrogen bonds with the backbone –NH and –C=O groups. The chains fold up to form a β sheet in (A) and an α helix in (B). The hydrogen bonds with water are replaced by hydrogen bonds formed within the secondary structural elements.

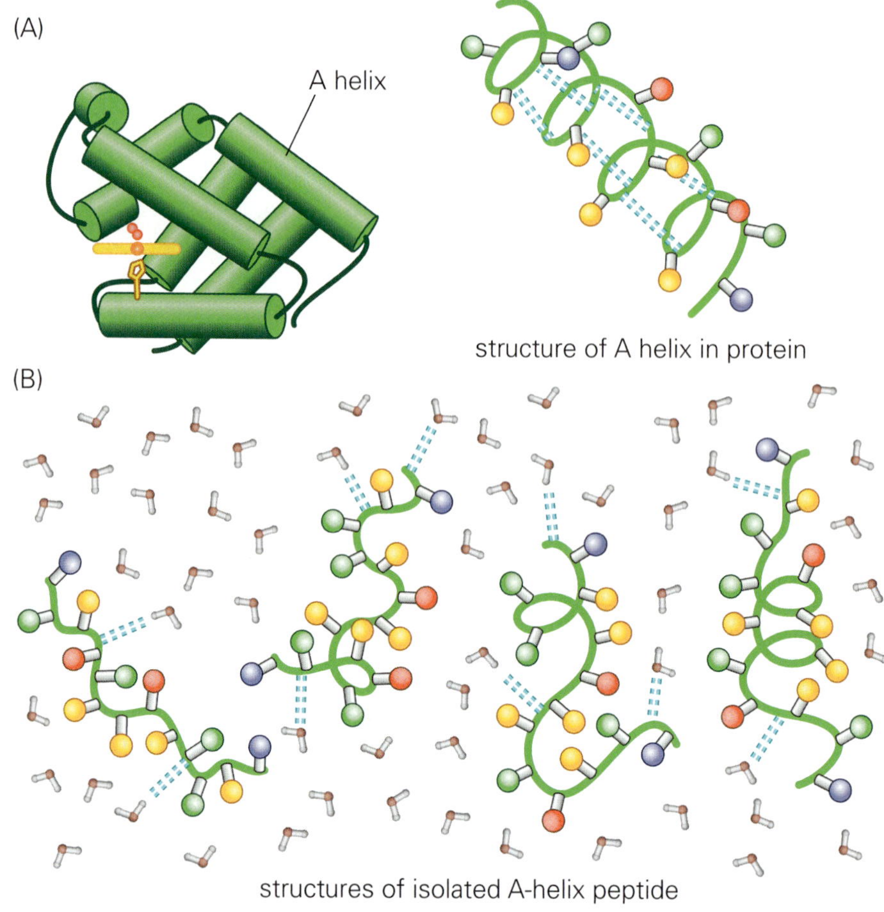


**Figure 4.9 Isolated fragments of proteins do not usually have stable structures.** (A) The structure of myoglobin is shown schematically, with the A helix indicated. The A helix is also shown separately, with the sidechains indicated by spheres, colored according to whether they are hydrophobic (*yellow*), polar (*green*), or charged (*red* or *blue*). (B) A short peptide with the same sequence as the A helix of myoglobin does not have a stable structure in solution, although it shows a tendency to form transient helical turns.

bonds with water being replaced by hydrogen bonds with other parts of the protein backbone. Because of this exchange, the net hydrogen-bonding energy of the backbone does not change much as the protein folds up.

One consequence of the importance of the hydrophobic effect in protein folding is that the individual secondary structural elements do not usually have a very stable structure by themselves when the protein is unfolded. Consider, for example, the structure of an α-helical protein, such as myoglobin, which is illustrated in Figure 4.1. Small fragments of the protein, such as a peptide spanning one of the α helices, do not have a well-defined structure when they are isolated from the rest of the protein (**Figure 4.9**). An isolated peptide segment corresponding to one of the helices in myoglobin cannot form a hydrophobic core because the elements that it normally packs against are missing.

## 4.4 The formation of α helices and β sheets satisfies the hydrogen-bonding requirements of the protein backbone

The hydrogen bonds formed by α helices and β sheets may not contribute much to the overall stability of a folded protein but, nevertheless, the formation of these hydrogen bonds is critical for protein folding. To resolve the apparent contradiction in this statement, consider what would happen if a backbone –NH or –C=O group failed to form a hydrogen bond when the protein folds up. In the unfolded state, the backbone forms hydrogen bonds with water (see Figure 4.8). Water is generally excluded from the interior of the protein and if, upon folding, a compensating hydrogen bond is not formed, then there will be an energetic penalty against folding. The construction of proteins from α helices and β sheets ensures that essentially all hydrogen-bonding groups that are excluded from water are able to form hydrogen bonds with each other.

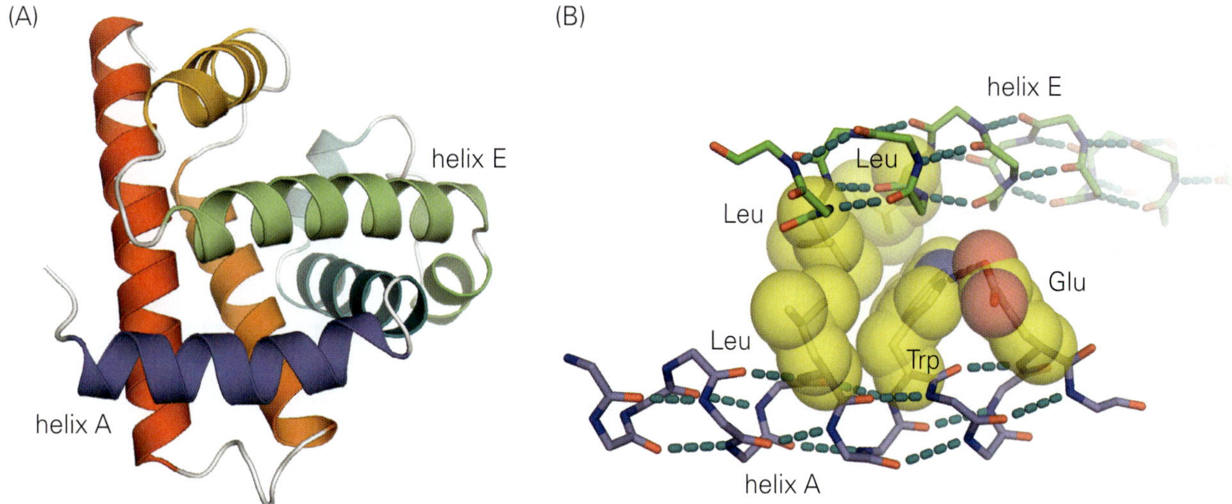


**Figure 4.10 Folded protein structures satisfy hydrogen-bonding requirements.** (A) The structure of myoglobin is shown. Two α helices, denoted A and E, pack together in the folded structure. An expanded view of the two helices is shown in (B), illustrating how hydrophobic sidechains (that is, three leucine residues and one tryptophan residue) are brought close together by the packing of the helices. Hydrogen bonds are indicated by dashed lines. Notice that the hydrogen-bonding capability of the backbone –NH and –C=O groups are satisfied by the formation of the two α helices. The tryptophan sidechain has a nitrogen atom that can form a hydrogen bond, and a glutamate sidechain is positioned so that it can form a hydrogen bond with the tryptophan. The glutamate sidechain is exposed to water, with which it also forms hydrogen bonds (not shown here). (PDB code: 1A6M.)

This principle is illustrated in **Figure 4.10**, which shows how two α helices pack together in myoglobin. The backbone groups form hydrogen bonds in a regular pattern, thereby compensating for the loss of hydrogen bonds with water, which is excluded from the interface between the two helices. Water molecules can sometimes be trapped within the folded structure, but this is relatively rare. The sidechains that are located at the interface are hydrophobic. One of these, a tryptophan, can form a hydrogen bond. The protein structure provides a glutamate sidechain that is positioned so that it can form a hydrogen bond with the tryptophan. The glutamate sidechain can form more than one hydrogen bond, but it is located at the edge of interface, and it is able to interact with water in order to satisfy its hydrogen-bonding capability.

# B. BACKBONE CONFORMATION

## 4.5 Protein folding involves conformational changes in the peptide backbone

Protein molecules consist of a single unbranched covalent backbone that extends from the N-terminus to the C-terminus. The three-dimensional fold of the protein molecule can be described in terms of torsional rotations about covalent bonds in the peptide backbone (**Figure 4.11**).

The covalent backbone has many possible conformations, and interconversions between one conformation and another occur readily. A **conformational change** is a change in the structure of a molecule that occurs due to rotations of parts of the molecule about covalent bonds (**Figure 4.12**). Such rotations are referred to as rotations about **torsion angles** or **dihedral angles**. The two structures denoted I and II in Figure 4.12 are simply two conformations of the same molecule, related by a rotation about the torsion angle denoted by $\phi$ in the illustration.

No covalent bonds need to be broken or remade in order to convert structure I into structure II, which are called **conformational isomers** of the molecule.

**Conformational change**

A change in structure that arises solely from rotations about covalent bonds is called a conformational change. Conformational rearrangements do not involve breaking and forming covalent bonds, and can often occur readily at room temperature.

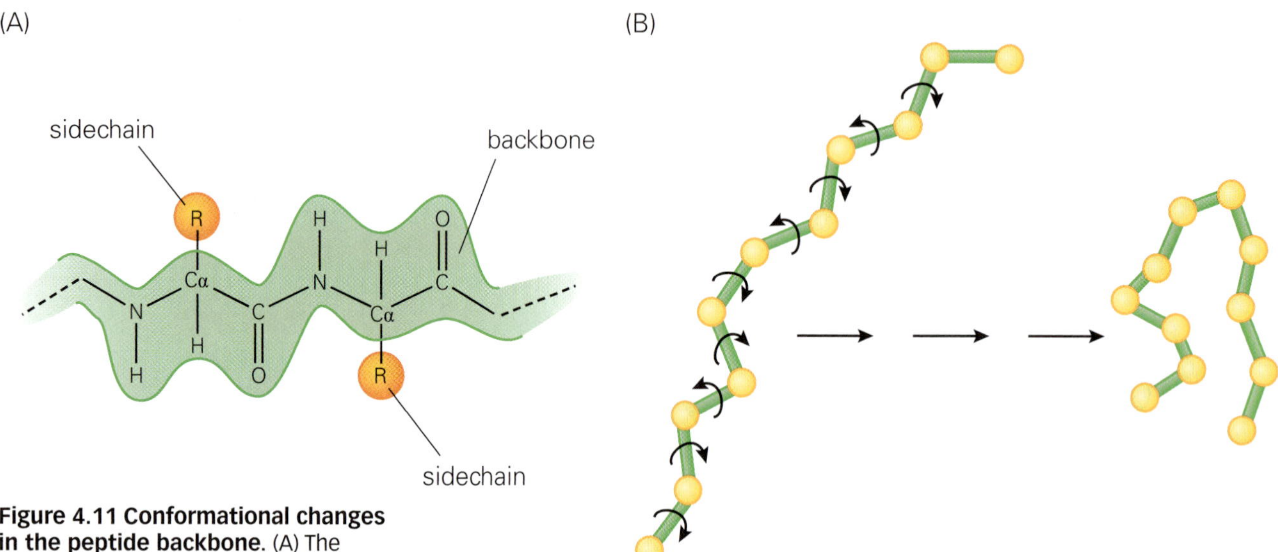


**Figure 4.11 Conformational changes in the peptide backbone.** (A) The chemical structure of a polypeptide (protein) chain, indicating the backbone. (B) A schematic diagram showing the folding of a protein chain into a compact three-dimensional structure. Each amino acid residue is represented by a sphere. The folding process is primarily a conformational change in the protein backbone.

Consider, instead, the structure denoted III in Figure 4.12. This structure cannot be obtained from the other two structures without breaking and remaking covalent bonds. Structure III is a **stereoisomer** of the other two structures.

## 4.6 Amino acids are chiral and only the L form stereoisomer is found in genetically encoded proteins

Molecules that form stereoisomers have special atoms known as **chiral centers**, which are bonded to inequivalent groups of atoms. The $C_\alpha$ atom of each amino acid, except for glycine, is a chiral center because it is bonded to the N atom of the amide, the C atom of the carbonyl, a hydrogen atom, and the R group (**Figure 4.13**). Amino acids, except for glycine, have two stereoisomers, known as the **L form** and the **D form**, which are illustrated in Figure 4.13. Only the L forms of the amino acids are found in genetically encoded proteins. Note that L amino acids

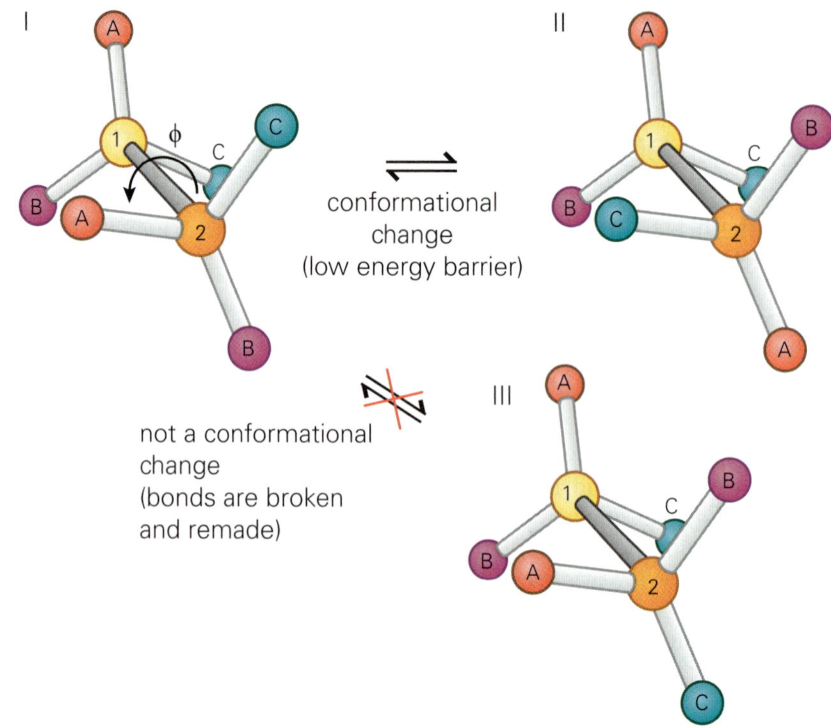


**Figure 4.12 Conformational changes and stereoisomers.** A simple molecule is illustrated here. Two of its atoms, labeled 1 and 2 are bonded together, and each has three substituents, labeled A, B, and C. A conformational change would simply rotate one set of substituents with respect to the other. Any conversion that requires making or breaking bonds is not a conformational change. For example, the molecular structure labeled III, which is a stereoisomer of molecule I, cannot be obtained from the structure labeled I without breaking and remaking covalent bonds.

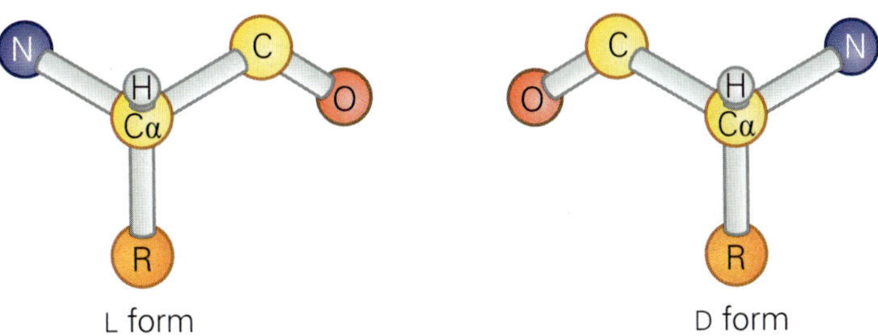


**Figure 4.13 The chirality of amino acids.** The $C_\alpha$ atom of each amino acid, except for glycine, is a chiral center. Only the L form of the amino acids, defined here on the left, is found in genetically encoded proteins.

cannot be converted to D amino acids by rotations about covalent bonds. The D-form amino acids are used in certain situations, such as in the construction of bacterial cell walls, but their utilization occurs as a result of the action of specialized enzymes and not as part of the genetic decoding process.

One consequence of the chirality of amino acids is that the structural elements in proteins have a definite handedness. α helices in naturally occurring proteins, for example, are right-handed. Chemists are able to synthesize artificial proteins in which all the amino acids are of the D form instead of the L form. The resulting protein structures are mirror images of their naturally occurring counterparts, with left-handed α helices instead of right-handed ones.

## 4.7 The peptide bond has partial double bond character, so rotations about it are hindered

Depending on the nature of the bond about which the rotation occurs, and the nature of the substituent groups, the energy barriers to a conformational change may be very small or insurmountably large. Rotations are strongly hindered when the covalent bond about which the rotations are being considered is a double bond, or has partial double bond character. High-resolution x-ray crystallographic studies have provided us with detailed structural information about the peptide group (**Figure 4.14**). Based on these studies, we know that the peptide bond (that is, the bond between the C atom in the first residue and the N atom in the second residue) and its four substituents are essentially coplanar. The four coplanar atoms of the peptide group (C, O, N, and H) form the **amide plane**, which is indicated by shading in Figure 4.14A.

**Amide plane**

The junction between two amino acid residues in a protein is formed by the C=O and N–H groups of the first and second residues, respectively. The four atoms in the peptide group are coplanar, and define the amide plane, which is illustrated in Figure 4.14.

A hint regarding the origin of the planarity of the peptide group is provided by looking at Figure 4.14A, in which the lengths of the various covalent bonds in the peptide group are given. The C–N bond length (1.33 Å) is shorter than the N–$C_\alpha$ bond length (1.46 Å). This 0.13-Å difference in bond length arises because the C–N bond has partial double bond character, which can be thought of as arising from resonance between the two structures shown in Figure 4.14B. There is, as a consequence, a very large energy penalty for distorting the peptide group away from planarity.

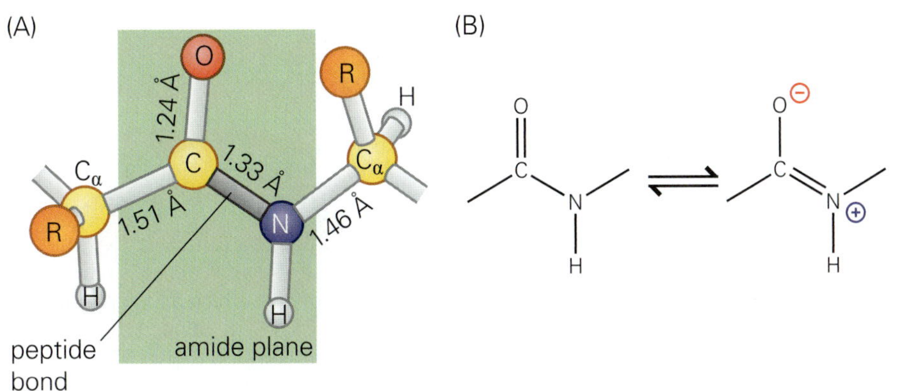


**Figure 4.14 The planarity of the peptide group.** (A) The structure of the peptide group is shown, with bond lengths indicated. The amide plane, which is restricted to planarity because of the partial double bond character of the C–N bond, is shaded. (B) Resonance structures for the peptide group. The nitrogen atom in the resonance structure on the right is shown with a positive charge. In an actual peptide group, the nitrogen has partial negative charge because it withdraws electrons from the hydrogen. (A, adapted from R.E. Marsh and J. Donohue, *Adv. Protein Chem.* 22: 235–256, 1967.)

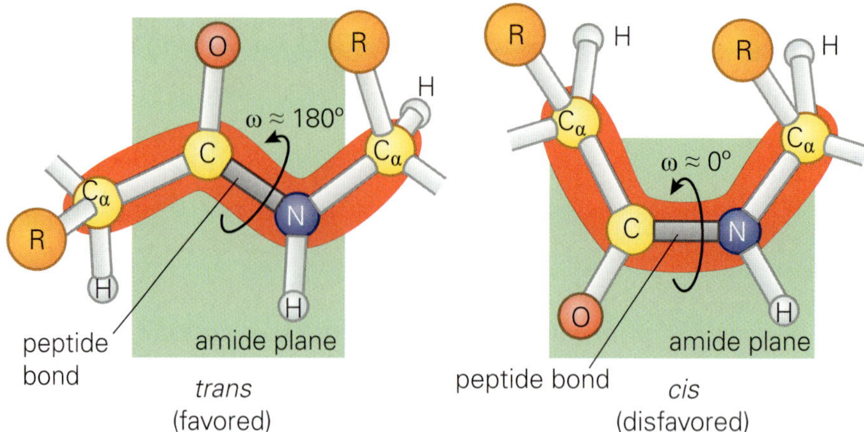


**Figure 4.15 The *trans* and *cis* peptide groups.** The torsion angle that defines rotation around the peptide bond is denoted ω. It has a value close to 180° for a *trans* peptide group and close to zero for a *cis* peptide group. The *cis* conformation results in collisions between adjacent $C_\alpha$ atoms, and is rarely seen.

## 4.8 Peptide groups can be in *cis* or *trans* conformations

There are two alternative conformations that maintain the planarity of the peptide group, shown in **Figure 4.15**. When the two $C_\alpha$ atoms are on opposite sides of the peptide bond, the peptide group is said to be in the ***trans*** conformation. The alternative, in which the $C_\alpha$ atoms are on the same side of the peptide bond, is referred to as the ***cis*** conformation. Peptide groups in proteins are rarely in the *cis* conformation because it brings the $C_\alpha$ groups into close contact.

Although rare, the *cis* conformation of the peptide group can be important for imposing a sharp bend in the direction of the peptide chain. **Figure 4.16** illustrates a *cis* peptide group formed by a proline residue and compares it with the *trans* peptide group normally formed by other amino acid residues. Proline residues are uniquely suited for forming *cis* peptide bonds because the sidechain is fused to the backbone, which reduces the energy difference between *cis* and *trans* forms (Figure 4.16B). Notice the sharp bend in the protein backbone at the location of the *cis* peptide bond, compared with the more extended conformation around a *trans* peptide bond (Figure 4.16C).

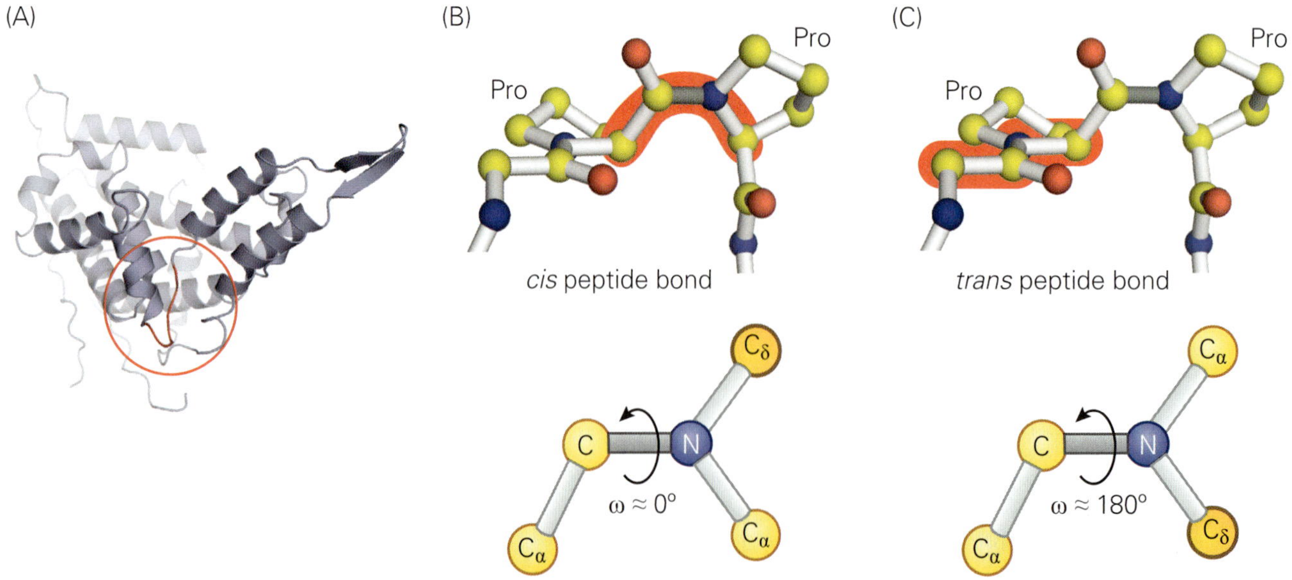


**Figure 4.16 Proline residues can form *cis* peptide bonds.** (A) A portion of a protein structure is shown in which a loop region (*red*) is strongly bent. Two adjacent residues within this loop are shown in expanded views in (B) and (C). (B) The bend in the peptide backbone is stabilized by the formation of a *cis* peptide bond at a proline residue. (C) The second proline has the more common *trans* peptide bond, for which the value of ω is close to 180°. (PDB code:2IIO.)

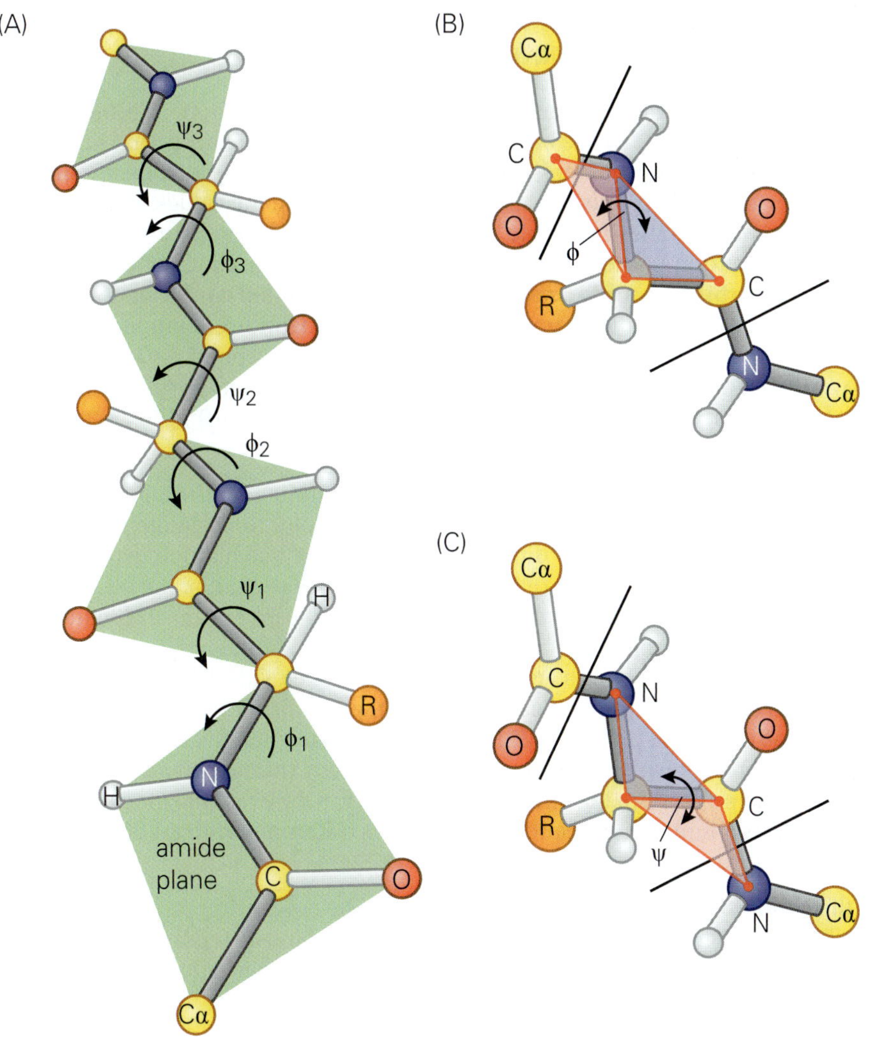


**Figure 4.17 The backbone torsion angles ϕ and ψ.** (A) This diagram illustrates the swivel points of the peptide backbone. (B, C) The definition of the backbone torsion angles ϕ and ψ. The ϕ torsion angle corresponds to rotation about the N–$C_\alpha$ bond. The conformation shown here corresponds to a value of ϕ = 180° and has the C atom of the carbonyl group *trans* with respect to the C atom of the carbonyl group of the previous residue. A value of ϕ = 0° would have these C atoms eclipsed. The ψ torsion angle corresponds to rotation about the $C_\alpha$–C bond. The conformation shown here is for ψ = 180° and has the backbone N atom *trans* with respect to the backbone N atom of the following residue. A value of ψ = 0° would have these N atoms eclipsed. The black lines indicate the boundaries between the residues. (Adapted from J.S. Richardson, *Adv. Protein Chem.* 34: 167–339, 1981.)

## 4.9 The backbone torsion angles ϕ (phi) and ψ (psi) determine the conformation of the protein chain

Because of the planarity of the peptide groups, a protein backbone only has freedom to rotate at the points where the peptide groups meet (**Figure 4.17**A). These swivel points are at each $C_\alpha$ atom of the peptide chain. The $C_\alpha$ atom is connected by single bonds to the amide nitrogen (N) and the carbonyl carbon (C) of the same residue. Rotation about the N–$C_\alpha$ and $C_\alpha$–C bonds define two torsion angles, denoted ϕ and ψ, respectively (Figure 4.17B). By convention, positive rotation in ϕ and ψ corresponds to right-handed rotations about the N–$C_\alpha$ and $C_\alpha$–C bonds. You can picture the sense of rotation by imagining your right hand grasping a stick that corresponds to these bonds, with the N-terminal end of the stick closest to you, as shown in **Figure 4.18**. The direction in which your fingers curl gives the sense of positive rotations in ϕ and ψ.

The values of ϕ and ψ for each residue in the polypeptide chain suffice to completely specify the three-dimensional structure of the peptide backbone. This is so because of the uniform stereochemistry of the building blocks of proteins. The covalent bond geometry and stereochemistry of each of the amino acids stays essentially constant at standard values, regardless of what protein the amino acid is in or in what sequence position. Given these standard stereochemical parameters, the only significant degrees of freedom in the peptide backbone are the ϕ and ψ torsion angles.

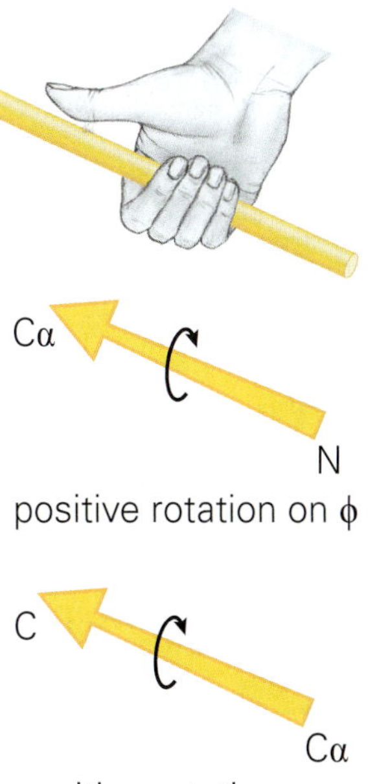


**Figure 4.18 Right-hand rule for determining the sense of ϕ and ψ.** Imagine grasping a rod representing the protein backbone with the fingers of your right hand. The thumb of your hand is aligned with the direction of the N–$C_\alpha$ bond (for ϕ) or the $C_\alpha$–C bond (for ψ). The direction in which your fingers are curled gives the sense of positive rotation in ϕ and ψ.

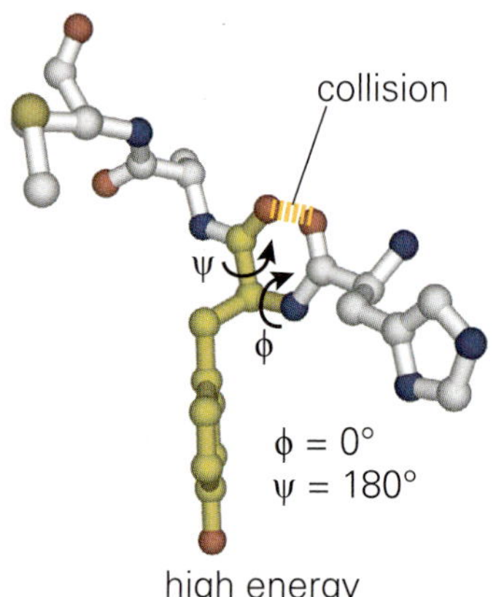


**Figure 4.19 A forbidden combination of ϕ and ψ.** This particular combination of backbone torsion angles, with ϕ = 0° and ψ = 180° for the tyrosine residue, results in a collision between the carbonyl oxygen of the tyrosine and the carbonyl oxygen of the histidine preceding it. This combination of ϕ and ψ is therefore disallowed.

## 4.10 The Ramachandran diagram defines the restrictions on backbone conformation

Not all values of ϕ and ψ are accessible to the peptide backbone. Certain values of ϕ and ψ bring atoms of the backbone into such close proximity that the repulsive component of the van der Waals energy prevents the atoms from coming any closer. van der Waals repulsions are so strong that, once the values of ϕ and ψ are in this regime, no other favorable energetic interaction (for example, hydrogen bonding) can compensate for the repulsion. Such combinations of ϕ and ψ correspond to **forbidden conformations**. One such forbidden conformation is illustrated in **Figure 4.19**, which shows the structure of a peptide containing a tyrosine residue, for which the values of ϕ and ψ are 0° and 180°, respectively. These values of ϕ and ψ lead to a collision between the carbonyl oxygen of the tyrosine and the carbonyl oxygen of the preceding peptide group. As a result, (0°, 180°) is a forbidden combination of ϕ and ψ values.

The van der Waals repulsions place very strong restraints on the allowed conformations of the polypeptide chain. This makes it useful to calculate the allowed values of ϕ and ψ for a simple dipeptide composed of two alanine residues (that is, Ala-Ala). This calculation is of broad generality because all of the other amino acid sidechains are larger than that of alanine (with the exception of glycine). This means that ϕ, ψ combinations that are forbidden for alanine will also be forbidden for all of the other amino acids (except for glycine). G.N. Ramachandran and colleagues carried out the first calculation of the conformational energy map for the alanine dipeptide (**Figure 4.20**). These earliest calculations treated the atoms as hard spheres, in which close contacts were strictly disallowed and no account was made for any stabilizing interaction. Such a two-dimensional map of molecular energy as a function of ϕ and ψ is known as a **Ramachandran diagram** (see Figure 4.20).

In the simplest Ramachandran diagram, the values of ϕ and ψ are grouped into two kinds of regions: allowed and disallowed. As you can see from Figure 4.20, most combinations of ϕ and ψ are disallowed due to interatomic collisions. There

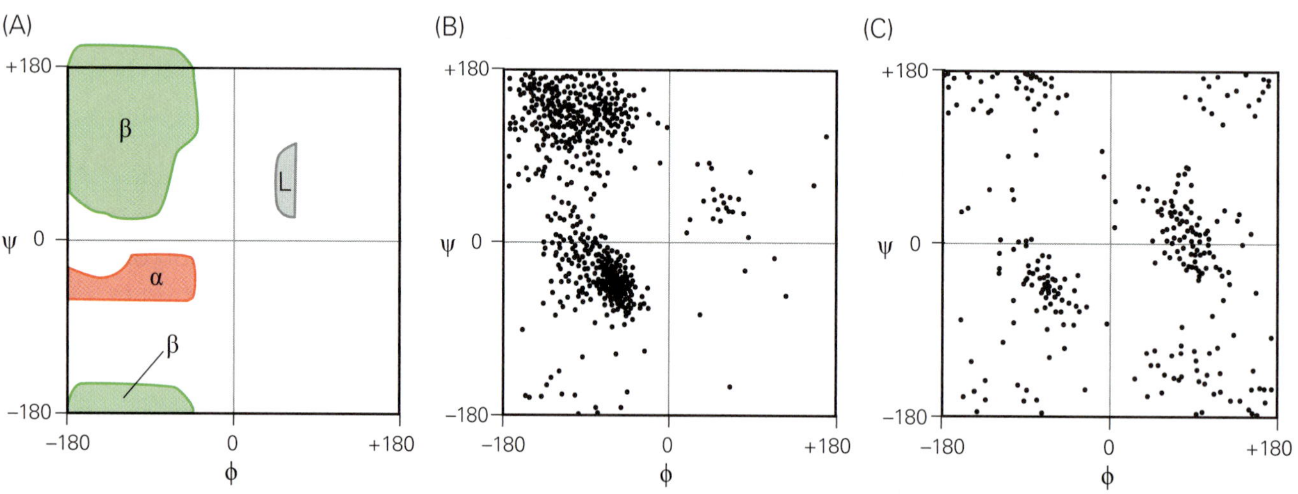


**Figure 4.20 The Ramachandran diagram.** (A) The values of ϕ and ψ for an alanine dipeptide are given by the two axes of the diagram. For each combination of ϕ and ψ, the structure of the dipeptide is calculated. If the structure has interatomic collisions, the conformation (that is, the combination of ϕ and ψ) is marked forbidden (the white regions in this diagram). Regions of allowed combinations of ϕ and ψ are shown in color and correspond to α-helical conformations (*red*), β strands (*green*), and left-handed helices (L). Notice that the top and right areas of the diagram are related to the bottom and left areas, respectively, by circular symmetry. The β region, therefore, appears at the top and the bottom of the diagram. (B) The actual values of ϕ and ψ from the crystal structures of several proteins are plotted. Each dot represents the values of ϕ and ψ for a residue in one of the proteins. (C) Values of ϕ and ψ for glycine residues in several different proteins. Glycine residues can access conformations that are forbidden to the other amino acids. (Adapted from C. Brändén and J. Tooze, Introduction to Protein Structure, 2nd ed. New York: Garland Science, 1999, and also J.S. Richardson, *Adv. Protein Chem.* 34: 167–339, 1981.)

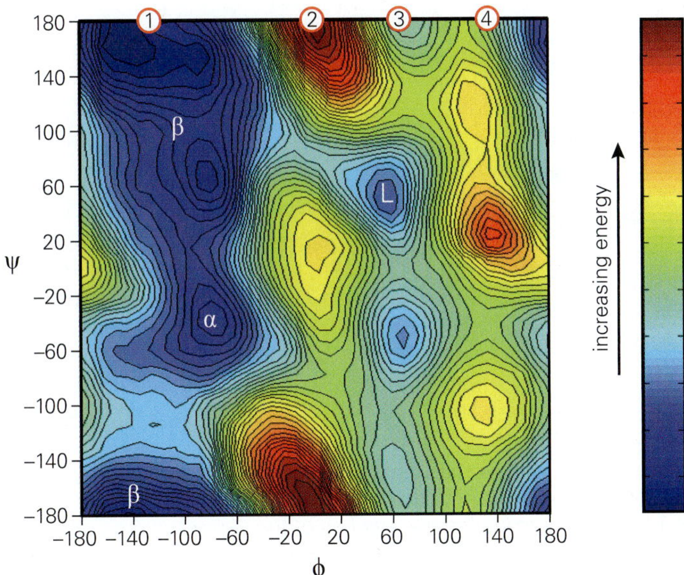


**Figure 4.21 A sophisticated version of the Ramachandran diagram.** Shown here is a Ramachandran diagram for the alanine dipeptide that uses a sophisticated evaluation of the energy of each conformation. Hydrogen bonding is accounted for, as is the interaction of the peptide with water molecules. Low- and high-energy regions are in *blue* and *red*, respectively. Notice that the lowest energy regions of this diagram are the same as those in Figure 4.20A. The conformations of a peptide with ϕ, ψ values corresponding to the points labeled 1–4 are shown in Figure 4.22. (Adapted from D.S. Chekmarev, T. Ishida, and R.M. Levy, *J. Phys. Chem. B* 108: 19487–19495, 2004. With permission from the American Chemical Society.)

are three allowed regions in the Ramachandran diagram, and these are labeled α, β, and L in Figure 4.20. The α region corresponds to values of ϕ and ψ that generate an α helix if repeated sequentially down the chain. Likewise, the region labeled β corresponds to ϕ and ψ combinations that generate a β strand if repeated sequentially. The L region generates a left-handed helical structure, which is not common in proteins.

It is remarkable that the simple hard-sphere calculation underlying the Ramachandran diagram restricts the conformation of the backbone to the α and β (plus L) regions. If we look at the observed values of ϕ and ψ in real protein structures, we find that these generally occur in the allowed regions of the Ramachandran diagram (Figure 4.20B). The one exception is glycine, which only has a hydrogen atom for an R group and is much less restricted in its conformation (Figure 4.20C).

**Ramachandran diagram**

This diagram shows the energy of an alanine dipeptide as a function of the backbone torsion angles ϕ and ψ. Allowed combinations of ϕ and ψ are those for which the backbone does not run into itself.

More sophisticated versions of the Ramachandran diagram use accurate descriptions of the van der Waals energy and include hydrogen-bonding effects and the effects of water molecules (**Figure 4.21**). Despite the inclusion of more interactions, the essential features of the restriction of ϕ, ψ combinations remain essentially the same. Low-energy regions of the diagram correspond to regions in which the backbone atoms do not collide with each other, whereas backbone collisions increase the energy of other conformations.

The Ramachandran maps shown in Figures 4.20 and 4.21 are calculated for an alanine dipeptide, but they help explain which conformations are disallowed for peptides containing other amino acid residues (except for glycine). This is because a conformation that is disallowed for alanine will also be disallowed for amino acids with bulkier sidechains. Four different conformations for the peptide illustrated in Figure 4.19 are shown in **Figure 4.22**. These conformations have different values of ϕ and ψ for the central tyrosine residues. Two of the conformations have very high energy because of collisions between backbone atoms, and these correspond to disallowed regions of the Ramachandran diagram for alanine dipeptides (see Figure 4.21).

## 4.11 α helices and β strands are formed when consecutive residues adopt similar values of ϕ and ψ

The α helix was first described in 1951 by Linus Pauling, who predicted that it was a structure that would be stable and energetically favorable in proteins. He made this prediction on the basis of accurate geometrical parameters that he had derived for the peptide unit from the results of crystallographic analyses of

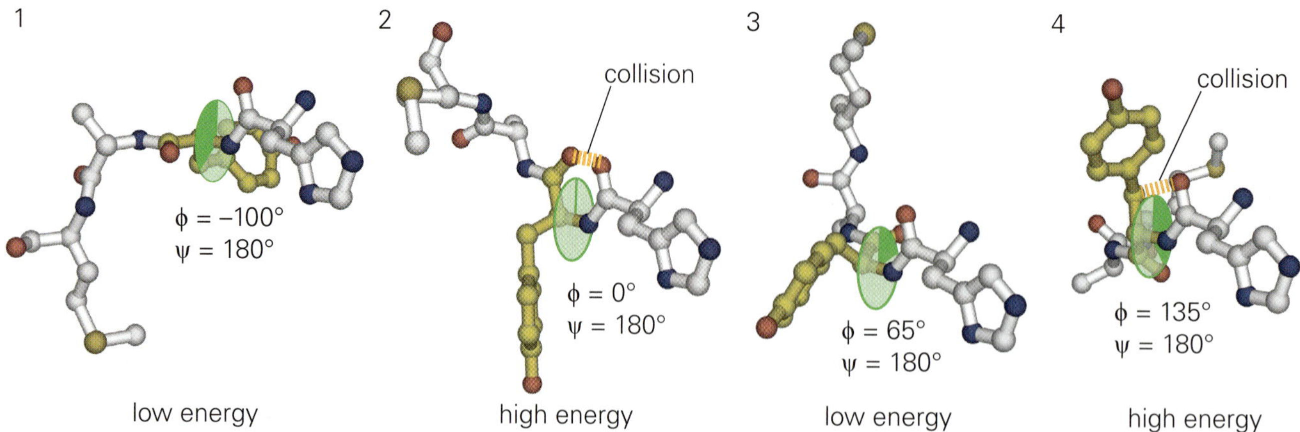


**Figure 4.22 Backbone collisions determine the high-energy regions of the Ramachandran diagram.** A peptide with a tyrosine residue (*yellow*) is shown here (see Figure 4.19). Four conformations of the peptide are illustrated, corresponding to the $\phi$, $\psi$ values labeled 1–4 in Figure 4.21. The $\phi$, $\psi$ combinations labeled 1 and 3 are in low-energy regions of the Ramachandran map, while those labeled 2 and 4 are in high-energy regions because of collisions between atoms in the peptide backbone. All four conformations have the same value of $\psi$. Changes in the value of $\phi$ are indicated by the *green* circle.

the structures of a range of small molecules. This prediction almost immediately received strong experimental support from x-ray data obtained by Max Perutz in Cambridge, UK, from hemoglobin crystals and keratin fibers. It was completely verified by John Kendrew's high-resolution structure of myoglobin, where all secondary structure is helical (see Figure 4.1).

The protein backbone forms an α helix when a stretch of consecutive residues all have similar values of $\phi$ and $\psi$, around −60° and −40°, respectively, corresponding to an allowed region in the bottom left quadrant of the Ramachandran diagram (**Figure 4.23**). Recall from Chapter 1 that an ideal α helix has 3.6 residues per turn, with hydrogen bonds between the C=O group of one residue (denoted $n$) and the NH of the residue that is four positions ahead in the chain (denoted $n + 4$) (Figure 1.36). Not all α helices in proteins obey this ideal rule. Note that the allowed region in the Ramachandran diagram is considerably larger than occupied by the $\phi$ and $\psi$ values for the near-ideal α helix depicted in Figure 4.23. There is, as a consequence, some variation in the precise conformations of α helices.

Variations on the α helix in which the chain is either more loosely or more tightly coiled, with hydrogen bonds to residues $n + 5$ or $n + 3$ instead of $n + 4$, are called the π helix and $3_{10}$ helix, respectively. The $3_{10}$ helix has three residues per turn and contains 10 atoms between the hydrogen bond donor and acceptor, hence its name. Both the π helix and the $3_{10}$ helix occur rarely and usually only at the ends of α helices or as single-turn helices. They are not energetically favorable, because the backbone atoms are too tightly packed in the $3_{10}$ helix and so loosely packed in the π helix that there is a hole through the middle. Only in the α helix are the backbone atoms properly packed to provide a stable structure.

An α helix can in theory be either right-handed or left-handed, depending on the screw direction of the chain. A left-handed α helix is not, however, a favorable conformation for L amino acids due to the close approach of the sidechains and the C=O group. Thus, the α helix that is observed in proteins is almost always right-handed. Short regions of left-handed α helices (3–5 residues) do occur in protein structures, but only occasionally.

As for the α helix, β strands are formed when a series of consecutive residues have similar values of $\phi$ and $\psi$, except that in this case these are clustered around −120° and +120°, respectively (see Figure 4.23). Recall from Chapter 1 that β strands associate to form either parallel, antiparallel, or mixed β sheets, depending on

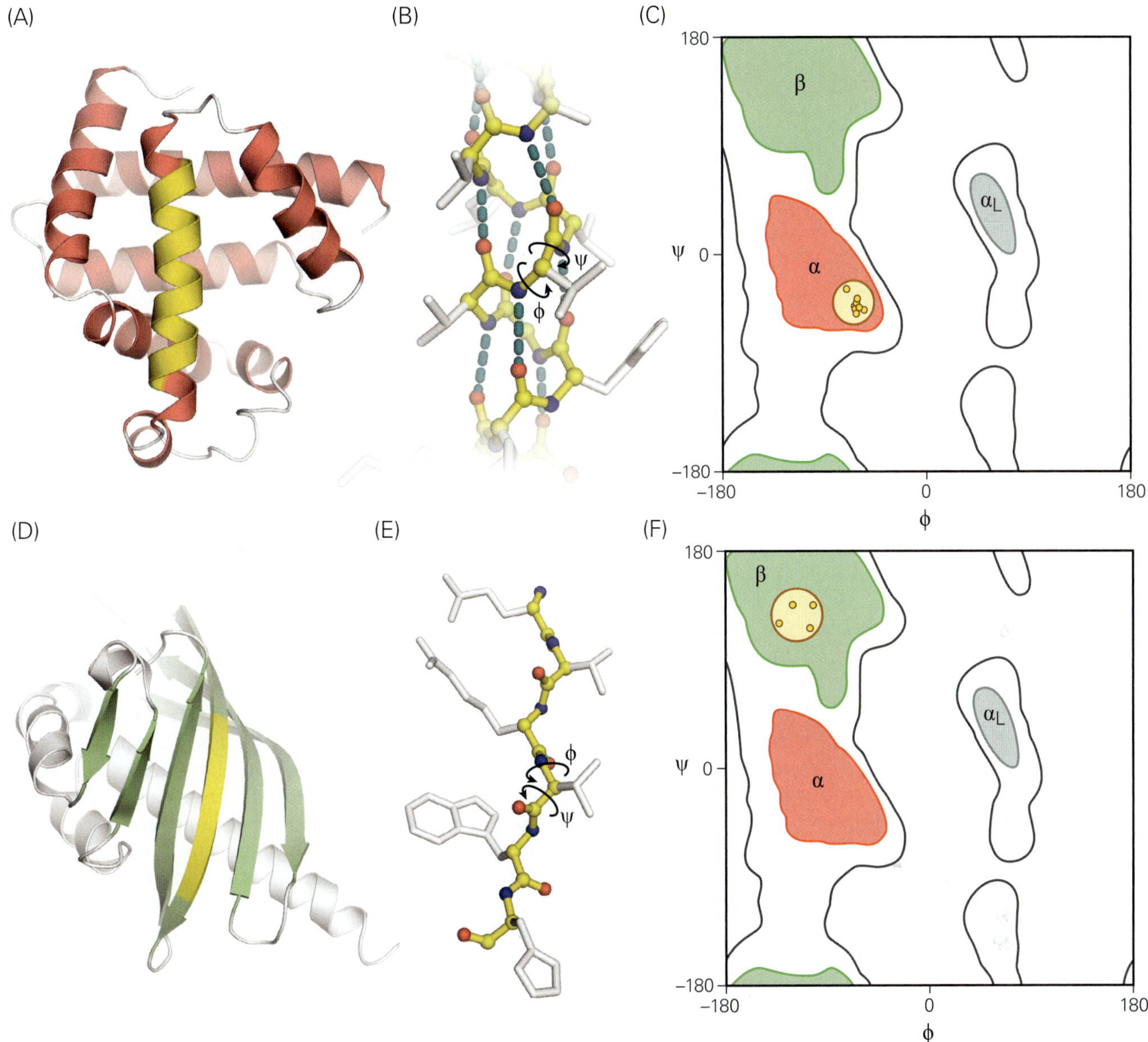


**Figure 4.23 Ramachandran diagrams for α helices and β strands.** (A–C) An α helix is formed when values of ϕ and ψ close to −60° and −50°, respectively, are repeated along the backbone. (A) The structure of myoglobin is shown, with one helix highlighted in *yellow*. (B) An expanded view of the structure of this helix, with ϕ and ψ indicated for a valine residue in the middle of the helix. (C) Values of ϕ and ψ for all the residues in the helix are plotted on a Ramachandran diagram. Note that these values all cluster close together in the α region of the diagram (*yellow* circle). (D–F) A β strand is formed when consecutive residues in the polypeptide chain have values of ϕ and ψ close to −120° and +120°, respectively. (D) The structure of a protein containing an antiparallel β sheet is shown, with one strand highlighted in *yellow*. (E) An expanded view of the structure of this strand, with ϕ and ψ indicated for a valine residue in the middle of the strand. (F) Values of ϕ and ψ for four residues in the strand are plotted on a Ramachandran diagram. These values are clustered in the β region of the diagram (*yellow* circle). (Ramachandran diagrams were generated using the program MolProbity: I.W. Davis et al., and D.C. Richardson, *Nuc. Acids Res.* 35: W375–W383, 2007; PDB codes: A and B, 1A6M; D and E, 1HKX.)

whether the strands all run in the same direction or not. The backbone conformations of parallel and antiparallel β sheets are slightly different, but in both cases the values of ϕ and ψ are clustered in the upper left quadrant of the Ramachandran diagram. The values of ϕ and ψ for a strand in an antiparallel β sheet are illustrated in Figure 4.23.

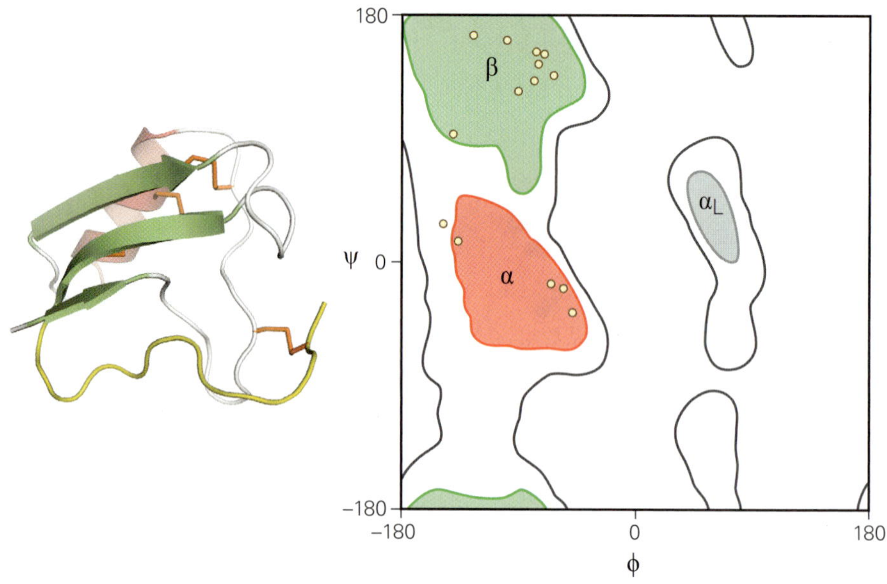


**Figure 4.24 Lack of clustering of ϕ and ψ values in a loop segment.** The structure of a scorpion toxin is shown on the left with a loop segment colored *yellow*. The ϕ and ψ values for the residues in the loop are shown in the Ramachandran diagram on the right. Note that these values are scattered across the α and β sectors of the diagram. This protein has four disulfide bonds, which are shown in *orange*. (Ramachandran diagram generated using the program MolProbity: I.W. Davis et al., and D.C. Richardson, *Nucleic Acids Res.* 35: W375–W383, 2007; PDB code: 1PTX.)

## 4.12 Loop segments have residues with very different values of ϕ and ψ

The loop regions that connect α helices and β strands, or form the N- and C-terminal tails of the protein, are usually at the surface of the molecule. The backbone C=O and NH groups of the loop regions are exposed to water, and so are not required to form hydrogen bonds with other polar groups in the protein. This feature relaxes the necessity to form a repeating structure that satisfies backbone hydrogen-bonding requirements, as in α helices and β sheets. As a consequence, the values of ϕ and ψ in consecutive residues in a loop segment can be quite different. This is illustrated in **Figure 4.24**, which shows the values of ϕ and ψ in a long loop region in a scorpion toxin. Note that the values of ϕ and ψ are scattered around the low-energy regions of the diagram corresponding to both the α and the β conformation. This example illustrates the fact that although the values of ϕ and ψ for a residue may lie in the α or β region of the Ramachandran diagram, the residue is not necessarily in an α helix or a β strand. Helices and strands are formed only when a series of consecutive residues all have ϕ and ψ values in either the α or the β region.

The scorpion toxin structure illustrated in Figure 4.24 has unusually long loop regions, with the loop highlighted in yellow being 14 residues long. We have emphasized that protein structures are stabilized by the packing of hydrophobic sidechains at the interfaces between helices and strands, which raises the question as to how the very long loops in the toxin structure are stabilized. The answer is that, in addition to the hydrophobic core, this particular protein fold is held together by four disulfide bonds that provide covalent linkages between different parts of the protein backbone (see Figure 4.24). Disulfide bonds are commonly found in secreted proteins. Disulfide bonds are unstable inside the cell because of the presence of molecules known as reducing agents, which break disulfide bonds.

Although they are irregular in conformation, loop structures in proteins quite often have specific conformations that are found in many different proteins. We shall not discuss the variety of loop conformations in detail, but simply illustrate one kind of loop—namely, that connecting adjacent antiparallel β strands. These are known as **β hairpin loops**, or **reverse turns**. As shown in **Figure 4.25**, such loops are usually quite short, and typically contain only four to six residues. **Figure 4.26** shows two of the most frequently occurring turns—the type I turn and the type II turn—which are distinguished by the ϕ, ψ values of the two central residues

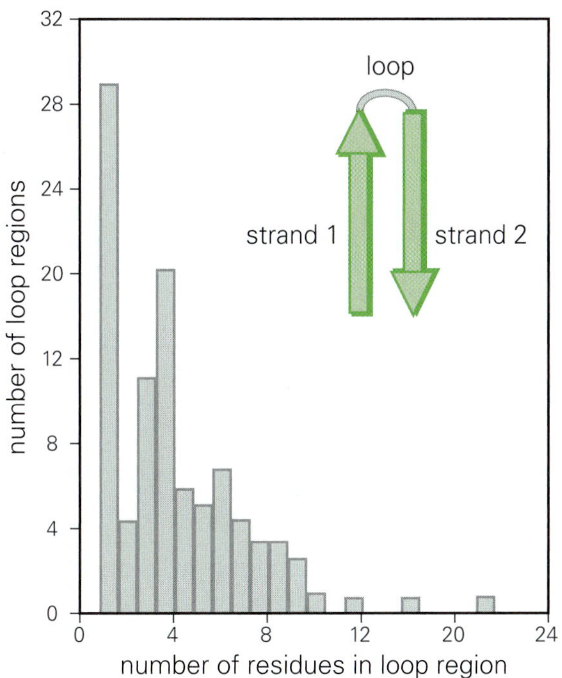


**Figure 4.25 Adjacent antiparallel β strands are often joined by hairpin loops.** This diagram shows the frequency with which loops of different lengths are found linking two β strands. Such loops are usually short. (Adapted from B.L. Sibanda and J.M. Thornton, *Nature* 316: 170–174, 1985. With permission from Macmillan Publishers Ltd.)

of the turn. The common occurrence of these two specific types of turn structures is a consequence of the fact that the polypeptide backbone is not arbitrarily flexible. The conformation of each residue in the turn has to be in one of the allowed regions of the Ramachandran diagram while allowing the direction of the chain to bend back on itself.

## 4.13 α helices and β strands are often amphipathic

The most common location for an α helix in a protein structure is along the outside of the protein, with one side of the helix facing the solution and the other side facing the hydrophobic interior of the protein. Therefore, with 3.6 residues per turn,

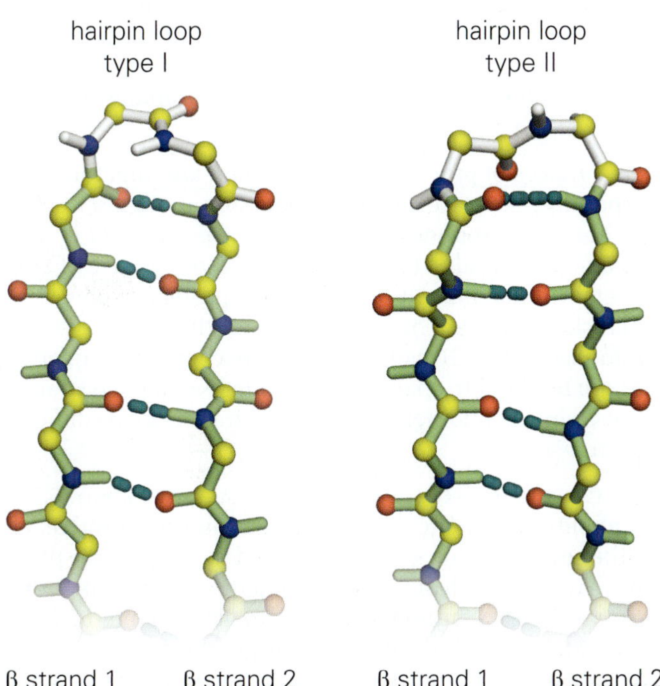


**Figure 4.26 Two kinds of β hairpin loops.** The diagram shows the two most frequently occurring kinds of two-residue hairpin loops: the type I turn (left) and the type II turn (right). Bonds within the hairpin loop are shown in *white*. (Adapted from C. Brändén and J. Tooze, Introduction to Protein Structure, 2nd ed. New York: Garland Science, 1999.)

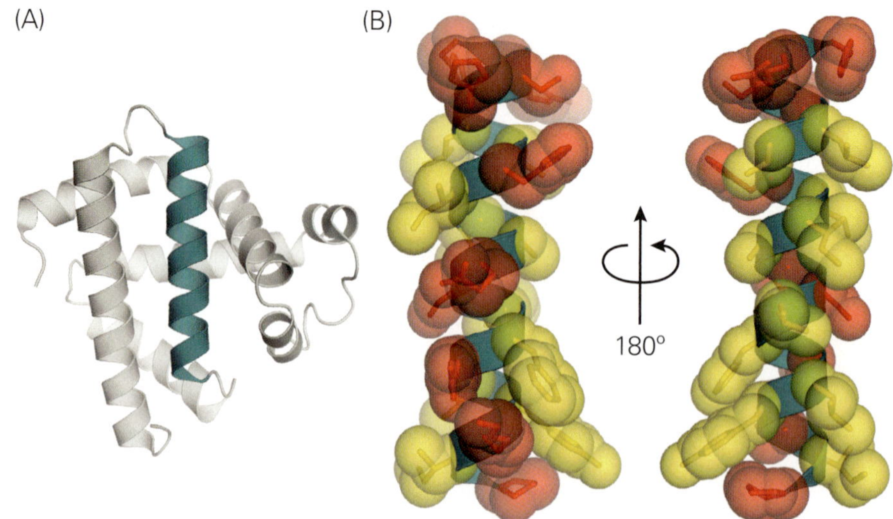


**Figure 4.27 An amphipathic α helix.** The structure of myoglobin is shown in (A), with one α helix colored *blue*. (B) This α helix is shown in an expanded view. The view on the left is in the same orientation as in (A). The view on the right is rotated by 180° with respect to the first one, revealing the face of the helix that packs against the rest of the protein. Hydrophobic sidechains are colored *yellow* and polar and charged sidechains are colored *red*.

there is a tendency for sidechains to change from hydrophobic to hydrophilic with a periodicity of three to four residues. α helices that have one hydrophobic face and one hydrophilic face are known as **amphipathic helices** (**Figure 4.27**).

Not all α helices in globular proteins are amphipathic, because residues that face the solution can sometimes be hydrophobic and, furthermore, α helices can be either completely buried within the protein or completely exposed. **Figure 4.28** shows the structures of α helices in three different proteins, with their amino acid sequences. One of the helices is totally buried, and most of the residues in this helix are hydrophobic (Figure 4.28A). A partially exposed α helix is shown in Figure 4.28B. The hydrophobic residues in this helix are arranged in an $i$, $i + 3$, or $i$, $i + 4$ pattern, where $i$ refers to the position of one residue in the sequence. Figure 4.28C shows the structure of a completely exposed α helix, and most of the residues in this helix are polar. Completely buried or completely exposed α helices are relatively rare in proteins.

**Amphipathic α helices and β sheets**

Secondary structural elements that have distinctive faces, one hydrophobic and one hydrophilic, are referred to as amphipathic. The hydrophobic faces of amphipathic sheets and helices can pack against each other to form a hydrophobic core, leaving the hydrophilic faces to interact favorably with water. This is a central principle in the architecture of most globular proteins.

A convenient way to illustrate the amino acid sequences in helices is the helical wheel or spiral. Because one turn in an α helix spans 3.6 residues, each residue can be plotted every 360/3.6 = 100° around a circle or a spiral, as shown in Figure 4.28. Such a plot shows the projection of the position of the residues onto a plane perpendicular to the helical axis. Residues on one face of the helix cluster together on one side of the spiral. It is immediately obvious from the helical wheels that one side of the helix illustrated in Figure 4.28B is hydrophilic and the other side hydrophobic—that is, this helix is amphipathic.

β sheets on the surfaces of proteins usually contain amphipathic β strands. One such strand, spanning seven residues, is illustrated in **Figure 4.29**. This strand contains three buried hydrophobic residues that are arranged in an $i$, $i + 2$ pattern (that is, every other residue is hydrophobic). The presence of amphipathic β strands can often be quite difficult to discern in a protein sequence. Some of the surface exposed sidechains are hydrophobic, as is the case for the valine in the strand shown in Figure 4.29. Because the peptide chain is completely extended, fewer residues are required to span a protein domain with a β strand than is the case for α helices, which is another reason that β strands can be difficult to discern in sequence patterns.

The distance between alternate $C_\alpha$ atoms in a β strand is about 6.5 Å (that is, a translation of about 3.2 Å along the strand direction per residue, as shown in **Figure 4.30**). The seven-residue strand illustrated in Figure 4.29 spans a distance of 20 Å, which is a typical width of a protein domain, and so the pattern of alternating hydrophobic residues is limited to just three such residues in this case. Recall from Chapter 1 that the rise along the helix axis per residue is ~1.5 Å in an α helix, and so a 14-residue α helix spans the same distance (see Figure 4.30).

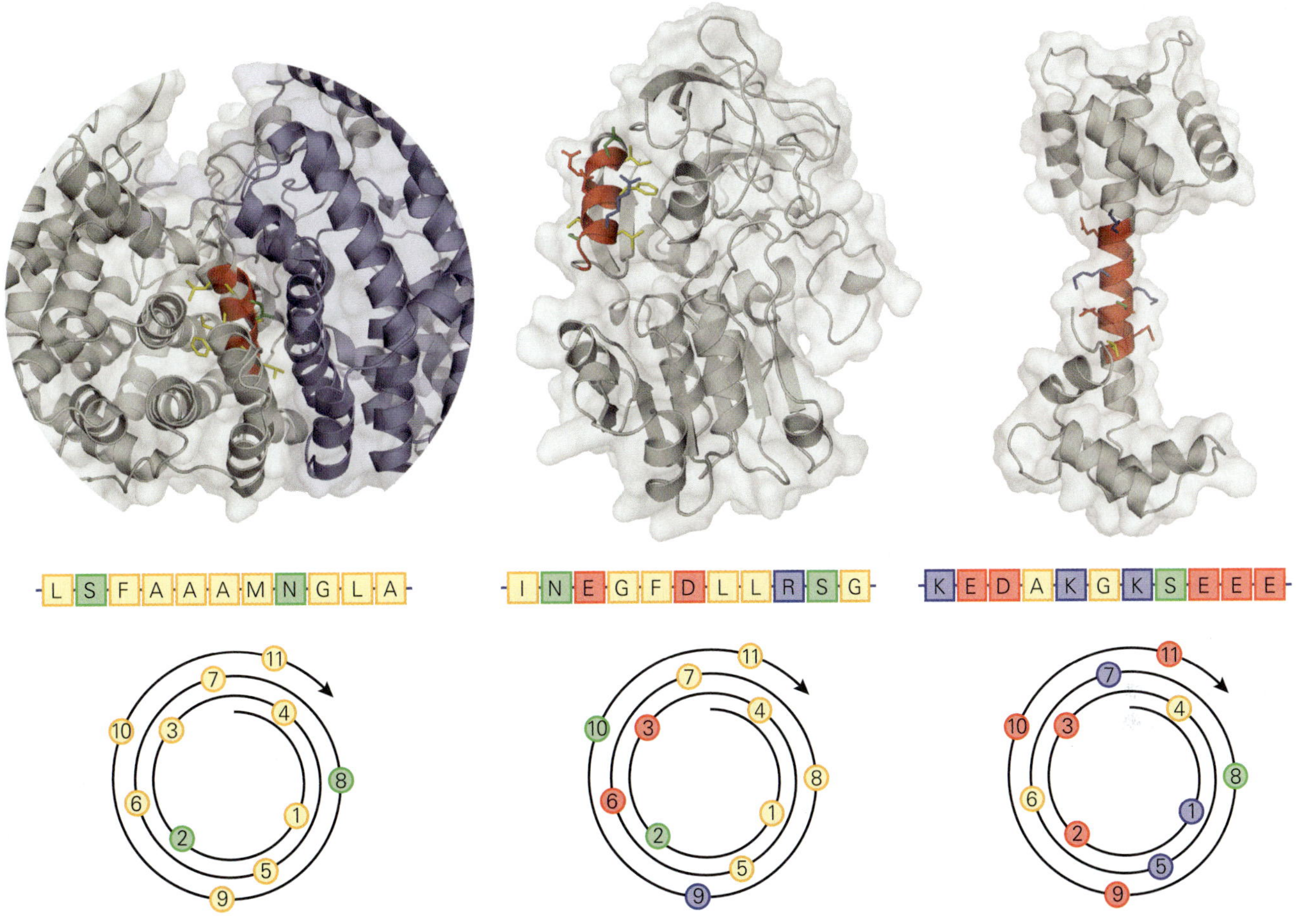

**Figure 4.28 Examples of α helices in three different proteins.** (A) A completely buried α helix (*red*) from the enzyme citrate synthase. (B) A partially exposed helix from the enzyme alcohol dehydrogenase. (C) A completely exposed helix from troponin-C. The amino acid sequences of the helices are shown below the structures, with charged residues in *red* and *blue*, polar resides in *green*, and hydrophobic residues in *yellow*. The sequences of the α helices are also shown on helical wheels or spirals, in which amino acid residues are plotted every 100° around the spiral. (Adapted from C. Brändén and J. Tooze, Introduction to Protein Structure, 2nd ed. New York: Garland Science, 1999; PDB codes: A, 1CTS; B, 1A71; and C, 5TNC.)

## 4.14 Some amino acids are preferred over others in α helices

The amino acids have different preferences either for or against being in α helices, and a ranked list of such preferences is called a helix propensity scale (**Figure 4.31**). These preferences arise due to differences in energy of the contacts made by sidechain atoms with the backbone. Alanine, which has the smallest sidechain,

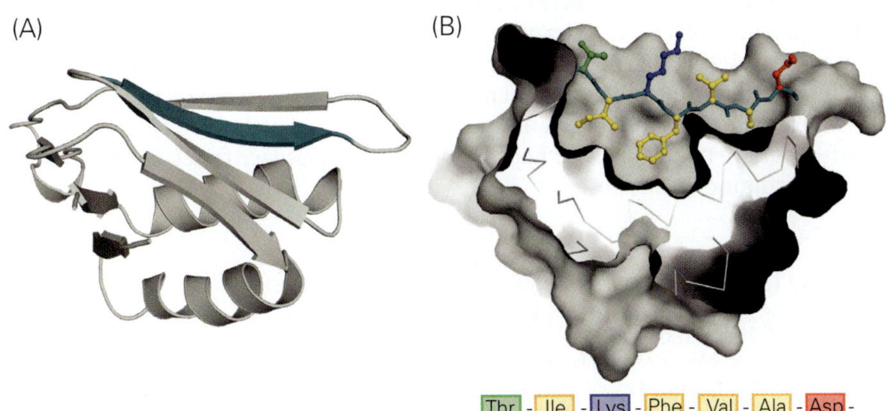


**Figure 4.29 An amphipathic β strand.** (A) A protein structure is shown in which one β strand is colored *teal*. (B) A cross section through the protein, showing packing of the same β strand against the rest of the protein. The amino acid sequence of the strand is shown below. Notice that the buried hydrophobic residues (isoleucine, phenylalanine, and alanine) occur in an $i$, $i + 2$ pattern. (PDB code: 1PLQ.)

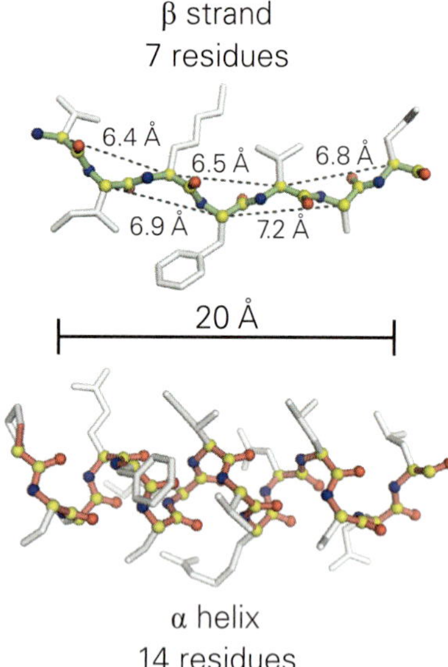


**Figure 4.30 Comparing the dimensions of α helices and β strands.** A distance of 20 Å is spanned by seven residues in a β strand and 14 residues in an α helix.

is the residue that is most preferred in an α helix. Leucine, which does not have a branched $C_\beta$ atom, is better accommodated in a helix than valine, in which the two methyl groups attached to the $C_\beta$ atom make closer contact with the backbone atoms. The ordering of the preferences of the other amino acids is harder to understand without considering the results of complicated energy calculations.

The list of helix preferences does not provide a hard and fast set of rules for the extent to which particular amino acids stabilize α helices because interactions between residues in the helix and those in other parts of the protein can also weaken or strengthen an α helix. They do, however, provide a useful guide in protein engineering efforts to alter the stability of proteins or to design artificial proteins.

Glycine and proline tend to destabilize α helices. The last atom of the proline sidechain is bonded to the backbone N atom, thus forming a ring structure, $C_\alpha$–$CH_2$–$CH_2$–$CH_2$–N (see Figure 1.29). This prevents the N atom from participating in hydrogen bonding and also provides some steric hindrance to the α-helical conformation. Proline fits very well in the first turn of an α helix, but it usually produces a significant bend if it is anywhere else in the helix, and so is a helix-breaking residue. Glycine is rarely found in the central regions of α helices. The backbone conformations of glycine are much less restricted than for other amino acids (see Figure 4.20), and it reduces the stability of the secondary structure into which it is incorporated.

There are also preferences for particular amino acids to be found at the ends of helices. These residues can help satisfy the hydrogen-bonding requirements of the -NH and -C=O groups at the ends of helices. The flexibility of glycine makes it a residue that is commonly found at the C-terminal ends of α helices. By taking on conformations that are energetically unfavorable for residues with larger sidechains, glycine can cap an α helix and induce a bend in the backbone. This is illustrated in **Figure 4.32**, which also shows how a serine sidechain can provide a cap for the N-terminal end of a helix.

The propensity of different amino acids to be found in β sheets is less clearly delineated than for α helices, because local interactions between neighboring sidechains play a stronger role in the stability of β-sheet structure.

## C. STRUCTURAL MOTIFS AND DOMAINS IN SOLUBLE PROTEINS

### 4.15 Secondary structure elements are connected to form simple motifs

Simple combinations of a few secondary structure elements with a specific geometric arrangement have been found to occur frequently in protein structures. These units are called **structural motifs**, and are formed by packing sidechains from adjacent α helices or β strands close to each other. Some of these motifs can be associated with a particular function, such as binding to DNA, metal ions, or small molecules; others have no specific biological function alone, but are part of larger structural and functional assemblies. There is a wide variety of structural motifs that have been observed in protein structures, and we describe just a few of the more common ones here.

Ala
Leu
Met
Arg
Lys
Gln
Glu
Ile
Trp
Ser
Tyr
Phe
Val
Thr
His
Cys
Asn
Asp
⋮
Gly
Pro

**Figure 4.31 A helix propensity scale.** The amino acids are listed in decreasing likelihood of being present in an α helix, from top to bottom (*blue*: helix stabilizing; *red*: helix destabilizing). This particular scale is based on experimentally observed frequencies of amino acids found in the middle portions of α helices exposed to solvent. (Adapted from C.N. Pace and J.M. Scholtz, *Biophys. J.* 75: 422–427, 1998.)

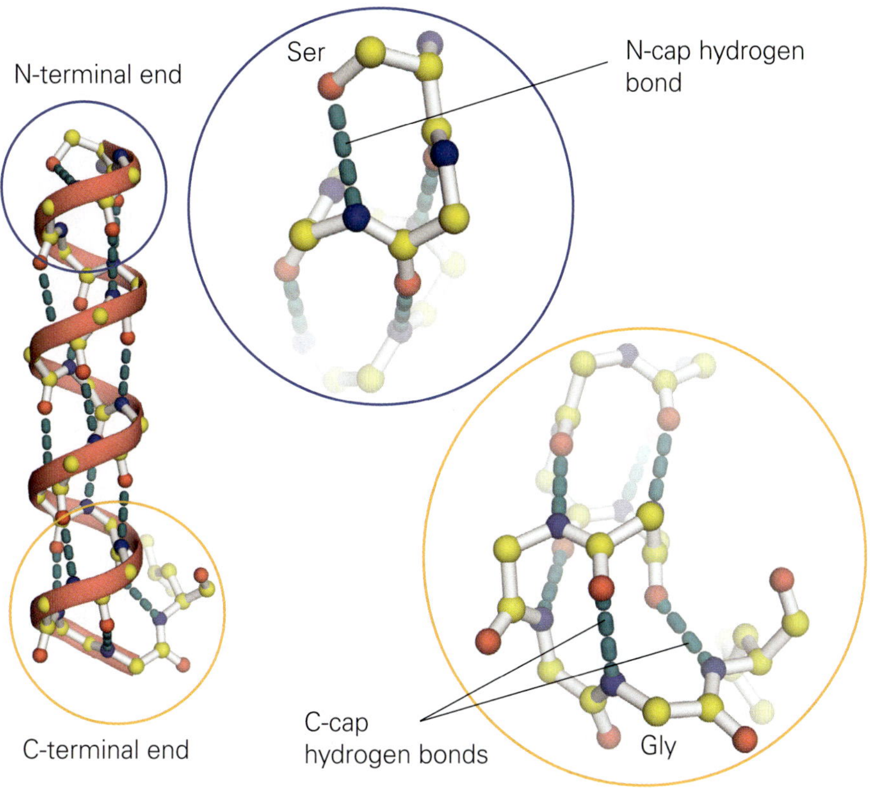


**Figure 4.32 Helix cap residues.** The structure of an α helix is shown here with the N- and C-terminal regions expanded. A serine sidechain at the N-terminus of the helix caps the helix by forming a hydrogen bond (denoted "N-cap") with a backbone nitrogen atom. A glycine residue at the other end of the helix causes a bend in the backbone, which allows the backbone nitrogen atoms of the glycine and the next residue to form hydrogen bonds (denoted "C-cap") with the backbone carbonyl groups of preceding residues. (PDB code: 1TIM.)

**Structural motif**

A three-dimensional arrangement of two or more secondary structural elements that is commonly found in many proteins is called a structural motif. Motifs are typically components of larger domains, and more than one motif may be found in a protein domain.

A particularly simple motif consists of two α helices joined by a short turn region, called a **helix-turn-helix motif** (**Figure 4.33**A). Helix-turn-helix motifs are commonly found in proteins that recognize specific sequences in DNA. One of the helices is inserted into the major groove of DNA, where sidechains emanating from the helix make sequence-specific contacts with the bases in the major groove (**Figure 4.34**).

A similar motif is the **helix-loop-helix motif** (see Figure 4.33B), in which the connection between the two helices is longer, as in the protein calmodulin. Calmodulin is a small protein that responds to changes in calcium levels in the cell by changing its structure. The helix-loop-helix motif appears four times in the structure of calmodulin, in each case forming a calcium-binding site. **Figure 4.35** shows this motif, which is called an **EF hand** because the helices of this motif resemble the thumb and forefinger of a right hand, and were labeled E and F in the structure for which the motif was first described.

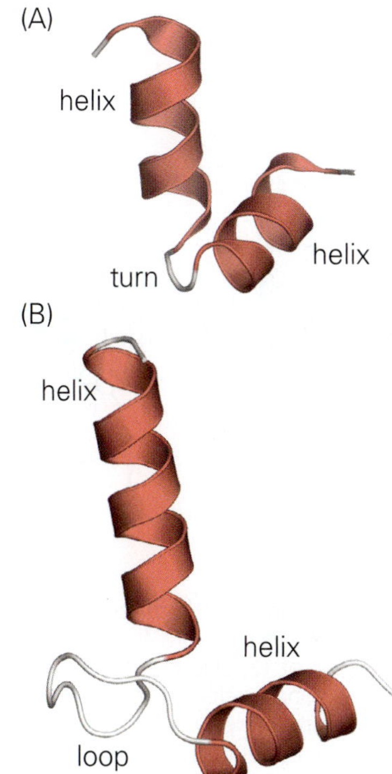


**Figure 4.33 Simple helical motifs.** (A) Two α helices that are connected by a short turn region constitute a helix-turn-helix motif. A DNA binding motif, common to many transcription factors, is illustrated (see Figure 4.34). (B) A helix-loop-helix motif involved in calcium binding is present in many proteins whose function is regulated by calcium (see Figure 4.35). (Adapted from C. Brändén and J. Tooze, Introduction to Protein Structure, 2nd ed. New York: Garland Science, 1999; PDB codes: A, 1LLI and B, 1CM1.)

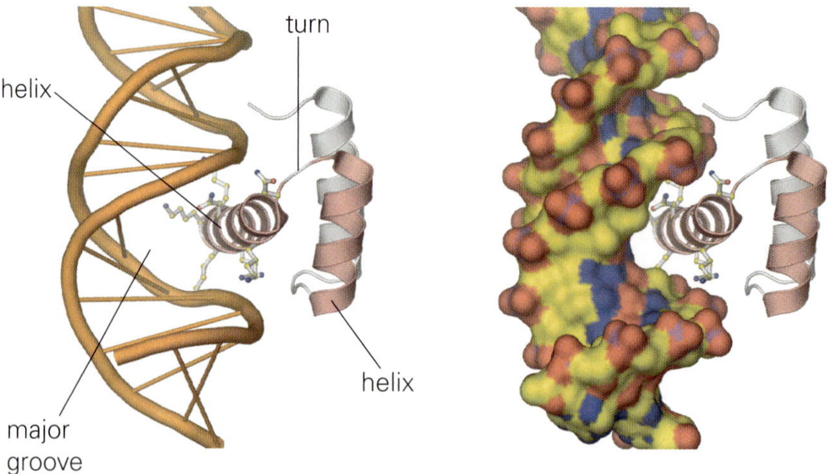


**Figure 4.34 Interaction between DNA and the helix-turn-helix motif of a DNA binding protein.** (PDB code: 1DU0.)

The loop region between the two α helices in an EF hand motif binds the calcium atom. Carboxyl sidechains from Asp and Glu, backbone carbonyl groups, and water molecules form the ligands to the metal ion (see Figure 4.35). Thus, both the specific backbone conformation of the loop and specific sidechains are required to provide the function of this calcium-binding motif. The helix-loop-helix motif provides a scaffold that holds the calcium ligands in the proper position to bind and release calcium. When calcium is bound to all four EF hands, calmodulin adopts a structure that enables it to bind to many target proteins, as illustrated in Figure 4.35. The ability to bind to these proteins is usually lost when $Ca^{2+}$ dissociates from calmodulin.

Antiparallel β strands are connected together by turns or hairpins (see Figure 4.26). For example, four adjacent antiparallel β strands are frequently arranged in a pattern similar to the repeating unit of an ornamental pattern, or fret, used in ancient Greece, which is now called a Greek key. In proteins, the motif is therefore called a **Greek key motif**, and **Figure 4.36** shows an example of such a motif. The Greek key motif is not associated with any specialized function, but it occurs frequently in protein structures.

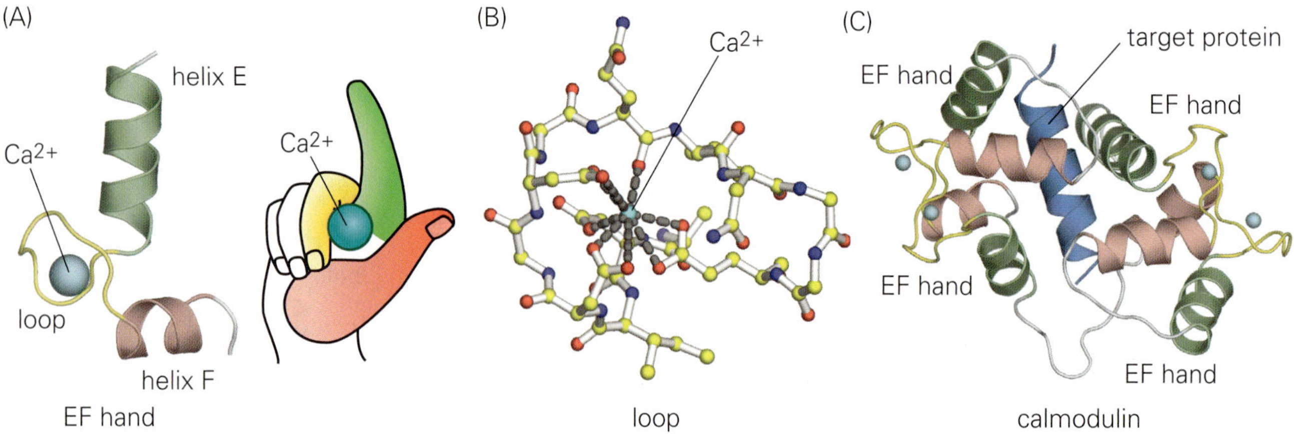


**Figure 4.35 Schematic diagrams of the calcium-binding EF hand motif.** (A) The calcium-binding motif is symbolized by a right hand. Helix E (*green*) runs from the tip to the base of the forefinger (also *green*). The flexed middle finger (*yellow*) corresponds to the loop region of 12 residues that binds calcium (*blue*). Helix F (*red*) runs to the end of the thumb. (B) The calcium ion is bound to one of the EF hand motifs in the protein calmodulin through several oxygen atoms (*red*) of the protein. (C) The structure of calmodulin is built up from four EF hand motifs. With four $Ca^{2+}$ ions bound to it, calmodulin adopts a structure that enables it to bind to helices in many of its target proteins. One such helix is shown in *blue*. (Adapted from C. Brändén and J. Tooze, Introduction to Protein Structure, 2nd ed. New York: Garland Science, 1999; PDB code: 1CM1.)

The hairpin motif is a simple and frequently used way to connect two antiparallel β strands, since the connected ends of the β strands are close together at the same edge of the β sheet. How are parallel β strands connected? If two adjacent strands are consecutive in the amino acid sequence, the two ends that must be joined are at opposite edges of the β sheet. The polypeptide chain must cross the β sheet from one edge to the other and connect the next β strand close to the point where the first β strand started. Such crossover connections are frequently made by α helices. The polypeptide chain must turn twice using loop regions, and the motif that is formed is thus a β strand followed by a loop, an α helix, another loop, and, finally, the second β strand.

This motif is called a **β-α-β motif** (**Figure 4.37**) and is found as part of almost every protein structure that has a parallel β sheet. For example, the enzyme triosephosphate isomerase is entirely built up by repeated combinations of this motif, where two successive motifs share one β strand (the structure of this enzyme is discussed in Section 4.22). The α helix in the β-α-β motif connects the carboxyl end of one β strand with the amino end of the next β strand (see Figure 4.37) and is usually oriented so that the helical axis is approximately parallel to the β strands. The α helix packs against the β strands and thus shields the hydrophobic residues of the β strands from the solvent.

**Figure 4.36 The Greek key motif.** (A) An antiparallel β sheet with the strands arranged in a Greek key motif. (B) A topology diagram for the Greek key motif. (PDB code: A, 1SNC.)

## 4.16 Amphipathic α helices can form dimeric structures called coiled coils

Despite its frequent occurrence in proteins, an isolated α helix is only marginally stable in solution (see Figure 4.9). α helices are stabilized in proteins by being packed together through hydrophobic sidechains. The simplest way to achieve such stabilization is to pack two α helices together. As early as 1953, Francis Crick reasoned that the sidechain interactions are optimized if the two α helices are not straight rods but are wound around each other in a supercoil, a so-called **coiled-coil** arrangement (**Figure 4.38**). Coiled coils are the basis for the structures of many fibrous proteins, and coiled coils in fibers can extend over many hundreds of amino acid residues to produce long, flexible structures that contribute to the strength and flexibility of the fibers (see Figure 4.38A). Much shorter coiled coils are used in a variety of proteins to promote formation of homo- and heterodimers (see Figure 4.38B).

The coiled-coil arrangement of α helices is called a **supercoil**, or **superhelix**. These terms reflect the fact that the α helix is itself a coil, and the two α helices coil around each other (see Figure 4.38). Coiled coils can either be parallel, as is the

**Coiled coil**

Two α helices that coil around each other are referred to as a coiled coil. These can either be parallel (when the two protein chains run in the same direction) or antiparallel (when the chains run in opposite directions). Residues at the *i*, *i* + 3, and *i* + 7 positions in a coiled coil face each other at the interface between the helices. These residues are usually hydrophobic.

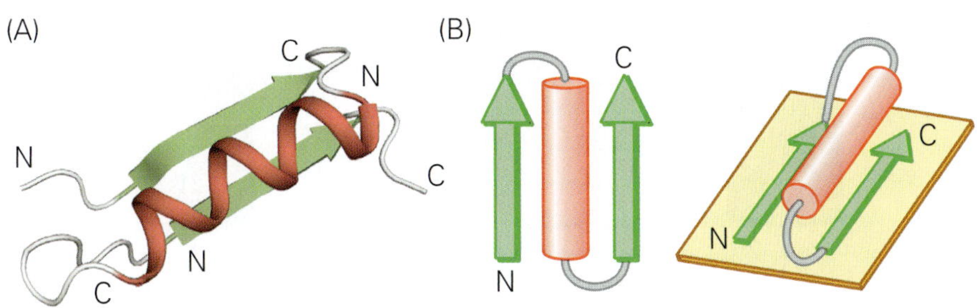


**Figure 4.37 β-α-β motifs.** (A) Two adjacent parallel β strands are usually connected by an α helix from the C-terminus of strand 1 to the N-terminus of strand 2. Most protein structures that contain parallel β sheets are built up from combinations of such β-α-β motifs. (B) Topological diagrams of the β-α-β motif. (Adapted from J.S. Richardson, *Adv. Protein Chem.* 34: 167–339, 1981.)

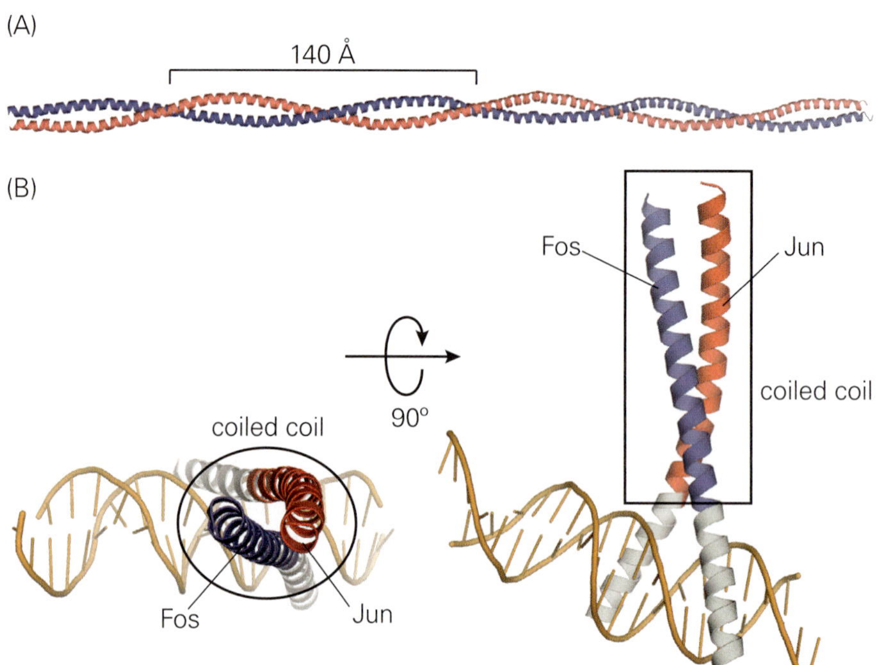


**Figure 4.38 Coiled-coil structures.** (A) A long dimeric coiled coil is shown, found in proteins that make up muscle fiber. The two helices coil around each other to form a superhelix, with a repeating distance of about 140 Å. (B) Proteins that form dimers often do so by using much shorter segments that form coiled coils. Shown here are two orthogonal views of the DNA-binding domains of two transcription factors known as Fos and Jun. These proteins heterodimerize by forming a coiled coil, and then splay apart to recognize the base pairs of DNA in the major groove. Note that the two proteins coil around each other in a left-handed sense. (PDB code: 1C1G and 1A02.)

case for the structure shown in **Figure 4.39**, with the chains running in the same direction, or antiparallel (see below). Parallel coiled coils are more common than antiparallel ones.

**Supercoil**

When the axis of a helix is coiled rather than straight, the resulting structure is called a supercoil or a superhelix.

For a coiled-coil structure to be stable, hydrophobic residues have to be brought into register at the interface between the helices. In order to achieve this, a coiled coil must have a left-handed superhelical structure. To see why this is so, consider the straight α helix illustrated in Figure 4.39A. If one residue is aligned with a line parallel to the helix axis, the residue two turns up from it will be to the *left* of this line. This is because the position exactly two turns up from a residue at position $i$ will be $3.6 \times 2 = 7.2$ residues along the helix backbone—that is, ahead of the position of the residue at the $i + 7$ position. If two α helices were to pack together, as shown in Figure 4.39B and C, the hydrophobic residues would not be in register unless the helices were supercoiled in a left-handed sense.

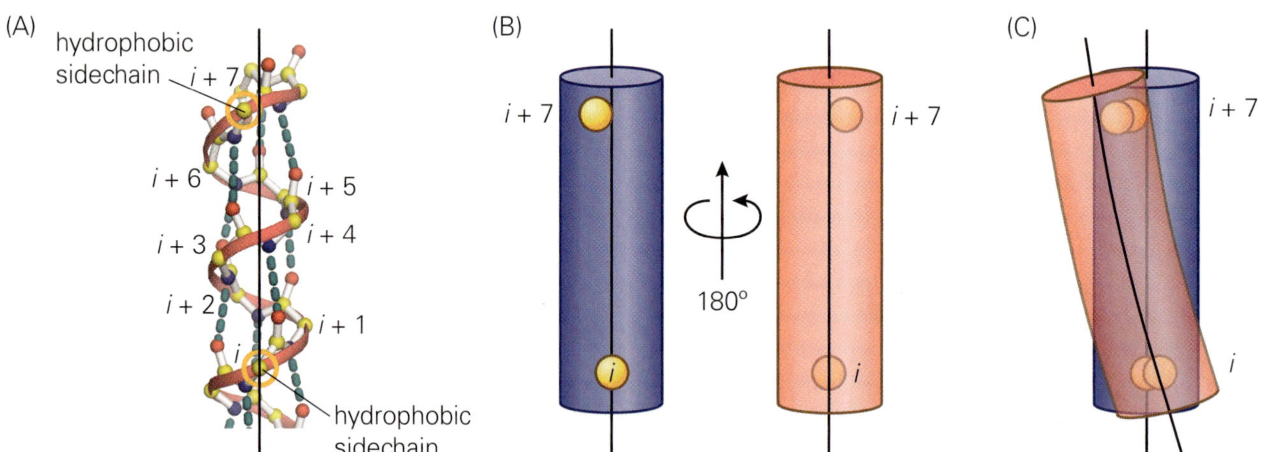


**Figure 4.39 Schematic diagram of the coiled-coil structure.** (A) An α helix is depicted with the $C_\alpha$ atoms of the residues at the $i$ and the $i + 7$ positions circled. A vertical line is drawn parallel to the helix axis and running through the residue at the $i^{th}$ position. Note that the residue at the $i + 7$ position is to the *left* of this line. (B) Schematic representation of two α helices, with one rotated by 180° with respect to the other. Hydrophobic residues at $i$ and $i + 7$ are shown in *yellow*. (C) Packing of the two helices depicted in (C). In order for the hydrophobic residues at $i$ and $i + 7$ to be aligned, the *red* helix is shown curling to the left. In real coiled coils, such as the ones shown in Figure 4.38, both helices are curled in a left-handed sense and so the extent of curling for each helix is less than that shown here for the *red* helix.

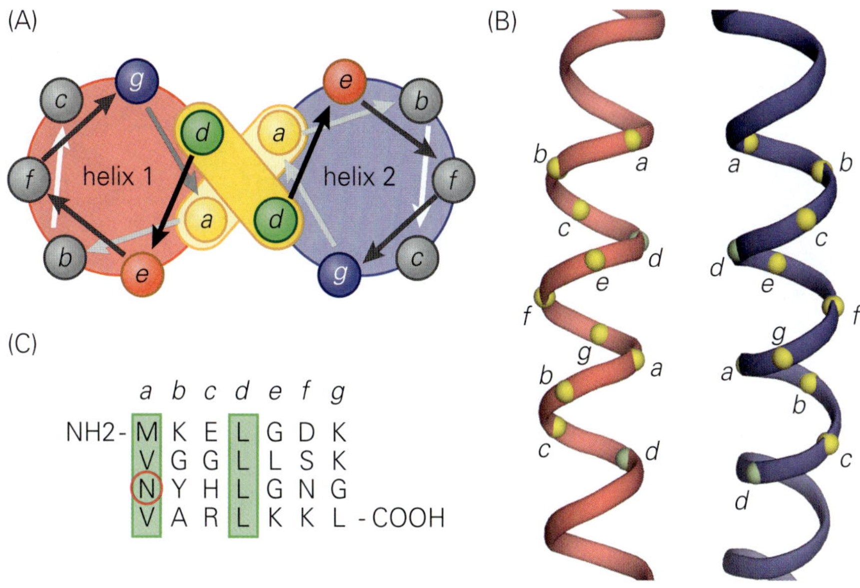


**Figure 4.40 Pattern of amino acids in a parallel coiled coil.** (A) Helical wheels corresponding to two helices in a parallel coiled coil. A heptad of residues is shown for each helix. The residues are shown in the order *d*, *e*, *f*, *g*, *a*, *b*, *c*, from the top to the bottom. The residues at the *d* and *a* positions interact with the corresponding residues in the other helix. (B) Diagram showing the backbone of the two polypeptide chains. (C) The amino acid sequence of the transcription factor GCN4, showing four heptad repeats. Within each heptad the residues at the *a* and *d* positions are hydrophobic, except for an asparagine at the *a* position in the third repeat (circled in *red*). The asparagine makes hydrogen bonds with the equivalent sidechain in the other helix (not shown). (Adapted from C. Brändén and J. Tooze, Introduction to Protein Structure, 2nd ed. New York: Garland Science, 1999; PDB code: 1KDD.)

## 4.17 Hydrophobic sidechains in coiled coils are repeated in a heptad pattern

A left-handed supercoil of two right-handed α helices reduces the effective number of residues per turn in each helix from 3.6 to 3.5, so the pattern of sidechain interactions between the helices repeats precisely every seven residues—that is, after two turns (**Figure 4.40**). As a consequence, the pattern of hydrophobic residues repeats every seven residues in the sequences of proteins that form coiled coils. This pattern is known as a **heptad repeat**. The amino acid residues within one such heptad repeat are usually labeled *a*–*g* (see Figure 4.40), and of these, the *a* and *d* residues are nearly always hydrophobic (usually a leucine, valine, or isoleucine).

When two α helices form a parallel coiled-coil structure, the sidechains of the *d* residues pack against each other every second turn of the α helices (see Figure 4.40). The hydrophobic interface between the α helices is completed by the *a* residues, which are also usually hydrophobic and also pack against each other (**Figure 4.41**). Polar residues that can form hydrogen bonds across the interface between the two helices are occasionally seen at the *a* and *d* positions (see, for example, the third heptad repeat in Figure 4.40C, which contains an asparagine sidechain at the *a* position). These polar residues determine the register of the two helices because the polar sidechains must be positioned across from each other. Residues *e* and *g*, which border the hydrophobic core (see Figure 4.40), are often charged residues. The sidechains of these residues can provide ionic interactions (salt bridges) between the α helices that also help define the register of the helices (**Figure 4.42**).

Ion pairing interactions are also important for determining whether a coiled coil is parallel or antiparallel. An antiparallel coiled coil is illustrated in **Figure 4.43**, in which an ion pair is formed between a lysine sidechain in one helix and a glutamate sidechain in the other. In antiparallel coiled coils, the residue at the *a* and *d* positions in one helix interact with the residues at the *d* and *a* positions in the other helix, respectively. This contrasts with the interactions seen in parallel coiled coils, in which the residues at the *a* position in one helix interact with the residues at the *a* position in the other helix, and likewise for the residues at the *d* positions. This means that charged residues that form ion pairs across the helices can only interact in one orientation and not the other.

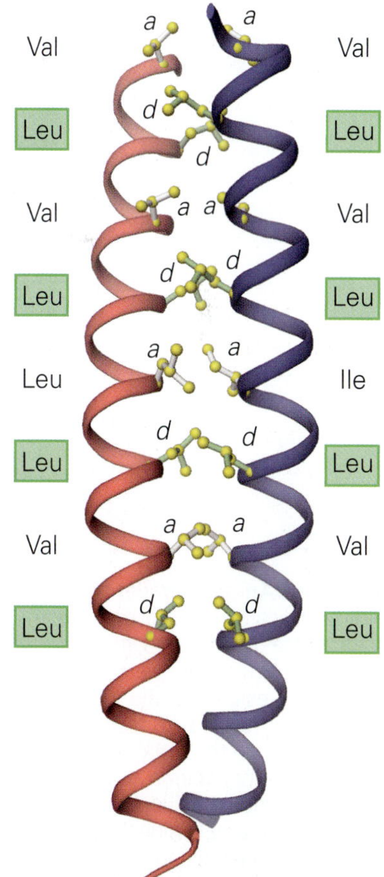


**Figure 4.41 The packing of hydrophobic sidechains in a coiled-coil structure.** The structure of a parallel coiled coil is shown, with the sidechains at the *a* and *d* positions indicated. (PDB code: 1KDD.)

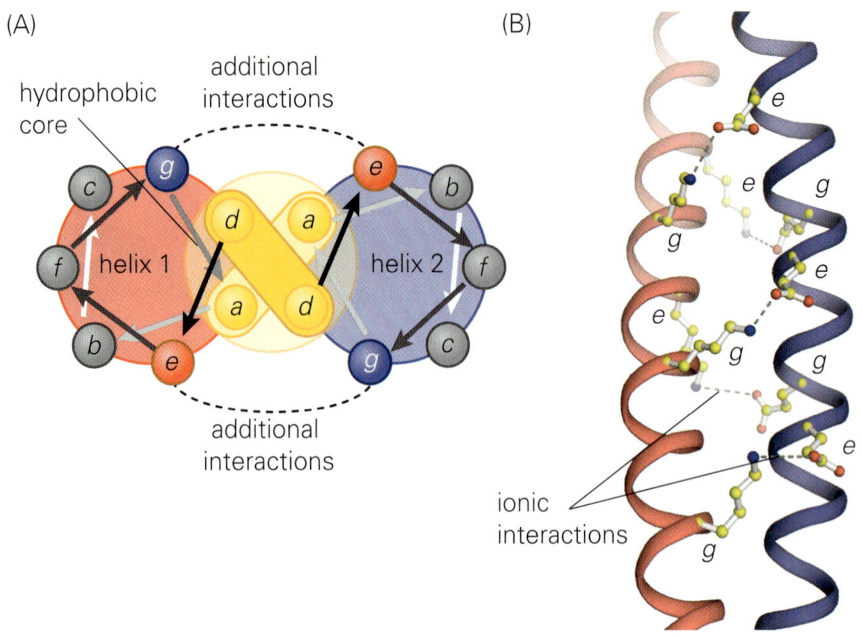


**Figure 4.42 Ionic interactions can stabilize coiled-coil structures.** (A) Helical wheel representation, as in Figure 4.40A. (B) View from the side of the coiled-coil structure. The residues labeled *e* and *g* in the heptad sequence can form salt bridges between the two α helices—the *e* residue in one helix with the *g* residue in the second and vice versa. (PDB code: 1KDD.)

**Heptad repeat**

In coiled-coil structures, every seventh residue occupies an identical position along the helix. This leads to a pattern of hydrophobic residues that repeats in groups of seven, known as the heptad repeat. Residues in each repeat are labeled *a* to *g*, and the residues at the *a* and *d* positions are usually hydrophobic and interact with the corresponding residues in the partner helix.

## 4.18 α helices that are integrated into complex protein structures do not usually form coiled coils

Coiled coils are typically formed by two α helices that are isolated from other structural elements. In more complex protein structures, the helices interact with many different structural elements and usually do not form coiled coils. The simplest and most frequent α-helical domain consists of four α helices arranged in a bundle with the helical axes almost parallel to each other. A schematic representation of the structure of a four-helix bundle is shown in **Figure 4.44**A. Each helix is relatively straight, and the sidechains of each helix in the four-helix bundle are arranged so that hydrophobic sidechains are buried between the helices and hydrophilic sidechains are on the outer surface of the bundle (Figure 4.44B). This arrangement creates a hydrophobic core in the middle of the bundle along

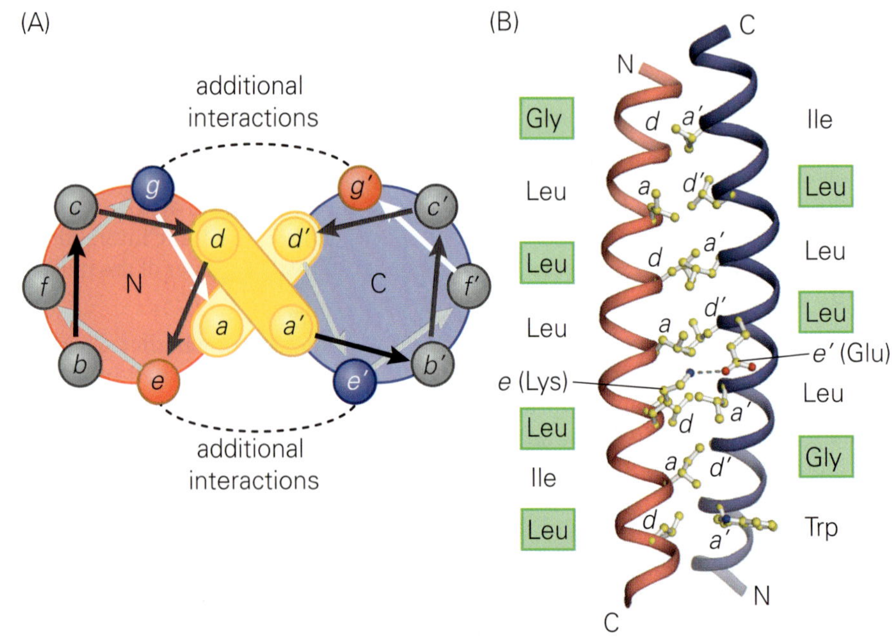


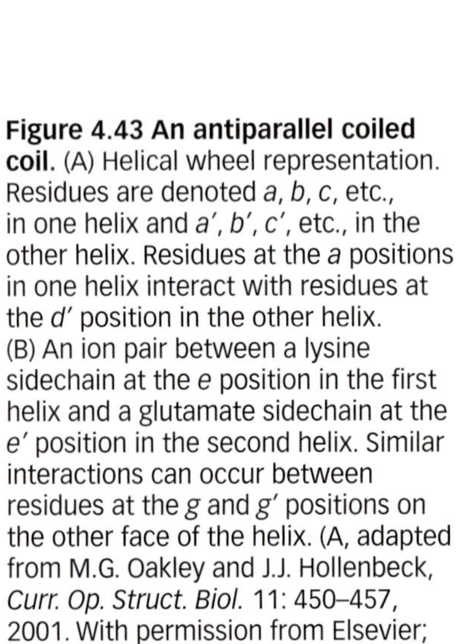

**Figure 4.43 An antiparallel coiled coil.** (A) Helical wheel representation. Residues are denoted *a*, *b*, *c*, etc., in one helix and *a′*, *b′*, *c′*, etc., in the other helix. Residues at the *a* positions in one helix interact with residues at the *d′* position in the other helix. (B) An ion pair between a lysine sidechain at the *e* position in the first helix and a glutamate sidechain at the *e′* position in the second helix. Similar interactions can occur between residues at the *g* and *g′* positions on the other face of the helix. (A, adapted from M.G. Oakley and J.J. Hollenbeck, *Curr. Op. Struct. Biol.* 11: 450–457, 2001. With permission from Elsevier; PDB code: 1A92.)

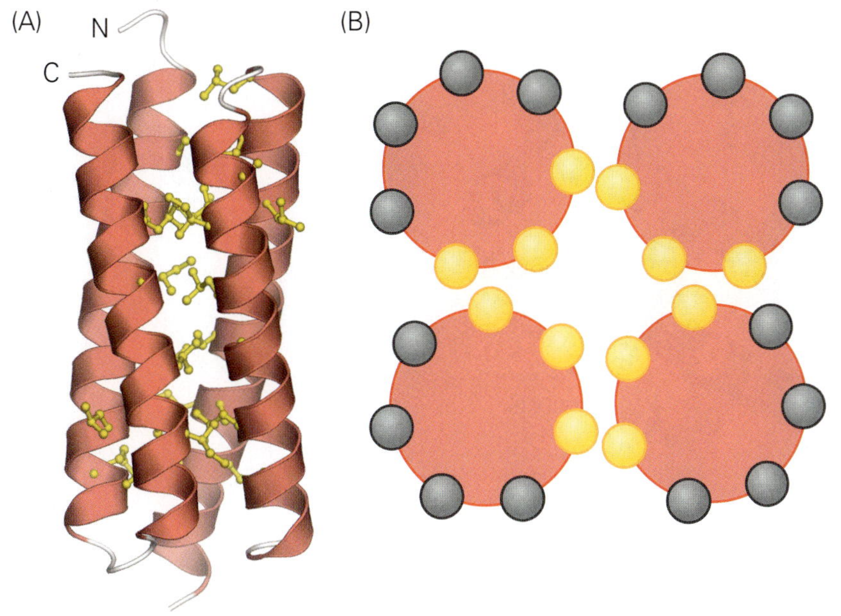


**Figure 4.44 A four-helix bundle protein.** (A) Structure of a four-helix bundle domain. Sidechains that form the hydrophobic core are shown in *yellow*. (B) Schematic view of a projection down the bundle axis. Large circles represent the α helices; small circles are sidechains. *Yellow* circles are the buried hydrophobic sidechains. (Adapted from C. Brändén and J. Tooze, Introduction to Protein Structure, 2nd ed. New York: Garland Science, 1999; PDB code: 2GTO.)

**Ridges into grooves packing of α helices**

Helical proteins that do not have coiled coils have relatively straight α helices. The sidechains of these α helices form ridges and grooves that are arranged at characteristic angles with respect to the helix axis. α helices pack against each other so that a ridge on one helix inserts into a groove on the other. This results in a limited set of crossing angles between two helices.

its length, where the sidechains are so closely packed that water is excluded. Many variations of the four-helix bundle are found in nature, including some that involve parallel helices.

Larger helical domains can have several α helices packed together in a complex pattern to form a globular domain. The globin fold, for example, found in hemoglobin, contains several α helices that form a box around the oxygen binding site (Figure 4.1). These helices pack against each other at various angles. An example of a larger helical protein, known as muramidase, is shown in **Figure 4.45**. This protein has 450 residues that form 27 α helices arranged in a two-layered ring.

## 4.19 The sidechains of α helices form ridges and grooves

When we compare the arrangements of the α helices in coiled-coil structures (Figure 4.41), a four-helix-bundle structure (Figure 4.44), the globin fold (Figure 4.1), and muramidase (Figure 4.45), the geometry of α-helical packing appears to be quite variable. Nevertheless, there are some constraints on the manner in which two helices interact with each other. α-helical structures that do not form coiled coils pack their α helices according to a **ridges into grooves** pattern. In the four-helix bundle, the α helices pack almost parallel, or antiparallel, to each other, with an angle of about 20° between the helical axes. In the globin fold, the angles between the helical axes are usually larger, in most cases around 50°. These packing motifs are dictated by the geometry of the surfaces of the α helices.

Because the sidechains of an α helix are arranged in a helical row along the surface of the helix, they form ridges separated by shallow furrows, or grooves, on the surface (**Figure 4.46**). α helices pack with the ridges on one helix packing into the grooves of the other and vice versa. The ridges and grooves are formed by amino acids that are usually three or four residues apart, as illustrated in **Figure 4.47**.

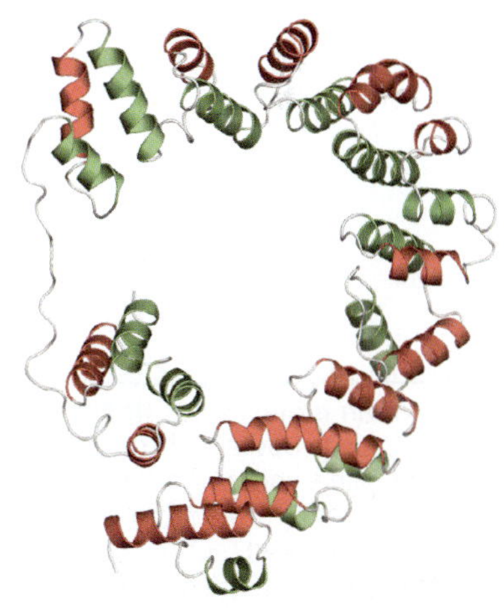

**Figure 4.45 Schematic diagram of the structure of one domain of a bacterial muramidase.** The structure, which comprises ~450 residues, is built up from 27 α helices arranged in a two-layered ring. The ring has a large central hole, like a doughnut, with a diameter of about 30 Å. (Adapted from C. Brändén and J. Tooze, Introduction to Protein Structure, 2nd ed. New York: Garland Science, 1999; PDB code: 1SLY.)

## 4.20 α helices pack against each other with a limited set of crossing angles

The most common way of packing α helices is by fitting the ridges formed by a row of residues separated in sequence by four in one helix into the same type of grooves in the other helix (see Figure 4.47). In this case, the ridges and grooves form an angle of about 25° to the helical axis. These ridges and grooves are

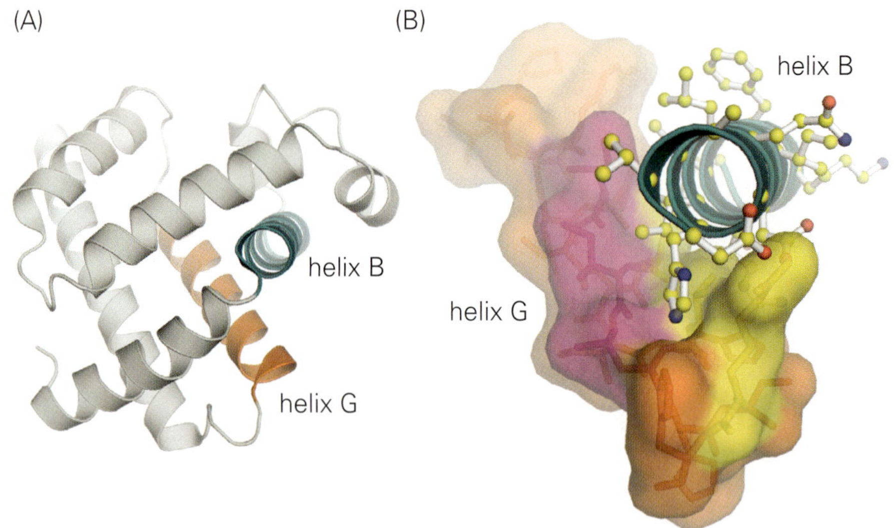


**Figure 4.46 Ridges into grooves packing of two helices.** (A) The structure of myoglobin, with two helices, B and G, colored *cyan* and *orange*, respectively. (B) The packing of helix B onto helix G. Helix G is shown in a similar orientation to that in (A), with two adjacent ridges colored *magenta* and *yellow*. The sidechains of helix B are inserted into the groove formed by these residues in helix G. (PDB code: 1JP6.)

depicted schematically for two helices in **Figure 4.48**. In order to pack the two helices shown in Figure 4.48A (red and blue) against each other, one of these (the blue one in Figure 4.48A) must be turned around 180° out of the plane of the paper and placed on top of the other (red). In the interface between the two α helices, the directions of the ridges and grooves are then on opposite sides of the vertical axis, as illustrated in Figure 4.48. The blue α helix must then be rotated by about 50° (25° + 25°) in order for the ridges of this helix to fit into the grooves of the other.

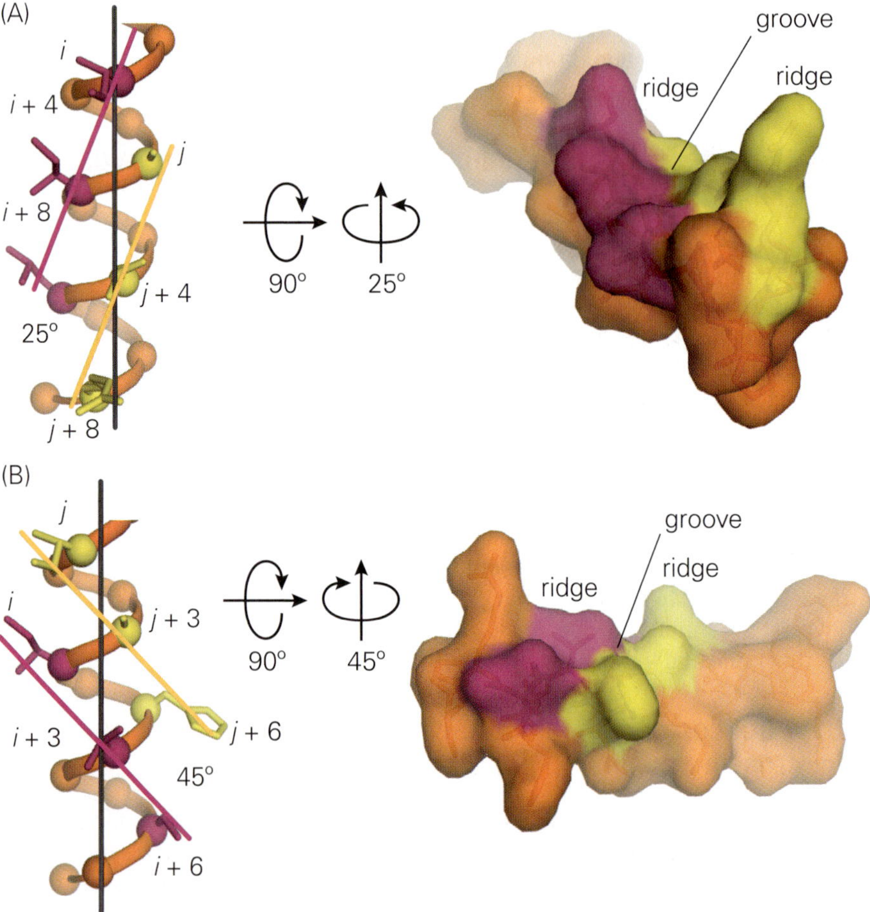


**Figure 4.47 The sidechains on the surface of an α helix form ridges separated by grooves.** (A) The left side of the panel shows α helix G in myoglobin with ridges formed by sidechains spaced four residues apart colored *magenta* and *yellow*. The ridges make an angle of ~+25° with the helix axis. The right side of the panel shows a rotated view with the surface of the α helix, similar to the view of this helix shown in Figure 4.46B. (B) As in (A), for ridges formed by sidechains spaced three residues apart. These ridges make an angle of ~−45° with the helix axis. The depth and shape of the ridges and grooves are determined by the nature of the sidechains on the ridges. (PDB code: 1JP6.)

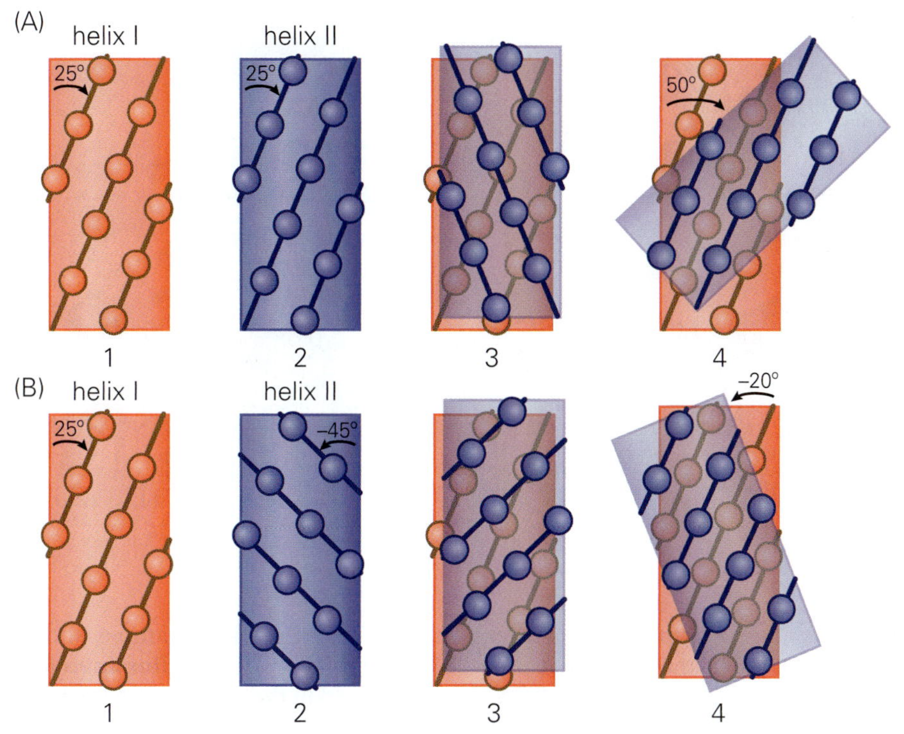


**Figure 4.48 Crossing angles for α helices.** By fitting the ridges of sidechains from one helix into the grooves between sidechains of the other helix and vice versa, α helices pack against each other. (A) Two α helices, I and II, with ridges from sidechains separated by four residues marked in *red* and *blue*, respectively. Panels 1 and 2 are the same view of the two α helices. In panel 3, the *blue* α helix has been rotated by 180° about a vertical axis and placed on top of the *red* helix. In panel 4, the ridges of one α helix are packed into the grooves of the other by rotating the *blue* helix by 50° about an axis coming out of the plane of the page. (B) In the *red* α helix, the ridges are formed by sidechains separated by four residues, and in the *blue* α helix, by three residues. The *blue* helix is flipped and rotated −20° in order to pack ridges into grooves. (Adapted from C. Brändén and J. Tooze, Introduction to Protein Structure, 2nd ed. New York: Garland Science, 1999, and also C. Chothia, M. Levitt, and D. Richardson, *Proc. Natl. Acad. Sci. USA* 74: 4130–4134, 1977.)

In another frequently occurring packing mode, the ridges formed by amino acids three residues apart fit into the grooves of amino acids four residues apart and vice versa. The direction of the first type of ridge forms an angle of about 45° to the helical axis, whereas the other type makes an angle of about 25° to the axis in the opposite direction (Figure 4.48B). In the interface, however, after one helix has been rotated 180°, these directions are on the same side of the helical axis. Thus, an inclination of about -20° (that is, 25° – 45°) between the two α helices will fit these ridges and grooves into each other.

These rules for fitting ridges into grooves are quite general: they explain the geometrical arrangements of adjacent α helices observed in many protein structures. The resulting crossing angles are determined by the backbone of the protein chain, and so do not change much when the sequence of the protein changes through evolution. This feature underlies the conservation of structure in helical proteins even when the sequence changes substantially, a point we shall discuss in more detail in Chapter 5.

## 4.21 Structures with alternating α helices and β strands are very common

Many protein domains contain alternating α helices and β strands, which form a central parallel or mixed β sheet surrounded by α helices, known as **α/β structures**. All the glycolytic enzymes, for example, are α/β structures, as are many other enzymes, as well as proteins that bind and transport metabolites. In α/β domains, binding crevices are formed by loop regions. These regions do not contribute to the structural stability of the fold, but participate instead in binding and catalytic action.

There are two major classes of α/β proteins and several other variant forms. In the first class, there is a core of twisted parallel β strands in a closed arrangement, like the staves of a barrel. The α helices that connect the parallel β strands are on the outside of this barrel (**Figure 4.49**A). This domain structure is often called the TIM barrel from the structure of the enzyme triosephosphate isomerase (usually abbreviated as TIM), where it was first observed. The second class contains an

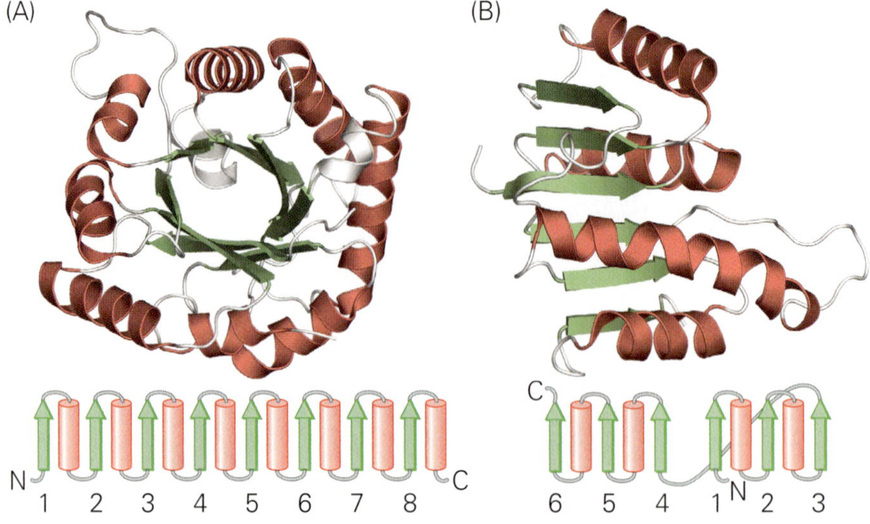


**Figure 4.49 α/β domains are found in many proteins.** Two classes of α/β domains are shown here. (A) A closed β barrel, exemplified by schematic and topological diagrams of the enzyme triosephosphate isomerase. (B) An open twisted sheet with helices on both sides. (Adapted from J.S. Richardson, *Adv. Protein Chem.* 34: 167–339, 1981; PDB codes: A, 1TIM and B, 1LDM.)

open β sheet surrounded by α helices on both sides. A typical example, shown in Figure 4.49B, is that of a nucleotide-binding domain called the **Rossman fold** after Michael Rossman, who first discovered this fold in the enzyme lactate dehydrogenase in 1970 (Rossman folds are discussed in more detail in Chapter 5).

The barrel and open-sheet structures described above are both built up from β-α-β motifs. To illustrate how they differ, let us consider two β-α-β motifs: $\beta_1$-$\alpha_{1,2}$-$\beta_2$ and $\beta_3$-$\alpha_{3,4}$-$\beta_4$ that are linked together by a helix, $\alpha_{2,3}$. There are two fundamentally different ways these two motifs can be connected into a β sheet of four parallel strands, as shown in **Figure 4.50**. Strand $\beta_3$ can be aligned adjacent either to strand $\beta_2$, giving the strand order 1 2 3 4, or to strand $\beta_1$, giving the strand order 4 3 1 2.

In the first case, with strand order 1 2 3 4, the two β-α-β motifs are joined with the same orientation. It turns out that virtually all β-α-β motifs have right-handed

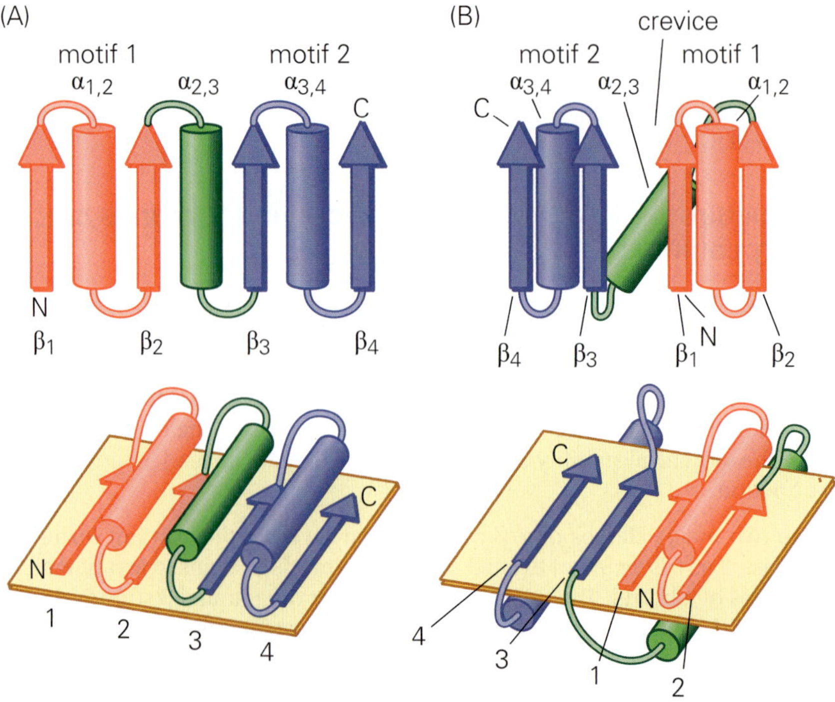


**Figure 4.50 α/β motifs.** Two such motifs can be joined into a four-stranded parallel β sheet in two different ways. They can be aligned with the α helices either on the same side of the β sheet (A) or on opposite sides (B). In case (A), the strand order is 1 2 3 4. The motifs are aligned in this way in barrel structures (see Figure 4.49). In (B), the first β strands of both motifs are adjacent, giving the strand order 4 3 1 2. Open twisted sheets contain at least one motif alignment of this kind. In both cases the motifs are joined by an α helix (*green*). (Adapted from C. Brändén and J. Tooze, Introduction to Protein Structure, 2nd ed. New York: Garland Science, 1999, and also J.S. Richardson, *Adv. Protein Chem.* 34: 167–339, 1981.)

topology (**Figure 4.51**). This is an empirical rule that almost always applies, although no convincing explanation has been found for why it is highly preferred. Because the β-α-β unit is almost always a right-handed structure, all three α helices (one from each motif and the joining helix) are on the same side, above the β sheet (Figure 4.50A). In barrel structures, the β-α-β motifs are linked in this way and consist of consecutive β-α-β units, all in the same orientation.

In the second case, with strand order 4 3 1 2, the second motif must be turned around in order to align β strands 1 and 3. As a result of the right-handed structure of the β-α-β motif, the α helix connecting strands 3 and 4 is on the other side of the β sheet (Figure 4.50B). In open twisted β-sheet structures, there are always one or more such alignments and, therefore, there are α helices on both sides of the β sheet.

**Figure 4.51 α/β structures are usually right-handed.** If the palm of a right hand is aligned along the direction of the β strands, the α helices run in the direction of the curled fingers.

## 4.22 α/β barrels occur in many different enzymes

In α/β structures where the strand order is 1 2 3 4, all connections are on the same side of the β sheet. An open twisted β sheet of this sort with four or more parallel β strands would leave one side of the parallel β sheet exposed to the solvent and the other side shielded by the α helices. Such a domain structure is rarely observed, except as part of more complex structures where loop regions, extra α helices, or additional β sheets cover the exposed side of the β sheet.

More commonly, a closed barrel of twisted β strands is formed with all the connecting α helices on the outside of the barrel, as shown in **Figure 4.52**. More than four β strands are needed to provide enough staves to form a closed barrel, and almost all the closed α/β barrels observed to date have eight parallel β strands. These are arranged such that β strand 8 is adjacent and hydrogen bonded to β strand 1. The interior of the barrel is filled with hydrophobic sidechains that form the core of the protein (see Figure 4.52).

In all these α/β-barrel domains, the active site is in a very similar position. It is situated in the bottom of a funnel-shaped pocket created by the eight loops that

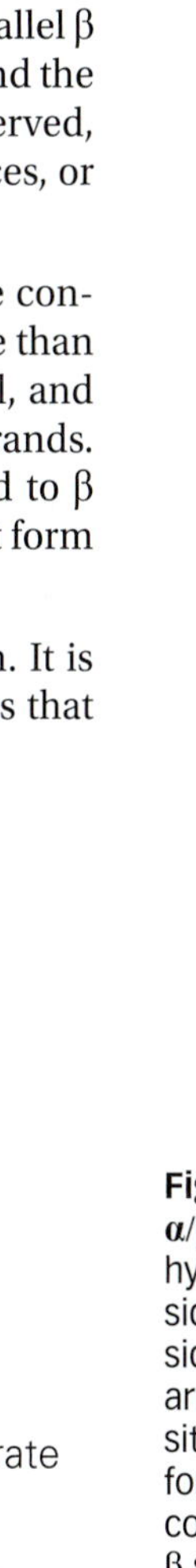

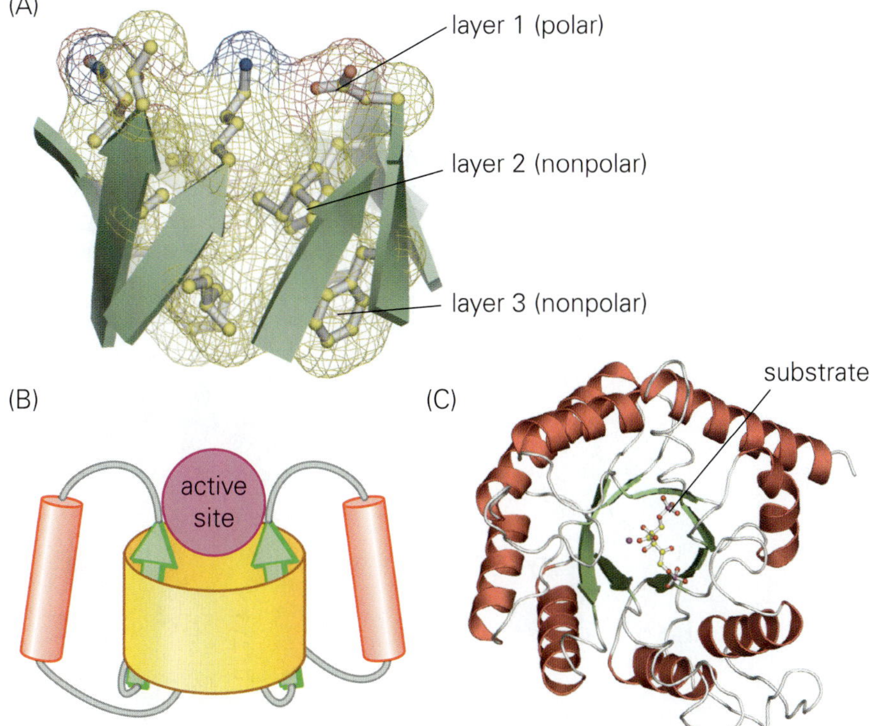


**Figure 4.52 Architecture of an α/β-barrel.** (A) A tightly packed hydrophobic core is formed by sidechains from the β strands. The sidechains within the barrel are arranged in three layers. (B) The active site in all α/β barrels is in a pocket formed by the loop regions that connect the carboxyl ends of the β strands with the adjacent α helices. (C) A view from the top of the barrel of the active site of the enzyme RuBisCo (ribulose bisphosphate carboxylase), which is involved in $CO_2$ fixation in plants. A substrate analog (shown in stick representation) binds across the barrel. (Adapted from C. Brändén and J. Tooze, Introduction to Protein Structure, 2nd ed. New York: Garland Science, 1999; PDB codes: A, 1GOX and C, 8RUC.)

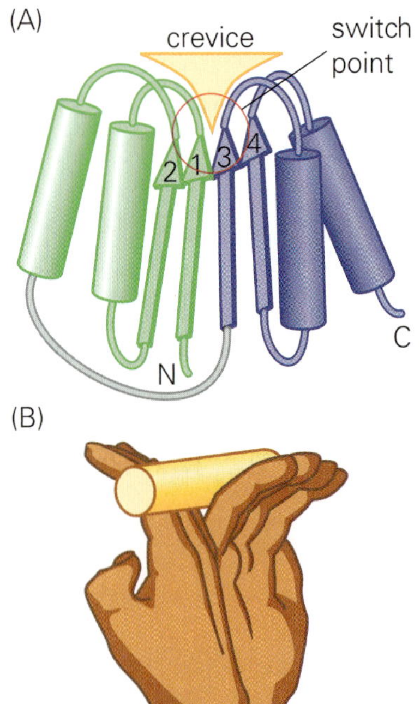


**Figure 4.53 α/β structures with helices on both sides of the β sheet.** (A) The active site in open twisted α/β domains is in a crevice outside the carboxyl ends of the β strands. The crevice is situated at a switch point in the structure, with the adjacent loops going in. This is illustrated by the curled fingers of two right hands (B), where the top halves of the fingers represent loop regions and the bottom halves represent the β strands. The rod represents a bound molecule in the binding crevice. (Adapted from C. Brändén and J. Tooze, Introduction to Protein Structure, 2nd ed. New York: Garland Science, 1999.)

connect the carboxyl end of the β strands with the amino end of the α helices (see Figure 4.52). Residues that participate in binding and catalysis are in these loop regions.

## 4.23 α/β open-sheet structures contain α helices on both sides of the β sheet

In the other major class of α/β structures, there are α helices on both sides of the β sheet (see Figure 4.49B). This has at least three important consequences. First, a closed barrel cannot be formed unless the β strands completely enclose the α helices on one side of the β sheet. Such structures have never been found and are very unlikely to occur, because a large number of β strands would be required to enclose even a single α helix. Instead, the β strands are arranged into an open twisted β sheet such as that shown in Figure 4.49B.

There are always two adjacent β strands (for example, $\beta_1$ and $\beta_3$ in Figure 4.50B) in the interior of the β sheet whose connections to the flanking β strands are on opposite sides of the β sheet. The loop from one of these two central β strands goes above the β sheet, whereas the other loop goes below. This switch point creates a crevice outside the edge of the β sheet between these two loops (**Figure 4.53**). Almost all ligand binding sites in this class of α/β proteins are located in crevices of this type at the carboxyl edge of the β sheet (we define the carboxyl edge of the sheet as the edge that is formed by the carboxyl ends of the parallel β strands in the sheet).

Finally, in open-sheet structures, the α helices are packed against both sides of the β sheet. Each β strand thus contributes hydrophobic sidechains to pack against α helices in two similar hydrophobic core regions, one on each side of the β sheet.

## 4.24 Proteins with antiparallel β sheets often form structures called β barrels

Antiparallel β structures comprise a large and diverse group of protein domains. The group includes enzymes, transport proteins, antibodies, cell surface proteins, and virus coat proteins. The cores of these domains are built up by β strands that can vary in number from four or five to over 10. The β strands are arranged in a predominantly antiparallel fashion and usually in such a way that they form two β sheets that are joined together and packed against each other.

The β sheets have the usual right-handed twist and, when two such twisted β sheets are packed together, they form a barrel-like structure, which is illustrated for the enzyme superoxide dismutase in **Figure 4.54**. Superoxide dismutase has a β structure comprising eight antiparallel β strands. In addition, the enzyme has two metal atoms, Cu and Zn, that help catalyze the conversion of a superoxide

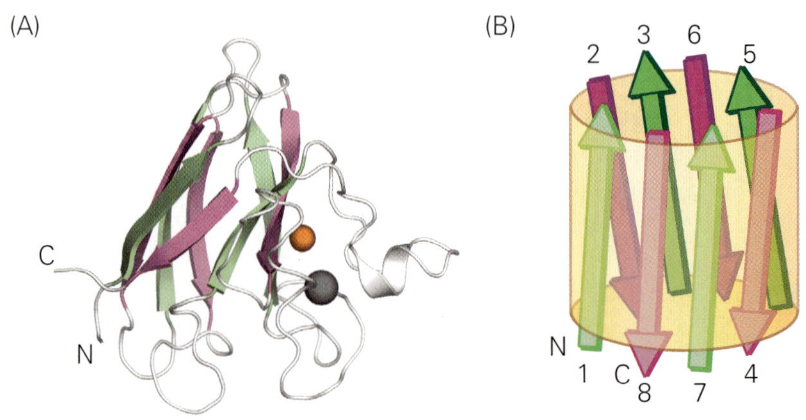


**Figure 4.54 The enzyme superoxide dismutase.** (A) This enzyme has a β structure comprising eight antiparallel β strands. In addition, the enzyme has two metal atoms, Cu and Zn (*orange* and *gray* spheres), that participate in the catalytic action: the conversion of a superoxide radical to hydrogen peroxide and oxygen. The eight β strands are arranged around the surface of a barrel, which is viewed perpendicular to the barrel axis in (B). (Adapted from C. Brändén and J. Tooze, Introduction to Protein Structure, 2nd ed. New York: Garland Science, 1999; PDB code: 2SOD.)

radical to hydrogen peroxide and oxygen (superoxide dismutase is discussed further in Chapter 16). Antiparallel β structures, such as that seen in superoxide dismutase, have a core of hydrophobic sidechains inside the barrel provided by residues in the β strands.

## 4.25 Up-and-down β barrels have a simple topology

The β structure with the simplest topology is obtained if each successive β strand is added adjacent to the previous strand until the last strand is joined by hydrogen bonds to the first strand and the barrel is closed (**Figure 4.55**). These are called up-and-down β sheets or barrels. The arrangement of β strands is similar to that in the α/β barrel structures we have just described, except that here the strands are antiparallel and all the connections are hairpins.

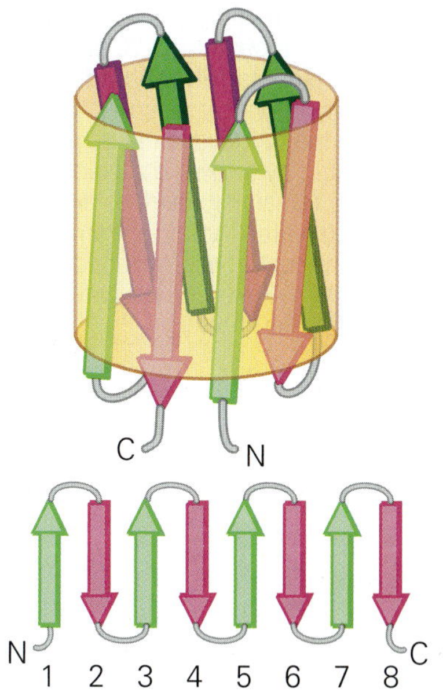


**Figure 4.55 An up-and-down β barrel.** The eight β strands are all antiparallel to each other and are connected by hairpin loops. (Adapted from C. Brändén and J. Tooze, Introduction to Protein Structure, 2nd ed. New York: Garland Science, 1999.)

An example of an up-and-down β barrel is the retinol-binding protein, which is a single polypeptide chain of ~180 amino acid residues (**Figure 4.56**). This protein is responsible for transporting the lipid alcohol vitamin A (retinol) from its storage site in the liver to the various vitamin-A-dependent tissues.

The hydrophobic retinol molecule is packed against hydrophobic sidechains from the β strands in the barrel's core (Figure 4.56B). Every other sidechain in the core of the barrel is exposed to water. As a consequence, hydrophobic residues alternate with polar or charged hydrophilic residues in the amino acid sequences of the β strands. This structure provides a nice illustration of amphipathic β strands (see Section 4.13), as illustrated in **Figure 4.57** for strands 2, 3, and 4 of retinol-binding protein.

## 4.26 Up-and-down β sheets can form repetitive structures

Another example of up-and-down β sheets is found in proteins that do not form a simple barrel, but instead form several small sheets, each with a small number of β strands, which are arranged like the blades of a propeller. Loop regions between the β strands form the active site in the middle of one side of the propeller. In related structures, there are different numbers of the same motif arranged like propellers with different numbers of blades.

The β-propeller architecture is illustrated by the structure of a portion of an enzyme known as neuraminidase. This portion consists of a single domain built up from six closely packed, similarly folded motifs. The motif is a simple up-and-down antiparallel β sheet of four strands (**Figure 4.58**). The strands have a rather

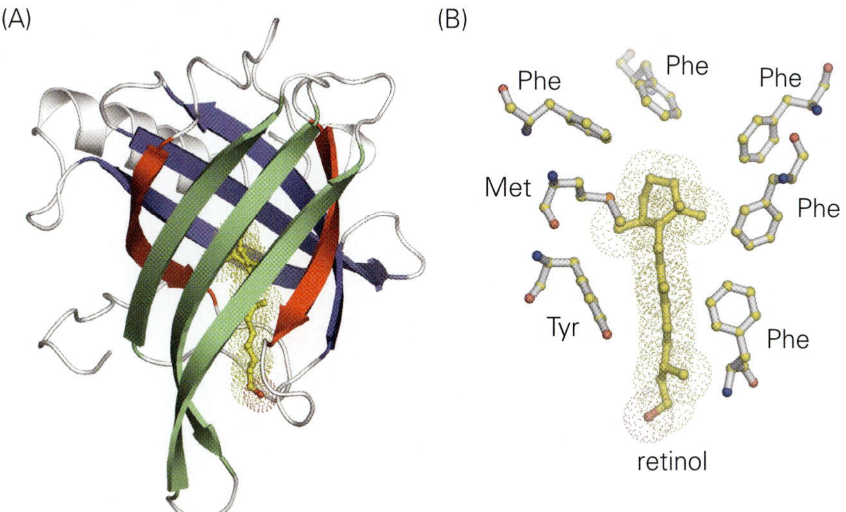


**Figure 4.56 Structure of retinol-binding protein, an up-and-down β barrel.** (A) The structure can be regarded as two sheets (*green* and *blue*) packed against each other. Some of the twisted β strands (*red*) participate in both β sheets. A retinol molecule (vitamin A, *yellow*), is bound inside the barrel. (B) The binding site for retinol inside the RBP barrel is lined with hydrophobic residues. (Adapted from C. Brändén and J. Tooze, Introduction to Protein Structure, 2nd ed. New York: Garland Science, 1999; PDB code: 1BRP).

| strand number | residue number | amino acid sequence | | | | | | | |
|---|---|---|---|---|---|---|---|---|---|
| | | ↓ | | ↓ | | ↓ | | ↓ | |
| 2 | 41–48 | Ile | Val | Ala | Glu | Phe | Ser | Val | Asp |
| 3 | 53–60 | Met | Ser | Ala | Thr | Ala | Lys | Gly | Arg |
| 4 | 71–78 | Ala | Asp | Met | Val | Gly | Thr | Phe | Thr |

**Figure 4.57 The amino acid sequence of β strands 2, 3, and 4 in human plasma retinol-binding protein.** Hydrophobic residues that point into the barrel (see Figure 4.56) are indicated by arrows and colored *green*. The remaining residues are exposed to water. (Adapted from C. Brändén and J. Tooze, Introduction to Protein Structure, 2nd ed. New York: Garland Science, 1999.)

large twist such that the directions of the first and the fourth strands differ by 90°. This allows the six up-and-down β motifs to be arranged with an approximate six-fold symmetry around an axis through the center of the subunit (see Figure 4.58A). These six β sheets are arranged like six blades of a propeller.

## 4.27 Greek key motifs occur frequently in antiparallel β structures

The Greek key motif, discussed in Section 4.15 (see Figure 4.36), provides a simple way to connect antiparallel β strands that are on opposite sides of a barrel structure. Assume that we have eight antiparallel β strands arranged in a barrel structure, as illustrated in **Figure 4.59**. We decide that we want to connect strand number $n$ to an antiparallel strand at the same end of the barrel. We do not want to

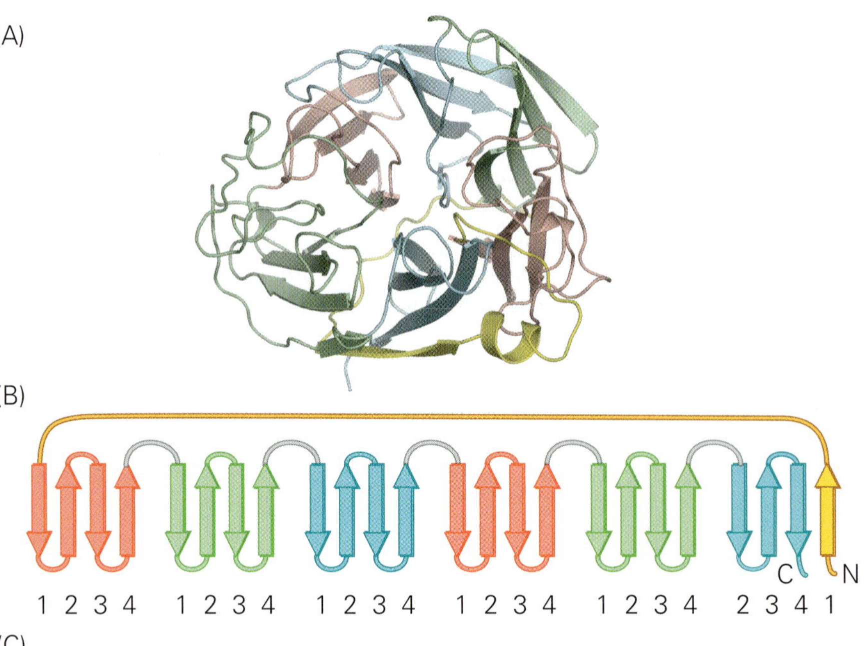


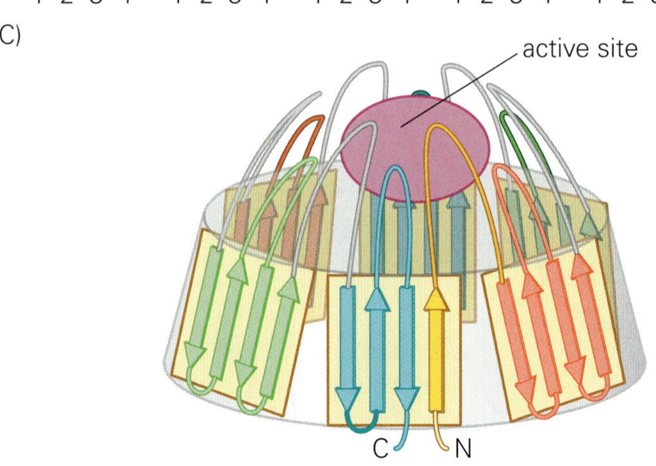


**Figure 4.58 A protein with β propeller architecture.** The neuraminidase domain of influenza virus is built up from six similar, consecutive motifs of four up-and-down antiparallel β strands. (B) Topological diagram, showing the six blades of the propeller. (C) The loop regions that connect the propellers (*gray* in B) in combination with the loops that connect strands 2 and 3 within each propeller form a wide funnel-shaped active site pocket. (Adapted from C. Brändén and J. Tooze, Introduction to Protein Structure, 2nd ed. New York: Garland Science, 1999; PDB code: 1NN2.)

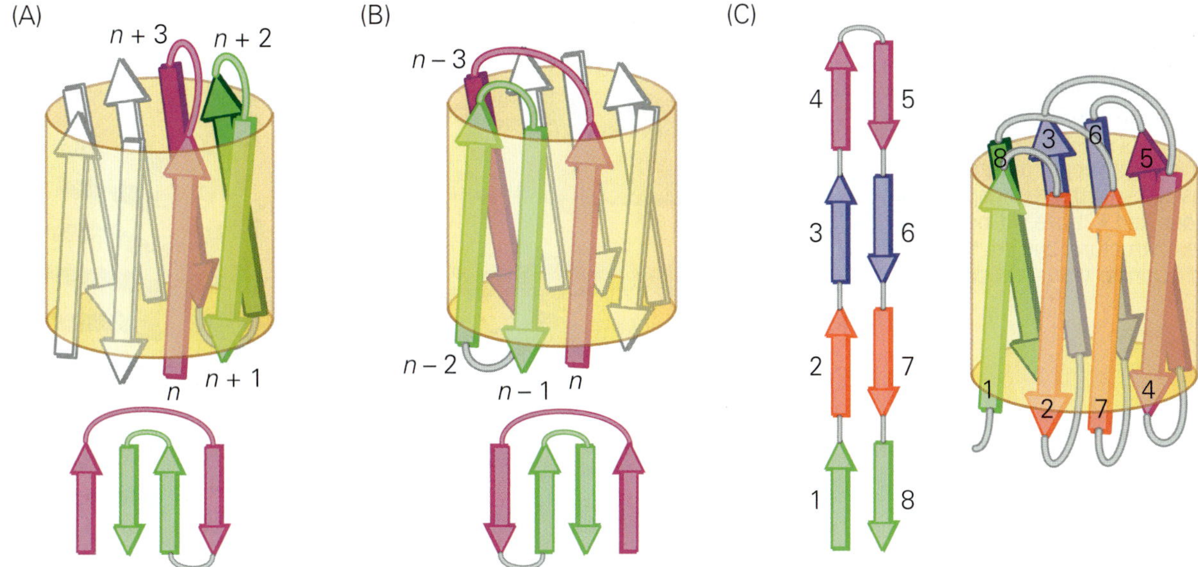


**Figure 4.59 Idealized diagrams of the Greek key motif.** This motif is formed when one of the connections of four antiparallel β strands is not a hairpin connection. The motif occurs when strand number *n* is connected to strand *n* + 3 (A) or *n* – 3 (B), instead of *n* + 1 or *n* – 1 in an eight-stranded antiparallel β sheet or barrel. The two different possible connections give two different hands of the Greek key motif. (C) A simple illustration of the way eight β strands are arranged in a jelly-roll motif. Imagine that the eight β strands are drawn as arrows along two edges of a strip of paper. The strands are arranged such that strand 1 is opposite strand 8, etc. The β strands are separated by loop regions. Now, imagine that the strip of paper is wrapped around a barrel such that the β strands follow the surface of the barrel and the loop regions (*gray*) provide the connections at both ends of the barrel. The β strands are now arranged in a jelly-roll motif. (Adapted from C. Brändén and J. Tooze, Introduction to Protein Structure, 2nd ed. New York: Garland Science, 1999.)

connect it to strand number $n + 1$, as in the up-and-down barrels just described, nor do we want to connect it to strand number $n - 1$, which is equivalent to turning the up-and-down barrel in Figure 4.59 upside down. What alternatives remain?

Figure 4.59 shows that there are only two alternatives. We can connect it either to strand number $n + 3$ or to $n - 3$. Both cases require only short loop regions that traverse the end of the barrel. How do we now continue the connections? The simplest way to connect the strands that were skipped over is to join them by up-and-down connections, as illustrated in Figure 4.59.

We have now connected four adjacent strands of the barrel in a simple and logical fashion, requiring only short loop regions. The result is the Greek key motif described previously (see Figure 4.36), which is found in the large majority of antiparallel β structures. The two cases represent the two possible different hands, but the hand that corresponds to the case where β strand $n$ is linked to β strand $n + 3$, as in Figure 4.59A, is the one commonly found in nature.

The remaining four strands of the barrel can be joined either by up-and-down connections before and after the motif or by another Greek key motif. An example of a more complicated arrangement of Greek key motifs, known as the "jelly-roll" motif, is shown in Figure 4.59C.

## 4.28 Certain structural motifs can be repeated almost endlessly to form elongated structures

The protein structures we have discussed so far in this part of the chapter have compact domains that surround discrete hydrophobic cores. One exception is the coiled-coil structure, in which the heptad pattern of hydrophobic residues can be repeated almost endlessly, leading to very long structures (see Figure 4.38). There are several other structural motifs that are not as symmetrical as the coiled coil, but nevertheless can be repeated to yield elongated structures. These structures

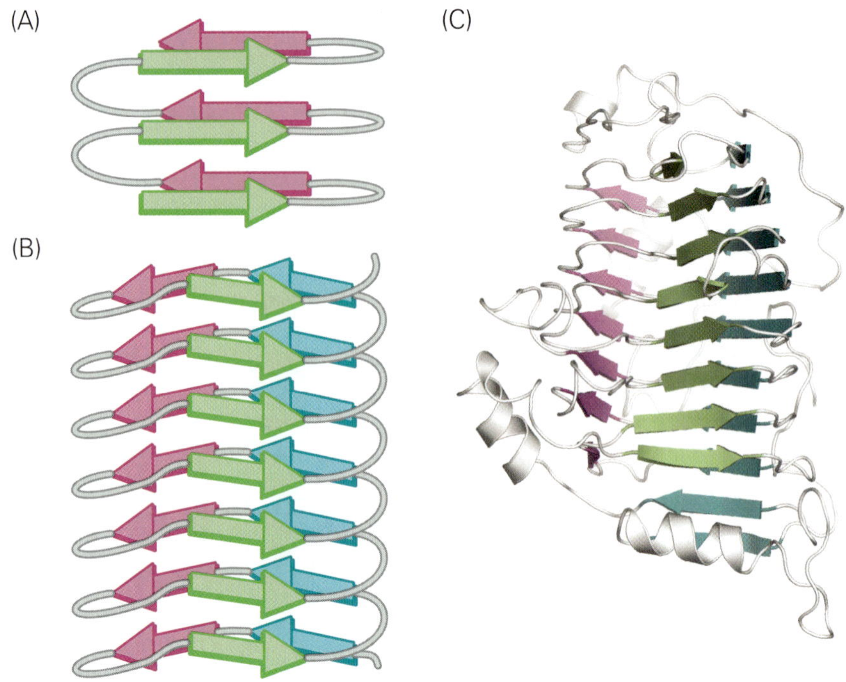


**Figure 4.60 Repetitive β helix structures.** (A) Three coils of a two-sheet helix. (B) A three-sheet β helix. (C) A three-sheet β helix in a protein structure. Each β sheet is composed of 7–10 parallel β strands with an average length of three to five residues in each strand. (Adapted from C. Brändén and J. Tooze, Introduction to Protein Structure, 2nd ed. New York: Garland Science, 1999; PDB code: 2PEC.)

are often part of larger assemblies, where they serve as protein docking sites or scaffolds for organizing assemblies of proteins.

One such structural motif is the β helix, which has the polypeptide chain coiled into a wide helix, formed by β strands separated by loop regions. In the simplest form, the two-sheet β helix, each turn of the helix comprises two β strands and two loop regions (**Figure 4.60**A). This structural unit is repeated several times to form a right-handed coiled structure, which comprises two adjacent parallel β sheets with a hydrophobic core in between. In a more complex β helix, each turn of the helix contains three short strands, each with three to five residues, connected by loop regions (Figure 4.60B and C). The β helix, therefore, comprises three parallel β sheets roughly arranged as the three sides of a prism.

There are also several α helical repeating motifs. A structure formed by one such motif, known as the armadillo motif, is shown in **Figure 4.61**. Each armadillo

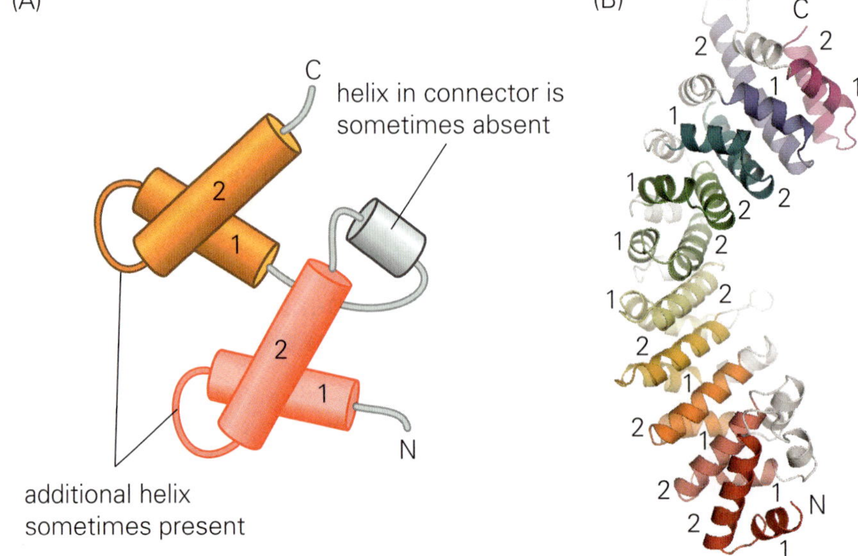


**Figure 4.61 A repeating helical structure known as the armadillo motif.** (A) A schematic diagram of two armadillo motifs is shown. Each motif consists of two α helices. (B) The arrangement of 10 armadillo motifs in a protein known as karyopherin, a protein that transports other proteins into the nucleus. (PDB code: 1BK5.)

motif consists of two α helices, connected by a loop or another α helix. Several armadillo motifs can be strung together to form a large protein, as shown in Figure 4.61B for a protein containing 10 armadillo motifs.

## 4.29 Catalytic sites are usually located within core elements of protein folds

A protein domain usually has a catalytic or binding function that is preserved during evolution. This fundamental function is modulated as the sequence of the protein evolves, resulting in changes in the particular small molecules or proteins that are recognized by the domain. This change in specificity comes about through changes in the residues that surround the functional or **active site**, and such changes are accumulated without the protein losing its ability to fold into a stable three-dimensional structure. In the case of enzymes, such changes in sequence can also result in fine tuning of the speed with which the domain catalyzes a chemical reaction.

This ability of protein domains to acquire new specificity and functional characteristics is a key aspect of variation and natural selection (see Chapter 5). Protein domains usually have their active sites located within a core region of the three-dimensional fold, with radical evolutionary changes occurring in amino acids of the periphery. These changes are able to alter the specificity of the protein for its substrates without compromising the integrity of the active site. In the α/β-barrel proteins, such as triosephosphate isomerase, the active site is located at the base of the mouth of the barrel, with the specificity-determining loops surrounding it on the periphery (see Figure 4.52). This principle is a general one, as we illustrate for an enzyme, carboxypeptidase, with a completely different structural scaffold.

**Active site**

The region of an enzyme where substrate binds and a chemical reaction is catalyzed.

Carboxypeptidases are zinc-containing enzymes that catalyze the hydrolysis of polypeptides at the C-terminal peptide bond. The mammalian enzyme is a monomeric protein comprising ~300 amino acid residues. The zinc atom is essential for catalysis because it binds to the carbonyl oxygen of the substrate. This binding weakens the C=O bond by withdrawing electrons from the carbon atom and thus facilitates cleavage of the adjacent peptide bond. Carboxypeptidase is a large single-domain structure comprising a mixed β sheet of eight β strands (**Figure 4.62**) with α helices on both sides. Some of the loop regions are very long and curl around the central core of the structure.

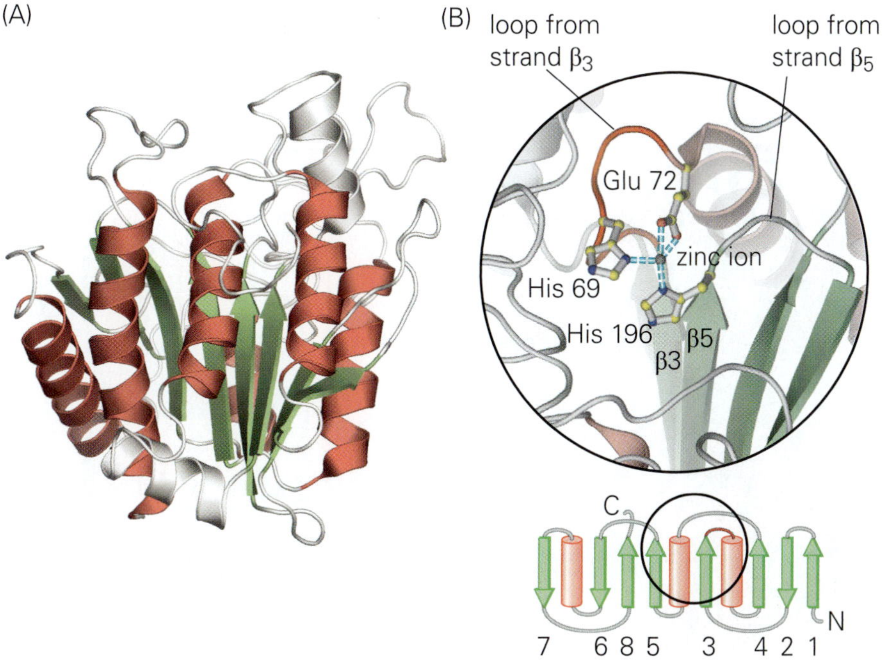

(A)

(B) loop from strand $\beta_3$ · loop from strand $\beta_5$ · Glu 72 · zinc ion · His 69 · His 196 · β5 · β3 · C · N · 7 6 8 5 3 4 2 1

**Figure 4.62 The active site in carboxypeptidase.** (A) The structure of carboxypeptidase. (B) Detailed view of the zinc environment in carboxypeptidase. (Adapted from C. Brändén and J. Tooze, Introduction to Protein Structure, 2nd ed. New York: Garland Science, 1999; PDB code: 5CPA.)

Even though carboxypeptidase is a large and complicated enzyme, the catalytic center containing the zinc atom is not located at an arbitrary position. Rather, a small structural core, with a key functional attribute (in this case, zinc binding) contains the catalytic center, which is elaborated on by the addition of further structural elements.

## 4.30 Binding sites are often located at the interfaces between domains

Another structural feature that enables the evolution of new function in proteins is the location of binding sites at the interfaces between domains. The structure of an individual domain is subject to the constraint that changes in the sequence have to be consistent with the folding of the protein. As a result of this constraint, the central core regions of proteins remain highly conserved during evolution (see Chapter 5). The relative orientation between domains in a multidomain protein and the nature of the interfacial residues are much less constrained because each domain can be thought of as an individual folding unit. Alterations in the residues that border the interfaces between domains are less likely to affect the overall stability of the protein than mutations that disturb the structural scaffold within each domain. Interdomain regions also provide deep invaginations within which small molecules can bind. As a consequence, the active sites of multidomain proteins are usually located at the interdomain interfaces.

An example of a protein with an interdomain binding site is the arabinose-binding protein, which is one of a family of related proteins that occur in the periplasmic space between the inner and outer cell membranes of Gram-negative bacteria such as *E. coli.* These proteins are components of active transport systems for various sugars, amino acids, and ions and, as its name implies, arabinose-binding protein is involved in the transport of arabinose sugars. It is a single polypeptide chain of ~300 amino acids folded into two domains of similar structure and topology (**Figure 4.63**). Arabinose-binding protein is one of several sugar-binding proteins, and we discuss the function of the closely related maltose-binding protein in Section 4.43.

The architecture of arabinose-binding protein consists of two α/β open-sheet domains formed from a single polypeptide chain. This is a common architecture in proteins and, in almost all these cases, the active sites are found in cleft regions between the two domains. In arabinose-binding protein, the domains are

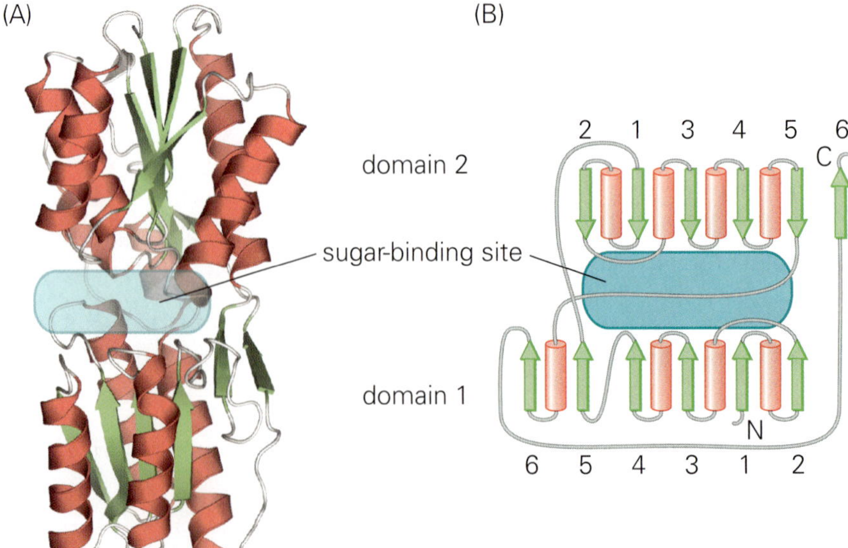


**Figure 4.63 Location of a binding site at the interface between two domains.** (A) The polypeptide chain of the arabinose-binding protein in *E. coli* contains two open twisted α/β domains of similar structure. (B) A schematic diagram of the protein. The two domains are oriented such that the ends of the β strands face each other on opposite sides of a crevice in which the sugar molecule binds (*blue* shading). (Adapted from C. Brändén and J. Tooze, Introduction to Protein Structure, 2nd ed. New York: Garland Science, 1999; PDB code: 5ABP.)

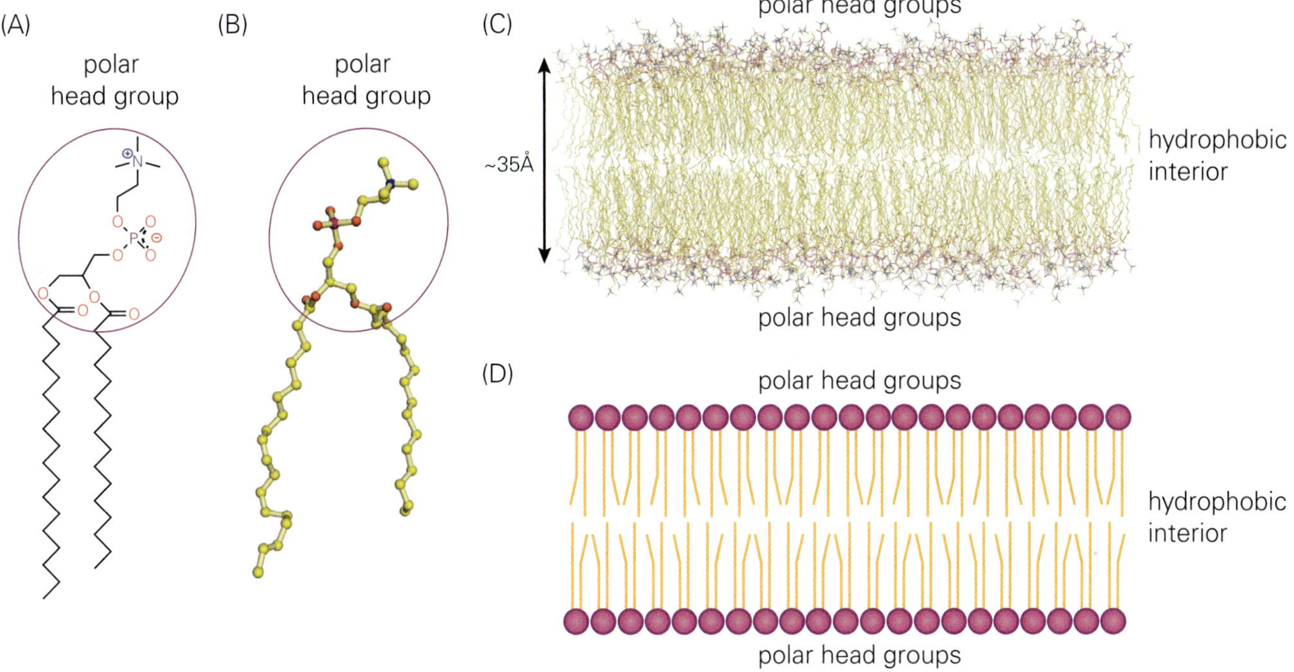


**Figure 4.64 A lipid bilayer.** (A) Chemical structure of a phospholipid. (B) Molecular model of a phospholipid. (C) Molecular structure of a lipid bilayer. (D) Schematic diagram of a lipid bilayer.

oriented in such a way that the carboxyl edges of both β sheets point toward the active site. Notice, in Figure 4.63, that one of the domains is inserted as a unit into the connection between a strand and a helix in the other. Loop regions adjacent to the switch points (see Figure 4.53) of both domains participate in forming the active site. In enzymatic reactions where two different substrates participate, they could be bound to different domains and brought together for catalytic reactions by the orientation of these domains. In other proteins, the two domains bind different regions of the same ligand. The bacterial arabinose-binding protein is an example of this second case.

# D. STRUCTURAL PRINCIPLES OF MEMBRANE PROTEINS

## 4.31 Lipid bilayers form barriers that are nearly impermeable to polar molecules

Cells and the organelles within them are bounded by membranes, which are extremely thin (~35 Å) films of lipids (see Chapter 3). The principal lipid components of membranes are **phospholipids**, in which a phosphorylated polar or charged scaffold, known as the **head group**, is attached to two hydrocarbon chains. Cell membranes are composed of two sheets of phospholipids that are packed against each other to form a **lipid bilayer**, in which the polar head groups are on the outside and the hydrophobic chains are on the inside (**Figure 4.64**).

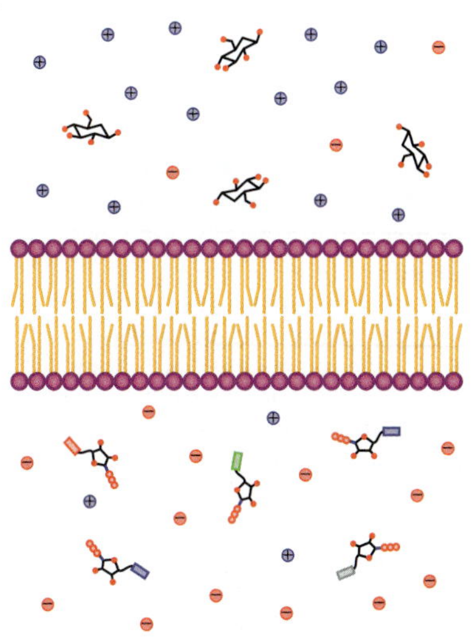

**Figure 4.65 Phospholipid bilayers are impermeable to polar molecules.** The hydrophobic center of the lipid bilayer cannot form hydrogen bonds, and so polar molecules do not readily pass through membranes.

A biological membrane functions as a permeability barrier that impedes the movement of polar or charged molecules across the membrane (**Figure 4.65**). The interior of the bilayer is hydrophobic, with no hydrogen-bonding capability. If a polar molecule were to enter the lipid bilayer, it would lose hydrogen bonds with water without gaining compensatory hydrogen bonds from the lipid. The loss of hydrogen bonding imposes a large energy penalty that opposes the movement of polar molecules into the bilayer.

Lipid bilayers establish discrete compartments within the cell and prevent the random mixing of the contents of one compartment with those of another. But,

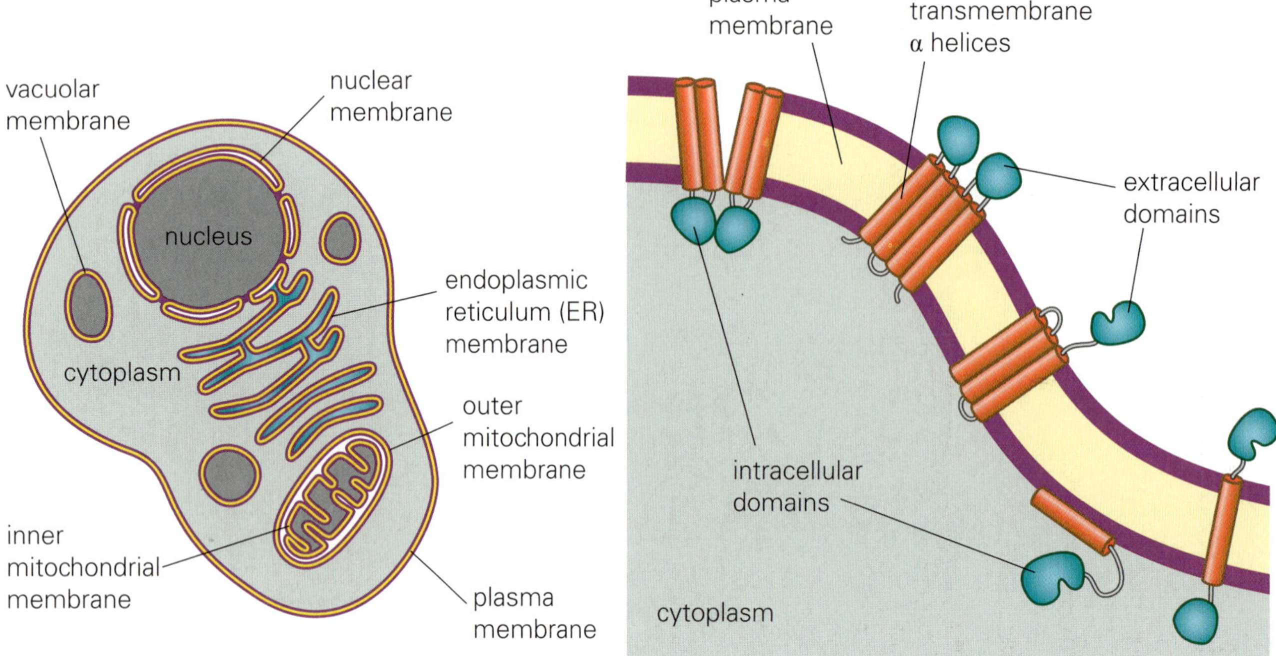


**Figure 4.66 Cellular membranes.** A schematic diagram of a eukaryotic cell is shown, with various cellular compartments and membranes indicated. The drawing is not to scale, and the width of the membranes with respect to the size of the cell is exaggerated. The exploded view shows a small segment of the plasma membrane along with several membrane proteins.

a living cell depends on the exchange of molecules across these membranes, without which the cell could not maintain its energy level or synthesize its molecular components. Thus, the transport of polar molecules across the membrane is facilitated by proteins embedded in the lipid bilayer.

A eukaryotic cell is illustrated schematically in **Figure 4.66**, in which various cellular compartments are shown along with the membranes that encapsulate them. The plasma membrane, for example, is the outer membrane of the cell. Other membranes, such as that of the endoplasmic reticulum, serve to separate internal compartments from the cytoplasm. All of these membranes are decorated with a dense array of proteins, known as **integral membrane proteins**, as shown in Figure 4.66.

Integral membrane proteins serve as mediators between the cell and its environment or the interior of an organelle and the cytoplasm. They catalyze the specific transport of metabolites and ions across the membrane barriers. They convert the energy of sunlight into chemical and electrical energy, and they couple the flow of electrons to the synthesis of ATP. Some membrane proteins act as signal receptors and transduce signals across the membrane. The signals can be, for example, neurotransmitters, growth factors, hormones, light, or chemotactic stimuli. Membrane proteins of the plasma membrane are also involved in cell–cell recognition.

**Integral membrane proteins**

Proteins that are associated with the membrane for the entirety of their lifespan are known as integral membrane proteins. The three-dimensional structure of these proteins depends on interactions with the lipids in the membrane. Proteins with at least one peptide segment that crosses the membrane bilayer are transmembrane proteins.

## 4.32 Membrane proteins have distinct regions that interact with the lipid bilayer

Membrane proteins with the simplest architecture have three distinct regions: one segment that crosses the membrane, known as the **transmembrane segment**, and two segments, one on each side of the membrane, that are exposed to water (**Figure 4.67**A). The amino acid residues within the transmembrane segment are almost exclusively hydrophobic, a point which we discuss in more detail in Section 4.34. The segments outside the membrane have amino acid distributions that are similar to those found in water-soluble proteins, and these segments often form folded structures with a hydrophobic core and a hydrophilic exterior.

The polypeptide chains of other kinds of transmembrane proteins pass through the membrane several times, usually as α helices (Figure 4.67B). Recall from

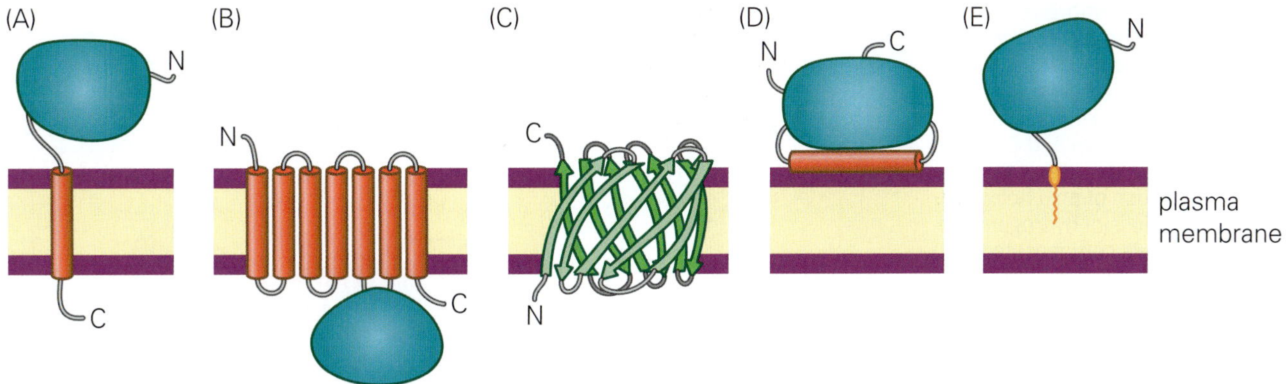


**Figure 4.67 Five different ways in which protein molecules may be bound to a membrane.** From left to right are (A) a protein whose polypeptide chain traverses the membrane once as an α helix, (B) a protein that forms several transmembrane α helices connected by hydrophilic loop regions, (C) a protein with several β strands that forms a channel through the membrane, (D) a peripheral membrane protein that is anchored to the membrane by one α helix that is parallel to the plane of the membrane and interacts with it, and (E) a peripheral membrane protein that is covalently linked to a lipid. (Adapted from C. Brändén and J. Tooze, Introduction to Protein Structure, 2nd ed. New York: Garland Science, 1999.)

Chapter 1 that the rise along the helix axis is 1.5 Å per residue in an α helix, and so ~25 residues (six to seven turns, with 3.6 residues per turn) are required to span a bilayer that is ~35 Å thick. Some transmembrane proteins use β strands to cross the membrane (Figure 4.67C). In these cases, the hydrophilic regions on either side of the membrane are the termini of the chain and the loops between the membrane-spanning parts. Some membrane-associated proteins do not traverse the membrane, but are instead attached to one side, either through α helices that lie parallel to the membrane surface (Figure 4.67D) or by fatty acids, covalently linked to the protein, that intercalate in the lipid bilayer of the membrane (Figure 4.67E). These are known as **peripheral membrane proteins**.

## 4.33 The hydrophobicity of the lipid bilayer requires the formation of regular secondary structure within the membrane

We had noted, in Section 4.4, that when the peptide backbone of a water-soluble protein travels into the hydrophobic core, it is required to form regular secondary structure (either α helices or β sheets; see Figure 4.8). This is a consequence of the hydrogen-bonding requirements of the peptide backbone. Because the sidechains in the hydrophobic core are nonpolar and do not form hydrogen bonds, the peptide backbone must hydrogen bond with itself. α helices and β sheets can be thought of as efficient mechanisms for completely satisfying the hydrogen-bonding requirements of the peptide backbone.

The hydrophobic interior of a lipid bilayer, like the hydrophobic core of a protein, provides no hydrogen-bond donors or acceptors. The energetic penalty for removing a polar group from water, with which it can form strong hydrogen bonds, into the lipid bilayer can be very substantial unless the hydrogen-bonding capacity of the group is satisfied. An α helix is completely self-contained in terms of its backbone hydrogen bonding, and proteins that cross the membrane only once always do so by forming a membrane-spanning α helix (**Figure 4.68**).

One corollary of the need to satisfy the hydrogen-bonding requirement of the backbone is that membrane-spanning α helices are highly stable on their own, in contrast to isolated α helices in soluble proteins. As discussed in Section 4.4, water forms hydrogen bonds with the peptide backbone when it is unfolded, thereby weakening the α-helical conformation in soluble proteins. The situation within the membrane is quite different because of the absence of water, and isolated membrane-spanning α helices do not unfold easily within the membrane.

All of the membrane-spanning segments of proteins with multiple transmembrane segments form α helices (**Figure 4.69**A) or, in rarer cases, β sheets (Figure 4.69B). Proteins that use β sheets to span the membrane must ensure that there are no loose edges to the sheet with uncompensated hydrogen bonds. β sheets that span the membrane are always seen to form closed barrels (see Figure 4.69B).

**Peripheral membrane proteins**

Proteins that are tightly associated with membranes, but do not traverse the membrane, are known as peripheral membrane proteins. Such proteins may be bound to the membrane by noncovalent interactions or covalently attached to a membrane lipid.

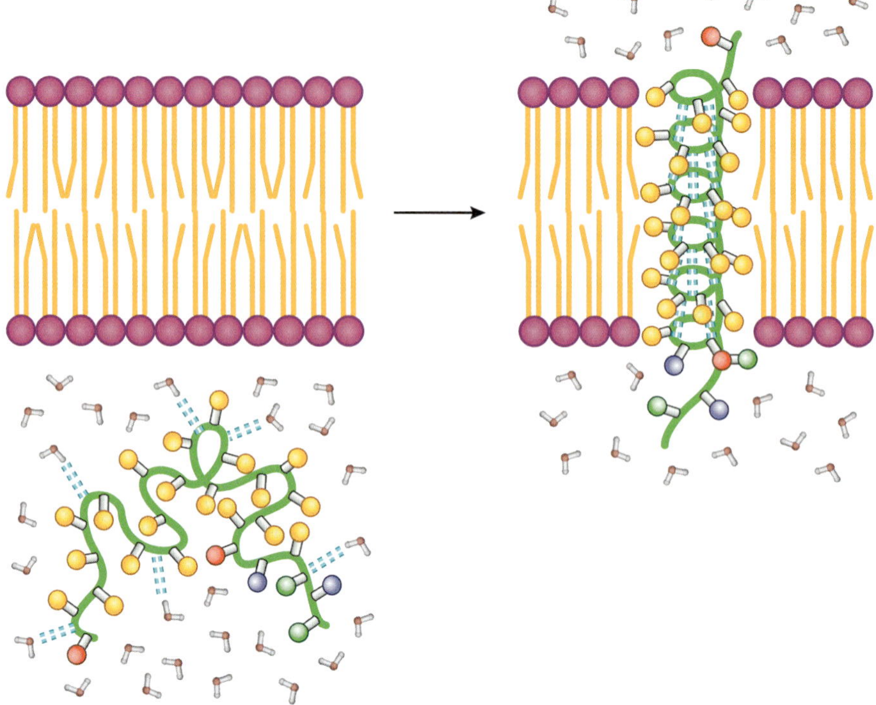

**Figure 4.68 Transmembrane α helices satisfy the hydrogen-bonding requirements of the peptide backbone.** When a transmembrane helix enters a lipid bilayer, the backbone NH and C=O groups lose hydrogen bonds with water. The formation of an α helix ensures that the loss of hydrogen bonds with water is compensated for. Roughly 25 residues are required to span the bilayer, and these form six to seven turns of helix.

**Bacteriorhodopsin**

Bacteriorhodopsin was the first integral membrane protein to have its three-dimensional structure determined. Bacteriorhodopsin is a proton pump found in certain "light-harvesting" bacteria, and it couples light energy to the generation of a proton gradient across the cell membrane. Bacteriorhodopsin has seven transmembrane helices, and its structure is reminiscent of that of a large family of transmembrane proteins in eukaryotic cells known as G-protein-coupled receptors (GPCRs). GPCRs transduce signals across the membrane. Rhodopsin, for example, is a GPCR that converts the detection of light photons into a neuronal signal.

## 4.34 The more polar sidechains are rarely found within membrane-spanning α helices, except when they are required for specific functions

As we shall see in Section 4.35, the amino acids can be ranked in terms of their hydrophobicity or polarity. Membrane proteins very rarely insert the more polar or charged sidechains into the hydrophobic environment of the membrane. When a polar sidechain is inserted into the membrane, it usually has an interaction partner in the same protein or in an interacting protein or small molecule.

One might imagine that membrane proteins might be constructed in an "inside-out" fashion, with hydrophobic residues on the outer membrane-interacting face and polar sidechains in the interior. This is not the case, however, for α-helical membrane proteins, as illustrated in **Figure 4.70** for two of the helices in **bacteriorhodopsin**, an integral membrane protein with seven transmembrane helices (some β barrel proteins provide an exception, as discussed in Section 4.37). The interactions between transmembrane helices within the same protein rely mainly on hydrophobic residues.

One reason for the general exclusion of polar residues at these interfaces is that hydrogen-bonding interactions, in contrast to hydrophobic interactions, require

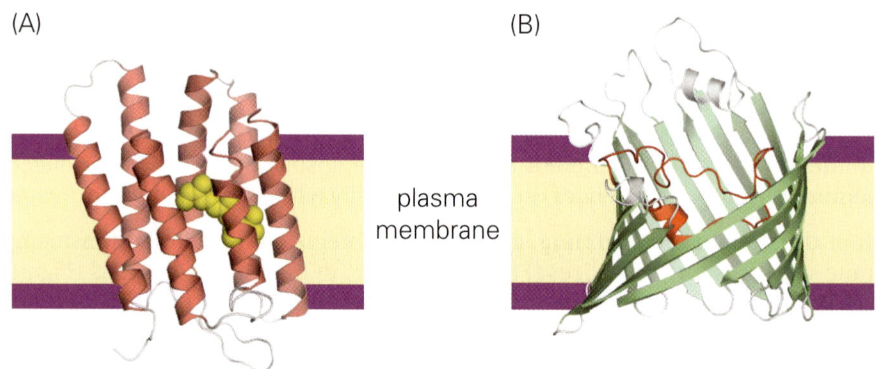


**Figure 4.69 α helices and β sheets in membrane proteins.** (A) Structure of bacteriorhodopsin, which consists of seven transmembrane helices. The protein contains a light-sensitive chromophore known as retinal, shown in *yellow*. (B) Structure of a porin, in which 16 β strands form an antiparallel β barrel that traverses the membrane. A long loop between two β strands (*red*) constricts the channel of the barrel. The porin protein is part of a trimer, and the two other proteins in the trimer are not shown. The β strands in the front of the molecule are shorter because they pack against other molecules in the trimer rather than the membrane. (PDB codes: A, 1C8S and B, 2POR.)

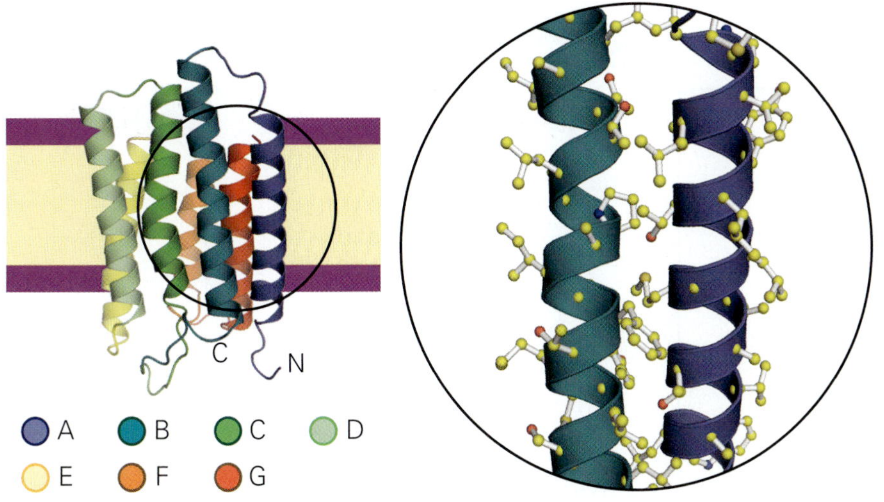


**Figure 4.70 Hydrophobic residues in the helices of bacteriorhodopsin.** The structure of bacteriorhodopsin is shown on the left, with the A and B helices shown in an expanded view on the right. Note that most of the residues at the interfaces between the helices are hydrophobic, as are the residues on the outer surface that faces the membrane.

the interacting groups to be in quite specific geometries. The proper coordination of one polar residue by another in a hydrophobic environment (where water cannot help) may be difficult to achieve.

Another reason for the dearth of interacting polar sidechains within the membrane could arise from the sensitivity to mutation of interactions between polar sidechains. The accidental loss of one polar residue in such a circumstance is likely to destabilize the protein, because it would leave its hydrogen-bonding partner unsatisfied. Most single-base mutations in codons for hydrophobic residues result in a codon for another hydrophobic residue (see Figure 1.48), making hydrophobic interactions less sensitive to the effects of mutations.

Some polar residues, particularly charged ones, are critical for the proper functioning of the protein and are found within the membrane-spanning segments. **Figure 4.71** shows the distribution of amino acid residues in bacteriorhodopsin. Several

**Figure 4.71 Amino acid sequence of bacteriorhodopsin.** The seven transmembrane helices of bacteriorhodopsin are shown as cylinders. Positively and negatively charged residues are indicated in *blue* and *red*, respectively. Notice that there are several of them within the membrane-spanning regions of the helices. These residues have important functional roles, which accounts for their presence within the membrane. (Adapted with permission from D.M. Engelman et al., and B.A. Wallace, *Proc. Natl. Acad. Sci. USA* 77: 2023–2027, 1980.)

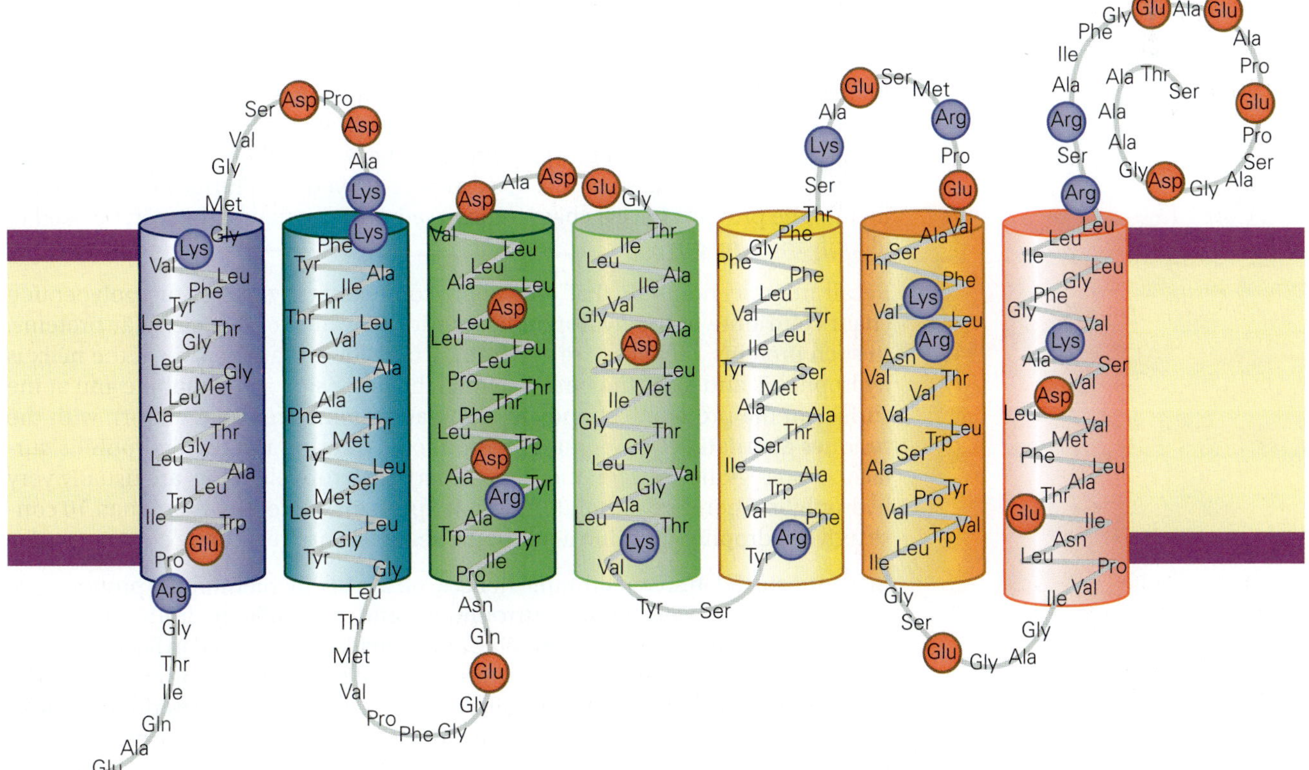

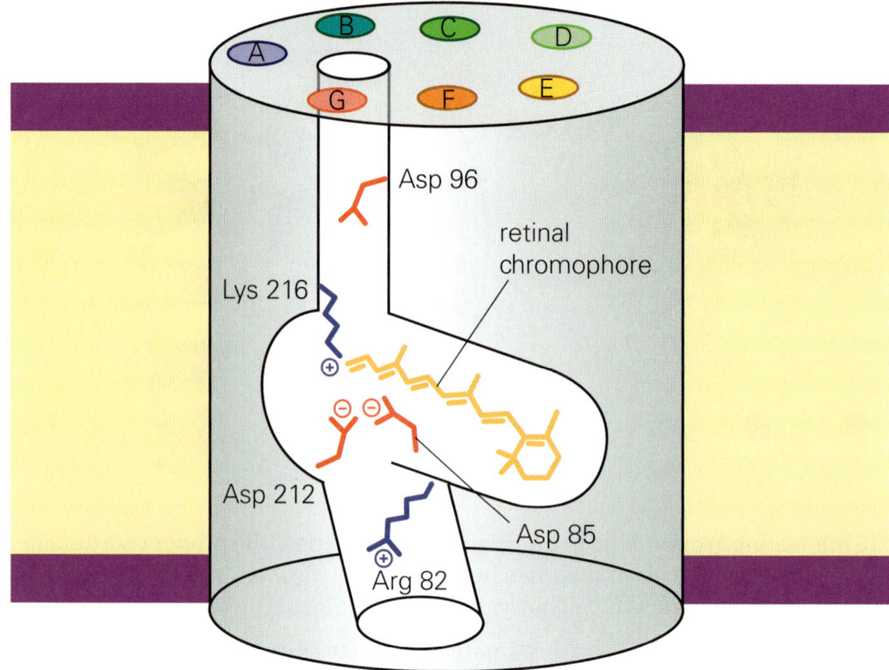


**Figure 4.72 Charged residues in bacteriorhodopsin.** A proton conducting conduit within the protein is indicated. Charged residues line this pathway and interact with the retinal chromophore. (Adapted from C. Brändén and J. Tooze, Introduction to Protein Structure, 2nd ed. New York: Garland Science, 1999, and R. Henderson et al., and K.H. Downing, *J. Mol. Biol.* 213: 899–929, 1990. With permission from Elsevier.)

charged residues are located within the transmembrane segments. Bacteriorhodopsin contains an internal network of residues that is part of the mechanism by which it functions as a light-driven proton pump, and the retinal chromophore is part of this conduit. Most of the charged residues either line this conduit, where they facilitate the movement of protons, or interact with retinal (**Figure 4.72**). We discuss the mechanism of bacteriorhodopsin in more detail in Sections 4.38–4.40, where the important role played by some of these charged residues is explained.

## 4.35 Transmembrane α helices can be predicted from amino acid sequences

Analysis of the genomes of various organisms indicates that roughly 20–30% of all proteins are integral membrane proteins with multiple transmembrane segments. Relatively few membrane proteins have had their three-dimensional structures determined, so how is this estimate arrived at? It turns out that integral membrane proteins have a characteristic sequence signature that can be used to identify them readily.

Recall from Section 4.13 that only short contiguous regions of the polypeptide chain contribute to the hydrophobic interior of water-soluble globular proteins. In such proteins, α helices are generally arranged so that one side of the helix is hydrophobic and faces the interior, while the other side is hydrophilic and at the surface of the protein. β strands in globular proteins are usually short, with the residues alternating between the hydrophobic interior and the hydrophilic surface. Loop regions between these secondary structure elements are usually very hydrophilic. Therefore, in soluble globular proteins, regions of more than 10 consecutive hydrophobic amino acids in the sequence are very rarely encountered.

In contrast, the transmembrane helices observed in membrane proteins are embedded in a hydrophobic surrounding and are built up from contiguous regions of predominantly hydrophobic amino acids. As we noted in Section 4.32, a minimum of about 20 residues is required to span the bilayer using an α helix. The α helices in typical membrane proteins comprise about 25 to 30 residues. Some of these residues are in segments of the helices that extend outside the hydrophobic part of the membrane.

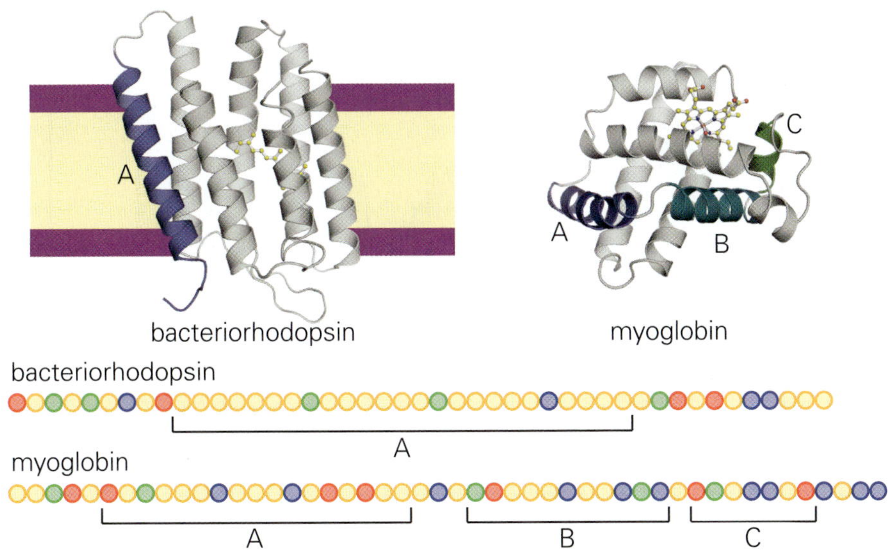


**Figure 4.73 Helices in a membrane protein and a soluble protein.** The structures of bacteriorhodopsin (left) and myoglobin (right) are compared. The first helix (A) in bacteriorhodopsin is colored *blue*. The first three helices in myoglobin (A, B, and C) are colored *blue*, *cyan*, and *green*. The amino acid sequences of the N-terminal portions of the two proteins are depicted schematically in the diagrams below. Each amino acid is indicated by a circle, colored *yellow*, *green*, *red*, and *blue* for hydrophobic, polar, negatively charged, and positively charged residues, respectively. (PDB codes: 1C8S and 1MBC.)

The distinction between the helices in integral membrane proteins and in soluble proteins is illustrated in **Figure 4.73**, which compares the nature of the amino acids found in the N-terminal regions of bacteriorhodopsin and a soluble protein, myoglobin. There is one α helix in this region of bacteriorhodopsin (helix A), which is about 25 residues long and consists mainly of hydrophobic residues. There are three shorter helices in a corresponding stretch in myoglobin, ranging from ~8 to ~17 residues long. Each of these helices is amphipathic, and contains no more than three hydrophobic residues in an uninterrupted stretch. This comparison suggests that the presence of transmembrane helices can be detected by looking for long stretches of hydrophobic residues in the sequences of proteins.

Naively, one might assume that it should be possible to merely scan the sequence and pick out regions with about 20 consecutive hydrophobic amino acids. In practice, such regions can sometimes be difficult to identify by eye because there are occasional hydrophilic sidechains (which are often important for function) among the hydrophobic sidechains (as seen in Figure 4.71). Hydrophobic residues, however, are in a clear majority in transmembrane helices, and such residues occur less frequently in other contiguous regions of the polypeptide chain.

**Hydrophobicity scale**

A hydrophobicity scale assigns a numerical value to each amino acid based on its hydrophobicity. There are many different hydrophobicity scales, which use different criteria for assigning a relative value for the hydrophobicity of the amino acids. One commonly used scale relies on the partitioning of amino acids between water and octanol to assign a value for the hydrophobicity.

## 4.36 Hydrophobicity scales are used to identify transmembrane helices

Each amino acid residue has a different degree of hydrophobicity associated with its sidechain. It is easy to state that residues such as Phe, Ile, and Leu are the most hydrophobic and that charged residues such as Arg and Asp are at the other end of the scale. However, to order all sidechains according to hydrophobicity and to assign actual numbers that represent their degree of hydrophobicity is not trivial. A **hydrophobicity scale** assigns a numerical value to each amino acid based on its hydrophobicity, allowing us to calculate the mean value of the hydrophobicity for any particular stretch of protein sequence. Many such hydrophobicity scales have been developed, on the basis of solubility measurements of the amino acids in different solvents, vapor pressures of sidechain analogs, analysis of sidechain distributions within soluble proteins, and theoretical energy calculations.

One of these hydrophobicity scales is shown in **Figure 4.74**. This particular scale is based on the propensity of each amino acid residue to partition between water

| Trp | Phe | Leu | Ile | Met | Tyr | Val | Cys | Pro | His$^0$ | Thr | Ser | Ala | Gln | Asn | Gly | Arg | His+ | Lys | Glu | Asp |
|---|---|---|---|---|---|---|---|---|---|---|---|---|---|---|---|---|---|---|---|---|
| −8.8 | −7.1 | −5.0 | −4.5 | −2.9 | −2.9 | −2.1 | 0.0 | +0.4 | +0.4 | +0.8 | +2.1 | +2.1 | +3.3 | +3.8 | +4.5 | +7.5 | +9.5 | +11.7 | +15.0 | +15.0 |

**Figure 4.74 A hydrophobicity scale.** The amino acids are sorted from left to right based on the extent to which the residue partitions into octanol, a hydrophobic solvent. The numbers in the scale are the free energies of transfer from water to octanol, in units of kJ•mol$^{-1}$. The more negative the transfer free energy, the more readily the amino acid partitions into octanol. The most hydrophobic residues are on the left (*blue*) and the most polar residues are on the right (*red*). (Adapted from S.H. White and W.C. Wimley, *Annu. Rev. Biophys. Biomol. Struct.* 28: 319–365, 1999.)

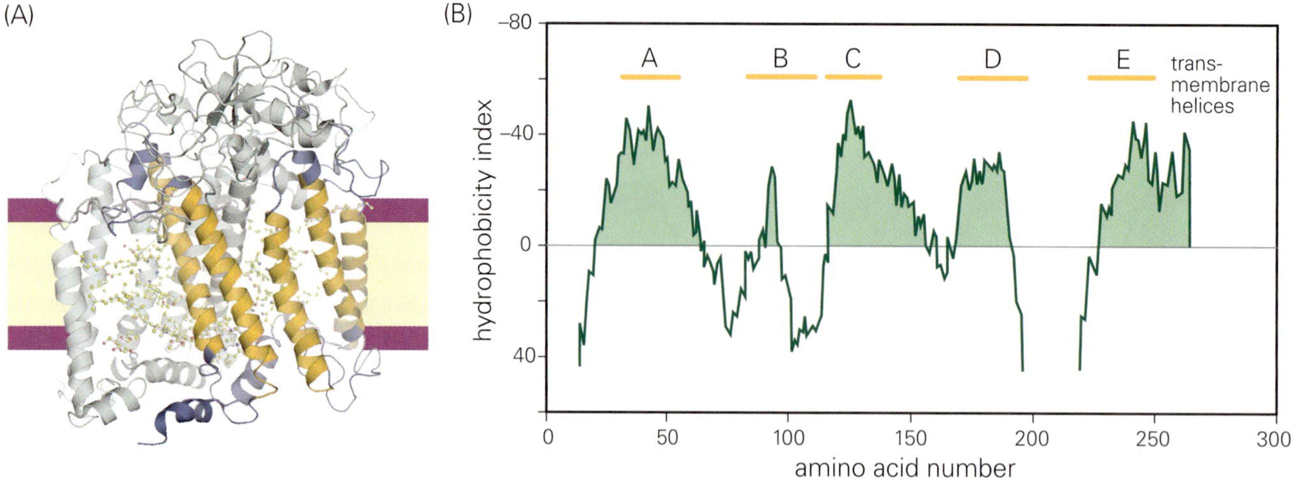


**Figure 4.75 Transmembrane helices and the hydrophobicity index.** (A) The structure of the photosynthetic reaction center from *Rhodobacter sphaeroides*. The backbone of one of the subunits of the reaction center is shown in color, with the transmembrane segments in *yellow* and the rest of the protein in *blue*. There are five transmembrane helices, labeled A–E. The protein contains several cofactors involved in the conversion of light energy to chemical energy, and these are shown in stick representation. (B) The hydrophobicity index as a function of residue number for the subunit of the reaction center shown in color in (A). The hydrophobicity index at any position along the sequence is the aggregate value of the water/octanol transfer free energies for 19 contiguous residues in the sequence, centered on the residue in question. Clustered regions of negative hydrophobicity index indicate the presence of transmembrane α helices. Note that all five transmembrane helices are identified by this criterion. (Adapted from S.H. White and W.C. Wimley, *Annu. Rev. Biophys. Biomol. Struct.* 28: 319–365, 1999; PDB code: 4RCR.)

and the hydrophobic solvent octanol. The value assigned to each amino acid in the scale is known as the transfer free energy, and reflects the ease with which the amino acid partitions into octanol from water. The more hydrophobic amino acids have lower values on this scale. Note that tryptophan is the most hydrophobic amino acid according to the scale shown in Figure 4.74. We had classified tryptophan as having a polar group in Figure 1.29, but you should recognize that the two fused aromatic rings also make it very hydrophobic.

Hydrophobicity scales, such as the one shown in Figure 4.74, are used to identify those segments of the amino acid sequence of a protein that have hydrophobic properties consistent with a transmembrane helix. For each position in the sequence, a mean value of the hydrophobicity is calculated. The hydrophobic index is the mean value of the hydrophobicity of the amino acids within a "window," usually 19 residues long, around each position. The size of the window corresponds to the expected length of a transmembrane helix, and so a residue in the middle of such a helix is expected to have a high hydrophobic index. When the hydrophobic indices are plotted against residue numbers, the resulting curves, called hydrophobicity plots, identify possible transmembrane helices as broad peaks with highly negative values for the hydrophobicity index, corresponding to a favorable predicted free energy of transfer into membranes.

The application of the hydrophobicity index to the identification of transmembrane helices is illustrated in **Figure 4.75**, which shows the structure of an integral membrane protein known as a photosynthetic reaction center. The protein contains two membrane-spanning subunits, one of which is shown in color in Figure 4.75A. This subunit consists of five transmembrane helices, labeled A–E. The hydrophobicity index as a function of residue number is calculated from the sequence of the protein and is graphed as a function of residue number in Figure 4.75B.

The hydrophobicity index is calculated from the amino acid sequence alone, without knowledge of the three-dimensional structure. Comparison with the known structure shows that the transmembrane helices are identified by clustered regions of the sequence in which the hydrophobicity index is negative (see Figure 4.75B). The exact ends of transmembrane helices cannot be predicted accurately, because they are usually inserted within the polar head groups of the membrane lipids and therefore contain charged and polar residues. All of the transmembrane helices in this particular subunit have a segment of at least 19 consecutive amino acids that contains no charged sidechains. The majority of residues in these segments have aliphatic sidechains, but there are a number of polar residues, such as Ser, Thr, Tyr, and Trp, among them.

## 4.37 Integral membrane proteins are stabilized by van der Waals contacts and hydrogen bonds

The folding of soluble proteins is driven by the removal of hydrophobic sidechains from water to form a hydrophobic core, as we have emphasized in Section 4.3. The hydrophobic effect cannot be dominant in the folding of membrane proteins because, once in the membrane, hydrophobic sidechains can either interact with the nonpolar aliphatic chains of lipids or with other hydrophobic sidechains. It turns out that the folded structures of membrane proteins are stabilized by two kinds of interactions, van der Waals forces and hydrogen bonds, which are more important in stabilizing membrane proteins than soluble proteins.

Even though all the faces of transmembrane α helices are hydrophobic, these helices pack together in specific ways due to complementarity in their shape. This is demonstrated by an experiment in which bacteriorhodopsin was reconstituted from two pieces of the protein that were made separately. The two fragments consist of the first five helices and the last two helices, respectively (**Figure 4.76**). When these two pieces are introduced together into a lipid bilayer, they interact with each other and with retinal to form a fully functional bacteriorhodopsin molecule that functions properly as a light-driven proton pump.

The complementarity in the shapes of the two pieces of bacteriorhodopsin results in numerous contacts between them, each one contributing a small amount of stabilizing energy due to van der Waals forces. In addition, if there are hydrogen-bonding interactions between helices (**Figure 4.77**), these hydrogen bonds are a powerful driving force for the formation of the correct structure, because

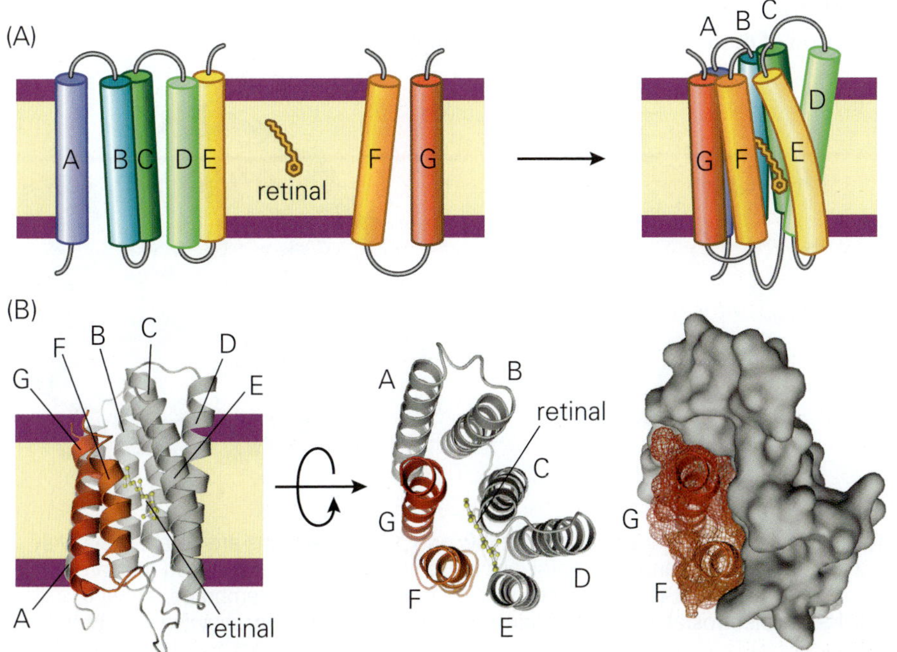


**Figure 4.76 Self-assembly of fragments of bacteriorhodopsin.** (A) Two separate fragments of bacteriorhodopsin are made and inserted into membranes. They assemble spontaneously into the correct three-dimensional structure that is active as a proton pump. (B) The helices of bacteriorhodopsin are packed tightly against each other. As a consequence, the two fragments (one containing helices A–E and the other helices F and G) have mutually complementary shapes. The diagram on the right shows the molecular surfaces of the two fragments. The surface of the first fragment is opaque, while that of the second fragment is shown as a mesh. Notice the tight fit between the two surfaces. (PDB code: 1C8S.)

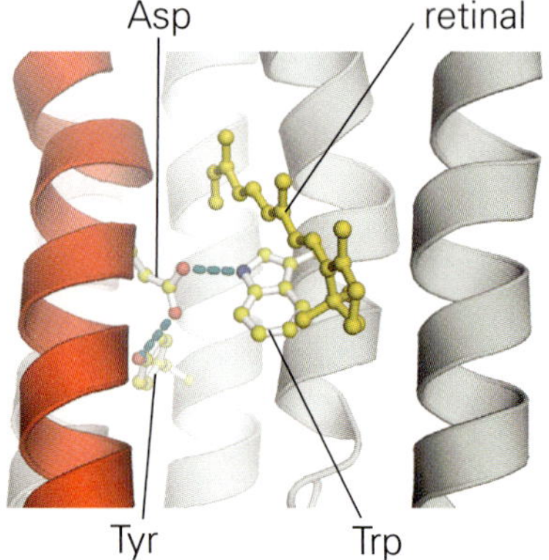


**Figure 4.77 Internal hydrogen bonds in bacteriorhodopsin.** An aspartate sidechain presented by one α helix forms hydrogen bonds with a tyrosine and a tryptophan presented by two other helices. The formation of hydrogen bonds such as these are important for the stability of the protein. (PDB code: 1C8S.)

conformations that do not have the hydrogen-bonding partners correctly paired will be very high in energy. This contrasts with the situation in water-soluble proteins, where hydrogen bonds with water weaken the effective strength of hydrogen bonds between protein groups.

## 4.38 Porins contain β barrels that form transmembrane channels

Our discussion has focused so far on α helical proteins, which constitute the vast majority of integral membrane proteins. There is, however, an important class of membrane proteins in bacteria that have a β-barrel architecture. Gram-negative bacteria are surrounded by two membranes, an inner plasma membrane and an outer membrane. These are separated by a periplasmic space. The outer membrane contains proteins known as porins that form open water-filled channels that allow the passive diffusion of nutrients and waste products across the outer membrane. These channels are restricted in size, and this excludes larger, potentially toxic compounds from entering the cell. Porins are among the most abundant proteins in bacteria. Each *E. coli* cell contains about 100,000 copies of porin molecules in its outer membrane.

A porin molecule from the bacterium *Rhodobacter capsulatus* is shown in **Figure 4.78**. The porin folds into a 16-stranded up-and-down antiparallel β barrel in which all β strands form hydrogen bonds to their neighbors. In contrast to the β barrels in soluble proteins, which have a tightly packed hydrophobic core, the porin barrels contain a central channel because of the large number of β strands. The channel, however, is partially blocked by a long loop region between two β strands that projects into the channel. This arrangement creates a narrow region in the middle of the channel, about 9 Å long and 8 Å in diameter, which defines the size of solute molecules that can traverse the channel. The entrance to the channel is lined almost exclusively with positively and negatively charged groups that are arranged on opposite sides of the channel, causing a transverse electric field across the pore. This asymmetric arrangement of charges contributes to the selection of molecules that can pass through the channel.

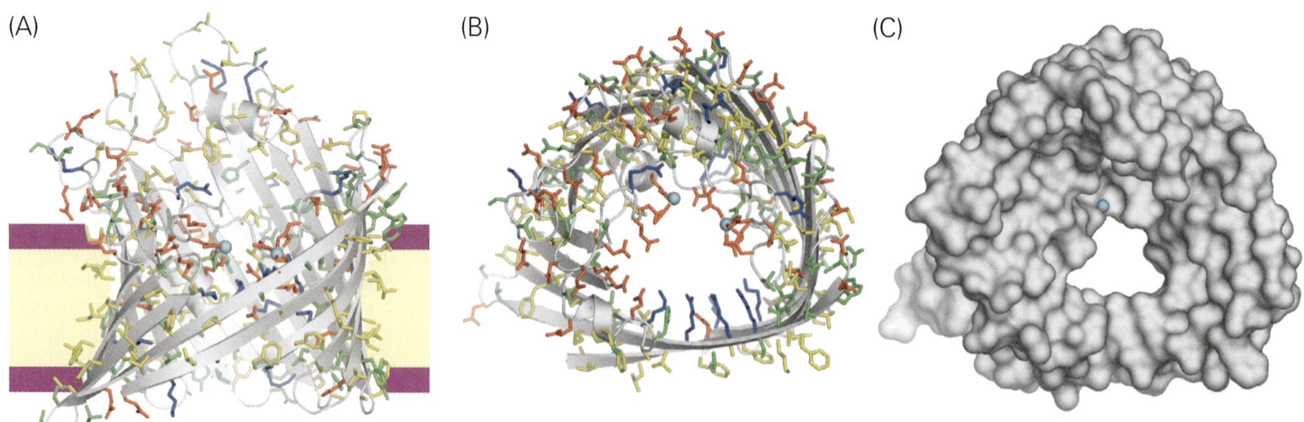


**Figure 4.78 Structure of a porin subunit.** Hydrophobic sidechains are colored *yellow* and polar sidechains are *green*. Negatively and positively charged sidechains are *blue* and *red*, respectively, and two metal ions bound to the protein are shown as *cyan* spheres. (A) Side view, showing how the protein is embedded in the membrane. (B) View from above the membrane, showing the charged residues that line the eyelet and act as an electrostatic filter. (C) Molecular surface of the porin, showing the central channel. (PDB code: 2POR.)

Since the outside of the barrel faces the hydrophobic lipids of the membrane and the inside forms the solvent-exposed channel, one would expect the β strands to contain alternating hydrophobic and hydrophilic sidechains. This requirement is not strict, however, because internal residues can be hydrophobic if they are in contact with hydrophobic residues from other regions of the protein. Thus, in contrast to α-helical membrane proteins, the sequences of porins do not exhibit readily identifiable patterns.

## 4.39 Pumps and transporters use energy to move molecules across the membrane

We shall end our discussion of membrane proteins in this chapter by looking at the mechanism of two specialized transmembrane proteins that facilitate the movement of specific molecules or ions across the lipid bilayer. Both of these are examples of **active transporters**, a large class of membrane proteins that actively drive the transport of molecules or ions against a concentration gradient (**Figure 4.79**). We have already been introduced to the structure of one, bacteriorhodopsin, which pumps protons across the membrane by using light energy. The other, the maltose transporter, uses ATP as a fuel to move molecules of the sugar maltose into bacterial cells.

Active transporters are quite distinct from another class of membrane proteins that allow molecules to transit from one side of the membrane to the other in accordance with the concentration gradient (that is, the net flow of the transiting molecules is from a region of high concentration to a region of low concentration). Such a process is known as passive transport, and it is facilitated by channel proteins that simply provide a tunnel through the membrane, within which molecules can transit from one side of the membrane to the other. The porin proteins found in the outer membranes of bacteria are one kind of passive transporter, but they are rather nonspecific in terms of their selectivity (see Section 4.38).

Most membrane channels are extraordinarily specific, and this specificity is crucial for maintaining the proper chemical composition of the cell. For example, water molecules move across the membrane by transiting through proteins known as aquaporins, which allow water molecules to pass, but do not permit the leakage of protons. Other channels are specialized for the conductance of particular metal ions, such as $K^+$, $Na^+$, or $Ca^{2+}$. Potassium channels, discussed in Chapter 11, allow the rapid movement of $K^+$ ions through them, but are essentially impermeable to sodium ions.

**Active and passive transporters**

Proteins that form channels in the membrane and facilitate the movement of specific molecules from one side of the membrane to the other are known as transporters. Active transporters are membrane proteins that use energy to move molecules against a concentration gradient. Passive transporters do not use energy to control the flow, and the net flow of molecules is simply from the side of the membrane where they are at higher concentration to the side where they are at lower concentration.

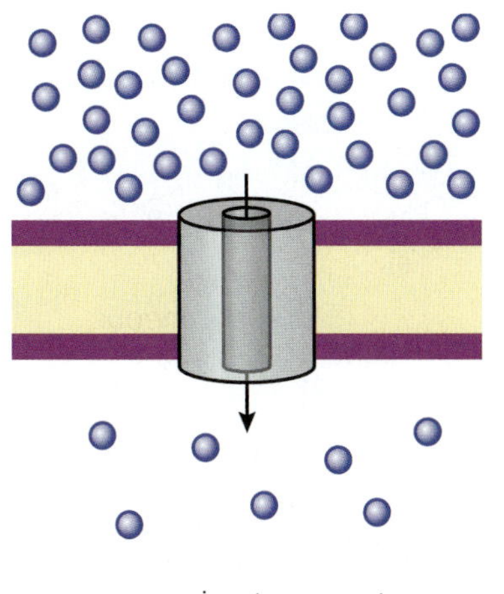


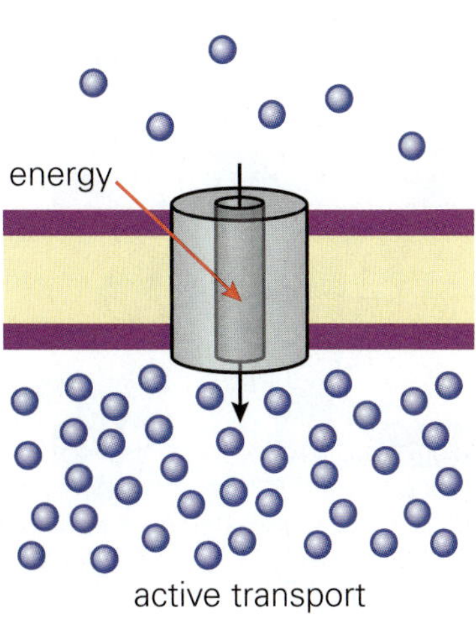


**Figure 4.79 Passive and active transporters.** A passive transporter is a protein that provides a channel that allows molecules to cross from one side of the membrane to the other. The direction of flow is determined by the concentrations of the transported molecule on the two sides of the membrane. Active transporters move molecules in one direction, regardless of the concentration differences. Active transporters require an input of energy.

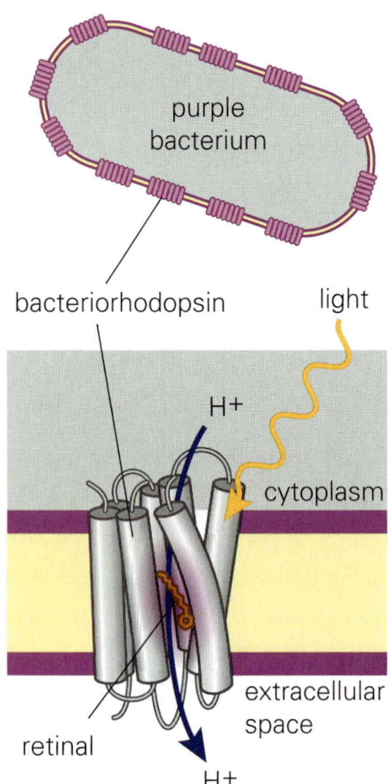


**Figure 4.80 Bacteriorhodopsin is a light-driven proton pump.** The membrane of the purple bacterium, *Halobacterium halobium*, is densely packed with molecules of bacteriorhodopsin. Light causes isomerization of the retinal chromophore bound to bacteriorhodopsin, which is coupled to the movement of protons out of the cell.

## 4.40 Bacteriorhodopsin uses light energy to pump protons across the membrane

Particularly important among active transporters are those that are involved in generating a chemical gradient that can be coupled to the production of ATP. Here we provide a brief description of how bacteriorhodopsin utilizes light energy to generate a gradient in proton concentration across the membrane (**Figure 4.80**).

Bacteriorhodopsin is found in the membrane of *Halobacterium halobium*, a bacterium that gives the salt flats in which it lives a characteristic purple color. Bacteriorhodopsin binds a photosensitive pigment known as retinal (**Figure 4.81**), which is closely related to the pigment that is used to capture light in our eyes. Bacteriorhodopsin uses the energy of light to pump protons across the membrane and is part of a particularly simple biological system for the conversion of light energy to chemical energy.

When retinal bound to bacteriorhodopsin absorbs a photon, it undergoes an isomerization that involves rotation about a double bond. Excitation by light switches retinal from a conformation in which all of the double bonds are in the *trans* conformation (referred to as all-*trans* retinal) to a conformation where the bond between carbon atom 13 and carbon atom 14 is in a *cis* conformation (13-*cis* retinal; see Figure 4.81). This light-driven conformational change in the retinal molecule is coupled to the movement of protons from the cytosol to the extracellular space, creating a proton gradient. This gradient is used to generate ATP and to transport ions and molecules across the membrane.

## 4.41 A hydrogen-bonded chain of water molecules can serve as a proton conducting "wire"

The mechanism for proton pumping by bacteriorhodopsin relies on the inability of charged species, such as protons, to transit the hydrophobic interior of the lipid bilayer with an appreciable rate. By spanning the bilayer, proteins such as bacteriorhodopsin can couple conformational changes, such as those induced by light, to the movement of protons through the protein. These mechanisms rely ultimately on the ability of the sidechains of glutamate, aspartate, or histidine to pick up or lose a proton by interconverting between neutral and charged forms (see Chapter 10 for a detailed discussion of the titration properties of these sidechains).

To understand how bacteriorhodopsin works, let us first consider a hypothetical protein that allows protons to move through it from one side of the membrane to the other (**Figure 4.82**). The protein contains a channel that is filled with water molecules, and two aspartic acid sidechains, labeled A and B, are located at each end of the channel. A chain of three water molecules connects the two aspartic acid sidechains.

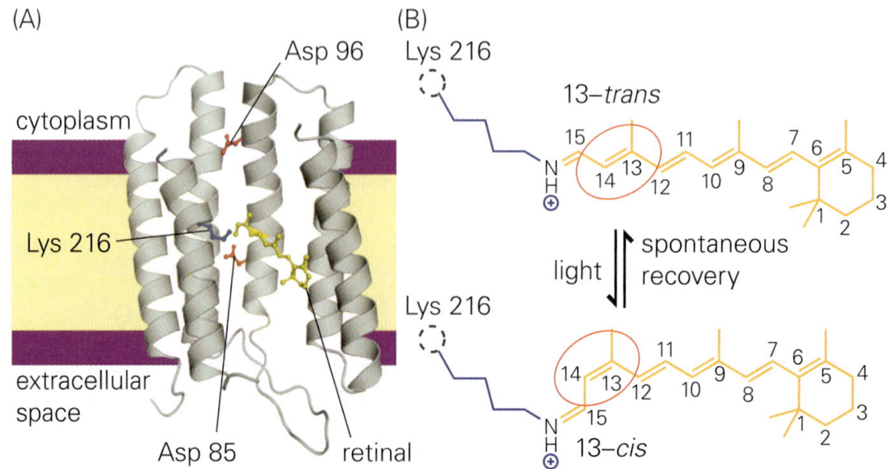


**Figure 4.81 The photoisomerization of retinal.** (A) The structure of bacteriorhodopsin, with one of the helices removed to show retinal bound in the interior. The retinal molecule forms a Schiff base with Lys 216 (*blue*). Two aspartic acid residues that are important for the proton-pumping process are shown in *red*. (B) The chemical structure of retinal in the Schiff base form, showing the light-induced *trans–cis* isomerization about the bond between carbon atoms 13 and 14.

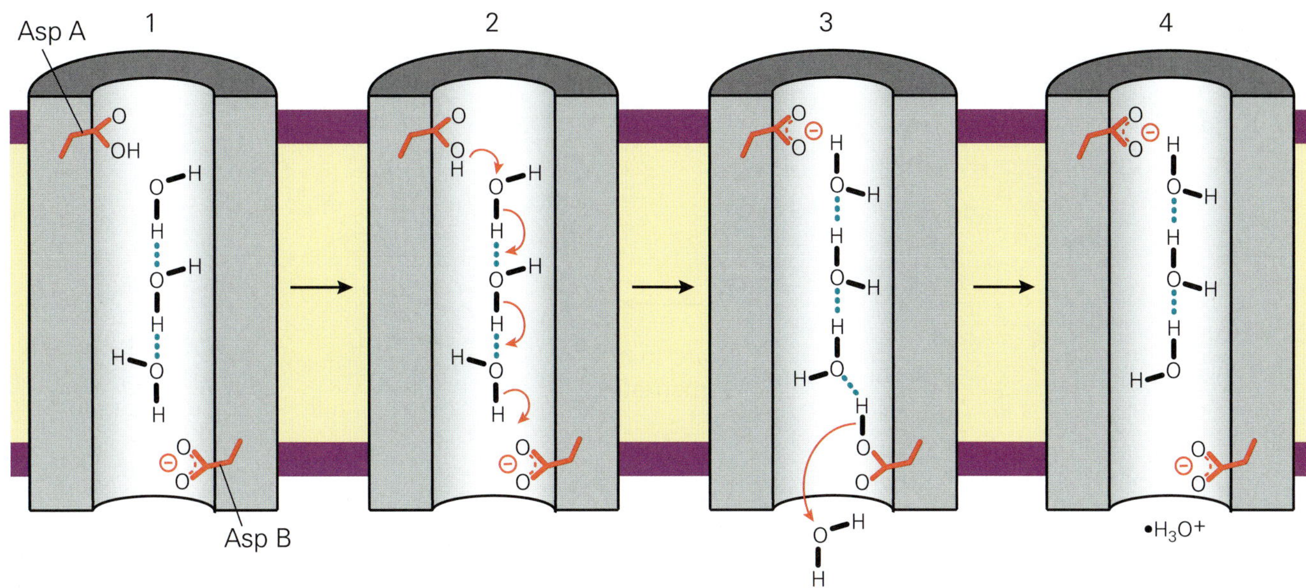


**Figure 4.82 Proton conduction by water molecules through a hypothetical protein.** The protein contains a channel in which water molecules are bound. There are two aspartic acid sidechains, labeled A and B, at either end of the channel. In the process shown here, aspartic acid A starts off protonated and ends up transferring a proton to aspartic acid B and, ultimately, to the other side of the membrane. Each *red* arrow indicates the transfer of a proton from one oxygen atom to an adjacent one, and should not be confused with electron transfer. The proton transfer involves a domino-like transfer of protons along the water network. The mechanism, as shown here, has no intrinsic directionality to it and could just as well run in the reverse direction.

To start with, aspartic acid A is protonated. It then releases a proton and becomes negatively charged. The released proton is picked up by the water molecule located next to the aspartic acid sidechain, which in turn donates a proton to the next water molecule, and so on down the chain. The last water molecule in the chain is located next to aspartic acid B, which is negatively charged to begin with, but becomes neutral when it picks up a proton from the last water molecule. Finally, aspartic acid B releases a proton to a water molecule outside the membrane, converting it to a hydronium ion ($H_3O^+$). The net effect of this series of events is to move a proton from one side of the membrane to the other, using the water network as a proton transfer "wire."

## 4.42 Conformational changes in retinal impose directionality to proton flow in bacteriorhodopsin

Like the hypothetical protein depicted in Figure 4.82, bacteriorhodopsin has two critical aspartic acid residues, one closer to the cytoplasm (Asp 96) and the other closer to the extracellular space (Asp 85; see Figure 4.81). The proton transfer pathway for the hypothetical protein, drawn in Figure 4.82, has no directionality to it. Proton transfer could just as well run in the reverse direction, moving a proton from aspartic acid B to aspartic acid A. If such a protein were inserted into a membrane, it would simply cause protons to leak from one side of the membrane to the other, with the direction of net flow determined by which side has a higher proton concentration. The trick to bacteriorhodopsin is that the retinal chromophore is inserted between the two aspartic acid residues, as shown in Figure 4.81. The conformation of the retinal controls the direction of proton transfer and imposes directionality on the process regardless of the concentration of protons on the two sides of the membrane.

Retinal is bound in a pocket of bacteriorhodopsin about equidistant from the two sides of the membrane (see Figure 4.81). The pigment forms a Schiff base with a lysine residue, Lys 216; in other words, it is covalently linked to the nitrogen atom of the lysine sidechain that is protonated and therefore has a positive charge (see Figure 4.81). In the *trans* state of retinal, before light is absorbed, Asp 85 is close to the positive charge of the Schiff base (**Figure 4.83**A). The structural change of the retinal molecule due to the *trans*-to-*cis* photoisomerization causes the Schiff base to change its position relative to Asp 85, which induces transfer of the Schiff-base proton to the aspartate group (Figure 4.83B). Once the Schiff base–Asp 85 ion pair is converted to a neutral pair by this proton transfer, the protein undergoes a

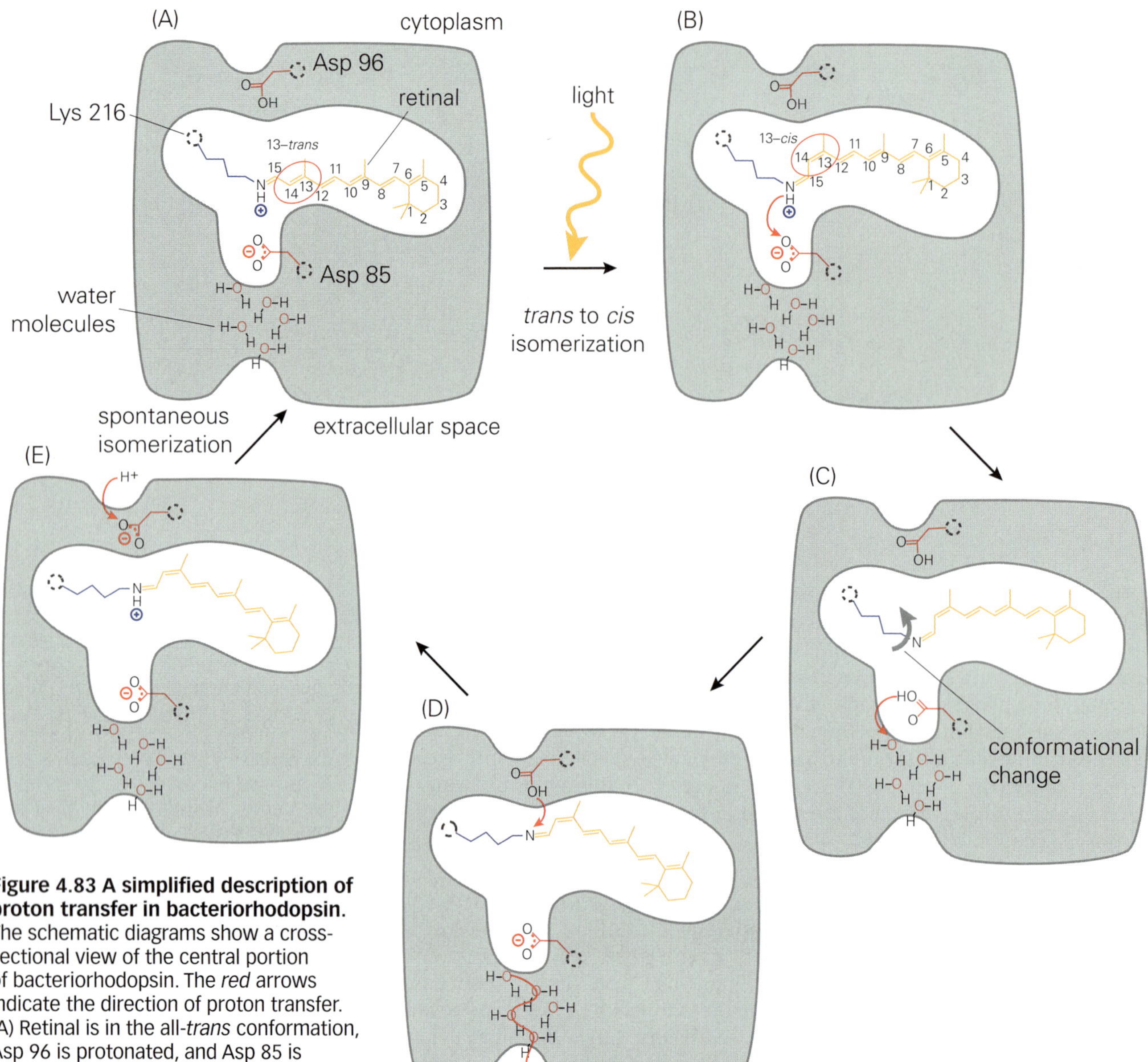


**Figure 4.83 A simplified description of proton transfer in bacteriorhodopsin.** The schematic diagrams show a cross-sectional view of the central portion of bacteriorhodopsin. The *red* arrows indicate the direction of proton transfer. (A) Retinal is in the all-*trans* conformation, Asp 96 is protonated, and Asp 85 is deprotonated. (B) Retinal absorbs a photon and goes from all-*trans* to 13-*cis*. This moves the protonated nitrogen of the Schiff base close to Asp 85, which picks up the proton. (C) Asp 85 donates a proton to a water molecule that is part of a hydrogen-bonded network, leading eventually to the ejection of a proton at the extracellular side. (D) Strain in the retinal molecule causes another conformational change, which moves the unprotonated Schiff-base nitrogen closer to the protonated Asp 96 sidechain. This allows Asp 96 to donate a proton to the Schiff-base nitrogen. (E) Asp 96 picks up a proton from the cytoplasm. The 13-*cis* conformation of retinal eventually relaxes to the lower-energy all-*trans* conformation, resetting the system for another cycle. (Adapted from C. Brändén and J. Tooze, Introduction to Protein Structure, 2nd ed. New York: Garland Science, 1999.)

conformational change that involves a reorganization of some of the transmembrane helices that bind retinal, with the consequence that the Schiff base is moved away from Asp 85 and towards Asp 96 (Figure 4.83C). Asp 85 then delivers a proton through a water network to the extracellular space (Figure 4.83D) and Asp 96 reprotonates the Schiff base (Figure 4.83E). Retinal subsequently relaxes back to the *trans* state and the protein is ready for another cycle of photoisomerization-induced proton transfer.

The essential aspect of the bacteriorhodopsin mechanism is that light causes a chemical change at the active site that alters the conformation of the protein, which in turn drives protons from the cytosolic side of the membrane to the extracellular side. The critical role played by the transmembrane protein is to provide a conduit across the hydrophobic span of the membrane for the transit of charged species. This general principle underlies the mechanism of the much more complicated membrane protein assemblies, known as photosynthetic reaction centers (see Figure 4.75), that couples light to chemical energy in plants and photosynthetic bacteria.

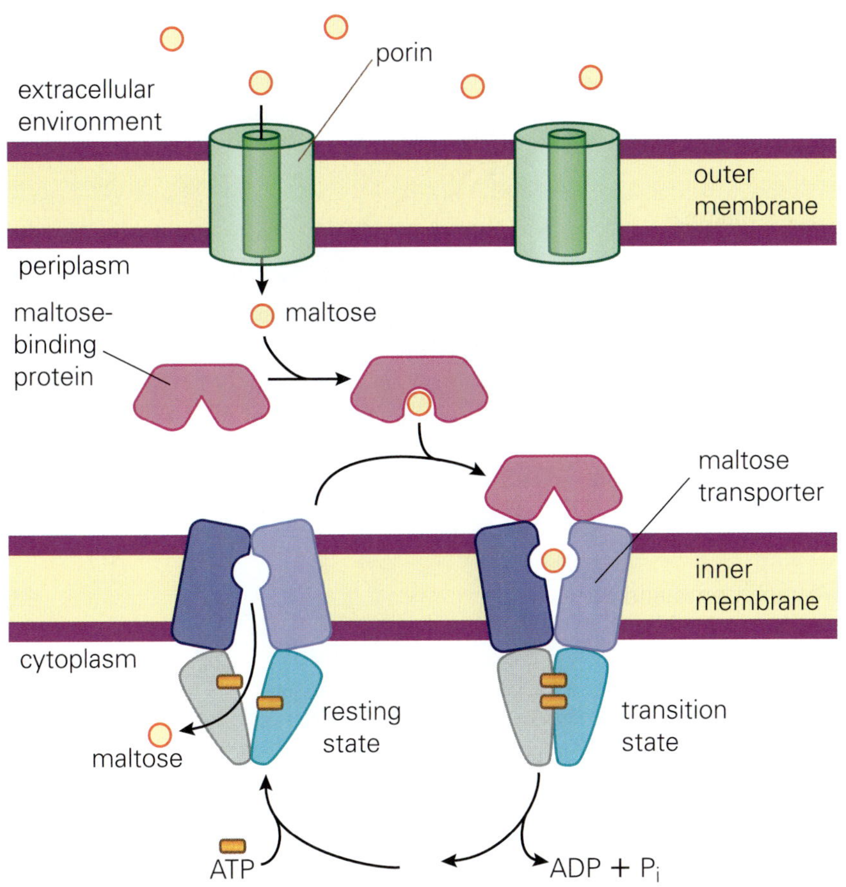


**Figure 4.84 Mechanism of a bacterial maltose transporter.** Porin channels in the outer membrane allow entry of maltose sugars into the periplasmic space. Here they are captured by maltose-binding protein, which then docks onto the maltose transporter. The conformation of the transporter changes from a cytoplasm-facing one (the resting state) to a periplasm-facing one (the transition state) upon the docking of the binding protein. This is accompanied by closure of the ATP binding sites, which results in the hydrolysis of ATP. The loss of ATP destabilizes this conformation of the transporter, which switches back to the resting state and releases maltose into the cytoplasm. (Adapted from M.L. Oldham et al., and J. Chen, *Nature* 450: 515–521, 2007. With permission from Macmillan Publishers Ltd.)

## 4.43 Active transporters cycle between conformations that are open to the interior or the exterior of the cell

In bacteriorhodopsin, light triggers a conformational change that enforces directionality in the proton transport process, allowing protons to be pumped in a direction that is energetically uphill. Most active transporters, with the exception of those involved in photosynthesis, utilize ATP hydrolysis as the source of energy to move molecules against a concentration gradient. As an example of an ATP-driven transport system, we describe the maltose uptake system, by which bacteria import malto-oligosaccharides, sugars consisting of several glucose units (see Chapter 3), into the cell.

Recall from Section 4.38 that Gram-negative bacteria, such as *E. coli*, have two membranes, separated by a region known as the periplasmic space. The outer membrane contains a large number of porins (Figure 4.78), which allow molecules such as sugars to enter the periplasmic space without active pumping. In the periplasm these molecules are bound by specific binding proteins (the structure of one such binding protein is shown in Figure 4.63), and the binding protein, in turn, binds to the periplasmic face of a transmembrane transporter in the inner membrane. In the case of maltose sugars, these are captured in the periplasmic space by maltose-binding protein, as shown schematically in **Figure 4.84**.

A key aspect of the transport mechanism is the cycling of the membrane-spanning portion of the transporter between two conformations. In one conformation, the interior binding site for maltose is open towards the periplasm, but closed towards the cytoplasm (**Figure 4.85**). In the other conformation, the binding site faces the cytoplasm, allowing maltose to exit into the cell. This **alternating access mechanism** is common to many active transporters, which are otherwise quite

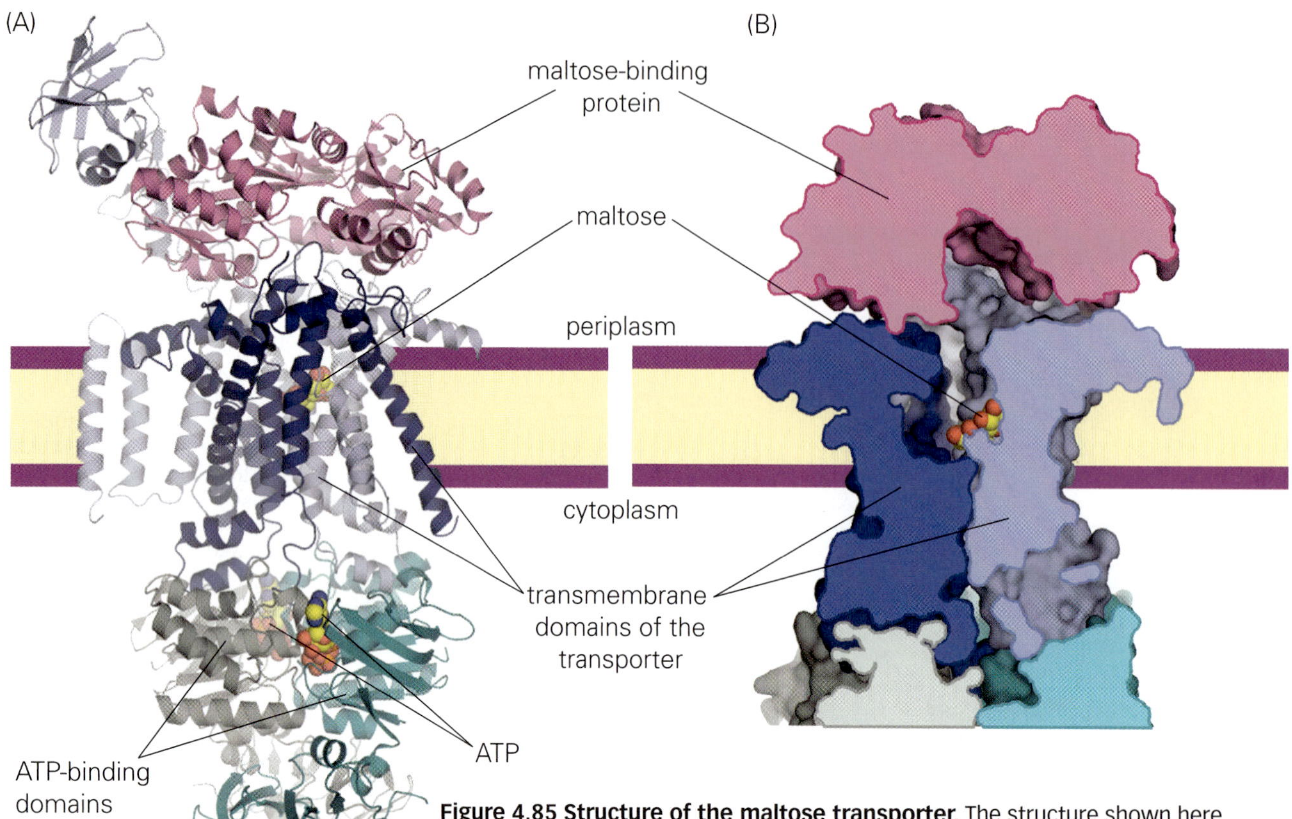


**Figure 4.85 Structure of the maltose transporter.** The structure shown here corresponds to the transition state shown in Figure 4.84, with maltose-binding protein bound to the transporter. Maltose has been released from the binding protein and has moved deep into the transmembrane domains of the transporter. (A) Overall view of the structure. (B) A close-up view of the maltose-binding protein and the transmembrane domains, showing their molecular surfaces. Note that the transmembrane segments are open towards the periplasmic side, where the maltose-binding protein is docked. This conformation causes the hydrolysis of ATP bound to the cytoplasmic side of the transporter, which causes the transmembrane domains to release the binding protein and open towards the cytoplasmic side (see Figure 4.84). (Adapted from M.L. Oldham et al., and J. Chen, *Nature* 450: 515–521, 2007.)

different in terms of their three-dimensional structures. In order to ensure that the transport of molecules occurs in only one direction, the conformation of the transmembrane segments is coupled to the presence of the cargo molecule. In the case of the maltose transporter, the transmembrane segments open to the periplasm only when maltose-binding protein, loaded with maltose, binds to the transporter (see Figure 4.84).

## 4.44 ATP binding and hydrolysis provides the driving force for the transport of sugars into the cell

The docking of the maltose-binding protein to the periplasmic side of the transporter is coordinated with the binding of ATP to the cytoplasmic side. The transporter contains two ATP-binding domains that are separated when the transporter is open to the cytoplasm (denoted the resting state in Figure 4.84) but are pulled together when the maltose-binding protein is docked. Each molecule of ATP is coordinated by both ATP-binding domains, as shown in **Figure 4.86**. The resulting binding energy enables the transporter to open up maltose-binding protein, so that maltose is released into the interior of the transporter.

The ATP-binding domains of the transporter are not just passive structures that provide a docking site for ATP. They are actually members of a family of enzymes known as ATPases that catalyze the hydrolysis of ATP to yield ADP and phosphate ion ($P_i$), as shown in Figure 4.86A. Each ATP binding site is a catalytic center in which the proper configuration of amino acid sidechains results in the cleavage of the γ phosphate group of ATP. In the resting state (see Figure 4.84), the separation between the two ATP-binding domains is such that the catalytic activity is switched off.

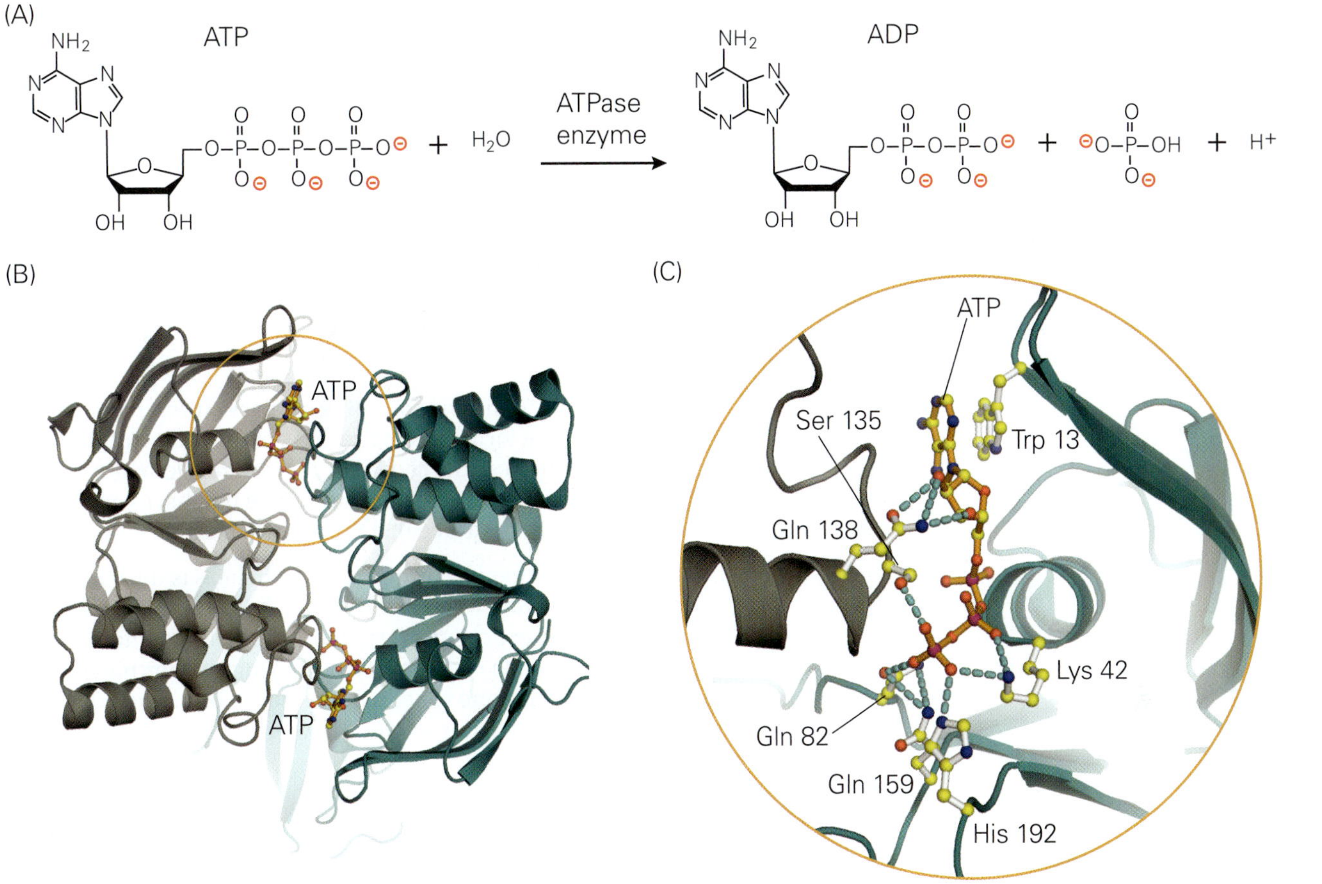

When the transporter switches to the transition state, the closure of the interface between the two ATP-binding domains completes the configuration of sidechains at each active site. Note that the γ phosphate of ATP makes several hydrogen bonds with the protein, and thus is critical to holding the two ATP-binding domains together. The γ phosphate is lost upon the hydrolysis of ATP, and the loss of the interactions made by it destabilizes the transporter, which switches back to the resting state. In this state, the membrane-spanning segments are open towards the cytoplasm, and maltose exits the transporter. The essential element of this mechanism, in which ATP binding results in a conformational change that causes a required action, but also triggers ATP hydrolysis, is common to many crucial biological mechanisms, particular molecular motors.

**Figure 4.86 ATP binding and hydrolysis by the maltose transporter.** (A) ATP hydrolysis is catalyzed by enzymes known as ATPases. The action of these enzymes causes ATP to break down to ADP and phosphate ion. (B) In the transition state of the transporter (see Figure 4.84), the two ATP-binding domains are close together, with two molecules of ATP bound at the interface. (C) Each molecule of ATP forms hydrogen bonds with sidechains from both domains. These sidechains are in a configuration that stimulates ATP hydrolysis. (Adapted from M.L. Oldham et al., and J. Chen, *Nature* 450: 515–521, 2007.)

## Summary

Protein structures are organized in a hierarchical fashion, with secondary structural elements (α helices and β strands) organized into structural motifs that assemble into protein domains with characteristic three-dimensional folds. One or more domains constitute the tertiary structure of the protein. Many proteins consist of more than one subunit, and the organization of subunits in such proteins is referred to as the quaternary structure.

The folding of water-soluble proteins is driven by the hydrophobic effect. The interiors of protein molecules contain mainly hydrophobic sidechains. The protein backbone in the interior is arranged in secondary structures so as to satisfy the hydrogen-bonding requirements of the protein backbone. The formation of such hydrogen bonds is critical for stability, because the backbone NH and C=O groups form hydrogen bonds with water when the protein is unfolded. Because of the exchange with water, the formation of hydrogen bonds in the folded protein does not contribute a large amount of stabilization energy.

The polypeptide backbone of the protein has two degrees of freedom per residue. These are the backbone torsion angles $\phi$ and $\psi$, which correspond to rotations about the bonds that connect each $C_\alpha$ atom to the rigid peptide groups on either side. Interatomic collisions restrict the values of $\phi$ and $\psi$ to certain combinations, described by the Ramachandran diagram. These correspond to the backbone conformations seen in β strands and α helices.

Protein structures are built up by combinations of secondary structural elements, known as structural motifs. These form the core regions of the molecule and they are connected by loops at the surface. The secondary structural elements in these motifs are usually amphipathic, with one hydrophobic face that packs against the rest of the protein and one hydrophilic face that is exposed to water.

An important structural motif formed by amphipathic α helices is the coiled coil, in which α helices coil around each other. Hydrophobic residues in a coiled coil interact in a pattern that repeats every seven residues. Helices also interact without forming coiled coils, and in such cases the helix packing is determined by fitting the ridges of sidechains along one α helix into grooves between sidechains of the other helix. This limits the crossing angles between α helices to the certain values.

The β-α-β motif, which consists of two parallel β strands joined by an α helix, occurs in almost all structures that have a parallel β sheet. Four antiparallel β strands that are arranged in a specific way comprise the Greek key motif, which is frequently found in structures with antiparallel β sheets.

Many enzymes have α/β barrel structures and all have their active sites in very similar positions, at the bottom of a funnel-shaped pocket created by the loops that connect the carboxyl end of the β strands with the amino end of the α helices. Proteins consisting of only β strands are often organized into barrel-like structures as well. The barrels may be quite flattened to form sandwich-like structures with two distinct β sheets.

In general, for proteins with different architectures, the active sites are located within the core elements of the fold. As the sequence of the protein changes during evolution, the core elements stay relatively unchanged, while the peripheral elements can alter dramatically. In multidomain proteins, the active sites are often located at the interfaces between domains, which form deep clefts into which small molecules can bind.

Membrane proteins are quite distinct from soluble proteins because the outer surfaces of their transmembrane segments are almost exclusively hydrophobic. The hydrophobic interior of a lipid bilayer has no hydrogen-bonding capability. As a consequence, a membrane-spanning protein segment has to adopt a regular secondary structure, most commonly an α helix, so as to satisfy the hydrogen-bonding requirements of the protein backbone. Some membrane proteins are constructed of β strands, in which case the transmembrane segments have to form a closed barrel so as to not leave any backbone polar groups exposed to the hydrophobic region of the membrane. Such proteins are usually found only in the outer membranes of bacteria.

The width of a lipid bilayer is typically ~35 Å, which means that about 25 residues in an α helix are required to span the membrane. The presence of transmembrane α helices in a protein sequence can be recognized by the occurrence of 20–25 hydrophobic residues in a contiguous stretch with very few polar residues. Such long stretches of hydrophobic residues are not seen in water-soluble proteins. The more polar residues are excluded from membrane-spanning segments unless they have a specific function to perform. The identification of transmembrane segments is aided by hydrophobicity scales, which rank the amino acids in terms of their ability to partition into a hydrophobic environment.

The lipid bilayer is impermeable to ions and polar molecules. Energy production by cells relies on the generation of concentration gradients across the membrane, particularly of ions such as protons. Membrane proteins play a crucial role in the generation of these concentration gradients. The protein bacteriorhodopsin, for example, couples light energy to pump protons across the membrane. The absorption of light by a chromophore, retinal, causes conformational changes that result in the unidirectional movement of protons through the protein, from one side of the membrane to the other. The transport of small molecules such as sugars also relies on conformational changes in membrane proteins known as transporters. In this case, the energy required to move these molecules across a concentration gradient is provided by the hydrolysis of ATP.

## Key Concepts

### A. GENERAL PRINCIPLES

- Proteins are classified into two main groups: soluble proteins and membrane proteins.
- Protein domains are the fundamental units of tertiary structure.
- The folding of soluble proteins is driven by the formation of a hydrophobic core.
- The formation of α helices and β strands satisfies the hydrogen-bonding requirements of the protein backbone.

### B. BACKBONE CONFORMATION

- The peptide bond has partial double bond character, so rotations about it are hindered.
- Peptide groups can be in *cis* or *trans* conformations, but *cis* conformations are rare.
- The backbone torsion angles $\phi$ and $\psi$ determine the conformation of the protein chain.
- The Ramachandran diagram indicates the restrictions on backbone conformation.
- α helices and β strands are formed when consecutive residues adopt similar values of $\phi$ and $\psi$.
- α helices and β strands are often amphipathic.
- Some amino acids are preferred over others in α helices.

### C. STRUCTURAL MOTIFS AND DOMAINS IN SOLUBLE PROTEINS

- Secondary structure elements are connected to form simple motifs.
- Amphipathic α helices can form dimeric structures called coiled coils.
- Hydrophobic sidechains in coiled coils are repeated in a heptad pattern.
- The sidechains of α helices form ridges and grooves that pack against each other.
- α helices pack against each other with a limited set of crossing angles.
- Structures with alternating α helices and β strands are very common.
- Proteins with antiparallel β sheets often form barrels.
- Active sites are located within core elements of protein folds and at the interfaces between domains.

### D. STRUCTURAL PRINCIPLES OF MEMBRANE PROTEINS

- Lipid bilayers form barriers that are nearly impermeable to polar molecules.
- Protein segments that cross the membrane do so as α helices or closed β barrels.
- The more polar sidechains are rarely found within membrane-spanning α helices.
- Transmembrane α helices can be predicted from amino acid sequences by using hydrophobicity scales.
- Transmembrane proteins are stabilized by van der Waals contacts and hydrogen bonds.
- Pumps and transporters use energy to move molecules across the membrane.

## Problems

### True/False and Multiple Choice

1. Globular proteins are generally embedded in the interior of a lipid bilayer.

   True/False

2. The secondary structure of a protein refers to the extent and order of its $\alpha$ helices and $\beta$ sheets.

   True/False

3. Which of the following statements regarding protein domains is NOT true?

   a. Secondary structural elements of a domain generally pack so that a hydrophobic core is formed.

   b. A protein domain normally contains 50–200 residues.

   c. Protein domains are units of tertiary protein structure.

   d. Proteins are only comprised of one protein domain.

4. Match the following proteins with their quaternary structure:

| | |
|---|---|
| a. hemoglobin | i. one subunit |
| b. RNA polymerase | ii. many structurally similar subunits |
| c. myoglobin | iii. many subunits of varied structure |

5. Most protein conformational changes involve breaking and reforming several covalent bonds along the polypeptide chain.

   True/False

6. The only genetically encoded amino acid without a stereoisomer is:

   a. alanine
   b. tryptophan
   c. glycine
   d. proline
   e. lysine

7. The least restricted $\phi$ and $\psi$ angles are found in polypeptides in which class of secondary structure?

   a. right-handed $\alpha$ helix
   b. $\beta$ sheet
   c. left-handed $\alpha$ helix
   d. loop

8. The active site in open twisted $\alpha/\beta$ domains is in a crevice outside the carboxyl ends of the $\beta$ strands.

   True/False

### Fill in the Blank

9. Soluble proteins have mostly ________ sidechains on the inside and mostly ________ sidechains on the outside.

10. The organization of the protein subunits in multimeric proteins is known as the __________ structure.

11. SH2 domains bind to ________ and SH3 domains bind to ________.

12. ______ residues form *cis* peptide bonds in proteins with significant frequency.

13. $\alpha$ helices that have one hydrophobic face and one hydrophilic face are known as ________ helices.

### Quantitative/Essay

14. A group of scientists isolate a novel strain of bacteria. Although the bacteria have normal nucleic acids and proteins, the strain has an abnormal membrane. In this strain, the lipid bilayer is twice as thick (70 Å) as that of a normal strain of bacteria. This extra thickness is entirely due to longer hydrophobic tails and not due to changes in the charged head groups of the phospholipids that comprise the membrane. If you isolated a single transmembrane helix from a protein from this strain, how long would you expect it to be?

15. After sequencing the genome of the bacteria in Problem 14, the scientists want to predict all of the membrane proteins.

    a. Describe the hydrophobicity index procedure for predicting $\alpha$ helical membrane proteins.

    b. How would the hydrophobicity index calculation differ for this new strain compared to normal bacteria?

16. Using the hydrophobicity scale in Figure 4.74, calculate the hydrophobicity index for the 19 contiguous residue window defined by the following sequence:
    Pro-Gly-Ala-Val-Val-Ile-Trp-Phe-Val-Val-Met-Ser-Ala-Ile-Ile-Phe-Tyr-Ala-Thr

    Could this segment be part of a transmembrane helix?

17. Why are isolated secondary structural elements not usually stable in isolation, even though all backbone hydrogen bonds are satisfied?

18. Draw an alanine-alanine dipeptide. Indicate the peptide bond. What factors restrict the rotation about the peptide bond?

19. The simplest form of the Ramachandran diagram (shown in Figure 4.20) is calculated by ignoring

hydrogen bonding, interactions with water, and the hydrophobic effect. Why is this simple form of the diagram still an effective predictor of protein backbone conformation?

**20.** How are the large loop elements of scorpion toxin stabilized without participating in the hydrophobic core?

**21.** Draw a helical wheel for the following sequences:

a. Leu-Asp-Lys-Ile-Val-Arg-Phe-Leu-Gln-Ser-Tyr

b. Leu-Asp-Leu-Lys-Arg-Ser-Glu-Leu-Asn-Tyr-Asn

For each, highlight the hydrophobic residues by marking them with an asterix (*). Do these sequences form amphipathic helices?

**22.** The ridges and grooves of an α helix form an angle of ~25° or ~45° to the helical axis. What are the spacings between the residues that line these two types of grooves? How do these geometric considerations constrain the packing of pairs of helices at relative angles of 50° and 20°?

**23.** Why are binding sites in proteins often located at interdomain boundaries?

**24.** In contrast to globular proteins and regions exposed to the cytoplasm, the integral membrane portions of membrane proteins contain very few residues not ordered into secondary structure elements such as β sheets and α helices. Why is satisfying backbone hydrogen bonds by forming secondary structure elements more important for protein folding in the membrane than in solution?

**25.** A scientist isolates a membrane protein that transports sodium ions from inside the bacterium (where it is at low concentration) into solution (where the sodium ion concentration is high).

a. Does this protein use active or passive transport?

b. What are two energy sources for accomplishing this transport?

**26.** What is the conformational change that occurs when bacteriorhodopsin absorbs light?

**27.** Two mutated bacteriorhodopsin proteins have a Val or Glu at the normal Asp 85 position.

a. Do you predict that either of these mutated proteins will transport protons?

b. If not, at what stage of the proton transport "wire" will the proton transport be blocked compared to wild type?

## Further Reading

### General


Brändén C-I & Tooze J (1999) Introduction to Protein Structure, 2nd ed. New York: Garland Science. Parts of this chapter are adapted, with permission, from Introduction to Protein Structure.

Creighton TE (1993) Proteins: Structures and Molecular Properties, 2nd ed. New York: W.H. Freeman.

Petsko GA & Ringe D (2004) Protein Structure and Function. New York: Oxford University Press.

Richardson JS (1981) The anatomy and taxonomy of protein structure. *Adv. Protein Chem.* 34, 167–339. The following website is an updated online version of this article: http://kinemage.biochem.duke.edu/teaching/anatax/index.html

Tanford C & Reynolds JA (2001) Nature's Robots: A History of Proteins. New York: Oxford University Press.

Voet D & Voet JG (2004) Biochemistry, 3rd ed. New York: John Wiley & Sons.

Williamson M (2011) How Proteins Work. New York: Garland Science.


### References

#### A. General Principles


Bashton M & Chothia C (2007) The generation of new protein functions by the combination of domains. *Structure* 15, 85–99.

Dyson HJ & Wright PE (2004) Unfolded proteins and protein folding studied by NMR. *Chem. Rev.* 104, 3607–3622.

Pawson T (2004) Specificity in signal transduction: from phosphotyrosine-SH2 domain interactions to complex cellular systems. *Cell* 116, 191–203.

Tanford C (1978) The hydrophobic effect and the organization of living matter. *Science* 200, 1012–1018.


#### B. Backbone Conformation


Pauling L & Corey RB (1953) Stable configurations of polypeptide chains. *Proc. R. Soc. Lond. B Biol. Sci.* 141, 21–33.

Ramachandran GN & Sasisekharan V (1968) Conformation of polypeptides and proteins. *Adv. Protein Chem.* 23, 283–438.


#### C. Structural Motifs and Domains in Soluble Proteins


Chothia C, Hubbard T, Brenner S, Barns H & Murzin A (1997) Protein folds in the all-β and all-α classes. *Annu. Rev. Biophys. Biomol. Struct.* 26, 597–627.

Chothia C, Levitt M & Richardson D (1977) Structure of proteins: Packing of α-helices and pleated sheets. *Proc. Natl. Acad. Sci. USA* 74, 4130–4134.

Crick FH (1952) Is α-keratin a coiled coil? *Nature* 170, 882–883.

Woolfson DN (2005) The design of coiled-coil structures and assemblies. *Adv. Protein Chem.* 70, 79–112.

### D. Structural Principles of Membrane Proteins

Khorana HG (1988) Bacteriorhodopsin, a membrane protein that uses light to translocate protons. *J. Biol. Chem.* 263, 7439–7442.

Kyte J & Doolittle RF (1982) A simple method for displaying the hydropathic character of a protein. *J. Mol. Biol.* 157, 105–132.

Lanyi JK (2004) Bacteriorhodopsin. *Annu. Rev. Physiol.* 66, 665–688.

Subramaniam S & Henderson R (2000) Crystallographic analysis of protein conformational changes in the bacteriorhodopsin photocycle. *Biochim. et Biophys. Acta* 1460, 157–165.

White SH & Wimley WC (1999) Membrane protein folding and stability: Physical principles. *Annu. Rev. Biophys. Biomol. Struct.* 28, 319–365.

# Evolutionary Variation in Proteins

Variation and natural selection, the two centrally important tools of evolution, operate ultimately at the level of protein and RNA function. Changes in the nucleotide sequences of genes, brought about by errors in replication, by genetic recombination, or by damage inflicted by the environment, find expression in the altered sequences of proteins and RNA. In this chapter, we shall study how variation in the sequences of related proteins is reflected in their three-dimensional structures. We first discuss a very important principle, which is that all the information that is required to fold a protein into a specific three-dimensional structure is contained within the sequence of the protein. Despite this, the general shape of folded proteins is remarkably tolerant of changes in the amino acid sequence. This tolerance allows sequence variation to accumulate without disrupting the ability of proteins to fold and function. These changes eventually alter the properties of proteins and thereby provide the functional variation that natural selection draws upon. We next discuss how the analysis of the potentially limitless variety of proteins in nature is simplified in practice by the fact that proteins are constructed in a modular fashion from combinations of domains, from a large but limited repertoire. Finally, we illustrate the versatility with which nature creates complicated and highly specialized proteins by exploiting variation in the sequences of these domains.

## A. THE THERMODYNAMIC HYPOTHESIS

### 5.1 The structure of a protein is determined by its sequence

In the flow of information from genes to proteins, an essential step is the spontaneous adoption by protein molecules of the folded structures that enable them to function. The interactions between the atoms in a protein control the folding of the protein molecule into a well-defined structure, which is known as the **native structure** of the protein. This concept is sometimes referred to as the **thermodynamic hypothesis in protein folding**, reflecting our belief that the native structures of proteins arise because polypeptide chains are driven to adopt structures that optimize intrinsic molecular properties, without requiring an external template. To understand what is meant by this idea, imagine that we have a test tube containing unfolded protein molecules in solution. As the test tube comes to thermodynamic equilibrium, the thermodynamic hypothesis states that the protein molecules will fold spontaneously into their functional shapes, without the help of external agents. That is, the sequence of the protein determines its structure (**Figure 5.1**).

**Native structure**

The native structure of a protein is the conformation that it adopts under normal physiological conditions.

**Thermodynamic hypothesis in protein folding**

The thermodynamic hypothesis refers to the idea that the native structure of a protein is determined solely by the properties of the protein and is not the result of an external template.

A crucial aspect of the thermodynamic hypothesis is that each protein molecule is competent to adopt its native structure spontaneously. In the case of protein complexes that contain several different tightly associated proteins, all of the components of the assembly may need to work together to form the native structure. Some proteins can only fold after post-translational modification, such as glycosylation or proteolytic cleavage, or after a cofactor (for example, heme) has

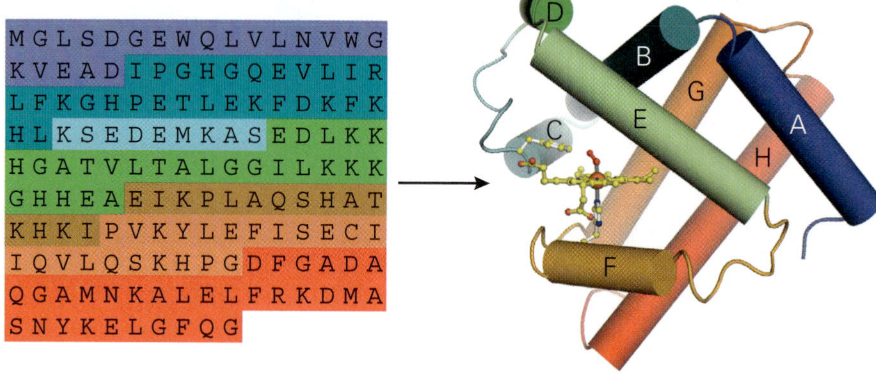


**Figure 5.1 Sequence determines structure**. Shown on the left is the amino acid sequence of the human protein myoglobin, an oxygen storage protein. The sequence and the structure are both colored in a spectrum from *blue* to *red,* from the N-terminus to the C-terminus. You can access the sequence of this and other proteins by going to the online genome databank at the US National Library of Medicine: http://www.nlm.nih.gov.

On the right is shown the three-dimensional structure of the myoglobin protein. The protein has a protoporphyrin IX group (an iron-containing heme group; shown in stick representation) bound to it. (PDB code: 1MBC.)

bound. But even in these more complex cases, the final structure is adopted in accordance with the thermodynamic hypothesis.

Spontaneous assembly is essential for the enormous diversity of structures that are adopted by protein molecules. Imagine for a moment that protein molecules were not able to self-assemble, and depended instead on a system of "assembly machines" that guided each protein molecule into its correct conformation as folding occurred. It is difficult to see how such assembly machines could have evolved to be general enough to impose specific structures upon the enormous variety of protein molecules in the cell. Instead, because proteins operate under the thermodynamic principle, the information required to assemble protein structures is inherent in the polypeptide chains themselves.

Although the folding of proteins occurs without specific guidance from other molecules, we shall see in Chapter 18 that there are proteins that operate in a general way to assist the folding of many different proteins. These proteins are known as **molecular chaperones**, and their primary role is to prevent the aggregation of unfolded proteins into nonspecific clumps. Aggregation can occur when hydrophobic groups, normally in the interior of the protein, are exposed during the folding process. Chaperones prevent aggregation by repeatedly binding to and releasing protein chains until they are folded properly and no longer expose these hydrophobic groups. Although chaperones are crucial for the folding of many proteins, the acquisition of specific structure by a particular protein is still a function of the sequence of the protein itself.

**Molecular chaperones**

Molecular chaperones are protein molecules that help other proteins to fold up properly. Chaperones do this by preventing the nonspecific aggregation of protein chains.

## 5.2 The thermodynamic hypothesis was first established for an enzyme known as ribonuclease-A, which can be unfolded and folded reversibly

That a protein can fold into its native structure without any external assistance was first demonstrated conclusively by Christian Anfinsen in the early 1960s. Anfinsen's experiments, which won him the Nobel Prize in 1972, were carried out using a small enzyme known as ribonuclease-A, which catalyzes the breakdown of RNA into ribonucleotides. What is particularly impressive about these experiments is that they were done without recourse to sophisticated analytical instrumentation, such as NMR spectroscopy or x-ray crystallography, which had only just been developed for application to biological molecules at that time. Indeed, the three-dimensional structure of ribonuclease-A had not yet been determined when these experiments were carried out. How was Anfinsen able to draw such a broad conclusion about the fundamental mechanism of protein folding from simple experiments and without knowledge of the structure of ribonuclease-A?

The catalytic activity of ribonuclease-A purified from natural sources can be measured and used as a standard to determine the amount of active and properly folded enzyme in any solution containing ribonuclease-A. Anfinsen assumed that ribonuclease-A can only achieve catalytic activity when it folds up into its proper native structure, which is shown in **Figure 5.2**. Unless all the catalytic groups come together into a precise configuration, it is unlikely that the enzyme would be able to cleave RNA. By measuring the catalytic activity of ribonuclease-A, Anfinsen could determine readily the extent to which the native structure of the enzyme was present in a sample.

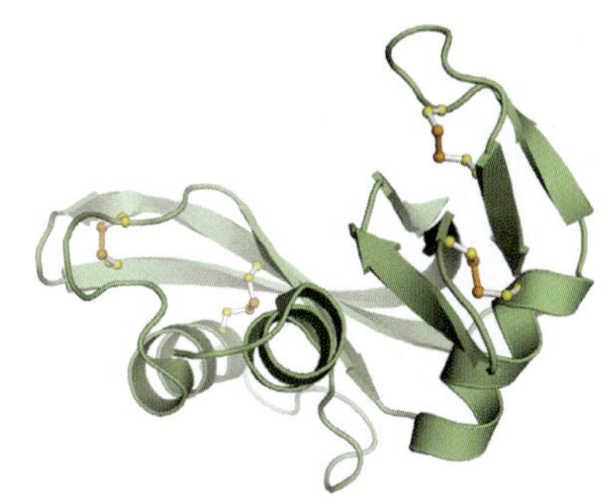

**Figure 5.2 Ribonuclease-A.** The structure of the ribonuclease enzyme used in Anfinsen's experiments is shown here. This protein contains four disulfide bonds (*orange*). (PDB code: 5RSA.)

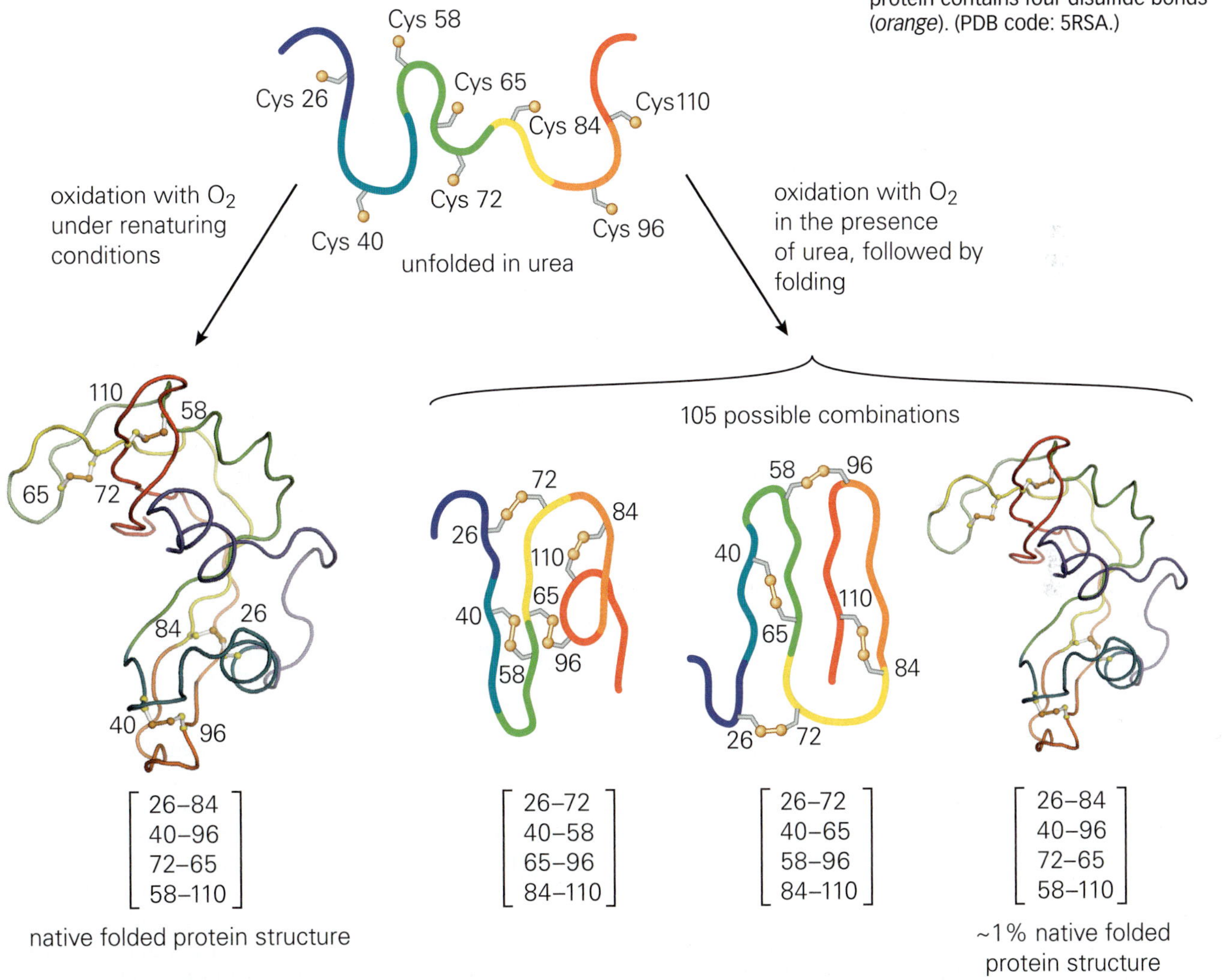

Eight of the 124 amino acids in ribonuclease-A are cysteine residues, and in the native form of the protein all eight are oxidized to form four disulfide bonds (see Figure 5.2). Anfinsen also relied on the disulfide bonds of ribonuclease-A to serve as indicators of the extent to which the native structure of ribonuclease is formed by the population of molecules in a solution. If we assume that ribonuclease-A can be active catalytically only when the polypeptide chain adopts the proper native structure, then it is likely that only one set of all the possible pairings between cysteine sidechains is consistent with active ribonuclease-A (**Figure 5.3**).

Cys 26, Cys 40, Cys 58, Cys 65, Cys 72, Cys 84, Cys 96, Cys110
unfolded in urea
oxidation with $O_2$ under renaturing conditions
oxidation with $O_2$ in the presence of urea, followed by folding
105 possible combinations
[26–84, 40–96, 72–65, 58–110] native folded protein structure
[26–72, 40–58, 65–96, 84–110]
[26–72, 40–65, 58–96, 84–110]
[26–84, 40–96, 72–65, 58–110] ~1% native folded protein structure

**Figure 5.3 A schematic representation of Anfinsen's experiment on ribonuclease-A.** The polypeptide chain of ribonuclease-A is shown at the top of the diagram, colored from *blue* to *red* to indicate progression from the N-terminus to the C-terminus. The chain is shown in a reduced and unfolded conformation. The eight cysteine residues in the chain are indicated, with their thiol groups shown as *orange* spheres. When the reduced form of ribonuclease-A (with no disulfide bonds formed) is allowed to oxidize in the absence of urea, the native folded structure is obtained. This form of the structure, shown on the left, is fully active and has a specific set of connections between cysteine residues that are paired in disulfide linkages, indicated in square brackets. However, when the protein is first oxidized in urea and then allowed to fold in the absence of urea, the disulfides linkages become scrambled. In the Anfinsen experiment, the sample was oxidized extensively, so that no cysteine residues were left unpaired. We assume that each cysteine in the unfolded chain is equally likely to pair with any other cysteine. Under these circumstances, only one of the 105 possible combinations of disulfide pairings corresponds to the correct native structure of the protein. (Adapted from D. Voet and J.G. Voet, Biochemistry, 3rd ed. New York: John Wiley & Sons, 2004.)

**Figure 5.4 The chemical structures of urea and the guanidinium ion, which are potent denaturants of proteins.**

urea

guanidinium ion

The disulfide bonds in ribonuclease-A are readily broken by the addition of reducing agents such as mercaptoethanol, a molecule that contains a reduced thiol group. Reduced ribonuclease-A is then easily unfolded by denaturing agents such as urea (**Figure 5.4**). The addition of two to five molecules of urea or guanidinium ion per 50 molecules of water can result in the unfolding (denaturation) of protein molecules. These denaturing agents are very polar, and they perturb the hydrogen-bonded structure of water. This alters the balance of forces that underlies the hydrophobic effect, and the stability of the folded protein structure is lost.

Anfinsen found that ribonuclease-A denatured in this way has no catalytic activity. The enzyme can be refolded by removing the urea in the presence of oxygen, which causes the disulfide bonds to form again (see Figure 5.3). When this is done, the native enzyme is recovered almost completely, with essentially no loss of catalytic activity. This process is reversible, and the protein can be unfolded and folded repeatedly in a test tube, demonstrating that all the information required for adopting the particular three-dimensional structure that is necessary for the function of ribonuclease-A is encoded entirely in its amino acid sequence.

## 5.3 By counting the number of possible rearrangements of disulfide bonds, we can confirm that ribonuclease-A is completely unfolded by urea and reducing agents

The interpretation of the Anfinsen experiment in terms of the thermodynamic principle of protein folding hinges on the assumption that the reduced and denatured protein corresponds to an essentially random conformation of the polypeptide chain, with no memory of the folded structure. How can we be sure that the addition of mercaptoethanol and urea completely unfolded the protein molecule? One way to do this is to consider how many different ways the eight cysteine residues in ribonuclease can form four disulfide linkages. When the protein molecule is folded into the native structure, one specific set of disulfide pairings occurs. There are, however, 105 different ways in which to form four sets of linkages between eight cysteine residues, 104 of which correspond to incorrect linkages (**Box 5.1**).

We shall make the simplifying assumption that each of the cysteines can pair with any other cysteine with equal probability when ribonuclease-A folds. In reality, cysteines that are closer together in sequence will be more likely to form disulfide bonds because of the entropy of the protein chain (a property that is explained in Chapter 10). Nevertheless, a simple calculation illustrates how Anfinsen arrived at his conclusions. If ribonuclease-A adopts random conformations when it is transferred to urea, and if the cysteine residues in the protein are oxidized to form disulfide linkages when the protein is unfolded, then the correct set of disulfide linkages would be expected to occur only about 1% of the time (1 in 105, as explained in Box 5.1).

Anfinsen showed that when the disulfide bonds in ribonuclease-A are allowed to form in the presence of urea (when the protein is denatured), only about 1% of maximal enzymatic activity is recovered when this oxidized protein is eventually renatured by the removal of urea (**Figure 5.5**). This suggests that the addition of urea randomizes the conformation of the polypeptide chain sufficiently so that the pairing between cysteine residues when disulfide bonds are formed is essentially random. If the urea is removed, the protein begins to fold and, if the disulfide bonds form at this stage, they do so correctly and the protein regains full activity.

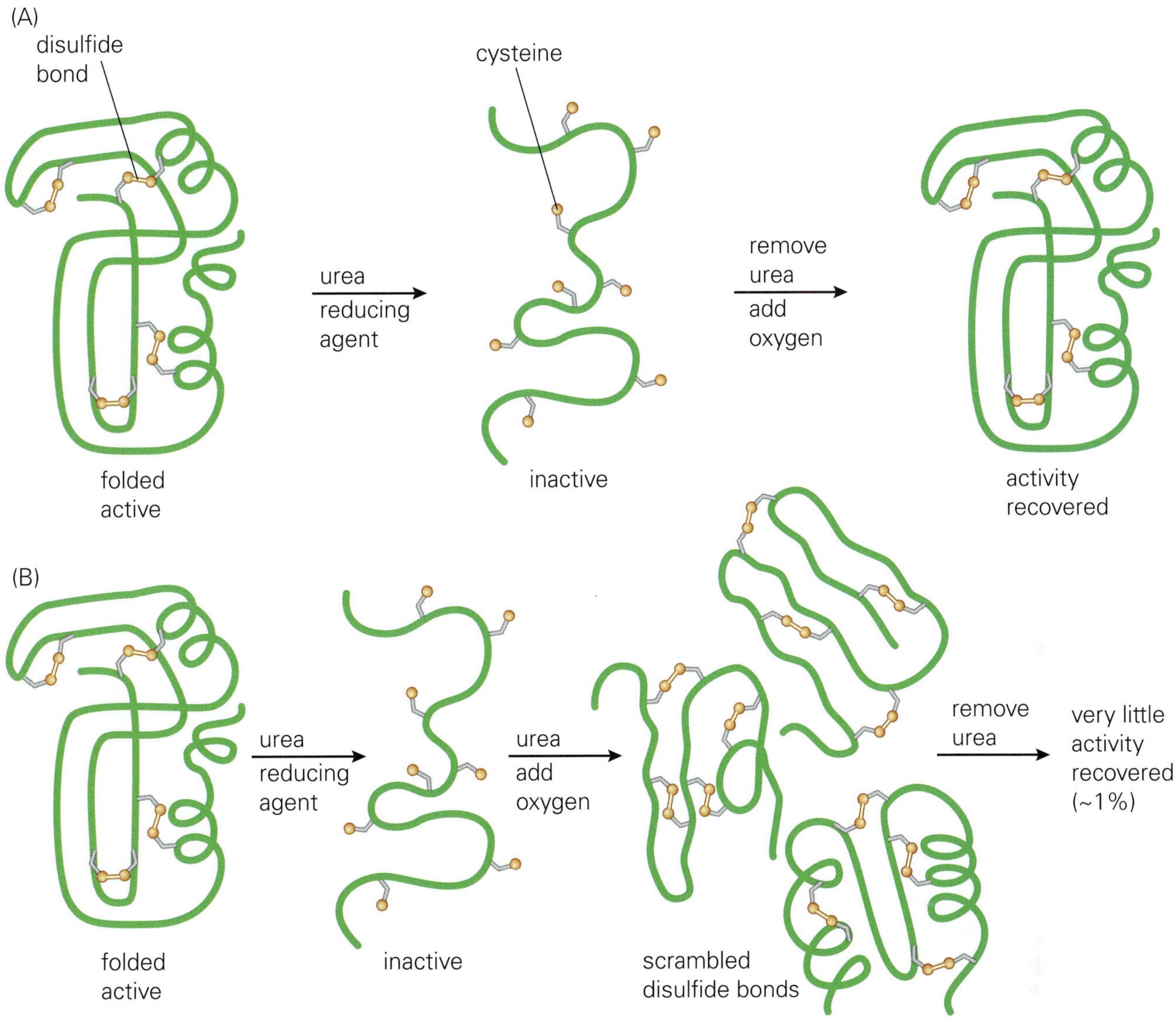


**Figure 5.5 Refolding of ribonuclease-A in the Anfinsen experiment.** (A) When refolding and disulfide bond formation occur together, the activity that is lost upon denaturation is regained. (B) When the disulfide bonds are allowed to form in the presence of urea, very little activity is regained upon subsequent removal of urea. This is interpreted to mean that, in the presence of urea, the protein chain adopts random conformations that allow the disulfide bonds to be scrambled.

# B. SEQUENCE COMPARISONS AND THE BLOSUM MATRIX

## 5.4 Protein structure is conserved during evolution while amino acid sequences vary

The very first protein structures to be determined, in the 1960s, were those of two oxygen-binding proteins known as hemoglobin and myoglobin (Chapter 4). **Hemoglobin** is the oxygen-carrying protein in blood that gives it its red color, and **myoglobin** is a related protein that is found in muscle tissue. The structures of myoglobin and hemoglobin were the first to illustrate an important consequence of the way that proteins evolve: there are families of proteins that have very similar three-dimensional architectures, but which differ considerably in their amino acid sequences.

Myoglobin and hemoglobin are both members of a family of proteins known as the globins, all of which bind to small gaseous molecules such $O_2$ and CO. Globins encapsulate a large heterocyclic ring system known as protoporphyrin-IX or heme (**Figure 5.6**). An iron atom in the heme group binds oxygen molecules reversibly, and most members of the globin family are involved in either transporting or storing oxygen.

## Box 5.1 Counting disulfide bond rearrangements

Ribonuclease-A has eight cysteine residues (**Figure 5.1.1**). The protein is unfolded and the disulfide bonds are broken. The protein is then oxidized under denaturing conditions so that all the cysteine residues form disulfide bonds with other cysteine residues. If we assume that the protein chain is completely flexible, and that each cysteine can pair with any other cysteine, how many different disulfide bonds are possible?

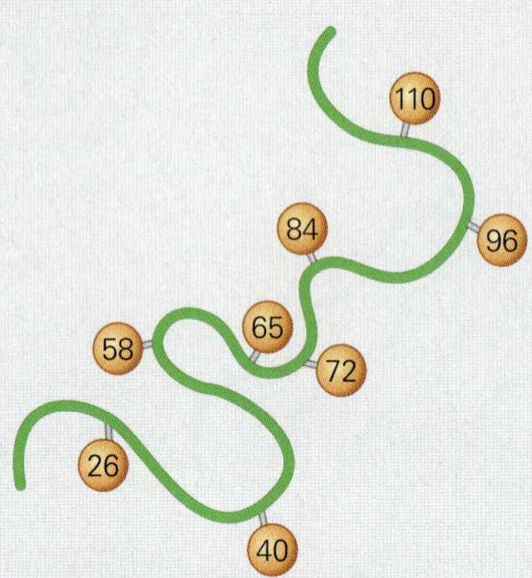


**Figure 5.1.1 The eight cysteine residues of ribonuclease-A, the enzyme used in Anfinsen's experiments.**

One possible set of cysteine pairings to form four disulfide bonds is shown in **Figure 5.1.2**. Note that the order of cysteine residues within a disulfide bond is not important, nor is the vertical position of a disulfide bond in the diagram. In counting disulfide bonds, we have to correctly account for such alternate ways of drawing the same disulfide bonded pattern.

We can count the possible number of disulfide bonds as follows. We draw four pairs of circles, each with two Cys residues representing one disulfide bond (**Figure 5.1.3**).

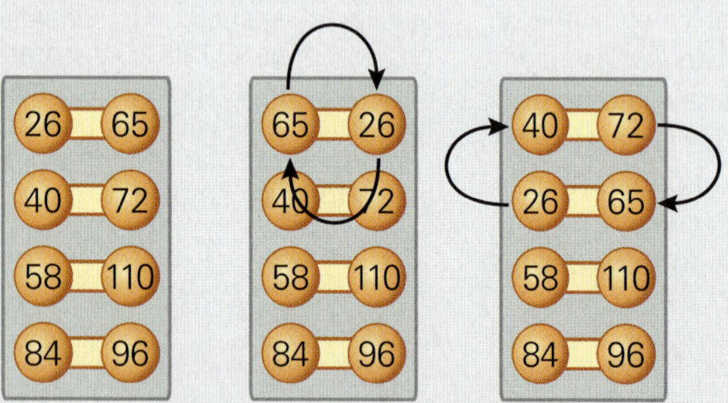


**Figure 5.1.2 One possible set of cysteine pairings.** The three sets of cysteine pairings shown here are all equivalent and should be counted only once.

There are eight ways of choosing a cysteine to fill the first circle, but once we have chosen the first one, there are only seven ways of picking the second one, six ways of picking the third one, and so on. This gives a total of $8 \times 7 \times 6 \ldots \times 1 = 8! = 40{,}320$ combinations.

This number overcounts the actual number of combinations, so we have to correct it as follows. First, the order of the cysteine residues in each pair of circles does not matter. As shown in Figure 5.1.2, 26–65 is the same as 65–26; both refer to a disulfide bond between Cys 26 and Cys 65. To account for this, we need to divide the total count by a factor of 2 for each pair of circles:

$$\frac{40{,}320}{2 \times 2 \times 2 \times 2} = 2520$$

A second correction is that the order of the pair of circles does not matter. That is, if the 26–65 pairing appears first, this is the same situation as when it appears second (or at any other position; see Figure 5.1.2). To correct for this, we have to divide the total count by the number of ways of rearranging four pairs of circles, which is $4 \times 3 \times 2 \times 1 = 4! = 24$. Thus, the number of ways of forming four disulfide bonds between eight cysteine residues is:

$$\frac{2520}{24} = 105$$

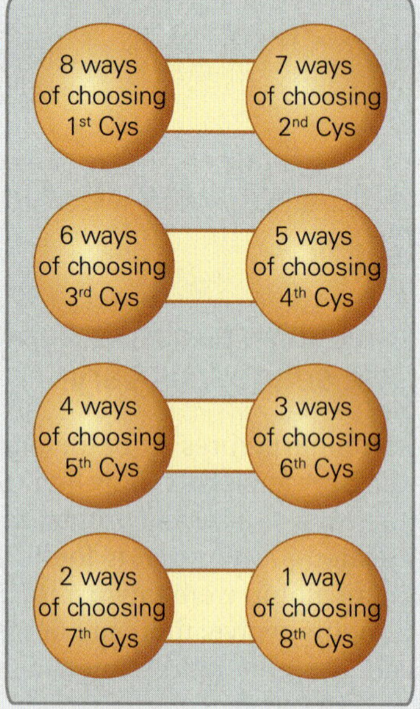


**Figure 5.1.3 Counting the ways of forming disulfide bonds.** After we count the number of ways of filling the circles, we have to correct for vertical and horizontal rearrangements.

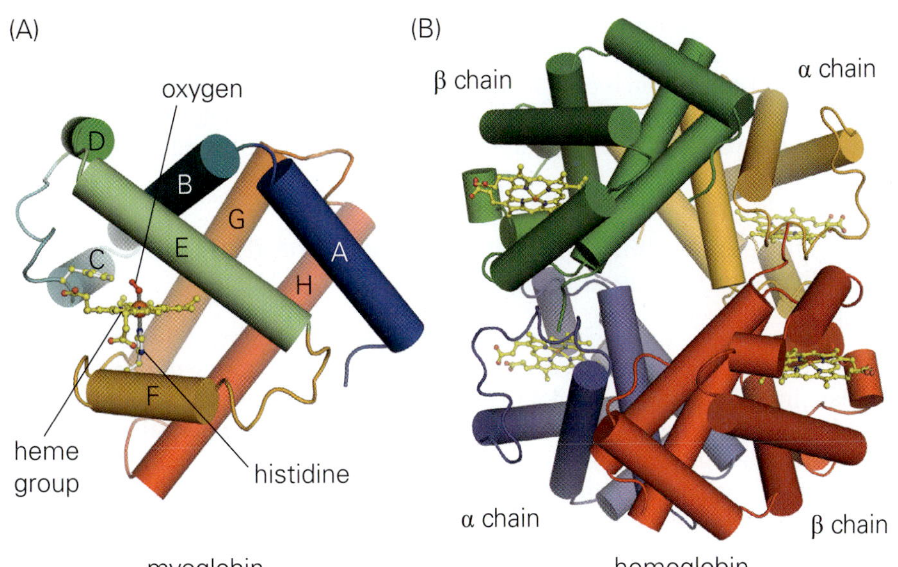


**Figure 5.6 Structures of (A) human myoglobin and (B) hemoglobin.** The heme groups are drawn in a ball-and-stick representation. The histidine sidechain that links the heme iron atom (shown as a large sphere in the center of the heme group) is shown in myoglobin, as is a molecule of oxygen that is bound to the iron atom. The protein structure of myoglobin, shared by each of the subunits of hemoglobin, is known as the globin fold. For myoglobin, each of the helices is shown in a different color, whereas for hemoglobin, each subunit is shown in a different color. (PDB codes: A, 1MBC; B, 1A00.)

Myoglobin is a monomer with a single heme group and one oxygen-binding site (see Figure 5.6A). Hemoglobin is a tetramer in which two each of two kinds of subunits (named α and β) are assembled into a unit that can bind up to four molecules of oxygen (see Figure 5.6B). The assembly of hemoglobin into tetramers allows it to function efficiently as an oxygen-transport protein in blood because the structures of the four subunits of hemoglobin in a tetramer are coupled. The binding of oxygen to any one subunit in the tetramer causes a conformational change that is transmitted to the other subunits, increasing their affinity for oxygen. The phenomenon whereby the binding of a molecule at one site changes the conformation of a protein at another site is known as allostery. We discuss the allostery of hemoglobin in detail in Chapter 14, where we will see how it influences the ability of hemoglobin to release oxygen upon demand. For now, we focus on the similarities and differences in the three-dimensional structures of hemoglobin and myoglobin, and how these relate to their sequences.

The polypeptide chain of myoglobin forms eight α helices (labeled A to H in Figure 5.6A), and the three-dimensional arrangement of these helices is known as the **globin fold.** The polypeptide chains of both the α and the β subunits of hemoglobin also adopt globin folds, and both subunits look strikingly similar to myoglobin (see Figure 5.6).

An alignment of the sequences of human myoglobin and hemoglobin (α and β subunits) is shown in **Figure 5.7**, with the residues that are identical in all three sequences highlighted in green. Since the three-dimensional structure of a protein is determined by its amino acid sequence, we might expect that the sequences of myoglobin and the α and β subunits of hemoglobin would be very similar. It is surprising, then, that only 28 of the ~150 residues of myoglobin are identical in both the α and β subunits of hemoglobin (Figure 5.7). If we compare the sequences in

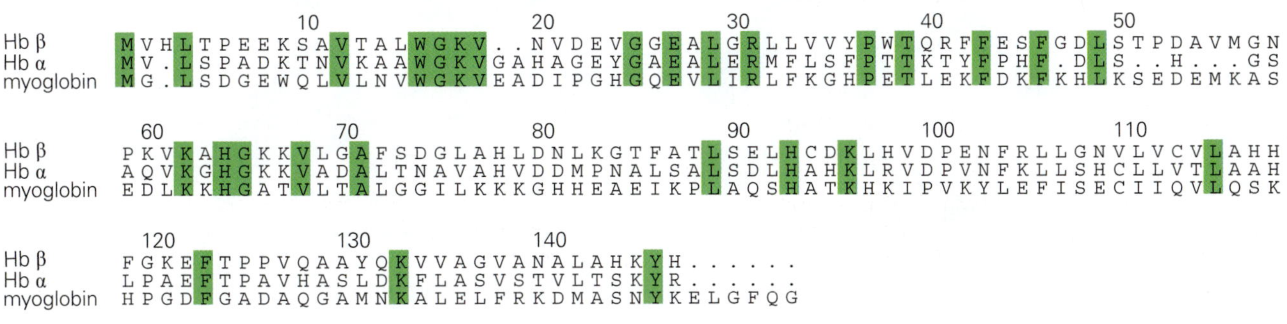


**Figure 5.7 Alignments of the sequences of human myoglobin and hemoglobin.** Identical residues are marked in *green*. The three sequences are identical at only 28 positions out of ~150. In the sequence alignments, gaps are introduced at various positions to improve the match between the sequences. These are indicated by dots. The α and β chains of hemoglobin are abbreviated as Hbα and Hbβ, respectively.

a pair-wise manner, we find that while 45% of the residues are identical between the α and β subunits of hemoglobin, the level of sequence identity between these subunits and myoglobin is only 30% and 26%, respectively. This is a rather striking observation: the close correlations in the fold of the polypeptide chains of hemoglobin and myoglobin are maintained despite very significant differences in the sequence.

## 5.5 The globin fold is preserved in proteins that share very little sequence similarity

The tendency of natural selection to preserve the general shape of the protein fold while allowing sequences to drift is illustrated by comparing the structures of human myoglobin and hemoglobin to those of very distantly related members of the globin family. These include hemoglobins from invertebrates, such as clams, insects, and earthworms, as well as hemoglobin-like oxygen-binding proteins found in the root nodules of certain plants. The three-dimensional structures of these globins are shown in **Figure 5.8**. The polypeptide chains of each of these

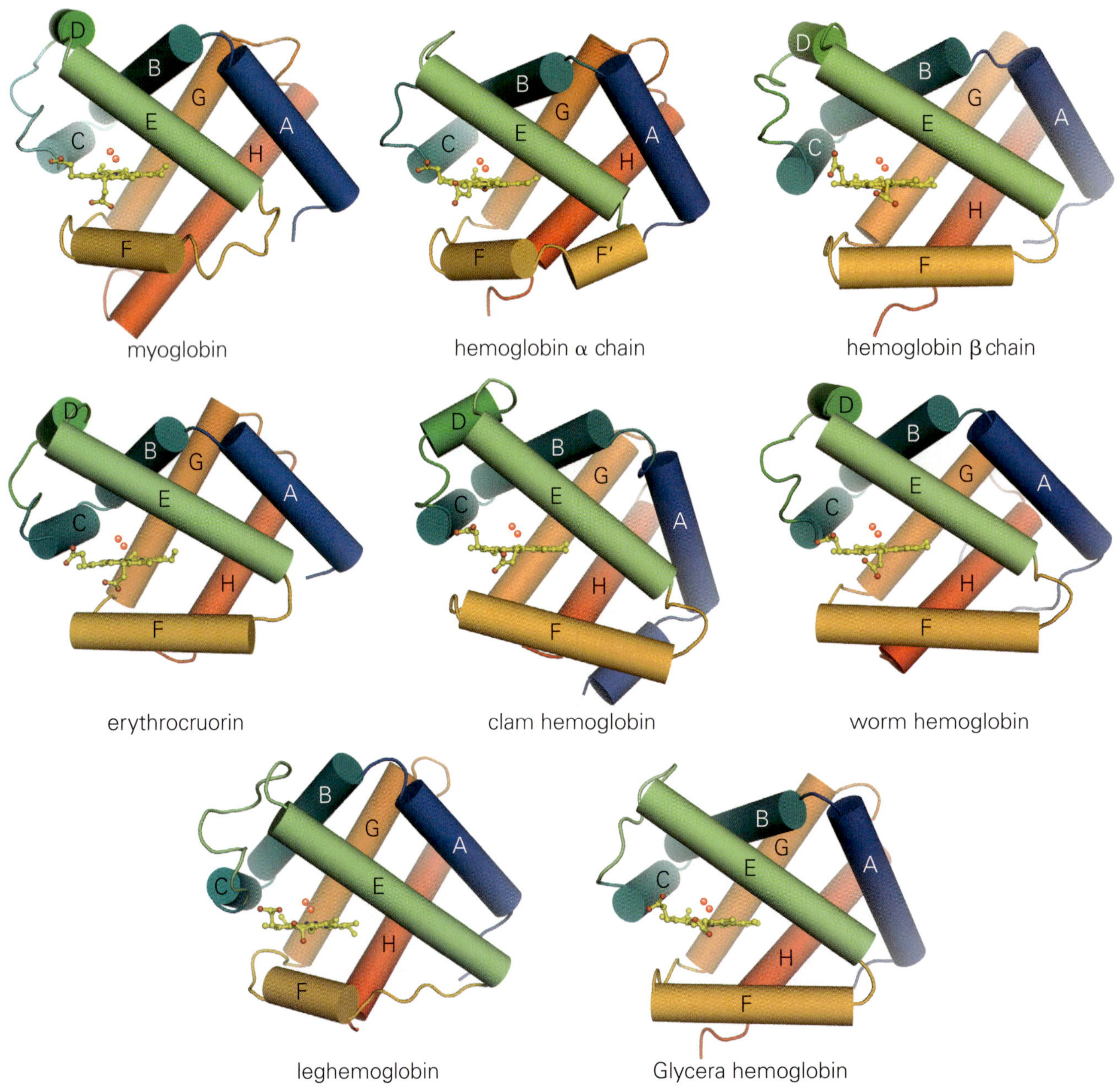


**Figure 5.8 Various examples of proteins with globin folds.** The top row shows myoglobin (PDB code: 1MBC) and the two subunits of human hemoglobin (PDB code: 1A00). The other structures are of oxygen-binding proteins from an insect (erythrocruorin; PDB code: 1ECA), a clam (PDB code: 1HBI), a worm (PDB code: 1BIN), a plant (leghemoglobin; PDB code: 2LHB), and a marine worm (*Glycera dibranchiata*; PDB code: 1HGB).

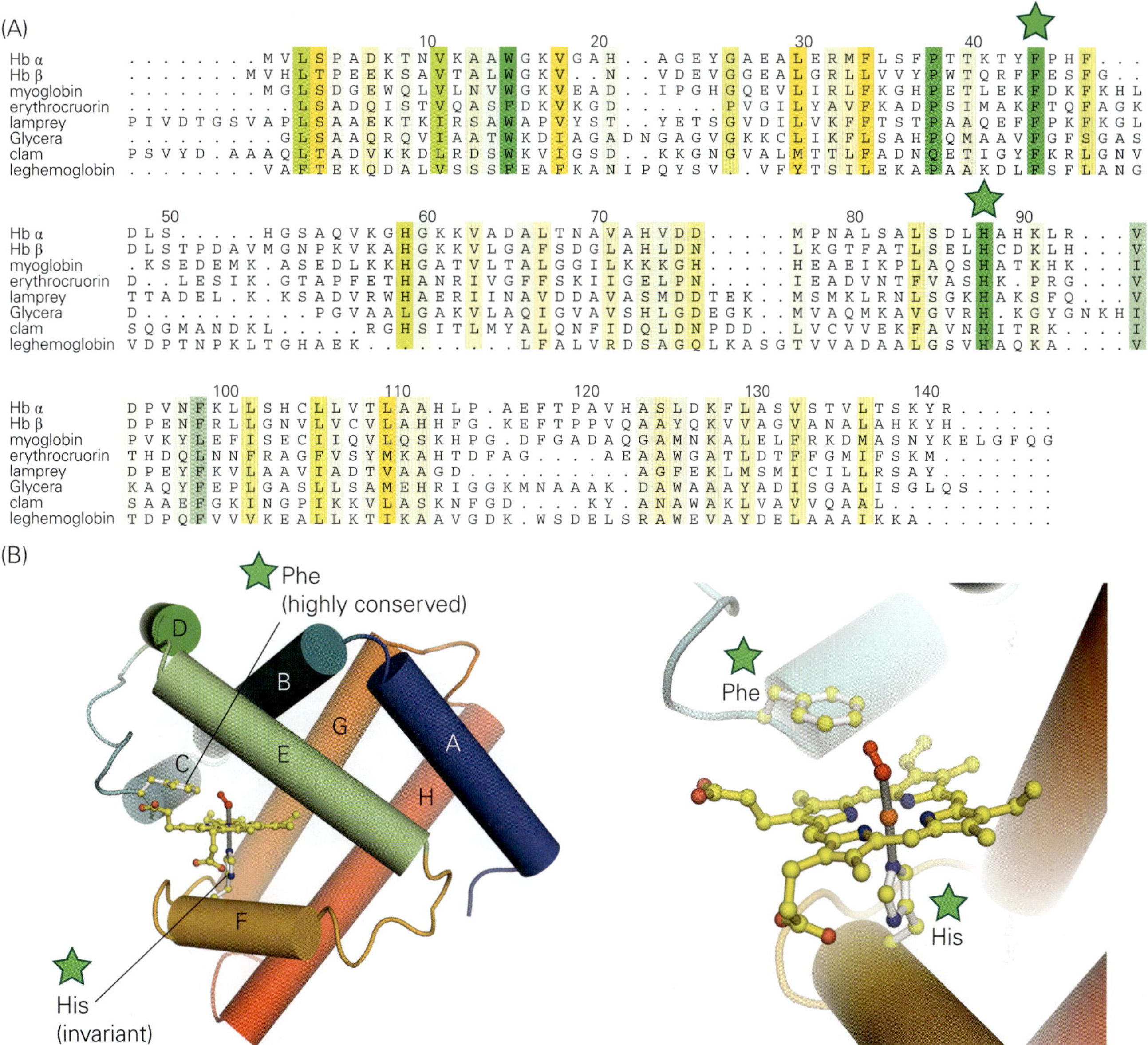


proteins adopts the globin fold, within which a heme group is bound just as in myoglobin and hemoglobin. The general features of the chain fold are strikingly similar to those we have seen before in human myoglobin and hemoglobin, and include the formation of seven or eight α helices.

An alignment of the sequences of several globins is given in **Figure 5.9**. This alignment is colored according to the degree of similarity across different proteins of each of the positions in the polypeptide chain, with highly variable positions left uncolored, and positions where similar amino acids are found being colored from yellow to green as the degree of similarity increases. Just how the degree of similarity is calculated is discussed a little later in this chapter.

Despite the structural conservation, comparison of the sequences of these globins yields a remarkable result. Of the ~150 amino acid residues in each of these proteins, only two residues are preserved without substitution at equivalent positions in the polypeptide chains (see Figure 5.9). These two residues are a histidine in helix F and a phenylalanine located just after helix C. The histidine sidechain forms a covalent bond with the iron atom of the heme group and is therefore necessary for the attachment of the heme group to the protein. The phenylalanine

**Figure 5.9 Alignment of globin sequences.** (A) Sequence similarity in the globins. The sequences of several globin proteins are shown, with the level of sequence similiarity indicated by color (*green* = most similiar; *yellow* to *white* = decreasing similarity). Two residues that are invariant across these sequences are indicated by stars. The histidine is invariant in all globins. The phenylalanine is sometimes replaced by another residue in globins that are not shown here. (B) The structure of myoglobin, showing the locations of the invariant phenylalanine and histidine residues. The degree of similarity between residues at equivalent positions in different proteins is determined by the BLOSUM matrix, as explained in Section 5.7. Hb, hemoglobin. (PDB code: 1MBC.)

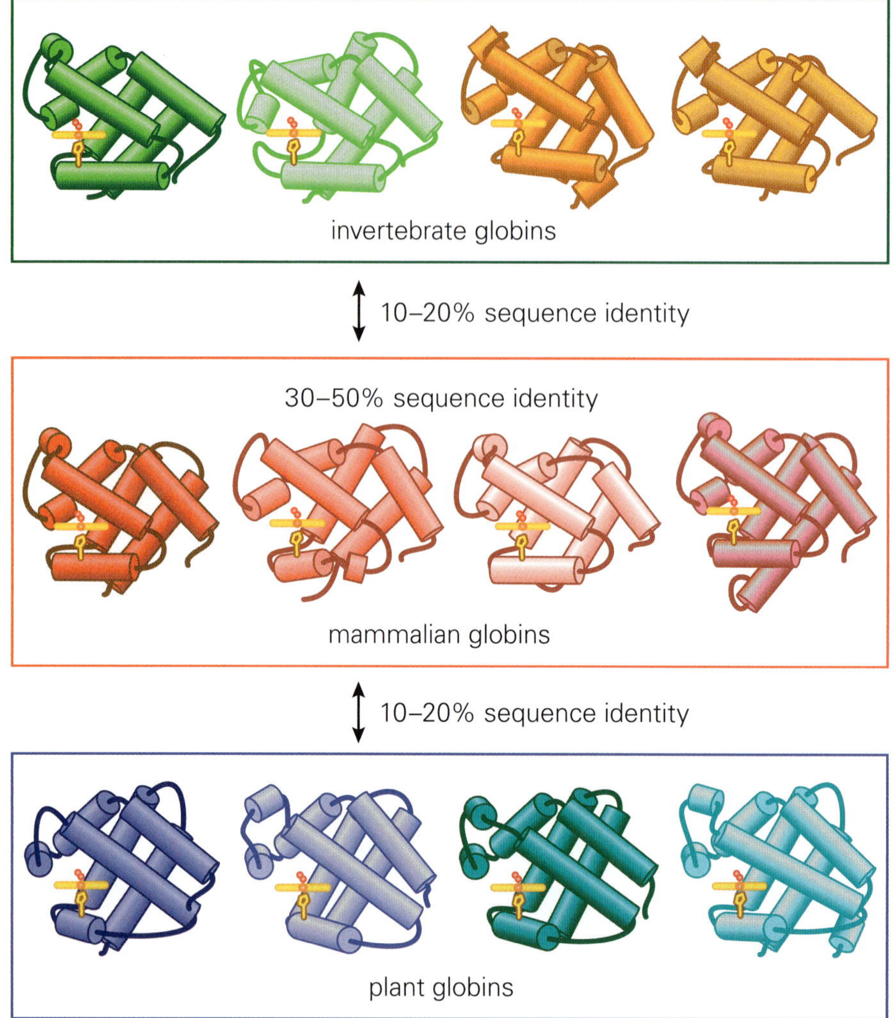


**Figure 5.10 Conservation of the globin fold.** Proteins with globin folds are found in all eukaryotes. Shown here are schematic representations of globin proteins from mammals (center) as well as from invertebrates (top) and plants (bottom). The mammalian proteins are closely related in sequence to each other, but share very little sequence identity with invertebrate or plant globins.

sidechain packs against the flat surface of the heme group and presumably helps stabilize the orientation of the heme within the protein. The sequence requirements for a functional globin protein are apparently so broad that only two amino acids are absolutely required to be invariant across the sequences that are shown in Figure 5.9.

The similarity in the overall chain fold of each of the members of the globin family is apparent to the eye from the structures in Figure 5.8. In contrast, the amino acid sequences of the more distantly related pairs of globins cannot be aligned reliably without referring to either the three-dimensional structures or the sequences of other members of the globin family. For example, the sequence identity between human hemoglobin and a plant globin known as leghemoglobin is only 14% (**Figure 5.10**).

The extensive variation in sequence among the globins illustrates a deep asymmetry in the relationship between protein sequence and three-dimensional structure. The sequence of amino acids in a protein dictates the specific three-dimensional structure that the protein will fold into. Nevertheless, apparently unrelated sequences of amino acids can generate similar three-dimensional structures, as shown in Figure 5.10. Such relationships, in which one property is not uniquely defined by another, are described as degenerate. The relationship between three-dimensional structure and protein sequence is degenerate in that many different protein sequences correspond to the same protein fold.

## 5.6 Similarities in protein sequences can be quantified by considering the frequencies with which amino acids are substituted for each other in related proteins

In the previous section, we used the percentage level of sequence identity as a measure of the similarity between two sequences. A problem with sequence identity as a metric for sequence comparison is that it fails to take into account the properties of the amino acids at positions where substitutions have occurred. Substitution of one amino acid with another that has similar properties is likely to have relatively little effect on the thermodynamics of protein folding. For example, the serine residue (S) at the fourth position in the polypeptide chain of myoglobin is replaced by a threonine residue (T) in the β subunit of hemoglobin (see Figure 5.7). Serine and threonine are rather similar in their chemical properties (see Chapter 1), so we refer to this kind of substitution as a **conservative substitution** or replacement. On the other hand, myoglobin has a glycine residue (G) at the seventh position in the chain, whereas the β subunit of hemoglobin has a glutamate residue (E) at the corresponding position (see Figure 5.7). The negatively charged glutamate residue has quite different properties from the small glycine residue, so this is an example of a **nonconservative substitution**.

**Conservative substitution**

The replacement of an amino acid residue in one protein by one with similar chemical properties in another protein is called a conservative substitution. The replacement of serine by threonine, or leucine by isoleucine, are examples of conservative substitutions.

To what extent are each of the substitutions in the globin sequences conservative or nonconservative? To answer this question, we need a quantitative measure of the similarities and differences between amino acids that substitute for each other. We could, perhaps, base a conservation score on how similar the chemical structures of the two amino acid sidechains are. A leucine sidechain is similar in structure to an isoleucine sidechain, for example, but how similar is it to other hydrophobic sidechains, such as those of phenylalanine or proline? It is difficult to assign numbers to these chemical similarities or differences in a meaningful way.

Instead of considering the chemical structure of the sidechains, a useful metric for comparing amino acids can be derived by taking a statistical approach and using alignments of the sequences of proteins that are known to be related evolutionarily (for example, the sequences of globin proteins). Since we know that the proteins whose sequences we have aligned are related to each other, we can assume that the frequency with which amino acids are substituted for each other in the alignment reflects their similarity. Another way of thinking about this is that, if two related proteins carry out the same or very similar function in two different organisms, then changes in the amino acid sequences have already been filtered through the process of natural selection and evolution.

## 5.7 The BLOSUM matrix is a commonly used set of amino acid substitution scores

A block of aligned sequences for a set of related proteins is used to generate an **amino acid substitution matrix**. One commonly used amino acid substitution matrix is known as BLOSUM, for "block substitution matrix" (**Figure 5.11**). The rows and columns of the BLOSUM matrix are labeled $i$ and $j$, respectively, and there are 20 rows and columns corresponding to each of the 20 amino acids. Each entry in the matrix is called a **substitution score**, $S_{ij}$, and the value of $S_{ij}$ is related to the frequency with which the $i^{\text{th}}$ type of amino acid is replaced by the $j^{\text{th}}$ type of amino acid in the alignments of related proteins, relative to the probability that this substitution could occur by random chance, given the abundance of each of the amino acids.

If the value of $S_{ij}$ is positive, then the $i^{\text{th}}$ type of amino acid is substituted by the $j^{\text{th}}$ type of amino acid more often than expected by random chance. The BLOSUM substitution score for the leucine–isoleucine pair is positive (+2), for example, because isoleucine is found to substitute for leucine more commonly than expected by chance—that is, such substitutions tend to be favored by evolution.

**Amino acid substitution matrix**

A matrix in which each row and column corresponds to one of the 20 amino acids. Each entry in the matrix is related to the probability that one amino acid is replaced by the other in proteins that are related evolutionarily.

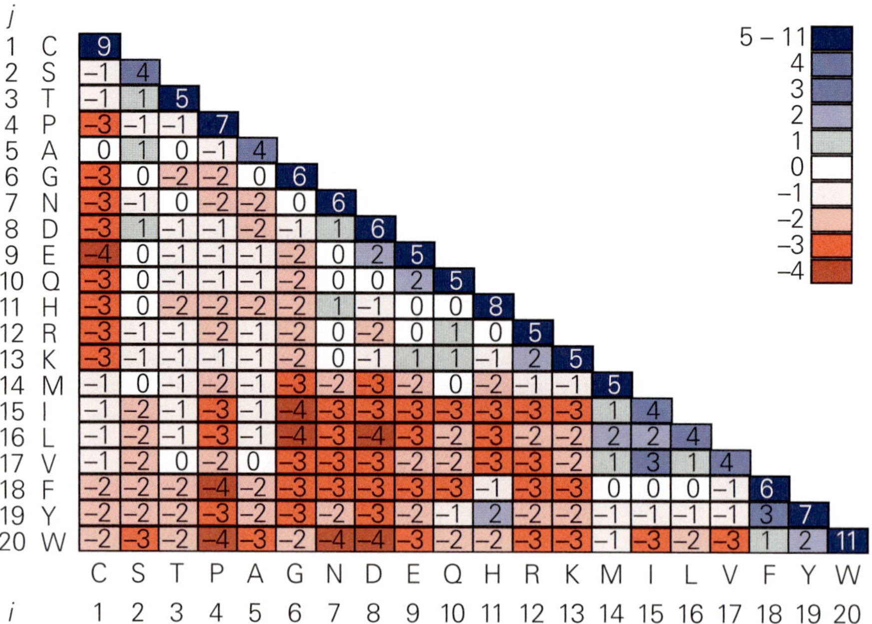


**Figure 5.11 The BLOSUM substitution matrix.** The BLOSUM score, $S_{ij}$, for a substitution of the $i^{th}$ residue by the $j^{th}$ one is color coded. The numerical values for each matrix element are shown. The color bar ranges from *dark blue* for a BLOSUM score of +5 or greater to *dark red* for a BLOSUM score of –4. The particular matrix shown here is referred to as the BLOSUM-62 matrix because it is based on sequence alignments in which no pair of sequences have greater than 62% sequence identity. (The BLOSUM-62 matrix values are taken from S. Henikoff and J.G. Henikoff, *Proc. Natl. Acad. Sci. USA* 89: 10915–10919, 1992.)

In contrast, the substitution score for the leucine–aspartate pair is negative (–4), consistent with our expectation that the replacement of a leucine by an aspartic acid, or vice versa, would be more detrimental to function and is therefore usually eliminated during evolution.

## 5.8 The first step in deriving substitution scores is to determine the frequencies of amino acid substitutions and correct for amino acid abundances

The values in the BLOSUM matrix are derived by first generating sets of alignments of the sequences of proteins that are related to each other. The sequences of globins from different species might provide one such set of aligned sequences. Likewise, the sequences of triosephosphate isomerase enzymes might provide another set of aligned sequences, and so on. Each set of aligned sequences is broken into smaller blocks that are uninterrupted by insertions or deletions. The frequencies of amino acid substitutions are calculated for each block and then averaged over the whole set of aligned sequences. One such block is illustrated in **Figure 5.12**, which shows the alignment of a short segment of eight globin sequences. In actual practice, the sequence blocks that are used to derive the BLOSUM matrix include hundreds of sequences from a very diverse range of species.

It is important that the sequences within an aligned block not be too closely related to each other. To see why this is so, consider the fact that the sequences of hemoglobin subunits from mammalian species are nearly identical to each other because they have not diverged much on an evolutionary timescale. If many mammalian hemoglobin sequences are included in a block of aligned

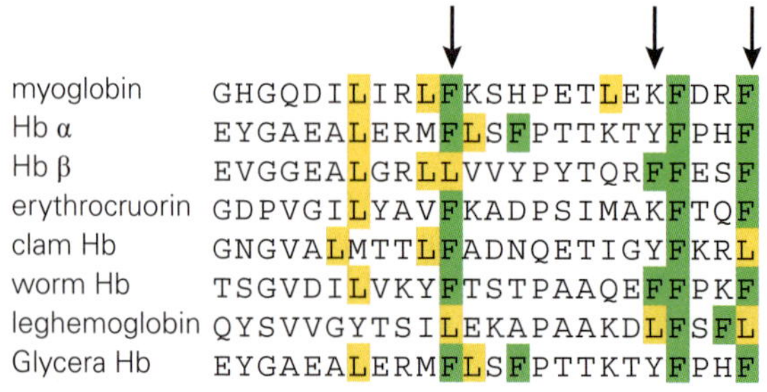


**Figure 5.12 A sequence block.** A block of aligned sequences with no insertions or deletions is shown, consisting of eight different globin sequences ($N$ = 8), each consisting of 24 residues ($M$ = 24). Phe (F) and Leu (L) residues are highlighted in *green* and *yellow*, respectively. There are three columns, indicated by arrows, where there are both Leu and Phe residues. Hemoglobin is abbreviated as Hb.

globin sequences, it would skew the statistical analysis of sequence conservation within the block, leading to the erroneous conclusion that most positions within globin sequences are highly conserved. To avoid this problem, sequences that share a higher degree of sequence identity than a chosen cutoff value are replaced by a single example, so that they do not introduce undue bias in the statistics. A threshold level of 62% sequence identity for inclusion of related sequences in the aligned blocks has been found empirically to produce a BLOSUM matrix that is particularly effective at aligning the sequences of distantly related proteins. The corresponding matrix is known as BLOSUM-62 and is shown in Figure 5.11.

To generate the BLOSUM matrix, one calculates the frequency with which particular pairs of amino acids (for example, F and L) occur at corresponding positions in the aligned blocks (that is, in the same column). One cannot simply stop there, however, because the 20 amino acids do not occur with the same abundance in proteins. If one amino acid (for example, L) is much more abundant than another (for example, C), then L–F pairings should occur more frequently by random chance than C–F pairings, because of the greater abundance of leucine in proteins when compared to cysteine. We need to see whether the L–F and C–F pairings occur more or less frequently than that expected by random chance, given the relative abundances of the different amino acids.

In order to account for amino acid abundances, we define the **substitution likelihood ratio**, $L_{ij}$, of the $i^{\text{th}}$ kind of amino acid being substituted by the $j^{\text{th}}$ kind of amino acid in a related set of proteins as follows:

$$L_{ij} = \frac{\text{frequency of an } i\text{–}j \text{ substitution in the alignment block of related proteins}}{\text{frequency of an } i\text{–}j \text{ substitution in the same block, but with the positions of all the amino acids scrambled randomly}}$$

$$= \frac{f_{ij}}{p_{ij}} \tag{5.1}$$

If $L_{ij} > 1$, then the $i^{\text{th}}$ and $j^{\text{th}}$ amino acids are found at the same position in related sequences more frequently than would be expected by random chance. The $i$–$j$ substitution is then considered to be conservative, and $L_{ij}$ is a quantitative measure of the degree of conservation. If $L_{ij} < 1$ then the $i$–$j$ substitution is found less frequently than expected by random chance. Evolution has selected against this substitution, which is considered to be nonconservative.

The numerator in Equation 5.1, denoted $f_{ij}$, is the observed frequency of $i, j$ pairs in the alignment block. This frequency is given by:

$$f_{ij} = \frac{\text{number of times the } i^{\text{th}} \text{ type and the } j^{\text{th}} \text{ type occur in the same column}}{\text{total number of amino acid pairs in the sequence block}} \tag{5.2}$$

The denominator in Equation 5.1, denoted $p_{ij}$, is the frequency of $i$–$j$ substitutions that would be expected if the amino acids in the block were distributed randomly (see **Box 5.2** for an explanation of how $f_{ij}$ and $p_{ij}$ are calculated for the block shown in Figure 5.12). This term corresponds to the substitution frequencies expected for unrelated proteins, and it corrects for the relative abundances of the individual amino acids in the block. To calculate $p_{ij}$, we start by calculating the probability $p_i$ of finding the $i^{\text{th}}$ type of amino acid anywhere in the sequence block:

$$p_i = \frac{\text{number of occurences of the } i^{\text{th}} \text{ amino acid}}{\text{number of positions in the alignment}}$$

$$= \frac{n_i}{N \times M} \tag{5.3}$$

In Equation 5.3, $n_i$ is the number of times the $i^{\text{th}}$ type of amino acid (for example, L) is found in the block, with $N$ sequences and $M$ residues in each sequence.

## Box 5.2 Calculation of amino acid substitution frequencies

To see how the substitution frequencies are derived, we shall use the sequence block in Figure 5.12 as an example. Based on the sequences in this block, we shall calculate the value of $L_{LF}$—that is, the likelihood of leucine-phenylalanine substitutions relative to random substitutions. According to Equation 5.1, $L_{ij}$ is given by ratio of $f_{ij}$, the frequency of *i-j* substitutions observed in the sequence, to $p_{ij}$, the probability of *i-j* substitutions in the same sequence block, but with the amino acids scrambled randomly:

$$L_{ij} = \frac{f_{ij}}{p_{ij}} \tag{5.1}$$

We shall first calculate the value of $p_{ij}$, which is given by Equation 5.4:

$$p_{ij} = 2 \times p_i \times p_j = 2 \times \frac{n_i \times n_j}{(N \times M)^2} \tag{5.4}$$

Here, *i* stands for leucine and *j* stands for phenylalanine. There are 18 occurrences of L ($n_i$ = 18) and 25 occurrences of F ($n_j$ = 25). There are eight sequences aligned ($N$ = 8) and 24 positions within each sequence ($M$ = 24).

Hence, from Equation 5.3, the probability $p_i$ of finding the $i^{\text{th}}$ residue anywhere in the sequence block is given by:

$$p_i = \frac{18}{8 \times 24} = 0.09375 \quad (i = \text{leucine}) \tag{5.2.1}$$

Likewise,

$$p_j = \frac{25}{8 \times 24} = 0.1302 \quad (j = \text{phenylalanine}) \tag{5.2.2}$$

The joint probability of finding L and F at two specified positions (for example, at two positions in the same column) is (see Equation 5.4):

$$p_{ij} = 2 \times p_i \times p_j = 2 \times 0.09375 \times 0.1302 = 0.0244 \tag{5.2.3}$$

Next, we shall calculate the value of $f_{ij}$ using Equation 5.2:

$$f_{ij} = \frac{\text{number of times the } i^{\text{th}} \text{ type and the } j^{\text{th}} \text{ type occur in the same column}}{\text{total number of amino acid pairs in the sequence block}} \tag{5.2.4}$$

We begin by counting the number of times F and L occur together in the same column. If there are $n_L$ leucine residues in a column and $n_F$ phenylalanine residues in the same column, then the number of possible pairs is given by

$$n_{FL} = \text{number of pairs} = n_L \times n_F \tag{5.2.5}$$

There are three columns in the sequence block where F and L are both present (indicated by arrows in Figure 5.12). For the first column, there are six Phe residues ($n_F$ = 6) and two Leu residues ($n_L$ = 2) and so:

$$n_{FL} = 6 \times 2 = 12 \quad (\text{for first column})$$

For the second column, $n_F$ = 2 and $n_L$ = 1, so:

$$n_{FL} = 2 \times 1 = 2 \quad (\text{for second column})$$

For the third column, $n_F$ = 6 and $n_L$ = 2, so:

$$n_{FL} = 6 \text{ x } 2 = 12 \quad (\text{for third column})$$

This probability, $p_i$, lets us calculate the frequency of finding an *i*, *j* pair if the amino acids were distributed randomly in the sequence block. For a random distribution, we assume that the probability of finding the $i^{\text{th}}$ type of amino acid anywhere is independent of the probability of finding the $j^{\text{th}}$ type of amino acid somewhere else. The probability of finding the $i^{\text{th}}$ type of amino acid at one position and the $j^{\text{th}}$ type of amino acid at another position in the same column is given by the product of the individual probabilities:

$$p_{ij} = 2 \times p_i \times p_j = 2 \times \frac{n_i \times n_j}{(N \times M)^2} \tag{5.4}$$

The multiplicative factor of 2 in Equation 5.4 is necessary because we do not consider the order to be important: we consider an L–F pair to be the same as an F–L pair. Once we have determined $f_{ij}$ and $p_{ij}$, the substitution likelihood ratio, $L_{ij}$, for the *i-j* substitution is determined readily by taking the ratio of these two values (see Equation 5.1).

## 5.9 The substitution score is defined in terms of the logarithm of the substitution likelihood

For computational convenience, the BLOSUM substitution score, $S_{ij}$, for the *i*, *j* pairing is defined in terms of the base-2 logarithm of the substitution likelihood as follows:

The total number of F–L pairings is therefore:

$12 + 2 + 12 = 26$

Now we need to calculate the total number of amino acid pairings, without regard to sequence, in the entire alignment. In any particular column, there are eight amino acid residues. There are eight possible ways of choosing the first amino acid in the pair, and seven possible ways of choosing the second one. There are, in this way, 8 × 7 = 56 possible pairings in each column. Since there are 24 columns, the total number of pairings is 56 × 24 = 1,344. The order of the pairings does not matter, so we have to divide this count by two, giving us 1344/2 = 672 possible pairings.

We can now calculate the value of $f_{ij}$, by using Equation 5.2:

$$f_{ij} = \frac{26}{672} = 0.0387 \quad (5.2.6)$$

The value of $p_{ij}$ (0.0244; Equation 5.2.3), where $p_{ij}$ is the expected probability of finding a Leu and a Phe in the same column by random chance, is lower than the value of $f_{ij}$ (0.0387), the observed frequency of finding Phe and Leu in the same column in the given sequence alignment. From this we conclude that the Phe–Leu substitution is a conservative one (that is, it has been selected for in evolution). The likelihood ratio for the Phe–Leu substitution is given by

$$L_{ij} = \frac{f_{ij}}{p_{ij}} = \frac{0.0387}{0.0244} = 1.59$$

The log-likelihood substitution score, $S_{ij}$, is given by:

$$S_{ij} = 2\log_2 L_{ij}$$

If your calculator does not have the $\log_2$ function, you can calculate $\log_2 L_{ij}$ as follows:

If $\log_2 L_{ij} = x$, then:

$$2^x = L_{ij} = 1.59$$

$$\Rightarrow \left(10^{0.301}\right)^x = 10^{0.301x} = 1.59$$

$$\Rightarrow 0.301x = \log_{10} 1.59 = 0.201$$

$$\Rightarrow x = \log_2 L_{ij} = \frac{0.201}{0.301} = 0.668$$

$$S_{ij} = 2 \times 0.668 = 1.336$$

The value of $S_{ij}$ that we have calculated here is not the same as the actual value of $S_{ij}$ for the Phe–Leu substitution in the BLOSUM substitution matrix. In the BLOSUM substitution matrix, $S_{ij}$ = 0 for Phe–Leu, where a value of $S_{ij}$ = 0 indicates a neutral substitution—that is, it is neither favored nor disfavored by natural selection. The discrepancy between our calculated score and the one that is in the BLOSUM matrix arises from the poor sampling statistics in our small sequence alignment block (see Figure 5.12). In actual practice, the values of $S_{ij}$ are derived from the analysis of hundreds of sequence alignment blocks by automated computer procedures.

$$S_{ij} = 2 \times \log_2 L_{ij} \quad (5.5)$$

Given this definition, the substitution likelihood ratio, $L_{ij}$, can be calculated from the substitution score, $S_{ij}$, by taking the exponent of both sides of Equation 5.5:

$$L_{ij} = 2^{\frac{S_{ij}}{2}} \quad (5.6)$$

The substitution score, $S_{ij}$, defined in this way is additive when considering sequence variation at multiple positions. The multiplication by 2 is arbitrary, and it sets the scale for the scores so that a likelihood ratio of 2 corresponds to a substitution score, $S_{ij}$, of 2. It is very common to take the logarithm of probabilities or frequencies when considering statistics, because of the resulting additivity. In Chapter 7 we shall see that this concept underlies the definition of entropy. The choice of the base-2 logarithm in the definition of the substitution matrix is a matter of computational convenience—binary computers handle base-2 logarithms much faster than natural or base-10 logarithms.

To understand why the BLOSUM substitution score is additive, consider the example shown in **Figure 5.13**. In this example, a segment of a protein (Protein 1) is being compared with segments from two other proteins (Proteins 2 and 3). Proteins 1 and 2 have only one identical residue over the span being aligned, whereas Proteins 1 and 3 have three identical residues. Although Proteins 1 and 3 have a

| | 1 | 2 | 3 | 4 | 5 | 6 | 7 | 8 | 9 | 10 | 11 | |
|---|---|---|---|---|---|---|---|---|---|---|---|---|
| $i$ | A | V | N | G | S | T | L | L | A | W | R | protein 1 |
| | | | | | | | | | | • | | |
| $j$ | M | L | S | Q | D | P | F | I | Q | W | K | protein 2 |
| $L_{ij}$ | $L_{AM}$ | $L_{VL}$ | $L_{NS}$ | $L_{GQ}$ | $L_{SD}$ | $L_{TP}$ | $L_{LF}$ | $L_{LI}$ | $L_{AQ}$ | $L_{WW}$ | $L_{RK}$ | $L = L_{AM} \times L_{VL} \times \ldots = 91.4$ |
| | 0.71 | 1.41 | 0.71 | 1.00 | 1.41 | 0.71 | 1.00 | 2.00 | 0.71 | 45.25 | 2.00 | |
| $S_{ij}$ | $S_{AM}$ | $S_{VL}$ | $S_{NS}$ | $S_{GQ}$ | $S_{SD}$ | $S_{TP}$ | $S_{LF}$ | $S_{LI}$ | $S_{AQ}$ | $S_{WW}$ | $S_{RK}$ | $S = S_{AM} + S_{VL} + \ldots = 13$ |
| | −1 | +1 | −1 | 0 | +1 | −1 | 0 | +2 | −1 | +11 | +2 | |

| | 1 | 2 | 3 | 4 | 5 | 6 | 7 | 8 | 9 | 10 | 11 | |
|---|---|---|---|---|---|---|---|---|---|---|---|---|
| $i$ | A | V | N | G | S | T | L | L | A | W | R | protein 1 |
| | | | | • | • | • | | | | | | |
| $j$ | D | Q | R | G | S | T | Y | P | K | C | L | protein 3 |
| $L_{ij}$ | $L_{AD}$ | $L_{VQ}$ | $L_{NR}$ | $L_{GG}$ | $L_{SS}$ | $L_{TT}$ | $L_{LY}$ | $L_{LP}$ | $L_{AK}$ | $L_{WC}$ | $L_{RL}$ | $L = L_{AD} \times L_{VQ} \times \ldots = 2.0$ |
| | 0.50 | 0.50 | 1.00 | 8.00 | 4.00 | 5.66 | 0.71 | 0.35 | 0.71 | 0.5 | 0.5 | |
| $S_{ij}$ | $S_{AD}$ | $S_{VQ}$ | $S_{NR}$ | $S_{GG}$ | $S_{SS}$ | $S_{TT}$ | $S_{LY}$ | $S_{LP}$ | $S_{AK}$ | $S_{WC}$ | $S_{RL}$ | $S = S_{AD} + S_{VQ} + \ldots = 2$ |
| | −2 | −2 | 0 | +6 | +4 | +5 | −1 | −3 | −1 | −2 | −2 | |

**Figure 5.13 Comparing substitution likelihood ratios, $L_{ij}$, and substitution scores, $S_{ij}$.** A segment of a protein (Protein 1) is aligned with segments of two other proteins (Proteins 2 and 3). The substitution likelihood ratios, $L_{ij}$, at each position in the two alignments are shown, as are the substitution scores, $S_{ij}$. We assume that the probabilities of substitutions at each position are independent, and so the aggregate likelihood ratio, $L$, for each alignment is the product of the substitution likelihoods at each position. Because the substitution score, $S_{ij}$, is defined in terms of the logarithm of the likelihood ratios (see Equation 5.5), the substitution score for the whole alignment, $S$, is the sum of the individual substitution scores. When the alignments involve large numbers of residues, it is more convenient to use the logarithmic substitution score because it is additive.

higher level of sequence identity, a glance at the alignments shows that the substitutions between Proteins 1 and 2 are more conservative than those between Proteins 1 and 3. Is Protein 1 more closely related to Protein 2 or to Protein 3? To answer this question quantitatively, we calculate the substitution likelihood ratios for the two alignments shown in Figure 5.13.

First, we calculate the likelihood that Proteins 1 and 2 are related. This is given by the likelihood ratio, $L$:

$$L = \frac{\text{probability that the alignment is of related proteins}}{\text{probability that the alignment is of unrelated proteins}} \tag{5.7}$$

We assume that the substitutions at each position are independent and so the joint likelihood is the product of the individual likelihoods:

$$L = \left( \frac{\begin{array}{c}\text{probability that an A–M substitution}\\ \text{occurs in related proteins}\end{array}}{\begin{array}{c}\text{probability that an A–M substitution}\\ \text{occurs in unrelated proteins}\end{array}} \right) \times \left( \frac{\begin{array}{c}\text{probability that a V–L substitution}\\ \text{occurs in related proteins}\end{array}}{\begin{array}{c}\text{probability that a V–L substitution}\\ \text{occurs in unrelated proteins}\end{array}} \right)$$
$$\times \left( \frac{\begin{array}{c}\text{probability that an N–S substitution}\\ \text{occurs in related proteins}\end{array}}{\begin{array}{c}\text{probability that an N–S substitution}\\ \text{occurs in unrelated proteins}\end{array}} \right) \times \ldots \tag{5.8}$$

To relate Equation 5.8 to the substitution likelihood ratios derived from the aligned sequence blocks, we reason as follows. The probability that a particular substitution (for example, A–M) occurs in related proteins is given by the frequency of that pairing in the aligned sequence blocks. Also, the probability that the same substitution occurs in unrelated proteins is given by the probability of finding this pairing in the scrambled sequence block. That is, for the A–M substitution:

$$\frac{\begin{array}{c}\text{probability that an A–M substitution}\\ \text{occurs in related proteins}\end{array}}{\begin{array}{c}\text{probability that an A–M substitution}\\ \text{occurs in unrelated proteins}\end{array}} = \frac{f_{AM}}{p_{AM}} = L_{AM} \tag{5.9}$$

This tells us that the substitution likelihood ratio for the alignment of Protein 1 to Protein 2 is simply given by the product of the likelihood ratios of the individual substitutions, which are shown in Figure 5.13 :

$$L = L_{AM} \times L_{VL} \times L_{NS} \times \ldots = 91.4 \tag{5.10}$$

According to Equation 5.10, the alignment between Proteins 1 and 2 is 91.4 times more likely to have occurred in related proteins than by random chance. A similar calculation for the second alignment shows us that it is only twice as likely to have arisen from related proteins relative to random chance (Figure 5.13). Hence, even though the number of identical residues is greater in the second alignment, we can conclude that Protein 1 is more likely to be related to Protein 2 than to Protein 3.

For much larger alignments, where thousands of positions are being compared by computer, it is faster to use a score that is additive rather than multiplicative. Recall that the BLOSUM substitution score is defined in terms of the base-2 logarithm of the likelihood of substitution ($S_{ij} = 2 \times \log_2 L_{ij}$; see Equation 5.5). Taking the base-2 logarithm of both sides of Equation 5.10, we get:

$$\begin{aligned}\log_2(L) &= \log_2(L_{AM} \times L_{VL} \times L_{NS} \times \ldots) \\ &= \log_2(L_{AM}) + \log_2(L_{VL}) + \log_2(L_{NS}) + \ldots \end{aligned} \tag{5.11}$$

Recall from Section 5.9 that the score defined in this way is multiplied by 2 to give the actual score. This is done so that a score of 2 corresponds to a likelihood ratio of 2. Multiplying both sides of Equation 5.11 by 2 gives:

$$\begin{aligned}&2 \times \log_2 L = 2 \times \log_2(L_{AM}) + 2 \times \log_2(L_{VL}) + 2 \times \log_2(L_{NS}) + \ldots \\ &\Rightarrow S = S_{AM} + S_{VL} + S_{NS} + \ldots \end{aligned} \tag{5.12}$$

Equation 5.12 confirms that the BLOSUM substitution scores defined according to Equation 5.5 are indeed additive. Thus, to compare two alignments, we can simply add up the substitution scores at each position. Alignments with greater aggregate scores indicate closer evolutionary relationships than those with lower aggregate scores, as indicated in Figure 5.13.

## 5.10 The BLOSUM substitution scores reflect the chemical properties of the amino acids

To understand how to interpret the substitution scores, let us look at the elements of the BLOSUM matrix for two residues, tryptophan (W) and arginine (R) (**Figure 5.14**). Entries on the diagonal in the BLOSUM matrix tell us how strongly conserved each of the amino acids tends to be. For tryptophan (W), the diagonal score, $S_{WW}$, is 11. The corresponding likelihood ratio, $L_{WW}$, is 45 (see Equation 5.6). This means that in the alignment blocks used to derive the BLOSUM matrix, a W–W match is seen 45 times more often than would be expected based on random chance alone.

Tryptophan is a particularly difficult amino acid to replace because no other amino acid resembles tryptophan closely in size or shape. The large aromatic ring

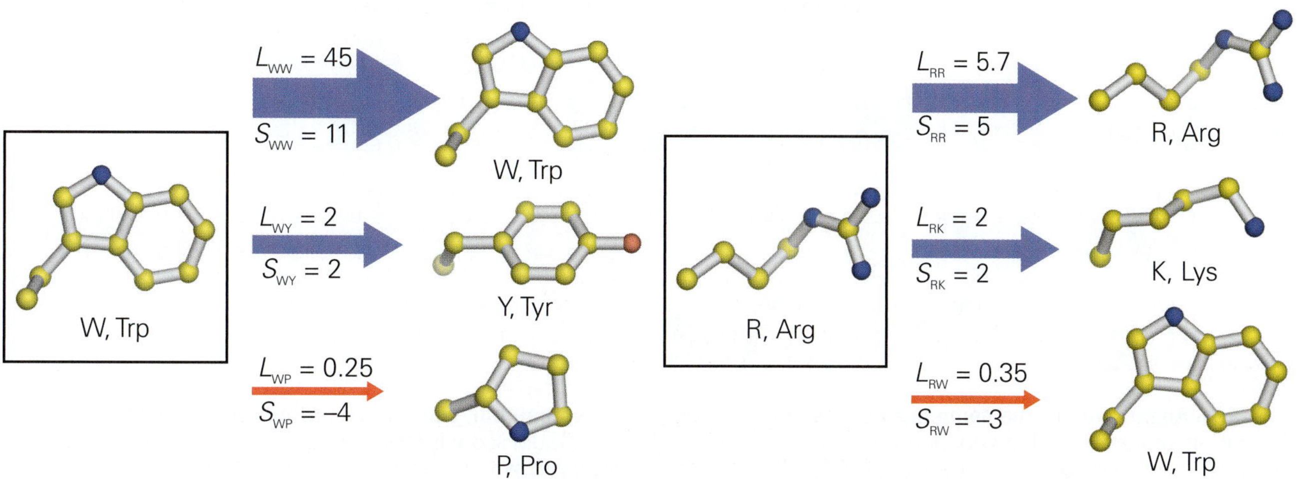


**Figure 5.14 Examples of substitution scores for tryptophan and arginine.** Positive substitution scores, such as for Trp being replaced by Trp or Tyr, indicate conservation. Negative substitution scores, such as for Trp being replaced by Pro or Arg being replaced by Trp, indicate negative selection.

in the tryptophan sidechain is often an important component of the hydrophobic cores of proteins, and the high score for W–W matches in the BLOSUM matrix reflects the tendency of tryptophan residues to be strongly conserved, because substitutions would be likely to destabilize the protein. Tryptophan also has positive scores when substituted by tyrosine ($S_{WY} = 2$) and phenylalanine ($S_{WF} = 1$). The likelihood ratios corresponding to these scores are $L_{WY} = 2.0$ and $L_{WF} = 1.4$, respectively, indicating a bias towards these chemically reasonable substitutions, both of which have aromatic sidechains.

The scores for all other substitutions for tryptophan are negative. For example, $S_{ij} = -4$ for tryptophan substituted by proline, asparagine, and aspartic acid. The corresponding likelihood ratio, $L_{ij}$ is 0.25 in each case, indicating that these substitutions are seen four times less frequently than would be expected based on random chance. This means there is selective pressure against such substitutions during evolution.

Arginine has a lower diagonal score ($S_{RR} = 5$) than tryptophan. This means that arginine–arginine matches are only 5.7 times more likely to be seen than expected from random probability, in contrast to the 45-fold enhancement of the W–W match probability. Arginine residues tend to be on the surfaces of proteins, where they can often be replaced by other polar residues without functional consequence. In the BLOSUM matrix, the most likely substitution for arginine is lysine ($S_{RK} = 2$, $L_{RK} = 2$), which is also positively charged, followed by several other polar residues. Arginine is rarely substituted by tryptophan ($S_{RW} = -3$, $L_{RW} = 0.35$), reflecting the tendencies of these two residues to be found in polar and hydrophobic environments within the protein, respectively.

## 5.11 Substitution scores are used to align sequences and to detect similarities between proteins

In Section 5.4, we emphasized the fact that the sequences of myoglobin and the two subunits of hemoglobin have surprisingly few residues in common. The alignment in Figure 5.7 shows that the residues that are identical between the three sequences are sparsely distributed. A much more informative version of this alignment is shown in **Figure 5.15**, where each position is colored based on the BLOSUM substitution scores (a similar color coding is also used in Figure 5.9). We can see in Figure 5.15 that most of the substitutions between these three proteins are in fact conservative, consistent with their close structural and functional relationship. The least conservative substitutions tend to occur on the surface of the protein, where they are less likely to affect the stability of the protein.

The BLOSUM substitution scores are also helpful in aligning the sequences of proteins when the level of sequence identity is low. Computer programs such as BLAST, which is available on the website of the U.S. National Library of Medicine, enable extremely rapid sequence alignment. Many alternative alignments

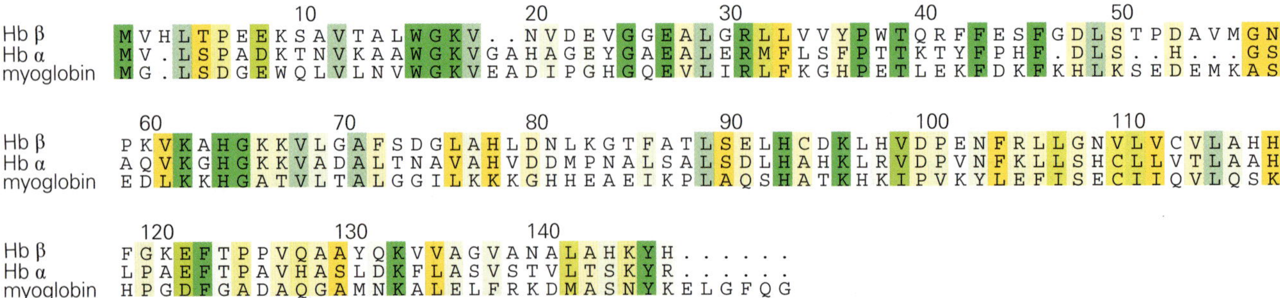


**Figure 5.15 Alignments of the sequences of human myoglobin and hemoglobin, colored according the BLOSUM substitution scores at each position.** Positions at which the averaged BLOSUM score is less than zero are not colored. Positions at which the averaged BLOSUM score is greater than zero are colored from *yellow* to *green* with increasing degree of conservation.

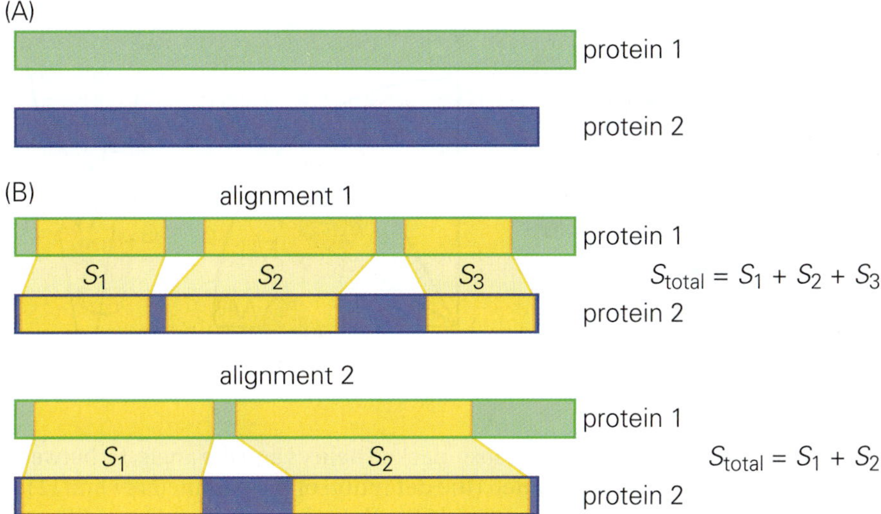


**Figure 5.16 BLOSUM substitution scores are used to optimize the alignment of sequences.** (A) Schematic representations of the sequences of two proteins, 1 and 2, which are distantly related. (B) Sequence alignment programs, such as BLAST, break up each protein into segments of different length and generate various possible matches between segments in the two proteins. Two such alignments are shown here. Segments within each protein that have high substitution scores ($S_1$, $S_2$, etc.) when matched with segments in the other protein are indicated by *yellow* shading. Many such alignments are generated by computer, although only two are shown in the diagram. The alignment with the highest total substitution score ($S_{total}$) is taken as the optimal sequence alignment between the two proteins.

are evaluated by these programs, as indicated schematically in **Figure 5.16**. The alignment that yields the highest aggregate substitution score is reported back to the user as the optimal alignment. This process is also used to detect relationships between any protein of interest (for example, a new gene that has just been sequenced) and the sequences of all known proteins in genomic databases (**Figure 5.17**). This is done by aligning the sequence of interest to all other protein sequences. For each alignment, the probability that the two sequences are related is evaluated using the BLOSUM scores, and the most likely matches are reported.

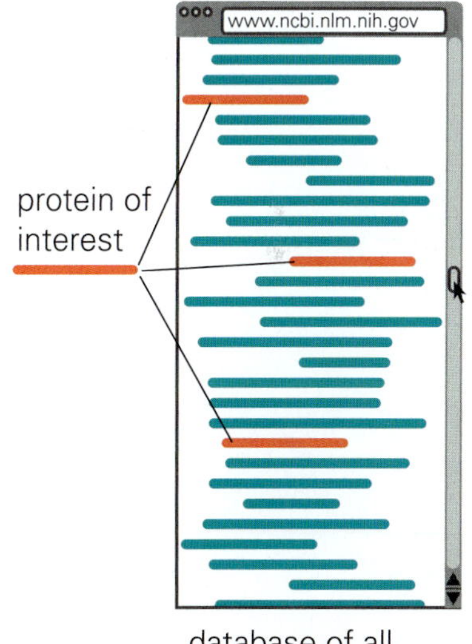


**Figure 5.17 Detecting similarities between a protein of interest and all known protein sequences.** The protein of interest (*red* line on the left) is aligned against all known proteins in a genomic database at the U.S. National Library of Medicine website, and the probability of each alignment being due to related proteins is evaluated using the BLOSUM scores. The alignments that are most likely to be due to evolutionary relatedness are reported (*red* lines on the right).

# C. STRUCTURAL VARIATION IN PROTEINS

## 5.12 Small but significant differences in protein structures arise from differences in sequences

Why are the structures of globins so stable over evolutionary time? If we superimpose any two of these structures, we find that there are differences in the relative orientations of the helices and in the distances of the helices from each other (**Figure 5.18**). Other differences are apparent when we look at the structures closely. Helix D is absent in leghemoglobin (a plant globin), for example, and the lengths and precise structures of the interhelical loops are quite variable from globin to globin. But, overall, these structural differences are relatively minor and do not affect in a substantial way the architecture of the core scaffold of the protein.

The most likely explanation for the stability of the globin fold is that secondary structural elements have to pack against each other in a way that optimizes sidechain interaction, and there are a limited number of ways of packing α helices against each other. Recall from Chapter 4 that α helices are packed such that the knobs or ridges on one helix fit into grooves on the other helix (see Figure 4.46). Thus, two helices pack together in a small number of orientations (Figure 4.48 illustrates two such orientations). Each pair of close-packed helices in the globin fold adopts one of these stable interhelical geometries. For example, the G and H helices come close together such that the helical axes are about 9 Å apart. The angle between the helical axes is ~25°. Changes in the globin sequence that survive natural selection tend to preserve this stable orientation or "click state," with relatively small structural changes (see Figure 5.18).

The range of variation in interhelical geometry in globins is illustrated in Figure 5.18B. For each individual globin (for example, human myoglobin), the distance, *r*,

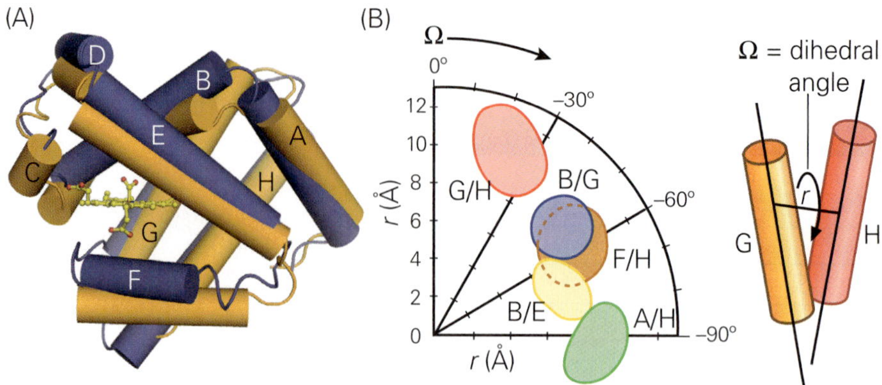


**Figure 5.18 Comparison between the structures of individual globins.** (A) Two globins are shown superimposed on each other, one colored *yellow* and one *blue*. (B) The range of interhelix geometries for several globins is shown. The definition of the parameters *r*, the distance between two helices, and Ω, the angle between helices, is shown on the right. The values of *r* and Ω are different for the G–H pair of helices in different globins. This range of values of *r* and Ω for different pairs of close-packed helices is indicated by the colored areas in the graph in the middle. (B, adapted from A.M. Lesk and C. Chothia, *J. Mol. Biol.* 136: 225–270, 1980. With permission from Elsevier.)

between the centers of each pair of close-packed helices and the angle, Ω, between the axes of these helices is computed (the definition of the parameters *r* and Ω for the G–H helix pair is shown in Figure 5.18B). The values of *r* and Ω are different for the G–H pair of helices in different globins, but lie within a rather restricted range. This range of values is indicated by the red area in the graph, labeled G/H. The graph indicates that the distance, *r*, between the G and H helices ranges from ~8 Å to ~12 Å in different globins. The angle, Ω, between these two helices varies between ~–10° and ~–35° in different globins. These changes in interhelical geometry correspond to a small degree of variation within a stable ridges-into-grooves packing arrangement, which tends to preserve the overall nature of the globin fold (see Section 4.20).

This idea, that there is a very large but nevertheless limited number of stable arrangements of secondary structural elements in protein folds, appears to be quite general. While maintaining a stable fold, evolutionary variation on the surfaces of proteins affects their ability to interact with other proteins, and dictates whether they are monomers or organized into higher-order assemblies. Changes in the interior of the proteins result in subtle alterations in the details of the fold and, combined with changes on the surface, allow the acquisition of new function. This variation in functional properties is a general feature of families of proteins and has allowed for the fine-tuning of many different cellular functions while maintaining the ability of protein molecules to fold spontaneously.

## 5.13 Proteins retain a common structural core as their sequences diverge

The extent to which the structures of two proteins differ depends on how different their sequences are. At one level, this states the obvious: two proteins that share a high level of sequence similarity can be expected to have very similar structures. On other hand, what can we say about proteins that are not so similar in their sequences? When comparing the structures of two related proteins, we can partition the structures into two regions. In one region, the two proteins have similar chain fold and, within this region, we can discuss the magnitude of structural differences in a meaningful manner. This region is known as the **common structural core** (**Figure 5.19**). In the other region, the chain fold is so different that it is not meaningful to quantify the extent to which the structures differ. The distinction between these two regions is not sharply defined, but we could consider polypeptide chains to have different folds when the positions of corresponding $C_\alpha$ atoms in this region deviate by more than ~3 Å. The $C_\alpha$–$C_\alpha$ distance in a polypeptide chain is about 3.7 Å, so a 3-Å cutoff in choosing the common core allows residues within the common core of one protein to be aligned unambiguously with a residue in the common core of the other.

**Common structural core**

When comparing the structures of two proteins, the region of the structure that is similar between the two proteins is called the common structural core (see Figure 5.19). The positions of $C_\alpha$ atoms within the common structural core are closely aligned in the two proteins (that is, the $C_\alpha$ positions deviate by less than ~3 Å).

The size of the common core, expressed as a fraction of the size of either of the proteins being compared, decreases as the sequences of proteins diverge

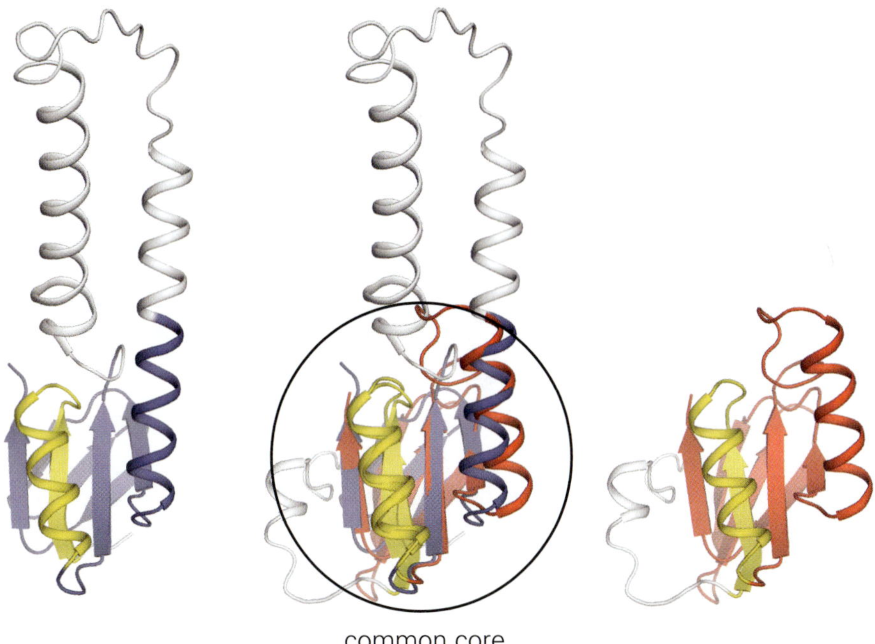


**Figure 5.19 The common structural core for two proteins.** Two domains from different proteins are shown on the left and on the right. In the middle, their structures are superimposed, and the common structural core is outlined and shown in color. These are nucleotide-binding proteins, and a central structural element that is important in binding nucleotide is colored *yellow* in both proteins (see Figure 5.41). (PDB codes: 1GET and 1TDE).

(**Figure 5.20**). For proteins that share greater than 50% sequence identity, the common core is very large (90% or more of the structure). As the sequence identity drops below ~20%, however, the size of the common core becomes variable and can range from as much as 80–90% to as little as 40% of the protein structure. Helical proteins tend to retain a larger common structural core as their sequences diverge, because of the relatively inflexible nature of α-helical structure. On the other hand, structures that contain a large degree of β sheet tend to lose structural correspondence more quickly as the sequences of related proteins diverge.

## 5.14 Structural overlap within the common core decreases as the sequences of proteins diverge

A measure of the degree of structural overlap between two proteins is the root mean square (rms) deviation in the positions of $C_\alpha$ atoms in the common core between the two structures, as explained in **Figure 5.21.** The pairs of structures being compared are aligned so as to minimize the rms deviation between $C_\alpha$ positions, and the rms deviation for this optimal alignment is reported. As discussed in the last section, for the comparison to be meaningful, the rms deviation is calculated using residues within the common core of the two proteins being compared.

The rms deviation in backbone atoms within the common structural core for several pairs of related proteins is shown in **Figure 5.22.** Proteins that are related by ~50% or more sequence identity have common cores that are structurally very similar, with rms deviations that are less than ~1 Å. The rms deviation in backbone positions rises steadily as the level of sequence identity drops, and the deviations can exceed 2 Å when the two proteins being compared share less than 20% sequence identity.

What are the consequences of the changes in protein structure that result from a drift in amino acid sequence? Genetic changes and natural selection result in structural divergence between members of a family of related proteins, but the common core regions undergo the least alteration in structure. The catalytic centers and binding sites of enzymes are usually located within such structural cores. The disposition of residues that are crucial for catalysis is subject to the constraint that enzymes would cease to function if the catalytic center were distorted greatly.

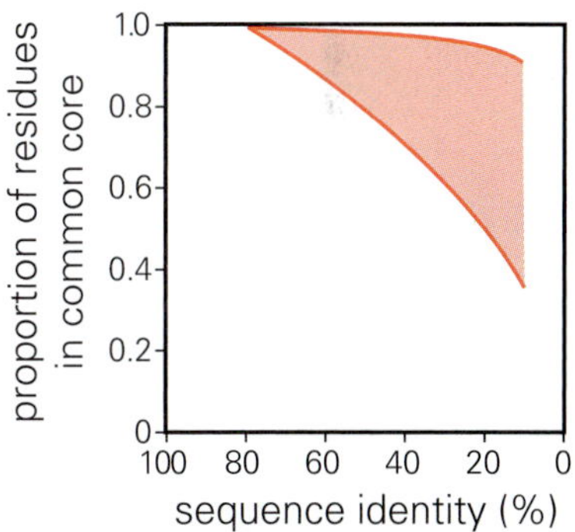


**Figure 5.20 The proportion of residues in the structural core that is common to two proteins decreases as their sequences diverge.** The sequence identity between two proteins being compared is plotted on the horizontal axis, from high to low sequence identity. The fraction of the structure that is in the common core is plotted on the vertical axis. These fractions fall within the range of values indicated by the shading. (Adapted from C. Chothia and A.M. Lesk, *EMBO J.* 5: 823–826, 1986. With permission from Macmillan Publishers Ltd.)

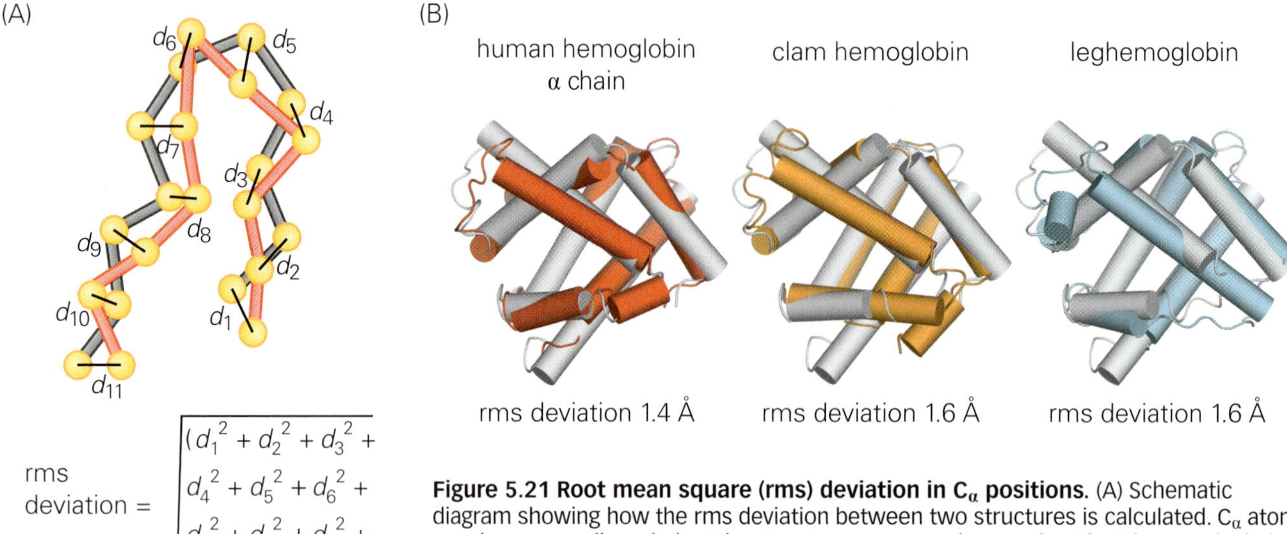


$$\text{rms deviation} = \sqrt{\frac{(d_1^2 + d_2^2 + d_3^2 + d_4^2 + d_5^2 + d_6^2 + d_7^2 + d_8^2 + d_9^2 + d_{10}^2 + d_{11}^2)}{11}}$$

**Figure 5.21 Root mean square (rms) deviation in $C_\alpha$ positions.** (A) Schematic diagram showing how the rms deviation between two structures is calculated. $C_\alpha$ atoms are shown as *yellow* circles. The structures are superimposed so that the rms deviation between $C_\alpha$ positions is minimized, and the resulting rms deviation is reported. (B) Structural overlap between myoglobin (*gray*) and three other globin proteins. The rms deviation in $C_\alpha$ positions for residues in the common core is shown below the structures. (PDB codes: 1MBC, 1A00, 1HB1, and 1BIN.)

As a consequence, residues that are important for catalysis stay relatively fixed in space. Nevertheless, the structure of the protein can vary considerably in regions surrounding the central structural core, and the process of evolution can lead readily to the acquisition of new substrate recognition properties (see Sections 4.29 and 4.30).

## 5.15 Sequence comparisons alone are insufficient to establish structural similarity between distantly related proteins

Consider two proteins whose sequences are not very closely related (for example, the sequence identity between them is 30% or less). Can we say anything about whether or not their structures are related? This is a very important question because the number of known protein sequences is several orders of magnitude greater than the number of proteins whose structures have been determined experimentally. Because structure is closely related to function, in order to understand how a protein works, we need to know its three-dimensional structure.

If we could establish reliable relationships between proteins of known structure and proteins whose sequences are known but whose structures have not yet been determined, then we would greatly increase our ability to model the structures of a large number of proteins, based on the more limited set of known protein structures. Unfortunately, when the sequences are very divergent, comparisons of sequences alone can fail to establish the presence of relationships in structure.

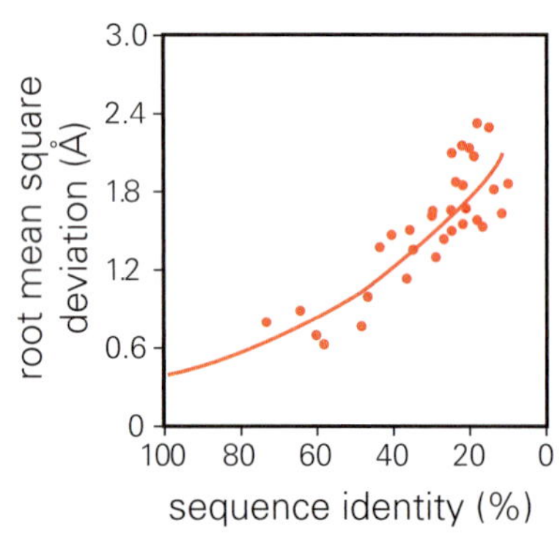


**Figure 5.22 Structural variation as a function of sequence identity.** The dependence of rms deviation in backbone atom positions as a function of sequence identity for the common structural core of various pairs of proteins (not just globins). (Adapted from C. Chothia and A.M. Lesk, *EMBO J.* 5: 823–826, 1986. With permission from Macmillan Publishers Ltd.)

The limitation of sequence comparisons as a probe for structural similarity is demonstrated by a systematic comparison of a very large number of proteins of known structure. This was done by using computer programs to align the sequences of all proteins with known structures with each other, in blocks of varying length. For each such alignment, the level of sequence identity was computed. Since the structures of both proteins in all the pair-wise comparisons are known, it is easy to establish whether the sequences being compared correspond to proteins of similar or different three-dimensional structures. The results of this analysis are shown in **Figure 5.23**. When the lengths of the sequences being compared are small (that is, 10–20 residues), then no level of sequence identity is sufficient to

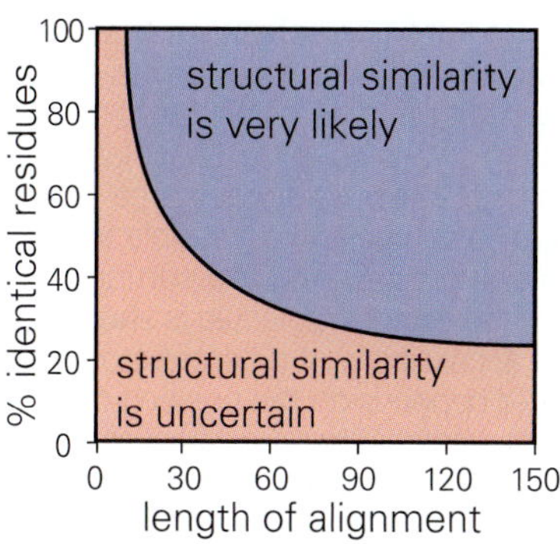


**Figure 5.23 Threshold levels in sequence identity for the reliable assignment of structural similarity.** The solid line separating the *blue* and *red* regions is the threshold value of sequence identity (the vertical axis), above which the structures of the two proteins being compared are highly likely to be similar. Below this value, the structures may be similar or different. (Adapted from C. Sander and R. Schneider, *Proteins* 9: 56–68, 1991.)

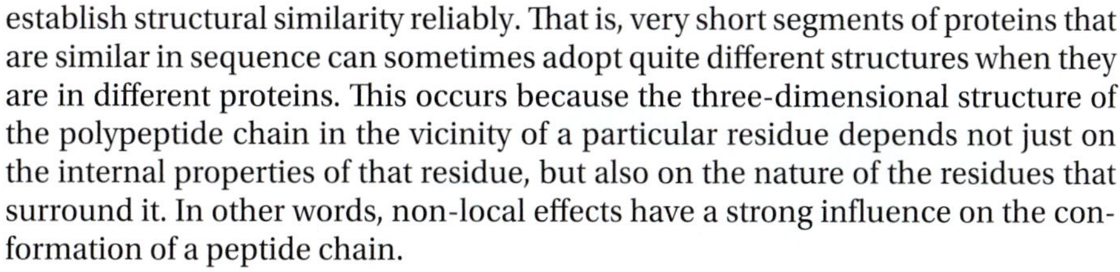

establish structural similarity reliably. That is, very short segments of proteins that are similar in sequence can sometimes adopt quite different structures when they are in different proteins. This occurs because the three-dimensional structure of the polypeptide chain in the vicinity of a particular residue depends not just on the internal properties of that residue, but also on the nature of the residues that surround it. In other words, non-local effects have a strong influence on the conformation of a peptide chain.

Sequence comparisons become more informative when the lengths of the segments being compared are ~50 residues or greater. Polypeptide segments of this length encompass secondary structural elements that are folded against each other in proteins. Figure 5.23 shows that structural similarity is virtually assured when the sequence identity between longer segments is greater than ~25%. Figure 5.23 also shows that proteins that are related by less than ~25% sequence identity can have significantly different three-dimensional structures.

There are several advanced methods for detecting structural relationships between distantly related proteins that go beyond simply matching sequences. One class of methods relies on sophisticated statistical procedures and the availability of very large numbers of potentially related sequences to establish evolutionary relationships, without necessarily using information about the three-dimensional structure. We shall not discuss these methods here, but they are readily accessible on websites that provide access to genomic databases (for example, the alignment procedure known as PSI-BLAST on the National Library of Medicine genome database). Another class of methods works only when the structure of at least one member of a family of related proteins has been determined experimentally. One such method is described in detail in the following sections because it helps us better understand the relationship between sequence and three-dimensional structure.

## 5.16 The amino acids have preferences for different environments in folded proteins

As we move from position to position along the polypeptide chain of a folded protein, the chemical environment surrounding the residues in the chain varies considerably. A key parameter in describing the environment of an atom is the solvent accessibility of the atom (**Figure 5.24**). Atoms in different positions within a protein structure have different degrees of accessibility to water molecules. An atom located at the surface of a protein would have most of its surface area accessible to water molecules. In contrast, atoms within the hydrophobic core may be completely inaccessible to water. Atoms in more polar groups prefer water-accessible positions, whereas atoms in hydrophobic groups prefer positions that are inaccessible to water.

A residue at the amino terminal end of the polypeptide chain is often in an exposed environment, where it can form hydrogen bonds with water. Polar sidechains are likely to be favored at such exposed positions. As the polypeptide backbone moves into the interior of the protein, the environment becomes more hydrophobic, and sidechains at these positions are likely to pack against the sidechains of nonpolar residues such as leucine, isoleucine and phenylalanine. Charged and

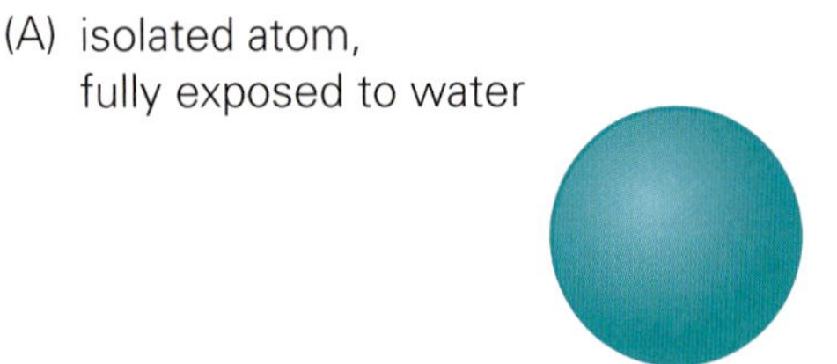


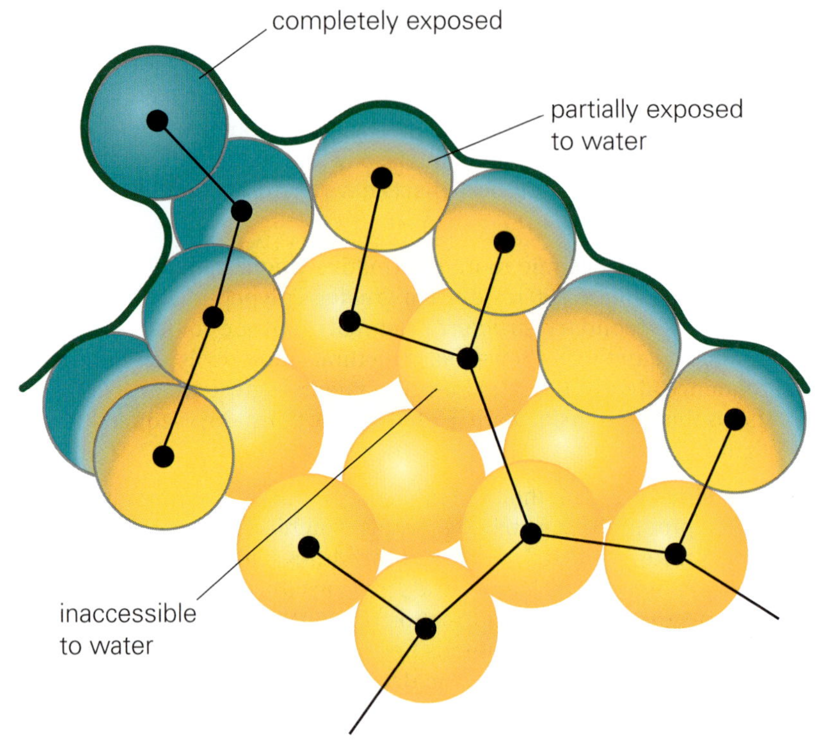


**Figure 5.24 The solvent accessibility of atoms.** The atoms in a protein have different accessibilities to water molecules. In (A), an isolated atom is shown. Its surface area (*blue*) is completely accessible to water molecules. In (B), a portion of a folded protein molecule is shown. The accessibility of different atoms to water depends on the position of the atom. Regions that are inaccessible to water are colored *yellow*. Atoms in more polar groups prefer water-accessible positions.

polar sidechains would not be favored at such positions. At other positions in the structure, the sidechains might encounter hydrogen-bonding donors and acceptors, or charged groups, and only certain sidechains (for example, asparagine, histidine, or serine) might be able to interact favorably with these groups. The secondary structure (that is, α helix, β strand, or neither) is also an attribute of the environment, because the amino acids have different preferences for being in one secondary structural element over another (Section 4.14). The environments of three residues in myoglobin are shown in **Figure 5.25**.

**Fold-recognition or threading algorithms**

Fold-recognition or threading algorithms are computational procedures that help predict the three-dimensional structures of proteins of unknown structure whose sequences are not very closely related to proteins that have had their structures determined. These methods take advantage of the fact that any particular protein fold places constraints on the kinds of amino acids that can be accommodated by the fold—that is, they explicitly use knowledge of the three-dimensional structures of proteins.

## 5.17 Fold-recognition algorithms evaluate the probability that the sequence of a protein is compatible with a known three-dimensional structure

There are various methods for establishing whether the sequence of a protein of unknown structure is compatible with the particular pattern of chemical environments provided by the chain fold of a protein whose structure is known. These methods are often referred to as **fold-recognition algorithms**, because they seek to establish possible connections between known folds and sequences. They are also known as **threading algorithms**, because they can be thought of as "threading" the sequence to be tested along the course of the known chain fold, and then determining whether the sequence is compatible with the particular fold (**Figure 5.26**).

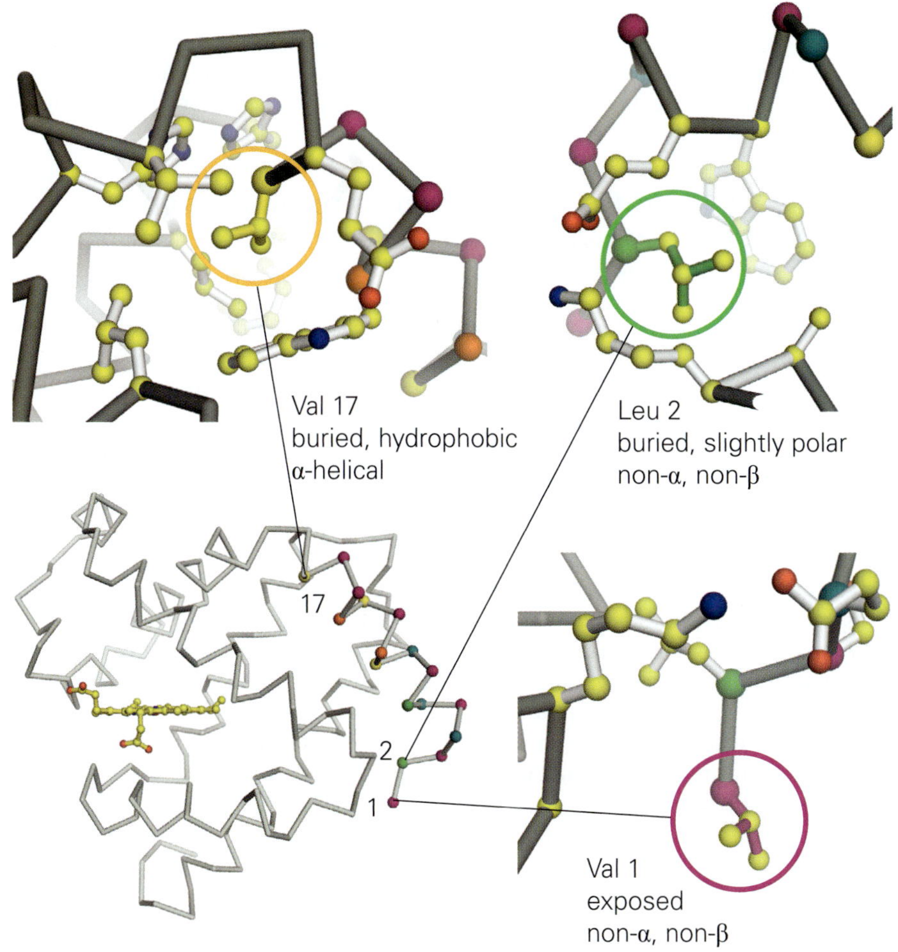


**Figure 5.25 Environmental classes for three residues of myoglobin.** The structure of myoglobin is shown in the lower left, with the $C_\alpha$ atoms of the first 17 residues indicated by colored spheres. The colors of these spheres correspond to the 18 environmental classes of the 3D-1D profile method (see Figure 5.27). The actual environment of the sidechains is shown in expanded views for residues 1, 2, and 17.

Conceptually, these computer algorithms work by first guessing at a potential alignment between the positions along the chain of the known structure and the sequence of amino acids in the protein of unknown structure. This alignment is then improved iteratively so as to overcome errors introduced by incorrect alignments. Given an alignment, the amino acids corresponding to the sequence of the unknown structure are mapped onto the corresponding positions along the polypeptide chain of the known structure. The likelihood that such a positioning is correct is then evaluated in terms of what we know about the properties of

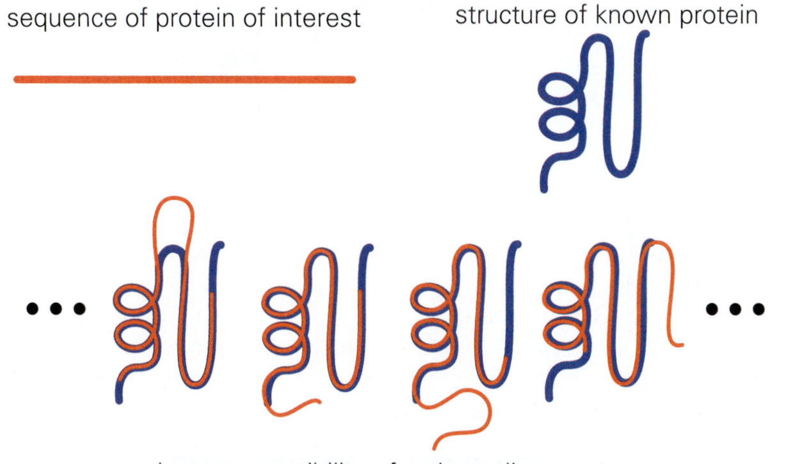


**Figure 5.26 Fold recognition.** Fold-recognition or threading algorithms test the compatibility of the sequence of a protein of interest (*red*) with the known structure of another protein (*blue*). Different alignments of the sequence onto the structure are generated, and each one is scored for compatibility.

amino acids in protein structures. The various methods available today differ in how they carry out this evaluation, and all of them are imperfect in their ability to model protein structures and to estimate their stability. It is also possible that the protein of interest has a completely novel structure. In such cases, fold-recognition algorithms are bound to fail.

## 5.18 The 3D-1D profile method maps three-dimensional structural information onto a one-dimensional set of environmental descriptors

One fold-recognition algorithm, known as the **3D-1D profile method**, converts information about the chemical and structural environments at each position of the chain of the known three-dimensional structure (the "3D" information) into a one-dimensional list that provides one environmental descriptor per residue (the "1D" list; hence, 3D-1D profile) (**Figure 5.27**). This list is known as the **environmental profile** of the protein. The mapping of three-dimensional structural information into a one-dimensional list reduces the complexity of the fold-recognition problem.

Given a three-dimensional protein structure, the environment at each position along the chain is characterized in three different ways in order to create the environmental profile. First, the extent to which the residue at a particular position is buried within the known structure is calculated. For simplicity, the extent of burial is classified as exposed, partially buried, or completely buried (see Figure 5.27). Next, the nature of the environment is calculated by measuring the extent to which polar atoms are in contact with the residue in question. Buried residues are grouped into three classes ($B_1$, $B_2$, and $B_3$), depending on the extent to which their environment is polar. Partially buried residues are grouped into just two classes

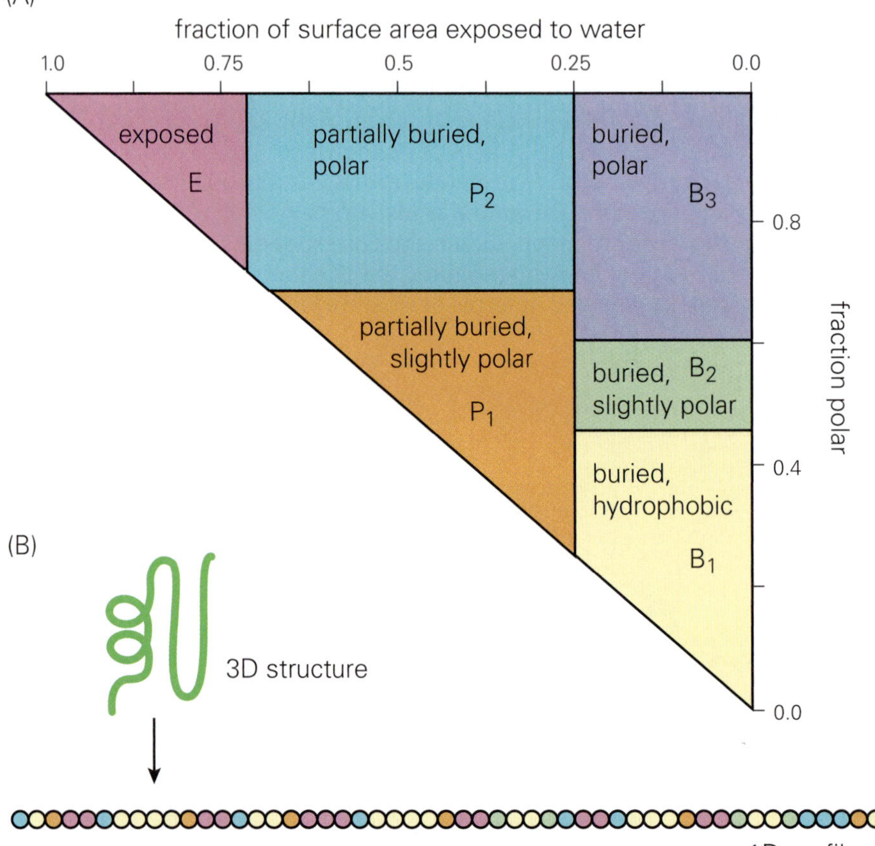


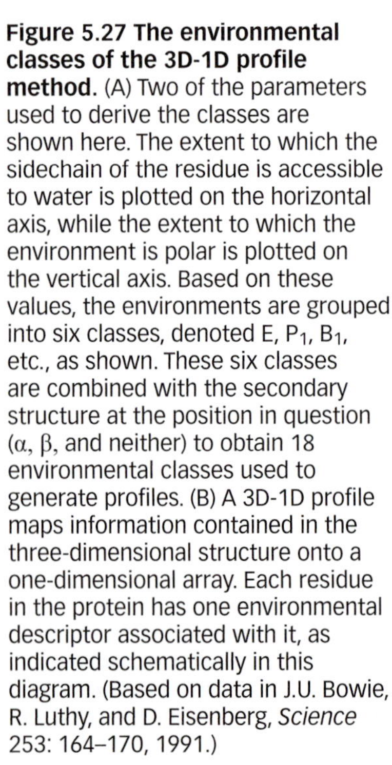

**Figure 5.27 The environmental classes of the 3D-1D profile method.** (A) Two of the parameters used to derive the classes are shown here. The extent to which the sidechain of the residue is accessible to water is plotted on the horizontal axis, while the extent to which the environment is polar is plotted on the vertical axis. Based on these values, the environments are grouped into six classes, denoted E, $P_1$, $B_1$, etc., as shown. These six classes are combined with the secondary structure at the position in question (α, β, and neither) to obtain 18 environmental classes used to generate profiles. (B) A 3D-1D profile maps information contained in the three-dimensional structure onto a one-dimensional array. Each residue in the protein has one environmental descriptor associated with it, as indicated schematically in this diagram. (Based on data in J.U. Bowie, R. Luthy, and D. Eisenberg, *Science* 253: 164–170, 1991.)

($P_1$ and $P_2$) based on the polarity of their environment, and exposed residues are assumed to be in a polar environment (E). Additionally, the secondary structure at each position along the chain is noted as either α helical, β strand, or neither.

The combination of these parameters results in a total of 18 different **environmental classes** in which a particular residue might find itself (see Figure 5.27). The partitioning of the environments into just 18 different classes is arbitrary and represents a compromise between the complexity of a more complete description and the speed with which sequences can be tested using a computer. Every known protein structure is mapped in this way into a one-dimensional array that lists the environmental class at each position in the polypeptide chain, as indicated schematically in Figure 5.27B. These arrays are collected in a database against which the sequences of proteins of unknown structure can be tested.

## 5.19 The database of known protein structures is used to generate a scoring matrix that gives the likelihood of finding each amino acid in a particular environmental class

The profile matching method relies on the fact that different amino acids have markedly different preferences for being found in the various environmental classes (**Figure 5.28**). These preferences are established by analyzing the database of known protein structures and determining the frequency with which the 20 amino acids are found in each of the 18 environmental classes in all the known protein structures. The result is a matrix of **environmental scores** that gives the likelihood that any of the amino acids is found in a particular environment class (see Figure 5.28).

The environmental score is similar in concept to the BLOSUM substitution score described in Section 5.7, except that it scores matches between amino acids and environmental descriptors, rather than other amino acids. The environmental score, $S_{ij}$, for finding amino acid $i$ in environment $j$ is related to the logarithm of the probability, $P_{ij}$, of finding amino acid $i$ in environment $j$: $S_{ij} = \log(P_{ij})$. We shall not discuss the details of how these scores are derived, but simply note that the definition of the scores in terms of the logarithm of a probability means that the scores are additive.

To see how the environmental scores are utilized, consider the first 10 residues in the structure of myoglobin, which are shown in **Figure 5.29**. The environmental

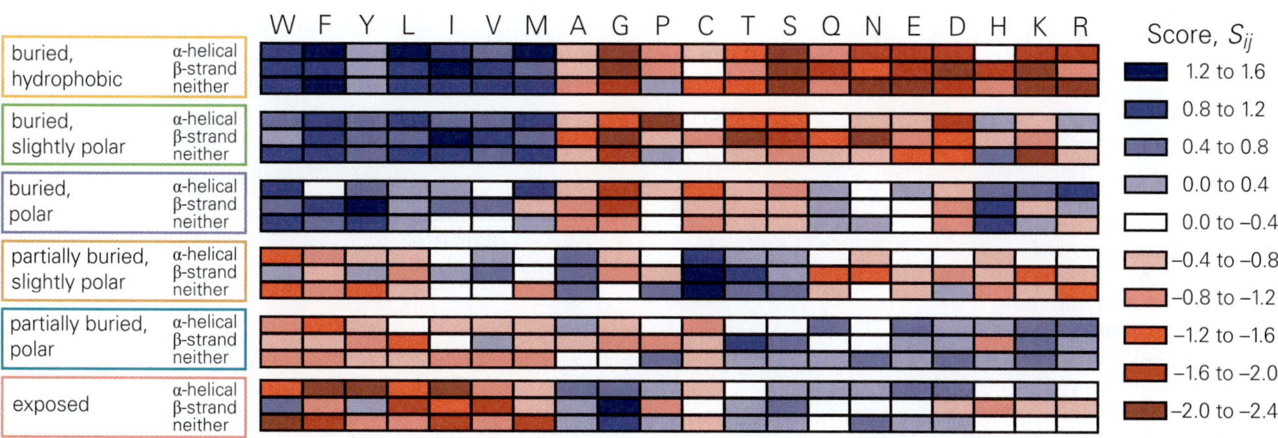


**Figure 5.28 The preferences of each of the amino acids for the 18 environmental classes.** The 18 classes are indicated vertically on the left. The 20 amino acids are shown at the top, using their single-letter codes. The environmental score for each of the amino acids being in one of the environmental classes is color coded. (Adapted from J.U. Bowie, R. Luthy, and D. Eisenberg, *Science* 253: 164–170, 1991.)

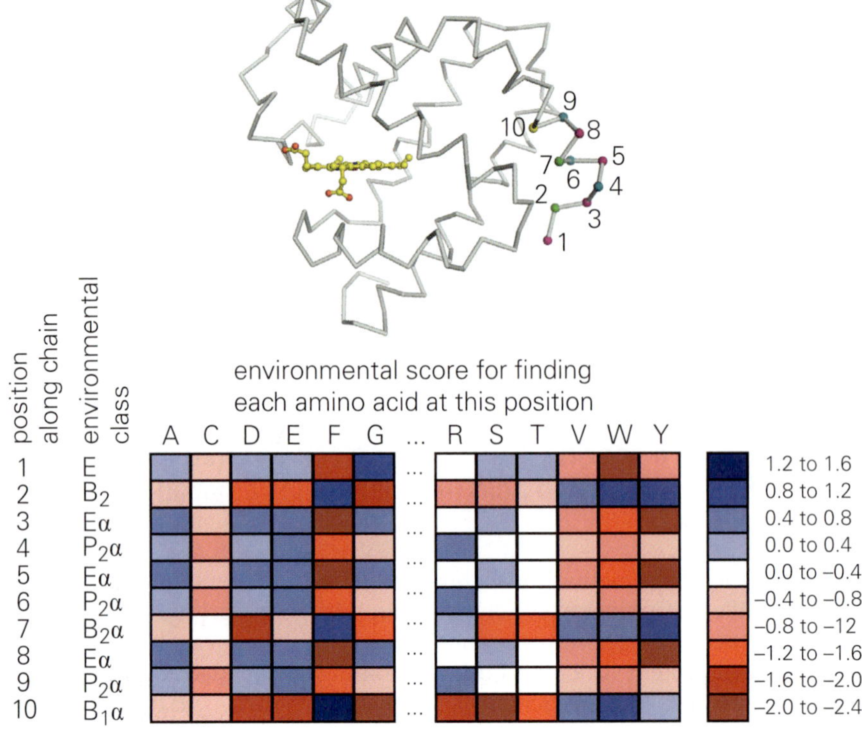


**Figure 5.29 3D-1D scores for finding different amino acids at each of the first 10 positions in the polypeptide chain of myoglobin.** For each position, the environmental class is given, using the notation explained in Figure 5.27. At each position, the environmental score for various amino acids (taken from the matrix in Figure 5.28) is color coded. Not all the amino acids are shown here. (Adapted from J.U. Bowie, R. Luthy, and D. Eisenberg, *Science* 253: 164–170, 1991.)

class for each of these positions is evaluated as discussed in Section 5.15. Given the environmental class, each of the 20 amino acids is favored to a greater or lesser extent at each position along the chain, according to the scores shown in Figure 5.28. For example, the first position in myoglobin is in an exposed environment that is in neither an α helix nor a β strand. Nonpolar residues, such as phenylalanine (F) and tryptophan (W), are disfavored at this position. Phenylalanine is instead favored at positions 2, 7, and 10, which are buried.

## 5.20 The 3D-1D profile method matches sequences with structures

One way to use the 3D-1D method is to take the sequence of a protein of unknown structure and align it with the environmental profile of every protein with known structure (**Figure 5.30**). To do this, the test sequence is aligned with each of the environmental profiles in the structural database, much as it might be aligned against all of the known sequences in genomic databases (compare Figure 5.30 with Figure 5.17). The difference between the BLOSUM-based alignment method

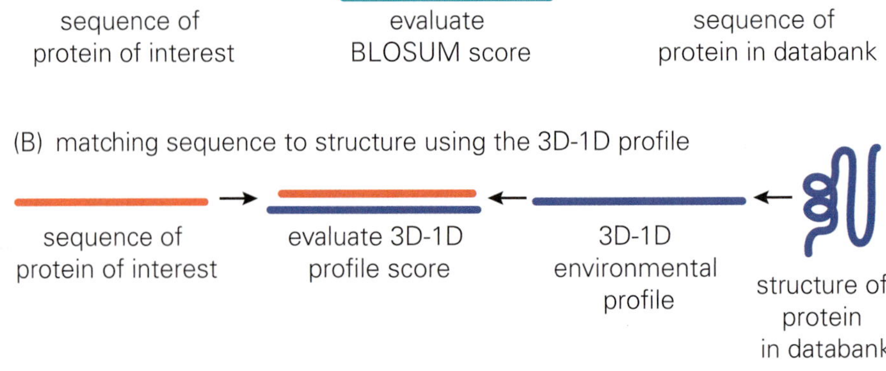


**Figure 5.30 Comparison of sequence alignment using the BLOSUM matrix and matching sequence to structure using the 3D-1D profile.** (A) The BLOSUM matrix is used to evaluate the probability that one sequence is related to another. (B) The 3D-1D profile method is used to evaluate the probability that the sequence of a protein of unknown structure (for example, a newly discovered protein) is compatible with a known three-dimensional structure.

and the 3D-1D profile method is that in the latter case the alignment embeds within it information that is derived from known three-dimensional structures (see Figure 5.30).

When using the 3D-1D profile method, an aggregate environmental score, *S*, for that particular alignment is calculated, where *S* is the sum of the environmental scores at each position of the alignment. Protein structures that return the highest aggregate environmental scores are the most likely to have structures that are related to the protein of interest.

Another way to use the 3D-1D profile method is to search the database of protein sequences with a particular structure and to look for proteins that might share the same structure. The results of such a calculation using the structure of myoglobin are shown in **Figure 5.31**. As expected, the 3D-1D profile for myoglobin pulls out the sequences of myoglobins with very high scores. It also pulls out the sequences of a number of globins that do not share significant sequence similarity with myoglobin, but nevertheless are likely to adopt the same chain fold. Note, however, that a few sequences that are known to be those of globins are not in fact differentiated from the bulk of the sequences in the databank. This highlights the fact that fold-recognition methods have limitations that arise from the approximations that they use. Despite these limitations, these methods are constantly being improved and their ability to discriminate between related and unrelated proteins is becoming increasingly powerful.

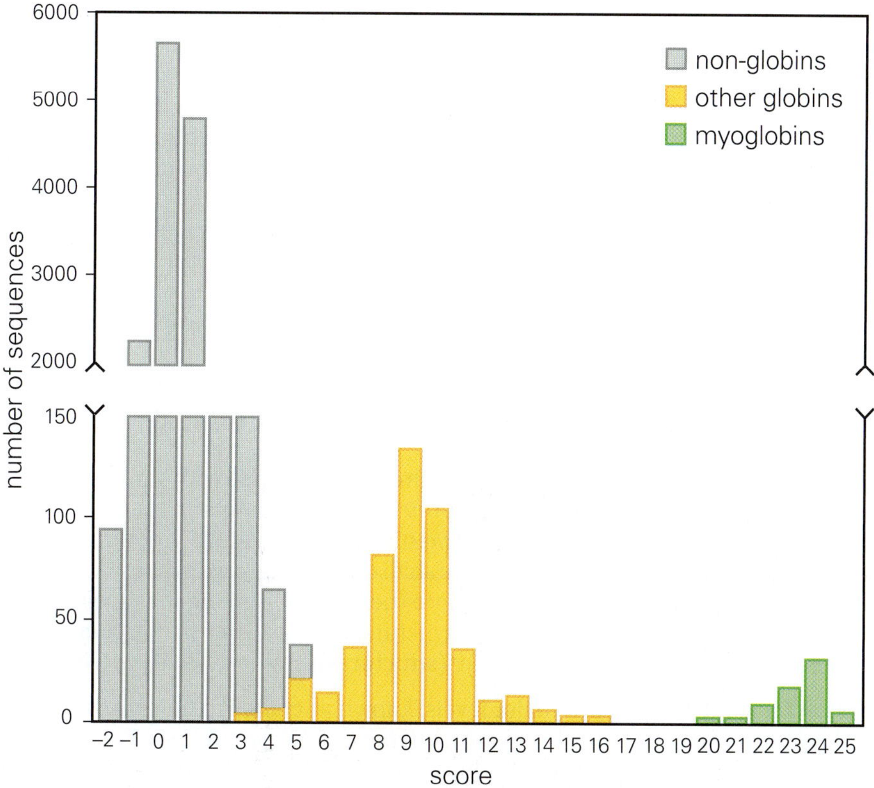


**Figure 5.31 Searching for globin folds in a database of sequences.** Shown here is the distribution of 3D-1D scores for a large number of sequences when matched against the environmental profile of myoglobin. The score for each alignment is plotted on the horizontal axis, and the number of sequences with a particular score is shown on the vertical axis. The great majority of sequences have low scores—that is, they are unlikely to be compatible with the environments of residues in the myoglobin structure. Sequences corresponding to various myoglobins have high scores (*green*). Hemoglobin and other known globins are indicated in *yellow*. The scores shown here are normalized by the standard deviation of the distribution. (Adapted from J.U. Bowie, R. Luthy, and D. Eisenberg, *Science* 253: 164–170, 1991. With permission from the AAAS.)

# D. THE EVOLUTION OF MODULAR DOMAINS

## 5.21 Domains are the fundamental unit of protein evolution

The problem of understanding protein structure and function is greatly simplified by the recognition that protein molecules are modular. We have introduced, in Chapter 4, the idea that large protein molecules are usually constructed from several domains, through processes of genetic recombination and gene duplication, in different combinations in different proteins (Section 4.2). As a consequence, there are far fewer kinds of domains than there are kinds of protein molecules. Examples of 20 different kinds of protein domains are shown in **Figure 5.32**. These domains are compact structures that can, in principle, be inserted into different proteins without affecting their internal structure.

An example of a protein made up of two different domains is shown in **Figure 5.33**. This protein, known as alkaline protease (proteases are enzymes that cleave other proteins), contains an N-terminal domain with an $\alpha/\beta$ structure and a C-terminal domain that is constructed almost entirely of $\beta$ strands. The N-terminal domain of alkaline protease contains the enzymatic active site, which contains a zinc ion that is required for catalyzing the cleavage of peptide bonds. The C-terminal domain adopts a structure known as a $\beta$ solenoid (see Figure 5.33), in which $\beta$ strands spiral to form a cylindrical stalk.

The N-terminal domain has its own tightly packed hydrophobic core, shown in Figure 5.33A, and shares a structural core with certain other zinc-containing protease enzymes that do not have the $\beta$ solenoid stalk domain. For example, a protease known as astacin, which is found in crayfish, shares a ~100-residue structural core with alkaline protease (Figure 5.33B). A comparison of the structures of the N-terminal domain of alkaline protease and astacin shows how the catalytic center (marked by the zinc ion in Figure 5.33) is encompassed within a structural core that is decorated by quite different loop structures in the two proteins. Within the structural core, the sequence similarity between the N-terminal domain of alkaline protease and astacin is quite recognizable, with a sequence identity of ~25%. On looking at the amino acid conservation scores in the alignment, there is sufficient sequence conservation for us to surmise that these domains have evolved from a common ancestor (Figure 5.33C).

## 5.22 Domains can be organized into families with similar folds

Each of the component domains of a protein can be grouped with similar domains in other proteins. The classification of protein domains into structural families is rather subjective, and there is more than one way to organize groupings of protein domains with similar structures. We shall discuss one organizational strategy, known as CATH, to provide a glimpse into the kind of thinking that goes into the development of a hierarchy of protein structural families. Another popular organizational database for proteins is SCOP (structural classification of proteins).

The CATH database has four major levels, known as class, architecture, topology, and homologous superfamily (hence the name CATH). **Figure 5.34** illustrates the CATH hierarchy. There is also an additional level of classification within CATH, known as sequence family, in which proteins that are particularly similar in sequence are grouped together.

The CATH organization starts by separating proteins into classes based on their secondary structure composition. There are four main classes, one for proteins that are mainly $\alpha$ helical, one for proteins that are constructed mainly of $\beta$ strands, one for proteins that contain a mixture of $\alpha$ helices and $\beta$ strands, and one for proteins that do not have regular secondary structure. Examples of proteins in the first three classes are shown in Figure 5.34.

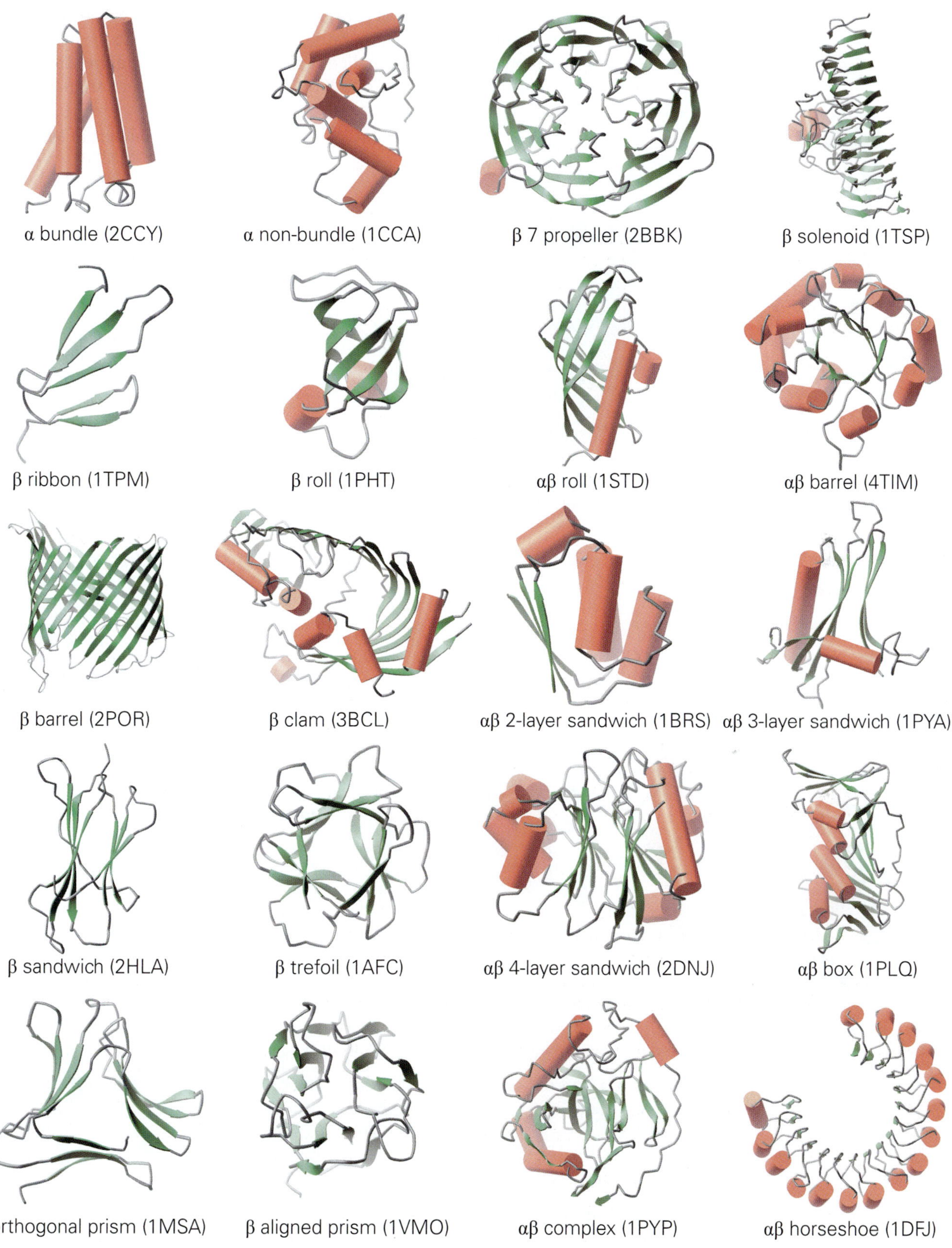


**Figure 5.32 Examples of 20 different protein domains.** The protein databank codes are shown in parenthesis for each structure. (Adapted from C.A. Orengo et al., and J.M. Thornton, *Structure* 5: 1093–1108, 1997. With permission from Cell Press.)

The next level down in the CATH organization groups together structures that have similar architectures. The term "architecture" is used here to mean the nature of the general organization of α helices and β strands in three dimensions, without referring to the connectivity of the polypeptide chain. There are ~40 distinct

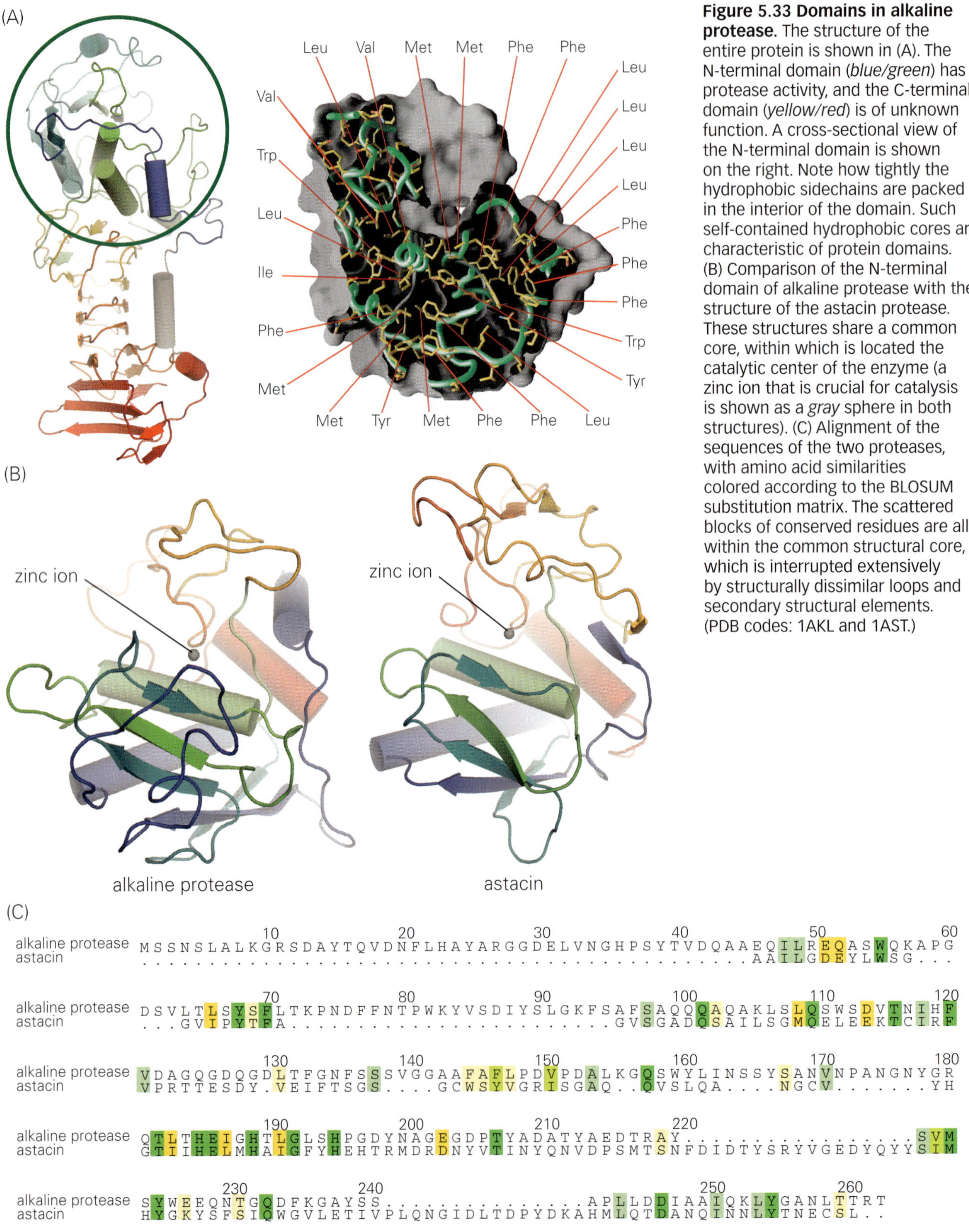

(A)

Leu Val Met Met Phe Phe
Val
Trp
Leu
Ile
Phe
Met
Met Tyr Met Phe Phe Leu
Leu
Leu
Leu
Leu
Phe
Phe
Phe
Trp
Tyr

(B)

zinc ion

zinc ion

alkaline protease

astacin

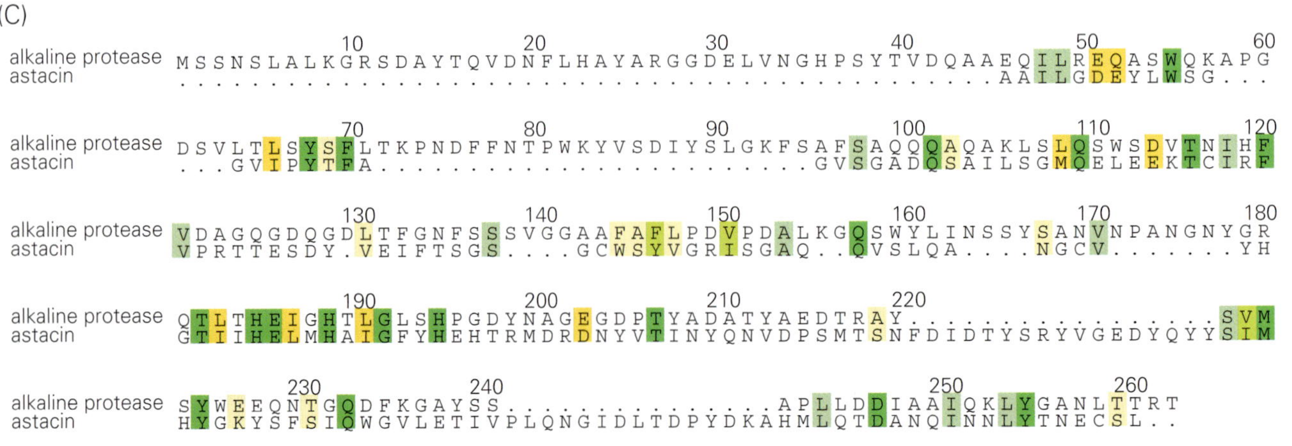


**Figure 5.33 Domains in alkaline protease.** The structure of the entire protein is shown in (A). The N-terminal domain (*blue/green*) has protease activity, and the C-terminal domain (*yellow/red*) is of unknown function. A cross-sectional view of the N-terminal domain is shown on the right. Note how tightly the hydrophobic sidechains are packed in the interior of the domain. Such self-contained hydrophobic cores are characteristic of protein domains. (B) Comparison of the N-terminal domain of alkaline protease with the structure of the astacin protease. These structures share a common core, within which is located the catalytic center of the enzyme (a zinc ion that is crucial for catalysis is shown as a *gray* sphere in both structures). (C) Alignment of the sequences of the two proteases, with amino acid similarities colored according to the BLOSUM substitution matrix. The scattered blocks of conserved residues are all within the common structural core, which is interrupted extensively by structurally dissimilar loops and secondary structural elements. (PDB codes: 1AKL and 1AST.)

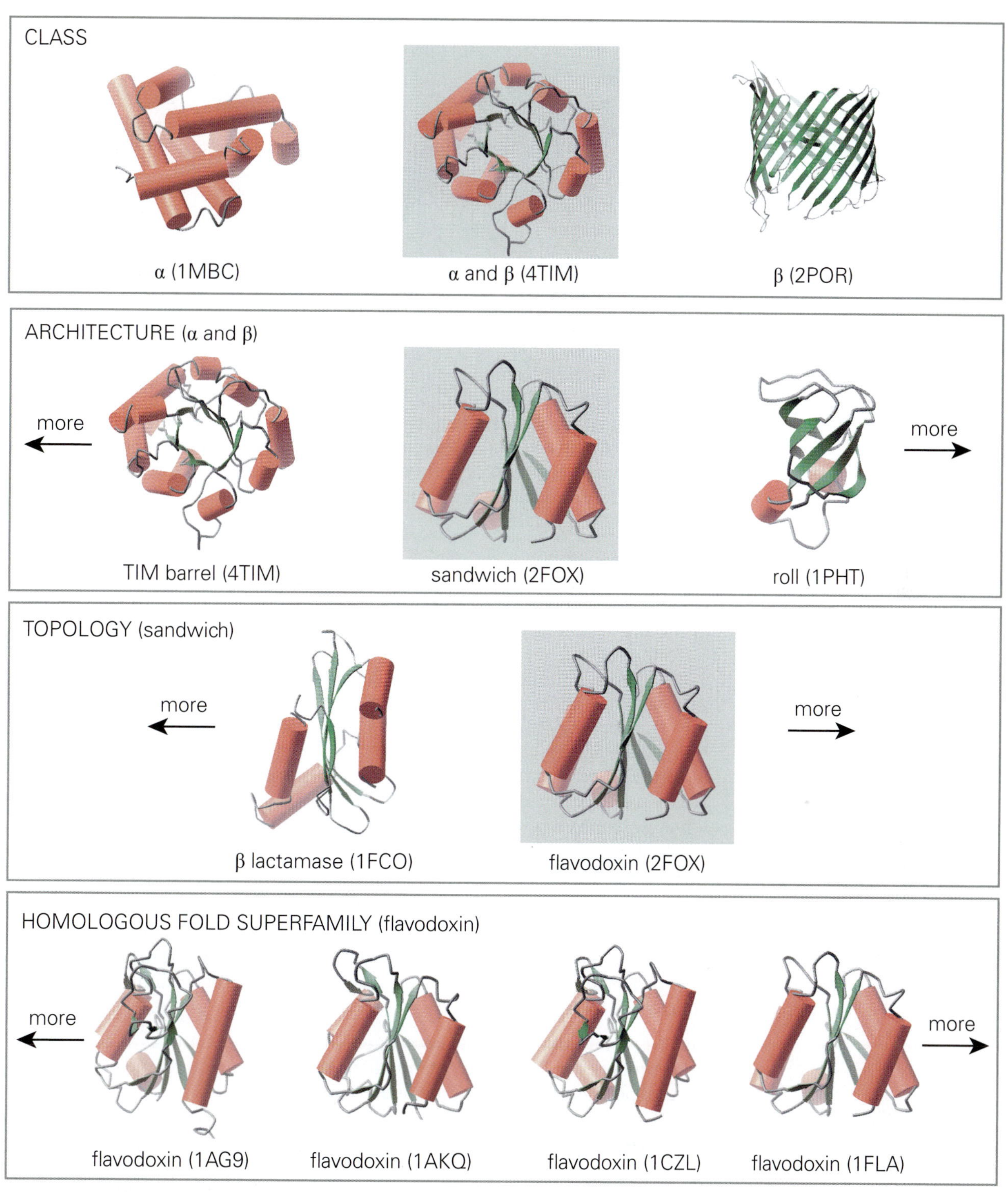


**Figure 5.34 The organization of the CATH database.** The hierarchical grouping of protein domains is illustrated with representative structures. (Adapted from C.A. Orengo et al., and J.M. Thornton, *Structure* 5: 1093–1108, 1997. With permission from Cell Press; PDB codes are shown in parentheses.)

architectures that are recognized in the CATH classification at present, and the number of architectures may grow as more structures are determined. Some of the architectures are given historical names in CATH, such as the "TIM barrel" shown in Figure 5.34. Other architectures are given names in CATH that are evocative of their shapes or structures, such as the "sandwich" and "roll" architectures illustrated in Figure 5.34.

Proteins with similar architectures are next divided into groupings with similar topology (the "T" in CATH). Proteins that have the same topology have their secondary structural elements connected in the same order. The term "topology" is

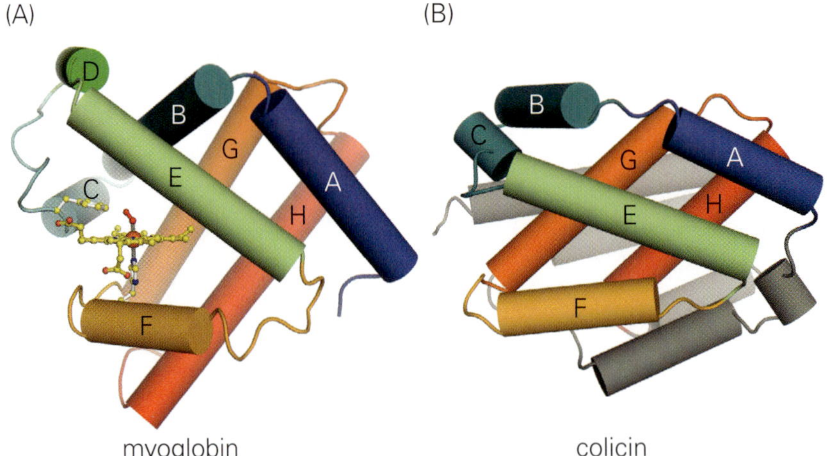


**Figure 5.35 Comparison of the structures of myoglobin and colicin.** Both proteins belong to the same topology grouping within CATH. (PDB codes: 1MBC and 1COL.)

used loosely in the same sense as the term "protein fold," and does not have a very precisely defined meaning in this context. Proteins within the same topology class have the same general fold, and need not share significant sequence similarity. All of the globins illustrated in Figure 5.8 belong in the same topology grouping within the CATH database.

The topology grouping is broader than that implied by the comparison of the globins, because proteins that share no functional similarity can still be grouped in the same topology classification. For example, colicins are proteins involved in the disruption of bacterial membranes and are not heme-binding proteins. Nevertheless, one form of colicin has the same chain fold as the globins (**Figure 5.35**). The organization and sequential connectivity of helices in the common structural core of these proteins are very similar, so they are grouped together into the same topology (fold) family in the CATH database. Colicin shares no significant sequence similarity with myoglobin, does not bind heme, and is involved in membrane rupturing rather than oxygen transport. Colicin and myoglobin are therefore grouped into different homologous superfamilies within the same topology family. It is unclear if this colicin is related evolutionarily to the globins, or whether the similarities in their structures are accidental, due to their having independently evolved the same solution to the folding problem.

The grouping of proteins with similar topology is further broken down into groups of proteins that have both high structural similarity and similarities in their functions. Such groupings are referred to as "homologous superfamilies." The true globins are all grouped together into one homologous superfamily, whereas the colicins are treated separately.

Finally, proteins that share strong sequence identity (> 35%) are grouped together into sequence families. Of the mammalian globins, myoglobins and hemoglobins are grouped into two separate sequence families.

## 5.23 The number of distinct fold families is likely to be limited

The rate at which new chain folds are being discovered is much lower than the rate at which structures are being determined. This suggests that the total number of protein folds is very much smaller than the total number of protein sequences. **Table 5.1** lists the total number of proteins for a number of microorganisms whose genomes have been sequenced completely. The table also lists the number of homologous families, as well as the expected number of different folds (at the level of domains) in each of the organisms, and this reveals an interesting trend. In very simple organisms such as *M. genitalium*, which has only 480 proteins, most of the proteins tend to be unique (the domains in these proteins fall into an estimated 400–550 different families, and the number of unique folds is expected to be between 240 and 352).

**Table 5.1 Number of genes, families, and folds in different microorganisms.**

| Species | Number of proteins in the genome | Total number of families | Number of structurally characterized families | Predicted number of folds |
|---|---|---|---|---|
| Aggregate | | 4000–7000 | 1000 | 900–1300 |
| *M. genitalium* | 480 | 400–600 | 70 | 250–350 |
| *R. prowazekii* | 834 | 750–950 | 122 | 350–500 |
| *A. aeolicus* | 1522 | 950–1100 | 154 | 400–550 |
| *M. jannaschii* | 1715 | 850–950 | 74 | 300–400 |
| *A. pernix* | 1760 | 950–1000 | 62 | 300–450 |
| *Synechocystis* sp. | 3169 | 1700–2200 | 220 | 450–650 |
| *M. tuberculosis* | 3900 | 1500–2000 | 200 | 450–700 |
| *B. subtilis* | 4100 | 1800–2100 | 260 | 450–700 |
| *E. coli* | 4289 | 2000–2600 | 353 | 550–800 |
| *S. cerevisiae* | 6530 | 2400–4500 | 234 | 500–720 |

(Adapted from Y.I. Wolf, N.V. Grishin, and E.V. Koonin, *J. Mol. Biol.* 299: 897–905, 2000. With permission from Elsevier.)

In more complicated organisms, the increase in the size of the genome does not result in a proportional increase in the number of families or folds. For example, the genome of the yeast *S. cerevisiae* has 6,530 proteins, 13.6 times as many as *M. genitalium*. However, the upper estimate for the number of families in yeast is only eight times larger than the estimate for *M. genitalium*. The growth in the number of folds is even smaller, with yeast expected to contain only twice as many different folds as *M. genitalium*, using the upper estimate. An analysis of several completely sequenced genomes has led to an estimate of only a few thousand different kinds of folds that may be found in nature.

Why is the number of protein folds so limited? Given the essentially infinite degree to which amino acids sequences can vary, we might have expected that the total number of protein folds would be extremely large. We do not understand the reasons for this limitation, but natural selection will favor protein sequences that can fold into stable structures. One remarkable feature of commonly found folds is their resistance to the potentially deleterious effects of mutation, a point that we discuss in the next section.

Among the possible folds, some are more likely to be populated than others due to their particularly optimal folding and stability properties. **Figure 5.36** shows the relative sizes of the topology and homologous superfamily groupings in the CATH organization of protein domains. Certain homologous superfamilies have more representatives in the protein databank than others. The most highly populated family is known as the **Rossmann fold**, an $\alpha/\beta$ structure that is common to many proteins that bind to nucleotides and is named after its discoverer, Michael Rossmann. We discuss the Rossmann fold in more detail in Section 5.26.

**Rossmann fold**

A very commonly occurring protein fold, found in nucleotide-binding domains. The Rossmann fold was the first modular protein domain to be identified.

## 5.24 Protein domains are remarkably tolerant of changes in amino acid sequence, even in the hydrophobic core

Protein structures retain the essential character of their chain fold, even though genetic variation and natural selection lead to changes in amino acid sequence. Residues in the hydrophobic cores of proteins are very closely packed, and substitutions in the core can be expected to disturb the stability of the protein.

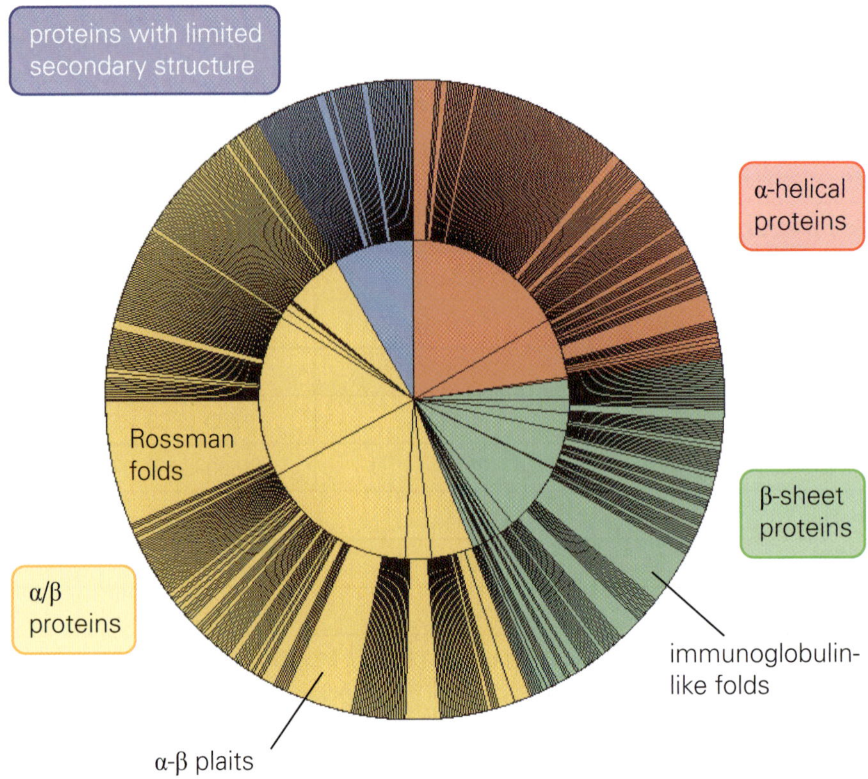


**Figure 5.36 A pie chart showing the population of the first three levels of the CATH database.** The class grouping is shown in the center, with α-helical structures in *red*, β-sheet structures in *green*, α/β structures in *yellow*, and structures without significant secondary structural elements in *blue*. The size of each segment represents the number of different groupings that are present within each major grouping. Topology families are shown in the outer wheel. (Adapted from F. Pearl et al., and C. Orengo, *Nuc. Acids Res.* 33: D247–D251, 2005. With permission from Oxford University Press.)

Nevertheless, experiments in which mutations are introduced into proteins show that proteins can function even when residues that are tightly packed in the core are replaced, particularly when these substitutions are nonpolar.

This ability of proteins to tolerate mutations is a requirement for evolution to proceed. If mutations in proteins were generally so damaging that the proteins ceased to function, then the sequences of proteins would be "frozen" because of selection against mutations leading to a loss of function. One reason for the great diversity of proteins generated by natural selection is that proteins can accumulate many mutations without critical damage to their functions. How is it though, that proteins are so tolerant of mutation? We shall address this question by summarizing the results of experiments on a protein known as lambda (λ) repressor.

Lambda is the name of a bacteriophage—that is, a virus that infects bacteria. Lambda repressor is a transcriptional regulator encoded by the genome of bacteriophage λ, and it controls the expression of genes that are involved in the determination of the life cycle of the bacteriophage. Lambda repressor works by recognizing specific DNA sequences at two adjacent sites in the major groove of DNA (**Figure 5.37**). The ability to recognize DNA depends upon the formation of a specific three-dimensional structure, since changes in the structure would disrupt the pattern of interactions between the protein and the DNA.

The sensitivity of the structure of λ repressor to changes in the nature of the amino acids that make up its hydrophobic core was examined by modifying a strain of *E. coli* so that the bacterium was rendered completely dependent for life on the presence of active λ repressor protein. The gene for λ repressor was then modified so as to randomly substitute other amino acids for seven residues in the hydrophobic core of the repressor (these residues are indicated in yellow in Figure 5.37). The modified λ repressor genes were then introduced into the engineered *E. coli* cells, and the gene for λ repressor was isolated from surviving cells and sequenced. In many cases it was found that mutations had been introduced into the hydrophobic core of λ repressor (**Figure 5.38**).

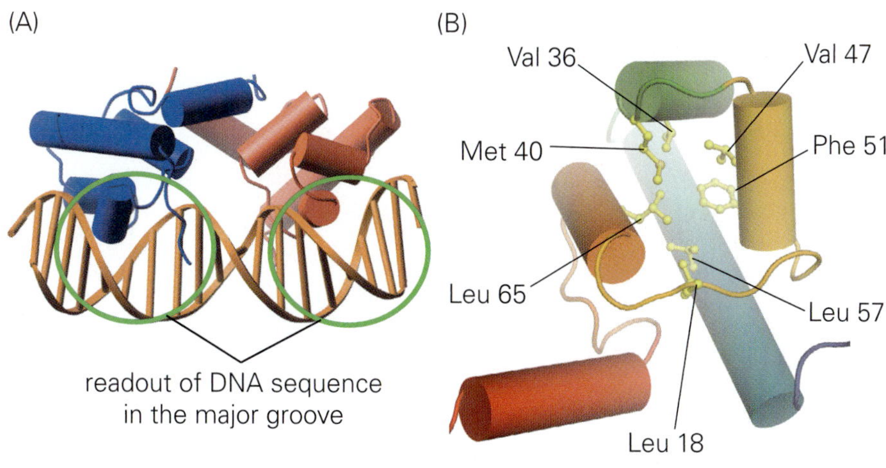


**Figure 5.37 The structure of λ repressor.** (A) Lambda repressor recognizes DNA as a dimer, and the structure of the complex with DNA is shown here. The precise positioning of the two proteins with respect to the DNA is crucial for the proper control of transcription. (B) The hydrophobic core of λ repressor. The seven residues that were altered in a random manner in the selection experiment are indicated in *yellow*. (PDB code: 1LMB.)

Since the cells that survived are required to have a functional λ repressor that can properly recognize its target, the inference is that the protein structure is able to tolerate the amino acid substitutions shown in Figure 5.38. The substitutions include the replacement of the bulkier phenylalanine sidechain with the smaller leucine sidechain, or the smaller valine and alanine sidechains with the larger leucine sidechain. In other cases, the shapes of the sidechains have been altered, such as in the substitution of methionine or isoleucine for leucine. Some of the active variants of λ repressor contain four substitutions at the seven positions in the hydrophobic core at which mutations were introduced.

## 5.25 Structural plasticity in protein domains increases the tolerance to mutation

The λ repressor experiment reveals a considerable plasticity in the hydrophobic core of this small protein. How does the protein structure cope with changes in the size and shape of the residues that pack together to form the hydrophobic core? The answer lies in the fact that protein molecules are dynamic systems in which the positions of the atoms fluctuate considerably. Rather than being a rigid interlocking set of sidechains, in which a shape mismatch would prevent proper folding, the protein structure is constantly fluctuating about a mean conformation (**Figure 5.39**).

Computer simulations of the motions of atoms in proteins indicate that the α helices and β strands in proteins move back and forth to a considerable extent. Transient fluctuations in the structures of proteins are estimated to be as large as the

**Figure 5.38 Variation in the sequences of the active forms of λ repressor.** The seven residues in the hydrophobic core that were varied randomly are shown in *yellow*. Alterations in the sequence of λ repressor molecules that were mutated but still retained activity are indicated in *green*. (Adapted from W.A. Lim and R.T. Sauer, *Nature* 339: 31–36, 1989. With permission from Macmillan Publishers Ltd.)

**Figure 5.39 Fluctuations in the hydrophobic core of a protein as seen in a computer simulation of the motions of atoms.** Ten structures are superimposed, separated from each other by 1 nanosecond during a simulation of the motions of atoms over 10 nanoseconds, at room temperature. Notice the large excursions in the structure of the hydrophobic residues. Such flexibility in the hydrophobic core makes it possible to accommodate amino acid substitutions.

differences between the static average structures of proteins with similar, but not identical, sequences. Because the structures of protein molecules are constantly fluctuating, proteins can accommodate alterations in their hydrophobic cores without falling apart. Not all substitutions are equally likely to be tolerated, hence the relative rarity of finding polar sidechains in the hydrophobic core. But, as the preceding discussion on λ repressor demonstrated, the flexibility of the protein structure allows the substitution of sidechains with quite different shapes within the hydrophobic core, without loss of function.

The introduction of a small number of mutations does not usually affect the function of a protein to a significant extent (unless the mutation is at a crucial position, such as at the position of the invariant histidine in myoglobin; see Figure 5.9). Eventually, as more mutations accumulate, the properties of the protein begin to change, and natural selection then determines whether the gene corresponding to the protein with altered properties continues to be transmitted to progeny or not. If the structural cores of proteins were intolerant of mutations, then evolution would be impossible.

## 5.26 The Rossmann fold is found in many nucleotide binding proteins

In order to get an idea of the power of natural selection acting on random variation to produce highly differentiated proteins, we shall take a closer look at proteins containing **nucleotide-binding domains**. Nucleotide-binding domains were the first fold family to be identified, and we shall see how this fold has evolved to participate in a wide diversity of cellular functions through variation in domain structure and combining with other domains.

We shall look at nucleotide-binding domains that bind NADPH (nicotinamide adenine dinucleotide phosphate) and FAD (flavin adenine dinucleotide) (**Figure 5.40**). Dinucleotides, such as NADPH and FADH, are known as **coenzymes** because they provide chemical functionalities that are not available within the protein components of the enzymes that utilize them. NADPH (or a variant with one less phosphate group, known as NADH) is an electron donor that transfers a hydride ion (two electrons and a proton) from the nicotinamide group to its reaction partners, becoming oxidized to $NADP^+$ in the process. The flavin ring system of FAD can accept the hydride ion from NADPH, becoming reduced to $FADH_2$. The reduced flavin can then transfer the electrons and protons to various substrates. Electron transfer reactions such as these are discussed in more detail in Chapter 11.

Enzymes that use FAD or NADPH have in common protein domains that are specialized for binding to the central portion of the dinucleotides, in which two negatively charged phosphate groups require neutralization. This dinucleotide-binding

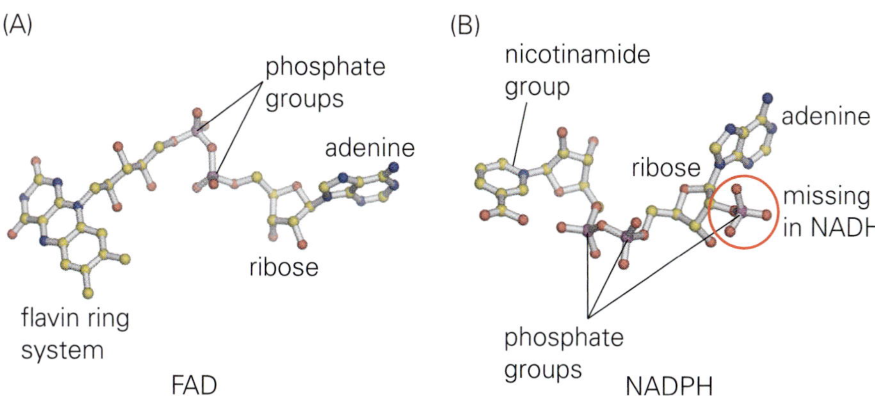


**Figure 5.40 The structures of (A) flavin adenine dinucleotide (FAD) and (B) nicotinamide adenine dinucleotide phosphate (NADPH).** A phosphate group that is present in NADPH but absent in NADH is indicated.

domain consists of a set of parallel β strands, to which are connected two or more α helices in a right-handed manner (**Figure 5.41**). These kinds of domains are referred to as Rossmann folds, and we noted earlier that these are the most commonly occurring domain folds in the CATH database (see Figure 5.35). The most important elements of the Rossman fold are a loop that interacts closely with the phosphate groups, known as the "P loop," for phosphate-binding loop, and the α helix connected to this loop. These elements are colored yellow in Figure 5.41.

Recall from Chapter 1 (Section 1.4) that amide and carbonyl groups of the peptide linkage have dipoles, and that these dipoles are parallel to each other (see Figure 1.11A). In an α helix all of the peptide linkages are parallel to each other, as shown in Figure 1.36. As a result, an α helix behaves as a single electrostatic dipole, known as a **helix dipole**, due to the additive effect of each of the individual dipole moments of the carbonyl and amide groups. The positive pole of the helix dipole is at the N-terminal end of the helix, and in the Rossmann fold the α helix connected to the P-loop is positioned so that its N-terminal end is near the phosphate groups of the nucleotide. In this way, the electrostatic dipole moment of the helix (indicated by the arrow in Figure 5.41) provides a favorable interaction with the negatively charged phosphate groups.

**Helix dipole**

An α helix behaves as if it is an electrostatic dipole, with the positive and negative poles at the N-terminal and C-terminal ends of the helix, respectively. The dipole moment of the helix is a consequence of the fact that the peptide linkages are all arranged parallel to the axis of the helix, and their individual dipole moments reinforce each other.

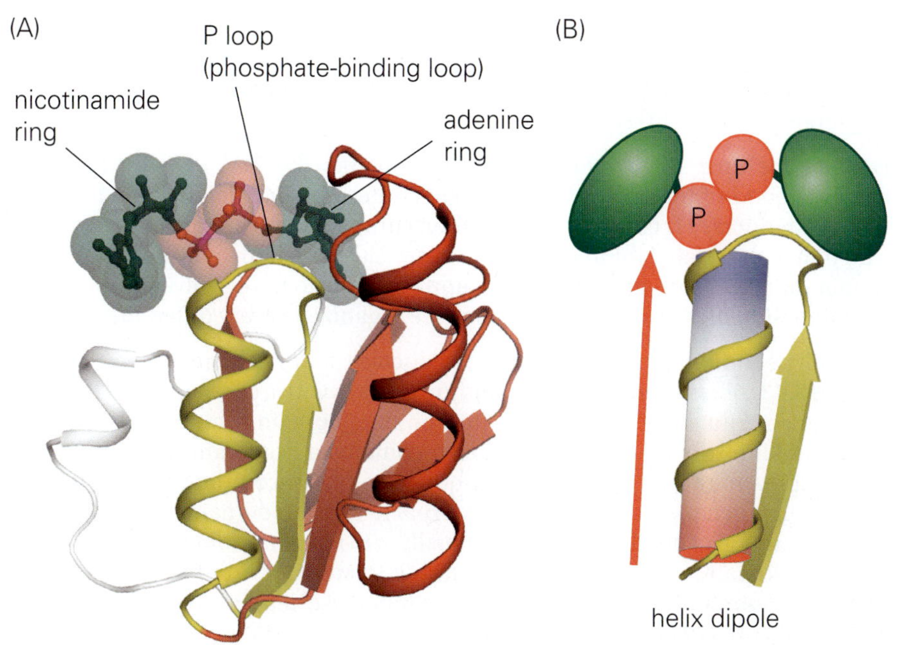


**Figure 5.41 The Rossmann fold.** (A) NADH bound to a Rossmann fold, which is one component of a large enzyme. (B) The most important elements of the Rossmann fold are the loop that interacts closely with the phosphate groups, which is known as the "P loop," for a phosphate-binding loop, and the α helix connected to this loop. The electrostatic dipole moment of the helix (indicated by the arrow) stabilizes the negatively charged phosphate groups. The positive pole of the helix dipole is at the N-terminal end of the helix, and is shaded *blue*. The negative pole of the helix is shaded *red*. (PDB code: 1GET.)

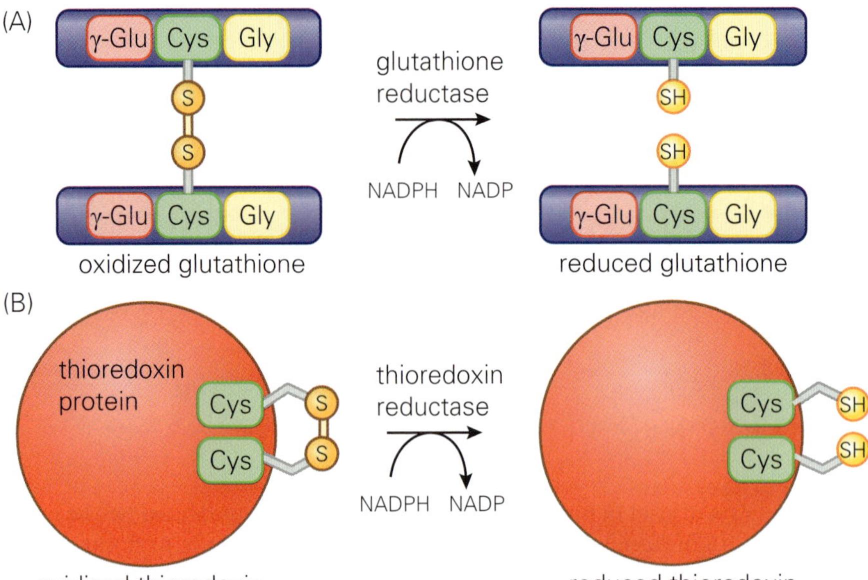


**Figure 5.42 Schematic representation of the reactions catalyzed by (A) glutathione reductase and (B) thioredoxin reductase.** (A) Glutathione is a tripeptide and, in the oxidized form, two molecules of glutathione are linked by a disulfide bond (the bond between the two *orange* circles). The enzyme glutathione reductase uses the reducing power of NADPH to break the disulfide bond, releasing oxidized NADP and reduced glutathione. The first residue in glutathione is a glutamate residue that is linked to the cysteine through its sidechain carboxyl group (denoted γ-Glu). (B) Thioredoxin reductase catalyzes the same reaction, but the disulfide bond that it breaks is within a protein, thioredoxin, that is about 100 residues long.

## 5.27 Thioredoxin reductase and glutathione reductase are enzymes that diverged from a common ancestor, but their active sites arose through convergent evolution

We shall now look at one example of how specialized functions arise from combining domains into larger proteins. The proteins we shall discuss are members of a class of enzymes known as disulfide reductases. The general reaction catalyzed by these enzymes is the transfer of electrons from the reduced nicotinamide ring of NADPH to the flavin ring of FAD. The reduced flavin ($FADH_2$) then transfers the electrons to a disulfide bond at the active site of the enzyme, which is consequently broken, releasing two cysteine residues. Finally, the reduced cysteines are utilized by the enzyme to transfer electrons to a disulfide bond in a substrate molecule, resulting in the reduction and breakage of the disulfide bond (see Chapter 11 for an explanation of oxidation and reduction). The two substrates we shall consider are glutathione, a small tripeptide, and thioredoxin, a protein (**Figure 5.42**). Both are reducing agents that help to maintain the redox balance of the cell, and disulfide reductase enzymes convert them from the oxidized to the reduced state. Thus, both enzymes serve as conduits for facilitated transfer of electrons from the NADPH to a disulfide bond in their respective substrates. The interesting thing here is the way in which evolution has come up with two very different arrangements of related domains in order to accommodate small (glutathione) and large (thioredoxin) substrates.

Each of the enzymes contains four components: the NADPH, the FAD, the disulfide bond of the enzyme, and the substrate. We shall first discuss how glutathione reductase is constructed, and then compare the structure of this enzyme with that of the related thioredoxin reductase. Although both enzymes utilize very similar dinucleotide-binding domains and are likely to have evolved from a common ancestor, the manner in which they are organized is very different.

The polypeptide chain of glutathione reductase is folded into three domains (**Figure 5.43**). First, there is an FAD-binding domain with a central core that adopts a Rossmann fold. A second dinucleotide-binding domain, which binds NADPH, is inserted into the first domain and also contains a Rossmann fold. Insertions of one domain into the structure of another are not uncommon and, as long as the insertion occurs within a flexible external loop, both domains are able to fold properly and carry out their functions. Finally, a C-terminal domain, known as the interface domain, is involved in mediating interactions with the corresponding domain in another molecule of glutathione reductase, leading to the formation of a dimer (**Figure 5.44**).

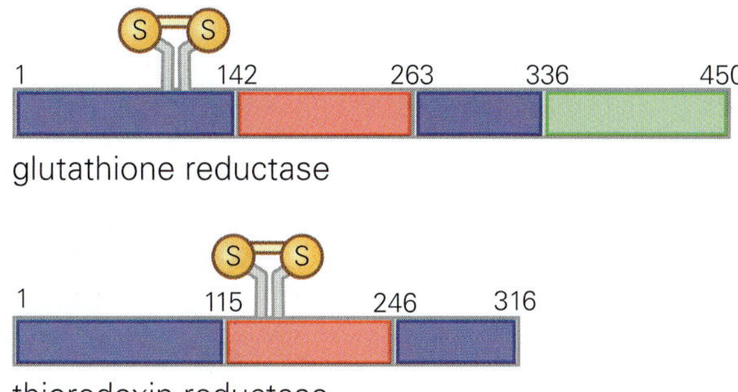


**Figure 5.43 Schematic representation of the domain organization of glutathione reductase (top) and thioredoxin reductase (bottom).** Each protein has one FAD-binding domain (*blue*) and one NADPH-binding domain (*red*), with the NADPH domain inserted into a loop in the FAD domain. Glutathione reductase contains an interface domain (*green*) that is required for the formation of dimers of that enzyme. Both enzymes contain a reactive disulfide bond (*orange* spheres), which plays a key role in the electron transfer reaction. The disulfide bond is located within the FAD domain in glutathione reductase, but found instead in the NADPH domain in thioredoxin reductase.

The glutathione binding site is located at the interface between two molecules of the dimer, within a deep pocket. One side of the substrate binding site is formed by residues from the interface domain of one of the monomers in the dimer, and the other side is composed of residues from the NADPH- and FAD-binding domains of the other monomer (see Figure 5.44). The enzyme has two active sites, which act independently of each other. The substrate binding site is narrow and deep, and well suited to accommodate the relatively small glutathione molecule. Glutathione reductase cannot utilize the much larger protein thioredoxin as a substrate, simply because the protein could not fit into the narrow channel in which glutathione binds.

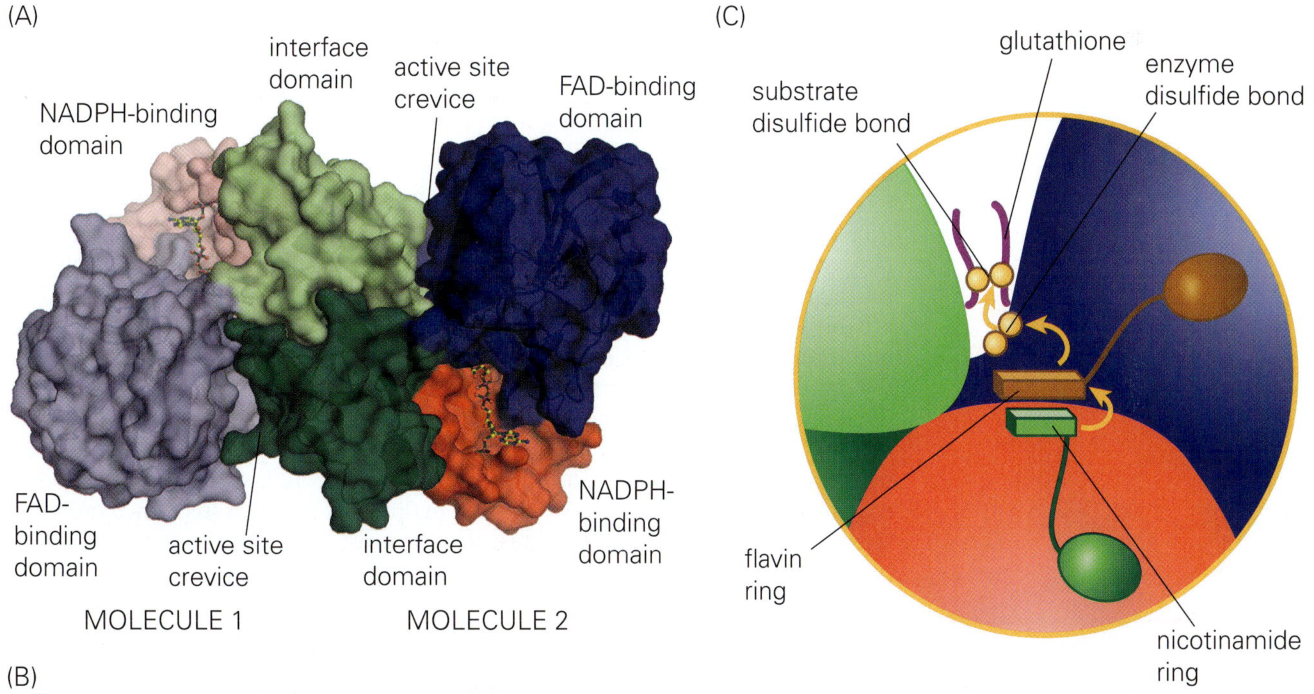


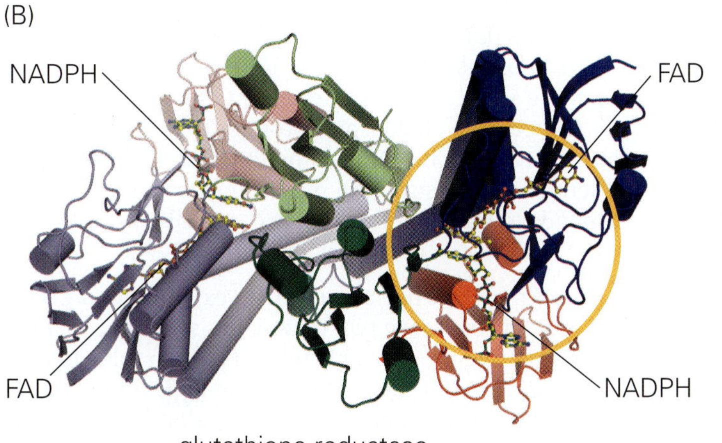


**Figure 5.44 The structure of glutathione reductase.** (A) The molecular surface. Two binding sites for glutathione are formed at the interface between the two molecules in the dimer. The FAD and NADPH domains are colored *blue* and *red*, respectively. (B) The polypeptide backbone of the protein. One active site is circled. (C) A schematic representation of one active site. The flow of electrons from the reduced nicotinamide ring to the flavin ring, from there to the disulfide bond of the enzyme and ultimately to the substrate disulfide bond is indicated by the *orange* arrows. (PDB code: 1GET.)

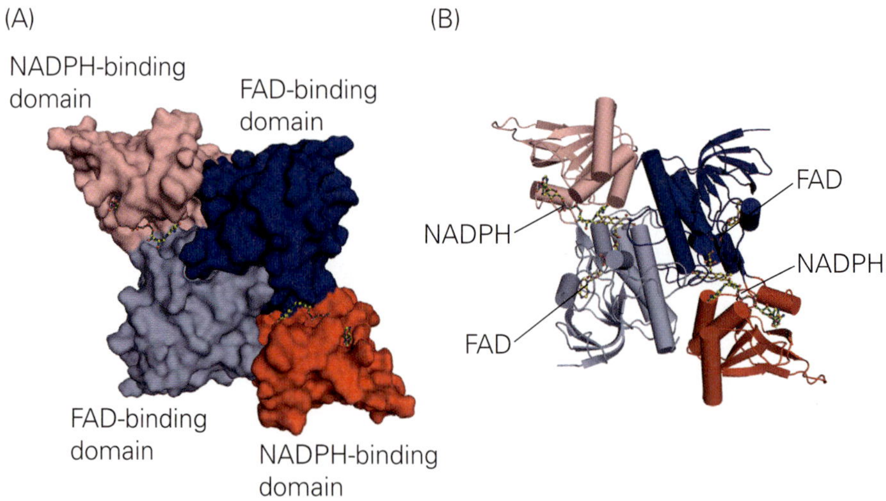


**Figure 5.45 The structure of thioredoxin reductase.** (A) The molecular surface of the dimer, with the FAD and NADPH domains colored *blue* and *red*, respectively. Thioredoxin binds to the relatively flat surface at the interfaces of the two domains. (B) The chain fold of the two molecules. Note the difference between the assembly of this dimer and that of glutathione reductase (shown in Figure 5.44). (PDB code: 1TDF.)

Thioredoxin reductase utilizes a completely different quaternary architecture in order to recognize its substrate, thioredoxin. Like glutathione reductase, it contains an NADPH-binding domain and an FAD-binding domain, and it forms a dimer (**Figure 5.45**). But thioredoxin reductase lacks the interface domains that mediate dimer formation in glutathione reductase, and so it assembles into a dimer quite differently (see Figure 5.45). The dimer of thioredoxin reductase has a relatively flat surface without any crevice corresponding to the glutathione-binding site of glutathione reductase.

Similarities between the sequences of the nucleotide-binding domains of glutathione reductase and thioredoxin reductase indicate that the two proteins diverged from a common ancestor (**Figure 5.46**). First, through a process of gene duplication and fusion, an ancestral nucleotide-binding domain must have developed into a protein with the ability to bind both FAD and NADPH. This two-domain unit is likely to have emerged before the divergence of separate glutathione reductase and thioredoxin reductase enzymes. This is suggested by the similarity between the corresponding domains of these enzymes, and the fact that the NADPH-binding domain is inserted into the FAD-binding domain in essentially the same way in both. The two-domain unit then has developed into two different dimeric forms by processes of divergent evolution. As you can see in Figure 5.43, a disulfide bond that mediates electron transfer is located in different domains in the two proteins. The acquisition of disulfide reductase activity must therefore have occurred independently in the two cases, through a process of convergent evolution.

## Summary

One guiding principle in biology is that the three-dimensional structures of proteins are determined by their amino acid sequences. In this chapter, we have seen that many alterations in the sequences of proteins do not change the overall nature of the fold of the polypeptide chains, and proteins that share very little sequence similarity can nevertheless have the same three-dimensional fold. The relationship between sequence and structure is therefore not symmetrical: while one sequence corresponds to one fold, that fold is compatible with a large number of sequences.

The number of protein structures that have been determined experimentally is far smaller than the number of proteins that have had their genes sequenced. Many of these proteins of unknown structure are probably related in their folds to proteins whose structures are already known. Nevertheless, the degeneracy in the relationship between protein structure and sequence means that it can be difficult to know whether two proteins that share limited sequence similarity also

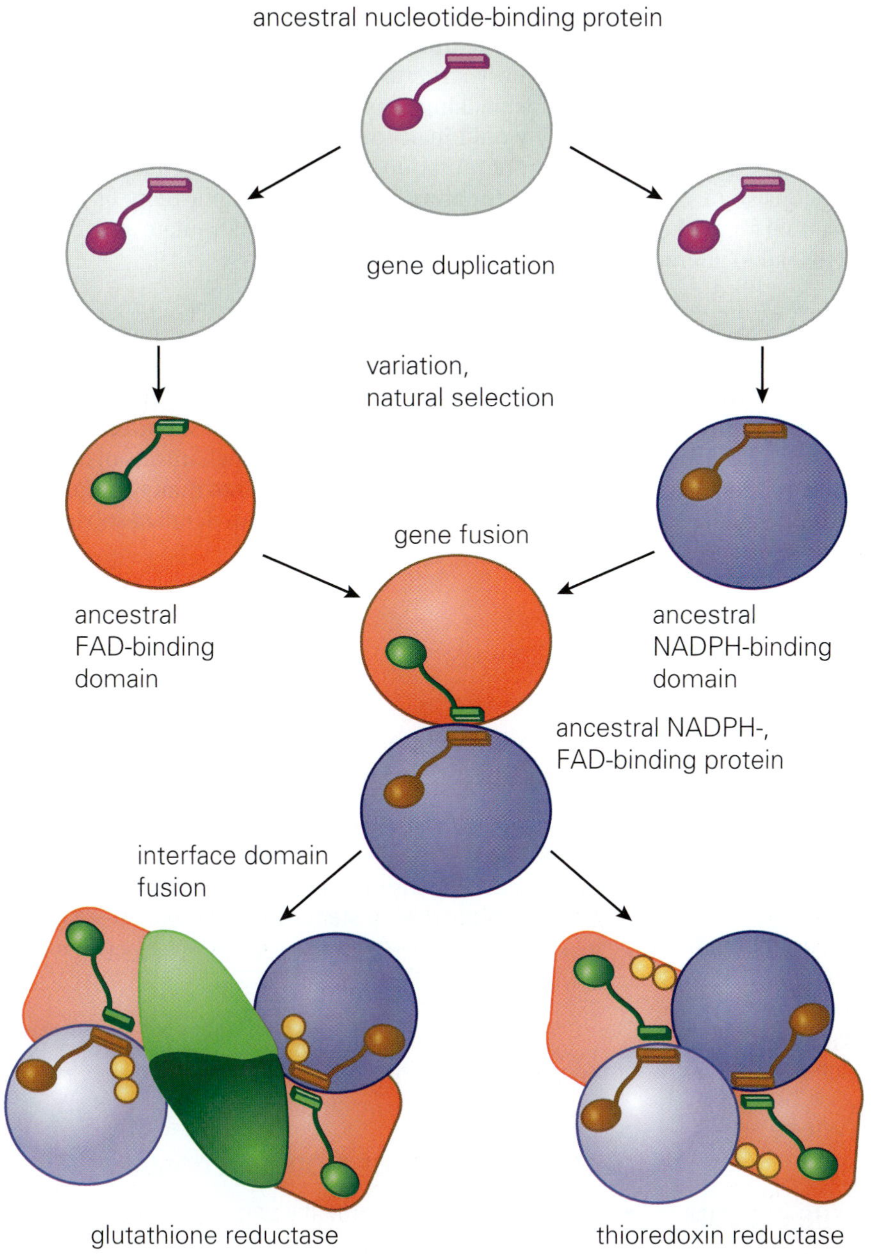

**Figure 5.46 A schematic illustration of a possible pathway by which the disulfide reductase enzymes evolved from an ancestral nucleotide-binding protein.** An ancestral nucleotide-binding domain is assumed to have undergone gene duplication, with the two resulting domains evolving, independently, the ability to bind to NADPH and FAD, respectively. Gene fusion then resulted in both domains becoming part of the same protein. This ancestral NADPH- and FAD-binding module was further duplicated, leading eventually to the evolution of glutathione reductase and thioredoxin reductase. The cysteine residues that undergo oxidation/reduction are shown as *yellow* spheres.

have the same structure. The detection of similarities in distantly related proteins relies on the use of amino acid substitution scores, such as those embodied in the block substitution matrix (BLOSUM). These substitution scores are derived from aligned blocks of sequences of related proteins, and they indicate the likelihood that one amino acid is substituted for another in proteins that are evolutionarily related. Substitution scores are used to align protein sequences with each other in a way that maximizes the likelihood that the sequences being aligned are related.

Very distantly related proteins can have the same fold, but the relationship between them may not be detectable by methods that align amino acid sequences. Fold-recognition methods exploit the knowledge of the three-dimensional environments of the amino acids at each position of a known structure to find out which sequences in genomic databases might be compatible with the given fold. These methods are helpful in relating proteins in the much larger sequence database to proteins in the smaller structural database. Since the active sites of proteins tend to be conserved through evolution, such connections provide valuable information about the functions of proteins that are poorly characterized at present.

Protein domains retain a common structural core as their sequences diverge during evolution. The stability of the structural core arises because there are particularly stable ways in which secondary structural elements can pack against each other. Mutations that disturb this packing drastically are likely to be eliminated by natural selection, since they would lead to malfunctioning proteins. Protein domains are remarkably tolerant of mutation, with many different substitutions being consistent with folding and function. This structural plasticity of proteins is essential for evolution, because it allows mutations to accumulate and, eventually, to lead to altered function.

Complex proteins are constructed by combining a relatively small number of folds in different ways. Proteins with very similar domains can be assembled quite differently and can operate using radically different mechanisms. This makes it very difficult to understand the workings of complicated protein machineries without first determining the structures of the intact functioning units. Fortunately, the concerted efforts of biologists to understand the detailed mechanisms of different cellular systems have provided us with structural insights into many complicated macromolecular assemblies, some of which are discussed in detail in subsequent chapters.

## Key Concepts

### A. THE THERMODYNAMIC HYPOTHESIS

- The structure of a protein is determined by its sequence and does not require an external template. This is known as the thermodynamic hypothesis of protein folding.
- The thermodynamic hypothesis was first established for the enzyme ribonuclease-A, which can be folded and unfolded reversibly.
- Ribonuclease-A, like all enzymes, needs to adopt a well-defined structure in order to function. This corresponds to one unique set of disulfide bonds among 105 possible pairings of eight cysteine residues.

### B. SEQUENCE COMPARISONS AND THE BLOSUM MATRIX

- Protein folds are conserved while amino acid sequences vary during evolution.
- Similarities in protein sequences are analyzed by evaluating the likelihood of amino acid substitutions, based on evolutionarily related sequences.
- Block substitution matrices (for example, BLOSUM-62) tell us the likelihood that one amino acid is replaced by another in two evolutionarily related proteins.
- The BLOSUM substitution score is defined in terms of the logarithm of the likelihood that one amino acid in a protein is replaced by another during evolution.
- Substitution scores are used to align sequences and to detect similarities between proteins.

### C. STRUCTURAL VARIATION IN PROTEINS

- Proteins retain a common structural core as their sequences diverge during evolution.
- Sequence comparisons can fail to detect structural similarities between distantly related proteins.
- The amino acids have preferences for different environments in folded proteins.
- Fold-recognition algorithms, also known as threading algorithms, match patterns of structural environments in known protein structures with the sequences of proteins with unknown structure.

### D. THE EVOLUTION OF MODULAR DOMAINS

- Domains are the fundamental unit of protein evolution.
- Protein domains can be grouped into hierarchical classes based on their secondary structural elements, their folds, and their similarity to other sequences.
- Protein structures are remarkably tolerant of mutations, even in the hydrophobic core. This property of proteins is essential for evolution.
- The number of distinct folds in proteins is likely to be limited.
- The Rossmann fold is found in many nucleotide-binding proteins and illustrates the versatility with which evolution generates new functions while retaining a protein fold.

## Problems

### True/False and Multiple Choice

1. The BLOSUM scoring matrix gives a measure of how conservative a mutation is. For substitutions of aspartic acid (Asp, D), which of the following orderings correctly places the amino acids from most conservative to least conservative?
   a. K,L,A,C,E,S
   b. E,S,K,A,C,L
   c. L,C,A,K,S,E
   d. A,C,E,K,L,S

2. An environment profile in the 3D-1D profile method compares:
   I. the stability of the amino acid in varying solvents
   II. the burial of each amino acid in the structure
   III. the hydrophobicity surrounding each amino acid
   IV. the type of secondary structure element containing each amino acid
   a. I, II, III, and IV
   b. I and IV
   c. II, III, and IV
   d. II and IV

3. Protein domains can be assembled together in many different ways, because surface sidechains can be mutated easily without losing protein stability.
   True/False

4. In contrast to ribonuclease, some proteins cannot fold without the assistance of proteins known as molecular chaperones. This means the thermodynamic hypothesis of protein folding does not apply to these proteins.
   True/False

5. Two proteins that share more than 50% sequence identity over a 100-residue stretch are likely to have the same three-dimensional fold.
   True/False

### Fill in the Blank

6. Two common chemical denaturants of proteins are guanidinium and _______.

7. The core of a protein generally contains residues from the ________ class of amino acids.

8. Globin proteins bind the iron-containing ________ cofactor.

9. A disulfide bond links two _______ residues.

10. According to the BLOSUM substitution matrix, the most conservative mutation from tryptophan (W), other than to itself, is to ______, which has a score of ______.

### Quantitative/Essay

11. Many soluble human proteins can be expressed in the *E. coli* bacteria or using an *in vitro* translation system. How can these proteins fold without the cellular machinery present in human cells?

12. What level of activity (1, 10, or 100%) is predicted for ribonuclease-A when it is subject to each of the following stepwise procedures?
    a. i. denatured, then
       ii. reduced, then
       iii. exposed to oxygen, then
       iv. refolded by removing urea
    b. i. denatured, then
       ii. reduced, then
       iii. refolded by removing urea while exposed to oxygen
    c. i. denatured, then
       ii. refolded by removing urea

    Rationalize the predictions based on the effect of denaturation and reduction–oxidation of the ribonuclease-A cysteine residues. Assume that all possible unfolded conformations are equally likely and that folding is faster than oxidation.

13. A folded protein structure contains six ion pairs between lysine and glutamate. There are no other possible ion pairs in the protein. A chemical cross-linker forms a covalent bond between ion-paired lysine and glutamate sidechains. By analogy to the Anfinsen experiment, the following experiments are done:

    a. folded protein → unfold with urea → remove urea to refold → add cross-linker → remove excess cross-linker → measure activity

    b. folded protein → unfold with urea → add cross-linker → remove excess cross-linker → remove urea to refold → measure activity

    The cross-linker does not by itself alter the activity of the protein when the correct ion pairs are formed. A protein with incorrect ion pairs cross-linked would be inactive. The activity measured at the beginning and end of experiment (a) is 100%. What percentage of the activity do you expect to observe at the end of experiment (b)? Assume that all unfolded conformations are equally likely.

14. Use the BLOSUM substitution matrix (Figure 5.11) to compute the sum of the substitution scores ($S_{ij}$) and the overall likelihood ratio ($L$) of the following short alignments:
    a.
    ```
    PADKTN
    PEEKSA
    ```
    b.
    ```
    KFLASV
    ATWDPE
    ```

**15.** Based on the BLOSUM matrix, how much more likely is it that:

a. tryptophan is substituted by a tyrosine than a tryptophan is substituted by a cysteine?
b. sequence (i) DPKRFL is related to sequence (ii) EPKRFI than sequence (i) is related to sequence (iii) KGKRYA?

To answer this question, you must calculate the ratio of the likelihood ratios for each case. Explain the significance of higher likelihood.

**16.** Proteins known as cyclophilins catalyze proline *cis-trans* isomerization. A catalytic arginine residue is invariant in all cyclophilins. All other positions change residue identity in different cyclophilins despite the fact the variant proteins have the same overall fold and general catalytic activity. Explain, given the relationship between protein sequence and structure, how catalytic activity is retained even though most residues can change.

**17.** Why is the sequence similarity generally higher when comparing two globins from mammals than when comparing a globin from a mammal and a globin from a plant?

**18.** How might the tolerated variation in the hydrophobic core of the lambda repressor change if the hydrophobic core of wild type lambda repressor were more tightly packed?

**19.** What characteristics define a protein domain?

**20.** The diagram below shows the size distribution for globular proteins produced by the bacterium *E. coli*. Explain why the distribution of protein sizes has the periodicity that is seen in the diagram and estimate a value for *x*.

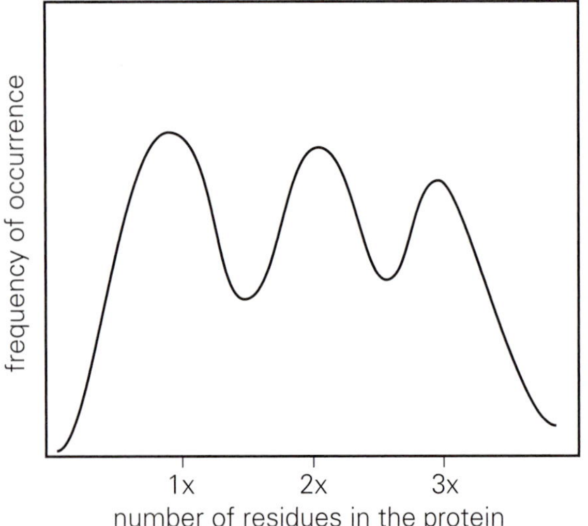


**21.** The number of distinct protein folds is limited. Why might this be so? Approximately how many folds are there (hundreds, thousands, millions, or billions)?

**22.** How many folds are represented below? Describe what CATH class each fold belongs to and how the secondary structure is arranged in each fold.

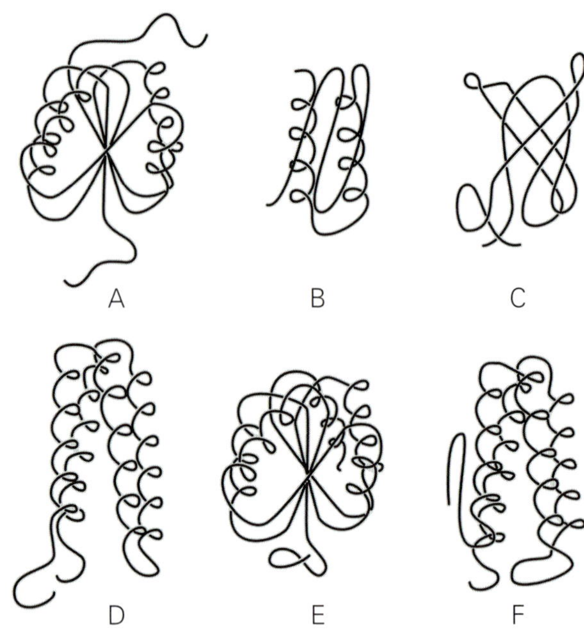


**23.** A threading program is used and it gives two possible predicted folds for a particular sequence. They differ mainly in the placement of a single helix. In predicted fold (A), the helix is entirely within the hydrophobic core. In predicted fold (B), the residues of the helix are mostly exposed to solvent. The sequence of the helix is LIVFLAIL. Explain how the 3D-1D profile method could be used to distinguish between the two possible folds.

**24.** What structural features of the Rossmann domain enable it to bind nucleotides?

**25.** Both thioredoxin reductase and glutathione reductase use fused FAD- and NADPH-binding domains and dimerization to accomplish their cellular functions. Which structural feature likely evolved first, the fused domains or dimerization?

# Further Reading

## General


Creighton TE (1993) Proteins: Structures and Molecular Properties, 2nd ed. New York: W.H. Freeman.

Dickerson RE & Geis I (1983) Hemoglobin: Structure, Function, Evolution, and Pathology. Menlo Park, CA: Benjamin Cummings.

Gu J & Bourne PE (2009) Structural Bioinformatics, 2nd ed. Hoboken, NJ: John Wiley & Sons.

Lesk AM (2008) Introduction to Bioinformatics, 3rd ed. New York: Oxford University Press.


## References

### A. The Thermodynamic Hypothesis


Anfinsen CB (1972) Nobel Lecture: Studies on the Principles that Govern the Folding of Protein Chains. T. Frängsmyr, ed. Stockholm: The Nobel Foundation.

Epstein CJ, Goldberger RF & Anfinsen CB (1963) The genetic control of tertiary protein structure: Studies with model systems. *Cold Spring Harb. Symp. Quant. Biol.* 28, 439–449.

Sela M & Lifson S (1959) The reformation of disulfide bridges in proteins. *Biochim. Biophys. Acta* 36, 471–478.


### B. Sequence Comparisons and the BLOSUM Matrix


Altschul SF, Madden TL, Schaffer AA, Zhang J, Zhang Z, Miller W & Lipman DJ (1997) Gapped BLAST and PSI-BLAST: A new generation of protein database search programs. *Nuc. Acids Res.* 25, 3389–3402.

BLAST: Basic Local Alignment Search Tool. http://blast.ncbi.nlm.nih.gov/Blast.cgi

Henikoff S & Henikoff JG (2000) Amino acid substitution matrices. *Adv. Protein Chem.* 54, 73–97.

Krishnamurthy N & Sjolander KV (2005) Basic protein sequence analysis. *Curr. Protoc. Mol. Biol.* Chapter 19, Unit 19.15.


### C. Structural Variation in Proteins


Bowie JU, Luthy R & Eisenberg D (1991) A method to identify protein sequences that fold into a known three-dimensional structure. *Science* 253, 164–170.

Holm L & Sander C (1996) Mapping the protein universe. *Science* 273, 595–602.

Jones DT, Taylor WR & Thornton JM (1992) A new approach to protein fold recognition. *Nature* 358, 86–89.

Marti-Renom MA, Madhusudhan MS & Sali A (2004) Alignment of protein sequences by their profiles. *Prot. Sci.* 13, 1071–1087.

Richardson JS (1981) The anatomy and taxonomy of protein structure. *Adv. Protein Chem.* 34, 167–339.


### D. The Evolution of Modular Domains


Andreeva A, Howorth D, Brenner SE, Hubbard TJ, Chothia C & Murzin AG (2004) SCOP database in 2004: Refinements integrate structure and sequence family data. *Nuc. Acids Res*. 32, D226–D229.

Buehner M, Ford GC, Moras D, Olsen KW & Rossmann MG (1973) D-glyceraldehyde-3-phosphate dehydrogenase: Three-dimensional structure and evolutionary significance. *Proc. Natl. Acad. Sci. USA* 70, 3052–3054.

Chothia C & Gough J (2009) Genomic and structural aspects of protein evolution. *Biochem. J.* 419, 15–28.

Karplus M & McCammon JA (2002) Molecular dynamics simulations of biomolecules. *Nat. Struct. Biol.* 9, 646–652.

Koonin EV, Wolf YI & Karev GP (2002) The structure of the protein universe and genome evolution. *Nature* 420, 218–223.

Laskowski RA & Thornton JM (2008) Understanding the molecular machinery of genetics through 3D structures. *Nat. Rev. Genet.* 9, 141–151.

Levitt M (2009) Nature of the protein universe. *Proc. Natl. Acad. Sci. USA* 106, 11079–11084.

Orengo CA, Michie AD, Jones S, Jones DT, Swindells MB & Thornton JM (1997) CATH—a hierarchic classification of protein domain structures. *Structure* 5, 1093–1108.

Wolf YI, Grishin NV & Koonin EV (2000) Estimating the number of protein folds and families from complete genome data. *J. Mol. Biol.* 299, 897–905.

Xia Y & Levitt M (2004) Simulating protein evolution in sequence and structure space. *Curr. Opin. Struct. Biol.* 14, 202–207.

# PART II

## ENERGY AND ENTROPY

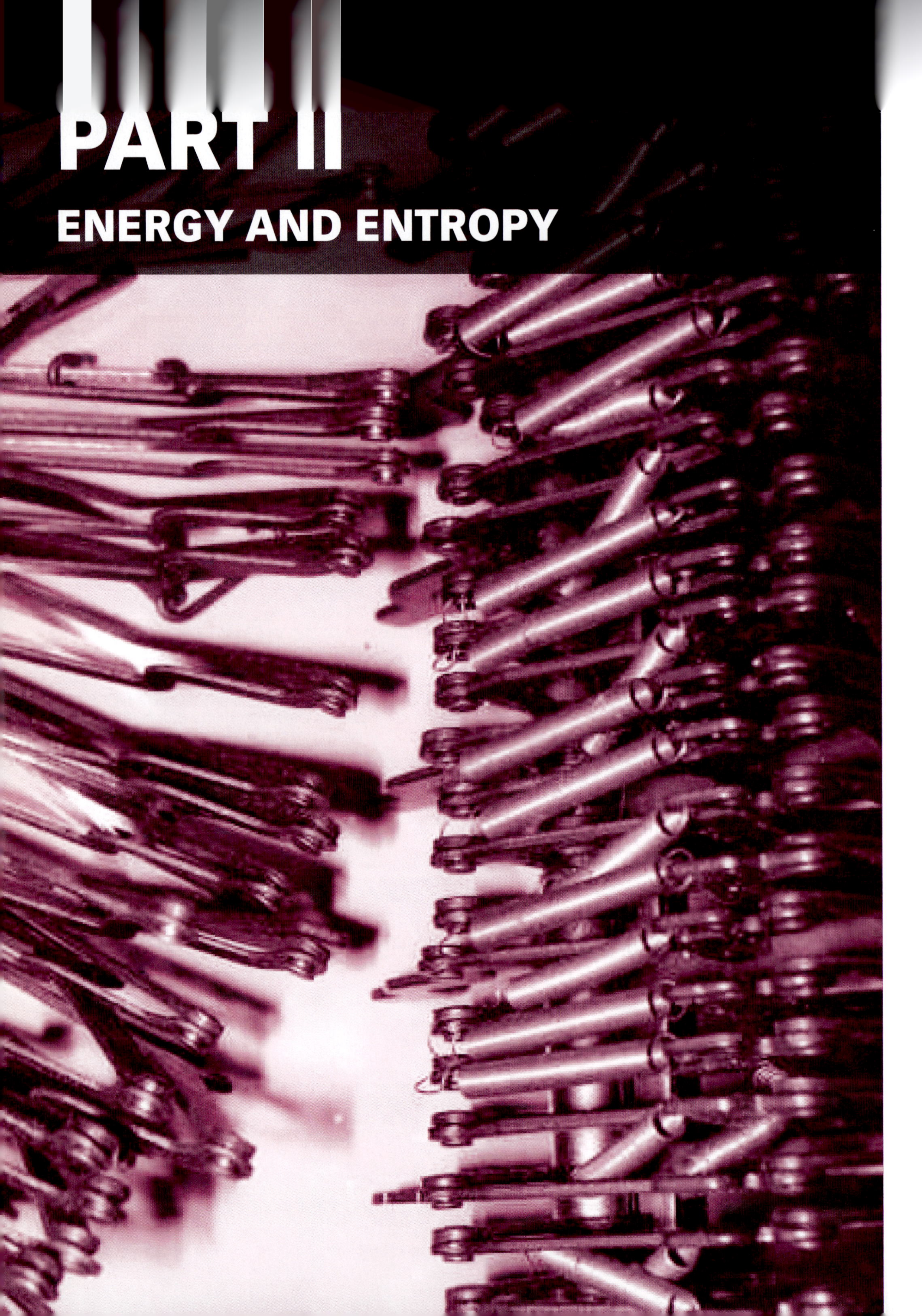

# Energy and Intermolecular Forces

The functional activity of proteins, DNA, and RNA depends ultimately on the forces between atoms in these molecules. These forces depend, in turn, on the energetics of interactions between atoms, the details of which are the focus of this chapter. In the study of biological processes, we are often interested in how energy is stored in molecular systems, and how it is released when needed. Light energy, for example, is captured in plant cells and coupled to the production of carbohydrates such as glucose. Glucose is an "energy-rich" molecule, and the oxidation of glucose releases energy that is utilized by cells for the conversion of ADP to ATP.

We are familiar with the concepts of force and energy from classical Newtonian mechanics, which describes the mechanical translation and rotation of objects, the energy of compressing springs, and the movement of charges in electric fields. These concepts are all directly applicable to molecular interactions, but we need to first make a link to the underlying quantum mechanical description of molecular energies. In addition, we have to consider the fact that molecules undergo chemical reactions—that is, they make or break bonds, which can take up or release energy. Chemical reactions are not the only transformations in which changes in energy are important. Molecular recognition events, such as the binding of a transcription factor to its target DNA site, or of a hormone to its receptor, can also take up or release energy.

The spontaneous nature of certain chemical reactions, such as the hydrolysis of ATP by enzymes, is used to drive processes that would not otherwise occur spontaneously in the cell (**Figure 6.1**). As an example, consider how a molecular motor known as kinesin works. Kinesins transport cargo, such as vesicles, inside cells by moving along tracks known as microtubules. Kinesin forms dimers consisting of two ATP-binding motor domains, which bind to the microtubule track and are linked to the cargo through a coiled coil. The orientation of the connector between the coiled coil and the motor domain changes drastically when ATP is hydrolyzed by kinesin to form ADP (Figure 6.1B). The terminal phosphate of ATP forms hydrogen bonds and other interactions with the protein, and these are lost when ATP is converted to ADP. The changes in the conformation of the connector cause kinesin to "walk" along microtubules, as shown schematically in Figure 6.1C. The mechanism of kinesin is discussed in more detail in Chapter 17 (Section 17.26).

In this chapter, we study three different aspects of molecular energy. First, we look at heat transfer between a region of interest, such as a test tube, and its surroundings. Heat transfer is the most obvious manifestation of changes in molecular energy and the most readily measurable. We use heat transfer to introduce the concept of energy conservation, known as the **first law of thermodynamics**. The second part of the chapter deals with **heat capacity**, which is the amount of heat required to bring about a specific temperature increase in a material or solution. Heat capacity measurements provide insight into the nature of forces between molecules. The first two sections of this chapter are rather abstract, but they introduce several ideas that are critical to the discussion of the nature of spontaneous change in subsequent chapters. Finally, the third part of the chapter discusses

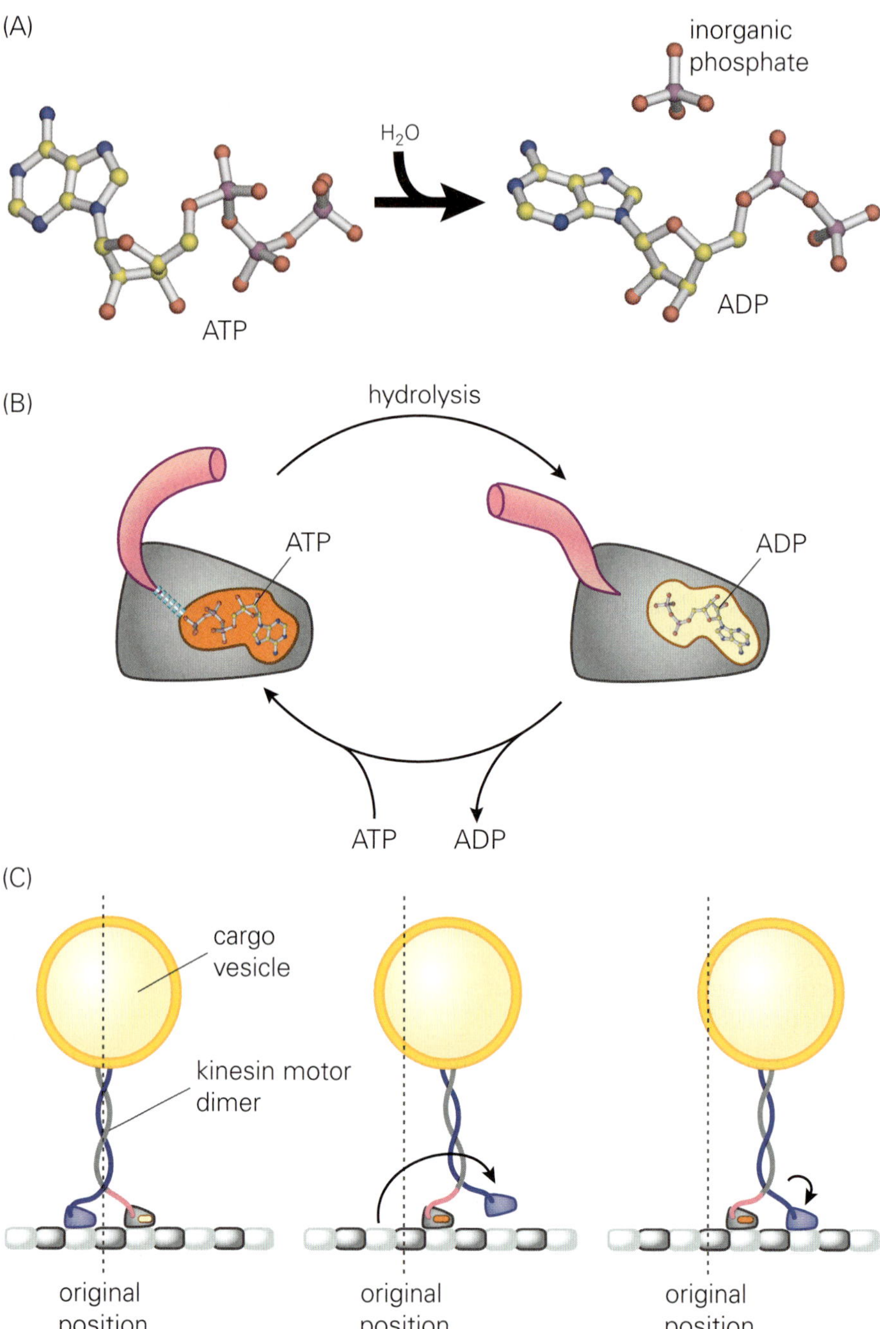


**Figure 6.1 The coupling of ATP hydrolysis to mechanical motion.** (A) ATP loses its terminal phosphate group upon hydrolysis. This reaction occurs rapidly in the forward direction when catalyzed by enzymes. (B) A schematic diagram of the motor domain of kinesin. ATP binds to the motor domain and, by forming hydrogen bonds and other interactions, provides the energy to hold a connector domain (*pink*) in a particular conformation (shown curving to the right in the diagram). ATP hydrolysis leads to a loss of these interactions, and the connector domain relaxes to another conformation (pointing to the left). (C) Kinesin "walks" along microtubule tracks. The microtubule proteins are shown in *gray*. The kinesin motor domains are in *blue* and *gray*. One of the kinesin motor domains (*gray*) is initially in the ADP bound state (with the connector, in *pink*, pointing to the left). ATP binding drives the connector to the right, which results in the other motor domain (*blue*) flipping over to the right. The process is repeated by ATP binding to the *blue* kinesin motor domain, and the cargo vesicle (*yellow*) moves to the right. (C, adapted from A. Gennerich and R.D. Vale, *Curr. Opin. Cell Biol.* 21(1): 59–67, 2009.)

how we can calculate molecular energies from knowledge of the three-dimensional structures of molecules such as proteins and RNA.

# A. THERMODYNAMICS OF HEAT TRANSFER

## 6.1 In order to keep track of changes in energy, we define the region of interest as the "system"

One of the simplest ways to study energetic changes in biochemical processes is to carry out the reaction (for example, a chemical reaction or binding event) in a test tube and to measure the heat released or taken up as the reaction proceeds. If

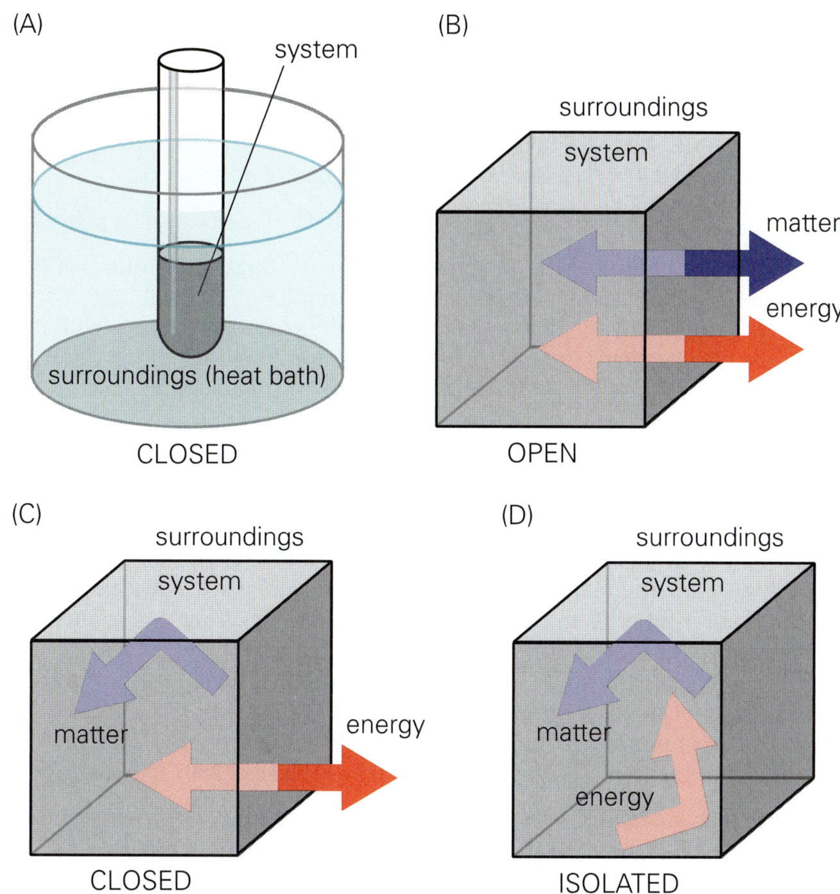


**Figure 6.2 Definition of the system.** We may be interested in processes occurring inside a test tube, as shown in (A), or in a cell. In all cases we need to define the system in order to keep track of the flow of energy in a meaningful way. Open, closed, and isolated systems are illustrated in B, C, and D, respectively.

a reaction releases energy, then this energy is usually dissipated as heat, and the amount of heat released can be measured quite accurately by instruments known as **calorimeters**. Calorimetry is discussed further in Chapters 10 (see Section 10.25) and 12 (Section 12.23).

In order to keep track of how much energy is released when a particular process occurs, we need to define the nature of the experimental system with respect to the transfer of energy and matter (that is, atoms and molecules). Consider a test tube or reaction chamber that we are observing (**Figure 6.2**). We divide the field of observation into two conceptual regions. A central region, referred to as the **system**, contains everything that we are interested in studying directly. The definition of the system is arbitrary, but once we have decided on a definition, we have to use it consistently when keeping track of the flow of matter and energy. Everything else that is not part of the system is referred to as the **surroundings**. Molecules are either in the system or in the surroundings and, in the most general case, both matter and energy can exchange between the system and the surroundings.

Depending on the nature of the boundary between the system and the surroundings, we have three kinds of systems. In an **open system**, molecules and energy can move in and out the system (Figure 6.2B). In this case, the "system" is purely a conceptualization that focuses attention on a particular region of space within which we are interested in observing molecular properties. Open systems are useful for studying diffusion and other molecular transport phenomena, which we shall return to in Chapter 17. In a **closed system**, matter (molecules) stays inside the system, whereas energy is exchanged with the surroundings (Figure 6.2C). We usually study chemical reactions in closed systems, such as a test tube that is immersed in a water bath (Figure 6.2A). Finally, in an **isolated system** (Figure 6.2D), matter and energy both stay inside the system.

## 6.2 Energy released by chemical reactions is converted to heat and work

When an **exothermic** reaction or process occurs inside the system, energy is released. For example, the combustion of glucose ($C_6H_{12}O_6$) releases 2801 $kJ \bullet mol^{-1}$ of energy at 298 K (see **Box 6.1** for a discussion of units):

$$C_6H_{12}O_6 + 6O_2 \longrightarrow 6CO_2 + 6H_2O \qquad \Delta U = -2801 \text{ kJ} \bullet \text{mol}^{-1}$$

Here $U$ denotes the energy of the system and $\Delta U = U$ (final) - $U$ (initial) is the change in energy when the reaction proceeds to completion (we will use "$U$" to denote energy rather than "$E$," so as not to confuse the symbol for energy with the symbol for the electric field). In an exothermic reaction, the final energy (that is, the energy of the products) is lower than the energy of the reactants, and $\Delta U$ is negative.

**Extensive and intensive properties**

An extensive property of a system depends on the size of the system. *Energy* is an extensive property, as is the *volume*. Intensive properties do not depend on the size of the system. The *temperature*, *density*, and *pressure* of the system are all intensive properties.

Energy is an **extensive property** of the system. Extensive properties depend on the size of the system, in contrast, to **intensive properties**, such as temperature, which are independent of the size of the system. If we double the size of the system, keeping everything else the same, the amount of energy in the system also doubles (**Figure 6.3**). The magnitude of the change in energy depends, therefore, on the number of molecules that participate in the reaction. To avoid confusion, we usually specify the amount of material under consideration when discussing energy changes, most commonly by considering a mole of reactants. In the case of the combustion of glucose, when we say that $\Delta U = -2801 \text{ kJ} \bullet \text{mol}^{-1}$ we mean that the conversion of one mole of glucose and six moles of oxygen into six moles of carbon dioxide and six moles of water results in the release of 2801 kJ of energy.

When glucose reacts with oxygen and releases energy, where does this energy go? In a system that exchanges energy with the surroundings, energy transfer between the system and the surroundings can occur either as **heat** or **work**. When energy is

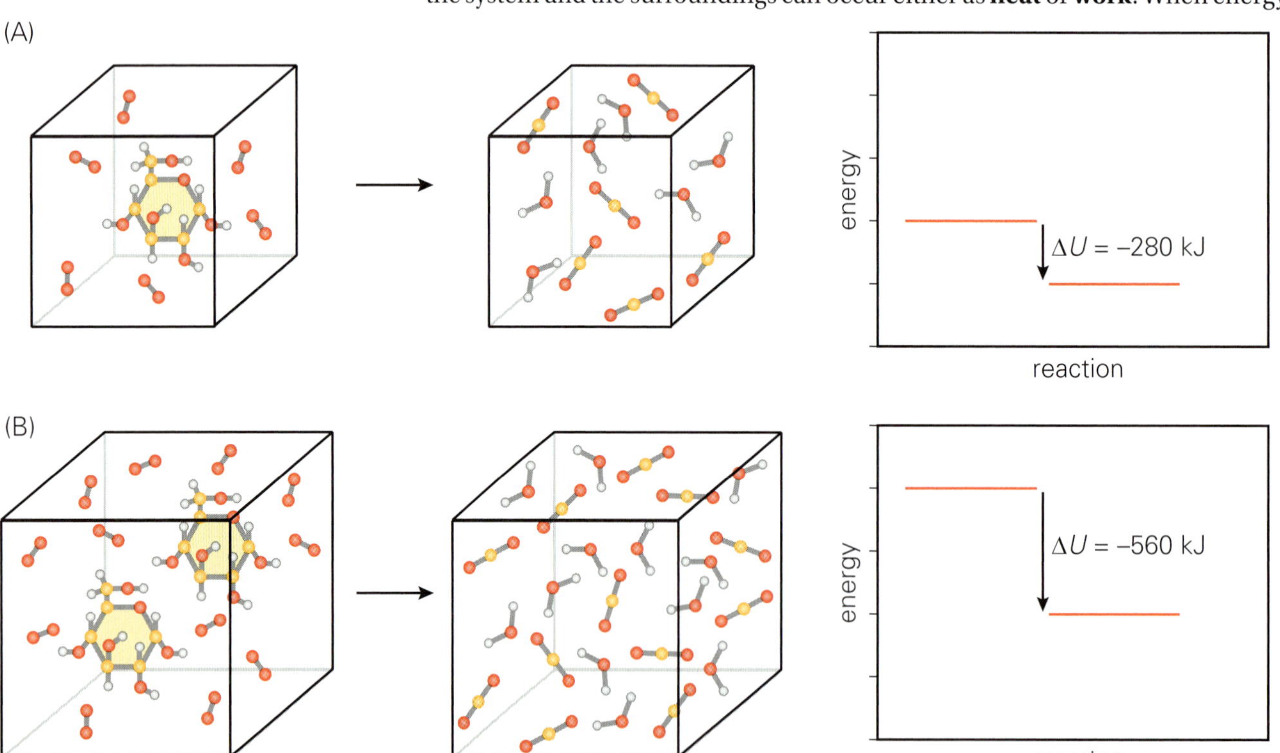


**Figure 6.3 Energy is an extensive property.** Shown here is a schematic representation of the reaction of glucose with oxygen to give $CO_2$ and water. The graph on the right shows the change in energy as the reaction goes from reactants to products. The total energy of a system is proportional to the size of the system, assuming that the density of the molecules stays the same. The system in B is twice as large as in A. Thus, the total energy change is twice as large in B as it is in A. We account for this by reporting energy changes per mole of reactants (that is, we use units of $J \bullet mol^{-1}$ or $kJ \bullet mol^{-1}$).

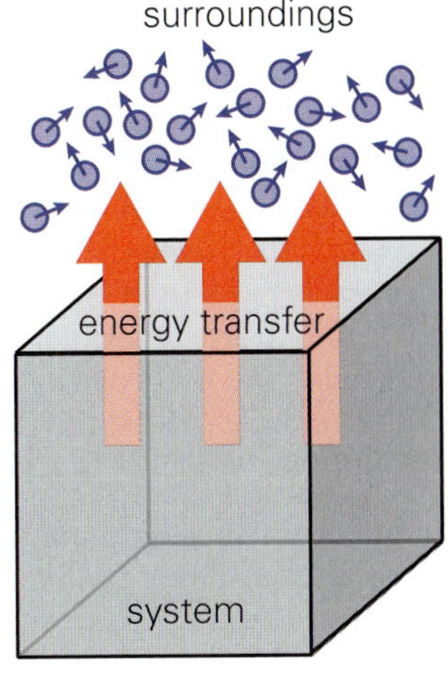


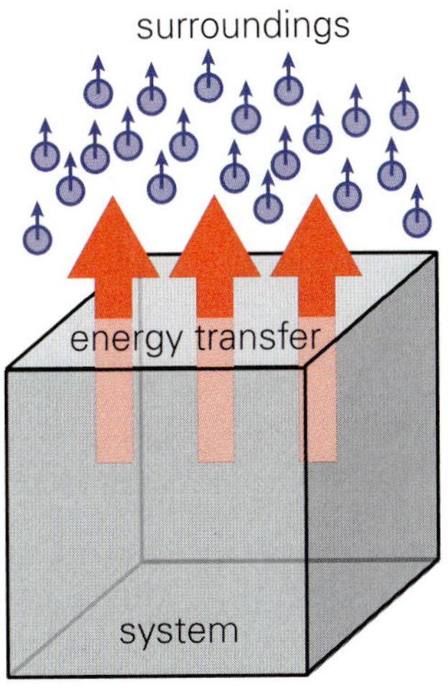


**Figure 6.4 Energy exchange between the system and the surroundings can take place as heat (A) or work (B).**

**Heat and work**

Energy transfer to and from the system occurs in the form of heat and work. Heat transfer results in increased random motion of molecules. When energy is transferred as work, it results in the ordered movement of some component of the system or the surroundings, such as the movement of a piston.

transferred to the surroundings as heat, this stimulates increased random motion of molecules in the surroundings (**Figure 6.4**). In contrast, when the system does **mechanical work** on the surroundings, it causes the ordered movement of some part of the surroundings. A simple example of mechanical work done by a chemical reaction is when one of the products of the reaction is a gas that causes the volume of the system to expand (**Figure 6.5**).

The essential difference between heat and work is that energy transferred as heat cannot easily be brought back into the system. Energy that is transferred out of the system as work, in contrast, can be "stored" and then may be readily available for coupling to another process. Energy transfer as work is not restricted to mechanical work, and can also be done in the form of chemical, electrical, or other kinds of work (see Chapter 9, Part C).

**The first law of thermodynamics**

Also referred to as the law of conservation of energy, the first law states that energy is neither created nor destroyed in a physical or chemical process.

## 6.3 The first law of thermodynamics states that energy is conserved

In all physical and chemical processes, the total energy of the system and the surroundings stays constant. This statement is known as the **first law of thermodynamics**, which is written mathematically (in the language of calculus—that is, differentials) as follows:

$$dU_{\text{total}} = dU_{\text{system}} + dU_{\text{surroundings}} = 0 \tag{6.1}$$

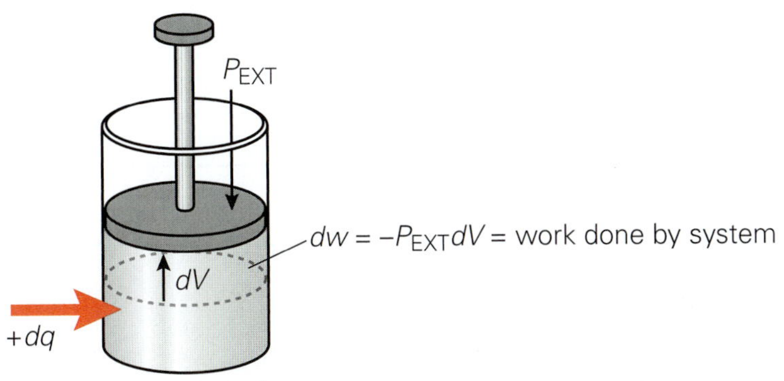


**Figure 6.5 Convention for the signs of work and heat.** A very small change in the volume of the system upon expansion is illustrated here (exaggerated for clarity). We follow the convention that the work done or heat transferred have positive (+) signs when the energy of the system increases. In the example shown here, the system expands against an external pressure, and loses energy as it expands. Hence, a negative sign is associated with the work done ($-P_{\text{EXT}}dV$). If the system were compressed, the sign of the work done would be positive. An amount of heat, $dq$, delivered *to* the system by the surroundings has a positive sign, because it increases the energy of the system.

## Box 6.1 Units

In this book we shall generally use the international system of units, known as SI units. In this metric system, the unit of length is the meter (m), the unit of mass is the kilogram (kg), the unit of energy is the joule (J), and the unit of temperature is the kelvin (K) (Table 6.1.1). In a few cases we will use centimeters (cm) and grams (g) instead to fit with other commonly used units.

The SI units allow us to make direct connections to measurements in physics but it is common, in chemistry and biology, to use non-SI units to describe energy, length, and mass. Biochemists usually report energy in units of calories or kilocalories (cal or kcal), and not joules (J) or kilojoules (kJ). One calorie is equal to 4.184 J. Since the energy change in a process depends on the amount of material involved in the process, confusion is avoided by usually reporting energy changes in units of J•mol$^{-1}$, kJ•mol$^{-1}$, cal•mol$^{-1}$, or kcal•mol$^{-1}$. The use of these units specifies that the amount of energy being reported applies to an Avogadro's number of molecules (that is, $6.022 \times 10^{23}$ molecules). Other nonstandard units commonly used in biochemistry are the Å ($10^{-10}$ m) to describe length and the dalton (Da) to describe molecular mass. A carbon–hydrogen bond length is ~1 Å, and so the Å unit is the natural one to use when describing molecular dimensions. Likewise, the mass of a hydrogen atom is approximately 1 Da, making it a natural unit for molecular mass.

When calculating the value of a physical quantity, it is crucial to pay attention to the units being used in order to avoid obtaining meaningless results. A process called **dimensional analysis** is often helpful in this regard. Dimensional analysis is the process of checking that the units on both sides of any equation linking physical variables are balanced. Consider, for example, Newton's law:

$$F = m \times a \tag{6.1.1}$$

**Table 6.1.1 Important units.**

| Quantity | Unit | Other commonly used units |
|---|---|---|
| Length | meter (m) | angstrom (Å) (1Å = $10^{-10}$ m) |
| Mass | kilogram (kg) | dalton (Da) (1 Da = 1.00 g•mol$^{-1}$)<br>pound (lb) (1 kg = 2.2 lb) |
| Volume | cubic meter (m$^3$) | liter (L) (1 L = $10^{-3}$ m$^3$) |
| Time | seconds (sec) | |
| Pressure | pascal (Pa) | atmosphere (atm)<br>(1 atm = 1.0133 × $10^5$ Pa) |
| Energy | joule (J) (1 J = 1 kg•m$^2$•sec$^{-2}$)<br>joule•mol$^{-1}$ | calorie (1 cal = 4.184 J) |
| Temperature | kelvin (K) | degrees Celsius (°C)<br>0°C = 273 K |
| Force | joule•m$^{-1}$<br>newton (N) (1 N = 1 kg•m•sec$^{-2}$) | |
| Electrical current | ampere (A) | |
| Electric charge | coulomb (C)<br>(1 C = 1 A•sec) | |
| Electrical potential difference | volt (V) (1 V = 1 kg•m$^2$•sec$^{-3}$•A$^{-1}$) | |
| Amount of matter | mole (mol)<br>(1 mol = 6.022 × $10^{23}$ molecules) | |

**Table 6.1.2 Important physical constants.**

| Name | Symbol | Value |
|---|---|---|
| Boltzmann's constant | $k_B$ | $1.3807 \times 10^{-23}$ J•K$^{-1}$ |
| Gas constant (Boltzmann's constant per mole) | $R$ | 8.3145 J•K$^{-1}$•mol$^{-1}$<br>1.987 cal•K$^{-1}$•mol$^{-1}$<br>0.0825 L•atm•K$^{-1}$•mol$^{-1}$ |
| Avogadro's number | $N_A$ | $6.022 \times 10^{23}$ mol$^{-1}$ |
| Faraday constant (magnitude of the charge on a mole of electrons) | $F$ | 96486.3 C•mol$^{-1}$ |

Equation 6.1.1 states that the force ($F$) acting on an object is equal to the mass times acceleration ($m \times a$). The acceleration is the rate of change of velocity, $v$:

$$a = \frac{dv}{dt} \tag{6.1.2}$$

The velocity, in turn, is the rate of change of position, $x$:

$$v = \frac{dx}{dt} \tag{6.1.3}$$

From Equation 6.1.3, we can infer that the unit of velocity is m•sec$^{-1}$ [the right-hand side of the equation has (increment in position—that is, distance)/(increment in time)]. Combining this information with Equation 6.1.2 tells us that the unit of acceleration is m•sec$^{-2}$. Equation 6.1.1, then, tells us that the units of force must be kg•m•sec$^{-2}$.

The standard unit of force is the newton (Table 6.1.1), and it follows that if you calculate mass in kilograms (kg), length in meters (m), and time in seconds (sec), you obtain the force in newtons. If you choose, instead, to measure mass in grams, your calculated value of the force will no longer be in newtons.

If an equation has a dimensionless (that is, unitless) quantity on one side, then all the units on the other side must cancel out in order for the equation to make sense. For example, consider Equation 6.32 in this chapter, which relates the ratio of the numbers of molecules in two energy levels of a molecular system to the energy difference between the levels (the Boltzmann distribution):

$$\frac{N_2}{N_1} = e^{-\Delta U / k_B T} \tag{6.1.4}$$

The left-hand side of the equation is a ratio of two numbers, and so it is a pure unitless number. This means that the quantity $\left(e^{-\Delta U / k_B T}\right)$ is also a pure unitless number. The only way this can be true is if the quantity $\left(\frac{-\Delta U}{k_B T}\right)$ is also a pure number (it is unitless). The numerator of this term, $-\Delta U$, is the difference in energy between two energy levels and therefore has units of energy. In the denominator, $T$ is the temperature and $k_B$ is the Boltzmann constant. In order for $\left(\frac{-\Delta U}{k_B T}\right)$ to be unitless, $k_B T$ must have units of energy (for example, J). Hence, if we report $\Delta U$ in units of joules (J), and the temperature, $T$, in units of kelvin, then it follows that the units of $k_B$ must be J•K$^{-1}$. **Table 6.1.2** lists the value of $k_B$, which is $1.3807 \times 10^{-23}$ J•K$^{-1}$. If we choose to report energy in units of J•mol$^{-1}$, as is commonly done in biochemistry, instead of J, then we need to multiply the value of $k_B$ by $6.022 \times 10^{23}$ (Avogadro's number), which gives us $k_B = 8.3145$ J•K$^{-1}$•mol$^{-1}$. The Boltzmann constant in units of J•K$^{-1}$•mol$^{-1}$ is also known as $R$, the gas constant (see Table 6.1.2).

$dU$ is a very small (actually, an infinitesimally small) change in the total energy of the system and the surroundings, and the value of $dU_{total}$ is set to zero because the total energy is constant. The first equality in Equation 6.1 reflects the fact that the total energy, $U_{total}$, is the sum of the energy of the system, $U_{system}$, and the energy of the surroundings, $U_{surroundings}$. Equation 6.1 can be rewritten as follows:

$$dU_{system} = -\, dU_{surroundings} \qquad (6.2)$$

Equation 6.2 is simply a statement of the fact that if the energy of the system increases, then the surroundings lose an equivalent amount of energy.

When the energy of the system changes, this results in the energy of the surroundings changing in two ways: the system can release heat to the surroundings, and the system can do work on the surroundings. The amount of heat transferred during an infinitesimally small step in a process is written as $dq$ and the amount of work done during such a step is denoted $dw$ (**Figure 6.6**). If no work is done during the process, then the change in the energy of the system is equal to the heat transferred:

$$dU = dq \qquad \text{(no work done)} \qquad (6.3)$$

In Equation 6.3, the change in energy refers to the system, but we have dropped the subscript, for convenience. In the rest of this chapter we shall use unsubscripted variables to refer to the system.

Consider, instead, an isolated system (one that exchanges no heat with the surroundings). In this case, if the process involves work being done, then the change in energy of the system is given by:

$$dU = dw \qquad \text{(no heat exchange)} \qquad (6.4)$$

For the general case, where heat transfer occurs in addition to work being done, we combine Equations 6.3 and 6.4 to get a mathematical restatement of the first law:

$$dU = dq + dw \qquad (6.5)$$

In keeping track of energy changes, we need to pay attention to the sign associated with the quantities of the heat and work in the process. Since our focus is on the system, we associate a positive sign with changes that *increase* the energy of the system (see Figure 6.6). Thus, if a quantity of heat, $dq$, is transferred *to* the system from the surroundings, then the energy of the system goes up and the quantity of heat transferred has a positive sign. If the process involves work being done *on* the system by the surroundings (if, for example, the system is compressed), the quantity of work done has a positive sign. If, instead, the system undergoes a change that results in work being done *by* the system *on* the surroundings (as would be case if the system expands), then the energy of the system is reduced (because the system uses energy to do work) and the quantity of work done has a negative sign.

The signs associated with energy, work, and heat can sometimes be confusing, but we can always check that the signs we are using make sense in terms of conservation of energy (Equation 6.1). For example, if 10 units of energy are delivered to a system in the form of heat, and the system does work on the surroundings equivalent to 5 units, then the energy of the system must increase by 5 units.

## 6.4 For a process occurring under constant pressure conditions, the heat transferred is equal to the change in the enthalpy of the system

Biological reactions typically occur under conditions of constant pressure (see **Box 6.2** for a discussion of pressure). When measuring heat transfer under these conditions, we need to be alert as to whether there are significant changes in the volume of the system during the process. Changes in volume would result in

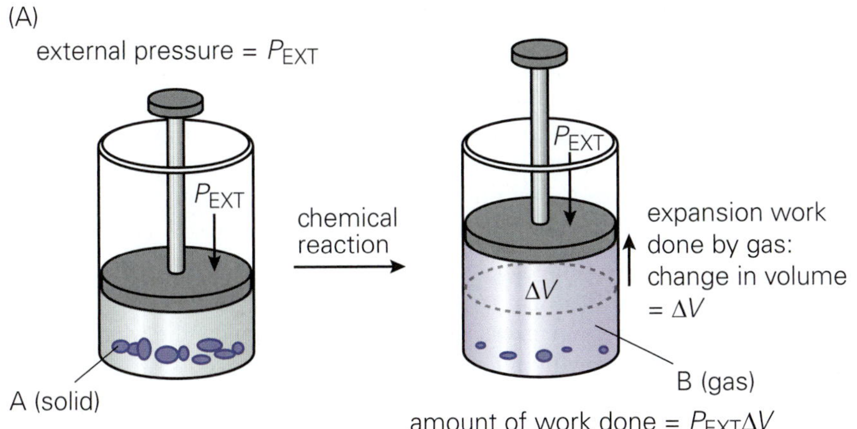


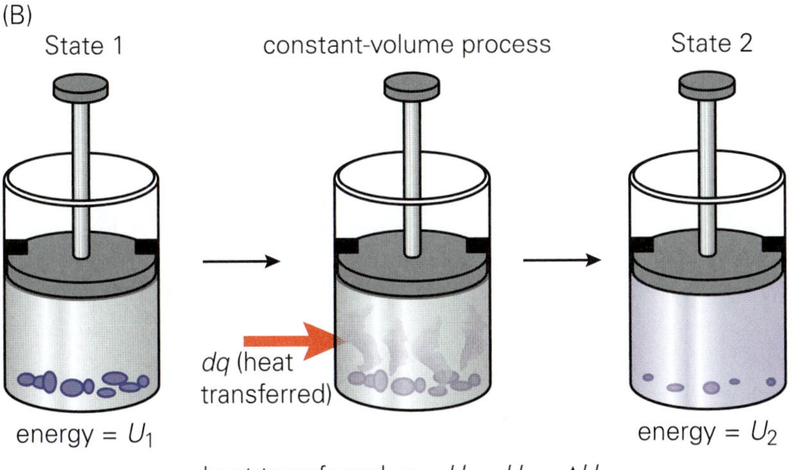


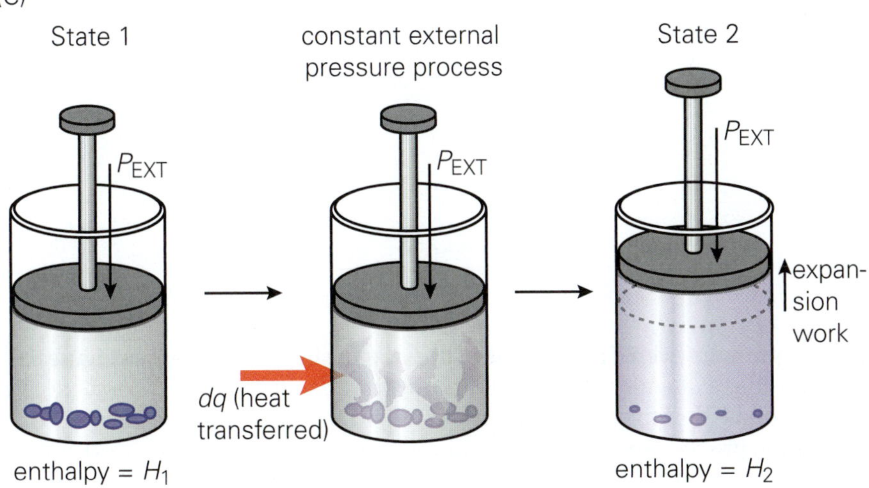


**Figure 6.6 Energy and Enthalpy.** (A) Under conditions of constant external pressure, reactions or processes that change the volume of the system ($\Delta V$) cause the system to do mechanical work. In this case, converting solid A to gaseous B increases the volume. (B) For a process occurring under conditions of constant volume, the heat transferred to the system is given by the change in energy, $U$, of the system. (C) For a process occurring under constant pressure conditions, the heat transferred is given by the change in enthalpy, $H$, of the system.

expansion work being done by the system against the atmosphere, which would change the amount of heat transferred from the surroundings.

To see how expansion work affects the amount of heat transferred, consider a test tube in which a chemical reaction is taking place. The test tube is capped by a frictionless piston, which can move up and down, thereby maintaining

## Box 6.2 Pressure

The pressure exerted on a system is the force applied per unit area:

$$\text{pressure} = \frac{\text{force}}{\text{area}} \tag{6.2.1}$$

Force has units of kg•m•sec$^{-2}$ (see Box 6.1), so it follows that the units of pressure are kg•m$^{-1}$•sec$^{-2}$ = N•m$^{-2}$ (where N = 1 newton = 1 kg•m•sec$^{-2}$). In the standard system of units, 1 N•m$^{-2}$ is defined as the **pascal** (Pa). The pressure exerted by the atmosphere at sea level is ~10$^5$ Pa, which is defined as 1 atmosphere of pressure, or 1 atm.

To get a sense of these units, consider a box with lid of cross-sectional area 0.01 m$^2$ (the dimensions of the edges of the box are 10 cm by 10 cm; see **Figure 6.2.1**).

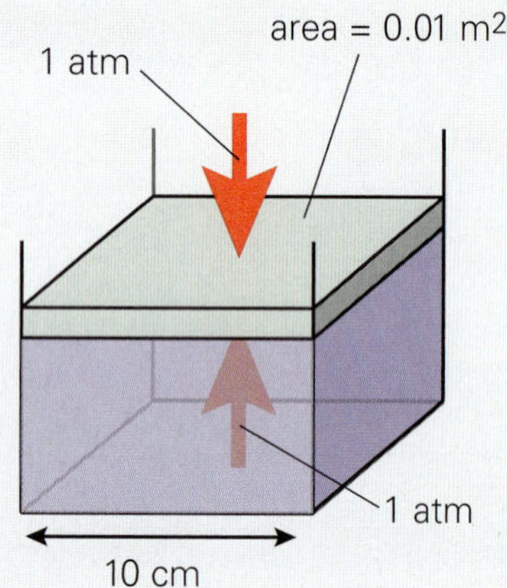


**Figure 6.2.1 A box containing air at 1 atmosphere pressure, balanced by the pressure of the atmosphere.**

The box contains air at 1 atm pressure, which is balanced by the external pressure of 1 atm. If we moved the box into a vacuum chamber, how much of a weight would we have to put on the lid of the box in order to prevent it from flying open?

Consider a weight of $W$ kg that is just enough to keep the lid from flying open in a vacuum. This means that the pressure exerted on the lid of the box by the weight balances the internal pressure of the gas (1 atm). What is the magnitude of the weight, $W$? We can calculate this as follows:

$$\text{force due to weight} = W \times g \tag{6.2.2}$$

where $g$ is the acceleration due to gravity (9.8 m•sec$^{-2}$). Using Equation 6.2.1, we get:

$$\begin{aligned}\text{pressure due to weight} &= \frac{\text{force}}{\text{area}} \\ &= \frac{9.8\ \text{m}\bullet\text{sec}^{-2}}{0.01\ \text{m}^2} \times W\ \text{kg} \\ &= 980 \times W\ \text{kg}\bullet\text{m}^{-1}\bullet\text{sec}^{-2} \\ &= 980 \times W\ \text{Pa}\end{aligned} \tag{6.2.3}$$

The pressure exerted by the weight has to balance the internal pressure of 1 atm (= 10$^5$ Pa), and so:

$$\begin{aligned}10^5\ \text{Pa} &= 980 \times W\ \text{Pa} \\ \Rightarrow\ W &= \frac{10^5}{980} \approx 100\end{aligned} \tag{6.2.4}$$

Equation 6.2.4 tells us that we would have to put a weight of ~100 kg (~220 lb) on the lid in order to prevent it from opening in vacuum. Air exerts considerable force at 1 atm pressure.

constant pressure even if the volume of the system should change (see Figure 6.6). First, consider a chemical reaction that converts a mole of a solid (A) to a mole of another solid (B). If the molar volumes of A and B are essentially the same, then the reaction proceeds with no change in volume. Thus, no expansion work is done by the system on the surroundings. Under these conditions, the heat transferred to the system, $dq$, is given by the first law (Equation 6.5), with the work term set to zero:

$$dq = dU \tag{6.6}$$

Equation 6.6 is a useful result, because it tells us that by measuring the heat taken up (or released) during the reaction, we can determine the difference in energy between A and B.

Now consider, instead, what happens if the volume of the system changes as A is converted to B (the volume change would be quite large if A were a solid, for example, and B were a gas). In this case, the piston would rise as the reaction proceeds in order to maintain constant pressure, and the system does work on the surroundings. The magnitude of the expansion work done in an infinitesimally small step of the process is $P_{EXT}dV$, where $dV$ is the change in volume of the system and $P_{EXT}$ is the (constant) external pressure. Because work is being done *by* the system, the quantity of work done has a negative sign ($dw = -P_{EXT}dV$). By rearranging Equation 6.5 we can see that the heat transferred to the system is given by:

$$dq = dU - dw = dU - (-P_{EXT}dV) = dU + P_{EXT}dV \qquad (6.7)$$

The term $dq$ on the left side of Equation 6.7 is the amount of heat that is transferred to the system. This amount of heat can do two things: (1) it can raise the energy of the system (this is denoted by the term $dU$), and (2) it can cause the system to do work. In this particular case, we are considering only the mechanical work of gas expansion, and the $+P_{EXT}dV$ term represents the amount of heat that goes towards the work done by the system.

Equation 6.7 tells us that, in situations where the volume of the system changes during the reaction, the heat transferred to the system is no longer equal to the change in energy. Nevertheless, the heat absorbed or released is an important experimental property (because it is easy to measure) and we consider it a critical property of the system. We define a variable of the system known as the **enthalpy**, $H$, as follows:

$$H = U + PV \qquad (6.8)$$

At any instant the energy depends only on the present conditions, such as pressure and temperature, and the interaction between all the atoms in the system. The energy of the system is therefore completely determined by the present state of the system and not by its history. Energy is therefore referred to as a **state variable** or **state function**. Likewise, the pressure and volume are also state variables (they depend only on present conditions). As you can see from Equation 6.8, the enthalpy, $H$, is completely defined in terms of state variables, and so it too is a state variable: it is completely defined by the present state of the system and no knowledge of the past history is required to specify the enthalpy.

To understand the meaning of the term "state variable," it might be useful to consider a simple analogy. The amount of money in your pocket is a state variable. We can determine the amount of money in your pocket at any time by just looking in your pocket and counting. We refer to the amount of money in your pocket as a "variable" because it can change from moment to moment as you spend or earn money. But, at any moment, we can easily determine the value of the variable. In contrast, the amount of money you have earned over your lifetime is not a state variable. It is impossible to assess the value of this variable from your current assets without a detailed knowledge of your income and spending history.

An infinitesimally small change in the enthalpy of the system is related to infinitesimally small changes in the other variables through the following differential equation:

$$dH = dU + PdV + VdP \qquad (6.9)$$

If the process occurs under constant pressure (for example system pressure $P = P_{EXT} = 1$ atmosphere throughout the process), then the $VdP$ term in Equation 6.9 is zero. Therefore, under this constant pressure condition:

$$dH = dU + P_{EXT}dV \qquad (6.10)$$

Changes in enthalpy are the focus of much attention when considering biochemical reactions because they are equal in magnitude to the amount of heat transferred under constant-pressure conditions. Comparing Equation 6.10 with Equation 6.7, we see that the change in enthalpy of the system is simply the heat taken up by the system under conditions of constant pressure (see Figure 6.6):

$$dH = dq \quad \text{(at constant pressure)} \qquad (6.11)$$

**Enthalpy**

The change in enthalpy of a system during a process carried out under constant pressure conditions is equal to the heat taken up by the system during the process.

Enthalpy is defined as follows:

$$H = U + PV$$

Since biochemical processes usually occur under constant pressure conditions (one atmosphere ambient pressure), the change in enthalpy is readily determined if the amount of heat transferred is measured ($\Delta H = q$). If the change in volume is negligible, then the change in enthalpy is essentially equal to the change in energy.

**State variables or functions**

Properties of the system that are independent of the history of the system are known as state variables or functions. The values of these variables can be calculated from knowledge of the present state of the system, without knowing its history. Energy, volume, temperature, and pressure are state variables. The heat transferred to a system is not a state variable, because its magnitude depends on what has been done to the system in the past.

Enthalpy is not a term that is familiar to us from daily use; we do not wake up each morning and ask, "How is my enthalpy today?" Nevertheless, we can recognize from Equation 6.10 that if no gases are involved in the process of interest, then the volume changes in biochemical reactions are negligible and the corresponding change in enthalpy is essentially the change in the energy of the system. Thus, the terms "energy" and "enthalpy" are often used in an interchangeable sense when discussing biological systems. We do need, however, to remain alert to the fact that when the reaction under consideration involves a change in gas stoichiometry (for example, in the process of respiration), then the magnitudes of the changes in energy and enthalpy are not necessarily the same.

## 6.5 Changes in energy do not always indicate the direction of spontaneous change

One of the things we are most interested in when studying biochemical reactions is the direction of spontaneous change. If we mix two molecules, A and B, will they react and convert to new molecular states, or will they remain stably as they are? Most chemical reactions have kinetic barriers that slow down the reaction rate and, in the present discussion, we are not considering the effects of these kinetic barriers, which will be discussed in Chapter 15. Rather, we are asking what would happen to the system if we waited long enough for equilibrium to establish itself.

Let us suppose that A and B could interact and convert to two new molecules, C and D. If the energy of C and D is lower than that of A and B, then the reaction is exothermic:

$$A + B \rightarrow C + D \quad \Delta U = \text{negative} \quad \text{(energy released)}$$

Our experience with everyday objects tells us that if energy is stored in a system (consider, for example, an elastic bungee cord that is held stretched out), then the system will relax spontaneously to a state of lower energy when the constraints are released (the bungee cord will immediately collapse into a more compact state when we let go of it) (**Figure 6.7**). Thus, regarding the outcome of mixing A molecules with B molecules, our intuition may guide us as follows: A and B are high-energy molecules and, when allowed to react with each other, they would convert to C and D molecules because C and D are of lower energy.

It turns out, unfortunately, that our everyday experience with macroscopic objects is not always a reliable guide to the behavior of molecular systems. The essential

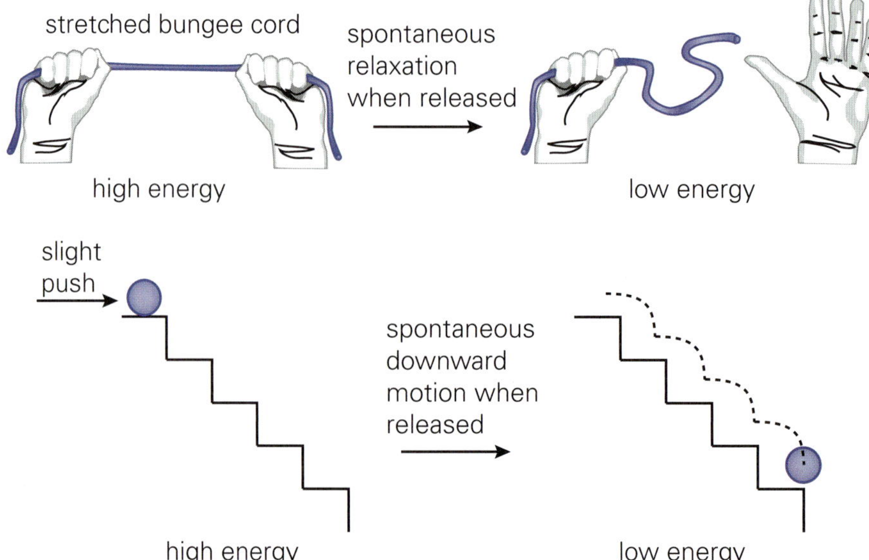


**Figure 6.7 The direction of spontaneous change for macroscopic objects is given by the direction of decreasing energy.** A spring obeying Hooke's law, illustrated here by an elastic bungee cord that is stretched, will relax spontaneously to its original length when released. A ball that is at the top of a staircase will roll down the steps spontaneously if pushed slightly.

difference is that when we work with macroscopic objects, such as a spring that obeys Hooke's law or a ball rolling down a staircase, we are usually dealing with one, two, or a small number of objects, and we ignore the microscopic molecular structure of the objects (see Figure 6.7). In such cases, the energy of the macroscopic object is a good indicator of the direction of spontaneous change. A stretched spring will relax spontaneously when released, just as a ball on a higher step of a staircase will roll down the steps when pushed (see Figure 6.7).

The fundamental difference between these kinds of macroscopic systems and the behavior of a system of molecules studied at the microscopic level is that in the latter case we are typically observing a very large number of molecules (for example, ~$10^{23}$ molecules in a mole of material). Such large numbers of molecules exhibit behavior that is controlled not only by the energy of the system, but also by another collective property of the system known as the **entropy**. The entropy of a system is a measure of its disorder, as we shall see in Chapter 7. A familiar example of the effect of entropy is shown in **Figure 6.8**, which illustrates what happens when a drop of colored dye is introduced into a beaker of water. The dye will diffuse until it is spread evenly throughout the beaker. This process is driven by increasing entropy, and will occur spontaneously, even if there is no energetic driving force (see Chapter 7).

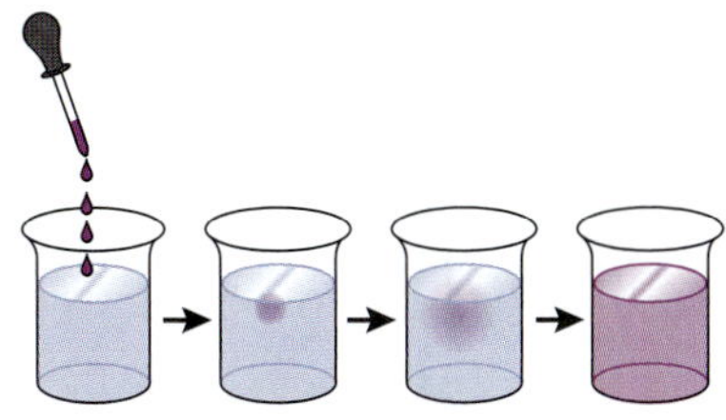

**Figure 6.8 Spontaneous diffusion of a dye in water.** When a drop of a purple dye is introduced into a beaker of water, the dye spreads throughout the beaker. This will happen spontaneously, even if there is no energetic driving force—that is, even if the molecules of the dye do not have a lower energy when they are dispersed. This is an example of a process that is driven by entropy.

## 6.6 The isothermal expansion of an ideal gas occurs spontaneously even though the energy of the gas does not change

Another simple experiment that demonstrates that energy changes are not always a reliable indicator of the direction of spontaneous change is the isothermal expansion of an **ideal monatomic gas**. The gas is composed of atoms, such as argon, in which interatomic interactions are so weak that they can be neglected. If we push the piston down (that is, temporarily adding a force to the external pressure), we compress the gas and, as a consequence, the work done is converted into energy stored in the gas atoms in the piston (**Figure 6.9**). The internal pressure of the gas changes from an initial value, which is the same as the external pressure, $P_{EXT}$, to a value we shall denote as $P_1$ ($P_1 > P_{EXT}$).

Now, consider what happens when we release the piston in two different ways. In the first case, we insulate the piston from the surrounding so that no heat exchange can occur. This is referred to as an **adiabatic** process (that is, without heat transfer). In the second case, we immerse the piston in a large water bath that is maintained at constant temperature. In this case, the system can exchange energy with the heat bath, the presence of which ensures that the expansion occurs at constant temperature (referred to as an **isothermal** process).

The piston expands spontaneously when released in both the adiabatic and the isothermal processes because the internal pressure, $P_1$, is greater than the external pressure, $P_{EXT}$. The outcome, however, is different in each case. In both cases the system does work on the surroundings by moving the piston against an external pressure, so that the final internal pressure at equilibrium is $P_{EXT}$. In the case of the isolated system, the energy to do this work must come from the internal energy of the gas (there is no other source of energy). From the first law (Equation 6.5), we know that the change in internal energy of the gas, $dU$, is given by:

$$dU = dq + dw \tag{6.12}$$

where $dq$ is the heat added to the system, which is zero for an adiabatic process. Hence, the change in energy is simply equal to the quantity of work done:

$$dU = dw \tag{6.13}$$

For an ideal monatomic gas, the energy of the gas is equal to the **kinetic energy**, $(3/2)nRT$, where $n$ is the number of moles of the gas, $R$ is the gas constant, and

**Ideal monatomic gas**

An ideal monatomic gas consists of atoms that do not interact with each other. An ideal gas obeys the ideal gas laws. For example, the pressure, $P$, volume, $V$, and the temperature, $T$, of an ideal gas are related by the following equation:

$$PV = nRT$$

where $n$ is the number of moles of the gas and $R$ is the gas constant. Noble gases, such as argon, are very close to ideal in their behavior.

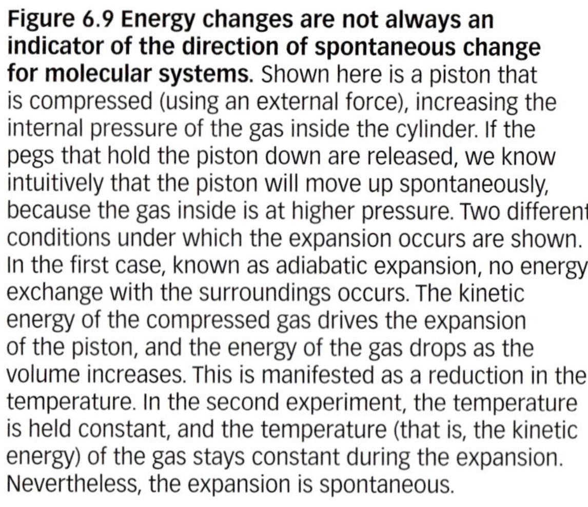

**Figure 6.9 Energy changes are not always an indicator of the direction of spontaneous change for molecular systems.** Shown here is a piston that is compressed (using an external force), increasing the internal pressure of the gas inside the cylinder. If the pegs that hold the piston down are released, we know intuitively that the piston will move up spontaneously, because the gas inside is at higher pressure. Two different conditions under which the expansion occurs are shown. In the first case, known as adiabatic expansion, no energy exchange with the surroundings occurs. The kinetic energy of the compressed gas drives the expansion of the piston, and the energy of the gas drops as the volume increases. This is manifested as a reduction in the temperature. In the second experiment, the temperature is held constant, and the temperature (that is, the kinetic energy) of the gas stays constant during the expansion. Nevertheless, the expansion is spontaneous.

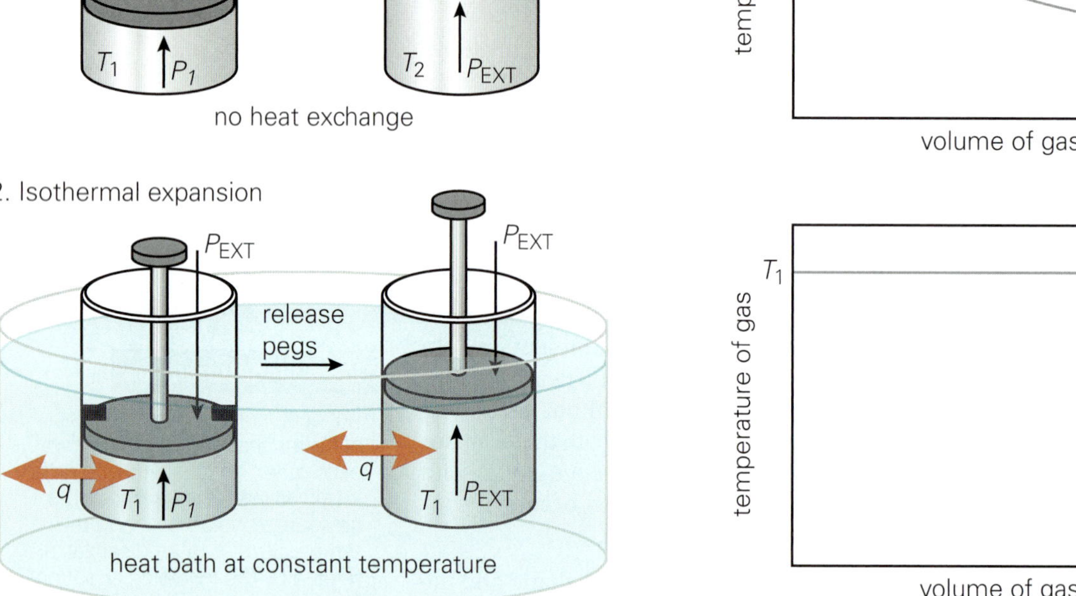


$T$ is the absolute temperature (see **Box 6.3** for a justification of this statement). The system does work against the surroundings, and so the quantity of work done has a negative sign, and so $dw$ is negative, in which case $dU$ is negative. The energy of the gas decreases as the piston expands, and this means that the temperature of the gas decreases as the gas expands (see Figure 6.9). Thus, the expansion of the piston is "fueled" by the kinetic energy of the gas molecules, and the kinetic energy of the gas molecules decreases as the piston expands.

In the case of isothermal expansion, the temperature of the system is kept constant because it is coupled to a heat bath. For an ideal monatomic gas, the energy depends only on the temperature (see Box 6.3), and so the energy of the gas stays constant during the expansion. We know intuitively that when we release the piston it will move up spontaneously, because the gas inside the piston is at high pressure. But the energy of the gas does not change during the expansion, and so what provides the energy to drive the piston? Again, the answer comes from looking at the first law and noting that $dU$ is zero for an ideal gas at constant temperature:

$$dU = dq + dw = 0 \quad \text{(constant temperature)} \tag{6.14}$$

Thus,

$$dq = -dw \tag{6.15}$$

According to Equation 6.15, the magnitude of work done by the system is equal to the magnitude of the heat transferred to the piston by the water bath. Thus, although the energy of the gas stays constant during the expansion process, heat comes into the system from the surroundings (the heat bath), and this energy goes back into the surroundings in the form of work (the piston is part of the surroundings).

The energy of the system does not change during the isothermal expansion of an ideal gas, and yet the process is spontaneous. It turns out, instead, that the entropy of the system is the property that tells us that the piston will move spontaneously when released. We defer a discussion of entropy until Chapter 7, in which we shall reexamine the process of isothermal gas expansion.

# B. HEAT CAPACITIES AND THE BOLTZMANN DISTRIBUTION

Heat transfer and changes in temperature do not seem like very interesting things to measure, and for an ideal gas that is certainly true. The situation is quite different, however, when we heat solutions containing biological molecules and measure how quickly the temperature changes with respect to heat input. Measurements like this can give us some interesting insights into the energetics of these molecules. We begin this part of the chapter by considering heat transfer to a piston containing an ideal gas, and then point out that solutions of protein or nucleic acid molecules behave very differently. The effect of temperature on the energetics of molecules can be explained by the **Boltzmann distribution**, which gives the probability of finding molecules with different energies. The ideas introduced in this part of the chapter are explored further in Chapters 18 (for protein folding) and 19 (for the melting of double-stranded DNA).

## 6.7 The heat capacity of an ideal monatomic gas is constant

Consider an ideal gas in a piston, with initial pressure, temperature, and volume given by $P_1$, $V_1$, and $T_1$, respectively. Let us assume that the piston is isolated, except for a heat source which can deliver measured amounts of heat into the system (**Figure 6.10**). As we start adding heat to the system, the temperature will rise. The **heat capacity** of the system is defined as the amount of heat required to increase the temperature of the system by 1 kelvin. If the heat transfer is carried out at constant volume, then the heat capacity is referred to as $C_V$. The heat transferred to the system under constant-volume conditions (when no work is done) is equal to the increase in internal energy ($\Delta U$) of the system, and so we can write:

$$C_V = \Delta U \tag{6.16}$$

for an increase in temperature of 1 K.

We can convert this to a differential form by recognizing that $C_V$ is the rate of change of energy with respect to temperature at constant volume:

$$C_V = \left(\frac{\partial U}{\partial T}\right)_V \tag{6.17}$$

**Heat capacity**

The heat capacity of a system is the amount of heat required to increase the temperature of the system by 1 kelvin. The magnitude of the heat capacity depends on the conditions under which the system is heated. If the heating is carried out under constant-volume conditions, the heat capacity is denoted $C_V$. If, instead, the heating occurs under conditions of constant pressure, the heat capacity is denoted $C_P$.

## Box 6.3 The relationship between temperature and kinetic energy for an ideal monatomic gas

We now justify the statement in the main text that the total kinetic energy of an ideal gas is given by (3/2) $nRT$, where $n$ is the number of moles of the gas, $R$ is the gas constant, and $T$ is the temperature. To do this, we use ideas from the **kinetic theory of gases** that were first introduced in the nineteenth century by scientists such as James Clerk Maxwell and Ludwig Boltzmann to understand the molecular basis for the laws of thermodynamics. While we will not pursue this level of detailed analysis of the kinetic theory of molecules in this book, we do return, in Chapter 17, to the idea that the velocities of individual molecules can be related to macroscopic phenomena.

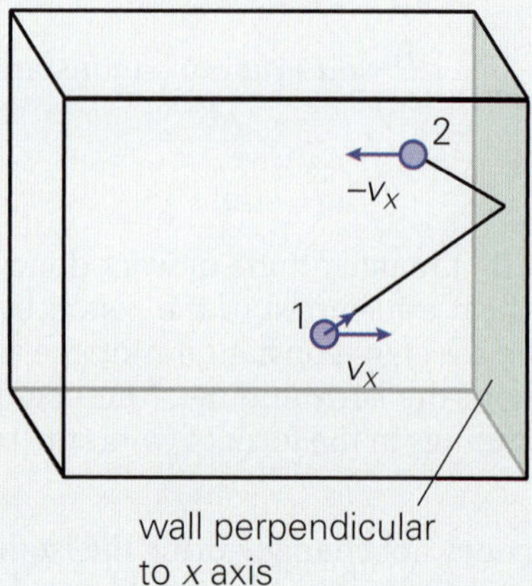


**Figure 6.3.2** When an atom collides with a wall elastically, the component of the velocity that is orthogonal to the wall is exactly reversed by the collision.

We start with the ideal gas law:

$$PV = nRT \tag{6.3.1}$$

where $P$ is the pressure, $V$ is the volume, $n$ is the number of moles of gas atoms, $T$ is the absolute temperature, and $R$ is the gas constant.

We begin by deriving a relationship between the kinetic energy of the gas and the pressure. The pressure is defined by:

$$\text{pressure} = \frac{\text{force}}{\text{area}} \tag{6.3.2}$$

The force on the walls of the container arises due to collisions of the atoms with the wall. We assume that these collisions are perfectly elastic—that is, the atoms lose no energy and only change their direction upon collision.

Consider an atom moving with a velocity vector, $\vec{v}$. The velocity vector has three components, $\vec{v}_x$, $\vec{v}_y$, and $\vec{v}_z$, which can be considered independently (**Figure 6.3.1**). The magnitudes of the three velocity components satisfy the following relationship:

$$v^2 = v_x^2 + v_y^2 + v_z^2 \tag{6.3.3}$$

The momentum of the molecule is given by:

$$\text{momentum} = \text{mass} \times \text{velocity} = m\vec{v} \tag{6.3.4}$$

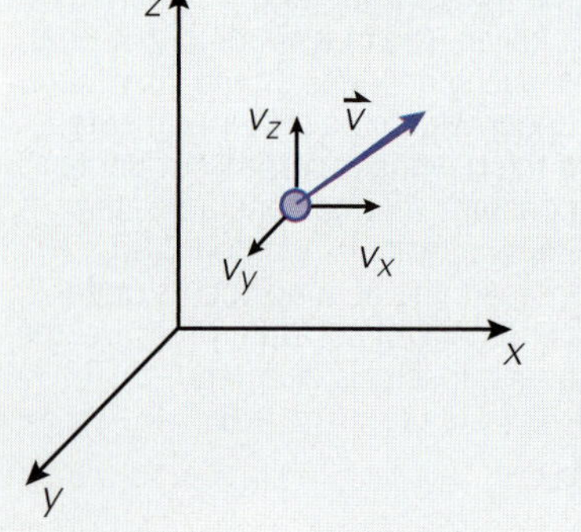


**Figure 6.3.1** The velocity vector, $\vec{v}$, of an atom has three orthogonal components, denoted $v_x$, $v_y$, and $v_z$.

where $m$ is the mass of the atom. We can relate the force to the rate of change of momentum by using Newton's laws of motion:

$$\text{force} = m\vec{a} = m\frac{d\vec{v}}{dt} = \frac{d(m\vec{v})}{dt} \tag{6.3.5}$$

where $\vec{a}$ is the acceleration.

Consider a wall perpendicular to the $x$ axis. When an atom with $x$-component velocity $v_x$ hits the wall and rebounds elastically, the $x$-component of the momentum changes from $+mv_x$ to $-mv_x$. Because the collision between the atom and the wall is assumed to be perfectly elastic, the velocity is exactly reversed (**Figure 6.3.2**). This allows us to calculate the change in momentum:

$$\text{change in momentum} = mv_x - (-mv_x) = 2m|v_x| \tag{6.3.6}$$

Consider a small time interval, $\Delta t$. The number of atoms with velocity $v_x$ that hit the wall during this time is given by Equation 6.3.7:

number of atoms to hit the wall in time interval $\Delta t$

$$= \left\{|v_x|\frac{\Delta t}{2}\right\}\frac{AN}{V} \tag{6.3.7}$$

where $A$ is the area of wall, $N$ is number of atoms in the box, and $V$ is the volume of the box.

How do we arrive at Equation 6.3.7? We start by considering the volume within the box that contains all the atoms with velocity $v_x$ that will hit the wall within the time interval $\Delta t$ (**Figure 6.3.3**). The number of atoms in this volume is given by:

$$\text{number of atoms in shaded volume} = \frac{N}{V}(A)|v_x|\Delta t \tag{6.3.8}$$

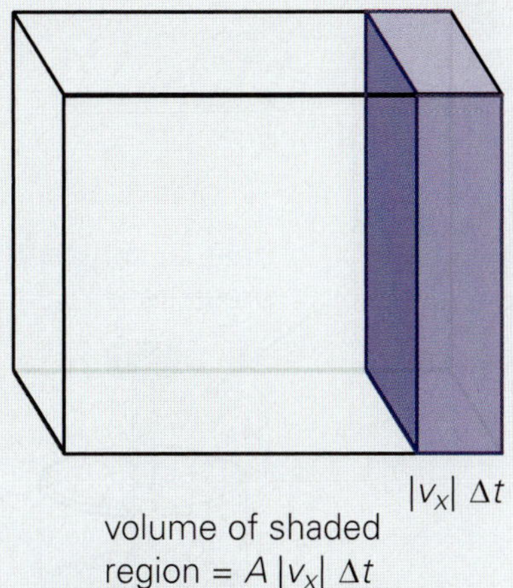


**Figure 6.3.3 The number of atoms with velocity $v_x$ that collide with the wall in a time interval $\Delta t$.** An atom with velocity $v_x$ in the $x$ direction travels a distance $|v_x|\Delta t$ in time $\Delta t$. Hence, all atoms that are in the shaded volume and moving to the right will collide with the wall in the time interval $\Delta t$.

where $N$ is the number of particles in the box and $V$ is the total volume of the box (that is, $N/V$ is the number density of particles). On average, half the molecules will have $v_x < 0$ (that is, they are moving away from the wall), and so the number of atoms moving towards the wall is given by the right-hand side of Equation 6.3.8 divided by 2. This gives us Equation 6.3.7.

Thus, the total change in momentum over a time period, $\Delta t$, is given by combining Equation 6.3.6, which gives us the change in momentum for one atom, with Equation 6.3.7, which gives us the number of atoms that collide with the wall in the given period of time:

total change in momentum over time period $\Delta t$

$$= 2m\,|\,v_x\,|\,\frac{1}{2}\frac{N}{V}A|v_x|\Delta t = mv_x^2\frac{NA}{V}\Delta t \qquad (6.3.9)$$

Equation 6.3.9 allows us to calculate the rate of change of momentum, which is equal to the force on the wall:

force = rate of change momentum =

$$mv_x^2\frac{NA}{V}\frac{\Delta t}{\Delta t} = mv_x^2\frac{NA}{V} \qquad (6.3.10)$$

Since the pressure is the force per unit area (Equation 6.3.2), we can use Equation 6.3.10 to calculate the pressure on the wall:

$$\text{pressure, } P = mv_x^2\frac{N}{V} \qquad (6.3.11)$$

The argument so far has focused only on atoms with a particular velocity along the $x$ coordinate, denoted $v_x$. But not all atoms will have the same velocity and, for a box containing a large number of atoms, there will be a distribution of atoms with different velocities. The net pressure will arise from the average of all the collisions involving atoms with different velocities. We therefore replace the square velocity, $v_x^2$ , by the mean square value, $<v_x^2>$:

$$P = m<v_x^2>\frac{N}{V} \qquad (6.3.12)$$

Equation 6.3.12 relates the pressure to the mean square value of the $x$-component of the velocity. We can relate the pressure to the mean square value of the total velocity, $<v^2>$, by taking the average value of both sides of Equation 6.3.3:

$$<v^2> = <v_x^2> + <v_y^2> + <v_z^2> \qquad (6.3.13)$$

We assume that the box is isotropic—that is, we assume that all three directions are equivalent in terms of the mean square velocities:

$$<v_x^2> = <v_y^2> = <v_z^2> \qquad (6.3.14)$$

Substituting Equation 6.3.14 into Equation 6.3.13, we get:

$$<v^2> = 3<v_x^2> = 3<v_y^2> = 3<v_z^2> \qquad (6.3.15)$$

Using the value of $<v_x^2>$ from Equation 6.3.15 in Equation 6.3.12 and rearranging, we get:

$$PV = \frac{1}{3}m<v^2>N \qquad (6.3.16)$$

Using the ideal gas law (Equation 6.3.1), we can modify Equation 6.3.16 as follows:

$$PV = nRT = \frac{1}{3}m<v^2>N \qquad (6.3.17)$$

The mean kinetic energy per molecule in the box is $(1/2)m<v^2>$, and the total kinetic energy (K.E.) of all the molecules is $(N/2)m<v^2>$, and so Equation 6.3.17 can be rewritten as:

$$nRT = \frac{1}{3}m<v^2>N = \frac{2}{3}\frac{1}{2}m<v^2>N = \frac{2}{3}K.E. \qquad (6.3.18)$$

Equation 6.3.18 can be rearranged to give the expression for the total kinetic energy of an ideal gas:

$$\text{K.E.} = \frac{3}{2}nRT \qquad (6.3.19)$$

Temperature is, in general, an indicator of the extent of molecular motion in a system. We shall consider a more precise definition of the temperature in Chapter 8.

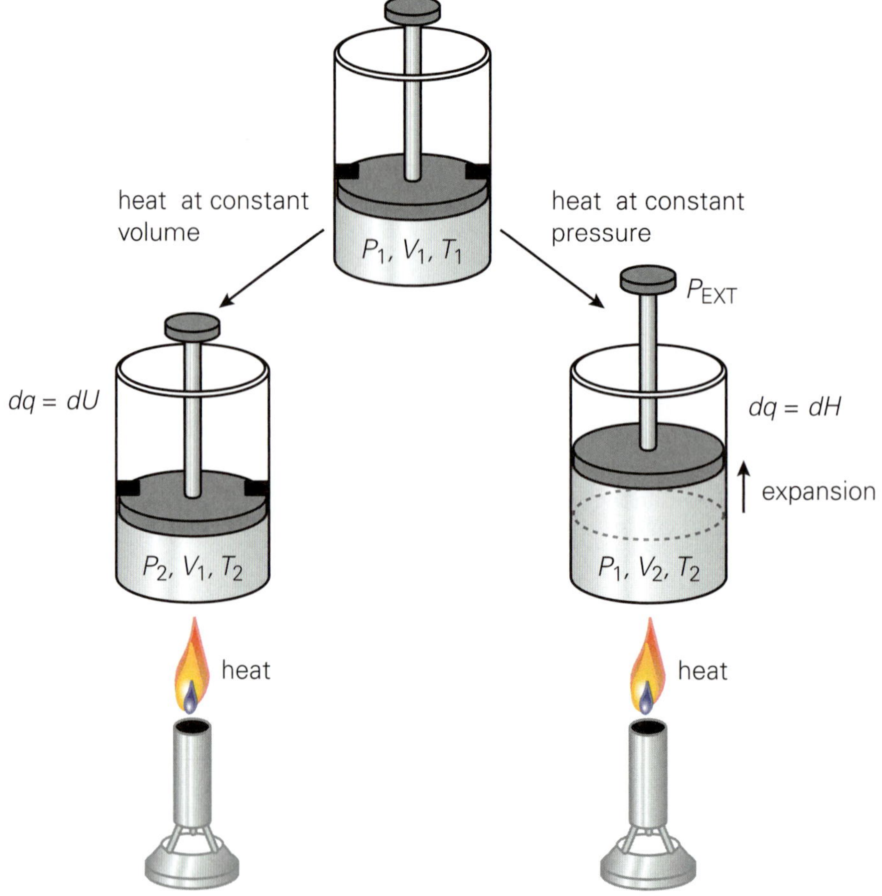


**Figure 6.10 Heating a system under conditions of constant volume (left) and constant pressure (right).** The heat taken up is equal to the change in energy when the volume is constant and no work is done (left). The heat transferred is equal to the change in enthalpy when the process occurs at constant pressure (right).

For an ideal monatomic gas, the energy, $U$, is the same as the kinetic energy, which is given by:

$$U = \frac{3}{2} nRT$$

(see Equation 6.3.19, Box 6.3), where $n$ is the number of moles of gas molecules in the system. Therefore,

$$C_V = \frac{\partial}{\partial T}\left(\frac{3}{2} nRT\right) = \frac{3}{2} nR \tag{6.18}$$

for an ideal gas.

Thus, the heat capacity of an ideal monatomic gas is a constant that does not depend on the nature of the gas. This because the heat that is delivered to the ideal gas can do only one thing, which is to increase the kinetic energy of the gas molecules, and thereby increase the temperature.

If we heat the gas at constant volume until the temperature reaches $T_2$, the total heat delivered to the system is:

$$q = \int_{T_1}^{T_2} C_V \, dT = C_V (T_2 - T_1) \tag{6.19}$$

The right-hand side of Equation 6.19 reflects the fact that $C_V$ is constant (it does not change with temperature) for an ideal monatomic gas. Equation 6.17 can be rearranged to express an infinitesimal change in energy, $dU$, in terms of an infinitesimal change in temperature, $dT$:

$$dU = C_V dT \tag{6.20}$$

According to Equation 6.20, the heat capacity of an ideal gas is a proportionality constant relating changes in temperature to changes in energy. For substances that are not ideal gases, $C_V$ is itself a function of temperature, and the relationship between changes in energy and changes in temperature is more complicated.

Now let us repeat the same experiment, but this time we allow the piston to move, so that the pressure stays constant. The heat capacity at constant pressure, $C_P$, is defined as the amount of heat required to increase the temperature of the system by 1 kelvin under conditions of constant pressure. Under such conditions, the amount of heat transferred is equal to the change in enthalpy of the system, which leads to the following definition of the heat capacity:

$$C_P = \left(\frac{\partial H}{\partial T}\right)_P \tag{6.21}$$

Equation 6.21 can be rearranged to show that $C_P$ is the proportionality constant relating infinitesimal changes in enthalpy to infinitesimal changes in temperature:

$$dH = C_P dT \tag{6.22}$$

By combining Equations 6.20 and 6.22, we can calculate the difference between the heat capacities at constant pressure and constant volume for an ideal gas:

$$(C_P - C_V)dT = dH - dU \tag{6.23}$$

Using the relationship between enthalpy and energy (Equation 6.8) and the ideal gas equation, we can rewrite the right-hand side of Equation 6.23:

$$H = U + PV = U + nRT$$

$$\Rightarrow dH = dU + nRdT$$

$$\Rightarrow dH - dU = nRdT \tag{6.24}$$

Comparing Equations 6.23 and 6.24, we can see that:

$$C_P - C_V = nR \tag{6.25}$$

for an ideal gas.

Equation 6.25 tells us that the heat capacity of an ideal gas at constant pressure is always greater than the heat capacity at constant volume. The atoms of an ideal gas do not interact with each other in a way that changes the energy. As heat is added to the system at constant volume, all that happens is that the kinetic energy of the system increases, and this results in the heat capacity being constant with temperature. When heat is added under constant-pressure conditions, some of the heat goes into the expansion work done by the gas, and so not all of it is available to increase the temperature of the gas. Thus, more heat is required to bring about the same increase in temperature.

## 6.8 The heat capacity of a macromolecular solution increases and then decreases with temperature

Now let us see what happens when we measure the heat capacity of a macromolecular solution. **Figure 6.11** shows the heat capacity as a function of temperature for a solution of a protein molecule. The protein solution is heated gradually, and the change in temperature is measured as heat is added. At first, the temperature rises by the same amount for the addition of a particular amount of heat—that is the heat capacity is roughly constant. The biochemical unit of energy, the calorie (1 cal = 4.184 J) is defined by the heat capacity of water, and 1 cal is the amount of heat required to raise the temperature of 1 g of water by 1 K at 287.5 K and 1 atm pressure. If the dissolved protein or nucleic acid molecules do not perturb the structure of water very much, then the measured heat capacity of the solution might be close to that expected for pure water.

As we continue to heat the solution of protein or RNA/DNA molecules, something interesting happens. At some point, the heat capacity of the solution rises

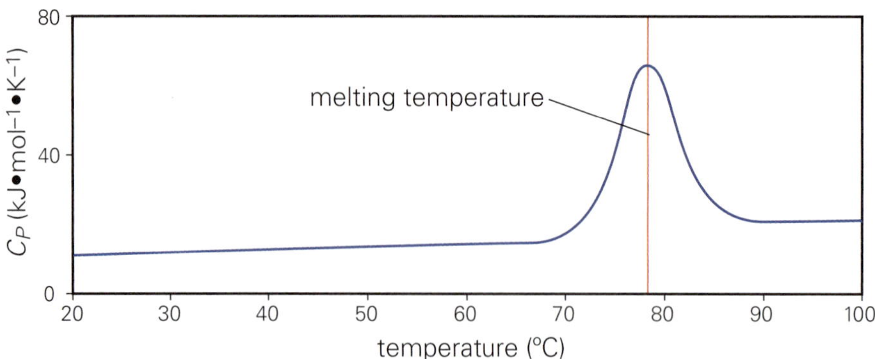


**Figure 6.11 Heat capacity of a protein as a function of temperature.** The graph shows the heat capacity at constant pressure ($C_P$) as a function of temperature. The heat capacity rises gradually at first, and then increases sharply before decreasing. The temperature at the peak value of the heat capacity is known as the melting temperature of the protein and is indicated by a *red* line. See Chapter 10 for a detailed analysis of this graph.

sharply, reaches a maximum value, and then falls sharply to a new plateau value (see Figure 6.11). A sample of pure water does not exhibit this behavior, which is characteristic of protein or DNA molecules present in the solution. As discussed in Chapter 10, what happens when heat is added to the system is that the protein molecules, which are folded to begin with, begin to unfold. The heat capacity reaches a maximum value when half of the protein molecules are unfolded. The temperature corresponding to this situation is known as the **melting temperature** of the protein.

Why does the heat capacity increase and then decrease as the protein molecules unfold? To understand why this happens we need to consider how the energy of a molecule changes with temperature. For any kind of molecule more complex than an ideal monatomic gas, delivery of heat into the system changes the energy of the system in two ways (**Figure 6.12**). Some of the heat goes into increasing the kinetic energy of the molecules, which manifests itself as an increase in the temperature of the system. In addition, some of the heat goes into exciting molecules from lower energy states into higher ones, thereby increasing their **potential energy**. In the example shown in Figure 6.12B, one of the molecules has undergone a conformational change from a *trans* conformation to a *cis* conformation. The conformational change takes up energy, and so the increase in kinetic energy,

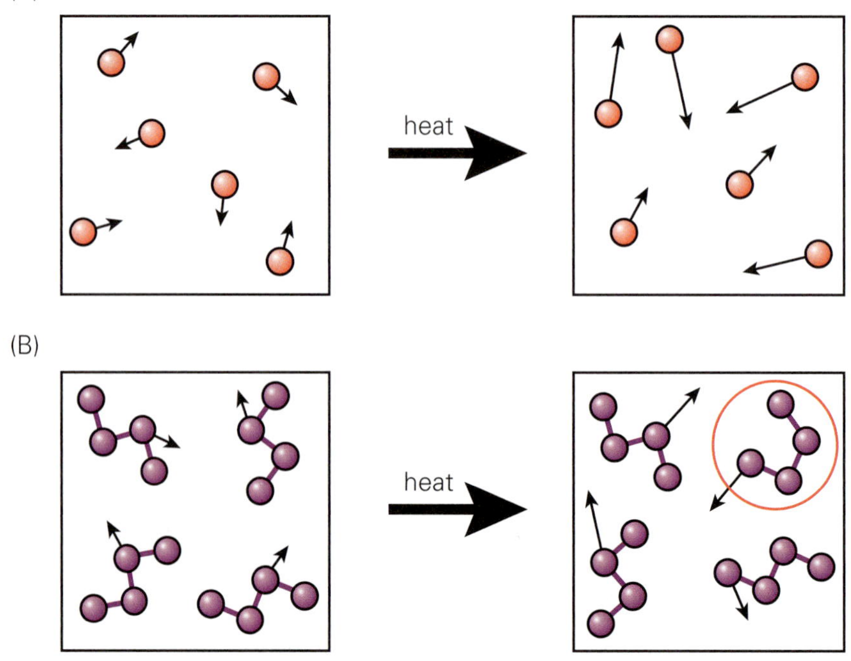


**Figure 6.12 More complex molecules take up more energy for the same increment in temperature.** (A) A schematic representation of an ideal monatomic gas. The lengths of the small arrows indicate the velocities of the atoms. As heat is delivered to the system, the atoms move faster, and the kinetic energy increases (see Box 6.3). (B) A more complicated molecule, with four atoms, is shown here. At low temperature, all the molecules are in one conformation. As heat is delivered, the molecules start to move faster, but they can also undergo a conformational change to a higher-energy conformation, indicated by the *red* circle.

which is detected as temperature, is lower than for the monatomic gas shown in Figure 6.12A. The more ways that a molecule has of taking up energy, the higher the heat capacity.

In order to understand the different modes by which molecules increase their potential energy, we need to consider the probability that a molecule can take up energy and transit from a low-energy state to one of higher energy. We start by defining the potential energy of a molecule in the next section.

## 6.9 The potential energy of a molecular system is the energy stored in molecules and their interactions

An important concept in understanding molecular interactions is that of **potential energy**, which is that part of the total molecular energy not due to atomic motion. Potential energy is energy stored in the molecule (for example, in the chemical bonds) or in the interactions between atoms in the same or different molecules.

In the physics of macroscopic objects, the potential energy is given by the work done to bring the system to a particular state. For example, **Figure 6.13** shows a person lifting a weight of mass $m$ to a height $h$. The quantity of work done, $w$, is given by

$$w = \text{force due to gravity} \times \text{distance} = m \times g \times h \tag{6.26}$$

where $g$ is the acceleration due to gravity. The work done increases the potential energy of the weight by an amount equal to $m \times g \times h$. If the weight is allowed to fall under the influence of gravity, then the maximum work that can be done by the falling weight is given by the potential energy, $mgh$.

Now consider the system of two atoms shown in **Figure 6.14**. Imagine that one of them is fixed at the origin and that the other one can move. When the atoms are very far apart, they neither attract nor repel each other. At this point, a negligible amount of work needs to be done to move the second atom a little closer to or further away from the first, because the atoms exert essentially no force on each other. The situation begins to change, however, when the atoms get closer together. When the attraction or repulsion between the atoms becomes significant, the atoms exert forces on each other, and these forces determine the amount of work required to move the second atom relative to the first.

If the interatomic distance is such that the atoms feel a mutually attractive force, then some amount of work needs to be done to increase the separation between the atoms. Alternatively, if the atoms feel a repulsive force, then work needs to be

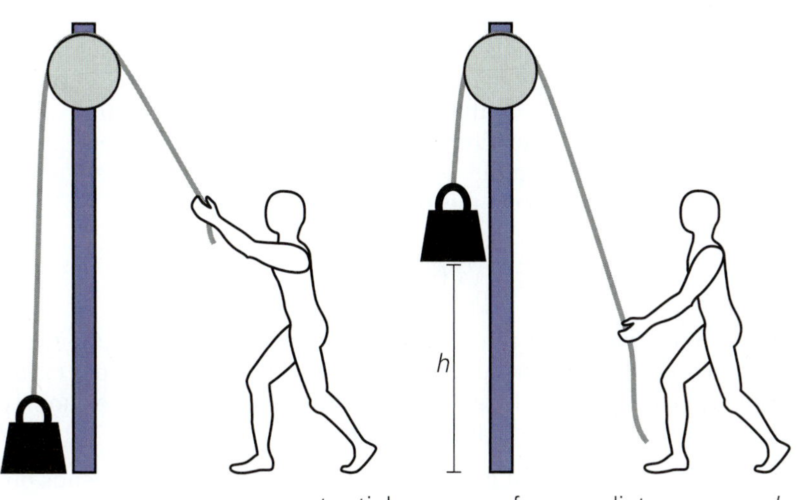


**Figure 6.13 The potential energy for a weight lifted by a pulley.** A weight of mass $m$ is lifted to a height $h$. The potential energy of the weight is equal to its capacity to do work. In this case the potential energy is $mgh$, where $g$ is the acceleration due to gravity. The maximum amount of work that can be done by the falling weight, if the pulley is released, is $mgh$.

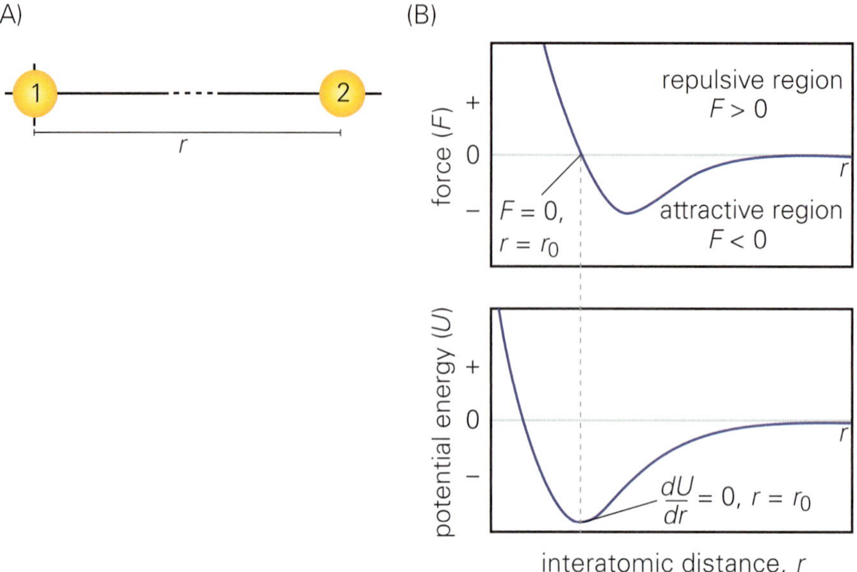


**Figure 6.14 The potential energy and force for two interacting atoms.** (A) Two atoms, denoted 1 and 2, are separated by a distance $r$. (B) The graphs show the energy and the force as a function of distance. The force is zero is when the potential energy is at a minimum (when the interatomic distance has an optimal value, denoted $r_0$).

done to decrease their separation. We relate an infinitesimally small amount of work done, $dw$, to the force, $F(r)$ at a distance $r$, exerted on each atom as follows:

$$dw = -F(r)dr \tag{6.27}$$

We take the sign of $dw$, the work done, to be positive if the external agent works against the force to effect the change, thereby increasing the distance when the atoms attract each other, or decreasing it when they repel each other. As before, work done *on* the system (which increases the energy of the system) has a positive sign.

The potential energy, $U_{pot}$, of a system of atoms is the energy "stored" in the system. For the simple case of the two atoms considered above, $U_{pot}$ is a function of $r$, the interatomic distance, and is defined as the work done in moving the second atom from a distance of infinite separation from the first ($r = \infty$) to a particular interatomic separation, $r$. The work done is the negative of the integral of force over the distance:

$$U_{pot}(r) = -\int_{\infty}^{r} F(r^*)dr^* \tag{6.28}$$

In Equation 6.28, $r^*$ is the integration variable representing the interatomic distance. Note the negative signs in Equations 6.27 and 6.28. The force is a vector, with both magnitude and direction. The force vector points in the direction of decreasing energy: if we move in the direction of the force vector, the energy decreases. This relationship implies that the force is the negative of the first derivative of the potential energy with respect to the interatomic separation, $r$:

$$F(r) = -\frac{dU_{pot}}{dr} \tag{6.29}$$

Figure 6.14B illustrates how the force and the potential energy of the two atoms change as a function of the separation, $r$, between the atoms. Because $r$ is the distance between the atoms, it depends only on the relative positions of atoms and is unaffected by the net movements of the atoms.

For a collection of more than two atoms, the potential energy depends on their internal configuration in a complicated way, depending on the nature of the bonding interactions between the atoms. We discuss the various components of the potential energy of molecules in later sections of this chapter. For now we simply consider that, when heat is added to a molecular system, the heat increases the potential energy as well as the kinetic energy, and that molecules that have a greater capacity for increasing their potential energy have a higher heat capacity.

## 6.10 The Boltzmann distribution describes the population of molecules in different energy levels

The principles of quantum mechanics tell us that only certain values of molecular energies are allowed—that is, energy is quantized. The potential energy levels of an isolated atom are shown schematically in **Figure 6.15**A. Each level in the energy diagram corresponds to a different electronic excitation state. The atom can increase its total energy by moving faster (increasing its translational kinetic energy, not shown in Figure 6.15) or by converting to an excited electronic state (that is, increasing its potential energy). Figure 6.15B shows the energy levels for a triatomic molecule. In addition to increasing its translational kinetic energy and electronic potential energy, the triatomic molecule can also rotate faster (that is, increase its rotational angular momentum, which is a form of kinetic energy). The bonds between atoms have vibrational energy, and the triatomic molecule can also pick up energy by moving to states of higher vibrational energy.

In Section 6.7, where we discussed the heat capacity of a monatomic gas, we ignored the electronic energy levels of the atoms. It turns out that at room temperature there simply is not enough energy in the collisions between atoms to excite them from the electronic ground state to the first excited state. The energy required for such a transition is on the order of ~1,000 kJ•mol$^{-1}$. To appreciate why this energy gap is too large to be surmounted at room temperature, we need to consider the probability that an individual atom or molecule in a large population acquires a certain amount of energy from collisions. This probability is given by a mathematical function known as the **Boltzmann distribution**. The Boltzmann distribution, which is named after Ludwig Boltzmann, is explained in more detail in Chapter 8, but we provide a brief introduction here in order to understand how molecules are distributed among different energy levels.

**Boltzmann distribution**

The Boltzmann distribution is a mathematical expression that tells us the probability of finding molecules in different energy levels at a given temperature and at equilibrium. According to the Boltzmann distribution, the number of molecules, $N$, in a particular energy level decreases exponentially with the energy, $U$, of the energy level:

$$N \propto e^{-\frac{U}{k_B T}}$$

The exponential term is known as the Boltzmann factor, $k_B$ is the Boltzmann constant and $T$ is the absolute temperature.

Suppose there are $N$ atoms or molecules in a system. The number of molecules $N_i$ that will be found in an energy level with energy $U_i$ at equilibrium is given by the Boltzmann distribution:

$$N_i = \frac{Ne^{-U_i/k_B T}}{Q} \qquad Q = \sum_i e^{-U_i/k_B T} \tag{6.30}$$

where $k_B$ is the **Boltzmann constant** ($k_B = 1.4 \times 10^{-23}$ J•K$^{-1}$), $T$ is the absolute temperature, and $Q$ is called the partition function.

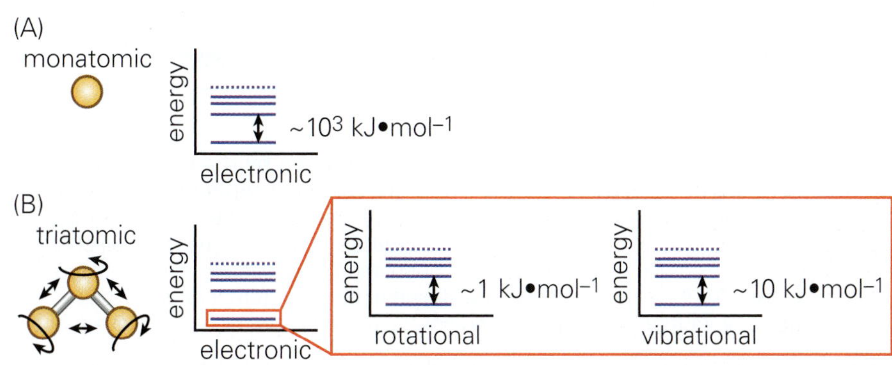


**Figure 6.15 Energy levels for an isolated atom and for a triatomic molecule.** (A) An isolated atom has translational kinetic energy and electronic energy levels (potential energy). Only the electronic energy levels are shown here. (B) A triatomic molecule has rotational energy and vibrational energy in addition to translational kinetic energy and electronic energy. For each electronic energy level, there are different energy levels corresponding to the rotational energy and to the vibrational energy.

The partition function, $Q$, is a normalization factor that ensures that the probability of finding a molecule in any one of the energy levels is 1.0, as will be explained in Chapter 8. The value of $Q$ is a constant at a given temperature (that is, we assume that the energies, $U_i$, of the different levels do not change with temperature), and so we can write a simpler form of the Boltzmann distribution as follows:

$$N_i \propto e^{-U_i/k_B T} \tag{6.31}$$

The exponential term on the right-hand side of Equation 6.31 is known as the **Boltzmann factor**, and it determines the relative population of an energy level, as explained below. Note that the term $k_B T$ has units of energy, and that it is the ratio of the energy to the value of $k_B T$ that enters into the Boltzmann factor.

The Boltzmann distribution helps us understand which energy levels will be populated to a significant extent at a certain temperature at equilibrium (**Figure 6.16**). For example, consider any two energy levels for a molecule. Call the lower energy level 1, and call the higher energy level 2. The ratio of the number of molecules in level 2, $N_2$, to the number of molecules in level 1, $N_1$, is given by the Boltzmann distribution as follows:

$$\frac{N_2}{N_1} = e^{-\Delta U/k_B T} \tag{6.32}$$

where $\Delta U = U_2 - U_1$.

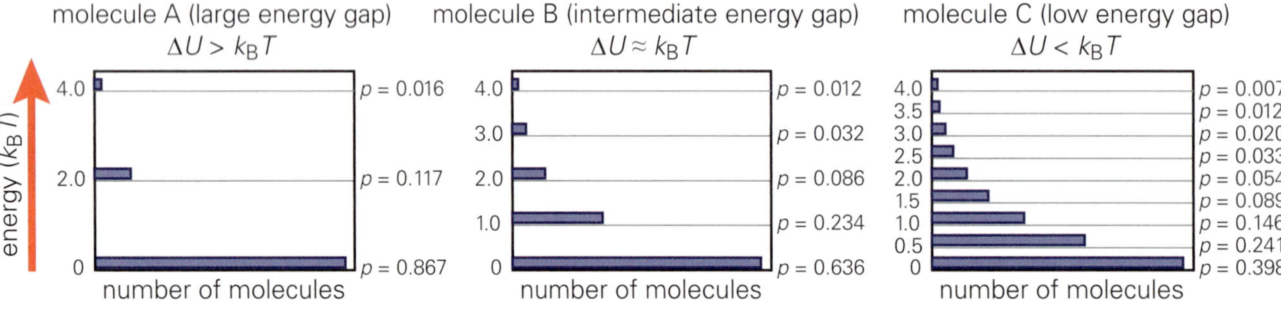


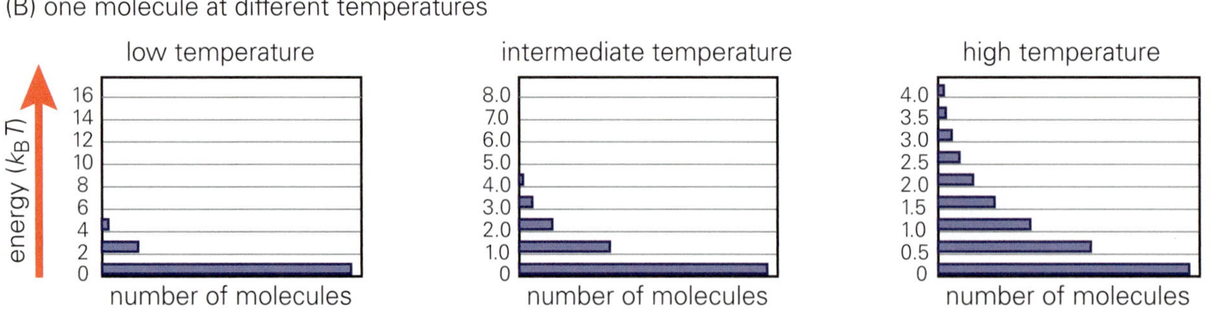


**Figure 6.16 Population of energy levels as a function of the energy gap and temperature, as given by the Boltzmann distribution.** (A) The energy spacings for three different kinds of molecules are shown, all at the same temperature. The energy of each level is shown in multiples of $k_B T$. Molecule A has a large energy gap ($\Delta U > k_B T$). The population of the upper levels is small. Molecule B has an intermediate energy gap ($\Delta U \approx k_B T$). The population of the upper levels increases. Molecule C has a small energy gap ($\Delta U < k_B T$), and the population of the upper levels becomes much larger. The relative population, $p$, of each level is indicated on the right of each diagram. (B) Population of energy levels for the same molecule are shown at different temperatures. The population of the upper levels increases with temperature. Note that real molecules do not have equally spaced energy levels, and the examples shown here are merely illustrative.

At 300 K, the value of $k_B T$ is $4.2 \times 10^{-21}$ J. This is an awkwardly small number, and it is more convenient to switch the units of energy to kJ•mol$^{-1}$. Since 1 mol = 6.022 $\times 10^{23}$ molecules,

$$k_B T = 4.2 \times 10^{-21} \times 6.022 \times 10^{23} \text{ J} \bullet \text{mol}^{-1}$$
$$= 25.29 \times 10^{2} \text{ J} \bullet \text{mol}^{-1}$$
$$= 2.529 \text{ kJ} \bullet \text{mol}^{-1} \tag{6.33}$$

Thus,

$$\frac{N_2}{N_1} = e^{-\Delta U / 2.529}$$

at 300 K, where $\Delta U$ is the difference in energy between the first and second levels in units of kJ•mol$^{-1}$.

**Table 6.1** expresses the Boltzmann factor $\left(e^{-\Delta U / k_B T}\right)$ as a function of energy in multiples of the value of $k_B T$. Based on these values, we can see that when the difference in energy between two levels is much greater than $k_B T$, the higher energy level is not populated to a significant extent (see Figure 6.15). For example, if the energy difference between two levels is $2k_B T$, then the population of the upper level is ~13% that of the lower level. If the energy difference is $10k_B T$, then the population of the upper level decreases to 0.0045%.

The Boltzmann factor explains why we ignored electronic excitations when considering the contributions to the heat capacity of an ideal gas. Because we are concerned primarily with biological systems, we consider energy changes that occur near room temperature—that is, $T \approx 300$ K and $k_B T \approx 2.5$ kJ•mol$^{-1}$. The energy difference between the lowest electronic energy level and the next highest one is typically ~1000 kJ•mol$^{-1}$—that is, $\Delta U >> k_B T$ for electronic energy levels at room temperature. The much smaller thermal energy at room temperature (~2.5 kJ•mol$^{-1}$) is very unlikely to excite molecules to higher electronic energy levels. We can recognize that this is the case by considering that ultraviolet radiation can excite electrons into higher energy levels. The wavelength of ultraviolet radiation is ~100 nm, corresponding to an energy of ~1,200 kJ•mol$^{-1}$.

When the Boltzmann constant ($k_B = 1.4 \times 10^{-23}$ J•K$^{-1}$) is expressed in units of J•mol$^{-1}$, it is equivalent to the gas constant, $R$. That is, the gas constant is simply Boltzmann's constant multiplied by Avogadro's number ($N_A$):

$$R \text{ (Gas constant)} = N_A \times k_B = 6.022 \times 10^{23} \times 1.38 \times 10^{-23} = 8.31 \text{ J} \bullet \text{mol}^{-1} \bullet \text{K}^{-1} \tag{6.34}$$

What this means is that if we use units of J•mol$^{-1}$ for the energy, then instead of $k_B T$ we simply use $RT$ to calculate the Boltzmann factor.

**Table 6.1 The Boltzmann factor as a function of energy.**

| $\Delta U$ (kJ•mol$^{-1}$) | $\frac{\Delta U}{k_B T}$ ($T$ = 300 K, $k_B T \approx 2.5$ kJ•mol$^{-1}$) | $e^{\frac{-\Delta U}{k_B T}}$ |
|---|---|---|
| 0.25 | 0.1 | 0.90 |
| 0.50 | 0.5 | 0.61 |
| 2.50 | 1.0 | 0.37 |
| 5.00 | 2.0 | 0.13 |
| 15.00 | 6.0 | 0.00067 |
| 25.00 | 10.0 | 0.0000045 |

## 6.11 The energy required to break interatomic interactions in folded macromolecules gives rise to the peak in heat capacity

The energy available from collisions at room temperature can excite molecular vibrations and conformational changes, which refer to all the different kinds of fluctuations in the internal structure of the molecules. We can estimate the energy required to excite the vibrations of individual covalent bonds because we know that these are excited by infrared radiation, with wavelengths around 2000 nm. This corresponds to ~60 kJ•mol$^{-1}$ (~24 $k_BT$ at 300 K). The energy difference between the lowest energy level and the first excited state for bond vibrations must be comparable. Thus, covalent bonds are very stiff, and their vibrations are unlikely to be excited at room temperature.

The key to understanding the heat capacity changes in complex molecules like proteins and DNA is to recognize that they are held together by numerous weak interactions, such as hydrogen bonds and van der Waals interactions. These interactions stabilize the folded structure collectively, but each individual interaction is easily broken. The energy required to break each interaction contributes to the increase in heat capacity as the protein unfolds.

We can now understand the origin of the characteristic heat capacity changes of proteins and DNA as follows. At low temperature, the molecules are folded into stable three-dimensional structures. As the temperature is increased, the structures undergo cooperative unfolding transitions, when the many weak interactions that hold the structure together begin to break. The effect of temperature on the unfolding of a small part of a protein is shown in **Figure 6.17**, in which states A and B denote the fully folded conformation and a partially unfolded conformation of the protein. We assume that the energy of the protein is 5 kJ•mol$^{-1}$ higher when the interactions made by this small region are disrupted in state B. The population of state B increases from ~10% to 20% as the temperature increases from 300 K to 350 K. The increase in population of state B requires energy, which manifests itself as an increase in the heat capacity of the protein solution.

Although the increase in population of the partially unfolded state shown in Figure 6.17 may seem small, there are many such interactions that hold the protein together. Once a small number start to break, other interactions rapidly become unstable, and the protein unfolds cooperatively. Once all of the interactions are broken (that is, the protein is completely unfolded), energy no longer goes into the unfolding process, and the heat capacity is reduced. The maximum value of the heat capacity occurs at the melting temperature, when the folding process is halfway towards completion (**Figure 6.18**).

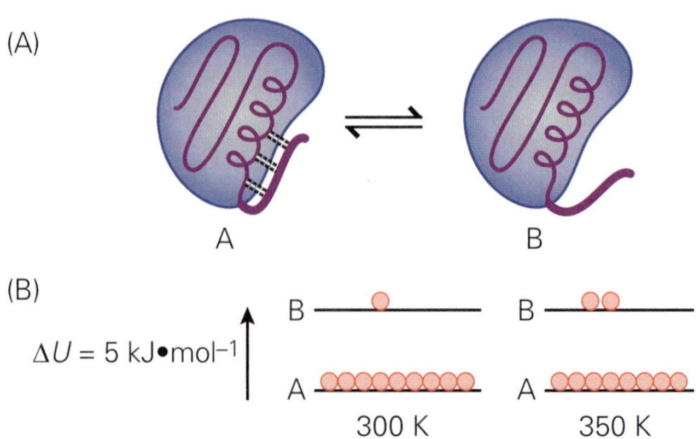


**Figure 6.17 Protein molecules take up energy as they unfold.** (A) A schematic drawing of a protein molecule in two states, denoted A and B. A corresponds to the fully folded conformation and B corresponds to a partially unfolded one. Interactions that are made in state A but broken in state B are indicated by dashed lines. (B) An energy diagram showing the relative populations of the two states at 300 K and at 350 K. State B is higher in energy by 5 kJ•mol$^{-1}$ than state A. The circles indicate the relative populations of the two states at the two temperatures, as given by the Boltzmann distribution.

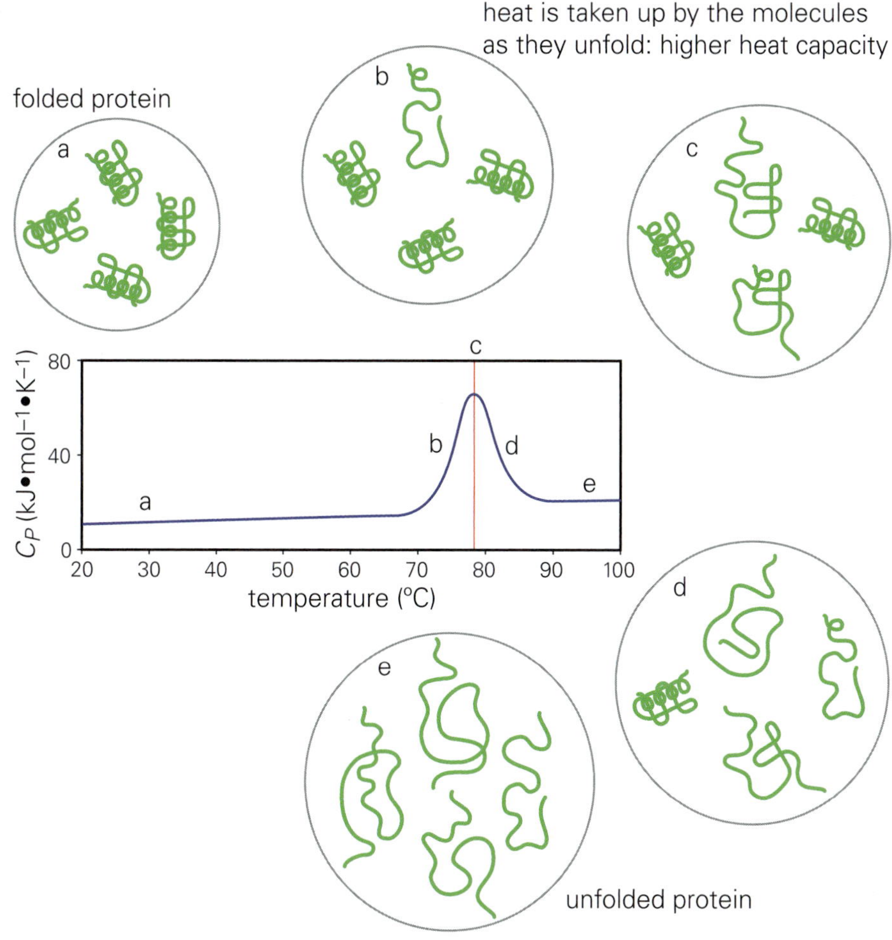


**Figure 6.18 Heat capacity and protein unfolding.** The heat capacity curve shown in Figure 6.11 is redrawn here, along with an interpretation of the molecular behavior underlying the changes in the heat capacity. At lower temperatures (< 40°C), the protein molecule is mainly folded. As the temperature is increased the protein begins to unfold and energy goes into breaking the numerous interactions that stabilize the folded structure. This manifests itself as an increased heat capacity. Once most of the protein molecules have unfolded the heat capacity decreases.

# C. ENERGETICS OF INTERMOLECULAR INTERACTIONS

## 6.12 Simplified energy functions are used to calculate molecular potential energies

In order to understand what happens to molecules as they take up and release energy, we need to be able to relate the structure of a molecule to its potential energy. Molecular potential energies can be calculated from first principles if we use quantum mechanics, but these calculations require an enormous amount of computer time. A typical protein or RNA molecule contains thousands of atoms and, even with the fastest computers available today, it is difficult to use quantum mechanical methods for the accurate calculation of potential energies for molecules of this size. In practice what is done instead is to use mathematical expressions known as **empirical potential energy functions**, which are highly simplified functions that allow approximate molecular energies to be calculated from knowledge of the three-dimensional coordinates of all the atoms in a molecule. The energy functions are also used to calculate forces, which are obtained by computing the first derivative of the potential energy with respect to atomic positions.

**Empirical potential energy functions**

These are relatively simple mathematical expressions that allow us to calculate the potential energy of a molecule or a collection of molecules, given the conformation (that is, the internal structure) of each molecule and their relative positions. Empirical potential energy functions are approximations to the true quantum mechanical energy of the system.

In quantum mechanical calculations, all of the electrons and nuclei in a molecule are considered explicitly, and probability functions that describe the distribution of electrons around nuclear positions are obtained by solving Schrödinger's equation on a computer. All the properties of the molecule, including its covalent bonded structure and the potential energy, can be derived accurately from the electronic wave functions that correspond to different configurations of the nuclei. This provides the most detailed and accurate information about the structure and electronic properties of molecules and forms the basis for all descriptions of molecular energies.

A major limitation of quantum mechanical calculations is the time required to carry out these calculations, even with the fastest computers available today. The calculation of molecular energies by quantum mechanics proceeds by considering a particular configuration of atomic nuclei and then generating the electronic wave functions appropriate for that nuclear configuration. The computer time required for the most accurate quantum mechanical calculations carried out today increases roughly as the seventh power of the number of electrons in the molecule. This means that a calculation that might take a minute for a benzene molecule (~40 electrons) would take ~$4 \times 10^{14}$ minutes for a very small protein molecule with ~5000 electrons, an impossibly long time. While less accurate quantum mechanical calculations on an organic compound of ~100 electrons take 1–2 hours to complete on a modern computer, and scale less rapidly with size than do the most accurate calculations, the increased computer power required to treat protein molecules with these less accurate methods still makes them impractical. While quantum mechanical calculations can be simplified by breaking the protein into smaller pieces and treating each piece separately, a general way to recombine the results accurately has not yet been developed. An additional complication arises because the electronic wave functions have to be recalculated when the nuclear positions change due to molecular motions. The need to calculate energies for a wide range of conformations, not just for a few, makes the routine application of quantum mechanical calculations impractical at present for molecules the size of proteins.

## 6.13 Empirical potential energy functions enable rapid calculation of molecular energies

Instead of using quantum mechanics, the rapid calculation of molecular energies for large molecules, such as proteins and nucleic acids, relies on highly simplified descriptions of the relationship between molecular geometry and potential energy. These calculations do not treat electrons explicitly (the effects of the electrons are embedded in the form of the energy functions), and consider molecular structure only in terms of nuclear positions. These relationships are called empirical potential energy functions because they are not derived from first principles, but are instead approximate representations of the empirically observed behavior of molecules in terms of simple mathematical expressions that can be very quickly calculated using computers. The approximations introduced have a price, which is that the gain in speed is offset by a loss in accuracy. The highly accurate calculation of the energies of protein molecules remains a problem that has no really satisfactory solution at the present time. Despite these limitations, empirical energy functions are the only practical way to describe the energies of macromolecules and the forces within them.

While quantum mechanics describes the properties of matter in terms of wave functions, classical or Newtonian mechanics considers the movement of particles or larger objects in terms of potential energies and forces, which depend on the distances between the particles (**Figure 6.19**). In principle, quantum mechanical calculations on small pieces of protein molecules tell us how their energy changes as a function of the distances between the various atomic nuclei. Such calculations can then be used to derive the form of simple energy functions that approximate the energy as a function of nuclear positions. These energy functions are then used within the framework of classical mechanics to calculate structural and dynamical properties of the molecule. We also ignore the quantization of energy that is a fundamental aspect of quantum mechanics. Because we are treating low-energy vibrations and interactions, this turns out not to be a bad approximation. Thus, instead of considering discrete energy levels as we did in discussing the

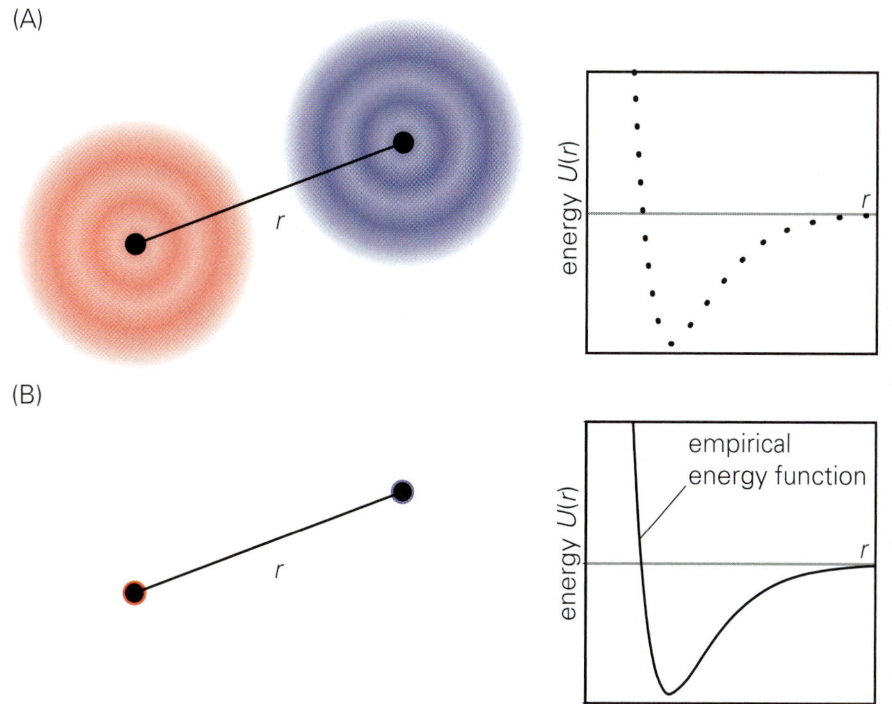


**Figure 6.19 Calculation of energy in quantum and classical mechanics.** (A) Two interacting atoms are shown, and their wave functions are illustrated schematically. Values of the ground-state energy, calculated using quantum mechanics for specific values of the interatomic distance, *r*, are shown as dots in the graph. (B) The empirical energy function for the interaction between the two atoms is graphed. The empirical energy function is a continuous mathematical function that approximates the quantum mechanical energy.

Boltzmann distribution, we assume that all values of the energy are allowed in principle. The Boltzmann distribution then gives us the relative populations of states with different energies.

There is no single empirical energy function that best describes the energetics of proteins, DNA, and RNA. Instead, the detailed mathematical formulation of these functions reflects choices made by the scientists who derive these functions. The simplest empirical energy functions typically include mathematical terms that describe the change in the energy of a protein molecule as its covalent bonds and angles are deformed, as the molecule changes its conformation by rotating about its covalent bonds, and as clusters of atoms arranged in particular geometries, such as planar groups, are distorted from their equilibrium configurations. In addition to these terms, which depend on interatomic distances between covalently bonded clusters of atoms, the energy functions also include **noncovalent energy terms** that describe changes in the energy as atoms that are not covalently bonded approach each other (**Figure 6.20**). These include the van der Waals interactions between atoms and electrostatic interactions, such as ion pairs and hydrogen bonds.

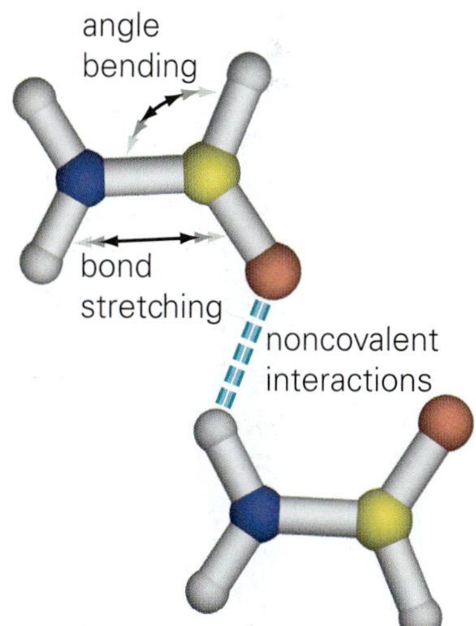


**Figure 6.20 Covalent and noncovalent interactions.** The deformations of covalently bonded atoms (such as bond stretching) have a high energetic penalty and, as a consequence, the geometry stays close to the optimal values at room temperature. Noncovalent interactions between groups that are not covalently bonded are readily broken.

## 6.14 The energies of covalent bonds are approximated by functions such as the Morse potential

To illustrate how empirical energy functions are formulated, consider two atoms, labeled A and B, that are covalently bonded to each other (**Figure 6.21**). We can calculate the energy of such a pair of A and B atoms using quantum mechanics. These calculations show us that the atoms experience a repulsive force when the atoms approach each other closely, due to overlap between their electronic orbitals. The repulsive force decreases rapidly when the distance between the atoms is increased. The interatom force becomes zero at a distance corresponding to the optimal bond length, $r_0$. If the distance between the atoms is increased beyond $r_0$, the atoms experience an attractive force until the bond between them is broken and the force between them again becomes zero. In summary, the potential energy of this two-atom system is zero for large interatomic separations, becomes negative for closer distances that correspond to covalent bond formation, and then rises sharply as the atoms move closer together.

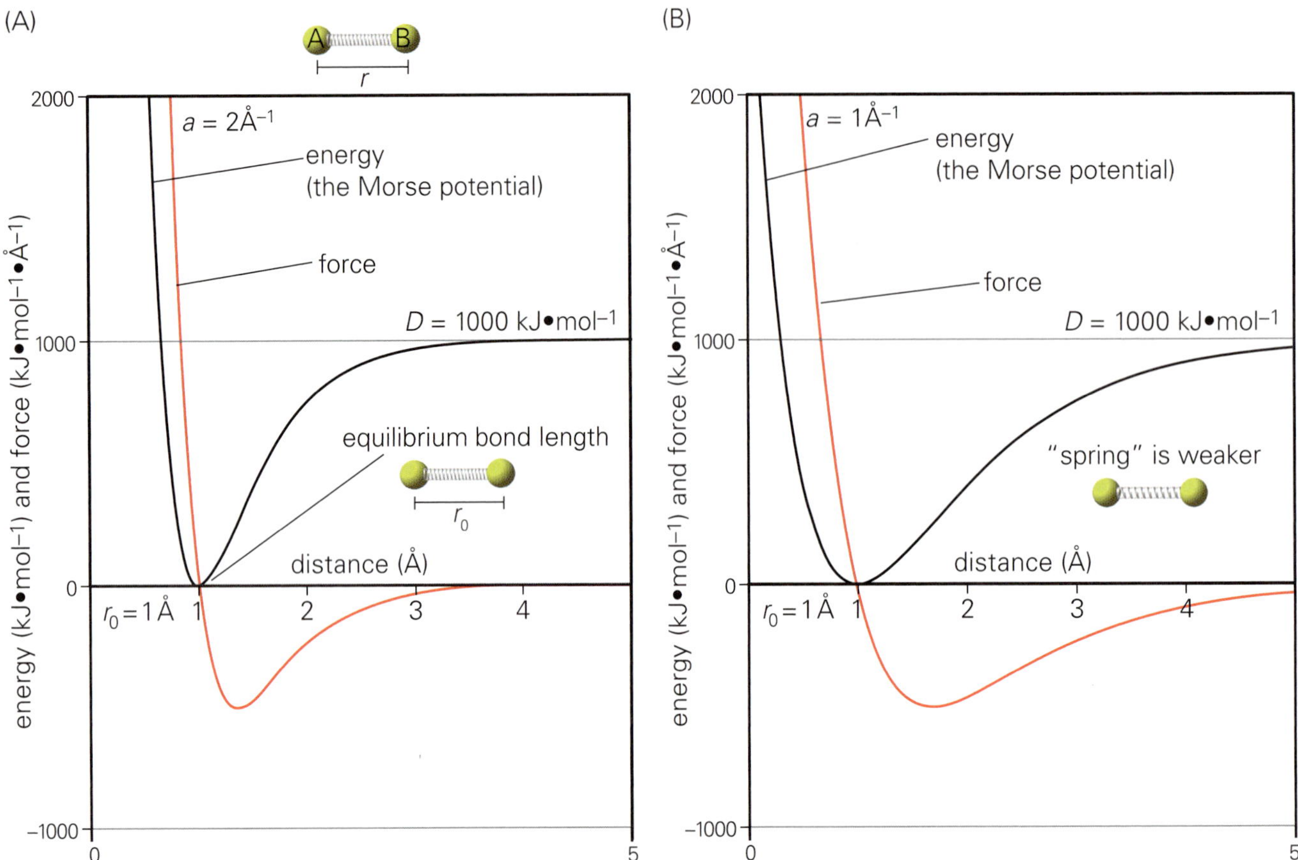


**Figure 6.21 The energy and the force on each atom in an A–B covalently bonded pair as a function of the interatomic distance**. The energy and force are calculated using the Morse potential (Equation 6.35). For this example, the equilibrium (optimal) bond length, $r_0$, is 1.0 Å. (A) The values of the parameters *a* and *D* are 2 Å$^{-1}$ and 1000 kJ•mol$^{-1}$, respectively. The parameter *D* is the stabilization energy. (B) The effect of the parameter *a* on the energy and force is illustrated by keeping all other parameters the same as in (A), but changing the value of the parameter *a* to 1.0 Å$^{-1}$. This results in a "softening" of the spring (the energy rises less sharply with interatomic distance), while the net stabilization of the covalent bond does not change. Note that the force is negative (that is, attractive) for distances larger than the optimal distance (1.0 Å) and positive (that is, repulsive) for distances shorter than the optimal distance.

The behavior of the interatomic force is modeled by the following simple expression for the potential energy that is based on an exponential function of the interatomic distance, $r$:

$$U_{\text{bond}}(r) = D\left[1 - e^{-a(r-r_0)}\right]^2 \tag{6.35}$$

In Equation 6.35, $U_{\text{bond}}$ is the energy of the covalent bond when the interatomic separation is $r$.

This expression, known as the **Morse potential** (see Figure 6.21), has three adjustable parameters—$D$, $a$, and $r_0$—that are characteristic of the nature of the covalent bond between A and B. These three parameters are related to the chemical properties of the molecule. For example, when the interatomic separation, $r$, is equal to $r_0$, then:

$$U_{\text{bond}}(r_0) = D\left[1 - e^{-a(r-r_0)}\right]^2 = D\left[1 - e^{0}\right] = 0 \tag{6.36}$$

Thus, the energy is zero when $r = r_0$, and so $r_0$ is called the **optimal bond length**.

When the atoms are infinitely far apart ($r = \infty$),

$$U_{\text{bond}}(r = \infty) = D\left[1 - e^{-\infty}\right]^2 = D \tag{6.37}$$

The parameter $D$ is the amount by which the potential energy of the A–B pair is reduced when the covalent bond is formed. $D$ is therefore related to the enthalpy of making the A–B bond, which is the heat released when A and B atoms fuse to form A–B, an experimentally measurable quantity. It may seem strange that the energy has a high value ($D$) when the atoms are infinitely far apart. You should recognize, however, that we are only interested in differences in potential energy, and so the zero of the energy scale is arbitrary. Our analysis would be unchanged if we were to subtract $D$ from the energy, giving us an energy of zero when the atoms are far apart and $-D$ when their separation is equal to the optimal value.

The parameter $a$ describes how sharply the energy rises when the interatomic separation changes from $r_0$ (see Figure 6.21). This parameter is related to the vibrational frequency of the A–B bond: the sharper the variation in energy, the higher the vibrational frequency. Vibrational frequencies of covalent bonds can be measured using techniques such as infrared spectroscopy, allowing us to estimate the values of $a$ experimentally for different kinds of covalent bonds.

Although the Morse potential (Equation 6.35) is a particularly simple form of an energy function for covalent bonds, we can simplify this even further for situations where covalent bonds are not broken. This is usually the case when we consider protein function, because most covalent bonds are stable at room temperature and, under normal biological conditions, are only made or broken at the active sites of enzymes (bonds between hydrogen atoms and oxygen or nitrogen atoms are an exception). Also, covalent bonds are very stiff, and at room temperature the bond lengths do not fluctuate much more than 0.1 Å away from the optimal values.

If we ignore chemical reactions and chemical groups that can release or accept protons, we can replace the Morse potential with a simpler form that describes the energy and force associated with stretching the bond in terms of a Hooke's law spring:

$$U_{\text{bond}}(r) = \frac{1}{2}K_b(r - r_0)^2 \tag{6.38}$$

and

$$F(r) = -K_b(r - r_0) \tag{6.39}$$

In these equations, $r_0$ is the optimal bond length and $K_b$ is known as the **force constant**. We can relate these Hooke's law equations to the Morse potential by choosing a value for the force constant, $K_b$, such that the Hooke's law expression matches the Morse potential as well as possible for distances that are close to the optimal bond length (**Figure 6.22**). Values for optimal bond lengths are derived

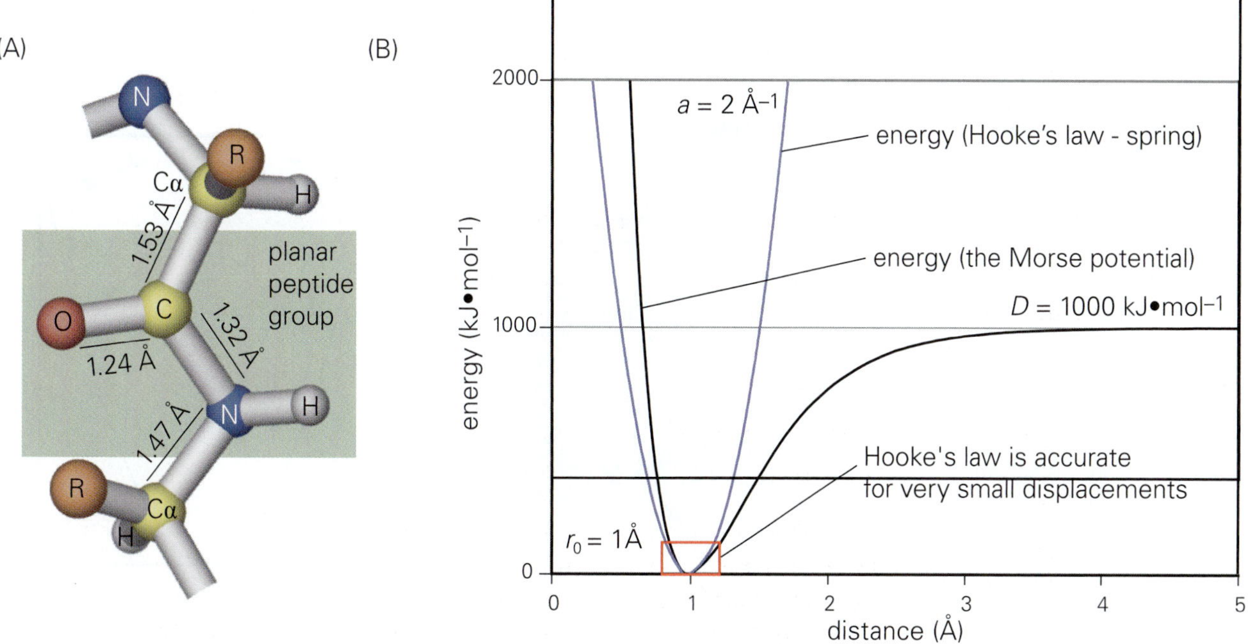


**Figure 6.22 Covalent bonds in the peptide backbone.** (A) A section of protein chain is shown, with amino acid sidechains in *orange* and indicated by "R." The peptide linkage between two adjacent amino acid residues is indicated by the *green* box. The lengths of covalent bonds between the nonhydrogen atoms in the peptide backbone are indicated. (B) A Hooke's law function that matches the Morse potential for small displacements about the optimal bond length is shown. Note that the energy rises sharply without limit as the interatomic separation increases, so this function does not model covalent bond breakage. (A, adapted from R.E. Marsh and J. Donohue, *Adv. Protein Chem.* 22: 235–256, 1967.)

from crystal structures of amino acids and peptides, as shown in Figure 6.22A for the peptide group. The values of $K_b$ for covalent bonds are typically in the range of ~400–2000 kJ•mol$^{-1}$•Å$^{-2}$.

## 6.15 Other terms in the energy function describe torsion angles and the deformations in the angles between covalent bonds

There are at least two other energetic terms associated with covalent bonding that need to be considered when calculating the potential energy. One term describes changes in the energy of the molecule that arise from angular distortions in the covalently bonded structure. Two adjacent covalent bonds in a molecule are constrained to an angular separation, $\theta_0$, known as the optimal bond angle (**Figure 6.23**). If the bond angle is distorted, a spring-like force tends to restore the angular separation between the bonds back to the optimal value. The energy for bond deformations is given by:

$$U_{\text{angle}} = \frac{1}{2} K_\theta (\theta - \theta_0)^2 \tag{6.40}$$

where $\theta$ is the value of the bond angle and $K_\theta$ is the force constant for deformation of the angle. Like the covalent bonds, the bond angles in a protein are also stiff. The values of $K_\theta$ range from 50 to 200 J•mol$^{-1}$•degree$^{-2}$. Again, as for bond vibrations, the Boltzmann distribution tells us that only very small displacements of bond angles are likely to occur at room temperature. Like the lengths of covalent bonds, the angles between covalent bonds do not change very much when a protein molecule alters its conformation or binds to other molecules.

Another term in the energy function concerns the relative rotation or torsion of two segments of the molecule about a covalent bond. Such a rotation is known as a **dihedral angle rotation**, because its magnitude is given by the angle between two planes formed by adjacent pairs of covalent bonds (a "dihedral" is the angle between two planes; **Figure 6.24**). Dihedral angle rotations about single bonds are "soft," in that the barrier to rotation about these bonds is usually not very high relative to the value of $k_BT$. Rotations about the $\phi$ and $\psi$ angles in the backbone of the polypeptide chain correspond to such "soft" dihedral rotations (see Figure 6.24). For soft dihedral angles, the dependence of the energy on the dihedral angle, $\chi$, is approximated by a term involving cos $(n\chi - \delta)$, where $\delta$ is the offset angle of the dihedral energy term and $n$ is an integer whose value determines the periodicity of the energy term:

$$U_{\text{dihedral}} = \frac{1}{2} K_\chi [1 + \cos(n\chi - \delta)] \tag{6.41}$$

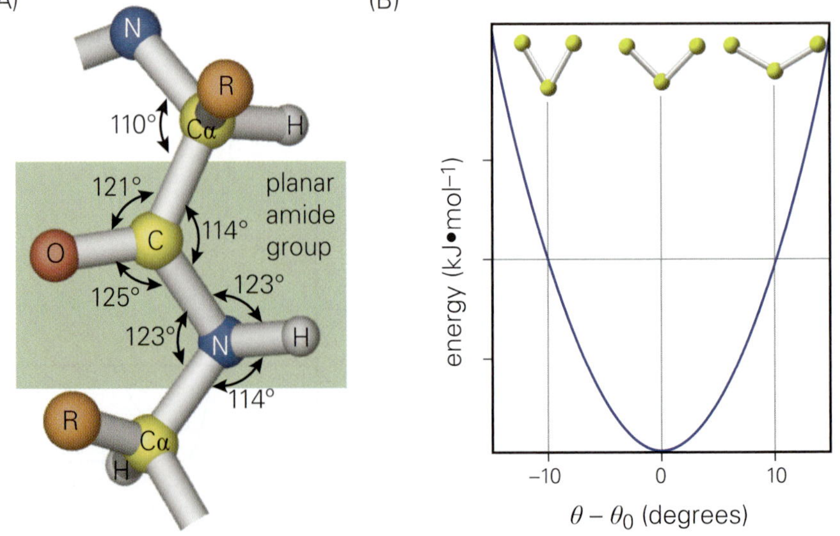


**Figure 6.23 Deformations in angles between covalent bonds.** (A) The angles between covalent bonds in the peptide chain. The same section of peptide chain that is illustrated in Figure 6.20A is shown here. The optimal values of the angles between adjacent covalent bonds is shown. (B) The energy of deforming an angle between covalent bonds is a quadratic function of the deformation angle and is plotted in this diagram. (A, adapted from R.E. Marsh and J. Donohue, *Adv. Protein Chem*. 22:235–256, 1967.)

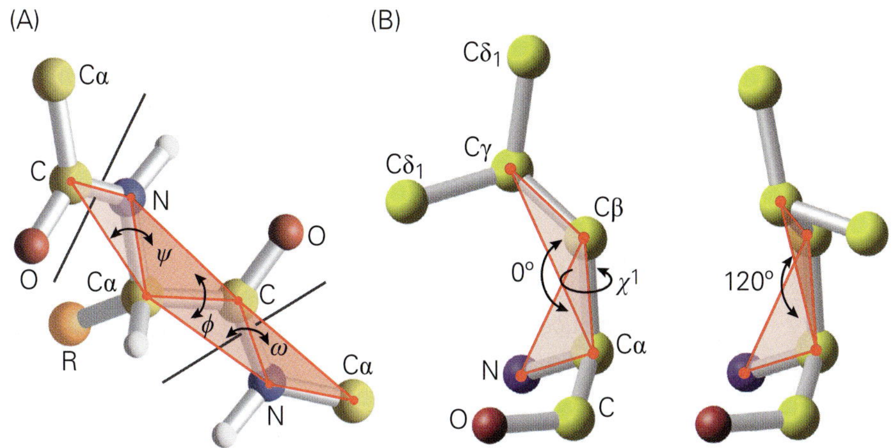


**Figure 6.24 Dihedral angles.** (A) Dihedral angles describe rotations about bonds in the structure and can be thought of as the angles between two planes, which are indicated in *red*. As discussed in Chapter 4, the peptide backbone has two dihedral angles that rotate easily, $\phi$ and $\psi$. The dihedral angle, $\omega$, is not free to rotate very much. (B) A dihedral angle, referred to as $\chi^1$, in the sidechain of leucine.

The application of Equation 6.41 is illustrated in **Figure 6.25**, which shows how the value of $U_{\text{dihedral}}$ changes as a sidechain dihedral angle in leucine, denoted $\chi^1$, is varied from 0° to 360°. For this dihedral angle, which corresponds to rotation about the $C_\alpha$–$C_\beta$ bond, the value of $\delta$ is zero, and so the energy has a maximum value when the value of $\chi^1$ is zero. The value of $n$ is 3, and so the energy is maximal for three values of $\chi^1$ (0°, 120°, 240°). These correspond to conformations in which the substituents of the $C_\alpha$ and $C_\beta$ atoms are in eclipsed configurations, and so these values of $\chi^1$ are hindered sterically. There are also three values of $\chi^1$ for which the energy has minimal values (60°, 180° and 300°). The substituents of the $C_\alpha$ and $C_\beta$ atoms are in staggered configurations for these values of $\chi^1$ and so there is minimal steric overlap. You can work out that the value of the parameter $K_\chi$ corresponds to the difference in energy between the eclipsed and staggered conformations. The value of $K_\chi$ is in the range of 1–20 kJ•mol$^{-1}$ for dihedral angles in protein sidechains.

As you can see from this example, when $n > 1$, the energy function given by Equation 6.41 has more than one value of the dihedral angle, $\chi$, with the same value for the potential energy. According to the Boltzmann distribution, each of the alternative values of the dihedral angles is equally likely, and the molecule will hop between these different allowed conformations. This is the primary source of flexibility in protein molecules, and we have already seen its consequences in the Ramachandran diagram (see Chapter 4).

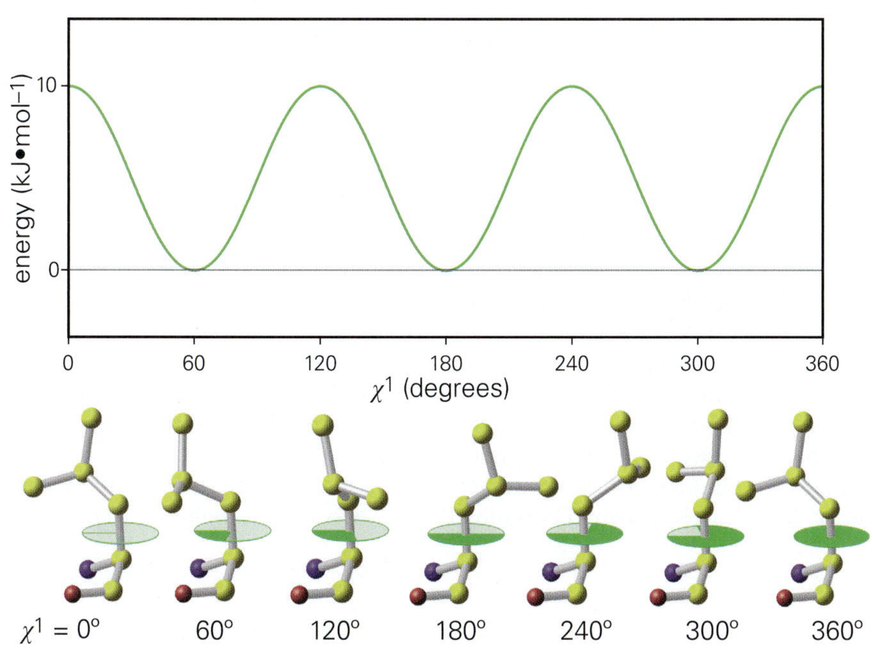


**Figure 6.25 The change in energy as a dihedral angle is rotated.** Various conformations of a leucine residue are shown at the bottom, and these differ only in the value of the sidechain dihedral angle, $\chi^1$. The extent of dihedral angle rotation is illustrated schematically by the *green* disk, in which the *dark green* sector corresponds to the extent of rotation. The dihedral angle energy as a function of $\chi^1$ is plotted above.

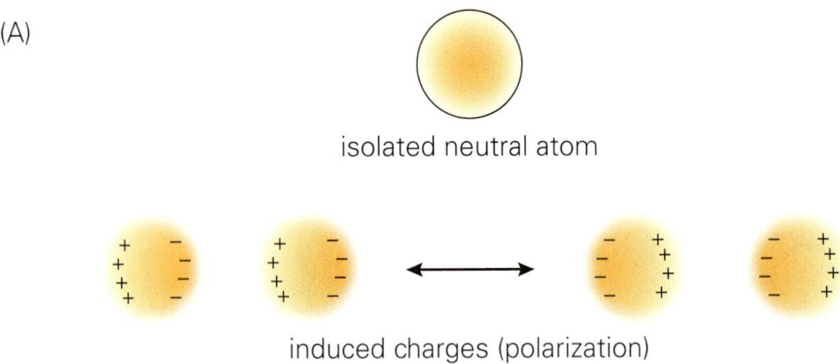


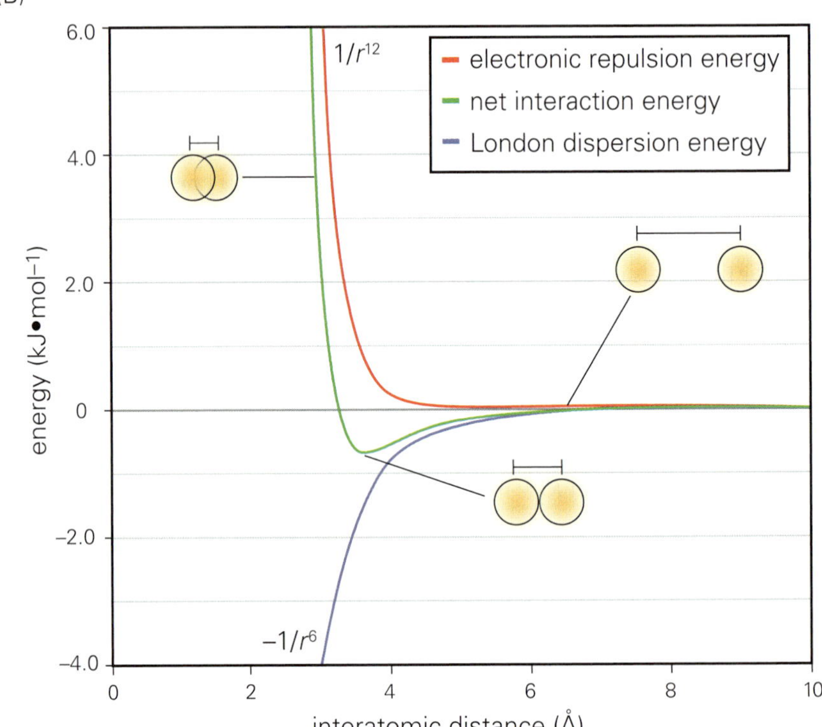


**Figure 6.26 van der Waals interactions between atoms.** (A) The attractive part of the van der Waals interaction is a consequence of induced dipoles in atoms. (B) The diagram shows how the van der Waals interaction energy between two atoms (shown as *yellow* spheres) changes as a function of interatomic distance. The van der Waals energy consists of two terms, a repulsive component that varies as ($1/r^{12}$) and an attractive component that varies at ($-1/r^6$). These two components are plotted as *red* and *blue* lines in the diagram, and the total van der Waals energy is indicated by the *green* line. Note that the energy scale here is ~1000 times smaller than that shown for the bond energy in Figure 6.22.

## 6.16 The van der Waals energy term describes weak attractions and strong repulsions between atoms

As two nonbonded atoms approach each other, the interaction energy between them becomes favorable until it reaches a minimum at a particular distance. This attraction between the two atoms is a consequence of the polarization of the electronic cloud of each atom by the other and is known as the **London dispersion force** or the **van der Waals attraction** (**Figure 6.26**). As the atoms move even closer to each other, the energy begins to rise sharply due to the interpenetration of the electronic shells of the atoms. The dependence of the interaction energy on the distance between two nonbonded atoms is described reasonably well by a function that treats the attractive component as varying as the inverse sixth power of the distance, and the repulsive component as the inverse 12th power of the distance (see Figure 6.26). This interaction term is known as the **van der Waals energy**, $U_{vdw}$, and is shown in Equation 6.42 for two atoms labeled $i$ and $j$:

$$U_{vdw} = \varepsilon_{ij}\left[\left(\frac{R_{\min_{ij}}}{r_{ij}}\right)^{12} - 2\left(\frac{R_{\min_{ij}}}{r_{ij}}\right)^{6}\right] \tag{6.42}$$

The energy function described by Equation 6.42 is also known as the **Lennard-Jones potential**.

The factor $\varepsilon_{ij}$ is the van der Waals stabilization energy or "well depth," which is the reduction in potential energy that occurs when two nonbonded atoms, labeled *i* and *j*, move closer together from an infinite separation until they are at their optimal interatomic separation, $R_{min}$ (see Figure 6.26). When the two atoms are of the same chemical type, $R_{min}/2.0$ is known as the **van der Waals radius** of that particular type of atom (see Table 1.1). The van der Waals radii of different atoms are used in computer graphics programs to visualize how well atoms pack together in protein or nucleic acid structures (**Figure 6.27**).

The maximal van der Waals attraction between two atoms corresponds to a stabilization energy, $\varepsilon_{ij}$, of about 1 kJ•mol$^{-1}$. From considerations of the Boltzmann distribution, we can see that this amount of stabilization is insufficient to hold

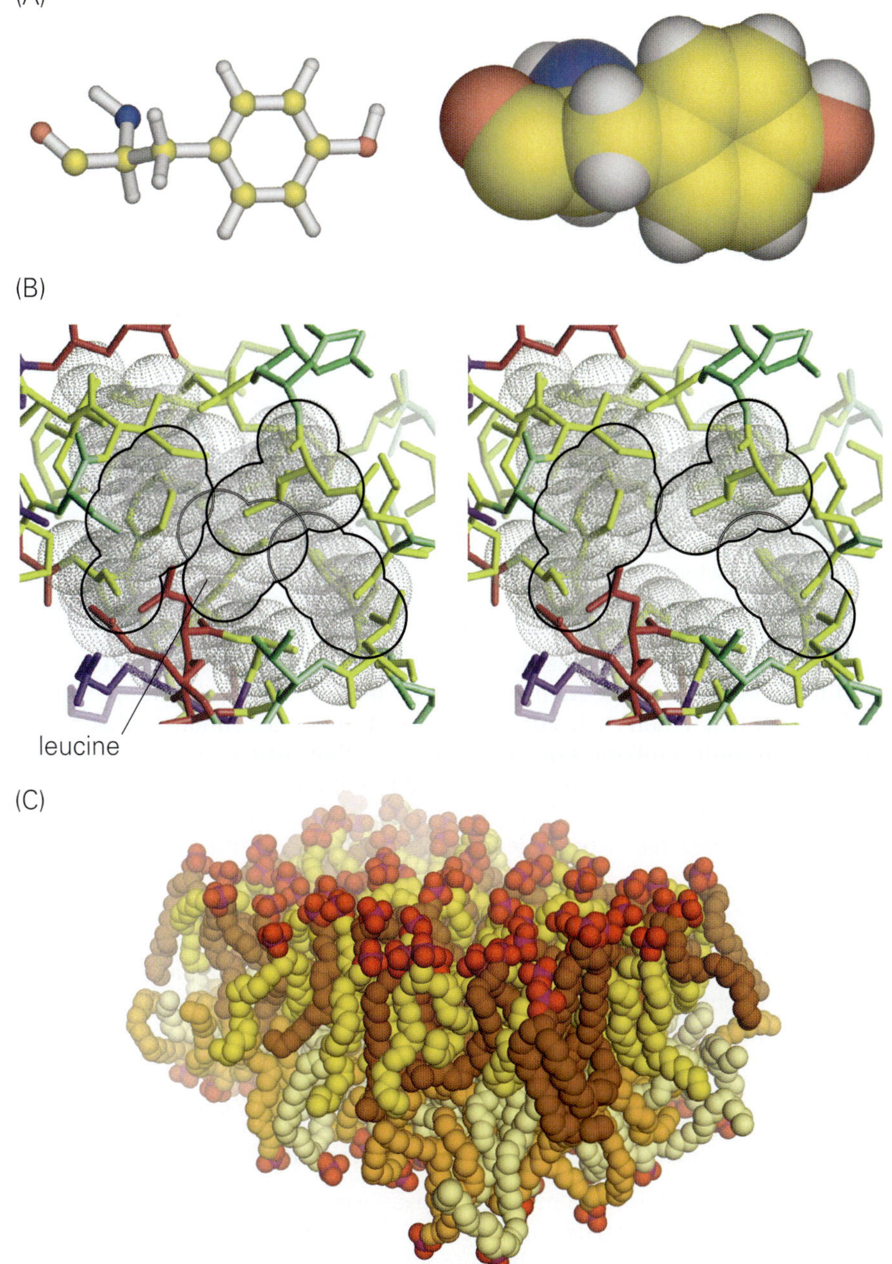


**Figure 6.27 van der Waals interactions.** (A) A tyrosine residue is shown, with spheres with radii corresponding to the van der Waals radii of the atoms. (B) A number of hydrophobic sidechains in the core of a protein are shown in this diagram. Dotted spheres are shown around each atom, and the radii of the spheres correspond to the van der Waals radii of the atoms. The effect of removing a leucine sidechain is shown in the panel on the right. The leucine sidechain and three of its neighbors are outlined in *black*. The atoms in these sidechains are in van der Waals contact. The projection view shown here makes it appear that the sidechains overlap, but in reality they merely touch each other. (C) The structure of a typical lipid bilayer. The phosphate and oxygen atoms of the head groups are shown in *purple* and *red*, respectively. Carbon and hydrogen atoms are shown in various colors, and these are the only atoms that are present in the interior of the bilayer.

two atoms together—thermal fluctuations can easily move them apart ($\varepsilon_{ij} < k_BT$ at room temperature). Any individual van der Waals contact provides only a small degree of stabilization, but van der Waals interactions can add up to provide a significant net stabilization to the structures of proteins, because the interiors of folded proteins contains hundreds of atoms that are very tightly packed.

The importance of van der Waals interactions in stabilizing protein structures can be appreciated by analyzing the effects of replacing residues that have large sidechains, such as leucine, with alanine. If such a mutation is introduced within the hydrophobic core, the protein usually becomes less stable (see Figure 6.27). Most of the reduction in stability is due to a reduction in the hydrophobic effect (as you can see in Figure 4.74, alanine is less hydrophobic than leucine and, consequently, contributes less to the stability of the hydrophobic core). One component of the destabilization does arise, however, from the loss of van der Waals interactions due to the creation of packing defects or cavities around the smaller alanine sidechain in the mutant protein. Sidechains that are adjacent to the cavity participate in fewer van der Waals interactions in the mutant protein than in the wild-type one. The extent to which the creation of an internal cavity destabilizes the protein has been estimated to be about 90 J•mol$^{-1}$ per 1 Å$^3$ of cavity volume. The typical cavity introduced by such a mutation has a volume of 10 to 100 Å$^3$, and so the packing of each leucine sidechain in the hydrophobic core contributes approximately 0.9 to 9 kJ•mol$^{-1}$ in stabilization energy through van der Waals interactions.

The strength of the van der Waals interaction between two atoms depends only on their relative separation and not on their direction of approach. This means that atoms that interact through van der Waals interactions alone are unlikely to be constrained to sharply restricted configurations. Such atoms are able to slide past each other, resulting in "greasy" structures that are flexible. An example of an organized biological structure dominated in its interior by van der Waals interactions is the lipid bilayer (Figures 6.27C and 3.39). Within the central region of the bilayer, the aliphatic chains interact with each other through van der Waals interactions alone. Measurements of molecular motion within the lipid bilayer show that these aliphatic chains move around considerably and are quite fluid in their general properties, as discussed in Section 3.18.

In contrast to membrane lipids, the aliphatic sidechains of amino acid residues inside folded proteins have more restricted flexibility. The key to imposing structural order on the otherwise slippery van der Waals interactions is the electrostatic interactions between charged atoms, which are distributed throughout the protein. The importance of these interactions has been made clear by the studies of artificial proteins with sequences that have been designed to fold up into particular structures. When such a designed protein sequence contains a hydrophobic core with only aliphatic sidechains, the resulting structures are very flexible and not well defined. The formation of a well-defined structure usually involves sidechains with hydrogen-bonding capabilities, which can interact in a more directional manner than is possible with the van der Waals interactions alone.

## 6.17 Atoms in proteins and nucleic acids are partially charged

Most of the atoms in a protein or nucleic acid molecule have an associated electric charge. In chemistry and biochemistry, we usually express charge in terms of the magnitude of the charge on the electron, $e$, which is known as the **elementary charge**. That is, when we say that an atom has a +1 charge, we mean that the positive charge is equivalent in magnitude to that of an electron—namely, $1.602 \times 10^{-19}$ coulombs (C). An atom may be partially charged, which means that it bears a charge that is a fraction of that of an electron (for example, +0.5 $e$ or –0.25 $e$).

The charges on the atoms in a molecule are determined by quantum mechanical calculations of the electron density. The results of such a calculation are shown in **Figure 6.28** for a small molecule that mimics an amino acid. The electron density

**Elementary charge**

The magnitude of the charge on the electron is $1.602 \times 10^{-19}$ coulombs (C). This is known as the elementary charge. The charges on atoms are described in terms of the elementary charge rather than coulombs. Thus, an atom with a single positive charge is said to have a +1 charge, which is understood to be $1.602 \times 10^{-19}$ C.

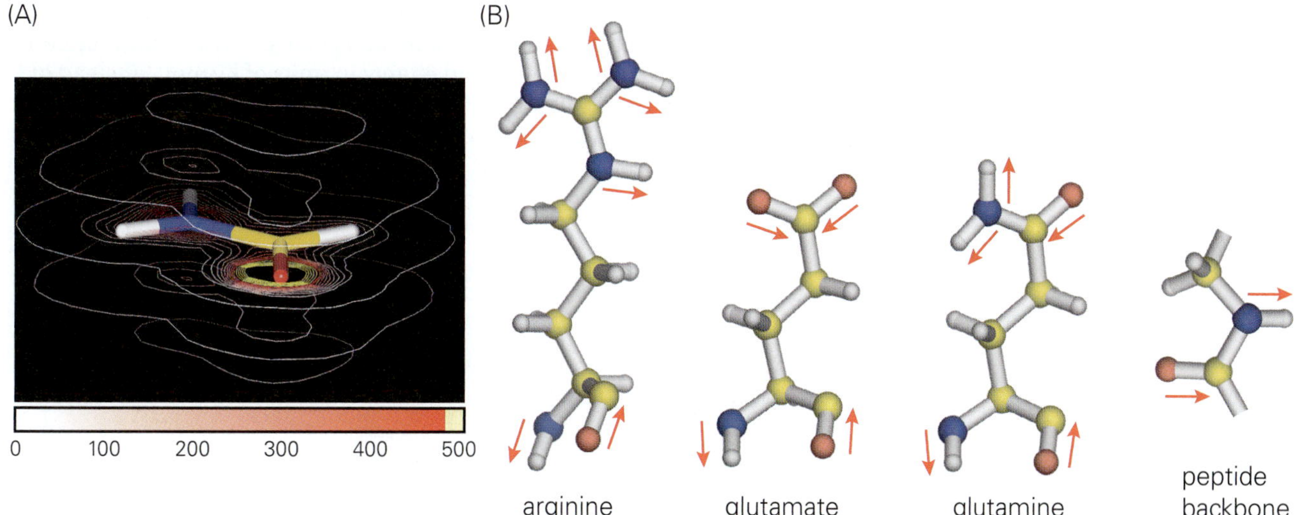


**Figure 6.28 Partial charges on atoms.** (A) The results of quantum mechanical calculations of electron density are shown for a small molecule that mimics an amino acid. The electron density is indicated by contour lines, with regions of excess electron density colored *red* and *yellow*. (B) Partial charges in amino acid residues. The partitioning of charge between adjacent atoms leads to the formation of electrostatic dipoles, indicated by arrows that point from the atom with partial negative charge to the one with partial positive charge. All of the atoms in the molecules have some degree of partial charge on them, but only the largest dipoles are shown. Neutral amino acids, such as glutamine, have no net charge, but they too have partial charges on their atoms. (A, adapted from A.R. Leach, Molecular Modelling: Principles and Applications. Upper Saddle River, NJ: Prentice Hall, 2001.)

is indicated by contour lines, and regions of excess electron density (that is, regions where the electron density is greater than would be expected for a neutral atom) are indicated by red and yellow contours. Notice that the nitrogen and oxygen atoms in the molecule are in regions of excess electron density, and so these atoms behave as if they have negative charge. This is consistent with the greater electronegativity of nitrogen and oxygen compared to the carbon and hydrogen atoms that they are bonded to (the electronegativity scale of the atoms is shown in Table 1.1). This withdrawal of electrons towards the more electronegative atoms is balanced by the development of positive charge on the other atoms.

The results of quantum mechanical calculations, such as this one, are used to determine the values of the partial charges assigned to each atom in amino acid residues and the nucleotides in RNA or DNA. In a charged residue, such as arginine or glutamate, the net charge is distributed over several atoms in the sidechain, with increased negative charge on the more electronegative atoms. This partitioning of charge leads to the formation of small dipoles, which are indicated by the arrows in Figure 6.28. The atoms in the backbone of the amino acid residue also have partial charges, leading to two dipoles, one associated with the carbonyl group, and one with the nitrogen and hydrogen. Neutral amino acids, such as glutamine, have no net charge, but they too have partial charges on their atoms (see Figure 6.28). This polarization results in electron density being concentrated in the vicinity of atoms that are more electronegative than others in the residue, and which consequently have net negative partial charges (see Chapter 2).

## 6.18 Electrostatic interactions are governed by Coulomb's law

The electrostatic energy of interaction, $U_{\text{electrostatic}}$, between two atoms, $i$ and $j$, is described by Coulomb's law:

$$U_{\text{electrostatic}} = \left(\frac{1}{4\pi\varepsilon_0}\right)\left(\frac{q_i q_j}{r_{ij}}\right) \tag{6.43}$$

where $q_i$ and $q_j$ are the charges on the two atoms, and $r_{ij}$ is the distance separating them. The parameter $\varepsilon_0$ is known as the permittivity of vacuum and is given by:

$$\varepsilon_0 = 8.854 \times 10^{-12}\ \text{C}^2 \bullet \text{N}^{-1} \bullet \text{m}^{-2} \tag{6.44}$$

Comparing Equations 6.43 and 6.44 we can see that if the charges are expressed in coulombs (C) and the interatomic distance in meters (m), then Equation 6.43 yields the energy in joules (1 J = 1 N•m). The permittivity of water is ~ 80 times higher than that of vacuum, and the effects of this on the interactions between atoms will be discussed in Section 6.23.

A form of Coulomb's law that is more convenient for biochemical calculations is obtained if we express charge in terms of elementary charges instead of coulombs, distance in Å units instead of meters, and energy in units of kJ•mol$^{-1}$ instead of J. We start by expressing charge in units of the elementary charge, using the fact that an elementary charge is 1.602 × 10$^{-19}$ coulombs:

$$U = \left(\frac{1}{4\pi\varepsilon_0}\right)\left(\frac{q_i q_j}{r_{ij}}\right)\left(1.602\times10^{-19}\right)^2 \text{J}$$
$$= \left(\frac{1}{4\pi\varepsilon_0}\right)\left(\frac{q_i q_j}{r_{ij}}\right)\times 2.57\times10^{-38}\ \text{J} \tag{6.45}$$

For convenience, we have dropped the subscript for $U$ in Equation 6.45. Substituting the value of the vacuum permittivity, $\varepsilon_0$, from Equation 6.44 into Equation 6.45, we get:

$$U = \left(\frac{q_i q_j}{r_{ij}}\right)\times\frac{2.57\times10^{-38}}{111.3\times10^{-12}} = \left(\frac{q_i q_j}{r_{ij}}\right)\times 0.0231\times10^{-26}\,\text{J} \tag{6.46}$$

Expressing distance in Å units rather than meters, and using the fact that 1 Å = 10$^{-10}$ meters, we get:

$$U = \left(\frac{q_i q_j}{r_{ij}}\right)\times 0.0231\times10^{-26}\times10^{10} = \left(\frac{q_i q_j}{r_{ij}}\right)\times 0.0231\times10^{-16}\ \text{J} \tag{6.47}$$

The energy in Equation 6.47 is in units of joules (J). If we wish to express the energy in units of kJ•mol$^{-1}$, instead, Equation 6.47 becomes:

$$U = \left(\frac{q_i q_j}{r_{ij}}\right)\times 0.0231\times10^{-16}\times 6.022\times10^{23}\times10^{-3}$$
$$= \left(\frac{q_i q_j}{r_{ij}}\right)\times 1391\ \text{kJ}\bullet\text{mol}^{-1} \tag{6.48}$$

In Equation 6.48, we have multiplied the right-hand side by Avogadro's number (6.022 × 10$^{23}$) in order to convert to units of J•mol$^{-1}$, and then by 10$^{-3}$ to convert J•mol$^{-1}$ to kJ•mol$^{-1}$. In Equation 6.48, the two charges are expressed in terms of elementary charges, and the interatomic distance is in Å units.

To see how to use this equation, consider an ion with a unit positive charge that is 4.0 Å from an ion with a unit negative charge. We use Equation 6.48 to calculate the energy as follows:

$$U = \frac{-1\times1}{4}\times1391 = -347.8\ \text{kJ}\bullet\text{mol}^{-1} \tag{6.49}$$

This is a very strong energy of interaction, much greater than the value of $k_BT$ at room temperature (~2.5 kJ•mol$^{-1}$). This energy is calculated for the two charges interacting in a vacuum. As discussed in Section 6.23, water weakens the strength of electrostatic interactions by about 80-fold. Thus, the same two ions interacting in water will have an interaction energy of only ~–4.34 kJ•mol$^{-1}$, which is less than $2k_BT$. Thus, interactions between charges are easily broken in water.

The strongest electrostatic interactions arise between atoms in residues that have a net charge, because atoms in these residues bear the largest individual charges (**Figure 6.29**). At neutral pH, lysine and arginine residues each have a net positive charge of +1, and aspartate and glutamate are negatively charged (–1). Histidine, which has a p$K_a$ near 7, may be positively charged or neutral, depending on its environment and the pH of the solution (the concept of the p$K_a$ of an amino acid sidechain is explained in Section 10.17). Interactions between oppositely charged

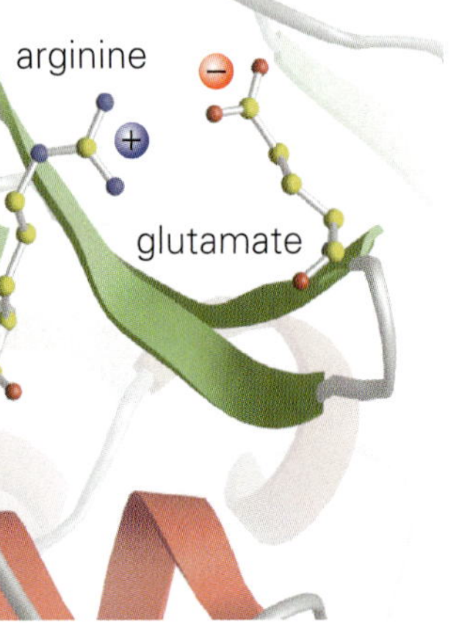


**Figure 6.29 Ion-pairing interactions.** When oppositely charged residues interact closely, as shown here for sidechains on the surface of a protein, they are said to form an ion pair, or a salt bridge.

sidechains are known as ionic interactions, and the particularly favorable interactions between oppositely charged sidechains are referred to as **ion-pairing interactions** or **salt bridges** (see Figure 6.29).

Electrostatic interactions may also be important for neutral residues because of partial charges on the atoms, or because of polarization of the electron density that can occur for any atom in amino acid residues or nucleotides (see Figure 6.28).

## 6.19 Hydrogen bonds are an important class of electrostatic interactions

According to Coulomb's law, the electrostatic interaction between two charged atoms falls off slowly with distance (that is, with a $1/r$ dependence, where $r$ is the distance between the charges, as shown in **Figure 6.30**). In contrast, when clusters of atoms that have a total charge of zero interact with each other, the total energy for the interaction between the clusters falls off much more quickly with distance than $1/r$. Since the net charge on each cluster is zero, at large distances the interactions between individual charged atoms in the two clusters cancel out, leaving no net attraction. When the clusters are closer together, the distances between various pairs of atoms are not the same, and there is a net electrostatic interaction. For example, the energy of interaction between favorably aligned dipoles falls off as $-1/r^3$, where $r$ is the distance between the interacting dipoles. This function decays to zero much more quickly than the $-1/r$ fall-off in the interaction energy between two opposite charges (see Figure 6.30).

The interaction energy for two dipoles depends strongly on their relative orientation. The most favorable interaction occurs when the dipoles approach each other co-linearly, with the positive pole of one dipole pointing towards the negative pole of the other one. If the orientation of one of the dipoles is flipped, this attractive interaction is converted to a repulsive one. The strongly directional nature of the short-range interactions between the dipoles or larger clusters of partially charged atoms provides the structural glue that holds specific protein conformations in place.

A particularly important electrostatic interaction in biology occurs between oxygen or nitrogen atoms bearing hydrogen (known as the donor atoms), and oxygen or nitrogen atoms bearing lone pairs of electrons (the acceptor atoms, **Figure 6.31**). The hydrogen atom "bridges" the donor and acceptor atoms, and such an interaction is known as a **hydrogen bond** (see Section 1.4). The stabilization that results from the formation of a hydrogen bond can be understood in terms of electrostatics. The hydrogen attached to each of the donor atoms has a small net positive charge and the more electronegative donor atom has a small net negative charge, forming a dipole (see Figure 6.31A). Likewise, the acceptor atom bearing

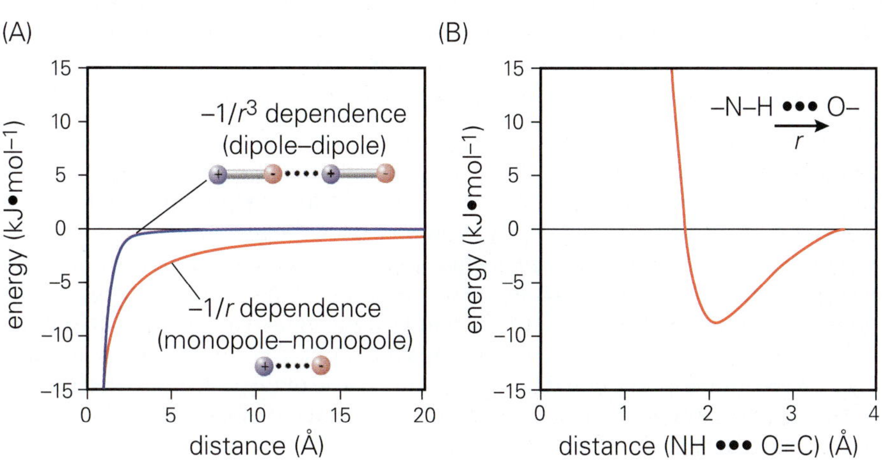


**Figure 6.30 Electrostatic energy and hydrogen bonds.** (A) The dependence of electrostatic energy on distance. The interactions between two opposite charges falls off with distance as ($-1/r$), where $r$ is the distance between the charges. The interaction between two pairs of charges (dipoles) falls off much more quickly with distance and has a ($-1/r^3$) dependence. (B) The change in energy as the distance between the hydrogen and oxygen atoms in a NH···O=C hydrogen bond is increased. The energy shown here is the sum of the van der Waals and electrostatic energy terms.

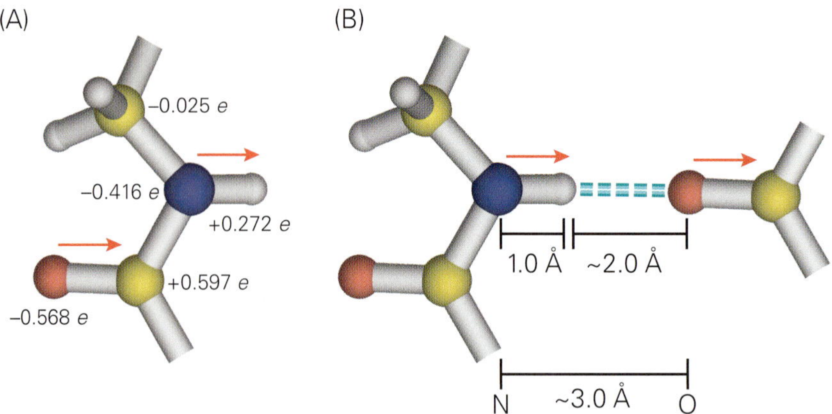


**Figure 6.31 A hydrogen bond between two peptide groups.** (A) Partial charges on the atoms in the backbone of a peptide, shown as multiples of an elementary charge, *e*. (B) The amide (N–H) and carbonyl (C=O) groups in the backbone interact by forming hydrogen bonds that align the dipoles (indicated by arrows that point towards the positively charged atom in the pair). The formation of this kind of backbone hydrogen bond underlies the stability of α helices and β sheets in proteins.

the lone pair of electrons has a small net negative charge and the atom that it is covalently bonded to has a small net positive charge, forming another dipole (see Figure 6.31B). In this simplest kind of hydrogen bond, the electrostatic interaction between two groups of atoms can be described as a dipole–dipole interaction, embedded within the van der Waals interactions that determine the distance of closest approach between the various atoms. Particularly tight hydrogen bonding interactions have the hydrogen and oxygen atoms separated by about 2 Å.

As we have seen already, hydrogen bonds are very important in determining the conformations of proteins and nucleic acids. In proteins, the amide NH and carbonyl C=O groups of the polypeptide backbone interact with each other to form arrays of hydrogen bonds in α helices and β sheets (**Figure 6.32**; see also Figures 1.36–1.38). The minimum energy configuration of interacting dipoles is one in which the dipoles are co-linear. α helices and β sheets are arrays of dipoles and, in particularly stable and well-formed elements of secondary structure, the N–H··· O–C interaction tends to be linear. However, hydrogen-bonding geometry is not sharply restricted, as can be seen in the variation of hydrogen-bonding angles in the parallel β sheet shown in Figure 6.32C.

**Figure 6.33** shows the results of a survey of high-resolution protein structures, indicating the directional tendencies of the hydrogen bonds formed by protein sidechains. As for the backbone hydrogen bonds, the spread of orientations in the alignments of the hydrogen-bonding groups in these sidechains shows that these interactions are not tightly locked into completely stiff geometrical constraints—rather, hydrogen bonds resemble the hooks and eyes of Velcro straps, with a degree of individual flexibility but accumulating cooperative strength at interfaces.

If you examine Figure 6.33 carefully you will see that the simple dipole-based model for hydrogen bonding that we have used in this chapter does not explain some of the structural details of the geometries of hydrogen bonds that are found in proteins. Take, for example, the location of hydrogen bond donors around the oxygen atoms of the sidechains of residues such as Asp or Asn. Instead of lying along the direction defined by the C–O bond, as we might expect based on dipole-dipole interactions, you can see that groups that donate hydrogen bonds to these oxygens are found in two clusters, one on either side of the C–O bond vector. This is because the orbitals associated with the lone pair of electrons of the oxygen atoms point in these directions.

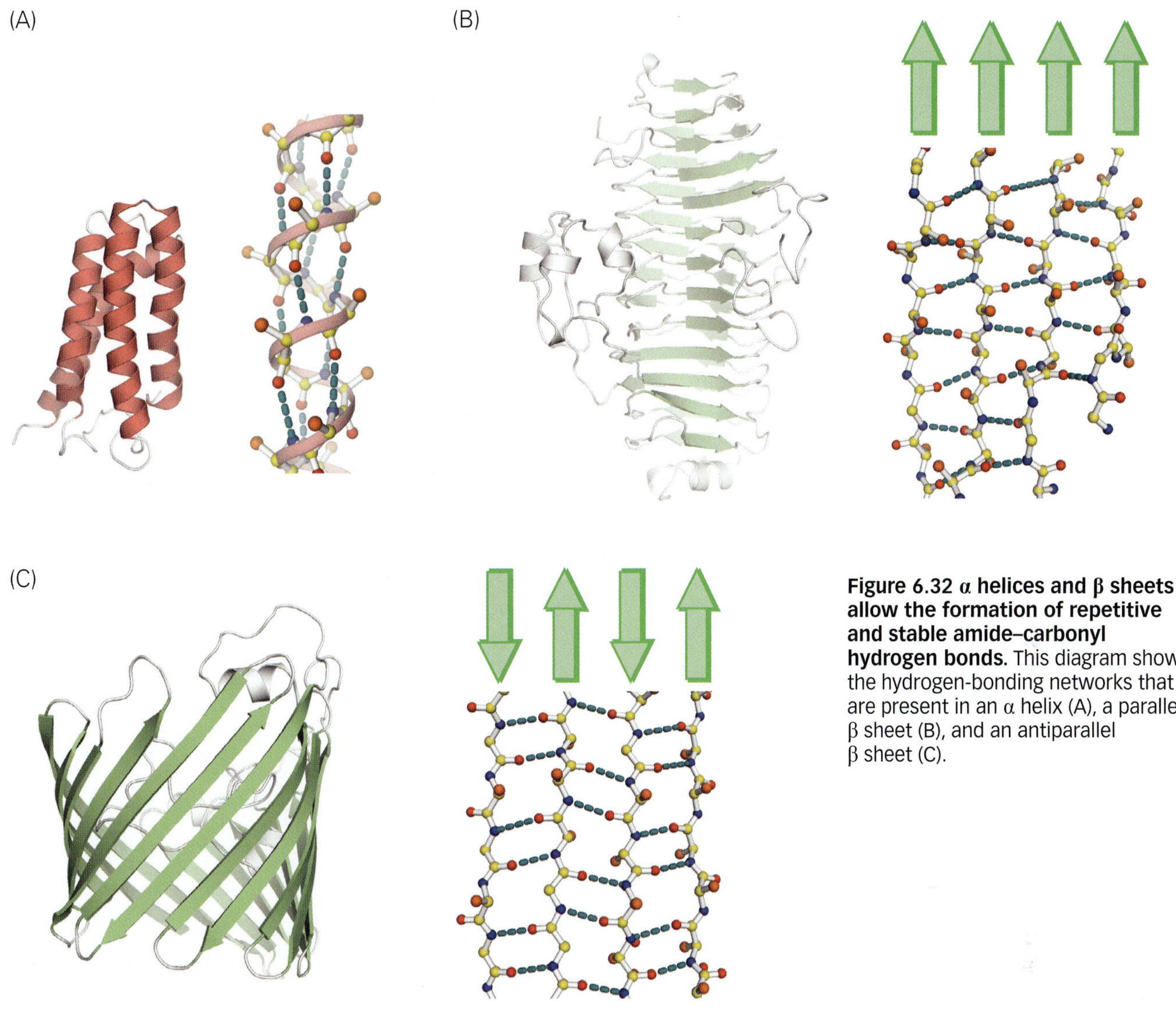


**Figure 6.32 α helices and β sheets allow the formation of repetitive and stable amide–carbonyl hydrogen bonds.** This diagram shows the hydrogen-bonding networks that are present in an α helix (A), a parallel β sheet (B), and an antiparallel β sheet (C).

## 6.20 Empirical energy functions are used in computer programs to calculate molecular energies

Given a three-dimensional structure for a protein or nucleic acid molecule, empirical energy functions can be used to calculate the energy of the molecule. There are two key assumptions that are made when doing these calculations: (1) the component energy terms are assumed to be **additive** and (2) the values of the parameters, such as optimal bond lengths, force constants, and partial charges, are assumed to be **transferable** from one macromolecule to another.

By saying that the energy is additive, we mean that the total energy for a configuration of atoms in a molecule is simply the sum of the component energy terms. For example, consider three atoms, A, B, and C, that are covalently bonded as shown in **Figure 6.34**, and that interact noncovalently with a fourth atom, D. To calculate the total energy, we simply add up the component energy terms (see Figure 6.34), treating each component as if the other components were not present. What this means is that we assume, for example, that the strength of the van der Waals interaction between the A and D atoms is not influenced by the positions of the B and C atoms. This is a reasonable assumption for the van der Waals energy term, but can introduce errors for the electrostatic energy. In the latter case, the polarization of

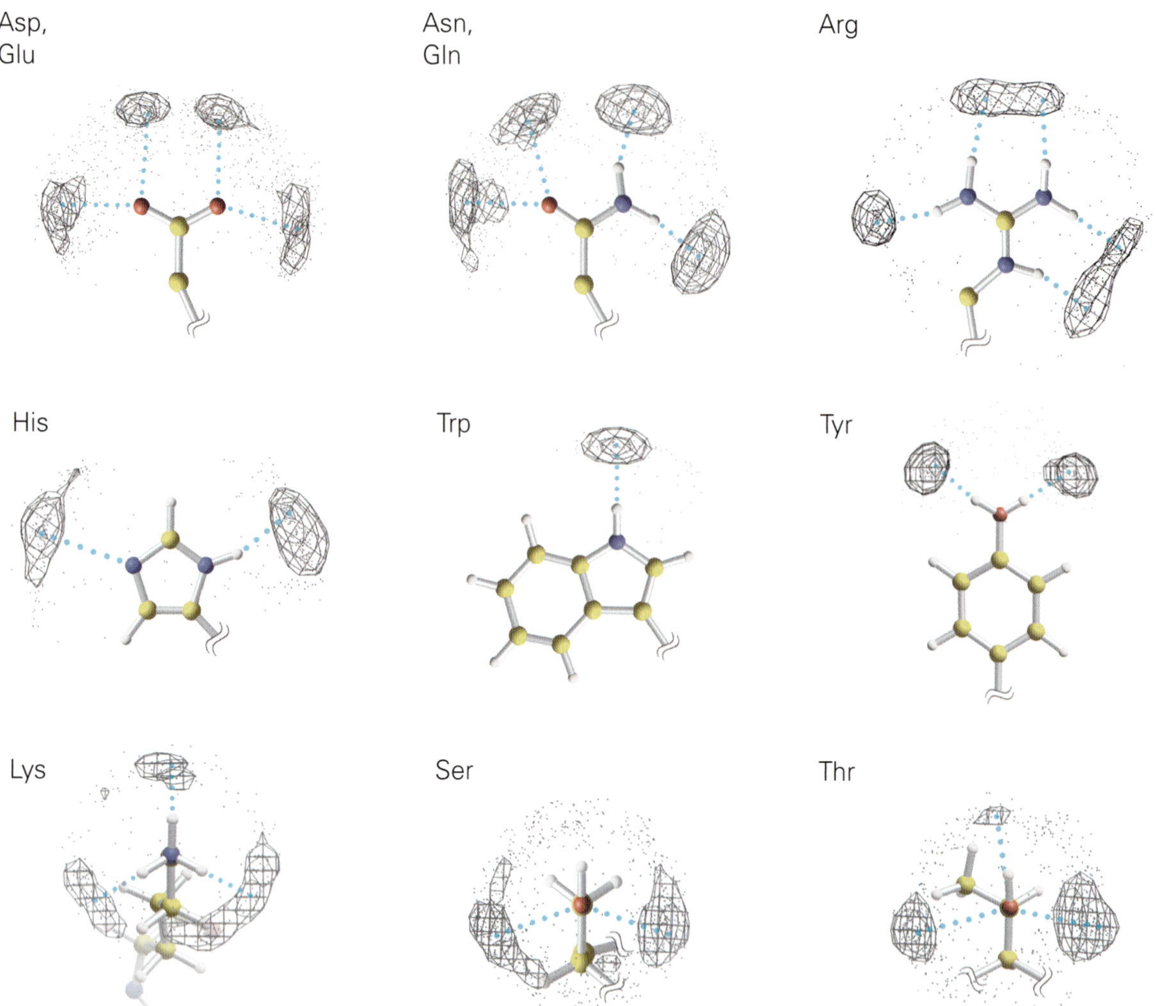


**Figure 6.33 The hydrogen-bonding capabilities of protein sidechains.** The sidechains that can form hydrogen bonds are shown in each of the panels of the figure. For each sidechain, the location of the nitrogen or oxygen atom in a hydrogen-bonding partner was determined from a large number of protein structures in the protein databank, and these locations are indicated by dots. Regions of high density of partner atoms are encompassed by meshes, and these represent regions of most favorable hydrogen-bond formation. For Tyr there are two equivalent positions for the hydroxyl proton, both of which are indicated. For Ser and Thr the $C_\beta$-$O_\gamma$ bond is pointed directly out of the plane of the page, and the hydroxyl proton is pointed up (the other protons are from $C_\beta$). (Adapted from J.A. Ippolito et al., and D. Christianson, *J. Mol. Biol*. 215(3): 457–471, 1990.)

atoms can be affected by the presence of other atoms, thereby changing the partial charges of the atoms and the electrostatic energy. Despite this complication, the assumption of additivity is still necessitated by limitations imposed by the speed of present-day computers. The calculation of energy would simply take too long if we had to recalculate the atomic polarization for each configuration of atoms.

The assumption of parameter transferability is another simplification that is introduced in order to make the energy calculation computationally tractable. We assume that parameters in the energy function can be defined once for all the component amino acids and nucleic acids, and that these parameters can be transferred without modification for use in energy calculations for any particular protein or oligonucleotide. For example, the force constant ($K$) and the optimal bond length ($r_0$) for the $C_\alpha$–$C_\beta$ bond in a leucine residue is assumed to be the same for all leucine residues in all proteins (see Equation 6.36). This is likely to be a good assumption for all energy terms, except for electrostatics once again, because polarization effects can make the transferability of electrostatic parameters problematic.

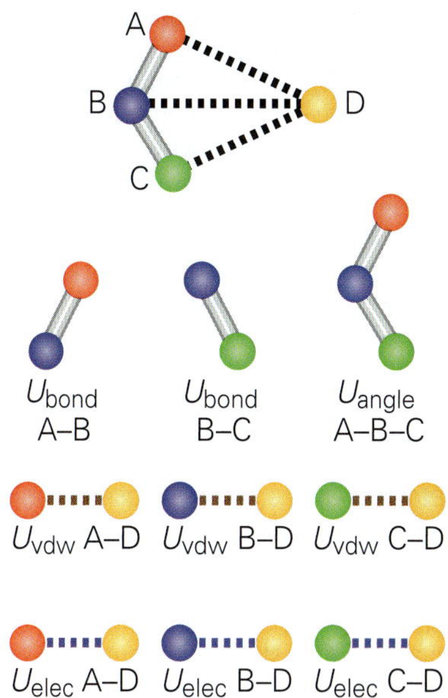


**Figure 6.34 The additivity of energetic terms.** A simplification that is made in order to calculate molecular energies quickly is to assume that the individual energy terms are additive. Shown here is the interaction of three covalently bonded atoms, A, B and C, that interact noncovalently with a fourth atom, D. The total energy of the interacting system is the sum of the individual energy terms, which are shown below.

## 6.21 Interactions with water weaken the effective strengths of hydrogen bonds in proteins

The discussion so far has centered on an isolated protein molecule, as if it were present in a vacuum. This is never the case, and realistic calculations of molecular energies need to account for the fact that the protein molecule is surrounded by water molecules and salt ions. One important consequence of the presence of water molecules is the reduction in the apparent strength of hydrogen bonds and ion-pairing interactions.

Measurement of the energies of the hydrogen bonds that form between small molecules in the gas phase indicates that, for atoms that have no net charge, the stabilization energy of hydrogen bonds is about 10 to 20 kJ•mol$^{-1}$. Hydrogen bonds in which one or more of the interacting partners is charged are stronger, with stabilization energies as high as 40 kJ•mol$^{-1}$. These values for the stabilization energy of the hydrogen bond are much larger than $k_BT$ (~2.5 kJ•mol$^{-1}$ at 300 K), and so we would expect such hydrogen bonds to be rarely broken at room temperature. Figure 6.30 shows the variation in energy as the distance between the N-H···O=C hydrogen-bonded groups is increased for two peptide units. The figure shows that the stabilization energy for this hydrogen bond is ~10 kJ•mol$^{-1}$.

We know, however, that most hydrogen bonds in a protein are constantly in a state of flux, as the protein molecule folds and undergoes structural fluctuations. This must mean that the stabilization energies of hydrogen bonds are comparable to $k_BT$. Indeed, measurements of the hydrogen-bond strengths in proteins show that, when the hydrogen-bonding partners are neutral, the net stabilization energy gained by forming a hydrogen bond is only in the range of ~2 to ~4 kJ•mol$^{-1}$. Hydrogen bonds in which one or more of the partners is charged are also weaker in proteins than between small molecules in the gas phase, contributing ~4 to ~8 kJ•mol$^{-1}$ of stabilization energy.

The crucial difference between the strengths of hydrogen bonds in proteins and those formed by small molecules in the gas phase is that the energy of hydrogen-bond formation in proteins is strongly influenced by water (**Figure 6.35**). Water is particularly good at forming hydrogen bonds with the backbone and sidechains of the amino acid residues in a protein. The change in energy when a hydrogen bond within a protein is broken is, consequently, a result of the difference in energy between the internal hydrogen bond and the hydrogen bonds that the protein residue can make with water. This results in the relatively small net energies associated with the formation of hydrogen bonds in proteins.

We can gain some appreciation for the effect of water on hydrogen-bond strengths by studying the behavior of a small organic compound, acetamide (see Figure 6.35C). Acetamide contains hydrogen-bonding groups that are similar to those found in proteins. When acetamide is dissolved in chloroform, a nonpolar solvent that cannot form hydrogen bonds, it can only form hydrogen bonds with itself. In this case, the net enthalpic stabilization of these hydrogen bonds has been measured experimentally to be –17 kJ•mol$^{-1}$. When acetamide is dissolved in water, with which it can form hydrogen bonds, the net enthalpic stabilization due to hydrogen bonds between molecules of acetamide becomes essentially zero. This is explained by the fact that dimer formation by acetamide now requires breaking hydrogen bonds to water (see Figure 6.35C).

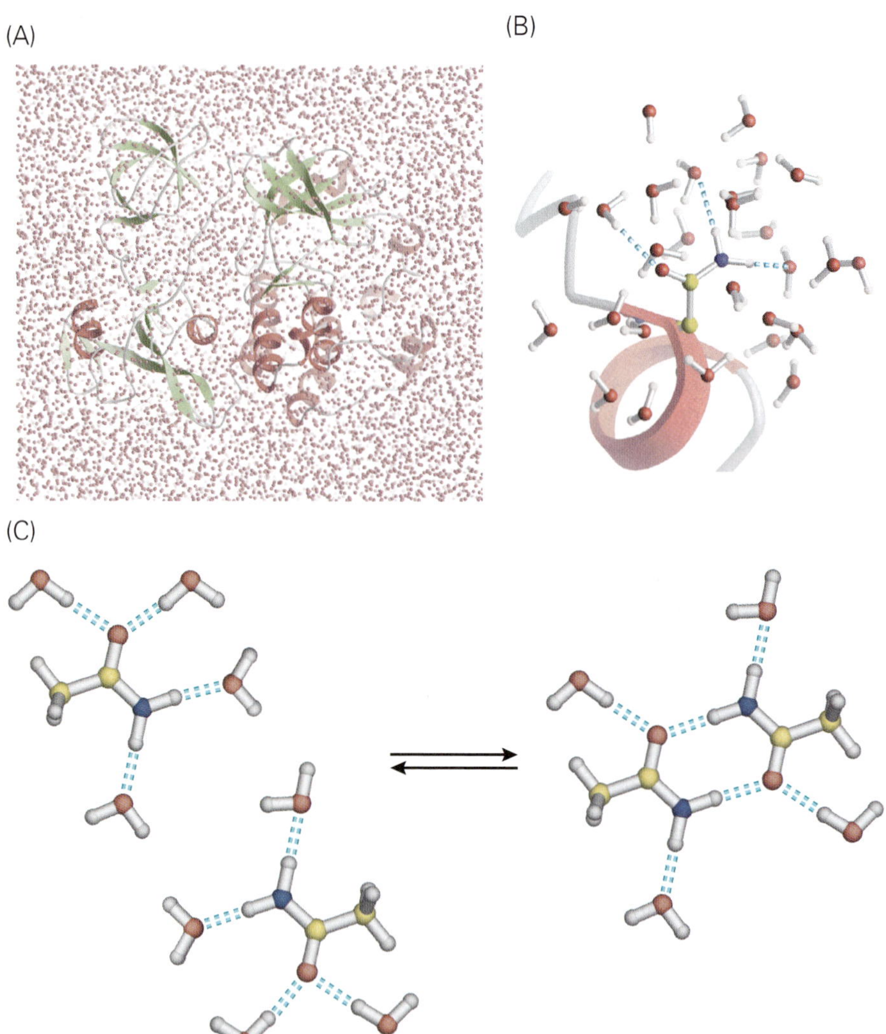


**Figure 6.35 Protein molecules are surrounded by water.** (A) The oxygen atoms of water molecules surrounding a protein are shown as *red* spheres. These water molecules are not easily observable by experimental methods because they are constantly moving around. The picture shown here was generated using a computer simulation of the structure of water around the protein. (B) Polar sidechains on the surface of the protein form hydrogen bonds with the water. An asparagine sidechain is shown, along with the water molecules that are close to it. Hydrogen bonds between the sidechain and water are shown by dotted lines. This picture is an instantaneous snapshot of the dynamic water structure around the protein. (C) Two molecules of acetamide are shown interacting with water, but not with each other. The picture on the right shows the two molecules forming hydrogen bonds with each other, which results in the loss of hydrogen bonds to water. This weakens the effective strength of the hydrogen bonds between molecules of acetamide.

## 6.22 The presence of hydrogen-bonding groups in a protein is important for solubility and specificity

If the competition with water makes hydrogen bonds only marginally stabilizing, why does the protein have so many hydrogen-bonding groups placed throughout its structure? One reason is that the polar groups of the protein provide for solubility and specificity. Were a protein to lack polar groups, it would be completely insoluble in water and would aggregate rather than fold. In addition, a polymer chain composed only of hydrophobic atoms would be unlikely to fold into a specific structure because the nondirectional van der Waals interactions would make it difficult to discriminate one condensed conformation of the chain from another. The constraints on conformation that are imposed by the need to maintain hydrogen-bonding complementarity is probably the single most important factor that determines the ability of protein chains to fold up into specific and well-defined three-dimensional structures.

Although the tendency of water to compete for hydrogen bonds with the protein ends up reducing the stability of the folded protein structure, this competition with water is crucial for protein function. As we shall see throughout this book, the ability of protein molecules to change their conformations in response to external cues is a necessary aspect of their function. Water molecules allow hydrogen bonds to be broken readily and remade without major expenditures of energy. Had the hydrogen bonds retained the full strength that is latent in them, then protein molecules would be rigidly locked into particular conformations and would be functionally inert.

Perhaps the most important conclusion to be drawn from the analysis of hydrogen bonding in proteins is that it is not so much the energy associated with a particular hydrogen bond that is important, as much as the fact that the hydrogen bond is made. If a sidechain with hydrogen-bonding capability, such as a tyrosine sidechain, were to be folded into the protein structure in such a way that the sidechain could not form a hydrogen bond, then the folded protein would be destabilized by the withdrawal of such a sidechain away from its interactions with water. However, by providing compensating hydrogen bonds within the folded structure, the protein can maintain the balance of interactions. The secondary structural elements of the protein are particularly good at satisfying the hydrogen-bonding requirements of the backbone. Whenever polar sidechains are found inside the protein structure, they are almost invariably associated with one or more suitable hydrogen-bonding partners.

Amino acid sidechains that bear a full charge interact particularly strongly with water, even when they form an ion-pairing interaction with another sidechain (see Figure 6.29 for an illustration of an ion pair). Removal of a charged sidechain from water consequently has a large energetic penalty, known as the **desolvation energy**. The desolvation energy of charged sidechains is so high that they are rarely found in the interiors of proteins.

## 6.23 The water surrounding protein molecules strongly influences electrostatic interactions

Water molecules are polarizable, and they respond to the presence of charged atoms near them. The presence of adjacent charges increases the dipole moment of water so as to balance the charge. This tends to reduce the electrostatic interaction energy between two charges in water. The attenuation of the interaction energy between two charges in a polarizable medium, such as water, is accounted for by modifying Coulomb's law. If the polarizability of the medium is uniform throughout space, then the energy of interaction between two charges is attenuated by a factor, ε, known as the **dielectric constant** of the medium. This leads to the following modified form of Coulomb's law:

$$U = \frac{1}{\varepsilon}\left(\frac{q_i q_j}{r_{ij}\text{Å}}\right) \times 1390 \text{ kJ}\bullet\text{mol}^{-1} \tag{6.50}$$

Equation 6.50 is derived from Equation 6.48 (Coulomb's law for charges interacting in a vacuum, with $q_i$ and $q_j$ being elementary charges, and the interatomic distance being expressed in Å units). This modified form of Coulomb's law can be used to calculate the interaction between two charges in water if we ignore the molecular details of the structure of water and simply assume that the strength of the interaction between the charges is reduced by an appropriate amount. The effective dielectric constant of water is 80, and this relatively large value reflects the ability of water to attenuate electrostatic interactions strongly. The effective interaction energy between the two charges in water is then given by Coulomb's law, with the dielectric constant set to 80 (**Figure 6.36**).

Two charged atoms in water do not actually see a medium of uniform dielectric around them, but are instead surrounded by discrete molecules of water (see Figure 6.36B). Each charged atom in water is surrounded and stabilized by the dipoles of water, and consequently interacts much less strongly with other charged atoms in the solution. The structure of the surrounding water fluctuates constantly due to thermal collisions (see Figure 6.36C) and, if we average over a long time and over many configurations of the water molecules, we find that the effective or averaged interaction energy between the two charges is reduced by a factor of 80 (the effective dielectric constant of water) relative to their interaction in a vacuum.

**Dielectric constant**

The dielectric constant, denoted ε, is a scale factor that reduces the magnitude of the electrostatic energy as calculated using Coulomb's law. The dielectric constant accounts for the effect of the environment in weakening the interaction between charges. The dielectric constant of bulk water is 80. The interior of a protein, which is slightly polar, has a much smaller dielectric constant (~2).

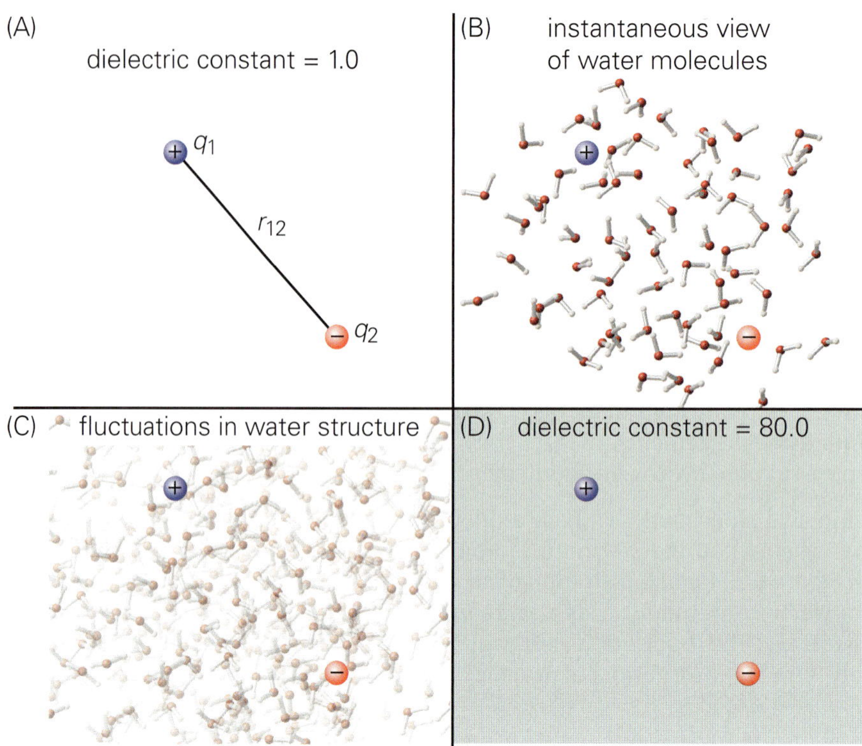


**Figure 6.36 The electrostatic interaction between charged atoms that are surrounded by water molecules is attenuated.** In (A), two ions, one positively charged (*blue*) and one negatively charged (*red*) are shown in a vacuum. The electrostatic energy is given by Coulomb's law. (B) The two ions are shown surrounded by water molecules. For this configuration of waters and ions, we can still calculate the energy using Coulomb's law, but we need to sum over all the atoms in the system, including the oxygen and hydrogen atoms of the water molecules. (C) All the atoms in the system are continually moving, and the net electrostatic energy of the system is the average over all instantaneous configurations. Over time, the effect of the water molecules on the ions is blurred out, and the detailed structure of water becomes less important. (D) We can approximate the effect of water molecules on the interactions between the ions by reducing the strength of the electrostatic energy by a factor known as the dielectric constant. This affords us a great simplification, because we are back to considering just the positions of the two ions.

When the molecular environment surrounding two interacting charges is less polarizable than water, the attenuation of the electrostatic interaction is correspondingly smaller. For example, a charge inside a protein molecule is surrounded by many chemical groups that are not very polarizable (aliphatic groups in sidechains, for example), and by some groups that are fairly polarizable (such as the amide and carbonyl groups of the backbone). In contrast to water, which has a very dynamic structure and can readily reorient to interact with charges, atoms in the interior of a protein are relatively rigid and are therefore limited in their ability to attenuate electrostatic interactions. The effective dielectric constant for the interior of a protein molecule turns out to be very low and is between 2 and 4, depending on the particular environment.

An electrostatic calculation that approximates the effect of polarizable groups by a choice of suitable dielectric constants for the medium is referred to as a **continuum dielectric model** for electrostatics. The application of continuum dielectric models to the consideration of electrostatic effects in proteins is a very powerful tool in understanding protein function, because it allows us to visualize effects that would be otherwise impractical to calculate if we had to account for each individual water molecule. Without the use of the continuum dielectric approximation, we would have to average the electrostatic energies over a large number of water configurations, which would make such calculations extremely time consuming.

In the continuum dielectric approach to electrostatics, individual solvent molecules are replaced by a region of uniform dielectric constant outside the protein, which represents the averaged effects of the solvent. The calculation of electrostatic effects in a protein involves the consideration of at least two regions with different dielectric constants (**Figure 6.37**). The region outside the protein, corresponding to the bulk solvent, has an effective dielectric constant of 80, that of water. Inside the protein, the structural environment is quite inhomogeneous, but in the simplest approximation this is treated as a region with a uniformly low dielectric constant (typically this is set to a value between 2 and 4). Because of the nonuniformity of the dielectric medium, Coulomb's law is no longer applicable.

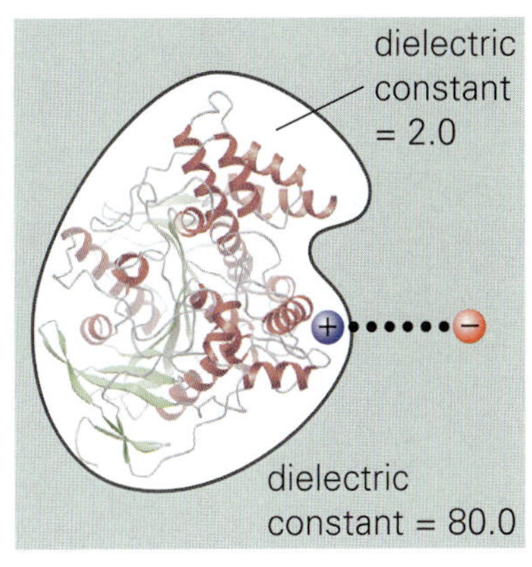


**Figure 6.37 A protein surrounded by water creates dielectric boundaries.** The difficulty in calculating electrostatic energies for a protein molecule immersed in water is illustrated. The region within the protein (colored *white*) has a low dielectric constant ($\varepsilon = 2$). The region outside the protein, filled with water molecules (not shown), has a high dielectric constant ($\varepsilon = 80$). Consider two charged atoms, one inside the protein (*blue*) and one outside (*red*). The interaction energy between them cannot be calculated using Coulomb's formula, because the space between them has different dielectric constants in different regions. Instead, the Poisson equation has to be solved in order to calculate the spatial distribution of the electrostatic potential.

Instead, the fundamental equation describing the electrostatic potential generated by the set of charges in the protein is a differential equation known as the **Poisson equation**, which we shall not discuss in detail.

The solution to the Poisson equation relates the spatial distribution of charges within the protein to the electrostatic potential, $\phi(\vec{r})$, at any point in space, $\vec{r}$. The value of the electrostatic potential at any point is the energy required to move a unit positive charge from infinity to that point (this concept is discussed in more detail in Section 11.10). Solution of the Poisson equation yields a three-dimensional map of the electrostatic potential as a function of position in space (**Figure 6.38**). This map allows us to calculate the interaction energy between charged atoms and the protein.

Protein molecules normally function in environments with appreciable ionic strength, corresponding typically to a concentration of 100–150 mM NaCl. The effect of ionic strength on the electrostatic potential generated by the protein can be calculated by using a modified form of the Poisson equation, known as the **Poisson–Boltzmann equation**. The Poisson–Boltzmann equation is derived by assuming that the ions in the solution are distributed throughout the solvent region as given by a Boltzmann distribution based on the electrostatic potential. The Poisson–Boltzmann equation takes account of the fact that the ions will be distributed more towards regions of electrostatic potential that are energetically favorable, and this results in a screening of electrostatic effects by the ions.

## 6.24 The shapes of proteins change the electrostatic fields generated by charges within the protein

The results of solving the Poisson–Boltzmann equation depend on the shape of the protein under consideration. The shape of the protein determines the shape

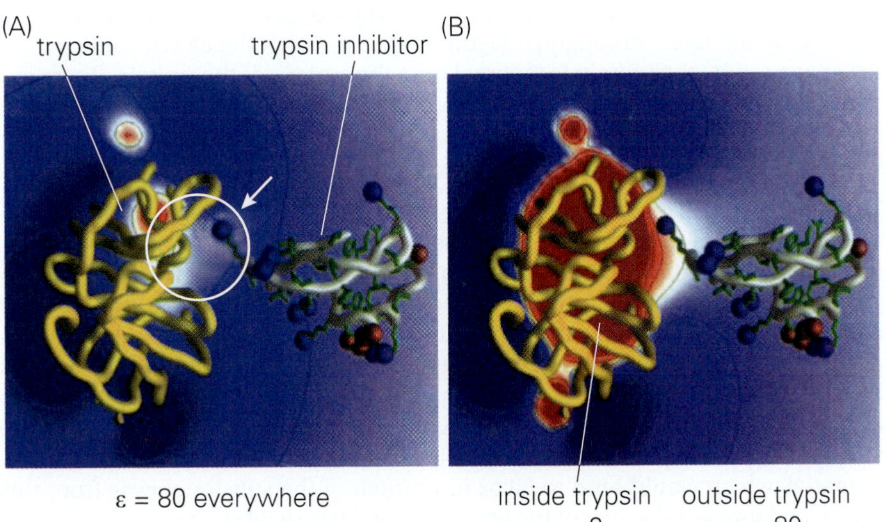


**Figure 6.38 The electrostatic potential generated by a protein.** (A) The structure of trypsin, a protease enzyme, is shown in *yellow*. The electrostatic potential generated by the charges in trypsin is illustrated in color for a plane passing through the active site of trypsin (*white* circle). *Blue* and *red* indicate regions of positive and negative electrostatic potential, respectively. The dielectric constant is assumed to be 80 throughout space. An inhibitor protein that binds to trypsin in also shown, slightly separated from it. The inhibitor protein was not included in the electrostatic calculation. A lysine residue in the inhibitor binds to trypsin (*white* arrow). The positively charged lysine residue is in a region of positive electrostatic potential, which would not favor binding to trypsin. (B) The electrostatic calculation was repeated, but this time the dielectric constant was set to 2 inside trypsin and 80 outside. Even though the charges that generate the electrostatic potential are the same as in (A), the map of the potential looks very different. Note that the active site of trypsin is now in a region of negative electrostatic potential, which favors binding of the inhibitor. (From B. Honig and A. Nicholls, *Science* 268: 1144–1149, 1995. With permission from the AAAS.)

of the region of low dielectric, which in turn results in the generation of features in the map of electrostatic potentials that are not apparent from consideration of the three-dimensional structure of the protein alone. This feature is illustrated in Figure 6.38, which shows maps of the electrostatic potential around trypsin, a protease enzyme.

The electrostatic potential is calculated in two ways. First, the dielectric constant is assumed to be 80 everywhere in space. In this case, the electrostatic potential at any point is given by Coulomb's law, and it depends only on the location of the charges in the trypsin molecule. Trypsin has more positively charged residues than negatively charged ones. As a result, the electrostatic potential is generally positive, even within the active site. This would hinder the binding of a protein that inhibits trypsin, because the inhibitor is also positively charged (see Figure 6.38A).

The results of a different calculation of the electrostatic potential around trypsin is shown in Figure 6.38B. For this calculation it was assumed that the value of the dielectric constant was 2 inside trypsin, and 80 outside, and the Poisson–Boltzmann equation was used to calculate the electrostatic potential generated by trypsin. The shape of the low dielectric region changes the nature of the electrostatic map dramatically. Even though the same charges are located at the same positions in this calculation as in the first one, there now is a large region of negative potential that leads into the active site. This change in the electrostatic potential arises because the low dielectric medium inside the protein does not attenuate the electric field generated by a few negatively charged residues that are in strategic positions around the active site of trypsin. A positively charged group, such as the lysine residue in the inhibitor protein that is highlighted in Figure 6.38B, would be drawn into the region of negative electrostatic potential at the active site. This is an example of an **electrostatic focusing effect** due to the shape of a protein.

The relevance of electrostatic focusing effects for proteins can be further illustrated by considering the enzyme acetylcholine esterase. Acetylcholine esterase functions at neuronal synapses, where it helps to terminate neuronal signaling by hydrolyzing the messenger molecule acetylcholine. Acetylcholine esterase works extremely fast, and the speed of this enzyme is limited only by how fast the substrate can get to it. There are several features of the electrostatic potential generated by this enzyme that speed up the binding of the positively charged acetylcholine molecule to the enzyme.

The structure of acetylcholine esterase is shown in **Figure 6.39**. As might be expected for an enzyme that binds to a positively charged substrate, there is a clustering of negatively charged residues on the surface of the enzyme bordering the active site. The nature of the electrostatic attraction between the substrate and the enzyme is illustrated strikingly in Figure 6.39C, in which the electrostatic potential generated by the protein is mapped onto the surface of the protein. There is a region of intense negative electrostatic potential in the vicinity of the binding site for acetylcholine. The favorable electrostatic potential is enhanced in this case by the focusing effect of the shape of the enzyme binding site.

The electrostatic potential generated by the protein also enhances the ability of the substrate to find the active site of the enzyme. This is illustrated in two ways in Figure 6.39. Electrostatic field lines emanating from regions of negative electrostatic potential in the vicinity of the active site are shown in Figure 6.39D. Positive charges would move along these field lines, which act as an electrostatic funnel. Another way in which the enzyme guides the substrate into the active site is by making the backside of the enzyme positively charged. When we look at the enzyme from an orthogonal view (see Figure 6.39E), we see that the surface of the enzyme that is distal to the active site is within a region of positive electrostatic potential, which would tend to favor movement of the substrate away from this region and towards the side of the enzyme (see Figure 6.39F).

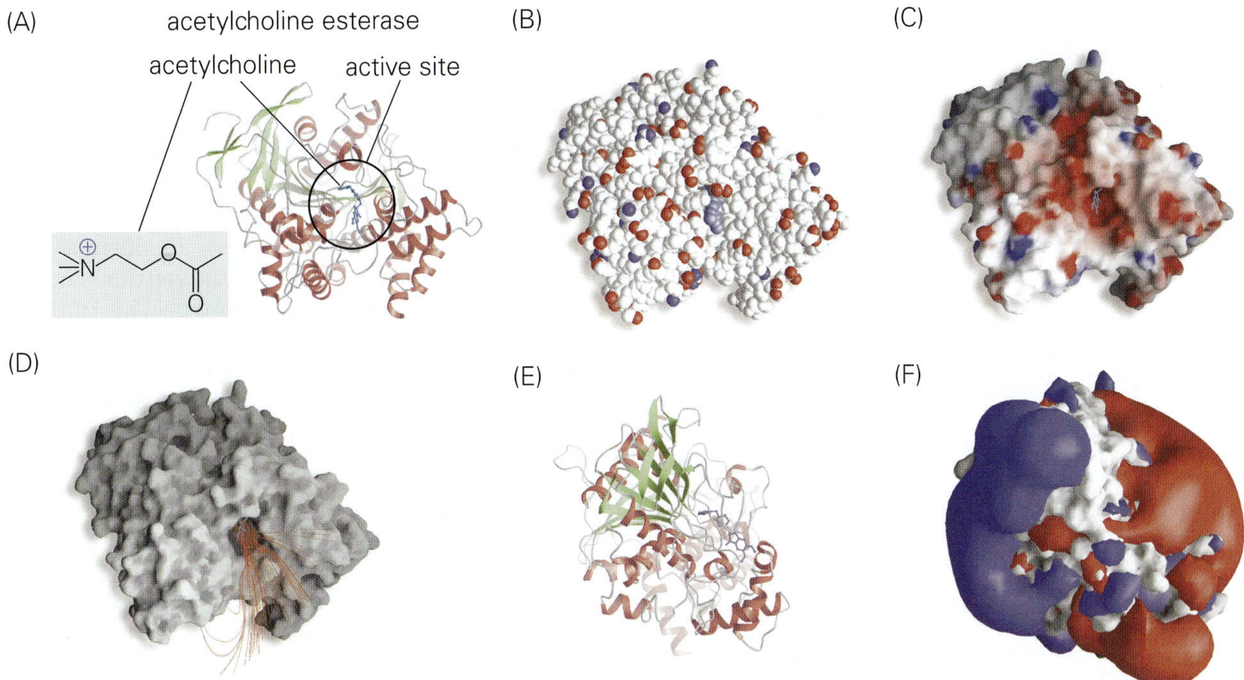


**Figure 6.39 Electrostatic calculations using continuum dielectric models for proteins.** (A) and (B) show the structure of the enzyme acetylcholine esterase. In (B), the oxygen and nitrogen atoms of charged sidechains (Asp, Glu, Arg, and Lys) are shown in *red* and *blue*, respectively. The electrostatic potential map surrounding the molecule is calculated by solving the Poisson–Boltzmann equation, and the results of the calculation are shown in different ways. In (C), the molecular surface of the enzyme is shown, and the surface is colored according to the electrostatic potential. The redder the surface, the more attractive (that is, lower energy) is that region of the surface for a positive charge being placed there. Thus *red* regions have negative electrostatic potential. The *blue* regions have positive electrostatic potential. In (D), some of the field lines corresponding to the potential map are shown. At any point on the map, the force on a unit positive charge is calculated. Each of the *red* lines indicates the path along which a positive charge would move due to the electrostatic force. The direction of the force lines are such that a positive charge would be drawn into the center of the active site. (E) and (F) show views of the enzyme that are rotated by ~90° with respect to the views in A–D. In (F), the *red* and *blue* surfaces are isopotential surfaces, on which a charged particle would have the same electrostatic energy anywhere (*red*, negative potential; *blue* positive potential). The electrostatic potential generated by the enzyme biases the substrate, which is positively charged, away from the left side and towards the right, where the active site is located. (PDB code: 1ACJ.)

## Summary

This chapter began with an introduction to a simple concept in thermodynamics—namely, the law of conservation of energy, also known as the first law of thermodynamics. The first law of thermodynamics helps us keep track of energy flow during a process by requiring us to account for how much of the change in energy is due to heat added to or removed from the system and how much is utilized to do work.

Calorimetry, which measures how much heat is taken up when the system undergoes a transformation, is a very important tool in the experimental measurement of molecular energies. Consideration of the amount of heat taken up by a system leads to the concept of heat capacity, which is the amount of heat required to raise the temperature of the system by one kelvin. The heat capacity of an ideal monatomic gas does not change with temperature, but more complex molecules, including proteins and DNA, have more ways to take up heat by increasing their potential energy rather than increasing the temperature. Thus, more complex molecules often have higher heat capacities than simpler ones.

In order to see whether a particular energetic transition is likely to happen, we refer to the Boltzmann distribution, which tells us that if the change in energy during a transition, $\Delta U$, is very much greater than the value of $k_B T$, then that transition is unlikely to happen at the given temperature. This means that electronic transitions are unlikely to occur at room temperature. Instead, molecules take up energy by undergoing conformational fluctuations, such as oscillating about covalent bonds (torsional fluctuations) or making and reforming van der Waals or hydrogen-bonding interactions. The availability of such low-energy transitions ($\Delta U \approx k_B T$) increases the heat capacity of the molecule.

We use mathematical expressions known as energy functions to calculate the energy of a protein or nucleic acid molecule. Of the many kinds of terms in these energy functions, those describing noncovalent interactions are the most important for understanding protein and nucleic acid function. The van der Waals interactions underlie the stabilization that results from the close packing of atoms in the interiors of proteins. The van der Waals interactions are also what gives shape to the component parts of the protein, which fit together like pieces of a jigsaw puzzle in a folded protein structure. Van der Waals interactions are not directional, and it is the inclusion of electrostatic interactions between atoms that provides specificity and locks in a particular configuration of atoms. Electrostatic interactions include a wide range of effects, such as the formation of ion pairs and hydrogen bonds.

Although the nature of electrostatic interactions is simple to understand in principle, in practice it is very difficult to estimate the strength of electrostatic interactions in proteins because of interference from water molecules. Water molecules weaken the effective strength of hydrogen bonds in proteins by competing for hydrogen-bond donors and acceptors. The benefit of this is that protein structures are not too rigid—the ability of water to form hydrogen bonds with the protein facilitates conformational changes that otherwise might not occur if hydrogen bonds were too strong. Water molecules also attenuate long-range electrostatic interactions. Insight into how water modulates the electrostatic field generated by proteins is obtained by using continuum dielectric approximations, which do not consider water molecules explicitly, and instead mimic the effects of water by surrounding the protein with a region of uniformly high dielectric constant. These kinds of calculations have shown us that the ability of proteins to locate and bind their substrates is strongly augmented by the electrostatic fields generated by charges within the protein.

## Key Concepts

### A. THERMODYNAMICS OF HEAT TRANSFER

- Energy released by chemical reactions is converted into heat and work.
- Heat is energy released in the form of disordered molecular motion, while mechanical (pressure-volume) work involves a displacement of the system or surroundings, reflecting energy that can be stored and later released.
- The total energy of the system and the surroundings is conserved. This statement is known as the first law of thermodynamics.
- The heat transferred to a system under conditions of constant pressure is equal to the change in enthalpy of the system.
- If the change in volume is small, then the change in enthalpy is essentially the change in energy.
- Changes in energy do not always indicate the direction of spontaneous change.

### B. HEAT CAPACITIES AND THE BOLTZMANN DISTRIBUTION

- The heat capacity of an ideal monatomic gas is constant with temperature.
- The heat capacity of a macromolecular solution increases and then decreases with temperature as the macromolecule unfolds.
- The potential energy of a molecular system is the energy that is stored in molecules and their interactions. It is that part of the energy that is *not* due to molecular motions (that is, not kinetic energy).
- The Boltzmann distribution describes the populations of molecules in different energy levels. Energy levels corresponding to energies much greater than $k_B T$ above the lowest energy level are not highly populated.
- The energy required to break interatomic interactions in folded macromolecules gives rise to the peak in heat capacity when the temperature is increased.

### C. ENERGETICS OF INTERMOLECULAR INTERACTIONS

- Simplified energy functions are used to calculate molecular potential energies very rapidly.
- The energies of covalent bonds are approximated by functions such as the Morse potential, which accounts for covalent bond breakage, or Hooke's law, which does not.
- The van der Waals energy function describes weak attractions and strong repulsions between atoms.
- The atoms in proteins and nucleic acids are partially charged.
- The interactions between charged atoms is governed by Coulomb's law.
- Hydrogen bonds are an important class of electrostatic interactions.
- Interactions with water weaken the effective strengths of hydrogen bonds and ionic interactions in proteins.
- Hydrogen bonds are important for solubility and specificity.
- The shapes of proteins influence the electrostatic fields generated by charges within the proteins.

## Problems

### True/False and Multiple Choice

1. Which of the following properties are extensive (choose all that apply):

   a. temperature
   b. pressure
   c. amount of heat released
   d. density
   e. energy
   f. molarity
   g. number of moles
   h. mass
   i. volume

2. A piston containing an ideal gas expands isothermally from 7 atm pressure to 2 atm pressure. The energy of the system (that is, the contents of the piston) remains constant during this process.

   True/False

3. A protein is negatively charged, but it binds a negatively charged small molecule faster than it binds a positively charged small molecule. The most reasonable explanation for this phenomenon is:

   a. Even though the protein is negatively charged overall, electrostatic focusing effects provide a pathway for a negatively charged molecule to enter the active site.

   b. The presence of water molecules screen the electrostatic effects.

   c. The negatively charged molecule makes stronger hydrogen bonds than the positively charged molecule.

   d. Charged proteins normally bind substrates with the same overall charge.

4. The enthalpy change for a process is equal to the heat transferred to the system under which of the following conditions?

   a. constant volume
   b. constant pressure
   c. constant temperature
   d. reversible expansion
   e. expansion at a constant pressure followed by reversible compression

5. Energy is a good indicator of the direction of spontaneous change for macroscopic mechanical objects but not for molecular processes.

   True/False

6. Which of the following statements about potential energy is true?

   a. The potential energy of an ideal gas is equal to its kinetic energy.

   b. The potential energy of an atom is the work done in moving the atom from infinity to its present position.

   c. Potential energy is always absolutely conserved.

   d. The potential energy of a system always increases.

   e. Potential energy is an intensive function.

7. The type of function that best describes the energy of a hydrogen bond as the distance between the hydrogen atom and the acceptor atom varies between 1.5 and 2.5 Å, is:

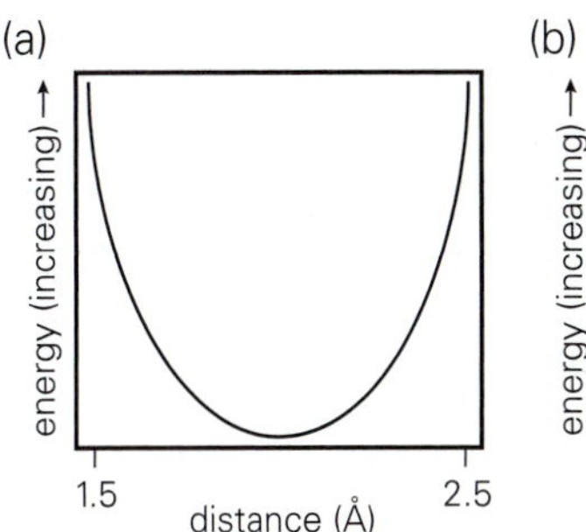


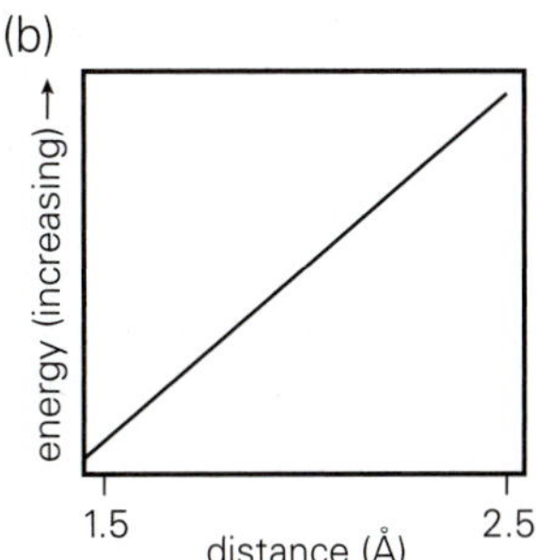

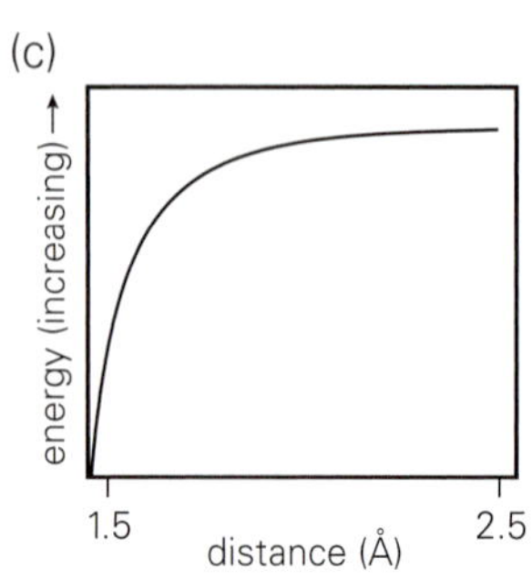


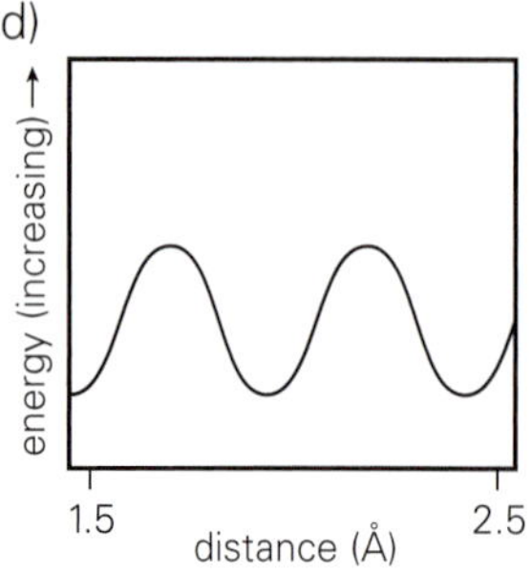


## Fill in the Blank

8. A reaction that releases energy is called an _______ reaction.

9. The first law of thermodynamics states that the total energy of the system and surroundings remains _________.

10. Electrostatic interactions are governed by _________'s law

11. An ion pair with lysine can be formed by the sidechain of ________ or __________.

12. The energy for van der Waals interactions approaches _____ as the distance between atoms goes to zero and approaches _____ at infinite distance.

13. The force arising from a noncovalent interaction between two atoms is always ______ at the energy minimum.

## Quantitative/Essay

14. One mole of an ideal monatomic gas is initially at 300 K and 1010 $J \cdot L^{-1}$ pressure inside a cylinder with a frictionless piston. (Note that 1 $J \cdot L^{-1} = 10^3$ Pa.)

    The gas expands until the pressure is 101 $J \cdot L^{-1}$. Calculate $\Delta U$ (the change in energy), $q$ (the heat transferred to the system), and $w$ (the work done by the system) when the expansion is isothermal and reversible.

15. One mole of an ideal monatomic gas is initially at 300 K and 1010 $J \cdot L^{-1}$ pressure inside a cylinder with a frictionless piston. The cylinder is an isolated (adiabatic) system. The gas expands against zero external pressure. When the volume expands to 24.7 L, a peg stops the piston. Calculate $\Delta U$ (the change in energy), $q$ (the heat transferred to the system), and $w$ (the work done by the system) when the expansion is isothermal and reversible.

16. One mole of a different ideal monatomic gas is initially at 300 K and 1010 $J \cdot L^{-1}$ pressure inside a cylinder with a frictionless piston . Calculate $\Delta U$ (the change in energy), $q$ (the heat transferred to the system), and $w$ (the work done by the system) when the expansion is isothermal and against a constant pressure of 101 $J \cdot L^{-1}$.

17. The complete oxidation of one mole of a sugar produces carbon dioxide and water. 2000 kJ of heat is transferred from the system to the surroundings. The rearrangement of bonds as 0.5 moles of the sugar are oxidized generates heat in an open test tube (101 $J \cdot L^{-1}$ pressure and 300 K temperature). What is the change in internal energy of the system ($\Delta U$)? What is the change in enthalpy of the system($\Delta H$)?

18. A system is maintained at thermal equilibrium (at the same temperature) with its surroundings and has an enthalpy of 50 kJ. It has 100 kJ of heat transferred to it, which causes it to expand against a constant pressure of 1 atm. It is then compressed back to its initial volume. All steps are at constant temperature. What is its final enthalpy?

19. How much kinetic energy does a system containing 3 moles of an ideal gas at 300 K possess? What is the heat capacity at constant volume? How much heat would need to be transferred to the system to raise the temperature by 15°C?

20. Below are two graphs of heat capacity changes in a differential scanning calorimeter experiment. In the experiment, protein molecules in solution are unfolded by increasing the temperature. In both graphs, temperature increases from left to right. Which graph correctly characterizes the heat capacity changes of a protein as it is heated? Which characteristics in the graph lead to this conclusion?

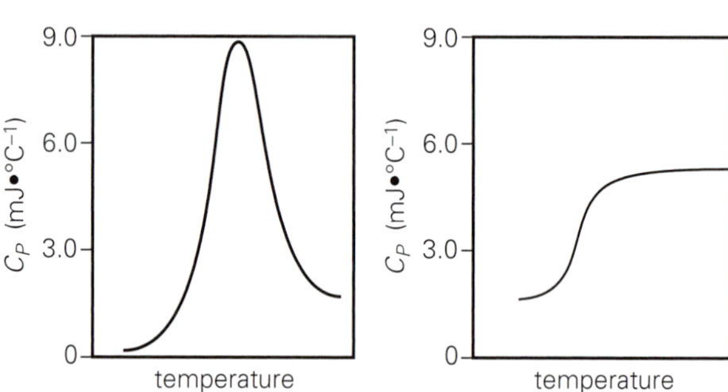


21. Consider a protein with different energy levels for electronic transitions and molecular vibrations. What are the approximate energy differences between the levels for each type? To what extent do the populations of levels within each type explain why the heat capacity of a protein is highest at the unfolding point?

22. The only two accessible conformations of a protein differ by 2 $kJ \cdot mol^{-1}$. What percentage of protein molecules will be in the higher energy state:
    a. at 350 K?
    b. at 270 K?

23. Consider a ribosome translating a codon to an amino acid. Compared to a perfect pairing, imagine that a codon–anticodon pairing with one incorrect base pair is 10 $kJ \cdot mol^{-1}$ higher in energy, two incorrect base pairs is 20 $kJ \cdot mol^{-1}$ higher, and three incorrect base pairs is 30 $kJ \cdot mol^{-1}$ higher.

How many ways are there of making a one-, two-, or three-base-pair mismatch? What is the partition function of the codon–anticodon system at 300 K? If energy difference is the only consideration, what percent of decoding events occur without an error (no mismatches) at 300 K?

Hint: There are three incorrect single-base mismatches in each of the three positions (A, B, and C)(3 × 3 = 9); 3 × 3 = 9 incorrect two-base mismatches in three possible combinations (AB, AC, and BC) of positions (9 × 3 = 27); and 3 × 3 × 3 = 27 incorrect three-base mismatches.

**24.** The energy of a covalent bond as a function of interatomic distance, $r$, is described by the following equation:

$$U(r) = 2000\,(r - 1.5)^2$$

In this equation, energy is expressed in units of kJ•mol$^{-1}$ and distance in Å. What is the equation that describes the force with respect to position? What is the energy when the interatomic distance is 2 Å? At the same value of the interatomic distance, what is the force and is the force pushing the bonded atoms closer together or further apart?

**25.** A salt bridge between an arginine and a glutamic acid has the two bridging atoms 3.5 Å apart. Assume that these bridging atoms have elementary charges of +1 and –1, respectively.

What is the energy if the residues are a) on the protein surface ($\varepsilon = 80$) and b) in the core of the protein ($\varepsilon = 2$)?

**26.** A particular sidechain has a dihedral angle for which the energy is governed by the following equation:

$$U_{dihedral} = 2 \times \cos(3 \times \text{angle})\ \text{kJ•mol}^{-1}$$

a. What is the energy of a sidechain with an angle of 60°?
b. With an angle of 90°?
c. What is the relative population of sidechains with an angle of 60° versus 90° at 300 K?

**27.** For two atoms, separated by distances between 0 and 4 Å, the van der Waals energy is well described by the following equation:

$$U(r) = \varepsilon\left(\frac{r_{min}}{r}\right)^2$$

where $\varepsilon = 2$ kJ•mol$^{-1}$. Calculate the value of $r_{min}$ if the energy at 3 Å is 1.5 kJ•mol$^{-1}$.

**28.** A calorimeter is used to measure the heat capacity of two molecules, A and B. Molecule A is a simple small molecule, whereas molecule B is a complex polymer (like DNA). Define heat capacity and explain why molecule B is likely to have a higher heat capacity.

**29.** Water forms hydrogen bonds with proteins. How might these hydrogen bonds alter the ability of a protein to undergo conformational changes in water versus in the gas phase?

# Further Reading

## General

Atkins PW & De Paula J (2006) Atkins' Physical Chemistry. New York: Oxford University Press.

Chang R (2005) Physical Chemistry for the Biosciences. Sausalito, CA: University Science Books.

Eisenberg DS & Crothers DM (1979) Physical Chemistry: With Applications to the Life Sciences. Menlo Park, CA: Benjamin Cummings.

Halliday D, Resnick R & Walker J (2001) Fundamentals of Physics, 6th ed. New York: John Wiley & Sons.

Haynie DT (2008) Biological Thermodynamics. New York: Cambridge University Press.

Voet D & Voet JG (2004) Biochemistry, 3rd ed. New York: John Wiley & Sons.

## References

### C. Energetics of Intermolecular Interactions

Baker EN & Hubbard RE (1984) Hydrogen-bonding in globular proteins. *Prog. Biophys. Mol. Biol.* 44, 97–179.

Honig B & Nicholls A (1995) Classical electrostatics in biology and chemistry. *Science* 268, 1144–1149.

Karplus M & McCammon JA (2002) Molecular dynamics simulations of biomolecules. *Nat. Struct. Biol.* 9, 646–652.

Leach AR (2001) Molecular Modeling: Principles and Applications. Upper Saddle River, NJ: Prentice Hall.

Matthews BW & Liu L (2009) A review about nothing: Are apolar cavities in proteins really empty? *Protein Sci.* 18, 494–502.

Schlick T (2002) Molecular Modeling and Simulation: An Interdisciplinary Guide. New York: Springer.

Wang W, Donini O, Reyes CM & Kollman PA (2001) Biomolecular simulations: Recent developments in force fields, simulations of enzyme catalysis, protein-ligand, protein-protein, and protein-nucleic acid noncovalent interactions. *Annu. Rev. Biophys. Biomol. Struct.* 30, 211–243.

# Entropy

In our attempt to understand the functioning of biological molecules, we need to relate the conformations and energies of individual molecules to the bulk properties of the system that are readily amenable to experimental measurement. At first glance, this might seem like an impossibly difficult problem because of the very large number of molecules contained within even a small volume of a sample. How could we possibly calculate the aggregate behavior of such a large number of molecules, given the essentially random movement and complex behavior of individual molecules? Although difficult, this problem is not as hopeless as it might appear at first glance because of a simplification that occurs when we consider the aggregate statistical properties of very large numbers of molecules.

If we look at only a small number of molecules (for example, 5, 10, or 100), it is difficult to predict their behavior with a great deal of certainty because each molecule does indeed behave erratically due to the stochastic nature of thermal motion. If we consider much larger numbers of molecules (for example, 100,000 or more), however, it turns out that we can relate many statistical properties of the entire system (such as the density, the mean energy, or the mean square displacement of molecules) to the properties of individual molecules. Although we still cannot be sure what any individual molecule is doing at any given time, the aggregate or bulk properties of collections of molecules are subject to certain statistical laws that are relatively straightforward to understand.

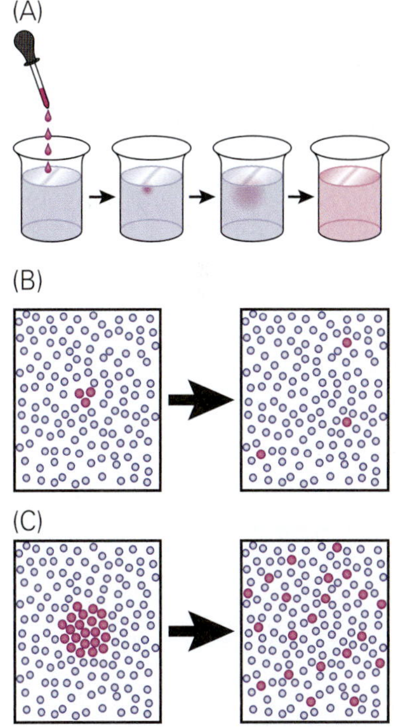


**Figure 7.1 Molecular diffusion.** (A) When a drop of dye is introduced into a beaker of water, the dye spreads uniformly throughout the beaker. (B, C) Schematic representations of the diffusion process at the molecular level. Water molecules are shown as *blue* dots and dye molecules are *purple*. In B, only three dye molecules are introduced. In C, a larger number of dye molecules is introduced.

The idea that the aggregate behavior of large numbers of molecules can be predicted with accuracy is illustrated by considering molecular diffusion (**Figure 7.1**). We had noted in Chapter 6 that dye molecules will spread throughout a container, even if there is no energetic driving force underlying their diffusion (see Figure 6.8). This process is illustrated schematically at the molecular level in Figure 7.1, which shows that if only a small number of dye molecules are inserted into a container, then their positions at a later time cannot be predicted. When many dye molecules are introduced, the position of any individual dye molecule still cannot be predicted. But, when the number of dye molecules is very large, we can say with certainty that the density of dye molecules will be uniform throughout the container at equilibrium. In this case, the density of the dye molecules is an aggregate property that is amenable to statistical analysis.

The application of statistical methods of analysis to the properties of molecules is based on a concept known as the **entropy**, which is related to the number of equivalent arrangements or configurations of the system. The study of the connection between molecular properties and thermodynamic properties using statistical treatments is called **statistical thermodynamics**. The discussion of statistical thermodynamics in this and the subsequent chapter is influenced by Dill and Bromberg's *Molecular Driving Forces* (see Further Reading at the end of the chapter). Consult this excellent book for a more advanced and thorough treatment of the concepts introduced here.

We begin our introduction to the essential concepts of this field by studying coin tosses, where the outcome can either be heads or tails (**Figure 7.2**). Coin tosses are a sequence of trials with binary outcomes and, as we shall see, such sequences

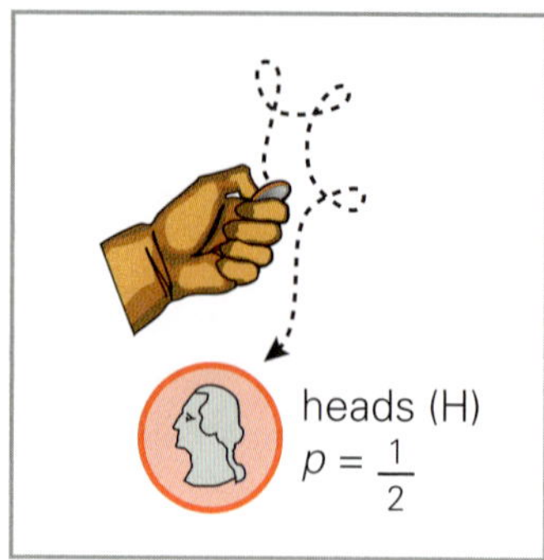


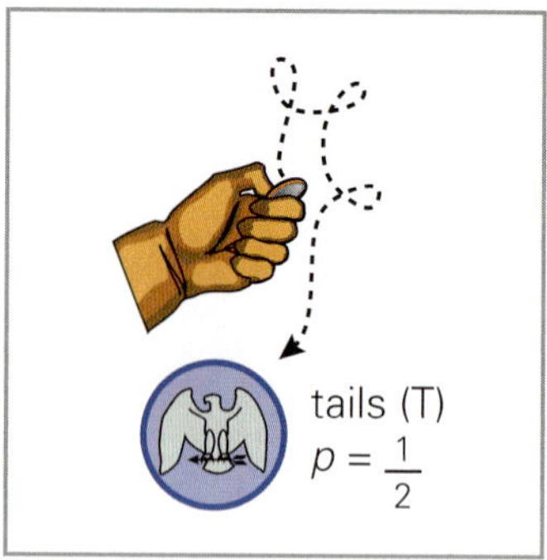


**Figure 7.2 The probability of the outcomes of coin tosses.** In an unbiased coin toss, the probability, $p$, of the outcome being heads (H) or tails (T) is equal to 1/2 in both cases.

can be used to analyze many situations of interest in biology, including molecular diffusion.

Although we cannot predict the outcome of any particular sequence of coin tosses, the statistical outcome of the aggregate number of heads and tails in a large number of coin tosses can be predicted with high precision. The common-sense concepts introduced by studying coin tosses allow us to introduce the statistical definition of entropy, which is related in a simple way to probability. A guiding principle that emerges is that the direction of spontaneous change in a molecular system that is not at equilibrium is always in the direction of increasing entropy. This principle, also known as the **second law of thermodynamics**, means that entropy is maximal at equilibrium. As a result, we evaluate whether or not a system and its surroundings are at equilibrium by computing the entropy and determining whether its value is maximal.

In this chapter, we consider changes in entropy that arise solely from changes in molecular positions, without changes in energy. In Chapter 8, we study the interplay between entropy and energy, which leads to the crucial concept of free energy, which is discussed in Chapter 9.

# A. COUNTING STATISTICS AND MULTIPLICITY

## 7.1 Different sequences of outcomes in a series of coin tosses have equal probabilities

Consider a series of coin tosses in which the coin is unbiased—that is, the probability of the toss yielding a head (H) is equal to the probability of the same toss yielding a tail (T) (see Figure 7.2). What is the probability ($p$) of obtaining first a head and then a tail in a series of two coin tosses? The probability of obtaining H in the first toss is 1/2 and, because the coin tosses are unbiased, the probability of obtaining T in the second toss is also 1/2. The combined probability of obtaining first a head and then a tail is the product of the individual probabilities:

$$p = \frac{1}{2} \times \frac{1}{2} = \frac{1}{4} \tag{7.1}$$

Likewise, in a series of four coin tosses, the probability ($p$) of obtaining the sequence HTHT is given by:

$$p = \frac{1}{2} \times \frac{1}{2} \times \frac{1}{2} \times \frac{1}{2} = \frac{1}{16} \tag{7.2}$$

In fact, the probability of obtaining *any* particular sequence of outcomes in a series of four coin tosses, such as HHHH or HTHH, is the same (1/16, as shown

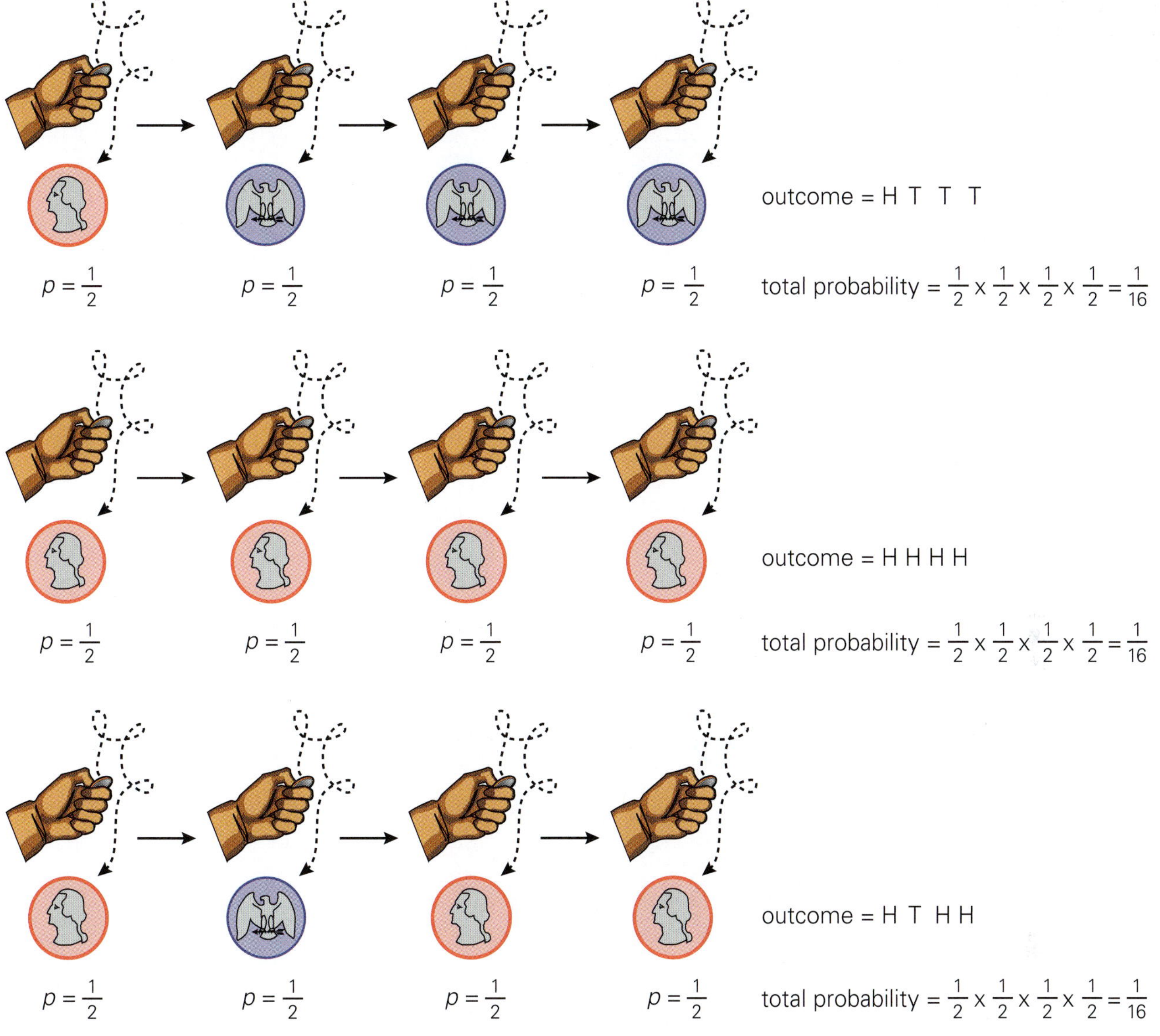


**Figure 7.3 The probability of an outcome in a series of coin tosses.** If a coin is tossed four times, the probability of any particular sequence of outcomes, such as the sequence H T T T, is $\frac{1}{2}\times\frac{1}{2}\times\frac{1}{2}\times\frac{1}{2}=\frac{1}{16}$.

in **Figure 7.3**). As the number of coin tosses in the series increases, the probability of obtaining any particular sequence of heads and tails becomes smaller and smaller. At the same time, the probability of obtaining any particular sequence is always the same as that for any other sequence that contains the same number of coin tosses.

## 7.2 When considering aggregate outcomes, the most likely result is the one that has maximum multiplicity

Now, instead of being concerned with the result of any individual coin toss, let us consider the *aggregate* or net result of the series of coin tosses (**Figure 7.4**). What, for example, is the probability of getting one H and three T's in a series of four coin tosses? This probability ($P$) is given by:

$$P = \frac{\text{number of desired outcomes}}{\text{total number of possible outcomes}} \tag{7.3}$$

where the number of desired outcomes (the number of different ways of achieving a particular outcome) is referred to as the **multiplicity** of the outcome. According

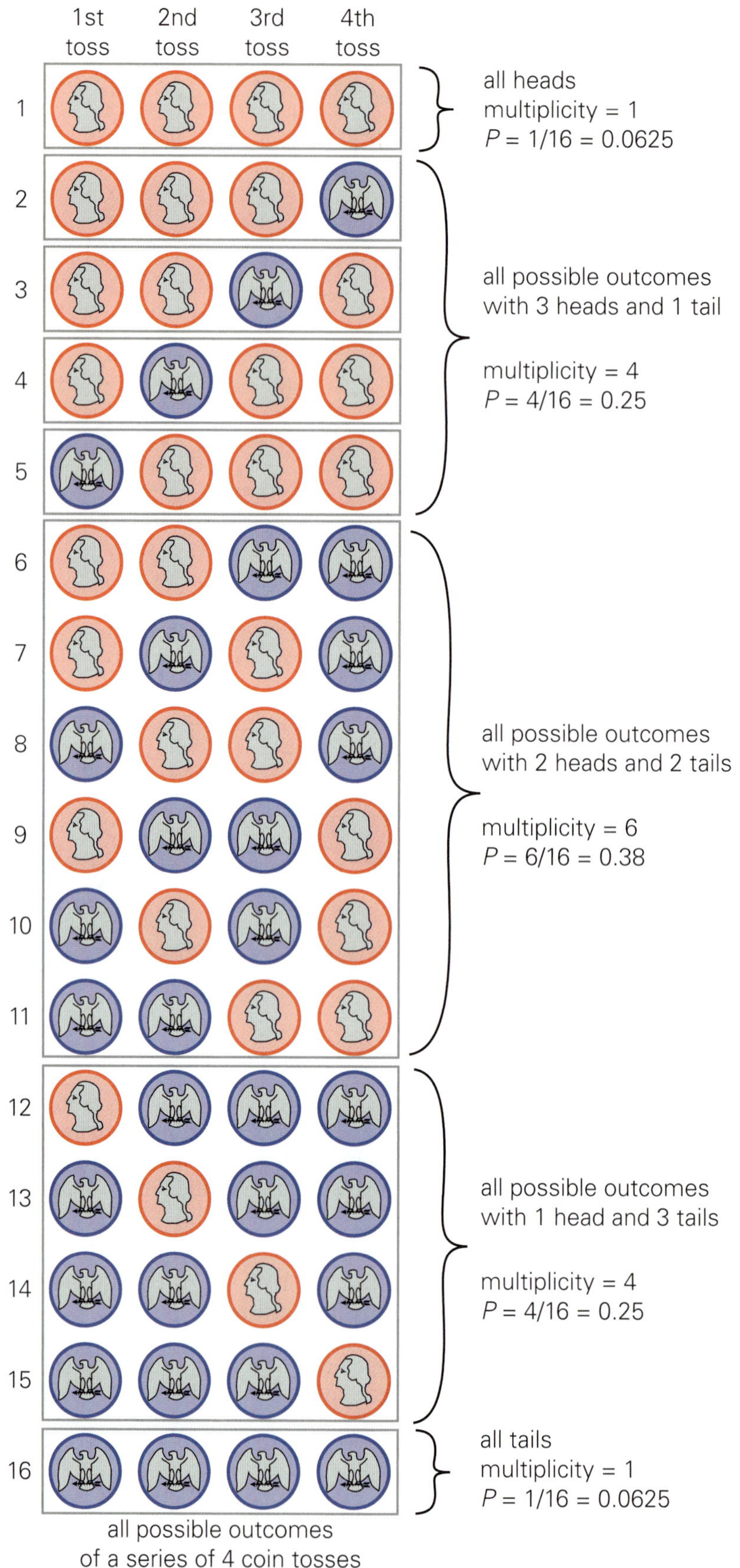


**Figure 7.4 The probability of aggregate outcomes.** All possible outcomes of a series of four coin tosses are shown. There are 16 possible outcomes, and each individual outcome has the same probability of occurring as any of the other ones. Aggregate outcomes, in which we consider only the total number of H's or T's, are grouped together. There are six ways of getting 2 H's and 2 T's (multiplicity = 6). There are four ways of getting 1 H and 3 T's or 3 H's and 1 T (multiplicity = 4). There is only one way each of getting all H's or all T's.

to Figure 7.4, there are four possible outcomes of a series of four coin tosses that correspond to the desired outcome of one H and three T's: HTTT, THTT, TTHT, and TTTH. The multiplicity of this outcome is therefore four. The total number of all possible outcomes, without regard to any pattern of H's and T's, is $2 \times 2 \times 2 \times 2 = 16$. According to Equation 7.3, therefore, the probability ($P$) of seeing one H and three T's in a series of four coin tosses is $P = 4/16 = 0.25$.

Because the coin is unbiased, the probability of getting two heads and two tails in a series of four coin tosses is greater than that for obtaining one heads and three tails. According to Figure 7.4, there are six outcomes that have an equal number of heads and tails:

HHTT THHT
HTHT THTH
HTTH TTHH

Thus, the multiplicity of getting two heads and two tails, in any order, in a series of four coin tosses, is six. The net probability of getting two heads and two tails in a series of four coin tosses is therefore $6/16 = 0.38$, which is indeed higher than the probability of getting one H and three T's in series of four tosses (where $P = 0.25$). Nevertheless, for such a small number of coin tosses, the probability of getting an outcome in which the numbers of heads and tails are unequal is significant, and we cannot have a high certainty of getting an equal number of heads and tails in a series of four coin tosses.

**Multiplicity**

The multiplicity of an outcome is the number of different ways in which that outcome can be achieved. For molecular systems, the multiplicity refers to the number of different molecular configurations that are consistent with the macroscopic parameters that define the system (for example, temperature or volume). The multiplicity is denoted by *W* in this book.

## 7.3 The multiplicity of an outcome of coin tosses can be calculated using a simple formula involving factorials

We can determine the multiplicity for any particular outcome of a series of four coin tosses by referring to the set of all possible outcomes shown in Figure 7.4 and identifying those that correspond to the desired outcome. This procedure becomes very tedious for larger numbers of coin tosses, though, so we turn instead to the formula in Equation 7.4, where $W(M, N)$ is the multiplicity, $M$ is the number of coin tosses in the series, and $N$ is the number of heads obtained (in any order):

$$W(M,N) = \frac{M!}{N!(M-N)!} \tag{7.4}$$

Thus, the multiplicity ($W$) of getting two heads ($N = 2$) and two tails, in any order, in a series of four coin tosses ($M = 4$) is given by:

$$W(4,2) = \frac{4!}{2!(4-2)!} = \frac{24}{2 \times 2} = 6 \tag{7.5}$$

The answer given by Equation 7.5 exactly matches the situation that is illustrated in Figure 7.4.

How do we arrive at the formula in Equation 7.4? Imagine that we have $M$ coins in a bag (**Figure 7.5**). Imagine, as well, that each coin in the bag is labeled distinctly (for example 1, 2, 3, ..., $M$). We reach into the bag and randomly pull out one coin, then randomly pull out another coin and place it next to the first one, and so on until we have removed all $M$ coins from the bag. In how many different ways can these $M$ coins be pulled from the bag?

There are $M$ possibilities for choosing the first coin, but once the first coin is selected and removed, there are only $(M-1)$ choices for the second coin. Because each independent choice of the first coin can be combined with each independent choice of the second coin to get a different possible outcome, there are $M \times (M - 1)$ different ways in which the first two coins can be chosen. Likewise, there are $M \times (M - 1) \times (M - 2)$ different ways in which the first three coins can be chosen and, by extension, the number of different ways in which all the coins can be chosen is given by the **factorial** of $M$:

$$M! = M \times (M-1) \times (M-2) \times \ldots \times 3 \times 2 \times 1 \tag{7.6}$$

**Factorials**

The value of $M!$ (pronounced "$M$ factorial") is given by:

$M! = M \times (M - 1) \times (M - 2) \times \ldots \times 2 \times 1$

For example, the value of 5! is given by:

$5! = 5 \times 4 \times 3 \times 2 \times 1$
$= 120$

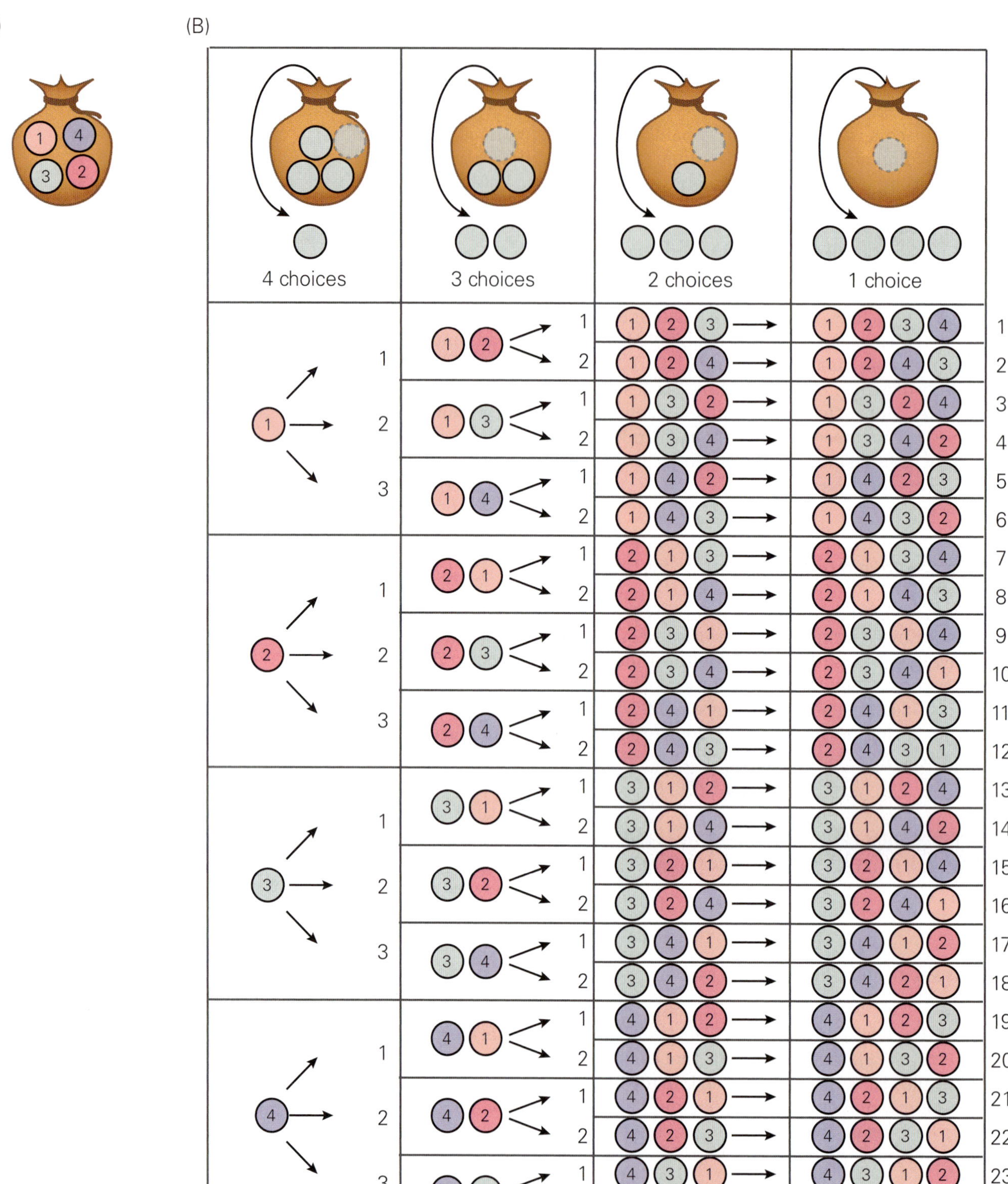


**Figure 7.5 The number of ways of picking four distinct coins in random order.** (A) Imagine a bag containing four different coins, labeled 1, 2, 3, and 4. A coin is selected by reaching into the bag and randomly picking a coin. (B) All possible outcomes of randomly picking four coins from the bag in sequence are shown. There are four ways of picking the first coin, three ways of picking the second coin, two ways of picking the third coin, and one way of picking the last coin in each sequence. There are a total of $4 \times 3 \times 2 \times 1 = 4!$ different possible arrangements of the coins.

Up until now, we have ignored whether the coins have landed heads up or tails up when we pick them. Let us assume, however, that $N$ of the coins in the bag are labeled "H" (for "heads"), and $M - N$ are labeled "T" for "tails." That is, the coins are labeled $H_1, H_2, ..., H_N, T_{N+1}, T_{N+2}, ..., T_M$ (**Figure 7.6**).

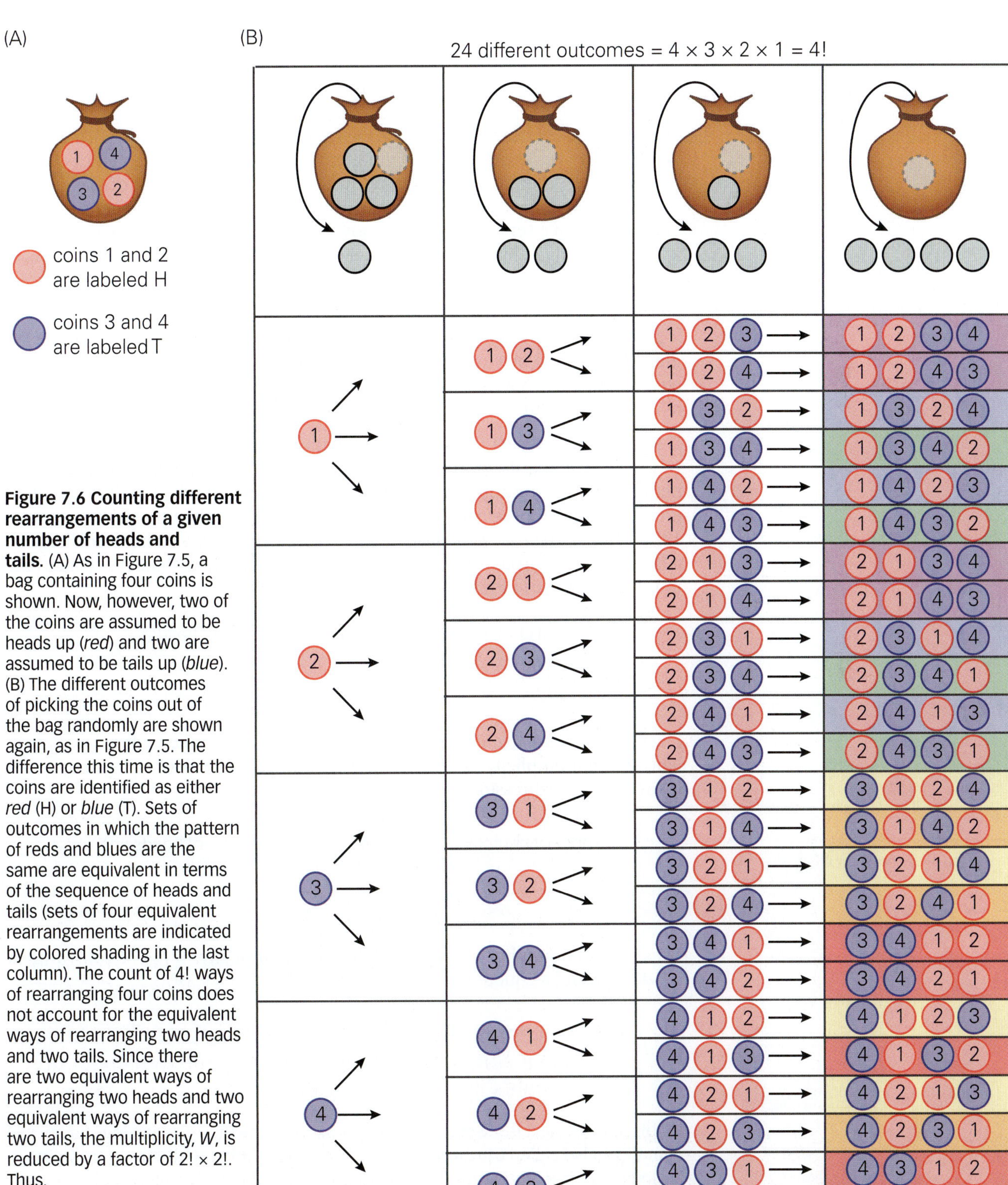


**Figure 7.6 Counting different rearrangements of a given number of heads and tails.** (A) As in Figure 7.5, a bag containing four coins is shown. Now, however, two of the coins are assumed to be heads up (*red*) and two are assumed to be tails up (*blue*). (B) The different outcomes of picking the coins out of the bag randomly are shown again, as in Figure 7.5. The difference this time is that the coins are identified as either *red* (H) or *blue* (T). Sets of outcomes in which the pattern of reds and blues are the same are equivalent in terms of the sequence of heads and tails (sets of four equivalent rearrangements are indicated by colored shading in the last column). The count of 4! ways of rearranging four coins does not account for the equivalent ways of rearranging two heads and two tails. Since there are two equivalent ways of rearranging two heads and two equivalent ways of rearranging two tails, the multiplicity, $W$, is reduced by a factor of $2! \times 2!$. Thus,

$$W = \frac{4!}{2!\,2!} = \frac{24}{4} = 6.$$

In the $M!$ different arrangements of the coins, there are many sequences of heads and tails in which the labels (numbers) on the coins are different, but the overall pattern of heads and tails is the same. For example, the following outcomes from a series of 10 coin tosses have identical patterns of heads and tails:

$$H_1 H_2 T_6 T_7 H_3 H_4 T_8 T_9 T_{10} H_5$$

$$H_2 H_1 T_6 T_7 H_3 H_5 T_8 T_9 T_{10} H_4$$

$$H_5 H_3 T_6 T_7 H_1 H_4 T_8 T_9 T_{10} H_2$$

The sequences of heads and tails are the same, but the labels on the H's are different. In the $M!$ ways of choosing $M$ different coins, each of these sequences is considered to be different, and each contributes to the total number of choices (see Figure 7.5). When we count the number of ways in which we can get five heads and five tails in a series of 10 coin tosses, however, each of these three sequences is treated as equivalent. We therefore have to divide $M!$ by the number of ways of rearranging $N$ heads, which is $N!$.

There are five heads in the three 10-coin sequences, so there are $5! = 5 \times 4 \times 3 \times 2 \times 1 = 120$ equivalent arrangements of these sequences, as long as all the H's are considered to be equivalent. All rearrangements of the $(M - N)$ tails are also equivalent, so we have to correct the total count by a factor of $(M - N)!$ to prevent overcounting all the permutations of tails in a given sequence.

If we consider each of the 10 coins to be unique, then there are $10! = 3{,}628{,}800$ ways of arranging them. If five of these coins are considered to be equivalent because they correspond to outcomes of H, then there are $5! = 120$ equivalent arrangements of these coins. For every arrangement of the other coins, we have 120 equivalent rearrangements of the five H's, so the original count is too large by a factor of 120. Correcting for the overcounting of the equivalent H's reduces the multiplicity to the following:

$$\text{number of choices (corrected for rearrangements of heads)} = \frac{3{,}628{,}800}{120} = 30{,}240$$

Likewise, there are five T's in each sequence that have to be considered as equivalent, so the multiplicity has to be divided by another factor of 120, giving us the following final expression:

$$\text{number of choices (corrected for rearrangements of heads and tails)} = \frac{30{,}240}{120} = 252$$

Thus, the total number of arrangements (the multiplicity, $W$) of $M$ coins with $N$ heads is given by Equation 7.4.

## 7.4 The concept of multiplicity is broadly applicable in biology because a series of coin flips is analogous to a collection of molecules in alternative states

In this and the next three sections, we discuss how the binary counting statistics of coin flips, also called **binomial statistics**, can be applied to different problems in biology. This is a digression from the main theme of this chapter, which is to connect the idea of multiplicity to entropy. You may, if you wish, skip ahead to Section 7.8, in which the connection to entropy begins to be made.

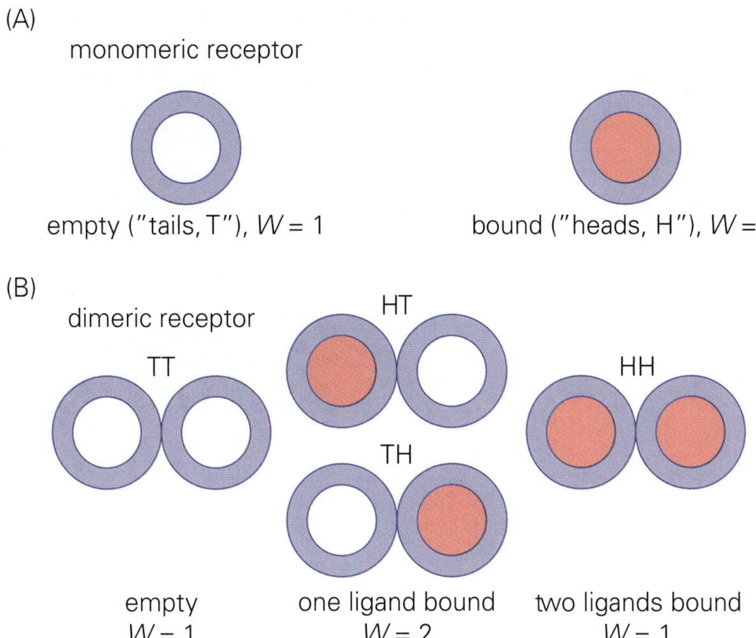


**Figure 7.7 An analogy between receptor–ligand interactions and coin tosses.** In these diagrams, the receptor is shown as a *blue* circle and the ligand as a smaller *red* circle. Two cases are shown here. In (A), the receptor is a monomer and there are two possible outcomes: either the ligand is not bound ("tails" or T) or the ligand is bound ("heads" or H). In (B), the receptor is a dimer. Now there are three possible sets of outcomes: no ligand bound, one ligand bound, and two ligands bound. As for coin flips, the multiplicity, *W*, for each outcome is the number equivalent rearrangements.

The simple counting statistics of coin flips apply to any series of events in which individual events have binary outcomes. Quite often, when considering a molecular mechanism in biology, we can characterize molecules as being in one of two states, such as *on* or *off* or *bound* or *unbound*. Consider, for example, the binding of two molecules to each other. One of the molecules, referred to as the **ligand**, may be a metabolite or a drug. The other molecule may be a protein or RNA molecule that has an affinity for the ligand and is referred to as the **receptor**. The receptor molecule could be a multimer with multiple subunits, each with a binding site for the ligand.

**Figure 7.7** illustrates the analogy between ligand binding and a series of coin tosses. A binding site on the receptor is either bound to the ligand ("heads") or it is not ("tails"). Thus, each of the binding sites in a multimeric receptor can be treated as a trial with a binary outcome.

We shall analyze situations in which the outcomes of binding events (for example, whether or not a binding site is occupied by a ligand) are biased—that is, one outcome is favored over another. In the case of ligand binding, the bias results from the strength of the interaction between the receptor and the ligand, and the concentration of the ligand. This corresponds to a series of coin tosses in which the coin is not fair (that is, the probability of getting heads is different from the probability of getting tails).

## 7.5 The binding of ligands to a receptor can be monitored by fluorescence microscopy

We can observe the binding of ligands to receptors by tagging the ligands with a fluorescent molecule and by using a microscope to detect the fluorescence from the ligand. One way to do this is to immobilize the receptor on a microscope slide and to flow a solution containing fluorescently tagged ligand over the immobilized receptor. This allows ligand molecules that are bound to the receptor to be observed directly in a fluorescence microscope, as shown in **Figure 7.8**.

For the purposes of our discussion, we assume that the receptors form multisubunit complexes, with more than one binding site for the ligand. If we look at a

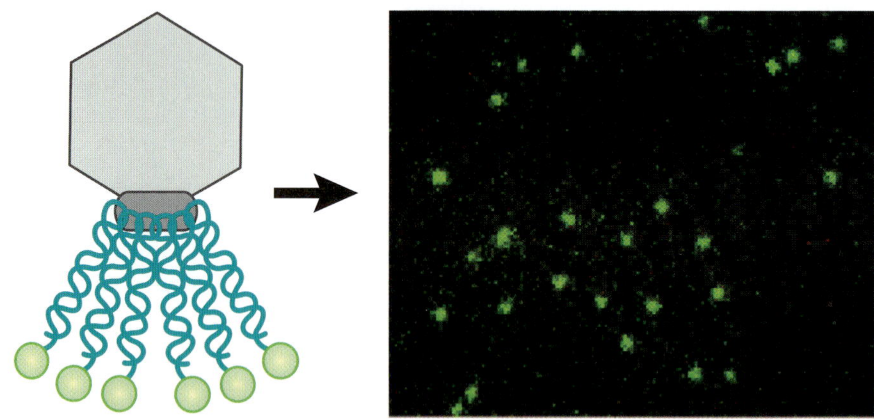

**Figure 7.8 Imaging fluorescently tagged molecules.** In this experiment, a virus particle is the receptor. The virus contains six RNA molecules (*blue* structures) that are bound by dye molecules. Up to six dye molecules can bind to one virus headpiece. When viewed through a fluorescence microscope, each virus particle is observed as a spot of green fluorescence. (Adapted from H. Zhang et al., and P. Guo, *RNA* 13: 1793–1802, 2007. With permission from Cold Spring Harbor Press.)

number of receptor–ligand complexes in the microscope and ask how many ligand molecules are bound to each receptor complex, what kind of distribution do we expect to see?

Individual receptor complexes can be identified in the microscope if the spatial separation between receptor complexes is large enough (Figure 7.8). The resolving power of light microscopes is not high enough for individual fluorescent molecules within one complex to be detected individually. Nevertheless, the number of fluorescent molecules bound to each receptor can be counted by observing the stepwise reduction in fluorescence intensity as individual molecules become bleached by the light, as indicated in **Figure 7.9**. In the example shown in Figure 7.9, we infer that there are two fluorescent molecules within the field of view. This is because the fluorescence intensity is reduced in two steps to the background level.

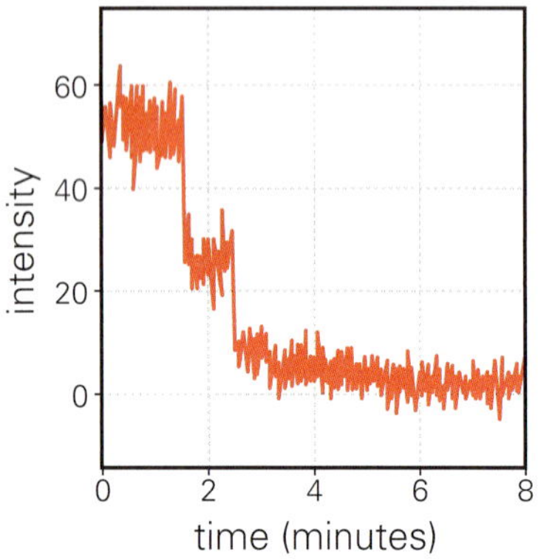

**Figure 7.9 Counting the number of fluorescent molecules in a region of fluorescence.** One of the fluorescent spots in the sample shown in Figure 7.8 is observed in a fluorescence microscope, and the intensity of fluorescence is recorded as a function of time. As the molecules in the region absorb light, they eventually become bleached (that is, they no longer fluoresce), and the intensity is reduced. The observation of two stepwise reductions in intensity in this region shows that there are two fluorescent molecules in the region. (Adapted from H. Zhang et al., and P. Guo, *RNA* 13: 1793–1802, 2007. With permission from Cold Spring Harbor Press.)

## 7.6 Pascal's triangle describes the multiplicity of outcomes for a series of binary events

Consider a multimeric receptor complex with $M$ binding sites for a ligand. The receptor complex can be bound to no ligand, or it can be bound to 1, 2, 3, ..., $M$ ligand molecules. If we assume that each of the ligand binding sites in the receptor is independent of the others, then the probability of seeing a certain number of ligands bound to a receptor complex is proportional to the number of ways in which the bound ligands can be rearranged among the binding sites. This is given by the multiplicity $W(M, N)$, where $N$ is the number of ligand molecules bound (number of positive outcomes) and $M$ is the total number of binding sites in the receptor complex (total number of events).

Analogous to the number of ways of obtaining $N$ heads in a series of $M$ coin tosses, the multiplicity $W(M, N)$ is given by:

$$W(M,N) = \frac{M!}{N!(M-N)!} \tag{7.7}$$

We can use Equation 7.7 to write down the multiplicities for various numbers of ligand molecules bound ($N$, the number of positive outcomes) to receptor complexes with different numbers of binding sites ($M$, the number of events). The multiplicities for receptors containing 1 to 10 binding sites (or events) are shown schematically in **Figure 7.10**. This diagram is known as **Pascal's triangle**. Each row in Pascal's triangle corresponds to a series of events with $M$ trials, and the entries in the row are the multiplicities for $N$ positive outcomes ($N = 0, 1, 2, ..., M$). The numbers in Pascal's triangle are known as **binomial coefficients**. The binomial coefficients exhibit many fascinating numerical relationships. For example, each binomial coefficient is the sum of the two numbers immediately above it, except for the numbers on the edges, which are all 1.

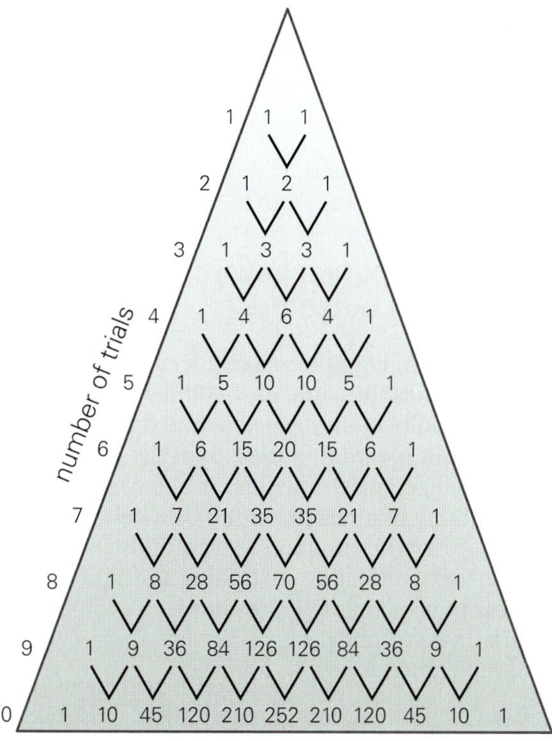


**Figure 7.10 Pascal's triangle.** The numbers in this triangle correspond to the multiplicities of different outcomes in a series of trials, where each event in the trial has two possible outcomes, such as heads or tails for a series of coin tosses. Each row in the triangle corresponds to an increasing number of trials (for example, *M* coin tosses). Within each row, the numbers from left to right represent the multiplicity of outcomes where there are *N* of one kind (for example, *N* heads). The triangle has 1 along the left and right edges. All other numbers are generated by taking the sum of the two numbers diagonally above it, as indicated by the slanted lines. The numbers in the triangle are referred to as binomial coefficients.

**Binomial coefficients**

The values of the multiplicity $W(M, N)$ are known as binomial coefficients. A simple diagrammatic representation of binomial coefficients for different numbers of trials is given by Pascal's triangle, named after the French mathematician Blaise Pascal.

As shown in **Figure 7.11**, the first row in Pascal's triangle corresponds to a receptor with one binding site (or a trial with one event). There are two entries for this row, the first corresponding to no ligand bound (no positive outcome) and the second to one ligand bound (one positive outcome). The multiplicity for both cases is 1. Likewise, the second row corresponds to multiplicities for receptors with two binding sites (two events), and the entries in this row correspond to zero, one, and two ligands bound (or positive outcomes). For a receptor with three binding sites, there are four possible outcomes, with multiplicities of 1, 3, 3, and 1.

We can verify that each entry in the triangle does indeed correspond to the multiplicities that are given by Equation 7.4. For example, for the case of a receptor

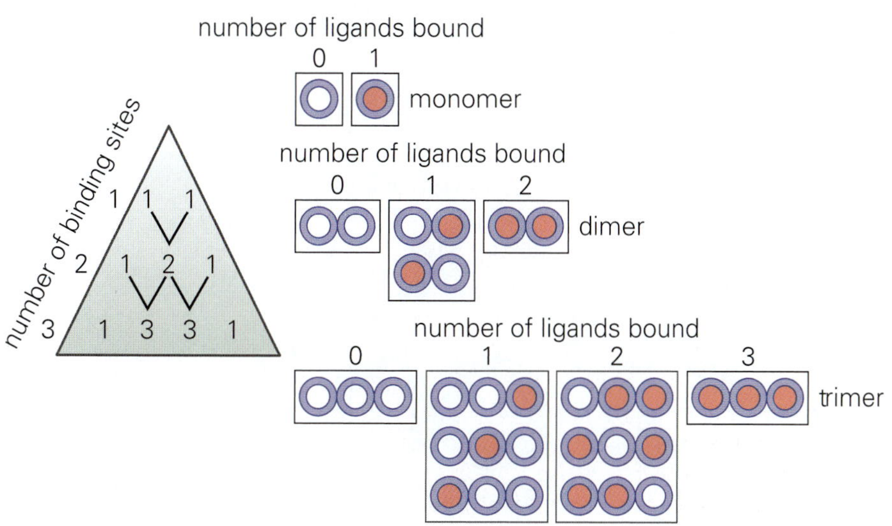


**Figure 7.11 An application of Pascal's triangle to receptor complexes.** The first three rows of Pascal's triangle are shown on the left. In this case, each trial refers to whether a binding site on the receptor is bound to ligand or not. Each receptor molecule is represented by a circle, with the binding site represented by a smaller circle within the larger circle. The binding site is either occupied by ligand (*red* circles) or empty (*white* circles). The first three rows of Pascal's triangle correspond to ligand binding by receptors that form monomers, dimers, and trimers, respectively. Note that the numbers in Pascal's triangle (the binominal coefficients) correspond to the number of different ways in which ligand can bind to the receptor complexes.

complex with seven binding sites or events, the multiplicity for having four ligands bound (or four positive outcomes) is given by:

$$W(7,4)=\frac{7!}{4!3!}=35 \tag{7.8}$$

Note that the fifth entry in the seventh row of Pascal's triangle, corresponding to four ligands bound, is indeed 35 (see Figure 7.10).

## 7.7 The binomial distribution governs the probability of events with binary outcomes

Recall from our discussion of coin flips that, if a coin is unbiased, then any particular series of outcomes has the same probability as any other series. If we ask, instead, about the probability of an *aggregate* outcome (for example, two heads in four coin tosses, without regard to order), then the probability is proportional to the multiplicity of the aggregate outcome. In the case of ligand binding, if we assume that the probability of a receptor molecule being bound to ligand is the same as it being empty, then the situation corresponds exactly to a series of unbiased coin tosses. What is the probability of finding a certain number of ligand molecules bound to a receptor complex? For example, if each complex is a tetramer with four binding sites, what is the probability of seeing receptor complexes with one, two, three, or four ligand molecules bound?

**Binomial distribution**

The probability distribution $P(M, N)$ that determines the probability of obtaining $N$ positive outcomes in a trial with $M$ binary events is known as the binomial distribution. When the number of trials, $M$, is large, the binomial distribution is well approximated by a Gaussian distribution, as explained in Section 7.12.

If we assume that the ligands bind independently of each other, then the probability $P(M, N)$ of seeing $N$ receptor molecules bound to ligand within a complex with $M$ binding sites is given by:

$$P(M,N)=W(M,N)\times\left(\frac{1}{2}\right)^{N}\times\left(\frac{1}{2}\right)^{M-N} \tag{7.9}$$

Equation 7.9 is known as the **binomial distribution**. The first term on the right-hand side of Equation 7.9 is the multiplicity, $W$, which accounts for the number of equivalent rearrangements of the bound and empty receptors, as illustrated for a dimeric receptor in **Figure 7.12**. The second term accounts for the probability of seeing $N$ ligands bound to the receptor complex, which is equivalent to seeing $N$ heads in a series of coin tosses, which is $(1/2)^N$. If there are $N$ ligands bound, then there are $M - N$ empty receptors, and the probability for this occurring is $(1/2)^{M-N}$, which is given by the last term.

We do not expect, in general, that the probability of a binding site being empty will be the same as that of the site being occupied. The probabilities will depend on the interaction energy between the receptor and the ligand, with stronger interactions resulting in higher occupancy. The probability of binding will also depend on the concentration of the ligand—namely, the higher the concentration, the more likely that a receptor molecule will be bound rather than empty.

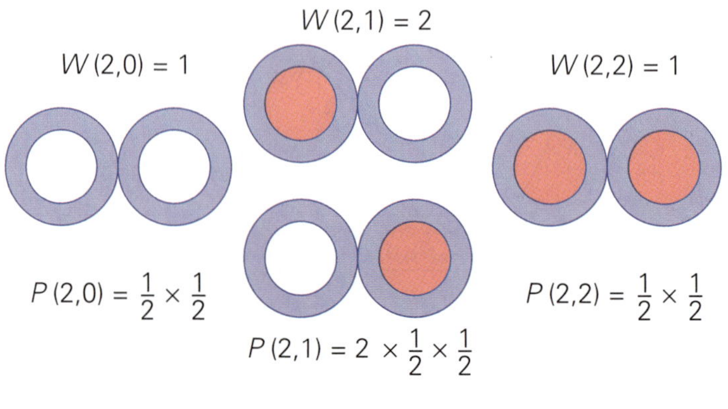

$W(2,0) = 1$

$W(2,1) = 2$

$W(2,2) = 1$

$P(2,0) = \frac{1}{2} \times \frac{1}{2}$

$P(2,1) = 2 \times \frac{1}{2} \times \frac{1}{2}$

$P(2,2) = \frac{1}{2} \times \frac{1}{2}$

**Figure 7.12 The probability of observing different numbers of ligands bound to a dimeric receptor.** The diagram shows a dimeric receptor with no ligands bound (left) and with one or two ligands bounds (middle and right). The probability of any receptor molecule being bound or empty is the same and is equal to 1/2. Receptor binding is assumed to be independent, and so the joint probability for a dimer is the product of the independent probabilities for each site. Each of the four configurations shown here has the same probability of occurring, but we are twice as likely to see receptor dimers with two ligands bound than with one or none because there are two ways in which one ligand can bind to a dimeric receptor.

A detailed analysis of binding equilibria is provided in Chapter 12. For now we shall assume that the probability of an individual receptor being bound is known, and simply focus on the distribution of bound and empty receptor molecules, given this probability.

If the probability of a binding site being occupied by ligand is $p$, then the probability that the site is unoccupied is $(1 - p)$. The probability $P(M, N)$ of finding $N$ binding sites occupied in a receptor complex with $M$ binding sites is then given by:

$$P(M,N) = W(M,N) \times p^{N} \times (1-p)^{M-N} \tag{7.10}$$

Equation 7.10 is a modified form of the binomial distribution that applies to any situation where there is a bias in a binary outcome, as in coin flips involving a weighted coin.

To illustrate the application of Equation 7.10, we return to the experiment described in Section 7.5, in which virus particles are observed using a fluorescence microscope (see Figure 7.8). Each virus particle can bind up to six molecules of a green fluorescent dye. Individual virus particles are identified in the microscope, and the number of dye molecules bound to each one is counted.

The experimental conditions are such that the probability of a dye molecule being bound to any one binding site is known to be 0.7. We can calculate the expected distribution of dye molecules per virus particle by using the binomial distribution, as shown in **Table 7.1**. In this case, the probability, $p$, of a positive outcome (that is, binding site occupied) is 0.7 and that of a negative outcome, $1 - p$, is 0.3. The total number of binding sites, $M$, is 6, and the possible values of $N$ (the number of ligands bound) range from 0 to 6.

The observed frequency distribution for the number of dye molecules per virus particle is shown in **Figure 7.13**, which compares these results with those calculated in Table 7.1. Note that the distribution peaks at $N = 4$—that is, the most probable outcome is that each virus particle is bound to four dye molecules. The agreement between the experimental data and the predictions from the binomial distribution is quite good for three to six ligand molecules bound. For one or two ligand molecules bound, the experimental results are smaller than the predicted values. This is because of the difficulty in identifying receptor complexes with small numbers of ligands bound, because such complexes exhibit lower levels of fluorescence and are therefore more difficult to count accurately.

**Table 7.1 Probability of seeing different numbers of ligand molecules bound to a receptor with six binding sites.**

| Number bound ($N$) | Multiplicity ($W$) | $p^N (1-p)^{M-N}$ | $P = W \times p^N (1-p)^{M-N}$ |
|---|---|---|---|
| 0 | 1 | $(0.3)^6 = 0.00073$ | 0.00073 |
| 1 | 6 | $(0.7) \times (0.3)^5 = 0.00170$ | 0.0102 |
| 2 | 15 | $(0.7)^2 \times (0.3)^4 = 0.00397$ | 0.0596 |
| 3 | 20 | $(0.7)^3 \times (0.3)^3 = 0.00926$ | 0.1852 |
| 4 | 15 | $(0.7)^4 \times (0.3)^2 = 0.02161$ | 0.3241 |
| 5 | 6 | $(0.7)^5 \times (0.3)^1 = 0.05042$ | 0.3025 |
| 6 | 1 | $(0.7)^6 = 0.11765$ | 0.1176 |

In this example, the probability, $p$, of a binding site being occupied is 0.7. The entries in the table show how the probability $P(M, N)$ for different numbers of ligand bound is calculated using Equation 7.10. Here $M$, the total number of binding sites, is 6. $N$, the number of ligands bound, varies from 0 to 6.

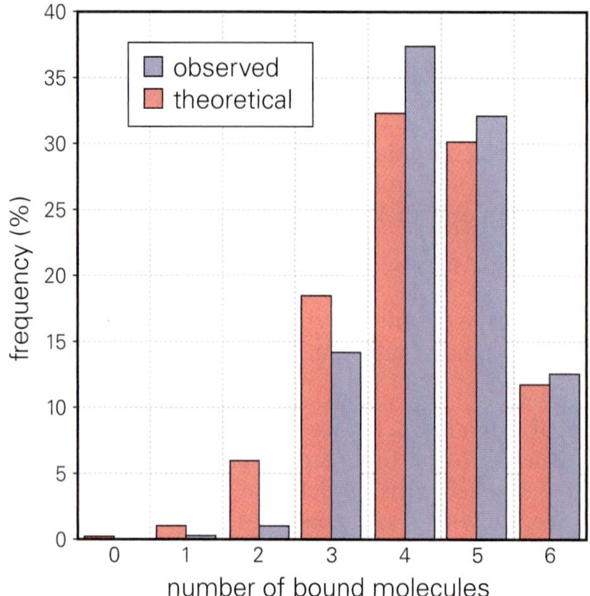


**Figure 7.13 The distribution of dye molecules bound to a virus with six binding sites for the dye.** The data shown here correspond to the experiment depicted in Figure 7.8. A green fluorescent dye binds to a virus that has six binding sites for the dye. The probability of a dye molecule binding to one of the binding sites is 0.7 under the conditions of the experiment. The number of dye molecules bound to individual viruses was counted, and the resulting frequency distribution is shown in *blue*. The *red* distribution is the expected frequency based on the binomial distribution (see Table 7.1). (Experimental data are from H. Zhang et al., and P. Guo, *RNA* 13: 1793–1802, 2007.)

## 7.8 When the number of events is large, Stirling's approximation simplifies the calculation of the multiplicity

Equation 7.4, which expresses the multiplicity, $W$, in terms of factorials, is difficult to apply when the number of events is very large. Consider, for example, a series of 1000 coin tosses. According to Equation 7.4, the multiplicity for the set of outcomes with 500 heads and 500 tails is given by:

$$W(1000,500) = \frac{1000!}{500!500!} \tag{7.11}$$

If you try computing the value of $W$ directly using Equation 7.11, you may find that your calculator cannot handle the factorial terms. Fortunately, we can turn to a useful result in mathematics known as **Stirling's approximation**, which holds true for large integers, $n$:

$$\ln n! \approx n \ln n - n \tag{7.12}$$

How large does $n$ need to be in order for us to be able to use Equation 7.12 to calculate reliable values for $\ln n!$? A more accurate version of Stirling's approximation is given by the following expression:

$$\ln n! \approx n \ln n - n + \tfrac{1}{2}\ln n + \tfrac{1}{2}\ln(2\pi) \tag{7.13}$$

**Stirling's approximation**

Calculations involving the factorials of large numbers are simplified by using Stirling's approximation:

$\ln n! = n \ln n - n$

Stirling's approximation can be used when $n$ is larger than about 100, and it becomes more accurate as $n$ becomes larger.

We are justified in using the simpler expression for $\ln n!$ given in Equation 7.12 if the value of $n$ is large enough that the last two terms in Equation 7.13 can be neglected. As shown in **Table 7.2**, the use of Equation 7.12 instead of Equation 7.13 introduces a small (~2%) error in the value of $\ln n!$ when $n$ is 50. The error is negligible when $n$ is 1000 or greater.

We can now use Stirling's approximation to calculate the value of $W$ when $M$ = 1000 and $N$ = 500. We begin by taking the logarithm of both sides of Equation 7.11:

$$\begin{aligned} \ln W(1000,500) &= \ln\left(\frac{1000!}{500!500!}\right) \\ &= \ln 1000! - (\ln 500! + \ln 500!) \\ &= \ln 1000! - 2 \times \ln 500! \end{aligned} \tag{7.14}$$

We then use Stirling's approximation to evaluate each of the terms in Equation 7.14:

$$\ln 1000! \approx 1000 \ln 1000 - 1000 \tag{7.15}$$

**Table 7.2 Error in using the simpler form of Stirling's approximation to calculate ln*n*!.**

| $n$ | $n \ln n - n$ | $\frac{1}{2} \ln n + \frac{1}{2} \ln (2\pi)$ | Percent error |
|---|---|---|---|
| 50 | 145.6 | 2.87 | 2.0 |
| 100 | 360.5 | 3.22 | 0.9 |
| 1000 | 5907.8 | 4.37 | 0.1 |
| 10,000 | 82,103.4 | 5.52 | <0.0001 |

The entries in the table compare the values of the first two terms in Equation 7.13 ($n \ln n - n$) and the values of the second two terms [$\frac{1}{2} \ln n + \frac{1}{2} \ln (2\pi)$], for different values of $n$. The last column shows the percent error in the value of $\ln n!$ if Equation 7.12 is used to calculate $\ln n!$ instead of Equation 7.13.

and

$$\ln 500! \approx 500 \ln 500 - 500 \tag{7.16}$$

Substituting these expressions for ln 1000! and ln 500! in Equation 7.14, we get:

$$\begin{aligned}\ln W(1000{,}500) &\approx 1000 \ln 1000 - 1000 - 2 \times (500 \ln 500 - 500) \\ &= 6907.755 - 1000 - 2 \times (3107.304 - 500) \\ &= 693.15\end{aligned} \tag{7.17}$$

Now that we know the value of ln $W$, we can calculate the value of $W$ by using the fact that if $\ln W = x$, then $W = e^x$. Thus,

$$W(1000{,}500) = e^{693.15} \tag{7.18}$$

To convert this value to a more recognizable form, use the fact that the value of $e$ is 2.718, which can be written as $10^{0.43429}$. We can therefore express Equation 7.18 in the following way:

$$W(1000{,}500) = e^{693.15} = \left(10^{0.43429}\right)^{693.15} \approx 10^{301} \tag{7.19}$$

Equation 7.19 tells us that, for a series of 1000 coin tosses, the multiplicity of outcomes with 500 heads and 500 tails is approximately $10^{301}$, an enormously large number.

## 7.9 The relative probability of two outcomes is given by the ratios of their multiplicities

What is the probability of observing some other distribution of heads and tails, such as 600 heads ($N = 600$) in a series of 1000 coin tosses? To calculate the relative probabilities of two aggregate outcomes (for example, $N = 500$ and $N = 600$), we calculate the ratio of their multiplicities. Let us start by calculating the multiplicity for outcomes with 600 heads in a series of 1000 coin tosses:

$$W(1000{,}600) = \frac{1000!}{600!\,400!} \tag{7.20}$$

Taking the logarithm of both sides of Equation 7.20 and using Stirling's approximation:

$$\begin{aligned}\ln W(1000{,}600) &\approx 1000 \ln 1000 - 1000 - \\ &\quad (600 \ln 600 - 600) - (400 \ln 400 - 400) \\ &= 6908 - 1000 - (3838 - 600) - (2397 - 400) \\ &= 673\end{aligned} \tag{7.21}$$

Thus,

$$W(1000{,}600) = e^{673} = \left(10^{0.434}\right)^{673} \approx 10^{292} \tag{7.22}$$

**Probability and multiplicity**

The probability of an aggregate outcome that corresponds to equivalent individual events (for example, seeing a particular configuration of molecules) is proportional to the multiplicity of that outcome. Thus, the relative probabilities of two outcomes (A and B) are given by the ratio of their multiplicities:

$$\frac{P_A}{P_B} = \frac{W_A}{W_B}$$

This number ($10^{292}$) is so large that its significance is difficult to grasp. To help you understand what this number means, we shall calculate the probability that a series of 1000 coin tosses results in 600 heads and compare it with the probability of obtaining an equal number of heads and tails (that is, 500 each) for the same number of coin tosses. The probability of obtaining 600 heads in 1000 coin tosses, denoted $P(1000,600)$, is given by:

$$P(1000,600) = \frac{W(1000,600)}{\text{total number of outcomes}} = \frac{W(1000,600)}{2^{1000}} \tag{7.23}$$

How much more likely are we to get 500 heads in a series of 1000 coin tosses than 600 heads? The probability of observing 500 heads in a series of 1000 coin tosses is given by:

$$P(1000,500) = \frac{W(1000,500)}{2^{1000}}$$

The ratio of the two probabilities is therefore given by:

$$\frac{P(1000,500)}{P(1000,600)} = \frac{W(1000,500)}{W(1000,600)} \tag{7.24}$$

We begin by calculating the logarithm of the ratios of the multiplicities:

$$\ln\left[\frac{W(1000,500)}{W(1000,600)}\right] = \ln\left[W(1000,500)\right] - \left[\ln W(1000,600)\right] \tag{7.25}$$

Substituting the values for $\ln W(1000, 500)$ and $\ln W(1000, 600)$ from Equations 7.17 and 7.21, we get:

$$\ln\left[W(1000,500)\right] - \left[\ln W(1000,600)\right] = 693 - 673$$

$$\Rightarrow \ln\left[\frac{W(1000,500)}{W(1000,600)}\right] = 20 \tag{7.26}$$

$$\Rightarrow \frac{W(1000,500)}{W(1000,600)} = e^{20} = \left(10^{0.434}\right)^{20} \approx 10^{8.68} = 4.8\times 10^{8} \tag{7.27}$$

In a series of 1000 coin tosses, therefore, an outcome with an equal number of heads and tails is about 480 million times more likely than an outcome with 60% heads and 40% tails. In practical terms, the odds of obtaining a sequence with 60% heads is very nearly zero compared to the odds of getting an equal number of heads and tails.

## 7.10 As the number of events increases, the less likely outcomes become increasingly rare

We can use Equation 7.4 or Pascal's triangle (see Figure 7.10) to calculate the value of $W$ for all possible outcomes of tossing a coin four times, as shown in **Table 7.3**. As shown in that table, the most probable outcome is the one in which there is an equal number of heads and tails. But, for such a small number of coin tosses, there is a reasonable probability that some other outcome will be obtained, such as one with one tail and three heads.

The data in Table 7.3 are graphed in **Figure 7.14**, which shows the multiplicity $[W(M, N)]$ as a function of different numbers of heads ($N$) for four coin tosses ($M = 4$). The maximum value of $W$ is 6 and it is obtained for $N = 2$ (an equal number of heads and tails); the minimal value of $W$ is 1 and it is obtained for either $N = 0$ or $N = 4$.

As the number of coin tosses ($M$) increases, the distribution of $W$ becomes more and more sharply peaked (Figure 7.14B and C). Assuming that the coin is fair, the distribution is always centered at the value of $N$ corresponding to an equal

**Table 7.3 Multiplicity of various outcomes of four coin tosses.**

| Outcome | Example | $W = \frac{M!}{N!(M-N)!}$ <br> $M$ = number of tosses <br> $N$ = number of heads |
|---|---|---|
| No heads | TTTT | $W = \frac{4!}{0!\,4!} = 1$ |
| 1 head | HTTT | $W = \frac{4!}{1!\,3!} = 4$ |
| 2 heads | THTH | $W = \frac{4!}{2!\,2!} = 6$ |
| 3 heads | THHH | $W = \frac{4!}{3!\,1!} = 4$ |
| 4 heads | HHHH | $W = \frac{4!}{4!\,0!} = 1$ |

number of heads and tails and, as $M$ gets larger, the values of $W$ for other values of $N$ that are away from the mean value become vanishingly small.

Consider, for example, the ratio of the maximal and minimal values of $W$ for a series of 10 coin tosses ($M = 10$). In this case, the multiplicity, $W$, for an outcome with five heads ($N = 5$) is given by:

$$W(10,5) = \frac{10!}{5!\,5!} = \frac{3{,}628{,}800}{120 \times 120} = 252 \quad (7.28)$$

The minimum values of $W$ are obtained for $N = 0$ and $N = 10$ [$W(10, 1) = W(10, 10) = 1$]. This means that there are 252 ways of getting five heads and five tails in a series of 10 coin tosses, compared to only one way of getting all tails ($N = 0$) or only one way of getting all heads ($N = 10$).

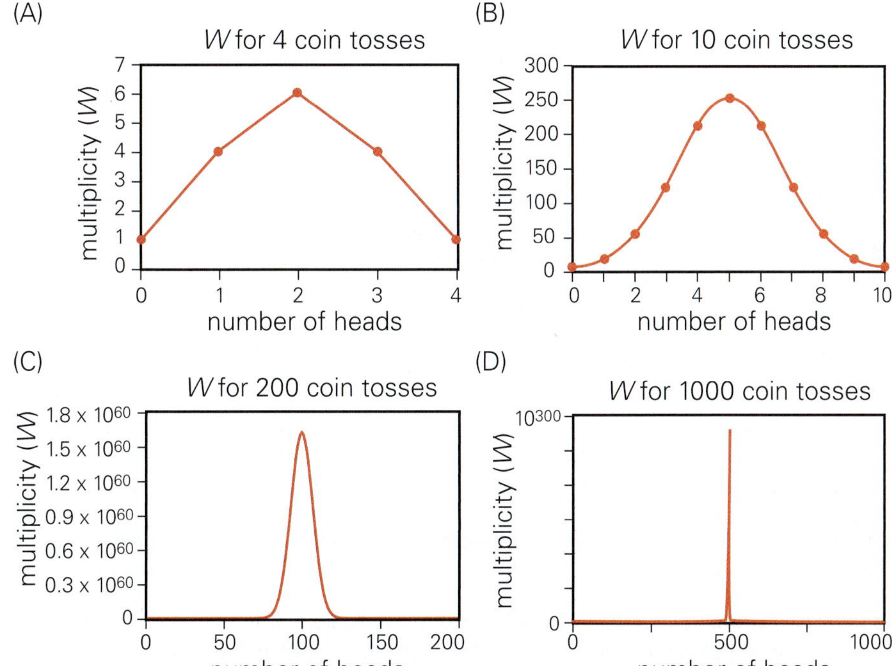


**Figure 7.14 The multiplicity of different outcomes for a series of coin tosses.** (A) The multiplicity for outcomes with 0, 1, 2, 3, and 4 heads in a series of four coin tosses is shown (see Table 7.1). The multiplicity as a function of the number of heads is shown for 10, 200, and 1000 coin tosses in B, C, and D, respectively. As the number of coin tosses increases, the distribution of multiplicities becomes sharply peaked, until the multiplicity of the most likely result is enormously greater than the multiplicities for other results.

Figure 7.14 also graphs the multiplicity when $M = 200$ and when $M = 1000$. For the large values of $M$, such as 200 or 1000, the value of $W$ very near the center of the distribution is vastly larger than for outcomes away from the center.

## 7.11 For coin tosses, outcomes with equal numbers of heads and tails have maximal multiplicity

For 1000 coin tosses, the outcome with $N = 500$ (that is, with an equal number of heads and tails) is most likely because there are more ways of achieving this outcome than any other. That is, the outcome with an equal number of heads and tails has a higher multiplicity than any other outcome—the *multiplicity is maximal.*

To show mathematically that the multiplicity is maximal, we ask, "What is the value of $N$ (that is, the number of heads), given a value of $M$ (that is, the number of coin tosses), for which the multiplicity, $W$, is maximal?" Because $W$ is a function of $N$ (Equation 7.4), one condition for $W$ to be maximal is:

$$\frac{dW}{dN} = 0 \tag{7.29}$$

Equation 7.29 also holds for values of $N$ for which $W$ is at a minimum (**Figure 7.15A**). The minimal value of $W$ occurs for $N = 0$ or $N = M$, so we exclude these two values from consideration.

If $W$ is maximal for a particular value of $N$, then $\ln W$ is also maximal, as shown in Figure 7.15. This occurs because the value of $\ln W$ rises and falls as the value of $N$ rises and falls. Therefore, another condition for $W$ to be maximal is:

$$\frac{d \ln W}{dN} = 0 \tag{7.30}$$

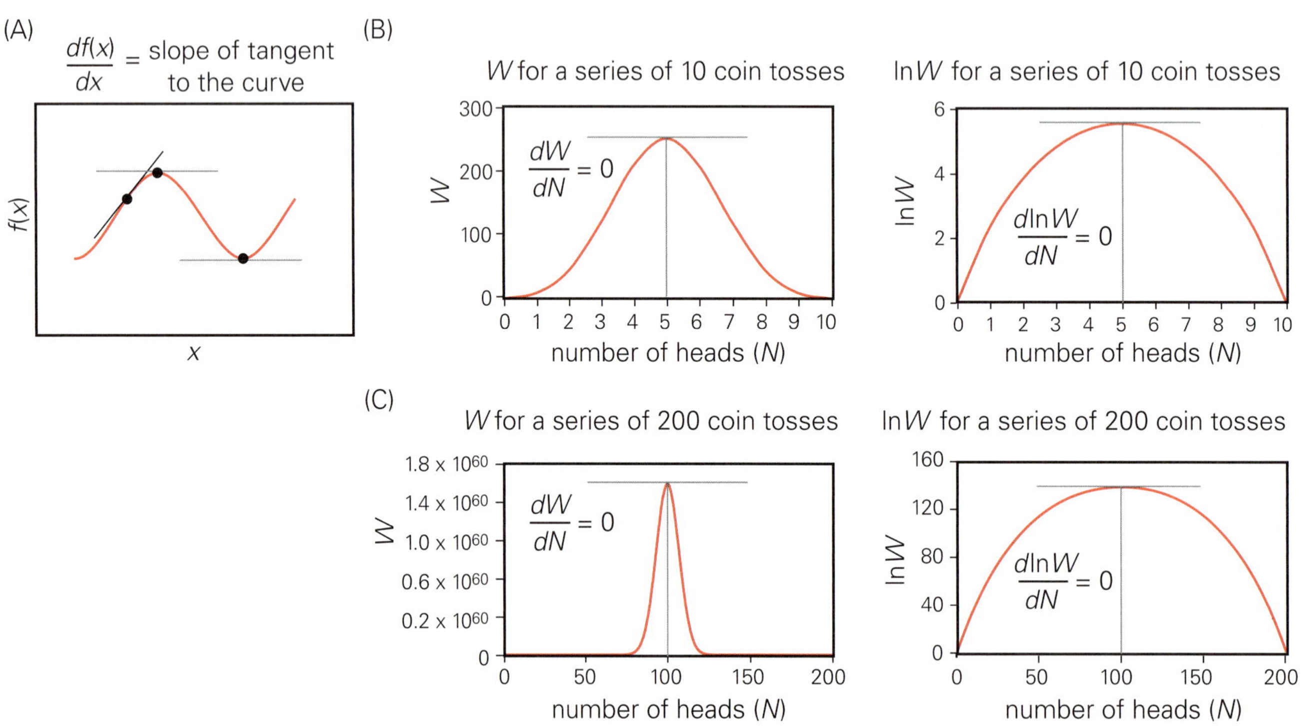


**Figure 7.15 Determining the value of *N* (number of heads) for which the multiplicity, *W*, is maximal.** (A) The derivative of a function of $x$, $f(x)$, with respect to $x$, is the slope of the line that is tangent to the function at $x$. When the function has maximal or minimal value, the slope is zero (the tangent line is horizontal). In (B), the multiplicity, $W$, and the logarithm of the multiplicity, $\ln W$, are shown as a function of $N$ for 10 coin tosses. The maximum multiplicity corresponds to $N = 5$ (equal number of heads and tails). The derivatives of both $W$ and $\ln W$ are zero for $N = 5$. (C) As in (B), but for 200 coin tosses. The derivatives of $W$ and $\ln W$ are zero when $N = 100$.

Equation 7.30 is useful because we can apply Stirling's approximation (Equation 7.12) to evaluate the derivative. If we combine Equation 7.30 with Equation 7.4 and then apply Stirling's approximation, it can be shown that:

$$\frac{d\ln W}{dN}=\frac{d}{dN}\ln\left[\frac{M!}{N!(M-N)!}\right]=-\ln N+\ln(M-N) \tag{7.31}$$

Equation 7.31 holds true when $M$ and $N$ are both large numbers.

For $W$ to be maximal, the left-hand side of Equation 7.31 must be zero. Thus,

$$0=-\ln N+\ln(M-N) \tag{7.32}$$

$$\Rightarrow \ln N=\ln(M-N)$$

$$\Rightarrow N=M-N$$

$$\Rightarrow N=M/2 \tag{7.33}$$

Equation 7.33 confirms our intuitive expectation that the outcome that maximizes the multiplicity ($W$) is one in which the number of heads ($N$) is equal to the number of tails.

As the number of coin tosses ($M$) increases, the ratio of the multiplicity for the most likely outcome (that is, equal numbers of heads and tails) to any other outcome increases sharply. For very large values of $M$, the only outcomes that will be observed in practical terms are ones in which a very nearly equal number of heads and tails are obtained (see Figure 7.14). This result, obtained here for the simple case of coin tosses, is broadly general: the observed outcome of an experiment involving a large number of trials (whether it is coin tosses or reacting molecules) is the one that corresponds to the maximum multiplicity.

Much of thermodynamics is concerned with defining the conditions for which the multiplicity of the system and its surroundings (or more practically, $\ln W$) is maximal.

## 7.12 When the number of events is very large, the probability distribution is well approximated by a Gaussian distribution

If the number of events in series of trials with binary outcomes is large, we have seen that Stirling's approximation can be used to simplify the calculation of the multiplicity. We now show that when this is done, an additional simplification occurs, which is that the binomial probability distribution is well approximated by a function of continuous variables known as the **Gaussian distribution**.

If the number of events in a series is $M$ and the probability of a positive outcome is $p$, then the probability of obtaining $N$ positive outcomes in a series of $M$ events is given by:

$$P(M,N)=W(M,N)\times p^N\times(1-p)^{M-N}$$

$$=\frac{M!}{N!(M-N)!}\times p^N\times(1-p)^{M-N} \tag{7.34}$$

Equation 7.34 is the same as Equation 7.10, and you can refer to Section 7.7 for an explanation of the various terms in the equation. $N$ is a discrete integer variable in Equation 7.34 (that is, $N$ can take on the values 1, 2, 3, ..., $M$). When $M$ is very large, we can consider $N$ to be a continuous rather than a discrete variable because the interval between consecutive values is small compared to the range of $N$ (**Figure 7.16**). In order to emphasize the switch from an integer variable to a continuous one, we rewrite Equation 7.34 by using $x$ instead of $N$ to represent the number of positive outcomes. Now, for a series of $M$ events, the probability of finding

**Gaussian distribution**

A probability distribution for the values of a variable, $x$, of the form:

$P(x) = Ce^{-ax^2}$

is known as a Gaussian distribution. The parameters $a$ and $C$ are constants that define the width of the distribution and ensure that it is normalized.

$x$ positive outcomes is given by:

$$P(M,x)=\frac{M!}{x!(M-x)!}\times p^x\times(1-p)^{M-x} \quad (7.35)$$

When $M$ is large, we expect the probability, $P(M, x)$, to be small when $x$ is small. This is because, when $M$ is large, the probability for obtaining only a small number of positive outcomes will also be small. We can, therefore, restrict the analysis to values of $x$ that are also large. When $M$ and $x$ are both large, we can apply Stirling's approximation to Equation 7.35, as explained in **Box 7.1**. The probability distribution is then well approximated by:

$$P(x)=\frac{1}{\sqrt{2\pi}\times\sigma}\exp\left[\frac{-(x-\mu)^2}{2\sigma^2}\right] \quad (7.36)$$

The expression on the right-hand side of Equation 7.36 is known as a **Gaussian function**, and the probability distribution expressed in this way is known as a Gaussian distribution or a **normal distribution**. The two parameters $\mu$ and $\sigma$ are the **mean** and **standard deviation** of the distribution, respectively. The mean, $\mu$, is the center of the Gaussian distribution (for example, $x = 50$ in Figure 7.16C), whereas the standard deviation, $\sigma$, is related to the width of the distribution (see below).

## 7.13 The Gaussian distribution is centered at the mean value and has a width that is proportional to the standard deviation

A discrete distribution, such as the one shown in Figure 7.16B, is characterized by the parameters $M$ (the number of trials) and $p$ (the probability of a favorable outcome in any individual trial). To convert the discrete distribution to a Gaussian distribution, we use the following relationships between $M$ and $p$ (for the discrete distribution) and $\mu$ and $\sigma$ (for the corresponding Gaussian distribution), as explained in Box 7.1:

$$\mu = M\,p \quad (7.37)$$

$$\sigma^2 = M\,p(1-p) \quad (7.38)$$

The Gaussian distribution has a maximum value when $x = \mu$ (the mean value)—that is, when the exponent is zero. Because of the squared term in the exponent, the Gaussian distribution is symmetrical about the mean value. One parameter of interest is the width of the distribution at half-maximal height, which tells us

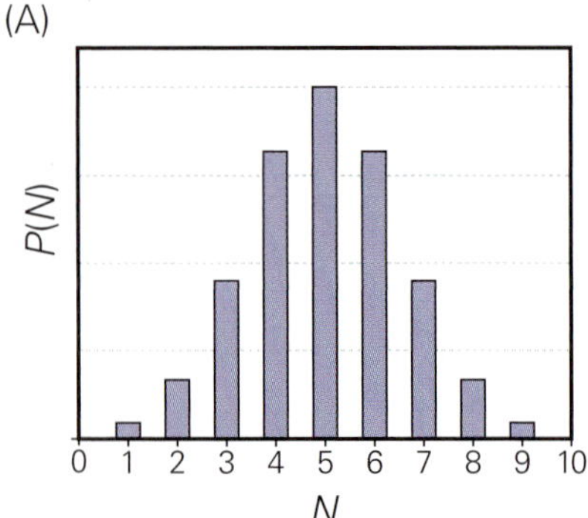


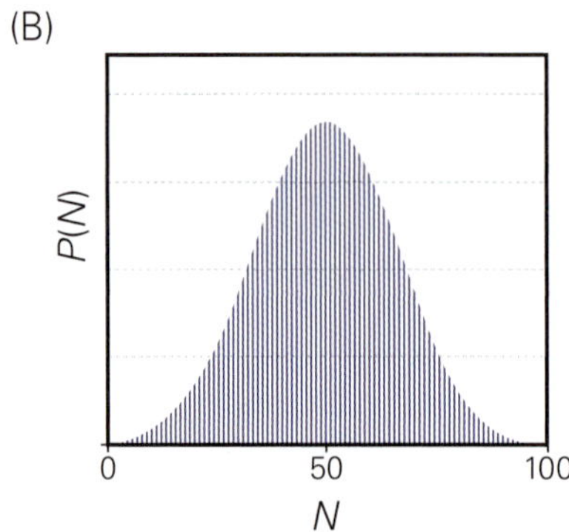


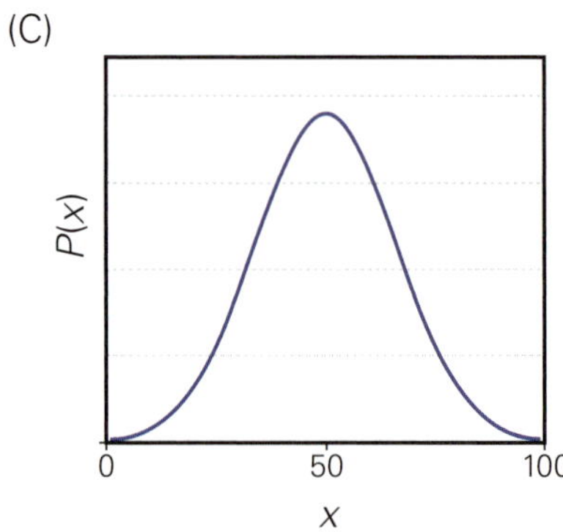


**Figure 7.16 Switching from discrete to continuous variables.** (A) The probability distribution, $P(N)$, for a series of 10 events. Here $N$, the number of positive outcomes, has discrete integer values ranging from 0 to 10. (B) The probability distribution for a series of 100 events. $N$ has integer values ranging from 0 to 100, but the increment in $N$ is small compared to the range, so the values of $N$ are spaced closely together. (C) The same probability distribution as in (B), but we have replaced $N$ with a continuous variable $x$.

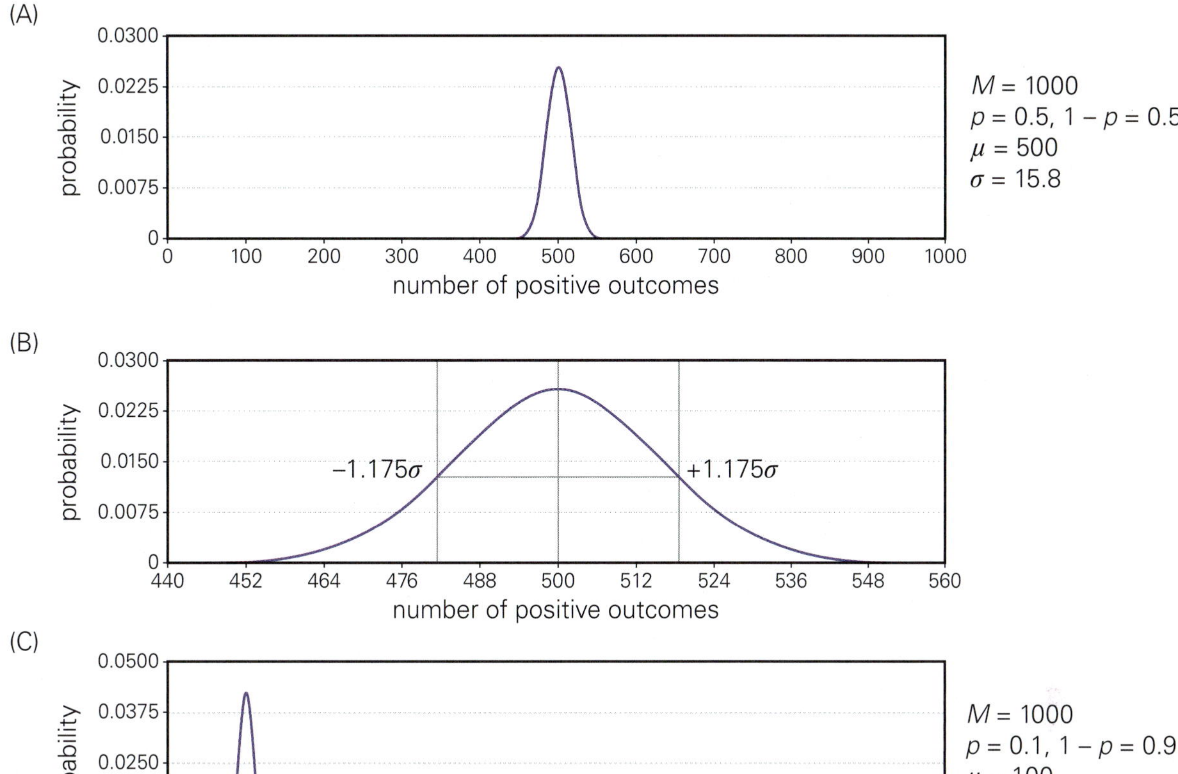


**Figure 7.17 Gaussian distributions.** When the number of events in a series of trials, $M$, is large, then the probability of having different numbers of positive outcomes is given by a Gaussian distribution. (A) A Gaussian distribution for a series of 1000 events ($M$ = 1000). The probability of a positive outcome is the same as that for a negative outcome ($p$ = 0.5, $1 - p$ = 0.5). (B) An expanded view of the distribution shown in (A), centered around the mean number of positive outcomes ($\mu$ = 500). The vertical lines on either side of the mean value correspond to numbers of positive outcomes that are 1.175 standard deviations away from the mean value. The distance between these values, indicated by the horizontal line, is the width of the distribution at half maximal height. (C) A Gaussian distribution for a series of 1000 binary events. The outcome for each event is biased, with positive outcomes disfavored (that is, $p$ = 0.1, $1 - p$ = 0.9). The mean number of positive outcomes is 100.

something about the range of values of $x$ for which the probability is appreciably large (**Figure 7.17**).

To obtain an expression for the width of the distribution at half of the maximum height, we calculate the value of $P(x_{1/2}) = {}^1/_2 P(\mu)$ and solve for $x_{1/2}$, where $x_{1/2}$ is the value of $x$ when $P$ is half maximal:

$$P(x_{1/2}) = \frac{1}{\sqrt{2\pi}\times\sigma}\exp\left[\frac{-(x_{1/2}-\mu)^2}{2\sigma^2}\right]$$

$$= \frac{1}{2}\frac{1}{\sqrt{2\pi}\times\sigma}\exp\left[\frac{-(\mu-\mu)^2}{2\sigma^2}\right]$$

$$\Rightarrow \exp\left[\frac{-(x_{1/2}-\mu)^2}{2\sigma^2}\right] = \frac{1}{2}$$

$$\Rightarrow \frac{-(x_{1/2}-\mu)^2}{2\sigma^2} = \ln\left(\frac{1}{2}\right) = -0.69$$

$$\Rightarrow (x_{1/2}-\mu) = \pm\sqrt{0.69\times 2\sigma^2} = \pm 1.175\sigma \qquad (7.39)$$

According to Equation 7.39, the half-width of the distribution is $1.175\sigma$, and so the full width of the distribution is $2(x_{1/2} - \mu) \approx 2.35\sigma$. This means that the width of the distribution is proportional to the standard deviation and is proportional, therefore, to $M^{1/2}$. In other words, the width of the distribution increases as the square root of the number of events (for example, coin flips). If 100 times as many events are considered, the width of the distribution will increase only 10-fold. This

## Box 7.1 Converting from a discrete binomial distribution to a continuous Gaussian distribution

Here we explain the relationship between the binomial distribution and the Gaussian distribution for the probabilities of outcomes of a series of events. We assume that the events are binary—that is, each individual event has one of two outcomes (for example, heads or tails in a series of coin flips). If the probability of each outcome is the same (that is, the coin is "fair"), then the probability of either outcome is 1/2. For a series of $M$ events, the probability for the $N$ positive outcomes is:

$$P(M,N)=W(M,N)\frac{1}{2^M} = \frac{M!}{N!(M-N)!}\,\frac{1}{2^M} \quad (7.1.1)$$

This is a **binomial distribution**, which is a discrete function of the variables $M$ and $N$, and is defined for integer values of $M$ and $N$ (where $N \leq M$).

In the more general case, the probabilities of positive and negative outcomes are not equal. If we define $p$ to be the probability of a positive outcome (for example, heads), then $(1 - p)$ is the probability of the other result (for example, tails). The probability for a specific combination of results from $M$-independent trials is just the product of the individual probabilities. A more general form for the probability for obtaining $N$ positive results specified by probability $p$ and $(M - N)$ negative results with probability $1 - p$ is then:

$$P(N;M,p)=\frac{M!}{N!(M-N)!}p^N(1-p)^{M-N} \quad (7.1.2)$$

This remains a discrete function in which $M$ and $N$ are integers, although the probability $p$ can take on any value between 0 and 1. As the numbers of events ($M$) and positive outcomes ($N$) get large, the factorial terms become intractable, and so Sterling's approximation for the natural logarithm of factorials is used. As discussed in Section 7.12, when $M$ is large we can switch to a continuous variable $x$ that replaces $N$ as the number of positive outcomes (see Figure 7.16).

A brief derivation will follow to show that the probability function $P(N;\, M,\, p)$ is very well described by:

$$P(x)=\frac{1}{\sqrt{2\pi}\times\sigma}\exp\left[\frac{-(x-\mu)^2}{2\sigma^2}\right] \quad (7.1.3)$$

In Equation 7.1.3, $\mu$ is the **mean value** of $x$, with $\mu = Mp$, and $\sigma^2$ is the **variance** of $x$, with $\sigma^2 = Mp(1 - p)$. The standard deviation of $x$ is the the square root of the variance ($\sigma$).

$P(x)$ is called a **Gaussian function**, which is defined by the exponential of the negative square of the variable. Based on the definitions of $\mu$ and $\sigma$, $P(x)$ is still an implicit function of $M$ and $p$.

### The Derivation

As noted above, the equivalence between the discrete and continuous distributions is derived by using Stirling's approximation. Using fair coin flips as an example, we use $H$ as the number of heads, $T$ as the number of tails, and $M$ as the total number of coin flips. $M = H + T$, and $D$ is the difference between heads and tails, $D = H - T$. The result obtained using these variables is actually completely general, because we can just consider $H$ to be the number of occurrences of any outcome of interest and replace $^1/_2$ with $p$ to get the more general form.

The mean number of heads will be $\mu = {}^M/_2$. These definitions can be combined to give $H = (M + D)/2$ and $T = (M - D)/2$, or equivalently, $D/2 = H - \mu$ and $T/2 = H + \mu$.

With $M$ and $H$ specified, the values of $T$ and $D$ are also defined. Using these values, we now cast the probability in terms of the variables $M$ and $D$:

$$P(H)=\frac{M!}{H!\;T!\;2^M}=\frac{M!}{\left[\frac{1}{2}(M+D)\right]!\;\left[\frac{1}{2}(M-D)\right]!\;2^M}$$

We then use Stirling's approximation with the extra terms that make it more accurate (see Equation 7.13):

$$\ln(x!)=\tfrac{1}{2}\ln(2\pi)+(x+\tfrac{1}{2})\ln x-x$$

Applying this to the factorial terms in $P(H)$ above, rearranging, and combining terms gives:

$$\ln\left[P(H)\right]=\ln\left[\tfrac{2}{\pi M}\right]-\tfrac{1}{2}(M+D+1)\ln\left[1+\tfrac{D}{M}\right]-\tfrac{1}{2}(M-D+1)\ln\left[1-\tfrac{D}{M}\right]$$

Then we need one more approximation—namely, from series expansions when $x$ is small (that is, $M >> D$):

$$\ln(1+x)\approx x-\tfrac{1}{2}x^2 \quad\text{and}\quad \ln(1-x)\approx -x-\tfrac{1}{2}x^2$$

Applying these gives:

$$\ln\left[P(H)\right]\approx\tfrac{1}{2}\ln\left(\frac{2}{\pi M}\right)-\frac{D^2}{2M}$$

Then, taking the exponential of each side:

$$P(H)=\left(\frac{2}{\pi M}\right)^{\frac{1}{2}}e^{\frac{-D^2}{2M}}$$

Using the definitions given initially, this can be rewritten in terms of $\mu$, specifically substituting $D = 2(H - \mu)$ and $\sigma^2 = M/4$. And then, to emphasize the conversion to a continuous variable, we substitute $x$ for $H$, which together give the result previously obtained:

$$P(x)=\left(\frac{1}{2\pi}\right)^{\frac{1}{2}}\frac{1}{\sigma}\,e^{\frac{-(x-\mu)^2}{2\sigma^2}}$$

The numbers in front of the actual Gaussian function keep it normalized so that adding the probabilities of all possible values of $x$ gives 1.

In spite of the approximations used that require that $M$ and $N$ both be large, the Gaussian function is remarkably good, even for modest values of these variables. The graph in **Figure 7.1.1** compares the true binomial distribution for 12 coin flips with the distribution predicted by the Gaussian function. Overall the two functions are very close.

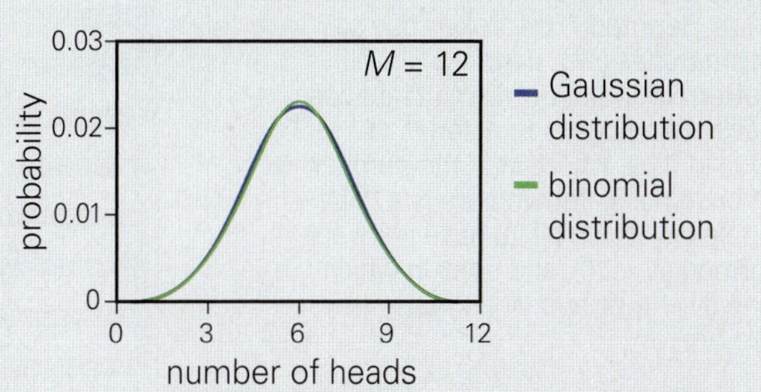


**Figure 7.1.1 Comparison of the true binomial distribution for 12 coin flips (*green*) and the predicted values from the appropriate Gaussian function (*blue*).** For larger values of $M$, the agreement is even better.

means that as the range of values increases, the distribution becomes more and more sharp relative to the number of events.

## 7.14 Application of the Gaussian distribution enables statistical analysis of a series of binary outcomes

An advantage of using the Gaussian distribution is that it is much easier to use for calculations as a function of $x$, the number of positive outcomes, than the discrete version. Because the Gaussian distribution is a function of a continuous variable, one can also calculate the derivatives of the function and do other mathematical manipulations.

As an example, consider a series of 1000 flips of a fair coin, meaning that $p = 1/2$. Using Equations 7.37 and 7.38, we find that the mean value for the number of heads in such a series of coin tosses is given by $\mu = 1000 \times 1/2 = 500$, and the standard deviation is given by $\sigma = (1000 \times 1/2 \times 1/2)^{1/2} = 15.8$. The width of the distribution is $2.355 \times 15.8 \approx 37$. This means that if sets of 1000 coin flips are repeated many times, then on the average we will obtain 500 heads and 500 tails. But, for any one particular set of 1000 coin flips, results of $500 \pm 19$ heads are reasonably probable (with a probability within 50% of the most probable result). The probability of larger deviations from the average drops off rapidly. The distribution of probabilities, calculated using the Gaussian function, is shown for this case in **Figure 7.18**. Integrating this curve shows that the probability of being within $1\sigma$ of the mean is 68%, within $2\sigma$ of the mean is 95.5%, and within $3\sigma$ of the mean is 99.7% (see Figure 7.18).

For a biologically relevant example, consider DNA double helices in which mismatches occur at random base pairs due to errors in the DNA replication process. The analogy to coin flips is illustrated in **Figure 7.19**. Each DNA duplex is

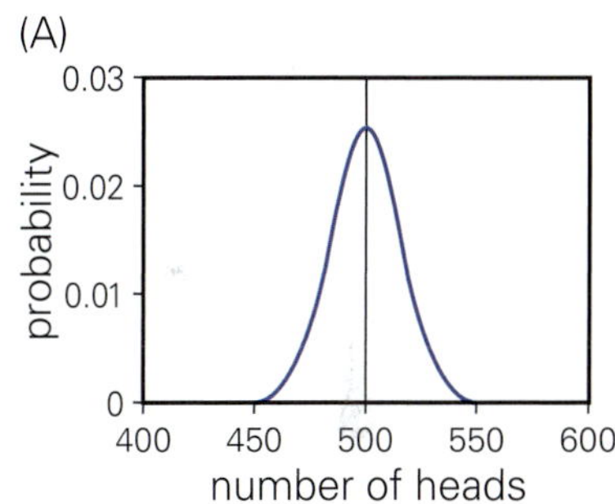


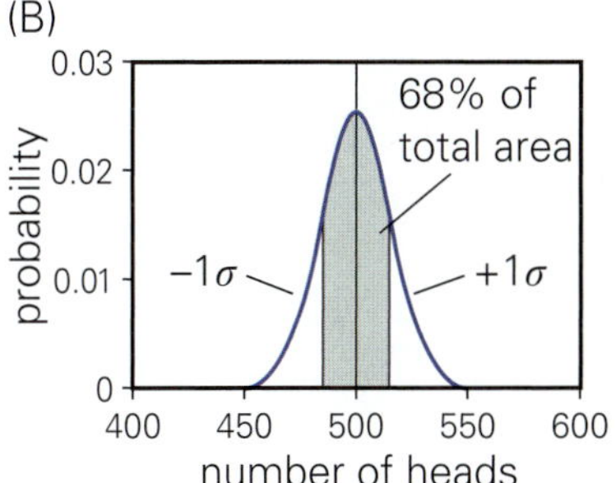


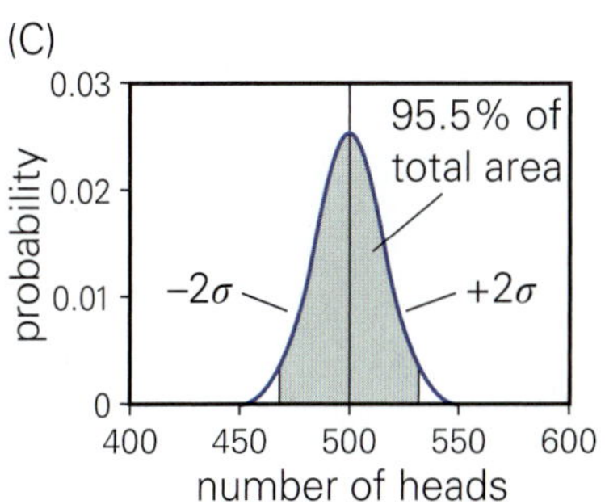


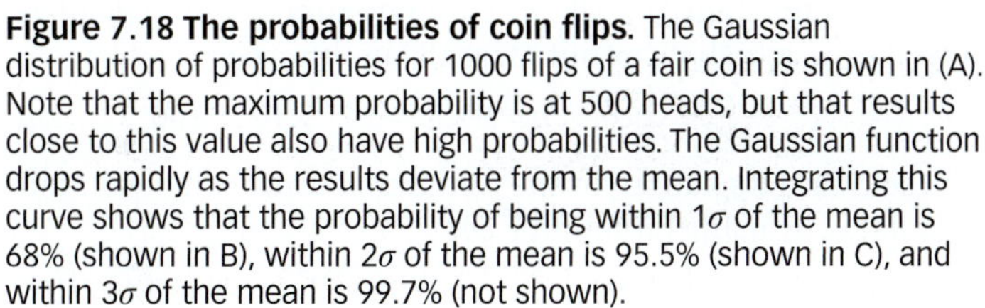

**Figure 7.18 The probabilities of coin flips.** The Gaussian distribution of probabilities for 1000 flips of a fair coin is shown in (A). Note that the maximum probability is at 500 heads, but that results close to this value also have high probabilities. The Gaussian function drops rapidly as the results deviate from the mean. Integrating this curve shows that the probability of being within $1\sigma$ of the mean is 68% (shown in B), within $2\sigma$ of the mean is 95.5% (shown in C), and within $3\sigma$ of the mean is 99.7% (not shown).

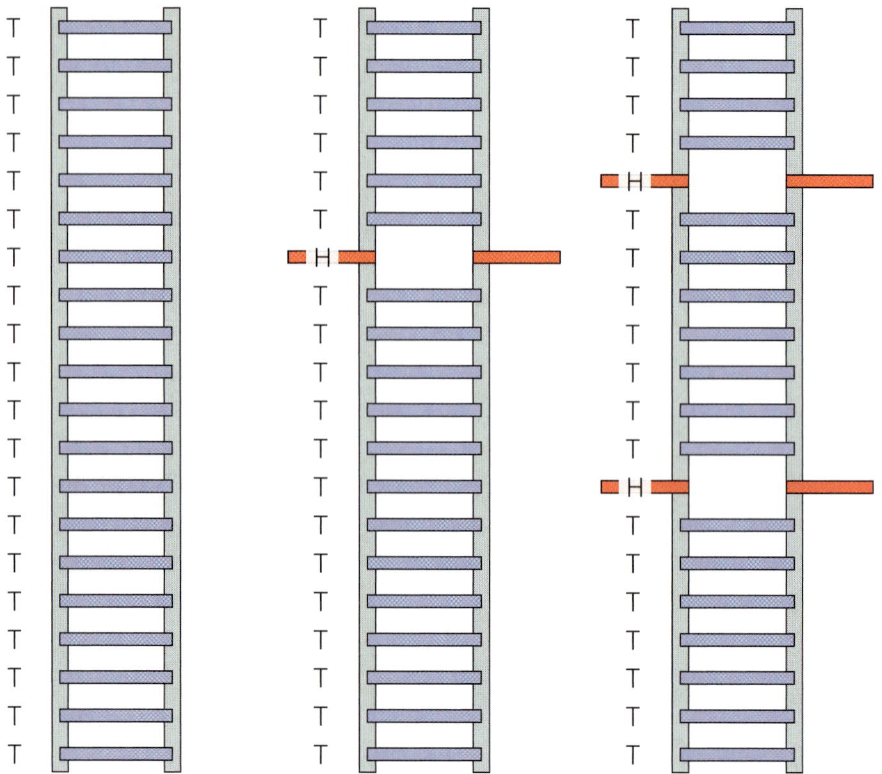


**Figure 7.19 An analogy between mismatches in DNA and coin flips.** A schematic representation of DNA double helices is shown here, with base pairs either formed (*blue*, denoted T) or broken due to mismatches (*red*, denoted H). The pattern of intact or mismatched base pairs is analogous to a series of $M$ coin flips, where $M$ is the number of base pairs. Shown here is a DNA duplex in which all 20 base pairs are formed ($N$ = 20), and ones in which one ($N$ = 1) or two ($N$ = 2) base pairs are broken.

considered to be a series of trials in which each base pair is an "event." Since our focus is on mismatches, we shall consider a mismatched base pair to be a positive outcome (that is, a "heads"). If the base pair is correctly formed, we shall consider it to be a negative outcome (that is, a "tails").

As discussed in Chapter 19, DNA is replicated by DNA polymerase with remarkably high accuracy, with errors being introduced in the newly synthesized strand with a very low frequency (for example, one mistake being made per million nucleotides copied). Most of these errors are corrected by a number of repair processes in the cell and so, at the end of the replication process, there are remarkably few errors. In such situations, the number of positive outcomes (mismatches) may be so small that we cannot apply the Gaussian distribution to calculate the probabilities of mismatches.

There are other situations, however, in which the replication process can be quite error-prone. This can occur naturally under conditions where the organism is being stressed, or in a test tube when an experimenter wishes to introduce variation in a population of DNA molecules. As long as there is a reasonable number of mismatches in the DNA molecule (for example, more than 50), the Gaussian distribution can be used to tell us what to expect. To do this we shall assume that the probability of an error being introduced is the same at each base pair, and that these probabilities are independent of each other.

We denote the probability of a mismatch being introduced as $p$. If, on average, the replication process makes one error per 10,000 nucleotides replicated, then $p = 10^{-4}$. If we consider a DNA molecule with one million nucleotides, as in a bacterial chromosome, then $M = 10^6$. We use Equation 7.24 to calculate the mean value of mistakes, $\mu$, which turns out to be 100 in this case:

$$\mu = p \times M = 10^{-4} \times 10^{6} = 100 \tag{7.40}$$

The standard deviation for the number of errors incorporated in the DNA can be calculated by using Equation 7.38:

$$\sigma = \sqrt{p \times (1 - p) \times M} = \sqrt{10^{-4} \times (1 - 10^{-4}) \times 10^{6}} = \sqrt{100} = 10 \tag{7.41}$$

With the values of the mean and standard deviation in hand, we can calculate the probability of finding a specific number of mismatches in the chromosome. As an example, let us calculate the probabilities of finding 100, 110, and 150 mismatches in a DNA duplex, given this error rate. First, for 100 mismatches:

$$P(100) = \frac{1}{\sqrt{2\pi} \times \sigma} e^{-(100-\mu)^2/2\sigma^2} = \frac{1}{25.1} e^0 = \frac{1}{25.1} = 0.0398 \tag{7.42}$$

Now, for 110 mismatches (that is, one standard deviation away from the mean value):

$$P(110) = \frac{1}{\sqrt{2\pi} \times \sigma} e^{-(110-\mu)^2/2\sigma^2} = \frac{1}{25.1} e^{-10^2/2\times 100} = \frac{0.606}{25.1} = 0.0241 \tag{7.43}$$

Finally, for 150 mismatches (that is, five standard deviations away from the mean value):

$$\begin{aligned} P(150) &= \frac{1}{\sqrt{2\pi} \times \sigma} e^{-(150-\mu)^2/2\sigma^2} = \frac{1}{25.1} e^{-50^2/2\times 100} \\ &= \frac{3.7\times 10^{-6}}{25.1} = 0.147 \times 10^{-6} \end{aligned} \tag{7.44}$$

As we can see from Equation 7.44, the probability of obtaining an aggregate number of mismatches more than a few standard deviations away from the mean value is very close to zero.

# B. ENTROPY

## 7.15 The logarithm of the multiplicity (ln *W*) is related to the entropy

The simple idea of multiplicity that we have introduced for a series of events with binary outcomes makes it possible to analyze the direction of spontaneous change for the isothermal (constant temperature) expansion of a gas. Recall from Chapter 6 that if a gas behaves ideally (that is, if the atoms of the gas do not attract or repel each other), then its energy does not change during an isothermal expansion. Nevertheless, if we compress a piston containing an ideal gas and then release the piston, it will expand spontaneously even under isothermal conditions (**Figure 7.20**).

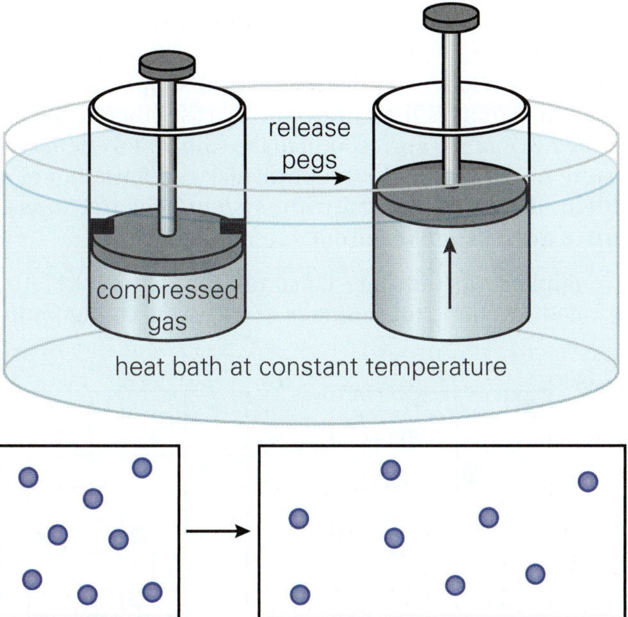


**Figure 7.20 Isothermal expansion of a gas.** If a piston compressing a gas is released, the gas will expand spontaneously. To analyze gas expansion, we will represent the piston as a box containing atoms, as shown in the lower diagram. The volume of the box increases as the piston expands.

The energy does not indicate the direction of spontaneous change in this case. In order to relate some other property of the molecules of the gas to the direction of spontaneous change (that is, expansion), we need a property of the system that will explain why the gas expands spontaneously even though the energy does not change. For an ideal gas, we can describe the physics of the random motion of individual gas molecules, from which we can deduce that the gas molecules exert a pressure on the walls of the container. If the pressure is unbalanced, then the container will expand (if the pressure inside is greater than the pressure outside) or contract (if the pressure outside is greater than the pressure inside) (see Box 6.3).

**Entropy**

The entropy of a system of molecules is a measure of the disorder in the system. The greater the number of equivalent rearrangements of the molecules, the greater the value of the entropy.

For systems that are more complicated than an ideal gas, it becomes very difficult to calculate the aggregate behavior of molecules by considering the physics of individual molecules. Instead, we turn to a statistical approach, which is based on the **entropy** of the system. Entropy is sometimes described as the randomness of a system (that is, the more ordered the system, the lower the entropy). We will define the entropy of a system as the logarithm of the multiplicity ($\ln W$) of the system.

For the expansion of an ideal gas, the introduction of entropy may seem like a needless complication because the behavior of the system is intrinsically simple. It turns out, though, that the statistical concepts that underlie entropy provide powerful tools for the analysis of much more complicated systems. We shall use the isothermal expansion of an ideal gas as a first step towards a more general statistical treatment of entropy.

## 7.16 The multiplicity of a molecular system is the number of equivalent configurations of the molecules (microstates)

Understanding multiplicity is crucial to understanding the direction of spontaneous change in molecular systems. In a sequence of coin tosses, the multiplicity is the number of different sequences that have the same aggregate outcome (that is, the same number of heads or tails). The multiplicity of a molecular system is defined as the number of different configurations or conformations of the component atoms or molecules that are equivalent.

To better understand this definition, consult **Figure 7.21**, which shows two distinct configurations of the six atoms of an ideal gas. Only the positions of the atoms are shown in Figure 7.21, but to characterize the system completely we would also have to specify the velocities of the atoms, which determine their kinetic energies. Each particular spatial configuration of the atoms corresponds to a multitude of different possible velocities of the atoms. For simplicity, we shall ignore the velocites of the atoms in the following discussion and simply assume that the total kinetic energy of the atoms is the same in every case (that is, the configurations shown in Figures 7.21A and B correspond to the same total energy and, therefore, to the same **state** of the system). Each particular configuration of atoms that is consistent with the definition of a state of the system is referred to as a **microstate**. We shall return to a discussion of the energies in Chapter 8.

What is the multiplicity of the state illustrated in Figure 7.21? To answer this question, we need to count the number of different configurations of atoms

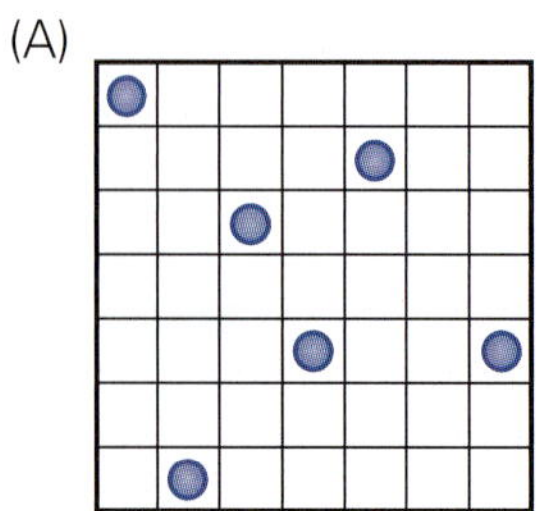


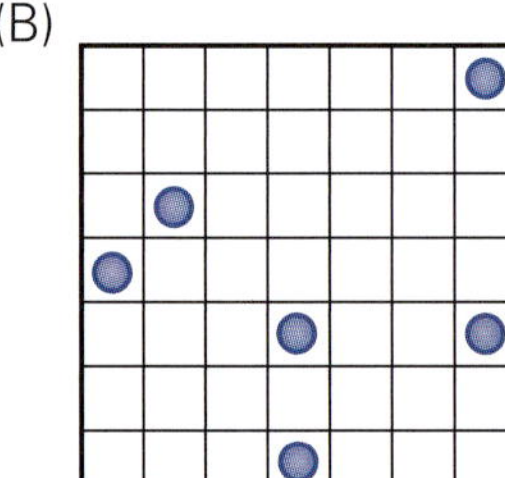


**Figure 7.21 Two configurations of atoms of an ideal gas.** All realistic considerations of a volume of an ideal gas would involve a very large number atoms (~$10^{23}$), but for simplicity only six atoms are shown in this figure. The container is divided up into grid boxes (49 are shown in the figure). The size of each grid box is the same, and they allow us to identify the location of each atom and thereby define the configuration. Realistically, the grid boxes need to be small enough to uniquely define the positions of the atoms, but for illustration we show a coarse grid of 7 × 7 = 49 squares. The actual number of boxes into which the space is divided is arbitrary, but in comparing different systems, we need to use the same size of grid boxes.

(microstates) that are possible. To do this, we must divide up the space within the system into a set of $M$ grid boxes, as shown in Figure 7.21. The number of grid boxes ($M$) is arbitrary (it is $7 \times 7 = 49$ in Figure 7.21), but is chosen to be sufficiently large that the positions of the atoms can be described at an adequate level of accuracy.

How many distinct configurations of atoms (microstates) are possible, given the six atoms and the 7 × 7 grid in Figure 7.21? This number is the multiplicity of the system, which we denote by $W$. The value of $W$ is given by:

$$W(M,N) = \frac{M!}{N!(M-N)!} \tag{7.45}$$

where $M$ is the number of grid boxes into which the system is divided and $N$ is the number of atoms. Equation 7.45 is justified in **Box 7.2**, but it is essentially the same as Equation 7.4 for the multiplicity of coin tosses, if you treat the total number of grid boxes as the number of coin tosses, the number of atoms (filled boxes) as the number of heads, and the number of empty boxes as the number of tails. For the example in Figure 7.21, where $M = 49$ and $N = 6$, $W$ is given by:

$$W(49,6) = \frac{49!}{6!(49-6)!} = \frac{6.08 \times 10^{62}}{720 \times 6.04 \times 10^{52}} \approx 1.4 \times 10^{7} \tag{7.46}$$

Thus, even for this very simple example of six atoms and 49 grid boxes, there are ~10 million different configurations or microstates, each of which is equally probable.

**States and microstates**

A state of a system is characterized by the global properties of the system, such as the temperature, pressure, and number of molecules. A microstate is a specific configuration of molecules that is consistent with the state. Each state corresponds to many different microstates.

## 7.17 The multiplicity of a system increases as the volume increases

What happens when we double the volume of the system while keeping the number of atoms the same, as shown in **Figure 7.22**? To do this, we increase the number of grid boxes from 49 to 98, while keeping the volume of each grid box the same as before (see Figure 7.22). Now $M = 98$ and $N = 6$, so the multiplicity ($W$) is given by:

$$W(98,6) = \frac{98!}{6!(98-6)!} \tag{7.47}$$

In this case, the numbers are becoming large enough that it is more convenient to switch to $\ln W$:

$$\ln W(98,6) = \ln 98! - \ln 6! - \ln 92! \tag{7.48}$$

Stirling's approximation ($\ln n! \approx n \ln n - n$) only holds for large numbers, so we can use it to calculate the values of $\ln 98!$ and $\ln 92!$, but not that of $\ln 6!$, which we must calculate directly.

$$\begin{aligned} \ln W(98,6) &= 98 \times \ln(98) - 98 - \ln(720) - (92 \times \ln 92 - 92) \\ &= 351.3 - 6.58 - 324.0 \\ &= 20.72 \end{aligned} \tag{7.49}$$

$$\Rightarrow W(98,6) = e^{20.72} = (10^{0.434})^{20.72} \approx 10^{9} \tag{7.50}$$

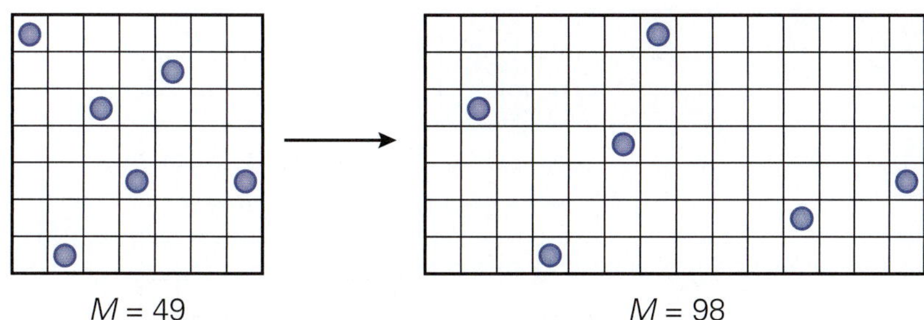


**Figure 7.22 Calculating the change in multiplicity when the volume of the system expands.** On the left is shown a system of six atoms (*blue* circles) in a volume divided into 49 (7 × 7) grid boxes. To see what happens when the system expands to twice the volume, we keep the size of each grid box the same but double the number of grid boxes to 98.

## Box 7.2 Multiplicity of atomic configurations in grid boxes (Equation 7.45)

The multiplicity, $W$, for $N$ atoms distributed among $M$ grid boxes is given by Equation 7.45:

$$W = \frac{M!}{N!(M-N)!}$$

To justify this equation, let us start by considering a set of $M$ atoms, one for each grid box. There are $M!$ ways of rearranging $M$ different atoms in $M$ grid boxes (**Figure 7.2.1**). The multiplicity of the system is equal to $M!$ in this case. Now consider the situation where we separate the $M$ different grid boxes into two groups: $N$ of them have atoms in them and $(M - N)$ correspond to grid boxes that are empty (**Figure 7.2.2**). We consider the $N$ grid boxes that contain atoms to all be equivalent, and do not count rearrangements of these grid boxes as separate configurations. Likewise, we also consider the $M - N$ empty grid boxes to be equivalent.

In the example shown in Figure 7.2.2, the atoms are in grid boxes labeled 1 to 6, and the unoccupied boxes correspond to the boxes labeled 7 to 49. When we count the total number of rearrangements of the $M$ labels, we overcount the number of configurations corresponding to the atoms by 6!. Likewise, we overcount the number of configurations corresponding to rearrangements of the empty grid boxes by $(M - N)! = 43!$. Thus, the multiplicity, $W$, is given by:

$$W = \frac{M!}{\left(\begin{array}{l}\text{number of ways of}\\ \text{rearranging grid boxes}\\ \text{containing atoms}\end{array}\right)\left(\begin{array}{l}\text{number of ways of}\\ \text{rearranging empty grid boxes}\end{array}\right)} = \frac{M!}{N!(M-N)!} \quad (7.2.1)$$

This argument is essentially the same as that used to derive Equation 7.4 for the multiplicity of a series of coin tosses, and the two problems are conceptually equivalent. We can see this by treating each of the grid boxes as one coin toss. If the grid box is filled with an atom, we call it "heads," and if it is empty, we call it "tails."

This argument can be extended to situations where we have more than one kind of atom. Suppose we have $M$ grid boxes, $N_1$ atoms of type 1, $N_2$ atoms of type 2, and so on. The total number of all kinds of atoms is $N_T = N_1 + N_2 + \ldots N_t$, where there are $t$ different kinds of atoms. The

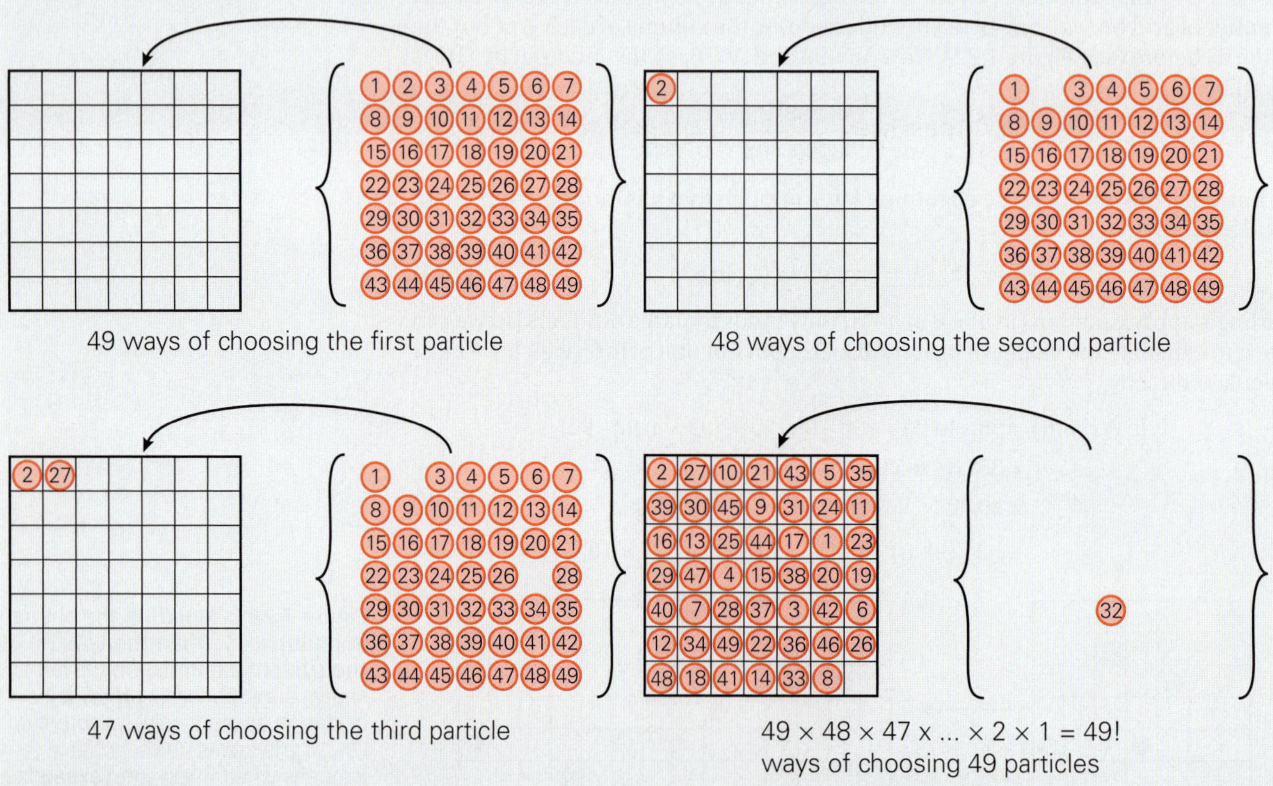


**Figure 7.2.1 There are $M!$ = 49! ways of filling a set of 49 grid boxes with 49 different particles.**

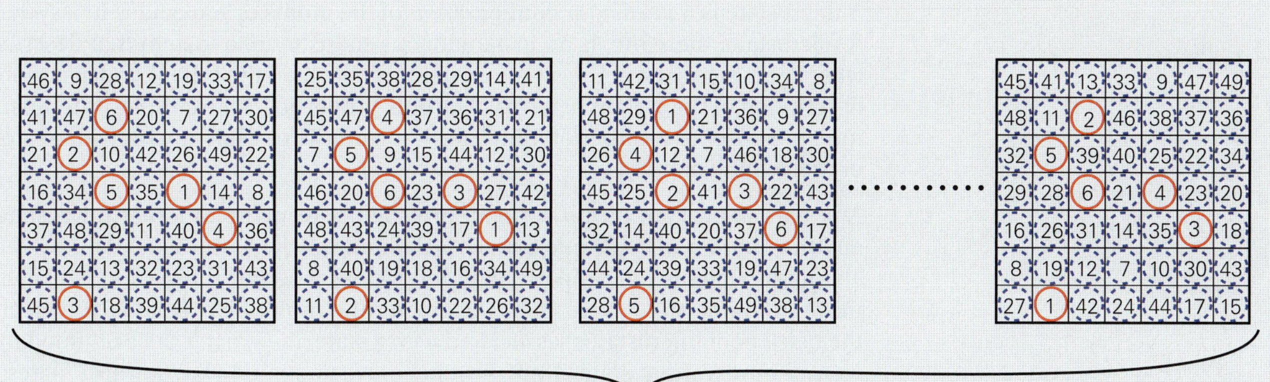

number of rearrangements of red atoms = 6!

number of rearrangements of empty boxes = 43!

total number of rearrangements = 6! × 43!

**Figure 7.2.2 All rearrangements of filled and empty boxes are considered to be equivalent.**

number of empty grid boxes is now $M - N_T$. The multiplicity of such as system is given by:

$$W = \frac{M!}{N_1!N_2!\ldots N_t!(M - N_T)!} \qquad (7.2.2)$$

Thus, when comparing the values of $W(49, 6)$ and $W(98, 6)$ in Equations 7.46 and 7.50, we see that doubling the volume of this simple system from 49 to 98 grid boxes increases the multiplicity 100-fold, from ~10 million to ~1 billion.

What does this result mean? The atoms are not subject to any external energy field (that is, there is no force directing them to any particular region of the container), so every possible configuration of the atoms is equally likely. The configuration (microstate) shown in **Figure 7.23**A, in which all of the atoms are in the left half of the box, is just as likely as the one shown in Figure 7.23B, in which the atoms are distributed evenly throughout the box. There are, however, ~100 times more configurations that correspond to distributions that are not restricted to the left half of the box than there are ones in which all of the atoms are on the left. When we look at the box, therefore, we are 100 times more likely to see the atoms on both sides of the box than we are to see them localized in the left half.

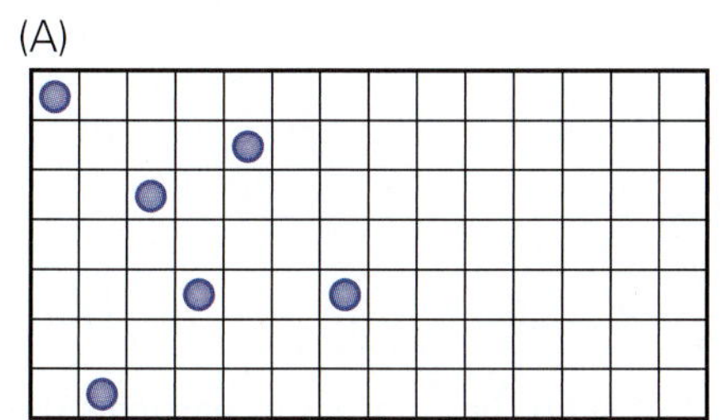


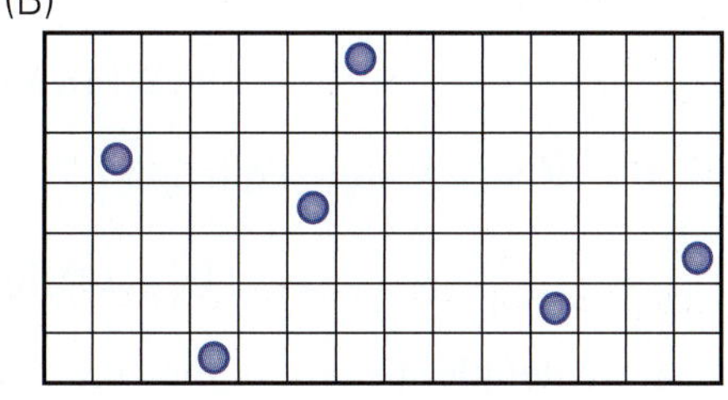


**Figure 7.23 All configurations of atoms have equal probability.** The configuration shown in (A), in which all the atoms are on the left-hand side, is just as likely as the configuration shown in (B), in which the atoms are more evenly distributed.

## 7.18 For a large number of atoms, the state with maximal multiplicity is the state that is observed at equilibrium

A microstate is a *particular* configuration of the atoms or molecules in a system. A microstate, therefore, is an instantaneous snapshot of a system that describes what every molecule in the system is doing. A state, on the other hand, is a global description of the system in terms of its bulk properties (for example, its temperature, pressure, or number of molecules), regardless of what any individual atom or molecule is doing.

Any particular state of the system corresponds to many different microstates, and the number of these microstates is the multiplicity of the state of the system. The multiplicity is the property that determines whether the system is likely to move from one state to another. The greater the multiplicity (that is, the greater the number of microstates) associated with a state, the greater the likelihood that we will observe that state. Thus, systems move spontaneously towards states of increasing multiplicity that are consistent with conservation of matter and energy.

Consider, for example, the sudden release of a piston that had been keeping a volume of gas compressed, as shown in **Figure 7.24**. Initially all the atoms of the gas are on the left-hand side of the piston chamber, corresponding to the original volume of the piston. If there are $10^4$ atoms in the piston chamber, then the number of atoms on the left (denoted $N_{left}$) is initially $10^4$, and the number of atoms on the right (denoted $N_{right}$) is initially 0. Over time, atoms move away from their initial positions, so $N_{left}$ will decrease and $N_{right}$ will increase (Figure 7.24B). The system is not at equilibrium with respect to these two variables initially, because they change with time. Eventually, however, the atoms fill up the entire chamber, at which point $N_{left}$ and $N_{right}$ stop changing because they have both reached a value of $0.5 \times 10^4$. At this point the system is at equilibrium with respect to these two variables.

To use statistical arguments to determine the equilibrium point of the system, we compute the multiplicity of the system for different values of the variables of interest (in this case, $N_{left}$ and $N_{right}$). Each possible value of $N_{left}$ (or $N_{right}$) corresponds to a possible state of the system, and states with higher values of the multiplicity are more likely to be observed than those with lower multiplicity. If the system can convert from a state with low multiplicity to a state with higher multiplicity, it will do so with a probability given by the ratios of the multiplicities of the two states.

When we consider a small number of atoms (just as when we consider a small number of coin tosses), there is a significant probability that we will observe outcomes that deviate from the expected even distribution. In the example presented in Figure 7.23, in which we consider only six atoms, there is a 1 in 100 chance that we will find all the atoms in the left half of the box. What happens when we consider a much larger number of atoms, such as 10,000?

We first compute the multiplicity ($W_1$) for $10^4$ atoms restricted to the left half of the container, which is divided up into a large number of grid boxes (we shall choose $10^8$ grid boxes for this example, so $N_{left} = 10^4$ and $M = 10^8$). The value of $M$ is arbitrary, as long as we have a very large number of grid boxes and we use the same grid size to compare different states of the same system. When $M$ and $N$ are both large numbers, and $M >> N$, the following approximation applies (see **Box 7.3** for the justification):

$$\ln W(M,N) \approx N \ln {}^{M}\!/\!_{N} \tag{7.51}$$

For $M = 10^8$ and $N = 10^4$, Equation 7.51 yields:

$$\ln W_1 = \left(10^4\right) \ln \frac{10^8}{10^4} = 9.2 \times 10^4 \tag{7.52}$$

Next, we double the size of the container, but keep the number of particles the

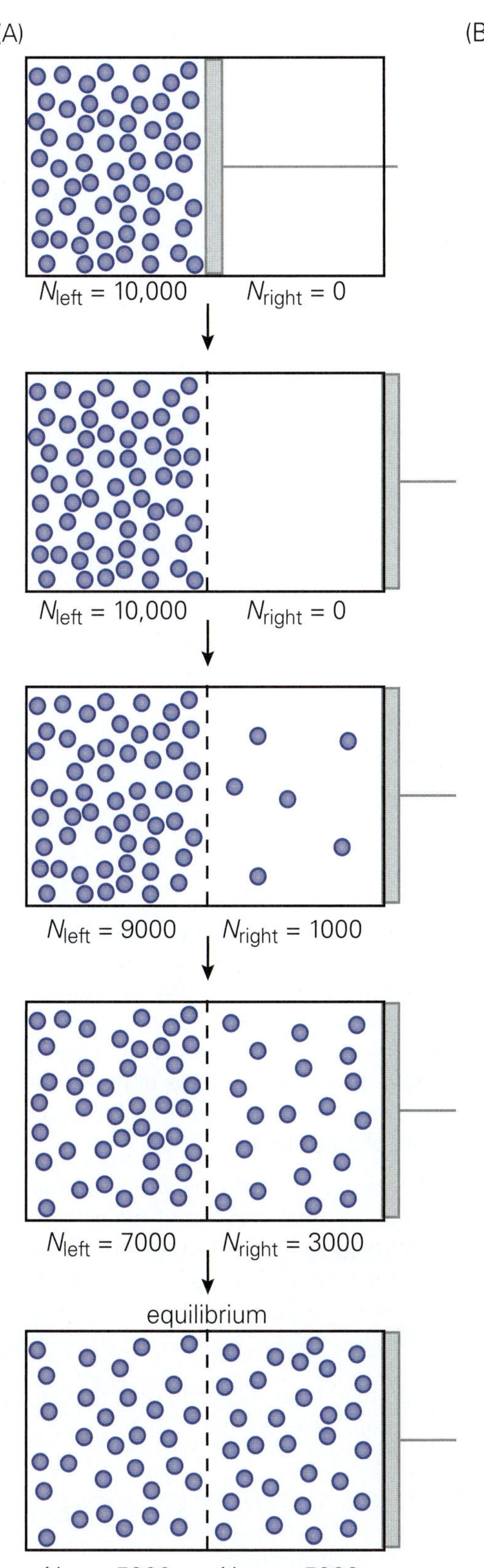


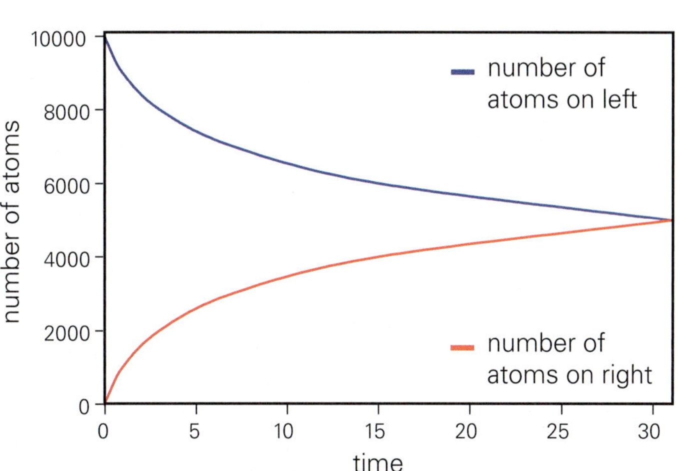


**Figure 7.24 Changes in the state of a system as a piston is suddenly released.** (A) A piston containing atoms of a gas is shown schematically. The piston is initially compressed so that all the atoms are in the left half of the chamber. The piston is suddenly released. The atoms move around until they completely fill the chamber. (B) As the system reaches equilibrium, the average number of atoms in the left half of the chamber becomes equal to that in the right half.

same. Thus, $M = 2 \times 10^8$ and $N = 10^4$. The new multiplicity ($W_2$) is given by:

$$\ln W_2 = \left(10^4\right) \ln \frac{2 \times 10^8}{10^4} = 9.9 \times 10^4 \tag{7.53}$$

The ratio of the two multiplicities is calculated by combining Equations 7.52 and 7.53:

$$\ln W_2 - \ln W_1 = (9.9 - 9.2) \times 10^4 = 7000 \tag{7.54}$$

$$\Rightarrow \frac{W_2}{W_1} = e^{7000} = \left(10^{0.434}\right)^{7000} \approx 10^{3000} \tag{7.55}$$

Equation 7.55 tells us that the probability of observing all the atoms in a volume corresponding to the left side of the box (proportional to $W_1$) compared to an unrestricted distribution (proportional to $W_2$) is 1 in $10^{3000}$, which is essentially zero.

The multiplicity of the system effectively acts as a driving force for expansion. Even a small increase in the volume (that is, a small increase in the number of grid boxes) increases the multiplicity dramatically and makes it highly unlikely that we will observe the atoms within the restricted space corresponding to the

## Box 7.3 A simplified expression for the multiplicity that is applicable when the number of atoms is much smaller than the size of the system

The expression for $\ln W$ simplifies when the number of molecules, $N$, is much smaller than the number of grid boxes, $M$, into which the system is divided:

$$\ln W \approx N \ln \frac{M}{N} \tag{7.51}$$

where $W$ is the multiplicity of a system of $N$ molecules arranged in $M$ grid boxes and $N$ is much smaller than $M$. We assume that $M$ and $N$ are sufficiently large so that we can use Stirling's approximation.

We start with the expression for the multiplicity, $W$ (see Box 7.2):

$$W = \frac{M!}{(M-N)!N!} \tag{7.3.1}$$

$$\Rightarrow \ln W = \ln M! - \ln N! - \ln (M-N)! \tag{7.3.2}$$

To simplify Equation 7.3.2, we use use Stirling's approximation:

$$\ln n! = n \ln n - n$$

where $n$ is a large number. Thus, Equation 7.3.2 becomes:

$$\begin{aligned}\Rightarrow \ln W &= M \ln M - M - (N \ln N - N) \\ &\quad - [(M-N)\ln(M-N) - M + N]\end{aligned} \tag{7.3.3}$$

The $+M$ and $-M$ terms in Equation 7.3.3 cancel, as do the $+N$ and $-N$ terms:

$$\begin{aligned}\Rightarrow \ln W &= M \ln M - N \ln N - M \ln(M-N) \\ &\quad + N \ln(M-N)\end{aligned} \tag{7.3.4}$$

In general it is reasonable to assume that $M >> N$. This is because $M$ is the number of grid boxes into which the space is divided and $M$ can be made arbitrarily large by dividing the space into smaller and smaller volumes. Hence,

$$M - N \approx M \tag{7.3.5}$$

We use Equation 7.3.5 to simplify Equation 7.3.4 and thus obtain:

$$\begin{aligned}\ln W &\approx M \ln(M) - N \ln(N) - M \ln(M) + N \ln(M) \\ &= N \ln(M) - N \ln(N)\end{aligned} \tag{7.3.6}$$

From Equation 7.3.6 it follows that:

$$\ln W \approx N \ln\left(\frac{M}{N}\right) \tag{7.3.7}$$

This justifies Equation 7.51, which is applicable when there is only one type of molecule in the system. When there is more than one type of molecule, the value of $\ln W$ can be calculated separately for each molecule and the results added together. We shall justify this for the case of a system with two different kinds of molecules, but the result can be easily generalized to collections of more than two types of molecules.

Suppose we have $N_A$ molecules of type A and $N_B$ molecules of type B, and the total number of molecules is $N_T$. The multiplicity is given by (see Equation 7.1.2 in Box 7.1):

$$W = \frac{M!}{N_A! N_B! (M - N_T)!} \tag{7.3.8}$$

By taking the logarithm of both sides and then applying Stirling's approximation, it can be shown that:

$$\ln W = N_A \ln\left(\frac{M}{N_A}\right) + N_B \ln\left(\frac{M}{N_B}\right) \tag{7.3.9}$$

From Equation 7.3.9 we can see that:

$$\ln W = \ln W_A + \ln W_B$$

where $W_A$ and $W_B$ are the multiplicities of the A and B molecules, respectively, calculated independently.

initial volume. Molecular systems tend naturally towards states of increased multiplicity, which is one statement of the **second law of thermodynamics**. We shall discuss other statements of the second law later in the chapter, but keep in mind that spontaneous change is driven by a tendency to increase multiplicity.

## 7.19 The Boltzmann constant, $k_B$, is a proportionality constant linking entropy to the logarithm of the multiplicity (ln *W*)

As we saw in the previous section, the value of the multiplicity ($W$) becomes unmanageably large when we consider more than a small number of atoms. The logarithm of the multiplicity ($\ln W$), on the other hand, increases much more slowly with the number of atoms, and so the value of $\ln W$ is easier to work with than the value of $W$. We have already noted, in Section 7.11, that $\ln W$ increases or decreases when $W$ increases or decreases and that $\ln W$ has a maximal value when $W$ is maximal.

Another convenient property of $\ln W$ is that it is an **extensive** and additive property of the system. To see why this is so, consider the two systems, A and B, shown in **Figure 7.25**. If the multiplicity of A is $W_A$ and the multiplicity of B is $W_B$, then the multiplicity of the combined system ($W_{A+B}$) is $W_A \times W_B$.

$$W_{A+B} = W_A \times W_B \tag{7.56}$$

$$\Rightarrow \ln W_{A+B} = \ln W_A + \ln W_B \tag{7.57}$$

Equation 7.57 demonstrates that $\ln W$ is an additive property of the system, but it also shows that $\ln W$ scales with the size of the system. That is, $\ln W$ is an *extensive* property that is also additive (**Figure 7.26**). Moreover, $\ln W$ is a state function,

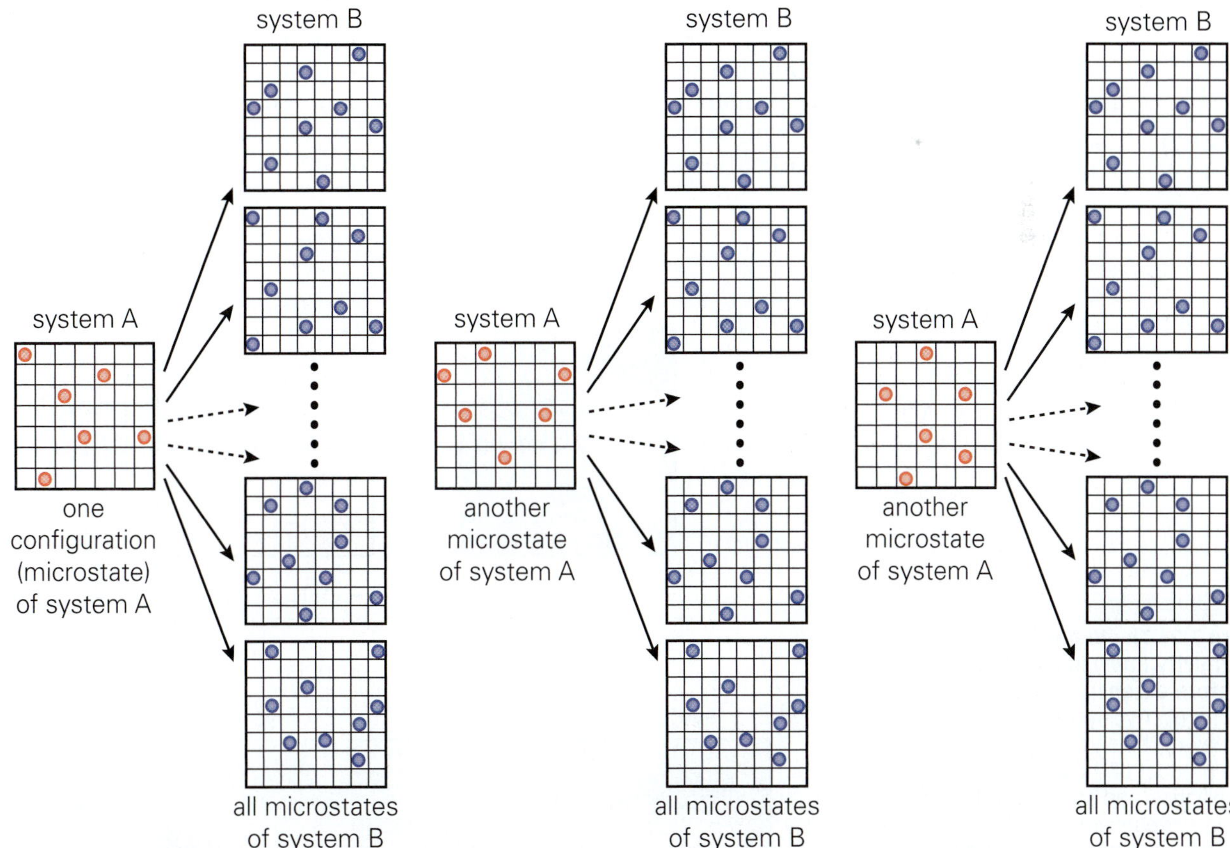


**Figure 7.25 The multiplicity of a combined system is the product of the multiplicities of the subsystems.** Two systems are shown, one containing *red* atoms and one containing *blue* atoms. Each unique configuration of *red* atoms can be paired with each unique configuration of *blue* atoms. The total multiplicity is therefore the product of the individual multiplicities.

because it depends only on the parameters that define the present state of the system, not on its history (that is, not on the path used to achieve its present state; see Section 6.4). Based on these properties of ln $W$, a new state function of the system, known as the **entropy** ($S$), is defined as follows:

**Statistical definition of entropy**

The Boltzmann constant, $k_B$, relates the multiplicity, $W$, to entropy, $S$, through the statistical definition of entropy:

$S = k_B \ln W$.

$$S = k_B \ln W \tag{7.58}$$

where $k_B$ is the **Boltzmann constant**, which has a value of $1.38 \times 10^{-23}$ J•K$^{-1}$. Equation 7.45 is known as the **statistical definition of entropy**. The entropy defined in this way has the same units as the Boltzmann constant—that is, energy/temperature.

We have already encountered the Boltzmann constant when we discussed the Boltzmann distribution in Section 6.10. Recall that the Boltzmann constant is simply the gas constant ($R$) divided by Avogadro's number ($N_A$):

$$k_B = \frac{R}{N_A} = \frac{8.314 \text{ J}\bullet\text{K}^{-1}}{6.023 \times 10^{23}} = 1.38 \times 10^{-23} \text{ J}\bullet\text{K}^{-1} \tag{7.59}$$

## 7.20 The change in entropy is related to the heat transferred during a process

The definition of entropy given in Equation 7.58, in which a pure number (ln $W$) is multiplied by a constant ($k_B$) with units of energy over temperature, arose due to historical reasons. In the nineteenth century, the concept of entropy was derived from the study of heat engines, and the change in entropy during a process, $\Delta S$, was defined as follows:

$$\Delta S = \frac{q_{rev}}{T} \tag{7.60}$$

where $q_{rev}$ is the heat transferred to the system during a **reversible** or **near-equilibrium** transition (discussed in greater detail in Section 7.21) from one state to another, and $T$ is the absolute temperature. Equation 7.60 is known as the **thermodynamic definition of entropy**. The two definitions of entropy are made consistent by multiplying ln $W$ by the Boltzmann constant, $k_B$. As we discuss

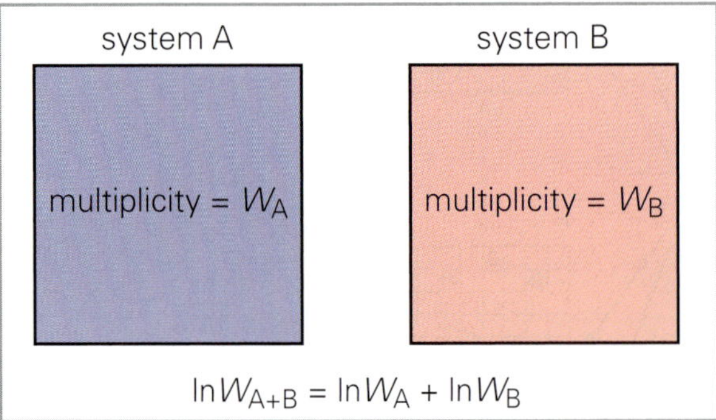


**Figure 7.26 The logarithm of the multiplicity (ln W) is an additive and therefore extensive property of the system.** (A) Two systems, A and B, are shown. The combined multiplicity is the product of the individual multiplicities, and so ln $W$ is additive. (B) Two systems are shown that are identical in terms of their composition, except that one is twice the size of the other. The value of ln $W$ for the larger system is twice the value of ln $W$ for the smaller system.

further in Part C of Chapter 8, the value of the Boltzmann constant depends on the units of energy and temperature because of a fundamental relationship that links changes in the energy and the entropy to the temperature.

The thermodynamic definition of the entropy (Equation 7.60) is not particularly intuitive if we are interested in molecular properties rather than heat engines. It is of enormous practical importance, however, because the amount of heat transferred to a system during a process, such as a chemical reaction, is an experimentally measurable quantity. We can use these experimental results to understand molecular properties, because we will show that Equations 7.58 and 7.60 are strictly equivalent definitions of entropy, first in Section 7.23 for the case of an isothermal gas expansion, and in Chapter 8 for a general case.

**Thermodynamic definition of entropy**

A historical definition of entropy based on heat transfer is known as the thermodynamic definition of entropy:

$$\Delta S = \frac{q_{rev}}{T}$$

where $\Delta S$ is the change in entropy for a reversible (near-equilibrium) process, $q_{rev}$ is the heat transferred, and $T$ is the absolute temperature. This definition of the entropy turns out to be equivalent to the statistical definition.

## 7.21 The work done in a near-equilibrium process is greater than for a nonequilibrium process

The thermodynamic definition of the entropy is cast in terms of heat transfer during a process that occurs near equilibrium, also called a reversible process. The importance of a near-equilibrium process is that, because the interaction between the surroundings and the system is nearly in balance, one can gain information about system variables in a way that is not possible for nonequilibrium processes. A simple analogy might help explain why this is the case. Imagine that you engaged in an arm wrestling contest with a friend, and that you wished to use this contest to gauge just how strong your friend is. How would you do this? If you pushed too hard against your friend's arm, you might only realize that you are stronger than her. If you pushed back only weakly, then you would simply lose the contest and not really get a sense of how strong your friend is. Now imagine that you pushed back against your friend with almost as much force as she exerted on you, and then increased the force very gradually to push her arm back very slowly. By pushing back in this way, you can, if you are strong enough, figure out precisely how much resistance your friend is capable of exerting on your arm. This would be an example of a near-equilibrium process. Realize also that you would be *working hardest* during a near-equilibrium arm wrestling contest.

No real process proceeds so slowly that it is strictly near-equilibrium. Nevertheless, the importance of the concept of a near-equilibrium process is that the work done during such a process is directly related to the change in a state function, as is the heat transferred during such a process. The amount of work done in a near-equilibrium process is also the *maximum* amount of work that can be extracted from the system during a change from one state to another. While this statement is true in general, we shall illustrate it using the particularly simple case of gas expansion.

**Near-equilibrium or reversible process**

A process is referred to as near-equilibrium or reversible if it is carried out so slowly that the external driving force is always nearly in balance with the internal forces. When a system moves from one state to another, the amount of heat absorbed and the amount of work done depend on how exactly the transformation occurred. By specifying a near-equilibrium process, we can calculate the heat and work precisely. For such a process, the work done is maximal, given the initial and final states of the system.

In the case of the isothermal expansion of an ideal monatomic gas, imagine a process in which the pressure on the piston (the external pressure, $P_{EXT}$) is reduced very gradually, so that the expansion occurs in a series of very small *near-equilibrium* steps, as shown in **Figure 7.27**. In one such small step, the work done by the piston when it expands against a constant external pressure ($P_{EXT}$) is given by:

$$w = -P_{EXT}\,\Delta V \tag{7.61}$$

The sign is negative because work done *by* the system is negative (see Section 6.3).

According to Equation 7.61, the magnitude of the work done by the system increases as the external pressure increases. The maximum amount of work is done when $P_{EXT}$ is set to be just less than the internal pressure of the piston ($P_{INT}$)—that is, when $P_{EXT}$ and $P_{INT}$ are almost balanced—and the process occurs reversibly.

The piston must expand for the system to do work, so $P_{EXT}$ must be less than $P_{INT}$. If $P_{EXT}$ were greater than $P_{INT}$, the piston would compress rather than expand. The maximum value of $P_{EXT}$ for which work can still be extracted from the system is a value that is just slightly less than $P_{INT}$, which corresponds to a near-equilibrium process.

The graph in Figure 7.27C shows that the amount of work done by the system is considerably less when the piston expands suddenly due to a sharp decrease in external pressure. This kind of expansion is *nonequilibrium* in the thermodynamic sense—that is, the system is very far from equilibrium as the process occurs. The amount of work extracted from a system undergoing a nonequilibrium process is always less than the amount of work extracted from the same system undergoing a near-equilibrium process.

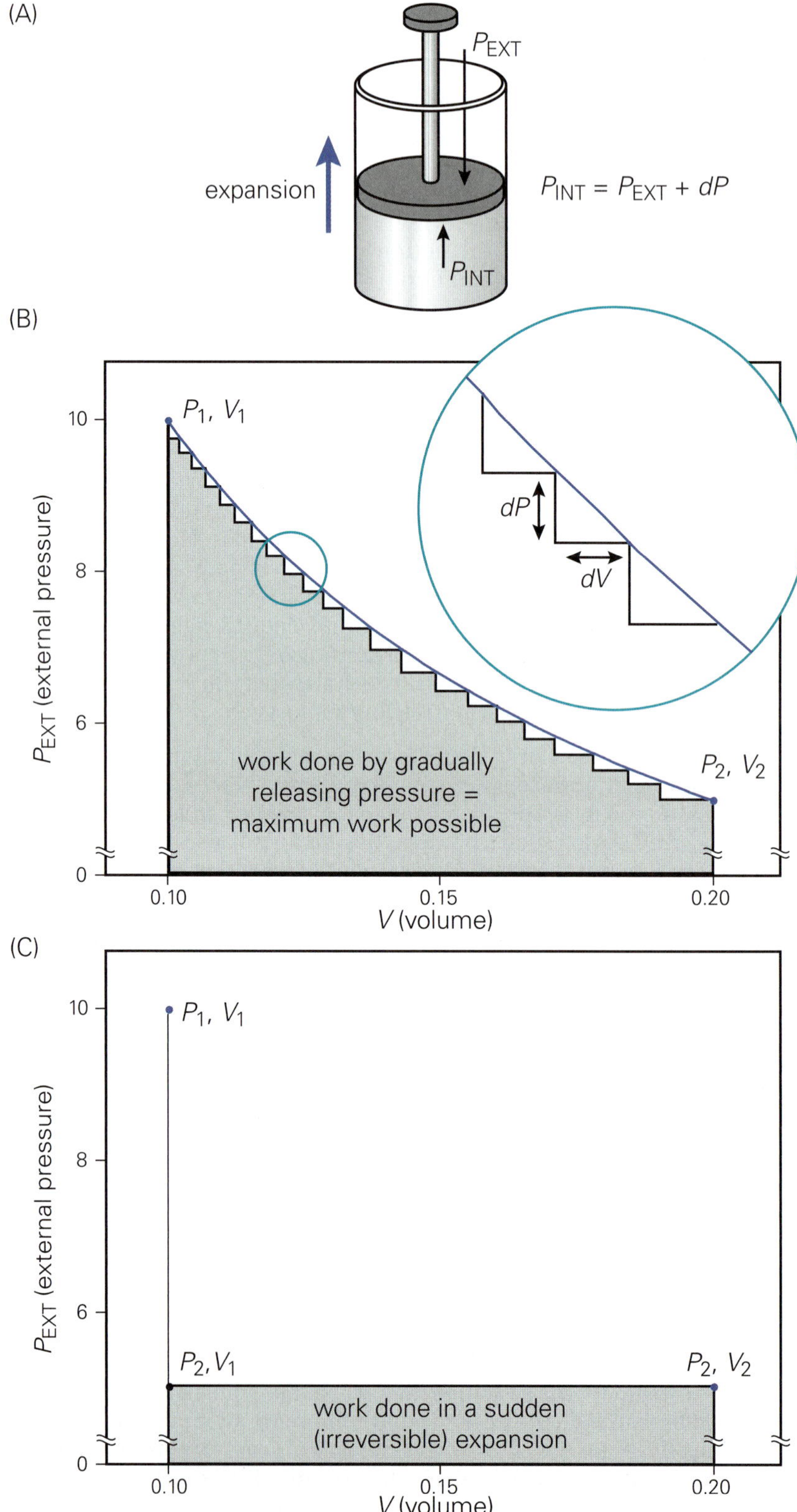


**Figure 7.27 The near-equilibrium expansion of a gas.** (A) A schematic diagram of a piston. When the expansion is carried out in a near-equilibrium (reversible) manner, the external pressure, $P_{EXT}$, is gradually reduced so that the internal pressure, $P_{INT}$, is only greater than $P_{EXT}$ by a small amount, $dP$. (B) A pressure–volume diagram (*P–V* diagram) for a near-equilibrium expansion. The area under the pressure–volume curve corresponds to the work done by the system, Section 7.22. (C) A *P–V* diagram for a sudden and nonequilibrium expansion.

## 7.22 The work done in a near-equilibrium process is related to the change in entropy

If the expansion occurs in a series of infinitesimally small steps, in which the external pressure is reduced by an infinitesimally small amount, $dP$, we can write:

$$P_{EXT} = P_{INT} - dP \tag{7.62}$$

The work done by the system during such an infinitesimal step, $dw$, is given by:

$$dw = -P_{EXT}\,dV = -(P_{INT} - dP)\,dV \tag{7.63}$$

Hence, $w_{rev}$, the work done over the entire course of the near-equilibrium or reversible expansion, from an initial volume $V_1$ to a final volume $V_2$, is given by:

$$w_{rev} = -\int_{V_1}^{V_2} (P_{INT} - dP)dV \tag{7.64}$$

$$\Rightarrow w_{rev} = -\int_{V_1}^{V_2} P_{INT}\,dV + \int_{V_1}^{V_2} dP dV \tag{7.65}$$

Notice that, according to Equation 7.65, the work done depends only on the temperature and pressure of the system. You should realize, however, that the work done by the system is always against an external force ($P_{EXT}$ in this example). It is only because we have specified that the process is near equilibrium (Equation 7.62) that we are able to express work in terms of system properties.

We can ignore the second term in Equation 7.65 because it is the product of two infinitesimally small quantities and will therefore be extremely small compared to the first term. Next, we can apply the ideal gas law to the internal pressure:

$$P_{INT} = nRT/V \tag{7.66}$$

where $n$ is the number of moles of gas in the system, $R$ is the gas constant, and $T$ is the absolute temperature.

Combining Equations 7.65 and 7.66 leads to the following expression for the work done in a near-equilibrium process:

$$w_{rev} = -\int_{V_1}^{V_2} \frac{nRT}{V} dV = -nRT \ln \frac{V_2}{V_1} \tag{7.67}$$

Recall that the change in energy for an ideal monoatomic gas is zero for an isothermal process (Section 6.6). Hence, in accordance with the first law, the heat taken up by the system in a near-equilibrium, reversible, isothermal expansion is:

$$q_{rev} = -w_{rev} = +nRT \ln \frac{V_2}{V_1} \tag{7.68}$$

Next, by substituting the expression in Equation 7.68 for $q_{rev}$ in the thermodynamic definition of the change in entropy for the process (Equation 7.60), we get the following expression for the change in entropy for the isothermal expansion of an ideal gas:

$$\Delta S = \frac{q_{rev}}{T} = nR \ln \frac{V_2}{V_1} \tag{7.69}$$

Finally, by combining Equations 7.67 and 7.69, we get the following relationship between the reversible (or maximum) work extracted from the system and the entropy change:

$$w_{rev} = -q_{rev} = -T\Delta S \tag{7.70}$$

This is a profound result because it demonstrates that the change in entropy is related to the maximum work that can be extracted from the system. Moreover, the maximum work that can be extracted from the isothermal expansion of an ideal monatomic gas is given by the product of the temperature and the change in entropy, and does *not* depend on the energy. In short, there is no change in energy and the work done is due solely to an increase in the entropy of the system.

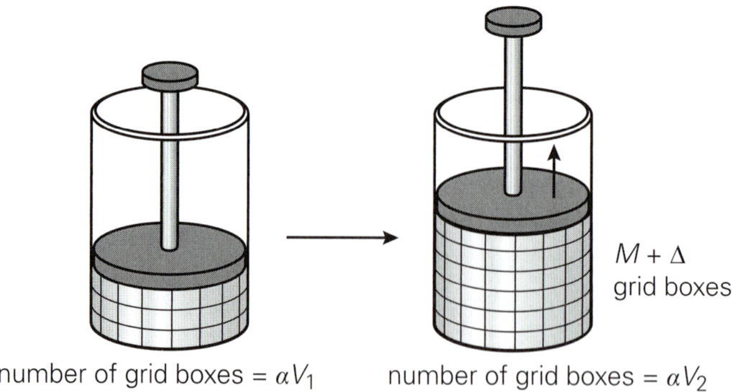


**Figure 7.28 Linking volume changes to changes in multiplicity during gas expansion.** We divide the volume of the piston into a number of grid boxes such that the number of grid boxes per unit volume is α. When the piston expands, the number of grid boxes increases proportionally to the volume.

In this particularly simple case, the only kind of work that the system is capable of doing is expansion work against an external pressure. In Chapter 10, we shall see that in a general process, in which the energy and the entropy both change, the maximum work extractable from the process, in addition to expansion work, is given by the change in a quantity known as the "free energy," which is a combination of both the change in energy and $T\Delta S$.

## 7.23 The statistical and thermodynamic definitions of entropy are equivalent

Equation 7.69 relates the change in entropy, using the thermodynamic definition of entropy, to the initial and final volumes of the system for the expansion of an ideal gas. We can now compare this expression to that obtained using the statistical definition of the entropy (that is, $S = k_B \ln W$; Equation 7.58). First, we must divide up the volume of the piston into $M$ grid boxes. As the piston expands, the number of grid boxes increases from $M$ to $M + \Delta$ (**Figure 7.28**). We can compute the change in multiplicity ($W$) that arises due to the increase in volume (that is, due to the increase in the number of grid boxes).

When considering a large number of atoms, $N$, we can use Stirling's approximation (Equation 7.12), so the expression for $\ln W$ simplifies to:

$$\ln W(M,N) = \ln\left[\frac{M!}{N!(M-N)!}\right] \approx N \ln \frac{M}{N} \tag{7.71}$$

as shown previously in Equation 7.51 and explained in Box 7.3. Equation 7.71, which is connected to the statistical definition of the entropy, can be related to Equation 7.69, which is connected to the thermodynamic definition. To do this we need to relate the volume of the piston to the number of grid boxes ($M$) in Equation 7.71.

The value of $M$ for a given state of the system is arbitrary (as long as each grid box is small enough to depict the positions of the atoms accurately), but the number of grid boxes in the system has to be strictly proportional to the volume of the system in order for the calculation of the multiplicity to remain consistent as the volume changes.

The volume ($V$) is related to the number of grid boxes ($M$) through a proportionality constant $\alpha$:

$$M = \alpha V \tag{7.72}$$

Thus, $M_1 = \alpha V_1$ and $M_2 = \alpha V_2$ for the piston before and after an isothermal expansion of an ideal gas (see Figure 7.28).

If we increase the volume of the system from $V_1$ to $V_2$, what is the corresponding change in $\ln W$ (which changes from $\ln W_1$ to $\ln W_2$)? To begin, substitute the expression for $M$ from Equation 7.72 into the expression for $\ln W$ in Equation 7.71:

$$\ln W_1 = N \ln\left(\frac{M_1}{N}\right) = N \ln\left(\frac{\alpha V_1}{N}\right) \tag{7.73}$$

Similarly,

$$\ln W_2 = N \ln\left(\frac{M_2}{N}\right) = N \ln\left(\frac{\alpha V_2}{N}\right) \quad (7.74)$$

Then,

$$\begin{aligned} \ln W_2 - \ln W_1 &= \ln\left(\frac{W_2}{W_1}\right) \\ &= N \ln\left(\frac{\alpha V_2}{N} \times \frac{N}{\alpha V_1}\right) \\ &= N \ln \frac{V_2}{V_1} \end{aligned} \quad (7.75)$$

Using the statistical definition of the entropy ($S = k_B \ln W$), we get:

$$\frac{(S_2 - S_1)}{k_B} = \frac{\Delta S}{k_B} = \ln W_2 - \ln W_1 = N \ln \frac{V_2}{V_1} \quad (7.76)$$

where $S_1$ and $S_2$ are the values of the entropy when the volume of the system is $V_1$ and $V_2$, respectively.

The number of atoms ($N$) can be expressed in terms of the number of moles ($n$) because $N = nN_A$, where $N_A$ is Avogadro's number. Thus,

$$\Delta S = nN_A k_B \ln \frac{V_2}{V_1} \quad (7.77)$$

The product of the Boltzmann constant ($k_B$) and Avogadro's number ($N_A$) is the gas constant, $R$, so Equation 7.77 can be rewritten as:

$$\Delta S = nR \ln \frac{V_2}{V_1} \quad (7.78)$$

Equation 7.78 (derived from the statistical definition of the entropy) is identical to Equation 7.69 (derived from the thermodynamic definition of the entropy) for the isothermal expansion of an ideal monatomic gas. The statistical and thermodynamic definitions of the entropy given in Equations 7.58 and 7.60 are therefore equivalent, provided that we multiply $\ln W$ by the Boltzmann constant, $k_B$.

## 7.24 The second law of thermodynamics states that spontaneous change occurs in the direction of increasing entropy

The increase in the entropy of a system indicates the direction of spontaneous change. An isolated molecular system will *not* convert spontaneously from a state of high entropy (high multiplicity) to one of lower entropy (lower multiplicity). The only way to achieve that kind of conversion is to couple the system to an external agent (that is, something in the surroundings) whose gain in entropy is great enough to bring about the change. In other words, for a process to occur spontaneously, the combined entropy of the system and the surroundings (which includes all external forces), must increase. This statement is another expression of the **second law of thermodynamics**.

**Second law of thermodynamics**

One statement of the second law is that the combined entropy of the system and the surroundings always increases for a spontaneous process. This is equivalent to saying that the entropy of the system and the surroundings has a maximum value at equilibrium, which is referred to as the maximal entropy principle.

If we denote the entropy of the system by $S_{sys}$ and the entropy of the surroundings by $S_{surr}$, then at equilibrium the second law can be stated as follows:

$$S_{sys} + S_{surr} = \text{maximal} \quad \text{(at equilibrium)} \quad (7.79)$$

Equation 7.79 is known as the **maximal entropy principle**. We commonly express Equation 7.79 in terms of the differential terms, $dS_{sys}$ and $dS_{surr}$, which are infinitesimally small changes in the values of $S_{sys}$ and $S_{surr}$, respectively:

$$dS_{sys} + dS_{surr} = 0 \text{ (at equilibrium)} \quad (7.80)$$

Equation 7.80 makes it possible to define the conditions under which a system is at equilibrium. If a system is initially *not* at equilibrium, it will rearrange over time so that the entropy of the system and its surroundings increases until it is maximal. An increase in the entropy corresponds to an increase in the number of equally probable microstates that are all consistent with the definition of the system. Thus, as the entropy increases, the system takes on more and more alternative configurations, thereby becoming less and less ordered. The second law of thermodynamics implies that spontaneous change always brings about an increase in the disorder of the system and its surroundings.

We return to a detailed analysis of the consequences of Equation 7.80 in Chapter 8, when we consider the interplay between energy and entropy. For now, though, we will study one more example to see how the maximal entropy principle (Equation 7.79) determines the outcome of molecular events.

## 7.25 Diffusion across a semipermeable membrane can lead to unequal numbers of molecules on the two sides of the membrane

To illustrate the application of the maximal entropy principle, we shall consider a highly simplified conceptual model for a process known as **osmosis**. Osmosis occurs when a semipermeable membrane separates two parts of a chamber, restricting the movement of one kind of molecule, but allowing free diffusion of another kind of molecule (particularly the solvent). Osmosis is important in biology because all living cells are contained within semipermeable membranes that allow certain molecules and ions to pass through freely, while blocking the transport of other molecules and ions.

A chamber that is partitioned by a semipermeable barrier is illustrated in **Figure 7.29**. There are two kinds of molecules within the chamber, type A (*blue*) and type B (*red*). The blue A molecules pass freely through the barrier, but the red B molecules cannot. Assume that the A and B molecules are inert (that is, they do not interact energetically with each other).

To keep things simple, we shall assume that the number of type A molecules is the same as the number of type B molecules. Let us suppose that initially we have $N$ molecules of type A and no molecules of type B on the right side of the barrier and $N$ molecules of type B and no molecules of type A on the left.

Is this situation, which is depicted in Figure 7.29A, at equilibrium? If we had, for example, only 1000 type A molecules in the system, then a situation in which there were 500 type A molecules on the left side and 500 molecules on the right side would certainly be at equilibrium. We are now considering something special, though, in which the 500 type B molecules on the left side are prevented from crossing the partition. Since type A molecules can freely cross the barrier, will there be a net flow of type A molecules from the right side to the left side? If so, what would be the situation at equilibrium? Would we have an equal number of type A molecules on both sides of the partition, or would we have some other distribution?

We can analyze problems such as this by using a statistical approach. We simply calculate the entropy of various possible states of the system, with the expectation that the state of the system with the highest entropy is the one that will be observed experimentally at equilibrium.

We begin by dividing up each half of the chamber into $M$ grid boxes, with $M >> N$ (we are assuming that the density of the molecules is low). The value of the entropy for the left side of the chamber, $S_{left}$, can be calculated as follows:

$$\frac{S_{left}}{k_B} = \ln W_{left} = N \ln\left(\frac{M}{N}\right) \tag{7.81}$$

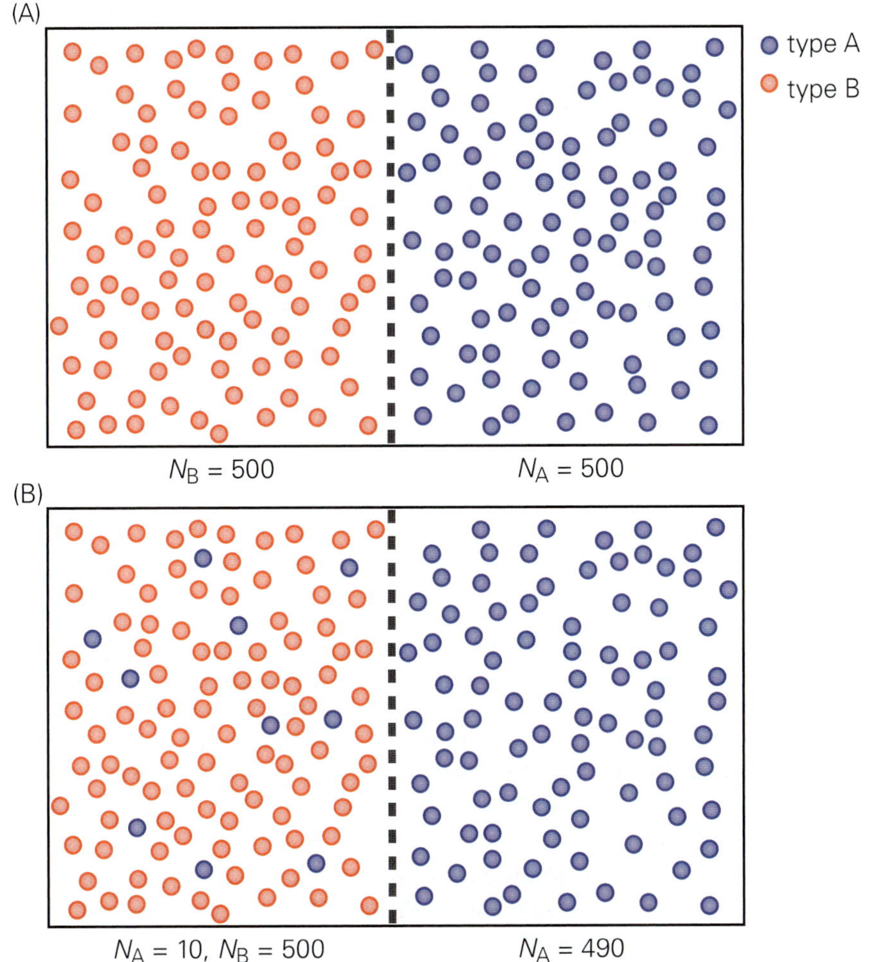


**Figure 7.29 Diffusion across a semipermeable barrier.** The system consists of a chamber, which is divided into two halves (left and right) by a semipermeable barrier. The barrier allows type A molecules (*blue*) to pass freely, but blocks the passage of type B molecules (*red*). Initially there is an equal number of molecules on both sides of the barrier (A). In (B), 10 type A molecules have moved from the right side to the left side.

In Equation 7.81, we have used the fact that when $M$ and $N$ both large and $M >> N$, then the expression for the multiplicity simplifies to (see Box 7.3):

$$\ln W(M,N) \approx N \ln\left(\frac{M}{N}\right)$$

Because we have the same number of B-type molecules on the left as we have A-type molecules on the right, the initial total entropy is:

$$\frac{S_{\text{total}}}{k_B} = \frac{S_{\text{left}}}{k_B} + \frac{S_{\text{right}}}{k_B} = 2\frac{S_{\text{left}}}{k_B} = 2N \ln\left(\frac{M}{N}\right) \tag{7.82}$$

Let us calculate how the entropy changes when a fraction of the A-type molecules move from the right side to the left. Denoting the fraction of A-type molecules that have moved across the membrane as $x$, the left side of the chamber will have $N$ B-type molecules (these have not crossed the partition) as well as $xN$ A-type molecules.

How do we calculate the entropy when we have two kinds of molecules present? As explained in Box 7.2, when $M >> N$, the two kinds of molecules behave independently, and we can simply add their individual contributions to the entropy together:

$$\begin{aligned}\frac{S_{\text{left}}}{k_B} &= N \ln\left(\frac{M}{N}\right) + xN \ln\left(\frac{M}{xN}\right) \\ &= N \ln\left(\frac{M}{N}\right) + xN \ln\left(\frac{M}{N}\right) - xN \ln x \\ &= (1+x) N \ln\left(\frac{M}{N}\right) - xN \ln x\end{aligned} \tag{7.83}$$

The right-hand side of the chamber now has $(1 - x)N$ molecules of type A, so the entropy for this region is given by:

$$\frac{S_{\text{right}}}{k_B} = (1-x)N \ln\left(\frac{M}{(1-x)N}\right)$$
$$= (1-x)N \ln\left(\frac{M}{N}\right) - (1-x)N \ln(1-x) \qquad (7.84)$$

The total entropy ($S_{\text{total}}$) is therefore given by:

$$\frac{S_{\text{total}}}{k_B} = \frac{S_{\text{left}} + S_{\text{right}}}{k_B} = 2N\ln\left(\frac{M}{N}\right) - N\left[x\ln x + (1-x)\ln(1-x)\right] \qquad (7.85)$$

By subtracting Equation 7.82 from Equation 7.85 we get an expression for the change in entropy as A-type molecules move from the right side to the left side of the partition:

$$\frac{\Delta S}{k_B} = -N\left[x\ln x + (1-x)\ln(1-x)\right] \qquad (7.86)$$

We can express the entropy change on a per-molecule basis as follows:

$$\frac{\Delta S}{Nk_B} = -\left[x\ln x + (1-x)\ln(1-x)\right] \qquad (7.87)$$

The graph in **Figure 7.30** shows how the entropy of the system changes as type A molecules are moved from the right side to the left. The total entropy of the system increases as an increasing number of type A molecules move to the left, and it reaches a maximum value when 50% of the A molecules have crossed over from the right side to the left. As more A molecules move from the right side to the left, the total entropy begins to decrease.

In this situation, the B-type molecules in the left-hand chamber contribute a constant value to the total entropy, and so the calculation is essentially that of the A-type molecules equilibrating to fill the entire chamber as if the B-type molecules were not present. This simplification arises because we are considering a very low density of molecules. It might also appear that the calculation applies only to gases, but a little thought will show that the results also hold for molecules in dilute solution. To see why this is the case, imagine that both sides of the

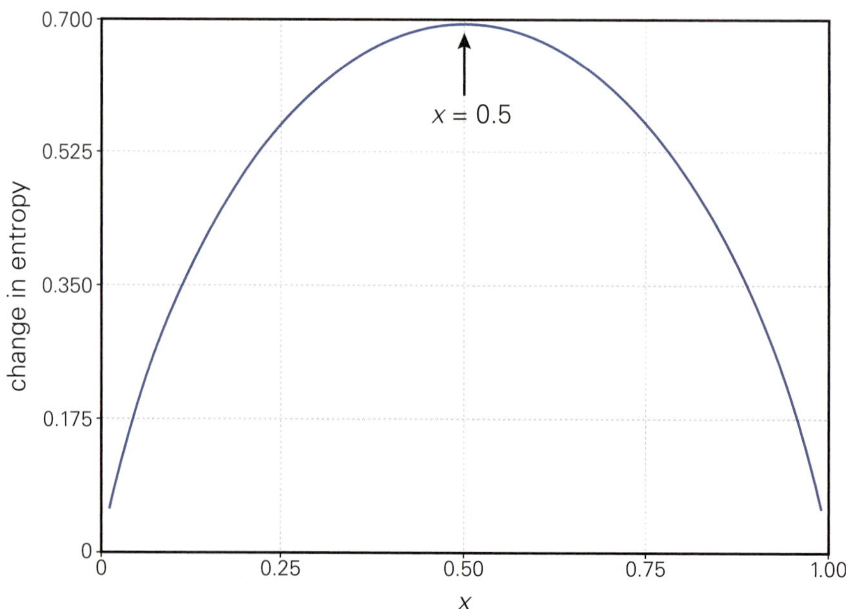


**Figure 7.30 Entropy changes for diffusion across a semipermeable membrane.** The system that is analyzed is illustrated in Figure 7.29. The value of

$$\frac{\Delta S}{Nk_B} = -\left[x\ln x + (1-x)\ln(1-x)\right]$$

(see Equation 7.87) is graphed as a function of $x$, where $x$ is the fraction of A molecules that have moved from the right side of the barrier to the left side, and $\Delta S$ is the change in entropy relative to the situation when no A molecules have crossed the barrier. The entropy is maximal when half of the A-type molecules have moved to the left.

chamber contain C-type molecules (for example, water). If the number of C-type molecules is much larger than that of the A- and B-type molecules (that is, the A and B molecules are solutes), then the contribution of the C-type molecules to the entropy is constant, and Equation 7.87 will apply.

The volume of the system is held constant in our simple model calculation. As the A molecules move into the left-hand chamber, the total number of molecules in the left-hand chamber increases because of the constant number of B molecules—that is, the density of the left-hand side of the chamber increases. A more realistic calculation would show that the pressure in the left-hand chamber increases as A molecules move in, and so the left-hand side of the chamber would swell if it were free to do so. This phenomenon is known as **osmotic pressure**. A simple manifestation of osmotic pressure is the observation that if cells are transferred to a solution with low solute concentration (for example, pure water), the cells burst. Most of the solutes within the cell are prevented by the cell membrane from moving out of the cell, but water molecules enter the cell through special channels, as discussed in Section 11.17. This causes the osmotic pressure in the cell to rise, eventually leading the cell membrane to rupture.

## Summary

The multiplicity of a state of a system is the number of different configurations or microstates of the system that are consistent with the parameters that define the given state. Any particular microstate is just as likely to be observed as any other microstate with the same total energy. As a consequence, the instantaneous configuration of a large collection of molecules is impossible to predict. Nevertheless, when we consider the aggregate properties that define a particular state of the system, we see that not all states are equally likely. States with a higher multiplicity (corresponding to a larger number of microstates) will be observed more often than states with low multiplicity.

For a series of $M$ events with binary outcomes, the multiplicity $W(M, N)$ associated with $N$ positive outcomes is given by the following expression:

$$W(M,N)=\frac{M!}{N!(M-N)!}$$

The values of $W(M, N)$ are known as binomial coefficients, and these are tabulated in a diagram known as Pascal's triangle. The corresponding probability distribution for different numbers of positive outcomes is known as the binomial distribution. When the number of trials becomes very large, the number of positive outcomes, $N$, can be represented by a continuous rather than a discrete variable and the corresponding probability distribution is known as a Gaussian distribution.

As the number of events becomes large, the probability distribution becomes sharply peaked around the most probable outcome. Outcomes more than a few standard deviations away from the most likely outcome have vanishingly small probability. Likewise, when considering molecular properties, the only realistic outcomes are ones that maximize the multiplicity. If a molecular system can convert to a state of higher multiplicity, it will do so spontaneously. This principle allows us to account for the spontaneous expansion of a gas even when there is no net input of energy.

The entropy ($S$) is a state variable that can be defined in terms of the multiplicity:

$$S=k_B \ln W$$

where $k_B$ is the Boltzmann constant and $W$ is the multiplicity. The entropy ($S$) is a more convenient quantity to compute and analyze than the multiplicity ($W$) because $W$ becomes astronomically large for systems with large numbers of molecules, whereas the value of $S$ remains manageable. The value of the Boltzmann constant ($1.38 \times 10^{-23}$ J•K$^{-1}$) can be computed by dividing the gas constant

($R = 8.314\ \text{J•mol}^{-1}\text{•K}^{-1}$) by Avogadro's number ($N_A = 6.022 \times 10^{23}\ \text{mol}^{-1}$). The Boltzmann constant is needed in the statistical definition of the entropy in order to make it consistent with the thermodynamic definition of the entropy—namely,

$$\Delta S = \frac{q_{rev}}{T}$$

where $q_{rev}$ is the heat transferred to the system during a near-equilibrium (reversible) process and $T$ is the absolute temperature. The statistical and thermodynamic definitions of the entropy are equivalent, a fact that is established in Chapter 8.

According to the second law of thermodynamics, all spontaneous transformations involve an increase in the entropy of the system and its surroundings. The second law can be expressed mathematically as follows:

$$dS_{sys} + dS_{surr} > 0$$

for a spontaneous process, where $dS_{sys}$ and $dS_{surr}$ are infinitesimal changes in the entropies of the system and the surroundings, respectively. This is known as the maximal entropy principle. If a system is at equilibrium, which means that it is stable and not subject to spontaneous change, it follows that:

$$dS_{sys} + dS_{surr} = 0$$

## Key Concepts

### A. COUNTING STATISTICS AND MULTIPLICITY

- The probability of obtaining a particular outcome (that is, a given number of heads) in a series of coin tosses is proportional to the multiplicity of the outcome.
- The multiplicity, $W$, for an outcome corresponding to $N$ heads in a series of $M$ coin tosses is given by Equation 7.4:

$$W(M,N) = \frac{M!}{N!(M-N)!}$$

- When the number of events is large, Stirling's approximation (Equation 7.12) simplifies the calculation of the multiplicity. Stirling's approximation states that when $n$ is a large integer, then:

$$\ln n! \approx n \ln n - n$$

- For a large number of coin tosses, the only observable outcomes will be ones that are close to the outcome with maximal multiplicity. This corresponds to an equal number of heads and tails when the coin is unbiased.
- The binomial distribution governs the probability of events with binary outcomes.
- When the number of events is large, the probability distribution is given by a Gaussian function (Equation 7.36):

$$P(x) = \frac{1}{\sqrt{2\pi} \times \sigma} \exp\left[\frac{-(x-\mu)^2}{2\sigma^2}\right]$$

- The multiplicity ($W$) of a molecular system is the number of equivalent configurations of the molecules (microstates) corresponding to the state of the system. If there are $N$ equivalent configurations of the molecules and, if the space available to the molecules is divided up into $M$ grid boxes, then the multiplicity is given by Equation 7.45:

$$W(M,N) = \frac{M!}{N!(M-N)!}$$

- The relative probability of two outcomes is given by the ratios of their multiplicities.
- When the number of molecules is large, the distribution of multiplicities is very sharply peaked. As a consequence, we are unlikely to observe states of the system that are more than a few standard deviations away from the state with maximum multiplicity.

### B. ENTROPY

- The logarithm of the multiplicity, $\ln W$, is an additive and therefore extensive property of the system, and it is more convenient to work with than $W$ because its numerical value remains manageable as the number of molecules in the system increases. The logarithm, $\ln W$, is maximal when $W$ is maximal.
- The entropy, $S$, of the system is given by Equation 7.58:

$$S = k_B \ln W$$

Equation 7.58 is known as the statistical definition of the entropy. The Boltzmann constant, $k_B$, is a proportionality constant linking the entropy to $\ln W$.
- The change in entropy on moving from one state to another is related to the heat transferred when the process is carried out near-equilibrium, or reversibly ($q_{rev}$), as shown in Equation 7.60:

$$\Delta S = \frac{q_{rev}}{T}$$

Equation 7.60 is known as the thermodynamic definition of the entropy.

- The statistical and thermodynamic definitions of the entropy are equivalent.
- A reversible process is one that is always near-equilibrium. The work done during a near-equilibrium process is always greater than the work done during an irreversible process, and it is related to the change in entropy.
- The direction of spontaneous change is given by the direction of increasing entropy. Entropy is maximal when the system is at equilibrium. This is known as the second law of thermodynamics.

## Problems

### True/False and Multiple Choice

1. It is easier to predict the bulk behavior of a small number of molecules than a large number of molecules.

   True/False

2. Two sets of molecules are mixed into a system at time zero. After 10 seconds, equilibrium has been reached. At what time was the entropy of the system maximal?

   a. 10 seconds (equilibrium)

   b. 5 seconds (half the time it takes to reach equilibrium)

   c. immediately upon mixing of the two sets of molecules

   d. prior to mixing of the two sets of molecules

3. Consider a coin with two sides (H = heads; T = tails). The probability of observing HHHHHTTTTT is equal to the probability of observing HTHTHTHTHT.

   True/False

4. On a 10-sided die, with a side for each number from 1 to 10, the probability of rolling a 5 is:

   a. 5/10

   b. 1/10

   c. 1/9

   d. 5/51

   e. 1/6

5. An isolated molecular system exists in two states of equal energy. State A has high multiplicity, whereas State B has low multiplicity. Without any external agents, a system in State B will spontaneously convert to State A.

   True/False

6. A system is divided into two halves separated by a removable divider. Initially, with the divider in place, the left half has only red molecules and the right half has only blue molecules. The divider is removed and equilibrium is reached. Which of the following statements correctly characterizes the system?

   a. It contains only blue molecules on the left side.

   b. It contains only red molecules on the left side.

   c. It contains a mixture of red and blue molecules throughout the system.

   d. It contains only blue molecules at the bottom of the system.

   e. It has an equal number of red and blue molecules on each side.

7. A state corresponds to many different microstates.

   True/False

### Fill in the Blank

8. The work done in a near-equilibrium expansion of an ideal gas is ____________ than for a nonequilibrium expansion.

9. When the volume of a system increases, its multiplicity ________.

10. The log of the multiplicity of the system is an __________________ property of the system.

11. The combined entropy of the system and surroundings always ______________ for a spontaneous process.

12. A drop of dye is added to a container of solvent. When equilibrium is reached the concentration of dye will be ___________ within the container.

### Quantitative/Essay

13. A coin is weighted deliberately so that the probability of tossing heads is twice the probability of tossing tails. What is the probability of tossing three heads in a row? What is the probability of tossing three tails

in a row? What is the relative probability of tossing three heads versus three tails?

**14.** Shown below is a portion of Pascal's triangle, with the tenth row filled in.

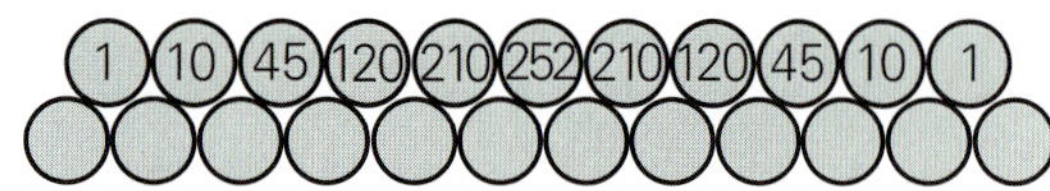


a. Fill in the values for the eleventh row. What simple rule can you use to fill in these values?

b. Using Pascal's triangle, calculate how much more likely it is to get five heads and six tails in a series of 11 coin tosses than getting four heads and seven tails.

**15.** Consider the following three cases of grid boxes. Molecule X can move between and occupy any box. Calculate the multiplicity and entropy for each case.

a.

| X | X |  | X |  | X |
|---|---|---|---|---|---|

b.

| X |  |  | X |  |  |
|---|---|---|---|---|---|

c.

| X |  | X | X |  |  |
|---|---|---|---|---|---|

**16.** Calculate the entropy of the system depicted below. There are four types of molecules (that is, X, O, Y, and Z) that can be arranged in any way in the available boxes. (Hint: Use Equation 7.1.2.)

| X |  |  |  |  |
|---|---|---|---|---|
| Y |  |  | X | Y |
| Z | O |  | Y | O |
|  | O |  | Z |  |
|  | O |  |  |  |

**17.** Consider the systems below consisting of different numbers of identical molecules (indicated by the symbol X) and equal-sized grid boxes that are either empty or occupied by a molecule. Which system has a higher entropy? What is the difference in entropy between A and B?

**A.**

| X |  |  |  |
|---|---|---|---|
|  | X |  |  |
|  |  | X |  |
|  |  |  | X |

**B.**

| X |  |  | X |
|---|---|---|---|
|  | X |  |  |
|  |  | X |  |
|  | X |  | X |

**18.** A bag contains 20 coins, each marked by the one-letter code for an amino acid. What is the probability of drawing each of the three large amino acids (Y, W, and F) once, in any order, without returning the coins to the bag after each draw? What is the probability if coins are returned to the bag after each draw?

What is the probability of drawing the three letters out in exactly the order of hydrophobicity (a unique ordering) without returning any coins to the bag?

**19.** What is the multiplicity of the following system?

| X | X | O | O |  |
|---|---|---|---|---|
|  | X |  |  |  |
|  |  | O |  | X |
|  |  |  | O |  |
| O |  |  |  |  |

If an additional five empty grid boxes are added, does the multiplicity of the system increase or decrease?

**20.** System A has a multiplicity of 15, whereas System B has a multiplicity of 12. What is the total entropy of Systems A and B?

**21.** In a near-equilibrium expansion, 15 kJ of work are done by a system at a constant temperature of 300 K. To reach the same state, 5 kJ of work are done in a nonequilibrium process. What is the entropy change for the two processes?

22. Considering the following system at two time points, A and B.

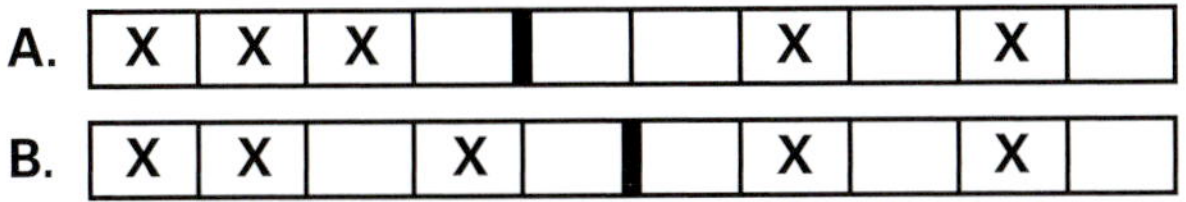

The system is divided by a movable partition. The molecules (X) cannot move across the partition, but they can move between the grid boxes on the same side of the partition. At which time point does the system have the higher multiplicity? At which time point is the system closer to equilibrium?

23. Consider a coin that is weighted so that it is twice as likely to come up heads as tails. Answer the following questions using the Gaussian distribution description for the results of a series of coin tosses:

a. What is the most likely outcome of a series of 2000 coin tosses using this biased coin?

b. What is the relative probability of getting 1500 heads versus 1000 heads in a series of 2000 coin tosses using this coin? Express your answer as a power of 10.

24. A system consists of molecules that convert between two colors (green and yellow). There are spaces for 10,000 molecules, but there are only 6000 molecules in the system. The system starts in a state with 2000 green molecules and 4000 yellow molecules. What is the entropy of the system? (Hint: Use Stirling's approximation.)

25. The system in Problem 24 converts to a final state with 6000 yellow molecules and no green molecules. Assuming no energy difference between yellow and green molecules, would this conversion occur spontaneously?

26. How much work is done in compressing one mole of an ideal gas from a starting volume of 1 L to a final volume of 250 mL at a constant temperature of 293 K? What is the change in entropy? Assume that the process occurs near-equilibrium (reversibly).

27. If the increase in entropy indicates the direction of spontaneous change, how can a system ever undergo a process that results in a decrease in the entropy of the system?

28. In our calculations, why do we work with the natural log of the multiplicity?

## Further Reading

### General

Atkins PW & De Paula J (2009) Atkins' Physical Chemistry, 9th ed. New York: Oxford University Press.

Dill KA & Bromberg S (2010) Molecular Driving Forces: Statistical Thermodynamics in Biology, Chemistry, Physics, and Nanoscience, 2nd ed. New York: Garland Science. The discussion of entropy in this chapter was guided by the treatment in this book, which is at a more advanced level.

Eisenberg DS & Crothers DM (1979) Physical Chemistry: With Applications to the Life Sciences. Menlo Park, CA: Benjamin Cummings.

McQuarrie DA (2000) Statistical Mechanics. Sausalito, CA: University Science Books.

# Linking Energy and Entropy: The Boltzmann Distribution

Energy and entropy are linked in fundamental ways that impact our understanding of numerous processes in biology, including the end points of biochemical reactions, the binding of drugs to their targets, the coupling of solar energy to food production, and the generation of electrical signals in nerve cells. The coupling between energy and entropy results in two concepts that are particularly useful in biochemistry—namely, the **Boltzmann distribution**, which governs the probability of finding molecules with particular energies, and the **free energy**, which determines whether a system is at equilibrium and defines the driving force to react if not at equilibrium.

In this chapter, we shall learn how to calculate the multiplicity, and therefore the entropy, of systems in which the molecules can be in different energy levels. We also introduce a new expression for the entropy, which is based on the probability of observing molecules in different energy levels. This probabilistic definition of the entropy is simpler to use when calculating the entropy of large numbers of molecules. We then rationalize why the Boltzmann distribution describes the population of molecules in different energy levels at equilibrium. Finally, we discuss the relationship between temperature and changes in entropy, which is crucial for the development of free energy in Chapter 9.

Despite the apparent complexity of some of the topics discussed in this chapter, the essential concepts are a direct extension of the principles of the conservation of energy and the tendency of entropy to always increase. Thus, much of this chapter is concerned with linkages between the first and second laws of thermodynamics. The treatment of entropy in this chapter, as in Chapter 7, is influenced by Dill and Bromberg's "Molecular Driving Forces" (see Further Reading at the end of the chapter), which provides a more thorough treatment of the concepts introduced here.

## A. ENERGY DISTRIBUTIONS AND ENTROPY

### 8.1 The thermodynamic definition of the entropy provides a link to experimental observations

In Chapter 7 we introduced two definitions of the entropy ($S$) of a system. The **statistical definition of the entropy** is given by:

$$S = k_B \ln W \tag{8.1}$$

where $S$ is the entropy, $k_B$ is the Boltzmann constant, and $W$ is the multiplicity of the system (**Figure 8.1**). Defining the entropy in terms of the multiplicity follows naturally from the idea that a system will evolve from less probable states to more probable ones. As we saw in Chapter 7, the greater the multiplicity (that is, the greater the number of equivalent configurations or microstates) of a state of the system, the more probable it is that the state will be observed. Equation 8.1 defines the entropy in terms of the logarithm of the multiplicity rather than the multiplicity itself, making it easier to deal with large numerical values of the multiplicity.

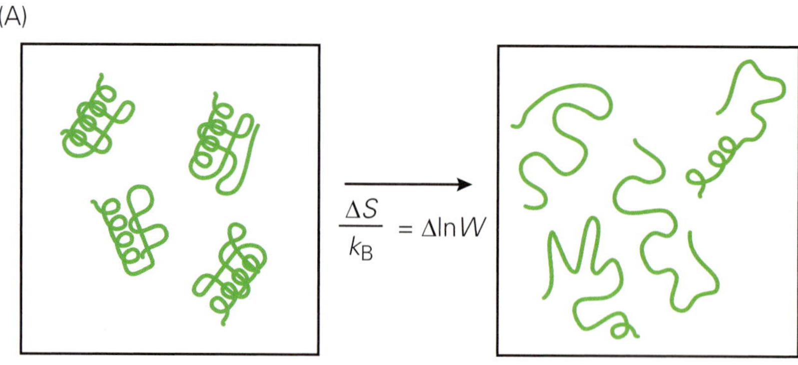


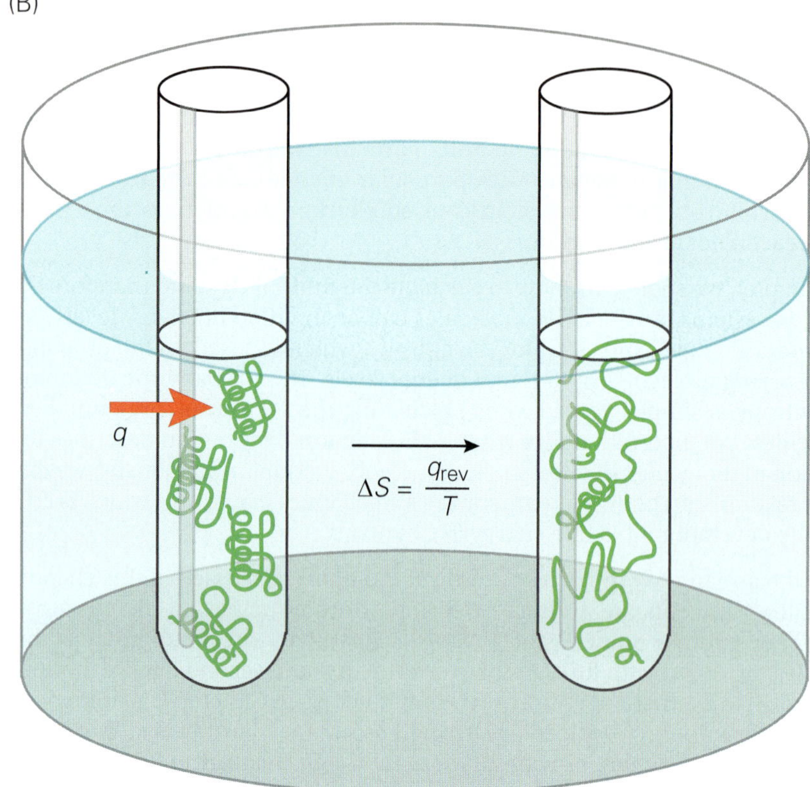


**Figure 8.1 Statistical and thermodynamic definitions of entropy.** (A) The statistical definition of entropy is based on the concept of the multiplicity of the system. The greater the multiplicity, the greater the entropy. The change in entropy for a process, such as protein unfolding, is proportional to the change in the logarithm of the multiplicity. (B) The thermodynamic definition of the entropy relates the change in entropy for a process to the heat delivered to the system while the process is carried out under reversible (that is, slow and near-equilibrium) conditions.

Chapter 7 also introduced the following definition of the entropy, which appears at first glance to be unrelated to Equation 8.1:

$$\Delta S = \frac{q_{\text{rev}}}{T} \tag{8.2}$$

where $\Delta S$ is the change in entropy of a system upon transforming from one state to another, $q_{\text{rev}}$ is the heat (that is, energy in the form of random molecular motion) transferred to the system during the transformation, using a reversible process (that is, a near-equilibrium process; see Section 7.21), and $T$ is the temperature (which is held constant during the transformation and is measured on the absolute scale). Recall from Chapter 7 that Equation 8.2 is referred to as the **thermodynamic definition of the entropy**, and that it arose historically from studying the direction of spontaneous change in heat transfer processes in heat engines.

The multiplicity, $W$, which underlies the definition of the entropy in Equation 8.1, is an abstract concept. Although we can understand readily what $W$ means, we cannot compute the value of $W$ easily for systems of even moderate complexity, and there is no straightforward way to measure the value of $W$ experimentally. Equation 8.2, on the other hand, tells us that if we carry out a transformation from

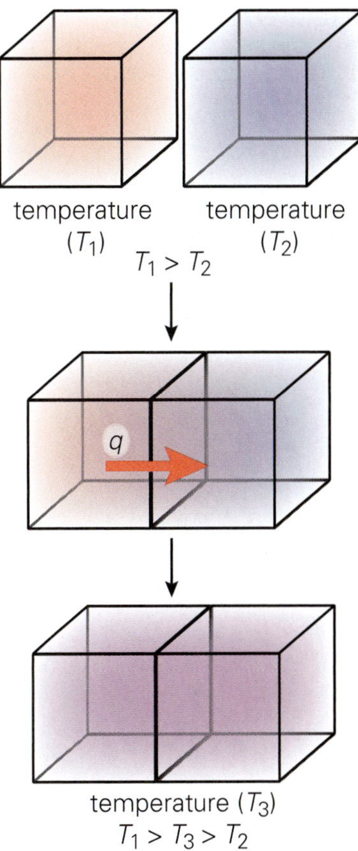


**Figure 8.2 Heat exchange between objects at different temperatures.** Two objects, at temperatures $T_1$ and $T_2$, respectively (where $T_1 > T_2$), are brought into thermal contact. There is net energy transfer in the form of heat from the object at higher temperature to the object at lower temperature until the temperatures of the two objects become equal.

one state of the system to another using a near-equilibrium and isothermal process, then the heat taken up by the system during the process is a direct measure of the entropy change between the two end states. Equation 8.2, therefore, provides a link between experiments we carry out in the laboratory and the value of the entropy.

Recall also from Chapter 7 that the statistical and thermodynamic definitions of the entropy are equivalent for an ideal monatomic gas. In this chapter, we shall show that these two definitions are strictly equivalent for *all* systems. In order to do this, we need to find a relationship between $W$, the multiplicity (or $S$, the entropy), and $T$, the temperature of the system, and we need to link these two quantities to the amount of heat transferred in a reversible process.

## 8.2 The concept of temperature provides a connection between the statistical and thermodynamic definitions of entropy

The key to relating the two definitions of entropy embodied in Equations 8.1 and 8.2 turns out to be the concept of temperature. What exactly do we mean by the "temperature" of a system? One definition is to say that the temperature is a property of the system that determines whether or not the system will transfer heat to or from another system that it is in contact with. If two systems at different temperatures are brought into contact, there will be net transfer of heat from the system at higher temperature to the one at lower temperature, until equilibrium is established (**Figure 8.2**). At equilibrium, the two systems will have the same temperature.

Defining the temperature of a system in terms of heat flow seems related intuitively to the energy of the system: the higher the energy of the molecules in the system, the higher we expect the temperature to be. If we place an object in which the molecules are moving faster (that is, they have higher kinetic energy) in contact with one in which the molecules are moving slower, our intuition tells us that on average there will be transfer of kinetic energy from the molecules in the first object to the molecules in the second object, until the molecules in both objects have the same average kinetic energy (**Figure 8.3**). Why, then, don't we just define the temperature as equal to the kinetic energy of the system, or at least connect it in some simple way to the kinetic energy of the system?

A direct relationship between kinetic energy and temperature (kinetic energy = $3RT/2$) is easy to establish for an ideal gas, as we showed in Box 6.3. For more complex materials, however, an analogous relationship is difficult to establish. One problem is that the energy alone does not always predict the direction of spontaneous change for molecular systems (see Section 6.5). Just as in the case

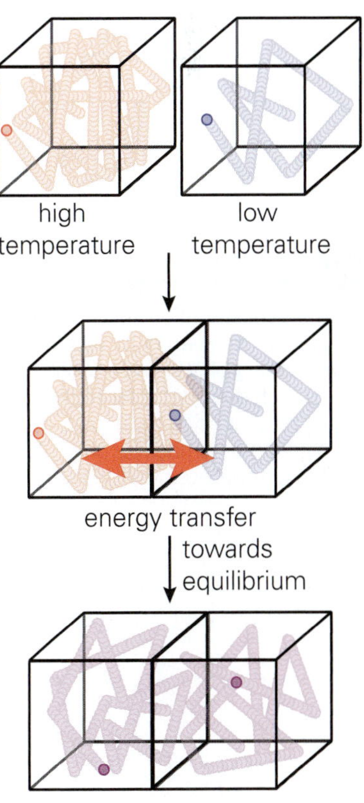


**Figure 8.3 Kinetic energy and temperature.** For a monatomic ideal gas, the kinetic energy is related to the temperature by the following equation:

$$\text{kinetic energy} = \frac{3}{2}RT$$

Thus, the higher the temperature, the higher the kinetic energy and the more the atoms move over the same time period. The object on the left is initially at higher temperature than the object on the right. When the two objects are brought into contact, energy is transferred until the kinetic energies become equal.

for isothermal gas expansion, where the spontaneous expansion process occurs without any net change in energy (as discussed in Chapters 6 and 7), this apparent paradox is resolved by considering the entropies of the systems rather than the energies. Energy transfer between two systems that are in contact but isolated from everything else always occurs in the direction of increasing entropy, which is what the second law of thermodynamics tells us should happen.

Because we are describing temperature in terms of the energy transfer between systems, we need to be able to write down a mathematical expression that relates the temperature to the entropy of the systems that are exchanging energy. In our treatment of entropy so far (that is, in Chapter 7), we have not considered energy explicitly (we have only looked at systems for which all configurations have the same energy). We begin our discussion of temperature by considering how we can calculate the multiplicity of a system in which the individual molecules have different energies.

## 8.3 Energy distributions describe the populations of molecules with different energies

Consider a set of molecules that do not interact with each other—that is, the energies of the molecule do not depend on their relative positions. Each molecule can exist in different energy levels, as governed by the quantum mechanics of the system. The molecules can collide with each other and exchange energy that way. We shall make a few simplifying assumptions in the following discussion of the relationship between energy and entropy. A more complicated treatment without these assumptions leads to the same conclusions, but we shall not pursue such an analysis here.

The first simplification is to assume that the molecules are all identical—that is, they have identical ladders of energy levels, which are shown schematically in **Figure 8.4**. We assume, furthermore, that the energy levels are evenly spaced and that the energy of the lowest level is $U_0 = 0$ (arbitrary energy units), that the energy of the second level is $U_1 = 1$, and so on. At each instant, any particular molecule in the system may be in any one of the energy levels that are accessible to it, given the total energy of the system (that is, no molecule can be in an energy level that is higher than the total energy of the system, and the total energy is conserved). Finally, we assume that each energy level is unique. This ignores the possibility that different levels have the same energy, a feature of real molecular energy levels that is known as degeneracy.

We assume that the motion of all the molecules is chaotic and that they are constantly colliding with each other and exchanging energy. Instead of tracking each molecule individually, which would be very difficult, we monitor only the manner in which the molecules are distributed among the different energy levels. That is, we ask how many molecules are in the lowest energy level, how many are in the next one up, and so on. The population of molecules in each energy level is called the **energy distribution**. Any particular energy distribution is illustrated by denoting the numbers of molecules in each level by an appropriate number of circles in each level. In **Figure 8.5**, for example, three circles are shown on the lowest rung in the energy diagram because the system consists of three molecules, each of which is in the lowest energy level.

**Energy distribution**

An energy distribution specifies the population (number) of molecules in each energy level. The energy distribution does not refer to the identities of specific molecules, just the aggregate number of molecules in each energy level.

## 8.4 The multiplicity of an energy distribution is the number of equivalent configurations of molecules that results in the same energy distribution

How do we calculate the multiplicity of a particular energy distribution? The multiplicity is simply the number of equivalent configurations of the molecules that is consistent with the definition of the energy distribution. Each equivalent configuration is a particular distribution of molecules among the energy levels, called a

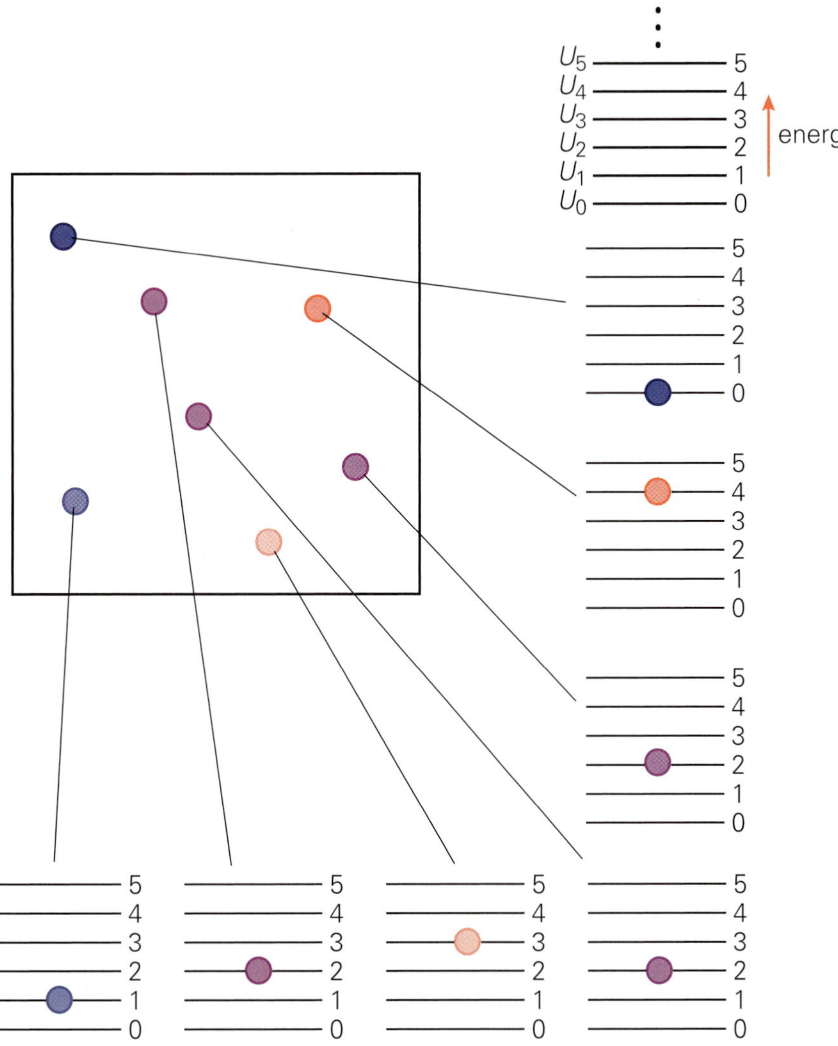


**Figure 8.4 Energy levels.** A system containing many identical molecules is shown schematically. Each molecule has a particular energy, indicated by different colors, and the value of the energy is indicated by a dot on the appropriate energy level.

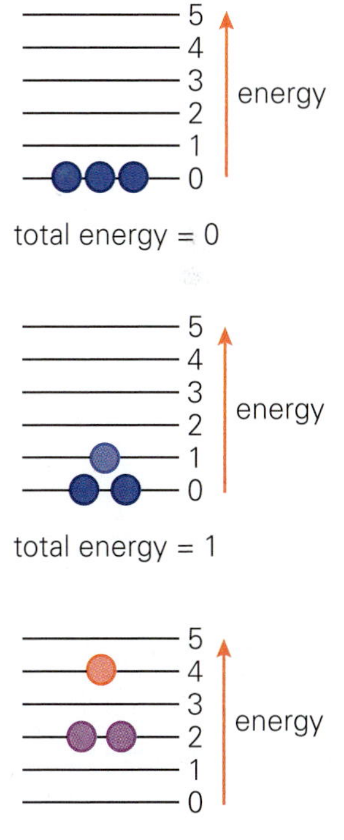


**Figure 8.5 Energy distributions.** Three different energy distributions for three molecules are shown schematically. The number of molecules in each level is represented by the number of circles in each energy level.

microstate. There usually are many different microstates that are consistent with a particular energy distribution, as shown in **Figure 8.6** for the three-molecule system from Figure 8.5. Thus, in order to calculate the multiplicity, we need a way to count all of the different microstates that are consistent with a particular energy distribution.

We define the overall state of the system by the volume ($V$), the number of particles ($N$), and the total energy ($U$). Given a certain number of molecules ($N$) in a volume ($V$), there is a component of the multiplicity that is associated with the number of different ways of rearranging the positions of the molecules within the volume of the system. In Chapter 7, we calculated this term by ignoring the energies of the molecules and simply focusing on the positional multiplicity of the system, which we denoted $W$. We shall now denote the positional multiplicity as $W_{\text{positional}}$.

In addition to the positional multiplicity, for every particular positional configuration of the molecules there are different ways in which the total energy ($U$) can be distributed among the molecules (**Figure 8.7**). The number of different ways in which the energy can be distributed among the molecules is a component of the multiplicity that is distinct from the positional multiplicity, and we denote it by $W_{\text{energy}}$. If we assume that the molecules do not interact with each other (that is, they do not influence each others' potential energy), then $W_{\text{energy}}$ is the same for every possible positional configuration of the molecules (that is, the energy does

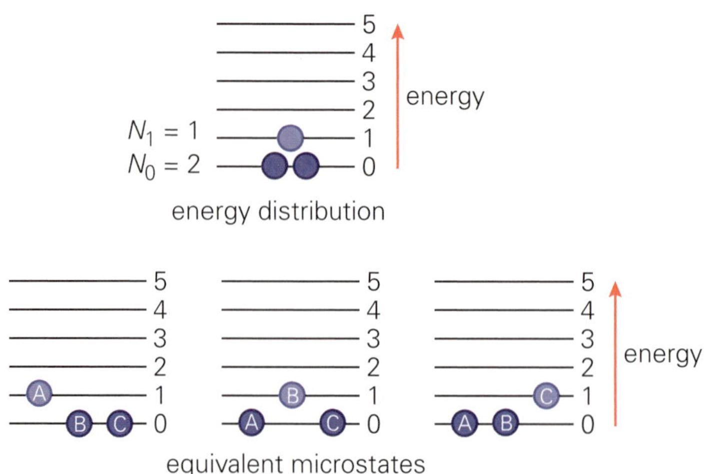


**Figure 8.6 Energy distributions and microstates.** An energy distribution with three molecules and a total energy of 1 energy unit is shown above. The term "energy distribution" refers to the populations of molecules in the different energy levels. In this case there are two molecules in the first level and one in the second. The energy distribution does not specify which molecule is in which level; instead, it simply reflects the aggregate probability of finding molecules in different energy levels. By specifying which individual molecules are in particular energy levels, we define the various microstates of the system that are consistent with a particular energy distribution. In the example shown here, there are three different microstates that correspond to the energy distribution shown above.

not depend on the positions of the molecules, each of which behaves independently). The total multiplicity ($W_{total}$) is therefore given by:

$$W_{total} = W_{positional} \times W_{energy} \tag{8.3}$$

Equation 8.3 makes it possible to separate the total entropy of the system, $S_{total}$, into two components, one due to positional rearrangements and one due to rearrangements of the energy. The total entropy is given by Equation 8.4:

$$S_{total} = k_B \ln W_{total} \tag{8.4}$$

Substituting the value of $W_{total}$ from Equation 8.3, we get:

$$S_{total} = k_B \ln(W_{positional} \times W_{energy}) = k_B \ln W_{positional} + k_B \ln W_{energy}$$
$$= S_{positional} + S_{energy} \tag{8.5}$$

where $S_{positional}$ is the entropy that arises from positional variations (sometimes called the **configurational entropy**) and $S_{energy}$ is the entropy due to the redistribution of the energy among different molecules. If the number of molecules and

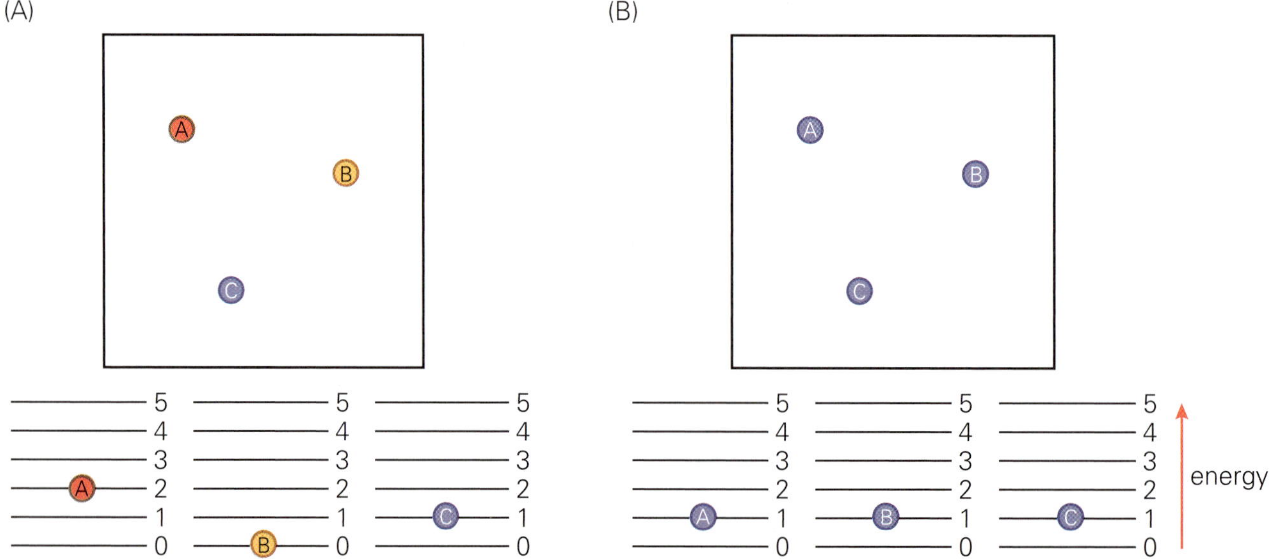


**Figure 8.7 Configurational multiplicity and the multiplicity of energy levels.** Shown here is a system with three molecules. There are many different positions (configurations) of these molecules, one of which is shown. For each configuration, there are different microstates corresponding to different values of energy for each molecule, subject to the constraint that the total energy is constant. Two alternative microstates for the same positional configuration are illustrated for a total energy of three energy units.

the volume of the system remain constant, then the $S_{positional}$ term is a constant, and the change in entropy for a change from state 1 to state 2 is given by:

$$\begin{aligned}\Delta S &= S_{total}(\text{state 2}) - S_{total}(\text{state 1}) \\ &= S_{energy}(\text{state 2}) + S_{positional}(\text{state 2}) \\ &\quad - S_{energy}(\text{state 1}) - S_{positional}(\text{state 1}) \\ &= S_{energy}(\text{state 2}) - S_{energy}(\text{state 1}) \\ \Rightarrow \Delta S &= \Delta S_{energy} \end{aligned} \tag{8.6}$$

(for a process occurring under conditions of constant volume and a fixed number of molecules).

By restricting the present discussion to a constant volume and a fixed number of molecules, we can ignore the positional entropy when calculating changes in the entropy, because it cancels out in Equation 8.6. We shall not discuss positional entropy further in this chapter, but we need to take it into consideration whenever there are changes in the volume of the system, or when the molecules interact with each other and it is no longer possible to separate the energy distribution from the positional distribution.

## 8.5 The multiplicity of a system with different energy levels can be calculated by counting the number of equivalent molecular rearrangements of energy

How do we calculate $W_{energy}$? First, we need to know what the total energy ($U$) of the system is, because all of the possible configurations of the system must satisfy the constraint that the total energy is a constant (that is, the first law of thermodynamics is always obeyed). The total energy ($U_{total}$) is given by:

$$U_{total} = \sum_i N_i U_i \tag{8.7}$$

where $N_i$ is the number of atoms in the $i^{th}$ energy level and $U_i$ is the energy of the $i^{th}$ level (**Figure 8.8**). The multiplicity of the system will depend on how many units of energy we put into the system.

Consider a process that randomly puts one unit of energy into a system consisting of three molecules that are initially all in the lowest energy level (that is, $U_{total} = 0$ initially, and $U_{total} = 1$ finally). There is only one possible energy distribution that results from this process, and it has one molecule in the first excited level and two molecules in the lowest energy level, as shown in **Figure 8.9**. Three different microstates correspond to this energy distribution, because any one of the three molecules could pick up the unit of energy by random chance. If we treat the one excited molecule in the energy level diagram (see Figure 8.9) as "heads" and the

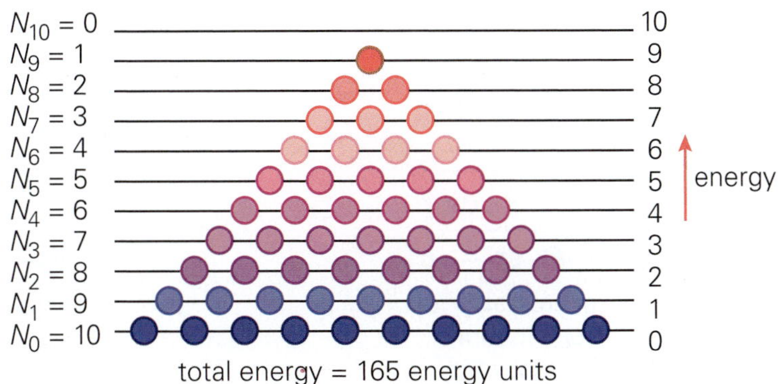


**Figure 8.8 Energy distributions and the conservation of energy.** In the distribution shown, there is a total of 55 molecules that are distributed among the lowest 10 energy levels. The number of molecules in the $i^{th}$ energy level is denoted by $N_i$. Although many different energy distributions are consistent with the total energy of this particular distribution, distributions that correspond to a different value of the total energy are not allowed if the system is isolated. For example, a distribution that is identical to the one shown here but in which $N_9 = 0$ and $N_{10} = 1$ would be disallowed because it corresponds to an increase in the total energy by 1 energy unit.

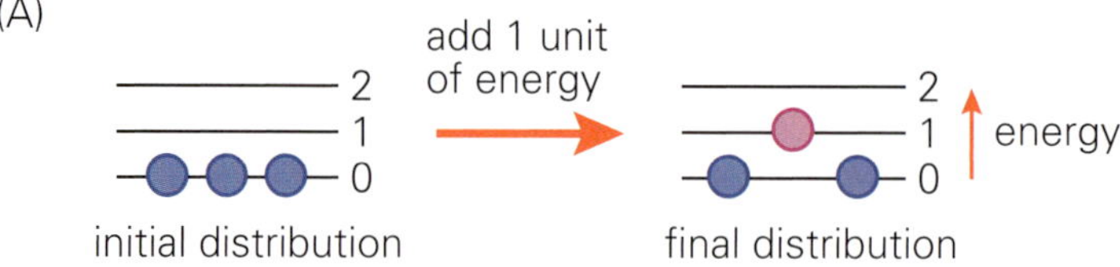


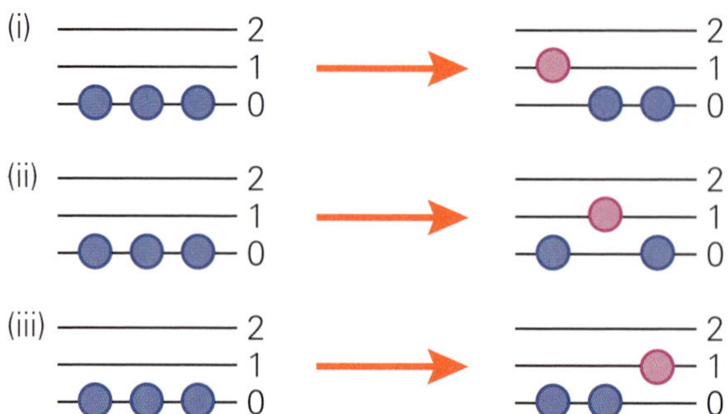


**Figure 8.9 Adding one unit of energy to a system at minimal energy.** (A) The process of adding one unit of energy to a system of three molecules is illustrated here. (B) By random chance, any one of the three molecules can pick up the unit of energy, and so there are three ways of achieving the end result. This corresponds to the multiplicity of the final energy distribution, which is 3 ($W$ = 3).

two molecules in the ground state as "tails," the multiplicity ($W_{\text{energy}}$) of the energy distribution is:

$$W_{\text{energy}} = \frac{3!}{2!1!} = 3 \tag{8.8}$$

Because the molecules are identical, we do not count different permutations of the molecules in the lowest energy rung separately. Thus, in general, if we have $N$ molecules distributed among a total of $t$ different energy levels, then the multiplicity is given by:

$$W = \frac{N!}{N_1!N_2!\ldots N_t!} \tag{8.9}$$

where $N_1$, $N_2$, … , $N_i$ are the numbers of molecules in the first, second, etc., energy level (up to the highest level, $t$). We have dropped the "energy" subscript in Equation 8.9, with the understanding that "$W$" refers to the number of different rearrangements of molecules that are possible among the energy levels.

Now consider a process that puts two units of energy into the system consisting of three molecules that are initially all in the lowest energy level (that is, $U = 0 \rightarrow U = 2$). There are *two* different energy distributions that are valid outcomes of this process. One energy distribution, shown in **Figure 8.10**A, puts two units of energy

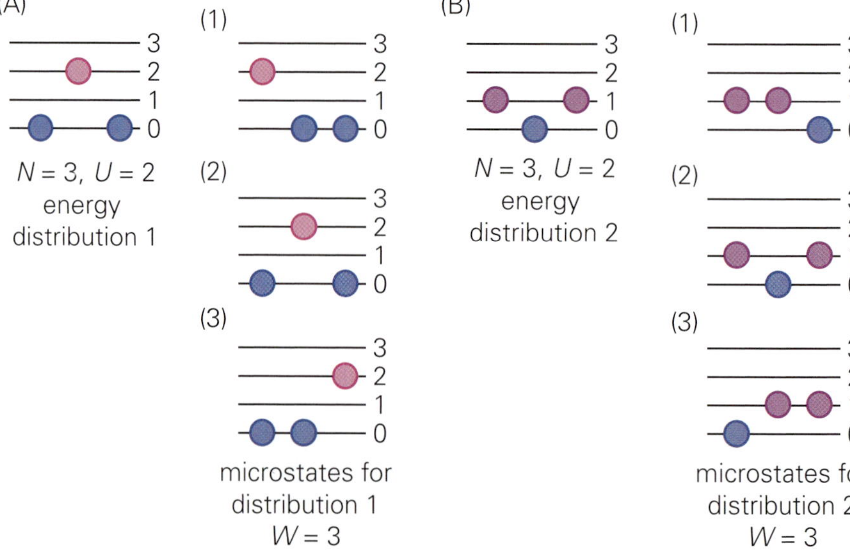


**Figure 8.10 Energy distributions and multiplicities for $N$ = 3, $U$ = 2.** Shown is a system with three molecules and a total of two energy units. (A) and (B) show the two possible energy distributions for this system. Each distribution corresponds to three equivalent microstates, so the multiplicity of each distribution is 3. (Adapted from K.A. Dill and S. Bromberg, Molecular Driving Forces: Statistical Thermodynamics in Biology, Chemistry, Physics, and Nanoscience, 2nd ed. New York: Garland Science, 2010.)

into one molecule and none into the other two. The multiplicity for this energy distribution can be calculated using Equation 8.8, just as it was for $U = 1$ (that is, one excited molecule and two molecules in the ground state). The multiplicity, therefore, is three. Another possible energy distribution has one unit of energy in each of two molecules and none in a third (Figure 8.10B). The multiplicity for this distribution is also 3.

Now suppose we repeatedly and randomly add two units of energy to systems that initially have zero energy. What is the energy distribution likely to be? Since both ways of adding two units of energy to the system have the same multiplicity, we are just as likely to obtain energy distributions in which one molecule has taken up two units of energy as we are to see energy distributions in which two molecules have taken up one unit of energy each. As with coin tosses and molecular diffusion (Chapter 7), calculating the multiplicity of the different possible energy distributions allows us to infer the probability of obtaining a particular energy distribution relative to that for another energy distribution. When calculating the multiplicity of a system with energy levels, an "outcome" is a particular distribution of molecules in energy levels, and the multiplicity of any particular distribution is the number of microstates that correspond to it.

The situation becomes a little more complicated when we put three units of energy into the system. As shown in **Figure 8.11**, there are three different energy distributions that correspond to this case. The energy distribution shown in Figure 8.11A

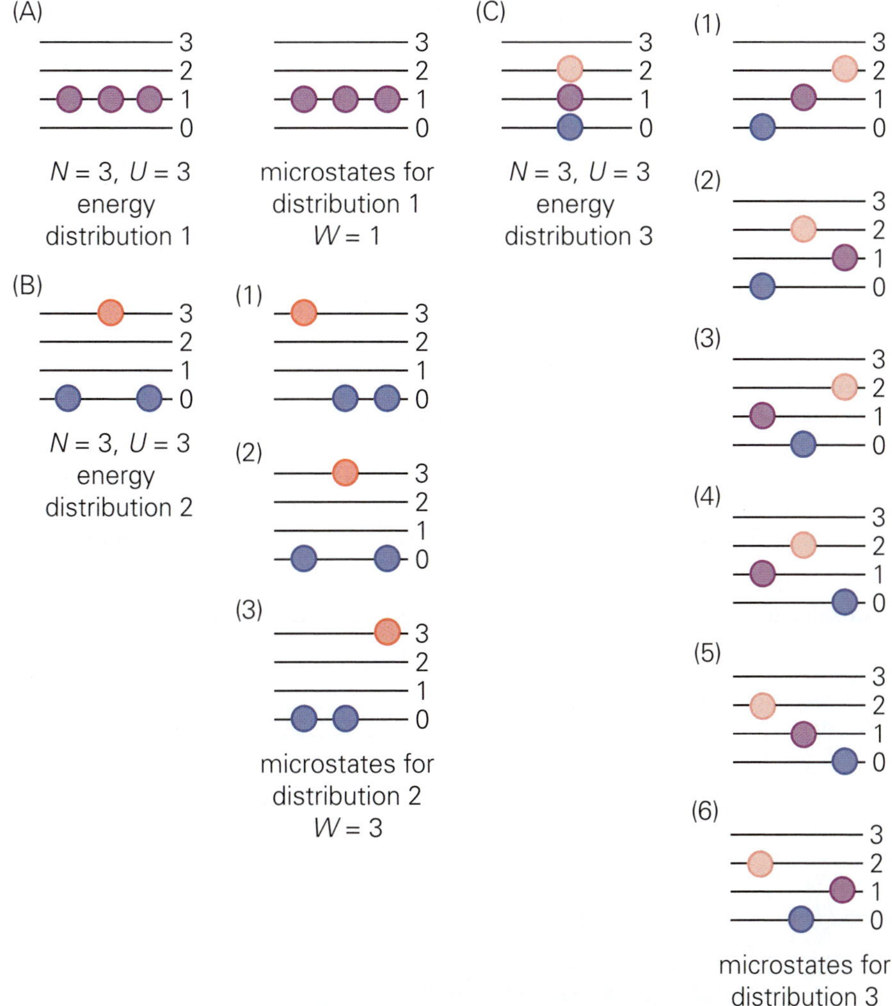


**Figure 8.11 Energy distributions and multiplicities for $N = 3$, $U = 3$.** There are three different energy distributions for a system with three molecules and a total energy of three units. These have multiplicities of 1, 3, and 6, as shown in (A), (B) and (C). (Adapted from K.A. Dill and S. Bromberg, Molecular Driving Forces: Statistical Thermodynamics in Biology, Chemistry, Physics, and Nanoscience, 2nd ed. New York: Garland Science, 2010.)

results from putting one unit of energy into each of the three molecules. In this case:

$$W = \frac{3!}{1!\,3!} = 1 \tag{8.10}$$

Alternatively, the distribution obtained from putting all of the energy into one molecule and none into the other two is shown in Figure 8.11B. In this case:

$$W = \frac{3!}{2!1!} = 3 \tag{8.11}$$

Finally, putting two units of energy into one molecule, one unit of energy into a second molecule, and none into the third yields the distribution shown in Figure 8.11C. The multiplicity (that is, the number of different microstates) is:

$$W = \frac{3!}{1!1!1!} = 6 \tag{8.12}$$

If we repeatedly and randomly add three units of energy to systems consisting of three molecules that are initially at zero energy, we are most likely to obtain the third situation, in which one molecule has taken up two units of energy, since this energy distribution has the largest value of the multiplicity. When we consider only a small number of molecules, though, such as the three in this example, there is a significant probability that we will observe energy distributions other than the one with the maximum multiplicity. In this case, for example, there is a one in 10 chance that all the molecules will be excited to the first energy level, with none in the higher energy levels.

# B. THE BOLTZMANN DISTRIBUTION

## 8.6 For large numbers of molecules, a probabilistic expression for the entropy is more convenient

When the number of molecules in a system is very large, then the statistical definition of the entropy (Equation 8.1) can be written in terms of the population of molecules in each level. This is known as the **probabilistic definition of entropy**, which is stated as follows:

$$S = -Nk_B \sum_{i=1}^{t} p_i \ln p_i \tag{8.13}$$

where $N$ is the total number of molecules in the system, $k_B$ is the Boltzmann constant, and $p_i$ is the *probability* that a molecule is found in the $i^{th}$ energy level. The probability, $p_i$, is the **fractional occupancy** of the $i^{th}$ level—that is, the fraction of the total number of molecules that are found in this level (**Figure 8.12**). In **Box 8.1**, we show that Equation 8.13 is just an alternative way of writing Equation 8.1, the statistical definition of the entropy, when the number of molecules is large.

We are introducing this new expression for the entropy because Equation 8.13 is simpler to use than Equation 8.1 when very large numbers of molecules are

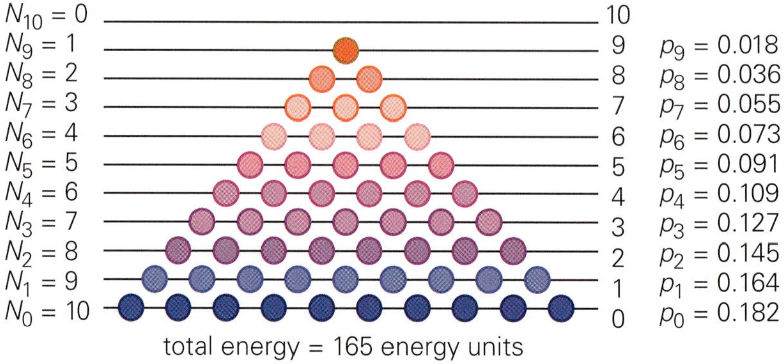


**Figure 8.12 The probabilities of finding molecules in energy levels.** The energy distribution shown originally in Figure 8.8 is shown again here, with the probabilities of finding molecules in each level indicated. The total number of atoms, $N$, is 55. The number of atoms in each level, $N_i$, is shown on the left next to each level. The probability, $p_i$, of finding a molecule in the $i^{th}$ level is given by the fractional occupancy of the level, $N_i/N$, and is shown on the right.

## Box 8.1 Derivation of the probabilistic definition of the entropy

Here we show that the probabilistic expression for the entropy (Equation 8.13) is equivalent to the statistical definition of entropy, when the number of molecules is large. That is, we need to show that if:

$$S = k_B \ln W \qquad (8.1.1)$$

then it must be true that

$$S = -Nk_B \sum_{i=1}^{t} p_i \ln p_i \qquad (8.1.2)$$

where $p_i = \frac{N_i}{N}$, $N_i$ is the number of molecules in the $i^{\text{th}}$ energy level, and $N$ is the total number of molecules.

Suppose we have a system in which the molecules can be in any one of 1 to $t$ energy levels. If there are $N$ molecules, and the number of molecules in the $i^{\text{th}}$ level is $N_i$, then the multiplicity of the system, $W$, is given by:

$$W = \frac{N!}{N_1!N_2!\ldots N_t!} \quad \text{(see Equation 8.9)}$$

In order to convert this to a probabilistic form, we use Stirling's approximation:

$$\ln N! \approx N \ln N - N \qquad (8.1.3)$$

We are now going to express Stirling's approximation in an alternative form. We start by exponentiating both sides of Equation 8.1.3:

$$e^{\ln N!} = e^{N \ln N - N} \qquad (8.1.4)$$

We simplify the right-hand side of Equation 8.1.4 as follows:

$$e^{N \ln N - N} = e^{N \ln N} e^{-N} = \frac{e^{N \ln N}}{e^N} = \frac{\left(e^{\ln N}\right)^N}{e^N} = \left(\frac{e^{\ln N}}{e}\right)^N \qquad (8.1.5)$$

This, now, looks very complicated, but we can use the fact that $e^{\ln x} = x$ to simplify it:

$$\Rightarrow \left(\frac{e^{\ln N}}{e}\right)^N = \left(\frac{N}{e}\right)^N \qquad (8.1.6)$$

Likewise,

$$e^{\ln N!} = N! \qquad (8.1.7)$$

$$\Rightarrow N! = \left(\frac{N}{e}\right)^N \qquad (8.1.8)$$

Equation 8.1.8 is an alternative form of Stirling's approximation.

Using Equation 8.1.8, we can rewrite the multiplicity as follows:

$$W = \frac{(N/e)^N}{(N_1/e)^{N_1} (N_2/e)^{N_2} \ldots (N_t/e)^{N_t}}$$

$$\Rightarrow W = \frac{N^N \left(\frac{1}{e}\right)^N}{N_1^{N_1} N_2^{N_2} \ldots N_t^{N_t} \left(\frac{1}{e}\right)^{N_1+N_2+N_3+\ldots N_t}} \qquad (8.1.9)$$

But,

$$N_1 + N_2 + N_3 + \ldots + N_t = N \qquad (8.1.10)$$

So,

$$\left(\frac{1}{e}\right)^{N_1+N_2+\ldots+N_t} = \left(\frac{1}{e}\right)^N \qquad (8.1.11)$$

According to Equation 8.1.11, the $\left(\frac{1}{e}\right)$ terms in Equation 8.1.9 cancel each other. Equation 8.1.9 therefore simplifies to:

$$W = \frac{N^N}{N_1^{N_1} N_2^{N_2} \ldots N_t^{N_t}} \qquad (8.1.12)$$

We can define the probability of finding a molecule at any energy level as the number of molecules in that level divided by the total number of molecules:

$$p_j = \frac{N_j}{N} \qquad (8.1.13)$$

where $p_j$ is the probability of the finding of a molecule in the $j^{\text{th}}$ energy level.

Going back to Equation 8.1.12, we can rewrite the expression for the multiplicity as:

$$W = \frac{N^{N_1} N^{N_2} N^{N_3} \ldots N^{N_t}}{N_1^{N_1} N_2^{N_2} N_3^{N_3} \ldots N_t^{N_t}} = \left(\frac{N}{N_1}\right)^{N_1} \left(\frac{N}{N_2}\right)^{N_2} \left(\frac{N}{N_3}\right)^{N_3} \ldots \left(\frac{N}{N_t}\right)^{N_t} \qquad (8.1.14)$$

Using Equation 8.1.13, we can now express the multiplicity in terms of the probabilities, $p_i$:

$$W = p_1^{-N_1} p_2^{-N_2} \ldots p_t^{-N_t} \qquad (8.1.15)$$

$$\Rightarrow \ln W = -\sum_{i=1}^{t} N_i \ln p_i \qquad (8.1.16)$$

Now divide both sides of Equation 8.1.16 by $N$:

$$\Rightarrow \frac{\ln W}{N} = -\frac{1}{N} \sum_{i=1}^{t} N_i \ln p_i = -\sum_{i=1}^{t} \frac{N_i}{N} \ln p_i = -\sum_{i=1}^{t} p_i \ln p_i \qquad (8.1.17)$$

$$\Rightarrow \ln W = -N \sum_{i=1}^{t} p_i \ln p_i \qquad (8.1.18)$$

Using the statistical definition of the entropy (Equation 8.1), and combining it with Equation 8.1.18, we get:

$$S = k_B \ln W = -Nk_B \sum_{i=1}^{t} p_i \ln p_i \qquad (8.1.19)$$

Equation 8.1.19 demonstrates that the statistical (Equation 8.1) and the probabilistic (Equation 8.1.13) definitions of the entropy are equivalent. Note that this argument relied on the application of Stirling's approximation (see Equation 8.1.9). Hence, the equivalence between the two definitions is only valid when the number of atoms in the systems is so large that Stirling's approximation holds true for all levels that are populated to a significant extent.

involved. The multiplicity ($W$) in Equation 8.1 must be calculated by using Equation 8.9, and it is tedious to calculate the factorial values that enter into Equation 8.9 when the number of molecules is large. The probabilistic definition of the entropy considers only the fractional occupancy of each energy level (or outcome) and these fractional occupancies (that is, what fraction of the molecules are in a particular energy level) are always less than 1.0.

By using Equation 8.13 to define the entropy, we are focusing attention on the **probability distribution** of the molecules, which specifies the probability of finding molecules in different energy levels. The probability of finding a molecule in a particular energy level is given by:

$$p_i = {}^{N_i}\!/_{N} \tag{8.14}$$

where $N_i$ is the number of molecules in the $i^{\text{th}}$ energy level and $N$ is the total number of molecules (see Figure 8.12). The probabilities defined in this way are said to be **normalized**, which means that the sum of all the individual probabilities is 1.0:

$$\begin{aligned} & p_1 + p_2 + p_3 + \ldots p_t \\ & = \frac{N_1}{N} + \frac{N_2}{N} + \ldots \frac{N_t}{N} = \frac{\sum_{i=1}^{t} N_i}{N} = \frac{N}{N} = 1 \end{aligned} \tag{8.15}$$

The molecule must be in *one* of the energy levels (without specifying which one), so the normalization of a probability distribution ensures that the probability of finding a molecule in some available energy level is unity. The normalization condition can be written as:

$$\sum_i p_i = 1 \tag{8.16}$$

The probability of being in any particular energy level is always less than or equal to 1.0, so the value of $\ln p_i$ is less than or equal to zero (the logarithm of a number that is less than 1.0 is negative):

$$\ln p_i \leq 0 \tag{8.17}$$

The negative sign in Equation 8.13 therefore ensures that the entropy is always positive.

To see how the probabilistic definition makes it easier to calculate the entropy, consider a collection of dice where each die represents a molecule that can be in one of six conformations. **Figure 8.13** compares the entropy of such a system when all six outcomes are equally likely with situations where the outcomes are biased. Bias in the probability of outcomes lowers the entropy.

**Figure 8.14** illustrates the link between probability and entropy with a molecular example. Imagine a system in which all 10,000 molecules of the system are divided equally between the two lowest energy levels of the system (see Figure 8.14A). Alternatively, imagine the 10,000 molecules are divided equally among the first five energy levels, as shown in Figure 8.13B. What are the values of the entropy in the two cases?

In the first case, the probability of being in either one of the first two levels is 0.5, and the probability of being in any other level is zero. According to Equation 8.13, the entropy is calculated as follows:

$$\begin{aligned} \frac{S}{Nk_B} &= -\sum_{i=1}^{t} p_i \ln p_i = -0.5 \times \ln 0.5 - 0.5 \times \ln 0.5 \\ &= 0.5 \times 0.693 + 0.5 \times 0.693 = 0.693 \end{aligned} \tag{8.18}$$

Note that the value obtained in Equation 8.18 is $S/(Nk_B)$ and not $S$. We shall often calculate $S/(Nk_B)$ rather than $S$ when we are only interested in comparing the values of the entropy in situations where the number of molecules is the same.

$S/(Nk_B)$ is a unitless number, because $N$ is a pure number and $k_B$ has the same units as entropy. $S/(Nk_B)$ is the entropy per molecule, expressed as a multiple of the Boltzmann constant.

In the second case (that is, Figure 8.14B), the probability of being in any of the five lowest levels is 0.2, so:

$$\frac{S}{N k_B} = 5 \times (-0.2 \times \ln 0.2) = 5 \times 0.2 \times 1.61 = 1.61 \tag{8.19}$$

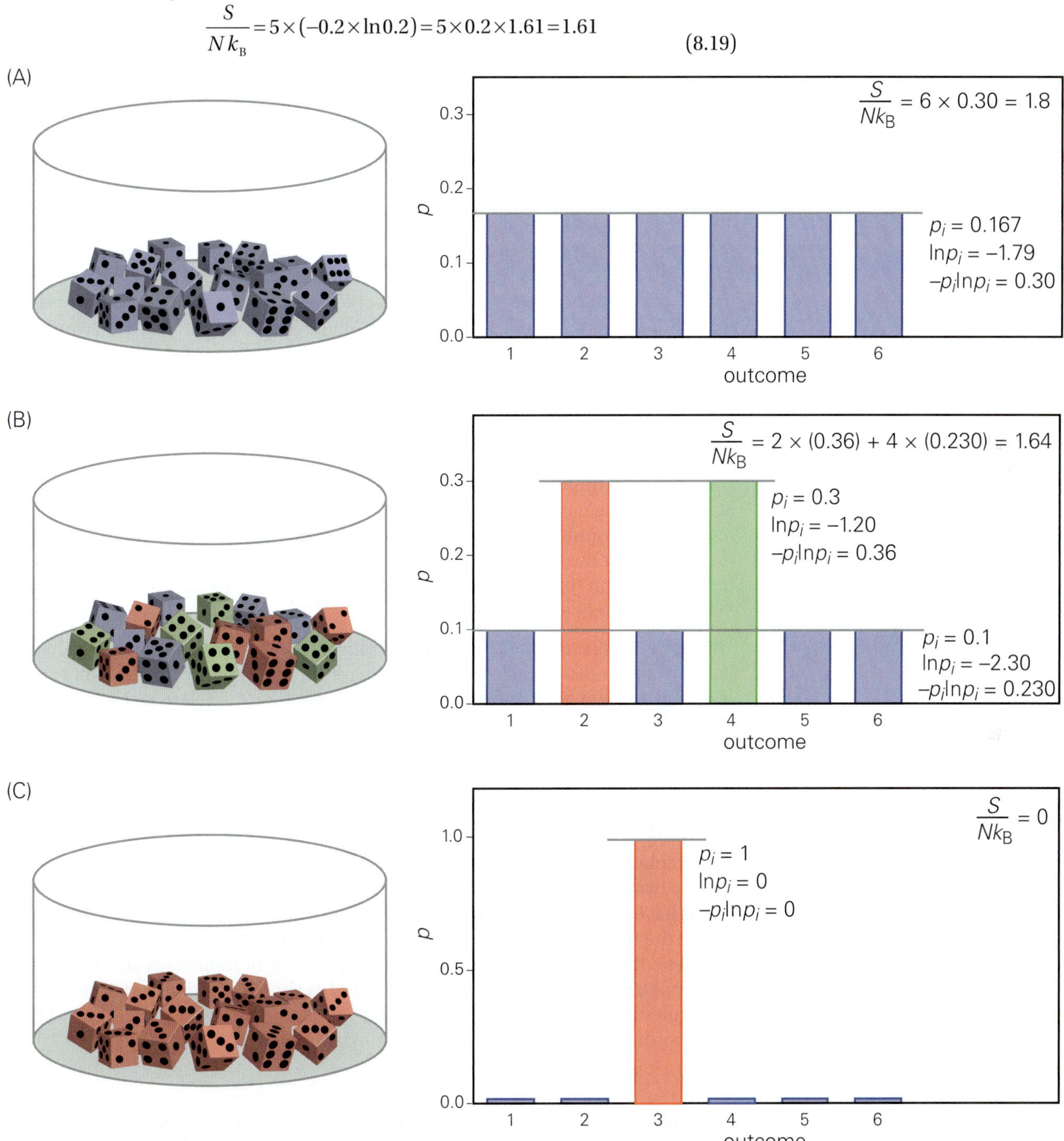


**Figure 8.13 Calculating the entropy of the outcome of dice throws using probability.** (A) The system consists of a large number, $N$, of unbiased dice. The probability for each possible outcome for an individual die is equal to 0.167 (1/6) for all outcomes. (B) The collection of dice in the system is "spiked" with two different kinds of biased dice; *red* dice are biased towards 2 and *green* dice are biased towards 4. The probabilities of the various outcomes is now different, as shown. The value of the entropy decreases. (C) The system consists of only one kind of dice, and these are all so heavily weighted that they invariably come up with only 3. The probability of all other outcomes is zero. The entropy for this situation is zero. These examples illustrate the general principle that the entropy of a system increases as the probability of alternative outcomes becomes more even: the entropy increases with "disorder."

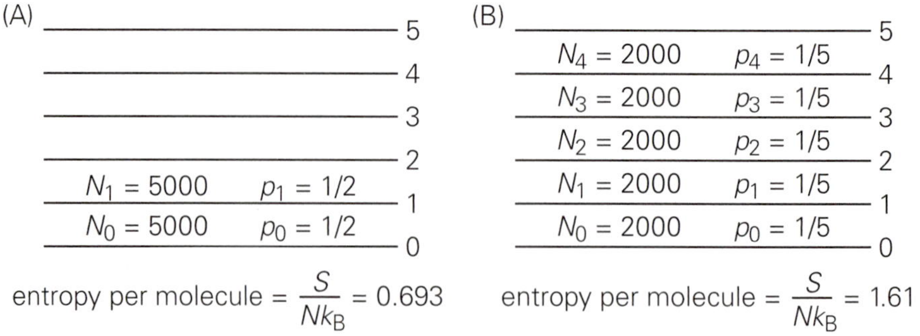

(A)

| Level | $N$ | $p$ |
|---|---|---|
| 5 | | |
| 4 | | |
| 3 | | |
| 2 | | |
| 1 | $N_1 = 5000$ | $p_1 = 1/2$ |
| 0 | $N_0 = 5000$ | $p_0 = 1/2$ |

entropy per molecule $= \frac{S}{Nk_B} = 0.693$

(B)

| Level | $N$ | $p$ |
|---|---|---|
| 5 | | |
| 4 | $N_4 = 2000$ | $p_4 = 1/5$ |
| 3 | $N_3 = 2000$ | $p_3 = 1/5$ |
| 2 | $N_2 = 2000$ | $p_2 = 1/5$ |
| 1 | $N_1 = 2000$ | $p_1 = 1/5$ |
| 0 | $N_0 = 2000$ | $p_0 = 1/5$ |

entropy per molecule $= \frac{S}{Nk_B} = 1.61$

**Figure 8.14 Comparing the entropy for narrower and broader distributions of molecular energies.** A system with 10,000 molecules is illustrated. In the first situation (A), half of the molecules are in the lowest energy level and half of the molecules are in the next level. In (B), the molecules are evenly distributed over the first five levels. The situation in (B) has higher entropy.

Thus, the entropy increases as the number of energy levels that are occupied increases. Similarly, systems that have molecules distributed over a larger number of energy levels (that is, Figure 8.14B) are more disordered, and increased disorder corresponds to increased entropy. We could have calculated the entropies using Equations 8.1 and 8.9, and we would have obtained the same answers. The calculations, however, would have been far more laborious.

## 8.7 The multiplicity of a system changes when energy is transferred between systems

How does the multiplicity of energy distributions change when two systems are able to exchange energy with each other? We assume that the atoms in each system move in a random and unpredictable manner, and exchange energy whenever they collide. At the boundary between the two systems, energy can be transferred from one system to the other in the form of heat (that is, in the disordered motion of the component atoms).

The atoms involved do not "know" anything about the direction in which the net energy transfer takes place between the two systems. Rather, every possible mode of energy transfer from one molecule to another is considered to be equally likely, as long as the total energy of the system is conserved (**Figure 8.15**). In the end, energy distributions that are associated with higher multiplicities are more likely to be observed than those with lower multiplicities. The logic used to analyze the outcomes of energy transfer is essentially the same as that used in Chapter 7 to analyze the outcomes of coin tosses or of molecular diffusion.

For simplicity, consider a system (system A) in which the molecules have access to only two energy levels, as shown in **Figure 8.16**. A realistic situation in which atoms have access to only two energy levels occurs in nuclear magnetic resonance, where the nuclear spin of a proton is either aligned with or against an external magnetic field. Suppose that the system consists of 10 molecules and has a total energy of $U = 2$. The molecules are distributed as shown in Figure 8.16A, with eight molecules in the lowest level ($U_0 = 0$) and two molecules in the higher level ($U_1 = 1$). According to Equation 8.9, the multiplicity of system A (that is, $W_A$) is:

$$W_A = \frac{10!}{2!8!} = 45 \tag{8.20}$$

Now consider a second system (system B), which also consists of 10 molecules that can only access the same two energy levels (Figure 8.16A). The total energy of system B is $U = 4$ and its multiplicity ($W_B$) is:

$$W_B = \frac{10!}{4!6!} = 210 \tag{8.21}$$

What happens when we put systems A and B into "thermal contact"—that is, A and B can exchange energy, but not molecules? The systems are isolated from everything else, so the total energy of the combined systems must remain constant.

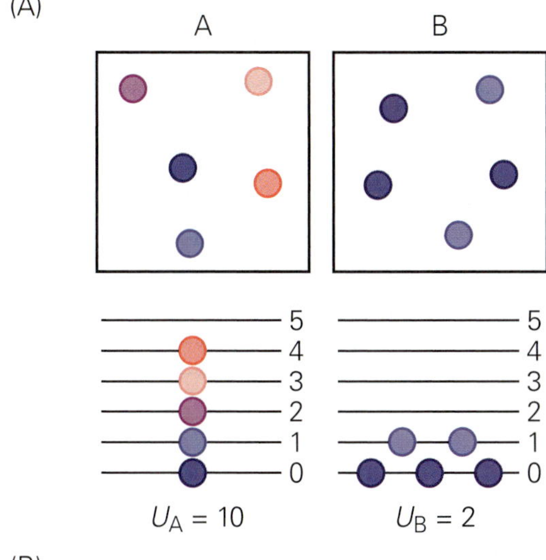


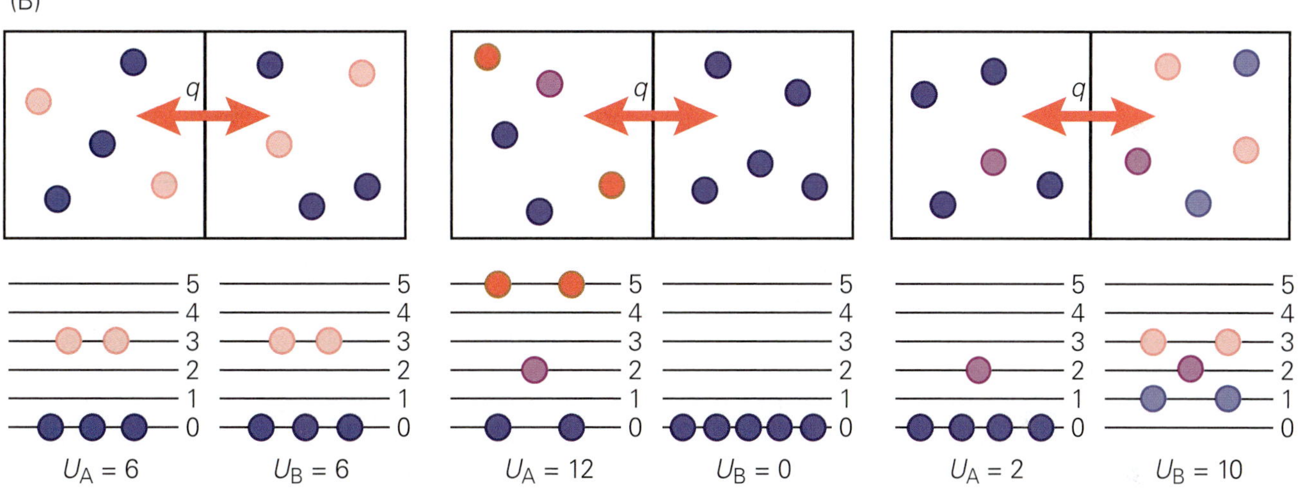


**Figure 8.15 Two systems in thermal contact exchange energy.** (A) Two systems, A and B, each contain five molecules. Initially, the energy of A is 10 units and that of B is two units. (B) The systems are brought into contact. We assume that all microstates corresponding to a total energy of 12 units are equally likely, and a few of these microstates are illustrated here. At equilibrium, the energy distribution corresponding to the maximum number of microstates (that is, the maximum multiplicity or entropy) is most likely to be observed.

The total multiplicity ($W_{A+B}$) of the two systems is the product of the individual multiplicities, so the initial value of $W_{A+B}$ is given by:

$$W_{A+B} = W_A \times W_B = 45 \times 210 = 9450 \tag{8.22}$$

We can now calculate the combined multiplicity ($W_{A+B}$) for every possible redistribution of the total energy that still conserves the total energy. For example, the energy could be redistributed so that system A has zero energy and system B has six units of energy (Figure 8.16B). In this case,

$$W_A = \frac{10!}{0!10!} = 1 \tag{8.23}$$

$$W_B = \frac{10!}{6!\,4!} = \frac{3{,}628{,}800}{720 \times 24} = 210 \tag{8.24}$$

The combined multiplicity is the product of $W_A$ and $W_B$:

$$W_{A+B} = W_A \times W_B = 210 \tag{8.25}$$

The combined multiplicity for the outcome depicted in Figure 8.16B (where all the energy is partitioned to system B) is lower than the combined multiplicity for Figure 8.16A (compare Equations 8.22 and 8.25). This means that the outcome shown in Figure 8.16B has a lower probability of occurring than that shown in Figure 8.16A.

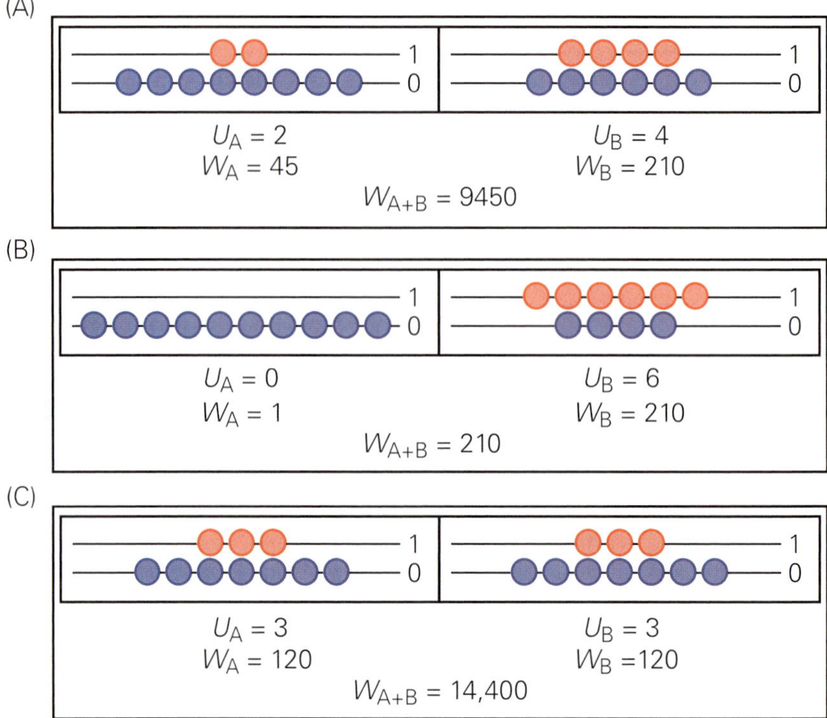


**Figure 8.16 Multiplicity for systems with two energy levels.** Two systems, A and B, are shown. The molecules in these systems have access to only two energy levels, with 0 and 1 units of energy, respectively. (A) System A initially has two units of energy and System B has four. (B) All of the energy in the system is transferred to System B. (C) The energy is distributed equally between the two systems. (Adapted from K.A. Dill and S. Bromberg, Molecular Driving Forces: Statistical Thermodynamics in Biology, Chemistry, Physics, and Nanoscience, 2nd ed. New York: Garland Science, 2010.)

How do these two outcomes compare with one in which the energy is distributed evenly, with both systems having three units of energy, as shown in Figure 8.16C? In this case,

$$W_A = W_B = \frac{10!}{3!7!} = \frac{3{,}628{,}800}{6 \times 5040} = 120 \tag{8.26}$$

and

$$W_{A+B} = W_A \times W_B = 14{,}400 \tag{8.27}$$

The combined multiplicity of the system in Figure 8.16C (14,400) is much higher than for the system in Figure 8.16A (9450). We are therefore more likely to observe outcomes in which the energy is distributed equally. The multiplicity of system B is reduced (from 210 to 120) upon the redistribution of energy, but this is more than offset by the increase in the multiplicity of system A (from 45 to 120).

When the multiplicity of the combined system ($W_{A+B}$) is calculated for all possible redistributions of the energy, the results shown in **Figure 8.17** are obtained. The maximum value of $W_{A+B}$ occurs for the state in which there is equal energy in both systems (that is, $U_A = U_B$). When the two systems are brought into thermal contact, therefore, the most likely outcome is one in which the energy is distributed equally between the two systems. Since system B starts off at higher energy, there is a net transfer of energy from B to A.

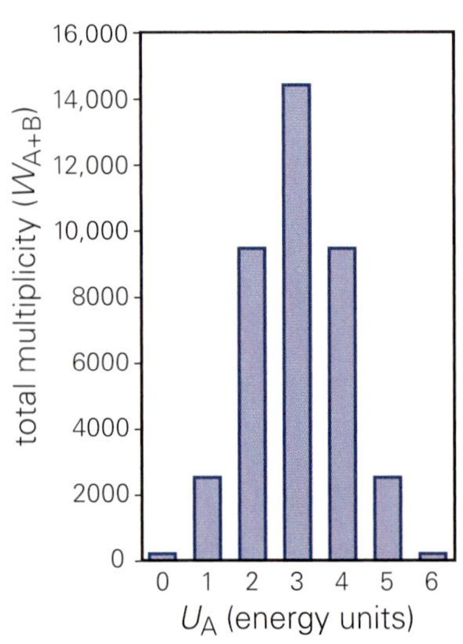


**Figure 8.17 Multiplicity as a function of energy redistribution.** The combined multiplicity, $W_{A+B}$, of the systems A and B described in Figure 8.16 is shown as a function of the energy of system A, $U_A$. The maximum value of the multiplicity occurs when $U_A = U_B = 3$ energy units.

## 8.8 Systems in thermal contact exchange heat until the combined entropy of the two systems is maximal

Recall from Chapter 7 that when we consider the positional multiplicity of only a small number of molecules, there is a significant probability that we will observe outcomes that have less than the maximal value of the multiplicity. If we look at a very small number of molecules in a box, for example, there is a significant probability that we will find all the molecules in the left half of the box and none in the right half.

The situation is similar for the redistribution of energy when we bring two systems into thermal contact. Consider, for example, the system discussed in Section 8.7, which included a total of 20 molecules. The graph in Figure 8.17 shows that the distribution with maximum multiplicity is the one in which the energy is distributed equally, so this is the state with the highest probability of being observed. There is significant probability, however, of observing outcomes in which the energy is distributed unequally. But, as we saw in Chapter 7 for positional entropy, the difference between the multiplicity of the most likely outcome and the multiplicities of all other outcomes becomes enormously large when the number of molecules increases. For large numbers of molecules, the only outcomes that have realistic chances of being observed are ones that are very close to the one with maximum multiplicity.

What happens if we scale up the size of the system discussed in Section 8.7 by a factor of 1000? That is, consider a larger system in which $N_A = N_B = 10{,}000$ and $U_A + U_B = 6000$, where $N_A$ and $N_B$ are the number of atoms in systems A and B, respectively, and $U_A$ and $U_B$ are the energies of systems A and B, respectively. We must now calculate $\ln W$ instead of $W$ because the number of molecules involved is so large, and we can do so for all possible values of $U_A$ and $U_B$, making sure that energy is conserved.

The value of $\ln W_{A+B}$ as a function of the energy of system A is shown in **Figure 8.18**. Again, the maximum value of $\ln W_{A+B}$ is obtained when the energy is distributed equally between systems A and B, with 3000 units of energy in system A and 3000 units in system B.

When we scale up the system, the much larger number of molecules involved results in the combined multiplicity being much more sensitive to deviations from the state of maximal multiplicity. To see why this is so, we redistribute 10% of the energy of system A to system B, so that system A now has 2700 units of energy and system B has 3300 units (Case 1). We compare the multiplicity for this situation with that when the energy is distributed equally between the two systems (Case 2).

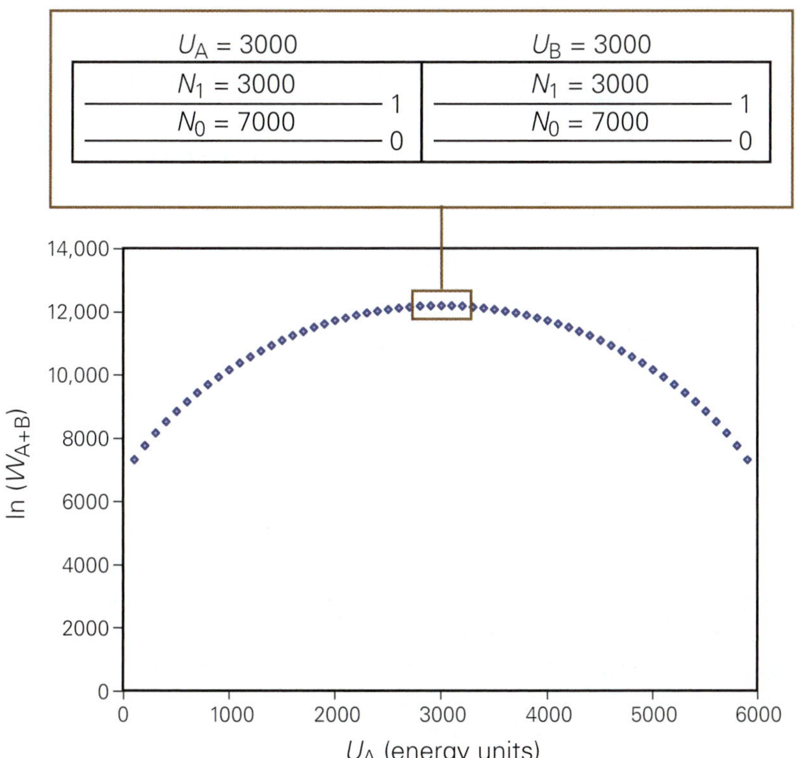


**Figure 8.18 The multiplicity for larger systems.** The logarithm of the combined multiplicity ($\ln W_{A+B}$) is shown for two systems, A and B. A and B are similar to the systems illustrated in Figures 8.16 and 8.17, except that each has 10,000 molecules. The maximum value of $\ln W_{A+B}$ occurs when the energy is distributed equally between A and B.

With such a large number of molecules, we can use the probabilistic definition of entropy (Equation 8.13) to calculate $\ln W$. We start by calculating the probabilities, $p_1$ and $p_2$, of observing molecules in the two energy levels for system A in Case 1 (unequal distribution of energy). Remember, though, that redistributing 300 units of energy from system A to system B means that 300 molecules in the $U_1$ energy level of system A will be back in the $U_0$ level. As a result, $N_0 = 7300$ and $N_1 = 2700$ for system A, and we calculate the probabilities as follows:

$$p_1 = \frac{7300}{10{,}000} = 0.73 \tag{8.28}$$

$$p_2 = \frac{2700}{10{,}000} = 0.27 \tag{8.29}$$

Given these values for the probabilities, we can calculate the entropy of system A by using Equation 8.13:

$$\begin{aligned}\frac{S_A}{Nk_B} &= -\sum_i p_i \ln p_i \\ &= -0.73\ln 0.73 - 0.27\ln 0.27 \\ &= 0.230 + 0.353 \\ &= 0.583\end{aligned} \tag{8.30}$$

Since $S_A = k_B \ln W_A$, it follows that:

$$\begin{aligned}\ln W_A &= S_A/k_B = -N\sum_i p_i \ln p_i \\ &= N \times 0.583 = 10{,}000 \times 0.583 = 5830\end{aligned} \tag{8.31}$$

Likewise, redistributing 300 units of energy into system B means that 300 molecules that were in the $U_0$ energy level will be in $U_1$ level instead. Thus, $N_0 = 6700$ and $N_1 = 3300$ for system B. The probabilities for finding molecules in the two levels of system B are therefore given by:

$$p_1 = \frac{6700}{10{,}000} = 0.67 \tag{8.32}$$

$$p_2 = \frac{3300}{10{,}000} = 0.33 \tag{8.33}$$

Using these values for the probabilities and Equation 8.13, we calculate the value of the entropy and $\ln W_B$ as follows:

$$\begin{aligned}\frac{S_B}{Nk_B} &= -\sum_i p_i \ln p_i = -0.67\ln 0.67 - 0.33\ln 0.33 \\ &= +0.268 + 0.366 \\ &= 0.634\end{aligned} \tag{8.34}$$

$$\ln W_B = S_B/k_B = N \times 0.634 = 10{,}000 \times 0.634 = 6340 \tag{8.35}$$

Equations 8.31 and 8.35 allow us to calculate the logarithm of the total multiplicity of the combined systems, $(W_{A+B})_{\text{Case 1}}$:

$$(\ln W_{A+B})_{\text{Case 1}} = \ln W_A + \ln W_B = 5830 + 6340 = 12170 \tag{8.36}$$

We can now compare this value of $\ln W_{A+B}$ with the one obtained when both systems have 3000 units of energy each (that is, Case 2). In this case, $\ln W_A = \ln W_B$, so we can do just one series of calculations to obtain either $\ln W_A$ or $\ln W_B$ and then simply double the resulting value to obtain $\ln W_{A+B}$:

We proceed as follows:

$$p_1 = \frac{7000}{10{,}000} = 0.7 \quad p_2 = \frac{3000}{10{,}000} = 0.3 \tag{8.37}$$

$$\frac{S_A}{k_B} = \frac{S_B}{k_B} = -\sum_i p_i \ln p_i = -0.7\ln 0.7 - 0.3\ln 0.3$$
$$= 0.25 + 0.361 = 0.611 \quad (8.38)$$

$$\ln W_A = \ln W_B = N \times 0.611 = 10{,}000 \times 0.611 = 6110 \quad (8.39)$$

$$(\ln W_{A+B})_{\text{Case 2}} = \ln W_A + \ln W_B = 6110 + 6110 = 12{,}220 \quad (8.40)$$

With the values of the combined multiplicities in hand (Equations 8.36 and 8.40), we can now evaluate the probability of observing a situation in which 10% of the energy has been redistributed from system A to system B (Case 1: non-maximal multiplicity) versus a situation where energy has not been redistributed (Case 2: maximal multiplicity). The answer is proportional to the ratios of the combined multiplicities, which can be calculated as follows:

$$\ln\left[\frac{(W_{A+B})_{\text{Case 1}}}{(W_{A+B})_{\text{Case 2}}}\right] = \ln(W_{A+B})_{\text{Case 1}} - \ln(W_{A+B})_{\text{Case 2}}$$
$$= 12{,}170 - 12{,}220 = -50 \quad (8.41)$$

We now take the exponent of both sides of Equation 8.41 to obtain the desired ratio of multiplicities:

$$\left[\frac{(W_{A+B})_{\text{Case1}}}{(W_{A+B})_{\text{Case2}}}\right] = e^{-50} = \left(10^{0.434}\right)^{-50} \approx 10^{-22} \quad (8.42)$$

Based on this calculation, the probability of observing outcomes in which the energy imbalance between the two systems is ~10% of the optimal energy of each system is essentially zero (because $10^{-22}$ is an extremely small number). The result in Equation 8.42, moreover, has been obtained for a system containing just 20,000 molecules, which is still very small when compared with realistic experimental systems, which might contain on the order of $10^{20}$ molecules. For systems with such large numbers of molecules, the probability that the energy fluctuates to any appreciable extent from the optimal distribution is minute.

To summarize the discussion so far, when two systems are brought into thermal contact, energy flows between the systems until the combined entropy of the two systems is maximized. This principle is a manifestation of the second law of thermodynamics, and we shall use it in Chapter 9 to derive the concept of the free energy of the system. For now, we turn our attention to the distribution of energy within one system, which is the Boltzmann distribution at equilibrium.

## 8.9 Many energy distributions are consistent with the total energy of a system, but some have higher multiplicity than others

In Chapter 6, we introduced the idea, without justification, that the energies of molecules at equilibrium obey a probability distribution known as the **Boltzmann distribution**. In Section 8.5, we discussed the fact that for a system with a fixed total energy, there are many different energy distributions that are consistent with the same total energy. In general, these distributions have different multiplicities, and it follows that energy distributions with higher multiplicities are more likely to be observed. This idea is the key to understanding the Boltzmann distribution.

Consider a system with a defined total energy, $U_{\text{total}}$, and a fixed number of atoms, $N$. As an illustration, **Figure 8.19** shows such a system. To simplify analysis, we assume that the energy levels are equally spaced, and that each energy level is one energy unit higher than the lower one. There are 30 atoms in the system, and the total energy is 27 units.

At equilibrium, all possible microstates of the system (that is, all configurations that satisfy the constraint on total energy) are equally likely. These microstates

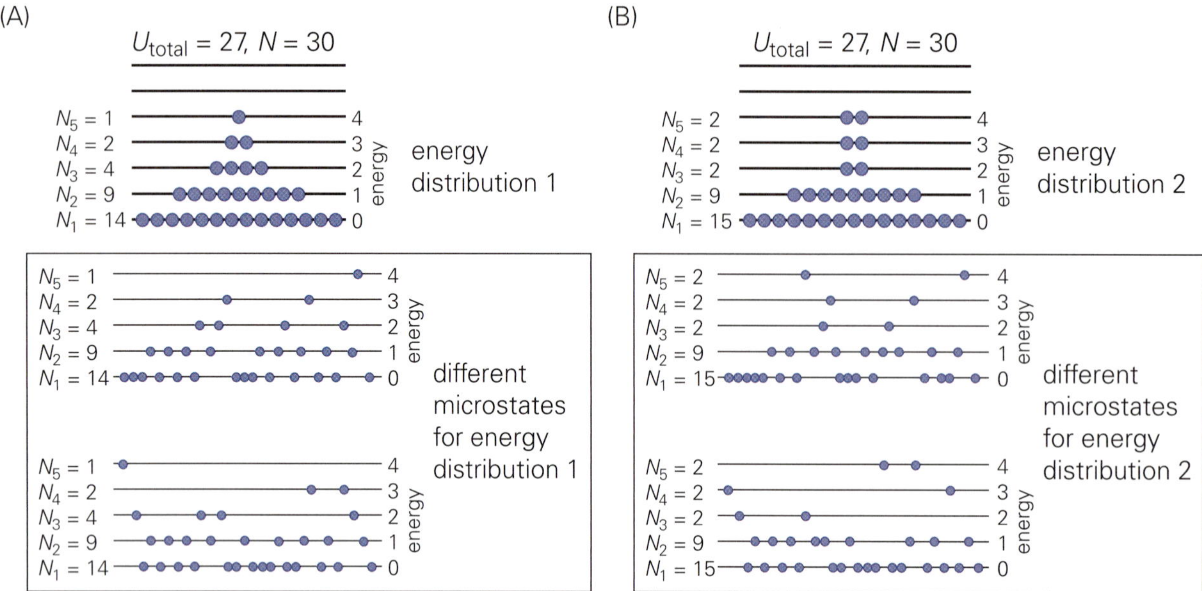


**Figure 8.19 Energy distributions and microstates for a system with fixed total energy and number of atoms.** The total energy of the system is 27 units, distributed among energy levels that are equally spaced in steps of one energy unit. Two of the many different energy distributions that are consistent with the total energy are shown in (A) and (B). Each energy distribution corresponds to many different microstates, only two of which are shown for each distribution. In the diagrams for each microstate, the horizontal positions identify specific atoms. This contrasts with the diagrams for the energy distributions, in which specific atoms are not identified, and the dots simply represent populations.

correspond to many different energy distributions, each of which is consistent with the number of atoms and the total energy of the system. Two such energy distributions are shown in Figure 8.19. Energy distribution 1 has 14 atoms in the lowest level (with zero units of energy each), nine in the second level (with one unit of energy each), and so on. Distribution 2 has a different number of atoms in each level, but the same total energy and the same number of atoms as distribution 1.

Each energy distribution has many different microstates associated with it, and the number of microstates associated with each distribution is the multiplicity (Figure 8.19 shows two possible microstates for each distribution). Because the system in the example has only a small number of atoms in it, we use Equation 8.9 to count the number of microstates associated with each distribution explicitly. That is, we cannot use the probabilistic expression for the entropy (Equation 8.13) to derive the multiplicity for this particular case.

For distribution 1, the multiplicity, $W_1$, is given by:

$$W_1 = \frac{30!}{1!2!4!9!14!} = 1.74 \times 10^{14} \tag{8.43}$$

For distribution 2, the multiplicity, $W_2$, is given by:

$$W_2 = \frac{30!}{2!2!2!9!15!} = 0.69 \times 10^{14} \tag{8.44}$$

According to Equations 8.43 and 8.44, distribution 1 is more likely than distribution 2 because it has ~2.5 times more microstates associated with it.

As the number of atoms in the system becomes very large, one particular energy distribution ends up dominating because the multiplicity associated with it is very much greater than for other distributions. The distribution with maximum multiplicity is the Boltzmann distribution.

## 8.10 The energy distribution at equilibrium must have an exponential form

Recall from Chapter 6 that the Boltzmann distribution is an exponential function of the energy:

$$p_j = Ae^{-\beta u_j} \tag{8.45}$$

where $p_j$ is the probability of finding a molecule in the $j^{\text{th}}$ energy level, $u_j$ is the energy of the $j^{\text{th}}$ energy level, and $A$ and $\beta$ are constants (that is, they do not depend on the energy levels). It turns out that the probability distribution function describing molecular energies has to be an exponential function, because of certain conditions that such a function must satisfy. We present an explanation of why this is so that is based on an analysis of the Boltzmann distribution in Tipler and Llewellyn's textbook on modern physics, which is listed under Further Reading, at the end of this chapter.

Let us start by assuming that we do not know the functional form of the probability distribution. We assume that this function depends on the energy—that is, given the energy of a level, the function returns the probability of finding a molecule in that level. If $N_j$ is the number of atoms in the $j^{\text{th}}$ energy level, and $N$ is the total number of atoms, then the probability distribution must be a function, $f(u_j)$, that is as yet unspecified but which satisfies the following equation:

$$p_j = \frac{N_j}{N} = A\,f(u_j) \tag{8.46}$$

The parameter $A$ is a proportionality constant that ensures that the probability function is normalized—that is, the sum of its value over all energy levels is 1.0. We discuss the significance of normalization in Section 8.11.

The special property of energy distribution functions that helps define the Boltzmann distribution becomes apparent when we consider the probability of finding two specific atoms in two specific energy levels. We then compare this with the probability of finding two atoms with the same total energy as in the first case, but without specifying which energy level each atom is in. This comparison reveals the special defining property that we are looking for.

First, we consider the probability, $p_{X,j}$, of finding a particular atom, labeled X, in the $j^{\text{th}}$ energy level of a system (**Figure 8.20**). According to Equation 8.46:

$$p_{X,j} = A\,f(u_j) \tag{8.47}$$

Likewise, the probability of finding another atom, labeled Y, in the $k^{\text{th}}$ energy level, with energy $u_k$, is given by:

$$p_{Y,k} = A\,f(u_k) \tag{8.48}$$

We assume that the atoms in the system are independent. Therefore, the joint probability, $p_{XY}$ of finding the X atom in the $j^{\text{th}}$ energy level and the Y atom in the

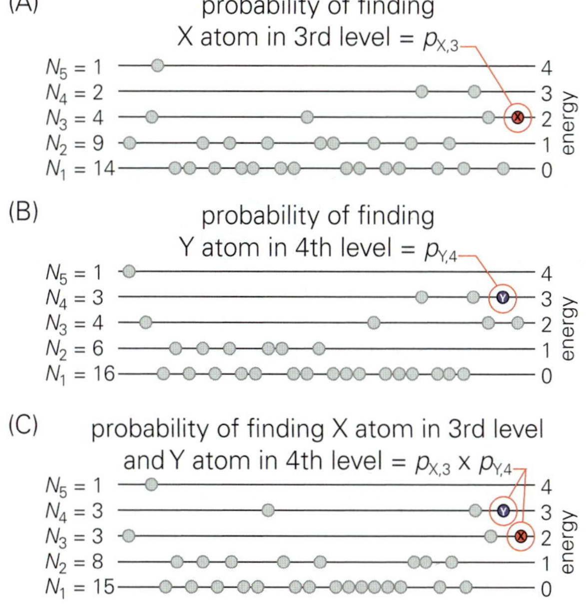


**Figure 8.20 Joint probability of finding two atoms in two energy levels.** (A) A particular microstate with an atom labeled "X" in energy level 3. The probability of finding atom X in level 3 is denoted $p_{X,3}$. (B) Similar to (A), but with a different atom, labeled "Y," in energy level 4. The probability of finding atom Y in level 4 is denoted $p_{Y,4}$. (C) A microstate with atom X in level 3 *and* atom Y in level 4. The probability of this happening is given by $p_{X,3} \times p_{Y,4}$.

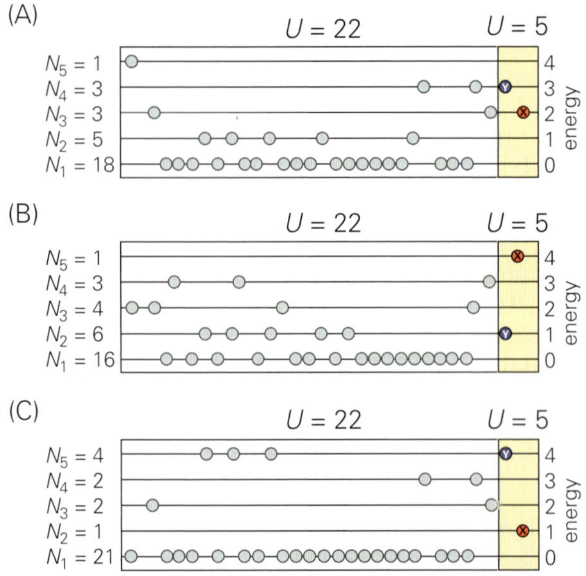


**Figure 8.21 Microstates of a system with X and Y atoms having the same total energy.** Three different microstates are shown. In each one, the total energy of the X and Y molecules is five units, but the five units are distributed differently between X and Y. The remainder of the system (28 atoms) have 22 units of energy in each case.

$k^{th}$ energy level is just the product of the individual probabilities:

$$p_{XY} = p_{X,j} \times p_{Y,k} = Af(u_j) \times Af(u_k) \tag{8.49}$$

The situation we are considering is illustrated in Figure 8.20, with one of the atoms labeled "X" (in panel A) and another labeled "Y" (in panel B). The probability, $p_{XY}$ of finding atom X in level 3 and atom Y in level 4 in the same microstate is given, according to Equation 8.49, by the product of the independent probabilities of finding the individual atoms in these levels:

$$p_{XY} = p_{X,3} \times p_{Y,4} = Af(3) \times Af(4) \tag{8.50}$$

Notice that the total energy of the X and Y atoms in the situation shown in Figure 18.20C is five energy units (atom X has two energy units and atom Y has three energy units).

Now consider the probability that the X and Y atoms have a total energy of $u_j + u_k$ (which is five in the example shown in Figure 18.20C). There are many distinct microstates in which the total energy of the X and Y atoms is five units, but in which the energy is distributed differently between the X and Y atoms, as shown in **Figure 8.21**. In each microstate where the energy of X and Y is $u_j + u_k$, the energy of the rest of the system is always $U_{total} - (u_j + u_k)$, where $U_{total}$ is the total energy of the system.

The important point that emerges is that all configurations of the X and Y atoms that have the same total energy for these two atoms (that is, $u_j + u_k$) have the same probability of occurring. To see why this is the case, recognize that the probability of a particular configuration of X and Y atoms is proportional to the multiplicity of the system with the X and Y atoms in the defined energy levels. This is given by the multiplicity of the rest of the system, excluding the X and Y atoms (the atoms in the rest of the system are shown shaded in light yellow in Figure 8.21). But in each case, the rest of the system has the same total energy (22 units in our example) and the same number of atoms (28 atoms in our example). So, the number of microstates, and therefore the multiplicity, for the rest of the system must be the same in each case.

By this reasoning, we conclude that the probability of finding the two atoms X and Y with total energy $u_j + u_k$ is given by an unspecified function $g(u_j + u_k)$, which depends on the sum of the energies of the two atoms. Thus:

$$p_{XY} = Bg(u_j + u_k) \tag{8.51}$$

The parameter $B$ in Equation 8.51 is a normalization constant.

Comparing Equations 8.49 and 8.51, we see that:

$$p_{XY} = A^2 f(u_j) \times f(u_k) = Bg(u_j + u_k)$$

$$\Rightarrow f(u_j) \times f(u_k) \propto g(u_j + u_k) \quad (8.52)$$

Equation 8.52 is the key result that reveals the nature of the probability distribution function for energy. According to the Equation 8.52, *f* with one argument multiplied by *f* with a second argument is proportional to a function whose argument is the sum of the two arguments of *f*. There is only one mathematical function that has this property, and that is the exponential function.

The exponential function satisfies Equation 8.52. If *f* and *g* are both exponential functions, then we can write:

$$f(u_j) = e^{-\beta u_j}$$

and

$$f(u_k) = e^{-\beta u_k}$$

Then:

$$f(u_j) \times f(u_k) = e^{-\beta u_j} \times e^{-\beta u_k} = e^{-\beta(u_k + u_j)} = g(u_j + u_k) \quad (8.53)$$

From this, we reason that the probability distribution must be given by:

$$p_j \propto Ae^{-\beta u_j} \quad (8.54)$$

In principle, we could also choose the probability function as $e^{+\beta u_j}$, which would also satisfy Equation 8.52. But, the probability of finding molecules at higher energy levels must be smaller than the probability of finding molecules in lower energy levels. Hence, we choose $e^{-\beta u_j}$ as the form of the probability function, because the value of this function decreases with increasing energy.

In **Box 8.2**, we provide a more rigorous derivation of the Boltzmann distribution, which starts from the statistical definition of the entropy and then applies the principles of energy conservation and entropy maximization. This analysis shows, for a system that exchanges heat with the surroundings, that the Boltzmann distribution is the one that maximizes the combined entropy of the system and the surroundings. The parameter β in Equation 8.54 is shown to be equal to $1/k_BT$. The complete form of the Boltzmann function is therefore:

$$p_j = Ae^{-u_j/k_BT} \quad (8.55)$$

## 8.11 The partition function indicates the accessibility of the higher energy levels of the system

As the system moves from a nonequilibrium state to equilibrium, its energy distribution keeps changing (**Figure 8.22**). The population of energy levels at

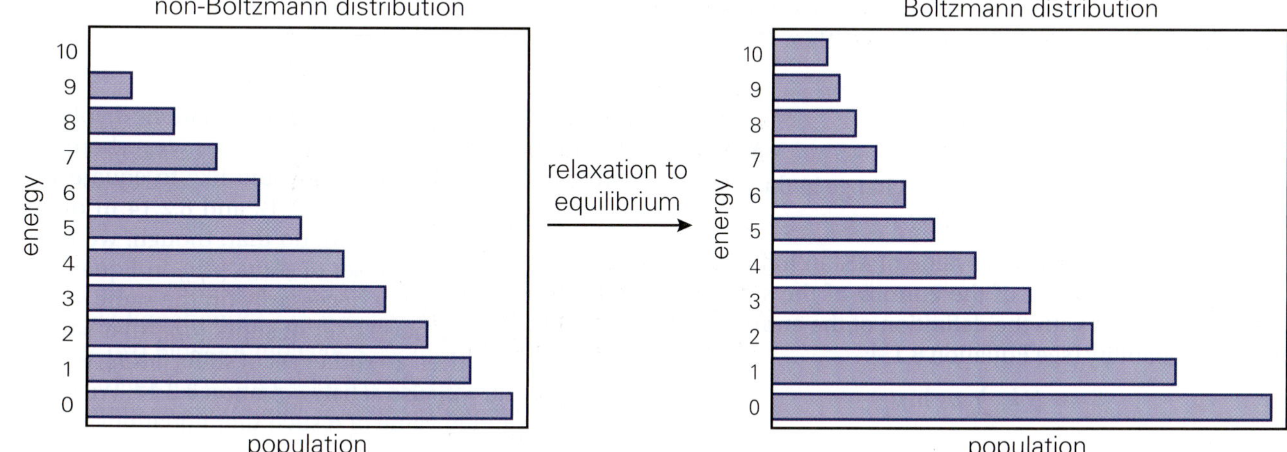


**Figure 8.22 The energy distribution tends towards the Boltzmann distribution as the system relaxes towards equilibrium.** Two molecular energy distributions are illustrated here. The vertical axis shows the energy, and the horizontal bars reflect the relative populations of the energy levels. The distribution at left does not correspond to an equilibrium situation because the populations of the levels do not obey the Boltzmann rule.

## Quantitative/Essay

Assume $k_B = 1\ \text{J}\bullet\text{K}^{-1}$ for all problems.

**13.** A system starts with a multiplicity of 2000. Two kJ of heat are transferred into the system reversibly at 298 K. What is the multiplicity now?

**14.** A system starts with a multiplicity of $10^{99}$. Heat is transferred out of the system reversibly at 298 K such that the final multiplicity is $10^5$. How much heat was lost?

**15.** A system of noninteracting atoms has a positional multiplicity of 120 and an energetic multiplicity of 5. What is the total multiplicity? What are the positional, energetic, and total entropies of the system?

**16.** A system with 1 mole of an ideal gas changes its volume from 10 L to 1 L by isothermal compression. What is the change in entropy of the system? What is the change in the entropy associated with the energy distribution? (Use $R = 8.31\ \text{J}\bullet\text{K}^{-1}\bullet\text{mol}^{-1}$.)

**17.** Consider a system with five molecules and three energy levels. The energy levels are such that a molecule at energy level 1 contributes 1 J of energy to $U_{TOTAL}$, a molecule at energy level 2 contributes 2 J of energy to $U_{TOTAL}$, and a molecule in energy level 3 contributes 3 J of energy to $U_{TOTAL}$. Which has more internal energy, state A or state B?

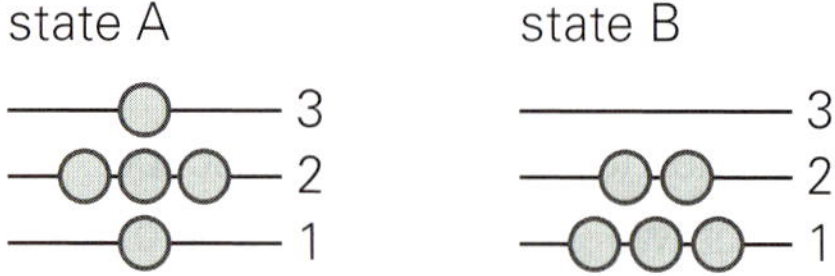


**18.** Consider the system and the two states from Problem 17. Which state has greater multiplicity?

**19.** A system has 100,000 molecules at energy level 1, 10,000 molecules at energy level 2, and 1000 molecules at energy level 3. What is the entropy of the system? (*Hint*: Use the probabilistic definition.)

**20.** A system with 100,000 molecules has two energy levels (A and B). At first, the two energy levels are populated equally. After a reversible process, energy level A is populated by 65% of the molecules and the system is at 293K.

a. What is the difference in energy between the two levels?
b. How much heat was added or removed from the system?
c. What is the change in entropy?

**21.** Two systems, A and B, are placed in thermal contact and isolated from the rest of the world. Shown next are three possible microstates for the combined system. Two of these are valid microstates for the combined system, and one is not. Identify the inconsistent microstate and explain why it does not belong with the other two. (*Hint*: Consider Figure 8.15.)

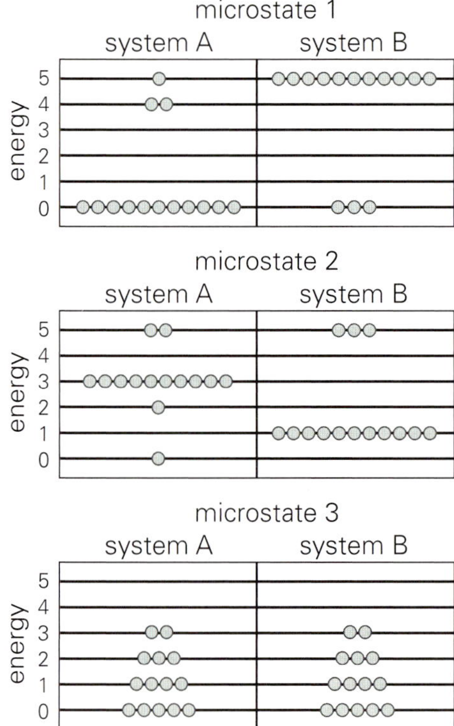


**22.** A system has two types of molecules (black and gray) that interconvert. The log of the multiplicity of energy of the two types of molecules individually versus the total system energy is plotted below. Plot the log of the total multiplicity of energy and indicate where the maximum value for the combined multiplicity occurs.

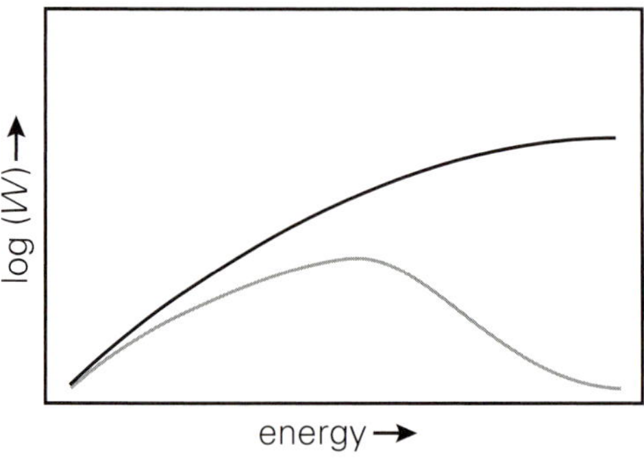


**23.** What is the entropy of the following energy distribution? What is the multiplicity?

$N_3 = 200$ molecules
$N_2 = 1100$ molecules
$N_1 = 10{,}500$ molecules

**24.** A system has a partition function of 1.3 at 293 K. At that temperature, an energy level is populated by 10% of the molecules. What is the energy of that level?

**25.** Explain how energy can move spontaneously from a system of low total energy to a system of higher total energy.

Comparing Equations 8.49 and 8.51, we see that:

$$p_{XY} = A^2 f(u_j) \times f(u_k) = Bg(u_j + u_k)$$

$$\Rightarrow f(u_j) \times f(u_k) \propto g(u_j + u_k) \tag{8.52}$$

Equation 8.52 is the key result that reveals the nature of the probability distribution function for energy. According to the Equation 8.52, $f$ with one argument multiplied by $f$ with a second argument is proportional to a function whose argument is the sum of the two arguments of $f$. There is only one mathematical function that has this property, and that is the exponential function.

The exponential function satisfies Equation 8.52. If $f$ and $g$ are both exponential functions, then we can write:

$$f(u_j) = e^{-\beta u_j}$$

and

$$f(u_k) = e^{-\beta u_k}$$

Then:

$$f(u_j) \times f(u_k) = e^{-\beta u_j} \times e^{-\beta u_k} = e^{-\beta(u_k + u_j)} = g(u_j + u_k) \tag{8.53}$$

From this, we reason that the probability distribution must be given by:

$$p_j \propto Ae^{-\beta u_j} \tag{8.54}$$

In principle, we could also choose the probability function as $e^{+\beta u_j}$, which would also satisfy Equation 8.52. But, the probability of finding molecules at higher energy levels must be smaller than the probability of finding molecules in lower energy levels. Hence, we choose $e^{-\beta u_j}$ as the form of the probability function, because the value of this function decreases with increasing energy.

In **Box 8.2**, we provide a more rigorous derivation of the Boltzmann distribution, which starts from the statistical definition of the entropy and then applies the principles of energy conservation and entropy maximization. This analysis shows, for a system that exchanges heat with the surroundings, that the Boltzmann distribution is the one that maximizes the combined entropy of the system and the surroundings. The parameter β in Equation 8.54 is shown to be equal to $1/k_BT$. The complete form of the Boltzmann function is therefore:

$$p_j = Ae^{-u_j/k_BT} \tag{8.55}$$

## 8.11 The partition function indicates the accessibility of the higher energy levels of the system

As the system moves from a nonequilibrium state to equilibrium, its energy distribution keeps changing (**Figure 8.22**). The population of energy levels at

**Figure 8.22 The energy distribution tends towards the Boltzmann distribution as the system relaxes towards equilibrium.** Two molecular energy distributions are illustrated here. The vertical axis shows the energy, and the horizontal bars reflect the relative populations of the energy levels. The distribution at left does not correspond to an equilibrium situation because the populations of the levels do not obey the Boltzmann rule.

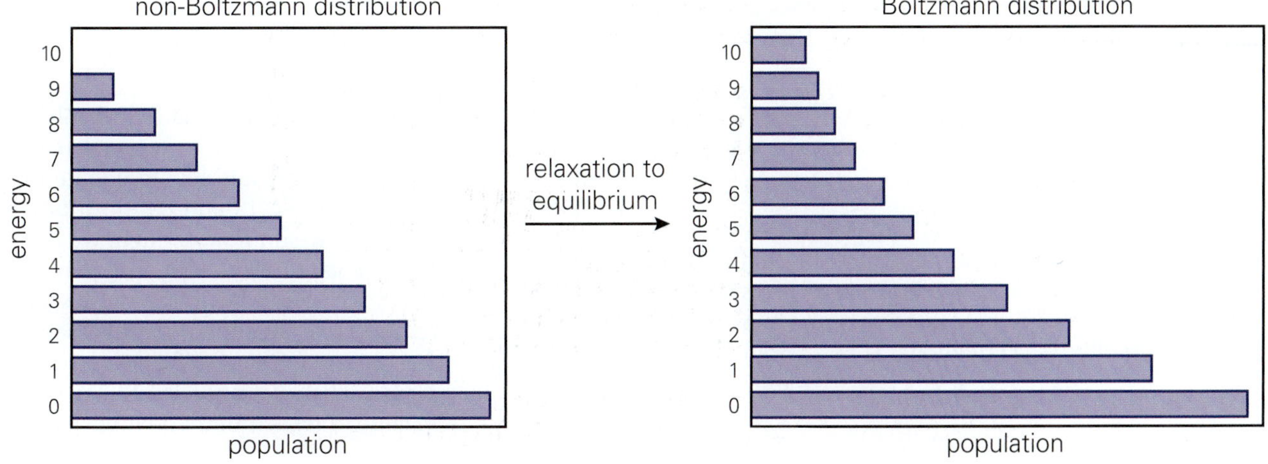

## Box 8.2 Derivation of the Boltzmann distribution from the statistical definition of entropy and the free energy

In Section 8.10, we used a conceptual argument to reason that the probability $p_j$ of finding a molecule with energy $u_j$ is given by:

$$p_j = A e^{-\beta u_j} \tag{8.2.1}$$

where β is a constant. We now show that the exponential form of the Boltzmann distribution arises naturally from the probabilistic definition of the entropy. We shall consider the general case of a system that is able to exchange heat with the surroundings. In such a situation, according to the second law, the condition for equilibrium is that the combined entropy of the system and the surroundings is maximal. In Chapter 9, we show that, for a system at constant volume, the requirement that the total entropy be maximal is equivalent to the free energy of the system being at a minimum. The free energy, denoted $A$, is given by:

$$A = U - TS \tag{8.2.2}$$

where $U$, $T$, and $S$ are the energy, temperature, and entropy of the system. By using the condition that the free energy of the system is at a minimum at equilibrium, we can also show that $\beta = 1/(k_B T)$, where $k_B$ is the Boltzmann constant and $T$ is the temperature.

When the temperature and volume are held constant, the condition for equilibrium is:

$$dA = dU - TdS = 0 \tag{8.2.3}$$

The internal energy, $U$, of the system depends on the probabilities, $p_j$, of the energy levels in the following way:

$$U = \sum_{j=1}^{t} N_j u_j = N \sum_{j=1}^{t} p_j u_j \tag{8.2.4}$$

where $N$ is the total number of atoms in the system, $N_j$ is the number of atoms in the $j^{th}$ level, and $u_j$ is the energy of the $j^{th}$ level. We can calculate the value of an infinitesimally small change in the energy as follows:

$$dU = N \sum_{j=1}^{t} p_j du_j + N \sum_{j=1}^{t} u_j dp_j \tag{8.2.5}$$

We assume that the values of the energy levels are constant—that is, the $u_j$ values do not change. Thus, the first term in Equation 8.2.5 is zero, and so:

$$dU = N \sum_{j=1}^{t} u_j dp_j \tag{8.2.6}$$

where $dp_j$ is an infinitesimally small change in the probability of finding molecules in level $j$.

We also need to calculate the value of $dS$, an infinitesimal change in the value of the entropy of the system. To do this, we start with the definition of the entropy in terms of probability (see Equation 8.13):

$$S = -N k_B \sum_{j} p_j \ln p_j \tag{8.2.7}$$

Take the derivative of $S$ with respect to $p_j$:

$$\begin{aligned}\left(\frac{\partial S}{\partial p_j}\right)_{\substack{p_i=\text{constant}\\ \text{for } i\neq j}} &= -N k_B \left[ p_j \left(\frac{\partial \ln p_j}{\partial p_j}\right) + \ln p_j \left(\frac{\partial p_j}{\partial p_j}\right)\right] \\ &= -N k_B \left(\frac{p_j}{p_j} + \ln p_j\right) \\ &= N k_B \left(1 + \ln p_j\right)\end{aligned} \tag{8.2.8}$$

The derivatives in Equation 8.2.8 are partial derivatives of the entropy with the probabilities associated with different energy levels (see Box 8.3 for a brief review of partial derivatives).

An infinitesimal change in $S$ due to an infinitesimal change in $p_j$ is therefore given by:

$$dS = -N k_B \left(1 + \ln p_j\right) dp_j \tag{8.2.9}$$

Considering changes in the probabilities of all the energy levels, we get the following expression for the total differential $dS$:

$$dS = -N k_B \sum_{j=1}^{t} \left(1 + \ln p_j\right) dp_j \tag{8.2.10}$$

Combining Equations 8.2.6 and 8.2.10, the condition for minimum free energy becomes:

$$dU - TdS = N \sum_{j=1}^{t} u_j dp_j + N k_B T \sum_{j=1}^{t} \left(1 + \ln p_j\right) dp_j = 0 \tag{8.2.11}$$

Dividing both sides of Equation 8.2.11 by $N$, we get:

$$\sum_{j=1}^{t} u_j dp_j + k_B T \sum_{j=1}^{t} \left(1 + \ln p_j\right) dp_j = 0 \tag{8.2.12}$$

Equation 8.2.12 is a restatement of the fact that the free energy is a minimum at equilibrium, with a focus on the molecular energy levels. There is one other condition that the system must satisfy, and this is the normalization condition:

$$\sum_{j=1}^{t} p_j = 1 \tag{8.2.13}$$

$$\Rightarrow \sum_{j=1}^{t} dp_j = 0 \tag{8.2.14}$$

Equations 8.2.12 and 8.2.14 are both equations involving infinitesimal changes in the probabilities, $dp_j$, where the right-hand sides are zero. We would like to equate the left-hand sides of Equations 8.2.12 and 8.2.14 in order to proceed further. Before we do that, though, we have to be a little careful. The set of shifts $\{dp_j\}$ in Equations 8.2.12 and 8.3.14 are not necessarily on the same scale. For example, if we have a set of shifts $\{dp_j\}$ that satisfy Equation 8.2.12, then any other set of shifts $\{\alpha dp_j\}$, where α is a constant, will also satisfy Equation 8.2.14. We need to find the unique value of α such that Equations 8.2.14

and 8.2.12 are both satisfied in a consistent way. To do this, we incorporate α, a constant whose value we have yet to determine, directly into an equation that combines Equations 8.2.12 and 8.2.14:

$$\sum_{j=1}^{t} u_j\, dp_j + k_B T \sum_{j=1}^{t} (1+\ln p_j)\, dp_j + \alpha \sum_{j=1}^{t} dp_j = 0 \qquad (8.2.15)$$

Grouping terms together, we get:

$$\sum_{j=1}^{t} \left[ u_j + k_B T(1+\ln p_j) + \alpha \right] dp_j = 0 \qquad (8.2.16)$$

The values of $dp_j$, although infinitesimally small, are not in general zero. Thus, the only way the left-hand side of Equation 8.2.16 can be guaranteed to be zero is if each of the individual terms in the brackets is always equal to zero. This yields the following condition for all values of $j$:

$$u_j + k_B T(1+\ln p_j) + \alpha = 0 \qquad (8.2.17)$$

$$\Rightarrow 1+\ln p_j = \frac{-\alpha - u_j}{k_B T}$$

$$\Rightarrow \ln p_j = \frac{-\alpha - u_j}{k_B T} - 1$$

$$\Rightarrow p_j = e^{-\alpha/k_B T} e^{-u_j/k_B T} e^{-1}$$

$$\Rightarrow p_j = \left[ e^{\left(-\alpha/k_B T\right) - 1} \right] e^{-u_j/k_B T} \qquad (8.2.18)$$

The term in square brackets is a constant (that is, it does not depend on the energy levels). We denote this term by $A$, and so we can write:

$$p_j = A e^{-u_j/k_B T} \qquad (8.2.19)$$

Equation 8.2.19 is in exactly the same form as Equation 8.45 in the main text, and comparing the two equations shows that $\beta = 1/k_B T$.

The analysis that led to Equation 8.2.19 shows that the exponential form of the probability distribution follows directly from the probabilistic definition of the entropy, which is in turn derived from the application of Stirling's approximation to the calculation of the multiplicity of the system. The value of the proportionality factor $A$ in Equation 8.2.19 is evaluated by using the normalization condition, as discussed in the main text.

equilibrium will be given by the Boltzmann distribution, but in order to calculate these populations, we need to evaluate the normalization constant, $A$, in Equation 8.55.

The value of the constant $A$ is chosen so that Equation 8.55 satisfies the normalization condition:

$$\sum_{j=1}^{t} p_i = 1 \Rightarrow \sum_{j=1}^{t} A e^{-\beta u_j} = A \sum_{j=1}^{t} e^{-\beta u_j} = 1$$

$$\Rightarrow A = \frac{1}{\sum_{j=1}^{t} e^{-\beta u_j}} = \frac{1}{Q}$$

$$\Rightarrow Q = \sum_{j=1}^{t} e^{-\beta u_j} \qquad (8.56)$$

The term $Q$ in Equation 8.56 known as the **partition function**. Combining Equations 8.55 and 8.56, we have the following expression for the Boltzmann distribution:

$$p_j = \frac{e^{-u_j/k_B T}}{Q} \qquad (8.57)$$

To see how to apply Equation 8.57 in practice, we shall work through a simple example. Consider two kinds of molecules, A and B, where the energy levels of A are more closely spaced together than the energy levels of molecule B (**Figure 8.23**). This difference in spacing could be due to differences in electronic structure or, more interestingly for proteins, to conformational restrictions that are tighter

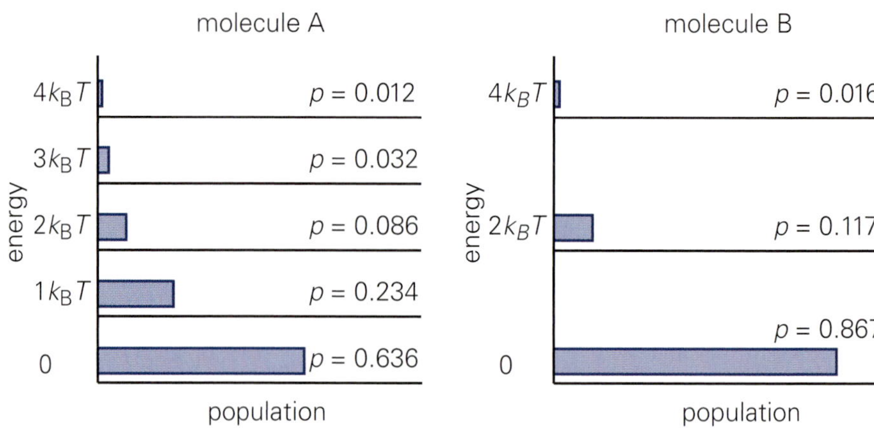


**Figure 8.23 Molecules with different energy gaps.** Molecules A and B differ in the separation between their energy levels. In molecule A, the energy levels are spaced more closely together and, as a consequence, the energy is distributed over more energy levels than for molecule B. The partition function for molecule A is larger than the partition function for molecule B.

in molecule B than in molecule A. Let us assume that the energy levels are spaced by $k_B T$ in molecule A and by $2k_B T$ in molecule B. Generally speaking, molecules do *not* have evenly spaced energy levels, and the actual spacings depend on the molecular structures in a complicated way, but scaling the energy in terms of $k_B T$ makes the calculations much easier to do for this example.

The partition function for molecule A is given by:

$$Q_A = \sum_{i=1}^{t} e^{-u_i/k_B T} \qquad \text{(see Equation 8.56)}$$

Expanding and evaluating the first few terms on the right-hand side of Equation 8.56, we get:

$$\begin{aligned} Q_A &= e^{-0/k_B T} + e^{-k_B T/k_B T} + e^{-2k_B T/k_B T} + \ldots \\ &= 1 + e^{-1} + e^{-2} + e^{-3} + \ldots \\ &= 1 + 0.368 + 0.135 + 0.050 + 0.018 + \ldots \\ &\approx 1.571 \end{aligned} \qquad (8.58)$$

Likewise, the partition function for molecule B is given by:

$$\begin{aligned} Q_B &= \sum_{i=1}^{t} e^{-u_i/k_B T} \\ &= e^{-0/k_B T} + e^{-2k_B T/k_B T} + e^{-4k_B T/k_B T} + e^{-6k_B T/k_B T} + \ldots \\ &= 1 + e^{-2} + e^{-4} + e^{-6} + \ldots \\ &= 1 + 0.135 + 0.018 + \ldots \\ &\approx 1.153 \end{aligned} \qquad (8.59)$$

Now that we have the partition functions of the two molecules, we can calculate the populations of molecules in each of the energy levels for the two systems by using Equation 8.57. The occupancies of the various levels, calculated in this way, are shown in Figure 8.23 for the two systems. Based on this figure, system A, which has the smaller energy gap, has a larger fractional occupancy in the higher levels than system B. This is reflected in the partition function of A, the value of which is greater than that of B.

Recall from Section 6.10 that the value of the Boltzmann constant, $k_B$, expressed in units of kJ•mol$^{-1}$ is the same as that of the gas constant, $R$. Recall also that the value of $k_B T$ (or $RT$) at 300 K is $8.314 \times 10^{-3} \times 300 = 2.49$ kJ•mol$^{-1}$. The value of $k_B T$ is a useful number to remember. As discussed in Section 6.10, if two states of a system are separated by less than a $k_B T$ of energy, then both will be significantly populated relative to each other (this is also reflected in Figure 8.23). The value of $k_B T$ at a particular temperature is sometimes referred to as "thermal energy." This represents a difference in energy that is readily surmounted by thermal fluctuations, in accordance with the Boltzmann distribution.

## 8.12 For large numbers of molecules, non-Boltzmann distributions of the energy are highly unlikely

When the number of molecules in the system is very large, the probability of observing molecules in any distribution other than the Boltzmann distribution becomes vanishingly small. We shall demonstrate this point with a simple numerical example in which we consider a system that is isolated (that is, with constant energy). As before, for computational convenience we choose the units of energy to be $k_BT$, so that the energy of the first excited level is $k_BT$, the energy of of the second excited level is $2k_BT$, and so on.

The number of molecules in each level for a distribution obeying the Boltzmann rule are shown in **Figure 8.24**, with 10,000 × $10^{20}$ molecules in the lowest energy level, 3679 × $10^{20}$ molecules in the next level, and so on. Additionally, the figure includes a distribution that does not obey the Boltzmann rule. In both cases, the total number of atoms is the same ($N$ = 15,780 × $10^{20}$) and the total energy is the same ($U$ = 8948 × $10^{20}$ $k_BT$). What is the relative probability of observing the Boltzmann distribution (distribution A in Figure 8.24) relative to the alternative distribution (B in Figure 8.24)? We can answer this question by calculating the entropies of the two distributions.

The probabilities of finding molecules in each of the energy levels for the two distributions are shown in Figure 8.24. For distribution A (the Boltzmann distribution), the entropy ($S_A$) is given by:

$$\frac{S_A}{Nk_B} = -\sum_{i=1}^{t} p_i \ln p_i \tag{8.60}$$

Expanding Equation 8.60 and evaluating the terms by referring to Figure 8.24, we get:

$$\begin{aligned}\frac{S_A}{Nk_B} &= -0.6337\ln 0.6337 - 0.2331\ln 0.2331 - 0.08574\ln 0.08574 - 0.03155\ln 0.03155\\ &\quad -0.01161\ln 0.01161 - 0.004270\ln 0.004270\\ &= 0.2891 + 0.3395 + 0.2106 + 0.1090 + 0.05173 + 0.02330\\ &= 1.023\end{aligned} \tag{8.61}$$

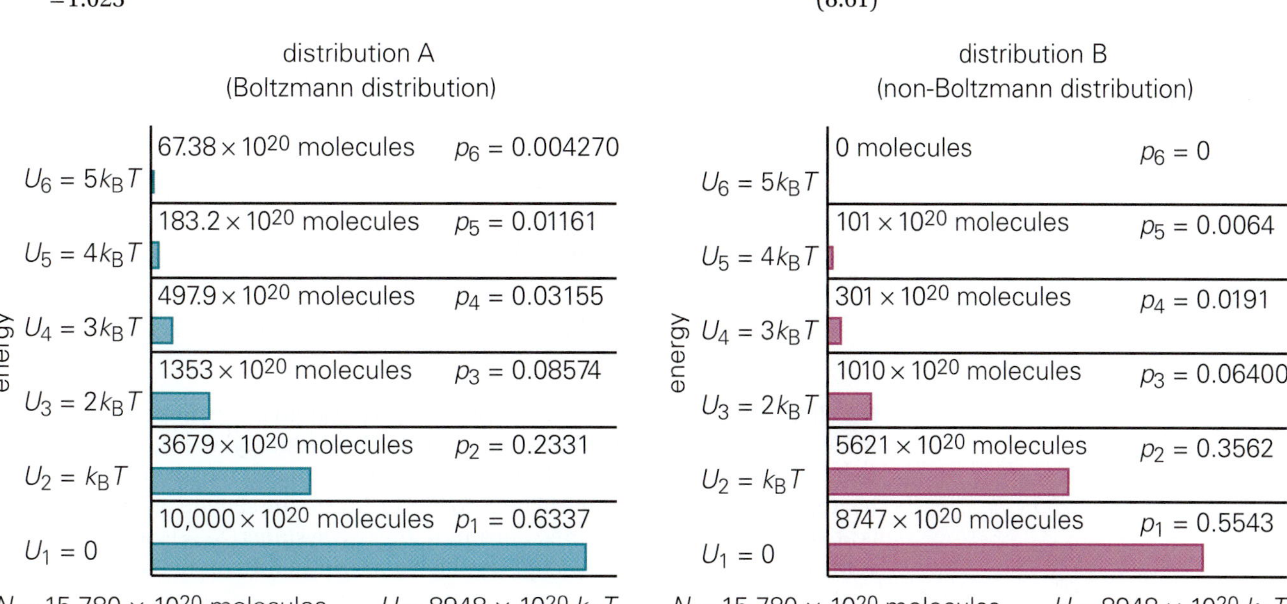


**Figure 8.24 Populations of energy levels in Boltzmann and non-Boltzmann distributions.** In this example, we consider a system with 15,780 × $10^{20}$ molecules and a total energy of 8948 × $10^{20}$ $k_BT$. To make the calculation simple we do not specify the value of $k_BT$, and we assume that the energy levels are spaced equally apart, with a separation of $k_BT$. The energy of each level, $U_i$, and the probability of finding a molecule in each level, $p_i$, are shown for both distributions.

Likewise, for system B (the non-Boltzmann distribution):

$$\begin{aligned}\frac{S_B}{Nk_B} &= -0.5543\ln 0.5543 - 0.3562\ln 0.3562 - 0.06400\ln 0.06400 \\ &\quad -0.01910\ln 0.01910 - 0.0064\ln 0.0064 \\ &= 0.3271 + 0.3677 + 0.1759 + 0.07560 + 0.03233 \\ &= 0.9786 \end{aligned} \tag{8.62}$$

How much more likely is it that we will observe the Boltzmann distribution (A) rather than the non-Boltzmann distribution (B)? This value can be calculated from the ratio of $W_A$ (the multiplicity of distribution A) to $W_B$ (the multiplicity of distribution B). To obtain the multiplicities, we follow the same procedure as in Section 8.8. For distribution A, we use Equation 8.61:

$$\ln W_A = {}^{S_A}\!\big/_{k_B} = N \times 1.023 \tag{8.63}$$

where $N$ is the total number of atoms in the system.

Likewise, for distribution B, we use Equation 8.62:

$$\ln W_B = \frac{S_B}{k_B} = N \times 0.9786 \tag{8.64}$$

Combining Equations 8.63 and 8.64, we obtain the ratios of the multiplicities as follows:

$$\begin{aligned}\ln\frac{W_A}{W_B} &= \ln W_A - \ln W_B = N \times (1.023 - 0.9786) \\ &= N \times 0.0444 \end{aligned} \tag{8.65}$$

Taking the exponent of both sides of Equation 8.65, we get the desired result:

$$\frac{W_A}{W_B} = e^{N \times 0.0444} \tag{8.66}$$

According to Equation 8.66, the ratio of the multiplicities for the Boltzmann distribution versus the non-Boltzmann distribution depends on the total number of molecules in the system. For any reasonably large number of molecules (for example, $N > 1000$), the ratio is a truly enormous number. For the particular example discussed here, $N = 15{,}780 \times 10^{20}$, the number of configurations corresponding to the Boltzmann distribution is astronomically higher than the number of configurations corresponding to the non-Boltzmann distribution. As a consequence, we are much more likely to observe the molecules arranged in the Boltzmann distribution than in this particular non-Boltzmann distribution at equilibrium.

# C. ENTROPY AND TEMPERATURE

## 8.13 The rate of change of entropy with respect to energy is related to the temperature

Recall from Section 8.1 that the temperature of a system is the property that determines whether or not the system will exchange heat (energy) with another system. In other words, if two systems (A and B) are at the same temperature, then the net heat transfer between them will be zero. According to the second law, equilibrium is achieved when the combined entropy of the two systems ($S_{A+B}$) has a maximum value (**Figure 8.25**). This concept can be expressed in differential form, which at equilibrium is:

$$dS_{A+B} = 0 \tag{8.67}$$

Analyzing the consequences of Equation 8.67 leads to an understanding of the connection between entropy and temperature.

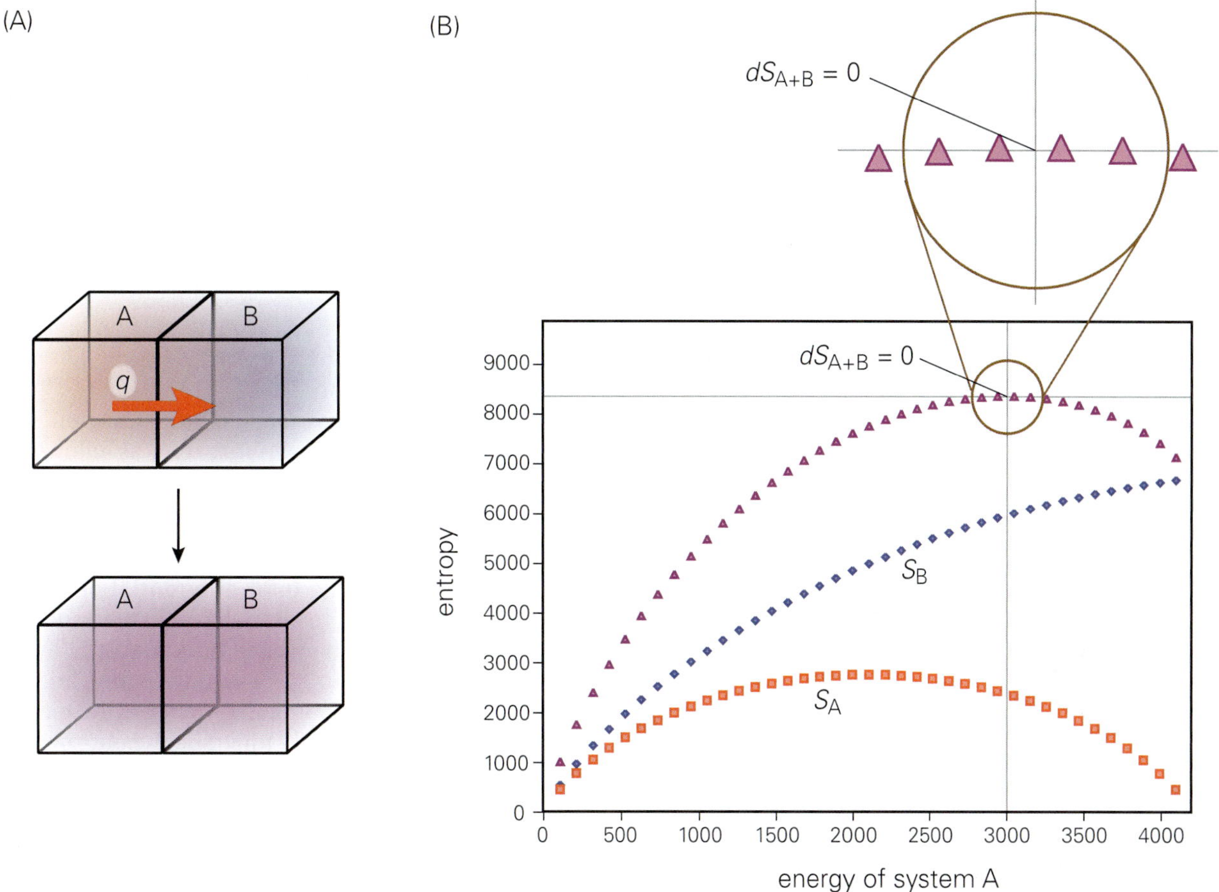


**Figure 8.25 Equilibrium is determined by the combined entropy of two systems that are isolated from the rest of the world.** (A) Two systems, A and B, are isolated from the rest of the world. Heat is transferred from system A to system B until equilibrium is reached. (B) The variation in entropy as a function of a changing variable (for example, energy of system A) is graphed in the diagram. $S_A$ and $S_B$ are the entropies of the individual systems. The equilibrium position (circled) corresponds to the position of maximum combined entropy of the two systems ($S_{A+B}$). (A, adapted from K.A. Dill and S. Bromberg, Molecular Driving Forces: Statistical Thermodynamics in Biology, Chemistry, Physics, and Nanoscience, 2nd ed. New York: Garland Science, 2010.)

The total entropy ($S_{A+B}$) of the combined systems is given by the sum of the individual entropies:

$$S_{A+B} = S_A + S_B \tag{8.68}$$

where $S_A$ and $S_B$ are the entropies of systems A and B, respectively.

Combining Equations 8.67 and 8.68, we obtain the following condition on the entropy at equilibrium:

$$dS_{A+B} = dS_A + dS_B = 0 \tag{8.69}$$

We must consider all the ways in which the entropy of system A ($S_A$) can change. The entropy is an extensive property of the system, and a function of various other extensive properties of the system, such as the volume ($V_A$), the energy ($U_A$), and the number of molecules ($N_A$). The dependence of $S_A$ on $V_A$, $U_A$, and $N_A$ can be written as follows:

$$S_A = f(V_A, U_A, N_A) \tag{8.70}$$

Equation 8.70 simply states that if the volume, energy, or number of particles of the system changes, then the entropy changes, too. Using the principles of differential calculus (**Box 8.3**), we can write down the following expression for an infinitesimal change in the entropy ($dS_A$) in terms of infinitesimal changes in the variables $V_A$, $U_A$, and $N_A$:

$$dS_A = \left(\frac{\partial S_A}{\partial V_A}\right)_{U_A,N_A} dV_A + \left(\frac{\partial S_A}{\partial U_A}\right)_{V_A,N_A} dU_A + \left(\frac{\partial S_A}{\partial N_A}\right)_{U_A,V_A} dN_A \tag{8.71}$$

## Box 8.3 Partial derivatives of a function

In relating changes in the entropy to changes in other variables (for example, energy or volume), we have to consider functions of many variables. The concept of a partial derivative of a function is critical to understanding the derivation of the Boltzmann distribution in Box 8.2 and the discussion of temperature in part C of this chapter.

The concept of a partial derivative is a simple extension of the derivative of a function of a single variable. Consider a function:

$$y = f(x) \tag{8.3.1}$$

The function $f(x)$ assigns a number, $y$, for each value of $x$. Here, $f(x)$ is a function of a single variable, such as $f(x) = x^2$ or $f(x) = \log(x - 14)$.

Now consider a function of two variables:

$$z = f(x,y) \tag{8.3.2}$$

In this case, the function $f(x,y)$ assigns a number (that is, a value) to $z$ for pairs of $x$, $y$ values (**Figure 8.3.1**). The function defines a surface, rather than a curve, and is known as a **multivariable function**.

Partial derivatives are derivatives of a multivariable function taken with respect to only one variable, keeping the other variables fixed (**Figure 8.3.2**):

The partial derivative of the function $f(x,y)$ with respect to $x$ is defined as follows:

$$\left(\frac{\partial f}{\partial x}\right)_y = \lim_{\Delta x \to 0}\left[\frac{f(x+\Delta x, y) - f(x,y)}{\Delta x}\right] \tag{8.3.3}$$

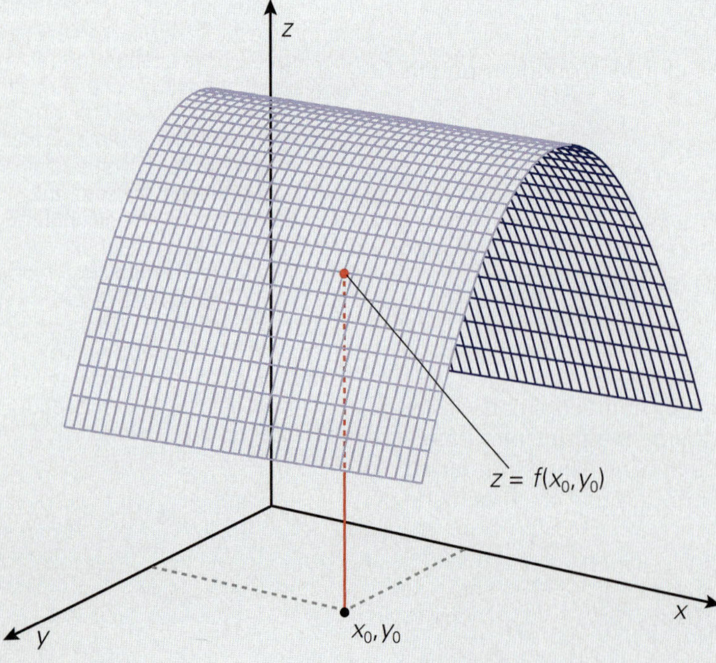


Note that this is essentially the same as the definition of the derivative of the single variable function $f(x)$ with respect to $x$, except for the complication that one variable ($y$) is kept fixed as the derivative is computed.

Likewise, the partial derivative of $f(x,y)$ with respect to $y$ is defined as follows:

$$\left(\frac{\partial f}{\partial y}\right)_x = \lim_{\Delta y \to 0}\left[\frac{f(x, y+\Delta y) - f(x,y)}{\Delta y}\right] \tag{8.3.4}$$

A small change in the value of the function, $df$, is expressed in terms of changes in $x$ and $y$ as follows:

$$df = \left(\frac{\partial f}{\partial x}\right)_y dx + \left(\frac{\partial f}{\partial y}\right)_x dy \tag{8.3.5}$$

The term $df$ is called the **total differential** of the function $f$. Equation 8.3.5 is very important in thermodynamics because it relates changes in the variables $x$ and $y$ to changes in the function $f$. For example, the variables $x$ and $y$ might be system parameters such as the temperature and the pressure, and $f$ might be the entropy of the system.

For a function with $t$ variables (for example, $x_1, x_2, \ldots, x_t$), the total differential is expressed as:

$$df = \sum_{i=1}^{t}\left(\frac{\partial f}{\partial x_i}\right)_{x_{j\neq i}} dx_i \tag{8.3.6}$$

Here, the subscript $x_{j\neq i}$ indicates that each derivative is calculated with all other variables kept fixed.

To find a minimum, or a maximum, of a multivariable function, we set the total differential to be zero. For example, for the function $f(x, y)$, the condition for the function to be at an extremum (that is, at either a minimum or a maximum) is:

$$df = \left(\frac{\partial f}{\partial x}\right)_y dx + \left(\frac{\partial f}{\partial y}\right)_x dy = 0 \tag{8.3.7}$$

**Figure 8.3.1 A function of two variables.** The function $f(x, y)$ is a function of two variables, $x$ and $y$. For every pair of values of $x$ and $y$, the function assigns a value for $z$. (Adapted from K.A. Dill and S. Bromberg, Molecular Driving Forces: Statistical Thermodynamics in Biology, Chemistry, Physics, and Nanoscience, 2nd ed. New York: Garland Science, 2010.)

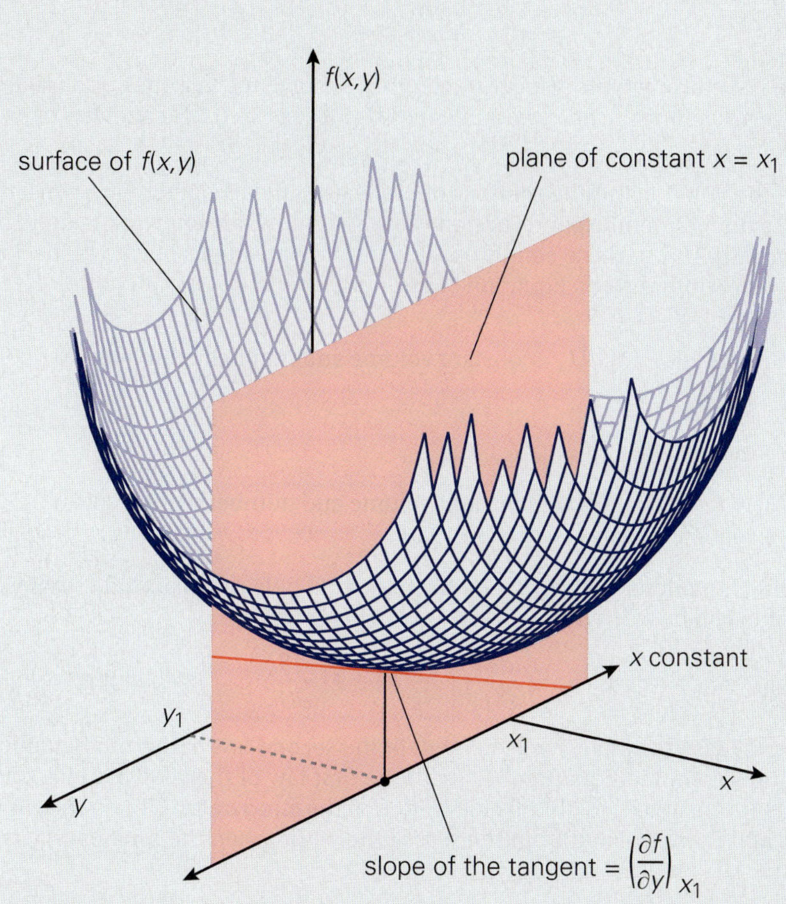


**Figure 8.3.2 Partial derivatives of a function.** The function $f$ is a function of two variables, $x$ and $y$. The partial derivative of $f$ with respect to $y$ is the derivative of $f$ with respect to $y$, keeping $x$ constant (fixed) at a particular value. In this illustration, the value of $x$ is fixed at $x_1$. The partial derivative at the point $x_1$, $y_1$ is the tangent to the curve defined by the intersection of the surface of $f(x, y)$, shown as a *blue* mesh, and the plane of constant $x$ (*red*). The *red* tangent line lies entirely within the *red* plane. (Adapted from K.A. Dill and S. Bromberg, Molecular Driving Forces: Statistical Thermodynamics in Biology, Chemistry, Physics, and Nanoscience, 2nd ed. New York: Garland Science, 2010.)

Since $dx$ and $dy$ are independent variables, both terms must be zero in order for the condition to be satisfied (**Figure 8.3.3**):

$$\left(\frac{\partial f}{\partial x}\right)_y = 0 \text{ and } \left(\frac{\partial f}{\partial y}\right)_x = 0 \tag{8.3.8}$$

Equation 8.3.8 is the condition for $f(x,y)$ to be at an extremum (maximum or minimum).

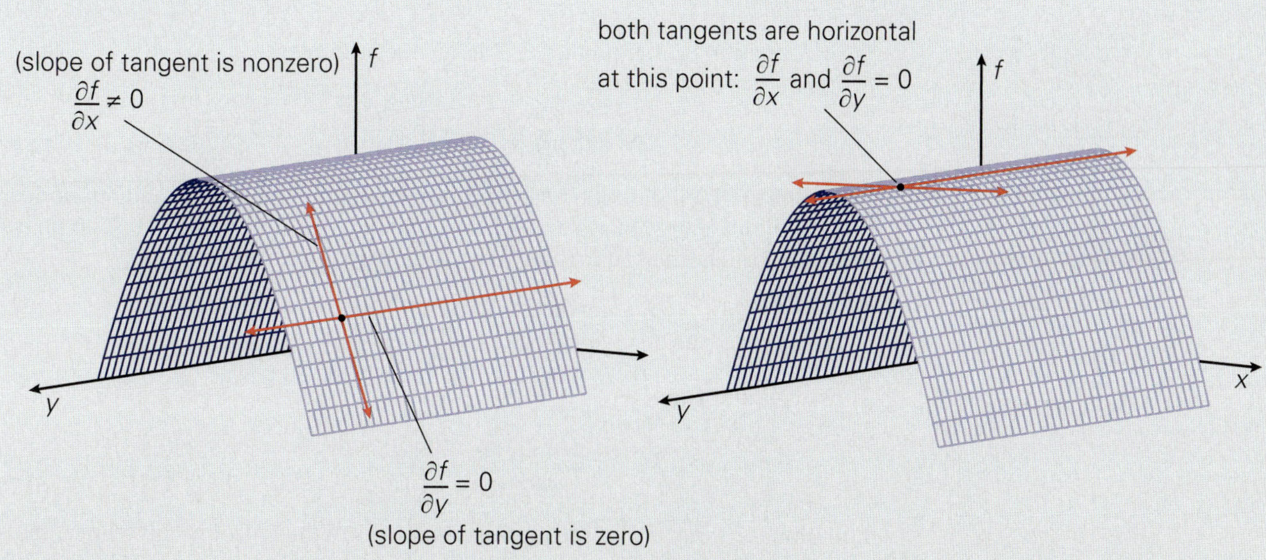


**Figure 8.3.3 Determining an extremum (maximal or minimal) point for a function of two variables.** (Adapted from K.A. Dill and S. Bromberg, Molecular Driving Forces: Statistical Thermodynamics in Biology, Chemistry, Physics, and Nanoscience, 2nd ed. New York: Garland Science, 2010.)

The partial derivatives $\left(\frac{\partial S_A}{\partial V_A}\right)_{U_A,N_A}$, $\left(\frac{\partial S_A}{\partial U_A}\right)_{V_A,N_A}$, and $\left(\frac{\partial S_A}{\partial N_A}\right)_{U_A,V_A}$ are the rates of change of entropy with respect to volume, energy, and number of particles, respectively, keeping the other variables constant (see Box 8.3 for a review of partial derivatives).

Suppose the volume and the number of molecules in each system are constant (that is, neither system expands or contracts and no expansion work is done by either system). Under these conditions, the partial derivatives of with respect to $V_A$ and $N_A$ are both zero, so Equation 8.71 reduces to:

$$dS_A = \left(\frac{\partial S_A}{\partial U_A}\right)_{V_A,N_A} dU_A \quad \text{(constant volume and number of molecules)} \tag{8.72}$$

Likewise:

$$dS_B = \left(\frac{\partial S_B}{\partial U_B}\right)_{V_B,N_B} dU_B \quad \text{(constant volume and number of molecules)} \tag{8.73}$$

Combining Equations 8.69, 8.72, and 8.73, the principle of maximum entropy at equilibrium can be written as:

$$dS_{A+B} = dS_A + dS_B = \left(\frac{\partial S_A}{\partial U_A}\right)_{V_A,N_A} dU_A + \left(\frac{\partial S_B}{\partial U_B}\right)_{V_B,N_B} dU_B = 0 \tag{8.74}$$

Equation 8.74 is essentially a restatement of the second law of thermodynamics. We can now combine Equation 8.74 with the first law of thermodynamics (that is, energy is conserved) to obtain a definition of the temperature. To do this, note that if A and B are isolated from the rest of the world, then the total energy is a constant:

$$U_{A+B} = U_A + U_B = \text{constant} \tag{8.75}$$

From Equation 8.75, it follows that:

$$\begin{aligned} dU_{A+B} &= 0 = dU_A + dU_B \\ &\Rightarrow dU_A = -dU_B \end{aligned} \tag{8.76}$$

Using Equation 8.76, Equation 8.74 can be rewritten as:

$$dS_{A+B} = \left[\left(\frac{\partial S_A}{\partial U_A}\right)_{V_A,N_A} - \left(\frac{\partial S_B}{\partial U_B}\right)_{V_B,N_B}\right] dU_A = 0 \tag{8.77}$$

The value of $dU_A$ in Equation 8.77 represents an infinitesimally small but nonzero change in the energy of system A. Thus, the only way for Equation 8.77 to be true is for the term in the square brackets to be zero at equilibrium:

$$\left[\left(\frac{\partial S_A}{\partial U_A}\right)_{V_A,N_A} - \left(\frac{\partial S_B}{\partial U_B}\right)_{V_B,N_B}\right] = 0 \tag{8.78}$$

Thus, at equilibrium:

$$\left(\frac{\partial S_A}{\partial U_A}\right)_{V_A,N_A} = \left(\frac{\partial S_B}{\partial U_B}\right)_{V_B,N_B} \tag{8.79}$$

Equation 8.79 shows that the partial derivatives of the entropy with respect to the energy for the two systems *are equal* when the two systems are in thermal equilibrium (see Figure 8.24). But if systems A and B are in thermal equilibrium, then the temperatures of the two systems are the same by definition. Equation 8.79 identifies a property (the derivative of entropy with respect to energy) that is equalized when systems that are in contact come to equilibrium. Thus, Equation 8.79

provides a way to define the temperature of a system. There is one complication, however. If it is true that:

$$\left(\frac{\partial S_A}{\partial U_A}\right)_{V_A,N_A}=\left(\frac{\partial S_B}{\partial U_B}\right)_{V_B,N_B} \quad \text{(from Equation 8.79)}$$

then it is also true that:

$$\left(\frac{\partial U_A}{\partial S_A}\right)_{V_A,N_A}=\left(\frac{\partial U_B}{\partial S_B}\right)_{V_B,N_B} \quad (8.80)$$

We therefore have two different choices for defining the temperature. We could define the temperature as the rate of change of entropy with respect to energy:

$$\left(\frac{\partial S_A}{\partial U_B}\right)_{V_A,N_A}=\left(\frac{\partial S_B}{\partial U_B}\right)_{V_B,N_B}\stackrel{?}{=}T \quad (8.81)$$

Or, alternatively, we could define the temperature as the rate of change of energy with respect to entropy:

$$\left(\frac{\partial U_A}{\partial S_A}\right)_{V_A,N_A}=\left(\frac{\partial U_B}{\partial S_B}\right)_{V_B,N_B}\stackrel{?}{=}T \quad (8.82)$$

Both definitions of the temperature are valid in principle, but the second definition is more in line with our intuitive feeling that increasing energy is correlated with increased entropy and temperature (**Box 8.4**), so we define temperature as:

$$T\equiv\left(\frac{\partial U_A}{\partial S_A}\right)_{V_A,N_A}=\left(\frac{\partial S_A}{\partial U_A}\right)_{V_A,N_A}^{-1} \quad (8.83)$$

A graphical explanation of the meaning of temperature as defined in Equation 8.83 is provided in **Figure 8.26**.

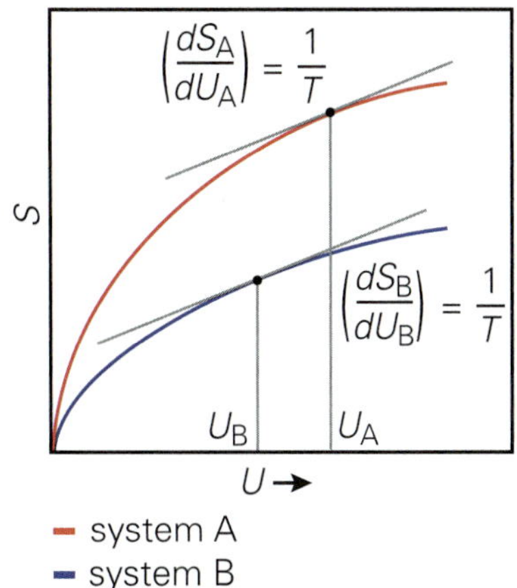


**Figure 8.26 Dependence of entropy upon energy for two systems.** Two systems, A and B, are in thermal contact. In general, the entropy of each system increases with increasing energy. The points at which the tangents to the entropy (*S*) vs. energy (*U*) curves have the same slope are shown. At equilibrium, the slopes of *S* vs. *U* are equal at $U_A$ and $U_B$, but $U_A$ is not equal, necessarily, to $U_B$. (Adapted from K.A. Dill and S. Bromberg, Molecular Driving Forces: Statistical Thermodynamics in Biology, Chemistry, Physics, and Nanoscience, 2nd ed. New York: Garland Science, 2010.)

What are the units of temperature? We can rewrite Equation 8.83 as:

$$T\approx\frac{\Delta U}{\Delta S} \quad (8.84)$$

where $\Delta S$ is a very small change in the entropy of the system and $\Delta U$ is the corresponding change in the energy (see Box 8.4 for an example of how Equation 8.84 can be applied in practice). Thus, the units of temperature depend on the units chosen for the energy and the entropy. According to Equation 8.1, the units of entropy are the same as that of the Boltzmann constant, which has units of energy/temperature. Therefore, the choice of units for the Boltzmann constant and the temperature scale have to be consistent.

By convention, we use the **absolute scale temperature**, in which the freezing point of water is 273.15 K (degrees kelvin). The value of $k_B$ is 1.38066 × $10^{-23}$ J•K$^{-1}$ in SI units, and so we must report the temperature on the absolute Kelvin scale if we use this value of $k_B$.

For a system with unlimited energy levels, increasing the entropy of the system should also increase the energy of the system (**Figure 8.27**). That is, if $\Delta S$ is positive, then $\Delta U$ should also be positive. Equation 8.84 then implies that the value of the temperature is always positive. The zero of the absolute scale would be

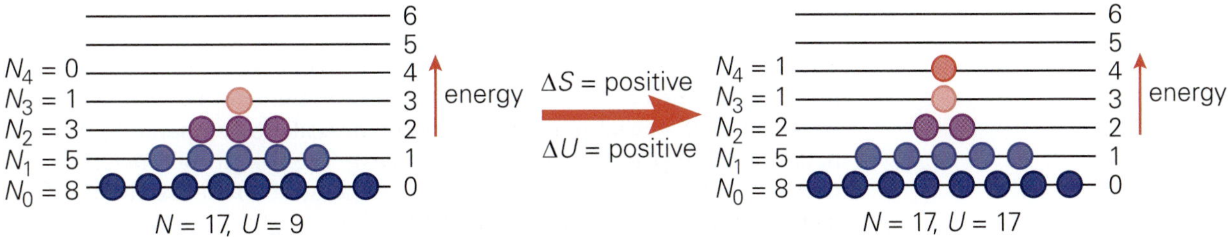


**Figure 8.27 An increase in energy usually corresponds to an increase in entropy.** For a system with unbounded energy levels, an increase in energy results in more energy levels being populated, which results in an increase in entropy.

## Box 8.4 A thermodynamic definition of the temperature

Consider a system with $N$ molecules and two energy levels, as shown in **Figure 8.4.1**. We compare two states of the system, state A and state B. In state A, most of the molecules are in the lowest energy level ($p_1 = 0.999$, $p_2 = 0.001$). In state B, which has a higher energy, 40% of the molecules are in the higher energy level ($p_1 = 0.600$, $p_2 = 0.400$). We would say intuitively that state A is at lower temperature than state B because it has lower entropy and lower energy, and we would like to define the temperature so as to be consistent with this definition.

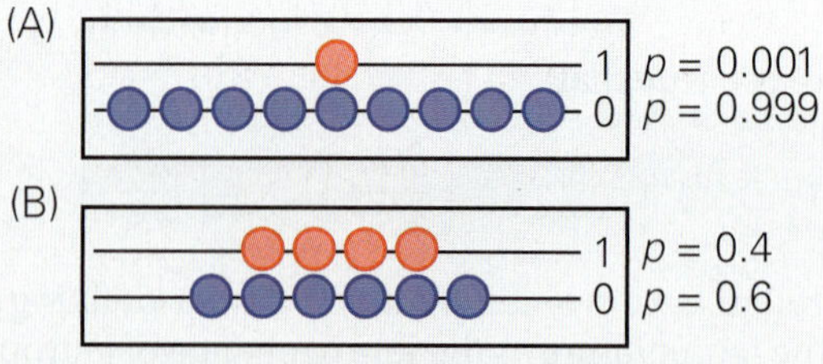


**Figure 8.4.1 A system with two energy levels and two different energy distributions.**

Let us compare the values of the derivative $\left(\frac{\partial S}{\partial U}\right)_{N,V}$ for the two states of the system. We do this by approximating the derivative in the following way:

$$\left(\frac{\partial S}{\partial U}\right)_{N,V} \approx \frac{\Delta S}{\Delta U} \tag{8.4.1}$$

where $\Delta S$ is the change in entropy when we increase the energy by a small amount, $\Delta U$, while keeping the number of molecules constant. For state A, the initial value of the entropy is given by:

$$\begin{aligned}\frac{S_1}{Nk_B} &= -p_1 \ln p_1 - p_2 \ln p_2 \\ &= -0.001 \ln 0.001 - 0.999 \ln 0.999 \\ &= 0.00691 + 0.00100 \\ &= 0.00791\end{aligned} \tag{8.4.2}$$

Now let us add a small amount of energy, equal to $0.002N$ energy units, to the system. What we mean by this is that we promote $0.002N$ molecules from the lowest energy level to the higher level. Thus, $p_1$ is now 0.997 and $p_2$ is 0.003. The entropy is now given by:

$$\begin{aligned}\frac{S_2}{Nk_B} &= -0.003 \ln 0.003 - 0.997 \ln 0.997 \\ &= 0.01743 + 0.00299 \\ &= 0.02042\end{aligned} \tag{8.4.3}$$

Therefore:

$$\frac{\Delta S}{Nk_B} = \frac{S_2 - S_1}{Nk_B} = 0.02042 - 0.00791 = 0.01251 \tag{8.4.4}$$

And so,

$$\frac{\Delta S}{\Delta U} = \frac{0.01251 \times Nk_B}{0.002 \times N} = 6.25 k_B \tag{8.4.5}$$

Let us now add the same amount of energy ($0.002N$) to the system in state B, and calculate the ratio of $\Delta S$ to $\Delta U$.

Initially, $p_1 = 0.600$ and $p_2 = 0.400$ and so:

$$\frac{S_1}{Nk_B} = -0.4 \ln 0.4 - 0.6 \ln 0.6 = 0.3665 + 0.3065 = 0.6730 \tag{8.4.6}$$

Finally, $p_1 = 0.598$ and $p_2 = 0.402$, and so:

$$\frac{S_2}{Nk_B} = -0.402 \ln 0.402 - 0.598 \ln 0.598 = 0.6738 \tag{8.4.7}$$

$$\Rightarrow \frac{\Delta S}{Nk_B} = 0.0008 \tag{8.4.8}$$

$$\Rightarrow \frac{\Delta S}{\Delta U} = \frac{0.0008 \times Nk_B}{0.002 \times N} = 0.4 k_B \tag{8.4.9}$$

Thus, the value of $\frac{\Delta S}{\Delta U}$ is ~15 times as large for state A when compared to state B. This reflects a general phenomenon. States with lower initial entropy, such as state A, undergo larger changes in entropy when their energies are increased. Thus, the value of $\left(\frac{\partial S}{\partial U}\right)_{N,V}$ is expected to be larger for states with lower entropy. Since we expect that in general the entropy will increase with temperature, we choose to define the temperature in terms of $\left(\frac{\partial S}{\partial U}\right)_{N,V}^{-1}$ rather than $\left(\frac{\partial S}{\partial U}\right)_{N,V}$.

reached if all molecular motion could be frozen out and the energy and entropy were both zero. The fahrenheit and centigrade scales of temperature, on the other hand, allow temperature values to be negative. This would make no sense in Equation 8.84, so you should always convert temperature to the absolute scale before carrying out a thermodynamic calculation.

## 8.14 The statistical and thermodynamic definitions of the entropy are equivalent

The thermodynamic definition of the entropy (Equation 8.2) includes $q_{rev}$, the quantity of heat transferred in a near-equilibrium (reversible) process. The preceding discussion helps us understand why heat transfer is related to the entropy. A logical place to begin to see how this heat term arises is the first law of thermodynamics, which is stated in differential form as:

$$dq = dU - dw \tag{8.85}$$

Recall that if the system does work *on* the surroundings, then the sign of the work term is *negative* and this is reflected in the negative sign in front of $dw$ in Equation 8.85. We have written the first law equation in this way to emphasize that an infinitesimal amount of heat ($dq$) that is added to the system can be used to increase the energy of the system by $dU$ and to do an amount of work, $dw$. We shall now analyze the interconversion between heat and work as a process that occurs within a system. But, in contrast to the analysis in Sections 7.21 and 7.22, we shall not assume that the system contains an ideal gas.

Rearranging Equation 8.85 we get:

$$dU = dq + dw \tag{8.86}$$

Equation 8.86 is always true, but when we consider a process that goes from an initial state (state 1) to a final state (state 2), the amount of heat taken up by the system ($\int dq = q$) and the work done by the system $\left(\int dw = w\right)$ will depend on the path taken to go from state 1 to state 2.

If the conversion in volume is brought about by a sudden decrease in the external pressure, for example, then the work done by the system is small, and the heat taken up by the system is small (**Figure 8.28**). Because $dw$ is path dependent, $dq$ is also path dependent, and cannot in general be related directly to the change in a state function such as the entropy, which is independent of path. We can make such a connection, however, if we specify the path taken during the process. In particular, consider how the energy changes when the system undergoes a near-equilibrium (reversible) conversion from state 1 to state 2.

The energy of the system ($U$) is an extensive state function. Just as we did for the entropy in Equation 8.71, we can express the dependence of energy on other extensive variables as follows:

$$dU = \left(\frac{\partial U}{\partial S}\right)_{V,N} dS + \left(\frac{\partial U}{\partial V}\right)_{S,N} dV + \left(\frac{\partial U}{\partial N}\right)_{V,S} dN \tag{8.87}$$

If we restrict ourselves to situations in which the number of molecules, $N$, is fixed, then Equation 8.87 reduces to:

$$dU = \left(\frac{\partial U}{\partial S}\right)_{V,N} dS + \left(\frac{\partial U}{\partial V}\right)_{S,N} dV \tag{8.88}$$

Using Equation 8.83, which defines the temperature, we get:

$$dU = TdS + \left(\frac{\partial U}{\partial V}\right)_{S,N} dV \tag{8.89}$$

A little consideration tells us that the partial derivative of energy with respect to volume, $\left(\frac{\partial U}{\partial V}\right)_{S,N}$, must be equal to the *pressure*, $P$, of the system. The second term in Equation 8.89 describes how the energy of the system changes due to a change in volume. It has the form:

$$dU = XdV \tag{8.90}$$

The term on the right-hand side of Equation 8.90 is essentially the magnitude of the mechanical work done by the system. If the system expands against an

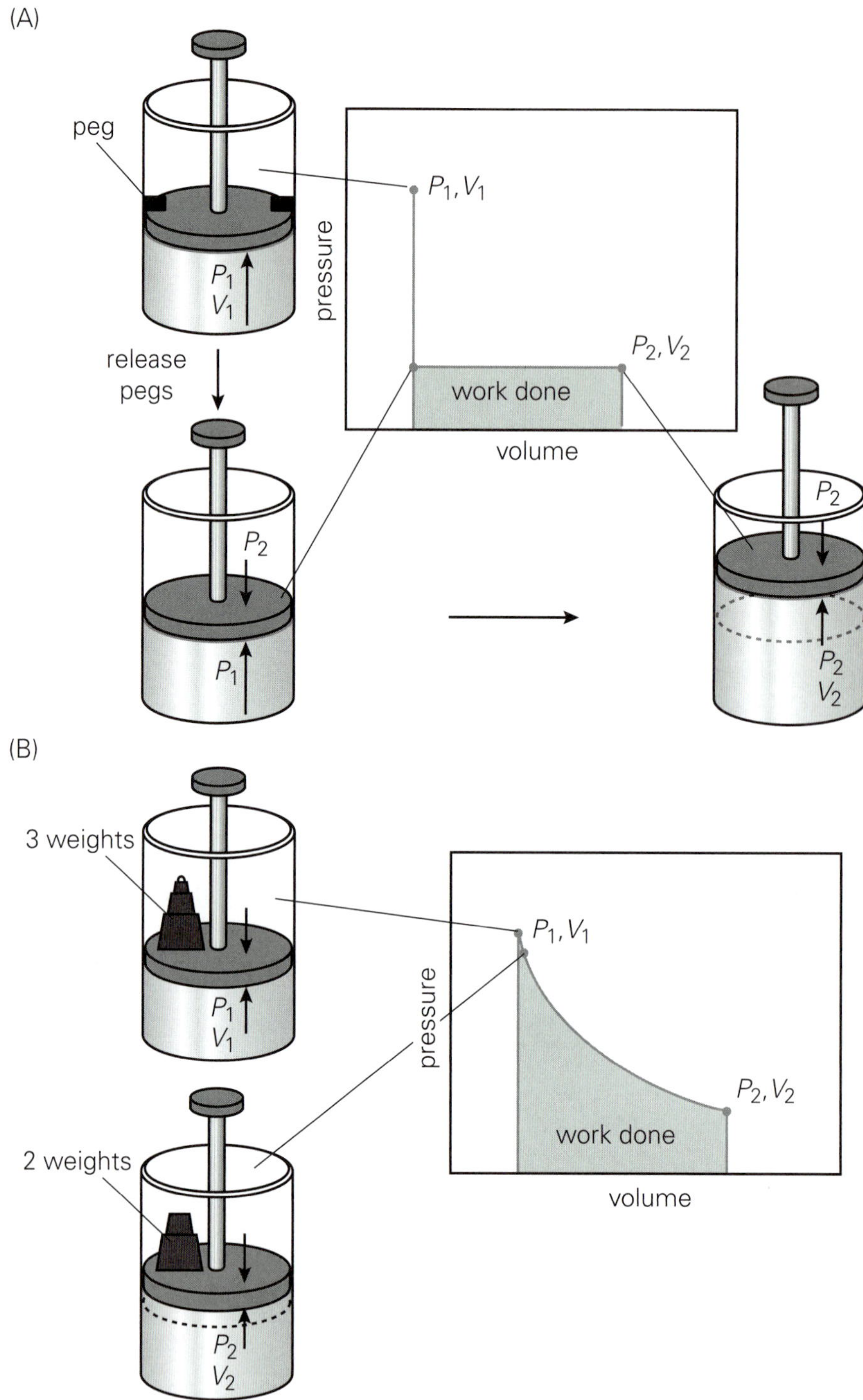


**Figure 8.28 The work done in a process depends on the details of how the process is carried out.** This example illustrates a system in which a gas expands, but we do not assume that the gas is an ideal gas. (A) and (B) show processes that go from the same initial state to the same final state. In the first process, (A), the pressure on the piston is suddenly released, and the piston expands against a constant external pressure. The magnitude of the work done by the piston is given by the shaded area in the pressure-volume diagram. In the second process, (B), the external pressure is reduced so slowly that the internal and external pressures are always nearly in balance. The amount of work done is much greater in this case, as indicated by the shaded area.

external pressure, $P_{EXT}$, (that is, if its volume increases), then the energy change due to the expansion is given by:

$$dU = dw = -P_{EXT}\, dV$$

(change in energy due to work done) (8.91)

In Equation 8.91, the term $dw$ is the amount of work done by the system in an infinitesimal expansion step.

If the expansion occurs in a near-equilibrium manner, then the internal pressure ($P$) of the system is always very close to the external pressure ($P_{EXT}$). In this case, the pressure ($P$) of the system is given by:

$$P = P_{EXT} + dP \tag{8.92}$$

where $dP$ is a small increment in the pressure. Hence, from Equation 8.91:

$$dU = -(P - dP)dV = -PdV \tag{8.93}$$

where we have neglected the product of two infinitesimally small terms ($dP\, dV$).

Comparing Equations 8.88 and 8.93, we can identify the partial derivative of energy with respect to volume as the negative of the pressure:

$$\left(\frac{\partial U}{\partial V}\right)_{S,N} = -P \tag{8.94}$$

Substituting Equation 8.94 into Equation 8.89, we get:

$$dU = TdS - PdV \tag{8.95}$$

The term $-PdV$ in Equation 8.95 is the work done by the system because for the reversible process we can equate $P$ and $P_{EXT}$ (Equation 8.92). For a nonreversible process we cannot make this equivalence. Hence, Equation 8.95 can be written as:

$$dU = TdS + dw \tag{8.96}$$

Comparing Equation 8.96 with Equation 8.86 (that is, $dU = dq + dw$), then (for a near-equilibrium process) it turns out that:

$$dq = TdS \;\Rightarrow\; dS = \frac{(dq)_{\text{rev}}}{T} \tag{8.97}$$

where $(dq)_{\text{rev}}$ is the heat transferred to the system during an infinitesimal step in a near-equilibrium process. The change in entropy that results from the entire process of going from an initial state 1 to a final state 2 can be obtained by integrating both sides of Equation 8.97:

$$\Delta S = \int_1^2 dS = \int_1^2 \frac{(dq)_{\text{rev}}}{T} = \frac{q_{\text{rev}}}{T} \tag{8.98}$$

where $q_{\text{rev}}$ is the total amount of heat taken up during the reversible process, which is assumed to be occurring at constant temperature. Equations 8.97 and 8.98 are equivalent statements of the thermodynamic definition of the entropy.

## Summary

A system with molecules in different energy levels can be characterized by an energy distribution, which describes how many molecules are in each energy level. The multiplicity of the energy distribution ($W$) is the number of microstates that correspond to the distribution. Using similar arguments to the ones we used in Chapter 7 to derive the expression for the multiplicity of positional configurations, we obtain the following expression for the multiplicity of an energy distribution:

$$W = \frac{N!}{N_1!N_2!\ldots N_t!} \tag{8.9}$$

where $N$ is the total number of molecules in the system, and $N_1$, $N_2$, ..., $N_t$, are the numbers of molecules in each energy level. When the number of molecules is large, we can use Stirling's approximation to convert the statistical definition of entropy ($S = k_B \ln W$) to the following alternative expression in terms of probabilities:

$$S = -Nk_B \sum_{i=1}^{t} p_i \ln p_i \tag{8.13}$$

Equation 8.13 is easier to use in practice than Equation 8.9 because the values of $p_i$, the probabilities of finding molecules in the different energy levels, are always less than 1.0. Equation 8.13 is known as the probabilistic definition of the entropy.

At the microscopic level, the principle that the entropy of a system and its surroundings is maximal at equilibrium translates into the Boltzmann distribution, which specifies the probabilities of finding molecules in different energy levels at equilibrium. The Boltzmann distribution is given by:

$$p_j = \frac{e^{-u_j/k_B T}}{Q} \tag{8.57}$$

where $p_j$ is the probability of finding molecules in the $j^{\text{th}}$ level, $u_j$ is the energy of the $j^{\text{th}}$ level, $k_B$ is the Boltzmann constant, and $T$ is the absolute temperature. $Q$ is the partition function of the system, which indicates the extent to which the molecules of the system are distributed over different energy levels. The partition function, $Q$, is given by:

$$Q = \sum_{j=1}^{t} e^{-\beta u_j} \tag{8.56}$$

As the value of $Q$ increases, so does the accessibility of the higher energy levels of the system at a given temperature. Thus, the value of $k_B T$, which is also known as the "thermal energy" at a given temperature, is an important parameter. If levels have an energy $< k_B T$ above that of the lowest level then they will be occupied appreciably. If the energy is significantly more than $k_B T$ above the energy of the lowest level, then the occupancy becomes very low.

In order to connect the probabilistic or statistical definition of entropy to the amount of heat transferred during a process, we need to relate the entropy to the temperature. The temperature of a system is that property of the system that determines whether net heat transfer occurs when the system is placed in thermal contact with another system. When two systems are in thermal equilibrium, there is no net heat transfer between them, and their combined entropies are maximal (assuming that the two systems are isolated from everything else). This leads to the following definition of the temperature:

$$T \equiv \left(\frac{\partial U_A}{\partial S_A}\right)_{V_A, N_A} = \left(\frac{\partial S_A}{\partial U_A}\right)_{V_A, N_A}^{-1} \tag{8.83}$$

Based on Equation 8.83, systems at lower temperatures undergo a larger increase in entropy for the same energy input than systems at higher temperatures.

The first law of thermodynamics relates the change in energy of a system, the heat transferred to a system, and the work done by the system. For a reversible or near-equilibrium process, in which the work done is readily related to system properties, the first law of thermodynamics and Equation 8.83 together lead to the conclusion that:

$$\Delta S = \frac{q_{\text{rev}}}{T} \tag{8.98}$$

where $\Delta S$ is the change in entropy between two states of a system, $T$ is the temperature, and $q_{\text{rev}}$ is the heat delivered to the system when the transformation is carried out under reversible (near-equilibrium) conditions. Equation 8.98 is known as the thermodynamic definition of the entropy, and in this chapter we have shown that it is equivalent to the statistical definition of the entropy in a completely general way.

## Key Concepts

### A. ENERGY DISTRIBUTIONS AND ENTROPY

- The thermodynamic definition of the entropy provides a link to experimental observations because it relates heat transfer to changes in the entropy.
- The concept of temperature, which is the thermodynamic parameter that determines the direction of heat flow, provides a connection between the statistical and thermodynamic definitions of entropy.
- Energy distributions describe the populations of molecules with different energies.
- The multiplicity of an energy distribution is the number of equivalent configurations of molecules that results in the same energy distribution.

- The multiplicity of a system with different energy levels, $W$, can be calculated by counting the number of equivalent molecular rearrangements in Equation 8.9:

$$W = \frac{N!}{N_1!\,N_2!\ldots N_t!}$$

**B. THE BOLTZMANN DISTRIBUTION**

- For large numbers of molecules, a probabilistic expression for the entropy is more convenient as shown in Equation 8.13:

$$S = -Nk_B \sum_{i=1}^{t} p_i \ln p_i$$

- Systems in thermal contact transfer heat between each other until the combined entropy of the two systems is maximal.
- Many energy distributions are consistent with the total energy of a system, but some have higher multiplicities than others.
- The energy distribution at equilibrium is given by the Boltzmann distribution, Equation 8.57:

$$p_j = \frac{e^{-u_j/k_BT}}{Q}$$

- For large numbers of molecules, the Boltzmann distribution has maximum multiplicity and non-Boltzmann distributions of the energy are highly unlikely.

**C. ENTROPY AND TEMPERATURE**

- The rate of change of the entropy with respect to energy is related to the temperature, given by Equation 8.83:

$$T \equiv \left(\frac{\partial U_A}{\partial S_A}\right)_{V_A,N_A} = \left(\frac{\partial S_A}{\partial U_A}\right)^{-1}_{V_A,N_A}$$

- The statistical and thermodynamic definitions of the entropy are equivalent in a general way, not just for monatomic ideal gases.

## Problems

### True/False and Multiple Choice

1. For a system converting from state 1 to state 2: $(k_B \ln W_2 - k_B \ln W_1) = q_{rev}/T$.
   True/False

2. If the second-lowest energy level is separated from the ground state by $0.5k_BT$, then the second-lowest energy level will not be occupied appreciably.
   True/False

3. There are many equivalent microstates corresponding to a particular energy distribution.
   True/False

4. Which of the following definitions of entropy are equivalent for a large system:
   a. probabilistic definition
   b. thermodynamic definition
   c. statistical definition
   d. all of the above

5. The probabilistic definition of entropy is more accurate than the statistical definition for small systems.
   True/False

6. After spontaneous heat transfer between systems, the overall multiplicity is:
   a. lower than it was before the transfer of heat
   b. zero
   c. maximized
   d. minimized

7. The partition function ($Q$) is needed to describe:
   a. The volume of two gases on either side of a partition.
   b. The extent to which many different energy levels are occupied.
   c. How the statistical definition of entropy is equivalent to the probabilistic definition of entropy.
   d. The relationship between work and pressure.

### Fill in the Blank

8. The Boltzmann distribution describes the energy of molecules at ________________.

9. The Boltzmann constant is the gas constant ($R$) divided by ___________ .

10. Kinetic energy is due to the _________ of atoms. Potential energy is due to the __________ of atoms.

11. The direction of spontaneous change is governed by an increase in ___________, thus explaining how energy can be transferred "uphill," that is from a system of lower total energy to one with higher total energy.

12. If two systems of different temperatures are brought together, there will be a net transfer of heat from the system of _________ temperature to the system of ________ temperature.

## Quantitative/Essay

Assume $k_B = 1$ J•K$^{-1}$ for all problems.

**13.** A system starts with a multiplicity of 2000. Two kJ of heat are transferred into the system reversibly at 298 K. What is the multiplicity now?

**14.** A system starts with a multiplicity of $10^{99}$. Heat is transferred out of the system reversibly at 298 K such that the final multiplicity is $10^5$. How much heat was lost?

**15.** A system of noninteracting atoms has a positional multiplicity of 120 and an energetic multiplicity of 5. What is the total multiplicity? What are the positional, energetic, and total entropies of the system?

**16.** A system with 1 mole of an ideal gas changes its volume from 10 L to 1 L by isothermal compression. What is the change in entropy of the system? What is the change in the entropy associated with the energy distribution? (Use $R = 8.31$ J•K$^{-1}$•mol$^{-1}$.)

**17.** Consider a system with five molecules and three energy levels. The energy levels are such that a molecule at energy level 1 contributes 1 J of energy to $U_{TOTAL}$, a molecule at energy level 2 contributes 2 J of energy to $U_{TOTAL}$, and a molecule in energy level 3 contributes 3 J of energy to $U_{TOTAL}$. Which has more internal energy, state A or state B?

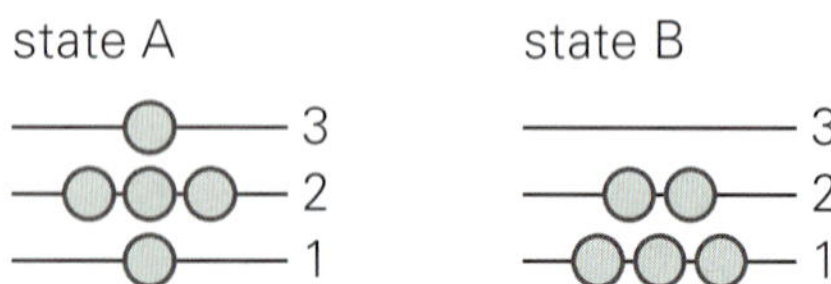


**18.** Consider the system and the two states from Problem 17. Which state has greater multiplicity?

**19.** A system has 100,000 molecules at energy level 1, 10,000 molecules at energy level 2, and 1000 molecules at energy level 3. What is the entropy of the system? (*Hint*: Use the probabilistic definition.)

**20.** A system with 100,000 molecules has two energy levels (A and B). At first, the two energy levels are populated equally. After a reversible process, energy level A is populated by 65% of the molecules and the system is at 293K.

a. What is the difference in energy between the two levels?
b. How much heat was added or removed from the system?
c. What is the change in entropy?

**21.** Two systems, A and B, are placed in thermal contact and isolated from the rest of the world. Shown next are three possible microstates for the combined system. Two of these are valid microstates for the combined system, and one is not. Identify the inconsistent microstate and explain why it does not belong with the other two. (*Hint*: Consider Figure 8.15.)

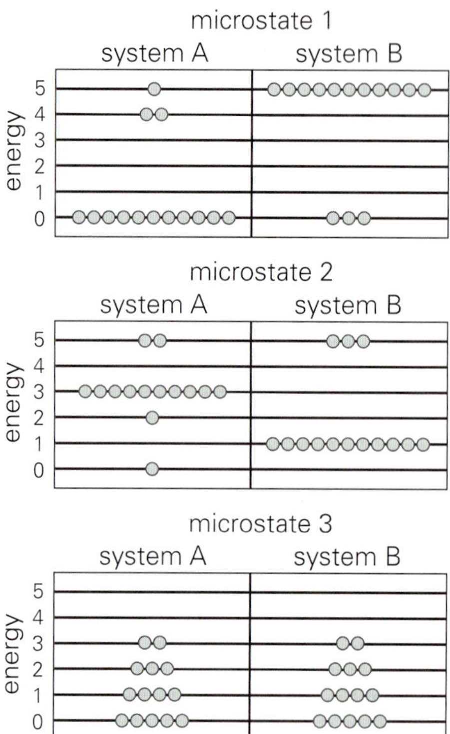


**22.** A system has two types of molecules (black and gray) that interconvert. The log of the multiplicity of energy of the two types of molecules individually versus the total system energy is plotted below. Plot the log of the total multiplicity of energy and indicate where the maximum value for the combined multiplicity occurs.

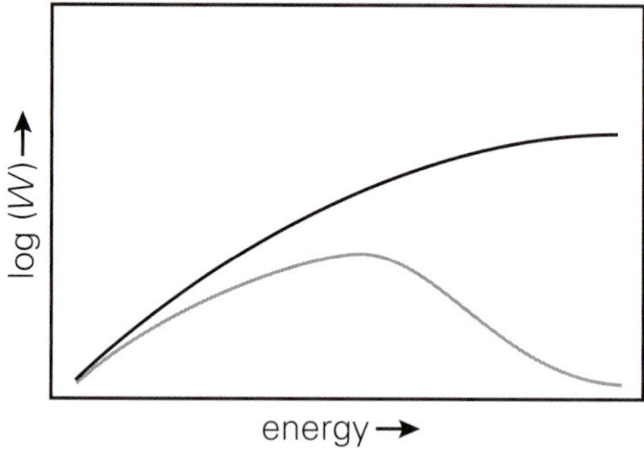


**23.** What is the entropy of the following energy distribution? What is the multiplicity?

$N_3 = 200$ molecules
$N_2 = 1100$ molecules
$N_1 = 10{,}500$ molecules

**24.** A system has a partition function of 1.3 at 293 K. At that temperature, an energy level is populated by 10% of the molecules. What is the energy of that level?

**25.** Explain how energy can move spontaneously from a system of low total energy to a system of higher total energy.

26. Why is it more common to think of temperature as being related to the rate of change of energy with respect to entropy than it is to think of it as being related to the rate of change of entropy with respect to energy?

## Further Reading

### General

Atkins PW & De Paula J (2010) Atkins' Physical Chemistry, 9th ed. New York: Oxford University Press.

Dill KA & Bromberg S (2010) Molecular Driving Forces: Statistical Thermodynamics in Chemistry, Biology, Physics, and Nanoscience, 2nd ed. New York: Garland Science. The discussion of entropy and partial derivatives in this chapter was guided by the treatment in this book, which is at a more advanced level.

Eisenberg DS & Crothers DM (1979) Physical Chemistry: With Applications to the Life Sciences. Menlo Park, CA: Benjamin/Cummings.

McQuarrie DA (2000) Statistical Mechanics. Sausalito, CA: University Science Books.

Tipler PA & Llewellyn RA (2007) Modern Physics, 5th ed. New York: W.H. Freeman & Co.

# PART III

## FREE ENERGY

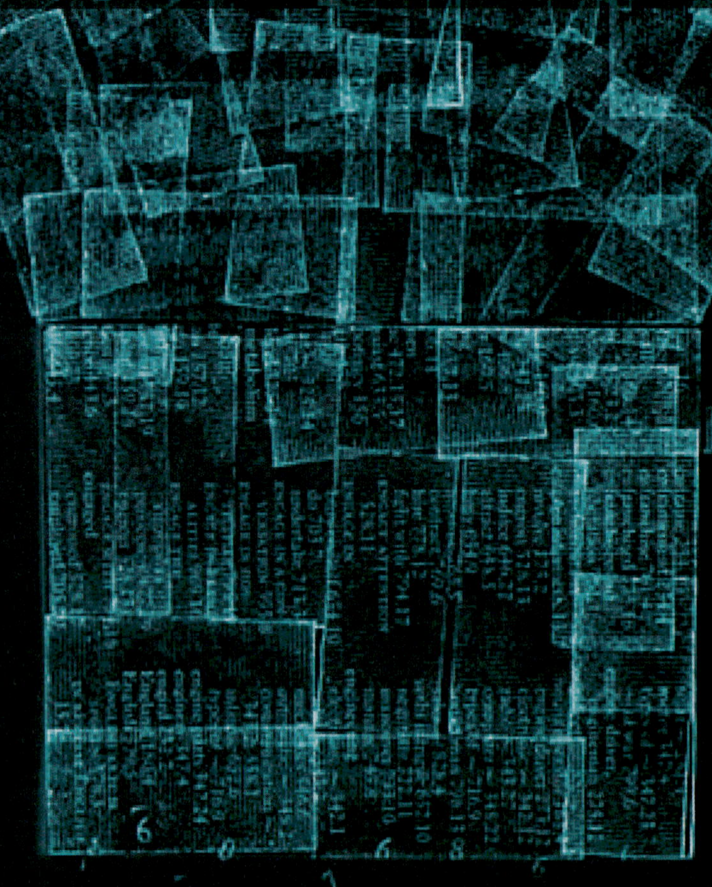

# CHAPTER 9

# Free Energy

How do we know when a chemical or biological process is at equilibrium? Various macroscopic properties of the system, such as the total energy, the temperature, or the entropy, are stable and unchanging at equilibrium (**Figure 9.1**). But molecular systems are always far from being static at a microscopic level, with each of the molecules moving along chaotic trajectories and exchanging energy as they collide. Nevertheless, for systems with large numbers of molecules, there is no net change in the global properties of the system at equilibrium.

If we perturb a system that is at equilibrium, perhaps by changing the temperature or by adding more molecules, then the system will no longer be at equilibrium. The macroscopic variables of the system will respond to the changes until the system reaches a new equilibrium, defined by the new set of environmental conditions (**Figure 9.2**). As the system moves towards a new equilibrium, what is the direction of spontaneous change? Will the temperature rise or fall? Will the concentrations of particular molecules increase or decrease? What will the values of these parameters be after equilibrium has been reestablished?

The first and second laws of thermodynamics provide powerful guiding principles that help answer these questions. We will find it helpful to define a new property of the system, known as the **free energy**, which incorporates both the energy of the system (which is constrained by the first law) and its entropy (constrained

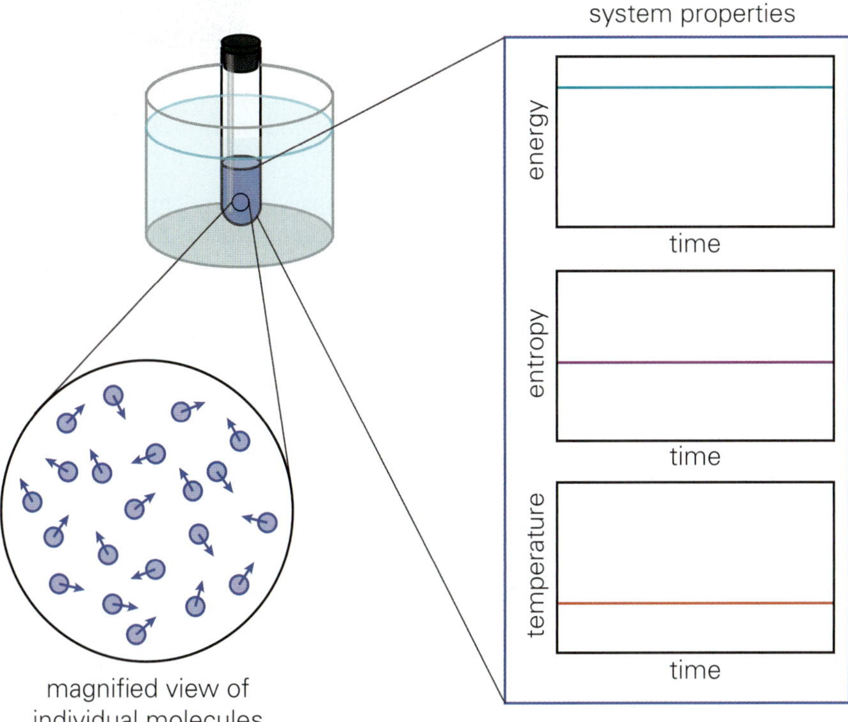


**Figure 9.1 A system at equilibrium.** A system (the test tube and its contents) is shown immersed in a water bath (surroundings). The movements of the individual molecules within the system are chaotic and unpredictable, but the global properties of the entire system, such as the temperature and the total energy, are stable with time.

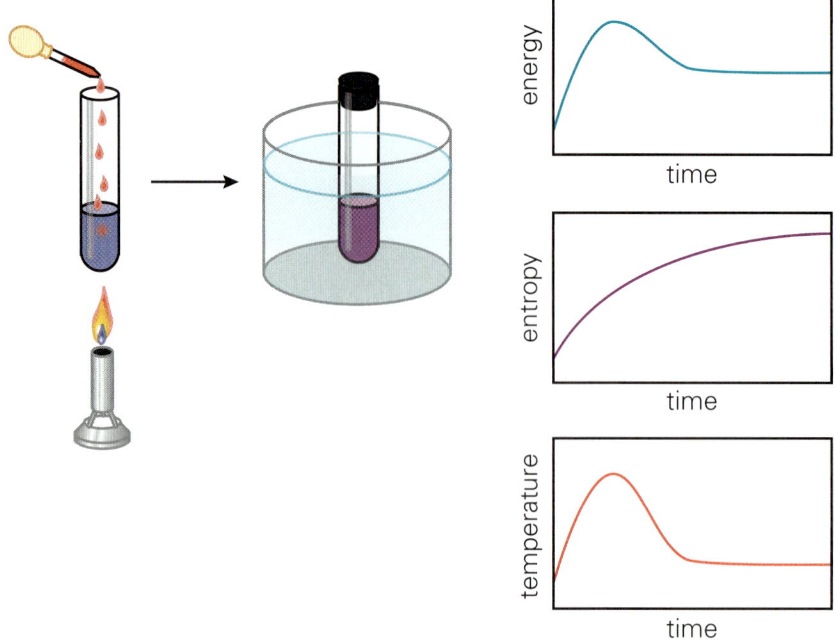


**Figure 9.2 A perturbed system moves to a new equilibrium state after the perturbation.** A system (test tube) is perturbed by adding more molecules to it (*red*) and by heating it. The system is then placed in a water bath and not perturbed further. The global properties of the system relax to new equilibrium values, which are stable as long as the system is not perturbed again.

by the second law). The free energy of a system always *decreases* when a process occurs spontaneously, and is at a minimum when the system is at equilibrium. As we shall see, the change in free energy that occurs during a process is equal to the maximum amount of work that can be extracted from the process.

# A. FREE ENERGY

## 9.1 The combined entropy of the system and the surroundings increases for a spontaneous process

Recall from our study of entropy (Chapters 7 and 8) that spontaneous change in a system occurs in the direction that increases the combined entropy of the system *and* the surroundings (**Figure 9.3**). To develop the concept of free energy, we shall convert this condition on the combined entropy of the system and the surroundings into a new condition for equilibrium that depends *only* on the properties of the system.

Most biochemical processes occur under conditions of constant pressure and temperature, and so our analysis emphasizes a reaction or process that occurs under these conditions, as depicted in **Figure 9.4**. The reaction chamber is immersed in a bath that is so large that the system is essentially held at constant temperature. The reaction chamber and the heat bath are isolated from everything else, so there

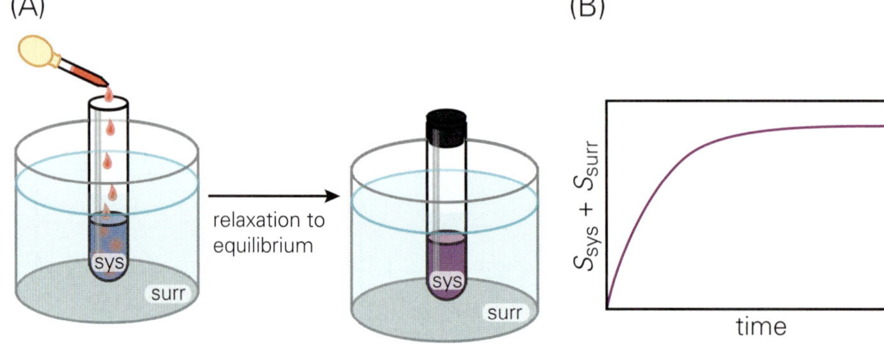


**Figure 9.3 The combined entropy of the system and the surroundings increases in a spontaneous process.** (A) A system, denoted *sys*, is perturbed from equilibrium as new material is added to it. The system relaxes to equilibrium after the perturbation. (B) The graph shows the combined entropy of the system and the surroundings (denoted *surr*) as a function of time. The relaxation towards equilibrium is accompanied by an increase in the combined entropy of the system and the surroundings, in accordance with the second law of thermodynamics.

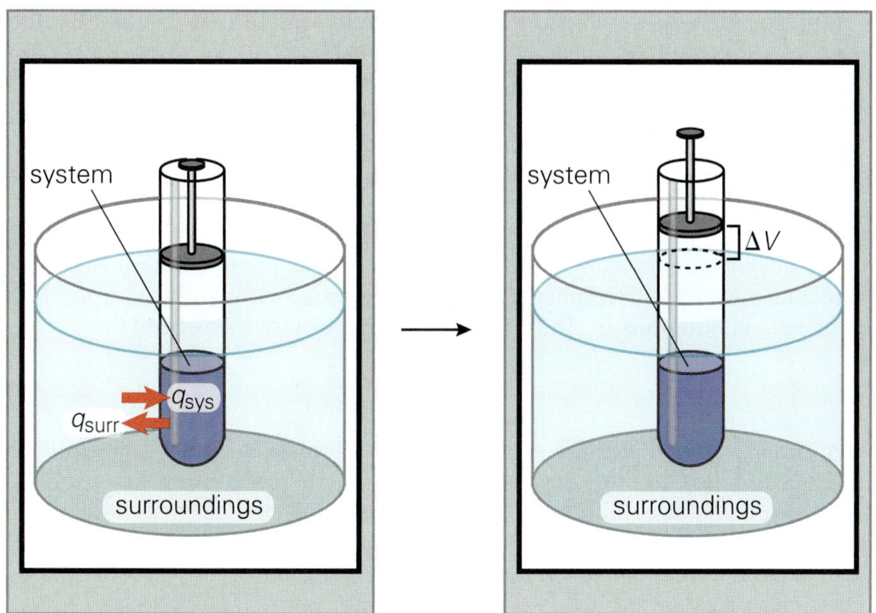


**Figure 9.4 An idealization of a system at constant temperature and pressure.** The system is shown schematically as a test tube that is capped by a frictionless piston. The diagram shows a process occurring, during which heat is exchanged between the system and the surroundings. The system expands, and the piston adjusts so as to maintain constant pressure.

is no heat transfer into or out of the combination of the chamber plus the bath. The chamber is capped by a frictionless piston, which adjusts itself so as to maintain the system at constant pressure.

The piston in the system shown in Figure 9.4 is a conceptual device that lets us calculate the expansion work done by the system during a process. In a more realistic situation, there would be no piston, and the system would be open to the atmosphere. The expansion work done by the system would still be given by the change in volume of the system multiplied by the pressure, and would correspond to the work done to push against the atmosphere.

If we denote the experimental set-up (the reaction chamber and its contents) as the "system," and everything else (the heat bath) as the "surroundings," then the condition for equilibrium is given by the second law of thermodynamics, which states that the total entropy of the system and the surroundings, $S_{total}$, has a maximal value at equilibrium (**Figure 9.5**). This means is that if the system is at

**Figure 9.5 The combined entropy of the system and the surroundings is maximal at equilibrium.** The graph shows the dependence of the total entropy, $S_{total} = S_{sys} + S_{surr}$, on the energy of the system ($U_{sys}$). At equilibrium, a small change in $U_{sys}$, denoted here by Δ, produces essentially no change in $S_{total}$. The expanded view shows the variation of $S_{total}$ for values of $U_{sys}$ that are very near the equilibrium value. The slope of the horizontal line, which is tangent to the curve describing the entropy, represents the derivative of the entropy with respect to the energy.

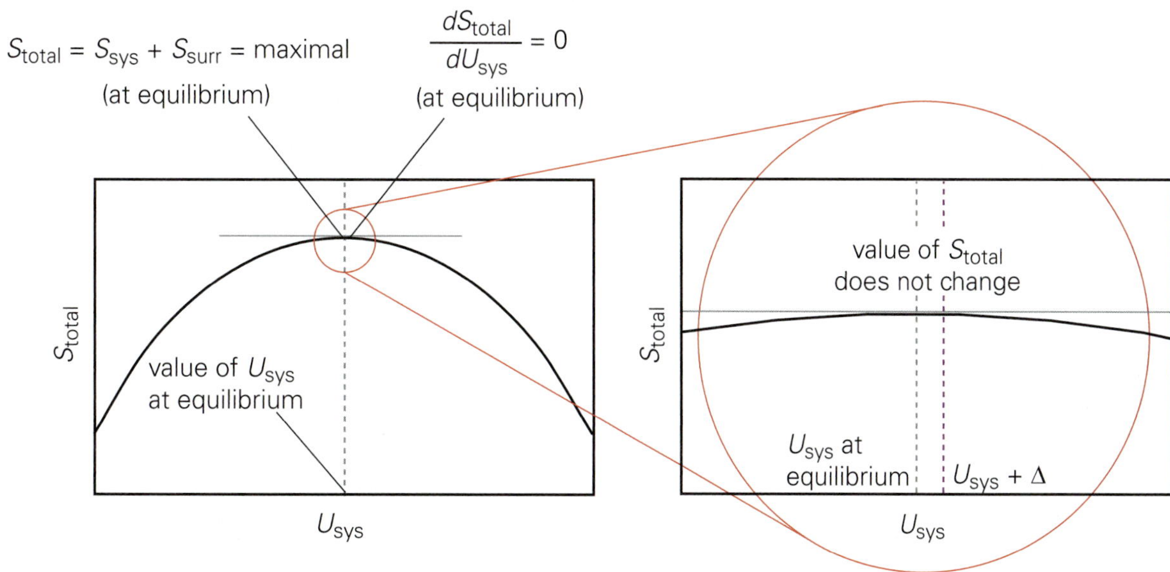

equilibrium, then the derivative of the total entropy with respect to any variable of the system (for example, the energy, $U_{sys}$) is zero:

$$\frac{dS_{total}}{dU_{sys}} = 0 \text{ (at equilibrium)} \tag{9.1}$$

This condition is illustrated in Figure 9.5, which shows that the slope of the graph of $S_{total}$ as a function of $U_{sys}$ is zero at the equilibrium point. Although the derivative shown in Equation 9.1 is that of the total entropy with respect to the energy of the system, you should recognize that, according to the second law, the derivative of the total entropy with respect to *any* variable of the system or the surroundings will be zero at equilibrium. Thus, a more general form of the second law is given by:

$$dS_{total} = 0 \text{ (at equilibrium)} \tag{9.2}$$

Equation 9.2 states that if the system is at equilibrium, then any infinitesimally small perturbation to the system will result in no change in the total entropy. If the system is *not* at equilibrium, then a spontaneous change in the system will *increase* the total entropy. This allows us to express a completely general condition on spontaneous changes in the total entropy as follows:

$$dS_{total} \geq 0 \tag{9.3}$$

If the system and the surroundings are at equilibrium, then the combined entropy is at a maximum and the right-hand side of Equation 9.3 is set to zero. If the system and its surroundings are *not* at equilibrium, then a spontaneous change will increase the combined entropies of the system and its surroundings—that is, $dS_{total} > 0$ for that change.

## 9.2 The change in entropy of the surroundings is related to the change in energy and volume of the system

Equation 9.3 defines the condition for spontaneous change in terms of the total entropy—that is, the entropy of the system *and* that of the surroundings. It would be more convenient, instead, if we could rewrite Equation 9.3 in a way that only involves system parameters. We begin by rewriting Equation 9.3 in terms of infinitesimal changes in the entropies of the system and the surroundings:

$$dS_{total} = dS_{sys} + dS_{surr} \geq 0 \tag{9.4}$$

We would like to rearrange Equation 9.4 in some way that focuses attention only on the properties of the system—that is, we want to get rid of the $dS_{surr}$ term because it is only the system that we are directly interested in. We start by equating the heat transferred to the system with the heat lost by the surroundings:

$$dq_{sys} = -dq_{surr} \tag{9.5}$$

Using the first law (that is, $dU = dq + dw$; see Equation 6.5) gives:

$$dU_{sys} - dw_{sys} = -(dU_{surr} - dw_{surr}) \tag{9.6}$$

If we assume that the expansion work done by the surroundings is equal in magnitude but opposite in sign to the work done by the system—that is, if we ignore heat dissipation, then the work terms in Equation 9.6 cancel, and we get:

$$dU_{sys} = -dU_{surr} \tag{9.7}$$

Recall from Equation 8.87 that changes in the energy, $U$, of a system can be expressed as a function of changes in other variables as follows:

$$dU = \left(\frac{\partial U}{\partial S}\right)_{V,N} dS + \left(\frac{\partial U}{\partial V}\right)_{S,N} dV + \left(\frac{\partial U}{\partial N}\right)_{V,S} dN \tag{9.8}$$

Recall also from Equation 8.83 that the rate of change of energy, $U$, with respect to entropy, $S$, is the temperature, $T$:

$$\left(\frac{\partial U}{\partial S}\right)_{V,N} = T \tag{9.9}$$

Likewise, from Equation 8.94, the rate of change of energy with respect to volume, $V$, is the negative of the pressure, $P$:

$$\left(\frac{\partial U}{\partial V}\right)_{S,N} = -P \tag{9.10}$$

If the number of molecules does not change (that is, if $dN = 0$), then by combining Equations 9.8, 9.9, and 9.10, we get:

$$dU = TdS - PdV \tag{9.11}$$

Equation 9.11 applies separately both to the system and to the surroundings. By rearranging Equation 9.11, we obtain the following expression for an infinitesimal change in the entropy of the surroundings:

$$dS_{surr} = \frac{1}{T}dU_{surr} + \frac{P}{T}dV_{surr} \tag{9.12}$$

Substituting $-dU_{sys}$ for $dU_{surr}$ (see Equation 9.7), we get:

$$dS_{surr} = \frac{-1}{T}dU_{sys} + \frac{P}{T}dV_{surr} \tag{9.13}$$

The change in the volume of the surroundings ($dV_{surr}$) has to be equal in magnitude but opposite in sign to the change in volume of the system ($dV_{sys}$):

$$dV_{surr} = -dV_{sys} \tag{9.14}$$

Substituting Equation 9.14 in Equation 9.13, we see that changes in the entropy of the surroundings can be expressed in terms of changes in the energy and the volume of the system:

$$dS_{surr} = -\frac{1}{T}dU_{sys} - \frac{P}{T}dV_{sys} \tag{9.15}$$

## 9.3 The Gibbs free energy ($G$) of the system always decreases in a spontaneous process occurring at constant pressure and temperature

Recall from Chapter 6 that the enthalpy of the system, $H_{sys}$, is given by:

$$H_{sys} = U_{sys} + PV_{sys} \tag{9.16}$$

According to Equation 9.16, an infinitesimal change in the enthalpy at constant pressure, $dH_{sys}$, is given by:

$$dH_{sys} = dU_{sys} + PdV_{sys} + V_{sys}dP = dU_{sys} + PdV_{sys} \tag{9.17}$$

The second equality in Equation 9.17 reflects the fact that under constant-pressure conditions, $dP = 0$. By combining Equations 9.15 and 9.17, we get the following expression for an infinitesimal change in the entropy of the surroundings, under constant pressure:

$$dS_{surr} = -\frac{1}{T}dH_{sys} \tag{9.18}$$

Equation 9.18 is equivalent to the thermodynamic definition of the entropy, because the change in enthalpy under constant pressure is equal to the heat transferred (Equation 7.60). By combining Equations 9.4 and 9.18, the general condition that must be satisfied for a process to occur spontaneously at constant temperature and pressure can be written as:

$$dS_{sys} + dS_{surr} = dS_{sys} - \frac{dH_{sys}}{T} \geq 0$$

$$\Rightarrow dH_{sys} - TdS_{sys} \leq 0 \tag{9.19}$$

We now define a new state function of the system, which we call the **Gibbs free energy** ($G$):

$$G = H - TS \qquad (9.20)$$

**Gibbs free energy**

The Gibbs free energy, $G$, of a system is given by:

$$G = H - TS$$

For a system at constant pressure and temperature, the value of $G$ always decreases in a spontaneous process.

All of the variables in Equation 9.20 refer to the system alone, and so we no longer use subscripts to distinguish them. The enthalpy ($H$), the temperature ($T$), and the entropy ($S$) in Equation 9.20 are all state variables, and so $G$ is also a state function. The Gibbs free energy is named after Josiah Willard Gibbs, who first introduced the ideas that led to the definition given in Equation 9.20. As we discuss below, in part C of this chapter, the change in Gibbs free energy during a process is equal to the maximum amount of non-expansion work that can be extracted from the process. The change in Gibbs free energy is therefore the amount of energy (or heat) that is "free" to be converted to work (the rest is bound up in entropy).

An infinitesimally small change in $G$ (that is, $dG$) is given by:

$$dG = dH - TdS - SdT$$

At constant temperature the value of $dT$ is zero, and so this equation reduces to:

$$dG = dH - TdS \qquad (9.21)$$

Substituting Equation 9.21 into Equation 9.19 yields:

$$dG \leq 0 \text{ (constant pressure and temperature)} \qquad (9.22)$$

Thus, a spontaneous process at constant temperature and pressure always involves a *decrease* in the Gibbs free energy of the system (that is, $dG < 0$). It follows, then, that the value of the Gibbs free energy is at a minimum (that is, $dG = 0$) at equilibrium, as shown in **Figure 9.6**.

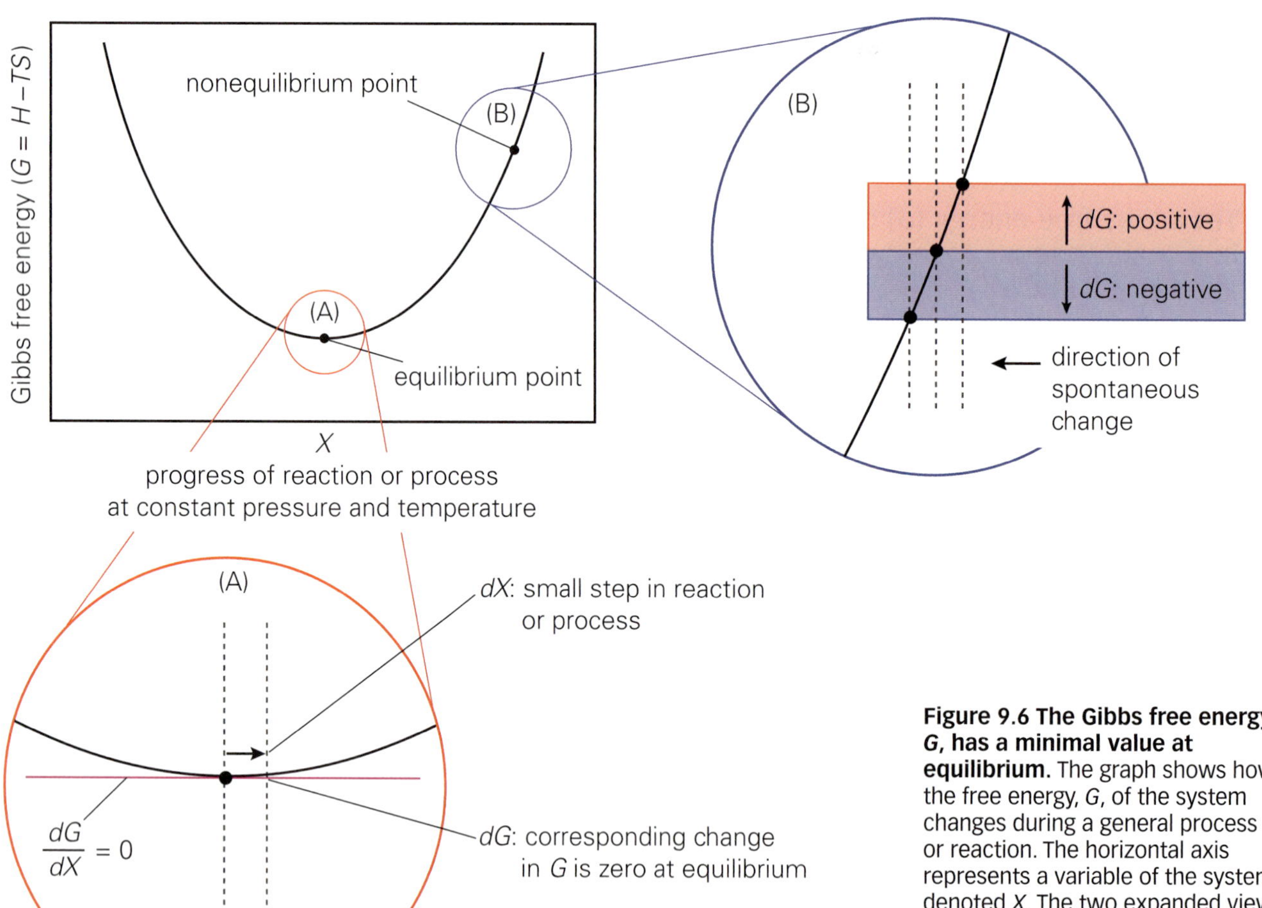


**Figure 9.6 The Gibbs free energy, $G$, has a minimal value at equilibrium.** The graph shows how the free energy, $G$, of the system changes during a general process or reaction. The horizontal axis represents a variable of the system, denoted $X$. The two expanded views show the variation of the free energy (A) when $X$ is close to the equilibrium value and (B) when $X$ is far from equilibrium.

The graph in Figure 9.6 shows how the free energy, $G$, of a system changes during a general process or reaction. The horizontal axis of the graph represents a change in the state of the system, such as during a chemical reaction or an energetic relaxation after a jump in the temperature. The equilibrium point is reached when small changes in any system variable, denoted $X$, lead to essentially no change in the free energy. That is, the derivative of the free energy, $G$, with respect to the variable $X$ is zero at equilibrium:

$$\frac{dG}{dX}=0 \tag{9.23}$$

The condition given by Equation 9.23 defines an extremum point of the function $G$—that is, the free energy is either at a maximum or a minimum when this condition holds true (**Figure 9.7**). Only the points of minimum free energy are stable. When the free energy is at a minimum, small fluctuations in the system parameters will increase the free energy and the system will relax back to the equilibrium point spontaneously.

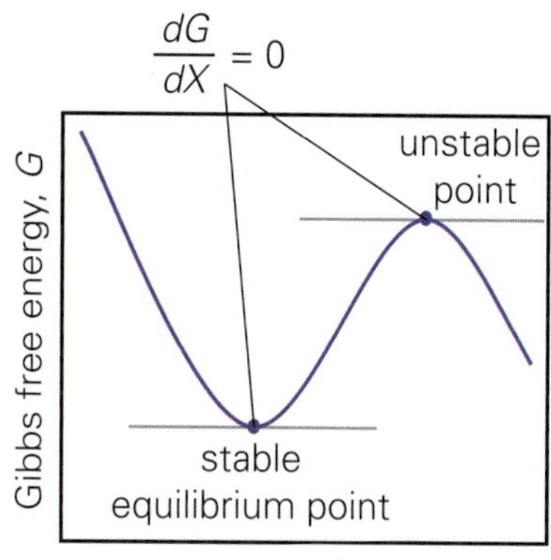


**Figure 9.7 The Gibbs free energy ($G$) is at a minimum at equilibrium.** The diagram shows the change in $G$ as a function of a system variable, $X$. The derivative of $G$ with respect to $X$ is zero when $G$ is either maximal or minimal. Values of $X$ for which the value of $G$ is maximal are unstable because small fluctuations in the system will cause the system to move away from that point.

## 9.4 The Helmholtz free energy ($A$) determines the direction of spontaneous change when the volume is constant

A slightly different expression for the free energy is obtained when a process occurs with no change in volume, as illustrated in **Figure 9.8**. Equation 9.15, which relates an infinitesimal change in the entropy of the surroundings to variables of the system, is modified as follows under constant-volume conditions:

$$dS_{\text{surr}}=-\frac{1}{T}dU_{\text{sys}}-\frac{P}{T}dV_{\text{sys}}=-\frac{1}{T}dU_{\text{sys}} \text{ (constant volume)} \tag{9.24}$$

The condition on changes in the total entropy (Equation 9.4) then becomes:

$$dS_{\text{sys}}+dS_{\text{surr}}=dS_{\text{sys}}-\frac{1}{T}dU_{\text{sys}}\geq 0 \tag{9.25}$$

Multiplying both sides of Equation 9.25 by the temperature ($T$) yields:

$$TdS_{\text{sys}}-dU_{\text{sys}}\geq 0 \tag{9.26}$$

$$\Rightarrow dU_{\text{sys}}-TdS_{\text{sys}}\leq 0 \tag{9.27}$$

We now define a new state function of the system called the **Helmholtz free energy**, named after Hermann von Helmholtz. The Helmholtz free energy is denoted $A$ (after *arbeit*, the German word for work) and is defined as follows:

$$A=U-TS \tag{9.28}$$

At constant temperature (that is, $dT=0$), an infinitesimal change in the Helmholtz free energy, $dA$, is given by:

$$dA=dU-TdS-SdT=dU-TdS \tag{9.29}$$

Comparing Equation 9.29 with Equation 9.27, we see that the change in $A$ must satisfy the following condition for a process to occur spontaneously at constant volume:

$$dA<0 \tag{9.30}$$

And, for a system at equilibrium, the value of the function $A$ must be at a minimum:

$$dA=0 \tag{9.31}$$

According to Equations 9.30 and 9.31, changes in system parameters that result in a reduction in the value of $A$ will occur spontaneously under conditions of constant volume and temperature.

Biochemists usually use the Gibbs free energy rather than the Helmholtz free energy, because biochemical reactions almost always occur under conditions of

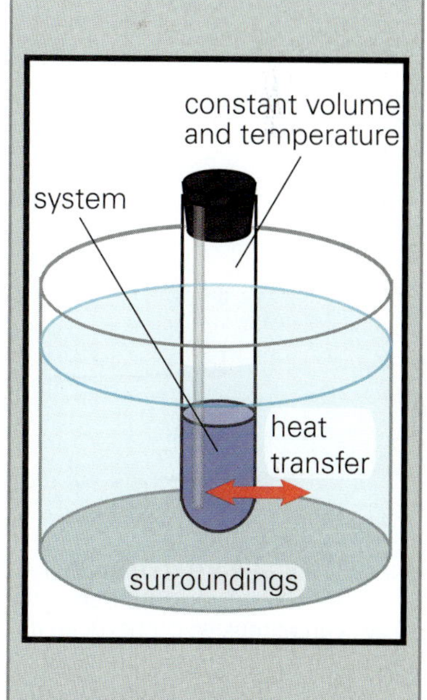


**Figure 9.8 A process occurring under conditions of constant volume and temperature.** The system is shown here schematically as a stoppered test tube. The system exchanges heat with the surroundings and is maintained at constant temperature.

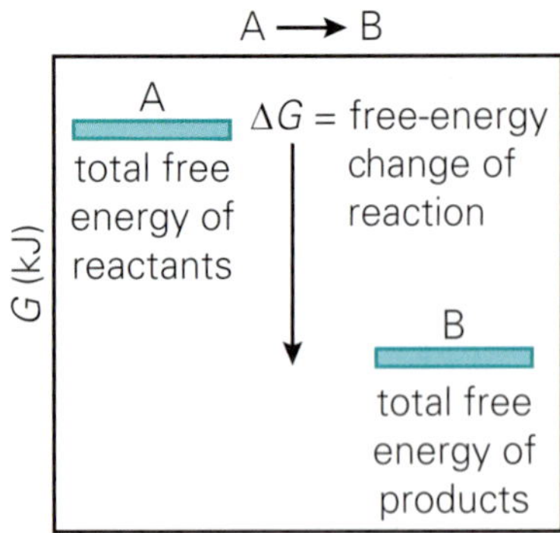


**Figure 9.9 Free-energy change of a reaction.** When type A molecules are converted to type B molecules, the free-energy change of the reaction, $\Delta G$, is the difference between the free energies of the B and A molecules.

constant pressure. For reactions that do not involve changes in the number of gas molecules, the change in volume is often negligible, and under such conditions the two forms of the free energy can be used interchangeably to determine the direction of spontaneous change.

# B. STANDARD FREE-ENERGY CHANGES

## 9.5 Standard free-energy changes are defined with reference to defined standard states

Now that we have a prescription for identifying the direction of spontaneous change in terms of the free energy (that is, $dG < 0$), we are in a position to use it to determine the direction in which a chemical reaction will proceed under given conditions. We shall defer a discussion of how to determine the equilibrium points of chemical reactions until Chapter 10, when we discuss the concept of the equilibrium constant. Here we simply introduce some bookkeeping conventions so that we can calculate and compare free-energy values correctly.

**Helmholtz free energy**

The Helmholtz free energy, *A*, of a system is given by:

$$A = U - TS$$

For a system at constant volume and temperature, the value of *A* always decreases in a spontaneous process.

Consider the hydrolysis of ATP:

$$\text{ATP} + \text{H}_2\text{O} \rightarrow \text{ADP} + \text{P}_\text{i} \tag{9.32}$$

where $P_i$ represents the phosphate ion, $[H(PO_4)^{2-}/H_2(PO_4)^{-}]$. In order to determine whether the reaction will proceed spontaneously from left to right, we need to know the value of the total change in free energy ($\Delta G$) for the reaction (**Figure 9.9**):

$$\Delta G = \int_{\text{reactants}}^{\text{products}} dG = G(\text{products}) - G(\text{reactants}) \tag{9.33}$$

The integral in Equation 9.33 indicates that we are summing over all the infinitesimal changes in free energy as the reactants are converted to products, and the value of the integral is just the difference between the free energies of the products, *G* (products), and the reactants, *G* (reactants). Writing out the free energies of the individual molecules that constitute the reactants and the products explicitly, we get:

$$\Delta G = G(\text{ADP} + \text{P}_\text{i}) - G(\text{ATP} + \text{H}_2\text{O}) \tag{9.34}$$

The free energy is an extensive property of the system because it depends on enthalpy and entropy, which are both extensive properties. This means that the values of each of the individual terms in Equation 9.34 will depend on how much ATP and water enter into the reaction, as explained in **Figure 9.10**. In order to

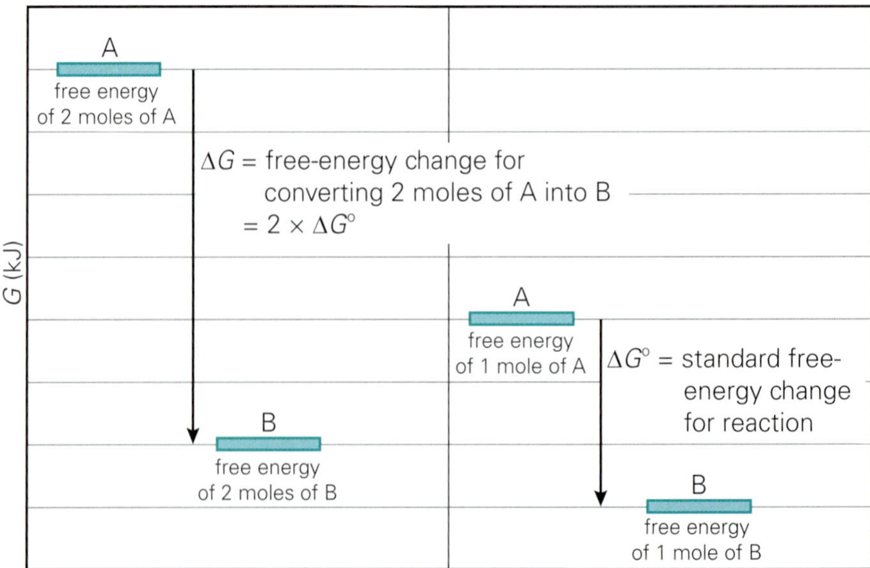


**Figure 9.10 Standard free-energy change for a reaction.** The free energy is an extensive property, so it depends on the amount of material in the system. The free energy of two moles of A-type molecules is twice that of one mole of A-type molecules. The standard free-energy change, $\Delta G^\circ$, for the reaction, A → B, is the difference in free energy between one mole of B-type molecules and one mole of A-type molecules, under standard conditions of temperature, pressure, and concentration.

provide a standard reference value for the free-energy change, we define the **molar free energy** of a molecule as the free energy of one mole of that molecule. Since the free energy of a molecule changes with temperature, pressure, and whether it is pure or in a mixture, we need to know the conditions for which a free-energy change is being reported.

By convention, free-energy changes are quoted for the **standard state**, which refers to a condition of defined concentration and pressure. Most biochemical reactions occur in aqueous solution, and the standard state is set to be a one molar (M) solution of the molecules in water and 1 atm pressure. The free-energy change that occurs upon converting a stoichiometric equivalent of reactant molecules into the stoichiometric equivalent of product molecules, all under standard conditions, is known as the **standard free-energy change**, $\Delta G^o$, of the reaction (see Figure 9.10). Standard free-energy changes are usually reported at room temperature (298 K). The value of $\Delta G^o$ is temperature dependent, and so if we are interested in the standard free-energy change at a different temperature, then we have to account for the change in $\Delta G^o$ with temperature. This is discussed further in Chapter 10.

**Standard state**

The standard state is a condition of defined molecular state, concentration, and pressure that is used as a reference for reporting free energy values. For water soluble molecules, the standard state is usually a 1 M solution of the molecule at 1 atm pressure. The standard state for a solid material is the pure solid. The standard state of water is pure water (55 M). The concentration of water is assumed to be constant at 55 M, and the concentration of $H^+$ is $10^{-7}$ M (that is, pH = 7.0). This definition of the standard state is the biochemical standard state. In other branches of chemistry, the standard state proton concentration is 1 M (that is, pH 0).

There are some important special cases in the definition of standard states. The standard state for water is pure water (55 M). For a solid material the standard state is the pure solid. The natural standard state for protons in water is that of an aqueous solution at pH 7 ($[H^+] = 10^{-7}$ M). This convention, common in biochemistry, defines the **biochemical standard state**, and is different from the convention in other branches of chemistry, where the standard states are set to be 1 M for *all* solutions, including $H^+$ in water. To remind us of this difference, standard free-energy changes in biochemistry are often denoted as $\Delta G^{o\prime}$ rather than $\Delta G^o$. We shall always use the biochemical definition of the standard state in this book, unless stated otherwise, and will therefore drop the "prime" denotation.

**Standard free-energy change, $\Delta G^o$**

$\Delta G^o$ is the change in free energy when a molar equivalent of reactants are converted into products under standard conditions of concentration. If the biochemical standard state is used (that is, pH 7.0), the standard free-energy change is denoted $\Delta G^{o\prime}$.

The standard free-energy change ($\Delta G^o$) for the hydrolysis of ATP is –28 kJ•mol$^{-1}$. The value of $\Delta G^o$ for the hydrolysis of ATP depends on the concentration of $Mg^{2+}$ ions in the solution, as well as the pH. The value quoted here is for pH 7 and a magnesium ion concentration of 100 mM. The negative value of the free-energy change means that the reaction will proceed spontaneously to the right for a 1 M solution of pure ATP in water, because the free energy of ATP and water (the reactants) is higher than the free energy of ADP and $P_i$ (the products).

The standard free-energy change is a hypothetical concept, corresponding to the complete conversion of one mole of ATP (as a 1 M solution) into one mole of ADP (as a 1 M solution). In reality, if we start the reaction with a 1 M solution of ATP, the reaction will *not* proceed to completion, but will instead come to an equilibrium point at which the concentration of ATP is not precisely zero (this effect arises from the concentration dependence of free energy, as we shall see in Section 10.8).

## 9.6 The zero point of the free-energy scale is set by the free energy of the elements in their most stable forms

If we know the molar free energies of the reactants and products, we can readily calculate the molar free-energy change of a reaction by using Equation 9.34, but how do we know these values in the first place? By convention, the molar free energy of a molecule is the standard free-energy change that results from converting stoichiometric amounts of the pure elements that are its constituents into one mole of the molecule of interest under standard conditions. This is referred to as the **standard free energy of formation** of the molecule ($\Delta_f G^o$).

We are usually interested only in differences in the values of free energies rather than the absolute values of the free energies. It therefore suffices to define a specific zero point for the free-energy scale, and to then measure all free-energy

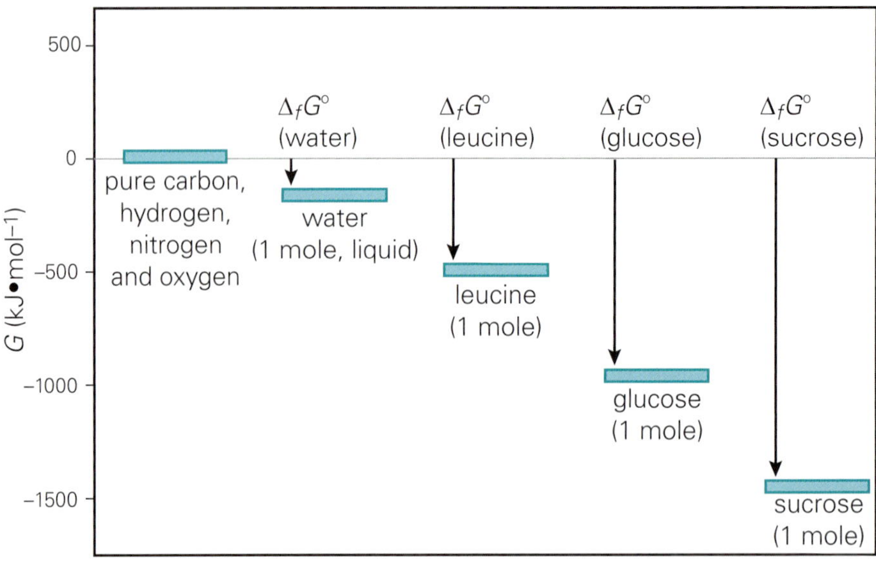


**Figure 9.11 Standard free energies of formation.** By convention, the molar free energies of elemental molecules in their standard states are set to zero. The standard free-energy change for converting the pure elemental molecules into more complex molecules, in molar stoichiometry, is known as the standard free energy of formation, $\Delta_f G^o$, for the molecule.

**Table 9.1 Standard free energies of formation of some biochemical compounds (1 atm, 298 K).**

| Compound | $\Delta G^{o\prime}$ (kJ•mol$^{-1}$) |
|---|---|
| acetate$^-$ | −369.2 |
| $CO_2$ (gas) | −394.4 |
| $CO_2$ (aqueous solution) | −386.2 |
| carbonate ion | −587.1 |
| ethanol | −181.5 |
| fructose | −915.4 |
| fructose-6-phosphate$^{2-}$ | −1758.3 |
| α-D-glucose | −917.2 |
| glucose-6-phosphate$^{2-}$ | −1760.2 |
| $H^+$ (aqueous solution) | 0.0 |
| $H_2$ (gas) | 0.0 |
| $H_2O$ (liquid) | −237.2 |
| isocitrate$^{3-}$ | −1160.0 |
| lactate$^-$ | −516.6 |
| $OH^-$ | −157.3 |
| pyruvate$^-$ | −474.5 |
| succinate$^{2-}$ | −690.2 |

(From D. Voet and J.G. Voet, Biochemistry, 3rd ed. New York: John Wiley & Sons, 2004; D.E. Metzler, Biochemistry, The Chemical Reactions of Living Cells. New York: Academic Press, 1977.)

changes relative to the zero values of this scale. By convention, the free energy of an element (for example, $N_2$, or $O_2$) in the most stable form of that element (for example, oxygen gas, $O_2$, and not ozone, $O_3$) has its free energy set to zero under standard conditions. The free energies of all other molecular forms or phases are measured relative to this set point for the zero on the free-energy scale (**Figure 9.11**). Using this convention, the free-energy change for any reaction under standard conditions is given by:

$$\Delta G^o = \sum_{\substack{\text{all}\\ \text{products}}} \Delta_f G^o(\text{product}) - \sum_{\substack{\text{all}\\ \text{reactants}}} \Delta_f G^o(\text{reactant}) \qquad (9.35)$$

The free energies of formation of a few compounds that are important in biochemistry are given in **Table 9.1**. A more extensive list of such compounds may be found in biochemistry textbooks, or in compendia such as the *Handbook of Biochemistry and Molecular Biology* (see Further Reading at the end of this chapter).

## 9.7 Thermodynamic cycles allow the determination of the free energies of formation of complex molecules from simpler ones

It is usually very difficult, if not impossible, to measure the free energy of formation of a complex molecule by converting elemental molecules directly into the complex molecule in one step. The free energy is a state function, however, and so the change in free energy for a process is independent of the path, and can be obtained by summing over the free-energy changes for any particular stepwise pathway that links reactants and products. We can break up the formation of a complex molecule into a series of intermediate reactions involving less complex molecules. If the values of $\Delta_f G^o$ for these less complex molecules can be determined experimentally, these values can then be combined to yield the value of $\Delta_f G^o$ for the more complex molecule.

We first illustrate this idea in an abstract way by considering the free energy of formation of a hypothetical molecule, denoted Z. Then, in the subsequent sections, we make these ideas concrete by discussing how the free energy of formation of a particular molecule, glucose, is determined experimentally.

Imagine that the molecule Z is formed from the elements A, B, C, and D:

$$A + B + C + D \xrightarrow{\Delta_f G^o(Z)} Z \qquad (9.36)$$

The standard free-energy change for this reaction is the standard free energy of formation of Z, $\Delta_f G^o(Z)$. The reaction depicted in Equation 9.36 may be impossible to carry out in one step in a test tube, but imagine that we can carry out the following reactions without difficulty:

$$A + B \xrightarrow{\Delta_f G^\circ(X)} X \tag{9.37}$$

$$C + D \xrightarrow{\Delta_f G^\circ(Y)} Y \tag{9.38}$$

$$X + Y \xrightarrow{\Delta G^\circ(X+Y \to Z)} Z \tag{9.39}$$

The standard free-energy changes for the reactions depicted in Equations 9.37–9.39 are noted above the arrows. Since A, B, C, and D are elements (for example, carbon or hydrogen), the standard free-energy changes for the first two reactions are the free energies of formation ($\Delta_f G^o$) of X and Y.

The free-energy changes along two different pathways in going from the elements A, B, C, and D to the complex molecule Z are illustrated using the **thermodynamic cycle** in **Figure 9.12**. A thermodynamic cycle is a set of reactions in which reactants and products are connected by two different pathways. The total free-energy change along either pathway is the same, but one pathway may be more accessible to experimental measurement than the other. Thermodynamic cycles allow free-energy changes to be determined for processes that are conceptually important but difficult to study directly. In Chapter 13, for example, we use thermodynamic cycles to determine the free-energy difference between correctly and incorrectly paired nucleotides in double-helical DNA.

**Thermodynamic cycle**

Two different pathways that connect the same set of reactants and products, with the same starting and ending conditions, define a thermodynamic cycle. The change in any state variable, such as the free energy, enthalpy, or entropy, must the same following either pathway. Thermodynamic cycles are useful when one of the pathways is experimentally accessible, but the other one is not. Experimental measurements on one pathway provide information about the other pathway. See Figure 9.12 for an example of a thermodynamic cycle.

The free-energy change for the pathway denoted 1 in Figure 9.12 (*red* arrow) is the same as that for the set of transformations denoted 2 (*green* arrow). Hence:

$$\Delta_f G^\circ(Z) = \Delta_f G^\circ(X) + \Delta_f G^\circ(Y) + \Delta G^\circ(X+Y \to Z) \tag{9.40}$$

Thus, for this simple case, we can work out the free energy of formation of Z by adding up the standard free-energy changes for the three reactions shown in Equations 9.37–9.39 (**Figure 9.13**).

Over the years, scientists have painstakingly measured and tabulated the standard free energies of formation of most of the important biological molecules (a few of these are shown in Table 9.1). From these data, it is usually possible to work out the free-energy change for a particular reaction of interest under standard conditions.

We have said nothing so far about how we actually determine the values of the free-energy changes for reactions such as those that comprise pathway 2 in Figure 9.12. To provide some idea of how this is done, we discuss the experimental determination of the free energy of formation of glucose.

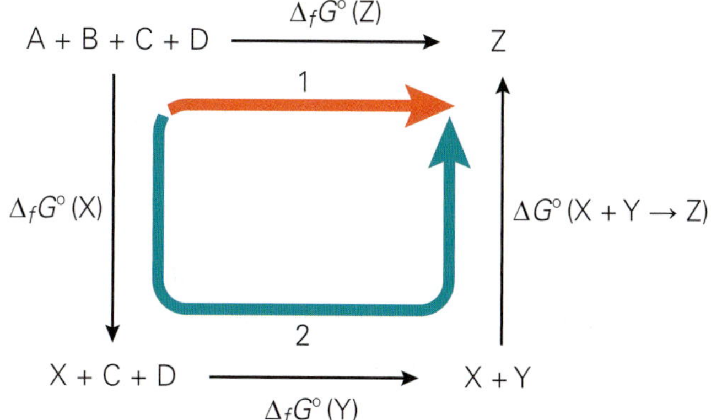


**Figure 9.12 A thermodynamic cycle representating how free energies of formation are derived.** The reaction depicted in Equations 9.36–9.39 are shown here. Compare this diagram with Figure 9.13, which shows the same free-energy changes in a different way.

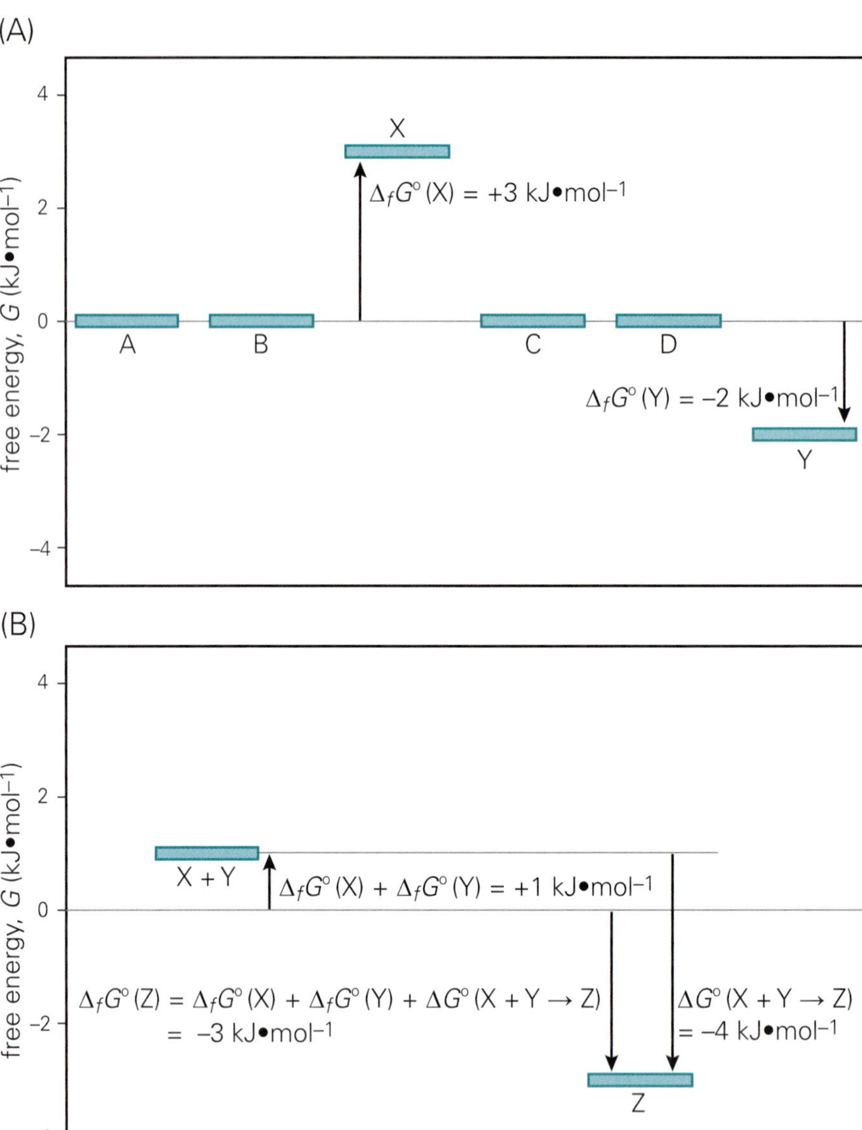


**Figure 9.13 The standard free energy of formation ($\Delta_f G^o$) of a complex molecule, Z.** (A) Z is composed of the elementary molecules A, B, C, and D. A and B react to form X, and C and D react to form Y. (B) The standard free energy of formation of Z, $\Delta_f G^o(Z)$ is the sum of the free energies of formation of X and Y and the standard free-energy change of the reaction X + Y → Z. The additivity of the free energies is illustrated in the form of a thermodynamic cycle in Figure 9.12.

## 9.8 The free energy of formation of glucose is obtained by considering three combustion reactions

Recall from Section 3.1 that glucose has the chemical formula $C_6H_{12}O_6$. The free energy of formation of glucose is therefore given by the standard free-energy change of the following reaction:

$$6\,C + 3\,O_2 + 6\,H_2 \xrightarrow{\Delta_f G^\circ\,(\text{glucose})} C_6H_{12}O_6 \tag{9.41}$$

In this reaction, graphite (the most stable form of carbon) is combined with gaseous oxygen and hydrogen to form glucose. This is not a reaction that can be carried out in the laboratory, and so we consider, instead, the following three reactions that yield glucose from the elements:

$$\begin{array}{c} 6\times\left(C + O_2 \xrightarrow{\Delta_f G^\circ\,(CO_2)} CO_2\right) \\ 3\times\left(2\,H_2 + O_2 \xrightarrow{\Delta_f G^\circ\,(H_2O)} 2\,H_2O\right) \\ 6\,CO_2 + 6H_2O \xrightarrow{-\Delta G^\circ\,(\text{glucose combustion})} C_6H_{12}O_6 + 6O_2 \\ \hline 6\,C + 3\,O_2 + 6\,H_2 \xrightarrow{\Delta_f G^\circ\,(\text{glucose})} C_6H_{12}O_6 \end{array} \tag{9.42}$$

In the first reaction, elemental carbon (graphite) combines with oxygen to yield carbon dioxide, and the standard free-energy change for this reaction is the free energy of formation of carbon dioxide, $\Delta_f G^o(CO_2)$. This is a **combustion reaction**, involving the burning of a compound in oxygen. The second reaction is the combustion of hydrogen to produce water, with a standard free-energy change equal to the free energy of formation of water, $\Delta_f G^o(H_2O)$. The third reaction is the reverse of the combustion of glucose to produce carbon dioxide and water. The free-energy change of this reaction is the negative of the standard free-energy change for glucose combustion, $\Delta G^o$.

## 9.9 Enthalpies and entropies of formation can be combined to give the free energy of formation

**Figure 9.14** shows the thermodynamic cycle formed by the chemical reactions depicted in Equation 9.42. Notice that all three reactions in pathway 2 are combustion reactions. Combustion reactions are difficult to initiate, but once initiated, they proceed essentially to completion. As we discuss in Chapter 10, one common way to determine the value of the standard free-energy change of a reaction is to determine the equilibrium constant by measuring the concentrations of the reactions and products at equilibrium. This is impractical for combustion reactions, because the concentrations of the reactants are negligible after the combustion takes place. Instead, the changes in enthalpy and entropy for the reactions are determined separately, and then combined to yield the change in free energy.

The **standard change in enthalpy** for a reaction is denoted $\Delta H^o$, and this is the change in enthalpy when one mole of reactants is converted to one mole of products under standard conditions. The **standard change in entropy** is defined similarly, and is denoted $\Delta S^o$. The **enthalpy of formation** of a compound, denoted $\Delta_f H^o$, is the difference in the enthalpy of one mole of the compound in the standard state and stoichiometric equivalents of the corresponding elements. The **entropy of formation** ($\Delta_f S^o$) is defined similarly.

From the definition of the free energy (Equation 9.20), the standard free-energy change for a reaction, $\Delta G^o$, can be written in terms of the standard changes in enthalpy and entropy as:

$$\Delta G^o = \Delta H^o - T\Delta S^o \tag{9.43}$$

The enthalpy and the entropy are state functions, so changes in their values do *not* depend on the path followed between an initial state and a final state. Just as we did for the free energy, we can construct thermodynamic cycles for the changes in enthalpy and entropy. These cycles allow us to choose the most convenient experimental route to determine the values of the changes in enthalpy and entropy.

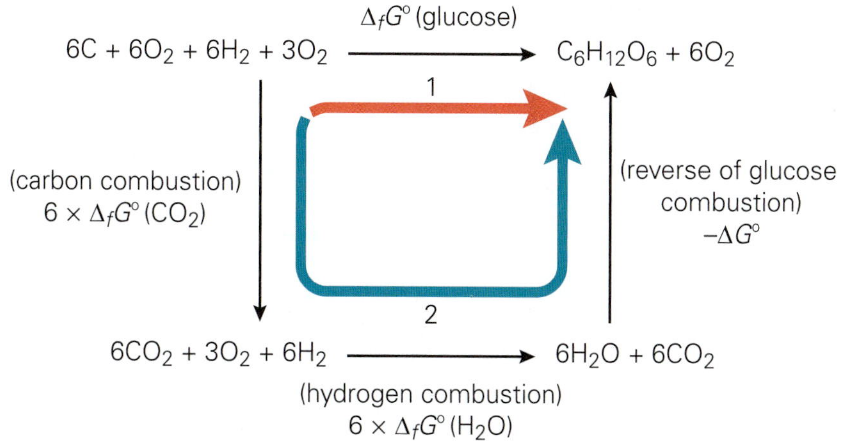


**Figure 9.14 Thermodynamic cycle for the free energy of formation of glucose.** Pathway 1 shows the formation of glucose from the elements. Pathway 2 involves three combustion reactions (one in reverse) that are combined to produce glucose from the elements.

## 9.10 Calorimetric measurements yield the standard enthalpy changes associated with combustion reactions

A thermodynamic cycle for enthalpy changes in the formation of glucose is shown in **Figure 9.15**. The figure also shows the experimentally determined values for the enthalpy changes for the three combustion steps. These enthalpy changes are combined as shown in Figure 9.15 to obtain the standard enthalpy of formation of glucose at 25°C, which turns out to be –1266 kJ•mol$^{-1}$.

The enthalpy changes shown in Figure 9.15 are obtained by using calorimeters to measure the heat released by the reaction. Recall that the change in enthalpy for a process occurring at constant pressure is equal to the heat transferred during the process (Section 6.4). An instrument known as a **bomb calorimeter** is used to determine the amount of heat released when a defined amount of a compound undergoes complete combustion. A bomb calorimeter is simply a chamber enclosed by a thick metal casing. The compound of interest is inserted into the chamber, which is then sealed and pressurized with oxygen. Combustion of the compound is initiated by an electric discharge, and the heat released is measured. Appropriate corrections for the changes in pressure that occur during the reaction are made, and the standard change in enthalpy for the reaction is obtained.

## 9.11 The entropy of formation of a compound is derived from heat capacity measurements

The standard entropy of formation of glucose, $\Delta_f S^o$, is given by:

$$\Delta_f S^o(\text{glucose}) = S^o(\text{glucose}) - \left(6 \times S^o(\text{carbon}) - 12 \times S^o(\text{hydrogen}) - 6 \times S^o(\text{carbon})\right) \quad (9.44)$$

where the $S^o$ values are the standard molar entropies of glucose and the elements. It turns out that these values have been determined experimentally, and so the entropy of formation of glucose can be calculated directly, without going around a thermodynamic cycle as we did for the enthalpy.

The thermodynamic definition of the entropy (Equation 7.60) tells us that the heat transferred during an infinitesimally small step in a reversible process, $dq_{\text{rev}}$, is related to the change in entropy, $dS$, in the following way:

$$dS = \frac{dq_{\text{rev}}}{T} \quad (9.45)$$

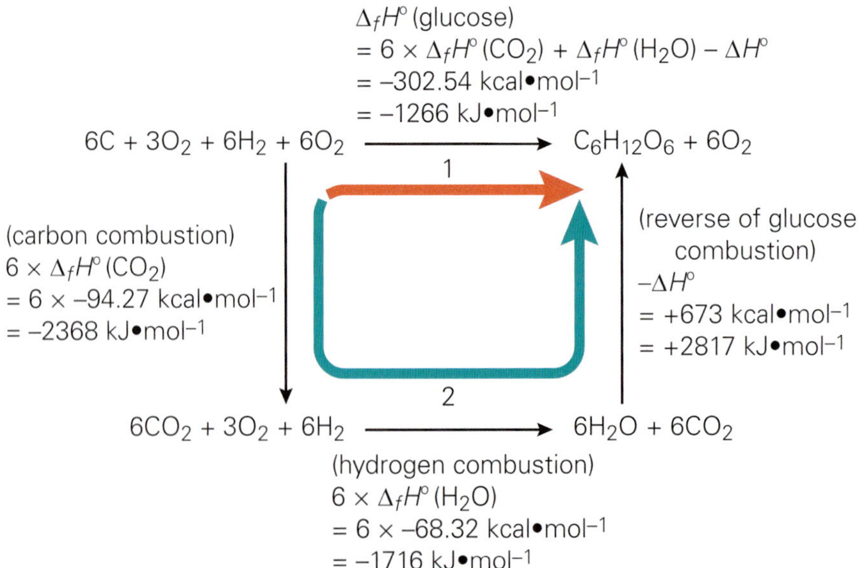


**Figure 9.15 Thermodynamic cycle for the standard enthalpy of formation of glucose at 298 K.** Experimentally determined values for the standard enthalpy changes for the combustion of carbon, hydrogen, and glucose (in reverse) are shown. The standard enthalpy of formation of glucose is obtained by combining these values in the proper stoichiometric ratios. (The data shown here are as reported by G.S. Parks, K.K. Kelley, and H.M. Huffman, *J. Am. Chem. Soc.* 51: 1969–1973, 1929. This paper uses the so-called "15°C calorie," which is based on the heat capacity of water at 15°C and is equal to 4.186 J.)

If the process is carried out at constant pressure, then the heat transferred is simply the change in enthalpy of the system, and is independent of the path connecting the initial and final states of the system:

$$dS = \frac{dH}{T} \text{ (for a process at constant pressure)} \quad (9.46)$$

We can integrate Equation 9.46 to obtain the change in entropy as the temperature is changed from an initial value, $T_1$, to a final value, $T_2$.

$$\Delta S^o = S^o(T_2) - S^o(T_1) = \int_{T_1}^{T_2} \frac{dH^o}{T} = \int_{T_1}^{T_2} \frac{C_P^o}{T} dT \text{ (at constant pressure)} \quad (9.47)$$

In Equation 9.47, the superscripts refer to the fact that we are considering one mole of the substance. We have used the fact that the change in enthalpy, $dH^o$, is equal to $C_P^o dT$, where $C_P^o$ is the molar heat capacity of the system at constant pressure (see Equation 6.22). Thus, if we measure the heat capacity at constant pressure at a number of temperatures, we can use Equation 9.47 to calculate the change in entropy between the two temperatures.

It is assumed that the entropy of a pure substance is zero when the temperature is absolute zero on the Kelvin scale. This principle is sometimes referred to as the **third law of thermodynamics** and, as a consequence, the integral from 0 K to any temperature, $T_2$, is equal to the entropy of the substance at the temperature $T_2$:

$$\Delta S = S(T_2) - S(T = 0\,\text{K}) = S(T_2) = \int_0^{T_2} \frac{C_P}{T} dT \text{ (at constant pressure)} \quad (9.48)$$

**Figure 9.16** shows experimental measurements of the molar heat capacity of an aliphatic substance ($C_P^o$; that is, the heat capacity per mole) between 20 K and 120 K. The experimentally determined data points are fit by a polynomial function, which is used to extrapolate the values of the heat capacity down to 0 K. The integral of $C_P^o/T$ from 0 K to 298 K (25°C), using a combination of extrapolated values and actual data, yields the molar entropy of the substance at 298 K. Heat capacity measurements, such as the ones shown in Figure 9.16, have been used to determine that the standard entropy of formation of glucose is –1.180 kJ•K$^{-1}$•mol$^{-1}$ at 298 K. If the substance undergoes a phase transition at temperature $T_{pt}$, then the associated entropy change calculated from the heat associated with the transition, $q_{pt}$, must also be added ($\Delta S_{pt} = q_{pt}/T_{pt}$).

Once the enthalpy and entropy of formation are both known, Equation 9.43 can be used to derive the free energy of formation of the substance. In the case of glucose at 298 K, $\Delta_f H^o$ = –1266 kJ•mol$^{-1}$ (see Figure 9.15) and $\Delta_f S^o$ = –1.180 kJ•K$^{-1}$•mol$^{-1}$ (**Figure 9.17**), and so:

$$\Delta_f G^o = -1266 - 298 \times (-1.18) = -914.4 \text{ kJ•mol}^{-1} \quad (9.49)$$

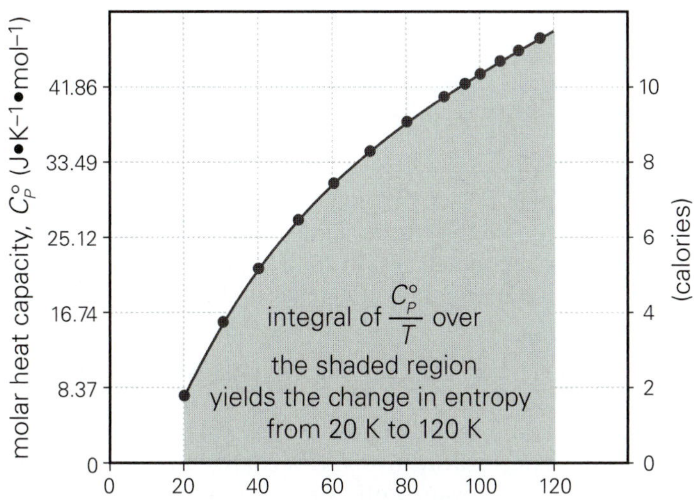


**Figure 9.16 The heat capacity of an aliphatic compound.** Shown here are experimental measurements (circles) of the molar heat capacity of a substance ($C_P^o$; that is, the heat capacity per mole) between 20 K and 120 K. The *black* line is a polynomial fit of the data, which is used to extrapolate the value of the heat capacity down to 0 K (the extrapolation is not shown here). The integral of $\frac{C_P^o}{T}$ over a temperature range (for example, the shaded region) gives the change in entropy over that temperature range. As in Figure 9.15, the calorie unit shown here is the so-called "15°C calorie," which is equal to 4.186 J. (Adapted from G.S. Parks, K.K. Kelley, and H.M. Huffman, *J. Phys. Chem.* 33: 1802–1805, 1929.)

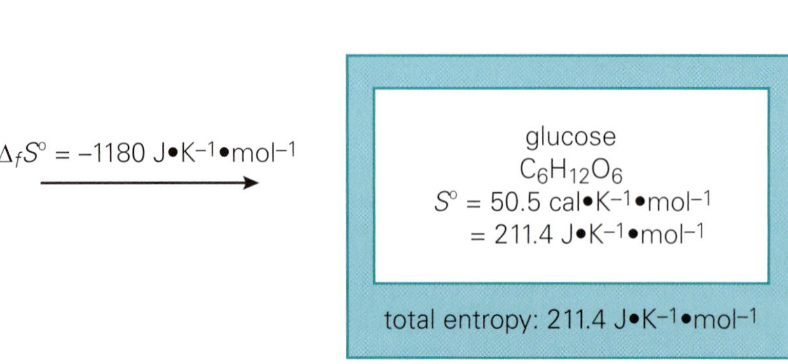


**Figure 9.17 The standard entropy of formation of glucose at 298 K, derived from heat capacity measurements such as those shown in Figure 9.16.** The standard molar entropies of the elements from which glucose is formed are shown on the left, multiplied by the appropriate stoichiometries. The molar entropy of glucose is shown on the right. The standard entropy of formation of glucose is the difference between the entropy of glucose and the total entropy of the elements. The original data are reported in 15°C calorie units (shown here, and see legend to Figure 9.15). (Data from G.S. Parks, K.K. Kelley and H.M. Huffman, *J. Am. Chem. Soc.* 51: 1969–1973, 1929.)

Note that this value for the standard free energy of formation of glucose (–914.4 kJ•mol$^{-1}$) is for pure, solid glucose. If one considers, instead, the free energy of formation of glucose in aqueous solution, one has to take into account the free-energy change that occurs upon dissolving glucose in water. The value of this free-energy change can be obtained by measuring the solubility of glucose in water, and turns out to be very small (about –2 kJ•mol$^{-1}$ for a 1 M solution of glucose). Thus, the standard free energy of formation of a standard state solution of glucose is –917 kJ•mol$^{-1}$, which is the value reported in Table 9.1. Most of the free energy of glucose is "stored" in the covalent bonds and, in comparison, the free energy of its interaction with water is very small.

# C. FREE ENERGY AND WORK

## 9.12 Expansion work is not the only kind of work that can be done by a system

In Section 6.2 we introduced the idea that energy can enter or leave a system in two ways, as *heat* and *work*. Recall that heat involves a chaotic movement of molecules in the system and the surroundings, with energy being transferred by random collisions between molecules. In contrast, energy transfer through work involves the directional movement of molecules in the system and the surroundings, as illustrated in Figure 6.4.

All of our discussion so far has focused exclusively on **expansion work**. If the system undergoes a change in volume during a process, then work is done against the external pressure, as shown in **Figure 9.18**. The system can, however, do other kinds of work, without necessarily involving a change in volume. **Table 9.2** lists some examples of non-expansion work that are important in molecular processes. These include **electrical work**, in which charges move against a gradient of

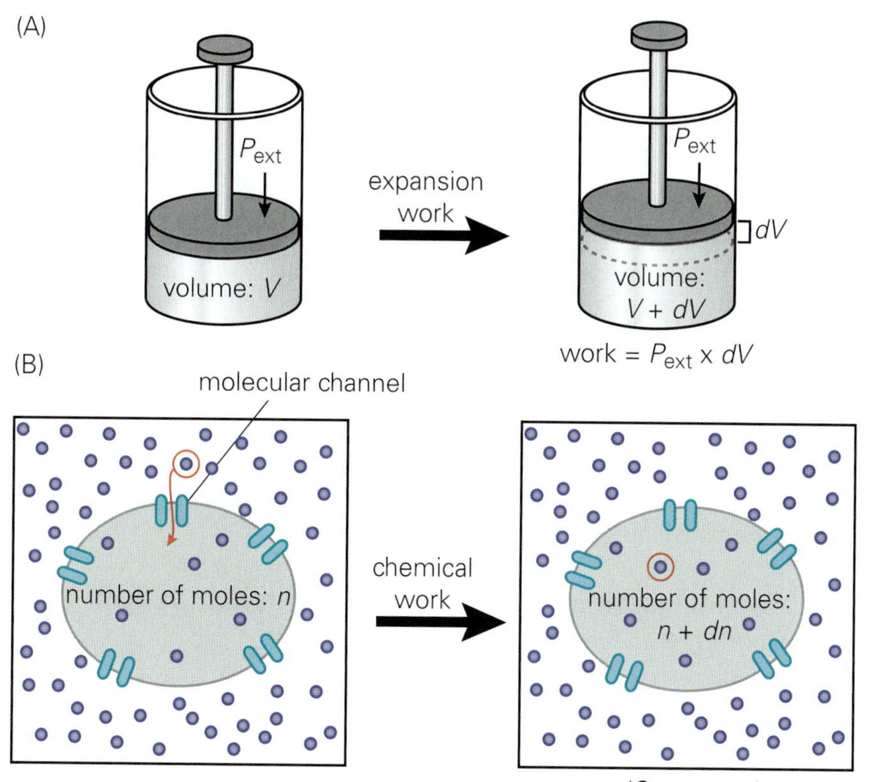


**Figure 9.18 Comparison of expansion and chemical work.** (A) The expansion of a piston. The work done in an infinitesimal step during the expansion is given by $P_{ext} \times dV$. (B) An example of chemical work, which involves the movement of molecules into a cell, along a concentration gradient. The amount of chemical work is given by the change in free energy, $dG$, for the movement of a small number of molecules into the cell. This change in free energy is given by $\Delta\mu \times dn$. As explained in Chapter 10, the term $\Delta\mu$ is the difference in chemical potential for molecules inside and outside the cell, and $dn$ is an infinitesimal change in the number of moles of molecules inside the cell.

electrical potential (see Chapter 11), and **chemical work**, which involves changes in the free energy of the system that result from changes in the number of molecules.

The general definition of work involves the product of force and displacement. As explained in Section 6.9, the work, $dw$, done in an infinitesimal step of a process in which there is a displacement, $dr$, against a force, $F$, is given by:

$$dw = F\,dr \tag{9.50}$$

Note that the sign of the force will depend on whether the movement is against the force or aligned with it. The total work done during a process is obtained by

**Table 9.2 Different kinds of work that can be done by a system.**

| Type of work | Intensive variable | Extensive differential | Expression for work |
|---|---|---|---|
| General | force, $F$ | change in distance, $dr$ | $w = \int F\ dr$ |
| Expansion | pressure, $P$ | change in volume, $dV$ | $w = \int P\ dV$ |
| Electrical | voltage difference, $\Delta E$ | change in charge, $dq$ | $w = \int \Delta E\ dq$ |
| Surface | surface tension, $\gamma$ | change in surface area, $dA$ | $w = \int \gamma\ dA$ |
| Stretching | tension, $\tau$ | change in length, $dl$ | $w = \int \tau\ dl$ |
| Chemical | chemical potential difference, $\Delta\mu$ | change in number of moles of the molecule, $dn$ | $w = \int \Delta\mu\ dn$ |

(Adapted from D.S. Eisenberg and D.M. Crothers, Physical Chemistry: With Applications to the Life Sciences. Menlo Park, CA: Benjamin/Cummings, 1979. With permission from Benjamin/Cummings.)

integrating Equation 9.50:

$$w = \int F \, dr \tag{9.51}$$

As you can see, this is the expression given for the general form of work in the first entry in Table 9.2. The expression for the work involves the product of an intensive variable (force) and the differential of an extensive variable (distance). Each of the examples of different kinds of work listed in Table 9.2 also involves the product of an intensive variable (for example, the pressure or the electrical potential) and the differential of an extensive variable (for example, volume or amount of charge). The change in the extensive variable defines the directionality of the energy transfer underlying the work (for example, whether a rubber band stretches or contracts).

The first three examples of non-expansion work in Table 9.2 are familiar to us from the physics of macroscopic objects (that is, electric potential, surface tension, and stretching), and can be thought of as particular instances of the general idea of displacement against a force. Chemical work is quite different, however, and can be understood only in molecular terms.

## 9.13 Chemical work involves changes in the numbers of molecules

To better understand what is meant by chemical work, study Figure 9.18B. Chemical work involves changes in the numbers of molecules of a certain species, such as in a chemical reaction or during transport across a concentration gradient. The latter process is illustrated in Figure 9.18B, in which the system consists of a cell with molecular channels in the membrane. In this example, chemical work refers to the movement of molecules into the cell through these channels, along a concentration gradient (there are more molecules outside the cell than inside).

Now consider what happens to the free energy when a small number of molecules move from outside the cell to the inside. The number of moles of molecules inside the cell changes from $n$ initially to $n + dn$ after the movement. The variable corresponding to a force that enters into the expression for the chemical work associated with this movement is the difference in chemical potential, $\Delta\mu$, between molecules outside the cell and inside. We defer a detailed explanation of chemical potential to Chapter 10, and for the present it is sufficient to understand that the chemical potential of a type of molecule is simply the free energy of one mole of these molecules under the specified conditions. The change in free energy, $dG$, associated with the movement of $dn$ moles of these molecules is therefore $\Delta\mu \times dn$. The net change in free energy as molecules move into the cell is given by:

$$\Delta G = \int \Delta\mu \, dn \tag{9.52}$$

As we shall see in the following sections, there is a close relationship between the work done during a process and the change in free energy associated with the process. Because of the similarity between Equation 9.52 and the general expression for work (Equation 9.51), the change in free energy associated with changes in the numbers of molecules is commonly referred to as **chemical work**. You should be aware, however, that chemical work is not the same as the other kinds of work described in Table 9.2, because it does not involve the movement of molecules against a physical force.

## 9.14 The decrease in the Gibbs free energy for a process is the maximum amount of non-expansion work that the system is capable of doing under constant pressure and temperature

In our earlier derivation of the condition that the free energy must decrease for a spontaneous process, we assumed that the system is only capable of expansion work (see Section 9.3). Now let us see what condition is imposed on the free energy when the system is capable of other forms of work, such as chemical work. To keep things simple, we shall consider a process that occurs reversibly, and under constant temperature and pressure.

According to the second law of thermodynamics, the general condition that the process must satisfy is:

$$dS_{sys} + dS_{surr} \geq 0 \qquad \text{(see Equation 9.4)}$$

For a reversible process, the change in entropy of the surroundings can be expressed in terms of the heat transferred to the surroundings:

$$dS_{surr} = \frac{dq_{surr}}{T} = -\frac{dq_{sys}}{T} \tag{9.53}$$

We use the first law of thermodynamics to express the heat transferred to the system in terms of the change in energy and the work done by the system (see Equation 6.5):

$$\begin{aligned} dq_{sys} &= dU_{sys} - dw_{sys} \\ &= dU_{sys} - (-PdV_{sys}) - dw_{non\text{-}exp} \\ &= dU_{sys} + PdV_{sys} - dw_{non\text{-}exp} \\ &= dH_{sys} - dw_{non\text{-}exp} \end{aligned} \tag{9.54}$$

In Equation 9.54, we have introduced the possibility that the system does both expansion work ($-PdV$) and non-expansion work ($dw_{non\text{-}exp}$). Recall from Chapter 6 (see Figure 6.5) that if the system does work *on* the surroundings (that is, if $\Delta V$ is positive), then the work has a negative sign. Likewise, for non-expansion work, the negative sign in front of $dw_{non\text{-}exp}$ means that the second term in Equation 9.54 is positive if the system does work (some of the heat taken up by the system goes towards the work done). We have also equated $dU + PdV$ to an infinitesimal change in the enthalpy, $dH$ (see Equation 9.17 and note that pressure is constant).

Combining Equations 9.53 and 9.54, we get:

$$dS_{surr} = -\left(\frac{1}{T}\right)\left(dH_{sys} - dw_{non\text{-}exp}\right) \tag{9.55}$$

Substituting this expression for $dS_{surr}$ into the expression for the second law of thermodynamics (Equation 9.4) gives:

$$dS_{sys} + dS_{surr} = dS_{sys} - \left(\frac{1}{T}\right)\left(dH_{sys} - dw_{non\text{-}exp}\right) \geq 0 \tag{9.56}$$

Multiplying both sides of Equation 9.56 by the temperature, $T$, we get:

$$\begin{aligned} &TdS - \left(dH - dw_{non\text{-}exp}\right) \geq 0 \\ &\Rightarrow TdS - dH + dw_{non\text{-}exp} \geq 0 \\ &\Rightarrow dH - TdS - dw_{non\text{-}exp} \geq 0 \\ &\Rightarrow dG - dw_{non\text{-}exp} \leq 0 \end{aligned} \tag{9.57}$$

In the steps leading to Equation 9.57 all of the variables refer to the system. Adding $dw_{non\text{-}exp}$ to both sides of Equation 9.57, we obtain the following relationship between the change in free energy and the non-expansion work done:

$$dG \leq dw_{non\text{-}exp} \tag{9.58}$$

At first glance it looks as if Equation 9.58 is telling us that the work done by the system ($dw$) is greater than the change in the free energy ($dG$). But, as we discussed earlier, if the system does work *on* the surroundings (that is, if work is extracted from the system), then $dw$ is a *negative* number (**Figure 9.19**A). This must mean that the absolute magnitude of $dw$ is less than or equal to the magnitude of $dG$ when work is extracted from the system. Integrating over the whole process, the total work done, $w$, and the net change in free energy, $\Delta G$, must satisfy the following relationship for a spontaneous process:

$$|w_{non\text{-}exp}| \leq |\Delta G| \tag{9.59}$$

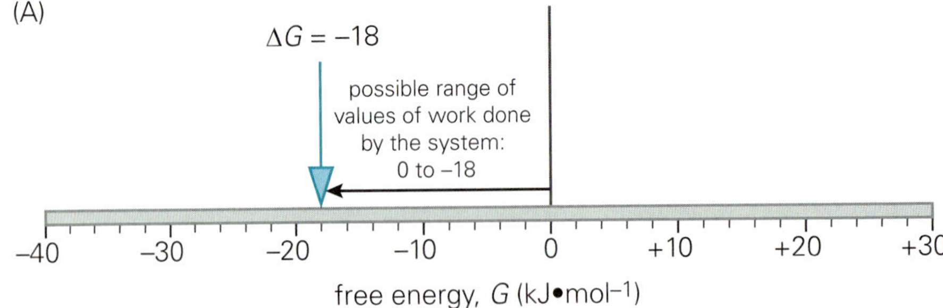


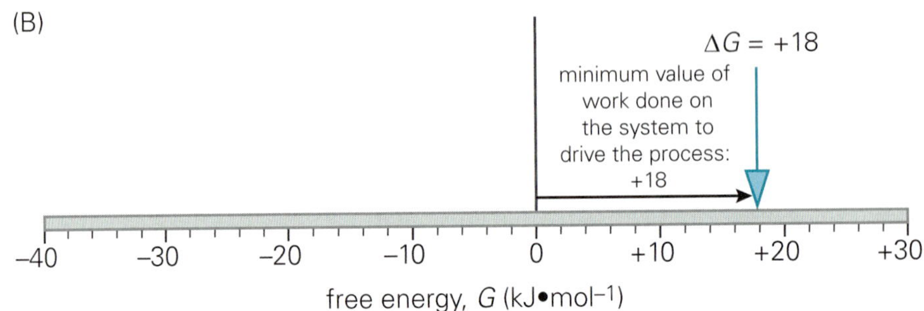


**Figure 9.19 Comparing the values of the free-energy change, ΔG, and the non-expansion work done.** (A) The work done is driven by a process with a negative change in free energy ($\Delta G = -18$ kJ•mol$^{-1}$). The possible values of the work done by the process range from 0 to −18 kJ•mol$^{-1}$. (B) A process occurs with an unfavorable change in free energy ($\Delta G = +18$ kJ•mol$^{-1}$). Although the process will not happen spontaneously, it can be made to happen if at least 18 kJ•mol$^{-1}$ of work is done on the system by the surroundings.

Thus, the maximum possible non-expansion work that can be extracted from the system (that is, when the system does work *on* the surroundings) is given by the magnitude of the change in free energy, $\Delta G$, for the process driving the work. Work can be extracted from the system only if the free-energy change associated with the process is negative. If the free-energy change for a process is *greater* than zero, it will not happen spontaneously, and work cannot be extracted from it. The process can, however, be made to happen if work is done on the system by the surroundings, as illustrated schematically in Figure 9.19B. We provide specific examples of these two situations in the next two sections.

## 9.15 The coupling of ATP hydrolysis to work underlies many processes in biology

From the discussion in Section 9.14, you should now understand that the change in the free energy of a process is equal to the maximum amount of work that can be extracted from that process. Physical work, such as the compression of a piston, or the directed movement of an object, is visualized readily and the amount of work done can be calculated using the equations given in Table 9.2. One example of physical work at the cellular level is the action of molecular motors. Kinesin, which was introduced at the beginning of Chapter 6 and is discussed further in Section 17.26, is one such molecular motor, and it moves vesicles along microtubule tracks inside a cell (**Figure 9.20**). The work done by kinesin is given by the displacement ($\Delta r$) multiplied by the resistive force ($F$) due to friction and viscosity, as shown in Figure 9.20A. The movement of the vesicle is powered by the hydrolysis of ATP within the motor domain of the kinesin protein, with a change in free energy, $\Delta G_{ATP}$, that depends on the conditions under which the motor is operating (see Figure 9.20B). The amount of work done is limited by the value of $\Delta G_{ATP}$.

The generation of chemical work is particularly important for cellular function. In Section 4.39, we had discussed how proteins known as **active transporters** are able to move molecules across the cell membrane and against a concentration gradient. For example, nutrients are concentrated within cells by proteins, such as the maltose transporter, which hydrolyzes two molecules of ATP for every molecule of maltose moved across the membrane (see Section 4.43).

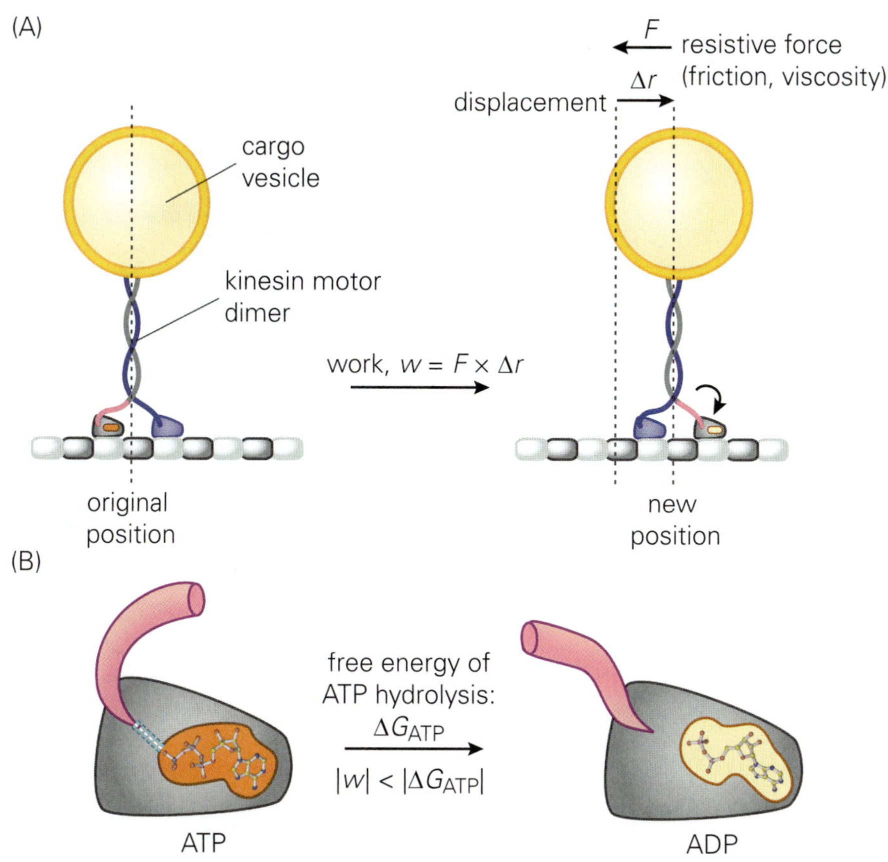


**Figure 9.20 An example of physical work.** (A) A kinesin motor protein moves a cargo vesicle along microtubule track (refer to Figure 6.1 for more information). The work done is given by the displacement ($\Delta r$) multiplied by the resistive force due to friction and viscosity ($F$). (B) The work done is coupled to the hydrolysis of ATP by the motor domain. The value of the free-energy change for hydrolysis of ATP, $\Delta G_{ATP}$, sets a limit on the amount of work that can be done.

A schematic representation of ATP-coupled active transport is shown in **Figure 9.21**, in which a cell is indicated by a purple oval. A number of nutrient molecules are also shown, both inside and outside the cell. The active transport process moves nutrient molecules from outside the cell to the inside, generating a concentration gradient across the cell membrane. The net movement of molecules is in one direction, from inside to outside, which is sometimes referred to **vectorial transport**. As we discuss in Section 10.4, the movement of molecules into the cell increases the chemical potential of these molecules inside the cell. Even

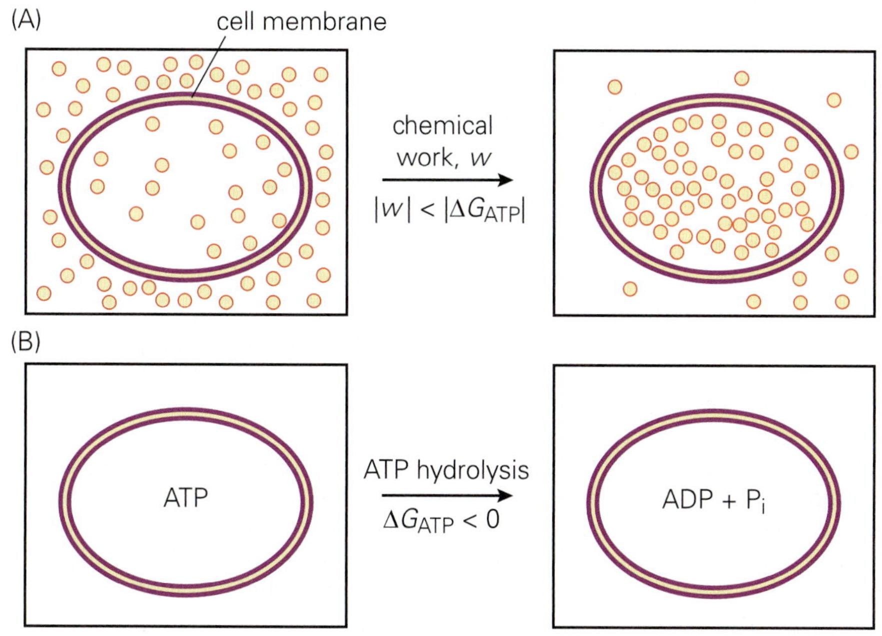


**Figure 9.21 The coupling of chemical work to ATP hydrolysis.** (A) The process of concentrating molecules, such as nutrients, inside the cell is an example of chemical work. (B) The work done is coupled to the hydrolysis of ATP by transporter proteins, such as the one illustrated in Figure 9.22. The free-energy change upon ATP hydrolysis, $\Delta G_{ATP}$, limits the extent to which molecules can be moved against a concentration gradient.

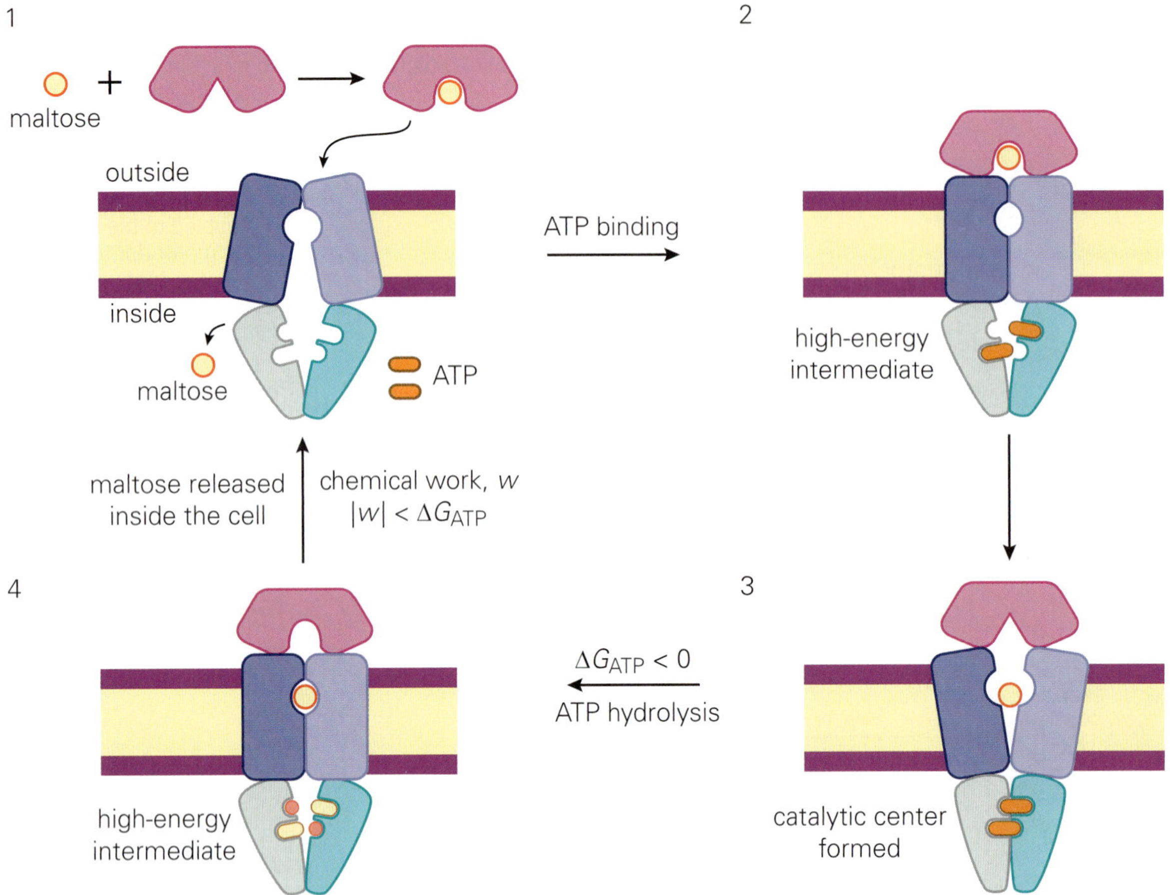


**Figure 9.22 The movement of maltose molecules into a cell by the maltose transporter.** Consult Figures 4.84, 4.85, and 4.86 for a description of the structure of this transporter. The maltose transporter (*blue*) initially has the maltose binding site empty and open to the interior of the cell (state 1). ATP binds to the cytoplasmic ATPase domains, and a maltose-loaded binding protein (*magenta*) binds to the exterior face of the transporter. This switches the transporter to state 2, which relaxes to state 3. The maltose molecule is now released from the binding protein into the interior of the transporter, and the catalytic center of the ATPase domains are properly configured for catalysis. ATP hydrolysis then occurs, which destabilizes state 3, converting the complex to state 4, which then releases ADP and maltose as it returns to state 1. (Adapted from B.H. Shilton, *Biochim. Biophys. Acta* 1778: 1772–1780, 2008.)

though the molecules do not move against a physical force, the vectorial nature and the increase in chemical potential during transport provides an analogy to work against a physical force.

The chemical work done in moving the nutrient molecules into the cell is driven by the hydrolysis of ATP by the transporter proteins. The hydrolysis of ATP must be tightly coupled to the transport process, or else the energy stored in ATP will be dissipated. In the case of the maltose transporter, this coupling is accomplished by a series of conformational changes triggered by ATP and the maltose-binding protein, as discussed in Sections 4.43 and 4.44, and as illustrated in **Figure 9.22**.

The cycle of maltose transport begins with the empty transporter (state 1 in Figure 9.22). The first step in the cycle is the binding of ATP and the maltose-loaded binding protein to the inner and outer surfaces of the cell membrane, respectively. These interactions help stabilize a high-energy conformation of the transporter (denoted 2 in Figure 9.22), which relaxes subsequently to a conformation in which maltose moves into the interior of the transporter. This movement of maltose is coupled to the structural changes that result in the formation of a properly configured catalytic center in the ATP-binding domains (state 3; compare Figure 4.86C). Although ATP is bound prior to this step, the active site of the catalytic center is not properly formed, and ATP hydrolysis is prevented. Once the transporter reaches state 3, it can catalyze ATP hydrolysis, resulting in the formation of ADP. The loss of interactions with ATP destabilizes the state 3 conformation, which converts to state 4. This conformation is also unstable, and it relaxes by releasing maltose to the inside of the cell and the empty binding protein to the outside. The transporter is now back at the initial state, so it can bind ATP again and transport more maltose molecules.

The ATP hydrolysis step is critical for driving the cycle in one direction (the forward direction shown in Figure 9.22), because the transporter cannot easily go back from state 4 to state 3 once ATP is hydrolyzed, and ADP and maltose are released. But the cycle cannot proceed without limit. At some point, the concentration of maltose inside the cell become so high that it is not readily released in state 4. If this is the case, then the cycle will tend to move backwards from state 4 to state 3, and the active transport process will stop. The limiting condition is reached when the work done in moving maltose into the cell exceeds the value of $\Delta G_{ATP}$.

Many important transporters in the cell couple the movement of ions across the membrane to the hydrolysis of ATP. Such transporters do electrical work in addition to chemical work and are discussed in Chapter 11.

## 9.16 The synthesis of ATP is coupled to the movement of ions across the membrane, down a concentration gradient

If we study the cycle shown in Figure 9.22, it is reasonable to suppose that, if one began with an extremely high concentration of maltose inside the cell, then we might be able to run the cycle backwards, thereby converting the maltose transporter into an ATP synthesis machine. While this is conceptually reasonable, most transporters do not run in reverse for a number of practical reasons. The concentration of maltose required to reverse the cycle, for example, might exceed the solubility of maltose. Instead, cells utilize a specialized molecular machine for the synthesis of ATP, known as $F_o$-$F_1$ ATP synthase, or simply ATP synthase. ATP synthase couples the transport of protons down a concentration gradient to the synthesis of ATP from ADP, and is a complicated molecular assembly. We shall not provide a detailed description of ATP synthase here, but instead introduce some simple ideas concerning how ATP might be synthesized by such a machine.

To begin with, consider the hydrolysis of ATP:

$$\mathrm{ATP} \rightleftharpoons \mathrm{ADP} + \mathrm{P_i} \tag{9.60}$$

In the next chapter we shall show that the concentrations of ATP, ADP, and $P_i$ must obey the following relationship at equilibrium:

$$K = \frac{[\mathrm{ADP}][\mathrm{P_i}]}{[\mathrm{ATP}]} \tag{9.61}$$

where $K$ is a constant, known as the equilibrium constant, and it has a fixed value under constant temperature conditions.

An important consequence of Equation 9.61 is that it tells us that we could, in principle, run the ATP hydrolysis reaction backwards by increasing the concentrations of ADP and $P_i$. If we pump more molecules of ADP and $P_i$ into the cell, then at a high enough concentration the molecules will react to produce ATP, in order to maintain a constant value of $K$ at equilibrium. Such a reversal of the ATP hydrolysis reaction illustrates **Le Châtelier's principle**, which states that if the concentration of one or more molecules in a reaction are changed, the system will adjust itself and reach a new equilibrium so as to balance the change.

Biological systems are unable to drive the progress of a normally uphill chemical reaction, such as ATP synthesis, by changing the concentrations of the reactants over very large ranges in value. In particular, a cell cannot drive the production of ATP by increasing the concentration of ADP or phosphate ions drastically, because the interior of a living cell cannot tolerate very large changes in the concentrations of phosphate ion.

Instead, the ATP synthesis process begins by utilizing a separate set of proteins to generate a concentration gradient of protons across the membrane (see Section 11.6). Cells have evolved so that they can tolerate large differences in proton concentrations across the membrane, and this concentration gradient is used to drive conformational changes in the ATP synthesis machine. The energetic cost of forcing these conformational changes is paid for by the work done to generate the concentration gradient.

Let us denote a highly simplified and hypothetical ATP synthesizing machine by F (**Figure 9.23**). The protein F binds to ADP and $P_i$ to form a complex denoted (ADP, $P_i$)•F. The protein can undergo a conformational change to another form that we denote F*. The structure of F* is such that it forces the ADP and $P_i$ molecules so close together that they unite to form ATP:

$$(\mathrm{ADP}, \mathrm{P_i})\bullet \mathrm{F} \rightleftharpoons (\mathrm{ATP})\bullet \mathrm{F}^* \quad (9.62)$$

Because the formation of ATP is uphill energetically, the equilibrium point for the reaction shown in Equation 9.62 lies far to the left under normal cellular concentrations of ADP and $P_i$ (see Figure 9.23A). On its own, the protein F cannot do anything to force the production of ATP from ADP. This is where nature plays a trick, by coupling ATP synthesis to another process, one that is downhill in terms of free energy.

Imagine that the protein F has a binding site for a proton, $H^+$. The binding of $H^+$ to F is strongly favored by the conversion of F to the $F^*$ form, which we write as follows:

$$(\mathrm{ADP}, \mathrm{P_i})\bullet \mathrm{F} + \mathrm{H}^+ \rightleftharpoons (\mathrm{ATP})\bullet \mathrm{F}^* \bullet \mathrm{H}^+ \quad (9.63)$$

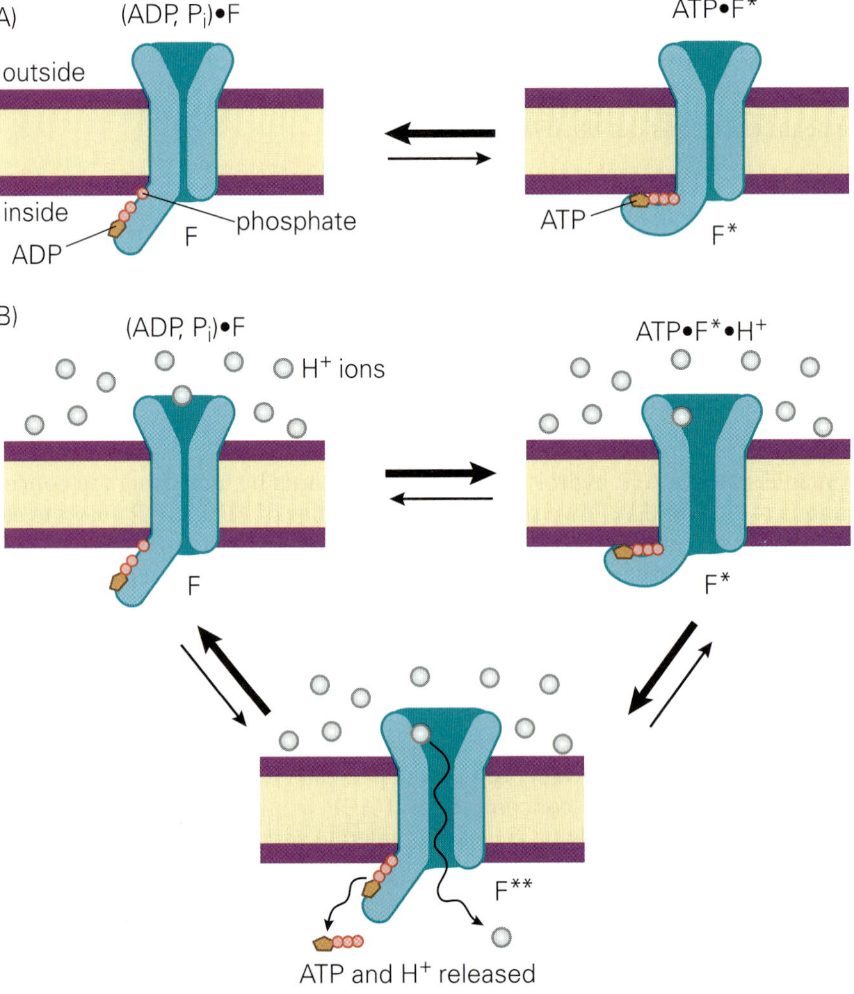


**Figure 9.23 A simplified conceptualization of the ATP synthesis process.** (A) A transmembrane protein, denoted F, binds to ADP and $P_i$ and undergoes a conformational change to a form $F^*$. $F^*$ favors the binding of ATP over ADP, but the equilibrium is shifted strongly to the left. (B) The production of ATP is coupled to the transport of a proton down a concentration gradient. The proton binds preferentially to the $F^*$ form and, because protons are at high concentration outside the cell, they drive the production of ATP inside the cell. The binding of the proton to F causes the channel to open, causing the proton to move into the cell. The release of the proton from $F^*$ causes the system to cycle, first converting $F^*$ to a third form ($F^{**}$), which releases ATP, then back to F. The actual mechanism of ATP synthesis by $F_1$ ATP synthase is far more complicated (see Figure 9.24).

The reaction depicted in Equation 9.63 is made more favorable in the right-hand direction by increasing the concentration of protons. This occurs because the equilibrium for the reaction is governed by the following equilibrium constant:

$$K = \frac{[(\text{ATP})\bullet \text{F}^* \bullet \text{H}^+]}{[(\text{ADP}, \text{P}_\text{i})\bullet \text{F}][\text{H}^+]} \tag{9.64}$$

If the proton concentration is increased, then the concentration of (ATP)•F$^*$•H$^+$ will increase and the concentration of (ADP, $P_i$)•F will decrease in order to maintain the ratio at the value of *K*.

If the concentration of protons outside the cell is high, then ATP and the proton will remain bound to the protein F. ATP needs to be released from the protein in order to allow the process to cycle. A key aspect to making the process cyclic is the fact that the F protein is a membrane-spanning channel that has a conduit for the passage of the protons, which opens when ATP is synthesized. The opening of the channel results in the proton passing through the channel and into the cell. This causes a further conformational change in the F protein, converting it from the F$^*$ form to a third form (F$^{**}$) in a way that releases ATP from F. This recloses the channel and the cycle repeats as long as the concentration of protons outside the cell is high enough to drive the system, generating one molecule of ATP in each cycle.

Although this simplified and hypothetical discussion captures some of the essential aspects of how ATP is synthesized, it is not an accurate description of the far more complicated mechanism by which the actual ATP synthase enzyme works. The structure of ATP synthase, which is a complex of several different proteins, is shown in **Figure 9.24**. The catalytic portion of the complex consists of six subunits, with three active sites where ATP synthesis occurs. The catalytic assembly encircles a protein that acts as a shaft. The shaft is forced to rotate as protons move through a channel in the membrane-spanning portion of the complex (see Figure 9.24). The shaft has an asymmetric shape and, as it rotates, it induces the catalytic subunits to adopt three distinct conformations, depending on their position with respect to the shaft.

The changes in the conformations of the catalytic subunits as ATP is synthesized are shown schematically in **Figure 9.25**. One of the catalytic subunits binds ADP and $P_i$ loosely, the next one forces them together to form ATP, and the third one binds and then releases an ATP molecule synthesized in the previous cycle. As the

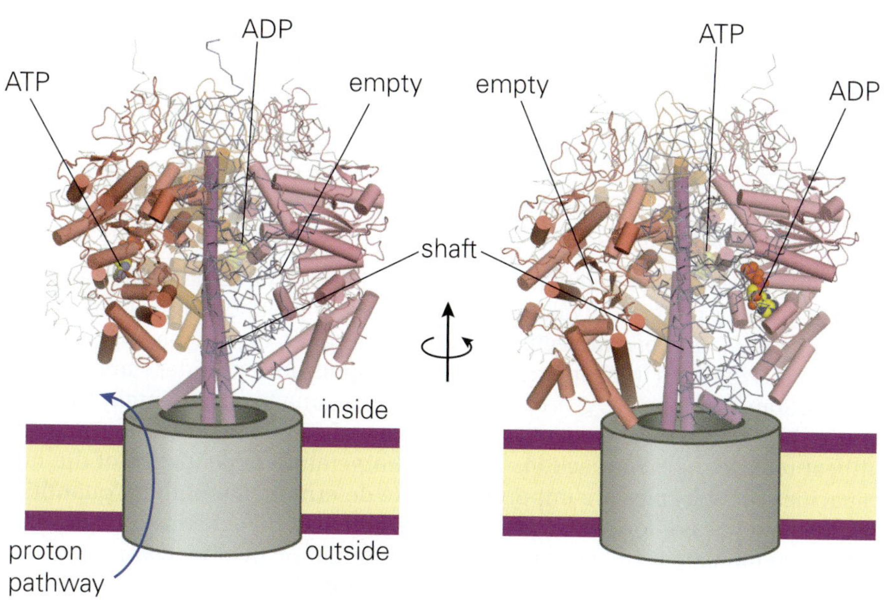


**Figure 9.24 Structure of the $F_0$-$F_1$ ATP synthase.** The intracellular portion of the assembly is formed by six subunits that form a shell around a central shaft (*purple*). The three subunits that contain the catalytic centers are shown with their helices and strands represented as cylinders and arrows. One of them is bound to ATP, one to ADP, and one is empty. The central shaft is forced to rotate when protons move through the transmembrane portion of the ATP synthase (*gray*). The rotation of the shaft causes conformational changes in the catalytic subunits, as shown by comparing the two diagrams. The subunit bound to ADP in the left drawing has converted it to ATP in the right drawing, with a molecule of ATP being released from the subunit on the left hand side of the molecule. Not shown here are additional components of the ATP synthase that hold the catalytic subunits in place while the shaft rotates. (PDB codes: 1H8E and 1H8H.)

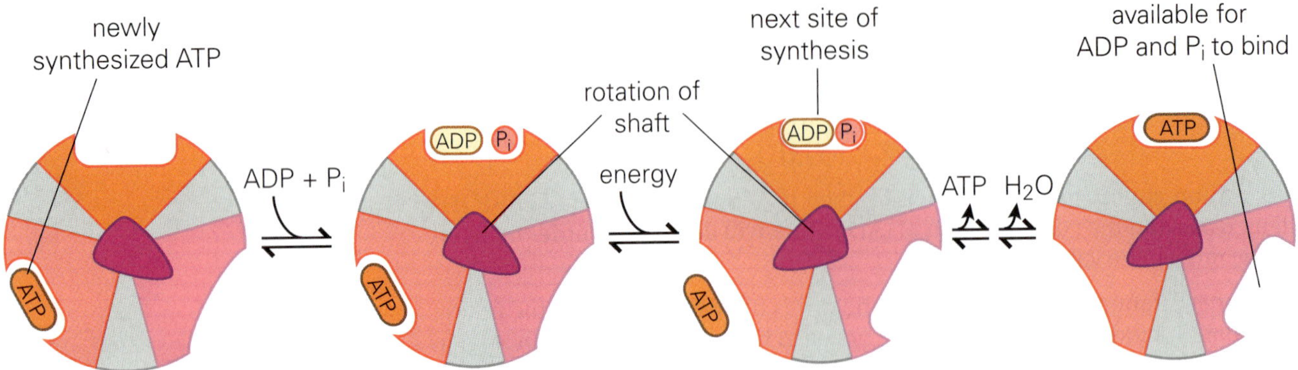


**Figure 9.25 The binding change mechanism for ATP synthesis.** Shown here schematically are the six subunits of the catalytic complex of the ATP synthase, in cross section. The rotating shaft is in the middle. The conformational changes driven by the rotation of the shaft are indicated by changes in the shapes of the three colored subunits and by whether ATP, ADP, or nothing is bound to them. (Adapted from P.D. Boyer, *Biochim. Biophys. Acta* 1140: 215–250, 1993. With permission from Elsevier.)

shaft rotates, it forces each catalytic center to adopt one of these three conformations in sequence. This leads to the generation and release of ATP at each catalytic center in turn, as shown in Figure 9.25. This mechanism, in which the binding of ATP to the catalytic subunits is changed by rotation of the central shaft, is known as the **binding change mechanism** for ATP synthesis. The ATP synthase machine and its mechanism are conserved in all the branches of life, reflecting its central role in generating the chemical power that drives cellular processes.

## Summary

The second law of thermodynamics states that the combined entropy of the system and the surroundings increases for a spontaneous process and is maximal at equilibrium. At constant temperature and pressure, the constraint imposed by the second law, which involves properties of both the system and the surroundings, can be converted to one that depends only on the properties of the system. This condition is that the Gibbs free energy of the system decreases for a spontaneous process, and is minimal at equilibrium. The Gibbs free energy, $G$, incorporates the enthalpy of the system, $H$, its entropy, $S$, and the temperature, $T$, in the following way:

$$G = H - TS$$

If a constant temperature process involves no changes in volume, then the Helmholtz free energy, $A$, of the system decreases for a spontaneous process. The Helmholtz free energy depends on the energy, $U$, the entropy, $S$, and the temperature, $T$:

$$A = U - TS$$

For a reaction at constant pressure and temperature, the difference in free energy between the reactants and the products is denoted $\Delta G$. If the value of $\Delta G$ is negative, the reaction can occur spontaneously. The free energy of a system is an extensive property, and so the magnitude of $\Delta G$ depends on how many molecules are involved in the reaction. To make comparisons meaningful, we use the standard free-energy change of the reaction, $\Delta G^{o}$, which is the change in free energy when one mole of reactants is converted to one mole of products under standard conditions of concentration and 1 atm pressure.

The value of $\Delta G^{o}$ can be calculated as the difference in the standard free energies of formation of molar equivalents of reactants and products. The standard free energy of formation of a molecule is the free-energy change in making one mole of the molecule from its elements in their most stable forms, under standard conditions. Standard free energies of formation are obtained experimentally by determining the standard enthalpies and entropies of formation. The measurement of heat released during reactions and the temperature dependence of the heat capacities of molecules are important for the determination of these quantities. The free energies of formation of many compounds of biochemical importance have been tabulated, and the free energies of formation of other compounds can be determined by the analysis of thermodynamic cycles.

In addition to expansion work against an external pressure, molecular systems can also do other kinds of work, such as chemical or electrical work. The maximum amount of such non-expansion work that can be done by a process is given by the free-energy change of the process.

## Key Concepts

### A. FREE ENERGY

- The combined entropy of the system *and* the surroundings increases in a spontaneous process.
- By linking the entropy of the surroundings to a system property, we obtain the free energy of the system as the indicator of spontaneous change.
- For a process occurring at constant pressure and temperature, the Gibbs free energy of the system, $G = H - TS$, always decreases in the direction of spontaneous change. For a system at equilibrium, the Gibbs free energy is at a minimum.
- The Helmholtz free energy of the system, $A = U - TS$, decreases for a spontaneous process occurring at constant volume and temperature.

### B. STANDARD FREE-ENERGY CHANGES

- The zero point of the free-energy scale is set by the free energy of the most stable form of the elements.
- Thermodynamic cycles make it possible to determine the free energies of formation of complex molecules when the free energies of formation of simpler molecules are known.
- The standard free-energy change for a reaction, $\Delta G^o$, is the change in molar free energy upon stoichiometric conversion of reactants to products under standard conditions.
- Enthalpies and entropies of formation can be combined to give the free energy of formation.
- Calorimetric measurements yield the standard enthalpy changes associated with combustion reactions.
- The entropy of formation of a compound can be derived from heat capacity measurements.

### C. FREE ENERGY AND WORK

- Expansion work is not the only kind of work that can be done by a system.
- Chemical work involves changes in the numbers or concentrations of molecules.
- The decrease in the Gibbs free energy during a process is the maximum amount of non-expansion work that the system is capable of doing under constant pressure.
- The coupling of ATP hydrolysis to work underlies many processes in biology.
- The synthesis of ATP is coupled to the movement of ions across the membrane, down a concentration gradient.

## Problems

### True/False and Multiple Choice

1. For a system at equilibrium, Gibbs free energy is maximized.

   True/False

2. The work done by biological systems is most commonly which type of work?

   a. pressure
   b. chemical
   c. home
   d. volume
   e. heat

3. According to Le Châtelier's principle:

   a. A reaction always favors the formation of as much product as possible.
   b. Reactants will only react upon the addition of external heat or pressure.
   c. The system will adjust, if more reactants are added, and a new equilibrium will be established to balance the change.
   d. The Gibbs free energy is always greater than the amount of work done plus the Helmholtz free energy.

4. The value of $\Delta_f G^o$ for elemental oxygen is:

   a. $-15$ kJ•mol$^{-1}$.
   b. 0 kJ•mol$^{-1}$.
   c. Much larger than the free energy of elemental hydrogen and much less than the free energy of elemental uranium.
   d. Equal to the free energy of elemental tungsten.
   e. Both b and d.

5. The cell generates ATP from ADP and $P_i$ by coupling the synthesis to an energetically unfavorable process.

   True/False

6. At equilibrium:
   a. There is no change in temperature over time.
   b. The free energy of the system is minimized.
   c. The combined entropy of the system and surroundings is maximized.
   d. Only b and c.
   e. a, b, and c.

## Fill in the Blank

7. Gibbs free energy is used to describe systems with constant __________, while Helmholtz free energy is used to describe systems with constant __________.

8. The standard state has a pressure of __________.

9. The change in free energy of a process is equal to the __________ amount of work extracted from that process.

10. In biochemistry, the standard state for water is ____ M. Excepting $H^+$, for all other solutions, the standard state is _____ M.

## Quantitative/Essay Problems

11. A system is at constant volume and 29°C. The internal energy is –40 kJ and the entropy is 31 J•K$^{-1}$. What is the value of the Helmholz free energy?

12. Consider the distribution of molecules in energy levels in the diagram below, where each "X" represents one molecule. A molecule in energy level 1 contributes one unit of energy to the total internal energy, a molecule in energy level 2 contributes two units of energy to the total internal energy, etc. The system is at 273 K.

| | |
|---|---|
| Energy level 4 | X |
| Energy level 3 | XX |
| Energy level 2 | XXXXX X |
| Energy level 1 | XXXXX XXXXX XXXXX |

What are the values of the (a) internal energy (*U*), (b) entropy (*S*), and (c) Helmholtz free energy (*A*) of the system? Assume that $k_B$ = 1 energy unit•K$^{-1}$.

13. A system at 275 K in state A has an enthalpy of –25 kJ and an entropy of 2 J•K$^{-1}$. In state B, it has an enthalpy of –20 kJ and an entropy of 10 J•K$^{-1}$. Will state A convert spontaneously to state B?

14. Assume that entropy and enthalpy changes are independent of temperature. A system in state A has an enthalpy of –22 kJ and an entropy of 7 J•K$^{-1}$. In state B, it has an enthalpy of –12 kJ and an entropy of 15 J•K$^{-1}$. At what temperatures will state B be favored?

15. A reaction has an enthalpy change of 200 kJ and an entropy change of 250 J•K$^{-1}$. Assume that entropy and enthalpy changes are independent of temperature. At what temperature will the Gibbs free energy be zero?

16. What is the expression for the equilibrium constant for the reaction A $\rightleftharpoons$ B + C?

17. Consider the reaction A + 2B → C. Values of $\Delta_f G^o$ for A, B, and C are as follows:

    A = –34 kJ•mol$^{-1}$
    B = 84 kJ•mol$^{-1}$
    C = –112 kJ•mol$^{-1}$

    What is the value of $\Delta G^o$ for the reaction?

18. Consider the following reactions in which A, B, C, and D are elements in their most stable states:

    A + B → Y
    C + D → Z
    Y + Z → J

    For each reaction, $\Delta G^o = 0$. Using the following table of $\Delta G^o$ values, what is the value of $\Delta G^o$ for the formation for J?

| Reaction | $\Delta G^o$ (kJ•mol$^{-1}$) |
|---|---|
| A + B → Y | –70 |
| C + D → Z | 23 |
| Y + Z → J | 15 |

19. Consider the reaction, 2I + J → K, and the following table of enthalpies and entropies of formation at 298 K:

| | I | J | K |
|---|---|---|---|
| $\Delta_f H^o$ (kJ•mol$^{-1}$) | 0 | –30 | –300 |
| $\Delta_f S^o$ (J•K$^{-1}$•mol$^{-1}$) | 120 | 200 | 60 |

What is the standard free-energy change ($\Delta G^o$) for the reaction?

20. Consider the same reaction as in Problem 19. Assume that enthalpy and entropy do not vary with temperature. At what temperature will the reaction begin to proceed spontaneously in the opposite direction (that is, at what temperature will $\Delta G^o_{reactants} < \Delta G^o_{products}$)?

21. The value of $\Delta G^o$ for forming water from elemental oxygen and hydrogen is –237 kJ•mol$^{-1}$.

$$2H_2 + O_2 \rightarrow 2H_2O$$

Calculate the standard enthalpy of formation of water using the entropies below. (*Hint*: Elemental oxygen and hydrogen have the same enthalpy of formation.)

| | $O_2$ | $H_2$ | $H_2O$ |
|---|---|---|---|
| $S^o$ (J•K$^{-1}$•mol$^{-1}$) | 205 | 130 | 70 |

22. A chemist wants to develop a fuel by converting water back to elemental hydrogen and oxygen using coupled ATP hydrolysis to drive the reaction. Given that the value of $\Delta_f G^o$ for water is –237 kJ•mol$^{-1}$ and that one mole of ATP hydrolyzed to ADP yields –30 kJ•mol$^{-1}$, how much ATP is needed to yield three moles of $H_2$ gas?

**23.** A molecular motor moves along a microtubule track in steps of 100 Å displacements. The motor hydrolyzes one molecule of ATP per step. The motor operates under conditions where the free-energy change for ATP hydrolysis is –60 kJ•mol$^{-1}$. What is the maximum resistive force against which the motor can move cargo?

**24.** Explain why the absolute value of the work done by a process at constant pressure can never be greater than the absolute value of the Gibbs free-energy change for the process.

**25.** The equilibrium constant for a chemical system is 1. More reactants are added to the system. Explain what will happen to the system as it reestablishes equilibrium.

## Further Reading

### General

Atkins PW & De Paula J (2010) Atkins' Physical Chemistry, 9th ed. New York: Oxford University Press.

Chang R (2005) Physical Chemistry for the Biosciences. Sausalito, CA: University Science Books.

Dill KA & Bromberg S (2010) Molecular Driving Forces: Statistical Thermodynamics in Chemistry, Biology, Physics, and Nanoscience, 2nd ed. New York: Garland Science.

Eisenberg DS & Crothers DM (1979) Physical Chemistry: With Applications to the Life Sciences. Menlo Park, CA: Benjamin/Cummings.

Halliday D, Resnick R & Walker J (2010) Fundamentals of Physics, 9th ed. New York: John Wiley & Sons.

Haynie DT (2008) Biological Thermodynamics, 2nd ed. New York: Cambridge University Press.

Lundblad RL & Macdonald F (2012) Handbook of Biochemistry and Molecular Biology, 4th ed. Boca Raton, FL: CRC Press.

Voet D & Voet JG (2011) Biochemistry, 4th ed. New York: John Wiley & Sons.

### C. Free Energy and Work

Boyer PD (1993) The binding change mechanism for ATP synthase—some probabilities and possibilities. *Biochim. Biophys. Acta* 1140, 215–250.

Gennerich A & Vale RD (2009) Walking the walk: How kinesin and dynein coordinate their steps. *Curr. Opin. Cell Biol.* 21, 59–67.

Shilton BH (2008) The dynamics of the MBP-MalFGK(2) interaction: A prototype for binding protein dependent ABC-transporter systems. *Biochim. Biophys. Acta* 1778, 1772–1780.

Tinoco I Jr, Li PT & Bustamante C (2006) Determination of thermodynamics and kinetics of RNA reactions by force. *Q. Rev. Biophys.* 39, 325–360.

Walker JE (1997) Nobel Prize in Chemistry Lecture. Nobelprize.org. 21 Nov 2010 http://nobelprize.org/nobel_prizes/chemistry/laureates/1997/walker-lecture.html

# CHAPTER 10

# Chemical Potential and the Drive to Equilibrium

In Chapter 9, we introduced the concept of the free energy of a system and showed that it is at a minimum when the system is at equilibrium. We are, however, often interested in systems that are *not* at equilibrium. Living systems, for example, are constantly in a state of flux and never reach equilibrium. But, given any nonequilibrium situation, there is a drive towards equilibrium that determines what changes in the system will occur spontaneously. By understanding how the free energy changes with system parameters, we can understand the direction of spontaneous change and the capacity of the system to do work.

The most important of the parameters that control changes in the free energy in biological systems are the concentrations of molecular species. In part A of this chapter, we introduce the concept of the **chemical potential**, which is the dependence of free energy on the number of molecules present of a particular chemical species. The chemical potential can be understood as the free energy per molecule or the molar free energy of a chemical species. In part B, we describe a relationship between the free energy and the concentrations of molecules at equilibrium, which introduces the concept of the **equilibrium constant**, $K$.

The equilibrium constant tells us the ratios of concentrations of reactants and products at equilibrium. We are often also interested in whether a particular reaction will go forward or backward under given nonequilibrium conditions of concentration. Knowledge of the equilibrium constant, combined with a related parameter known as the **reaction quotient**, which depends on the actual concentrations of the reactants and products, allows us to calculate the reaction free-energy change, $\Delta G$, which determines the driving force for the reaction.

The value of $K$ is sometimes readily measured, and it is linked closely to the changes in the enthalpy and entropy for binding or chemical reactions. To illustrate the application of equilibrium constants, we discuss, in part C, the titration properties of acids and bases.

The drive towards equilibrium involves a combination of changes in energy and entropy, which often offset one another. Part D of this chapter provides a specific example of how free energy and the equilibrium constant determine the direction in which a particular reaction, protein folding, progresses. We first consider a conceptual model for protein folding and show how energy and entropy are interlinked in the folding process. We next describe how the free-energy change, and the associated changes in entropy and enthalpy, are determined experimentally for a protein-folding reaction.

## A. CHEMICAL POTENTIAL

In this part of the chapter, we consider how the free energy per molecule changes with the concentration of molecules. We obtain a result that is central to biochemistry, which is that the chemical potential (that is, the free energy per molecule) varies as the logarithm of the concentration.

## 10.1 The chemical potential of a molecular species is the molar free energy of that species

Suppose we have a system that is maintained at constant temperature and pressure, containing $N_i$ molecules of type $i$. Now suppose that we add one additional molecule of type $i$ to the system, keeping the temperature and pressure constant. The change in free energy is given by:

$$\Delta G = G_{\text{final}} - G_{\text{initial}} \tag{10.1}$$

Because we have added only one molecule, though, the change in free energy is essentially the free energy of one molecule in the system:

$$\Delta G = G \text{ (for one molecule)} \tag{10.2}$$

The **molecular chemical potential** ($\mu_i$) is defined as the rate of change of the free energy with respect to the number of molecules:

$$\mu_i = \left(\frac{\partial G}{\partial N_i}\right)_{T,P,N_{i\neq j}} \approx \frac{\Delta G}{\Delta N_i} = G \text{ (for one molecule)} \tag{10.3}$$

According to Equation 10.3, the chemical potential for the $i^{\text{th}}$ species is essentially the free energy of one molecule of type $i$ in the system. The chemical potential defined in this way has units of energy per molecule (for example, J•molecule$^{-1}$).

When studying chemical reactions, we usually consider the number of moles rather than the number of molecules. One mole is just an Avogadro's number of molecules ($N_A = 6.022 \times 10^{23}$ molecules). We denote the number of moles of the $i^{\text{th}}$ kind of molecule by $n_i$. We can define the molar chemical potential in terms of the number of moles as follows:

$$\bar{\mu}_i = \left(\frac{\partial G}{\partial n_i}\right)_{T,P,n_{i\neq j}} \tag{10.4}$$

The chemical potential, defined as in Equation 10.4, has units of energy•mol$^{-1}$ (for example, J•mol$^{-1}$).

Because the number of molecules of type $i$ in the system ($N_i$) is given by $N_i = N_A n_i$, we can write:

$$\left(\frac{\partial G}{\partial n_i}\right)_{T,P,n_{i\neq j}} = \left(\frac{\partial G}{\partial N_i}\right)_{T,P,N_{i\neq j}} \left(\frac{\partial N_i}{\partial n_i}\right)_{T,P,n_{i\neq j}} = N_A \left(\frac{\partial G}{\partial N_i}\right)_{T,P,N_{i\neq j}} \tag{10.5}$$

According to Equation 10.3, $\mu_i$ is the free energy of one molecule of type $i$ in the system. According to Equation 10.5, therefore, the chemical potential $\bar{\mu}_i$ (in terms of moles) is the free energy of an Avogadro's number of molecules, which is the **molar free energy** for that species of molecule. In general, when we discuss chemical potential, it will be in molar terms, and the bar above the symbol is generally dropped because the meaning is clear.

**Chemical potential, $\mu$**

The molecular chemical potential is the rate of change of free energy with respect to the number of molecules at constant temperature, pressure, and number of other molecules:

$$\mu_i = \left(\frac{\partial G}{\partial N_i}\right)_{T,P,N_{i\neq j}}$$

$\mu_i$ is the free energy of one $i$-type molecule, or the molar free energy, $G_i$.

## 10.2 Molecules move spontaneously from regions of high chemical potential to regions of low chemical potential

The manner in which the free energy of a system changes with the numbers of molecules of various species is important for understanding how chemical reactions come to equilibrium. Before we study chemical reactions, though, we need to consider a simpler process involving just one type of molecule. One example of such a process is diffusion, in which molecules move spontaneously from one region of a system to another (diffusion in discussed in more detail in Chapter 17).

Consider a chamber containing a solution of only one kind of solute (which we will call A). Here we take A to be uncharged, the effects of electrical charges and electric fields will be addressed in Chapter 11. The surroundings of the chamber also contain a solution of A molecules. A higher concentration of A molecules is placed

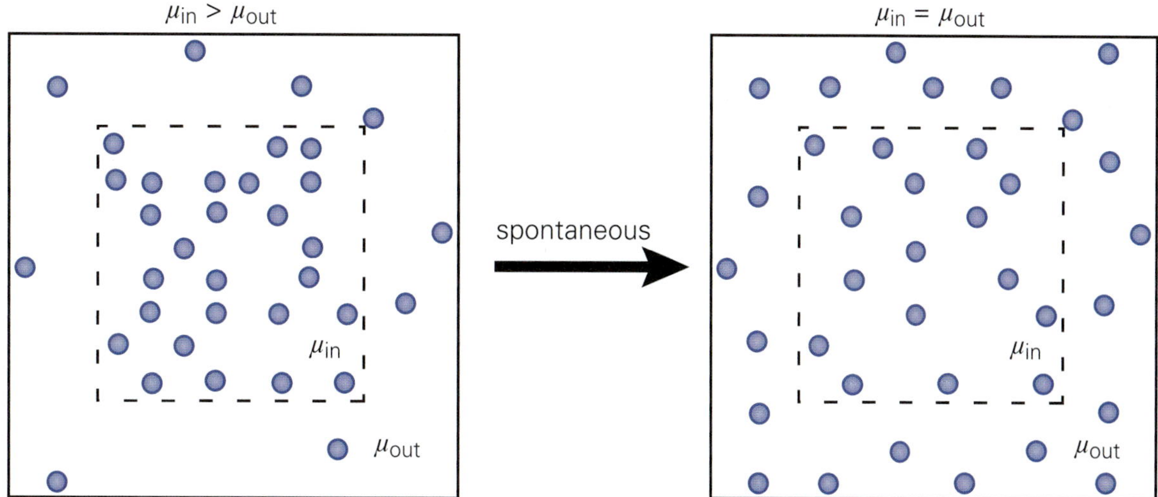


**Figure 10.1 Schematic representation of a concentration differential.** The central region, indicated by dashed lines, initially has a higher concentration of molecules than the outside. The chemical potential, $\mu$, of the molecules is also initially higher inside than outside. At equilibrium, $\mu_{in} = \mu_{out}$.

inside the chamber than outside the chamber, as shown in **Figure 10.1**. The walls of the chamber are pierced with holes, so that the molecules can move freely in and out of the chamber. What will happen when the system reaches equilibrium?

The condition for equilibrium is that the total free energy of the system, $G$, is at a minimum (that is, $dG = 0$), so at constant temperature ($dT = 0$) and pressure ($dP = 0$) we use Equation 10.4 to get:

$$dG = \mu_{in}\, dN_{in} + \mu_{out}\, dN_{out} = 0 \tag{10.6}$$

where $N_{in}$ and $N_{out}$ are the numbers of A molecules inside and outside the chamber, respectively, and $\mu_{in}$ and $\mu_{out}$ are the chemical potentials inside and outside the chamber, respectively. If you are unfamiliar with the operations in differential calculus that lead to Equation 10.6, Box 8.3 provides a review of the essential concepts.

Because $N_{in}$ and $N_{out}$ both refer to the numbers of A molecules, they must change in a coupled manner:

$$dN_{in} = -dN_{out} \tag{10.7}$$

Thus, if the number of A molecules *outside* the chamber *increases*, then the number of A molecules *inside* the chamber must *decrease* by an equivalent amount.

Using Equation 10.7, we can substitute $-dN_{in}$ for $dN_{out}$ in Equation 10.6 to get:

$$dG = \mu_{in}\, dN_{in} - \mu_{out}\, dN_{in} = 0 \tag{10.8}$$

Therefore,

$$(\mu_{in} - \mu_{out})\, dN_{in} = 0 \qquad \text{(at equilibrium)} \tag{10.9}$$

Since the infinitesimal displacements $dN_{in}$ can be nonzero, the only way in which Equation 10.9 can always be true is if the first term on the left-hand side is zero:

$$\mu_{in} - \mu_{out} = 0 \qquad \text{(at equilibrium)} \tag{10.10}$$

Thus,

$$\mu_{in} = \mu_{out} \qquad \text{(at equilibrium)} \tag{10.11}$$

According to Equation 10.11, the chemical potential of the molecules will be the same inside and outside the chamber at equilibrium. This condition, that chemical potentials become equal, is equivalent to the statement that equilibrium is reached.

What happens when the system is not at equilibrium? We know that a spontaneous change must always decrease the free energy, and the free energy is minimized at equilibrium, so:

$$dG < 0 \qquad \text{(for a change towards equilibrium)} \tag{10.12}$$

Substituting the value of $dG$ from Equation 10.8 into Equation 10.12, we get:

$$dG = (\mu_{in} - \mu_{out})\, dN_{in} < 0 \qquad (10.13)$$

For $dG$ to be less than zero, the sign of $dN_{in}$ must be opposite to the sign of $(\mu_{in} - \mu_{out})$. Therefore, if $N_{in}$ decreases spontaneously (that is, if $dN_{in} < 0$), then $(\mu_{in} - \mu_{out}) > 0$. It follows that:

$$\mu_{in} > \mu_{out} \quad \text{(if } N_{in} \text{ decreases spontaneously)} \qquad (10.14)$$

According to Equation 10.14, molecules move spontaneously from regions of high chemical potential to regions of low chemical potential.

## 10.3 Biochemical reactions are assumed to occur in ideal and dilute solutions, which simplifies the calculation of the chemical potential

One of the most important conclusions of this chapter, discussed in the next section, is that the chemical potential of a solute is related to the logarithm of the concentration of the solute. The reason this relationship is so important is that it underlies the discussion of chemical equilibrium that follows. The logarithmic dependence of chemical potential on solute concentration comes about as a consequence of the positional entropy of the solute.

It is much easier to calculate the entropy of the solute when the solution is very dilute. Most biochemical reactions occur in aqueous solution, where the concentration of water is 55 M. The physiological concentrations of proteins and small molecules such as ATP are in the millimolar ($10^{-3}$ M) range or below, so the concentrations of these solutes are low compared to the concentration of water (that is, the solutions are very dilute).

**Ideal dilute solution**

A solution in which the concentration of the solute is so low that individual solute molecules do not influence each other is referred to as an ideal dilute solution.

If a solution is so dilute that the solute molecules do not interact with each other, the solution is referred to as an **ideal dilute solution** (Figure 10.2). What this also means is that, although the solute molecules do interact with the solvent molecules in an ideal dilute solution, the energy of interaction between any individual solute molecule with the solvent molecules does not change with solute concentration. The molar energy (or enthalpy) of the solute is independent of the concentration of the solute in an ideal dilute solution. As we shall see, changes in the chemical potential of an ideal solution as a function of concentration are due solely to changes in the entropy of the solute.

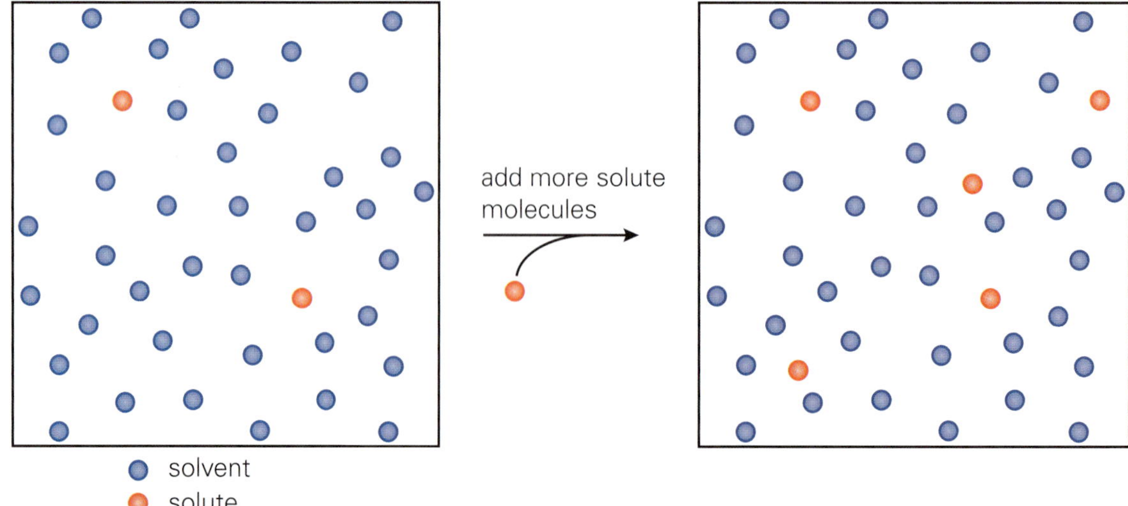


**Figure 10.2 An ideal dilute solution.** An ideal dilute solution is one in which the solute molecules do not influence each other energetically. In the example shown schematically here, the addition of more solute molecules is assumed to result in no change in the average energy (or energy per molecule) of the other solute molecules.

region 1

● Solute: $N_{A1}$, $\mu_{A1}$
● Solvent: $N_B$, $\mu_B$
$M$ grid boxes

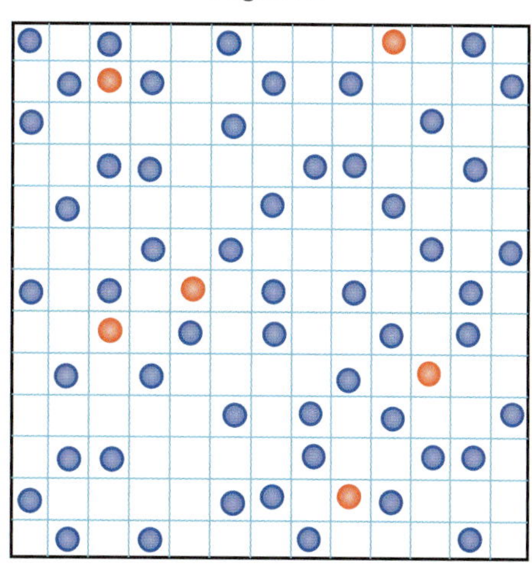


**Figure 10.3 The relationship between chemical potential and concentration.** Shown are two regions of an ideal dilute solution, with solute (A, *red*) and solvent (B, *blue*) molecules. For an ideal dilute solution, the difference in solute chemical potential ($\Delta\mu_A$) is given by:

$$\Delta\mu_A = \mu_{A2} - \mu_{A1} = RT\ln\left(\frac{C_2}{C_1}\right)$$

where $C_1$ and $C_2$ are the concentrations of A in regions 1 and 2, respectively.

## 10.4 The chemical potential is proportional to the logarithm of the concentration

We know intuitively that spontaneous diffusion drives molecules from regions of higher concentration to regions of lower concentration. We have also seen that the chemical potential must be greater at higher concentration. What, then, is the mathematical relationship between concentration and chemical potential? For an ideal, dilute solution of a solute in a solvent, the relationship between the chemical potential at two concentrations ($C_1$ and $C_2$) of the solute is as follows:

$$\Delta\mu = \mu_2 - \mu_1 = RT\ln\left(\frac{C_2}{C_1}\right) \tag{10.15}$$

In Equation 10.15, $\mu_1$ is the chemical potential (that is, the free energy per molecule of solute) when the concentration of the solute is $C_1$ and $\mu_2$ is the chemical potential when the concentration of the solute is $C_2$. Equation 10.15 is central to much of biochemical analysis, and in this section we discuss how we arrive at this equation.

To better understand Equation 10.15, let us consider two regions of equal volume, both containing a solution of A molecules (the solute) in B molecules (the solvent; for example, water). The concentration of A is higher in region 1 than in region 2. Thus, the number of A molecules in region 1 ($N_{A1}$) is greater than the number of A molecules in region 2 ($N_{A2}$), as shown in **Figure 10.3**. Since both solutions are dilute, we assume that the number of B (solvent) molecules in both regions ($N_B$) is much larger than the number of A molecules in either region (that is, $N_B >> N_{A1}$ and $N_B >> N_{A2}$). Also, we assume that $N_B$ is the same in both regions because both regions have the same volume. That is, we assume that the concentration of the solvent does not change when the concentration of the solute changes. This is true only for very dilute solutions.

The free energies of the A molecules in the two regions, $G_1$ and $G_2$, are given by:

$$G_1 = H_1 - TS_1 \tag{10.16}$$

and

$$G_2 = H_2 - TS_2 \tag{10.17}$$

Hence,

$$\mu_{A1} = \left(\frac{\partial G_1}{\partial N_{A1}}\right)_{T,P,N_B} = \left(\frac{\partial H_1}{\partial N_{A1}}\right)_{T,P,N_B} - T\left(\frac{\partial S_1}{\partial N_{A1}}\right)_{T,P,N_B} \tag{10.18}$$

where $\mu_{A1}$ is the chemical potential of the A molecules in region 1. Likewise, the chemical potential of the A molecules in region 2 is given by:

$$\mu_{A2} = \left(\frac{\partial H_2}{\partial N_{A2}}\right)_{T,P,N_B} - T\left(\frac{\partial S_2}{\partial N_{A2}}\right)_{T,P,N_B} \tag{10.19}$$

As discussed previously, for an ideal solution, the energy and enthalpy per molecule (or per mole) is assumed to be independent of solute concentration. Thus, $\left(\frac{\partial H}{\partial N_i}\right)_{T,P,N_{j\neq i}}$, which is the enthalpy per molecule of solute, is independent of concentration. Since the only difference between regions 1 and 2 is the concentration of the solute (A), it follows that the first terms on the right-hand sides of Equations 10.18 and 10.19 have the same value—that is, $\left(\frac{\partial H_1}{\partial N_{A1}}\right)_{T,P,N_B} = \left(\frac{\partial H_2}{\partial N_{A2}}\right)_{T,P,N_B}$.

As a consequence, the **difference in chemical potential**, $\Delta\mu$, between the two regions is given by:

$$\Delta\mu = \mu_{A2} - \mu_{A1} = -T\left(\frac{\partial S_2}{\partial N_{A2}} - \frac{\partial S_1}{\partial N_{A1}}\right)_{T,P,N_B} \tag{10.20}$$

where we have subtracted both sides of Equation 10.19 from Equation 10.18.

In order to evaluate the derivatives in Equation 10.20, we need to write down expressions for the entropy in regions 1 and 2 as a function of the number of A molecules in them. To do this, we partition each region into $M$ grid boxes (recall that both regions have the same volume, so $M$ must be the same for both regions in order to obtain meaningful results; see Figure 10.3).

It will be easier to calculate the entropy if we use the probabilistic definition (see Section 8.6). In Chapter 8, we used the probabilistic definition of the entropy for calculations involving molecules in different energy levels. In order to apply the probabilistic definition of the entropy to the calculation of the positional entropy, we use the fact that each grid box has to be in one of the three states: (1) the grid box is empty, (2) the grid box is occupied by an A molecule, or (3) the grid box is occupied by a B molecule. The calculation of the multiplicity is essentially a calculation of the number of all possible rearrangements of the states of the grid boxes (**Figure 10.4**). We can cast this in the form of probabilities as follows.

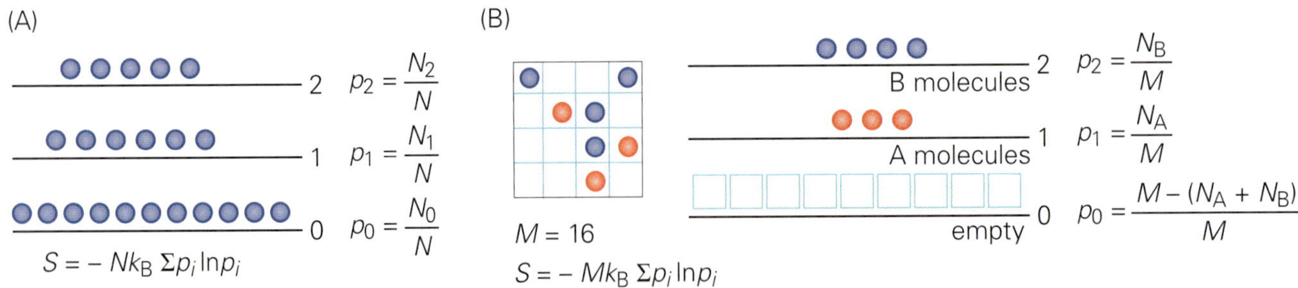


**Figure 10.4 Calculating positional entropy by using the statistical definition of the entropy.** (A) An energy distribution is shown, and the entropy can be calculated from the probabilities of finding molecules in each of the energy levels, as indicated. (B) The positional entropy can also be calculated in this way by realizing that the multiplicity is the number of all possible rearrangements of the states of $M$ grid boxes. For two kinds of molecules in $M$ grid boxes, each of the grid boxes can be considered to be in one of three states (that is, empty, containing an A molecule, or containing a B molecule). We can calculate the probabilities of each of these states and thereby calculate the entropy.

In the first part of the calculation, the probability ($p_1$) that a grid box is empty is:

$$p_1 = \frac{\text{number of empty gridboxes}}{\text{total number of gridboxes}}$$

$$p_1 = \frac{M - (N_{A1} + N_B)}{M} \tag{10.21}$$

In the second part of the calculation, the probability, $p_2$, that a grid box is occupied by an A molecule is:

$$p_2 = \frac{N_{A1}}{M} \tag{10.22}$$

Finally, in the third part of the calculation, the probability, $p_3$, that a grid box is occupied by a B molecule is:

$$p_3 = \frac{N_B}{M} \tag{10.23}$$

Note that in Equations 10.21, 10.22, and 10.23 we are assuming that each of the probabilities is independent, which is a reasonable assumption if the number of grid boxes is very large compared to the number of molecules.

Using the probabilistic definition of entropy (see Equation 8.13), the entropy in region 1, $S_1$, is given by:

$$S_1 = -M_B k \sum_{i=1}^{3} p_i \ln p_i \tag{10.24}$$

In order to compute the derivative of $S_1$ with respect to $N_{A1}$, we need consider only $p_1$ and $p_2$, because $p_3$ does not depend on $N_{A1}$ (see Equation 10.23). Thus,

$$\left(\frac{\partial S_1}{\partial N_{A1}}\right)_{T,P,N_B} = -M k_B \left[\frac{\partial}{\partial N_{A1}}(p_1 \ln p_1 + p_2 \ln p_2)\right] \tag{10.25}$$

And, for region 2,

$$\left(\frac{\partial S_2}{\partial N_{A2}}\right)_{T,P,N_B} = -M k_B \left[\frac{\partial}{\partial N_{A2}}(p_1 \ln p_1 + p_2 \ln p_2)\right] \tag{10.26}$$

Combining Equations 10.20, 10.25, and 10.26, we get:

$$\Delta\mu = \mu_{A2} - \mu_{A1} = k_B T\left(\ln\frac{N_{A2}}{M} - \ln\frac{N_{A1}}{M}\right) \tag{10.27}$$

The steps leading to Equation 10.27 are explained in **Box 10.1**. The concentration of A molecules in region 1 is $C_1$ and in region 2 is $C_2$, where $C_1$ and $C_2$ are given by:

$$C_1 = \frac{N_{A1}}{V} \quad \text{and} \quad C_2 = \frac{N_{A2}}{V} \tag{10.28}$$

Although the value of $M$ is arbitrary, it has to be proportional to the volume, so we can express $M$ in terms of the volume as follows:

$$M = \alpha V \tag{10.29}$$

where $\alpha$ is a constant of proportionality.

Substituting the values of $C_1$, $C_2$, and $M$ in Equation 10.27, we get:

$$\Delta\mu = k_B T\left(\ln\frac{C_2 V}{\alpha V} - \ln\frac{C_1 V}{\alpha V}\right) \tag{10.30}$$

$$\Rightarrow \Delta\mu = k_B T \ln\left(\frac{C_2}{C_1}\right) \tag{10.31}$$

Equation 10.31 describes the fundamental relationship between the difference in the chemical potential between two regions and the concentration of a solute in the two regions. In order to obtain this equation we defined the concentration

## Box 10.1 The relationship between differences in chemical potential and differences in concentration

Consider two equal-volume regions of a solution of A molecules (the solute) in B molecules (the solvent). The number of A molecules in regions 1 and 2 are $N_{A1}$ and $N_{A2}$, respectively. Assume that the solution is dilute, so the number of B molecules (the solvent) is much greater than the number of A molecules (that is, $N_B >> N_{A1}, N_{A2}$).

For region 1,

$$S_1 = -M k_B \sum_{i=1}^{3} p_i \ln p_i \quad \text{(see Equation 10.24)} \qquad (10.1.1)$$

where $M$ is the number of grid boxes, $k_B$ is Boltzmann's constant, and $p_i$ is probability that a grid box (1) is empty, (2) contains an A molecule, or (3) contains a B molecule.

The probability of finding an empty box, $p_1$, is:

$$p_1 = \frac{M - (N_B + N_{A1})}{M} \qquad (10.1.2)$$

The probability of finding a box with an A molecule, $p_2$, is:

$$p_2 = \frac{N_{A1}}{M} \qquad (10.1.3)$$

The probability of finding a box with a B molecule, $p_3$, is:

$$p_3 = \frac{N_B}{M} \qquad (10.1.4)$$

There is a similar set of terms for the other region. Only $p_1$ and $p_2$ depend on the number of A molecules, so the chemical potential is given by the derivative of the first two terms with respect to the number of A molecules ($N_A$).

The first derivative to consider is:

$$\frac{\partial}{\partial N_{A1}}(-M k_B p_1 \ln p_1)$$

$$= -M k_B \frac{\partial}{\partial N_{A1}}\left\{\frac{M-(N_B+N_{A1})}{M}\ln\frac{[M-(N_B+N_{A1})]}{M}\right\}$$

$$= -M k_B \left\{-\frac{1}{M}\ln\frac{[M-(N_B+N_{A1})]}{M} + \frac{M-(N_B+N_{A1})}{M}\frac{\partial}{\partial N_{A1}}\ln\frac{[M-(N_B+N_{A1})]}{M}\right\} \qquad (10.1.5)$$

We evaluate the derivative in Equation 10.1.5 as follows, using the chain rule.

Define a variable, $X$, as:

$$X = \frac{M-(N_B+N_{A1})}{M} \qquad (10.1.6)$$

Then,

$$\frac{\partial \ln X}{\partial N_{A1}} = \frac{\partial \ln X}{\partial X}\frac{\partial X}{\partial N_{A1}} = \frac{1}{X}\frac{\partial X}{\partial N_{A1}}$$

$$= \frac{M}{[M-(N_B+N_{A1})]}\left(\frac{-1}{M}\right)$$

Using this result, the second term in Equation 10.1.5 simplifies as follows:

$$\frac{M-(N_B+N_{A1})}{M}\frac{\partial}{\partial N_{A1}}\ln\left\{\frac{[M-(N_B+N_{A1})]}{M}\right\}$$

$$= \frac{M-(N_B+N_{A1})}{M}\frac{M}{[M-(N_B+N_{A1})]}\left(\frac{-1}{M}\right)$$

$$= -\left(\frac{1}{M}\right)$$

Therefore:

$$\frac{\partial}{\partial N_{A1}}(-M k_B p_1 \ln p_1) = -M k_B\left\{-\frac{1}{M}\ln\left[\frac{M-(N_B+N_{A1})}{M}\right] - \frac{1}{M}\right\}$$

$$= k_B \ln\left[\frac{M-(N_B+N_{A1})}{M}\right] + k_B \qquad (10.1.7)$$

The second term in the expression for the chemical potential is:

$$\frac{\partial}{\partial N_{A1}}(-M k_B p_2 \ln p_2) = -M k_B \frac{\partial}{\partial N_{A1}}\left(\frac{N_{A1}}{M}\ln\frac{N_{A1}}{M}\right)$$

$$= -M k_B\left[\frac{1}{M}\ln\frac{N_{A1}}{M} + \left(\frac{N_{A1}}{M}\frac{M}{N_{A1}}\right)\frac{1}{M}\right]$$

$$= -M k_B\left(\frac{1}{M}\ln\frac{N_{A1}}{M} + \frac{1}{M}\right)$$

$$= -k_B \ln\frac{N_{A1}}{M} - k_B \qquad (10.1.8)$$

Hence, the chemical potential is given by combining Equations 10.1.7 and 10.1.8:

$$\mu_{A1} = -T\left(\frac{\partial S_1}{\partial N_{A1}}\right)$$

$$= -T\left\{-k_B \ln\frac{N_{A1}}{M} - k_B + k_B \ln\left[\frac{M-(N_B+N_{A1})}{M}\right] + k_B\right\}$$

$$= k_B T \ln\frac{N_{A1}}{M} - k_B T \ln\left[\frac{M-(N_B+N_{A1})}{M}\right] \qquad (10.1.9)$$

Since $N_B >> N_{A1}$ (that is, since the number of solvent molecules is very much greater than the number of solute molecules):

$$M-(N_B+N_{A1}) \approx M - N_B \qquad (10.1.10)$$

Substituting Equation 10.1.10 into Equation 10.1.9, we get:

$$\mu_{A1} = +k_B T \ln \frac{N_{A1}}{M} - k_B T \ln \frac{(M - N_B)}{M} \quad (10.1.11)$$

The second term in Equation 10.1.11 does not depend on $N_{A1}$. When we consider the difference in chemical potential between the two regions, we use Equation 10.1.11 to calculate the chemical potential in each region:

$$\Delta\mu_1 = \mu_{A2} - \mu_{A1} = k_B T \left( \ln \frac{N_{A2}}{M} - \ln \frac{N_{A1}}{M} \right) \quad (10.1.12)$$

Note that the $k_B \ln \left[ \frac{(M - N_B)}{M} \right]$ term is the same in both regions, so it cancels out in Equation 10.1.12. Equation 10.1.12 is the same as Equation 10.27 in the main text.

in terms of the number of molecules per unit volume (see Equations 10.28 and 10.29). If, instead, we define concentration as the number of moles per unit volume (that is, in molar units), then we must multiply the right-hand side of Equation 10.31 by Avogadro's number, $N_A$. This gives us Equation 10.15.

Suppose $C_1 > C_2$—that is, suppose region 1 has a higher concentration of the solute (A) than region 2 (this is the situation depicted in Figure 10.3), then $\ln\left(\frac{C_2}{C_1}\right)$ is negative, and $\Delta\mu$ is negative, too. Since $\Delta\mu = \mu_2 - \mu_1$, this means that the A molecules have a higher chemical potential in region 1 than in region 2. This makes sense, because we expect molecules to move spontaneously from region 1 (high concentration) to region 2 (low concentration), which is in the direction of decreasing chemical potential.

## 10.5 Chemical potentials at arbitrary concentrations are calculated with reference to standard concentrations

The discussion so far has treated the chemical potential as the free energy per molecule, whereas it is more convenient to treat the chemical potential as the *molar* free energy when studying biochemical reactions. Switching to molar units rather than numbers of molecules, Equation 10.31 can be rewritten as:

$$\Delta\mu = RT \ln\left(\frac{C_2}{C_1}\right) \quad (10.32)$$

where $C_1$ and $C_2$ are now concentrations in molar units instead of numbers of molecules per unit volume, $R$ is the gas constant, $T$ is the absolute temperature, and the chemical potential is now the free energy per mole rather than per molecule.

The chemical potential of a molecule under standard conditions is denoted by $\mu^o$. Equation 10.32 allows us to calculate the chemical potential of the molecule at nonstandard concentrations, if we know the value of $\mu^o$. When the concentration of the molecule is some value, $C$, other than the standard concentration, $C^o$, the chemical potential is given by:

$$\mu = \mu^\circ + RT \ln\left(\frac{C}{C^\circ}\right) = \mu^\circ + RT \ln C \quad (10.33)$$

The second equality in Equation 10.33 assumes that $C^o = 1$ M, as is true for most situations. Recall from Section 9.5 that one exception is water, for which the standard condition is pure water, with $[H_2O] = 55$ M. In biochemical calculations, the standard concentration of protons is taken to be that in pure water at 25°C, which is $[H^+] = 10^{-7}$ (pH 7).

# B. EQUILIBRIUM CONSTANTS

When a reaction comes to equilibrium, the concentrations of reactants and products are related to each other by the equilibrium constant. In this section, we define the equilibrium constant and show how it is related to the standard free energy change for the reaction ($\Delta G^{o}$).

## 10.6 The chemical potentials of the reactants and products are balanced at equilibrium

Consider a general chemical reaction:

$$\upsilon_A A + \upsilon_B B \rightleftharpoons \upsilon_C C + \upsilon_D D \tag{10.34}$$

where $\upsilon_A$, $\upsilon_B$, $\upsilon_C$, and $\upsilon_D$ are stoichiometric coefficients for the balanced chemical reaction. According to Equation 10.34, $\upsilon_A$ molecules of A combine with $\upsilon_B$ molecules of B to yield $\upsilon_C$ molecules of C and $\upsilon_D$ molecules of D.

The change in the free energy, $G$, as the reaction progresses is given by:

$$dG = \mu_A \, dn_A + \mu_B \, dn_B + \mu_C \, dn_C + \mu_D \, dn_D \tag{10.35}$$

Equation 10.35 follows directly from the definition of the chemical potential as the derivative of the free energy with respect to the numbers of moles of a molecule. Consult Box 8.3 for a brief review of differential calculus.

The changes in the numbers of moles, $dn_A$, $dn_B$, $dn_C$, and $dn_D$, are coupled to each other. They do not vary independently because changes in the numbers of A and B molecules must result in correlated changes in the numbers of C and D molecules.

**Reaction progress variable, $\xi$**

$\xi$ (Greek letter xi) measures how far a reaction has proceeded. If the stoichiometric coefficient for a reactant $i$ is $\upsilon_i$, then the amount of this reactant that has been consumed is given by $\upsilon_i d\xi$.

In order to account for the coupled nature of the changes in the number of molecules, we define a **reaction progress variable**, $\xi$ (Greek letter xi). $\xi$ is simply a measure of how far the reaction has progressed from the left to the right (**Figure 10.5**). When $\xi = 0$, the reaction has not occurred at all and when $\xi = 1$, all reactants have been converted to products. Suppose we start with an initial population of molecules denoted $n_A(0)$, $n_B(0)$, $n_C(0)$, and $n_D(0)$. As the reaction progresses, the value of $\xi$ increases and the population of molecules changes as follows:

$$dn_A = -\upsilon_A \, d\xi \tag{10.36}$$

$$dn_B = -\upsilon_B \, d\xi \tag{10.37}$$

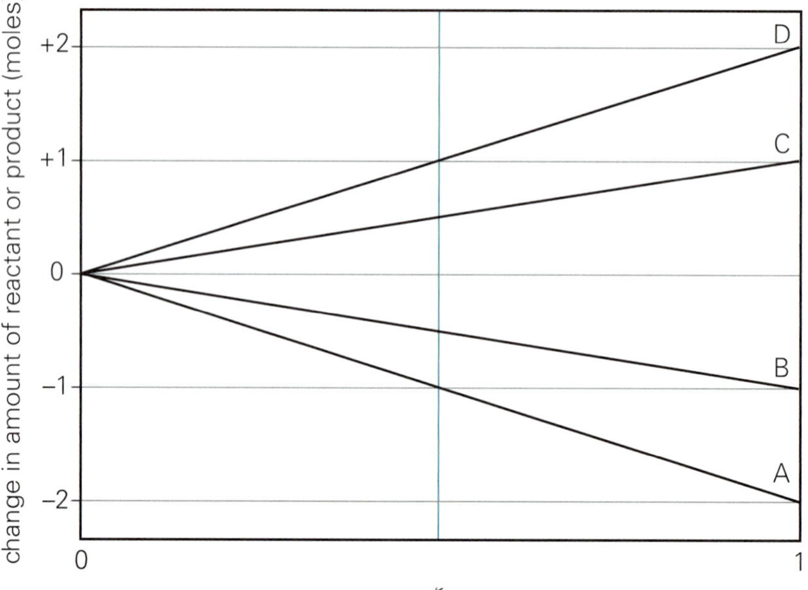


**Figure 10.5 The reaction progress variable $\xi$ keeps track of the balance in consumption and generation of reactants and products as a chemical reaction progresses.** For a given value of $\xi$, the change in the amount of a molecule (for example, A) is given by $\upsilon_A\xi$, where $\upsilon_A$ is the stoichiometric coefficient for A. For example, consider the following reaction:

$$2A + B \rightarrow C + 2D$$

Here, $\upsilon_A = 2$, $\upsilon_B = 1$, $\upsilon_C = 1$, and $\upsilon_D = 2$. The changes in the amounts of A, B, C, and D as $\xi$ changes are shown in the diagram.

$$dn_C = +\upsilon_C \, d\xi \tag{10.38}$$

$$dn_D = +\upsilon_D \, d\xi \tag{10.39}$$

In these equations, $dn_A$, $dn_B$, $dn_C$, and $dn_D$ are the changes in the number of moles of A, B, C, and D, respectively. The signs in Equations 10.36–10.39 reflect the fact that, as the reaction progresses, A and B molecules are consumed and C and D molecules are produced. Substituting these equations into Equation 10.35 we get:

$$dG = -\upsilon_A \mu_A \, d\xi - \upsilon_B \mu_B \, d\xi + \upsilon_C \mu_C \, d\xi + \upsilon_D \mu_D \, d\xi$$

or

$$dG = (-\upsilon_A \mu_A - \upsilon_B \mu_B + \upsilon_C \mu_C + \upsilon_D \mu_D) d\xi \tag{10.40}$$

When the reaction progresses to the equilibrium point, then the value of the free energy will be at a minimum (that is $dG = 0$). Imposing this condition on Equation 10.40, we get:

$$dG = 0 = (-\upsilon_A \mu_A - \upsilon_B \mu_B + \upsilon_C \mu_C + \upsilon_D \mu_D) d\xi \tag{10.41}$$

Since $d\xi$ can be nonzero, the only way for Equation 10.41 to always be true is for the following condition to hold:

$$-\upsilon_A \mu_A - \upsilon_B \mu_B + \upsilon_C \mu_C + \upsilon_D \mu_D = 0 \tag{10.42}$$

or

$$\upsilon_A \mu_A + \upsilon_B \mu_B = \upsilon_C \mu_C + \upsilon_D \mu_D \quad \text{(at equilibrium)} \tag{10.43}$$

Equation 10.43 tells us that at equilibrium the chemical potentials of reactants and products, multiplied by the stoichiometric coefficients, are balanced (**Figure 10.6**).

(A)

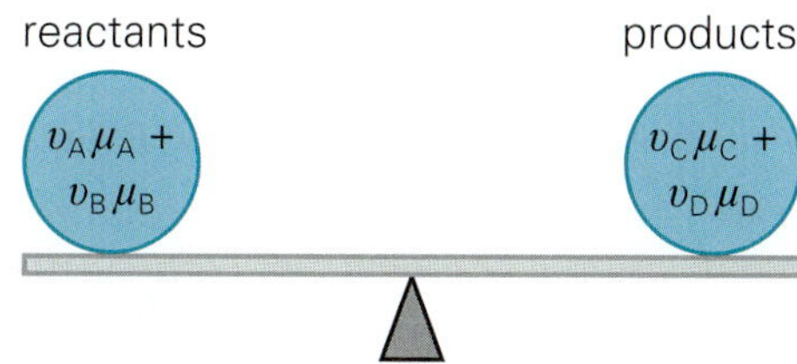


(B)

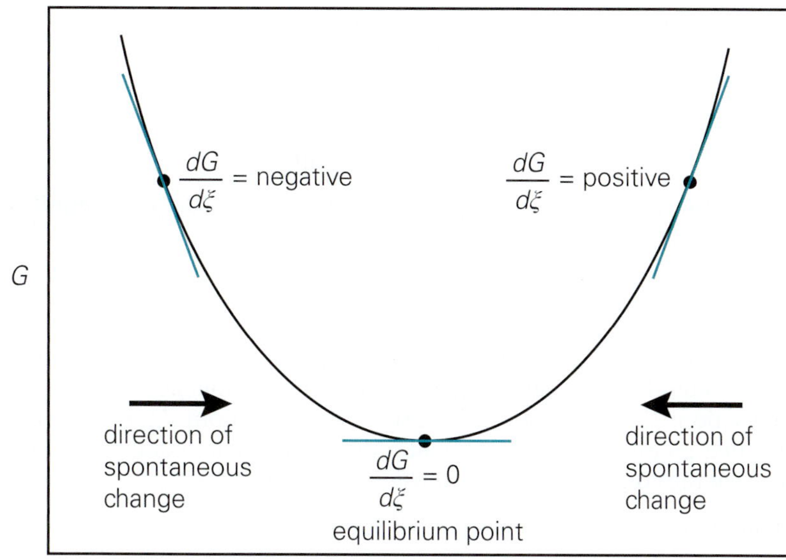


**Figure 10.6 Chemical potentials of reactants and products.** (A) The chemical potentials of the reactants and the products are balanced at equilibrium. (B) Spontaneous change in the progress of a reaction. The value of $dG/d\xi$ determines whether a chemical reaction will proceed to the right or to the left under a specified set of conditions.

## 10.7 The concentrations of reactants and products at equilibrium define the equilibrium constant (*K*), which is related to the standard free energy change ($\Delta G^o$) for the reaction

When the reaction comes to equilibrium, let the concentrations of A, B, C, and D be denoted by $[A]_{eq}$, $[B]_{eq}$, $[C]_{eq}$, and $[D]_{eq}$, respectively. The chemical potentials in Equation 10.43 refer to these equilibrium concentrations. According to Equation 10.33, the values of $\mu_A$, $\mu_B$, $\mu_C$, and $\mu_D$ are related to the standard state values of the chemical potentials ($\mu_A^o$, $\mu_B^o$, $\mu_C^o$, and $\mu_D^o$) by:

$$\mu_A = \mu_A^o + RT\ln\left(\frac{[A]}{[A]^o}\right) \tag{10.44}$$

$$\mu_B = \mu_B^o + RT\ln\left(\frac{[B]}{[B]^o}\right) \tag{10.45}$$

$$\mu_C = \mu_C^o + RT\ln\left(\frac{[C]}{[C]^o}\right) \tag{10.46}$$

$$\mu_D = \mu_D^o + RT\ln\left(\frac{[D]}{[D]^o}\right) \tag{10.47}$$

**Equilibrium constant**

The equilibrium constant, $K_{eq}$, for a reaction relates the concentrations of reactants and products at equilibrium. The value of $K_{eq}$ is given by:

$$K_{eq} = \frac{[C]_{eq}^{\upsilon_C}[D]_{eq}^{\upsilon_D}}{[A]_{eq}^{\upsilon_A}[B]_{eq}^{\upsilon_B}}$$

for a simple reaction involving A, B, C, and D. $[A]_{eq}$, $[B]_{eq}$, $[C]_{eq}$, and $[D]_{eq}$ are the concentrations at equilibrium, and $\upsilon_A$, $\upsilon_B$, $\upsilon_C$, and $\upsilon_D$, are the stoichiometric coefficients of the reaction.

In Equations 10.44–10.47, the terms $[A]^o$, $[B]^o$, etc., are the standard state concentrations. For most solutions, the standard state corresponds to a 1 M solution, and so these terms usually disappear. They have been retained in Equations 10.44–10.47 to remind us that the arguments of the logarithms are dimensionless numbers, and do not have units of concentration.

Substituting Equations 10.44–10.47 into Equation 10.43 with the equilibrium concentrations, we get:

$$\upsilon_A\mu_A^o + \upsilon_A RT\ln[A]_{eq} + \upsilon_B\mu_B^o + \upsilon_B RT\ln[B]_{eq}$$

$$= \upsilon_C\mu_C^o + \upsilon_C RT\ln[C]_{eq} + \upsilon_D\mu_D^o + \upsilon_D RT\ln[D]_{eq} \tag{10.48}$$

Rearranging terms, we get:

$$\upsilon_C\mu_C^o + \upsilon_C\mu_D^o - \upsilon_A\mu_A^o - \upsilon_B\mu_B^o = -RT\ln\frac{[C]_{eq}^{\upsilon_C}[D]_{eq}^{\upsilon_D}}{[A]_{eq}^{\upsilon_A}[B]_{eq}^{\upsilon_B}} \tag{10.49}$$

The terms on the left-hand side of the equation represent the free-energy change upon a complete conversion of stoichiometric equivalents of reactants to products under standard conditions—that is, the left-hand side is equal to the value of $\Delta G^o$ for the reaction. Thus, Equation 10.49 can be rewritten as:

$$\Delta G^o = -RT\ln\frac{[C]_{eq}^{\upsilon_C}[D]_{eq}^{\upsilon_D}}{[A]_{eq}^{\upsilon_A}[B]_{eq}^{\upsilon_B}} \tag{10.50}$$

We define the ratio of concentrations in Equation 10.50 as the **equilibrium constant**, $K_{eq}$, for the reaction. Thus,

$$K_{eq} = \frac{[C]_{eq}^{\upsilon_C}[D]_{eq}^{\upsilon_D}}{[A]_{eq}^{\upsilon_A}[B]_{eq}^{\upsilon_B}} \tag{10.51}$$

where $[A]_{eq}$, $[B]_{eq}$, $[C]_{eq}$, and $[D]_{eq}$ are the concentrations of the reactants and products at equilibrium, and $\upsilon_C$, $\upsilon_D$, $\upsilon_A$, and $\upsilon_B$ are the stoichiometric coefficients of the reaction. The subscript "eq" is usually dropped from $K_{eq}$ and the concentrations that go into it, but you should note that using an equilibrium constant implies that the concentrations are those at equilibrium.

The standard free energy change ($\Delta G^o$) is given by:

$$\Delta G^o = -RT\ln K_{eq} \tag{10.52}$$

It follows, from Equation 10.52, that the value of $K_{eq}$ is given by:

$$K_{eq} = e^{-\left(\frac{\Delta G^{o}}{RT}\right)} \tag{10.53}$$

The equilibrium constant, $K_{eq}$, must of necessity be a unitless number. The left-hand side of Equation 10.52, $\Delta G^{o}$, has units of energy (J•mol$^{-1}$). On the right-hand side, the $RT$ term has units of J•mol$^{-1}$, so $\ln K_{eq}$ (and therefore $K_{eq}$) must be dimensionless. If the stoichiometric coefficients ($\upsilon_C$, $\upsilon_D$, $\upsilon_A$, and $\upsilon_B$) are not equal to each other, then the right-hand side of Equation 10.51 might not appear to be a dimensionless number (that is, if $\upsilon_A$, $\upsilon_B$, $\upsilon_C$, and $\upsilon_D$ are all different, then all of the concentration units do not cancel out). Nevertheless, $K$ is always a unitless number because it should really be written as follows:

$$K_{eq} = \frac{\left(\frac{[C]_{eq}}{[C]^{o}}\right)^{\upsilon_C}\left(\frac{[D]_{eq}}{[D]^{o}}\right)^{\upsilon_D}}{\left(\frac{[A]_{eq}}{[A]^{o}}\right)^{\upsilon_A}\left(\frac{[B]_{eq}}{[B]^{o}}\right)^{\upsilon_B}} \tag{10.54}$$

where $[A]^{o}$, $[B]^{o}$, $[C]^{o}$, and $[D]^{o}$ are the concentrations of the reactants and products in their standard states. Because the values of these standard state concentrations are typically 1 M, they are usually not written out explicitly, but we must remember that $K_{eq}$ is a unitless number. In spite of this, equilibrium constants are often presented or spoken of as if they had concentration units, most commonly when describing the equilibrium constants for dissociation reactions, as we shall discuss in Chapter 12.

## 10.8 Equilibrium constants can be used to calculate the extent of reaction at equilibrium

Recall from Section 9.5 that the standard free-energy change for ATP hydrolysis, at pH 7.0 and 0.01 M $Mg^{2+}$ ion, is –28 kJ•mol$^{-1}$:

$$\text{ATP} + \text{H}_2\text{O} \rightarrow \text{ADP} + \text{P}_i \qquad \Delta G^{o} = -28\ \text{kJ•mol}^{-1} \tag{10.55}$$

This means that 1 mole of ADP and $P_i$ in a one molar solution is lower in free energy by 28 kJ•mol$^{-1}$ than 1 mole of ATP and 1 mole of water, under the given set of conditions. Does this mean that if ATP is allowed to react with water all of the ATP molecules will be converted to ADP? Let us analyze the extent to which this reaction does actually go to completion.

We start by calculating the value of the equilibrium constant, $K$:

$$K = \frac{[\text{ADP}][\text{P}_i]}{[\text{ATP}][\text{H}_2\text{O}]} \tag{10.56}$$

In Equation 10.56 we have written $K$ without a subscript, with the understanding that the concentrations in the equation are equilibrium values. Recall that each of the concentrations in the definition of $K$ is divided by the concentration in the standard state. For water, the standard state concentration is 55 M, and the effective concentration of water does not change when all the solutes are very dilute. The ratio of the concentration of water to the standard state concentration is therefore always very close to 1, in which case water can be dropped from Equation 10.56. The equilibrium constant is therefore rewritten as:

$$K = \frac{[\text{ADP}][\text{P}_i]}{[\text{ATP}]} \tag{10.57}$$

At room temperature the value of $RT$ is 8.314 × 298 = 2478 J•mol$^{-1}$, or 2.478 kJ•mol$^{-1}$. It follows from Equation 10.53 that the value of $K$ is:

$$K = e^{+28/2.478} = e^{+11.3} \approx \left(10^{0.43}\right)^{11.3} \approx 10^{5} \tag{10.58}$$

The concentration of phosphate in a living cell is maintained in the range of 1–10 mM. Assuming that the phosphate ion concentration is held fixed at a concentration of 10 mM ($10^{-2}$ M), the ratio of ADP to ATP at equilibrium is given by:

$$\frac{[\mathrm{ADP}]}{[\mathrm{ATP}]} \times 10^{-2} = 10^{5}$$

and so:

$$\frac{[\mathrm{ADP}]}{[\mathrm{ATP}]} = 10^{7} \tag{10.59}$$

Equation 10.59 tells us that at equilibrium almost all the ATP will indeed have converted to ADP, except for roughly 1 part per 10 million. The substantial free energy difference between ATP and its hydrolysis products does indeed drive the reaction nearly to completion.

For reactions with less negative standard free-energy changes, the extent of reaction at equilibrium will be smaller. Consider, for example, the hydrolysis of glycerol-1-phosphate:

$$\text{glycerol-1-phosphate} + \mathrm{H_2O} \longrightarrow \text{glycerol} + \mathrm{P_i}$$

$$\Delta G^{o} = -9.2\ \mathrm{kJ{\bullet}mol^{-1}} \tag{10.60}$$

In this case, the equilibrium constant, $K$, is given by:

$$K = \frac{[\text{glycerol}][\mathrm{P_i}]}{[\text{ glycerol-1-phosphate }]} \tag{10.61}$$

where:

$$K = e^{-\Delta G^{o}/RT} = e^{9.2/2.478} = e^{3.7} \approx \left(10^{0.43}\right)^{3.7} \approx 10^{1.6} \tag{10.62}$$

Again, assuming that the phosphate concentration remains constant at 10 mM,

$$\frac{[\text{glycerol}]}{[\text{ glycerol-1-phosphate }]} \times 10^{-2} = 10^{1.6}$$

$$\Rightarrow \frac{[\text{glycerol}]}{[\text{ glycerol-1-phosphate }]} = 10^{-3.6} \approx 3980 \tag{10.63}$$

In this case, the reaction goes to ~99.97% completion. As the value of $\Delta G^{o}$ becomes less negative, the equilibrium point of the reaction moves further and further away from completion.

## 10.9 The free-energy change for the reaction ($\Delta G$), not the standard free-energy change ($\Delta G^{o}$), determines the direction of spontaneous change

Let us return to the chemical reaction depicted in Equation 10.34:

$$\upsilon_{A}\mathrm{A} + \upsilon_{B}\mathrm{B} \rightleftharpoons \upsilon_{C}\mathrm{C} + \upsilon_{D}\mathrm{D}$$

In general, the concentrations of A, B, C, and D are arbitrary (for example, these chemicals can be added into the reaction chamber in arbitrary amounts), and the reaction is not at equilibrium. In particular, the concentrations of key metabolite molecules (for example, ATP) are held at levels that are far from equilibrium in a living cell.

Returning to Equation 10.40, the value of $dG$ is nonzero under nonequilibrium conditions:

$$dG = \left(-\upsilon_{A}\mu_{A} - \upsilon_{B}\mu_{B} + \upsilon_{C}\mu_{C} + \upsilon_{D}\mu_{D}\right)d\xi \neq 0 \tag{10.64}$$

**Reaction free energy, $\Delta G$**

$\Delta G$ is the molar free-energy change upon stoichiometric conversion of reactants to products, given their actual nonequilibrium concentrations.

We can rewrite Equation 10.64 as:

$$\frac{dG}{d\xi} = \upsilon_{C}\mu_{C} + \upsilon_{D}\mu_{D} - \upsilon_{A}\mu_{A} - \upsilon_{B}\mu_{B} = \Delta G \neq 0 \tag{10.65}$$

The second term in Equation 10.65 is the **free-energy change for the reaction** or the **reaction free energy** ($\Delta G$). It is the change in free energy upon carrying out a stoichiometric conversion of molar equivalent reactants to products, given the specified (that is, not necessarily equilibrium or standard state) concentrations of each of the molecular species.

$\Delta G$, and not $\Delta G^o$, determines the direction of spontaneous change. If $\Delta G$ is negative, the reaction will proceed as written, to the right. If $\Delta G$ is positive, then the reaction will proceed to the left (see Figure 10.6B).

## 10.10 The ratio of the reaction quotient (*Q*) to the equilibrium constant (*K*) determines the thermodynamic drive of a reaction

Now let us consider how to determine the value of $\Delta G$ when the concentrations of A, B, C, and D are not at equilibrium. We denote these concentrations as $[A]_{obs}$, $[B]_{obs}$, $[C]_{obs}$, and $[D]_{obs}$, respectively where the subscript denotes "observed" concentrations, as determined, perhaps, by measuring the actual concentrations of these molecules in a cell in the nonequilibrium condition.

The free-energy change for the reaction can be rewritten as:

$$\Delta G = \left[ \upsilon_C \mu_C^o + \upsilon_D \mu_D^o - \upsilon_A \mu_A^o - \upsilon_B \mu_B^o \right] + RT \ln \frac{[C]_{obs}^{\upsilon_C} [D]_{obs}^{\upsilon_D}}{[A]_{obs}^{\upsilon_A} [B]_{obs}^{\upsilon_B}}$$

$$\Rightarrow \Delta G = \Delta G^o + RT \ln Q \tag{10.66}$$

where $Q$ is defined by:

$$Q = \frac{[C]_{obs}^{\upsilon_C} [D]_{obs}^{\upsilon_D}}{[A]_{obs}^{\upsilon_A} [B]_{obs}^{\upsilon_B}}$$

$Q$ is known as the **reaction quotient**. Note that $Q$ is not the same as the equilibrium constant because $Q$ involves arbitrary nonequilibrium values of the concentrations of the reactants and products. If we know the values of $[A]_{obs}$, $[B]_{obs}$, $[C]_{obs}$, and $[D]_{obs}$, then we can calculate $Q$ and thereby determine if the reaction is at equilibrium or not.

**Reaction quotient**

Denoted $Q$, the definition of the reaction quotient looks similar to that of the equilibrium constant:

$$Q = \frac{[C]^{\upsilon_C} [D]^{\upsilon_D}}{[A]^{\upsilon_A} [B]^{\upsilon_B}}$$

The difference is that the concentrations in the definition of $Q$ are not equilibrium concentrations. Rather, they are the concentrations under the present observed conditions. The value of $Q$ helps us determine if the reaction will proceed to the left or to the right.

Combining Equations 10.52 and 10.66, we get:

$$\Delta G = \Delta G^o + RT \ln Q = -RT \ln K + RT \ln Q$$

$$\Rightarrow \Delta G = +RT \ln \frac{Q}{K} = +2.3RT \log_{10} \frac{Q}{K} \tag{10.67}$$

Equation 10.67 makes it clear that the capacity of a reaction to run in the forward direction is determined by the ratio of $Q$ to $K$, known as the **mass action ratio.** When $Q$ is less than $K$, (that is, the mass action ratio is less than 1), $\Delta G$ is negative, and the reaction progresses in the forward direction. On the other hand, when $Q$ is greater than $K$, the reaction will go backwards.

**Mass action ratio**

The ratio of the reaction quotient, $Q$, to the equilibrium constant, $K$, is known as the mass action ratio, $Q/K$. A reaction will proceed in the forward reaction if the mass action ratio is less than 1.0.

**Figure 10.7** shows the dependence of the free energy ($G$) on the ratio of $Q$ to $K$ for a simple reaction, $A \rightleftharpoons B$. The slope of the free-energy curve at any point gives the value of $\Delta G$. The reaction is at equilibrium when the mass action ratio is 1.0. For every 10-fold increase in the mass action ratio beyond 1.0, the value of $\Delta G$ increases by $2.3RT$ (see Equation 10.71), which corresponds to an increase of 5.8 $kJ \cdot mol^{-1}$ at a temperature of 300 K. Likewise, for every 10-fold decrease in the mass action ration below 1.0, the value of $\Delta G$ decreases by $2.3RT$.

## 10.11 ATP concentrations are maintained at high levels in cells, thereby increasing the driving force for ATP hydrolysis

We shall illustrate the importance of the mass action ratio by returning to the hydrolysis of ATP to yield ADP. The analysis in Section 10.9 demonstrates that if the ATP concentration in the cell is allowed to reach equilibrium, then there will

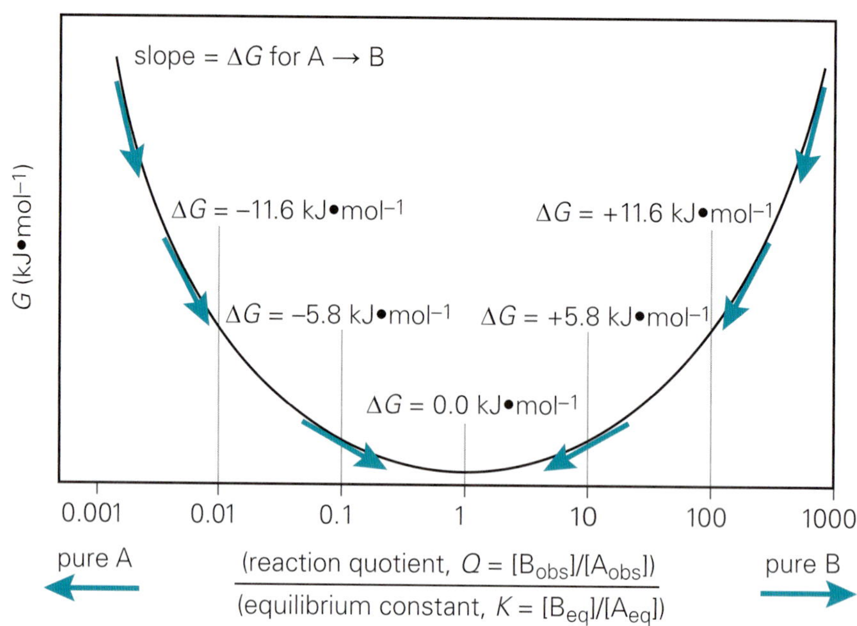


**Figure 10.7 Relationship between the free-energy change ($\Delta G$) and the mass action ratio.** The free energy ($G$) is indicated by a *black* line, and the direction of spontaneous change is indicated by arrows at selected points. When the mass action ratio is 1.0, the value of the free energy ($G$) is at a minimum and the value of the slope, which is $\Delta G$, is zero. For every factor of 10 increase in the value of the mass action ratio ($Q/K$) beyond the equilibrium value, the value of $\Delta G$ increases by 2.3$RT$ (see Equation 10.67), which corresponds to an increase of 5.8 kJ•M$^{-1}$ at a temperature of 300 K. For every 10-fold decrease below the equilibrium value, the value of $\Delta G$ decreases by the same amount. Values of $\Delta G$ for a temperature of 300 K ($RT \approx 2.5$ kJ•mol$^{-1}$) are indicated for selected values of the mass action ratio. (Adapted from D.G. Nicholls and S.J. Ferguson, Bioenergetics, 3rd ed. San Diego: Academic Press, 2002. With permission from Elsevier.)

eventually be very little ATP left in the cell (almost all of it will convert to ADP). In living cells, the ratio of ATP to ADP is held nearly constant, however, by the action of metabolic reactions that consume sugars and other sources of energy used to synthesize ATP. **Table 10.1** shows the value of the free energy change ($\Delta G$) for ATP hydrolysis for various values of the mass action ratio, assuming that the concentrations of phosphate and $Mg^{2+}$ ions are held constant at 10 mM, at pH 7.

ATP concentrations in the cytoplasm are held so that [ATP]/[ADP] is ~1000. When the mass action ratio is clamped at this level, the value of $\Delta G$ for the ATP hydrolysis reaction is −57 kJ•mol$^{-1}$—that is, substantially more favorable than the value of $\Delta G^{o}$ (−28 kJ•mol$^{-1}$). Under these conditions, the ATP hydrolysis reaction can be coupled to other reactions that are substantially uphill in terms of free energy, making such transformations possible. On the other hand, if the ATP concentration is allowed to dissipate, the free-energy change for ATP hydrolysis is drastically reduced and such energetic couplings will not be possible.

## C. ACID–BASE EQUILIBRIA

All of biochemistry involves reactions that do not go to completion, with reversible interconversion between reactants and products. Here we discuss an important class of such reactions, involving weak acids and bases.

**Table 10.1 Relationship between the free energy change ($\Delta G$) for ATP hydrolysis and the reaction quotient.**

| $Q$ | $Q/K$ | $\Delta G$ (kJ•mol$^{-1}$) | [ATP]/[ADP] | Relevant condition |
|---|---|---|---|---|
| $10^5$ | 1 | 0 | $10^{-7}$ | equilibrium |
| $10^3$ | $10^{-2}$ | −11 | $10^{-5}$ | |
| 1 | $10^{-5}$ | −28 | $10^{-2}$ | standard condition |
| $10^{-3}$ | $10^{-8}$ | −46 | $10^1$ | mitochondrial matrix |
| $10^{-5}$ | $10^{-10}$ | −57 | $10^3$ | cytoplasm |

(Adapted from D.G. Nicholls and S.J. Ferguson, Bioenergetics, 3rd ed. San Diego: Academic Press, 2002. With permission from Elsevier.)

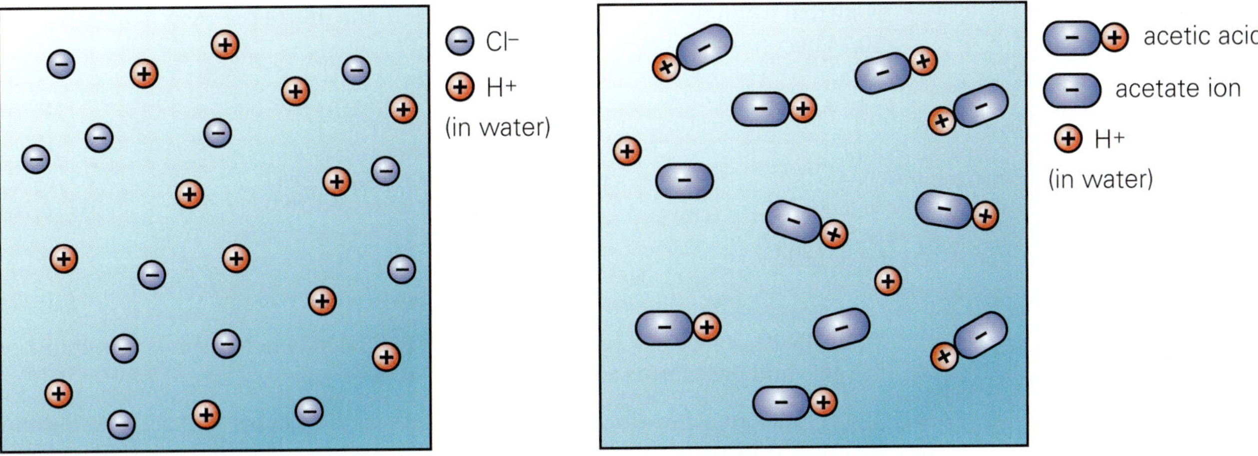


**Figure 10.8 Strong and weak acids.** A strong acid, such as HCl (left), dissociates completely in water. The resulting proton concentration is equal to the total concentration of the added acid. A weak acid, such as acetic acid (right), dissociates very little in water. The resulting proton concentration is very low compared to the total concentration of the weak acid.

## 10.12 The Henderson–Hasselbalch equation relates the pH of a solution of a weak acid to the concentrations of the acid and its conjugate base

An acid is a molecule that dissociates in water to release protons:

$$HA \rightleftharpoons H^+ + A^- \tag{10.68}$$

In Equation 10.68, HA refers to the acid. Dissociation of the acid releases a proton ($H^+$) and a base ($A^-$), which is referred to as the **conjugate base** of the acid.

A strong acid (such as HCl) dissociates completely in water (**Figure 10.8**). A weak acid, in contrast, dissociates only to a small extent. Acetic acid, $CH_3COOH$, which dissociates to yield a proton and the acetate ion, is a weak acid:

$$CH_3COOH \rightleftharpoons CH_3COO^- + H^+ \tag{10.69}$$

The equilibrium constant for the reaction depicted in Equation 10.68 is denoted $K_a$ (for **acid dissociation constant**):

$$K_a = \frac{[H^+][A^-]}{[HA]} \tag{10.70}$$

An important relationship between $K_a$ and the pH of a solution emerges if we take the logarithm of both sides of Equation 10.70:

$$\log_{10} K_a = \log_{10}[H^+] + \log_{10}\left(\frac{[A^-]}{[HA]}\right) \tag{10.71}$$

Recall that $pH = -\log_{10}([H^+])$, and so Equation 10.71 can be rewritten as:

$$\log_{10} K_a = -pH + \log_{10}\left(\frac{[A^-]}{[HA]}\right) \tag{10.72}$$

or

$$pH = -\log_{10} K_a + \log_{10}\left(\frac{[A^-]}{[HA]}\right) \tag{10.73}$$

The term $-\log_{10} K_a$ is defined as the **p$K_a$ value** of the weak acid, HA. Thus,

$$pH = pK_a + \log_{10}\left(\frac{[A^-]}{[HA]}\right) \tag{10.74}$$

Equation 10.74 is known as the **Henderson–Hasselbalch Equation.**

**p$K_a$ value**

The p$K_a$ value for a weak acid is the negative logarithm of the acid dissociation constant ($K_a$):

$$pK_a = -\log_{10} K_a$$

The p$K_a$ is the pH at which the concentration of the acid and its conjugate base are equal.

**Henderson–Hasselbalch equation**

The Henderson–Hasselbalch equation relates the pH of a solution to the ratio of the concentrations of a weak acid and its conjugate base:

$$pH = pK_a + \log_{10}\left(\frac{[A^-]}{[HA]}\right)$$

Note that when the pH of a solution is equal to the $pK_a$ of the weak acid, Equation 10.74 becomes:

$$pK_a = pK_a - \log_{10}\left(\frac{[A^-]}{[HA]}\right) \quad (10.75)$$

and

$$\log_{10}\left(\frac{[A^-]}{[HA]}\right) = 0 \text{ (for pH = } pK_a) \quad (10.76)$$

As a result,

$$[A^-] = [HA] \text{ (when pH = } pK_a) \quad (10.77)$$

Thus, the $pK_a$ of a weak acid is the value of the pH at which the concentrations of the acid and base forms are equal.

The Henderson–Hasselbalch equation allows us to calculate the pH of a solution that results from mixing the acid and base forms of a weak acid. We shall explore how this is done in the following sections.

## 10.13 The proton concentration ($[H^+]$) in pure water at room temperature corresponds to a pH value of 7.0

Water is itself a weak acid and can dissociate to release protons and hydroxide ions:

$$H_2O \rightleftharpoons H^+ + OH^- \quad (10.78)$$

Protons in water are present in a hydrated form (for example, $H_3O^+$), but to keep things simple we shall simply denote an aqueous proton as $H^+$. The equilibrium constant, $K_w$, for the dissociation of water is:

$$K_w = \frac{[H^+][OH^-]}{[H_2O]} \quad (10.79)$$

Water dissociates only to a small extent, so the concentration of undissociated water ($[H_2O]$) does not change significantly from its standard state value of 55 M (recall the discussion in Section 10.8). The ratio $[H_2O]/[H_2O]^o$ is very close to 1, so the concentration of $H_2O$ can be omitted in Equation 10.79:

$$K_w = [H^+][OH^-] = 1.0 \times 10^{-14} \text{ (at 25°C)} \quad (10.80)$$

The equilibrium constant for water dissociation ($K_w$) is also referred to as the **ion product of water**. At 25°C, the standard free-energy change for the reaction ($\Delta G^o$) is:

$$\Delta G^o = -RT \ln K_w = -2.5 \times -32.2 = +79.9 \text{ kJ•mol}^{-1} \quad (10.81)$$

Thus, the dissociation of water is highly unfavorable. Nevertheless, pure water does dissociate to a small extent, leading to a measurable concentration of protons.

**Ion product of water**

The dissociation constant for water, $K_w$, is also called the ion product of water.

$$K_w = [H^+][OH^-] = 1.0 \times 10^{-14}$$

at 25°C. The pH of pure water is:

$$-\log_{10}\sqrt{K_w}.$$

Thus, the pH of pure water is 7.0 at 25°C.

To calculate the pH of pure water, we use the fact that $[H^+] = [OH^-]$. This is true because there is no other source of protons or hydroxide ions. At 25°C, therefore,

$$K_w = [H^+][OH^-] = [H^+]^2$$

so

$$[H^+] = \sqrt{K_w} = 10^{-7} \text{ M} \quad (10.82)$$

and

$$pH = -\log[H^+] = 7.0$$

According to Equation 10.82, the pH of pure water is 7.0 at 25°C. This pH value is referred to as **neutral pH**. If the proton concentration in a solution is greater than $10^{-7}$ M (that is, the pH is lower than 7.0), the solution is referred to as **acidic**. Likewise, if the proton concentration is lower than $10^{-7}$ M (that is, the pH is greater than 7.0), the solution is referred to as **basic**.

## 10.14 The temperature dependence of the equilibrium constant allows us to determine the values of $\Delta H^o$ and $\Delta S^o$

The equilibrium "constant" for a reaction is only constant at a particular temperature, because its value changes if the temperature changes. To understand why this is so, consider that the standard change in enthalpy ($\Delta H^o$) and the standard change in entropy ($\Delta S^o$) are related to the equilibrium constant, $K$, by Equation 10.83:

$$\Delta G^o = \Delta H^o - T\Delta S^o = -RT\ln K \quad (10.83)$$

$$\ln K = -\frac{\Delta H^o}{RT} + \frac{\Delta S^o}{R} \quad (10.84)$$

where $T$ is the absolute temperature. In addition to the explicit dependence on temperature shown in Equation 10.84, the standard changes in enthalpy and entropy for reaction can also be temperature dependent.

We shall illustrate the temperature dependence of equilibrium constants by considering the ion product of water ($K_W$). As we shall see, the values of $\Delta H^o$ and $\Delta S^o$ for the dissociation of water are nearly temperature independent over a significant temperature range. Nevertheless, as Equation 10.84 makes clear, the value of $K_W$ is still temperature dependent because of the $1/T$ term.

Equation 10.84 is known as the **van't Hoff equation**, and it states that the logarithm of the equilibrium constant (ln $K_W$, in the case of water dissociation) varies linearly with $(1/T)$, provided that $\Delta H^o$ and $\Delta S^o$ are independent of temperature. Note that $T$ in Equation 10.84 is the absolute temperature, which must always be used when calculating thermodynamic quantities. Equation 10.84 will not hold if we express temperature on the Celsius or Fahrenheit scales.

**van't Hoff equation**

The van't Hoff equation gives the temperature dependence of the equilibrium constant, relating the logarithm of $K$ to the values of $\Delta H^o$ and $\Delta S^o$:

$$\ln K = -\frac{\Delta H^o}{RT} + \frac{\Delta S^o}{R}$$

Experimentally measured values of p$K_W$ (that is, $-\log_{10} K_W$) and pH for pure water are given in **Table 10.2**. Note that the pH drops from 7.47 to 6.51 as the temperature increases from 0°C to 60°C. The data in Table 10.2 allow us to check that the enthalpy and entropy changes for water dissociation are really independent of temperature. A graph of ln $K_W$ versus the inverse of the absolute temperature is shown in **Figure 10.9**. Notice that the temperature dependence of ln $K_W$ is well described by a straight line, which is consistent with the van't Hoff equation (that is, Equation 10.84), and confirms that $\Delta H^o$ and $\Delta S^o$ are indeed independent of temperature.

The equation for the straight line that best fits the data shown in Figure 10.9 is:

$$\ln K_w = -6723\left(\frac{1}{T}\right) - 9.718 \quad (10.85)$$

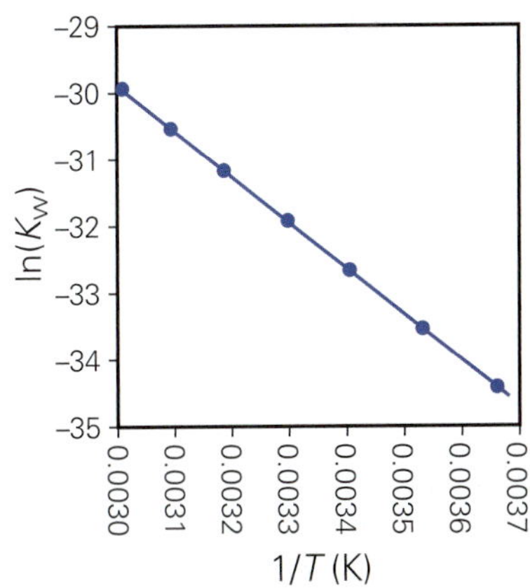


**Figure 10.9 The dissociation constant of water ($K_W$) varies with temperature.** Shown here is the dependence of ln($K_W$) as a function of $1/T$ based on the data given in Table 10.2. As predicted by the van't Hoff equation (that is, Equation 10.84), the graph of ln($K_W$) with respect to $1/T$ is a straight line. The slope of the line is −6723, and corresponds to the value of $-\left(\Delta H^o/R\right)$. The intercept of the line is −9.718, which corresponds to the value of $\Delta S^o/R$.

**Table 10.2 The pH and p$K_W$ of water as a function of temperature.**

| Temperature (°C) | p$K_W$ | pH (p$K_W$/2) |
|---|---|---|
| 0 | 14.9435 | 7.47 |
| 10 | 14.5346 | 7.27 |
| 20 | 14.1669 | 7.08 |
| 30 | 13.8330 | 6.92 |
| 40 | 13.5348 | 6.77 |
| 50 | 13.2617 | 6.63 |
| 60 | 13.0171 | 6.51 |

(Values from Handbook of Chemistry and Physics, 91st ed. Boca Raton: CRC Press, 2010.)

Comparing Equation 10.85 with the van't Hoff equation (Equation 10.84), we see that:

$$\frac{-\Delta H^o}{R} = -6723 \tag{10.86}$$

Thus,

$$\Delta H^o = R \times 6723 \text{ J•mol}^{-1} = +55.9 \text{ kJ•mol}^{-1} \tag{10.87}$$

Also,

$$\frac{\Delta S^o}{R} = -9.718$$

so

$$\Delta S^o = -R \times 9.718 = -80.8 \text{ J•K}^{-1}\text{•mol}^{-1} \tag{10.88}$$

The results of the van't Hoff analysis are consistent with the observation that the dissociation of water increases with increasing temperature (that is, the pH decreases with temperature). Because $\Delta H^o$ is positive, energy is required for the dissociation of water. As the temperature increases, more energy is available and the extent of dissociation increases.

The van't Hoff equation provides a convenient way to obtain experimental values for $\Delta H^o$ and $\Delta S^o$ for reactions where these terms are temperature independent. We shall use it in Chapter 18 to analyze the stability of duplex DNA. It should be noted, however, that in cases where the hydrophobic effect is substantial, the enthalpy changes are often temperature dependent. This is true, for example, when considering protein folding, as we discuss in part D of this chapter.

## 10.15 Weak acids, such as acetic acid, dissociate very little in water

Now let us see what happens when we dissolve a weak acid in water at room temperature (25°C). For example, consider a solution of 100 mM (0.1 M) acetic acid in water. The p$K_a$ value of acetic acid is 4.76. What is the pH of this solution? To solve this problem, we start by writing down several things we know about the system.

*Constraint 1:* The total amount of acetic acid (HA) and acetate ion ($A^-$) is fixed and is determined by the amount of acetic acid added:

$$[HA] + [A^-] = [AH]_{total} = 0.1 \text{ M} \tag{10.89}$$

*Constraint 2:* The solution always stays electrically neutral. There are two negatively charged species in the solution: acetate ion ($A^-$) and hydroxide ion ($OH^-$) from the dissociation of water. There is one positively charged species ($H^+$). Thus,

$$[A^-] + [OH^-] = [H^+] \tag{10.90}$$

*Constraint 3:* The ion product of water is constant at 25°C:

$$[H^+][OH^-] = 10^{-14} \tag{10.91}$$

*Constraint 4:* The p$K_a$ value for acetic acid is 4.76, so the equilibrium constant for the dissociation of acetic acid is:

$$K_a = \frac{[A^-][H^+]}{[HA]} = 10^{-4.76} = 1.74 \times 10^{-5} \tag{10.92}$$

We expect a solution of acetic acid in water to be acidic; that is, $[H^+] > 10^{-7}$. From constraint 3 we know that $[OH^-] = \frac{10^{-14}}{[H^+]}$. Thus, if the resulting pH is 6.0, $[OH^-] = 10^{-8}$. If the resulting pH is 5.0, then $[OH^-] = 10^{-9}$, and so on. As long as the resulting pH is one or more pH units below neutrality, we expect $[OH^-]$ to be at least 100-fold smaller than $[H^+]$. We can express this as follows:

$$[H^+] >> [OH^-] \tag{10.93}$$

Equation 10.93 tells us that we can neglect $[OH^-]$ in Equation 10.90, which yields:

$$[A^-] \approx [H^+] \tag{10.94}$$

To determine the value of $[H^+]$, we substitute the value of $[A^-]$ from Equation 10.94 in Equation 10.92:

$$\frac{[H^+]^2}{[HA]} = 1.74 \times 10^{-5} \tag{10.95}$$

Combining Equations 10.89 and 10.94, we get:

$$[HA] = 0.1 - [H^+] \tag{10.96}$$

Combining Equations 10.95 and 10.96, we get:

$$\frac{[H^+]^2}{0.1 - [H^+]} = 1.74 \times 10^{-5} \tag{10.97}$$

To solve this equation, denote $[H^+]$ by $x$ and rearrange to get:

$$x^2 + 1.74 \times 10^{-5}(x - 0.1) = 0 \tag{10.98}$$

We can solve this quadratic equation for $x$, obtaining:

$$x = [H^+] = 1.31 \times 10^{-3} \tag{10.99}$$

Thus, $\text{pH} = -\log_{10}([H^+]) = 2.88$.

These results allow us to calculate the degree of dissociation of acetic acid in water. Since, $[H^+] \approx [A^-]$ we see that

$$\frac{[A^-]}{[HA]_{total}} = \frac{1.31 \times 10^{-3}}{0.1} = 1.31 \times 10^{-2} \tag{10.100}$$

Thus, ~99% of the acetic acid is in the undissociated form (that is, $CH_3COOH$), in which case the concentration of the acid, [HA], is essentially unchanged from the total amount added. This property allows acetic acid to buffer the pH of a solution when combined with a source of acetate ion, as we see in the next section.

## 10.16 Solutions of weak acids and their conjugate bases act as buffers

The conjugate base of acetic acid is the acetate ion ($CH_3COO^-$). If we dissolve a salt of the acetate ion (for example, sodium acetate) in water, the salt dissociates completely:

$$Na^+CH_3COO^- \xrightarrow{\text{water}} Na^+ + CH_3COO^- \tag{10.101}$$

Now consider what happens if we mix an equimolar amount of acetic acid and sodium acetate. As we reasoned in the previous section, a solution of acetic acid in water does not dissociate appreciably. When we add an equivalent of sodium acetate to this solution, very little of the added acetate ion can combine with protons to form acetic acid. As a result:

$$[A^-] \approx [HA] \tag{10.102}$$

We can now calculate the pH of the resulting solution by using the Henderson–Hasselbalch equation (Equation 10.74):

$$\begin{aligned} \text{pH} &= \text{p}K_a + \log\left(\frac{[A^-]}{[HA]}\right) \\ &= \text{p}K_a + \log(1) \\ \Rightarrow \text{pH} &= \text{p}K_a = 4.76 \end{aligned} \tag{10.103}$$

In general, the pH of an equimolar solution of a weak acid and the salt of its conjugate base is the same as the $\text{p}K_a$ value of the weak acid.

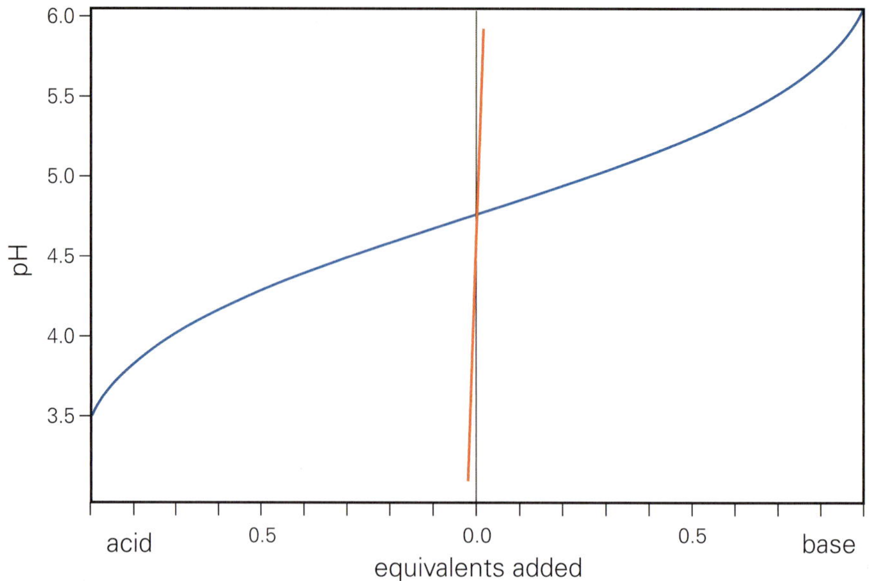


**Figure 10.10 A solution of a weak acid and its conjugate base acts as a buffer against changes in pH.** Shown here as a blue curve are the pH values, calculated using the Henderson–Hasselbalch equation, as a result of adding strong acid or strong base to a solution containing equimolar 0.1 M acetic acid and sodium acetate. The *red* line represents the pH change that would occur in the absence of buffer.

Now consider what happens when we add a strong acid to a solution of acetic acid and acetate ion (**Figure 10.10**). For example, consider a solution in which acetic acid and acetate ion are both at 5 mM, to which 1 mM HCl is added. The HCl dissociates completely, releasing 1 mM $H^+$ ions into solution. These excess protons will react with acetate ion to form acetic acid, which alters the ratio of [HA] to [$A^-$]. We can now use these altered concentrations to calculate the new pH value:

$$\begin{aligned} \mathrm{pH} &= \mathrm{p}K_\mathrm{a} + \log\left(\frac{[\mathrm{A}^-]}{[\mathrm{HA}]}\right) \\ &= 4.76 + \log\left(\frac{5-1}{5+1}\right) = 4.76 + \log(0.67) \\ \mathrm{pH} &= 4.58 \end{aligned} \tag{10.104}$$

The addition of 1 mM HCl to this solution has lowered the pH by only 0.18 pH units, from 4.76 to 4.58.

Consider, instead, what would happen if 1 mM HCl were added to a solution of pure water. In this case, the resulting [$H^+$] concentration would be 1 mM ($10^{-3}$ M). The resulting pH is therefore 3.0, a drop of four pH units from that of pure water. Thus, the presence of the acetic acid/acetate acid–base pair in solution dramatically reduces the change in pH upon adding a strong acid.

A similar effect is seen upon adding a strong base (for example, NaOH) to the solution containing acetic acid and acetate ion (see Figure 10.10). In this case, the presence of $OH^-$ ions in solution causes acetic acid to dissociate, yielding acetate ion and $H^+$. The protons then react with $OH^-$ to generate water. Again, interconversion between the acid and base forms of the weak acid stabilizes (that is, buffers) the pH of the solution.

We can change the reference pH of the buffer solution by altering the molar ratios of the acid and base forms. For example, if the solution contains 10 equivalents of acetic acid to one equivalent of acetate ion, then the Henderson–Hasselbalch equation tells us that:

$$\mathrm{pH} = 4.76 + \log(0.1) = 3.76 \tag{10.105}$$

If, instead, we add one equivalent of acetic acid to 10 equivalents of acetate ion, then the resulting pH is:

$$\mathrm{pH} = 4.76 + \log(10) = 5.76 \tag{10.106}$$

The pH within a eukaryotic cell is held between 7.0 and 7.4. Nature uses two buffers to control pH in cells and tissues. The first buffer involves hydrogen phosphates:

$$H_2PO_4^- \rightleftharpoons H^+ + HPO_4^{2-} \; (pK_a = 7.2) \tag{10.107}$$

The other is carbon dioxide/bicarbonate:

$$H_2O + CO_2 \rightleftharpoons H_2CO_3 \rightleftharpoons HCO_3^- + H^+ \; (pK_a = 6.36) \tag{10.108}$$

## 10.17 The charges on biological macromolecules are affected by the pH

Equilibria involving weak acids and bases are of great importance in biology. Many of the amino acid sidechains, as well as the N-terminal amino group and the C-terminal carboxyl group of the protein backbone, can pick up or lose protons, depending on the pH of the solution (**Figure 10.11**). This alters the charge

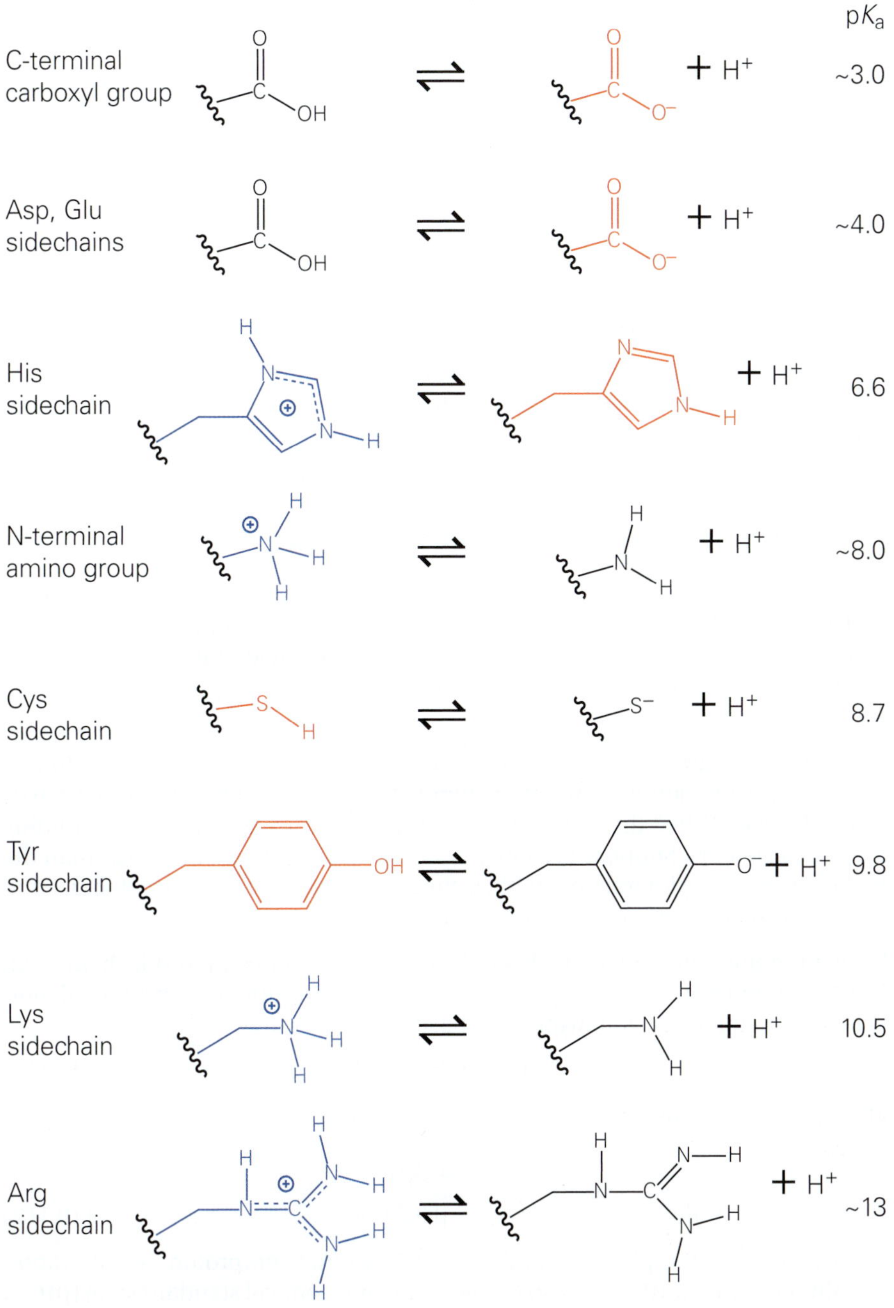

**Figure 10.11 Acid–base equilibria in protein groups.** Shown here are sidechains, as well as the terminal amino and carboxy groups of the backbone, that can exist in acidic or basic forms, along with their p$K_a$ values. The acid or base form that is prevalent at neutral pH is drawn in color. The p$K_a$ values given here are those for isolated amino acids, or unfolded protein chains. Interactions between sidechains can dramatically alter the p$K_a$ values of these groups, as discussed in the main text. (Adapted from J. Kyte, Structure in Protein Chemistry, 2nd ed. New York: Garland Science, 2007.)

on the sidechain, which can have profound consequences for the function of the protein. DNA bases can also become charged upon protonation. In Chapter 2, for example, we discussed the formation of Hoogsteen base pairs between a positively charged cytosine and a guanine (see Figure 2.37). Phosphate groups in the backbone of DNA and in molecules such as ATP also participate in acid–base equilibria. Weak acids and bases are critical for buffering the pH, both inside cells and in the extracellular environment within tissues.

To illustrate this phenomenon, we shall consider the histidine sidechain, which can exist in either a neutral form (His) or a protonated form ($His^+$; see Figure 10.11). The $pK_a$ value of the histidine sidechain in an isolated amino acid is ~6.0. If the pH of a solution is 7.0, what fraction of the histidine molecules are expected to be charged? We can calculate this ratio by using the Henderson–Hasselbalch equation:

$$\log\left(\frac{[\text{His}]}{[\text{His}^+]}\right) = \text{pH} - \text{pK}_a = 7-6=1$$
$$\text{so} \quad \frac{[\text{His}]}{[\text{His}^+]} = 10 \text{ (when pH} = 7) \tag{10.109}$$

Thus, roughly 10% of the histidine sidechains will be protonated at pH 7. On the other hand, if the pH is reduced to 5, then:

$$\log\left(\frac{[\text{His}]}{[\text{His}^+]}\right) = 5-6=-1$$
$$\text{and so} \quad \frac{[\text{His}]}{[\text{His}^+]} = 0.1 \tag{10.110}$$

This analysis tells us that a change in pH of a single unit above or below the $pK_a$ value of the histidine sidechain can lead to a drastic change in the relative populations of the charged and uncharged forms of the sidechain.

## 10.18 The charge on an amino acid sidechain can be altered by interactions in the folded protein

In Section 10.17, we have seen that because the $pK_a$ value of a histidine sidechain is ~6.0 at neutral pH, only ~10% of the molecules will carry a positive charge on the sidechain. This is true for an isolated histidine amino acid, but the situation can be quite different in a folded protein molecule.

Consider a histidine sidechain that is placed next to a negatively charged aspartic acid sidechain in a folded protein (**Figure 10.12**). When the histidine is charged, a strong ion-pairing interaction is formed between it and the aspartic acid sidechain. When the histidine is neutral, a much weaker hydrogen-bonding interaction occurs. Suppose the ion pair is ~35 kJ•mol$^{-1}$ lower in energy than the hydrogen-bonding interaction. What is the consequence of this extra stabilization for the $pK_a$ of this particular histidine sidechain?

To analyze this situation, let us begin by considering a histidine sidechain in an isolated amino acid—that is, one that is not interacting with another charged molecule. It undergoes the following acid–base equilibrium:

$$\text{His}^+ \rightleftharpoons \text{His} + \text{H}^+ \tag{10.111}$$

The equilibrium constant for this reaction is given by:

$$K = \frac{[\text{His}][\text{H}^+]}{[\text{His}^+]} \tag{10.112}$$

We assume that the pH is buffered at pH 7, so that the proton concentration remains constant at $10^{-7}$. We also consider the biochemical standard state (pH 7),

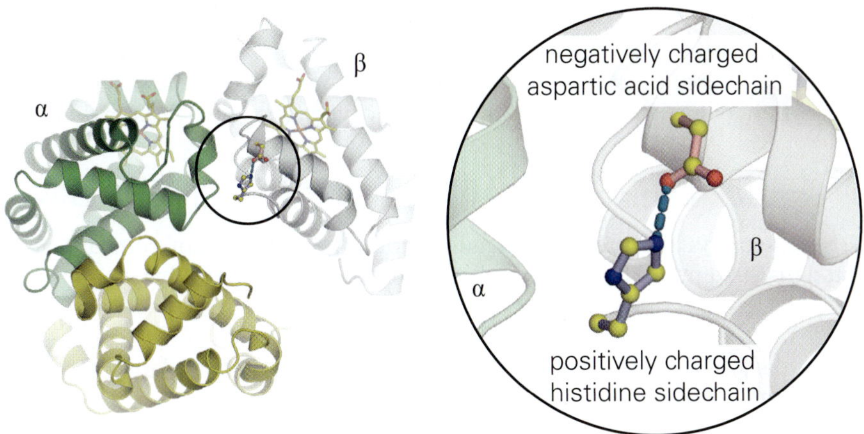


**Figure 10.12 A histidine sidechain interacting with an aspartic acid sidechain in hemoglobin.** The proximity of the negatively charged aspartic acid group will favor the protonated and positively charged histidine sidechain, because of the stronger ion pair that results. The aspartic acid sidechain increases the p$K_a$ value of this histidine sidechain. The importance of this interaction for the function of hemoglobin is discussed in Chapter 14 (see Figure 14.25).

so that the proton concentration can be neglected in Equation 10.112 (recall that the equilibrium constant expression always involves dividing the concentrations by the standard state values). Thus, the expression for $K$ can be written as:

$$K = \frac{[\text{His}]}{[\text{His}^+]} \text{ (biochemical standard state)} \tag{10.113}$$

In Section 10.17, we found that the ratio of [His] to [His$^+$] is 10 at neutral pH, so $K$ = 10. The standard free-energy change for the reaction at 300 K is therefore:

$$\Delta G^{o\prime} = -RT \ln K = -8.314 \times 300 \times \ln 10 = -5.7 \text{ kJ}\bullet\text{mol}^{-1} \tag{10.114}$$

According to Equation 10.114, the neutral form of the histidine is more stable at pH 7, as expected. In Equation 10.114, the primed symbol, $\Delta G^{o\prime}$ indicates the biochemical standard state (pH 7).

In the folded protein, the presence of the aspartic acid sidechain stabilizes His$^+$ relative to His by 35 kJ•mol$^{-1}$ (we are ignoring any changes in the entropy). As a consequence, the value of $\Delta G^{o\prime}$ for the deprotonation reaction becomes +29.3 kJ•mol$^{-1}$ (that is, -5.7 + 35 = +29.3). We can now calculate the value of the new equilibrium constant:

$$K = e^{-\Delta G^{o}/RT} = e^{\frac{-29{,}300}{8.314 \times 300}} = 7.9 \times 10^{-6} \tag{10.115}$$

Thus,

$$\frac{[\text{His}]}{[\text{His}^+]} = 7.9 \times 10^{-6} \text{ (in the folded protein)} \tag{10.116}$$

We can now calculate the p$K_a$ value for the histidine sidechain in the folded protein by rearranging the Henderson–Hasselbalch equation:

$$\begin{aligned} pK_a &= \text{pH} - \log\left(\frac{[\text{His}]}{[\text{His}^+]}\right) \\ &= 7.0 - \log(7.9 \times 10^{-6}) \\ &= 7.0 + 5.1 \\ &= 12.1 \end{aligned} \tag{10.117}$$

Equation 10.117 presents us with a remarkable result, which is that the p$K_a$ of the histidine sidechain has increased by ~6 pH units, to ~12 from ~6. An isolated histidine sidechain is half deprotonated at pH 6. In contrast, the presence of the aspartic acid sidechain stabilizes the protonated form of histidine, and as a consequence the proton concentration has to be reduced a million-fold before this sidechain becomes neutral. At neutral pH, this particular histidine sidechain will be essentially fully protonated.

The alteration of the $pK_a$ values of amino acid sidechains in folded proteins is often critical to protein function. In Chapter 14, we shall see that increases in the $pK_a$ values of histidine sidechains, due to interactions with acidic sidechains, is an important element of the high efficiency of oxygen transport by hemoglobin.

# D. FREE-ENERGY CHANGES IN PROTEIN FOLDING

We have seen, in the preceding sections, that the equilibrium constant, $K$, is related to the standard free-energy change for the reaction, $\Delta G^o$, by:

$$K = e^{-\Delta G^o/RT} \tag{10.118}$$

The standard free-energy change depends, in turn, on the standard changes in enthalpy (essentially, the energy) and entropy:

$$\Delta G^o = \Delta H^o - T\Delta S^o \tag{10.119}$$

Thus, in order to understand the molecular basis for the value of the equilibrium constant for a reaction, we need to know something about the differences in energy and entropy between reactants and products. Here we use protein folding as an example in order to see how the energy and the entropy change during a reaction, and how these changes can be measured.

## 10.19 The protein folding reaction is simplified by ignoring intermediate conformations

We first consider a highly simplified model for protein folding that allows us to calculate the energy and entropy changes for the folding reaction. The point of this analysis is to get a qualitative understanding of the terms that enter into the free energy of a system. In order to do this in a straightforward way, we shall introduce several approximations that help make the calculation particularly simple. A realistic calculation of the free energy of protein folding would require us to calculate the energy of the protein and the surrounding water molecules accurately. In addition, we would have to calculate the entropy of the protein and the water molecules, in both the folded and the unfolded states, by calculating the probabilities of observing different conformations of the protein and the surrounding water. Such an analysis would require very extensive computer-based calculations. The simple treatment presented here captures some of the essential concepts underlying protein folding, particularly that of the hydrophobic effect, without requiring computers. But, keep in mind that the treatment has been highly simplified.

The protein folding reaction is the conversion of unfolded protein chains, which do not have unique conformations, to folded chains with well-defined conformations. Protein folding is discussed in more detail in Chapter 18, where we shall see that small protein domains fold cooperatively—that is, once a chain starts to fold, it very quickly transitions into the folded state. Because of the high degree of cooperativity in the reaction, at equilibrium there is typically a very low population of structures that have an intermediate degree of folding. We shall ignore intermediate conformations and simply assume that a protein molecule is either completely folded or completely unfolded (**Figure 10.13**).

If we denoted the folded protein as F and the unfolded molecule as U, the protein folding reaction can be written as:

$$\mathrm{U} \rightleftharpoons \mathrm{F} \tag{10.120}$$

The equilibrium constant, $K$:

$$K_{\text{folding}} = \frac{[\mathrm{F}]}{[\mathrm{U}]} \text{ (at equilibrium)} \tag{10.121}$$

The subscript for $K$ in Equation 10.121 denotes that we are considering the folding reaction. In subsequent sections, when we discuss experimental measurements, we consider the unfolding reaction, instead:

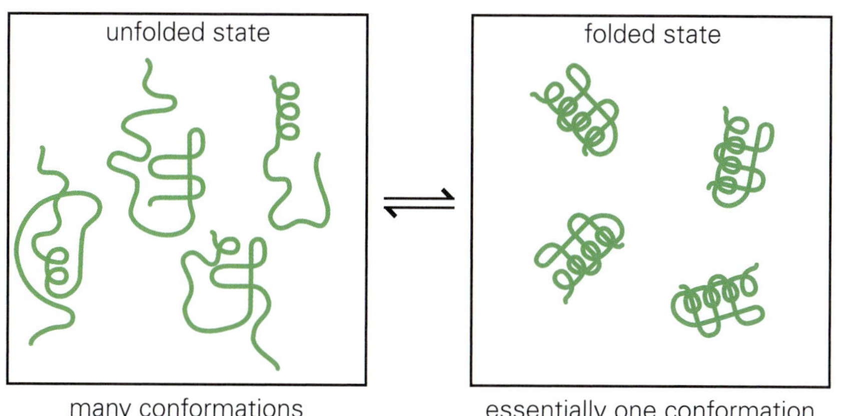


**Figure 10.13 A schematic representation for a simplified version of a protein-folding reaction.** The unfolded state is represented on the left, consisting of random conformations of the protein backbone. Each of the unfolded conformations is assumed to have the same energy, which is a simplification. On the right is shown the folded conformation, which is assumed to be unique.

$$\mathrm{F} \rightleftharpoons \mathrm{U} \tag{10.122}$$

The equilibrium constant for this reaction is given by:

$$K_{\text{unfolding}} = \frac{[\mathrm{U}]}{[\mathrm{F}]} = K_{\text{folding}}^{-1} \tag{10.123}$$

Note that Equations 10.120—10.123 hold for a monomeric protein. Proteins that form dimers or higher-order multimers exhibit more complex equilibria.

## 10.20 Protein folding results from a balance between energy and entropy

There are many different conformations that are accessible to the protein in the unfolded state (see Figure 10.13). When the protein folds up, however, it becomes conformationally restricted, and the folded state corresponds to far fewer distinct conformations. The entropy of the protein molecules must decrease substantially upon converting from the unfolded state to the folded state.

The *decrease* in entropy of the protein chain opposes protein folding because the value of $-T\Delta S$ is positive when $\Delta S$ is negative (see Equation 10.119). Nevertheless, protein molecules do fold spontaneously, so the overall free energy change upon folding must be negative.

It is natural to think that protein folding might be driven by more favorable energetic interactions in the folded protein. For example, there are numerous hydrogen bonds in the folded protein, most of which are disrupted in the unfolded protein (**Figure 10.14**). The close packing of atoms in the interiors of folded proteins also contributes to stabilization, as a consequence of favorable van der Waals interaction energies (**Figure 10.15**).

The extent to which these factors actually stabilize the folded structure is reduced significantly, however, by the additional interactions that the protein makes with water molecules in the unfolded state. Water molecules form strong hydrogen bonds with all of the exposed polar groups in an unfolded protein (see Figure 10.14). Thus, when a protein unfolds, it exchanges intramolecular hydrogen bonds with hydrogen bonds to water. As a consequence, the energetic contribution of hydrogen bonding ends up being small, although it generally favors the formation of the folded structure. Likewise, the extent to which van der Waals interactions stabilize the folded structure is important but relatively small (see Figure 10.15).

It turns out that a major driving "force" underlying protein folding is the hydrophobic effect, which is the tendency of nonpolar groups to sequester themselves away from water. The origins of the hydrophobic effect are complex and are still not completely understood. One important component of the hydrophobic effect is the decrease in the entropy of water molecules when nonpolar groups are

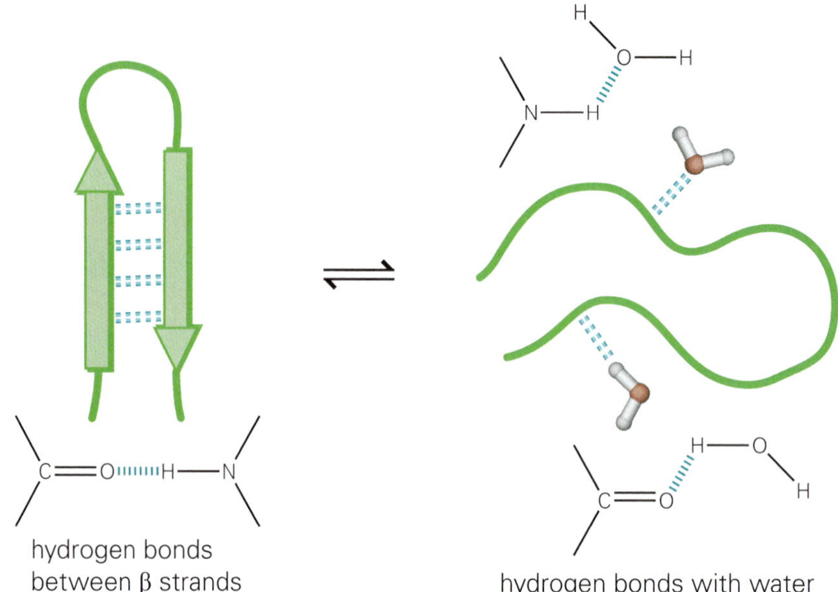


**Figure 10.14 Hydrogen bonds in protein folding.** In a folded protein structure, most hydrogen-bonding groups interact with other groups within the protein. When the protein unfolds, these groups interact with water instead. The difference in hydrogen-bonding energy between the two situations is close to zero, but the hydrogen bonds in the folded protein might have slightly lower energy if the groups are favorably oriented.

introduced into water. In other words, the entropy of water molecules *increases* when nonpolar groups are removed from water, in particular by clustering inside the folded protein.

## 10.21 The entropy of the unfolded protein chain is proportional to the logarithm of the number of conformations of the chain

The entropy, $S$, of a system is given by (see Equation 8.13):

$$S = -Nk_B \sum_i p_i \ln p_i \tag{10.124}$$

where the sum is over all the different energy levels or configurations of the molecules, $N$ is the number of molecules, and $p_i$ is the probability of observing a molecule in the $i^{\text{th}}$ configuration (that is, the relative population of molecules in that configuration).

We shall make the grossly simplifying assumption that all unfolded conformations of the protein that are significantly populated have the same energy (**Figure 10.16**). Let us assume that there are $N_C$ possible unfolded conformations. Each unfolded conformation is equally probable, and any particular protein molecule

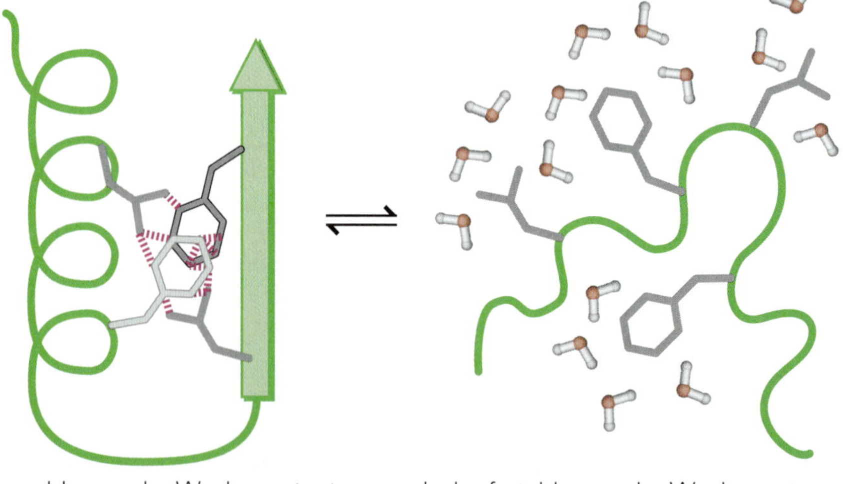


**Figure 10.15 van der Waals contacts in protein folding.** In the folded protein structure, shown on the left, atoms in the core of the protein make numerous contacts that are favorable in terms of the van der Waals energy. Although the contribution of each individual contact to the total energy is small, the large number of such contacts can lower the energy of the folded protein significantly. In the unfolded protein, contacts between atoms is fleeting, and the protein is not stabilized to the same extent by van der Waals interactions.

can be in one of $N_C$ conformations (**Figure 10.17**). If there are a total of $N$ different protein molecules in the system, then each of these protein molecules can be in one of the $N_C$ conformations. The entropy of the unfolded state, $S_{unfolded}$, is then given by:

$$S_{unfolded} = -Nk_B \sum_i^{N_C} p_i \ln p_i \tag{10.125}$$

The summation in Equation 10.129 runs over all of the $N_C$ conformations. The probability, $p_i$, is the probability of a protein adopting the $i^{th}$ conformation. Since all of the unfolded conformations have the same energy, they are all equally likely, and:

$$p_i = \frac{1}{N_C} \tag{10.126}$$

$$S_{unfolded} = -Nk_B \sum_i^{N_C} \frac{1}{N_C} \ln \frac{1}{N_C} = +Nk_B \frac{N_C}{N_C} \ln N_C = Nk_B \ln N_C \tag{10.127}$$

Hence, the entropy per molecule is:

$$\frac{S_{unfolded}}{N} = k_B \ln N_C \tag{10.128}$$

We will switch to using units of J•mol$^{-1}$ for energy, and J•mol$^{-1}$•K$^{-1}$ for entropy. In order to do that, we have to consider the entropy per mole of molecules, and so we multiply both sides of Equation 10.128 by Avogadro's number, $N_A$, to obtain the following expression for the entropy of a mole of protein molecules. Assuming standard conditions, the molar entropy is denoted:

$$S^o_{unfolded} = N_A \times \frac{S_{unfolded}}{N} = N_A k_B \ln N_C = R \ln N_C \tag{10.129}$$

Equation 10.129 presents a very simple result, which is that the molar entropy of the unfolded chain is proportional to the logarithm of the number of conformations of the unfolded chain. Strictly speaking, Equation 10.129 applies only to the case where all of the conformations have the same energy. Nevertheless, the logarithmic relationship between the number of conformations and the entropy is very generally applicable. Higher-energy conformations do exist. But, as the Boltzmann distribution tells us, they have lower probability of occurring than low-energy conformations, and the probability decreases exponentially as the energy of the conformation increases. As a consequence, high-energy conformations make a much smaller contribution to the entropy.

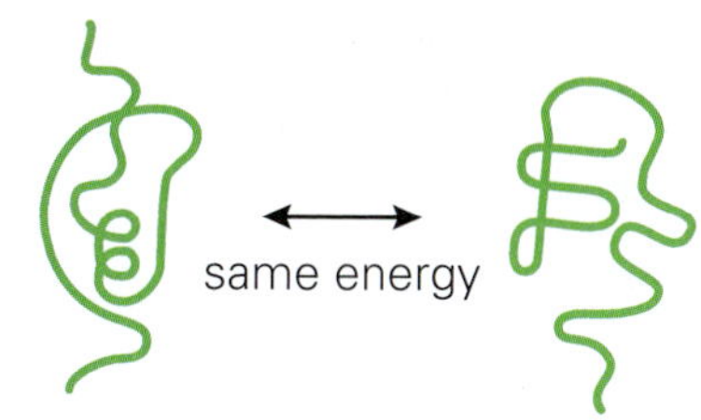


**Figure 10.16 Energy of unfolded conformations.** In order to simplify the calculation of the entropy, we shall assume that all possible conformations of the protein interact similarly with the water molecules and have the same energy.

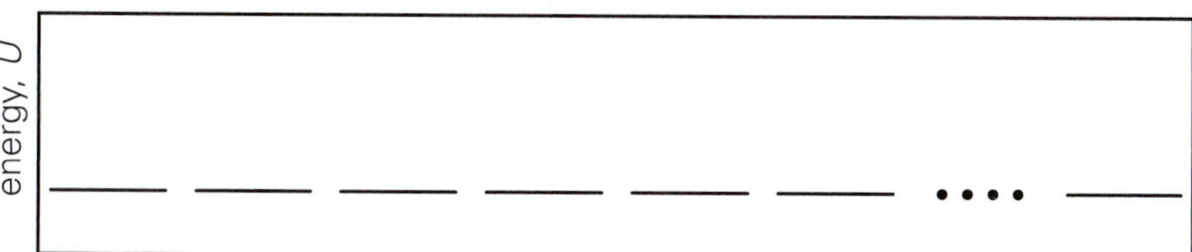


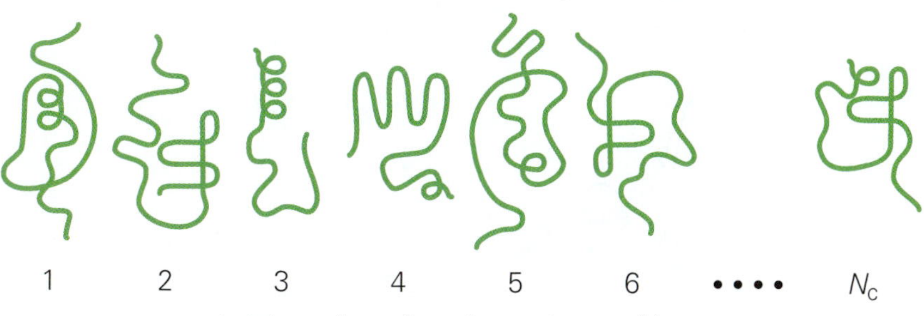


**Figure 10.17 All unfolded conformations are equally probable.** We assume that there are $N_C$ unfolded conformations. The energy of each conformation is shown in this diagram, along with the structure. Because all of the energies are the same, the probability of finding any of these conformations is the same.

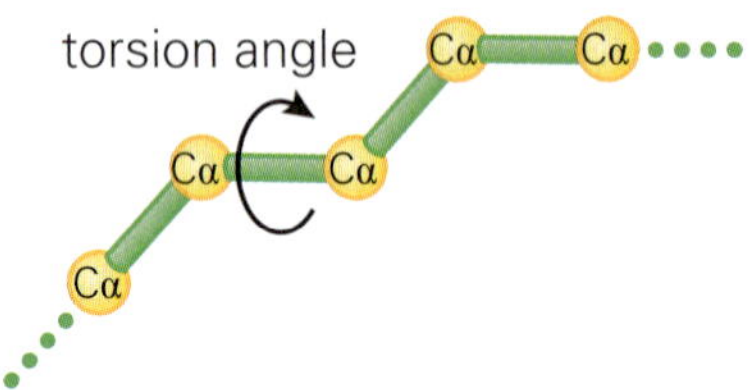


**Figure 10.18 Simplified representation of the protein backbone.** The backbone atoms of each residue are represented by a single sphere at the $C_\alpha$ position. Alternative conformations of the protein chain are generated by rotations about the bonds linking these spheres, as illustrated for one such torsion angle.

## 10.22 The number of conformations of the unfolded chain can be estimated by counting the number of low-energy torsional isomers

Although we have reduced the problem of calculating the entropy of the protein chain to one of counting the number of low-energy conformations of the chain, this is still a very difficult problem. As discussed in Section 4.9, alternative conformations of a protein molecule that have low energy arise from changes in the torsion angles (dihedral angles) between groups of $\phi$ and $\psi$. In addition, each sidechain typically has one or more dihedral angles that can also vary. This leads to an enormous number of possible combinations of torsion angles, many of which will be disallowed (that is, have very high energy) due to collisions between atoms. The Ramachandran diagram, for example, describes which combinations of $\phi$ and $\psi$ are disallowed due to collisions between atoms (see Figure 4.20).

The complexity of torsion angle changes in proteins means that we cannot accurately estimate the number of low-energy conformations that can be adopted by a protein chain without using very powerful computers. Nevertheless, we shall simplify matters by considering a very simple model for the protein chain, one that allows us to count the number of possible conformations.

First, we represent the polypeptide backbone by considering only the $C_\alpha$ atoms—that is, the backbone atoms of each residue are replaced by a single sphere (**Figure 10.18**). Each sphere is connected to the adjacent ones by bonds, and rotations about these bonds (torsional rotations) alter the conformation of the protein. Next, we assume that the torsion angle at each linkage can only be in one of three stable states, as illustrated in **Figure 10.19**. That is, we assume that there are only three possible conformations at each residue–residue linkage. This corresponds to assuming that only the three lowest energy combinations of $\phi$ and $\psi$ in the Ramachandran diagram have a high probability of occurring (see Figure 4.21).

We also need to consider the conformation of the sidechains. For simplicity, we shall assume that each sidechain is in one of two stable conformational states, as illustrated in **Figure 10.20**.

The combination of backbone and sidechain conformations gives a total of six conformations per residue (3 × 2). We denote the number of amino acid residues in the protein chain by $N_{res}$. The total number of conformations adopted by the chain, $N_C$, is then given by:

$$N_C = 6 \times 6 \times 6 \times \ldots \times 6 \ (N_{res} \text{ times}) = 6^{N_{res}} \tag{10.130}$$

Some of these conformations will not be allowed (that is, they will have impossibly high energy) because certain combinations of torsion angles will cause the chain to intersect itself, although most do not. We will ignore this complication

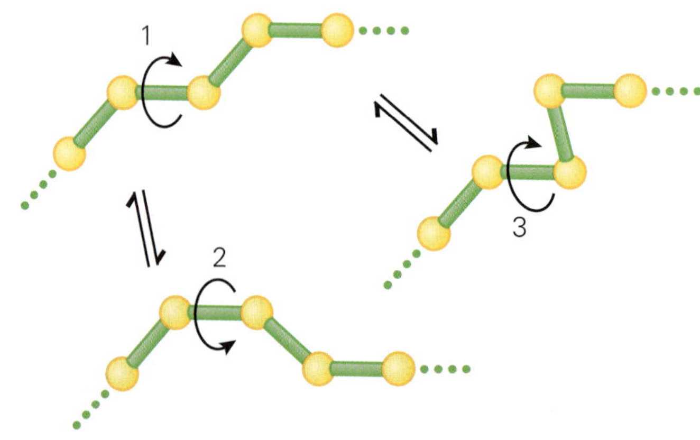


**Figure 10.19 Allowed conformations of the peptide backbone in the simplified model.** We assume that the torsion angle describing rotations about the bond connecting adjacent $C_\alpha$ atoms in the simplified model can take on only three different values. This $C_\alpha$–$C_\alpha$ bond is an imaginary bond, and rotations about it correspond to different combinations of $\phi$ and $\psi$, which are the actual torsion angles governing backbone conformation.

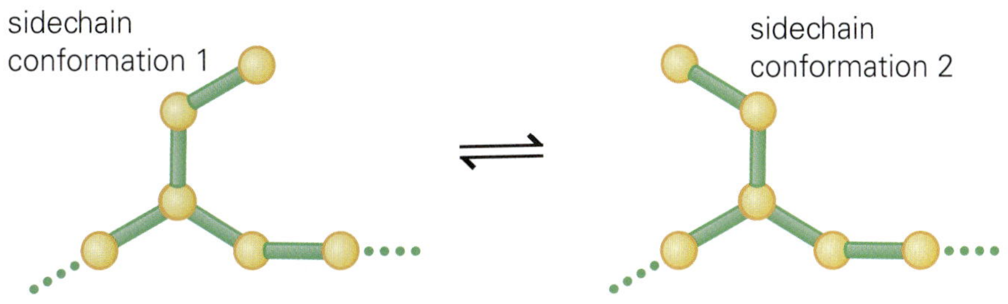


**Figure 10.20 Sidechain conformations.** For simplicity, we assume that each sidechain can be in one of two conformations, of equal energy.

and assume that the number of conformations of the protein chain, $N_C$, is given by $6^{N_{res}}$, where $N_{res}$ is the number of residues in the protein.

By combining Equations 10.129 and 10.130, we see that the molar entropy, $S^o_{unfolded}$, of the unfolded state is then given by:

$$S^o_{unfolded} = R\ln 6^{N_{res}} = RN_{res}\ln 6 \quad (10.131)$$

Thus, for a small protein containing 50 residues (that is, $N_{res} = 50$),

$$S^o_{unfolded} = R \times 50 \times 1.79 = R \times 89.5 \quad (10.132)$$

## 10.23 The free-energy change opposes protein folding if the entropy of water molecules is not considered

In the simple model that we are considering, the entropy of the folded chain is zero, because it is assumed to have only one conformation. Thus, according to Equation 10.132, the change in entropy upon folding for a protein with 50 amino acid residues is given by:

$$\Delta S^o = 0 - S^o_{unfolded} = -S^o_{unfolded} = -R \times 89.5 \quad (10.133)$$

We also need to calculate the change in enthalpy, $\Delta H^o$. If we assume that changes in the volume of the protein solution are negligible, $\Delta H^o$ is equal to the change in energy ($\Delta U^o$). Again, we take a very simple approach. Each residue in the folded protein actually makes a complicated set of interactions, but we shall ignore all of the detail and simply assume that each residue makes one favorable interaction, on balance. As we have argued previously, because of the loss of interactions with water, the net energetic stabilization of the protein is expected to be quite small. We shall assume that the molar change in energy of the chain upon forming this interaction is 3 kJ•mol$^{-1}$. This value is simply a guess, but is about that expected for the strength of a hydrogen bond after removal of the interacting groups from water. Thus, the molar change in enthalpy (energy) upon folding, $\Delta H^o$, is given by:

$$\Delta H^o = -N_{res} \times 3 = -150 \text{ kJ}\bullet\text{mol}^{-1} \quad (10.134)$$

This value is in agreement with the experimental data discussed in Section 10.28.

At 300 K, the standard change in Gibbs free energy, $\Delta G^o$, is given by:

$$\Delta G^o = \Delta H^o - T\Delta S^o \quad (10.135)$$

Using Equation 10.133 for the value of $\Delta S^o$, we get:

$$T\Delta S^o = +300 \times 89.5 \times 8.314 \text{ J} = 223 \text{ kJ}\bullet\text{mol}^{-1}$$

so

$$\Delta G^o = \Delta H^o - T\Delta S^o = -150 + 223 = +73 \text{ kJ}\bullet\text{mol}^{-1} \quad (10.136)$$

According to Equation 10.136, the change in free energy is positive, and so the protein will *not* fold spontaneously. This is because the reduction in entropy of the protein chain opposes folding. In other words, the value of $-T\Delta S^o$ (that is, +223 kJ•mol$^{-1}$) is positive and greater in magnitude than the value of the $\Delta H^o$ term (that is, −150 kJ•mol$^{-1}$).

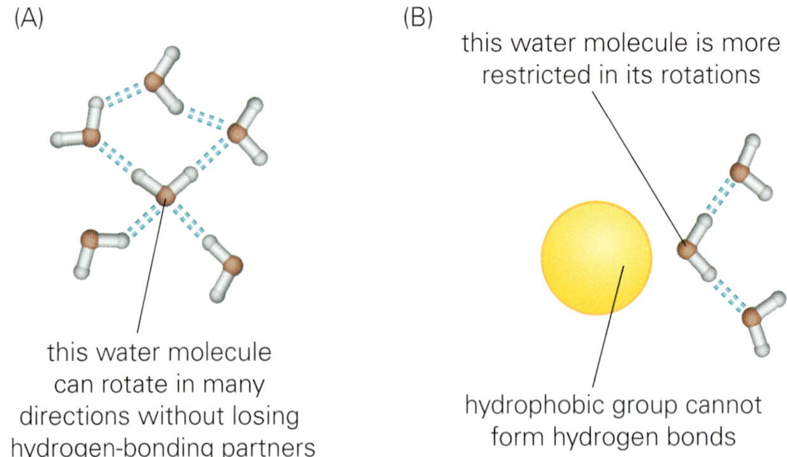


**Figure 10.21 Hydrophobic groups restrict the motional freedom (entropy) of water.** A water molecule in bulk water can rotate in many directions and find suitable hydrogen-bonding partners in other water molecules. If a nonpolar group is present in water, then water molecules near it are restricted.

So, what then drives protein folding? To understand this, we need to consider the entropy of water, which we have ignored up to now.

## 10.24 Protein folding is driven by an increase in water entropy

The protein chain consists of two kinds of groups, those that like to interact with water (hydrophilic) and those that do not (hydrophobic). Hydrophobic groups restrict the rotation of water molecules because water cannot form hydrogen bonds with them. To see why this is so, consider a water molecule that is in bulk water (that is, it is surrounded only by other water molecules). As this water molecule tumbles, it can form hydrogen bonds with other water molecules while changing its orientation (**Figure 10.21**). Compare this situation with one in which the water molecule is adjacent to a nonpolar (hydrophobic) group. If the water molecule were to tumble into an orientation so that it pointed its hydrogen atoms or lone pair of electrons towards the nonpolar group, then its ability to engage in hydrogen-bonding interactions is reduced (see Figure 10.21B). As a consequence, the energy goes up when the water molecule breaks hydrogen bonds with other water molecules and moves into orientations where it cannot form as many hydrogen bonds.

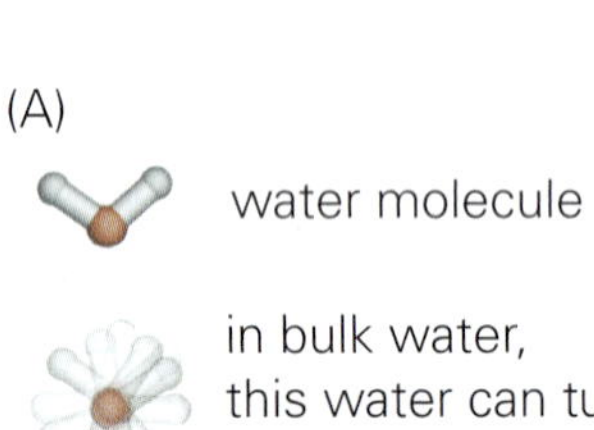


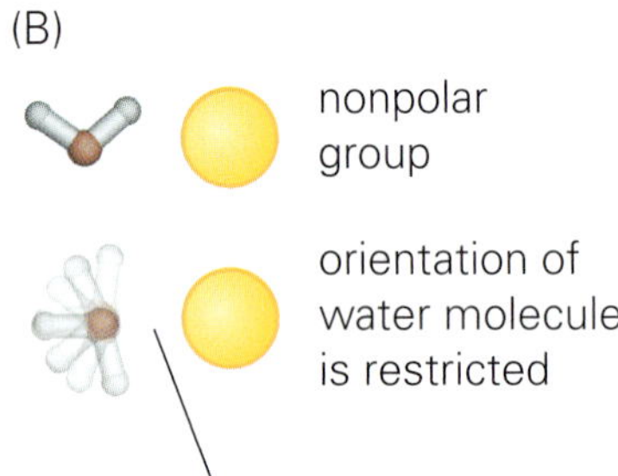


**Figure 10.22 Nonpolar groups restrict the orientations of water molecules.** (A) A water molecule in bulk water is shown, with no restrictions on its orientation. (B) A water molecule near a nonpolar group. The orientations of this water molecule are restricted because the energy goes up if it positions hydrogen-bonding groups near the nonpolar group.

A realistic treatment of the entropy of water is very complicated and requires the calculation of the probabilities of many different configurations of a large number of water molecules. Instead, let us assume that the number of low-energy rotational conformations for the central water molecule goes from six in the first case (when it is surrounded only by water molecules) to two in the second case (when a nonpolar group displaces some of the water molecules). In reality, water molecules are not restricted to a specific set of orientations. Rather, this reduction in allowed configurations is meant to reflect a restriction in the allowed orientations of the water, as illustrated in **Figure 10.22**.

Assume that half the residues in the protein are hydrophobic (25 out of 50), and that each interacts with two water molecules in the unfolded state, as shown in **Figure 10.23**. Considering only the waters that are affected by the hydrophobic groups, we can now calculate the change in the entropy of the water molecules when the protein folds. We shall do this by using Equation 10.129—that is, by relating the entropy to the logarithm of the number of conformations.

When the protein is unfolded, there are $2N_{res}/2$ water molecules that are affected (that is, half of the residues are hydrophobic, and these affect two water molecules each). Each of these water molecules is in one of two conformations, of equal energy. Thus, the total number of water conformations is $2 \times 2 \times 2 \times \ldots$ ($N_{res}$ times, because there are $2N_{res}/2 = N_{res}$ water molecules affected), which is $2^{N_{res}}$. Thus, using Equation 10.129:

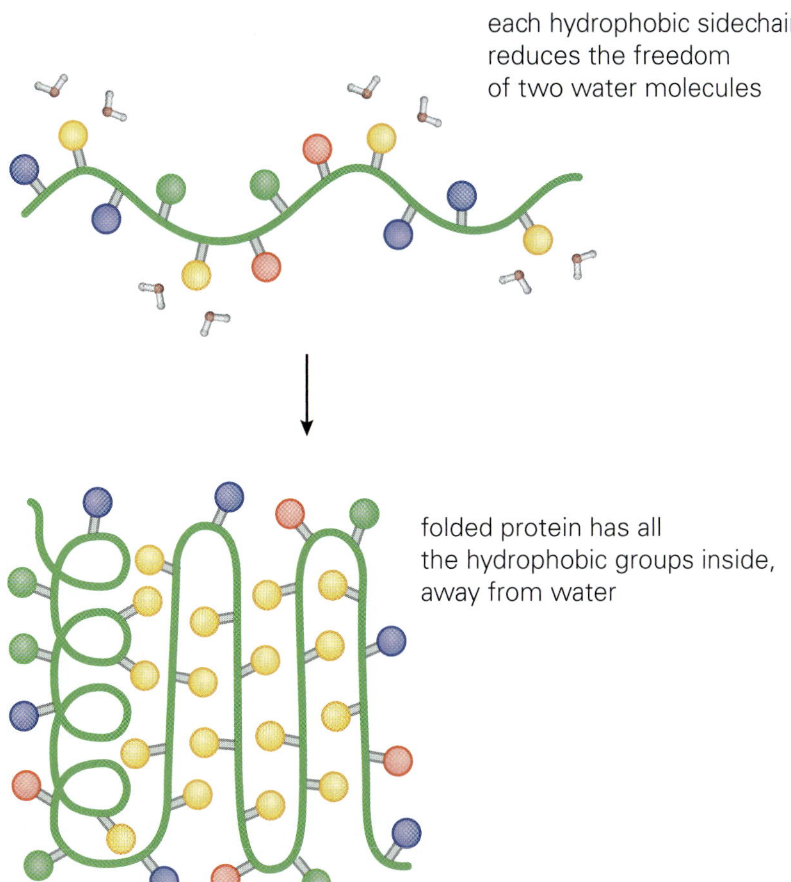


**Figure 10.23 Hydrophobic groups in protein folding.** In our simple model, we assume that half of the protein sidechains are hydrophobic, and that all of them are buried in the folded protein. In the unfolded state, each hydrophobic sidechain interacts with two water molecules. We assume that these water molecules are the only ones affected by the unfolded protein.

$$S^{o}_{\text{water,unfolded}} = R\ln 2^{N_{\text{res}}} = R \times N_{\text{res}} \ln 2 \qquad (10.137)$$

where $S^{o}_{\text{water, unfolded}}$ is the molar entropy of the water molecules that are affected by the hydrophobic groups when the protein molecule is unfolded. When the protein folds up, each of these water molecules is released, and the total number of conformations for these water molecules is now 6 × 6 × 6 × ... ($N_{\text{res}}$ times), which is $6^{N_{\text{res}}}$. Thus,

$$S^{o}_{\text{water,folded}} = R\ln 6^{N_{\text{res}}} = R \times N_{\text{res}} \ln 6 \qquad (10.138)$$

Using Equations 10.137 and 10.138, the standard change in entropy of the water is calculated to be:

$$\Delta S^{o}(\text{water}) = S^{o}_{\text{water,folded}} - S^{o}_{\text{water,unfolded}} = R \times N_{\text{res}}(\ln 6 - \ln 2) = R \times N_{\text{res}} \times 1.1 \qquad (10.139)$$

If we assume that $\Delta H^{o}$ is zero for the water molecules (the number and strengths of hydrogen bonds made by water do not change), then

$$\Delta G^{o}\,(\text{water}) = -T \times \Delta S^{o}\,(\text{water}) = -300 \times R \times 55 = -137 \text{ kJ}\bullet\text{mol}^{-1} \qquad (10.140)$$

The net change in free energy for the folding reaction is given by:

$$\Delta G^{\circ}(\text{total}) = \Delta G^{\circ}(\text{protein}) + \Delta G^{\circ}(\text{water}) \qquad (10.141)$$

Substituting the values of $\Delta G^{o}$(protein) from Equation 10.136 and $\Delta G^{o}$(water) from Equation 10.140, we get:

$$\Delta G^{o}\,(\text{total}) = +73 - 137 = -64 \text{ kJ}\bullet\text{mol}^{-1} \qquad (10.142)$$

Thus, the overall value of the standard free-energy change for folding is negative. The water molecules make a critical contribution to the protein folding reaction through the hydrophobic effect, without which folding would not be spontaneous.

The value of $\Delta G^{\circ}$ calculated above is indeed roughly the amount by which the free energy changes when a small, very stable protein molecule folds, as we shall see in the next few sections. The calculation we have done is very crude and contains many arbitrary assumptions, but it captures the importance of the hydrophobic effect in protein folding.

## 10.25 Calorimetric measurements allow the experimental determination of the free energy of protein folding

We now consider how the changes in energy and entropy that occur during protein folding are determined experimentally. Calorimetry, an experimental technique that measures how much heat is absorbed by a system, allows us to determine the values of the standard enthalpy change ($\Delta H^{\circ}$) and the standard entropy change ($\Delta S^{\circ}$) for protein folding reactions.

A schematic diagram of an instrument known as a **differential scanning calorimeter** is shown in the **Figure 10.24**. The instrument consists of two identical chambers, denoted A and B in Figure 10.24, which are in a thermally isolated container. Chamber A contains a solution of the protein of interest, at a known concentration and buffered at a fixed pH. Chamber B contains everything in chamber A (for example, the buffer solution), but without the protein molecules. By using heating elements and electronic controls, the temperature of the solutions in both chambers is gradually increased in lockstep, so that the temperature of chamber A is always the same as that of chamber B. This gradual increase in temperature is referred to as a "temperature scan."

Because chamber A contains protein molecules and chamber B does not, more heat is transferred to chamber A than to chamber B in order to increase the temperature by the same amount in each. In other words, the heat capacity of chamber A is greater than that of chamber B (see Section 6.8). By analyzing the difference in heat capacity between the two chambers as the temperature is scanned, we can determine the heat capacity of the protein, and how it changes as the temperature increases.

## 10.26 The heat capacity of a protein solution depends on the relative population of folded and unfolded molecules, and on the energy required to unfold the protein

The differential heat capacity as a function of temperature for the small protein lysozyme is shown in **Figure 10.25**. The differential heat capacity is the excess heat capacity of the protein molecules (and associated water molecules and ions) with respect to buffer solution, and we shall simply refer to it as the heat capacity of the protein. The heat capacity of the protein illustrated in Figure 10.25 increases

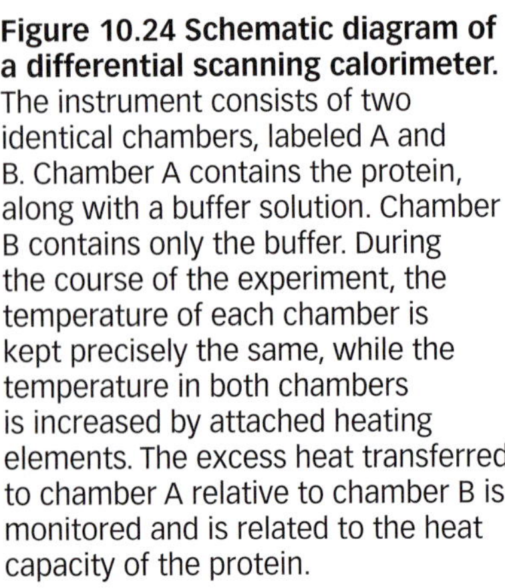

**Figure 10.24 Schematic diagram of a differential scanning calorimeter.** The instrument consists of two identical chambers, labeled A and B. Chamber A contains the protein, along with a buffer solution. Chamber B contains only the buffer. During the course of the experiment, the temperature of each chamber is kept precisely the same, while the temperature in both chambers is increased by attached heating elements. The excess heat transferred to chamber A relative to chamber B is monitored and is related to the heat capacity of the protein.

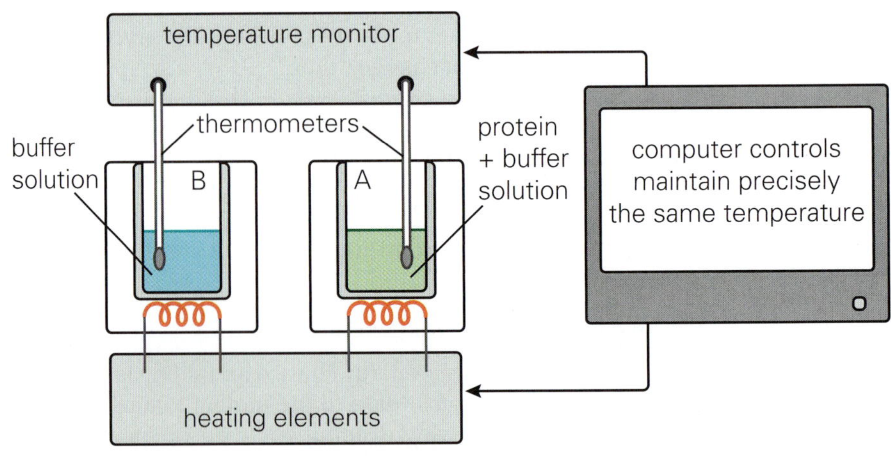

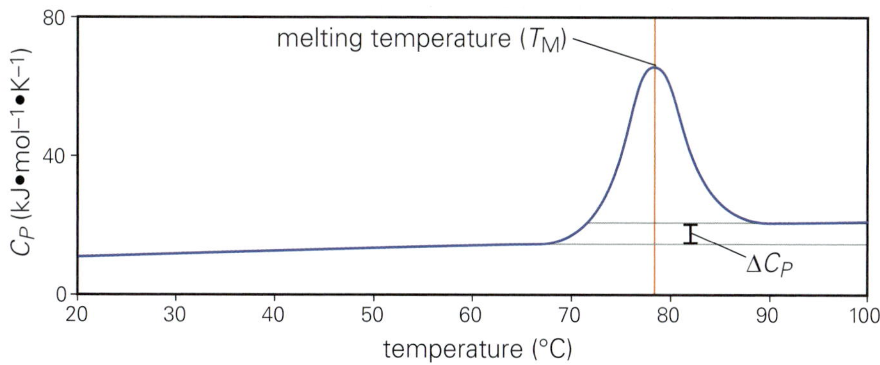


**Figure 10.25 Heat capacity ($C_P$) of a small protein (lysozyme) as a function of temperature.** The difference between the value of $C_P$ at high temperature (when the protein is fully unfolded) and low temperature (when the protein is fully folded) is denoted $\Delta C_P$. The peak in the heat capacity curve occurs at the melting temperature ($T_M$). (Adapted from W. Pfeil and P.L. Privalov, *Biophys. Chem.* 4: 23–32, 1976.)

very gradually as the temperature is increased from 20°C to 70°C. At that point, the heat capacity increases sharply, reaches a maximum value at ~80°C and then decreases rapidly. At temperatures above ~85°C, the heat capacity reaches a stable value, but one that is higher than that of the protein solution at lower temperature. The variation in the heat capacity of the protein as a function of temperature is known as a **melting curve**.

The unfolding of the protein can be described as an equilibrium between two states, the folded state (denoted F) and the unfolded state (denoted U)—see Equations 10.120–10.123 and Figure 10.13. For the calorimetric experiment, we start with folded protein molecules and gradually unfold them as the temperature is increased. It is therefore convenient to consider the unfolding reaction (Equation 10.122).

The standard free-energy change for the unfolding reaction is given by:

$$\Delta G^{o}_{unfolding} = \Delta H^{o}_{unfolding} - T\Delta S^{o}_{unfolding} \quad (10.143)$$

As the temperature increases, more and more of the protein unfolds. At a certain temperature, the value of $\Delta G^o$ becomes equal to zero, and the population of folded and unfolded molecules is equal. This temperature is known as the **melting temperature of a protein** ($T_M$), and it corresponds to the peak value of the heat capacity (see Figure 10.25).

**Melting temperature of a protein**

At the melting temperature, $T_M$, half the protein molecules in a solution are unfolded. The standard free-energy change for folding, $\Delta G^o$, is zero at $T = T_M$ for a monomeric protein. Melting temperatures of different proteins vary over a wide range.

Recall from Chapter 6 that the heat capacity of the folded protein is lower than that of the unfolded protein, as can also be seen in Figure 10.25. Thus, at the lower end of the temperature scale, where the folded protein is more stable, the measured heat capacity is essentially that of the folded protein. Since the experiment is carried out at constant pressure, the heat capacity is denoted $C_P^F$, where the superscript "F" refers to the folded protein and the subscript "$P$" refers to constant pressure. Likewise, at the higher end of the temperature scale, most of the protein molecules are unfolded and the measured heat capacity corresponds to that of the unfolded protein ($C_P^U$).

$C_P^F$ and $C_P^U$ are related to the ability of folded and unfolded protein molecules, as well as associated solvent molecules, to take up energy by increasing their vibrations. It turns out that $C_P^U$ and $C_P^F$ depend weakly on temperature, and their difference is essentially independent of temperature. As the temperature increases, the fraction ($f$) of the molecules that are folded decreases, and the fraction of molecules that are unfolded ($1 - f$) increases (**Figure 10.26**). Because folding and unfolding are highly cooperative, the value of $f$ decreases very slowly until the temperature is just below $T_M$, at which point the value of $f$ decreases sharply.

The measured heat capacity includes contributions from $C_P^F$ and $C_P^U$, weighted by the fraction of each state present at any temperature:

$$C_P^{F+U} = f\,C_P^F + (1-f)\,C_P^U \quad (10.144)$$

where $f$ is the fraction of protein in the folded state. $C_P^{F+U}$ is the combined heat capacity of the folded and unfolded proteins.

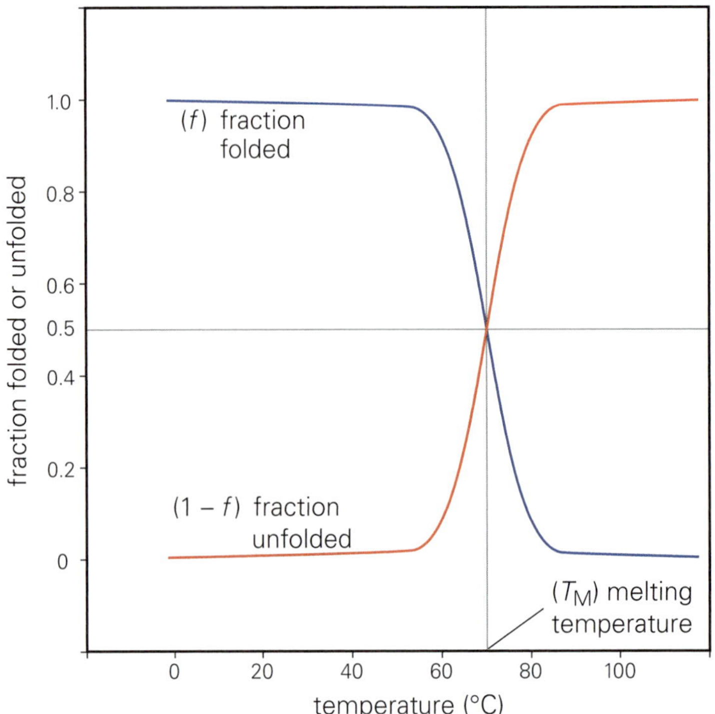


**Figure 10.26 The fraction of protein molecules that are folded or unfolded, as a function of temperature.** There is generally a sharp transition from the fully folded to the fully unfolded state. The melting temperature ($T_M$) is the temperature at which half the protein molecules are folded.

The expected behavior of $C_P^{F+U}$ is shown in **Figure 10.27**. The change in $C_P^{F+U}$ with temperature does not explain the prominent peak in the heat capacity, centered at the melting temperature ($T_M$). This peak arises from an additional process that takes up energy as the temperature is increased, which is the conversion of protein molecules from the folded form to the unfolded one. It takes energy to break the interactions that stabilize the folded protein and this leads to an additional heat capacity term, denoted $C_P^{\text{unfolding}}$ (**Figure 10.28**). As the temperature approaches $T_M$, more and more protein molecules take up heat as they unfold, leading to increased heat capacity (see Figure 10.28). Once the temperature crosses $T_M$, most of the protein molecules are already unfolded, and the heat capacity decreases. Hence, the observed heat capacity is a sum of three terms:

$$C_P^{\text{observed}} = C_P^{F+U} + C_P^{\text{unfolding}} = f\ C_P^F + (1-f)\ C_P^U + C_P^{\text{unfolding}} \tag{10.145}$$

Although we have focused on the protein molecules in this discussion, it should be stressed that the measurements correspond directly to differences in enthalpies of all interactions, including those between the protein and the solvent.

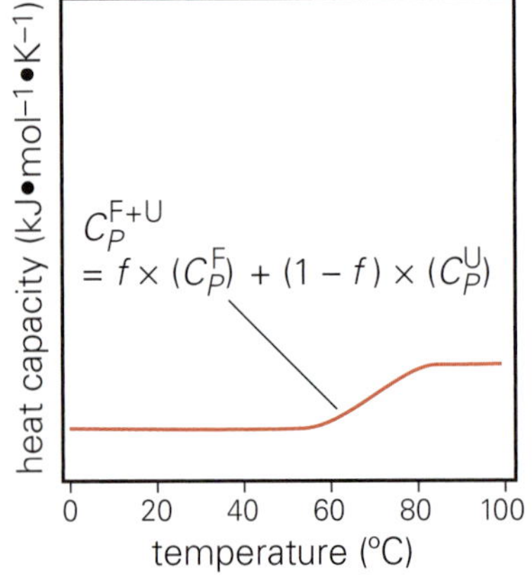


**Figure 10.27 Contributions of $C_P^F$ and $C_P^U$ to the heat capacity of a protein.** The *red* line shows the heat capacity due to the folded and unfolded proteins. It does not include the contribution that arises from the heat required to convert the folded form to the unfolded form.

## 10.27 The area under the peak in the melting curve is the enthalpy change for unfolding at the melting temperature

If we integrate the value of $C_P^{\text{unfolding}}$ over the narrow temperature range that spans the peak in the melting curve, we can determine the total amount of heat taken up as the folded protein is converted to the unfolded form:

$$\int C_P^{\text{unfolding}}\, dT = H_{\text{unfolded}} - H_{\text{folded}} = \Delta H_{\text{unfolding}} \tag{10.146}$$

$\Delta H_{\text{unfolding}}$ is essentially the change in enthalpy for the conversion of all of the folded protein to the unfolded form. Since we know the concentration of the protein, we can readily calculate the enthalpy change per mole of protein ($\Delta H^o$). Since most of the protein molecules undergo the unfolding process over a narrow range of temperature near $T_M$, we equate $\Delta H^o$ with the enthalpy change for unfolding at the melting temperature, $\Delta H^o_{\text{unfolding}}(T_M)$.

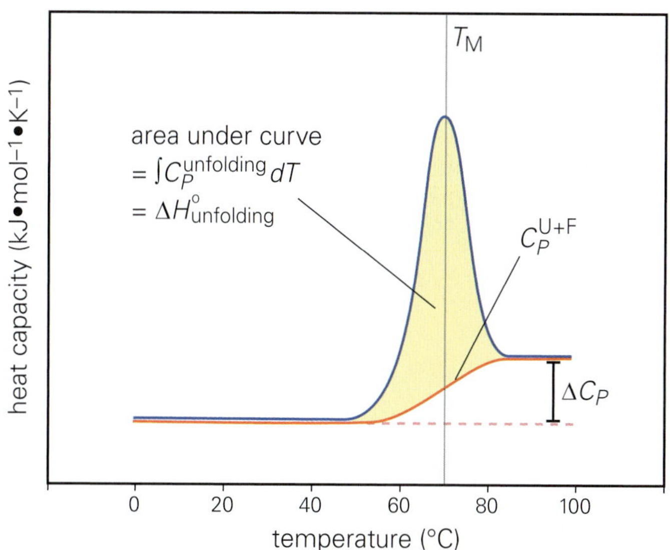


**Figure 10.28 Determination of the standard enthalpy change of unfolding for a protein.** Integrating the melting curve and subtracting the heat capacity of the unfolded and folded proteins yields the enthalpy change for unfolding (*yellow* shaded area).

At the melting temperature:

$$K(\text{at } T_M) = \frac{[U]}{[F]} = 1$$

so

$$\Delta G^o(T_M) = -RT\ln K(T_M) = 0$$

for a monomeric protein. Hence:

$$\Delta G^o_{unfolding}\ (\text{at } T_M) = \Delta H^o_{unfolding} - T_M \Delta S^o_{unfolding} = 0$$

$$\Rightarrow \Delta S^o_{unfolding} = \frac{\Delta H^o_{unfolding}}{T_M} \qquad (10.147)$$

Thus, the calorimetric experiment gives us the values of $\Delta H^o$ and $\Delta S^o$ for the unfolding reaction at the melting temperature.

## 10.28 The heat capacities of the folded and unfolded protein allow the determination of $\Delta H^o$ and $\Delta S^o$ for unfolding at any temperature

When we considered the dissociation of water, we found that the values of $\Delta H^o$ and $\Delta S^o$ for that process were independent of temperature (see Section 10.14). But, as we noted, reactions that involve the hydrophobic effect, such as protein folding, do not behave this way. Although the calorimetric experiment gives us the values of $\Delta H^o$ and $\Delta S^o$ at the melting temperature, how do we determine these values at some other temperature (for example 25°C)? Because $\Delta H^o$ and $\Delta S^o$ are temperature dependent, we cannot simply apply the van't Hoff analysis (Section 10.14). Fortunately, the calorimetric data also give us the value of the difference between heat capacities of the folded and unfolded proteins ($\Delta C_P = C_P^U - C_P^F$), and this allows us to calculate how both $\Delta H^o$ and $\Delta S^o$ depend on temperature (**Box 10.2** for details):

$$\Delta H^o_{unfolding}(T) = \Delta H^o_{unfolding}(T_M) + \Delta C_P (T - T_M) \qquad (10.148)$$

$$\Delta S^o_{unfolding}(T) = \Delta S^o_{unfolding}(T_M) + \Delta C_P \ln\left(\frac{T}{T_M}\right) \qquad (10.149)$$

These equations tell us that if $\Delta C_P$ were zero (that is, if the folded and unfolded proteins had the same heat capacity), then $\Delta H^o_{unfolding}$ and $\Delta S^o_{unfolding}$ would be independent of temperature. But, in fact, experimental measurements of protein

## Box 10.2 Determining the enthalpy and entropy changes at a general temperature, *T*, from heat capacity measurements

Consider the following thermodynamic cycle:

$$\begin{array}{ccc} F(T_M) & \xrightarrow[3]{\Delta H_3^o} & U(T_M) \\ \Delta H_2^o \uparrow 2 & & 4 \downarrow \Delta H_4^o \\ F(T) & \xrightarrow[\Delta H_1^o]{1} & U(T) \end{array}$$

Here, the process labeled 1 is the one we are interested in, where the folded protein is converted to the unfolded form at temperature $T$. Process 2 corresponds to changing the temperature of the folded protein from $T$ to $T_M$. Process 3 is the unfolding of the protein at the melting temperature, $T_M$. Finally, process 4 involves changing the temperature of the unfolded protein from $T_M$ back to $T$.

Because enthalpy is a state function, we can write:

$$\Delta H^o_{\text{unfolding}}(T) = \Delta H_1^o = \Delta H_2^o + \Delta H_3^o + \Delta H_4^o \tag{10.2.1}$$

$\Delta H_2^o$ and $\Delta H_4^o$ can be readily determined from the measured heat capacities of the folded and unfolded proteins, respectively:

$$\Delta H_2^o = \int_T^{T_M} C_P^F dT = C_P^F(T_M - T) \tag{10.2.2}$$

and

$$\Delta H_4^o = \int_{T_M}^{T} C_P^U dT = C_P^U(T - T_M) \tag{10.2.3}$$

$\Delta H_3^o$ is the enthalpy of unfolding at the melting temperature, which is measured experimentally:

$$\Delta H_3^o = \Delta H^o_{\text{unfolding}}(T_M) \tag{10.2.4}$$

By substituting these values for $\Delta H_2^o$, $\Delta H_3^o$, and $\Delta H_4^o$ into Equation 10.2.1, we get:

$$\begin{aligned} \Delta H^o_{\text{unfolding}}(T) &= \Delta H^o_{\text{unfolding}}(T_M) + C_P^U(T - T_M) + C_P^F(T_M - T) \\ &= \Delta H^o_{\text{unfolding}}(T_M) + (C_P^U - C_P^F)(T - T_M) \end{aligned}$$

$$\Delta H^o_{\text{unfolding}}(T) = \Delta H^o_{\text{unfolding}}(T_M) + \Delta C_P(T - T_M) \tag{10.2.5}$$

where $\Delta C_P$ is the difference between the heat capacities of the unfolded and folded proteins.

The entropy change of the unfolding reaction can also be obtained by using a similar thermodynamic cycle:

$$\begin{array}{ccc} F(T_M) & \xrightarrow[3]{\Delta S_3^o} & U(T_M) \\ \Delta S_2^o \uparrow 2 & & 4 \downarrow \Delta S_4^o \\ F(T) & \xrightarrow[\Delta S_1^o]{1} & U(T) \end{array}$$

Here:

$$\Delta S_1^o = \Delta S^o_{\text{unfolding}}(T) = \Delta S_2^o + \Delta S_3^o + \Delta S_4^o \tag{10.2.6}$$

$\Delta S_2^o$ is the entropy change on increasing the temperature of the folded protein from $T$ to $T_M$. Recall from Chapter 7 that the change in entropy is related to $dq$, the heat taken up during the process that is carried out very slowly (reversibly):

$$dS = \frac{dq}{T} \tag{10.2.7}$$

Since $dq = C_P dT$, we can write:

$$dS = \frac{C_P dT}{T} \Rightarrow \Delta S = \int \frac{C_P dT}{T} = C_P \int \frac{dT}{T} \tag{10.2.8}$$

Thus,

$$\Delta S_2^o = C_P^F \int_T^{T_M} \frac{dT}{T} = C_P^F \ln\left(T_M/T\right) \tag{10.2.9}$$

Likewise,

$$\Delta S_4^o = C_P^U \ln\left(T/T_M\right)$$

$\Delta S_3^o$ is the entropy change of unfolding at the melting temperature, denoted as $\Delta S^o_{\text{unfolding}}(T_M)$, which is determined experimentally. Combining these equations we get:

$$\Delta S^o_{\text{unfolding}}(T) = \Delta S^o_{\text{unfolding}}(T_M) + \Delta C_P \ln\left(T/T_M\right) \tag{10.2.10}$$

The standard free-energy change for unfolding at a temperature $T$ is therefore given by:

$$\begin{aligned} \Delta G^o_{\text{unfolding}}(T) &= \Delta H^o_{\text{unfolding}}(T_M) - T\Delta S^o_{\text{unfolding}}(T_M) \\ &+ \Delta C_P\left[T - T_M - T\ln\left(T/T_M\right)\right] \end{aligned} \tag{10.2.11}$$

unfolding demonstrate that the value of $\Delta C_P$ always has a positive value for proteins, as can be seen for lysozyme in Figure 10.25.

Combining Equations 10.148 and 10.149, we can calculate the standard free-energy change of unfolding ($\Delta G^o_{\text{unfolding}}$) at any temperature, $T$:

$$\begin{aligned} \Delta G^o_{\text{unfolding}}(T) &= \Delta H^o_{\text{unfolding}}(T) - T\Delta S^o_{\text{unfolding}}(T) \\ \Delta G^o_{\text{unfolding}}(T) &= \Delta H^o_{\text{unfolding}}(T_M) + \Delta C_P(T - T_M) \\ &- T\left(\Delta S^o_{\text{unfolding}}(T_M) + \Delta C_P \ln\left(\frac{T}{T_M}\right)\right) \end{aligned} \tag{10.150}$$

This important equation describes the stability of the folded protein at any temperature of interest.

Experimentally determined values of $\Delta C_P$, as well as values for $\Delta H^o$ and $\Delta S^o$ at specified temperatures, are shown in **Table 10.3** for several proteins. These proteins are of different sizes, and so have different values of $\Delta H^o$, $\Delta S^o$ and $\Delta C_P$. In order to make comparisons easier, these values are divided by the number of residues in the protein.

One striking aspect of the data in Table 10.3 is that this very diverse set of proteins all have quite similar thermodynamic parameters on a per residue basis. This justifies to some extent our very simple computational analysis of proteins in the previous sections, where we simply assumed that each residue makes a similar contribution to the enthalpy and entropy of folding.

Given the data in Table 10.3, the values of $\Delta H^o$ and $\Delta S^o$ can be calculated as a function of temperature using Equations 10.148 and 10.149, respectively. The results for one particular small protein (protein G-B1) are graphed in **Figure 10.29**. The value of $\Delta C_P$ for this protein is 53 $J{\bullet}K^{-1}\ mol^{-1}$ per residue (see Table 10.3). Protein G-B1 has 56 residues, and so the value of $\Delta C_P$ for the whole protein is $56 \times 53 \approx 3000\ J{\bullet}K^{-1}{\bullet}mol^{-1} = 3\ kJ{\bullet}K^{-1}{\bullet}mol^{-1}$. Consider a change in temperature from 20°C (293 K) to 30°C (313 K). According to Equation 10.148, the value of $\Delta H^o_{unfolding}$ would then change as follows:

$$\Delta H^o_{unfolding}(T{=}313K) - \Delta H^o_{unfolding}(T{=}293K) =$$
$$\Delta C_p(313-293) = 3.0\times10 = 30\,kJ{\bullet}\,mol^{-1} \qquad (10.151)$$

Indeed, $\Delta H^o$ increases by 30 $kJ{\bullet}mol^{-1}$ over this temperature range for protein G-B1, as reflected in the graph in Figure 10.29. Note that $\Delta H^o$and $T\Delta S^o$ change much more rapidly than their difference, which is $\Delta G^o$ (**Figure 10.30**).

How do these experimental results compare with our simple computational model for folding? Figure 10.30 shows the value of $\Delta G^o_{unfolding}$ as a function of temperature for G-B1. At room temperature (25°C, 298 K), the experimentally determined value of $\Delta G^o_{unfolding}$ is +28 $kJ{\bullet}mol^{-1}$ and $\Delta H^o_{unfolding}$ is $\approx$ +78 $kJ{\bullet}mol^{-1}$. In terms of the

**Table 10.3 Thermodynamic parameters for unfolding of various proteins.**

| Protein | Molecular weight | $\Delta H^o$ (25°C) [$kJ{\bullet}mol^{-1}$ per residue] | $\Delta S^o$ (110°C) [$J{\bullet}mol^{-1}$ per residue] | $\Delta C_P$ [$J{\bullet}K^{-1}{\bullet}mol^{-1}$ per residue] |
|---|---|---|---|---|
| Protein G-B1 | 7200 | 1.4 | 16.1 | 53 |
| Parvalbumin | 11,500 | 1.4 | 16.8 | 46 |
| Cytochrome *c* | 12,400 | 0.64 | 17.8 | 67 |
| Ribonuclease A | 13,600 | 2.4 | 17.8 | 44 |
| Hen lysozyme | 14,300 | 2.0 | 17.6 | 52 |
| Staph. nuclease | 16,800 | 0.85 | 17.5 | 61 |
| Myoglobin | 17,900 | 0.04 | 17.9 | 75 |
| Papain | 23,400 | 0.93 | 17.0 | 60 |
| β-Papain | 23,800 | 1.3 | 17.9 | 58 |
| α-Chymotrypsin | 25,200 | 1.1 | 18.0 | 58 |
| Average | | 1.2 ± 0.7 | 17.4 ± 0.6 | 57 ± 9 |

This table presents values of $\Delta H^o$, $\Delta S^o$, and $\Delta C_P$ for several proteins. The temperatures for which $\Delta H^o$ and $\Delta S^o$ are calculated were chosen to reflect the enthalpy and entropy of the protein chain itself, without the hydrophobic interaction (see Baldwin, 1986, in Further Reading). The values scale with the size of the protein; in order to make comparison easier, the values have been divided by the number of residues in the protein to give a "per residue" value. (Adapted from P. Alexander et al., and P. Bryan, *Biochemistry* 31: 3597–3603, 1992. With permission from the American Chemical Society.)

folding reaction, $\Delta G^o_{folding}$ is therefore –28 kJ•mol$^{-1}$ and $\Delta H^o_{folding}$ is –78 kJ•mol$^{-1}$. These values are comparable to the values of $\Delta G^o_{folding}$ and $\Delta H^o_{folding}$ we calculated for our highly simplified model:

$$\Delta G^o_{folding} = -64 \text{ kJ•mol}^{-1}, \quad \Delta H^o_{folding} = -150 \text{ kJ•mol}^{-1}$$

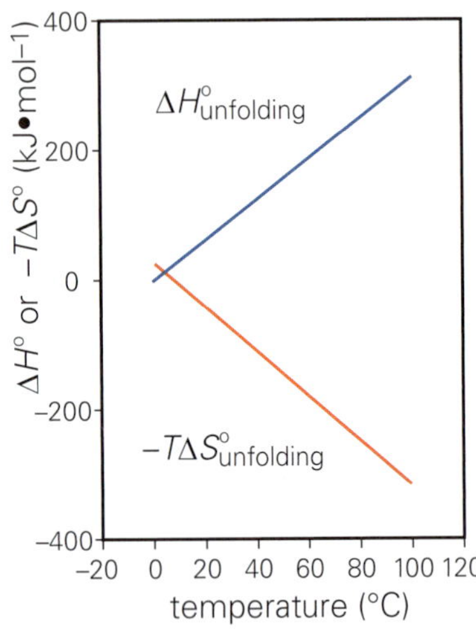


**Figure 10.29 Temperature dependence of the enthalpy and entropy components of the free energy of unfolding.** (Based on data for protein G-B1 in Table 10.3.)

## 10.29 Folded proteins become unstable at very low temperature because of changes in $\Delta H^o$ and $\Delta S^o$

There is one fundamental way in which the experimental measurements of protein folding differ from the result of our simple computational model. Note that the energy and entropy changes in the simple model are both *independent* of temperature (see Equations 10.133 and 10.134). The free energy of folding is given by:

$$\Delta G^o_{folding} = \Delta H^o_{folding} - T\Delta S^o_{folding}$$

Recall that in our model calculation, $\Delta S^o_{folding}$ is positive, which predicts that the stability of the folded protein decreases linearly with temperature, shown by the blue line in Figure 10.30. This contrasts markedly with the experimental situation, where the stability of the folded protein is maximal at a particular temperature and decreases as the temperature either increases or decreases with respect to this temperature. The temperature of maximum stability varies considerably for different proteins within the same organism. In addition, as you might expect, proteins from thermophilic organisms (ones that grow at elevated temperatures) remain folded at temperatures that would lead to the unfolding of most proteins from normal organisms.

To better understand why proteins lose stability at low temperatures, we will examine the values of $\Delta H^o_{unfolding}$ and $\Delta S^o_{unfolding}$ for protein G-B1. First, let us use the data in Table 10.3 to figure out what these values are. From Table 10.3, the value of $\Delta H^o_{unfolding}$ at 25°C (298 K) is +1.4 kJ•mol$^{-1}$ per residue. Since this protein has 56 residues, $\Delta H^o_{unfolding}$ = +1.4 × 56 = +78.4 kJ•mol$^{-1}$. From the graph of $\Delta G^o_{unfolding}$ versus temperature (see Figure 10.30), we see that $T_M$ = 85°C for protein G-B1. Using Equation 10.148 and the known value of $\Delta C_P$, we can calculate the value of $\Delta H^o_{unfolding}$ ($T_M$), which turns out to be +258 kJ•mol$^{-1}$. Recalling Equation 10.147, at the melting temperature:

$$\Delta S^o_{unfolding} = \frac{\Delta H^o_{unfolding}}{T_M} \tag{10.152}$$

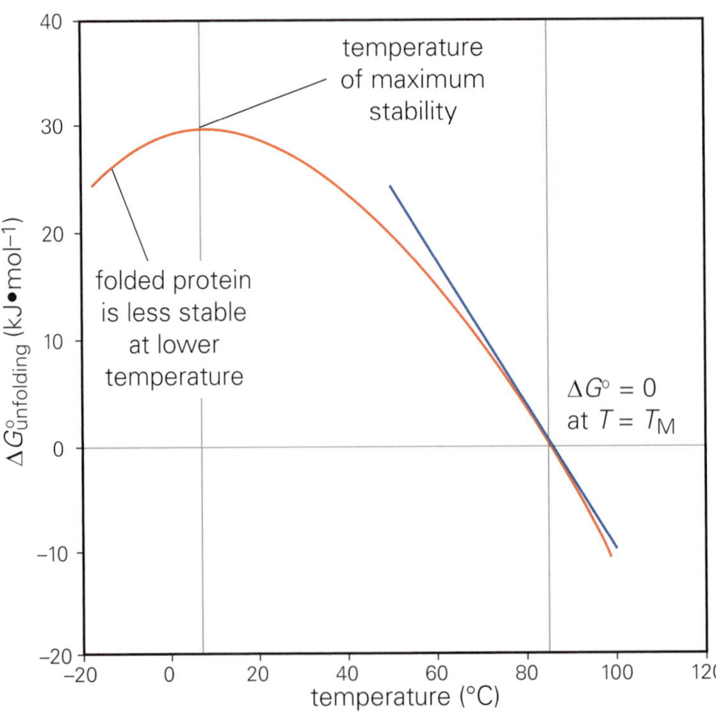


**Figure 10.30 The change in the free energy of $\Delta G^o_{unfolding}$, as a function of temperature for a small protein (B-G1).** This "protein stability curve" is based on the data in Table 10.3 and Equation 10.150. This curve shows that this protein has a maximum stability at about 7°C. The *blue* line reflects what the protein stability would be if $\Delta H$ and $\Delta S$ were temperature independent, and equal to their values at the melting temperature.

Substituting the values of $\Delta H^{\mathrm{o}}_{\mathrm{unfolding}}$ and $T_{\mathrm{M}}$, we find that $\Delta S^{\mathrm{o}}_{\mathrm{unfolding}}$ is 0.72 kJ•K$^{-1}$ mol$^{-1}$ for protein G-B1, at the melting temperature.

We can now readily calculate the value of $\Delta G^{\mathrm{o}}_{\mathrm{unfolding}}$ for protein G-B1 at any temperature by using these values, assuming either that $\Delta H^{\mathrm{o}}_{\mathrm{unfolding}}$ and $\Delta S^{\mathrm{o}}_{\mathrm{unfolding}}$ are temperature independent, or that they vary with temperature due to $\Delta C_P$. In the latter case, we have to use Equation 10.150 to relate $\Delta G^{\mathrm{o}}_{\mathrm{unfolding}}$ to $\Delta H^{\mathrm{o}}_{\mathrm{unfolding}}$ and $\Delta S^{\mathrm{o}}_{\mathrm{unfolding}}$. The results of both calculations are graphed in Figure 10.30. When the $\Delta C_P$ term is included, we see that $\Delta G^{\mathrm{o}}_{\mathrm{unfolding}}$ has a maximum value at ~7°C and decreases at *both* higher *and* lower temperatures. In contrast, using temperature-independent values for the changes in enthalpy and entropy, the value of $\Delta G^{\mathrm{o}}_{\mathrm{unfolding}}$ continues to increase as temperature gets lower.

The loss of protein stability at lower temperature is known as **cold denaturation**, and it runs counter to intuition. We can see that the curvature in the protein stability curve (free energy vs. temperature) arises as a consequence of the terms involving $\Delta C_P$ in Equation 10.150. These terms, which describe the increased heat capacity of the unfolded protein, are not explained by our simple model for protein folding. $\Delta C_P$ is observed to be positive for protein-unfolding reactions and this behavior arises because of changes in the energetic interactions between hydrophobic residues in the protein and water molecules (neglected in our simple treatment). The origin of the $\Delta C_P$ term is still not completely understood, but it is found to be correlated empirically with the increased exposure of hydrophobic sidechains in the unfolded protein. In proteins with low stability and large values of $\Delta C_P$, the free-energy curve can drop sufficiently at low temperature that it crosses $\Delta G^{\mathrm{o}}_{\mathrm{unfolding}} = 0$, the point at which the folded protein becomes unstable and unfolds. Such proteins do not fold at low temperature.

**Cold denaturation**

Cold denaturation refers to the unfolding of proteins induced at low temperature. This arises from the $\Delta C_P$ term in the protein stability equation, which correlates with the burial of hydrophobic residues in the folded state of the protein.

## Summary

This chapter introduced the concept of the chemical potential, $\mu$, which is the rate of change of free energy with respect to the number of molecules. The chemical potential is essentially the free energy per molecule, or per mole. For ideal dilute solutions, a key result is that the chemical potential is related to the logarithm of the concentration. If the concentrations of a molecule in two solutions are $C_1$ and $C_2$, then the difference in chemical potential for that molecule is given by:

$$\Delta\mu = \mu_2 - \mu_1 = RT\ln\left(\frac{C_2}{C_1}\right) \tag{10.32}$$

The chemical potential for a molecule under standard conditions (typically, 1 M solution) is denoted $\mu^{\mathrm{o}}$. Equation 10.32 allows us to calculate the chemical potential of the molecule at a nonstandard concentration ($C$):

$$\mu = \mu^{\circ} + RT\ln\left(\frac{C}{C^{\circ}}\right) = \mu^{\circ} + RT\ln C \tag{10.33}$$

In Equation 10.33, we have assumed that the standard concentration, $C^{\mathrm{o}}$, is 1 M, and have omitted it in the final expression.

For a general chemical reaction of the form:

$$\upsilon_{\mathrm{A}}\,\mathrm{A} + \upsilon_{\mathrm{B}}\,\mathrm{B} \rightarrow \upsilon_{\mathrm{C}}\,\mathrm{C} + \upsilon_{\mathrm{D}}\,\mathrm{D}$$

where $\upsilon_{\mathrm{A}}$, $\upsilon_{\mathrm{B}}$, $\upsilon_{\mathrm{C}}$, and $\upsilon_{\mathrm{D}}$ are the stoichiometric coefficients, the value of the standard free-energy change for the reaction, $\Delta G^{\mathrm{o}}$, is given by:

$$\Delta G^{\mathrm{o}} = \upsilon_{\mathrm{C}}\mu^{\mathrm{o}}_{\mathrm{C}} + \upsilon_{\mathrm{D}}\mu^{\mathrm{o}}_{\mathrm{D}} - \upsilon_{\mathrm{A}}\mu^{\mathrm{o}}_{\mathrm{A}} - \upsilon_{\mathrm{B}}\mu^{\mathrm{o}}_{\mathrm{B}} \tag{10.54}$$

where $\mu^{\mathrm{o}}_{\mathrm{A}}$, $\mu^{\mathrm{o}}_{\mathrm{B}}$, $\mu^{\mathrm{o}}_{\mathrm{C}}$, and $\mu^{\mathrm{o}}_{\mathrm{D}}$, are the chemical potentials (molar free energies) of the A, B, C, and D molecules, respectively, under standard conditions (typically 1 M solutions). The equilibrium point for the reaction is defined by the following condition:

$$\Delta G^{\mathrm{o}} = -RT\ln K_{\mathrm{eq}} \tag{10.52}$$

where $K$ is the equilibrium constant for the reaction. The value of $K$ is related to the concentrations, [A], [B], [C], and [D] at equilibrium as follows:

$$K = \frac{[\mathrm{C}]^{\upsilon_{\mathrm{C}}}\,[\mathrm{D}]^{\upsilon_{\mathrm{D}}}}{[\mathrm{A}]^{\upsilon_{\mathrm{A}}}\,[\mathrm{B}]^{\upsilon_{\mathrm{B}}}} \tag{10.51}$$

The free-energy change for the reaction, $\Delta G$, is given by:

$$\Delta G = \Delta G^{o} + RT\ln Q \tag{10.66}$$

where $Q$, the reaction quotient, is given by:

$$Q = \frac{[C]_{obs}^{\upsilon_C}\,[D]_{obs}^{\upsilon_D}}{[A]_{obs}^{\upsilon_A}\,[B]_{obs}^{\upsilon_B}}$$

Here $[A]_{obs}$, etc., refer to the actual (observed) concentrations of the reactants and products under reaction conditions of interest, rather than the equilibrium values. The ratio of $Q$ to $K$ ($Q/K$) is known as the mass action ratio and is related to the free-energy change ($\Delta G$) as follows:

$$\Delta G = +RT\ln\frac{Q}{K} = +2.3RT\log_{10}\frac{Q}{K} \tag{10.67}$$

It is the value of $\Delta G$, rather than $\Delta G^{o}$, that determines whether a reaction is driven to the left or to the right under specific conditions.

Biochemical reactions typically do not go to completion, and the extent of reaction is determined by the relevant equilibrium constant and the concentration of reactants and products. An important class of such reactions involve weak acids and bases. The pH of a solution of a weak acid (HA) and its conjugate base ($A^-$) is given by:

$$\text{pH} = \text{p}K_a + \log_{10}\left(\frac{[A^-]}{[HA]}\right) \tag{10.74}$$

Equation 10.78 is known as the Henderson–Hasselbalch equation, and it explains the ability of weak acids and bases to stabilize or buffer the pH of solutions. This property is critical for the proper functioning of biological systems, because the charges on molecules such as proteins, DNA, and RNA change with pH.

Equilibrium constants have fixed values at a fixed temperature, but their values change when the temperature changes. Water itself is a weak acid and dissociates to yield $H^+$ and $OH^-$. The pH of pure water is 7.0, but the pH decreases as the temperature increases, because the equilibrium constant for the dissociation of water changes in value.

The net free-energy change in biological reactions is the result of a balance between changes in energy (strictly speaking, the enthalpy) and the entropy. We have illustrated this balance by considering a simple model for protein folding, which shows why protein folding is spontaneous, even though the entropy of the protein chain is reduced substantially upon folding. The hydrophobic effect, which is largely a consequence of the reduction in water entropy when nonpolar groups are exposed to water, is a critical driving force in protein folding. By sequestering the nonpolar groups in the interior, the protein folding process results in an increase in the entropy of water. This is a key driving force for protein folding, as important as energetic considerations.

The principal concept that allows us to estimate the entropy of the protein chain, and of water molecules that are affected by it, is a relationship between the entropy and the logarithm of the number of conformations accessible to the system. If we assume that all of the conformations have the same energy, then the molar entropy, $S$, of a system containing molecules that can adopt $N_C$ different conformations is given by:

$$\Delta S = R\ln N_C \tag{10.129}$$

The values of the thermodynamic parameters $\Delta G^{o}$, $\Delta H^{o}$, and $\Delta S^{o}$ can be determined experimentally using calorimetry. A key result is that the values of $\Delta H^{o}$ and $\Delta S^{o}$ for protein folding vary with temperature. This temperature dependence arises because the heat capacity of the unfolded protein $\left(\Delta C_P^{U}\right)$ is higher than that of the folded protein $\left(\Delta C_P^{F}\right)$. The origin of this difference in heat capacity ($\Delta C_P$) is not well understood, but it is found to be correlated with the extent of interaction of hydrophobic groups with water.

## Key Concepts

### A. CHEMICAL POTENTIAL

- The chemical potential, $\mu$, of a molecular species is the derivative of the free energy with respect to the number of molecules of that species. For dilute solutions, the chemical potential is equivalent to the molar free energy of that species.
- Molecules move spontaneously from regions of high chemical potential to regions of low chemical potential.
- Biochemical reactions are assumed to occur in ideal and dilute solutions, which simplifies the calculation of the chemical potential.
- The chemical potential is proportional to the logarithm of the concentration. The difference in chemical potential between two solutions, one at concentration $C_1$ and the other at concentration $C_2$, is given by Equation 10.15:

$$\Delta\mu = \mu_2 - \mu_1 = RT\ln\left(\frac{C_2}{C_1}\right)$$

- Chemical potentials at arbitrary concentrations are calculated with reference to standard concentrations. If $\mu^{\text{o}}$ is the chemical potential of the standard state solution (with concentration $C^{\text{o}}$), then the chemical potential for a solution at an arbitrary concentration $C$ is given by:

$$\mu = \mu^{\text{o}} + RT\ln\left(\frac{C}{C^{\text{o}}}\right)$$

In most cases, the standard state concentration is 1 M, and so the expression for the chemical potential simplifies to Equation 10.33:

$$\mu = \mu^{\text{o}} + RT\ln C$$

### B. EQUILIBRIUM CONSTANTS

- The concentrations of reactants and products at equilibrium define the equilibrium constant ($K$), which is related to the standard free-energy change ($\Delta G^{\text{o}}$) for the reaction: $\Delta G^{\text{o}} = -RT\ln K$.

Here $K$ is the equilibrium constant. Consider a chemical reaction in which $\upsilon_A$ molecules of A and $\upsilon_B$ molecules of B react to produce $\upsilon_C$ molecules of C and $\upsilon_D$ molecules of D:

$$\upsilon_A\text{A} + \upsilon_B\text{B} \rightleftharpoons \upsilon_C\text{C} + \upsilon_D\text{D}$$

The equilibrium constant is defined by Equation 10.51:

$$K = \frac{[\text{C}]^{\upsilon_C}\,[\text{D}]^{\upsilon_D}}{[\text{A}]^{\upsilon_A}\,[\text{B}]^{\upsilon_B}}$$

In this expression, the concentrations of A, B, C, and D are the equilibrium concentrations.

- The free-energy change for the reaction ($\Delta G$), not the standard free-energy change ($\Delta G^{\text{o}}$), determines the direction of spontaneous change. The free-energy change for a reaction with reactants and products at arbitrary (that is, nonequilibrium) concentrations is given by Equation 10.70:

$$\Delta G = \Delta G^{\text{o}} + RT\ln Q$$

In this equation, $Q$ is the reaction quotient and is given by:

$$Q = \frac{[\text{C}]_{\text{obs}}^{\upsilon_C}\,[\text{D}]_{\text{obs}}^{\upsilon_D}}{[\text{A}]_{\text{obs}}^{\upsilon_A}\,[\text{B}]_{\text{obs}}^{\upsilon_B}}$$

- The ratio of the reaction quotient ($Q$) to the equilibrium constant ($K$) determines the thermodynamic drive of a reaction. The reaction free energy, $\Delta G$, is given by Equation 10.71:

$$\Delta G = +RT\ln\frac{Q}{K} = +2.3RT\log_{10}\frac{Q}{K}$$

If $\Delta G$ is negative, the reaction will proceed spontaneously in the forward direction, from reactants to products.

- ATP concentrations are maintained at high levels and ADP at low levels in cells, thereby increasing the driving force for ATP hydrolysis.

### C. ACID–BASE EQUILIBRIA

- The Henderson–Hasselbalch equation (Equation 10.74) relates the pH of a solution of a weak acid to the concentrations of the acid and its conjugate base:

$$\text{pH} = \text{p}K_a + \log_{10}\left(\frac{[\text{A}^-]}{[\text{HA}]}\right)$$

- The proton concentration ($[\text{H}^+]$) in pure water at room temperature corresponds to a pH value of 7.0.
- The temperature dependence of the equilibrium constant allows us to determine $\Delta H^{\text{o}}$ and $\Delta S^{\text{o}}$ through the van't Hoff equation (Equation 10.84):

$$\ln K = -\frac{\Delta H^{\text{o}}}{RT} + \frac{\Delta S^{\text{o}}}{R}$$

- Weak acids, such as acetic acid, dissociate very little in water.
- Solutions of weak acids and their conjugate bases act as buffers.
- The charges on biological macromolecules are affected by the pH and by interactions within the macromolecules.

### D. FREE-ENERGY CHANGES IN PROTEIN FOLDING

- Protein folding results from a balance between energy and entropy.
- The entropy, $S_{\text{unfolded}}$, of the unfolded protein chain is proportional to the logarithm of the number of conformations of the chain:

$$\frac{S_{\text{unfolded}}}{N} = R\ln N_C$$

In this equation, $N_C$ is the number of conformations of the unfolded chain and $N$ is the number of protein molecules.

- The number of conformations of the unfolded chain can be estimated by counting the number of low-energy torsional isomers.
- Protein folding is driven by an increase in water entropy upon displacement from hydrophobic sidechains.
- Calorimetric measurements make it possible to experimentally determine the free energy of protein folding.
- The heat capacity of a protein solution depends on the relative population of folded and unfolded molecules, and on the energy required to unfold the protein.
- Measurement of the difference in the heat capacities of the folded and unfolded protein ($\Delta C_P$) makes it possible to determine the value of $\Delta H^o$ and $\Delta S^o$ for unfolding at any temperature.
- Folded proteins may be unstable at very low temperature because of the way in which $\Delta H^o$ and $\Delta S^o$ change with temperature.

## Problems

### True/False and Multiple Choice

1. Molecules move spontaneously from regions of low chemical potential to regions of high concentration.
   True/False
2. The difference in chemical potential for a region with 500 mM of molecule B and a region with 1 M of molecule B is equal to:
   a. $k_BT \ln(0.5)$
   b. 0
   c. 1
   d. $k_BT \ln(500)$
   e. $k_BT \ln(1)$
3. The proton concentration in pure water at standard state (298 K) is:
   a. equal to 14
   b. always less than –7
   c. the square root of the ion product
   d. $10^7$
   e. 81 kJ•mol$^{-1}$
4. The melting temperature ($T_M$) is the temperature at which 100% of the protein molecules are unfolded.
   True/False
5. Which of the following must be independent of temperature when properly applying the van't Hoff equation?
   a. $K_{eq}$
   b. $\Delta S^o$
   c. $1/T$
   d. pH
6. The p$K_a$ of a protein sidechain depends only on the chemical identity of the sidechain, not on the surrounding environment.
   True/False
7. During protein folding, the entropy of water:
   a. increases
   b. decreases
   c. is equal to the protein entropy change
   d. is zero

### Fill in the Blank

8. A region with a high chemical potential for molecule A has a _______ concentration of molecule A than a region with low chemical potential.
9. To make a buffer, add a weak acid to its conjugate _______.
10. _____, ______, and ______ sidechains generally have a p$K_a$ less than 7.0. _______, ________, and ______ sidechains have a p$K_a$ greater than 9.
11. The integral of the melting curve of heat capacity versus temperature yields the ____________ change of protein unfolding.
12. The ______________ change for a reaction determines the direction of spontaneous change.

### Quantitative/Essay

13. Two regions of an ideal dilute solution have a difference in concentration of potassium ions ($K^+$). At 293 K, what is the difference in chemical potential between region 1, with a concentration of 0.5 M $K^+$, and region 2, which has a concentration of 2 mM?
14. The difference in chemical potential for a particular molecule between two regions of an ideal dilute solution is 5 kJ•mol$^{-1}$. The region with the higher chemical potential has a concentration of 200 mM. What is the concentration of the molecule in the other region at 293 K?
15. A cell with an internal calcium ion ($Ca^{2+}$) concentration of 20 μM is placed in media with a $Ca^{2+}$ concentration of 70 mM. What is the difference in chemical potential for $Ca^{2+}$ ions between the inside and outside of the cell at 310 K?
16. At equilibrium, in a test tube, the concentration of GDP is 1 M, of GTP is 20 μM, and of $P_i$ is 1 M. What is the equilibrium constant of the reaction, GTP $\rightleftharpoons$ GDP + $P_i$?

17. The reaction, A + 2B $\rightleftharpoons$ C, has an equilibrium constant of 2000. During a reaction, the concentration of A is 0.01 M, of B is 0.2 M, and of C is 0.5 M.
    a. What is the reaction quotient (*Q*)?
    b. In what direction will the reaction proceed?

18. The arginine-tRNA synthetase enzyme catalyzes the reaction that charges a tRNA with the amino acid arginine:

    $$\text{ATP} + \text{arginine} + \text{tRNA} \rightleftharpoons \text{AMP} + \text{PP}_i + \text{arginyl-tRNA}$$

    The value of the equilibrium constant is 1.13.

    At equilibrium, following an *in vitro* reaction, the concentration of ATP is 2 μM, of arginine is 500 mM, and of arginyl-tRNA is 10 μM. The concentrations of AMP and of $PP_i$ are 500 μM. What is the concentration of arginyl-tRNA?

19. The $pK_a$ of a weak acid is 5. What is the pH when the concentration of the acid form is 0.5 M and the concentration of the conjugate base form is 0.05 M?

20. The pH of a 0.15 M propionic acid/0.1 M sodium propionate buffer is 4.71. What is the $pK_a$ of propionic acid?

21. Consider a protein with a surface-exposed histidine residue in a pH 4 solution. What is the fraction of protein molecules in which this histidine residue is charged? (Assume that the $pK_a$ is 6.0.)

22. For a protein with a surface-exposed aspartic acid, at what pH will this residue be neutral in 75% of the protein molecules? (Assume that the $pK_a$ is 4.0.)

23. A histidine is involved in an interaction with a glutamic acid that stabilizes the charged form of the histidine, such that the value of $\Delta G^o$ for deprotonation is 15 kJ•mol$^{-1}$ at pH 7.0 and 293 K (calculated using the biochemical standard state). What is the $pK_a$ of this histidine?

24. At the $T_M$ of a protein (55°C), the value of $\Delta H^o_{unfolding}$ is 15 kJ•mol$^{-1}$. What is the value of $\Delta S^o_{unfolding}$?

25. A protein has a $\Delta H^o_{unfolding}$ value of 140 kJ•mol$^{-1}$ at 25°C. The value of $\Delta C_P$ is 7.5 kJ•K$^{-1}$•mol$^{-1}$. The value of $\Delta H^o_{unfolding}$ at the $T_M$ is 230 kJ•mol$^{-1}$. What is the value of $T_M$?

26. A lysine sidechain has four torsion angles, each of which can take on three different values (60°, –60°, and 180°). Each unique combination of angles is called a rotamer. For example, a lysine residue where the first, second, third, and fourth torsion angles are all 60° is one unique rotamer, whereas a residue where the first, second, and third torsion angles are 60° and the fourth torsion angle is 180° is a second rotamer.

    In contrast, a serine sidechain has only one torsion angle, which can take on three different values.

    Assume that all possible dihedral angles are allowed at each angle for residues at this surface-exposed position.

    a. What is the difference in molar entropy between a protein with a surface-exposed lysine and an otherwise identical protein with a serine mutation at that position?
    b. Why might the simplification that each lysine torsion angle is able to adopt any of the three staggered positions, independent of the conformation at other torsion angles, lead to an overestimate of the number of low-energy conformations that lysine can adopt?

27. In the hydrophobic core of a folded protein, there are three alanine and five phenylalanine residues that are buried, and do not interact with water. Assume:
    - In solution, waters can take on seven energetically equal states.
    - Two waters are ordered around each alanine in the unfolded state.
    - Six waters are ordered around each phenylalanine in the unfolded state.
    - In the unfolded state, waters are ordered around alanine or phenylalanine residues and can take on only two energetically equal states.

    What is the difference in the entropy of the water due to the burying of these residues as this protein folds?

28. Why do proteins denature at cold temperatures?

29. How do hydrophobic interactions provide favorable entropy for protein folding?

## Further Reading

### General

Dill KA & Bromberg S (2010) Molecular Driving Forces: Statistical Thermodynamics in Chemistry, Biology, Physics, and Nanoscience, 2nd ed. New York: Garland Science.

#### B. Equilibrium Constants

Nicholls DG & Ferguson SJ (2002) Bioenergetics, 3rd ed. San Diego, CA: Academic Press.

#### D. Free-Energy Changes in Protein Folding

Baldwin RL (1986) Temperature dependence of the hydrophobic interaction in protein folding. *Proc. Natl. Acad. Sci. USA* 83, 8069–8072.

Becktel WJ & Schellman JA (1987) Protein stability curves. *Biopolymers* 26, 1859–1877.

Pfeil W & Privalov PL (1976) Thermodynamic investigations of proteins. I. Standard functions for proteins with lysozyme as an example. *Biophys. Chem.* 4, 23–32. (*This is the first of four related papers*.)

Privalov PL & Gill SJ (1988) Stability of protein structure and hydrophobic interaction. *Adv. Protein Chem.* 39, 191–234.

# Voltages and Free Energy

In this chapter we study two different kinds of processes in which electrical potential differences (voltages) govern how a system responds to perturbations. We begin by studying chemical reactions in which an electron is transferred from one molecule to another, a process referred to as an oxidation–reduction reaction. In the first two parts of this chapter, we discuss molecules that undergo oxidation–reduction reactions in biological systems and examine how the free-energy changes associated with these reactions are related to the voltages generated by electrochemical cells. In the last two parts of this chapter, we study the voltages that develop across biological membranes due to an asymmetric charge distribution. The propagation of these voltage differences, known as action potentials, underlies the transmission of signals in neurons.

**Redox reactions**

Chemical reactions that involve the transfer of an electron from one molecule to another are referred to as redox reactions. For example, consider the reaction in which a molecule (M) loses an electron:

$$M \rightleftharpoons M^+ + e^-$$

When the reaction proceeds from left to right, the M molecule is said to be oxidized. When the reaction proceeds in the reverse direction the $M^+$ ion is said to be reduced.

## A. OXIDATION–REDUCTION REACTIONS IN BIOLOGY

Chemical reactions that involve the transfer of electrons from one molecule to another are crucial in biology, particularly for processes that involve the trapping and utilization of energy. Here we introduce some basic principles concerning electron transfer reactions and provide some examples of their importance for energy capture and utilization.

### 11.1 Reactions involving the transfer of electrons are referred to as oxidation–reduction reactions

Metal ions and certain organic molecules that can exist in more than one stable oxidation state can participate in electron transfer reactions. In such a reaction, the molecule (or ion) that accepts the electron is said to undergo a **reduction**, whereas the molecule that loses the electron is said to be **oxidized**. Such reactions are referred to as oxidation–reduction reactions or **redox reactions**. Redox reactions typically occur in pairs, where one involves the release of electrons from one molecule (the oxidation half-reaction) and the other involves the acceptance of the electrons by the other molecule (the reduction half-reaction).

Consider what happens, for example, when a piece of zinc metal is immersed in a solution containing cupric ion ($Cu^{2+}$). As we discuss in Section 11.8, the zinc metal dissolves to form zinc ion ($Zn^{2+}$), while the cupric ion is converted to metallic copper (**Figure 11.1**). The net reaction is the sum of two half reactions, as shown in Equations 11.1 and 11.2. Both reactions proceed to the right as written. Equation 11.1 is the oxidation half-reaction (the zinc atom loses electrons and is oxidized), and Equation 11.2 is the reduction half-reaction (the cupric ion accepts electrons and is reduced).

$$Zn \rightleftharpoons Zn^{2+} + 2e^- \qquad (11.1)$$

$$Cu^{2+} + 2e^- \rightleftharpoons Cu \qquad (11.2)$$

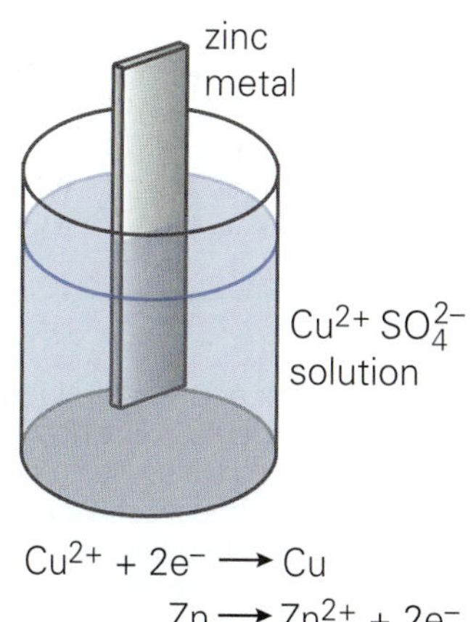


**Figure 11.1 A simple redox reaction.** A piece of zinc metal is immersed in a solution of copper sulfate. The zinc metal dissolves while transferring two electrons to a $Cu^{2+}$ ion, converting it to metallic copper, which is deposited on the zinc metal. See Equations 11.1 and 11.2.

The oxidized and reduced forms of a molecule are known as a **redox couple**. In Equation 11.1, the zinc atom and the $Zn^{2+}$ ion form a redox couple, as do the copper atom and the $Cu^{2+}$ ion in Equation 11.2.

## 11.2 Biologically important redox-active metals are bound to proteins

Metal ions involved in biologically important redox reactions do not diffuse freely in solution, but are bound instead to protein molecules, as illustrated in **Figure 11.2**. The protein molecules are often organized into large complexes that bring different redox active centers into close proximity, allowing electrons to be transferred from one protein-bound metal ion to another. Such electron transfer complexes are important in the conversion of light energy to ATP during photosynthesis and in the process of **oxidative phosphorylation**, which couples the oxidation of fuel molecules, such as glucose, to ATP production.

One common means of attaching metal ions to proteins is to have the metal coordinated to a large organic molecule, such as a heme group. The heme-bound metal is coordinated, in turn, by one or two sidechains from the protein, as illustrated for a protein known as a cytochrome in Figure 11.2. Another common means of attaching a metal to a protein is through direct coordination of the metal through the sulfur atoms of cysteine or methionine residues, as in the iron–sulfur cluster of a protein known as rubredoxin (Figure 11.2B). The metal ions in proteins such as cytochrome *c* and rubredoxin can gain or lose electrons and can therefore form a redox couple. Electrons can be transferred to and from these metal ions, resulting in their oxidation or reduction.

The coordination by the protein also modifies the properties of the metal ion, altering the free energy of the reduced form with respect to the oxidized form. This tuning of the relative propensity to be oxidized or reduced is critical for setting up a cascade of electron transfer events, in which the direction of electron transfer is determined by the relative free energies of the oxidized and reduced forms of different protein-bound metals. This tuning is critical for photosynthesis and oxidative phosphorylation, where electrons are driven from one center to another in an energetically downhill process. Proteins also protect the metal ions from chemical reactions that might prevent the metal from participating in these kinds of electron transfer reactions.

## 11.3 Nicotinamide adenine dinucleotide ($NAD^+$) is an important mediator of redox reactions in biology

In addition to metal ions, many organic compounds also play important roles in biological redox reactions. Nicotinamide adenine dinucleotide ($NAD^+$) is an organic compound that undergoes redox reactions and is extremely important

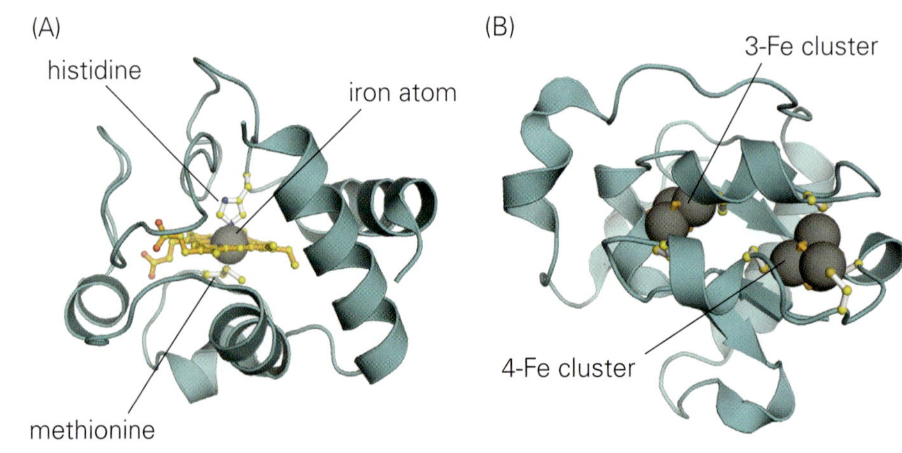


**Figure 11.2 Redox active metal ions in proteins.** (A) A single iron atom (*gray* sphere) bound to a heme group and coordinated by two sidechains (a histidine and a methionine) in the protein cytochrome *c*. (B) Two multi-iron clusters (*gray* spheres) in the protein rubredoxin. One cluster contains three iron atoms, whereas the other contains four iron atoms (only three of these are visible). The iron atoms in each cluster are cross-linked by sulfur atoms, so these clusters are called iron–sulfur clusters. Each iron–sulfur cluster is also coordinated to sulfur atoms of cysteine residues in the protein. (PDB codes: 5CYT and 1FER.)

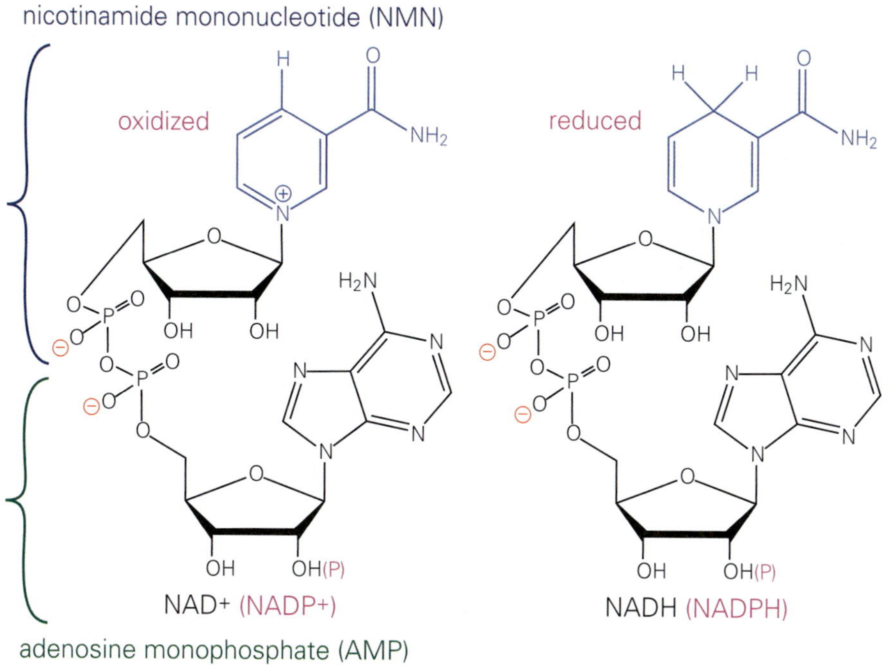

**Figure 11.3 The redox couple NAD⁺/NADH.** Nicotinamide adenine dinucleotide consists of two nucleotides. The nucleotides are linked together by two phosphate groups that are attached to their 5′ hydroxyl groups. The redox active element is the nicotinamide base (shown in *blue*).

in biological energy transduction (**Figure 11.3**). $NAD^+$ can undergo the following redox half-reaction, in which two electrons and a proton are involved in the reduction of $NAD^+$ to generate NADH:

$$NAD^+ + H^+ + 2e^- \rightleftharpoons NADH \tag{11.3}$$

The proton and the two electrons do not exist as separate species, but rather are considered to form a so-called "hydride ion" ($H:^-$).

$NAD^+$/NADH binds to specialized nucleotide binding domains, introduced in Section 5.26. These domains are components of larger proteins or multiprotein complexes, and they present the redox-active nicotimamide group in a manner that allows it to be oxidized or reduced efficiently by other compounds. The adenosine monophosphate component of $NAD^+$ plays no role in the redox reaction, but it anchors the cofactor to the protein. The 2′ hydroxyl group of the AMP portion of $NAD^+$/NADH can be phosphorylated, as shown in Figure 11.3. This modified redox couple, known as $NADP^+$/NADPH, binds to a separate class of enzymes, allowing the cell to maintain two separate pools of these redox-active molecules.

To illustrate how $NAD^+$ can oxidize substrates, consider the mechanism of alcohol dehydrogenase. This enzyme catalyzes the oxidation of alcohols to aldehydes, but it can also operate in the reverse direction, converting acetaldehyde to ethanol. **Figure 11.4** shows the structure of alcohol dehydrogenase bound to benzyl alcohol, which is converted to benzaldehyde upon oxidation by $NAD^+$ (the details of the reaction are shown in **Figure 11.5**). The substrate molecule (benzyl alcohol) binds near the nicotinic acid group of $NAD^+$. The nicotinic acid group abstracts a hydride ion from the substrate, thus oxidizing the substrate and releasing a proton.

## 11.4 Flavins and quinones can undergo oxidation or reduction in two steps of one electron each

NADH is the most important source of reducing power in the cell, and the generation of NADH by the reduction of $NAD^+$ is a key step in harnessing light energy in photosynthesis and utilizing the energy stored in food. The reducing power of NADH is harnessed to generate ATP, as we describe below. As shown in Equation 11.3, the reduction of $NAD^+$ involves the transfer of two electrons per molecule

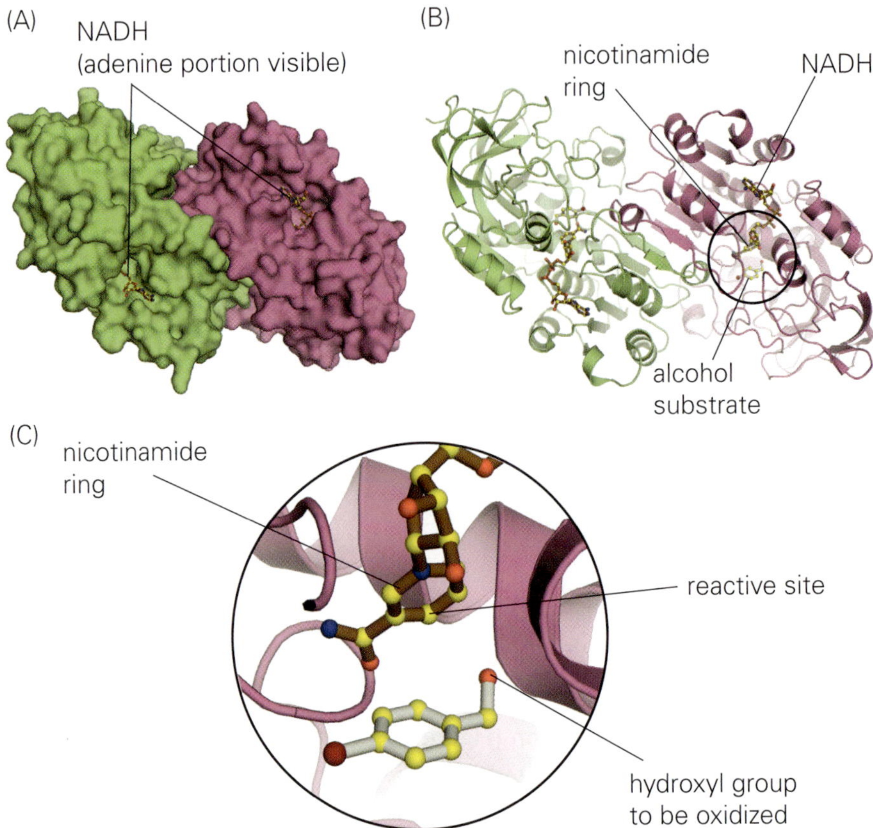


**Figure 11.4 NADH and alcohol dehydrogenase.** The structure of mammalian liver alcohol dehydrogenase is shown. (A) The surface of the enzyme, which has two identical subunits. $NAD^+$ is bound to the active sites of both subunits, and the adenine portion of NADH is visible at the surface. The two subunits are shown in *green* and *purple*. (B) The two subunits are shown in ribbon representation. The active site of the *purple* subunit is shown in an expanded view in (C), which shows the relative orientation of the nicotinamide ring and the alcohol substrate (benzyl alcohol). $NAD^+$ oxidizes the alcohol substrate to an aldehyde, as shown in Figure 11.5. (PDB code: 1HLD.)

of $NAD^+$. In many processes, such as the capture of light energy in photosynthesis, the fundamental steps involve the transfer of single electrons. Two families of redox-active organic compounds that are important in biology—namely, the flavins and quinones—are able to take up either one or two electrons and are therefore able to couple processes that transfer single electrons to the reduction of $NAD^+$, which requires two electrons.

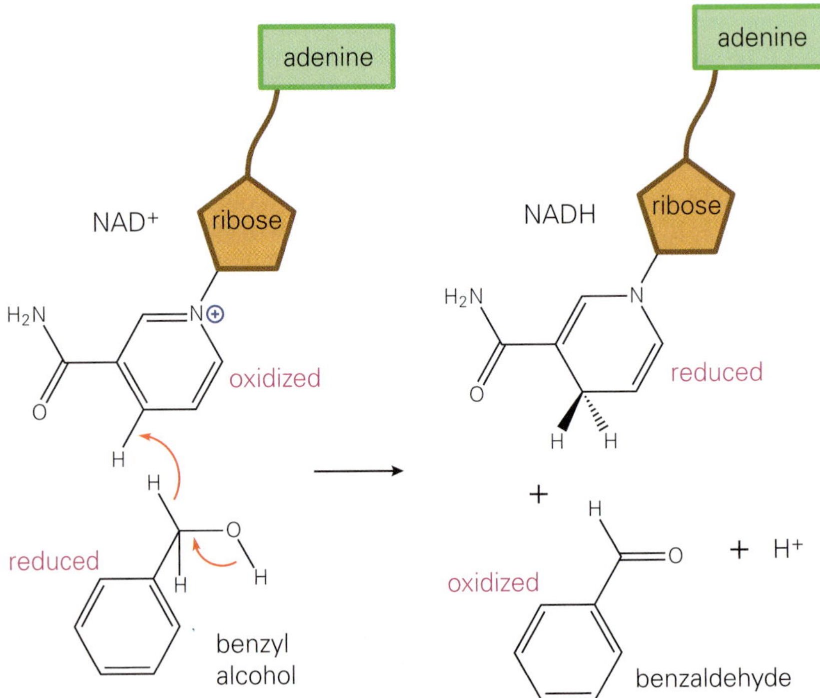


**Figure 11.5 The reaction catalyzed by alcohol dehydrogenase.** The enzyme binds to $NAD^+$/NADH and to the substrate (benzyl alcohol in this case) as shown in Figure 11.4. $NAD^+$ abstracts a hydride ion ($H^+ + 2e^-$) from the substrate, which also loses a proton. The net effect is the oxidation of the alcohol to the aldehyde. The enzyme will also catalyze the reaction in the reverse direction, depending on the concentrations of the alcohol, the aldehyde, and the relative ratio of $NAD^+$ to NADH.

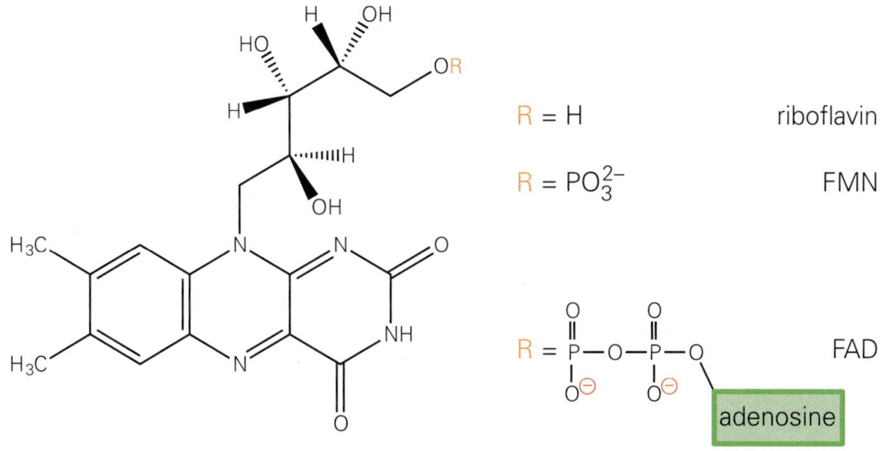

R = H riboflavin

R = $PO_3^{2-}$ FMN

R = adenosine diphosphate FAD

**Figure 11.6 Chemical structure of the flavins.** The heterocyclic ring system is known as an isoalloxazine group. The structure shown here is essentially that of riboflavin (vitamin $B_2$). In flavin mononucleotide (FMN), R represents a phosphate group. In flavin adenine dinucleotide, R is adenosine diphosphate. (Adapted from C. Walsh, *Acc. Chem. Res.* 13: 148–155, 1980. With permission from the American Chemical Society.)

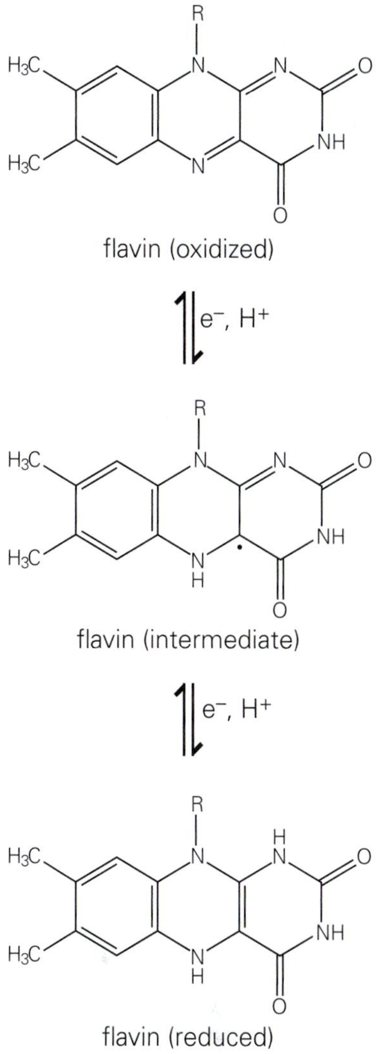

The flavins are derivatives of riboflavin (vitamin $B_2$), in which the isoalloxazine ring system of riboflavin is phosphorylated to give flavin mononucleotide (FMN) and then adenylated to give flavin adenine dinucleotide (FAD) (**Figure 11.6**). In contrast to NADH/NADPH and $NAD^+$/$NADP^+$, which are soluble in the cytoplasm when not bound proteins, the flavin cofactor is always tightly bound to proteins. Like $NAD^+$ or $NADP^+$, the flavins can undergo two-electron reductions, but they can also transfer or take up electrons in single-electron steps, as shown in **Figure 11.7**. Likewise, ubiquinones (one form of which is known as coenzyme Q, illustrated in **Figure 11.8**) can also be reduced in two steps involving one electron each. Ubiquinones play an important role in mediating electron transfer between membrane-bound redox centers, because they are soluble in the lipid bilayer.

flavin (oxidized)

$e^-$, $H^+$

flavin (intermediate)

$e^-$, $H^+$

flavin (reduced)

**Figure 11.7 Redox states of the flavin group.** The three states shown here are connected by single electron transfers.

## 11.5 The oxidation of glucose is coupled to the generation of NADH and $FADH_2$

Electron transfer reactions are fundamental to cellular processes that convert the energy stored in food into ATP, which is then utilized by the molecular machinery of the cell to drive chemical reactions against a free-energy gradient. The complete oxidation of glucose releases a very large amount of free energy:

$$C_6H_{12}O_6 + 6O_2 \rightarrow 6CO_2 + 6H_2O \; (\Delta G^o = -2823 \text{ kJ} \bullet \text{mol}^{-1}) \qquad (11.4)$$

ubiquinone $\xrightarrow{e^-, H^+}$ semiquinone $\xrightarrow{e^-, H^+}$ ubiquinol

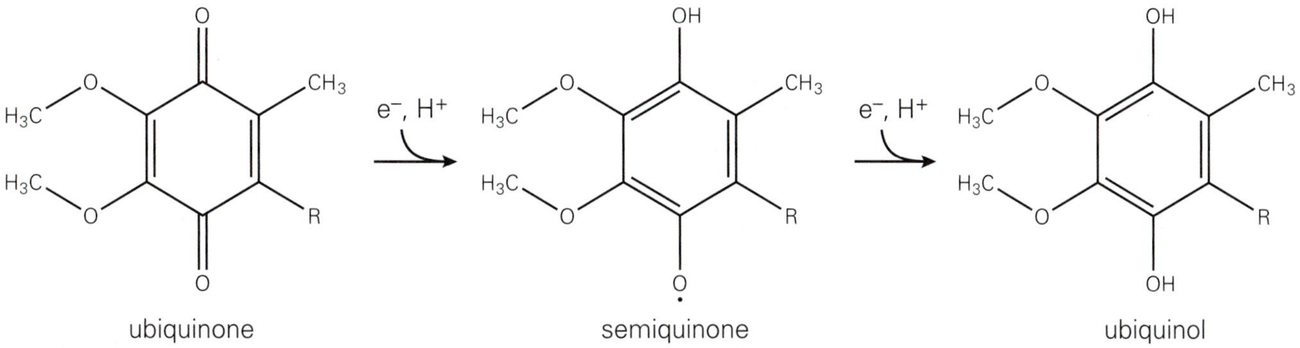

**Figure 11.8 Redox reactions involving ubiquinone.** Ubiqinone (also known as coenzyme Q) contains a quinone head group (shown here) to which a long hydrocarbon tail (indicated here by R) is attached. Ubiquinone can undergo two reductions, involving one electron at each step, to generate the semiquinone and ubiquinol forms. The presence of the aliphatic R group makes all three forms soluble in the membrane.

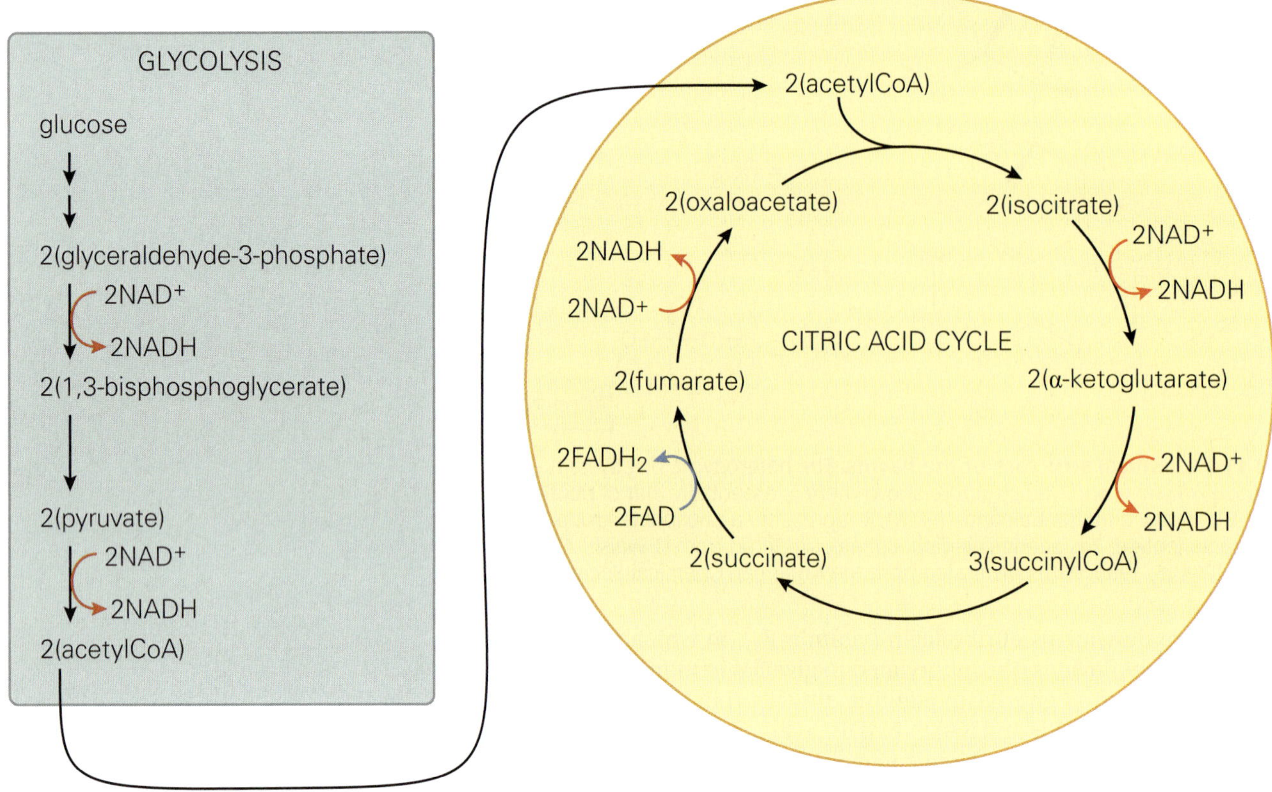


**Figure 11.9 Schematic representation of the steps that lead to NADH and $FADH_2$ in glycolysis and the citric acid cycle.** The stepwise oxidation of glucose by a series of enzyme-catalyzed reactions results in the reduction of $NAD^+$ and FAD to NADH and $FADH_2$, respectively. The free energy stored in NADH and $FADH_2$ is used subsequently to generate a proton gradient across the mitochondrial membrane. The diagram does not show all of the intermediate molecules in glycolysis and the citric acid cycle. (Adapted from D. Voet and J.G. Voet, Biochemistry, 3rd ed. New York: John Wiley & Sons, 2004.)

Although it is not apparent that this reaction involves electron transfer, the reaction can be written as two half reactions that show the transfer of electrons explicitly:

$$C_6H_{12}O_6 + 6H_2O \rightarrow 6CO_2 + 24H^+ + 24e^-$$
$$6O_2 + 24H^+ + 24e^- \rightarrow 12H_2O \tag{11.5}$$

The electrons and protons are not liberated and recaptured in one step. Instead, they are transferred to $NAD^+$ and FAD, resulting in the generation of 10 NADH molecules and two molecules of $FADH_2$. The chemical steps that underlie this conversion are referred to as glycolysis and the citric acid cycle, which are illustrated schematically in **Figure 11.9**.

One advantage of the stepwise oxidation of glucose is that it greatly increases the efficiency by which the free energy that is liberated is harnessed for the generation of ATP (**Figure 11.10**). A thermodynamic analysis shows that biochemical systems can harness up to 50% of the free energy stored in glucose by converting it to ATP. This process is considerably more efficient than combustion reactions in man-made engines. If a fuel, such as gasoline, is burned, then an explosive reaction occurs and a great amount of energy is released all at once. Only about 20% of this energy, though, is converted to work by combustion engines.

The efficiency of a combustion engine depends on the ratio of the temperature in the reaction chamber and the exterior heat sink, with greater temperature ratios yielding greater efficiency. Biological systems, which are fragile, would be damaged by the generation of too much heat. Instead, the combustion (oxidation) of biological fuels, such as glucose or fat, occurs in a series of stepwise reactions, with the release of small amounts of free energy at each step. These reactions store the free energy that is released by increasing the reducing power of the cell—that is, by increasing the concentrations of NADH and protein-bound $FADH_2$. The increased reducing power is harnessed to generate ATP, as described in the next section.

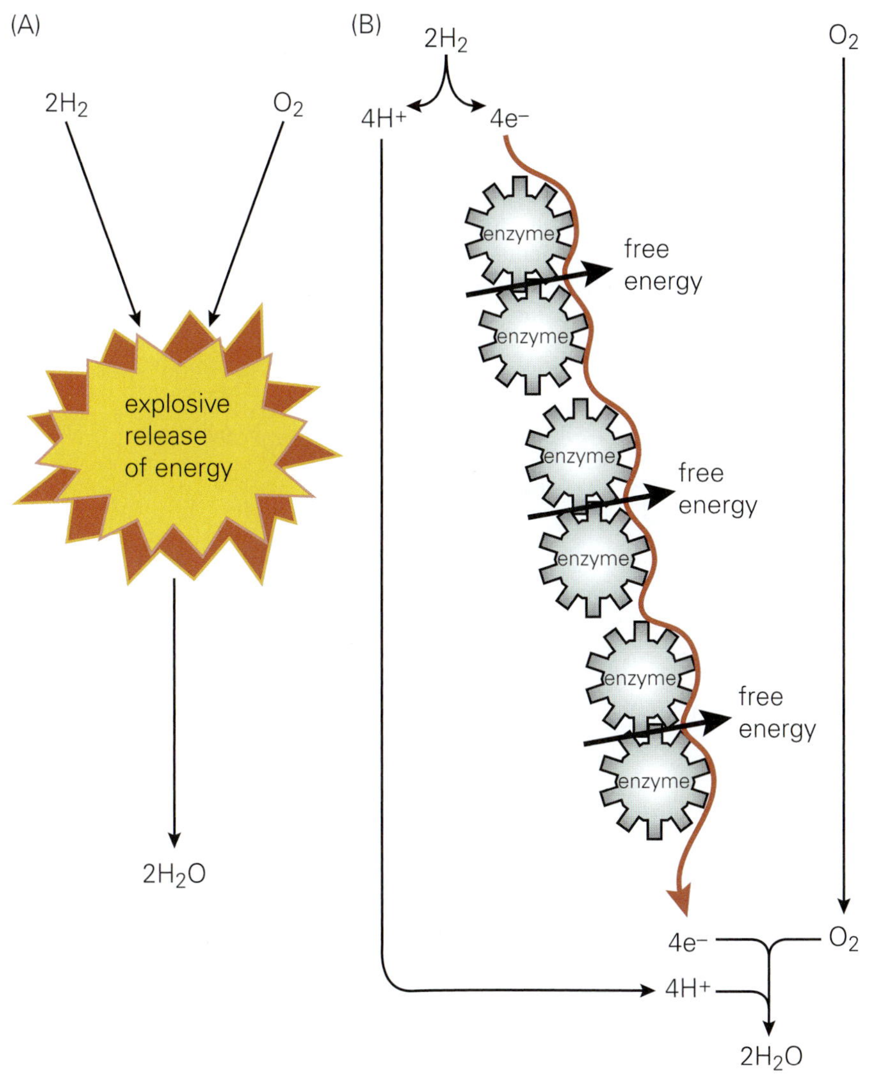


**Figure 11.10 Biological oxidation allows energy to be harnessed in a series of controlled steps.** In (A), the combustion of hydrogen, which leads to the explosive release of energy, is illustrated schematically. This is contrasted with the oxidation of molecules such as glucose in cells, where the abstraction of protons ($H^+$) and electrons ($e^-$) occurs in a stepwise fashion, with the storage of energy (more correctly, free energy) at each step. Each of the cogs in the wheel in this schematic diagram represents an enzyme that reduces $NAD^+$ or FAD. (Adapted from B. Alberts et al., Molecular Biology of the Cell, 5th ed. New York: Garland Science, 2008.)

## 11.6 Mitochondria are cellular compartments in which NADH and $FADH_2$ are used to generate ATP

The sites of ATP production in eukaryotic cells are the **mitochondria**, which are specialized compartments with their own membranes (**Figure 11.11**). One key feature that distinguishes a mitochondrion from other cellular organelles is that it contains two membranes, one encapsulated entirely within the other. Each membrane consists of a lipid bilayer. The inner membrane forms numerous invaginations, known as cristae, so that it has an extremely large surface area. The surface

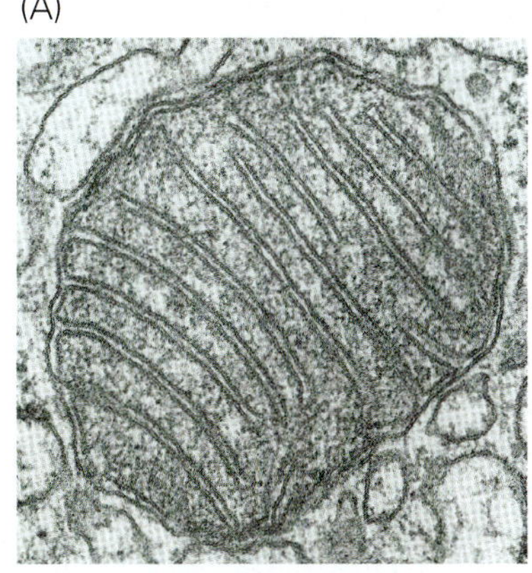


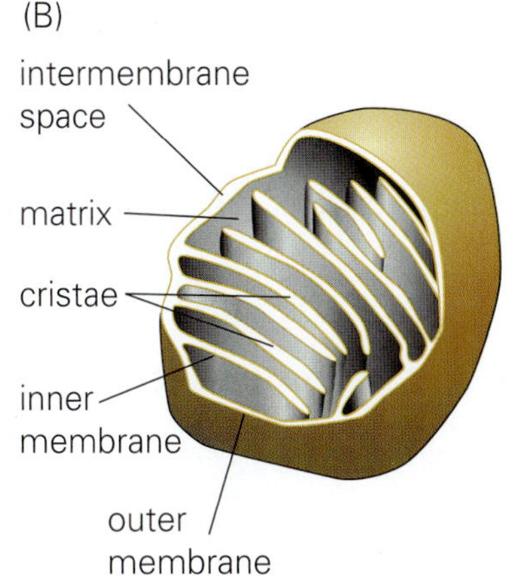


**Figure 11.11 The structure of mitochondria.** (A) An electron micrograph of a mitochondrion, which is a specialized cellular compartment known as an organelle. (B) A schematic diagram of a mitochondrion. Mitochondria contain two membranes, called inner and outer. The inner membrane, which is highly invaginated, encloses an internal space known as the matrix. The reactions of the citric acid cycle (see Figure 11.9) take place here. The outer membrane seals the organelle. The space between the inner and outer membranes (*white*) is the region where the proton concentration increases in order to drive ATP synthesis. Proton pumps and the ATP synthesis machinery are embedded in the inner membrane. (A, courtesy of Daniel S. Friend; B, adapted from B. Alberts et al., Molecular Biology of the Cell, 4th ed. New York: Garland Science, 2002.)

of this inner membrane is studded with the protein complexes that pump protons out of the inner mitochondrial space and into the intermembrane space (this is illustrated schematically in **Figure 11.12**).

Pyruvate molecules, which are generated from glucose by glycolysis (see Figure 11.9), are transported into mitochondria, where they are converted to acetyl coenzyme A (acetyl CoA). Acetyl CoA enters the citric acid cycle, which takes place entirely within the mitochondria. Other fuel molecules that can also be converted into acetyl CoA, such as fatty acids, are also transported into the mitochondria. The reducing power thus generated in the mitochondria, where these reactions occur, is used, in turn, to reduce components of a set of integral membrane

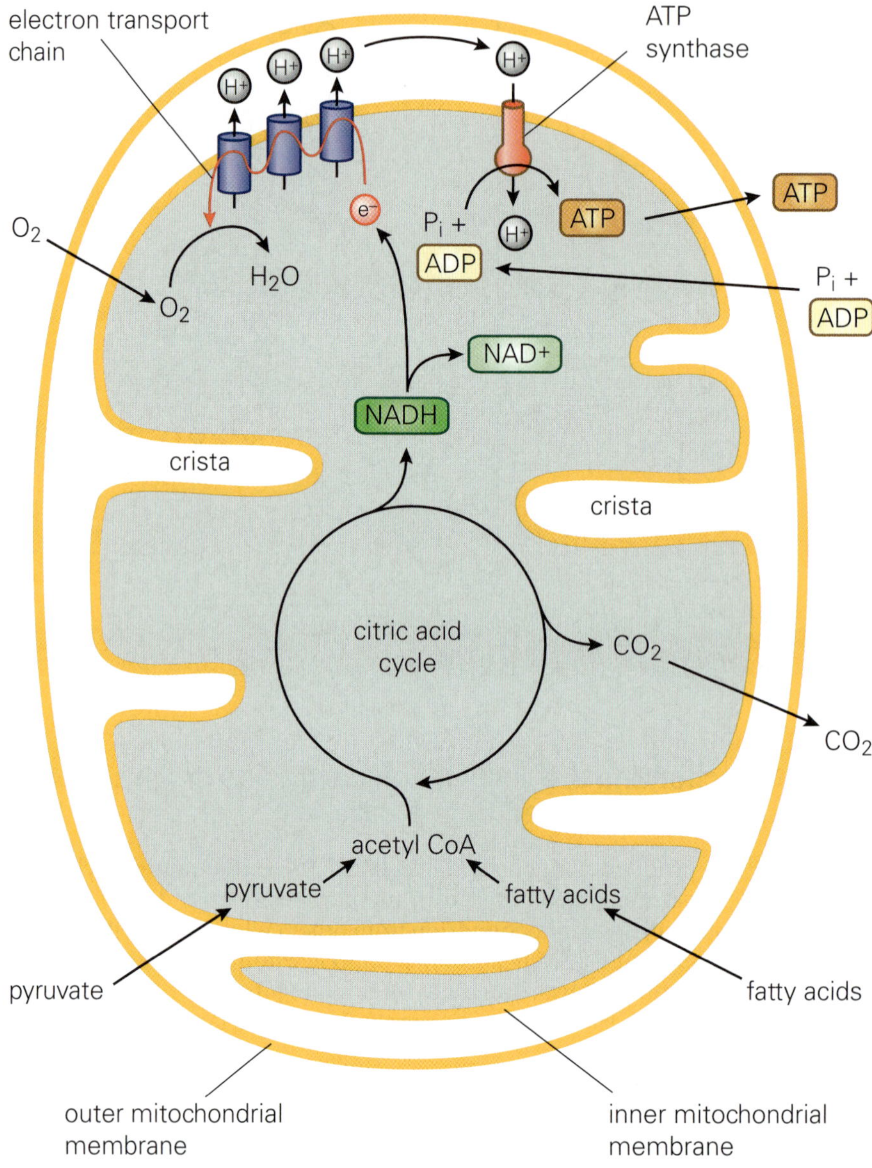


**Figure 11.12 Schematic diagram of a mitochondrion.** The organelle consists of inner and outer membranes. The inner membrane contains integral membrane proteins involved in energy transduction. These are the electron transport chain, which uses the reducing power of NADH (generated in the inner mitochondrial space by the citric acid cycle) to pump protons into the intermembrane space. This generates a gradient in proton concentration across the inner membrane. This gradient drives the synthesis of ATP by the ATP synthase enzyme, which is also integral to the inner membrane. (Adapted from B. Alberts et al., Molecular Biology of the Cell, 5th ed. New York: Garland Science, 2008.)

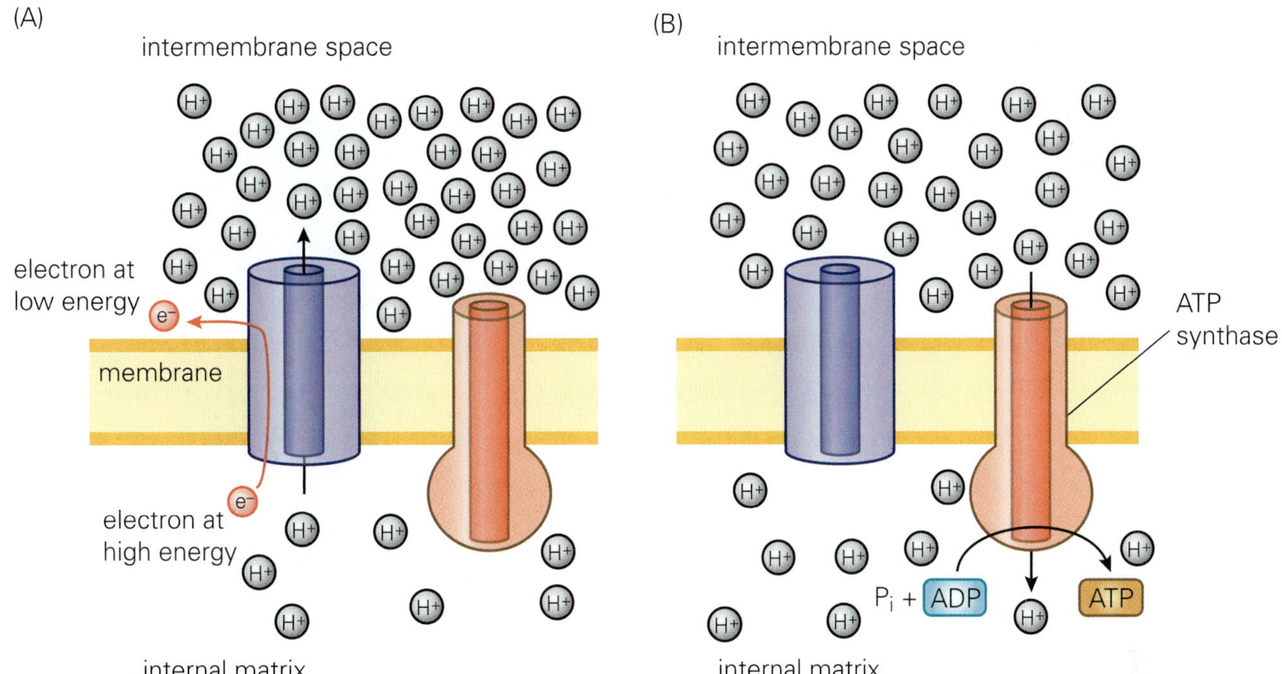


**Figure 11.13 Proton gradient generation and ATP synthesis.** In (A), the electron transfer chain couples the high free energy of electrons in NADH to the transport of the protons against a proton gradient. The free energy of NADH is therefore stored in the reduced pH of the intermembrane space. This concentration gradient is then used, as shown in (B), by the ATP synthase protein to generate ATP from ADP. (Adapted from B. Alberts et al., Molecular Biology of the Cell, 5th ed. New York: Garland Science, 2008.)

proteins known as the electron transport chain. The high free energy of NADH and $FADH_2$ is used to move protons against a concentration gradient, thus building up a high proton concentration in the intermembrane space. This, in turn, is coupled to rotational movements within another integral membrane protein, ATP synthase, which generates ATP from ADP (see Section 9.16). This process is shown schematically in **Figure 11.13**.

## 11.7 Absorption of light creates molecules with high reducing power in photosynthesis

The ultimate source of energy for life on Earth is the sun. Plants convert solar energy into biological fuel (food) in two steps. First, light energy is captured in specialized compartments within plant cells, known as **chloroplasts**. The capture of light energy is used to drive the generation of a proton gradient across a membrane. As in the case of energy production by mitochondria, the proton gradient provides the free-energy drive that powers an ATP synthase enzyme (very similar to the one used in mitochondria) to make ATP. In the second step, the free energy of ATP is used to synthesize glucose and other carbohydrates from carbon dioxide and other inputs in a process known as carbon fixation. This process is also driven by NADPH, which is another product of photosynthesis.

Like mitochondria, chloroplasts are organelles that have an outer and an inner cell membrane (**Figure 11.14**). In contrast to the mitochondria, the inner membrane of chloroplasts is not highly invaginated. Chloroplasts, however, are distinguished by having a third internal membrane that is highly folded up to form a large number of pouch-like structures. This membrane, which is known as the thylakoid (from the Greek word meaning "pouch like"), encloses a separate internal compartment known as the thylakoid space. The photosynthetic machinery, which consists of several integral membrane proteins, is embedded in the thylakoid membrane.

The critical first step in photosynthesis is the capture of a photon of light by a chromophore known as chlorophyll. Chlorophyll consists of a porphyrin ring system, similar to that seen in the heme group (**Figure 11.15**). In contrast to heme, though, where an iron atom is bound to the porphyrin ring, the metal atom in chlorophyll is a magnesium. As a result of the excitation by light, an electron is

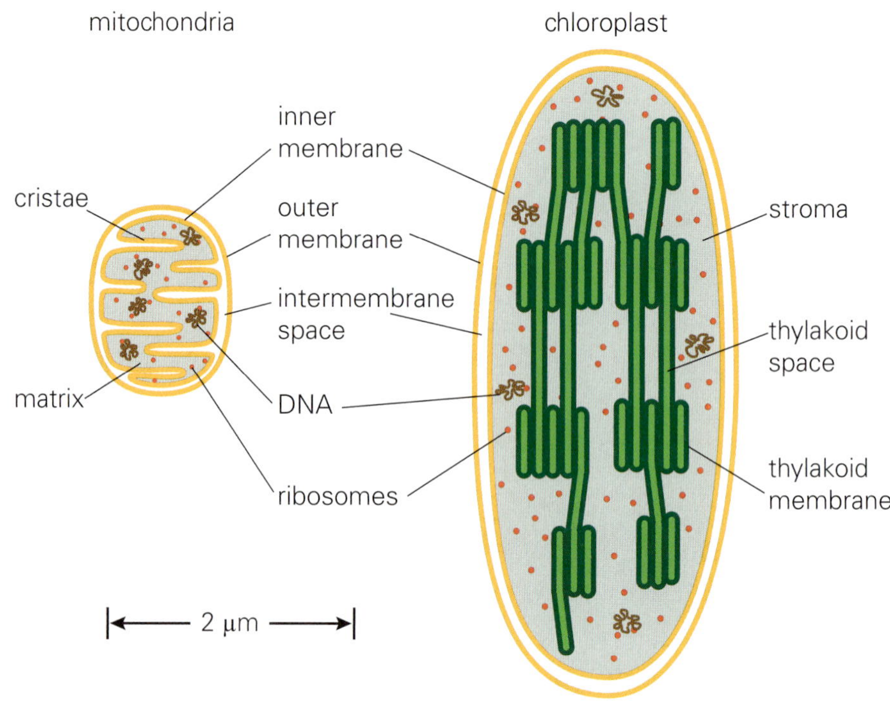


**Figure 11.14 A comparison of a mitochondrion and a chloroplast.** Both have an outer membrane and an inner membrane. Chloroplasts also have a third internal membrane, known as the thylakoid membrane. This membrane encloses a separate space, known as the thylakoid space. The photosynthetic apparatus is embedded within the thylakoid membrane. (Adapted from B. Alberts et al., Molecular Biology of the Cell, 5th ed. New York: Garland Science, 2008.)

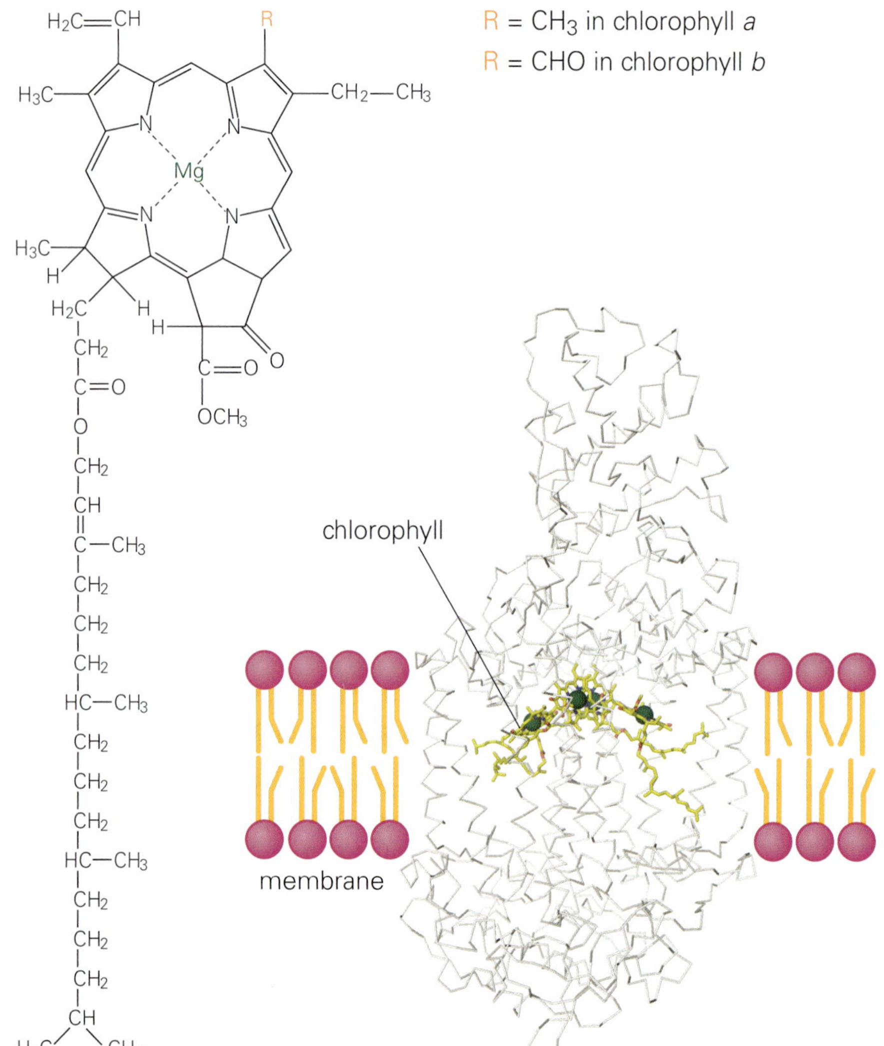


**Figure 11.15 Structure of a photosynthetic reaction center**. The structure of a chlorophyll molecule is shown on the left. On the right is shown the crystal structure of a bacterial photosynthetic reaction center, whose structure and mechanistic principles are similar to those of the photosynthetic complexes found in plants. Four of the chlorophyll molecules are shown in *yellow*, with their magnesium atoms indicated as *green* spheres. The initial absorption of light energy occurs at one of the two central chlorophyll molecules. A number of other metal ions and redox-active molecules are present within the reaction center, but these are not shown in this diagram. (PDB code: 4RCR.)

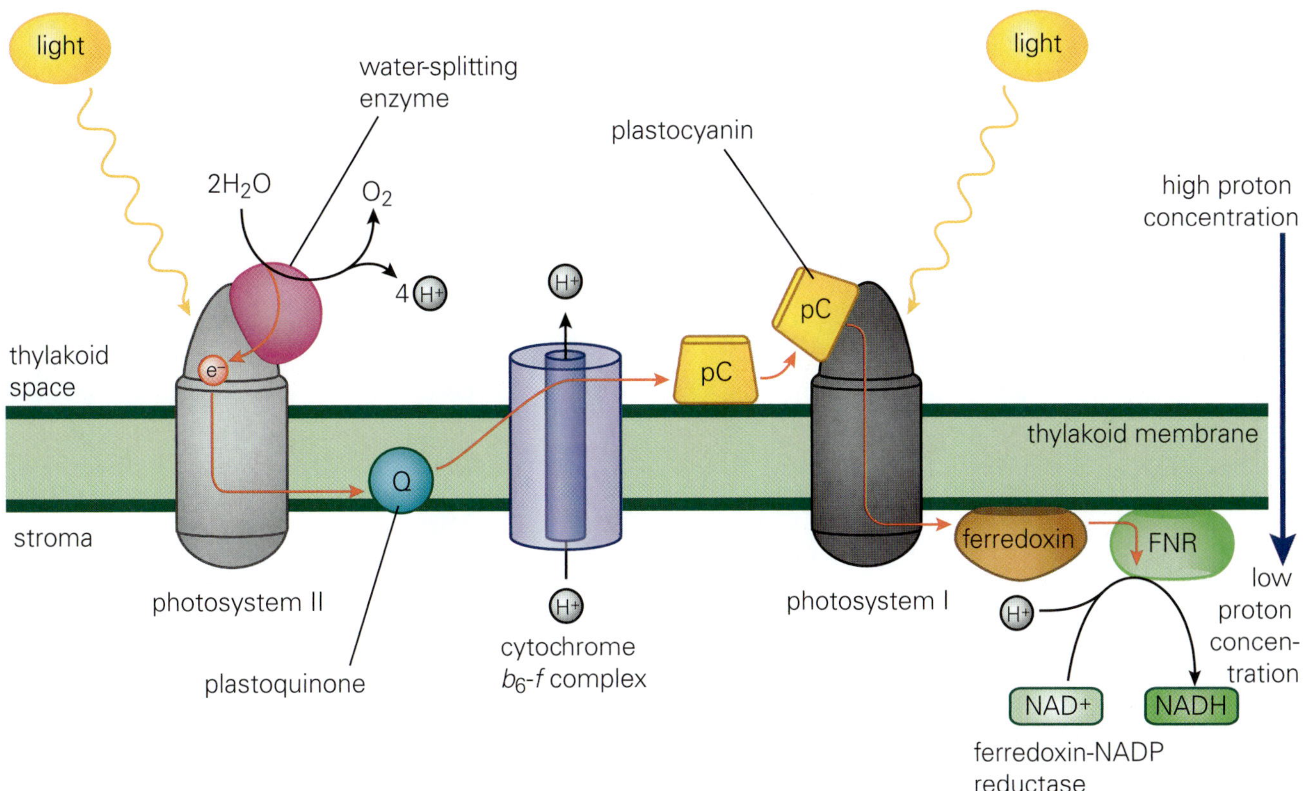


**Figure 11.16 The photosynthetic complexes of the thylakoid membrane in plants.** There are three integral membrane protein complexes (photosystems I and II and the cytochrome $b_6$-$f$ complex). Plastoquinone is a redox-active quinone that is within the membrane (see Figure 11.8). Plastocyanin and ferredoxin are proteins that are associated with the membrane and serve as electron shuttles. The net effect of the capture of photons is the generation of a proton gradient and the reduction of $NAD^+$ to NADH by ferredoxin nucleotide reductase (FNR). (Adapted from B. Alberts, et al., Molecular Biology of the Cell, 5th ed. New York: Garland Science, 2008.)

ejected from the metal ion in one of the chlorophylls, leading to charge separation. The ejected electron has very high reducing power (that is, very high free energy), and it is rapidly transferred from redox center to redox center in a series of membrane-bound proteins, known as the photosynthetic reaction center (see Figure 11.15) and the photosystems (**Figure 11.16**). The rapid transfer of the electron minimizes the unproductive back reaction in which it simply recombines with the chlorophyll from which it had been ejected.

As the free energy of the electron is reduced in a step-by-step process, much of the free energy is recovered in two important forms—namely, protons are pumped across the membrane, up the concentration gradient, and $NAD^+$ molecules are reduced to form NADH, which is also used to pump additional protons (**Figure 11.17**). The proton gradient that is generated by photosynthesis results in an increased proton concentration (lower pH) in the thylakoid space. This gradient is used by ATP synthase enzymes, located within the thylakoid membrane, to make ATP. Thus, understanding how this remarkable system works requires us to understand the free-energy changes underlying redox reactions, which is the focus of Part B of this chapter.

## B. REDUCTION POTENTIALS AND FREE ENERGY

Molecules that form a redox couple can be used to generate current in an electrical circuit. In this part of the chapter we study the relationship between the voltage generated by such a circuit and the free-energy change of the underlying redox reactions.

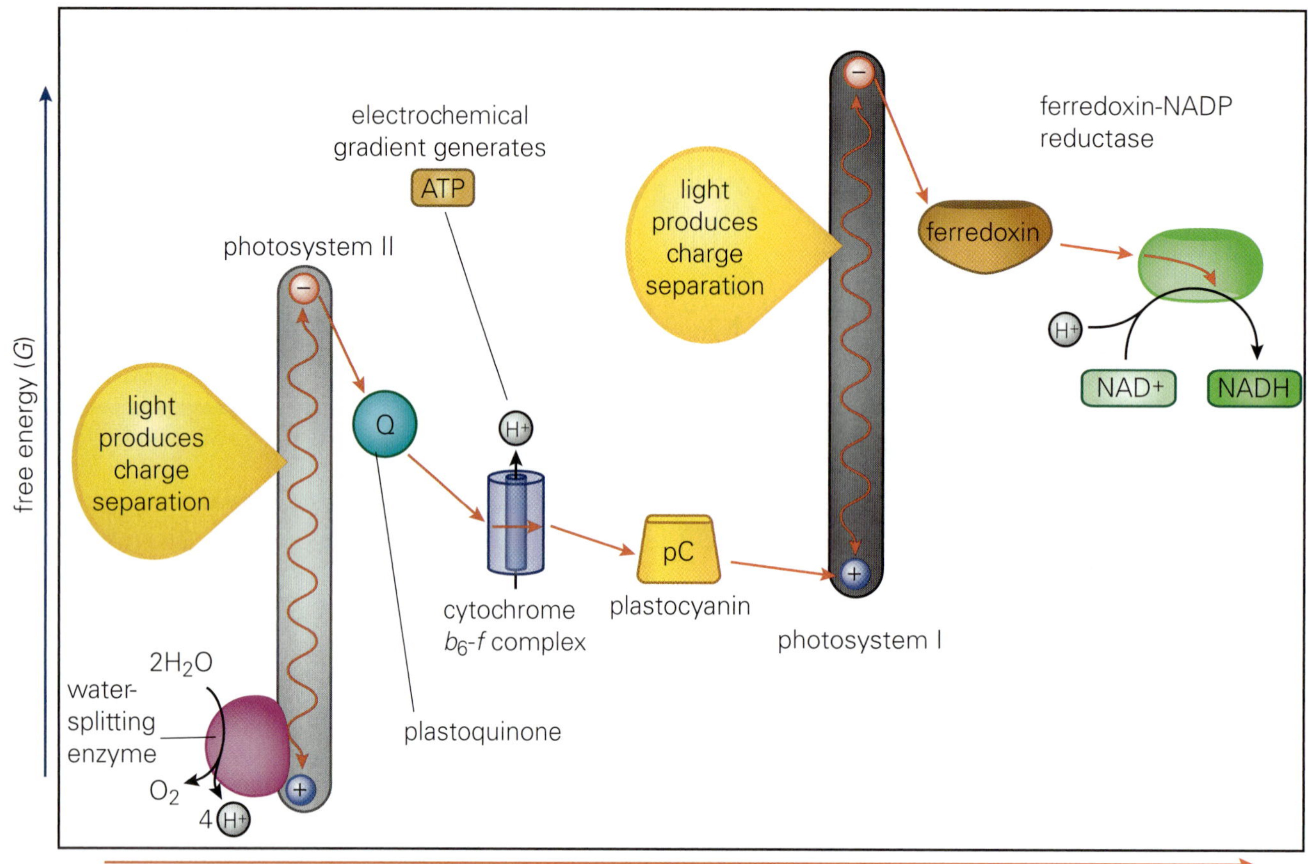


**Figure 11.17 Free energy and electron flow in the photosynthetic complexes.** This diagram indicates the free energies of the various intermediates in the processes shown in Figure 11.16. The first step involves the ejection of a high-energy electron from a metal in photosystem II, which is coupled to the splitting of water. The excited electron is transferred to photosystem I in a series of steps that are downhill in free energy. This transfer is coupled to the pumping of protons across the membrane to generate an electrochemical gradient. Photosystem I is excited by another photon, and the resulting high-energy electron is transferred via ferredoxin and FNR to $NAD^+$ to generate NADH. (Adapted from B. Alberts et al., Molecular Biology of the Cell, 5th ed. New York: Garland Science, 2008.)

## 11.8 Electrochemical cells can be constructed by linking two redox couples

Suppose we have two molecules, A and B. If A donates electrons to B spontaneously, becoming oxidized in the process, then it follows that the oxidized form of A is lower in free energy than the oxidized form of B. Because an electron transfer is involved in this reaction, it turns out that the free-energy differences can be related in a straightforward way to the voltage in an **electrochemical cell** constructed using the A and B molecules. Consideration of such electrochemical cells leads to the concept of the **reduction potential** of a molecule, a parameter that determines the direction of electron flow in coupled oxidation–reduction reactions. We begin the study of reduction potentials by first examining how current can be made to flow in electrochemical cells. In subsequent sections, we relate the voltage driving current flow to reduction potentials and to changes in free energy.

**Electrochemical cells**

An electrochemical cell is a battery that is powered by connecting two redox reactions into a circuit. Consider the following two half reactions:

$$A \rightarrow A^+ + e^-$$

and

$$B^+ + e^- \rightarrow B$$

If the free energy of B is lower than that of A, a current can be made to flow between a chamber containing A and another chamber containing $B^+$. Such a system is called an electrochemical cell.

The concept of an electrochemical cell is most straightforward to understand by considering reactions involving metal ions, although the idea can be readily generalized to other molecules. Let us start by considering a piece of zinc metal (Zn, solid) that is placed in an aqueous solution of $CuSO_4$ ($Cu^{2+}$, solution; $SO_4^{2-}$, solution), as shown schematically in Figure 11.1. We will refer to such a piece of metal

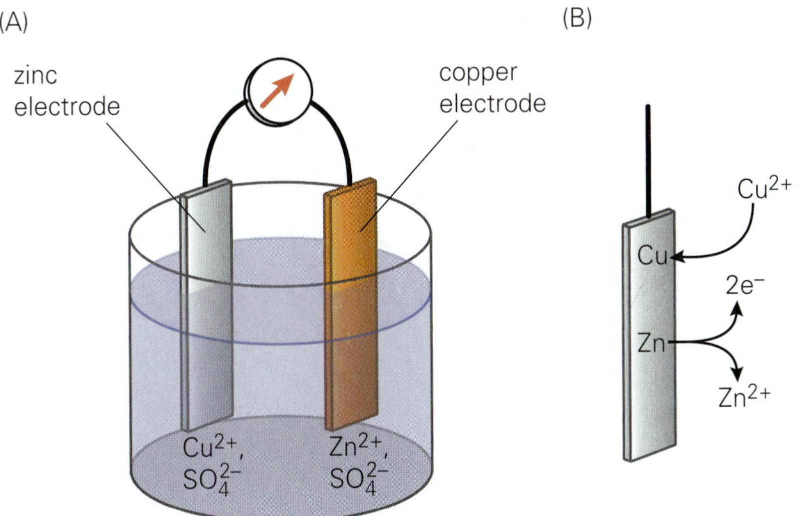


**Figure 11.18 Zinc and copper electrodes connected by a wire.** (A) A piece of copper metal and a piece of zinc metal are immersed in a solution containing copper sulfate and zinc sulfate. The metals are referred to as electrodes and are connected by wires to a voltmeter to measure current flow. (B) Schematic representation of what happens at the zinc electrode. Although the zinc will dissolve spontaneously, releasing electrons, these electrons can be captured by copper ions in solution, and so no current will flow through the external wire.

as an **electrode**, anticipating the fact that current can be made to flow through it under certain circumstances. As discussed in Section 11.1, the zinc electrode dissolves and copper metal gets deposited on the zinc electrode. These reactions can be written as follows:

$$\text{Zn (solid)} \rightarrow \text{Zn}^{2+}\text{ (solution)} + 2\text{e}^- \quad (11.6)$$

$$2\text{e}^- + \text{Cu}^{2+}\text{ (solution)} \rightarrow \text{Cu (solid)} \quad (11.7)$$

This combination of reactions is spontaneous, but if we reverse the situation, by placing a copper electrode in a solution of zinc sulfate, nothing happens. That is, the following two coupled half reactions are not spontaneous:

$$\text{Cu (solid)} \rightarrow \text{Cu}^{2+}\text{ (solution)} + 2\text{e}^- \quad (11.8)$$

$$2\text{e}^- + \text{Zn}^{2+}\text{ (solution)} \rightarrow \text{Zn (solid)} \quad (11.9)$$

From these observations we can conclude that copper "attracts" electrons more than zinc does. That is, copper metal (Cu) has lower free energy relative to its ion than zinc metal (Zn) does.

Let us now see how we can construct an electrochemical cell (battery) whereby we can cause a current to flow as a consequence of the spontaneous nature of the two half reactions denoted in Equations 11.6 and 11.7. First, consider the situation depicted in **Figure 11.18** in which a copper electrode and a zinc electrode are immersed in a solution containing copper sulfate and zinc sulfate. If we connect the two metal electrodes by a conducting wire, do we expect current to flow through the wire? To answer this question, think about what happens at the zinc electrode. According to Equation 11.6, the zinc metal dissolves, liberating two electrons. But, because there are $Cu^{2+}$ ions in the solution, the electrons can immediately reduce these $Cu^{2+}$ ions, without going through the circuit. Thus, no current is expected to flow in a cell that is set up as illustrated in Figure 11.18, although copper metal will be deposited on the zinc electrode.

Now suppose we construct two separate cells, one containing the zinc electrode (shown on the left in **Figure 11.19**A) and one containing the copper electrode (shown on the right in Figure 11.19A). What happens when we connect the two electrodes with a wire and measure current flow? In the left-hand electrode, zinc atoms can dissolve, releasing zinc ions ($Zn^{2+}$) and two electrons each. Because there are no copper ions in the solution, the electrons could flow through the wire, entering the right-hand cell. There, the two electrons could reduce copper ions ($Cu^{2+}$), resulting in the deposition of copper metal on the electrode. But, if such a reaction were to happen, it would lead to the build up of positive charge in the

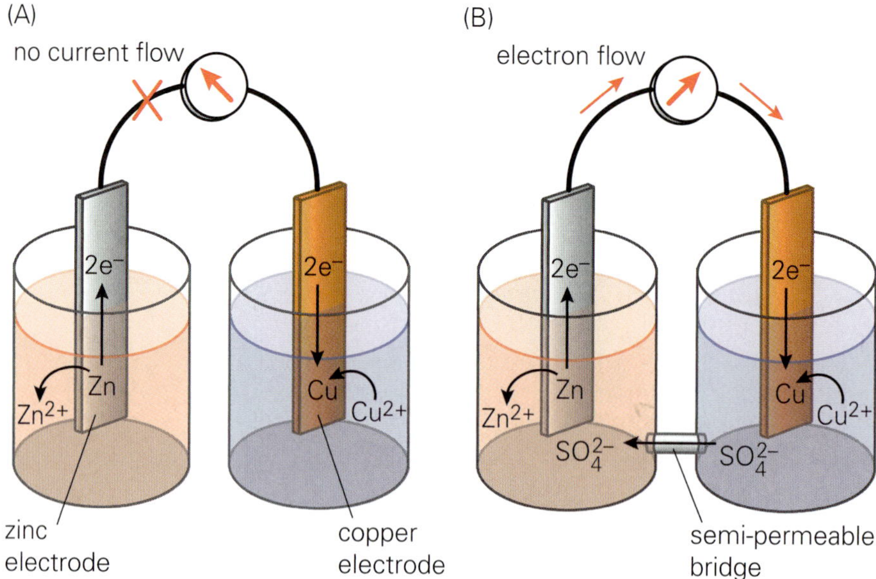


**Figure 11.19 An electrochemical cell.** (A) Two separate half-cells are shown, one containing a zinc electrode and one containing a copper electrode. No current sustained flows when the electrodes are connected, because the circuit is incomplete. (B) The circuit is completed by connecting the two half-cells by a semipermeable bridge. The bridge allows the flow of sulfate ions, but prevents metal ions from passing through. If copper ions could pass through the bridge, then they would directly pick up electrons from the zinc electrode. The semipermeable nature of the bridge prevents this from happening.

left-hand cell and the depletion of positive charge in the right-hand cell. In fact, no sustained current will flow unless the circuit is completed in some way.

How can we complete the circuit? We cannot simply make a completely permeable connection between two cells, because then we will be back at the situation illustrated in Figure 11.18, and this will result in a short circuit with no current flow through the external wire. What is done instead is to place a semipermeable bridge between the two electrodes, one that impedes the flow of copper and zinc ions, but allows the flow of sulfate ions. Such bridges can be made from certain ceramics or gels that repel positively charged metal ions, but allow the flow of the negatively charged sulfate ions, as illustrated in Figure 11.19B. If the two cells are connected by such a semipermeable bridge, then current will flow through the external wire.

In the electrochemical cell depicted in Figure 11.19B, zinc ions release electrons at the left electrode:

$$\text{Zn (solid)} \rightarrow \text{Zn}^{2+} + 2\text{e}^- \qquad (11.10)$$

These electrons move along the wire conductor until they reach the copper electrode, where they react with copper ions:

$$\text{Cu}^{2+}\text{ (solution)} + 2\text{e}^- \rightarrow \text{Cu (solid)} \qquad (11.11)$$

From the point of view of the external circuit, the zinc electrode behaves as if it is a source of electrons and is therefore designated as the negative electrode, or **anode** (**Figure 11.20**). The copper electrode is referred to as the **cathode**, or positive electrode. By convention, electrochemical cells are drawn with the anode on the left and the cathode on the right—that is, electron flow occurs from left to right in this conventional view. A shorthand notation is sometimes used to depict the underlying chemical reactions. For the Zn/Cu electrochemical cell, the two chemical reactions given in Equations 11.10 and 11.11 are written in shorthand form as:

$$\text{Zn }(s)\,|\,\text{Zn}^{2+}\,||\,\text{Cu}^{2+}\,|\,\text{Cu }(s) \qquad (11.12)$$

In this notation, the vertical double bars separate the reaction into two half reactions, one corresponding to the cell on the left and one to the cell on the right. In each half reaction, the spontaneous direction is indicated by reading from left to right. Thus, zinc (solid) is converted to zinc ion in solution ($ZnSO_4$), and copper ion in solution ($CuSO_4$) is converted to solid copper.

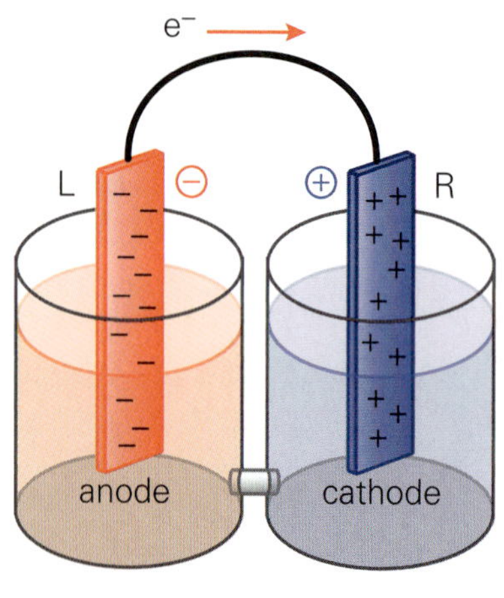


**Figure 11.20** The polarity of an electrochemical cell. Electrons in the external circuit flow from the anode (left) to the cathode (right). Electrochemical cells are, by convention, drawn with the anode on the left.

## 11.9 The voltage generated by an electrochemical cell with the reactants at standard conditions is known as the standard cell potential

Imagine that the circuit shown in Figure 11.19B includes a voltmeter, which registers a voltage difference between the two electrodes. As soon as current starts to flow, the concentration of sulfate ions in both cells will change. Likewise, the concentration of zinc ions in the left-hand cell and the concentration of copper ions in the right-hand cell will also change. Since the driving force for current flow is given by the differences in free energy between the reactants and products in Equations 11.6 and 11.7, which are concentration dependent, the current (and the corresponding voltage) will change as the reaction progresses. Thus, the cell depicted in Figure 11.19B is not at equilibrium and, once the electrodes are connected, the cell will run down. Eventually, when equilibrium is reached, there will be no current flow.

The voltage generated by the cell depends, therefore, on the concentrations of the reactants in each half-cell. We can, by modifying the apparatus, measure the voltage that corresponds to a defined concentration of reactants in the two half-cells. In order to do this, we set up each half-cell with a specified initial set of concentrations (for example, 1 M solutions of zinc sulfate and copper sulfate). The zinc and copper electrodes are then connected to an external circuit, but the voltage is balanced by an opposing voltage generated by an external power source, as depicted in **Figure 11.21**, so that there is no current flow. The voltage generated by the electrochemical cell is established by knowing the voltage generated by the external power source that just balances the electrochemical cell so that there is no current flow.

When such a measurement is carried out using standard state concentrations of the reactants (that is, 1 M solution concentrations), the resulting voltage is known as the **standard potential** of the cell. The standard state of the metal electrodes is assumed to be the pure metal and is unchanged as the reactions in the electrochemical cell proceed. When the zinc–copper electrochemical cell is set up with 1 M solutions of zinc sulfate and copper sulfate in each half-cell at 298 K, the measured voltage is 1.104 V. This is the value of the standard cell potential for the reaction depicted in Equation 11.12, in which cupric ion and zinc metal are converted to copper metal and zinc ion:

$$Zn\,(s)\,|\,Zn^{2+}\,(1\text{ M})\,||\,Cu^{2+}\,(1\text{ M})\,|\,Cu\,(s)$$

**Standard cell potential**

The voltage generated by an electrochemical cell in which the reactants are all at standard concentrations (1 M solution, solid metals) is known as the standard cell potential. The voltage generated by such a cell depends on the temperature, so standard potentials are usually given for 298 K.

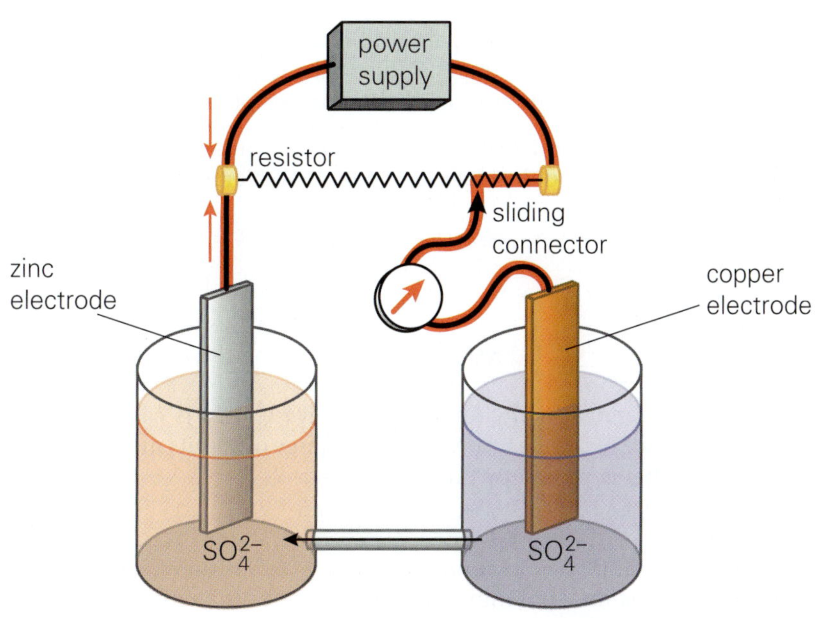


**Figure 11.21 Measuring the voltage generated by an electrochemical cell at a defined concentration of reactants.** The electrochemical cell is connected to an external power supply that puts out a constant DC voltage. The circuit is completed through a resistor that is connected to the electrochemical cell through a slider, which allows the external voltage to be matched to the voltage of the cell so that there is no current flow. The external portion of the resulting circuit is highlighted in *red*.

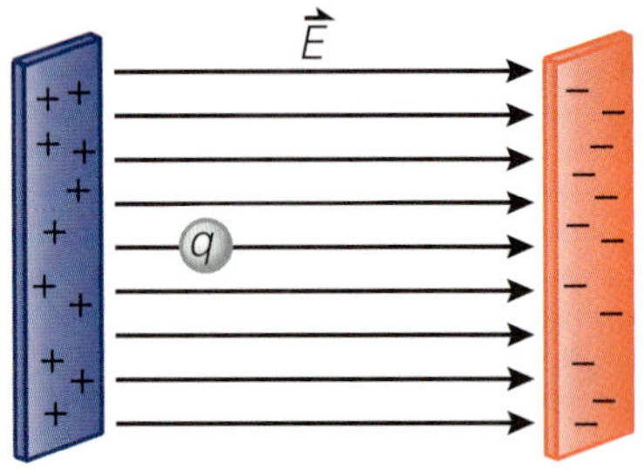


**Figure 11.22 An electric field.** Shown here is a schematic representation of two metal plates, one positively charged and one negatively charged. Electric field lines extend from the positively charged plate to the negatively charged one. A charged particle with charge $q$ is shown. If $q$ is positive, then the particle will move along the field lines, towards the negative plate.

## 11.10 The electric potential difference (voltage) between two points is the work done in moving a unit charge between the two points

The standard cell potential for a reaction is related in a fundamental way to the standard free-energy change for the reaction underlying the electrochemical cell. This can be understood by considering that the voltage generated by the cell is related to the electrical work that can be done by the cell. From the discussion in part C of Chapter 9, we can appreciate that the electrical work is powered by the underlying chemical reaction, for which the free-energy change gives the maximum amount of work that can be extracted from the reaction. Thus, in order to relate the standard reduction potential to the free-energy change, we simply calculate the amount of work that the cell can do under standard conditions.

In this section, we review some basic principles of electric fields and electrical work as a prelude to discussing the link between cell potentials and free-energy changes. The concepts introduced here will also be helpful in understanding parts C and D of this chapter, where we discuss the work done in moving charges across biological cell membranes.

Consider an electric field, such as that generated by two metal plates held at different voltages (**Figure 11.22**). The electric field lines are vectors denoted by $\vec{E}$ in Figure 11.22, and they represent the directions along which a positive charge moves in response to the electric field. The force, $\vec{F}$, experienced by a charged particle, with charge, $q$, is given by:

$$\vec{F} = q\vec{E} \tag{11.13}$$

To hold the particle at a specific point and move it slowly a force of the same magnitude but opposite sign must be applied. Now suppose we move the charged particle by an infinitesimal distance, $d\vec{r}$. The work, $w$, done on the charge particle is given by:

$$dw = -\vec{F} \bullet d\vec{r} = -q\vec{E} \bullet d\vec{r} \tag{11.14}$$

Note that the negative sign in front of the force term in Equation 11.14 arises from the fact that the work is done in the context of the holding force, that is of opposite sign to the electrical force.

The work done on the charge changes the potential energy of the charged particle. If we ignore heat dissipation, an infinitesimal change in potential energy, $dU$, is given by the work done, $dw$, on the charge:

$$dw = dU = -q\vec{E} \bullet d\vec{r} \tag{11.15}$$

According to Equation 11.15, if the particle has a positive charge, and it moves in the same direction as the force, then its potential energy decreases. If the particle is moved against the force, on the other hand, then its potential energy increases.

On moving the charged particle by a finite distance, from position 1 to position 2, the total work done, $w$, and the change in potential energy, $\Delta U$, is given by integrating Equation 11.15:

$$w = \Delta U = \int_1^2 -q\vec{E} \bullet d\vec{r} \tag{11.16}$$

For a unit point charge (that is, a +1 charge),

$$\Delta U = \int_1^2 -\vec{E} \bullet d\vec{r} = \Delta E \tag{11.17}$$

This change in potential energy for a unit positive charge is called the **electric potential difference** of the electric field. For a charge of $q$ units moving from point 1 to point 2, the change in energy, $\Delta U$, is given by:

$$\Delta U = q \times \Delta E \tag{11.18}$$

The standard unit for electric potential difference ($\Delta E$) is the **volt** (V). The standard unit of charge is the coulomb (C). If a positive unit charge of 1 coulomb is moved across a potential difference of 1 volt, then the work done is equal to 1 J (joule). Thus,

$$1\ \text{V} = 1\frac{\text{joule}}{\text{coulomb}} = 1\ \text{J}\bullet\text{C}^{-1} \tag{11.19}$$

In biochemistry, a natural unit of charge is the magnitude of the charge on the electron, $e$, ($e = 1.602 \times 10^{-19}$ C), which is called the elementary charge. The magnitude of the charge on 1 mole of electrons is known as the Faraday constant, $\mathcal{F}$ (we use this special script symbol to differentiate the Faraday constant from the unit of capacitance, the farad, F; see Section 11.26). The value of the Faraday constant is given by:

$$\begin{aligned} 1\ \mathcal{F} &= N_A \times e = (6.022 \times 10^{23})(1.602 \times 10^{-19}\ \text{C}) \\ &= 9.647 \times 10^4\ \text{C} \approx 96{,}500\ \text{C} \end{aligned} \tag{11.20}$$

We will usually use the approximate value of 96,500 C for 1 faraday.

What is the work done in moving electrons across an electric potential difference of 1 V? If the potential difference is +1 V, then it means that the final position of the electrons is at higher electric potential (that is, more favorable for the electron). The work done is negative, because the energy of the electron decreases, and is –96,500 J•mol$^{-1}$, or –96.5 kJ•mol$^{-1}$. Conversely, if a proton (with a single positive charge) is moved across a +1 V potential difference, then the work done is +96.5 kJ•mol$^{-1}$. As we shall see in part D of this chapter, the voltage difference across many biological membranes is approximately 100 mV (0.1 V). Thus, the energy required to move a proton from the negative side of the membrane to the positive side is 9.65 kJ•mol$^{-1}$, or approximately 10 kJ•mol$^{-1}$.

## 11.11 Standard reduction potentials are related to the standard free-energy change of the redox reaction underlying the electrochemical cell

We now return to a discussion of electrochemical cells and consider how to relate the measured voltages to the standard free-energy changes ($\Delta G^{\text{o}}$) of the underlying reactions.

To begin, let us denote the electric potential of an electrode by $E$. The electric potential of the electrode on the left is denoted $E_L$ and that of the electrode on the right by $E_R$. The voltage generated by the cell, $\Delta E$, is simply the difference between the electric potentials of the two electrodes. By convention, as shown in **Figure 11.23**, the value of $\Delta E$ is given by:

$$\Delta E = E_R - E_L \tag{11.21}$$

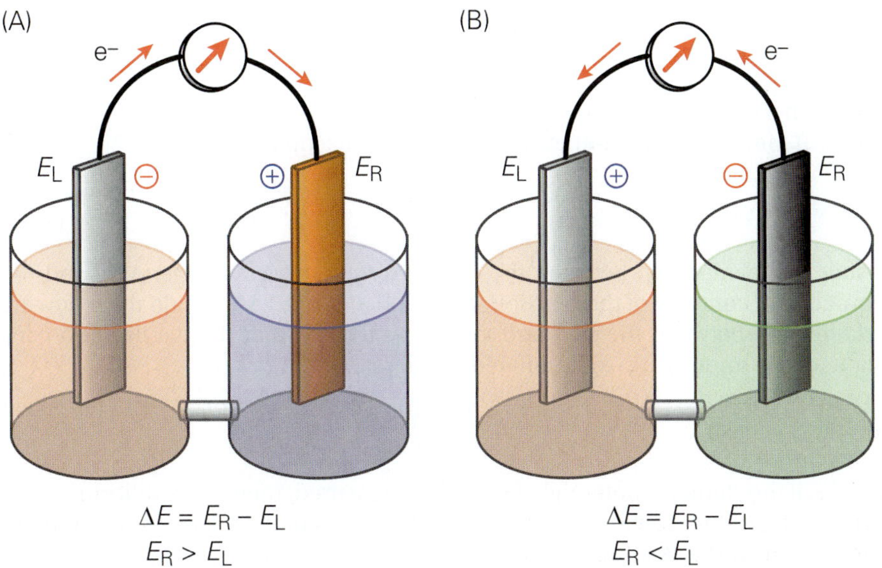


**Figure 11.23 Conventions regarding electric potential differences.** In the electrochemical cell shown in (A), the potential of the left electrode is $E_L$ and that of the right electrode is $E_R$. Because $E_L$ is less than $E_R$, electrons flow spontaneously to the right (electrons move towards regions with higher electric potential). This corresponds to the situation for the Cu–Zn cell, written as follows: Zn (*s*) | $Zn^{2+}$ (1 M) || $Cu^{2+}$ (1 M) | Cu (*s*). (B) In this case, the potential of the left electrode is *higher* than that of the right electrode. Electrons now flow spontaneously from the right cell to the left. This would happen, for example, if the copper half-cell were replaced with a manganese half-cell: $Mn^{2+}$ (1 M) | Mn (*s*) || Zn (*s*) | $Zn^{2+}$ (1 M).

Note that the spontaneous flow of electrons is from regions of low electric potential to regions of high electric potential. Thus, if $\Delta E$ is positive, then electrons flow spontaneously from the left cell (lower electric potential) to the right cell (higher electric potential). This can be confusing because we are accustomed to thinking of spontaneous processes occurring in the direction of lower free energy. The confusion arises because historically the flow of current has been defined in terms of the movement of *positive* charge, which moves downward in terms of electric potential. That is, positive charge moves from high electric potential (more positive) to low electric potential (more negative). Electrons move spontaneously in the opposite direction.

As discussed in Section 11.9, if the concentrations of the reactants in the half-cells are at their standard state concentrations (that is, 1 M solution), then the voltage or electric potential difference across the electrodes is known as the standard cell potential, which is denoted $\Delta E^{\text{o}}$, or sometimes $E^{\text{o}}$. $E^{\text{o}}$ is also called the **standard voltage** of the cell. It follows from Equation 11.21 that, under standard conditions $\Delta E^{\text{o}}$ is given by:

$$\Delta E^{\text{o}} = E^{\text{o}}_{\text{R}} - E^{\text{o}}_{\text{L}} \tag{11.22}$$

where $E^{\text{o}}_{\text{R}}$ and $E^{\text{o}}_{\text{L}}$ are the **standard electrode potentials** of the right and left half-cells, respectively. Standard electrode potentials are discussed further in Section 11.12, where we discuss how they are determined.

How do we relate the cell potentials to the free-energy change of the reaction? Under conditions of constant temperature and pressure, the free-energy change, $\Delta G$, is the maximum reversible work that can be extracted from the reaction (see Section 9.14). Because this is an electrical cell, the magnitude of the work, $w$, done by the cell is given by the following (see Equation 11.14):

$$\text{work} = (\text{charge moved}) \times (\text{voltage})$$

For a reaction in which $\upsilon$ moles of electrons are transferred:

$$\text{work} = -\upsilon\mathcal{F}\Delta E \tag{11.23}$$

where $\mathcal{F}$ is the magnitude of the charge on 1 mole of electrons (see Equation 11.20), and $\Delta E$ is the voltage of the cell. This is the maximum amount of work that can be extracted from the cell, assuming no dissipative losses to heat. We therefore equate the work to the free-energy change of the reaction:

$$\text{work} = \Delta G = -\upsilon\mathcal{F}\Delta E \tag{11.24}$$

If the two half-cells are under standard conditions, then $\Delta E = \Delta E^{\text{o}}$ (the standard reduction potential or standard voltage) and $\Delta G = \Delta G^{\text{o}}$ (the standard free energy change). In this case:

$$\Delta G^{\text{o}} = -\upsilon\mathcal{F}\Delta E^{\text{o}} \tag{11.25}$$

Equation 11.25 can be rearranged to express the standard voltage of the cell in terms of the standard free-energy change of the reaction:

$$\Delta E^{\text{o}} = \frac{-\Delta G^{\text{o}}}{\upsilon\mathcal{F}} \tag{11.26}$$

Using the relationship between the standard free-energy change, $\Delta G^{\text{o}}$, and the equilibrium constant of the reaction, $K$, ($\Delta G^{\text{o}} = -RT \ln K$), we obtain the following relationship between the standard voltage of the cell and the equilibrium constant for the underlying oxidation–reduction reaction:

$$\Delta E^{\text{o}} = \frac{+RT \ln K}{\upsilon\mathcal{F}} \tag{11.27}$$

Hence, if the standard potential of the cell is measured, then the equilibrium constant of the redox reaction is derived readily. This is one powerful application of the concept of the electrochemical potential.

## 11.12 Electrode potentials are measured relative to a standard hydrogen electrode

According to Equation 11.22, the standard potential or voltage of a cell is simply the difference between the standard electrode potentials of the right and left half-cells. But, when we measure the voltage of a particular electrochemical cell, we do not get any information regarding the individual electrode potentials: only the difference in electrical potential is measured. Nevertheless, standard electrode potentials are extremely useful because all electrode potentials can be determined relative to a standard reference electrode. Once this is done, it becomes straightforward to determine the relative electrode potentials of the half-cells from the measurement of the standard potentials of the combined cells.

By convention all electrode voltages are specified relative to the **standard hydrogen electrode**. The hydrogen electrode, illustrated schematically in **Figure 11.24**, consists of a solution of hydrogen chloride (1 M concentration), which is in equilibrium with hydrogen gas (at 1 atmosphere pressure, at room temperature). The hydrogen electrode forms a half-cell that needs to be connected to another cell in order to get a flow of current. Depending on the cell that it is connected to, the following reaction runs in either the forward or backward direction:

$$H^+ \text{ (1 M)} + e^- \rightleftharpoons \tfrac{1}{2}H_2 \text{(gas, 1 atm pressure)} \qquad E^o \equiv 0 \text{ V} \qquad (11.28)$$

The electrons that are either released or taken up by the half-cell are delivered by a platinum wire that is inserted into chamber of the hydrogen electrode. The platinum simply serves as an inert metal that does not directly contribute to the oxidation–reduction reactions that are being monitored.

The electrode potential of the hydrogen electrode is defined to be zero volts—that is, $E^o$ (hydrogen) = 0 V. Thus, if we connect the standard hydrogen electrode to any other electrode under standard conditions, the measured voltage is called the standard reduction potential of the other electrode ($E^o$).

The use of the hydrogen electrode to determine standard electrode potentials is illustrated in **Figure 11.25** for a silver electrode. The hydrogen electrode (on the left) is connected to a silver electrode that is immersed in a 1 M solution of silver nitrate. When the half-cells are connected, the measured voltage is +0.8 V. Thus, the standard electrode potential ($E^o$) for the silver electrode ($Ag/Ag^+$) is +0.8 V:

$$Ag^+ + e^- \rightarrow Ag \qquad E^o = +0.8 \text{ V}$$

The positive value of $E^o$ implies that the reaction proceeds spontaneously to the right as written, so the silver electrode must consume electrons (that is, it is the

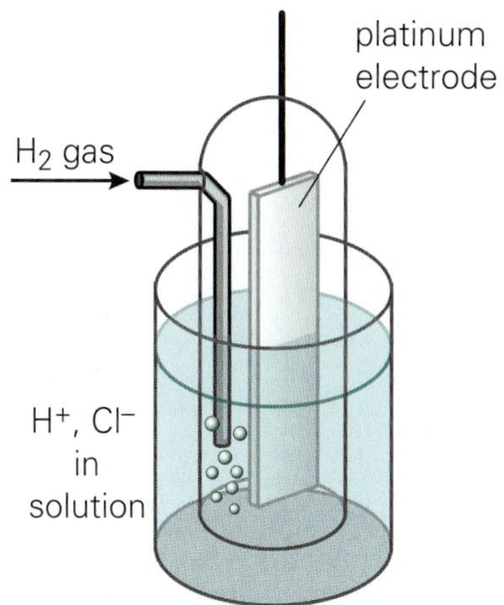


**Figure 11.24 The standard hydrogen electrode.** This half-cell consists of a solution of 1 M HCl, into which hydrogen gas at 1 atmosphere pressure is bubbled. A glass chamber encloses hydrogen gas in equilibrium with the solution. An inert platinum electrode serves as the conduit for electrons, but does not participate directly in the oxidation–reduction reactions.

**Standard hydrogen electrode**

The electrode potentials for various electrodes are referenced to a standard hydrogen electrode. This electrode is illustrated in Figure 11.24, and it reflects the oxidation of hydrogen gas at 1 atm pressure to yield protons and electrons under conditions where the concentration of protons is 1 M at room temperature.

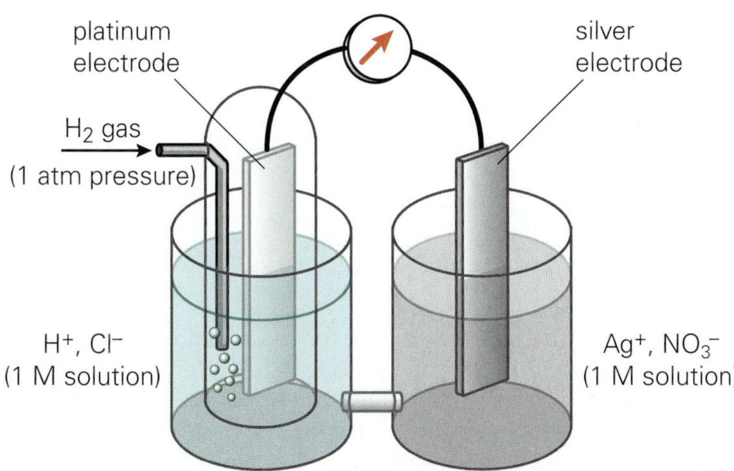


**Figure 11.25 Measurement of the standard electrode potential of silver.** A standard hydrogen electrode forms the left half-cell. The right half-cell consists of a silver electrode immersed in a 1 M solution of silver nitrate. The measured voltage of this cell is +0.8 V, which is the standard electrode potential of the $Ag/Ag^+$ half-cell. Since the voltage of the cell is positive, electrons flow from the left (the hydrogen electrode) to the right (the silver electrode). Silver ion ($Ag^+$) is reduced to form metallic silver, while hydrogen gas is oxidized to form $H^+$ ions.

cathode). Therefore, the hydrogen electrode must be the anode that generates electrons, with the following half reaction:

$$\tfrac{1}{2}H_2 \rightarrow H^+ + e^- \qquad E^o = 0\text{ V} \qquad (11.29)$$

Thus, the net reaction is:

$$Ag^+ + \tfrac{1}{2}H_2 \rightarrow Ag\,(\text{solid}) + H^+ \qquad \Delta E^o = +0.8\text{ V} \qquad (11.30)$$

In practice, relatively few electrode potentials are obtained by directly combining the half-cell of interest with the standard hydrogen electrode. Instead, most values of the standard electrode potentials are obtained by measuring the standard voltages of various combinations of half-cells and deducing the electrode potentials by appropriate subtraction.

**Table 11.1 Standard reduction potentials (electrode potentials) with respect to the hydrogen electrode (pH 0) for various half reactions.**

| Half reaction | Standard reduction potential ($E^o$) in volts (V) |
|---|---|
| $F_2 + 2e^- \rightarrow 2F^-$ | +2.87 |
| $Co^{3+} + e^- \rightarrow Co^{2+}$ | +1.80 |
| $Cl_2 + 2e^- \rightarrow 2Cl^-$ | +1.36 |
| $Ag^+ + e^- \rightarrow Ag$ | +0.80 |
| $Fe^{3+} + e^- \rightarrow Fe^{2+}$ | +0.77 |
| $I_2 + 2e^- \rightarrow 2I^-$ | +0.54 |
| $Cu^{2+} + 2e^- \rightarrow Cu$ | +0.34 |
| $Sn^{4+} + 2e^- \rightarrow Sn^{2+}$ | +0.15 |
| $2H^+ + 2e^- \rightarrow H_2$ | 0.0 |
| $Fe^{3+} + 3e^- \rightarrow Fe$ | −0.04 |
| $Pb^{2+} + 2e^- \rightarrow Pb$ | −0.13 |
| $Sn^{2+} + 2e^- \rightarrow Sn$ | −0.14 |
| $Ni^{2+} + 2e^- \rightarrow Ni$ | −0.25 |
| $Co^{2+} + 2e^- \rightarrow Co$ | −0.29 |
| $Cd^{2+} + 2e^- \rightarrow Cd$ | −0.40 |
| $Fe^{2+} + 2e^- \rightarrow Fe$ | −0.41 |
| $Cr^{3+} + 3e^- \rightarrow Cr$ | −0.74 |
| $Zn^{2+} + 2e^- \rightarrow Zn$ | −0.76 |
| $Mn^{2+} + 2e^- \rightarrow Mn$ | −1.18 |
| $Al^{3+} + 3e^- \rightarrow Al$ | −1.66 |
| $Mg^{2+} + 2e^- \rightarrow Mg$ | −2.37 |

(Adapted from R. Chang, Physical Chemistry for the Chemical and Biological Sciences, 3rd ed. Sausalito, CA: University Science Books, 2000.)

Standard electrode potentials (or standard reduction potentials) for various half reactions are listed in **Table 11.1**. Notice that each reaction is written as a reduction (for example, $Zn^{2+} + 2e^- \rightarrow Zn$; $E^o = -0.76$ V). The standard electrode potential of the oxidation half reaction is given simply by taking the negative of the electrode potential for the reduction reaction (that is, $Zn \rightarrow Zn^{2+} + 2e^-$ ; $E^o = +0.76$ V).

All of the standard electrode potentials in Table 11.1 are referenced to the standard state for the hydrogen electrode, in which hydrogen gas is in equilibrium with a solution at pH 0 (that is, $[H^+] = 1$ M, the chemical standard state). Recall from Section 9.5 that in biochemistry the standard state is defined as pH 7 ($[H^+] = 10^{-7}$ M). Standard electrode potentials that are referenced to a hydrogen electrode at pH 7 are denoted $E^{o\prime}$ (this is analogous to the distinction between $\Delta G^o$ and $\Delta G^{o\prime}$). We shall look at how electrode potentials change with concentration in Section 11.14, where we shall discuss how to covert between the chemical and biochemical standard electrode potentials.

## 11.13 Tabulated values of standard electrode potentials allow ready calculation of the standard potential of an electrochemical cell

We began this chapter by stating that an electrochemical cell that couples copper and zinc will run in the direction of copper reduction and zinc oxidation. We can now see why this is so, given the electrode potentials for the copper and zinc half-cells. From Table 11.1, we have the following standard electrode potentials:

$$2e^- + Cu^{2+} \rightarrow Cu \qquad E^o = +0.340\text{ V} \qquad (11.31)$$

$$2e^- + Zn^{2+} \rightarrow Zn \qquad E^o = -0.762\text{ V} \qquad (11.32)$$

The electrode potential for copper reduction is more positive than it is for zinc reduction. This means that the free-energy change for copper reduction is more negative than it is for zinc reduction (see Equation 11.25). Therefore, under standard conditions, the copper will be reduced and the zinc will be oxidized. Under nonstandard conditions of concentration, however, the opposite could, in principle, occur. This is discussed in Section 11.14.

When we reverse the reaction governing the zinc half-cell, the sign of the electrode potential changes. The combined reaction then becomes:

$$2e^- + Cu^{2+} \rightarrow Cu \qquad E^o = +0.340\text{ V} \qquad (11.33)$$

$$Zn \rightarrow Zn^{2+} + 2e^- \qquad E^o = +0.762\text{ V} \qquad (11.34)$$

$$Cu^{2+} + Zn \rightarrow Zn^{2+} + Cu \qquad \Delta E^o = +1.102\text{ V} \qquad (11.35)$$

What is the standard free-energy change for the reaction shown in Equation 11.35? Recall from Equation 11.25 that $\Delta G^o = -\upsilon \mathcal{F} \Delta E^o$. Here, $\upsilon$ is the stoichiometric number of electrons transferred in the reaction, which is 2 in this case, and $\mathcal{F}$ is the Faraday constant (96,500 C•mol$^{-1}$). Hence, the value of $\Delta G^o$ for this reaction is −(2)(96,500)(1.102) = −212,686 J•mol$^{-1}$ = −212.7 kJ•mol$^{-1}$.

In the case of the copper/zinc reactions, we simply added the values of the standard electrode potentials to get the standard reduction potential for the combined reaction (Equations 11.33–11.35). This is correct, because the stoichiometry of the electrons is the same in the two half reactions. If this is not the case, then the electrode potentials are not generally additive. In such cases, it is can be useful to consider the standard free-energy changes, which are additive, instead of the electrode potentials, as illustrated by two examples.

The first example involves zinc and silver electrodes. The standard electrode potentials for the two half reactions in this case are:

$$Zn\ (s) \rightarrow Zn^{2+} + 2\ e^- \qquad E^o = +0.76 \qquad (11.36)$$

$$Ag^+ + e^- \rightarrow Ag\ (s) \qquad E^o = +0.80\ V \qquad (11.37)$$

The combined reaction is:

$$Zn\ (s) + 2Ag^+ \rightarrow Zn^{2+} + 2Ag\ (s) \qquad (11.38)$$

We shall calculate the standard reduction potential for the combined reaction by first calculating the standard free-energy changes for the two half reactions, and then combining them. For the zinc half reaction, two electrons are produced and so:

$$\Delta G^o\ (\text{zinc oxidation}) = -2E^o\mathcal{F} = -145.9\ kJ \bullet mol^{-1} \qquad (11.39)$$

For the silver half-reaction, only one electron is produced, and so:

$$\Delta G^o\ (\text{silver reduction}) = -E^o\mathcal{F} = -77.2\ kJ \bullet mol^{-1} \qquad (11.40)$$

The combined reaction has a standard free-energy change of:

$$\Delta G^o = \Delta G^o(Zn) + 2\Delta G^o(Ag) = -300.3\ kJ \bullet mol^{-1}$$

We have to multiply the molar free-energy change of silver (Ag) by 2 because 2 moles of silver participate per mole of zinc (Zn).

The standard voltage of the combined cell, $\Delta E^o$, is given by:

$$\Delta E^o = \frac{-\Delta G^o}{\upsilon\mathcal{F}} = \frac{-(300.3 \times 10^3)}{(2)\,(96{,}500)} = 1.56\ V \qquad (11.41)$$

Note in this case that the standard voltage of the reaction is, once again, simply obtained by adding the electrode potentials of the half reactions. When the stoichiometries are more complicated, the electrode potentials need not be additive, as the next example illustrates.

Consider the reactions:

$$Fe^{2+} + 2e^- \rightarrow Fe\ (s) \qquad E_1^o = -0.447 \qquad (11.42)$$

$$Fe^{3+} + e^- \rightarrow Fe^{2+} \qquad E_2^o = 0.771\ V \qquad (11.43)$$

What, then, is the standard reduction potential for the following reaction?

$$Fe^{3+} + 3e^- \rightarrow Fe\ (s) \qquad E_3^o = ? \qquad (11.44)$$

The third reaction is the sum of the first two reactions, but its standard reduction potential is not the sum of the potentials of the first two reactions. To see why this is the case, consider the relationship between free energies of each of the reactions:

$$\Delta G_3^o = \Delta G_1^o + \Delta G_2^o \qquad (11.45)$$

or

$$-\upsilon_3\mathcal{F}E_3^o = -\upsilon_1\mathcal{F}E_1^o - \upsilon_2\mathcal{F}E_2^o \qquad (11.46)$$

This means that:

$$E_3^o = \frac{\upsilon_1 E_1^o + \upsilon_2 E_2^o}{\upsilon_3} \qquad (11.47)$$

where

$$\upsilon_1 = 2, \upsilon_2 = 1, \text{ and } \upsilon_3 = 3.$$

According to Equation 11.47, the standard reduction potential for the combined reaction is –0.041 V, which is not the sum of the two electrode potentials in the half reactions. In all circumstances, however, the standard electrode potential can be calculated correctly by considering the standard free-energy change for the reaction, as shown in this example.

## 11.14 The Nernst equation describes how the cell potential changes with the concentrations of the redox reactants

Consider a general reaction of the form:

$$a\text{A} + b\text{B} \rightleftharpoons c\text{C} + d\text{D} \tag{11.48}$$

where $a$, $b$, $c$, and $d$ are stoichiometric coefficients. Equation 11.48 represents the net chemical reaction and does not show the electron transfer steps. For example, suppose that there are intermediate steps in which electron transfer occurs:

$$a\text{A} \rightleftharpoons c\text{C} + \upsilon\text{e}^-$$

$$\upsilon\text{e}^- + b\text{B} \rightleftharpoons d\text{D}$$

The stoichiometric reactions involve the transfer of electrons, but these do not enter into the final balanced chemical reaction.

When we move away from standard conditions, the free-energy change for the reaction is given by:

$$\Delta G = \Delta G^\circ + RT \ln \frac{[\text{C}]^c[\text{D}]^d}{[\text{A}]^a[\text{B}]^b} \tag{11.49}$$

where the argument of the logarithm is the reaction quotient or mass action ratio, $Q$, and the concentrations are not the equilibrium concentrations (see Section 10.10). We can calculate the reduction potential or voltage of the electrochemical cell corresponding to these reactions by using Equation 11.24 in combination with Equation 11.49:

$$\Delta E = \frac{-\Delta G}{\upsilon \mathcal{F}} = \frac{-\Delta G^\circ}{\upsilon \mathcal{F}} - \frac{RT}{\upsilon \mathcal{F}} \ln \frac{[\text{C}]^c[\text{D}]^d}{[\text{A}]^a[\text{B}]^b}$$

Since, according to Equation 11.26, $\Delta E^\circ = \frac{-\Delta G^\circ}{\upsilon \mathcal{F}}$, this means that:

$$\Delta E = \Delta E^\circ - \frac{RT}{\upsilon \mathcal{F}} \ln \frac{[\text{C}]^c[\text{D}]^d}{[\text{A}]^a[\text{B}]^b} \tag{11.50}$$

Equation 11.50 is one form of the **Nernst equation**. The Nernst equation is very important because it allows us to calculate the change in voltage of an electrochemical cell as the concentrations of the reacting species change. Another form of the Nernst equation is used in calculating the voltage change across biological membranes that arises due to differences in the concentrations of ionic species (see Section 11.25).

## 11.15 The standard state for reduction potentials in biochemistry is pH 7

For a reaction that consumes or releases protons, biochemists consider the standard state to be a solution at pH 7 (that is, the standard proton concentration, $[\text{H}^+]$, is taken to be $10^{-7}$ M, not pH 0, where $[\text{H}^+] = 1$ M). It is straightforward to convert between reduction potentials using the pH 0 standard to the biochemical standard. To see how this is done, consider the following electron transfer reaction, in which the stoichiometry of the proton is $x$ (as in Equation 11.48, the electrons do not appear in this combined equation):

$$A + B \rightarrow C + xH^+ \tag{11.51}$$

The biochemical standard reduction potential for this reaction is denoted $\Delta E^{o\prime}$, and its relationship to the standard reduction potential ($\Delta E^o$) can be calculated using the Nernst equation (Equation 11.50):

$$\Delta E = \Delta E^o - \frac{RT}{\upsilon \mathcal{F}} \ln \frac{[C][H^+]^x}{[A][B]} \tag{11.52}$$

For the biochemical standard state, the concentrations of A, B, and C are all 1 M, and $[H^+]$ is $10^{-7}$ M. Thus,

$$\begin{aligned}\Delta E^{o\prime} &= \Delta E^o - \frac{RT}{\upsilon \mathcal{F}} \ln (10^{-7})^x \\ &= \Delta E^o - \left(\frac{RTx}{\upsilon \mathcal{F}}\right) \ln (10^{-7}) = \Delta E^o + 16.12 \left(\frac{RTx}{\upsilon \mathcal{F}}\right)\end{aligned} \tag{11.53}$$

In Equation 11.53, $\upsilon$ is the stoichiometry of the electrons that are involved in the reaction (its value will depend on the chemical details of the reaction). At room temperature (that is, $T = 298$ K), this simplifies to:

$$\Delta E^{o\prime} = \Delta E^o + \frac{0.414x}{\upsilon} \tag{11.54}$$

Many biological electron transfer reactions involve protons, and for such reactions we use tabulated values of the biochemical standard reduction potentials (**Table 11.2**). These values can be used to calculate the standard reduction potentials of combined reactions exactly as was done previously in Section 11.13 for reactions involving metals. For example, consider the reduction of oxygen by NADH, which underlies the production of ATP in oxidative phosphorylation. The two reductive half-reactions are:

$$\begin{aligned} &NAD^+ + H^+ + 2e^- \rightarrow NADH && E^{o\prime} = -0.320 \text{ V} \\ &\tfrac{1}{2}\, O_2 + 2H^+ + 2e^- \rightarrow H_2O && E^{o\prime} = 0.816 \text{ V} \end{aligned} \tag{11.55}$$

The reduction of oxygen has a much more positive electrode potential than the reduction of $NAD^+$. In other words, oxygen is a much stronger oxidizing agent than $NAD^+$, and the first reaction will run in the reverse direction, giving the following overall reaction:

$$NADH + H^+ + \tfrac{1}{2}\, O_2 \rightarrow NAD^+ + H_2O \qquad E^{o\prime} = 1.136 \tag{11.56}$$

The standard free-energy change for the reaction is given by $\Delta G^{o\prime} = -\upsilon \mathcal{F} E^{o\prime}$, where $\upsilon$ is the number of electrons transferred in the balanced chemical reaction. Recall that the oxidation of each NADH molecule generates two electrons, and so $\upsilon = 2$ in this case:

$$\begin{aligned}\Delta G^{o\prime} &= -(2)(96{,}500 \text{ C}\bullet\text{mol}^{-1})\,(1.136 \text{ V}) \\ &= -219 \text{ kJ}\bullet\text{mol}^{-1}\end{aligned} \tag{11.57}$$

This is far more free energy than can be stored in one molecule of ATP ($\Delta G^{o\prime}$ for ATP synthesis is ~30 kJ•mol$^{-1}$), but the stepwise harnessing of energy in the electron transport chain makes efficient use of this energy by producing several ATP molecules per NADH oxidized.

## C. ION PUMPS AND CHANNELS IN NEURONS

In the first part of this chapter we saw that the generation of electrochemical gradients is a key mechanism for harnessing energy from light and from fuel molecules and coupling it to the production of ATP and reducing power. Thus, the ability of cells to utilize electrochemical gradients in this way must have occurred very early in the evolution of life. In the remaining two parts of this chapter we study how

**Table 11.2 Biochemical standard reduction potentials (electrode potentials).**

| Half reaction | $E^{0\prime}$ (V) |
|---|---|
| $\frac{1}{2}O_2 + 2H^+ + 2e^- \rightarrow H_2O$ | 0.816 |
| $SO_4^{2-} + 2H^+ + 2e^- \rightarrow SO_3^{2-} + H_2O$ | 0.480 |
| $NO_3^- + 2H^+ + 2e^- \rightarrow NO_2^- + H_2O$ | 0.420 |
| $O_2 + 2H^+ + 2e^- \rightarrow H_2O_2$ | 0.295 |
| cytochrome-*c* ($Fe^{3+}$) + $e^-$ → cytochrome-*c* ($Fe^{2+}$) | 0.235 |
| ubiquinone + $2H^+$ + $2e^-$ → ubiquinol | 0.045 |
| fumarate + $2H^+$ + $2e^-$ → succinate | 0.031 |
| FAD + $2H^+$ + $2e^-$ → $FADH_2$ (in typical enzymes) | ~−0.0 |
| oxaloacetate + $2H^+$ + $2e^-$ → malate | −0.166 |
| pyruvate + $2H^+$ + $2e^-$ → lactate | −0.185 |
| actetaldehyde + $2H^+$ + $2e^-$ → ethanol | −0.197 |
| FAD + $2H^+$ + $2e^-$ → $FADH_2$ (free in solution) | −0.219 |
| S + $2H^+$ + $2e^-$ → $H_2S$ | −0.230 |
| $NADP^+$ + $H^+$ + $2e^-$ → NADPH | −0.320 |
| ubiquinone + $2H^+$ + $2e^-$ → ubiquinol | −0.340 |
| acetate + $2H^+$ + $2e^-$ → acetaldehyde + $H_2O$ | −0.581 |

For reactions that do not involve a proton, these values are the same as the standard reduction potentials (for example, those in Table 11.1). For reactions involving a proton, the values shown here are related to the standard reduction potentials by Equation 11.54. (Data from R. Chang, Physical Chemistry for the Chemical and Biological Sciences, 3rd ed. Sausalito, CA: University Science Books, 2000 and H.A. Sober, Handbook of Biochemistry, (ed). Cleveland, OH: The Chemical Rubber Company, 1968. FAD values from C. Walsh, *Acc. Chem. Res.* 13: 148–155, 1980.)

higher organisms have adapted electrochemical gradients to create a mechanism for the transmission of electrical signals through neurons. This process underlies all sensory and cognitive function in animals, but is connected in a fundamental way to the very basic mechanisms of energy transduction in all types of cells.

## 11.16 Neuronal cells use electrical signals to transmit information

Neurons are highly specialized cells in animals that convey information from one part of the body to the other. Although there are many different types of neurons that serve specialized functions, all neurons share a similar general structure. The cell body or soma of the neuron contains the nucleus (**Figure 11.26**A). The cell body is usually located at one end of the neuron, not in the center. There are two kinds of structures emanating from the cell body. Dendrites are relatively short branched protrusions, shown above the cell body in Figure 11.26A. A single axon also emerges from the cell body and leads away from it. The axonal trunk has occasional branches emanating from it. Both the main axonal trunk and the axonal branches sprout synaptic terminals at their ends (Figure 11.26A).

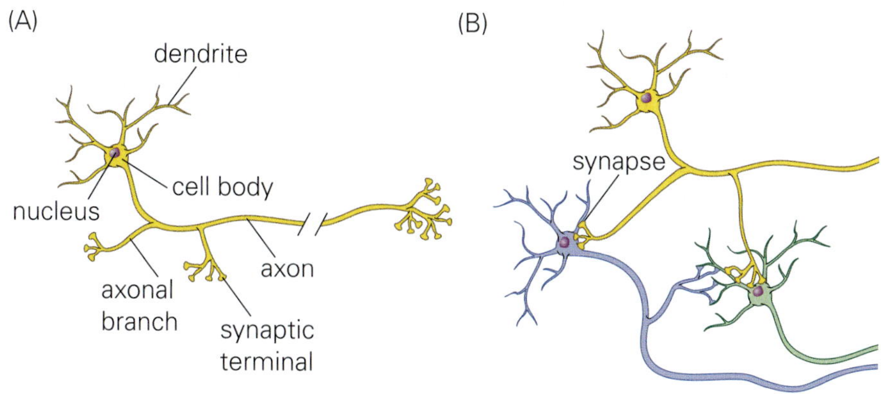


**Figure 11.26 Structures of neurons and neuronal junctions.** (A) A single neuronal cell. The neuron consists of a cell body with a nucleus located at one end of the neuron. There are two kinds of structures emanating from the cell body. Dendrites are the shorter branched protrusions extending above the cell body in this diagram. A single axon emerges from the cell body and leads away from it. The axon can be very long compared to the width of the cell body. Signals enter the neuron through the dendrites and the cell body, and travel through the trunk of the axon and to the axonal branches. The synaptic terminals are responsible for transmitting the signal to other neurons. (B) Three interconnected neurons are shown here. The neuron at the top (*yellow*) makes connections, referred to as synapses, with the dendrites and cell body of the neurons colored *blue* and *green*, as does the *blue* neuron with the *green* neuron. (Adapted from C.F. Stevens, Neurophysiology: A Primer. New York: John Wiley & Sons, 1966.)

The dendrites in a neuron receive signals from the synaptic terminals of other neurons (Figure 11.26B). This information takes the form of electrical pulses, but neurons are not like metal wires that conduct electrons. Instead, the transmission of an electrical signal through a neuron involves highly localized changes in the electrical potential difference across the plasma membrane of the cell. The electrical potential inside the dendrite is initially negative with respect to the outside. Upon stimulation, the electrical potential inside the dendrite becomes less negative with respect to the outside over a period of several milliseconds and then reverts back to its initial value. This change in electrical potential is propagated down the dendrite and into the cell body.

Depending on the nature of the input, the cell body then generates a voltage spike at the base of the axon, in a region known as the axon hillock (**Figure 11.27**). This voltage spike is known as an **action potential**, and it is propagated down the length of the axon to the synaptic terminals. Like the signals that travel down the dendrites, action potentials involve transient changes in the distribution of charges across the plasma membrane. Action potentials are remarkable, however, because they are transmitted essentially without attenuation down the axon (see Figure 11.27). The arrival of action potentials activates the synaptic terminals, resulting in the stimulation of other neurons that are connected to these terminals. In the rest of this chapter, we shall study how the action potential is generated and how the neuron enables it to propagate without attenuation.

**Action potential**

A transient change in the voltage difference across the plasma membrane of an axon is called an action potential, also called a *nerve impulse*. Action potentials move along the axon with essentially no attenuation. They are the currency of signal transmission through the axon.

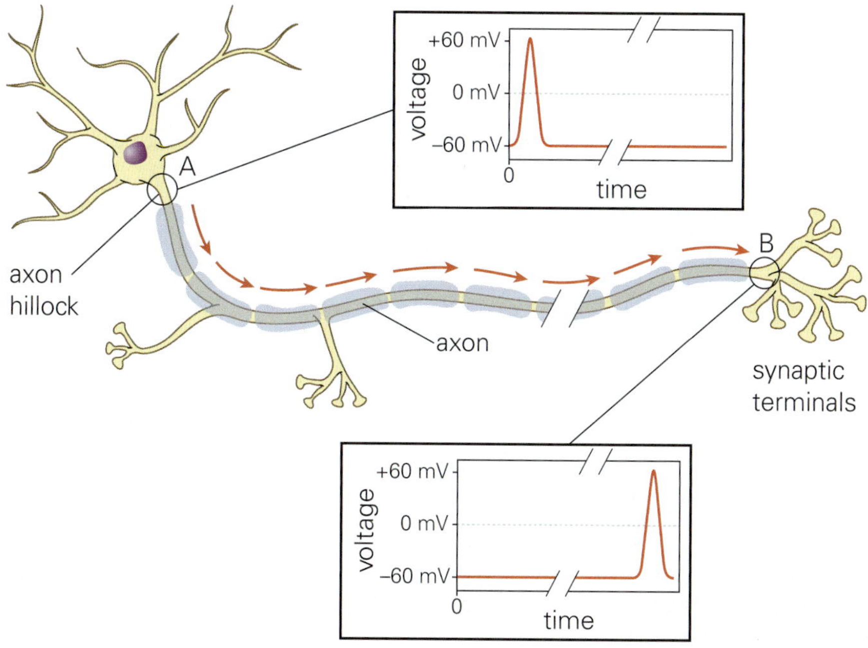


**Figure 11.27 Transmission of an action potential down an axon.** An action potential, which is a localized spike in the voltage difference across the membrane of the nerve cell, originates at the axon hillock, in the region marked A. The graph shows the electrical potential difference, or voltage, across the membrane as a function of time in region A. The arrows depict the transmission of the action potential along the axon. The lower graph shows the action potential near the synaptic terminal, marked B. The action potential is transmitted essentially without attenuation.

## 11.17 An electrical potential difference across the membrane is essential for the functioning of all cells

There is an electric potential difference across the cell membrane in all living cells, not just neurons—that is, the net charge on one side of the membrane is *not* the same as on the other side (**Figure 11.28**A). This voltage difference across the cell membrane is called the **membrane potential**, and is denoted $E_m$ or $\Delta\Psi$. By convention, the membrane potential is defined with reference to the external electric potential (Figure 11.28B). That is, if the values of the electrical potential at the inner and outer surfaces of the membrane are denoted $E_{in}$ and $E_{out}$, respectively, then the membrane potential is given by:

$$E_m = E_{in} - E_{out} \tag{11.58}$$

**Membrane potential**

The electrical potential difference across the cell membrane (that is, the difference between the potential inside the cell and the potential outside the cell) is known as the membrane potential. Resting mammalian cells have a membrane potential that is approximately –70 mV, with the interior of the cell at a negative potential with respect to the outside.

Membrane potentials are measured by using two electrodes that are connected to an oscilloscope or voltmeter that records the voltage differences between the two electrodes. The tip of one electrode is immersed in the solution outside the cell, whereas the tip of the other electrode is inserted into the cytoplasm of the cell (the electrode is thin enough to penetrate the cell membrane without disrupting it).

A resting mammalian cell maintains an electrical potential difference of about –70 mV across the cell membrane—that is, the interior surface of the membrane is negatively charged with respect to the outside. This membrane potential is set by the action of an ATP-driven pump in the plasma membrane, which couples the movement of three $Na^+$ ions out of the cell to the inward movement of two $K^+$ ions (the sodium-potassium pump; see Section 11.18). The cell membrane also contains ion channels that are highly specific for the flow of $Na^+$, $K^+$, or $Cl^-$ ions. In contrast to the sodium-potassium pump, which moves ions against their concentration gradient, the ion channels simply allow ions to move through them in either direction, with the direction of net flow determined by the free energy of the ions. The combination of these two features generates the membrane potential.

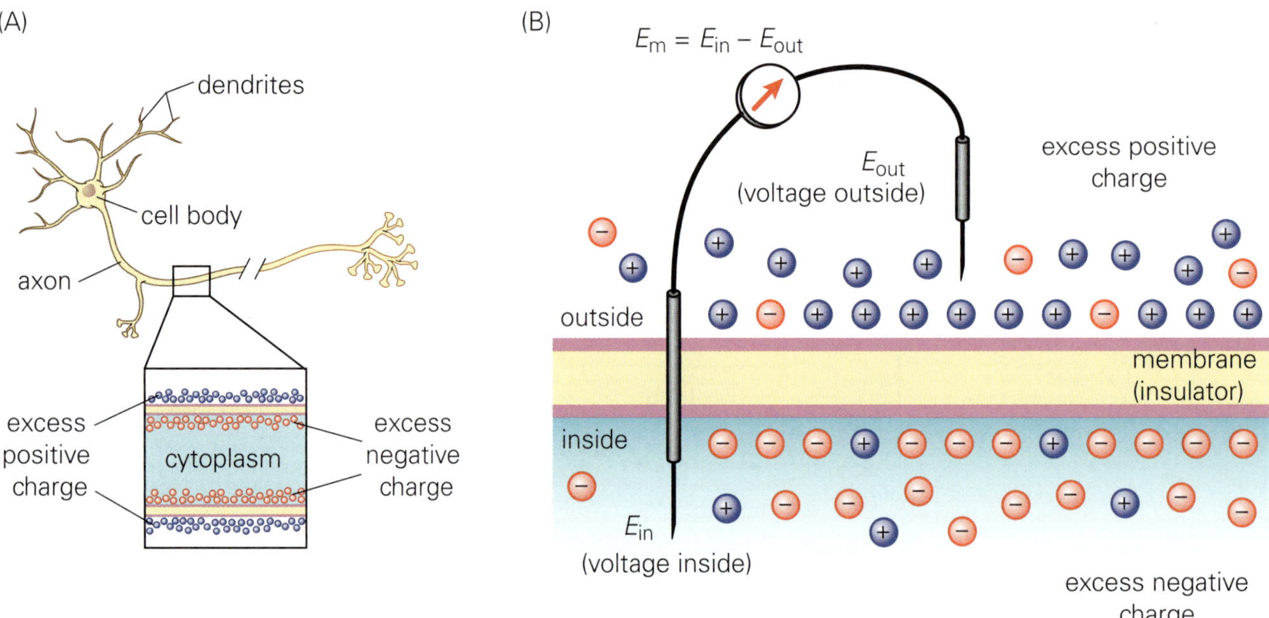


**Figure 11.28 Membrane potential.** (A) A schematic diagram of a neuron. A cross section of a region of the axon is show in the expanded view. There are more negative ions inside the cell than positive ions, and the excess negative ions are shown as *red* spheres. The situation is reversed outside the cell, where there is excess positive charge (*blue* spheres). (B) There is an electrical potential difference across the membrane, due to the asymmetric distribution of charge across the membrane. The potential difference across the membrane is referred to as the membrane potential, denoted $E_m$, and is given by the difference between the internal and external electrical potential: $E_m = E_{in} - E_{out}$. Membrane potentials are measured by inserting electrodes on both sides of the membrane and measuring the voltage across the electrodes.

**Table 11.3** shows the concentrations of $Na^+$ and $K^+$ inside the giant axon from squids (an experimental system in which the basic principles of neuronal conduction were worked out) and in a human cell. In both cases, the concentration of $K^+$ inside the cell is about 10 times greater than it is outside the cell. The situation is the opposite for $Na^+$, the concentration of which is about 10 times lower inside the cell than it is outside.

This difference in concentrations of $Na^+$ and $K^+$ inside and outside the cell is an example of an **electrochemical gradient** across the cell membrane. The electrochemical gradient is a free-energy gradient arising from a combination of the differences in electrical potential and the chemical potential across the membrane. The electrochemical gradient is used by cells to move nutrients into the cell against their concentration gradient. The membrane potential is also an essential aspect of the ability of animal cells to sequester nucleotides, proteins, sugars and various metabolites within them without bursting due to osmotic pressure.

Recall from Section 7.25 that cells swell up and burst if they are moved into pure water. This occurs because water moves in and out of the cell through specific channels called aquaporins, whereas most other molecules inside the cell cannot cross the membrane (**Figure 11.29**A). This leads to the inward movement of water and an increase in the internal pressure. Animal cells do not have a cell wall, and the plasma membrane cannot withstand this osmotic pressure.

The generation of the membrane potential by the sodium-potassium pump provides a mechanism by which cells compensate for the osmotic pressure. A key aspect of this mechanism is the presence of chloride channels in the cell membrane (Figure 11.29B). Because the sodium-potassium pump maintains the interior of the cell at a negative potential, chloride ions move out of the cell through chloride channels. The outward movement of chloride ions reduces their

**Table 11.3 Approximate concentrations of sodium and potassium ions in some cells and environments.**

| | $[K^+]$ (mM) | $[Na^+]$ (mM) |
|---|---|---|
| Sea water | 10 | 460 |
| Squid giant axon | 400 | 50 |
| Human plasma | 5 | 140 |
| Human cell | 150 | 10 |

Notice that the sodium and potassium concentrations inside the cells are inverted with reference to their respective values outside. The concentrations shown here are only indicative—the actual concentrations of these ions varies with cell type. (Data from A.C. Giese, Cell Physiology, 4th ed. Philadelphia, PA: W.B. Saunders, 1973, and M.J. Ackerman and D.E. Clapham, *N. Engl. J. Med.* 336(22): 1575–1586, 1997.)

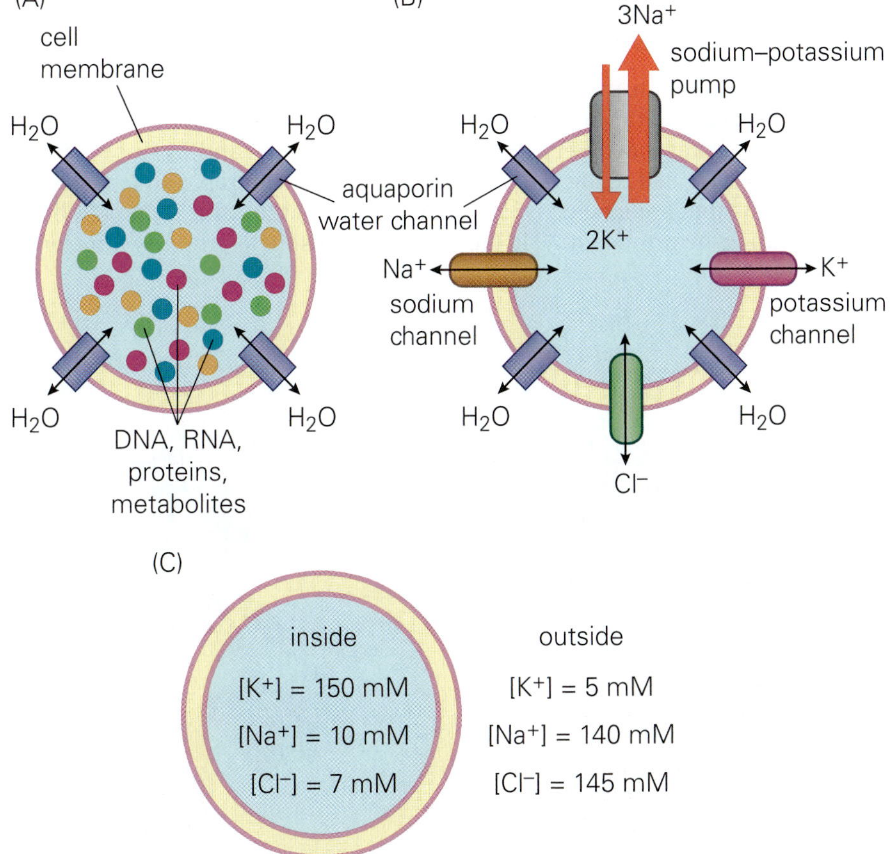


**Figure 11.29 The role of membrane potential in compensating for osmotic pressure.** (A) A schematic diagram of an animal cell, indicating the high concentration of molecules within it that cannot cross the cell membrane. The presence of water channels in the cell membrane, known as aquaporins, allows water to equilibrate rapidly across the membrane. This leads to the development of an osmotic pressure across the membrane. (B) The cell contains sodium–potassium pumps that maintain the interior of the cell at a negative electrical potential with respect to the outside. They do this by pumping three $Na^+$ ions out of the cell for every two $K^+$ ions that they drive in. The membrane also contains various ion channels, including chloride channels. (C) Due to the negative electrical potential inside the cell, there is a net outflow of chloride ions, leading to a reduction in its internal concentration by ~140 mM. This balances the osmotic pressure generated by the molecules that cannot cross the membrane. (Adapted from C.M. Armstrong, *Proc. Natl. Acad. Sci. USA* 100: 6257–6262, 2003.)

concentration inside the cell substantially (~7 mM inside compared with 150 mM outside; see Figure 11.29C). This internal deficit in chloride ion concentration provide osmotic "room" for the essential molecules that the cell needs to sequester within it.

Neuronal cells have evolved to exploit the free energy stored in the electrochemical gradient to transmit electrical signals down the axon without attenuation. The rest of this chapter focuses on the mechanism of axonal transmission.

## 11.18 The sodium–potassium pump hydrolyzes ATP to move $Na^+$ ions out of the cell with the coupled movement of $K^+$ ions into the cell

The electrochemical gradient across the neuronal membrane is maintained by an integral membrane protein known as the sodium–potassium ($Na^+/K^+$) pump, the structure of which is shown in **Figure 11.30**. These proteins are also referred to as $Na^+/K^+$-ATPases, because they couple the free energy released upon ATP hydrolysis to the pumping of $Na^+$ and $K^+$ ions against their respective concentration gradients. For each molecule of ATP that is hydrolyzed, two potassium ions get pumped into the cell and three sodium ions get pumped out (**Figure 11.31**).

The pump is an active transporter, and its mechanism is similar in general terms to that of the maltose transporter described in Sections 4.43 and 9.15, although its structure is quite different. Like the maltose transporter, the pump cycles between two states, one open to the exterior of the cell and one open to the interior (see Figure 11.31). When the pump is open to the interior (state 1 in Figure 11.31), it binds to $Na^+$ ions with higher affinity than to $K^+$ ions. The binding of ATP then closes the pump, and the subsequent hydrolysis of ATP and the release of ADP converts the pump into a state where it is open to the exterior. The pump now has higher affinity for $K^+$ than for $Na^+$, so it releases the $Na^+$ ions into the external solution and picks up $K^+$ ions. Upon release of the phosphate ion (which came from the hydrolysis of ATP), it switches back to a conformation that is open to the interior, which results in the transport of $K^+$ ions into the cell.

The sodium–potassium pump consists of a transmembrane segment that provides the pathway for ions to transit from one side of the membrane to the other. The most important machinery that couples ATP binding and hydrolysis to ion translocation is located in the cytoplasmic portion of the pump, and it consists of three domains that are labeled N, P, and A in Figure 11.30. The N domain binds ATP and, upon hydrolysis of ATP, the terminal phosphate group is transferred

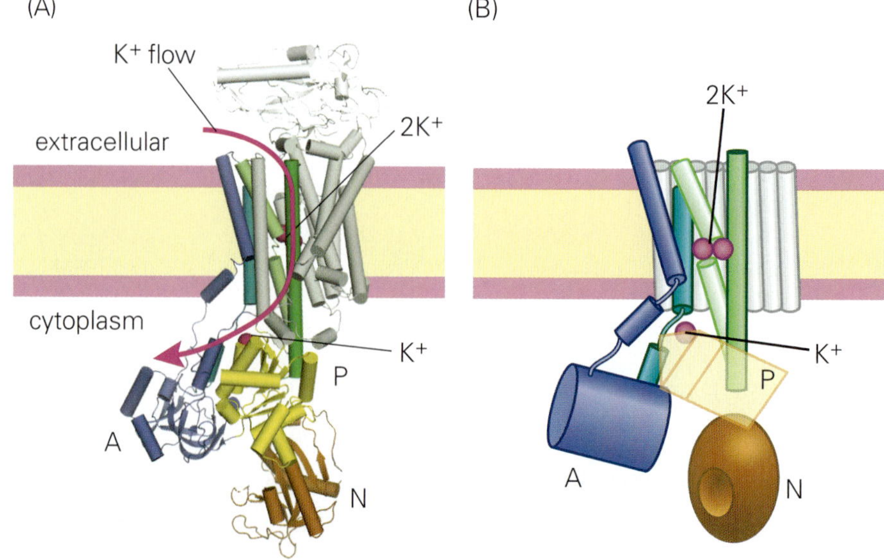


**Figure 11.30 Structure of a $Na^+/K^+$ ATPase pump.** (A) Crystal structure of a $Na^+/K^+$ pump. The pump consists of two subunits, one shown colored and one in *light gray*. There are three cytoplasmic domains, labeled A, P, and N. N is the nucleotide- (ATP-) binding domain, and the P domain receives the phosphate group from ATP hydrolysis. The A domain is an actuator that undergoes large conformational changes that are coupled to changes in the structure of the transmembrane segments (see Figure 11.32). There are three $K^+$ ions bound to this structure. Two of these ions are in an intermediate stage of translocation and are sequestered deep within the membrane-spanning section of the protein in this conformation, and one is bound to the P domain, from which it can easily escape. The direction of potassium ion flow is indicated by an arrow. (B) Highly schematized representation of the structure of the $Na^+$-$K^+$ pump. The drawing is not to scale, and the subunit shown in *light gray* in (A) is not shown in this figure. This schematic diagram is the basis for the ATPase cycle shown in Figure 11.32. (A, PDB code: 2ZXE; B, adapted from C. Toyoshima, *Biochim. Biophys. Acta*, 1793: 941–946, 2009.)

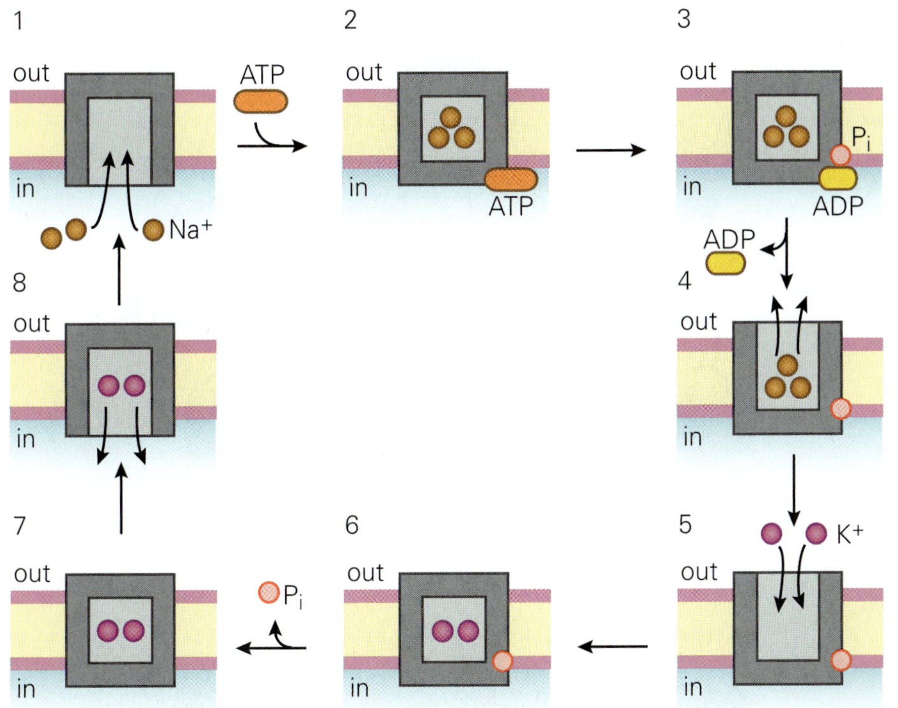


**Figure 11.31 Schematic diagram showing how ATP hydrolysis drives the $Na^+/K^+$-ATPase pump.** In one conformation of the pump (shown in states 1 and 8 in the diagram), an internal ion binding site is open to the cytoplasm (the side of the membrane labeled "in"). This conformation has higher affinity for $Na^+$ than for $K^+$. In another conformation, shown in states 4 and 5, the binding site is open to the extracellular space (labeled "out"). This conformation has higher affinity for $K^+$ than for $Na^+$. The cycle is driven by the binding of ATP, followed by hydrolysis and release of first ADP and then phosphate ion ($P_i$). (Adapted from J.P. Morth et al., and P. Nissen, *Nature* 450: 1043–1049, 2007.)

to an aspartic acid sidechain in the P domain. The A domain is an actuator that undergoes large conformational changes during the cycle and is the domain that is responsible for coupling ATP binding and hydrolysis to structural changes in the transmembrane segment.

A considerable amount of information is available about the structural changes that occur in the pump during the cycle. This information comes from studies on the $Ca^{2+}/H^+$ pump, also known as the $Ca^{2+}$-ATPase, which pumps calcium ions out of cells with the coupled movement of protons into the cytoplasm. The $Ca^{2+}$-ATPase is closely related in structure to the sodium-potassium pump, and its action on $Ca^{2+}$ and $H^+$ is analogous to the action of the sodium-potassium pump on $Na^+$ and $K^+$, respectively.

The crystal structure of the $Ca^{2+}$-ATPase has been determined in a large number of states along the reaction cycle by Chikashi Toyoshima and colleagues (four of these structures are shown schematically in **Figure 11.32**). We shall not describe these structures in detail, but simply note that the general idea is that the chemical changes brought about by the hydrolysis of ATP result in conformational changes that open and close the ion binding sites within the transmembrane segment. The thermodynamic implications of such a mechanism have been described conceptually in Section 9.15.

## 11.19 Sodium and potassium channels allow ions to move quickly across the membrane

The $Na^+$-$K^+$ pump is rather slow, because it has to hydrolyze ATP and undergo major conformational changes while switching between states that are open alternatively to the inside and the outside of the cell. The slowest step in the cycle is the conversion between the states labeled 3 and 4 in Figure 11.31, and this step occurs at a maximal rate of ~50 per second. At this rate, each pump will move ~100 $K^+$ and ~150 $Na^+$ ions into and out of the cell per second. Even though there are ~1000 $Na^+$-$K^+$ pumps per square micron of axonal surface, this is much too slow for the purpose of nerve conduction.

Ion channels that are specific for $Na^+$ and $K^+$ allow ions to move through them at extraordinary speeds: open potassium channels conduct ten million to a hundred

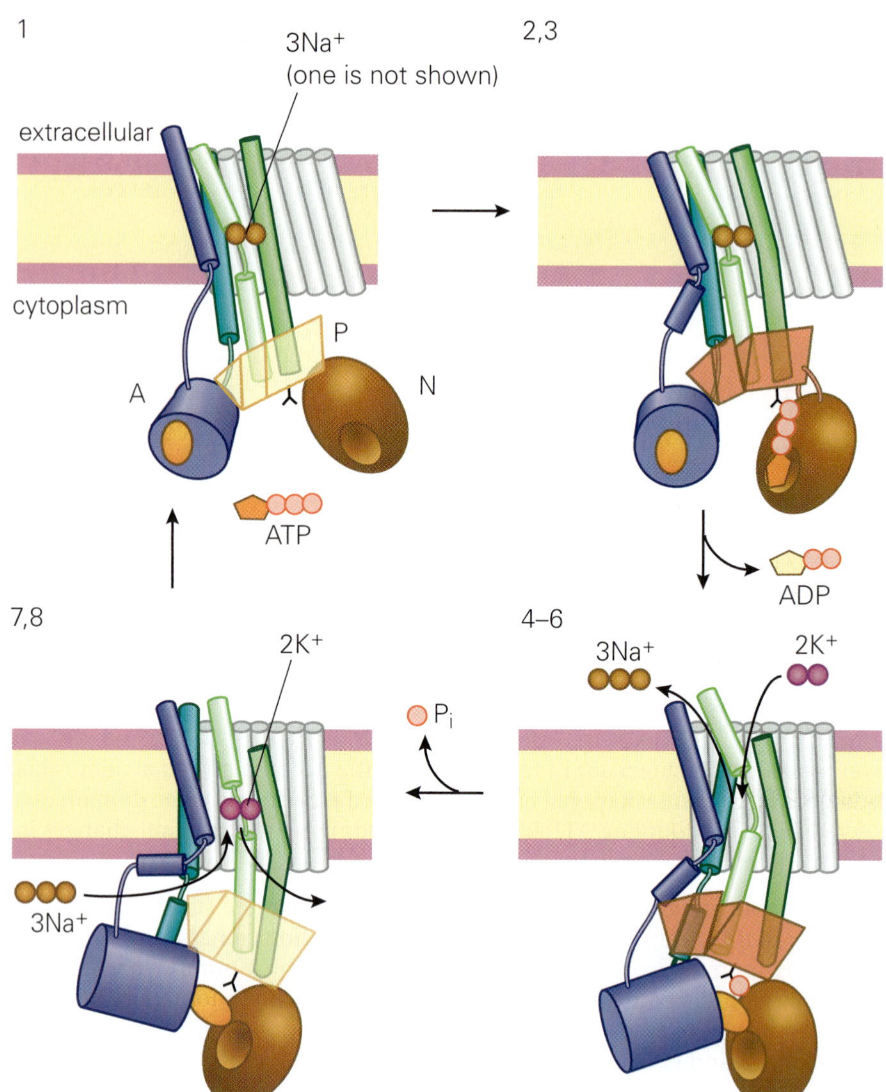


**Figure 11.32 Structural basis for the ATPase cycle of the Na$^+$-K$^+$ pump.** This schematic should be compared with the more abstract version shown in Figure 11.31 and the structural drawings in Figure 11.30. The numbers next to each stage in the cycle refer to the corresponding states in Figure 11.31. Only two of the three Na$^+$ ions that are pumped in each cycle are shown. A loop in the A domain that interacts with the N domain in two of the steps in the cycle is shown as an *orange* oval. This diagram is based on an ATPase that pumps Ca$^{2+}$ and H$^+$, instead of K$^+$ and Na$^+$, for which crystal structures in different steps of the cycle are available. The Ca$^{2+}$/H$^+$-ATPase is very closely related to the Na$^+$/K$^+$-ATPase. (Adapted from C. Toyoshima, *Biochim. Biophys. Acta* 1793: 941-946, 2009.)

million ions per second—that is, about 10,000 times faster than the pumps can move ions. To appreciate just how fast these rates are, we can compare them with the speed at which ions diffuse freely through water. An ion moving in water undergoes Brownian (random) diffusion in three dimensions, but to compare this with movement through the ion channel, we shall imagine that the ion undergoes a one-dimensional random walk in water.

Computer simulations have shown that the **diffusion constant** for a sodium ion in water is roughly $1.0 \times 10^{-5}$ cm$^2$•sec$^{-1}$. The meaning of the diffusion constant is explained in Chapter 17, but for the present discussion it is sufficient to know that the diffusion constant, $D$, relates the root mean square displacement in a one-dimensional random walk to the time elapsed, $t$:

$$\text{root mean square displacement} = \sqrt{2Dt} \tag{11.59}$$

The span of the potassium channel is ~40 Å ($4 \times 10^{-7}$ cm, the thickness of the membrane). We assume that the time needed for an ion to cover this distance in a random walk is comparable to the time required for the root mean square displacement to grow to this value. By using Equation 11.59, we can calculate that this would take $8 \times 10^{-9}$ seconds (that is, 8 nanoseconds). If the flow rate through an ion channel is $10^8$ ions per second, then the average transit time for one ion is 10 nanoseconds, which is close to the diffusion time in water. Remarkably, such high transit rates through the channel are accomplished while retaining specificity for a particular kind of ion.

The very high speed with which ion channels pass ions in just one direction is only possible because the $Na^+$-$K^+$ pumps maintain the electrochemical gradient that drives ions through the channels. It has been estimated that about 30% of the energy consumption of a neuron goes towards running the pumps that power this gradient. These pumps run all the time, and even though they are very slow, they keep up with the demands of the axon because the action potentials are very short lived and, as we discuss below, because they involve the movement of only a relatively small number of sodium or potassium ions at a time.

How do ion channels allow such rapid transit of ions while maintaining specificity for a particular ion? A breakthrough in our understanding occurred with the determination of the crystal structure of a potassium channel by Roderick MacKinnon and colleagues, which provided answers to both questions.

## 11.20 Sodium and potassium channels contain a conserved tetrameric pore domain

The core segment of potassium channels forms a tetrameric structure with an ion-conducting pore running through the middle of the tetramer. The pore-forming domains have similar structures in sodium and potassium channels, and the potassium channel structure serves as a good model for that of sodium channels as well.

The pore-forming domain consists of two transmembrane helices that are connected by a segment that inserts partway through the middle of the structure and forms a filter that selects a specific ion (**Figure 11.33**A). Four pore-forming domains come together to form a tetramer with the **selectivity filter** in the middle,

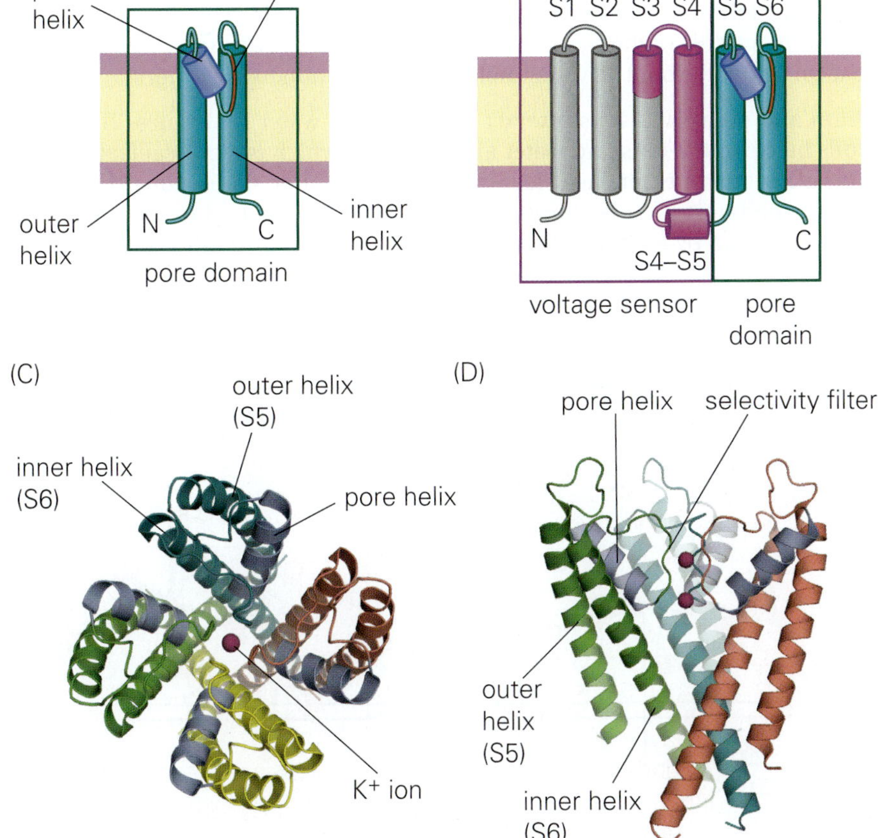


**Figure 11.33 The structure of the pore-forming domain of potassium channels.** (A) $K^+$, $Na^+$, and $Ca^{2+}$ channels have in common a pore domain of very similar structure that consists of two membrane-spanning helices, a pore helix, and a selectivity filter. (B) Schematic diagram of a voltage-gated $K^+$ channel, which opens and closes in response to changes in the membrane voltage (see Section 11.39 and Figure 11.62). S1 to S4: helices of the voltage sensor; S5 and S6: helices of the pore domain. (C) Crystal structure of the tetrameric pore domain of a bacterial $K^+$ channel, viewed from above the plane of the membrane. A $K^+$ ion is shown in the center of the pore. The four pore helices are in *light blue*. (D) Side view of the bacterial $K^+$ channel, with the ion permeation pathway vertical and in the center of the structure. The subunit colored *yellow* in (C) has been removed. (PDB code: 1K4C.)

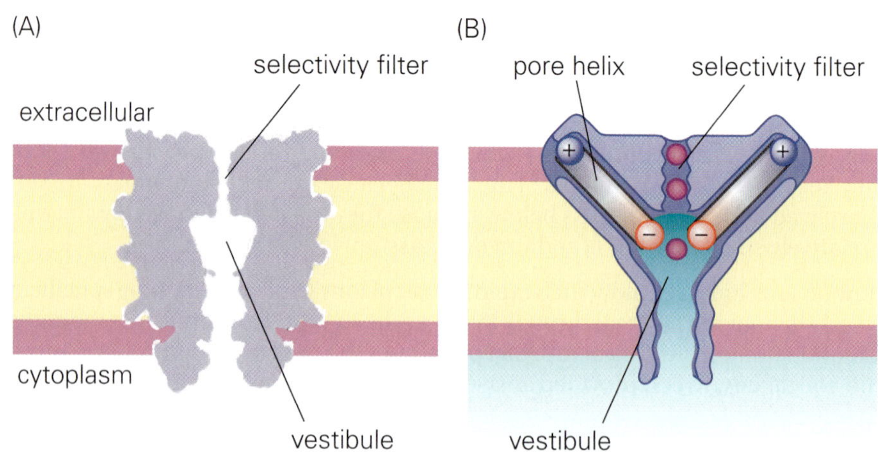


**Figure 11.34 The ion conduit through the potassium channel.** (A) Cross section through the crystal structure of the channel, showing the cavity in the interior, known as the vestibule. (B) Schematic diagram of the channel, showing a cut-away view. The vestibule is wide, allowing ions to remain solvated in this region. The most constricted part of the channel is formed by the selectivity filter, which only allows desolvated ions to pass through. Two of the four pore helices in the channel are shown. The negative poles of the helix dipoles of these helices are pointed towards the center of the vestibule, which may stabilize an ion before it enters the selectivity filter. (Adapted from D.A. Doyle et al., and R. MacKinnon, *Science* 280: 69–76, 1998. With permission from the AAAS; PDB code: 1K4C.)

generating a channel through which the ions move (Figure 11.33B and C). This selectivity filter is polar, thereby facilitating the movement of ions through it. The selectivity filter is highly conserved in potassium channels from all the branches of life, but its sequence is different in the sodium channels.

In addition to the pore domain, these channels contain elements that gate the flow of ions through the pore. Most pertinent for our subsequent discussion are voltage sensor domains, which open and close these channels in response to changes in the membrane potential. We discuss voltage sensors in Section 11.40.

## 11.21 A large vestibule within the channel reduces the distance over which ions have to move without associated water molecules

The width of the conduit through the channel is not uniform (**Figure 11.34**). The cytoplasmic entrance is relatively wide, and it opens up into an even wider vestibule in the middle of the transmembrane portion of the channel. This vestibule can accommodate about 50 water molecules in addition to the $K^+$ ion.

Sodium and potassium ions interact strongly with water molecules. The interactions between the ions and water are not static, which is a critical feature that enables the ions to enter the selectivity filter. Computer simulations show that, although the interaction is strong, one water molecule bound to an ion can be rapidly displaced by another when these bonds break transiently (**Figure 11.35**).

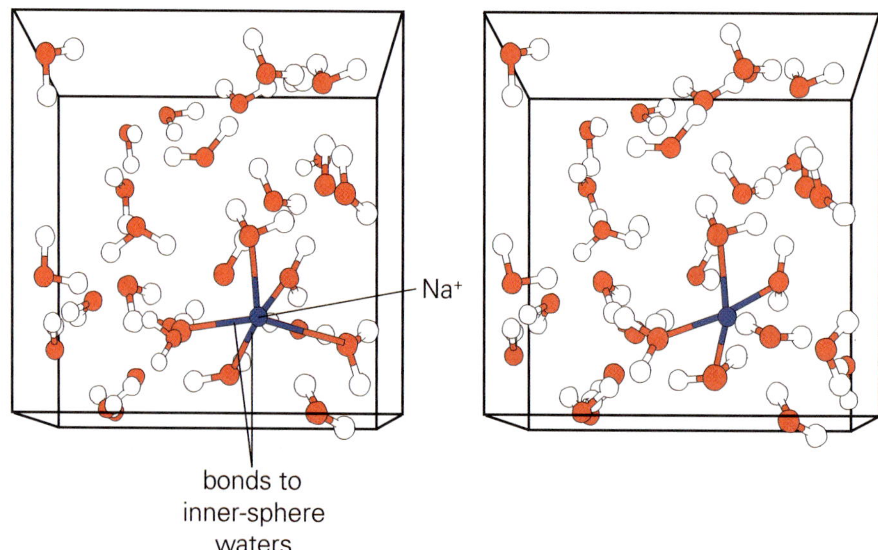


**Figure 11.35 Coordination of sodium ions by water molecules.** Two instantaneous snapshots from a molecular dynamics simulation are shown, separated in time by 70 femtoseconds ($10^{-15}$ sec). The sodium ion (*blue*) has five inner-sphere waters in the first frame and four in the next. (Adapted from S.B. Rempe and L.R. Pratt, *Fluid Phase Equilibria* 183–184: 121–132, 2001. With permission from Elsevier.)

On average, the ion interacts closely with four to five water molecules at any one time. The construction of the potassium channel allows the ion to remain solvated as it transits into the vestibule. The ion is also stabilized by the negative ends of helix dipoles that point into the center of the vestibule (see Figure 11.34B).

The channel narrows considerably at the point at which the selectivity filter begins (see Figure 11.34A). The pore through the selectivity filter is so constricted that the ions have to pass through it in single file, with all their inner-shell water molecules stripped away (**Figure 11.36**). However, the architecture of the potassium channel reduces the distance over which the desolvated ions have to move to less than half the span of the transmembrane segment.

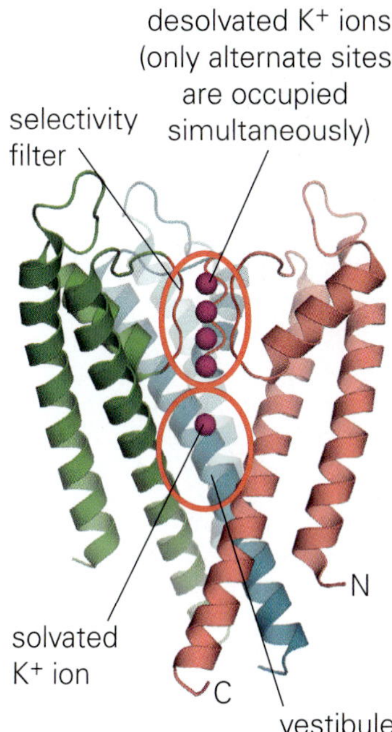


**Figure 11.36 $K^+$ ions pass through the selectivity filter in single file.** Shown here is the crystal structure of the potassium channel, with the positions of ions indicated by *purple* spheres. Four ion binding sites are shown in the selectivity filter, but not all four are occupied at the same time. (Adapted from J.H. Morais-Cabral, Y. Zhou, and R. MacKinnon, *Nature* 414: 37–42, 2001. With permission from Macmillan Publishers Ltd; PDB code: 1K4C.)

## 11.22 Carbonyl groups in the selectivity filter provide specificity for $K^+$ ions by substituting for the inner-sphere waters

The selectivity filter consists of an extended strand spanning five residues in which two glycine residues are invariant across all potassium channels (**Figure 11.37**A). The presence of these glycine residues allows the strand to adopt a conformation that would be forbidden otherwise, in which all of the carbonyl groups point in the same direction (in a normal β strand, alternate carbonyl groups would point in opposite directions, as explained in Chapter 3).

Recall from Section 2.23 that metal ions have a set of inner-sphere ligands that are bound directly to the ions. The number and geometry of the inner-sphere

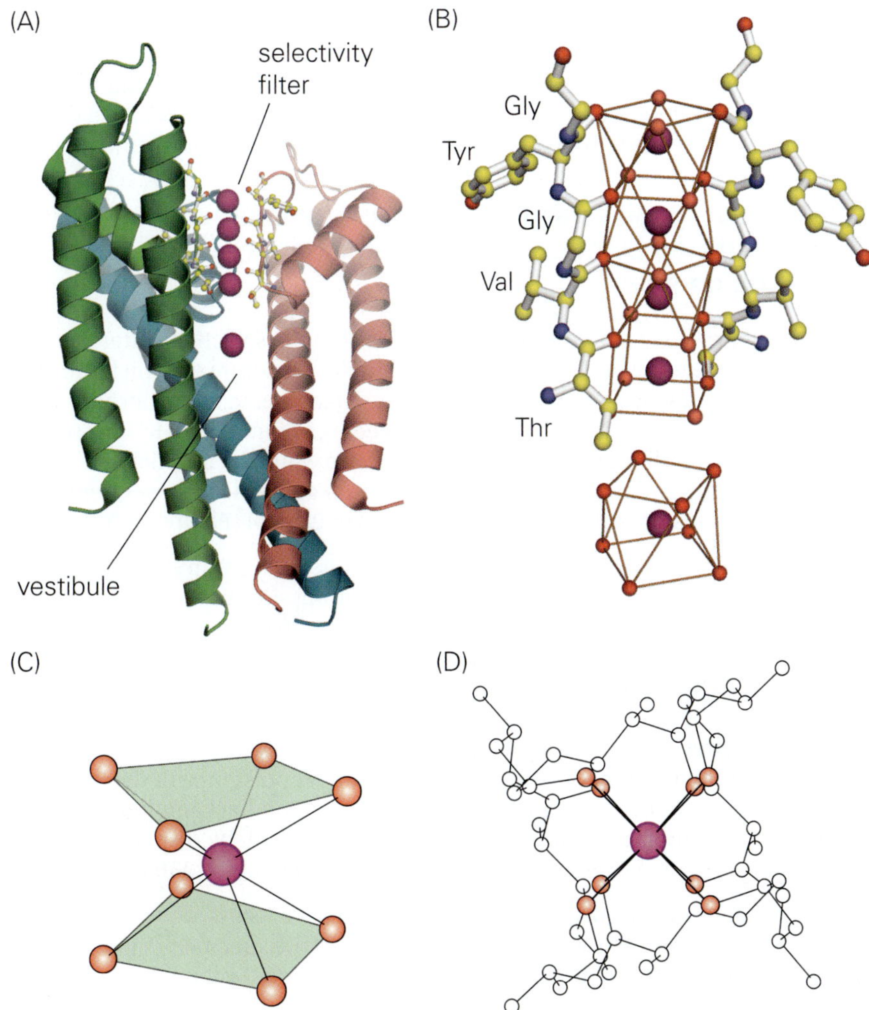


**Figure 11.37 Coordination of potassium ions in the selectivity filter.** (A) Structure of the potassium channel, with one subunit removed. $K^+$ ions are shown as *purple* spheres. Residues in the selectivity filters of the subunits are shown as sticks. (B) Expanded view of four $K^+$ ions in the selectivity filter and one in the vestibule. Oxygen atoms (*red*) from the protein coordinate the $K^+$ ions in the selectivity filter, and oxygen atoms in water molecules (*red*) coordinate the $K^+$ ion in the vestibule. Although four ions are shown here, at any one time no more than two $K^+$ ions can occupy the selectivity filter. Only two of the four protein strands that make up the selectivity filter are shown. (C) Square antiprism coordination of a $K^+$ ion by eight oxygen atoms in the selectivity filter of the $K^+$ channel. (D) Structure of the antibiotic nonactin bound to a $K^+$ ion (*purple*). Eight oxygen atoms (*red*) in nonactin coordinate the $K^+$ ion. (A and B, PDB code: 1K4C; B, adapted from data in Y. Zhu et al., and R. MacKinnon, *Nature* 414: 43–48, 2001; D, adapted from M. Dobler, J.D. Dunitz, and B.T. Kilbourn, *Helv. Chim. Acta* 52: 2573–2583, 1969.)

ligands are distinctive features of each metal ion. The oxygen atoms of the carbonyl groups in the selectivity filter are positioned to satisfy the inner-sphere coordination of $K^+$ ions with near optimal geometry. This allows the selectivity filter to capture $K^+$ ions from the vestibule or the extracellular space without a large energetic penalty.

The coordination of $K^+$ ions within the selectivity filter is illustrated in Figure 11.37B. Notice that each of the four ions has eight oxygens coordinating it, and that each of these oxygens is provided by the backbone of the selectivity filter. A $K^+$ ion in the vestibule is also shown in the figure, and this ion is coordinated by eight water molecules. The oxygen atoms in the selectivity filter mimic the arrangement of water molecules around the solvated ion in the vestibule. This observation suggests that the selectivity filter serves as a surrogate for the natural hydration of $K^+$ ions.

The eight oxygen atoms surrounding the $K^+$ ion form a square antiprism (see Figure 11.37C). This geometrical arrangement is a distorted cube in which two of the faces are planar and rotated with respect to each other. A similar arrangement of eight oxygen atoms surrounding a $K^+$ ion is seen in the structure of the antibiotic nonactin (see Figure 11.37D). Nonactin binds preferentially to $K^+$ ions over $Na^+$ or $Ca^{2+}$ ions, consistent with the idea that the geometric arrangement of oxygen atoms within the selectivity filter of the potassium channel is optimized for accommodating $K^+$ ions.

The backbone atoms of the selectivity filter are relatively rigid, and this feature will tend to maintain the geometry of the carbonyl groups despite fluctuations in the protein structure. The sodium ion is smaller than the potassium ion (the atomic radius of $Na^+$ is 1.9 Å, versus 2.43 Å for $K^+$), and one explanation for the specificity of the potassium channel is that the selectivity filter cannot properly satisfy the inner-shell coordination of the sodium ion, which would therefore pay a much larger energetic cost if it entered the filter. It is also thought that the carbonyl groups interact preferentially with $K^+$ ions compared to $Na^+$ ions because of their electronic properties, but this aspect of the interaction is still poorly understood.

## 11.23 Rapid transit of K⁺ ions through the channel is facilitated by hopping between isoenergetic binding sites

The apparent presence of four potassium ions within the selectivity filter (see Figure 11.36) is misleading because the crystallographic structure on which the illustration is based provides a time-averaged view. There are, in fact, alternate states in which only every other binding site is occupied by $K^+$ ions. If two $K^+$ ions were to occupy adjacent binding sites within the filter, there would be a sufficiently strong electrostatic repulsion between them to force them apart. As a consequence, the channel has $K^+$ ions bound at either sites 1 and 3 or 2 and 4 (**Figure 11.38**).

Analysis of the crystallographic data, as well as results from computer simulations, suggest that $K^+$ ions bound to each of the four binding sites have nearly the same energy and are also close in energy to ions that are just about to enter or leave the channel. This suggests that the rapid movement of ions through the channel proceeds through a "knock-on" mechanism, as illustrated in Figure 11.38. Imagine that there are two ions bound to the selectivity filter. These can hop back and forth between sites 1 and 3 or 2 and 4. If site 4 is vacant, a $K^+$ ion can enter from the vestibule and bind to it. This will lead to electrostatic repulsion with the ion at site 3, which will either move outward or displace the new ion back into the vestibule. If the ion at site 3 moves to site 2, then the ion at site 1 will be displaced into the extracellular space due to electrostatic repulsion (see Figure 11.38B and C).

The channel does not impose any directionality on the process of ion conduction. A similar knock-on process will occur if an ion enters the specificity filter from the extracellular side, and the direction of the net flux of ions will be determined by the electrochemical gradient.

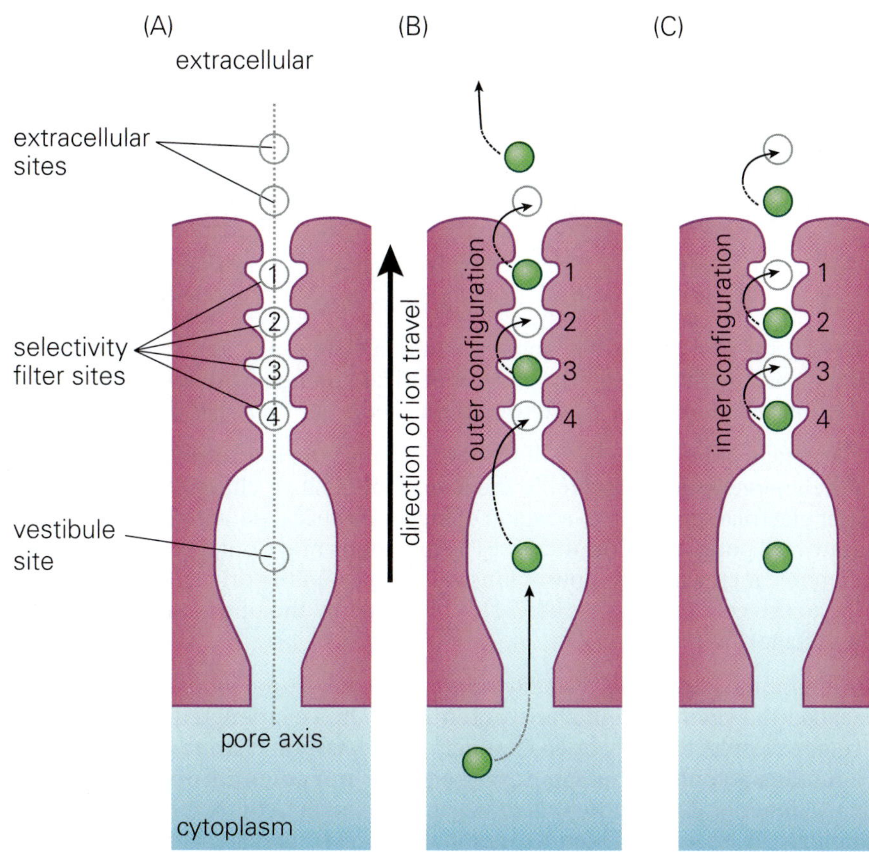


**Figure 11.38 Model for the transit of ions through the selectivity filter.** (A) The four binding sites in the selectivity filter are indicated, as well as three additional sites where solvated $K^+$ ions are bound. (B and C) The model specifies that only alternate sites within the specificity filter are occupied simultaneously by $K^+$ ions. The arrival of a third ion in the specificity filter results in the expulsion of an ion at the other end, as shown here, or the ejection of the third ion (not shown). (Adapted from C. Miller, *Nature* 414: 23–24, 2001. With permission from Macmillan Publishers Ltd.)

# D. THE TRANSMISSION OF ACTION POTENTIALS IN NEURONS

In this part of the chapter we focus on the mechanisms underlying the transmission of action potentials down the axon of the neuron without attenuation. First, we consider what happens to a voltage spike when it travels through the neuron without being regenerated. We shall see that voltage spikes decay within a short distance of their point of origin, because of various dissipative processes. For neuronal signaling to be effective, there has to be a way to regenerate the voltage spike periodically. This leads to the understanding that voltage-gated sodium and potassium channels respond to voltage spikes by opening and closing transiently, thereby regenerating the spike.

## 11.24 The asymmetric distribution of ions across the cell membrane generates an equilibrium membrane potential

To understand the origin of the membrane potential, let us study a hypothetical situation in which we consider just the concentration of $K^+$ ions within a cell and in its environment (**Figure 11.39**). To begin with, let us assume that the concentration of $K^+$ ions is high inside the cell (for example, 150 mM; see Figure 11.29) and low outside (for example, 5 mM). Now imagine that the cell membrane contains some open $K^+$ channels, but no other open ion channels.

The concentration gradient provides a driving force for the movement of $K^+$ ions from inside the cell to the outside through these channels. But, as $K^+$ ions move out, positive charge is depleted inside the cell and builds up outside the cell. In

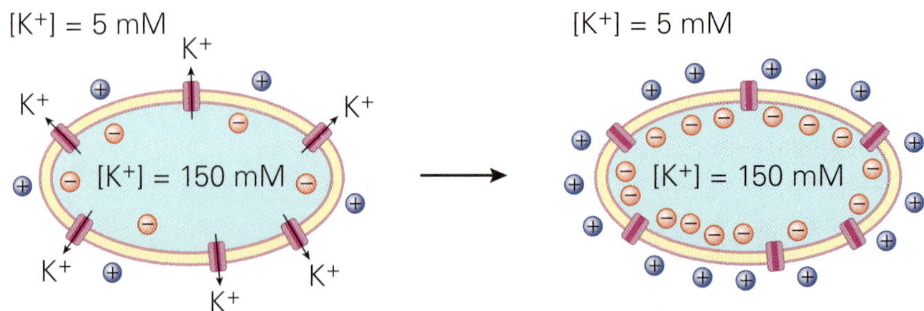


**Figure 11.39 Generation of an equilibrium membrane potential.** A schematic diagram of the cross section of a cell is shown here, with open potassium channels in the membrane. The concentration of $K^+$ ions is higher inside the cell than outside, so the open potassium channels leak $K^+$ ions from inside to outside. The outward movement of $K^+$ ions leads to polarization of the membrane, with an increased negative charge inside the cell and an increased positive charge outside the cell. This electrical gradient opposes the flow of $K^+$ ions. The net flux of $K^+$ ions stops when the concentration gradient is balanced by the electrical gradient.

this way, the movement of $K^+$ ions generates a voltage across the cell membrane, with the inside of the cell at negative electrical potential with respect to the outside. The cell membrane is said to be **polarized**.

This voltage *opposes* the flow of $K^+$ ions, because the outward flow corresponds to the energetically unfavorable movement of positive charge from a region of lower electrical potential to a region of higher electrical potential. The opposition to outward potassium flow increases as more ions move out, because the voltage difference increases. The flow of ions will stop when the drive to move outwards (due to the concentration gradient) is balanced by the opposition generated by the voltage difference.

**Resting membrane potential**

The membrane potential of a cell at steady state, with no net flux through ion channels, is known as the resting membrane potential, denoted $E^*_m$. The value of $E^*_m$ for a mammalian cell is usually around –70 mV.

The membrane potential at steady state (that is, when there is no net flow of ions) is called the **resting membrane potential** of the cell, denoted $E^*_m$. In this particular example, the only open channels are potassium channels and the resting membrane potential is the same as the **equilibrium potential of potassium**, $E_K$—that is, the membrane potential at which there is no net flux through potassium channels, given a certain concentration differential. The equilibrium potential of potassium is also called the **Nernst potential** of potassium, for reasons that will become clear in Section 11.25.

Does the leakage of $K^+$ ions through the open channels affect the concentration gradient across the membrane? As explained in Section 11.28, a change in the membrane potential of ~100 mV corresponds to the movement of a very small number of ions across the membrane. As a consequence, the membrane potential is established without a substantial change in the intracellular concentration of $K^+$ ions.

**Nernst potential of an ion**

The membrane potential at which there is no net flux through channels for a particular ion (for example, $K^+$) is known as the Nernst potential for that ion. The Nernst potential depends on the concentrations of the ion inside and outside the cell.

We have not considered the behavior of other ions, such as $Na^+$, $Cl^-$, or $Ca^{2+}$, in this discussion. A more general treatment for the case when more than one kind of ion channel is present is given in Section 11.31. It turns out that, in resting neurons, the leakage of ions across the membrane occurs primarily through potassium channels, and so the resting membrane potential, $E^*_m$, is very close to the Nernst potential of potassium, $E_K$.

## 11.25 The Nernst equation relates the equilibrium membrane potential to the concentrations of ions inside and outside the cell

To calculate the value of the equilibrium membrane potential of potassium, we begin by recognizing that the driving force for potassium moving out of the cell is given by the difference in the chemical potential of potassium across the membrane. This driving force is opposed by the voltage that develops across the membrane (**Figure 11.40**). We can calculate the equilibrium membrane potential for potassium by balancing the electrical potential difference against the chemical potential difference.

If the membrane potential is $E_m$, then the work done in moving one mole of a single positive charge from inside to outside is given by:

$$\text{work} = -E_m \times \mathcal{F} \tag{11.60}$$

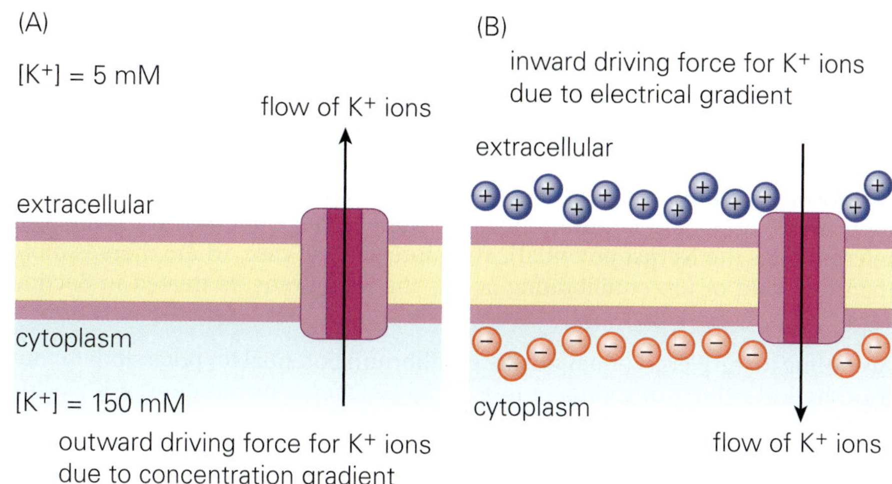


**Figure 11.40 Opposing driving forces control the movement of K+ ions.** (A) The outward movement of K+ ions is driven by the concentration gradient across the cell membrane. (B) As K+ ions leave the cell, excess negative and positive charges build up inside and outside the cell, respectively. This generates a voltage across the cell membrane, which opposes the outward movement of K+ ions.

where $\mathcal{F}$ is the Faraday constant (the magnitude of the charge on a mole of electrons; see Section 11.10). The work done is equal to the change in free energy, assuming that the process is carried out reversibly. We denote the change in free energy by $\Delta G_{elec}$, where the subscript indicates that we are referring to electrical work:

$$\text{electrical work} = \Delta G_{elec} = -E_m \times \mathcal{F} \tag{11.61}$$

Note the significance of the negative sign in Equation 11.61. If the membrane potential ($E_m$) is negative, then the free-energy change in moving a positive charge from inside the cell to the outside has a net positive sign. This is consistent with the fact that the energy increases when the charge moves against the electrical potential gradient.

The change in free energy due to electrical work is balanced by the difference in free energy due to the difference in concentration, which we denote by $\Delta G_{conc}$:

$$\begin{aligned}\Delta G_{conc} = G_{out} - G_{in} &= RT\ln([K^+]_{out}) - RT\ln([K^+]_{in}) \\ &= RT\ln\left(\frac{[K^+]_{out}}{[K^+]_{in}}\right)\end{aligned} \tag{11.62}$$

This term has a negative value because the concentration of potassium outside the cells is lower than the concentration inside, leading to a negative value for the logarithmic term. The negative value indicates that the outward flow of potassium is driven by the concentration difference.

The free-energy differences given by Equations 11.61 and 11.62 are equal and opposite at equilibrium—that is, net flow of potassium stops when the total free-energy change is zero:

$$\begin{aligned}\Delta G\,(\text{at equilibrium}) = 0 &= \Delta G_{conc} + \Delta G_{elec} \\ &= RT\ln\left(\frac{[K^+]_{out}}{[K^+]_{in}}\right) - E_m\mathcal{F}\end{aligned} \tag{11.63}$$

The resting membrane potential, $E^*_m$, is therefore given by:

$$E^*_m = E_K = \frac{RT}{\mathcal{F}}\ln\left(\frac{[K^+]_{out}}{[K^+]_{in}}\right) = (0.059)\log\left(\frac{[K^+]_{out}}{[K^+]_{in}}\right)$$

The value of $E^*_m$ given by this equation has units of volts (V). If we wish to express $E^*_m$ in units of millivolts (mV), we must modify the equation as follows:

$$E^*_m(\text{in mV}) = E_K = (59)\log\left(\frac{[K^+]_{out}}{[K^+]_{in}}\right) \tag{11.64}$$

In Equation 11.64 we have used the fact that at room temperature (298 K) the value of $RT/\mathscr{F}$ is 0.0257 V.

Equation 11.64 is a form of the Nernst equation, which we first encountered in Section 11.14 (Equation 11.50). Note that in Equation 11.64 the resting membrane potential, $E^*_m$, is the same as the equilibrium potential for potassium ($E_K$) because there are no other ions contributing to the membrane potential (this is why $E_K$ is referred to as the Nernst potential). The more general case, where there is more than one kind of ion equilibrating across the membrane, is treated in Section 11.31.

According to the Nernst equation, the equilibrium potential for potassium for our hypothetical cell (corresponding to a 30-fold higher concentration of $K^+$ inside relative to outside) would be given by:

$$E_K(\text{mV}) = (59)\log\left(\frac{[K^+]_{out}}{[K^+]_{in}}\right) = (59)\log\left(\frac{1}{30}\right)$$
$$= -87\text{ mV} \tag{11.65}$$

If we were to consider the equilibrium potential due to a doubly charged ion, such as $Mg^{2+}$, we would have to modify the Nernst equation to account for the increased charge on the ion. In general, if the charge on an ion M is $\upsilon$, then the equilibrium potential for M, $E_M$, is given by a more general form of the Nernst equation, where we denote the metal ion as M:

$$E_M = \frac{RT}{\upsilon\mathscr{F}}\ln\left(\frac{[M^{\upsilon+}]_{out}}{[M^{\upsilon+}]_{in}}\right) = \frac{59}{\upsilon}\log\left(\frac{[M^{\upsilon+}]_{out}}{[M^{\upsilon+}]_{in}}\right) \tag{11.66}$$

In order to derive the Nernst equation, we began with a certain initial concentration difference for $K^+$ outside and inside the cell, and then we deduced the membrane potential that balanced the driving force due to the concentration difference. The movement of ions to generate the membrane potential changes the concentration of ions inside and outside the cell. It turns out that the number of potassium ions that have to move across the membrane is very small compared to the total concentration of ions, as shown below in Section 11.28, and so we can assume that the effective values of $[K^+]_{out}$ and $[K^+]_{in}$ remain essentially constant as the membrane potential develops. This means that the resting potential can be calculated directly from the initial concentrations of potassium inside and outside the membrane.

## 11.26 Cell membranes act as electrical capacitors

In order to calculate the number of charges that have to move across the membrane to establish the equilibrium membrane potential, we shall make an analogy between the development of the membrane potential and the charging of an electrical **capacitor**. The analogy to electrical capacitors is also helpful in understanding how the action potential is propagated.

In the simplest case, a capacitor consists of a pair of conducting plates separated by an insulating material, as illustrated in **Figure 11.41**. If an uncharged capacitor

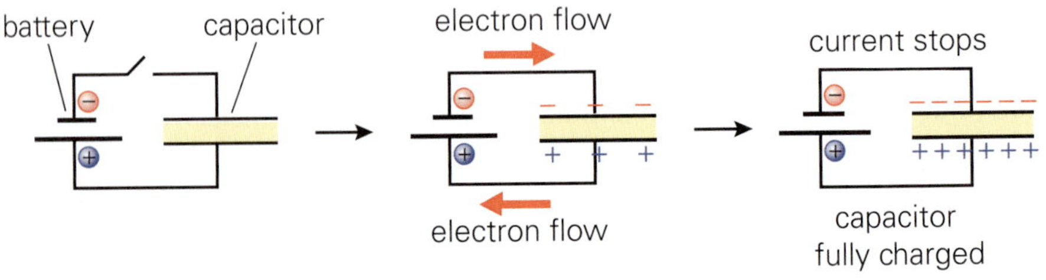


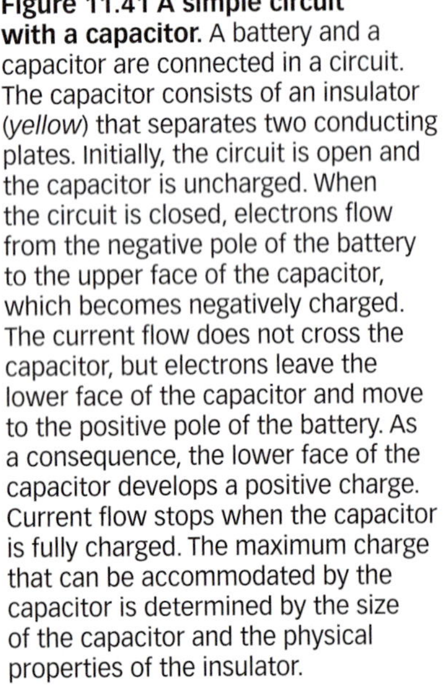
**Figure 11.41 A simple circuit with a capacitor.** A battery and a capacitor are connected in a circuit. The capacitor consists of an insulator (*yellow*) that separates two conducting plates. Initially, the circuit is open and the capacitor is uncharged. When the circuit is closed, electrons flow from the negative pole of the battery to the upper face of the capacitor, which becomes negatively charged. The current flow does not cross the capacitor, but electrons leave the lower face of the capacitor and move to the positive pole of the battery. As a consequence, the lower face of the capacitor develops a positive charge. Current flow stops when the capacitor is fully charged. The maximum charge that can be accommodated by the capacitor is determined by the size of the capacitor and the physical properties of the insulator.

is connected to a battery, then negative charge accumulates on the side of capacitor that is connected to the negative pole of the battery and an equal amount of positive charge develops on the other face of the capacitor. This results in a flow of current in the circuit, which is known as the **capacitive current**. The capacitive current is transient, and it stops when the capacitor is fully charged (see Figure 11.41).

The positive charge on one face of a charged capacitor is balanced by an equal amount of negative charge on the other face. The magnitude of this charge, $q$, is referred to as the charge on the capacitor. Even though the capacitor has both positive and negative charge on it, the value of $q$ is always taken to have a positive sign.

The maximum charge that can be held by the capacitor is determined by the size of the capacitor—the larger the surface area of the conducting plates of the capacitor, the more charge it can hold. The maximum charge also depends on the nature and thickness of the insulating material that separates the two conducting plates. The thinner the insulating material, the greater the amount of charge that can be held by the capacitor. When the capacitor comes to equilibrium, the ratio of the charge, $q$, on the capacitor and the magnitude of the voltage, $V$, across the capacitor is known as the **capacitance**, $C$, of the capacitor:

$$C = \frac{q}{V} \tag{11.67}$$

The capacitance is defined to always be positive and is simply the capacity of the capacitor to hold charge, given a certain voltage across the capacitor. The capacitance, $C$, should not be confused with the unit of charge (coulombs, C, not italicized). The capacitance ($C$, italicized) has units of coulombs × volt$^{-1}$. This unit is known as the **farad**:

$$1 \text{ farad (unit of capacitance)} = \frac{1 \text{ coulomb}}{1 \text{ volt}} \tag{11.68}$$

To see why the cell membrane behaves as a capacitor, consider what happens to the flow of current and to the membrane voltage when the ionic concentration outside the cell changes. Imagine a cell with just one kind of channel, specific to potassium ions (**Figure 11.42**). The cell is immersed in a solution in which the

**Figure 11.42 The cell membrane as a capacitor.** (A) In this schematic diagram the membrane of a cell is shown in *yellow*, with potassium channels indicated in *purple*. The concentration of potassium ions is initially the same inside and outside the cell, and there is no ion flow and no membrane potential. The cell is then moved to a solution in which the concentration of potassium is 10-fold lower (middle panel). Potassium ions move out of the cell, generating excess negative charge inside the cell and excess positive charge outside the cell. The excess charges attract each other across the insulating material of the cell membrane and are therefore localized near the cell membrane. (B) The cell membrane acts as a capacitor, which is charged by a "battery" in which the electromotive force is given by the chemical potential difference between the potassium ions inside and outside the cell. Electrodes inserted into the cell measure the voltage difference across the membrane (the membrane potential) and the current flow through the membrane, and these are graphed as a function of time. The transient capacitive current stops when the membrane potential reaches its equilibrium value.

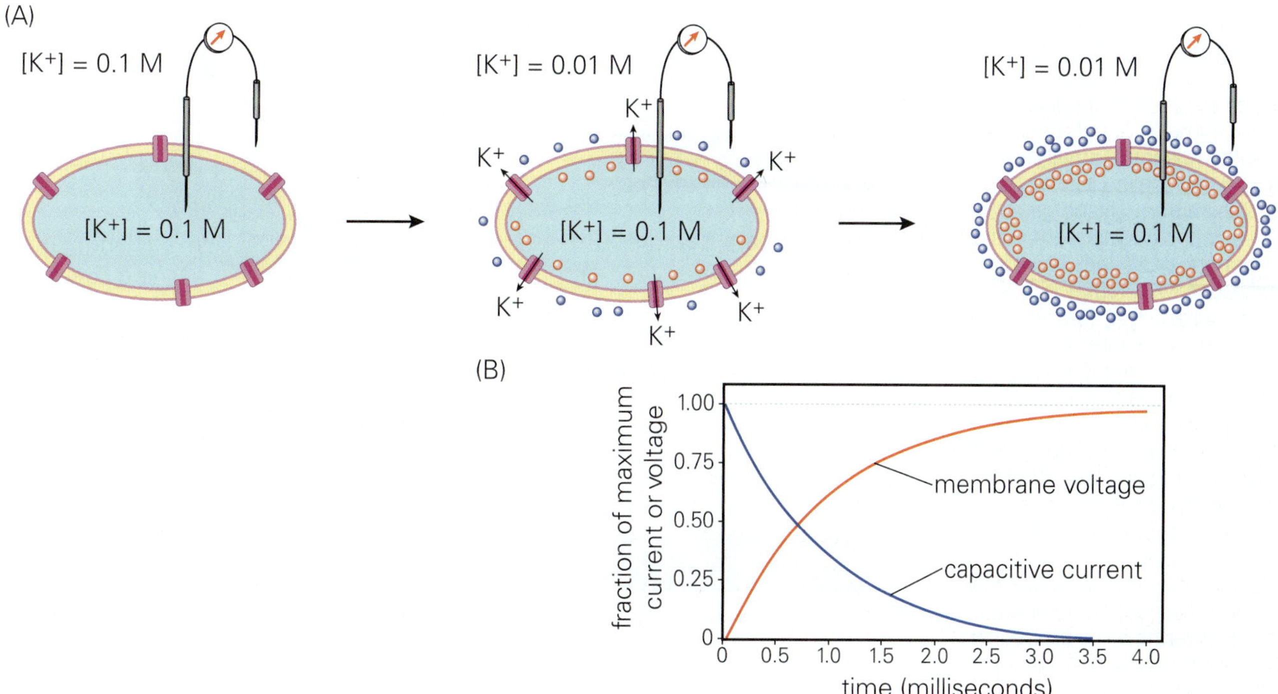

**Capacitance**

An insulator sandwiched between two conductors is known as a capacitor. A capacitor builds up charge on its surfaces when the conductors are connected to an electrical circuit. The ratio of the maximum charge to the voltage across the conductors is known as the capacitance of the conductor. The capacitance, $C$, is defined by Equation 11.67.

**Depolarization of the cell membrane**

The membrane potential is usually negative—that is, the cytoplasmic surface of the cell membrane is usually at lower electrical potential than the extracellular surface. If there is an influx of positively charged ions, then the inner surface becomes less negative, or may even become positive with respect to the electrical potential of the outer surface. This is referred to as depolarization. Alternatively, if there is a net outflow of positive ions, then the membrane becomes hyperpolarized.

concentration of potassium ions is the same as that inside the cell, and all other ionic concentrations are also equal on both sides of the membrane. Because there is no concentration gradient across the membrane, there will be no net current flow and no membrane potential.

Now imagine what happens when the cell is moved into a solution where the concentration of potassium ions is much lower than inside the cell. Initially, the number of positive charges outside the cell is balanced by an equal number of negative charges, and the same is true inside the cell. Thus, initially, the voltage across the membrane is zero. But, because of the difference in potassium ion concentration across the membrane, potassium ions flow out of the cell through the channels. This leads to the accumulation of excess negative charge inside the cell and excess positive charge outside. These charges attract each other across the cell membrane, so they will be located on either side of the cell membrane, as shown in Figure 11.42. The net movement of potassium out of the cell stops when the membrane potential is equal to the equilibrium potential for potassium, as given by the Nernst equation.

Comparison of the processes shown in Figures 11.41 and 11.42 suggests that the development of the membrane potential is analogous to the charging of a capacitor (the cell membrane) by a battery (the chemical potential difference of potassium ions inside and outside the cell). The flow of current that occurs when the cell is moved to the solution with lower potassium ion concentration can be modeled by an electrical circuit in which a battery is connected in parallel to a resistor and a capacitor, as discussed further in Section 11.30.

## 11.27 The depolarization of the membrane is a key step in initiating a neuronal signal

The transmission of signals between neurons occurs at synaptic junctions, where one neuron triggers the flow of ions into another (**Figure 11.43**). The initial trigger is provided by the release of small molecules known as neurotransmitters by

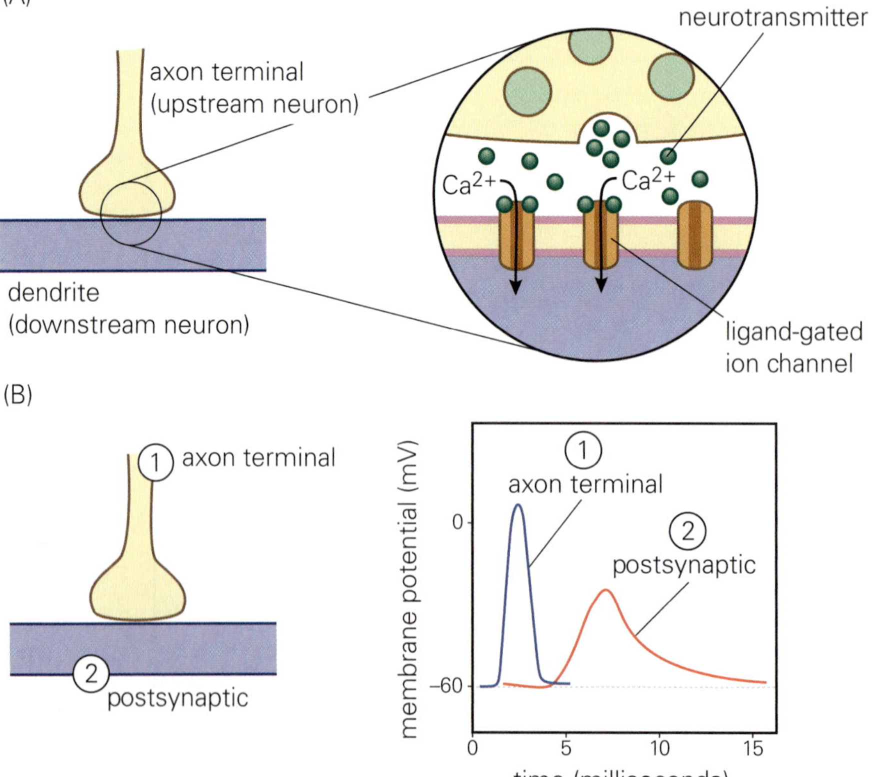


**Figure 11.43 Signal transmission across a neural synapse.** (A) Shown here is an axon terminal from one neuron forming a synapse with a dendrite of another neuron. The arrival of an action potential at the synaptic terminus of the first neuron causes the release of small molecules such as acetylcholine, known as neurotransmitters. Ligand-gated ion channels in the downstream neuron, shown in the expanded view of the synapse, open when they bind to the neurotransmitter, causing the influx of ions (for example, $Ca^{2+}$) into the dendrite. (B) The influx of ions changes the membrane potential of the downstream neuron in the vicinity of the synapse, known as the postsynaptic region. The diagram shows the results of measuring the membrane potential in the axon terminal (labeled 1) and in the postsynaptic region (labeled 2). The arrival of an action potential at the axon terminal is followed by depolarization of the postsynaptic region a short time later.

the upstream neuron. The release of neurotransmitters from the first neuron is initiated by the arrival of an action potential at the axonal terminal. The neurotransmitter molecules diffuse across the space between the two cells and bind to ligand-activated ion channels at the cell surface at the neuron receiving the signal. These are ion channels that open when the neurotransmitters bind to the extracellular portions of the receptors, resulting in the flow of $Ca^{2+}$ and $Na^{+}$ ions into the cell.

Before the arrival of the signal, the membrane of the second cell in the vicinity of the synaptic junction is polarized in the usual way, with the inside negative and the outside positive. The injection of positively charged ions (for example, $Ca^{2+}$) into the cell results in a **depolarization** of the neuron, as illustrated in **Figure 11.44**. Because of the excess positive charge on the cytoplasmic face of the membrane in the vicinity of the activated ion channels, the membrane potential is more positive than normal in this region. If the channels open and then close transiently, the inflow of positive ions generates a **voltage spike** in the dendrite or cell body. More than one such voltage spike may be integrated before an action potential is generated at the base of the axon.

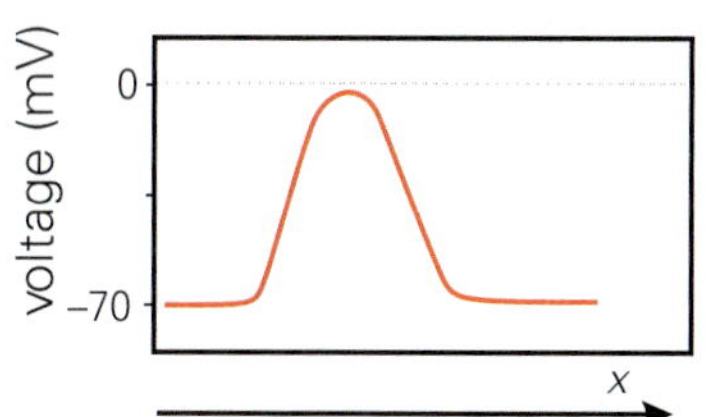


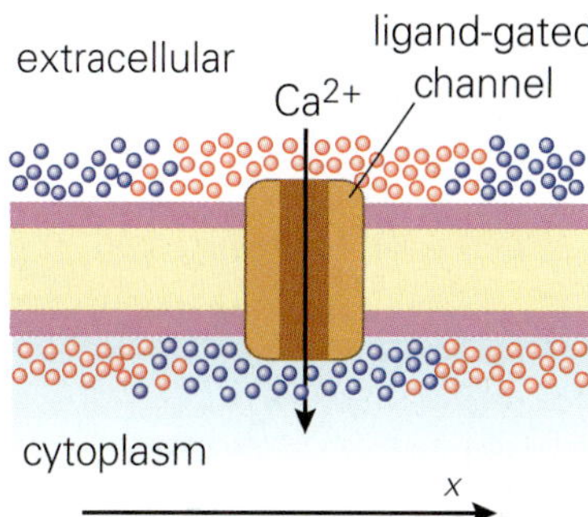


**Figure 11.44 Depolarization of the membrane in the vicinity of a ligand-activated channel.** The transient influx of positive ions (for example, $Ca^{2+}$) inverts the membrane potential so that there is excess positive charge inside and excess negative charge outside. The membrane potential as a function of longitudinal position *x* is shown in the graph.

## 11.28 Membrane potentials are altered by the movement of relatively few ions, enabling rapid axonal transmission

Measurement shows that the capacitance of a typical cell membrane is around 1 microfarad (μF) per square centimeter. This allows us to calculate how many ions move across the membrane per unit area for a given change in membrane potential. This calculation reveals that the number of ions that move across the membrane is extremely small compared to the total number of ions in the cell and, as a consequence, the polarization or depolarization of the membrane does not lead to an appreciable change in the cellular concentrations of ions.

We shall use the the giant axon of the squid to analyze how many ions move across the membrane for a 100 mV change in membrane potential. This axon is particularly large, with a cross-sectional diameter of about 500 μm. Consider a thin slice, 1 μm thick, of this axon (**Figure 11.45**). Expressing the surface area of the membrane within this slice in units of square centimeters, we get:

$$\begin{aligned}\text{surface area} &= \text{circumference} \times \text{thickness}\\ &= 2\pi r(1\times10^{-6}\text{ m}) = (2)(3.14)(250\times10^{-6})(10^{-6})\text{ m}^2\\ &= 1570\times10^{-12}\text{ m}^2 \approx 2\times10^{-9}\text{ m}^2 = 2\times10^{-5}\text{ cm}^2 \end{aligned}\quad(11.69)$$

Using the value of 1 μF ($1\times10^{-6}$ F) per square centimeter for the capacitance per unit area, the total capacitance of this slice of the axon, *C*, is given by:

$$C=(1\times10^{-6})(2\times10^{-5})=2\times10^{-11}\text{ F}\quad(11.70)$$

Assume that the resting membrane potential is –0.1 V. We can use Equation 11.67 to calculate the charge on the membrane capacitor at this voltage:

$$q=C\times V=(2\times10^{-11}\text{ F})(0.1\text{ V})=2\times10^{-12}\text{ C}\quad(11.71)$$

Thus, the charge on either side of the membrane of this slice of the axon is 2 picocoulombs. Note that in Equation 11.71 we use the magnitude of the membrane voltage rather than its actual value because the charge on a capacitor is always defined to be positive.

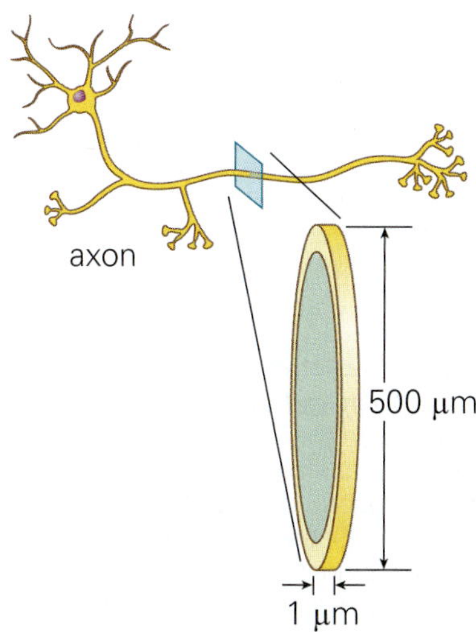


**Figure 11.45 A thin slice of a giant axon from a squid.** The membrane of the entire cell acts as a capacitor, but it is convenient for calculations to focus on small regions, such as this slice.

Now consider what happens when the membrane depolarizes during the passage of an action potential. As we discuss in Section 11.38, depolarization of the membrane occurs because of the influx of $Na^{+}$ ions through transiently open sodium channels. If the membrane potential switches from –100 mV to 0 mV, then 2 picocoulombs of charge will have to move across the membrane. To calculate the number of $Na^{+}$ ions that move, we use the fact that the charge corresponding to one mole of unit charge is 1 faraday ($1\,\mathcal{F} \approx$ 96,500 coulombs; see Section 11.10):

$$\begin{aligned}&\text{number of sodium ions that cross the membrane}\\&= \frac{2\times10^{-12}\text{C}}{96{,}500\ \text{C}\bullet\text{mol}^{-1}}\times \text{Avogadro's number}\\&\approx(2\times10^{-17})(6.022\times10^{23})\approx12\times10^{6}\end{aligned} \tag{11.72}$$

Thus, approximately 12 million sodium ions must move from one side of the membrane to the other in order to change the membrane potential by 0.1 V. How does this compare with the total number of sodium ions in this slice of the axon? If we assume that the sodium ion concentration inside the axon is 0.01 M, then:

$$\begin{aligned}&\text{number of sodium ions in slice of axon}\\&=[\text{volume of slice (in liters)}](\text{molarity})(\text{Avogadro's number})\\&=(\text{cross-sectional area})(\text{thickness})(\text{molarity})(\text{Avogadro's number})\\&\approx(2\times10^{-13}\ \text{m}^3)(1000\ \text{L}\bullet\text{m}^{-3})(0.01\ \text{mol}\bullet\text{L}^{-1})(6.023\times10^{23})\\&\approx1.2\times10^{12}\end{aligned} \tag{11.73}$$

Comparing Equations 11.72 and 11.73, we see that the number of potassium ions that move outwards in order to change the membrane potential by 100 mV reduces the number of ions inside the slice by only 10 parts per million.

A rough estimate of the surface density of sodium channels in the squid giant axon is ~100 per square micron. Thus, given the surface area of the slice (~$2 \times 10^{-9}$ $m^2$ or $2 \times 10^3$ $\mu m^2$; see Equation 11.69), we can expect that there are roughly 20,000 sodium channels on the surface of the slice. Therefore, on average, only about 60 ions move through each channel. Similar conclusions can be drawn about the potassium channels in the axon. Thus, alterations of the membrane potential are brought about by a very small number of ions moving through each ion channel. This occurs very rapidly, as we have discussed in Section 11.23, and underlies the very fast transmission of voltage changes down the axon.

## 11.29 The propagation of voltage changes can be understood by treating the axon as an electrical circuit

To understand how the action potential is propagated, we need to know how perturbations in the membrane potential in one region of the axon spread down the axonal body. We expect that the voltage spike is attenuated as it moves away from the site of initiation. The signal dissipates over the length of the neuron due to two effects, as illustrated in **Figure 11.46**. First, the diffusion of ions away from the point of initiation of the voltage spike dilutes the excess charge. In addition, because the membrane potential within the region of the spike is more positive than the Nernst potential for potassium, open potassium channels enable $K^+$ ions to leak out of the cell, further dissipating the signal.

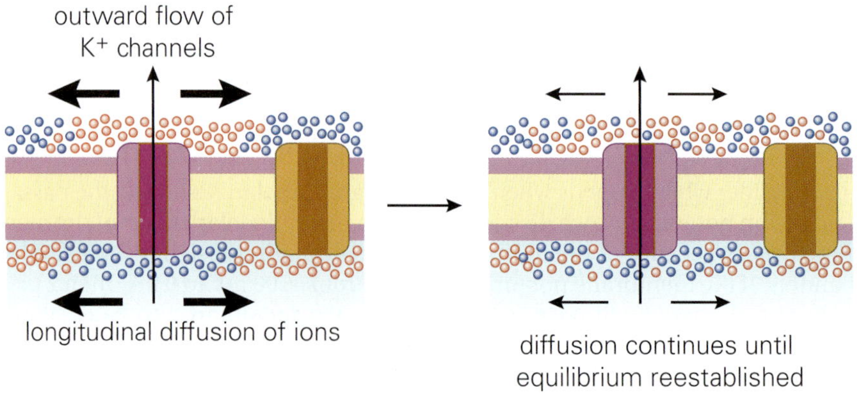


**Figure 11.46 Return to equilibrium after the ligand-gated channels close.** Once the influx of positive ions ends, the membrane potential returns to its equilibrium value through two processes. One is the longitudinal diffusion of ions away from the region of depolarization, and the other is the leakage of $K^+$ ions through open potassium channels (*purple*). These two processes lead to attenuation of the membrane depolarization at the original position, but also to a transient outward spread of membrane polarization.

In the 1950s, Alan Hodgkin and Andrew Huxley carried out landmark measurements on the propagation of voltage spikes in neuronal axons. They discovered that a voltage spike induced in a neuronal cell is not dissipated, but is instead regenerated as it moves along the axon. The regeneration of the voltage spike relies on the presence within the membrane of voltage-dependent potassium and sodium channels, which open and close as the membrane depolarizes and repolarizes. Hodgkin and Huxley won the Nobel Prize in 1963 for their discoveries concerning the mechanism of neuronal signal transmission.

A key to the discoveries made by Hodgkin and Huxley was the realization that the propagation of signals in neuronal axons can be very well described in terms of electrical circuits involving **batteries**, **resistors** and **capacitors**. We have already introduced the concept of the membrane as a capacitor, in Section 11.26, where we also referred to the chemical potential difference of metal ions inside and outside the cell acting as a battery or source of electromotive force. The movement of ions through channels in the membrane generates a current, which is characterized by the **conductance** of the channel (that is, the inverse of the electrical **resistance**). Current can also flow longitudinally down the axon of the neuron, which also has an associated conductance or resistance. Using these concepts, we can construct electrical circuits that are analogous to elements of the neuron. We can analyze the properties of such circuits using rules that may be familiar from physics. One such rule is **Ohm's law**, which states that the voltage across a resistor is given by the product of the current and the resistance.

**Conductance and resistance**

The conductance of a channel is the inverse of its electrical resistance, denoted $R$. When describing electrical circuits, it is more common to use resistance because we think of the impedance to current flow due to an element of the circuit. When discussing ion channels, however, it is more intuitive to describe their facilitation of ion flow in terms of the conductance, denoted $g$. The units of resistance and conductance are ohms (Ω) and siemens (S), respectively.

It is not obvious why it is justifiable to treat a neuron as an electrical circuit to which the rules governing macroscopic circuits, such as Ohm's law, are applicable. Ohm's law is not a fundamental law of nature; instead, it is simply an empirical observation that applies to many macroscopic conductors. The movement of charge in neurons is quite different from that in macroscopic conductors, involving as it does the passage of relatively small numbers of ions through highly constricted pathways in ion channels. Nevertheless, it is a fact that the circuit analysis of neurons, as originally developed by Hodgkin and Huxley, goes a long way toward explaining the action potential. We shall simply use it here without making any attempt to bridge the gap between our molecular understanding of ion channels and the application of circuit analysis to the movement of charge. Our treatment is inspired by the simplified and highly readable description of the circuit analysis of neurons provided by Charles Stevens in his book "Neurophysiology: A Primer" (see reading list at the end of this chapter).

## 11.30 The propagation of changes in membrane potential in the axon are described by the cable equation

To develop an analogy between a neuronal axon and a circuit, we shall first assume that the axon is radially symmetric, so there are only two spatial dimensions to consider in analyzing current flow (**Figure 11.47**A). We denote one dimension as $x$, and it represents the longitudinal position along the direction of the central axis of the axon. The other dimension is denoted $y$, and it represents the radial distance from the central axis of the axon.

We are interested in how the membrane potential, $E_m(x,t)$, changes with time, $t$, and $x$, the longitudinal position along the axon. Changes in the membrane potential can be related to two kinds of currents, an axial or longitudinal current, $I_{ax}$, and a membrane current, $I_m$, which is along $y$ (Figure 11.47B).

**Ohm's law**

**Ohm's law** states that the voltage difference, $E$, across a resistor is equal to product of the current, $I$, and the resistance, $R$:

$$E = I \times R$$

Ohm's law can also be stated in terms of the conductance, $g$:

$$I = g \times E$$

The membrane current can be thought of as being confined to planes that are perpendicular to the longitudinal axis of the axon (**Figure 11.48**). A useful conceptualization is to consider the axon as being divided into a series of thin slices or disks (see Figure 11.48 and **Figure 11.49**A). There are two components to the membrane current within each such disk. First, there is the flow of potassium and sodium ions through the ion channels that are specific for each ion. These

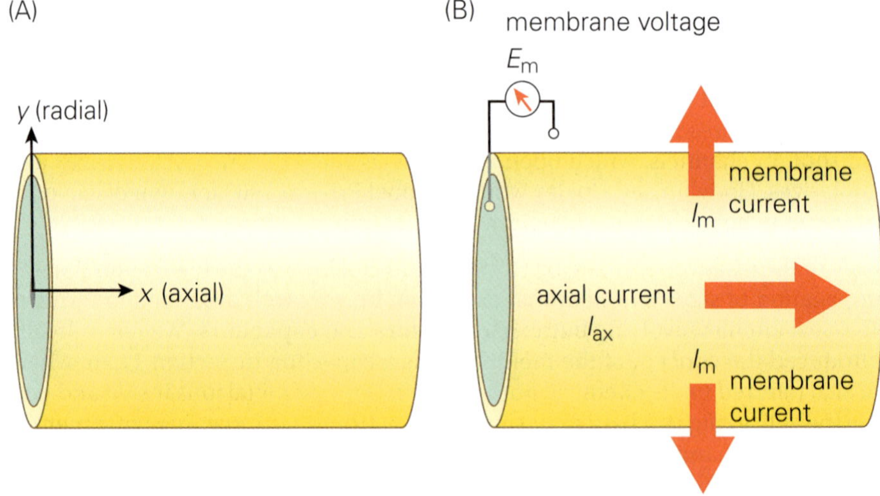


**Figure 11.47 Analysis of currents in the axon.** (A) Coordinate system used in the analysis. The *x* axis represents the axial distance along the axon. The *y* axis represents the radial distance from the axis of the axon. (B) Decomposition of axonal currents into the membrane current, $I_m$, and the axial current, $I_{ax}$. Both currents, as well as the membrane voltage, $E_m$, depend on the axial position, *x*.

currents are referred to as **ionic currents**. Although there are channels that are specific for several other kinds of ions in cells (for example, $Ca^{2+}$ or $Cl^-$), the propagation of action potentials depends only on sodium and potassium current, so we shall ignore all other ions in the subsequent discussion. In addition to the ionic current, there are transient **capacitive currents**, which occur whenever the membrane potential changes. These are due to changes in the number of charges localized at the membrane.

These two types of membrane current within each disk can be described by the electrical circuit shown in Figure 11.49. This **equivalent circuit**, so named because it represents the behavior of the axon, consists of two resistors and two batteries connected in parallel to a capacitor (see Figure 11.49B). The resistor on the left represents all of the potassium channels in the slice of the axon being considered and has a conductance of $g_K$. This resistor is connected in series to a battery that represents the electromotive force generated by the chemical potential difference between potassium ions inside and outside the cell. The positive pole of this battery is towards the outside of the cell, indicating that the potassium ions move outwards spontaneously. The resistor and battery on the right represent the sodium channels, with conductance $g_{Na}$, and the chemical gradient of sodium.

Note that the positive pole of the battery corresponding to the sodium ion flow is switched with respect to that for potassium ions. This assumes that the sodium concentration gradient is inverted with respect to that for potassium, with higher sodium concentrations outside the cell than inside. The capacitor represents the charging of the membrane as ions move through the channels.

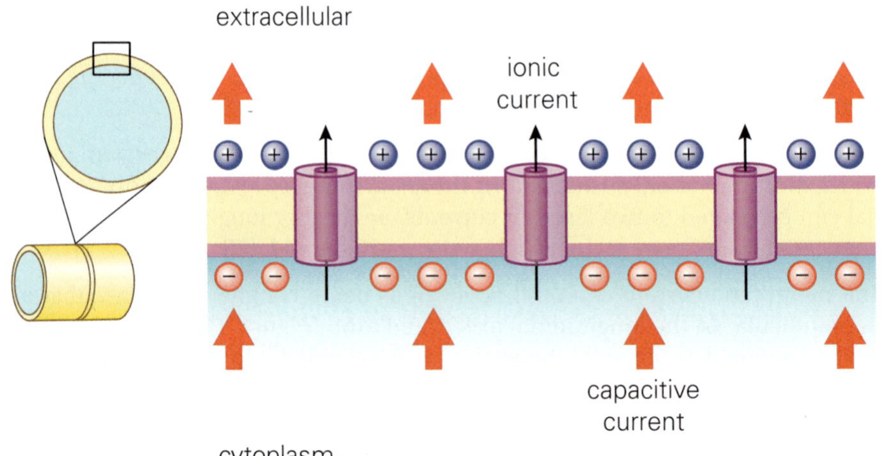


**Figure 11.48 Two types of membrane current in an axon.** The first is an ionic current, indicated by *black* arrows, which refers to the movement of ions through channels (for example, potassium ions through potassium channels). The ionic current occurs when the membrane potential is different from the equilibrium potential for the ion. The second current is a transient capacitive current, indicated by *red* arrows, which arises only when the membrane potential changes. By convention, currents are drawn in the direction of positive charge movement. Because the outer surface of the membrane becomes more positively charged when the membrane depolarizes, the capacitive current behaves as if it is moving from the inside to the outside.

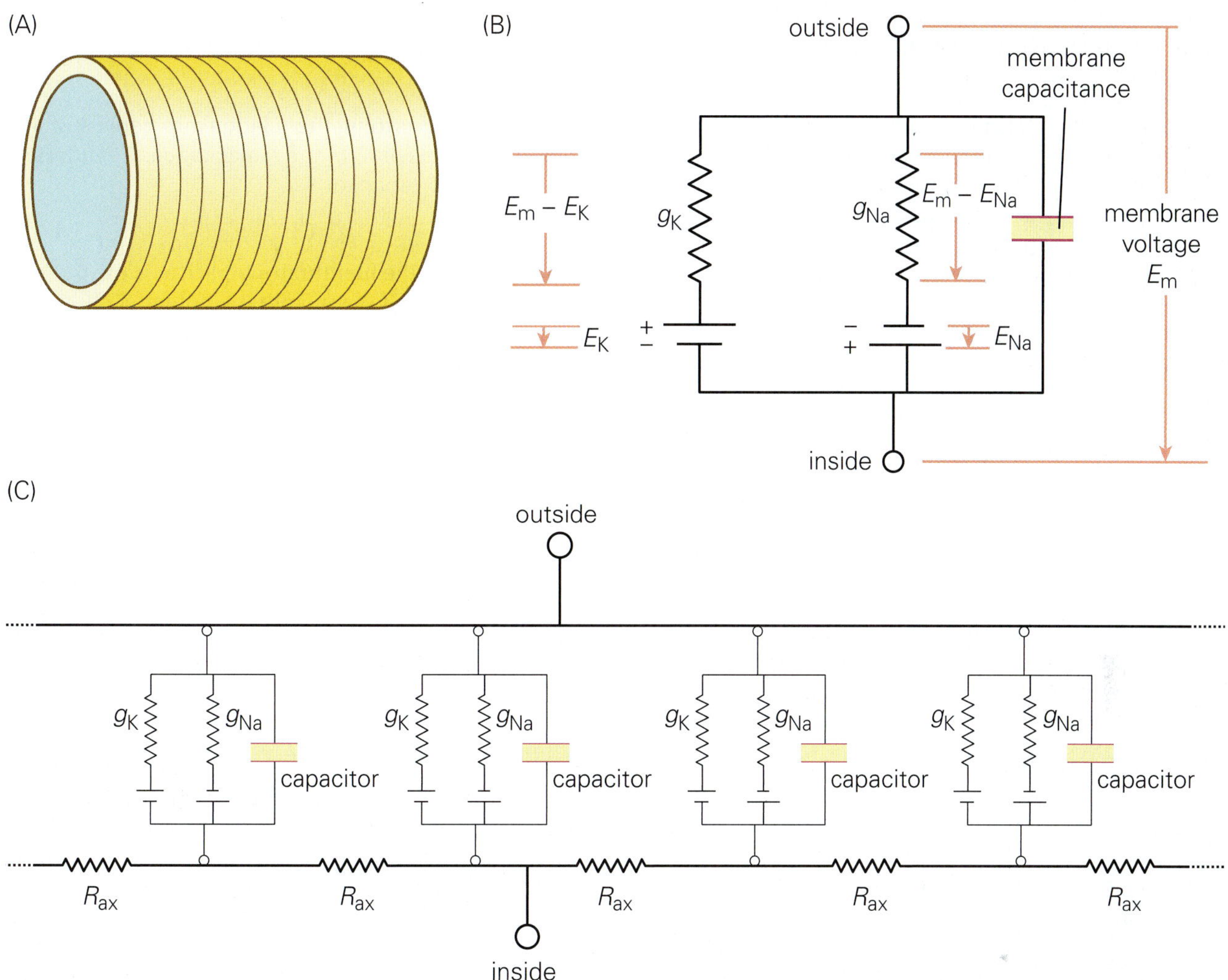


**Figure 11.49 Equivalent circuit for axonal current.** (A) The axon is divided conceptually into a series of adjacent and infinitesimally thin disks. (B) The electrical circuit corresponding to one disk. The circuit consists of two resistors and two batteries, connected in parallel to a capacitor. The resistors on the left and the right represent all of the potassium and sodium channels within one disk, respectively. Each resistor is connected to a battery that represents the electromotive force generated by the chemical potential difference between potassium and sodium ions inside and outside the cell. The capacitance of the membrane is indicated in *yellow*. The voltage across each element of the circuit is indicated. (C) The circuit corresponding to the entire axon is constructed by joining together the circuits for each individual disk. The resistance to axial flow of current is denoted by resistors connecting adjacent circuits, with resistance $R_{ax}$.

The flow of current through the entire axon can by analyzed by joining together a series of such individual circuits, as shown in Figure 11.49C. As explained in **Box 11.1**, analysis of this circuit shows that the membrane current, $I_m$, satisfies the following relationship:

$$\begin{aligned} I_{\mathrm{m}} &= \frac{1}{R_{ax}} \frac{\partial^2 E_{\mathrm{m}}}{\partial x^2} \\ &= C\frac{\partial E_{\mathrm{m}}}{\partial t} + g_{\mathrm{Na}}(E_{\mathrm{m}} - E_{\mathrm{Na}}) + g_{\mathrm{K}}(E_{\mathrm{m}} - E_{\mathrm{K}}) \end{aligned} \quad (11.74)$$

Equation 11.74 is known as the **cable equation**, so-named because it was developed originally to describe macroscopic electrical cables, such as those used to transmit telephone signals. When applied to axons, this second-order differential equation relates the membrane potential, $E_m$, to the conductances of the sodium and potassium channels ($g_{Na}$ and $g_K$) and to their Nernst potentials ($E_{Na}$ and $E_K$).

The conductances, $g_K$ and $g_{Na}$, are the most important parameters in the propagation of action potentials. If $g_K$ and $g_{Na}$ were constant (that is, if the ability of

## Box 11.1 Derivation of the cable equation

We begin by considering an infinitesimally thin disk-like section of the axon, as indicated in Figure 11.48. Within this section, the total membrane current has an ionic component, due to flux through ion channels, and a capacitive component (**Figure 11.1.1**). The ionic currents due to $K^+$ and $Na^+$ are denoted $I_K$ and $I_{Na}$, respectively. The ionic current through the potassium channels is characterized by a resistance, $R_K$, which is the inverse of the aggregate conductance of all of the potassium channels within this disk, $g_K$. Using Ohm's law, we can write:

$$I_K = \frac{[E_m(x,t) - E_K]}{R_K} = g_K[E_m(x,t) - E_K] \quad (11.1.1)$$

The term $[E_m(x, t) - E_K]$ is the difference between the actual membrane potential and the Nernst potential for potassium, and is therefore the electrical driving force for potassium movement across the membrane. The membrane potential, $E_m(x, t)$, depends on both position (it is different from disk to disk) and time, but for simplicity we shall just write it as $E_m$.

The sodium current is given by:

$$I_{Na} = \frac{(E_m - E_{Na})}{R_{Na}} = g_{Na}(E_m - E_{Na}) \quad (11.1.2)$$

The total ionic current, $I_{ionic}$, is the sum of the sodium and potassium currents:

$$\begin{aligned} I_{ionic} &= I_{Na} + I_K \\ &= g_K(E_m - E_K) + g_{Na}(E_m - E_{Na}) \end{aligned} \quad (11.1.3)$$

We also need to consider the transient current that is due to the capacitive charging of the cell membrane when the membrane voltage changes (see Figure 11.42). To understand this term, start with a rearranged form of Equation 11.67 describing the relationship between the charge on the capacitor, $q$, the capacitance, $C$, and the voltage across the capacitor, $V$:

$$q = CV \quad (11.1.4)$$

The voltage across the membrane capacitor is the membrane potential, $E_m$. If the membrane potential changes with time, then the charge on the membrane will also change with time, leading to the capacitive current. Because the capacitance of the membrane does not change, we can write the following expression for $I_{cap}$, the capacitive current:

$$I_{cap} = \frac{\partial q}{\partial t} = \frac{\partial CV}{\partial t} = C\frac{\partial E_m}{\partial t} \quad (11.1.5)$$

The total membrane current, $I_m$, in this thin slice of the axon is then given by combining Equations 11.1.3 and 11.1.5:

$$I_m = C\frac{\partial E_m}{\partial t} + g_{Na}(E_m - E_{Na}) + g_K(E_m - E_K) \quad (11.1.6)$$

Equation 11.1.6 describes the currents in the circuit shown in Figure 11.1.1, which represents one thin slice of the neuron. The current through the entirety of the neuron can be described by linking together a series of such circuits, as shown in Figure 11.48C. Each subcircuit is coupled to the next one by a resistor with resistance $R_{ax}$, which represents the resistance to axial current flow per unit length of axon.

Since $R_{ax}$ is the axial resistance per unit length, the resistance to axial current flow over an infinitesimal distance $dx$ is $R_{ax}dx$. The voltage change over $dx$ is the rate of change of voltage along $x$ multiplied by the distance, which is given by:

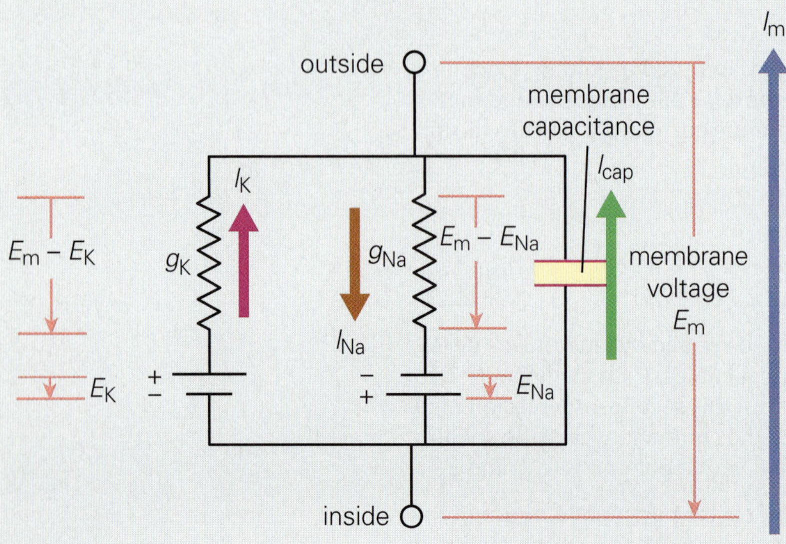


**Figure 11.1.1 Components of the membrane current.** The circuit shown here represents one thin disk in the axon (see Figure 11.48). The total current passing through the membrane, $I_m$, has three components. Two of these, $I_K$ and $I_{Na}$, are the currents passing through the potassium and sodium channels, respectively. The third component, $I_{cap}$, is the capacitive current that results from charging the membrane. The voltages across each of the components of the circuit are indicated in *red*.

$$\text{voltage change over } dx = \left(\frac{\partial E_m}{\partial x}\right) dx \tag{11.1.7}$$

We now apply Ohm's law to calculate the axial current:

$$I_{ax} = \frac{\text{voltage}}{\text{resistance}} = \frac{1}{(R_{ax}dx)}\left(\frac{\partial E_m}{\partial x}\right) dx = \frac{1}{R_{ax}}\left(\frac{\partial E_m}{\partial x}\right) \tag{11.1.8}$$

The axial current, $I_{ax}$, can be related to the membrane current, $I_m$, by noting that all current flow must be accounted for by the sum of the axial current and the membrane current. Thus, if the axial current changes at any point, then the rate of change in the axial current must be equal to the membrane current (**Figure 11.1.2**). This gives us the following relationship, where we have used Equation 11.1.8:

$$I_m = \frac{\partial I_{ax}}{\partial x} = \frac{1}{R_{ax}} \frac{\partial^2 E_m}{\partial x^2} \tag{11.1.9}$$

Combining Equations 11.1.6 and 11.1.9, and explicitly noting the time and space dependence of $E_m$, we get:

$$\begin{aligned} I_m &= \frac{1}{R_{ax}} \frac{\partial^2 E_m(x,t)}{\partial x^2} \\ &= C\frac{\partial E_m(x,t)}{\partial t} + g_{Na}(E_m - E_{Na}) + g_K(E_m - E_K) \end{aligned} \tag{11.1.10}$$

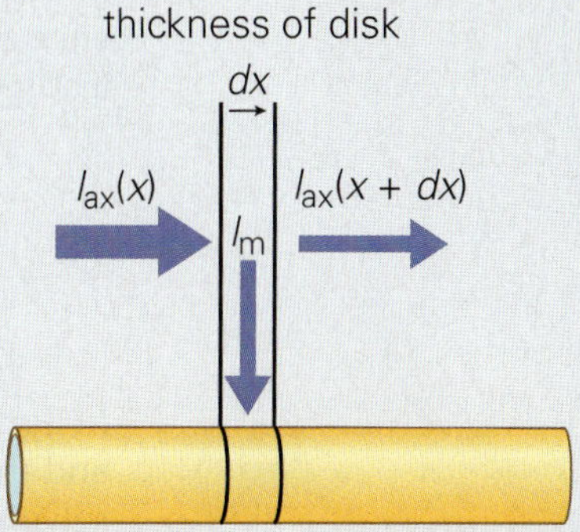

**Figure 11.1.2 Relationship between axial current, $I_{ax}$, and membrane current, $I_m$.** The axial and membrane currents together account for all of the current in the axon. The diagram shows a reduction in the axial current as it crosses an infinitesimally small axial slice of the neuron, of width $dx$. The decrease in axial current per unit distance is the derivative of $I_{ax}$ with respect to $x$:

$$\frac{\partial I_{ax}}{\partial x}$$

This derivative is equal to the membrane current.

Equation 11.1.10 is a general form of the **cable equation**, which is a second-order partial differential equation that describes how the membrane potential varies over the length of the axon and with time.

the channels to allow passage of ions were unchanging), then the axon would be nothing more than a leaky electrical cable in which charge transport occurs via ions rather than electrons. Instead, as we shall see, the conductances of the membrane undergo dramatic changes as a function of the membrane voltage, cycling between states of low conduction and high conduction.

One can model the changes in membrane potential mathematically, as a function of different initial conditions and parameter values, by using the cable equation. A remarkable achievement of Hodgkin and Huxley in the 1950s was their accurate recapitulation of the propagation of action potentials by applying Equation 11.74. We shall not go through such an exercise here, which would require extensive computation, but shall look instead at certain limiting conditions under which the equation can be applied, which will allow us to understand the general behavior predicted by the equation.

## 11.31 The resting membrane potential is determined by a combination of the basal conductances of potassium and sodium channels

In our earlier discussion of the resting membrane potential (see Section 11.25), we considered a cell with only potassium channels. In that case, the resting potential of the cell is simply the Nernst potential of potassium (see Equation 11.62). In reality, the resting potential of the cell is determined by the effect of *all* of the different kinds of open ion channels and the concentration gradients of the corresponding ions. We shall now use the cable equation to calculate the resting membrane potential for a cell containing both sodium and potassium channels.

When the cell is at rest (that is, when there are no voltage spikes), the sodium and potassium conductances do not vary with time. There are always some sodium and potassium channels that are open, and these set the basal or resting membrane potential. In the absence of a stimulating potential, both the time dependence and space dependence of the membrane potential vanish in Equation 11.74, leading to the following simplified relationship:

$$\frac{1}{R_{ax}}\frac{\partial^2 E_m}{\partial x^2} = 0 = g^*_{Na}(E^*_m - E_{Na}) + g^*_K(E^*_m - E_K) \tag{11.75}$$

Thus,

$$g^*_{Na}(E^*_m - E_{Na}) = -g^*_K(E^*_m - E_K) \tag{11.76}$$

In Equation 11.75 we have used asterisks to denote the fact that the conductances and the membrane potentials correspond to the resting state. Rearranging and solving for $E^*_m$, we get:

$$E^*_m = \left(\frac{g^*_{Na}}{g^*_{Na} + g^*_K}\right)E_{Na} + \left(\frac{g^*_K}{g^*_{Na} + g^*_K}\right)E_K \tag{11.77}$$

It turns out that the potassium conductivity of resting neurons is about 100 times greater than the sodium conductivity. This is because there are potassium channels in the cell membrane that are not voltage gated and are constitutively open (there are very few such sodium channels). Thus, according to Equation 11.77, the resting membrane potential is essentially equal to the equilibrium potential for potassium, which we estimated to be about –90 mV (see Equation 11.65).

## 11.32 The propagation of a voltage spike without triggering voltage-gated ion channels is known as passive spread

Consider what happens when a voltage spike is propagated without activation of voltage-gated ion channels. As we shall see in the following sections, the cable equation predicts that the perturbation in the membrane potential induced by the voltage spike will be attenuated as it travels away from the point of initiation (**Figure 11.50**). The attenuation of the voltage spike is due to the diffusive spread of the excess positive charge and the leakage of ions through ion channels, as illustrated in Figure 11.46. This process, in which the voltage spike is not regenerated as it traverses the axon, is known as **passive spread**.

**Passive spread**

The propagation of a voltage spike without regeneration is known as passive spread. Passive spread occurs through diffusion, and the amplitude of the action potential decreases during this process.

Because no voltage-dependent channels are activated during passive spread, the sodium and potassium conductances of the membrane stay constant during the process. Under these conditions, the form of the cable equation can be simplified somewhat.

It is convenient to define the membrane voltage as the deviation from the resting potential, $E^*_m$, as follows:

$$V(x,t) = E_m(x,t) - E^*_m \tag{11.78}$$

Here $V(x,t)$ is a perturbation in the voltage—that is, the extent to which the membrane voltage differs from the resting value. Note that both the membrane voltage, $E_m(x,t)$, and the perturbed voltage, $V(x,t)$, vary with position and time, but we shall omit this dependence for simplicity.

From Equation 11.78, it follows that:

$$E_m = V + E^*_m \tag{11.79}$$

Substituting this expression for the membrane potential into the cable equation (11.74), we get:

$$\frac{1}{R_{ax}}\frac{\partial^2 V}{\partial x^2} = C\frac{\partial V}{\partial t} + (g^*_{Na} + g^*_K)V + (g^*_{Na} + g^*_K)E^*_m - (g^*_{Na}E_{Na} + g^*_K E_K) \tag{11.80}$$

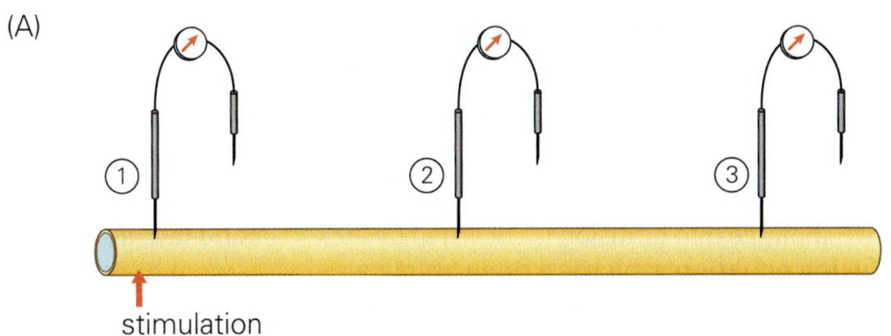


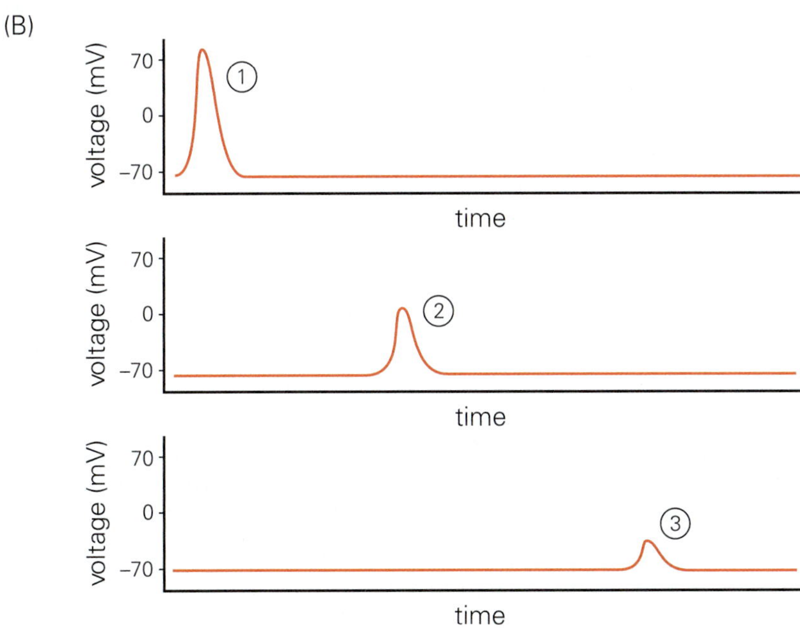


**Figure 11.50 Passive spread of a voltage spike.** (A) A section of an axon is shown, with the point at which a voltage spike is generated indicated by a *red* arrow. The membrane voltage is measured at three positions along the axon, marked 1, 2, and 3. There are no voltage-dependent channels in this section of the axon. (B) Graphs of the membrane voltage as a function of time at each position.

By rearranging Equation 11.77 we obtain:

$$(g^*_{\text{Na}} + g^*_{\text{K}})E^*_{\text{m}} = g^*_{\text{Na}}E_{\text{Na}} + g^*_{\text{K}}E_{\text{K}} \tag{11.81}$$

According to Equation 11.81, then, the last two terms in Equation 11.80 cancel each other, leaving us with the following simplified form of the cable equation:

$$\begin{aligned}\frac{1}{R_{\text{ax}}}\frac{\partial^2 V(x,t)}{\partial x^2} &= C\frac{\partial V(x,t)}{\partial t} + (g^*_{\text{Na}} + g^*_{\text{K}})V(x,t)\\ &= C\frac{\partial V(x,t)}{\partial t} + \frac{1}{R^*_{\text{m}}}V(x,t)\end{aligned} \tag{11.82}$$

To obtain the second equality in Equation 11.82, we have used the fact that the resistance of the membrane at equilibrium, $R^*_{\text{m}}$, is the inverse of the sum of the sodium and potassium conductances, $g^*_{\text{Na}} + g^*_{\text{K}}$.

In order to derive Equation 11.82, we have assumed that the sodium and potassium conductances are equal to their resting values (that is, $g^*_{\text{Na}}$ and $g^*_{\text{K}}$). As a consequence, Equation 11.82 cannot be used to analyze the effect of the activation of voltage-dependent ion channels. If these are activated, then the original form of the cable equation must be used (Equation 11.74), with the conductances $g_{\text{K}}$ and $g_{\text{Na}}$ taking on time-dependent forms.

## 11.33 If membrane currents are neglected, then the cable equation is analogous to a diffusion equation

The first extreme condition we shall examine is to see what happens if we neglect the ionic currents through the sodium and potassium channels. In this case, the

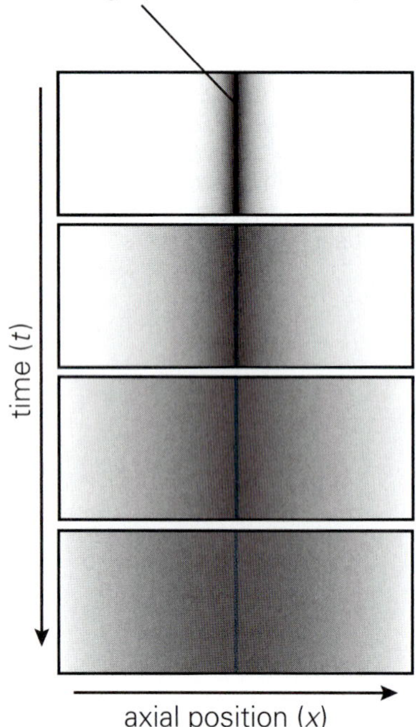


**Figure 11.51 Analogy between molecular diffusion and the spread of a perturbation in the membrane voltage.** Shown here is a series of snapshots at different times of a chamber or, equivalently, a section of an axon. The shading represents the concentration of solute molecules, $c(x,t)$, or the perturbation in the membrane voltage, $V(x,t)$. As time progresses, the concentration gradient dissipates, as does the voltage perturbation. If we ignore the presence of ion channels in the membrane, then the dissipation of the voltage perturbation is governed by the diffusion equation.

second term on the right-hand side of Equation 11.82 is omitted, and we obtain the following modified form of the cable equation:

$$\frac{\partial^2 V(x,t)}{\partial x^2} = R_{ax} C \frac{\partial V(x,t)}{\partial t} \tag{11.83}$$

In Chapter 17 we discuss the diffusion of molecules along concentration gradients, and we shall see that diffusion is governed by **Fick's second law**, also known as the **diffusion equation**, which is written as follows:

$$\frac{\partial c(x,t)}{\partial t} = D \frac{\partial^2 c(x,t)}{\partial x^2} \tag{11.84}$$

Here, $c(x,t)$ is the concentration of a solute at position $x$ and time $t$. As explained in Chapter 17, $D$ is the diffusion constant of the solute (we have encountered the diffusion constant in Section 11.19), and it is related to how fast the solute molecules diffuse through the solvent. Rearranging Equation 11.84, we get:

$$\frac{\partial^2 c(x,t)}{\partial x^2} = \frac{1}{D} \frac{\partial c(x,t)}{\partial t} \tag{11.85}$$

Comparing Equations 11.83 and 11.85, we can see that they have the same form, with the voltage, $V$, in Equation 11.83 replaced by the concentration of the solute, $c$, in Equation 11.85. That the two equations are connected makes sense, because the perturbation in the membrane voltage, $V$, is determined by the excess concentration of positively charged ions. We can also see that the term $R_{ax}C$ in the cable equation, which is the product of the axial resistance of the neuron and the membrane capacitance, is analogous to the inverse of the diffusion constant, $D$. This point is important in understanding how the structure of neurons is optimized to speed the transmission of action potentials.

The diffusion equation describes how the concentration of a solute changes with position and time. Chapter 17 explains that the solution to the diffusion equation is given by:

$$c(x,t) = \frac{\alpha}{\sqrt{Dt}} \exp\left(\frac{-x^2}{4Dt}\right) \tag{11.86}$$

Here, $\alpha$ is a constant that ensures that the solution to the diffusion equation conserves the mass of the solute—the details need not concern us here, and are gone into more fully in Chapter 17. The general behavior of the function given in Equation 11.86 is illustrated in **Figure 11.51**. If we start with a thin band of concentrated solute in the center of a chamber, the solute molecules spread away from the center with time, as indicated by the shading in the drawing. By analogy, if the rectangle represents a section of an axon, with a voltage perturbation or spike represented by the shading, then the voltage perturbation will attenuate in the center with time and spread through the rest of the neuron.

The rate of spread of solute molecules is determined by the diffusion constant. Likewise, the rate of spread of a voltage perturbation is determined by the value of $(R_{ax}C)^{-1}$, which plays a role analogous to the diffusion constant. This is illustrated in **Figure 11.52**, which shows how the voltage perturbation spreads with time for two different values of $(R_{ax}C)^{-1}$. The graphs in Figure 11.52 are simply one-dimensional representations of the distributions shown in Figure 11.51. Initially, the voltage perturbation is localized close to the initial position, but with time it spreads out. If the value of $(R_{ax}C)^{-1}$ is increased by a factor of five, the voltage perturbation is seen to spread much more rapidly. Notice that when the value of $(R_{ax}C)^{-1}$ is 1.0 (using arbitrary units), there is very little change in the voltage at a position three units away from the center at any of the time points that are graphed. In contrast, when the value of $(R_{ax}C)^{-1}$ is 5.0, there is an even more appreciable increase in the voltage perturbation at this position for the second two time points.

The important point of this analysis is that in order to increase the speed with which an action potential or voltage perturbation travels through a neuron the

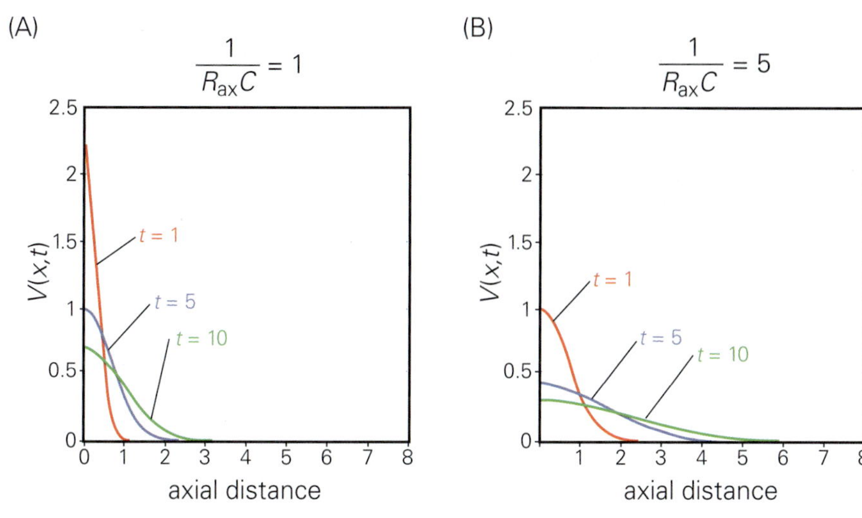


**Figure 11.52 The effect of the "diffusion constant" on the rate of spread of the voltage perturbation.** The dissipation of a voltage perturbation with time is graphed here, as predicted by the solution to the diffusion equation and using arbitrary units (Equation 11.85). The factor ($1/R_{ax}C$) is analogous to the diffusion constant, $D$, in Equation 11.85. (A) The value of $1/R_{ax}C$ is set to 1.0. (B) The value of $1/R_{ax}C$ is set to 5.0. Note the much faster spread of the voltage perturbation in this case.

value of $(R_{ax}C)^{-1}$ should be made as large as possible. Thus, for optimal transmission, the axial resistance and the membrane capacitance should both be as low as possible.

## 11.34 Leakage through open ion channels limits the spread of a voltage perturbation

The second extreme situation we consider is one in which current flux through the axon achieves a steady state, with no time dependence in the membrane voltage. Imagine that a stimulating electrode delivers a constant current at one point along the axon, as illustrated in **Figure 11.53**A. As soon as the current is switched on, the membrane will depolarize, and the depolarization will spread through the axon. If the current is maintained at a constant level, then eventually a steady-state situation will be reached, where the input current is balanced by the outward flux of ions through open ion channels, and the membrane potential at each position along the axon will be constant with time.

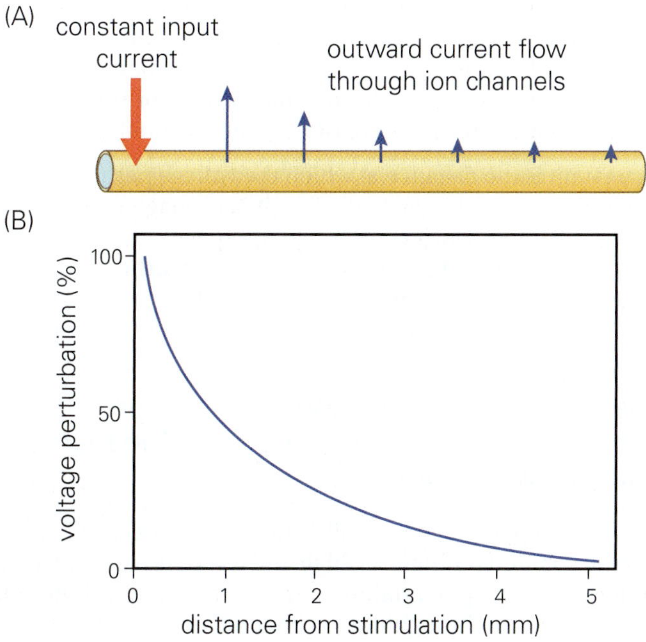


**Figure 11.53 Effect of stimulation with constant input current.** (A) The *red* arrow indicates an electrode delivering a constant input current. Because the membrane depolarizes, open ion channels in the membrane will allow fluxes of ions (ionic currents) out of the membrane, indicated by *blue* arrows. (B) According to the cable equation, the voltage perturbation decreases exponentially with distance from the point of stimulation, and so does the ionic current. (B, adapted from C.F. Stevens, Neurophysiology: A Primer. New York: John Wiley & Sons, 1966. With permission from John Wiley & Sons, Inc.)

A simple analogy might help explain why a steady-state situation is reached under these circumstances. Imagine a very long garden hose connected to a tap at one end and closed at the other end. The hose has numerous small holes along its length, through which water can flow out of the hose. When the tap is turned on, water flows into the hose at a constant rate, and pressure builds up within the hose. The pressure is highest close to the tap, but dissipates along the length of the hose because water flows out of the holes in the hose. A steady state is achieved in a short time after turning on the tap, with water leaking out of the hose at a high rate near the tap and with decreasing rate as the distance from the tap increases.

Because the membrane potential in the axon does not vary with time under the steady-state condition, we can set the time derivative to zero in the cable equation to obtain the following expression:

$$\frac{1}{R_{\text{ax}}}\frac{\partial^2 V(x)}{\partial x^2}=\frac{1}{R_{\text{m}}^*}V(x,t) \tag{11.87}$$

This second-order differential equation can be solved by imposing the boundary condition that the voltage perturbation, $V(x,t)$, goes to zero at very large distances from the point of stimulation. The solution to the equation is then given by:

$$V(x)=V(0)\exp\left(\frac{-x}{\alpha}\right) \tag{11.88}$$

where $V(0)$ is the voltage perturbation at the site of stimulation, $\alpha$ is a scale constant, and $x$ is the distance from the site of stimulation ($x$ is assumed to be always positive). According to Equation 11.88, the voltage perturbation decays exponentially from the site of stimulation, as shown graphically in Figure 11.53B.

The parameter $\alpha$ is known as the **space constant** of the axon, because it determines the distance over which the voltage perturbation decays. The value of $\alpha$ is given by the square root of the ratio of the equilibrium membrane resistance and the axial resistance:

$$\alpha=\sqrt{\frac{R_{\text{m}}^*}{R_{\text{ax}}}} \tag{11.89}$$

Equations 11.88 and 11.89 can be verified by taking the second derivative of $V(x)$ and seeing that it satisfies Equation 11.87.

Equation 11.88 tells us that the greater the value of $\alpha$, the greater the spread of the voltage perturbation. Thus, given Equation 11.89, the transmission of the signal can be increased in two ways. Decreases in the axial resistance of the axon, $R_{\text{ax}}$, will increase the distance over which the signal is transmitted. Likewise, the signal will also increase with an increase in the equilibrium membrane resistance, $R_{\text{m}}^*$, which corresponds to a decrease in the conductivity of sodium and potassium through the membrane (that is, decreased leakage of ions).

The space constant is the distance at which the voltage perturbation decreases to ~37% of the initial value (that is, $1/e$) when it is propagated by passive diffusion alone. Measurements on neurons show that the space constant is in the range of a few millimeters, as indicated in Figure 11.53. Most neuronal axons in the human body are much longer than a few millimeters, and so passive spread cannot, by itself, account for the transmission of action potentials in these axons.

## 11.35 The time taken to develop a membrane potential is determined by the conductance of the membrane and its capacitance

Consider the first step in the process illustrated in Figure 11.42, in which a cell is moved between solutions containing high and low potassium ion concentrations. Because of the chemical potential difference across the cell membrane, potassium ions begin to move through the ion channels, as illustrated in **Figure 11.54**.

As ions flow out of the cell, the membrane capacitor begins to charge because negative and positive ions move towards the inner and outer faces of the membrane, respectively (see Figure 11.54). The process stops when the membrane can no longer accommodate additional ions in its vicinity, due to the charge repulsion between the ions. The total charge, $q$, held by the membrane capacitor when equilibrium is reached is given by $q = C \times V$, where $C$ is the capacitance of the membrane and $V$ is the voltage across it (see Equation 11.67).

The rate at which ions flow through these channels is determined by the conductance of the channel. We denote the aggregate conductance of all of the potassium channels in a region of the membrane by $g_K$, which is the inverse of the resistance of the membrane to potassium ion flow, $R_K$. The time taken to develop the membrane potential will depend on the resistance, $R_K$, and also on the total number of charges that have to move across the membrane to establish the equilibrium

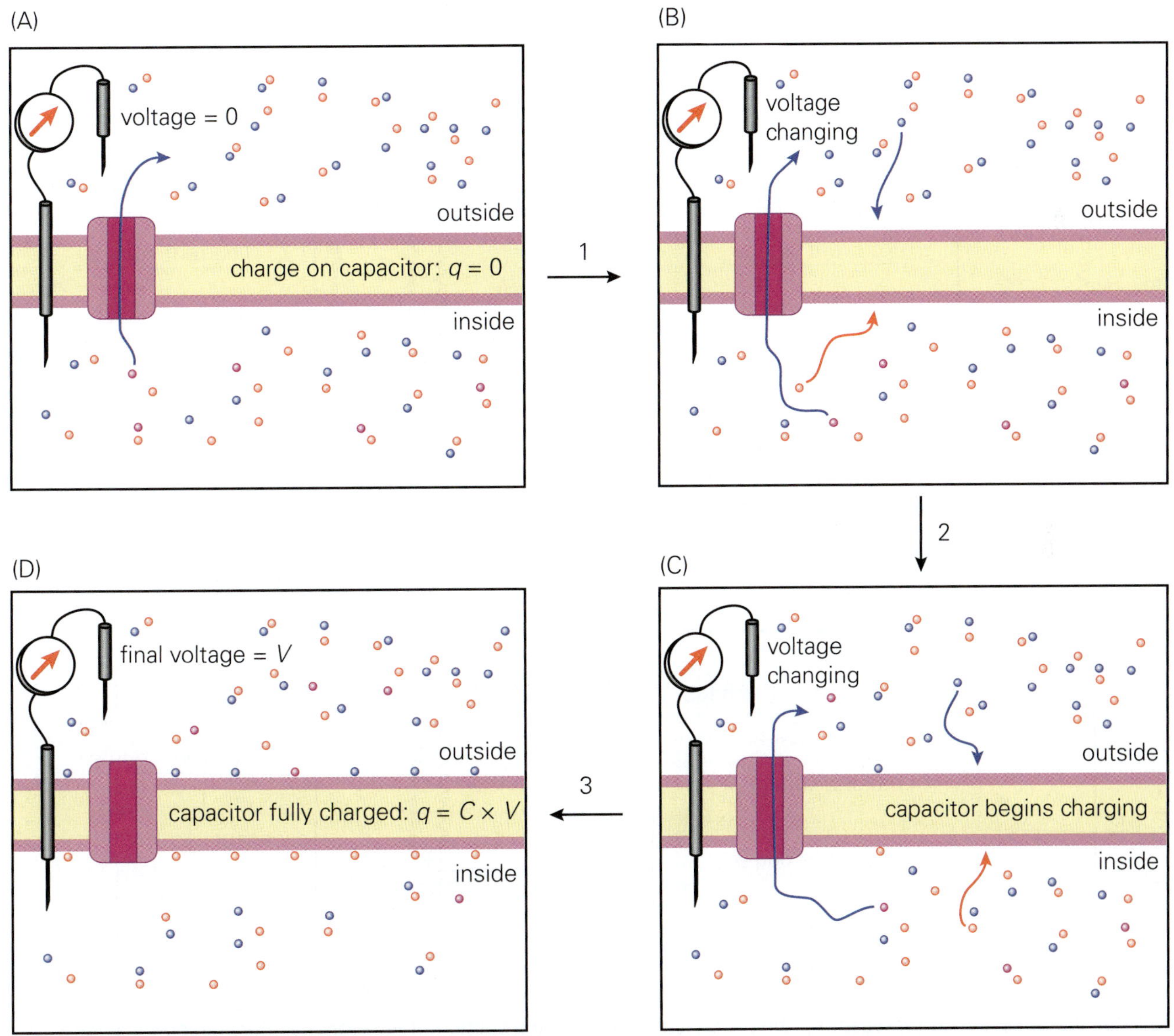


**Figure 11.54 The coupling between ion flux and capacitive charging of the membrane.** A small section of the membrane of a cell is shown in *yellow*, with a potassium channel shown in *purple*. The $K^+$ concentration is higher inside than outside. (A) The membrane potential is initially zero, and $K^+$ ions (*purple*) begin to move out. (B and C) The outward movement of $K^+$ ions results in excess negative (*red*) and positive (*blue*) ions inside and outside, respectively. These ions attract each other across the membrane and are localized near it, charging the membrane capacitor. (D) The capacitor is fully charged when no more ions can be accommodated due to electrostatic repulsion, so net ion flow stops.

voltage. The higher the capacitance of the membrane, the greater the number of charges that have to move, and the longer the charging process will take.

This process can be represented by the circuit shown in **Figure 11.55**, which consists of a resistor and a battery connected in parallel to a capacitor. One experimental technique to study membrane capacitance is to introduce an electrode that runs axially through the neuron, in order to make the electric potential uniform along the axis of the neuron. In this case, there is no axial current, and the membrane current is equal to the current that is applied through the external electrode, $I_{app}$. The cable equation (Equation 11.74) can then be written as:

$$I_m = I_{app} = C\frac{\partial V(t)}{\partial t} + \frac{1}{R_K^*}V(t) \tag{11.90}$$

In Equation 11.90 $R_K^*$ is the aggregate resistance of the potassium channels in the membrane. We are assuming that the flow through the sodium channels can be ignored, as long as voltage-gated channels are not activated, and so $R_K^*$ is the same as the membrane resistance, $R_m^*$. Now consider what happens if the neuron is

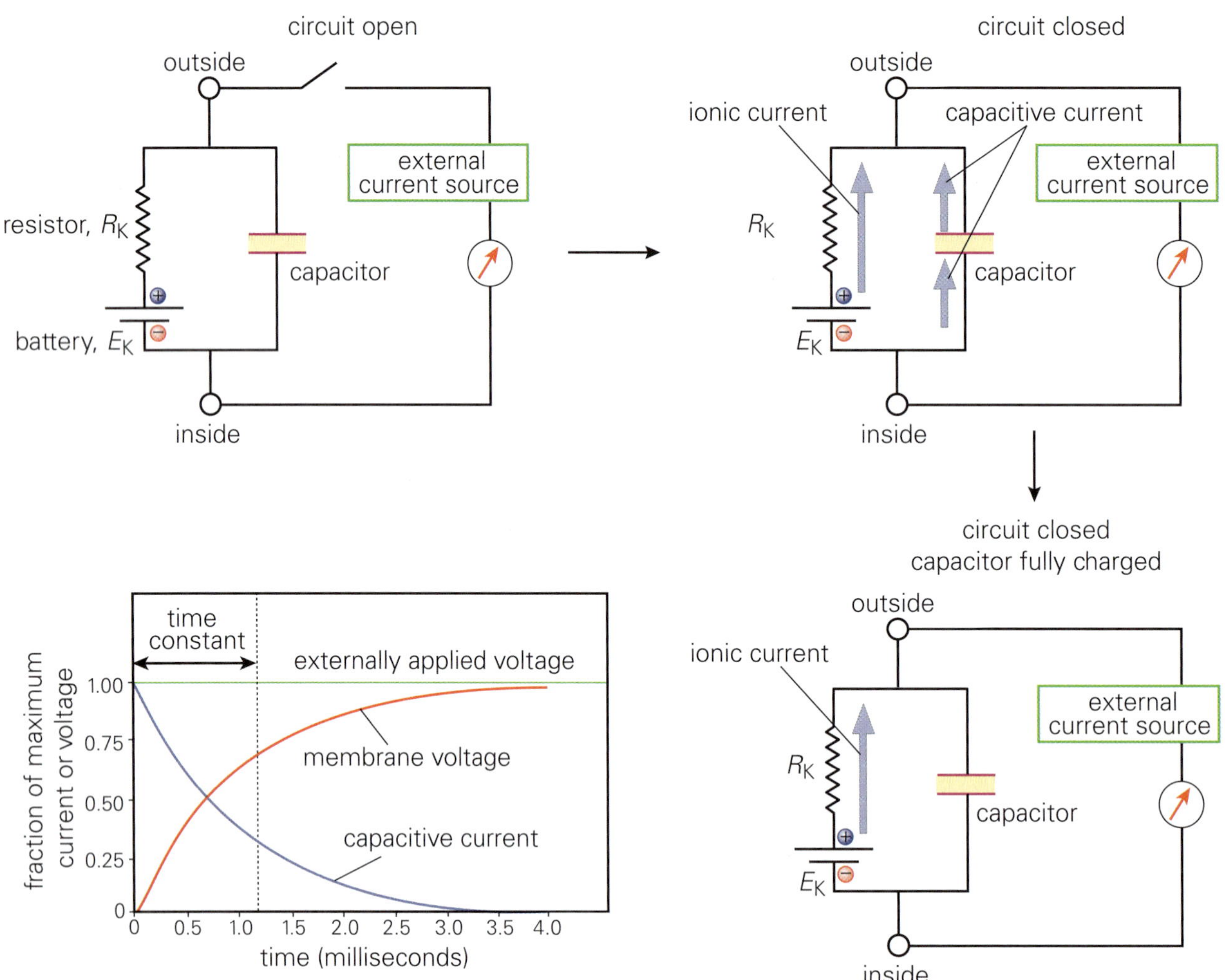


**Figure 11.55 Electrical circuit representing the membrane potential in a section of an axon.** A battery representing the electromotive force generated by the potassium concentration gradient ($E_K$) is connected in series to a resistor representing all of the potassium channels in the membrane ($R_K$), as shown in the upper left. A capacitor representing the membrane is connected in parallel to the resistor and the battery. An external current source and voltmeter are part of an experimental apparatus probing the axon and are shown on the right. When the circuit is closed, a capacitive current is observed until the capacitor is fully charged. The graph shows the time-dependent changes in the voltage across the membrane and the capacitive current, in response to an externally applied voltage. The membrane voltage increases from its initial value to a final value equal to the applied voltage. The time constant of the membrane is the time required for the capacitive current to decrease to ~37% (1/*e*) of its initial value.

stimulated by a pulse in which the applied current is changed and held at a constant value. What will be the response of the membrane potential? When current is first applied to this circuit, it flows through both branches of the circuit until the capacitor is fully charged (see Figure 11.55). At that point, current flows only through the resistor and the final voltage, $V_f$, is given by Ohm's law:

$$V_f = I_m \times R_K^* = \frac{I_m}{g_K^*} \qquad (11.91)$$

If the applied current is constant, you can verify that the solution to Equation 11.90 is given by:

$$V(t) = V_f\left[1 - \exp\left(\frac{-t}{R_K^* C}\right)\right] \qquad (11.92)$$

Thus, the voltage across the capacitor (that is, the membrane voltage) takes some time to reach its final value. The term $R_K^*C$ is known as the **time constant** of the circuit, and it is the time taken for the voltage to reach ~63% of its final value, or for the capacitive current to decrease to 37% of its initial value. Resting cells have time constants that range from 10 microseconds to 1 second.

The time constant for a typical neuron is on the order of milliseconds. To see why this is so, recall from Section 11.19 that open potassium or sodium channels conduct ions at the rate of $10^7$–$10^8$ ions per second. Let us denote the number of potassium channels in a square centimeter of membrane by $N$. If these are the only channels that are open, then the membrane current per square centimeter, $I_m$, is given by:

$$\begin{aligned} I_m &= \text{charge moved per second} \\ &= \frac{(\text{number of ions moved per second per channel})(\text{Faraday constant})(\text{number of ion channels})}{\text{Avogadro's number}} \\ &= \frac{(10^8)(96{,}500)(N)}{6.023\times10^{23}} = (1.6\times10^{-11})(N)\,\text{C}\bullet\text{sec}^{-1} \end{aligned} \qquad (11.93)$$

In Equation 11.93 we have assumed that each channel conducts $10^8$ ions per second.

According to Ohm's law, the conductance of the membrane, $g_K$, is the current divided by the membrane potential, $E_m$. Assuming that the membrane potential is 0.1 V, then the conductance is given by:

$$g_K^* = \frac{I_m}{E_m} = \frac{(1.6\times10^{-11})(N)}{0.1} = (1.6\times10^{-10})(N)\ \text{siemens} \qquad (11.94)$$

If there are 100,000 open ion channels per square centimeter of membrane (that is, if $N = 10^5$), then according to Equation 11.94, the conductance is $1.6 \times 10^{-5}$ siemens•cm$^{-2}$. The resistance of the membrane, $R_K^*$ is simply the inverse of this and is, therefore, 60,000 Ω•cm$^{-2}$. Using the typical value for the capacitance of the membrane (that is, 1 microfarad per square centimeter: $1 \times 10^{-6}$ F •cm$^{-2}$), the time constant of the membrane is given by:

$$\tau = R_K^*C = (60{,}000)(1\times10^{-6}) = 0.06\ \text{sec} = 60\ \text{msec} \qquad (11.95)$$

In order to speed up the response of the neuron, the time constant should be made smaller. This can be achieved by decreasing the membrane resistance, which speeds up the flow of ions across the membrane, or by decreasing the capacitance, which decreases the number of charges that need to move across the membrane to establish a given membrane potential.

## 11.36 Myelination of mammalian neurons facilitates the transmission of action potentials

The preceding analysis of the cable equation tells us that there are several ways in which the transmission of voltage spikes can be increased during the process of passive spread. One is to decrease the axial resistance of the neuron, and the other is to decrease the capacitance. If the membrane resistance is increased, then this will increase the spatial spread of the signal, although it will also increase the time required to charge the membrane capacitance.

The simplest way to decrease the axial resistance of the axon is to make it thicker, because the resistance of a conductor is inversely proportional to its cross-sectional area. This is the solution achieved by the neurons of simple organisms—we had noted earlier that the giant axon of the squid is roughly 500 μm in diameter (see Section 11.28). This would be impractical, however, for a complex organism, such as a mammal, whose body plan requires that a great many neurons be packed closely together. The human brain, for example, is estimated to have as many as 100 billion neurons in it, and these have to be as thin as possible while still allowing rapid signaling.

As we discuss in Section 11.37, a key mechanism for the transmission of action potentials is their active regeneration as they proceed down the axon. But this process consumes a great deal of energy, and there is an advantage to maximizing the extent to which action potentials are transmitted by passive spread before they are regenerated. The solution that nature has found to is to reduce the capacitance of the neuronal membrane, as well as its conductance to sodium and potassium. This is accomplished by coating the axon in a highly insulating layer that is generated by wrapping a sheath known as **myelin** around it (**Figure 11.56**). The axon is divided into millimeter long segments that are wrapped in myelin sheaths. Between these segments are short stretches of unmyelinated axon, known as **nodes of Ranvier**, named after the French scientist Louis-Antoine Ranvier, who discovered these interruptions in the myelin sheath in the nineteenth century. The nodes of Ranvier contain voltage-dependent sodium and potassium channels that are crucial for regenerating the action potential, as we discuss in Section 11.37.

The capacitance of the surface of a myelinated axon is about 1000 times smaller than that of an unmyelinated neuron, thereby increasing the effective diffusion constant ($R_{ax}C$) for propagation of the voltage spike by 1000-fold (see Section 11.32). Because the myelin sheath is an insulator, the conductance of the membrane to sodium and potassium is also decreased, which has the effect of increasing the space constant for passive spread (see Section 11.34). Both of these factors help increase the speed and extent of the transmission of a voltage spike.

## 11.37 Action potentials are regenerated periodically as they travel down the axon

Even with the boost provided by myelination, action potentials cannot be transmitted very far by passive spread and will die out within a few millimeters of their point of initiation. The transmission of an action potential in a myelinated neuron is illustrated in **Figure 11.57**. An action potential generated in the first node of Ranvier, shown on the left, enters a myelinated section of the axon, within which it is transmitted through passive spread. The strength of the action potential is attenuated as it moves through the myelinated section. When it enters the second

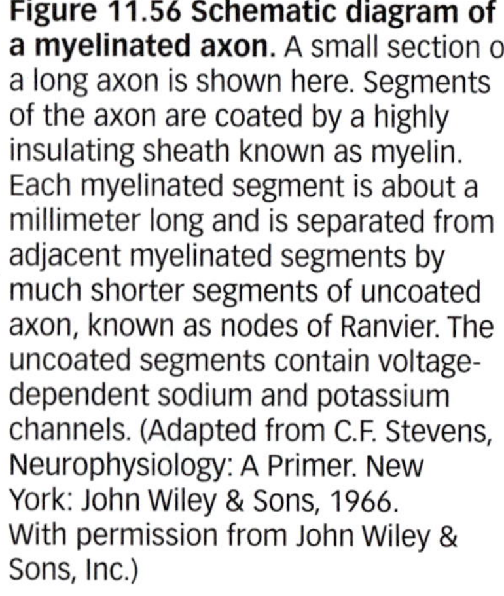

**Figure 11.56 Schematic diagram of a myelinated axon.** A small section of a long axon is shown here. Segments of the axon are coated by a highly insulating sheath known as myelin. Each myelinated segment is about a millimeter long and is separated from adjacent myelinated segments by much shorter segments of uncoated axon, known as nodes of Ranvier. The uncoated segments contain voltage-dependent sodium and potassium channels. (Adapted from C.F. Stevens, Neurophysiology: A Primer. New York: John Wiley & Sons, 1966. With permission from John Wiley & Sons, Inc.)

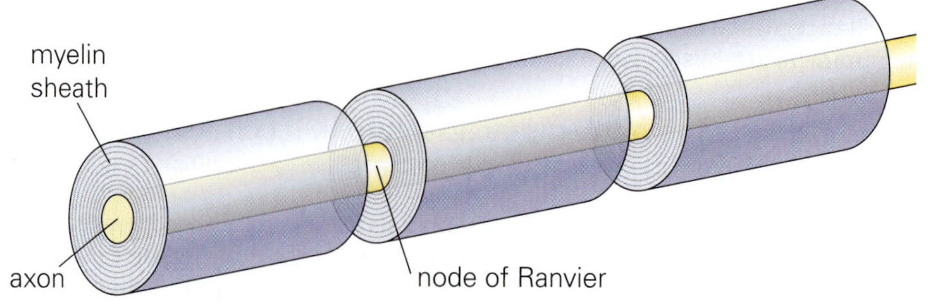

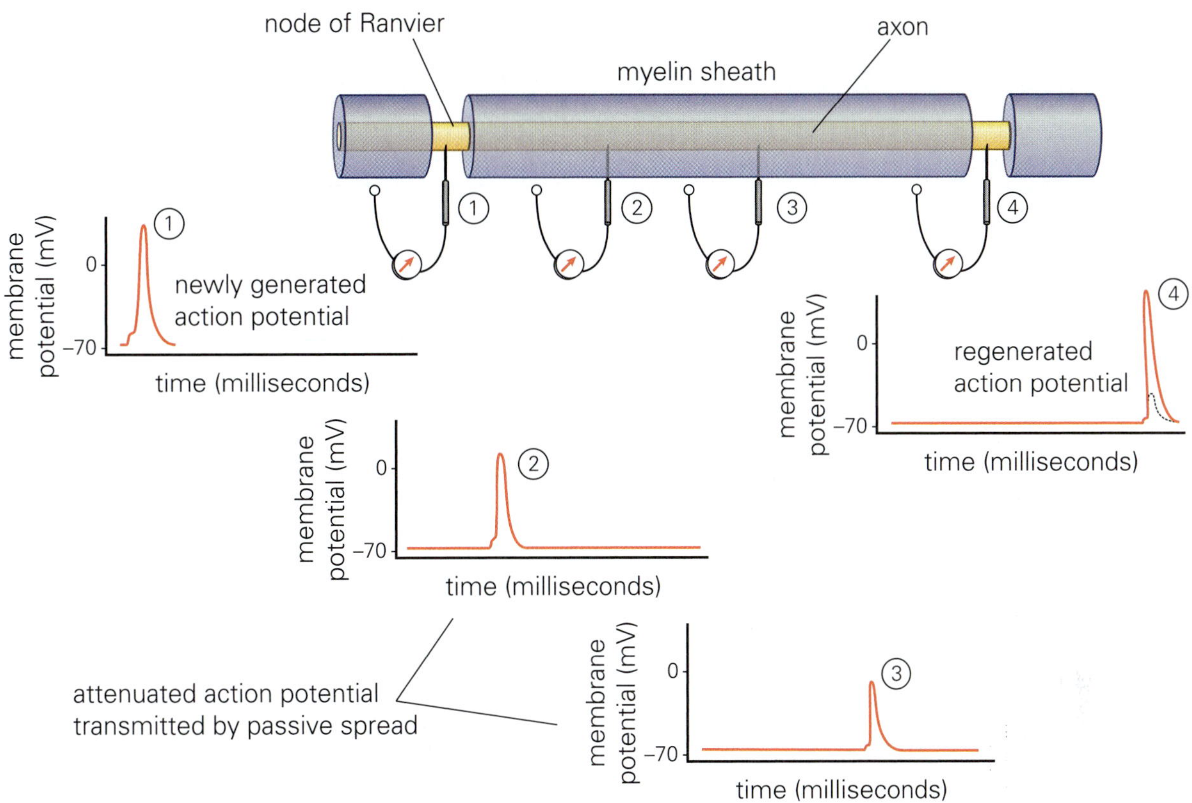


node of Ranvier, a remarkable thing happens, which is that the action potential is regenerated so that it regains full strength. The regeneration of the action potential is due to the opening of voltage-gated sodium and potassium channels, as we discuss below. This process is repeated as the action potential transits through each myelinated segment.

**Figure 11.57 The transmission of action potentials.** A section of a myelinated axon is shown here. Electrodes are inserted into the axon at various points to measure the progress of an action potential that is moving from left to right. A newly generated action potential is formed at the first node of Ranvier (at the position marked 1). It is transmitted through the myelinated section of the neuron by passive spread, during which it is attenuated (positions 2 and 3). At the next node of Ranvier the attenuated action potential triggers the opening of voltage-gated sodium and potassium channels, which regenerate the action potential (position 4). In the last graph, the *gray* line indicates the expected strength of the action potential if it had not been regenerated. (Adapted from C.F. Stevens, Neurophysiology: A Primer. New York: John Wiley & Sons, 1966. With permission from John Wiley & Sons, Inc.)

The repeated renewal of the action potential results in the signal being transmitted without any reduction in strength as it moves down the axon. The transmission of action potentials in unmyelinated neurons occurs through a similar process of regular regeneration, except that the voltage-gated ion channels are distributed more evenly throughout the axon.

Experiments in which an axon is stimulated by an electrode inserted at a specific point reveal that the response of the axon is different depending on whether the stimulus results in hyperpolarization (that is, the membrane potential becomes more negative than normal) or depolarization (the membrane potential becomes less negative). The basic set-up of the experiment is illustrated in **Figure 11.58**A and consists of a stimulating electrode and an electrode that reads out the membrane potential at a point close to the point of stimulations.

When the stimulus is hyperpolarizing, the membrane potential becomes more negative than the resting level, and the response is proportional to the strength of the stimulation (Figure 11.58B). This is consistent with a passive spread of the voltage perturbation from the point of stimulation to the point of measurement.

Something quite different happens when the stimulations are depolarizing (Figure 11.58C). In this case, when the stimulus intensity is small, the response is again proportional to the strength of the stimulus. But, when the stimulus intensity crosses a threshold level, there is a strong response from the axon, and membrane potential becomes greatly depolarized (that is, positive). If a series of stimuli are delivered that are all above the threshold level, then the response is the same in each case, with the extent of depolarization of the membrane being

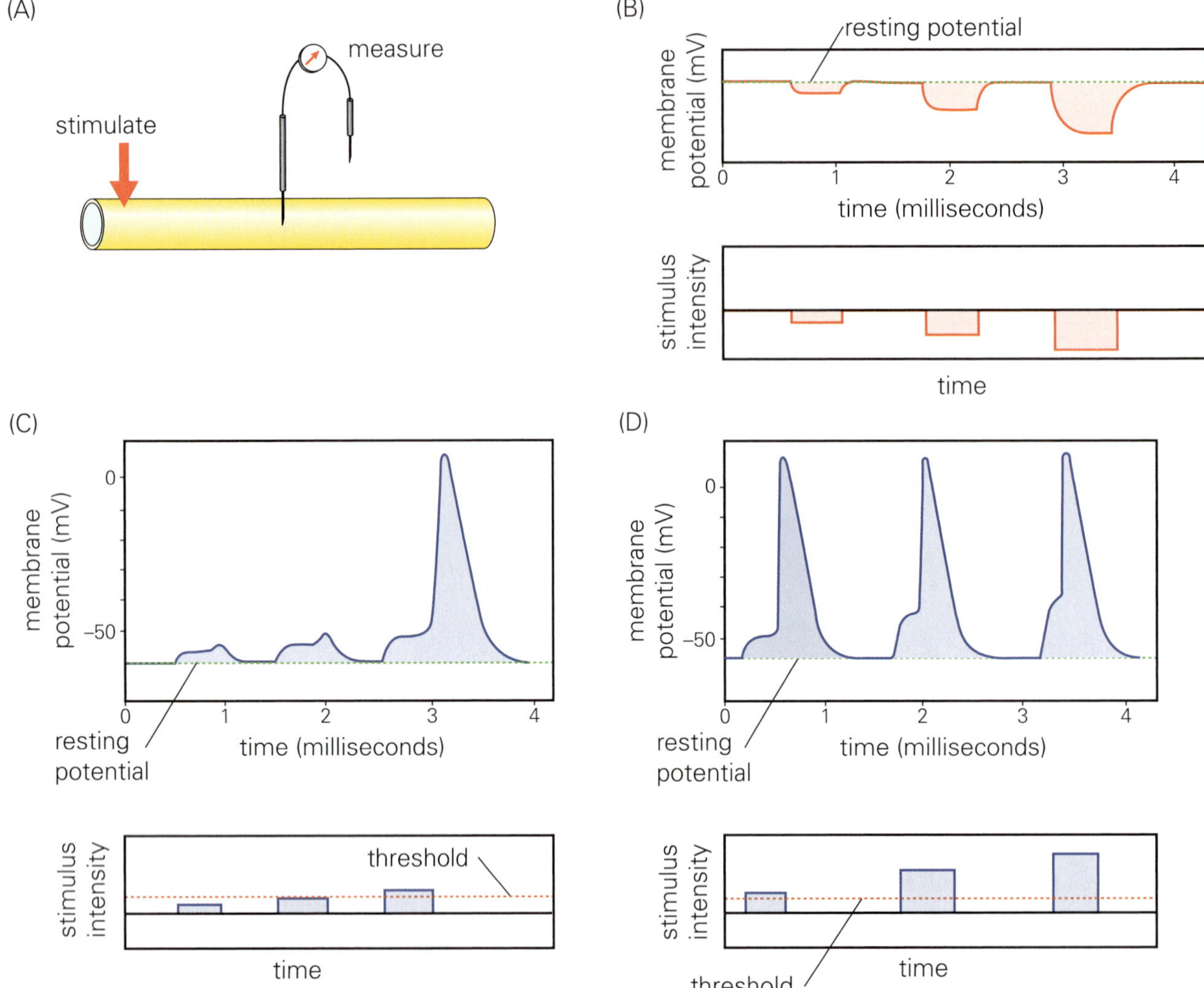


**Figure 11.58 Response of an axon to hyperpolarizing and depolarizing pulses.** (A) In this experiment a series of electrical pulses are delivered to the axon at one point (*red* arrow), and the resulting membrane potential at a point nearby is measured using another electrode. (B) When a series of hyperpolarizing (negative) pulses are delivered, the membrane potential changes in a way that is proportional to the strength of the stimulus. (C) The response to depolarizing (positive) pulses is different. Below a threshold level of stimulus, the response is proportional to the strength of the stimulus. Above a threshold level, a strong depolarization of the membrane occurs. (D) The response to depolarizing pulses is independent of the strength of the stimulus once the threshold level is crossed. (B, C and D, adapted from C.F. Stevens, Neurophysiology: A Primer. New York: John Wiley & Sons, 1966. With permission from John Wiley & Sons, Inc.)

independent of the strength of the signal (Figure 11.58D). Thus, the ability of the axon to respond strongly to depolarizing pulses is all or nothing—weak pulses are transmitted by passive spread, but strong depolarizing pulses result in the same amplified response from the axon.

Hodgkin and Huxley found that if the sodium in the external solution bathing the neuron is replaced by choline, then the axonal membrane does not depolarize when stimulated by an electrode. Like $Na^+$, choline is an ion with a single positive charge, but it does not pass through ion channels in the membrane. Thus, they inferred that the rapid depolarization of the membrane when the stimulus is above the threshold level is due to the transient opening of sodium channels.

The process by which the action potential is regenerated is shown schematically in **Figure 11.59**, where the membrane potential is graphed as a function of time at a point in the axon as an action potential passes through. The action potential first arrives by passive spread, much attenuated from its maximal value (indicated by the bump in the graph labeled 1 in Figure 11.59). At this point voltage-gated $Na^+$ channels are closed (see the schematic diagram marked 1 in Figure 11.59). The attenuated action potential is still strong enough to trigger the opening of these channels, leading to a strong inward $Na^+$ current, and the membrane potential becomes depolarized (diagram 2 in Figure 11.59).

Voltage-gated ion channels contain an inactivation domain that is attached to the main body of the channel by a flexible tether, much like a ball on a chain. When the channel is open, one of the four inactivation domains is able to bind to the

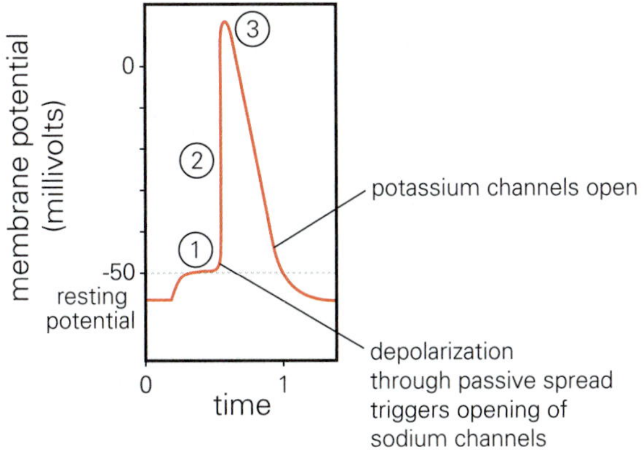


**Figure 11.59 Mechanism by which an action potential is regenerated.** The graph at the top shows the membrane potential as a function of time at a point in the axon. The diagrams below show a schematic representation of conformational changes in a voltage-gated sodium channel at the points labeled 1, 2, and 3 in the graph. The channel has four symmetrical subunits, and conformational changes in it are driven by movements of a voltage sensor. Only one of the four voltage sensors is shown.

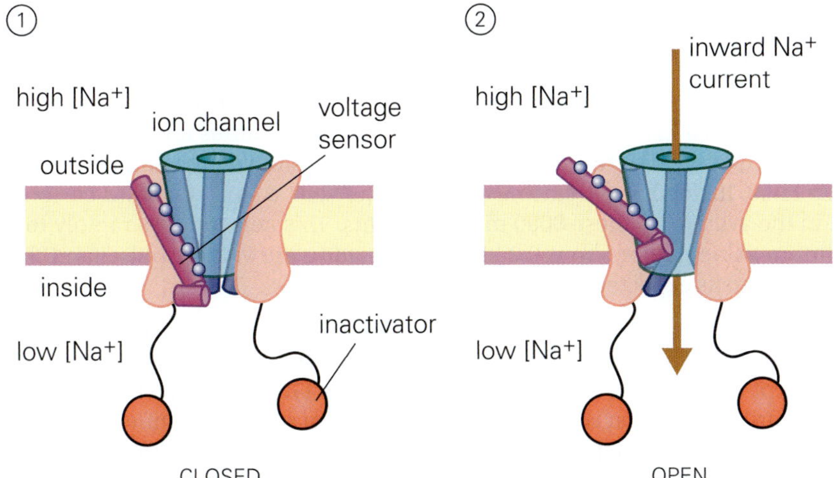


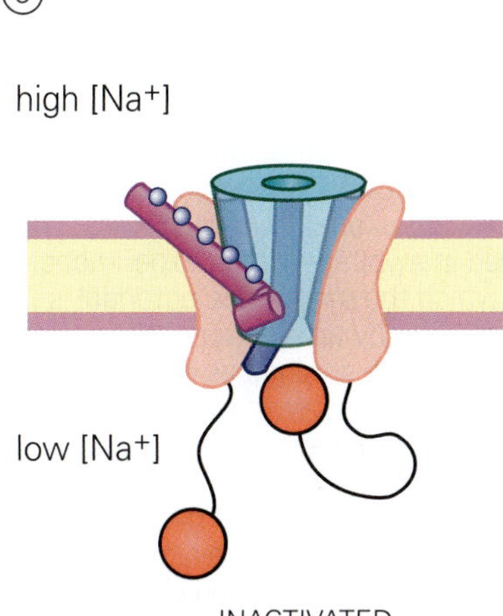


base of the channel and block further ion flow (diagram 3 in Figure 11.59). This stops the $Na^+$ current. At about the same time, voltage-gated $K^+$ channels open and cause an outward $K^+$ current that restores the polarization of the membrane (not shown in Figure 11.59). The net effect of these processes is the regeneration of the strength of the action potential.

The blockage of ion channels by the inactivation domain (diagram 3 in Figure 11.59) is slow to release. This feature gives directionality to the propagation of the action potential because the activation of downstream ion channels does not reactivate ion channels that have already been activated and inactivated. This increases the efficiency of the transmission of the signal.

## 11.38 A positively charged sensor in voltage-gated ion channels moves across the membrane upon depolarization

Hodgkin and Huxley had reasoned, based on their early experiments in the 1950s, that the opening of ion channels in response to depolarization of the membrane required some element in the membrane that moved in response to changes in voltage. Some 20 years later, Clay Armstrong and Francisco Bezanilla carried out technically challenging experiments that actually measured a small current, referred to as a **gating current**, that results from a conformational change in the voltage-dependent channels when the membrane is depolarized (**Figure 11.60**).

Armstrong and Bezanilla used the squid giant axon under conditions where sodium and potassium were removed both inside and outside the axon. This was done by replacing the external sodium with the buffer Tris, and perfusing the interior of the axon with cesium ions. Neither Tris nor cesium move through the ion

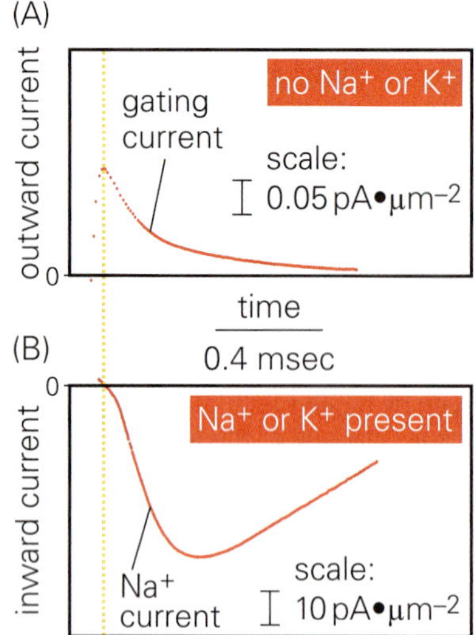


**Figure 11.60 Experimental observation of the gating current.** The giant axon of the squid was used in a voltage clamp experiment, in which the membrane potential is controlled by an electrode. The current measured is graphed as a function of time. (A) The average current observed upon alternating voltage pulses of +70 mV and –70 mV relative to the resting potential of –70 mV. $Na^+$ and $K^+$ ions were replaced by ions that do not move through ion channels. A small outward current is observed that is associated with the gating current (see Figure 11.61). (B) Current measured during an experiment in which $Na^+$ and $K^+$ ions were present and the membrane potential was depolarized by 70 mV. A large inward sodium current is observed. Note that the current scale in (A) is magnified 200-fold with respect to the scale in (B), as indicated by the vertical scale bars. (Adapted from C.M. Armstrong and F. Bezanilla, *Nature* 242: 459–461, 1973.)

channels, but they maintain the ionic strength that is required for the integrity of the axon. They used an apparatus known as a **voltage clamp**, in which the membrane potential is under the control of an electrode that runs axially through the interior of the axon. The membrane potential could then be stepped through various values and the currents measured.

In absence of $K^+$ or $Na^+$, there should be no ionic currents when the membrane potential changes. There are, however, large capacitive currents of the sort depicted in Figure 11.55, which are generated by the response of the ionic environment of the membrane to the changes in the voltage. The capacitive currents should have the same magnitude but opposite sign if the sign of the voltage perturbation is flipped. For example, if the membrane is held at the resting potential of –70 mV, and the voltage is hyperpolarized by –70 mV (to a final membrane potential of –140 mV), then an inward capacitive current is observed. If, instead, the membrane potential starts at –70 mV and is then depolarized by +70 mV (to a final membrane potential of 0 mV), then we would expect an outward capacitive current of the same magnitude as before.

What Armstrong and Bezanilla did was to subject the axon to alternating voltage pulses of +70 mV and –70 mV, repeating this 4000 times and averaging the measured currents. The result of this experiment are shown in the upper graph in Figure 11.60. Instead of the currents averaging to zero, a small outward current is seen in the average of these 8000 measurements. This current rises rapidly to its maximum value within about 0.2 msec of initiating the voltage pulse, and then decays to zero as expected for a capacitive current.

The outcome of this experiment can be compared with the result of measuring the current generated in response to a +70 mV depolarization of the membrane in the presence of $Na^+$ and $K^+$ (the lower graph in Figure 11.60). In this case, there is a large inward current that is caused by the opening of sodium channels, as identified originally by Hodgkin and Huxley (potassium channels are slower to open and do not contribute much to the current on the timescale shown in this graph). Notice, however, that the small capacitive current in the first experiment peaks *before* the sodium current builds up. Based on this and other observations, the current seen in the first experiment is identified as the gating current associated with a conformational change in the voltage-gated sodium channel that triggers its opening.

Integration of the current over time, coupled with knowledge of the density of sodium channels in the membrane, allows one to calculate the total number of charges that move per channel during the gating process. Modern measurements show that between 12 and 16 positive charges move across the membrane per tetrameric channel. How might the movement of these charges be coupled to the opening of the channel? We do not know for sure because we only have crystal structures for the open form of the channel (see Section 11.40). But we can surmise, based on many different lines of evidence, that the switching process may occur in a conceptual sense as shown in **Figure 11.61**.

A schematic representation of a voltage-gated ion channel is shown in Figure 11.61. The channel has four subunits with four-fold symmetry, but the voltage sensor is shown for only one subunit (**Figure 11.62**). A helix in the sensor (the S4 helix; see Section 11.40) contains five or six positively charged residues, which are lined up on one face of the sensor. When the membrane is polarized (negative inside), the interaction of these positive charges with the electric field pulls the helix towards the cytoplasmic face of the membrane in such a way that it interacts with the central helices of the channel, causing closure of the channel (see diagram 1 in Figure 11.61). When the membrane is depolarized (more positive inside), the altered electric field imposes an upward force on the helix (see diagram 2 in Figure 11.61). This causes an upward movement of the sensor helix and the opening of the channel (see diagram 3 in Figure 11.61). The movements of the charges in the voltage sensor are detected as the gating current in the experiment described in Figure 11.60.

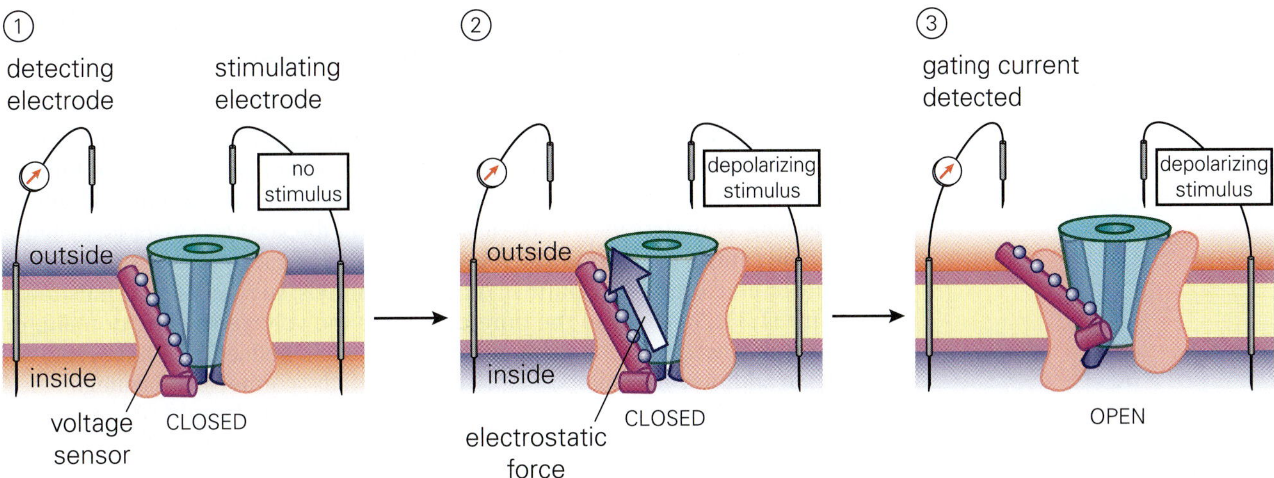


**Figure 11.61 Conceptual basis for the origin of the gating current.** The diagrams show a highly schematic representation of a voltage-gated ion channel. The channel has four subunits with four-fold symmetry, but the voltage sensor is shown for only one subunit (see Figure 11.62). Positively charged residues in the voltage sensor are shown as *blue* circles. Two electrodes are shown. One measures current and the other is used to set the membrane voltage in the voltage clamp experiment described in Section 11.38. (1) In the resting state, the membrane is polarized so that the inner face of the membrane is at a more negative electric potential compared to the outer face (indicated by *red* and *blue*). The voltage sensor is pulled downward by the electric field and interacts with the central helices of the channel in such a way that the channel is closed. (2) The membrane is depolarized (that is, is more positive inside). The electric field tends to pull the voltage sensor upwards. (3) The voltage sensor moves up, and the movement of the positive charges borne by the sensor is detected as the gating current. The movement results in a conformational change that opens the channel.

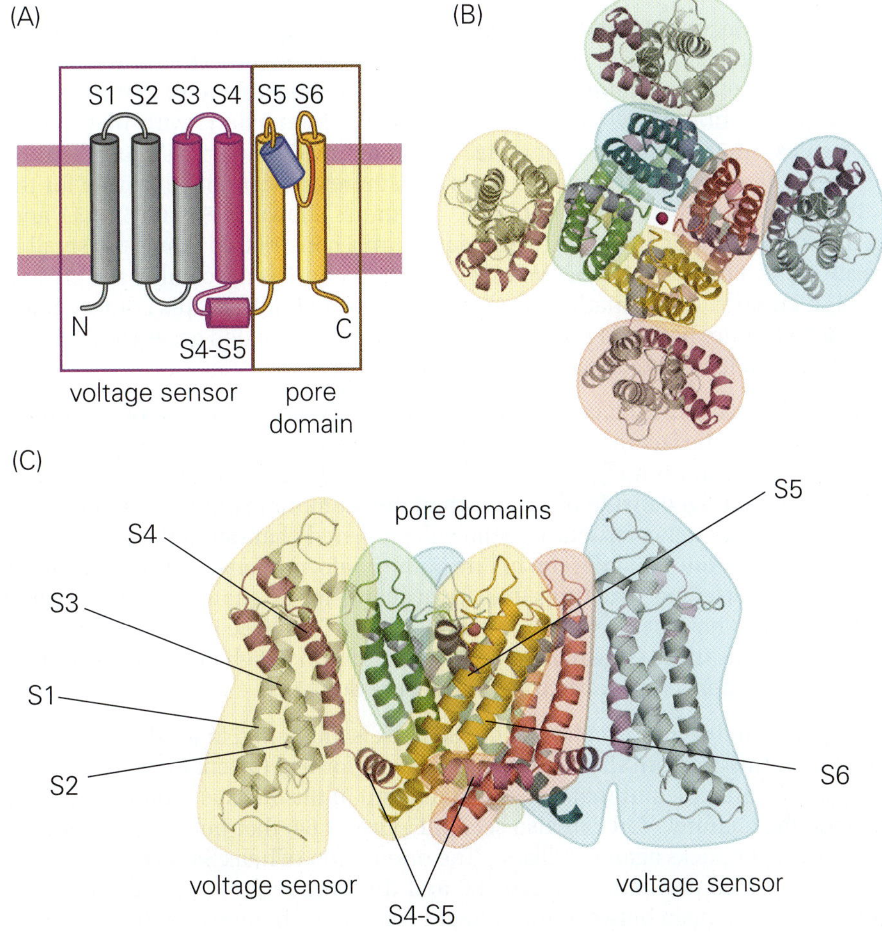


**Figure 11.62 Structure of a voltage-gated K⁺ channel.** (A) Schematic diagram of the secondary structure of voltage-gated channels (compare with Figure 11.33). (B) Crystal structure of a voltage-gated K⁺ channel. The tetrameric pore domains are colored as in Figure 11.32, with potassium ions within the selectivity filter in *purple*. The voltage sensors are shown in *magenta* and *gray*—the segments colored *magenta* are particularly important for voltage sensing. (C) Side view of the channel. For clarity, the voltage sensor in the front (connected to the pore domain colored *red*) is not shown, except for the S4-S5 helix. (B and C, adapted from X. Tao et al., and R. MacKinnon, *Science* 328: 67–73, 2010; PDB code: 2R9R.)

## 11.39 The structures of voltage-gated $K^+$ channels show that the voltage sensors form paddle-like structures that surround the core of the channel

Roughly 50 years after Hodgkin and Huxley's first identification of the role for voltage-gated ion channels in the transmission of action potentials, MacKinnon and colleagues determined the structure of a voltage-gated $K^+$ channel, providing an essentially complete view of the architecture of the voltage sensor (see Figure 11.62). The central tetrameric assembly of the pore domains is very similar in its general architecture to that found in other $K^+$ channels, such as the one illustrated in Figure 11.33. Attached to the pore domain is the voltage-sensing domain, or voltage sensor, which contains four transmembrane α helices denoted S1, S2, S3, and S4. In addition, a smaller α helix (the S4-S5 helix) connects the voltage-sensing domain to the two transmembrane helices of the pore domain (that is, the S5 and S6 helices).

The voltage sensor forms a four-helix bundle that is loosely attached to the pore domain and is shaped like a paddle. The key element of the voltage-sensing domain is helix S4. As explained in Section 11.41, this helix contains several positively charged residues, mostly arginines, that move in response to changes in the membrane potential (it is the movement of these residues that is detected as the gating current that we described in Section 11.38). Even though S4 is a transmembrane helix, the positively charged residues are accommodated within the membrane-spanning segment because of interactions with negatively charged residues provided by other helices in the voltage sensor. Note that each voltage sensor packs against the pore domain of another subunit and not the one to which it is covalently attached. This hand-over of the sensor across subunits facilitates an iris-like opening and closing of the entry to the vestibule in response to changes in membrane voltage.

Recall from Section 11.21 that $K^+$ and $Na^+$ ions in solution are always bound strongly to at least four or five water molecules (see Figure 11.35). In order for an ion to enter the vestibule from solution, it has to be able to pass through the lower constriction of the channel without losing interactions with these water molecules—this region of the channel does not provide the precise coordination of the ion that the specificity filter does and, therefore, cannot substitute for the water molecules. A closed form of a $K^+$ channel is shown in **Figure 11.63**A and, in this form, the four S6 helices of the pore form a constriction that is too small to easily admit a hydrated $K^+$ ion. Voltage-gated $K^+$ channels have only been crystallized in an open conformation, as shown in Figure 11.63B. In this conformation the S6 helices are bent and rotated, and they allow unimpeded access to the inner vestibule that leads into the selectivity filter.

That the voltage-gated channels have only been crystallized in the open conformation reflects the fact that the channel does not experience a voltage gradient in the crystal. Recall from Figure 11.61 that the gate is closed when the membrane is negatively charged inside but open when the membrane potential is zero, as in the crystal. Although a structure of the voltage-gated channel in the closed form is not available, one can infer the nature of the conformational change that closes the channel based on structures that are available for $K^+$ channels that are gated differently and have been crystallized in the closed form (see Figure 11.63). These models, such as those shown in Figure 11.63C, rely on a high degree of conservation in the sequences and structures of the pore domains.

The change in the conformation of the S6 helix is illustrated in a side view in **Figure 11.64**. This diagram shows that, in the closed form of the channel, the S6 helix is relatively straight, and that a substantial bend in this helix is required in order to open the channel. You can also see, in Figure 11.64, that helix S4-S5 from the voltage sensor packs against helix S6. The orientation of helix S4-S5 can therefore influence the conformation of helix S6 and this is thought to provide the main coupling mechanism between the voltage sensor and channel opening.

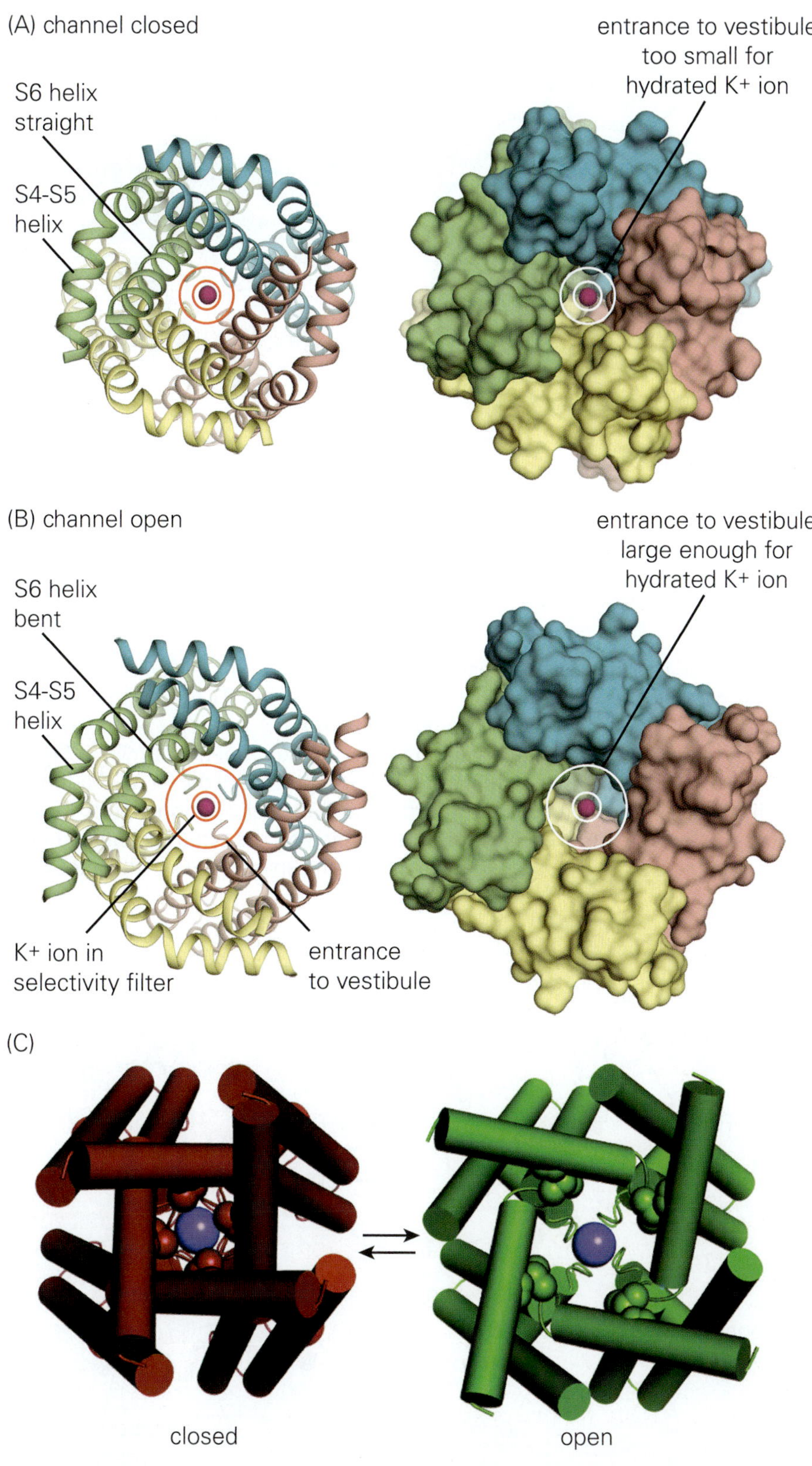


**Figure 11.63 Open and closed forms of $K^+$ channels.** (A) Structure of a closed form of a bacterial $K^+$ channel. The entry to the vestibule, indicated by the larger of the two concentric circles, is too constricted to allow easy passage of a hydrated $K^+$ ion. The inner circle (*red*) indicates the selectivity filter, with an ion bound within it. (B) Structure of the open form of a voltage-gated $K^+$ channel. A conformational change in the channel, particularly the bending of helix S6, opens the entrance to the vestibule so that a hydrated $K^+$ ion can pass. The transition to the open form is coupled to movements of the S4-S5 helix of the voltage sensor (shaded). (C) Models of a $K^+$ channel, showing the iris-like opening of the vestibule.
A conformational change in the channel, particularly the bending of helix S6, opens the entrance to the vestibule so that a hydrated $K^+$ ion (*blue*) can pass. Hydrophobic residues that line the opening are shown as spheres. The bacterial channel does not have an S4-S5 helix, and its position in (A) is modeled as described in X. Tao et al. and R. MacKinnon, *Science* 328: 67–73, 2010. (A, B: PDB codes: 1K4C and 2R9R. C, from S.K. Upadhyay, P. Nagarajan, and M.K. Mathew, *J. Physiol.* 587: 3851–3868, 2009.)

## 11.40 The crystal structure of a voltage-gated $K^+$ channel suggests how the voltage sensor opens and closes the channel

The S4 helix bears six residues that are positively charged (**Figure 11.65**). These are labeled R0, R1, R2, R3, R4, and K5, based on residues that are present in the first potassium channel to be cloned, known as the Shaker channel because of

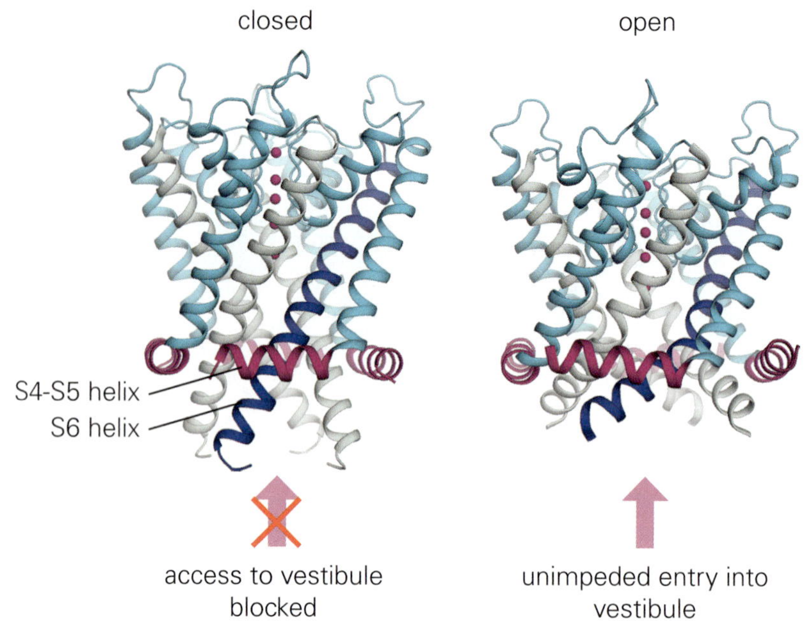


**Figure 11.64 Structural differences between open and closed forms of the K⁺ channel.** The structures shown here are the same as those illustrated in Figure 11.63. Each subunit of the channel is colored the same way, except that three S6 helices are in *gray* and one is in *blue*. The S4-S5 helices are in *magenta*. The structure on the left is closed—the four S6 helices block access to the vestibule. The structure on the right is open—the S6 helices are bent and they provide access to the vestibule. (Adapted from X. Tao et al., and R. MacKinnon, *Science* 328: 67–73, 2010; PDB codes: 1K4C and 2R9R.)

the phenotype in fruit flies that bear a mutation in this channel (in the version of this channel that was crystallized the residue at position R1 is glutamine rather than arginine). The charged residues repeat in an $i$, $i + 3$ pattern, with two non-polar residues between each charged one (refer to the sequence shown in Figure 11.65A). The $i$, $i + 3$ pattern is significant because a portion of the S4 helix adopts a $3_{10}$ helical rather than an α helical conformation. A $3_{10}$ helix is more tightly wound than an α helix, and has every third residue aligned on the same face of the helix. A $3_{10}$ helix is also more extended than an α helix and spans a greater distance with

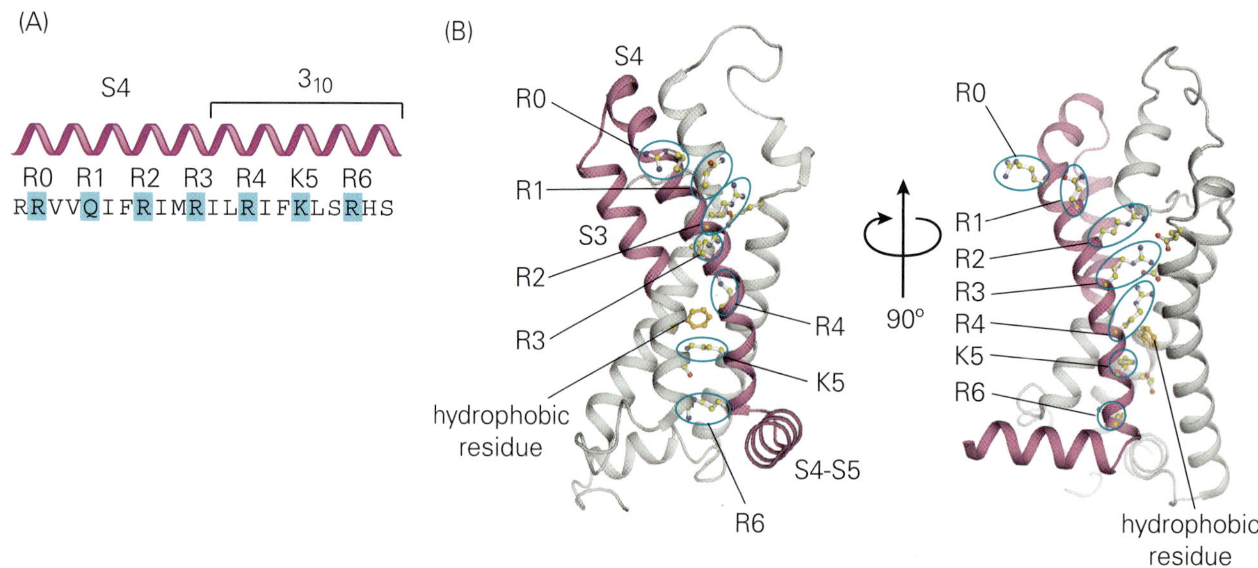


**Figure 11.65 Structure of the voltage sensor in the open state of the channel.** (A) Sequence of the S4 helix and its secondary structure. The notation used for the charged residues is based on the Shaker K⁺ channel. In the channel that was crystallized, the residue at the R1 position is glutamine rather than arginine. (B) Structure of the voltage sensor. The positively charged residues are highlighted. Ion-pairing interactions with acidic residues are shown, and a hydrophobic sidechain that serves as a barrier to proton leakage is indicated. The orientation of the voltage sensor on the left is similar to that in the subunit shaded in *yellow* in Figure 11.62. (Adapted from X. Tao et al., and R. Mackinnon, *Science* 328: 67–73, 2010; PDB code: 2R9R.)

the same number of residues. Each of these features of the $3_{10}$ helix comes into play in explaining how a series of charged residues are accommodated within the structure of a transmembrane protein.

Residues R0, R1, and R2 are completely exposed to water at the extracellular face of the membrane, and this is because helix S4 is splayed out from helices S1 and S2. Residues R3, R4, K5, and R6 are part of the $3_{10}$ helix, so they are all presented on one surface of S4—namely, that facing the S1 and S2 helices. These helices provide two clusters of acidic residues that engage the charged residues provided by S4. One cluster is towards the extracellular side and engages R3 and R4, while the other cluster is towards the intracellular side and engages K5 and R6 (see Figure 11.64). The two clusters are separated by a phenylalanine residue, which is thought to provide a barrier to the leakage of protons through the interface between S4 and the S1 and S2 helices.

Note that the formation of a $3_{10}$ helix by the C-terminal portion of the S4 helix extends the distance spanned by the S4 segment—if this segment were entirely α-helical, then the R0, R1, and R2 residues would have to interact with the rest of the protein instead of being exposed to water. Because of the $i$, $i + 3$ spacing between the charged residues, these would have to circle around an α helix, requiring many more elements of the structure to be brought into play in order to neutralize the charges.

The formation of a $3_{10}$ helix may also be critical for the switch to the closed state of the channel. There is evidence that the R1 residue interacts with the lower acidic cluster, near the intracellular face, when the channel is in the closed conformation. This would require a dramatic conformational change relative to the structure shown in Figure 11.64, where the R1 residue is exposed to water on the extracellular side. One possibility is that the N-terminal half of the S4 helix switches to a $3_{10}$ helix in the closed state and then somehow moves down to present R1 to the lower acidic cluster.

A speculative model for the closed form is shown in **Figure 11.66.** The essential idea in this model is that, when the membrane is polarized, negative charges on the cytoplasmic face of the membrane pull the S4 helix downwards, bringing the R1 and R2 residues into a position where they can interact with the lower electrostatic cluster. This causes a downward movement of the S4-S5 helix in each of the four voltage sensors, which pushes down on the S6 helices of the pore domains of the adjacent subunits of the channel, causing them to straighten. The straightened forms of the S6 helix block the opening to the vestibule, thus preventing the flow of ions.

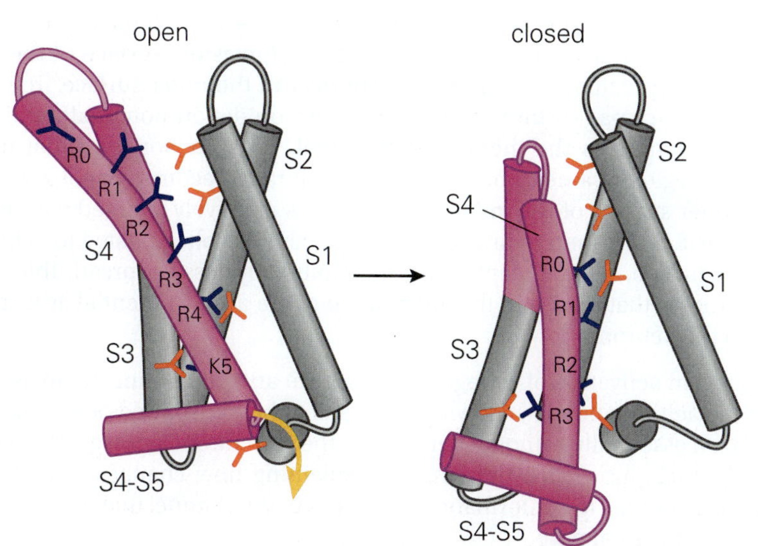


**Figure 11.66 Model for how a voltage-gated channel might switch from open to closed.** A schematic diagram for the voltage sensor in the open form is shown on the left (see Figure 11.65). The R0 and R1 sidechains interact with water at the exterior face of the channel. A hypothetical structure for the closed form is shown on the right, in which the R0 and R1 residues have moved down and interact with the lower electrostatic cluster. As shown in Figure 11.64, the S4-S5 helix packs against the S6 helix of the pore domain. The movement shown here for the voltage sensor changes the position of the S4-S5 helix in such a way that it forces helix S6 to straighten, thereby closing the channel. (Adapted from S.B. Long et al., and R. MacKinnon, *Nature* 450: 376–382, 2007.)

## Summary

In this chapter we have studied two quite different types of processes that depend on electrical potential differences: oxidation–reduction reactions and signal transmission in neurons. Oxidation–reduction reactions are at the heart of the mechanisms by which light energy emanating from the sun is captured by photosynthetic cells, thereby providing free energy for the generation of ATP. The processes that allow us to respond to our environment and to achieve consciousness rely on the very rapid transmission of electrical signals in neurons, which are made possible by the action of ion channels. The action potentials generated by these neurons are propagated without attenuation because they are regenerated by the ability of the cell to tap into the free energy stored in the form of concentration gradients in potassium and sodium ions. These concentration gradients are established by the action of ATP-dependent ion pumps.

Living systems uses a variety of redox-active molecules that differ in their ability to take up or release electrons. Some molecules are easily oxidized (that is, they give up electrons readily). Other molecules are more easily reduced, (that is, they take up electrons more readily). These differences in their redox potentials allow nature to control the release of large amounts of energy from processes that would otherwise be dissipated as heat. A pair of molecules or atoms that can mutually reduce and oxidize each other is known as a redox couple. Electrochemical cells can be generated by connecting two chambers containing different redox couples in an electrical circuit. The voltage generated across the two chambers is related to the difference in the values of the free-energy, $\Delta G$, for the chemical reactions corresponding to the two redox couples. The voltages generated by electrochemical cells are measured readily, providing an experimentally straightforward route to measuring free-energy changes associated with redox reactions.

Living cells maintain an asymmetric distribution of ions across the cell membrane, most notably for sodium and potassium. Mammalian cells, for example, have high concentrations of potassium inside the cell compared to the extracellular environment, while the reverse is true for sodium ions. Cell membranes contain ion channels, which allow specific ions to pass through while preventing the flow of other kinds of ions. Some of these ion channels are always open, and the flux of ions through these channels establishes an equilibrium membrane potential, which is the voltage across the cell membrane that balances the flow of ions due to the concentration gradients. The structures of potassium channels explain how very high rates of transit of ions are enabled while maintaining specificity for particular ions.

The Nernst equation allows us to calculate the equilibrium membrane potential, given the concentrations of various ionic species inside and outside of the cell. Mammalian cells are polarized such that the cytoplasmic surface of the membrane is at approximately –70 mV with respect to the outer surface. The transmission of information through neurons relies on action potentials, which are transient changes in the membrane potential. As an action potential transits through a region of a neuronal axon, the membrane becomes depolarized (that is, the inner surface of the membrane becomes positively charged with respect to the outside) and then returns to its resting value. If voltage-gated ion channels are not activated, action potentials are transmitted by passive spread. This is a diffusive process that results in the attenuation of the action potential as it spreads through the neuron.

The transient activation of voltage-gated sodium and potassium channels regenerates the action potential, allowing the signal to be transmitted down the length of the axon essentially without attenuation. The structures of voltage-gated potassium channels have been determined, explaining how changes in membrane voltage are coupled to conformational changes in the channel that open and close access to the central pore.

## Key Concepts

### A. OXIDATION REACTIONS IN BIOLOGY

- Reactions involving the transfer of electrons are referred to as oxidation–reduction reactions.
- Metal ions bound to proteins are one class of biologically important redox-active species.
- Nicotinamide adenine dinucleotide, flavins, and quinones are some of the important nonmetal mediators of redox reactions in biology.
- The oxidation of glucose is coupled to the generation of NADH and $FADH_2$.
- Mitochondria are cellular compartments in which NADH and $FADH_2$ are used to generate ATP.
- Photosynthesis uses energy from light absorption to drive synthesis of reduced compounds, and oxidation of water to oxygen.

### B. REDUCTION POTENTIALS AND FREE ENERGY

- Electrochemical cells are generated by connecting two chambers containing molecules that undergo redox reactions.
- The voltage generated by an electrochemical cell with the reactants at standard conditions is known as the standard potential for the reaction.
- Standard potentials are related to the standard free-energy change of the redox reaction underlying the electrochemical cell.
- Standard reduction potentials are measured relative to a standard hydrogen electrode.
- Tabulated values of standard electrode reduction potentials allow ready calculation of the standard potential of an electrochemical cell.
- One form of the Nernst equation describes how the cell potential changes with the concentrations of the redox reactants.
- Standard reduction potentials in biochemistry refer to a standard state of pH 7.

### C. ION PUMPS AND CHANNELS IN NEURONS

- Neuronal cells use electrical signals to transmit information.
- Cell membranes have an electrical potential difference between the inside and the outside of the cell.
- Sodium–potassium pumps maintain the equilibrium membrane potential by hydrolyzing ATP to move three $Na^+$ ions out of the cell in every cycle, with the coupled movement of two $K^+$ ions into the cell.
- Potassium and sodium channels allow ions to move quickly across the membrane, in response to the thermodynamic gradients.
- Potassium and sodium channels contain a conserved tetrameric pore domain that provides high selectivity for a specific ion.
- Very rapid transit of $K^+$ ions through the channel is facilitated by hopping between isoenergetic binding sites.

### D. THE TRANSMISSION OF ACTION POTENTIALS IN NEURONS

- The asymmetric distribution of ions across the cell membrane generates an equilibrium membrane potential.
- The Nernst equation relates the equilibrium membrane potential to the concentrations of ions inside and outside the membrane.
- Cell membranes act as electrical capacitors.
- Membrane potentials are altered by the movement of relatively few ions, enabling rapid neuronal transmission.
- Action potentials are initiated by activation of ligand-gated ion channels in the cell body or the dendrites.
- The propagation of voltage changes in neurons can be understood by treating the neuron as an electrical circuit.
- The propagation of a voltage spike without regeneration is known as passive spread.
- The time taken to develop a membrane potential is determined by the conductance of the membrane and its capacitance.
- Myelination of mammalian neurons facilitates the transmission of action potentials.
- The opening and closing of voltage-gated $Na^+$ and $K^+$ channels regenerates action potentials, ensuring that long range transmission occurs without attenuation.
- A positively charged sensor in voltage-gated ion channels moves across the membrane upon depolarization.
- The crystal structure of a voltage-gated $K^+$ channel suggests a mechanism for how the conformation of the voltage sensor is coupled to the opening and closing of the channel.

## Problems

### True/False and Multiple Choice

1. An oxidative half-reaction involves the release of electrons from a molecule.

   True/False

2. Which of the following is a common consequence of a protein coordinating a metal:

   a. The free energy of the reduced and oxidized form of the metal are altered.
   b. The metal is protected from some chemical reactions.
   c. The protein undergoes a conformational change.
   d. It enables the transfer of electrons over long distances.
   e. All of the above.

3. Standard reduction potentials in biochemistry correspond to solutions with $[H^+] = 1$ M.

   True/False

4. Which of the following conformational changes is not thought to occur as a $K^+$ channel closes due to a voltage change:

   a. Helix S6 becomes straighter.
   b. Helix S4 moves downward.
   c. Charged residues on helix S4 form new interactions.
   d. The selectivity filter changes size.
   e. The C-terminus of helix S4 forms a $3_{10}$ helix.

5. Most ion channels have little selectivity because ions are small relative to the size of the channel pore.

   True/False

6. Which feature allows mitochondria and chloroplasts, unique from other organelles, to maintain large proton gradients:

   a. They contain membrane spanning proteins.
   b. They have a second internal membrane.
   c. There are only one of each per cell.
   d. Their membranes consists of a lipid bilayer.

7. Which of the following statements about the passive spread of a voltage perturbation in a neuron are true:

   i The voltage spike is not regenerated.
   ii The potassium conductance of the membrane changes locally.
   iii The sodium conductance of the membrane remains constant.
   iv It only occurs in non-myelinated axons.

   a. None of the above.
   b. All of the above.
   c. (i) and (iii).
   d. (i), (ii) and (iii).
   e. (ii) and (iv).

### Fill in the Blank

8. The oxidized and reduced forms of a molecule are known as a ___________.

9. The energy from _______ is used to produce reduced compounds in photosynthesis.

10. Sodium and potassium pumps hydrolyze ________ to move ions in and out of the cell.

11. By decreasing the capacitance of the surface of the axon, ______________ greatly facilitates the transmission of action potentials.

12. A _________________ can build up charge on two conducting surfaces with an insulator sandwiched between them.

### Quantitative/Essay

13. How does phosphorylation contribute to increasing the diversity of redox active molecules in the cell?

14. A new microorganism is isolated from a lake and is placed into a solution of KCl. The voltage difference across its membrane is measured at 120 mV. How much energy is required to move a proton from the negative side of the membrane to the positive side?

15. An electrochemical cell couples manganese and copper, each in the presence of its $2^+$ ion.

    a. What is the combined redox reaction?
    b. What is the the value of $\Delta G^o$ for the reaction?

16. A researcher assembles an electrochemical cell with silver and an unknown metal in a 1 M solution of its ion. She measures the free energy of the combined reaction under standard conditions to be $-150$ kJ•mol$^{-1}$. What is the potential of the oxidative half reaction if one electron is transferred by the unknown metal to silver under standard conditions?

17. A new ATP-producing protein is discovered that couples ATP production to the oxidation of NADPH by oxidative phosphorylation. Assume that the value of $\Delta G^o$ for ATP synthesis is 30 kJ•mol$^{-1}$. If this protein only produces 1 molecule of ATP per reaction that consumes one NADPH:

    a. How much free energy is wasted, under standard conditions?
    b. How many more ATP molecules could be created by a perfectly efficient electron transport chain from one NADPH?

18. Consider the structure of the potassium channel and answer the following questions.

    a. Given that the four $K^+$ binding sites are roughly isoenergetic, why are usually only two of these sites occupied at any given time?

b. Based on your answer to (a), how does the "knock-on" model explain the very high conductance of potassium channels?

c. What establishes the directionality of $K^+$ ion flow through a potassium channel?

**19.** A scientist measures the potential across the membrane of a cell. At room temperature, the pH outside the cell is 7.4 and the pH inside the cell is 7.1. What is the membrane potential for protons across the bilayer?

**20.** A squid axon is immersed in seawater during a laboratory experiment, and the resting potential across the axonal membrane is –76 mV, at room temperature. The concentrations of $Na^+$ and $K^+$ in the seawater and inside the squid axon are given in the table below:

| | $[K^+]$ (mM) | $[Na^+]$ (mM) |
|---|---|---|
| inside squid axon | 150 | 30 |
| seawater | 5 | 200 |

a. What are the membrane potentials for $Na^+$ and $K^+$ under these conditions?

b. Do you expect the resting membrane potential to be equal to the sum of the membrane potentials for $Na^+$ and $K^+$? Explain your answer.

**21.** A scientist grows a tiny synthetic nerve cell with a surface area of $3.1 \times 10^{-7}$ cm$^2$. In its culture medium, the cell has a resting membrane potential of 75 mV. During an action potential that depolarizes the cell to 0 mV, $6.2 \times 10^{-12}$ C of charge move across the membrane. What is the capacitance of the cell per unit area of the cell?

**22.** A cell has the intracellular and extracellular concentrations of $K^+$ and $Na^+$ as shown below, at room temperature. If the resting conductance of its $K^+$ channels is 30 times greater than its $Na^+$ channels, what is the resting potential of the cell (assuming no other ions contribute significantly)?

| | $[K^+]$ (mM) | $[Na^+]$ (mM) |
|---|---|---|
| intracellular | 200 | 15 |
| extracellular | 12 | 360 |

**23.** A region of an axon has a time constant of 120 msec. The capacitance of the membrane is 1 μF•cm$^{-2}$ and there are 25,000 potassium channels per cm$^2$ of surface area of the membrane. How many ions does a typical potassium channel conduct per second when the membrane potential is 100 mV?

**24.** Immediately prior to and subsequent to an action potential, there is little conductance through sodium channels; however, conductance is blocked through different mechanisms before and after the action potential passes. Explain the structural mechanisms underlying the closed states of the channel before and after the action potential passes.

## Further Reading

### General

Atkins PW & De Paula J (2006) Atkins' Physical Chemistry. Oxford, UK: Oxford University Press.

Chang R (2000) Physical Chemistry for the Chemical and Biological Sciences, 3rd ed. Sausalito, CA: University Science Books.

Halliday D, Resnick R & Walker J (2001) Fundamentals of Physics, 6th ed. New York: John Wiley & Sons.

Kandel ER, Schwartz J & Jessell T (2000) Principles of Neural Science, 4th ed. New York: McGraw-Hill.

Serway RA & Jewett JW (2004) Physics for Scientists and Engineers. Belmont, CA: Thomson-Brooks/Cole.

Voet D & Voet JG (2004) Biochemistry, 3rd ed. New York: John Wiley & Sons.

#### A and B. Oxidation–Reduction Reactions

Cramer WA & Knaff DB (1990) Energy Transduction in Biological Membranes: A Textbook of Bioenergetics. New York: Springer-Verlag.

Haynie DT (2008) Biological Thermodynamics. Cambridge, UK: Cambridge University Press.

Nicholls DG & Ferguson SJ (2002) Bioenergetics 3, 3rd ed. San Diego, CA: Academic Press.

#### C. Ion Pumps and Channels in Neurons

Ackerman MJ & Clapham DE (1997) Ion channels—basic science and clinical disease. *N. Engl. J .Med.* 336, 1575–1586.

Armstrong CM (2003) The Na/K pump, Cl ion, and osmotic stabilization of cells. *Proc. Natl. Acad. Sci. U.S.A.* 100, 6257–6262.

Choe S (2002) Potassium channel structures. *Nat. Rev. Neurosci.* 3, 115–121.

Hille B (2001) Ion channels of Excitable Membranes, 3rd ed. Sunderland, MA: Sinauer Associates.

MacKinnon R (2003) Potassium channels. *FEBS Lett.* 555, 62–65.

MacKinnon R (2003) Nobel Prize Lecture. http://nobelprize.org/nobel_prizes/chemistry/laureates/2003/mackinnon-lecture.html

Miller C (2001) See potassium run. *Nature* 414, 23–24.

Roux B (2005) Ion conduction and selectivity in K(+) channels. *Annu. Rev. Biophys. Biomol. Struct.* 34, 153–171.

Toyoshima C (2009) How $Ca^{2+}$-ATPase pumps ions across the sarcoplasmic reticulum membrane. *Biochim. Biophys. Acta* 1793, 941–946.

### D. The Transmission of Action Potentials in Neurons

Hodgkin and Huxley won the Nobel Prize in Physiology or Medicine in 1963 for discovering how action potentials are propagated. The following two articles summarize their findings:

Hodgkin AL (1964) The ionic basis of nervous conduction. *Science* 145, 1148–1154.

Huxley AF (1964) Excitation and conduction in nerve: quantitative analysis. *Science* 145, 1154–1159.

*The following book provides one of the best introductions to the electrical properties of nerve cells. Much of the discussion of the propagation of action potentials is based on the treatment provided in this book:*

Stevens CF (1966) Neurophysiology: A Primer. New York: John Wiley & Sons.

Tombola F, Pathak MM & Isacoff EY (2006) How does voltage open an ion channel? *Annu. Rev. Cell Dev. Biol.* 22, 23–52.

# PART IV

## MOLECULAR INTERACTIONS

# CHAPTER 12

# Molecular Recognition: The Thermodynamics of Binding

In Chapter 10, we introduced the concept of the equilibrium constant, *K*, which governs the concentrations of reactants and products in a reaction that has reached equilibrium. In this chapter, we focus on the analysis of equilibrium constants for a particularly important subset of molecular reactions—namely, those involving the binding of one molecule to another. These **molecular recognition** events (**Figure 12.1**) underlie all of the critical processes in biology, including the recognition of proper substrates by enzymes, the transmission of cellular signals, the recognition of one cell by another, the control of transcription and translation, and the fidelity of DNA replication.

We begin by analyzing the thermodynamics of binding interactions in which two molecules form a **noncovalent complex**. Such complexes are held together by ionic, hydrogen-bonding, or hydrophobic interactions, which are much weaker than covalent bonds. Noncovalent complexes usually dissociate to an appreciable extent at room temperature, leading to a mixture of unbound and bound molecules at equilibrium. By measuring the concentration of the free and associated species at equilibrium, we can calculate the strength of the molecular interaction.

We focus on noncovalent interactions between proteins and their **ligands**, a term that is typically used to describe a smaller molecule that binds to a larger one. The ligand might be a drug molecule or a substrate for an enzyme, but more generally it could also be another protein molecule, or any other kind of macromolecule, such as DNA or RNA (see Figure 12.1). Although we focus in this chapter on protein molecules as the receptors for the ligands, the receptors could also be DNA or RNA molecules.

In Part A of this chapter we describe the thermodynamics of the simplest molecular interactions, which involve a single receptor interacting with a single ligand. An important class of interactions in biology involves more than one ligand molecule binding to a receptor. When the binding of one of these molecules alters the affinity of the other molecules for the receptor, the interactions are referred to as **allosteric** (**Figure 12.2**). Allosteric proteins, which are discussed in Chapter 14, are very important in biology, because the outputs that they generate are much more sensitive to changes in input levels.

**Ligand**

We use the term "ligand" in a general sense to mean any molecule that binds to another molecule. In biochemistry, "ligand" usually refers to a small molecule, such as an organic compound, which binds to a macromolecule, such as a protein. The macromolecule is often referred to as the receptor for the ligand.

In Part B we apply the principles developed in the first part of this chapter to a discussion of one class of binding interactions—namely, those of proteins with drugs. We shall see that molecular recognition in biology involves a trade-off between **affinity** and **specificity**. High-affinity binding is often achieved by increasing the hydrophobicity of the ligand, while specificity relies on hydrogen-bonding interactions. Affinity and specificity in protein–protein and protein–nucleic acid interactions are the focus of Chapter 13.

## A. THERMODYNAMICS OF MOLECULAR INTERACTIONS

In this part of the chapter we discuss some fundamental concepts concerning the equilibrium constants for noncovalent interactions. We shall focus on interactions

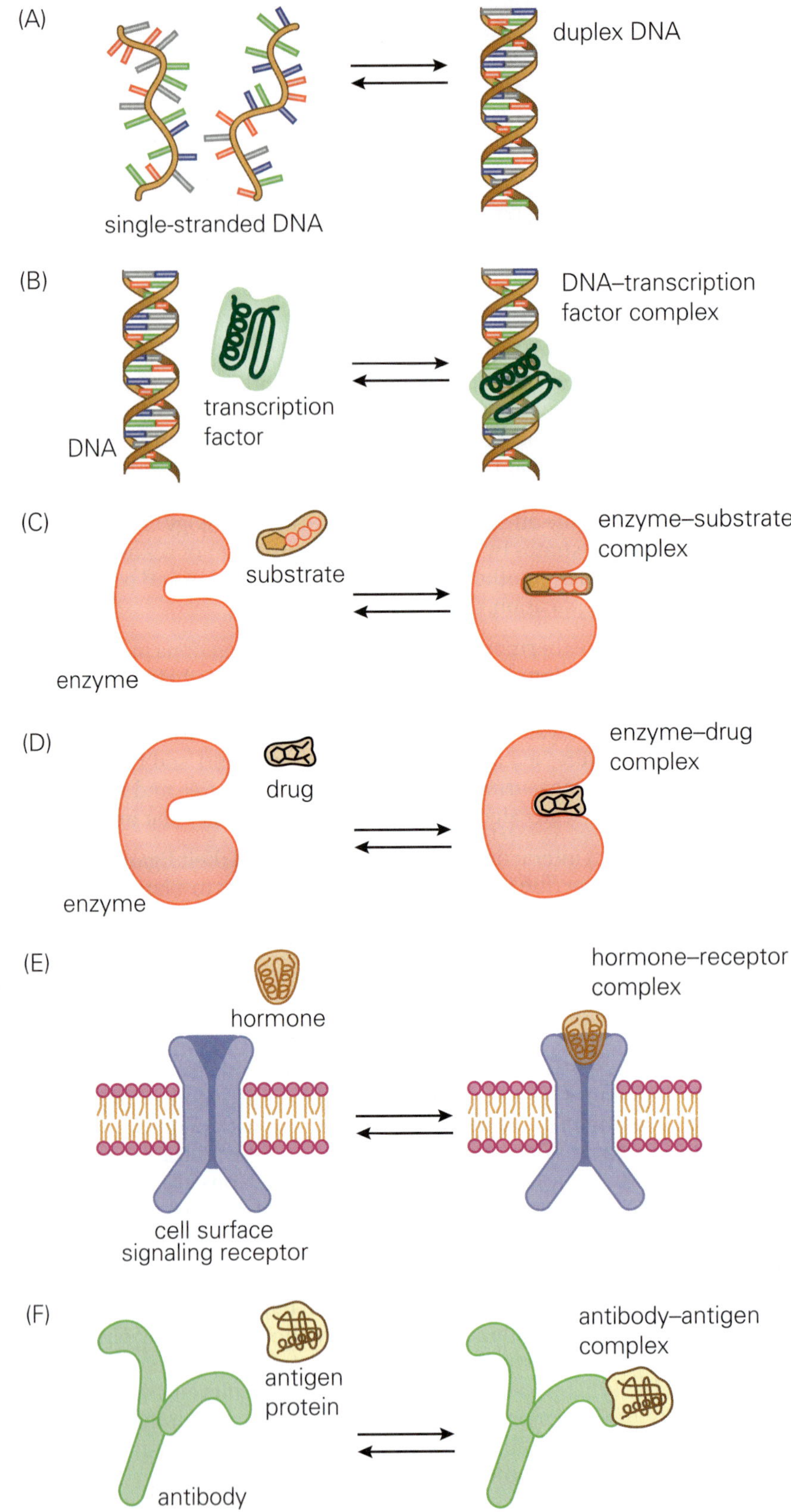


**Figure 12.1 Molecular recognition.** Shown here are some examples of the many kinds of noncovalent interactions that are important in biology. In each case, the interactions between molecules are strong enough for the biological function, but sufficiently weak that there is an equilibrium between bound and unbound molecules.

between proteins and small molecule ligands, but the concepts introduced here are quite general. We shall extend these ideas to protein–protein interactions, protein–DNA interactions, and protein–RNA interactions in Chapter 13.

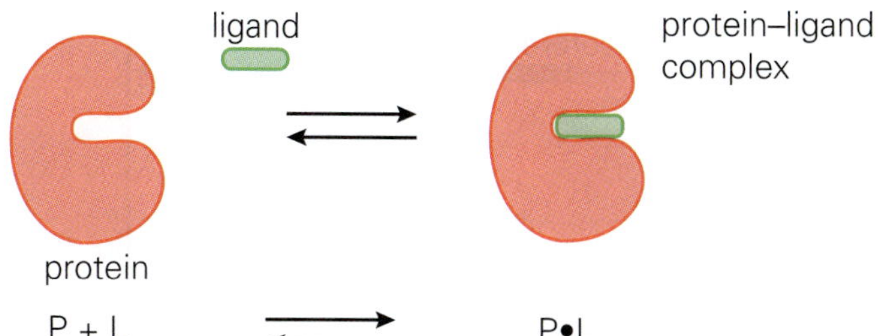


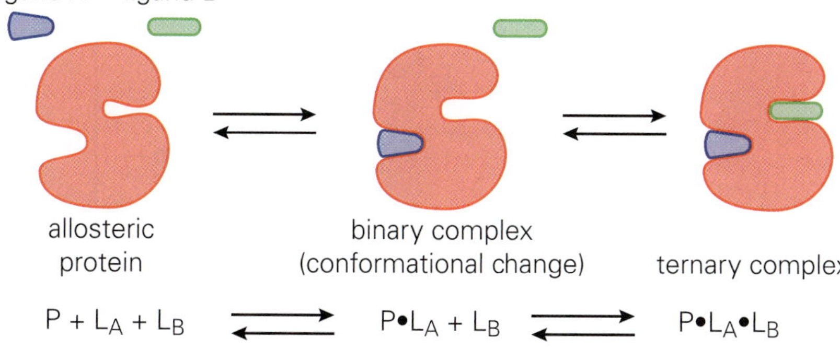


**Figure 12.2 Allosteric and non-allosteric binding interactions.** (A) In a simple binding interaction, each encounter between the protein molecule and a ligand molecule is independent of other binding interactions. (B) An allosteric protein has more than one ligand binding site, and the binding of a ligand to one site influences the affinity of the other site for its ligand. Allosteric binding sites are often symmetry related sites in oligomeric proteins. Allosteric interactions are discussed in Chapter 14.

## 12.1 The affinity of a protein for a ligand is characterized by the dissociation constant, $K_D$

We begin our analysis of noncovalent complexes by restating some thermodynamic relationships that are familiar to us from Chapter 10, but which we now place explicitly in the context of a ligand, L, binding noncovalently to a protein, P.

The general binding equilibrium for the interaction of a protein, P, with a ligand, L, can be written as follows:

$$P + L \rightleftharpoons P \bullet L \tag{12.1}$$

In Equation 12.1, P•L represents the noncovalent protein–ligand complex. The equilibrium constant, $K$, for the reaction shown in Equation 12.1, is given by Equation 12.2 (see Equation 10.51):

$$K = \frac{[P \bullet L]}{[P][L]} \tag{12.2}$$

In Equation 12.2, [P•L] is the concentration of the liganded protein, [P] is the concentration of the free protein, and [L] is the concentration of the free ligand. Because the binding reaction (Equation 12.1), as read from left to right, is in the direction of association, the equilibrium constant as defined in Equation 12.2 is referred to as the **association constant**, $K_A$:

$$K_A = \frac{[P \bullet L]}{[P][L]} \tag{12.3}$$

The standard free-energy change, $\Delta G^o$, for the binding reaction is given by Equation 12.4 (see Equation 10.52):

$$\Delta G^o = -RT \ln K_A \tag{12.4}$$

Recall that $\Delta G^o$ is the change in free energy upon converting one mole of reactants into a stoichiometric equivalent of products (**Figure 12.3**). In this case, $\Delta G^o$ is the change in free energy when 1 mole of protein binds to 1 mole of ligand under

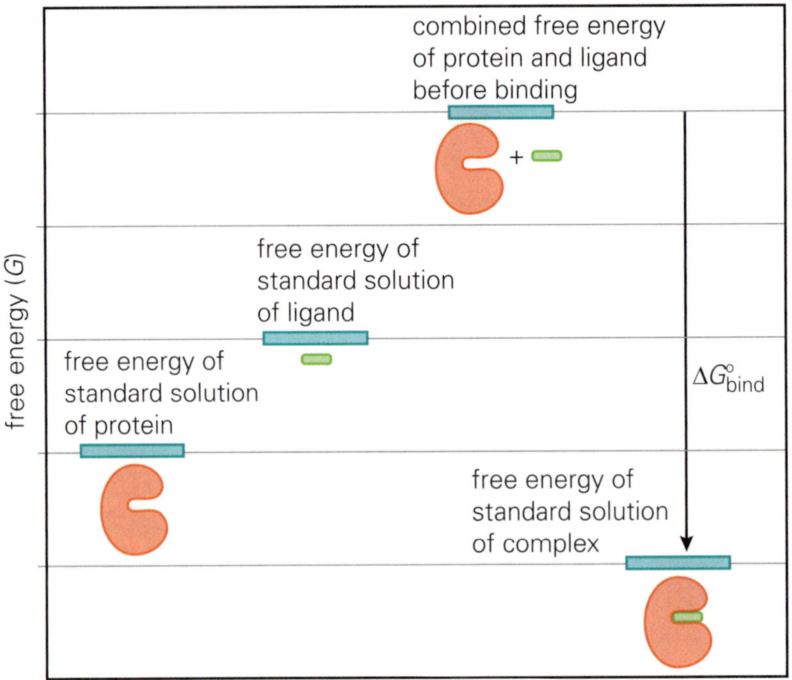


**Figure 12.3 Schematic diagram showing the changes in free energy upon ligand binding.** The standard free-energy change for binding, $\Delta G^o_{bind}$, refers to the conversion of one mole of protein and one mole of ligand to a complex under standard conditions.

standard conditions (1 molar solution of each). The standard free-energy change upon complex formation is called the **binding free-energy change** or, more simply, just the **binding free-energy**, $\Delta G^o_{bind}$:

$$\Delta G^o_{bind} = -RT \ln K_A \tag{12.5}$$

The value of $\Delta G^o_{bind}$ is a measure of the **affinity** of the interaction, that is, how strongly the molecules bind to each other. It is common practice to characterize the affinity of a binding interaction in terms of the equilibrium constant for the dissociation reaction, $K_D$, rather than the association constant, $K_A$. The dissociation reaction is simply the reverse of the association reaction:

$$P \bullet L \rightleftharpoons P + L \tag{12.6}$$

As a result, the **dissociation constant**, $K_D$, is the inverse of the association constant:

$$K_D = \frac{[P][L]}{[P \bullet L]} = \frac{1}{K_A} \tag{12.7}$$

It follows from Equations 12.5 and 12.7, then, that the binding free-energy is given by:

$$\Delta G^o_{bind} = +RT \ln K_D \tag{12.8}$$

Although the dissociation constant is a dimensionless number, it is usually discussed as if it has molar units of concentration (see Section 12.3). Biologically important nonconvalent interactions have dissociation constants that range from picomolar to nanomolar (that is, $10^{-12}$–$10^{-9}$ M) for the tightest interactions, to millimolar (that is, $10^{-3}$ M) for the weakest ones (**Table 12.1**). These correspond to standard free-energy changes upon binding of approximately −50 kJ•mol$^{-1}$ for the tightest interactions to approximately −17 kJ•mol$^{-1}$ for the weaker ones. Small-molecule drugs usually bind very tightly to their target proteins, with dissociation constants in the nanomolar ($10^{-9}$ M) to picomolar ($10^{-12}$ M) range. The value of the dissociation constant is sometimes simply referred to by biochemists as the "affinity" of an interaction; an interaction with a dissociation constant of 1 nanomolar is described as having a *nanomolar affinity*.

**Affinity**

The affinity of a molecular interaction refers to its strength. The greater the decrease in free energy upon binding, the greater the affinity. Another important concept is the specificity of the interaction, which refers to the relative strength of the interactions made between one protein and alternative ligands. Biologically relevant interactions are usually highly specific, as discussed in more detail in Chapter 13.

**Table 12.1 Typical strengths of different kinds of interactions.**

| Type of interaction | $K_D$ (molar) | $\Delta G^\circ_{bind}$ (at 300 K) (kJ•mol$^{-1}$) |
|---|---|---|
| Enzyme–ATP | ~1 × $10^{-3}$ to ~1 × $10^{-6}$ (millimolar to micromolar) | −17 to −35 |
| Signaling protein binding to a target | ~1 × $10^{-6}$ (micromolar) | −35 |
| Sequence-specific recognition of DNA by a transcription factor | ~1 × $10^{-9}$ (nanomolar) | −52 |
| Small molecule inhibitors of proteins (drugs) | ~1 × $10^{-9}$ to ~1 × $10^{-12}$ (nanomolar to picomolar) | −52 to −69 |
| Biotin binding to avidin protein (one of the strongest known noncovalent interactions) | ~1 × $10^{-15}$ (femtomolar) | −86 |

## 12.2 The value of $K_D$ corresponds to the concentration of free ligand at which the protein is half saturated

The reason that the dissociation constant, $K_D$, is more commonly referred to than the association constant, $K_A$, is that the value of $K_D$ is equal in magnitude to the concentration of free ligand at which half the protein molecules are bound to ligand (and half are unliganded) at equilibrium (**Figure 12.4**). The value of $K_D$ is therefore determined readily if we have some way of measuring the fraction of protein molecules that are bound to ligand.

It is straightforward to see why the value of $K_D$ corresponds to the ligand concentration at which the protein is half saturated. Let us define a parameter, *f*, which is the **fractional saturation** or **fractional occupancy** of the ligand binding sites in the protein molecules. If we assume that each protein molecule can bind to one ligand molecule, then *f* is the ratio of the number of protein molecules that have ligand bound to them to the total number of protein molecules (see Figure 12.4). In terms of concentrations, *f* can be expressed as:

$$f = \frac{\text{concentration of protein with ligand bound}}{\text{total protein concentration}} = \frac{[\text{P}\bullet\text{L}]}{[\text{P}]+[\text{P}\bullet\text{L}]} \qquad (12.9)$$

Using Equation 12.7, we can relate [P•L] to the dissociation constant as follows:

$$[\text{P}\bullet\text{L}] = \frac{[\text{P}][\text{L}]}{K_\text{D}} \qquad (12.10)$$

Substituting the expression for [P•L] from Equation 12.10 into Equation 12.9, we get:

$$f = \frac{[\text{P}][\text{L}]}{K_\text{D}\left([\text{P}]+\frac{[\text{P}][\text{L}]}{K_\text{D}}\right)}$$

$$\Rightarrow f = \frac{[\text{L}]}{K_\text{D}\left(1+\frac{[\text{L}]}{K_\text{D}}\right)} = \frac{[\text{L}]}{K_\text{D}+[\text{L}]} = \frac{\frac{[\text{L}]}{K_\text{D}}}{1+\frac{[\text{L}]}{K_\text{D}}} \qquad (12.11)$$

By using Equation 12.11, we can calculate the value of the fractional saturation, *f*, when the ligand concentration is equal in magnitude to the value of the dissociation constant. That is, if

$$f = \frac{[\text{L}]}{[\text{L}]+K_\text{D}}$$

**Fractional saturation, *f***

The fractional saturation is the extent to which the binding sites on a protein are filled with ligand. For a protein with a single ligand binding site, the value of *f* is given by the ratio of the concentration of the protein with ligand bound to the total protein concentration. The fractional saturation is an important parameter, because experimentally measurable responses to ligand binding usually depend directly on the fractional saturation.

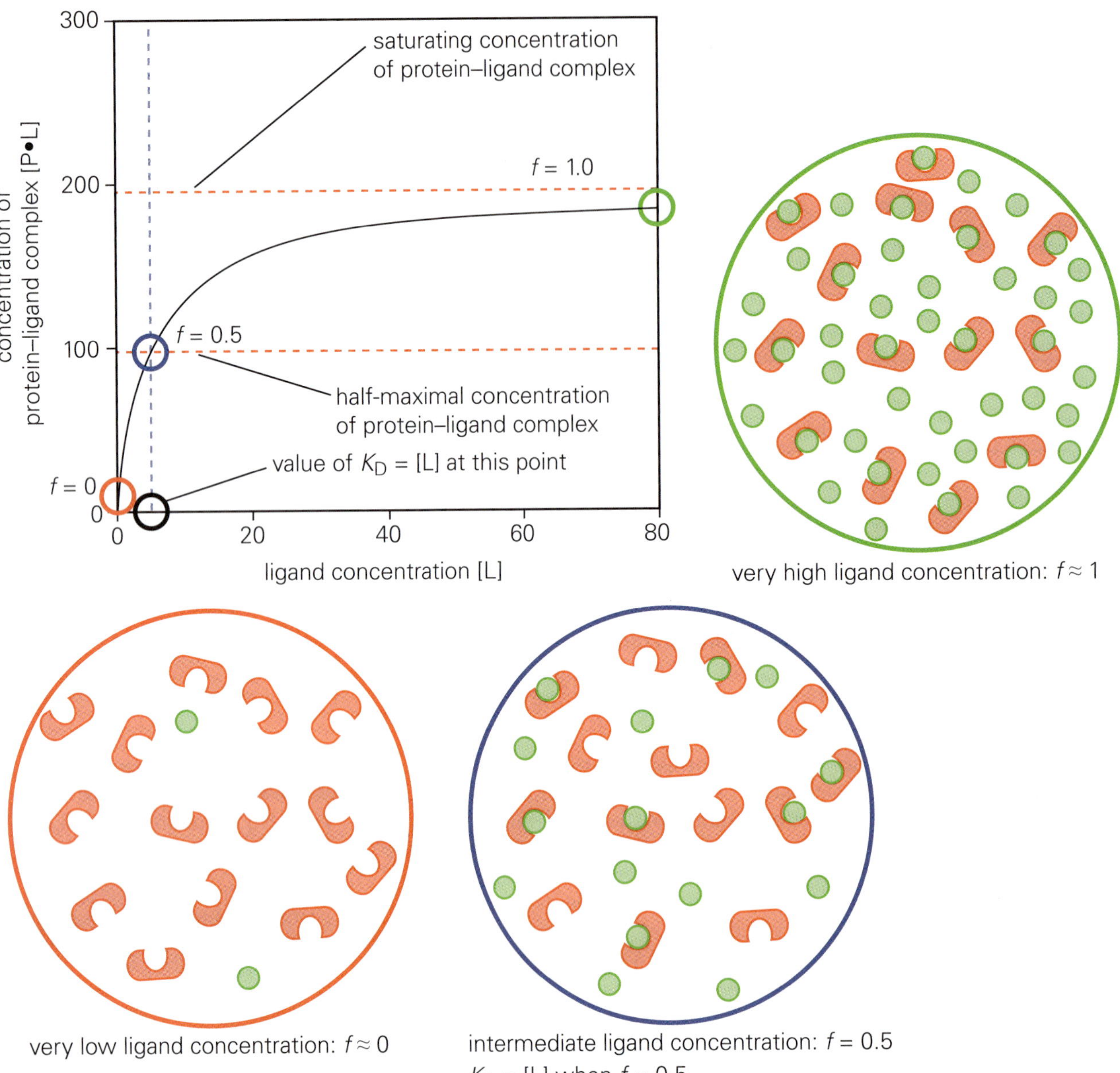


**Figure 12.4 The fractional saturation, *f*, as a function of free ligand concentration, [L].** The value of *f* ranges from zero (no binding) to 1.0 (all protein molecules are bound to ligand). The graph of *f* versus [L] shown here is known as a binding isotherm. The shape of the binding curve is that of a rectangular hyperbola.

then when [L] = $K_D$

$$f = \frac{K_D}{K_D + K_D} = \frac{1}{2} \tag{12.12}$$

According to Equation 12.12, when the protein is half saturated (that is, when half the protein molecules in the solution have ligand bound to them), then the value of the ligand concentration is equal to the dissociation constant (see Figure 12.4).

A plot of fractional saturation, *f*, as a function of ligand concentration, measured at constant temperature, is known as a **binding isotherm** or **binding curve**. The term "isotherm" refers to the fact that all of the measurements have to be made at a constant temperature in order for the binding curve to be meaningful. Note that the fractional occupancy, *f*, at a given concentration of free ligand, [L], depends on the dissociation constant, $K_D$ (Equation 12.7). The value of $K_D$ depends, in turn, on the temperature (see Equation 12.8). That is, the dissociation "constant" is a constant only if the temperature is maintained at a constant value. If the temperature is allowed to fluctuate while a series of binding measurements are made then the results will make little sense.

The shape of the binding isotherm shown in Figure 12.4 is referred to as a *rectangular hyperbola*, which is a curve traced out by a cone when it intersects a plane. The binding isotherm for the simple equilibrium represented by Equation 12.1 is sometimes referred to as a **hyperbolic binding isotherm**.

**Binding isotherm**

A series of measurements of the extent of binding or the fractional saturation, *f*, as a function of ligand concentration is known as a binding isotherm. Such a measurement can be analyzed to yield the dissociation constant only if all of the measurements are made at the same temperature and, hence, the series of measurements is called an *isotherm*.

## 12.3 The dissociation constant is a dimensionless number, but is commonly referred to in concentration units

The dissociation constant, like all equilibrium constants, is a dimensionless number. This has to be true, as we can see by considering Equation 12.8:

$$\Delta G^{o}_{bind} = RT \ln K_D$$

$\Delta G^{o}_{bind}$ has units of energy (for example, kJ•mol$^{-1}$), as does $RT$ on the right-hand side of the equation. Hence, for the units to balance, $K_D$ must be a pure number. If $K_D$ is dimensionless, how is it that we equate $K_D$ with ligand concentration in Equation 12.12? The apparent discrepancy in the units of $K_D$ arises because we customarily omit the values of the standard state concentrations in the definition of the equilibrium constants (see Section 10.7).

If we write out the complete expression for the dissociation constant, then we have the following expression (see Equation 10.54):

$$K_D = \frac{\dfrac{[P]}{[P]^o}\dfrac{[L]}{[L]^o}}{\dfrac{[P\bullet L]}{[P\bullet L]^o}} \tag{12.13}$$

where $[P]^\circ$, $[L]^\circ$, and $[P\bullet L]^\circ$ are standard state concentrations and are numerically equal to 1 M, and are therefore usually not written out explicitly.

**Hyperbolic binding curve or isotherm**

The simple non-allosteric binding of a ligand to a protein results in a hyperbolic relationship between the fractional saturation, *f*, and the ligand concentration [L]. Because of this relationship, a simple ligand-binding equilibrium is referred to as hyperbolic binding. Deviation from the hyperbolic shape of the binding curve is evidence for more complicated phenomena, such as allostery or multiple binding sites with different affinities.

We can rewrite equation 12.13 as:

$$K_D = \left(\frac{[P\bullet L]^o}{[P]^o[L]^o}\right)\frac{[P][L]}{[P\bullet L]} = \left(\frac{[P\bullet L]^o}{[P]^o[L]^o}\right)K_D^* \tag{12.14}$$

where $K_D^*$, a pseudo equilibrium constant, is given by:

$$K_D^* = \left(\frac{[P][L]}{[P\bullet L]}\right) \tag{12.15}$$

$K_D^*$ has units of concentration, and its value is equal to the ligand concentration at which the protein is half saturated. Because the value of the term $\left(\dfrac{[P\bullet L]^o}{[P]^o[L]^o}\right)$ in Equation 12.14 is 1.0, the numerical values of $K_D$ and $K_D^*$ are the same, even though they have different units. We use $K_D$ and $K_D^*$ interchangeably in practice, and will often use molar units when referring to $K_D$.

## 12.4 Dissociation constants are determined experimentally using binding assays

Dissociation constants are derived experimentally from binding isotherms, which rely on methods for measuring the amount of ligand bound to the protein. There are many different ways of making such a measurement, known as a **binding assay**. Exactly how a binding assay is carried out depends on the details of the interaction being monitored and the ingenuity of the biochemical investigator. Here we discuss an example in which radioactivity is used to monitor the amount of ligand bound to the receptor for the hormone estrogen (**Figure 12.5**).

Estrogen is a hormone in females, and its receptor is a site-specific DNA binding protein. The estrogen receptor belongs to a large family of closely related transcription factors known as the **nuclear** or **steroid hormone receptors**. The estrogen receptor consists of two important domains, one that binds to the hormone

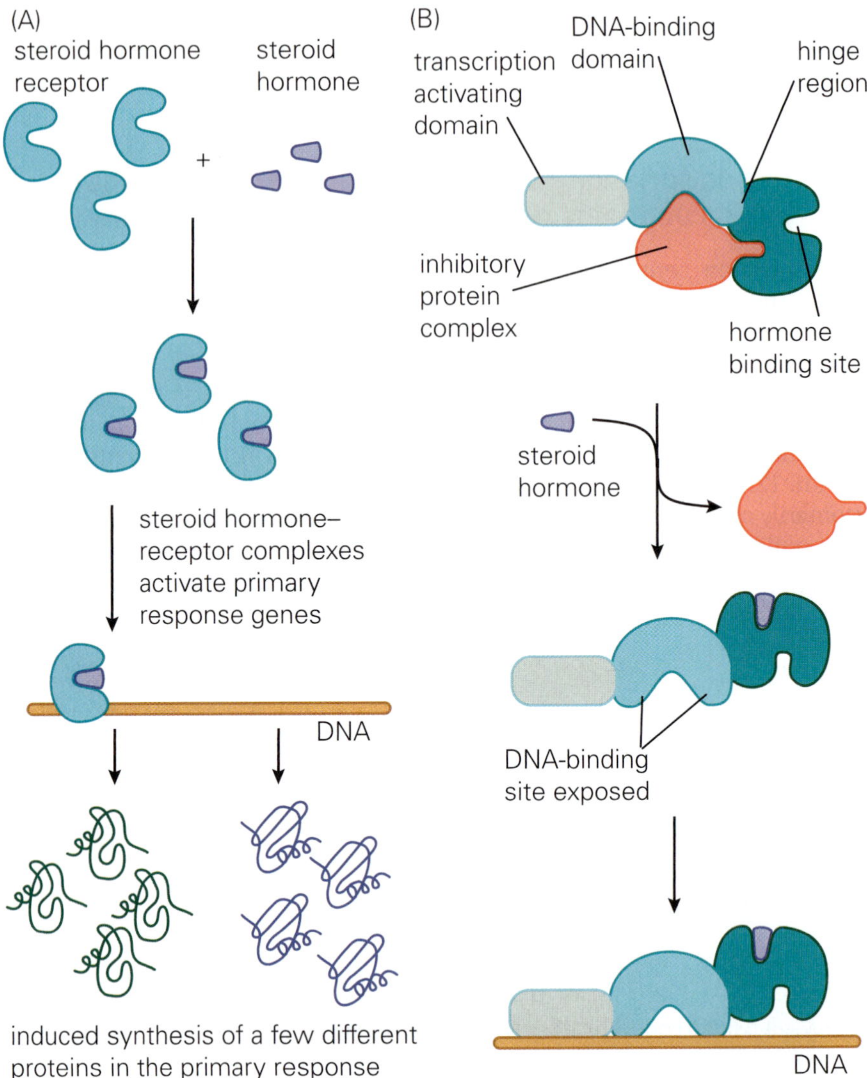


**Figure 12.5 Mechanism of steroid receptors, such as the estrogen receptor.** (A) Simplified functional diagram: these receptors bind to specific sites on DNA and activate transcription, but only when bound to their specific ligand (for example, estrogen). (B) Structural mechanism: the binding of the hormone to its receptor causes a conformational change, exposing the DNA-binding domain of the receptor, allowing it to interact with target DNA sequences. The active conformation is actually a dimer (second molecule not shown). The activated receptor also binds to proteins that are responsible for recruiting the transcriptional machinery (not shown here). In this complicated pathway, the amount of signaling can be modified by many other factors, such as phosphorylation of the receptor. (Adapted from B. Alberts et al. Molecular Biology of the Cell, 5th ed. New York: Garland Science, 2008.)

and one that binds to DNA (see Figure 12.5). When estrogen binds to the receptor, it promotes the dimerization of the receptor, which facilitates the binding of the receptor to sites on DNA that contain specific recognition sequences. Binding of estrogen to the receptor also induces a conformational change in the ligand-binding domain, which results in the recruitment of proteins known as transcriptional co-activators to the receptors. The co-activator proteins are responsible for turning on transcription from the gene.

In the binding assay shown in **Figure 12.6**, the estrogen sample contains a known amount of radioactively labeled estrogen that has been synthesized separately and mixed in with the normal estrogen (see Figure 12.6C). It is assumed that the presence of the radioactive isotope in the labeled estrogen molecule does not affect its ability to bind to the receptor. This allows us to assume that the amount of radioactivity that remains associated with the receptor after the free ligand is removed is proportional to the total amount of ligand bound by the receptor.

In order to measure the amount of ligand bound by the receptor, we need a way to separate the bound ligand from the unbound ligand. In the experiment shown in Figure 12.6, a negatively charged resin is added to the solution. The estrogen receptor binds to this resin, and a centrifugation step separates the bound from

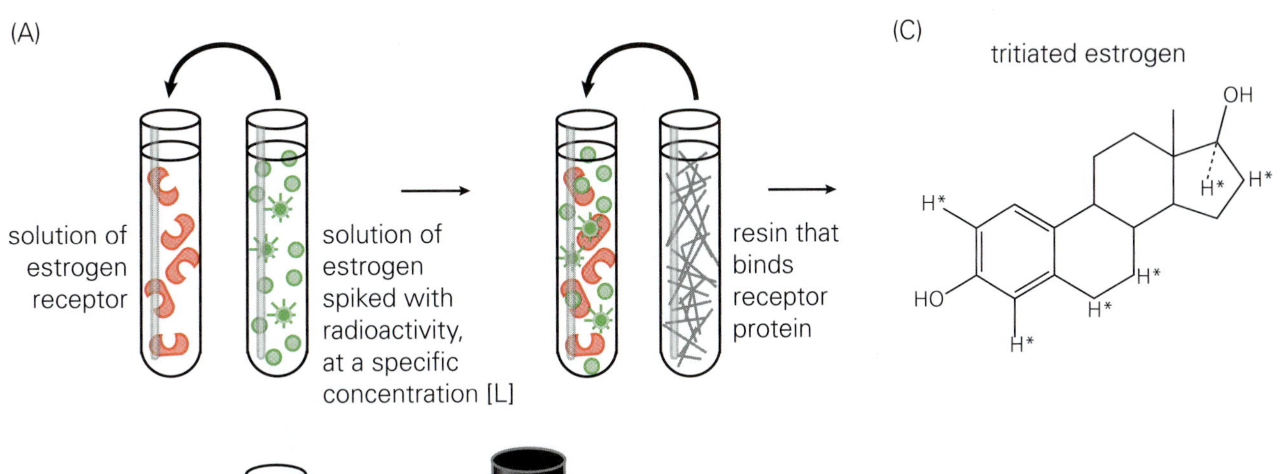


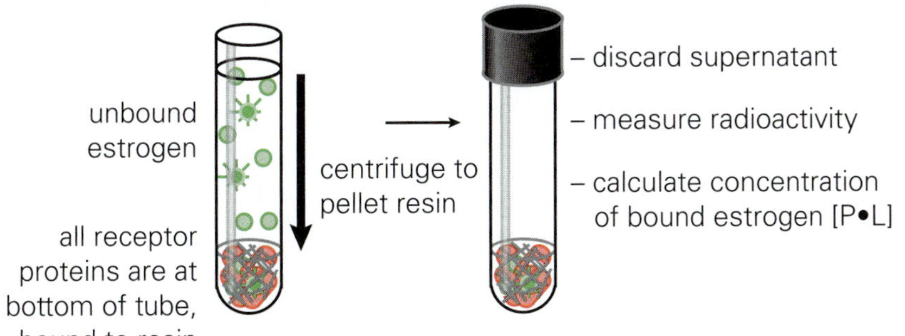


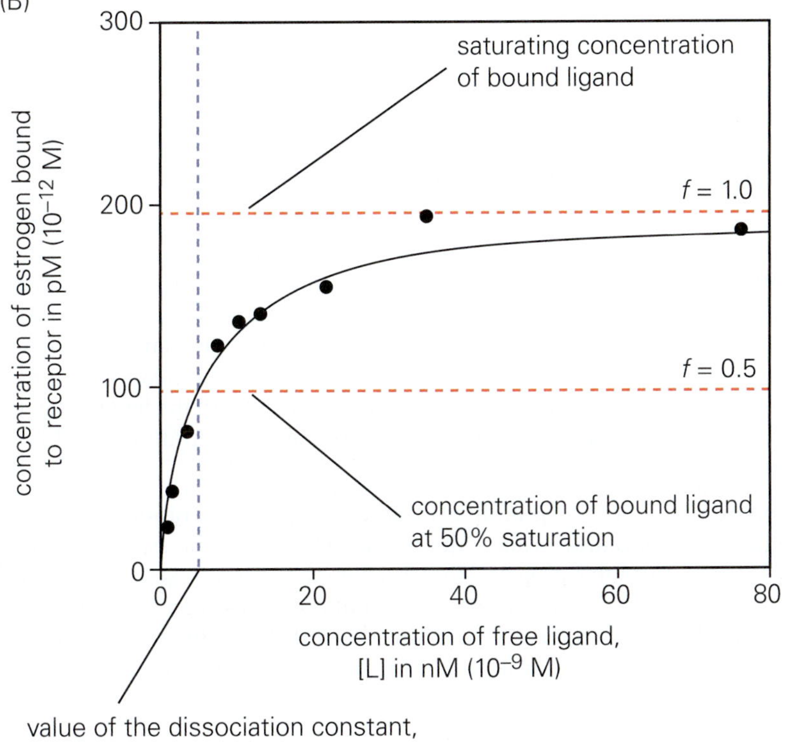


**Figure 12.6 Binding isotherm for estrogen binding to its receptor.** (A) In the particular assay shown here, the solution of estrogen contains a known fraction of radioactively labeled estrogen. Separation of the receptor–estrogen complex from unbound estrogen makes it possible to determine the bound ligand concentration ([P•L]) by measuring the radioactivity. (B) The binding isotherm, generated by plotting [P•L] as a function of [L], reaches a plateau value, for which $f = 1.0$. The value of the dissociation constant, $K_D$, is given by the ligand concentration at the half-maximal value of $f$, the fractional saturation. (C) The chemical structure of 17-β estradiol, the particular estrogen used in this experiment. Sites where hydrogen is replaced by tritium ($^3H$) are indicated by asterisks. (B, adapted from D.N. Petersen et al., and T.A. Brown, *Endocrinology* 139: 1082–1092, 1998. With permission from The Endocrine Society.)

the unbound ligand. The pelleted resin with the protein is transferred to a vial, and the total amount of radioactivity in it is estimated by using a liquid scintillation counter. Another common way to separate the bound ligand from the free ligand is to pass the solution through a filter that allows the solution to flow through but

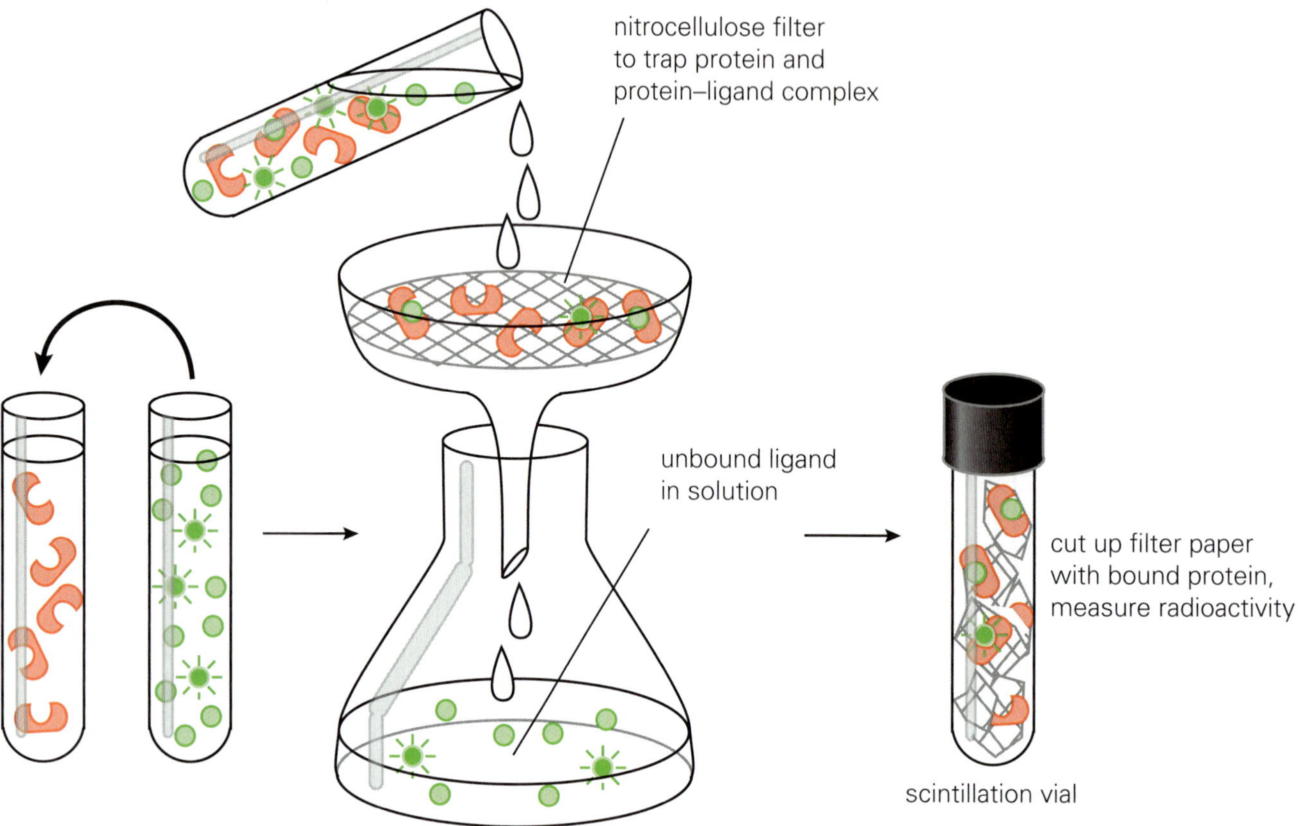


**Figure 12.7 Filter binding assays.** Proteins stick to filters made of material such as nitrocellulose, which allows ligand bound to the protein to be separated from unbound ligand. The presence of ligand bound to protein is then detected using a readout such as radioactivity.

to which the protein molecules (and the ligands bound to them) adhere. Filters made of nitrocellulose are commonly used for this purpose, and such a **filter binding assay** is illustrated in **Figure 12.7**.

The amount of bound ligand is plotted as a function of the total ligand concentration (usually assumed to be equal to the free ligand concentration, see Section 12.6) to obtain a binding isotherm, as shown in Figure 12.6B In this experiment, the radioactively labeled estrogen contains tritium atoms instead of normal hydrogens at several positions (Figure 12.6C). The tritium atoms emit β particles (that is, high-velocity electrons), which are detected by the scintillation counter. The amount of estrogen bound to the protein is proportional to the level of radioactivity detected and is plotted as a function of estrogen concentration to yield the binding isotherm.

Notice in Figure 12.6B that the amount of bound estrogen in the binding isotherm reaches a maximum or plateau value. This occurs when all of the estrogen receptor molecules are bound to estrogen—that is, when the receptor is saturated. The concentration of estrogen at which the amount of bound estrogen is half that of the saturating value gives us the dissociation constant. The value of $K_D$ for estrogen binding to estrogen receptor is ~5 nM ($5 \times 10^{-9}$ M), according to the data in Figure 12.6.

## 12.5 Binding isotherms plotted with logarithmic axes are commonly used to determine the dissociation constant

A binding curve or isotherm, such as the one shown in **Figure 12.8**A, is a graph of the value of $f$, the fractional saturation of the protein, as a function of the ligand concentration, [L]. The value of $K_D$ can be estimated by reading off the concentration at which the value of $f$ is equal to 0.5. As we can see in Figure 12.8A, the range of concentrations over which the value of $f$ changes from low to high is relatively narrow, and the most informative data points are crowded together on the left side of the graph. This can make it difficult to estimate the value of $f$ by visual inspection of the binding isotherms. We could expand this region of the graph to more easily read out the value of $K_D$. Alternatively, we could switch to a graph with logarithmic axes, which would spread the data out more conveniently.

One such logarithmic graph involving fractional saturation and ligand concentration is shown in Figure 12.8B. This kind of graph turns out to be particularly useful for analyzing allostery in binding, and so we introduce it here and apply it to the analysis of allostery in Chapter 14.

To understand the nature of the graph in Figure 12.8B, we start with the expression for the fraction, $f$, of the protein that is bound to the ligand (that is, Equation 12.11):

$$f = \frac{[\mathrm{L}]}{[\mathrm{L}] + K_\mathrm{D}}$$

The fraction of the protein that is *not* bound to ligand is given by $1 - f$:

$$1 - f = 1 - \frac{[\mathrm{L}]}{[\mathrm{L}] + K_\mathrm{D}} = \frac{[\mathrm{L}] + K_\mathrm{D} - [\mathrm{L}]}{[\mathrm{L}] + K_\mathrm{D}} = \frac{K_\mathrm{D}}{[\mathrm{L}] + K_\mathrm{D}} \qquad (12.16)$$

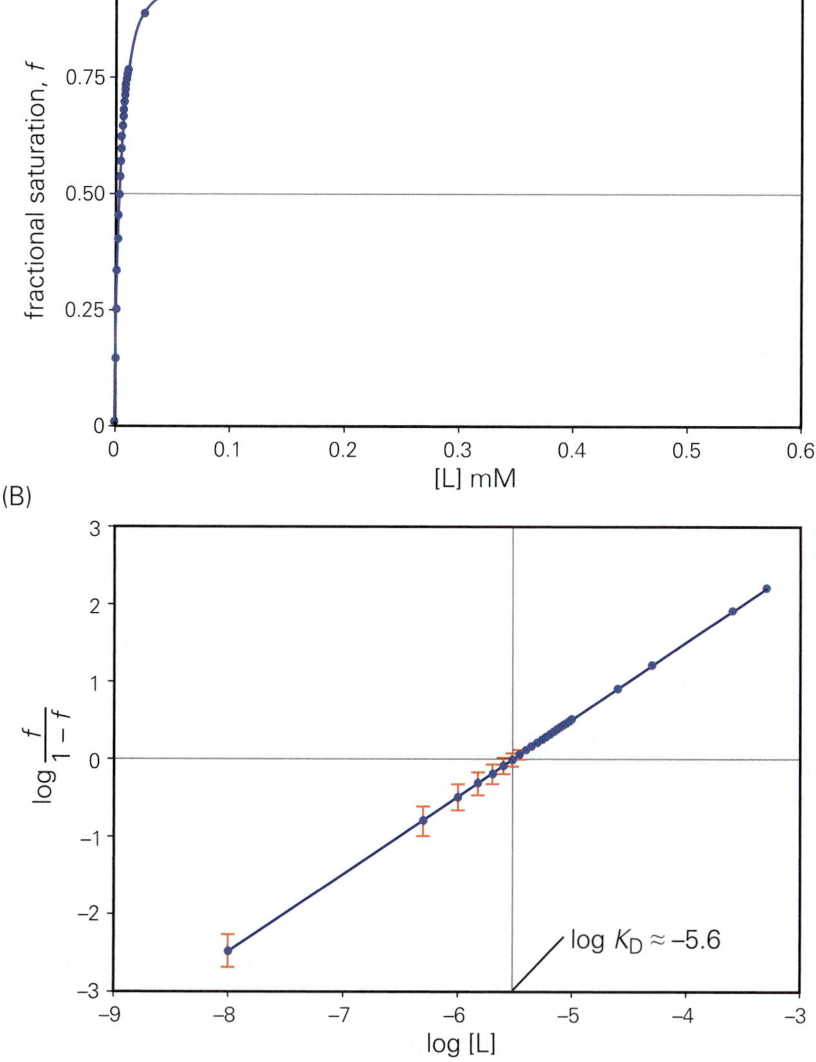


**Figure 12.8. Binding isotherms with logarithmic axes.** (A) A normal binding isotherm. The binding of ligand to protein is measured over a wide range of ligand concentration, which leads to some of the data being compressed into a small region of the graph. (B) The data shown in (A) are graphed using logarithmic axes. A linear plot is obtained when $\log\left(\frac{f}{1-f}\right)$—that is, when $\log\left(\frac{\text{fraction bound}}{\text{fraction unbound}}\right)$—is graphed versus log [L]. Note that the values of $\log\left(\frac{f}{1-f}\right)$ associated with low values of log [L] often have large errors associated with them because of errors in measurement when the ligand concentration is very low.

Using Equations 12.11 and 12.16, we can calculate the ratio of the fraction of protein that is bound to the fraction that is unbound:

$$\frac{\text{fraction bound}}{\text{fraction unbound}} = \frac{f}{1-f} = \frac{[\mathrm{L}]}{[\mathrm{L}]+K_\mathrm{D}} \frac{[\mathrm{L}]+K_\mathrm{D}}{K_\mathrm{D}} = \frac{[\mathrm{L}]}{K_\mathrm{D}} \tag{12.17}$$

Taking the logarithm of both sides of Equation 12.17, we get:

$$\log\left(\frac{f}{1-f}\right) = \log\left(\frac{[\mathrm{L}]}{K_\mathrm{D}}\right) = \log[\mathrm{L}] - \log K_\mathrm{D} \tag{12.18}$$

As shown in Figure 12.8B, graphing the value of $\log\left(\frac{f}{1-f}\right)$ as a function of log [L] yields a straight line. When the protein is half saturated—that is, when $\left(\frac{f}{1-f}\right)=1$, then $\log\left(\frac{f}{1-f}\right)$ is zero. The intercept of the line on the horizontal axis is therefore equal to log $K_\mathrm{D}$.

A logarithmic graph such as the one shown in Figure 12.8B is a convenient way of checking the assumption that the protein binds to the ligand in a simple way, as described by Equation 12.1. As we shall see in Chapter 14, if the actual binding isotherm is not linear, or has a slope that is not unity, then this could be an indication that the actual binding process is more complex, and might include factors such as allostery.

The slope of the binding isotherm and the value of its intercept on the horizontal axis can be difficult to determine accurately if the binding data have errors in them. Values of the fractional saturation determined at low ligand concentration are particularly error prone, because the detection signal (for example, fluorescence or radioactivity) is correspondingly weak at low ligand concentration. Care must therefore be taken to ensure that errors associated with very weak signals do not unduly bias the analysis of the binding isotherm.

## 12.6 When the ligand is in great excess over the protein, the free ligand concentration, [L], is essentially equal to the total ligand concentration

In Figure 12.8, the critical parameter is the free ligand concentration, [L]—that is, the concentration of the ligand that is not bound to the protein. In most situations we are more concerned with the total ligand concentration, $[\mathrm{L}]_{total}$, because this is something we know directly from the total amount of ligand added to the system under study. For example, if a patient takes a pill that contains 500 mg of a drug, the total concentration of the drug in the blood can be estimated by knowing the volume of blood in a typical human body (~5 liters). The free ligand concentration, [L], is a different matter, and can only be determined, in principle, by making a measurement.

In many biochemical applications, including the study of drug binding, a simplification occurs because the number (or concentration) of protein molecules is usually very small compared to that of the drug (**Figure 12.9**). The maximum concentration of bound ligand, $[\mathrm{L}]_{bound}$, is therefore very small compared to the total ligand concentration, $[\mathrm{L}]_{total}$, if protein concentration is very low compared to total ligand concentration:

$$[\mathrm{L}]_{bound} << [\mathrm{L}]_{total} \tag{12.19}$$

Because the total ligand concentration is the sum of the free ligand concentration, [L], and the bound ligand concentration, $[\mathrm{L}]_{bound}$, it follows that the free ligand concentration is essentially the same as the total ligand concentration when $[\mathrm{L}]_{bound} << [\mathrm{L}]_{total}$:

$$[\mathrm{L}]_{total} = [\mathrm{L}] + [\mathrm{L}]_{bound} \approx [\mathrm{L}] \tag{12.20}$$

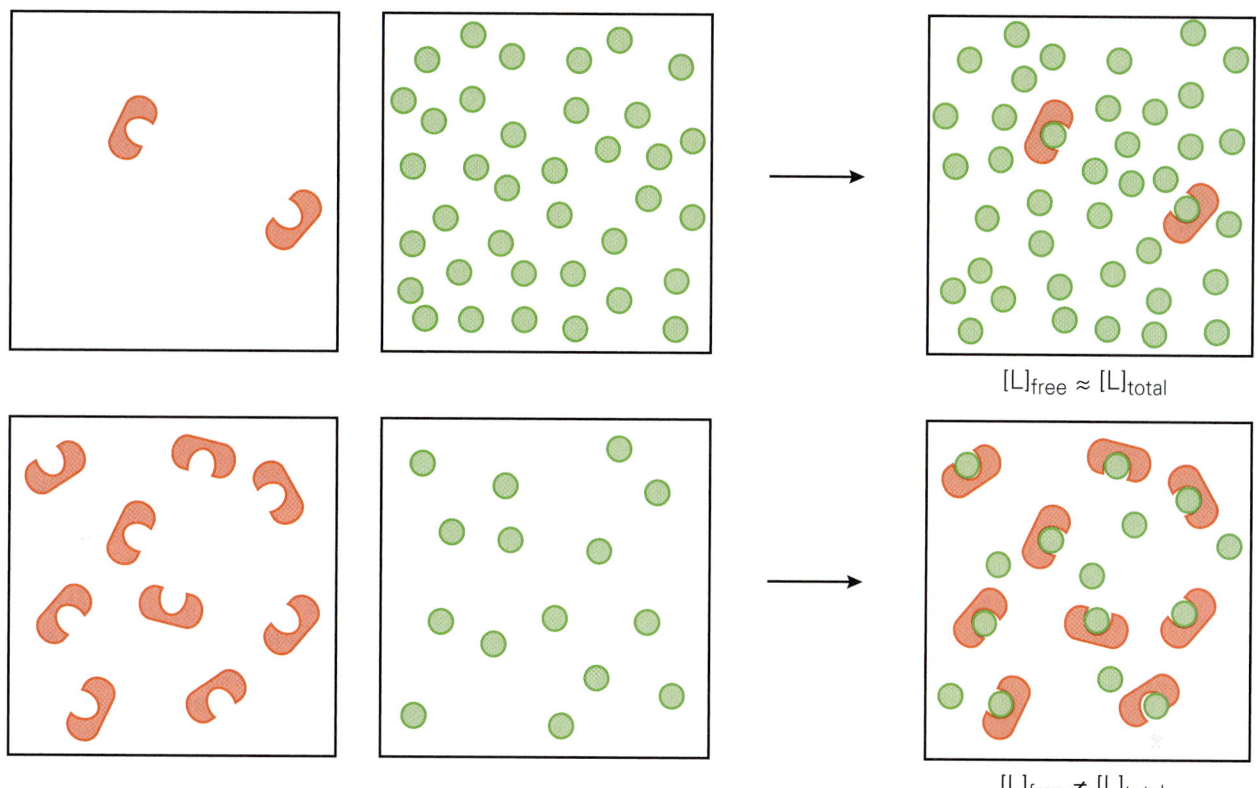


**Figure 12.9 The free ligand concentration.** (A) When the concentration of ligand is much greater than that of the protein, then the concentration of the unbound (free) ligand, [L], does not change much as ligand binds to protein. (B) When the concentrations of ligand and protein are comparable, the free ligand concentration is affected by how much ligand is bound to protein.

In calculations involving the saturation of protein binding sites by a ligand, we often assume that the amount of bound ligand is very small compared to the total amount of ligand available, and in such cases we use the free ligand concentration and the total ligand concentration interchangeably.

## 12.7 Scatchard analysis makes it possible to estimate the value of $K_D$ when the concentration of the receptor is unknown

In the preceding analysis of binding isotherms, we assumed that both the protein and ligand concentrations were known. Without knowing the protein concentration, we cannot calculate the fractional saturation, *f*, knowledge of which is critical for determining the value of $K_D$. There are many situations in biology where it is straightforward to determine the concentration of bound and unbound ligand, but the protein concentration is not directly measurable. This is the case, for example, when we study the binding of a ligand to a cellular protein, without fractionation or purification.

When the assumption of single-site binding is valid, a method known as **Scatchard analysis**, named after physical chemist George Scatchard, allows us to determine the dissociation constant as well as the total protein concentration.

Let us return to the definition of the dissociation constant, $K_D$, given originally in Equation 12.7:

$$K_D = \frac{[P][L]}{[P\bullet L]}$$

**Scatchard analysis**

A simple binding equilibrium between a protein and a ligand results in the hyperbolic binding curve shown in Figure 12.9B. Deviations from the hyperbolic curve, however, can be difficult to detect visually. Scatchard analysis involves rearranging the basic binding equation to yield the following form, known as the Scatchard equation:

$$\frac{[L]_{bound}}{[L]} = -\frac{1}{K_D}[L]_{bound} + \frac{[P]_{total}}{K_D}$$

The Scatchard equation, which is an alternative form of the hyperbolic binding equation, tells us that the ratio of bound to free ligand concentration is related linearly to the bound ligand concentration (Figure 12.10C).

The concentration of the protein–ligand complex, [P•L], is the same as the concentration of the bound ligand, $[L]_{bound}$, assuming that the ligand does not bind to anything else. We can therefore express the concentration of the free protein, [P], as follows:

$$[P]=[P]_{total}-[P\bullet L]=[P]_{total}-[L]_{bound} \quad (12.21)$$

Hence,

$$K_D=\frac{\left([P]_{total}-[L]_{bound}\right)[L]}{[L]_{bound}} \quad (12.22)$$

Rearranging Equation 12.22, we get:

$$\frac{[L]_{bound}}{[L]}=\frac{[P]_{total}-[L]_{bound}}{K_D} \quad (12.23)$$

Or,

$$\frac{\text{concentration of bound ligand}}{\text{concentration of free ligand}}=\frac{[L]_{bound}}{[L]}$$

$$=-\frac{1}{K_D}[L]_{bound}+\frac{[P]_{total}}{K_D} \quad (12.24)$$

Equation 12.24 is known as the **Scatchard equation**. It tells us that, for a simple binding equilibrium, the ratio of the concentrations of the free and bound ligands is related linearly to the concentration of the bound ligand. An example of a Scatchard plot is shown in **Figure 12.10**C. The slope of the line is related inversely to the dissociation constant, and knowing the protein concentration is not required to derive this value. The total protein concentration, $[P]_{total}$, is in fact obtained from the analysis, because the intercept of the Scatchard plot on the vertical axis is the ratio of the values of the protein concentration and the dissociation constant.

## 12.8 Scatchard analysis can be applied to unpurified proteins

As an example of the application of the Scatchard equation, we shall look at the binding of the hormone retinoic acid to its receptor, the retinoic acid receptor. Retinoic acid is a derivative of vitamin A and is a very important signaling molecule in mammalian development. Many of the effects of retinoic acid are transduced by the retinoic acid receptor, which is a relative of the estrogen receptor discussed in Section 12.4.

Scatchard analysis lets us measure the binding properties of the retinoic acid receptor in cellular extracts without going to the trouble of purifying the receptor. Cells expressing the receptor are lysed, and the cell lysate containing the receptors is used for a series of binding measurements, each one corresponding to a different total ligand (retinoic acid) concentration, but with everything else kept the same. As for the determination of the binding isotherm (see Figure 12.5), samples containing normal retinoic acid are spiked with retinoic acid that contains the radioactive isotope tritium ($^3H$). For each binding measurement, the receptor-containing lysate and the $^3H$-labeled retinoic acid are mixed together for several hours to ensure that the mixture reaches equilibrium.

For the case of the receptor–retinoic acid interaction, advantage is taken of the fact that the free retinoic acid binds to charcoal, whereas the retinoic acid bound to the receptor does not (see Figure 12.10). Note that this procedure for separating bound ligand is different from that used to separate bound estrogen in the experiment discussed earlier (see Figure 12.5). It is often the case that a distinctive binding assay needs to be developed to suit the needs of the particular system under study.

Pellets of charcoal are added to the reaction mixture, and the charcoal is separated from the main solution by centrifugation. It is assumed that the rate at which the retinoic acid dissociates from the charcoal is slow enough that this procedure

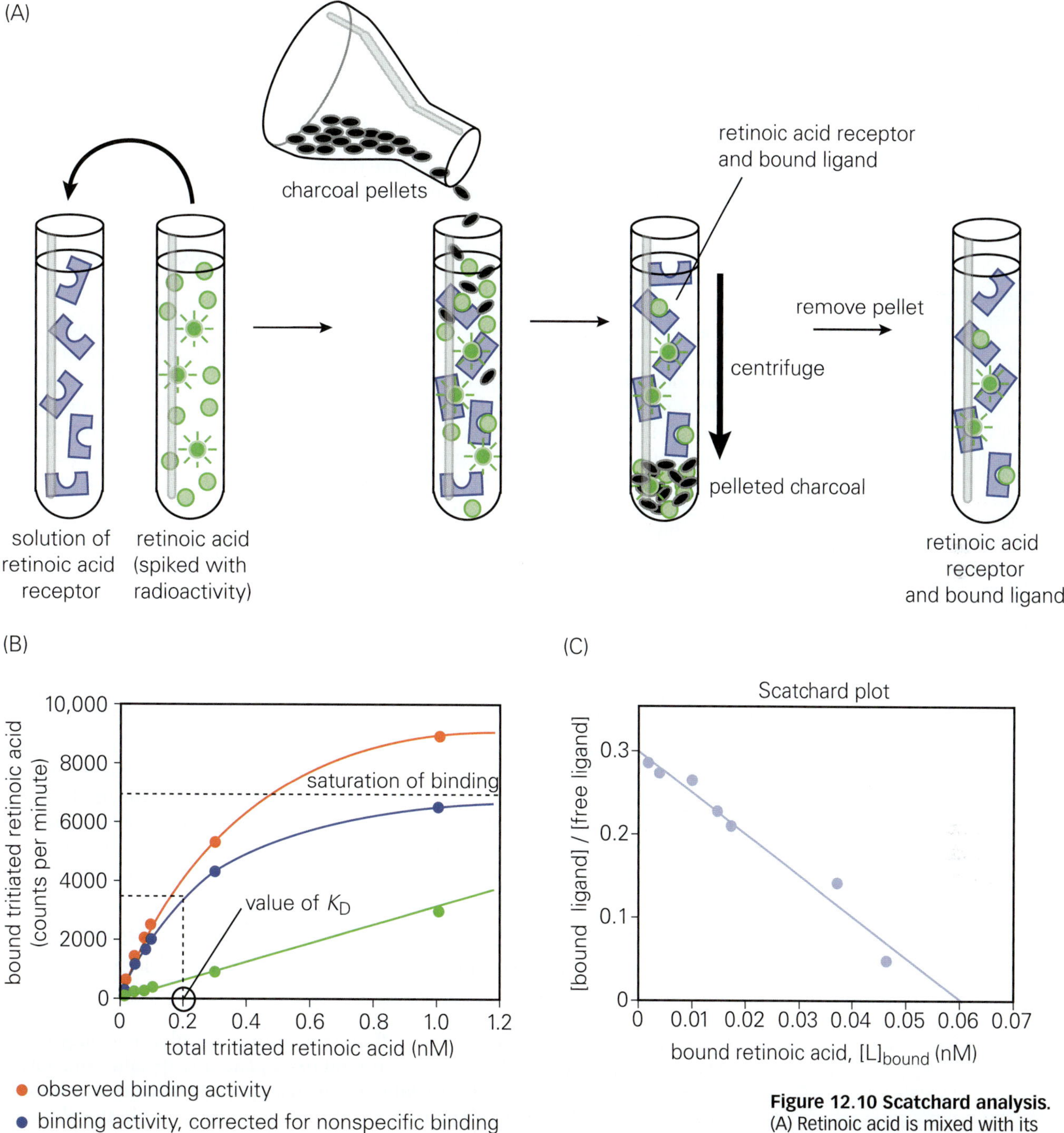


**Figure 12.10 Scatchard analysis.** (A) Retinoic acid is mixed with its receptor and the unbound retinoic acid is separated by binding it to charcoal. (B) The observed binding isotherm is shown in *red*. The observed binding data contain a contribution from nonspecific binding to material such as the plastic in the test tube (*green*). The corrected binding isotherm (*blue*) is used for further analysis. (C) The Scatchard plot of the corrected binding isotherm. (B and C, adapted from N. Yang et al., and R.M. Evans, *Proc. Natl. Acad. Sci. USA* 88: 3559–3563, 1991. With permission from the National Academy of Sciences.)

leads to a faithful separation of the receptor-bound and free retinoic acid (see Figure 12.10A). Once the bound retinoic acid is separated from the unbound portion, the total amount of $^3$H-labeled retinoic acid in both fractions can be readily measured using a scintillation counter to determine the amount of radioactive material in the samples. The amount of bound retinoic acid as a function of the concentration of the free retinoic acid concentration can be graphed as shown in Figure 12.10B.

According to Equation 12.24, if we plot the ratio of bound ligand to free ligand as a function of bound ligand concentration, then we should get a straight line. This is indeed the case for the binding of retinoic acid to its receptor, as shown in Figure

12.10C. The slope of the line is equal to $-\frac{1}{K_D}$, making it possible to determine the value of the dissociation constant, which is ~0.2 nM in this case. The intercept of the line on the vertical axis is $\frac{[P]_{total}}{K_D}$. The Scatchard analysis, therefore, allows us to determine the concentration of the receptor protein in the cell lysates, even though the receptor protein was not purified.

## 12.9 Saturable binding is a hallmark of specific binding interactions

**Saturable binding**

In a situation where a ligand binds to a single binding site on a protein and no other, all of the protein molecules are bound to ligand at very high ligand concentrations. Increasing the ligand concentration beyond this point does not lead to any increase in protein binding. Such a binding interaction is referred to as being saturable. When a binding isotherm does *not* saturate, even at very high ligand concentrations, then it usually indicates that the ligand is also binding to things other than the protein of interest.

In any binding measurement, there is usually a need to correct the data for systematic effects that lead to distortions in the apparent values of the amount of ligand bound to the protein. In the case of the retinoic acid receptor, for instance, it turns out that there is a significant amount of nonspecific binding of retinoic acid to something other than its receptor. This can be seen by adding a 100-fold excess of unlabeled retinoic acid to each of the binding reactions (see Figure 12.10B). We would now expect all of the receptor molecules to bind predominantly to the unlabeled retinoic acid. Nevertheless, when this is done, a certain amount of labeled retinoic acid is still seen to be retained in the fraction left behind after the charcoal is removed (see Figure 12.10B). This binding occurs, presumably, to other proteins in the cell lysate or to the material (such as plastic) that makes up the reaction chamber. Whatever the nonspecific target may be, it offers so many binding sites that the addition of 100-fold excess of unlabeled retinoic acid still leaves sufficient binding sites to capture some labeled retinoic acid. If we subtract this nonspecific binding from the total binding measured in the absence of unlabeled retinoic acid, then we get a corrected binding isotherm (see Figure 12.10B).

Notice that the corrected binding isotherm in Figure 12.10B shows saturation of the binding—that is, the amount of bound retinoic acid reaches a maximum plateau value and then does not increase further. A plateau value in the binding isotherm, referred to as **saturable binding**, is a hallmark of the binding of a ligand to a defined binding site on a specific protein whose availability is limited. In contrast, nonspecific binding often shows no evidence of reaching a plateau value.

## 12.10 The value of the dissociation constant, $K_D$, defines the ligand concentration range over which the protein switches from unbound to bound

A question we are often concerned with in biochemistry or pharmacology is the extent to which a particular protein is bound to a ligand at a specific concentration of the ligand. For example, if a patient is to take a pill that delivers an inhibitor for an enzyme, what should be the concentration of the inhibitor in the blood in order to have most of the enzyme bound to the inhibitor? A related question concerns the specificity of the interaction. Most ligands will bind to more than one protein in the cell. Can we choose a ligand concentration such that one protein is bound to the ligand and another is not?

For a simple binding equilibrium involving one ligand and one protein, it is straightforward to estimate what the ligand concentration has to be in order to saturate the protein. The protein goes from having very little ligand bound to being almost saturated within a concentration range that extends from ~$0.1K_D$ to ~$10K_D$—that is, over a concentration range that spans two orders of magnitude. For example, if the concentration of the free ligand, [L], is 10 times the value of $K_D$, then the value of the fractional saturation, $f$, is given by:

$$f = \frac{[L]}{[L]+K_D} = \frac{10K_D}{10K_D+K_D} = \frac{10}{11} = 0.91 \quad (12.25)$$

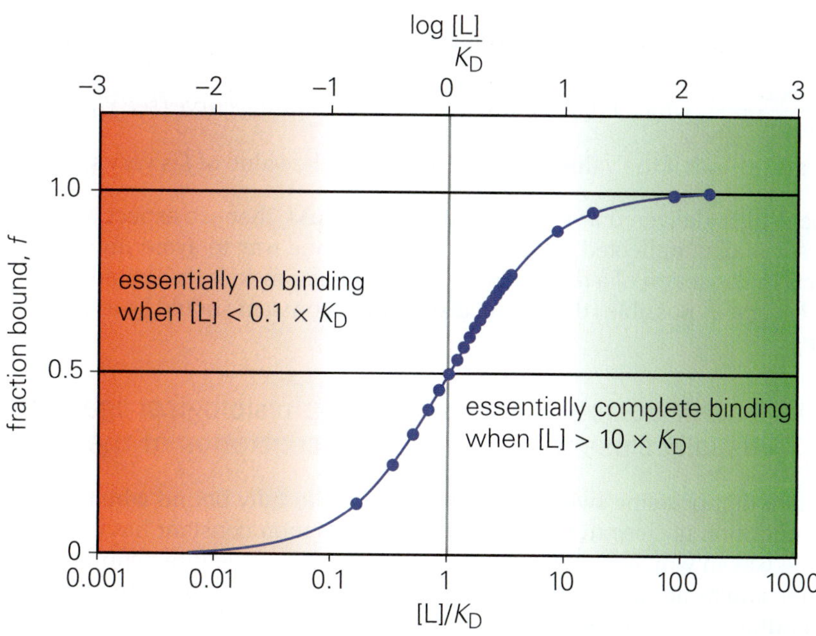


**Figure 12.11 A "universal" binding isotherm.** This graph expresses concentration in terms of the dimensionless number $\frac{[\mathrm{L}]}{K_\mathrm{D}}$. This graph is "universal" in the sense that it applies to any simple binding equilibrium.

Hence, the protein is 91% saturated when the ligand concentration is 10 times greater than the value of $K_\mathrm{D}$. Likewise, when $[\mathrm{L}] = 0.1K_\mathrm{D}$, then the fractional saturation is given by:

$$f = \frac{0.1K_\mathrm{D}}{K_\mathrm{D} + 0.1K_\mathrm{D}} = \frac{0.1}{1.1} = 9\% \tag{12.26}$$

At this lower concentration, only 9% of the protein is bound to the ligand. Thus, the ligand concentration range that is within a factor of 10 on either side of the value of the dissociation constant is the range in which the protein switches from being essentially unbound to nearly completely bound.

We can appreciate the way in which the population of protein molecules switches from bound to unbound by plotting the ligand concentration on a logarithmic scale, since in practice the concentrations of ligand under consideration span several orders of magnitude [for example, picomolar (that is, $10^{-12}$ M) to millimolar (that is, $10^{-3}$ M)]. It is particularly useful to plot the fractional saturation, $f$, as a function of $\log\left(\frac{[\mathrm{L}]}{K_\mathrm{D}}\right)$. By expressing the ligand concentration in terms of the dissociation constant, we get a "universal" binding curve (**Figure 12.11**) that is helpful in the discussion of ligands binding to alternative target proteins (see Chapter 13).

As an example, consider a drug that binds to a target protein, A, with a dissociation constant of 1 nM ($10^{-9}$ M). The interaction between the drug and protein A is critical for treatment of a disease. Now imagine that the drug also binds to another protein, B, and that this interaction has undesirable side effects. Let us suppose that the binding of the drug to protein B occurs with a dissociation constant of 10 μM ($10^{-5}$ M). How do we determine a concentration at which to deliver the drug so that protein A is essentially shut down by the drug, while protein B is essentially unaffected?

Using the universal binding curve in Figure 12.11, we look for a concentration range within which binding to A is maximal while binding to B is minimal. As the value of $\frac{[\mathrm{L}]}{K_\mathrm{D}}$ approaches 100, the value of $f$ approaches 1.0. Since the value of $K_\mathrm{D}$ for protein A is $1{\times}10^{-9}$ M (0.001 μM), protein A will be essentially saturated if the drug is delivered at a concentration of 0.1 μM (note that we assume that Equation

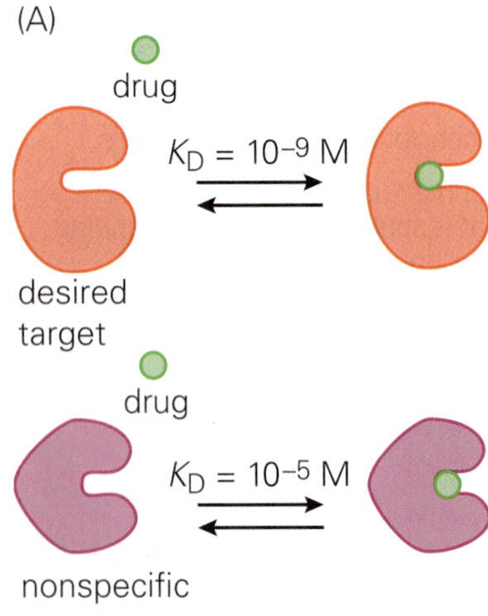


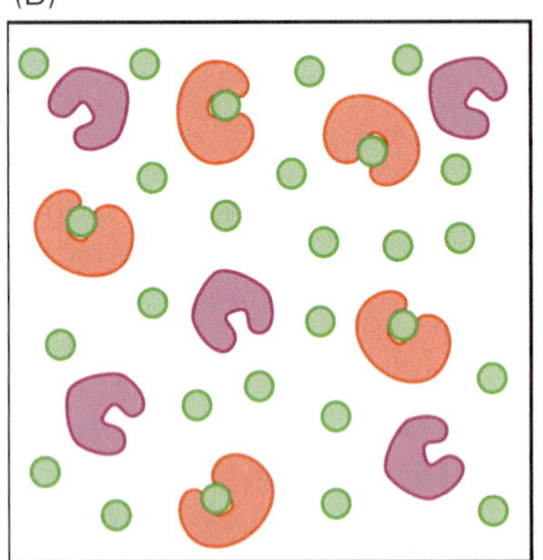


**Figure 12.12 Affinity and specificity in drug binding.** (A) A drug binds tightly to a desired protein and weakly to an undesired target. (B) The drug is delivered at a concentration that is much below the value of $K_D$ for the undesired target. Very little binding to the undesired target occurs. Binding to the desired target is maintained if the concentration is above the value of $K_D$ for that target.

**Evolutionary relationship between dissociation constants and physiological concentrations of ligands**

The dissociation constant of a protein for a naturally occurring ligand is usually close to the physiological concentration of the ligand. Much tighter interactions are unnecessary, because the protein is 99% saturated when the ligand concentration is 100 times greater than the dissociation constant.

12.20 holds true). At this concentration of the drug, the value of $\frac{[L]}{K_D}$ for protein B is $\frac{0.1\times10^{-6}\,M}{1\times10^{-5}\,M}$, which is 0.01. From the universal binding curve (see Figure 12.11), we can see that if the value of $\frac{[L]}{K_D}$ is 0.01, then the value of $f$ is very small. Thus, if the drug is delivered at a concentration of 0.1 μM, then we expect protein B to be essentially unaffected (**Figure 12.12**). Thus, one way to avoid unwanted side effects in the action of a drug is to make its interaction with its desired target protein as tight as possible (that is, the dissociation constant should be as low as possible).

## 12.11 The dissociation constant for a physiological ligand is usually close to the natural concentration of the ligand

The fact that proteins switch from being empty to fully bound when the ligand concentration is close to the value of the dissociation constant has implications for the way in which evolution "tunes" the strength of the interaction between a protein and its natural ligands. In most cases, the dissociation constant for a natural binding interaction is lower by no more than a factor of 10–100 than the physiological concentration of the ligand. For example, the concentration of ATP in the cell is approximately 1 mM ($10^{-3}$ M). Later in the chapter we discuss enzymes known as protein kinases, which bind to ATP and transfer the terminal phosphate group to the sidechains of proteins. The dissociation constant of ATP for protein kinases is typically ~10 μM ($10^{-5}$ M)—that is, approximately one-hundredth that of the physiological ATP concentration. Certain motor proteins known as kinesins, which utilize ATP as a fuel to power the movement of organelles and other objects inside the cell, also bind to ATP with a similar dissociation constant, even though kinesins are completely unrelated to the protein kinases in terms of structure and mechanism.

It is easy to understand why the dissociation constant of a protein for its natural ligand is relatively close to the physiological concentration of the ligand. Suppose that a protein accumulates mutations that lead to an increased affinity for the ligand, such that the dissociation constant is much smaller than the natural concentration of the ligand. If a value of $\frac{[L]}{K_D} = 100$ is reached, then the saturation ($f$) is given by (see Equation 12.11):

$$f = \frac{\frac{[L]}{K_D}}{1+\frac{[L]}{K_D}} = \frac{100}{101} = 99\%$$

At this point the protein is essentially saturated, and further increases in affinity will not lead to any appreciable increases in ligand binding to the protein. As a consequence, mutations that do lead to higher affinity will not have an evolutionary advantage, and will likely disappear due to evolutionary drift. Furthermore, if the dissociation constant becomes too small, the protein would always have ligand bound. Signaling systems in cells use ligand binding to proteins as a way to turn proteins on and off, and always having the ligand bound would prevent normal "on"–"off" signaling in such systems. Even if the ligand concentration drops to very low levels, the tighter binding may make the rate of dissociation of the ligand very slow (see Chapter 15), which may interfere with function.

Mutations that weaken the binding, so that the value of $K_D$ increases much beyond the physiological level of the ligand, will result in a failure of the protein to bind to the ligand. Because of the loss of function, such mutations will also be selected against.

# B. DRUG BINDING BY PROTEINS

The concepts introduced in the first section of this chapter find practical application in the process of drug development. In this section we discuss some of the important principles governing the action of small molecules that block protein targets.

## 12.12 Most drugs are developed by optimizing the inhibition of protein targets

The development of drugs for a specific disease often begins with the identification of the proteins that are involved in steps critical to disease progression, and then proceeds to the design or discovery of molecules that inhibit the function of these proteins. In the initial steps of this process, the molecules that are first discovered to inhibit the target protein usually do not have all of the properties that are desirable in a drug. Such molecules are called **lead compounds**. Lead compounds are sometimes discovered serendipitously, as might happen through the isolation and identification of active ingredients in medicinal plants. Alternatively, lead compounds are discovered through screening a large collection of organic compounds (called a "library") or by designing molecules that would be expected, based on chemical knowledge, to fit into the active site of the protein target.

**Lead compound**

Small molecules that inhibit a target protein but do not have high affinity or other desirable properties are called lead compounds. Drug development often involves the systematic modification and optimization of lead compounds.

An example of a lead compound that has been designed to inhibit a protein is shown in **Figure 12.13**. The target protein is an enzyme known as a **protein kinase**. Protein kinases are enzymes that transfer the terminal phosphate of ATP to the hydroxyl groups of serine, threonine, or tyrosine residues on proteins. The malfunctioning of one such kinase, known as Abl, causes **chronic myelogenous leukemia**, a cancer in which white blood cells proliferate without control. Pharmaceutical chemists have developed an inhibitor of Abl known as imatinib (marketed as Gleevec™), which blocks the action of Abl and is an effective treatment for the leukemia.

**Protein kinases**

These proteins carry out cellular signaling by phosphorylating serine, threonine, or tyrosines in specific sets of target proteins. The human genome contains approximately 500 different protein kinases, all of which are very closely related in their catalytic domain (known as the kinase domain), but which respond to different input signals using additional domains with different functions. Bacterial cells also utilize kinases that phosphorylate histidine and aspartate residues, but these proteins form a distinct family that is unrelated to the protein kinases found in animals.

How was imatinib developed? The Abl kinase cannot function unless it can bind to ATP, and so molecules that bind very tightly to the ATP-binding site of the kinase will block the action of the enzyme. By studying how ATP binds to kinase proteins, chemists designed a small organic compound (shown in Figure 12.13C) to bind to the ATP-binding site of protein kinases. The compound is designed to form hydrogen bonds with the protein in a manner that mimics the adenine group of ATP, and it also mimics some aspects of the planar aromatic nature of the adenine group. The chemists chose to add substituents at three positions on the scaffold provided by this organic compound, leading to the development of lead compounds that inhibited the Abl kinase.

These lead compounds bound only weakly to the Abl kinase, and they also inhibited many other protein kinases. The chemists then tested a large number of compounds in an iterative series of steps, checking for increased affinity for Abl and decreased cross-reactivity at each step (the affinities of some of these compounds for Abl are shown in **Figure 12.14**). This process of **lead optimization** eventually led to imatinib, which is a highly successful cancer drug (see Figure 12.14).

## 12.13 Signaling molecules are protein targets in cancer drug development

The design of imatinib to block the Abl kinase is an example of the development of a drug to treat a particular cancer—chronic myelogenous leukemia. The Abl kinase happens to be a **tyrosine kinase** that catalyzes the phosphorylation of specific tyrosine residues in proteins that control signal transmission between cells. Tyrosine phosphorylation is a key signaling switch in animal cells, and it results in the activation of signaling pathways that control cell growth and differentiation. As you might imagine, there are many other forms of cancer that are caused by the aberrant behavior of signaling proteins such as the Abl kinase.

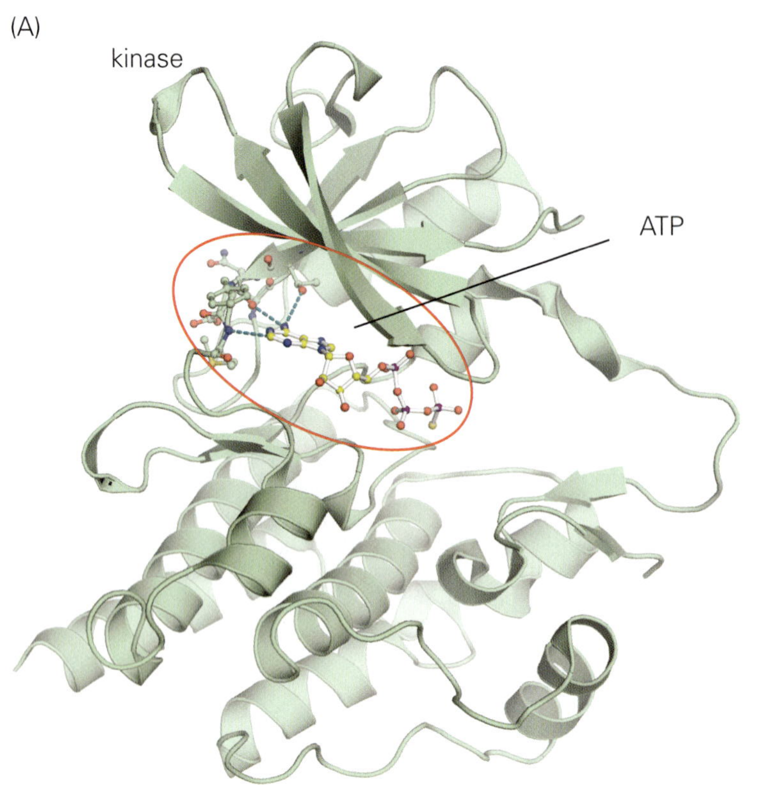


**Figure 12.13 Design of a lead compound to inhibit a protein kinase.** (A) The structure of the kinase, with ATP bound to it. Three of the many hydrogen bonds made by the adenine group of ATP are shown. (B) Expanded view of a part of the ATP-binding site, showing hydrogen bonds involving the adenine group, with flanking hydrophobic and polar pockets indicated. (C) A chemical scaffold for the development of lead compounds. The scaffold was designed to form two of the hydrogen bonds made by the adenine group. Substituents at three positions, marked $R_1$, $R_2$, and $R_3$, were used to develop lead compounds. (PDB code: 2G1T.)

Another example of a protein kinase that is an important target in cancer therapy is a cell surface protein known as the epidermal growth factor (EGF) receptor. EGF is a small protein hormone that conveys messages between cells and works by binding to and activating EGF receptors that are displayed by cells that it encounters (**Figure 12.15**). Like the Abl kinase, the EGF receptor is a protein tyrosine kinase.

Tyrosine kinases such as the EGF receptor are normally kept off, and their catalytic activity is released only when external signals received by the cell require them to turn on. Malfunctions in the control of the EGF receptor causes some

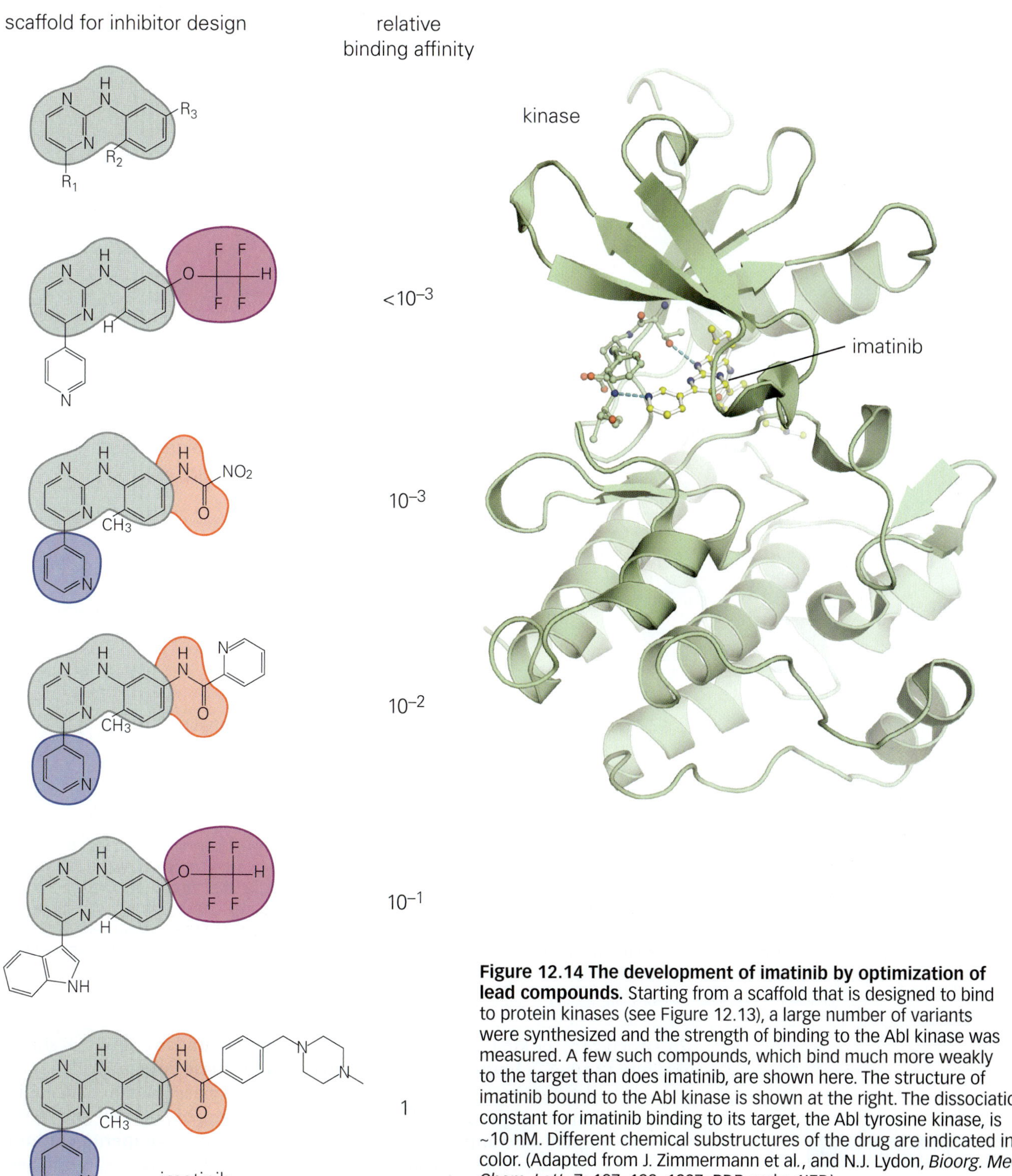


**Figure 12.14 The development of imatinib by optimization of lead compounds.** Starting from a scaffold that is designed to bind to protein kinases (see Figure 12.13), a large number of variants were synthesized and the strength of binding to the Abl kinase was measured. A few such compounds, which bind much more weakly to the target than does imatinib, are shown here. The structure of imatinib bound to the Abl kinase is shown at the right. The dissociation constant for imatinib binding to its target, the Abl tyrosine kinase, is ~10 nM. Different chemical substructures of the drug are indicated in color. (Adapted from J. Zimmermann et al., and N.J. Lydon, *Bioorg. Med. Chem. Lett.* 7: 187–192, 1997; PDB code: 1IEP.)

forms of breast and lung cancers, as well as some brain tumors known as glioblastomas. Two drugs, trastuzumab, and erlotinib are used to treat some of these cancers because they bind to and shut down the signaling activity of the EGF receptor (**Figure 12.16**). They are very different from each other, however, because trastuzumab is an antibody (a protein) while erlotinib is a small organic compound that is synthesized chemically (see Figure 12.16). Trastuzumab binds to the extracellular portion of the EGF receptor, whereas erlotinib binds to the intracellular kinase domain.

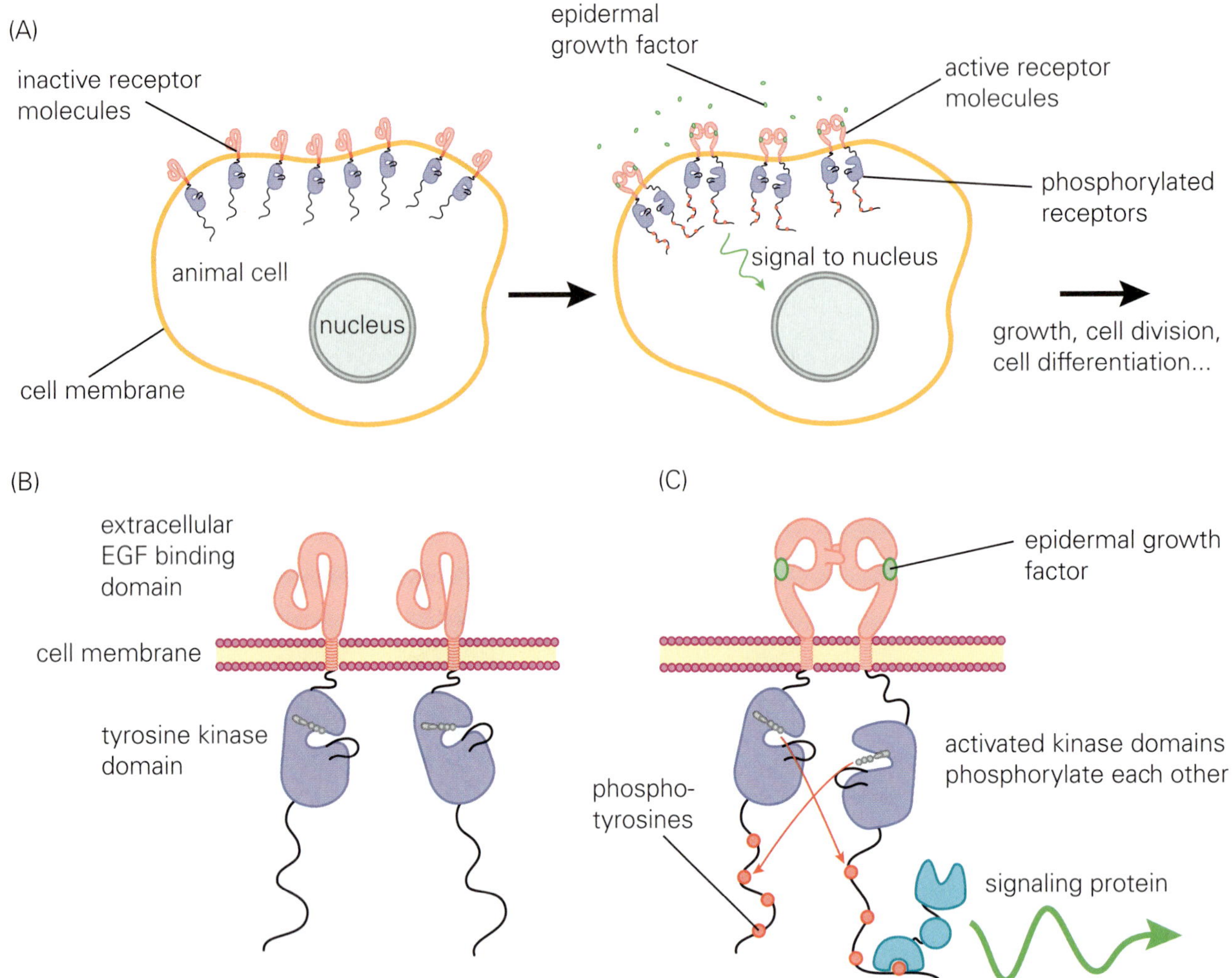


**Figure 12.15 The epidermal growth factor (EGF) receptor.** (A) The EGF receptor is normally in an inactive state on the surfaces of cells. When activated by the arrival of the EGF hormone (a small protein) outside the cell, the receptor sends signals to the nucleus and to various components of the cellular machinery. (B) The receptor contains an extracellular hormone (EGF) binding domain and an intracellular tyrosine kinase domain. (C) Upon activation, the tyrosine kinase domain phosphorylates the tail of the receptor, leading to the recruitment of other signaling proteins that then transmit the message downstream. In several cancers, the EGF receptor is activated improperly, leading to phosphorylation in the absence of an input signal.

The structure of the kinase domain of the EGF receptor bound to erlotinib is shown in Figure 12.16. ATP binds within a deep cavity in the kinase domain, which is the active site of the enzyme. When erlotinib binds to the kinase domain, it occupies much of the space required for ATP binding. The binding of erlotinib and ATP is therefore mutually incompatible, and erlotinib is known as a **competitive inhibitor** of the kinase domain with respect to ATP as is imatinib. As we shall see in Section 12.18, erlotinib binds much more tightly to the EGF receptor than does ATP, and the presence of even a small amount of erlotinib is sufficient to shut down the tyrosine kinase activity of the EGF receptor.

**Competitive inhibitor**

A molecule that blocks the functioning of a protein by displacing a naturally occurring ligand of the protein is known as a competitive inhibitor.
A noncompetitive inhibitor is one that binds to some site on the protein other than the binding site of the natural ligand and exerts its influence through an allosteric mechanism.

## 12.14 Most small molecule drugs work by displacing a natural ligand for a protein

Most drugs that are small molecules work by binding to and blocking naturally occurring ligand binding sites on proteins, as do imatinib and erlotinib. To illustrate this principle further, we describe two classes of drugs that are effective in the treatment of acquired immune deficiency syndrome (AIDS), both of which block the function of proteins produced by the human immunodeficiency virus (HIV) (**Figure 12.17**). The HIV viral genome is made up of RNA, and the target of one class of HIV drugs is the enzyme reverse transcriptase, an RNA-dependent DNA polymerase that converts viral genetic information from RNA to DNA (**Figure 12.18**). The DNA then feeds into the cellular transcriptional machinery, thereby directing the production of proteins that are essential for the replication of the virus. HIV drugs known as nucleotide analogs work by mimicking the structure of nucleotide triphosphates, which results in their binding tightly to the reverse transcriptase enzyme. The nucleotide analogs are so designed that they cannot be incorporated into a growing DNA chain, and they therefore prevent further replication of the viral genome (see Figure 12.18).

Another critical step in the life cycle of the HIV virus is the cleavage of large precursor protein molecules, produced by translation of the HIV genome, into smaller fragments that are the properly functional protein units (see Figure 12.17). This cleavage is carried out by a **protease** (a protein-cleaving enzyme), known as HIV protease, which is itself encoded by the viral genome. A second class of HIV drugs,

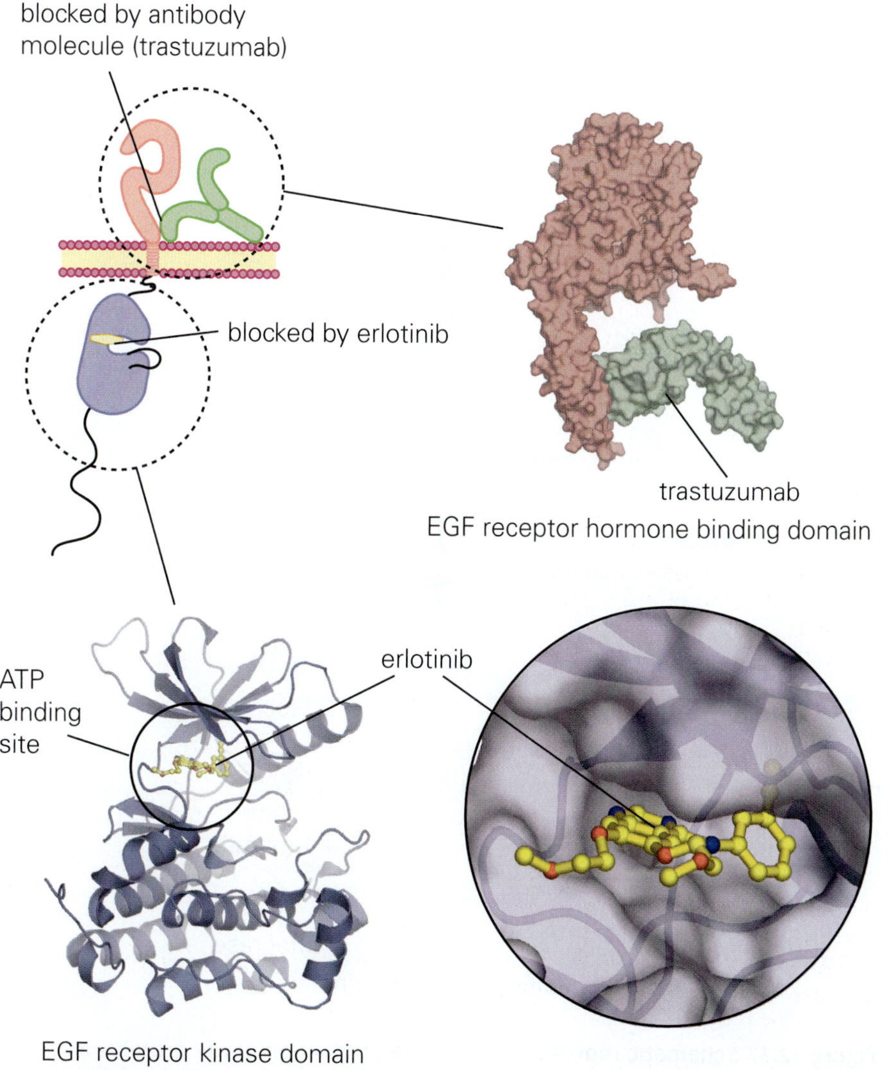


**Figure 12.16 The cancer drugs trastuzumab (Herceptin™) and erlotinib (Tarceva™) work by blocking the EGF receptor.** Trastuzumab is a protein antibody that binds to the external portion of the EGF receptor. Erlotinib is a small organic compound that enters cells and displaces ATP from the kinase domain of the EGF receptor. (PDB codes: 1M17, 1M14.)

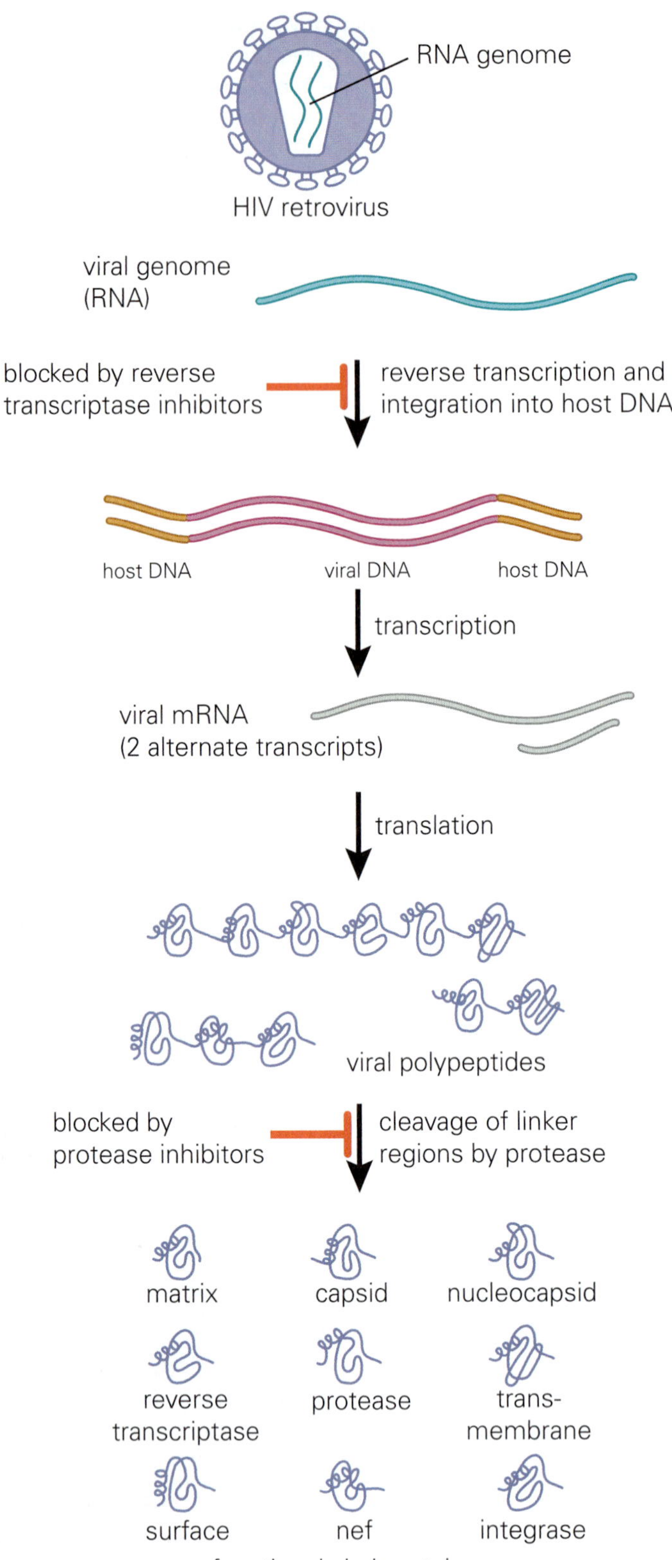


**Figure 12.17 Schematic representation of the production of HIV proteins in an infected cell.** The viral genomic RNA is reverse transcribed into DNA, which is integrated into the host DNA. The reverse transcriptase and integrase enzymes are necessary for this process and are targets of anti-HIV therapy. Once integrated, viral genes are translated into large polyproteins that need to be cleaved in order to generate the functioning forms of the individual viral proteins. This cleavage is catalyzed by HIV protease, another enzyme that is targeted by inhibitors in HIV therapy.

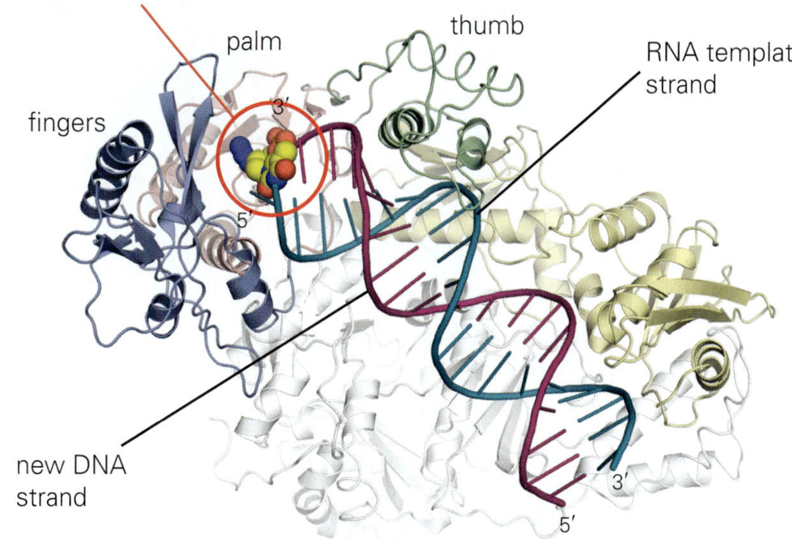


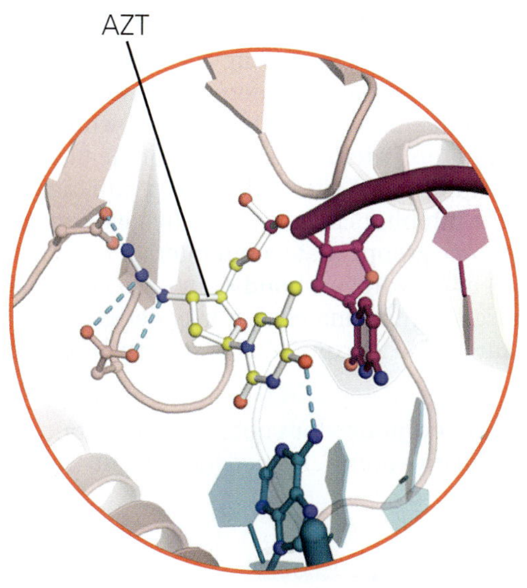


**Figure 12.18 Inhibition of HIV reverse transcriptase.** Small molecules that mimic the structures of nucleotides bind to the active site of HIV reverse transcriptase, thereby blocking the conversion of the viral genome into DNA. Shown here is an RNA-DNA hybrid double helix bound to the reverse transcriptase enzyme. An HIV drug known as AZT (3′-azido-3′-deoxythymidine) is bound to the enzyme, as shown in detail in the expanded view of the active site. AZT mimics a nucleotide, thus shutting down the enzyme. (PDB code: 1N5Y.)

known as protease inhibitors, work by binding to the active site of HIV protease and displacing its normal substrates (**Figure 12.19**). This prevents HIV protease from working, and the virus is then unable to produce the basic machinery that is required for its replication, thus stopping the viral infection.

It is difficult to develop small-molecule inhibitors that are effective against proteins that do not bind to naturally occurring small molecules, peptides, or nucleotides during their normal function. This is because, as we shall see, the high affinity of drugs for their targets usually arises from hydrophobic effects. Proteins such as enzymes and signaling switches contain cavities or invaginations where small molecules are bound naturally, as shown in **Figure 12.20** for a protein kinase. Such cavities provide opportunities for generating the hydrophobic interactions that can drive drug binding.

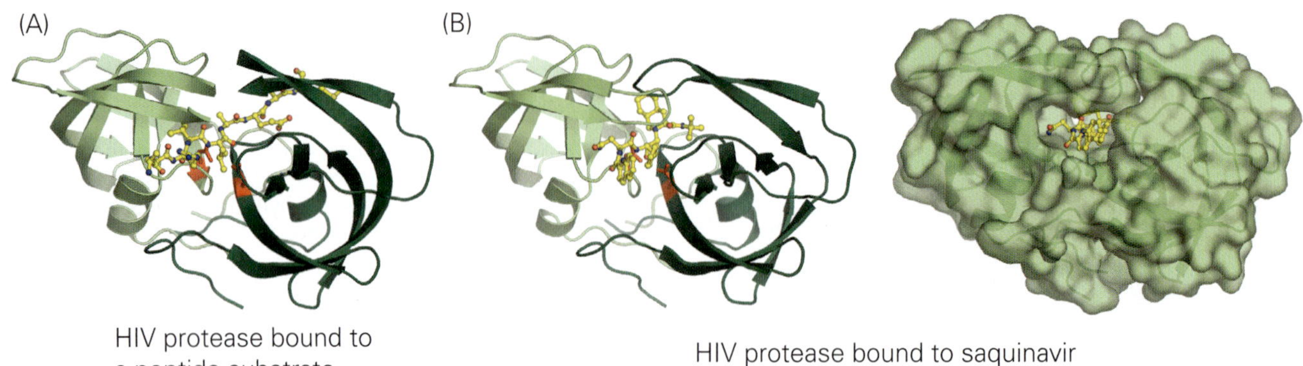


**Figure 12.19 HIV protease inhibitors.** (A) The structure of HIV protease bound to a peptide substrate. Residues at the active site of the enzyme are shown in *red*. (B) A drug known as saquinavir binds to the active site of HIV protease and displaces the substrate. Saquinavir binds to a deep channel in the protease that is normally occupied by the peptide substrate. (PDB codes: 1F7A, 1FB7.)

## 12.15 The binding of drugs to their target proteins often results in conformational changes in the protein

Proteins usually have very high specificity for a particular ligand while ignoring very closely related molecules. In 1894, long before the structural principles of proteins were understood, the insightful organic chemist Emil Fischer proposed the **lock-and-key model** for how a substrate binds to an enzyme. Fischer viewed the protein as a rigid "lock" and the ligand or substrate as a "key," as illustrated in **Figure 12.21**A. Only a key designed to fit a specific lock would fit well enough to allow binding and reaction. There can be many locks, each with a specific key that fits and is allowed to react, thus explaining the specificity of proteins.

The static pictures of protein structures, including the many illustrations in this textbook, lend themselves naturally to an interpretation of ligand binding in terms of the lock-and-key model. Based on this model, it would seem fairly easy to design high-affinity inhibitors, when the structure of the target protein is known, by placing atoms of the inhibitor in sites that overlap with the binding site of a natural ligand. Atoms of the inhibitor can be chosen to have complementary functionality, such as hydrophobic groups that would interact with hydrophobic regions of the protein surface, or a hydrogen-bond donor selected at the site of an acceptor on the protein target. Unfortunately for drug design, Fischer's lock-and-key model is not particularly good at explaining how ligands actually bind to proteins.

The problem is that the lock-and-key model does not take into account the intrinsic flexibility of proteins. An alternative model, termed **induced fit**, was introduced

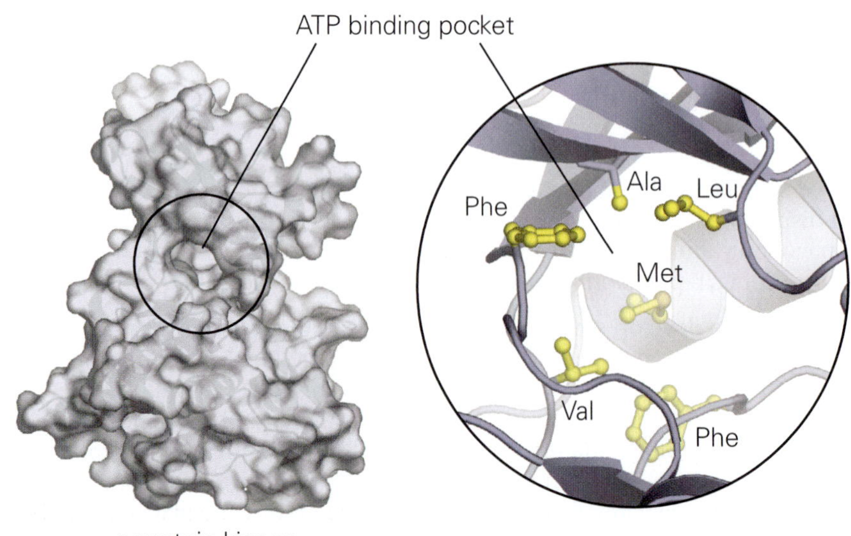


**Figure 12.20 The ATP binding pocket of a protein kinase.** Hydrophobic residues that line pockets such as this one provide opportunities for tight drug binding. The structure shown here is that of the tyrosine kinase Abl. (PDB code: 1F4J.)

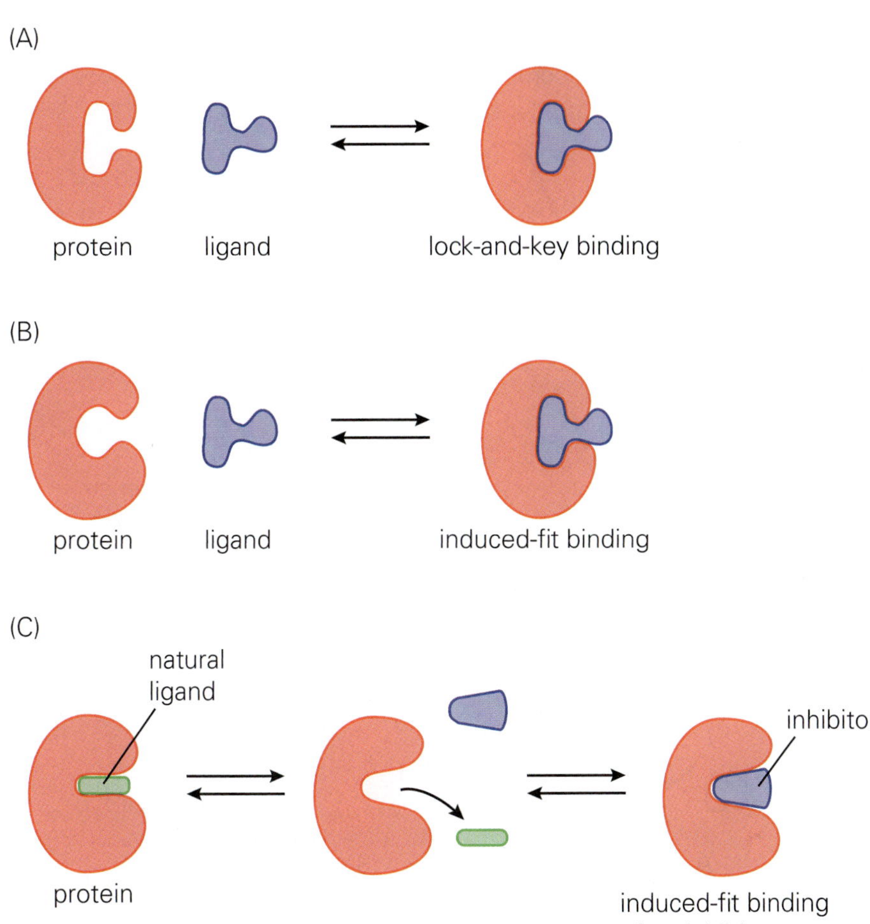


**Figure 12.21 Lock-and-key versus induced-fit interactions.** (A) In a lock-and-key interaction, the protein binds to the ligand without undergoing conformational changes. Only ligands with the proper shape can bind to the protein. (B) In an induced-fit interaction, the protein changes its shape in order to bind the ligand. (C) Fluctuations in the structure of a protein allow inhibitors to trap the protein in a conformation that may not normally be populated significantly. In this example of induced-fit binding, the shape of the protein when bound to the natural ligand is different from its shape when bound to the inhibitor.

by Daniel Koshland in 1958. In this model, the binding site is considered to have some plasticity, allowing it to change somewhat to accommodate binding of the ligand or substrate, like a hand fitting into a stretchy glove (Figure 12.21B). We now know that proteins are quite flexible (recall the discussion in Section 5.25), and can undergo conformational changes upon binding of ligands. This greatly complicates the process of computational inhibitor design because the inhibitor may bind to a conformation of the protein that is different from the one that binds to the natural ligand (Figure 12.21C). It is difficult to predict such conformational changes and, hence, to evaluate whether a potential ligand fits well to the target.

## 12.16 Induced-fit binding occurs through selection by the ligand of one among many preexisting conformations of the protein

When we look at the illustrations of protein structures in this book we "see" one conformation. This conformation is often the one with the lowest free energy, under the conditions of the experimental structure determination. It is important to recognize, however, that there are many other conformations that are not too much higher in free energy. Such states will be populated according to equilibrium constants determined by their free energies, following the ideas developed in Chapters 9 and 10. As we discuss in Chapter 18, most proteins are only marginally stable, and so the unfolded conformation is always populated to some extent. In addition, there are many folded conformations that differ to varying degrees from the one with lowest free energy.

The term "induced fit" implies that the conformational change is induced by the binding of the ligand and would not occur in the absence of ligand. But, in reality,

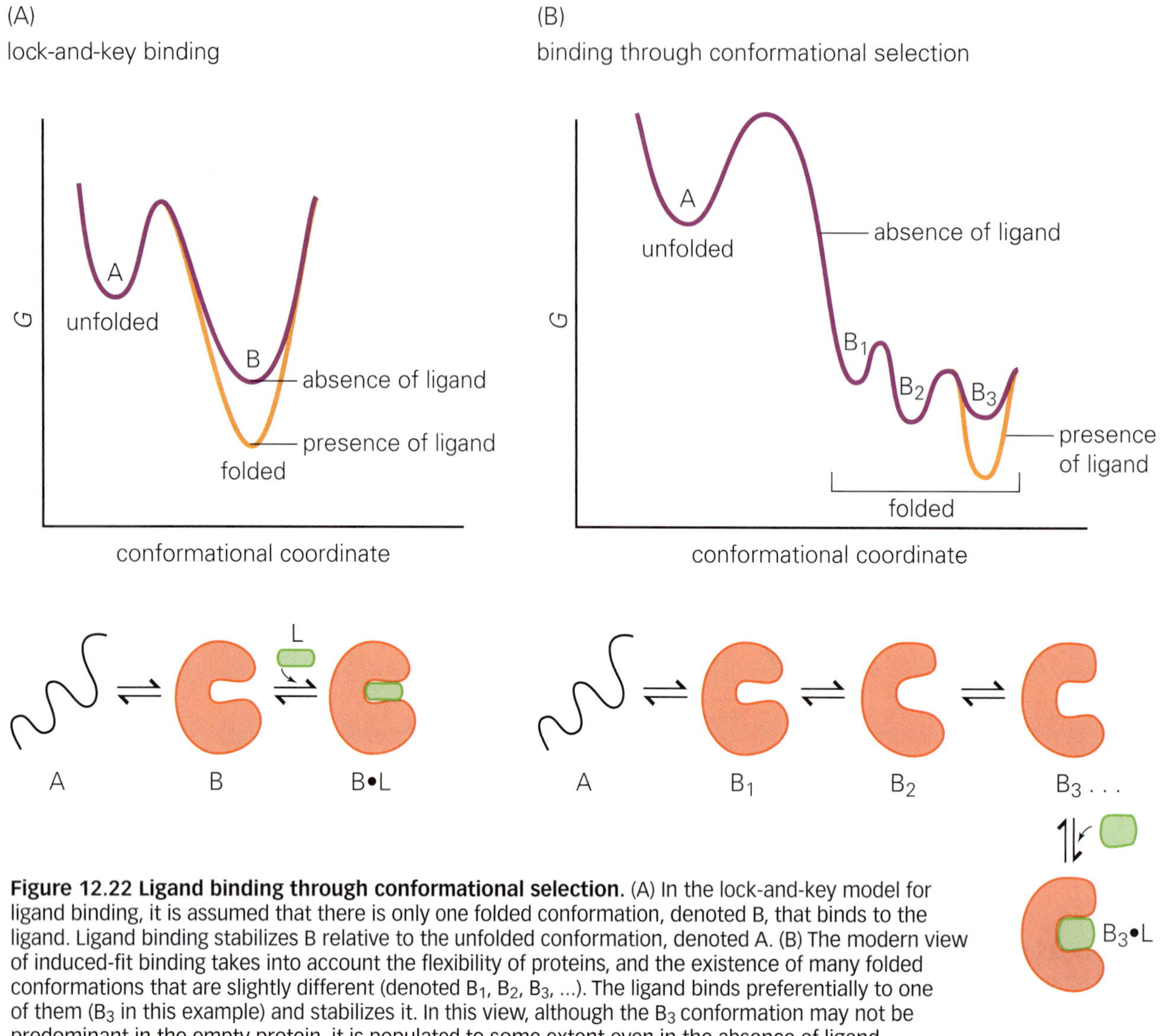


**Figure 12.22 Ligand binding through conformational selection.** (A) In the lock-and-key model for ligand binding, it is assumed that there is only one folded conformation, denoted B, that binds to the ligand. Ligand binding stabilizes B relative to the unfolded conformation, denoted A. (B) The modern view of induced-fit binding takes into account the flexibility of proteins, and the existence of many folded conformations that are slightly different (denoted $B_1$, $B_2$, $B_3$, ...). The ligand binds preferentially to one of them ($B_3$ in this example) and stabilizes it. In this view, although the $B_3$ conformation may not be predominant in the empty protein, it is populated to some extent even in the absence of ligand.

the free energy of the induced conformation must not be too high, or the binding of the ligand would be weakened (this idea will be made quantitative in Section 12.18). It is now believed that induced-fit binding involves the binding of the ligand to conformations of the protein that are populated, although at a low level, even in the absence of ligand (**Figure 12.22**). This is referred to as **conformational selection** by the ligand. Regardless of the precise term used, the idea of an adaptive fit during complex formation is very important, not just for the recognition of small-molecule ligands, but also for the interactions between macromolecules (see Chapter 13).

An example in which the requirement for conformational change becomes obvious is the interaction of ATP with the EGF receptor. Looking at a space-filling model, shown in **Figure 12.23**, the ATP is so deeply buried in the enzyme that it is not easy to see how it could enter or exit the site without a conformational change of the kinase domain. It is found experimentally, however, that ATP binds and dissociates very rapidly, many times per second. This occurs through movements of the lobes of the kinase domain, opening with respect to each other. Binding of other molecules, such as inhibitors, also occurs through such states, as discussed in the next section.

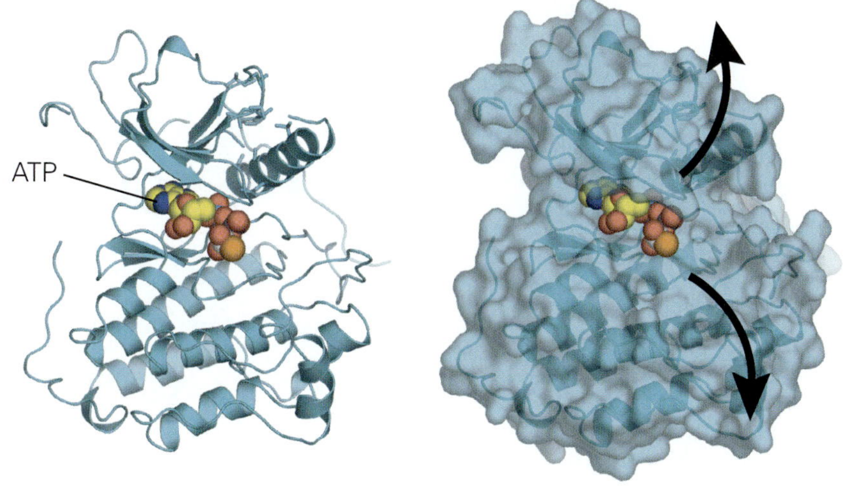


**Figure 12.23 The binding of ligands usually requires proteins to open and close.** The structure of ATP bound to the EGF receptor kinase domain is shown here. A surface representation of the kinase domain, shown on the right, indicates that the drug cannot enter or exit the binding site without breathing motions in the kinase domain that provide access. (PDB code: 2GS6.)

## 12.17 Conformational changes in the protein underlie the specificity of a cancer drug known as imatinib

An example of the importance of conformational changes in enabling the binding of a drug to a kinase is provided by the binding of imatinib to the Abl kinase. The structure of the Abl kinase when bound to imatinib is different from that seen when the kinase is bound to its natural substrate, ATP. As shown in **Figure 12.24**, imatinib penetrates deeper into the body of the protein kinase than does ATP (how ATP binds to a kinase is illustrated in Figures 12.13 and 12.23). This deep penetration by the drug is made possible by a change in the conformation of a centrally located structural element in the protein kinase, known as the activation loop.

Because tyrosine kinases are signaling switches, they have evolved to cycle between active and inactive conformations in response to external stimuli. The activation loop is a critical part of this switching mechanism because it converts from an inactive conformation that blocks substrate binding to an active and open conformation upon being phosphorylated. The inactive conformation of the activation loop recognized by imatinib is likely to be relevant to this fundamental switching mechanism of tyrosine kinases.

There are several dozen different tyrosine kinases in human cells, all of which are very closely related to each other in terms of sequence. The tyrosine kinases, in turn, are closely related in sequence to several hundred different protein kinases that phosphorylate serine or threonine residues instead of tyrosine. All protein kinases look very similar when they turn on, because they all catalyze the same chemical reaction (phosphate transfer from ATP to a hydroxyl group on the substrate protein) and they have to satisfy the dictates of chemistry. Despite this similarity, each kinase responds to a unique set of activating signals, and they often look quite different when they are switched off.

Imatinib is capable of distinguishing between very closely related targets on the basis of differences in their inactive conformations (imatinib does not bind tightly to the active conformation). Shown in Figure 12.24D is the structure of a tyrosine

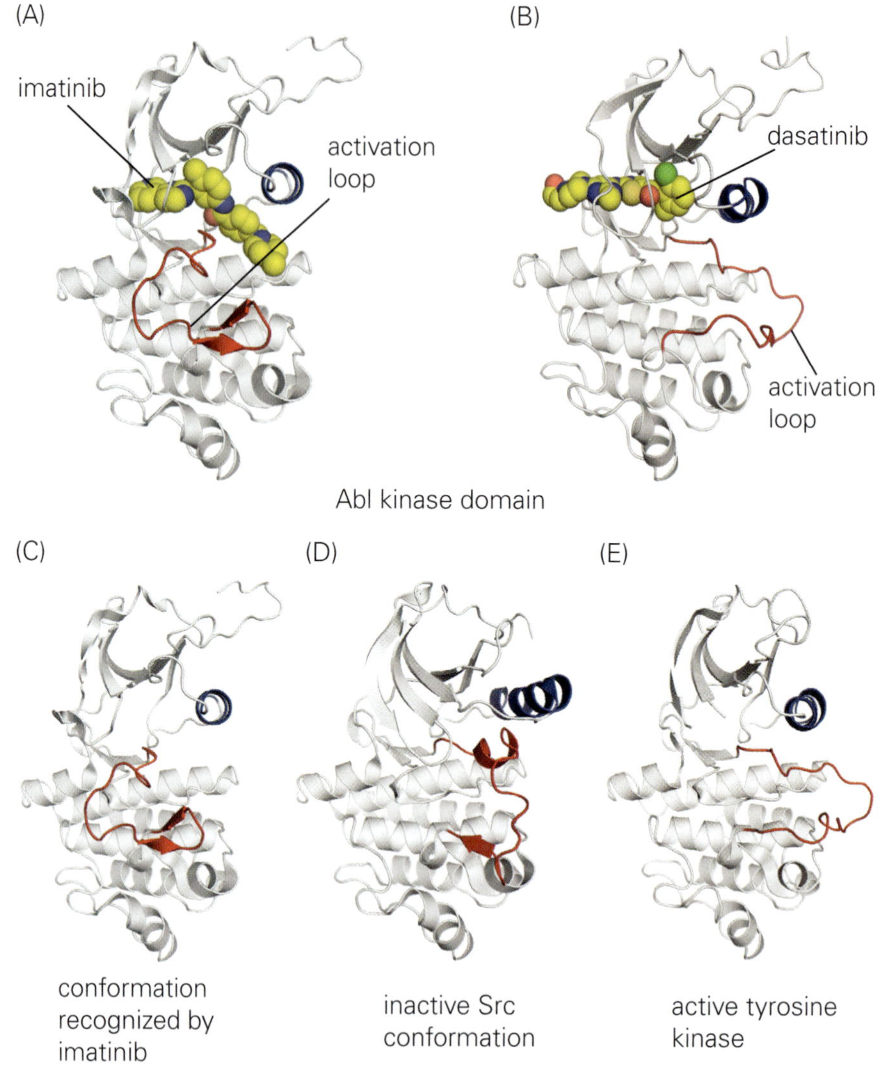


**Figure 12.24 The specificity of imatinib.** (A) Structure of imatinib bound to the kinase domain of Abl. The drug penetrates deeply into the kinase domain and cannot bind tightly unless the kinase is in the specific inactive conformation shown here and also in (C). (B) Dasatinib, shown on the right, does not penetrate so much and binds to the active form of the kinase. (C) Conformation of Abl, as bound to imatinib, with the drug removed. (D) Conformation of inactive Src kinase. Differences in the structure of a centrally located "activation loop" prevent the binding of imatinib. (E) The conformation of the active kinase domain, to which imatinib also cannot bind. (PDB codes: 1IEP, 2GQG, 1QCF, 2G2I.)

kinase known as the sarcoma kinase (Src) in an inactive state, the conformation of which is incompatible with the binding of imatinib. This kind of specificity in a drug such as imatinib is important for clinical efficacy, but because it relies on conformational changes in the target protein, it can be very difficult to predict or design from first principles.

## 12.18 Conformational changes in the target protein can weaken the affinity of an inhibitor

The existence of alternative conformations of the protein complicates the analysis of binding equilibria. For example, the simple expression that relates the fractional saturation of the protein, $f$, to the dissociation constant, $K_D$, and the free ligand concentration, [L], was derived by assuming that the protein partitions between only two states, free and ligand-bound (see Equation 12.1):

$$P + L \xrightleftharpoons{\Delta G^\circ_{\text{bind}}} P \bullet L$$

If, instead, the protein exists in two populations, only one of which is competent to bind the ligand, then this equilibrium has to be modified as follows:

$$P^* + L \xrightleftharpoons{\Delta G^\circ_1} P + L \xrightleftharpoons{\Delta G^\circ_2} P \bullet L \tag{12.27}$$

Here P* refers to a population of the protein to which the ligand does not bind, $\Delta G_1^\circ$ is the standard free-energy change for converting P* to P, and $\Delta G_2^\circ$ is the free energy for L binding to P. If the value of $\Delta G_1^\circ$ is much greater than the value of $RT$ (~2.5 kJ•mol$^{-1}$ at room temperature), then P* is the predominant conformation of the protein in the absence of ligand. Some of the intrinsic binding free energy of the ligand, $\Delta G_2^\circ$, then goes towards converting P* to P, which weakens the effective binding free energy (**Figure 12.25**).

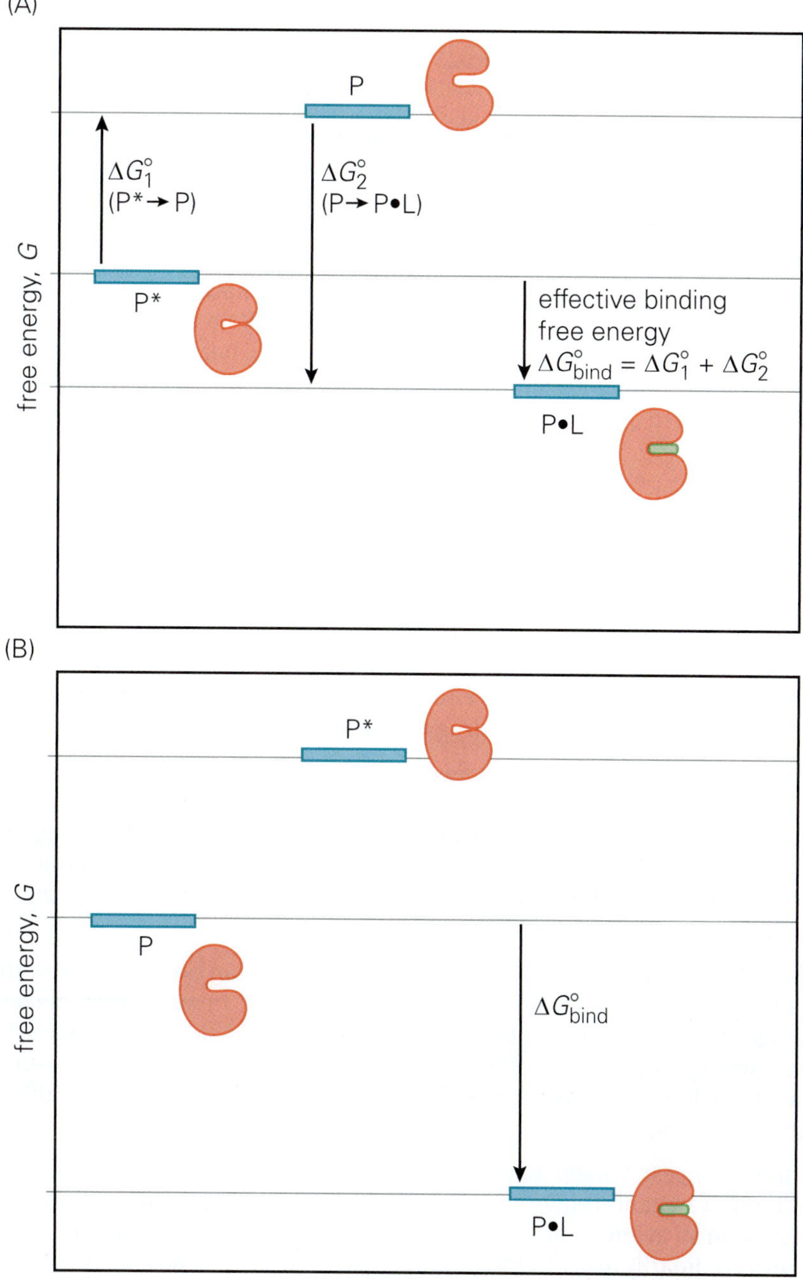


**Figure 12.25 Effect of conformational changes on $\Delta G_{bind}^\circ$.** A protein is assumed to exist in two conformations, P (open) and P* (closed). (A) If the free energy of P* is lower than that of P so that P* is the dominant form in solution, then the effective binding free energy is reduced. (B) If the free energy of P is much lower than that of P*, then the binding free energy is unaffected because the population of molecules in the P* conformation is small.

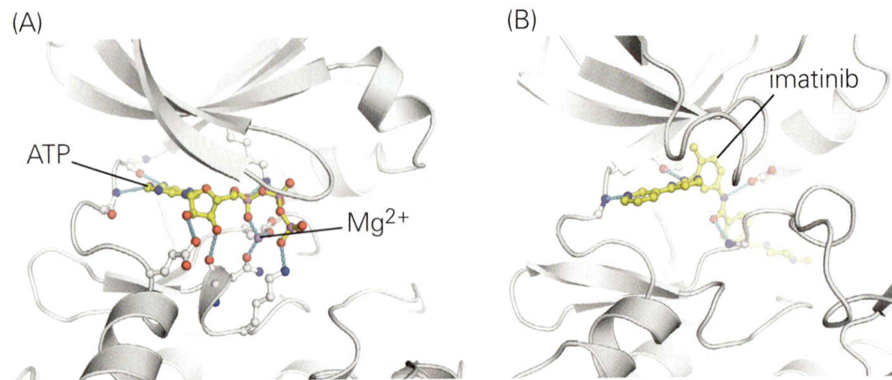


**Figure 12.26 Comparison of ATP and imatinib binding to protein kinases.** (A) The ATP–kinase complex forms many hydrogen bonds (*blue* dotted lines), but it is a much weaker complex than the imatinib–kinase complex (B), which is stabilized primarily by hydrophobic interactions. (PDB codes: 1ATP and 1OPJ.)

The importance of this effect can be appreciated by considering another drug that is effective against chronic myelogenous leukemia, known as dasatinib (see Figure 12.24B). Dasatinib is ~350 times more potent an inhibitor of the Abl kinase than imatinib. Dasatinib recognizes the active conformation of the kinase domain, which is likely to be the predominant form in cancer cells. Thus, one reason dasatinib has a higher affinity for Abl is that it does not have to stabilize a less-populated conformation. The price paid for higher affinity, in this case, is lower specificity. Dasatinib inhibits both the Src kinases and Abl, because the structure of the active kinase domain is similar in both cases (compare Panels B and E of Figure 12.24).

## 12.19 The strength of noncovalent interactions usually correlates with hydrophobic interactions

As discussed in Section 12.11, the affinity of proteins for their natural ligands is usually related to the physiological concentration of the ligand. Because ATP is quite abundant in cells ([ATP] ≈ 1 mM), the affinity of protein kinases for ATP is relatively weak ($K_D \approx 10$ μM; $\Delta G^o \approx -29$ kJ•mol$^{-1}$). Drugs that are kinase inhibitors, in contrast, bind very tightly ($K_D \approx 10$ nM; $\Delta G^o \approx -47$ kJ•mol$^{-1}$). What features account for the higher affinity of the drug for the kinase?

A detailed view of the interactions between ATP and a protein kinase is shown in **Figure 12.26**A. The ATP molecule is quite polar, with several hydrogen-bond donors and acceptors. In addition, the three phosphate groups are highly charged. The protein forms seven or eight hydrogen bonds with ATP. The protein compensates for the charge of ATP by coordinating one or two magnesium ions ($Mg^{2+}$) that coordinate, in turn, the phosphate groups.

The highly polar interactions between ATP and the kinase can be contrasted with how imatinib binds to the Abl kinase (see Figure 12.26B). Imatinib only makes four hydrogen bonds with the protein. On the other hand, imatinib forms a much more extensive hydrophobic interface with the protein than does ATP. This suggests that the hydrophobic interactions are what make the kinase-imatinib affinity particularly strong.

One of the strongest known noncovalent interactions involving a protein is between biotin (a vitamin) and a family of proteins known as avidins (because they bind biotin so avidly). The dissociation constant for the avidin-biotin interaction is extraordinarily low, at ~1 femtomolar ($10^{-15}$ M; $\Delta G^o \approx -86$ kJ•mol$^{-1}$). The structure of the complex of an avidin-like protein with biotin is shown in **Figure 12.27.** Biotin does form five hydrogen bonds with the protein, but these hydrogen bonds are not the dominant factor in the affinity of biotin for the protein. Instead, the strength of the binding arises from the large number of nonpolar contacts between biotin atoms and atoms of the protein. Hydrophobic contacts formed within an invagination of the protein surface are a characteristic feature of the high-affinity interactions of proteins with small-molecule ligands.

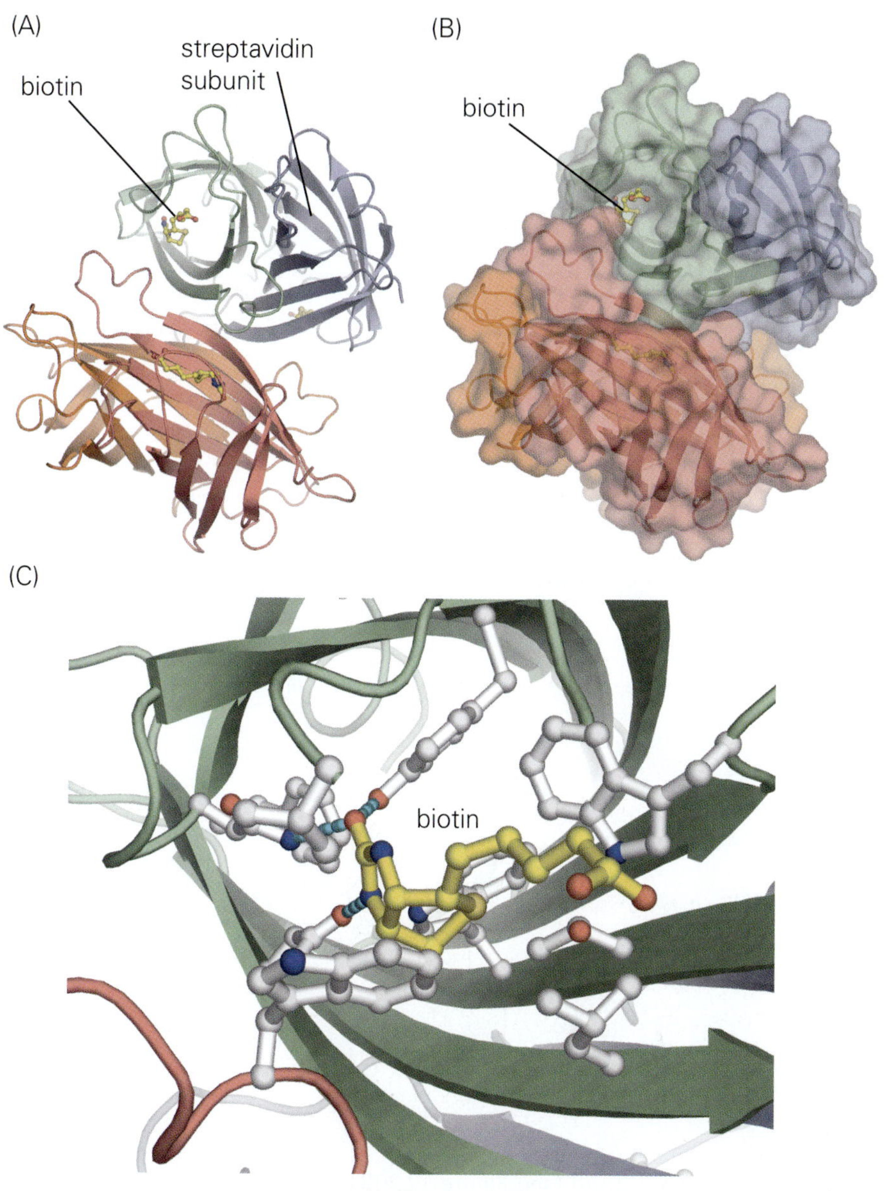


**Figure 12.27 Structure of biotin bound to streptavidin, one of the tightest known noncovalent interactions, with $K_D \approx 10^{-15}$ M.** (A) Streptavidin is a tetrameric protein of the avidin family and each subunit binds one molecule of biotin. In this diagram, two of the streptavidin subunits are in *blue*, and two are in *red*. (B) Biotin is almost completely engulfed by the protein at binding sites that are formed at the interfaces between protein subunits. (C) Although biotin forms several hydrogen bonds with the protein, it is the nonpolar contacts with protein sidechains that are the most important factor in the high affinity of the interaction. (PDB code: 1STP.)

## 12.20 Cholesterol-lowering drugs known as statins take advantage of hydrophobic interactions to block their target enzyme

To further illustrate the importance of hydrophobicity, we consider a class of drugs known as the **statins**. Statins are widely used to reduce cholesterol levels in people who produce too much of this steroid lipid (**Figure 12.28**). High levels of cholesterol in the blood is linked to blockage of the arteries, which can lead to heart disease. Reducing the dietary intake of cholesterol is one way to control the levels of this molecule in the blood, but for many people the cellular production of cholesterol remains a serious problem. Statins shut down the activity of an enzyme known as 3-hydroxy-3-methyl glutaryl coenzyme A reductase (HMG-CoA reductase), which catalyzes a key step in the cellular synthesis of cholesterol (**Figure 12.29**). Combined with control of dietary cholesterol intake, the use of statins has proven to be an effective therapy for the prevention of heart disease.

**Statins**

Statins are a class of small-molecule drugs that are effective in lowering cholesterol levels in humans. Statins inhibit an enzyme known as HMG-CoA reductase, which catalyzes a key step in the synthesis of cholesterol.

The structure of HMG-CoA reductase bound to its natural substrate, HMG-CoA, is shown in Figure 12.29B. HMG-CoA consists of two components, HMG (the 3-hydroxyl-3-methylglutaryl group) and the coenzyme A group, which are linked

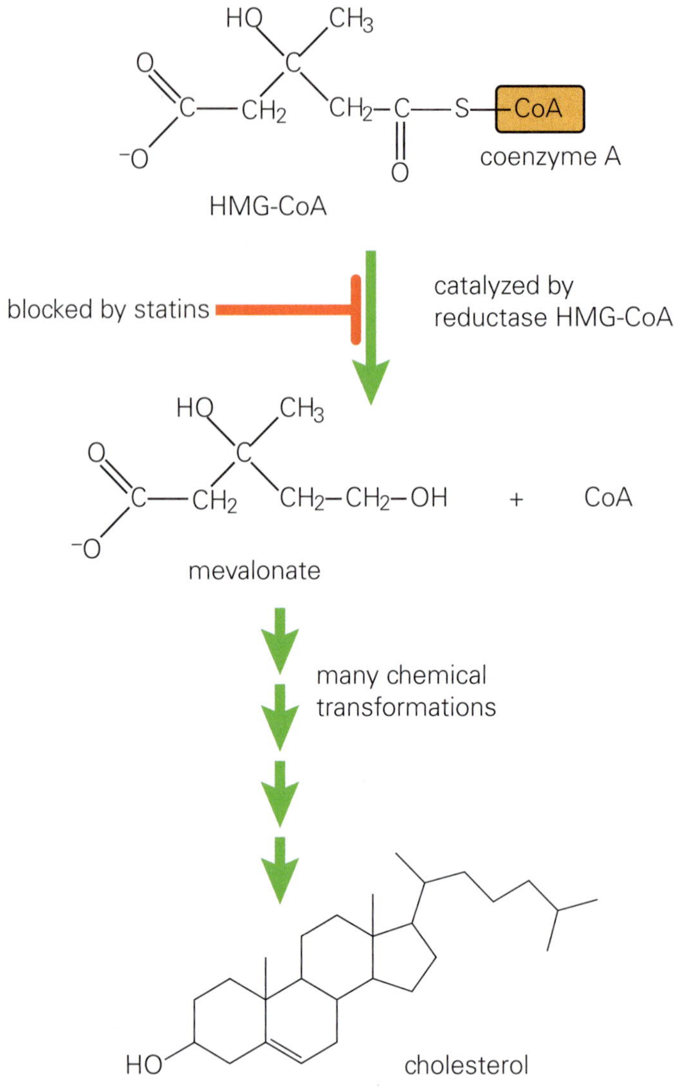

**Figure 12.28 Statins and HMG-CoA reductase.** A schematic diagram showing the synthesis of cholesterol and the role played by the enzyme HMG-CoA reductase, which is blocked by statins.



together by a thioester bond (see Figure 12.29B). The HMG group is bound at a polar binding site, where it forms several hydrogen bonds with protein sidechains. The CoA group is long and skinny, and it runs along the surface of HMG-CoA reductase.

Figure 12.29C shows the structure of HMG-CoA reductase bound to a statin drug, atorvastatin (marketed under the name Lipitor™). One portion of the drug resembles HMG and is bound to the enzyme at the same site as HMG, where it makes a similar set of hydrogen bonds with the protein. The rest of the drug molecule bears no resemblance to the natural substrate and consists of several aromatic rings that are attached to a central heterocyclic ring. These aromatic rings make the drug significantly bulkier and more hydrophobic than the CoA portion of HMG-CoA.

There are many different statin drugs available in the clinic, and the structures of many of these bound to HMG-CoA reductase have been determined (**Figure 12.30**). All of these drugs have in common a polar portion that is structurally similar to HMG. Attached to this polar head group is a large aromatic component, which is chemically different in each drug. Whereas the binding of the polar component to the protein is the same in each, with maintenance of roughly the same hydrogen bonds as seen for HMG, the hydrophobic portions interact differently with the protein because of differences in their structure.

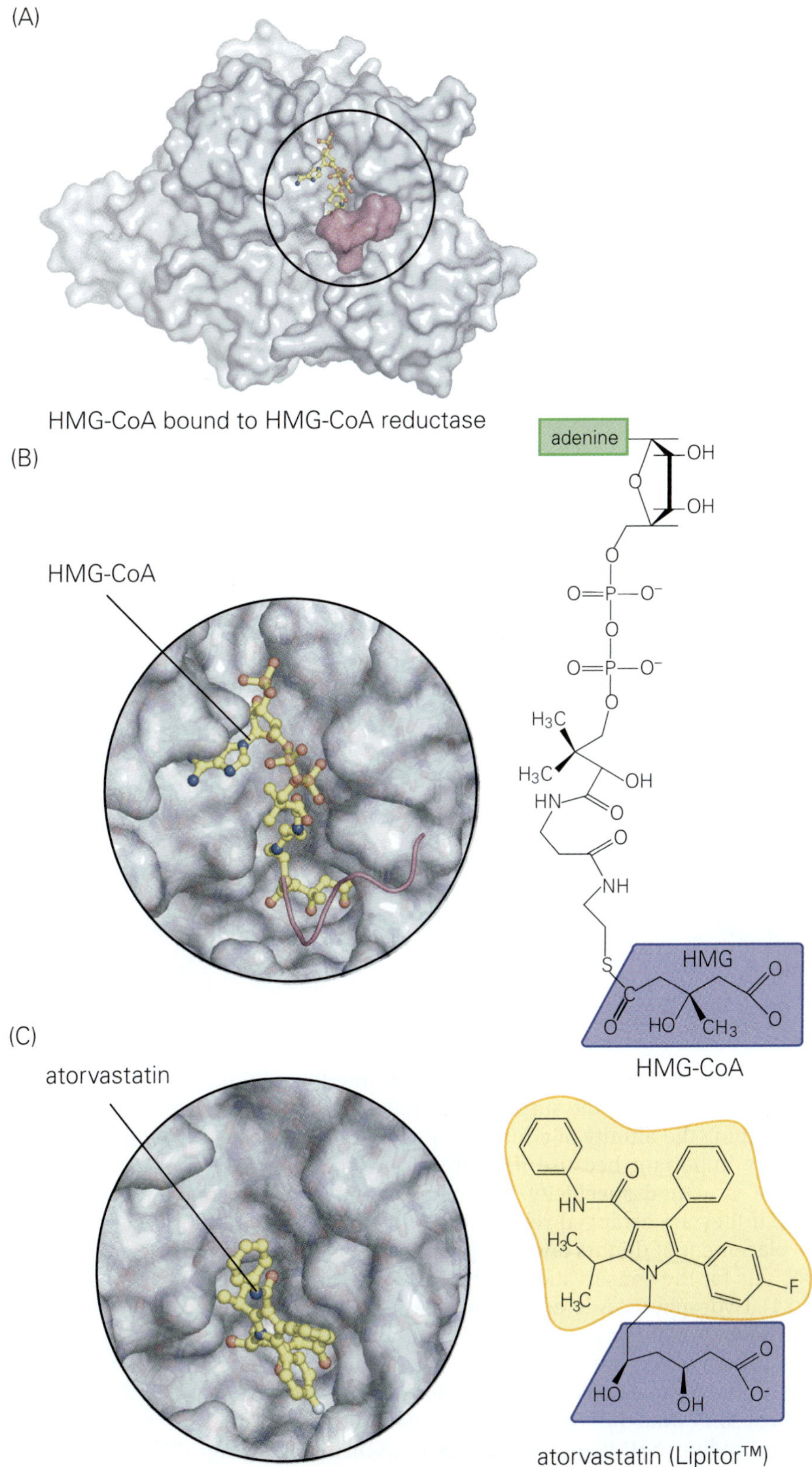


**Figure 12.29 Structure of HMG-CoA reductase**. (A) The structure of the enzyme bound to HMG-CoA reductase is shown. (B) Expanded view of the active site, showing the binding of HMG-CoA, whose chemical structure is shown on the right. A helical segment (*purple*) that covers HMG-CoA is shown as a tube so as to not obscure the view of the substrate. (C) Structure of atorvastatin (Lipitor™) bound to the active site of the enzyme. The helical segment (*purple* in B) is disordered in the structures of drug complexes and is not shown. (PDB codes: 1DQ9, 1HWK.)

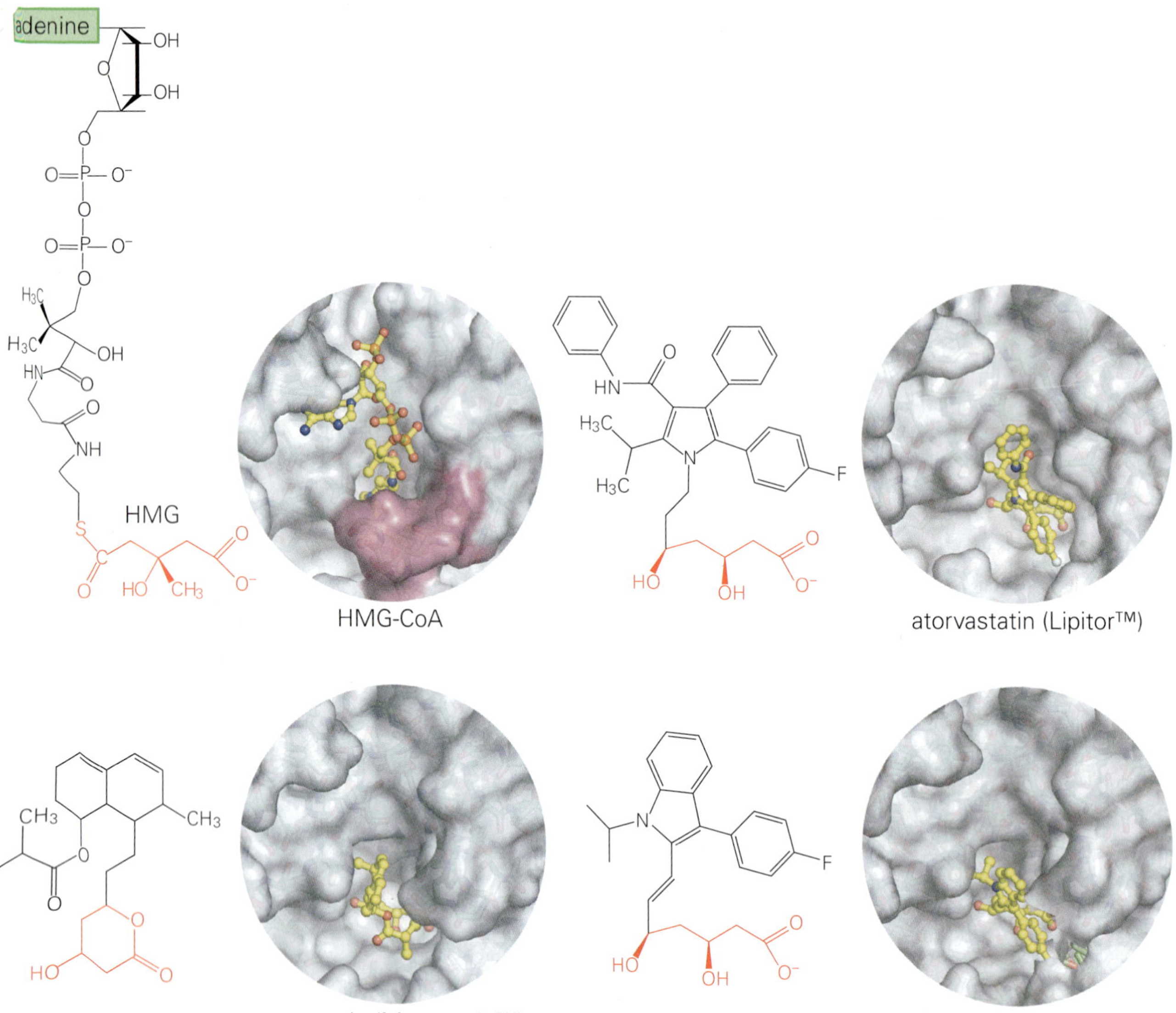


**Figure 12.30 Structures of HMG-CoA and various statins bound to HMG-CoA reductase.** (PDB codes: 1DQ9, 1HWK, 1HW8, 1HWI.)

The similarities and differences between the binding of the statins to HMG-CoA reductase underscore the surprisingly limited role played by hydrogen bonds in determining the affinity of a binding interaction. Nevertheless, hydrogen bonds are very important because they are directional and, as a consequence, they impose structural specificity in molecular interactions. Polar groups in proteins, if they are arranged relatively rigidly, impose geometrical constraints on the placement of donors and acceptors in the ligands that bind near them. This accounts for the conservation of the HMG-like portion in all the statins (**Figure 12.31**). Hydrophobic interactions, on the other hand, are not strongly directional. The predominant constraint is the sequestration of the interacting groups away from water (see Figure 12.31B), and the precise interdigitation of the groups is of secondary importance. The various statins achieve hydrophobic stabilization through somewhat different intermolecular interactions. The thermodynamics of statins binding to HMG-CoA reductase is discussed in Section 12.23.

## 12.21 The apparent affinity of a competitive inhibitor for a protein is reduced by the presence of the natural ligand

A binding isotherm for the interaction of a purified protein with an inhibitor yields a measure of the dissociation constant directly, as we have seen in Section 12.2. A complication arises if the natural ligand for the protein is present in abundance

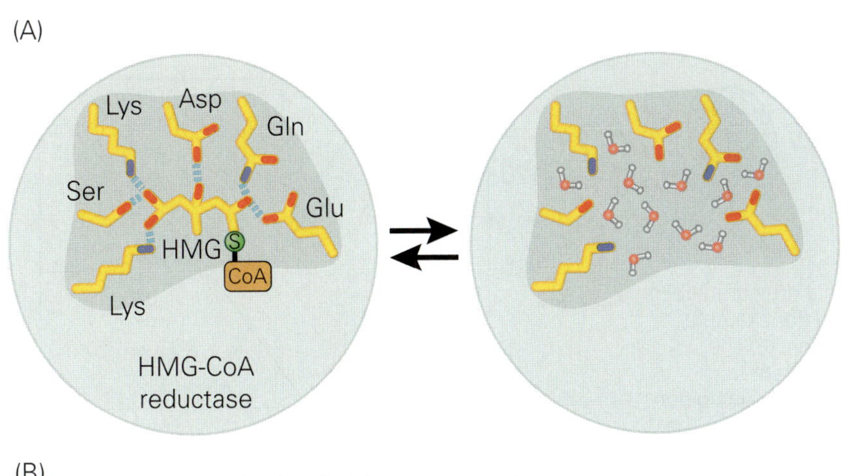


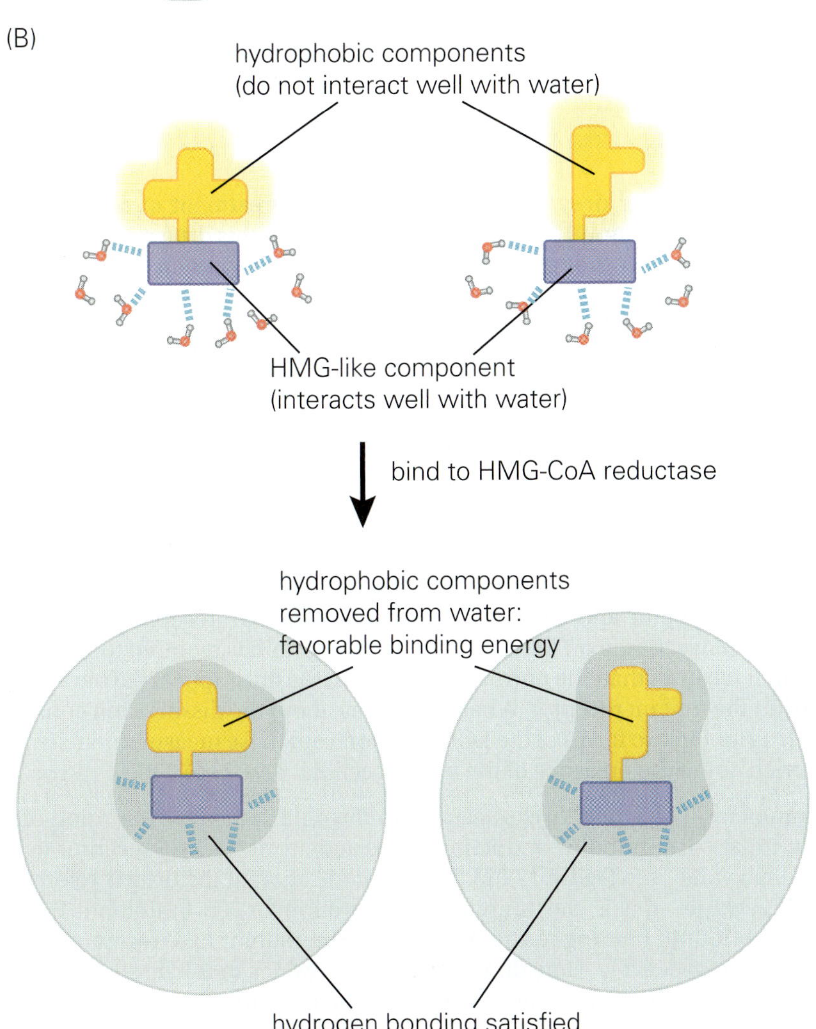


**Figure 12.31 Hydrophilic and hydrophobic components of the statins.** (A) The HMG portion of HMG-CoA makes several hydrogen bonds with the protein. When HMG-CoA is not bound, these hydrogen bonds are replaced by hydrogen bonds to water. (B) Drugs that bind to HMG-CoA reductase have one component that satisfies the hydrogen-bonding requirements of the active site and one component that is hydrophobic.

during the determination of the isotherm, as would be the case if the measurement is made directly in the cell or by using cellular extracts. When measuring the affinity of a drug for a protein kinase, for example, ATP is present at high concentrations (~1 mM) in the cell and protein kinases are normally saturated with ATP. If ATP is present, we have to modify the analysis of the binding isotherm to account for the fact that the inhibitor has to compete with ATP for access to the binding site on the protein (**Figure 12.32**A).

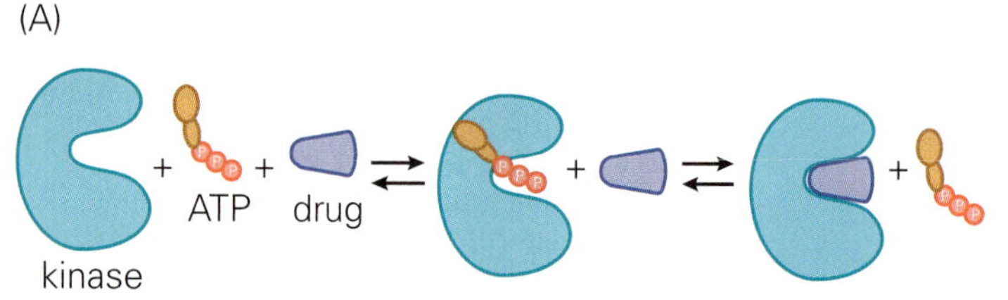


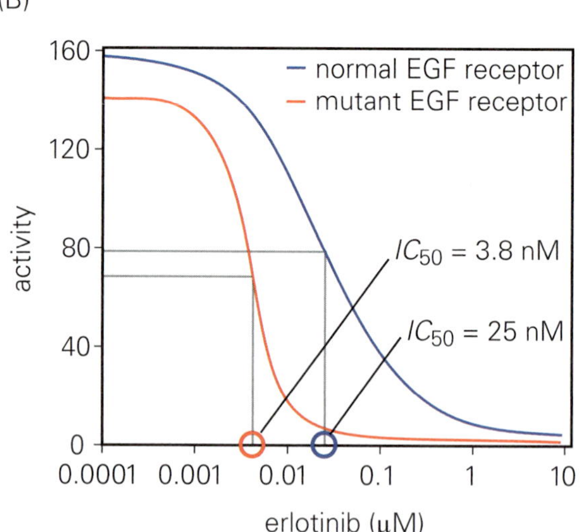


**Figure 12.32 Competitive inhibition.** (A) When a drug or inhibitor binds to the same binding site as a natural ligand (for example, ATP), the process is known as competitive inhibition. (B) The apparent affinity of the inhibitor is reduced to an extent that depends on the concentration of the natural ligand. The inhibition of normal and mutant EGF receptors by erlotinib is shown in the presence of ATP. The concentration of the drug at which the activity of the protein is reduced to half the maximal value is known as the $IC_{50}$ value. (B, adapted from K.D. Carey et al., and M.X. Sliwkowski, *Cancer Res.* 66: 8163–8171, 2006.)

Recall that the cancer drug erlotinib is used in the treatment of certain breast and lung cancers (Section 12.13). Not all patients given this drug respond equally well to the treatment, and it was discovered that patients who responded more favorably had particular mutations in the EGF receptor, while the others did not. The affinity of erlotinib for EGF receptors in cells isolated from these patients was studied by measuring the activity of the receptor (that is, the ability of the receptor to phosphorylate tyrosine residues) in the presence of increasing concentrations of the drug (see Figure 12.32B). The activity of the receptor is high when there is no inhibitor present, and it decreases to essentially zero when increasing concentrations of erlotinib are added to the cells. The activity of the receptor is proportional to the fraction of the protein that is not bound to the drug, and so the data in Figure 12.32 can be interpreted as a binding isotherm.

The measurements shown in Figure 12.32B were made using cells containing the normal EGF receptor and also with cells containing EGF receptors that have a mutant sequence. Based on the data, much less erlotinib is required to shut down the mutant receptor than the normal one—that is, the drug appears to have higher affinity for the mutant receptor. What can we say about the dissociation constants of the drug for the two forms of the EGF receptor from these measurements, which are carried out in the presence of the normal cellular levels of ATP?

The parameter that is readily extracted from the data is the concentration of the inhibitor that corresponds to reduction of the activity of the EGF receptor to half its maximal value (see Figure 12.32). This concentration of the drug is referred to as the ***$IC_{50}$* value**—that is, the inhibitor concentration for 50% inhibition. The $IC_{50}$ values for erlotinib binding to the normal and the mutant EGF receptors are 25 and 3.8 nM, respectively. How do we relate these values to the dissociation constants for the drug binding to the two forms of the receptor?

**$IC_{50}$ value for an inhibitor**

The concentration of inhibitor that reduces the activity of a protein to half the maximal value (that is, the value seen in the absence of the inhibitor) is known as the $IC_{50}$ value.

We refer to the dissociation constant for the inhibitor binding to the protein as $\boldsymbol{K_I}$ in order to distinguish it from the dissociation constant, $K_D$, for ATP binding to the protein. If there were no ATP present during the measurement, then the $IC_{50}$ value would simply be equal to the value of the dissociation constant for the inhibitor–drug interaction (that is, $K_I$). ATP is, however, necessarily present when the enzyme activity is measured. Because of the competition with ATP, we have to modify the relationship between $IC_{50}$ and the inhibitor dissociation constant, $K_I$, to include contributions from $K_D$ (for ATP) and the concentration of ATP, [L], as follows:

$$K_I = (IC_{50})\left(\frac{K_D}{K_D + [L]}\right) \tag{12.28}$$

See **Box 12.1** for an explanation of how we arrive at Equation 12.28. If we assume that the concentration of ATP in the cell is ~1 mM, and that the dissociation constant for ATP binding to the kinase domains of both the normal and the mutant EGR receptor is ~10 μM ($1 \times 10^{-2}$ mM; a typical value for protein kinases), then Equation 12.28 tells us that the derived values of $K_I$ for the normal and mutant receptors are 100 times lower than the $IC_{50}$ (that is, 0.25 nM and 0.038 nM, respectively). A more detailed analysis shows, however, that the mutant protein binds ATP roughly 50 times more weakly than does the normal protein. The derived values of $K_I$ are therefore more nearly equal, and the increased ability of the drug to inhibit the mutant protein arises at least in part from an alteration in the interaction between ATP and the protein. These issues point to the complexities that often arise in relating the thermodynamic binding data for drugs to the actual efficacies that are observed when the drugs are delivered to patients.

## 12.22 Entropy lost by drug molecules upon binding is regained through the hydrophobic effect and the release of protein-bound water molecules

When a ligand molecule that is tumbling freely in solution binds to a protein, there is a significant loss of entropy (**Figure 12.33**). Three translational and three rotational degrees of freedom of the ligand are lost, and this acts as a barrier to

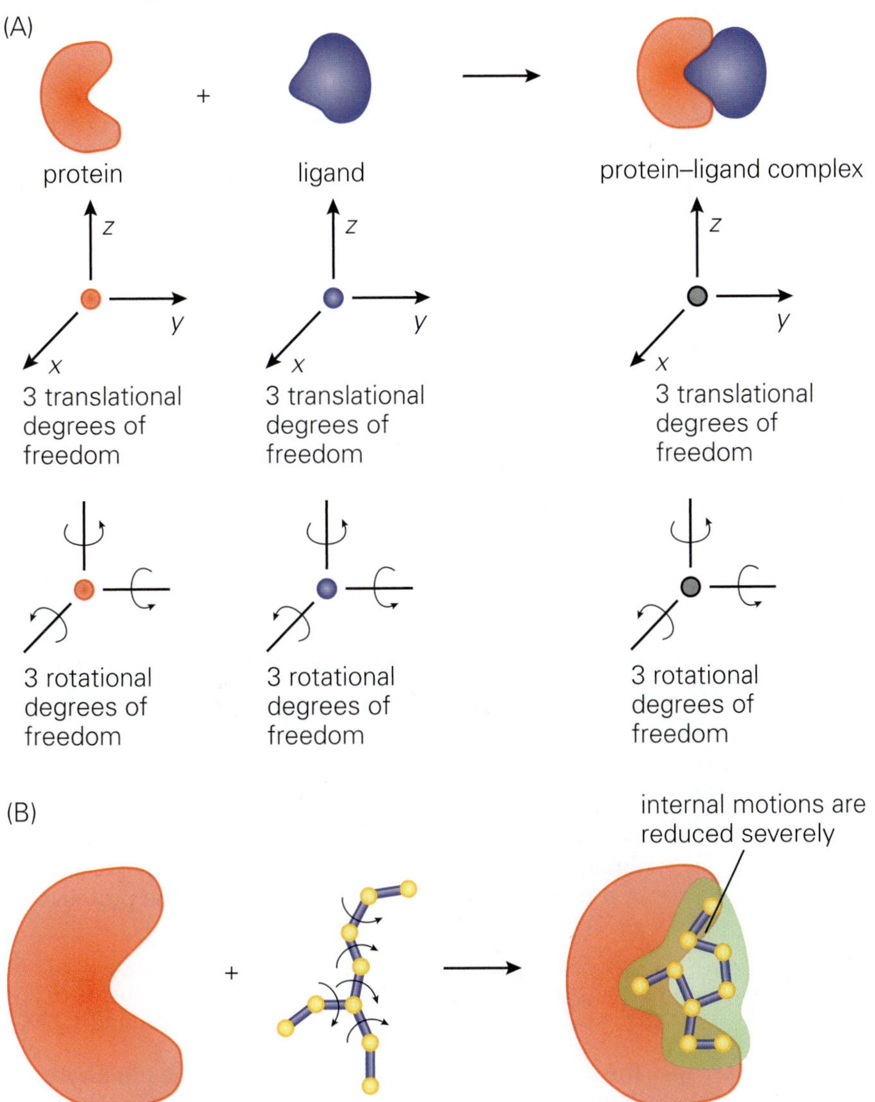


**Figure 12.33 Entropy changes upon binding.** (A) Reduction in translational and rotational degrees of freedom. (B) Reduction in internal degrees of freedom.

## Box 12.1 The relationship between the inhibitor concentration for half-maximal activity, $IC_{50}$, and $K_I$, the dissociation constant for a competitive inhibitor

In Section 12.20 we stated that the dissociation constant for the binding of a competitive inhibitor, $K_I$, is related to the value of $IC_{50}$, the concentration of the inhibitor at which the activity of the protein is reduced to half the maximal value, by Equation 12.28:

$$K_I = (IC_{50})\left(\frac{K_D}{K_D + [L]}\right)$$

We now show why Equation 12.28 is valid for the case where a protein with one binding site binds either to the ligand, L, or an inhibitor, I.

We assume that the activity of the protein is proportional to the concentration of the protein bound to the ligand:

$$\text{activity} = C[P \bullet L] \quad (12.1.1)$$

where $C$ is held constant. We now need to find an expression for [P•L] in the presence of the inhibitor. The reaction scheme that describes competitive inhibition of P by an inhibitor I is as follows:

$$P \bullet I + L \overset{K_I}{\rightleftharpoons} P + L + I \overset{K_D}{\rightleftharpoons} I + P \bullet L \quad (12.1.2)$$

When the reaction described by Equation 12.1.2 comes to equilibrium, the concentrations of the two different protein complexes, the free protein and the free ligand and the inhibitor, will all obey the equations that define the two dissociation constants:

$$K_D = \frac{[P][L]}{[P \bullet L]};\ K_I = \frac{[P][I]}{[P \bullet I]} \quad \text{(at equilibrium)} \quad (12.1.3)$$

From Equation 12.1.3, we can write down an expression for [P•L] as follows:

$$[P \bullet L] = \frac{[P][L]}{K_D} \quad (12.1.4)$$

We use the definition of $K_I$ in Equation 12.1.3 to substitute for [P] in Equation 12.1.4:

$$[P] = \frac{K_I[P \bullet I]}{[I]} \Rightarrow [P \bullet L] = \left(\frac{K_I[P \bullet I]}{[I]}\right)\left(\frac{[L]}{K_D}\right) \quad (12.1.5)$$

The total protein concentration, $[P]_{total}$, is now given by the sum of the free protein concentration and the concentrations of the protein bound to the ligand and to the inhibitor:

$$[P]_{total} = [P] + [P \bullet I] + [P \bullet L] \Rightarrow [P \bullet I] = [P]_{total} - [P] + [P \bullet L] \quad (12.1.6)$$

Substituting the expression for [P•I] from Equation 12.1.6 into Equation 12.1.5, we get:

$$[P \bullet L] = \left(\frac{K_I}{[I]}\right)\left(\frac{[L]}{K_D}\right)([P]_{total} - [P \bullet L] - [P]) \quad (12.1.7)$$

Rearranging terms, we get:

$$\left(\frac{[I]}{K_I}\right)[P \bullet L] + \left(\frac{[L]}{K_D}\right)[P \bullet L] + \frac{[L][P]}{K_D} = \frac{[L]}{K_D}[P]_{total} \quad (12.1.8)$$

The last term on the left-hand side is equal to [P•L] (see Equation 12.1.3), and so Equation 12.1.8 can be rewritten as:

$$\left[\left(\frac{[I]}{K_I}\right) + 1 + \left(\frac{[L]}{K_D}\right)\right][P \bullet L] = \frac{[L]}{K_D}[P]_{total} \quad (12.1.9)$$

Rearranging Equation 12.1.9, we get the following expression for [P•L]:

$$[P \bullet L] = ([P]_{total})\frac{[L]}{K_D\left(1 + \frac{[I]}{K_I}\right) + [L]} \quad (12.1.10)$$

Denoting the term $\left(1 + \frac{[I]}{K_I}\right)$ as α, Equation 12.1.10 is rewritten as:

$$[P \bullet L] = ([P]_{total})\frac{[L]}{\alpha K_D + [L]} \quad (12.1.11)$$

If there were no inhibitor present, then the concentration of the protein bound to the ligand would be given by the product of the total protein concentration and the fractional saturation, $f$:

$$[P \bullet L] = ([P]_{total})(f) \quad (12.1.12)$$

where

$$f = \left(\frac{[L]}{K_D + [L]}\right)$$

Comparing Equations 12.1.11 and 12.1.12, we see that the effect of the inhibitor is to scale the dissociation constant, $K_D$, by the parameter, α. The value of α is always greater than 1.0, so the effect of the inhibitor is to weaken the apparent strength of the interaction between the protein and the ligand, as expected.

When the concentration of the inhibitor is very low, the $\frac{[I]}{K_I}$ term in the definition of α can be neglected. In that case, $\alpha \approx 1$ and the expression for [P•L] reduces to Equation 12.1.12.

Returning to the analysis of the $IC_{50}$ value, combining Equations 12.1.1 and 12.1.11 shows that, in the presence of a competitive inhibitor, the activity of the protein is given by:

$$\text{activity} = C[P \bullet L] = (C[P]_{total})\left(\frac{[L]}{\alpha K_D + [L]}\right) \quad (12.1.13)$$

The ratio of the activity in the absence of the inhibitor to that observed in the presence of the inhibitor is given by:

$$\frac{(\text{activity})_{\text{inhibitor}}}{(\text{activity})_{\text{no inhibitor}}} = \frac{(C[\text{P}]_{\text{total}})\left(\dfrac{[\text{L}]}{\alpha K_{\text{D}}+[\text{L}]}\right)}{(C[\text{P}]_{\text{total}})\left(\dfrac{[\text{L}]}{K_{\text{D}}+[\text{L}]}\right)} = \frac{\left(\dfrac{[\text{L}]}{\alpha K_{\text{D}}+[\text{L}]}\right)}{\left(\dfrac{[\text{L}]}{K_{\text{D}}+[\text{L}]}\right)} \quad (12.1.14)$$

When the concentration of the inhibitor, [I], is equal to the value of the $IC_{50}$, then this ratio is equal to 0.5, by definition, and so:

$$\frac{\left(\dfrac{[\text{L}]}{\alpha K_{\text{D}}+[\text{L}]}\right)}{\left(\dfrac{[\text{L}]}{K_{\text{D}}+[\text{L}]}\right)} = \frac{1}{2} \quad (12.1.15)$$

When the inhibitor concentration is equal to $IC_{50}$, the scaling parameter α is given by:

$$\alpha = 1 + \frac{IC_{50}}{K_{\text{I}}} \quad (12.1.16)$$

Substituting this expression for α into Equation 12.1.15 and, with some simple algebraic manipulation, we finally get the desired relationship between $K_I$ and $IC_{50}$:

$$K_{\text{I}} = IC_{50}\left(\frac{K_{\text{D}}}{K_{\text{D}}+[\text{L}]}\right) \quad \text{(Equation 12.28)}$$

binding. In addition, certain internal motions of the ligand, such as rotations about single bonds, may also be frozen out, again an unfavorable contribution to the free-energy change associated with ligand binding.

**Table 12.2** lists the expected values of the entropy change for the various degrees of freedom lost by a small molecule when it binds to a protein and the corresponding contribution to the free energy at room temperature. The loss of translational and rotational entropy can together account for a contribution of ~+50 kJ•mol$^{-1}$ to $\Delta G^{\circ}_{\text{bind}}$, which is a significant penalty. Each torsion angle in the ligand that is frozen into a single conformation contributes an additional ~4–6 kJ•mol$^{-1}$ to the binding free energy.

Natural ligands, such as ATP, are often very polar, and such molecules can compensate for the unfavorable binding entropy by forming networks of hydrogen-bonding interactions. But what about a molecule such as imatinib, which binds with high affinity but makes only four hydrogen bonds to the protein (see Figure 12.26)? The answer is that the hydrophobic effect plays a role in stabilizing the interactions between drugs such as imatinib and their targets.

An estimate of the potential contribution of the hydrophobic effect can be made by looking at the change in free energy upon moving small nonpolar molecules from water to a nonpolar solvent, such as carbon tetrachloride. **Table 12.3** lists the

**Table 12.2 Changes in entropy upon freezing out the motions of a small molecule.**

| | Entropy change (J•mol$^{-1}$•K$^{-1}$) | Contributions to $\Delta G^{\circ}$ at 300 K (kJ•mol$^{-1}$) |
|---|---|---|
| Loss of translational entropy for a small molecule | 125–150 | 37.5–45.0 |
| Loss of rotational entropy for a small molecule (for example, propane) | 90 | 27 |
| Rotation around internal bonds (per bond) | 12.5–21 | 3.75–6.3 |

(Data from M.I. Page and W.P. Jencks, *Proc. Natl. Acad. Sci. USA* 68: 1678–1683, 1971.)

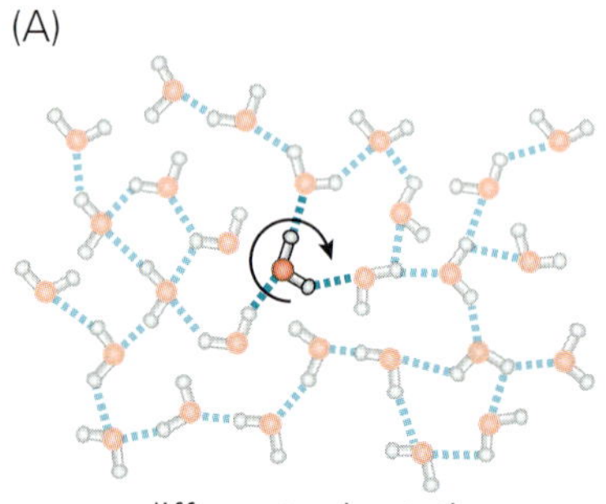


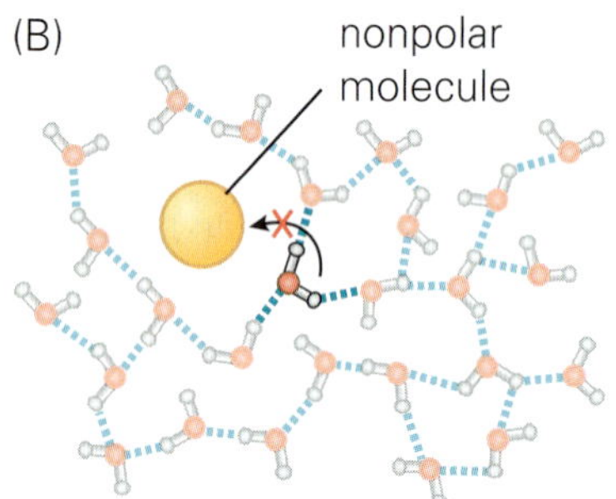


**Figure 12.34 The hydrophobic effect.** (A) A water molecule in bulk water can rotate in many directions. (B) The presence of a nearby nonpolar group restricts the movement of the water, reducing the entropy. The water molecules are released when the nonpolar molecule binds to a protein.

**Table 12.3 Transfer free energies from water to carbon tetrachloride for methane and propane, at 300 K.**

| | $\Delta H^o$ ($kJ \bullet mol^{-1}$) | $\Delta S^o$ ($J \bullet mol^{-1} K^{-1}$) | $\Delta G^o$ ($kJ \bullet mol^{-1}$) |
|---|---|---|---|
| $CH_4$ (in $H_2O$) → $CH_4$ (in $CCl_4$) | +10.8 | +50.2 | –4.3 |
| $C_3H_8$ (in $H_2O$) → $C_3H_8$ (in $CCl_4$) | +8.1 | +93.8 | –20.0 |

(Adapted from data in M. H. Abraham, *J. Am. Chem. Soc.* 104: 2085–2094, 1982.)

thermodynamic parameters for the transfer of methane and propane from water to carbon tetrachloride. The process is strongly favored by an increase in entropy, with an overall favorable value for $\Delta G^o$ of –4.1 and –20.0 $kJ \bullet mol^{-1}$ for methane and propane, respectively. The favorable entropy change is due to the hydrophobic effect and can be understood as arising from restrictions on the movement of the water molecules in the presence of nonpolar molecules (**Figure 12.34**). The magnitude of the favorable entropy changes in these two cases makes it clear that a drug molecule that is sufficiently hydrophobic can overcome the entropic penalty that arises from the loss of its own motion.

In Section 12.10, we had noted that the higher the affinity of a drug for its target, the lower the concentration at which it could be delivered, thereby avoiding side effects. A ready route towards increasing the strength of the interaction is to make the drug more hydrophobic. Unfortunately, increased hydrophobicity brings with it many attendant problems, such as decreased solubility and an increased tendency to adhere to various cellular and extracellular components.

An additional favorable entropy term can also play a role in promoting drug binding if the active site of the protein traps water molecules, as illustrated in **Figure 12.35.** The presence of polar groups on the surface of the protein often leads to the relatively tight association of water molecules with surface groups on the protein. These bound water molecules have a lower entropy than water molecules in the bulk solution. If the binding of a drug causes the release of some of these water molecules, then a favorable entropy change can end up benefiting the process.

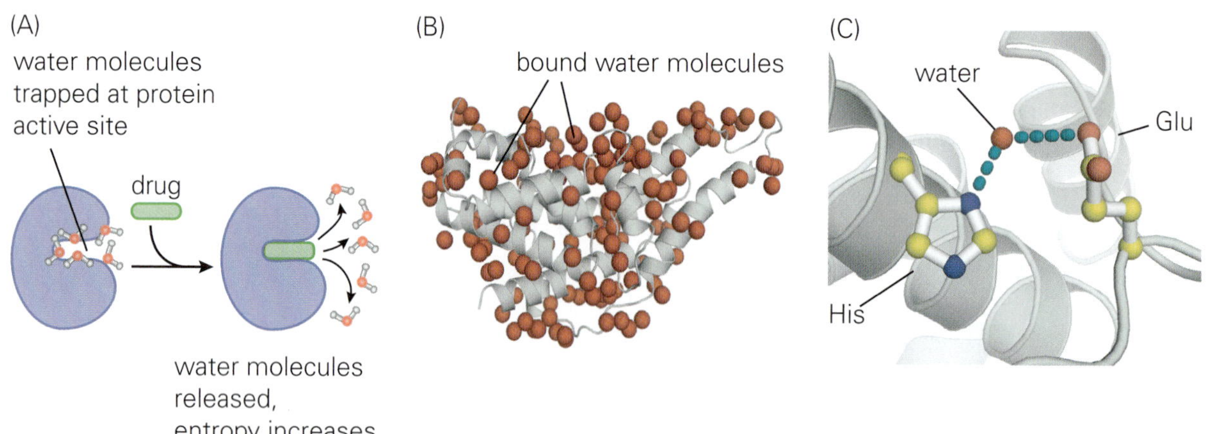


**Figure 12.35 Release of trapped water molecules.** (A) Polar binding sites in proteins can trap water molecules. Release of these waters upon binding also increases the entropy, although this effect is distinct from the hydrophobic effect. (B) Water molecules localized to a protein surface. The surface of a protein molecule is typically covered with ~100 or more trapped water molecules. (C) These water molecules participate in hydrogen-bonded networks and, depending on how tightly they are bound, their entropy is reduced with respect to bulk water. (PDB code: 2II0.)

## 12.23 Isothermal titration calorimetry allows us to determine the enthalpic and entropic components of the binding free energy

As we have emphasized previously, the net enthalpy change upon binding is a balance between interactions made with water and with the protein. We now see that the entropy change is also a balance involving water. A complete understanding of the origins of the affinity between a small molecule and a protein therefore requires knowledge of both the enthalpy change and the entropy change upon binding. This important information is not obtained from the analysis of binding isotherms, which provide only the value of the net free-energy change, $\Delta G^o$.

There are two commonly used methods to obtain the enthalpy and entropy changes for a binding reaction. One of them is known as **isothermal titration calorimetry**, and the principle underlying this technique is described in **Box 12.2**. The second method is to measure the equilibrium constant at a series of temperatures, and thereby derive the binding enthalpies and entropies. This process is called van't Hoff analysis, and its application to the fidelity of DNA base pairing is described in Chapter 19.

The binding of various statin drugs to HMG CoA-reductase has been analyzed by isothermal titration calorimetry (**Figure 12.36**; see Box 12.2). The top part of panel A in Figure 12.36 shows the heat released upon each injection of ligand solution into the protein solution. Integrating the area under the curve for each injection gives the total amount of heat released for that injection. The lower part of the

**Figure 12.36 Thermodynamics of statin binding to HMG-CoA reductase.** (A) Representative data from isothermal titration calorimetry (see Box 12.2). (B) Contributions of $\Delta H^o$ and $T\Delta S^o$ to the free energy of binding ($\Delta G^o$) for various statins. Note that the value of $\Delta H$ for fluvastatin is zero, and so the value of $\Delta G$ is derived from enzyme inhibition rather than calorimetry. (C) Structures of the statins for which thermodynamic data are reported. (Adapted from T. Carbonell and E. Freire, *Biochemistry* 44: 11741–11748, 2005. With permission from The American Chemical Society.)

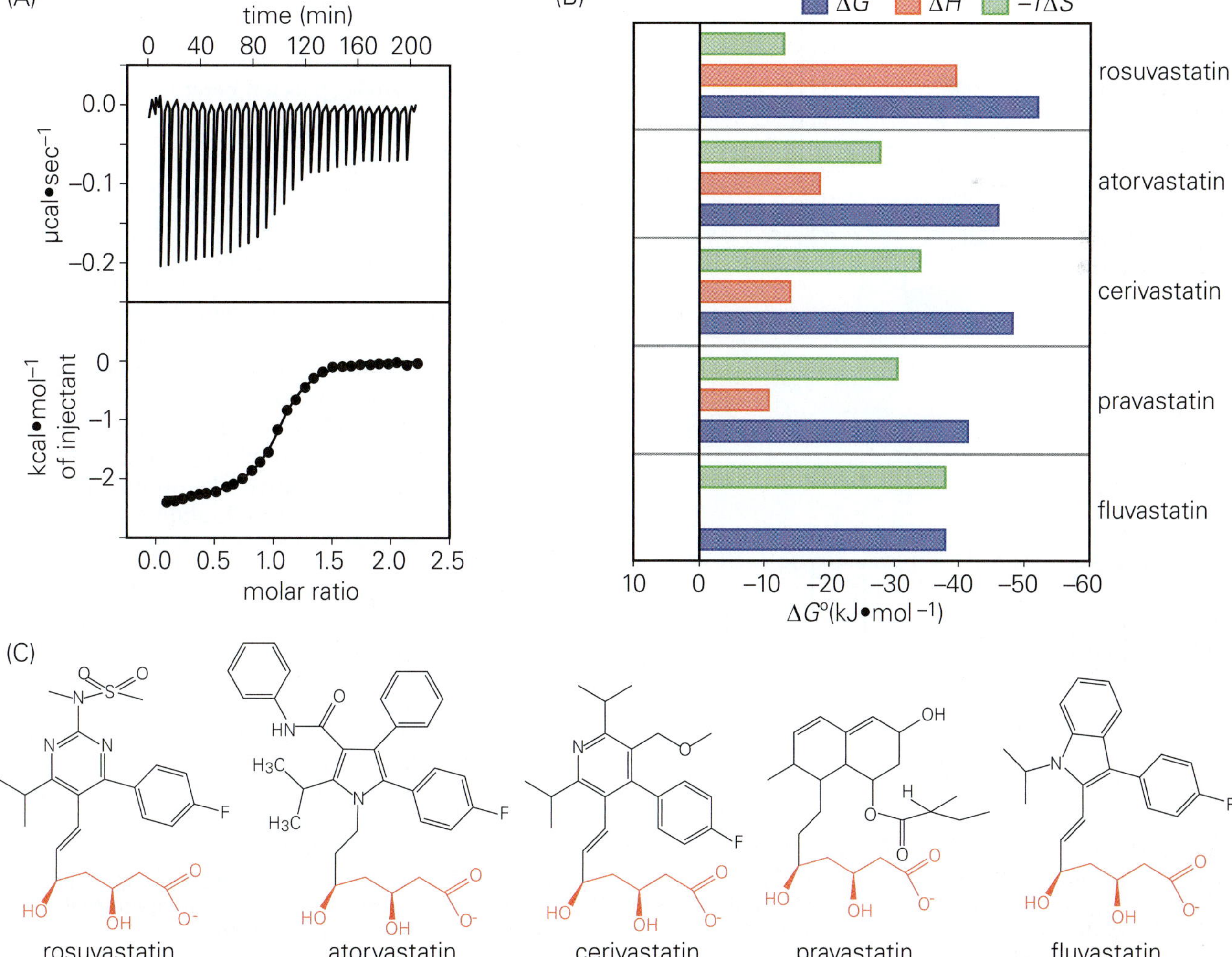

## Box 12.2 Isothermal titration calorimetry

Isothermal titration calorimetry is particularly useful for the analysis of the thermodynamics of binding interactions because it provides a way to obtain not only the dissociation constant, $K_D$, (or, equivalently, the binding free energy, $\Delta G^o$), but also the standard enthalpy and entropy changes upon binding ($\Delta H^o$ and $\Delta S^o$, respectively). In contrast to most other methods for studying binding, which rely on monitoring changes in properties of the protein or the ligand upon binding, titration calorimetry relies on direct measurement of the heat released upon binding. The measurement of heat released (that is, a calorimetric measurement) is carried out at a constant temperature while adding the ligand to the protein drop by drop (hence the term *isothermal titration*).

Not surprisingly, titration calorimetry gives us a direct measure of the enthalpy change of the binding reaction. The value of the dissociation constant ($K_D$) is obtained indirectly, by analyzing the manner in which the amount of heat released during the titration of ligand into the protein changes with ligand concentration, as discussed below. This gives us the standard free-energy change, $\Delta G^o$, of the reaction. The standard entropy change is then derived using the following equation:

$$\Delta S^o = -\frac{\Delta G^o - \Delta H^o}{T}$$

The isothermal titration calorimeter consists of a thermally insulated chamber, within which are two cells containing solutions (**Figure 12.2.1**). During the course of the experiment, the two cells are kept at the same constant temperature. One cell serves as the reference and typically contains everything except the ligand. The other cell is the one in which the binding reaction is carried out and is connected to a syringe that delivers the ligand solution drop by drop during the course of the titration. The amount of heat released (or taken up) during the injection of a drop of ligand solution is measured by removing or adding known amounts of heat to the sample cell so as to keep its temperature constant.

Let us analyze the outcome of an isothermal titration calorimetric experiment for a simple situation in which a protein, P, interacts with a ligand, L. The protein has only one binding site for the ligand. For every drop of ligand added to the protein, the reaction cell either releases heat (if the reaction is exothermic) or takes up heat (if it is endothermic). In rare cases, when the binding reaction proceeds with no change in enthalpy, the reaction cell will neither take up nor release heat. Isothermal titration calorimetry cannot be used in such situations.

The amount of heat transferred rises sharply upon addition of the first drops of the ligand and then tapers off more slowly as the system reaches equilibrium (**Figure 12.2.2**). With each drop of ligand that is added, more and more protein molecules are bound to the ligand at equilibrium, so there are

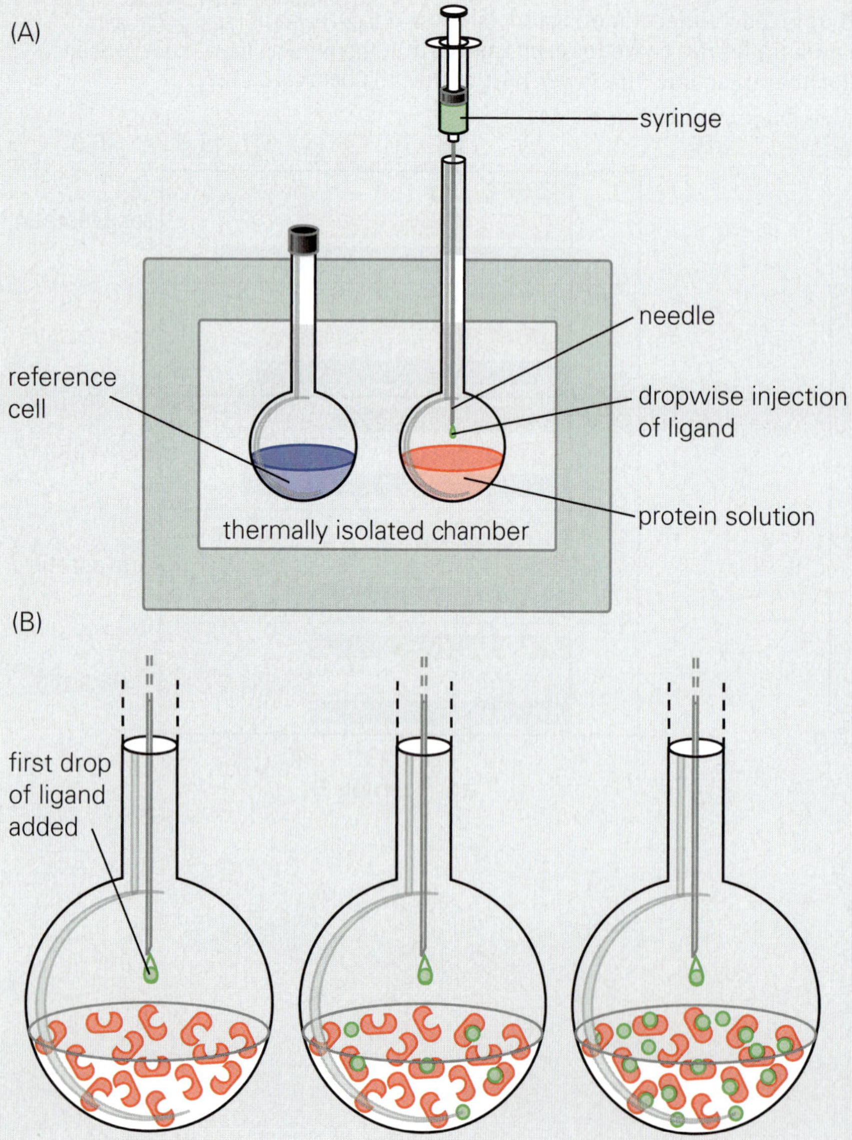

**Figure 12.2.1 Schematic diagram of an instrument used for isothermal titration calorimetry.**

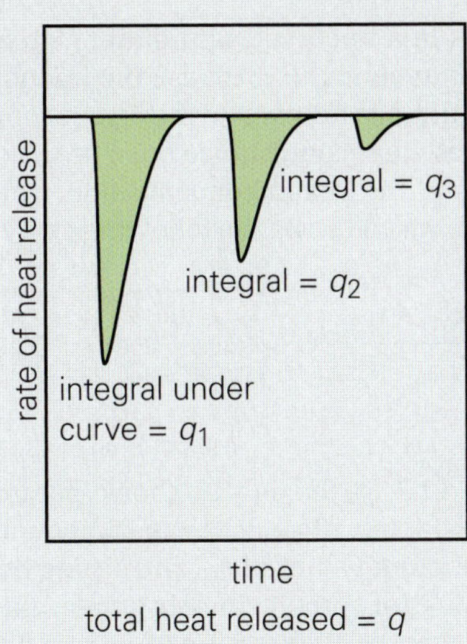


**Figure 12.2.2 Data acquired during an isothermal titration calorimetric experiment.** The ligand is injected into the protein sample, and the heat released is measured drop by drop. The total amount of heat released up to any point in the titration is calculated by integrating the heat released for each injection.

fewer unliganded protein molecules in the population to bind to the ligand molecules. The amount of heat delivered (or released) decreases, therefore, with each subsequent injection, finally reaching zero when the protein is saturated with ligand. The amount of heat transferred for each injection of ligand is obtained by integrating the area under the heat transfer curve for each injection (see Figure 12.2.2).

At any point in the titration, the total heat released up to that point, $q$, is directly related to the fraction of the protein that is bound to ligand, $f$, at that point in the titration:

$$q = (f)([\mathrm{P}]_{\mathrm{TOTAL}})(V_o)(\Delta H^o) \tag{12.2.1}$$

In Equation 12.2.1, $\Delta H^o$ is the standard enthalpy change of binding, which is equal to the heat released when one mole of protein binds to one mole of ligand; $q$ is the heat released up to this point in the titration, and it is given by the product of $\Delta H^o$ and the number of moles of protein that have bound to ligand. The term $(f)([\mathrm{P}]_{\mathrm{total}})$ is the same as [P•L]—that is, the concentration of the bound protein. Multiplying [P•L] by the volume of the cell ($V_o$) gives the number of moles of protein bound to ligand.

The value of $f$ at any point in the titration can be related to the dissociation constant, $K_D$, as follows (see Equation 12.11):

$$f = \frac{\frac{[\mathrm{L}]}{K_D}}{1 + \frac{[\mathrm{L}]}{K_D}} \tag{12.2.2}$$

In Equation 12.2.2, [L] is the concentration of the free ligand. In the main text we have usually simplified matters by ignoring the difference between the free ligand concentration and the total ligand concentration when analyzing binding data. We cannot make this approximation when interpreting calorimetric data because the output (heat absorbed) reflects the changes in the free ligand concentration directly. We therefore need to account for the difference between the total ligand concentration, $[\mathrm{L}]_{\mathrm{total}}$, which is the value we control, and the free ligand concentration, [L].

We start by writing down the following expression, which accounts for all of the states of the ligand:

$$[\mathrm{L}]_{\mathrm{total}} = [\mathrm{L}] + [\mathrm{P}\bullet\mathrm{L}] = [\mathrm{L}] + (f)([\mathrm{P}]_{\mathrm{total}}) \tag{12.2.3}$$

$$\Rightarrow [\mathrm{L}] = [\mathrm{L}]_{\mathrm{total}} - (f)([\mathrm{P}]_{\mathrm{total}}) \tag{12.2.4}$$

Substituting the expression for [L] from Equation 12.2.4 into Equation 12.2.2, we get:

$$f = \frac{\frac{[\mathrm{L}]}{K_D}}{1 + \frac{[\mathrm{L}]}{K_D}} = \frac{\frac{[\mathrm{L}]_{\mathrm{total}}}{K_D} - (f)\frac{[\mathrm{P}]_{\mathrm{total}}}{K_D}}{1 + \left[\frac{[\mathrm{L}]_{\mathrm{total}}}{K_D} - (f)\frac{[\mathrm{P}]_{\mathrm{total}}}{K_D}\right]} \tag{12.2.5}$$

$$\Rightarrow f + (f)\frac{[\mathrm{L}]_{\mathrm{total}}}{K_D} - (f^2)\frac{[\mathrm{P}]_{\mathrm{total}}}{K_D} = \frac{[\mathrm{L}]_{\mathrm{total}}}{K_D} - (f)\frac{[\mathrm{P}]_{\mathrm{total}}}{K_D} \tag{12.2.6}$$

Rearranging and grouping terms, we get:

$$(f^2)\frac{[\mathrm{P}]_{\mathrm{total}}}{K_D} - (f)\left(1 + \frac{[\mathrm{L}]_{\mathrm{total}}}{K_D} + \frac{[\mathrm{P}]_{\mathrm{total}}}{K_D}\right) + \frac{[\mathrm{L}]_{\mathrm{total}}}{K_D} = 0 \tag{12.2.7}$$

Multiplying by $\frac{K_D}{[\mathrm{P}]_{\mathrm{total}}}$, we get:

$$f^2 - (f)\left(1 + \frac{[\mathrm{L}]_{\mathrm{total}}}{[\mathrm{P}]_{\mathrm{total}}} + \frac{K_D}{[\mathrm{P}]_{\mathrm{total}}}\right) + \frac{[\mathrm{L}]_{\mathrm{total}}}{[\mathrm{P}]_{\mathrm{total}}} = 0 \tag{12.2.8}$$

This is a quadratic equation in $f$ for which the two possible solutions are given by:

$$f = \frac{\left(1 + \frac{K_D}{[\mathrm{P}]_{\mathrm{total}}} + \frac{[\mathrm{L}]_{\mathrm{total}}}{[\mathrm{P}]_{\mathrm{total}}}\right)}{2} \pm \frac{\sqrt{\left(1 + \frac{K_D}{[\mathrm{P}]_{\mathrm{total}}} + \frac{[\mathrm{L}]_{\mathrm{total}}}{[\mathrm{P}]_{\mathrm{total}}}\right)^2 - 4\frac{[\mathrm{L}]_{\mathrm{total}}}{[\mathrm{P}]_{\mathrm{total}}}}}{2} \tag{12.2.9}$$

In order to choose between the two solutions, we rewrite the expression for $f$ in terms of two parameters, $a$ and $b$:

$$\left.\begin{aligned} f &= \frac{a}{2} + \frac{\sqrt{a^2 + b}}{2} \\ f &= \frac{a}{2} - \frac{\sqrt{a^2 + b}}{2} \end{aligned}\right\} \text{2 choices} \tag{12.2.10}$$

The expressions for $a$ and $b$ in Equation 12.2.10 are given by:

$$a = 1 + \frac{K_D}{[P]_{total}} + \frac{[L]_{total}}{[P]_{total}} \quad (12.2.11)$$

and

$$b = -(4)\frac{[L]_{total}}{[P]_{total}} \quad (12.2.12)$$

The minimum value of $f$ is obtained when the ligand concentration is zero—that is, when $b = 0$:

$$\left.\begin{aligned} f &= \frac{1}{2}(a + \sqrt{a^2}) \\ f &= \frac{1}{2}(a - \sqrt{a^2}) \end{aligned}\right\} (\text{when } b = 0) \quad (12.2.13)$$

The physically correct value of $f$ is obtained only for the second choice, which gives $f = 0$. The first choice corresponds to a fractional saturation, $f$, greater than 1, which is not meaningful. Because the fractional saturation is a continuous function of the ligand concentration, it follows that the second choice must be the correct one at all values of the ligand concentration, not just when [L] = 0. Thus, we can write the following expression for $f$:

$$f = \frac{1}{2}\left[\left(1 + \frac{K_D}{[P]_{total}} + \frac{[L]_{total}}{[P]_{total}}\right) - \sqrt{\left(1 + \frac{K_D}{[P]_{total}} + \frac{[L]_{total}}{[P]_{total}}\right)^2 - (4)\frac{[L]_{total}}{[P]_{total}}}\right] \quad (12.2.14)$$

In Equation 12.2.14, the only unknown parameter is $K_D$, because $[L]_{total}$ and $[P]_{total}$ are experimentally fixed. By choosing different values of $K_D$ and seeing how well the resulting values of $f$ predict the observed value of $q$ (that is, the heat absorbed), the value of $K_D$ can be estimated empirically.

panel gives the amount of heat associated with binding (on a molar basis) for each point in the titration. The amount of heat released is related to the fractional saturation, $f$, at that point, as explained in Box 12.2. By fitting the value of $f$ over the course of the titration, the value of $K_D$ (and, therefore, $\Delta G^o$) for the reaction is derived. The value of $\Delta H^o$ is obtained directly from the total amount of heat released.

The statins all bind tightly to their target enzyme ($\Delta G^o_{bind}$ = –38 kJ•mol$^{-1}$ to –52 kJ•mol$^{-1}$), but the balance between enthalpy and entropy is different in each case (see Figure 12.34). The binding of fluvastatin to HMG CoA-reductase is entirely entropy-driven, with $\Delta H^o \approx 0$ and $-T\Delta S^o$ = –38 kJ•mol$^{-1}$ at 25°C. At the other extreme is rosuvastatin, with $\Delta H^o$ = –39 kJ•mol$^{-1}$ and $-T\Delta S^o$ = –12.5 kJ•mol$^{-1}$. These differences are consistent with the higher polarity of rosuvastatin. Note that the entropy of binding is favorable in all cases, despite the loss of the translational and rotational entropy of the drug.

## Summary

The cell is an intricate network of interactions between protein molecules, nucleic acids, sugars, lipids, and small molecules. The affinity or strength, of these interactions is critical for the proper functioning of the cellular machinery, and the focus of this chapter has been on understanding how we measure binding strengths and the factors that modulate specificity. These concepts are also important for the development of drugs that are targeted against proteins that are malfunctioning in disease.

The strength of a noncovalent interaction between two molecules is characterized by the dissociation constant, $K_D$, which is related to the standard binding free energy, $\Delta G^o$ through the following equation:

$$\Delta G^o_{bind} = +RT \ln K_D$$

The values of the dissociation constant and the binding free energy for a noncovalent interaction are obtained by measuring the amount of complex that forms as the concentration of the ligand is increased. An important parameter in this regard is the fractional saturation, *f*, and a graph of the fractional saturation as a function of ligand concentration is known as a binding isotherm. The fractional concentration is related to the dissociation constant and the ligand concentration by the following equation:

$$f = \frac{[L]}{K_D + [L]}$$

According to this equation, the value of the dissociation constant is equal to the ligand concentration at which the protein is half saturated with ligand. The protein switches from being essentially unbound to almost completely bound within a concentration range of ligand that spans two orders of magnitude around the value of the dissociation constant. One consequence of this is that the dissociation constants of proteins for their natural ligands are usually close to the natural abundance of the ligand. An implication for drug development is that the dissociation constant for the drug binding to its desired target should be as low as possible and that for undesired targets should be as high as possible.

Drug molecules usually work by binding to the active sites of proteins and displacing their natural ligands. Such drugs are known as competitive inhibitors, and their effective strengths depend on the concentration of the natural ligand, such as ATP. The dissociation constant for an inhibitor binding to a protein is referred to as $K_I$. If the binding of the inhibitor to the drug is monitored by measuring the activity of the protein in the presence of the inhibitor and the natural ligand, then the concentration of the inhibitor that results in a reduction of activity to half the maximal value is known as the $IC_{50}$ value. The $IC_{50}$ value is related to the dissociation constant of the inhibitor, $K_I$, by the following equation:

$$K_I = (IC_{50})\left(\frac{K_D}{K_D + [L]}\right)$$

where $K_D$ is the dissociation constant of the natural ligand and [L] is the concentration of the natural ligand.

Small molecules, whether they are natural ligands or inhibitors, lose a considerable amount of entropy when they bind to proteins, and this serves as a substantial barrier to binding. Nevertheless, the process of binding many protein–ligand interactions has a net favorable entropy change, which is a consequence of the release of water molecules from hydrophobic groups on the ligand or from the protein. Although hydrogen bonding is a very important determinant of specificity in molecular recognition, it often does not contribute substantially to the net free-energy change. This is because water molecules form strong hydrogen bonds with polar groups on the ligand and the protein, and so the free energy gained upon complex formation represents a balance between the energy of hydrogen bonding in the complex and the hydrogen bonds to water that are broken when the intermolecular interface is formed.

As a result of these considerations, the hydrophobic effect usually provides the most important single determinant of binding free energy in the tightest intermolecular interactions. Unfortunately, it is impractical to increase the affinity of a ligand for its target by arbitrarily increasing its hydrophobicity, because the interactions between nonpolar groups is relatively nonspecific. As a consequence, molecules that are extremely hydrophobic are likely to adhere to many different components of the cell and are unlikely to be effective as specific inhibitors. An additional problem is that very hydrophobic molecules are insoluble in water and cannot be readily delivered into cells. The development of an effective drug therefore involves balancing the contributions of polar groups with nonpolar ones, and this makes the optimization of lead compounds a rather difficult task.

## Key Concepts

### A. THERMODYNAMICS OF MOLECULAR INTERACTIONS

- The affinity of a protein for a ligand is characterized by the dissociation constant, $K_D$.
- The value of the dissociation constant, $K_D$, for a binding interaction is the ligand concentration at which half the receptors are bound to ligand.
- The dissociation constant is commonly expressed in concentration units because we ignore the standard state concentrations (1 M).
- For a simple binding reaction (that is, without allostery), the binding isotherm is hyperbolic.
- The Scatchard equation recasts the equation governing the binding isotherm into a linear form.
- Saturable binding is a hallmark of specific binding interactions.
- The value of the dissociation constant determines the concentration range of the ligand over which the receptor switches from unbound to bound.
- The dissociation constant for a physiological ligand is usually close to the natural concentration of the ligand.

### B. DRUG BINDING BY PROTEINS

- Many drugs are developed by optimizing the binding of small molecules to protein targets.
- Most small-molecule drugs work by displacing a natural ligand from a protein.
- The binding of drugs to their target proteins often results in conformational changes in the protein.
- Induced-fit binding occurs through selection by the ligand of one among many preexisting conformations of the protein.
- Conformational changes in the protein underlie the specificity of a cancer drug known as imatinib.
- Conformational changes in a protein can weaken the apparent affinity of a ligand.
- The strength of binding interactions is often correlated with the hydrophobicity of the interaction.
- Cholesterol-lowering drugs known as statins take advantage of hydrophobic interactions to block their target enzyme.
- Polar (for example, hydrogen-bonding) interactions confer specificity in binding.
- The apparent affinity of an inhibitor is reduced by the presence of the natural ligand.
- Entropy lost by drug molecules upon binding is regained through the hydrophobic effect and the release of protein-bound water molecules.
- Isothermal titration calorimetry allows us to determine the enthalpic and entropic components of the binding free energy.

## Problems

### True/False and Multiple Choice

1. Which of the following is not a process governed by molecular recognition?
   a. The fidelity of DNA replication.
   b. Passive diffusion.
   c. Active transport.
   d. Translation by the ribosome.
   e. Transcription by RNA polymerase.

2. Noncovalent interactions, such as ionic and hydrophobic interactions, are generally much weaker than covalent interactions.

   True/False

3. A protein binds to the DNA sequence AAAAA with a 10-nM affinity (that is, the value of the dissociation constant is 10 nM). The same protein binds to the RNA sequence AUAAUA with a 15-nM affinity and to a lipid with a 10-mM affinity. Which of the following statements is true:
   a. The protein binds to RNA and DNA with high affinity, but low specificity.
   b. The protein has high specificity for lipids over RNA and DNA.
   c. The protein binds with low affinity for RNA and DNA, but high specificity.
   d. The protein binds with low affinity and low specificity, for all three targets.

4. The value of $K_D$ corresponds to:
   a. The reciprocal of $K_A$.
   b. The concentration of ligand at half saturation of receptor.
   c. $e^{\Delta G^\circ / RT}$.
   d. [P][L]/[P•L].
   e. All of the above.

5. A linear ligand binding curve indicates a simple binding equilibrium, whereas a hyperbolic binding curve

indicates that the system contains a receptor that is allosteric or contains multiple binding sites with different affinities.

True/False

6. Some inhibitors of HIV reverse transcriptase are nucleotide analogs that displace the natural nucleotide substrates in the active site.

True/False

7. Which of the following would lead to a decrease in binding affinity?

a. Releasing more protein-bound water molecules upon binding.
b. Decreasing the number of rotational degrees of freedom of the ligand upon binding.
c. Adding more hydrophobic interactions to the ligand–protein interface.
d. Increasing the number of hydrogen bonds between the ligand and protein.

## Fill in the Blank

8. A _______ is a small molecule that binds to a macromolecule receptor.

9. In general, higher affinity is achieved by increasing the _________ of a ligand, while increasing specificity relies on increasing ______________.

10. A typical drug has a dissociation constant for its receptor in the _________ to _________ range.

11. Because they bind to the same site as ATP, many kinase inhibitors are _________ with respect to ATP.

12. Without __________, some drugs cannot enter or exit the active site of a kinase domain.

## Quantitative/Essay

13. The binding between the drug cyclosporin and the protein cyclophilin is measured, giving a $K_D$ value of 1.5 nM. What is the the value of $\Delta G^o_{bind}$ at 298 K?

14. Shown below is a binding isotherm for an enzyme called a phosphatase and a lead compound known as JF99.

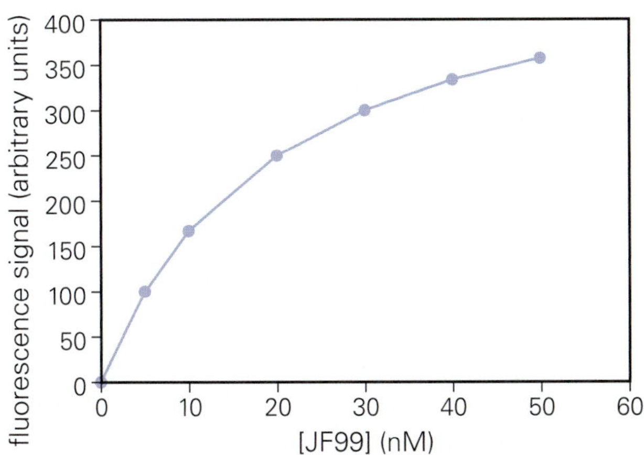


There is a tryptophan near the binding site, and the fluorescence from this residue increases upon binding the compound (shown on the *y* axis). At the highest concentration of JF99 tested (1 μM), the fluorescence signal is 490 units (not shown on the graph). Estimate the value of $K_D$.

15. What is the value of $\Delta G^o_{bind}$ for the interaction between the phosphatase and the inhibitor, based on the data shown in previous question, at 298 K?

16. At 298 K a variant of the protein streptavidin binds to biotin with a dissociation constant of 150 pM. What are the values of $K_A$ and $\Delta G^o_{bind}$?

17. Enough estrogen is dissolved into a solution containing the estrogen receptor that the total concentration of estrogen is 3 nM. The value of $K_D$ for the interaction between estrogen and its receptor is 5 nM. What fraction of the protein is bound to estrogen? Assume that the concentration of estrogen is much greater than that of the protein.

18. A binding assay is performed with a DNA-binding protein and a small DNA duplex. When the concentration of the free DNA duplex is 10 nM, 8% of the protein is bound to DNA. Assume the protein concentration is much lower than that of DNA. What is the value of $K_D$ for the interaction? What are the units of $K_D$? Explain your answer.

19. Using a purified sample of the protein cyclophilin, the Scatchard plot shown below is obtained when a form of cyclosporin is added. What is the value of $K_D$?

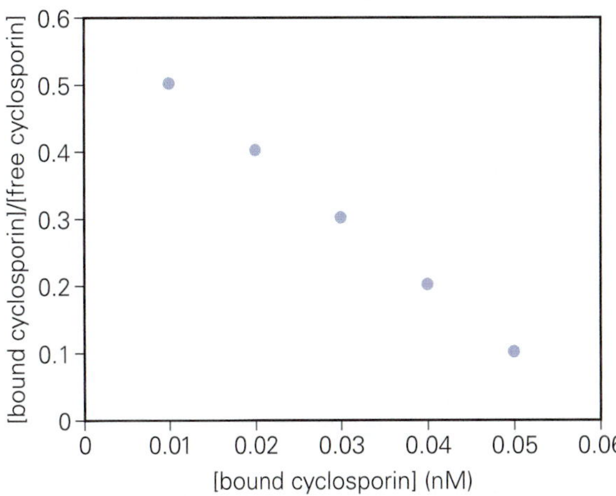


20. The drug jafrasitor (molecular weight 540 daltons) binds the histone deacetylase enzyme Sir2 with a dissociation constant of 0.1 nM. What mass of jafrasitor should be administered to a patient with a blood volume of 5.5 L such that Sir2 is at least 91% inhibited?

21. How do the conformational changes during kinase activation enable imatinib to distinguish between Abl and the hundreds of other related kinases in the cell?

22. Using 1 mL of partially purified cell lysate, which contains 1 mg of total protein, the following Scatchard

plot is obtained for the binding of the protein FKBP (molecular weight 20,000 daltons) and the drug rapamycin. What is the purity of the lysate with respect to FKBP?

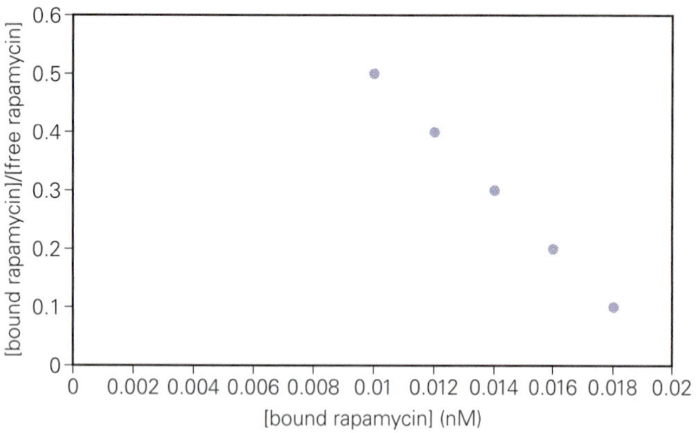


23. The $IC_{50}$ value for a drug called razundib for a Src kinase is 15 nM. The Src kinase is known to bind ATP with a dissociation constant of 0.05 mM. If the concentration of ATP in the cell is 0.75 mM, what is the value of $K_I$ for razundib with respect to the Src kinase?

24. An NMR analysis of the the drug fluvastatin indicates that it loses rotational and translational degrees of freedom upon binding to HMG-CoA reductase. Additionally, many internal bond rotations are restricted. Yet, an isothermal titration calorimetry experiment indicates that the binding reaction has a favorable change in entropy. What is the likely source of the favorable binding entropy change?

## Further Reading

### B. Drug Binding by Proteins

Carbonell T & Freire E (2005) Binding thermodynamics of statins to HMG-CoA reductase. *Biochemistry* 44, 11741–11748.

Cheng Y & Prusoff WH (1973) Relationship between the inhibition constant ($K_I$) and the concentration of inhibitor which causes 50 per cent inhibition ($I_{50}$) of an enzymatic reaction. *Biochem. Pharmacol.* 22, 3099–3108.

Congreve M, Murray CW & Blundell TL (2005) Structural biology and drug discovery. *Drug Discov. Today* 10, 895–907.

Istvan ES & Deisenhofer J (2001) Structural mechanism for statin inhibition of HMG-CoA reductase. *Science* 292, 1160–1164.

Jorgensen WL (2004) The many roles of computation in drug discovery. *Science* 303, 1813–1818.

Jura N, Zhang X, Endres NF, Seeliger MA, Schindler T & Kuriyan J (2011) Catalytic control in the EGF receptor and its connection to general kinase regulatory mechanisms. *Mol. Cell* 42, 9–22.

Kawasaki Y & Freire E (2011) Finding a better path to drug selectivity. *Drug Discov. Today* 16, 985–990.

Noble MEM, Endicott JA & Johnson LA (2004) Protein kinase inhibitors: insights into drug design from structure. *Science* 303, 1800–1805.

Richman DD (2001) HIV chemotherapy. *Nature* 410, 995–1001.

Teague SJ (2003) Implications of protein flexibility for drug discovery. *Nat. Rev. Drug Discov.* 2, 527–541.

Wells JA & McClendon CL (2007) Reaching for high-hanging fruit in drug discovery at protein-protein interfaces. *Nature* 450, 1001–1009.

# CHAPTER 13

# Specificity of Macromolecular Recognition

Almost every process in a living cell involves the dynamic interactions of biological macromolecules with one another. Cellular signaling, the control of metabolic activity, and the transport of material from one location in the cell to another are just a few of the essential processes that rely on protein–protein interactions. Both protein–protein and protein–DNA interactions are critical for the assembly of the required sets of transcription factors that initiate transcription of specific genes. Once mRNA is synthesized, its modification, including splicing, requires further sets of very specific protein–RNA interactions.

In order to understand how protein molecules recognize each other, and how proteins recognize specific nucleic acid sequences, we need to understand how residues arranged on protein surfaces interact with target molecules to provide both affinity and specificity. For the simplified analysis of binding in Chapter 12, we assumed that just two kinds of interacting molecules were present (the receptor and its ligand). In real systems there are many possible binding partners for any ligand, and the relative probabilities of forming different complexes determine the functional outcome. Thus, in addition to considering the affinity of a ligand for one receptor, we must also develop the concept of specificity—that is, the preference of a ligand for one particular target over others.

In this chapter, we begin by formally defining specificity in intermolecular interactions, and then discuss the molecular basis by which affinity and specificity are generated in protein–protein, protein–DNA, and protein–RNA interactions (**Figure 13.1**).

## A. AFFINITY AND SPECIFICITY

### 13.1 Both affinity and specificity are important in intermolecular interactions

The **affinity** of an intermolecular interaction reflects its strength. The affinity of two molecules for each other is given by the dissociation constant, $K_D$, for the complex, with lower values of $K_D$ corresponding to higher affinity. Alternatively, we can describe the affinity in terms of the free energy of formation of the complex, $\Delta G^o$ (where the more negative the value of $\Delta G^o$, the higher the affinity; see Section 12.1). If a complex is functionally important, then the interaction of the ligand and its target receptor must be sufficiently strong that they bind one another at their physiologically relevant concentrations (see Section 12.10).

**Affinity and specificity**

The *affinity* of a particular molecular interaction refers to its strength. The *specificity* of an interaction refers to the preference for a molecule to bind one particular target relative to all others. Interactions between biological macromolecules usually have high specificity.

High affinity between a molecule and its target is by itself insufficient to ensure activity *in vivo*. If a ligand were simply "sticky" and bound tightly to any receptor, then most of it would be bound to off-target receptors and not to the real target. In order to achieve a particular activity through binding, the ligand must have a preference for binding to a particular receptor—that is, the interaction needs to be specific.

The **specificity** of an interaction reflects the affinity of a ligand for one particular receptor relative to all other possible kinds of receptors, as described

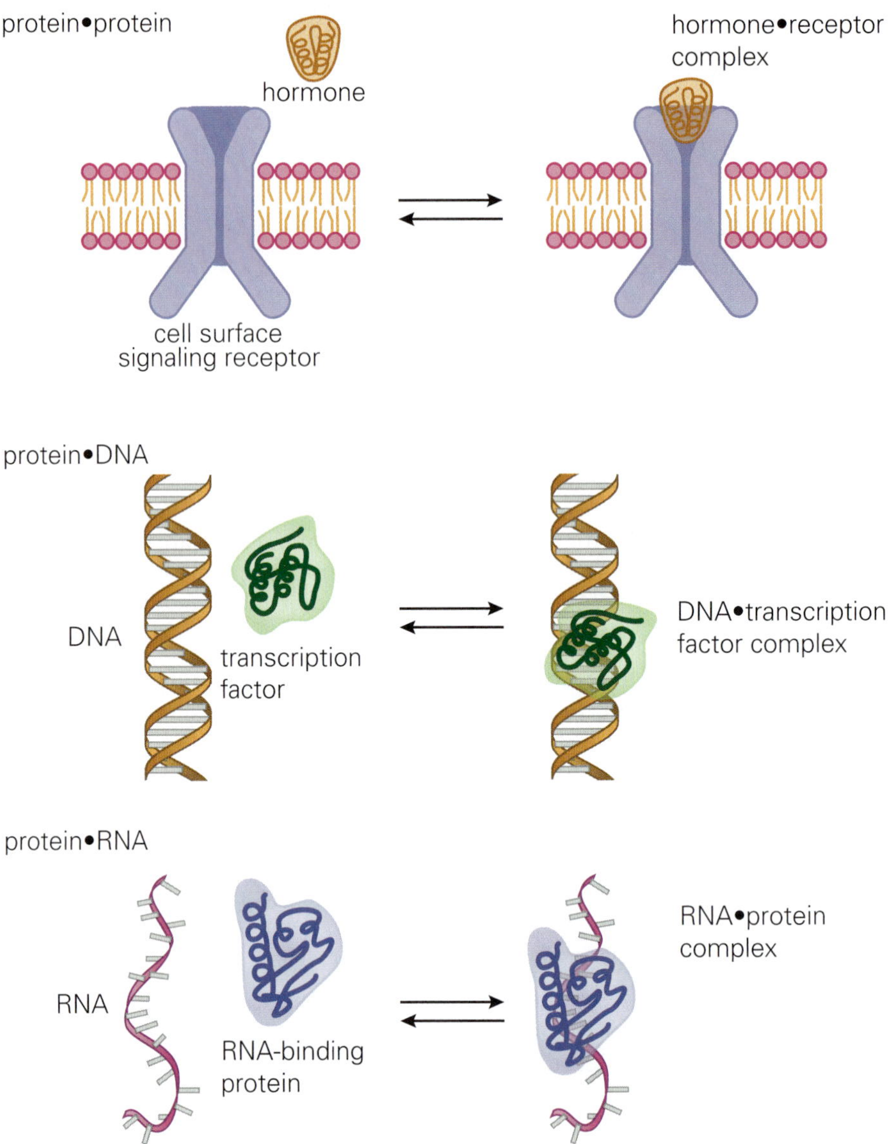


**Figure 13.1 Molecular recognition.** Examples of the three kinds of macromolecular interactions discussed in this chapter.

mathematically below. For many interactions in cells, the **affinity** of a ligand for one particular receptor is higher than for binding to other receptors (that is, functionally important binding usually has high specificity). The examples shown in Figure 13.1 are interactions with high specificity.

## 13.2 Proteins often have to choose between several closely related targets

The specificity of the interaction between a protein and its receptor is particularly important when the protein has to choose between many closely related targets in the cell. This is very common during cell signaling, when protein hormones can activate multiple receptors, leading to different outputs from each receptor. One such example is provided by a family of protein hormones known as the fibroblast growth factors (FGFs) and their receptors. The FGFs constitute a family of closely related proteins that control processes such as the growth of new blood vessels, wound healing, and embryonic development (**Figure 13.2**). The primary targets of these hormones are the fibroblast growth factor receptors (FGFRs), as shown in **Figure 13.3**. The FGF receptors are transmembrane tyrosine kinases with general similarity to the epidermal growth factor receptor that was described in Chapter 12 (see Figure 12.15).

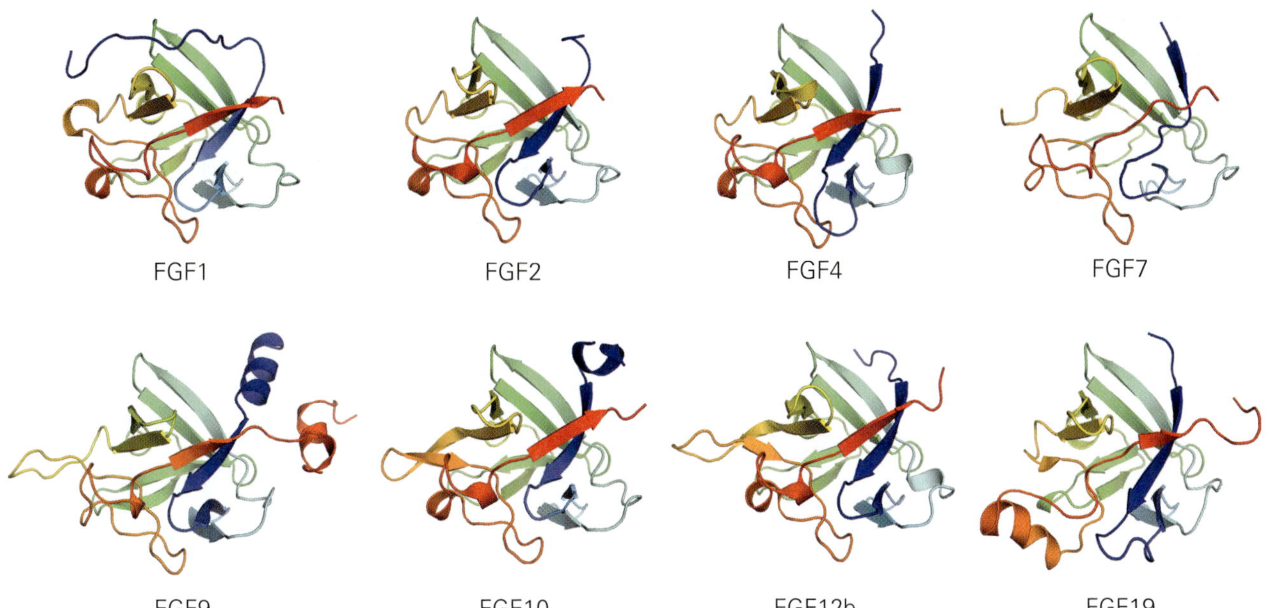


**Figure 13.2 Fibroblast growth factors.** The structures of eight related mammalian fibroblast growth factors (FGFs) are shown. They share the same core fold, with some differences in peripheral structures. The chains are shaded in the color of the rainbow, from N-terminus to C-terminus. (Adapted from M. Mohammadi, S.K. Olsen, and O.A. Ibrahimi, *Cytokine Growth Factor Rev.* 16: 107–137, 2005; PDB codes: 1RG8, 1CVS, 1IJT, 1IHK, 1QQK, 1NUN, 1Q1U, and 2P23.)

There are 18 known FGFs in humans, which are related in sequence and structure (some of these are shown schematically in Figure 13.2). There are four genes for FGF receptors in humans, but the resulting mRNA can be spliced alternatively, giving three more variants. In principle, any FGF can bind to any FGF receptor, so there are 126 possible FGF–FGFR combinations.

In order for this signaling system to have the required selective responses to FGF signals, a few of the interactions are strong (corresponding to the primary signaling pathways), while many others are very weak, as indicated in **Figure 13.4**. A given tissue may express only one or two of the receptors, and the binding affinity may be modified by glycosylation of the receptor (see Section 3.11), as well as binding to heparin sulfate on cell surfaces.

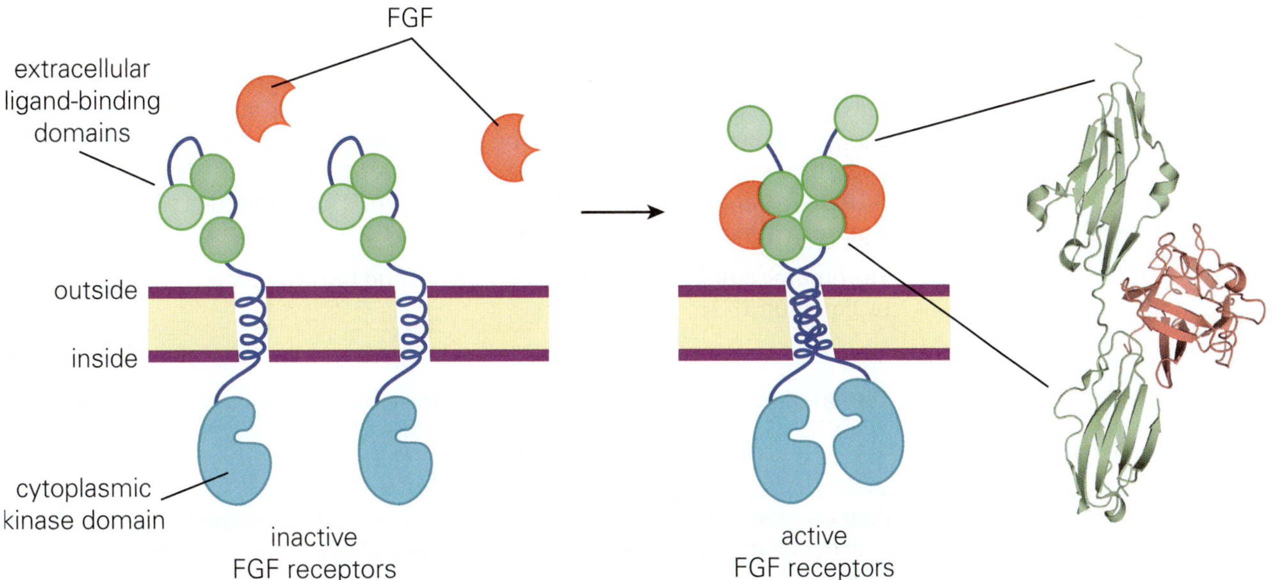


**Figure 13.3 Interaction between fibroblast growth factor (FGF) and its receptor.** FGF binding displaces an autoinhibition domain in the receptor, which blocks dimerization in the absence of FGF. The kinase domains then phosphorylate each other, which activates the receptor. To the right, the crystal structure of the complex of FGF1 with two domains of the FGF1 receptor is shown. Not shown in this diagram is heparin, a sulfated glycan that promotes receptor dimerization by bridging two FGF molecules. (PDB code: 1DJS.)

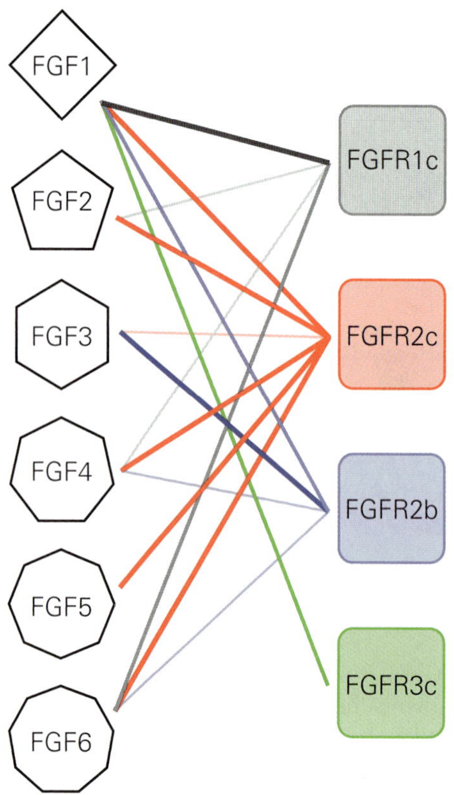


**Figure 13.4 Schematic representation of some FGF–FGFR interactions.** Each color reflects a different receptor, and the intensity of the line from each FGF reflects the affinity (relative to the tightest interaction for each). Missing lines correspond to interactions with very low affinity.

To understand the specificity of FGF signaling, it is necessary to understand how the occupancies of the various receptors change in response to altered levels of FGFs. We will return to a quantitative analysis of the FGF system after first defining more precisely what we mean by specificity.

## 13.3 Specificity is defined in terms of ratios of dissociation constants

In Section 12.10 we discussed the idea that substantial binding of a ligand to a receptor occurs when the concentration of the ligand exceeds the value of the dissociation constant. For the binding to be specific, two conditions must be satisfied. First, the dissociation constant for the primary target receptor (we will denote this receptor by $R_0$) must be smaller than the ligand concentration. Second, the dissociation constants for secondary or off-target receptors, denoted by $R_i$, must be larger than the ligand concentration. We use this idea to come up with a numerical value for the specificity of binding.

The binding equilibrium between the ligand and the primary receptor, $R_0$, is given by Equation 13.1:

$$R_0 \bullet L \rightleftharpoons R_0 + L \tag{13.1}$$

The strength of this interaction is characterized by the dissociation constant, $K_{D,R_0}$:

$$K_{D,R_0} = \frac{[R_0][L]}{[R_0 \bullet L]} \tag{13.2}$$

Suppose that the ligand can bind to a set of $N$ secondary receptors, $R_i$, in addition to the primary one. Each of these interactions is governed by a binding equilibrium:

$$R_i \bullet L \rightleftharpoons R_i + L \tag{13.3}$$

The dissociation constant for each of these binding equilibria is given by:

$$K_{D,R_i} = \frac{[R_i][L]}{[R_i \bullet L]} \tag{13.4}$$

Thus, when both $R_0$ and $R_i$ are present in the same sample, the concentration of free ligand, [L], will be identical in Equations 13.2 and 13.4, which must both be satisfied simultaneously.

We will define the **specificity factor**, $\alpha$, as the ratio of the concentration of ligand bound to the specific receptor, $R_0$, divided by the total concentration of ligand bound to all of the secondary receptors, $R_i$:

$$\alpha = \frac{[R_0 \bullet L]}{\sum_{i=1}^{N}[R_i \bullet L]} \tag{13.5}$$

It is useful to rewrite Equation 13.5 in terms of the dissociation constants and the total ligand concentration. To do this, we use the following set of relationships, where $R_i$ refers to any one of the several receptors:

**Specificity factor**

The ratio of the concentration of ligand bound to the target receptor to the total concentration of ligand bound to all other receptor is known as the specificity factor, $\alpha$.

$$[R_i]_{total} = [R_i] + [R_i \bullet L]$$

so:

$$[R_i] = [R_i]_{total} - [R_i \bullet L] \tag{13.6}$$

From Equation 13.4, the value of $[R_i \bullet L]$ is given by:

$$[R_i \bullet L] = \frac{[R_i][L]}{K_{D,R_i}}$$

Using Equation 13.6 we now get:

$$[R_i \bullet L] = \frac{([R_i]_{total} - [R_i \bullet L])[L]}{K_{D,R_i}} \tag{13.7}$$

Rearranging Equation 13.7 we obtain the following relationship:

$$[R_i \bullet L]\left(1+\frac{[L]}{K_D}\right) = \frac{[R_i]_{total}[L]}{K_{D,R_i}}$$

and so:

$$[R_i \bullet L] = \frac{[R_i]_{total}}{1+\frac{K_{D,R_i}}{[L]}} \tag{13.8}$$

Equation 13.8 holds for each of the receptor–ligand interactions separately. Putting these relationships into the numerator and denominator of Equation 13.5, we get:

$$\alpha = \frac{\dfrac{[R_0]_{total}}{1+\frac{K_{D,R_0}}{[L]}}}{\sum_{i=1}^{N}\left(\dfrac{[R_i]_{total}}{1+\frac{K_{D,R_i}}{[L]}}\right)} \tag{13.9}$$

Equation 13.9 allows us to calculate the specificity, $\alpha$ (that is, how much of the ligand is bound to the specific receptor), if we know the dissociation constants and the ligand concentration. A high value of $\alpha$ means that the primary target receptor is preferentially bound over the secondary receptors, shown schematically in **Figure 13.5**.

## 13.4 The specificity of binding depends on the concentration of ligand

What conditions favor high specificity in binding? From Equation 13.9, the value of $\alpha$ is large when the numerator is large but the denominator small. If we assume that the target and secondary receptors are at similar concentrations, $\alpha$ will be large when $K_{D,R_0} \leq [L] \leq K_{D,R_i}$ (see Figure 13.5A). In this range the target receptor is occupied significantly, but the secondary receptors are not.

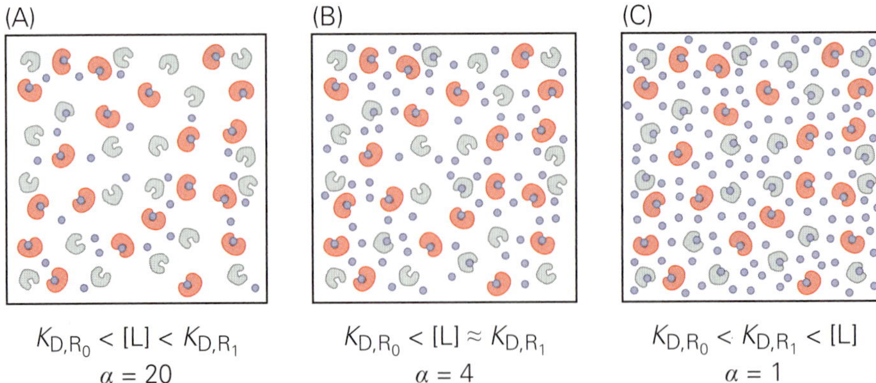


**Figure 13.5 Specificity of a ligand for two receptors.** The two receptors are denoted $R_0$ for a primary, high-affinity, receptor (*red*) and $R_1$ for a secondary one (*gray*). (A) When ligand is present at concentrations greater than $K_{D,R_0}$ but less than $K_{D,R_1}$, most of the ligand will be associated with $R_0$. The binding is highly specific under these circumstances. If [L] equals (B) or exceeds (C) $K_{D,R_1}$, then both receptors are occupied by ligand, thus lowering specificity.

If the ligand concentration is increased so that $[L] \approx K_{D,R_i}$ for some of the secondary receptors, then the value of $\alpha$ starts to decrease (see Figure 13.5B). When free ligand is present at high concentration, meaning that $[L]/K_{D,R_0} \gg 1$ and $[L]/K_{D,R_i} \gg 1$, then the value of $\alpha$ is small. If we assume that the primary target and secondary receptors are present at the same concentration, $[R_0]_{total} = [R_i]_{total}$, then the value of $\alpha$ approaches $1/N$, where $N$ is the number of receptor types. Under these conditions, both the target and the secondary receptors are fully occupied and the specificity factor is just the ratio of the number of target receptors to the total number of alternative receptors.

The value of $\alpha$ as a function of ligand concentration is graphed in **Figure 13.6** for the simple case when there are just two kinds of receptors, $R_0$ and $R_1$. At low ligand concentration, the primary receptor "wins," and the specificity is high. But, as the ligand concentration increases, an increasing amount of ligand binds to the secondary receptor, and the specificity decreases. Two situations are shown in this figure. In one case, the affinity of the primary receptor is much higher than that of the secondary receptor. In the other case, the difference between the affinities is not so large. When the values of $K_{D,R_0}$ and $K_{D,R_i}$ are very similar, then it is not possible to have high specificity (unless off-target receptors are only present at very low concentration), as can be seen in Figure 13.6.

This analysis also helps us understand the potential effects of a structural change in the ligand, L. If L is a protein whose sequence is altered through a mutation, for example, then binding to both target and secondary receptors may be modified. If the altered residue improves the steric fit to the primary receptor, $R_0$, or enhances the charge complementarity to it, then the affinity of the ligand for the $R_0$ receptor will increase (that is, $K_{D,R_0}$ becomes smaller). Such a change may result in an increase in specificity if the altered interactions do not also increase the affinity of the ligand for the secondary receptors by an equal amount. Mutations that widen the gap between $K_{D,R_0}$ and $K_{D,R_i}$ enhance specificity and are more likely to be kept during natural selection, while those that reduce this gap lead to lower specificity and are likely to be discarded during evolution.

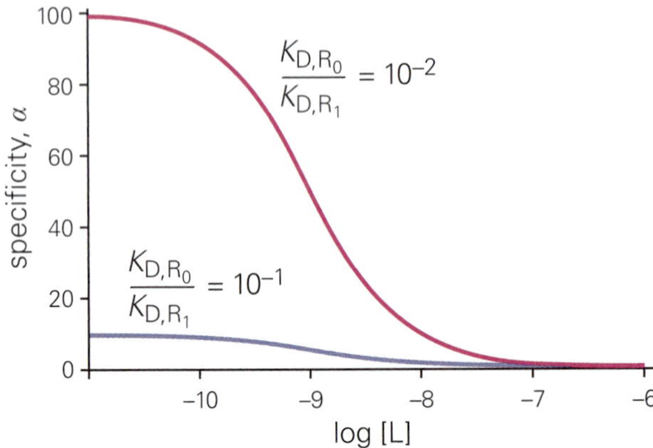


**Figure 13.6 Specificity as a function of ligand concentration.** In this example, there are two receptors, $R_0$ and $R_1$. The specificity, $\alpha$, is shown as a function of ligand concentration for dissociation constants of $10^{-9}$ for $R_0$ and $10^{-7}$ for $R_1$ (upper curve) and $10^{-9}$ for $R_0$ and $10^{-8}$ for $R_1$ (lower curve). In both cases, the specificity decreases with increasing ligand concentration, reaching a value of 1 at high ligand concentration.

## 13.5 Fractional occupancy and specificity are important for activities resulting from binding

The activity of a receptor is usually determined by whether a ligand is bound to it or not. It is therefore useful to describe the activity of a receptor in terms of fractional occupancy, a concept that we developed in Section 12.2. In particular, with the ligand–receptor notation used in this chapter, the fraction of receptors occupied by ligand, $f_{R_i}$, is:

$$f_{R_i} = \frac{[R_i \bullet L]}{[R_i]+[R_i \bullet L]} = \frac{[R_i \bullet L]}{[R_i]_{total}} \tag{13.10}$$

Hence, we can rewrite Equation 13.5 (the expression for the specificity, $\alpha$) in terms of the fractional occupancies:

$$\alpha = \frac{f_{R_0}[R_0]_{total}}{\sum_i f_{R_i}[R_i]_{total}} \tag{13.11}$$

As a result, a large fractional occupancy of the target receptor, $R_0$, and a low occupancy of off-target receptors, $R_i$, corresponds to high specificity.

It can be instructive to look at a simple case in which the ligand has just two different affinities for receptors, high for the primary receptor and 100 times weaker for all of the secondary receptors. For example, we assume that all receptors are present at equal total concentration, $[R_0]_{total} = [R_i]_{total}$, but with 10 different secondary receptors. In this case,

$$\alpha = \frac{f_{R_0}[R_0]_{total}}{10 f_{R_i}[R_i]_{total}} = \frac{f_{R_0}}{10 f_{R_i}} \tag{13.12}$$

As makes intuitive sense, the specificity depends on the ratio of the fractional occupancies (**Figure 13.7**). When ligand concentration is low (that is, when $[L]/K_{D,R_0} \ll 1$), you can see from Equation 13.9 that:

$$\alpha = \frac{\dfrac{[R_0]_{total}}{1+\dfrac{K_{D,R_0}}{[L]}}}{\sum_{i=1}^{N} \dfrac{[R_i]_{total}}{1+\dfrac{K_{D,R_i}}{[L]}}} \approx \frac{\dfrac{[R_0]_{total}[L]}{K_{D,R_0}}}{\sum_{i=1}^{10} \dfrac{[R_0]_{total}[L]}{100 K_{D,R_0}}} = 10 \tag{13.13}$$

As the concentration of ligand increases, the occupancy of $R_i$ increases and the specificity will decrease, reaching a limiting value of 0.1.

## 13.6 Most macromolecular interactions are a compromise between affinity and specificity

The requirements for generating high affinity and high specificity can often be conflicting for interactions between macromolecules. As we discussed in Section

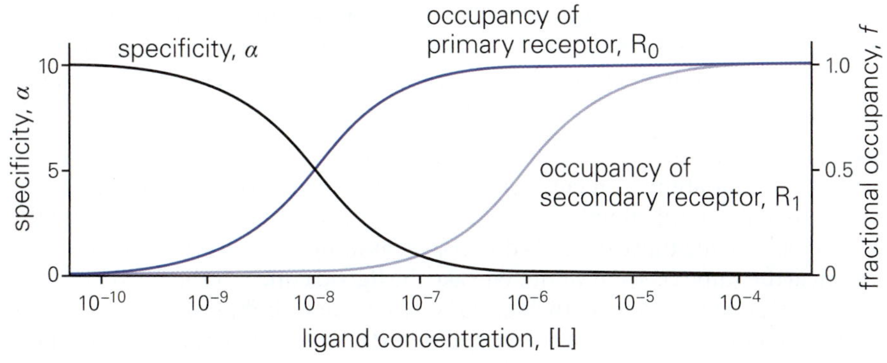


**Figure 13.7 Graph of specificity and receptor occupancies as a function of ligand concentration.** Curves are shown for the occupancy of both the primary receptor ($K_D = 10^{-8}$) and 10 off-target receptors ($K_D = 10^{-6}$), and the specificity, $\alpha$. Near $[L] = 10^{-7}$, there will be equal occupancy of the primary receptor and the set of secondary receptors; hence, $\alpha \approx 1$ at that point.

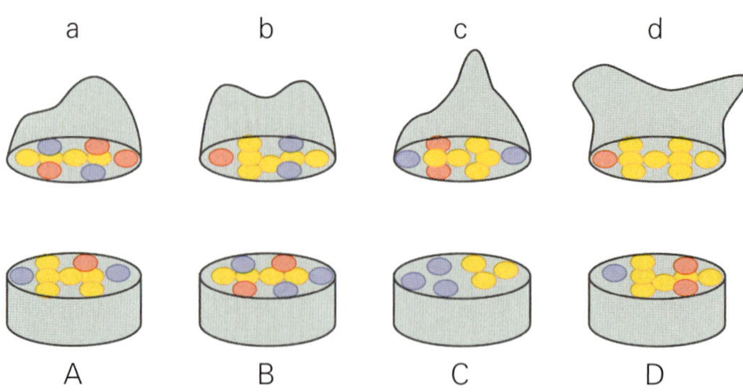


**Figure 13.8 Charge interactions and specificity.** Polar and electrostatic interactions, even if they do not provide net stabilization, can provide specificity by destabilizing off-target complexes. In this schematic, the ligands a, b, c, and d could potentially pair with receptors A, B, C, and D. The *yellow*, *blue*, and *red* regions are hydrophobic, positively charged, and negatively charged, respectively. The pairing b–D will be very specific.

12.19, the affinity of a noncovalent interaction is most easily increased by making the interface more hydrophobic. Hydrophobic interactions have lower specificity, however, because there are many hydrophobic amino acids, and many orientations of each one, that can make up a hydrophobic contact surface.

Polar or charged residues at an interface will only contribute favorably to the free energy of binding if there are complementary functional groups at specific positions on the opposite surface (**Figure 13.8**). For example, bringing a negatively charged residue into contact with a hydrophobic surface is unfavorable because the strong solvent interactions of the charged group are lost. But, if the contacted surface has an appropriately positioned positive charge, then the favorable electrostatic interaction compensates for the loss of solvent interactions. Many factors have entered into the evolution of specific interactions, including shape complementarity of surface hydrophobic patches, hydrogen bonding, and electrostatics.

The difference between on-target versus off-target affinities is the important factor for specificity, as shown in Figure 13.6. Interactions that lead to specificity quite often destabilize off-target complexes as often as they stabilize on-target ones. Because of this aspect of specificity, nonpolar contacts can play important roles in determining specificity, even when highly charged molecules such as nucleic acids are being recognized.

## 13.7 Fibroblast growth factors vary considerably in their affinities for receptors

Having now developed a precise definition of specificity, we return to the system of fibroblast growth factors (FGFs) and their receptors. As described in Section 13.2, there are many possible ligand–receptor pairings that can occur. The binding of several FGFs to different receptors has been studied quantitatively using purified ligands and receptors. Some of the dissociation constants that were determined are listed in **Table 13.1**. In addition to the natural receptors, folded RNA molecules have been developed that bind to FGFs with high affinity (RNAs do not normally bind to FGFs – these RNAs are artificial receptors for the FGFs). Binding of an FGF to the RNAs prevents binding to a real receptor, and hence blocks signaling. Data for the binding of one particular RNA to the FGFs are also given in Table 13.1.

Based on the data in this table, some FGFs, such as FGF1, bind to several receptors with similar affinity. In contrast, other FGFs are very selective (for example, FGF5). To make this distinction more quantitative, we can calculate values for the specificity, $\alpha$, using these data. To do this, we assume that all of the receptors are present at the same concentration. We take the ligand concentration to be 100 nM, which is high enough to occupy high-affinity receptor binding sites. Considering FGF1 interacting with the natural receptors, we take the highest affinity receptor,

FGFR2c, as the primary receptor ($R_0$) and find:

$$\alpha_{\text{FGFR2c}} = \frac{\dfrac{1}{1+\dfrac{K_{D,R_0}}{[L]}}}{\sum_i \left(\dfrac{1}{1+\dfrac{K_{D,R_i}}{[L]}}\right)}$$

$$= \frac{\dfrac{1}{1+\dfrac{0.94\times10^{-7}}{10^{-7}}}}{\dfrac{1}{1+\dfrac{1.4\times10^{-7}}{10^{-7}}}+\dfrac{1}{1+\dfrac{1.6\times10^{-7}}{10^{-7}}}+\dfrac{1}{1+\dfrac{2.3\times10^{-7}}{10^{-7}}}}$$

$$=0.47 \tag{13.14}$$

The value of $\alpha$, 0.47, is quite low, reflecting the fact that there is a similar level of binding to several other receptors. This result means that FGF1 probably stimulates multiple receptors when acting *in vivo*. But, for FGF5, again considering the natural receptors, with FGFR2c as the target,

$$\alpha_{\text{FGFR2c}} = \frac{\dfrac{1}{1+\dfrac{5.2\times10^{-7}}{10^{-7}}}}{\dfrac{1}{1+\dfrac{10^{-4}}{10^{-7}}}+\dfrac{1}{1+\dfrac{10^{-4}}{10^{-7}}}+\dfrac{1}{1+\dfrac{10^{-4}}{10^{-7}}}} = 54 \tag{13.15}$$

In this case, the specificity is much higher, even though the binding of FGF5 to the target receptor is actually weaker than for FGF1 (see Table 13.1). This value for the specificity indicates that FGF5 is likely to be bound to FGFR2c ~50 times more than to all of the other receptors combined. Note that the effective specificity could actually be higher, because the lower limit estimate for the dissociation constant, $10^{-4}$, was used for the off-target receptors for which binding was not detected. Also, because many tissues only express a subset of receptors, the specificity *in vivo* might be even higher than our estimate, because all receptors are not present.

**Table 13.1 Specificity of FGF ligands for FGF receptors.**

| Receptor ligand | FGFR1c | FGFR2c | FGFR2b | FGFR3c | FGF-binding RNA |
|---|---|---|---|---|---|
| FGF1 | $1.4 \times 10^{-7}$ | $9.4 \times 10^{-8}$ | $1.6 \times 10^{-7}$ | $2.3 \times 10^{-7}$ | $9.7 \times 10^{-7}$ |
| FGF2 | $6.2 \times 10^{-8}$ | $1.0 \times 10^{-8}$ | n.d. | n.d. | $3.5 \times 10^{-10}$ |
| FGF3 | n.d. | $1.2 \times 10^{-6}$ | $3.6 \times 10^{-7}$ | n.d. | n.d. |
| FGF4 | $1.7 \times 10^{-7}$ | $2.7 \times 10^{-8}$ | $5.3 \times 10^{-7}$ | n.d. | $5.6 \times 10^{-7}$ |
| FGF5 | n.d. | $5.2 \times 10^{-7}$ | n.d. | n.d. | $8.5 \times 10^{-9}$ |
| FGF6 | $1.0 \times 10^{-7}$ | $3.7 \times 10^{-8}$ | $6.6 \times 10^{-7}$ | n.d. | $6.1 \times 10^{-7}$ |

Dissociation constants for various combinations of FGF–receptor complexes are shown, in molar units. Also included are values for an artificial "receptor," which is a folded RNA molecule. n.d. indicates binding not detected, which means $K_D > 10^{-4}$. (Data from M. Mohammadi, S.K. Olsen, and O.A. Ibrahimi, *Cytokine Growth Fac. Rev.* 16, 107–137, 2005.)

We can also reverse the roles and consider the specificity of different ligands for one receptor. In this case, to apply the equations, we consider the FGFR as the fixed "ligand" and the different FGFs as the receptors. With this point of view, we see that FGFR3c is very specific for FGF1 (with $\alpha > 60$), while FGFR1c and FGFR2c have low specificity ($\alpha \approx 1$). The specificity for the RNA ($\alpha \approx 24$) is high, but in spite of very tight binding to FGF2, it is clear that specificity is lowered primarily by binding to one particular off-target hormone, FGF5. If the RNA were to be used as a drug, then this off-target binding might have to be reduced by altering the RNA sequence to avoid side effects.

## 13.8 The recognition of DNA by transcription factors involves discrimination between a very large number of off-target binding sites

The number of possible pairings of FGFs interacting with their receptors is moderate (126 possible combinations of ligand and receptor; see Section 13.2). For some other binding events, however, the number of off-target receptors is very large. One good example of this is the interaction of sequence-specific DNA-binding proteins with their target sequences in the context of genomic DNA, in which any set of base pairs can constitute a weak, nonspecific binding site.

One such case that has been studied extensively involves the lactose (lac) repressor from *E. coli*. This protein binds to specific DNA sequences (known as **operator sequences**) that control the expression of genes involved in lactose metabolism. In the absence of lactose, these enzymes are not needed, and the repressor binds tightly (with a dissociation constant $K_{D,Op} \approx 10^{-12}$ M) to the 12-base-pair operator DNA, blocking access of polymerase and preventing transcription of the genes, as shown in **Figure 13.9**. When lactose is present, it binds the repressor, reducing its affinity for operator by about 1000-fold ($K_{D,Op} \approx 10^{-9}$ M). This has the effect of dissociating the repressor-operator complex, leading to transcription of genes encoding enzymes involved in lactose metabolism.

Mechanisms by which transcription factors such as the lac repressor achieve specificity for particular DNA sequences are discussed in part C of this chapter. For now, we note that some of the protein–DNA contacts involve electrostatic attraction between positively charged groups on the protein and the negatively charged

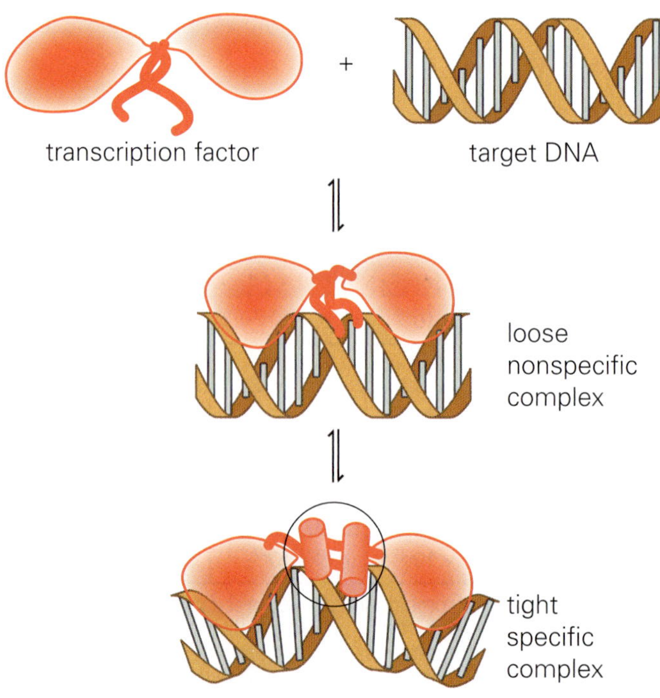


**Figure 13.9 Lac repressor DNA binding.** The lac repressor binds to operator DNA with high affinity, but can bind other DNAs, as well, with much lower affinity. Recognition of the specific operator DNA sequence is accompanied by structural rearrangement that increases interactions and hence stability. A region of secondary structure formation that occurs only in the specific complex is circled at the bottom. These figures show just the DNA-binding domains. (Adapted from C.G. Kalodimos et al., and R. Kaptein, *Science* 305: 386–389, 2004. With permission from AAAS.)

phosphate backbone of the DNA, and do not depend on the DNA sequence. Therefore, the lac repressor can also bind weakly to essentially any non-operator DNA sequence (the value of the dissociation constant for binding to a non-operator site, $K_{D,nonOp}$, is ~$10^{-4}$ M). The difference in affinity between sequence-specific and nonspecific binding is very large, making it seem that only specific binding would occur. But, there are about $5 \times 10^6$ base pairs in the *E. coli* genome, and if any contiguous set of 12 base pairs is considered to be a nontarget site, then there are about $5 \times 10^6$ such sites (the first base of the 12 can be at any base pair of the genome; **Figure 13.10**).

We can calculate the specificity of the interaction between lac repressor and its operator sequence by using Equation 13.13 for the highly simplified case that all non-operator DNA sequences are equivalent. This converts the sum in the denominator into just the number of non-operator sites, $N_{nonOp}$, multiplied by the dissociation constant per site. The specificity for operator DNA is then:

$$\alpha = \frac{\left(\dfrac{\dfrac{[L]}{K_{D,Op}}}{1+\dfrac{[L]}{K_{D,Op}}}\right)}{\sum_i \left(\dfrac{\dfrac{[L]}{K_{D,nonOp_i}}}{1+\dfrac{[L]}{K_{D,nonOp_i}}}\right)} = \frac{\dfrac{1}{1+\dfrac{K_{D,Op}}{[L]}}}{\dfrac{N_{nonOp}}{1+\dfrac{K_{D,nonOp}}{[L]}}} \tag{13.16}$$

Putting in the binding constants and the number of non-operator sites gives $\alpha \approx 0.1$, a very modest value in spite of the large difference between specific and nonspecific binding constants. This means that most of the repressor (~90%) is actually bound at nonspecific sites. In this case, the issue for function is whether the specific operator site is occupied or not and, to answer this question, we have to do a different calculation.

## 13.9 Lowering the affinity of lac repressor for the operator switches on transcription

To understand how lactose binding to lac repressor enables transcription, we need to consider the concentrations of the components inside the cell. A single *E. coli* cell has an internal volume of about $10^{-15}$ L (the shape of the cell can be approximated by a cylinder with a radius of 0.5 μm and a length of 2 μm). There

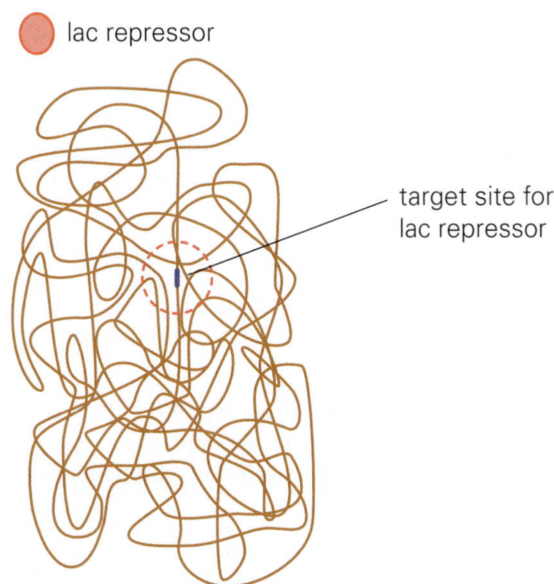


**Figure 13.10 Lac repressor–operator binding.** Lac repressor, top left (greatly enlarged for visibility), must find a specific site (the small *blue speck*, in the circled region) in a sea of background sites (the *brown line*).

are about 10 lac repressor tetramers (the form that binds DNA), 1 operator, and 5 $\times 10^6$ total base pairs per cell. These numbers imply that the total concentration of the lac repressor in the cell ($[\text{lacR}]_{\text{total}}$) is given by:

$$[\text{lacR}]_{\text{total}} = \frac{10 \text{ molecules}}{6\times 10^{23} \text{molecules}\bullet\text{mol}^{-1}} \times \frac{1}{10^{-15}\text{L}}$$
$$\approx 2\times 10^{-8} \text{ M} \qquad (13.17)$$

The concentration of the single operator site in the cell, $[\text{Op}]_{\text{total}}$, is given by:

$$[\text{Op}]_{\text{total}} = \frac{1 \text{ site}}{6\times 10^{23} \text{ sites}\bullet\text{mol}^{-1}} \times \frac{1}{10^{-15} \text{ L}}$$
$$\approx 2\times 10^{-9} \text{ M} \qquad (13.18)$$

If we assume that all the non-operator sites (denoted nonOp) are equivalent, and the number of such sites is the same as the number of base pairs, the concentration of non-operator sites ([nonOp]) is calculated from the number of operator sites as follows:

$$[\text{nonOp}]_{\text{total}} = \left(2\times 10^{-9}\right)\left(5\times 10^{6}\right)\text{M}$$
$$\approx 10^{-2} \text{ M} \qquad (13.19)$$

Because the concentration of non-operator sites is so high, very few can be occupied, and so we can assume that $[\text{nonOp}]_{\text{free}} = [\text{nonOp}]_{\text{total}}$. Then, we can calculate the fraction of repressor that is free:

$$K_{\text{D}} = 10^{-4} = \frac{[\text{lacR}][\text{nonOp}]}{[\text{lacR}\bullet\text{nonOp}]}$$

and so, by using Equation 13.19:

$$\frac{10^{-4}}{10^{-2}} = \frac{[\text{lacR}]}{[\text{lacR}\bullet\text{nonOp}]} = 10^{-2} \qquad (13.20)$$

According to Equation 13.20, 1% of the repressor is free and 99% is bound to non-operator DNA. Thus, given that $[\text{lacR}]_{\text{total}} = 2 \times 10^{-8}$ M, $[\text{lacR}]_{\text{free}} = 2 \times 10^{-10}$. Given this concentration of free repressor, determined by nonspecific binding, what will the occupancy of the operator site be? We apply the same idea, but use $K_{\text{D}} = 10^{-12}$ M for the operator complex in the *absence* of lactose:

$$K_{\text{D}} = 10^{-12} = \frac{[\text{lacR}]_{\text{free}}[\text{Op}]}{[\text{lacR}\bullet\text{Op}]}$$

or

$$\frac{[\text{Op}]}{[\text{lacR}\bullet\text{Op}]} = \frac{10^{-12}}{2\times 10^{-10}} \qquad (13.21)$$

In the non-operator case, the binding sites are in huge excess with respect to the repressor, and so we did not have to consider the depletion of binding sites due to repressor binding. But when we consider specific binding, the operator is present at low concentration, so we have to include another equation, which states that the total concentration of the operator site is the sum of the concentrations of the free and bound operator sites:

$$[\text{Op}]_{\text{total}} = [\text{Op}] + [\text{lacR}\bullet\text{Op}] = 2 \times 10^{-9} \text{ M} \qquad (13.22)$$

By combining Equations 13.21 and 13.22, we find that $[\text{Op}] = 10^{-11}$ M. Since $[\text{Op}]_{\text{total}} = 2 \times 10^{-9}$, only 0.5% of the operator is free. Functionally, then, transcription of the lactose related genes is off.

If we repeat the calculation, but this time with $K_{\text{D}} = 10^{-9}$ M, corresponding to repressor in the *presence* of lactose, then we find $[\text{Op}] = 1.66 \times 10^{-9}$ M—that is, 80% of the operator is free, allowing transcription. The switch for transcription has been thrown by the presence of lactose.

If the equilibrium constants are used to calculate the binding of repressor to operator in the *absence* of *any* competing nonspecific sites, then we find that with no lactose, 99.99% of operator would be bound, but in the presence of lactose, there would still be 95% bound, and transcription would remain low. This calculation shows that the binding constants have been tuned by evolution to undergo functional switching in the presence of nonspecific binding. It also shows that the specificity factor, $\alpha$, does not always provide the information needed to deduce when a ligand will function.

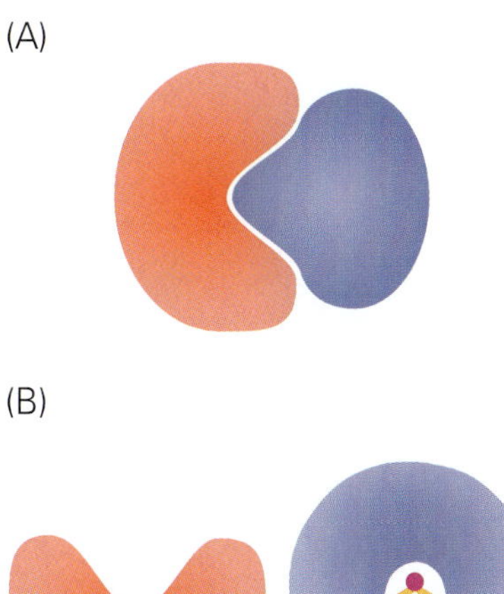


**Figure 13.11 Protein–protein interfaces.** (A) An interface between two folded domains. (B) Recognition of a peptide segment in a protein (*red*) by a peptide recognition domain (*blue*).

## B. PROTEIN–PROTEIN INTERACTIONS

In this part of the chapter we discuss some aspects of how proteins that interact with each other achieve specificity. Our focus is on interactions that are transient—in such cases the proteins that interact are usually capable of folding on their own into stable structures, but can form specific complexes when they find their appropriate binding partners. In other cases, which we do not discuss here, a protein molecule might require its binding partner in order to fold into a stable structure. For example, proteins that form coiled-coils (see Figure 4.38) may not form stable α helices in the absence of their binding partners. The principles that govern the formation of such constitutive protein complexes are somewhat different from those underlying the more transient interactions described here.

### 13.10 Protein–protein complexes involve interfaces between two folded domains or between a domain and a peptide segment

There are two major classes of protein–protein interactions, and the distinction between them is illustrated in **Figure 13.11**. In one class of interactions, the surfaces of two folded protein domains make extensive contact with each other (see Figure 13.11A). The residues that interact across this interface may come from several different structural elements within the two proteins. For example, in the complex of the protease enzyme trypsin with an inhibitor (discussed in Section 16.22), there are two discontinuous segments within the inhibitor that make contact with five segments from the enzyme, as shown in **Figure 13.12**.

The second class of interactions are mediated by specialized peptide recognition domains, which we discussed briefly in Section 4.2. These domains recognize short peptide segments in other proteins, often just five to 10 residues long (see

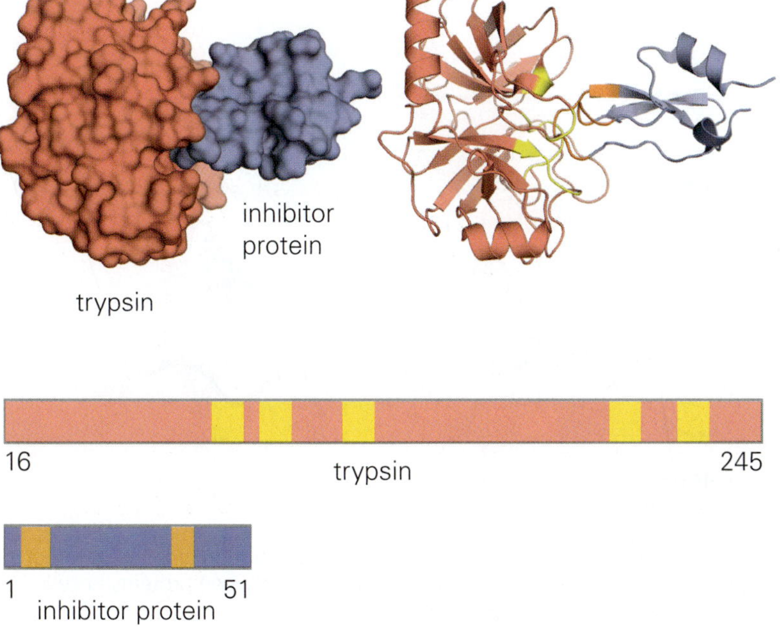


**Figure 13.12 Structure of a protease–inhibitor complex.** The structure of a protease enzyme (trypsin) bound to an inhibitor protein is shown. The regions of both proteins that form the interaction surface (the binding epitopes) are indicated in *yellow* and *orange*. These binding epitopes are from noncontiguous regions of the polypeptide chains, as shown in the bar diagrams that represent the chains. (PDB code: 3FP6.)

Figure 13.11B). Two examples of peptide recognition domains are the Src homology domains 2 and 3 (SH2 and SH3) that we discussed briefly in Chapter 4 (see Figure 4.4).

Interactions between folded protein domains, such as that between trypsin and its inhibitor, can be very tight and therefore very specific. Such complexes exhibit a wide range of dissociation constants, ranging from micromolar ($10^{-6}$ M) at the weaker end to nanomolar or picomolar ($10^{-9}$–$10^{-12}$ M) at the tighter end. The interactions between peptide recognition domains and their targets are weaker, with dissociation constants ranging from ~10 micromolar to ~100 nanomolar.

Many different kinds of peptide-recognition domains have now been identified. These have different folds, a few of which are illustrated in **Figure 13.13**, and they bind to a variety of peptide motifs. SH2 domains, for example, bind to peptide segments containing phosphorylated tyrosine residues, and SH3 domains recognize peptide segments that adopt a conformation known as a polyproline type II helix (usually containing Pro-X-X-Pro sequence motifs, where X may be any amino acid).

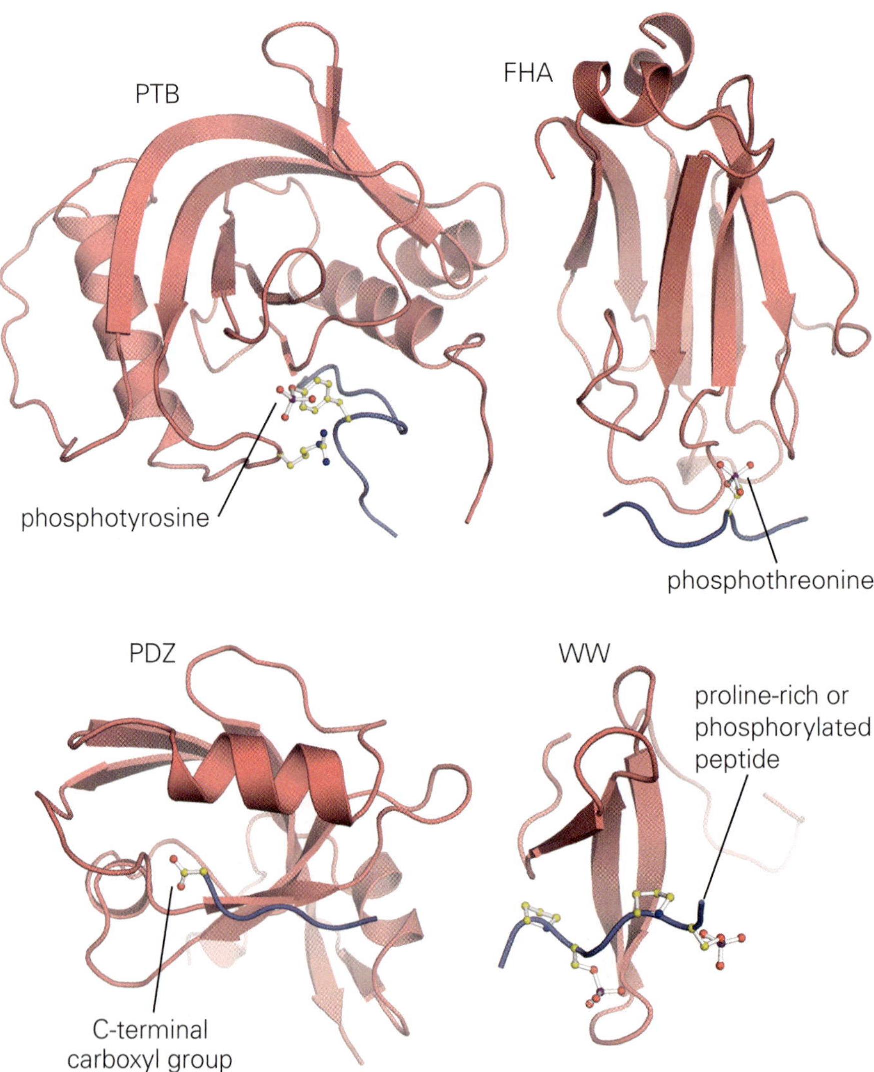


**Figure 13.13** Peptide interaction domains. Shown are four domains that interact with specific peptides to provide important functional links between proteins in cells. (PDB codes: 1SHC, 1G6G, 1BE9 and 1F8A.)

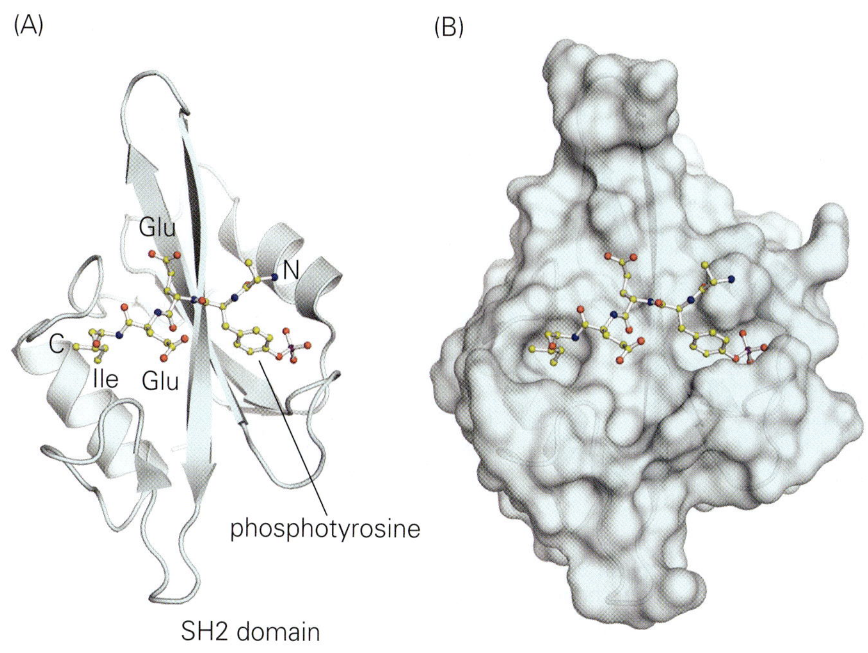


**Figure 13.14 Structure of an SH2 domain.** (A) The backbone of the Src SH2 domain is shown as a ribbon, with a bound target peptide shown in stick representation. The sequence of the peptide is P-Tyr-Glu-Glu-Ile, where P-Tyr stands for phosphotyrosine. (B) The surface of the SH2 domain is shown. Notice that the P-Tyr and Ile sidechains fit into socket-like depressions on the surface. (PDB code: 1SPS.)

## 13.11 SH2 domains are specific for peptides containing phosphotyrosine

The SH2 domain has a compact three-dimensional structure comprising about 100 amino acid residues (**Figure 13.14**). The core of the domain is a four-stranded antiparallel β sheet, flanked by two α helices. Phosphotyrosyl peptides bind to the SH2 domain in an extended conformation, using a binding surface that runs perpendicular to the central β sheet in the domain.

The binding surface of an SH2 domain is located on the side of the domain that is opposite to the N-terminal and C-terminal segments of the protein, which are located close together. It is this feature that allows these domains to function in a manner that is relatively independent of the larger proteins in which they are found. Evolution has clearly exploited the modularity in the design of these domains, as witnessed by the remarkable variation in the number and placement of such domains in signaling proteins (see Figure 4.5).

A crucial aspect of the specificity of an SH2 domain is its ability to distinguish between phosphorylated and unphosphorylated tyrosine residues in its targets. When phosphotyrosine is present in a peptide that has a proper sequence for interaction with the Src SH2 domain (for example, the peptide shown in Figure 13.14), the dissociation constant ($K_D$) is ~0.1 μM. The corresponding value of the standard free energy of binding, $\Delta G^o$, is –40 kJ•mol$^{-1}$ at 300 K (see Table 12.1).

If the tyrosine residue in the peptide is not phosphorylated, then the peptide fails to bind to the SH2 domain (the binding is below the limit of detection in the experiment, and the affinity is reduced by at least 10,000-fold). This difference corresponds to a reduction in the free energy of binding of at least ~22 kJ•mol$^{-1}$. Thus, interactions with the phosphotyrosine residue must contribute more than half of the free energy of binding of peptides to SH2 domains.

The key to recognition of phosphotyrosine is its interaction with an arginine sidechain that is invariant in all SH2 domains (**Figure 13.15**). The peptide lies along the surface of the SH2 domain, and the negatively charged phosphotyrosine sidechain reaches down from the surface to form hydrogen bonds with the positively charged arginine sidechain, which rises up to meet it from the interior of the domain.

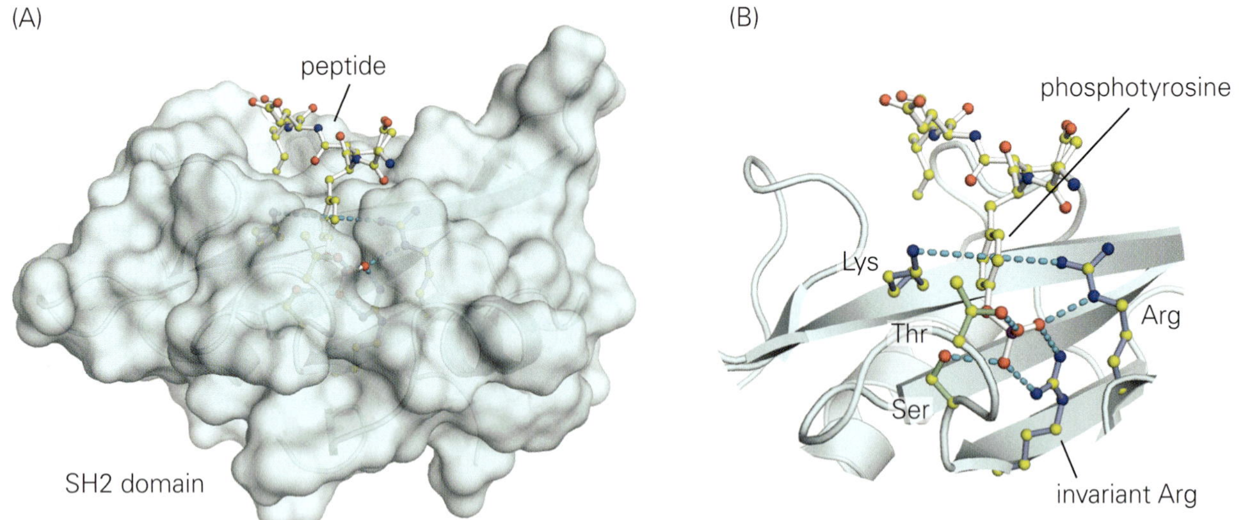


**Figure 13.15 Recognition of phosphotyrosine by an SH2 domain.** (A) The Src SH2 domain, bound to the same peptide shown in Figure 13.14. (B) An expanded view showing interactions with the phosphotyrosine residue. An invariant arginine sidechain that is critical for recognition is at the base of the phosphotyrosine binding pocket. Hydrogen bonds are indicated by dotted lines. The dotted line passing through the tyrosine ring indicates favorable interaction between a lysine and arginine and the electrons of the aromatic ring. (PDB code: 1SPS.)

The location of the invariant arginine is such that, in a fully extended conformation, the sidechain is just long enough to interact with the phosphate group of a fully extended phosphotyrosine sidechain, as you can see in Figure 13.15. In this way, the interaction provides a stereochemical "ruler" that can select for phosphotyrosine, because all other kinds of negatively charged groups in proteins are not long enough to engage the arginine.

The hydrogen-bonding potential of the phosphotyrosine group is completely satisfied by other interactions with the SH2 domain. In the Src SH2 domain, shown in Figure 13.15, a second arginine sidechain coordinates the phosphate group, with additional hydrogen bonds also provided by threonine and serine sidechains. The second arginine and a lysine sandwich the aromatic ring of the phosphotyrosine. This leads to favorable electrostatic interactions between these positively charged residues and the electrons of the aromatic ring. These **amino–aromatic interactions** also contribute to the specificity of the SH2 domain for phosphotyrosine.

## 13.12 Individual SH2 domains cannot discriminate sharply between different phosphotyrosine-containing sequences

Each SH2 domain has a preference for particular residues at three to six residues following the phosphotyrosine. The Src SH2 domain binds with particularly high affinity ($K_D \approx 0.1$ μM, $\Delta G^o \approx -40$ kJ•mol$^{-1}$) to peptides with the sequence phosphotyrosine-glutamate-glutamate-isoleucine (see Figures 13.14 and 13.15). The structure of the complex resembles a two-pronged plug (the peptide) engaging a two-holed socket (the SH2 domain) (see Figure 13.14). One of the sockets binds the phosphotyrosine sidechain, as discussed earlier. The second socket in the Src SH2 domain is lined by hydrophobic sidechains, and the isoleucine residue of the peptide is bound there. The two glutamate residues that are between the phosphotyrosine and the isoleucine lie along the surface of the SH2 domain, and are in the general vicinity of basic residues, which tend to disfavor hydrophobic or basic residues at these positions.

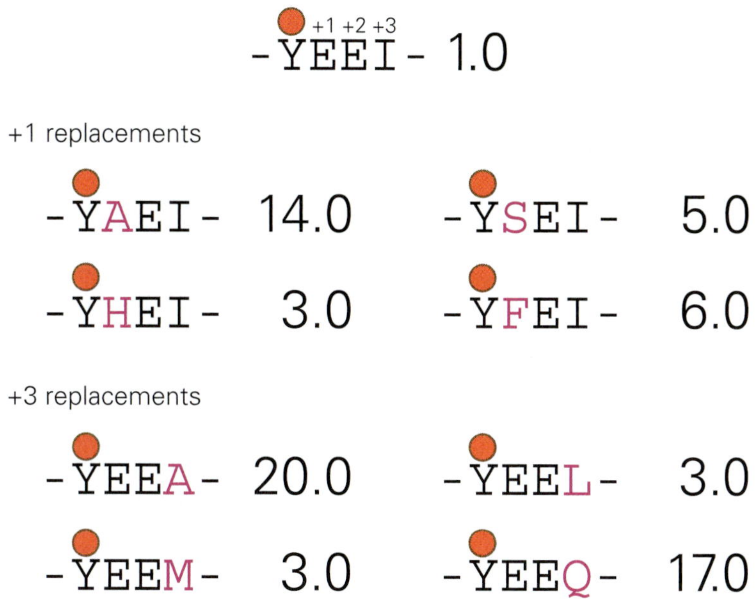

| | | | |
|---|---|---|---|
| -pYEEI- (+1 +2 +3) | 1.0 | | |
| **+1 replacements** | | | |
| -pYAEI- | 14.0 | -pYSEI- | 5.0 |
| -pYHEI- | 3.0 | -pYFEI- | 6.0 |
| **+3 replacements** | | | |
| -pYEEA- | 20.0 | -pYEEL- | 3.0 |
| -pYEEM- | 3.0 | -pYEEQ- | 17.0 |

**Figure 13.16 Sensitivity of an SH2 domain to changes in target sequence.** Shown here are the sequences of various peptides and the relative values of their dissociation constants for the Src SH2 domain. The structure of the SH2 domain bound to the optimal peptide is shown in Figure 13.14. (Based on data in T. Gilmer et al., and J. Berman, *J. Biol. Chem.* 269: 31711–31719, 1994.)

Despite the ability of SH2 domains to favor certain sequence motifs over others, it turns out that SH2 domains are unable to discriminate sharply between sets of peptides with different sequences flanking a phosphotyrosine residue. To appreciate why this is so, consider that, for an optimal peptide binding to the Src SH2 domain, at least –22 kJ•mol$^{-1}$ of the total binding energy of –40 kJ•mol$^{-1}$ comes from the phosphotyrosine (see the previous section). Now consider a non-optimal peptide in which the residues adjacent to the phosphotyrosine do not interfere with binding—that is, their contribution to the binding free-energy is zero. In that case, the free energy of binding will be due solely to the phosphotyrosine, which we estimate to be at least –22 kJ•mol$^{-1}$. This corresponds to a dissociation constant of 100 μM (refer to Section 12.1)—that is, 1000-fold weaker than the affinity of the optimal peptide.

The problem is that it is quite likely that some other sequences will, by chance, gain additional binding energy through the formation of one or two hydrogen bonds or hydrophobic interactions. If a peptide with a different sequence forms two additional hydrogen bonds, each worth 2 $k_BT$ (~5.0 kJ•mol$^{-1}$ at 300 K), then the dissociation constant of the peptide would be very close to that of the optimal peptide. To better appreciate why SH2 domains are not very specific, consider the data shown in **Figure 13.16** for the Src SH2 domain. Substitution of the residues at the first or third position in the target peptide by alanine can reduce the affinity for the SH2 domain by more than 10-fold, and combining these substitutions will reduce the affinity even further. But note that other changes, such as replacing the glutamate at the first position by histidine or the isoleucine at the third position by leucine or methionine, reduces the affinity only three-fold.

It is likely that there is a wide range of sequences that will bind to the Src SH2 domain with sufficient affinity to compromise specificity. This conclusion is true for peptide recognition domains in general: these domains do not achieve very high specificity on their own.

## 13.13 Combinations of peptide recognition domains have higher specificity than individual domains

As we have emphasized in the first part of this chapter, macromolecules usually need to interact with their targets with high specificity in order to carry out their biological function. If the interactions between peptide binding domains and their targets are not very specific, how do the proteins that contain these modules

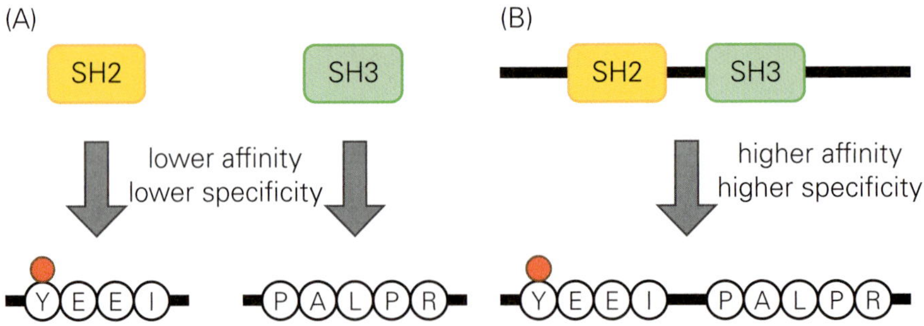


**Figure 13.17 Increased specificity of combinations of domains.** (A) Individual SH2 and SH3 domains bind their targets with relatively low affinity and specificity. (B) A protein that contains both domains will bind with much higher affinity and specificity to a segment containing both motifs with the proper order and spacing.

achieve specificity? The answer is that nature uses combinations of these domains in a single protein so that the specificity of the final interaction depends on the specificities of two or more of the component domains being satisfied simultaneously, as shown schematically in **Figure 13.17**. To understand this idea better, we shall take a closer look at the specificity of tandem arrangements of SH2 domains.

A tyrosine kinase known as ZAP-70, which plays an important role in the activation of T-cells in the immune system, contains two SH2 domains (**Figure 13.18**A). The kinase activity of ZAP-70 is switched on when the SH2 domains bind to peptides containing two phosphotyrosine residues that are spaced 10 or 11 residues apart, with appropriate sequences around them. The dissociation constant for ZAP-70 binding to such peptides is 2 nM, about 100-fold tighter than the interaction of the Src SH2 domain with a peptide containing an optimal sequence motif (see Figure 13.18B). But, if a peptide contains two phosphorylated tyrosines that do not have the correct spacing and sequence, the affinity is reduced more than 1000-fold (see Figure 13.18B).

Like ZAP-70, a tyrosine phosphatase known as SHP2 also contains two SH2 domains and is also switched on by peptides containing two phosphotyrosine residues. SHP2, however, prefers peptides with a different spacing between the two phosphotyrosine residues. SHP2 binds to its preferred target with a dissociation constant of 8 nM, but its dissociation constant for the ZAP-70 target is at least 100,000-fold greater.

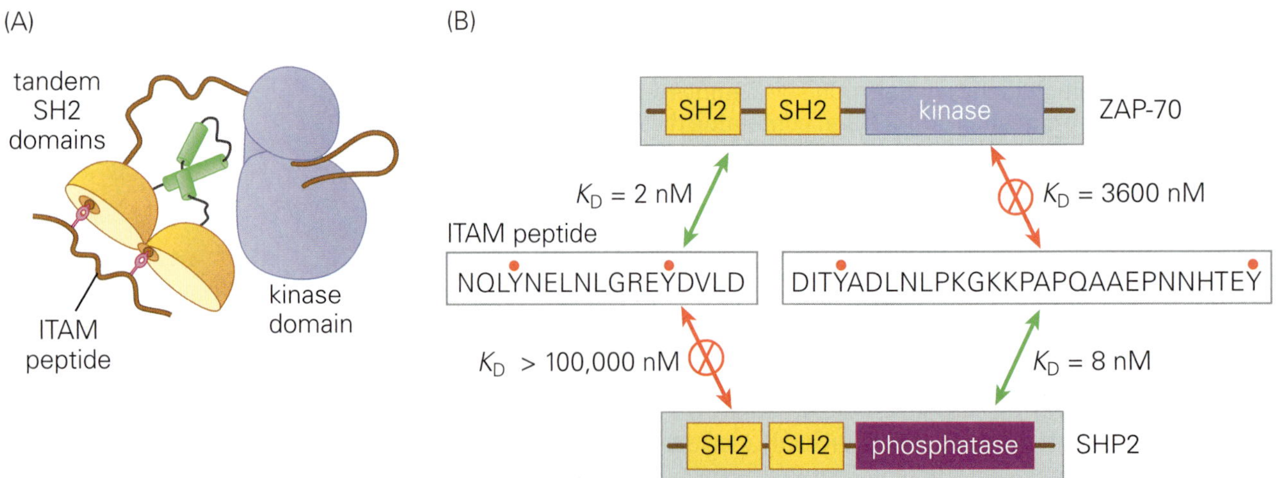


**Figure 13.18 Specificity of proteins with tandem SH2 domains.** (A) Schematic diagram of the structure of the ZAP-70 kinase. The binding of a doubly phosphorylated peptide, known as an ITAM peptide, to the SH2 domains activates the kinase. (B) Comparison of the peptide-binding affinities of ZAP-70 and the tyrosine phosphatase SHP2, which also contains two SH2 domains, but with different specificity. (Based on data from E.A. Ottinger, M.C. Botfield, and S.E. Shoelson, *J. Biol. Chem.* 273: 729–735, 1998.)

ZAP-70 and SHP2 have opposing actions because the former adds phosphate groups to proteins and the latter removes them. If a peptide activates ZAP-70 but also cross-reacts with SHP2, then the kinase action of ZAP-70 will be undone by SHP2. Given the data in Figure 13.18, we can calculate the specificity, $\alpha$, of ZAP-70 for its activating peptide in the presence of SHP2. If we assume that the target peptide is at a concentration of 0.1 μM and that the concentrations of ZAP-70 and SH2 are the same, then we can apply Equation 13.9:

$$\alpha = \frac{\dfrac{1}{1+\dfrac{K_{D1}}{[L]}}}{\dfrac{1}{1+\dfrac{K_{D2}}{[L]}}} = \frac{\dfrac{1}{1+\dfrac{2\times 10^{-9}}{10^{-7}}}}{\dfrac{1}{1+\dfrac{3600\times 10^{-9}}{10^{-7}}}} \approx \frac{0.980}{0.027} = 36 \tag{13.23}$$

The result in Equation 13.32 tells us that a peptide that activates ZAP-70 will activate SHP2 36-fold less strongly. This provides sufficient specificity for the kinase signal to be transmitted without attenuation by the phosphatase.

## 13.14 Protein–protein interfaces usually have a small hydrophobic core

What kinds of amino acids participate in the formation of a protein–protein interface? In contrast to the recognition of peptides by peptide-recognition domains, the interfaces between folded proteins do not usually contain any particular kind of sequence motifs or structural patterns. Residues at the interfaces between proteins interact with each other in much the same way as do residues within the same protein. There is packing of hydrophobic residues against each other, as well as hydrogen bonding and charge–charge interactions across the interface.

The amino acid composition of residues involved in forming interfaces is not very different from the surfaces of proteins in general. There is, however, an increased likelihood of finding aromatic and hydrophobic sidechains at interfaces (these are ~2.5 times and 1.5 times more likely to be found at interfaces than elsewhere on the surface, respectively). That aromatic and hydrophobic sidechains are only found slightly more frequently at interfaces reflects a balance between the importance of hydrophobic interactions for gaining affinity and the need to maintain the protein in a soluble form that does not interact indiscriminately with other proteins. A typical protein–protein interface has a small core of hydrophobic sidechains surrounded by several polar residues, as shown in **Figure 13.19**.

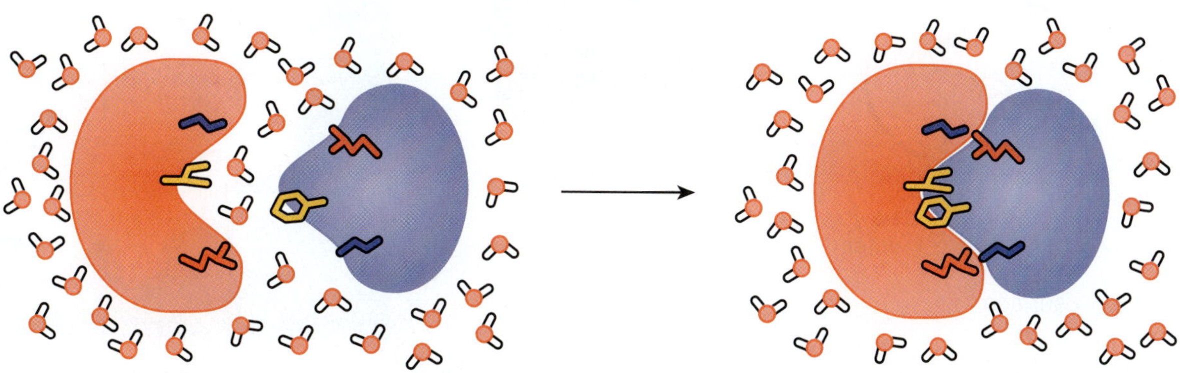

**Figure 13.19 Hydrophobic interactions at a protein–protein interface.** Most interfaces have a small hydrophobic core, as shown by Phe and Ile residues at the center of this schematic, which makes an important contribution to the binding affinity as a consequence of the release of water from the interface.

## 13.15 A typical protein–protein interface buries about 700 to 800 Å$^2$ of surface area on each protein

**Buried surface area**

Protein–protein interfaces are characterized by their buried surface area. This refers to the surface area on the interacting proteins that is accessible to water before the complex is formed, but is inaccessible in the complex.

When two proteins form a complex, they bring together regions of their surfaces, making a tight junction. How large is a typical interface between two proteins? One way to characterize the extent of an interface is to calculate the surface area that is buried at the interface. As described in Section 5.16, the accessible surface area of a protein is defined as the area of the surface that is mapped out by a spherical probe that rolls over the entire surface of the protein (see Figure 5.24). Typically, a probe with a radius corresponding to that of a water molecule (~1.4 Å) is used in such a calculation (**Figure 13.20**). By calculating the accessible surface areas of two proteins before and after they form a complex, we can determine the **buried surface area**, that is, the extent of the surface that is buried at the interface.

How much surface area do we expect to see buried at the interfaces between proteins? We can gain an appreciation for the magnitude of the change in surface area that occurs upon complex formation by considering the solvent exposure of the amino acid sidechains in hypothetical peptides of the form Ala-X-Ala, where X is any amino acid. We can construct models for such tripeptides in an extended conformation and compute the total solvent-accessible surface area for each of the sidechains placed at the "X" position. This gives us an estimate of the maximum possible accessible surface area for each of the sidechains. The solvent-accessible surface areas for a few sidechains are shown in **Figure 13.21**. Taking leucine as a typical sidechain, we can see that each sidechain that is fully buried at an interface will contribute ~100 Å$^2$, on average, to the area buried by the complex.

A histogram of the buried surface areas for many different protein complexes is shown in **Figure 13.22**. We see that these interfaces typically involve the burial of ~1300–1700 Å$^2$ of total surface area. Thus, the contribution from each protein is roughly 700–800 Å$^2$. A typical protein domain of ~150 residues has a radius of ~20 Å. If we assume that the domain has a spherical shape, then this corresponds to

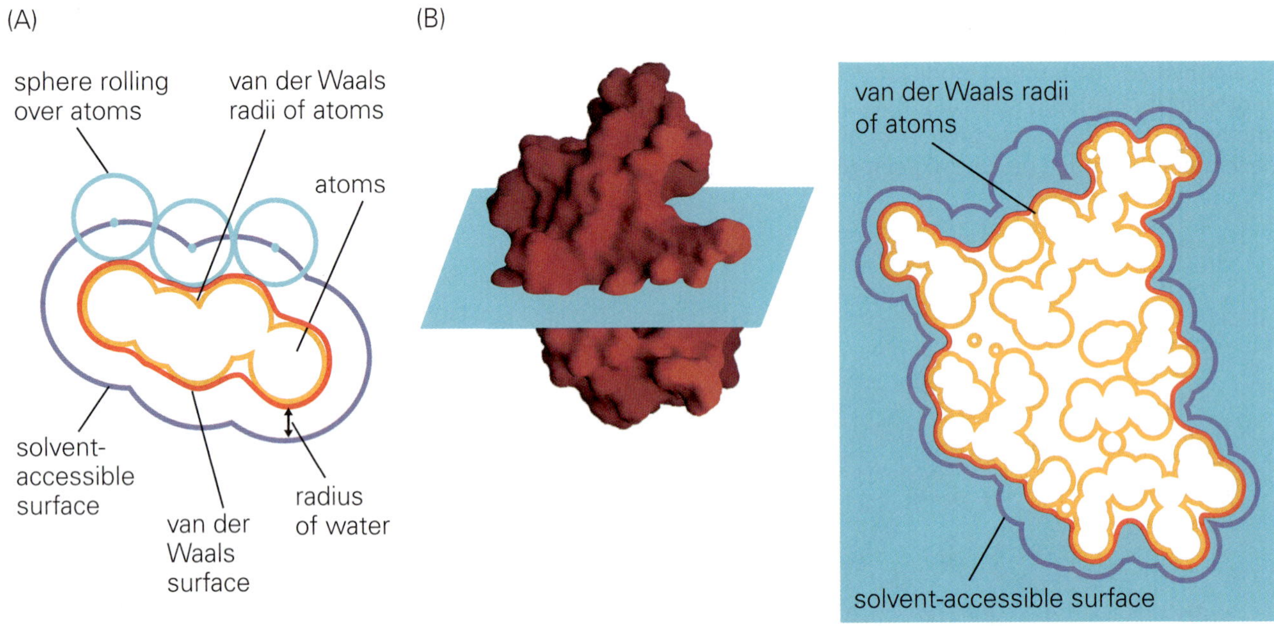


**Figure 13.20 The solvent accessible surface of a protein.** (A) A cross section through a small molecule is shown. A sphere corresponding in radius to a water molecule (radius of ~1.4 Å) is rolled around the molecule, using a computer. The surface traced out by the sphere is known as the solvent-accessible surface. (B) A cross-sectional slice through a protein (*red*) is shown in *light green*. The atoms in this slice are shown on the right, along with the outline of the solvent-accessible surface (*blue* line). (Adapted from B. Lee and F.M. Richards, *J. Mol. Biol.* 55: 379, 1971.)

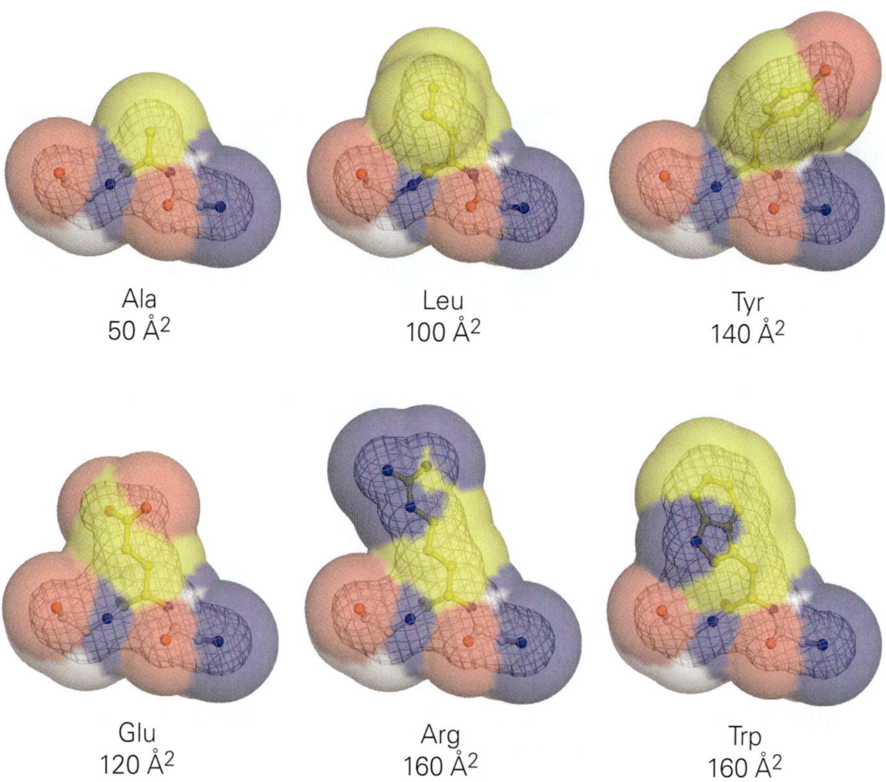


**Figure 13.21 The solvent-accessible surface areas of sidechains.** Shown here are six sidechains, with the van der Waals surfaces indicated by a mesh. The solid surface is the solvent-accessible surface area, which is the set of points traced out by a sphere of radius 1.4 Å, representing water, rolling over the van der Waals surface. The accessible surface area associated with each sidechain is shown below each structural diagram.

a surface area of ~5000 Å$^2$, and so a typical protein–protein interface extends over ~10% of the surface area of each protein. This corresponds to the two proteins touching each other side by side, as shown for the trypsin–inhibitor complex in Figure 13.12.

A typical protein–protein interface involves about ~30 residues at the contact surface. The total surface area buried (~1500 Å$^2$) is about half of what we might expect if all 30 residues at the interface went from being completely exposed to completely buried, assuming a surface area per sidechain of ~100 Å$^2$ (see Figure 13.21). This is because the interface is not perfectly complementary in shape and many residues, particularly at the edges of the interface, are not completely buried.

## 13.16 Water molecules form hydrogen-bonded networks at protein–protein interfaces

A typical protein–protein interface only has about 10 hydrogen bonds between protein atoms, suggesting that much of the stabilization comes from interactions between nonpolar residues. Although this can be true, the number of

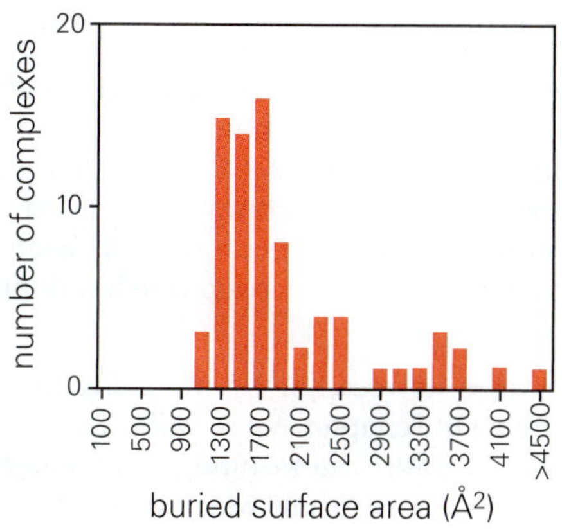


**Figure 13.22 Histogram of surface area buried at a protein–protein interface.** The number of complexes observed is plotted as a function of the size of the interface. (Adapted from L. Lo Conte, C. Chothia, and J. Janin, *J. Mol. Biol.* 285: 2177–2198, 1999. With permission from Elsevier.)

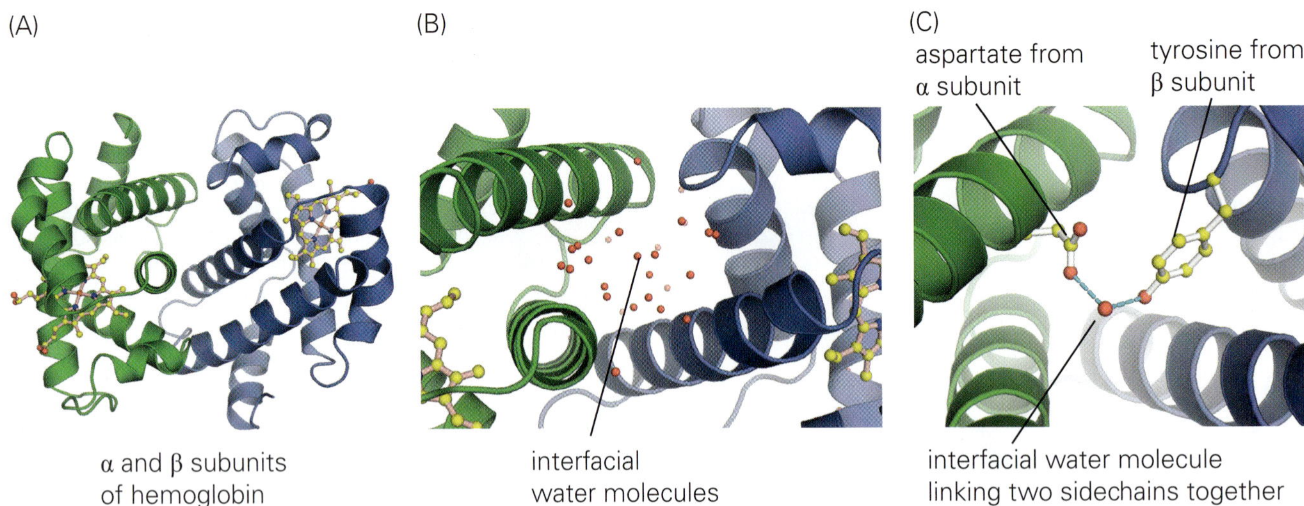


**Figure 13.23 Interfacial water molecules.** (A) Two subunits from human hemoglobin are shown. (B) A close-up view of the interface, showing interfacial water molecules as *red* spheres. (C) Interfacial water molecules, such as the one shown here, form hydrogen bonds that link the two subunits together. (PDB code: 2DN1.)

interfacial hydrogen bonds increases greatly if one includes water molecules that are localized between the two proteins. The average number of ordered water molecules at protein–protein interfaces is ~20. Most of these water molecules form hydrogen bonds with protein atoms on both sides of the interface, and so the number of water-mediated hydrogen bonds is typically larger than the number of direct hydrogen bonds between protein residues. Water molecules that are localized at the interface between two subunits of hemoglobin are shown in **Figure 13.23**.

Water molecules at the interfaces between proteins, known as **interfacial waters**, have quite different properties when compared to water molecules that are not close to the protein (referred to as **bulk water**). The interfacial water molecules are rather precisely localized on the protein surface, because they form hydrogen bonds with residues on the protein surface, as shown in Figure 13.23C. Recall from Section 12.22 that the release of protein-bound water molecules upon drug binding can contribute to a favorable change in entropy upon binding. The water molecules that remain bound at a protein–protein interface do not contribute to the entropy of the interaction, but they increase the specificity of the complex. The localized water molecules impose hydrogen-bonding requirements on protein atoms in the vicinity. An analysis of the packing density of atoms at the interface shows that inclusion of water molecules optimizes the packing—that is, increases the area of the contact surface. Both of these factors help ensure that the correct complex is selected for.

## 13.17 The interaction between growth hormone and its receptor is a model for understanding protein–protein interactions

A good case in which to examine how the demands of affinity and specificity are balanced in protein–protein interactions is the binding of human growth hormone to its receptor (**Figure 13.24**). Like many other growth factors, human growth hormone activates signaling pathways in cells by turning on tyrosine kinases. Growth hormone is so named because its deficiency leads to dwarfism in humans, whereas people who produce too much of it are prone to excessive growth, or gigantism.

Growth hormone activates its receptor by triggering its dimerization, as was described for FGF earlier in the chapter. A molecule of growth hormone first binds to one molecule of the receptor. The resulting 1:1 hormone–receptor complex then recruits a second receptor molecule, which interacts with both the hormone

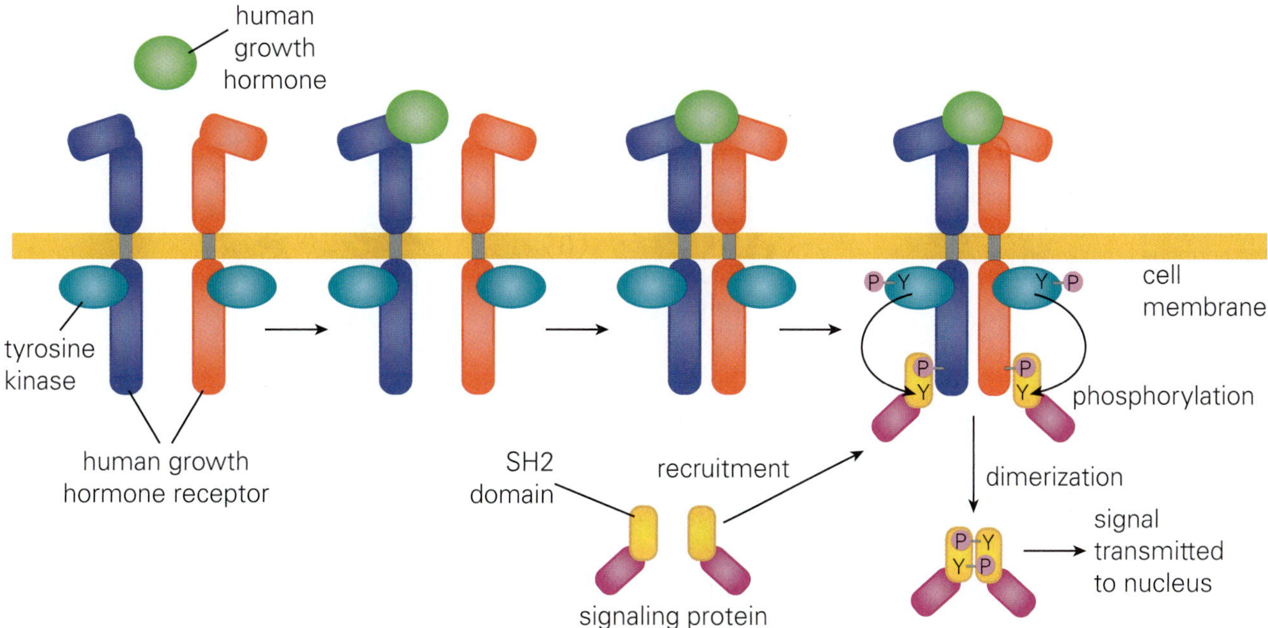


**Figure 13.24 Signaling by growth hormone**. The binding of growth hormone to its receptor induces dimerization of the receptor. The growth hormone receptor does not itself contain a tyrosine kinase domain. Instead, tyrosine kinases that are bound to the receptor promote phosphorylation of the receptor, which leads, in turn, to the recruitment of signaling proteins, known as STATs, that contain SH2 domains.

and the first receptor (see Figure 13.24). The active complex consists of one molecule of growth hormone bound to two molecules of the receptor, which is different from FGF, which forms a 2:2 complex with the receptor.

The two regions of interaction between growth hormone and its receptor have fairly typical size interfaces (**Figure 13.25**). One interface buries a total of 1230 $Å^2$ of surface area on the receptor and the hormone. The second interface is smaller, burying 900 $Å^2$ (see Figure 13.25B). The larger interface corresponds to the hormone-receptor complex that is of higher affinity and is the one that is likely to be formed when the hormone first encounters a receptor molecule. A 1:1 complex of the hormone and the receptor, corresponding to the diagram shown on the left-hand side of Figure 13.25B, has a dissociation constant, $K_D$, of 0.3 nM. This corresponds to a free energy of binding of $-55$ kJ•mol$^{-1}$ at 300 K. This a relatively high-affinity protein-protein complex, with a net stability that is comparable to that of many drug-protein interactions. In the following sections, we analyze the energetics of this 1:1 complex in more detail.

## 13.18 The major growth hormone–receptor interface contains many types of interactions

The interface between growth hormone and the receptor involves interactions between more than 30 residues from each protein in the tighter 1:1 complex. Some of these interactions are shown in **Figure 13.26**. Fifteen of these residues are involved in the formation of nine hydrogen bonds, which is typical for protein-protein interfaces. In addition, there are many nonpolar interactions at the interface, including contacts between hydrophobic residues.

A natural question is how the binding free energy of $-55$ kJ•mol$^{-1}$ is distributed among the interactions at the interface—that is, whether interactions between particular residues are more important than others. To begin this analysis, let us look at the interactions made by three residues on the receptor—namely, Trp 104, Glu 120, and Glu 127—and ask whether any of these sets of interactions appear to be stronger than the others (these interactions are shown in Figure 13.26).

Trp 104 is completely buried at the heart of the interface, and it forms a hydrogen bond with a residue on growth hormone and packs against hydrophobic groups

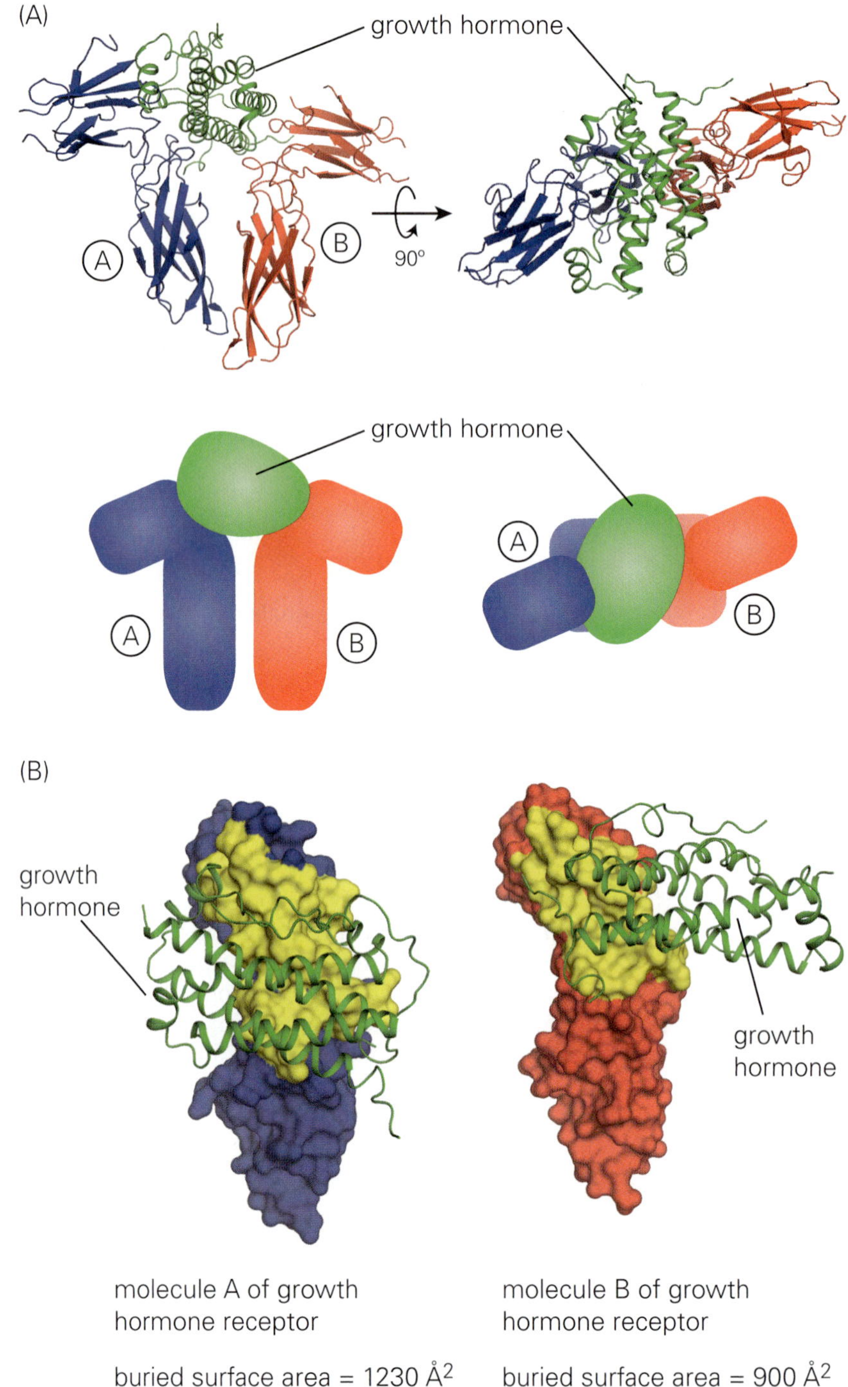


**Figure 13.25 Structure of the growth hormone–receptor complex.** (A) One molecule of growth hormone (*green*) binds to two molecules of the receptor. The extracellular domains of the receptor are shown in *red* and *blue*. (B) Footprint of the hormone on the two receptor molecules, labeled A and B. The regions that are in contact with the hormone are colored *yellow*. (PDB code: 3HHR.)

at the interface (see Figure 13.26). Trp 104 is thus well integrated into a network of interactions that might be expected to stabilize the protein complex significantly. Glu 127 is located more towards the periphery of the interface, but it forms hydrogen bonds with two positively charged residues from growth hormone, Lys 41 and Arg 167. Hydrogen-bonding interactions between oppositely charged residues are known to be particularly strong, and so we might expect this interaction to also stabilize the complex between hormone and receptor. Finally, Glu 120 is located right at the edge of the interface, but it reaches out to form a hydrogen bond with the sidechain of Gln 46, a neutral, polar residue of the hormone. Does this interaction also contribute towards the stability of the complex?

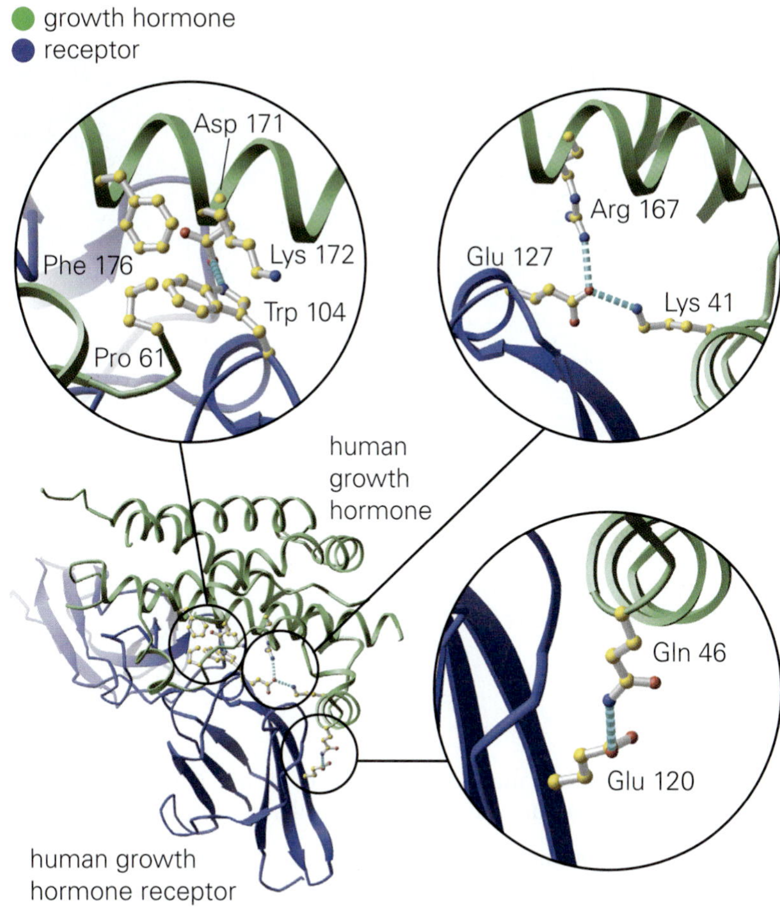


**Figure 13.26 Interactions between growth hormone and its receptor.** Shown here is the structure of the tighter 1:1 complex between the two proteins, which is depicted on the left in Figure 13.25B. The expanded views show contacts between residues in three regions of the interface. (PDB code: 3HHR.)

## 13.19 The interface between growth hormone and its receptor contains hot spots of binding affinity, which dominate the interaction

The relative contributions of different residues to the binding affinity between growth hormone and receptor were determined by mutating residues at the interface to alanine and measuring the change in the free energy of binding (**Figure 13.27**). If the binding free energy of the complex were distributed evenly among all of the residues at the interface, we would expect to see a small reduction (~1.8 kJ•mol$^{-1}$; that is, –55/30) in the binding free energy resulting from each mutation. In reality, however, the mutation of certain residues on the receptor, such as Trp 104 and Trp 169, to alanine reduces the binding energy substantially, by more than ~20 kJ•mol$^{-1}$, whereas mutating more than half of the other residues at the interface individually to alanine does not cause a significant change in binding free energy (see Figure 13.27). Large changes in binding affinity were also obtained when certain other hydrophobic residues at the interface were mutated (Trp 107, Ile 103, Ile 105, Pro 106, and Ile 165). In contrast, smaller changes in binding were observed when polar residues where mutated. Among the polar residues at the interface, five (Arg 43, Glu 44, Asp 126, Glu 127, and Asp 164) make smaller but still significant changes to the binding free energy (~5 kJ•mol$^{-1}$).

Going back to the three residues selected for examination (Trp 104, Glu120, and Glu 127), we see that the buried location of Trp 104 and its hydrophobic interactions do indeed correlate with a large change in the free energy of binding (~20 kJ•mol$^{-1}$) when this residue is mutated. The charged hydrogen bonds that Glu 127 makes with lysine and arginine sidechains on growth hormone also contribute favorably to binding, but not as much as the interactions with Trp 104—just

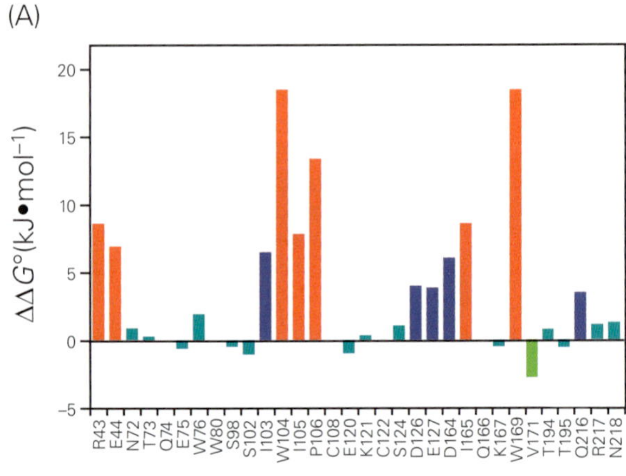


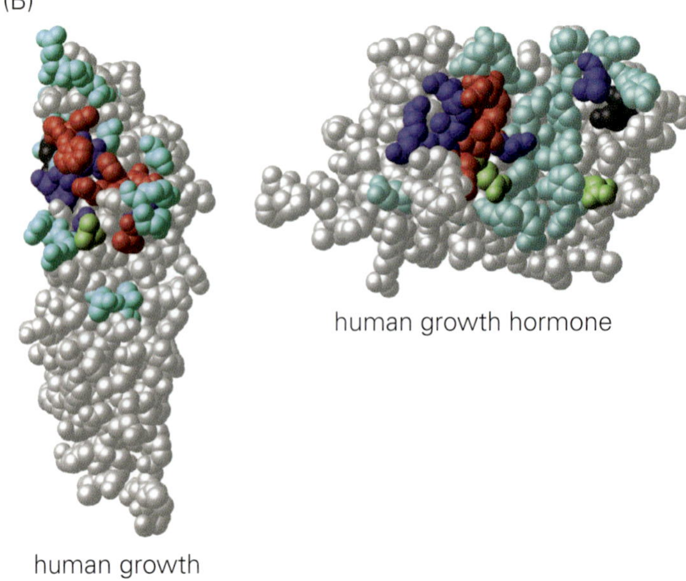


**Figure 13.27 Hot spots of binding free energy in the growth hormone–receptor interface.** (A) Residues at the interface in the tighter 1:1 growth hormone–receptor complex were mutated individually to alanine and the change in binding free energy ($\Delta\Delta G^o$) was measured for each protein. The values of $\Delta\Delta G^o$ are color coded (*red*, largest increase in value; *dark blue*, intermediate; *light blue*, small; and *green*, decrease in value). (B) The surface of growth hormone and its receptor are shown. Residues at the interface are colored according to the value of $\Delta\Delta G^o$, as in (A). (Adapted from T. Clackson and J.A. Wells, *Science* 267: 383–386, 1995. With permission from the AAAS; PDB code: 3HHR.)

4 kJ•mol$^{-1}$. The interaction made by Glu 120 does not stabilize the complex, because mutating this residue to alanine changes the binding affinity only slightly.

Similar results were obtained when the effects of mutations on the surface of growth hormone were analyzed (see Figure 13.27B). Most of the binding free energy is accounted for by about 25% of the total number of residues of growth hormone that are at the interface, and more than half the residues located at the interface make no significant contribution to binding free energy. The residues that are important on growth hormone and the receptor are complementary, in that for an interaction across the interface, the residues on both proteins are important. Thus, a cross section through the interface shows that the important, predominantly hydrophobic residues are at the center and are surrounded by polar residues. In some ways this pattern mimics the cross section through the interior of a folded protein.

## 13.20 Residues that do not contribute to binding affinity may be important for specificity

If so few residues contribute to binding affinity, then why is the interfacial region so large? One possible explanation is that the polar residues that surround the important residues impose hydrogen-bonding requirements on potential binding partners, thereby making an important contribution to the specificity of the interface.

This explanation is supported by the observation that converting eight residues in the interface to alanine, seven of which were charged or polar, yielded a protein with 50-fold higher affinity for the receptor. The observation further suggests that it should be possible to increase the affinity of the interaction beyond that seen with the naturally occurring proteins. Indeed, the substitution of 15 residues in the hormone by various residues, not just alanine, increased the affinity for the receptor by a factor of ~400 (a selection process was used to arrive at this optimized sequence). Clearly, the interface between growth hormone and its receptor has not evolved towards maximum affinity, and it appears that the demands of specificity have required that the affinity of the interface be lower than maximally possible.

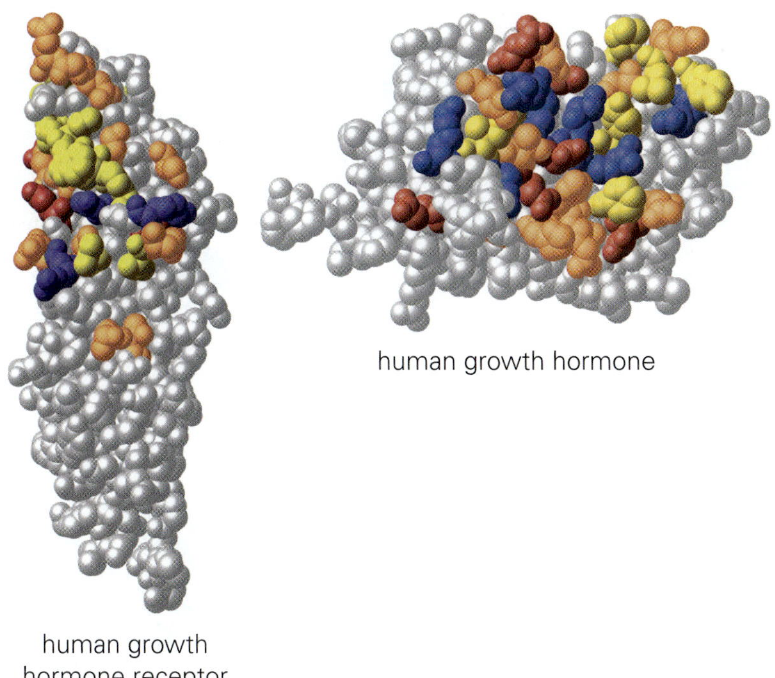


**Figure 13.28 Charged and polar groups dominate the growth hormone–receptor interface.** Interfacial residues in the complex are colored *red* and *blue* for positively charged and negatively charged sidechains, respectively. Residues with polar sidechains are colored *orange* and those that are hydrophobic are *yellow*. (PDB code: 3HHR.)

The analysis of the structures of many protein-protein complexes has shown that charged and polar groups are prevalent at these interfaces, as illustrated for the growth hormone complex in **Figure 13.28**. The strength of the interactions between the polar groups depends on their precise placement and relative orientation, because the energy of a hydrogen bond depends strongly on its geometry (see Section 6.19).

## 13.21 The desolvation of polar groups at interfaces makes a large contribution to the free energy of binding

Charged and polar amino acid sidechains interact strongly with water molecules and with dissolved ions when they are not interacting with each other. The favorable interactions at a protein-protein interface may be insufficient to overcome the energetic penalty of desolvating charged and polar groups. Thus, although hydrogen bond and ionic interactions at an interface may appear favorable, the cost of desolvating polar groups can disfavor the formation of protein-protein interfaces.

It is informative to look at some examples of specific protein-protein interfaces and to examine the extent to which electrostatic interactions help to stabilize them. We shall do this for four different protein complexes, shown in **Figure 13.29**, one of which is the complex between growth hormone and its receptor discussed in the preceding section. The other complexes are those formed by a ribonuclease known as barnase and an inhibitor of it (barstar), a neuraminidase (see Figure 4.58) bound to an antibody, and a signaling protein (Rap1A) bound to a part of a protein kinase, referred to as RapBD (for RaplA binding domain).

The experimentally determined binding free energies for the four complexes cover a range of values, from –34 kJ•mol$^{-1}$ (at room temperature) for the weakest complex (Rap1A–RapBD) to –80 kJ•mol$^{-1}$ for the strongest (barnase-barstar) (**Table 13.2**). A qualitative examination of the structures of these complexes does not allow us to predict which have highest affinity. Surface area buried in the complex correlates poorly with the strength of interaction of the proteins (see Table 13.2). For example, the barnase-barstar complex is ~30 kJ•mol$^{-1}$ more stable than the

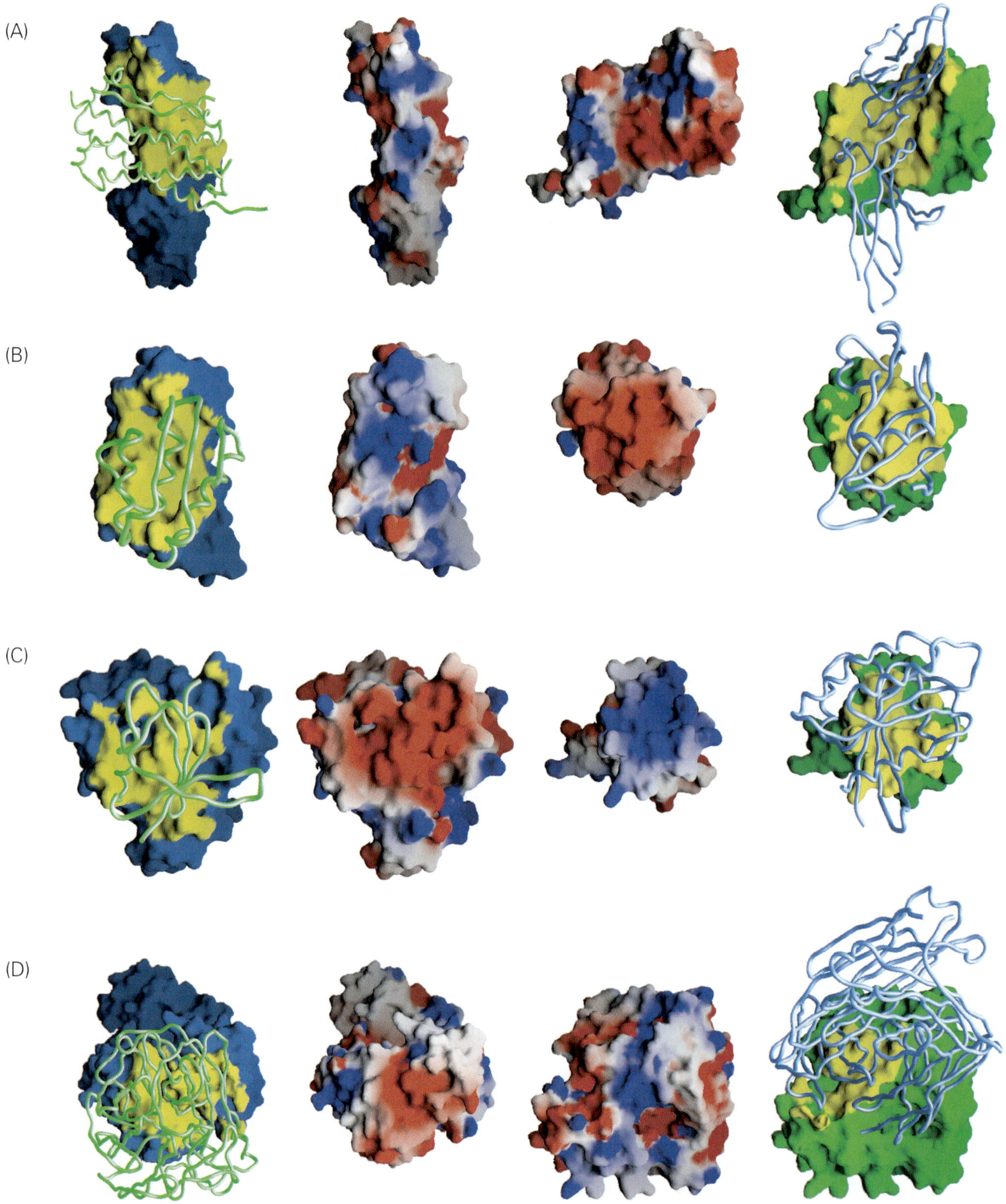

(A)

(B)

(C)

(D)

**Figure 13.29 The role of electrostatics in the stabilization of protein–protein complexes**. Two orthogonal views of each complex are shown on the extreme left and right of each panel. In each of these drawings the backbone of one of the proteins is shown as a ribbon and the molecular surface of the other protein is displayed. The surface buried at the interface is colored *yellow*. The two diagrams in the center of each panel show the electrostatic potential of the interacting proteins, calculated in the absence of the other. Regions of positive and negative electrostatic potential are colored *blue* and *red*, respectively. (A) Growth hormone–receptor (PDB code: 3HHR). (B) Barnase–barstar (PDB code: 1BGS). (C) Rap1A–RapBD (PDB code: 1C1Y). (D) Neuraminidase–antibody (PDB code: 1NCA).

**Table 13.2 Interaction energies and electrostatic contributions for some protein–protein complexes.**

| Complex | Experimental dissociation constant $K_D$ (M) | Binding free energy $\Delta G^o_{bind}$ at 300 K (kJ•mol$^{-1}$) | Electrostatic contribution to binding energy (kJ•mol$^{-1}$) $\Delta G_{elec} = \Delta G_{desolv} + \Delta G_{intra} + \Delta G_{inter}$ | Total buried surface area at interface (Å$^2$) | Surface area at interface from hydrophobic sidechains (Å$^2$) |
|---|---|---|---|---|---|
| Barnase–barstar | $10^{-14}$ | –80 | +15 = 769 – 97 – 657 | 1699 | 276 |
| Growth hormone–receptor | $0.3 \times 10^{-9}$ | –55 | +177 = 853 – 139 – 537 | 3021 | 648 |
| Neuraminidase–antibody | $0.25 \times 10^{-8}$ | –50 | +148 = 591 – 17 – 426 | 2208 | 569 |
| Rap1A–RapBD | $1.2 \times 10^{-6}$ | –34 | –51 = 374 + 31 – 456 | 1414 | 286 |

desolv = desolvation; intra = intramolecular; inter = intermolecular

(From F.B. Sheinerman and B. Honig, *J. Mol. Biol.* 318: 161–177, 2002. With permission from Elsevier.)

complexes formed between growth hormone and its receptor and between neuraminidase and an antibody, but the surface area buried between barnase and barstar is much less (1700 Å$^2$) than for the others (3000 Å$^2$ and 2200 Å$^2$, respectively). Once again, this comparison emphasizes the fact that not all of the interactions that occur at an interface between two proteins contribute favorably to the binding free energy.

Figure 13.29 shows the electrostatic potential generated by each of the interacting proteins, mapped to the molecular surface. Whereas the barnase-barstar and Rap1A-RapBD complexes involve interacting partners with electrostatically complementary surfaces, electrostatic complementarity in the growth hormone-receptor complex or in the neuraminidase-antibody complex is less clear.

This observation is reflected in the quantitative analysis of the total electrostatic contribution to the binding energy for each of these complexes, shown in Table 13.2. The electrostatic contribution to the binding energy was calculated for each of these complexes, using the method of continuum electrostatics described in Section 6.23. In this method, water molecules surrounding the protein are replaced by a uniform medium with a dielectric constant, ε, that is appropriate for water (typically, ε = 80). The protein interior is treated as a region of uniform, low dielectric (ε = 2). The electrostatic potential around the protein is then calculated by solving the Poisson-Boltzmann equation, using a salt concentration of the solution corresponding to physiological ionic strength. Electrostatic interactions are seen to be quite unfavorable for the growth hormone-receptor and neuraminidase-antibody complexes, with $\Delta G^o_{elec}$ values of +178 and +149 kJ•mol$^{-1}$, respectively. The barnase-barstar complex slightly disfavored by electrostatics, with a $\Delta G^o_{elec}$ value of +15.0 kJ•mol$^{-1}$. The Rap1A-RapBD complex is found to be significantly stabilized by the electrostatic contributions, with a $\Delta G^o_{elec}$ value of -51 kJ•mol$^{-1}$.

Further insight into the electrostatic interactions is gained by dissecting the total electrostatic contribution to the binding free energy, $\Delta G^o_{elec}$, into three components, $\Delta G^o_{desolv}$, $\Delta G^o_{inter}$, and $\Delta G^o_{intra}$ (see Table 13.2). $\Delta G^o_{desolv}$ is the free-energy change associated with desolvating the charged groups at the interface—that is, removing them from water. There are very large, unfavorable desolvation energies for each of the protein-protein complexes. $\Delta G^o_{inter}$ is the free-energy change associated with *inter*molecular electrostatic interactions across the interface, including hydrogen bonding and ion-pair formation. This term is favorable in all cases, but only in the Rap1A-RapBD complex is the value of $\Delta G^o_{inter}$ large enough to compensate for the unfavorable desolvation energy. The final term, $\Delta G^o_{intra}$, reflects changes in the free energy of *intra*molecular electrostatic interactions—that is, within the same protein upon complex formation. This term is usually small, and it arises because of changes in the dielectric environment of charged groups in both proteins upon formation of the complex.

The results shown in Table 13.2 for these specific systems are actually of broad generality. Charged sidechains and, to a lesser extent, polar sidechains, interact so favorably with water that their burial at an interface comes with a large desolvation penalty. Even though the overall electrostatic contribution to the binding free energy may end up being destabilizing, electrostatic interactions are critical because they help ensure the specificity of the interaction. Because of the large desolvation term in the binding free energy, only interfaces that offset this penalty to a substantial extent with extensive electrostatic complementarity will be stable. The strategic placement of charged and polar sidechains is the most important mechanism for preventing the formation of protein–protein complexes with inappropriate partners, just as we saw for protein–drug interactions (see Figure 12.31, for example).

# C. RECOGNITION OF NUCLEIC ACIDS BY PROTEINS

In this part of the chapter we look at some examples of how proteins recognize DNA and RNA. Recall from Section 2.11 that DNA is predominantly in the B-form under cellular conditions and that the wide major groove allows close approach of protein sidechains, which can hydrogen bond to bases and make sequence-specific contacts. Much of DNA recognition occurs through the major groove, but the minor groove is sometimes used as well. RNA displays a much wider spectrum of structures than does double-stranded DNA. Some RNA-binding proteins recognize unpaired bases primarily, while others recognize various aspects of double-helical structure. We end the chapter with some examples of how proteins recognize single-stranded RNA.

## 13.22 Complementarity in both electrostatics and shape is an important aspect of the recognition of double-helical DNA and RNA

Double helices formed by both DNA and RNA have the charged phosphate backbone on the outside, thereby presenting regions of high negative charge density for recognition at the surface. The molecular surfaces of A-form and B-form DNA are shown in **Figure 13.30**. The surfaces are colored according to the electrostatic potential, which provides a vivid illustration of just how negatively charged the surfaces of double-helical structures are. The two structures shown in Figure 13.30 have sequences in which a G-C dinucleotide pattern is repeated. The charge distribution within the grooves will be different for double helices with different nucleotide sequences, but the charge distribution of the backbone is essentially unaffected by differences in sequence.

As a result, proteins that bind to DNA or RNA typically do so with surfaces that are enhanced in positive charge. Protein surfaces in contact with DNA or RNA are enriched in arginine and lysine, and depleted in aspartic acid and glutamic acid. The arginine and lysine sidechains, as well as other polar sidechains, make hydrogen bonds to the phosphate groups, and these interactions account for ~60% of the hydrogen bonds between the protein and the DNA. For RNA, the presence of the 2′ hydroxyl group of the ribose provides additional hydrogen-bonding opportunities with the backbone that are absent in DNA. About one in four hydrogen bonds between proteins and RNA involve the 2′ hydroxyl group.

When the oppositely charged surfaces of the protein and the DNA or RNA are in contact, there is a generally favorable electrostatic interaction and surface

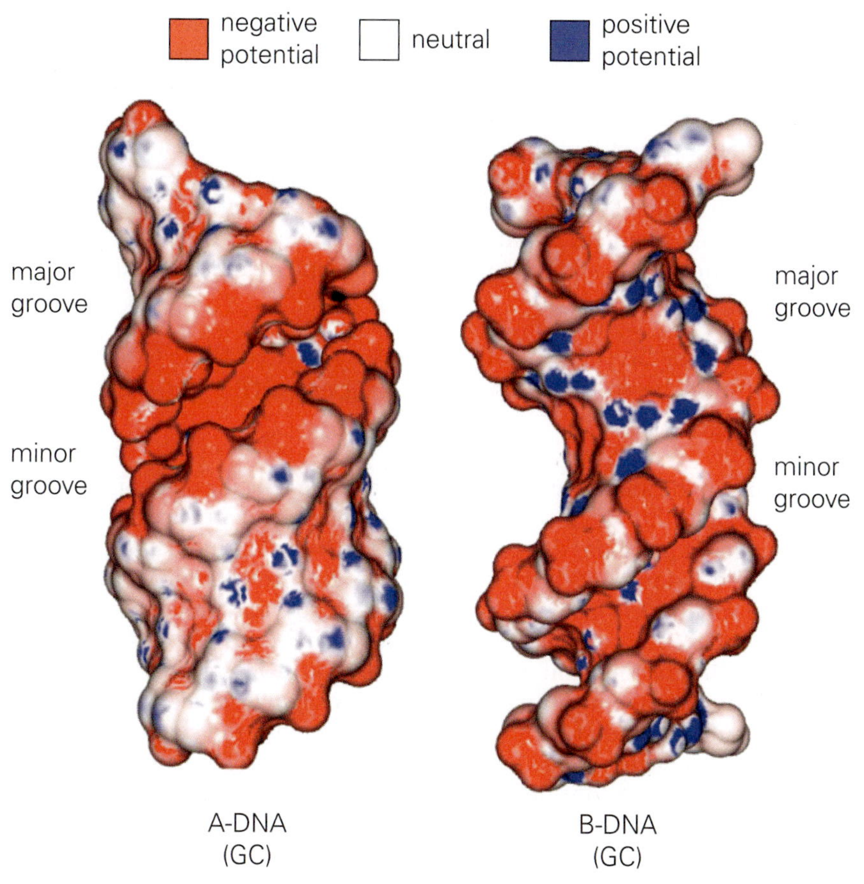


**Figure 13.30 The electrostatic potential of A-form and B-form DNA, calculated using the Poisson–Boltzmann equation in the absence of localized metal ions**. The structure and electrostatic potential of double helical RNA resembles that of A-form DNA. (Adapted from R. Rohs et al., and R.S. Mann, *Annu. Rev. Biochem.* 79: 233–269, 2010. With permission from Annual Reviews.)

complementarity (**Figure 13.31**). But, as we saw for protein–protein interactions in Section 13.21, it can be difficult to estimate the precise energetic balance because of desolvation effects. These effects can be particularly strong for nucleic acids because of the displacement of metal ions by the protein.

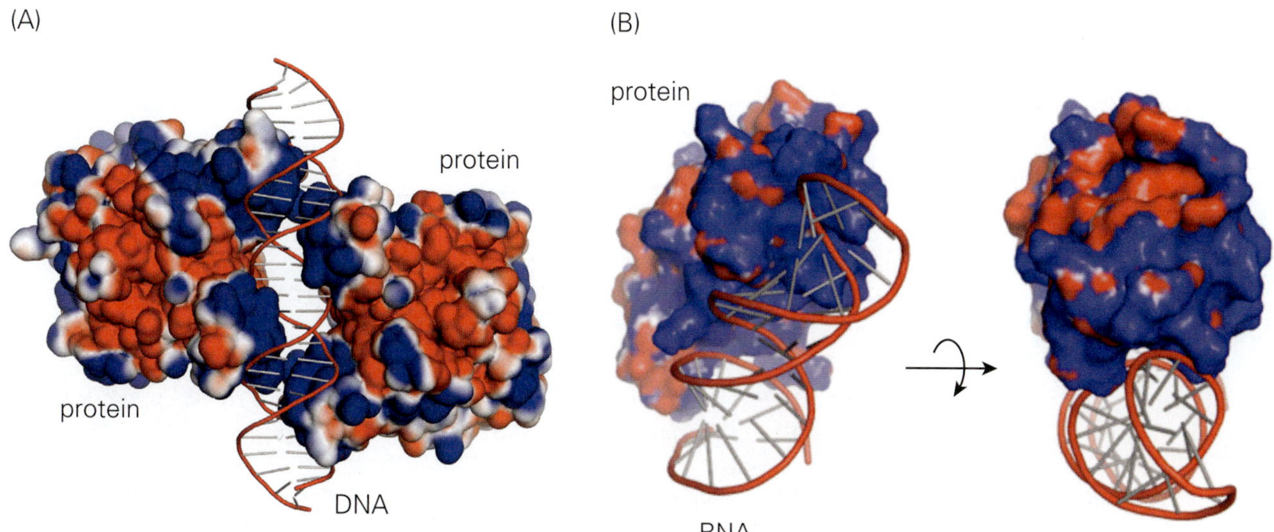


**Figure 13.31 Complementarity in shape and charge between proteins and their DNA or RNA targets**. The molecular surfaces of the proteins are colored according to electrostatic potential, calculated using the Poisson–Boltzmann equation. *Blue* and *red* indicate regions of positive and negative potential, respectively. (A) The structure of two dimers of a transcriptional repressor, QacR, bound to DNA. (B) Ribosomal protein S8 in complex with a stem-loop structure in the ribosomal RNA. (PDB codes: A, 1JT0; B, 1I6U.)

The structure of two dimers of a transcriptional repressor protein bound to DNA are shown in Figure 13.31A, along with a rendering of the electrostatic potential. Notice how certain segments of the protein fit into the major grooves of DNA and how these segments are positively charged. An example of the recognition of the shape and charge of double-helical RNA by a protein is provided in Figure 13.31B, which shows the ribosomal protein S8 in complex with a 37-nucleotide stem-loop fragment of ribosomal RNA. The protein has a noticeable concave surface and is strongly positively charged, as indicated by the blue color. The RNA fills most of the concave blue protein surface, to which it is complementary in both shape and charge.

## 13.23 Proteins distinguish between DNA and RNA double helices by recognizing differences in the geometry of the grooves

As we discuss in the following sections, the sequence-specific recognition of DNA occurs principally through the interaction of α helices and other secondary structural elements of proteins with the edges of the bases in the major groove. The major groove of A-form RNA is narrow and deep, and prevents the ready entry of α helices (see Section 2.11). The minor groove is shallow and broad, but is partially occluded by the 2′ hydroxyl groups of the sugars and does not have the distinctive pattern of interacting groups that the major groove does. Instead of relying on recognition through the major groove, proteins recognize folded RNA structures, including double-stranded RNA helices, by reading out the shape of the molecule, including nuances such as loops, bulges, or features of their intricate tertiary structures. We shall not consider these details here, but simply look at one aspect of recognition, which is how proteins distinguish between double-stranded DNA and RNA.

Many proteins that bind to double-stranded RNA contain a module known as a double-stranded RNA-binding domain (dsRB domain). These domains consist of 65–70 residues and are found in proteins involved in RNA interference (a process in which certain RNAs inhibit gene expression), the interferon-controlled response of mammalian cells to RNAs produced by viral infection, and pre-mRNA editing.

A dsRB domain that interacts with a major groove and the flanking minor grooves of an RNA double helix is shown in **Figure 13.32**A. In order for the dsRB domain

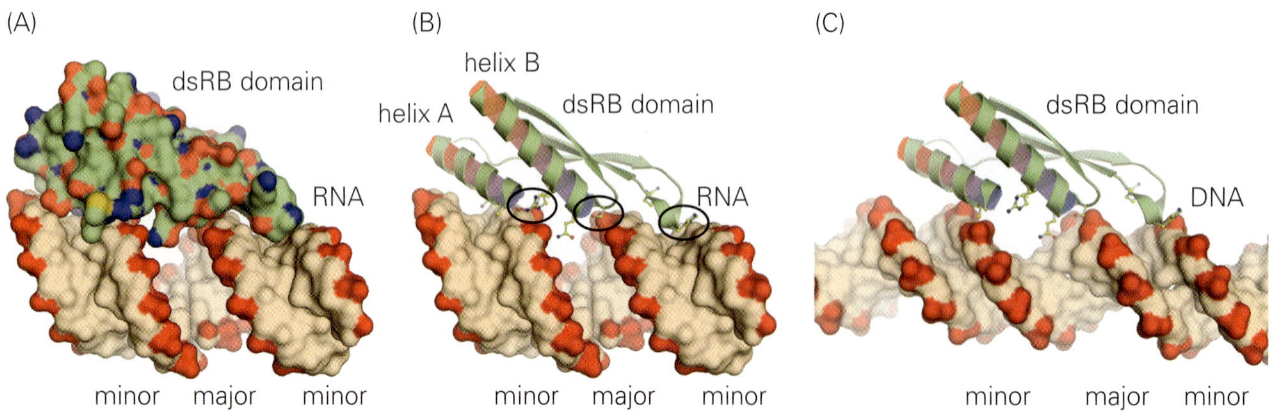


**Figure 13.32 Recognition of double-helical RNA by a double-stranded RNA-binding domain (dsRB domain) from frog RNA-binding protein A.** (A) Molecular surfaces of RNA and the dsRB domain, showing the complementarity in shape. (B) Structure of the complex with RNA. Interactions with the RNA backbone at the major groove and two flanking minor grooves are circled in *black*. The phosphate groups of the RNA are colored *red*. The helix dipoles of the two α helices are indicated by colored cylinders, with *blue* for the positive pole. (PDB code 1DI2.) (C) A hypothetical structure, generated by aligning the major groove of B-form DNA with that of the RNA in (A). The recognition elements of the protein no longer align properly with the phosphate groups.

to make contact with both the minor and major grooves of RNA, the RNA double helix must be long enough to present two adjacent faces of the minor groove to the protein. In this case, the N-terminal α helix (labeled helix A in Figure 13.32A) interacts with the RNA minor groove (shown on the left in Figure 13.32A). Residues presented by this helix are involved in direct interactions with two 2′-OH groups and a base in the minor groove of the RNA. The second helix interacts primarily with phosphate oxygens in the major groove of the RNA (helix B in Figure 13.32A). Both helices are aligned so that the positive poles of their helix dipoles are pointed towards the phosphate backbone. In addition, a helical turn in a loop connecting two β strands makes contact with the backbone of the other minor groove (on the right in Figure 13.32A).

Both the geometry of the interaction with RNA and the fact that there are many contacts with the 2′-OH groups that line the minor groove strongly favor binding to double-stranded RNA over double-stranded DNA. The ability of the dsRB domain to discriminate against double-stranded DNA is made obvious in Figure 13.32C, which shows B-form DNA that has been aligned so that the major groove matches that of the RNA in the structure of the actual complex (see Figure 13.32B). You can see that the recognition elements of the dsRB domain are appropriately configured for interaction with double-stranded RNA, but are misaligned with the backbone of the B-form DNA double helix.

## 13.24 Proteins recognize DNA sequences by both direct contacts and induced conformational changes in DNA

The sequence of a duplex DNA molecule can be read by examining the pattern of unique functional groups exposed in the major groove, as discussed in Chapter 2 (see Figure 2.13). When a specific pattern of hydrogen-bond donors, acceptors, and nonpolar groups on the protein is matched to a complementary pattern displayed on the DNA, the recognition is referred to as **direct readout** or **base readout** of the sequence, as illustrated in **Figure 13.33**. As we discuss in the next section, hydrogen-bonding interactions are the most important class of contacts between proteins and DNA. Although less common than hydrogen bonds, important contacts to DNA also involve nonpolar residues that interact with pyrimidines, particularly the methyl group of thymine (see Figure 13.33C).

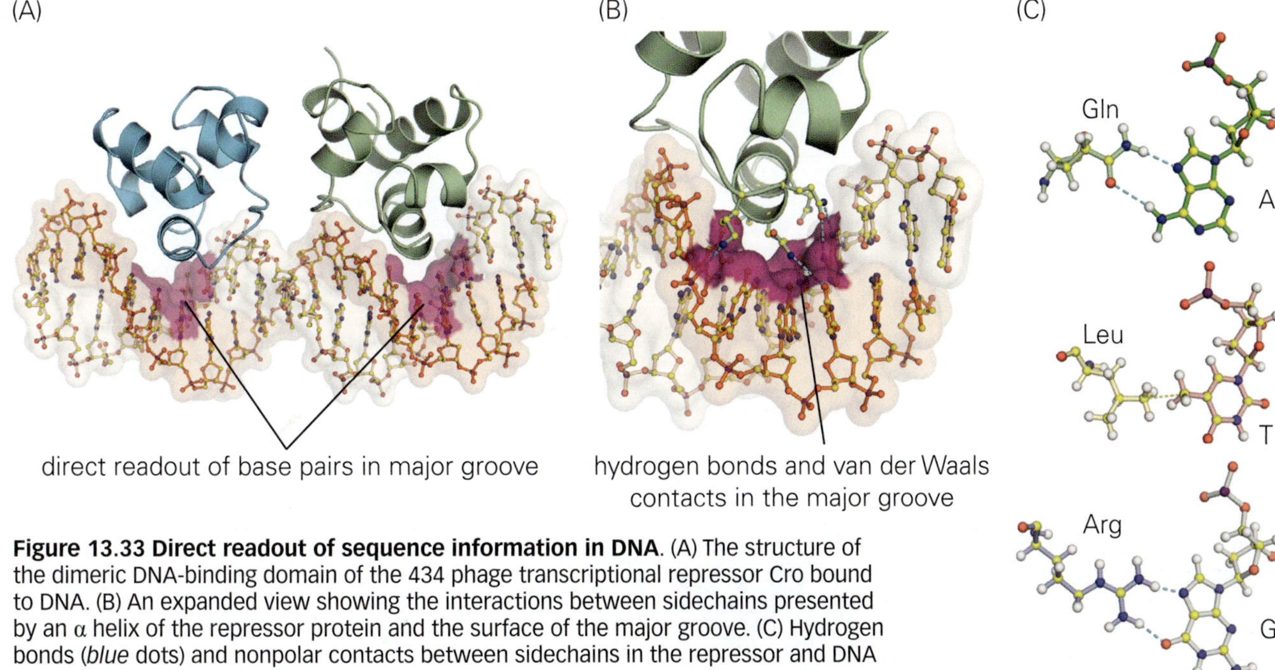


**Figure 13.33 Direct readout of sequence information in DNA.** (A) The structure of the dimeric DNA-binding domain of the 434 phage transcriptional repressor Cro bound to DNA. (B) An expanded view showing the interactions between sidechains presented by an α helix of the repressor protein and the surface of the major groove. (C) Hydrogen bonds (*blue* dots) and nonpolar contacts between sidechains in the repressor and DNA bases. (PDB code: 3CRO.)

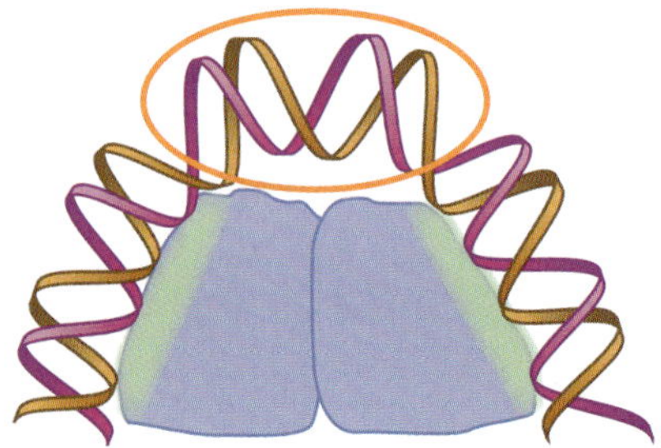

**Figure 13.34 Indirect readout of a DNA sequence.** The sequence of DNA within the *orange circle* can affect the energetics of binding, even though it does not make direct contact with the protein.

The specificity of a protein for a particular DNA sequence need not involve contact with the sequence that is recognized. The binding of a protein to DNA often involves a conformational change in the DNA. Whether a dramatic bend in the DNA, or a more subtle local shape change, the ability of DNA to undergo the required conformational change can depend on its sequence. When a protein recognizes the ability of a DNA sequence to undergo a particular conformational change, the process is referred to as **indirect readout** or **shape readout**. In the example shown schematically in **Figure 13.34**, the ability of the DNA to bend sharply will affect the ability of the protein to bind to it. The region of DNA that bends does not make contact with the protein, but sequences within this region that stiffen the DNA would lower binding affinity.

Indirect effects through DNA conformational changes are as important as the direct contacts for determining the specificity of DNA–protein interactions. Actual protein–DNA recognition utilizes a continuum of readout mechanisms that depend on the structural flexibility of both macromolecules.

## 13.25 Hydrogen bonding is a key determinant of specificity at DNA–protein interfaces

Given the richness of hydrogen-bond donor and acceptors at the edges of the base pairs, it is no surprise that many of the contacts between protein residues and DNA bases involve the formation of hydrogen bonds. Purine bases have more capacity for hydrogen bonding in the major groove than do pyrimidines (refer back to Figure 2.13). Guanine is most often observed to participate in hydrogen bonds, followed by adenine, cytosine, and thymine. The sidechain of arginine has five hydrogen-bond donors that can potentially interact with DNA bases in one or two base pairs, or with both DNA and another amino acid, as shown in **Figure 13.35** (the arginine sidechain shown in this figure forms only four out of the possible five hydrogen bonds). Such interactions are very restrictive in sequence because of the requirement for multiple correctly positioned hydrogen-bond acceptors in the DNA. The pairing of two hydrogen-bond donors in the arginine sidechain with the two acceptors on guanine is particularly energetically favorable and occurs frequently in DNA recognition.

A single hydrogen-bond donor or acceptor on the protein can often find an appropriate partner on the nucleic acid because both hydrogen-bond donors and acceptors are abundant on the DNA. For this reason, single hydrogen-bond donors or acceptors often do not discriminate between different sequences

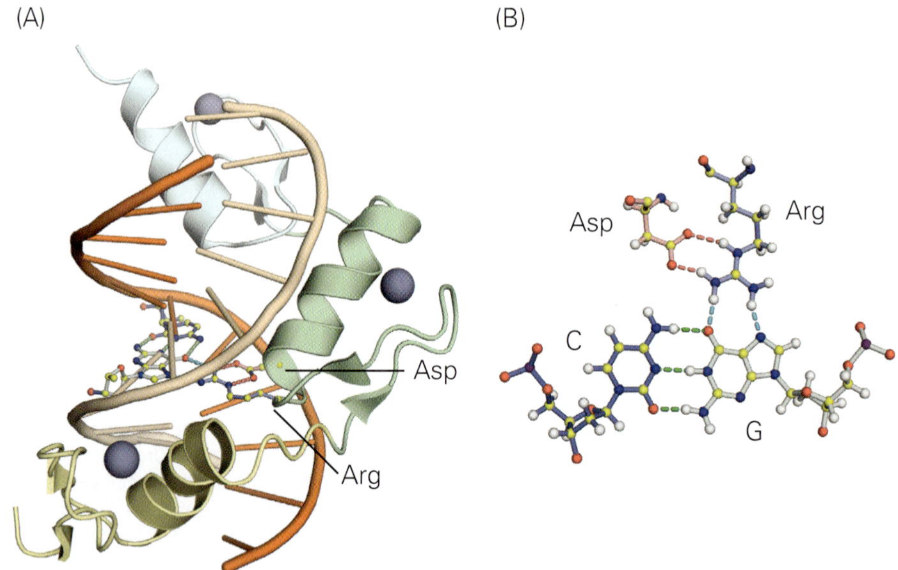


**Figure 13.35 Bidentate hydrogen bonds.** (A) Structure of three zinc finger modules bound to DNA (see Section 13.30). Two sidechains that form a hydrogen bond with bases in the major groove are indicated. (B) Expanded view of the hydrogen-bonded network. An aspartic acid hydrogen bonds to one edge of the arginine sidechain (*red* dashes), positioning it to interact optimally with a C-G base pair (*cyan* dashes; the Watson-Crick hydrogen bonds are *green* dashes). (PDB code: 1ZAA.)

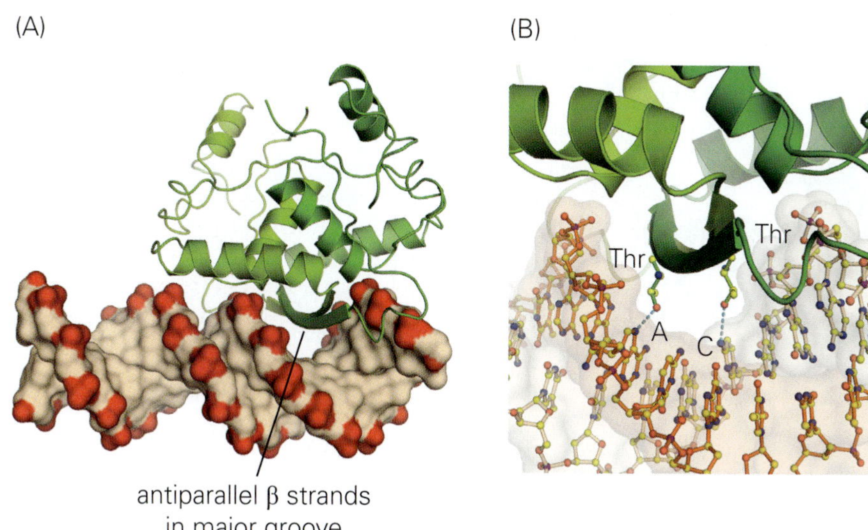


**Figure 13.36 Interaction of Met repressor with DNA**. (A) A Met repressor dimer recognizes DNA by inserting a pair of antiparallel β strands in the major groove. (B) Two threonine sidechains in the β strands form hydrogen bonds with an adenine from one DNA chain and a cytosine from the other. (PDB code: 1MJ2.)

available for binding. Bidentate hydrogen bonding, involving two different donor–acceptor pairs, are more specific because two pairs of complementary groups must be present and properly aligned to contribute fully to binding. An example of a bidentate hydrogen bond between an arginine sidechain and a guanine base is shown in Figure 13.35B. The arginine sidechain in this example is held in place by the formation of an additional set of bidentate hydrogen bonds with an aspartate sidechain.

Bidentate hydrogen bonds can be formed between a protein sidechain and one base, or with two bases in a base pair, two adjacent bases in one strand, or two bases in different base pairs and on opposite strands. Similar considerations apply to bifurcated hydrogen bonds (two hydrogen bonds that share the donor or acceptor). Flexible protein sidechains, such as lysine, that can move to find complementary groups on DNA are less likely to discriminate different sequences than sidechains that are fixed in position.

Neutral histidine, asparagine, glutamine, tyrosine, serine, and threonine sidechains have both donor and acceptor capabilities (Chapter 1). In spite of this, they are most often found involved in single hydrogen bonds with DNA, with context-dependent and variable contributions to specificity. The sidechains of serine and threonine are shorter, and this limits their ability to reach into the groove and make contact with the bases. Nevertheless, serine and threonine residues do make important contacts in some cases, as illustrated in **Figure 13.36** for the protein Met repressor. The Met repressor uses a pair of antiparallel β strands to interact with the major groove, highlighting the fact that proteins that interact specifically with DNA do so with many different structural motifs.

## 13.26 Water molecules can form specific hydrogen-bond bridges between protein and DNA

Earlier in the chapter we had discussed how water molecules are often a part of protein–protein interfaces, and can play an important role in specific recognition (see Section 13.16). The same is true of protein–nucleic acid complexes. In many protein–DNA complexes, water molecules form hydrogen bonds that bridge residues on the protein to bases in the DNA (**Figure 13.37**). Because there is still a requirement for specific functional groups on both protein and DNA, in specific orientations and with a specific spacing, these contacts can contribute to the ability of the protein to bind preferentially to the correct target sequences. Indeed, in some complexes of proteins with their cognate DNAs, essentially all of the hydrogen bonds linking the protein to the DNA are water-mediated.

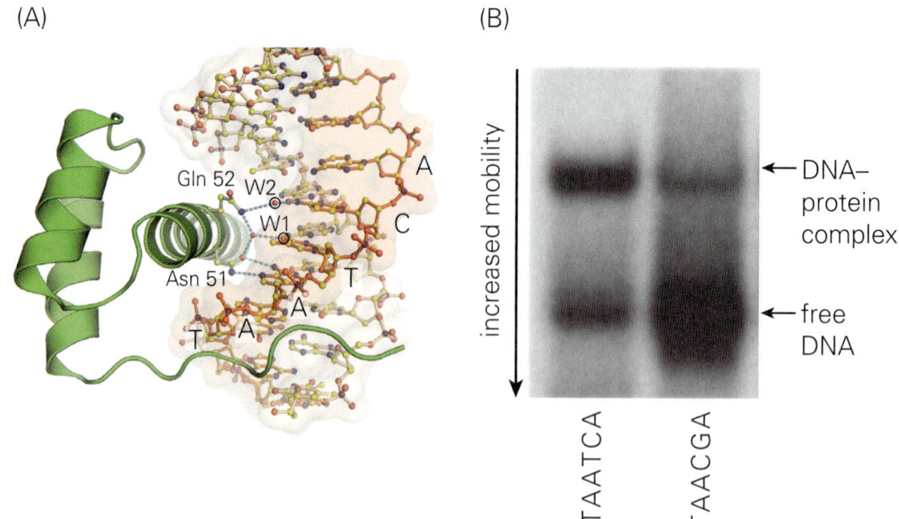


**Figure 13.37 Water molecules facilitate specific DNA recognition.** (A) Structure of a DNA-binding domain known as the Paired homeodomain bound to DNA. Two water molecules, labeled W1 and W2, form hydrogen bonds that bridge Asn 51 and Gln 52 of the protein to a T and C in the DNA. (B) A gel-shift experiment, showing that DNA in which the T and C are replaced by C and G binds with much lower affinity to the protein. (B, adapted from D.S. Wilson et al., and J. Kuriyan, *Cell* 82: 709–719, 1995; PDB code: 1FJL.)

The importance of water-mediated hydrogen bonds is illustrated in Figure 13.37, which shows how a transcription factor known as the Paired homeodomain forms a specific complex with DNA. The target DNA contains a TAATCA sequence motif. Notice that two sidechains from the protein, Asn 51 and Gln 52, form water-mediated hydrogen bonds with the second T and the C in the TAATCA motif. If the T and C are replaced by a C and G, the high affinity of the DNA for the protein is lost, even though the T and the C in the original motif only make contacts with the protein through water molecules. This is demonstrated in Figure 13.37B, which shows the results of a **gel-shift assay** on the normal and altered DNA.

Gel-shift assays are often used to determine the affinity of the interaction between proteins and DNA or RNA. As we shall see in Section 17.31, the mobility of molecules in electrophoresis gels is a function of both charge and shape, and hence when two molecules interact, the complex containing them will generally have a different mobility than the individual binding partners. Gel-shift assays are done by varying the concentrations of potentially interacting molecules and analyzing the extent to which complexes form by using electrophoresis to separate the species present (**Figure 13.38**). From the amounts of molecules in specific bands on the gel, the relative concentrations of the free molecules and complex can be estimated, and then, in turn, the binding constant is determined.

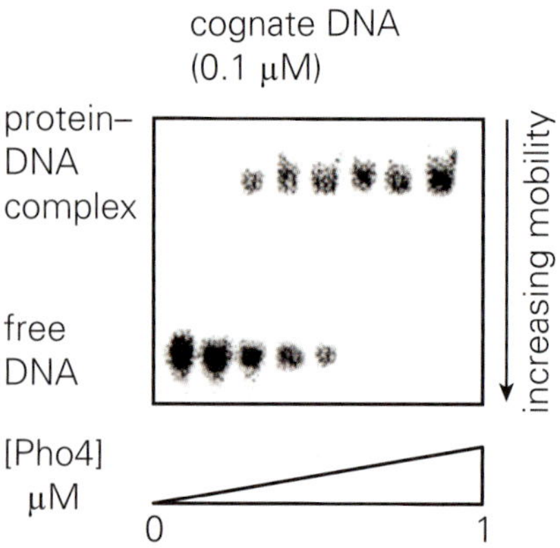

cognate DNA (0.1 μM)
protein–DNA complex
free DNA
increasing mobility
[Pho4] μM
0
1

**Figure 13.38 A gel-shift experiment.** A fixed amount of DNA containing the target sequence for a transcription factor, Pho4, is mixed with increasing concentrations of Pho4. The samples are electrophoresed in consecutive lanes of a gel. As the protein concentration increases, notice that the free DNA converts to the protein–DNA complex.

Binding of a protein to a nucleic acid usually reduces the mobility of the complex relative to the free nucleic acid. The position of the nucleic acid component can be identified by incorporating radioactive atoms into it and then detecting the decay events, or by addition of a fluorescent dye that binds to it and then detecting emitted light. In the case of the Paired homeodomain, you can see from the gel in Figure 13.37B that the amount of complex decreases substantially when the DNA sequence is modified, as judged by the increased amount of high mobility DNA.

## 13.27 Arginine interactions with the minor groove can provide sequence specificity through shape recognition

Recall from Chapter 2 that although the minor groove of DNA has both hydrogen-bond donors and acceptors, the spatial pattern in which they occur in different sequences is less distinct than for the major groove (see Figure 2.13). Nevertheless, sidechains that extend into the minor groove do contribute to sequence specificity. Much of the specificity can be explained through a shape readout mechanism that relies on the sequence-specific width and electrostatic potential of the minor groove. The minor groove tends to be narrower and deeper in segments of DNA with runs of 3-4 or more A-T or T-A base pairs. The narrower groove enhances the negative electrostatic potential generated by the phosphates along

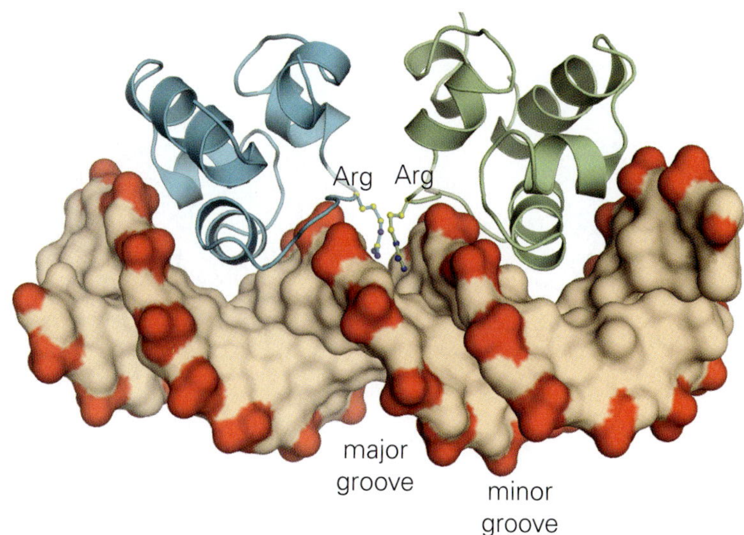


**Figure 13.39 Arginine sidechains in the minor groove.** In the phage 434 Cro protein–DNA complex, two arginines extend into a region of narrowed minor groove near the center of the complex (see also Figure 13.33). (PDB code: 3CRO.)

the groove, and positively charged sidechains interact more favorably with narrowed minor grooves than with normal minor grooves.

Arginine sidechains are most frequently inserted into regions of narrowed minor groove, as shown in **Figure 13.39**. The sequence specificity is not as high for this case as in most examples of direct readout because there are multiple A-T and T-A combinations that give rise to the narrowed minor groove.

The minor groove is normally too narrow to accommodate secondary structure elements within it and so segments of proteins that interact with the minor groove generally are not folded. Sidechains can penetrate quite deeply into the minor groove, allowing them to form hydrogen bonds and even some hydrophobic contacts (for example, between methylene groups in the sidechain and the edge of adenosine bases). If the DNA is strongly distorted by the protein, then α helices and β sheets might enter the minor groove.

## 13.28 DNA structural changes induced by binding vary widely

In some protein–DNA complexes, there is relatively little distortion of the structure of the protein or the DNA, as is the case for most of the examples we have discussed so far. In other protein–DNA complexes, much larger distortions of the DNA are found. We have looked at two such cases in Section 2.13 when we discussed the deformability of DNA. In the case of the nucleosome core particle, the basic packaging unit of chromosomal DNA, the double helix is wrapped around a protein core to form a solenoid (see Figure 2.23). The TATA-box binding protein, a key transcription factor, introduces a very sharp bend into DNA when it binds to promoter elements located upstream of genes (see Figure 2.24).

Another example of a protein that bends DNA is Lef-1, a member of the so-called high-mobility group (HMG) family of DNA-binding proteins (**Figure 13.40**). As is the case for the TATA-box binding protein, Lef-1 binds to DNA in the minor groove, with more hydrophobic contacts to A-T base pairs than hydrogen bonds to the bases. The minor groove is greatly expanded to accommodate the protein, and the DNA follows the shape of the protein surface, which induces a 90° bend toward the major groove. This distortion collapses the major groove and creates a region of high negative charge density within it. The tail of the protein wraps around the groove to position a sequence that is almost all Lys and Arg residues, so that their positive charges can interact favorably with the high negative electrostatic potential and stabilize the distorted DNA structure. If this tail is deleted, there is a very substantial drop in binding affinity. The structure of the folded part of the protein is affected relatively little by binding, and most of the structural change during complex formation is in the DNA.

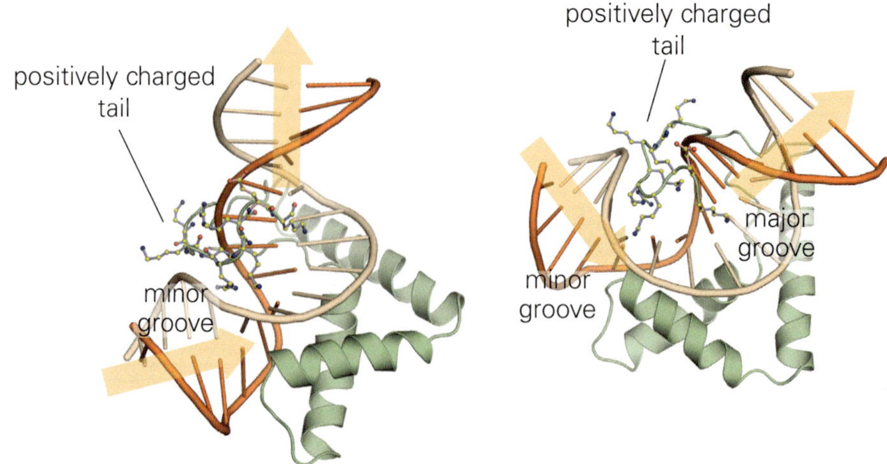


**Figure 13.40 Lef-1–DNA complex.** The structure of the Lef-1 protein in complex with a consensus target DNA sequence is shown, in two different views. Binding occurs from the minor groove, and a large distortion of the DNA is induced, with a bend of the helix axis (*orange* arrows) by almost 90°. (PDB code: 2LEF.)

## 13.29 Proteins that bind DNA as dimers do so with higher affinity than if they were monomers

Many DNA binding proteins, particularly in bacteria, occur as dimers and recognize **inverted repeat sequences**, or **palindromic sequences**. In such cases, each monomer recognizes the same features in each half of the inverted repeat DNA, allowing both higher affinity and sequence specificity to be achieved. An example of this is shown in **Figure 13.41**, with the sequence being recognized by the Cro repressor protein.

**Palindromic sequences**

A binding site in DNA that contains an inverted repeat (two copies of a sequence motif, with one running backwards) is a palindromic sequence. Dimeric proteins can bind to such sequences with two-fold symmetry.

If each half of the dimeric protein binds to each copy of the DNA sequence with the binding free energy of a monomer, then the dissociation constant of the dimer is the square of the dissociation constant of the monomer. To see why this is the case, begin with the fact that the standard free energy of binding, $\Delta G^o$, is given by Equation 12.8:

$$\Delta G^o = +RT\ln(K_D)$$

If the binding free energy is additive, then:

$$\Delta G^o_{dimer} = 2\Delta G^o_{monomer} = 2RT\ln(K_{D,monomer}) = RT\ln(K^2_{D,monomer}) \quad (13.24)$$

And so:

$$K_{D,dimer} = K^2_{D,monomer} \quad (13.25)$$

The multiplicative nature of the dissociation constants gives a dramatic enhancement in binding when the binding free energies are additive. If the dissociation constant for the monomer, $K_{D,monomer}$, is 1 μM ($10^{-6}$ M), then this analysis

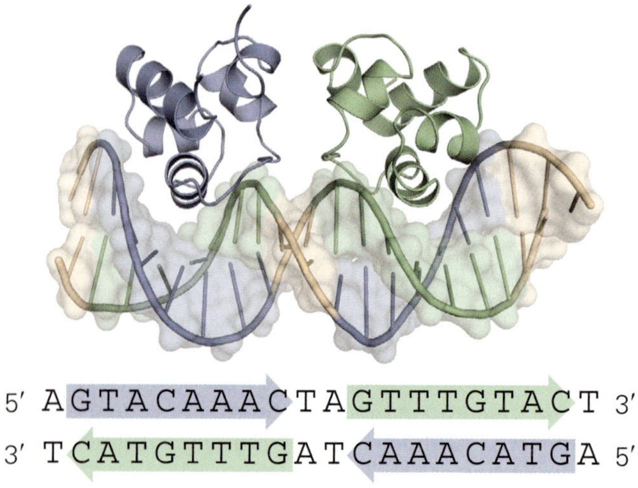

5′ AGTACAAACTAGTTTGTACT 3′
3′ TCATGTTTGATCAAACATGA 5′

**Figure 13.41 An inverted repeat, or palindrome, in a DNA-binding site.** The structure of the Cro repressor–DNA complex is shown. The two strands of DNA in the region that is recognized are colored blue and green, showing how the inverted repeat sequence makes a two-fold symmetric complex. The sequence of the DNA is shown below. (PDB code: 3CRO.)

predicts that the dissociation constant for the dimer, $K_{D,dimer}$, will be 1 pM ($10^{-12}$ M). Thus, two interactions that separately only have modest affinity for DNA can, together, make up one very strong one.

In real systems, the simple additivity of free energy rarely applies. Any free energy required to induce conformational changes in the DNA to accommodate dimer binding is subtracted from the binding free energy. In real systems, two monomers with dissociation constants around 1 μM often yield dissociation constants of 0.1 to 1 nM when binding to DNA as a dimer.

## 13.30 Linked DNA binding modules can increase binding affinity and specificity

How many bases must be recognized by a protein in order for it to interact specifically with a single sequence within a fairly large genome, such as that of humans? At any given site in a DNA, there are four possible bases, and the human genome has ~$3 \times 10^9$ base pairs. If all combinations of base pairs occurred for a set of 16 bases, then there would be $4.3 \times 10^9$ possible sequences. Although the sequence of human DNA is not really random (there are, for example, repeating sequences that reduce the randomness), this very rough calculation shows that approximately 16 base pairs would have to be recognized, and *each one* specifically, for a protein to target one site in the genome. There are few if any proteins known that recognize this length of DNA sequence by using a single folded domain.

As pointed out in the previous section, many bacterial proteins extend the region of contact by forming homodimers, with each half of the dimer interacting with one section of DNA. This strategy is somewhat restrictive because the sequences recognized by the two halves must be related. Bacterial genomes are considerably smaller than eukaryotic ones, with millions of base pairs rather than billions. With dimers, the bacterial proteins can make contact with 14–16 base pairs, which is sufficient for achieving good specificity within the genome (as discussed explicitly for lac repressor in part A of this chapter).

One mechanism that extends the sequences recognized by transcription factors is to combine several DNA recognition domains within the same protein. This strategy is analogous to that used by signaling proteins to achieve specificity, as we discussed in Section 13.13. A prime example of the combinatorial use of DNA binding modules is provided by the **zinc finger** proteins. The zinc finger is a small module, about 25 amino acids plus a short linker, with a fold stabilized by binding of a zinc ion, as shown in **Figure 13.42**. There are many different kinds of naturally

**Zinc fingers**

A class of small structural modules that are stabilized by the coordination of zinc by protein sidechains. There are many different kinds of zinc fingers, some of which bind DNA or RNA.

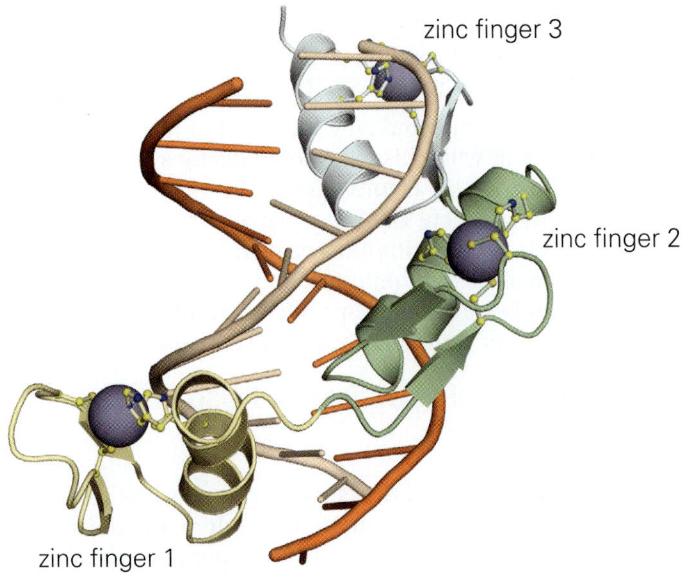


**Figure 13.42 Three zinc fingers in the protein Zif268 bound to DNA.** The three consecutive zinc fingers are shown bound to a cognate DNA sequence. The zinc atoms are shown as spheres, and the two histidine and two cysteine ligands for each zinc are shown. Contacts to DNA bases are made from the helix of each finger. See also Figure 13.32. (PDB code: 1ZAA.)

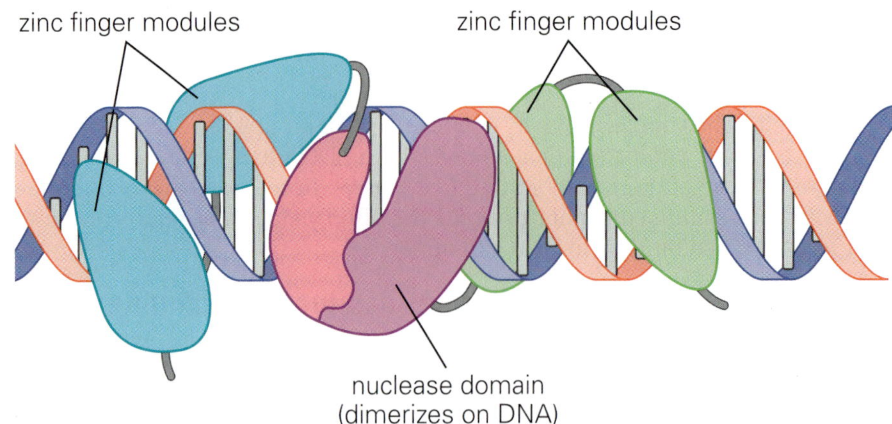


**Figure 13.43 Zinc finger nucleases.** Zinc fingers are selected from libraries for the desired sequence specificity and are combined in one polypeptide chain with a nuclease domain (*pink* and *purple*). The nuclease must be dimeric to be active, but only dimerizes very weakly on its own. If two different zinc finger nuclease hybrids are put into a cell, and bind close together in the genome, then the effective concentration of the nuclease will be high enough to form an active dimer. This can cut the DNA between the two zinc finger binding sites.

occurring zinc finger modules, with different folds and different patterns of zinc coordination, but the zinc fingers in the protein shown in Figure 13.42 are structurally similar.

Zinc fingers occur as tandem repeats in proteins, with three to six repeats appearing frequently. In fact, zinc fingers are among the most common modules in the human genome, and over 750 different proteins contain them. In addition to the DNA-binding domains, zinc-finger-containing proteins usually also have transactivation domains. These regions interact with components of the cellular transcription machinery, leading to the recruitment of polymerase complexes near the zinc finger binding sites, resulting in the initiation of transcription from neighboring genes.

The structure of a protein containing three zinc fingers bound to a cognate DNA shows that each zinc finger module recognizes three base pairs of DNA, as shown in Figure 13.42. The helix of each finger is inserted into the major groove, and sequence-specific contacts are made to each successive base. The spacer between modules is quite short, so there are no bases that are skipped, with the three modules reading out the sequence of nine successive base pairs. The linker between these modules allows them to adapt to the local DNA structure, making it possible for many repeats to be combined for recognition of long DNA sequences.

The finding that each zinc finger module reads out three successive base pairs stimulated work to create engineered versions of these that could be targeted to any desired DNA sequence. This was done by creating libraries of zinc fingers (a collection of variant forms of the protein with every possible amino acid at each of several designated positions within the protein sequence), and then using selection methods to find zinc fingers that bind a desired target DNA sequence. In this way, it has indeed been possible to target relatively long DNA elements of essentially arbitrary sequence by these synthetic zinc fingers.

One application of these engineered zinc fingers is to use them to cut genomic DNA at any desired site at will. This is done by fusing synthetic zinc fingers to nuclease domains, as shown in **Figure 13.43**. The nuclease domains in these applications must dimerize to cut DNA, but the dimer interface is very weak. To become active, two nuclease domains must be held in proximity by other attached targeting domains, such as the synthetic zinc fingers. It has been shown that these synthetic nucleases can be sufficiently specific to cut just one site in the human genome, into which another gene can be inserted for gene therapy applications.

## 13.31 Cooperative binding of proteins also enhances specificity

In Section 13.13, we had discussed the fact that proteins that contain more than one peptide binding module, such as multiple SH2 domains, are much more specific in binding to their targets than an individual peptide binding module. The

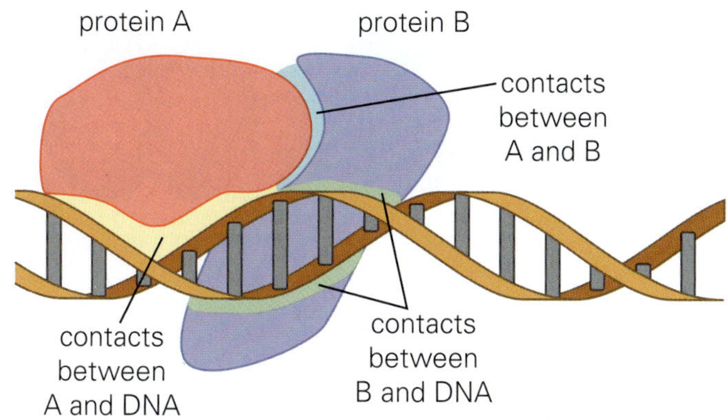


**Figure 13.44 Cooperative protein–DNA binding.** A schematic drawing is shown of a complex of two proteins with DNA. Each protein has an interface in contact with DNA, but when both are present, there is also a protein–protein interface, leading to cooperative binding.

same principle applies for DNA recognition. Combinations of covalently linked DNA-binding domains appear in many DNA-binding proteins. The zinc finger proteins discussed in Section 13.30 are just one example. The role of the multiple domains is primarily to enhance the sequence selectivity of binding, because high affinity to a DNA target can be achieved within single domains when needed.

Proteins that are not covalently linked to each other can also bind cooperatively to DNA and, in this way, increase their specificity. By **cooperative DNA binding** we mean that the binding of one protein to DNA increases the affinity of another protein for DNA. The quantitative principles of cooperative binding will be discussed in Chapter 14. For the present, we can just think of the binding site for one protein as comprising a DNA sequence and part of another protein, as shown schematically in **Figure 13.44**. As suggested by the shaded regions, which indicate the interfaces with DNA, the two proteins can bind to DNA independently. When both bind simultaneously, there is, in addition to the two protein–DNA interfaces, a protein–protein interface. Each of these interfaces contributes to the total free-energy change upon complex formation. Favorable interactions from the protein–protein interface, formed in association with DNA binding, enhances the affinity of the complex at DNA sites containing the cognate sequences for both proteins.

**Cooperative DNA binding**

When the binding of one protein to DNA increases the affinity of another protein for DNA, the two proteins are said to bind cooperatively. Cooperativity in binding is discussed in Chapter 14.

The cooperative binding of transcription factors to gene regulatory regions is critical for the regulation of gene expression, particularly in eukaryotic organisms. A good example is provided by the transcriptional regulation of interferons, which are protein hormones that are involved in cellular defenses against pathogens such as viruses. The production of interferon-β is triggered by the binding of multiple proteins to control regions in the DNA known as **enhancer elements**, located upstream of the gene. These proteins include transcription factors, such as nuclear factor κ-B (NF-κB) and c-Jun, that play multiple roles in the cell. NF-κB and c-Jun bind to the enhancer element of the interferon-β gene cooperatively with other transcription factors, including interferon response factors (IRFs) and ATF-2 and, when they do so, they recruit additional proteins that eventually bring RNA polymerase to the promoter, resulting in the initiation of transcription.

The structure of a complex of eight different transcription factors bound to the enhancer element of the interferon-β gene is shown in **Figure 13.45**. The complex of these proteins and the DNA is known as an **enhanceosome**, and its formation is the first step in the initiation of transcription. The proteins bound to DNA form a single contiguous unit that make uninterrupted contact with ~50 base pairs of DNA, allowing for very specific recognition of particular sites in the human genome that allow this assembly to form (refer to the discussion at the beginning of Section 13.30).

Why is the binding of the transcription factors to the enhancer element cooperative? Part of the answer is that each of the proteins in the enhanceosome makes contact with other components of the complex, thereby facilitating the assembly. But, as you can see in Figure 13.45, the contacts between the interferon response factors and c-Jun/AFT-2 and NF-κB are not particularly extensive. Another aspect

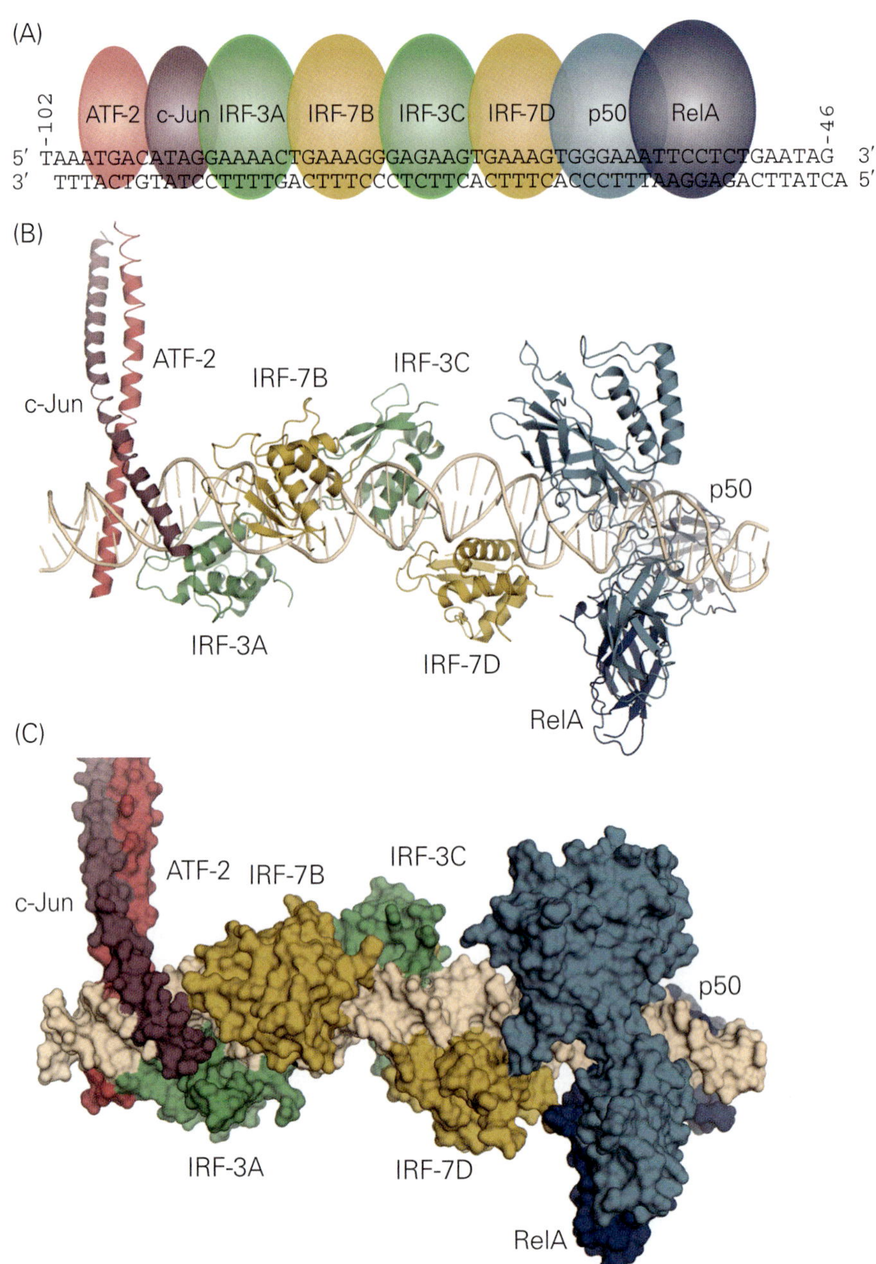


**Figure 13.45 Proteins binding to the interferon-β enhancer.** (A) Sequence of the enhancer element, showing the binding sites for eight proteins that bind to it. The numbers above the gene sequence indicate the distance from the transcription start site. (B) Structure of the enhanceosome, a protein–DNA complex in which multiple transcription factors are bound to the enhancer element. Four of the proteins are interferon response factors, denoted IRF. The transcription factor NF-κB is a heterodimer formed by the p50 and RelA proteins. Note the distortions induced in the DNA by the transcription factors. (C) Molecular surfaces of the proteins and DNA. [The structural models in (B) and (C) are generated using PDB entries 2O61, 1T2K, and 2PI0, as described in D. Panne, T. Maniatis, and S.C. Harrison, *Cell* 129: 1111–1123, 2007.] (A, adapted from D. Panne et al. With permission from Elsevier.)

of the complex that can increase the cooperativity of binding is the distortion in the structure of DNA that is induced by the binding of the transcription factors (this distortion is apparent to the eye in Figure 13.45B and C).

To understand why the distortion of DNA by the binding of one protein can enhance the binding of another protein, let us turn to a simpler example, that of the Paired homeodomain binding to palindromic sites in DNA (the Paired homeodomain was introduced in Figure 13.37). The Paired homeodomain binds cooperatively to DNA containing an inverted repeat of the TAATCA motif that we described in Section 13.26 as being optimal for the binding of this protein. The cooperativity of binding is demonstrated by a gel-shift assay shown in **Figure 13.46**. As the concentration of the protein is increased, the DNA switches from being unbound to being bound to two molecules of the homeodomain, with very little of the population in which only one homeodomain is bound. The Paired homeodomain is monomeric in solution, so we can infer that it is the interaction with DNA that promotes the cooperative binding of the homeodomain to DNA.

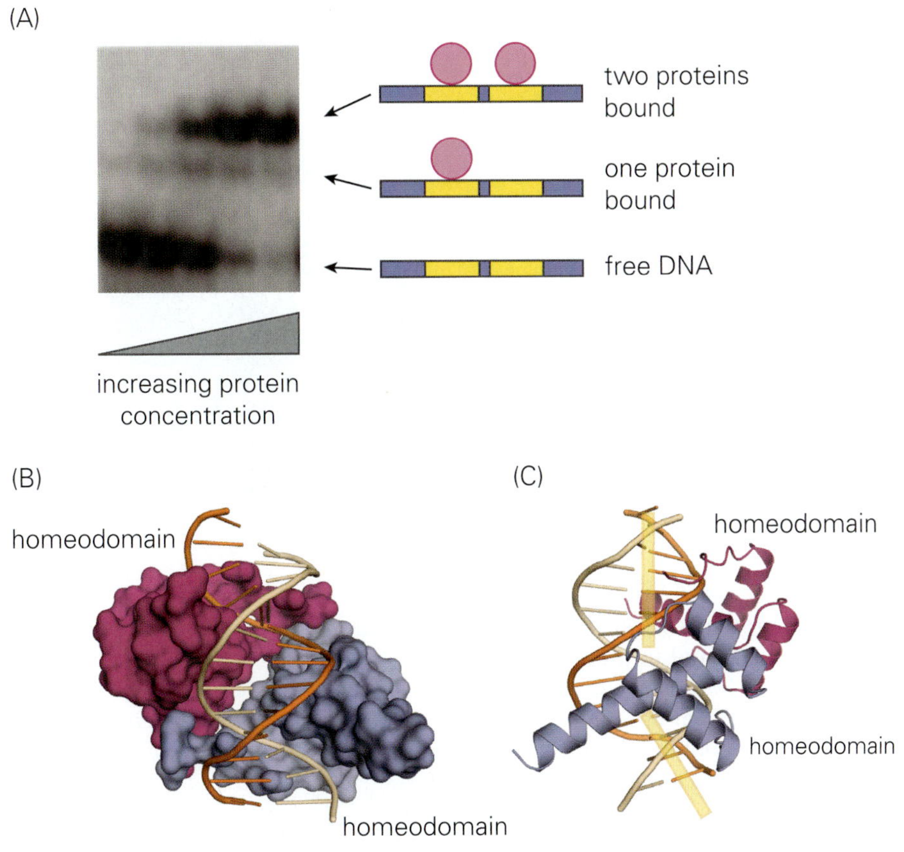


**Figure 13.46 Cooperative binding of the Paired homeodomain to DNA containing a palindromic binding site.** (A) Gel-shift assay, showing that as the protein concentration is increased, the DNA switches from being unbound to having two homeodomains bound, with little binding of monomeric protein. The Paired protein is a monomer in solution. (B) Structure of the Paired homeodomain bound to DNA (also shown in Figure 13.37). (C) An orthogonal view, showing the bend in DNA induced by Paired binding. The contacts between Paired molecules seen in (B) require that the DNA be bent. (A, adapted from D.S. Wilson et al., and J. Kuriyan, *Cell* 82: 709–719, 1995; PDB code: 1FJL.)

The structure of the Paired homeodomain bound to DNA is shown in Figure 13.46B and C. As you can see, the two proteins touch each other when bound to DNA. The contact between the two proteins requires that the DNA be bent by ~20° (the bending of the DNA is shown in Figure 13.46C). By comparing this structure with that of single homeodomains bound to DNA, it was deduced that the binding of the first homeodomain induces the bend in DNA, which then facilitates the binding of the second homeodomain by setting up the framework for forming interactions between them. If the DNA did not bend, then the homeodomains would not touch and, presumably, the cooperativity would be reduced or eliminated.

## 13.32 Proteins that recognize single-stranded RNA interact extensively with the bases

We shall now end our discussion of the interactions of nucleic acids with proteins by considering how single-stranded RNA is recognized specifically by proteins. One principal difference between the specific recognition of RNA versus DNA is that the interactions of proteins with RNA often involve single-stranded nucleotide segments that are not base paired. Proteins that recognize regions of RNA in which the bases are not paired typically do so by interacting with the flat surface of the nucleic acid bases, which are accessible instead of being sequestered within double helices. The protein surface that interacts with the nucleic acid bases is not predominately positively charged and, instead, is rather hydrophobic.

Proteins are quite versatile in their ability to recognize single-stranded RNA and utilize many different structural scaffolds to do so. For example, one class of RNA-binding modules known as KH domains bind RNA within a cleft, making contacts with both the bases and the phosphate backbone (**Figure 13.47**). KH domains are named for their sequence homology to the human heterogeneous nuclear

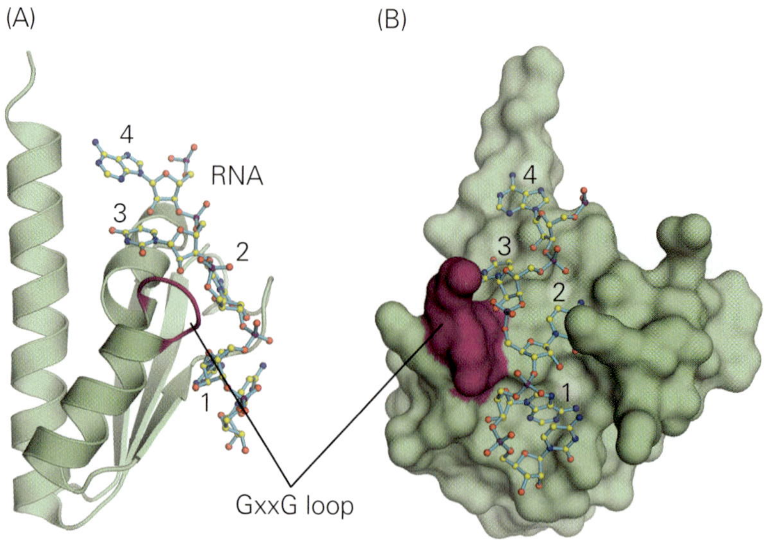


**Figure 13.47 A KH domain bound to RNA**. The RNA shown here is just a small segment of a larger RNA molecule. (A) Ribbon drawing of the structure, with the four nucleotides that make specific interactions with the KH domain numbered. The GxxG motif that is a characteristic feature of KH domains is in *purple*. (B) The molecular surface of the KH domain is shown, with the view rotated ~90° with respect to that in (A). (PDB code: 1EC6.)

ribonucleoprotein K (hnRNP K Homology). KH domains consist of about 70 residues and are found in a diverse set of proteins in eukaryotes, eubacteria, and archaea, often in multiple copies within the same protein. One form of the fragile X mental retardation syndrome, which is the most common form of inherited mental impairment in humans, is associated with loss of function of a particular KH domain, due to a single point mutation.

The structures of different KH domains with bound RNA reveal some common features. The RNA ligand is bound in a cleft on the surface of the protein, in an extended conformation (see Figure 13.47). Each KH domain interacts specifically with four nucleotides. The first three nucleotides are spread out on the surface of the protein, with the second and third lying in a hydrophobic pocket of the protein. The third nucleotide also makes backbone contacts with a conserved residue within the KH domain. Recognition of the RNA by the KH domain relies on hydrogen bonds between RNA bases and both the backbone and sidechains of the protein. A loop connecting two α helices contains a "GxxG" sequence motif, in which two glycine residues are separated by two variable residues (see Figure 13.47). The glycine residues in the loop enable the close approach of a phosphate group and the formation of a hydrogen bond with a backbone amide group of the protein.

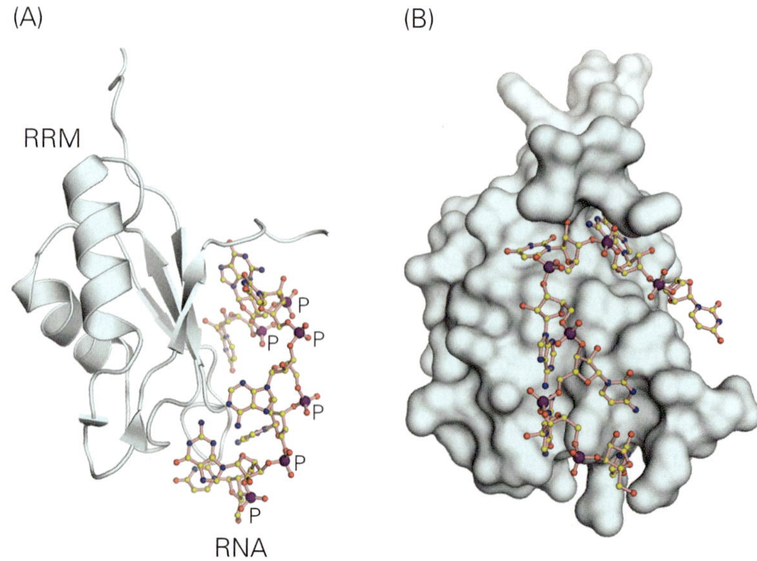


**Figure 13.48 Complex of an RRM domain with RNA.** (A) Structure of the Fox1 RRM domain bound to RNA. This view emphasizes the fact that the protein interacts with the bases and not the phosphate groups (P). (B) Orthogonal view, showing how the RNA bases tuck into crevices on the surface of the RRM domain. These crevices are formed by sidechains that stack against the bases (see Figure 13.49). (PDB code: 2ERR.)

## 13.33 Stacking interactions between amino acid sidechains and nucleotide bases are an important aspect of RNA recognition

We have discussed earlier, in Section 2.3, how the electronic polarization of the nucleotide bases leads to favorable stacking energy when one base lies flat on top of another base. A base that is swung out from the rest of the RNA molecule can interact favorably with the aromatic sidechains of proteins, and with the guanidinium group of arginine, by forming similar stacking interactions. Such stacking interactions between bases and protein sidechains are commonly seen in complexes of RNA with RNA-binding proteins.

Stacking interactions are the principal mechanism by which small domains known as RNA recognition motifs, or RRM domains, recognize RNA (**Figure 13.48** shows the RNA complex of an RRM domain of the protein Fox1). RRM domains, which typically contain about 90 residues, are present in about 1% of human genes, and proteins containing RRM domains participate in a variety of processes. RRM domains generally bind to RNA, although at least some seem to function by binding single-stranded DNA or other proteins.

Most RRM domains interact with their target RNA across a surface formed by a four-stranded β-sheet (see Figure 13.48). Two conserved aromatic sidechains, which are next to each other on the β sheet surface, stack with the bases of the nucleic acid (**Figure 13.49**A shows His 120 and Phe 160 of the Fox1 RRM domain stacking with RNA bases). In the Fox1 RRM domain, the sidechain of Phe 126 stacks between two bases (Figure 13.49B). Another highly conserved aromatic residue, Phe 158 in Fox1, is often inserted between sugar rings of the RNA (Figure 13.49C).

The stacking interactions seen in RRM–RNA complexes contribute substantially to the free energy of binding. The wild-type RRM domains bind to the RNA with a dissociation constant, $K_D$, of 1.1 nM (that is, $\Delta G^o = -52$ kJ•mol$^{-1}$ at 300 K). Mutation of Phe 126 to alanine increases the value of $K_D$ 1000-fold, to 1.6 μM ($\Delta G^o = -33$ kJ•mol$^{-1}$). This indicates that the stacking interactions between this aromatic residue and the RNA bases provides ~19 kJ•mol$^{-1}$, or 36%, of the binding free energy.

The principle that RNA bases can be recognized by their ability to stack against protein sidechains is also utilized by proteins containing PUF repeats

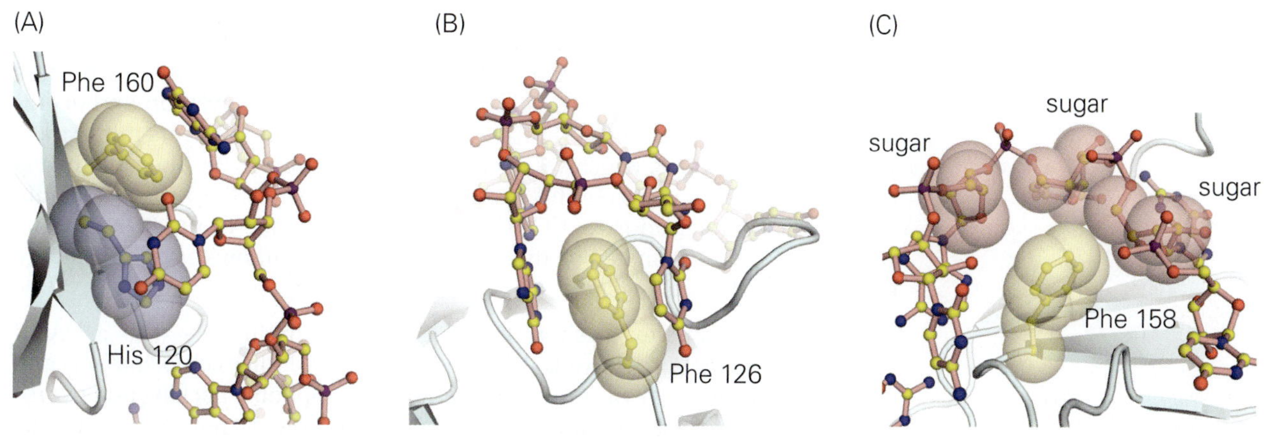


**Figure 13.49 Interaction of aromatic sidechains in an RRM domain with bases and sugars.** (A) Histidine and phenylalanine sidechains stack against the bases of a cytosine and a guanine, respectively. (B) A phenylalanine sidechain stacks against two bases. (C) A phenylalanine sidechain packs against three sugar groups, helping to orient the phosphate groups outwards, away from the surface of the domain. (PDB code: 2ERR.)

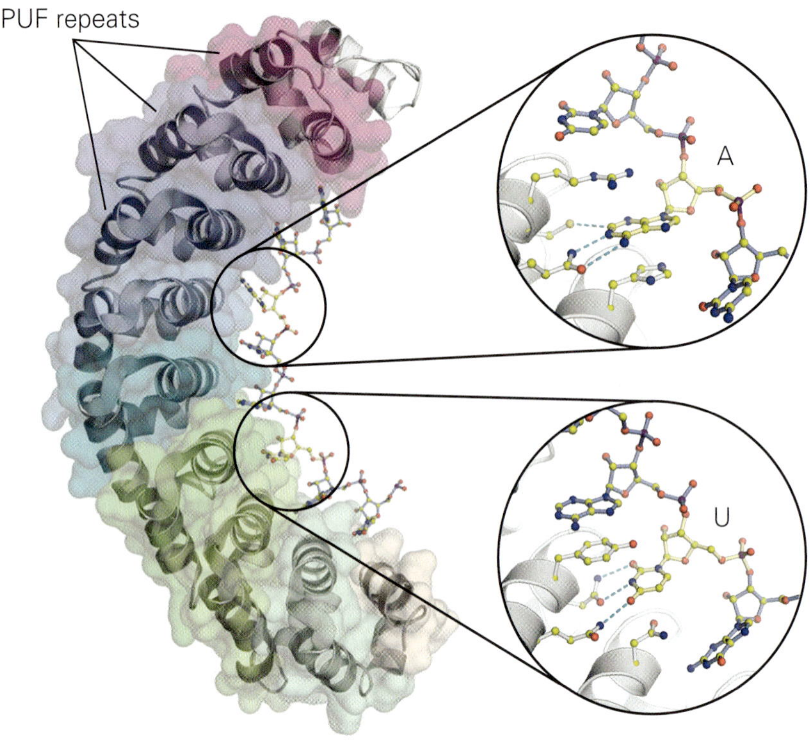


**Figure 13.50 A PUF repeat protein bound to RNA.** The protein consists of nine PUF repeats, which are indicated in rainbow colors. The bases of the RNA are recognized at the interfaces between PUF repeats, as shown in the expanded views for two bases. (PDB code: 1M8Y.)

(**Figure 13.50**). PUF repeats are helical modules that are chained together to construct larger units and are named for the proteins in which they were first identified. Proteins containing PUF repeats regulate translation and mRNA stability by binding to the 3′ untranslated regions of mRNAs.

Each repeat consists of three α helices, and a single RNA base is recognized at each of the interfaces between repeats (see Figure 13.50). In addition to stacking between protein sidechains and the bases, the edges of the bases form hydrogen bonds with the protein, leading to a highly specific interaction. The modular architecture of PUF proteins allows one to change the order of the repeat units within the protein. In this way, mutant PUF proteins have been created that recognize a diverse array of RNA molecules.

Base stacking is also a prominent feature of the mechanism by which zinc fingers in certain RNA-binding proteins recognize single-stranded RNA. The structure of a pair of zinc fingers from the protein TIS11d, bound to an AU-rich RNA, is shown in **Figure 13.51**. TIS11d is involved in mRNA deadenylation and degradation. The zinc fingers in TIS11d are distinct from the zinc fingers in the transcription factor Zif268, which we encountered in Section 13.30. The zinc fingers in Zif268 have two cysteine and two histidine ligands for the zinc (see Figure 13.42) and are referred to as CCHH zinc fingers. The zinc in the RNA-binding zinc fingers are coordinated by three cysteine ligands and one histidine and are known as CCCH zinc fingers.

In the TIS11d complex, each zinc finger interacts exclusively with a UAUU sequence in the RNA. The two UAUU sites adopt very similar conformations and bind each zinc finger in an equivalent manner. Hydrophobic packing and hydrogen bonding dominate the interface between the protein and RNA, especially through interactions between the sidechains of conserved aromatic amino acids (red), which stack with RNA bases (yellow) (see Figure 13.51). The specificity for RNA sequences comes from a network of intermolecular hydrogen bonds, predominantly between the Watson-Crick edges of the bases and amide and carbonyl groups of the protein backbone.

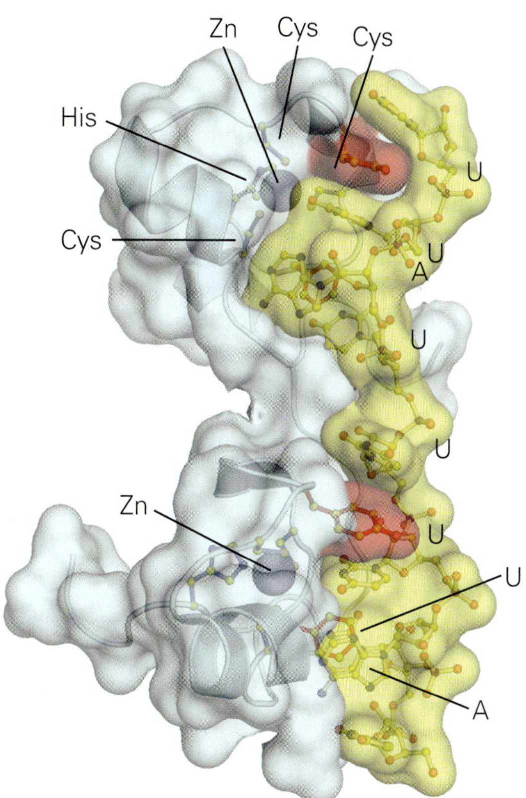


**Figure 13.51 Single-stranded RNA bound to a CCCH zinc protein (TIS11d).** The single-stranded RNA (*yellow*) fits the surface of the protein (*gray*). Aromatic protein residues (*red*) stack between the RNA bases. (PDB code: 1RGO.)

## Summary

For a biological ligand to function by binding to a target, the affinity for the target must be high enough to have a significant fractional occupancy and the specificity must be sufficiently high that the ligand is not tied up by other off-target receptors. Specificity is defined by the ratio of the concentration of the ligand bound to the target of interest divided by the sum of the concentrations of the ligand bound to all other targets. The number of off-target receptors and their affinities are important in determining selectivity. The energetic contributions to both affinity and specificity in intermolecular interactions are not spread evenly among residues at the intermolecular interface—rather, they are typically localized to a few hot spots. The residues contributing to affinity may be different from those that determine specificity.

Protein–protein interfaces typically remove over 1000 $Å^2$ of protein surface from contact with water, and often a substantial part of the contact surface is formed by hydrophobic residues. Water plays an important role in the hydrophobic effect that drives burial of such regions, but individual water molecules can also form hydrogen-bond bridges between proteins. Ionic interactions across interfaces can also contribute to stability and, even when they do not stabilize a complex substantially, they may still contribute to specificity by requiring complementary charges at specific sites in the binding partner.

Many proteins that bind DNA must recognize a specific set of sequences to have a selective effect (for example, in controlling transcription). Specific sequences in DNA can be read out by contacts to the hydrogen-bond donors and acceptors on the edges of the bases. This is often accompanied by distortion of the DNA (and sometimes the protein) in the process of recognition, and the energetic cost of this may be an important factor in determining specificity. Ionic contacts between the protein and the DNA can take many forms, from direct hydrogen bond-salt bridges to phosphates to the localization of sidechains in regions of negative electrostatic potential, such as in the minor groove. To recognize long DNA sequences, proteins often have multiple modules, each of which makes contact with a short sequence. Protein–protein interactions are also exploited to induce the assembly of multiprotein complexes at specific sites.

RNA is discriminated from DNA by the difference in helix geometry in base-paired regions, and by the presence of the 2′ hydroxyl group (absent in DNA) that is often involved in hydrogen bonds to proteins. Many RNA binding proteins recognize unpaired, unstructured regions (although there may be a requirement for flanking secondary structure), because RNA is generally not fully base paired. Stacking between bases and protein sidechains is a common mechanism for the recognition of unpaired bases, with the edges of the bases also being recognized by hydrogen bonding.

## Key Concepts

### A. AFFINITY AND SPECIFICITY

- To understand how intermolecular interactions lead to specific function, it is necessary to understand both the affinity and specificity of binding.
- Specificity is defined using the relative occupancy of a target binding site relative to the occupancy of all off-target binding sites.
- Specificity varies with the concentration of ligand, decreasing as ligand concentration increases.
- Many natural ligand–receptor systems have many ligands and many receptors, with highly variable specificity in different systems.
- The energetic value of inter-residue contacts is highly variable; a small number of "hot spots" often contribute most of the binding free energy.

### B. PROTEIN–PROTEIN INTERACTIONS

- Peptide recognition domains are modules that contribute to specific inter-protein interactions.
- The extent of surface area that becomes inaccessible to water upon complex formation is useful to characterize complex formation.
- Some water molecules remain in inter-protein interfaces and often bridge other hydrogen-bonding groups.
- Residues that do not contribute substantially to binding affinity may still be important for the specificity of an interaction.
- The relative contributions of hydrophobic, polar, and charged interactions to binding are highly variable in different complexes.

### C. RECOGNITION OF NUCLEIC ACIDS BY PROTEINS

- Proteins distinguish between double-helical DNA and RNA by the differences in groove shape and the functional groups displayed for interaction.
- Sequence-specific recognition of DNA by proteins involves both the interaction of correctly positioned complementary functional groups and the sequence-dependent distortion of DNA.
- Hydrogen bonding is important in contacts between protein and DNA; multiple contacts to a single residue may contribute more to specificity than single contacts.
- As for protein–protein interfaces, bridging water molecules are important in some protein–DNA complexes.
- In some protein–DNA complexes, major distortion of the DNA occurs and is used for specific recognition.
- Specificity can be increased by linking distinct DNA interaction modules in one polypeptide chain.
- The length of DNA recognized can be increased by cooperative interactions of multiple sequence-specific binding proteins at neighboring DNA sites, mediated by direct protein–protein contact and/or by modulation of DNA structure.
- In RNA, the 2′ OH plays an important role in contacts to proteins, in addition to bases and phosphates, which are also important in DNA.
- For recognizing unstructured regions of RNAs (those lacking base pairing), stacking between bases and protein sidechains contributes to binding affinity and specificity.

## Problems

### True/False and Multiple Choice

1. The affinity of a macromolecular interaction reflects the strength of an interaction for one receptor relative to all other possible receptors.

   True/False

2. Which of the following is an attribute of the FGF–FGFR family of interactions?

   a. Each FGF ligand has high specificity for an FGF receptor.
   b. Because there are 18 FGF and 4 FGFR genes, there are 72 potential interactions.
   c. Signaling specificity is enhanced through selective expression of only a few receptors per tissue type.
   d. FGFRs are soluble serine/threonine kinases.
   e. FGF proteins have highly diverse folds.

3. Specificity depends on the concentration of ligand, whereas affinity is independent of ligand concentration.

   True/False

4. Most of the binding energy for SH2 domain–peptide interactions is contributed by:

   a. Amino–aromatic interactions.
   b. The phosphorylation of the peptide tyrosine.
   c. The amino acid in the peptide + 1 position.
   d. The amino acid in the peptide + 3 position.

5. Which of the following statements regarding interfacial waters is true?

   a. They contribute favorably to the entropy of protein–protein recognition.
   b. They help to optimize the packing at the contact surface.
   c. They contribute only to affinity, but not to the specificity of protein–protein recognition.
   d. There are typically fewer water-mediated hydrogen bonds than direct hydrogen bonds at protein–protein interfaces.

6. Interface residues that do not contribute greatly to binding affinity are also generally unimportant for specificity.

   True/False

7. Which of the following is not an important aspect of protein–nucleic acid recognition?

   a. Insertion of arginine sidechains into the minor groove of the nucleic acid.
   b. Conformational changes of the protein and nucleic acid.
   c. Stacking interactions between serine sidechains and nucleic acid bases.
   d. Stacking interactions between tyrosine sidechains and nucleic acid bases.
   e. Water-mediated hydrogen bonds between protein and nucleic acid.

### Fill in the Blank

8. ____________ residues, identified by mutating residues to alanine, contribute a larger than expected energy to the interaction affinity of human growth hormone with human growth hormone receptor.

9. The kinase Zap-70 contains two _______ domains that bind to two __________ residues in its targets.

10. Proteins distinguish double-helical RNA and DNA by differences in _______ shape.

11. Protein–protein interfaces typically bury approximately ____________ in surface area.

12. Examining complexes of nucleic acids and their binding proteins reveals a high ____________ in both shape and charge.

### Quantitative/Essay

(Assume $T = 300$ K and $RT = 2.5$ kJ•mol$^{-1}$ for all questions.)

13. A single-molecule microscopy experiment is performed on a slide containing 1000 molecules each of green, red, cyan, and yellow fluorescent proteins (GFP, RFP, CFP, and YFP, respectively). The experiment measures the number of antibodies bound to each type of protein, finding that 900 molecules of GFP, 10 of RFP, 1 of CFP, and 3 of YFP are bound at 1 nM concentration of antibody.

    a. What is the specificity of the antibody for GFP?
    b. What is the $K_D$ of the GFP–antibody interaction?

14. A scientist wants to engineer an antibody to distinguish between two proteins (Cyclophilin A and Cyclophilin B) with a specificity of 500 at 1 nM concentration for each protein. Her starting material is an antibody that binds with 10 nM $K_D$ to both proteins. She finds that she can easily make mutations that decrease the affinity for Cyclophilin B without affecting the affinity for Cyclophilin A. When she achieves the desired specificity, what is the $K_D$ for Cyclophilin B?

15. A tetracycline repressor (TetR) protein, which is present in *E. coli* at $10^{-8}$ M concentration, binds to the tetO site with a $K_D$ of $10^{-10}$ in the absence of the drug tetracycline and a $K_D$ of $10^{-6}$ in the presence of tetracycline. There is one tetO site in the *E. coli* genome, giving a concentration of $10^{-9}$ M per cell. There are $2 \times 10^5$ nonmatching sites per cell (a concentration of $5 \times 10^{-2}$ M), to which TetR binds nonspecifically with a $K_D$ of $10^{-5}$.

    a. What is the specificity for the tetO site over the nonmatching sites when no tetracycline is present?
    b. What is the specificity for the tetO site over the nonmatching sites when tetracycline is present?
    c. How can TetR repressors switch transcription with such low specificity values?

**16.** A zinc finger protein is isolated from a yeast cell. The value of $K_D$ for its binding site is 3 μM. In the presence of glucose, the protein dimerizes and recognizes an inverted repeat binding site.

a. What is the expected value of $K_D$ if the binding is additive?

b. The dimeric $K_D$ is measured at 5 nM. Why does this value deviate from the expected $K_D$?

**17.** A tryptophan residue near the periphery of a protein–protein interface is mutated to alanine and changes the $K_D$ of binding from 1 nM to 40 μM at 300 K.

a. How much binding energy was contributed by that residue?

b. Explain whether or not the tryptophan residue is likely a hot spot residue.

**18.** A protein–protein interface has a 10 nM affinity at 300 K. A series of mutants are made in which each residue at the interface is replaced by alanine. A lysine residue at the center of the interface is mutated, and found to contribute 4 kJ•mol$^{-1}$ to the binding free energy.

a. What is the new $K_D$?

b. Explain whether or not the lysine residue is a hot spot residue.

**19.** The transcription factor FraJ binds a poly-A DNA sequence with a 10 nM $K_D$ and a poly-G DNA sequence with a 27 μM $K_D$. Mutation of a critical Phe residue to Ala results in a loss of 20 kJ•mol$^{-1}$ in binding free energy for the poly-A sequence, but only a loss of 4 kJ•mol$^{-1}$ on binding to the poly-G sequence. What is the change in specificity for the poly-A sequence over the poly-G sequence at $10^{-8}$ M concentration of FraJ at 300 K?

**20.** A protein–protein interface comprises 22 residues at the contact surface. From structures of the isolated proteins, it is expected that completely burying these residues would cause a surface area reduction of ~2000 Å$^2$. However, a structure of the interface reveals that only 1200 Å$^2$ of surface area is buried. Why is there a discrepancy between the expected and measured surface area reductions?

**21.** Consider the dsRB domain and its potential for interacting with DNA and RNA (see Figure 13.32). What is the predicted effect on the specificity and affinity of recognition for the two types of nucleic acids of:

a. An Arg to Ala mutation at the binding interface?

b. Insertion of loop residues that change the relative spacing of helix A and helix B?

**22.** A DNA-binding domain binds the sequence GATCGCAA**T**ATCGATCGATC with a 25 nM affinity. A mutation of an Arg to Ala in the protein or a mutation of the underlined "T" to "G" in the DNA sequence both result in a 9 kJ•mol$^{-1}$ loss of binding free energy. Simultaneous mutation of both the protein and the DNA also results in a 9 kJ•mol$^{-1}$ loss of binding free energy.

a. What is the effect on the $K_D$ of any of these mutations?

b. What does the double mutant result suggest about the structural basis for the protein–DNA interaction?

**23.** Each subunit of a homodimeric transcription factor can individually recognize a DNA half-site with a 5 μM $K_D$. The dimeric form of the transcription factor recognizes the full inverted repeat DNA site with a 50 nM $K_D$. How much free energy is used to induce the conformational changes of the protein and DNA during the binding of the dimeric transcription factor?

**24.** A complex of seven transcription factors binds a DNA enhancer element. The binding is cooperative. What are two molecular mechanisms that the transcription factors might use to achieve this cooperativity?

## Further Reading


Eaton BE, Gold L & Zichi DA (1995) Let's get specific: The relationship between specificity and affinity. *Chem. & Biol.* 2, 633–638.


## References

### A. Affinity and Specificity


Mohammadi M, Olsen SK & Ibrahimi OA (2005) Structural basis for fibroblast growth factor receptor activation. *Cyt. & Growth Fact. Rev.* 16, 107–137.


### B. Protein–protein Interactions


Clackson T & Wells JA (1995) A hot spot of binding energy in a hormone-receptor interface. *Science* 267, 383–386.

LoConte L, Chothia C & Janin J (1999) The atomic structure of protein–protein recognition sites. *J. Mol. Biol.* 285, 2177–2198.

Pearce KH, Cunningham BC, Fuh G, Teeri T & Wells JA (1999) Growth hormone binding affinity for its receptor surpasses the requirements for cellular activity. *Biochemistry* 38, 81–89.

Scott JD & Pawson T (2009) Cell signaling in space and time: where proteins come together and when they're apart. *Science* 326, 1220–1224.

Sheinerman FB & Honig B (2002) On the role of electrostatic interactions in the design of protein-protein interfaces. *J. Mol. Biol.* 318, 161–177.


### C. Recognition of Nucleic Acids by Proteins


Clery A, Blatter M & Allain FH (2008) RNA recognition motifs: boring? Not quite. *Curr. Op. Struct. Biol.* 18, 290–298.

Draper DE (1995) Protein–RNA recognition. *Annu. Rev. Biochem.* 64, 593–620.

Harrison SC (1991) A structural taxonomy of DNA-binding proteins. *Nature* 353, 715–719.

Kao-Huang Y, Revzin A, Butler AP, O'Conner P, Noble DW & von Hippel PH (1977) Nonspecific DNA binding to genome-regulating proteins as a biological control mechanism: Measurement of DNA-bound *Escherichia coli lac* repressor *in vivo*. *Proc. Natl. Acad. Sci. USA* 74, 4228–4232.

Klug A (2010) The discovery of zinc fingers and their applications in gene regulation and genome manipulation. *Ann. Rev. Biochem.* 79, 213–231.

Marmorstein R & Fitzgerald MX (2003) Modulation of DNA-binding domains for sequence specific DNA recognition. *Gene* 304, 1–12.

Murphy IV FV & Churchill MEA (2000) Nonsequence-specific DNA recognition: A structural perspective. *Structure* 8, R83–R89.

Patikoglou G & Burley SK (1997) Eukaryotic transcription factor–DNA complexes. *Annu. Rev. Biophys. Biomol. Struct.* 26, 289–325.

Rohs R, Jin X, West SM, Joshi R, Honig B & Mann RS (2010) Origins of specificity in protein-DNA recognition. *Ann. Rev. Biochem.* 79, 233–269.

Seeman NC, Rosenberg JM & Rich A (1976) Sequence-specific recognition of double helical nucleic acids by proteins. *Proc. Natl. Acad. Sci. USA* 73, 804–808.

Valverde R, Edwards L & Regen L (2008) Structure and function of KH domains. *FEBS J.* 275, 2712–2726.

# CHAPTER 14

# Allostery

Protein molecules that play critical cellular roles are usually switchable—that is, their function can be turned on or off. This modulation of activity can occur through covalent modification, such as phosphorylation, or through the binding of ligand molecules. In either case, the active site of the protein is altered in such a way that the activity of the protein changes. In the most general sense, such communication between two sites in a protein (the active site and the site of modification or binding) is referred to as **allostery**. Allosteric regulation involves the stabilization of one conformation of the protein with respect to alternative conformations, thereby altering the activity of the protein.

Allosteric proteins are critical for the responsiveness of cells to external signals. Allostery is also important for the regulation of metabolic pathways. For example, as the levels of a metabolite molecule build up in the cell, the molecule binds to enzymes involved in its synthesis and turns them off allosterically. Such systems are discussed in Chapter 16. Allostery was originally discovered in the oxygen transport protein hemoglobin, where it increases the efficiency of oxygen uptake and release. We shall discuss hemoglobin in detail in this chapter.

Before beginning a discussion of allostery in hemoglobin, we introduce the concept of ultrasensitivity, which refers to situations in which the response of a protein to an input is sharper than expected from the hyperbolic response that results from a simple binding equilibrium. Ultrasensitivity in molecular systems can arise from many different mechanisms, but allosteric proteins are usually at the heart of the mechanism.

We first illustrate the concept of ultrasensitivity by discussing how bacteria switch from directed movement to tumbling movements in response to the presence of chemical attractants or repellants. In another example, also drawn from signal transduction, we describe how sets of coupled protein kinases give rise to ultrasensitive responses in signaling systems. We then turn to the major focus of the chapter, which is to illustrate the basic principles of allostery using oxygen transport as an example. We shall see that hemoglobin is allosteric, and that its structure changes in response to the binding of oxygen. This structural change makes hemoglobin ultrasensitive to oxygen concentration and a much more efficient oxygen transporter than it would be otherwise.

## A. ULTRASENSITIVITY OF MOLECULAR RESPONSES

### 14.1 Molecular outputs that depend on independent binding events switch from on to off over a 100-fold range in input strength

Cells continually make decisions that affect how they respond to their environment and grow. The decision-making circuits in cells typically consist of proteins (or, in some cases, RNAs) that process "inputs" and generate "outputs," as illustrated in **Figure 14.1**. At the heart of each such processing unit is an interaction between a protein (or RNA) and another molecule that binds to it (a ligand). The output generated by such a unit depends on the strength of the interaction between the protein and the ligand. Hence, an important aspect of understanding how such processing units work is the thermodynamics of binding interactions, which was the focus of Chapters 12 and 13.

There are many examples of proteins that are components of decision-making circuits (see Figure 14.1). Every enzyme in the cell can be thought of in this way, because an enzyme binds to a substrate (the "input") and generates a product (the "output"). Cell-surface receptors, such as the epidermal growth factor receptor (see Section 12.13) are activated when they bind to hormones, and the outputs of the activated receptor cause changes in cellular behavior. Transcription factors that bind to DNA alter the rates at which genes are transcribed, depending on how strongly they are bound to their target DNA elements.

In the simplest case, the output generated by a protein is proportional to the amount of complex formed by the input ligand molecule and the protein. The rate of an enzyme-catalyzed reaction, for example, is proportional to the concentration of the enzyme–substrate complex (see Chapter 16). If the binding of one ligand molecule does not affect the binding of the others (that is, if the binding events are independent), then the formation of the complex as a function of ligand concentration is described by a hyperbolic binding isotherm (see Figure 12.4). In such a situation, the output as a function of input strength is also hyperbolic (**Figure 14.2A**).

**Graded response**

An output function that depends on the input in a hyperbolic fashion, as in a simple binding equilibrium, is known as a graded, or linear, response. In such a response, the output switches from ~10% to ~90% of the maximum response over a 100-fold change in input strength.

Recall from Section 12.10 that the fractional saturation, $f$, of a protein increases from 0.1 to 0.9 as the ligand concentration is increased ~100-fold, from ~$0.1K_D$ to $10K_D$, where $K_D$ is the dissociation constant for the interaction. This is illustrated in Figure 14.2B, in which the input strength (for example, concentration of the ligand) is shown on a logarithmic scale. As made clear in Figure 14.2B, if the binding events are independent, then the protein responds gradually to increases in the concentration of the ligand, switching from off (unbound) to on (bound) over two log units of concentration. Such a response is referred to as a **graded response**. Graded responses are also known as **linear responses** because the output is a linear function of the input when the input strength is low, as you can see by considering Equation 12.11.

## 14.2 The response of many biological systems is ultrasensitive, with the switch from off to on occurring over a less than 100-fold range in concentration

A graded response is much too sluggish to be useful for many biological systems. The ~100-fold increase in the input signal strength (for example, concentration of the ligand) that is required to switch the system from on to off has several undesirable features. First, in living systems, the concentrations of molecules do not usually change by 100-fold as processes occur. One example, which we study in more detail in part B of this chapter, is the function of hemoglobin in transporting oxygen. There is only a three-fold difference in dissolved oxygen concentration between the lungs and the tissues, and if the response of hemoglobin to oxygen concentration were graded, it would be a very inefficient oxygen transporter.

Second, cells have to be able to respond decisively to small changes in the concentrations of the molecules that they detect. A graded response is not very sensitive to small changes in concentration, and it does not produce a switch-like response to an input signal. The need for a sharp, switch-like response can be understood by imagining that we were in a house that was catching fire. Our noses are able to detect trace amounts of the molecules that are in smoke and our nervous system galvanizes us to make a rapid response as soon as smoke is detected. A switch-like response is essential for both the rapid response and for preventing too many false alarms. A graded response in our noses would lead to unfortunate outcomes.

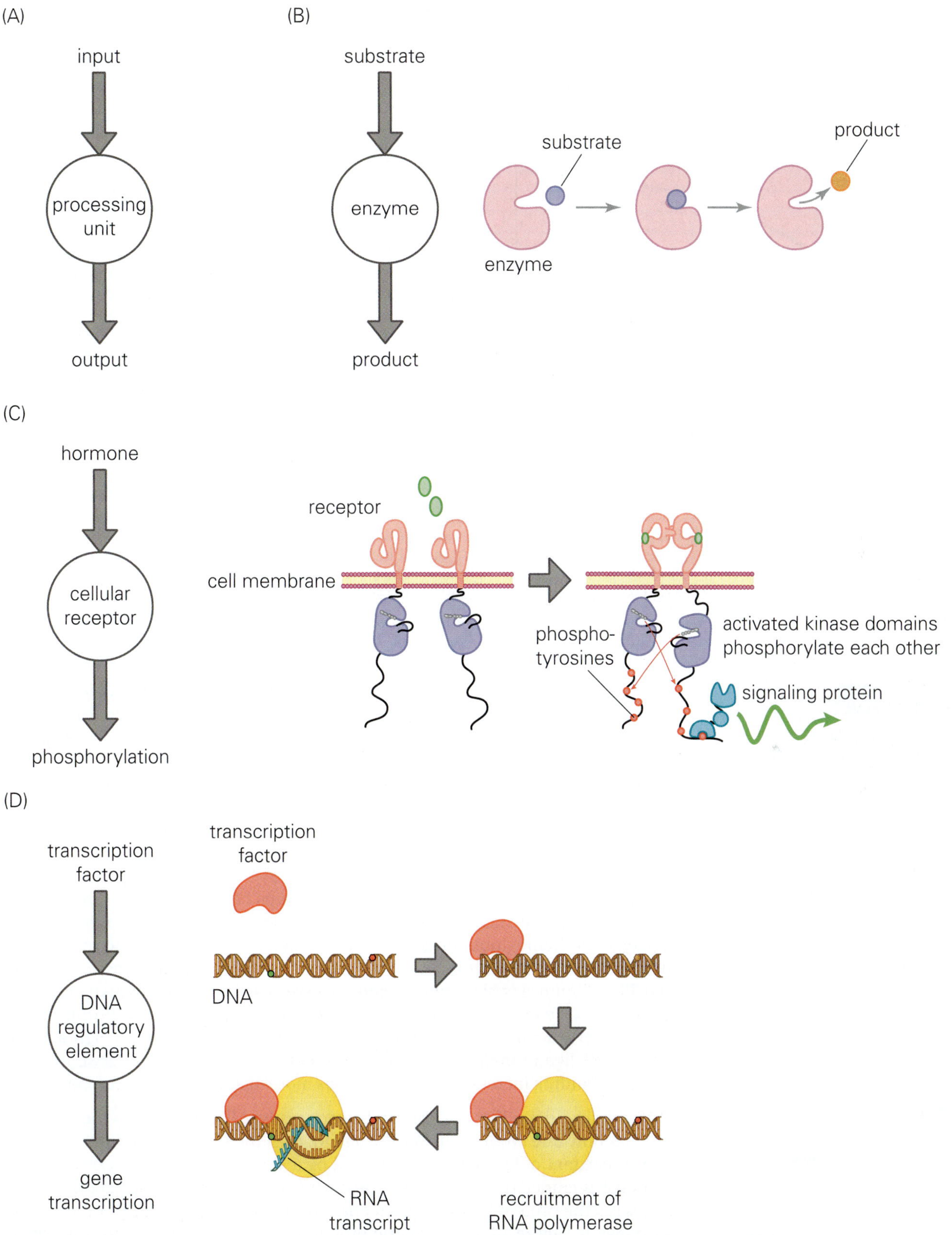


**Figure 14.1 Proteins as input processing units.** (A) Schematic representation of the conversion of inputs into outputs. (B) The substrates of enzymes can be thought of as inputs, with their products as the outputs. (C) Cell signaling systems respond to inputs, such as hormones, and generate responses, such as phosphorylation (refer to Figure 12.15 for a more detailed discussion of the receptor shown here). (D) Transcription factors bind to regulatory elements in DNA and help recruit RNA polymerase. This initiates transcription, as explained in Chapter 1. (The concepts illustrated here are based on a discussion in D. Bray, *Nature* 376: 307–312, 1995.)

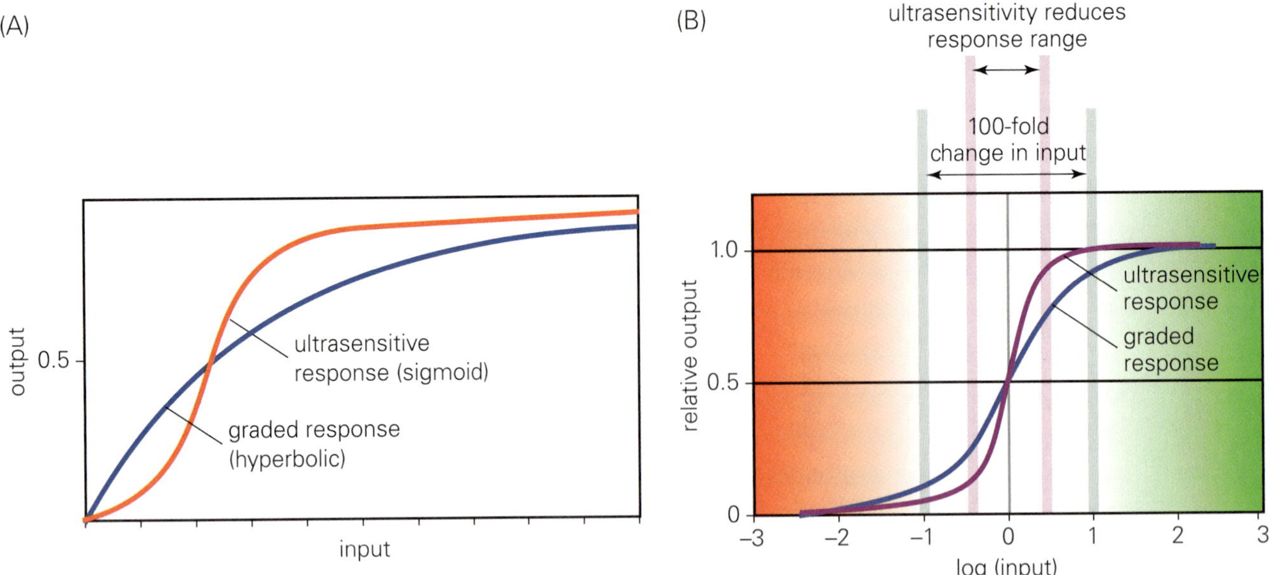


**Figure 14.2 Graded and ultrasensitive responses.** (A) A graded response results from independent binding events between input molecules and the protein. The result is a hyperbolic response curve (*blue*). An ultrasensitive response has a sigmoid rather than a hyperbolic binding curve (*red*) and results from cooperativity in binding events. (B) Graded and ultrasensitive response curves, with the input strength plotted on a logarithmic scale. A protein with a graded response switches from output levels that are at 10% of the maximum to 90% of the maximum over a ~100-fold span of input strength (see Figure 12.11). An ultrasensitive response is sharper, and switches from off to on over a compressed concentration range.

**Ultrasensitivity**

An ultrasensitive system is one in which the response to an input is sharper than expected from a simple binding equilibrium. For example, the response of a protein to a ligand can be defined as the fractional saturation, *f* (see Chapter 12). When *f* rises from 0.1 to 0.9 over a less than ~100-fold concentration range of the ligand, the system is said to be ultrasensitive.

**Cooperativity**

When the binding of a ligand to a protein is ultrasensitive, the binding is said to be cooperative. As more ligand molecules bind to the protein, the saturation of the protein increases more sharply than would be expected for a normal binding event, as if the ligand molecules "cooperate" with each other.

Most of the key systems in cells are **ultrasensitive**, rather than graded, in their response to input signals. An ultrasensitive system is one that switches from *off* to *on* over a less than ~100-fold change in the strength of the input signal. The response of an ultrasensitive system to changes in the input level is illustrated in Figure 14.2B.

## 14.3 Cooperativity and allostery are features of many ultrasensitive systems

There are many ways in which molecular systems can display ultrasensitivity. If the underlying mechanism is governed only by binding events, as in the binding of oxygen to hemoglobin or a transcription factor to DNA, then the response is determined by the binding isotherm for the interaction. In such cases, an ultrasensitive response is also called a **cooperative response**. Cooperativity refers to the fact that the extent of binding increases more sharply than for a normal binding isotherm, as the concentration of the ligand increases (refer to Figure 14.2). That is, the ligand molecules appear to "cooperate" with each other so that the extent of binding increases more sharply as more ligand molecules are bound.

How can ligand binding to a protein be cooperative? If the protein is monomeric, and has only one binding site for the ligand, then the system cannot be cooperative. The extent of binding is determined by a simple binding isotherm, and the response is graded (**Figure 14.3**A). The most common way in which cooperativity arises is if the protein has more than one binding site for the ligand. For example, as shown in Figure 14.3B, a dimeric protein has two binding sites for the ligand per dimeric unit. Now imagine that in the absence of ligand both binding sites are closed, so that it is difficult for ligand molecules to bind. The protein changes conformation when the ligand binds to it, thus opening up one binding site. If the dimer is so constructed that the conformational change that occurs when ligand binds to one subunit of the dimer is transmitted to the other subunit, then the second binding site opens up before a ligand has bound. This facilitates the binding of a second ligand molecule to the dimer, and cooperative binding results (see Figure 14.3B).

The transmission of the conformational change from one binding site to another is an example of **allostery**. An allosteric protein is one in which the activity of the protein is modulated by interactions that occur at a distance from the active site. An allosteric protein has a catalytic center or binding site ("active site"), and another site where a ligand can bind or where covalent modification of the protein occurs ("allosteric site"). The properties of the active site depend on the state of the allosteric site. Allosteric proteins are often oligomeric, as in the example shown in Figure 14.3B. In this case, each of the binding sites in the oligomer is also an allosteric site that can communicate to the other binding sites.

**Allostery**

An allosteric protein is one in which the activity of the protein is modulated by interactions that occur at a distance from the active site.

In the example shown in Figure 14.3B, cooperativity arises from conformational changes (that is, allosteric changes) in the structure of the protein. In principle, cooperative binding can occur without conformational changes in the protein or the ligand. Consider, for example, the binding of a transcription factor to a DNA segment with two binding sites for the transcription factor that are inverted with respect to each other (Figure 14.3C). The first protein to bind to the DNA does so weakly, but when the second protein binds, it can interact with the first protein as well as with the DNA. The additional interactions with the first protein increase

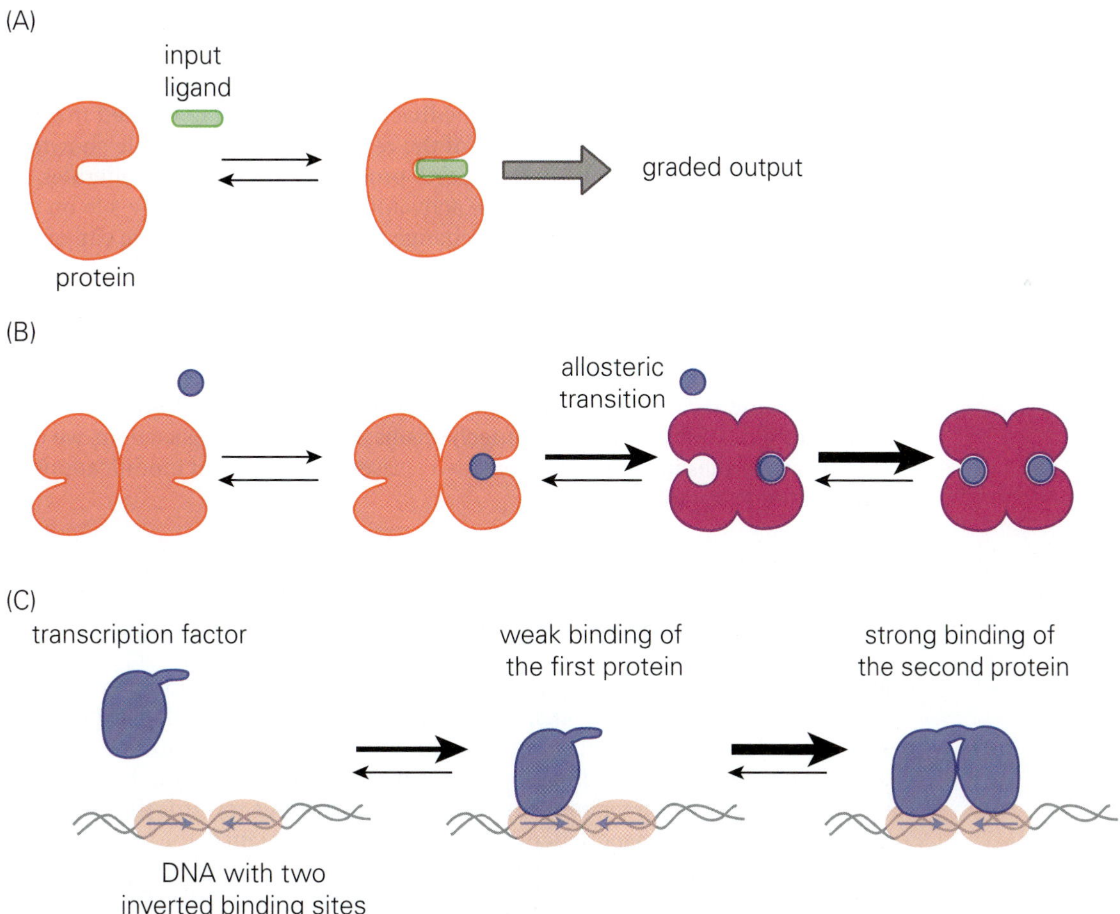


**Figure 14.3 Cooperativity and allostery.** (A) A monomeric protein with a single binding site cannot display ultrasensitivity (the output is graded). (B) A dimeric protein with two binding sites can display ultrasensitivity if the conformations of the two binding sites are coupled. Ligand binding to one site causes an allosteric change in the other, thereby increasing affinity for the ligand. Ligand binding to such a system is cooperative. (C) Cooperative binding can occur without allostery. In this example, a transcription factor has two closely spaced and inverted binding sites on DNA. If two molecules of the transcription factor interact with each other when bound to DNA, then cooperative binding results without necessitating a conformational change.

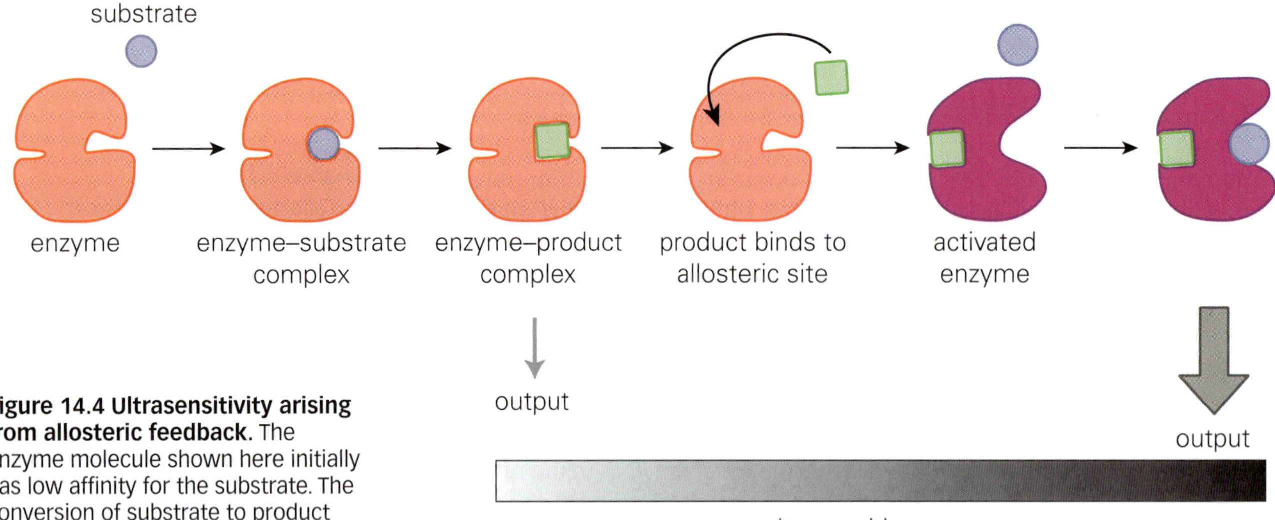


**Figure 14.4 Ultrasensitivity arising from allosteric feedback.** The enzyme molecule shown here initially has low affinity for the substrate. The conversion of substrate to product activates the enzyme because the product molecule can bind to an allosteric site on the enzyme, causing a conformational change that opens up the active site.

the affinity of the second protein for DNA, leading to cooperative binding. Cooperativity in this case need not be associated with allosteric changes in the protein or the DNA, although it usually is.

Finally, ultrasensitivity can also arise through mechanisms other than cooperative binding. Many different kinds of feedback mechanisms are utilized in cells, and these can increase the sharpness of the response. In the example shown in **Figure 14.4**, the conversion of substrate to product by an enzyme generates an allosteric activator of the enzyme. The activator binds to an allosteric site on the enzyme and causes a conformational change that opens up the active site of the enzyme, thereby speeding up the rate of catalysis (see Chapter 16).

To summarize, although the binding of ligands to proteins is a key step in cellular processes, the response that results is often sharper than expected from independent binding events. This sharpness in the response is known as ultrasensitivity, and it can arise from cooperativity in binding, from allosteric changes in the protein, and from feedback mechanisms. In the following sections we discuss two examples in which ultrasensitivity arises from covalent modifications of enzymes that alter their activity.

## 14.4 Bacterial movement towards attractants and away from repellants is governed by signaling proteins that bind to the flagellar motor

Many processes in biology involve the response of cellular machinery to changes in the concentrations of molecules. For example, bacteria exhibit a behavior known as **chemotaxis**, which is a directed movement towards sources of food and away from toxins. Bacteria normally move in an erratic fashion, but if a food source is placed in their vicinity, they are observed to move towards the food. This directed motion is a consequence of the correlated movement of the bacterial flagella, hair-like structures on the cell surface, which all rotate in a counterclockwise direction and propel the bacterium forward in a whip-like fashion (**Figure 14.5**A; flagella are also illustrated in Figure 1). If there is a chemical repellant present, instead, then the flagella start to rotate in a clockwise fashion and disengage from each other. The uncorrelated rotations of the disengaged flagella cause the bacteria to tumble and, eventually, to move away from the repellant (Figure 14.5B).

Each bacterial flagellum is driven by a motor protein that is embedded in the cell wall (Figure 14.5C). The direction of rotation of the flagellum is controlled by a

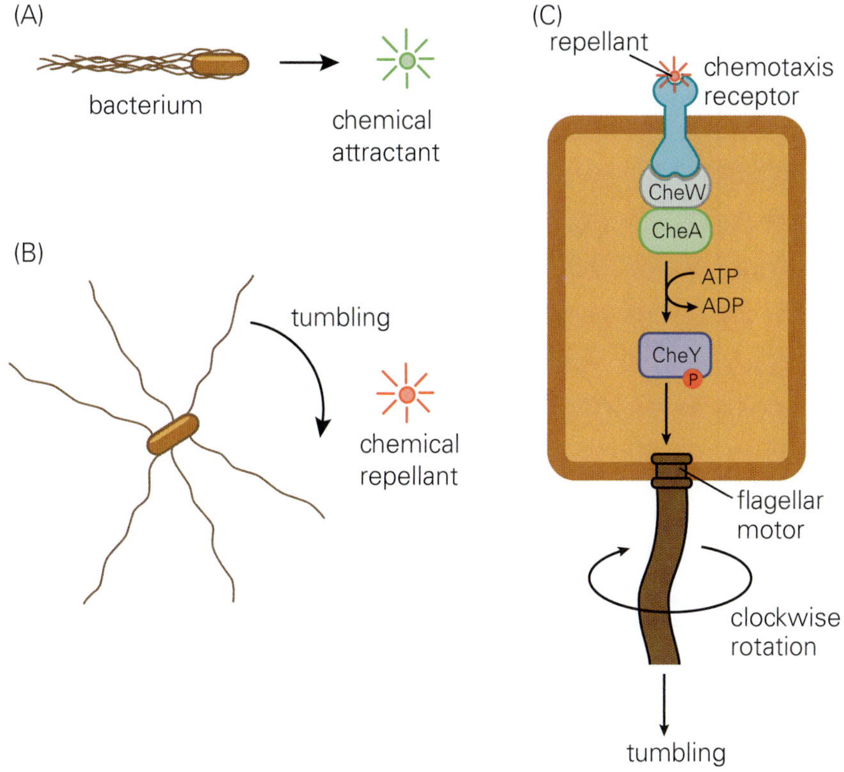


**Figure 14.5 The chemotaxis response in bacteria.** (A) When a chemical attractant (for example, a food source) is in the vicinity, bacteria swim towards the attractant. This motion is a consequence of the counterclockwise rotation of the bacterial flagella, which are hair-like structures on the surface of the cell. (B) In contrast, the presence of a repellant (for example, a toxin) causes the bacteria to tumble. This is because of clockwise and uncorrelated rotation of the flagella. After tumbling the bacteria swim forward again, but only continue to do so if moving away from the repellant. Moving toward repellant causes more frequent tumbling. This yields a net movement away from repellant. (C) Generation of the tumbling response in bacteria. Binding of a repellant molecule to a cell-surface receptor results in the phosphorylation of a protein known as CheY. Phosphorylated CheY binds to the flagellar motor protein of each flagellum and causes it to rotate in a clockwise fashion. This causes each flagellum to move apart from the others and results in tumbling. (Adapted from B. Alberts et al. Molecular Biology of the Cell, 5th ed. New York: Garland Science, 2008.)

protein known as CheY. When CheY is bound to the motor protein, the flagellum rotates in a clockwise fashion and tumbling results. When CheY is disengaged, the rotation is in a counterclockwise fashion. In order for CheY to bind to the motor protein, it has to be phosphorylated on an aspartate residue, and this phosphorylation is brought about as a consequence of a repellant molecule binding to a receptor on the surface of the cell. The signal transduction pathway connecting the chemotaxis receptor to the flagellar motor is shown in Figure 14.5C.

## 14.5 The flagellar motor switches to clockwise rotation when the concentration of CheY increases over a narrow range

When repellant molecules bind to the chemotaxis receptor, it results in the phosphorylation of CheY molecules in the cell. The concentration of phosphorylated CheY, [CheY-P], determines the probability that CheY is bound to the flagellar motor proteins, which determines, in turn, whether clockwise rotation is initiated or not.

The concentration of CheY-P in individual bacterial cells has been measured in a clever experiment. The gene for the CheY protein was fused to the gene for a green fluorescent protein (GFP), and the fused gene was inserted into cells that lack CheY. By measuring the amount of green fluorescence emanating from the cell, the concentration of CheY protein in the cell can be estimated. There is no easy way to distinguish CheY from CheY-P in individual cells, but it is known that when the repellant molecules bind to receptors, all of the CheY molecules are eventually converted to the phosphorylated form—that is, [CheY] = [CheY-P]. The CheY-GFP gene is under the control of an inducible promoter, and the levels of CheY-GFP protein can be varied by controlling the level of gene expression.

The experimentally observed relationship between [CheY-P] and the extent of clockwise rotation of the flagella is shown in **Figure 14.6**. There is very little clockwise rotation when [CheY-P] is less than ~2 μM. As the concentration increases

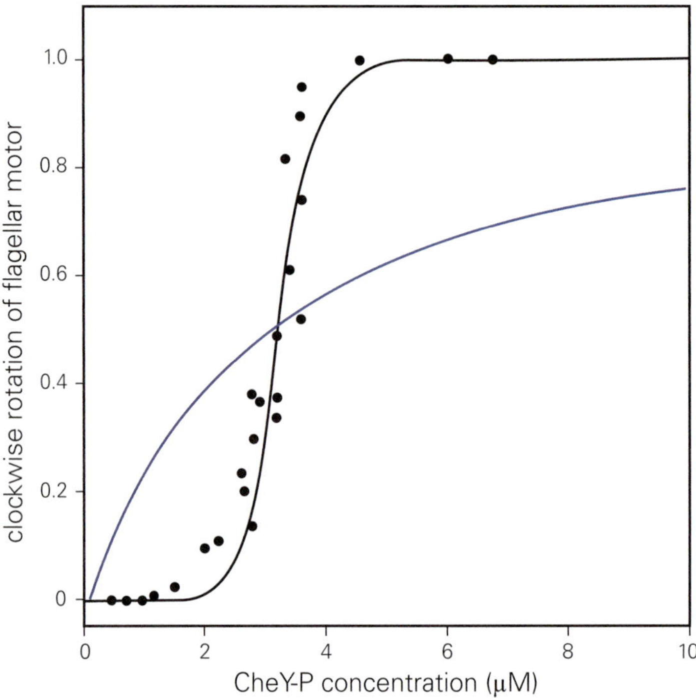


**Figure 14.6 Ultrasensitivity in bacterial chemotaxis.** Experimental measurements of the extent of clockwise flagellar rotation in individual bacteria are plotted against the concentration of the phosphorylated form of the signaling protein CheY (CheY-P). Notice the sharp switch-like response of the flagellar motor. The concentration of CheY-P at which the response is half-maximal is ~3 μM. The binding isotherm corresponding to a $K_D$ value of 3 μM is shown as a *blue line*. (Adapted from P. Cluzel, M. Surette, and S. Leibler, *Science* 287: 1652–1655, 2000. With permission from the AAAS.)

beyond this point, clockwise rotation switches on and reaches a maximal value when [CheY-P] is ~4 μM. The concentration of CheY-P at which the extent of clockwise rotation is half-maximal is ~3 μM. The transition from counterclockwise rotation of the motor is initiated and completed within a two-fold change in [CheY-P]—that is, the response is ultrasensitive.

## 14.6 The response of the flagellar motor to concentrations of CheY is ultrasensitive

Because the extent of clockwise rotation is governed by the binding of CheY-P to the flagellar motor, the response curve shown in Figure 14.6 can be interpreted in terms of a binding isotherm (see Chapter 12). The ligand in this case is CheY-P, and the receptor is the flagellar motor. The concentration of CheY-P at which the response is half-maximal ([CheY-P] = 3 μM) should then correspond to the value of the dissociation constant, $K_D$, for the binding interaction.

If the interaction between CheY-P and the motor protein can be explained by a simple binding equilibrium, then we can calculate the expected binding isotherm by knowing the value of $K_D$, as explained in Chapter 12 (see Equation 12.11):

$$f = \frac{[L]}{K_D + [L]} = \frac{[\text{CheY-P}]}{3.0 \times 10^{-6} + [\text{CheY-P}]} \tag{14.1}$$

Here $f$ is the fractional saturation of the motor protein and should be proportional to the extent of clockwise rotation. The values of $f$, calculated from Equation 14.1, are shown in Figure 14.6 as a function of [CheY-P] and display the expected hyperbolic behavior.

Note how poorly the calculated values of $f$ correlate with the experimentally observed response of the flagellar motors. Although both curves have their half-maximal values when [CheY-P] is equal to 3 μM, the calculated value of $f$ is higher than the response for concentrations below $K_D$, but lower than the actual response when the concentration is higher than $K_D$. The actual behavior of the flagellar motor when the concentration of CheY-P changes resembles a sharp, switch-like response and is obviously ultrasensitive. The motor does not turn on until concentration of CheY-P has reached a certain level, but then the motor turns on decisively, over a narrow concentration range.

Decisive molecular switches, such as the activation of clockwise rotation in the flagellar motor, are very important in biology because they are less sensitive to noise (small amounts of the input signal do not evoke the response) and are efficient (once a threshold concentration is reached, the system switches on the desired process).

## 14.7 The MAP kinase pathway involves the sequential activation of a set of three protein kinases

The discussion of the direction of rotation of the bacterial flagellar motor introduced the concept of ultrasensitivity, but did not provide a connection to allostery in proteins. The detailed molecular processes underlying ultrasensitivity in the flagellar motor system are still not completely understood. We shall now consider another ultrasensitive signaling system, one involving the coupled activation of protein kinases. This system, known as the MAP kinase pathway (for mitogen-activated protein kinase; a mitogen is a molecule that causes cells to divide) is a very important signaling device in all eukaryotic cells.

The MAP kinase pathway relies on three different but closely related protein kinases that phosphorylate serine or threonine residues in proteins. One of them, known simply as MAP kinase, operates at the lowest level of the pathway (**Figure 14.7**). MAP kinase phosphorylates proteins such as transcription factors, and thereby evokes a cellular response to an input signal.

MAP kinase is activated by phosphorylation. This phosphorylation is carried out by a second kinase in the pathway, known as MAP kinase kinase. MAP kinase kinase is itself turned on by a third kinase in the pathway, MAP kinase kinase

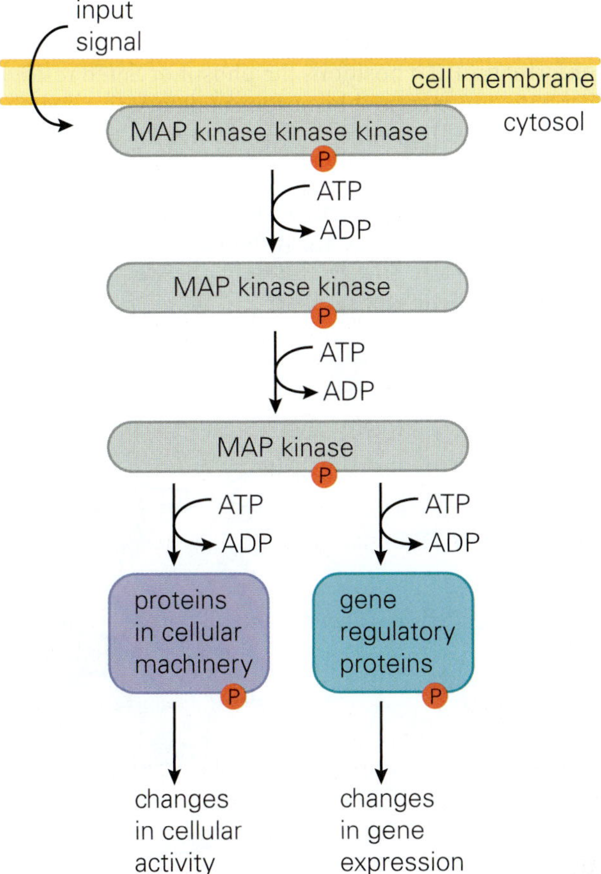


**Figure 14.7 The MAP kinase pathway.** There are three protein kinases in this pathway, known as MAP kinase, MAP kinase kinase, and MAP kinase kinase kinase. MAP kinase kinase kinase is activated first, followed by sequential activation of the other two. MAP kinase is lowest in the chain, and it phosphorylates cellular proteins, resulting in a response to the incoming signal.

kinase, which phosphorylates it. MAP kinase kinase kinase is at the top of the chain and is activated by the arrival of an input signal at the cell surface (see Figure 14.7).

## 14.8 Phosphorylation controls the activity of protein kinases by allosteric modulation of the structure of the active site

The MAP kinases are members of the large family of eukaryotic protein kinases, which phosphorylate proteins on serine, threonine, or tyrosine residues. There are about 500 protein kinases in the human genome, all of which share a highly conserved catalytic domain, known as the kinase domain (kinase domains were discussed in Sections 12.12 and 12.13). Many kinases, including the MAP kinases, are activated by phosphorylation, which typically occurs upon threonine, serine, or tyrosine residues in a loop known as the activation loop (**Figure 14.8**).

In the active conformation, the catalytic center of the kinase is open and the two substrates (ATP and a peptide chain) can bind to it. Kinases can have many different inactive conformations (see Figure 12.24). In one particular inactive conformation shown in Figure 14.8, access to the catalytic center is blocked by the activation loop. In addition, residues that are required for the catalysis of the phosphate transfer reaction are not in the correct place. For example, a glutamate sidechain that is critical (shown in red in Figure 14.8) is moved away from the catalytic center.

In the unphosphorylated state, the inactive conformation of the kinase domain is lower in free energy than the active conformation (**Figure 14.9**). Phosphorylation of one or more residues in the activation loop stabilizes the active conformation, because of interactions made by the negatively charged phosphate group and positively charged sidechains in the kinase domain. In the case of the kinase domain shown in Figure 14.9, a phosphorylated tyrosine residue in the activation loop interacts with two arginine sidechains in a manner that stabilizes the open conformation of the activation loop. Phosphorylation can also destabilize the inactive conformation, if it positions the phosphorylated residue in a location where the phosphate group cannot be accommodated. This switches the relative free energy of the active and inactive conformations, resulting in the activation of the kinase.

Phosphorylation is just one way in which a kinase can be switched from an inactive to an active state. Some kinases are switched on by interacting with another protein. In such cases the free energy of binding between the two proteins helps stabilize the active form of the kinase.

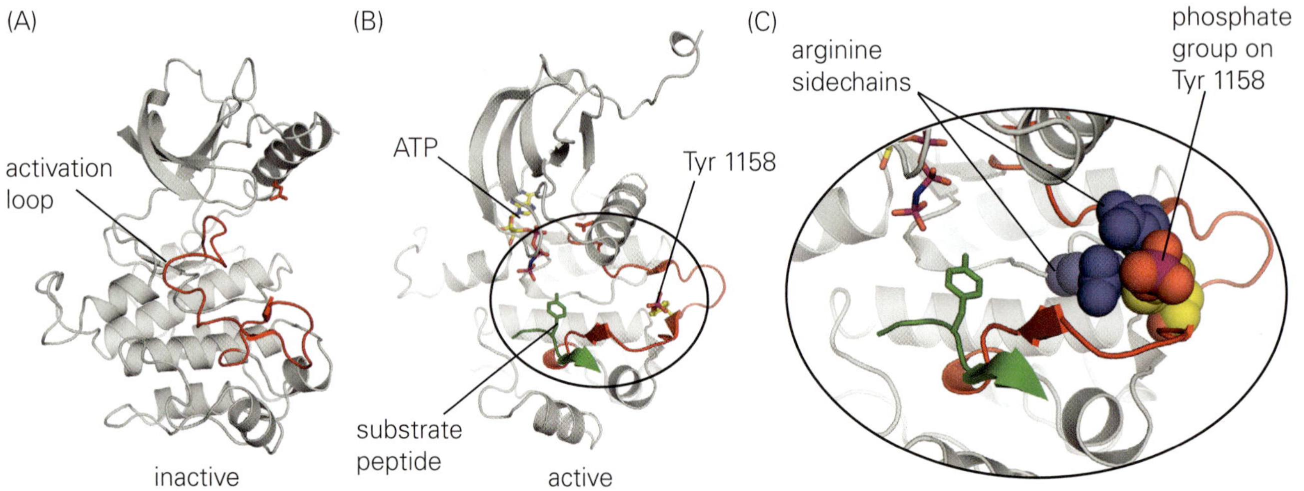


**Figure 14.8 Phosphorylation control of the conformation of a protein kinase.** The inactive (A) and active (B) conformations of a protein kinase are shown here. In the inactive conformation, a central loop (*red*), known as the activation loop, is folded into the active site of the kinase domain. This prevents the binding of the substrates (ATP and a peptide that is to be phosphorylated). In the active conformation, the activation loop is folded out, and supports ATP and peptide substrate binding. The particular kinase domain shown here is that of the insulin receptor tyrosine kinase, which is representative of the highly conserved structure of all eukaryotic protein kinases. (C) Phosphorylation of a tyrosine residue (Tyr 1158) stabilizes the active conformation, because two arginine sidechains in the protein interact with the negatively charged phosphate group. (PDB codes: A, 1IRK; B, 1IR3.)

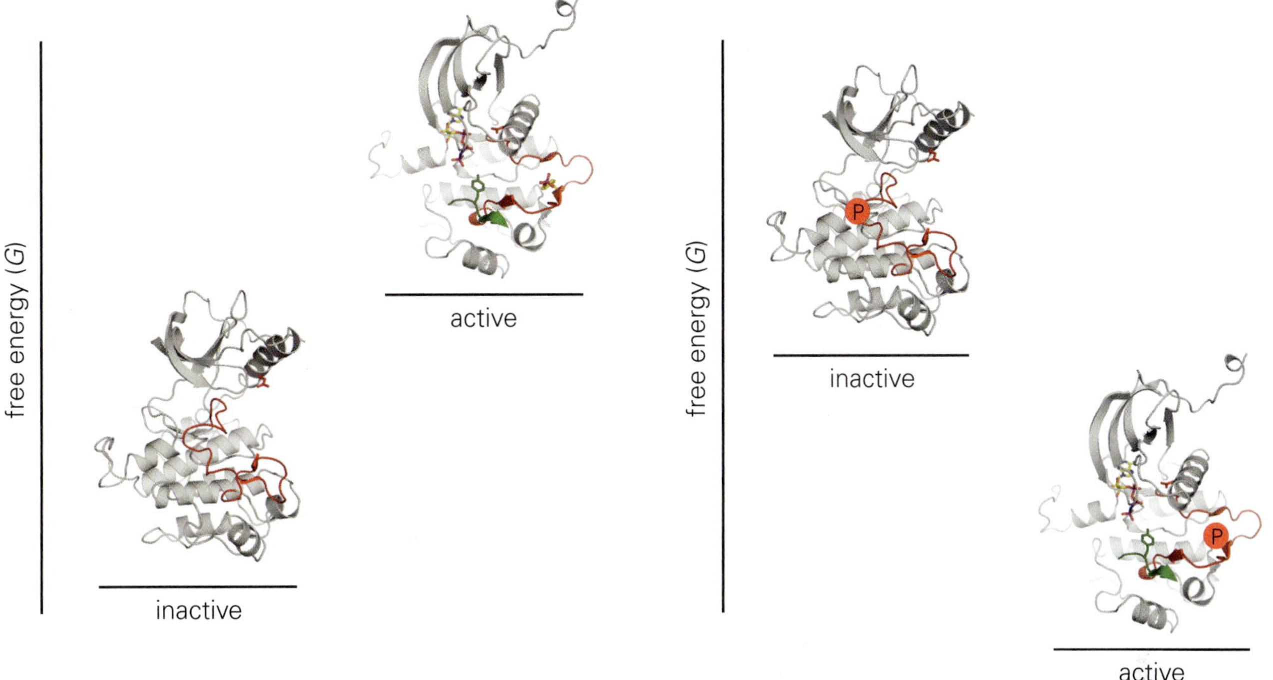


**Figure 14.9 Effect of phosphorylation on the conformations of a protein kinase.** In the unphosphorylated state (left), the inactive conformation of the protein kinase is lower in free energy and is therefore the predominant form. Upon phosphorylation of the activation loop, the active conformation is more stable and the inactive conformation may be destabilized. The net effect of phosphorylation is to increase the activity of the enzyme. (PDB codes: 1IR3 and 1IRK.)

## 14.9 The sequential phosphorylation of the MAP kinases leads to an ultrasensitive signaling switch

The switching behavior of the MAP kinase pathway has been studied in the development of frog eggs, where this pathway controls the transition of the oocyte from a resting state to a stage where meiotic cell division is initiated. The addition of the hormone progesterone initiates this transition by activating a MAP kinase kinase kinase. Progesterone invokes an all-or-none response in the MAP kinase pathway. As the concentration of progesterone in the medium increases, the cell transitions from a state in which MAP kinase is not activated to a state where it is essentially completely activated. States in which the level of MAP kinase activation is at an intermediate stage are rarely observed.

A quantitative analysis of the MAP kinase pathway is facilitated by the fact that it can be initiated artificially by taking cell extracts from resting oocytes and adding an activated MAP kinase kinase kinase. The levels of phosphorylated and unphosphorylated MAP kinase can then be detected using an antibody (**Figure 14.10**). The amount of phosphorylated MAP kinase generated in the cellular extracts upon addition of activated MAP kinase kinase kinase demonstrates the all-or-nothing response (see Figure 14.10). When the concentration of the activating enzyme is 0.02 μM or below, essentially all of the MAP kinase is in the unphosphorylated form. Strikingly, when the concentration of the activating enzyme increases to 0.05 μM, all of the MAP kinase is the phosphorylated form. This transition from an unactivated to an activated form over such a narrow concentration range (that is, a two-fold increase in concentration) is a hallmark of an ultrasensitive process.

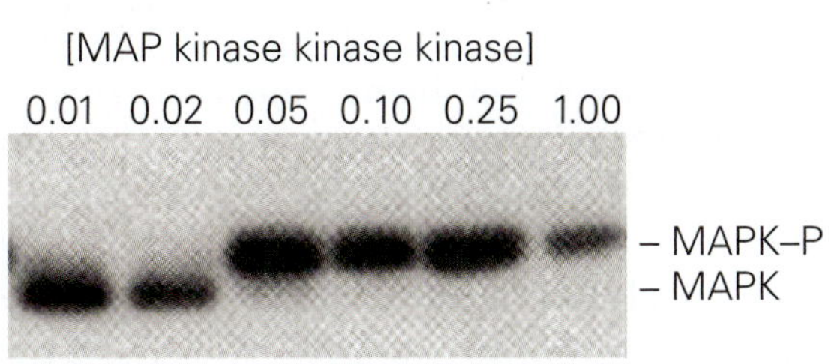


**Figure 14.10 Activation of MAP kinase in frog oocyte extracts by a MAP kinase kinase kinase.** The concentration of the activated MAP kinase kinase kinase is shown above each lane (in μM). The levels of MAP kinase are detected by immunoblotting. Phosphorylation of MAP kinase retards its electrophoretic mobility. Note the sharp transition from unphosphorylated to phosphorylated MAP kinase when the concentration of the activating enzyme is between 0.02 to 0.05 μM. (Adapted from C.Y. Huang and J.E. Ferrell, Jr., *Proc. Natl. Acad. Sci. USA* 93: 10078–10083, 1996. With permission from the National Academy of Sciences.)

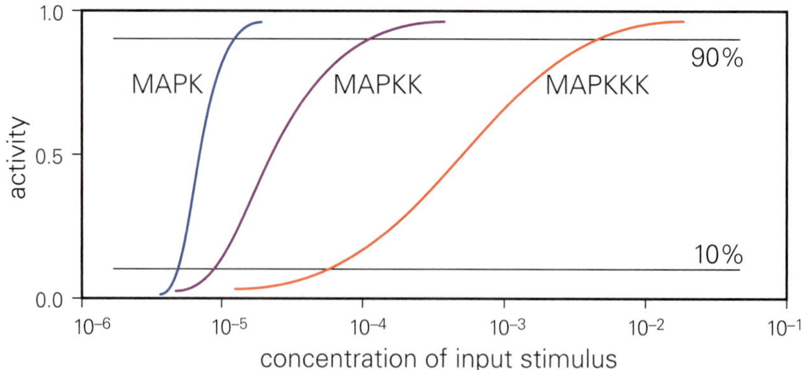


**Figure 14.11 Simulation results for the activation of the MAP kinase pathway.** The concentration of the activating molecule (for example, progesterone) is shown on a logarithmic scale. The calculated activation levels (phosphorylation levels) of the three MAP kinases are shown, with values between 1.0 and 0.0 for maximal and minimal levels of activity. The levels of activity corresponding to 10% and 90% of the maximal values are shown as horizontal lines. (Adapted from C.Y. Huang and J.E. Ferrell, Jr., *Proc. Natl. Acad. Sci. USA* 93: 10078–10083, 1996. With permission from the National Academy of Sciences.)

The origin of the ultrasensitivity can be analyzed by simulating the MAP kinase pathway and calculating the resulting relationship between the input signal and the activation state of MAP kinase. This has been done, and although we shall not discuss the details here, the results are interesting to consider (**Figure 14.11**). For a graded (linear) change in the input strength (for example, the concentration of progesterone), the activation state of the MAP kinase kinase kinase corresponds to a hyperbolic response curve (this curve has an "S" shape in Figure 14.11 because of the use of logarithmic axes; compare with the binding isotherm shown in Figure 14.2B).

The activity of the MAP kinase kinase kinase behaves as if it reflects a simple binding equilibrium between the input signaling molecule and the enzyme. As we shall see in Chapter 16, this kind of behavior is known as Michaelis–Menten kinetics and is a characteristic feature of non-allosteric systems governed by simple binding equilibria.

The response of MAP kinase kinase kinase to the concentration of the input signal is sluggish and requires an 81-fold increase in the signal to go from 10% of the maximal activity to 90%. To see why this is the case, we assume that the response is proportional to the fraction, $f$, of the enzyme that has responded to the signal (in the simple ligand binding case, this is the fraction bound). We use an alternative form of Equation 14.1 (see Equation 12.17):

$$\frac{f}{1-f} = \frac{[\mathrm{L}]}{K_\mathrm{D}} \tag{14.2}$$

For two different fractional saturations, $f_1$ and $f_2$, corresponding to ligand concentrations, $[\mathrm{L}]_1$ and $[\mathrm{L}]_2$, Equation 14.2 yields:

$$\frac{[\mathrm{L}]_1}{[\mathrm{L}]_2} = \left(\frac{f_1}{1-f_1}\right)\left(\frac{1-f_2}{f_2}\right) \tag{14.3}$$

If $f_1$ is 0.9 and $f_2$ is 0.1, then Equation 14.3 tells us that the ratio of the corresponding ligand concentrations is 81 for a system obeying a hyperbolic binding isotherm (in Section 14.2, we had more loosely said that the change in concentration is ~100-fold). This result gives us a definition of ultrasensitivity—namely, an ultrasensitive system is one in which a less than 81-fold (more loosely, ~100-fold) change in the concentration of the stimulus is sufficient to change the response from 10% of the maximal value to 90% of the maximal value.

The response of MAP kinase to the input stimulus is clearly ultrasensitive. Only a 2.5-fold increase in the concentration of the stimulus converts the level of MAP kinase activity from 10% to 90% of the maximal value (see Figures 14.10 and 14.11). MAP kinase kinase shows an intermediate level of ultrasensitivity, requiring a 13-fold increase in the concentration of the stimulus to be activated by the same amount.

The net result of the coupling of the three kinases in the MAP kinase pathway is that it converts a graded (linear) input signal into a sharply defined all-or-nothing switch. At the heart of this process is the control of the activity of the MAP kinases by phosphorylation. Just how the sequential coupling of the activation of the kinases results in ultrasensitivity is well understood, but complicated. We shall not discuss this further, but turn instead to allostery in hemoglobin, which is much simpler to analyze in detail.

# B. ALLOSTERY IN HEMOGLOBIN

## 14.10 Allosteric proteins exhibit positive or negative cooperativity

In Chapter 12, where we studied the thermodynamics of ligand binding, the discussion was centered on binding interactions in which the binding events are independent of each other. That is, the binding of one ligand molecule to a protein is assumed to have no influence on the strength of the interaction between other ligand molecules and their binding sites. Many molecular recognition events in biology, in contrast, are cooperative, with the binding of one ligand molecule to a protein resulting in alterations in the affinity of the protein for other ligand molecules. Such systems are by nature allosteric.

In the most general case, consider a protein, P, with two distinct binding sites (**Figure 14.12**). One site is specific for one kind of ligand, $L_A$, and the other site is specific for a different ligand, $L_B$. If P is allosteric, then the binding of $L_A$ to P affects the strength of the interaction with $L_B$ and vice versa. For example, as illustrated in Figure 14.12, the binding of $L_A$ in the absence of $L_B$ is characterized by a dissociation constant $K_{D1}$. When $L_B$ binds to P, it induces a conformational change such that the structure of the binding site for $L_A$ is altered. If this structural change results in tighter binding, then the dissociation constant for $L_A$ in the presence of $L_B$, $K_{D2}$, is smaller than $K_{D1}$. This phenomenon is known as **positive cooperativity**, because the two ligands cooperate to increase the strengths of the interactions with the protein. Alternatively, if the binding of $L_B$ to P decreases the strength of the interaction of $L_A$ with P, then the phenomenon is known as **negative cooperativity**. Positive cooperativity is much more common in nature than negative cooperativity.

**Positive and negative cooperativity**

When two or more ligands bind to a protein in such a way that they mutually reinforce each of their binding affinities, the phenomenon is called positive cooperativity. If the ligands make it more difficult for each other to bind, the phenomenon is called negative cooperativity.

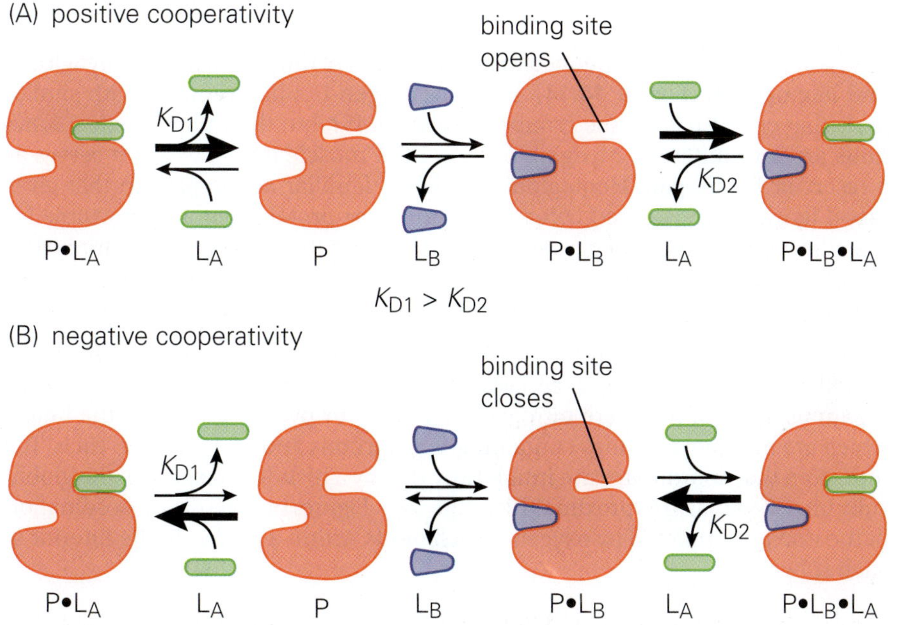


**Figure 14.12 Positive and negative cooperativity.** (A) Positive cooperativity occurs when two ligands, $L_A$ and $L_B$, mutually reinforce their affinities with a protein, P. In the binding scheme shown here, $L_B$ increases the affinity of $L_A$ for P. The reverse is also true, although not shown explicitly. (B) Negative cooperativity occurs when each of the ligands decreases the binding affinity of the other.

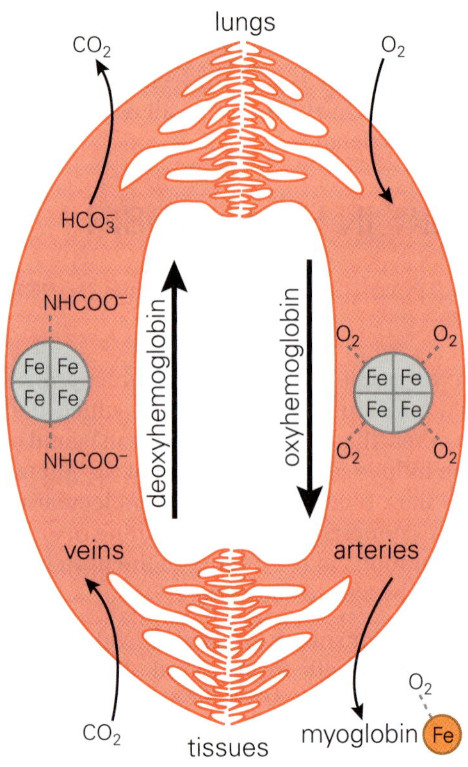


**Figure 14.13 Functions of hemoglobin.** Hemoglobin transports oxygen from the tissues to the lungs. Four oxygen atoms are bound per hemoglobin tetramer, at each iron-containing heme group. Hemoglobin also transports carbon dioxide, a by-product of respiration, from the tissues to the veins. (Adapted from R.E. Dickerson and I. Geis, Hemoglobin: Structure, Function, Evolution, and Pathology. Menlo Park, CA: Benjamin-Cummings, 1983. With permission from Pearson Education Ltd.)

As discussed in Section 14.3 (see Figure 14.3), a very common kind of allostery occurs when a protein or a protein assembly has more than one binding site for the same ligand. This occurs naturally if the protein is an oligomer (for example, a dimer, trimer, or tetramer). If such a multimeric protein is allosteric, then when the ligand binds to the empty protein a conformational change in the protein is induced such that the structures of the other binding sites are altered, and the dissociation constants for the next ligands to bind are different from the first one. The oxygen-transport protein hemoglobin (**Figure 14.13**) is an example of such a multimeric and allosteric system, and we analyze its mechanism in detail in the following sections.

## 14.11 The heme group in hemoglobin binds oxygen reversibly

In Chapter 5, we were introduced to two members of the globin family, myoglobin and hemoglobin (**Figure 14.14**). Both are oxygen-binding proteins with similar folds that encompass heme groups (**Figure 14.15**), but their physiological functions are very different. Myoglobin is found in muscle tissue, where it serves as a reservoir for oxygen. Hemoglobin, which is familiar as the protein that gives blood its characteristic red color, is the transport protein that is responsible for the efficient movement of oxygen from the lungs to the tissues (see Figure 14.13). Myoglobin is a monomer, with one oxygen-binding site, whereas hemoglobin is a tetramer, with four oxygen-binding sites per tetramer (see Figure 14.14B).

Functional hemoglobin and myoglobin have a ferrous ($Fe^{2+}$) iron atom in the middle of the heme group (see Figure 14.15). The iron atom is coordinated by six atoms. Four of these are nitrogen atoms of four pyrrole groups in the heme, which are heterocyclic rings containing four carbons and one nitrogen each. The fifth atom that coordinates the iron is provided by a histidine residue of the globin subunit, known as the proximal histidine (see Figure 14.14C). The coordination of the iron is completed by oxygen, which binds within a pocket in the interior of each subunit.

The ferrous iron atom binds to oxygen reversibly. This is a critical aspect of the function of hemoglobin, because it allows the heme groups to pick up oxygen in

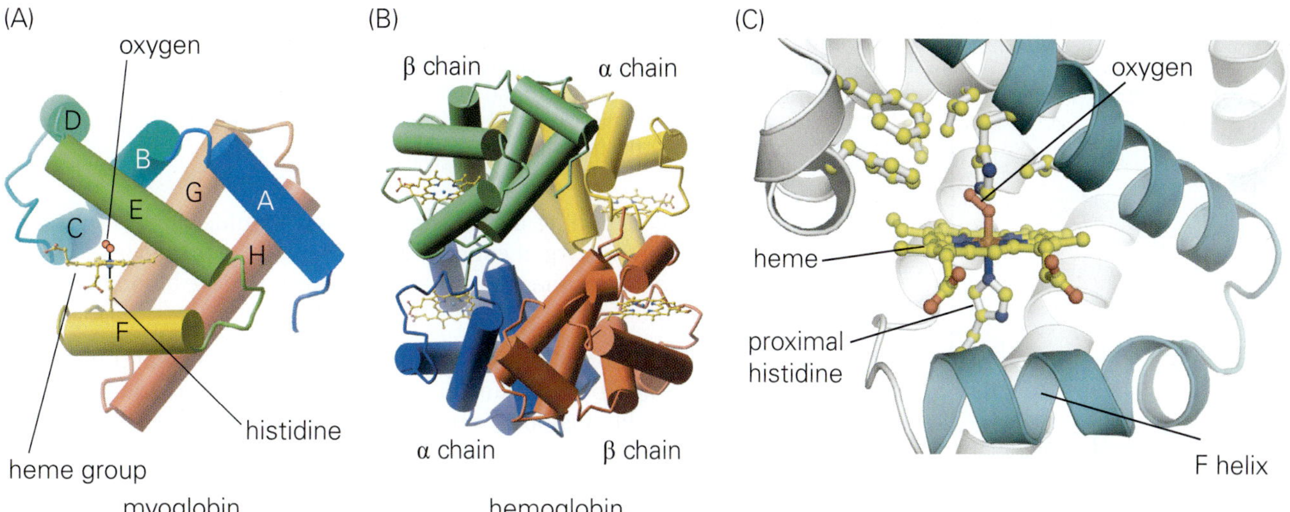


**Figure 14.14 Globin and heme structures.** (A) Structure of myoglobin. (B) Structure of human hemoglobin. (C) Close up of one of the four heme binding sites in hemoglobin, with oxygen bound to the iron atom. (PDB codes: A, 1MBC; B, 1A00.)

the lungs but also to let it go in the tissues. The heme iron is readily oxidized to $Fe^{3+}$ (ferric), and part of the function of the protein is to protect the heme group and to make the iron atom difficult to oxidize. Once oxidized, the iron atom no longer binds oxygen reversibly.

## 14.12 Hemoglobin increases the solubility of oxygen in blood and makes its transport to the tissues more efficient

Hemoglobin has at least two critical roles to play in our respiratory systems. First, it greatly increases the oxygen carrying capacity of the blood. Oxygen is not particularly soluble, and the dissolved oxygen concentration in blood is ~ 0.1 mM ($10^{-4}$ M). As we shall see, the dissolved oxygen concentration changes by a factor of three between arterial and venous blood, but 0.1 mM is a good reference value. The hemoglobin concentration in human blood is about 15 g deciliter$^{-1}$, using a common clinical unit. Since the molecular weight of a hemoglobin tetramer is ~70,000, this means that the concentration of hemoglobin in the blood is ~ 2 mM ($2 \times 10^{-3}$ M). Because each tetramer has four oxygen binding sites, the effective

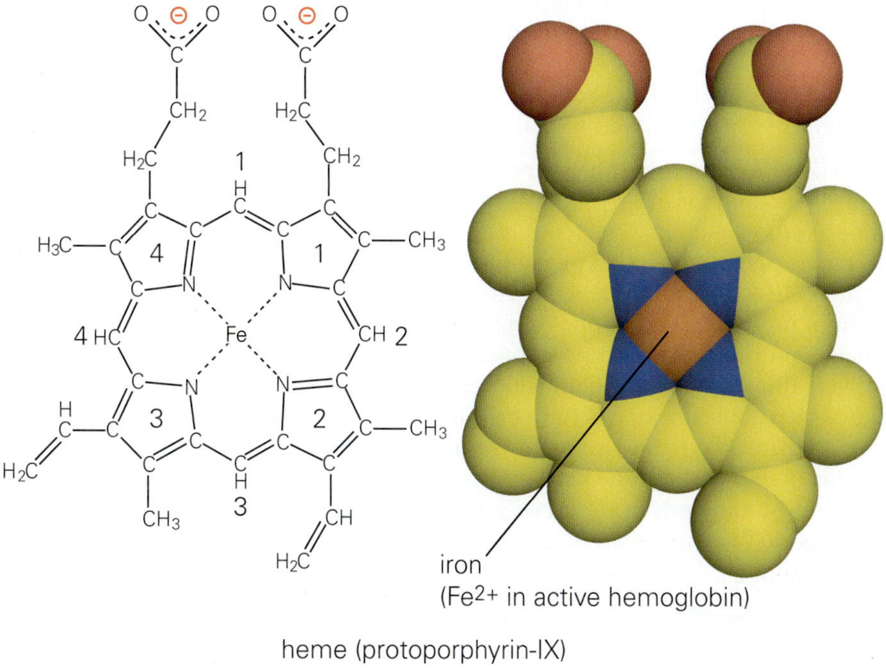

iron ($Fe^{2+}$ in active hemoglobin)

heme (protoporphyrin-IX)

**Figure 14.15 Structure of the heme group, also known as protoporphyrin-IX.**

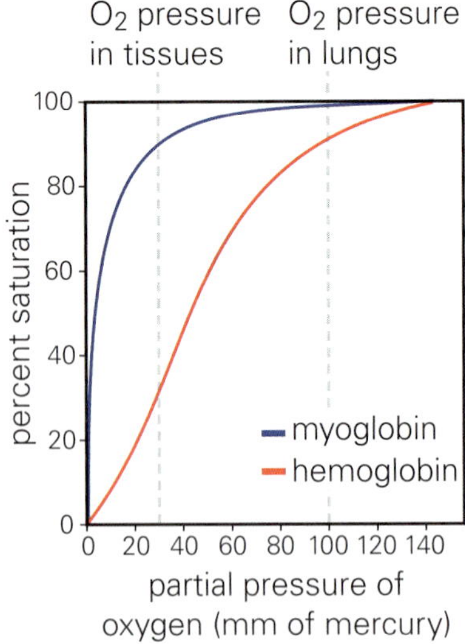


**Figure 14.16 Binding isotherms for myoglobin (*blue*) and hemoglobin (*red*).** The fractional saturation, *f*, is expressed as a percentage and is graphed as a function of dissolved oxygen concentration. The oxygen concentration is reported in terms of the partial pressure of oxygen. The dissolved oxygen concentration in blood, which is in the range of 0.1 mM, is proportional to the partial pressure of oxygen.

concentration of oxygen binding sites in hemoglobin is ~8 mM, which is ~80 times greater than the dissolved oxygen concentration. Hence, the presence of hemoglobin in the blood increases the oxygen carrying capacity of the blood by almost two orders of magnitude.

The second critical function of hemoglobin is to serve as an efficient transporter of oxygen from the lungs to the tissues. This is accomplished by alterations in the fractional saturation of hemoglobin as it moves from the arterial blood, with high dissolved oxygen concentration, to venous blood, with low dissolved oxygen concentration.

The oxygen concentration in blood, $[O_2]$, is usually expressed in pressure units, because the partial pressure of oxygen gas in equilibrium with dissolved oxygen is measured readily. The commonly used unit of partial pressure is the torr, with 1 torr corresponding to 1 mm of Hg at 20° C (760 torr = 1 atm of pressure). These nuances need not concern us here, because the concentration of oxygen in molar units is proportional to the partial pressure. The oxygen concentration of venous blood is ~30 torr, whereas it is ~100 torr in arterial blood.

The rather small difference (about three-fold) in dissolved oxygen concentration between venous and arterial blood means that hemoglobin would be a very inefficient transporter of oxygen if it did not have an ultrasensitive response to oxygen. To appreciate this, let us first look at the binding isotherm for hemoglobin, shown in **Figure 14.16**. The binding isotherm is clearly not hyperbolic. Instead, it switches between a concave and a convex shape, resembling a distorted form of the letter "S." Such a binding isotherm is referred to as a **sigmoid binding curve**. The sigmoid shape of the hemoglobin oxygen binding isotherm is characteristic of an allosteric system with positive cooperativity and is a feature that is essential for the proper delivery of oxygen to the tissues from the lungs.

**Sigmoid binding curve**

When the binding of ligands to a protein is cooperative, the shape of the binding curve (fractional saturation versus ligand concentration) is no longer hyperbolic. Instead, it resembles an "S" shape and is called sigmoid (for sigma, the Greek letter S).

When the oxygen concentration is 100 torr, the fractional saturation, *f*, is ~90%. That is, as the hemoglobin is carried through the lungs, it becomes nearly completely loaded with oxygen, with roughly 36 oxygen molecules bound for every 10 hemoglobin tetramers. When the dissolved oxygen concentration is ~30 torr (corresponding to the oxygen concentrations in tissues), the value of *f* drops to ~30%. This means that roughly 24 oxygen molecules will be released per 10 hemoglobin tetramers.

The sigmoid oxygen-binding curve of hemoglobin makes it a relatively efficient delivery system for oxygen, as can be appreciated by looking at the binding isotherm for myoglobin. Myoglobin is a monomeric protein, and so it has a graded response to oxygen. The oxygen-binding isotherm for myoglobin, also shown in Figure 14.16, is hyperbolic, consistent with the expected simplicity of an equilibrium process in which one oxygen molecule binds to the heme group of one myoglobin molecule with no cooperativity. Myoglobin has a somewhat higher oxygen affinity than does hemoglobin, and so it is nearly completely saturated if the oxygen concentration is 100 torr. When the oxygen concentration drops to ~30 torr, myoglobin still remains highly saturated, with a value of *f* near 90%. Hence, if hemoglobin had the same binding isotherm as myoglobin, only ~4 molecules of oxygen would be released for every 10 tetramers of hemoglobin that move from the lungs to the tissues.

The inefficiency of noncooperative proteins in ligand transport processes is a general phenomenon. Recall from Section 14.9 that a noncooperative protein switches from ~10% saturation to ~90% saturation over a 81-fold difference in

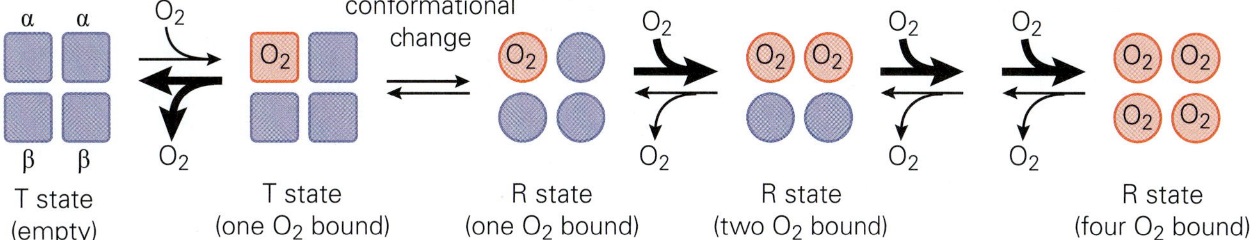


**Figure 14.17 Allosteric transitions in human hemoglobin.** Deoxyhemoglobin exists in a conformation known as the T state. The binding of oxygen triggers a change in quaternary structure to the R state, which has increased affinity for oxygen.

free ligand concentration (see Equation 14.3). The three-fold difference in oxygen concentration in arterial and venous blood is therefore insufficient to trigger substantial unloading if hemoglobin were not ultrasensitive to oxygen concentration.

## 14.13 Hemoglobin undergoes conformational changes as it binds to and releases oxygen

As we shall see, the sigmoid binding curve for hemoglobin arises from conformational changes that are brought about by oxygen binding. Empty hemoglobin has four oxygen-binding sites that all have low affinity for oxygen (**Figure 14.17**). This conformation of hemoglobin is referred to as the "tense" or T state of the system. The first oxygen to bind to such a hemoglobin tetramer does so with difficulty, but this initial binding event triggers a conformational change in the assembly such that the three remaining binding sites are switched to a conformation that has much higher affinity for oxygen. This conformation is called the "relaxed" or R state of the system. The last oxygen molecule to bind to R-state hemoglobin binds ~1000 times more tightly than does the first oxygen molecule to bind to T-state hemoglobin. It is this switch in binding affinity that allows hemoglobin to pick up oxygen with high affinity in the lungs, but then release it with alacrity in the tissues.

We shall study the origins of the sigmoid binding curve of hemoglobin in two stages. First, by analyzing a simpler system with just two binding sites, we shall understand why allosteric systems have sigmoid binding curves (Sections 14.14 and 14.15). Next, we shall look at the structural mechanisms that underlie the ability of hemoglobin to alter the structure of its oxygen-binding sites in response to oxygen binding at a distant site (Sections 14.16 and 14.17). Our current understanding of allostery has its roots in concepts first developed for hemoglobin by Jacques Monod, Jeffries Wyman, and Jean-Pierre Changeux, and placed within a structural framework by Max Perutz, a founder of protein crystallography.

## 14.14 The sigmoid binding isotherm for an allosteric protein arises from switching between low- and high-affinity binding isotherms

We shall now consider the consequences of allostery for a dimeric protein, P, with two equivalent binding sites for a ligand, L, that can bind to either site (**Figure 14.18**). Let us suppose that P exhibits positive cooperativity, so that the binding of the second ligand is tighter than the binding of the first one. P switches between two conformations, one in which all of the binding sites are in the low-affinity conformation (the T state), and one in which all of the binding sites are in the high-affinity conformation (the R state).

The binding isotherms for the T and R states are shown in **Figure 14.19**A. For positive cooperativity, the dissociation constant for the T state, $K_{D1}$, is larger than that for the R state, $K_{D2}$. Recall from Section 12.5 that the graph of the logarithm of the ratio of the fraction of protein bound to the fraction unbound, $\log[f/(1-f)]$, versus the logarithm of the concentration of the ligand, log[L], is a straight line. In the log-log plot shown in Figure 14.19A, the binding isotherms for the T and R states are straight lines that are parallel to each other, with the R-state line on the left.

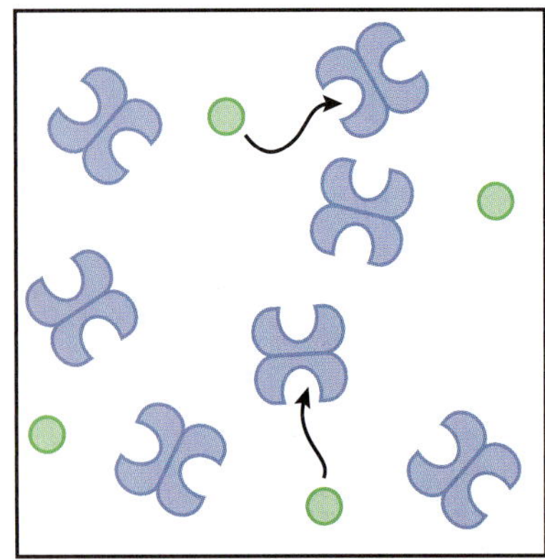


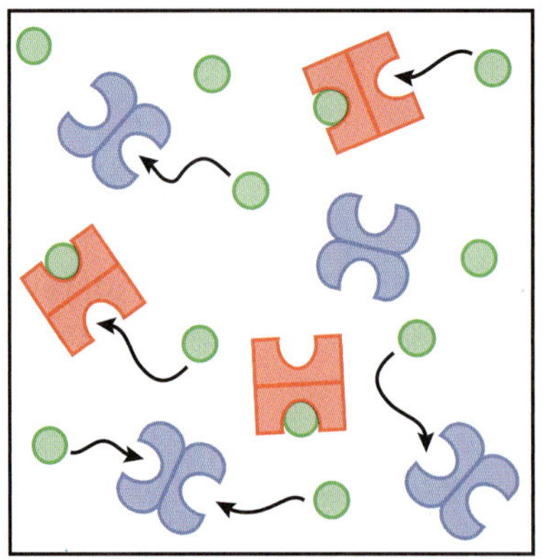


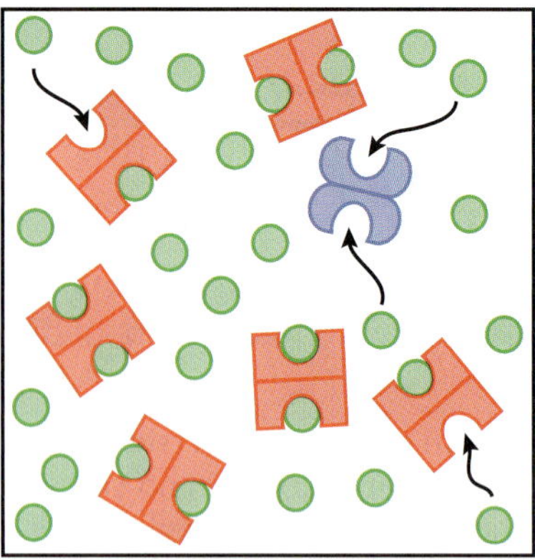


**Figure 14.18 A dimeric allosteric protein.** The protein switches between two conformations, denoted T (low affinity, *blue*) and R (high affinity, *red*). When the ligand concentration is low, all of the protein is in the T state, and ligand binds with difficulty. As the ligand concentration increases, the protein starts to switch to the R state as ligand binds, and ligand affinity increases. This is an example of positive cooperativity.

We assume that the structure of P is always symmetrical—that is, conformations in which the two subunits have different conformations are disallowed. A very common form of allostery in biological systems (including hemoglobin) involves symmetrical multisubunit molecules, and for such systems the condition of symmetry is a reasonable one to impose. Notice that the dimeric molecules depicted in Figure 14.18 are all symmetrical, with both subunits in the R-state conformation, or both in the T-state conformation.

First, consider what happens when the ligand concentration is very low. At this point, most of the protein molecules are empty, and all of the available binding sites correspond to the low-affinity binding site (T state, dissociation constant $K_{D1}$). At very low ligand concentrations, the binding reaction will be governed by the isotherm corresponding to the low-affinity T state, as shown in Figure 14.19B. As the ligand concentration increases, more and more of the protein molecules have one ligand bound and have switched to the R-state conformation, which has higher affinity for the ligand (dissociation constant $K_{D2}$). The population of proteins is now mixed, with some in the T state and some in the R state. Ligand molecules now have a choice of binding sites and the thermodynamics of binding cannot be understood without a detailed calculation, which we shall examine in Section 14.15. Finally, when the ligand concentration is very high, all of the protein molecules have at least one site occupied, and so essentially all of the available sites are the in the high-affinity R state. The binding of the ligand is now governed by the binding isotherm corresponding to the high-affinity site (see Figure 14.19B).

**Figure 14.20** shows a graph of the data using normal (that is, not logarithmic) axes. The binding isotherm switches from one hyperbolic binding curve at low concentration to another one at high concentration, resulting in the sigmoidal shape that is characteristic of positive cooperativity.

## 14.15 The degree of cooperativity between binding sites in an allosteric protein is characterized by the Hill coefficient

The log-log plot for the binding isotherm of a noncooperative protein is governed by the following equation (see Equation 12.18 in Section 12.5):

$$\log\left(\frac{f}{1-f}\right)=\log[\mathrm{L}]-\log K_{\mathrm{D}} \qquad (14.4)$$

According to Equation 14.4, the slope of the binding isotherm is 1.0. For an allosteric system with positive cooperativity, the slope of the binding isotherm is also ~1.0 at the two extremes of ligand concentration. At intermediate values of ligand

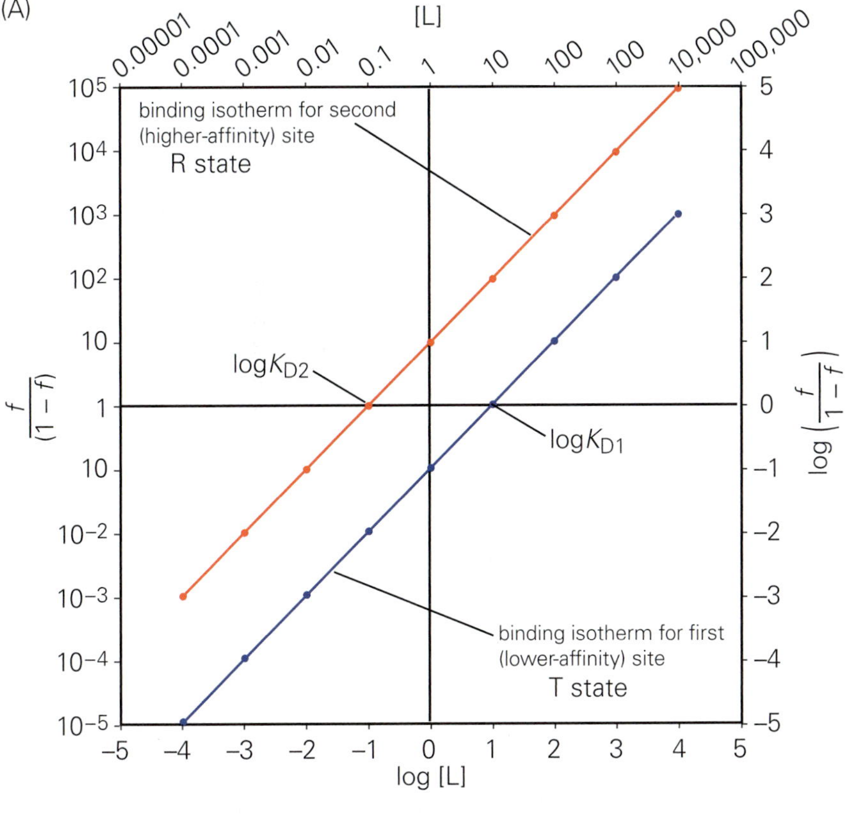


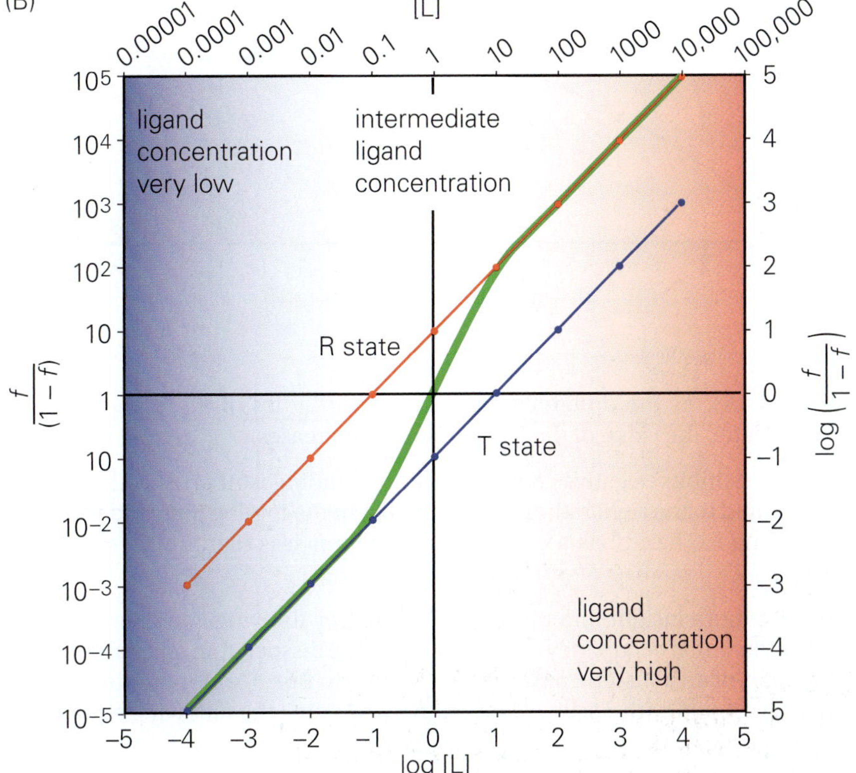


**Figure 14.19 Binding isotherm for a dimeric protein with positive cooperativity.** (A) Binding isotherms for the T (*blue*) and R (*red*) states are shown in a log-log plot (refer to Section 12.5). (B) As the ligand concentration increases, the binding isotherm switches from the T state isotherm to the R state isotherm. The resulting binding isotherm is no longer linear (*green*). The ligand concentration is expressed in micromolar units.

concentration, the binding isotherm curves up and has a slope that is greater than unity. The slope of the binding isotherm at the point where the protein is half saturated [that is, $\log\left(\frac{f}{1-f}\right)$ is zero] is known as the **Hill coefficient**, $n_H$

**Hill coefficient**

This parameter, named after the physiologist Archibald Hill, reflects the steepness of the log-log binding isotherm at the point when the protein is half saturated with ligand. A noncooperative system has a Hill coefficient of unity. Systems exhibiting positive and negative cooperativity have Hill coefficients that are greater than and less than unity, respectively.

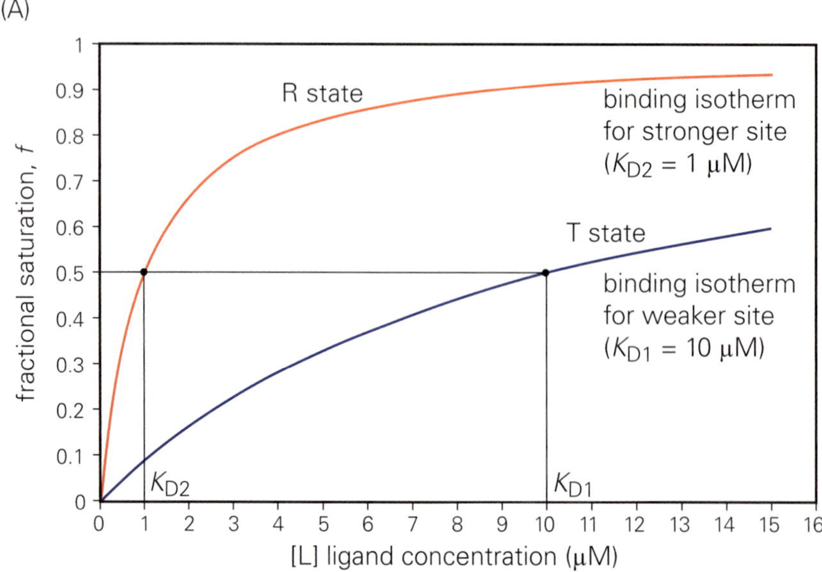


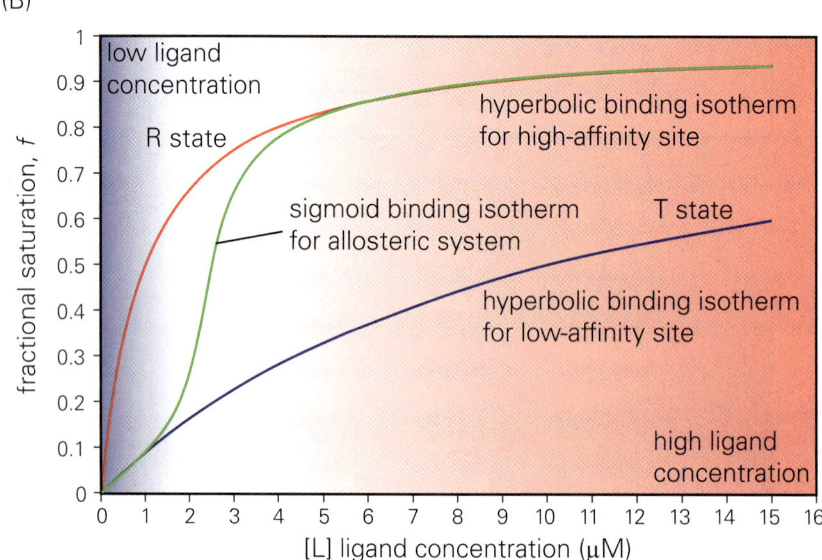


**Figure 14.20 Binding isotherm graphed with normal axes for a dimeric protein with positive cooperativity.** (A) Binding isotherms for the T state (*blue*) and R state (*red*). (B) The net binding isotherm (*green*) is a combination of the T- and R-state isotherms.

(**Figure 14.21**A). As the difference between the affinities of the T and R states becomes greater, the value of the Hill coefficient increases.

If the system exhibits negative cooperativity, the binding isotherm for the second ligand is shifted to the right rather than to the left in the log-log plot, and the composite binding isotherm has a Hill coefficient that is less than unity (see Figure 14.21).

We can gain more insight into the Hill coefficient by deriving an expression that relates the fractional occupancy of the protein to the dissociation constants for the R and T states of an allosteric protein. This is done in **Box 14.1** for the simple case of a dimeric protein with positive cooperativity, yielding the following equation:

$$\frac{f}{1-f}=\frac{\frac{[L]}{K_{D1}}+\left(\frac{[L]}{K_{D1}}\right)\left(\frac{[L]}{K_{D2}}\right)}{1+\frac{[L]}{K_{D1}}} \tag{14.5}$$

Equation 14.5 allows us to plot the binding isotherms as a function of different values of $K_{D1}$ and $K_{D2}$ (the graphs in Figure 14.21 are based on Equation 14.5). In the

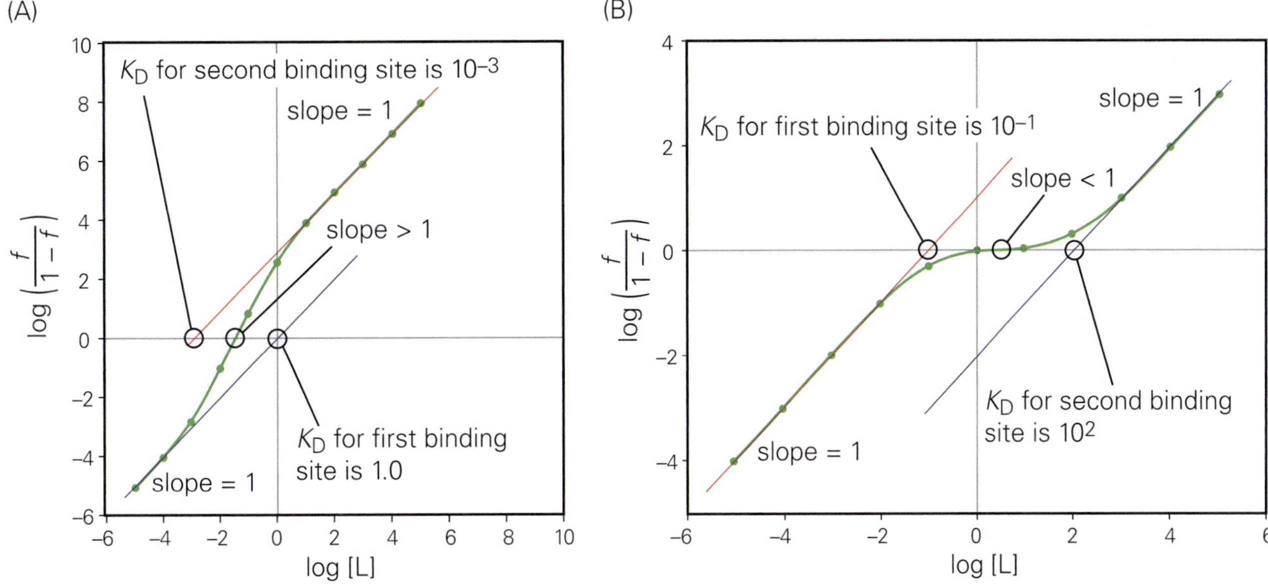


**Figure 14.21 Hill coefficients for positive and negative cooperativity.** (A) The binding isotherm for a dimeric protein is shown, calculated using Equation 14.5, as described in Box 14.1 The protein exhibits positive cooperativity, with the values of $K_D$ for the T and R states being 1.0 M and $10^{-3}$ M, respectively. At the point of half saturation, when $\log\left(\frac{f}{1-f}\right)$ is zero, the slope of the binding curve is greater than 1.0. The slope at this point is defined as the Hill coefficient, $n_H$, and its value is 1.94 (see Equation 14.6). (B) The binding isotherm for a dimeric protein with negative cooperativity. The T and R states have $K_D$ values of $10^{-1}$ M and $10^2$ M, respectively. In this case, the binding isotherm has a slope less than unity for intermediate concentrations of ligand. The value of $n_H$ is 0.061, according to Equation 14.6. The ligand concentration is expressed in micromolar units.

case of positive cooperativity, as the ratio of $\frac{K_{D1}}{K_{D2}}$ increases (that is, as the degree of cooperativity increases), the binding isotherm becomes more and more nonlinear. The slope of the Hill coefficient does not increase without limit, however, and has a maximum value of 2.0, which corresponds to the number of subunits in the protein. A relationship between the Hill coefficient, $n_H$, and the ratio of the two dissociation constants is derived in Box 14.1 and is as follows:

$$n_H = \frac{2}{1+\sqrt{\frac{K_{D2}}{K_{D1}}}} \tag{14.6}$$

According to Equation 14.6, if $K_{D1}$ and $K_{D2}$ are equal (that is, if the system is not cooperative), then the value of $n_H$ is 1.0, which is consistent with Equation 14.4. When the system exhibits an increasing degree of positive cooperativity, the ratio in the denominator becomes smaller as the affinity of the R state becomes tighter. The maximum value of $n_H$ is therefore 2.0.

Even for the simplest situation of a protein with two allosterically coupled binding sites, the derivation of the equation governing the fractional saturation is rather complicated (see Box 14.1). In practice, the binding isotherms for proteins with more than two allosterically coupled sites are generated using computer simulations. The analysis of such simulations reveals that the the maximum value of $n_H$ is given by the number of binding sites that are coupled allosterically. For hemoglobin, for example, the Hill coefficient is therefore expected to be 4.0 or less. The actual value of the Hill coefficient for human hemoglobin is in the range of ~2.5 to ~3.5, with the precise value depending on conditions such as the pH (see Sections 14.20 and 14.21).

Biochemists commonly use a simplified analysis of the Hill coefficient, in which intermediate states with the proteins partially saturated are ignored. This analysis is not physically meaningful but, because it is so commonly used, it is described in **Box 14.2**.

## 14.16 The tertiary structure of each hemoglobin subunit changes upon oxygen binding

Crystal structures of ligand-bound and ligand-free hemoglobin, first determined by Max Perutz, have led to the elucidation of the structural basis for the cooperativity in oxygen binding. The key feature of the mechanism is the change in the

## Box 14.1 The Hill coefficient for a protein with two allosterically coupled binding sites

Consider a protein, P, that has two binding sites for a ligand, L (see Figure 14.18). Most commonly such a situation would occur if the protein, P, forms a dimer, but we do not consider the oligomerization state of the protein explicitly here. The binding isotherm that characterizes the binding of L to P is shown in Figure 14.21, using log-log axes.

Let us denote the two conformations of the binding sites by R and T. When no ligand is bound to the protein, both binding sites are in conformation T. When one or two ligands are bound to the protein, both sites switch to conformation R. The dissociation constants for the binding of the ligand to the two conformations are denoted $K_{D1}$ and $K_{D2}$, respectively.

When $K_{D1}$ and $K_{D2}$ are not equal, the binding isotherm is no longer linear. The Hill coefficient, $n_H$, characterizes the steepness of the curvature of the binding isotherm. The slope of the binding curve is not uniform and, in order to be precise, we define the Hill coefficient as the slope of the binding isotherm at the point when the protein is 50% saturated with ligand—that is, when the value of $f$ is 0.5 and $\log[f/(1-f)]$ is zero (see Figure 14.21).

The value of the Hill coefficient, $n_H$, for a protein, P, with two allosterically coupled binding sites is given by:

$$n_H = \frac{2}{1+\sqrt{\dfrac{K_{D2}}{K_{D1}}}} \qquad (14.1.1)$$

Equation 14.1.1 is obtained by treating the binding of the ligand to the protein as if it were described by the following two sequential binding equilibria:

$$\text{(i)}\quad P + L \underset{K_{D1}}{\rightleftharpoons} P\bullet L \qquad (14.1.2)$$

$$\text{(ii)}\quad P\bullet L + L \underset{K_{D2}}{\rightleftharpoons} P\bullet L_2 \qquad (14.1.3)$$

In order to derive Equation 14.1.1, we proceed in two steps. In step 1, we relate the ratio of bound and unbound protein to the two dissociation constants and the free ligand concentration, [L], by the following equation:

$$\frac{f}{1-f} = \frac{\dfrac{[L]}{K_{D1}} + \left(\dfrac{[L]}{K_{D1}}\right)\left(\dfrac{[L]}{K_{D2}}\right)}{1+\dfrac{[L]}{K_{D1}}} \qquad (14.1.4)$$

Equation 14.1.4 allows us to graph the expected binding isotherm for different values of the dissociation constants, as illustrated in the main text.

In Step 2, we use Equation 14.1.4 to calculate the Hill coefficient, which is the slope of $\log_{10}[f/(1-f)]$ (that is, the derivative of $\log_{10}\left(\frac{f}{1-f}\right)$ with respect to $\log_{10}[L]$), evaluated at the ligand concentration when $f = 0.5$:

$$\text{Hill coefficient, } n_H = \left.\frac{d\log_{10}\left(\dfrac{f}{1-f}\right)}{d\log_{10}[L]}\right|_{\text{evaluated at } f=0.5} \qquad (14.1.5)$$

Evaluation of this derivative gives us the desired definition of the Hill coefficient, $n_H$ (Equation 14.1.1). Steps 1 and 2 are straightforward, but they involve rather tedious algebra and repeated application of the chain rule of differentiation. You can work out these relationships yourself by using the definitions of the dissociation constants and the fractional saturation. In practice, allosteric binding interactions are usually estimated empirically by numerical analysis of actual binding data, or by analyzing binding equilibria using the principles of statistical thermodynamics. You may, if you are interested, refer to the paper by Attila Szabo and Martin Karplus on a statistical thermodynamic model for hemoglobin, in the reading list at the end of this chapter.

state of electron spins of the iron atom of the heme group upon coordination by oxygen. In the absence of oxygen, the iron atom is coordinated by five nitrogen atoms, four provided by the heme group and one provided by the proximal histidine (Figure 14.14C). The five-coordinate iron atom has high electron spin (that is, the d-electrons are in different orbitals and are not paired) and, consequently, an ionic radius that is too large to be accommodated within the central cavity of the heme group.

The heme group is therefore buckled in the deoxy form, and the iron atom is popped out of the plane of the four nitrogen atoms of the heme group (**Figure 14.22**A). Oxygen binding converts the iron atom from the five-coordinate state with high electron spin to a six-coordinate state with low spin (pairing electrons in orbitals), which results in a reduction of the effective size of the iron atom. The iron atom can now be accommodated within a planar heme group, as seen in

Figure 14.22B. The heme group is a delocalized aromatic ring system and prefers, therefore, to adopt a planar structure. Strain induced in the heme group by the high-spin state of the five-coordinate iron atom is released upon oxygen binding.

## 14.17 Changes in the tertiary structure of each subunit are coupled to a change in the quaternary structure of hemoglobin

The changes in heme structure that occur upon oxygen binding are fundamentally a consequence of the chemical structures of the heme group, the iron atom, and the oxygen molecule, and are also seen in noncooperative globins, such as myoglobin. What is special about hemoglobin is that these changes in heme structure are transmitted to the other subunits of the tetramer, which are able to "sense" the presence or absence of oxygen at each of the other subunits. First, the structure of each individual subunit changes upon oxygen binding, as shown in Figure 14.22C. A principal component of this conformational change is

### Box 14.2 A commonly used but physically implausible analysis of the Hill coefficient

It is common practice in biochemistry to analyze Hill coefficients in a way that is different from the treatment in Box 14.1. This alternative analysis is easier to understand but, unfortunately, it is physically implausible. Nevertheless, we describe it here because it is very commonly used. The analysis begins by assuming that if $n$ ligand molecules bind cooperatively to a protein, then the binding reaction can be written as:

$$P + nL \rightleftharpoons P \bullet nL \qquad (14.2.1)$$

The dissociation constant for this reaction is given by:

$$K_D = \frac{[P][L]^n}{[P \bullet L_n]} \qquad (14.2.2)$$

The binding reaction represented by Equations 14.2.1 and 14.2.2 is physically implausible. These equations would be true for binding reactions in which all $n$ ligand molecules bind to the protein *simultaneously*, with no intermediate steps in the binding (an all or none model). Although this is unrealistic, let us proceed to see what the implications of these two equations are. We begin by writing down an expression for the fractional saturation of the protein, $f$, by using Equation 14.2.2:

$$f = \frac{[P \bullet L_n]}{[P \bullet L_n] + [P]} = \frac{\dfrac{[P][L]^n}{K_D}}{\dfrac{[P][L]^n}{K_D} + [P]} = \frac{\dfrac{[L]^n}{K_D}}{\dfrac{[L]^n}{K_D} + 1} \qquad (14.2.3)$$

Using Equation 14.2.3, we obtain the following expression for the fraction of protein that is unbound, $1 - f$:

$$1 - f = 1 - \frac{\dfrac{[L]^n}{K_D}}{\dfrac{[L]^n}{K_D} + 1} = \frac{1}{\dfrac{[L]^n}{K_D} + 1} \qquad (14.2.4)$$

By combining Equations 14.2.3 and 14.2.4, we obtain an expression for the ratio of the fraction bound to the fraction unbound:

$$\frac{f}{1-f} = \frac{[L]^n}{K_D} \qquad (14.2.5)$$

Notice that Equation 14.2.5 is very similar to the equation describing a simple binding equilibrium (see Equation 14.2 in the main text), except that the ligand concentration is raised to the exponent $n$. Taking the logarithm of both sides of Equation 14.2.5, we get:

$$\log\left(\frac{f}{1-f}\right) = n \log[L] - \log K_D \qquad (14.2.6)$$

According to Equation 14.2.6, a graph of $\log\left(\frac{f}{1-f}\right)$ versus log [L] yields a straight line with slope $n$. The exponent $n$ is commonly equated with the Hill coefficient (also called the **Hill slope**, because in this analysis $n$ is indeed the slope of the line). Although this analysis is very commonly used, you should recognize that it does not correspond to a realistic physical process, and you should consider the Hill coefficient derived in this way to simply reflect the steepness of the binding curve. It is inappropriate to equate the value of $n$ derived in this way with the number of molecules or binding sites involved in cooperative binding. A realistic analysis of Hill coefficients would take into account each of the steps in the complete set of binding reactions, as is done for a dimeric protein in Box 14.1.

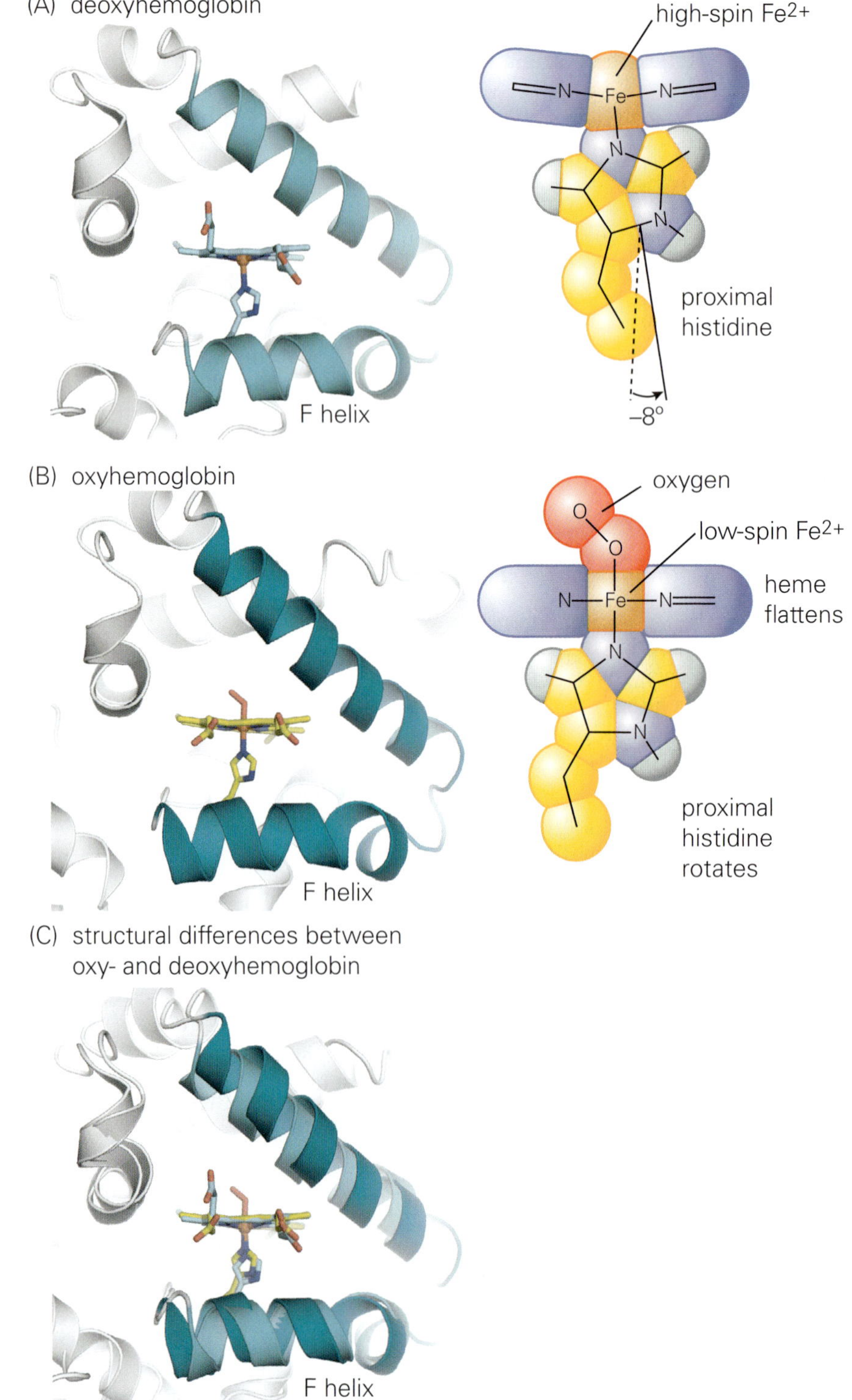


**Figure 14.22 Structural changes in hemoglobin induced by oxygen binding.** (A) Structure of deoxyhemoglobin. (B) The structure of oxyhemoglobin. The hexacoordinate iron atom is smaller and is pulled into the plane of the heme group by the oxygen atom. (C) Comparison of the structures of oxy- and deoxyhemoglobin. The structural changes that result from oxygen binding are propagated throughout the molecule. This is the basis for the allosteric mechanism. (PDB codes: A, 4HHB; B, 1HHO.)

the alteration of the position of the proximal histidine, which is presented by an α helix known as helix F. The upward movement of the iron atom upon oxygen binding (in the orientation seen in Figure 14.22) results in a coupled movement of helix F, which is propagated throughout the protein. This change in structure of an individual globin subunit is coupled to a change in the relative orientations of the four subunits of the hemoglobin tetramer.

As shown in Figure 14.14, the hemoglobin tetramer contains two each of two different subunits, named α and β. Each α subunit is paired tightly with one β subunit,

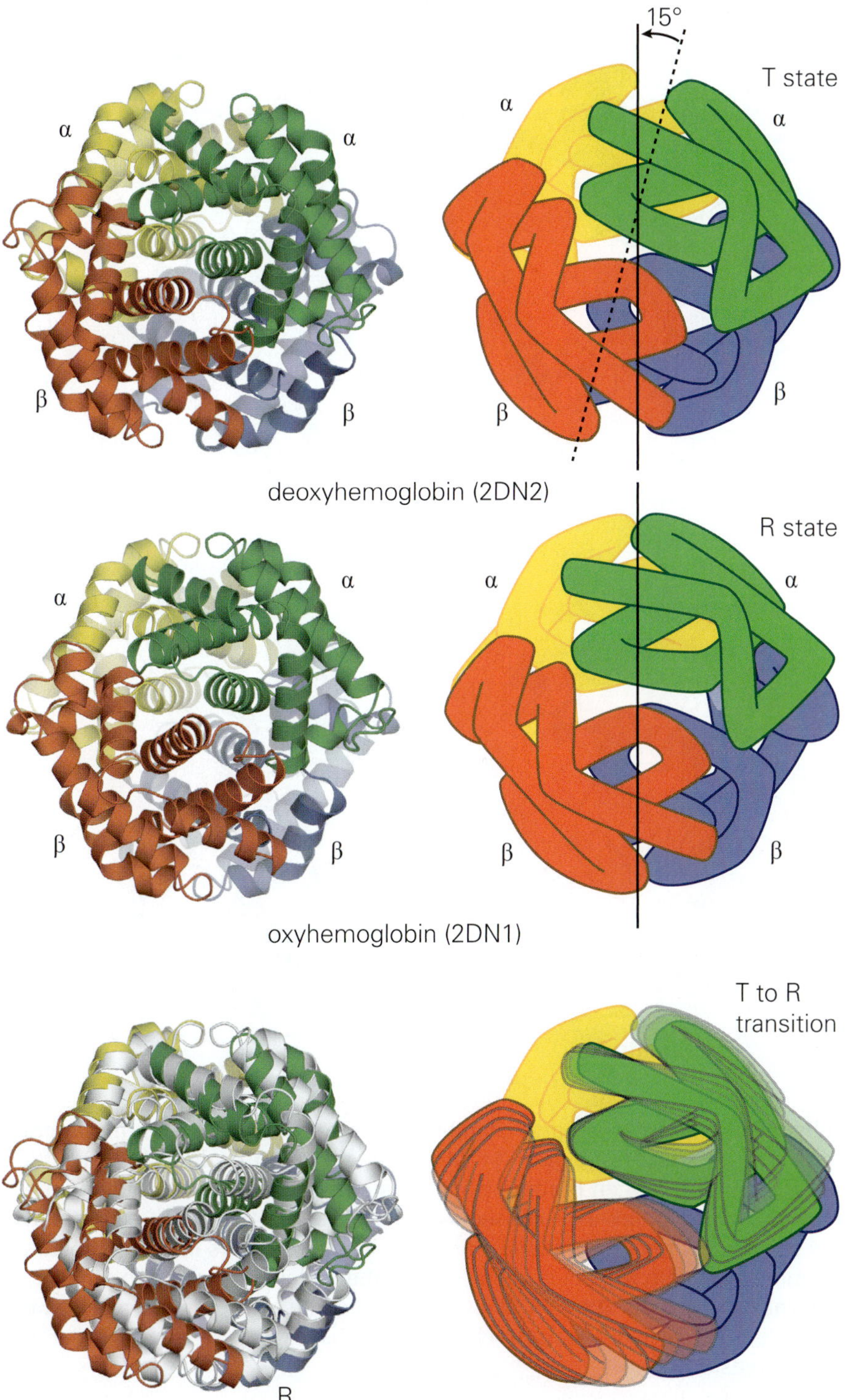


**Figure 14.23 The T → R transition in hemoglobin.** (A) Quaternary structure of deoxyhemoglobin, in the T state. All four heme groups are unliganded. (B) Quaternary structure of oxyhemoglobin, in the R state. All four heme groups have oxygen bound to them. (C) The major change in the quaternary structure is the rotation of one α/β dimer with respect to the other, by ~15°. (Schematic diagrams adapted from R.E. Dickerson and I. Geis, Hemoglobin: Structure, Function, Evolution, and Pathology. Menlo Park, CA: Benjamin-Cummings, 1983.) (PDB codes: 4HHB, 1HHO.)

and the structure can be thought of as a dimer of α/β heterodimers (**Figure 14.23** shows a schematic representation of the hemoglobin tetramer). The relative orientation of each α subunit with respect to its paired β does not change very much upon oxygen ligation. What does happen is that one α/β heterodimer rotates by ~15° with respect to the other one when oxygen binds, as shown in Figure 14.23C. Oxygen binding therefore triggers both a local conformational change by altering the position of helix F (see Figure 14.22C), and a global change in the organization of the quaternary structure of hemoglobin (see Figure 14.23C).

A close-up view of the structures of liganded and unliganded hemoglobin reveals how the binding of oxygen triggers the change in quaternary structure. The T-state

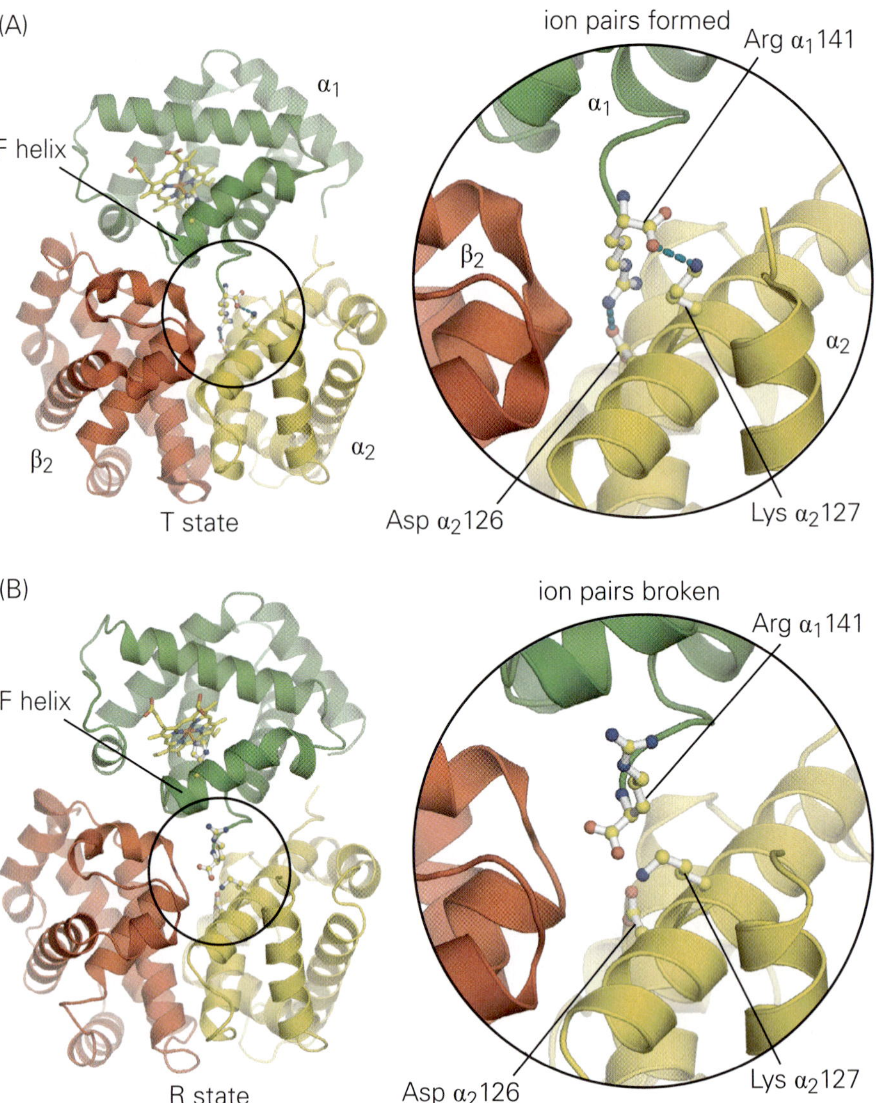


**Figure 14.24 Breakage of ion pairs in the T → R transition.** (A) The T state of deoxyhemoglobin. Two of the ion-pairing interactions at the interface between two α subunits are shown. The sidechain of the C-terminal arginine residue, Arg $\alpha_1$141, is paired with the sidechain of Asp $\alpha_2$126. (B) The R-state structure of oxyhemoglobin. The ion pairs shown in (A) are disrupted. (PDB codes: 4HHB, 1HHO.)

(unliganded) structure is partly stabilized by a network of ion-pairing interactions between subunits (**Figure 14.24**A). The alteration in the F helix position that occurs upon oxygen binding is inconsistent with the maintenance of these ion pairs, which break as a consequence (Figure 14.24B). The T-state quaternary structure is no longer stable, and the α/β heterodimers in the hemoglobin tetramer rotate with respect to each other as the molecules settle into the R-state quaternary structure. **Figure 14.25**A shows a schematic representation of how the interactions between the subunits change as the hemoglobin tetramer undergoes the R–T transition.

## 14.18 The hemoglobin tetramer is always in equilibrium between R and T states, and oxygen binding biases the equilibrium

One way to think about the hemoglobin mechanism is that the interfacial ion pairs in T-state hemoglobin "resist" the binding of oxygen, which has to pull apart the ion pairs as the iron atom moves into the plane of the heme group. Once the first oxygen atom binds, the hemoglobin tetramer is more likely to switch to the R-state quaternary structure, in which the ion pairs are broken, as shown in Figure 14.25A, and the three empty ligand binding sites can more readily bind to the next oxygen molecule that arrives.

More generally, we can think of the hemoglobin tetramer as always being in an equilibrium between the R and T states, whether or not oxygen is bound (Figure 14.25B). The binding of oxygen biases the equilibrium towards the R state, whereas the departure of oxygen shifts the equilibrium towards the T state. This model for hemoglobin allostery is known as the Monod–Wyman–Changeux–Perutz (MWC–Perutz) model, after the names of the scientists who first developed it (see Figure 14.25). As you can imagine, the analysis of the fractional saturation as a function of oxygen concentration in the MWC–Perutz model is extremely complicated because of the large number of dissociation constants involved and is best analyzed by computer simulations. Nevertheless, it is a conceptually straightforward extension of the simpler analysis of a dimeric allosteric protein that we have worked through in Sections 14.14 and 14.15 and in Box 14.1.

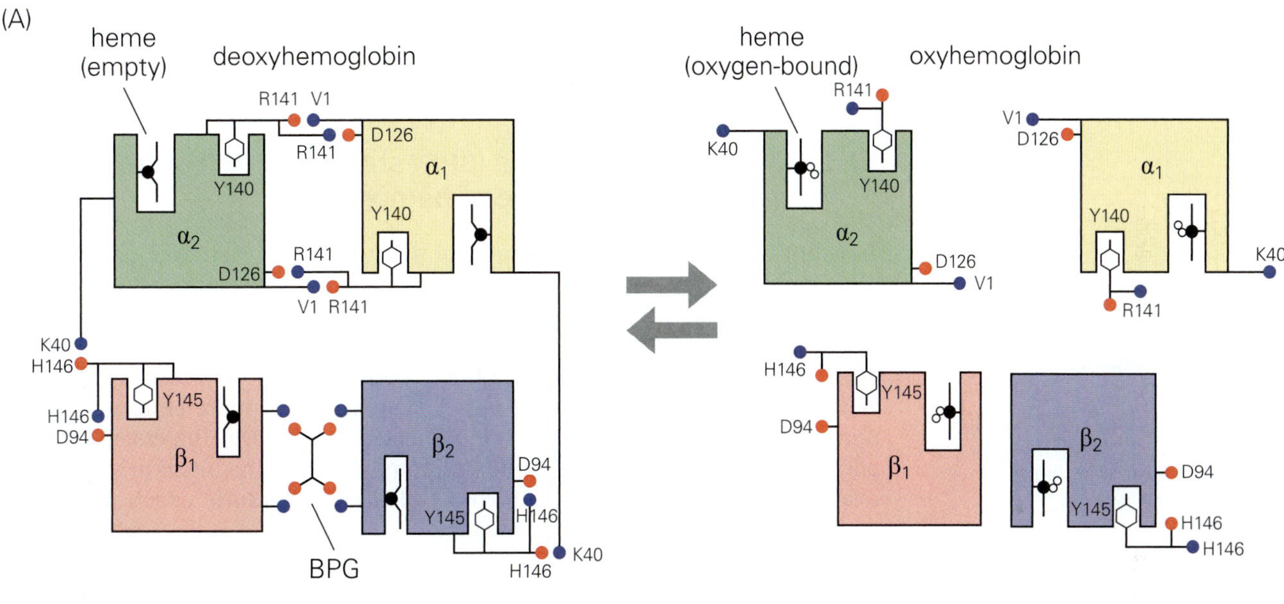


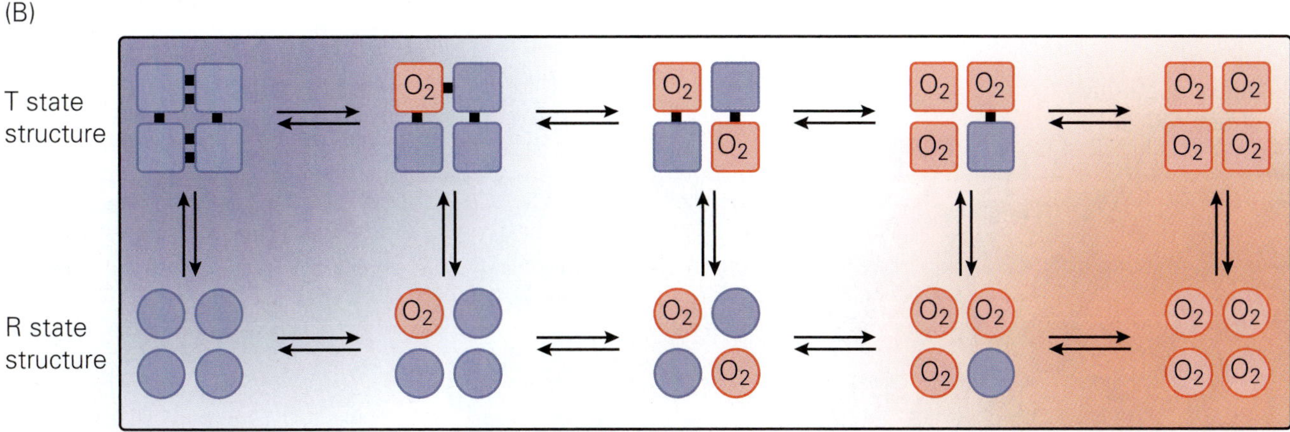


**Figure 14.25 The Monod–Wyman–Changeux–Perutz (MWC–Perutz) model for hemoglobin allostery.** This model assumes that the hemoglobin tetramer is in equilibrium between two quaternary states, T and R. Ligand binding shifts the equilibrium from T to R. (A) Schematic diagram showing changes in ion-pairing and other interactions between the T (deoxy) and R (oxy) states of hemoglobin. Positively and negatively charged groups are indicated by *blue* and *red* dots, respectively. The T and R states are both compact, but these diagrams have been distorted for clarity. The T and R states are shown here as empty and bound to oxygen, respectively, but you should think of these as alternate structures that the tetramer can adopt, whether or not oxygen is bound. (B) Schematic illustration of the MWC–Perutz model. Ion pairs are indicated by lines connecting the subunits (A, from M.F. Perutz, A.J. Wilkinson, M. Paoli, and G.G. Dodson, *Annu. Rev. Biophys. Biomol. Struct.* 27: 1–34, 1998. With permission from Annual Reviews; B, adapted from W.A. Eaton et al., and A. Mozzarelli, *Nat. Struct. Biol.* 6: 351–358, 1999. With permission from Macmillan Publishers Ltd.)

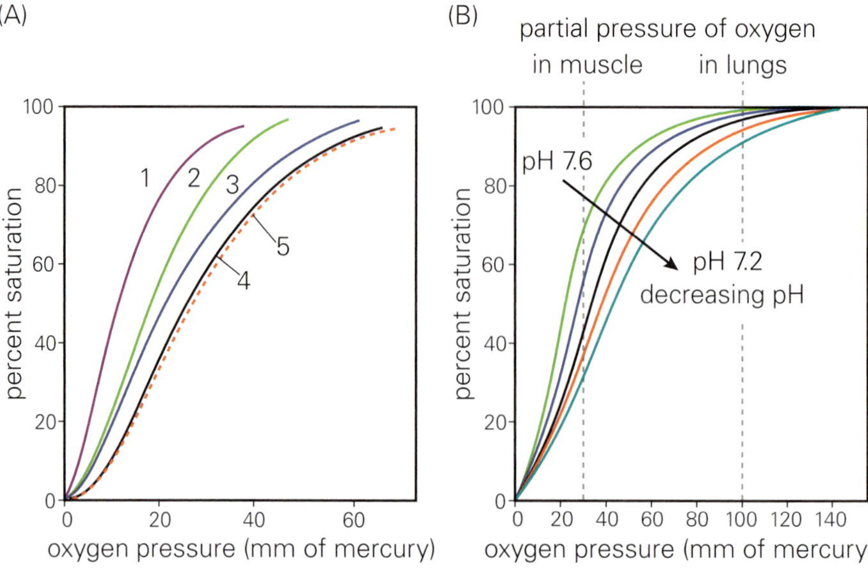


**Figure 14.26 Allosteric effectors.** The oxygen-binding isotherm for purified hemoglobin at pH 7.6 is labeled 1. Upon adding $CO_2$ or bisphosphoglycerate (BPG), the oxygen affinity of hemoglobin decreases (curves 2 and 3). $CO_2$ and BPG, added together (curve 4), make hemoglobin behave as it does in blood (curve 5, dotted). (B) Effect of pH on oxygen affinity (the Bohr effect). As the pH decreases (that is, as the proton concentration increases), the oxygen affinity of hemoglobin decreases. (A, adapted from J.V. Kilmartin and L. Rossi-Bernardi, *Physiol. Rev.* 53: 884, 1973. With permission from the American Physiological Society; B, adapted from R.E. Benesch and R. Benesch, *Adv. Prot. Chem.* 28: 212, 1974. With permission from Elsevier.))

## 14.19 Bisphosphoglycerate (BPG) stabilizes the T-state quaternary structure of hemoglobin

The shift between the R and T states of the hemoglobin tetramer is somewhat more complex than described in the previous sections. It turns out that when the binding isotherm for purified hemoglobin is measured at pH 7.6 or above, hemoglobin is essentially saturated at an oxygen pressure of ~100 torr. The saturation drops to only ~60% upon reducing the oxygen concentration to 30 torr (**Figure 14.26**). Hemoglobin in red blood cells is actually much more efficient at transporting oxygen than these figures would lead us to believe. The essential difference between purified hemoglobin and hemoglobin in the blood arises from its response to chemical factors known as **allosteric effectors**, which modify the oxygen-binding properties of hemoglobin.

**Allosteric effectors**

Allosteric effectors are molecules that bind to a protein at a site other than the active site and affect the binding of the main target or substrate of the protein.

One important allosteric effector is a small molecule known as bisphosphoglycerate (BPG). BPG binds to deoxyhemoglobin at an interfacial site between the two β subunits (**Figure 14.27**; also shown in Figure 14.25A), where positively charged

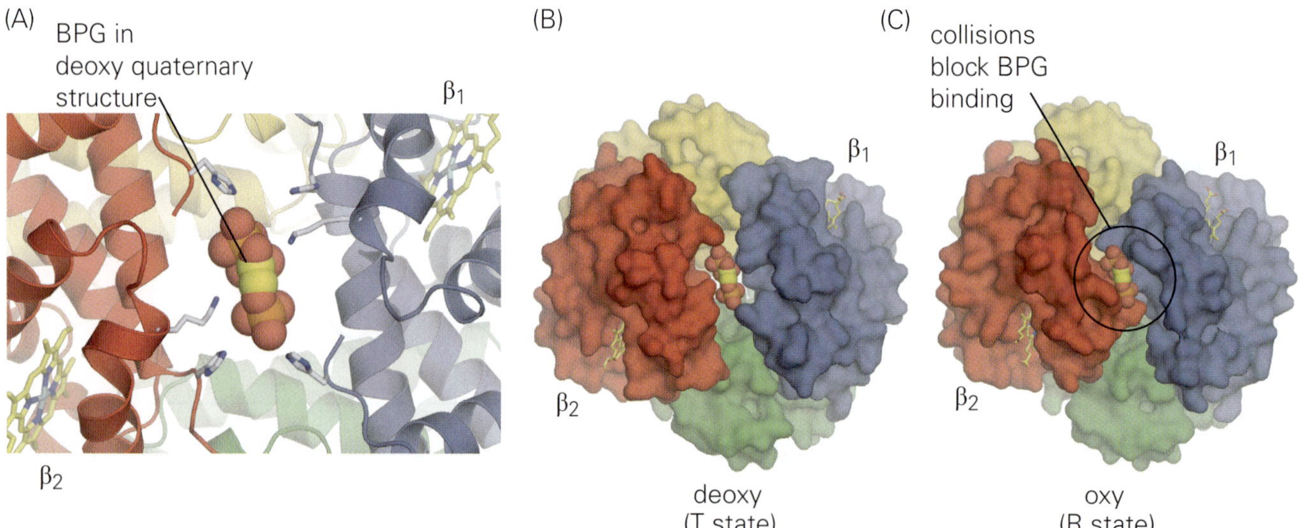


**Figure 14.27 Binding of BPG to hemoglobin.** (A) Bisphosphoglycerate (BPG) is a negatively charged molecule that binds in the vicinity of positively charged residues (histidines and lysines) in the T-state quaternary structure. This binding mode is facilitated by lower pH, which increases the protonation of the histidines. (B) The BPG molecule is accommodated within a cleft that is formed between the two β subunits in the T-state structure, as also shown schematically in Figure 14.25A. (C) This cleft closes in the R-state structure. BPG therefore opposes the binding of oxygen. (PDB codes: 4HHB, 1HHO.)

residues interact with the negatively charged phosphate groups of BPG. The rotation between the two α/β dimers that accompanies the T-to-R transition closes the cleft within which BPG binds (see Figure 14.27C). BPG, which therefore opposes the T-to-R transition, is present in high concentrations in red blood cells and, as a consequence, the oxygen affinity of hemoglobin in the blood is reduced relative to that of purified hemoglobin. The presence of BPG "tunes" the binding isotherm of hemoglobin to better match the difference in oxygen concentration between arterial and venous blood (see Figure 14.26A).

## 14.20 The low pH in venous blood stabilizes the T-state quaternary structure of hemoglobin

Another important allosteric effector of hemoglobin is carbon dioxide, $CO_2$. $CO_2$ affects the oxygen-binding properties of hemoglobin directly, by reacting with the N-terminal amino groups of the protein. The N-terminal regions of both the α and the β subunits are involved in interfacial interactions (**Figure 14.28**), and so the change in their physical properties alters the balance of the R and T states. $CO_2$ also affects the T-to-R transition indirectly by causing the release of protons and a concomitant drop in the pH of the blood. This proton release occurs upon the reaction of $CO_2$ with water to form bicarbonate, a reaction catalyzed in red blood cells by the enzyme carbonic anhydrase:

$$CO_2 + H_2O \rightleftharpoons H^+ + HCO_3^-$$

As with BPG, the oxygen affinity of hemoglobin decreases when the proton concentration increases (that is, when the pH decreases). This phenomenon, known as the **Bohr effect**, is particularly important because it maximizes the release of oxygen from hemoglobin in the tissues, where the $CO_2$ concentration is high and the pH is low.

**Bohr effect**

The reduction in the affinity of hemoglobin for oxygen that occurs at lower pH is known as the Bohr effect, named after the physiologist Christian Bohr. This phenomenon is important because respiration acidifies venous blood, and the Bohr effect facilitates the release of oxygen from hemoglobin.

(A)

α

β

ion pair with Asp β94 favored by low pH

Asp β94

Lys α40

F helix

β

α

His β146

(B)

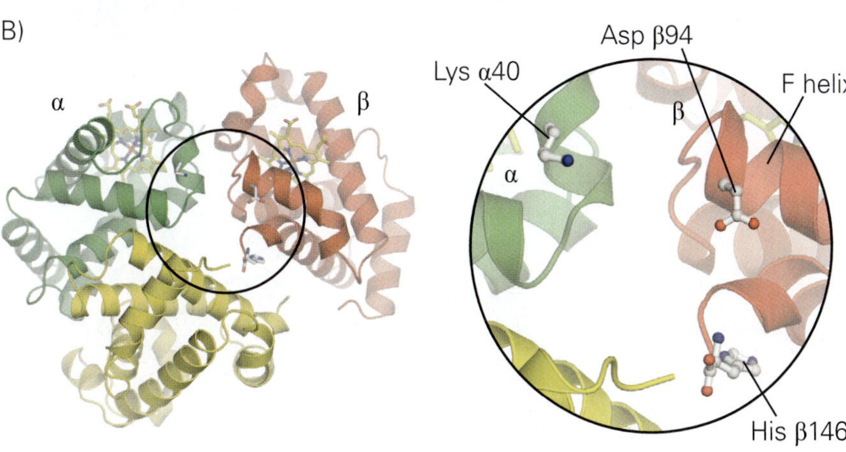


**Figure 14.28 Origin of the Bohr effect.** The Bohr effect arises from a pH-dependent change in the structure of hemoglobin such that the deoxy form (that is, the T state) is favored. (A) Structure of T-state hemoglobin, showing an ion-pairing network involving His β146. The interaction between the histidine sidechain and that of Asp β94 would be favored at low pH, when the histidine sidechain is protonated. (B) Structure of R-state hemoglobin. The network of interactions shown in (A) is disrupted. (PDB codes: 4HHB, 1HHO.)

The Bohr effect arises from interactions made by the amino-terminal groups of the peptide chains as well as certain histidine residues, which are stabilized in the charged form by interactions with negatively charged groups in the deoxy but not in the oxy structure. For example, Figure 14.28 shows a network of interactions that involves ion pairing between His β146 and Asp β94. His β146 is connected across an interface to the sidechain of Lys α40. The T-to-R transition leads to a substantial rearrangement of these residues, and this network of interactions is disrupted. Low pH will favor protonation of His β146, thereby stabilizing the interaction with Asp β94 and making the T-to-R transition more difficult. As a consequence, the oxygen-binding isotherm is right-shifted as the pH decreases (see Figure 14.26B). The combined effects of $CO_2$, the Bohr effect, and BPG make hemoglobin much more ultrasensitive to oxygen concentration than it would otherwise be.

## 14.21 Hemoglobins across evolution have acquired distinct allosteric mechanisms for achieving ultrasensitivity

As discussed in Section 14.12, hemoglobin is critical for respiration because it helps overcome the limited solubility of oxygen in blood and also increases the efficiency of oxygen release in tissues. Not surprisingly, other organisms with blood-based respiratory systems also utilize allosteric hemoglobin molecules to overcome these intrinsic limitations in oxygen uptake and transport. Recall from Chapter 5 that we looked at the structures of the globin subunits of hemoglobins from a diverse array of species and noted the striking conservation of three-dimensional structure despite the great extent of sequence drift amongst the globins. What is surprising, however, is that the allosteric mechanisms utilized by non-mammalian hemoglobins are often quite different from that described in the previous sections for human hemoglobin.

The oxygen-binding isotherm for a dimeric clam hemoglobin, drawn in a log-log plot, is compared with that for human hemoglobin in **Figure 14.29**A. Although the evidence for cooperativity is not as striking as for human hemoglobin, the slope of the binding isotherm for clam hemoglobin clearly deviates from the line of unit slope. The value of the Hill coefficient, $n_H$, is 1.4, corresponding to a ~5.4-fold increase in affinity for the second oxygen that binds, relative to the first one (see Equation 14.6).

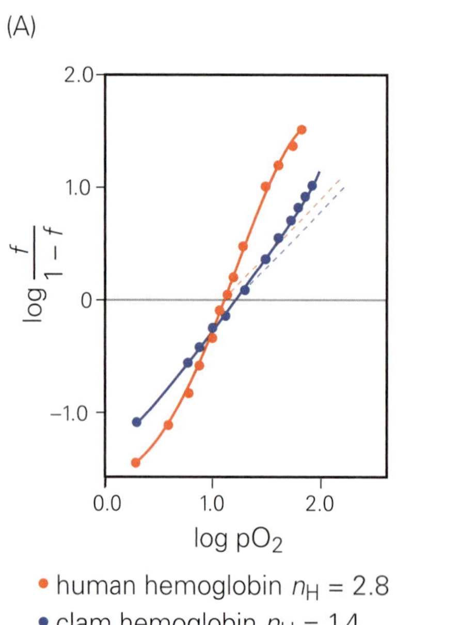


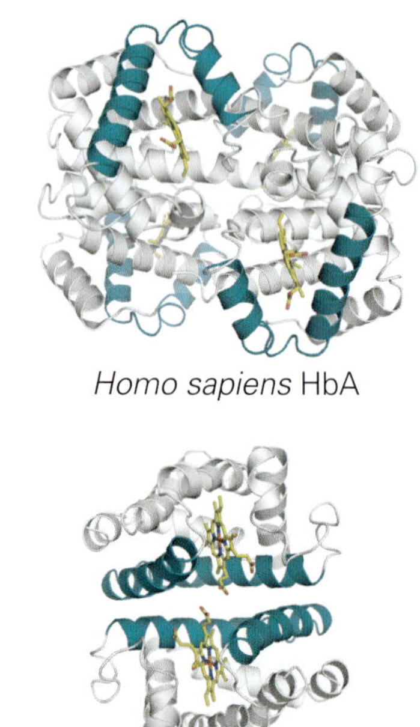


**Figure 14.29 Cooperativity and the structure of clam hemoglobins.** (A) Binding isotherm for clam hemoglobin compared with that of human hemoglobin. Straight lines of unit slope are shown as dashed lines. (B) The structure of clam hemoglobin compared with that of human hemoglobin. The F and G helices in each globin subunit are colored *blue*. Note that the relative orientation of globin subunits in clam hemoglobin is different from that in human hemoglobin. (A, adapted from R.C. San George and R.L. Nagel, *J. Biol. Chem.* 260: 4331–4337, 1985. With permission from American Society for Biochemistry and Molecular Biology; PDB codes: 2HHB and 1HLM.)

The MWC–Perutz mechanism for allostery in human hemoglobin has as its central element the rotation of two $\alpha/\beta$ dimers with respect to each other. Such a mechanism cannot be operative in the clam hemoglobins, which are dimers rather than tetramers. The structure of a dimeric hemoglobin from a clam is compared with that of human hemoglobin in Figure 14.29B. Note that the heme groups in the clam hemoglobin are located close each other in the dimer, whereas the heme groups in the human hemoglobin tetramer are located far apart from each other. It turns out that the allosteric mechanism in the clam hemoglobin involves a rather direct transmission of the effects of oxygen binding at one heme group to the adjacent one, with interfacial water molecules playing a key role in the signaling mechanism. Thus, nature has come up with an altogether different mechanism for achieving allostery in the clam hemoglobin.

The quaternary structures of hemoglobins from a diverse set of species are shown in **Figure 14.30**, revealing a striking diversity of assembly patterns. Some hemoglobins are dimeric, some are tetrameric, and others are organized into higher-order oligomers. The diversity of interfacial packing suggests that the details of the allosteric mechanisms are different in each case. The intricate series of molecular interactions seen in human hemoglobin, in which the change in size of the iron atom is transmitted through the proximal histidine and the F helix to the breakage of ion pairs at the inter-subunit interfaces, is apparently only one of several ways in which the globin fold can be adapted to yield a cooperative response to oxygen binding.

## 14.22 Allosteric mechanisms are likely to evolve by the accretion of random mutations in colocalized proteins

The allosteric mechanism of human hemoglobin is a marvel of molecular engineering, and yet the comparison with other hemoglobins shows that it is not a unique mechanism. This is a very common finding in biological systems, where it is often observed that phylogenetically related proteins have acquired allosteric mechanisms that appear to be unrelated in terms of molecular detail. This suggests that allostery evolves randomly, by the serial acquisition of mutations. When we consider the end result of such an evolutionary process, such as the series of linked interactions that constitute the allosteric response in human hemoglobin, it is difficult to imagine how such a remarkable mechanism was arrived at by the "blind" process of mutational change and natural selection.

One speculation regarding the evolution of allosteric mechanisms invokes the colocalization of protein subunits as a critical first step in the development of cooperative interactions. There are many different ways in which two proteins (or two domains of the same protein) could be colocalized in the cell (**Figure 14.31**). These could include binding to adjacent sites on DNA or on a scaffold protein, being expressed together as part of a larger protein as a consequence of gene fusion or duplication, or by being enclosed within the same restricted cellular compartment. The effect of such colocalization is to increase greatly the local concentration of the proteins, which serves to amplify the effects of random mutations in terms of their effects on neighboring domains or proteins.

Consider, for example, two protein domains that are diffusing in the cell in an uncorrelated fashion. A random mutation on the surface of one of the proteins is unlikely to lead to a dramatic change in the interaction (or lack thereof) between the two proteins. Such a mutation might add or remove a hydrogen-bonding group, or increase or decrease the size of a hydrophobic patch on the surface of one of the proteins. A single such potential interaction is probably not worth much more than about 1 to $2k_BT$ in terms of the free energy of binding and is unlikely to overcome the substantial entropic barrier to their interaction.

Now let us compare this situation to what happens when the two protein domains are localized together at the cell membrane, on DNA, or tethered together as part

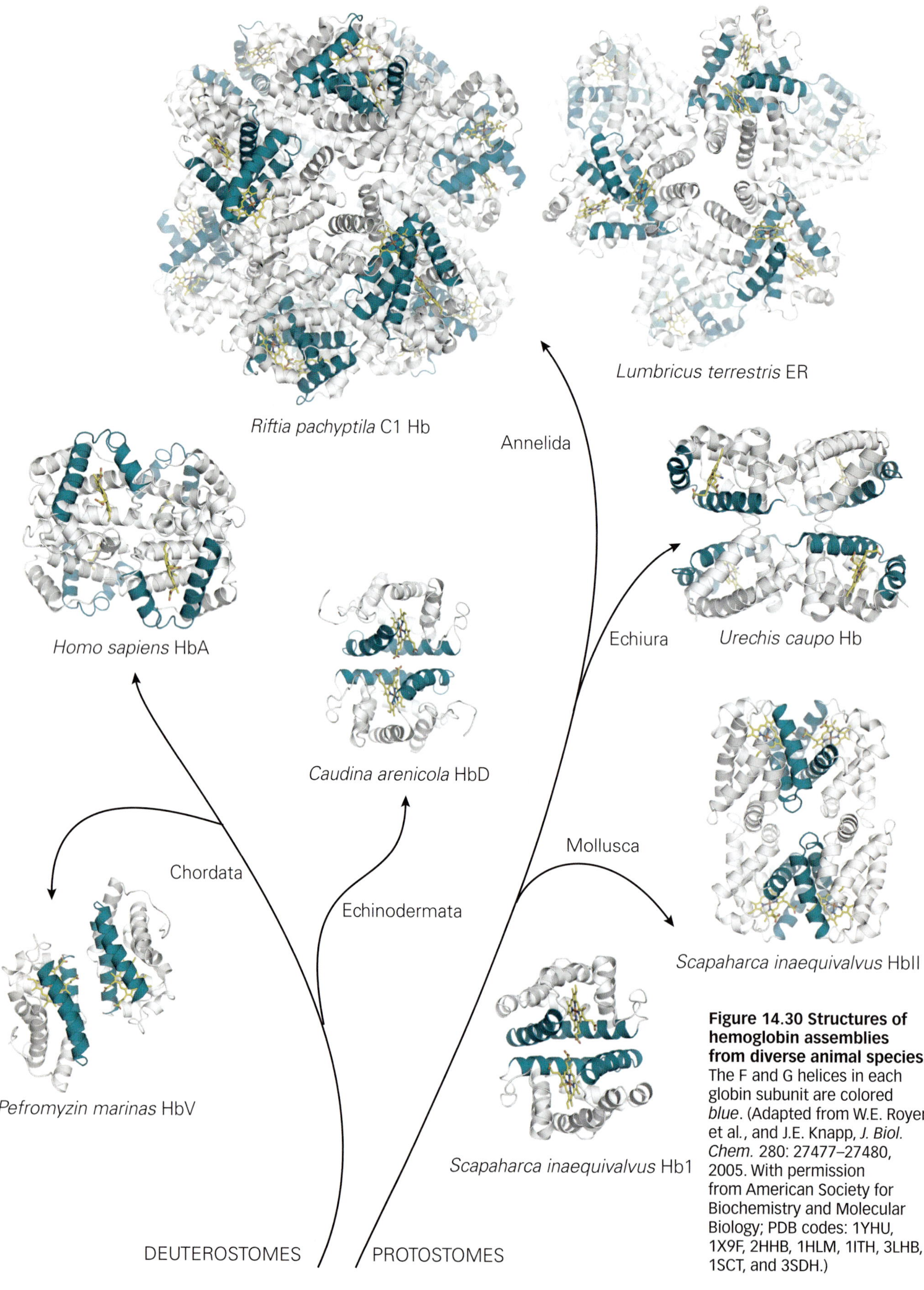


**Figure 14.30 Structures of hemoglobin assemblies from diverse animal species.** The F and G helices in each globin subunit are colored *blue*. (Adapted from W.E. Royer et al., and J.E. Knapp, *J. Biol. Chem.* 280: 27477–27480, 2005. With permission from American Society for Biochemistry and Molecular Biology; PDB codes: 1YHU, 1X9F, 2HHB, 1HLM, 1ITH, 3LHB, 1SCT, and 3SDH.)

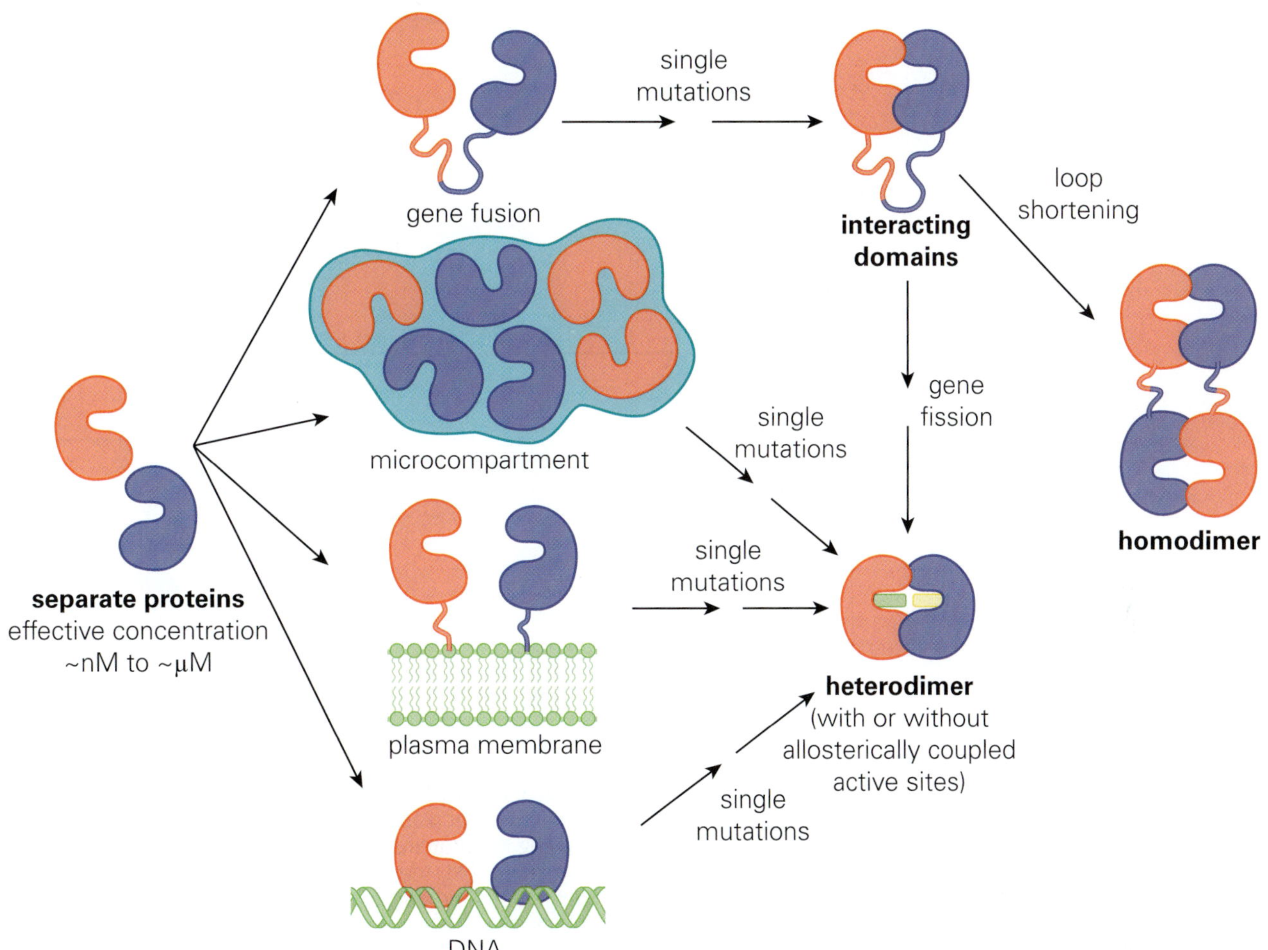


**Figure 14.31 A speculation regarding the evolution of allosteric portions.** Two proteins moving about freely in the cell are unlikely to change their interaction due to random mutations. If the two proteins are colocalized in some way, the increase in local concentration can amplify the effects of random mutations very significantly. If the effect of a mutation in one protein is to alter the behavior of its tethered partner in a favorable way, then the mutation will be selected for over evolutionary time. Additional mutations are then selected for, eventually leading to a highly dependent partnership. (Adapted from J. Kuriyan and D. Eisenberg, *Nature* 450: 983–990, 2007.)

of the same polypeptide chain. If we assume, for example, that the two proteins are localized in a spherical region of radius ~100 Å, then the local concentration of the two proteins within this region is in the millimolar range ($10^{-3}$ M). This is very much higher than the typical concentration of proteins within the cell, which is in the micromolar range ($10^{-6}$ M) or lower.

What is the effect on complex formation of increasing the local concentration of each protein by a factor of 1000 (that is, from micromolar to millimolar)? As shown in **Figure 14.32**, the effect is to move the interaction from the low-concentration end of the binding curve to the high-concentration one, which results in most of the molecules interacting with each other. Another way to think about the effect of a high local concentration is to consider the effect that it has on the free energy of binding. Recall that the reaction free energy, $\Delta G$, for a reaction under nonequilibrium conditions of concentrations of reactants and products, is given by:

$$\Delta G = \Delta G^{\circ} + RT \ln Q \tag{14.7}$$

where $Q$ is the reaction quotient (Section 10.10). For a binding equilibrium involving two molecules, A and B, that form a complex, $Q$ is related to the nonequilibrium concentrations of A, B, and the complex, A•B, by:

$$Q = \frac{[\text{A} \bullet \text{B}]}{[\text{A}][\text{B}]} \tag{14.8}$$

Now consider how the reaction free energy changes if we increase the concentrations of A and B by a factor of 1000. To do this in an approximate way, we ignore any

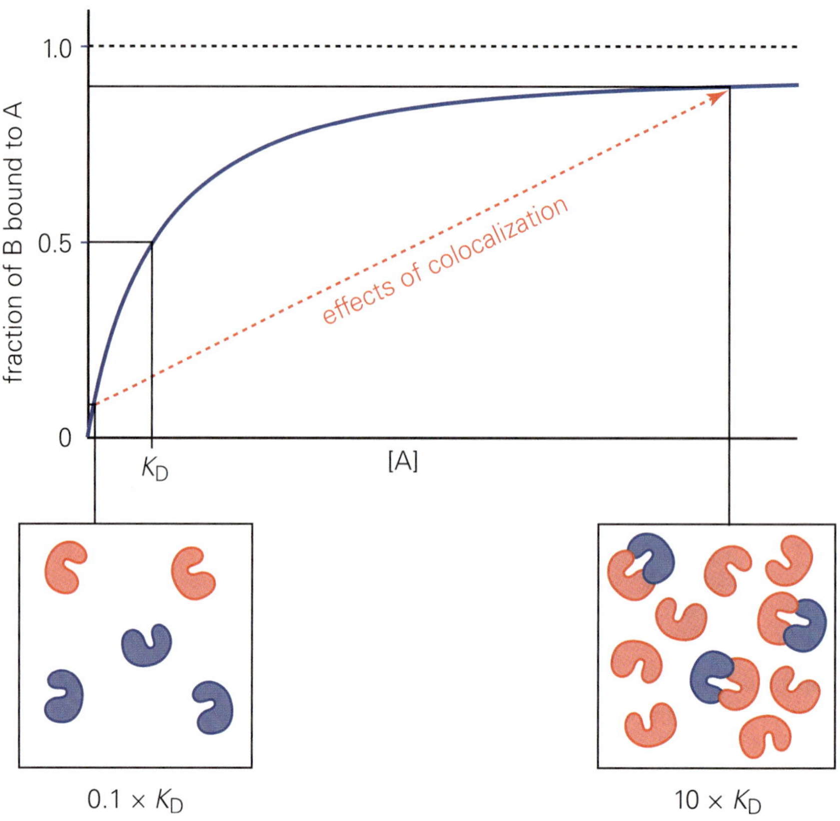


**Figure 14.32 The effect of colocalization on complex formation.** When two random proteins (A, *red*, and B, *blue*) are in the cytoplasm, their concentrations are low and they are unlikely to interact. Colocalizing the proteins, such as at the membrane, can increase their effective concentration by ~1000-fold. The effective concentration may then be above the dissociation constant, and the proteins can interact. (Adapted from J. Kuriyan and D. Eisenberg, *Nature* 450: 983–990, 2007.)

differences in the concentrations of the product, A•B, and calculate the change in reaction free energy as follows:

$$\Delta\Delta G = \Delta G\,(\text{higher concentration}) - \Delta G\,(\text{lower concentration})$$

$$= RT\ln\left[\frac{(10^{-6})(10^{-6})}{(10^{-3})(10^{-3})}\right] = RT\ln(10^{-6}) = -34.5\ \text{kJ mol}^{-1}\ (\text{at }300\ \text{K}) \tag{14.9}$$

Equation 14.9 tells us that an increase in the effective concentration of each partner in a binding reaction by a factor of 1000 is equivalent to providing an additional ~35 kJ•mol$^{-1}$ of favorable free energy for the binding reaction. That is, when proteins are in highly concentrated local environments, there is a very substantial boost in the effective free-energy change upon binding. Under these circumstances, random mutations that otherwise might have no effect on their mutual interactions could now potentially lead to a significant effect (see Figure 14.31). The two proteins might now begin to interact, and the orientation of one with respect to the other could alter the binding or catalytic properties of either protein. If these changes are unfavorable, natural selection will ensure that they are lost. If, on the other hand, the mutations lead to an increased fitness of the organism, then they will be selected for.

The accumulation of a series of such mutations can result eventually in the development of an interdependent set of allosteric interactions between the two domains. A corollary of this idea is that allostery will evolve in an opportunistic manner through the accretion of random mutations, with different members of a set of related proteins acquiring allostery in different ways. This is what is indeed seen, not just for the hemoglobins but also for many different proteins in the cell.

## Summary

A key to the regulation of cellular processes is the switching on and off of the activities of proteins. This is achieved either by covalent modification or by the binding of one molecule to another. Many of the key proteins in cellular regulation are ultrasensitive. The response of an ultrasensitive protein to an input signal (for example, the concentration of a molecule that affects its function) is sharper than predicted by a simple hyperbolic binding curve. For simple binding interactions, the ligand concentration has to change by ~100-fold (more precisely, by 81-fold) in order for the protein to go from 10% bound to 90% bound. If the protein is ultrasensitive to the ligand, a smaller increase in ligand concentration suffices to bring about the same change in occupancy.

One example of an ultrasensitive switch is seen in bacterial chemotaxis. A very small increase in the concentration of the phosphorylated form of the signaling protein CheY, in response to the binding of a repellant molecule, switches the direction of the bacterial flagellar motor from counterclockwise to clockwise, which causes the bacterium to tumble randomly. We also looked at a set of coupled protein kinases, known as the MAP kinases, which respond to signaling molecules such as progesterone. A 2.5-fold increase in progesterone concentration is sufficient to switch the signaling pathway from inactive to the completely activated state.

Proteins that have more than one binding site for ligands often exhibit cooperativity in the binding of these ligands, a phenomenon known as allostery. In the case of hemoglobin, allostery increases the efficiency of oxygen transport from the lungs to the tissue. Hemoglobin is a tetramer, with four binding sites for oxygen. Hemoglobin with no oxygen bound is in a conformation known as the T state, in which all four binding sites have low affinity for oxygen. As oxygen begins to bind, the conformation of the hemoglobin tetramer switches to the R state, in which the binding sites have high affinity for ligand. Conversely, when oxygen molecules dissociate in the tissues, the conformation switches to the T state, which facilitates efficient unloading of oxygen. The oxygen-binding properties of hemoglobin are further modulated by allosteric effectors, such as $CO_2$, protons, and bisphosphoglycerate (BPG), which optimize the loading and unloading properties of hemoglobin.

The binding isotherm for an allosteric protein is different in a characteristic way from that for a non-allosteric protein. A graph of the fractional saturation, $f$, versus ligand concentration is a hyperbolic curve for a non-allosteric protein, but exhibits a sigmoid shape for an allosteric one. A log-log plot, in which the logarithm of $(f/1 - f)$ is plotted versus the logarithm of the ligand concentration, is linear for a non-allosteric protein, with unit slope. The comparable graph for an allosteric protein is nonlinear, and its slope at the midpoint of saturation (when $f = 0.5$) is known as the Hill coefficient. The value of the Hill coefficient is greater than 1, but less than the number of allosterically coupled binding sites for positive cooperativity, and less than 1 for negative cooperativity.

Oxygen binding by hemoglobin is the only allosteric process we have discussed in this chapter, but allostery is a very important general phenomenon in molecular interactions. Although the molecular details of an allosteric system such as hemoglobin look complicated, such systems arise through the accumulation of random mutations that modulate a fundamental function (oxygen binding in the case of hemoglobin) and are selected for over evolutionary time. This principle is illustrated by the diversity of allosteric mechanisms found in hemoglobins from different animal species, all of which utilize the same globin fold and heme group to bind oxygen.

## Key Concepts

### A. ULTRASENSITIVITY OF MOLECULAR RESPONSES

- Molecular outputs that depend on independent binding events switch from on to off over a ~100-fold range in input strength.
- Very small changes in the concentration of an input signal can lead to a large change in the response of an ultrasensitive protein.
- The response of many biological systems is ultrasensitive, with the switch from off to on occurring over a less than ~100-fold range in concentration.
- Cooperativity and allostery are features of many ultrasensitive systems.
- During bacterial chemotaxis, the response of the flagellar motor to the concentration of the signaling protein CheY–P is ultrasensitive.
- The sequential phosphorylation of the MAP kinases leads to an ultrasensitive signaling switch.

### B. ALLOSTERY IN HEMOGLOBIN

- Allosteric proteins exhibit positive or negative cooperativity.
- The heme group in hemoglobin binds oxygen reversibly.
- The oxygen-binding isotherm of hemoglobin has a sigmoid shape, indicating ultrasensitivity.
- Hemoglobin increases the solubility of oxygen in blood and also increases the efficiency of its transport to the tissues.
- The sigmoid binding isotherm of hemoglobin arises from switching between conformations that have low and high affinity for oxygen, denoted T and R, respectively.
- The degree of cooperativity between binding sites is characterized by the Hill coefficient.
- Changes in the tertiary structure of hemoglobin upon oxygen binding are coupled to changes in quaternary structure.
- The hemoglobin tetramer is in equilibrium between the R and T states, and oxygen binding biases the equilibrium towards the R state.
- Allosteric effectors, such as BPG and protons, increase the cooperativity of hemoglobin.
- Hemoglobins from different evolutionary lineages have evolved different mechanism for ultrasensitivity.
- Allosteric mechanisms are likely to evolve by the accretion of random mutations in colocalized proteins.

## Problems

### True/False and Multiple Choice

1. Which of the following can result in an allosteric modulation of activity?
   a. Covalent modification such as phosphorylation or acetylation.
   b. Oligomerization.
   c. Binding of a ligand.
   d. Stabilizing an alternative conformation.
   e. All of the above.

2. The transcription of a gene is controlled by a transcription factor binding either glucose or lactose. When there is 0.2 mM of either glucose or lactose in the cell, the gene is transcribed at about 10% of the maximum. When there is more than 2 mM glucose in the cell, the gene is fully induced. However, when there is 2 mM lactose in the cell, the amount of transcription is approximately half of the maximum. At 40 mM of either glucose or lactose in the cell, the gene is fully induced. The transcriptional regulation is likely:
   a. Ultrasensitive with respect to both glucose and lactose.
   b. Ultrasensitive with respect to lactose but not glucose.
   c. Graded with respect to glucose.
   d. Ultrasensitive with respect to glucose but not lactose.
   e. Hyperbolic with respect to glucose but not lactose.

3. In allosteric proteins, ligand binding can only result in positive cooperativity.

   True/False

4. Which of the following is *not* a known allosteric effector of hemoglobin?
   a. oxygen
   b. bisphosphoglycerate
   c. low pH
   d. myoglobin
   e. $CO_2$

5. Bacterial chemotaxis involves only random, Brownian movement.

   True/False

6. Because of the importance of the F helix in oxygen binding, the allosteric hemoglobin proteins of all organisms are tetrameric assemblies.

   True/False

7. Which of the following statements about the Hill coefficient are true?

i. It is the steepness of the log-log binding isotherm at the half saturation point.
ii. Allosteric systems have a Hill coefficient of exactly 1.
iii. The maximum value for a dimeric protein is 2.

a. Only (i) is true.
b. Both (i) and (iii) are true.
c. Both (ii) and (iii) are true.
d. All of the statements are true.

## Fill in the Blank

8. When a system is ultrasensitive, the response to an input is ______ than expected from the graded, hyperbolic response.

9. Oxygen biases the equilibrium of hemoglobin towards the _____ state.

10. The extent of flagellar clockwise rotation is governed by the ________________ of the protein CheY.

11. The _________ histidine residue of hemoglobin senses the effective size of the heme iron atom.

12. An effect of colocalization is to greatly increase the local _________ of two proteins.

## Quantitative/Essay

13. A dimeric enzyme, glucokinase, has a binding site for glucose in each subunit. The $K_D$ for the first binding event is 1 mM and the $K_D$ for the second event is 10 μM.

a. What is the Hill coefficient?
b. Is this protein positively or negatively cooperative with respect to glucose binding?

14. A dimeric hemoglobin is isolated from a fish. Each subunit contains a binding site for a xenon gas atom. The $K_D$ for the first binding event is measured to be 23 nM. The $K_D$ for the second binding event is measured to be 3.5 μM.

a. What is the Hill coefficient?
b. Is this protein positively or negatively cooperative with respect to xenon binding?

15. A dimeric enzyme with two identical binding sites has a Hill coefficent of 1.3. If the $K_D$ of the first binding site is 200 nM, what is the $K_D$ of the second binding site?

16. What is the ratio of bound to unbound receptor $[f/(1-f)]$ for a dimeric protein with two binding sites ($K_{D1} = 20$ nM; $K_{D2} = 2$ nM) at 15 nM concentration of ligand?

17. A dimeric allosteric protein is isolated. A scientist determines that the value of $K_D$ for the first binding site is 25 nM and that the Hill coefficient is 1.6.

a. What fraction of the protein has ligand bound at 10 nM concentration of free ligand?
b. A single point mutation abolishes all cooperativity in the protein such that the protein binds its ligand with an apparent $K_D$ equal to 25 nM. What is the fraction of the protein that has ligand bound at 10 nM concentration of free ligand?

18. The first and second binding sites of a positively cooperative allosteric dimeric protein have $K_D$ values of 100 mM and 10 μM, respectively.

a. Sketch the binding isotherms as $\log[f/(1-f)]$ versus $\log([L])$.
b. What is the value of the Hill coefficient?

19. The first and second binding sites of a negatively cooperative allosteric dimeric protein have $K_D$ values of 100 μM and 10 mM, respectively.

a. Sketch the binding isotherms as $\log[f/(1-f)]$ versus $\log([L])$.
b. What is the value of the Hill coefficient?

20. Two homologs of a protein are isolated. Homolog A is a monomer that binds glucose with a $K_D$ of 4 mM. Homolog B is a positively cooperative dimer. The $K_D$ of the first binding site of homolog B is measured at 4 mM. At 1 mM concentration of glucose, it is found that homolog B binds twice as much glucose as homolog A. What is the Hill coefficient of homolog B?

21. A cyclist is interested in cheating in a race by delivering more oxygen to his muscles. The cyclist reasons that since bisphosphoglycerate (BPG) stabilizes the "T" state of hemoglobin, which reduces its affinity for oxygen, that reducing the BPG concentration in his blood cells should be good for his performance. How might removing BPG have a detrimental effect on the delivery of oxygen to his muscles?

22. Due to a fascinating coupling between folding and ligand binding, the dimeric acetyl transferase enzyme AAC displays positive cooperativity at low temperatures and negative cooperativity at high temperatures. At 10°C, AAC has $K_{D1} = 330$ μM and $K_{D2} = 130$ μM. At 37°C, AAC has $K_{D1} = 200$ μM and $K_{D2} = 230$ μM.

a. What is the Hill coefficient at each temperature?
b. What is the change in fraction bound at 50 μM ligand between 10°C and 37°C?
c. Describe a reasonable mechanism underlying this unusual behavior.

23. How does $CO_2$ directly and indirectly stabilize the "T" state of hemoglobin in venous blood?

24. Two proteins are modified by myristoylation, which targets them to the plasma membrane in a cell at 25°C. This changes their effective local concentration from 10 nM to 1 μM. Assume that any favorable mutation would decrease the value of $\Delta G^o$ for binding by 4 $kJ \bullet mol^{-1}$. How many favorable mutations would have to occur in the absence of colocalization to result in an equivalent effective affinity as observed when the two proteins are colocalized?

25. A scientist finds two pathways, both of which depend on distinct kinase activities, which activate a stress response in yeast. One pathway responds to elevated

levels of salt and the other responds to elevated levels of caffeine. Both pathways result in transcription of the chaperone Hsp90. Below is a table listing the relative concentrations of salt or caffeine and the measured transcription of Hsp90. Explain which pathway likely contains a single kinase and which pathway likely contains a kinase cascade similar to the MAP kinase pathway.

| [NaCl] (mM) | Hsp90 transcription |
|---|---|
| 1 | 0 |
| 2 | 1 |
| 3 | 1 |
| 4 | 40 |
| 5 | 100 |
| 10 | 100 |
| 100 | 100 |
| 1000 | 100 |

| [Caffeine] (mM) | Hsp90 transcription |
|---|---|
| 1 | 0 |
| 2 | 0 |
| 3 | 1 |
| 4 | 1.1 |
| 5 | 2 |
| 10 | 10 |
| 100 | 50 |
| 1000 | 100 |

## Further Reading


Changeux J-P (2011) 50th anniversary of the word "Allosteric". *Prot. Sci.* 20, 1119–1124.

Changeux J-P & Edelstein SJ (2005) Allosteric mechanisms of signal transduction. *Science* 308, 1424–1428.

Cui Q & Karplus M (2008) Allostery and cooperativity revisited. *Prot. Sci.* 17, 1295–1307.

Dickerson RE & Geis I (1983) Hemoglobin: Structure, Function, Evolution, and Pathology. Menlo Park, CA: Benjamin-Cummings.

Kuriyan J & Eisenberg D (2007) The origin of protein interactions and allostery in colocalization. *Nature* 450, 983–990.

Philips R, Kondev J & Theriot J (2008) Physical Biology of the Cell. New York: Garland Science.

Ptashne M & Gann A (2002) Genes & Signals. Cold Spring Harbor, NY: Cold Spring Harbor Laboratory Press.

Swain JF & Gierasch LM (2006) The changing landscape of protein allostery. *Curr. Op. Struct. Biol.* 16, 102–108.

Voet D & Voet J (2004) Biochemistry, 3rd ed. New York: John Wiley & Sons; vol. 1, Chapter 10.


## References

### A. Ultrasensitivity of Molecular Responses


Bray D (1995) Protein molecules as computational elements in living cells. *Nature* 376, 307–312.

Cluzel P, Surette M & Leibler S (2000) An ultrasensitive bacterial motor revealed by monitoring signaling proteins in single cells. *Science* 287, 1652–1655.

Ferrell Jr JE & Machleder EM (1998) The biochemical basis of an all-or-none cell fate switch in Xenopus oocytes. *Science* 280, 895–898.

Huang CY & Ferrell Jr JE (1996) Ultrasensitivity in the mitogen-activated protein kinase cascade. *Proc. Natl. Acad. Sci. USA* 93, 10078–10083.


### B. Allostery in Hemoglobin


Eaton WA, Henry ER, Hofrichter J & Mozzarelli A (1999) Is cooperative oxygen binding by hemoglobin really understood? *Nat. Struct. Biol.* 6, 351–358.

Monod J, Wyman J & Changeux J-P (1965) On the nature of allosteric transitions: a plausible model. *J. Mol. Biol.* 12, 88–118.

Perutz MF, Wilkinson MJ, Paoli M, & Dodson GG (1998) The stereochemistry of the cooperative effects in hemoglobin revisited. *Annu. Rev. Biophys. Biomol. Struct.* 27, 1–34.

Royer Jr WE, Zhu H, Gorr TA, Flores JF & Knapp JE (2005) Allosteric hemoglobin assembly: Diversity and similarity. *J. Biol. Chem.* 280, 27477–27480.

Szabo, A & Karplus, M (1972) A mathematical model for structure-function relationships in hemoglobin. *J. Mol. Biol.* 72, 163–197.

# PART V

## KINETICS AND CATALYSIS

# CHAPTER 15

# The Rates of Molecular Processes

In the previous chapters, we have dealt with principles of thermodynamics, which help us keep track of the changes in energy and entropy. By combining energy and entropy we obtained the Gibbs free energy, *G*, which is a parameter that enables us to predict whether reactions will occur spontaneously. With a knowledge of the standard free-energy change for a reaction, $\Delta G^{o}$, the value of the equilibrium constant can be determined. This gives us a precise prediction of how far a reaction will go before it stops.

The discussion so far has focused on the conditions that govern equilibrium. Consider, for example, a reaction in which two kinds of reactant molecules (A and B) combine with each other to generate a third kind of molecule (C), as shown in **Figure 15.1**A. Given the free energies of A, B, and C, we can calculate the concentrations of the reactants and products at equilibrium (Figure 15.1B). You should realize, however, that knowing the free energies of A, B, and C does not tell us anything about how *long* these molecules will take to reach equilibrium—that is, how fast the reaction will occur. If we mix the reactants and products together, the concentrations of A, B, and C will relax towards their equilibrium values, as shown in Figure 15.1B. But *time* is a variable that we ignored in our analysis of energy, entropy, and free energy and, without additional information, we cannot predict how long the reaction will take.

Our intuition may tell us that if a reaction is very favorable—that is, if it is associated with a large, negative $\Delta G^{o}$ value—then it will occur rapidly. While such reactions can sometimes be fast, there is no general correlation between how

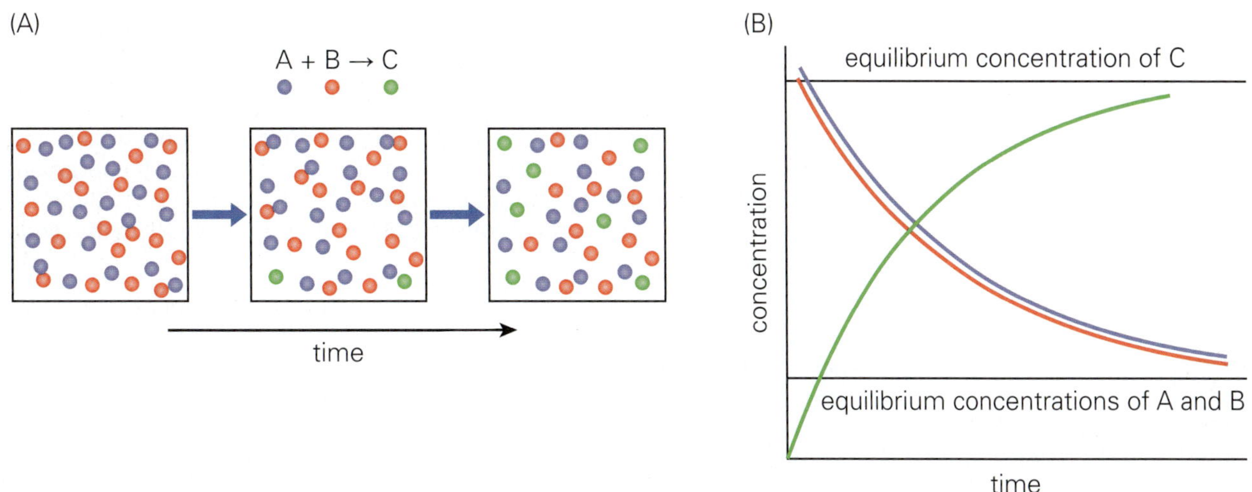


**Figure 15.1 Time as a variable in the progress of a reaction.** (A) Schematic representation of a chemical reaction in which A and B molecules react with each other to produce C molecules. (B) The time course of the reaction. The system is initially not at equilibrium, but with time, A and B molecules are converted to C molecules until equilibrium is reached. Knowing thermodynamic parameters, such as the free-energy change of the reaction, does not inform us about how fast the reaction will go.

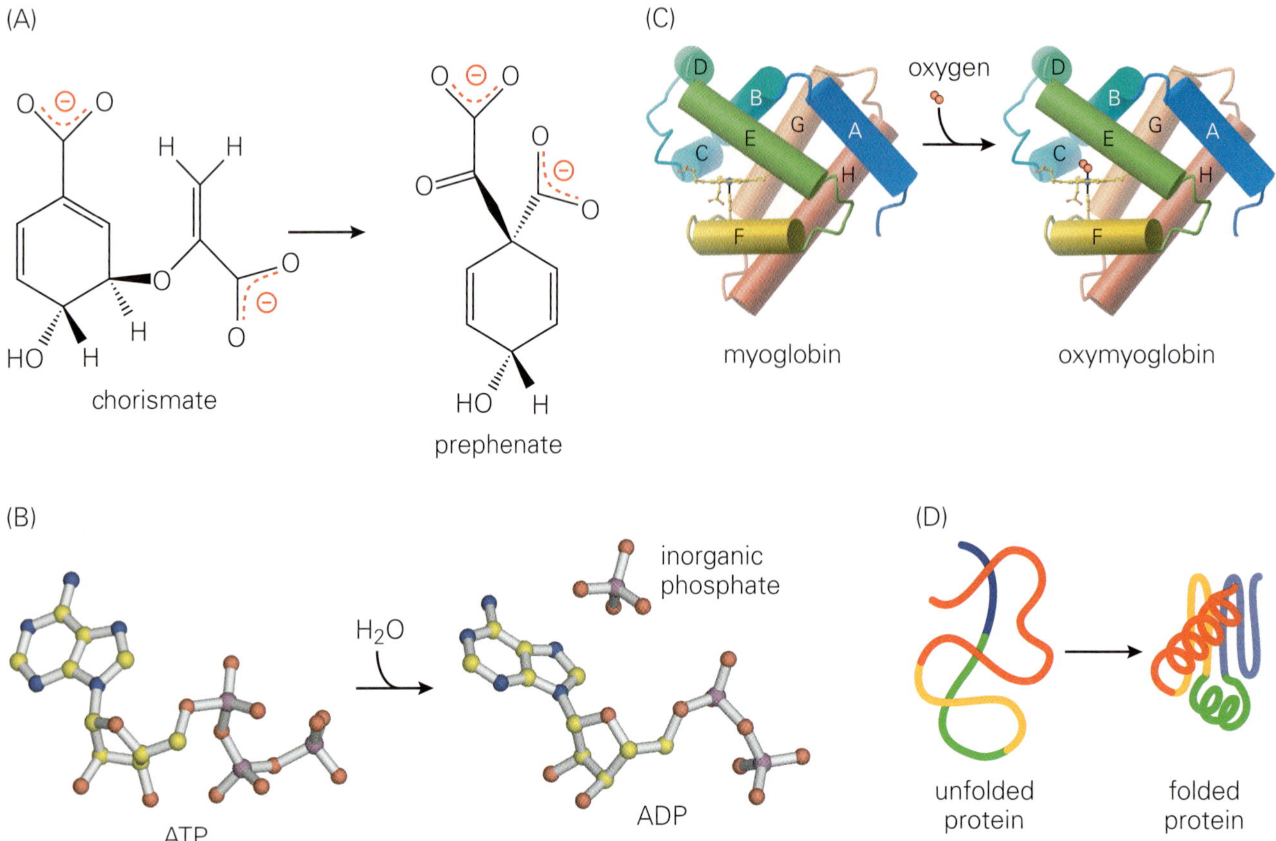


**Figure 15.2 Different kinds of reactions.** (A) The conversion of chorismate to prephenate, an intramolecular rearrangement. (B) The hydrolysis of ATP. (C) The binding of oxygen to myoglobin. (D) Protein folding.

favorable a reaction is in terms of the change in free energy and how fast it occurs. Additionally, the same reaction may occur at very different rates under different circumstances. This is indeed fortunate, because it allows reactions occurring in biological systems to be controlled. For example, the hydrolysis of ATP to give ADP and phosphate by addition of water is favorable in terms of $\Delta G^o$, as discussed in previous chapters, but the reaction occurs very slowly unless catalyzed by interactions with proteins in the cell. Thus, ATP is not rapidly hydrolyzed by water, enabling its use as an energy currency and reservoir in cells. Similarly, gasoline burns in oxygen, but the reaction essentially does not occur unless initiated by a spark, allowing us to use the reaction to power cars without worrying too much about exposing the gasoline to oxygen in air as we fill the tank of our car.

Living biological systems are not at equilibrium. While thermodynamics makes valuable predictions of when reactions can occur, and how much energy can be derived from them, the concentrations of most molecules in cells are determined by their **rates** of formation and consumption rather than simply by the equilibrium constants.

**Kinetics**

Kinetics is the study of the rates at which chemical and physical processes occur.

In this chapter and the next one, we describe the basic factors that determine the rates of chemical reactions and physical processes, a field of study generally known as **kinetics**. Kinetic principles are used to understand how fast molecular reactions occur. These reactions include chemical reactions in which covalent bonds are rearranged—that is, the reactants and products have the same number of atoms of each type, but some bonds (involving shared electrons) change between reactants and products. Other reactions may not involve bond breakage or formation. Such processes include the rotational isomerization of molecules, the association (binding) of one molecule to another, the folding of a protein, or the release of light from an excited molecule (fluorescence). A few examples of different kinds of reactions are shown in **Figure 15.2**.

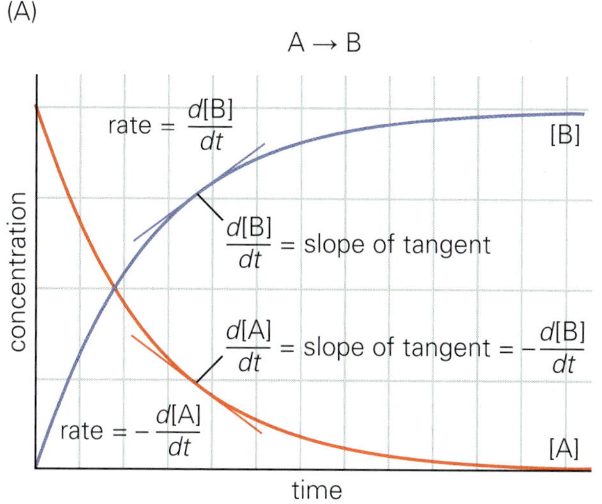


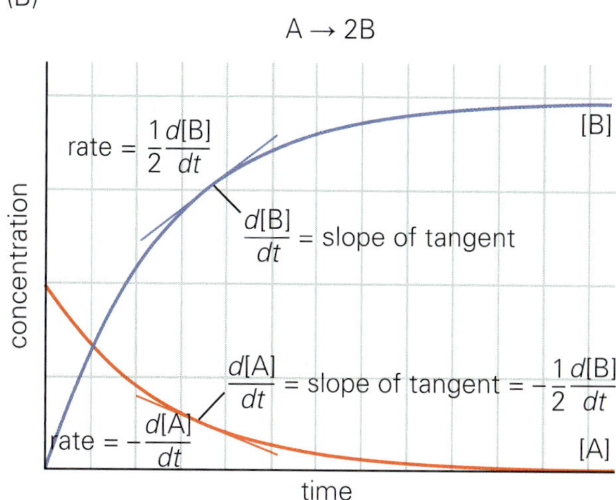


**Figure 15.3 The dependence of reaction rates on stoichiometry.** (A) The time dependence of the concentrations of reactants and products for the reaction A → B is shown. The rate of the reaction is the derivative of the concentration of product, [B], with time, which has a positive value. Because the stoichiometries of A and B are equal, the rate is equal to the negative of the derivative of [A] with time. (B) Concentrations of reactants and products for the reaction A → 2B. The concentration of B increases twice as rapidly as the decrease in concentration of A. The rate of the reaction is defined as $\frac{1}{2}\frac{d[\text{B}]}{dt}$, which is the same as $-\frac{d[\text{A}]}{dt}$.

# A. GENERAL KINETIC PRINCIPLES

## 15.1 The rate of reaction describes how fast concentrations change with time

A general chemical reaction can be written as follows:

$$a\text{A} + b\text{B} \rightarrow c\text{C} + d\text{D} \tag{15.1}$$

The uppercase letters represent the reactants and products. The lowercase letters are integers reflecting the stoichiometry (that is, $a$, $b$, $c$, and $d$ are stoichiometric coefficients whose values are such that the reaction is balanced and atoms are neither created nor destroyed). Note that the arrow in Equation 15.1 points in only one direction. In principle, any reaction can go forward or backward, but we first focus our analysis on the forward direction, for simplicity, and later extend it to include both directions. A general reaction may have any number of reactants and/or products, but the simple reaction depicted in Equation 15.1 suffices to introduce the concept of reaction rates.

The rate of the reaction is how fast the reactant changes to product with time. Usually we think in terms of the amount in a particular volume of interest—that is, the concentration, changing with time. Mathematically, the rate corresponds to the derivative of a concentration with respect to time. To provide an unambiguous definition, the rate is defined such that it does not matter which species is being followed (for the general reaction above):

$$\text{reaction rate} = -\frac{1}{a}\frac{d[\text{A}]}{dt} = -\frac{1}{b}\frac{d[\text{B}]}{dt} = \frac{1}{c}\frac{d[\text{C}]}{dt} = \frac{1}{d}\frac{d[\text{D}]}{dt} \tag{15.2}$$

The signs in Equation 15.2 reflect the fact that concentrations of the reactants decrease with time while the concentrations of the products increase. Thus, the rate, defined as in Equation 15.2, has a positive value as the reaction progresses.

To understand why we divide the time derivatives by the stoichiometric coefficients in Equation 15.2, study **Figure 15.3**, which illustrates the time dependence of the reactants and products for two particularly simple reactions. In the first example, the reaction involves the conversion of A molecules to B molecules. In this case, the rate is simply given by $-\frac{d[\text{A}]}{dt}$ or, equivalently, by $\frac{d[\text{B}]}{dt}$ (see Figure 15.3A).

In the second example, one molecule of A is converted to two molecules of B. This means that the concentration of B increases twice as rapidly as the concentration of A decreases. If we were to define the rate as simply the time derivative of one concentration (for example, $-\frac{d[A]}{dt}$ or $\frac{d[B]}{dt}$), then the value of the rate would depend on whether A or B was being followed. By dividing the concentration derivatives by the stoichiometric coefficients, as shown in Equation 15.2, the definition of the rate does not depend on which of the reactants or products is followed (see Figure 15.3B).

## 15.2 The rates of intermolecular reactions depend on the concentrations of the reactants

When we first learn about chemical reactions it is usually in the context of two different chemicals reacting (rearranging bonds, transferring atoms from one reactant to the other). A common example is a phosphate transfer reaction:

$$\text{glucose} + \text{ATP} \rightarrow \text{glucose-phosphate} + \text{ADP} \tag{15.3}$$

For such a process to occur, the reactants have to come close together (collide) in order to react, because bond rearrangements between two molecules require very close proximity. The probability of the reaction occurring per unit time should therefore be related to the rates of collisions. The rates at which molecules collide with each other will be examined in detail in Chapter 17, but for now we just need to know how the rates of reaction depend on concentration.

In a mixture of molecules A and B, the frequency with which A-B collisions occur is proportional to the concentrations [A] and [B]. We can understand why this is so by using the approach applied in Chapter 7 to calculate entropy. We divide space into boxes that may be occupied or not, as shown in **Figure 15.4**. The probability of a box being occupied by an A molecule is $P_A = N_A/n_{total}$, where $N_A$ is the number of A molecules in the box, and $n_{total}$ is the number of boxes. The concentration, [A], is proportional to $N_A/n_{total}$. Likewise, the probability of a box being occupied by a B molecule is $P_B = N_B/n_{total}$ and [B] is proportional to $N_B/n_{total}$. If we consider that a collision occurs when two neighboring boxes are occupied by A and by B, then the probability of this happening is $P_A P_B$ (**Figure 15.4**). The collision rate will therefore be proportional to the product of the concentrations, [A][B]. If the formation of the collision complex, A•B, is all that is required for the reaction to proceed, then the reaction rate will also be proportional to [A][B].

## 15.3 Rate laws define the relationship between the reaction rates and concentrations

The relationship of the rate of a reaction to the concentrations of the chemical species that are involved in the reaction is called the **rate law** or **rate equation**. Rate equations can be derived from knowing the fundamental steps that occur during a reaction, as discussed in subsequent sections of this chapter. For example, if two molecules, A and B, combine and react directly to yield a product molecule, C, then the reaction, $A + B \rightarrow C$, is called an **elementary reaction**. We would expect the rate of this reaction to be proportional to the product of the concentrations of the reactants, [A][B] (**Figure 15.5A**).

**Rate laws and rate constants**

A rate law or *rate equation* is a differential equation that specifies how the reaction rate depends on the concentrations of species present in the reaction mixture including, but not limited to, reactants and products. The rate constant, *k*, is a proportionality constant between the reaction rate and the concentrations of species that determine the reaction rate.

The rate law is the differential equation that describes the rate of change of concentration with time. For this elementary reaction, the rate law is:

$$\text{rate} = -\frac{d[A]}{dt} = \frac{d[C]}{dt} = k[A][B] \tag{15.4}$$

The proportionality constant, *k*, is called the **rate constant** of the reaction. What is the physical significance of the rate constant, *k*? The rate constant must be related to the frequency of collisions between the reactants, as discussed above. For example, conditions in which molecules move faster will lead to more frequent collisions and therefore faster reactions. The rates of movement of molecules, determined by factors like the viscosity, are discussed in Chapter 17 in some detail.

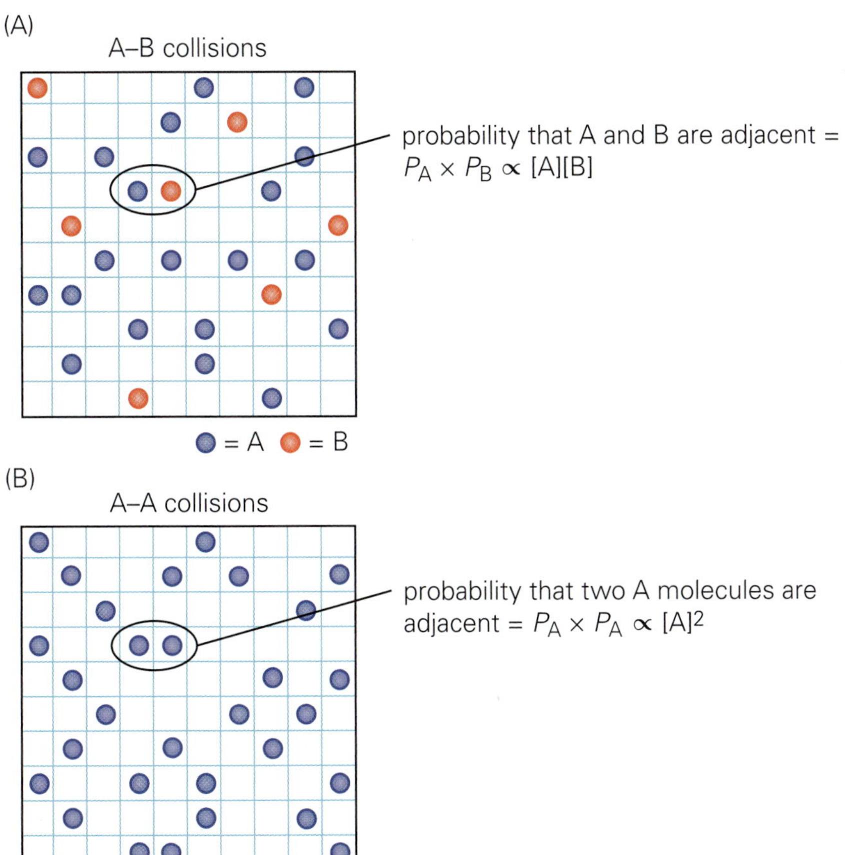


**Figure 15.4 Diagram for counting collisions.** (A) The frequency of collisions of B (*red molecules*) with A (*blue molecules*) is obtained by counting occurrences of A and B occupying neighboring boxes. The probability that a box is occupied by A is given by $P_A$, which is proportional to [A]. The probability that an adjacent box is occupied by B is given by $P_B$, which is proportional to [B]. The probability of A colliding with B is proportional to [A] × [B]. (B) If collisions of A with A lead to product, then the probability of collision is proportional to $[A]^2$.

The rate constant also reflects the fraction of collisions that actually lead to product formation, which is often very small. Most collisions are unproductive—that is, the two reactant molecules collide and then simply separate again without having undergone a reaction. The fraction of collisions that are productive depends on the chemical structure of the reactants and the energetics of the reaction. A more detailed analysis of the physical meaning of the rate constant is given in part C of this chapter.

Although it might be natural to think that the rate law would depend only on the concentrations of the reactants in the overall reaction, there are often other molecules present that do not appear in the overall reaction but still greatly affect the rate of reaction. For example, consider a reaction in which A molecules are converted to C molecules, but only if they collide with a catalyst, Z (see Figure 15.5B). Although the catalyst is unchanged by the reaction, it enters into the rate

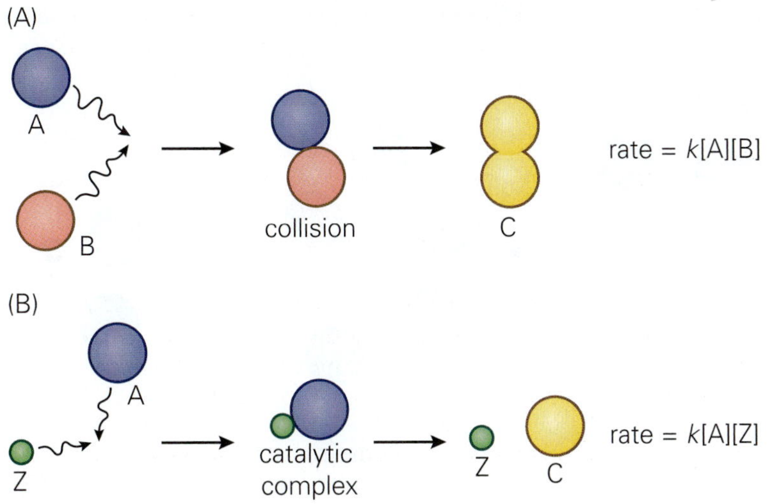

**Elementary reaction**

An elementary reaction represents the most basic step used to describe a reaction process. Many chemical mechanisms result from the combination of several elementary steps.

**Figure 15.5 Rate laws.** (A) Two molecules, A and B, collide to form a product molecule, C. The elementary reaction is A + B → C, and the rate law (shown on the right) depends on the concentrations of A and B. (B) The rate law can depend on the concentrations of molecules that do not enter into the overall reaction. In this example, an A molecule is converted to a C molecule, but only when it encounters a catalyst (Z). The rate law depends on the concentrations of both A and Z.

law because the reaction rate will be proportional to [A][Z]. This is particularly true for biological reactions, most of which are accelerated by catalysts that are proteins or RNA molecules.

## 15.4 The dependence of the rate law on the concentrations of reactants defines the order of the reaction

**Reaction order**

The order of the reaction is the sum of the concentration exponents in the rate law, which is the number of molecules colliding if it is an elementary reaction. An elementary bimolecular reaction has a reaction order of two.

The type of reaction depicted in Figure 15.5A, with two reactant molecules in the elementary reaction, is termed a **bimolecular reaction**. The sum of the exponents of the concentrations in a rate law is known as the **reaction order**. For the bimolecular reaction rate law, the order of the reaction is 1 + 1 = 2, and so bimolecular reactions are called **second-order** reactions. This concept of counting the number of required collision partners for the reaction to derive the order for elementary reactions is general, and schematic examples are shown in **Figure 15.6** for different orders of reaction.

The units of the rate constant depend on the order of the reaction. The rate of the reaction always has units of concentration per unit time. For second-order reactions, we can infer from Equation 15.4 that the units of $k$ are $M^{-1}\bullet sec^{-1}$. By convention, the rate constant is always taken to have a positive value. The sign on the

(A) first-order reaction

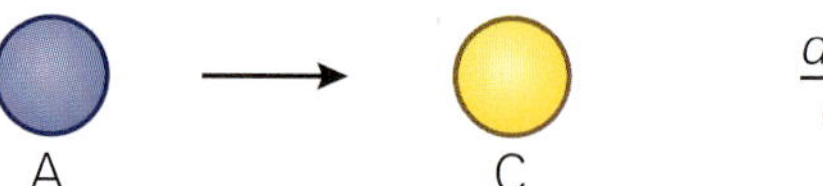


$$\frac{d[A]}{dt} = -k[A]$$

(B) second-order reaction

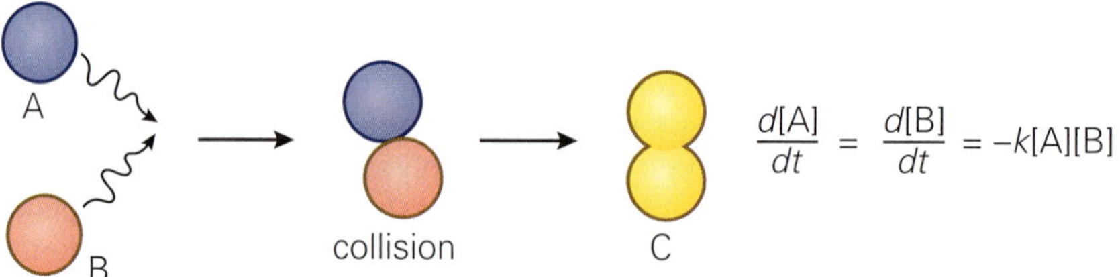


$$\frac{d[A]}{dt} = \frac{d[B]}{dt} = -k[A][B]$$

(C) third-order reaction

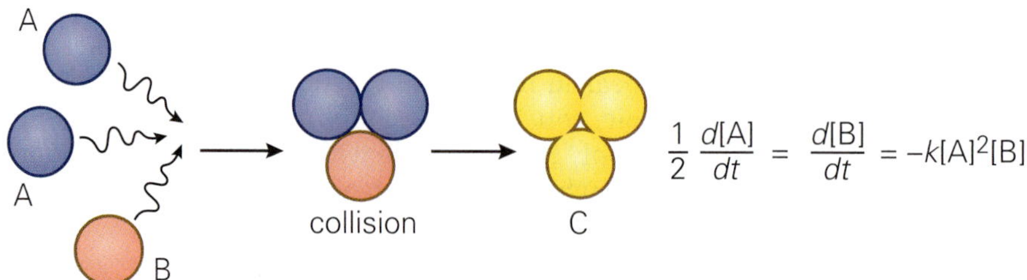


$$\frac{1}{2}\frac{d[A]}{dt} = \frac{d[B]}{dt} = -k[A]^2[B]$$

(D) zero-order reaction

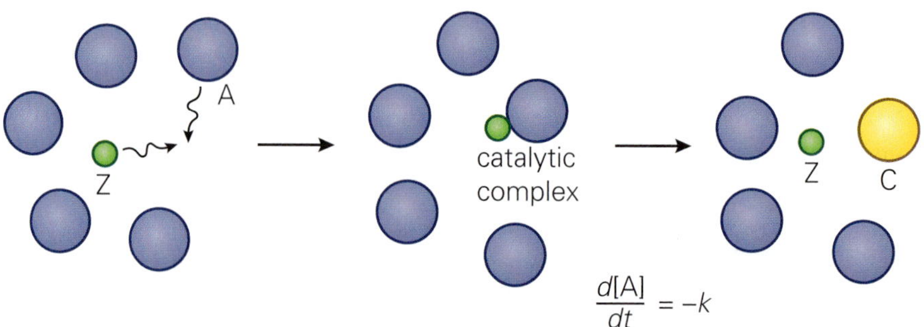


$$\frac{d[A]}{dt} = -k$$

(rate is independent of [A] when there are very few molecules of Z)

**Figure 15.6 Reaction order.** Schematic illustration of (A) first-, (B) second-, and (C) third-order reactions occurring as elementary processes. In (D), a zero-order reaction is shown, for which the overall reaction, A → C, is not an elementary process. The rate of Z colliding with A is fast compared to conversion of the catalytic complex to release product C. In this case, concentrations of reactants or products do not enter into the rate law because the rate-limiting step of the reaction involves collisions with the small number of Z molecules, and so the concentration of Z ends up controlling the reaction rate. Such behavior is common for enzymes with high concentrations of substrates, as discussed in detail in Chapter 16.

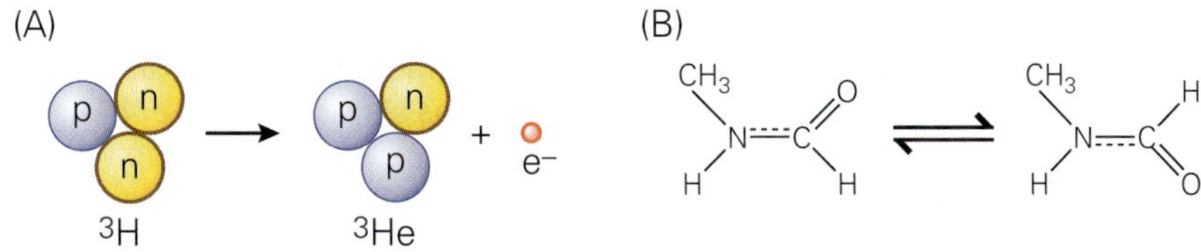


**Figure 15.7 Examples of first-order reactions.** (A) Radioactive decay. An example is the conversion of tritium (a form of hydrogen with one proton, p, and two neutrons, n) to helium (two protons, one neutron) that occurs by emission of an electron, a process known as β decay. (B) A bond isomerization reaction. A process such as rotation about a partial double bond in an amide (N-methyl formamide in the example shown) is a unimolecular reaction that occurs with first-order kinetics. This is very similar to the *cis–trans* isomerization of peptide bonds in proteins (see Figure 4.15).

right-hand side of Equation 15.2 reflects this convention, which is that reactant concentrations must decrease while product concentrations increase.

Second-order reactions may also involve the collision and reaction of two identical molecules. For example, if a reaction corresponds to A + A → C, where just collisions between two A molecules are required, then the rate law is:

$$\frac{d[\mathrm{A}]}{dt} = -k[\mathrm{A}][\mathrm{A}] = -k[\mathrm{A}]^2 \quad \text{or} \quad \frac{d[\mathrm{C}]}{dt} = +k[\mathrm{A}]^2 \tag{15.5}$$

Second-order reactions are the most common type of elementary reaction.

Elementary reactions that have three participating molecules are termed **termolecular** or **third order** (Figure 15.6C). For example, if two molecules of A and one of B are required to collide in order for the reaction to proceed, the rate law would be:

$$\frac{d[\mathrm{A}]}{dt} = -k[\mathrm{A}][\mathrm{A}][\mathrm{B}] = -k[\mathrm{A}]^2[\mathrm{B}] \tag{15.6}$$

The rate constant, $k$, for a third-order reaction has units of $M^{-2}{\bullet}sec^{-1}$. At low concentrations, a third-order reaction becomes very slow because the probability of three molecules being close enough to each other to collide simultaneously is very low. Fourth or higher-order reactions are virtually never observed.

It is also found experimentally that many reactions are **unimolecular**—that is, they depend linearly on the concentration of a single reactant. Such reactions obey **first-order kinetics**. For a first-order reaction of A → C:

$$\frac{d[\mathrm{A}]}{dt} = -k[\mathrm{A}] \quad \text{or} \quad \frac{d[\mathrm{C}]}{dt} = +k[\mathrm{A}] \tag{15.7}$$

The rate constant, $k$, for a first-order reaction has units of $sec^{-1}$.

Although not very common in chemical processes in which bonds are made and broken, there are many physical processes that obey first-order kinetics. The radioactive decay of a nucleus is one example and is illustrated in **Figure 15.7**A. Other examples are provided by the emission of light from a molecule in an electronic excited state or by rearrangement reactions such as isomerizations (Figure 15.7B).

It is also possible to have situations in which the rate for a process is actually independent of the concentration of the reactant. This might be the case when the reaction depends on a molecule, such as a catalyst, that is present in very low concentration compared to the concentrations of the reactants. In such a case, the reaction rate is determined by the concentration of the catalyst rather than that of the reactants (see Figure 15.6D). Such processes are referred to as **zero order**. The rate law for a zero-order process is given by:

$$\frac{d[\mathrm{A}]}{dt} = -k \quad \text{or} \quad \frac{d[\mathrm{C}]}{dt} = +k \tag{15.8}$$

In this case, the rate constant, $k$, has units of $M{\bullet}sec^{-1}$.

**Unimolecular reaction**

A unimolecular or first-order reaction is one in which the rate depends only on the concentration of the reactant to the first power.

## 15.5 The integration of rate equations predicts the time dependence of concentrations

The rate equations for the different kinds of reactions discussed above are all differential equations. In a kinetic measurement carried out in a laboratory it is usually the actual concentration of a reactant or product that is followed as a function

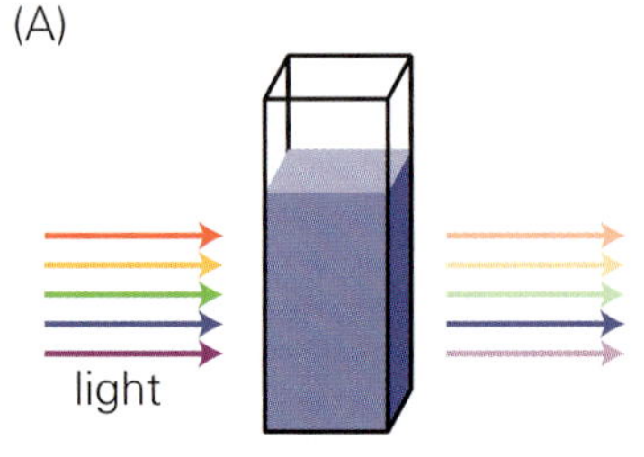


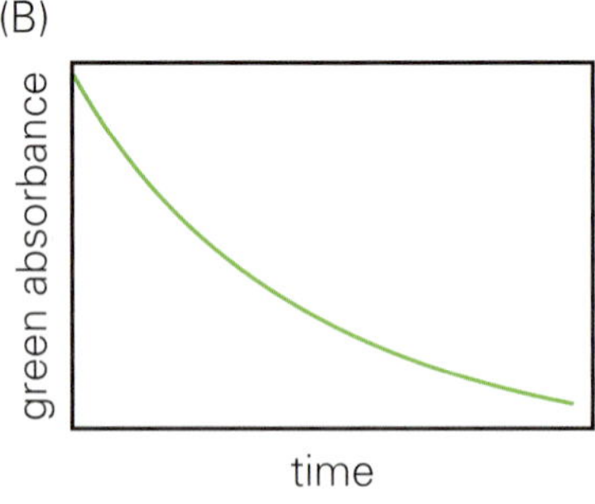


**Figure 15.8 Monitoring concentration during a reaction.** (A) Molecules absorb light at specific wavelengths, and this property allows their concentrations to be monitored during a reaction. In the example shown here, a reactant molecule absorbs some wavelengths more than others, leaving a blue color. (B) The change in concentration of reactants causes the amount of absorbance of green light, in this case, to decrease as the reaction progresses.

of time, rather than the time derivative. For example, we might monitor the concentration of a particular molecule in a reaction by measuring the absorption of visible or ultraviolet light at a specific wavelength, as shown in **Figure 15.8**. It is straightforward to convert a differential rate equation to a function of concentration versus time by specifying the initial conditions (concentrations of reactants at time = 0) and then integrating the rate equation. The time dependence of the concentration can be predicted in this way for each kind of elementary reaction. Comparison with experiment then helps determine the actual order of the reaction.

To set up the integration, we denote the initial concentrations of the various molecules by a subscript 0. For example, the initial concentration of a reactant, A, is denoted $[A]_0$. To integrate the equations, we separate the time and concentration variables, and then integrate from time = 0 to time = $t$. We now discuss reactions of various orders in turn.

## 15.6 Reactants disappear linearly with time for a zero-order reaction

Consider a reaction, A → C, that proceeds with zero-order kinetics. This means that the rate law does not have any concentration terms in it, as given by Equation 15.8. By rearranging Equation 15.8, we get:

$$d[A] = -k dt \tag{15.9}$$

We can now integrate both sides of Equation 15.9 from the beginning of the reaction ($t = 0$) to any time $t$. For the left-hand side of the equation, the limits on the concentration are from $[A]_0$ (at $t = 0$) to [A] (at $t = t$). This results in the following expression:

$$\int_{[A]_0}^{[A]} d[A] = -k \int_0^t dt \tag{15.10}$$

which gives:

$$[A] - [A]_0 = -kt \tag{15.11}$$

According to Equation 15.11, the reactant disappears linearly with time for a zero-order reaction, as shown in **Figure 15.9**A. For the reaction, A → C, the reactants are converted into a single product, C. It follows that the concentration of C increases linearly with time.

The concentration of a reactant can never be negative, and so Equation 15.11 applies only until the value of $kt$ becomes equal to $[A]_0$, the initial concentration of A. At subsequent times, the concentration of A is zero. Note that the rate constant, $k$, has units of $M \bullet sec^{-1}$ for a zero-order reaction.

## 15.7 The concentration of reactant decreases exponentially with time for a first-order reaction

Now consider a first-order reaction in which a single reactant is converted to a single product, A → C. The rate law for such a reaction is given by Equation 15.7:

$$\frac{d[A]}{dt} = -k[A]$$

By rearranging this equation, we get:

$$\frac{d[A]}{[A]} = -k dt \tag{15.12}$$

We now integrate both sides of Equation 15.12 to get:

$$\int_{[A]_0}^{[A]} \frac{d[A]}{[A]} = -k \int_0^t dt$$

This yields:

$$\ln[A] - \ln[A]_0 = -kt \tag{15.13}$$

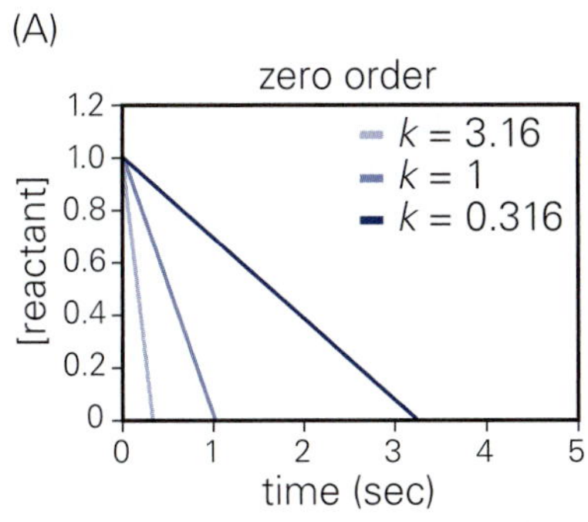


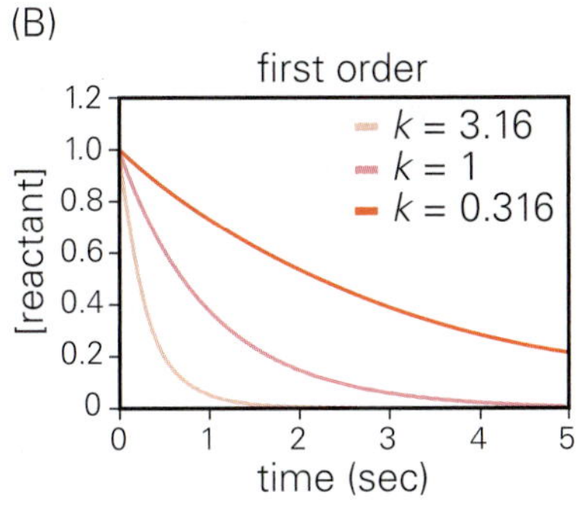


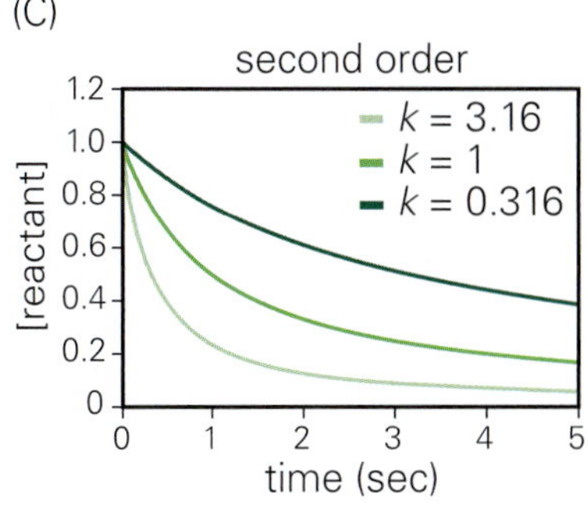


**Figure 15.9 Concentration of reactant versus time for zero-order, first-order, and second-order kinetics.** Graphs of concentration versus time are shown for three values of the rate constant, *k*, spanning a factor of 10 in rate (the values of *k* are 0.316, 1, and 3.16). (A) Zero-order kinetics. (B) First-order kinetics. (C) Second-order kinetics with only one type of reactant.

To get an expression for the concentration of A at time *t*, we take the exponential of both sides of Equation 15.13, which results in Equation 15.14:

$$[\mathrm{A}]=[\mathrm{A}]_0 e^{-kt} \tag{15.14}$$

According to Equation 15.14, the concentration of the reactant decays in a simple exponential fashion, as shown in Figure 15.9B. If the reactant is converted into a single product, C, then the concentration of C increases exponentially with time in the following way:

$$[\mathrm{C}]=[\mathrm{A}]_0-[\mathrm{A}]=[\mathrm{A}]_0(1-e^{-kt}) \tag{15.15}$$

## 15.8 The reactants decay more slowly in second-order reactions than in first-order reactions, but the details depend on the particular type of reaction and the conditions

Second-order reactions are more complicated to analyze than first-order reactions, and it is helpful to consider different limiting cases. To start with, consider a second-order reaction with just one kind of reactant, as shown in **Figure 15.10**. The rate law for this reaction is given by Equation 15.5:

$$\frac{d[\mathrm{A}]}{dt}=-k[\mathrm{A}]^2$$

Rearranging, we get:

$$\frac{d[\mathrm{A}]}{[\mathrm{A}]^2}=-kdt \tag{15.16}$$

We now integrate Equation 15.16 to get an expression for the change in the concentration of A with time:

$$\int_{[\mathrm{A}]_0}^{[\mathrm{A}]}\frac{d[\mathrm{A}]}{[\mathrm{A}]^2}=-k\int_0^t dt$$

$$\Rightarrow \frac{1}{[\mathrm{A}]_0}-\frac{1}{[\mathrm{A}]}=-kt$$

$$\Rightarrow \frac{1}{[\mathrm{A}]}=\frac{1}{[\mathrm{A}]_0}+kt=\frac{1+[\mathrm{A}]_0 kt}{[\mathrm{A}]_0}$$

$$\Rightarrow [\mathrm{A}]=\frac{[\mathrm{A}]_0}{1+[\mathrm{A}]_0 kt} \tag{15.17}$$

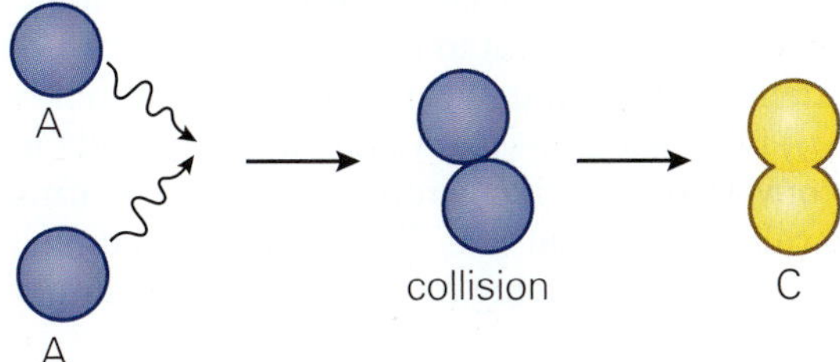


**Figure 15.10 A second-order reaction with just one type of reactant.** Two A molecules collide and react to form one C molecule.

Graphs of the concentration of A as a function of time for a second-order reaction, as given by Equation 15.17, are shown in Figure 15.9C. Because the rate depends on the square of the concentration of reactants, the reaction begins with the highest rate, and slows down as the concentration of reactants decreases with time and collisions become less frequent (see Figure 15.9C). The graph describing the time dependence of the concentration for the second-order reaction is qualitatively similar to the exponential decay for a first-order reaction (Figure 15.9B), but the former has a somewhat different shape because the reactant concentration decays more slowly at long times, as the concentration of reactants decreases.

For the bimolecular case in which there are two different kinds of reactant molecules (for example, $A + B \rightarrow C$), two cases must be considered. If $[A]_0 = [B]_0$, then because of the stoichiometry, we know that $[A] = [B]$ for all times, and mathematically the behavior is identical to the case in which two molecules of [A] react, as discussed above (that is, Equation 15.17 applies).

If the initial concentrations of A and B are different, then the integration of the rate equation is more complicated. Starting with the rate law (see Equation 15.6):

$$\frac{d[A]}{dt} = -k[A][B]$$

we obtain the the integrated form, as explained in **Box 15.1**:

$$\ln\left(\frac{[A]}{[B]}\right) - \ln\left(\frac{[A]_0}{[B]_0}\right) = ([A]_0 - [B]_0)kt \tag{15.18}$$

Although Equation 15.18 looks complicated, if $[A]_0 \gg [B]_0$ (or $[B]_0 \gg [A]_0$), then it reduces to the case of a first-order reaction. To see why this is so, consider what happens when the initial concentration of A is very much greater than that of B (that is, $[A]_0 \gg [B]_0$). As the reaction proceeds, A molecules react with B molecules, but because there are very few B molecules, the concentration of A is essentially unchanged and remains very close to $[A]_0$. In this case, the rate equation becomes:

$$\frac{d[A]}{dt} = -k[A][B] = -k[A]_0[B] = -k'[B] \tag{15.19}$$

By comparing Equation 15.19 with Equation 15.7, you can see that the reaction behaves as if it were a first-order reaction with a rate constant, $k'$. The rate constant, $k'$, in Equation 15.19 is called a **pseudo-first-order rate constant** because it applies to a second-order reaction under a limiting condition (that is, $[A]_0 \gg [B]_0$). The concept of a pseudo-first-order rate constant is an important one that we shall return to in Section 15.17, when we discuss the rates of reversible ligand binding to proteins.

When the initial concentrations of the reactants, $[A]_0$ and $[B]_0$, are comparable, but not equal, Equation 15.18 must be applied. Although we shall not analyze this situation in detail, the changes in concentration with time are intermediate between first- and second-order reactions with one reactant.

## 15.9 The half-life for a reaction provides a measure of the speed of the reaction

Although all the information about how fast a reaction will occur is present in the equations above, it is convenient to have a simple description of how long it takes for some extent of reaction to occur. The most commonly used descriptor is the **half-life** ($t_{1/2}$) for a reaction—that is, the time required for half of the initial reactant to have converted to product. From the integrated forms of the rate equations, it is easy to calculate the half-life. This is done by setting $[A] = [A]_0/2$ in the integrated equation and then solving for the value of the time corresponding to this concentration, which is $t_{1/2}$.

For a zero-order reaction, $t_{1/2} = [A]_0/2k$. Since the rate of reaction (that is, the number of moles per unit time) is constant, it makes sense that more reactant being present means it takes longer for half of it to react.

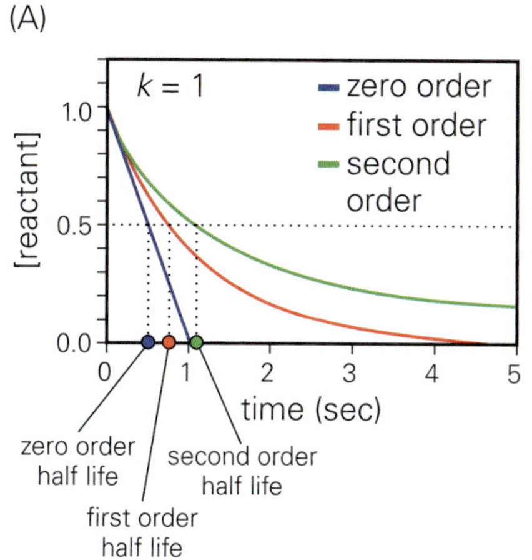


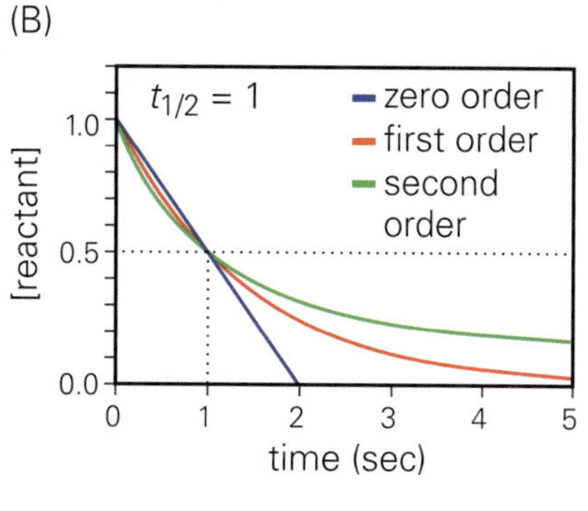


**Figure 15.11 Rate constants and half-lives.** The concentration of reactant as a function of time is graphed for zero-order, first-order, and second-order reactions (with identical molecules reacting). (A) The value of the rate constant, *k*, is the same in all three reactions. The half-life of a reaction, which is the time taken for the concentration of the reactant to decrease to 50% of its initial value, is also indicated. Dotted lines indicate the half-life for each reaction. (B) The value of the half-life is the same in all three cases.

For a first-order reaction, $t_{1/2} = \ln 2/k = 0.693/k$. In this case, the half-life does not depend on the amount of reactant initially present. The values of half-lives are very commonly encountered in descriptions of the decay of radioactive elements. The fact that the half-life is independent of the starting amount is a critical feature for using the decay of "radiocarbon," $^{14}C$, for determining the age of bones and other anthropological materials.

**Half-life and time constant**

The half-life of a reaction is the time for the concentration of a reactant to drop to half of its initial value. First-order processes are also commonly described by their time constants or lifetimes, which is the time required for the reactant to decay to ~37% (that is, $1/e$) of its initial value. The time constant for a first-order reaction is given by $1/k$, where $k$ is the rate constant.

For first-order reactions, the **time constant**, $\tau = 1/k$ (often referred to as the lifetime, especially for processes involving light, such as fluorescence), is used rather than half-life. The time constant is the time required for the concentration of the reactant to decay by a factor of $e^{-1}$ (that is, to decay to ~37% of its initial value). For a first-order reaction with a single product, the time constant is also the time taken for the concentration of the product to increase to ~63% of its final value.

For a second-order reaction with two identical molecules reacting, $t_{1/2} = 1/(k[A]_0)$. As for the zero-order case, the half-life depends on the initial starting concentration, but in the opposite way. In the second-order case, a higher concentration of reactant results in collisions being more likely, leading to a faster initial reaction and, hence, a shorter half-life. For the A + B reaction, a half-life can be defined, but it depends on both $[A]_0$ and $[B]_0$ in a complex way and, hence, is not particularly useful.

Examples of concentration versus time curves are shown in **Figure 15.11**A for the three common types of elementary reactions: zero, first, and second order. In all of these examples, the rate constant is 1 (with appropriate units), and the initial concentration of reactant is taken as 1. For comparison, graphs are also shown in which the half-life is the same for all three kinds of reactions (Figure 15.11B). The latter case emphasizes the similarity of first- and second-order behavior toward the beginning of each reaction, but shows how differently they behave at longer times.

It is worth noting that, for these elementary reactions, the half-life for a reaction depends on the initial concentration in a different way for each reaction order. This means that one can determine the order of reaction just by measuring the half-life for different initial concentrations and matching the observed dependence to the appropriate order of reaction. A related concept is discussed further in Section 15.23.

## 15.10 For reactions with intermediate steps, the slowest step determines the overall rate

The elementary reactions discussed above are sufficient to describe any individual step of a reaction, but many real reactions have multiple steps or processes

## Box 15.1 Integrated forms of rate equations

**Second-order reaction with unequal concentrations of reactants:** $A + B \rightarrow C$

For the second-order reaction with the condition that $[A]_0 \neq [B]_0$, we use the fact that the amounts of A and B that have reacted must be the same due to the stoichiometry:

$$[A]_0-[A]=[B]_0-[B] \quad \text{or} \quad [B]=[B]_0-[A]_0+[A] \tag{15.1.1}$$

For convenience, we define the following:

$$\Delta=[B]_0-[A]_0$$

so

$$[B]=\Delta+[A] \tag{15.1.2}$$

Using this, the basic kinetic equation becomes:

$$\frac{d[A]}{dt}=-k[A][B]=-k[A](\Delta+[A]) \tag{15.1.3}$$

Separating variables and then integrating Equation 15.1.3 from time 0 to $t$ and from $[A]_0$ to [A], we get:

$$\int_0^t dt=\int_{[A]_0}^{[A]}\frac{d[A]}{-k[A](\Delta+[A])}=\int_{[A]_0}^{[A]}\frac{d[A]}{-k\Delta[A]-k[A]^2} \tag{15.1.4}$$

Such polynomial integrals, of the form $dt = dx/(a + bx + cx^2)$, appear in standard mathematics tables. In this case, the general solution is:

$$t=\frac{1}{\sqrt{-q}}\log\left(\frac{2cx+b-\sqrt{-q}}{2cx+b+\sqrt{-q}}\right) \tag{15.1.5}$$

where $q = 4ac - b^2$. Then, using $x = [A]$, $a = 0$, $b =-k\Delta$, $c = -k$, and Equation 15.1.4, Equation 15.1.5 becomes:

$$t=\frac{1}{k\Delta}\log\left(\frac{[A]}{[A]+[B]_0-[A]_0}\right)-\frac{1}{k\Delta}\log\left(\frac{[A]_0}{[A]_0+[B]_0-[A]_0}\right)$$

$$=\frac{1}{k\Delta}\left[\log\left(\frac{[A]}{[B]}\right)-\log\left(\frac{[A]_0}{[B]_0}\right)\right] \tag{15.1.6}$$

We now insert the definition of $\Delta$ from Equation 15.1.2 and rearrange slightly to get Equation 15.18 in the main text:

$$\ln\left(\frac{[A]}{[B]}\right)-\ln\left(\frac{[A]_0}{[B]_0}\right)=([A]_0-[B]_0)kt \tag{15.1.7}$$

**Reversible reaction** $A \underset{k_r}{\overset{k_f}{\rightleftharpoons}} B$

Using $k_f$ and $k_r$ as the rate constants for the forward and reverse reactions, respectively, the differential rate equations for concentrations have a term for the disappearance of A to make B and for the formation of A from B. The basic rate equations are then:

$$\frac{d[A]}{dt}=-k_f[A]+k_r[B] \tag{15.1.8}$$

and

$$\frac{d[B]}{dt}=+k_f[A]-k_r[B] \tag{15.1.9}$$

Since A and B are the only molecules involved in the reaction, the sum of their concentrations must be constant. We can write this condition as $[A] + [B] = X$, where $X$ is a constant. Rearranging this gives $[B] = X - [A]$ and, substituting into the differential rate equation for [A] (Equation 15.1.8), we find:

$$\frac{d[A]}{dt}=-k_f[A]+k_r(X-[A])=-(k_f+k_r)[A]+k_rX \tag{15.1.10}$$

The variables [A] and $t$ can be separated again. This is most easily seen by substituting in a new variable:

$$Z=-(k_f+k_r)[A]+k_rX \tag{15.1.11}$$

Using this in Equation 15.1.10, and recognizing that $X$ is constant, we get:

$$\frac{dZ}{dt}=-(k_f+k_r)Z \tag{15.1.12}$$

Equation 15.1.12 can be integrated easily to give an exponential. To get back to [A] as the variable, substitute back in for $Z$ and use the value of $Z$ at time zero:

$$Z(0)=-(k_f+k_r)[A]_0+k_fX \tag{15.1.13}$$

Following these steps gives:

$$[A]=\left([A]_0-\frac{k_r}{k_f+k_r}X\right)e^{-(k_f+k_r)t}+\frac{k_r}{k_f+k_r}X \tag{15.1.14}$$

Note that the forward and reverse rates must be equal at equilibrium, so:

$$k_f[A]_{eq}=k_r[B]_{eq}=k_r(X-[A]_{eq})$$

Then,

$$[A]_{eq}=\frac{k_r}{k_f+k_r}X \tag{15.1.15}$$

and so

$$[A]=([A]_0-[A]_{eq})e^{-(k_f+k_r)t}+[A]_{eq} \tag{15.1.16}$$

Equation 15.1.16 is the same as Equation 15.31 in the main text.

**Steady-state situation for the series reaction**

$$A \xrightarrow{k_1} B \xrightarrow{k_2} C$$

We begin by assuming that under steady-state conditions the concentration of [A] is held constant. We set $[A] = [A]_0$ at all times, so:

$$\frac{d[B]}{dt}=+k_1[A]_0-k_2[B] \tag{15.1.17}$$

Since there is just one time-dependent variable now, [B], this is much easier to integrate than Equation 15.1.3. By substituting in $X = k_1[A]_0 - k_2[B]$ and then noting that $dX = -k_2\,d[B]$ or $d[B] = (-1/k_2)\,dX$, the integration can be written in the following simplified form:

$$\int_{X_0}^{X} \frac{-dX}{k_2 X} = \int_0^t dt \qquad (15.1.18)$$

Thus,

$$\ln\left(\frac{X}{X_0}\right) = -k_2 t \qquad (15.1.19)$$

Substituting back in for $X$ and noting that $X_0 = k_1[A]_0$ gives:

$$[B] = \frac{k_1}{k_2}[A]_0(1 - e^{-k_2 t}) \qquad (15.1.20)$$

These equations show that [B] increases, asymptotically approaching $[B]_{ss}$ (with a time constant $1/k_2$), and that at steady state:

$$[B]_{ss} = \frac{k_1}{k_2}[A]_0 \qquad (15.1.21)$$

Equation 15.1.21 is the same as Equation 15.42 in the main text.

between the reactants and final products. Consider, for example, a reaction in which an **intermediate**, B, is formed, that then goes on to product, C:

$$A \xrightarrow{k_1} B \xrightarrow{k_2} C \qquad (15.20)$$

The letter $k$ above the arrow for a step indicates the rate constant for that step in the reaction.

**Intermediate**

An intermediate is a chemical species that appears during the reaction but disappears when the reaction is complete.

Many chemical reactions have intermediates. Intermediates are also important for many other processes, including protein folding. As we discuss in Section 18.8, an unfolded protein often collapses to a partially folded intermediate, which then converts to the native and functional conformation of the protein (**Figure 15.12**).

If the elementary reactions of the two steps, A → B and B → C, are each first order, then there are two rate equations that govern the reaction. First, there is the rate equation for A converting to B. This is similar to the first-order rate equation given in Equation 15.7:

$$\frac{d[A]}{dt} = -k_1[A] \qquad (15.21)$$

The rate equation for the concentration of B is more complex, because the concentration of B can change in two ways. B is produced from A, but it is also converted to C. By considering both of these processes, we obtain the rate equation for B:

$$\frac{d[B]}{dt} = +k_1[A] - k_2[B] \qquad (15.22)$$

As for A, the rate equation for C is simple, because its concentration changes only when it is generated from or converted to B:

$$\frac{d[C]}{dt} = +k_2[B] \qquad (15.23)$$

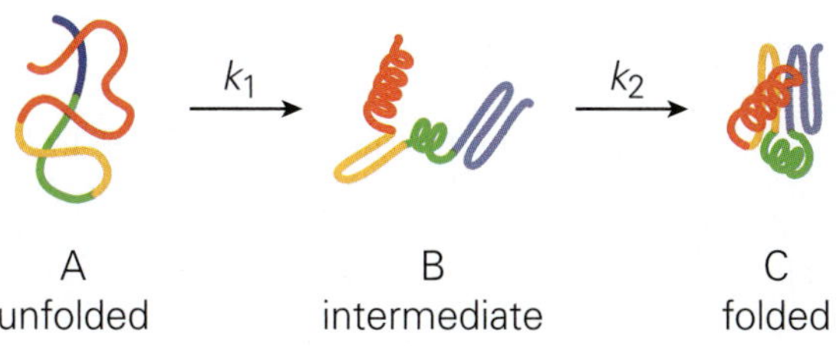


**Figure 15.12 A reaction with an intermediate.** A protein folding reaction is shown in which the unfolded protein first converts to an intermediate conformation that then progresses to the folded conformation. The rate constants associated with the two steps are denoted $k_1$ and $k_2$, respectively.

The integrated form of the equation for A is the same as if there was no reaction beyond B:

$$[\mathrm{A}]=[\mathrm{A}]_0 e^{-k_1 t} \tag{15.24}$$

To get the integrated equation for [B], we substitute the expression for [A] given in Equation 15.24 into Equation 15.22:

$$\frac{d[\mathrm{B}]}{dt}=+k_1[\mathrm{A}]-k_2[\mathrm{B}]=k_1[\mathrm{A}]_0 e^{-k_1 t}-k_2[\mathrm{B}] \tag{15.25}$$

As explained in **Box 15.2**, Equation 15.25 is integrated to yield the following expression for [B]:

$$[\mathrm{B}]=\frac{k_1[\mathrm{A}]_0}{k_2-k_1}\left(e^{-k_1 t}-e^{-k_2 t}\right) \tag{15.26}$$

To get the equation for the product, C, the solution for [B] is inserted into the differential equation, Equation 15.23, and the integration can again be performed with separation of variables to give:

$$[\mathrm{C}]=[\mathrm{A}]_0\left[1-\frac{1}{k_2-k_1}\left(k_2 e^{-k_1 t}-k_1 e^{-k_2 t}\right)\right] \tag{15.27}$$

The values of [A], [B], and [C] are plotted versus time in **Figure 15.13** for two different ratios of the rate constants, $k_1$ and $k_2$. In one case, shown in Figure 15.13A, the first step is the slow one. In the second case, the second step is the slow one (see Figure 15.13B). The behavior is intuitively what one expects. The reactant A decreases in a simple, first-order, exponential way—unaffected by what happens subsequently. The intermediate, B, increases with the same time constant as that for the decrease of A, and then B converts to C in a first-order manner with rate constant, $k_2$. If the production of B is much slower than its subsequent conversion to C (that is, if $k_1 \ll k_2$), then [B] never gets very high. On the other hand, if the first step is fast (that is, $k_1 \gg k_2$), then most of the A converts to B before it goes on to form C, as is seen in the graphs.

## Box 15.2 Integrating the rate equation for a reaction with intermediate steps

We seek to integrate Equation 15.25 in the main text:

$$\frac{d[\mathrm{B}]}{dt}=+k_1[\mathrm{A}]-k_2[\mathrm{B}]=k_1[\mathrm{A}]_0 e^{-k_1 t}-k_2[\mathrm{B}] \tag{15.25}$$

Equation 15.25 has just [B] and $t$ as variables, so it can be integrated. This integration is not straightforward, because the variables cannot easily be separated. However, the general form of Equation 15.25 derives from an application of the product rule for differentiation. Notice that Equation 15.25 is equivalent to a differential equation of the general form:

$$\frac{dy}{dx}+cy=g(x) \tag{15.2.1}$$

If we apply the product rule for differentiation to an exponential function, we get the following:

$$\frac{d}{dx}\left(e^{cx}y\right)=ce^{cx}y+e^{cx}\frac{dy}{dx}=e^{cx}\left(\frac{dy}{dx}+cy\right) \tag{15.2.2}$$

By combining Equations 15.2.1 and 15.2.2, we get:

$$\frac{d}{dx}\left(e^{cx}y\right)=e^{cx}g(x) \tag{15.2.3}$$

Equation 15.2.3 is then integrated to give:

$$\int\frac{d}{dx}\left(e^{cx}y\right)dx=e^{cx}y=\int e^{cx}g(x)dx \tag{15.2.4}$$

To apply Equation 15.2.4 to integrate the rate equation (Equation 15.25), we use the initial condition that $[\mathrm{B}]_0 = 0$ and substitute as follows:

$$x = t$$

$$g(x) = k_1[\mathrm{A}]_0\, exp(-k_1 t)$$

$$y = [\mathrm{B}] \text{ (which is a function of } x = t\text{)}$$

This gives us Equation 15.21 in the main text:

$$[\mathrm{B}]=\frac{k_1[\mathrm{A}]_0}{k_2-k_1}\left(e^{-k_1 t}-e^{-k_2 t}\right)$$

You can verify that, for [B] as defined in Equation 15.26, the derivative, $d[\mathrm{B}]/dt$, gives Equation 15.25, as it should.

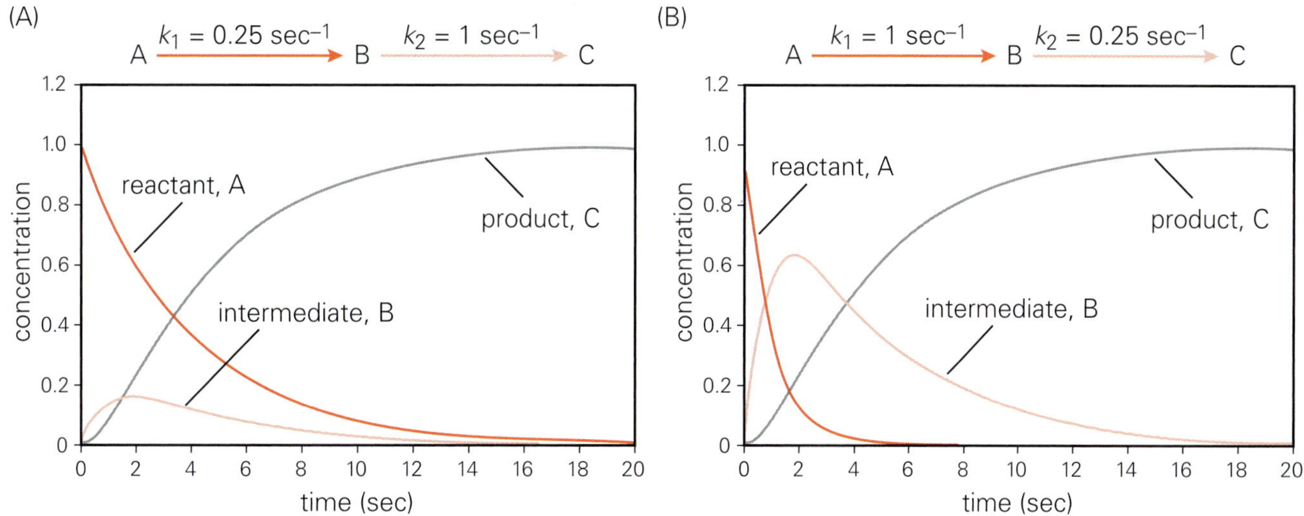


**Figure 15.13 Time dependence of concentrations for the reaction A → B → C.** (A) The rate constants for the conversion of A to B and of B to C are $k_1 = 0.25\ \text{sec}^{-1}$ and $k_2 = 1.0\ \text{sec}^{-1}$, respectively. The first step is slower than the second step and the intermediate does not build up very much, because it is rapidly converted to product. (B) The values of the rate constants, $k_1$ and $k_2$, are $1.0\ \text{sec}^{-1}$ and $0.25\ \text{sec}^{-1}$, respectively. The second step is slower than the first step. The intermediate builds up to a greater extent. The rate of final product formation is the same regardless of whether the first or the second step is slower, because the slower step has the same rate constant in both cases.

An important principle emerges when we compare Figures 15.13A and 15.13B. Note that the rate at which the product, C, builds up is the same in the two cases. This is because it is always the slowest step in a chain of reactions that limits the overall reaction rate. In the examples shown in Figure 15.13, the rate constant for the slower step is the same in the two cases, and so the rate of buildup of the product is the same. But, more generally, for any complex set of series reactions, the rate for the slowest step determines the rate of the overall process. That slowest step is called the **rate-determining step** for the reaction.

The graphs shown in Figure 15.13 are based on Equations 15.24, 15.26, and 15.27, which seem rather complicated. Nevertheless, the principle that the slowest step in a process limits the overall speed of the process is one that you can recognize from many aspects of everyday life. The rate at which you can move through a grocery store, for example, limits the speed at which you can exit the store (**Figure 15.14**). The rate at which cars move along a multilane highway is determined by

**Rate-determining step**

The rate-determining step in a series of reactions is the slowest step, which therefore ultimately limits the rate of formation of product.

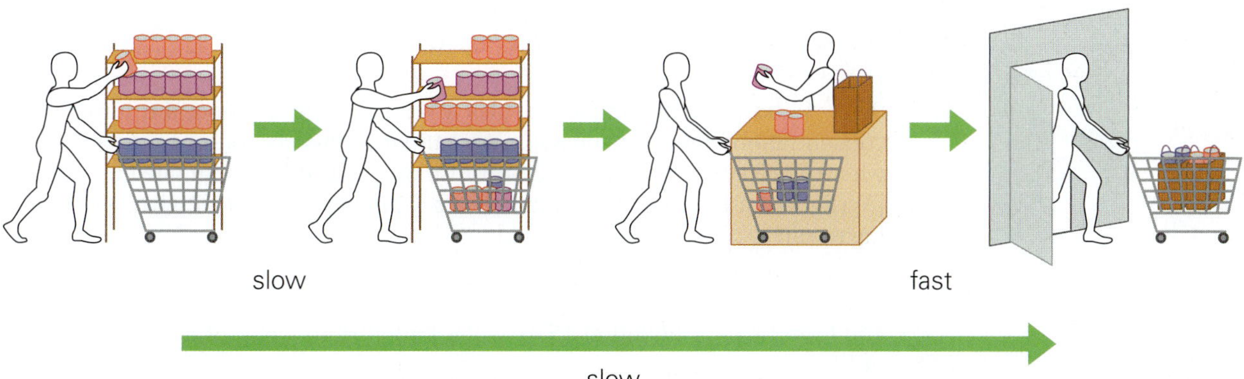


**Figure 15.14 The overall rate of a series of kinetic steps is determined by the rate of the slowest step.** This slowest step is called the rate-limiting step. In this illustration, the process of purchasing groceries at a store is depicted. The first step, loading the grocery cart, is assumed to be slow. The second step, paying for the groceries at the checkout counter, is assumed to be fast. The overall speed with which the groceries are purchased is then set by the speed at which the grocery cart is loaded.

the rate at which they transit a bottleneck, such as a toll booth. The concept of a rate-determining step finds many applications in biology. The overall throughput in metabolic pathways, for example, is often controlled by regulating the key rate-limiting steps in the pathway.

## B. REVERSIBLE REACTIONS, STEADY STATES, AND EQUILIBRIUM

In the first part of this chapter we have ignored reverse reactions, in which products convert back to reactants. This is a reasonable thing to do when the forward reaction is very favorable in free energy (that is, when the value of $\Delta G$ for the reaction is large and negative). In such cases, as we shall see, once products are formed, it is very unlikely that they will convert back to form the reactants. For many important biological processes, however, the magnitude of the free-energy change is not large, and both forward and reverse reactions need to be considered in the kinetic equations.

### 15.11 The forward and reverse rates must both be considered for a reversible reaction

A particularly simple example of a reversible reaction is a conformational rearrangement, such as the isomerization of a peptide bond where one of the residues is a proline. Recall from Section 4.8 that both the *cis* and *trans* conformers can be populated significantly if one the residues forming a peptide bond is proline.

A reversible unimolecular reaction in which molecules A and B interconvert can be written in the following way:

$$\mathrm{A} \underset{k_{\mathrm{r}}}{\overset{k_{\mathrm{f}}}{\rightleftarrows}} \mathrm{B} \tag{15.28}$$

where $k_f$ and $k_r$ are the rate constants for the forward and reverse reactions. The forward and reverse rate constants are also commonly denoted $k_1$ and $k_{-1}$, respectively.

We can work out the rate equations for concentrations as we did before, except that there will be a term that accounts for the the formation of A from B in addition to the term that describes the disappearance of A as it is converted to B. The basic rate equations are then:

$$\frac{d[\mathrm{A}]}{dt} = -k_{\mathrm{f}}[\mathrm{A}] + k_{\mathrm{r}}[\mathrm{B}] \tag{15.29}$$

and

$$\frac{d[\mathrm{B}]}{dt} = +k_{\mathrm{f}}[\mathrm{A}] - k_{\mathrm{r}}[\mathrm{B}] \tag{15.30}$$

Since A and B are the only molecules involved in the reaction, the sum of their concentrations must be constant. We can write this condition as $[\mathrm{A}] + [\mathrm{B}] = X$, where $X$ is a constant. Using this substitution in the differential rate equation for [A] (Equation 15.29) and, integrating it as explained in Box 15.1, gives the following result:

$$[\mathrm{A}] = ([\mathrm{A}]_0 - [\mathrm{A}]_{\mathrm{eq}})e^{-(k_{\mathrm{f}}+k_{\mathrm{r}})t} + [\mathrm{A}]_{\mathrm{eq}} \tag{15.31}$$

The term $[\mathrm{A}]_{\mathrm{eq}}$ in Equation 15.31 is the equilibrium value of the concentration of A.

There are two reactions occurring, one in the forward direction and one in reverse, and each of these reactions has a different rate constant associated with it. Nevertheless, the change in concentrations is described by a simple exponential function, as for a first-order reaction that proceeds in only one direction. The exponential relaxation has an effective rate constant that is the *sum* of the forward

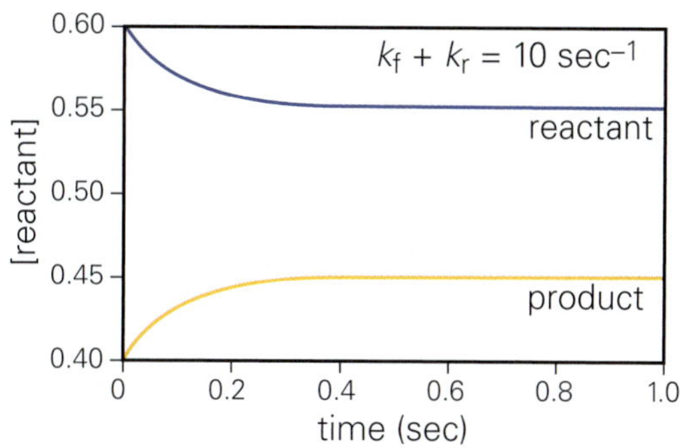


**Figure 15.15 Time dependence of reactants and products for a reversible unimolecular reaction.** In this case, the concentrations of reactant and product are 0.60 and 0.40 initially and then 0.45 and 0.55 at equilibrium. The sum of the forward and reverse rate constants is 10 $sec^{-1}$.

and reverse rate constants, as shown in **Figure 15.15**. From an experiment that measures only the time-dependence of the concentrations of reactants and products for such a reaction, it is not possible to determine the values of $k_f$ and $k_r$ individually.

It is important to note that, for reversible reactions, the concentration of reactant does not go to zero, but rather stops changing when the equilibrium concentration ($[A]_{eq}$) is reached. This leads to the extra terms in Equation 15.31 relative to Equation 15.14, which describes the concentration dependence for a first-order reaction that proceeds in only one direction.

## 15.12 The on and off rates of ligand binding can be measured by monitoring the approach to equilibrium

The analysis of forward and reverse rates for a reversible reaction finds an important practical application in determining how fast a ligand (for example, a drug molecule) binds to a protein. Such a binding interaction is a bimolecular reaction, while Equation 15.31 applies to unimolecular reactions. But the analysis is simplified by the fact that we can carry out the binding reaction under conditions where the concentration of the ligand is much greater than the concentration of the protein. In that case, the concentration of the ligand does not change much during the reaction. Under such circumstances, the reaction becomes pseudo-first order (see Section 15.8) because the ligand concentration is a constant. Recall that this is an assumption that we also made in the analysis of the equilibrium constants for ligand binding (see Section 12.6).

We write the binding reaction for protein, P, and ligand, L, as:

$$P + L \underset{k_{off}}{\overset{k_{on}}{\rightleftharpoons}} P{\bullet}L \qquad (15.32)$$

In Equation 15.32, $k_{on}$ and $k_{off}$ are the forward and reverse rate constants, respectively. The subscripts "on" and "off" are commonly used for rate constants involved in ligand binding.

The basic rate equations are:

$$\frac{d[P]}{dt} = -k_{on}\,[P][L] + k_{off}\,[P{\bullet}L] \qquad (15.33)$$

and

$$\frac{d[P{\bullet}L]}{dt} = +k_{on}\,[P][L] - k_{off}\,[P{\bullet}L] \qquad (15.34)$$

Because [L] is essentially constant (that is, we are considering conditions where $[L] \gg [P]$), these equations can be rewritten as:

$$\frac{d[P]}{dt} = -k'_{on}\,[P] + k_{off}\,[P{\bullet}L] \qquad (15.35)$$

and

$$\frac{d[P{\bullet}L]}{dt} = +k'_{on}\,[P] - k_{off}[P{\bullet}L] \qquad (15.36)$$

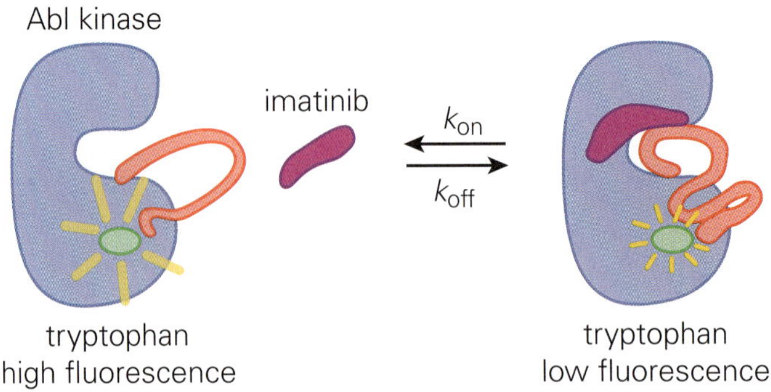


**Figure 15.16 Fluorescence changes can be used to detect the binding of a ligand.** The schematic diagram shows the binding of the cancer drug imatinib (*purple*) to the protein kinase Abl. A tryptophan sidechain in the protein (illustrated in Figure 15.23) has high fluorescence when no drug is bound and low fluorescence when imatinib is bound. This change in fluorescence can be used to detect the binding of imatinib to the protein kinase.

where $k'_{on} = k_{on}[L]$ is a pseudo-first-order rate constant that has a value that depends on the total ligand concentration, $[L]_0$ (that is, we assume that $[L] \approx [L]_0$). Equations 15.35 and 15.36 are the same as Equations 15.29 and 15.30, with P corresponding to A and P•L corresponding to B. Hence, we can use Equation 15.31 to write down an expression for the concentration of the product as a function of time:

$$[P] = \left([P]_0 - [P]_{eq}\right)e^{-(k'_{on} + k_{off})t} + [P]_{eq} \tag{15.37}$$

Here [P] is the time-dependent concentration of the free protein, $[P]_0$ is the initial concentration of the free protein, and $[P]_{eq}$ is the equilibrium concentration of the free protein in the presence of ligand.

Imagine that the binding of the ligand to the protein can be monitored instantaneously as it changes. This can be readily done if the fluorescence properties of the protein change upon ligand binding, as is often the case (**Figure 15.16**). Equation 15.37 then tells us that, if we mix together solutions containing the protein and ligand (the drug), then we should see an exponential relaxation of the concentration of the free protein from an initial value (when it is entirely unbound) to an equilibrium value (when some of it is bound to the ligand). The apparent rate constant for the process of ligand binding can be used to derive the actual on- and off-rates, with rate constants $k_{on}$ and $k_{off}$, respectively, by varying the ligand concentration as explained in the following example.

In Chapter 12, we discussed the binding of the cancer drug imatinib to the tyrosine kinase Abl. When imatinib binds to Abl, the intensity of fluorescence emission at 350 nm (upon excitation at 290 nm) is reduced (**Figure 15.17**). The change in fluorescence is due to changes in the environment of tryptophan residues in the protein upon drug binding. Thus, when imatinib and Abl are mixed together in a fluorescence spectrometer, the fluorescence signal is observed to decrease, as shown in Figure 15.17A. The change in the fluorescence signal is well described by an exponential curve, yielding a first-order rate constant, $k_{obs}$. According to Equation 15.37:

$$k_{obs} = k'_{on} + k_{off} \tag{15.38}$$

The apparent first-order forward rate constant, $k'_{on}$, is the product of the actual second-order forward rate constant and the ligand concentration ($k'_{on} = k_{on}[L]$; see the discussion following Equation 15.36). The value of $k_{obs}$ depends, therefore, on the ligand concentration as follows:

$$k_{obs} = k_{on}[L] + k_{off} \tag{15.39}$$

According to Equation 15.39, we should get a straight line if we measure the apparent rate constant for imatinib binding to Abl as a function of total imatinib concentration ([L]). This is indeed the case, as shown in Figure 15.17C. The slope of the line is the second-order forward rate constant for imatinib binding to Abl ($k_{on}$). If the line is extrapolated back to zero ligand concentration, the intercept of the line is the reverse rate constant ($k_{off}$, the off-rate).

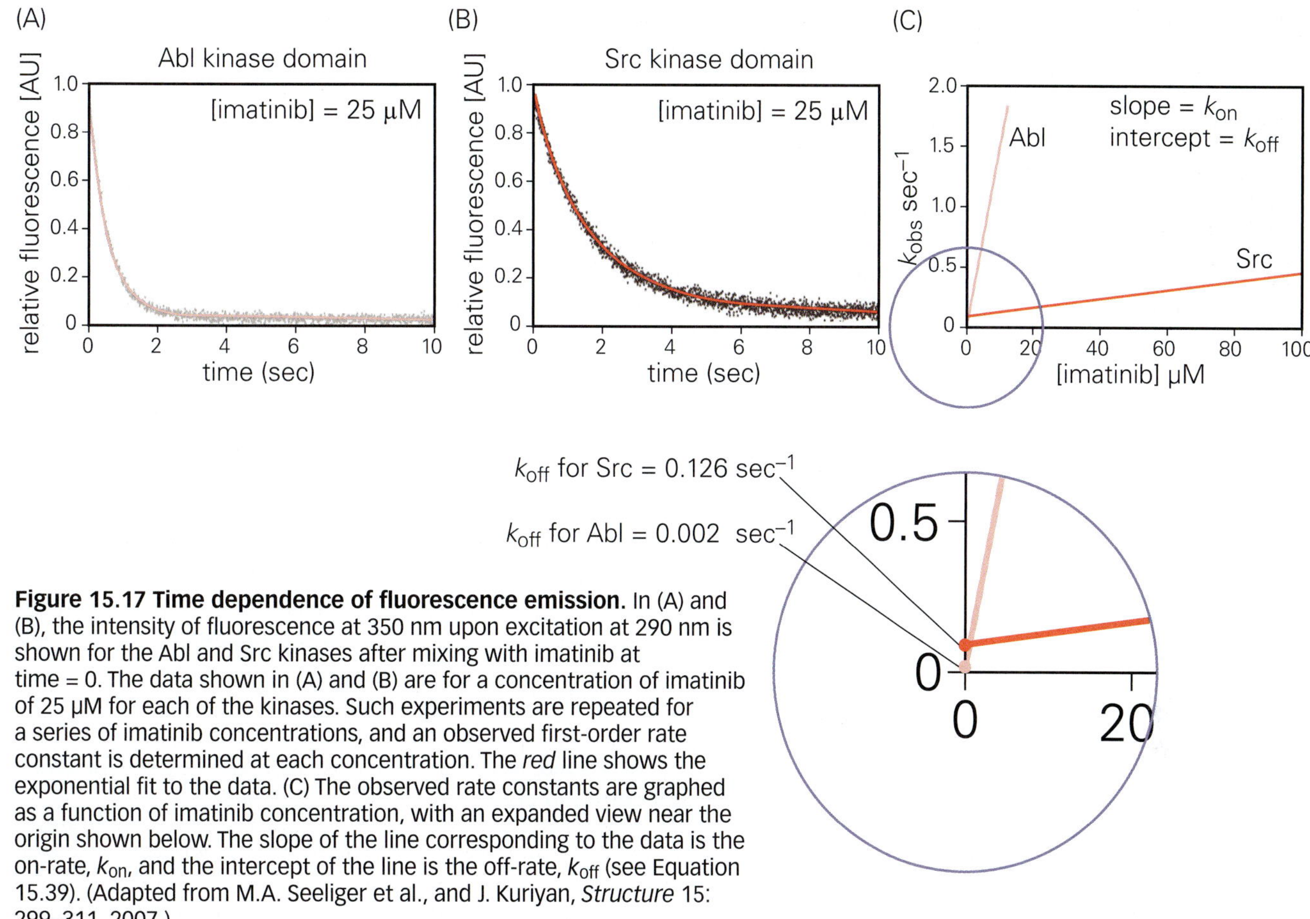


**Figure 15.17 Time dependence of fluorescence emission.** In (A) and (B), the intensity of fluorescence at 350 nm upon excitation at 290 nm is shown for the Abl and Src kinases after mixing with imatinib at time = 0. The data shown in (A) and (B) are for a concentration of imatinib of 25 μM for each of the kinases. Such experiments are repeated for a series of imatinib concentrations, and an observed first-order rate constant is determined at each concentration. The *red* line shows the exponential fit to the data. (C) The observed rate constants are graphed as a function of imatinib concentration, with an expanded view near the origin shown below. The slope of the line corresponding to the data is the on-rate, $k_{on}$, and the intercept of the line is the off-rate, $k_{off}$ (see Equation 15.39). (Adapted from M.A. Seeliger et al., and J. Kuriyan, *Structure* 15: 299–311, 2007.)

Analysis of the data shown in Figure 15.17 reveals that the on-rate for imatinib binding to Abl is $0.146 \times 10^6$ M$^{-1}$•sec$^{-1}$. Recall from Section 12.17 that imatinib is ineffective as an inhibitor of the tyrosine kinase Src. Data for imatinib binding to Src are shown in Figure 15.17B, for one concentration of the drug. The line corresponding to data measured over a range of imatinib concentrations has a much smaller slope than that for the Abl data because the on-rate for imatinib binding to Src ($k_{on} = 0.003 \times 10^6$ M$^{-1}$•sec$^{-1}$) is much slower than for Abl (Figure 15.17C). This makes sense, because the stable conformations of Src are incompatible with imatinib binding (see Section 12.17). Presumably, imatinib is slow to bind to Src because the protein is slow to convert to a conformation that can accommodate the drug.

The off-rate of imatinib from Abl is quite low, 0.00219 sec$^{-1}$. This corresponds to a half-life of $t_{1/2} = 0.693/k_{off} = 316$ sec. The dissociation of drug from Src is much faster, with $k_{off} = 0.126$ sec$^{-1}$, and so $t_{1/2} = 5.5$ sec.

## 15.13 Steady-state reactions are important in metabolism

As noted earlier, living organisms are never at equilibrium. It is usually the case, however, that the concentrations of many important metabolites in cells do not change much over significant periods of time, despite the fact that many different metabolic reactions are going on continuously. An example of a metabolic pathway that we have discussed earlier (Section 12.20), the production of cholesterol from acetyl coenzyme A (acetyl CoA), is shown in **Figure 15.18.** The rate-limiting step of this pathway is the conversion of 3-hydroxy-3-methylglutaryl-CoA (HMG-CoA) to mevalonate by the enzyme HMG-CoA reductase. Recall from Section 12.20 that HMG-CoA reductase is the target of cholesterol-lowering drugs. The

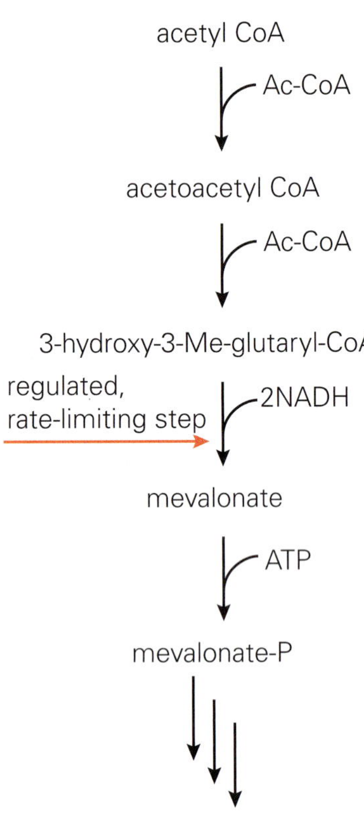


**Figure 15.18 A metabolic pathway.** Almost all metabolic pathways are series of reactions in which intermediates pass from one enzyme to another to either build up or degrade a product. A part of the cholesterol biosynthetic pathway is shown, with the side arrow showing a dominant, regulated rate-limiting step in this pathway (also discussed in Section 12.20 in the context of inhibitors).

HMG-CoA reductase enzyme is regulated by normal cellular processes so that the net flux through the pathway is held at an appropriate level, and the level of cholesterol in the cell is maintained.

The fact that metabolites, such as cholesterol, do not change much in concentration must mean that the rates of formation of these molecules equal the rates at which they are converted into other kinds of molecules. A series of reactions for which the concentrations of reactants and products do not change with time is said to have reached a **steady state**. As is discussed in subsequent sections, this is also true for a set of reactions that have reached equilibrium, but metabolic reactions can be far from equilibrium, yet still at a steady state.

Consider a part of a metabolic pathway, $A \xrightarrow{k_1} B \xrightarrow{k_2} C$, a series of two reactions of the sort that we discussed in Section 15.10. But now, instead of assuming that there is a fixed amount of A to begin the reaction, we will consider what happens if the concentration of A is held constant at all times. We will assume that both reactions are first order.

When the reaction begins, the concentration of B will start to increase, but as B increases, the rate of the reaction producing C will also increase. The rate of production of B from A and the rate of conversion of B to C eventually become equal, so the concentration of B no longer changes. To calculate the time dependence of [B], we set $[A] = [A]_0$ at all times, because we are assuming that the level of A is held constant. And so:

$$\frac{d[B]}{dt} = +k_1[A]_0 - k_2[B] \tag{15.40}$$

Equation 15.40 can be integrated, as explained in Box 15.1, yielding the following expression for the concentration of B:

$$[B] = \frac{k_1}{k_2}[A]_0(1 - e^{-k_2 t}) \tag{15.41}$$

Equation 15.41 shows that the concentration of [B] asymptotically approaches a final value with a time constant of $1/k_2$ (**Figure 15.19**; see Section 15.9 for an explanation of the time constant).

The asymptotic value of [B] is called the steady-state concentration, $[B]_{ss}$, which is obtained by setting $t$ to infinity in Equation 15.41 and noting that $e^{-\infty}$ is zero :

$$[B]_{ss} = \frac{k_1}{k_2}[A]_0 \tag{15.42}$$

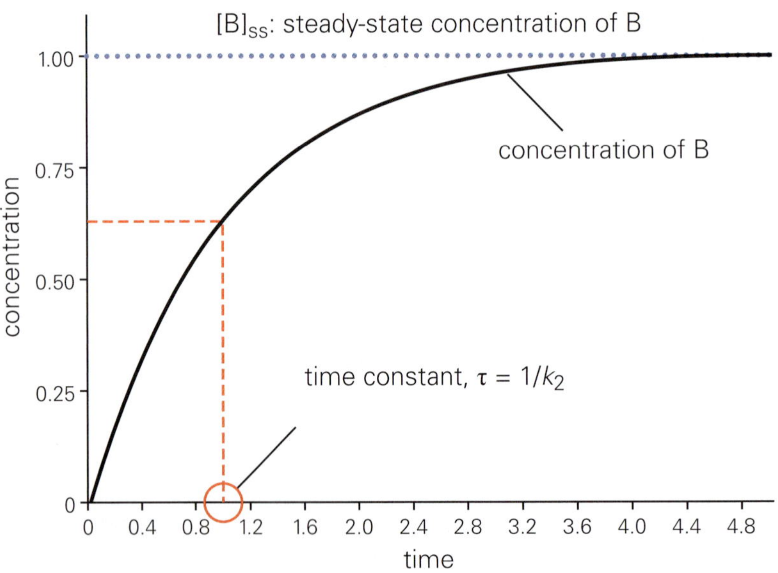


**Figure 15.19 Steady-state concentration.** Reactant A generates an intermediate, B, which is converted to a product, C. If the concentration of A is held fixed (that is, if it is constantly replenished), then the concentration of B will reach a steady-state value, $[B]_{SS}$. The graph shows the increase in the concentration of B with time, with a time constant equal to $1/k_2$, where $k_2$ is the rate constant for the conversion of B to C.

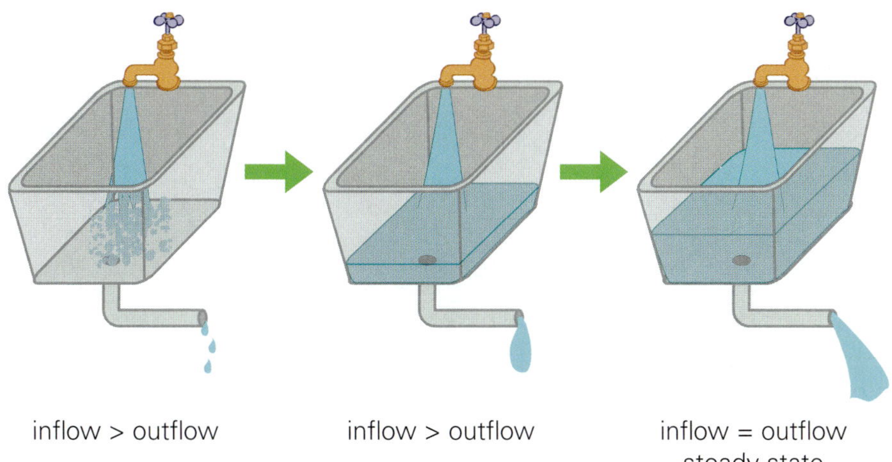


**Figure 15.20 Steady-state kinetics and a kitchen sink**. In the analogy to steady-state kinetics shown here, water flows into a sink and flows out of a drain. The rate at which water flows out of the drain is proportional to the pressure of the water, which depends on how much water is in the sink. As more water accumulates in the sink, a steady state is reached when the flow in from the faucet is matched by the flow out through the drain. The amount of water in the sink is constant under these conditions. Compare the buildup of water in the sink with the increase of the concentration of B in Figure 15.19.

If the rate constant for the production of B is larger than that for its conversion to C (that is, if $k_1 > k_2$), then $[B]_{ss}$ will be higher than $[A]_0$. Conversely, if $k_2 > k_1$, then $[B]_{ss}$ will be lower than $[A]_0$. You can understand why this is so by considering an analogy of a sink with a faucet and an open drain, as shown in **Figure 15.20**. Water entering the sink corresponds to the reaction making B from A, with the flow out of the drain depending on pressure (proportional to the amount of B present). The water level in the sink (the amount of B at steady state) depends on the relative flow in versus the flow out of the drain. If either the rate of flow in or the flow out of the drain changes, then the level of water ($[B]_{ss}$) will also change. In metabolic pathways, the rates of production and utilization of compounds are controlled by enzymes whose activities are modulated by interactions with regulatory factors, including the metabolites themselves (see Section 16.17).

**Steady state**

A set of reactions for which the concentrations of reactants and products do not change with time is said to be in a steady state. The reactants and products do not need to be at equilibrium for a steady state to occur. All that is required for a steady state is that the rates of production of molecules be balanced by the rates at which they are converted to other molecules.

Note that a whole pathway, ... $A \rightarrow B \rightarrow C \rightarrow D \rightarrow E$ ..., can come to steady state with each metabolite (A, B, C...) at a different concentration. The only requirement is that the rate at each step beyond the first must depend on the concentration of the metabolites, so that as the concentrations increase, the rate also increases. It is not necessary that the processes be first order, as was used in the example above (this case was chosen because it is easy to analyze). As noted in the previous section, the rate of product formation at the end the pathway will be determined by the rate of the slowest step. At steady state, the concentrations of metabolites before the slowest step will be higher than those after the slowest step.

## 15.14 For reactions with alternative products, the relative values of rate constants determine the distribution of products

We have, so far, only considered linear reaction schemes in which a reactant goes to only one kind of product. It is not uncommon, however, for reactions to be branched. Branched reactions have two or more pathways, with alternative intermediates and products generated from a reactant. One example of a branched reaction is the closure of a linear form of a sugar, which can go to either the α or the β anomer ring form (see Figure 3.4). Another branched reaction occurs when proteins go from their unfolded state to the correct, native conformation, or to one or more incorrectly folded conformations (**Figure 15.21**). An example that we shall discuss further in the next section is provided by fluorescence. Energy absorbed by a molecule from a photon of light can either be re-emitted as light or converted to heat.

A kinetic model for a reaction with two branches is given by:

$$A \xrightarrow{k_1} B \text{ and } A \xrightarrow{k_2} C$$

(A)

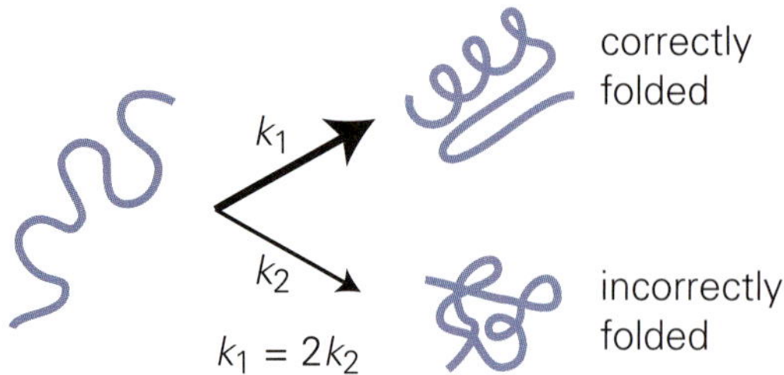


resulting population

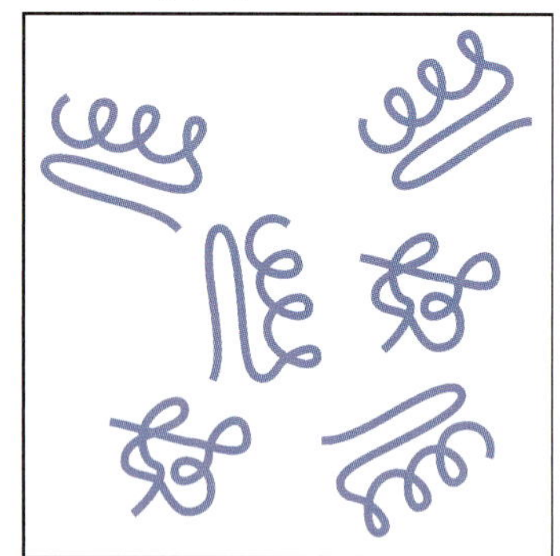

(B)

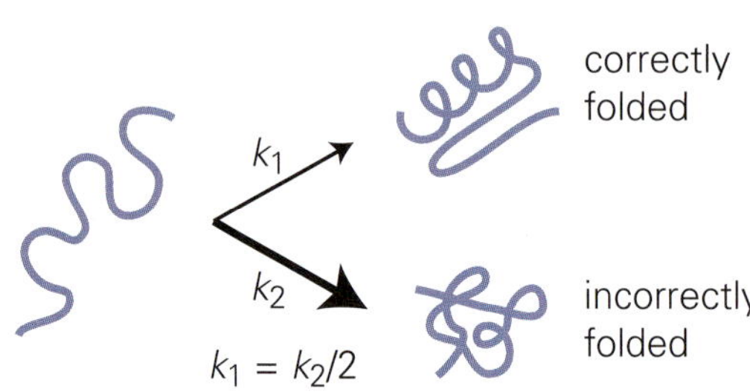


resulting population

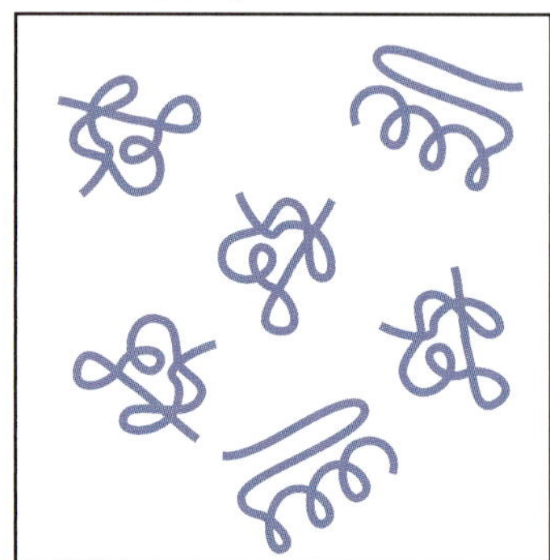

**Figure 15.21 Protein folding as an example of a branching reaction.** In this example, an unfolded protein molecule is assumed to be able to fold either into a correctly folded conformation, or an incorrectly folded one. (A) The rate constant for the formation of the correct structure, $k_1$, is assumed to be twice as large as the rate constant, $k_2$, for the formation of the incorrect structure. There will be twice as many correctly folded proteins formed as incorrectly folded ones. (B) If $k_1$ is half the value of $k_2$, then the opposite happens and there will be twice as many incorrectly folded proteins as correctly folded ones.

The rate equation for the reactant, A, must account for the two parallel reactions:

$$\frac{d[\mathrm{A}]}{dt}=-k_1[\mathrm{A}]-k_2[\mathrm{A}]=-(k_1+k_2)[\mathrm{A}] \tag{15.43}$$

There are two rate equations for the products:

$$\frac{d[\mathrm{B}]}{dt}=+k_1[\mathrm{A}] \text{ and } \frac{d[\mathrm{C}]}{dt}=+k_2[\mathrm{A}] \tag{15.44}$$

For the reactant, A, the result is just a simple first-order behavior, but with effective rate constant, $k_{\mathrm{eff}} = k_1+k_2$:

$$[\mathrm{A}]=[\mathrm{A}]_0 e^{-k_{\mathrm{eff}}t} \tag{15.45}$$

To solve for the concentration of the products ([B] and [C]), the expression for [A] is substituted into the differential equations for [B] and [C] (Equation 15.44) and integrated to give:

$$[\mathrm{B}]=\frac{k_1[\mathrm{A}]_0}{k_1+k_2}\left[1-e^{-(k_1+k_2)t}\right]=\frac{k_1[\mathrm{A}]_0}{k_{\mathrm{eff}}}\left(1-e^{-k_{\mathrm{eff}}t}\right) \tag{15.46}$$

and

$$[\mathrm{C}]=\frac{k_2[\mathrm{A}]_0}{k_1+k_2}\left[1-e^{-(k_1+k_2)t}\right]=\frac{k_2[\mathrm{A}]_0}{k_{\mathrm{eff}}}\left(1-e^{-k_{\mathrm{eff}}t}\right) \tag{15.47}$$

Each of these contains a ratio of rate constants, $[k_1/(k_1+k_2)]$ for B, the fraction of molecules that ultimately become B, and $[k_2/(k_1+k_2)]$ for those that become C. Equations 15.46 and 15.47 make it easy to calculate how much B is produced relative to C. The ratio of the concentrations of the two products is known as the **branching ratio**, [B]/[C]. Because the time dependence for each branch is identical, the branching ratio is constant at the value $k_1/k_2$ at all times.

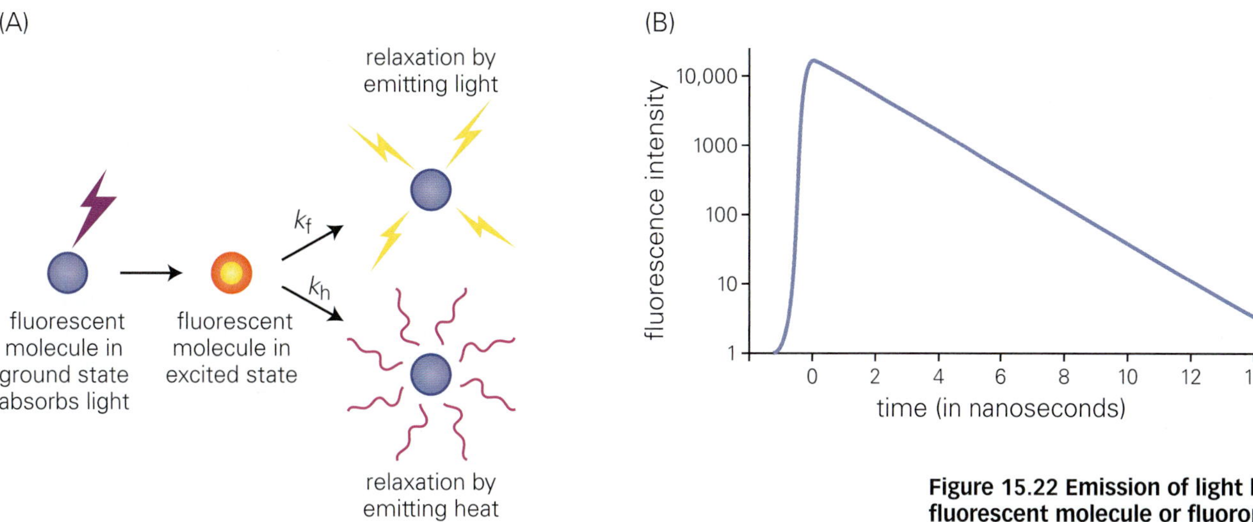


**Figure 15.22 Emission of light by a fluorescent molecule or fluorophore.** (A) Fluorophores can absorb light of an appropriate wavelength and convert to an excited electronic state (*red*). The fluorophore can then relax back to the ground state by two processes. In one process, known as fluorescence, the molecule emits light of longer wavelength than the incident light. In an alternative process, the molecule gives up its extra energy in the form of heat. The rate constants for the two processes, $k_f$ and $k_h$, vary depending on the environment of the fluorescent molecule. (B) A fluorescence decay curve. A logarithmic plot of fluorescence intensity versus time, $t$, is shown. The rise near $t = 0$ comes from the pulse of exciting light, and thereafter the intensity decays exponentially, giving a straight line in the log plot. Note that the fluorescence decays in a few nanoseconds, which is typical and reflects both the emission of light and the conversion to heat.

## 15.15 Measuring fluorescence provides an easy way to monitor kinetics

Fluorescence spectroscopy and imaging are important tools in the study of biological molecules and illustrate well a number of features of kinetic processes. Fluorescence results after a molecule, known as a **fluorophore**, is put into an electronically excited state by absorbing light (usually ultraviolet light). After excitation, the fluorophore can emit a photon of light (the process of fluorescence, in which the light emitted is at a longer wavelength than that absorbed), or convert the energy to vibrations (that is, releasing it as heat, as shown in **Figure 15.22**). The relative rates of these two processes depend on the characteristics of the fluorophore and the environment. The sensitivity to the environment makes fluorescence useful for following processes such as ligand binding (see Section 15.12) and protein folding (Section 18.5). We have also encountered the use of fluorescence to monitor the mobility of lipids (see Section 3.18).

To describe the kinetics of the fluorescence process, we start with some population of fluorophores, F, that have been excited, F*, and consider the two alternative reactions by which they return to the ground state:

$$\begin{aligned} &F^* \xrightarrow{k_f} F + \text{light} \\ &F^* \xrightarrow{k_h} F + \text{heat} \end{aligned} \tag{15.48}$$

with rate constants $k_f$ for fluorescence and $k_h$ for producing heat. Using Equation 15.45, we can obtain the following equation for the concentration of the excited fluorophores, $F^*$:

$$[F^*] = [F^*]_0 e^{-(k_f+k_h)t} \tag{15.49}$$

In a fluorescence experiment, what is actually measured is the the intensity of light emitted, $I_{light}$, as a function of time. $I_{light}$ is the fluorescence intensity, and has units of photons per second. The fluorescence intensity corresponds directly to the rate of production of light as a product—that is:

$$I_{light} = \frac{d[F]}{dt} = k_f[F^*] \tag{15.50}$$

In this case, the direct measurement is the *rate* of reaction, rather than the concentration of product.

Fluorophores can be excited using a very short pulse of light (from a flash lamp or a pulsed laser), creating the initial concentration of excited fluorophores, $[F^*]_0$. Using the last two equations together, the time dependence of fluorescence intensity, $I_{light}$, is:

$$I_{light} = k_f[F^*]_0 e^{-(k_f+k_h)t} \tag{15.51}$$

The time constant for the disappearance of fluorescence, known as the **fluorescence lifetime**, $\tau_f = 1/(k_f + k_h)$, is determined by the combination of the two processes that make excited molecules relax to their ground states, not just the rate of fluorescence itself. An example of a fluorescence decay curve is shown in Figure 15.22B. A change in either $k_f$ or $k_h$ will cause a change in fluorescence lifetime.

## 15.16 Fluorescence measurements can be carried out under steady-state conditions

Measurements of fluorescence use light for excitation. Because the excitation and the detection are done at different wavelengths, the exciting light and the fluorescence emission can be distinguished. That is, the fluorescent measurement can be carried out even if the exciting light is on. This allows measurements to be made under steady-state conditions, by continuously pumping molecules into the excited state. If we denote the rate constant for excitation of molecules as $k_e$, then the following kinetic scheme describes the process:

$$\mathrm{F} \xrightarrow{k_e} \mathrm{F}^* \xrightarrow{k_f} \mathrm{F} + \text{light}$$
$$\mathrm{F}^* \xrightarrow{k_h} \mathrm{F} + \text{heat} \tag{15.52}$$

Some short time after turning on the exciting light, the system will reach a steady state:

$$\frac{d[\mathrm{F}^*]_{ss}}{dt} = 0 = k_e[\mathrm{F}] - k_f[\mathrm{F}^*] - k_h[\mathrm{F}^*] \tag{15.53}$$

By rearranging Equation 15.53, we can see that the steady-state concentration of the excited fluorophore, $[\mathrm{F}^*]_{ss}$ is given by:

$$[\mathrm{F}^*]_{ss} = \frac{k_e}{k_f + k_h}[\mathrm{F}] \tag{15.54}$$

Recall from Equation 15.50 that the intensity of fluorescence emission is given by the product of the rate constant for fluorescence and the concentration of excited molecules. Combining this information with Equation 15.42, we get:

$$I_{\text{light}} = k_f[\mathrm{F}^*]_{ss} = \frac{k_e k_f}{k_f + k_h}[\mathrm{F}]_0 \tag{15.55}$$

To obtain Equation 15.55, we assume that the excitation is fairly weak, so that almost all fluorophores are in their ground state at any given time (that is, the concentration of unexcited fluorophores, [F], is essentially the same as the total fluorophore concentration, $[\mathrm{F}]_0$). Note that this assumption is similar to the assumption we made when analyzing ligand binding—namely, that the free ligand concentration is the same as the total ligand concentration (see Section 12.6). For the analysis of fluorescence, as for ligand binding, the advantage of working under such conditions is that the total concentration of fluorophore (or ligand, in the case of binding), a parameter that is known to the experimenter, is sufficient to calculate the rates or extent of binding.

Note that, according to Equation 15.55, the fluorescence intensity will change if any of the rate constants, $k_e$, $k_f$, or $k_h$, changes. These rate constants can change when the fluorescent group changes its environment. If the fluorescent group is attached to a protein, then conformational changes in the protein, or the binding of the ligand, can change the fluorescence intensity.

Tryptophan sidechains in proteins are fluorescent, and monitoring fluorescence from tryptophan residues is often a good way to detect the binding of a ligand, such as a drug molecule, to a protein. This is illustrated in Figure 15.17 for two drugs that inhibit the protein kinase as Abl, which was discussed earlier, in Section 15.12. The conformation of Abl is different when bound to these two drugs, dasatinib and imatinib (Section 12.17). A tryptophan sidechain in Abl is in quite different environments when these two drugs are bound, as shown in **Figure 15.23.** As a consequence, the fluorescence intensity is different in the two cases and is also different when no drug is bound.

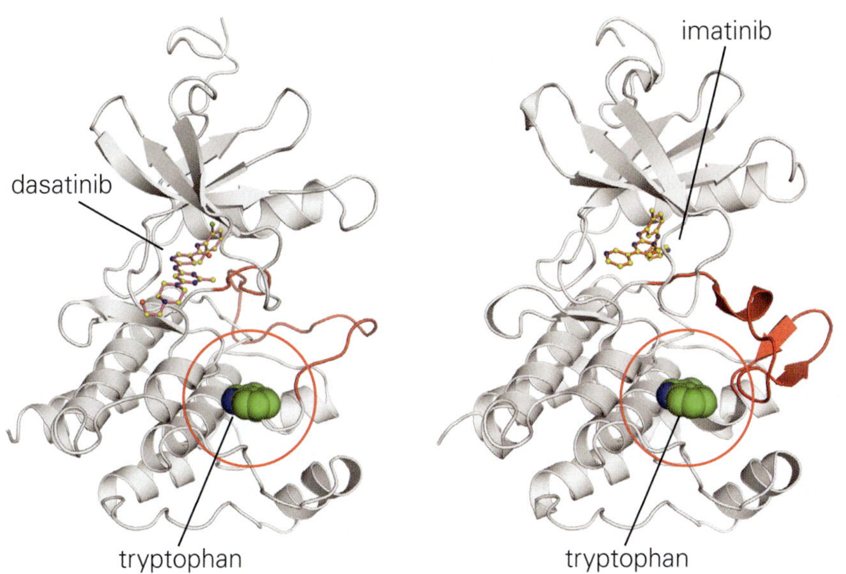


**Figure 15.23 The fluorescence of tryptophan residues is sensitive to protein conformation.** Shown here are the structures of the protein kinase Abl, bound to two different drugs, dasatinib and imatinib (these drugs are discussed in Chapter 12; see Figure 12.24). The conformation of a segment of the protein (colored *red*) is different depending on which drug is bound. The binding of the two drugs alters the fluorescence of this tryptophan sidechain differently because the conformation of the *red* segment is different when the two drugs are bound. The fluorescence of the tryptophan is also different when no drug is bound. (PDB codes: 2GQG and 1IEP)

Fluorescence from tryptophan sidechains is also commonly used both to follow the kinetics of unfolding of proteins and to determine the amount of unfolded protein at equilibrium, as discussed in Section 18.5.

## 15.17 Fluorescence quenchers provide a way to detect whether a fluorophore on a protein is accessible to the solvent

Another useful application of fluorescence comes about because it is easy to probe whether or not a fluorophore that is attached to a protein or RNA molecule is buried. This can be done by the addition of a **fluorescence quencher** to the solution. The quencher is a compound, such as acrylamide, to which the excitation energy is readily transferred from the fluorophore, and the quencher converts this energy rapidly to heat. In this way, fluorescence quenchers reduce the amount of light emitted by fluorescent molecules. The quencher must come into close contact with the fluorophore for energy to be transferred to it and, hence, the quencher will be ineffective if the fluorophore is buried.

For example, a tryptophan residue in the hydrophobic core of a protein would not have its fluorescence reduced efficiently by a quencher. But if the protein were to unfold, the tryptophan would be exposed to the solvent, and addition of a quencher would then drastically reduce its fluorescence.

Fluorescence quenching also provides a way to monitor the oligomerization of proteins. In the example shown in **Figure 15.24**, an enzyme, phosphofructokinase-2, switches between a monomeric and a tetrameric form depending on whether ligand is bound to it. In order to detect which state the enzyme is in, scientists attached a fluorophore to a cysteine residue that is exposed in the monomer but buried in the tetramer. In the absence of the ligand, fluorescence from the fluorophore is readily quenched by a quencher, Q. But, when the ligand is added to the enzyme, the amount of quenching is reduced substantially.

A kinetic model for this process just requires adding one more term to the scheme shown in Equation 15.52:

$$\begin{aligned} &\mathrm{F} \xrightarrow{k_e} \mathrm{F}^* \xrightarrow{k_f} \mathrm{F} + \text{light} \\ &\mathrm{F}^* \xrightarrow{k_h} \mathrm{F} + \text{heat} \\ &\mathrm{F}^* + \mathrm{Q} \xrightarrow{k_Q} \mathrm{F} + \mathrm{Q} + \text{heat} \end{aligned} \tag{15.56}$$

By using the ideas that we have discussed earlier, we can obtain expressions for the time dependence of the fluorescence intensity in the presence of the quencher, which is:

$$[\mathrm{F}^*]=[\mathrm{F}^*]_0 e^{-(k_\mathrm{f}+k_\mathrm{h}+k_\mathrm{Q}[\mathrm{Q}])t} \tag{15.57}$$

Likewise, the steady-state condition that applies in the case of continuous irradiation is:

$$I_{\mathrm{light}}=k_\mathrm{f}[\mathrm{F}^*]_{\mathrm{ss}}=\frac{k_\mathrm{e}k_\mathrm{f}}{k_\mathrm{f}+k_\mathrm{h}+k_\mathrm{Q}[\mathrm{Q}]}[\mathrm{F}]_0 \tag{15.58}$$

According to Equation 15.58, the lifetime will be shortened by the term $k_\mathrm{Q}[\mathrm{Q}]$. The steady-state fluorescence intensity will be reduced due to quenching if the fluorophore is accessible to the quencher, when $k_\mathrm{Q}$ has a large value. For a buried fluorophore, $k_\mathrm{Q}$ will be very small and little or no effect of the quencher will be observed. These effects are illustrated in Figure 15.24C for the enzyme phosphofructokinase-2. In this diagram, the ratio of the fluorescence intensity in the absence of the quencher to the intensity in the presence of the quencher is graphed as a function of the concentration of the quencher. By comparing Equations 15.55 and 15.58, we can see that this ratio is given by:

$$\frac{\text{fluorescence intensity in the absence of quencher}}{\text{fluorescence intensity in the presence of quencher}}$$

$$=1+\left(\frac{k_\mathrm{Q}}{k_\mathrm{f}+k_\mathrm{h}}\right)[\mathrm{Q}] \tag{15.59}$$

Equation 15.59 is known as the **Stern–Volmer equation**, and it allows us to determine the rate constant for fluorescence quenching, $k_\mathrm{Q}$. According to the

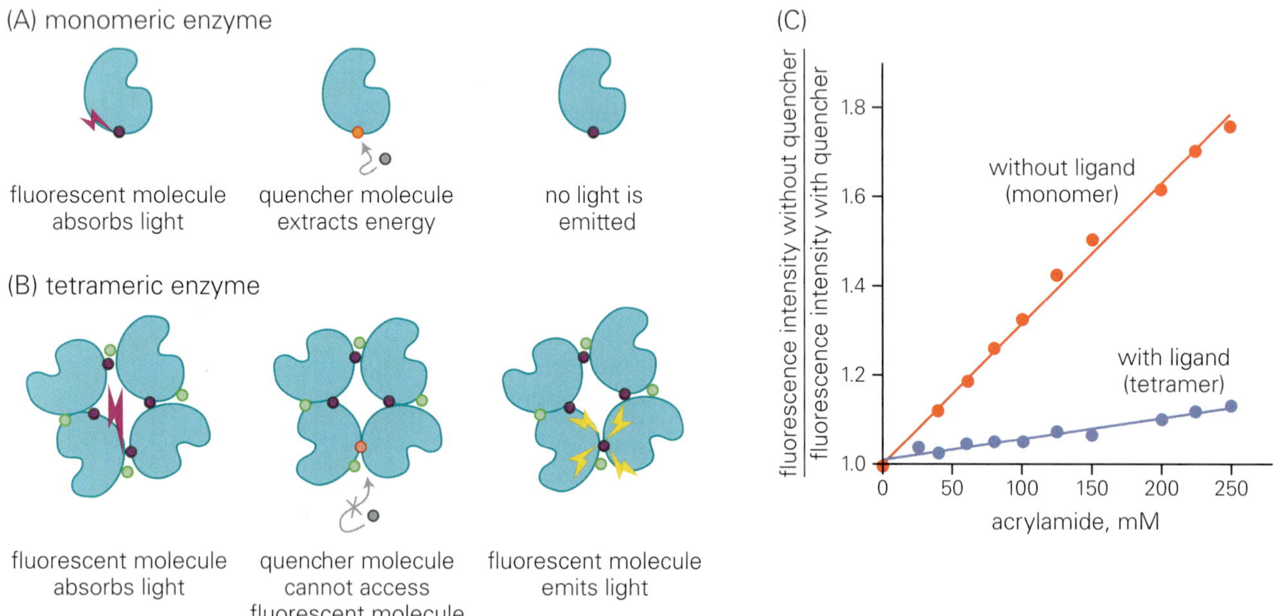


**Figure 15.24 Fluorescence quenching**. An enzyme that can exist in either a monomeric or a tetrameric form is shown schematically in (A) and (B). The surface of the enzyme is labeled with a fluorophore. (A) In the monomeric form, the fluorescent molecule is exposed. The fluorescent molecule absorbs light and converts to an excited electronic state (*red*). The presence of fluorescence quenchers in the solution results in the transfer of the excitation energy to the quenchers, and the amount of light emitted as fluorescence is reduced. (B) The enzyme converts to a tetrameric form upon binding to a ligand molecule (*green*). The fluorescent molecules are buried at the inter-protein interfaces of the tetramer and are inaccessible to the quencher molecules in the solution. When the fluorescent molecules are excited by incident light, they now emit light through fluorescence. (C) Graphs of the ratio of the fluorescence intensity without quencher (acrylamide) to that with quencher, in the presence (*blue*) and absence (*red*) of the ligand. The slopes of the lines give the values of $k_\mathrm{Q}/(k_\mathrm{f}+k_\mathrm{h})$, as shown in Equation 15.59 (the Stern–Volmer equation). (C, adapted from M. Baez et al., and V. Guixe, *Biochem. J.* 376: 277–283, 2003. With permisison from the Biochemical Society.)

Stern–Volmer equation, the slope of the graph of the fluorescence ratio as a function of [Q] yields the value of $k_Q/(k_f + k_h)$. Recall from Equation 15.51 that the term $(k_f + k_h)$ is the time constant for the fluorescence in the absence of the quencher and is therefore determined readily. In this way, the value of $k_Q$ can be derived from the data shown in Figure 15.24C. Note that the slope of the graph is much steeper in the absence of ligand, consistent with increased quenching due to the exposure of the fluorophore in the monomeric form of the enzyme.

## 15.18 The combination of forward and reverse rate constants is related to the equilibrium constant

When a reversible reaction reaches equilibrium, the concentrations of the reactants and products no longer change with time. For a reaction in which A and B interconvert, the condition for equilibrium is:

$$\frac{d[\mathrm{A}]}{dt} = 0 \text{ and } \frac{d[\mathrm{B}]}{dt} = 0 \qquad (15.60)$$

Substituting these expressions for the derivatives of the concentrations of A and B into the rate equation (Equation 15.29) give the following relationship between the rate constants and the equilibrium concentrations of A and B:

$$k_\mathrm{f}[\mathrm{A}]_\mathrm{eq} = k_\mathrm{r}[\mathrm{B}]_\mathrm{eq} \qquad (15.61)$$

Equation 15.61 can be rearranged to give:

$$\frac{k_\mathrm{f}}{k_\mathrm{r}} = \frac{[\mathrm{B}]_\mathrm{eq}}{[\mathrm{A}]_\mathrm{eq}} = K_\mathrm{eq} \qquad (15.62)$$

where $K_{eq}$ is the thermodynamic equilibrium constant. This shows that, although the equilibrium constant for a reaction does not define the values of $k_f$ or $k_r$, it does constrain their ratio. The actual values of $k_f$ and $k_r$ can be very large or very small, but the ratio of the two rate constants is always equal to the equilibrium constant.

The relationship between rate constants and equilibrium constant (Equation 15.62) holds for any order of reaction. For example, consider a second-order binding reaction between a protein, P, and a ligand, L. Recall that the dissociation constant ($K_D$) is related to the protein and ligand concentrations as follows:

$$K_\mathrm{D} = \frac{[\mathrm{P}][\mathrm{L}]}{[\mathrm{P{\bullet}L}]} \qquad (15.63)$$

At equilibrium,

$$\frac{d[\mathrm{P}]}{dt} = 0 \qquad (15.64)$$

Combining Equation 15.33 with Equation 15.64, we get:

$$\frac{d[\mathrm{P}]}{dt} = -k_\mathrm{on}[\mathrm{P}][\mathrm{L}] + k_\mathrm{off}[\mathrm{P{\bullet}L}] = 0 \qquad (15.65)$$

Thus,

$$-k_\mathrm{on}[\mathrm{P}][\mathrm{L}] = -k_\mathrm{off}[\mathrm{P{\bullet}L}] \qquad (15.66)$$

and so:

$$\frac{k_\mathrm{off}}{k_\mathrm{on}} = \frac{[\mathrm{P}][\mathrm{L}]}{[\mathrm{P{\bullet}L}]} = K_\mathrm{D} = \frac{1}{K_\mathrm{A}} \qquad (15.67)$$

where $K_A$ is the association constant. In Equations 15.63–15.67, the concentrations [P], [L] and [P•L] are all equilibrium concentrations.

Equation 15.67 allows us to link the on- and off-rates for imatinib binding, discussed in Section 15.12, to the equilibrium constants for imatinib binding to the Abl and Src kinases. For Abl:

$$K_\mathrm{D} = \frac{k_\mathrm{off}}{k_\mathrm{on}} = \frac{0.00219\ \mathrm{sec}^{-1}}{0.146\times10^{6}\ \mathrm{M}^{-1}{\bullet}\mathrm{sec}^{-1}} = 0.015\times10^{-6}\ \mathrm{M} \qquad (15.68)$$

For Src:

$$K_D = \frac{k_{off}}{k_{on}} = \frac{0.126\ \text{sec}^{-1}}{0.003 \times 10^6\ \text{M}^{-1} \bullet \text{sec}^{-1}} = 42 \times 10^{-6}\ \text{M} \tag{15.69}$$

Knowing the values of the dissociation constants from the kinetic measurements we can apply the ideas of specificity of binding that were developed in Chapter 13. The maximum specificity of imatinib, which will be at very low ligand concentrations, is given by the ratio of the dissociation constants:

$$\frac{(K_D)_{Src}}{(K_D)_{Abl}} = \frac{42 \times 10^{-6}}{0.015 \times 10^{-6}} = 2800 \tag{15.70}$$

While this specificity is very high, the relevant concentration range for therapeutic applications requires that most Abl molecules have imatinib bound while few Src molecules do. This requirement corresponds to:

$$(K_D)_{Abl} < [\text{imatinib}] < (K_D)_{Src}$$

With the actual $K_D$ values we can see that the typical drug concentration range in the body of 100 nM to 1 μM corresponds well to the range in which high specificity is maintained.

## 15.19 Relaxation methods provide a way to obtain rate constants for reversible reactions

The obvious way to determine the kinetic parameters governing a reaction is to mix the reactants together and monitor how quickly the products are formed. This is the approach underlying the analysis of drug binding to the Abl kinase, discussed in Section 15.12. There are many situations where it is inconvenient, or perhaps even impossible, to start with pure reactants and watch how fast the products are generated. Consider, for example, the dimerization of a protein. The monomeric and dimeric forms of the protein are always interconverting, and so it is not straightforward to separate monomers from dimers and then recombine them. How, then, might we learn about the rates of interconversion between monomers and dimers?

A different way of obtaining kinetic information is to start with a system at equilibrium, with both reactants and products present, and to perturb it so that it is no longer at equilibrium. By monitoring how the system returns to equilibrium, we can learn about the rates of interconversion between the molecules that define the system. Such approaches are known as **relaxation methods**. Relaxation methods provide a very powerful way to study kinetics, but they require that both the reactants and the products be present at measurable concentrations (that is, the reaction must have an equilibrium constant that is neither very large nor very small).

One way of perturbing the system is to change the temperature in a **temperature jump** experiment (**Figure 15.25**). This can be done within microseconds by using voltage pulses, or even within nanoseconds by using pulses of laser light. In this way, the kinetics of very fast processes can be measured.

**Relaxation method**

A relaxation method follows the rate of return to equilibrium after a sudden perturbation that disturbs the equilibrium.

Recall from Section 10.14 that the free energy, and hence the equilibrium constant, is temperature dependent. If the temperature is changed suddenly, then the reactants and products will no longer be at their equilibrium values at the new temperature. It turns out that, if the perturbation is small, the concentrations then relax to their new equilibrium values with simple exponential behavior, regardless of the order of the reaction (**Box 15.3**). For example, consider the following bimolecular reaction:

$$\text{A} + \text{B} \underset{k_r}{\overset{k_f}{\rightleftarrows}} \text{A}\bullet\text{B} \tag{15.71}$$

Here $k_f$ and $k_r$ are the rate constants for association and dissociation, respectively. If this reaction is perturbed slightly from equilibrium, then the concentrations will

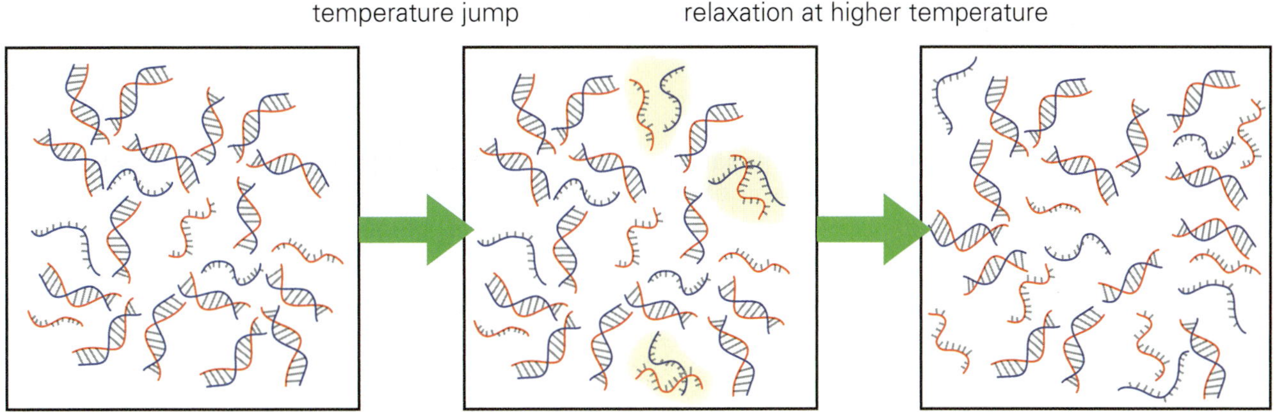


**Figure 15.25 An example of a temperature jump experiment.** A test tube containing two DNA molecules with complementary sequences (*red* and *blue*) is shown schematically. At equilibrium, most of the DNA molecules are in a double-helical form. The temperature is increased suddenly and some of the double helices begin to separate into individual strands of DNA (middle panel). The system relaxes to a new equilibrium at the higher temperature.

relax back to equilibrium with exponential kinetics, with a time constant, $\tau$, given by:

$$\tau = \frac{1}{k_{\text{apparent}}} = \frac{1}{k_r + k_f([A]_{eq} + [B]_{eq})} \qquad (15.72)$$

The time constant, $\tau$, is known as the **relaxation time**. The rate constant, $k_{\text{apparent}}$, is the apparent first-order rate constant ("apparent" because the reaction is not really first order). The subscript "eq" indicates that the concentrations in Equation 15.72 are the equilibrium concentrations after relaxation. An explanation of how Equation 15.72 comes about is given in Box 15.3.

## 15.20 Temperature jump experiments can be used to determine the association and dissociation rate constants for dimerization

Temperature jump experiments are particularly useful for studying the rates of dimerization of a protein (**Figure 15.26**). Consider a protein, P, that undergoes a monomer–dimer equilibrium:

$$2P \underset{k_r}{\overset{k_f}{\rightleftharpoons}} P_2 \qquad (15.73)$$

If the monomer and dimer forms are at equilibrium and the system is subject to a temperature jump, then the system will relax to a new equilibrium with the following time constant, $\tau$:

$$\tau = \frac{1}{k_{\text{apparent}}} = \frac{1}{4k_f[P]_{eq} + k_r} \qquad (15.74)$$

Equation 15.74 can be derived by following the same set of steps, explained in Box 15.3, that led to Equation 15.72 for the more general bimolecular reaction.

The term $[P]_{eq}$ in Equation 15.74 is the equilibrium concentration of the monomeric form of the protein. It is more convenient to express the relaxation time in terms of the total concentration of protein subunits, $[P]_{total}$, because that value is set by the experimenter when making up the protein solution. The total subunit concentration, $[P]_{total}$ is related to $[P]_{eq}$ in the following way:

$$[P]_{total} = [P] + 2[P_2] \qquad (15.75)$$

At equilibrium, the concentrations of the monomer and dimer forms are related to the rate constants as explained in Section 15.18:

$$[\mathrm{P}]_{\mathrm{eq}}^2 = \left(\frac{k_\mathrm{r}}{k_\mathrm{f}}\right)[\mathrm{P}_2]_{\mathrm{eq}} \tag{15.76}$$

By taking the square of Equation 15.74, and by combining Equations 15.75 and 15.76 and substituting, we get:

$$\tau^2 = \frac{1}{k_\mathrm{r}^2 + 8k_\mathrm{f}k_\mathrm{r}[\mathrm{P}]_{\mathrm{total}}} \tag{15.77}$$

## Box 15.3 First-order behavior after small perturbations in relaxation kinetics

Consider a second-order reaction in which two reactants form a single product, as discussed in Section 15.19 of the main text:

$$\mathrm{A} + \mathrm{B} \underset{k_\mathrm{r}}{\overset{k_\mathrm{f}}{\rightleftarrows}} \mathrm{C} \tag{15.3.1}$$

A, B and C are initially at equilibrium. The system is then subjected to a temperature jump and allowed to relax to a new equilibrium. We define the equilibrium concentrations *after* the perturbation to be $[\mathrm{A}]_{\mathrm{eq}}$, $[\mathrm{B}]_{\mathrm{eq}}$, and $[\mathrm{C}]_{\mathrm{eq}}$. The concentrations at any time during the change following the perturbation can be written as:

$$[\mathrm{A}] = [\mathrm{A}]_{\mathrm{eq}} + \Delta[\mathrm{A}] \tag{15.3.2}$$

$$[\mathrm{B}] = [\mathrm{B}]_{\mathrm{eq}} + \Delta[\mathrm{B}] \tag{15.3.3}$$

$$[\mathrm{C}] = [\mathrm{C}]_{\mathrm{eq}} + \Delta[\mathrm{C}] \tag{15.3.4}$$

The terms $\Delta[\mathrm{A}]$, $\Delta[\mathrm{B}]$, and $\Delta[\mathrm{C}]$ are the small deviations from equilibrium in the concentrations of A, B, and C, respectively, that are brought about by the temperature jump. Because C can only be produced from A and B, it follows that:

$$\Delta[\mathrm{A}] = \Delta[\mathrm{B}] = -\Delta[\mathrm{C}] \tag{15.3.5}$$

The equilibrium concentrations of the reactants and products are fixed. Hence, the expression for the rate of change in the concentration of the product, C, depends only on the perturbation in the concentration:

$$\frac{d[\mathrm{C}]}{dt} = \frac{d([\mathrm{C}]_{\mathrm{eq}} + \Delta[\mathrm{C}])}{dt} = \frac{d\Delta[\mathrm{C}]}{dt} \tag{15.3.6}$$

Combining Equation 15.3.6 with the rate equation for the second-order reaction depicted in Equation 15.3.1, we get:

$$\frac{d[\mathrm{C}]}{dt} = \frac{d\Delta[\mathrm{C}]}{dt} = k_\mathrm{f}[\mathrm{A}][\mathrm{B}] - k_\mathrm{r}[\mathrm{C}] = k_\mathrm{f}([\mathrm{A}]_{\mathrm{eq}} + \Delta[\mathrm{A}])([\mathrm{B}]_{\mathrm{eq}} + \Delta[\mathrm{B}]) - k_\mathrm{r}([\mathrm{C}]_{\mathrm{eq}} + \Delta[\mathrm{C}]) \tag{15.3.7}$$

Thus,

$$\frac{d\Delta[\mathrm{C}]}{dt} = k_f[\mathrm{A}]_{\mathrm{eq}}[\mathrm{B}]_{\mathrm{eq}} + k_\mathrm{f}[\mathrm{A}]_{\mathrm{eq}}\Delta[\mathrm{B}] + k_\mathrm{f}[\mathrm{B}]_{\mathrm{eq}}\Delta[\mathrm{A}] + k_\mathrm{f}\Delta[\mathrm{A}]\Delta[\mathrm{B}] - k_\mathrm{r}[\mathrm{C}]_{\mathrm{eq}} - k_\mathrm{r}\Delta[\mathrm{C}] \tag{15.3.8}$$

Recall from Section 15.18 that the ratio of the forward and reverse rates is equal to the equilibrium constant:

$$\frac{k_\mathrm{f}}{k_\mathrm{r}} = \frac{[\mathrm{A}]_{\mathrm{eq}}[\mathrm{B}]_{\mathrm{eq}}}{[\mathrm{C}]_{\mathrm{eq}}} \tag{15.3.9}$$

Rearranging Equation 15.3.9, we get:

$$k_\mathrm{f}[\mathrm{A}]_{\mathrm{eq}}[\mathrm{B}]_{\mathrm{eq}} = k_\mathrm{r}[\mathrm{C}]_{\mathrm{eq}} \tag{15.3.10}$$

Substituting Equation 15.3.10 in Equation 15.3.8, and by using Equation 15.3.5, we get:

$$\frac{d\Delta[\mathrm{C}]}{dt} = -k_\mathrm{f}[\mathrm{A}]_{\mathrm{eq}}\Delta[\mathrm{C}] - k_\mathrm{f}[\mathrm{B}]_{\mathrm{eq}}\Delta[\mathrm{C}] - k_\mathrm{r}\Delta[\mathrm{C}] + k_\mathrm{f}\Delta[\mathrm{C}]^2 \tag{15.3.11}$$

Now we use the fact that, if the perturbation is small, then the term $\Delta[\mathrm{C}]$ is small. We can therefore neglect $\Delta[\mathrm{C}]^2$, which is very small and can be ignored relative to the other terms. In this case, Equation 15.3.11 simplifies to:

$$\frac{d\Delta[\mathrm{C}]}{dt} = -\left[k_\mathrm{f}([\mathrm{A}]_{\mathrm{eq}} + [\mathrm{B}]_{\mathrm{eq}}) + k_\mathrm{r}\right]\Delta[\mathrm{C}] = -k_{\mathrm{eff}}\Delta[\mathrm{C}] \tag{15.3.12}$$

This is a simple first-order kinetic equation that gives rise to exponential time dependence, where the effective (or apparent) rate constant, $k_{\mathrm{eff}}$, is:

$$k_{\mathrm{eff}} = k_\mathrm{f}([\mathrm{A}]_{\mathrm{eq}} + [\mathrm{B}]_{\mathrm{eq}}) + k_\mathrm{r} \tag{15.3.13}$$

The associated time constant, or relaxation time, is:

$$\tau_{\mathrm{eff}} = \frac{1}{k_{\mathrm{eff}}} = \frac{1}{k_\mathrm{f}([\mathrm{A}]_{\mathrm{eq}} + [\mathrm{B}]_{\mathrm{eq}}) + k_\mathrm{r}} \tag{15.3.14}$$

The central point in this derivation, leading to a simple exponential dependence on time, is that only terms that are first order in $\Delta[\mathrm{X}]$ (where X is any of the chemical species) are kept. All higher-order terms (regardless of what they are in the actual kinetic mechanism) are dropped. This yields a first-order equation and thus gives an exponential approach to equilibrium. The relationship between the observed effective rate constant and the rate constants for specific steps in the reaction mechanism depends on the details of the mechanism. This relationship can be worked out through a process analogous to that presented above for the reaction of A and B forming C.

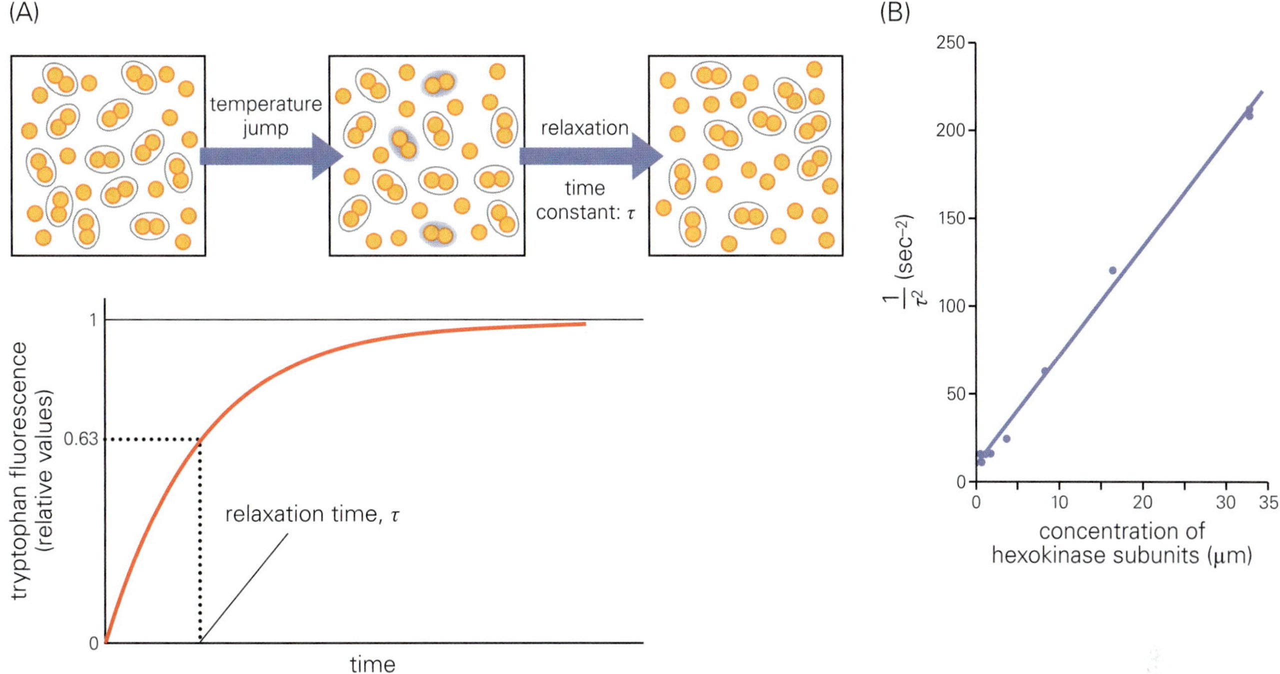


**Figure 15.26 Measurement of protein dimerization with a temperature jump experiment.** The fluorescence intensity from a sample of hexokinase, partly monomer and partly dimer (A), is followed as a function of time after a temperature jump. Repeating this process at several concentrations provides the data for (B), allowing determination of $k_f$ and $k_r$. (B, adapted from J. Hoggett and G. Kellett, *Biochem. J.* 287: 567–572, 1992. With permission from the Biochemical Society.)

To understand how this expression is used to derive the rates of association and dissociation, we take the inverse of both sides of Equation 15.77:

$$\frac{1}{\tau^2} = k_r^2 + 8k_f k_r [\mathrm{P}]_{\mathrm{TOTAL}} \tag{15.78}$$

According to Equation 15.78, the inverse of $\tau^2$ is related linearly to the total concentration of protein subunits. A graph of $1/\tau^2$ versus $[\mathrm{P}]_{\mathrm{total}}$ then yields a line. The intercept of the line is the square of the rate constant for dissociation, $k_r^2$. The slope of the line is equal to $8k_r k_f$ and, because $k_r$ is known from the intercept, the value of $k_f$ can be derived.

The application of Equation 15.83 to the analysis of the dimerization of a protein, hexokinase, is illustrated in Figure 15.26. Hexokinase equilibrates between monomers and dimers, and the extent of dimerization can be monitored by measuring the fluorescence from tryptophan sidechains in the protein. The monomer form has higher fluorescence than the dimer form, and so the tryptophan fluorescence increases if the equilibrium is shifted towards the monomer form. This is what is observed if a solution of hexokinase, initially at equilibrium, is subjected to a temperature jump (Figure 15.26A). The fluorescence signal relaxes exponentially to a new value when the temperature is increased, allowing the relaxation time, $\tau$, to be determined.

The temperature jump experiment is repeated with different concentrations of the hexokinase subunits, and the value of $1/\tau^2$ is graphed as a function of the total hexokinase subunit concentration, as shown in Figure 15.26B. As predicted by Equation 15.84, the graph of $1/\tau^2$ versus $[\mathrm{P}]_{\mathrm{total}}$ is a straight line. From these data on hexokinase, the association and dissociation rate constants were found to be $2.5 \times 10^5 \bullet \mathrm{M}^{-1}\ \mathrm{sec}^{-1}$ and $3.1\ \mathrm{sec}^{-1}$, respectively. The equilibrium association constant, $K_A$, is given by the ratio of the association rate to the dissociation rate, $k_f/k_r$, and is therefore $8 \times 10^4\ \mathrm{M}^{-1}$. In this way, the temperature jump experiment provides a comprehensive description of the kinetics and thermodynamics of dimer formation by hexokinase. In this particular case, an interesting finding is that glucose, the substrate of hexokinase, has a profound effect on the dimerization. In the presence of saturating amounts of glucose, the dimer form of the enzyme is weakened, and the dissociation rate increases about 10-fold.

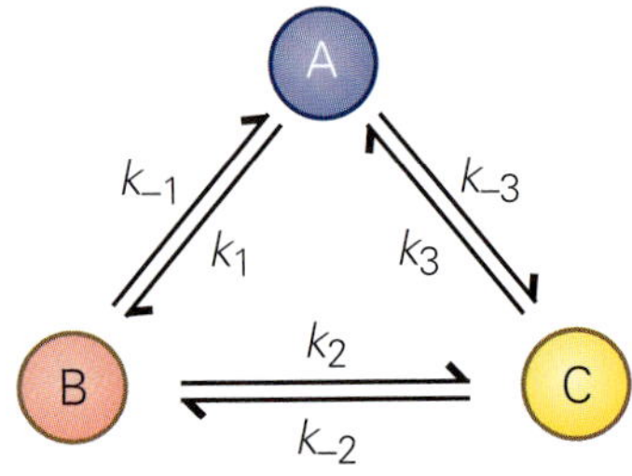


**Figure 15.27 A cyclic set of reactions.** At equilibrium, each of the individual reactions must also be at equilibrium and the forward and reverse fluxes at each step must be equal.

## 15.21 The rate constants for a cyclic set of reactions are coupled

In thinking about chemical reactions, an interesting case to consider is a cyclic set of reactions. An example of such a cyclic set would be three sequential reactions that lead back to the initial reactant. Schematically we can draw this as shown in **Figure 15.27**, in which all of the individual rate constants are identified. When such a system reaches equilibrium, there will be no further change in concentrations with time.

If the reaction was just the interconversion of A and B, then equilibrium has to occur through a balance of the forward and reverse reactions. In the cyclic case, it seems possible that the amount of A converted to B can be balanced by B converting to C and then C to A without requiring B to undergo the back reaction to A at all. But, in fact, each of these individual steps must be in equilibrium, and so the rate constants have to satisfy the following relationships:

$$\left.\begin{aligned} k_1[\text{A}] &= k_{-1}[\text{B}] \\ k_2[\text{B}] &= k_{-2}[\text{C}] \\ k_3[\text{C}] &= k_{-3}[\text{A}] \end{aligned}\right\} \qquad (15.79)$$

This rule is known as the **principle of microscopic reversibility**. This idea, which is also known as the principle of **detailed balance**, is a consequence of a principle in physics, which is that the equations of motion hold true when time is reversed. For any trajectory of atoms that obeys the laws of motion, a trajectory that follows the same path, but has the direction reversed, also obeys the laws of motion. An extension of this idea is that, for any molecular reaction at equilibrium, the forward rate must equal the reverse rate—that is, the forward and backward flux must be equal at every step. If this were not true, then it would be possible to create a kind of perpetual motion in which A converts to B, which converts to C, which then converts to A, always in a forward direction.

**The principle of microscopic reversibility**

This principle states that, for a set of cyclic reactions, the rates of the forward and backward reactions must be equal for every individual reaction in the cycle.

The principle of microscopic reversibility applies whether or not the reactions are unimolecular. An example is provided by the phosphorylation of glucose, which is depicted in **Figure 15.28**. Here, A is glucose + $P_i$ (phosphate ion); B is glucose-6-P (glucose phosphorylated at position 6); and C is glucose-1-P (glucose phosphorylated at the 1 position). Glucose-1-P and glucose-6-P are interconverted by the enzyme glucophosphomutase. Microscopic reversibility says that the amount of glucose + $P_i$ reacting to form glucose-1-P must equal the amount of glucose-1-P hydrolyzed in the same time, regardless of the rates of the other reactions.

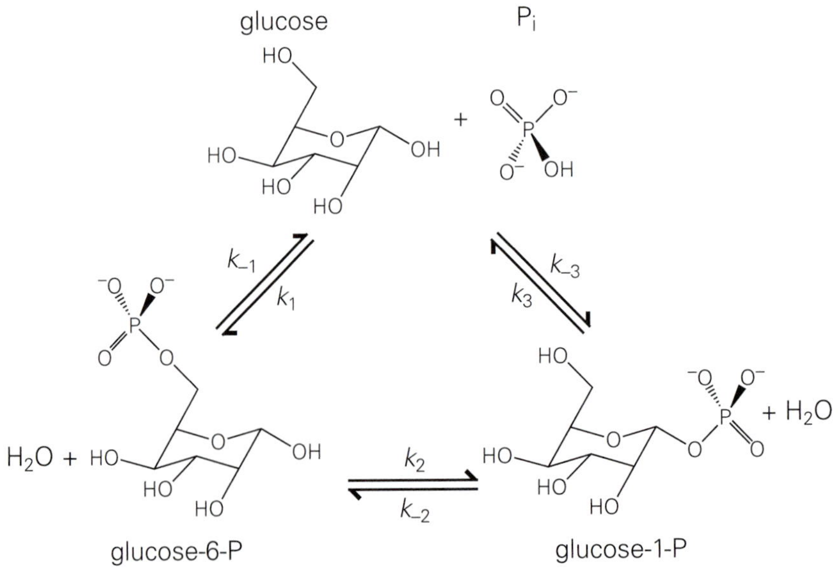

**Figure 15.28 The phosphorylation of glucose is an example of a cyclic set of reactions.** Glucose can be phosphorylated on either the 1 or 6 position, and these two isomers are interconverted by the enzyme glucophosphomutase.

For this set of cyclic reactions, the equilibrium constants are related to the rate constants as follows:

$$K_1 = \frac{[\text{glucose-6-P}]}{[\text{glucose}][\text{P}_\text{i}]} = \frac{k_{-1}}{k_1} \tag{15.80}$$

$$K_2 = \frac{[\text{glucose-1-P}]}{[\text{glucose-6-P}]} = \frac{k_{-2}}{k_2} \tag{15.81}$$

$$K_3 = \frac{[\text{glucose}][\text{P}_\text{i}]}{[\text{glucose-1-P}]} = \frac{k_{-3}}{k_3} \tag{15.82}$$

The concentrations in Equations 15.80–15.82 are all equilibrium concentrations. If we multiply the three equilibrium constants together, we get the following result:

$$K_1K_2K_3 = \left(\frac{[\text{glucose-6-P}]}{[\text{glucose}][\text{P}_\text{i}]}\right)\left(\frac{[\text{glucose-1-P}]}{[\text{glucose-6-P}]}\right)\left(\frac{[\text{glucose}][\text{P}_\text{i}]}{[\text{glucose-1-P}]}\right)$$

$$= 1 = \frac{k_{-1}k_{-2}k_{-3}}{k_1k_2k_3} \tag{15.83}$$

This means that:

$$k_1k_2k_3 = k_{-1}k_{-2}k_{-3} \tag{15.84}$$

Thus, the cyclic nature of the set of coupled reactions places an additional constraint on the values of the rate constants, which is that the product of the forward rate constants must equal the product of the reverse rate constants. This constraint is in addition to the requirements for them to agree with the individual equilibrium constants.

## C. FACTORS THAT AFFECT THE RATE CONSTANT

The rate constants specified in elementary reactions are "constant" in the sense that their values do not change with changes in concentrations of any reactants or products under a defined set of reaction conditions. They do change, however, when conditions change. For example, they are almost always strongly temperature dependent and often also depend on the solvent used. It is important to understand why the rate constants vary with such parameters and to learn how structural information about changes in the molecules during the reaction can be deduced from the behavior of the rate constants. In this part of the chapter, we discuss some of the factors that affect the observed rates of reactions.

### 15.22 Catalysts accelerate the rates of chemical reactions without being consumed in the process

A thermodynamic analysis of a chemical reaction depends only on the nature of the reactants and products and their concentrations. If there is another substance present that participates in terms of the reaction mechanism, but is neither consumed nor produced in the *overall* reaction, it does not enter into the thermodynamic analysis. For example, given the following overall reaction:

$$\text{A} + \text{B} \xrightarrow{k} \text{C} \tag{15.85}$$

a substance D could participate in the actual mechanism without appearing in the overall reaction:

$$\text{A} + \text{D} \xrightarrow{k_1} \text{E} \quad \text{and then} \quad \text{E} + \text{B} \xrightarrow{k_2} \text{C} + \text{D} \tag{15.86}$$

D is unchanged in the *overall net* reaction and E is created in the first step, but used in the second step. For these two reactions, the net equilibrium constant is:

$$K_{eq} = K_1 K_2 = \frac{[C]}{[A][B]} \tag{15.87}$$

with

$$K_1 = \frac{[E]}{[A][D]}$$

and

$$K_2 = \frac{[C][D]}{[E][B]}$$

Although D and E do not enter into the thermodynamic equilibrium constant, $K_{eq}$, the kinetics of the overall reaction can be affected enormously by the presence of D. For example, it is quite often the case that the process by which A collides with B to immediately form C can occur, but is very slow (that is, the value of the rate constant, $k$, is small). But, the overall rate of the reaction can be much faster if A collides rapidly with D to form E, which then collides rapidly with B to form C and release D (that is, the rate constants, $k_1$ and $k_2$, are much larger than $k$).

**Catalyst**

A catalyst is a substance that accelerates the rate of a reaction, but does not appear in the overall balanced reaction. Such substances participate in the reaction mechanism, but are regenerated in their original form in the course of the reaction.

Substances such as D, which accelerate reactions but do not appear in the overall balanced reaction, are called **catalysts** (**Figure 15.29**). Proteins (and some key RNAs) act as catalysts that accelerate the rates of many biochemical transformations so that these reactions can occur fast enough to sustain life. Protein and RNA catalysts are called **enzymes**, and are discussed in detail in Chapter 16.

## 15.23 Rate laws for reactions must be determined experimentally

Given the role of catalysts in affecting reaction rates, and the possibility of complex multistep mechanisms, it should now be apparent that it is often impossible to simply look at an overall reaction and write down the kinetic rate law that applies. If we know something about the chemistry of the reactions, a rational mechanism and rate law can be developed, but these must be tested experimentally to verify that the experimental concentration dependence is indeed in agreement with the proposed mechanism.

Even for complicated mechanisms, rate laws can usually be written as the products of concentrations of participating molecules:

$$\text{rate} = k[A]^a[B]^b[C]^c \tag{15.88}$$

For a complex reaction mechanism, some of the chemical species that appear in the rate law may not appear in the overall chemical reaction. This is the case for catalysts, for example. The exponents $a$, $b$, and $c$ need not have integer values. If one just picks initial concentrations, $[A]_0$, $[B]_0$, $[C]_0$, etc., with no idea of the overall

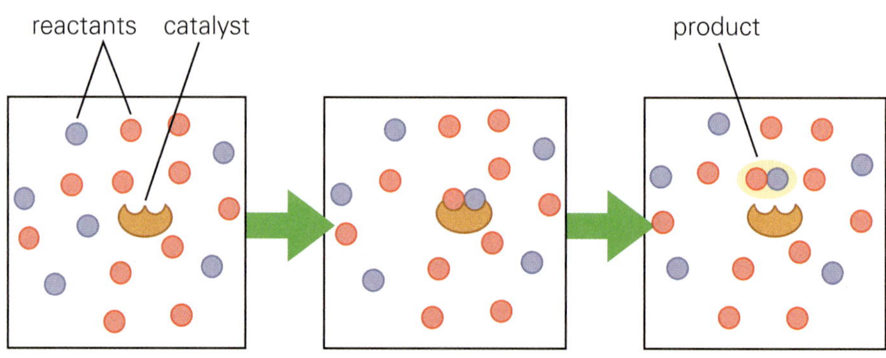


**Figure 15.29 A catalyst participates in a chemical reaction but is left unaltered by the reaction**. Shown here are two kinds of molecules (*red* and *blue*) that can react to form a covalent bond. A catalyst (*brown*) binds to both kinds of molecules and promotes their reaction by bringing them together. Once the reaction is completed the catalyst releases the product and can participate in more reactions.

mechanism, then it is not generally possible to fit measured concentrations as a function of time to an appropriate time-dependent concentration equation.

Fortunately, there are simple approaches that allow the exponents (and hence the rate law) to be determined experimentally. The most generally used approach is to measure the initial reaction rates (that is, the reaction rates are measured at times short enough that concentrations have not changed by a large fraction of their total value). Such measurements are made for different initial concentrations of one reactant. To determine the value of the exponent of [A], for example, measurements could be done using starting concentrations $[A]_0$ and then $2[A]_0$, and so on.

In terms of unknown exponents $a$, $b$, $c$, the rate is then written as:

$$\text{rate}_1 = k[A]_0^a[B]_0^b[C]_0^c \tag{15.89}$$

where $[A]_0$, $[B]_0$, and $[C]_0$ are the initial concentrations of A, B, and C. Equation 15.89 is valid if the measurement is done over a period of time that is short enough that the concentrations of the reactants remain close to their initial values (for example, $[A] \approx [A]_0$).

If we now double the initial concentration of A, from $[A]_0$ to $2[A]_0$, then the rate is given by:

$$\text{rate}_2 = k(2[A]_0)^a[B]_0^b[C]_0^c \tag{15.90}$$

The ratio of the two rates is:

$$\frac{\text{rate}_1}{\text{rate}_2} = \frac{k\,[A]_0^a[B]_0^b[C]_0^c}{k\,(2[A]_0)^a[B]_0^b[C]_0^c} = \left(\frac{1}{2}\right)^a \tag{15.91}$$

Taking the logarithm of both sides of Equation 15.91, we can see that the value of the exponent, $a$, is given by:

$$a = \frac{1}{\log 2}\log\left(\frac{\text{rate}_2}{\text{rate}_1}\right) \tag{15.92}$$

To determine the other exponents, $b$ and $c$, the same procedure would need to be repeated, varying the concentrations of the other reactants, one at a time.

Another approach to determine the value of the exponents is simply to begin with a large excess of all reagents except for one. For example, we keep $[A]_0$ small, but make $[B]_0$ and $[C]_0$ large. In this case, the concentrations of B and C essentially remain constant at $[B]_0$ and $[C]_0$ during the reaction, because only a small amount of B and C can react with the small amount of A that is present. This is the approach used to simplify the analysis of ligand binding to a protein in Section 15.12. In this case, the rate is given by:

$$\text{rate} = k'[A]^a \tag{15.93}$$

where

$$k' = k[B]_0^b[C]_0^c \tag{15.94}$$

Since there is only one reactant that is time dependent, the resulting rate (determined by measuring [A] as a function of time) can be fitted to give the value of the exponent, $a$.

## 15.24 The hydrolysis of sucrose provides an example of how a reaction mechanism is analyzed

An example of a reaction that demonstrates some aspects of the kinetic models that have been discussed is the hydrolysis of sucrose (ordinary table sugar). The chemical reaction is the hydrolysis of the disaccharide [sucrose = α-D-glucopyranosyl (1 → 2) β-D-fructofuranoside; see Figure 3.1] to give two monosaccharides, glucose and fructose. Written as a chemical equation, this is $C_{12}H_{22}O_{11} + H_2O \rightarrow C_6H_{12}O_6 + C_6H_{12}O_6$, but this makes much more sense with chemical structures, as shown in **Figure 15.30**.

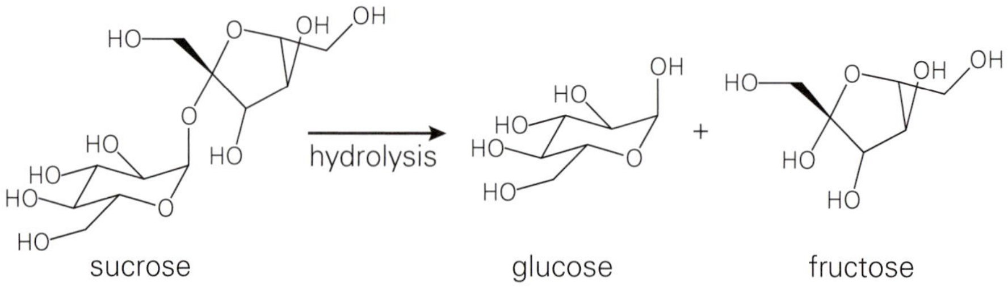

**Figure 15.30 Structures of sucrose and its hydrolysis products, glucose and fructose.** The process of sucrose hydrolysis is known as inversion because it was first followed by measuring how much the plane of polarization of light was rotated by passing through the solution. As the reaction proceeds in this case, the sense of rotation inverts for products relative to reactants.

The kinetics of sucrose hydrolysis was studied many years ago by following the reaction through observation of the rotation of polarized light by solutions containing sucrose. A solution of sucrose is dextrorotatory (that is, it rotates light in a counterclockwise fashion, as explained in Section 3.4). Glucose is dextrorotatory and fructose is levorotatory, but because fructose rotates the polarization of light more strongly than glucose, the net polarization of light by the products is levorotatory. Thus, as the reaction proceeds, the polarization of light by the solution switches from counterclockwise to clockwise (that is, the polarization inverts). This phenomenon led to the hydrolysis of sucrose being referred to as an **inversion reaction**.

The mechanism of sucrose hydrolysis can be deduced from three important observations concerning the kinetics of the reaction. The first observation is that although the hydrolysis reaction is very slow in pure water, the rate increases dramatically as the pH is lowered (that is, as the proton concentration increases). The nature of the acid used to lower the pH of the solution does not matter.

The second observation is that during the course of the reaction the pH of the solution does not change. Thus, although protons are clearly participating in the reaction (because the rate depends on the pH), the protons must be acting as catalysts.

The third observation is that the reaction is first order in proton concentration and also first order in sucrose. Thus, the rate law seems to be:

$$\frac{d[\text{sucrose}]}{dt} = k[\text{sucrose}][\text{H}^+] \tag{15.95}$$

We now combine these observations with some chemical knowledge to deduce the chemical mechanism. First, in reactions that break bonds, the stability of the leaving group is very important. From a study of other organic reactions, it is known that protonation of an oxygen can create a better leaving group (HOCH is a better leaving group than the unstable $^-$OCH group). In this case, it is the bridging oxygen in sucrose that must be protonated (**Figure 15.31**). But this is unfavorable, because ether oxygens (C–O–C) have a very low p$K_a$ value. Nevertheless,

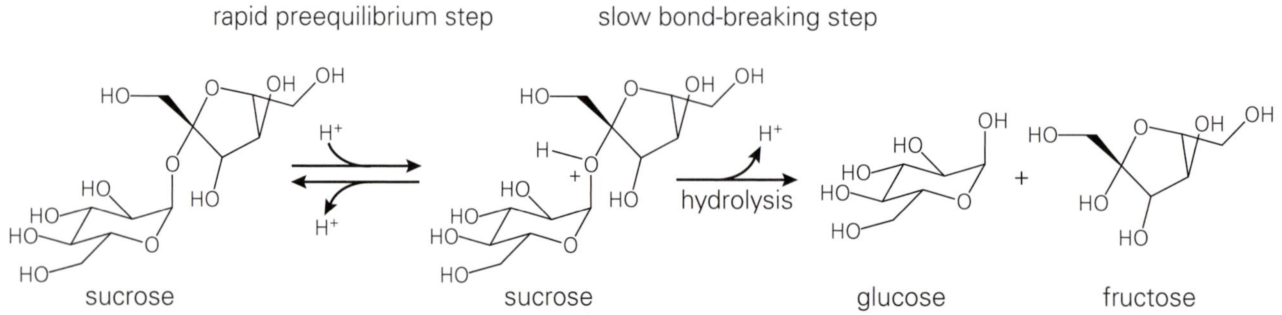

**Figure 15.31 Mechanism of sucrose hydrolysis.** A mechanism that is consistent with the observed rate law is shown. It involves reversible protonation followed by bond cleavage in the protonated form of sucrose to give glucose + fructose. The water molecule that reacts is not shown.

in solution we know that there will be a small amount of the protonated sugar present. This reaction is:

$$H^+ + \text{sucrose} \underset{k_{-1}}{\overset{k_1}{\rightleftharpoons}} \text{sucrose-H}^+ \tag{15.96}$$

and the rates for protonation/deprotonation reactions are generally rapid.

If we assume that the protonated sucrose reacts with water, then we can write:

$$\text{sucrose-H}^+ + H_2O \xrightarrow{k_2} \text{glucose} + \text{fructose} \tag{15.97}$$

The back reaction is not included because the forward reaction is thermodynamically favorable and the reaction goes essentially to completion (that is, all of the sucrose is converted to glucose and fructose). We can now develop a kinetic model from these two reactions.

First, note that product formation is an elementary reaction of sucrose-$H^+$ with $H_2O$, so the rate law is:

$$\frac{d[\text{glucose}]}{dt} = \frac{d[\text{fructose}]}{dt} = k_2[\text{sucrose-H}^+][H_2O] \tag{15.98}$$

To obtain an expression for [sucrose-$H^+$], we consider the reaction depicted in Equation 15.96. Protonation/deprotonation is much faster than the bond-breaking step that occurs in the reaction depicted in Equation 15.97. This means that the rate constants for the protonation/deprotonation step, $k_1$ and $k_{-1}$, are much greater than the rate constant for the bond-breaking step, $k_2$. This fact allows us to make a **"preequilibrium" approximation**—that is, we assume that the following is true:

$$[\text{sucrose-H}^+] \approx \frac{k_1}{k_{-1}}[\text{sucrose}][H^+] \tag{15.99}$$

**Preequilibrium approximation**

A preequilibrium approximation refers to a case in which there are forward and backward reactions that are both much faster than subsequent reaction steps. In this case, the reactants and products for this fast step always remain at relative concentrations determined as if they were fully at equilibrium.

Equation 15.99 is equivalent to stating that the protonated and deprotonated sucrose essentially remains at equilibrium throughout the reaction—that is, we are equating the ratio of the two rate constants to the equlibrium constant for the reaction (compare Equation 15.67). Then, by inserting Equation 15.99 into Equation 15.98, we find that the rate of product production is given by:

$$\begin{aligned}\frac{d[\text{glucose}]}{dt} &= k_2\frac{k_1}{k_{-1}}[\text{sucrose}][H^+][H_2O]\\ &= k_2'\frac{k_1}{k_{-1}}[\text{sucrose}][H^+]\end{aligned} \tag{15.100}$$

In the last step we have taken $k_2' = k_2[H_2O]$ because the reaction is in water (where $[H_2O] = 55\ \text{mol}\cdot\text{L}^{-1}$), and the concentration of water does not change significantly during the reaction.

This kinetic model, which is shown in Figure 15.31 and consists of a reversible protonation step followed by an essentially irreversible hydrolysis step, is consistent with experimental observation. The rate depends linearly on sucrose concentration and also on hydrogen ion concentration. It is important to note, however, that this agreement does not prove that this kinetic model is correct. It is often the case that there are other kinetic models that can give the same concentration dependence and, hence, would also be consistent with the kinetic observations. To distinguish among such possibilities we would need to carry out more complicated experiments, such as those that probe whether other intermediates are present.

## 15.25 The fastest possible bimolecular reaction rate is determined by the diffusion-limited rate of collision

In Section 15.2, we noted that the rate constant, $k$, for a bimolecular reaction must reflect the frequency of collisions per unit concentration. The rate constant also depends on the fraction of collisions that lead to formation of product. Both of these factors depend on reaction conditions, including temperature.

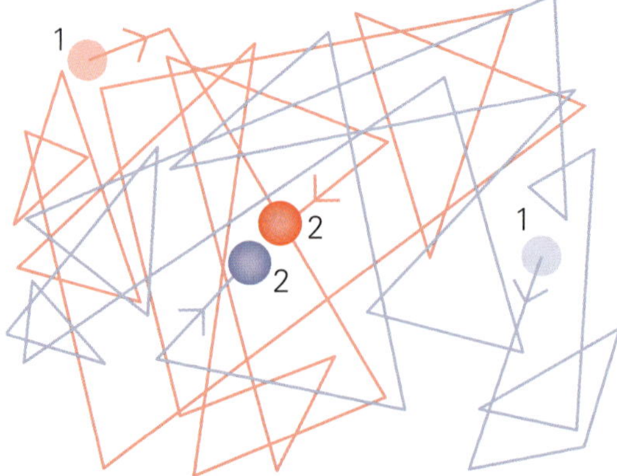


**Figure 15.32 The diffusion time limits the rate of reaction.** Two molecules in solution are shown here (*red* and *blue*). As they move around, they collide with water molecules (not shown), which changes their trajectories. Starting from the points indicated by 1, the molecules move randomly until they collide (indicated by the *red* and *blue* lines, ending in the positions marked 2). The frequency of such collisions is determined by the viscosity of the solution and the temperature.

For gases it is quite straightforward to calculate the rate at which collisions occur, $N_{\text{collision}}$. The rate of intermolecular collisions in air is given by:

$$N_{\text{collision}} = k_{\text{collision}}[\text{A}][\text{B}] \tag{15.101}$$

in which $k_{\text{collision}}$ is the rate constant for collisions. For molecules in air at 1 atm, the rate constant, $k_{\text{collision}}$, has a value of about $6 \times 10^{10}$ $M^{-1}{\bullet}sec^{-1}$. In spite of this high collision rate, molecules in gases move very significant distances between collisions (an average on the order of 500 molecular diameters).

For molecules moving in solution, as is the case for essentially all biological reactions, the movement of the molecules changes to that of much slower diffusion (see **Figure 15.32**), which is developed in detail in Chapter 17. It would seem that large molecules should collide more often, but they also move more slowly, which decreases the rate of collisions. For water at 25°C, the value of $k_{\text{collision}}$ is about $1 \times 10^{10}$ $M^{-1}{\bullet}sec^{-1}$, surprisingly similar to the value predicted for gas-phase collisions of molecules in air with each other.

If every collision between reactants in a solution leads to product, then $k_{\text{collision}}$ will be the value for the bimolecular rate constant. There are some reactions that do indeed reach this **diffusion-limited reaction** rate. An example of a reaction that occurs at the diffusion limit is that of iodine atoms reacting to form molecular iodine: $2I^{\bullet} \rightarrow I_2$. In this case, the delocalized electrons in the outermost orbitals on each atom must simply come close together to form the bond, and this can happen during any collision, with no orientational constraint. There are a few biological reactions that occur at rates close to this diffusion limit, but most do not.

**Diffusion-limited reaction**

A diffusion-limited reaction is one in which every collision between reactants leads to products. The rate of reaction in this case is limited by the rate of collisions, which is affected just by the rates of diffusion. The rate constant for collisions between molecules in water is about $1{\times}10^{10}$ $M^{-1}{\bullet}sec^{-1}$. A reaction that occurs with a rate constant close to this value is said to be diffusion-limited.

## 15.26 Most reactions occur more slowly than the diffusion-limited rate

Most observed reaction rate constants are much smaller than the diffusion-limited value. The reason is that most collisions of reactants do not lead to products—that is, after a collision the molecules usually remain as reactants (**Figure 15.33**). By comparing the observed rate constant for a bimolecular reaction, $k_{\text{obs}}$, to that for the expected rate constant for collisions ($\sim 10^{10}$ $M^{-1}{\bullet}sec^{-1}$), one can calculate the fraction of collisions that do lead to products. It turns out that very fast reactions have values of $k_{\text{obs}} = 10^6$ $M^{-1}{\bullet}sec^{-1}$, which means that only 1 collision in $10^4$ leads to products. For a slow reaction, with $k_{\text{obs}} = 1$ $M^{-1}{\bullet}sec^{-1}$, it is just 1 collision in $10^{10}$ that is productive.

There are two important factors that limit reaction rates. First, as shown in Figure 15.33, for the reaction to occur, the reactants must come together in a rather specific orientation. The importance of the correct orientation is illustrated in **Figure 15.34**, which shows how a hydroxide ion attacks ATP, resulting in the hydrolysis of ATP. The oxygen atom of the hydroxide ion must come in close and be in line with the P–O bond that will break. In this orientation, the attacking oxygen can get close enough to the phosphorous to begin to make a bond, forming a transient structure known as the **transition state**. The transition state is a point of commitment to the conversion of reactants to products. The energy of the reactants increases until the transition state is crossed, and then it decreases as conversion to product occurs. In the case of ATP hydrolysis, the O–P bond then breaks and the ADP departs. If the hydroxide comes into contact with any other part of the phosphate, then the transition state cannot form, and no reaction will occur. Thus, only collisions with "correct" relative positions and orientations form the transition state that leads to product.

**Transition state**

In a pathway connecting reactants to products, the potential energy increases until it has a maximum value at the transition state. Once the transition state is crossed, the reaction is downhill in terms of potential energy.

As the analysis of the ATP hydrolysis reaction makes clear, to get the number of collisions that can lead to product, the total rate of collisions must be multiplied by $f_p$, the fraction of collisions that can actually lead to product. This fraction is specific for the particular pair of reactants, differing widely depending on the structures involved and the nature of the chemical transformation. The collision rate, corrected for orientation, is given the symbol $A$ and is the rate of collisions multiplied by the fraction that are productive, $f_p$:

$$A = k_{collision} f_p \tag{15.102}$$

This corrected collision rate is called the **preexponential factor**, for reasons that are explained in Section 15.28.

The second factor limiting reaction rates is that the collision must provide enough energy for the reaction to occur, as discussed in the next section.

## 15.27 The activation energy is the minimum energy required to convert reactants to products during a collision between molecules

When the hydroxide ion approaches the ATP from the correct direction, the phosphate oxygens must be pushed away from their equilibrium tetrahedral geometry toward a planar position in order for the electronic orbitals of the hydroxide to overlap with those on the phosphorous to initiate bonding, and this requires energy. There is a minimum energy required to bring the reactants sufficiently close together to react, which is called the **activation energy** and is denoted $E_A$. This can be visualized by plotting the energy of the system as a function of a single **reaction coordinate**—a variable defined to describe whether colliding molecules are more like reactants or products, as shown in **Figure 15.35**. The activation

**Reaction coordinate**

A reaction coordinate is a variable that describes how far a reaction has progressed from separated reactants to final products. The reaction coordinate is usually some combination of distances and angles that change during the reaction.

(A)

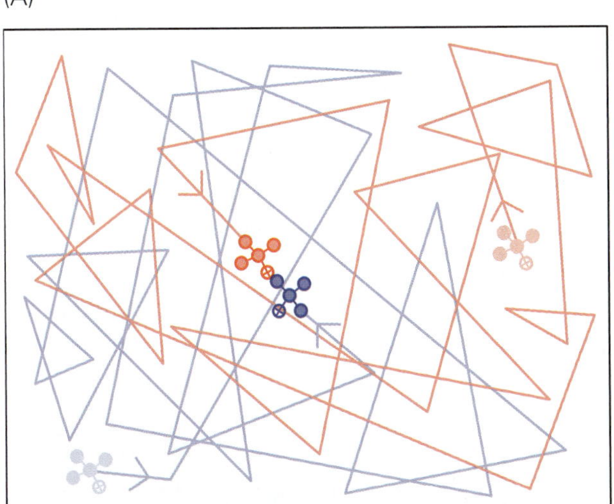

(B)

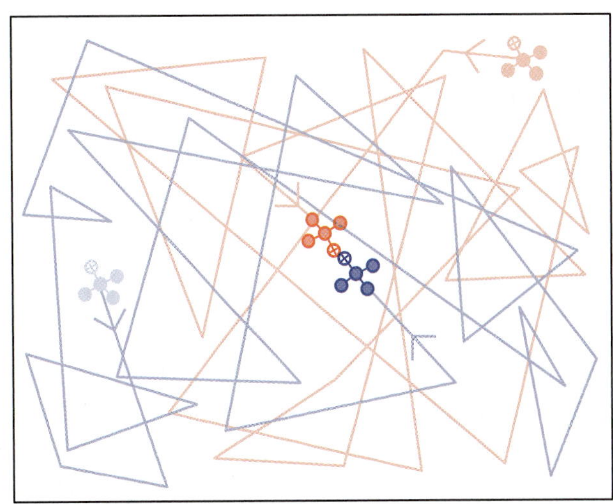

**Figure 15.33 Productive and nonproductive collisions.** Two molecules, each containing five atoms, are shown. Each molecule contains one reactive atom, indicated by a cross mark. A chemical reaction can occur only if the reactive atoms make contact. (A) The molecules collide, but the atoms that make contact are not both reactive. This is a nonproductive collision, because there is no possibility of a chemical reaction occurring. (B) Another collision is shown, but in this case the reactive atoms are the ones making contact. This is a productive collision, because a chemical reaction may occur, provided that there is enough energy in the collision.

energy is the difference in potential energy between the unreacted molecules and the transition state of the reaction.

Describing the reaction as occurring along just one coordinate is a great oversimplification, but this description provides useful insights into the fundamental aspects of the processes. A more complete diagram of energy versus distances for the exchange of a single atom with a diatomic molecule is also shown in Figure 15.35. Trajectories of the atoms as they go through a reaction correspond to lines on this potential energy surface and, in a complete description, would include vibrational motions as well as the relative translation.

**Activation energy**

The activation energy is the minimum increase in potential energy for reactants to be converted to products, and hence the minimum energy that reactants must have to be able to react. The requirement that the reactants have sufficient energy, $E_A$, to cross from reactants to products limits the rate of most reactions.

As molecules come together to undergo a chemical reaction, there is an increase in the potential energy, converted from kinetic energy of the reactants (which includes vibrational and rotational energy). The minimum increase in potential energy of the system along the reaction coordinate in order for the reaction to proceed is $E_A$, the activation energy. If the reacting molecules have sufficient energy to get past this point, then they can convert to products, releasing energy as they do. The value of $E_A$ is specific for the particular reactants, as is the shape of the energy versus reaction coordinate plot.

## 15.28 The reaction rate depends exponentially on the activation energy

Recall from part B of Chapter 8 that energy is distributed among molecules according to the Boltzmann distribution. The fraction of molecules that have energy equal to $E_A$ (expressed in J•mol$^{-1}$) is proportional to the Boltzmann factor:

$$e^{-\frac{E_A}{RT}}$$

When $E_A$ is much larger than $RT$, then this Boltzmann factor is very small and only occasionally do colliding molecules have sufficient energy to react. If, however, $E_A$ is much less than $RT$ (equivalent to saying $E_A \approx 0$), then essentially all collisions are reactive from the energy viewpoint and the rate becomes limited just by

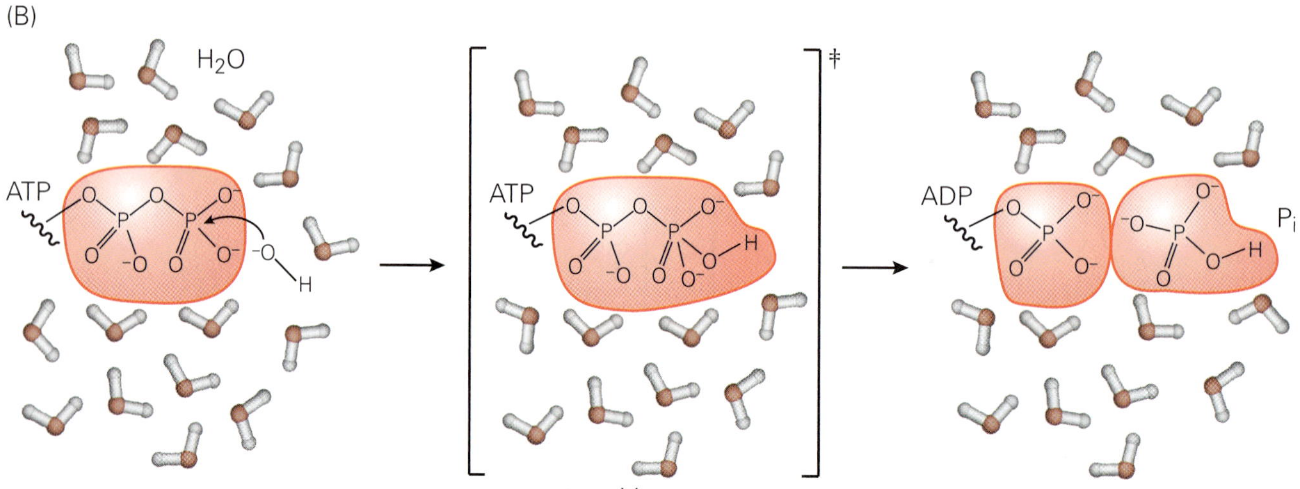


**Figure 15.34 Importance of the attacking group orientation in the hydrolysis of ATP.** (A) The structure of ATP. The terminal phosphate groups are highlighted. (B) Attack by a hydroxide group in water to give ADP and inorganic phosphate (P$_i$). Only the terminal phosphate groups of ATP are shown. The hydroxide ion attacks from a specific direction to form the transition state for the reaction, indicated by ‡. A few water molecules are shown, to stress the idea that the reaction occurs in solution and that there are interactions with water molecules that occur throughout the reaction.

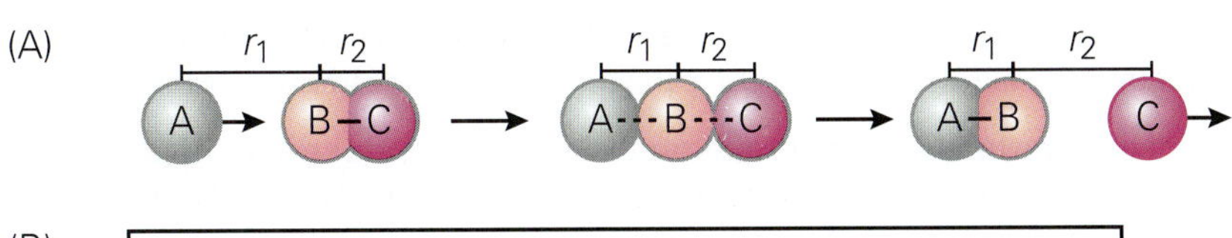


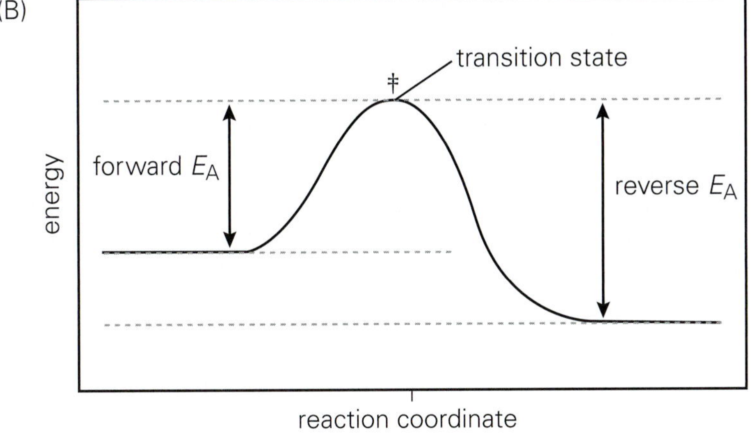


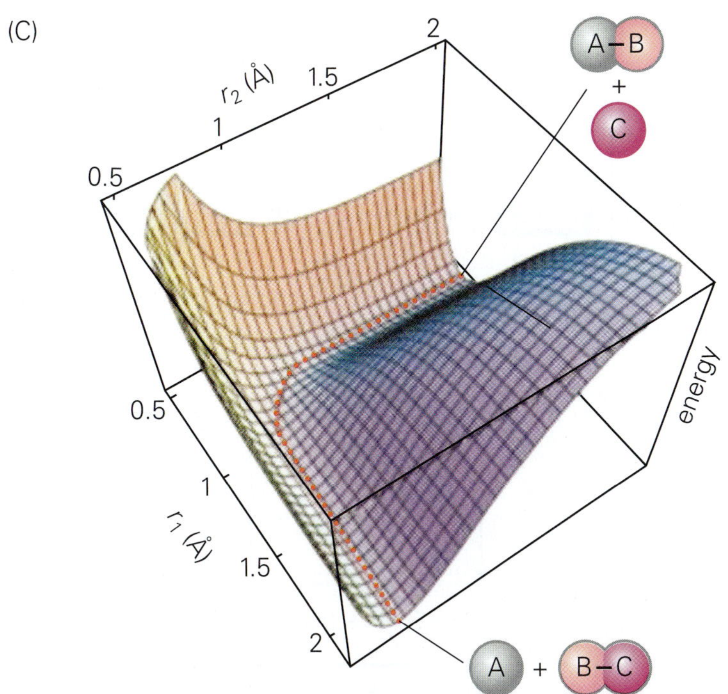


**Figure 15.35 Energy vs. reaction coordinate diagram.** (A) A simple atom exchange reaction is shown. The reaction coordinate is the difference in separation of atom 1 and 2 ($r_1$) and atoms 2 and 3 ($r_2$). (B) The energy as a function of the reaction coordinate. The maximum point on this curve is the transition state marked with ‡. The energy barriers that must be overcome to bring the atoms together in the forward and reverse directions are indicated by $E_A$. (C) A more complete energy diagram. The reaction pathway shown by the *red* dotted line represents the minimum energy path from reactants to products and, drawn as a function of a single variable, it would look like the curve in (B). (C, adapted from C.M. Dobson, A. Šali and M. Karplus, *Angew. Chemie* 37: 868–893, 1998. With permission from John Wiley & Sons.)

encounters through diffusion and the orientational factor. Note that the activation energies in the forward and reverse directions of reactions are different, and the orientational factors will also be different.

The expected reaction rate constant is a product of the two factors we have discussed. The first is the combination of the reaction collision rate and the orientation effect, $A$, as given in Equation 15.102. The second is the Boltzmann factor, and together these give the following expression for the rate constant:

$$k = Ae^{-\frac{E_A}{RT}} \tag{15.103}$$

An overall kinetic equation for a bimolecular reaction can therefore be written as follows:

$$\frac{d[\mathrm{C}]}{dt} = Ae^{-\frac{E_A}{RT}}[\mathrm{A}][\mathrm{B}] \tag{15.104}$$

The reason that the term $A$ (not to be confused with [A], which is the concentration of reactant, A) is referred to as the preexponential should now be obvious.

The exponential term gives rise to the typically strong temperature dependence of rate constants. If we take the logarithm of both sides of Equation 15.104, we see that $\ln(k)$ is linearly related to $1/T$:

$$\ln(k) = \left(-\frac{E_A}{R}\right)\frac{1}{T} + \ln(A) \tag{15.105}$$

This linear relationship between the logarithm of the rate constant and the inverse temperature is known as **Arrhenius rate behavior**. This relationship was noted initially by Jacobus van't Hoff in 1884, but the idea that this corresponded to a required minimum energy for the reaction came from Svante Arrhenius a few years later, and Equations 15.103 and 15.105 are now referred to as forms of the **Arrhenius equation**.

**Arrhenius rate behavior**

Arrhenius rate behavior gives an exponential dependence of the rate on temperature, $k \propto \exp(-E_A/RT)$.

The Arrhenius equation is commonly used to determine the activation energy of reactions. According to this equation, a graph of $\ln(k)$ versus $1/T$ should be a straight line, with the slope corresponding to $-E_A/R$, and the intercept corresponding to $\ln(A)$. The application of this type of analysis to the determination of the activation energy for the *cis–trans* isomerization of proline is shown in **Figure 15.36**.

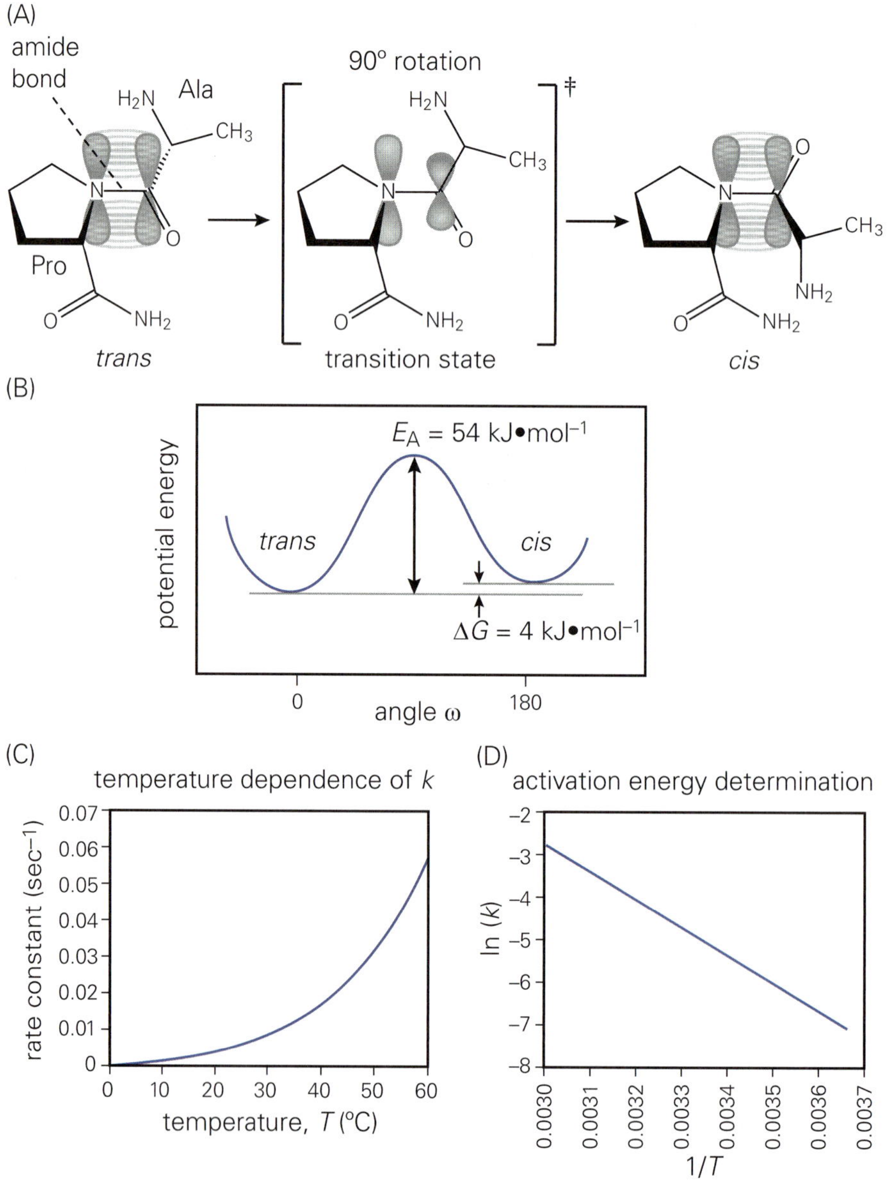

**Figure 15.36 Activation energy analysis for peptide bond isomerization.** (A) Rotation of the amide bond between alanine and proline is shown, with the orbitals contributing to the partial double bond between the amide nitrogen and carbonyl carbon indicated. (B) Energy as a function of the rotation angle for the Ala-Pro bond. (C) A plot of the rate constant for proline isomerization vs. temperature, $T$. (D) $\ln(k)$ vs. $1/T$. From the plot in part (D), the slope of the line gives the value of $(-E_A/R)$, with $E_A$ = 54 kJ•mol$^{-1}$ in this case.

In Section 4.8, we discussed the fact that the peptide bond can be in either the *cis* or *trans* conformation. Although the *cis* conformation is strongly disfavored for most amino acids, proline residues can adopt either conformation and can interconvert between the two conformations at room temperature.

The rotation of an amide bond between an alanine and a proline is shown in Figure 15.36A. As the bond rotates away from the optimal geometry, there is decreased overlap between the orbitals contributing to the partial double bond between the amide nitrogen and carbonyl carbon indicated. For a 90° rotation (center), there is no longer overlap of these *p* orbitals. This represents the transition state for this reaction. The energy as a function of the rotation angle for the Ala-Pro bond is shown in Figure 15.36B. The transition state, indicated by ‡, is about 54 kJ•mol$^{-1}$ higher in energy than the optimal planar geometry. This value of the of the activation energy is determined by measuring the temperature dependence of the rate constant for *cis–trans* isomerization, as shown in Figure 15.36C.

## 15.29 Transition state theory links kinetics to thermodynamic concepts

A somewhat different development for the analysis of the reaction rate constant, called **transition state theory**, is conceptually useful for understanding the contributions that determine the barrier to a reaction. In this view, there is a key point along the reaction trajectory for getting from the reactants to the products, which is the transition state, usually indicated by a "double dagger" symbol, ‡ (see Figures 15.34–15.36).

**Transition state theory**

Transition state theory uses the concept of an equilibrium between the ground state and the transition state for a reaction to predict the temperature dependence of the reaction rate.

Transition state theory assumes that the transition state is in equilibrium with the reactants, as explained in **Figure 15.37**. That is, the population of the transition state is assumed to be governed by the difference in free energy between it and the reactants. This provides the conceptual framework for the conversion of reactants to products:

$$\mathrm{A+B} \underset{}{\overset{K_{\mathrm{eq}}^{\ddagger}}{\rightleftarrows}} \mathrm{A{\bullet}B^{\ddagger}} \xrightarrow{k^{\ddagger}} \mathrm{C+D} \tag{15.106}$$

If A and B are taken to be in equilibrium with A•B$^{\ddagger}$, then we can write:

$$K_{\mathrm{eq}}^{\ddagger} = \frac{[\mathrm{A{\bullet}B^{\ddagger}}]}{[\mathrm{A}][\mathrm{B}]}$$

so

$$\frac{d[\mathrm{C}]}{dt} = k^{\ddagger}[\mathrm{A{\bullet}B^{\ddagger}}] = k^{\ddagger}K_{\mathrm{eq}}^{\ddagger}[\mathrm{A}][\mathrm{B}] \tag{15.107}$$

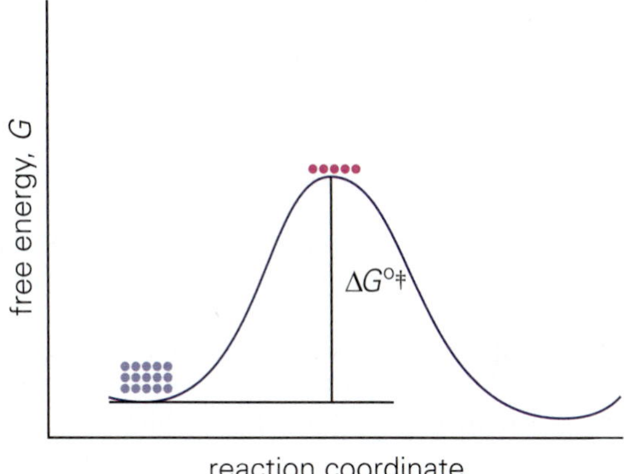


**Figure 15.37 Transition state theory.** In this theory, it is assumed that the number of molecules that are in the transition state for the reactions is determined by the equilibrium constant between the reactants and the transition state. (A) The graph shows the free energy as a function of the reaction coordinate for a reaction. The relative population of molecules in the reactant state and in the transition state is shown schematically by circles. (B) If the transition state has higher free energy, then the number of molecules in the transition state decreases.

In this analysis, we need to estimate $k^{\ddagger}$. Since the transition state is at a relative maximum in energy, it is intrinsically unstable, and any movement of atoms leads to a lower energy. Movements of atoms in molecules, even in this kind of transition state, correspond to vibrations. Because the bond being changed becomes weak in the transition state, the rate of crossing the transition state can be estimated as the rate of motion due to the vibration. The vibration frequency should be significantly lower than that typical of a stable, fully bonded molecule. This can be expressed as $k^{\ddagger} = \kappa\nu$, in which $\kappa$ is a transmission coefficient (the fraction of molecules that do not reform the original bond after crossing the barrier, normally a value near 1), and $\nu$ is the frequency of the relevant vibration. These ideas lead to the following expression for $k^{\ddagger}$:

$$k^{\ddagger} = \kappa \frac{k_B T}{h} \tag{15.108}$$

in which $k_B$ is Boltzmann's constant, $h$ is Planck's constant, $\kappa \approx 1$ as described above. Understanding the basis for Equation 15.108 requires a more complicated analysis than is possible here, and you can consult advanced textbooks on kinetics to see how this equation arises.

Using the thermodynamic concepts of an equilibrium constant that were developed earlier:

$$\Delta G^{\ddagger} = -RT \ln K^{\ddagger}_{eq} = \Delta H^{\ddagger} - T\Delta S^{\ddagger} \tag{15.109}$$

Or, in exponential form:

$$K^{\ddagger}_{eq} = e^{-\frac{\Delta G^{\ddagger}}{RT}} = e^{-\frac{\Delta H^{\ddagger}}{RT}} e^{\frac{\Delta S^{\ddagger}}{R}} \tag{15.110}$$

Thus, together these ideas give the following expression for the rate:

$$\frac{d[\mathrm{C}]}{dt} = \frac{k_B T}{h} e^{-\frac{\Delta H^{\ddagger}}{RT}} e^{\frac{\Delta S^{\ddagger}}{R}} [\mathrm{A}][\mathrm{B}] \tag{15.111}$$

By comparing this with the Arrhenius analysis for the reaction rate constant, it is easy to see a correspondence between $\Delta H^{\ddagger}$ and $E_A$. The transition state viewpoint predicts an extra temperature dependence from the linear $k_B T$ term, but this temperature dependence is very weak relative to that of the exponential, so that the correspondence between $\Delta H^{\ddagger}$ and $E_A$ in determining the overall temperature dependence is good.

The entropy change in getting to the transition state, $\Delta S^{\ddagger}$, and the restriction on angles that lead to products in the preexponential term, $A$, also have a clear connection. If the orientation for reaction is extremely restrictive, then there is a large loss of entropy in getting to the transition state ($\Delta S^{\ddagger}$ is negative and large because few of the available orientations are allowed, making $e^{\frac{\Delta S^{\ddagger}}{R}}$ very small and the reaction slow).

The transition state view provides somewhat clearer connections to thermodynamic ideas than the Arrhenius description, but either approach allows useful predictions to be made about factors affecting reaction rate constants.

## 15.30 Catalysts can work by decreasing the activation energy, by increasing the preexponential factor, or by completely altering the mechanism

With the ideas of the preceding section, we can now give a better description of the ways in which catalysts can accelerate a chemical reaction. The catalyst can decrease the activation energy for the reaction, $E_A$ (without significantly changing $A$), or can increase the "preexponential" coefficient, $A$ (without significantly affecting $E_A$), or can change both of these parameters. However, a catalyst can also completely change the mechanism of a reaction, in which case comparing the kinetic parameters is not very meaningful. Serine proteases, discussed in the following chapter, provide an example in which the mechanism for an enzyme-catalyzed hydrolysis reaction is changed from that which occurs in water alone.

In presenting the idea of an activation energy, we noted that the potential energy reaches a maximum at the transition state. One common way for biological catalysts to accelerate the rate of a reaction is to have favorable interactions with the transition state, thereby lowering its energy, without changing the energy of the reactants or products substantially. This lowers the activation energy, $E_A$, and thereby increases the rate of reaction, as shown in **Figure 15.38**. This effect can be quite dramatic, because the activation energy enters the rate equation in an exponent.

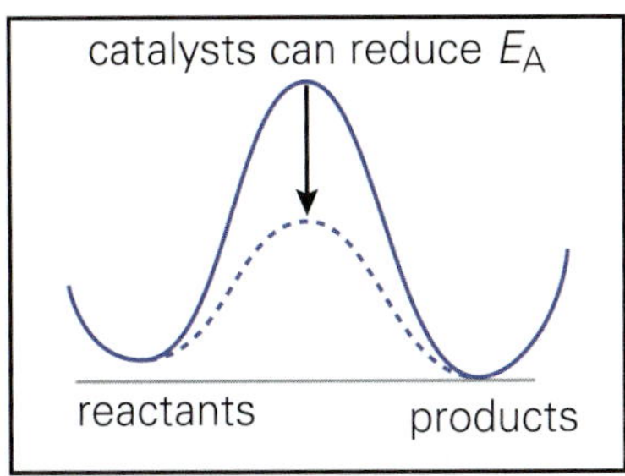


**Figure 15.38 Schematic drawing of the effect of a catalyst.** One effect of a catalyst can be to lower the activation energy. The arrow indicates the decreased barrier; the energy for the catalyzed reaction is shown as a dashed line.

The preexponential term, $A$, reflects a combination of the rates of collisions and the fraction of these that are in the correct orientation to actually react when sufficient energy is available. A catalyst can increase the rate of collisions or can cause more collisions to occur in a productive orientation by interacting with the reactants.

In biological systems, a vast majority of reactions are catalyzed by enzymes, which are discussed in the next chapter. Enzyme-catalyzed reactions in cells are critical for accelerating slow reactions sufficiently to meet the needs of the cells for specific compounds. The regulation of enzyme activity enables the rates of production of many metabolites to be controlled in response to the requirements of the cell.

## Summary

In this chapter, we have developed equations that describe the rates at which chemical reactions occur. Differential rate equations can be derived from the molecular mechanism of the reaction by using simple ideas predicted from the rates of collision for elementary reactions—for example, rate = $k[A][B]$ for the direct, bimolecular reaction of A with B. Differential rate laws for elementary reactions can be combined to explain the kinetics of complex, multistep reactions. Integration of these differential rate laws predicts how concentrations of reactants and products change with time during a reaction. Experimentally determined rate laws can be used to test the consistency of mechanisms with the experimental data, providing important insight into how the reaction occurs.

Situations giving rise to steady-state concentrations in series reactions can be easily understood with these ideas. Steady-state conditions are important in many metabolic processes and will be used in understanding enzyme reactions. In systems with equilibrium constants that are close to unity, both the forward and reverse reactions have to be considered to correctly predict kinetic behavior. At equilibrium, the rates of forward and reverse reactions must be equal, which constrains the ratio of the forward and reverse rate constants to be equal to the equilibrium constant. The values of the individual rate constants are not constrained by this relationship.

Reaction rates are limited by the rates of collision (ultimately by their rate of diffusion), by what range of relative orientations of the molecules bring the reactive parts of the molecule together to react, and by the height of the energy barrier (the activation energy) that must be surmounted to convert reactants to products. Thermodynamic principles tell us that the need for reactants to have at least a certain amount of energy should give rise to exponential dependence of the reaction rate constant on temperature, as is seen experimentally. The idea of a critical transition state that must be reached for a reaction to occur provides insights into how catalysts can act—that is, increasing the probability of reaching the transition state or lowering the activation energy so that a larger fraction of molecules have sufficient energy to cross it. However, catalysts can also alter the reaction mechanism, creating reaction mechanisms that cannot occur in their absence.

## Key Concepts

### A. GENERAL KINETIC PRINCIPLES

- The rate of a reaction describes how fast concentrations change with time.
- Reaction rates are often determined by the rate of collisions between molecules, which depends on the concentrations.
- Rate laws define the relationship between the reaction rates and concentrations.
- The dependence of the rate law on the concentrations of reactants defines the order of the reaction.
- The integration of rate equations predicts the time dependence of concentrations.
- The concentration of reactants decreases exponentially with time for a first-order reaction.
- The reactants decay more slowly in second-order reactions than in first-order reactions with the same rate constant, but the details depend on the particular type of reaction and the conditions.
- The half-life for a reaction provides a measure of the speed of the reaction.
- For a reaction with intermediate steps, the slowest step determines the overall rate.
- Multistep reactions have intermediates that build up as the reaction is initiated, but disappear as the reaction goes to completion.
- A steady-state condition in a reaction means that a concentration does not change with time although the reaction is occurring.
- Steady-state reactions are important in metabolism.
- For reactions with alternative products, the relative values of rate constants determine the distribution of products.
- Measurement of fluorescence provides an easy way to monitor kinetics.

### B. REVERSIBLE REACTIONS

- For reversible reactions (that is, ones for which the equilibrium constant is close to unity), both forward and reverse reactions must be considered in calculating the approach to equilibrium.
- The ratio of forward and reverse rate constants must equal the equilibrium constant.
- Relaxation methods provide a way to obtain rate constants for reversible reactions.
- Temperature jump experiments can be used to determine the association and dissociation rate constants for reactions such as dimerization.
- The rate constants for a cyclic set of reactions are coupled.

### C. FACTORS THAT AFFECT THE RATE CONSTANT

- Catalysts accelerate the rates of chemical reactions without being consumed in the process.
- Rate laws for reactions must be determined experimentally.
- The fastest possible reaction rate is determined by the diffusion-limited rate of collision.
- Most reactions occur more slowly than the diffusion-limited rate.
- The activation energy is the minimum energy required to convert reactants to products during a collision between molecules.
- The activation energy gives rise to an exponential dependence of rate on temperature through a Boltzmann factor.
- Transition state theory links kinetics to thermodynamic concepts.
- Catalysts for reactions can accelerate a reaction by lowering the activation energy or by affecting collision rates or orientation factors.

## Problems

### True/False and Multiple Choice

1. What is the order of this elementary reaction: $A + 2B \longrightarrow 1C$
   a. 2
   b. 7
   c. 4
   d. 3
   e. 8

2. The addition of a catalyst increases the rate of the reaction but not the equilibrium constant.
   True/False

3. For a reaction with a larger $\Delta G$ compared to a reaction with a smaller $\Delta G$,
   a. The reaction with the larger $\Delta G$ is always the faster reaction.
   b. The reaction with the larger $\Delta G$ is always the slower reaction.
   c. The two rates are equal.
   d. It is impossible to decide which reaction is faster.

4. For elementary reactions of zero and second order with a rate constant of 1 (with appropriate units),

the second-order reaction half-life is a shorter period of time.

True/False

5. The most basic step used to describe a reaction process is:

   a. The transition state.
   b. An elementary reaction.
   c. A unimolecular reaction.
   d. The steady state.
   e. The equilibrium rate.

6. The probability of a bimolecular reaction occurring is related to the rate of collisions between the two species of molecules.

   True/False

7. After excitation, the intensity of light emitted by a sample of fluorescent molecules drops exponentially with time.

   True/False

## Fill in the Blank

8. The symbol "‡" is used to represent the ____________ of the reaction.

9. The rate law gives the relationship between the rate of a reaction and the ________ of the chemical species involved in the reaction.

10. The rate constant for a _____-order elementary reaction is $M^{-1} \bullet sec^{-1}$.

11. The rate of product formation at the end of a pathway is determined by the rate of the ______ step.

12. Catalysts can work by changing the reaction mechanism, ________, or ________.

## Quantitative/Essay

13. How do the kinetics of hydrolysis of ATP to ADP and phosphate by water make it a good energy reservoir for the cell?

14. Below is a two-dimensional energy versus reaction coordinate diagram for the reaction AB + C → A + BC.

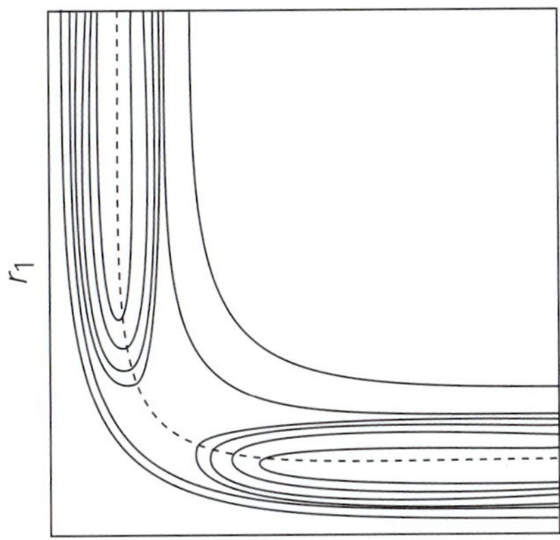


Circle the approximate area that represents the transition state and justify your answer. If $r_1$ corresponds to the A-B distance, and $r_2$ to the B-C distance, indicate what molecules would be present in the regions corresponding to the left hand and right hand sides of the reaction above.

15. Iodine-123 is important for medical imaging studies and follows first-order decay kinetics. A 15-μg sample of I-123 has decayed to 7.5 μg after 13 hours. After how much time will it decay to only 1.5 μg?

16. A reaction is half complete after 20 minutes. After 40 minutes the reaction is two-thirds complete. When will the reaction be 90% complete?

17. A protein (P) can either fold properly into the native state (N) or aggregate into a misfolded form (A). Both processes obey first-order kinetics. The branching ratio ([N]/[A]) is 9 and the effective rate constant, $k_{eff}$, is 15 $sec^{-1}$. What is the rate constant for native state folding?

18. Upon excitation, a modified green fluorescent protein emits photons that yield an initial intensity of 10,000 units in the fluorimeter. After 2 nanoseconds, the signal has decayed to 300 units. If the rate constant for fluorescent production of light ($k_f$) is 0.1 $nsec^{-1}$, what are the values of the rate constant for heat production ($k_h$) and the fluorescence lifetime ($\tau_f$)?

19. An experiment is performed to measure the affinity of the human proline isomerase hCypA to the inhibitor cyclosporin, analogous to the experiments described for imatinib in this chapter. At 25 μM cyclosporin, the value of $k_{obs}$ is measured to be 12.55 $sec^{-1}$. Given that the off-rate is 0.01 $sec^{-1}$, what is the on-rate?

20. A homologous proline isomerase, vCypA, is isolated from a deadly virus. The kinetics of cyclosporin binding to vCypA were measured. The off-rate is 0.1 $sec^{-1}$ and the on-rate is 0.05 $\mu M^{-1} \bullet sec^{-1}$. What concentration of cyclosporin was used to yield $k_{obs}$ = 120 $sec^{-1}$?

21. Several experiments indicate that the vCypA is essential for viral replication. Use the parameters calculated in Problems 19 and 20, and assume that the concentration of cyclosporin would be much larger than the concentration of either vCypA or hCypA. For each protein, calculate the value of $K_D$ for cyclosporin. Given the ratio of the two $K_D$ values, explain whether cyclosporin could be an effective treatment for this virus.

22. The activation energy for proline isomerization of a peptide depends on the identity of the preceding residue and obeys Arrhenius rate behavior. Experiments are conducted on the isomerization of an alanine-proline peptide. At 25°C (298 K) the observed rate constant is 0.05 $sec^{-1}$ and the value of $E_A$ is calculated to be 60 $kJ \bullet mol^{-1}$. What is the value of the preexponential factor (A)? Similar measurements are performed on a phenylalanine-proline peptide at 25°C, with a measured rate constant of 0.005 $sec^{-1}$. Assuming an identical preexponential factor as the alanine-proline peptide, what is the activation energy for this peptide?

**23.** A relaxation experiment probes the homodimerization of an oligonucleotide. At 10-nM oligonucleotide, the apparent rate constant ($k_{apparent}$) is 2.06 $sec^{-1}$. At 100-nM oligonucleotide, the apparent rate constant ($k_{apparent}$) is 6.34 $sec^{-1}$. What are the association and dissociation rate constants?

**24.** A protein folding reaction has two intermediate states, each of which individually obeys Arrhenius-type behavior. At low temperatures, forming the first intermediate is rate-limiting. At high temperatures, forming the second intermediate is rate-limiting.

a. Does the protein folding reaction obey Arrhenius-type behavior over all temperatures?

b. Forming which intermediate has a higher activation energy?

## Further Reading

Atkins PW & De Paula J (2006) Atkins' Physical Chemistry, 8th ed. Oxford, UK: Oxford University Press.

Eisenberg DS & Crothers DM (1979) Physical Chemistry: With Applications to the Life Sciences. Menlo Park, CA: Benjamin-Cummings.

Fersht A (1999) Structure and Mechanism in Protein Science: a Guide to Enzyme Catalysis and Protein Folding. San Francisco: W.H. Freeman.

Hammes GG (2000) Thermodynamics and Kinetics for the Biological Sciences. New York: John Wiley & Sons.

Steinfeld JI, Francisco JS & Hase WL (1999) Chemical Kinetics and Dynamics, 2nd ed. Upper Saddle River, NJ: Prentice Hall.

# CHAPTER 16

# Principles of Enzyme Catalysis

For two molecules to undergo a reaction, they must collide and be in a relative orientation that brings the reacting groups close together. The molecules must also have sufficient energy to overcome the energetic barrier to the reaction. We have noted, in Chapter 15, that catalysts can speed up the rates for reactions by bringing molecules together and by reducing the activation energy for the reaction.

Biologically important reactions are catalyzed by enzymes, which are composed of proteins or RNA. In this chapter, we will apply the general ideas concerning kinetics that were developed in Chapter 15 to understand how enzymes work. We will discuss a few specific enzymes in detail, describe how they catalyze the reactions that they have evolved to accelerate, and explain how their activity is regulated.

The kinetic analysis of enzyme reactions dates back over 100 years, to early studies by Victor Henri, Leonor Michaelis, and Maud Menten. The basic equations describing enzyme kinetics introduced by Henri, Michaelis, and Menten are still in use today, which is a tribute to these early workers, who obtained their insights long before the three-dimensional structures of proteins were known. The first structure of an enzyme was revealed only much later, in 1965, when David Phillips determined the crystal structure of the enzyme lysozyme, which hydrolyzes polysaccharides that are part of bacterial cell walls. Lysozyme has features that are common to many enzymes, including a cleft containing the residues responsible for catalysis, as shown in **Figure 16.1**.

Long after the first protein enzymes were characterized, it was shown that some RNAs in cells also have catalytic activity. This discovery caused a conceptual shift in biochemistry, because it expanded our definition of an enzyme to include RNA as well as protein catalysts. Thomas Cech first showed in 1982 that some RNAs containing introns and exons can catalyze a self-splicing reaction, excising the intron from pre-mRNA (**Figure 16.2**). At about the same time, Sidney Altman showed that the RNA component of an enzyme known as ribonuclease-P cleaves a precursor RNA molecule during the maturation of tRNA; Cech and Altman shared the Nobel Prize for their discoveries. In the last part of this chapter, we discuss the structure and mechanism of two kinds of catalytic RNAs, a self-cleaving RNA and a self-splicing one.

## A. MICHAELIS–MENTEN KINETICS

In this part of the chapter, we develop a very simple but extremely powerful conceptual framework for analyzing the rates of enzyme-catalyzed reactions. This scheme, known as Michaelis–Menten kinetics, is built on the idea that the substrate and the enzyme have to first bind and form a complex in order for the reaction to proceed. By using the Michaelis–Menten equations to model the measured rates of reactions under steady-state conditions, we can extract many useful parameters that are characteristic properties of enzymes.

(A)

A B C D enzymatic cleavage E F

GlcNAc MurNAc GlcNAc MurNAc GlcNAc MurNAc

(B)

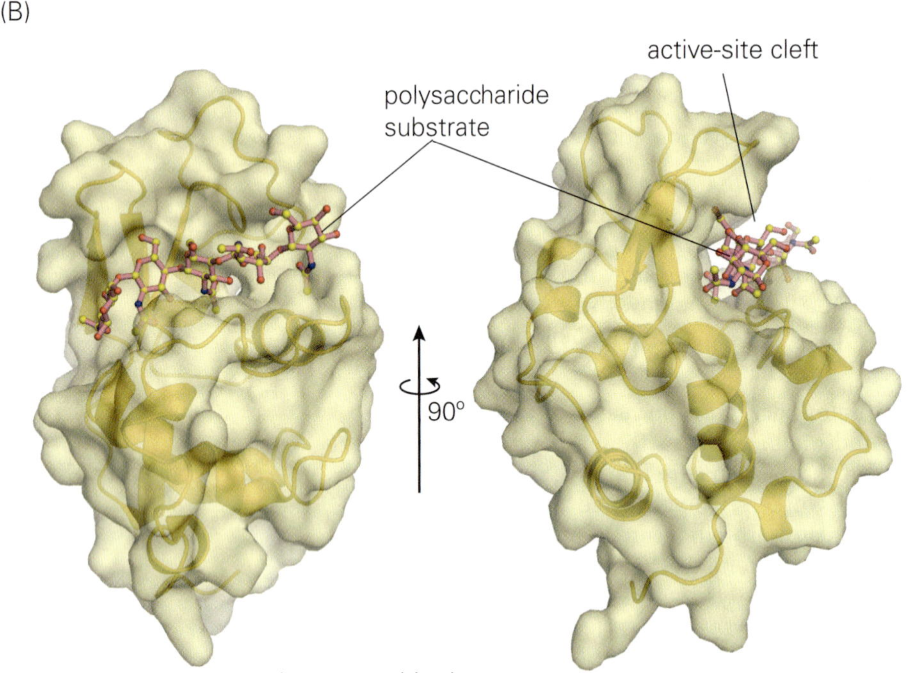


**Figure 16.1 A protein enzyme, lysozyme**. (A) Lysozyme catalyzes the cleavage of polysaccharides consisting of alternating GlcNAc and MurNAc residues (see Figure 3.9), such as the hexasaccharide shown here. (B) The structure of the enzyme hen egg white lysozyme is shown in complex with a polysaccharide substrate bound in the active site cleft. (A, adapted from N.C.J. Strynadka and M.N.G. James, *J. Mol. Biol.* 220: 401–424, 1991. With permission from Elsevier; PDB code: 1SFB.)

---

(A)

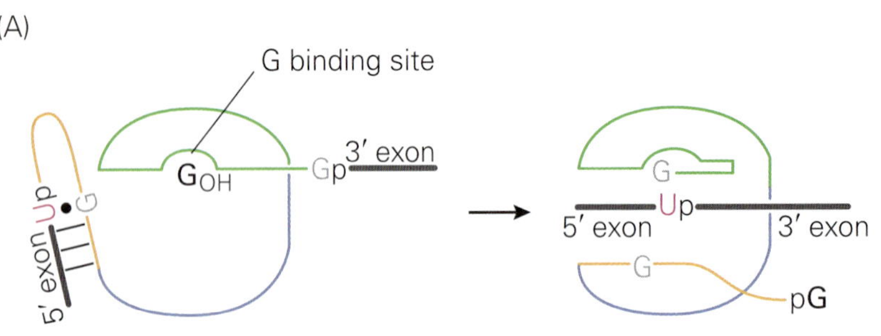


(B)

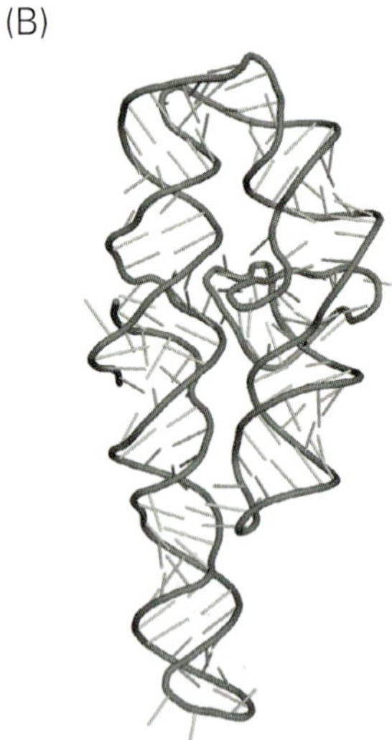


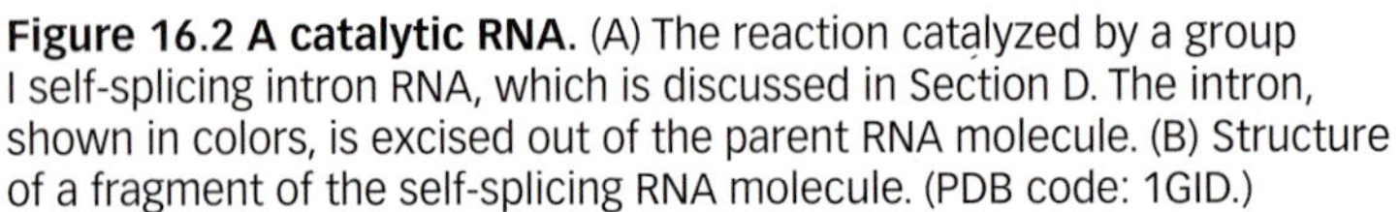

**Figure 16.2 A catalytic RNA.** (A) The reaction catalyzed by a group I self-splicing intron RNA, which is discussed in Section D. The intron, shown in colors, is excised out of the parent RNA molecule. (B) Structure of a fragment of the self-splicing RNA molecule. (PDB code: 1GID.)

## 16.1 Enzyme-catalyzed reactions can be described as a binding step followed by a catalytic step

For enzymatic reactions, the reactants are usually referred to as the **substrates** for the enzyme. Substrate, enzyme, and product will be indicated by S, E, and P, respectively. Within each enzyme there is an **active site**, which includes a region where the chemical reaction actually occurs (the **catalytic site**). The active site may also contain other regions that help hold and position the substrate at the correct location in the active site. Most of the essential elements of enzyme kinetic behavior are manifested when there is just a single substrate, and so that case will be considered first because it is the simplest to understand.

**Initial velocity**

When substrate is added to the enzyme to initiate a reaction, there is initially little or no product present and the amount of substrate has not decreased significantly. The rate of the reaction during this period is called the initial velocity, $v_0$.

For a single substrate, the overall enzyme-catalyzed reaction can be written as:

$$\mathrm{S}+\mathrm{E} \underset{k_r}{\overset{k_f}{\rightleftharpoons}} \mathrm{P}+\mathrm{E} \qquad (16.1)$$

We need, in general, to consider both the forward rate constant ($k_f$) and the rate constant for the reverse reaction ($k_r$). When analyzing the results of experiments, the situation is simplified if the reaction is initiated by adding the substrate to the enzyme, so that there are no product molecules initially. The rate of back reaction is often very slow, and hence it can be ignored in the analysis. The rate of product formation is then given by:

$$v = \frac{d[\mathrm{P}]}{dt} = k_f[\mathrm{S}][\mathrm{E}] \qquad (16.2)$$

The rate of product formation is called the **velocity of the reaction**, $v$. If the rate of the reaction is measured only during the initial period of the reaction (that is, before substrate is depleted and product builds up), then the rate is referred to as the **initial velocity**, $v_0$.

**Maximal velocity**

The maximal velocity, $V_{max}$, is the maximum rate of reaction catalyzed by an enzyme at very high substrate concentration.

Equation 16.2 predicts that the velocity of the reaction depends linearly on the concentrations of both substrate and enzyme. In actuality, if one measures the velocity of the reaction as a function of substrate concentration, holding enzyme concentration constant, results such as those shown in **Figure 16.3** are found. At very low substrate concentrations the response is indeed linear, but with increasing substrate the rate levels off, reaching a maximum value termed the **maximal velocity** ($V_{max}$). A second parameter, the **Michaelis constant** ($K_M$), specifies the concentration of substrate required to reach half of this maximum velocity. As a result, a more complex kinetic scheme than that described by Equation 16.1 is required to explain these data.

**Michaelis constant**

The Michaelis constant, $K_M$, is the concentration of substrate required for the reaction velocity to be $^1/_2$ $V_{max}$.

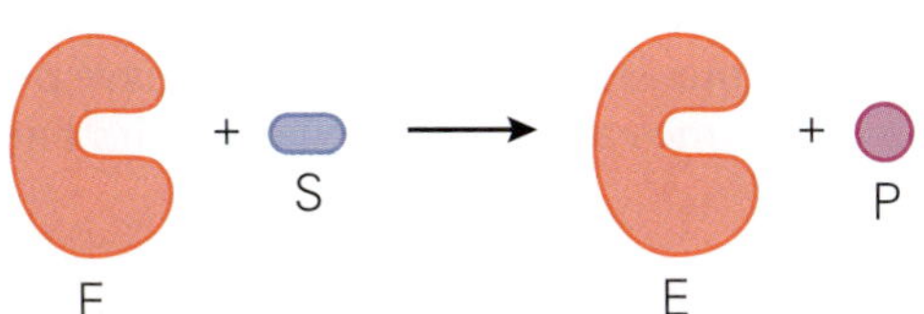


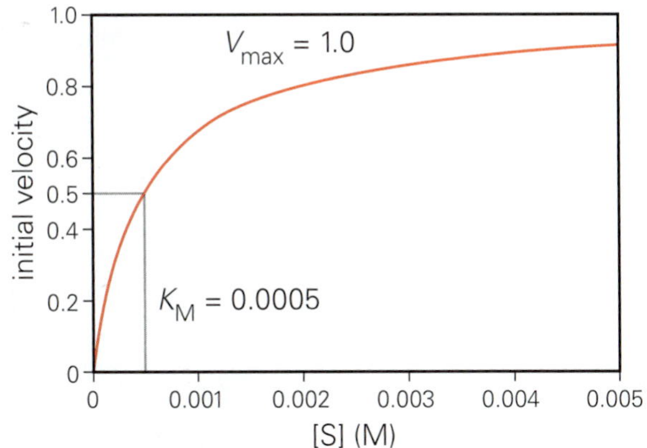


**Figure 16.3 A graph of the initial velocity of reaction for different concentrations of the substrate, S.** At very high concentration, the velocity approaches the value $V_{max}$, and it reaches half of that value when the substrate concentration is $K_M$, the Michaelis constant.

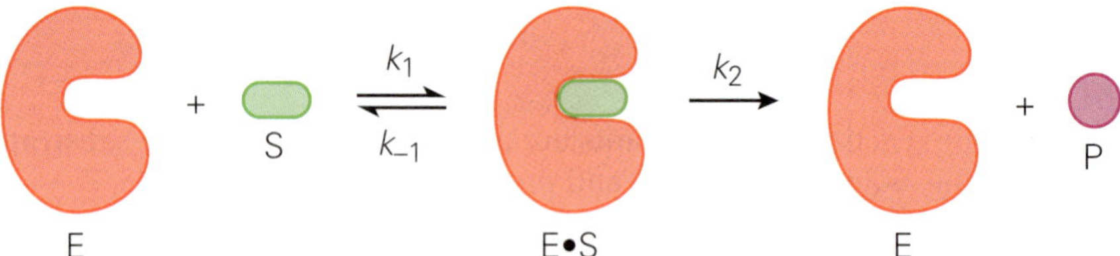


**Figure 16.4 A schematic drawing of the process of an enzyme binding substrate, which reacts and then is released as product.** The analysis of such a reaction scheme is known as Michaelis–Menten kinetics. An example of a real enzyme–substrate complex is shown in Figure 16.5.

The graph of reaction velocity as a function of substrate concentration (see Figure 16.3) is reminiscent of the hyperbolic binding isotherm for a ligand binding to a protein (see Figure 12.4). Recall that at low substrate concentration the fraction of the protein that is bound to ligand increases steeply with substrate concentration. But, at higher substrate concentration, as the protein becomes saturated, the binding curve levels off and approaches a maximum value asymptotically. There is, indeed, a close connection between the thermodynamics of binding and the kinetics of enzyme-catalyzed reactions, because the first step in catalysis is the binding of substrate to the enzyme. If the on- and off-rates for the substrate binding to the enzyme are fast compared to the catalytic step, then the binding can be considered to be a reversible event, as shown in **Figure 16.4**.

Considering binding and dissociation of the substrate as separate kinetic steps that come before the actual catalyzed reaction gives a more complicated kinetic model:

$$E + S \underset{k_{-1}}{\overset{k_1}{\rightleftharpoons}} E{\bullet}S \text{ and then } E{\bullet}S \xrightarrow{k_2} P + E \tag{16.3}$$

The rate constant $k_2$ is associated with the actual chemical step. As we discussed earlier, the reverse reaction is ignored because, if the measurements are made within a short time after mixing substrate and enzyme, then not enough product is generated to drive the reverse reaction. This description of kinetic behavior is called **Michaelis–Menten kinetics**, named for the biochemists who established this model in 1913, following the slightly earlier work of Henri.

One important consequence of the reaction scheme shown in Equation 16.3 is that the recognition of the substrate is separated from the catalysis of the chemical reaction. The specificity of the enzyme for the substrate often manifests itself in the first step, with the nature of the interactions between the substrate and the enzyme determining whether the substrate binds to the enzyme and whether it binds in an orientation appropriate for catalysis. This step is governed by the same set of principles that we have established in Chapters 12 and 13 for simple binding events. As an example, the structure of a protease enzyme bound to its substrate is shown in **Figure 16.5**. Proteases are enzymes that cleave peptide bonds. Some proteases are digestive enzymes and cleave peptides with hardly any specificity.

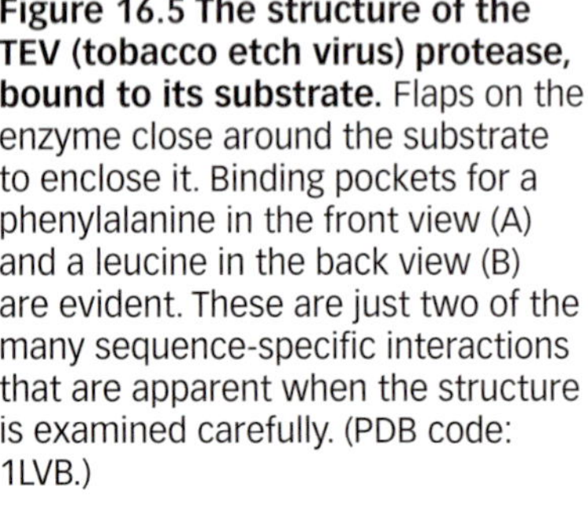

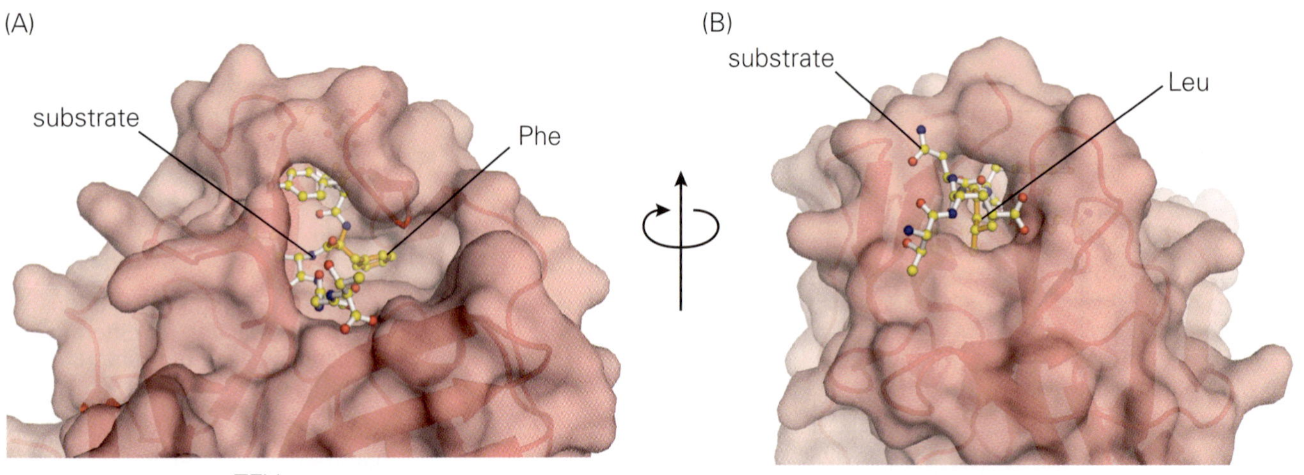


**Figure 16.5 The structure of the TEV (tobacco etch virus) protease, bound to its substrate.** Flaps on the enzyme close around the substrate to enclose it. Binding pockets for a phenylalanine in the front view (A) and a leucine in the back view (B) are evident. These are just two of the many sequence-specific interactions that are apparent when the structure is examined carefully. (PDB code: 1LVB.)

Others, such as the one shown in Figure 16.5, are very specific, and cleave only particular peptide bonds within a specific sequence context. As you can see from the structure shown in Figure 16.5, numerous interactions between the enzyme and the substrate ensure that only the correct target sequence binds with high affinity to the enzyme.

**Michaelis–Menten kinetics**

Michaelis–Menten kinetics derives from a model with reversible substrate binding to the enzyme, followed by the chemical transformation. It predicts the behavior of many enzymes very well.

## 16.2 The Michaelis–Menten equation describes the kinetics of the simplest enzyme-catalyzed reactions

For the reaction scheme shown in Equation 16.3, the initial velocity, $v_0$, is given by:

$$v_0 = \frac{d[\mathrm{P}]}{dt} = k_2[\mathrm{E \bullet S}] \tag{16.4}$$

If we initiate the reaction by mixing enzyme with substrate, the enzyme-substrate complex will begin to form, but can dissociate to release free enzyme. The complex can also generate product and then dissociate to release free enzyme. After some time, the rates of formation and dissociation of the enzyme complex will become equal, and so the concentrations of the free enzyme and the enzyme-substrate complex will reach constant values. Recall from Section 15.13 that this corresponds to a **steady-state** situation. We will use $[\mathrm{E{\bullet}S}]_{ss}$ to denote the steady state concentration of the enzyme-substrate complex. As long as there is a large amount of substrate present, $[\mathrm{E{\bullet}S}]_{ss}$ does not change with time and we can write:

$$\frac{d[\mathrm{E \bullet S}]_{ss}}{dt} = k_1[\mathrm{E}][\mathrm{S}] - k_{-1}[\mathrm{E \bullet S}]_{ss} - k_2[\mathrm{E \bullet S}]_{ss} = 0 \tag{16.5}$$

Rearranging this gives:

$$[\mathrm{E \bullet S}]_{ss} = \frac{k_1[\mathrm{E}][\mathrm{S}]}{k_{-1} + k_2} \tag{16.6}$$

Using Equation 16.6 in Equation 16.4, the initial velocity, $v_0$, is given by:

$$v_0 = k_2[\mathrm{E \bullet S}]_{ss} = \frac{k_1 k_2[\mathrm{E}][\mathrm{S}]}{k_{-1} + k_2} \tag{16.7}$$

In Equation 16.7, [E] is the concentration of free enzyme. The concentration of the free enzyme is not straightforward to determine experimentally or to calculate. [E] can be eliminated from the rate equations by rewriting them to use the total enzyme concentration present, $[\mathrm{E}]_0$:

$$[\mathrm{E}]_0 = [\mathrm{E}] + [\mathrm{E \bullet S}]_{ss} \quad \text{or} \quad [\mathrm{E}] = [\mathrm{E}]_0 - [\mathrm{E \bullet S}]_{ss} \tag{16.8}$$

The same idea can be applied to the substrate, but the substrate is usually present in large excess relative to enzyme. If the rates are measured only during a brief period immediately after initiating the reaction (again, by measuring the *initial velocity*, $v_0$), we can then assume that the concentration of the substrate ([S]) is essentially constant at its starting value $[\mathrm{S}]_0$:

$$[\mathrm{S}]_0 = [\mathrm{S}] + [\mathrm{E \bullet S}] \approx [\mathrm{S}] \tag{16.9}$$

Substituting the expression for [E] into the equation for the steady-state concentration of enzyme-substrate complex (that is, Equation 16.6) then gives:

$$[\mathrm{E \bullet S}]_{ss} = \frac{k_1([\mathrm{E}]_0 - [\mathrm{E \bullet S}]_{ss})[\mathrm{S}]}{k_{-1} + k_2} = \frac{k_1[\mathrm{E}]_0[\mathrm{S}]}{k_{-1} + k_2} - \frac{k_1[\mathrm{E \bullet S}]_{ss}[\mathrm{S}]}{k_{-1} + k_2} \tag{16.10}$$

Grouping together the terms containing $[\mathrm{E{\bullet}S}]_{ss}$ then gives:

$$[\mathrm{E \bullet S}]_{ss}\left(1 + \frac{k_1[\mathrm{S}]}{k_{-1} + k_2}\right) = \frac{k_1[\mathrm{E}]_0[\mathrm{S}]}{k_{-1} + k_2} \tag{16.11}$$

and again rearranging,

$$[\mathrm{E \bullet S}]_{ss} = \frac{[\mathrm{E}]_0}{\left(1 + \dfrac{k_{-1} + k_2}{k_1[\mathrm{S}]}\right)} \tag{16.12}$$

We define the Michaelis constant, $K_M$, as follows:

$$K_M = \frac{k_{-1} + k_2}{k_1} \tag{16.13}$$

Using this definition for $K_M$ in Equation 16.12, we get:

$$[E \bullet S]_{ss} = \frac{[E]_0}{\left(1 + \frac{K_M}{[S]}\right)} \tag{16.14}$$

We obtain an expression for the initial velocity by multiplying $[E{\bullet}S]_{ss}$ by $k_2$:

$$v_0 = k_2[E \bullet S]_{ss} = \frac{k_2[E]_0}{\left(1 + \frac{K_M}{[S]}\right)} \tag{16.15}$$

The maximum velocity, $V_{max}$, of the enzyme-catalyzed reaction occurs when all of the enzyme is bound to substrate—that is, when [E•S] is equal to $[E]_0$. And so the value of $V_{max}$ is given by:

$$V_{max} = k_2[E]_0 \tag{16.16}$$

Substituting this into Equation 16.15 gives:

$$v_0 = \frac{V_{max}}{\left(1 + \frac{K_M}{[S]}\right)} \tag{16.17}$$

Equation 16.17 is called the **Michaelis–Menten equation** and it predicts precisely the behavior shown in Figure 16.3 and **Figure 16.6**.

Note that when $[S] = K_M$, the velocity will be half of $V_{max}$, which is the functional definition of $K_M$ given originally in Section 16.1. The value of the Michaelis constant can be determined experimentally by using data such as those in Figure 16.6 and equating $K_M$ to the substrate concentration at which the velocity is half-maximal. The determination of the value of $K_M$ in this way is only possible if the rate of the reaction can in fact be measured at a sufficiently high substrate concentration that the plateau value is observed. This may not be possible if the substrate is not sufficiently soluble, and alternate methods for determining the value of $K_M$ are discussed below in Section 16.8.

**Michaelis–Menten equation**

For the simplest enzyme-catalyzed reaction, shown in Equation 16.3, the initial rate of the reaction under steady-state conditions is given by:

$$v_0 = \frac{V_{max}}{\left(1 + \frac{K_M}{[S]}\right)}$$

$V_{max}$ is the maximum rate of the reaction and $K_M$ is the Michaelis constant, which corresponds to the substrate concentration at which the rate is half-maximal.

## 16.3 The value of the Michaelis constant, $K_M$, is related to how much enzyme has substrate bound

The value of the Michaelis constant ($K_M$) determines how much of the enzyme is bound to the substrate, as can be seen from the expression for the concentration of the enzyme–substrate complex, $[E{\bullet}S]_{ss}$, given by Equation 16.14. $[E{\bullet}S]_{ss}$ can be expressed as the product of the fractional occupancy of the enzyme, $f$, multiplied by the total enzyme concentration, $[E]_0$:

$$[E \bullet S]_{ss} = f[E]_0 \tag{16.18}$$

where

$$f = \frac{1}{\left(1 + \frac{K_M}{[S]}\right)}$$

Based on the discussion in Chapter 12 (see Equation 12.11), we know that if the enzyme and the substrate are in equilibrium, then the fraction of enzyme bound to ligand, $f$, is given by:

$$f = \frac{[S]}{K_D + [S]} = \frac{1}{\left(1 + \frac{K_D}{[S]}\right)} \tag{16.19}$$

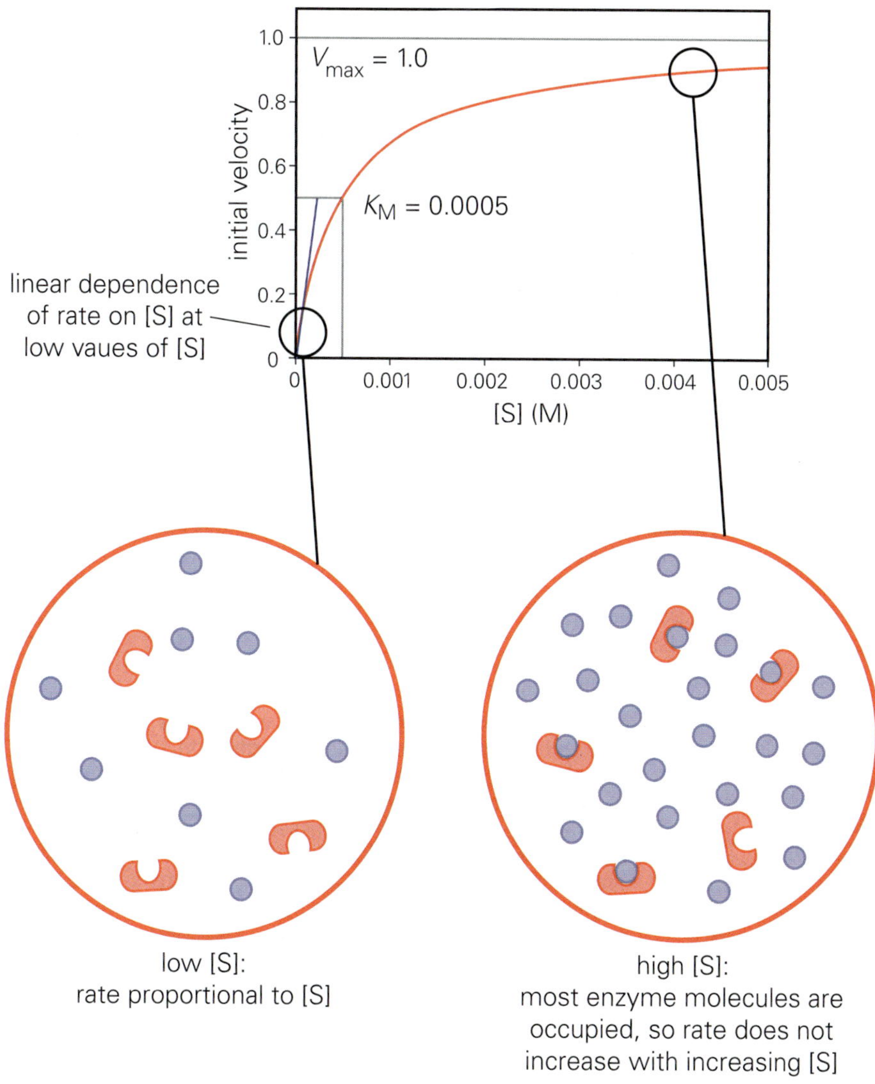


**Figure 16.6 The velocity of an enzyme reaction as a function of substrate.** The velocity is linear in [S] at very low values [S] (shown by the steep *blue* line intersecting the origin), because the reaction is bimolecular. The rate becomes independent of [S] at very high [S] because most of the enzyme molecules are occupied by substrate. The rate approaches a maximal value, denoted by the *gray* line, which is $V_{max}$. The value of $K_M$ for the enzyme is indicated by the substrate concentration corresponding to half-maximal rate.

It is apparent, from a comparison of Equations 16.18 and 16.19, that the Michaelis constant, $K_M$, plays a role analogous to that of the dissociation constant, $K_D$, in determining how much of the enzyme is bound to the substrate. Note, however, that $K_M$ and $K_D$ are fundamentally different. Equation 16.18 applies to the steady-state (nonequilibrium) situation, and the Michaelis constant is determined by a set of rate constants, including the rate constant for product formation (see Equation 16.13). Equation 16.19 applies only at equilibrium and the dissociation constant for the formation of the enzyme–substrate complex does not depend on the product at all, and is simply given by the ratio of rate constants for the dissociation of the enzyme–substrate complex and for its formation ( $K_D = \frac{k_{-1}}{k_1}$, as explained in Section 15.18).

Looking at the reaction scheme in Equation 16.3, if $k_2 \ll k_{-1}$ (that is, if the chemical catalysis step is slow compared to dissociation of the substrate), then the value of $K_M$ approaches the value of the dissociation constant for the enzyme–substrate complex (that is, $K_M \approx k_{-1}/k_1$). When $k_2$ is comparable to or larger than $k_{-1}$, then the occupancy of the active site decreases (as though the dissociation constant was higher, corresponding to apparently weaker binding) because substrate is converted to product and is released frequently, emptying the active site.

To make these ideas more concrete, let us look at an example in which we compare two situations. In one, there is a binding equilibrium between an enzyme and a substrate, with product formation blocked. In the other situation, the enzyme converts the substrate to product. Let us suppose that the substrate binds to the enzyme with a high rate constant (for example, $k_1 = 10^8\ \text{M}{\bullet}\text{sec}^{-1}$). Assume that the dissociation constant is 1 μM, which is a fairly typical value for binding equilibria in the cell. This means that the rate constant for dissociation, $k_{-1}$, must be:

$$k_{-1} = K_D k_1 = \left(10^{-6}\right)\left(10^{8}\right) = 100\ \text{sec}^{-1} \qquad (16.20)$$

We assume that the concentration of the substrate is equal to the value of the dissociation constant (for example, $[S] = K_D = 1$ μM; the concentration of the enzyme is assumed to be much lower). If the enzyme and the substrate are at equilibrium, then 50% of the enzyme will be bound to substrate, as shown in **Figure 16.7**A.

Now consider, instead, a steady-state situation in which the enzyme turns over the substrate to generate product. The occupancy of the enzyme will depend on the value of the catalytic rate constant, $k_2$, in addition to $k_{-1}$ and $k_1$. If the rate of catalysis is fast compared to the rate of dissociation of the enzyme–substrate complex ($k_{-1}$), then the value of $K_M$ will be larger than the value of $K_D$. For example, if the value of $k_2$ is 1000 sec$^{-1}$, then using Equation 16.13 to calculate the value of $K_M$, we get:

$$K_M = \frac{k_{-1} + k_2}{k_1} = \frac{100 + 1000}{10^8} = 11 \times 10^{-6} \qquad (16.21)$$



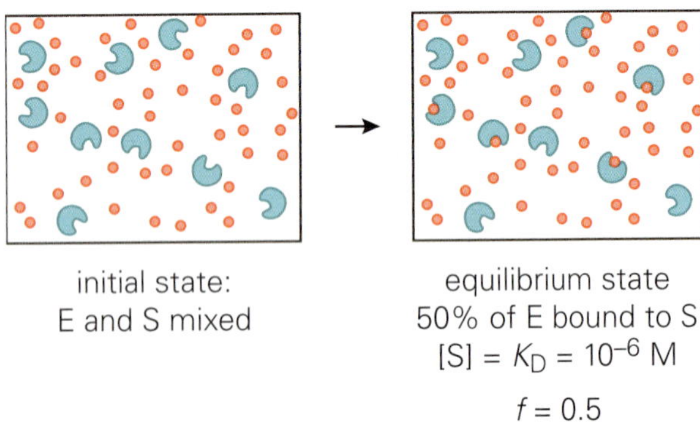




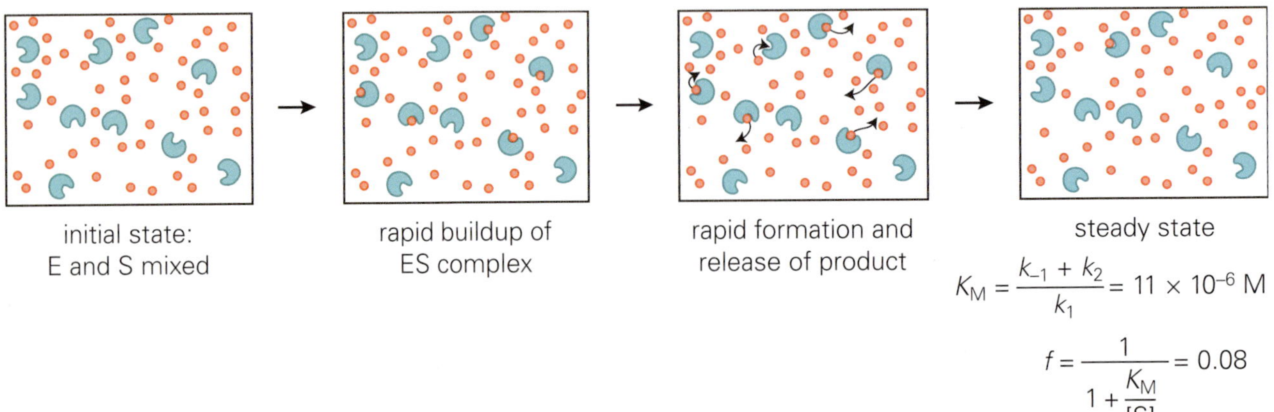


**Figure 16.7 The occupancy of an enzyme by substrate at equilibrium, with no product formation, and under steady-state conditions.** (A) Equilibrium. The catalytic step is assumed to be so slow that there is essentially no product formation. In the example discussed in the main text, the substrate concentration and the value of $K_D$ for the enzyme are both equal to 1 μM, and so 50% of the enzyme is bound to the substrate at equilibrium. (B) A steady-state situation, when the rate of product formation is fast compared to the rate of dissociation of the E•S complex. Only ~8% of the enzyme molecules are bound to substrate because of product formation and release.

(A)

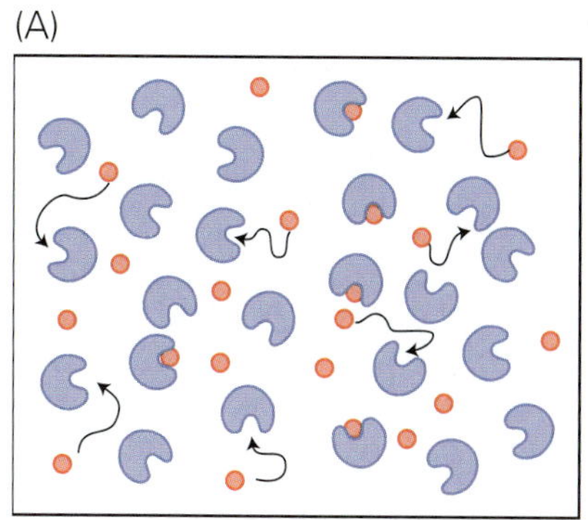

(B)

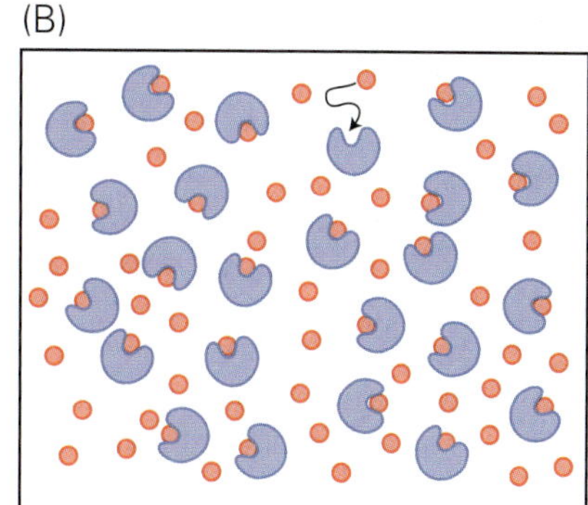

**Figure 16.8 Two extremes of substrate concentration.** (A) When the substrate concentration is low (that is, when [S] ≪ $K_M$), most of the enzyme molecules (*blue*) are empty. The rate is affected by the speed with which substrate molecules (*red*) bind to the enzyme. (B) When the substrate concentration is very high (that is, when [S] ≫ $K_M$), almost all of the enzyme molecules are bound to substrate. The rate of the reaction is then determined by the rate of the chemical catalysis step on the enzyme molecules. Further increases in the concentration of the substrate do not increase the rate because the additional substrate molecules do not have any empty enzyme molecules to which they can bind.

To calculate the occupancy of the enzyme, we use Equation 16.18:

$$f = \frac{1}{\left(1+\frac{K_M}{[S]}\right)} = \frac{1}{1+\frac{11\times10^{-6}}{1\times10^{-6}}} = \frac{1}{12} \approx 0.08 \tag{16.22}$$

Thus, because the enzyme turns over substrate and releases the product very quickly, only about 8% of the enzyme is occupied by substrate under steady-state conditions (Figure 16.7B).

## 16.4 Enzymes are characterized by their turnover numbers and their catalytic efficiencies

When the substrate concentration is low, as shown in **Figure 16.8**A, most of the enzyme molecules are not occupied by substrate. Under these conditions, the rate of the reaction should be proportional to substrate concentration. As more substrate molecules are added to the reaction, there are empty enzyme molecules to which they can bind, and the reaction proceeds faster. In particular, when [S] ≪ $K_M$, the 1 in the denominator of the Michaelis-Menten equation is small relative to $K_M$/[S] and can be ignored. At low substrate concentration, the Michaelis-Menten equation therefore takes a simpler form, in which the rate of the reaction is related linearly to substrate concentration:

$$v_0 = \frac{V_{max}}{1+\frac{K_M}{[S]}} \approx \frac{V_{max}}{\frac{K_M}{[S]}} = \left(\frac{k_2}{K_M}\right)[E]_0[S] \tag{16.23}$$

In Equation 16.23, we have used Equation 16.16 to replace $V_{max}$ with $k_2[E]_0$. Under such conditions (that is, when [S] ≪ $K_M$), most enzyme molecules are not occupied at any given time, so [E] ≈ $[E]_0$.

At high substrate concentration, specifically when [S] ≫ $K_M$, the enzyme becomes saturated with substrate (Figure 16.8B). Under such conditions, $K_M$/[S] ≪ 1, so $K_M$/[S] can be ignored in the Michaelis-Menten equation. Thus, as the substrate concentration is increased, the rate equation becomes independent of substrate concentration asymptotically:

$$v_0 = \frac{V_{max}}{1+\frac{K_M}{[S]}} \approx \frac{V_{max}}{1} = k_2[E]_0 \tag{16.24}$$

This asymptotic plateau in the value of the reaction rate is exactly the behavior observed experimentally, as shown in Figure 16.6.

The apparent rate constant for an enzyme reaction at high substrate concentration (that is, when the reaction is proceeding with maximal rate) is often referred to as the **catalytic rate constant**, $k_{cat}$. The value of $k_{cat}$ is also referred to as the **turnover number** for the enzyme, because it is the maximum number of reactions per second per mole of the enzyme. For a reaction obeying simple Michaelis-Menten kinetics, the value of $k_{cat}$ is the same as the value of $k_2$, the rate constant associated

**Catalytic rate constant, $k_{cat}$**

The apparent rate constant for an enzyme-catalyzed reaction operating at maximum rate is called the catalytic rate constant, $k_{cat}$. The maximum rate, $V_{max}$, occurs when the enzyme is saturated with substrate:

$V_{max} = k_{cat} [E]_0$

If the enzyme obeys Michaelis-Menten kinetics then $k_{cat}$ is the same as $k_2$, the rate constant for the catalytic step (see Equation 16.16).

**Catalytic efficiency**

The ratio of the catalytic rate constant to the Michaelis constant, $k_{cat}/K_M$, is known as the catalytic efficiency. This parameter is equivalent to the second-order rate constant for the enzyme-catalyzed reaction at low substrate concentration.

with the catalytic step. For more complicated reactions, such as those with multiple intermediates, $k_{cat}$ may not be related to any individual rate constant in a simple way.

From Equation 16.23 it is clear that, in this low-substrate limit, $k_2/K_M$ behaves as a second-order rate constant. The ratio $k_2/K_M$ is equivalent to $k_{cat}/K_M$, which is referred to as the **catalytic efficiency** of the enzyme. Enzymes with large values of $k_{cat}/K_M$, approaching values for rate constants of diffusion-limited second-order reactions (~$10^{10}$ $M^{-1}$•$sec^{-1}$; see Section 15.25), are very efficient. They bind substrates and carry out the catalytic step as rapidly as possible. The catalytic efficiency of a particular enzyme for different kinds of substrates is discussed in Section 16.6.

## 16.5 A "perfect" enzyme is one that catalyzes the chemical step of the reaction as fast as the substrate can get to the enzyme

The catalytic rate constant, $k_{cat}$, tells us how fast the enzyme catalyzes the chemical reaction once the substrate has arrived at the active site. Some enzymes need to be very fast in order to keep up with the demands of the cell. One such enzyme is triose phosphate isomerase, which catalyzes a reaction in glycolysis, the series of reactions that generates ATP by the conversion of glycogen to lactate (shown in **Figure 16.9** and also discussed in Section 11.5). This anaerobic process for the production of ATP is particularly critical for an animal to be able to mount a sudden and energetically costly effort, as in a "fight-or-flight" response to a threat.

As indicated schematically in Figure 16.9, during glycolysis glycogen is converted in a series of steps to fructose-1,6-bisphosphate, which is broken down to form dihydroxyacetone phosphate and glyceraldehyde-3-phosphate. Glyceraldehyde-3-phosphate is then converted to pyruvate, with the generation of ATP and reducing power (NADH). Pyruvate can enter the citric acid cycle (see Figure 11.9), which generates more ATP and NADH. Triose phosphate isomerase catalyzes the conversion of dihydroxyacetone phosphate to glyceraldehyde-3-phosphate. This doubles the amount of glyceraldehyde-3-phosphate that is converted to pyruvate and therefore increases the output of ATP and NADH from glycolysis and the citric acid cycle. If there were no triose phosphate isomerase, or if its catalytic action were to slow down, the cell would incur an energy deficit.

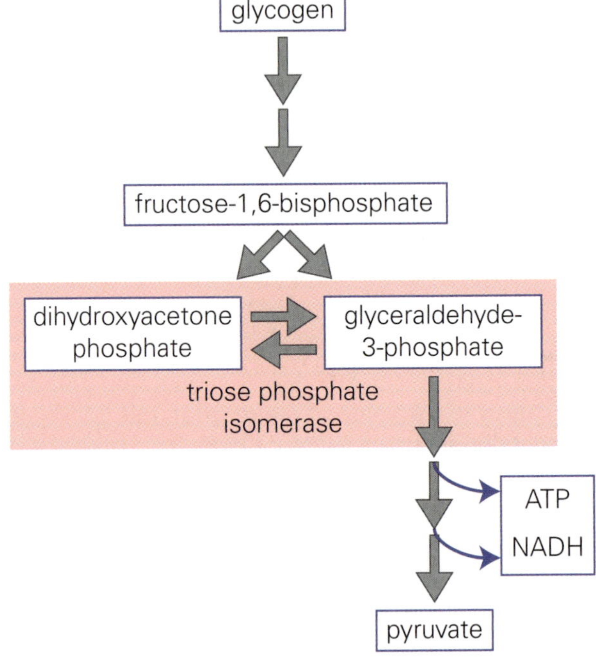


**Figure 16.9 Triose phosphate isomerase in glycolysis.** Shown here is a highly simplified diagram of the steps in glycolysis (you can also refer to Figure 11.9). Triose phosphate isomerase catalyzes the conversion of dihydroxyacetone phosphate to glyceraldehyde-3-phosphate, which is processed further to yield ATP and NADH. Without the action of this enzyme, only half of the breakdown products of fructose-1,6-bisphosphate can be used to generate ATP and NADH.

Triose phosphate isomerase is a "perfect" enzyme, in the sense that it catalyzes the conversion of dihydroxyacetone phosphate to glyceraldehyde-3-phosphate so fast that the reaction is diffusion-controlled. In other words, the rate of the reaction is determined by how fast the substrate collides with the active site of triose phosphate isomerase (see Section 15.25). A Michaelis–Menten analysis of the enzyme reveals that the value of the catalytic rate constant, $k_{cat}$, for triose phosphate isomerase is ~5000 $sec^{-1}$. Let us see what this tells us about how fast the reaction is catalyzed relative to the diffusion-limited rate.

In order to evaluate the reaction speed, we shall assume that the enzyme concentration is 1 nM ($10^{-9}$ M), and that the substrate is at 10 μM ($10^{-5}$ M), which is approximately the dihydroxyacetone phosphate concentration in a cell. The rate of the chemical step is then given by:

$$\text{rate} = k_{cat}[E]_0 = 5000 \times 10^{-9} = 50 \times 10^{-7} \text{ M} \bullet \text{sec}^{-1} \qquad (16.25)$$

Now let us calculate the diffusion-limited rate. Recall from Section 15.25 that the second-order rate constant, $k_{collision}$, for collision between two molecules in solution is ~$10^{10}$ $M^{-1} \bullet sec^{-1}$. In Section 15.26 we noted that not all collisions are productive, because the two molecules may not be oriented correctly for reaction. For an enzyme–substrate reaction, only collisions between the substrate and the enzyme that occur at the active site of the enzyme will be productive, and then only if the substrate is oriented correctly. This constraint reduces the effective rate constant for productive collisions. Measurement has shown that the values of the diffusional rate constant, $k_{diff}$, for enzyme–substrate reactions is in the range of $10^6$ to $10^8$ $M^{-1} \bullet sec^{-1}$ (that is, only 1 in 10,000 to 1 in 100 collisions satisfy the orientational constraint).

Thus, a reasonable upper limit for the rate of productive diffusional encounters between the substrate and triose phosphate isomerase is given by choosing $10^7$ $M^{-1} \bullet sec^{-1}$ as the value for the diffusional rate constant:

$$\text{rate} = k_{diff}[E]_0[S] = (10^7)(10^{-9})(10^{-5}) = 10^{-7} \text{ M} \bullet \text{sec}^{-1} \qquad (16.26)$$

You can see, by comparing the rate of the chemical step (Equation 16.25) and the rate of productive encounters (Equation 16.26), that triose phosphate isomerase is as fast a catalyst as it needs to be to keep up with the arrival of substrate at the active site. There will be no evolutionary pressure to make triose phosphate isomerase a faster enzyme: if it worked any faster, the reaction would be limited by the speed at which substrate collided with the active site.

Some enzymes actually work *faster* than the diffusion controlled limit. In Section 6.24 we discussed the fact that the electrostatic fields around proteins can be strongly polarized by the locations of charged groups and by the shape of the protein. This feature, called electrostatic focusing, is used by the enzyme acetylcholine esterase to guide positively charged substrates to the active site of the enzyme, so that they arrive at the active site faster than they would through normal diffusion (see Figure 6.39).

Another enzyme that uses electrostatic focusing is superoxide dismutase, an enzyme that converts the superoxide radical ($O_2^{\bullet -}$) into hydrogen peroxide and oxygen. The superoxide radical is highly reactive and can cause serious damage to DNA and other cellular components, something which is prevented by the action of the enzyme. Superoxide dismutase has an overall negative charge, which is counterintuitive because the substrate is also negatively charged. But, as shown in **Figure 16.10**, the negative electrostatic potential around most of the surface of the enzyme reduces the number of unproductive collisions because the substrate moves away from these regions. The active sites of the enzyme are in regions of positive electrostatic potential, and so the substrates are preferentially drawn into these regions. This is known as **electrostatic steering**, and this feature allows superoxide dismutase to catalyze the conversion of its substrate faster than the diffusion limit.

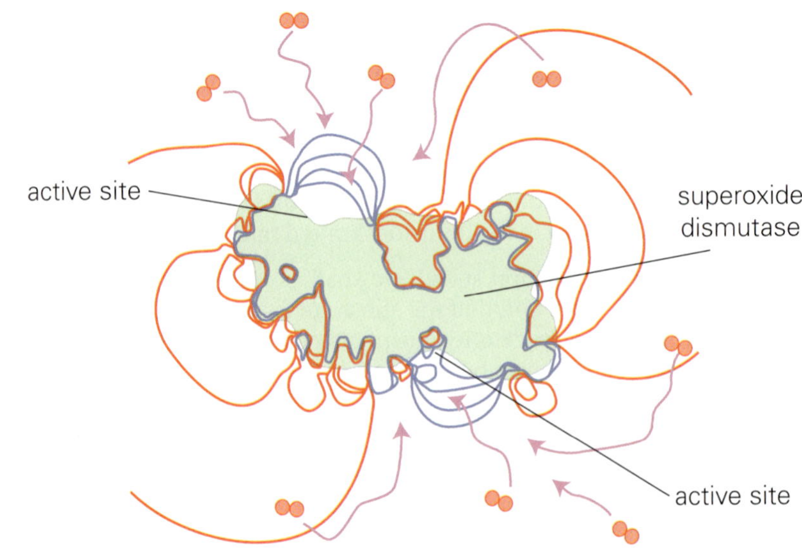


**Figure 16.10 The enzyme superoxide dismutase works faster than the diffusion limit.** The substrate for superoxide dismutase is the superoxide ion (*red* spheres), which is negatively charged. The electrostatic potential around the protein is largely negative (the *red* lines are contours of negative electrostatic potential), but has regions of positive electrostatic potential (*blue* contours) leading into the two active sites (the enzyme is dimeric). This increases the number of productive collisions between the substrate and the enzyme because the substrate is "steered" into the active sites. (Adapted from A.R. Leach, Molecular Modeling: Principles and Applications, 2nd ed. Upper Saddle River, NJ: Prentice Hall, 2001; based on data in D.E. McRee et al. and J.A. Tainer, *J. Biol. Chem.* 265: 14234–14241, 1990.)

## 16.6 In some cases the release of the product from the enzyme affects the rate of the reaction

Given the discussion of substrate binding to the protein as a discrete step in the Michaelis–Menten scheme, it is then natural to ask why product is directly released after being produced rather than being bound to the enzyme and released subsequently in a second step. In real systems, the dissociation of the product is indeed a separate step that is part of a complete description of the kinetics of the enzyme. The effects of this additional step, or of other reversible reactions, can also be added to the reaction scheme.

The following is a more complete reaction scheme for a simple enzyme-catalyzed reaction with one substrate:

$$\mathrm{E+S} \underset{k_{-1}}{\overset{k_1}{\rightleftarrows}} \mathrm{E\bullet S} \underset{k_{-2}}{\overset{k_2}{\rightleftarrows}} \mathrm{E\bullet P} \xrightarrow{k_3} \mathrm{E+P} \tag{16.27}$$

As before, we can apply exactly the same idea of a steady state for [E], [E•S], and [E•P], yielding a kinetic equation for the rate of formation of free product:

$$v_0 = \frac{V_{\max}}{\left(1+\dfrac{K'_{\mathrm{M}}}{[\mathrm{S}]}\right)} \tag{16.28}$$

where

$$V_{\max} = \frac{k_2 k_3 [\mathrm{E}]_0}{(k_2 + k_{-1} + k_3)}$$

and

$$K'_{\mathrm{M}} = \frac{(k_{-1}k_2 + k_{-1}k_3 + k_2 k_3)}{k_1(k_2 + k_{-2} + k_3)}$$

In Equation 16.28, $V_{\max}$ is the maximum velocity of the reaction and $K'_{\mathrm{M}}$ is a modified form of the Michaelis constant that includes terms involving the rate constant for product release, $k_3$. By comparing Equation 16.28 with the Michaelis–Menten equation (Equation 16.17), we see that, even when we need to consider product release, the same basic behavior appears in the overall reaction process. The difference is that, if product release is slow (that is, $k_3$ cannot be ignored), then the values of $V_{\max}$ and $K'_{\mathrm{M}}$ depend on the rate constants in a different way than in the original Michaelis–Menten equation.

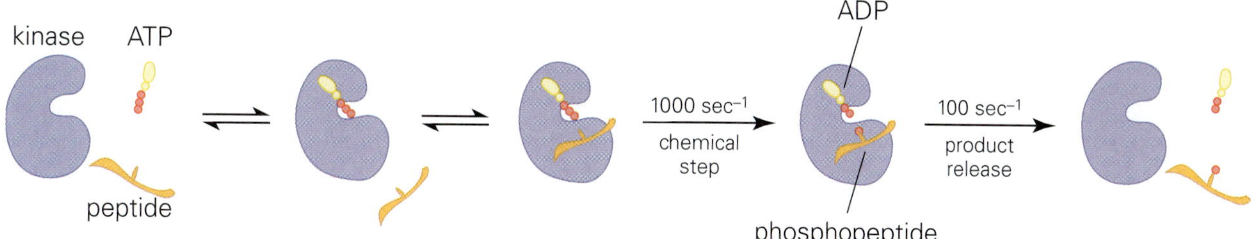


**Figure 16.11 A reaction for which product release is slow.** Shown here is the reaction catalyzed by a kinase enzyme, which transfers a phosphate group from ATP to a peptide substrate. The chemical step is fast, with a rate constant of 1000 sec$^{-1}$. Release of products is 10-fold slower, and this step dominates the overall rate of the reaction. (Adapted from J. Lew, S.S. Taylor, and J.A. Adams, *Biochemistry* 36: 6717–6724, 1997.)

Product dissociation, the step with rate constant $k_3$, is often very rapid (that is, $k_3$ is large relative to $k_{-1}$, $k_2$, and $k_{-2}$) and, in this limit, as one would expect, these equations return to the "simple" Michaelis–Menten form given previously. But, there are other cases where the overall rate of the reaction is controlled by the rate of release of products. One example of such a reaction is shown in **Figure 16.11**, which depicts peptide phosphorylation by a kinase enzyme (see Section 12.12 for a brief discussion of kinases). The chemical step, which is the transfer of a phosphate group from ATP to the peptide, is fast when catalyzed by the enzyme (the rate constant for this step is ~1000 sec$^{-1}$). In Chapter 12 we mentioned that the binding of ATP to the enzyme involves an induced fit and that the nucleotide cannot enter or leave the enzyme without a conformational change in the protein (see Figure 12.23). This makes the release of ADP very slow, and the overall rate of the reaction is much slower than the rate of the catalytic step.

## 16.7 The specificity of enzymes arises from both the rate of the chemical step and the value of $K_M$

For many enzymes, there are several different alternative natural substrates. Proteases, for example, cut the peptide backbone of different target proteins. A particular protease will only work on peptides with certain kinds of residues at the cleavage site and at nearby positions (**Figure 16.12**). Different peptides will have different values of the catalytic efficiency, $k_{cat}/K_M$ (or, equivalently, $k_2/K_M$ if the reaction follows simple Michaelis–Menten kinetics).

In Section 16.4, we explained that if the substrate concentration is low (that is, if [S] ≪ $K_M$), then $k_{cat}/K_M$ is the rate constant for the reaction between enzyme and substrate. That is, the rate of the reaction is directly proportional to $k_{cat}/K_M$

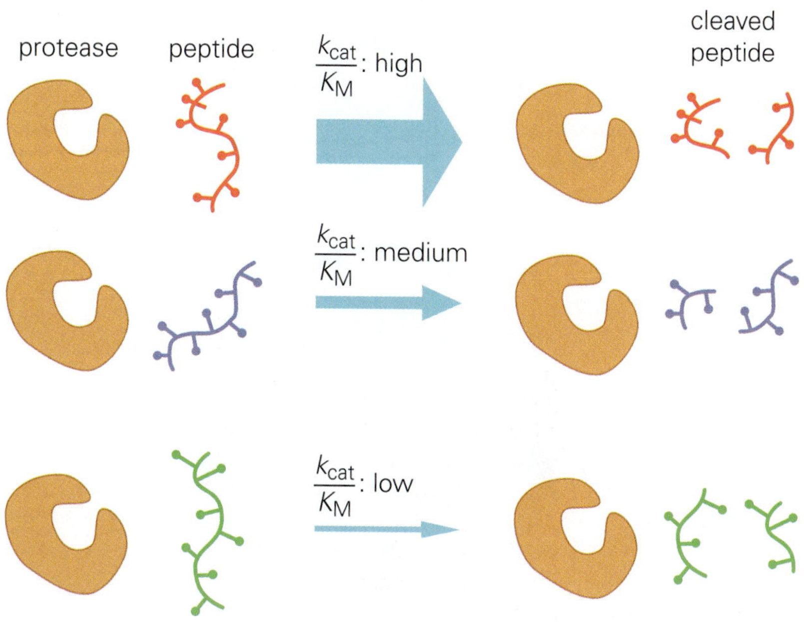

**Figure 16.12 Different substrates are turned over with different rates by the same enzyme.** Shown here is the action of a protease enzyme on three different substrates. At low substrate concentrations, the reaction rate is proportional to the catalytic efficiency, $k_{cat}/K_M$. The substrate with the highest value of $k_{cat}/K_M$ is degraded most rapidly. The ratio $k_{cat}/K_M$ is also called the specificity constant of the enzyme.

**Specificity constant**

The catalytic efficiency of an enzyme is the $k_{cat}/K_M$ ratio. For comparing reactions of different substrates with the same enzyme, the relative values of the catalytic efficiencies for different substrates predict the relative amounts of different products formed. For this reason, $k_{cat}/K_M$ is also referred to as the specificity constant.

(see Equation 16.22). For two different substrates, each present at the same low concentration, the relative amounts of product formed will be determined by the corresponding values of the $k_{cat}/K_M$ ratio and, hence, in this context, this ratio is also referred to as the **specificity constant** for the enzyme.

The rates of enzyme-catalyzed reactions vary tremendously, depending on the chemistry and the role of the enzyme in the cell. We mentioned earlier, in Section 16.5, that superoxide dismutase is essential because it degrades the very reactive superoxide ion. To minimize the levels of the superoxide, this enzyme works extremely fast, with a $k_{cat}/K_M$ value approaching $10^9$ $M^{-1}$•$sec^{-1}$. This rate constant is only a factor of 10 lower than the rate constant for collisions between molecules, without considering orientational effects (~$10^{10}$ $M^{-1}$•$sec^{-1}$; see Section 15.25). As explained above, superoxide dismutase is able to work so fast because the number of unproductive encounters between enzyme and substrate molecules is minimized by electrostatic steering (see Figure 16.10).

On the other hand, there are reactions for which control and specificity are much more important than speed. An example is the HIV protease (discussed further in Section 16.9), which is needed to cleave the viral polyprotein into the active individual components. This process requires precise cleavage at specific sites without significant cutting at other sites. This is accomplished by specific interactions of the substrate protein with binding pockets on the enzyme, which are also exploited for the binding of anti-HIV drugs (see Section 12.14). There are eight sites on the enzyme that interact with the peptide substrate, each accommodating one amino acid sidechain. These sites are designated P4, P3, P2, and P1 before the cleavage site and P1′, P2′, P3′, and P4′ after it.

Some of these sites have modest preference for particular amino acids. For example, the P4 site will accept 11 different residues with $k_{cat}/K_M$ varying from $45 \times 10^6$ $M^{-1}$•$sec^{-1}$ for serine (in V**S**QNY*PIVQ, cleaved at the *) to $0.2 \times 10^6$ $M^{-1}$•$sec^{-1}$ for leucine at the same site (**Figure 16.13**). One might expect the specificity of the protease to arise just from the binding (that is just from differences in $K_M$), but in fact $k_{cat}$ also varies by a factor of 100 among the amino acids for which cleavage occurs at a reasonable rate. The P3 site is more tolerant of variation, with $k_{cat}/K_M$ within a

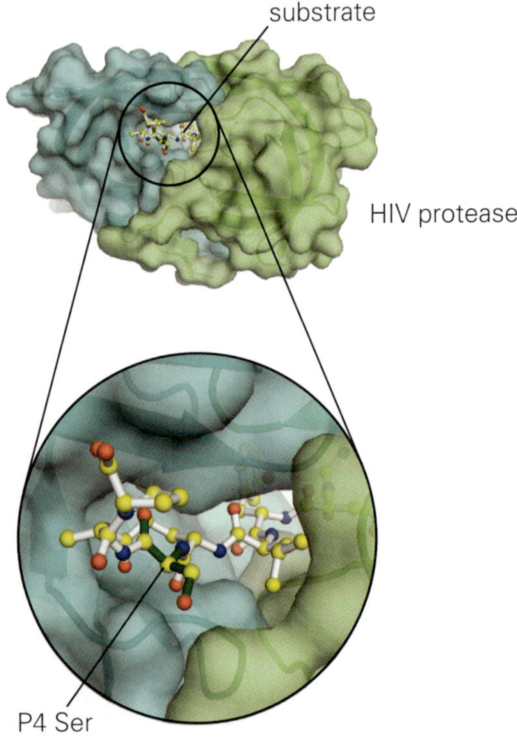


**Figure 16.13 The structure of the HIV protease.** The enzyme is a dimer, and the two subunits are colored differently. A peptide substrate bound to the enzyme is shown in sticks. The expanded view shows a serine residue at the P4 site. The sidechain of the serine is not enclosed in the binding pocket, but if it is replaced by residues that are too large, then steric clashes occur and the value of $k_{cat}/K_M$ is reduced. (PDB code: 3D3T.)

factor of 10 for 17 amino acids. On the other hand, sites P2 and P1 are much more restrictive. P1, for example, strongly prefers tyrosine or phenylalanine, with tryptophan reducing activity almost 10-fold, and leucine and methionine reducing it 20-fold. Sequences with other amino acids at this site are not cleaved. Thus, while cutting is quite efficient at optimally recognized sequences, effects on both $k_{cat}$ and $K_M$ reduce activity at other "incorrect" sites, giving the necessary specificity.

Protein kinases, which we discussed in Section 16.6, are regulatory enzymes that are themselves regulated in activity, in part by interactions with other proteins and in part by phosphorylation. Phosphorylation can be either autophosphorylation (that is, a particular kinase phosphorylating other enzyme molecules of the same kind) or transphosphorylation (that is, being phosphorylated by another different kinase). Uncontrolled (or improperly controlled) kinase signaling can create serious unbalances in processes in cells, leading to serious diseases such as cancer. Because of this, kinases are under tight control with specificity of activation being more important than the speed of the response. As such, many kinases have relatively low catalytic efficiencies, with $k_{cat}/K_M$ values of $10^3$ to $10^4$, a reduction of 100,000-fold from the diffusion-controlled rate.

## 16.8 Graphical analysis of enzyme kinetic data facilitates the estimation of kinetic parameters

The Michaelis-Menten equation provides a basis for graphical analysis of the kinetic data for an enzyme-catalyzed reaction. The observed dependence of $v_0$ on [S] can be fitted directly using a computer and least-squares approach to obtain the values of $V_{max}$ and $K_M$. However, in earlier days (that is, before computers were routinely available for such fitting), the data analysis was usually done by rearranging equations to give a linear form, for which fitting is easier and much of the analysis can be done graphically. Even now, such linear graphs provide a quick check that simple Michaelis-Menten behavior is being followed for the reaction under study.

One type of graphical analysis uses a **Lineweaver–Burk** plot (**Figure 16.14**). Both sides of the Michaelis-Menten equation (Equation 16.17) are inverted, giving:

$$\frac{1}{v_0} = \frac{1}{V_{max}} + \frac{K_M}{V_{max}}\frac{1}{[S]} \tag{16.29}$$

According to Equation 16.29, a graph of $\frac{1}{v_0}$ versus $\frac{1}{[S]}$ gives a line with slope $\frac{K_M}{V_{max}}$, a $y$-intercept of $\frac{1}{V_{max}}$, and an $x$-intercept of $-\frac{1}{K_M}$.

**Lineweaver–Burk**

A Lineweaver–Burk plot is a graph of $1/v_0$ vs. $1/[S]$. If the enzyme obeys Michaelis–Menten kinetics, then this gives a straight line with slope $K_M/V_{max}$ and an intercept of $1/V_{max}$.

A practical advantage of this type of analysis is that one need not measure rates at very high substrate concentrations in order to obtain the value of $V_{max}$. Practical considerations, such as the finite solubility of substrate, can limit what can be done experimentally in terms of measuring the rate of the reaction at very high substrate concentrations.

Another graphical approach is the **Eadie–Hofstee** plot, in which the rearrangement of the Michaelis-Menten equation is slightly different (see Figure 16.14). Multiplying both sides of the Michaelis-Menten equation by $(1 + K_M/[S])$ gives:

$$v_0 = -K_M\frac{v_0}{[S]} + V_{max} \tag{16.30}$$

Thus, a graph of $v_0$ versus $\frac{v_0}{[S]}$ gives a line with slope of $-K_M$ and a $y$-intercept of $V_{max}$.

**Eadie–Hofstee**

An Eadie–Hofstee plot is a graph of $v_0$ vs. $v_0/[S]$, which for Michaelis–Menten kinetics gives a straight line with slope $-K_M$ and an intercept of $V_{max}$.

The various ways of graphing enzyme kinetic data are shown in Figure 16.14. The observation of a straight line in either the Lineweaver-Burk plot or the

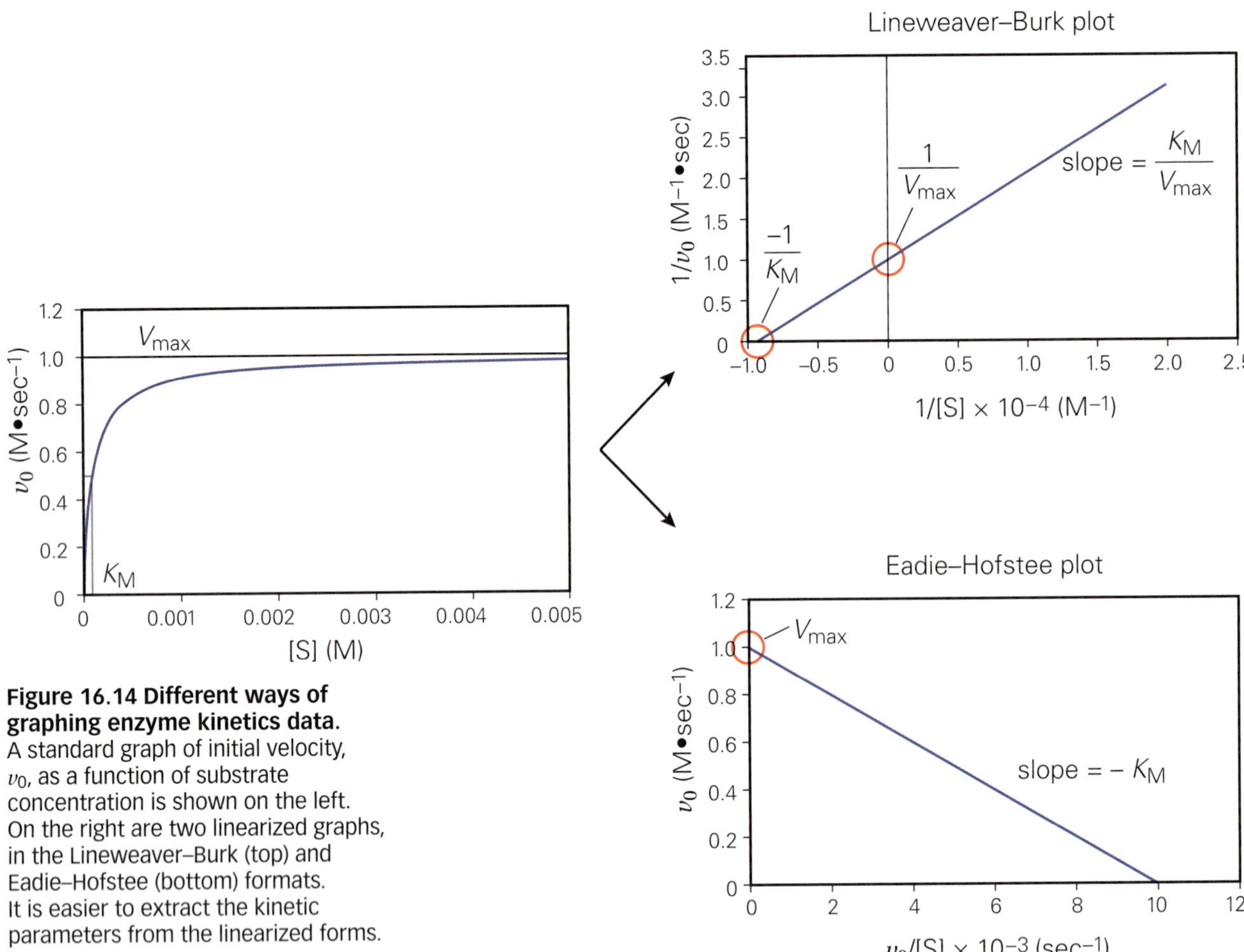


**Figure 16.14 Different ways of graphing enzyme kinetics data.** A standard graph of initial velocity, $v_0$, as a function of substrate concentration is shown on the left. On the right are two linearized graphs, in the Lineweaver–Burk (top) and Eadie–Hofstee (bottom) formats. It is easier to extract the kinetic parameters from the linearized forms.

Eadie-Hofstee plot is verification that "simple" Michaelis-Menten behavior is being followed. If the enzyme is oligomeric, then cooperative substrate-binding behavior is often observed, giving more complicated kinetic behavior and curvature when data are plotted in these linearized forms (discussed further below).

## B. INHIBITORS AND MORE COMPLEX REACTION SCHEMES

### 16.9 Competitive inhibitors block the active site of the enzyme in a reversible way

**Competitive inhibition**

Competitive inhibition occurs when the inhibiting compound can bind reversibly to the active site of an enzyme and block the binding of substrate, but the compound itself does not undergo a reaction.

There are both natural and man-made compounds that can act as inhibitors of enzymes, as discussed in Chapter 12. Such molecules can act to prevent enzyme catalysis in several different ways, and they are classified by how they interact with the enzyme, as shown schematically in **Figure 16.15**. The most common, and perhaps easiest to understand, is **competitive inhibition**. In this case, the inhibitor, I, binds to the enzyme in or near the active site in such a way that substrate cannot bind when the inhibitor is bound—that is, the substrate and inhibitor compete for the active site. The amount of enzyme available to bind to substrate is thereby reduced in the presence of inhibitor and, as one expects intuitively, the rate of product formation is reduced.

In competitive inhibition, the inhibitor does not undergo a chemical reaction on the enzyme due to differences in chemical structure relative to substrate. In some cases, the structure of the inhibitor may be rather similar to that of the substrate and in other cases completely different.

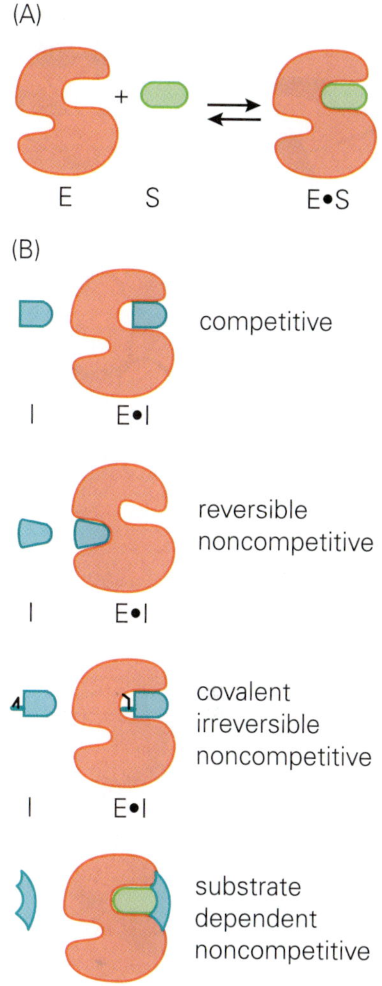


**Figure 16.15 An enzyme can have its activity decreased by binding inhibitors.** A schematic of an enzyme is shown with the active site and substrate at the top. In the lower panels, different binding modes that can lead to inhibition of the enzyme are shown. Many drugs that are important in health care act as inhibitors for specific enzymes in the body. Some examples have been discussed in Chapter 12 and others are presented in part B of this chapter.

Many natural inhibitors and drugs are competitive inhibitors of enzymes, and several examples of such inhibitors were discussed in Chapter 12. One such example is the HIV protease inhibitor, atazanavir, shown in **Figure 16.16.** Another competitive inhibitor of HIV protease, saquinavir, was discussed in Chapter 12 (see Figure 12.19). As you can see from Figure 16.16, substrate and inhibitor bind at the same site, competing for the enzyme. Although their binding sites overlap, the inhibitor and the substrate do not bind in precisely the same way because they do not have the same structure (see also Figure 16.16A and 16.16B). This has unfortunate consequences for therapy, because the HIV protease can acquire mutations that allow it to still bind the substrate and function, but which block the binding of the inhibitor. The positions of two such mutations are shown in Figure 16.16B. These mutations increase the value of the inhibitor constant, $K_I$ (see Section 12.21) for atazanavir such that the inhibitor is no longer effective at the concentrations used in therapy. Such a mutant enzyme is said to be resistant to the drug.

Because the resistance mutations have to preserve the ability of the protease to bind to the substrate, these mutations tend to occur near regions of the drug that do not overlap with atoms of the substrate, as shown in Figure 16.16C. Different drugs that block the HIV protease have differences in their shapes and, because of this, the mutations that confer resistance to these drugs are not the same. This makes it hopeful that a combination of drugs might be effective at evading resistance.

## 16.10 A competitive inhibitor does not affect the maximum velocity of the reaction, $V_{max}$, but it increases the Michaelis constant, $K_M$

The binding of the inhibitor is usually an equilibrium (reversible) process. To quantitatively predict the effect of the inhibitor, I, we use the same basic Michaelis–Menten kinetic model, but add to it a reaction corresponding to binding of inhibitor to the enzyme to form the inhibited complex, E•I:

$$E + S \underset{k_{-1}}{\overset{k_1}{\rightleftarrows}} E{\bullet}S \xrightarrow{k_2} E + P$$

$$E + I \underset{k_{-3}}{\overset{k_3}{\rightleftarrows}} E{\bullet}I \tag{16.31}$$

The dissociation constant for the enzyme–inhibitor complex is denoted $K_I$ (see Section 12.21) and is related to the rate constants and the equilibrium concentrations in the following way:

$$K_I = \frac{k_{-3}}{k_3} = \frac{[E][I]}{[E{\bullet}I]} \tag{16.32}$$

The kinetic analysis proceeds exactly as before, with a steady-state assumption for [E•S]. The expression for the total enzyme concentration, $[E]_0$, now has one extra term, which is the concentration of the enzyme–inhibitor complex:

$$[E]_0 = [E] + [E{\bullet}S]_{ss} + [E{\bullet}I] \tag{16.33}$$

Using Equation 16.6 for $[E{\bullet}S]_{ss}$ and combining with Equation 16.33, we get the following expression for the concentration of the free enzyme:

$$[E] = [E]_0 - [E{\bullet}S] - [E{\bullet}I] = [E]_0 - \frac{k_1[E][S]}{k_{-1}+k_2} - \frac{[E][I]}{K_I} \tag{16.34}$$

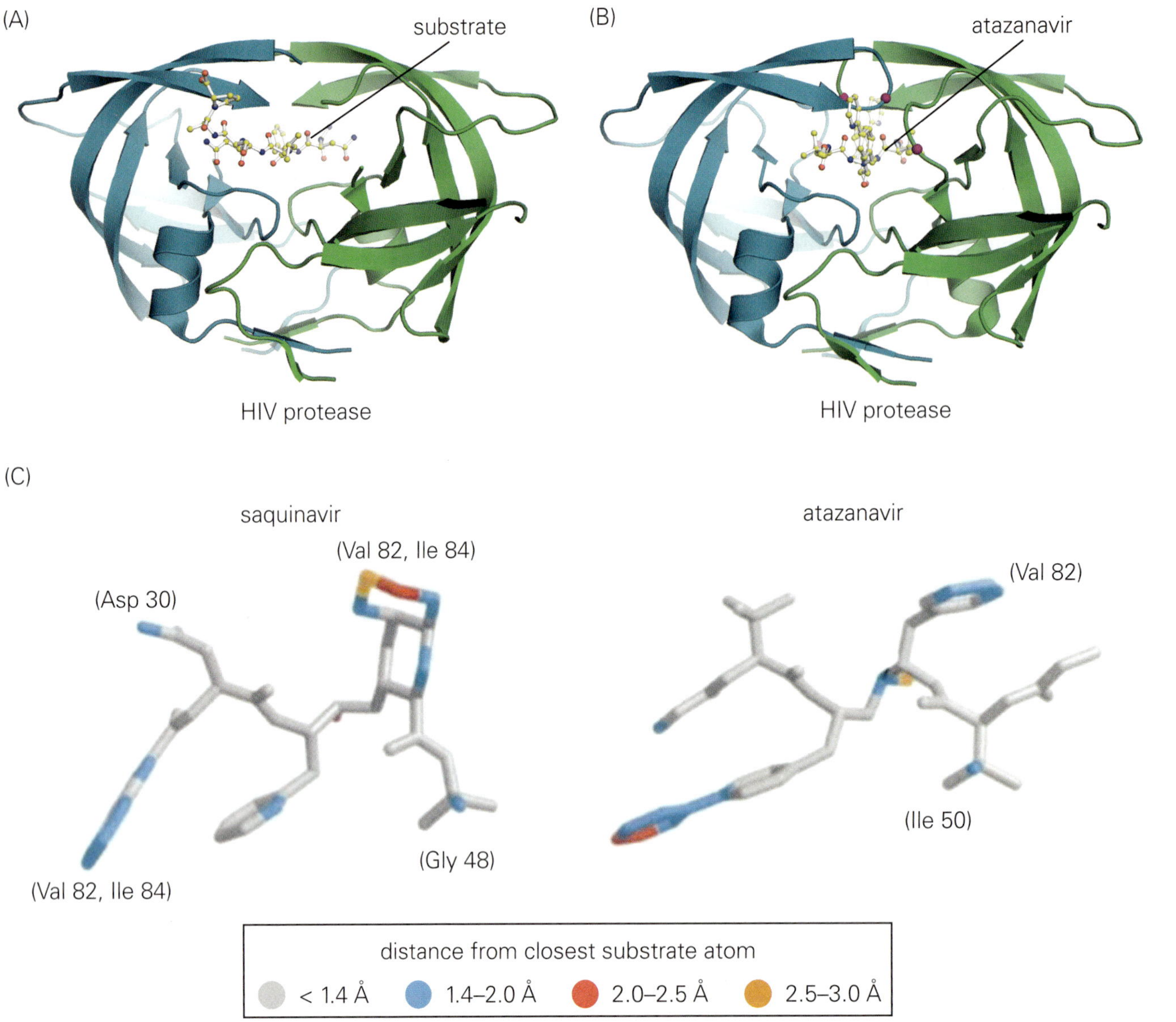

**Figure 16.16 Competitive inhibition of HIV protease.** (A) Structure of the substrate complex. (B) Structure of the complex with a competitive inhibitor, atazanavir, which is used in HIV therapy. The sites of two mutations that cause resistance to atazanavir are shown as *red* spheres (Ile 50 and Val 82). (C) The structures of saquinavir and atazanavir. The interaction of saquinavir with HIV protease is shown in Figure 12.19. The distances of atoms in the drug to the nearest atoms in the substrate [in the structure shown in (A)] are color coded. Some of the residues in the protease that confer resistance to the drug when mutated are indicated in parentheses, near the closest atom in the drug. (C, adapted from N.M. King et al. and C.A. Schiffer, *Chem. Biol.* 11: 1333–1338, 2004; PDB codes: A, 3D3T; B, 2AQU.)

Rearranging this to solve for [E] then gives:

$$[\mathrm{E}] = \frac{[\mathrm{E}]_0}{\left(1 + \frac{[\mathrm{S}]}{K_\mathrm{M}} + \frac{[\mathrm{I}]}{K_\mathrm{I}}\right)} \tag{16.35}$$

The initial velocity of the reaction, $v_0$, is given by:

$$v_0 = \frac{d[\mathrm{P}]}{dt} = \frac{k_1 k_2 [\mathrm{E}][\mathrm{S}]}{k_{-1} + k_2} = \frac{k_2[\mathrm{E}][\mathrm{S}]}{K_\mathrm{M}} \tag{16.36}$$

Substituting the expression for [E] given by Equation 16.35 into Equation 16.36 and rearranging gives:

$$v_0 = \frac{k_2[\mathrm{E}]_0}{\left[1 + \frac{K_\mathrm{M}}{[\mathrm{S}]}\left(1 + \frac{[\mathrm{I}]}{K_\mathrm{I}}\right)\right]} = \frac{V_{\max}}{\left(1 + \frac{K_\mathrm{M}^*}{[\mathrm{S}]}\right)} \tag{16.37}$$

where

$$K_\mathrm{M}^* = K_\mathrm{M}\left(1 + \frac{[\mathrm{I}]}{K_\mathrm{I}}\right)$$

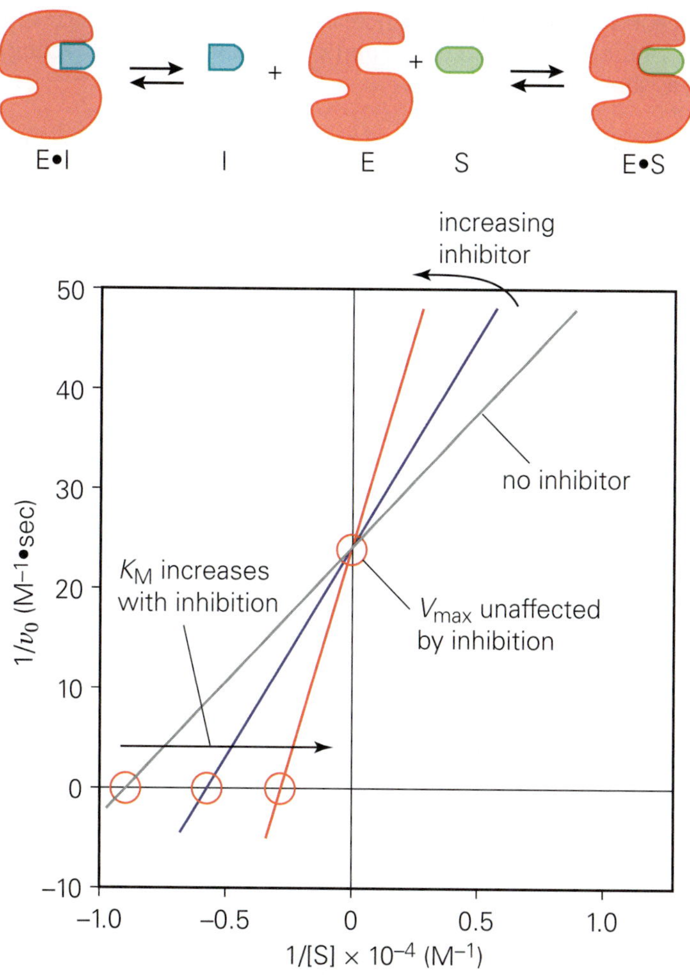


**Figure 16.17 A Lineweaver–Burk plot for an enzyme reaction in the presence of a competitive inhibitor.** As shown in the schematic diagram above the graph, the inhibitor and the substrate compete for the same site on the enzyme. As the substrate concentration increases, the inhibitor is displaced, and so the value of $v_{max}$ is unchanged by the inhibitor. Increasing concentrations of inhibitor increase the apparent value of $K_M$, and so the slope of the line increases.

This describes the kinetic behavior in the presence of a competitive inhibitor, as illustrated in **Figure 16.17**. At high substrate concentrations the rate of the reaction reaches a plateau value given by $V_{max}$, and this maximal rate is the same as for the uninhibited reaction. What has changed is the substrate concentration required for half-maximal rate, which is given by the apparent Michaelis constant, $K_M^*$. According to Equation 16.37, the value of $K_M^*$ is greater than the value of $K_M$, and so more substrate is required to get to the half-maximal rate.

As must be the case, if [I] = 0 or $K_I$ is large (that is, if inhibitor binds very weakly), then the enzyme will not have significant inhibitor bound and the equation reduces to exactly the original Michaelis–Menten behavior described above. If inhibitor is present and binds, then Michaelis–Menten behavior still applies, but with a modified $K_M$ value (denoted $K_M^*$). If $[I] \gg K_I$, then most of the enzyme is occupied by inhibitor. $K_M^*$ is then large and only at very high substrate concentration will the rate of the reaction go up to the uninhibited $V_{max}$ value. In the race between substrate and inhibitor to get to an open active site, a large excess of substrate can out-compete the inhibitor, even though the inhibitor may bind more tightly.

In a Lineweaver–Burk analysis for a competitively inhibited enzyme, a straight line will be observed for any given inhibitor concentration:

$$\frac{1}{v_0} = \frac{1}{V_{max}} + \frac{K_M^*}{V_{max}} \frac{1}{[S]} \tag{16.38}$$

where

$$K_M^* = K_M\left(1 + \frac{[I]}{K_I}\right)$$

As the concentration of inhibitor is increased, the *y*-intercept (corresponding to $V_{max}$) remains the same, but the slope and *x*-intercept change, reflecting different $K_M^*$ values, as shown in Figure 16.17.

## 16.11 Reversible noncompetitive inhibitors decrease the maximum velocity, $V_{max}$, without affecting the Michaelis constant, $K_M$

**Noncompetitive inhibitor**

A reversible noncompetitive inhibitor is a compound whose effects occur through binding to an enzyme somewhere other than the active site. Thus, a noncompetitive inhibitor prevents the chemical reaction from occurring, but does *not* do so by competing with substrate for the active site. Reversible noncompetitive inhibitors are allosteric inhibitors (see the definition of allostery in Section 14.3).

Another type of inhibition involves inhibitors that bind to the enzyme away from the active site, but somehow sabotage the enzyme mechanism. In the simplest case, the presence of substrate does not affect the binding of the inhibitor, and increasing the substrate concentration does not overcome the effect of the inhibitor. This kind of inhibition is called noncompetitive.

As an example, consider an enzyme that needs to close down around a substrate in order to position key amino acids in the active site for catalysis. The inhibitor could then be a compound that binds and prevents closure, without actually blocking the active site, as shown schematically in **Figure 16.18A**. An example of such an inhibitor is shown in Figure 16.18B. This is a compound that inhibits a protein kinase. The substrate may still bind to the active site (in this case, ATP is bound), but it cannot undergo the reaction to make product.

In the case of noncompetitive inhibition, to write a kinetic scheme, we must consider two classes of enzyme molecules, those that do not have inhibitor bound and hence are active, and those that do have inhibitor bound and are "dead." Those molecules with nothing bound are perfectly normal enzyme molecules that

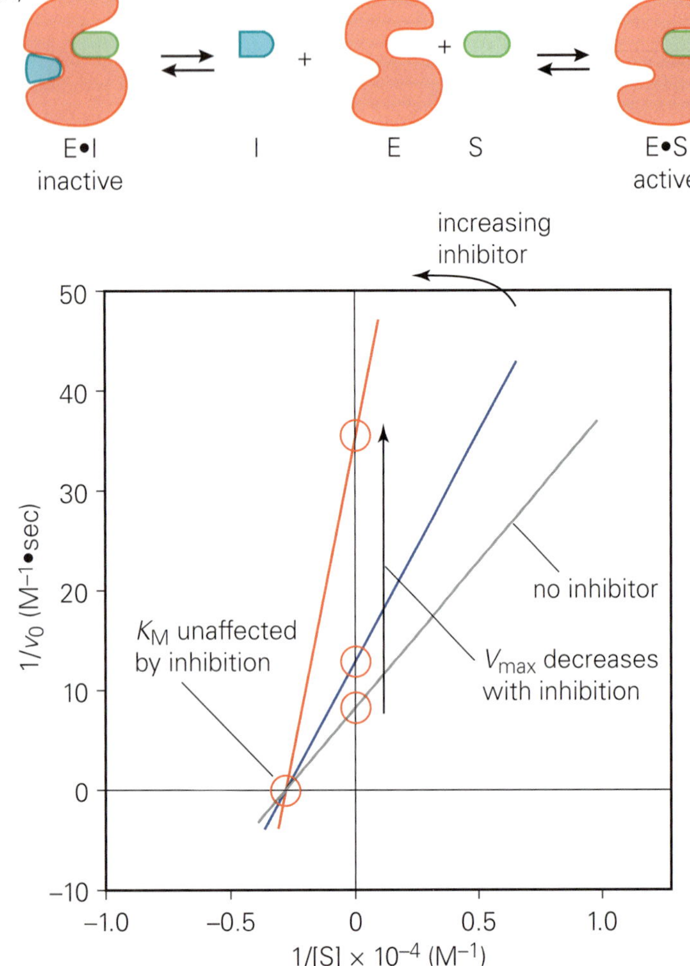


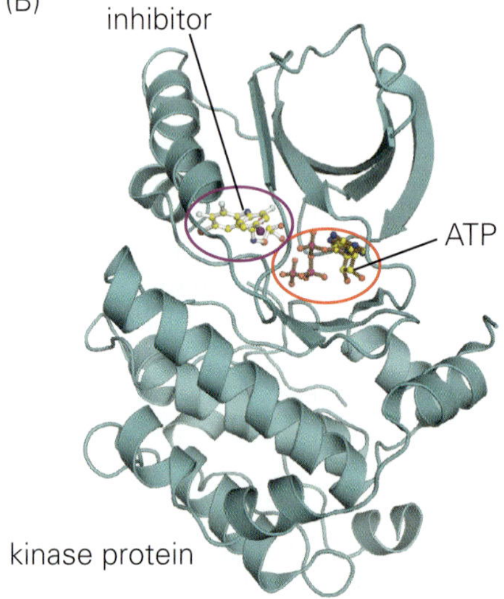


**Figure 16.18 Noncompetitive inhibition.** (A) A Lineweaver–Burk plot is shown for an enzyme reaction in the presence of a reversible noncompetitive inhibitor. As shown in the schematic diagram at the top, the inhibitor binds at a site on the enzyme other than the active site. Increasing concentrations of inhibitor reduce the value of $V_{max}$, but leave the value of $K_M$ unchanged. (B) An example of a reversible noncompetitive inhibitor bound to a protein kinase. The inhibitor binds near the ATP substrate, which is visible to the right of the inhibitor. The binding of the inhibitor and the substrate need not be mutually exclusive, but when the inhibitor is bound, the enzyme is inactive. Although the inhibitor does not compete with ATP or peptide substrates (not shown here) for their binding sites, it holds the kinase in an inactive conformation. (PDB code: 1S9J.)

have values of $K_M$ and $k_2$ that are the same as if no inhibitor was present. The effective concentration of enzyme ($[E]_0$), however, is reduced to $(1 - f_I)[E]_0$, where $f_I$ is the fraction of molecules that have inhibitor bound. The value of $f_I$ is calculated as a normal ligand binding equilibrium, as discussed in Section 12.2.

Noncompetitively inhibited enzymes will obey Michaelis–Menten behavior and will have the same $K_M$ value as the uninhibited solution, but they will have a modified value of $V_{max}$:

$$V^*_{max} = k_2(1 - f_I)[E]_0 \quad (16.39)$$

As more inhibitor is added, the value of $V_{max}$ will decrease, but $K_M$ will remain unchanged. In this case, linear Lineweaver–Burk plots are again observed:

$$\frac{1}{v_0} = \frac{1}{V^*_{max}} + \frac{K_M}{V^*_{max}} \frac{1}{[S]} \quad (16.40)$$

The slopes and $y$-intercepts of the lines change with the amount of inhibitor, while the $x$-intercept remains fixed (Figure 16.18). The difference between this behavior and what is observed for competitive inhibition makes them easy to distinguish just by looking at the qualitative behavior of the plots for reaction rates with different amounts of inhibitor present, as you can see by comparing Figures 16.17 and 16.18.

## 16.12 Substrate-dependent noncompetitive inhibitors only bind to the enzyme when the substrate is present

Within the class of noncompetitive inhibitors are compounds that bind only to the enzyme–substrate complex, E•S, but when bound prevent formation of product, as shown in **Figure 16.19**. Such compounds are relatively rare, but do occur. A specific example is shown in Figure 16.19B. This involves the enzyme catechol *O*-methyl transferase, to which the substrate must bind before the inhibitor can bind.

The kinetic behavior for such inhibitors is different from that seen for competitive inhibition because the equilibrium between E and E•S depends on [S], and thus binding of the inhibitor also depends on [S]. This behavior is called **substrate-dependent noncompetitive inhibition**. These inhibitors are sometimes called "uncompetitive" inhibitors, but this term can be confusing and we shall not use it.

To predict the behavior in this case, the binding of the inhibitor is described as an equilibrium analogous to the competitive case, but involving binding to E•S rather than E (hence the substrate dependence):

$$\text{E•S} + \text{I} \underset{k_{-3}}{\overset{k_3}{\rightleftharpoons}} \text{E•S•I} \quad (16.41)$$

**Substrate-dependent noncompetitive inhibitor**

A substrate-dependent noncompetitive inhibitor is a compound whose effects occur through binding to an enzyme–substrate complex, preventing completion of the enzyme reaction and release of the product.

To calculate the kinetic behavior, the same principle is used as was applied to the competitive inhibitor case. The concentration of free enzyme, [E], is calculated by keeping track of all forms of the enzyme: $[E]_0 = [E] + [ES] + [E•S•I]$. As the concentration of substrate increases, the amount of enzyme–substrate complex increases, which binds the inhibitor. Hence, even at high substrate concentrations the rate of the reaction is affected by changes in [S]. Going through a calculation such as that which led to Equation 16.37 gives the following equation for the velocity of reaction in the presence of substrate:

$$v_0 = \frac{V^*_{max}}{1 + \frac{K^*_M}{[S]}} \quad (16.42)$$

where

$$V^*_{max} = \frac{V_{max}}{1 + \frac{[I]}{K_I}}$$

and

$$K_M^* = \frac{K_M}{1+\frac{[I]}{K_I}}$$

Putting this in the Lineweaver–Burk format gives:

$$\frac{1}{v_0} = \frac{1}{V_{max}^*} + \frac{K_M^*}{V_{max}^*}\frac{1}{[S]} = \frac{1}{V_{max}}\left(1+\frac{[I]}{K_I}\right) + \frac{K_M}{V_{max}}\frac{1}{[S]} \tag{16.43}$$

From Equation 16.43 it is easy to see that the Lineweaver–Burk plot will still give a straight line. In this case, the slope, $\frac{K_M}{V_{max}}$, will be the same for different inhibitor concentrations because it only depends on the properties of the uninhibited enzyme. But, as the inhibitor concentration increases, both the *x*- and the *y*-intercepts will change, as shown in Figure 16.19.

Although it may not be possible to identify how an inhibitor will work from its structure alone, the descriptions above show that a systematic study of the kinetics of the enzyme-catalyzed reaction in the presence of variable amounts of inhibitor can be used to determine the mechanism of action. A comparison of the Lineweaver–Burk plots that derive from the different inhibition mechanisms depicted in **Figure 16.20** shows that the kinetic behavior for each case is distinct and is easily recognized by comparing the slopes and intercepts of the resulting lines.

## 16.13 Some noncompetitive inhibitors are linked irreversibly to the enzyme

In Section 16.11 we saw that reversible noncompetitive inhibitors block the activty of the enzyme in a way that cannot be overcome by the presence of a large amount

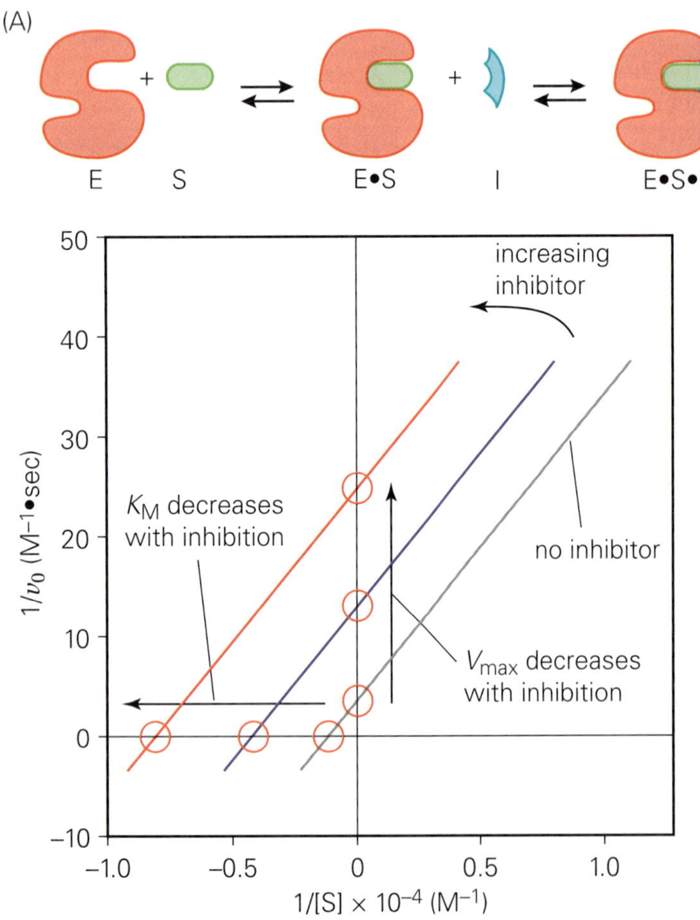


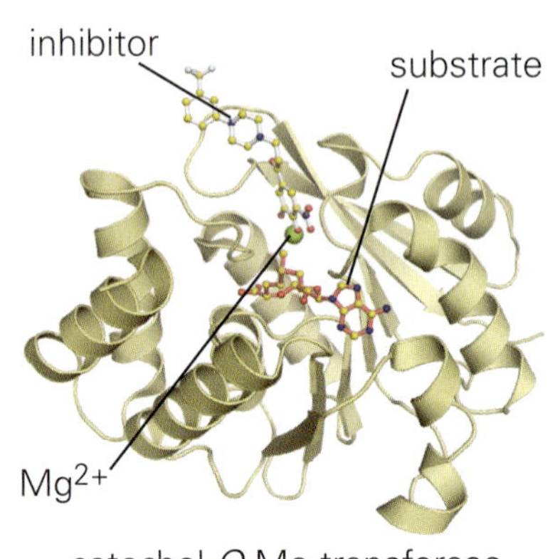


**Figure 16.19 Substrate-dependent noncompetitive inhibition.** (A) A Lineweaver–Burk plot for an enzyme reaction in the presence of a substrate-dependent noncompetitive inhibitor. A schematic representation of the action of the inhibitor is shown above the graph. (B) An example of a substrate-dependent noncompetitive reaction, involving the enzyme catechol *O*-Me transferase. A substrate molecule, *S*-adenosylmethionine, must bind to create the site at which the inhibitor (upper left) then binds. (PDB code: 1H1D.)

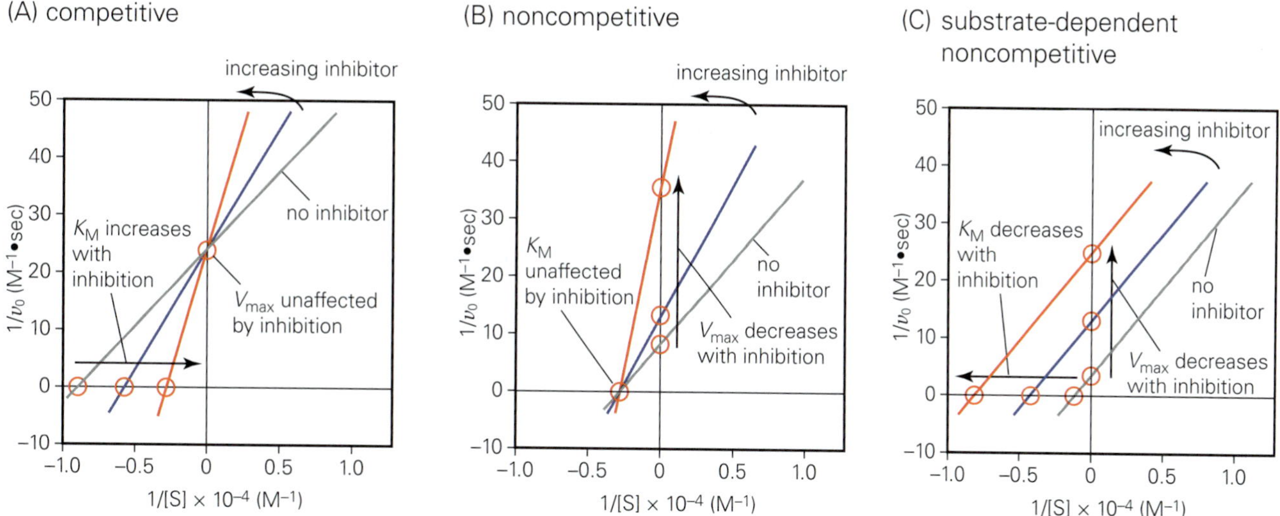


**Figure 16.20 Lineweaver–Burk plots are compared for the three inhibition mechanisms discussed.** (A) Competitive inhibition. (B) Noncompetitive inhibition. (C) Substrate-dependent noncompetitive inhibition. The distinct changes with added inhibitor allow the mechanisms to be easily distinguished.

of substrate. A second class of noncompetitive inhibitors are those that bind at the active site and react covalently with an amino acid in the active site. Such inhibitors are called **irreversible inhibitors** because they cannot dissociate and hence their effect is to block substrate binding permanently.

**Irreversible inhibitors**

A molecule that permanently damages the ability of the enzyme to catalyze reactions is called an irreversible inhibitor. Such inhibitors react covalently with the enzyme.

The structure of thymidylate synthase (which makes thymidine from uridine by the transfer of a methyl group) is shown in **Figure 16.21**, with a covalently attached inhibitor bound to the active site. The inhibitor, 5-fluoro-uridine monophosphate (5-F-UMP) is an analog of the substrate of the enzyme, uridine, and the inhibitor works by taking advantage of the catalytic mechanism of the enzyme. The inhibitor lets the first steps of the reaction occur, which is the attack by the sidechain of a cysteine residue from the enzyme at the C6 position of the uracil ring. But, the reaction is prevented from proceeding to completion because a proton in uracil, which normally needs to be abstracted, is replaced in the inhibitor by a fluorine. The resulting adduct of the enzyme with substrate analog 5-F-uridine and methylene tetrahydrofolate bound can no longer react and be released (Figure 16.19B).

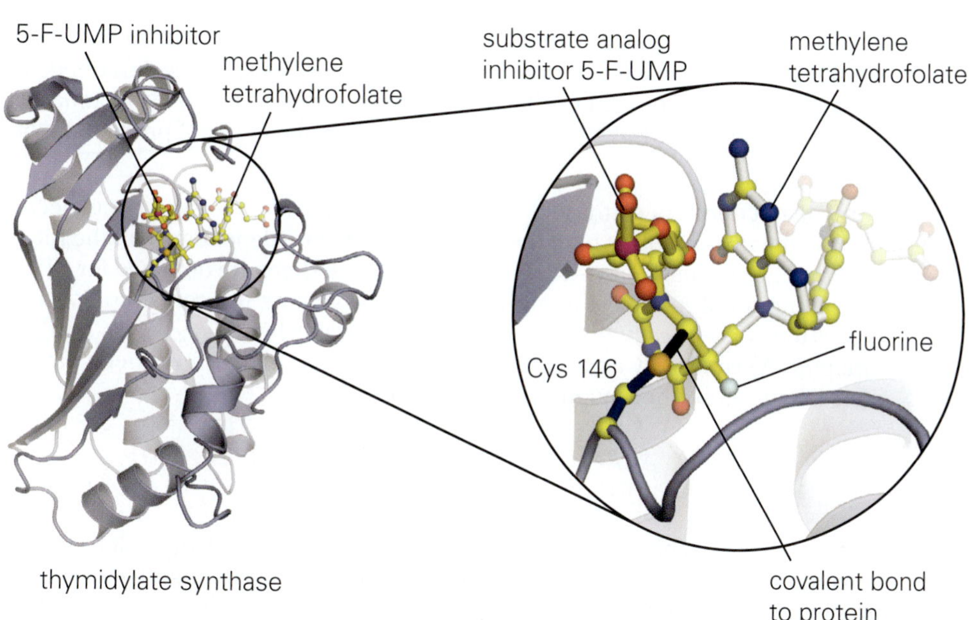


**Figure 16.21 The structure of thymidylate synthase (which makes thymidine from uridine by transfer of a methyl group) is shown with an irreversible (covalently attached) inhibitor bound.** This inhibitor is "mechanism-based," letting the first steps of the reaction occur (that is, attack of a Cys from the enzyme at the C6 position of the uracil ring), but then preventing completion of the reaction because a proton that needs to be abstracted is replaced by a fluorine. The resulting adduct of the enzyme with substrate analog 5-F-uridine and methylene tetrahydrofolate bound can no longer be released, and the enzyme is blocked permanently. (PDB code: 1TSN.)

Inhibitors such as 5-F-UMP are known as **suicide substrates** because they trigger the catalytic cycle just as a substrate does, but then they are trapped in a dead-end complex on the enzyme.

## 16.14 In a ping-pong mechanism the enzyme becomes modified temporarily during the reaction

The Michaelis–Menten equation described in the first part of this chapter is applicable when there is one substrate on which the enzyme acts. However, many important biological reactions have two (or more) reactants. For example, enzymes involved in the synthesis of biological polymers of any kind (such as DNA polymerase, discussed in detail in Chapter 19) bind the growing polymer and also the monomer that is to be added. Such reactions can take place in different ways. All of the basic ideas that have been developed so far apply also to the kinetic schemes describing such reactions, but because of multiple steps, the overall equations may become considerably more complicated.

One fairly common reaction scheme involves the transfer of a chemical group (for example, a phosphate) from one molecule to another, using the enzyme as an intermediate carrier of the group that is transferred. An example of this behavior is found in bacterial histidine kinases. These histidine kinases are unrelated to the protein kinases discussed earlier in this chapter (which phosphorylate serine, threonine, or tyrosine residues). The role of histidine kinases in bacterial chemotaxis was described briefly in Section 14.4.

For the histidine kinases, one substrate, denoted A, is ATP, from which the terminal phosphate group is transferred to a histidine residue on the enzyme and ADP is released. The kinase binds subsequently to a target protein, B (containing a receiver domain), and transfers the phosphate to an aspartate residue in the receiver domain, restoring the kinase to its original state. The phosphorylation of the receiver domain causes a change in conformation that affects its interaction with a target protein. Structures of the kinase CheA and receiver CheY involved in chemotactic signaling are shown in **Figure 16.22**.

The phosphate transfer from substrate A (ATP) to substrate B (the receiver domain, CheY) occurs in two distinct steps, with the overall process being referred to as a **"ping-pong" mechanism**. Schematically, this corresponds to:

$$\mathrm{E} + \mathrm{A} \underset{k_{-1}}{\overset{k_1}{\rightleftarrows}} \mathrm{E{\bullet}A} \xrightarrow{k_2} \mathrm{E{-}P} + \mathrm{Q}$$

$$\mathrm{E{-}P} + \mathrm{B} \underset{k_{-3}}{\overset{k_3}{\rightleftarrows}} \mathrm{E{-}P{\bullet}B} \xrightarrow{k_4} \mathrm{E} + \mathrm{P{-}B} \tag{16.44}$$

in which P is the fragment of substrate A that is transferred to B. Each of these steps is a basic enzyme reaction such as those that we have already described, but the two are coupled by the fact that the enzyme is the same and only the modified enzyme can react with the second substrate B to make the final product, **Figure 16.23**.

**"Ping-pong" mechanism**

A "ping-pong" mechanism is one in which there are two distinct steps required, the first of which produces a modified enzyme.

We shall not derive or go through the kinetic equations for the ping-pong mechanism in detail. We simply note that when the substrate dependence of the rate is analyzed in the Lineweaver–Burk approach, straight lines are obtained for a fixed concentration of A while varying substrate B. Changing the concentration of A moves the line vertically. This results in a set of parallel lines, $1/v_0$ versus $1/[\mathrm{B}]$, for different [A] values. By analyzing these data, one can obtain the kinetic parameters that describe the reaction.

## 16.15 For a reaction with multiple substrates, the order of binding can be random or sequential

With enzymes that bind two substrates, there is also the issue of the order in which substrates bind. For some enzymes the order of binding is random—that is, the binding of one substrate does not significantly affect the binding of the other.

In such a **random order**, sequential binding model, the kinetic equations must include two "branches":

$$\begin{array}{ccccc} & \overset{+A}{\nearrow\!\!\!\swarrow} & E \bullet A & \overset{+B}{\searrow\!\!\!\nwarrow} & \\ E & & & & E \bullet A \bullet B \rightleftharpoons E + A - B \\ & \underset{+B}{\searrow\!\!\!\nwarrow} & E \bullet B & \underset{+A}{\nearrow\!\!\!\swarrow} & \end{array} \tag{16.45}$$

The equation for the rate of product formation is complicated, but the ideas behind it are essentially the same as have been discussed. The general form of the rate equation is analogous to that for the Michaelis–Menten case, but the rate depends on both substrate concentrations.

If one substrate is present at high concentration (relative to its $K_M$ value), then the enzyme is essentially always bound to the high concentration substrate, and this complex can be considered as the enzymatic unit. We will use this approach in Chapter 19 when we analyze the DNA polymerase mechanism, which is shown

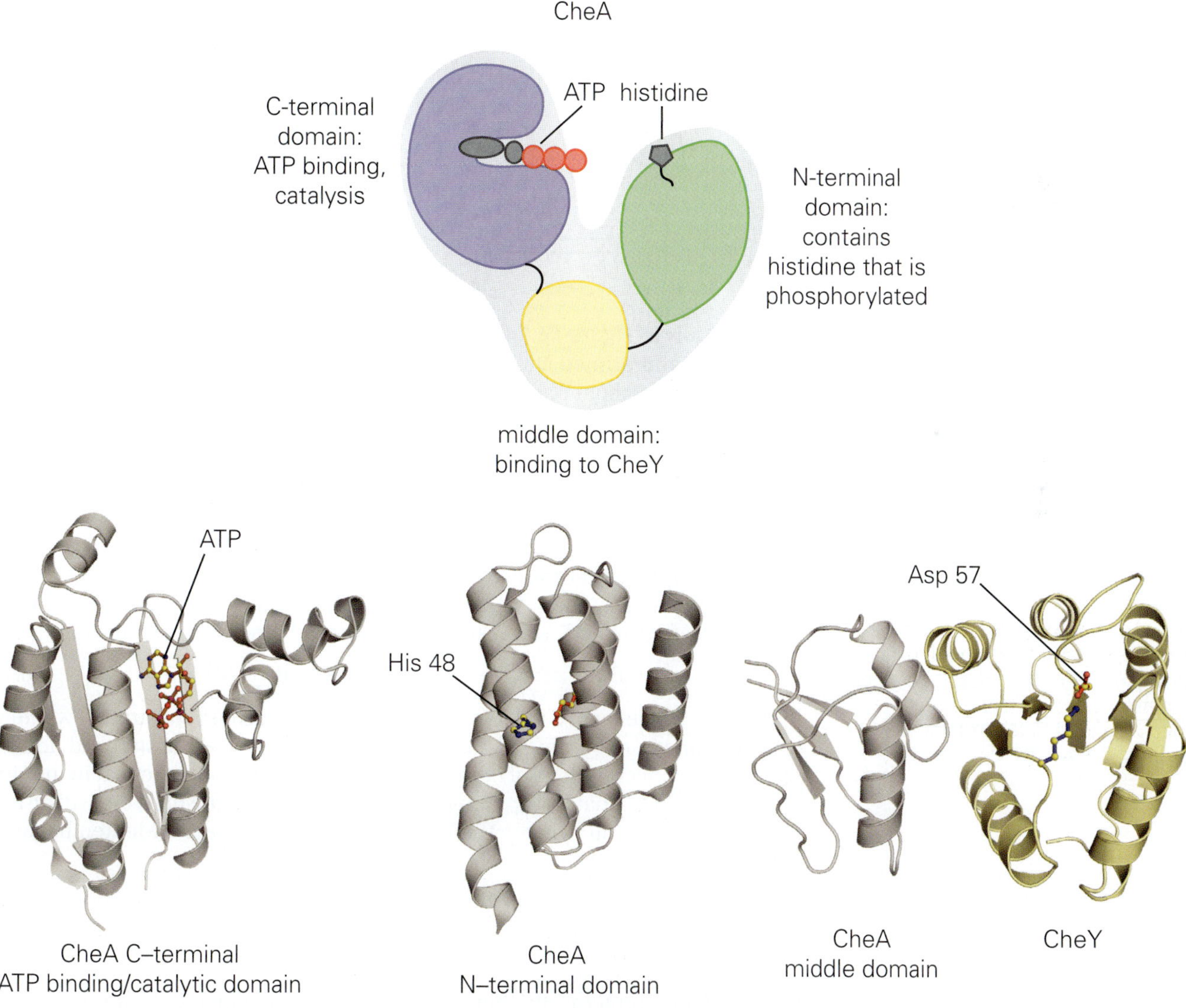


**Figure 16.22 The structure of CheA, a histidine kinase involved in bacterial chemotaxis.** A schematic representation of intact CheA is shown at the top. The structures of three domains of CheA are shown below. At the left is the C-terminal ATP binding/catalytic domain (with ATP shown) that transfers the terminal phosphate from ATP to His 48 (sidechain shown, with an important glutamate residue that is nearby) in the N-terminal domain (center panel). The target-binding domain of CheA (*gray*) in complex with the target (CheY; *light yellow*) is shown at the right. CheY gets phosphorylated on Asp57 (sidechain shown, together with a catalytically important lysine), and phospho-CheY binds to the flagellar motor to change its direction of rotation. (PDB codes: 1EAY, 1I5N, and 1I59.)

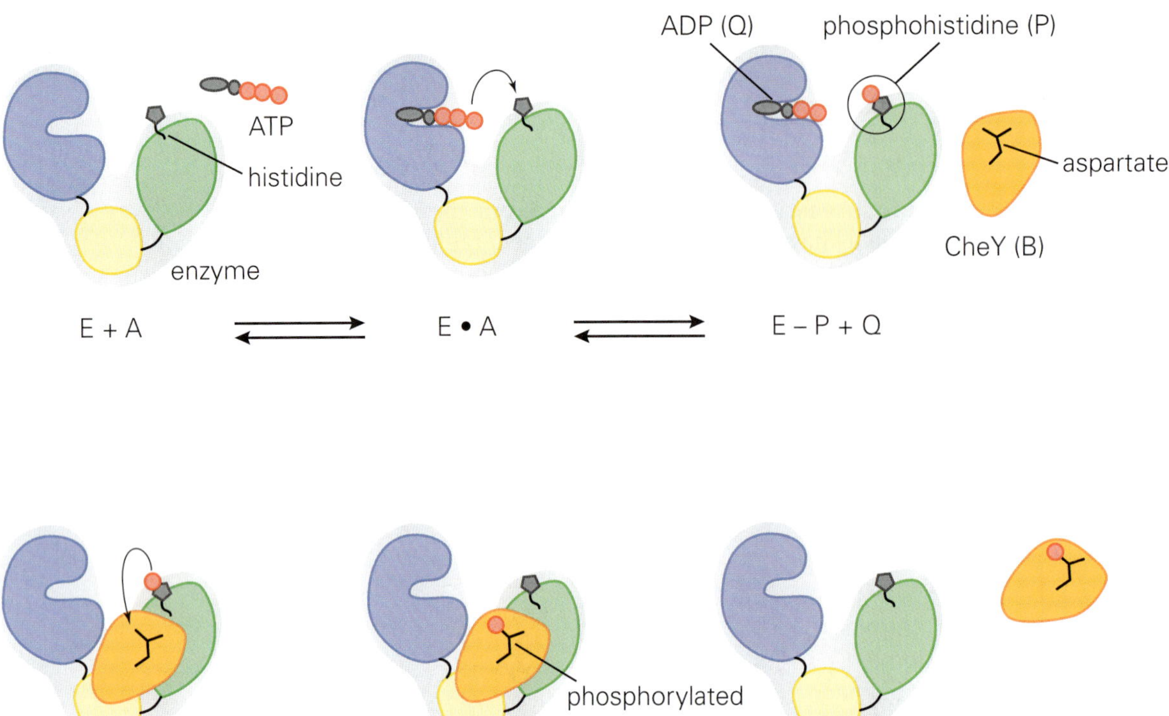


**Figure 16.23 A ping-pong mechanism.** The reaction on CheA with ATP and CheY is an example of an enzyme reaction that proceeds through a ping-pong mechanism. The CheA protein is the enzyme (E), and it first interacts with one substrate, ATP (designated A). The terminal phosphate group of ATP is transferred to a histidine residue on the N-terminal domain of CheA, forming CheA-P (E–P) and ADP (Q). The second substrate of the enzyme is the CheY protein (B), which binds to the middle domain of CheA. The phosphate group is transferred to an aspartate sidechain in the CheY protein, resulting in E + P–B.

schematically in **Figure 16.24**. The polymerase has two substrates, nucleotide triphosphates and primed DNA. Either substrate can bind first to the enzyme, and so the reaction scheme has two branches. Experimental analysis of the polymerization reaction is simplified by using high concentrations of nucleotide triphosphates. If this is done, then the empty enzyme is rarely present, and only one branch of the reaction needs to be considered (shown boxed in Figure 16.24).

For other enzymes, the binding of two substrates may be strongly coupled—that is, the binding of the first substrate can greatly enhance the binding of a second. This can happen either through direct interactions of the substrate or through a conformational change in the enzyme that is induced by the first substrate binding (an allosteric effect). This results in an **ordered binding** mechanism:

$$\mathrm{E} + \mathrm{A} \underset{k_{-1}}{\overset{k_1}{\rightleftharpoons}} \mathrm{E{\bullet}A} + \mathrm{B} \underset{k_{-2}}{\overset{k_2}{\rightleftharpoons}} \mathrm{E{\bullet}A{\bullet}B} \xrightarrow{k_3} \mathrm{E} + \mathrm{P} \qquad (16.46)$$

In this case, the effective $K_M$ for the enzyme with respect to substrate B will have contributions from binding to both A and B, while the catalytic step and limiting rate at high concentrations of both substrates will be equivalent to that in the simpler mechanism—that is, $V_{max} = k_3[\mathrm{E}]_0$. If [A] is high, then the first step will be fast, B binding to E•A will become the limiting step, and Michaelis–Menten behavior will again be followed.

## 16.16 Enzymes with multiple binding sites can display allosteric (cooperative) behavior

For oligomeric proteins, the binding of ligand to the individual subunits can be cooperative—that is, the binding of one ligand affects the conformation of the oligomer—thereby affecting the affinity of subsequent ligands that bind. A good example of positive cooperativity in binding is provided by oxygen binding to hemoglobin, which was discussed in Chapter 14.

The binding of substrate to an enzyme, before the chemical transformation occurs, is a process that is equivalent to ligand binding to a receptor. And so, just

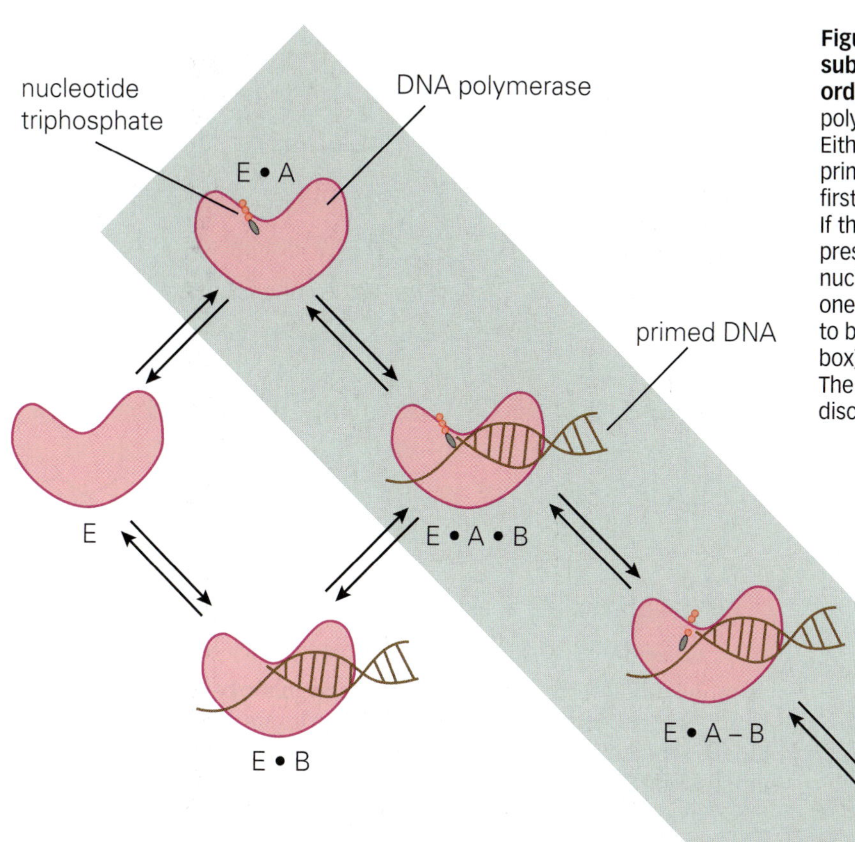


**Figure 16.24 A reaction with two substrates that bind in random order**. The reaction catalyzed by DNA polymerase is shown schematically. Either nucleotide triphosphate or primed DNA can bind to the enzyme first, followed by the other substrate. If the reaction is carried out in the presence of high concentrations of nucleotide triphosphates, then only one branch of the reaction needs to be considered (shown within the box), which simplifies the analysis. The kinetic analysis of this reaction is discussed in Chapter 19.

as ligand binding to a multimeric receptor can be cooperative, substrate binding can be cooperative if the enzyme has more than one binding site. Since the amount of substrate bound affects the rate of appearance of product, cooperativity in substrate binding is reflected in the rate versus substrate concentration profile of the enzyme, which is nonhyperbolic for an **allosteric enzyme**. Nature exploits this behavior to regulate the activity of the enzyme as conditions in the cell change (for example, the concentrations of metabolites). The activity of an enzyme might increase, for example, as its substrate becomes more abundant in the cell. Enzymes can also be affected allosterically by the products of the reaction, as well as by molecules that are structurally unrelated to either substrate or product.

Since cooperativity in binding has already been discussed extensively in Chapter 14, here we will just present one example that demonstrates several important features of allosteric enzymes. The enzyme we will consider is aspartate transcarbamylase (ATCase) from *E. coli*, which carries out a reaction early in the pathway for pyrimidine biosynthesis. The enzyme catalyzes the transfer of a carbamyl group from carbamylphosphate to aspartate, giving carbamylaspartate and phosphate, as shown in **Figure 16.25**. The product is subsequently cyclized and elaborated to generate pyrimidine nucleotides that are used in the synthesis of nucleic acids. To keep a proper balance between pyrimidine and purine bases, the activity of ATCase is increased by ATP but decreased by CTP (cytidine triphosphate). When the ratio of the concentrations of ATP and CTP moves from its "set point," the activity of ATCase is altered in the direction to reestablish it.

*E. coli* ATCase contains two types of protein subunits, one that is responsible for catalysis and the other for regulation (**Figure 16.26**). The ATCase assembly contains six regulatory subunits and six catalytic subunits—a multisubunit enzyme such as ATCase is known as a **holoenzyme** when fully assembled. The catalytic

**Allosteric enzyme**

An allosteric enzyme is one in which the activity of a catalytic center is altered by the binding of molecules at sites other than this catalytic center. Multimeric allosteric enzymes can exhibit cooperativity—that is, the binding of substrate to one active site increases the activity of another active site in the assembly.

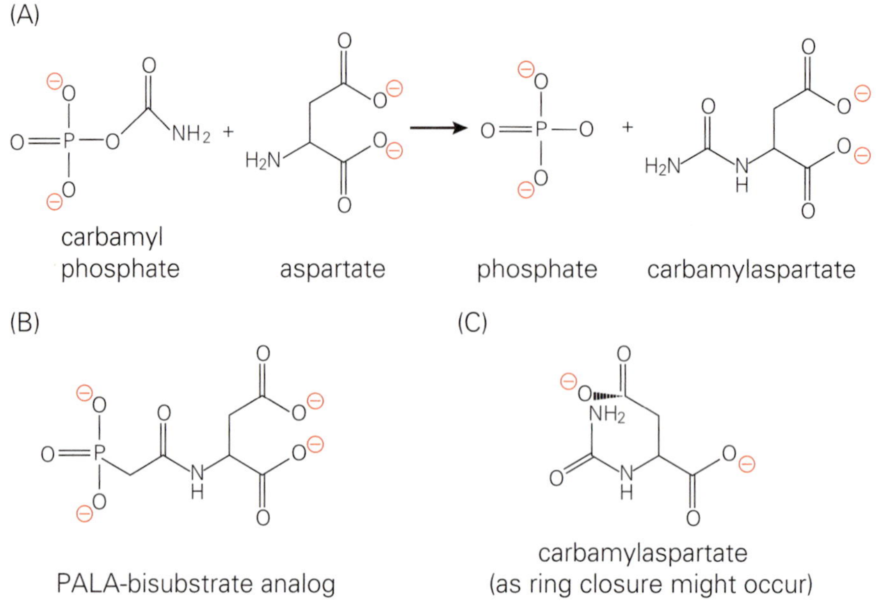

**Figure 16.25 The reaction catalyzed by aspartate transcarbamylase (ATCase).** (A) The chemistry of the aspartate transcarbamylase reaction is shown, with substrates carbamylphosphate and aspartate being converted to phosphate and carbamylaspartate. (B) The bi-substrate analog (mimicking both carbamylphosphate and aspartate) PALA is a good inhibitor of ATCase. (C) Carbamylaspartate is shown in a conformation that indicates how it will be cyclized to give the heterocyclic ring of the pyrimidine bases.

(A)

carbamyl phosphate + aspartate → phosphate + carbamylaspartate

(B)

PALA-bisubstrate analog

(C)

carbamylaspartate (as ring closure might occur)

subunits alone form trimers, while the regulatory subunits form dimers. In the intact holoenzyme, two of the catalytic trimers are bridged by three regulatory dimers. The catalytic sites are formed by the interfaces between the catalytic subunits, while the nucleotide binding sites are in the regulatory subunits, far from the active sites.

When the velocity of the reaction catalyzed by ATCase is measured as a function of the concentration of aspartate, at fixed carbamylphosphate concentration, results such as shown in **Figure 16.27** are obtained. The sigmoid curvature (introduced in Chapter 14) is inconsistent with a simple Michaelis–Menten mechanism and, if

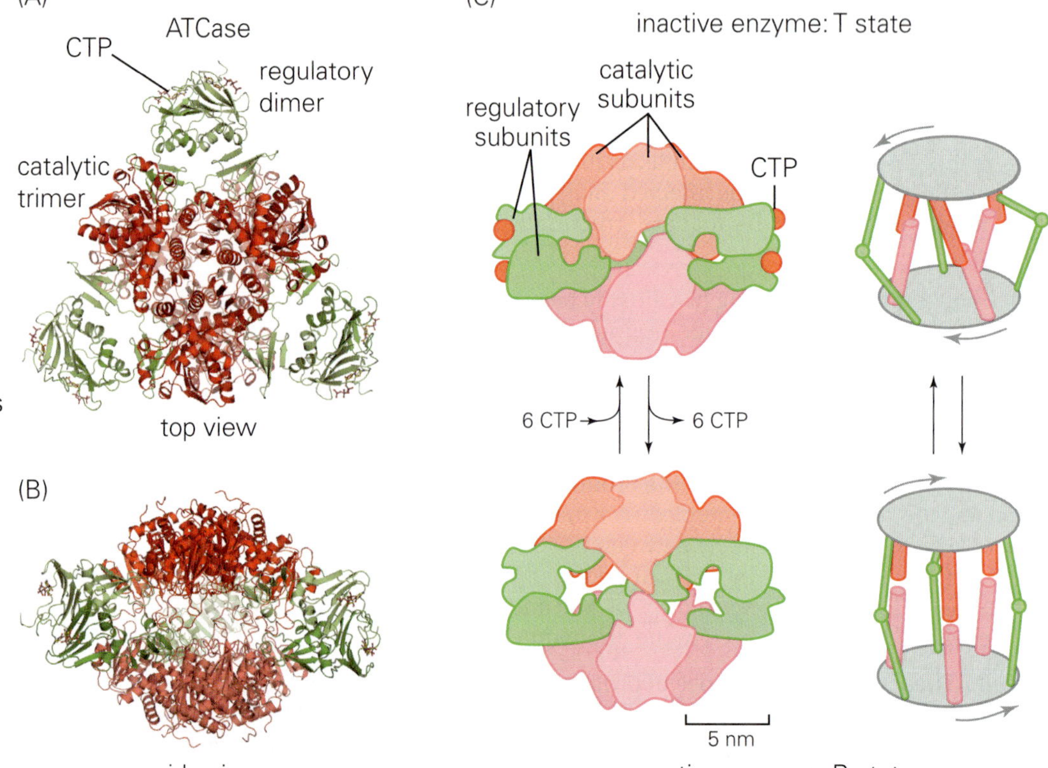


**Figure 16.26 Structure of the enzyme ATCase.** (A, B) Two orthogonal views of the holoenzyme, which is comprised of two catalytic trimers and three regulator dimers. The allosteric effector CTP is shown bound at the periphery of the regulatory domain. The structures shown here are of the inactive T state. (C) Schematic drawing of the allosteric transition in ATCase. The inactive T-state structure is shown at the top and corresponds to the structure shown in (A) and (B). The active R-state structure is shown at the bottom. The allosteric transition involves rotations of the subunits with respect to each other, as indicated in the diagrams on the right. (C, adapted from B. Alberts et al., Molecular Biology of the Cell, 5th ed. New York: Garland Science, 2008; PDB code: 1RAA.)

the data are plotted in the Lineweaver–Burk format, a curve rather than a straight line is found. Analogous measurements made by varying the concentration of carbamylphosphate while keeping the aspartate concentration fixed show the same behavior.

The sigmoidal shape observed is exactly what was described in Chapter 14 for cooperative binding of oxygen to hemoglobin, except that here it is the *rate of reaction* that is being monitored rather than a direct measurement of occupancy of the binding site. Since the measurement is activity rather than substrate binding, it is possible that the cooperativity arises through some modification in the rate of the catalytic step, rather than from cooperativity in substrate binding. Equilibrium binding studies with the bi-substrate analog inhibitor PALA (which binds both parts of the active site, thus preventing substrates from binding, but which cannot itself react; Figure 16.25B) showed cooperative binding with the same characteristics as the substrates, indicating that the cooperativity in velocity does indeed arise through cooperative substrate binding.

As for oxygen binding to hemoglobin, binding of substrate to ATCase induces a conformational change from a "tense" (T) to a "relaxed" (R) conformation, with the R conformation being the active one (Figure 16.26C). The first substrate molecule binding to the enzyme does not bind tightly because part of the binding energy goes into driving the conformational change that affects all active sites in the molecule. Subsequent substrates bind more tightly at other active sites that already have the high-affinity conformation.

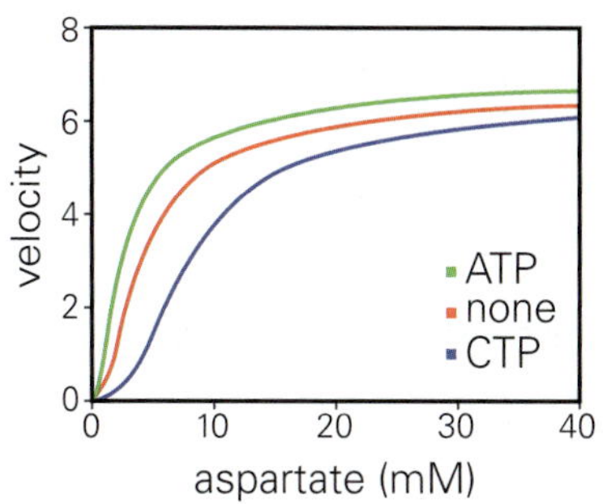


**Figure 16.27 The velocity of the reaction catalyzed by ATCase as a function of aspartate concentration for a fixed concentration of carbamylphosphate.** The velocity shows a sigmoid dependence on the substrate concentration. The allosteric effector ATP enhances the activity of the enzyme, while CTP reduces it relative to no nucleotide. The effects of substrates and nucleotides serve to regulate the activity of ATCase in cells. (Adapted from J.O. Newell, D.W. Markby, and H.K. Schachman, *J. Biol. Chem.* 264: 2476–2481, 1989. With permission from the American Society for Biochemistry and Molecular Biology.)

## 16.17 Product inhibition is a mechanism for regulating metabolite levels in cells

The metabolic pathway for the synthesis of pyrimidines is shown in **Figure 16.28**. The cooperative binding of substrate molecules to an enzyme, such as ATCase, enhances the reaction rate when the substrates are not at a high enough concentration to saturate the enzyme. However, if the end products of the pyrimidine biosynthetic pathway (for example, CTP) build up, it indicates that more pyrimidines are being produced than used and that the flux into that pathway can be reduced. This is accomplished by the binding of CTP to the regulatory subunit of ATCase, which increases the free-energy difference between the T and R conformations. This makes the free energy of substrate binding less favorable, making it more difficult for substrate to shift the conformation into the high-affinity state for subsequent molecules to bind, thereby lowering the rate of reaction and decreasing the flux of molecules through most of the pathway.

On the other hand, if ATP is more abundant than the pyrimidine nucleotides, then the flux through the pyrimidine path should be increased to balance the nucleotide pool. Initially it was thought that ATP binding had an analogous effect to CTP, but in the opposite sense—that is, decreasing the free energy between the tense and relaxed conformations and, hence, enhancing substrate binding. Both nucleotides bind to the regulatory domain, far from the active site, making this seem plausible. However, recent studies indicate that ATP affects the reaction in a different manner that is not yet completely understood, perhaps affecting the range of structural fluctuations in the enzyme.

# C. PROTEIN ENZYMES

There is a vast diversity of protein enzymes in nature that have evolved to catalyze many different kinds of processes. In this part of the chapter we review some of the most important aspects of the mechanisms of enzyme catalysis, focusing on proteins (RNA enzymes are described in the last part of this chapter). The discussion here is brief and far from complete. We discuss three principal mechanisms for rate acceleration by enzymes: transition state stabilization, acid-base catalysis, and co-localization. We do not cover many fascinating topics, such as the role

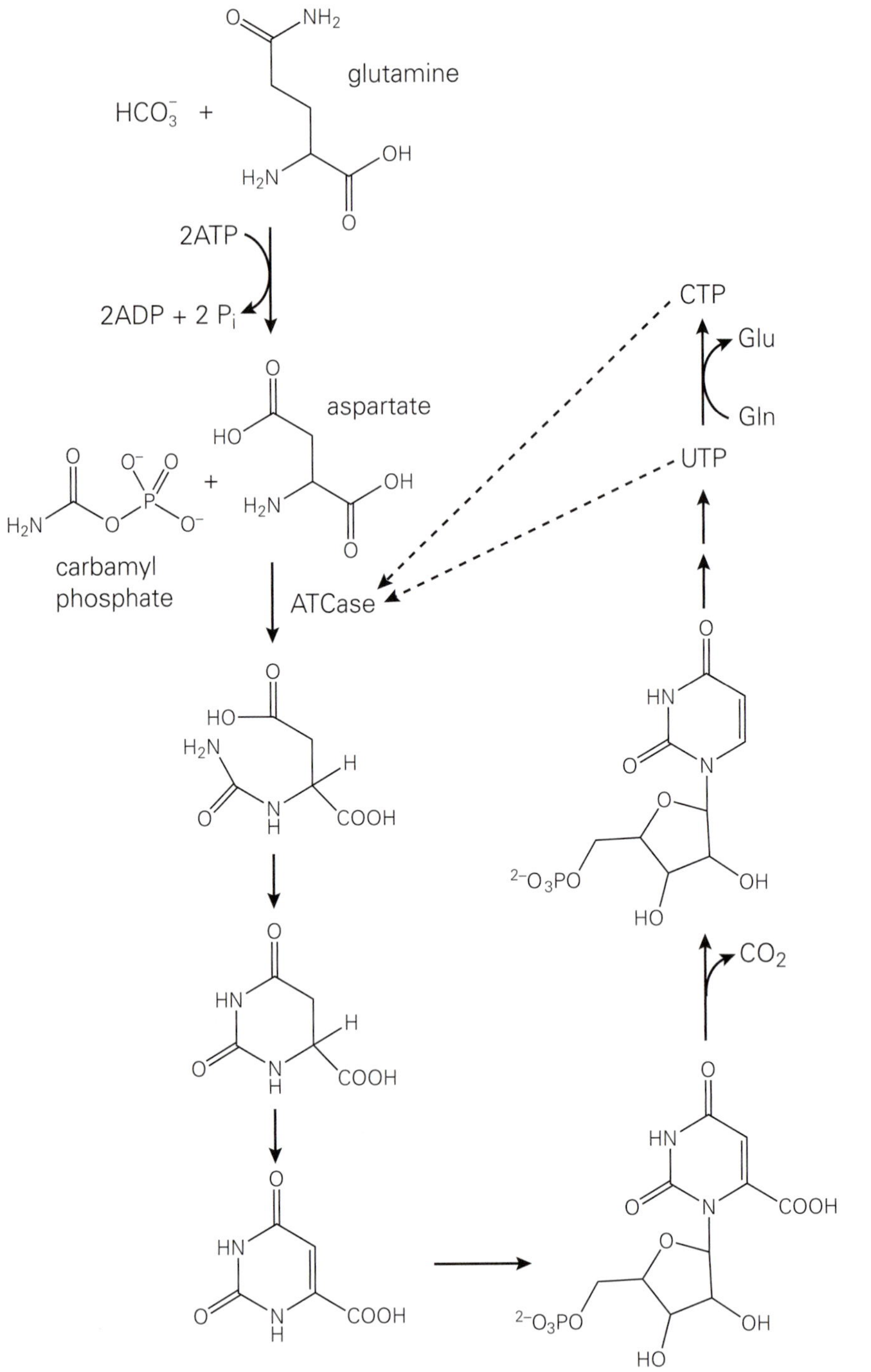

**Figure 16.28 The pathway for pyrimidine biosynthesis is shown, in which the reaction catalyzed by ATCase is the second step.** The end product of the pathway binds to and inhibits (noncompetitively) the activity of ATCase. UTP alone does not inhibit, but it does enhance the inhibition of CTP.



of dynamics in enzyme catalysis and the ability of enzymes to alter the chemical mechanisms by which the catalyzed reactions occur. We end this part of the chapter by discussing the mechanisms of a few specific enzymes.

## 16.18 Enzymes can accelerate reactions by large amounts

Enzymes do not just increase the rates of reaction, they also limit which molecules are converted into products. The balance between rate acceleration and specificity varies greatly for different enzymes. For some enzymes, particularly those that catalyze chemically difficult reactions, the ability to enhance the rate may completely dominate specificity in the evolution of function. For other enzymes, particularly those involved in regulation, the actual reaction rate is much less critical than the specificity for the correct substrate.

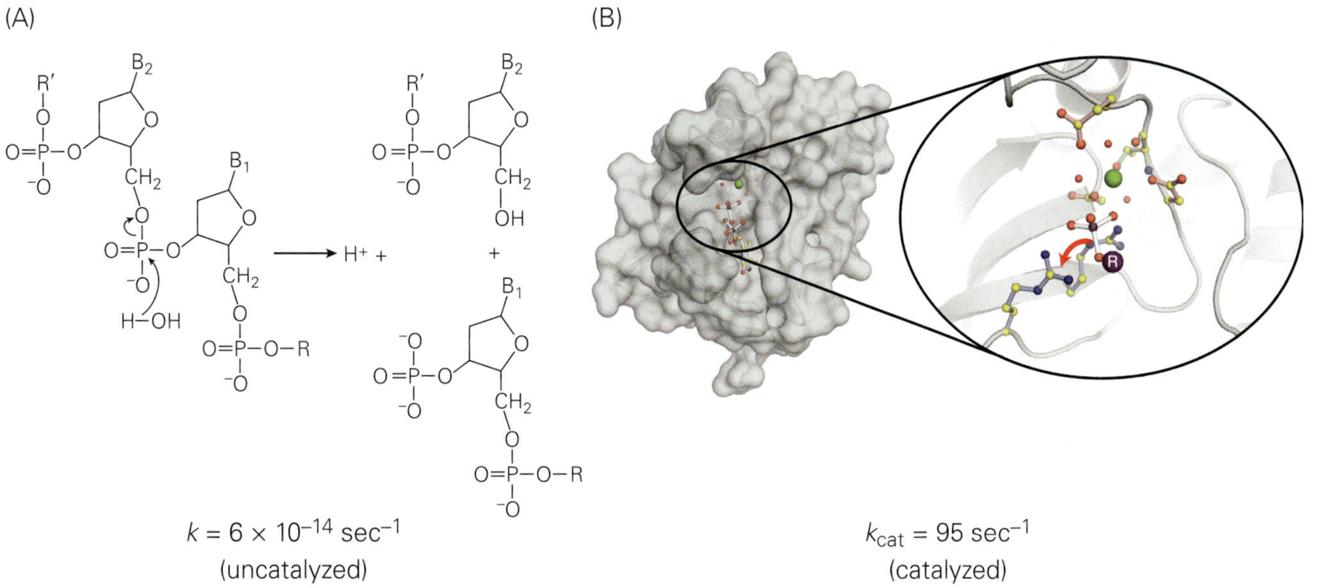

(A)

$k = 6 \times 10^{-14}$ sec$^{-1}$
(uncatalyzed)

(B)

$k_{cat} = 95$ sec$^{-1}$
(catalyzed)

**Figure 16.29 The acceleration of a reaction by an enzyme.** (A) The uncatalyzed cleavage of a nucleotide. (B) Cleavage of a nucleotide at the active site of the enzyme staphylococcus nuclease. In the expanded view an essential $Ca^{2+}$ ion is shown (*green* sphere), as are several sidechains of the protein. Only the phosphate group at the cleavage site in the substrate is shown, with the rest of the substrate indicated by R. The *red* arrow indicates the movement of electrons. (Adapted from E.H. Serpersu, D. Shortle, and A.S. Mildvan, *Biochemistry* 26: 1289–1300, 1987. With permission from the American Chemical Society; PDB code: 2SNS.)

As we discussed in Section 16.7, the catalytic efficiency ($k_{cat}/K_M$) is a useful parameter because it brings in both the rate of the reaction and the ability of the enzyme to recruit substrate when it is present at low concentration. Another measure of the catalytic power of an enzyme is to consider the rate acceleration provided by the enzyme relative to the rate of the equivalent uncatalyzed reaction. There are many reactions for which enzymes provide tremendous rate enhancements. For example, the hydrolysis of a peptide bond by water occurs more than $10^{10}$ times faster in the active site of a protease than in water alone. Other reactions are accelerated by even larger factors. An enzyme called staphylococcal nuclease, which cleaves nucleotides, is estimated to speed up the reaction so that it occurs $>10^{15}$-fold faster than the reaction in water alone (**Figure 16.29**).

How do enzymes generate such tremendous rate enhancements? The fundamental ideas were presented in Chapter 15 for catalysts in general: lower the activation energy, increase the preexponential factor, or change the reaction mechanism. Enzymes use all of these approaches, often exploiting the chemistry of bound cofactors to help. In fact, there are often so many factors contributing to rate acceleration that is it quite difficult to partition them and evaluate the contributions from each. In the following sections we give a brief discussion of major mechanisms by which enzymes catalyze reactions.

## 16.19 Transition state stabilization is a major contributor to rate enhancement by enzymes

Since the rate constant for a reaction depends exponentially on the activation energy, $E_A$, even a modest decrease in $E_A$ can lead to a substantial acceleration in the rate of product formation. As shown schematically in **Figure 16.30**, in an enzyme-catalyzed reaction, the rate-limiting step is usually the conversion of the enzyme–substrate complex to enzyme–product complex. This chemical transformation is rate limiting because of the high energy associated with the transition state, due to changes in geometry of the substrate during the reaction, often accompanied by changes in the distribution of charge in the molecule.

A good way to think about enzyme active sites and catalysis is the principle that optimum interaction occurs when the active site is complementary to the transition state in its properties. This includes matching the geometry of the active site to the transition state rather than to substrate or product, introducing charges (or partial charges) that make favorable electrostatic interactions with the transition state (particularly with atoms that change charge as the transition state is approached), and making hydrogen bonds that are optimized in the transition state.

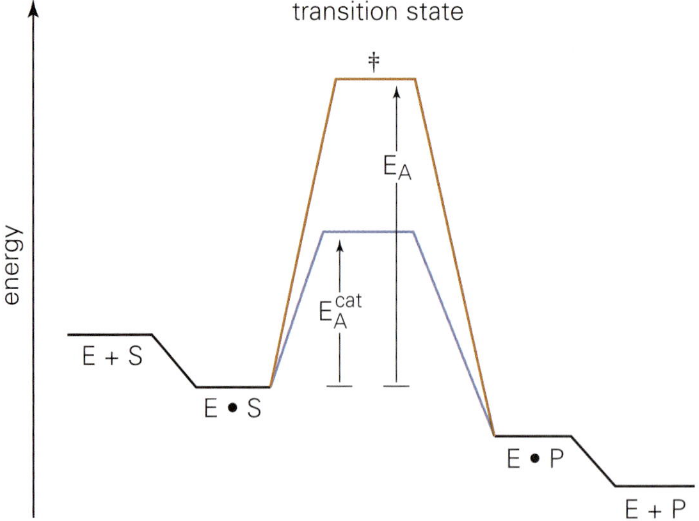


**Figure 16.30 An energy diagram for an enzyme reaction is shown.** The *blue* line reflects a change that leads to stabilization of the transition state, which lowers the activation energy.

The principle of transition state stabilization by enzymes is illustrated by considering the hydrolysis of ATP by the motor protein myosin (the use of ATP by motor proteins is discussed in Section 9.15). We briefly discussed the transition state for ATP hydrolysis in Chapter 15 (see Figure 15.34), and this is illustrated again in **Figure 16.31**A for the hydrolysis of ATP in water. Note that the reaction involves the attack of a hydroxide ion, generated from a water molecule on the terminal phosphate of ATP, leading to the formation of a transition state in which the terminal phosphate atom is bonded to five oxygen atoms instead of the normal four (see Figure 16.31A). Myosin binds to ATP at its active site and promotes the formation of the pentavalent phosphate structure that characterizes the transition state. The structure of the transition state complex of myosin and ATP is shown in Figure 16.31B. Once ATP binds to myosin, the formation of the transition state is facilitated by the fact that the enzyme is poised to make additional interactions with the pentavalent phosphate intermediate, as shown in Figure 16.31C. This assures an optimal interaction with the transition state, lowering its energy. Remember, however, that the transition state is still at a local maximum in free energy, so in spite of favorable interactions with the enzyme, the reaction still proceeds forward, downhill in free energy, to product.

The concept that the complementarity of enzyme and transition state lowers the free energy of the transition state is a very general one. The stabilization of the transition state by the enzyme virtually always contributes to enzyme catalysis, as will be seen in the examples in this chapter. Among the contributions to transition state stabilization, electrostatic effects are often particularly important. As discussed in Chapter 6, electrostatic interactions tend to be larger in magnitude

**Figure 16.31 Hydrolysis of ATP by myosin (*right*).** (A) The mechanism of ATP hydrolysis in water. The transition state involves a pentavalent phosphate group that is formed after the terminal phosphate group of ATP is attacked by a water molecule. (B) The structure of the myosin–transition state complex. An ATP molecule with a pentavalent terminal group is shown at the active site of the enzyme. The actual transition state is unstable and, in order to obtain this structure, a complex of ADP with vanadate is used (the vanadium atom is shown in *gray*). This mimics the true transition state. (C) Schematic diagram showing the interactions made between myosin and the substrate (left) and the transition state (right). In the transition state, the oxygen atom of the water molecule labeled $W_1$ makes a fifth bond to the phosphate atom of the terminal phosphate group of ATP. (C, adapted from H. Onishi et al. and M.F. Morales, *Proc. Natl. Acad. Sci. USA* 99: 15339–15344, 2002. With permission from the National Academy of Sciences; PDB code: 1VOM.)

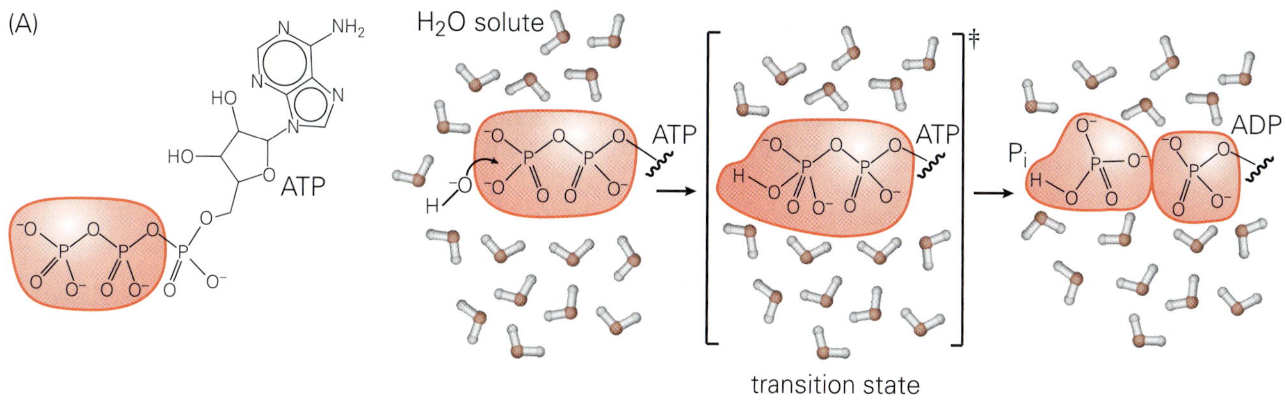
(A)
ATP
$H_2O$ solute
ATP
ATP
transition state
$P_i$
ADP

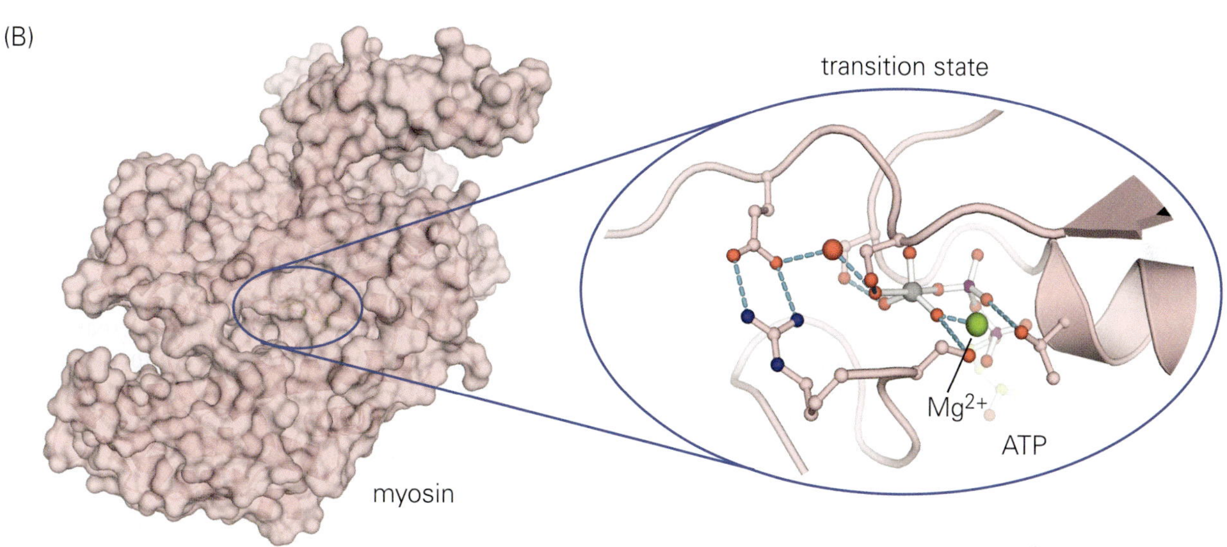
(B)
transition state
myosin
$Mg^{2+}$
ATP

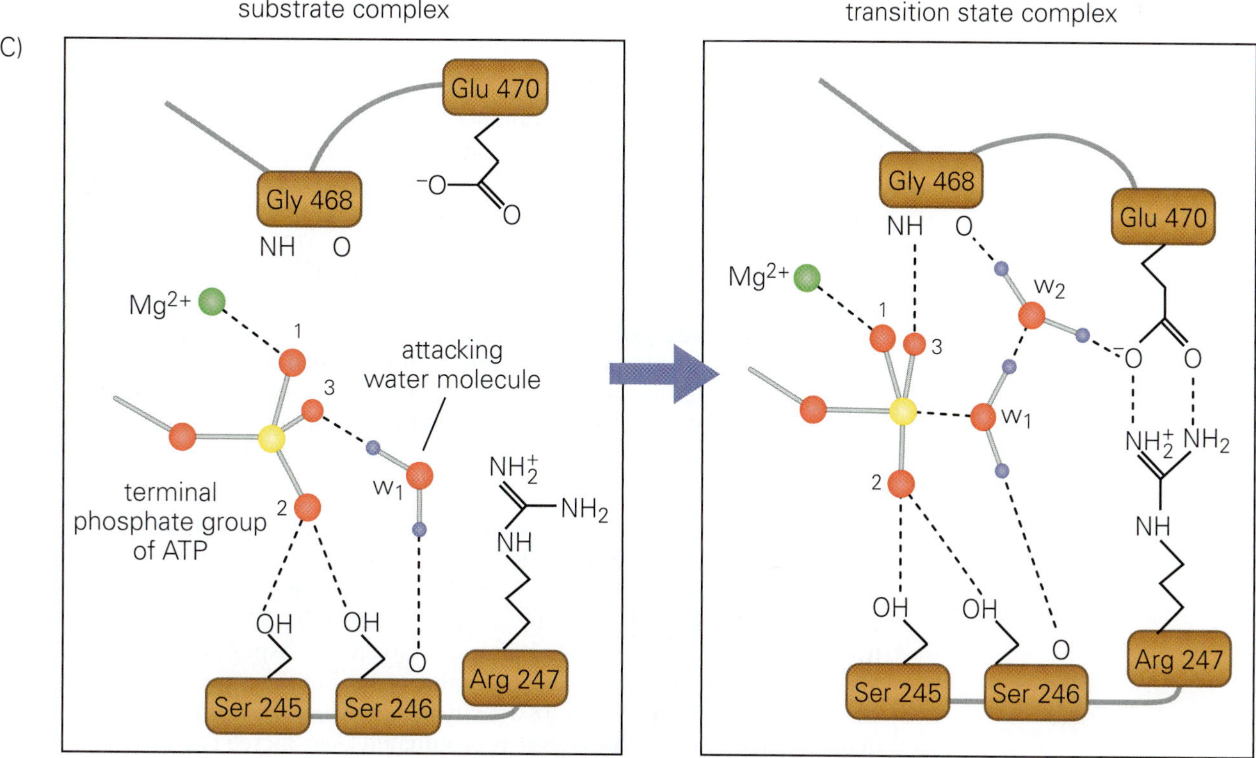
(C)
substrate complex
Glu 470
Gly 468
NH O
$Mg^{2+}$
attacking water molecule
terminal phosphate group of ATP
$w_1$
OH
OH
O
Ser 245
Ser 246
Arg 247
transition state complex
Gly 468
Glu 470
$w_2$
$w_1$
Ser 245
Ser 246
Arg 247

and longer range than most other interactions. This, together with the fact that many reactions cause changes in charge distribution in the substrate and/or enzyme during the reaction, explains the importance of electrostatics in catalysis. Although transition state stabilization is often the dominant contributor to catalysis, it is not always the most important factor.

## 16.20 Enzymes can act as acids or bases to enhance reaction rates

The hydrolysis of sucrose in water to give fructose and glucose was discussed in Section 15.24, and it was noted that the rate is pH-dependent, although hydrogen ions are neither consumed nor created during the overall reaction. The kinetics fit a model in which there is an equilibrium protonation of the oxygen bridging the two sugars, a process that creates a good leaving group and thereby facilitates the reaction.

Organisms cannot change the pH in the whole cell to accelerate a particular reaction. The way they get around this limitation is to use enzymes that can accomplish the same effect by exploiting residues in the active site that can act as general acids or bases. The important residues are generally ones that can undergo titrations in the normally accessible pH range, including Asp, Glu, His, Cys, Lys, sometimes the backbone N-terminal amine or C-terminal carboxylate, and on rare occasions, Tyr. Such resides can either donate (acting as an acid) or accept (acting as a base) hydrogen ions, depending on their initial state.

There are often several of these residues in the active sites of enzymes, creating an environment that can strongly shift the p$K_a$ value of the titrating group, priming it to act as an acid or a base during the reaction. Having a region of negative charge near an Asp or Glu sidechain will shift its p$K_a$ value from the normal value of about 4 up to 7 or more, so that it is ready to donate a proton. Similarly, a region of positive charge adjacent to a histidine will favor the neutral (unprotonated) form of that sidechain, holding it ready to accept a proton during a reaction. The degree of solvation of a sidechain by water can also affect its p$K_a$ value. Ionic forms are generally greatly stabilized by interactions with water, so the uncharged forms are favored in nonpolar environments. Once a reaction is completed and product is released, the acidic or basic residues in the enzyme reset to their original protonation state, ready to act again when the next substrate molecule binds.

These ideas are illustrated by the mechanism of ribonuclease-A, the enzyme used by Anfinsen to establish the spontaneous nature of protein folding (see Section 5.2). The structure of ribonuclease-A is shown in **Figure 16.32**. Like staphylococcus nuclease, which was discussed in Section 16.18 (see Figure 16.29), ribonuclease-A catalyzes the cleavage of ribonucleotides. Unlike staphylococcus nuclease, which uses a metal ion to activate a water molecule, ribonuclease-A uses two histidine residues (His 12 and His 119) for acid–base catalysis. First, as shown in Figure 16.32B, His 12 accepts a proton from the 2′-OH group of one nucleotide. The oxygen atom of the 2′ group attacks the phosphate group, leading to the formation of a cyclic phosphate intermediate and the transfer of a proton from His 119 to the leaving nucleotide. The reaction is completed in a second step in which the cyclic phosphate intermediate is cleaved by water to generate the product. The reaction is facilitated by the presence of Asp 121, which interacts with His 119 and stabilizes the protonated state of this histidine (see Figure 16.32A).

Measuring the rate of a reaction as a function of pH can provide evidence for the presence of titratable groups and, therefore, acid–base contributions to catalysis. It is usually the case that there is an optimal pH for the reaction, at which the rate is the highest. This is shown for lysozyme in **Figure 16.33**. In lysozyme there are two acidic residues involved in catalysis, Asp 35 and Asp 52, as shown in **Figure 16.34**. One of these has a normal p$K_a$ value of 3.9, but the p$K_a$ value of the other is shifted up to ~6.5. The optimum activity occurs when the normal one (Asp 52) is deprotonated, but the shifted one is protonated (Asp 35; see Figure 16.34). At low

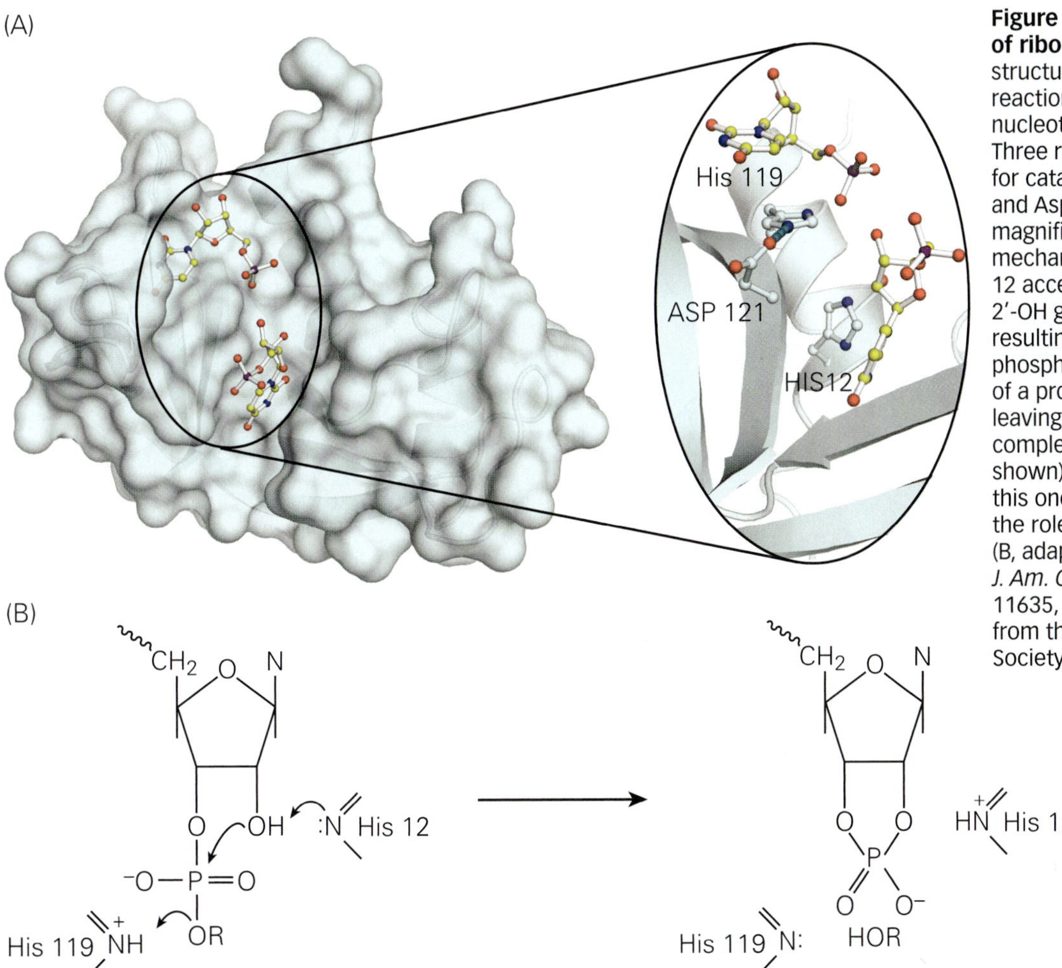


**Figure 16.32 The mechanism of ribonuclease-A**. (A) The structure of the enzyme, with the reaction products (two cleaved nucleotides) shown in sticks. Three residues that are important for catalysis (His 12, His 119, and Asp 121) are shown in the magnified view. (B) The reaction mechanism of ribonuclease. His 12 accepts a proton from the 2′-OH group of a nucleotide, resulting in cyclization of the phosphate group and the transfer of a proton from His 119 to the leaving group. The reaction is completed in a second step (not shown), which proceeds just like this one except that water plays the role of the 2′-OH group. (B, adapted from D. Herschlag, *J. Am. Chem. Soc.* 116: 11631–11635, 1994. With permission from the American Chemical Society; PDB code: 3DXG.)

pH, both become protonated and, at high pH, both are deprotonated, and so the activity is reduced at both low and high pH.

In discussions of catalysis, these effects are often referred to as "general acid–base catalysis." This term emphasizes the fact that often while the acid–base chemistry involves the transfer of a proton, there are cases in which the concept of a Lewis acid (which can accept a pair of electrons) or a Lewis base (which can donate a pair of electrons) is more relevant.

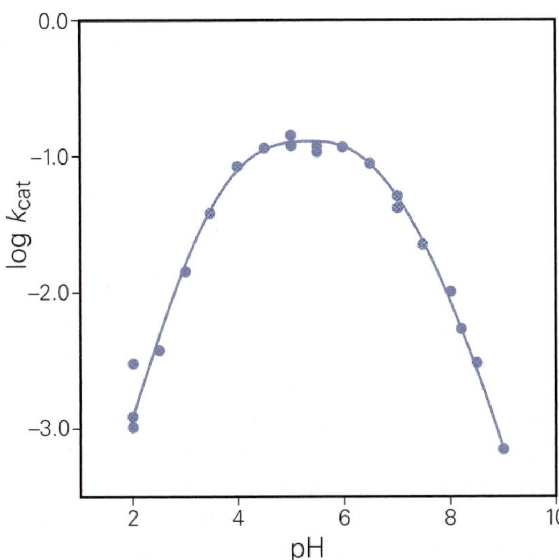


**Figure 16.33 The rate of hydrolysis of a substrate by the enzyme lysozyme**. (Adapted from S.K. Banerjee and J.A. Rupley, *J. Biol. Chem.* 250: 8267–8274, 1975. With permission from the American Society for Biochemistry and Molecular Biology.)

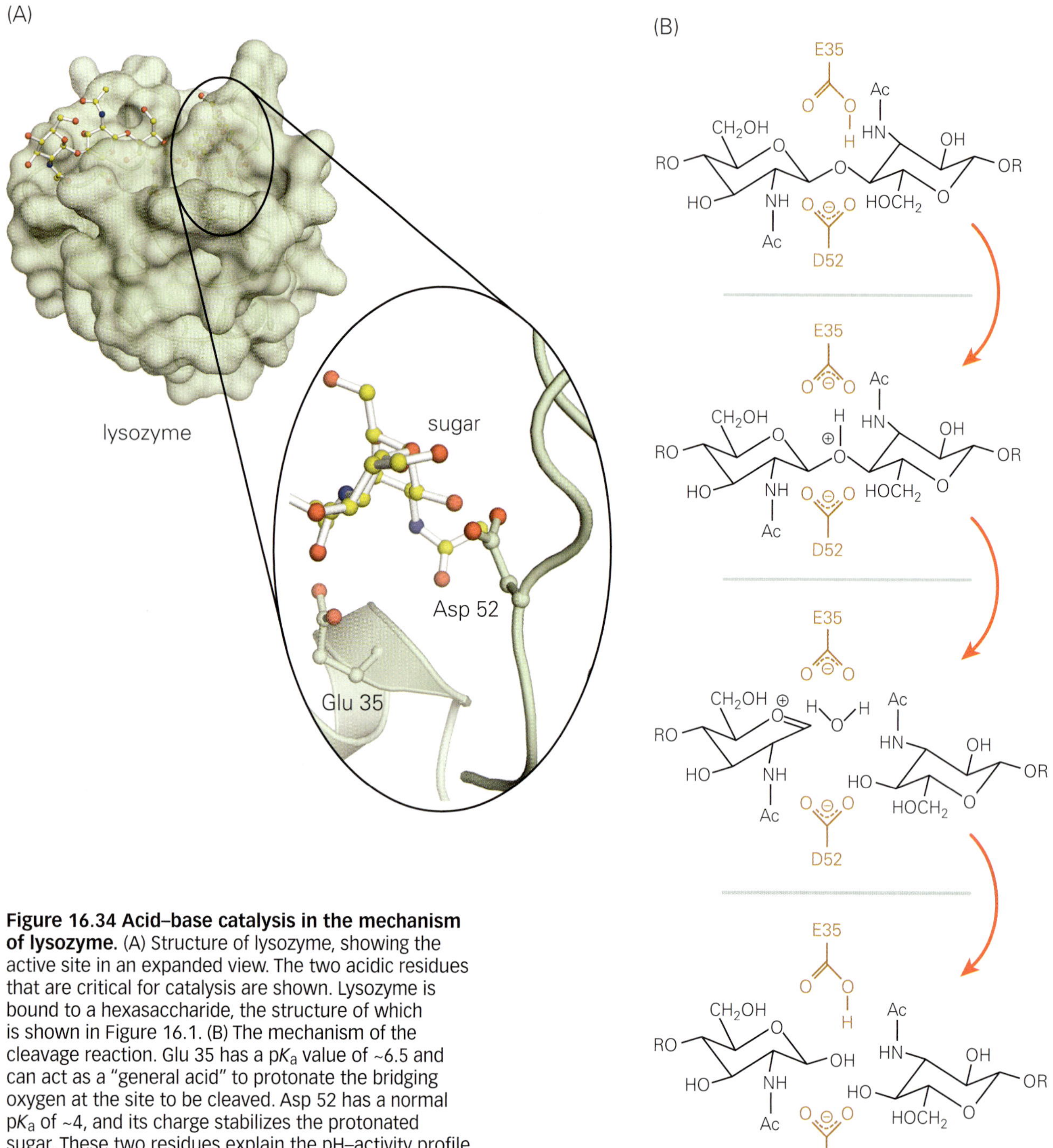


**Figure 16.34 Acid–base catalysis in the mechanism of lysozyme.** (A) Structure of lysozyme, showing the active site in an expanded view. The two acidic residues that are critical for catalysis are shown. Lysozyme is bound to a hexasaccharide, the structure of which is shown in Figure 16.1. (B) The mechanism of the cleavage reaction. Glu 35 has a p$K_a$ value of ~6.5 and can act as a "general acid" to protonate the bridging oxygen at the site to be cleaved. Asp 52 has a normal p$K_a$ of ~4, and its charge stabilizes the protonated sugar. These two residues explain the pH–activity profile shown in Figure 16.33. (PDB code: 1SFB.)

## 16.21 Proximity effects are important for many reactions

The interaction of an enzyme with substrate (that is holding substrate in proximity to active-site residues) is essential for the enzyme to effect catalysis. However, the term "proximity effect" is often used in the broader sense of bringing together two substrate molecules to react or to organize a substrate into a specific conformation that is optimal for an intramolecular reaction. Enzymes accomplish this by using the favorable free energy of binding to "pay" for the loss of translational and rotational freedom that occurs upon binding of one or more substrates to the enzyme. Proximity effects are basically entropic in origin, in contrast to the lowering of activation energy, which is enthalpic.

In Chapter 14, we discussed how the colocalization of proteins provides a considerable free-energy drive for the formation of molecular complexes (see Section 14.22). Colocalization is also a very important factor in enzyme catalysis, as indicated schematically in **Figure 16.35**. A specific example is shown in **Figure 16.36**, which illustrates the role of a DNA polymerase enzyme in adding nucleotides to primed DNA (see Section 1.16 for an introduction to the basic mechanism of DNA replication, which is discussed in detail in Chapter 19). While the overall reaction catalyzed by DNA polymerases is favorable in free energy, there are significant free-energy barriers that make the spontaneous condensation of nucleotides on to the primer DNA strand highly unlikely.

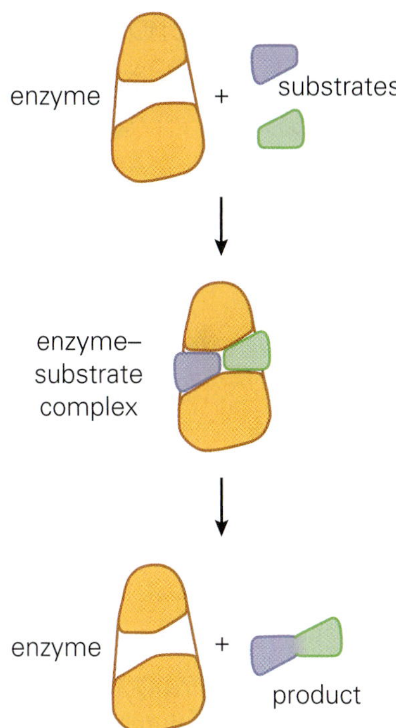


**Figure 16.35 A proximity effect for an enzyme can be as simple as holding together two substrates so that they can react rapidly to form a product**. In this schematic, the reaction that is catalyzed is the joining of two molecules to make one larger one, which occurs in many contexts, such as the synthesis of polysaccharides or DNA.

Electrostatic effects are one barrier to the spontaneous incorporation of new nucleotides into a growing DNA chain—the negative charge on the phosphates in the DNA's sugar-phosphate backbone will repel incoming nucleotides, which are also negatively charged (see Figure 16.36). Even if a nucleotide were to arrive in the vicinity of its target, the reaction can only proceed if stringent geometric requirements are met: the attacking 3′-hydroxyl group on the terminal nucleotide of the primer DNA must be positioned appropriately for nucleophilic attack on the α phosphate of the incoming nucleotide triphosphate. The need for spatial colocalization thus imposes a significant entropic barrier to the reaction. One function of the catalytic center in a DNA polymerase is to provide residues that stabilize and correctly steer the incoming nucleotide so as to facilitate the completion of each condensation step in the synthesis reaction. The mechanisms underlying this process are discussed in detail in Chapter 19.

One useful way to think about how much the binding of substrate to an enzyme affects a bimolecular reaction is to define an "effective concentration." This is the concentration in free solution that would be required to reach the same reaction rate as occurs on the enzyme after correcting for the change in activation energy. The effective concentrations of substrates and functional groups on the enzyme can be enormous, far above what is physically realistic (that is, molar concentrations beyond what the pure liquid substrate would be). To understand this, refer to Section 14.22, where we explained that if the concentration of interacting molecules is increased 1000-fold, then the free-energy drive for complex formation increases by ~35 kJ•mol$^{-1}$. You can work out that if the free energy of the transition state for a reaction is lowered by 35 kJ•mol$^{-1}$, then the reaction will proceed about a million times faster, at room temperature. This is comparable to the rate enhancement provided by many enzymes, and in order to achieve this without the enzyme the reactants would have to be present at concentrations that are 1000-fold higher than normal.

The control of relative orientation may also be important for controlling the stereochemical outcome of a reaction. For some types of substrate, there is the possibility of a reaction occurring from the "top" or the "bottom" (for example, addition across a double bond), as shown in **Figure 16.37**. For nonchiral substrates, symmetry requires that the rates of the reaction for the two possibilities be the same. On enzymes, however, which are always chiral, the binding sites for substrates will position them in one relative orientation only. Even if the reaction were not accelerated, this could change the racemic mixture of two chiral products from the solution reaction into one specific enantiomer for the enzyme product.

**Figure 16.36 The DNA polymerization reaction**. Nucleotide triphosphates are repelled from DNA because they are negatively charged. A DNA polymerase enzyme binds to primed DNA and provides a favorable electrostatic environment for the binding of the nucleotide triphosphate.

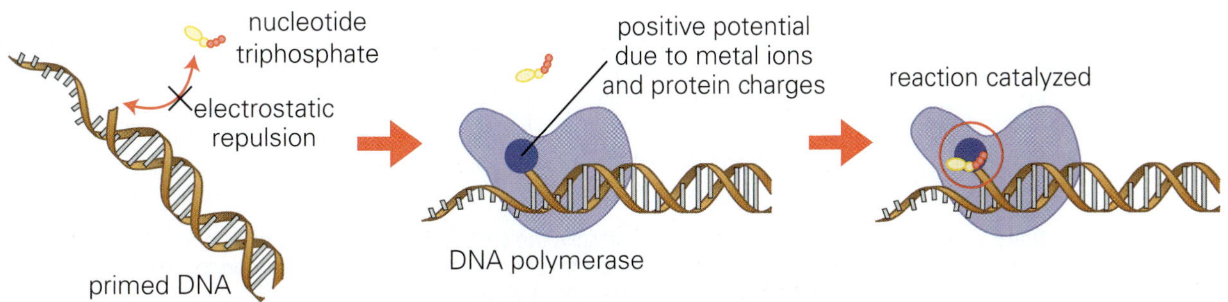

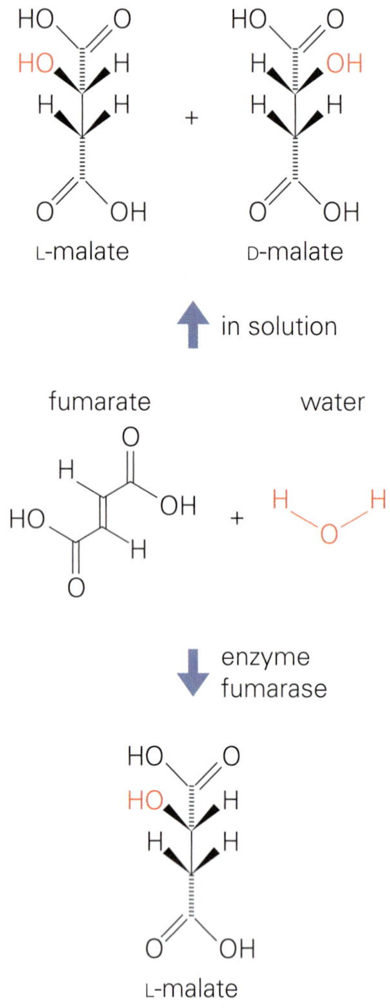

L-malate + D-malate

in solution

fumarate + water

enzyme fumarase

L-malate

**Figure 16.37 The enzyme fumarase adds water to fumarate, making only the L isomer of the product malate.** Addition of water in free solution would yield equal amounts of the D and L isomers.

## 16.22 The serine proteases are a large family of enzymes that contain a conserved Ser-His-Asp catalytic triad

The **serine proteases** form a large class of enzymes that break amide bonds between amino acid residues in polypeptides, producing two separate peptides by adding water across the bond. Cleavage of the polypeptide backbone can be part of a regulatory mechanism, inactivating the target protein by causing it to unfold, or activating it by removing an inhibitory pro-enzyme fragment. Proteolysis is also part of the machinery in cells that recycles components (in this case, the amino acids).

The mechanism by which peptide bonds are cleaved in water is shown in **Figure 16.38**. The free-energy change for the reaction is small but favorable under usual conditions in cells. The rate of cleavage of peptide bonds in pure water is very slow, due to a large activation energy and a low concentration of the required hydroxyl attacking group (see Figure 16.38A). The cleavage in water is also rather nonspecific, and peptide bonds between different pairs of amino acid residues are generally cleaved by water at a fairly similar rate.

The enzymes in the serine protease family accelerate the reaction by lowering the activation energy and by creating a high local concentration of an activated attacking nucleophile, and they also provide specificity for the amide bonds they cleave by selectively binding specific peptide sequences in a specific register. In order to accelerate the chemical steps, the enzymes replace the single-step direct cleavage that occurs in water by a two-step mechanism of the ping-pong type, which is described in detail in the next sections.

Trypsin is a much-studied member of the serine protease family, which has in its active site a "catalytic triad" of amino acids—namely, a serine, a histidine, and an aspartate—that are completely conserved in this family (**Figure 16.39**). These are arranged such that the serine (Ser 195 in bovine trypsin, shown in Figure 16.39) is at the edge of a groove on the surface of the enzyme that can interact with a substrate polypeptide. The hydroxyl group of the serine sidechain is hydrogen-bonded to the histidine (His 57 in trypsin), which hydrogen-bonds, in turn, to the aspartate (Asp 102 in trypsin) (see Figure 16.39).

## 16.23 Sidechain recognition positions the catalytic triad next to the peptide bond that is cleaved

The serine residue of the catalytic triad is positioned right next to the carbonyl group of the peptide bond that will be cleaved, as shown in Figure 16.39. As we shall see, the hydroxyl group of the serine residue plays a role analogous to that of the hydroxide ion in the uncatalyzed reaction. How does the serine "know" where it is to be positioned relative to the substrate? It turns out that the serine proteases recognize the sidechain of the residue that is at the N-terminal end of the bond that is being cleaved (that is, at the P1 position, as shown in Figure 16.39). This positions the site of cleavage precisely next to the serine residue of the catalytic triad.

P1 P1′ → [transition state (oxyanion)]‡ → P1 + P1′

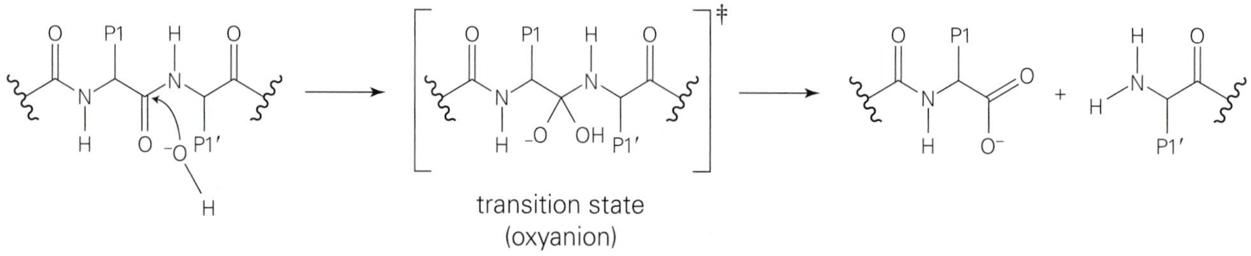

**Figure 16.38 Peptide bond cleavage in water.** A hydroxide ion attacks the carbonyl group of a peptide bond and forms an oxyanion transition state, with a negatively charged oxygen. This rearranges to yield the two product peptides. The uncatalyzed reaction is very slow because of the low concentration of hydroxide ions in water.

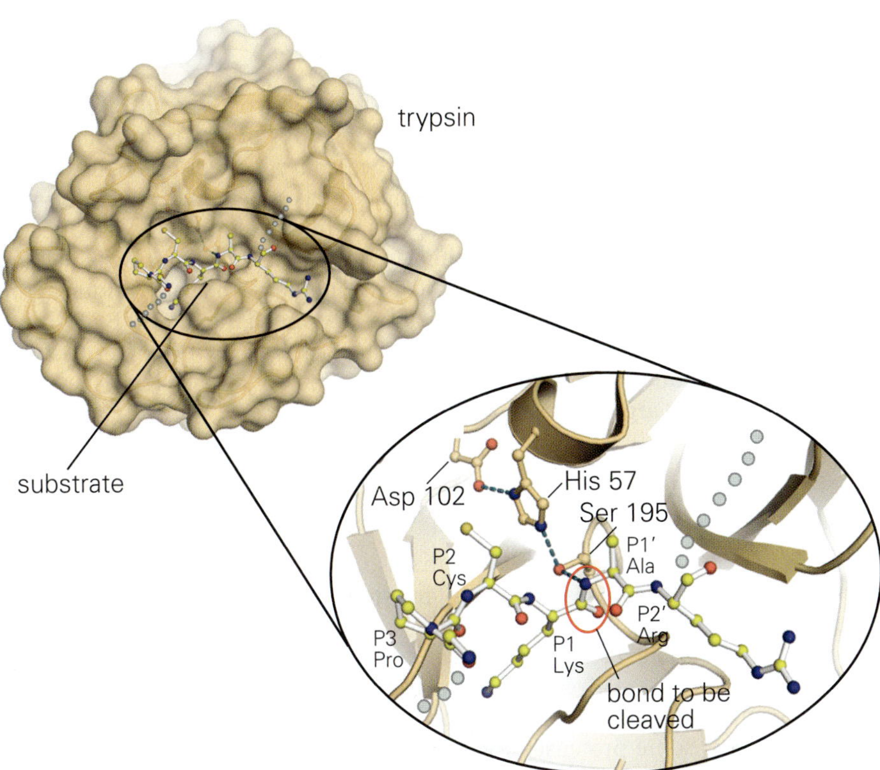


**Figure 16.39 The structure of trypsin.** The surface of trypsin is shown, with a portion of a substrate bound to it. The expanded view shows the active site region with the peptide substrate. The *gray dots* indicate the N- and C-terminal extensions of the peptide chain of the substrate. The three residues of the catalytic triad (Asp 102, His 57, and Ser 195) are shown. The residues that flank the peptide bond to be cleaved are denoted P1 and P1′, and the nomenclature used to identify the other residues in the peptide is indicated. The "substrate" shown in this structure is actually an inhibitor known as bovine pancreatic trypsin inhibitor (BPTI). Although it is actually a substrate, it binds to trypsin so tightly that it is not released rapidly, and in this way inhibits the enzyme. (PDB code: 3FP6.)

The precision with which the substrate is positioned on the enzyme can be appreciated by looking at the structure of a protein known as bovine pancreatic trypsin inhibitor (BPTI) bound to trypsin (**Figure 16.40**). BPTI is actually a substrate of trypsin, but it turns out that it binds too tightly to be released rapidly. This slows the cleavage reaction drastically, and BPTI is effectively an inhibitor rather than a substrate. The dissociation constant for BPTI binding to trypsin is about $5 \times 10^{-14}$ M, making this one of the tightest protein–protein interactions known.

**Serine proteases**

Serine proteases are enzymes that cleave peptide bonds. The catalytic center of these enzymes contains a catalytic triad: a serine, a histidine, and an aspartic acid. The hydroxyl group of the serine sidechain attacks the peptide bond that is to be cleaved.

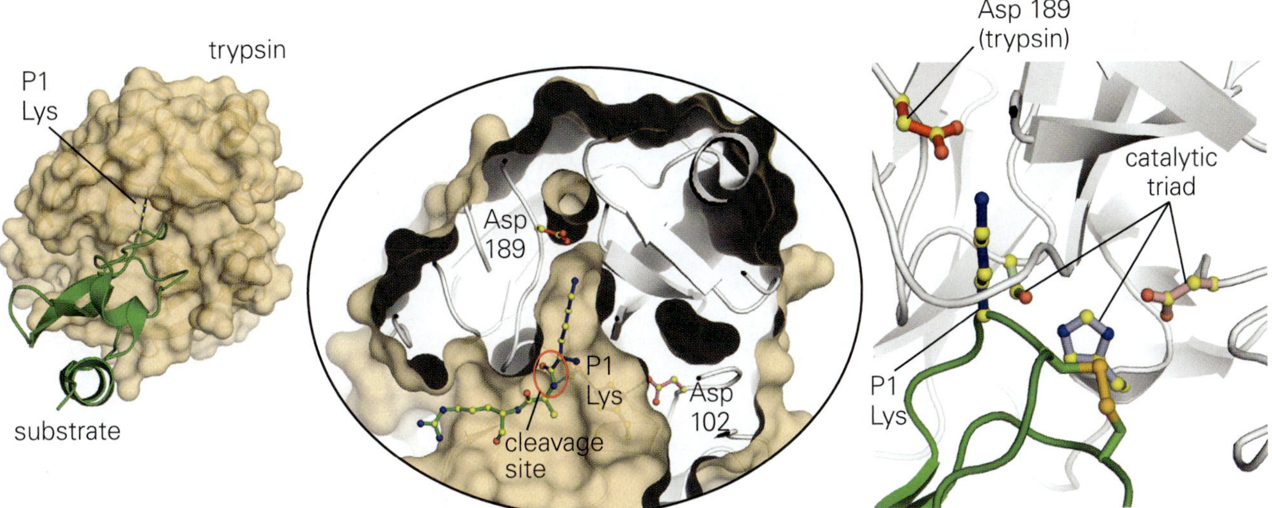


**Figure 16.40 The structure of trypsin in complex with the protein bovine pancreatic trypsin inhibitor (BPTI).** The diagram on the left shows the surface of trypsin, with BPTI as a ribbon. The lysine residue at the P1 position in BPTI (that is, the cleavage site) is indicated. The middle diagram shows an expanded view in which the front part of the complex has been cut away, showing how the P1 lysine residue is recognized by trypsin. The lysine fits into a deep pocket in trypsin (the specificity pocket) and forms a salt bridge to Asp 189 at the end of the pocket. The diagram on the right shows how the interaction between Asp 189 in the specificity pocket positions the backbone of the BPTI protein. Notice that the lysine is positioned so that the serine residue of the catalytic triad is directly below it. A disulfide bond joining loops of the inhibitor, stabilizing it and suppressing cleavage, can be seen just in front of the histidine. (PDB codes: 3BTK and 3FP6.)

The dissociation rate constant is estimated to be $5 \times 10^{-8}$ $sec^{-1}$, which corresponds to a bound half-life of about six months.

The surface of trypsin contains a groove within which the peptide substrate binds (see Figure 16.39). In the case of BPTI, one of the loops of the protein lies along this groove, as shown in Figure 16.40. The groove on trypsin has one deep pocket that is just past the amide bond that will be cleaved, as shown in the middle panel of Figure 16.40. The substrate is positioned so that the sidechain of the residue just before the amide bond to be cleaved (the P1 residue) fits into this pocket, which is relatively long, narrow, and hydrophobic, except for an aspartate residue at the very end of the pocket (Asp189). The P1 residue is lysine in the case of BPTI, but it can be either lysine or arginine in other substrates. The specificity pocket does not accommodate residues other than lysine and arginine, so trypsin cleaves proteins only after lysine and arginine residues.

## 16.24 The specificities of serine proteases vary considerably, but the catalytic triad is conserved

The overall structures of the serine proteases vary considerably (two examples are shown in **Figure 16.41**), as do the interactions leading to specificity. Chymotrypsin, which is closely related in structure to trypsin, has a somewhat wider and more hydrophobic specificity pocket than does trypsin, and it lacks the aspartate that in trypsin recognizes lysine or arginine residues in the substrate. As a consequence, chymotrypsin cuts the polypeptide backbone just after large hydrophobic residues. Some proteases, such as subtilisin (see Figure 16.41A) have no well-defined specificity pockets and, hence, little sequence preference for where they break amide bonds.

The catalytic triad is a completely conserved feature of all of the serine proteases. Some of the serine proteases, such as chymotrypsin and subtilisin, are closely related to trypsin and so it is not surprising that these catalytic residues are conserved. But other serine proteases, such as tripeptidyl peptidase, have distinct

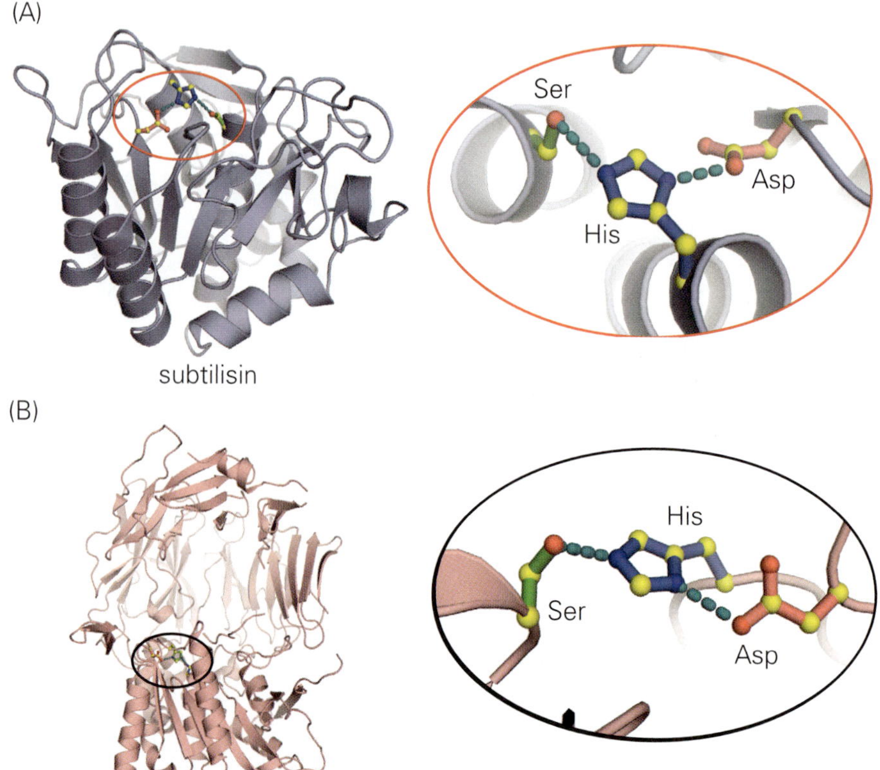


**Figure 16.41 Conservation of the catalytic triad in serine proteases.** (A) The structure of the broad specificity protease subtilisin is shown, with the catalytic triad shown in the expanded view. (B) The structure of tripeptidyl peptidase and its active site are shown. (PDB codes: A, 1C3L; B, 2D5L.)

folds and yet have a conserved catalytic triad. If you compare the structures of subtilisin and tripeptidyl peptidase, shown in Figure 16.41, you will see that the residues of the catalytic triad are presented by different secondary structural elements in the two enzymes. Clearly, evolution has found this arrangement of these three residues to be a good solution to the problem of cleaving the peptide bond.

## 16.25 Peptide cleavage in serine proteases proceeds via a ping-pong mechanism

The mechanism of peptide cleavage, as catalyzed by serine proteases, is shown in **Figure 16.42**. The reaction occurs in two steps. In the first step, with the substrate bound in the cleft of the enzyme, the amide bond that will be cleaved is right above the hydroxyl group of the serine of the catalytic triad. The hydrogen bond between this hydroxyl group and the histidine sidechain, which in turn forms a hydrogen bond with the aspartate (see Figure 16.39), makes the oxygen much more nucleophilic than it would normally be. Fluctuations in the protein structure drive the serine oxygen into the carbonyl, and these form a bond. This alters

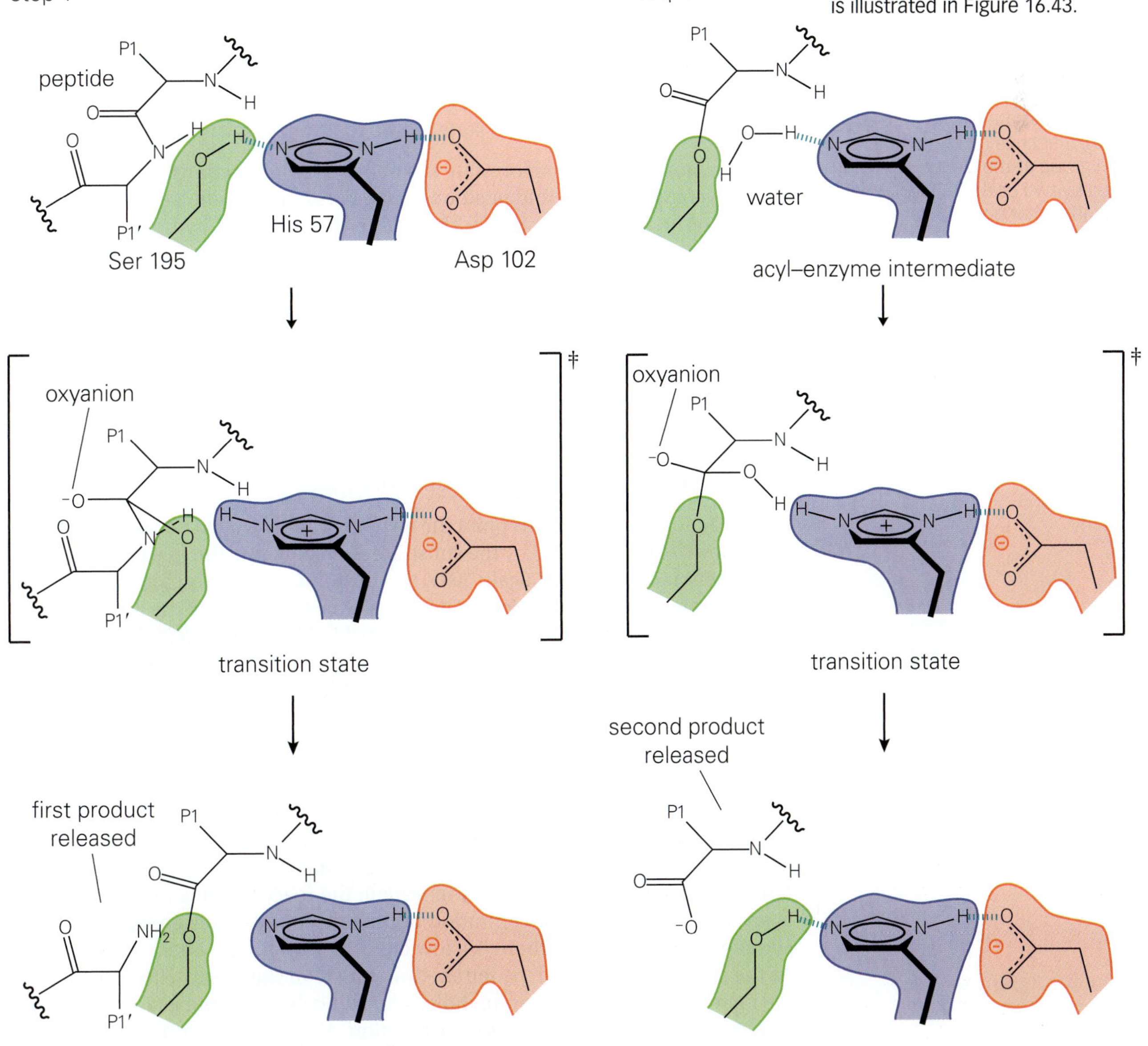


**Figure 16.42 Schematic drawing of peptide bond cleavage by a ping-pong mechanism, as catalyzed by a serine protease such as trypsin.** In the first step, the peptide bond is cleaved, and the C-terminal peptide product is released. At the end of this step, an intermediate is formed in which the remaining portion of the peptide is attached to the serine residue of the catalytic triad (the acyl–enzyme intermediate). The transition states for the reactions are shown in brackets with a ‡. Note that both transition states involve the formation of a negatively charged oxyanion. The oxyanion is stabilized by the "oxyanion hole" on the enzyme, which is illustrated in Figure 16.43.

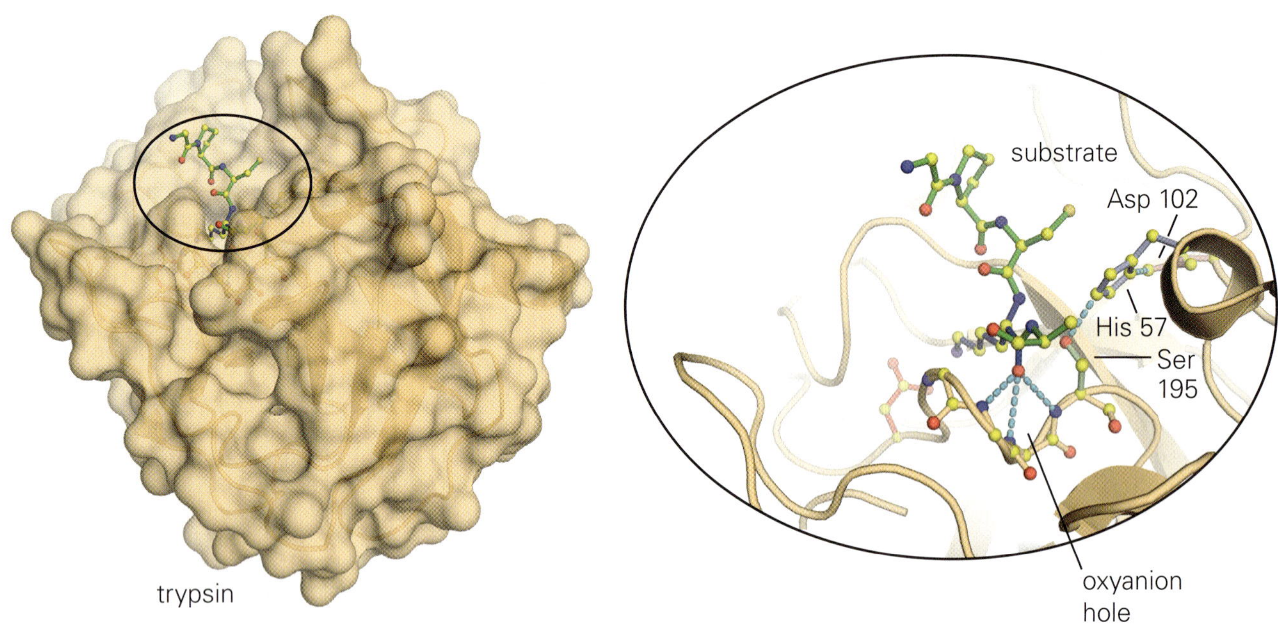


**Figure 16.43 The oxyanion hole in trypsin**. The structure of trypsin bound to a substrate is shown. The developing negative charge on the carbonyl group at the cleavage site is stabilized by two or three backbone amide groups that point towards it. (PDB code: 3FP6.)

the orbital configuration of the carbonyl from $sp^2$ to $sp^3$, which leaves a formal negative charge on one of the oxygens (an oxyanion). This is a high-energy configuration, and is the transition state for the first step of the reaction.

The oxyanion in the transition state is stabilized by interactions with the protein. The trypsin polypeptide fold positions three backbone amide N–H groups so that they point toward the negatively charged oxygen and can hydrogen bond to it. This feature in the structure is called an "oxyanion hole" and is illustrated in **Figure 16.43**. The interaction with the oxyanion hole lowers the energy of the transition state, and so helps to accelerate the reaction.

This process can reverse and go back to unmodified substrate. Alternatively, the proton that was associated originally with the serine hydroxyl can move to the amide and, rather than breaking the carbonyl–serine bond, it can break the carbonyl–amide nitrogen bond, releasing the C-terminal segment of the protein and leaving the N-terminal segment covalently bonded to the enzyme, forming an "acyl–enzyme" intermediate (see Figure 16.42).

If the reaction stopped at this point the enzyme would be dead because the active-site groove is still occupied by the substrate. To regenerate the free enzyme, a second reaction occurs. This again uses the histidine–aspartate pair, but this time they activate a water molecule that occupies a position close to the initial position of the serine sidechain. Through a process rather like the attack of the serine O–H on the amide bond, this water is made nucleophilic by the hydrogen bond to the histidine. The water molecule adds to the ester carbonyl group, making a tetrahedral transition state. This results in the elimination of the serine sidechain to give a free carboxyl at the end of the substrate peptide and regenerates the enzyme. The products diffuse away from trypsin, leaving it ready to act again.

There are several features of this process that contribute to the rate acceleration of the reaction by the enzyme relative to the reaction in solution. Substrate binding positions the amide bond to be cleaved very close to, and in the correct orientation relative to, the serine that will initiate the reaction. This means that there is a very high effective concentration of the nucleophile. This contrasts with the situation in water, where there is a low concentration of the attacking hydroxide. As the reaction occurs and the tetrahedral transition state forms, there is a favorable interaction with the oxyanion hole, stabilizing the transition state and thereby lowering the activation energy for the reaction. Although two steps are required in the reaction on the enzyme, rather than one in water, each of these is much faster on the enzyme, and so the overall process is accelerated greatly. The effective rate constant for the enzymatic reaction is as much as $10^{10}$-fold higher than that for the solution reaction. For trypsin, the value of $k_{cat}/K_M$ with small substrate peptides can be as high as $5 \times 10^5$ ($k_{cat}$ = 22 sec$^{-1}$, $K_M$ = 45 μM). This is a moderately high turnover rate for a reaction that is intrinsically difficult to catalyze.

## 16.26 Angiotensin-converting enzyme is a zinc-containing protease that is an important drug target

Earlier in this chapter we encountered two enzymes that both cleave nucleotides. One of these, staphylococcus nuclease, relies on a metal ion to activate a water molecule (see Figure 16.29). The other, ribonuclease-A, uses two histidines that engage in acid–base catalysis (see Figure 16.32). Protease enzymes are no different, and nature has come up with several different catalytic strategies to facilitate the cleavage of peptide bonds. The HIV protease, discussed in Section 16.9, uses two aspartic acid residues instead of a catalytic triad (HIV protease is an **aspartate protease**). Other proteases, known as **metalloproteases**, use metal ions to activate water molecules for catalysis.

One metalloprotease that is important in human biology is angiotensin-converting enzyme (ACE), a zinc-containing protease that cleaves an inactive 10 amino acid precursor peptide, angiotensin-I (with the amino acid sequence DRVYIHPFHL), to give mature, active angiotensin-II (DRVYIHPF), a potent vasoconstrictor that regulates blood pressure.

High blood pressure (hypertension) is a fairly common problem in middle aged or older people and, without treatment, it leads to heart disease. One successful approach to reducing blood pressure has been to selectively inhibit ACE with small organic molecules. These were selected for inhibitory activity against ACE from sets of compounds that were synthesized in laboratories. To be effective as a drug for treating humans, such compounds must be absorbed after being taken orally, must be fairly slowly excreted by the body, must not be toxic or have toxic metabolites, and must be quite selective for the target of interest (ACE in this case). Such compounds have been identified, tested in clinical trials, and are now widely used.

Although a knowledge of the enzyme structure is not required for developing a drug that targets it, having structural information can often help, particularly in optimizing high-affinity binding. The structure of human ACE with one of the clinically used ACE inhibitors, a drug called lisinopril, is shown in **Figure 16.44**.

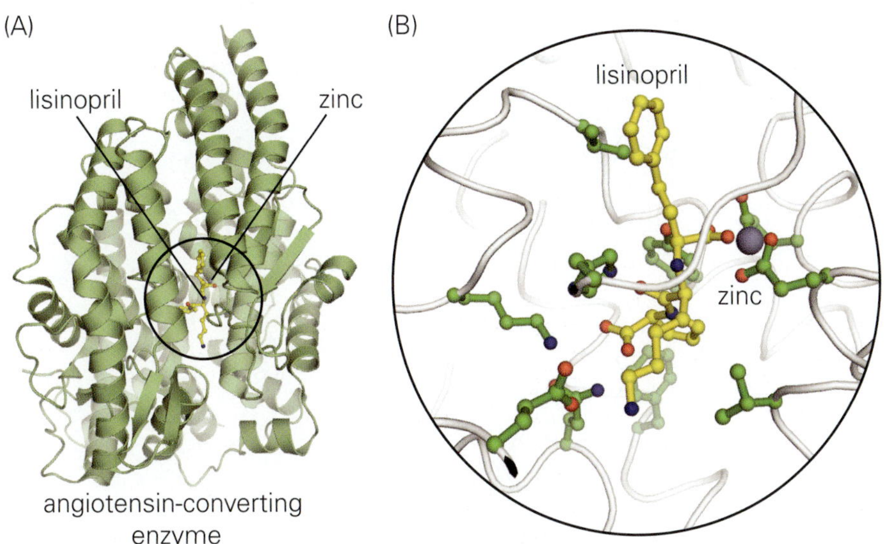


**Figure 16.44 The structure of angiotensin-converting enzyme (ACE), a zinc protease.** The active site is occupied by an inhibitor, lisinopril, an antihypertensive drug. The complementarity in shape and character of functional groups between enzyme and inhibitor is required to achieve the specificity needed for the application of an inhibitor as a drug. (PDB code: 1O86.)

The enzyme is substantially larger than trypsin and is mostly helical in secondary structure, with an interior cavity containing the zinc ion at the active site. The drug makes contact with many residues around the active site, which leads to its high specificity for ACE over the many other zinc proteases present in the human body. The drug-binding site is the same as that for substrate binding, leading to a competitive inhibition mechanism. The development of potent, selective enzyme inhibitors, such as lisinopril, is a major effort in the pharmaceutical industry.

## 16.27 Creatine kinase catalyzes phosphate transfer by stabilizing a planar phosphate intermediate

In Section 16.19, when discussing transition state stabilization as a general enzyme mechanism, we described how ATP hydrolysis by myosin involves a transition state with a planar phosphate group. The transfer of the terminal phosphate group of ATP, the most common phosphodonor in cells, to a variety of acceptors is generally important in biochemistry, and this reaction closely resembles the hydrolysis reaction. We have already encountered two different kinds of protein kinases (see Figures 16.11 and 16.22) that phosphorylate residues in proteins. Other kinases phosphorylate small molecules (for example, metabolites), thus generating key metabolic intermediates such as fructose-1,6,-bisphosphate (see Figure 16.9).

The enzyme creatine kinase carries out a reversible transfer of a phosphate group from ATP to the guanidine group of creatine, creating phosphocreatine and ADP, as shown in **Figure 16.45**. Phosphocreatine acts as an energy reservoir in cells such as muscle. When high energy demand begins to deplete ATP in such cells,

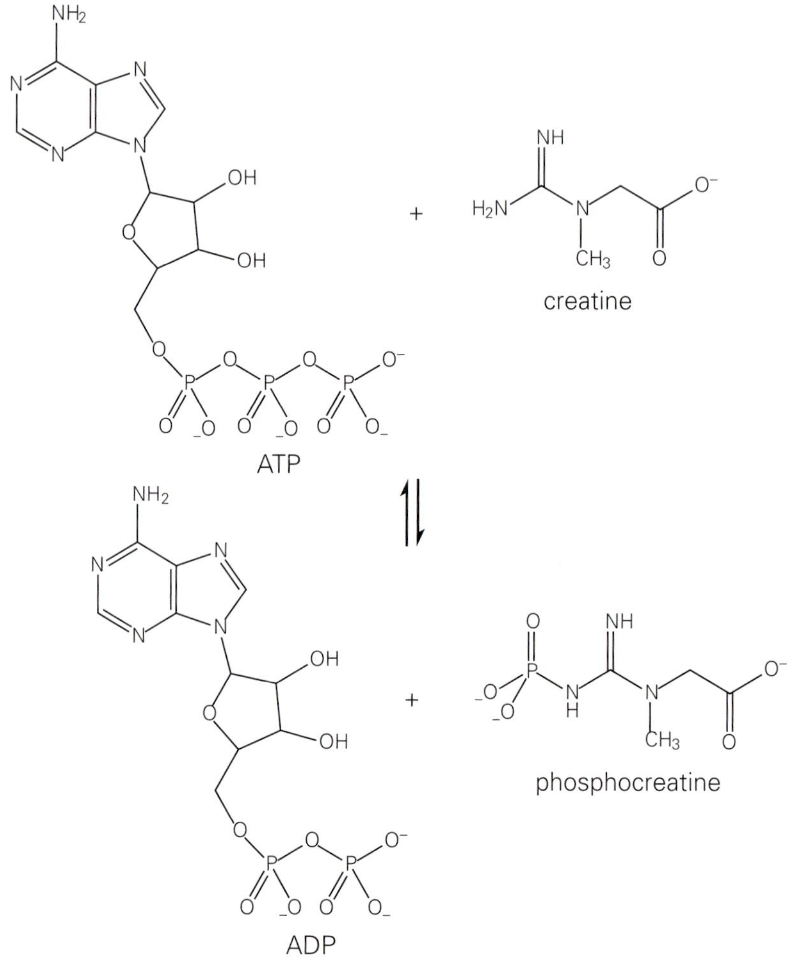

ATP + creatine ⇌ ADP + phosphocreatine

**Figure 16.45 The structures of reactants and products for the enzyme creatine kinase.** Phosphocreatine acts as an energy reserve for cells and can be quickly used to generate ATP through the action of this enzyme.

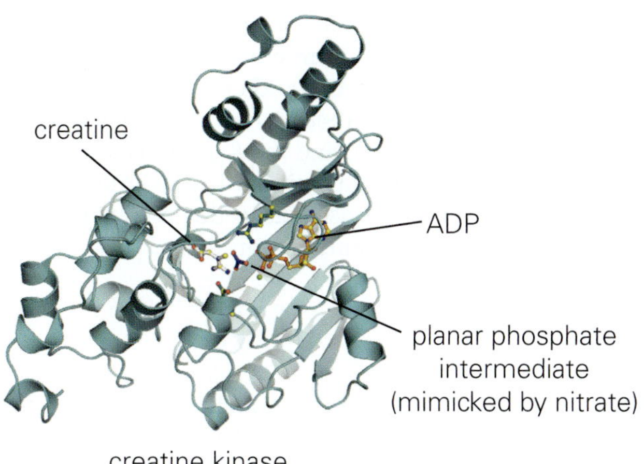


**Figure 16.46 The structure of creatine kinase.** The substrates bind in an active site cleft in the center of the enzyme. (PDB code: 1O86.)

the reverse transfer reaction occurs, with ADP being rephosphorylated to give ATP and creatine. This reversibility indicates that the free-energy difference between reactants and products under conditions in the cell must be quite small. A shift in concentration, such as when ATP decreases and ADP increases, alters the free energy enough that the reaction then reverses, and ATP is synthesized.

The two substrates bind in an interior cleft in the enzyme (**Figure 16.46**) in random order. As is true of many enzymes that use ATP, it is associated with a magnesium ion that bridges two of the phosphate groups, thus reducing the charge of the triphosphate. The two substrates are bound so that the receiving group of the creatine and the donor γ-phosphate group of the ATP are close together. The transfer of the phosphate is direct, with no covalent intermediate involving the enzyme.

The transition state of the reaction involves a distorted ATP molecule in which the terminal phosphate group has adopted a planar trigonal geometry. An actual transition state has a fleeting existence and, in order to visualize what it looks like, one has to use molecules that mimic essential aspects of the transition state but are stable and can therefore be subjected to structural analysis. In the structures shown in Figure 16.46 and **Figure 16.47**, the transition state of the reaction is mimicked by a complex of ADP and nitrate, a planar trigonal molecule. Because the central atom is nitrogen rather than phosphorus, it cannot actually react (recall that a vanadate group was used for the same purpose in the analysis of myosin; see Figure 16.31). The structure shown in Figure 16.47 supports the idea that the enzyme binds preferentially to the transition state, thereby stabilizing it. The stabilization of the transition state contributes significantly to catalysis by the protein.

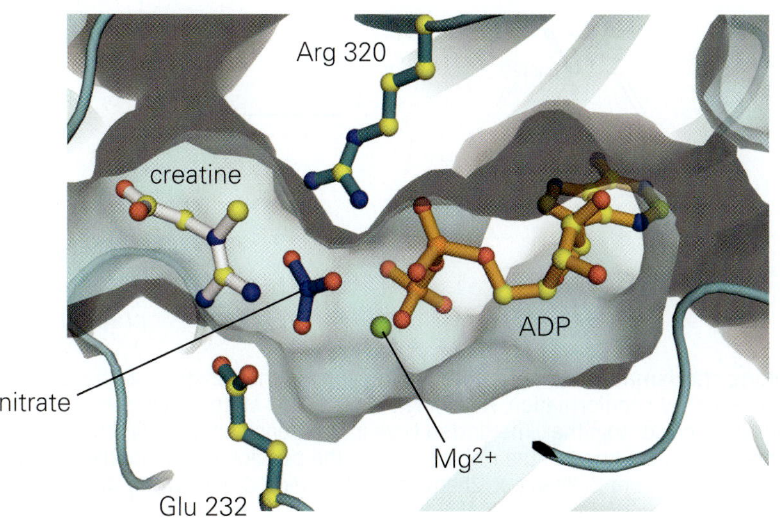


**Figure 16.47 The active site of creatine kinase.** A transition-state-like complex of creatine and ADP, with a planar nitrate ion ($NO_3^-$) mimicking the planar phosphate group of the true transition state. A few key sidechains from the enzyme, including an arginine stabilizing the transition state and a glutamate positioning the creatine substrate, are also shown. (PDB code: 1VRP.)

The phosphotransfer step is rate limiting in the reaction catalyzed by creatine kinase. For making phosphocreatine from creatine and ATP, the value of $k_{cat}$ is 160 $sec^{-1}$, and the $K_M$ values for Mg-ATP and creatine are 0.2 mM and 15 mM, respectively. In the reverse direction, the synthesis of ATP from phosphocreatine and ADP, the value of $k_{cat}$ is 470 $sec^{-1}$, and the $K_M$ values for Mg-ADP and phosphocreatine are 0.084 and 1.3 mM, respectively. The fact that these values for the forward and reverse reaction are so close tells us that the enzyme-catalyzed reaction is finely balanced between reactants and products.

## 16.28 Some enzymes work by populating disfavored conformations

One of the steps in the biosynthesis of the aromatic amino acids phenylalanine, tyrosine, and tryptophan is the isomerization of a complex intermediate, chorismate, to give prephenate, which is just a few chemical steps from the final amino acid products (this reaction was mentioned briefly in Chapter 15; see Figure 15.2). **Figure 16.48** shows the chemical transformation that occurs, a unimolecular reaction catalyzed by the enzyme chorismate mutase.

The transition state for the reaction requires rotation of the fragment that is being transferred, thus positioning the critical $\pi$ orbitals of the reacting double bonds in proximity. This conformation of the substrate can occur in solution, and indeed

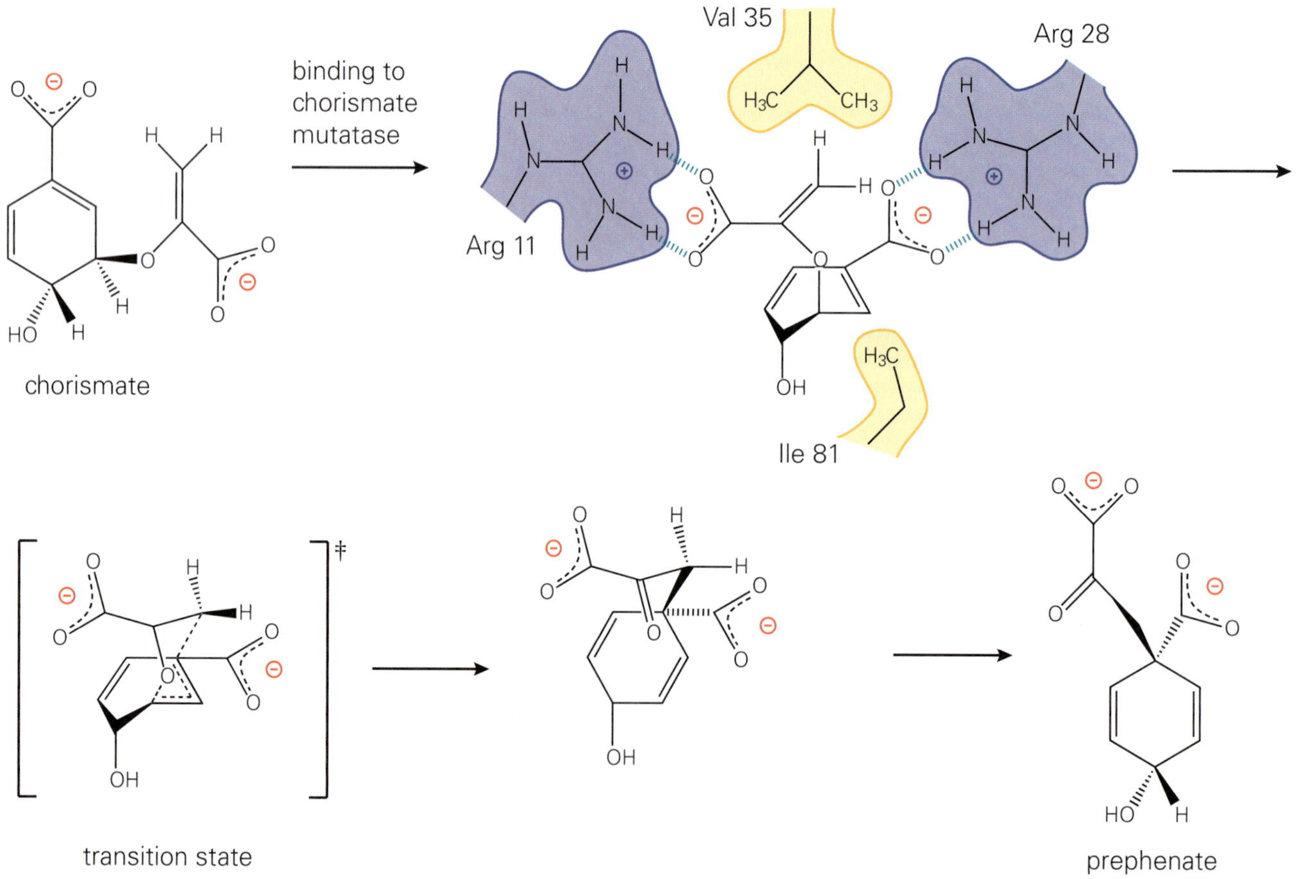

**Figure 16.48 The rearrangement reaction taking chorismate to prephenate.** Interactions of the substrate with the sidechains of the enzyme are shown, with the substrate in the reactive conformation. Reaching the transition state requires a further motion to bring the reacting orbitals of the double bonds even closer together. The dotted lines in the transition state (in brackets with a ‡) indicate bonds in which electrons are being rearranged as the bond to the oxygen breaks, the position of the double bond in the ring shifts, and the bond between the carbons forms. (Adapted from X. Zhang, X. Zhang, and T.C. Bruice, *Biochemistry* 44: 10443–10448, 2005. With permission from the American Chemical Society.)

the reaction does occur in water, but at a very slow rate. The enzyme speeds up the reaction by binding the substrate and holding it in the appropriate "near-attack" conformation shown in Figure 16.48. This conformation of the substrate is stabilized by specific contacts with sidechains of the enzyme, some of which are shown in Figure 16.48.

Even after the appropriate conformation is adopted by the substrate on the enzyme, the reaction requires a further structural fluctuation to occur that drives the orbitals together to overlap sufficiently for the new carbon–carbon bond to form and the old carbon–oxygen bond to break. It was found experimentally that the activation energy for the reaction is reduced from 84 kJ•mol$^{-1}$ in water to 50 kJ•mol$^{-1}$ on the enzyme. Computer calculations have shown that this 34 kJ•mol$^{-1}$ difference corresponds to the free energy cost of converting the chorismate from its low-energy states in water to the "near-attack" conformation required for the reaction. The energy required to alter the conformation comes from the energy of binding of the substrate to the enzyme. The remaining 50 kJ•mol$^{-1}$ of activation energy on the enzyme represents the energy required to drive the reacting orbitals together, and this energy is essentially the same whether it occurs in water or bound to the enzyme. Thus, catalysis in this enzyme system is achieved primarily through conformational restriction of the reacting substrate. The experimentally determined rate enhancement of the reaction is $2 \times 10^6$ relative to the reaction in solution. The $k_{cat}$ value for the enzyme-catalyzed reaction is 50 sec$^{-1}$ and the value of $K_M$ is 100 μM.

Chorismate mutase has a relatively simple fold, with three helices packed together with a second copy to make a dimer, as shown in **Figure 16.49**. The binding site for the chorismate is in a pocket in the center of the four-helix bundle formed through dimerization. There are positively charged sidechains that interact with the carboxylates of the substrate. There are also hydrogen bonds to heteroatoms of the substrate and some hydrophobic contacts that help induce the reactive conformation of the substrate when it is bound to the enzyme.

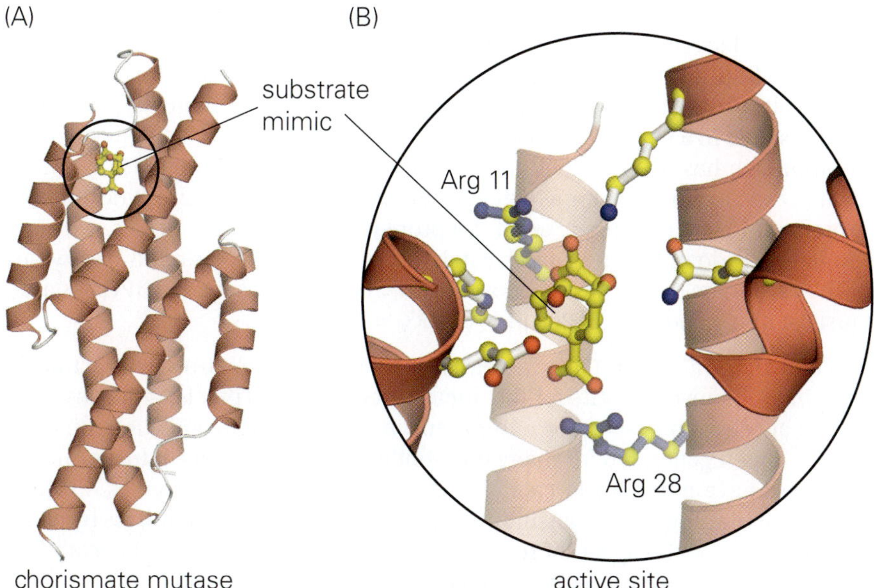


**Figure 16.49 The structure of a chorismate mutase.** (A) The enzyme is a dimer, with each subunit comprised of three helices that pack together in the dimer to form two four-helix bundles. The active sites in this structure are occupied by a substrate mimic (just one is shown) that has one of the key double bonds in the substrate reduced to a single bond, thus preventing the reorganization of the bonds that leads to product. (B) The expanded view of the active site shows the contacts between the enzyme and substrate or inhibitor. These contacts contribute to binding the substrate in the productive conformation, an important component of catalysis in this case. (PDB code: 1ECM.)

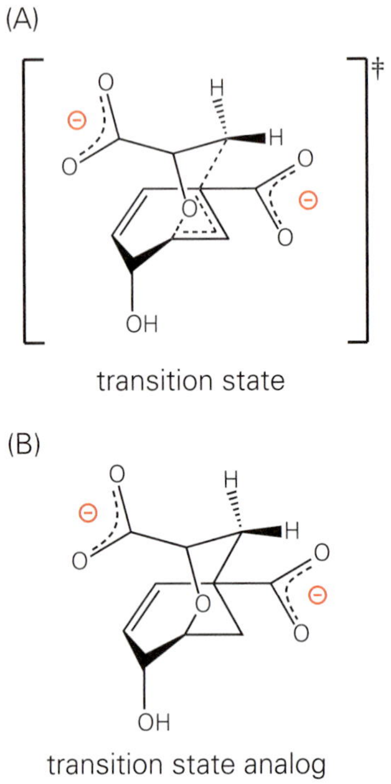

(A) transition state

(B) transition state analog

**Figure 16.50 The transition state for the reaction catalyzed by chorismate mutase.** Below the drawing of the transition state (A) is (B) the structure of the transition state analog that is bound to the enzyme in the structure shown in Figure 16.49.

The structure of chorismate mutase shown in Figure 16.49 has an inhibitor bound. This inhibitor is an example of a transition state analog—that is, a compound that looks like the transition state for the reaction in both composition and geometry, but is altered in some key way to prevent reaction. In this case, the reaction requires the presence of double bonds so that there can be a concerted shift in the bonding to give product (**Figure 16.50**). In the transition state analog, both the bonds to the oxygen and carbon are present, but the key ring double bond in substrate that shifts upon reaction is missing, so no reaction can occur. Because the geometry of the inhibitor is very similar to the transition state, with which the enzyme interacts strongly, the contacts to the enzyme lead to tight binding of the inhibitor.

## 16.29 Antibodies that bind transition state analogs can have catalytic activity

An interesting exploitation of the transition state stabilization concept is the development of antibodies (produced by the body to recognize "foreign" molecules) that are enzymes (that is, are catalytic). Catalytic antibodies are generated by first synthesizing molecules that are stable analogs of the transition state for the reaction to be catalyzed. Such analogs were mentioned in the previous section as very potent enzyme inhibitors, since they can bind tightly to the enzyme, which is optimized to stabilize the transition state. One example of such a compound is shown in Figure 16.50 for the chorismate mutase reaction.

The rationale behind catalytic antibodies is to turn this idea around and to find antibodies that bind the transition state analog tightly, which should then stabilize, in turn, the transition state itself. Such antibodies are generated by injecting the transition state analogs into animals (attached to a carrier protein to make them more immunogenic) to induce antibodies that recognize them. The antibodies are then screened for catalysis, using the substrate for which the analog mimicked the transition state.

For some transition state analogs, a significant fraction of the antibodies produced are in fact catalytic—that is, they act as enzymes for the desired reaction. One of the first reactions for which a catalytic antibody was developed is a simple hydrolysis of a carbonate. The transition state mimic contains a phosphate, which is stable in a tetrahedral geometry in place of the carbonate (**Figure 16.51**). This approach has been applied to many reactions for which there is no natural enzyme, but also to some for which natural enzymes exist. The catalytic antibodies produced have essentially all of the properties of natural enzymes, although they generally do not have as much catalytic power ($k_{cat}/K_M$) as natural enzymes. For the carbonate hydrolysis reaction shown, an antibody called MOPC has a $k_{cat}$ of just 0.4 $min^{-1}$ and a $K_M$ of 208 μM ($k_{cat}/K_M = 32$).

The catalytic antibody MOPC increases the rate of the hydrolysis reaction by about 1000-fold over the rate for the uncatalyzed reaction (that is, much less than the rate accelerations seen for natural enzymes). Although catalytic antibodies have now been generated for many different chemical reactions, it turns out that all of these have much lower catalytic efficiency that natural enzymes do. Thus, preferential binding to the transition state is just one of the mechanisms by which enzymes increase the speed of reactions and, by itself, transition state stabilization is not enough to explain the speed of enzyme-catalyzed reactions.

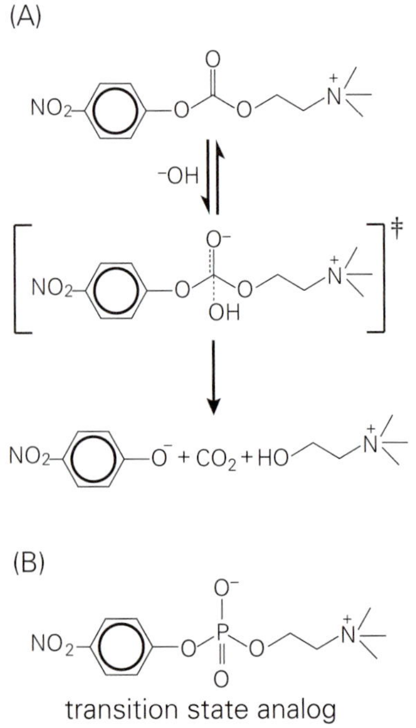

(A)

(B) transition state analog

**Figure 16.51 Transition state for hydrolysis of a carbonate.** (A) The reaction for hydrolysis of a carbonate by hydroxide is shown, with a tetrahedral intermediate transition state. (B) A phosphate analog that mimics the transition state with the same geometry and charge distribution, but is a stable molecule. Antibodies that recognize this molecule catalyze the hydrolysis reaction by stabilizing the transition state. (From S.J. Pollack, J.W. Jacobs, and P.G. Schultz, *Science* 234: 1570–1573, 1986. With permission from the AAAS.)

# D. RNA ENZYMES

Catalytic RNAs are known as **ribozymes**, and in this part of the chapter we will focus on ribozymes that can function, at least under some conditions, without the help of proteins. You should know, however, that most RNAs in the cell are bound to proteins in **ribonucleoprotein complexes**. Two such complexes (the ribosome and the spliceosome) catalyze two of the fundamental processes that maintain and express the genome—namely, protein synthesis and mRNA splicing. Other ribonucleoprotein complexes make essential contributions to RNA processing, protein translocation across membranes, gene silencing, RNA export, and the addition of telomeric DNA repeats to chromosomes ends.

**Ribozymes**

RNA molecules that catalyze chemical reactions are known as ribozymes.

The same principles that guide our understanding of protein enzymes—namely, the concept of the transition state for a reaction and the idea that a catalyst works in part by lowering the energy of activation—apply to the study of ribozymes. As for protein enzymes, RNA catalysts lower the energy of activation by interacting favorably with the transition state, thereby lowering its energy, without changing the energy of the reactants or products.

Most natural ribozymes catalyze phosphoryl transfer reactions, which require the activation of either the 2′-hydroxyl group in the ribose or a water molecule for nucleophilic attack of a phosphodiester bond. In the following sections, we describe two kinds of ribozymes: self-cleaving ribozymes and self-splicing introns.

## 16.30 Small self-cleaving ribozymes and ribonuclease proteins catalyze the same reaction

Small ribozymes that catalyze the cleavage of their own RNA backbone are called **self-cleaving ribozymes**. This is a large group of ribozymes that are commonly found in plant pathogens. All of these catalyze reversible phosphodiester cleavage reactions. In these reactions, a 2′-hydroxyl group in the ribozyme adjacent to the phosphodiester bond that is cleaved acts as the nucleophile (**Figure 16.52**) and the products are two oligonucleotides with new 5′ hydroxyl and 2′,3′-cyclic phosphate chain termini.

In these self-cleaving reactions, the reaction proceeds with an inversion of the stereochemical configuration of the nonbridging oxygen atoms that are bound to the phosphorus that is being attacked. This implies that the mechanism involves an $S_N2$-type reaction with a trigonal bipyramidal transition state, in which five oxygens form transient bonds with phosphorus (see Figure 16.52).

This internal phosphoryl transfer reaction is very similar to the reaction catalyzed by ribonuclease-A, a protein enzyme discussed in Section 16.20. The strand cleavage reaction catalyzed by ribonuclease-A and the self-cleaving ribozymes both involve the attack of the 2′ OH on the scissile phosphate (see Figure 16.52). Resolution of the transition state results in 2′,3′-cyclic phosphate and free 5′-OH products. Both the RNA catalysts and the ribonuclease protein use electrostatic stabilization, acid–base catalysis, and geometric constraints to promote this reaction.

## 16.31 Self-cleaving ribozymes use nucleotide bases for catalysis, even though these do not have p$K_a$ values well suited for proton transfer

While the concerted acid–base mechanism of the ribonuclease-A protein (see Figure 16.32) provides a framework for thinking about how self-cleaving ribozymes might work, the comparison breaks down because the chemical properties of amino acids and ribonucleotides are so different. In fact, the properties that make RNA particularly well suited for the storage and transmission of genetic information through complementary base pairing also make it less adept at acid–base catalysis.

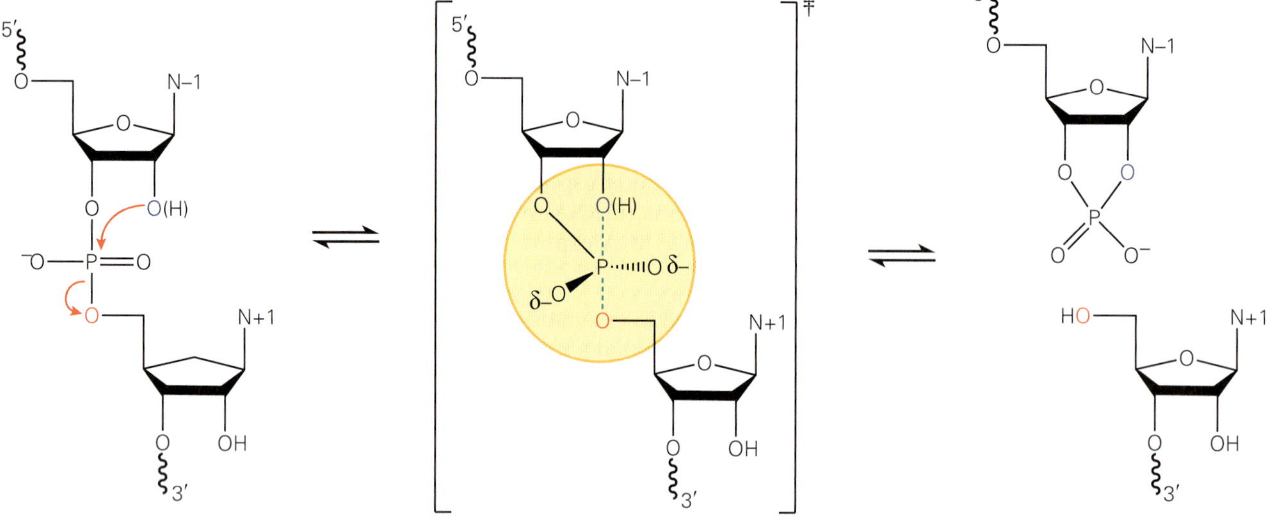


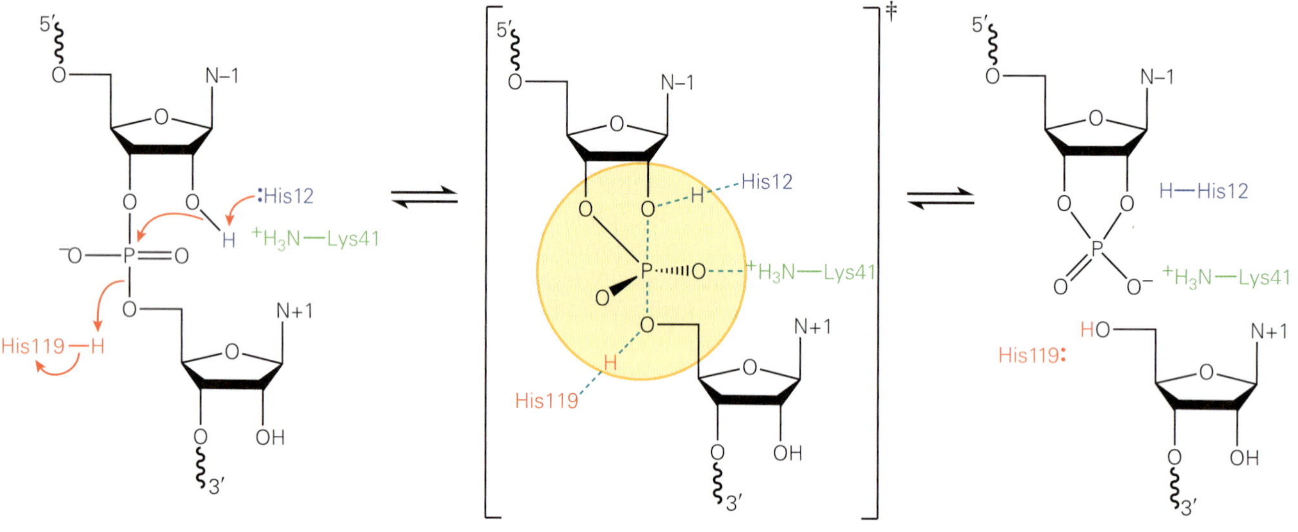


**Figure 16.52 Mechanisms of a self-cleaving ribozyme and the protein ribonuclease-A.** In both cases, the cyclic phosphate intermediate that is formed is converted to a free phosphate group by reaction with water. (A) The self-cleaving reaction mechanism. The reaction occurs with an inversion of the configuration of the phosphate group. This implies an $S_N2$-type in-line attack mechanism with a trigonal bipyramidal transition state (*yellow* shading). N–1 and N+1 refer to the nucleobases before and after the cleavage site, respectively. (B) Mechanism of ribonuclease-A, which catalyzes a similar reaction. See Figure 16.32 for more information on this enzyme.

If a ribonucleotide is to participate in acid-base catalysis, an unprotonated functional group, the base, must remove a proton and a protonated functional group, the acid, must donate a proton. This type of proton shuffling depends on a ribonucleotide being in the appropriate ionization state and being acidic or basic enough to donate or accept protons. RNA lacks a functional group that has a p$K_a$ value close to neutrality, making it less well suited to acid-base catalysis than proteins, which are able to use the histidine sidechain or, in some cases, the aspartate or glutamate sidechains.

In solution, ribonucleotide functional groups have ionization equilibria that are far outside the neutral pH range. For example, cytidine, adenosine, and guanosine have p$K_a$ values that are in the acidic range, and uridine and guanosine have p$K_a$ values in the basic range, as shown in **Figure 16.53.** Thus, only a small fraction of these nucleotide bases would be in the right ionization state to accept protons (act as a base) or donate protons (act as an acid) at neutral pH.

The expectation that no RNA functional groups would be available to function as an acid or a base at neutral pH led to the idea that all ribozymes would turn out to require metals for catalysis. Metal ions could, for example, stabilize negatively charged leaving groups or stabilize the negative charge that develops in the

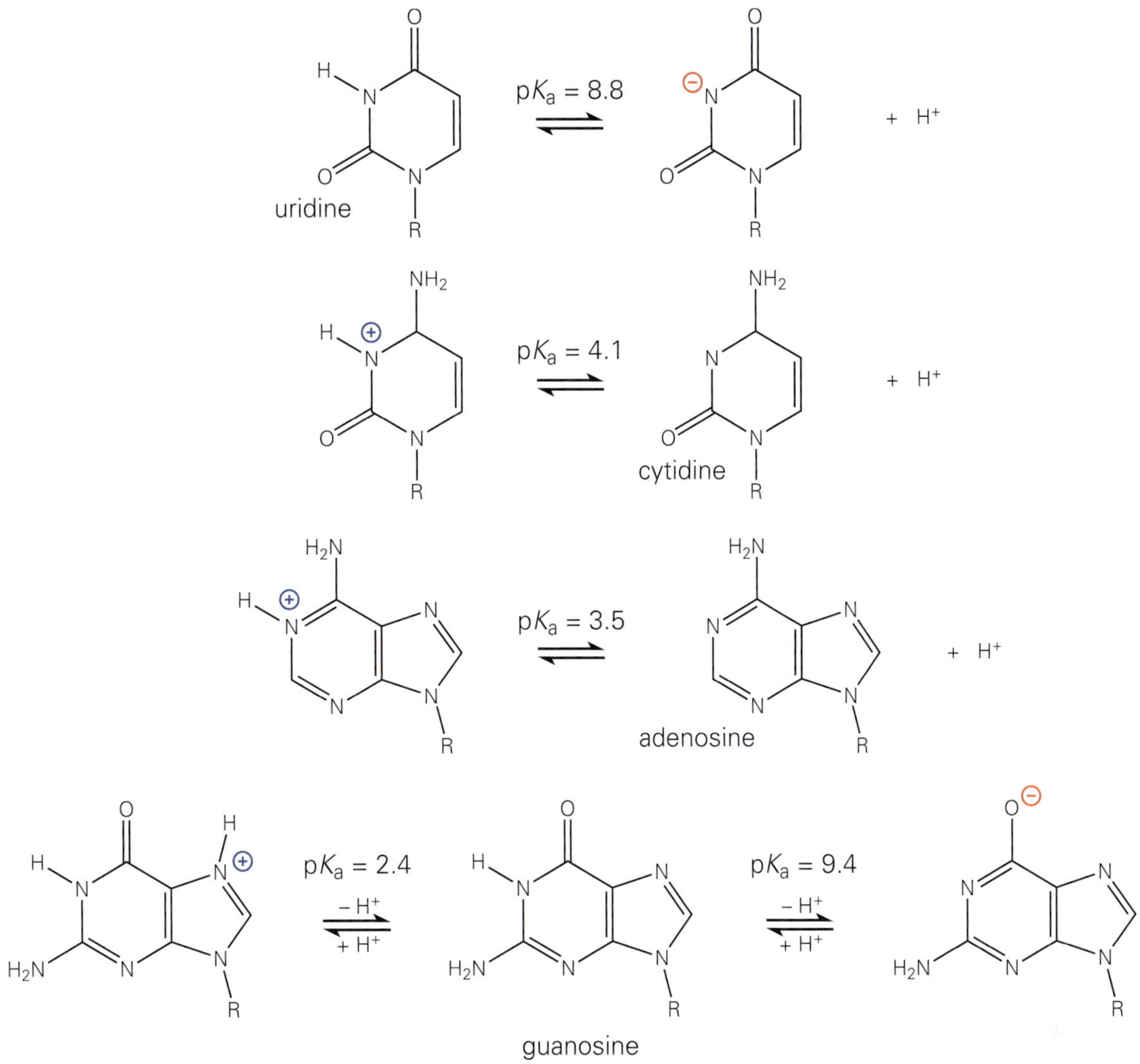

uridine

$pK_a = 8.8$

$+ H^+$

$pK_a = 4.1$

$+ H^+$

cytidine

$pK_a = 3.5$

$+ H^+$

adenosine

$pK_a = 2.4$

$- H^+$

$+ H^+$

guanosine

$pK_a = 9.4$

$- H^+$

$+ H^+$

**Figure 16.53 The p$K_a$ values of nucleotide bases.** The four nucleotides are shown in their typical tautomeric forms and the p$K_a$ values for the protonation of specific groups are shown. (Values for pyrimidines from J. Saurina et al. and A. Izquierdo-Ridorsa, *Anal. Chim. Acta* 408: 135–143, 2000; values for purines from S. Steenken, *Chem. Rev.* 89: 503–520, 1989.)

transition state. Metal-bound water could mediate general acid-base catalysis by donating protons to oxyanion leaving groups or by accepting protons from nucleophilic oxygens. Finally, the binding of metal ligands could help align reactants, thereby reducing entropic barriers to reaction.

While some ribozymes do require specific divalent metal ions for catalysis, self-cleaving ribozymes can promote catalysis in the absence of divalent cations by using only nucleotides. As discussed in Section 16.20, proteins can shift the p$K_a$ values of sidechains to optimize an amino acid for general acid-base catalysis and so can RNA. The presence of charged phosphate groups or cations in self-cleaving ribozymes results in appropriate shifts in the p$K_a$ values of the bases that participate in catalysis. In addition, active site organization and transition state electrostatic stabilization play a role in promoting catalysis.

## 16.32 Hairpin ribozymes optimize hydrogen bonds to the transition state rather than to the initial or final states

The self-cleaving ribozymes achieve specificity—that is, they cleave only at phosphodiester bonds within specific sequence contexts—by using base-pairing and other interactions to align the cleavage site within the RNA active site. We will consider the active site of one such ribozyme, called the hairpin ribozyme, in more detail. This ribozyme comes from the tobacco ringspot virus satellite RNA (satellite RNAs depend on their associated viruses for replication and encapsidation), where it plays an essential role in the processing of the viral RNA after replication. The active site of the ribozyme is in a cleft between two structural domains (denoted A and B) and includes nucleotides from helices in both domains (**Figure 16.54**).

Our understanding of the catalytic mechanism of the hairpin ribozymes is based on crystal structures of the state prior to catalysis, a state that mimics the transition

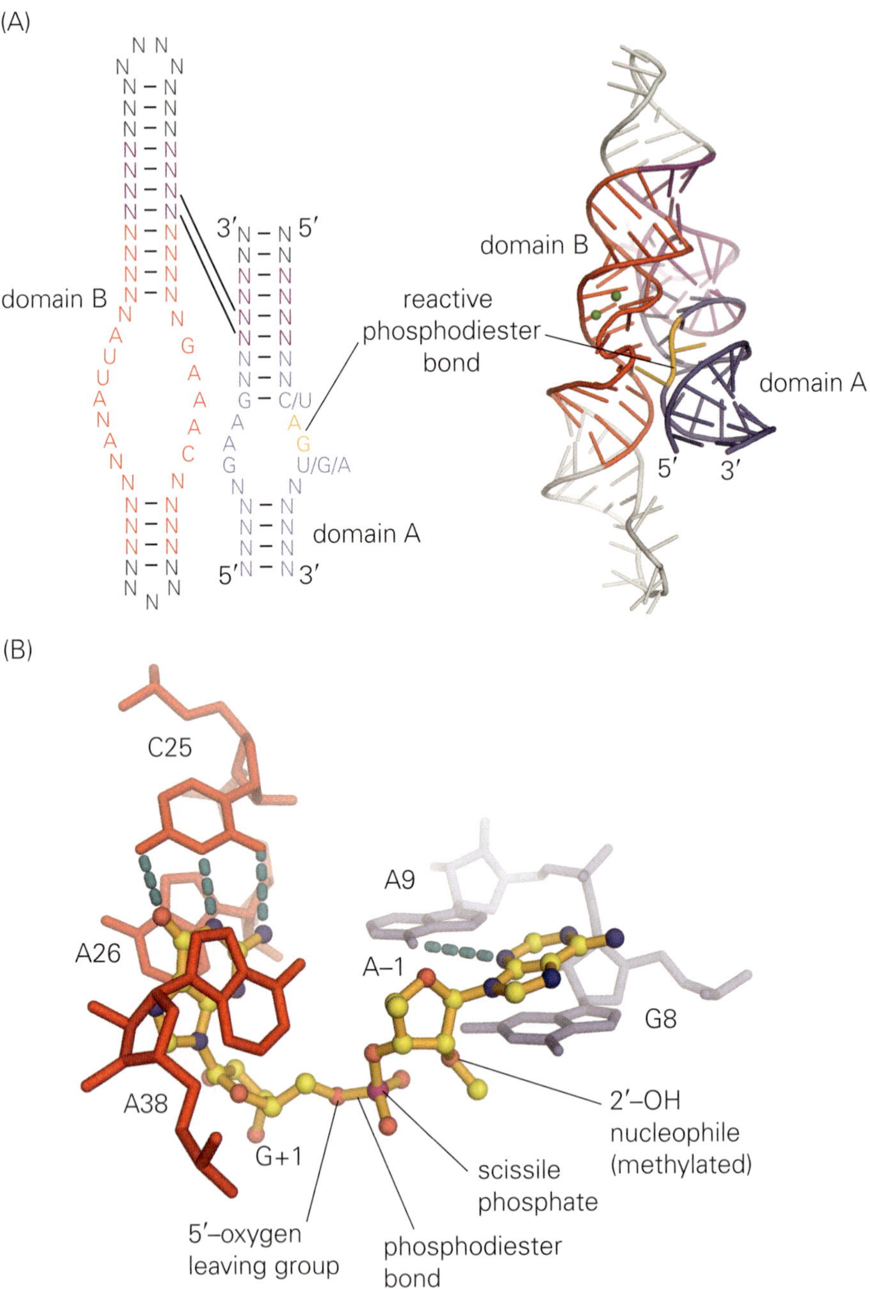


**Figure 16.54 Hairpin ribozyme.** (A) A secondary structure schematic of the hairpin ribozyme and the structure of the hairpin ribozyme complexed with a noncleavable substrate analog. The reactive phosphodiester bond is between the nucleotides colored yellow in both diagrams. The green spheres represent two bound calcium ions. (B) Active site of the hairpin ribozyme. The 2′-OH nucleophile (methylated to prevent reaction), the scissile phosphate, and 5′-oxygen leaving group are shown. Hydrogen bonds are shown as dashed lines. (A, adapted from P.B. Rupert and A.R. Ferre-D'Amare, *Nature* 410: 780–786, 2001; B, adapted from P.B. Rupert et al. and A.R. Ferre-D'Amare, *Science* 298: 1421–1424, 2002. With permission from the AAAS; PDB code: 1M5K.)

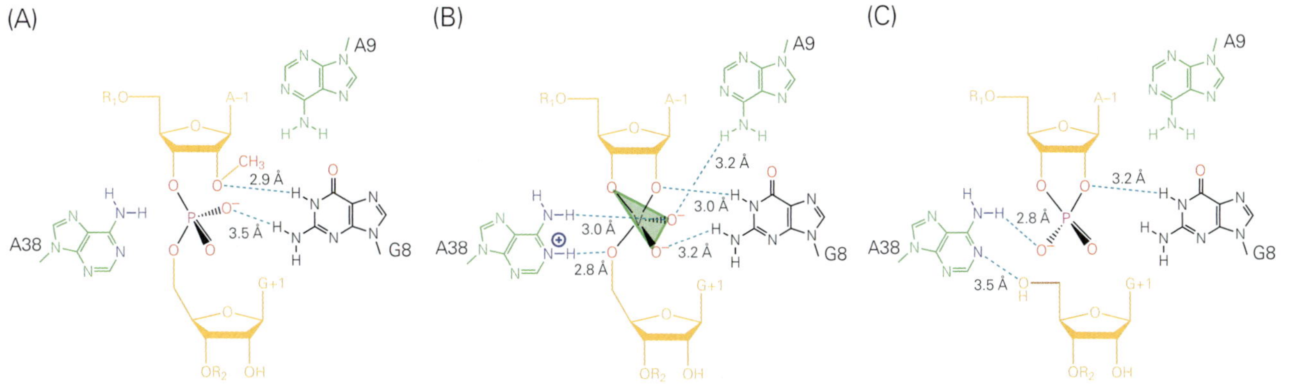

**Figure 16.55 Summary of active-site hydrogen bonds in the hairpin ribozyme.** (A) Precursor complex stabilized by two hydrogen bonds (*dashed* lines) from G8. (B) Transition state mimic complex stabilized by five hydrogen bonds. The *green* triangle connects the equatorial oxygens of the vanadate group. Protonation of A38 is inferred from the distance between N1 of A38 and the 5′-oxygen. (C) Product complex stabilized by three hydrogen bonds. (Adapted from P.B. Rupert et al. and A.R. Ferre-D'Amare, *Science* 298: 1421–1424, 2002. With permission from the AAAS; PDB codes: A, 1M5K; B, 1M5O; C, 1M5V.)

state, and the product state (**Figure 16.55**). The precursor state was observed by replacing the nucleophilic 2′-hydroxyl proton with a methyl group (see Figure 16.55A). A view of the transition state was obtained by using a close structural and chemical mimic of the trigonal bipyramidal phosphate transition state, the vanadate ion ($VO_4^{3-}$), as shown in Figure 16.55B. The product structure was observed by using an all-RNA substrate and trapping the cleaved form, with a 2′,3′-cyclic phosphate, in the crystals (see Figure 16.55C).

The number of hydrogen bonds seen in each of these structures suggests that this catalytic RNA binds most tightly to substrate in the transition state. In the precursor, the base of the G8 nucleotide makes two hydrogen bonds: one to the 2′-OH nucleophile and one to a phosphate oxygen. In the transition state mimic, five hydrogen bonds are formed between the bases of G8, A9, and A38 in the active site and oxygens of the vanadate ion. In the product, the bases of G8 and A38 make two hydrogen bonds to the cyclic phosphate and one bond to the 5′-OH leaving group. Thus, the hairpin ribozyme has evolved to maximize its hydrogen bonding interactions with the transition state, rather than with the starting or ending state.

## 16.33 There are at least two possible mechanisms for bond cleavage by the hairpin ribozyme

Structures of substrate and transition state complexes of the hairpin ribozyme place the G8 and A38 nucleotide bases close to the reactive phosphate (see Figure 16.55). The structures are equally consistent with two distinct mechanisms for how these two nucleotide bases participate in catalysis. But, regardless of the exact details, it is the active-site nucleotide bases that participate in catalysis, without the direct assistance of metal ions.

In the first mechanism, which is illustrated in **Figure 16.56**A, G8 and A38 accept and donate protons, respectively, and therefore act as general acid–base catalysts. In this mechanism, A38 must be protonated for cleavage to occur, allowing it to donate a proton to the 5′ oxygen. The guanosine base (G8) starts off negatively charged and withdraws a proton from the 2′-hydroxyl group, becoming neutral after cleavage. The two bases also make other interactions with the substrate. The N2 exocyclic amine of G8 interacts with a nonbridging phosphoryl oxygen and the exocyclic amine of A38 interacts with the other nonbridging phosphoryl oxygen.

In the second possible mechanism, shown in Figure 16.56B, the guanosine and the adenosine do not change their charge states. Instead, both G8 and A38 directly

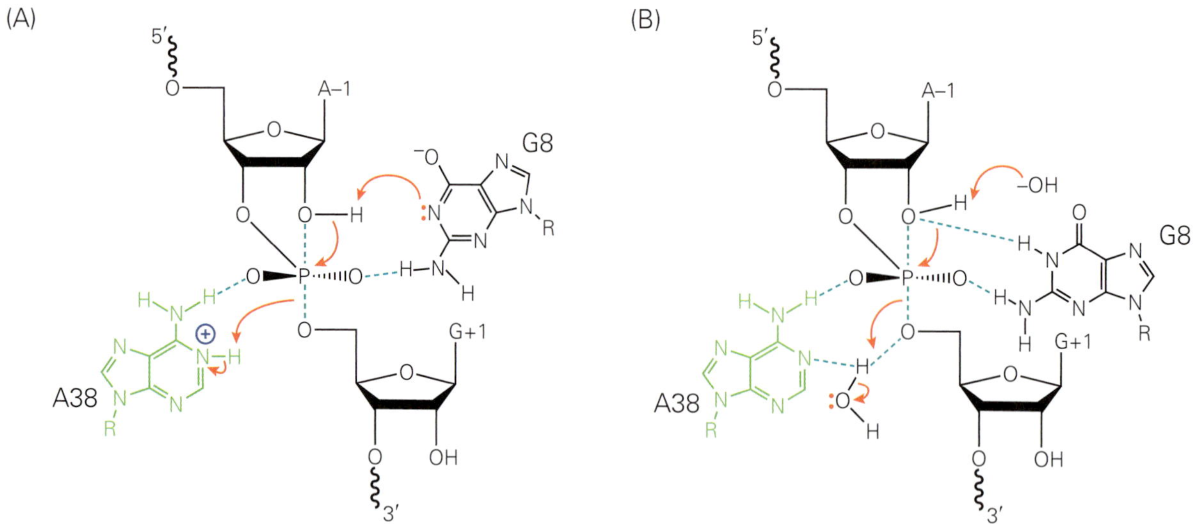


**Figure 16.56 Possible mechanisms for the hairpin ribozyme.** (A) G8 and A38 act as general acid–base catalysts. (B) Hydrogen-bond formation between the phosphate group and the bases positions the phosphate group for catalysis and provides electrostatic stabilization to the pentacovalent transition state, in which proton transfer occurs through specific acid–base catalysis, mediated by water molecules. (Adapted from M.J. Fedor, *Annu. Rev. Biophys.* 38: 271–299, 2009. With permission from Annual Reviews.)

stabilize the transition state and water molecules mediate the required proton transfer reactions. In this model, A38 accepts a hydrogen bond from a bridging 5′ oxygen that is protonated by water during cleavage. The amide group of G8, in its protonated form, donates hydrogen bonds to the 2′ and phosphoryl oxygens to provide electrostatic stabilization as a negative charge develops in the transition state. A water molecule observed near the 2′ OH of A1 (in the substrate) suggests that proton transfer could occur through specific acid-base catalysis.

## 16.34 The splicing reaction catalyzed by group I introns occurs in two steps

Group I introns are found in bacteria, lower eukaryotes, and higher plants. They occur as insertions into genes for rRNA, tRNA and mRNAs. These introns cleave themselves out of precursor RNAs, leaving behind a spliced RNA (see Section 1.18 for a general description of splicing carried out by the spliceosome, a ribonucleoprotein complex, through a different mechanism). All group I introns share a common core that consists of conserved base-paired elements designated as P1–P9, as shown schematically in **Figure 16.57** (the three-dimensional structure of a group I intron is shown in Figure 2.36). Group I introns were one of the first ribozymes to be discovered.

Group I intron self-splicing occurs in two steps, as illustrated in Figure 16.57, **Figure 16.58**, and **Figure 16.59**. In the unprocessed RNA two exons, called the 5′ exon and the 3′ exon, separated by the intron, are joined and the intron is excised. In the first step, which is shown in detail in Figure 16.58, there are two substrates—an exogenous guanosine (G) that is not a part of the self-splicing RNA, and a short intramolecular duplex that includes the 5′ exon. The 3′-OH nucleophile of the exogenous guanosine attacks the 5′-exon–intron junction. This cleaves the junction and adds guanosine to the 5′ end of the intron (see Figures 16.58 and 16.59A).

The first reaction is followed by a conformational change in which the exogenous guanosine leaves the site where it had been bound, and it is replaced by a conserved guanosine at the 3′ end of the intron (this is referred to as the ΩG

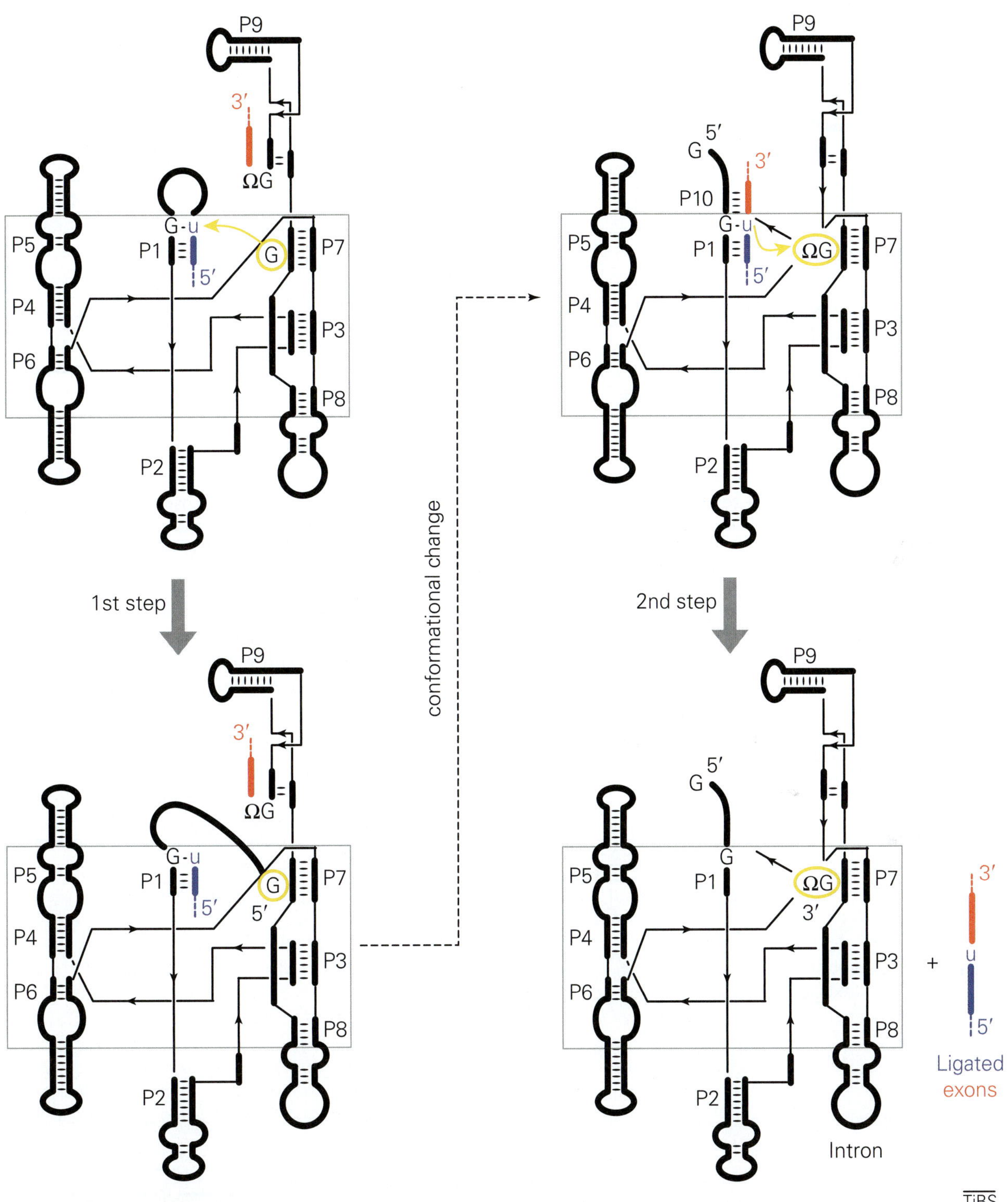


**Figure 16.57 Self-splicing in a group I intron**. The typical secondary structure of a group I intron consists of a number of conserved base-paired helices (P1–P9). The central catalytic core of the intron RNA is boxed in each of these diagrams. The reaction begins with the structure in the upper left, in which the 5′ end of the exon (*blue*) is base-paired with an internal guide sequence in the intron RNA (P1). In the first step of the reaction, an exogenous guanosine cofactor (G, circled in *yellow*) that is not part of the intron or the exon attacks the 5′ splice site (*yellow* arrow) in a nucleophilic reaction. A conformational change then occurs, in which the structure shown at the lower left converts to that shown on the upper right. During this transition, the exogenous guanosine is displaced by the G residue at the 3′ end of the intron (denoted ΩG, next to the intron strand in *red*). This reaction corresponds to the first reaction, but in the backward direction. The structure at the end, shown in the lower right, has the exogenous G covalently linked to the 5′ end of the intron, and the two parts of the exon (*red* and *blue*) are now ligated. (Adapted from Q. Vicens and T.R. Cech, *Trends Biochem. Sci*. 31: 41–51, 2006. With permission from Elsevier.)

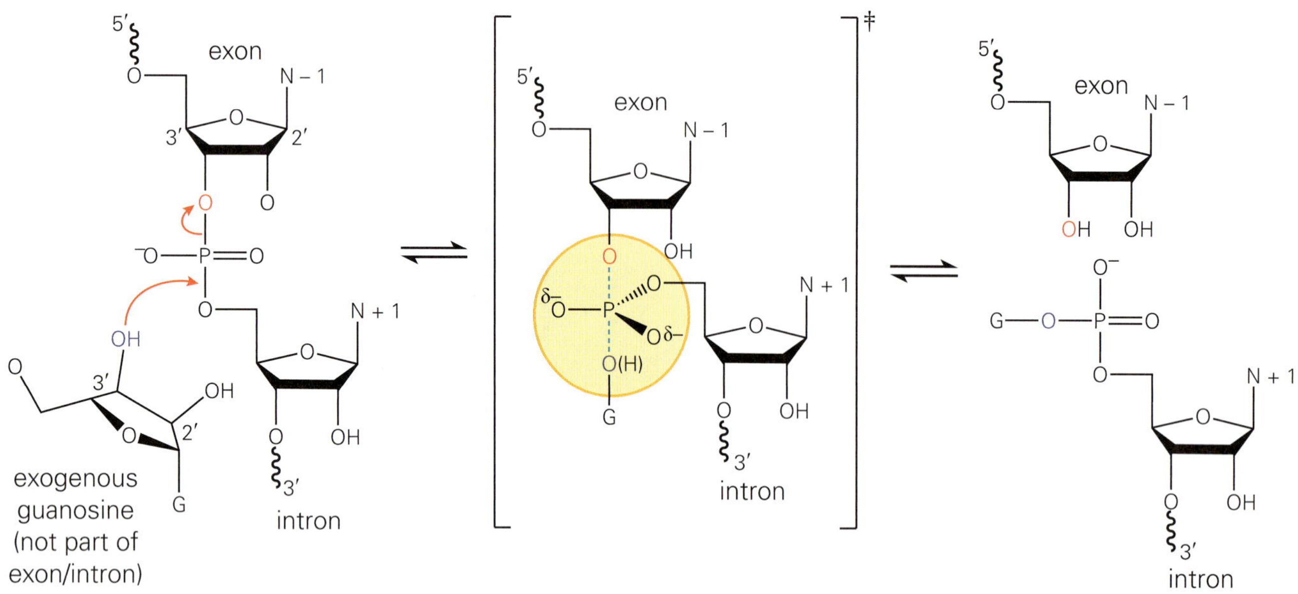


**Figure 16.58 The self-splicing reaction.** The 3′-hydroxyl group of an exogenous guanosine (*blue*) is the attacking nucleophile in the first step of group I self-splicing. The bridging 3′ oxygen (*red*) is the leaving group, and the new 3′ end of the 5′ exon becomes the attacking nucleophile during the second step of splicing (not shown). Refer to Figure 16.59 for a schematic view of the complete splicing reaction.

nucleotide, as indicated in Figures 16.57 and 16.59B). In the second reaction, the 3′ OH of the 5′ exon carries out nucleophilic attack on the ΩG nucleotide of the intron. This results in ligation of the exons and release of the linear intron (see Figure 16.59B). Chemically, the second step is the reverse of the first step.

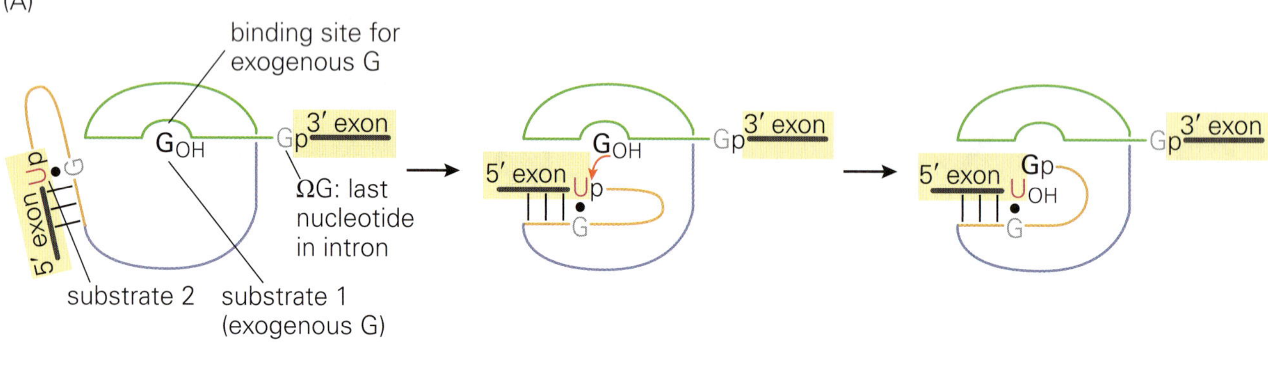


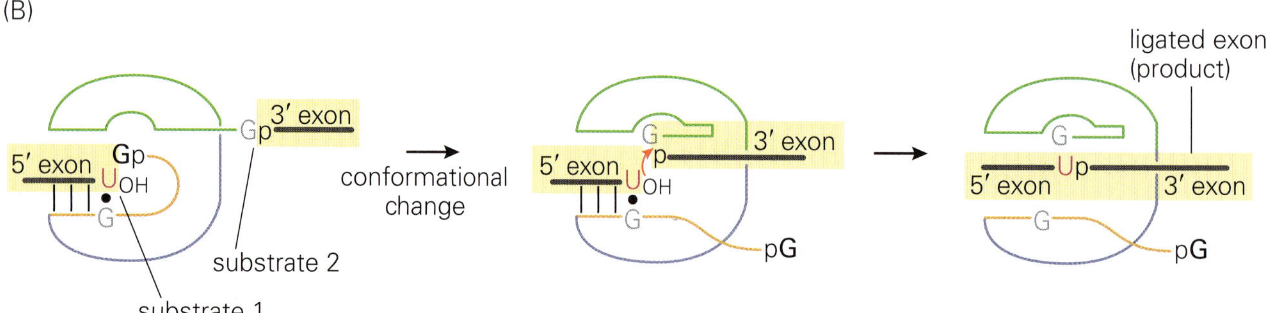


**Figure 16.59 The two steps of the group I intron self-splicing reaction.** These diagrams provide a highly simplified version of the reaction shown in Figure 16.57. (A) Step 1. There are two substrates, an exogenous guanosine nucleoside (G, in *black*) and a short duplex formed between the 5′ exon and the intron. The first reaction adds the guanosine nucleoside to the 5′ end of the intron and cleaves the 5′ exon–intron junction. (B) Step 2. The second reaction links the two exons and releases the intron. (Adapted from D.M.J. Lilley and F. Eckstein, Ribozymes and RNA Catalysis. Cambridge, UK: RSC Publishing, 2008. With permission from The Royal Society of Chemistry.)

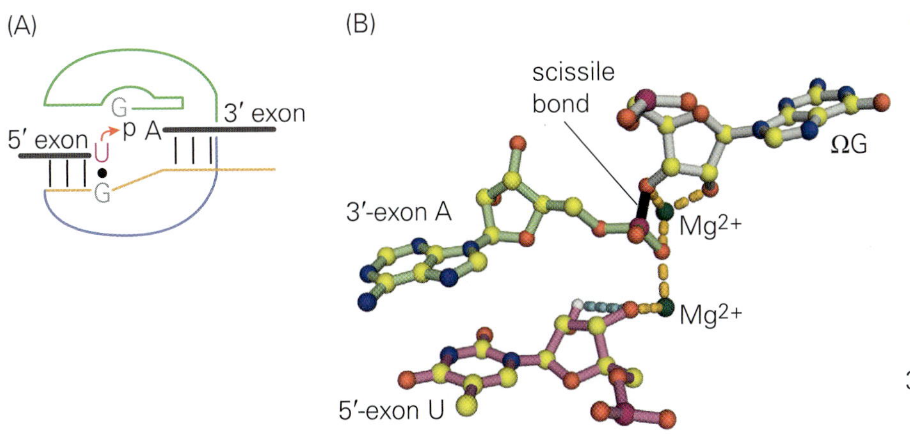


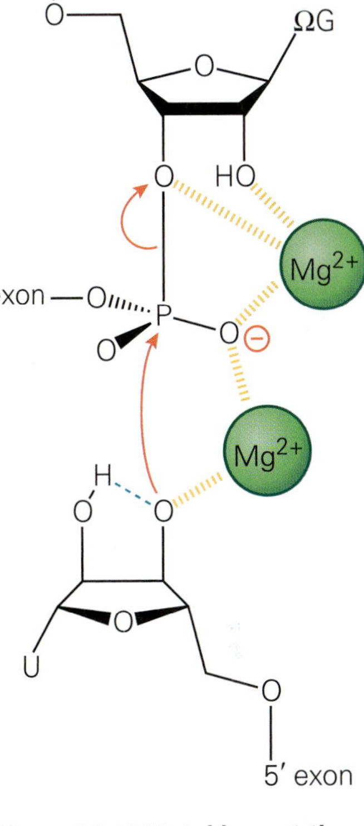


**Figure 16.60 Metal ions at the active site of a group I intron.** (A) Schematic diagram of the second step of the self-splicing reaction (see also Figure 16.59B). (B) Structure of the active site of a group I intron, at a state just before the chemical reaction. Two $Mg^{2+}$ ions that coordinate the phosphate linkage that will be cleaved are shown. (C) Role of $Mg^{2+}$ ions in stabilizing the transition state of the reaction. (PBD code: 1ZZN.)

## 16.35 Metal ions facilitate catalysis by group I introns

Earlier in this chapter, we have mentioned how protein enzymes use metal ions to activate water molecules for the hydrolytic cleavage of nucleotides or peptides (see Figures 16.29 and 16.47) or for the transfer of a phosphate group from ATP to a protein sidechain or a small molecule (Sections 16.14 and 16.27). In Chapter 19, we will discuss the mechanism of DNA polymerases, which use metal ions to facilitate the incorporation of nucleotides into a growing DNA chain (see Section 19.17). Like these protein enzymes, the group I introns use metal ions to catalyze the RNA cleavage and subsequent ligation step.

Metal ions at the active site of group I introns help activate the nucleophiles, stabilize leaving groups, and neutralize charge buildup in the transition state. The structure of the active site of a group I intron just before the second step of the reaction is shown in **Figure 16.60**. Two $Mg^{2+}$ ions are seen at the active site. One of these stabilizes the developing negative charge on the leaving group oxygen in the transition state. The other $Mg^{2+}$ ion helps deprotonate the 3′ oxygen of the G nucleophile.

The way in which these two metal ions assist catalysis in the group I intron is remarkably similar to the way in which metal ions participate in the DNA synthesis reaction catalyzed by DNA polymerase (these enzymes are discussed in Chapter 19). The group I intron RNA uses its phosphate backbone to form an active-site metal binding pocket (**Figure 16.61**A) in the same way that the polymerase uses the carboxylate groups of aspartate sidechains (Figure 16.61B).

## 16.36 Substitution of oxygen by sulfur in RNA helps identify metals that participate in catalysis

It is not straightforward to establish the importance of metal ions in catalysis by ribozymes. In the case of most protein enzymes, we can simply remove the metal ions and see if the enzyme is still active. Such a strategy works for most proteins because the metals usually play a specialized role in catalysis and are not required for the folding of the protein. But, as was discussed in Section 2.23, RNA folding depends upon the presence of metal ions that help overcome the substantial repulsion between the negatively charged phosphoryl groups of the RNA. As we shall see in Section 18.24, if we change the concentrations of metal ions, then we will affect the stability of folded RNA molecules drastically.

One way to distinguish those metal ions that stabilize RNA structures from those that directly participate in catalysis is to do so-called metal specificity-switch experiments. These experiments are designed to detect directly coordinated metal ions based on the different binding specificities of "hard" and "soft" metal ions for hard and soft ligands. **Hard metal ions**, such as $Na^+$, $K^+$, $Mg^{2+}$, and $Cr^{3+}$, generally favor interacting with oxygen ligands. **Soft metal ions**, on the other hand, such as

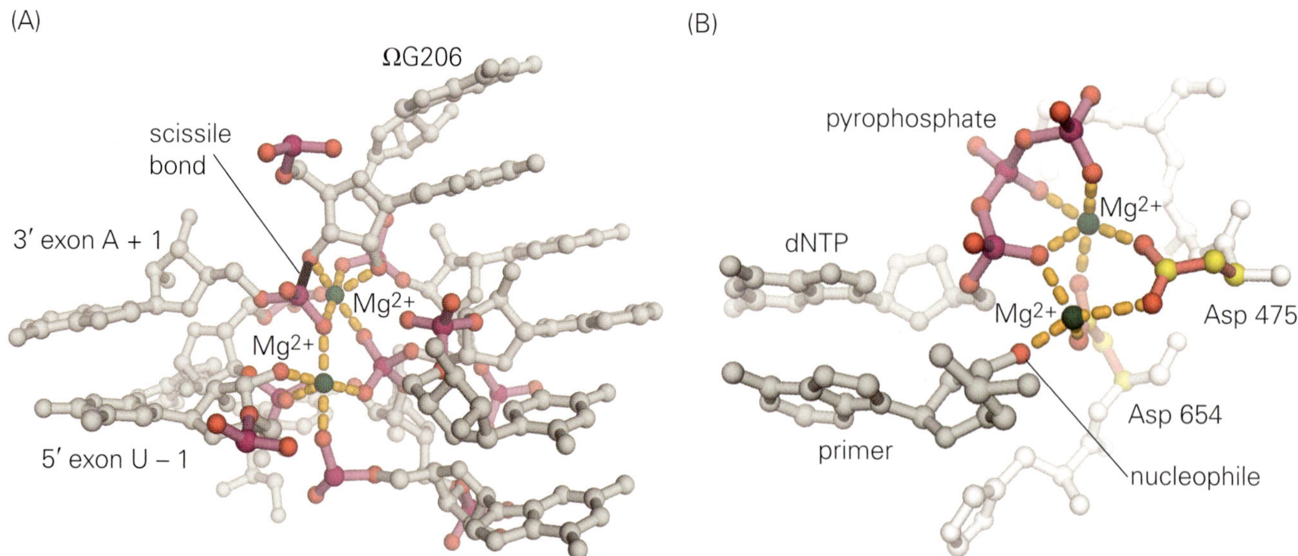


**Figure 16.61 Comparison of the active sites of a group I intron and a DNA polymerase.** (A) Group I intron structure. (B) T7 DNA polymerase. $Mg^{2+}$ ions are shown as *green* spheres. (Adapted from M.R. Stahley and S.A. Strobel *Curr. Opin. Struct. Biol.* 16: 319–326, 2006; PDB codes: A, 1ZZN; B, 1T7P.)

$Cu^{2+}$, $Au^{+}$, $Hg^{+}$, $Cd^{2+}$, $Pt^{2+}$, and $Mn^{2+}$, interact efficiently with sulfur or nitrogen ligands. They are also known as sulfur-loving or thiophilic metals (**Figure 16.62**).

If the interaction of $Mg^{2+}$ with a particular oxygen ligand in the RNA is important for catalysis, then substituting the oxygen with sulfur will decrease the activity of the enzyme in the presence of $Mg^{2+}$. The activity of the enzyme will be restored by adding soft metal ions such as $Mn^{2+}$ or $Cd^{2+}$ to the solution. Because we are only changing one of the several hundred or more metal–ligand interactions in the RNA, the effect on the stability of the RNA will be minimal. This type of rescue by thiophilic metals can confirm the importance of the $Mg^{2+}$ ion for catalysis.

We shall illustrate the utility of this approach using the group I intron as a model. The experiment involves the measurement of reaction rates as a function of metal ion concentration. This would be difficult to do for the normal form of the group I intron because the substrate is a part of the enzyme, and each ribozyme catalyzes the reaction only once and destroys itself in the process. These ribozymes can be converted into true multiple-turnover enzymes by separating the catalyst and substrate. This is done by removing the 5′ and 3′ exons (and the splice sites) from the group I intron, as shown in **Figure 16.63**.

By adding an oligonucleotide that corresponds to the 5′ exon–intron junction and the guanosine nucleoside (and is complementary to part of the intron so that the P1 helix can form), cleavage of the oligonucleotide is observed. The conversion to a *trans*-acting ribozyme allows the concentrations of substrate and catalytic components to be varied independently so that the kinetic parameters of the individual steps in the reaction can be determined. The concentrations of $Mg^{2+}$ and other metal ions, such as $Mn^{2+}$, can then be varied and their effects on catalysis established.

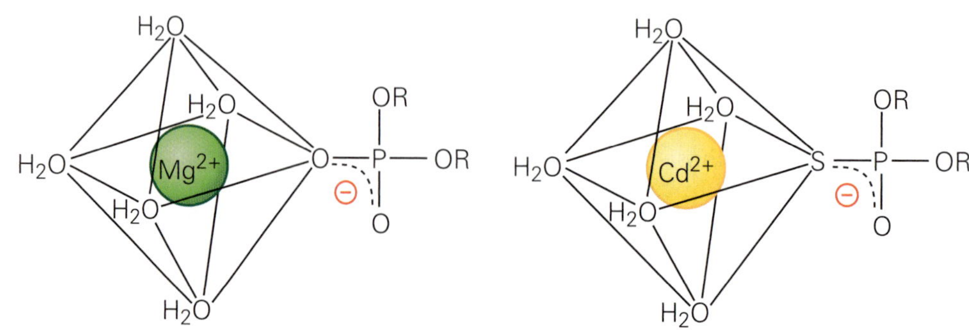


**Figure 16.62 Effect of sulfur substitution on metal ion specificity.** $Mg^{2+}$ is a hard metal ion that interacts efficiently with oxygen, which is a hard ligand (left panel). $Mg^{2+}$ interacts poorly with soft ligands, such as sulfur or nitrogen, and will not bind if the oxygen in a phosphate ligand is replaced by sulfur. Soft metal cations such as $Cd^{2+}$ or $Mn^{2+}$ can coordinate the modified ligand (right panel). (Adapted from M.J. Fedor and J.R. Williamson, *Nat. Rev. Mol. Cell Biol.* 6: 399–412, 2005. With permission from Macmillan Publishers Ltd.)

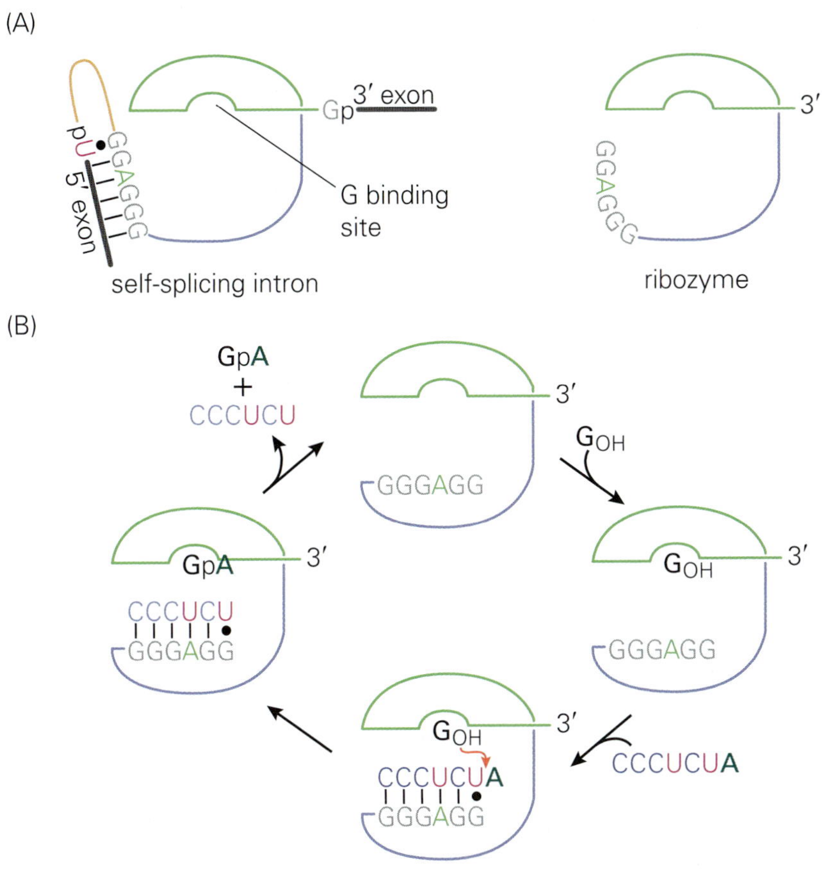


**Figure 16.63 Converting a group I self-splicing intron into a multiple turnover ribozyme.** (A) Group I self-splicing intron (left). The removal of the 5′ and 3′ exons (right) converts the group I intron into a ribozyme (right). (B) The basic catalytic cycle of the group I ribozyme. In this reaction, the guanosine nucleoside and the 5′-exon oligonucleotide are both added as exogenous substrates, and catalysis results in endonucleolytic cleavage of the 3′ nucleotide of the oligonucleotide and its addition to the end of the guanosine nucleotide. (Adapted from A.J. Zaug and T.R. Cech, *Biochem.* 25: 4478-82, 1986; D. Herschlag and T.R. Cech, *Biochem.* 29: 10159–10171, 1990.)

The metal ion interaction with the 3′ oxygen of the oligonucleotide substrate of a group I ribozyme was probed by replacing this oxygen with sulfur, as shown in **Figure 16.64**. The rate of ribozyme-catalyzed cleavage of the natural oligonucleotide

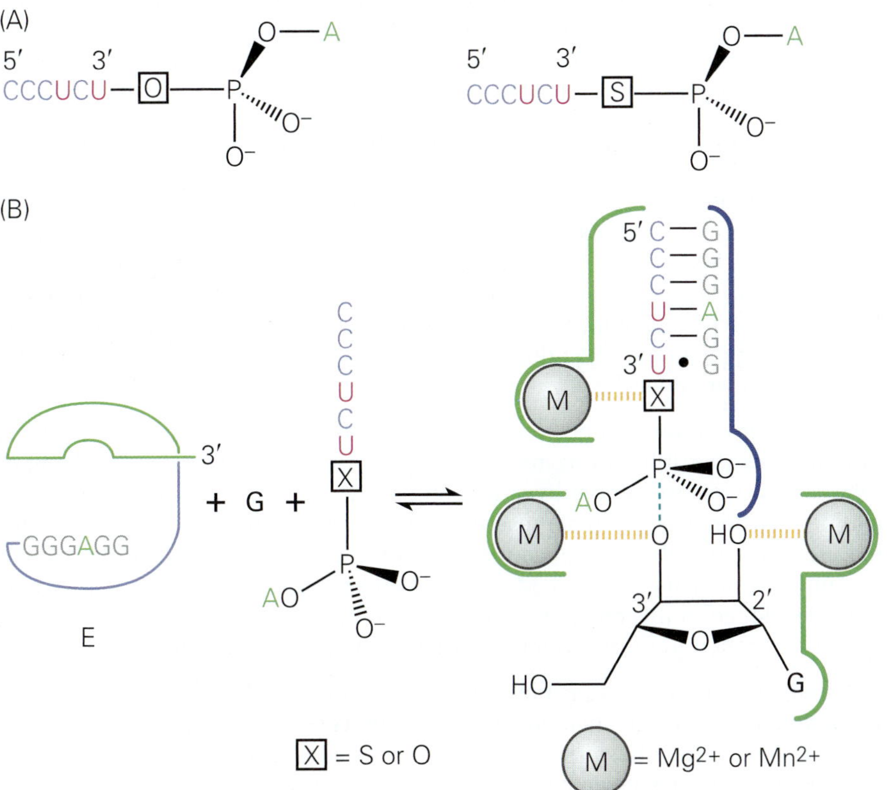


**Figure 16.64 Effect of hard and soft metals on ribozyme activity.** (A) The oligonucleotide substrate on the left contains a 3′ oxygen (boxed) while the substrate on the right contains a sulfur instead of oxygen. (B) The reaction between the ribozyme, denoted E, guanosine, and the oligonucleotide, in which X is either oxygen or sulfur. There are three metal-binding sites at the active site, denoted M. Only the metal next to the X position will be sensitive to the replacement of $Mg^{2+}$ by $Mn^{2+}$ when oxygen is replaced by sulfur there. Refer to Figure 16.58 to understand the details of the reaction depicted here. (Adapted from S. Shan et al. and D. Herschlag, *Proc. Natl. Acad. Sci. USA* 96: 12299–12304, 1999.)

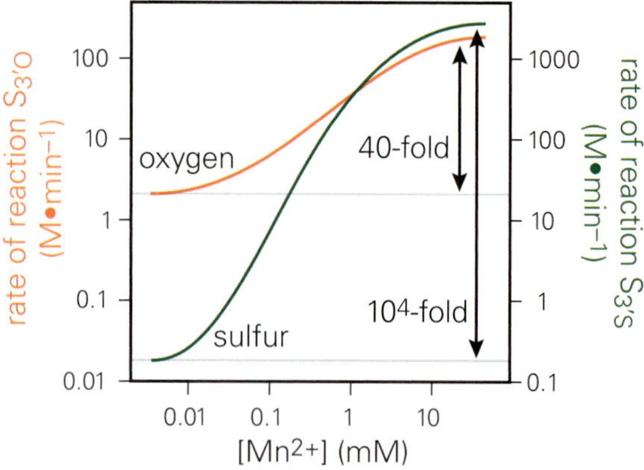


**Figure 16.65 The effect of $Mn^{2+}$ on the rate of the reaction with oligonucleotide substrates containing an oxygen replaced by sulfur.** The graph compares the rates of ribozyme-catalyzed reactions using the two substrates shown in Figure 16.64A. The concentration of $Mg^{2+}$ is held constant at 10 mM, and the concentration of $Mn^{2+}$ is varied. (Adapted from S. Shan et al. and D. Herschlag, *Proc. Natl. Acad. Sci. USA* 96: 12299–12304, 1999. With permission from the National Academy of Sciences.)

was compared with that for the sulfur-substituted one at different concentrations of $Mn^{2+}$, as shown in **Figure 16.65**. The concentration of $Mg^{2+}$ was held fixed at 10 mM in these experiments. Notice that, when sulfur replaces oxygen in the substrate, the enzyme is inactive when the concentration of $Mn^{2+}$ is very low, because $Mg^{2+}$ cannot coordinate the sulfur atom. The addition of $Mn^{2+}$ to the reaction mixture increases the rate of reaction $10^4$-fold, whereas the effect on the reactivity of the normal nucleotide containing oxygen at the 3′ position is only 40-fold. These experiments confirm the catalytic role of metal ion coordination by the substrate's 3′ oxygen.

## Summary

In this chapter we described the kinetic equations that govern the rates at which enzymatic reactions occur. A key aspect of the behavior of enzyme-catalyzed reactions is that a discrete, reversible binding of substrate to the enzyme is the first step. The enzyme, as a catalyst which is unchanged after a reaction cycle, reaches a steady-state distribution between free and substrate-bound forms. Using the steady-state value for the concentration of the enzyme–substrate complex leads to the Michaelis–Menten equation, which predicts that the velocity of the reaction reaches a plateau value at high substrate levels.

Two key parameters in the Michaelis–Menten equation are $K_M$, the Michaelis constant, and $k_2$, the rate constant associated with the catalytic step. The value of $K_M$ corresponds to the substrate concentration at which the enzyme-catalyzed reaction operates at half its maximal value. The rate constant $k_2$ is also referred to as $k_{cat}$, the catalytic rate constant. The maximum rate of the enzyme-catalyzed reaction, $V_{max}$, is given by the product of $k_{cat}$ and the total enzyme concentration, $[E]_0$. The catalytic efficiency $k_{cat}/K_M$, is particularly useful in comparing the ability of the enzyme to turn over different substrates, and this ratio is also called the specificity constant for the enzyme.

The Michaelis–Menten kinetic model also makes it possible to predict the effects of molecules that inhibit enzymes. Depending on how such compounds bind, they can give rise to competitive, noncompetitive, or substrate-dependent noncompetitive behavior. Each of these situations gives rise to characteristic changes in the rate versus substrate concentration profile, which allows the mechanism of the inhibitor to be determined. Many drugs are enzyme inhibitors, and the equations developed in this chapter allow the quantitative calculation of their effects at different concentrations, which relates to their efficacy in clinical applications.

Many oligomeric enzymes can bind substrates cooperatively, due to allosteric conformational changes. In such enzymes, the rate versus substrate curves become sigmoidal in shape. Allosteric coupling of active sites provides a mechanism for the regulation of enzyme activity, both by substrates and by other molecules.

Both the reaction mechanisms and the structures of many protein enzymes have been determined. The mechanisms show that stabilization of the transition state through specific interactions with the protein is a common mechanism of catalysis, although other factors, such as constraining the substrate conformation, can also play important roles. Catalysis of biological reactions can often be accomplished in several different ways, that is, enzymes that catalyze similar reactions may have very different mechanisms. The examples presented in this chapter illustrate the principles developed in understanding the enzyme reaction rates and the factors that affect them.

Like their protein counterparts, ribozymes use many different strategies to catalyze reactions. They can position the reactive groups in an optimal alignment; they can participate in general acid–base catalysis to activate a nucleophilic oxygen or to stabilize oxyanion leaving groups; they can stabilize a negative charge that accumulates in the transition state; or they can destabilize the ground state. Ribozymes are not limited to using only metal ions as functional groups in catalysis, but can also use nucleotide bases (as in self-cleaving ribozymes), the sugar hydroxyls, and even the phosphate backbone.

## Key Concepts

### A. MICHAELIS–MENTEN KINETICS

- Enzymes are biological molecules that catalyze most chemical reactions in cells.
- Enzyme-catalyzed reactions can be described as a binding step followed by a catalytic step.
- A steady-state analysis of the simplest scheme for such a reaction leads to the Michaelis–Menten equation, which relates the initial rate of reaction, $v_0$, to the substrate concentration, [S]:

$$v_0 = \frac{V_{max}}{\left(1 + \frac{K_M}{[S]}\right)}$$

- The Michaelis–Menten model, with parameters $K_M$ (the Michaelis constant) and $V_{max}$ (the maximum velocity) correctly predicts the initial velocities of many enzyme-catalyzed reactions under steady-state conditions.
- The value of the Michaelis constant, $K_M$, is related to how much enzyme is bound to substrate. Specifically, the $K_M$ value corresponds to the substrate concentration at which the rate is half-maximal.
- The rate constant associated with the catalytic step in the Michaelis–Menten scheme, $k_2$, is also called the catalytic rate constant, $k_{cat}$. This is also referred to as the turnover number, because it specifies how fast the enzyme can turnover the substrate, when the binding step is not rate-limiting.
- The ratio $k_{cat}/K_M$, is called the catalytic efficiency or the specificity constant. This ratio corresponds to the second-order rate constant for substrate binding to the enzyme when the substrate concentration is low.
- A "perfect" enzyme is one that catalyzes the chemical step of the reaction as fast as the substrate can get to the enzyme.
- In some cases, the rate of product release from the enzyme affects the overall rate of the reaction.
- The specificity of enzymes arises from both the rate of the chemical step and the value of $K_M$.
- Graphical analysis of enzyme kinetic data facilitates the estimation of kinetic parameters.

### B. INHIBITORS AND MORE COMPLEX REACTION SCHEMES

- Competitive inhibitors block the active site of the enzyme in a reversible way.
- A competitive inhibitor does not affect the maximum velocity of the reaction, $V_{max}$, but it increases the Michaelis constant, $K_M$.
- Reversible noncompetitive inhibitors decrease the maximum velocity, $V_{max}$, without affecting the Michaelis constant, $K_M$.
- Substrate-dependent noncompetitive inhibitors only bind to the enzyme when the substrate is present.
- Some noncompetitive inhibitors are linked irreversibly to the enzyme.
- In a ping-pong mechanism, the enzyme becomes modified temporarily during the reaction.
- For a reaction with multiple substrates, the order of binding can be random or sequential.
- Enzymes with multiple binding sites can display allosteric (cooperative) behavior.
- Product inhibition is a mechanism for regulating metabolite levels in cells.

### C. PROTEIN ENZYMES

- Enzymes can accelerate reactions by large amounts.
- Transition state stabilization is a major contributor to rate enhancement by enzymes.
- Protein sidechains or chemical groups in co-factors bound to the enzyme can act as acids or bases to enhance reaction rates.
- Proximity effects are important for many reactions.
- The serine proteases are a large family of enzymes that contain a conserved Ser-His-Asp catalytic triad.
- Sidechain recognition positions the catalytic triad next to the peptide bond that is cleaved.
- The specificities of serine proteases vary considerably, but the catalytic triad is conserved.
- Creatine kinase catalyzes phosphate transfer by stabilizing a planar phosphate intermediate.
- Some enzymes work by populating disfavored conformations.
- Antibodies that bind transition state analogs can have catalytic activity.

### D. RNA ENZYMES

- Self-cleaving ribozymes use nucleotide bases for catalysis, even though these do not have $pK_a$ values well suited for proton transfer.
- Hairpin ribozymes optimize hydrogen bonds to the transition state rather than to the initial or final states.
- The splicing reaction catalyzed by group I introns occurs in two steps, both involving nucleophilic attack on a bond that is cleaved.
- Metal ions facilitate catalysis by group I introns.
- Substitution of oxygen by sulfur in RNA helps identify metals that participate in catalysis.

## Problems

### True/False and Multiple Choice

1. The initial reaction velocity for an enzyme reaction reaches a maximum at high substrate concentration because the free enzyme can no longer regenerate at the end of each reaction cycle.

   True/False

2. The turnover number for an enzyme obeying Michaelis–Menten kinetics is:
   a. $k_2$.
   b. $k_{cat}/K_M$.
   c. $k_1/k_{-1}$.
   d. $(k_1 + k_2)$.
   e. $\Delta G^\ddagger$.

3. Catalytic antibodies are generally less efficient than natural enzymes that catalyze the same reactions.

   True/False

4. A metabolic enzyme generates the amino acid methionine. For a given substrate concentration, an experiment conducted in the presence of high initial concentrations of methionine generates less new methionine than an experiment conducted with no initial methionine present. This is likely an example of:
   a. A ping-pong mechanism of substrate binding.
   b. A proximity effect.
   c. Substrate strain.
   d. Product inhibition.
   e. A reaction intermediate.

5. Which of the following is *not* a commonly observed feature of proteases?
   a. The catalytic triad in the active site.
   b. Exclusively hydrophobic residues in the active site.
   c. A cysteine residue in the active site.
   d. Metal ions coordinated in the active site.
   e. A pair of acidic residues in the active site.

6. An enzyme inhibitor is observed to alter the $K_M$ but not the $V_{max}$ of a reaction. This inhibitor is most likely:
   a. A noncompetitive inhibitor.
   b. A competitive inhibitor.
   c. An allosteric inhibitor.
   d. A substrate-dependent noncompetitive inhibitor.
   e. A covalent inhibitor.

7. Due to its extremely slow dissociation kinetics, the protein bovine pancreatic trypsin inhibitor (BPTI) has broad specificity and inhibits more proteases than protease inhibitors that are small molecules.

   True/False

### Fill in the Blank

8. In the schemes for the catalyzed reactions considered in this chapter, S, E, and P refer to ______________, ______________, and ______________, respectively.

9. The specificity constant or catalytic efficiency is the ratio between _________ and ________.

10. In a plot of initial velocity versus substrate concentration, an allosteric enzyme displays a ___________ curve, whereas a non-allosteric enzyme that obeys Michaelis–Menten kinetics displays a ___________ curve.

11. The geometry of competitive inhibitors commonly mimics the ____________ of the reaction that the enzyme normally catalyzes.

12. Proteins are not the only polymers that act as catalysts. Catalytic ______ molecules are also essential for cells, including playing an essential role in protein synthesis.

## Quantitative/Essay

**13.** At 25°C, an enzyme accelerates a reaction by a factor of $10^5$ over the uncatalyzed reaction in water. If the effect of the enzyme is solely to reduce the energy of the transition state, by what amount does it reduce the energy of the transition state ($E_A$)?

**14.** Show that the equations plotted in Lineweaver–Burk and Eadie–Hofstee plots are equivalent.

**15.** In a reaction, $E + S \underset{k_{-1}}{\overset{k_1}{\rightleftharpoons}} E{\bullet}S \xrightarrow{k_2} E + P$, calculate the value of $K_M$ if the forward rate constant ($k_1$) for E•S formation is $4.3 \times 10^6\ sec^{-1}{\bullet}M^{-1}$, the reverse rate constant ($k_{-1}$) for E•S dissociation is $2.4 \times 10^2\ sec^{-1}$, and the turnover number ($k_2$) is $1.2 \times 10^3\ sec^{-1}$.

**16.** Presented below are Lineweaver–Burk plots for enzymatic reactions with (*red*) and without (*blue*) inhibitor. What type of inhibition is occurring in each case?

a.

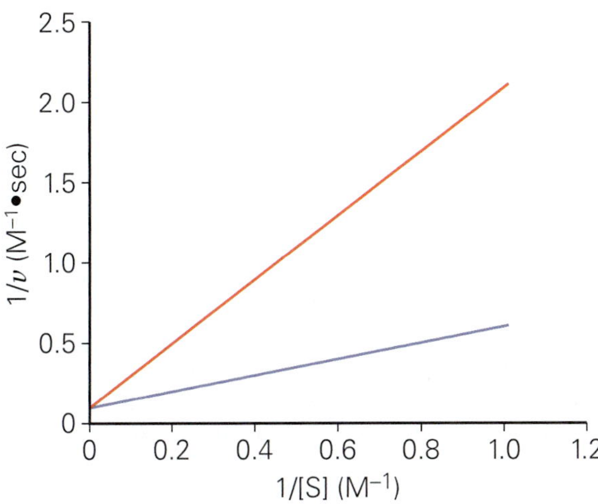


b.

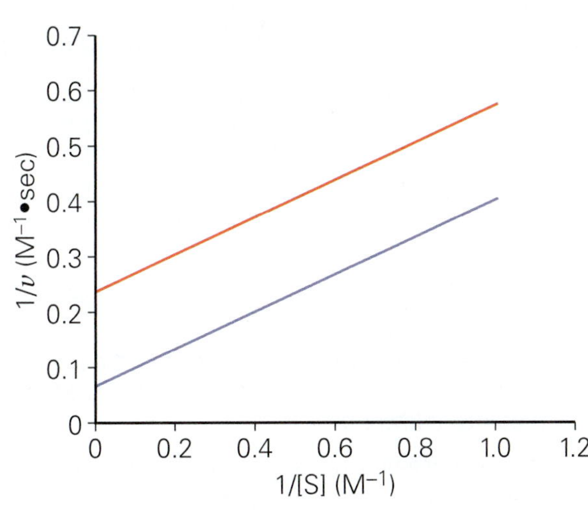


c.

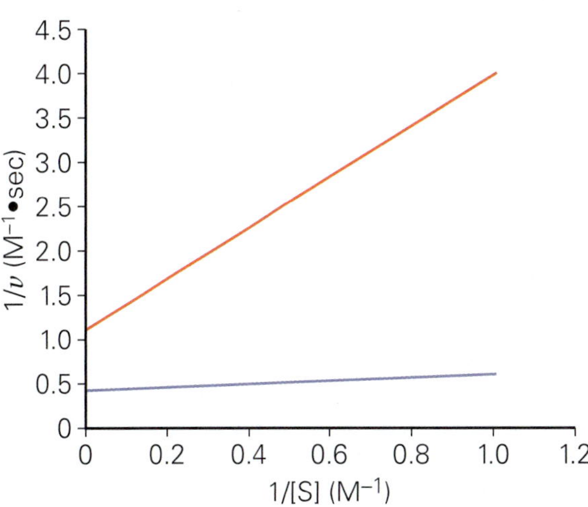


**17.** The table below lists initial velocities measured for an enzymatic reaction at different substrate concentrations in the presence and absence of an inhibitor. The enzyme concentration is identical in both reactions.

| [S] (mM) | *v*, no inhibitor ($mM{\bullet}sec^{-1}$) | *v*, with inhibitor ($mM{\bullet}sec^{-1}$) |
|---|---|---|
| 1 | 2.50 | 1.11 |
| 2 | 4.00 | 2.00 |
| 5 | 6.25 | 3.85 |
| 10 | 7.69 | 5.56 |
| 20 | 8.70 | 7.14 |

a. Graph a Lineweaver–Burk plot.
b. What are the apparent values of $V_{max}$ and $K_M$ for each experiment?
c. What is the inhibition mechanism?
d. If the concentration of inhibitor is 0.5 mM, what is the value of $K_I$?

**18.** The table below lists initial velocities measured for an enzymatic reaction at different substrate concentrations in the presence and absence of an inhibitor. The enzyme concentration is identical in both reactions.

| [S] (mM) | *v*, no inhibitor ($mM{\bullet}sec^{-1}$) | *v*, with inhibitor ($mM{\bullet}sec^{-1}$) |
|---|---|---|
| 1 | 1.000 | 0.923 |
| 5 | 1.154 | 1.053 |
| 10 | 1.176 | 1.071 |
| 50 | 1.195 | 1.087 |
| 100 | 1.198 | 1.089 |

a. Graph a Lineweaver–Burk plot for each set of data.
b. What are the apparent values of $V_{max}$ and $K_M$ for each experiment?
c. What is the inhibition mechanism?
d. If the concentration of inhibitor is 10 nM, what is the value of $K_I$?

**19.** The table below lists initial velocities measured for an enzymatic reaction at different substrate concentrations in the presence and absence of an inhibitor. The enzyme concentration is identical in both reactions.

| [S] (μM) | $v$, no inhibitor (μM•sec$^{-1}$) | $v$, with inhibitor (μM•sec$^{-1}$) |
|---|---|---|
| 10 | 8.93 | 6.94 |
| 20 | 10.42 | 8.93 |
| 30 | 11.03 | 9.87 |
| 100 | 12.02 | 11.57 |
| 200 | 12.25 | 12.02 |

a. Graph a Lineweaver–Burk plot for each set of data.
b. What are the values of $V_{max}$ and $K_M$ for each experiment?
c. What is the inhibition mechanism ?
d. If the concentration of inhibitor is 100 nM, what is the value of $K_I$?

**20.** An experiment with 10 nM of an enzyme obeying Michaelis-Menten kinetics yields a $V_{max}$ of $7 \times 10^{-3}$ M•sec$^{-1}$.

a. What is the turnover number ($k_2$)?
b. The experiment is repeated in the presence of a noncompetitive inhibitor and the $V_{max}$ is reduced to $5 \times 10^{-4}$ M•sec$^{-1}$. What fraction of the enzyme is bound to the inhibitor?

**21.** Given the following three data tables of substrate concentrations and initial velocities for enzymes that obey Michaelis–Menten kinetics, estimate $K_M$ for each enzyme in molar units.

a.

| [S] (mM) | $v_0$ (mM•sec$^{-1}$) |
|---|---|
| 1 | 266.7 |
| 3 | 553.8 |
| 5 | 705.9 |
| 50 | 1121.5 |
| 500 | 1191.7 |
| 5000 | 1199.2 |

b.

| [S] (nM) | $v_0$ (mM•min$^{-1}$) |
|---|---|
| 4 | 123.5 |
| 5 | 137.4 |
| 6 | 148.5 |
| 10 | 177.3 |
| 100 | 240.2 |
| 1000 | 249.0 |

c.

| [S] (mM) | $v_0$ (M•hour$^{-1}$) |
|---|---|
| 1 | 0.00 |
| 10 | 0.01 |
| 100 | 0.07 |
| 200 | 0.10 |
| 1000 | 0.17 |
| 5000 | 0.19 |

**22.** When the bi-substrate analog PALA is added to the enzyme ATCase at low concentration it increases the rate of reaction of aspartate and carbamylphosphate. However, at higher concentrations it decreases the reaction rate. How can PALA act as both an activator and inhibitor of ATCase?

**23.** In the search for the catalytic mechanism of an enzyme, three mutations of charged residues to alanine (which is uncharged) are made and compared with the wild type (WT) enzyme. At otherwise identical conditions and concentrations of enzymes, the following initial velocities (μM•sec$^{-1}$) are measured as a function of pH.

| pH | WT | Arg55Ala | Glu63Ala | Lys113Ala |
|---|---|---|---|---|
| 4 | $1.3 \times 10^3$ | $1.3 \times 10^3$ | $1.3 \times 10^2$ | $1.5 \times 10^3$ |
| 5 | $8.7 \times 10^5$ | $8.7 \times 10^5$ | $1.3 \times 10^2$ | $9.5 \times 10^5$ |
| 6 | $8.1 \times 10^4$ | $8.1 \times 10^4$ | $1.3 \times 10^2$ | $8.1 \times 10^4$ |
| 7 | $5.3 \times 10^3$ | $5.3 \times 10^3$ | $1.3 \times 10^2$ | $5.2 \times 10^3$ |

a. Explain which residue likely acts as a general acid/base during catalysis?
b. What is a possible mechanism for the slightly increased reaction velocities observed in the Lys113Ala mutants at lower pH?

**24.** Why is triose phosphate isomerase considered to be an example of a "perfect enzyme"?

# Further Reading

## General


Cook PF & Cleland WW (2007) Enzyme Kinetics and Mechanism. New York: Garland Science.

Copeland RA (2000) Enzymes: A Practical Introduction to Structure, Mechanism, and Data Analysis. New York: Wiley.

Fersht A (1998) Structure and Mechanism in Protein Science: A Guide to Enzyme Catalysis and Protein Folding. New York: W.H. Freeman.

Walsh C (1979) Enzymatic Reaction Mechanisms. New York: W.H. Freeman.


## References

### A. Michaelis–Menten kinetics


Albery WR & Knowles JR (1977) Perfection in enzyme catalysis: the energetics of triosephosphate isomerase. *Acc. Chem. Res.* 10, 105–111.

Dowd JE & Riggs DS (1965) A comparison of estimates of Michaelis–Menten kinetic constants from various linear transformations. *J. Biol. Chem.* 240, 863–869.

Hammes GG (2002) Multiple conformational changes in enzyme catalysis. *Biochemistry* 41, 8221–8228.

Kirsch JF (1973) Mechanism of enzyme action. *Annu. Rev. Biochem.* 42, 205–234.


### B. Inhibitors and more complex reaction schemes


Cleland WW (1963) The kinetics of enzyme-catalyzed reactions with two or more substrates or products: I. Nomenclature and rate equations. *Biochimica et Biophysica Acta* 67, 107–137.

Dixon M (1953) The determination of enzyme inhibitor constants. *Biochem. J.* 55, 170–171.

Wolfenden R (2006) Degrees of difficulty of water-consuming reactions in the absence of enzymes. *Chem. Rev.* 106, 3379–3396.


### C. Protein enzymes


Beck ZQ, Morris GM & Elder JH (2002) Defining HIV-1 protease substrate selectivity. *Curr. Drug Targ. Infect. Disorders* 2, 37–50.

Kraut J (1977) Serine proteases: structure and mechanism of catalysis. *Annu. Rev. Biochem.* 46, 331–358.

Pelmenschikov V & Siegbahn PE (2005) Copper-zinc superoxide dismutase: Theoretical insights into the catalytic mechanism. *Inorg. Chem.* 44, 3311–3320.

Pollack SJ, Jacobs JW & Schultz PG (1986) Selective chemical catalysis by an antibody. *Science* 234, 1570–1573.

Wang P-F, Flynn AJ, Naor MM, Jensen JH, Cui G, Merz KM, Kenyon GL & McLeish MJ (2006) Exploring the role of the active site cysteine in human muscle creatine kinase. *Biochemistry* 45, 11464–11472.

Zhang X, Zhang X & Bruice TC (2005) A definitive mechanism for chorismate mutase. *Biochemistry* 44, 10443–10448.


### D. RNA enzymes


Cochrane JC & Strobel SA (2008) Catalytic strategies of self-cleaving ribozymes. *Acc. Chem. Res.* 41, 1027–1035.

Fedor MJ (2009) Comparative enzymology and structural biology of RNA self-cleavage. *Annu. Rev. Biophys.* 38, 271–299.

Fedor MJ & Williamson JR (2005) The catalytic diversity of RNAs. *Nat. Rev. Mol. Cell. Biol.* 6, 399–412.

Lilley DMJ & Eckstein F (2008) Ribozymes and RNA Catalysis. Cambridge, UK: RSC Publishing.

Stahley MR & Strobel SA (2006) RNA splicing: group I intron crystal strcutures reveal the basis of splice site selection and metal ion catalysis. *Curr. Opin. Struct. Biol.* 16, 319–326.

Vicens Q & Cech TR (2006) Atomic level architecture of group I introns revealed. *Trends Biochem. Sci.* 31, 41–51.

# CHAPTER 17

# Diffusion and Transport

Throughout every cell the movement of molecules from one region to another is important to meet the demands of the processes that sustain life. A small molecule, such as ATP (the energy currency of cells), is synthesized by enzymes at the membranes of mitochondria, but is then needed in remote parts of the cell, where it is used for reactions such as powering chaperones for protein folding. Proteins themselves are synthesized by ribosomes in the cytoplasm, but may need to function elsewhere in the cell, such as by binding to DNA in the nucleus.

Much of the movement of molecules in cells is passive, requiring no energy source, and occurs in random directions. The random motions of molecules arise from constant collisions with water and other molecules. Such collisions cause transient accelerations in random directions and also cause the molecules to rotate. The resulting motions are maintained for only a very short time before further collisions change both the direction and the speed of the motion. Over time, these random motions cause molecules to move significant distances in a process referred to as **diffusion**. The same effects occur with larger particles, resulting in **Brownian motion** (in Brownian motion, the diffusing particles are large enough to see with a microscope). **Passive transport**, which occurs through diffusion alone, can be too slow to meet all of the needs of a cell. As a consequence, **active transport**, which requires an energy source, is also important. Some of the concepts that we shall develop in this chapter to describe diffusion are also useful for understanding aspects of active transport.

**Passive transport**

Passive transport is the movement of molecules through the process of diffusion. In contrast, the active transport of molecules in cells is driven by chemical energy—usually the hydrolysis of ATP or other nucleoside triphosphates.

In this chapter, we describe the quantitative relationships between molecular properties, such as mass and shape, and the rates of diffusion and other processes related to it. These relationships make it possible to deduce information about the molecular properties from measurements of rates of movement, and also to predict what processes in cells would be too slow if they occurred only through diffusion. Understanding diffusion also helps us to understand the maximum rate with which a chemical reaction can occur.

## A. RANDOM WALKS

### 17.1 Microscopic motion is well described by trajectories called random walks

To begin to understand the net movement of molecules undergoing random motions, let us first look at a slightly larger scale system, that of bacteria "swimming." The movement of many kinds of bacteria occurs as a series of discrete steps of directed movement, interspersed by tumbling that randomizes the direction for the subsequent step. The alternation of tumbling and directed motion makes the movement of bacteria similar to that of molecules, which move in a series of very many small steps in which the direction of movement changes randomly at each step.

Recall from the discussion of bacterial chemotaxis in Chapter 14 that many bacteria that undergo active movement have fibers, known as flagella, that extend from

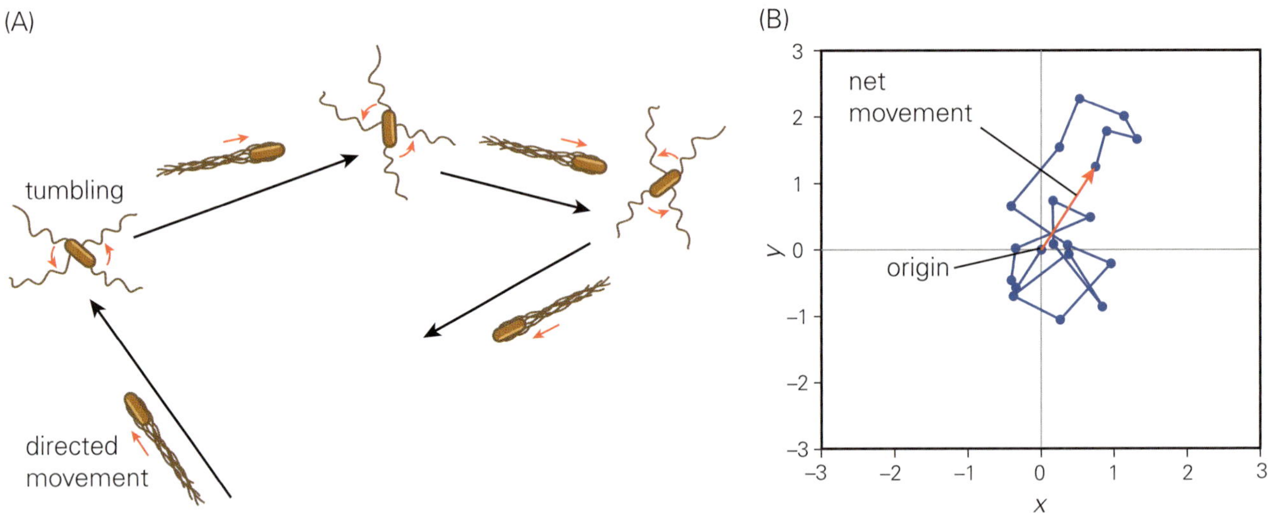


**Figure 17.1 Bacterial chemotaxis can be described as a random walk.** (A) In the absence of a chemical attractant or repellent, bacteria explore their environment by switching between directed movement and tumbling. (B) This movement resembles a two-dimensional random walk if the bacteria are growing in a thin film of water. The diagram shows 20 steps in such a random walk. The net movement or displacement (the vector sum of the steps) is shown by the *red* arrow. The net displacement after many steps is much smaller than the total distance traveled during all the steps.

the body of the cell (see Figure 14.5). In the absence of a chemical attractant or repellent, the bacteria alternate between straight swimming and tumbling (**Figure 17.1**A). During straight swimming, the counterclockwise rotation of the flagella causes them to form a bundle and work together like a miniature propeller. Tumbling occurs when the flagella rotate in a clockwise fashion, which causes them to separate and the directed movement to stop.

Each straight swim constitutes a step in a **random walk**, because the period of tumbling that follows the straight step causes the next straight step to occur in a random direction (Figure 17.1B). When bacteria are living in a thin film of water on top of a plate of agar in a Petri dish, as they commonly would be in a laboratory culture, their motion is basically constrained to two dimensions. If we start with a large number of bacteria in a single small spot on a plate, how far they will spread with time? By treating the movement of bacteria as a random walk, we can predict how quickly the clump of bacteria will spread out.

## 17.2 The analysis of bacterial movement is simplified by considering one-dimensional random walks with uniform step lengths and time intervals

To begin a quantitative analysis of the random motion of bacteria, we shall make a simplifying assumption. We will assume that each step has a uniform length and occurs over a fixed period of time. As we saw in the last section, this is not how bacteria actually move. Nevertheless, it turns out that after many steps, both the average size of the step and the average number of steps per unit time become precisely defined. As a consequence, the movement of bacteria over long times is well predicted by using this simpler approach.

**Random walk**

A series of discrete steps in which the direction of each step is uncorrelated with the direction of the previous step is known as a random walk.

We can simplify the process even further and consider just movement in one dimension, with each step going either to the left or to the right. We can then obtain information about movement in two or three dimensions by combining two or three one-dimensional random walks in orthogonal directions, as we discuss later.

The average displacement, $\delta$, from the original position after a total number of random steps, $N_{total}$, is given by the difference between the number of steps to the right, $N_R$, and the number of steps to the left, $N_L$:

$$\text{displacement, } \delta = (N_R - N_L) \tag{17.1}$$

We have assumed in Equation 17.1 that each step is of unit length. If the steps are not of unit length, then the actual displacement is obtained by multiplying the value of $\delta$ by the length of each step, $l_{step}$. Note that in Equation 17.1 $\delta$ refers to the distance between the initial point and the end point, and not to the summed length of all the steps, which is simply $N_{total}$.

Two one-dimensional random-walk trajectories of 30 steps each are illustrated in **Figure 17.2**, with each step separated vertically in time in the diagram. Both trajectories have uniform step lengths and start at the same point. One trajectory (shown in Figure 17.2A) consists of 15 steps to the left with 15 steps to the right mixed in randomly, and so the trajectory ends up where it began (the value of $\delta$ is zero). The second trajectory consists of 14 steps to the left and 16 steps to the right. This trajectory ends up at two steps to the right of the origin—that is, it has traveled a distance of two units from the origin ($\delta = +2$).

If steps to the right and to the left are equally likely, then we expect that, on average, the distance traveled will be zero. But, as shown in Figure 17.1B, trajectories with nonzero displacements are also possible. A trajectory would have maximum displacement if all of the steps went in one direction only, which is very unlikely, but not impossible.

## 17.3 The probability distribution for the number of moves in one direction is given by a Gaussian function

A one-dimensional random walk with uniform step size is analogous to a series of coin flips. In particular, we might consider a step to the right to correspond to a coin flip that comes up "heads" and a step to the left to one that comes up "tails." Any particular trajectory then corresponds to a series of coin tosses, with one particular pattern of heads and tails.

Recall, from the discussion of entropy in Chapter 7, that any particular sequence of heads and tails is equally likely (see Section 7.1). For example, the two trajectories shown in Figure 17.2 are equally likely. But, when we consider the probability of moving a certain distance from the origin (or obtaining a specified number of

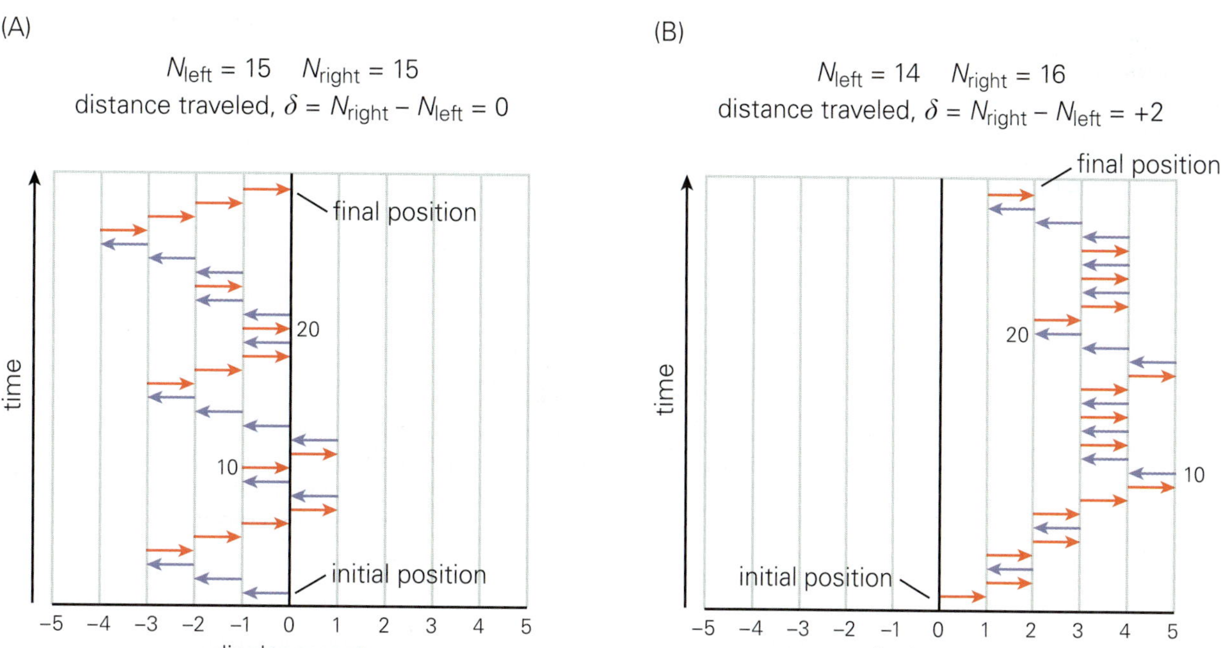


**Figure 17.2 One-dimensional random walks.** Two independent trajectories are shown, each with 20 uniform steps. Whether each step is to the left or the right is random and uncorrelated with the directions of other steps. Both trajectories start at the origin (0), but then propagate randomly. (A) The trajectory consists of 15 steps to the right and 15 steps to left, ending up at zero. (B) The trajectory consists of 16 steps to the right and 14 steps to the left, ending up at +2. In this illustration the steps are assumed to have unit length, and so the distance traveled is simply given by the difference between the number of steps to the right and to the left.

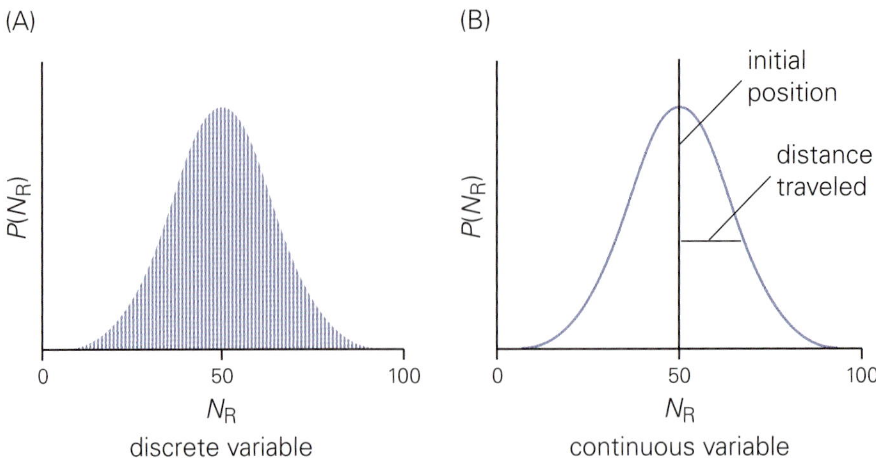


**Figure 17.3 Probability distributions for a one-dimensional random walk.** (A) Binomial distribution for the probability of observing different numbers of right-hand steps in a random walk with 100 steps. The number of steps to the right, $N_R$, is an integer and takes on values between 0 and 100. When averaged over many trajectories, the mean value of $N_R$ is 50. (B) If we assume that the number of steps to the right is a continuous variable, then the probability distribution is given by a Gaussian function. This is an accurate description of the probability distribution when the number of steps is very large (see Figure 7.16).

heads in a series of coin tosses), then the probability is given by the binomial distribution, and chance favors trajectories that do not move very far from the origin (that is, an even distribution of heads and tails is more likely; see Section 7.7).

The distance between the initial point and the end point in a random-walk trajectory is given by the difference between the number of steps to the right and to the left, as given by Equation 17.1 and illustrated in Figure 17.2. Different trajectories travel different distances from the origin, and the distribution of distances travelled has exactly the same form as the binomial distribution for the outcomes of coin flips. When the number of steps is large, the distribution is well described by a **Gaussian function** (see Section 7.12).

**Gaussian function**

A Gaussian function of a variable $x$ is an exponential of $-x^2$. Most distributions generated by random processes are Gaussian.

The conversion of a discrete binomial distribution to a Gaussian function of a continuous variable is illustrated in **Figure 17.3**. The expected distribution for $N_R$, the number of steps to the right, for a random walk with a total of 100 steps is shown in that figure, which is essentially the same as Figure 7.16. Figure 17.3A shows the binomial distribution, with a discrete integer variable $N_R$, and it is the expected bell-shaped curve peaked at the mean value, which is 50 in this case. Figure 17.3B shows the equivalent Gaussian function, in which we have replaced the integer variable $N_R$ with a continuous variable, also denoted $N_R$ (see the discussion in Section 7.12, where the continuous variable was denoted $x$).

With the Gaussian description, the probability of a trajectory having $N_R$ steps to the right is given by:

$$P(N_R) = \frac{1}{\sqrt{2\pi}\,\sigma} \exp\left(\frac{-(N_R - \mu)^2}{2\sigma^2}\right) \tag{17.2}$$

Equation 17.2 is the same as Equation 7.36, with the variable $x$ replaced by $N_R$. The term before the exponential is a normalization constant, which ensures that when the probability distribution is integrated over all possible values of $N_R$, the result is unity (**Box 17.1**). The mean value of the distribution, $\mu$, is given by:

$$\mu = \frac{N_{total}}{2} \tag{17.3}$$

The standard deviation, $\sigma$, is the square root of the variance of the distribution and is given by:

$$\sigma = \frac{\sqrt{N_{total}}}{2} \tag{17.4}$$

The standard deviation is the **root mean square displacement** (r.m.s. displacement) of the values in the distribution from the mean value (see Box 17.1).

Substituting these expressions for $\mu$ and $\sigma$ into Equation 17.2, we get an expression for the probability distribution that depends only on the total number of steps, $N_{\text{total}}$:

$$P(N_R) = \frac{2}{\sqrt{2\pi N_{\text{total}}}} \exp\left[\frac{-\left(N_R - \frac{N_{\text{total}}}{2}\right)^2}{\frac{N_{\text{total}}}{2}}\right] \tag{17.5}$$

## 17.4 The probability of moving a certain distance in a one-dimensional random walk is also given by a Gaussian function

In the preceding sections we have discussed the probability that a random walk trajectory has a certain number of steps in one direction (for example, $N_R$). We are usually more interested, however, in the probability that a trajectory ends up at a certain displacement, $\delta$, from the starting position. The value of $\delta$ is related to the number of steps in one direction (for example, $N_R$) in the following way:

$$\begin{aligned}\delta &= N_R - N_L = N_R - (N_{\text{total}} - N_R) \\ &= 2N_R - N_{\text{total}} = 2(N_R - \mu)\end{aligned} \tag{17.6}$$

How do we convert the known probability distribution for $N_R$ (Equation 17.5) into a probability distribution for $\delta$? Recognize that $\delta$ is defined by the value of $N_R$. The probability that the displacement is $\delta$ is given by the probability that there are $N_R$ steps to the right (that is, if the trajectory consists of $N_R$ steps to the right, then according to Equation 17.6, the displacement from the origin must be $\delta$). There is one subtlety, however, which is that the probability distribution for $\delta$, $P(\delta)$, will have a different normalization constant than does $P(N_R)$. This is because, as shown in **Figure 17.4**A, $N_R$ has a range from 0 to $N_{\text{total}}$, the total number of steps. The range of $\delta$, in contrast, is twice as large, and extends from $-N_{\text{total}}$, when all of the moves are to the left, to $+N_{\text{total}}$, when all of the moves are to the right (see Figure 17.4B).

As explained in **Box 17.2**, the expression for the probability distribution for the displacement, $\delta$, is given by:

$$P(\delta) = \frac{1}{\sqrt{2\pi N_{\text{total}}}} \exp\left(-\frac{\delta^2}{2N_{\text{total}}}\right) \tag{17.7}$$

The mean value of $\delta$ is zero because it is symmetric about the origin. The standard deviation of $\delta$ is given by $\sqrt{N_{\text{total}}}$, as you can surmise by comparing Equations 17.2 and 17.7. That the standard deviation is given by the square root of the number of

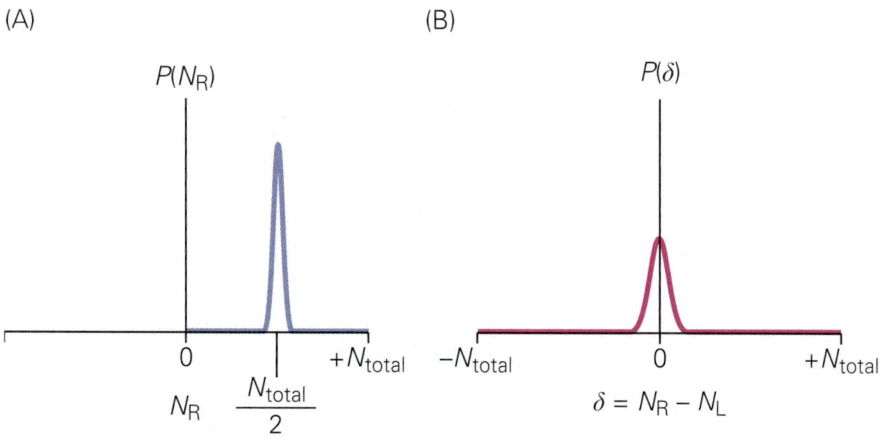


**Figure 17.4 Probability distributions for the number of steps and the distance traveled.** (A) Gaussian distribution for the number of steps to the right, $N_R$, in a one-dimensional random walk with $N_{\text{total}}$ steps. (B) Distribution of net displacements from the origin, $\delta$, in the same random walk. The distribution is also described by a Gaussian function, with larger width.

## Box 17.1 Gaussian distributions

Here we review a few properties of Gaussian functions that are useful for the discussion in the main text.

### Normalizing a Gaussian probability distribution

The first property of Gaussian functions is that the probability distribution must be normalized, which means that the integral of the probability distribution over the range of values is 1.0. This is because the total probability of obtaining *any* value of $N_R$ has to be unity:

$$\int_0^{N_{total}} P(N_R)dN_R = \frac{1}{\sqrt{2\pi}\,\sigma}\int_0^{N_{total}} \exp\left[-\frac{(N_R-\mu)^2}{2\sigma^2}\right]dN_R = 1 \quad (17.1.1)$$

It is instructive to verify that the probability distribution is indeed normalized, by evaluating the integral in Equation 17.1.1. To do this, we define a variable, $x$, such that $x = N_R - \mu$. This corresponds to simply shifting the function so that the mean value is zero, as shown in **Figure 17.1.1**. It follows from this definition that the differential elements of $x$ and $N_R$ are equal—that is, $dx = dN_R$. While the range of $N_R$ is from 0 to $N_{total}$, the range of $x$ is from $-N_{total}/2$ to $+N_{total}/2$. To evaluate the integral, we proceed as follows:

$$\int_0^{N_{total}} \exp\left(\frac{-(N_R-\mu)^2}{2\sigma^2}\right)dN_R = \int_{-\frac{N_{total}}{2}}^{+\frac{N_{total}}{2}} \exp\left(\frac{-x^2}{2\sigma^2}\right)dx = \int_{-\infty}^{+\infty} \exp\left(\frac{-x^2}{2\sigma^2}\right)dx \quad (17.1.2)$$

We have extended the range of the integral in Equation 17.1.2 from $-\infty$ to $+\infty$ because, as you can see from Figure 17.4B, the function being integrated falls off to zero at $\pm N_{total}/2$, and extending the range of the integration does not change the value of the integral. We can now use the following expression for the definite integral, given in **Table 17.1.1**, which you can also find in tables of common integrals:

$$\int_{-\infty}^{+\infty} \exp(-ax^2)\,dx = 2\int_0^{+\infty} \exp(-ax^2)\,dx = \sqrt{\frac{\pi}{a}} \quad (17.1.3)$$

By using Equation 17.1.3 to simplify Equation 17.1.2, we get:

$$\int_{-\infty}^{+\infty} \exp\left(\frac{-x^2}{2\sigma^2}\right)dx = \sqrt{\pi(2\sigma^2)} = \sigma\sqrt{2\pi} \quad (17.1.4)$$

**Table 17.1.1 Standard Gaussian integrals.**

| | |
|---|---|
| (1) | $\int_0^{\infty} e^{-ax^2}\,dx = \frac{1}{2}\sqrt{\frac{\pi}{a}}$ |
| (2) $(a > 0)$ | $\int_0^{\infty} xe^{-ax^2}\,dx = \frac{1}{2a}$ |
| (3) | $\int_0^{\infty} x^2e^{-ax^2}\,dx = \frac{1}{4}\sqrt{\frac{\pi}{a^3}}$ |
| (4) $(a > 0)$ | $\int_0^{\infty} x^3e^{-ax^2}\,dx = \frac{1}{2a^2}$ |
| (5) | $\int_0^{\infty} x^4e^{-ax^2}\,dx = \frac{6}{4}\sqrt{\frac{\pi}{a^5}}$ |

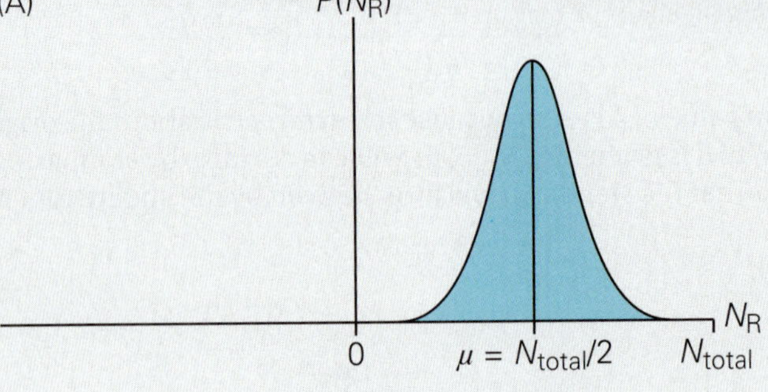


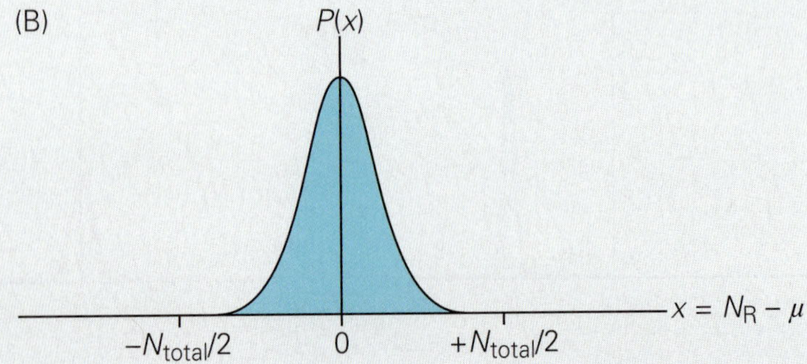


**Figure 17.1.1 Shifting a Gaussian distribution to the origin.** (A) Graph of a Gaussian distribution for the probability of steps to the right ($N_R$) in a random walk where the total number of steps is $N_{total}$. The integral of the probability distribution is the area under the curve, shaded *blue*. (B) The Gaussian distribution is shifted to the origin (that is, centered about zero) by switching to the variable $x$, where $x = N_R - \mu$. The range of the probability distribution is now from $-N_R/2$ to $+N_R/2$.

And, by combining Equations 17.1.1, 17.1.2, and 17.1.4, we verify that the probability distribution is indeed normalized:

$$\int_0^{N_{\text{total}}} P(N_R)\,dN_R = \frac{1}{\sigma\sqrt{2\pi}}\sigma\sqrt{2\pi} = 1 \qquad (17.1.5)$$

**The standard deviation of a Gaussian distribution**

It is also useful to review the significance of the **standard deviation, $\sigma$**. The standard deviation is the square root of the **variance** of the distribution. The variance is the mean value of the squared displacement from the mean:

$$\text{variance} = \langle (N_R - \mu)^2 \rangle \qquad (17.1.6)$$

The angular brackets in Equation 17.1.6 denote an average over the whole distribution. The average value of any variable is calculated by multiplying the variable by the probability distribution and integrating the result:

$$\langle (N_R - \mu)^2 \rangle = \frac{1}{\sqrt{2\pi}\,\sigma}\int_0^{N_{\text{total}}} (N_R - \mu)^2 \exp\left[-\frac{(N_R - \mu)^2}{2\sigma^2}\right] dN_R \qquad (17.1.7)$$

To evaluate the integral in Equation 17.1.7, we again use the shifted variable, $x$, which gives:

$$\frac{1}{\sigma\sqrt{2\pi}}\int_0^{N_{\text{total}}} (N_R - \mu)^2 \exp\left[-\frac{(N_R - \mu)^2}{2\sigma^2}\right] dN_R =$$

$$= \frac{1}{\sigma\sqrt{2\pi}}\int_{-\infty}^{+\infty} x^2 \exp\left(-\frac{x^2}{2\sigma^2}\right) dx \qquad (17.1.8)$$

By referring to a Table 17.1.1, you will find that:

$$\int_{-\infty}^{+\infty} x^2 \exp(-ax^2)\,dx = 2\int_0^{+\infty} x^2 \exp(-ax^2)\,dx = \frac{1}{2}\sqrt{\frac{\pi}{a^3}} \qquad (17.1.9)$$

Using Equation 17.1.9 in Equation 17.1.7, you will see that:

$$\langle (N_R - \mu)^2 \rangle = \left(\frac{1}{\sigma\sqrt{2\pi}}\right)\left(\frac{1}{2}\right)\sqrt{\pi 2^3 \sigma^6} = \sigma^2 \qquad (17.1.10)$$

It follows from Equation 17.1.9 that the standard deviation, $\sigma$, is indeed the square root of the variance. The standard deviation is also referred to the **root mean square (r.m.s.) displacement** from the mean.

steps can be appreciated by looking at **Figure 17.5**, which graphs the probability distributions for random-walk trajectories with different numbers of steps. The longest trajectory in Figure 17.5 has 10,000 steps, and is 100 times longer than the shortest one, which has only 100 steps. The width of the distribution for the longest trajectory, which is related to the standard deviation (see Section 17.5, below), is only 10 times larger than the width of the distribution for the shortest trajectory.

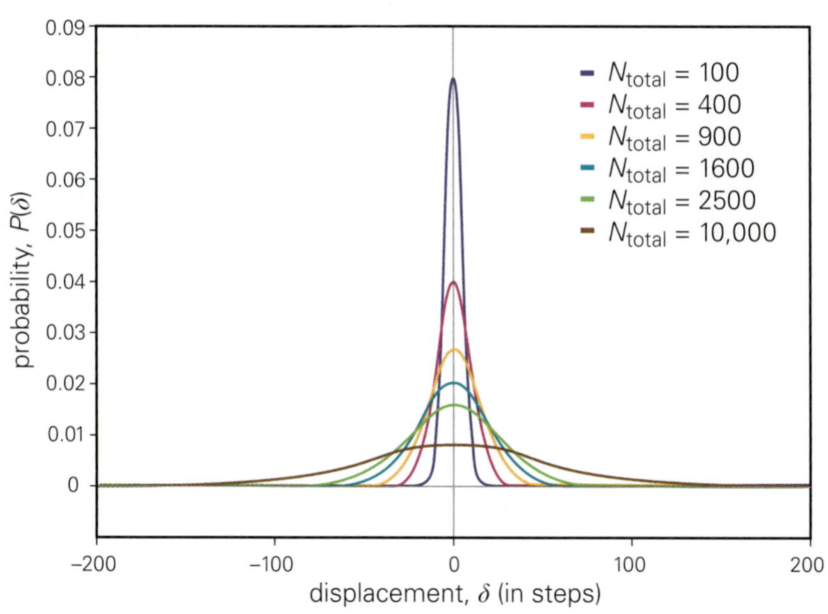


**Figure 17.5 Displacement probability for random walks.** The distribution of displacements, $\delta$, from the starting point is shown for increasing numbers of steps, $N_{\text{total}}$, for a one-dimensional random walk. These are Gaussian functions, defined in Equation 17.7.

## Box 17.2 Expressing the probability distribution for a random walk in terms of displacements from the mean position

Here we explain how we arrive at the expression for the probability of finding a displacement $\delta$ from the mean position for a one-dimensional random walk:

$$P(\delta)=\frac{1}{\sqrt{2\pi N_{total}}}\exp\left(-\frac{\delta^2}{2N_{total}}\right) \tag{17.7}$$

We begin with Equation 17.5 in the main text, which gives the probability of a trajectory moving $N_R$ steps to the right:

$$P(N_R)=\frac{2}{\sqrt{2\pi N_{total}}}\exp\left[\frac{-\left(N_R-\frac{N_{total}}{2}\right)^2}{\frac{N_{total}}{2}}\right] \tag{17.5}$$

As we noted in the main text, a trajectory with a displacement $\delta$ is equivalent to a trajectory that moves $N_R$ steps to the right, and so the probability of finding such a trajectory must be given by Equation 17.5. We need to express the exponent in Equation 17.5 in terms of $\delta$ rather than $N_R$. To do this, we express $N_R$ in terms of the displacement, $\delta$:

$$N_R=\frac{\delta}{2}+\frac{N_{total}}{2} \tag{17.2.1}$$

By substituting this expression in Equation 17.5, we get:

$$P(\delta)=K\exp\left[-\frac{\left(\frac{\delta}{2}\right)^2}{\frac{N_{total}}{2}}\right]=K\exp\left(-\frac{\delta^2}{2N_{total}}\right) \tag{17.2.2}$$

Here $K$ is the normalization constant. To determine its value, we integrate the probability distribution function over all values of $\delta$, and set the value of the integral to unity, as explained in Box 17.1.

$$\int_{-\infty}^{+\infty}P(\delta)d\delta=K\int_{-\infty}^{+\infty}\exp\left(-\frac{\delta^2}{2N_{total}}\right)d\delta=K\sqrt{2\pi N_{total}}=1 \tag{17.2.3}$$

And so:

$$K=\frac{1}{\sqrt{2\pi N_{total}}} \tag{17.2.4}$$

We can also determine the standard deviation directly by calculating the mean square displacement, $\langle\delta^2\rangle$, as follows:

$$\langle\delta^2\rangle=\int_{-\infty}^{+\infty}\delta^2P(\delta)d\delta=\frac{1}{\sqrt{2\pi N_{total}}}\int_{-\infty}^{+\infty}\delta^2\exp\left(-\frac{\delta^2}{2N_{total}}\right)d\delta \tag{17.2.5}$$

The integral in Equation 17.2.5 can be evaluated as explained in Box 17.1, yielding:

$$\langle\delta^2\rangle=\frac{1}{\sqrt{2\pi N_{total}}}\frac{\sqrt{\pi\left(2N_{total}\right)^3}}{2}=N_{total} \tag{17.2.6}$$

## 17.5 The width of the distribution of displacements increases with the square root of time for random walks

Now let us return to the question we posed at the end of Section 17.1, which was to ask how quickly a clump of bacteria would spread out from the spot where they are deposited initially. If we assume that each bacterium moves along a trajectory that is a random walk, then the spread of the bacterial population from the initial spot will be described by Equation 17.7. To understand why this is so, consider that there are a great many bacteria in the initial spot. They all start moving randomly and independently of each other. If we assume that the bacteria all take "steps" at a fixed rate, $R_{step}$, then the total number of steps, $N_{total} = R_{step}t$, increases linearly with time, $t$. At any particular time, the number of bacteria that have moved a certain displacement, $\delta$, from the origin is proportional to the probability that a random walk with $N_{total}$ steps ends up with that displacement, which is given by the Gaussian distribution shown in Equation 17.7 (**Figure 17.6**).

To see how quickly the bacteria spread out, it is useful to consider the width of the probability distribution. As we explain below, random movement in two and three dimensions can be described just as for one dimension, except that constant scale factors have to be included for the two cases, respectively. For now, we just

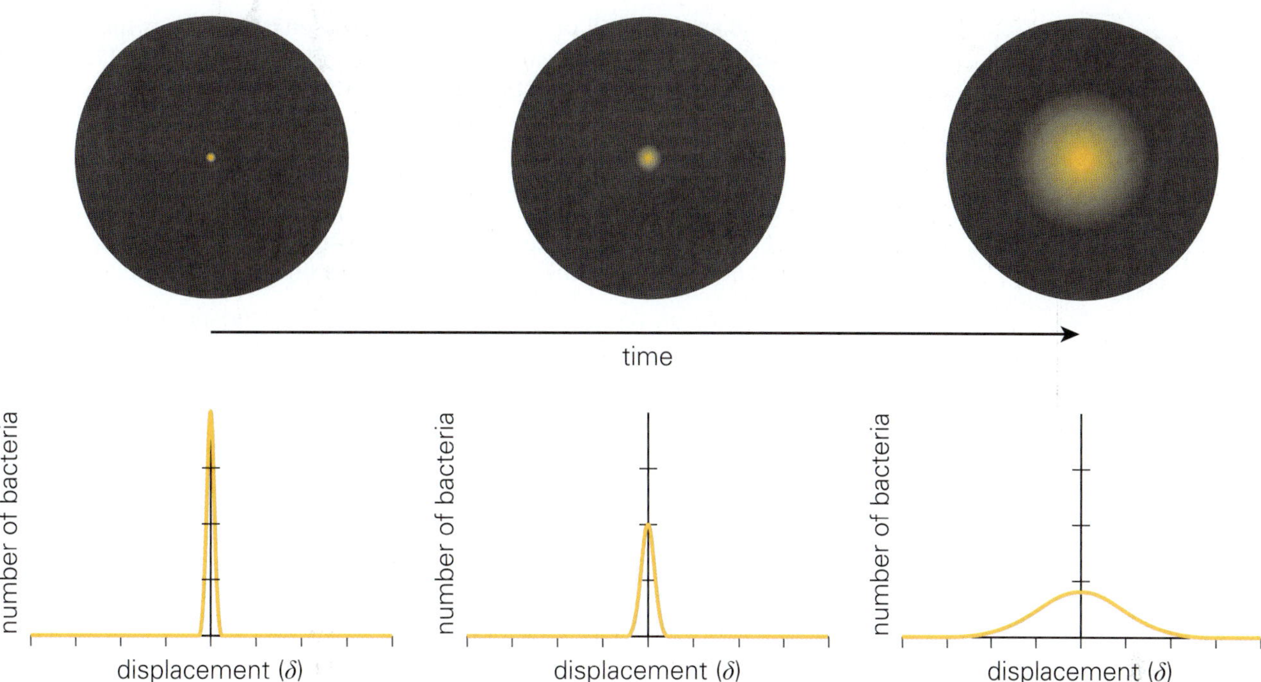


**Figure 17.6 Gaussian functions describe the outward spread of bacteria.** The diagram above shows that a clump of bacteria that is spotted on an agar plate spreads out with time. The graphs show the number of bacteria as a function of distance from the origin.

consider the probability distribution for one dimension and assume that the density of bacteria at a certain displacement from the origin will be proportional to the value of the probability distribution for that displacement.

How does the width of the probability distribution change with time? One practical definition of the width is simply the standard deviation of the distribution which, as we saw in the previous section, is $\sqrt{N_{\text{total}}}$. Recall from Section 7.14 that ~68% of the displacements will be within $1\sigma$ of the mean value (**Figure 17.7**A). Another measure of the width of the distribution is the displacement at which the probability is half of the maximum probability (see Section 7.13). The most probable location is the origin ($\delta = 0$) because steps to the left and to the right are equally probable. The displacement, $\delta_{1/2}$, for which the probability drops to half the value at the origin is called the **half-width at half height** (see Figure 17.7B). The population of bacteria starts dropping rapidly when the displacement from the origin is greater than the half-width. This gives us a rough measure of the edge of the spreading spot of bacteria.

As explained in Section 7.13, the half-width of the distribution, $\delta_{1/2}$ is given by:

$$\delta_{1/2} \approx 1.175\sigma = 1.175\sqrt{N_{\text{total}}} \tag{17.8}$$

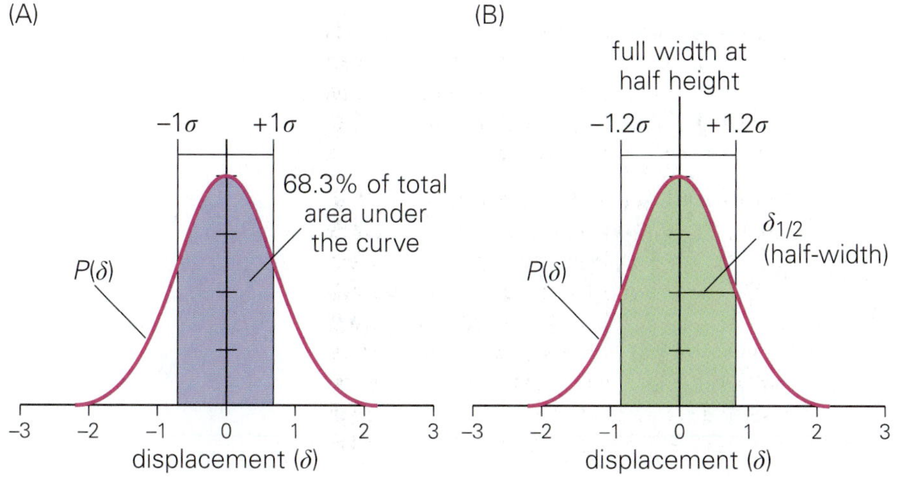


**Figure 17.7 Two measures of the width of a Gaussian distribution.** (A) The area within $1\sigma$ of the mean value covers ~68% of the distribution, and so $1\sigma$ can be considered the width of the distribution. (B) The half-width at half height is another measure of the width, and it is equal to ~$1.2\sigma$.

Equation 17.8 applies if all of the steps are of unit length, which is the assumption we made in order to derive the probability distribution (Equation 17.7). If the actual length of each step is $l_{step}$, then the half-width is given by:

$$\delta_{1/2} \approx \left(l_{step}\right)(1.175)\sqrt{N_{total}} \tag{17.9}$$

As you can see from Figure 17.7B, most of the bacteria will have displacements from the origin that are less than the half-width of the distribution. According to Equation 17.9, the region containing most of the bacteria will grow as the square root of time, because the number of steps increases linearly with time.

This dependence on the square root of time is a characteristic of random-walk processes (see Figure 17.5). Because steps to the left and to the right are equally probable, they tend to cancel each other out. As a consequence, even though the total length of each trajectory increases linearly with time, the spread of the distribution of displacements from the origin increases more slowly with time. This can be appreciated by referring back to Figure 17.4. If all of the cells began at the origin and take 10 steps per second, then the narrowest distribution shown in Figure 17.4 corresponds to the spread after 10 seconds and the widest distribution shows the spread after 1000 seconds. As we noted earlier, the spread between 10 and 1000 seconds has increased only 10-fold due to the $\sqrt{t}$ dependence. This is an indication that movement over substantial distances by random-walk processes is inefficient.

## 17.6 Random walks in two dimensions can be analyzed by combining two orthogonal one-dimensional random walks

Several trajectories for cells undergoing random walks in two dimensions are shown in **Figure 17.8**, in which each trajectory consists of 1000 steps. In these diagrams, each step is in a random direction with respect to the previous one and has a length that is distributed randomly between 0 and 1. For trajectories with such a large number of steps, it is clear from Figure 17.8 that the details of the direction

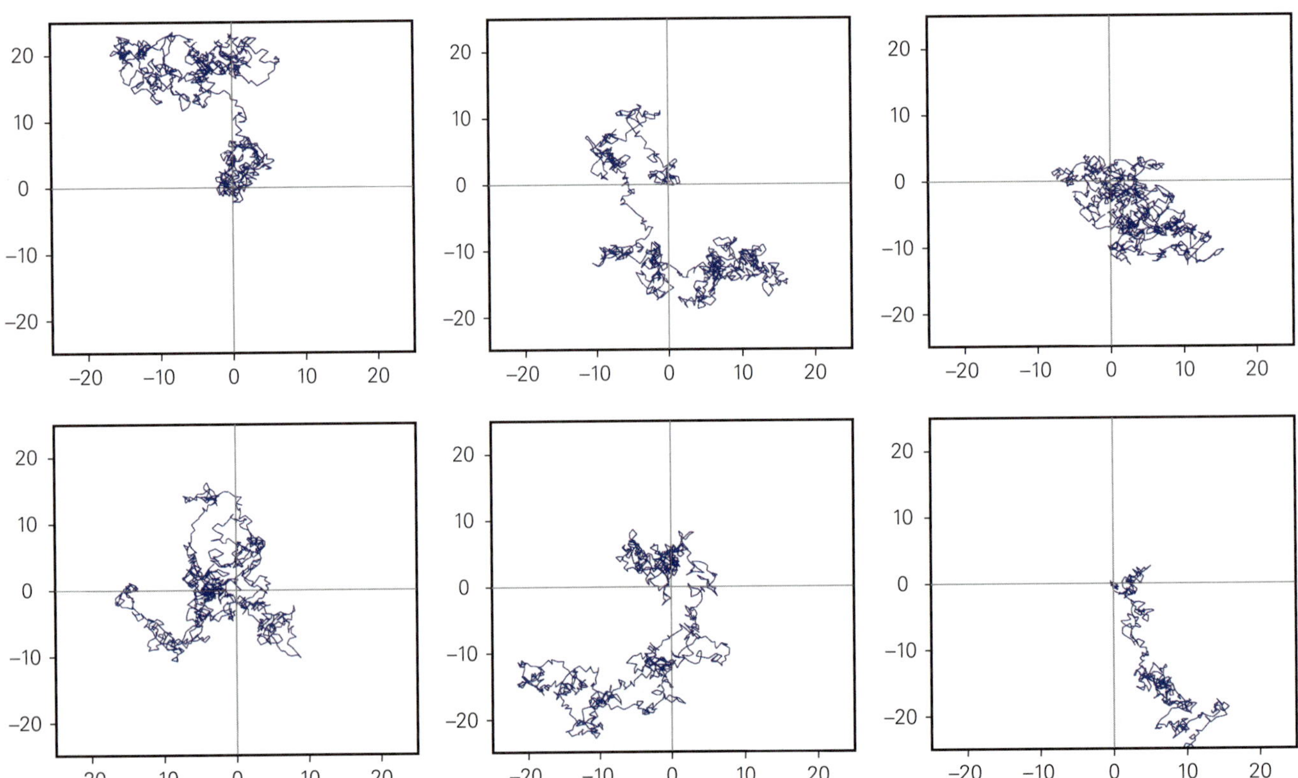


**Figure 17.8 Paths of individual cells in a two-dimensional random walk.** All paths begin at the origin and are 1000 steps in length in each case. For each step, the direction is random and the size of the step takes a random value between 0 and 1. The net displacement from the origin, averaged over a large number of cells, would give a two-dimensional Gaussian distribution with one-dimensional cross-sections that are equivalent for all directions away from the origin, such as those shown in Figure 17.6.

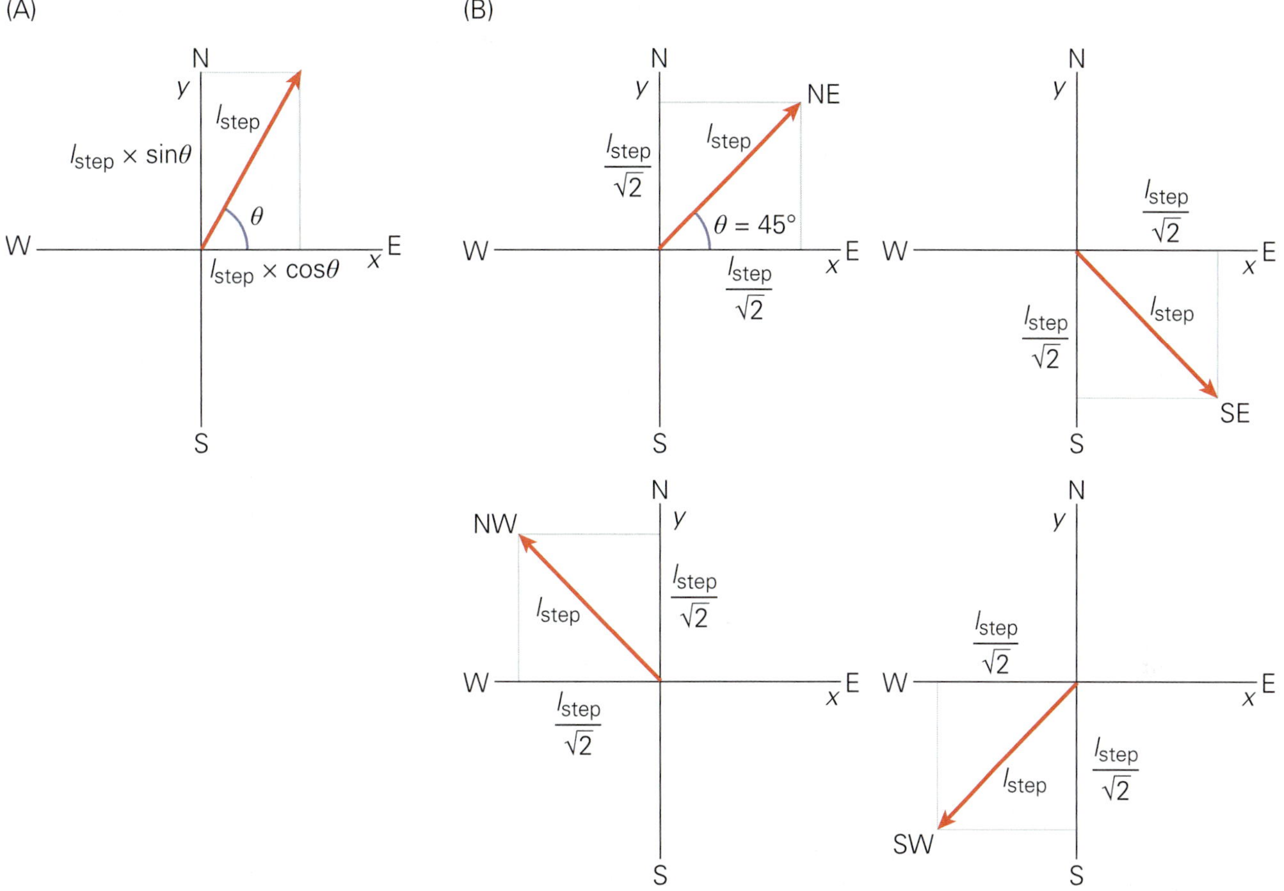


**Figure 17.9 The projection of a step in a two-dimensional random walk along two orthogonal directions.** (A) A single step, with length $l_{step}$, is shown, with random orientation. The angle between the $x$ axis and the vector corresponding to the step is denoted by $\theta$. The projections of the step on two orthogonal axes are given by $(l_{step})\cos\theta$ and $(l_{step})\sin\theta$, respectively. (B) As explained in Section 17.7, r.m.s. length of the projections on each axis, averaged over many steps in a two-dimensional random walk, is given by $\frac{l_{step}}{\sqrt{2}}$. If we work backwards and construct a two-dimensional random walk from two one-dimensional trajectories, each with this step size, then each step in the resulting two-dimensional random walk is restricted to one of four directions. These are indicated in the diagram using geographical directions (NE, northeast, etc.). See Figure 17.10 for a two-dimensional random walk built up in this way.

and length of each step cannot be made out easily compared to the scale of the overall motion. It turns out that, in the limit of a large number of steps, an accurate statistical description of the motion is obtained if we consider random walks with uniform step sizes corresponding to the average step size, just as we did for one-dimensional motion. But, in order to make the connection to Gaussian distributions, we need to make one more assumption.

We can describe movement in two dimensions by considering the projections of each displacement step along two orthogonal axes, denoted $x$ and $y$, as shown in **Figure 17.9**. We begin by assuming that each step in the two-dimensional random walk has a uniform displacement, $l_{step}$, that corresponds to the average step size of the entire random walk. The projections of a two-dimensional step on $x$ and $y$ are shown in Figure 17.9A.

We can see from this diagram that, even though the random walk consists of steps of uniform length in two dimensions, if the orientation of each step is random, then the lengths of projections on the $x$ and $y$ axes will not be uniform. Depending on the orientation of the two-dimensional step, the length of the step on each axis can vary between $+l_{step}$ to $-l_{step}$. Thus, the two-dimensional random walk can be described as a combination of two one-dimensional random walks, each with step sizes of variable length. For trajectories with large numbers of steps, we once again make the approximation that each one-dimensional random walk can be analyzed in terms of trajectories with steps of equal length. The length of each step is equal to the average step along that axis over the course of the random walk.

## 17.7 A two-dimensional random walk is described by two one-dimensional walks, but the effective step size for each is smaller by a factor of $\sqrt{2}$

What is the average step length along one axis in a two-dimensional random walk? As you can see from Figure 17.9, the projection of the step on an axis is ($l_{step}$) cos $\theta$, where $l_{step}$ is the length of the step and $\theta$ is the angle between the step vector and the axis. Because positive and negative displacements are equally likely, the average value of the projected displacement is zero. We shall calculate, instead, the root mean square displacement, which is defined as follows:

$$\text{r.m.s. step length along the } x \text{ axis} = \left\langle \left(l_{\text{step}}\right)^2 \cos^2\theta \right\rangle^{\frac{1}{2}} \tag{17.10}$$

The brackets indicate an average over all the steps of the trajectory. We assume that the orientation of each step is truly random—that is, all values of $\theta$ are equally likely and the trajectory is sufficiently long that these values are sampled uniformly. As explained in **Box 17.3**:

$$\text{r.m.s. step length along the } x \text{ axis} = \left\langle \left(l_{\text{step}}\right)^2 \cos^2\theta \right\rangle^{\frac{1}{2}} = \frac{l_{\text{step}}}{\sqrt{2}} \tag{17.11}$$

The statistical behavior of the trajectories along each axis (that is, $x$ and $y$) is the same as that described in Section 17.4 for a one-dimensional random walk, except that the step size is smaller by a factor of $\sqrt{2}$. This reduction in the effective step size occurs because the motion is broken down into two orthogonal and independent components. For each step, only part of the movement is in the radial direction, away from the origin. The time dependence of the outward movement (that is, the movement of the "edge" of the distribution) remains the same as for the one-dimensional distribution and scales as $\sqrt{t}$.

## 17.8 The assumption of uniform step lengths along each axis means that the random walk occurs on a grid

It turns out that, by assuming that we can treat movement along each axis in terms of a uniform step length given by Equation 17.21, we are assuming that each step in the two-dimensional random walk occurs on a grid, with only diagonal steps allowed. This is made clear in Figure 17.9B, which shows that, if the step along each axis is $\frac{l_{\text{step}}}{\sqrt{2}}$, then the actual step must lie exactly midway between each axis. Thus, only four kinds of steps are possible, corresponding to moves in the northwest, northeast, southeast, and southwest directions (assuming that $x$ and $y$ correspond to "east" and "north," respectively). This idea is illustrated in **Figure 17.10**, which shows how a two-dimensional random walk is broken down into two orthogonal one-dimensional random walks.

The random walk shown in Figure 17.10C is restricted to lie on the nodes of a two-dimensional grid or lattice. You can readily appreciate that it is straightforward to write computer algorithms that generate such random walks, given a small set of parameters such as the step size and the time constant. Indeed, such algorithms find powerful application in the analysis of a very diverse range of problems in which random steps influence some kind of aggregate behavior. In addition to the study of diffusion and Brownian motion, which is our particular focus here, random walks are used to study things as apparently unrelated as the structure of polymers and the variation in stock market prices.

## Box 17.3 Calculating the r.m.s. value of the projection along one axis of steps in random walks

### Two-dimensional random walks

Here we explain at how we arrive at Equation 17.11, which relates the r.m.s. value of the projection of a step on one axis in a two-dimensional random walk to the length of the step:

$$\text{r.m.s. step length along the } x \text{ axis} = \left\langle l_{\text{step}}^2 \cos^2\theta \right\rangle^{\frac{1}{2}} = \frac{l_{\text{step}}}{\sqrt{2}} \tag{17.11}$$

We assume that the orientation of each step is truly random—that is, all values of $\theta$ are equally likely and the trajectory is sufficiently long that these values are sampled uniformly. To calculate the average value, we express it in terms of integrals over values of $\theta$:

$$\left\langle l_{\text{step}}^2 \cos^2\theta \right\rangle = \left(l_{\text{step}}\right)^2 \left\langle \cos^2\theta \right\rangle = \left(l_{\text{step}}\right)^2 \frac{\int_0^{2\pi} \cos^2\theta \, d\theta}{\int_0^{2\pi} d\theta} \tag{17.3.1}$$

To evaluate Equation 17.3.1, we use the following expression for the integral of $\cos^2\theta$:

$$\int \cos^2\theta \, d\theta = \frac{\theta}{2} + \frac{\sin 2\theta}{4} + C \tag{17.3.2}$$

where $C$ is a constant of the integration.

Using Equation 17.3.2, we evaluate the definite integral in Equation 17.3.1 as follows:

$$\frac{\int_0^{2\pi} \cos^2\theta \, d\theta}{\int_0^{2\pi} d\theta} = \frac{\left[\frac{\theta}{2} + \frac{\sin 2\theta}{4}\right]_0^{2\pi}}{2\pi} = \frac{1}{2} \tag{17.3.3}$$

Although Equation 17.3.3 may look complicated, its meaning may be appreciated intuitively by looking at **Figure 17.3.1**, which shows a graph of $\cos^2\theta$ over the interval 0 to $2\pi$. The figure shows a horizontal red line that cuts the graph of the function exactly in half. This red line corresponds to the mean value of the function, which is 0.5 or 1/2, just as Equation 17.20 tells us.

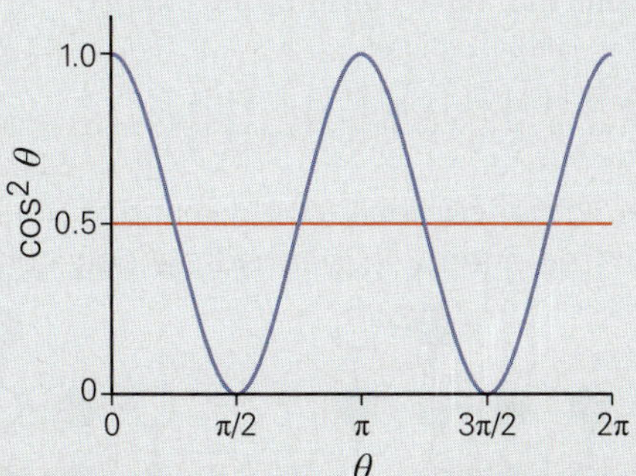


**Figure 17.3.1 The mean value of $\cos^2\theta$ over a full cycle is equal to 1/2.** The value of the function $\cos^2\theta$ is graphed in the interval 0 to $2\pi$. Notice that the horizontal *red* line cuts the function into two equal halves. The red line marks the mean value of the function over the interval, which is 1/2.

Combining Equations 17.3.1 and 17.3.3, we obtain an expression for the r.m.s. step length along the $x$ axis:

$$\text{r.m.s. step length along the } x \text{ axis} = \left\langle l_{\text{step}}^2 \cos^2\theta \right\rangle^{\frac{1}{2}} = \frac{l_{\text{step}}}{\sqrt{2}} \tag{17.3.4}$$

### Three-dimensional random walks

We now explain how we derived the expression for the projection along one axis of a step in a three-dimensional random walk. Denoting the chosen axis as $z$:

$$\text{mean square displacement along the } z \text{ axis} = \left\langle l_{\text{step}}^2 \cos^2\alpha \right\rangle = \frac{1}{3} l_{\text{step}}^2 \tag{17.12}$$

If you refer to Figure 17.12, you will see that the projection of the step along $z$ is $(l_{\text{step}})\cos\alpha$. Thus, the mean square value of the projection is given by:

$$\text{mean square value of step length along } z = \left\langle l_{\text{step}}^2 \cos^2\alpha \right\rangle \tag{17.3.5}$$

In calculating the mean square value, we have to account for the different weighting of vectors with different values of $\alpha$. This is because for each value of $\alpha$ there are a different number of vectors with the same value of $\alpha$. We do this by multiplying the projection of the vector by the circumference of the circle traced out by the tips of vectors with the same value of $\alpha$ (see Figure 17.12), which is $(2\pi)(l_{\text{step}})\sin\alpha$, and integrating over the range of $\alpha$:

$$\text{mean square value} = \frac{\int_0^{\pi} \left(l_{\text{step}}\right)^2 \left(\cos^2\alpha\right)(2\pi)\left(l_{\text{step}}\right)(\sin\alpha)\, d\alpha}{\int_0^{\pi} (2\pi)\left(l_{\text{step}}\right)(\sin\alpha)\, d\alpha} = \frac{l_{\text{step}}^2 \int_0^{\pi} \left(\cos^2\alpha\right)(\sin\alpha)\, d\alpha}{\int_0^{\pi} \sin\alpha \, d\alpha} \tag{17.3.6}$$

To evaluate this expression, we use the following integrals, which you can look up in a table of standard integrals:

$$\int \left(\cos^2\alpha\right)(\sin\alpha)\, d\alpha = -\frac{1}{3}\cos^3\alpha + C \tag{17.3.7}$$

$$\int \sin\alpha \, d\alpha = -\cos\alpha + C \tag{17.3.8}$$

By using these equations to evaluate the definite integrals in Equation 17.3.6, we get:

$$\text{mean square value} = \left\langle l_{\text{step}}^2 \cos^2\alpha \right\rangle = \frac{1}{3} l_{\text{step}}^2 \tag{17.3.9}$$

Thus, the r.m.s. value of the projection of the displacement on any one of the axes is given by:

$$\text{r.m.s. step length along any axis} = \frac{l_{\text{step}}}{\sqrt{3}} \tag{17.3.10}$$

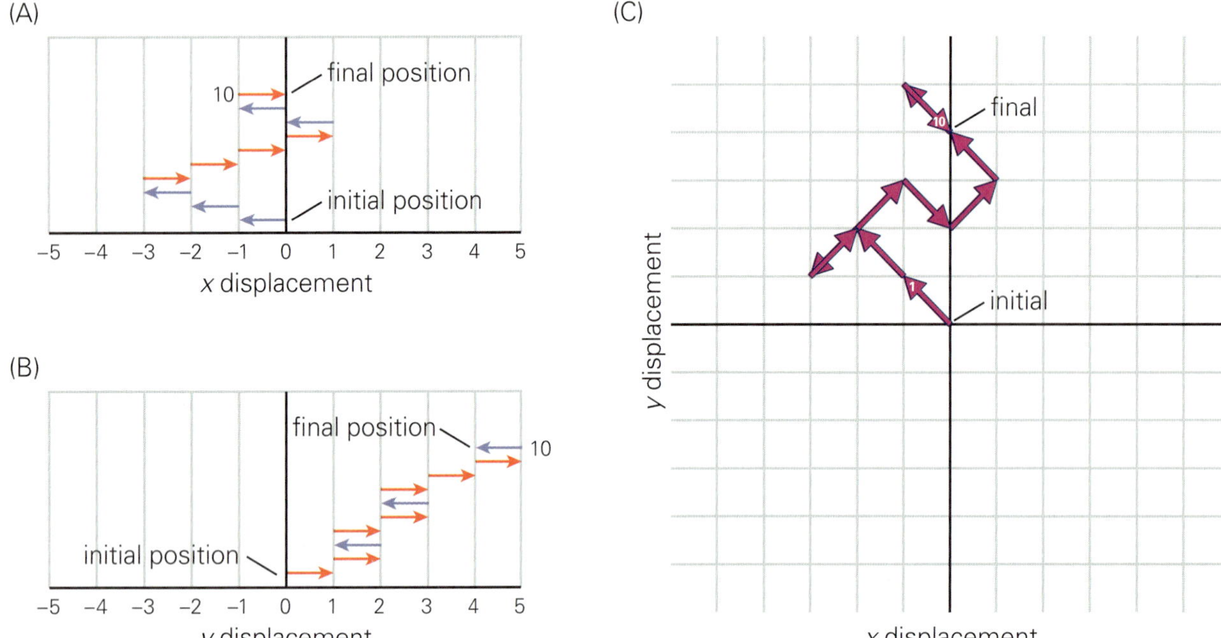


**Figure 17.10 Construction of a two-dimensional random walk from two orthogonal one-dimensional random walks.** (A, B) Displacements along the *x* and *y* directions during a 10-step random walk. Compare with Figure 17.2, and note that the trajectories for *x* and *y* displacements end up at 0 and +4, respectively. (C) A two-dimensional random walk generated by combining the sequential steps shown in (A) and (B). As expected, the trajectory ends up at $x = 0$ and $y = +4$.

As an example, consider the application of a two-dimensional random walk to the analysis of protein folding. One interesting question is whether there are particular patterns of amino acids that are capable of folding into a stable structure and others that are not. To study this question, scientists have analyzed the behavior of proteins that consist of just two kinds of amino acids, one hydrophobic and the other polar. A simple computational model for such a protein considers each residue to be a ball that is connected to two adjacent residues in a chain. Any particular conformation of the protein corresponds to a random-walk trajectory, in which the starting point is the first residue and the number of steps is simply the number of residues in the chain.

Three conformations of a 14-residue protein modeled in this way in two dimensions are shown in **Figure 17.11**. Using the random-walk algorithm, modified to prevent the chain from running into itself, all possible conformations of the protein can be generated very quickly on a computer. Different patterns of amino acids can then be placed on these chains, and the energies of the different structures can be evaluated using simple schemes. In the example shown in Figure 17.11, every time two hydrophobic residues are next to each other, the energy is deemed to be favorable. This kind of strategy has allowed scientists to understand what features allow proteins to adopt unique folded structures without being trapped in alternative conformations that have comparable energy.

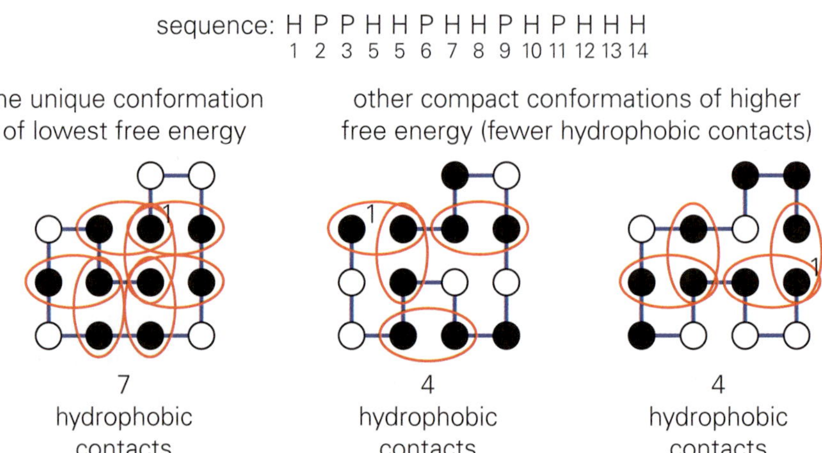


**Figure 17.11 Protein structure modeled by random walks.** A two-dimensional representation of a protein with 14 residues is shown here. The residues of the protein are constrained to lie on the points of a grid, and all possible conformations are generated by the trajectories of a two-dimensional random walk. The trajectories are self-avoiding, because two residues cannot lie on the same point. There are only two kinds of residues, hydrophobic (*black*) and polar (*white*). Favorable contacts between hydrophobic residues are outlined in *red*. Highly simplified protein models such as this help us to understand what kinds of sequence patterns lead to unique folded structures. (Adapted from K.A. Dill, *Biochemistry* 29: 7133–7155, 1990. With permission from Elsevier.)

## 17.9 A three-dimensional random walk is described by three orthogonal one-dimensional walks, and the effective step size for each is smaller by a factor of $\sqrt{3}$

The extension of our analysis of random walks to three dimensions is done in a way that is analogous to the extension to two dimensions, but with a further reduction in the rate of outward movement due to the fact that the steps can have components along the *x*, *y*, and *z* directions. This reduces the effective step size in each direction to $\frac{l_{step}}{\sqrt{3}}$. The spread of the distribution in each direction has the same time dependence as before, increasing with the square root of time.

To understand how the factor of $\sqrt{3}$ arises, consider a three-dimensional random walk with uniform step size, but with random direction at each step. The projection of such a step on to one of the three orthogonal axes is shown in **Figure 17.12**. The vector corresponding to the step makes an angle $\alpha$ with the vertical axis. The projection of the vector on to the vertical axis is given by $(l_{step})\cos\alpha$.

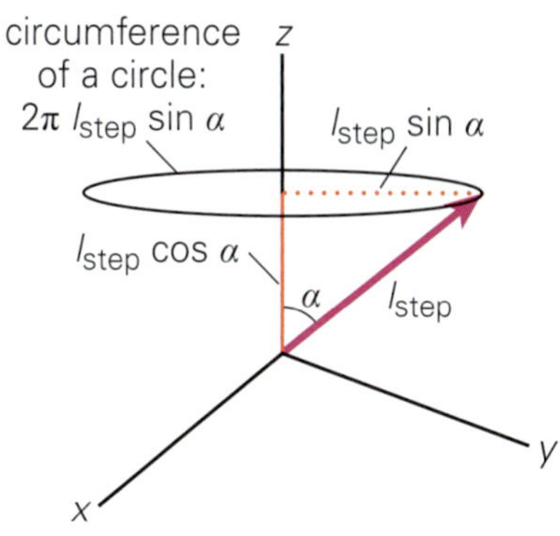


**Figure 17.12 Projection of a three-dimensional displacement step along one axis.** A step in a random direction is shown as a *purple* vector. The projection of the step on the vertical axis is shown in *red*. All vectors that have their tips on the circle have the same projection on the vertical axis. To calculate the mean square value of the projection, we need to average over all these vectors, and over all values of $\alpha$, the angle made by the vector with the vertical axis (see Box 17.3).

Each value of $\alpha$ defines a set of displacement vectors, the tips of which trace out a circle of radius $(l_{step})\sin\alpha$, as shown in Figure 17.12. The number of such vectors is proportional to sin $\alpha$, and therefore increases as $\alpha$ increases from 0 to 90 and then decreases again. The range of $\alpha$ is from 0° to 180° (2π). As explained in Box 17.3,

$$\text{mean square displacement along the } z \text{ axis} = \left\langle (l_{step})^2 \cos^2\alpha \right\rangle = \frac{1}{3} l_{step}^2 \quad (17.12)$$

Thus, the r.m.s. value of the projection of the displacement on any one of the axes is given by:

$$\text{r.m.s. step length along any axis} = \frac{l_{step}}{\sqrt{3}} \quad (17.13)$$

According to Equation 17.13, for a molecule diffusing in three dimensions, its rate of movement along any one direction is slower by a factor of $\sqrt{3}$ than its overall movement. We implicitly used this result earlier in Section 11.19, when we compared the rate of diffusion of potassium ions in water (that is, in three dimensions) to the rate at which they translocate through the potassium channel (in one dimension). The significance of this point will be made clear in Section 17.14, below, where we relate a parameter known as the diffusion constant to the r.m.s. displacement.

## 17.10 The movement of bacteria in the presence of attractants or repellents is described by biased random walks

The random-walk model provides a satisfactory description of the behavior of bacterial swimming. The actual movement of a single *E. coli* over a period of 30 seconds, in the absence of any chemical attractant (for example, nutrient molecules) or repellant, is shown in **Figure 17.13**. As we saw earlier, this random movement is a rather inefficient way for a bacterium to travel over substantial distances. When bacterial cells are really moving towards a nutrient source (or away from a toxin), this behavior is modified. The cells have receptors on the surface that respond to nutrients in the environment and, if the cell is swimming in a direction that increases the nutrient concentration, then the periods of straight swimming are continued for a longer time.

This means that the bacterium takes longer steps towards the nutrient than in other directions. On the other hand, if the concentration of the attractant decreases, then the cell tumbles and then sets off in a new direction to try its luck again. This leads to a biased random walk (still random in the sense that the new direction after a period of tumbling is always arbitrary), but with a larger step size in the direction of increasing nutrients. In the completely random walk, the center of the distribution always stays at the starting point, but in the biased case there is a net motion of the whole distribution in the favorable direction.

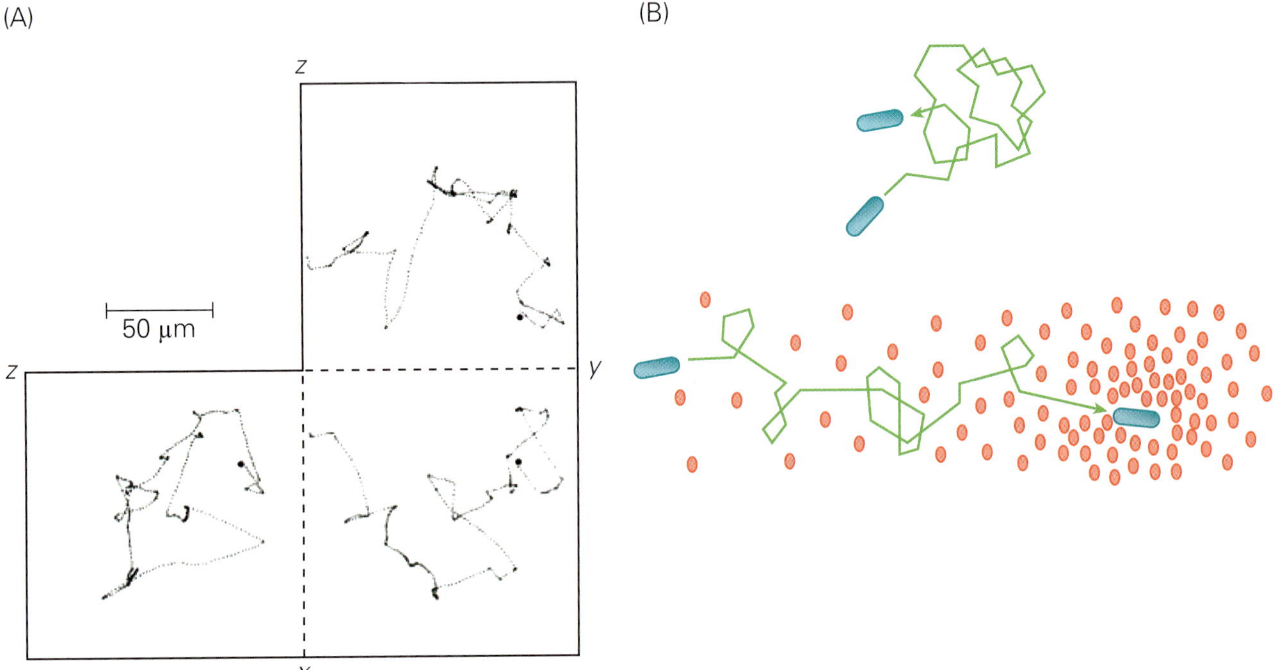


**Figure 17.13 Unbiased and biased random walks.** (A) Trajectory of an actual *E. coli* bacterium, observed over 30 seconds in a microscope. The diagram shows three orthogonal projections of the three-dimensional movement of the bacterium. There are no attractant or repellant molecules in the solution, and the random walk is unbiased. (B) Shown at the top is a schematic representation of an unbiased random walk, such as the trajectory depicted in (A), in which no particular direction is preferred. The diagram at the bottom shows what happens if there is a concentration gradient of attractant molecules (*red*). The steps that move the bacterium up the gradient are longer than steps in other directions, leading to a random walk that is biased towards the source of the attractant. (A, adapted from H.C. Berg and D.A. Brown, *Nature* 239: 500–504, 1972. With permission from Macmillan Publishers Ltd; B, adapted from B. Alberts et al., Molecular Biology of the Cell, 5th ed. New York: Garland Science, 2008.)

One can ask why the bacterial cells tumble at all if there is a nutrient gradient. Why don't they simply move directly towards the attractant? As discussed in following sections, molecules and cells are constantly buffeted by random collisions with their surroundings, and this means that the cells cannot maintain a straight line for very long and will invariably be diverted from the correct direction. They must, as a consequence, reset their direction of travel over and over again just to be sure that they are moving in the best direction. Imagine a small rowboat in strong winds and high waves—even a rower who could maintain a perfect line on calm water would have to turn to look for a landmark and correct the course of the boat to have a chance of reaching the target in rough conditions.

## B. MACROSCOPIC DESCRIPTION OF DIFFUSION

In the first part of this chapter we described the diffusive movement of bacteria with a focus on individual random trajectories, from which we built up a description of the aggregate behavior of many trajectories. In this part of the chapter we shall approach diffusion from a macroscopic perspective, instead, in which we do not consider individual trajectories at all.

### 17.11 Fick's first law states that the flux of molecules is proportional to the concentration gradient

A molecule in a solution is buffeted constantly by other molecules and therefore undergoes random motion. Instead of following the trajectories of individual molecules, we shall use differential equations, called **Fick's laws**, to describe changes in concentration due to molecular motion. These differential equations were developed by Adolf Fick, in the middle of the nineteenth century. If different parts of the solution have different concentrations of the solute, then the net movement of the solute molecules will be towards equalizing concentration throughout space. We first consider a case in which the concentration gradient is along one axis only, taken as the $x$ direction. The concentration profile along $x$ is defined by the function $c(x)$.

Imagine that the concentration decreases from left to right, as shown in **Figure 17.14**. The system is clearly not at equilibrium, and there will be a net movement of molecules from left to right, along the $x$ axis. The extent of movement is

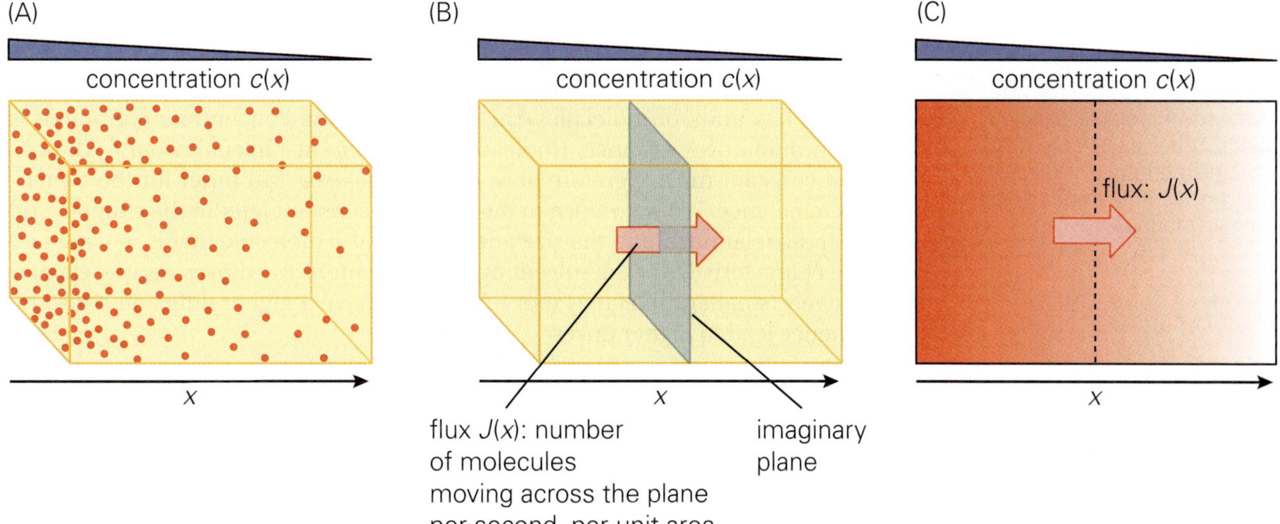


**Figure 17.14 Concentration gradient and flux of molecules in solution.** (A) A section of a three-dimensional volume of a solution is shown, with the solute in *red*. The concentration of the solute decreases from left to right, which is defined as the *x* axis. (B) Schematic representation of the flux of solute molecules across an imaginary plane that is perpendicular to the *x* axis. Because the concentration is higher on the left side of the plane, there is a net flux of molecules from left to right. (C) A one-dimensional projection of the concentration gradient. The imaginary plane is indicated by a *dashed* line.

characterized by the **molecular flux** along *x*, which is denoted $J(x)$. Consider an imaginary plane that is perpendicular to the *x* axis at position *x*, as shown in Figure 17.14B. The concentration of the solute is higher on the left side of the plane than the other, due to the gradient. The flux at position *x* is the net number of molecules crossing a unit area of the plane per unit time.

As molecules move randomly in the solution, there will be some that cross the imaginary plane. If the concentration is equal on each side of the plane, then the average number of solute molecules that cross from left to right will be equal to the number going from right to left, and there will be no net flux across the plane. When the concentration on one side is higher, however, more molecules will cross the plane from that side. There will be a net flux of molecules across the plane, and the difference in concentration will decrease with time.

**Fick's first law** simply states that the flux, $J(x)$, at position *x* is proportional to the concentration gradient at *x* (**Figure 17.15**). Fick's first law is stated mathematically as follows:

$$J(x) = -D\left[\frac{\partial c(x)}{\partial x}\right]_{y,z} \tag{17.14}$$

The derivative $\left(\frac{\partial c}{\partial x}\right)_{y,z}$ is the difference in concentration across the virtual plane at position *x* (that is, the concentration gradient along the *x* direction). The proportionality constant, *D*, is called the **diffusion constant**, and its value depends on the properties of the solute and the nature of its interactions with the solvent.

**Molecular flux, *J*(*x*)**

The flux of molecules in a direction *x* is the rate at which molecules cross a unit area of an imaginary plane perpendicular to the *x* axis.

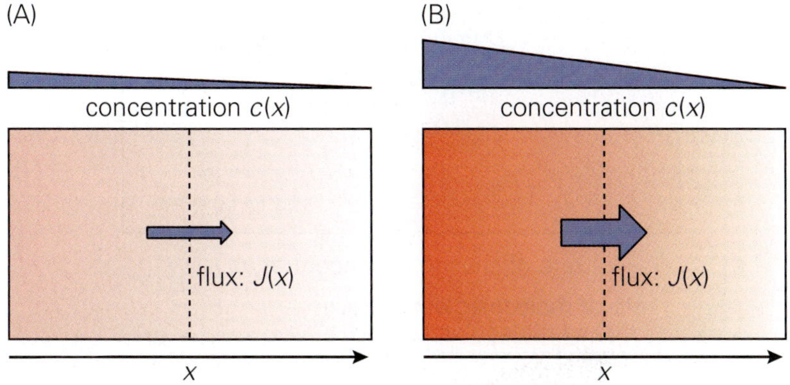

**Figure 17.15 Fick's first law.** A cross section of a three-dimensional volume is shown, with the variation in the concentration of solute molecules indicated, as in Figure 17.14C. The *dotted line* denotes an imaginary plane, and the magnitude of the flux across this plane is shown by the width of the arrow. According to Fick's first law, the flux is proportional to the concentration gradient. The gradient is shallower in (A) than in (B), and the flux is correspondingly smaller.

**Fick's laws**

Fick's laws are differential equations that describe the net movements of molecules in solutions. They can be solved to give the distribution of molecules after a specific time from a given starting distribution.

According to Equation 17.14, the greater the diffusion constant, the faster the solute will move down a concentration gradient. The negative sign in Equation 17.14 arises from the fact that molecules move *down* a concentration gradient.

The flux has units of molecules•cm$^{-2}$•sec$^{-1}$, and the concentration gradient is concentration over distance, (molecules•cm$^{-3}$)/cm or molecules•cm$^{-4}$. The diffusion constant must therefore have units of cm$^{2}$•sec$^{-1}$ in order for the dimensions to balance. We will return to the diffusion constant later in the chapter, and explain its relationship to the size and shape of the molecule that is moving and to the characteristics of the solvent in which the molecule moves. As one expects intuitively, smaller molecules move faster and have a higher diffusion constant, and hence lead to higher fluxes.

**Diffusion constant**

The diffusion constant, $D$, is a parameter that describes the average rate of motion of a molecule in solution. The units of the diffusion constant are cm$^{2}$•sec$^{-1}$.

## 17.12 Fick's second law describes the rate of change in concentration with time

Equation 17.14 shows that, when a concentration gradient is present, there will be a net flux of molecules from regions of high concentration to regions where the concentration is lower. To calculate how the concentration changes with time, we consider the fluxes going *into* and going *out of* an infinitesimally thin section in the sample, of width $dx$ (**Figure 17.16**). The rate of change of concentration with time, $\frac{\partial c(x,t)}{\partial t}$, is given by the difference in the rate at which molecules enter the thin section and the rate at which they leave, divided by the volume of section:

$$\frac{\partial c(x,t)}{\partial t} = \frac{\text{difference between numbers of molecules entering and leaving per unit time}}{\text{volume}} \quad (17.15)$$

If we consider molecules entering and leaving a unit cross-sectional area, then it follows from the definition of flux that the numerator is simply the difference between the flux at positions $x$ and $x + dx$. The volume of such a segment of the thin section is simply $dx$, as shown in Figure 17.16. Thus, Equation 17.15 becomes:

$$\frac{\partial c(x,t)}{\partial t} = \frac{J(x) - J(x+dx)}{dx} = -\frac{J(x+dx) - J(x)}{dx} \quad (17.16)$$

The right-hand side of Equation 17.16 is simply the derivative of $J(x)$ with respect to position, $x$. We can therefore rewrite Equation 17.16 in terms of the partial derivative of the flux with respect to $x$:

$$\frac{\partial c(x,t)}{\partial t} = -\frac{\partial J(x,t)}{\partial x} \quad (17.17)$$

In Equation 17.17, the flux is written as $J(x,t)$ to take explicit account of the time dependence.

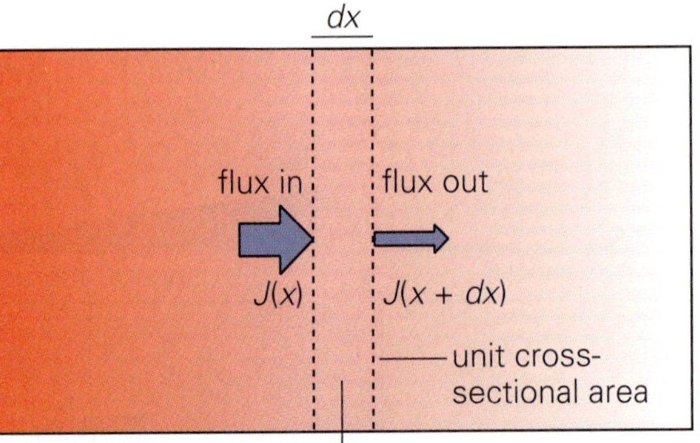


**Figure 17.16 Fick's second law.** As in Figure 17.14, the variation in the concentration of solute molecules is shown in two-dimensional projection. The space between *dashed lines* indicates an infinitesimally thin section, with thickness *dx* and unit cross-sectional area. The volume of this section is *dx*.

Next, we note that Equation 17.14, which defines Fick's first law, can be rearranged to give:

$$\frac{J(x,t)}{D}=-\frac{\partial c(x,t)}{\partial x} \tag{17.18}$$

Taking the partial derivative with respect to $x$ of both sides of Equation 17.31 gives us:

$$\frac{1}{D}\frac{\partial J(x,t)}{\partial x}=-\frac{\partial^2 c(x,t)}{\partial x^2} \tag{17.19}$$

By combining Equations 17.17 and 17.19, we get **Fick's second law**:

$$\frac{\partial c(x,t)}{\partial t}=D\frac{\partial^2 c(x,t)}{\partial x^2} \tag{17.20}$$

This equation describes how the concentration at each position in the sample (specified by picking a value for $x$) changes with time. Note that the right-hand side of Equation 17.20 involves the second derivative of concentration with respect to position. In a linear concentration gradient, this second derivative term is zero. The flux into and out of the slice is equal if the gradient is linear, and the concentration within the slice would not change with time.

The mathematical statement of Fick's second law (Equation 17.20) is also known as the **diffusion equation** because it allows us to calculate the net diffusive movement of molecules in response to concentration gradients. We encountered the diffusion equation in Chapter 11, during the discussion of ionic conduction in neurons. Recall from Section 11.33 that the passive spread of a voltage spike in a neuron is described by the diffusion equation when voltage-gated ion channels are not activated. This makes sense, because the dissipation of the voltage spike occurs due to the passive diffusion of ions down their concentration gradients, as long as the voltage-gated channels are closed and do not contribute to changes in the concentration of ions.

## 17.13 Integration of the diffusion equation allows us to calculate the change in concentration with time

The diffusion equation (Equation 17.33) allows us to calculate how the concentration changes with time due to diffusion. An instructive case is to start with a very narrow line of solute at high concentration at $x = 0$ and to calculate how the distribution of solute changes with time (**Figure 17.17**). In this simple case, diffusion occurs along only one direction, defined as the $x$ axis. There is no concentration gradient along the other directions (for example, the $y$ axis in Figure 17.17), and so the movement of molecules in those directions has no effect on the concentration profile and the system is fully described with just $x$ as a variable.

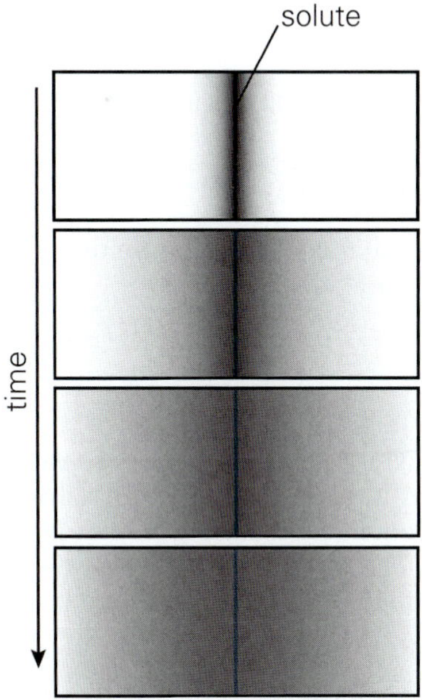


**Figure 17.17 The diffusion of molecules along one dimension is described by a Gaussian function.** The concentration of a solute is indicated by the density of *gray*, starting with the solute confined to a thin line in the center of the sample. The concentration profile at each time is given by the Gaussian function in Equation 17.21. The profiles of the concentration on a horizontal line look like the curves in Figure 17.5.

If we define the distance from the initial stripe of solute as $x$, then the concentration after diffusion has occurred for a time $t$ will be given by:

$$c(x,t)=\frac{n_0}{\sqrt{4\pi Dt}}\exp\left(-\frac{x^2}{4Dt}\right) \tag{17.21}$$

in which $n_0$ is the number of moles of solute present and $D$ is the diffusion constant. Equation 17.34 is the solution to the diffusion equation that satisfies the boundary conditions for the situation illustrated in Figure 17.17, as explained in **Box 17.4**.

The parameter $n_0$ in Equation 17.21 is the total number of moles of solute in the entire volume of the solution. The number of moles appears in Equation 17.21 because the total amount of substance present is conserved as it spreads in space. To understand this better, consider that the total number of moles is the integral of the concentration over the entire volume of the sample. If we assume that $c(x, t)$

is the concentration integrated over $y$ and $z$, then the integral of the concentration over the whole volume is given by:

$$n_0 = \int_{-\infty}^{+\infty} c(x,t)\,dx \tag{17.22}$$

In Equation 17.22, we have taken the limits of integration over $x$ to range from $-\infty$ to $+\infty$. Although the box has a finite length, we assume that the concentration of

## Box 17.4 Solving the diffusion equation

The consideration of fluxes of molecules gave rise to the differential equation that expresses how concentrations change in time due to diffusion. The basic differential equation relates the change in concentration with time at a point $x$ in space to the values of the second derivative of concentration gradient with position and the diffusion constant, given as Equation 17.20 (Fick's second law, or the diffusion equation):

$$\frac{\partial c(x,t)}{\partial t} = D\frac{\partial^2 c(x,t)}{\partial x^2} \tag{17.20}$$

This equation is deceptively simple in form, and it is not easy to solve. Fortunately, mathematicians have spent a great deal of time working out how to deal with such problems. The basic approach used for this type of differential equation is to write the concentration as a product of functions, one with the position variable, $x$, and the other with the time variable, $t$:

$$c(x,t) = C(x)T(t) \tag{17.4.1}$$

Substituting this into the diffusion equation gives:

$$C(x)\frac{dT(t)}{dt} = (D)T(t)\frac{d^2C(x)}{dx^2} \tag{17.4.2}$$

Equation 17.4.2 can be rearranged to separate the two variables:

$$\frac{1}{T(t)}\frac{dT(t)}{dt} = D\frac{1}{C(x)}\frac{d^2C(x)}{dx^2} \tag{17.4.3}$$

To solve the time part of the equation, recognize that what is on the right is a constant evaluated at the point $x$ in space. Likewise, for solving the spatial part of the equation, the left-hand side is a constant to be evaluated for some point in time. We can, therefore, find solutions to the following pair of equations:

$$\frac{1}{T(t)}\frac{dT(t)}{dt} = -D\omega^2 \quad \text{and} \quad D\frac{1}{C(x)}\frac{d^2C(x)}{dx^2} = -D\omega^2 \tag{17.4.4}$$

The general nature of these equations may be familiar, especially when rearranged to leave just the derivative on one side—that is:

$$\frac{dT(t)}{dt} = -D\omega^2 T(t) \quad \text{and} \quad \frac{d^2C(x)}{dx^2} = -\omega^2 C(x) \tag{17.4.5}$$

The forms of the solutions are probably also familiar—a function for which the derivative is the function itself is an exponential of the variable and, similarly, a function for which the second derivative gives back the function itself is a sine or cosine. The general solutions for the equations can then be written as sums (possibly of infinite numbers) of terms made up of the products of these:

$$\left(Ae^{-D\omega^2 t}\right)\sin\omega x \quad \text{and} \quad \left(Be^{-D\omega^2 t}\right)\cos\omega x$$

These still look quite far from the Gaussian function given as the solution to the diffusion equation (Equation 17.21). The next step is to apply boundary conditions that are specific for the problem we are studying. We know that (1) initially all of the solute is within a narrow strip at $x = 0$, (2) that the concentration is positive everywhere, and (3) that the total amount of solute is conserved. The correct specific solution for this problem must meet all three of these criteria and, in order to do so, the variables in the general solution can be restricted to specific values. For the actual solution for this case (and for many others), there are an infinite number of terms, but these can be recognized as an infinite series expansion for another function.

While we shall not work through a detailed conversion of the general solution to the specific one, it is quite easy to show that the Gaussian function specified in Equation 17.21 does satisfy the differential equation. Specifically, given:

$$c(x,t) = \frac{n_0}{\sqrt{4\pi Dt}}\exp\left(\frac{-x^2}{4Dt}\right) \tag{17.21}$$

then just applying the chain rule for derivatives,

$$\frac{\partial c(x,t)}{\partial t} = \frac{n_0}{\sqrt{4\pi D}}\left(-\frac{t^{-\frac{3}{2}}}{2} + \frac{x^2\,t^{-\frac{5}{2}}}{4D}\right)\exp\left(-\frac{x^2}{4Dt}\right) \tag{17.4.6}$$

and

$$\frac{\partial^2 c(x,t)}{\partial x^2} = \frac{\partial}{\partial x}\left[\frac{n_0}{\sqrt{4\pi D}}t^{-\frac{1}{2}}\left(-\frac{x}{2Dt}\right)\exp\left(-\frac{x^2}{4Dt}\right)\right] \tag{17.4.7}$$

then doing the second derivative, again with the chain rule,

$$\frac{\partial^2 c(x,t)}{\partial x^2} = \frac{n_0}{\sqrt{4\pi D}}\left(\frac{t^{-\frac{3}{2}}}{2D} + \frac{x^2\,t^{-\frac{5}{2}}}{4D^2}\right)\exp\left(-\frac{x^2}{4Dt}\right) \tag{17.4.8}$$

Comparing Equations 17.4.6 and 17.4.8, we can see that we get back the diffusion equation:

$$\frac{\partial c(x,t)}{\partial t} = D\frac{\partial^2 c(x,t)}{\partial x^2}$$

the solute decays to zero at the ends of the box. This simplifies the integration, because it allows us to use an expression for a standard Gaussian integral, as explained in Box 17.1. Substituting the expression for $c\ (x,\ t)$ given by Equation 17.21 into Equation 17.22, we get:

$$n_0 = \left(\frac{n_0}{\sqrt{4\pi Dt}}\right)\int_{-\infty}^{+\infty} \exp\left(\frac{-x^2}{4Dt}\right)dx = \frac{n_0}{\sqrt{4\pi Dt}}\sqrt{4\pi Dt} = n_0 \qquad (17.23)$$

See Box 17.1 to understand how the integral in Equation 17.23 is evaluated.

You should compare Equation 17.21, which is an expression describing the concentration, with Equation 17.7, which describes the probability of finding a displacement, $\delta$, after a certain number of steps in a random walk with discrete steps. The concentration of molecules at a certain displacement from the origin is proportional to the probability of finding molecules that have moved that far from the origin. The macroscopic treatment, in which we started with Fick's laws, has ended up with precisely the same Gaussian function that describes the spread of molecules (or bacteria) with distance or displacement.

## 17.14 The diffusion constant is related to the mean square displacement of molecules

The diffusion constant, $D$, tells us how fast molecules move through the solution, a point that we return to in the next section. Recall from the discussion in Section 17.5 that the number of steps, $N_{total}$, in a discrete random walk is given by $R_{step}t$, where $R_{step}$ is the rate at which the steps occur. By comparing Equations 17.7 and 17.21, you can see that $R_{step}t$ in the discrete random walk is replaced by $2Dt$ in the solution to the diffusion equation. By comparison with Equations 17.2 and 17.4 in Section 17.3, we can see that the standard deviation, or r.m.s. deviation for the distribution describing the spread of concentration (Equation 17.21), is given by:

$$\text{r.m.s. displacement (for 1-dimensional diffusion)} = \sqrt{2Dt} \qquad (17.24)$$

From Equation 17.24, it follows that the mean square displacement is given by $2Dt$. If the mean square displacement is expressed in units of $cm^2$, it follows that the units of the diffusion constant, $D$, are $cm^2{\bullet}sec^{-1}$, consistent with our earlier definition of $D$ in terms of Fick's law (see Section 17.11).

What about diffusion in two or three dimensions? For general cases the mathematics is not so easy. If one starts with an arbitrary concentration profile (such as that shown in **Figure 17.18**), then it is not always possible to write down an equation that specifies the concentration at each point in space as a function of time. For any sort of real application, the diffusion equation is solved numerically using a computer, by breaking up space into tiny volume elements. The concentration gradients and fluxes are calculated for each such element, and the change in concentration is propagated in time. In this way it is possible to generate very accurate descriptions of diffusion processes.

**Figure 17.18 The dissipation of a pattern through diffusion.** An initial concentration pattern blurs as diffusion of the molecules giving rise to it occurs. Time increases from the top to the bottom. Numerical methods must be used to calculate the evolution of concentrations in such complex cases.

There are a few special cases in which the solutions to the diffusion equation can be described by an exact expression. For example, by analogy to starting with a solute in a line, one can begin with the solute at a single point in a plane, from which it can diffuse outwards in two dimensions (**Figure 17.19**). The concentration gradient is along radial directions outward from the center, and the mathematical solution for the concentration is similar to Equation 17.7, except that the exponential $-x^2/4Dt$ is replaced with $-r^2/4Dt$, where $r$ is the distance from the starting point. As explained in **Box 17.5**, the concentration in this case is given by:

$$c(r,t) = \frac{n_0}{4\pi Dt}\exp\left(-\frac{r^2}{4Dt}\right) \qquad (17.25)$$

Box 17.5 also explains that the r.m.s. displacement of molecules from the origin is given by:

$$\text{r.m.s. displacement (for 2-dimensional diffusion)} = \sqrt{4Dt} \qquad (17.26)$$

## Box 17.5 Diffusion from a point in two and three dimensions

### Diffusion from a point in a plane

Any point in the plane has the polar coordinates $r$ and $\theta$, as shown in **Figure 17.5.1**, where we assume that the solute spreads outwards from the origin. The concentration gradient is along the radial direction characterized by $r$, and there is no gradient in the direction of the rotation, $\theta$. If we consider just the variation in concentration along one direction (for example, along the vector shown in Figure 17.5.1), then the concentration profile will look just like the one-dimensional case, except that the exponential $-x^2/4Dt$ is replaced with $-r^2/4Dt$. Thus, we can write the solution to the diffusion equation as:

$$c(r) = Ke^{-\frac{r^2}{4Dt}} \tag{17.5.1}$$

where $K$ is a normalization constant. To work out what $K$ is, we need to integrate $c(x)$ over the whole surface of the plane and set the integral equal to the number of moles of the solute, $n_0$. In order to do this, we must recognize that an infinitesimal surface element in two dimensions is given by $rd\theta dr$ (see Figure 17.5.1B). Thus:

$$n_0 = K\int_0^{2\pi}\int_0^{\infty} e^{-\frac{r^2}{4Dt}} rdrd\theta$$

$$= K\int_0^{2\pi} d\theta \int_0^{\infty} re^{-\frac{r^2}{4Dt}} dr$$

$$= 2\pi K\int_0^{\infty} re^{-\frac{r^2}{4Dt}} dr \tag{17.5.2}$$

Referring to the table of Gaussian integrals in Table 17.1.1 (in Box 17.1), we find that this simplifies to:

$$n_0 = K4\pi Dt$$

or

$$K = \frac{n_0}{4\pi Dt} \tag{17.5.3}$$

And so the concentration is given by the following equation, which is the same as Equation 17.38 in the main text.

$$c(r) = \frac{n_0}{4\pi Dt} e^{-\frac{r^2}{4Dt}} \tag{17.5.4}$$

Now we need an expression for the r.m.s. displacement of molecules from the origin as a function of time. The probability of finding a molecule at a distance $r$ from the origin is simply:

$$P(r) = \frac{1}{4\pi Dt} e^{-\frac{r^2}{4Dt}} \tag{17.5.5}$$

Equation 17.5.5 follows from the fact that the concentration of molecules in an infinitesimal surface element is given by the probability of finding molecules within that element multiplied by the total amount of molecules present, which is $n_0$. To calculate the mean square displacement, we average the value of $r^2$ over the distribution given by Equation 17.5.5:

$$\langle r^2 \rangle = \frac{1}{4\pi Dt}\int_0^{2\pi}\int_0^{\infty} r^2 e^{-\frac{r^2}{4Dt}} rdrd\theta \tag{17.5.6}$$

Again, we can evaluate the integral by turning to Table 17.1.1, which yields:

$$\langle r^2 \rangle = 4Dt \tag{17.5.7}$$

And so the r.m.s. displacement as a function of time is given by Equation 17.26 in the main text:

$$\text{r.m.s. displacement} = \sqrt{4Dt} \tag{17.26}$$

### Diffusion from a point in a three-dimensional volume

As for the two-dimensional case, the concentration is given by:

$$c(r) = Ke^{-\frac{r^2}{4Dt}} \tag{17.5.8}$$

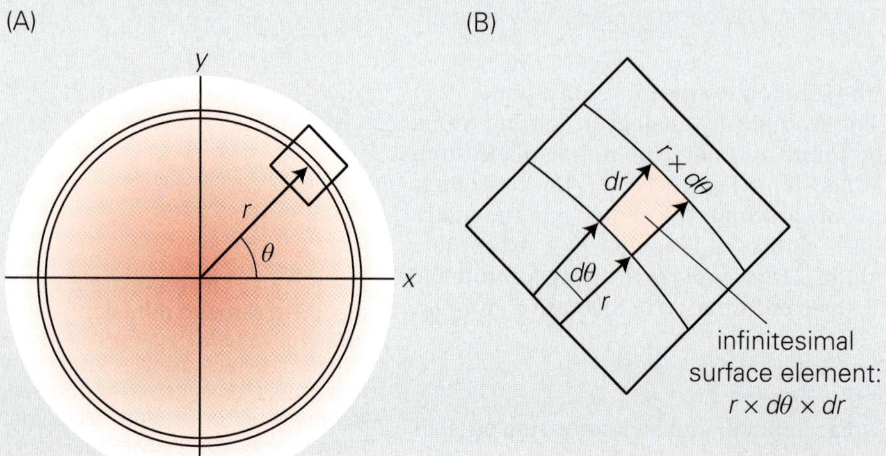


**Figure 17.5.1 Integration using polar coordinates.** (A) Any point on a two-dimensional surface has polar coordinates $r$ and $\theta$. (B) An infinitesimal surface element in polar coordinates is given by $r\,d\theta\,dr$.

We must now integrate over three spatial coordinates in order to evaluate the normalization constant, $K$. This is done using spherical polar coordinates, $r$, $\theta$, and $\phi$. As shown in **Figure 17.5.2**, an infinitesimal volume element in this coordinate system is given by $r^2 \sin\theta dr d\theta d\phi$. Once again, we evaluate the normalization condition by using Table 17.1.1 for the appropriate Gaussian integral:

$$n_0 = K\int_0^{\pi}\int_0^{2\pi}\int_0^{\infty} e^{-\frac{r^2}{4Dt}} r^2 \sin\theta \, dr d\phi d\theta$$

$$= K\int_0^{\pi} \sin\theta \, d\theta \int_0^{2\pi} d\phi \int_0^{\infty} r^2 e^{-\frac{r^2}{4Dt}} dr$$

$$= \frac{K}{4}(4\pi Dt)^{3/2} \qquad (17.5.8)$$

It follows, then, that the concentration is given by Equation 17.40 in the main text:

$$c(r) = \frac{4n_0}{(4\pi Dt)^{3/2}} e^{-\frac{r^2}{4Dt}} \qquad (17.40)$$

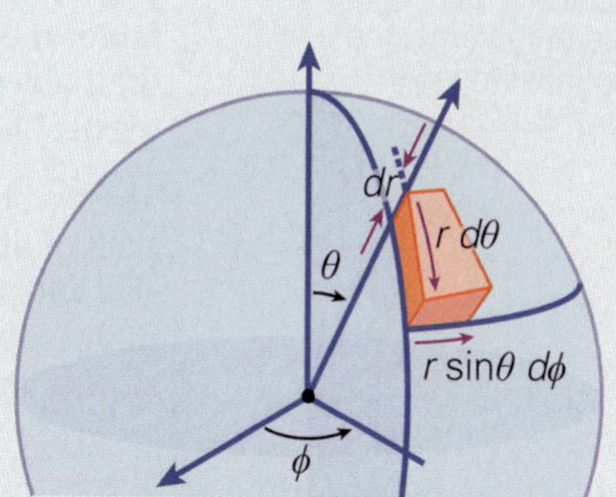


**Figure 17.5.2 Integration using spherical polar coordinates.** Any point in a three-dimensional volume has spherical polar coordinates $r$, $\theta$, and $\phi$. The value of $\phi$ ranges from 0 to $2\pi$, whereas that of $\theta$ ranges from 0 to $\pi$. An infinitesimal volume element is given by $r^2 \sin\theta dr d\theta d\phi$.

The mean square displacement is calculated as follows:

$$\langle r^2 \rangle = \frac{4n_0}{(4\pi Dt)^{3/2}} \int_0^{\pi}\int_0^{2\pi}\int_0^{\infty} e^{-\frac{r^2}{4Dt}} r^4 \sin\phi \, dr d\phi d\theta = 6Dt \qquad (17.5.9)$$

As before, the integrals in Equation 17.5.9 are evaluated by using one of the standard Gaussian integrals in Table 17.1.1.

Similar arguments apply for diffusion from a point in three dimensions (see Box 17.5). The concentration as a function of time depends only on the distance, $r$, from the origin, and is given by:

$$c(r,t) = \frac{4n_0}{(4\pi Dt)^{3/2}} \exp\left(-\frac{r^2}{4Dt}\right) \qquad (17.27)$$

The r.m.s. displacement of molecules in three dimensions, when diffusing outwards from a point, is given by:

$$\text{r.m.s. displacement (for 3-dimensional diffusion)} = \sqrt{6Dt} \qquad (17.28)$$

## 17.15 Diffusion constants depend on molecular properties such as size and shape

The diffusion constant, $D$, relates directly to how fast molecules move in solutions, such as in water, or in the cytoplasm of cells. The value of the diffusion constant

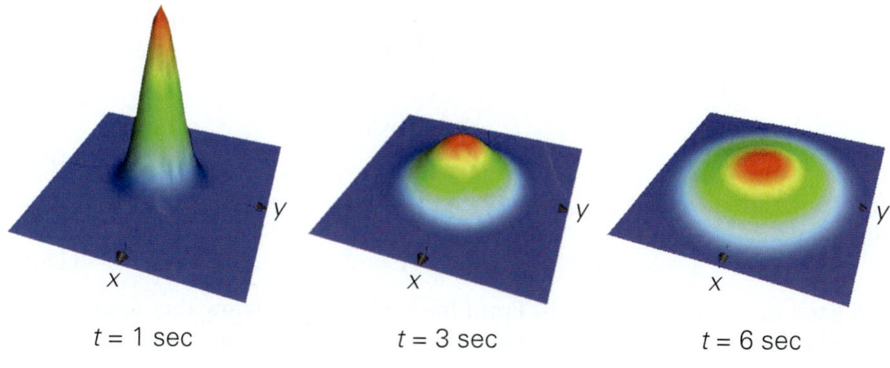


**Figure 17.19 Diffusion from a point in two dimensions.** The diagram shows the concentration of a solute, indicated by the height and color of the surface. The solution to the diffusion equation for three time points is shown.

**Friction factor**

The friction factor is a parameter that relates the size and shape of a molecule to the drag (the resistance to movement) it generates in a fluid.

depends upon molecular parameters, such as size and shape, and also the characteristics of the solution through which molecules move, particularly its viscosity.

Experimental determinations of diffusion constants show that, as one expects, larger molecules move more slowly than smaller ones. But the dependence of $D$ on molecular weight is not linear. In order to understand what the relationship is we need to consider the effects of friction on molecular movement.

The resistance to motion in a fluid is due to the need to move the fluid out of the path of the moving object. This results in a friction force that is dependent on the size and shape of the object, and on the velocity of motion and the characteristics of the solvent. An object moving rapidly meets more resistance than one moving slowly. The friction force, $F_{friction}$, depends on the velocity, $v$, as follows:

$$F_{friction} = -fv \tag{17.29}$$

The proportionality constant, $f$, is called the **friction factor**, and it incorporates the size, shape and solvent contributions. The negative sign on the right hand side reflects the fact that the friction force resists motion and hence its direction is opposite to that of the velocity. Given Equation 17.29 and the units for force (g•cm•sec$^{-2}$) and velocity (cm•sec$^{-1}$), the friction factor must have the nonintuitive units of g•sec$^{-1}$.

Under an applied force, such as gravity, an object will accelerate according to Newton's law ($F = ma$) until the friction force equals the applied force, at which point the object reaches a **terminal velocity**. The frictional force exactly cancels that causing acceleration when the terminal velocity is reached.

The dependence of frictional force on the shape of an object is well known in skydiving, an acrobatic feat in which the friction factor has a major practical role (**Figure 17.20**). A skydiver in free fall in a typical 'spread eagle' position reaches a terminal velocity of about 120 mph, while if the arms and legs are pulled in tightly the velocity goes up to about 200 mph—the difference coming just from the change in friction factor due to shape. When the skydiver opens a parachute the cross-section moving through the air increases enormously, and hence so does the friction factor—this slows the fall to a speed of about 12 mph, slow enough to make a safe landing.

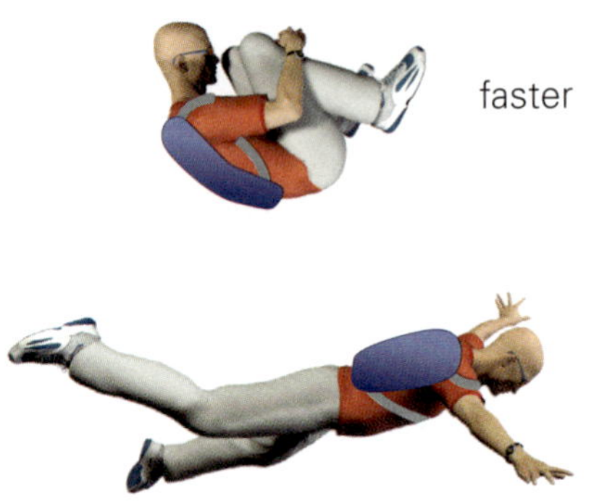

**Figure 17.20 Skydiving, an example of friction relating to speed.** The rate of the skydiver falling is greatly decreased by the parachute opening. This occurs because the open parachute creates a large amount of friction by requiring that a larger volume air be moved around it.

## 17.16 The diffusion constant is inversely related to the friction factor

Albert Einstein deduced that the random movement of very small particles in liquids can be understood in terms of frictionally damped motion. Such motion was first described for pollen granules by Robert Brown (and became known as Brownian motion). Einstein recognized that the motion of the molecules must be related to the driving force for diffusion.

In Equation 17.14 we defined $D$ as the proportionality constant between the flux of molecules through a plane, and the concentration gradient. To connect this with molecular properties we can think about the same principle in terms of motion of individual molecules. Again considering an imaginary plane passing through a sample, the flux of molecules through the plane will depend on the average velocity $v$ with which particles are moving, and on the number of particles near the plane, that is their concentration.

$$J(x) = v \times c(x) \tag{17.30}$$

Now we connect with the idea that molecules moving under a driving force, $F_{drive}$, reach a terminal velocity, $v_{terminal}$, when the driving force is balanced by the frictional force:

$$F_{friction} = -v_{terminal} f = -F_{drive} \tag{17.31}$$

We need to know what $F_{drive}$ is. From basic physics we know that forces correspond to the derivatives of potentials. The electrical forces on a charged particle, for example, arise from differences in electrical potential between two points in space, as discussed in Section 9.12. For diffusion the force arises from the difference in *chemical potential* at different points in space, due to a concentration difference.

The chemical potential was developed from the concept of free energy, and was discussed in detail in Chapter 10. For the present discussion we shall define the chemical potential on a per molecule basis rather than the per mole basis used in Chapter 10. This simply means that we need to switch from $R$ to $k_B$ in front of the logarithm of the concentration term in the equation for the chemical potential:

$$\mu = \mu^{\circ} + k_B T \ln\left(\frac{c}{1}\right) \tag{17.32}$$

In Equation 17.32, $\mu^{\circ}$ is the chemical potential of the standard state (1 M solution). If we consider a gradient in just the $x$ direction then the driving force for diffusion, $F_{\text{drive}}$, will be:

$$F_{\text{drive}} = \frac{-d\mu}{dx} = -k_B T \frac{d}{dx}\ln[c(x)] = -k_B T \frac{1}{c(x)}\frac{dc(x)}{dx} \tag{17.33}$$

We now combine Equation 17.33 with Fick's first law (Equation 17.27) and with Equation 17.32 to give:

$$J(x) = \frac{F_{\text{drive}}}{f}c(x) = \frac{-k_B T}{f}\frac{dc(x)}{dx} \tag{17.34}$$

The key is now to compare two equations that describe the flux of molecules (Equation 17.14 and 17.34):

$$J(x) = -D\frac{dc(x)}{dx} \quad \text{and} \quad J(x) = \frac{-k_B T}{f}\frac{dc(x)}{dx} \tag{17.35}$$

By comparing these two different descriptions of flux, which must correspond to the same thing and hence be equal, we obtain a relationship between the diffusion constant and the friction factor, $f$:

$$D = \frac{k_B T}{f} \tag{17.36}$$

While this equation shows that the rate of diffusion and the friction factor, $f$, are inversely related, it does not provide any insight into the relationship of either to molecular parameters. To take this next set step we need to consider the characteristics of the solution and examine how it interacts with a moving molecule.

## 17.17 Viscosity is a measure of the resistance to flow

To understand the behavior of solutions we need to understand how the motion of a moving particle is coupled to that of the solvent around it. A few solvent molecules will stick to the surface of the particle, and will be carried along with it. On the other hand, at a large distance away from the particle the solvent molecules will not be affected by the movement of the particle.

The dragging of solvent with the particle corresponds to transfer of momentum from the particle to the solvent. Recall from basic physics that the momentum, $p$, is the product of mass, $m$, and velocity, $v$: $p = mv$. When one part of a fluid is moving faster than another a gradient in momentum is created, as illustrated in **Figure 17.21**. The flux in momentum, $J_{p_x}$, is the momentum transferred along the $x$ direction per unit area per unit time. The **viscosity** of the solvent (given the symbol $\eta$) is the proportionality constant between the momentum flux and the velocity gradient along a direction $z$ orthogonal to $x$:

$$J_{p_x} = -\eta\frac{dv_x}{dz} \tag{17.37}$$

A 'sticky' solvent will transfer momentum over long distances, which corresponds to high viscosity. A 'thin' solvent will only transfer momentum over short distances, and has a low viscosity.

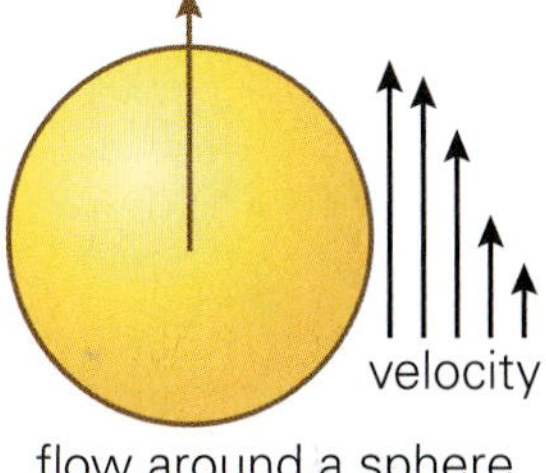


flow around a sphere

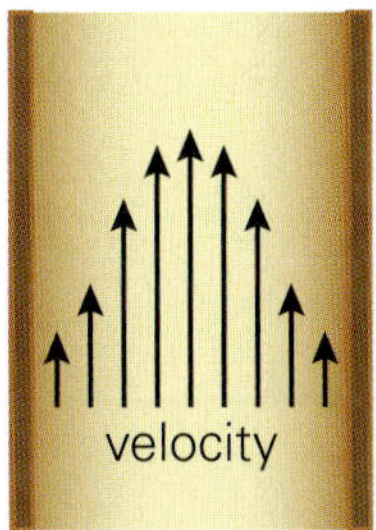


flow in a tube

**Figure 17.21 Gradients in momentum.** Flow around a sphere and in a tube are shown with vectors indicating the velocity of the flow at different positions relative to the boundary between the object and the fluid (solvent). For flow around a particle, the object is moving and the solvent is not, except near the particle, where interactions occur. For flow in a tube, the walls of the tube are fixed, while the fluid is moving. In both cases, the behavior is governed by the same rules and the same fundamental parameter—the viscosity of the solvent.

**Viscosity**

The viscosity of a solvent describes the coupling between movement of solute molecules and the solvent. In solvents with high viscosity, the movement of a solute disturbs the movement of the solvent over greater distances than for a solvent with lower viscosity. The higher the viscosity, the more the solvent resists the movement of the solute.

To understand the effect of viscosity on molecular movement note that if the viscosity is high then the layer of solvent molecules near the solute molecule affects other solvent molecules over a longer distance than when the viscosity is low. More solvent is dragged along with the solute molecule as it moves, making it more difficult for the solute molecule to move. Viscosity also determines the rate of flow of the solution as a whole. For example, if a solution is moving in a narrow tube the layer of molecules at the inner surface of the tube may tend to stick to the surface (see Figure 17.21). If the viscosity is high then the resistance to movement persists over a long distance, making flow of the solvent through the tube slow. If the viscosity is low then the effect of the layer at the wall does not propagate very far, and the center part will move with little resistance.

To determine the units of the viscosity, $\eta$, we look at the components of Equation 17.37. The momentum flux (momentum per unit area per unit time) will have units $(g{\bullet}cm{\bullet}sec^{-1})\,(cm^{-2})\,(sec^{-1}) = g{\bullet}cm^{-1}{\bullet}sec^{-2}$. The velocity gradient has units $(cm{\bullet}sec^{-1}{\bullet}cm^{-1}) = (sec^{-1})$, and hence the viscosity must have units of $g{\bullet}cm^{-1}{\bullet}sec^{-1}$ (the units of parameters such as this are not intuitively obvious, but can easily be determined from the defining equation, such as 17.36 in this case). The standard unit for viscosity is the **poise** ($1\ P = g{\bullet}cm^{-1}{\bullet}sec^{-1}$), but units of **centipoise** ($1\ cP = 10^{-2}\ P = 10^{-2}\ g{\bullet}cm^{-1}{\bullet}sec^{-1}$) are most commonly used for biological systems because the viscosity of water is approximately 1 cP.

## 17.18 Liquids with strong interactions between molecules have high viscosity

A large part of the value of the viscosity comes from the molecular characteristics of the solvent. Solvents with weak interactions between molecules (such as acetone, which is the principal component of nail polish remover) have low viscosity and they flow very easily. Acetone has a viscosity of 0.31 cP at 25°C. In water, the molecules form hydrogen bonds with other water molecules. This facilitates the transfer of momentum over longer distances, and so water has substantially higher viscosity than acetone, with a value of ~1 cP at 25°C. Glycerol ($HOCH_2CH(OH)CH_2OH$) has more extensive hydrogen bonding between molecules than does water, and a much higher viscosity, ~900 cP. Concentrated solutions of sugars have even higher viscosity. The viscosity of maple syrup, for example, is ~3200 cP.

Solutions of macromolecules have substantially higher viscosity than water. The fluid inside cells, the cytosol, is a rather concentrated solution of molecules and ions, in which proteins comprise 20–40% of the mass. Experimental measurements show that the viscosity of the cytosol in cells is approximately of 20 to 50 times that of water. As we shall see in the next section, the diffusion constant is inversely proportional to the viscosity. As a consequence, proteins move much slower in the cytoplasm than they do in water.

## 17.19 The Stokes–Einstein equation allows us to calculate the diffusion coefficients of molecules

A complicated set of differential equations, known as the Navier–Stokes equations, must be solved in order to describe fluid flow around an object as it moves. These equations include terms arising from the conservation of mass, energy, momentum, and angular momentum. For an object with an arbitrary shape, these equations cannot be solved analytically (that is, the solution cannot be written as an equation). Fortunately, as for the diffusion equation, the Navier–Stokes equations can be solved numerically on a computer for any case of interest.

For objects with very small masses, the forces arising from viscosity dominate the flow around the object (this is always the case for very small particles like individual molecules). In this limit the equations can be solved analytically for simple shapes, like a sphere. For a sphere, the friction factor for moving through liquid is:

$$f_{\text{sphere}} = 6\pi\eta r \qquad (17.38)$$

in which $\eta$ is the viscosity of the fluid and $r$ is the radius of the sphere. George Gabriel Stokes derived this relationship, and hence it is known as **Stokes' law**. This can then be combined with the relationship that Einstein derived, Equation 17.36, to give the **Stokes-Einstein equation**:

$$D_{\text{sphere}} = \frac{k_B T}{6\pi\eta r} \quad (17.39)$$

A good estimate of the rate of diffusion of nearly spherical molecules in solution can be made using the Stokes-Einstein equation, based just on the size of the molecule and the solvent viscosity. Note the inverse relationship between the diffusion constant and the viscosity, as expected from the discussion in the last section. High viscosity means high resistance to motion and hence slow diffusion.

To illustrate the utility of the Stokes-Einstein equation, we shall use it to calculate the diffusion constant in water for the protein myoglobin, familiar to us from Chapters 4 and 5 (see Figure 4.1). Myoglobin has a radius of ~20 Å, and is roughly spherical in shape. By convention, the diffusion constant has units of $cm^2{\bullet}sec^{-1}$, and so the radius of the myoglobin, $r$, must be expressed in centimeters for the dimensions to balance in Equation 17.39. Also, because the units of the viscosity, $\eta$, are $g{\bullet}cm^{-1}{\bullet}sec^{-1}$, we must express the Boltzmann constant ($k_B$) in consistent units. To begin with, we write down the values of each of the terms in Equation 17.39.

$$\begin{aligned} k_B &= 1.381\times10^{-23}\ m^2{\bullet}kg{\bullet}sec^{-2}{\bullet}K^{-1} \\ &= 1.381\times10^{-16}\ cm^2{\bullet}g{\bullet}sec^{-2}{\bullet}K^{-1} \end{aligned} \quad (17.40)$$

Assuming that the temperature is 300 K (approximately room temperature), we obtain the following value for $k_B T$:

$$k_B T = \left(1.381\times10^{-16}\right)(300) = 4.14\times10^{-14}\,cm^2{\bullet}g{\bullet}sec^{-2} \quad (17.41)$$

The values of the radius of myoglobin and the viscosity of water at room temperature are:

$$\begin{aligned} r &= 20\times10^{-8}\ cm \\ \eta &= 1\ cP = 0.01\ g{\bullet}cm^{-1}{\bullet}sec^{-1} \end{aligned} \quad (17.42)$$

Substituting all of these values into the Stokes-Einstein equation, we get:

$$D\,(\text{myoglobin, 300K}) = 1.1\times10^{-6}\ cm^2{\bullet}sec^{-1} \quad (17.43)$$

It turns out that this value for the diffusion constant is very close to the value measured experimentally for myoglobin in water. We can now use this value of $D$ to calculate the r.m.s. displacement of myoglobin molecules after 1 second of three-dimensional diffusion, by using Equation 17.28:

$$\text{r.m.s. displacement} = \sqrt{6Dt} = 2.6\times10^{-3}\ cm = 26\ \mu m \quad (17.44)$$

Thus, the typical distance that a myoglobin molecule will have moved from its starting point after 1 second of diffusion in water is about 26 µm. We can compare this rate of movement with that of a small metabolite molecule, such as sucrose (**Figure 17.22**). If we assume that the radius of sucrose is ~4 Å (that is, five times smaller than that of myoglobin), then the value of $D$ is five times larger than for myoglobin ($5.5\times10^{-6}\ cm^2{\bullet}sec^{-1}$). The time taken to travel a certain distance is inversely related to the diffusion constant, and so a sucrose molecule travels about five times faster than myoglobin. A small virus particle with a radius of ~140 Å, such as the one shown in Figure 17.22, travels about seven times slower than myoglobin.

Many eukaryotic cells are 10–50 µm across, and so our estimates of the speed of diffusion suggest that metabolites and small proteins can traverse the cell within a second. The actual movement of molecules inside a cell is much slower because, as we mentioned earlier, the viscosity of the cytoplasm is higher than that of water,

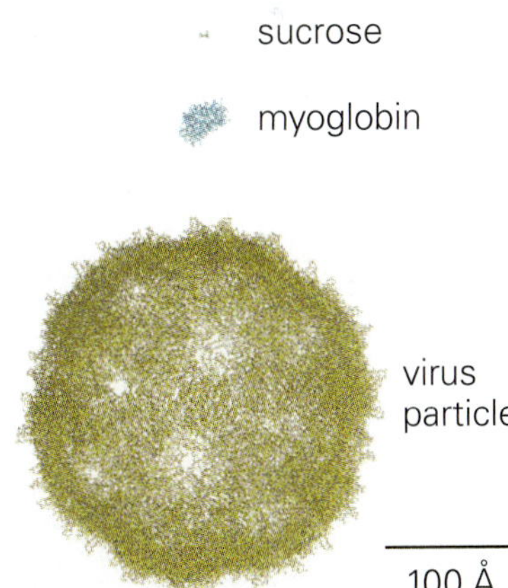


**Figure 17.22 Relative sizes of biological molecules.** Three rather different molecules are shown here. At the top, and barely visible, is sucrose (8 Å in diameter). In the middle is myoglobin (diameter of ~45 Å) and at the bottom is a virus known as MS2 (280 Å in diameter). The virus diffuses about 35 times more slowly in water than does sucrose.

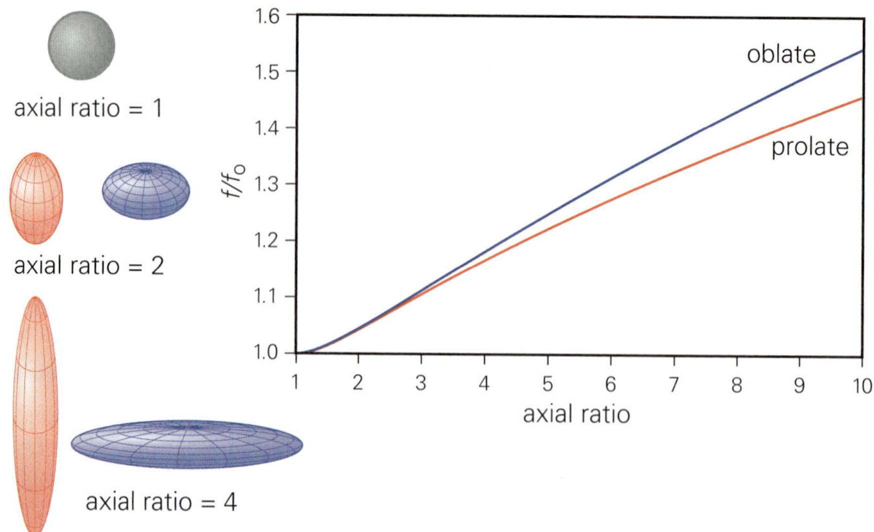


**Figure 17.23 Ratios of the friction factor for prolate and oblate shapes.** A sphere and prolate (*red*) and oblate (*blue*) shaped objects with axial ratios of 2 and 4 (all of which occupy the same volume of space—that is, correspond to the same molecular weight) are shown. At the right, a graph of the ratio of the friction factor to that for a sphere is shown for axial ratios between 1 and 10. Friction factors for the 2:1 ratio only differ from that of a sphere ($f_O$) by about 5%. The friction factors are generated by solving Navier–Stokes equations.

leading to diffusion constants for proteins the size of myoglobin that are in the range of $10^{-7}$ to $10^{-8}$ $cm^2 \bullet sec^{-1}$, compared to $10^{-6}$ in water. The apparent diffusion constants of proteins and other macromolecules can also be much slower than expected even after accounting for the viscosity of the cytoplasm. This is because macromolecules often interact with structural elements in the cell, instead of diffusing freely. As a consequence, cells often require active transport processes that consume energy to move certain molecules around at the speed that is required for essential processes (see Section 17.25).

All proteins are comprised of the same amino acids, which are all made from similar chemical groups, and so the density of a protein, $\rho_p$, is quite constant from one protein to another. Comparing spherical proteins, the mass (and hence the molecular weight, $M$) will be related to the volume of the protein:

$$M = \rho_p \frac{4}{3}\pi r^3 \quad \text{and} \quad r \propto \sqrt[3]{M} \tag{17.45}$$

Thus, for spherical molecules, the diffusion constant depends inversely on the cube root of the molecular weight.

## 17.20 The diffusion constants for nonspherical molecules are only slightly different from those calculated from the spherical approximation

For nonspherical molecules, the friction force is larger than for a spherical molecule with the same molecular weight. Although general equations cannot be written for the friction factors for arbitrary shapes, as noted above the values of the friction factors can be calculated with computers. By analyzing the results of such calculations we find that the values of the friction factor are only moderately sensitive to shape. For prolate spheroid shapes (like an American football), characterized by a long axis length $a$ and short axis length $b$, a ratio of $a/b = 2$ gives $f = 1.044 \times f_{sphere}$, where $f$ and $f_{sphere}$ are the friction factors for the two different shapes. A very asymmetric shape with $a/b = 10$ gives $f = 1.54 \times f_{sphere}$. For oblate, elipsoidal shapes (that is, disk-like), $a/b = 0.5$ gives $f = 1.042 \times f_{sphere}$ and $a/b = 0.1$ gives $f = 1.46 \times f_{sphere}$. A graph showing the $f/f_{sphere}$ value versus axial ratio is shown in **Figure 17.23.**

The increased friction, particularly for the prolate shape, seems counterintuitive because we might think that motion along the direction of the long axis would be facilitated, because of the narrower cross-section along this direction. While this is true, the direction of motion is actually random, and the molecule also frequently

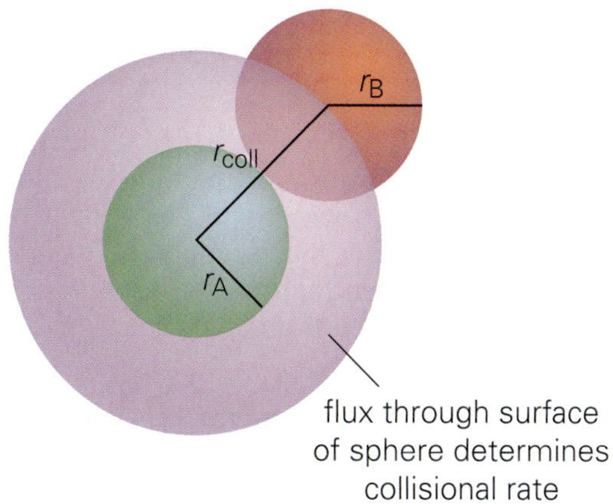


**Figure 17.24 Calculation of the diffusional collision rate.** Two spherical molecules, A and B, are shown here. The rate of collisions between them is obtained by calculating the flux of B molecules through the surface of a sphere around A, as explained in Box 17.6. The radius of this sphere, $r_{coll}$, is the sum of the radii of the two types of atoms.

moves in the directions of the short axes. The observed friction factors reflect an appropriately weighted average over motion in all directions. The viscous drag on molecules is larger than their momentum, and so they do not move continuously in the direction of least resistance the way macroscopic objects do.

## 17.21 Diffusion-limited reaction rate constants can be calculated from the diffusion constants of molecules

The idea that a rate constant cannot exceed the rate of collisions of the reactant molecules was discussed in Section 15.25. Reactions that occur at the rate of collisions are called **diffusion limited**, because the rate of collisions is determined by the rate of diffusion. The rate of collisions through diffusion can be calculated using Fick's laws. This is done by considering a sphere of radius $r_{coll}$ around a molecule, chosen such that, if the intermolecular distance reaches $r_{coll}$, then a collision occurs (**Figure 17.24**). For spherical molecules, this corresponds just to the sum of the radii of the molecules. The rate of collisions is obtained by calculating the flux of molecules that would go through the surface of this sphere.

The details of the rate calculations were worked out by Marian Smoluchowski and are explained in **Box 17.6**. The result is that the rate at which the A and B molecules collide with each other is given by:

$$\text{collision rate} = 4\pi r_{coll}\left(D_A + D_B\right)\bar{c}_A\bar{c}_B \tag{17.46}$$

Here, $D_A$ and $D_B$ are the diffusion constants for the A and B molecules. The variables $\bar{c}_A$ and $\bar{c}_B$ are the bulk concentrations of the A and B molecules, expressed as number of molecules per $cm^3$. The diffusion constants are in units of $cm^2{\bullet}sec^{-1}$ and, if the radius is expressed in units of cm, then the rate in Equation 17.46 is in units of $molecules{\bullet}sec^{-1}$. In Section 15.8, we described the more familiar rate equation in which concentrations are expressed in molar units:

$$\frac{d[AB]}{dt} = k_{coll}[A][B] \tag{17.47}$$

As explained in Box 17.6, the **collision rate constant**, $k_{coll}$, in Equation 17.47 is related to the diffusion constants of A and B in the following way:

$$k_{coll} = 4\pi r_{coll}\left(D_A + D_B\right)\left(6.022\times10^{20}\right) \tag{17.48}$$

The last term in Equation 17.48 is a conversion factor that has units of $M^{-1}{\bullet}cm^{-3}$, which ensures that the units of $k$ are $M^{-1}{\bullet}sec^{-1}$.

Let us now use Equation 17.48 to calculate what the collision rate constant is for sucrose molecules colliding with each other in water. In Section 17.19 we said that the radius of sucrose is ~4 Å, and so the value of $r_{coll}$ is 8 Å ($8 \times 10^{-8}$ cm). We calculated the diffusion constant for sucrose in water to be $5.5 \times 10^{-6}\ cm^2{\bullet}sec^{-1}$ (that is, five times greater than that of myoglobin). The collision rate constant for sucrose

## Box 17.6 Calculating diffusion-limited collision rates

The first step toward calculating the collision rate in solution, with molecules diffusing randomly, is to consider the coordinate system used for the calculation. For calculating the rate at which collisions occur, what matters is the just distance between the centers of the collision partners. The problem has inherent spherical symmetry. We initially discussed a concentration gradient along a single axis, so the derivative $\frac{\partial c}{\partial x}$ was sufficient to describe the position dependence of concentration. In a general case, the gradient is a vector quantity (that is, it has a direction) and is written in a more general form using the gradient operator (grad = $\nabla$) in Cartesian coordinates with unit vectors $\hat{x}, \hat{y}, \hat{z}$ :

$$\nabla c(x,y,z) = \left(\frac{\partial}{\partial x}\hat{x} + \frac{\partial}{\partial y}\hat{y} + \frac{\partial}{\partial z}\hat{z}\right)c(x,y,z) \quad (17.6.1)$$

This can be rewritten in polar coordinates. The second derivative, used for Fick's second law, looks more formidable in polar form, but is easier to use in cases of spherical symmetry:

$$\nabla^2 c(r,\theta,\phi) = \left[\frac{\partial}{\partial r^2} + \frac{2}{r}\frac{\partial}{\partial r} + \frac{1}{r^2}\left(\frac{1}{\sin^2\theta}\frac{\partial}{\partial \phi^2} + \frac{1}{\sin\theta}\frac{\partial}{\partial \theta}\sin\theta\frac{\partial}{\partial \theta}\right)\right]c(r,\theta,\phi) \quad (17.6.2)$$

To obtain the rate at which diffusive collisions occur, we set one atom (for example, A) at the origin, and then calculate the flux of B molecules that would come through a spherical surface surrounding it. The size of the sphere is chosen such that, when the center of a B atom crosses the surface of this sphere, a collision occurs. If the radii of A and B are $r_A$ and $r_B$, then the radius of the sphere is $r_{coll} = r_A + r_B$.

For a uniform, mixed solution in which a reaction is occurring, there is no concentration gradient, and so there is no time dependence of concentration from diffusion. Writing Fick's second law using the grad operator with this assumption then gives:

$$D\nabla^2 c_B = \frac{\partial c_B}{\partial t} = 0 \quad (17.6.3)$$

In polar form, using the fact that changes in angles $\theta$ and $\phi$ do not change distance and hence cannot cause collisions, the derivatives with them can be eliminated, giving:

$$\frac{\partial^2 c_B}{\partial r} + \frac{2}{r}\frac{\partial c_B}{\partial r} = 0 \quad (17.6.4)$$

A key step in then being able to calculate the actual collision rate (an approach originally used by Smoluchowski in 1917) is to consider that at any given instant it is very unlikely that a collision will actually be occurring, so the concentration of B at a distance $r_{coll}$ can be set to 0. This is equivalent to saying that the boundary at $r_{coll}$ acts as a sink. If a B molecule reaches it, then reaction with A occurs immediately, maintaining the concentration of B at zero at that distance with respect to A. At large values of distance, the concentration of B must reach the average bulk concentration, $\bar{c}_B$ , in solution. The solution to Equation 17.6.4 is evaluated with these boundary conditions:

$$\text{general solution: } c(r) = a + \frac{b}{r}$$

$$\text{specific case: } c(r) = \left(1 - \frac{r_{coll}}{r}\right)\bar{c} \quad (17.6.5)$$

In this approach, although there is not a macroscopic gradient, the assumption in the initial condition that no collision is occurring means that there is a gradient of collision partners as a function of distance from the molecule at the origin. The flux through the imaginary collision threshold sphere of radius $r_{coll}$ is calculated using Fick's first law:

$$J(r_{coll}) = D\left[\frac{\partial c_B(r)}{\partial r}\right]_{r=r_{coll}} \quad (17.6.6)$$

For the concentration as a function of distance calculated in Equation 17.6.5, this gives:

$$\frac{\partial c_B(r)}{\partial r} = \frac{r_{coll}\bar{c}_B}{r^2}\bigg|_{r=r_{coll}} = \frac{\bar{c}_B}{r_{coll}}$$

and so:

$$J(r_{coll}) = \frac{D\bar{c}_B}{r_{coll}} \quad (17.6.7)$$

Remember that the flux is defined as molecules crossing a unit area per unit time. To calculate the rate of collisions, then, the flux must be multiplied by the area of the surface, $A_{sphere} = 4\pi r_{coll}^2$, so:

$$\text{collision rate} = J(r_{coll})(\text{area}) = \frac{D\bar{c}}{r_{coll}}4\pi r_{coll}^2 = 4\pi r_{coll}D\bar{c} \quad (17.6.8)$$

Comparing this with the chemical kinetic version, rate = $k\bar{c}$, then gives:

$$k = 4\pi D r_{coll} \quad (17.6.9)$$

One further note is that $D$ is the diffusion constant describing the *relative motion* of the colliding molecules. We have taken one molecule at the origin and then considered the motion of the other one. This means that the effective diffusion constant, $D$, in the equations above, is really the sum of the diffusion constants of the two colliding molecules. We initially designated them A and B for considering the distance at which a collision occurs, $r_{coll} = r_A + r_B$. Using the same kind of notation, $D = D_A + D_B$.

is then given by substituting these values into Equation 17.48:

$$\begin{aligned} k_{\text{coll}} &= 4\pi\left(8\times10^{-8}\right)(2)\left(5.5\times10^{-6}\right)\left(6.022\times10^{20}\right) \\ &= 6659\times10^{6} \approx 1\times10^{10}\ \text{M}^{-1}\bullet\text{sec}^{-1} \end{aligned} \tag{17.49}$$

At first glance, looking at Equation 17.48, it would appear that the rate of collisions would be lower for larger molecules because $D$ decreases with the size of the molecule (see Equation 17.39). The value of $r_{\text{coll}}$, however, *increases* with the size of the molecule, which offsets the decrease in the diffusion constant and leaves the diffusion-limited rate of collisions essentially independent of the size of the molecule involved, with $k_{\text{coll}} \approx 10^{10}\ \text{M}^{-1}\bullet\text{sec}^{-1}$. Recall from Section 15.26 that the fastest chemical reactions occur considerably slower than the rate of diffusional collisions because not every collision leads to a productive reaction.

Equations 17.46 and 17.48 emphasize the relationship between the collision rate constant and the diffusion constants. Alternatively, we can express $k_{\text{coll}}$ in terms of the viscosity by substituting Equation 17.39 (the Stokes–Einstein equation) into Equation 17.48. For the case where there is only one kind of molecule colliding with itself:

$$\begin{aligned} k_{\text{coll}} &= 4\pi(2r)(2D)\left(6.022\times10^{20}\right) \\ &= 4\pi(2r)(2)\left(\frac{k_B T}{6\pi\ \eta\ r}\right)\left(6.022\times10^{20}\right) \\ &= \left(\frac{8}{3}\right)\left(\frac{k_B T}{\eta}\right)\left(6.022\times10^{20}\right) \end{aligned} \tag{17.50}$$

We have to be careful about units here. If we express $k_B T$ in units of J ($\text{kg}\bullet\text{m}^2\bullet\text{sec}^{-2}$), then we have to multiply by $10^7$ in order to match the units of the viscosity ($\text{g}\bullet\text{cm}^{-1}\bullet\text{sec}^{-1}$).

The expression for $k_{\text{coll}}$ then becomes:

$$\begin{aligned} k_{\text{coll}} &= \left(\frac{8}{3}\right)\left(\frac{k_B T}{\eta}\right)\left(6.022\times10^{20}\right)\left(10^{7}\right) \\ &= \frac{8RT}{3\eta}\times10^{4} \end{aligned} \tag{17.51}$$

where $R$ is the gas constant in units of $\text{J}\bullet\text{K}^{-1}$, $T$ is the absolute temperature, and $\eta$ is the viscosity in units of poise ($\text{g}\bullet\text{cm}^{-1}\bullet\text{sec}^{-1}$).

## 17.22 One-dimensional searches on DNA increase the rate at which transcription factors find their targets

In introducing random walks, we used the example of bacteria swimming in two dimensions—that is, on a surface. When considering the diffusion of biological macromolecules, we naturally think about their movement in three-dimensional space within the cell. There are circumstances, however, in which restricted diffusion (that is, in less than three dimensions) plays an important role in biology. One such example is provided by the binding of transcription factors to specific sites on DNA. To bind a particular target sequence, a protein must come into contact with that specific site (typically 10–20 base pairs in length) among all of the other genomic DNA that is present ($10^6$ to $10^{10}$ base pairs or more, depending on the organism).

A bacterial transcription factor known as lactose repressor (*lac*) binds to a specific sequence on DNA known as the operator (see Sections 13.8 and 13.9 for a discussion of how the lactose repressor recognizes specific sites on DNA). The rate constant for the bimolecular reaction of the repressor binding to its operator is ~1 × $10^{10}\ \text{M}^{-1}\bullet\text{sec}^{-1}$. This is essentially the same as the rate of diffusional collisions between two small molecules that we calculated in the last section, and it is very surprising that the lac repressor can find its operator so fast. Not every collision between the repressor and operator will lead to a proper interaction, because the repressor and the DNA will not be oriented correctly. Also, the operator is within a large DNA molecule (that is, the bacterial genome) that is relatively immobile. The high value for the observed rate constant indicates that something other than simple diffusion must underly the recognition process.

Most DNA-binding proteins have a positively charged face, which can interact loosely and nonspecifically with DNA. This observation leads to the idea that relatively weak, electrostatically driven binding could occur anywhere on the genomic DNA, followed by one-dimensional sliding of the protein along the DNA to find its proper recognition site (**Figure 17.25**). Such sliding substantially reduces the volume of space that must be searched, facilitating encounters with the proper sequence.

DNA is rarely naked in the cell, generally being covered with an assortment of proteins, including histones (or histone-like proteins in prokaryotic cells), as well as

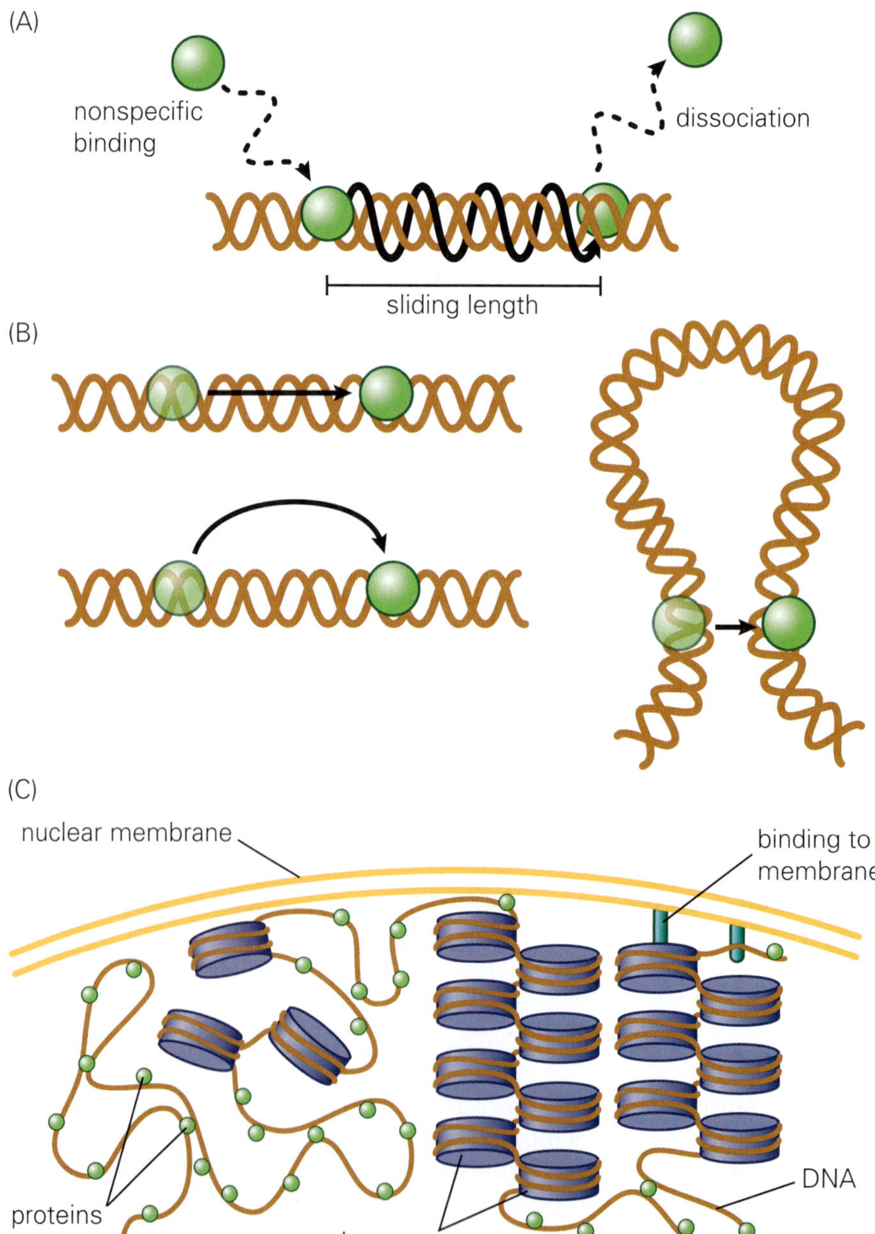


**Figure 17.25 Restricted diffusion on DNA.** The process of restricted diffusion occurs for proteins that can maintain loose, nonspecific contacts with the DNA duplex during sliding. (A) One-dimensional movement along the DNA. The *black* line indicates the movement of the protein (*green*) along the major groove of DNA. The movement corresponds to one-dimensional sliding along the groove. (B) Hops can occur by transiently breaking contact with the DNA and then binding nearby. The protein can also jump to a nearby strand, which is facilitated in oligomeric DNA-binding proteins that can bind more than one DNA region at a time. (C) In regions such as the nucleus, where DNA is organized by proteins such as histones, a combination of these restricted (or facilitated) diffusion modes contribute to proteins rapidly finding their cognate DNA sequences. (A and B, adapted from S.E. Halford and J.F. Marko, *Nucleic Acids Res.* 32: 3040–3052, 2004. With permission from Oxford University Press; C, adapted from M. Kampmann, *Mol. Microbiol.* 57: 889–899, 2005.)

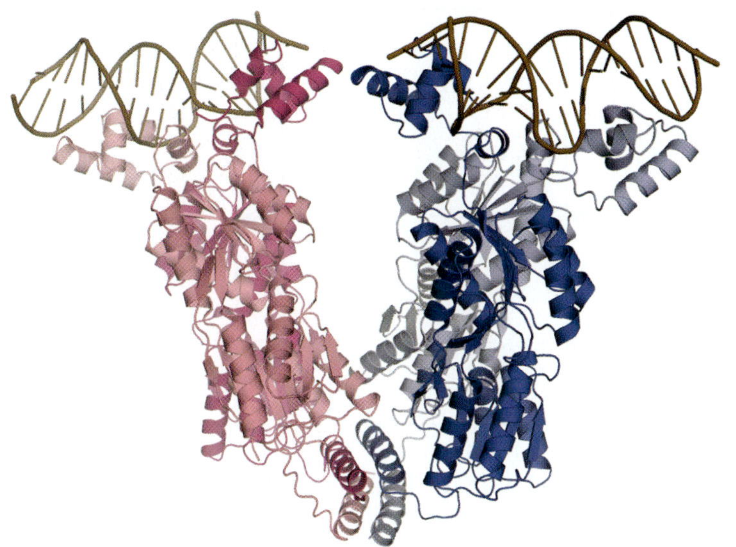


**Figure 17.26 Lac repressor tetramer bound to two DNA sites.** The structure shows two dimers within a tetramer, each of which can bind to target DNA sequences. This type of structure leads to the ability of the protein to "step" from one DNA strand to another. (Adapted from M. Lewis et al. and P. Lu, *Science* 271: 1247–1254, 1996; PDB code: 1LBG.)

many regulatory proteins bound at their particular target sequences. If diffusion were strictly one dimensional, then there would be major problems because the movement would be blocked by other proteins bound to DNA. It turns out that the loose electrostatic binding serves to hold the proteins on the DNA long enough to search significant lengths of sequence while allowing dissociation and rebinding, thereby allowing blockages to be bypassed.

The proteins that most efficiently find their target DNA sequences (such as lac repressor) have another important feature—namely, an oligomeric structure that makes possible simultaneous binding to two different DNA segments. During sliding encounters, transient looping of DNA can occur, as shown in Figure 17.25, which can lead to the protein binding at both ends, with a good chance of jumping to a new segment of DNA to search again by sliding. The structure of the lac repressor (**Figure 17.26**) suggests how the intermediate state might appear. In eukaryotic cells, the presence of histones and nucleosome structure brings many DNA segments together, further reducing the space that needs to be searched by proteins and facilitating transfer between segments.

## 17.23 Restricting diffusion to two-dimensional membranes can slow down the rate of encounter but still speed up reactions

Proteins that are anchored in membranes provide another context in which spatially restricted diffusion is important. As an example, consider cell surface receptors that are tyrosine kinases, which are activated by protein hormones such as insulin or the epidermal growth factor (see Section 12.13). One consequence of the activation of the receptor is that it switches on a key signaling protein known as Ras, which is associated with the cytoplasmic face of the cell membrane (**Figure 17.27**). Ras is a nucleotide-binding protein that is either bound to GDP or to GTP. Ras is active when bound to GTP, because it adopts a conformation that allows it to recruit other signaling proteins to the cell membrane, and these proteins transmit the signal further downstream. Once bound to Ras, GTP is slowly hydrolyzed and converts to GDP. GDP is tightly bound to Ras and does not dissociate easily, and thus Ras switches itself off.

How does the receptor activate Ras? When the receptor is activated by a hormone, it recruits a protein known as a nucleotide exchange factor to the membrane, as shown in Figure 17.27. The exchange factor binds to Ras, thereby causing the release of bound GDP and its replacement by GTP. Thus, Ras is converted from

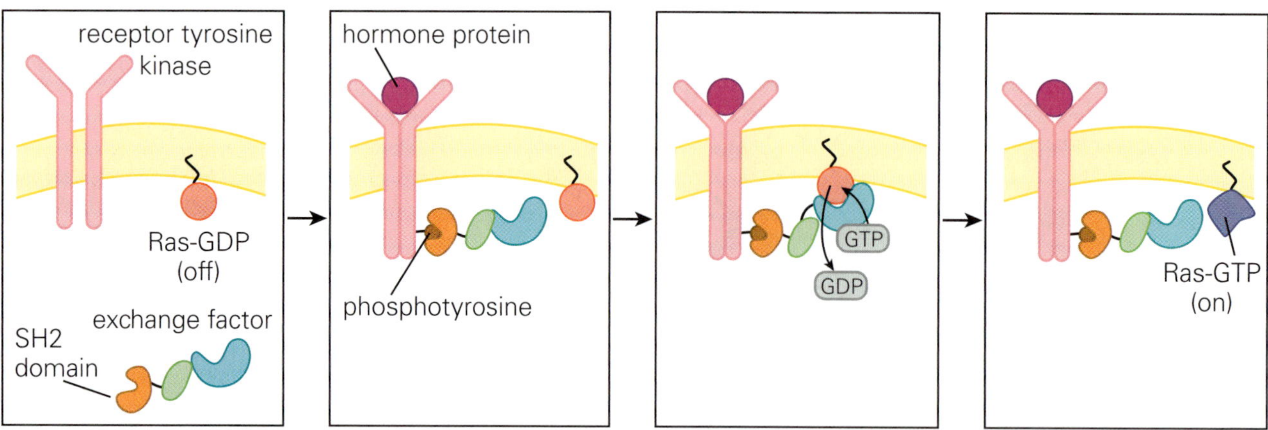


**Figure 17.27 Activation of Ras by a tyrosine kinase receptor.** When the receptor binds to a hormone, it becomes phosphorylated and recruits a nucleotide exchange factor to the membrane. Once anchored to the membrane by the receptor, the exchange factor encounters Ras and catalyzes the release of bound GDP and its replacement by GTP. Ras is active when loaded with GTP, and it sends a signal to other proteins at the membrane.

the inactive GDP-bound form to the active GTP-bound form. When the receptor is inactive, the exchange factor is in the cytoplasm and it does not activate Ras. When the receptor is activated, the exchange factor is brought to the membrane, where it then activates Ras.

Measurement of the rate of Ras activation by the exchange factor shows that localizing both Ras and the exchange factor to membranes speeds up the nucleotide exchange reaction by several hundred-fold relative to the rate when both proteins are in solution. Bringing the exchange factor to the membrane converts a three-dimensional search process for Ras into a two-dimensional one, and one might think that this speeds up the rate at which the exchange factor encounters Ras. It turns out, however, that the lateral diffusion of membrane-associated proteins can be very slow compared to three-dimensional diffusion in the cytoplasm, and so bringing the exchange factor to the membrane does not actually speed up the rate at which it encounters Ras. It is instructive to work though why this is the case.

We shall do this by calculating the time it takes for two molecules that are a certain distance apart to diffuse towards each other and collide. This is, in general, not an easy calculation to do, particularly if we consider the trajectories of individual molecules. We shall take a very simplified approach, in which we consider the populations of the two kinds of molecules and assume that the population distributions spread out from points that are separated by the initial distance between the two kinds of molecules, as shown in **Figure 17.28** for diffusion in two dimensions. The characteristic time for the two molecules to interact is then given by the time required for the r.m.s. displacements of the two distributions to equal half the displacement between them. We shall assume that there are equal numbers of the two kinds of molecules and that they have the same diffusion constants.

Let us first consider the case when the two interacting proteins (for example, Ras and the exchange factor) are distributed throughout the interior of the cell and collide with each other through three-dimensional diffusion. Imagine that the cell

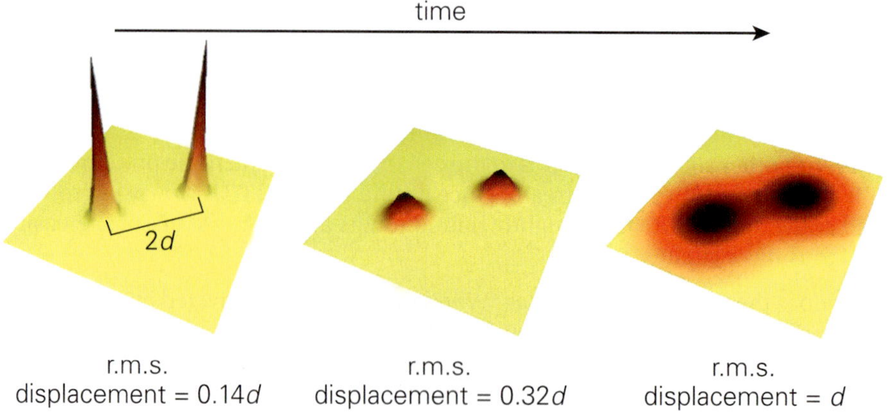


**Figure 17.28 Time taken for two molecules to collide through diffusion in two dimensions.** The two molecules are separated by a distance $2d$ initially. Shown here are the probability distributions of the molecules as a function of time. Initially, the likelihood of interaction is very small because the two distributions do not overlap. The r.m.s. displacement of the distributions increases with time and, when it is equal to $d$, the two distributions overlap substantially, with increased probability of collision.

is spherical and has a diameter of 100 μm. We assume that the two proteins are distributed evenly throughout the cell, as shown in **Figure 17.29**A, at a concentration of 1 nM each. This means that there are ~300,000 molecules of each protein per cell. The volume of the cell is ~500 × $10^{-15}$ $m^3$ and, because the molecules are evenly distributed, the volume per molecule of the two kinds of proteins is given by:

$$v = \frac{500\times10^{-15}}{(2)(300{,}000)} = 0.83\times10^{-18}\,\text{m}^3 \tag{17.52}$$

The average distance between the two kinds of proteins is estimated by taking the cube root of the volume, which is ~1 × $10^{-6}$ m, or 1 × $10^{-4}$ cm. Because both molecules are moving, we assume that a collision occurs when each molecule has moved half this distance (that is, 0.5 × $10^{-4}$ cm). We can now use Equation 17.28 to compute the characteristic time for collisions to occur. We do this by setting the root mean square displacement to be equal to half the estimated distance between proteins:

$$\text{r.m.s. displacement} = \sqrt{6Dt} = 0.5\times10^{-6} \tag{17.53}$$

The diffusion constant for a small protein in the cytoplasm is approximately $10^{-8}$ $cm^2$•$sec^{-1}$ (see Section 17.19). Using this value for the diffusion constant, we can calculate the time required for the displacement of a protein molecule to increase to the r.m.s. value:

$$t = \frac{(\text{r.m.s. displacement})^2}{6D} = \frac{(0.5\times10^{-4})^2}{(6)(1\times10^{-8})} = \sim 0.04\text{ sec} \tag{17.54}$$

Now let us study what happens when both proteins are restricted to the membrane, instead of being distributed evenly throughout the cell (Figure 17.29B). There are two key differences from the first situation that have a bearing on the calculation. The first is the obvious one, which is that the diffusion is now two dimensional, and we must use Equation 17.39 to calculate the time required. The second difference is that the diffusion constant for proteins moving on membranes is about a factor of 10 smaller than diffusion in the cytoplasm—that is, the value of $D$ is $10^{-9}$ $cm^2$•$sec^{-1}$. This is because the lipids in the membrane, although fluid, are tightly packed and restrain the movement of the proteins.

The surface area of the cell is 314 × $10^{-10}$ $m^2$, assuming the same diameter of 100 μm. As before, we assume that there are 300,000 molecules of each type on the membrane, and so the surface area per molecule is 52.3 × $10^{-15}$ $m^2$, which leads to an estimate of the average spacing between molecules of 0.0023 × $10^{-4}$ cm (that is, the spacing is about 20-fold smaller than before). We now use Equation 17.39 to relate the r.m.s. displacement of the proteins to the diffusion constant and the time:

$$\text{r.m.s. displacement} = \sqrt{4Dt} = 11.5\times10^{-6} \tag{17.55}$$

To calculate the time required, we use the value of $D$ = 1.0 × $10^{-9}$ $cm^2$•$sec^{-1}$ and solve Equation 17.68:

$$t = \frac{(\text{r.m.s. displacement})^2}{4D} = \frac{(11.5\times10^{-6})^2}{(4)(1\times10^{-9})} = \sim 0.05\text{ sec} \tag{17.56}$$

Thus, according to our calculation, restricting both proteins to the membrane actually *slows down* their rate of encounter. The principal cause of this effect is the lower value for the diffusion constant for movement on the membrane.

We began this section by stating that the reaction between Ras and the exchange factor occurs several hundred-fold faster when both proteins are on the membrane compared to when they are both diffusing in three dimensions. What might be the origin of this rate acceleration? The answer comes from considering the effective concentrations of the two molecules relative to each other. When the two molecules are both in the cytoplasm, they are at lower effective concentration than when they are both restricted to the membrane (compare panels A and B of Figure 17.29).

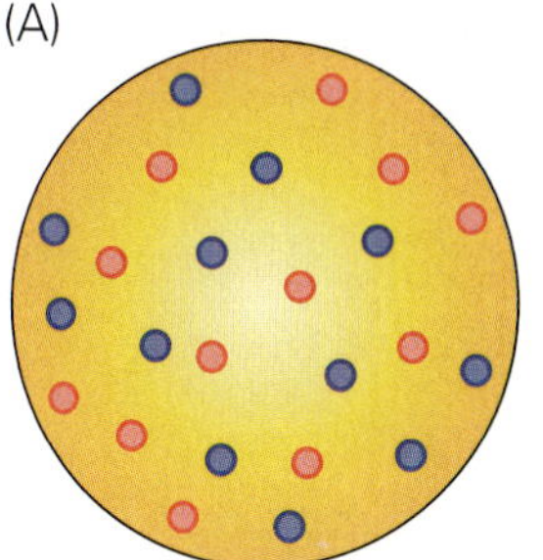


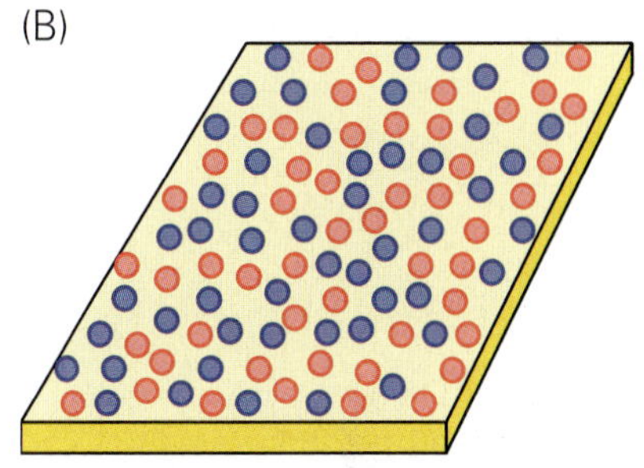


**Figure 17.29 Two kinds of molecules distributed throughout a spherical cell and the membrane surface.** (A) A cross section through a sphere is shown, with *red* and *blue* molecules distributed evenly. Only a small fraction of the total number of molecules within the sphere are shown in this planar cross section. (B) The same number of molecules are assumed to be restricted to the inner surface of the sphere. They are more densely packed, as shown here for a small portion of the surface.

To calculate the difference in effective concentration, we assume that when they are at the membrane, the proteins are confined to a shell of ~100 Å thickness. The volume of this shell is ~$300 \times 10^{-19}$ m$^3$ (surface area of the cell × 100 Å). This volume is ~10,000-fold smaller than the volume of the entire cell (~$500 \times 10^{-15}$ m$^3$). As we discussed in Section 14.22, this enhanced local concentration provides a considerable thermodynamic driving force that stabilizes the complex between the two proteins—that is, once the complex is formed, it is longer lived when both proteins are at the membrane. This facilitates the nucleotide exchange process by enabling the slow conformational changes that are necessary for the reaction to proceed.

## 17.24 Concentration gradients determine the outcomes of many biological processes

In real biological systems, molecules are being synthesized and degraded constantly through chemical and biochemical reactions. If the rates were uniform through a sample, as they would be *in vitro* in a test tube, then the reactions would not lead to concentration gradients. If, however, the rates vary in different regions in a cell, then concentration gradients are generated and may be present for extended periods of time. We have discussed how concentration gradients are used to drive the synthesis of ATP (Section 9.16) and to generate membrane potentials (Section 11.24). Concentration gradients also play important roles at a higher level of biological organization, such as in controlling the differentiation of cell types during embryonic development.

The first protein that was discovered to control differentiation through a concentration gradient occurs in the embryos of fruit flies (**Figure 17.30**). This protein, called bicoid, is necessary for the formation of distinct segments in the developing embryo. In the embryonic cell, the bicoid protein, together with others, activates the transcription of different sets of genes in different parts of the embryo. Ultimately these patterns of expressed genes determine the specific types of cells that emerge at that location—that is, what cells develop into what organs in the body.

The production of mRNA coding for bicoid is localized at the anterior end of the embryo, resulting from interactions of the embryo with the mother before the egg is even fertilized. The translation of bicoid mRNA into protein occurs at the anterior end, where the mRNA is localized, giving the newly expressed protein a high local concentration there. The protein can diffuse significantly before it is degraded. The combination of protein being synthesized primarily at one end of the cell and degraded throughout leads to a concentration gradient.

The shape of the bicoid gradient can be calculated using Fick's laws, as described above, but including terms that correspond to positive flux due to the synthesis of new molecules and negative flux due to their destruction, each with their appropriate dependence on position in the cell. Solving for the concentration as a function of time and space is done numerically with a computer. This model gave a

**Figure 17.30 Bicoid concentration profile in a fly embryo.** (A) A picture of a *Drosophila* embryo stained so that dark regions reflect high concentration of bicoid protein. (B) A graph of the concentration profile as a function of position along the long axis of two cells (100 is the left end; 0 is the right end). (Adapted from W. Driever and C. Nüsslein-Volhard, *Cell* 54: 95–104, 1988. With permission from Elsevier.)

(A)

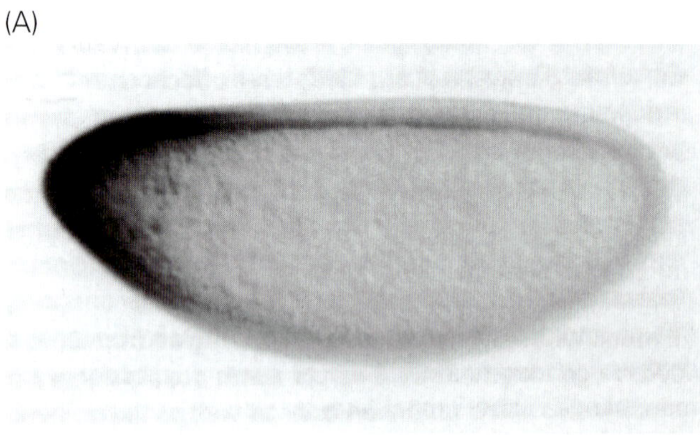

(B)

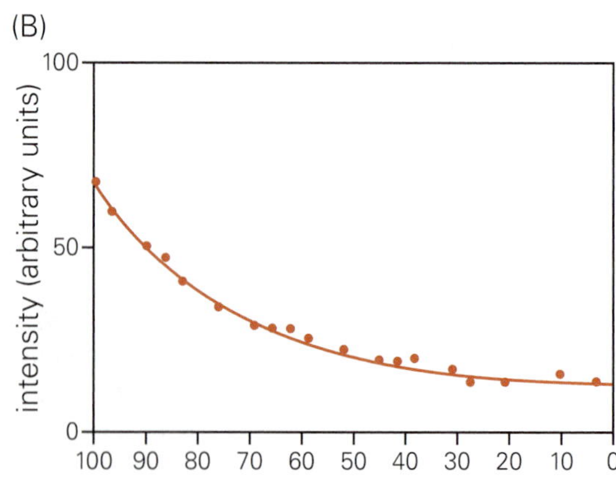

good prediction of the concentration gradient of bicoid protein. Very recent measurements of the rates of transport of the molecules involved, and more precise measurements of the mRNA in the cell, have shown that there is also a gradient of mRNA in the cell (decreasing in concentration with distance from the anterior end rather than near complete localization at the anterior end).

## 17.25 Cells use motor proteins to transport cargo over long distances and to specific locations

Simple diffusion allows molecules to move over micron-length scales within seconds (see Section 17.19). Because the r.m.s. displacement from a starting point increases only as the square root of the time, movement over greater distances requires much longer times, which becomes a problem for the transport of molecules in larger cells. Neuronal axons, for example, can extend for 1 mm (1000 μm) or more. Passive diffusion alone is insufficient to deliver proteins from the nucleus at one end of the cell to the axon terminus at the other end on the time scale of seconds to minutes. Even in smaller cells, diffusion is not fast enough to enable the necessary movement of large assemblies.

To overcome the limitations of passive transport (that is, just diffusion) and to be able to target specific proteins to particular regions, cells use chemical energy to actively move molecules and other large structures through the cytoplasm, a process called **active transport**. The term "active transport" is also used for moving ions across a membrane against a concentration gradient, also requiring energy. This was discussed in Section 9.15.

**Active transport**

Active transport is an energy-consuming process through which molecules are moved between different parts of cells. The term "active transport" is also used to mean movement of molecules across membranes against a concentration gradient, as was discussed in Chapter 11.

The general idea for how objects in cells are moved actively is very much analogous to what we routinely do at the macroscopic level: package what needs to be transported; connect it with a motor; and put it onto a track going in the desired direction. There are many variations of such transport systems in the cell, which are used in different contexts.

## 17.26 Vesicles are transported by kinesin motors that move along microtubule tracks

One of the major cellular transport systems uses a protein molecular motor called **kinesin**, which was described briefly in Chapter 6 (see Figure 6.1). The kinesin motor moves along protein filaments called **microtubules**. There are two essentially universal types of filaments inside cells, both comprised of noncovalently associated proteins. One type of filament is made from the protein actin and the other from tubulin, which makes microtubules (**Figure 17.31**). Actin filaments

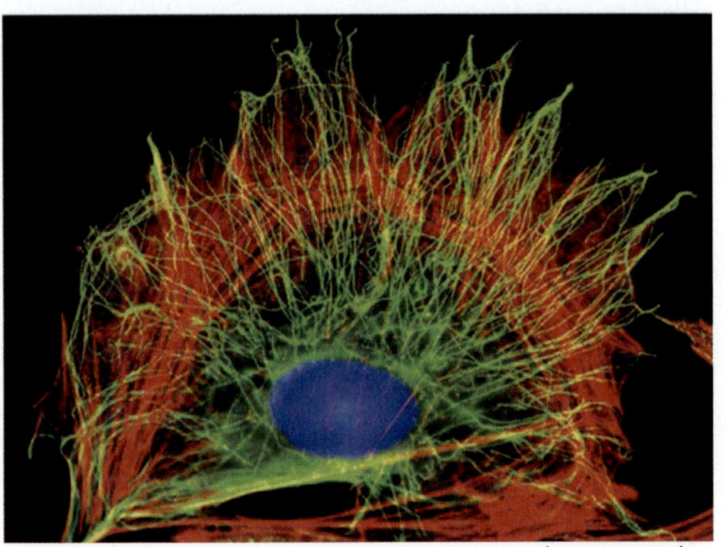


**Figure 17.31 Fluorescence micrograph showing filaments inside a cell.** Tubulin (stained *green*) and actin (stained *red*) are the two important components of cellular filaments. The tubulin filaments are called microtubules, and they act as tracks for active transport of materials within cells. DNA is stained *blue*. (Courtesy of Albert Tousson, University of Alabama at Birmingham, High-Resolution Imaging Facility.)

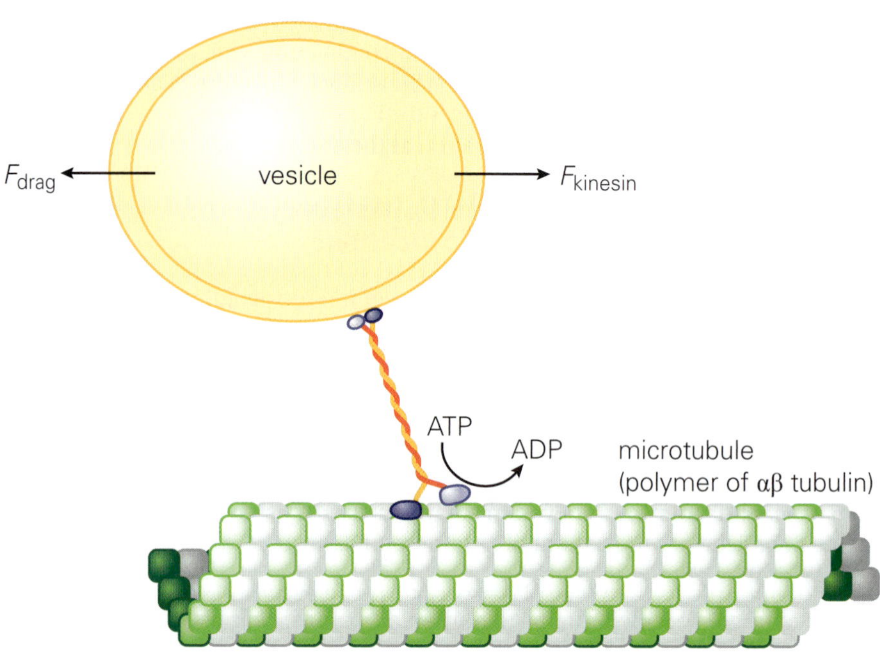


**Figure 17.32 Kinesin "walking" on tubulin.** A schematic drawing is shown of a kinesin molecule attached to a transport vesicle (*yellow*) moving along a microtubule (a polymer of αβ tubulin heterodimers that form a hollow tube). Kinesin dimers use the energy of ATP hydrolysis to move unidirectionally along the microtubule, dragging the vesicle along with it.

are primary determinants of the shape of the cell, and the growth of actin filaments underlies cell locomotion. Microtubules are involved in the positioning of organelles in the cells, and they are used for transport, by kinesins, of materials within the cell. Both types of filaments are dynamic, and they form and dissolve in response to the needs of the cell.

The packaging of the material to be transported is done by the formation of vesicles, which are small spheres of membrane that are pinched off the membranes in cellular compartments such as the Golgi (a central sorting house for processing cellular proteins). Vesicles vary in size, but are often in the range of 10–100 nm in diameter. The vesicles are surrounded by a membrane bilayer that contains proteins from the region from which it came, while the interior contains soluble proteins and small molecules.

Kinesin has an elongated structure, shown in **Figure 17.32**. At one end it has cargo-binding domains that adhere tightly to the vesicle to be transported. Attached to these by a long coiled coil (see Section 4.16) are a pair of ATPase domains (often called motor domains) that do work to pull the vesicle through the cell. The tracks that these motor domains move along are the microtubules, which are helical arrays of α,β-tubulin dimers.

The motor domains of kinesin bind to the tubulin proteins, and "walk" along them, as illustrated in **Figure 17.33** (see also Figure 6.1). One kinesin dimer can transport a large cargo vesicle, a few hundred nm in diameter, at a rate of 2 $\mu m \bullet sec^{-1}$, as shown by fluorescently labeling vesicles and then following the fluorescent spot under a microscope. This transport rate is much faster than passive diffusion of an object the size of the cargo vesicle. The motion is in one direction along the microtubule, rather than being random, which means that the distance traversed increases linearly with time rather than with $\sqrt{t}$ as for "random-walk" diffusion.

The ATPase domain of kinesin functions like other motor proteins and goes through a cycle of conformational changes as ATP is bound, hydrolyzed to ADP, and the ADP is released (see Figure 17.33). The two ATPase domains both bind to the microtubule at β subunits. When ATP binds to the leading ATPase subunit, already bound to a microtubule, there is a conformational change that exposes a high-affinity site for interaction with the linker, causing the linker to dock against the ATPase domain. This docking pulls on the connected coiled coil and the other ATPase domain, causing it to release from its binding site and rotate forward, traveling about 160 Å. Through random, diffusive motions it encounters another β-tubulin subunit and binds tightly to it. This binding leads to a conformational change in the now lagging ATPase subunit, causing it to hydrolyze its ATP to give ADP and $P_i$ (inorganic phosphate). The inorganic phosphate is then released, as is the ADP from the leading subunit. The system is then poised to again bind ATP to the leading ATPase, thereby initiating the next step along the microtubule.

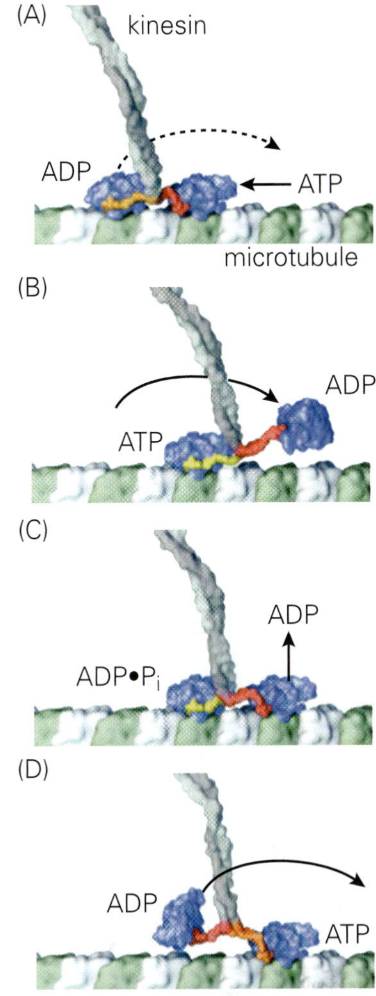


**Figure 17.33 Steps of kinesin on tubulin.** The processes involved in the individual steps taken as kinesin "walks" along a microtubule are shown. The individual motor domains are *blue*, and the colored linkers are associated with specific nucleotide states. (From R.D. Vale and R.A. Milligan, *Science* 288: 88–95, 2000. With permission from the AAAS.)

## 17.27 ATP hydrolysis provides a powerful driving force for kinesin movement

What is the efficiency of the kinesin motor in the thermodynamic sense? That is, what fraction of the free energy of hydrolysis is actually used to do the work? We can answer this question by calculating the work done by kinesin in moving cargo over a certain distance and comparing that value to the free energy released by the ATP molecules that are hydrolyzed during the movement.

In order to calculate the work done, we need to know the force against which the movement occurs. We shall assume that the cargo vesicle is a sphere of 100 nm and that kinesin moves it at 1.6 $\mu m{\bullet}sec^{-1}$ (this is about the maximum observed velocity). If you refer back to Equations 17.31 and 17.38, you will see that the drag force on a spherical object moving through a liquid is given by

$$\text{force} = \left(f_{\text{sphere}}\right)(v) = (6\pi r\eta)(v) \tag{17.57}$$

Here, $f_{\text{sphere}}$ is the friction factor for the sphere, $v$ and $r$ are the velocity and the radius of the sphere, respectively, and $\eta$ is the viscosity of the liquid (0.01 $g{\bullet}cm^{-1}{\bullet}sec^{-1}$ for water).

Using Equation 17.57 and substituting the values of $v$, $r$, and $\eta$ mentioned above, we get:

$$\begin{aligned}\text{force} &= (6\pi\eta r)(v) = (6)(3.14)(1\times10^{-2}\ g{\bullet}cm^{-1}{\bullet}sec^{-1})(10^{-5}\ cm)(1.6\times10^{-4}\ cm{\bullet}sec^{-1})\\ &= 3\times10^{-10}\ g{\bullet}cm{\bullet}sec^{-1}\end{aligned} \tag{17.58}$$

The work done against this force is given by the force multiplied by the distance, $x$, over which it acts. We can calculate the work done during 1 second of pulling as follows:

$$\begin{aligned}\text{work} = fx &= (3\times10^{-10}\ g{\bullet}cm{\bullet}sec^{-2})(1.6\times10^{-4}\ cm)\\ &= 5\times10^{-14}\ g{\bullet}cm^{2}{\bullet}sec^{-2}\end{aligned} \tag{17.59}$$

This motion is driven by ATP hydrolysis. If the kinesin motor moves at 1.6 $\mu m{\bullet}sec^{-1}$ in steps of 160 Å, there must be 100 steps per second of motion. This means that 100 ATP molecules are hydrolyzed per second by the kinesin motor. Under standard conditions, the free energy of hydrolysis of ATP is –31 $kJ{\bullet}mol^{-1}$. Under the real concentration conditions in the cell (that is, [ATP] > [ADP] and low [$P_i$]), the actual free energy of hydrolysis, and hence the work that can be derived, is somewhat larger. Nevertheless, for this calculation, we shall just use the value of $\Delta G^{o}$.

To compare the free energy released by ATP hydrolysis with the work done by kinesin (Equation 17.59), we must convert units, using 1 J = $10^7$ $g{\bullet}cm^2{\bullet}sec^{-2}$, and convert to molecules from moles using Avogadro's number ($N_A \approx 6\times10^{23}$ $molecules{\bullet}mol^{-1}$).

$$\begin{aligned}\text{free energy from 100 ATP} &= \left(10^2\text{molec}\right)\left(31\times10^3\ J{\bullet}mol^{-1}\right)\left(10^7 g{\bullet}cm^2{\bullet}sec^{-2}{\bullet}J^{-1}\right)\left(\frac{1\ \text{mol}}{6\times10^{23}\text{molec}}\right)\\ &= 5.2\times10^{-11}\ g{\bullet}cm^2{\bullet}sec^{-2}\end{aligned} \tag{17.60}$$

Comparing Equations 17.59 and 17.60 shows that there is far more energy available from ATP hydrolysis than is needed for kinesin to overcome the drag from the solution. Note that for this calculation we used the viscosity of pure water, but that the viscosity of the cytoplasm is higher by 20-fold or more. Even with the higher viscosity, the velocity is not limited by the free energy available. Instead, the speed of movement is probably limited by the properties of kinesin—that is, the limit is set by the rate at which the ATPase domains can complete their cycle of binding events and conformational changes.

# C. EXPERIMENTAL MEASUREMENT OF DIFFUSION

## 17.28 Diffusion constants can be measured experimentally in several ways

It might seem that a diffusion constant could be measured easily by simply watching how molecules mix when a drop of one solution is introduced into another. But, unfortunately, this kind of simple approach does not work in practice because the movement of molecules when solutions are mixed is usually dominated by convection effects. Even very small differences in density, such as those arising from temperature differences in parts of the sample, can cause such convective flow, in which the two solutions flow around each other on a macroscopic scale. This obscures the measurement of the microscopic diffusion effects.

**Dynamic light scattering**

Dynamic light scattering measures fluctuations in the number of scattering molecules in a small volume, through their effect on the scattered light intensity. The rate of light intensity fluctuation is directly related to the rate of diffusion of the scattering molecules.

A classical approach to determining a diffusion constant is to measure the rate of mixing of two solutions that are separated by a glass frit (a dense porous glass mesh that allows diffusion but prevents convection). A more modern approach uses the scattering of laser light from a sample. A method known as **dynamic light scattering** detects the fluctuations in light scattered by macromolecules diffusing through a small volume within the sample. Individual macromolecules are too small to be seen with a microscope, but the way in which they modulate the scattered light intensity can be detected and this allows the diffusion constant to be measured.

Macromolecules diffuse randomly through the illuminated volume, and variations in the number that are within that volume leads to fluctuations in the amount of scattered light. The rate at which molecules enter or leave the scattering volume is determined by their rate of diffusion, which is directly proportional to their diffusion constants. Hence, the rate of fluctuation of the light intensity is related to the rate of diffusion of the molecules doing the scattering.

These fluctuations are analyzed by calculating the time-correlation function, $C(\tau)$, for the intensity, $I(t)$:

$$C(\tau)=\langle[\Delta I(t)][\Delta I(t+\tau)]\rangle \tag{17.61}$$

Here, $\Delta I(t)$ and $\Delta I(t+\tau)$ are the fluctuations from the mean intensity at times $t$ and $t+\tau$. The angled brackets around the product of the intensity fluctuations indicate that $C(\tau)$ is the average value over all values of time, $t$.

The correlation function tells us how quickly the scattered intensities vary with time. If the intensity varies slowly, then $\Delta I(t)$ will be similar to $\Delta I(t+\tau)$, even when $\tau$ is a large offset in time, and the value of $C(\tau)$ will be large. In contrast, if the intensities vary very quickly with time, the correlation function will be small. This is because the fluctuations can have both negative and positive values and, as the intensities become uncorrelated, the mean value of the product tends to zero.

Scattered intensities and the corresponding correlation functions are shown as a function of time in **Figure 17.34**. Faster fluctuations in scattered intensity leads to a more rapid drop in the correlation function with time. Although we shall not discuss this in detail, it turns out that the correlation function decays exponentially with $\tau$, with a time constant that is linearly proportional to the diffusion constant. If two molecules of different size are present in the solution, then two components are seen in the correlation function, and the diffusion constants for each species can be determined. This is now a commonly used method to measure diffusion constants.

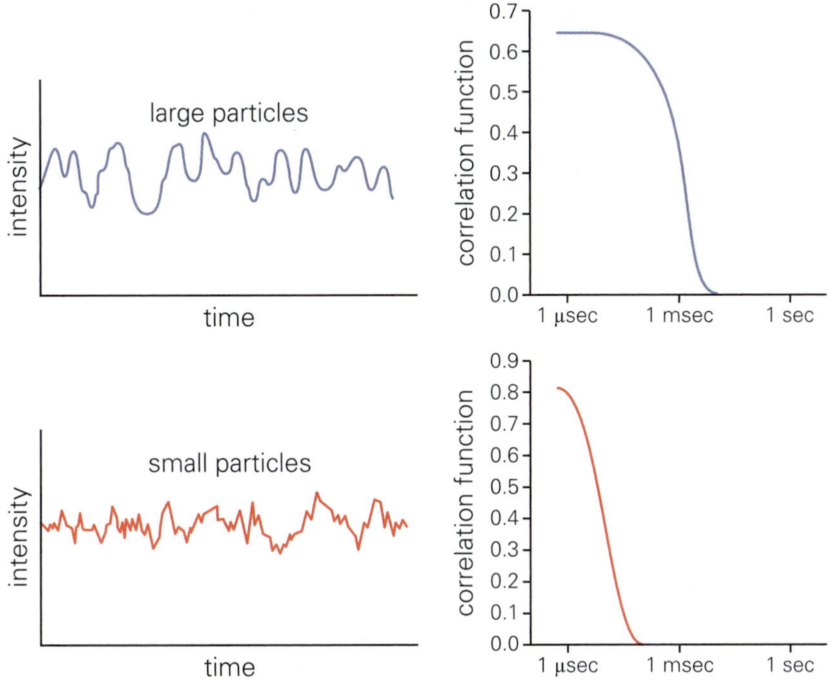


**Figure 17.34 Light scattering for diffusion constant measurements.** Fluctuations in the intensity of scattered light are measured in dynamic light scattering. Larger, slower moving molecules cause slow fluctuations (top left), while smaller, faster moving ones cause faster fluctuations (bottom left). The fluctuations are analyzed by calculating the correlation functions (right), which give a characteristic time constant for loss of correlation, related to the molecular size.

## 17.29 Movement of molecules in solution can be driven by centrifugal forces

A different approach to measuring the rate of motion of molecules is to create a continuous force that drives the motion in one direction. One way that a force can be created is through rotational motion in a **centrifuge**. The centrifugal force pushes the molecules away from the axis of rotation, as children learn from playground carousels.

**Centrifuge**

A centrifuge is an instrument that spins a sample very rapidly, making the molecules in it move under the centrifugal force.

From Newton's laws of motion, the centrifugal force, $F_{\text{cent}}$, is given by:

$$F_{\text{cent}} = \omega r^2 \tag{17.62}$$

where ω is the angular velocity (the rate of rotation in units of angle/time) and $r$ is the distance from the axis of rotation. The molecules in a spinning sample feel both the random Brownian forces and the centrifugal force, but the latter can be made quite large (up to more than $10^5$ times the force of gravity) and, hence, can affect the movement of the molecules substantially.

To apply a strong centrifugal force, the sample is put into a cell in a centrifuge rotor. The cell has windows on the top and bottom so that light can pass through and the concentration of the solute can be measured by determining the absorbance as a function of position in the cell while the rotor is spinning (**Figure 17.35**).

As the sample starts spinning, the molecules are accelerated toward the outside edge of the sample chamber (the molecules are usually denser than water). As they move, they will be subject to viscous drag and, like the skydiver that we described in Section 17.15, the molecules will quickly reach a velocity at which the centrifugal force and the frictional force become equal, and thereafter the molecules move at a constant (terminal) velocity.

To calculate the effect of the centrifugal force, the general relationship, $F = ma$ (force = mass × acceleration), must be modified because the molecules are in water. The water displaced by the molecules also feels the centrifugal force (buoyancy), and only the difference in mass of the protein and displaced solvent leads to the net force. The effective mass, $m_{\text{eff}}$, is the true mass of the macromolecule, $m_{\text{M}}$, scaled down by the ratio of the density of the protein to the density of the solvent:

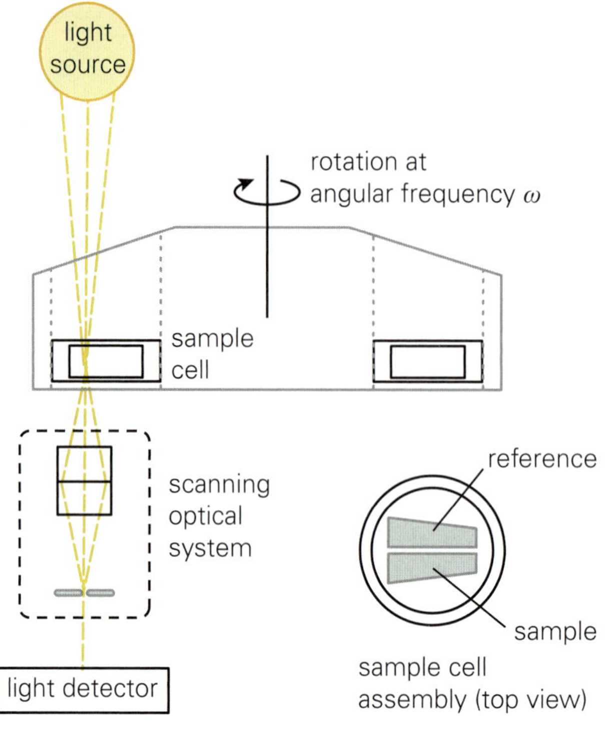


**Figure 17.35 Diagram of a centrifuge.** An analytical centrifuge spins samples inside a sample cell with windows so that the concentration can be measured as a function of position in the cell using absorption of light. The rotation rates in a laboratory centrifuge are as high as 40,000 rotations per minute.

$$m_{\text{eff}} = m_{\text{M}}\left(1 - \frac{\rho_{\text{solv}}}{\rho_{\text{macromol}}}\right) \tag{17.63}$$

where $\rho_{\text{solv}}$ and $\rho_{\text{macromol}}$ are the densities of the solvent and macromolecule, respectively. The density of water is near 1.0 g•cm$^{-3}$, while the protein density is usually between 1.33 and 1.43 g•cm$^{-3}$. Nucleic acids have densities between 2.0 and 2.5 g•cm$^{-3}$.

Under the centrifugal force from rotation, the molecules rapidly accelerate until the magnitude of the friction force, which is $fv$ (see Section 17.41), matches that of the centrifugal driving force at the terminal velocity, $v_{\text{term}}$:

$$m_{\text{eff}}\omega^2 r = f v_{\text{term}}$$

and so:

$$v_{\text{term}} = \frac{m_{\text{eff}}\omega^2 r}{f} \tag{17.64}$$

The terminal velocity depends on the force applied (and, hence, on the angular velocity), so it is usual to divide through by $\omega^2 r$ to give the **sedimentation coefficient**, $S$:

$$S \equiv \frac{v_{\text{term}}}{\omega^2 r} = \frac{m_{\text{eff}}}{f} \tag{17.65}$$

The sedimentation coefficient is a characteristic of the molecule that is related to its friction factor in solution.

**Sedimentation coefficient**

The sedimentation coefficient, $S$, of a macromolecule is inversely related to the friction factor, $f$, of the macromolecule (see Equation 17.65). Because the friction factor depends on the shape, the value of $S$ provides information on molecular shape.

The terminal velocity is measured experimentally by starting with a uniform solution in a centrifuge rotor, and then initiating spinning. As molecules throughout the cell begin to move to the bottom of the cell, a "boundary" develops, with the solution above the boundary being depleted of macromolecules. This boundary moves through the sample volume at a speed $v_{\text{term}}$. By measuring the concentration profile repeatedly, as shown in **Figure 17.36**, this boundary can be followed with time to determine $v_{\text{term}}$. Although all molecules feel the centrifugal force, individual molecules also experience random Brownian forces, which causes a spread in the boundary layer. Nevertheless, it is possible to determine the mean

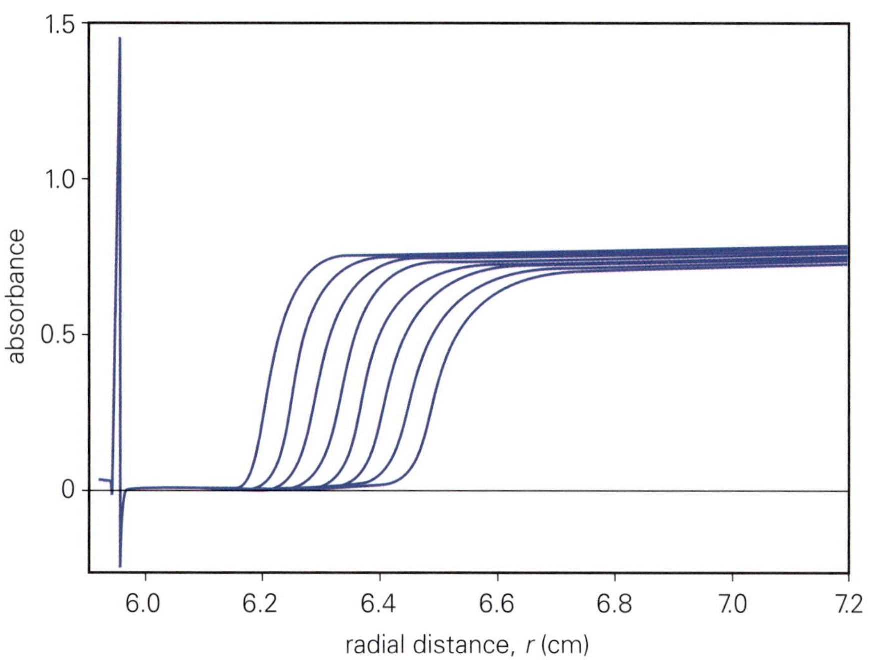


**Figure 17.36 Changes in protein concentration during centrifugation.** Concentration profiles are shown at different times for a sample undergoing centrifugation. The concentration at any position, $r$, within the sample cell is determined from the absorbance at 280 nm, which is linearly proportional to protein concentration. The boundary moves toward the outer edge of the rotor (right) with time. The rate of movement of the boundary allows the sedimentation coefficient to be determined.

position of the boundary with sufficient accuracy to enable calculation of the sedimentation coefficient.

Examining the units of $S$, $v_{term}$ is a velocity and must have units of $cm \bullet sec^{-1}$, while the units for $\omega^2 r$ are $cm \bullet sec^{-2}$, and so the units of the sedimentation constant will be sec. By using Equation 17.65 and parameter values that correspond to proteins, we find that the order of magnitude of $S$ is about $10^{-13}$ sec, which for convenience is defined as 1 Svedberg (1 S). This unit is named after Theodor Svedberg, who invented the high-speed ultracentrifuge for making such measurements.

Sedimentation velocities can be measured very accurately. For example, centrifugation was used to show that the enzyme aspartate transcarbamylase (discussed in Section 16.16) undergoes a conformational change in the presence of substrates, or the bisubstrate analog PALA. The conformational change results in a change in the friction factor, resulting in a small change in the sedimentation coefficient (~4%), but this small change can be measured accurately.

Large macromolecular assemblies are sometimes referred to by their sedimentation coefficients. For example, the two subunits of the bacterial ribosome have sedimentation coefficients of 50 S and 30 S, and these are called the 50S and 30S subunits, respectively. The intact ribosome, which is a complex of the large and small subunits plus a few other smaller RNA molecules, has a sedimentation coefficient of 70 S. The nonadditivity of the sedimentation coefficients of the subunits (that is, 30 S + 50 S ≠ 70 S) comes about because of different shape factors for the subunits alone or for the two bound together, as shown in **Figure 17.37**.

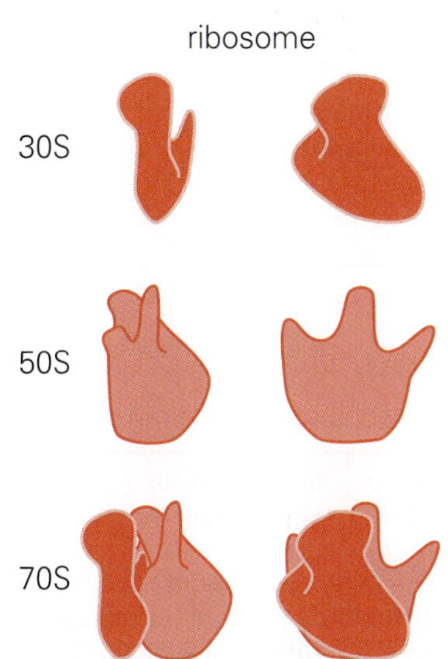


**Figure 17.37 Sedimentation coefficients for ribosome subunits.** Two views of the shapes of the 30S and 50S subunits of the bacterial ribosome are shown, together with that of the assembled 70S full ribosome.

## 17.30 Equilibrium centrifugation can be used to determine molecular weights

An equilibrium centrifugation experiment takes advantage of a combination of the motion under centrifugal force and diffusion in a concentration gradient. In this approach, after initiating centrifugation, the spinning is continued until the concentration profile stops changing, indicating that a concentration equilibrium has been reached. This can require a day or more, depending on the molecular weight of the dissolved macromolecule. The centrifugal force pushes all molecules toward the bottom, but as the concentration at the bottom increases, molecules will tend to diffuse back towards the top.

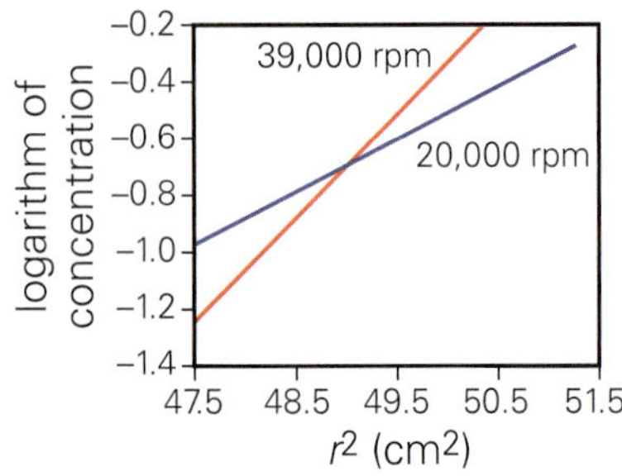


**Figure 17.38 Concentration profiles from equilibrium centrifugation experiments.** Results for two different spinning speeds are graphed. The linearity in the graph of ln(*c*) vs. $r^2$ indicates that there is just one effective molecular weight for the protein present (and hence just one oligomeric state).

The system reaches equilibrium when the flux due to the centrifugal force *exactly* equals the diffusive flux in the opposite direction. In thermodynamic terms, this means that the chemical potential (including both the energy due to centrifugation and free energy due to concentration) becomes uniform at each position in the cell. To calculate the requirements for this condition to be met, consider a plane through the sample, at a distance *r* from the rotation axis. The concentration at that position is $c(r)$, the concentration gradient is $dc(r)/dr$, and the area of the plane within the sample is *A*. Using Fick's law (Equation 17.14) and following the discussion in Section 17.29, the balance of the fluxes during a time, *dt*, required to meet the equilibrium condition, can be written as:

$$[c(r)](S)(\omega^2 r)(Adt)=(D)\left(\frac{dc(r)}{dr}\right)(Adt) \tag{17.66}$$

where *S* is the sedimentation coefficient and *D* is the diffusion constant. The left-hand side is the flux from centrifugation, while the right-hand side is that from diffusion in the concentration gradient. The terms *Adt* on each side cancel and, by inserting the values for *S* and *D* given by Equations 17.65 and 17.36, this becomes:

$$[c(r)]\left(\frac{m_{\text{eff}}}{f}\right)(\omega^2 r)=\left(\frac{k_B T}{f}\right)\left[\frac{dc(r)}{dr}\right] \tag{17.67}$$

One of the important things to note in this form of the equation is that *f* occurs in the denominator on *both* sides of the equation and hence cancels out. This leaves the dependence on $m_{\text{eff}}$, the effective mass of the centrifuging particle *without* the shape factor (see Equation 17.63). This equation can be further rearranged (converting to units on a per mole basis using $M_{\text{eff}} = N_A m_{\text{eff}}$ and $R = N_A k_B$) to give:

$$M_{\text{eff}}=M\left(1-\frac{\rho_{\text{solv}}}{\rho_{\text{macromol}}}\right)=\left(\frac{RT}{\omega^2}\right)\left(\frac{dc}{c}\right)\left(\frac{1}{r\,dr}\right) \tag{17.68}$$

Here *M* is the molecular weight of the macromolecule. Remembering that $d\ln(c) = dc/c$, and $dr^2 = 2rdr$, this can be reorganized again to give an expression for the molecular weight:

$$M=\frac{2RT}{\omega^2\left(1-\dfrac{\rho_{\text{solv}}}{\rho_{\text{macomol}}}\right)}\frac{d(\ln c)}{d(r^2)} \tag{17.69}$$

This equation indicates that a plot of ln(*c*) vs. $r^2$ should give a straight line, with slope related to *M* (**Figure 17.38**). By combining this information with the spinning angular velocity, $\omega$, the molecular weight, *M*, can be calculated.

If the graph of ln(*c*) versus $r^2$ produces a straight line, then this indicates that there is a single species of molecule in solution, with a defined molecular weight. If, instead, the molecule is in equilibrium between different oligomerization states (for example, monomer, dimer, trimer, etc.), then there will be deviations from linearity in the graph. This feature makes equilibrium centrifugation particularly valuable for determining the fractions of oligomeric species present. **Figure 17.39** shows an example of the profiles that would be expected for a monomer–tetramer equilibrium.

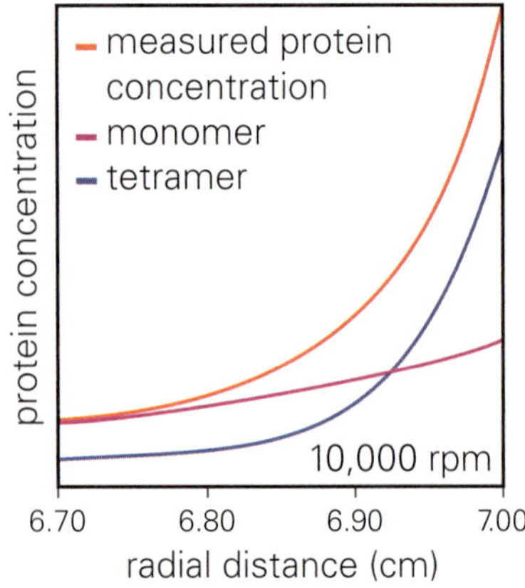


**Figure 17.39 Equilibrium centrifugation profiles for mixed oligomeric states.** Calculated curves are drawn for the expected total protein concentration as a function of position for a protein that has a monomer–tetramer equilibrium. To fit the full protein concentration profile, the concentration profiles for the monomeric and the tetrameric species must both be taken into consideration.

## 17.31 Electrophoresis provides an alternative method for driving molecular motion

An alternative to centrifugation is to induce the motion of molecules in solution through electrical forces—an approach called **electrophoresis**. The idea is quite simple. An electric field is applied to a sample by placing electrodes at each end and connecting the electrodes to an electrical source. An ion in the sample will be attracted to the oppositely charged electrode. A force $F_{\text{elec}} = ZeE$ is generated, where *Z* is the total net charge on the molecule, *e* is the magnitude of the charge of an electron, and *E* is the strength of the electric field (see Section 11.10). Under the

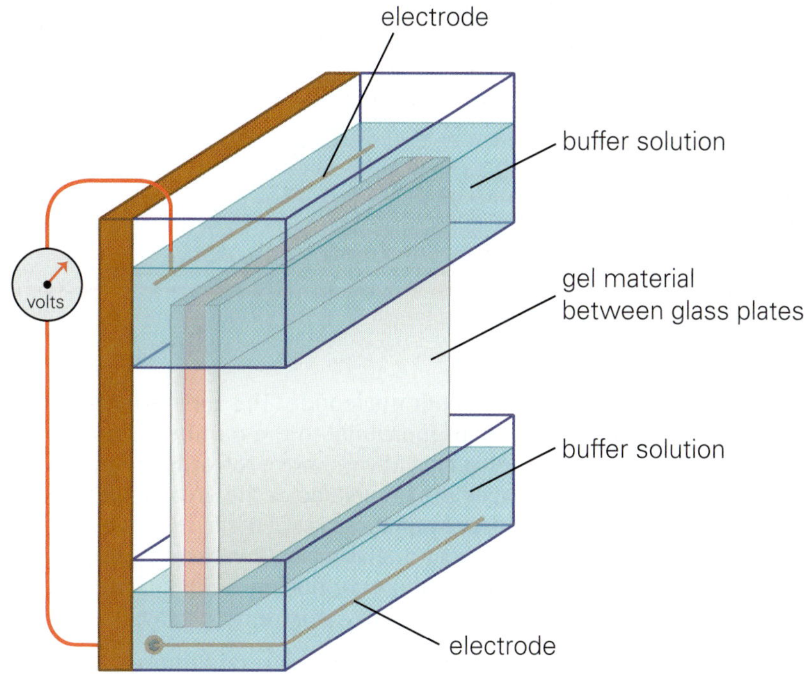


**Figure 17.40 A gel electrophoresis apparatus.** Electrophoresis experiments are commonly run in laboratories with the gel sandwiched between two vertical glass plates, each end immersed in a buffer solution. Wire electrodes are placed in the upper and lower buffer trays with a voltage source to provide the driving force to move the molecules from the top to the bottom. Samples are loaded into slots at the top of the gel so that many samples can be run in parallel (see Figures 17.42 and 17.43 for examples of such gels).

influence of this force, as in centrifugation, the molecules accelerate but quickly reach a terminal velocity at which the friction force, $fv_{\mathrm{term}}$, equals the electrical force:

$$fv_{\mathrm{term}} = ZeE$$

or

$$v_{\mathrm{term}} = \frac{ZeE}{f} \tag{17.70}$$

The **electrophoretic mobility** of the macromolecule is defined as the terminal velocity per unit electric field, $v_{\mathrm{term}}/E$:

$$\text{electrophoretic mobility} = \frac{v_{\mathrm{term}}}{E} = \frac{Ze}{f} \tag{17.71}$$

The electrophoretic mobility depends on the molecular size and shape through the friction factor, $f$, but also on the charge of the molecule.

**Electrophoresis**

Electrophoresis is the driven movement of charged molecules in an electric field, analogous to movement in a centrifugal field.

It is necessary to prevent convection in samples to be able to measure the real electrophoretic mobility. The common way to accomplish this is to add to the solution a substance that forms a gel. The gel is sandwiched between two glass plates, with buffer reservoirs above and below, each containing one of the electrodes, as shown in **Figure 17.40**. The two commonly used gel-forming substances are polyacrylamide and agarose, which is a linear, polymeric sugar that comes from red algae or seaweed (see Section 3.8).

The presence of the gel does affect the mobility of molecules moving in it (a higher concentration of gel polymer reduces mobility, as one would intuitively expect). As discussed in the next section, this generally means that the absolute mobility cannot be interpreted easily in terms of molecular parameters, so instead reference standards are run together with an experimental sample for comparison.

## 17.32 The electrophoretic mobility of nucleic acids decreases with size

As you can see from Equation 17.71, the electrophoretic mobility of molecules changes with size, shape, and charge. In nucleic acids, there is one phosphate

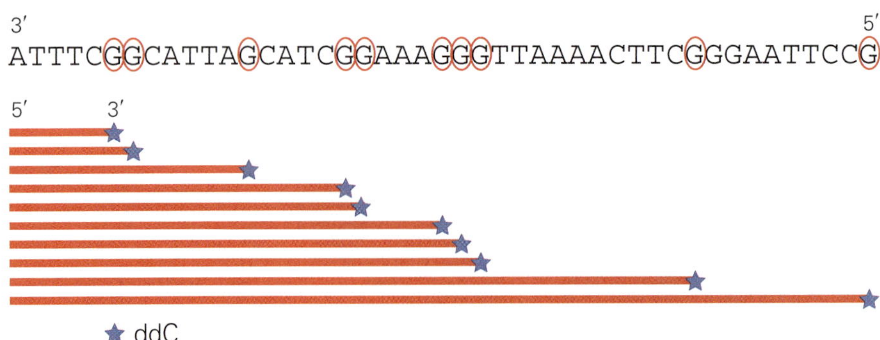


**Figure 17.41 A DNA sequencing reaction.** The DNA strand to be sequenced is shown at the top, and is used as a template to synthesize new strands of DNA. Small amounts of dideoxy cytidine (ddC, shown as a star) are added to the reaction. When ddC is incorporated opposite G in the template strand, further synthesis is terminated. The ddC nucleotides are labeled with a fluorophore, or with a radioactive isotope, so that the newly synthesized molecules can be detected on a gel, as shown in Figure 17.42. Additional reactions are run with ddA, ddG and ddT in order to provide complete information for sequencing.

group with one negative charge per nucleotide. This means that the total charge on a nucleic acid molecule is proportional to the number of nucleotides in the molecule. As a result, nucleic acids always move towards the positive pole during electrophoresis. The larger the nucleic acid, the greater the charge and the electrical force, but this is offset by an increase in the friction factor with size. The gels that are used for nucleic acid analysis also increase the friction factor. The net effect is that the observed mobility of nucleic acids during electrophoresis decreases slightly with increasing size, in a very uniform way.

For single-stranded oligonucleotides that are several hundred nucleotides long, the difference in mobility due to a single extra nucleotide can be detected in an electrophoresis gel. It is not possible to calculate the absolute mobility and derive the length of any particular nucleic acid in the sample, but a reference set of oligonucleotides of known lengths is run at the same time on the same gel in a separate lane, and the lengths of the unknown nucleotides are estimated by comparison with the known samples.

Electrophoresis is commonly used to sequence DNA. The DNA strand to be sequenced is used as a template for the synthesis of new strands by a DNA polymerase enzyme. Small amounts of the four types of dideoxynucleotides (in which the deoxyribose sugar lacks the 3′-hydroxyl group) are added along with the normal bases in four separate reactions. These dideoxynucleotides (ddA, ddT, ddC, and ddG) are incorporated into the newly synthesized strands, but they prevent further extension of the DNA strand. This leads to a ladder of DNA molecules that are terminated at different positions, as shown in **Figure 17.41.** In one sample, there is a series of DNA molecules of increasing length, all terminated by ddA, in another all the molecules are terminated by ddG, and so on.

A polyacrylamide gel is then used to separate the products of each reaction into individual bands of DNA, as shown in **Figure 17.42.** The dideoxynucleotides are labeled with a fluorescent tag or with a radioactive isotope so that they can be visualized in the gel. The sequence of the parent DNA strand can be determined by noting the order in which strands with different terminal dideoxynucleotides migrate down the gel. Modern electrophoresis for sequencing is generally done in a capillary rather than a gel, but the principle is the same.

## 17.33 Gel electrophoresis analysis of proteins is useful for size determination

We cannot readily interpret the electrophoretic mobility of proteins in terms of their size because only a few of the amino acids in a protein contribute to the

**Figure 17.42 A part of a DNA sequencing gel.** The label above each lane indicates that the reaction for that lane had a small amount of the dideoxynucleotide added. The shortest DNA strands are at the bottom of the gel. You can tell that the sequence of the first three positions in this segment of the sequence must be GGG because the three shortest DNAs are in the lane corresponding to ddC.

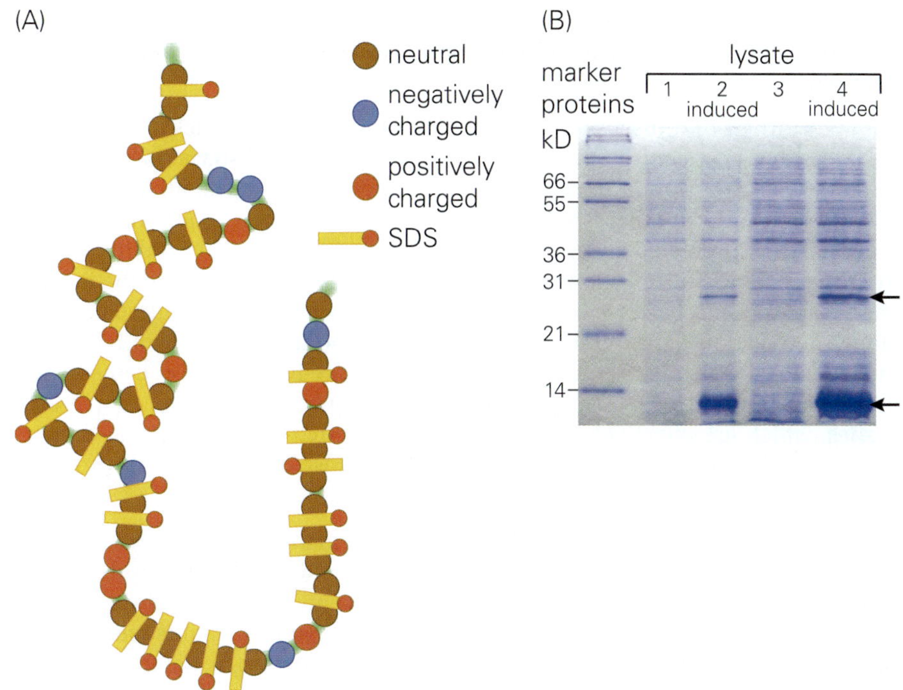


**Figure 17.43 SDS polyacrylamide gel electrophoresis (SDS-PAGE).** (A) The addition of SDS detergent unfolds proteins, as shown schematically here. On average one negatively charged SDS molecule is bound to the protein for every two amino acid residues, and so the net negative charge on the protein is proportional to the size of the protein. (B) An SDS-PAGE gel. The left lane contains a set of "marker" proteins of known molecular weight, visualized using a stain known as Coomasie blue. The lanes labeled 1 and 3 were loaded with a lysate of *E. coli* cells that contained a plasmid for expressing proteins of interest but in which expression was not induced. More lysate was added to lane 3 than to lane 1. The lanes marked 2 and 4 are the same amounts of lysate from cells in which the proteins of interest (marked with arrows) were induced to express.

overall charge of the molecule. The net charge may be positive, neutral, or negative, depending on the composition of the protein. For an unknown protein, it is not even possible to predict the direction of migration, much less the mobility, which also depends on size and shape.

The ambiguity in charge for proteins can be altered by the addition of a detergent that binds to proteins. The most commonly used detergent for this purpose is sodium dodecylsulfate, SDS (a 12-carbon hydrocarbon chain terminated by a negatively charged sulfate group; see Figure 3.37). The hydrocarbon tails of SDS cluster around the sidechains of the protein, particularly the hydrophobic ones, causing proteins to unfold, as shown in **Figure 17.43**A. One SDS molecule is bound on average for each two amino acids. Because each SDS carries one negative charge, its binding to the protein confers an overall net negative charge that is proportional to the number of amino acids in the protein, creating a situation analogous to that for nucleic acids.

The relationship between SDS-induced charge and size is only approximate because of amino acids in the protein that also have charges, but their contribution to the total charge is generally much smaller than the charge due to the SDS. When electrophoresis is carried out with SDS in the gel to maintain the denatured state, the proteins separate according to molecular weight, because the negative charge is roughly proportional to the molecular weight. The approximate molecular weight of the protein is estimated by comparisons to the mobility of known molecular weight reference proteins in the same gel (Figure 17.43B).

## Summary

The movement of molecules in cells is an integral part of their function. We began this chapter with an analysis of random walks, which provide a good description of the diffusive movements of molecules, and also that of small but macroscopic objects, such as bacteria. We saw that the probability distribution for displacements in a random walk is given by a Gaussian function, derived in much the same way as the binomial probability distribution for the outcomes of a series of coin tosses. An important conclusion is that the distance traveled by diffusing molecules increases with the square root of time. This means that biological molecules move micron-scale distances relatively quickly, but movement over longer distances is not very efficient. To facilitate movement over larger distances cells also use chemical energy to actively transport vesicles carrying needed molecules between different parts of the cell.

We then turned to a discussion of diffusion that began with Fick's laws, which describe how a concentration gradient changes in time and space. Just as for the random-walk description of diffusion, Fick's laws lead to a probability distribution for concentration that is a Gaussian function. The rate of random, diffusive motion can be predicted given the size and shape of the molecule and the viscosity of the solvent. Such motions are an intrinsic factor in determining the rates of chemical reactions, because collisions between molecules are a required first step along the reaction path. While we usually think about the motion of biological molecules as occurring in three dimensions, there are important processes, such as proteins binding to DNA and receptors becoming spatially organized in membranes, which are effectively of lower dimensionality. The reduction in dimensionality makes the search process faster, as in the case of DNA-binding proteins, or enhances reaction rates through local concentration effects.

The rates of molecular motion can be measured in many different ways, whether it occurs as a normal steady-state process (light scattering), or when the molecules are driven by a centrifugal or electrical field (centrifugation or electrophoresis). Measurements of molecular mobility allow diffusion constants and the size (radius and molecular weight) and shape of the molecules to be determined.

## Key Concepts

### A. RANDOM WALKS

- Random walks provide a good description of the microscopic movement of molecules occurring through random collisions.
- The probability distribution for the net number of moves in one direction is given by a Gaussian function.
- The distance traveled through a random walk increases as the square root of time.

### B. MACROSCOPIC DESCRIPTION OF DIFFUSION

- Fick's laws describe the average movement of molecules at the macroscopic level, such as the time evolution of concentration gradients due to diffusion.
- Integration of the diffusion equation (Fick's second law) allows us to calculate the change in concentration with time.
- Viscosity is the resistance to movement of molecules in solutions.
- The diffusion constant is related to the mean square displacement of molecules.
- The diffusion constant for a molecule is related to size and shape, and is inversely related to the viscosity of the solution.
- The Stokes–Einstein equation allows us to calculate the diffusion coefficients of molecules.
- Diffusion-limited reaction rate constants can be calculated from the diffusion constants of molecules.
- Concentration gradients play important roles in biology, including controlling aspects of development.
- For some biological processes, diffusion in one dimension (for example, proteins along DNA) or in two dimensions (for example, proteins anchored in membranes) is also important.
- Cells use motor proteins to transport cargo over long distances and to specific locations.

### C. EXPERIMENTAL MEASUREMENT OF DIFFUSION

- Diffusion rates can be determined experimentally by dynamic light scattering.
- The velocity of molecules moving during centrifugation is determined by both molecular mass and shape.
- In equilibrium centrifugation, the concentration profile is determined by just molecular mass, so it can be used to determine the oligomerization state when a protein's molecular mass is known from its sequence.
- Electrophoretic mobility is determined by charge and shape.
- Electrophoresis of proteins under denaturing conditions, and in the presence of sodium dodecylsulfate (SDS), can be used to estimate molecular weight.

# Problems

## True/False and Multiple Choice

1. Bacterial movement towards nutrients resembles a biased random walk.

   True/False

2. The concentration gradient of bicoid protein in *Drosophila* embryos is established by:

   a. passive diffusion of protein molecules.
   b. an mRNA gradient.
   c. spatially biased maternal deposition of mRNA.
   d. all of the above.

3. The distance moved in two-dimensional diffusion has a square-root dependence on time ( $\sqrt{t}$ ). In three dimensions, the distance covered through diffusion has a cubed-root dependence on time ( $\sqrt[3]{t}$ ).

   True/False

4. In passive transport with no barriers, if the concentration of molecules is high in one area but low in another area, then

   a. the molecules will have a net flux to the area of low concentration.
   b. the concentration will tend to be equal in both areas after a long period of time.
   c. the molecules will not move unless ATP is added.
   d. both a and b.
   e. none of the above.

5. A spherical virus with a radius of 30 nm will move more slowly by diffusion than a protein dimer with a radius of 5.5 nm.

   True/False

6. The rate of diffusion increases as friction increases.

   True/False

7. Water has a higher viscosity than acetone because

   a. it has a greater molecular mass.
   b. hydrogen bonds allow for the transfer of momentum.
   c. solute flux is higher in acetone.
   d. it is easier to concentrate water than it is acetone.

## Fill in the Blank

8. ________ transport moves molecules through diffusion, while __________ transport drives movement through the coupling to chemical energy.

9. The width of the Gaussian distribution describing a one-dimensional random walk ___________ with an increasing number of time steps.

10. Viscosity is a measure of a fluid's __________ to particles moving through it.

11. An oblate or prolate object will have increased friction relative to a _______ object that occupies the same volume.

12. The velocity at which friction balances applied force is called the ____________.

## Quantitative/Essay

13. The protein cyclophilin is a monomer with a diffusion constant of $1.2 \times 10^{-8}$ cm$^2$•sec$^{-1}$ in water at 25°C. Cyclophilin binds HIV capsid and the complex has a diffusion constant of $7 \times 10^{-9}$ cm$^2$•sec$^{-1}$. What is the difference in the typical (r.m.s.) distance that cyclophilin alone will travel versus the cyclophilin-capsid complex in 10 seconds?

14. There are approximately 250,000,000 hemoglobin molecules per red blood cell. How many collisions will an oxygen molecule ($O_2$; mass of 32 amu = $5.3 \times 10^{-26}$ kg) of radius 0.21 nm have with hemoglobin of radius 3 nm in a blood cell of volume $10^{-16}$ m$^3$ in 1 sec at 25°C? Assume both molecules are spheres, and that the viscosity is that of pure water.

15. What is the typical (r.m.s.) distance traveled by a spherical HIV virus (radius 120 nm) in 1 hour through the bloodstream (viscosity = 3 cP) at 37°C?

16. Why is the diffusion-limited rate of collisions essentially constant at $10^{10}$ M$^{-1}$•sec$^{-1}$—that is, why is it essentially independent of the size of the molecules involved?

17. What physical and chemical properties of DNA and the lac repressor account for the observation that the association rate is faster than the three-dimensional diffusion limit?

18. Below are traces from dynamic light scattering experiments over the same length of time using purified wild-type and mutant kinase protein. The wild-type protein exists primarily as a dimer, whereas the mutant protein is primarily a monomer. Explain which trace came from which protein.

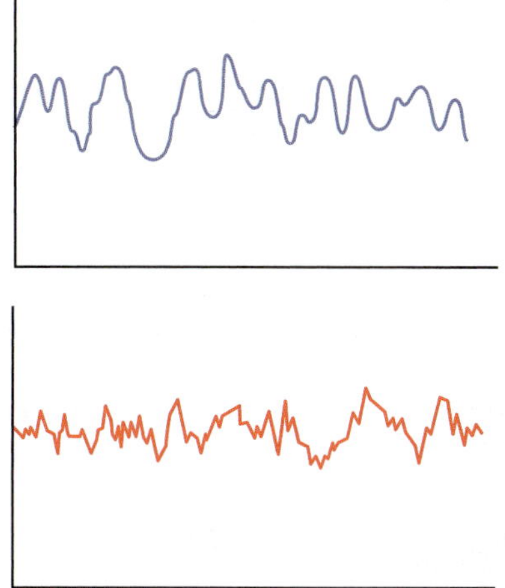

19. The same proteins from the previous problem are analyzed by equilibrium ultracentrifugation. The logarithm of the absorbance is plotted versus squared distance from the top of the measurement cell at 10,000 rpm for each protein.

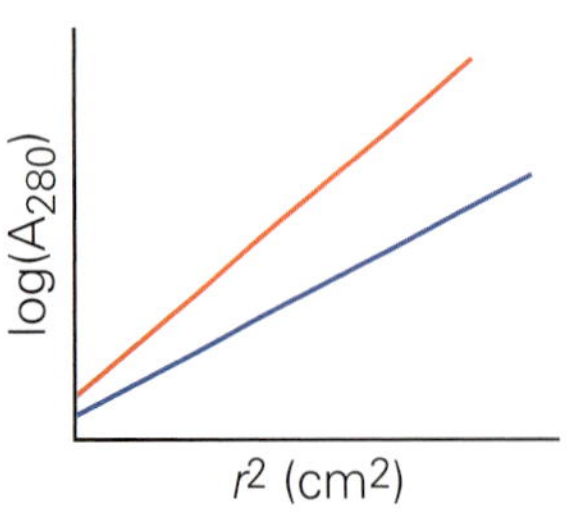


a. Explain which sample is the mutant and which is the wild type.
b. Is there monomer–dimer exchange in either the wild-type or mutant kinase?

20. Sedimentation coefficients (in Svedberg units) are often non-additive for macromolecular complexes. For example, the assembled ribosome and proteosome each have lower total sedimentation coefficients than one might expect given the constituents. How might a macromolecular assembly have a *higher* sedimentation coefficient than the sum of its subunits?

21. Why is it necessary to add agarose or polyacrylamide for electrophoresis experiments?

22. Why do proteins, but not nucleic acids, need to be covered in SDS to estimate the mass by gel electrophoresis?

23. Dynein is a cytoplasmic motor similar to kinesin, but it travels along microtubules in the opposite direction. A single dynein transports a vesicle 0.6 μm along an axon in 5 sec. Dynein steps use one cycle of ATP hydrolysis that move it 80 Å along a microtubule filament. Assuming all steps are forward along one filament, what is the ATP hydrolysis rate of dynein?

24. How much resistive force does a 50-nm vesicle experience if it is transported by dynein at 1 $\mu m \cdot sec^{-1}$ in the cytoplasm ($\eta = 0.2\ g \cdot cm^{-1} \cdot sec^{-1}$)?

## References


### A. Random walks

Adler J (1966) Chemotaxis in bacteria. *Science* 153, 708–716.

Berg HC (2004) *E. coli* in Motion. New York: Springer.

Berg HC (1993) Random Walks in Biology. Princeton, NJ: Princeton University Press.

### B. Macroscopic description of diffusion

Collins FC & Kimball GE (1949) Diffusion-controlled reaction rates. *J. Colloid Sci.* 4, 425–437.

Elf J, Li G-W & Xie XS (2007) Probing transcription factor dynamics at the single-molecule level in a living cell. *Science* 316, 1191–1194.

Elowitz MB, Surette MG, Wolf PE, Stock JB & Leibler S (1999) Protein mobility in the cytoplasm of *Escherichia coli*. *J. Bacteriol.* 181, 197–203.

Halford SE & Marko JF (2004) How do site-specific DNA-binding proteins find their targets? *Nucleic Acids Res.* 32, 3040–3052.

Kampmann M (2005) Facilitated diffusion in chromatin lattices: Mechanistic diversity and regulatory potential. *Mol. Micro*. 57, 889–899.

Kholodenko BN, Hoek JB & Westerhoff HV (2000) Why cytoplasmic signalling proteins should be recruited to cell membranes. *Trends Cell Biol.* 10, 173–178.

Spirov A, Fahmy K, Schneider M, Frei E, Noll M & Baumgartner S (2009) Formation of the bicoid morphogen gradient: An mRNA gradient dictates the protein gradient. *Development* 136, 605–614.

Vale RD & Milligan RA (2000) The way things move: Looking under the hood of molecular motors. *Science* 288, 88–95.

von Hippel PH & Berg OG (1989) Facilitated target location in biological systems. *J. Biol. Chem.* 264, 675–678.

### C. Experimental measurement of diffusion

Berne BJ & Pecora R (2000) Dynamic light scattering: with application to chemistry, biology, and physics. Mineola, NY: Dover.

Lebowitz J, Lewis MS & Schuck P (2002) Modern analytical ultracentrifugation in protein science: A tutorial review. *Protein Sci.* 11, 2067–2079.

# PART VI

## ASSEMBLY AND ACTIVITY

# CHAPTER 18

# Folding

Living cells continually transform the information stored in their genes into the architectural and mechanical assemblies and the enzymatic machinery needed to sustain life. As described in Chapter 1, the genetic code relates the sequence of nucleotide triplets, the codons, to the sequence of amino acid residues in the protein. With no information other than that implicit in the sequence of amino acids, proteins fold spontaneously into specific three-dimensional structures. In the first part of this chapter, we shall study how protein molecules fold.

Although many proteins can be folded and unfolded reversibly in a test tube, a serious problem that occurs with most proteins is that they may aggregate with other proteins during the process of folding. In the second part of this chapter, we discuss a class of proteins known as molecular chaperones, which are specialized to help other proteins fold properly.

RNA molecules that do not code for proteins play many different roles in the cell. The ability of RNA to function in these different processes derives in part from its ability to base pair with specific target sequences, but also from its capacity to form elaborate, folded three-dimensional structures. In the last part of this chapter, we briefly review a few aspects of how RNAs fold. The process of RNA folding has some fundamental differences from that of protein folding. The hydrophobic effect, which is the dominant driving force for protein folding (**Figure 18.1**A), is

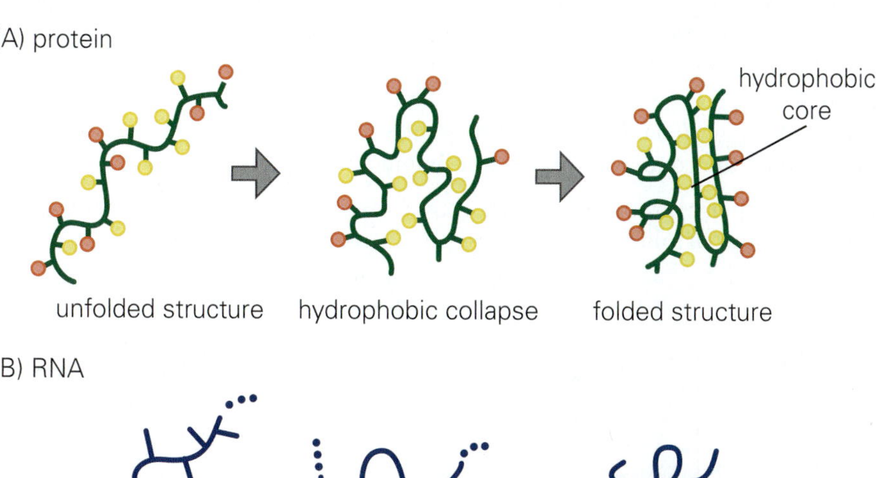


**Figure 18.1 Protein and RNA folding.** (A) Protein folding is driven by the hydrophobic effect and is very cooperative. In this schematic diagram, the protein backbone is in *green*, with hydrophobic and polar sidechains in *yellow* and *red*, respectively. (B) RNA folding is more hierarchical, with stable secondary structure formed first. Metal ions play an important role in RNA folding, but they are generally not very important in protein folding.

not so important in RNA folding. Instead, metal ions are crucial because they neutralize the charge of the negatively charged phosphate groups, allowing compact tertiary structures to form (Figure 18.1B).

# A. HOW PROTEINS FOLD

As we have discussed in Section 4.3 and part D of Chapter 10, the folding of water-soluble proteins is driven by the hydrophobic effect. The unfolded polypeptide chain collapses and becomes compact so as to exclude hydrophobic sidechains from water, creating a hydrophobic core (Figure 18.1A). The formation of secondary structural elements (α helices and β sheets) helps give the protein a defined three-dimensional shape. All of this can occur spontaneously, with the final structure determined by the amino acid sequence of the protein (see Chapter 5).

In this part of the chapter, we discuss aspects of the experimental study of the folding of water-soluble proteins that have well defined structures. We do not discuss membrane proteins or proteins that are intrinsically unstructured.

## 18.1 Protein folding is governed by thermodynamics

The idea that polypeptide chains have all of the information necessary to fold, described as the **thermodynamic hypothesis**, was introduced in Section 5.1. According to the thermodynamic hypothesis, the native structures of proteins correspond to conformations that are at a minimum in the free energy for the protein chain in water, even for very complicated protein folds (**Figure 18.2**). This is a critical feature, because if a specific assembly machine were necessary to fold each different protein, then cells would become impossibly complicated.

Anfinsen's experiment with ribonuclease-A, described in Chapter 5 (see Figure 5.3), showed that all of the information necessary to fold to the native state is present in the polypeptide chain alone. Since that time, many other proteins have been studied experimentally and shown to be capable of recovering their native structures after being unfolded.

An example of protein unfolding and refolding with temperature changes is shown in **Figure 18.3**. These data are for ribonuclease-H (an enzyme that cleaves the RNA strand of RNA-DNA hybrids). When the protein solution is heated, the value of $\Delta G^{\circ}_{folding}$ becomes less negative (see Figure 10.30), finally reaching a value of zero ($\Delta G^{\circ}_{folding} = 0$) at the melting temperature, $T_m$, which is ~53°C for ribonuclease-H under the conditions of the experiment illustrated in Figure 18.3.

For a monomeric protein, the equilibrium constant, $K_{folding}$, gives the ratio of the concentrations of folded and unfolded protein. At $T_m$:

$$K_{folding}(T_m) = \frac{[\text{folded}]}{[\text{unfolded}]} = 1 \tag{18.1}$$

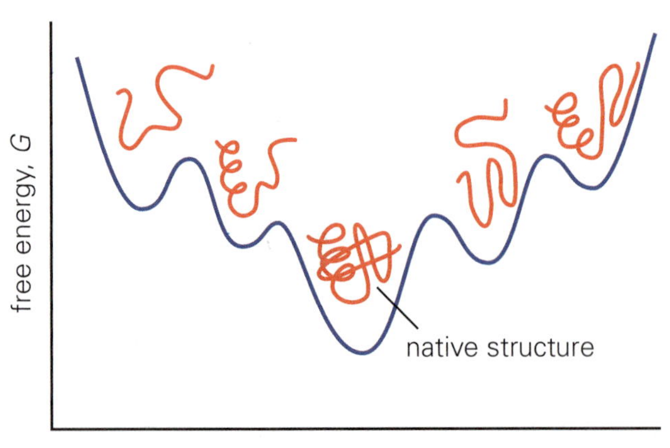


**Figure 18.2 Protein folding is governed by thermodynamics.** The thermodynamic hypothesis states that the native structure of a protein corresponds to a minimum in the free energy of the protein.

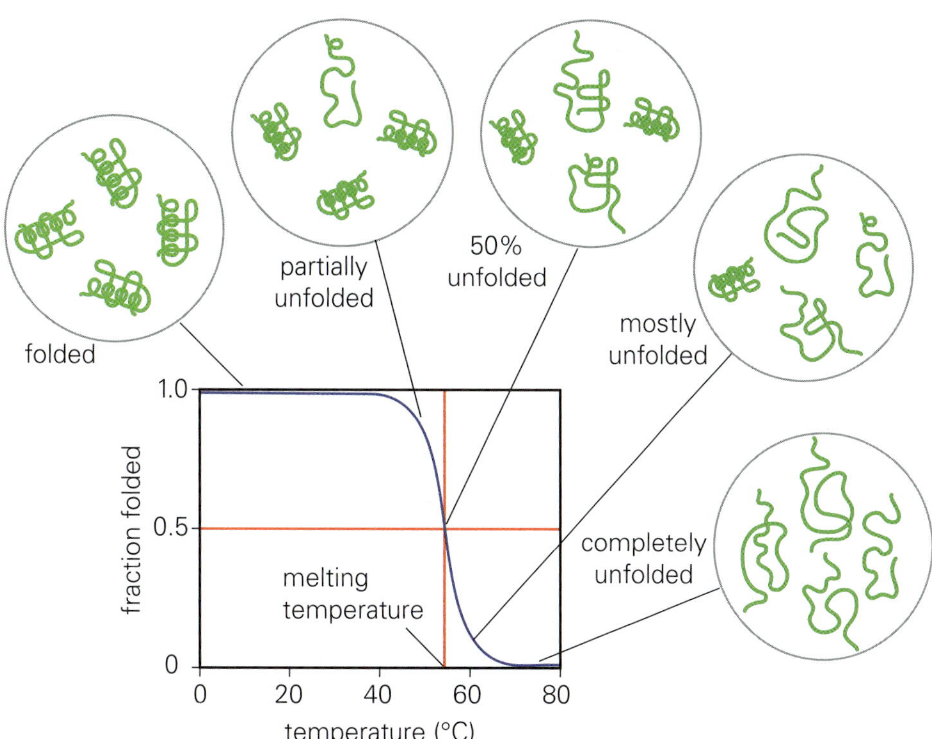


**Figure 18.3 Equilibrium unfolding curve for ribonuclease-H.** The curve shows the fraction of ribonuclease-H molecules that are folded as a function of temperature. The melting temperature (the midpoint of the unfolding curve where half of the protein is unfolded) is at ~53°C. (Adapted from J.M. Dabora and S. Marqusee, *Prot. Sci.* 3: 1401–1408, 1994.)

and so equal amounts of folded and unfolded protein are present at the melting temperature (review the discussion in part D of Chapter 10). Spectroscopic measurements can be used to follow changes in the concentrations of folded and unfolded protein present in a sample, from which the fraction folded (shown in Figure 18.3) can be calculated.

For many proteins, such as ribonuclease-H, unfolding and folding are reversible. After denaturation by heat, the protein refolds to the original conformation when the solution is cooled. The same effect can be achieved by the addition and removal of a denaturant, such as urea or guanidine. This is, as discussed in Chapter 5, the basis for Anfinsen's conclusion that all of the information needed for folding is contained in the primary amino acid sequence of a protein.

## 18.2 The reversibility of protein folding can also be demonstrated by manipulating single molecules

A particularly vivid way to demonstrate the reversibility of protein folding is to manipulate a single protein molecule and to unfold it by applying a pulling force across the two ends of the polypeptide chain. This kind of experiment has been made possible by the development of **optical tweezers**, which are beams of laser light that can be used to move microscopic objects with remarkable precision.

In order to pull on a single protein molecule by its two ends, the polypeptide chain is fused to a DNA molecule at each end. This is done is by introducing cysteine residues at the N- and C-termini of the protein chain. These cysteine residues react with thiol groups that are introduced artificially at the ends of the DNA molecules, thereby cross-linking the protein to the DNA by disulfide bonds. The two DNA molecules act as "handles" for manipulation of the protein. One handle is attached to a bead held on a small pipette, the other is attached to a bead that is captured by the laser beam of a molecular tweezer (**Figure 18.4**).

By moving the pipette, a pulling force is applied to the protein. Measuring the movement of the bead in the optical trap makes it possible to measure the force being applied. If the force is increased gradually, then the handles stretch, as

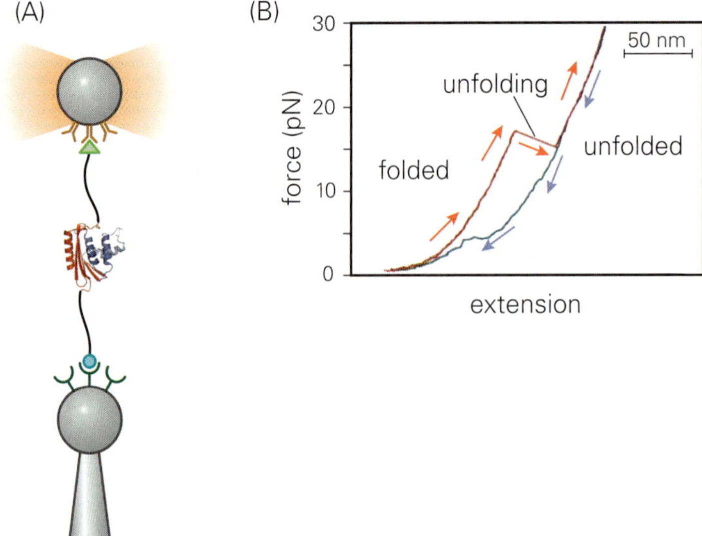


**Figure 18.4 Reversible unfolding of a protein by pulling.** (A) Two DNA handles (*black* lines) are linked to the protein, and the handles are bound to two beads. The lower bead is attached to a pipette and the the upper bead is held within an optical trap (*orange*). The optical trap allows the force being applied to be monitored very accurately. (B) Force exerted on the bead as a function of distance. *Red* trace, extension starting with folded protein. *Blue* trace, relaxation starting with unfolded protein. (Adapted from C. Cecconi et al. and S. Marqusee, *Science* 309: 2057–2060, 2005.)

shown by the first part of the red trace in Figure 18.4B. When the force becomes sufficiently high, the protein suddenly unfolds, allowing the distance between the beads to increase with no increase in force. If this is done very, very slowly (and, thus, reversibly in the thermodynamic sense), then the amount of work done by pulling to unfold the protein corresponds to the free energy of unfolding (see Section 9.14 for a discussion of the relationship between free energy and work).

If the beads are moved back together to remove the tension from the protein, it refolds (*blue trace* in Figure 18.4B). The process can be repeated over and over again with the same protein molecule, and the resulting traces are superimposable. The reversibility of the unfolding–folding cycles indicates that this individual protein molecule has all of the information needed to fold.

## 18.3 Unfolded states of proteins correspond to wide distributions of different conformations

Before we discuss the process of protein folding, it is important to understand what the unfolded protein is like. Most denatured proteins lack any significant regions of regular secondary or tertiary structure, with the backbone angles ($\phi$, $\psi$) for each residue taking on random values within the allowed regions in a Ramachandran diagram (see Figure 4.20). Sidechain dihedral angles are also highly variable. Because there are so many variable dihedral angles in even a small protein, essentially no two unfolded protein molecules will have exactly the same conformation at any given time.

To say a protein is in an unfolded "state" really means that the molecules populate an *ensemble* of unfolded conformations, commonly referred to as **random coil** conformations. Proteins fluctuate among these conformations very rapidly (as we shall see in following sections). The measured properties of unfolded proteins are averages of the whole ensemble.

**Random coil**

A random coil state of a polypeptide is one that lacks any well-defined secondary and tertiary structure. In such states polypeptides rapidly fluctuate between different, energetically equivalent conformations.

The fully folded states of proteins are very compact, with the interiors packed essentially as densely as possible. The unfolded states are substantially more expanded. One measure of the "size" for an unfolded state is the end-to-end distance, from the N-terminal amino acid to the C-terminal amino acid. A more practical measure of size is the **radius of gyration**, which can be obtained using a technique known as **small-angle x-ray scattering** (SAXS). SAXS, or solution x-ray scattering, involves measuring the scattering of x-rays by macromolecules in solution. Solution x-ray scattering data can be used to derive the size of the macromolecules (that is, the radius of gyration) as well their shapes.

The radius of gyration, $R_g$, is the r.m.s. displacement of the atoms in a macromolecule from the centroid. If the coordinates of the $i^{th}$ atom in a molecule with $N$ atoms is denoted by $\vec{r}_i$, then the centroid, $\vec{c}$, is defined as the average coordinate:

$$\vec{c} = \frac{1}{N}\sum_{i=1}^{N}\vec{r}_i \tag{18.2}$$

The radius of gyration, $R_g$, is given by:

$$R_g^2 = \frac{1}{N}\sum_{i=1}^{N}(\vec{r}_i - \vec{c})^2 \tag{18.3}$$

Each different unfolded conformer has a different size and, hence, it is useful to describe the distribution of interatomic distances, $P(r)$, which is related to the distribution of scattering density derived from solution x-ray scattering, shown in **Figure 18.5**.

To understand the nature of this distribution, we can think of vectors connecting successive amino acids in the protein chain as steps in a random walk with uniform step size, as discussed in Section 17.8. In this model, the end-to-end distance is the length of a random walk of $N$ steps in three dimensions, starting at the origin (N-terminus), with $N$ being the number of amino acids minus one. The end-to-end distance is then the displacement from the origin generated by the walk. The distribution of distances will be a Gaussian function, with a width related to the square root of the number of amino acids (see Section 17.4). The width of the distribution is also directly related to the radius of gyration. A more exact analysis, in which disallowed conformations (that is, ones in which two amino acids would occupy the same site) are excluded, gives a dependence on chain length of $N^{0.588}$ rather than $N^{0.5}$. This average end-to-end distance is much less than the length of the maximally extended chain, although such an extended chain is in principle a member of the conformational distribution. The agreement of experiment and the more exact theory is extremely good for proteins unfolded in denaturant, as shown in **Figure 18.6**.

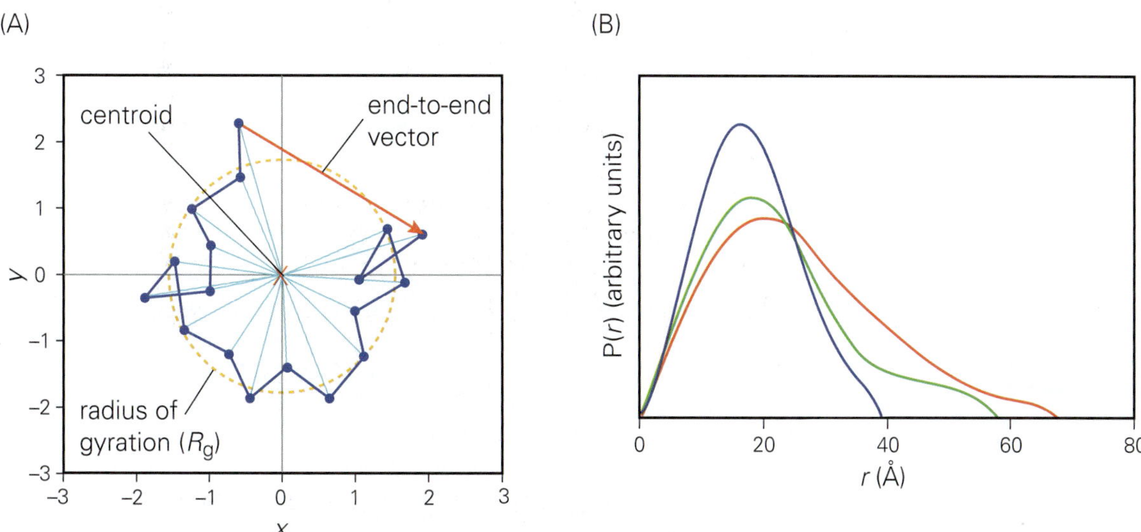


**Figure 18.5 Radius of gyration and radial distribution of interatomic distances in unfolded proteins.** (A) The radius of gyration, $R_g$, defines an average size of a molecule. (B) A graph showing the distribution of distances, P(r), between pairs of atoms in cytochrome *c* for the native state (*blue*) and the averages for partially folded intermediates (early in folding, *red;* later in folding, *green*) as measured by solution x-ray scattering. As expected, the distribution is wider in the intermediate states, but much less than that of the fully extended chain (not shown). (B, adapted from S. Akiyama et al. and T. Fujisawa, *Proc. Natl. Acad. Sci. USA* 99: 1329–1334, 2002. With permission from the National Academy of Sciences.)

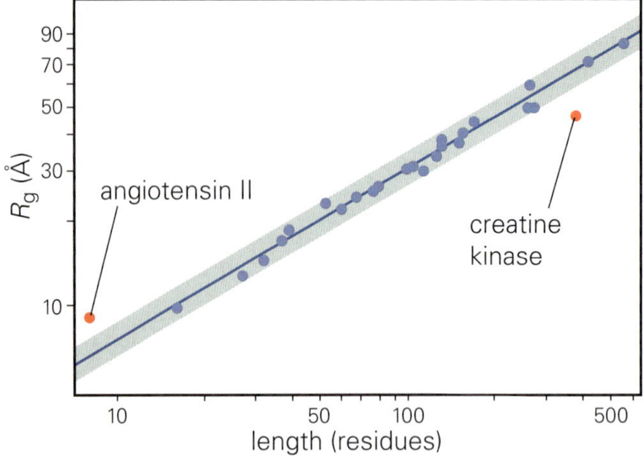


**Figure 18.6 Dependence of the radius of gyration, $R_g$, on the lengths of unfolded proteins.** The values of $R_g$ derived from solution x-ray scattering data for several denatured proteins is shown as a function of the length of the protein. Both axes are logarithmic. The straight line is a graph of the equation $R_g = R_0 N^{0.58}$, where $N$ is the length of the protein and $R_0$ is the intercept. Notice that most values of $R_g$ fall close to this line, indicating that the unfolded structures adopt random coil structures. Two outliers are shown in *red*. (Adapted from J.E. Kohn et al. K.W. Plaxco, *Proc. Natl. Acad. Sci. USA* 34: 12491–12496, 2004. With permission from the National Academy of Sciences.)

## 18.4 Protein folding cannot be explained by an exhaustive search of conformational space

Protein folding occurs as a series of conformational changes that connect the initial unfolded states to the final native state (see Figure 4.11). Conformations of unfolded proteins interconvert by rotations about single bonds. The energy barriers for such rotations are very low, and hence the rates of exchange are high (on the order of $10^{10}$ sec$^{-1}$ per rotation about a bond). Cyrus Levinthal pointed out that, in spite of the very rapid exchange among conformations, there are so many possible conformations that folding could not occur by having the protein sample all of them.

For a 100-amino-acid protein, assuming only 3 conformations allowed per amino acid, there would be $3^{100} = 5 \times 10^{47}$ unfolded conformations. This number of conformations is so large that it would take more than the age of the universe to sample them all. Hence, protein folding must occur by some directed process. Since the native conformation of each protein is specified by its amino acid sequence, the instructions for folding must also be encoded in the sequence, as noted in the introduction to this chapter. The rates of protein folding vary widely for different proteins of the same size, indicating that the amino acid sequence has important effects on the kinetics of folding as well on the final structure.

## 18.5 Many small proteins populate only fully unfolded and fully folded states

A variety of biophysical studies have shown that the folding of many small proteins seems to be **two-state**—that is, only fully unfolded and completely folded states are occupied with any significant population; there is no buildup of intermediates as folding occurs. Under native conditions, where the reverse reaction can be ignored, the folding of two-state proteins follows simple first-order kinetics described by a single rate constant, $k_f$ (**Figure 18.7**). This means that unfolded protein disappears exponentially with a time constant $\tau = 1/k_f$ (review the discussion of first-order processes in Section 15.7). There have been measurements of folding rates for many small proteins (fewer than 150 amino acids), with the finding that the rates differ by over five orders of magnitude—the time constants ranging from less than 10 μsec for the protein that folds the fastest to more than 1 second for the slower ones.

**Two-state folding**

Two-state folding means that only the unfolded protein and the fully folded protein occur at any significant concentration. Two-state folding kinetics can be fully described with a single rate constant.

What might be the limit to how fast a protein can fold? This issue has been addressed both theoretically and experimentally to show that the maximum rate constant for protein folding is $k_f \approx 100/N$ μsec$^{-1}$, where $N$ is the number of residues in the protein. For a protein of 100 amino acids, this means that the *minimum* time constant for folding should be about 1 μsec, only slightly less than the

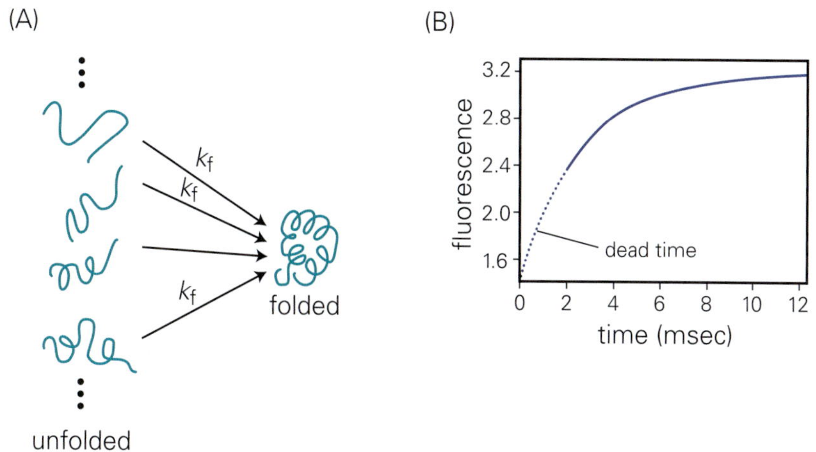


**Figure 18.7 Two-state folding of a protein.** (A) Schematic representation of the folding of a small protein. Each unfolded conformation of the protein converts to the folded form in a single step with a rate constant $k_f$. (B) The folding of a domain of λ repressor. The fluorescence of a tryptophan sidechain in the protein changes as the protein folds. The change in fluorescence is described by a single exponential, indicating a single rate process with a rate constant of ~500 $sec^{-1}$. The dotted line between 0 and 2 milliseconds indicates the "dead time" of the mixing required to initiate folding, during which data cannot be measured. (B, adapted from S. Ghaemmaghami et al. and T.G. Oas, *Biochemistry* 37: 9179–9185, 1998. With permission from the American Chemical Society.)

3 μsec fastest time experimentally observed. This "speed limit" is a consequence of the diffusive character of the motions of the protein chain—that is, small random rotations about single bonds change the relative positions of different segments of the protein as though the segments were diffusing. Just as collisions of molecules in solution cannot occur faster than allowed by diffusion, the collapse of the unfolded protein chain cannot happen arbitrarily fast, even when there is no specific free-energy barrier for the folding process.

## 18.6 The order in which secondary and tertiary interactions form can vary in different proteins

We know that the native states of proteins usually contain substantial amounts of secondary structure, and it seems reasonable to think that such structure would form early in the folding process. It is indeed the case that all types of local structure—irregular loops, helices, and β-hairpins—intrinsically *can* form fast relative to the time for the folding of the entire protein. To say that local structure *can form* fast means that the rate of local conformational changes is not rate limiting in this process; it does not mean that local structure actually *does form* fast.

This distinction is important because the actual rate of formation of secondary structural elements is sequence dependent. On the whole, the more stable the local structural element, the more rapidly it is formed. This is reasonable in the sense that formation of energetically favorable interactions offsets the loss of entropy in more folded states, making it less likely that partial structure formed through conformational fluctuations will be broken before the folding process is complete.

Consistent with this view, proteins have a range of folding behaviors that are distinguished by the speed of secondary structure formation. At one limit are proteins in which the secondary structure elements form quickly, which then assemble to form the native protein in the rate-limiting step. This is called a **diffusion collision** process. At the other end of behavior are proteins in which various distant segments of the polypeptide chain come together without forming secondary structure, stabilized instead by sidechain interactions, and which then reorganize locally to form the native secondary and tertiary structure; this is called **nucleation condensation**.

These two extreme models for how proteins fold are illustrated schematically in **Figure 18.8**. Proteins with more stable local structures, which may be single secondary structure elements or small groups of these, show folding patterns closer to the diffusion collision model. The transition state in this case is an ensemble of structures in which the secondary structural elements are formed, but which do not have the final compact tertiary structure. α-helical proteins tend to fold in

**Diffusion collision**

In the diffusion collision model for protein folding, secondary structural elements form first and undergo collisions with other secondary structure elements to form the native structure.

**Nucleation condensation**

In the nucleation condensation model for protein folding, a general compaction of the protein chain occurs first, leading to the nucleation of the native secondary and tertiary structure.

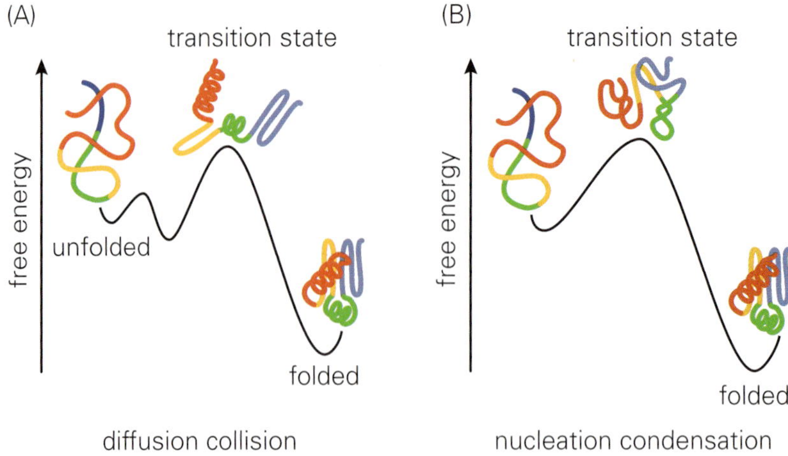


**Figure 18.8 Two models for how proteins fold.** Schematic representations of a diffusion collision process (A) and a nucleation condensation process (B). In the diffusion collision model, secondary structure forms first, and the rate-limiting process is the organization of the secondary structural elements into the fully folded structure. In the nucleation condensation model, the protein chain collapses into a compact form, followed by full formation of secondary structure. (Adapted from Y. Ivarsson et al. and S. Gianni, *Eur. Biophys. J.* 37: 721–728, 2008. With permission from Springer Science + Business Media.)

this way. Proteins that contain β sheets tend to fold differently, with the transition state corresponding to structures in which the chain has collapsed into a more compact form, prior to the formation of the final secondary and tertiary structure.

## 18.7 Folding rates are faster when residues close in sequence end up close together in the folded structure

The overall stability of a protein (that is, the total reduction in free energy upon folding) is not a good indicator of the rate at which the protein folds. Although helices tend to fold faster than β-sheet structures, the secondary structure of a protein also correlates poorly with its folding rate. A parameter that does correlate well (although not perfectly) with folding rate is the **contact order** of the structure. Contact order is a measure of the distance in the primary sequence between segments of the polypeptide chain that are brought into contact in the final folded structure (**Figure 18.9**). Hairpin structures (two helices or β strands with a short spacer between) have low contact order, while an N-terminal helix or strand packed with a C-terminal one will have maximal contact order.

**Contact order**

The contact order of a protein is a measure of how far apart in primary sequence are residues that make contact in the native folded structure, normalized for the length of the protein.

During the process of folding, collisions between nearby segments of the chain are likely to be more frequent than collisions between distant elements. If these local collisions can be productive (that is, lead to native structure), then the overall folding process should be faster than when distant elements are required to come together to stabilize a partially folded intermediate headed to the native structure. Thus, it seems reasonable that contact order should correlate inversely with folding rate, as seen experimentally (**Figure 18.10**). Note that there is considerable scatter in the correlation between contact order and folding rates for different proteins. For one value of the contact order, the folding rate constant can vary by nearly three log units (a factor of 1000). However, proteins with low contact order structures usually fold quickly, and those with high contact order structures usually fold slowly. Although the correlation is imperfect, measures related to the contact order are currently the best simple predictors for the folding rate of a protein that undergoes a two-state folding transition.

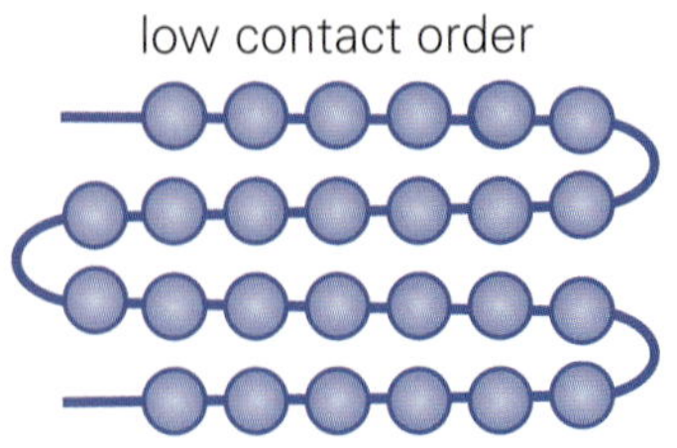


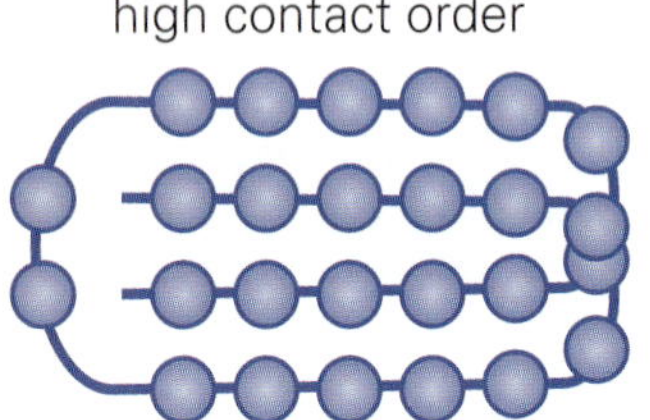


**Figure 18.9 Contact order of protein structures.** In structures with low contact order (top), residues that are close together in sequence are close together in the folded structure. Structures with high contact order (bottom) have residues that are spatially close but far apart in sequence. (Adapted from D. Baker, *Nature* 405: 39–42, 2000. With permission from Macmillan Publishers Ltd.)

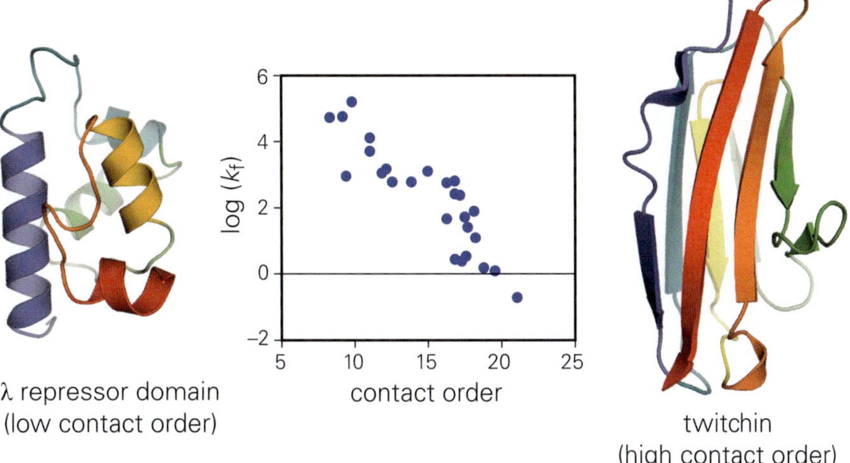


**Figure 18.10 The inverse relationship between contact order and folding rate.** The graph shows the logarithm of the folding rate ($k_f$) as a function of contact order. The structures shown on either side have low (left) and high (right) contact order. The sequence is color coded with the N-terminus *blue* and C-terminus *red*, following the order of colors in the rainbow between. (Adapted from K.W. Plaxco et al. and D. Baker, *Biochemistry* 39: 11177–11183, 2000. With permission from the American Chemical Society; PDB codes: 1LMB and 1WIT.)

## 18.8 The folding of some proteins involves the formation of transiently stable intermediates

The discussion so far has centered on proteins that fold in a two-state manner—that is, only unfolded and fully folded states are populated significantly (see Section 18.5). Most proteins, particularly larger ones, do not exhibit such simple folding behavior. Instead, folding involves intermediates, which are partially folded structures that are stable, although only transiently. The presence of such partially folded intermediates indicates that there is a local minimum in free energy along the folding pathway, with a barrier that must be overcome before folding is completed (**Figure 18.11**).

The formation of intermediates in folding can be detected by various techniques, including mass spectroscopy and NMR. The application of either technique to folding usually relies on a phenomenon known as **hydrogen exchange**. Certain groups in the protein, such as the backbone amide group, can readily exchange protons with the solvent (**Figure 18.12**). One way to follow folding is to investigate whether amide hydrogens are involved in hydrogen bonds, by determining how

**Hydrogen exchange**

Protons attached to certain groups, such as the backbone amide group, can readily exchange with protons in water. The rate of exchange slows down when these protons are involved in hydrogen bonds in folded structures. In this way, identification of segments of the protein that have slowly exchanging amides indicates which regions are likely to have secondary structure.

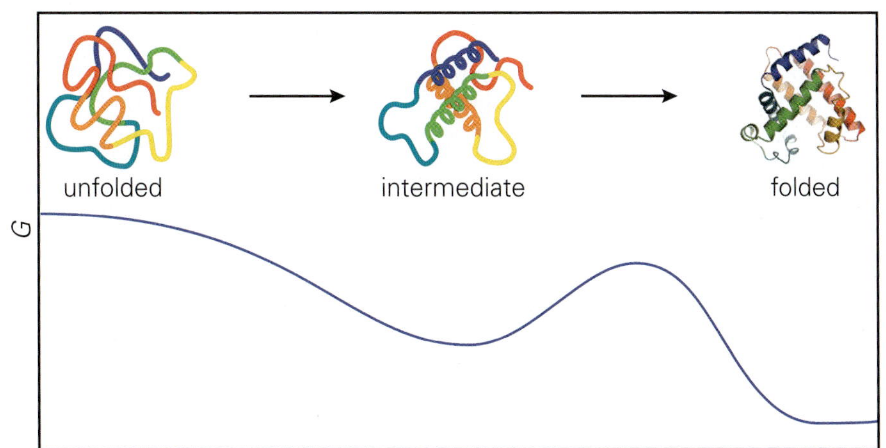


**Figure 18.11 A protein folding intermediate.** A schematic representation of a protein folding process, showing the change in free energy as a function of the extent of folding. The folding proceeds through an intermediate, which corresponds to a local minimum in the free energy.

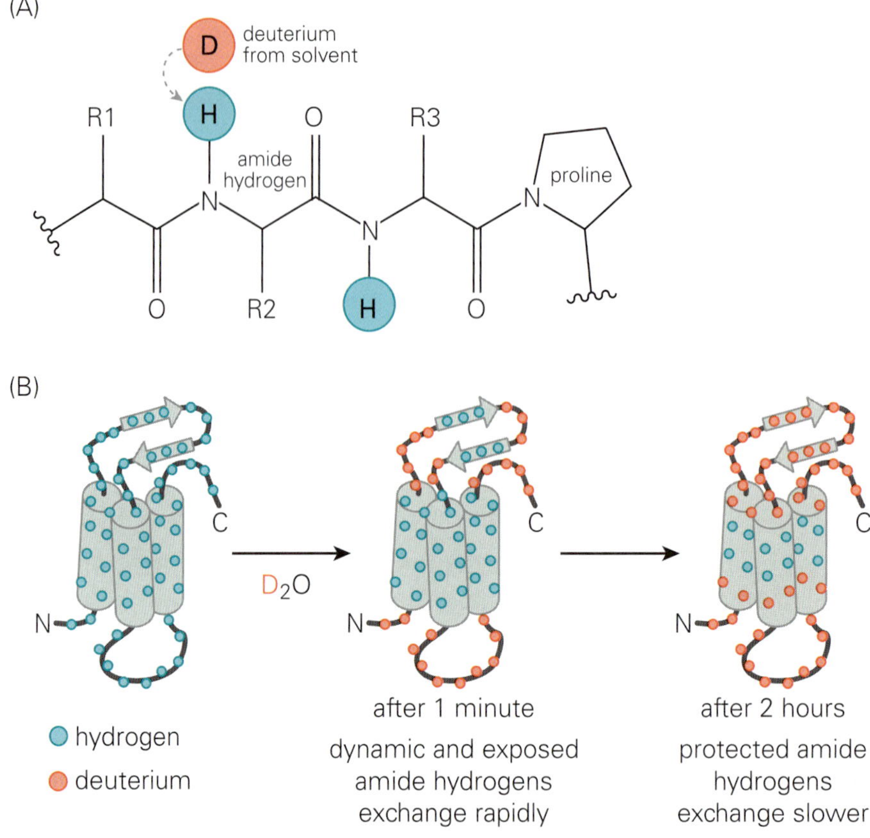


**Figure 18.12 The hydrogen exchange technique**. (A) The amide groups of all residues, except proline, can exchange protons (*blue*) with water. If the protein is transferred to heavy water, then the protons are replaced with deuterium (*red*), which is heavier. (B) The rate at which the amide protons exchange depends on the local environment. If the backbone forms a hydrogen bond, the rate of replacement by deuterium is slower. (Adapted from S.R. Marcsisin and J.R. Engen, *Anal. Bioanal. Chem.* 397: 967–972, 2010. With permission from Springer Science + Business Media.)

rapidly they can exchange with hydrogens from water. This can be detected by replacing water containing the normal hydrogen isotope ($^1$H with a mass of 1) with heavy water, containing deuterium ($^2$H with a mass of 2), and then "weighing" the protein molecules in a mass spectrometer to see how many normal hydrogens and how many deuteriums are present. If amide protons are involved in hydrogen bonds, and hence are not exposed to solvent, then they will not exchange. The solvent can be switched after letting the protein fold for different lengths of time to follow the successive formation of different secondary structure elements.

The detection of a folding intermediate for a globin using this hydrogen exchange approach is illustrated in **Figure 18.13**. Recall from Section 5.5 that the globins are helical proteins that contain a heme group. The particular globin for which hydrogen exchange data are shown in Figure 18.13 is a monomeric globin known as leghemoglobin, found in the root nodules of plants (the structure of this protein is compared with that of other globins in Figure 5.8). The data in Figure 18.13 are for the globin in the absence of the heme group, which is known as the **apo form** of the globin. The folding of apo-leghemoglobin proceeds through the formation of an intermediate, as shown in Figure 18.13A.

To detect the folding of the intermediate, the protein is first unfolded by transferring it to a solution of urea at high concentration (4.5 M; see Figure 5.4). Folding is initiated by transferring the unfolded protein from the denaturant solution to water. The protein is allowed to fold for a certain amount of time in normal water and is then transferred to heavy water (see Figure 18.13B). After folding is completed, the mass of the protein is determined by mass spectrometry, the results of which are shown in Figure 18.13C.

The rate at which exposed protons exchange with water is very fast compared to the time required for folding. Thus, when the protein is transferred to heavy water directly from the denaturant solution, hydrogen exchange occurs before folding and a maximal amount of deuterium is incorporated into the protein. The

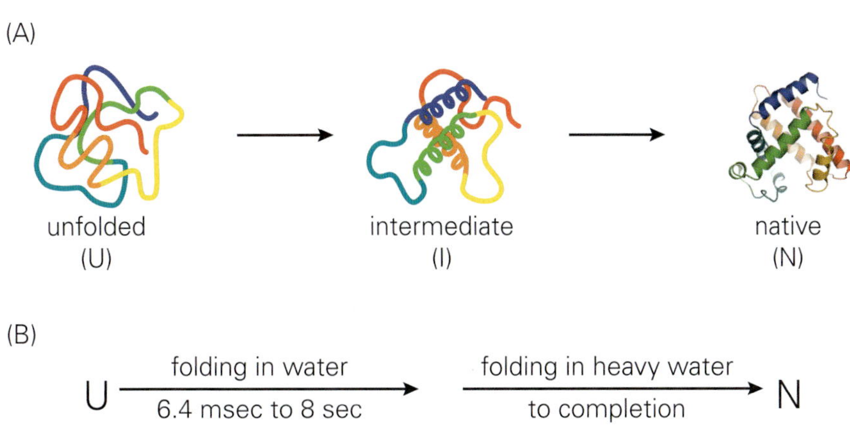


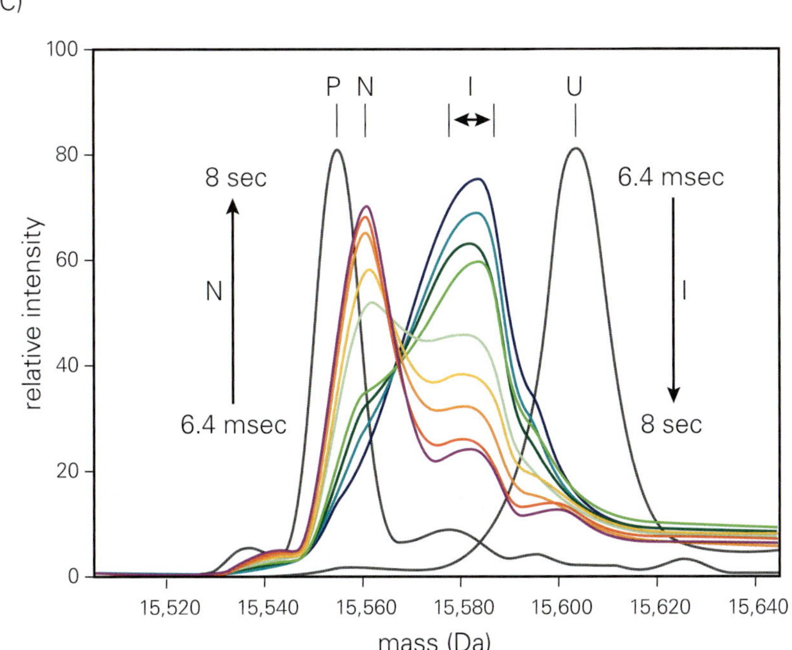


**Figure 18.13 Detection of intermediates in the folding of a globin.** (A) The conversion of the unfolded form of the globin, U, to the native structure, N, occurs through an intermediate, I. (B) Schematic representation of the hydrogen exchange experiment. (C) The graphs show the mass distributions for the globin for various times of folding in water before being transferred to heavy water. The colors indicate the time that folding occurs in water before the transfer to heavy water, with *blue* corresponding to shortest time (6.4 msec) and *purple* the longest time (8 sec). A reference standard, denoted P, is also shown. This is the mass distribution for protein that has never been exposed to deuterium. The broad mass distribution for P reflects the natural abundance of various isotopes. The native structure (N) has dynamic regions in which hydrogens exchange with deuterium to some extent, and so the mass distribution for N is shifted to the right relative to that for P. (Adapted from C. Nishimura et al., and P.E. Wright, *Nat. Struct. Biol.* 7: 679–686, 2000. With permission from Macmillan Publishers Ltd.)

corresponding mass distribution is indicated by unfolded, U, in Figure 18.13C. When the protein is allowed to fold in water for a short time before being transferred to heavy water, some secondary structure is formed, preventing exchange of the amides involved, giving a lighter mass. This is indicated by the *blue graph*, in which the mass distribution is shifted towards lower values than for the unfolded protein (intermediate, I, in Figure 18.13C). For long folding times, all regions of secondary structure are formed, giving maximal protection of amides, indicated by the *red graph* (which is close to that for the native protein, N). Note that in the native structure some amides are not hydrogen bonded, and hence can exchange, giving a mass heavier than the fully protonated form, P, that has never been exposed to heavy water.

The specific amino acids that form the secondary structural elements can be identified by using NMR spectroscopy. This method is much more time consuming than the mass spectrometric analysis shown in Figure 18.13, but it is capable of identifying each individual proton that is exchanged with deuterium. The analysis of hydrogen exchange in myoglobin by NMR showed that segments of the protein that correspond to helices A, G and H, and part of B, in the folded structure have hydrogen bonds formed in the intermediate, whereas segments corresponding to the other helices do not (see Figure 5.1 for the notation used to identify helices in the globin fold).

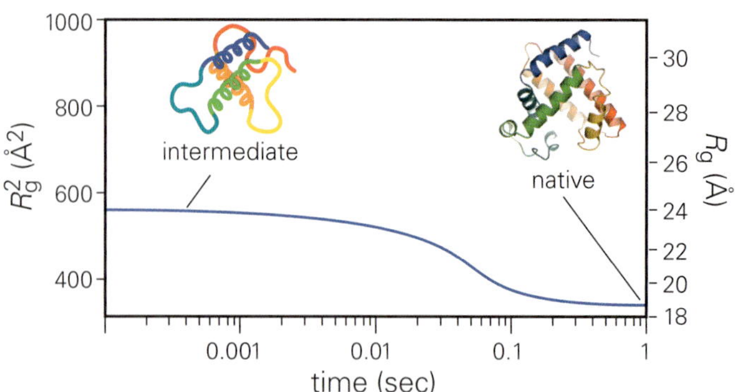


**Figure 18.14 Decrease in the radius of gyration of a globin during folding.** The radius of gyration of the protein (derived from solution x-ray scattering) is shown as a function of folding time. An initial very rapid collapse (with the value of $R_g$ decreasing from ~30 Å to ~24 Å) is not visible in this graph. This is followed by slower collapse to the more compact and fully folded form. (Adapted from T. Uzawa et al. and T. Fujisawa, *Proc. Natl. Acad. Sci. USA* 101: 1171–1176, 2004. With permission from the National Academy of Sciences.)

Further investigation using short peptides corresponding to the individual helices in the native structure showed that the G and H helices are particularly stable, and hence probably form very rapidly and nucleate structure of the intermediate in a diffusion collision mechanism. This intermediate is more compact than the fully unfolded protein, as shown by solution x-ray scattering (**Figure 18.14**), but less so than the native protein because parts of the chain remain in random coil conformations in the intermediate. The other helices form and pack to make the native structure in later steps. All of these steps are relatively rapid and myoglobin is fully folded within a few seconds.

## 18.9 Folding pathways can have multiple intermediates

The folding pathways of many proteins have more than one intermediate. Intestinal fatty acid binding protein, for example, has a structure composed primarily of β sheet and its folding involves the formation of distinct intermediates. The structure is a β sandwich with four- and six-stranded sheets packed against one another and a small helical hairpin (**Figure 18.15**A). All of the strands are parts of β hairpins, so the overall structure has relatively low contact order—that is, folding should occur rapidly (see Section 18.7).

Studies of the folding of the fatty acid binding protein using fluorescence have shown that there are three resolvable kinetic phases in its folding (Figure 18.15B). This reflects the presence of two intermediates, which are shown in Figure 18.15C. The distinct steps in folding correspond to: (1) a rapid collapse around a hydrophobic core (with a rate constant, $k_1 > 10{,}000$ sec$^{-1}$); (2) formation of β strands B–G, mostly in the top sheet ($k_2 \approx 1500$ sec$^{-1}$); and (3) formation of the last three β strands and completion of the hydrogen-bonding network of the backbone ($k_3 \approx 5$ sec$^{-1}$). The time dependence of the second and third steps is apparent in the graph of fluorescence versus time shown in Figure 18.15B. The first step is too fast to be seen in this graph.

The formation of intermediates is a critical step in enabling larger proteins to fold efficiently, but this is also a step at which things can go seriously wrong with the folding process. As we discuss in the next part of this chapter, there are various ways in which proteins end up misfolded. Some instances of misfolding, such as the formation of amyloid plaques, are the cause of several diseases in humans. Intermediates in folding pathways are not very stable, and they can be particularly prone to progression along pathways that lead to aggregates or other misfolded structures.

## 18.10 Changes in the sequence of a protein at certain positions can affect folding rates substantially

Protein folding is a highly cooperative process that involves conformational changes throughout the protein. Nevertheless, certain residues can affect folding rates more strongly than others. To illustrate this concept, we shall discuss some

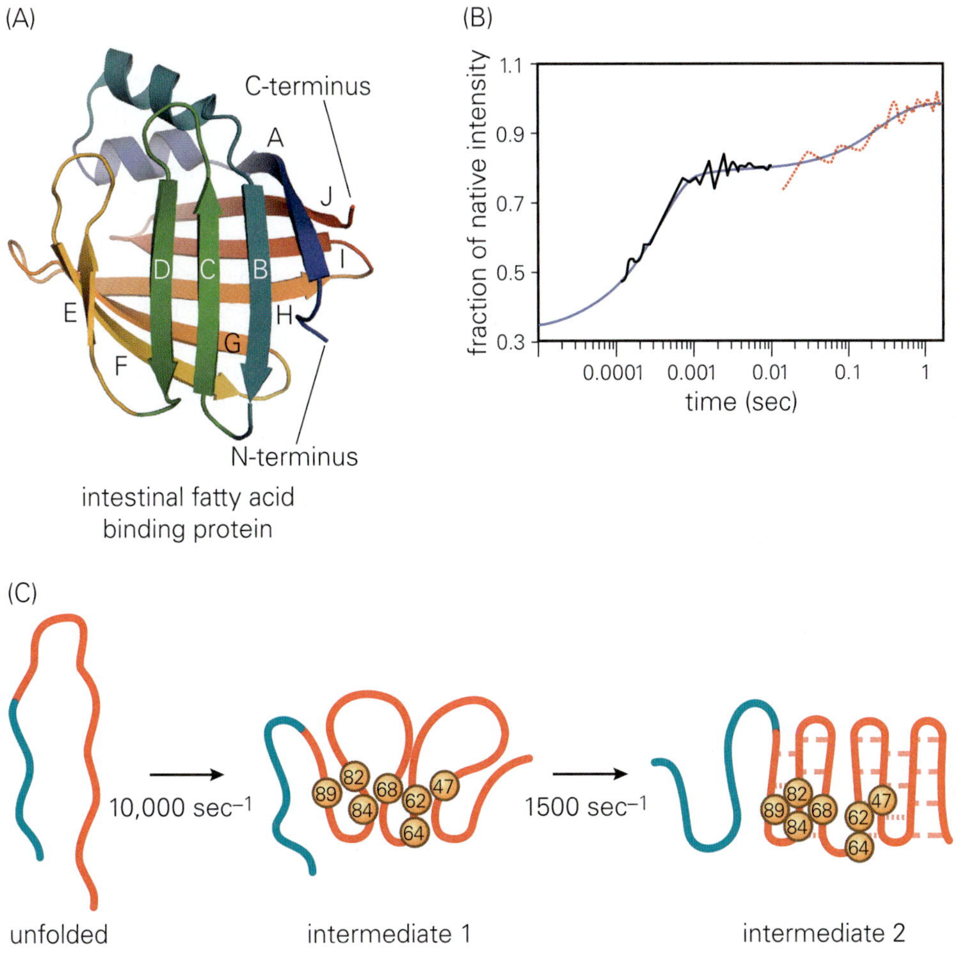

**Figure 18.15 A folding pathway with more than one intermediate**. (A) Structure of intestinal fatty acid binding protein, with the letters assigned to the β strands shown. (B) A graph of fluorescence changes following the second and third phases of folding for this protein. (C) Folding pathway of intestinal fatty acid binding protein. Steps in the folding process are shown schematically. Intermediate 1 is collapsed but has no secondary structure. Residues that are important for this initial collapse are shown as circles with the residue number. Secondary structure is formed in intermediate 2 and then is completed in the final folding step. (B and C, adapted from S.-R. Yeh, I.J. Ropson, and D.L. Rousseau, *Biochemistry* 40: 4205–4210, 2001. With permission from the American Chemical Society; PDB code: 1DC9.)

experiments with the DNA-binding domain of the λ repressor, which is shown in **Figure 18.16** (the λ repressor was described in Section 5.24).

The wild-type λ repressor domain takes about 250 μsec to fold, and it shows two-state folding behavior. One of the helices in the middle of the protein is unusual in that it contains two glycine residues (Gly 46 and Gly 48; see Figure 18.16A). Because glycine is less conformationally restricted than any other amino acid, the presence of glycine is likely to destabilize this helix (see Figure 4.31). Indeed, replacement of these glycine residues with alanine, a residue that strongly favors

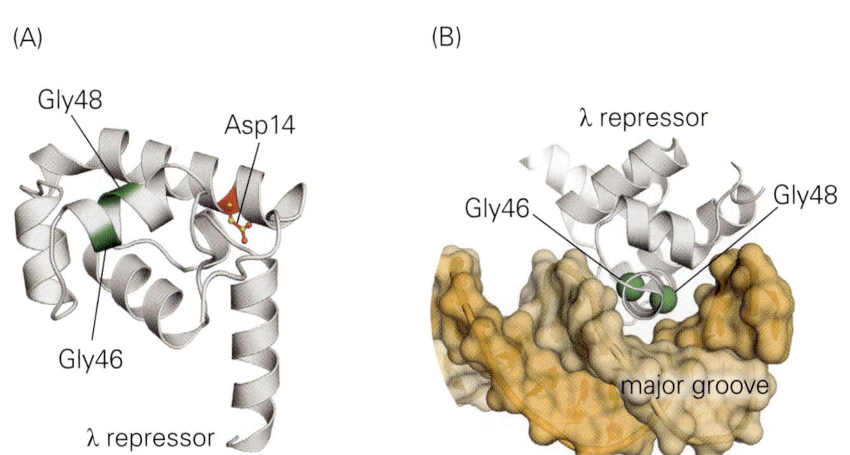


**Figure 18.16 Residues in λ repressor that slow down folding**. (A) The sites of glycine to alanine mutations that accelerate the folding reaction are in *green*. An aspartate residue whose replacement by alanine also accelerates the reaction is in *red*. (B) The structure of the complex of λ repressor with DNA is shown here, with the $C_\alpha$ atoms of Gly 46 and Gly 48 shown as *green spheres*. (A, based on data in R.E. Burton et al. and T.G. Oas, *Nat. Struct. Biol.* 4: 305–310, 1997; PDB code: 1LMB.)

helix formation, accelerates the rate of folding by over 10-fold. Additionally, changing Asp 14 to Ala (removing possible electrostatic interactions that impede folding) further accelerates folding by another 25%. A variant protein in which these three changes are combined folds with a time constant of 18 μsec, compared to 250 μsec for the wild type. The final folded structure is essentially the same for this variant as for the wild type, and so the principal effect of the mutations must be to remove kinetic barriers to folding.

Why does the wild-type protein have a sequence that is less than optimal for folding? The answer becomes apparent when we consider the function of the λ repressor, which is to bind specifically to certain DNA sequences (see Figure 5.37). As shown in Figure 18.16B, the two glycine residues that slow down folding are located at the heart of the interface between the protein and the major groove of DNA. There is not enough room at the interface to accommodate larger sidechains, and so evolution appears to have selected glycine at these two positions so that the helix can enter the major groove in a way that allows the formation of other important contacts. Because the protein must satisfy this functional constraint, evolution may have selected a sequence that is not optimal in terms of the folding rate.

Glycine residues can also help speed up the folding rate, instead of slowing it down, depending on where they are located. In the intestinal fatty acid binding protein, discussed in Section 18.9, there are glycine residues in the turns that connect strands of the β hairpins and the flexibility of glycine is important for the formation of the turn. Gly 121 is located in the turn connecting the last two β strands, and if this glycine is replaced by valine, the overall free energy of folding becomes less favorable and the rate of folding is lowered 1000-fold.

## 18.11 The nature of the transition state can be identified by mapping the effect of mutations on the folding and unfolding rates

Recall from Section 15.29 that the transition state is a point along the reaction trajectory at which the value of the free energy is at a local maximum. For protein folding, the transition state is not one well-defined state, but rather an ensemble of conformations. In the example shown in **Figure 18.17**, the formation of the first

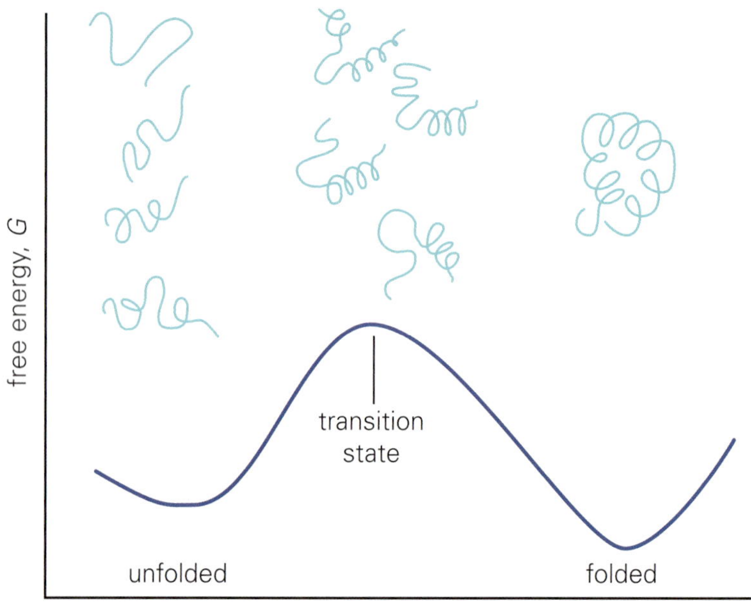


**Figure 18.17 The transition state for a folding reaction**. The free energy of a protein is shown as a function of the progress of the folding reaction. In this particular case, the formation of the first helix is rate limiting. The transition state corresponds to an ensemble of structures in which this helix alone is formed.

α helix in a protein is rate limiting for the folding reaction. The transition state is a collection of structures in which this helix is formed, with the rest of the structure unfolded. Folding is completed rapidly (that is, without a free-energy barrier), once this helix is formed.

The nature of the transition state for folding is distinct from that of intermediates that might occur as folding proceeds. The difference is apparent if you compare Figure 18.17 with Figure 18.11, which depicts an intermediate. Intermediates are conformations for which the free energy is locally at a minimum, and so they are stable to some extent. The transition state, in contrast, is very unstable. Once formed, the reaction proceeds rapidly in either the forward or the reverse direction.

Because any specific molecule exists at its transition state only for a short time during a reaction, transition states are very difficult to observe by direct measurement. To gain some understanding of the nature of the transition state, Alan Fersht and co-workers developed an approach, known as **$\phi$-value analysis**, for identifying the interactions that are important in the transition state for protein folding and unfolding. The method relies on the fact that the rate constant for a reaction is uniquely sensitive to interactions made in the transition state, because the ability to reach the transition state limits the rate of formation of product.

**$\phi$-value analysis**

Different residues in a protein affect the free energy of the transition state for folding to different extents. We can measure the relative contributions of different residues to the transition state by determining a parameter known as the phi–value, which captures the effects of mutations on the folding and unfolding rates (see Equation 18.9).

The approach used by Fersht and co-workers is to introduce mutations into the protein, and to compare the rates of folding and unfolding for the mutant and the wild-type proteins. It is assumed that amino acid residues do not interact in the unfolded state, and so the free energy of the unfolded state is assumed to be the same for wild-type and mutant proteins. The folding and unfolding rate constants, $k_f$ and $k_u$, respectively, for the wild-type and mutated proteins are then determined. These are used to calculate changes in the free energy of the folded and transition states.

First, consider the difference in free energy between the unfolded state and the transition state for the wild-type protein: $\Delta G_{U\rightarrow TS}$ . Recall from Section 15.29 that the rate of a reaction is related to the difference in free energy between the initial state and the transition state. According to transition state theory, the rate constant for the folding reaction, $k_f$, is related to $\Delta G_{U\rightarrow TS}$ in the following way:

$$k_f = Ae^{-\left(\frac{\Delta G_{U\rightarrow TS}}{RT}\right)} \tag{18.4}$$

The term before the exponential Boltzmann factor is a proportionality constant. You should consult **Figure 18.18** and review the discussion in Section 15.29 to understand how this equation arises.

By taking the logarithm of both sides of Equation 18.4, we obtain equations for the values of $\Delta G_{U\rightarrow TS}$ for the wild-type protein (WT) and the mutant (mut):

$$\Delta G_{U\rightarrow TS}(\text{wild type}) = -RT\ln\left(k_f^{WT}\right) + RT\ln A \tag{18.5}$$

$$\Delta G_{U\rightarrow TS}(\text{mutant}) = -RT\ln\left(k_f^{mut}\right) + RT\ln A \tag{18.6}$$

By subtracting these two equations, we get:

$$\begin{aligned}\Delta\Delta G_{U\rightarrow TS} &= \Delta G_{U\rightarrow TS}(\text{mutant}) - \Delta G_{U\rightarrow TS}(\text{wild type})\\ &= -RT\ln\left(\frac{k_f^{mut}}{k_f^{WT}}\right)\end{aligned} \tag{18.7}$$

According to Equation 18.7, by measuring the ratio of the rate constants for the folding reaction, we can figure out how much the mutation alters the free energy of the transition state (**Figure 18.19**A). By a similar line of reasoning (Figure 18.19B), we can determine the extent to which the mutation alters the free energy of the transition state relative to the folded state:

$$\Delta\Delta G_{F\rightarrow TS} = -RT\ln\left(\frac{k_u^{mut}}{k_u^{WT}}\right) \tag{18.8}$$

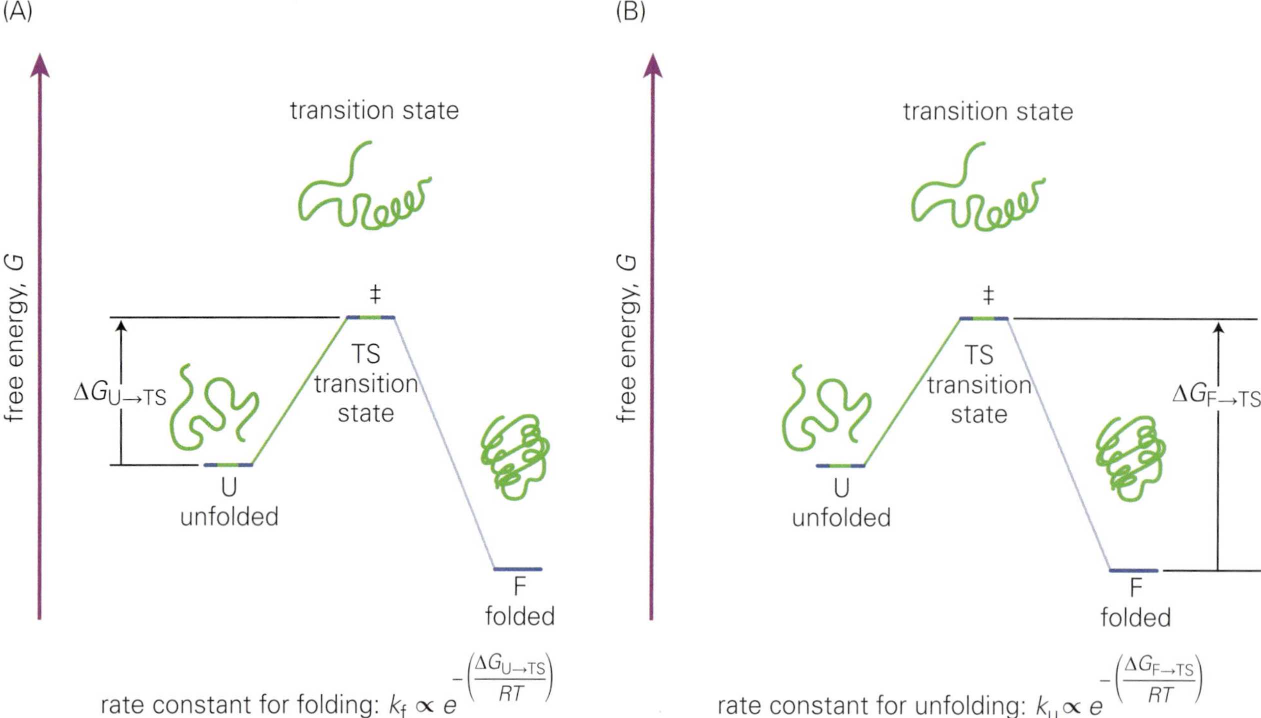


**Figure 18.18 Relationship between the free energy of the transition state and the rate constants for folding and unfolding.** (A) Free-energy diagram for a protein folding reaction, showing the unfolded (U), folded (F), and transition states (TS). The rate constant for folding, $k_f$, is proportional to a Boltzmann factor involving the difference in free energy between the unfolded state and the transition state. (B) Corresponding diagram for the unfolding reaction.

Consider what happens if a particular mutation destabilizes the folded state and the transition state by the same amount—this would be the case if the sidechain makes the same contacts in the transition state as it does in the native state (Figure 18.19C). The mutant protein will fold more slowly because the activation free energy for getting to the transition state from the unfolded state is increased. The rate of unfolding should be unaffected by the mutation because the activation free energy for unfolding is unchanged (see Figure 18.19C).

In contrast, if the mutation destabilizes the folded state, but does not affect the transition state, then the rate of folding will be unaffected because the activation free energy for going from the unfolded state to the transition state is unchanged (see Figure 18.19D). The mutant protein, however, will be faster at unfolding because the folded state is destabilized with respect to the transition state.

The degree to which a mutation affects the transition state is characterized by a parameter, $\phi_{\text{folding}}$, which is the ratio of how much the mutation changes the free energy of the transition state to the change in the folded state:

$$\phi_{\text{folding}} = \frac{\Delta G_{\text{TS}}}{\Delta G_{\text{F}}} = \frac{\Delta\Delta G_{\text{U}\to\text{TS}}}{\Delta\Delta G_{\text{U}\to\text{TS}} - \Delta\Delta G_{\text{F}\to\text{TS}}} \tag{18.9}$$

To understand Equation 18.9 you should study Figure 18.19, which explain how the various free energy differences are related to each other.

The value of $\phi_{\text{folding}}$ indicates the importance of a residue in the transition state for unfolding. $\phi_{\text{folding}} = 1$ means that the residue being tested (that is, the one that was mutated) has the same interactions in the folding transition state as occur in the fully folded state, and $\phi_{\text{folding}} = 0$ means that the transition state has the same interactions as in the fully unfolded state. Fractional $\phi_{\text{folding}}$ values indicate a partial interaction in the transition state and cannot be interpreted in detail.

By determining the $\phi_{\text{folding}}$ values for many residues, one can determine which residues in the protein are important for the formation of the transition state. The $\phi_{\text{folding}}$ values by themselves do not tell us what the transition state looks like. By combining this information with structural information from experimental measurements or computer simulations, we can piece together a picture of the structure of the transition state. For example, computer simulations were used to generate unfolding trajectories for a small protein that inhibits the protease chymotrypsin (**Figure 18.20**). The protein was also subjected to a $\phi_{\text{folding}}$ analysis, which identified those residues that are likely to be involved in energetically

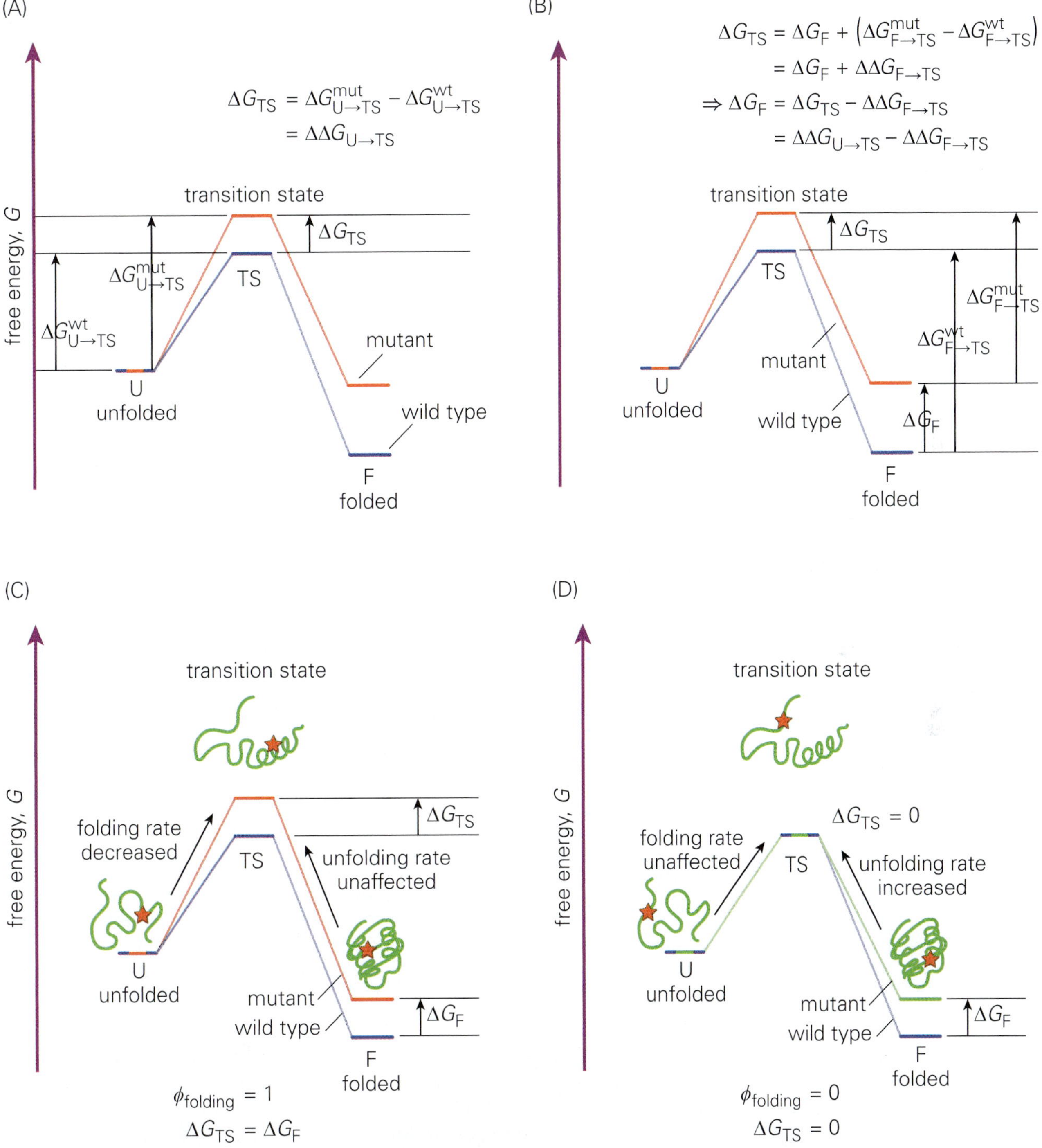


**Figure 18.19 Phi-value analysis of the effects of mutations on folding**. (A) Free energy diagrams for the wild type protein (*blue*) and a mutant (*red*). The relationship between the difference in the free energy of the transition state for the mutant versus the wild type ($\Delta G_{TS}$) and the change in the barrier for folding ($\Delta\Delta G_{U\to TS}$) is shown. (B) The energy diagram shown here is the same as in (A). The change in the free energy of the folded state ($\Delta G_F$) is related to the changes in the barriers for folding ($\Delta\Delta G_{U\to TS}$) and unfolding ($\Delta\Delta G_{F\to TS}$). (C) The effect of a mutation (*red* star) on the free energies of the folded state and the transition state are shown. The mutation destabilizes both states by an equal amount, and the value of $\phi_{\text{folding}}$ is 1.0. (D) A mutation at another location does not affect the transition state, although it destabilizes the folded state. The value of $\phi_{\text{folding}}$ is zero.

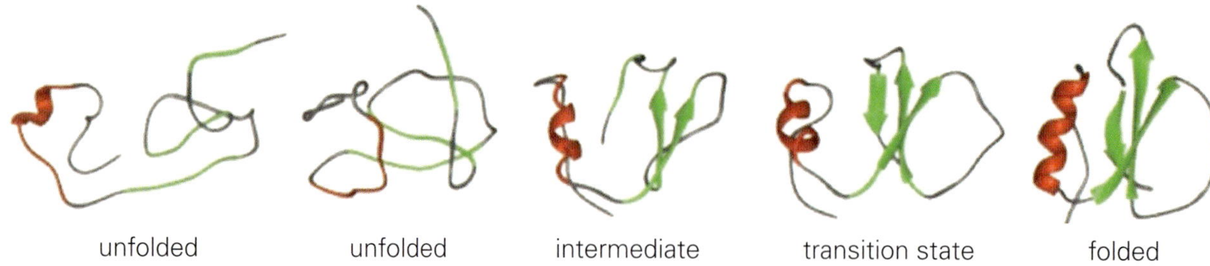


**Figure 18.20 Identification of transition states by comparing $\phi$-value analysis with the results of simulations.** The unfolding of a small protease inhibitor known as CI2 was simulated, resulting in many conformations of the protein. A few of these structures are shown here. Information from $\phi$-value analysis was used to validate the identification of the transition state. (Adapted from V. Daggett and A. Fersht, *Nat. Rev. Mol. Cell Biol.* 4: 497–502, 2003. With permission from Macmillan Publishers Ltd.)

**Figure 18.21 Hypothetical free-energy landscapes for protein folding.** These diagrams show free energy (vertical axis) as a function of combined conformational variables (two horizontal axes), for three different proteins. (A) There is a well-defined pathway of conformational transition from the unfolded state to the final folded state (F), with no free-energy barrier. (B) There is no defined pathway at all. All molecules can move downhill in free energy from their starting conformation to reach the folded state, again with no barrier. (C) The free-energy landscape is rugged; proteins starting in different conformations encounter different barriers. There are many local minima in which molecules could be trapped temporarily. (Adapted from K.A. Dill and H.S. Chan, *Nat Struct. Biol.* 4: 10–19, 1997. With permission from Macmillan Publishers Ltd.)

favorable interactions in the transition state. A consistent picture of the folding pathway has emerged by combining information from the simulations and the $\phi_{\text{folding}}$ analysis. We shall not go into the details of these results here, but simply note that the results of computer simulations, when correlated with experimental measurements, have allowed scientists to obtain detailed pictures of the folding process for many proteins.

## 18.12 The process of protein folding can be described as funneled movement on a multidimensional free-energy landscape

In our discussion of protein folding so far, we have used the term "folding pathway" as if folding proceeds along a defined route. It is clear, however, that this simple description cannot be completely correct because, as we have discussed, the unfolded state of a protein is an ensemble and does not correspond to just one conformation. In fact, there are so many conformations available to an unfolded protein that essentially every unfolded protein molecule in a solution would have a different conformation. As the molecules begin to fold, they move along different trajectories. The initial part of this process must be different for different molecules, but at some point the partially folded molecules begin to resemble each other. Eventually, the different trajectories converge on the folded state. The folded state should also be represented as an ensemble, but when the members are tightly clustered, it is usually easier to think of it as one structure.

Protein folding can be visualized as movement on a many-dimensional surface, the dimensions of the surface being the different conformational variables, including the backbone torsion angles, $\phi$ and $\psi$, as well as sets of sidechain torsion angles. This multidimensional surface is called a **free-energy landscape**, in which the value of the free energy for a particular combination of conformational variables is analogous to elevation in a conventional landscape (**Figure 18.21**).

It is impossible to visualize the >150-dimensional space that is required to describe the conformational variables for even a small protein of 50 residues. To represent the qualitative features of such surfaces, the conformational variables are projected in some way on to two dimensions, so that just the $x$ and $y$ dimensions reflect the conformation of the whole protein, and the $z$ axis is the free energy.

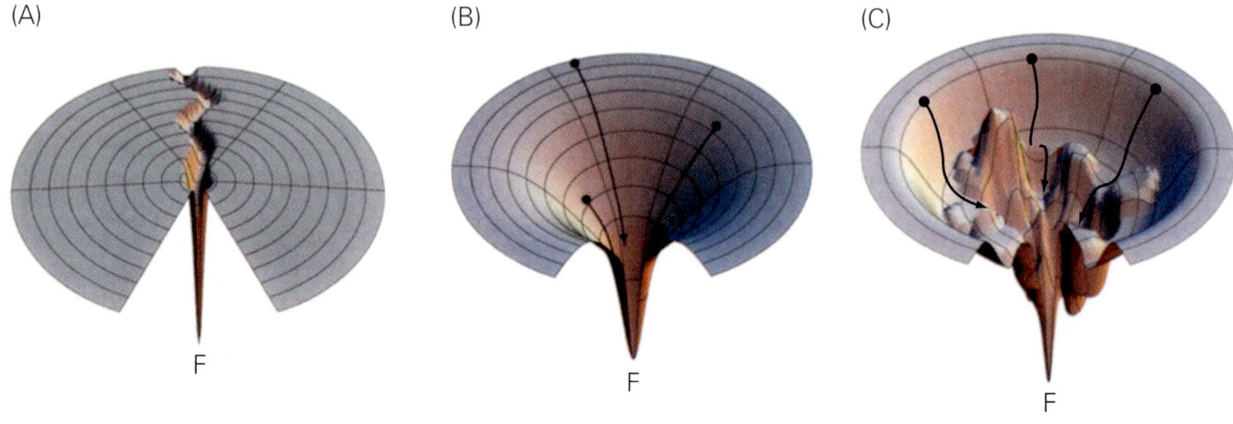

While grossly oversimplified, this free-energy landscape can display features that affect folding. The examples shown in Figure 18.21 have different characteristics, ranging from a simple defined low free-energy folding pathway to a rugged pathway that indicates that proteins folding from different starting conditions face different free-energy barriers.

A folding intermediate is also an ensemble of structures, in which parts of the chain have a reasonably well-defined conformation, with other parts in relatively random conformations. This means that with increasing time after the onset of folding, the set of conformations sampled by the ensemble converges, and occupies a much narrower range of conformations than seen in the unfolded state. That is, the pathways that individual proteins take from the unfolded to the folded state converge in some region of their conformation space. This feature of protein folding is referred to as a **folding funnel**. The presence of these funnels in the free energy landscape allows proteins to find the folded conformation in a reasonable amount of time, without having to search through all possible structures.

**Folding funnel**

The unfolded state of a protein corresponds to molecules in many different conformations, each of which begins to fold along a different pathway. Eventually, as the folding process continues, these pathways begin to converge towards the native structure. This convergence in folding pathways is called a folding funnel.

# B. CHAPERONES FOR PROTEIN FOLDING

Anyone who works with proteins realizes very soon that the ability of ribonuclease-A to fold and unfold reversibly, a feature that was central to Anfinsen's experiments, is far from universal behavior for proteins. The rapidly and reversibly folding proteins discussed in part A of this chapter are, in fact, atypical of the majority of larger proteins in cells, particularly those that have multiple domains. Such proteins fold much more slowly on their own, if at all, even as a pure, dilute solution. In this part of the chapter, we discuss how a class of proteins known as molecular chaperones help other proteins to fold.

## 18.13 Many proteins tend to aggregate rather than fold

Proteins are sensitive to perturbations that disrupt their native structure and, once they are unfolded, it is often difficult, if not impossible, to induce them to fold into the native structure again. A familiar example of this is the effect of heat (cooking) on egg whites (**Figure 18.22**). Individual protein molecules are small compared with the wavelength of light, and hence scatter light very weakly. Even concentrated solutions of native, folded proteins like egg white are essentially colorless. When the size of molecules (or aggregates) is comparable to the wavelength of visible light (around 600 nm), scattering becomes very strong and the solution appears cloudy or opaque. As egg white is heated, the proteins in it denature and become highly aggregated, scattering light much more strongly, making the "egg whites" white and opaque. A similar thing happens when solutions of purified proteins are heated in a test tube—that is, the protein molecules in the solution unfold and become aggregated, making the solution cloudy. Most of the protein precipitates into large clumps, a process that is essentially irreversible.

Why do protein molecules aggregate when they are heated? As the temperature of a solution of protein molecules is raised, the population of unfolded protein

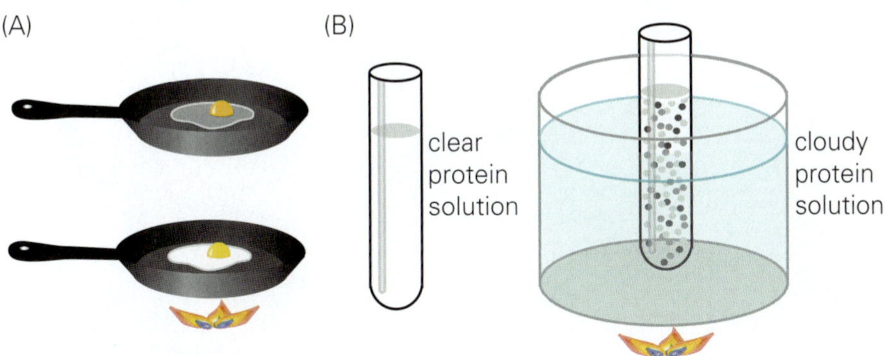


**Figure 18.22 Protein denaturation by heating.** (A) Egg white, a 15% protein solution, is translucent when the proteins are natively folded. When eggs are cooked, the proteins unfold and become aggregated. The large size of the protein aggregates makes them scatter light. (B) If a test tube containing a solution of protein is warmed, the solution turns cloudy as the protein aggregates. As in the case of cooking food, the size of protein aggregates is detected by eye as cloudiness.

(A) protein at low concentration

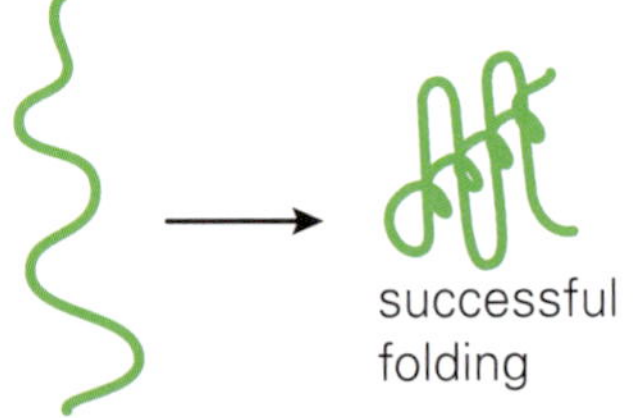


(B) protein at high concentration

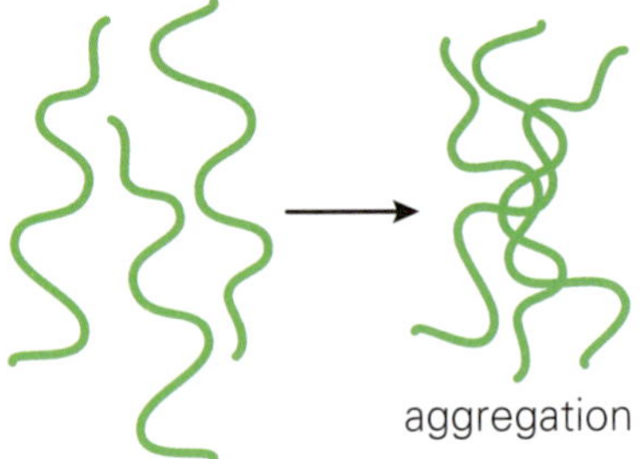


**Figure 18.23 Proteins at high concentration aggregate when unfolded**. (A) At low concentrations, proteins can fold before encountering other unfolded proteins. (B) At high concentrations, the hydrophobic effect leads to nonspecific aggregation.

molecules increases (see Chapter 10, part D), exposing hydrophobic sidechains that are normally sequestered in the interior of the protein. If the concentration of proteins is high, the hydrophobic sidechains from one partially unfolded protein molecule come into contact with exposed hydrophobic sidechains from another protein molecule and the chains stick together (**Figure 18.23**). If the protein molecules are at a sufficiently high concentration, the clumping becomes so extensive that the entire mass of protein precipitates.

## 18.14 The high concentration of macromolecules inside the cell makes the problem of aggregation particularly acute

The tendency of protein molecules to clump together when they are not completely folded prevents Anfinsen's thermodynamic principle from functioning smoothly *in vivo*. There is surprisingly little free space in the cytosol of a cell (**Figure 18.24**). Protein molecules and RNAs are so tightly packed inside the cell that the effective macromolecular concentration is ~300–400 mg•mL$^{-1}$ (for comparison, the density of pure water is 1000 mg•mL$^{-1}$). Schematic diagrams of the cell that show the cytosol as an expanse of uncluttered space are highly misleading, and it is much more appropriate to think of the cellular interior as resembling the crowded interior of a subway train at rush hour—that is, molecules are tightly packed together and jostling one another constantly.

Since much of the free space inside the cell is occupied by various macromolecules, two proteins that encounter each other in the cell have very little room in which to avoid interacting with each other. This phenomenon is known as **macromolecular crowding**, and it has the effect of increasing the apparent association constants between interacting molecules (this concept was discussed in the context of the evolution of allostery in Section 14.21). Crowding can make the effective interaction strengths between proteins in the cytosol 1000-fold higher than

**Macromolecular crowding**

Macromolecular crowding is the effect of macromolecules occupying a significant fraction of the volume of a cell. This increases the effective concentration of proteins. Macromolecular crowding appears to increase the strength of interaction between any particular pair of proteins because the actual space available for these proteins is much less than the volume of the cell.

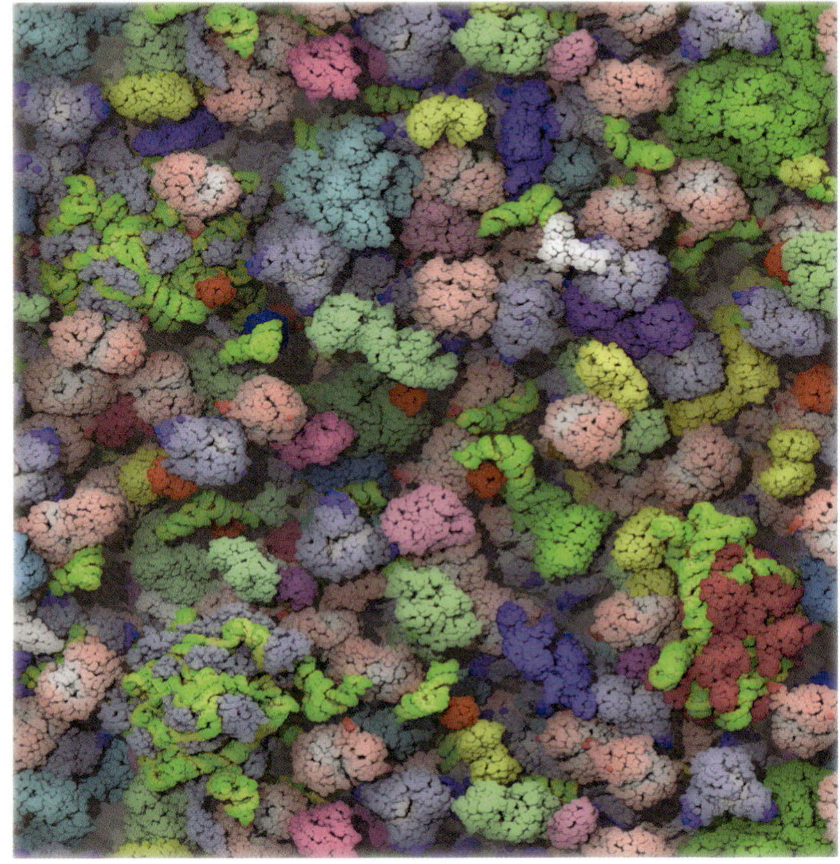

**Figure 18.24 Macromolecular crowding in the interior of a cell**. Proteins and RNA molecules inside a bacterial cell are shown here, in a snapshot from a computer simulation. The structures of the individual macromolecules are based on crystal structures. See also Figure 1, which is an artist's rendition of macromolecules in the interior of a bacterial cell. (From S.R. McGuffee and A.H. Elcock, *PLoS Comput. Biol.* 6: e1000694, 2010.)

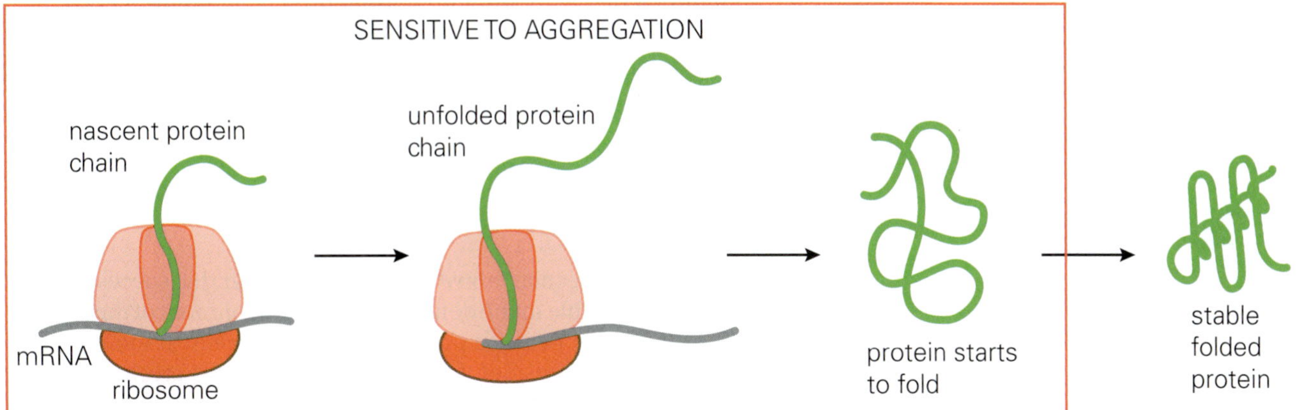


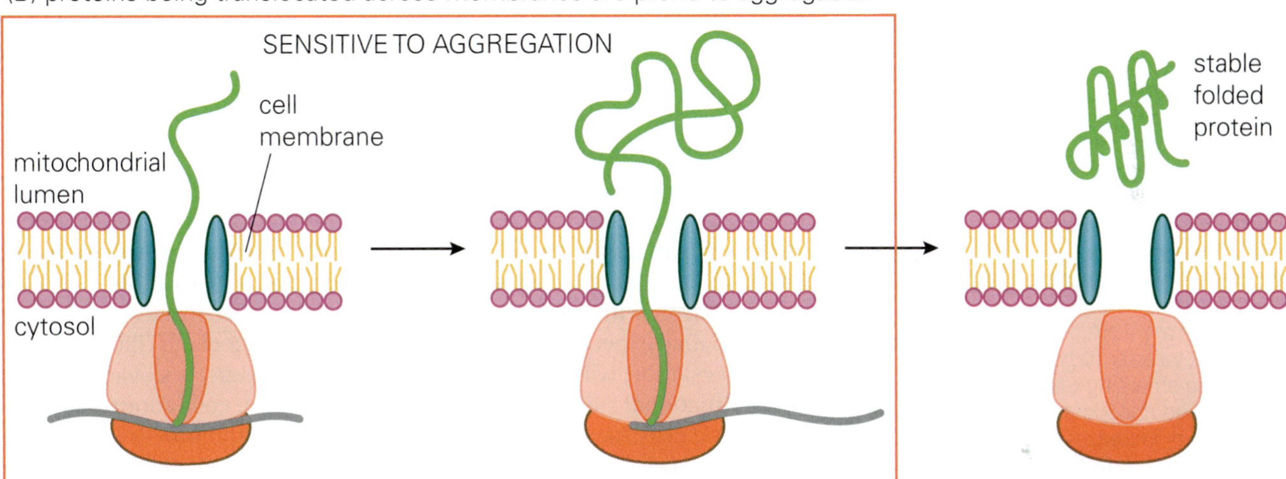


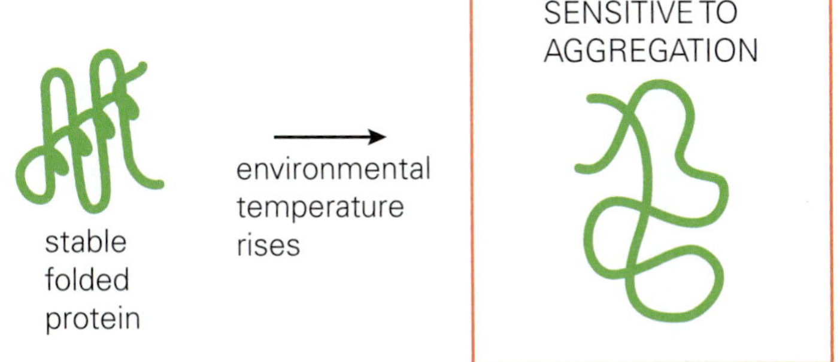


**Figure 18.25 Situations where proteins are prone to aggregate.** Partially folded forms of proteins expose hydrophobic sidechains and increase the chances of aggregation. (A) As a newly synthesized polypeptide emerges from a ribosome, it cannot fold into a stable structure until an entire domain has been exposed. (B) Proteins that are translocated across membranes do so in an unfolded state, and they cannot fold until the entire protein chain (or at least a complete domain) is within the new compartment. (C) When cells are subjected to heat shock (an increase in temperature), proteins begin to unfold.

when measured in dilute solutions (the typical conditions used in biochemistry experiments). Another consequence of macromolecular crowding is that the diffusion of protein molecules inside the cell is slower, making encounters last much longer than in dilute solutions. While crowding can increase the rate of folding in this way, it can also increase the probability that a newly synthesized polypeptide will encounter other proteins before it can fold.

In addition to the relatively short period during protein synthesis when polypeptide chains are susceptible to aggregation (**Figure 18.25**A), there are several other situations when aggregation is a potentially serious problem. Proteins that are transported across vesicular membranes must unfold to do so, and they have to refold into their proper conformation when they enter their final compartment (Figure 18.25B). For example, some proteins are synthesized in the cytoplasm, but are then transported into the mitochondria. Proteins are also prone to aggregate when cells undergo increases in ambient temperature, a condition known as a **heat shock**. A substantially increased temperature raises the amounts of completely or partially unfolded protein present (Figure 18.25C). Cells subject to such temperature excursions are at increased risk of losing important functional proteins due to aggregation.

## 18.15 Proteins inside the cell usually fold into a functional form rapidly

With the effects of macromolecular crowding, it would appear that successful protein folding is all but impossible in the densely packed environment of the cytosol. This is not the case, however, because proteins are very efficient at folding when synthesized inside the cell. Consider, for example, *E. coli* cells growing at 37°C in an aerobic glucose medium. The time it takes for the cells in such a culture to double in number (the doubling time) is about 40 minutes. From measurements of the total amount of protein in such cultures, we know that there are about 2.4 $\times 10^6$ polypeptide chains, with about 2000 different sequences, in each cell. On average this means that each cell is synthesizing about 60,000 protein molecules per minute. It is estimated that each *E. coli* cell contains 35,000 ribosomes, which means that on average each ribosome takes about 35 seconds to produce a new protein molecule.

Protein synthesis occurs rapidly, even in cells that are not dividing. Eukaryotic cells, for example, can respond to the arrival of messenger molecules such as hormones at the cell surface by changing the pattern of proteins that are produced inside the cell in a matter of minutes. The rapid production of new proteins inside cells contrasts sharply with the rate at which most proteins fold in a test tube. Even in Anfinsen's original experiments, he found that the time it took for ribonuclease-A to fold in the test tube, which can be as long as several hours, is slower than the rate at which functional ribonuclease-A is produced in the cell (about two minutes). Thus, living cells have the capacity to somehow aid the process of protein folding so that the speed at which the folded, functional forms of new proteins are produced is consistent with the requirements of the cell.

Proteins inside the cell are synthesized by ribosomes that are often part of large clusters called polyribosomes (**Figure 18.26**). Protein chains emerging from these

(A)

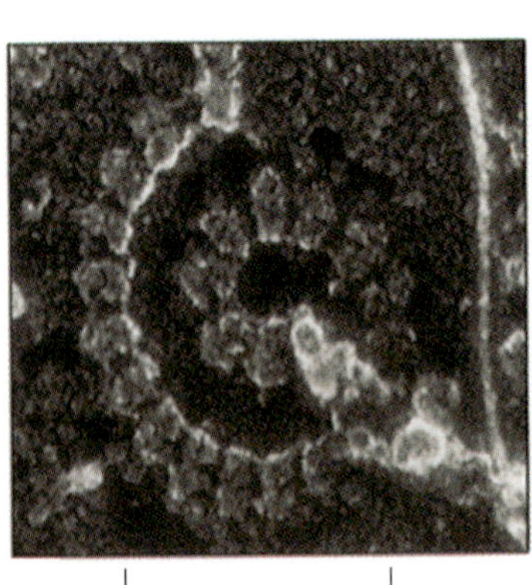


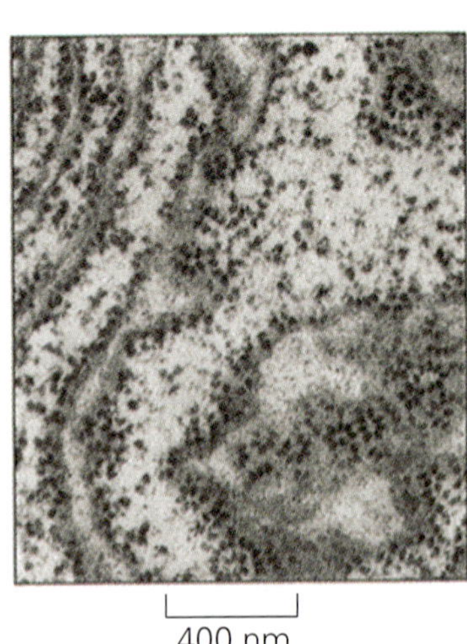


(B)

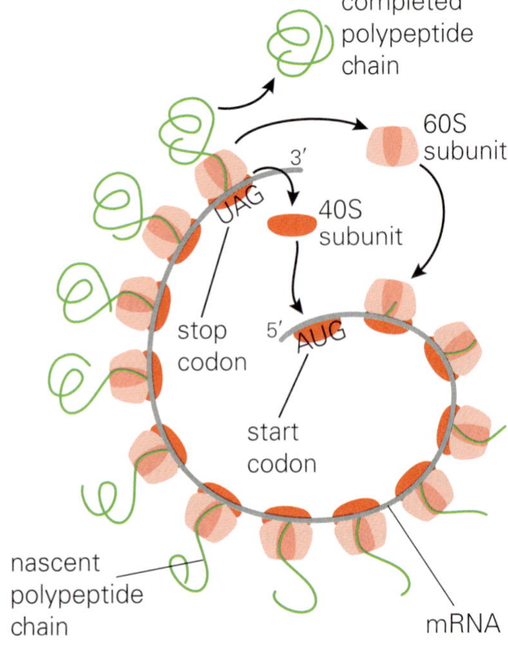


**Figure 18.26 Protein synthesis by polyribosomes.** Ribosomes inside the cell are tightly clustered into superstructures called polyribosomes, in which a single mRNA molecule is translated sequentially by ribosomes bound to it. (A) Electron micrographs of polyribosomes are shown at two different magnifications. (B) A schematic representation of a polyribosome that is translating a single piece of mRNA. Protein molecules at various stages of synthesis are emerging from these densely packed ribosomes, and their close proximity increases the chances of aggregation. (A, left, courtesy of John Heuser; right, courtesy of George Palade; B, adapted from B. Alberts et al., Molecular Biology of the Cell, 5th ed. New York: Garland Science, 2008.)

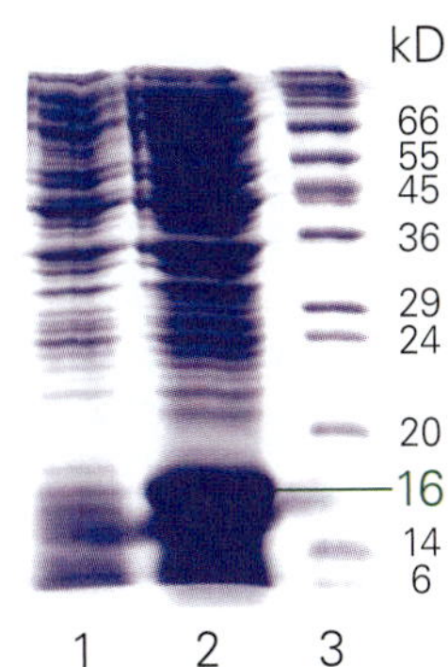


**Figure 18.27 Overexpression can produce insoluble protein.** A small soluble mammalian protein known as bone morphogen protein (BMP, 16 kD), was expressed from an engineered plasmid in *E. coli*. Cells expressing this protein were lysed, and the cell lysate was separated into soluble and insoluble fractions, which were analyzed on an SDS polyacrylamide gel (see Figure 17.43). The soluble proteins are shown in lane 1, and the insoluble proteins, which include membrane proteins, are in lane 2. Lane 3 contains molecular weight markers. BMP (indicated by the *green* line) is expressed at high levels, but is seen only in the insoluble fraction. (Courtesy of M. Nidanie Henderson.)

ribosomes are likely to encounter other protein chains that are in the process of folding. If the newly emergent protein chains were to aggregate irreversibly, the consequences for the cell would be catastrophic. Indeed, if cells are forced to produce unnaturally large amounts of a particular protein (for example, from a gene in a plasmid vector that has a high transcription level), it is common that the proteins are produced in an insoluble and nonfunctional form (**Figure 18.27**).

How do cells avoid such a breakdown in the final translation of the genetic code under normal circumstances? The answer is that cells contain many different kinds of molecular chaperones that are specialized for the task of helping proteins to fold efficiently. But even so, the aggregation of proteins cannot always be prevented, and this can sometimes have disastrous consequences, as discussed in the next section.

## 18.16 Some proteins form irreversible aggregates that are toxic to cells

A serious problem that occurs with eukaryotic cells is the aggregation of proteins to form ordered fibrils, known collectively as **amyloid** (**Figure 18.28**). Some of these fibrils are associated with neurodegenerative diseases, such as **Alzheimer's disease** and **spongiform neuroencephalopathies** (for example, scrapie in sheep and mad cow disease). Most proteins will form amyloid under certain conditions, particularly those that are partially denaturing, but these are not in general associated with disease.

**Amyloid**

The staining of fibrillar protein aggregates is similar to starch, and hence they became known as amyloid deposits (a misnomer that is now in widespread use). The corresponding medical disorders are called amyloid diseases.

Proteins that form amyloid vary widely in native structure, in function, and in just about any other characteristic one might consider. The single property that they all have in common is the formation of amyloid fibrils. These fibrils are made up of β strands, packed so that the long axis of the fibrils runs in the plane of the β sheet, with the individual strands running perpendicular to this axis (**Figure 18.29**). This kind of structure is called a **cross-β spine** (the term "cross-β" refers to a characteristic feature of the x-ray diffraction pattern of these fibrils).

(A)

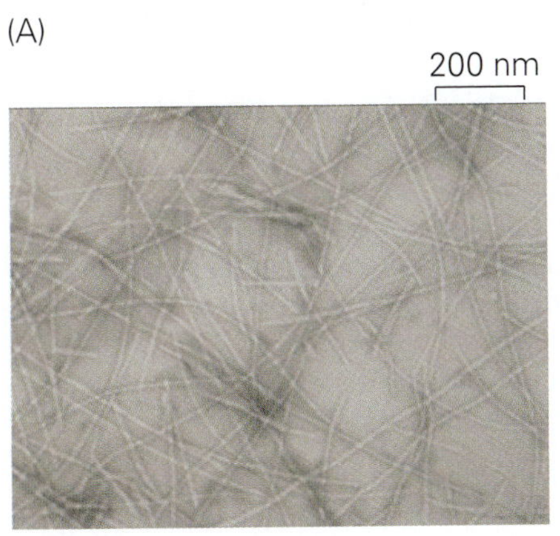


(B)

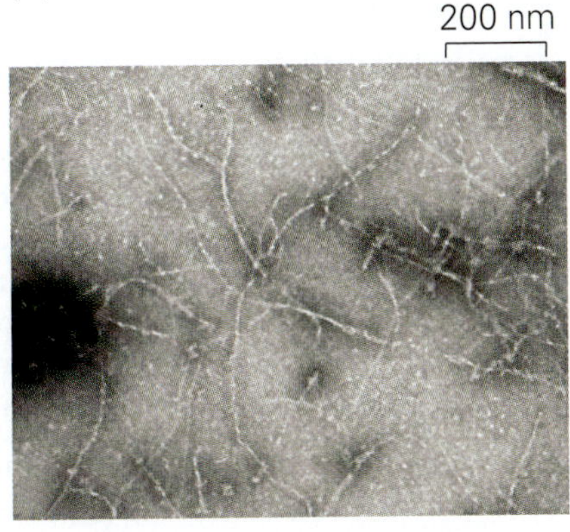


**Figure 18.28 Electron microscopic images of protein fibrils.** (A) Fibrils formed by the Alzheimer's β peptide (Aβ). (B) Fibrils of a fragment of the mouse prion protein. (A, from O.S. Makin and L.C. Serpell, *J. Mol. Biol.* 335: 1279, 2004. With permission from Elsevier; B, courtesy of the laboratory of David Wemmer.)

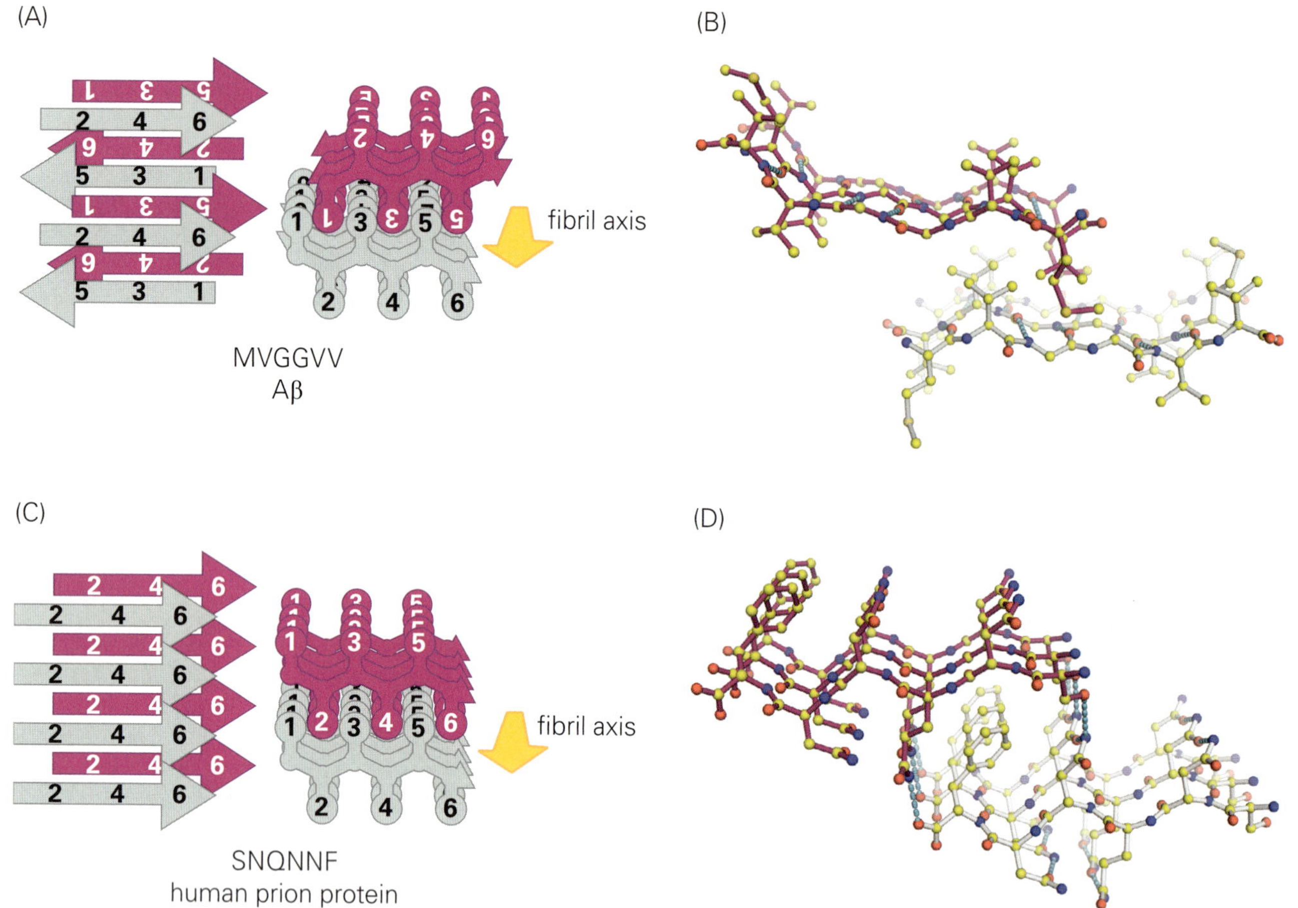


**Figure 18.29 Cross-β spine structures in fibrils.** (A) Schematic diagrams of the arrangement of β strands and sidechains in a cross-β spine formed by a short segment from the Aβ peptide. (B) Crystal structure of the peptide. (C, D) As in (A) and (B), but for a short segment from the human prion protein. (A and C, adapted from M.R. Sawaya et al. and D. Eisenberg, *Nature* 447: 453–457, 2007. With permission from Macmillan Publishers Ltd; PDB codes: B, 2ONA; D, 2OKZ.)

Every protein has the potential for forming fibrils with cross-β spine architectures because β sheets are intrinsically open structures that can be extended by adding additional strands to the edges, essentially without limit (**Figure 18.30**). In a folded protein, other tertiary interactions cap the edges of β sheets, preventing uncontrolled oligomerization. But if it so happens by chance that a short peptide segment within a protein has a sequence that stabilizes the formation of oligomeric β sheets, then fibrils can form if the protein is unable to fold into an alternative structure quickly enough.

For a cross-β spine to nucleate and grow, the β sheets have to pack against each other in a stable way. The two sheets that form the cross-β spine are held together by hydrogen bonds or hydrophobic interactions that are orthogonal to the fibril axis. For some peptides, the cross-β spine has strands arranged in antiparallel β sheets, as shown in Figure 18.29A, while in others the strands pack in a parallel arrangement (Figure 18.29C). There is considerable variation in the details of these interactions.

It was recognized many years ago that the brains of patients suffering from Alzheimer's disease undergo a slow degeneration, with associated fibrillar deposits of protein. Analysis of the deposits showed that they are composed primarily of a peptide 39 to 42 residues in length, known as **Alzheimer's beta peptide** (abbreviated Aβ). This was later shown to be a fragment from a large, membrane-associated protein, the **Alzheimer's precursor protein**. Aβ is produced primarily in the brain, and so its effects are localized there. These deposits are associated with development of disease, although the mechanisms by which they, or other aggregates of the peptide, lead to cell death are not yet clear.

The Aβ peptide is not a globular, folded protein, but instead forms only transient local structures. When Aβ peptide solutions are allowed to stand, they form

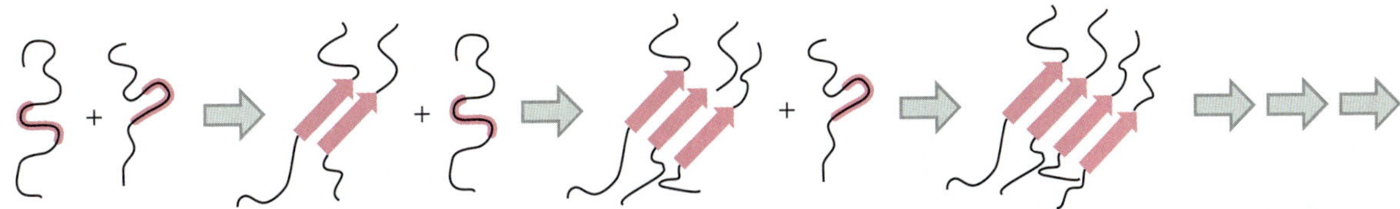

**Figure 18.30 Open-ended polymerization of peptide segments to form β sheets.** The sequence of a short segment of a protein, *pink*, stabilizes the formation of a β sheet by interacting with the same segment from another molecule. Because the hydrogen-bonding capabilities at the edges of the sheet are not capped, the sheet can grow by incorporating additional proteins into it. In actual cross-β spines, two such sheets pack against each other, as shown in Figure 18.29.

aggregates spontaneously, which become the long, slender fibrils that you can see in Figure 18.28A. The time required for fibrils to form is quite variable, because the rate for fibril growth is limited by nucleation. This means that an initial event, forming a "seed" from which the fibril grows, is the slow step.

A schematic representation of the cross-β spine formed by a short segment of the Aβ peptide, with sequence MVGGVV, is shown in Figure 18.29A. For this cross-β spine, the packing of peptides occurs through the sidechains of the hydrophobic residues, not unlike packing in the core of a native protein (Figure 18.29B).

Another example of an amyloid-forming protein is the **prion protein**, PrP, which causes transmissible spongiform encephalopathies. There are several diseases that cause a slow, irreversible degeneration of brain tissue, with development of vacuoles that make the tissue look like a sponge (hence the name spongiform encephalopathies for these diseases). One infected animal can transmit the disease to other normal, healthy animals. Many years of work, largely by Stanley Prusiner and co-workers, resulted in the remarkable discovery that a protein alone is responsible for the transmission of the disease, without the need for the transfer of genetic material. The term "prion" is used to describe the proteinaceous infectious particle, which is generated from the **prion protein** (PrP).

The key to the infectivity of prion diseases is the ability of fibrils or related smaller aggregates to nucleate the conversion of properly folded forms of the PrP protein into the amyloid form. The structure of the cross-β spine formed by a short segment of the PrP protein is shown in Figure 18.29C and D. The sequence of this peptide is SNQNNF. The serine, asparagine, and glutamine sidechains form hydrogen bonds with other sidechains and the backbone, both within the same β sheet and across β sheets. The interactions that hold together these fibrils are therefore quite different from the hydrophobic contacts that stabilize the fibrils formed by the Aβ peptide. The ability of glutamine and asparagine sidechains to form these zipper-like hydrogen bonds underlies the enrichment of these amino acids in peptide segments that tend to form fibrils.

## 18.17 Molecular chaperones are proteins that prevent protein aggregation

**Molecular chaperones** are proteins that assist in the folding of macromolecules or in the assembly of large macromolecular complexes. Chaperones are present in all cells, and many are required for cell viability. There are many different types of chaperones that belong to unrelated protein families. The expression of many chaperones is increased if the ambient temperature is increased, and so these proteins were originally named **heat-shock proteins**.

**Molecular chaperone**

Molecular chaperones are a group of unrelated protein families whose roles include preventing protein aggregation, assisting in correct folding and assembly of proteins, or disassembling complexes or aggregates.

A few representative chaperones are shown schematically in **Figure 18.31**. The fact that many different kinds of chaperones exist within a single organism and that particular chaperones are conserved across species points to their critical importance, even for normal cellular growth, differentiation, and development.

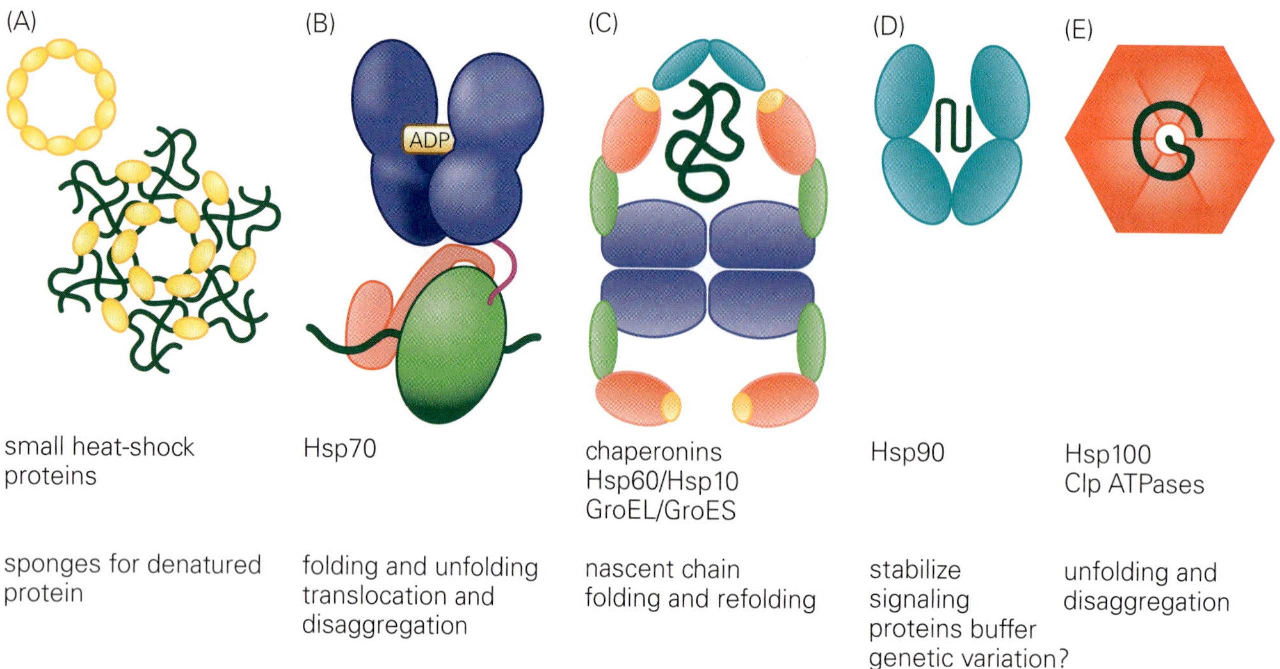


**Figure 18.31 Different kinds of chaperones.** (A) Various kinds of small heat-shock proteins bind to unfolded proteins. Substrate proteins are shown in *green* in each of the panels. (B) The Hsp70s are involved in protein folding and protein translocation across membranes. (C) The chaperonins participate in the folding and refolding of newly synthesized polypeptides in the cytosol. (D) Hsp90 binds to and stabilizes many proteins, particularly signaling proteins such as steroid receptors and protein kinases. (E) The Hsp100 family of chaperones are involved in protein unfolding, disaggregation, and the disassembly of protein complexes. For a discussion of the structures of various kinds of chaperones, see H.R. Saibil, *Curr. Op. Struct. Biol.* 18: 35–42 (2008).

The coexistence of different chaperones with related activities suggests that over the course of evolution they have become specialized to carry out specific tasks in the cell.

Molecular chaperones bind to partly folded or unfolded polypeptide chains and prevent aggregation that could irreversibly prevent folding (**Figure 18.32**). Although their mechanisms differ in detail, a common aspect to how they work is that they provide hydrophobic surfaces that interact with unfolded proteins in a reversible way. This reversibility is crucial, and most chaperones cycle between two different states. In one state they bind tightly to unfolded proteins, but then switch to a state that has low affinity for the polypeptide chain, causing release of the protein. The ability of molecular chaperones to bind and release unfolded protein chains repeatedly is driven by ATP binding and hydrolysis. When the proteins are released from the chaperones, they are free to fold into their thermodynamically favored native state. If they fail to do so, they bind to the chaperones again, with the cycle continuing until successful folding is achieved.

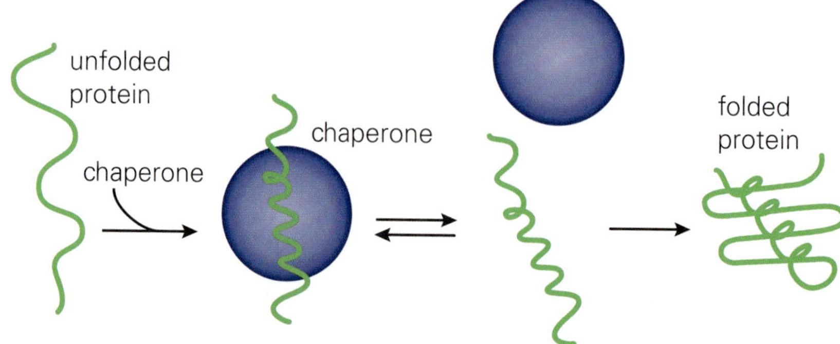


**Figure 18.32 Chaperones prevent aggregation by binding to and releasing unfolded proteins repeatedly**. A chaperone, indicated here by a *blue* sphere, binds to exposed hydrophobic segments, which are common in unfolded proteins. The chaperones then release the proteins, which can either fold or rebind to a chaperone.

We shall not discuss the many different kinds of chaperones in detail but focus, instead, on members of two particularly important families. One of these is **heat-shock protein 70** (Hsp70), so named because its molecular weight is ~70 kDa. Hsp70 binds to short hydrophobic segments in proteins (Figure 18.31B). Hsp70 is found in all organisms and is highly conserved in sequence. Another family of chaperones, known as the **chaperonins**, form large, cylindrical structures (Figure 18.31C). Chaperonins create compartments inside them that serve as "folding cages" in which a complete protein can fold while being sequestered from the cytosol. The bacterial chaperonin has two subunits, known as **GroEL** and **GroES**. The term "GroE" comes from the name of a bacterial gene that was discovered because it is essential for the growth of bacteriophage λ; the "L" and "S" in GroEL and GroES indicate the larger and the smaller of the two subunits. The molecular weight of GroEL is ~60 kDa and so it is also referred to as Hsp60.

Hsp70 and GroEL-ES cooperate to ensure the folding of proteins (**Figure 18.33**). As a newly synthesized protein chain emerges from the ribosome, it is coated by Hsp70. When the chain is released from the ribosome, the protein can fold or rebind Hsp70. Some proteins are unable to fold properly after release from Hsp70. These can interact with the GroEL-ES chaperonin system and thereby proceed to fold. Experimental studies of protein folding in *E. coli* have shown that ~30% of the cytoplasmic proteins in this bacterium are dependent on GroEL-ES for proper folding. The largest class of proteins that require help from chaperonins are those that have an α/β barrel topology, such as triose phosphate isomerase. As you can see in Figure 4.37, these proteins are constructed from a repeating pattern that can easily be scrambled if the wrong sets of β strands align with each other. This pseudo-symmetry in the structure might make these proteins more prone to aggregation.

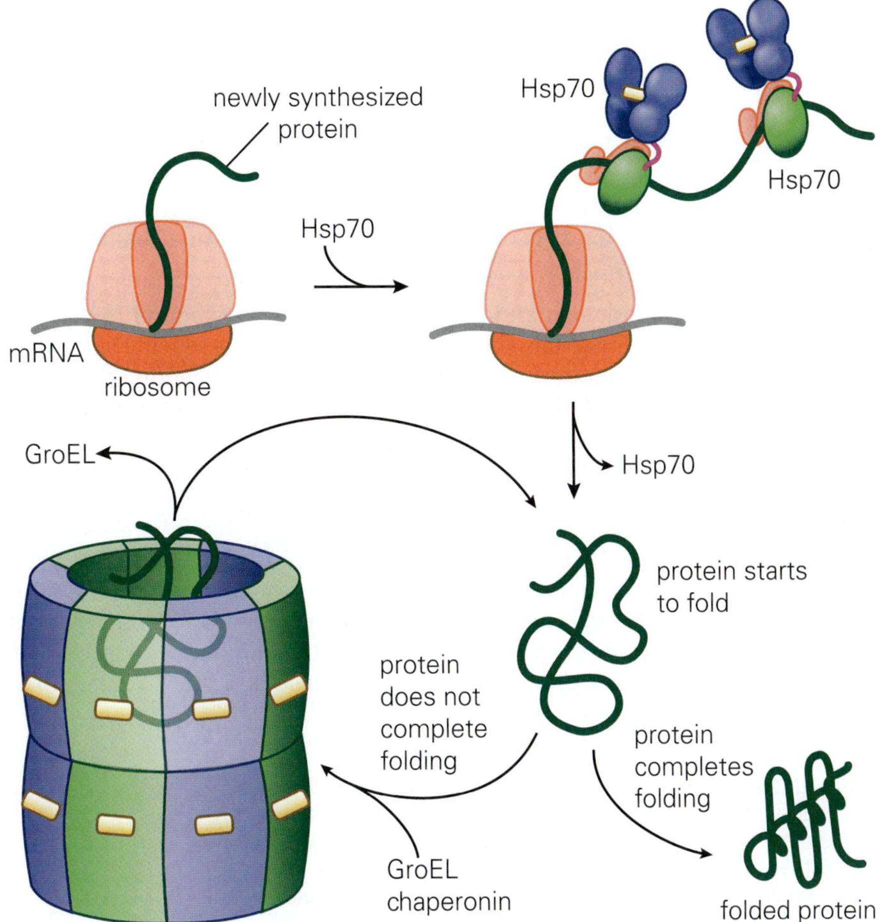


**Figure 18.33 Hsp70 and GroEL cooperate in protein folding.** As a newly synthesized protein emerges from the ribosome, it undergoes cycles of binding to and release from Hsp70 and GroEL until it is folded properly. The smaller subunit of the chaperonin, GroES, is not shown in this diagram. ATP binding sites are indicated by small *yellow* rectangles.

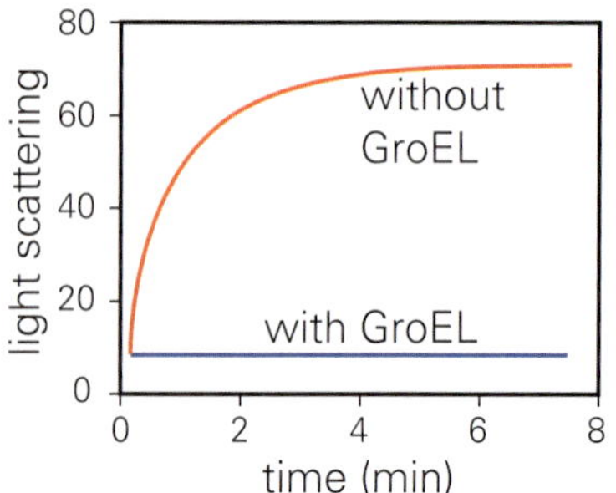


**Figure 18.34 Effect of GroEL on protein folding.** The aggregation of the enzyme citrate synthase is monitored by light scattering after transfer from a denaturing condition to one that favors folding. On its own, the enzyme aggregates instead of folding, as indicated by the increase in the intensity of scattered light. The presence of the chaperone GroEL prevents aggregation and promotes folding. (Adapted from J. Buchner, *FASEB J.* 10: 10–19, 1996. With permission from the Federation of American Societies for Experimental Biology.)

Hsp70 and GroEL (without Gro-ES) can also prevent aggregation by acting on their own. This is demonstrated in **Figure 18.34**, which shows what happens when the enzyme citrate synthase is transferred from denaturing conditions to conditions that favor folding. On its own, this protein aggregates instead of folding properly. Aggregation is monitored by light scattering, which increases steadily with time. If GroEL is added to the protein solution, the amount of scattered light remains low, showing that GroEL prevents aggregation of citrate synthase.

## 18.18 Hsp70 recognizes short peptides with sequences that are characteristic of the interior segments of proteins

The Hsp70 chaperone is able to distinguish between folded and unfolded proteins, and this property is crucial to its ability to bind and protect unfolded polypeptide chains. How does the Hsp70 protein achieve this remarkable discrimination in a completely general manner, without regard to the specific characteristics of the proteins that it must operate on? The answer to this question was arrived at by analyzing the kinds of peptides that are able to bind to Hsp70.

This was done by generating peptide libraries, which are mixtures of peptides that are essentially random in sequence. The peptide library was then separated into two pools, one that binds to Hsp70 and one that does not (**Figure 18.35**). The peptides in the pool that binds to Hsp70 were then sequenced. These experiments showed that Hsp70 recognizes relatively short sequence motifs, about seven residues in length. Peptides that bind to Hsp70 are enriched in hydrophobic residues, with sequences that would be typical for peptides that form the hydrophobic cores of folded proteins (see Figures 4.28 and 4.29 for examples of such segments). Thus, Hsp70 binds to segments of proteins that are usually exposed only in unfolded proteins.

The structure of a peptide containing three leucine residues bound to the peptide-binding domain of Hsp70 is shown in **Figure 18.36**. The leucine sidechains are sequestered from water by the interaction with Hsp70, preventing them from interacting with other hydrophobic peptides and potentially causing aggregation.

## 18.19 Hsp70 binds and releases protein chains in a cycle that is coupled to ATP binding and hydrolysis

The key to the proper functioning of Hsp70 as a chaperone, rather than as a sink for unfolded proteins, is its ability to release proteins after they have bound to it. Hsp70 can do this because its two domains are coupled: its ATPase domain changes conformation upon hydrolysis of ATP and the affinity for peptides of its peptide-binding domain is altered in response to cues from the ATPase domain (**Figure 18.37**).

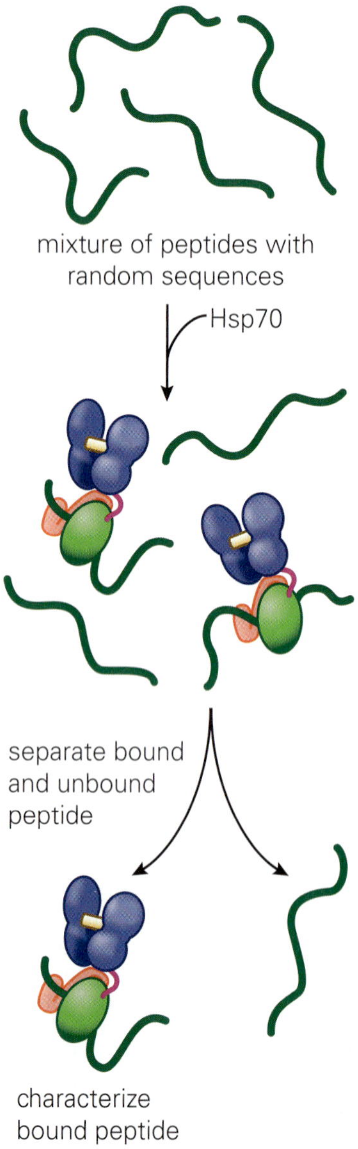

mixture of peptides with random sequences
Hsp70
separate bound and unbound peptide
characterize bound peptide

**Figure 18.35 Identifying peptides that bind to Hsp70.** Using chemical synthesis, a mixture of peptides with random sequences is produced. When Hsp70 is added to this mixture, some peptides bind to the chaperone and others do not. Those that bind can be separated and then their lengths and sequences can be determined. (For more details on this experiment, see G.C. Flynn, et al. and J.E. Rothman, *Nature* 353: 726–730, 1991.)

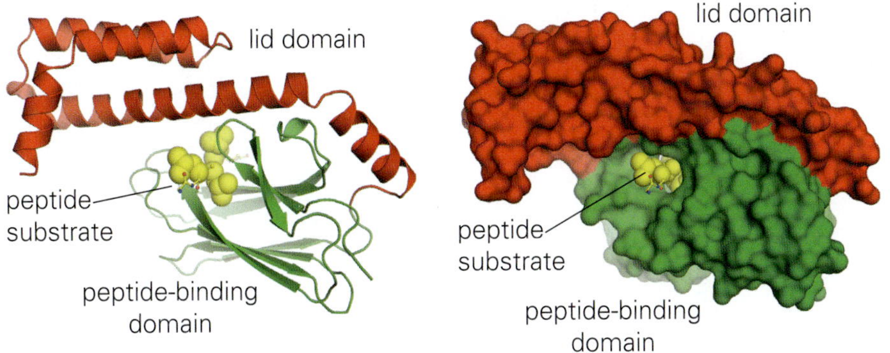


**Figure 18.36 Hsp70 sequesters hydrophobic sidechains in its target peptides.** The structure of the peptide-binding domain of Hsp70 is shown, complexed to a substrate peptide that contains three leucine residues (*yellow* spheres). The leucine residues are buried within a channel formed by the closure of the lid domain over the peptide-binding site. (PDB code: 1DKX.)

When ATP is bound to Hsp70, the affinity of the peptide for the substrate-binding domain is low, and peptide substrates are released rapidly. Substrate binding to the peptide-binding domain stimulates ATP hydrolysis and, with ADP bound, the substrates are trapped, exchanging very slowly. When ADP dissociates and is replaced by ATP, the peptide-binding site opens, and substrate can again exchange rapidly.

The determination of the structure of a relative of Hsp70 in yeast, known as Sse1, together with other structural and functional studies, revealed the mechanism by which the ATPase domain controls the affinity of the peptide-binding domain. While the ATP-binding and peptide-binding sites are far apart (they are separated by ~40 Å), there is an intricate network of hydrogen bonds that connects the ATPase domain to the peptide-binding domain. Upon ATP binding, the two lobes of the ATPase domain rotate against one another by ~25°. This is coupled to a conformational change in the peptide-binding domain: the helical lid of the peptide-binding domain lifts off the peptide-binding site, allowing release of the bound peptide (**Figure 18.38**). A new peptide can bind to Hsp70 once the first one

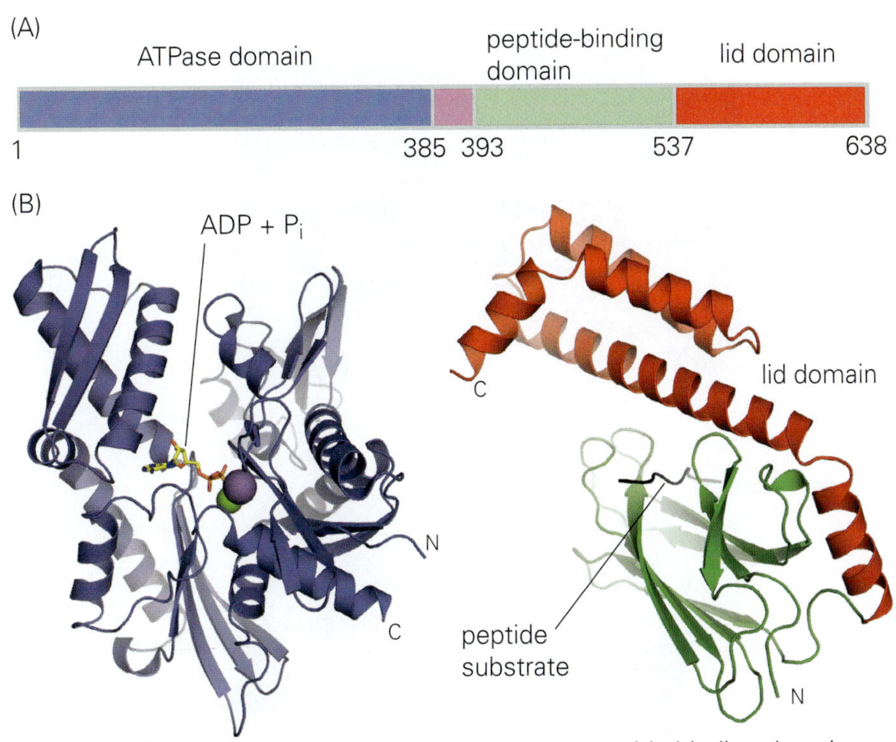


**Figure 18.37 The structure of Hsp70.** (A) A schematic representation of the domain boundaries of bacterial Hsp70. (B) The structure of the ATPase domain and the peptide-binding domain, determined separately by x-ray crystallography. The ATPase domain shown here is from mammalian Hsp70, which is very similar in both structure and sequence to the bacterial protein. (Adapted from Q. Liu and W.A. Hendrickson, *Cell* 131: 106–120, 2007; PDB codes: 1S3X and 1DKX.)

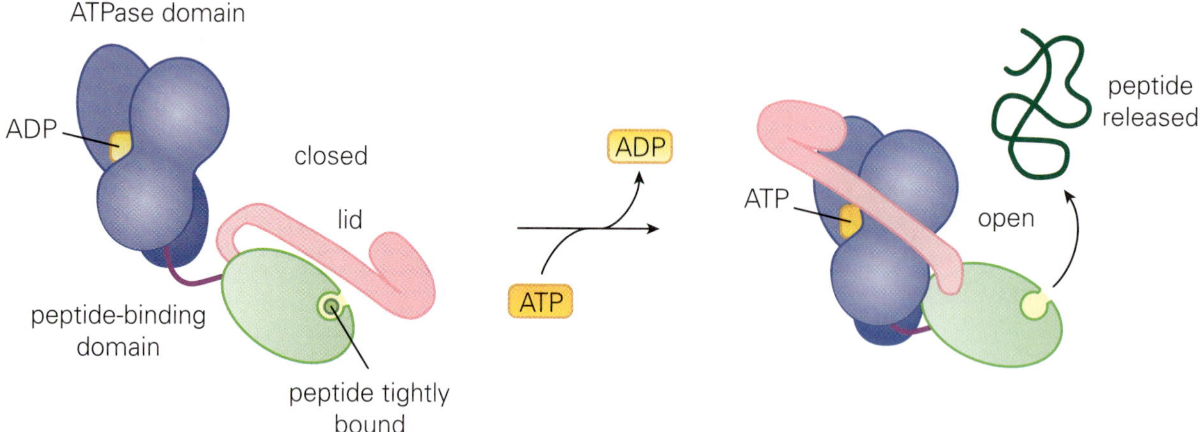


**Figure 18.38 Coupling between ATP binding and peptide release in Hsp70.** Hsp70 switches between two states. When ADP is bound to the ATPase domain, the peptide is bound tightly to the peptide-binding domain because the lid is closed. When ADP is released and replaced with ATP, the lid opens and peptide is released. Another peptide can bind at this stage, and the cycle is reset by ATP hydrolysis. (Adapted from Q. Liu and W.A. Hendrickson, *Cell* 131: 106–120, 2007.)

is released. At this stage, ATP hydrolysis causes the system to cycle back to the original state, with the lid closed and a new peptide bound to the peptide-binding domain.

## 18.20 The GroEL chaperonin forms a hollow double-ring structure within which protein molecules can fold

The bacterial chaperonins are double-ring structures made up of two heptameric assemblies of GroEL (see Figure 18.31C). The GroEL rings are sometimes capped by a lid formed by a heptamer of the smaller protein known as GroES (colored *green* in Figure 18.31C). While the bacterial chaperonins are assemblies of two seven-membered rings, the eukaryotic chaperonins, known as TriC, are assemblies of two rings formed by eight subunits. However, the general architecture of the eukaryotic chaperonins appears to be similar to that of GroEL.

The structure of the GroEL chaperonin is quite remarkable. Our first views of this complex came from electron microscopic images of GroEL alone and of GroEL complexed to GroES (**Figure 18.39**). These images show that GroEL forms

(A)

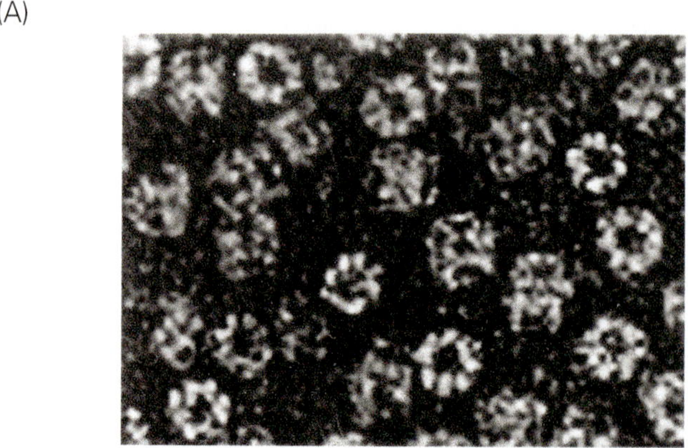

(B)

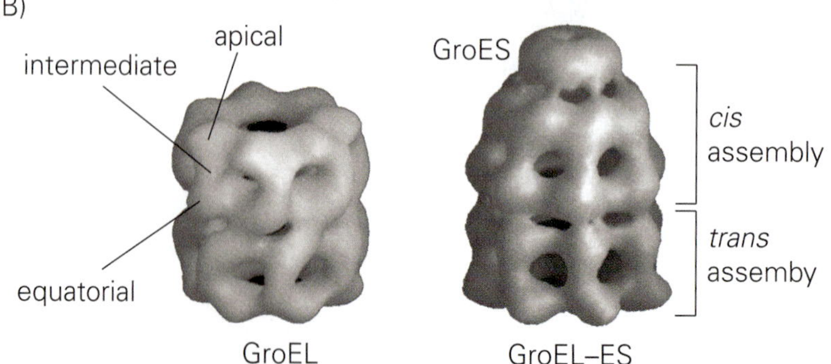


**Figure 18.39 Electron microscopic images of the GroEL chaperonin system.** (A) A view of individual chaperonin complexes, distributed on an electron microscope grid. (B) Many hundred individual images of these molecular assemblies are averaged in order to reconstruct the shape of the molecular assemblies shown here. Apical, intermediate, and equatorial are terms used for the domains of GroEL. GroES binds to one face of the GroEL assembly known as the *cis* face. (Adapted from A.M. Roseman et al. and H. Saibil, *Cell* 87: 241–251, 1996. With permission from Elsevier.)

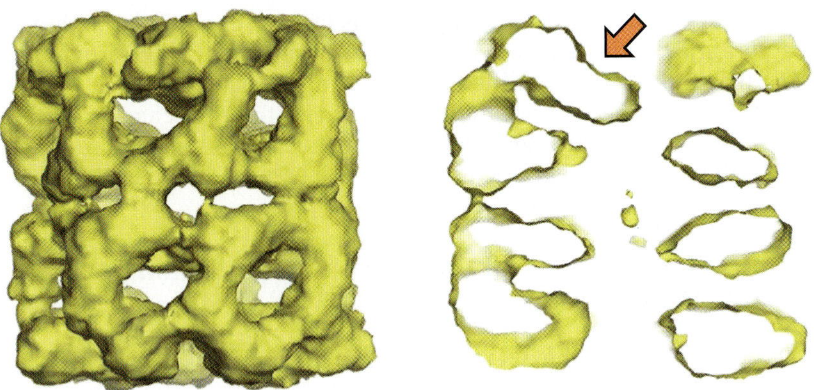

**Figure 18.40 Substrate binding in GroEL.** Side and cross-sectional views are shown from electron microscopic reconstructions of GroEL with the substrate malate dehydrogenase bound. The presence of additional density from substrate bound to GroEL is indicated by an arrow. (Adapted from N. Elad et al. and H.R. Saibil, *Mol. Cell* 26: 415–426, 2007. With permission from Elsevier.)

cage-like structures composed of highly symmetrical double-ring cylinders. Each ring is composed of seven identical copies of a molecule of GroEL (58 kDa each). The rings are assembled in a symmetric back-to-back fashion. Crystal structures of the 14-subunit GroEL assembly, determined subsequently at high resolution, are essentially identical in overall shape to the structures derived from electron microscopy.

Each subunit of GroEL consists of the three distinct structural domains, known as the equatorial, intermediate, and apical domains, which can be discerned in the electron microscope images (see Figure 18.39). The apical domain contains binding sites for unfolded proteins and also for GroES. The equatorial domain contains the ATP-binding site.

The GroEL cylinders are hollow, except for a thin partition between the equatorial domains of the two rings, which separates the internal space of the cylinder into two distinct cage-like chambers. When GroEL is mixed with unfolded protein chains and the molecules in the mixture are analyzed by electron microscopy, it is seen that the unfolded proteins enter the chambers within the GroEL cylinder and appear to be bound there in a stable manner (**Figure 18.40**).

The symmetrical double basket formed by GroEL undergoes a dramatic structural transition when GroES and ATP are added to it (**Figure 18.41**). In the absence of

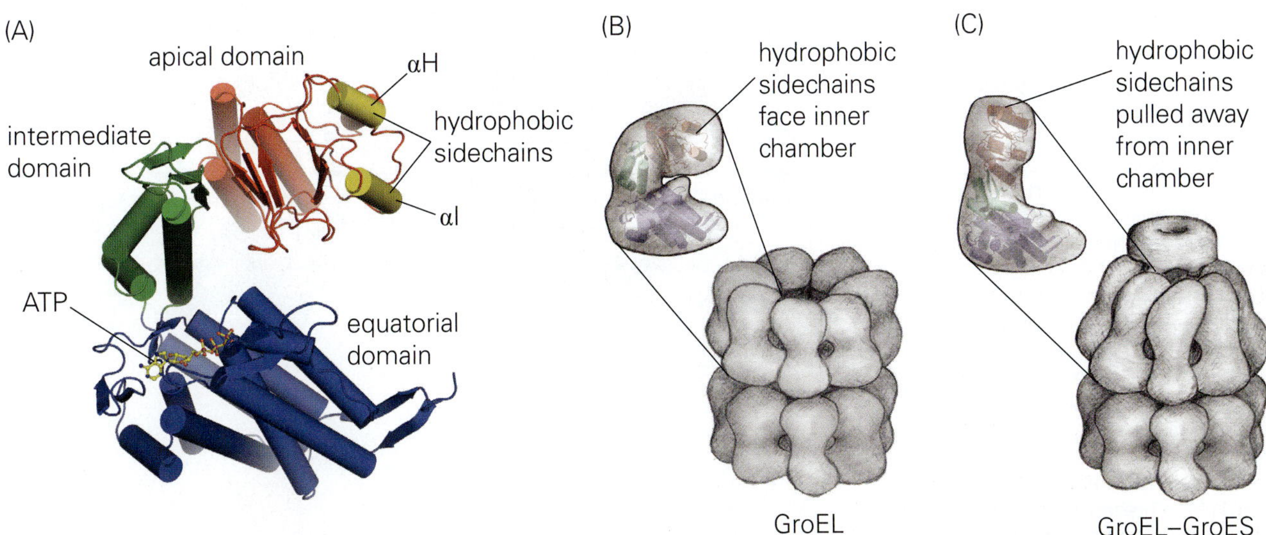


**Figure 18.41 Conformational changes in GroEL/GroES.** (A) The structure of a single subunit of GroEL in the absence of GroES. The two helices in the apical domain that are colored *yellow* present hydrophobic sidechains towards the central chamber of the assembled complex. (B) The structure of the GroEL assembly in the absence of GroES. (C) The structure of the GroEL–GroES assembly. (PDB codes: 1KP8 and 1AON.)

GroES, the inner apical surface of each ring is lined with nonpolar amino acids that capture and tightly bind protein folding intermediates. GroES forms a dome-like ring that caps one end of the GroEL cylinder, leaving the other open (see Figure 18.41). The molecules of GroEL that interact with GroES change their structure radically: they become elongated, and the internal volume that they encapsulate together with GroES is much larger than the size of the cavity in the GroEL cylinders.

## 18.21 GroEL works like a two-stroke engine, binding and releasing proteins

In the absence of GroES, the internal surface of the cavity is largely hydrophobic and provides good binding sites for unfolded protein chains. The large structural change in GroEL that occurs upon the binding of ATP and GroES results in a conversion of the internal surface so that it becomes mainly hydrophilic in character. Proteins bound within the cavity are released from the surface when GroES binds, and begin to fold within the GroEL–GroES cage.

Each subunit of GroEL in the assembly undergoes a large conformational change when GroES binds to the GroEL cylinder (see Figure 18.41 and **Figure 18.42**). The structural change arises from rotations of the intermediate and apical domains with respect to the equatorial domain. The intermediate domain swings towards the equatorial domain by ~25°, and the apical domain swings away from the center of the cylinder by ~60° and also rotates about its long axis by ~90° (see Figure 18.38). There are two major consequences of these *en bloc* movements. The intermediate domain closes over the nucleotide-binding site on the equatorial domain, thereby positioning a sidechain that acts as a catalytic base for ATP hydrolysis. In addition, as mentioned above, the rotations of the apical domain move the hydrophobic

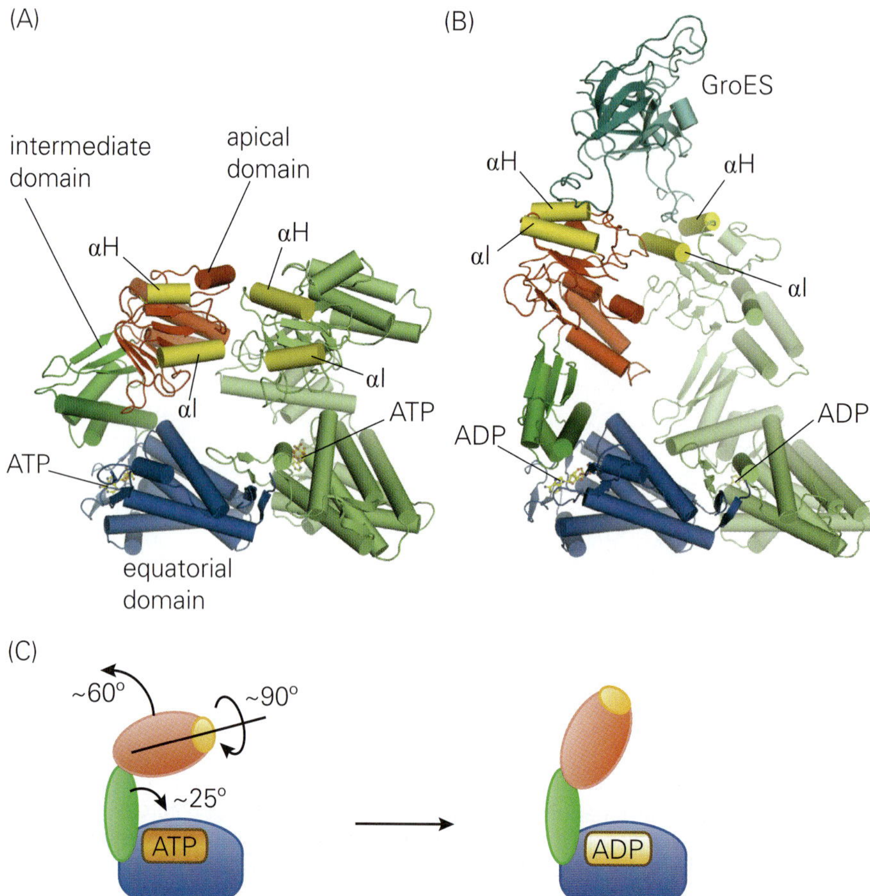


**Figure 18.42 Detailed view of the conformational changes in GroEL upon the binding of GroES.** (A) The structures of two adjacent molecules of GroEL are shown. Two helices that are important for recognizing substrates are colored *yellow*, and these face into the central channel in the GroEL assembly in this conformation. (B) The same two molecules of GroEL are shown in the structure of the GroEL–GroES assembly. A large conformational change has occurred in the GroEL molecules, moving the two helices from an internal position to an external one. These two helices now bind to a loop presented by GroES. (C) A schematic representation of the conformational change undergone by GroEL upon binding to GroES. ATP is required for the binding of GroES to GroEL, but is hydrolyzed after some time. (PDB codes: A, 1KP8; B, 1AON.)

sidechains that are important for interaction with unfolded substrates away from the central channel of the cylinder. The conformational changes of the GroEL subunits when GroES binds have the effect of releasing unfolded substrate proteins that are bound to GroEL in its open ATP-associated state.

In the state without ATP and GroES bound, hydrophobic residues line an interior cavity within each half of GroEL that has a volume of ~85,000 Å$^3$. Unfolded proteins take up more room than folded ones, but can extend out of the end of the GroEL chamber, which is open in the absence of GroES. When GroES binds to GroEL, the outward movements of the apical domain (Figure 18.42B) result in a considerable enlargement of the chamber, even though it is capped by GroES. The volume of the chamber is then ~175,000 Å$^3$, which is large enough to accommodate proteins up to the 60–70 kD range, even when they are unfolded.

These chambers in GroEL are referred to as **Anfinsen cages**, to emphasize the fact that they allow protein folding to occur according to the dictates of thermodynamics, without interference from other polypeptides. The two ends of GroEL bind and hydrolyze ATP in alternate cycles, thereby functioning like a two-stroke engine, as shown in **Figure 18.43**.

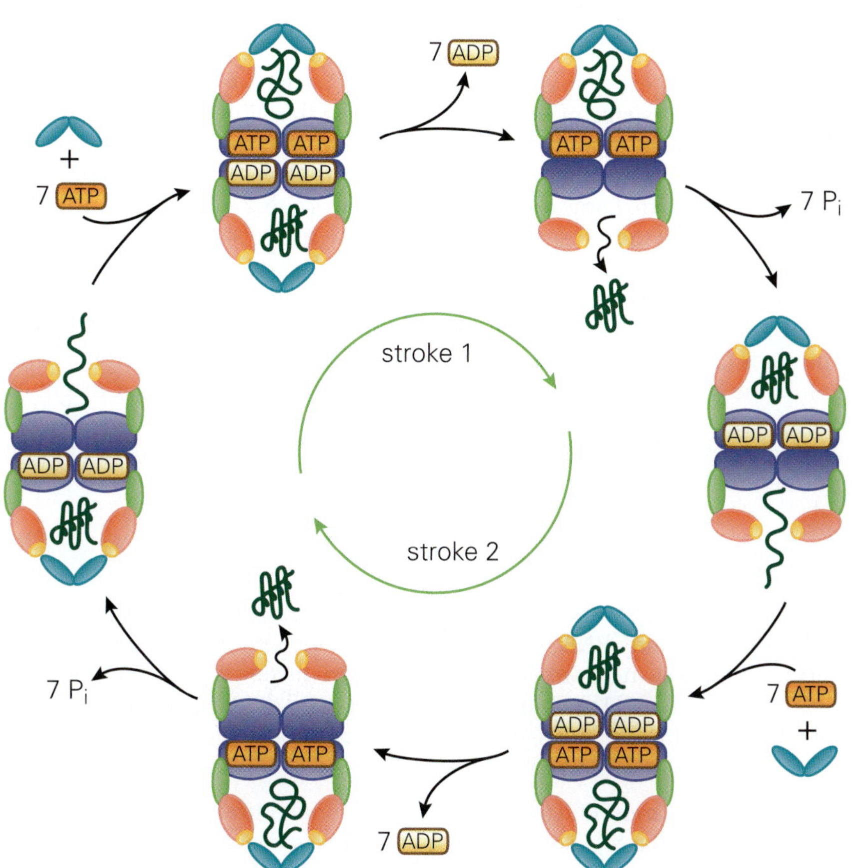


**Figure 18.43 GroEL–GroES works like a two-stroke engine.** The functional cycle of GroEL–GroES, illustrated schematically here, involves binding and discharging proteins from the two chambers in alternate cycles. Each stroke of the motor involves the hydrolysis of seven molecules of ATP, and results in the "charging" of the system with ATP so that it is ready to accept another substrate protein. Each ring contains a large central cavity in which folding takes place, one ring functioning as a folding compartment at a time. (For more details on this process, see A.L. Horwich, *Nat. Med.* 17: 1211–1216, 2011 and F.U. Hartl, *Nat. Med.* 17: 1206–1210, 2011.)

## 18.22 GroEL–GroES can accelerate the folding of proteins through passive and active mechanisms

GroEL, and its close relatives in other organisms, can help a wide variety of proteins to fold. Proteins that require GroEL are typically large (>20 kDa), are slow to fold on their own, and are aggregation prone. In principle, GroEL could promote productive folding by blocking aggregation (referred to as a passive mechanism) or by accelerating productive folding (an active mechanism). In the passive folding mechanism, GroEL has no direct effect on the conformation of the nonnative protein, and it simply serves to block interaction with other proteins, preventing aggregation. In contrast, in the active model, GroEL directly modifies the structure of the substrate protein, thereby affecting its folding.

In the passive model, the GroEL cavity provides a protected environment so that folding takes place effectively at "infinite dilution" (each GroEL-associated protein is separated physically from other aggregation-prone proteins). The GroEL ring shifts between states with high and low affinity for unfolded proteins, which can bind within the cavity and then are released for folding. This transition is driven by ATP hydrolysis and GroES binding, as described above.

The idea that the GroEL–GroES complex could play a more active role in the folding of at least some substrates comes from evidence that GroEL can in fact unfold proteins that are already partially or completely folded. Moreover, the folding of some proteins is accelerated within the GroEL–GroES cage, compared to the rate of folding outside the cage. What properties of the enclosed GroEL–GroES cavity can account for the experimental observation of accelerated folding? The cavity defines a restricted space of finite volume. While the dimensions of the GroEL cavity expand ~2-fold upon binding of GroES and ATP, the confinement within the GroEL–GroES cavity must impose serious constraints on the conformational freedom of the protein. Large regions of the protein's energy landscape are thereby excluded, confining the subsequent search for the native state to a smaller range of conformers, accelerating the folding process. If the GroEL–GroES complex can unfold substrate proteins, then it could also restart the folding of proteins that have become trapped in local free-energy minima in the conformational space of the proteins (nonproductive intermediates), using the energy of ATP hydrolysis to unfold the intermediates and then release them to allow folding. Current evidence suggests that GroEL–GroES works by a combination of protein encapsulation, which prevents aggregation, unfolding from unproductive local minima, and conformational restriction to accelerate protein folding.

# C. RNA FOLDING

An analysis of the sequence of the human genome indicates that there are potentially 10 times as many RNA molecules that are large enough to adopt complex tertiary structures as there are proteins encoded in the human genome. The function of most of these RNAs remains poorly understood, but if even a small fraction of them have specific functions, then the number of well-folded RNA molecules in the cell may well equal that of protein molecules.

Like proteins, RNAs must fold to attain their functional conformations, undergoing compaction from a large disordered array of states to specific three-dimensional structures. There are major differences, however, between RNAs and proteins that may make it more difficult for RNA molecules to get to a unique folded state. RNA has a lower information density than protein, with four nucleotides instead of the 20 naturally occurring amino acids. The driving force for protein folding is the exclusion of hydrophobic sidechains from water, leading to the initial collapse of the chain. The hydrophobic effect is not a dominant factor in RNA folding and, moreover, repulsions between the negatively charged phosphate groups of the RNA backbone present an electrostatic barrier to the formation of a compact, fully folded, functional structure. RNA secondary structures (primarily base-paired

A-form double helices) are more stable on their own than are protein secondary structures, and their formation can present kinetic "traps" during a conformational search for the native tertiary structure. Ultimately, RNA helices do interact to form tertiary structures, in a step that requires the presence of metal ions (usually $Mg^{2+}$; see Chapter 2).

## 18.23 The electrostatic field around RNA leads to the diffuse localization of metal ions

Metal ions or other positively charged ionic molecules that interact with RNA, referred to as **counterions**, neutralize the backbone charge of RNA. Their presence screens the repulsion between the phosphate groups so that tertiary structures can form. Virtually any positively charged ion can promote RNA folding, although monovalent ions (for example, $Na^+$ or $K^+$) are less efficient in this role than divalent metal ions (for example, $Mg^{2+}$ or $Ca^{2+}$).

The negatively charged backbone must be at least partially neutralized before an RNA molecule can begin to fold into a compact form. This is accomplished by an ionic "cloud" of monovalent and divalent ions that surround the RNA (see Section 2.23). Metal ions become more densely localized around the RNA as it folds, in a process known as **counterion condensation** (**Figure 18.44**A). Recall from Chapter 2 that these are known as diffuse ions (see Figure 2.41). The behavior of diffuse ions is determined by electrostatic interactions with the RNA that occur over much longer ranges of distances than for specific contacts (these metal ions may be located as much as 10–15 Å away from a phosphate group while still influencing the RNA molecule).

**Counterion condensation**

Counterion condensation refers to the increase in the local concentration of counterions (cations in the case of RNA) in regions of high electrostatic potential around a polymer. These ions are loosely bound but play an important role in stabilizing structures such as helices.

The concentration of diffuse ions in a particular region is proportional to the magnitude of the electrostatic potential in that region, and can be predicted by calculating the electrostatic field around RNA (**Figure 18.45**). In addition, specifically bound metal ions are also a crucial aspect of RNA tertiary structure (Figure 18.44B and Section 2.23).

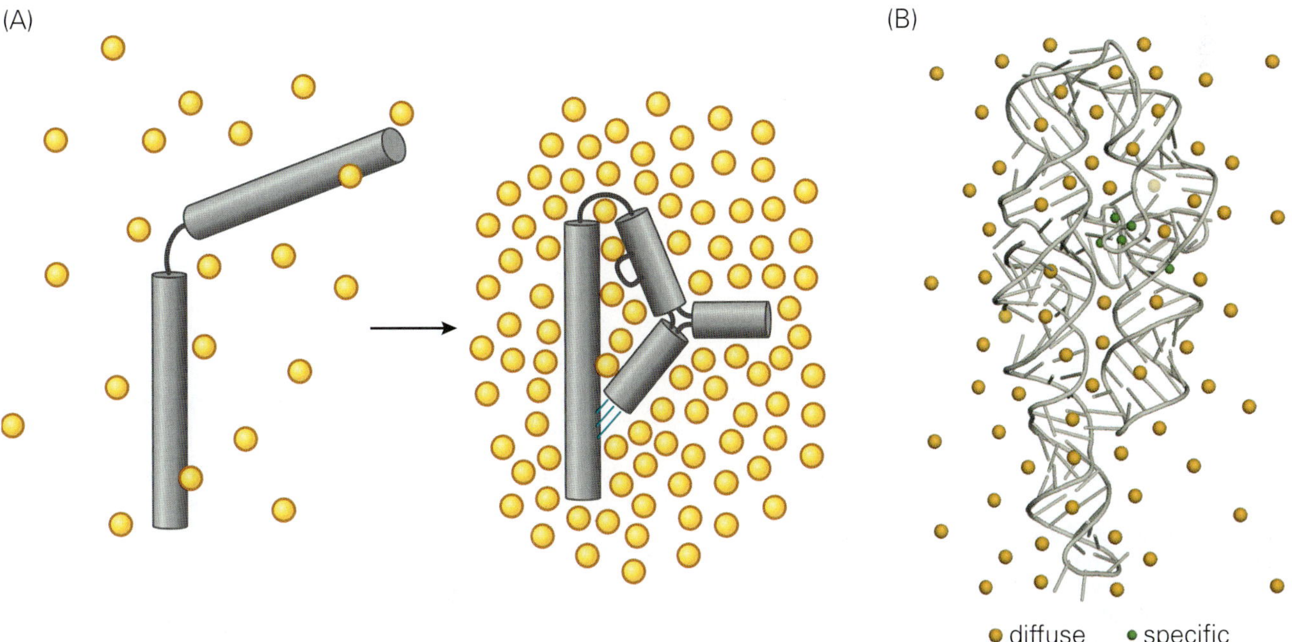


**Figure 18.44 Metal ions are required for RNA folding.** Electrostatic repulsion between phosphate groups prevents the formation of a compact tertiary structure unless metal ions neutralize the charge. (A) Diffuse metal ions condense around RNA as it folds. RNA helices are depicted as cylinders in this diagram and metal ions are *yellow*. (B) Folded RNA structures also have specifically bound metal ions (*green*). (PDB code 1GID.)

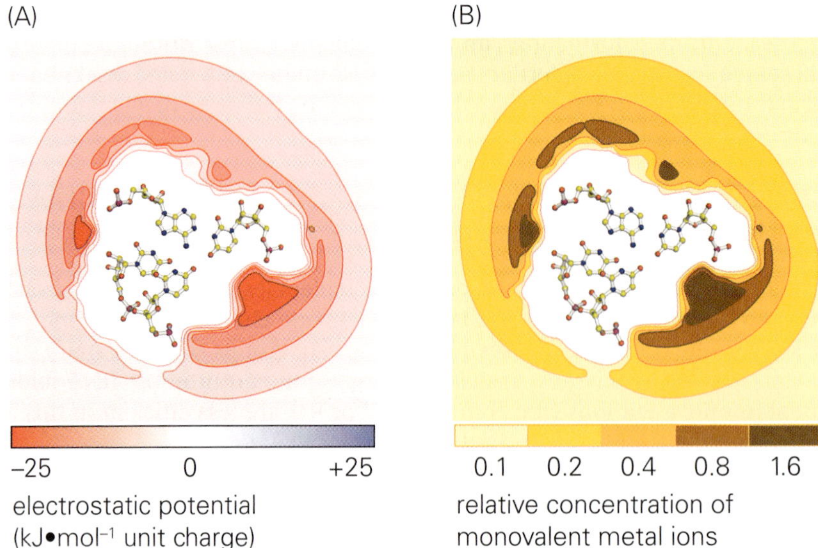


**Figure 18.45 The electrostatic potential and the concentration of diffuse ions around an RNA helix.** (A) Contours indicate the strength of the electrostatic potential, given in energy units for an interacting point charge. (B) Contour map of the concentration of cations resulting from the effects of the electrostatic field. Concentrations range from 1.6 M (*brown*) to 0.1 M (*light yellow*). The bulk monovalent salt concentration is 0.1 M. (Adapted from C. Garcia-Garcia and D.E. Draper, *J. Mol. Biol.* 331: 75–88, 2003. With permission from Elsevier.)

## 18.24 RNA folding can be driven by increasing the concentration of metal ions

Folded RNA structures are more compact than unfolded ones, and hence have a higher spatial density of negative charge. As a consequence, folded RNAs bind cations more tightly than unfolded RNAs. This principle applies even to simple structures, such as A-form RNA helices (and also B-form DNA helices), which are usually stable even at very low salt concentrations, but are always more stable in the presence of higher concentrations of salts.

As salts are added to a solution of an RNA, it is common to see a folding transition that corresponds to the development of tertiary structure. Because of their higher charge, divalent ions induce the same degree of compaction in RNA at much lower concentrations than do monovalent ions. This is primarily because the localization of a single divalent ion near to the RNA is entropically less costly than localizing two monovalent ions. This can be readily observed by comparing the metal-ion concentration dependence of RNA folding for a variety of metal ions (**Figure 18.46**).

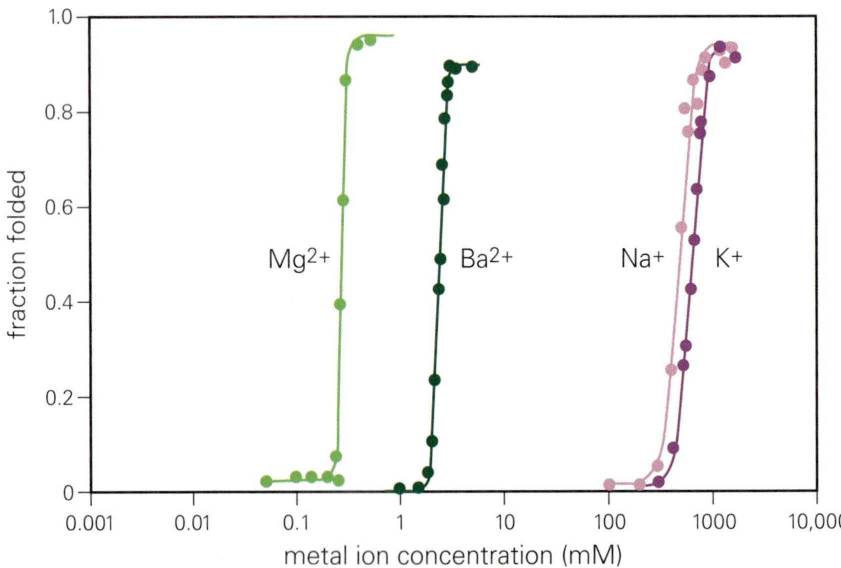


**Figure 18.46 Folding of a group I intron ribozyme in the presence of different metal ions at 30°C.** The fraction of ribozyme folded is plotted on the *y* axis and the metal ion concentration is on the *x* axis. (Adapted from S.L. Heilman-Miller, D. Thirumalai, and S.A. Woodson, *J. Mol. Biol.* 306: 1157–1166, 2001. With permission from Elsevier.)

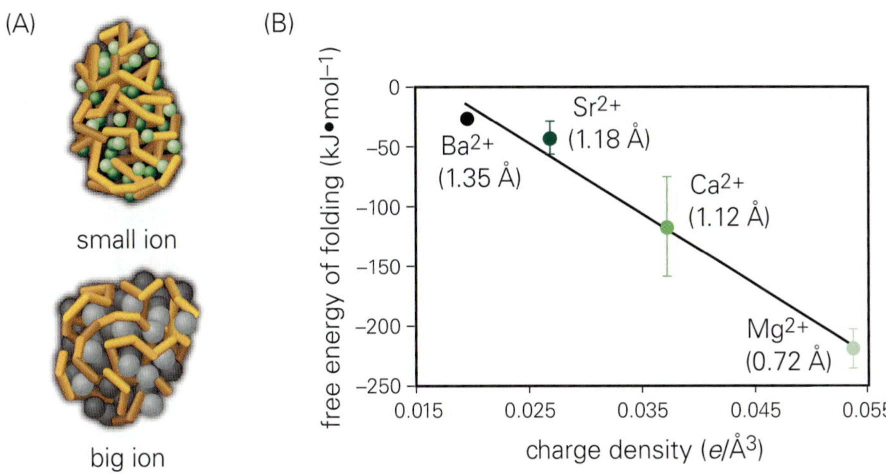


**Figure 18.47 The stability of a folded intron ribozyme increases with the charge density of metal ions.** (A) Schematic drawing of small and large ions bound to RNA. (B) Folding free energy of the RNA as a function of the cation charge density. The atomic radius of each ion is given in parentheses. The cation charge density is the charge of the ion divided by its volume. (Adapted from E. Koculi et al. and S.A. Woodson, *J. Am. Chem. Soc.* 129: 2676–2682, 2007. With permission from the American Chemical Society.)

Recall from Section 14.14 that, when the binding of a ligand to a receptor is cooperative, the binding isotherm displays a characteristic sigmoid shape. This means that the transition from unbound to fully bound is sharper than when ligand molecules bind independently. The transitions between unfolded and folded states of the RNA shown in Figure 18.46 indicate a very sharp change in the fraction folded as the ion concentration increases, particularly for $Mg^{2+}$. If a Hill coefficient is calculated from the $Mg^{2+}$ data (as described in Section 14.15), a value of ~20 is obtained. The large value of the Hill coefficient indicates that many ions (at least 20) become associated with the RNA when folding occurs.

Why is the binding of metal ions to RNA so cooperative during the folding process? One way to think about this is that ions are only very loosely associated with the RNA before tertiary structure forms. As the RNA structure fluctuates toward the folded state, regions of higher negative electrostatic potential are generated by the compaction of the structure (see Figure 18.45, which shows how regions of strongly negative electrostatic potential develop as the base pairs come together). These regions are more effective at localizing metal ions, which stabilizes this folded structure and can then cause yet more ions to become localized. The regions of high charge density can be extensive for a large RNA, and the cooperativity of folding is reflected in the cooperativity of ion binding.

Notice in Figure 18.46 that $Mg^{2+}$ ions are better than $Ba^{2+}$ ions at stimulating the folding of the RNA, even though both these ions are divalent. This is because the $Mg^{2+}$ ions are smaller than the $Ba^{2+}$ ions and so have a higher charge density. The charge density of an ion determines the closest distance between an ion and the RNA and also affects ion–ion distances and, thus, the strength of electrostatic interactions.

To determine the extent to which the size of metal ions changes the stability of RNA tertiary structure, the folding equilibrium of a ribozyme was compared in solutions of $Mg^{2+}$, $Ca^{2+}$, $Sr^{2+}$, and $Ba^{2+}$ (all as $Cl^-$ salts) (**Figure 18.47**). It is clear from these data that metal ions with a higher charge density are more effective at lowering the free energy of the folded form of RNA.

## 18.25 RNAs form stable secondary structural elements, which increases their tendency to misfold

The intrinsic stability of RNA secondary structures generally increases the likelihood that large RNAs will fold along multiple pathways with different intermediates. Some of these intermediates are misfolded or off-pathway intermediates, while others are native-like or on-pathway intermediates that are on their way to forming a unique native structure (**Figure 18.48**). This is in contrast to the folding of many small proteins, where isolated elements of secondary structure are usually unstable and folding occurs with apparent two-state kinetics (see Section 18.5).

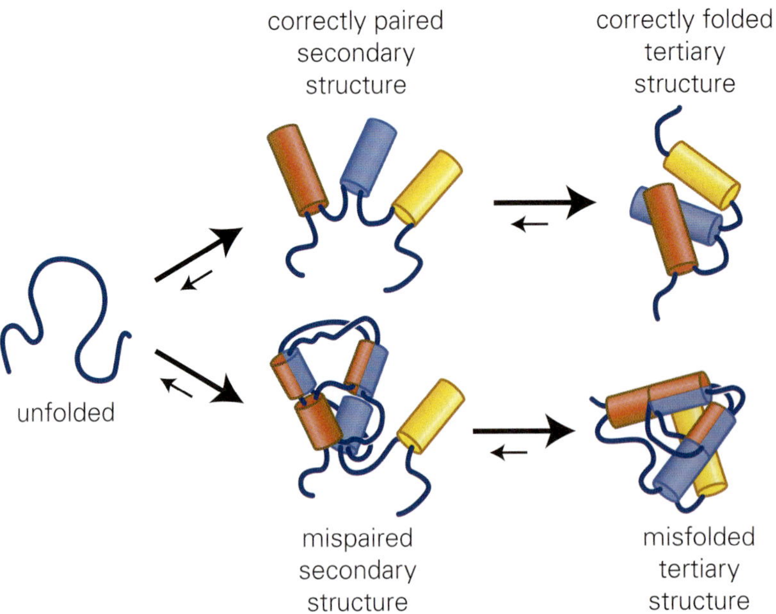


**Figure 18.48 RNA misfolding.** As an RNA molecule folds, the formation of alternate stable secondary structures (double helices, indicated in this schematic diagram as cylinders) can lead to misfolded structures.

The propensity of RNAs to easily misfold and become trapped in inactive conformations poses challenges for RNA folding. There are two distinct challenges—a thermodynamic one and a kinetic one. The thermodynamic challenge is that the nucleotide sequence of an RNA molecule must specify a single native structure that is more stable than all other possible structures at equilibrium (**Figure 18.49**). The fact that RNA secondary structure is highly stable (Chapter 2) suggests that it may be difficult for the sequences of RNAs to specify unique structures.

The kinetic problem is that the RNA molecule must be able to quickly form the native structure that allows it to function. This requires that the free-energy barrier to forming the folded structure ($\Delta G^{\ddagger}$) be sufficiently small, but also that the barriers to forming misfolded structures be higher (**Figure 18.50**). Again, because of similarities in the energetics of forming correctly and incorrectly folded structures, it is likely that the heights of the free-energy barriers will be similar in both cases. Indeed, many experiments have shown that RNAs have a tendency to form long-lived misfolded conformations. Many RNAs fold very slowly *in vitro*, requiring times on the order of minutes or longer. The conversion of the misfolded, non-native intermediates to the native state is slow because the RNA has to at least partially unfold before it can try to fold again.

The rate at which RNAs can fold will ultimately determine how quickly they can execute their biological functions. As for proteins, there are molecular chaperones that are specialized for the folding of RNA. Some of these are ATP-dependent motors, known as helicases, that can unwind double helices and cause incorrectly folded RNAs to unfold and attempt folding again.

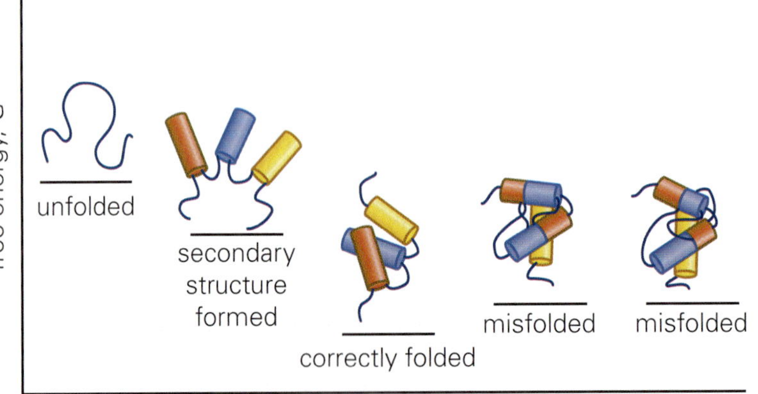


**Figure 18.49 The thermodynamic challenge in RNA folding.** In order for an RNA molecule to fold correctly, the free energy of the correctly folded, native state must be lower than that of the misfolded states. Some misfolded states have similar double-helical structures and stabilities to those of the correctly folded, native state, and therefore compete with the native state.

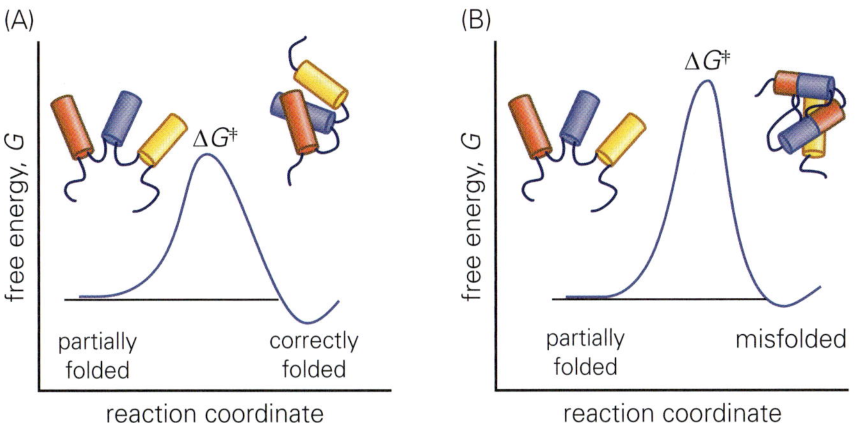


**Figure 18.50 The kinetic challenge in RNA folding.** (A) In order for RNA molecules to fold rapidly, the free-energy barrier ($\Delta G^{\ddagger}$) between the partially folded state and the folded state should be low. (B) At the same time, the barriers to the formation of misfolded states should be higher. In practice this can be difficult to achieve.

## 18.26 RNA folding is hierarchical with multiple stable intermediates

That RNA folding is hierarchical, with multiple folding pathways and distinct intermediates, has been established by probing RNA structure during folding. One method that provides information on structural changes in RNA is time-resolved **hydroxyl radical footprinting** (**Figure 18.51**).

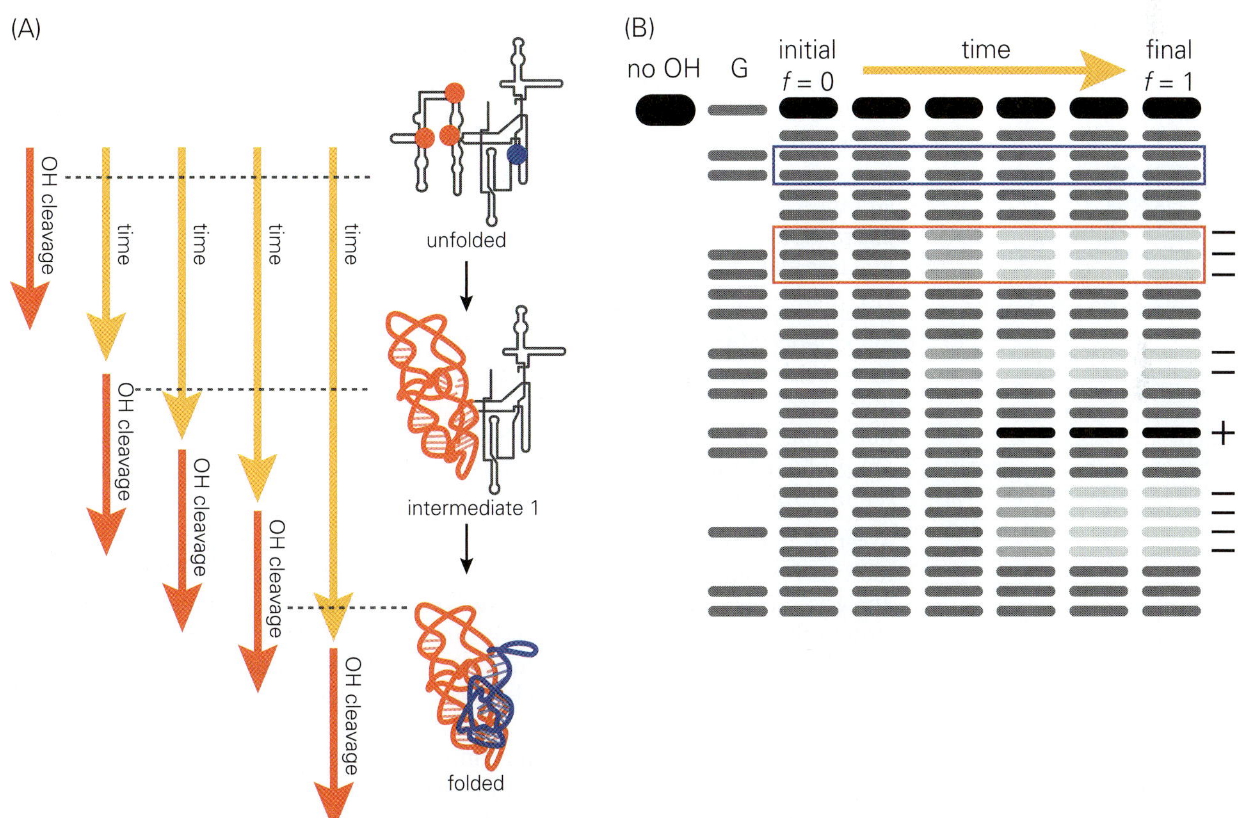


**Figure 18.51 Monitoring RNA folding by hydroxyl radical footprinting.** (A) A schematic representation of folding reactions that are allowed to proceed for different lengths of time, followed by a short burst of hydroxyl radical cleavage of the RNA. Hydroxyl radicals cleave the nucleic acid backbone where it is accessible to solvent. As RNA folding proceeds with time, some RNA backbone positions change their solvent accessibility (*red* and *blue* circles on the schematic representation of the RNA secondary structure). (B) The changes in solvent accessibility are visualized by high-resolution denaturing gel electrophoresis as increases (shown by a plus sign) or decreases (shown by minus signs) in band intensity with increased reaction time. The lane marked "initial" represents no cleavage. The "G" lane provides information on the RNA sequence. The lane marked $f = 0$ is the solvent accessibility of the RNA in its initial state (fraction folded = 0). The next four lanes report on the solvent accessibility of the backbone positions of the RNA as a function of time. The lane marked as "final, $f = 1$" reports on the solvent accessibility of the fully folded sample. (Adapted from I. Shcherbakova and M. Brenowitz, *Nat. Protoc.* 3: 288–302, 2008. With permission from Macmillan Publishers Ltd.)

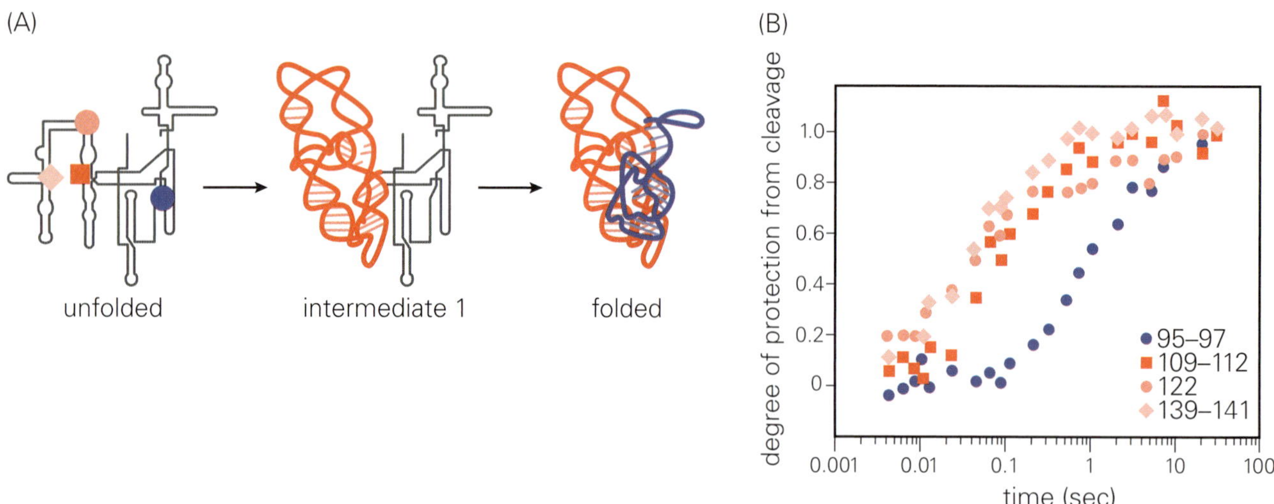


**Figure 18.52 Kinetics of RNA folding as measured by hydroxyl radical footprinting.** (A) Schematic representation of the folding process of a ribozyme. The sites of rapid nucleotide protection are in the *red* clusters, whereas the *blue* cluster of nucleotides are protected more slowly. The nucleotides within the *blue* cluster form the catalytic core of the ribozyme. (B) Time course of increase in protection from cleavage for various regions. The residues being monitored are indicated within the graph. (Adapted from I. Shcherbakova and M. Brenowitz, *Nat. Protoc.* 3: 288–302, 2008. With permission from Macmillan Publishers Ltd.)

A footprinting assay monitors changes in the sensitivity of individual residues to modification or cleavage. The hydroxyl radical is a particularly good footprinting reagent because it is small and it can cleave every exposed residue of a macromolecule, without regard to sequence (see Figure 18.51). Single- and double-stranded forms of RNA are susceptible to cleavage, while those involved in tertiary interactions are more likely to be inaccessible to hydroxyl radicals, and are therefore less likely to be cleaved.

**Hydroxyl radical footprinting**

Using this method it is possible to distinguish solvent-exposed regions of the RNA from those that are buried in the folded structure. Strand cleavage of exposed segments of the RNA backbone occurs rapidly in the presence of hydroxyl radical (•OH), while residues that are on the interior of a folded RNA are protected from cleavage.

Using time-resolved hydroxyl radical cleavage, it is possible to determine the folding kinetics of the *Tetrahymena* group I ribozyme by quantifying the changes in the sensitivity of individual sites and regions to cleavage by hydroxyl radicals, as a function of time (see Figure 18.51). After initiating folding of the RNA by the addition of $Mg^{2+}$, specific nucleotides within a region of the RNA (*red* in **Figure 18.52**) are protected most rapidly. In contrast, the catalytic core (*blue* in Figure 18.52) requires minutes to become fully protected.

A set of folding pathways deduced from the regions protected from hydroxyl radical cleavage are shown for the *Tetrahymena* ribozyme in **Figure 18.53**. This analysis reveals several parallel folding pathways that are populated with three structurally distinct intermediates. It turns out that the catalytic core (the *blue* domain) folds more slowly than the other domains because some of the helices in it become paired incorrectly. At least some mispaired interactions must be broken and reformed correctly before the native structure can form. The intermediates detected in these experiments contain secondary structural elements (that may be native or nonnative) and they also contain some tertiary structure, as evidenced by the regions that are protected from cleavage by hydroxyl radicals.

## 18.27 Collapse is an early event in the folding of RNA

The hydroxyl radical footprinting analysis discussed in the previous section does not provide a direct view of the structure of RNA molecules as they fold. Recall from Section 18.3 that solution x-ray scattering makes it possible to follow the size and shape of macromolecules in solution. The pattern of scattering of x-rays by molecules can be used to generate low-resolution information on their shapes, and this can be done as a function of time after initiating folding.

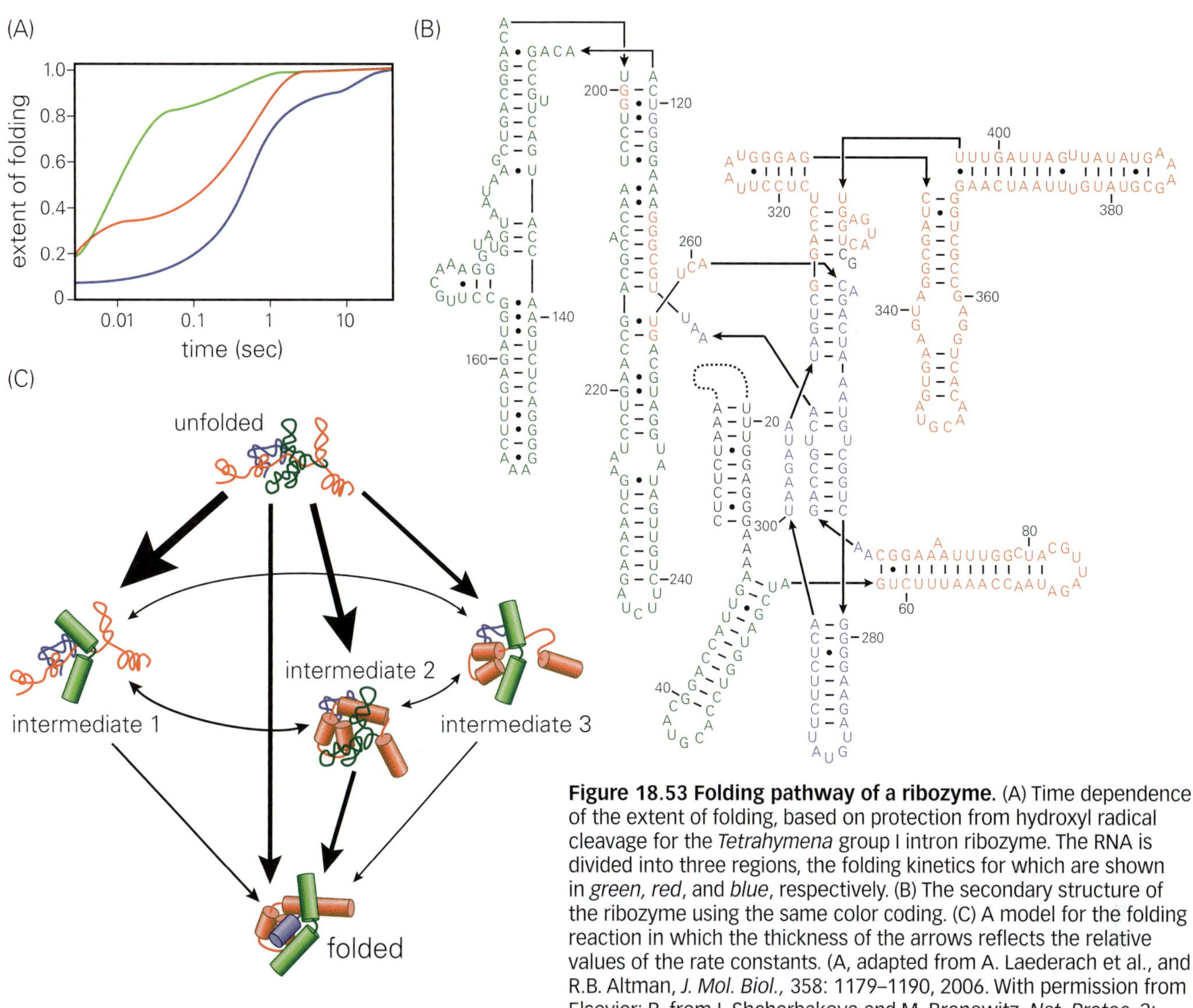


**Figure 18.53 Folding pathway of a ribozyme.** (A) Time dependence of the extent of folding, based on protection from hydroxyl radical cleavage for the *Tetrahymena* group I intron ribozyme. The RNA is divided into three regions, the folding kinetics for which are shown in *green, red*, and *blue*, respectively. (B) The secondary structure of the ribozyme using the same color coding. (C) A model for the folding reaction in which the thickness of the arrows reflects the relative values of the rate constants. (A, adapted from A. Laederach et al., and R.B. Altman, *J. Mol. Biol.*, 358: 1179–1190, 2006. With permission from Elsevier; B, from I. Shcherbakova and M. Brenowitz, *Nat. Protoc.* 3: 288–302, 2008. With permission from Macmillan Publishers Ltd.)

The folding of the *Tetrahymena* group I ribozyme was monitored by solution x-ray scattering as a function of time, yielding the model shown in **Figure 18.54**. The process of RNA collapse starts from the fully denatured state devoid of base pairing. Folding is initiated by adding metal ions. Some structural elements are in close proximity at 44 msec after folding is initiated and the shape of the RNA starts to resemble the final shape at 500 msec. The fully folded ribozyme is only slightly more compact and ordered than the structure seen at 500 msec. Thus, as folding begins, a substantial compaction in the low millisecond timescale was observed, with overall compaction and global shape changes largely complete within one second. The earliest detected tertiary structure is formed at least five-fold more slowly. The results suggest that the RNA rapidly forms a collapsed intermediate, which forms before there is significant formation of specific tertiary structure.

These experiments lead to a scheme for the folding of RNA, which is shown in **Figure 18.55**. The process begins with fully denatured or unfolded RNA completely devoid of base pairing (U). Association of diffuse metal ions with the unfolded RNA neutralizes most of the phosphate charge, allowing the formation of base pairs and local secondary structures. This is referred to as the pre-collapsed state (U*). The molecule then collapses into more compact conformations in which most of the secondary structures have formed (intermediate state, I). The collapse can be on the pathway towards folding, with native-like contacts, or off-pathway,

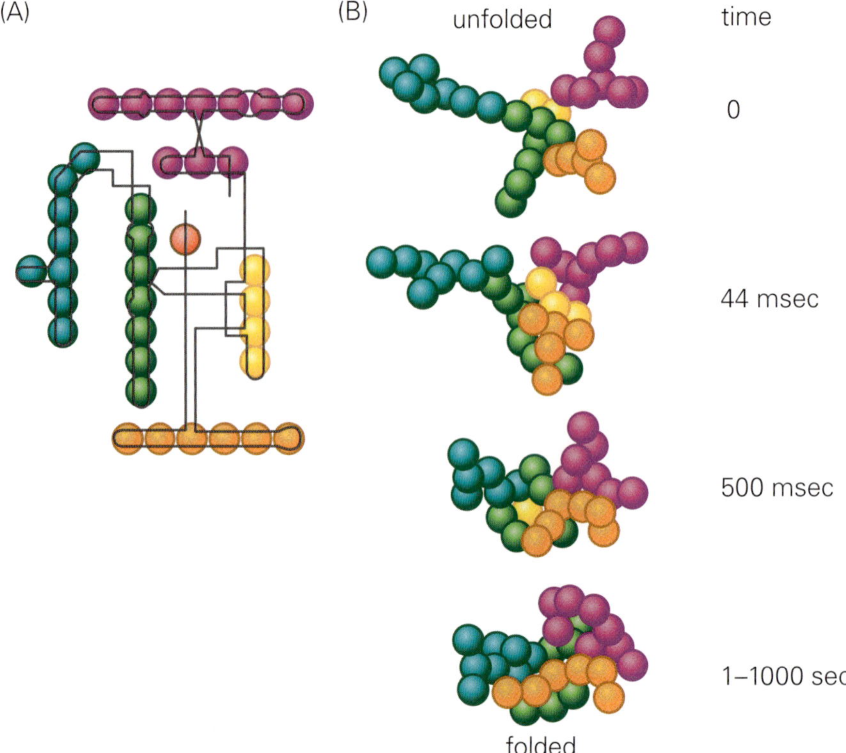


**Figure 18.54 Monitoring RNA folding by small-angle x-ray scattering**. (A) Secondary structure of the ribozyme. Each group of approximately five base pairs or five residues of a single-stranded region is represented as a colored sphere. (B) Time course of shape changes in ribozyme folding, based on solution x-ray scattering data. The spheres are as defined in (A). (Adapted from R. Russell et al. and L. Pollack, *Proc. Natl. Acad. Sci. USA* 99: 4266–4271, 2002. With permission from the National Academy of Sciences.)

resulting in a collection of nonnative intermediates. Additional rearrangements lead to the native tertiary structure (N), in which all of the secondary and tertiary structural elements have formed. The final native structure is stabilized by the binding of metal ions at specific sites.

## 18.28 RNA folding landscapes are highly rugged

In Section 18.12, we described the process of protein folding as movement on an energy landscape (see Figure 18.21). A similar conceptualization can be applied to RNA folding. Studies on the folding landscape of the group I intron ribozyme have revealed that the energy landscape is rugged (**Figure 18.56**). By "rugged" we mean that the energy landscape is composed of a series of crests and troughs of different heights and depths. In addition, the landscape contains a number of discrete minima. These minima are separated from each other by large energy barriers that make interconversion between them difficult. These local minima represent kinetic traps that increase the folding time of the RNA.

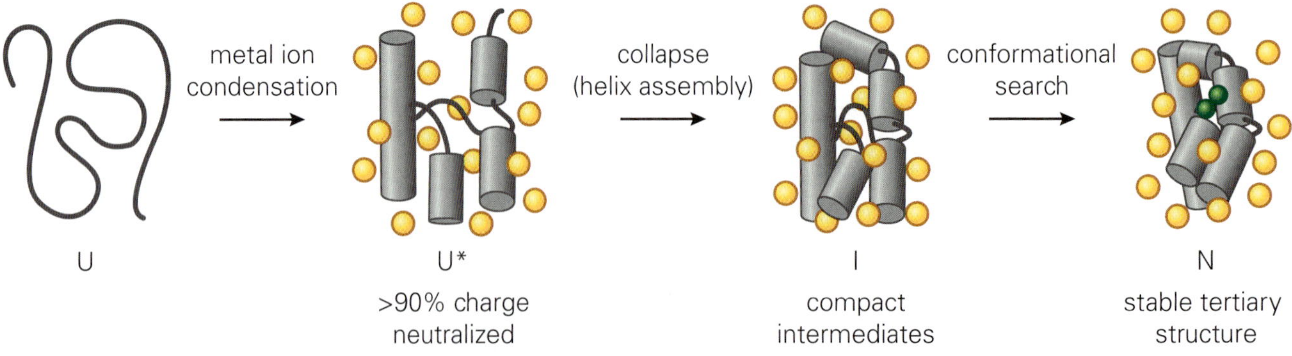


**Figure 18.55 Intermediates in the folding of RNA**. As folding begins, metal ions (*yellow*) associate nonspecifically with the unfolded RNA (U), resulting in more compact conformations (I). Subsequent conformational rearrangements lead to the native tertiary structure (N), which includes metals bound at specific sites (*green*). (Adapted from S.A. Woodson, *Curr. Opin. Chem. Biol.* 9: 104–109, 2005. With permission from Elsevier.)

Earlier in this chapter, in Section 18.1, we discussed how optical tweezers can be used to demonstrate the reversibility of protein folding (see Figure 18.4). We can do the same kind of experiment with RNA molecules, and directly watch a single RNA molecule hopping between different conformations as it unfolds and folds, again in response to pulling forces (**Figure 18.57**). These kinds of experiments provide direct information on the nature of the energy landscape for RNA folding.

Recall that this technique relies on a tightly focused laser beam (or beams) to trap a single molecule in three dimensions and, through movement of the beams, the molecules can be manipulated. Since biological molecules such as RNAs are often too small to be manipulated directly, the RNA is fused to duplex DNA "handles" that are attached to micron-sized beads. The beads are held between two optical traps or between one trap and a fixed mechanical support.

An experiment in which optical tweezers are used to pull on a single RNA hairpin is illustrated in Figure 18.57. In this experiment, the RNA hairpin is attached to DNA handles, each of which is tethered to a bead. One bead is held by a force-measuring laser trap, and the other is held by a micropipette. By moving the micropipette up or down, a force is exerted on the molecule, and the change in extension is measured. RNAs were stretched and relaxed repeatedly at fixed rates of increase of force (pN•sec$^{-1}$). A rip that increases the extension indicates unfolding, whereas folding is indicated by a "zip" that shortens the extension.

Once the RNA is unfolded in a single step (*blue* curve in Figure 18.57B; rip), if the hairpin is allowed to refold while the pulling force changes at rates of <1 pN•sec$^{-1}$, then a single transition (*red* curve in Figure 18.57B; zip) is observed (Figure 18.57B; single step). This is consistent with a single-step refolding process from an unfolded to a native structure. In contrast, when the RNA is allowed to

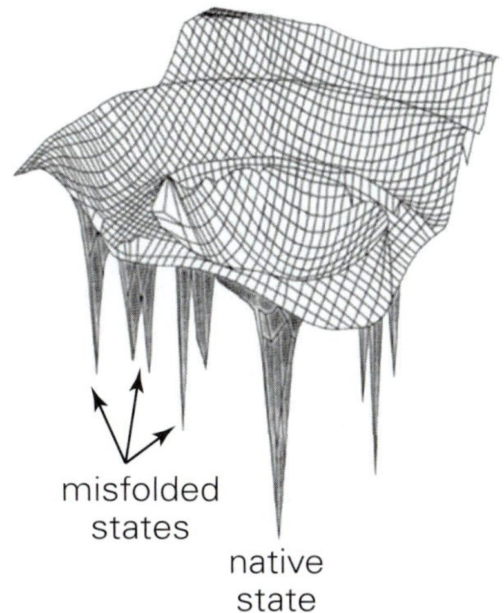


**Figure 18.56 An RNA folding landscape.** This diagram represents the energy of molecules as a function of their conformation (compare with Figure 18.21). Various local minima represent misfolded states or intermediates. The native state is at the global minimum. (From D. Thirumalai and S.A. Woodson, *Acc. Chem. Res.* 29: 433–443, 1996. With permission from the American Chemical Society.)

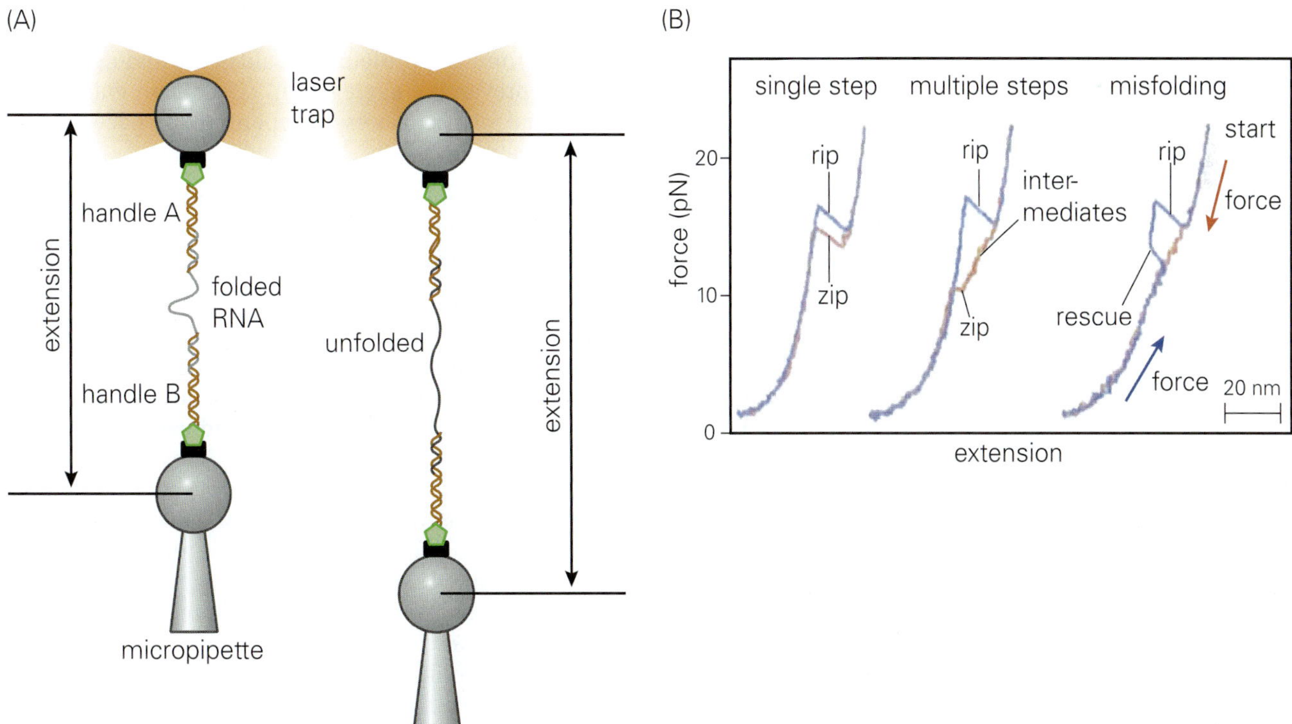


**Figure 18.57 Pulling on single molecules of RNA using optical tweezers**. (A) An RNA molecule is attached by two DNA "handles" bound to beads and manipulated using two optical traps. The RNA is stretched between the traps by an applied force. As RNA unfolds or refolds, the resulting change in extension is monitored. (B) Three types of force-extension curves for a hairpin RNA are observed. In each experiment, force was relaxed from 20 to 1 pN (*red*) and then raised back to 20 pN (*blue*). The increase in extension, a rip, indicates unfolding of the RNA. Folding and rescue are indicated by zips that shorten the extension. (Adapted from P.T. Li, J. Vieregg, and I. Tinoco, Jr., *Annu. Rev. Biochem*. 77: 77–100, 2008. With permission from Annual Reviews.)

refold while the pulling forces relax at a rate of 1.5 pN•sec$^{-1}$, most of the refolding curves show multiple-step refolding and small back and forth oscillations (Figure 18.57B; *red* curve, multiple steps) followed by a small zip. These oscillations most likely represent successive refolding and unfolding events as the RNA moves in and out of competing folding pathways. If the RNA forms an intermediate with native contacts, then it forms the remaining native structure cooperatively and the zip is observed.

Finally, when the RNA refolds while the pulling force is reduced at rates >2 pN sec$^{-1}$, a force extension curve with a transition known as a rescue is observed (Figure 18.57B; misfolding). This transition likely represents refolding of misfolded RNA molecules, allowing them to refold correctly to form the native hairpin RNA. This refolding seems to occur through multiple small unfolding/refolding steps and not through the complete unfolding of the RNA.

## Summary

Protein folding can be broadly understood in terms of thermodynamic principles operating on the interactions of different parts of the polypeptide chain in aqueous solution. Many proteins, particularly small ones, can fold rapidly to their native conformation without help. This process occurs through fluctuations in conformation, with consequent favorable interactions between different parts of the polypeptide chain. The rate at which this process is completed has been measured for purified proteins, and it has been found to be highly variable and correlated with the contact order (a measure of whether distant parts of the sequence must come together during folding).

When biologists first began to recognize the importance of molecular chaperones, which are required for the process of folding some proteins, there was some concern that a kind of "vital principle" was being introduced that made protein folding immune to simple reductionist thinking. As we have emphasized in this chapter, it is now clear that chaperones are proteins that aid in the folding of other proteins, with a primary role of preventing aggregation, rather than guiding the substrate protein during the actual folding process. Chaperones are necessary for the folding of a wide range of proteins in cells.

In spite of the action of chaperone proteins, protein aggregation can occur *in vivo*. This process seems to occur more frequently with increasing age, for reasons that are not yet understood, and can lead to formation of amyloid aggregates that are associated with cellular toxicity and disorders such as Alzheimer's disease.

The two chaperones that are best understood at present are the Hsp70 and GroEL systems. These two protein assemblies are unrelated in terms of their architecture and mechanism, but they have two key properties in common. Both systems have peptide-binding domains that are able to recognize and bind extended protein chains, principally through interactions with exposed hydrophobic sidechains in unfolded proteins. In both cases, the conformation of the peptide-binding domain undergoes cyclical conformational changes that release the substrate protein and then allow rebinding to occur. These conformational changes are driven by the binding and hydrolysis of ATP, and the cell expends many ATP molecules for every cycle of substrate binding and release by these chaperones. There is, however, one crucial difference between how Hsp70 and GroEL operate: in addition to preventing aggregation, GroEL provides an "Anfinsen cage" within which the process of folding can be completed in isolation from other proteins.

The ability of RNA to function in various biological processes in the cell depends on its ability to adopt elaborate and specific structures. This depends, in turn, on its ability to fold properly, with metal ions playing important roles in RNA folding. Much of the stabilization of an RNA is provided by the diffuse metal ions that are captured by the electrostatic field of the RNA. Because of the large electrostatic repulsion between helices (due to the polyelectrolyte nature of RNA), the ions in the "ion atmosphere" that surrounds the RNA provides a critical electrostatic contribution to the folding energetics, usually far greater than the contribution from ions bound to specific sites with the RNA.

RNA folding is a hierarchical process in which independently stable secondary structures form before tertiary structure. Thus, RNA folding typically begins with collapse from the expanded, unfolded state into a compact conformation, followed by additional conformational rearrangements that lead to the native tertiary structure. Large multidomain RNAs typically fold along multiple pathways through a rough or rugged landscape, where local energy minima, corresponding to intermediate states, act as traps that slow down the folding rate. These intermediates can be converted to the native state, but the transition is slow, because at least some interactions must be broken and reformed correctly in order to form the native structure.

## Key Concepts

### A. HOW PROTEINS FOLD

- The unfolded states of proteins correspond to wide distributions of different conformations.
- Protein folding is often fast and cannot be explained by an exhaustive search of conformational space.
- Many small proteins populate only fully unfolded and fully folded states.
- Folding rates are faster when residues close in sequence end up close together in the folded structure. That is, there is an inverse relationship between the folding rate and the contact order of a protein.
- The folding of some proteins involves the formation of transiently stable intermediates.
- The nature of the transition state for folding can be identified by mapping the effect of mutations on the folding and unfolding rates. This process is known as the $\phi$-value analysis.
- The process of protein folding can be described as movement on a multidimensional free-energy surface, with a funnel leading to the folded structure.

### B. CHAPERONES FOR PROTEIN FOLDING

- Many proteins tend to aggregate faster than they fold, particularly in a cellular environment.
- Some proteins form irreversible aggregates, nucleated around a cross-β spine core and known as amyloids, that are toxic to cells.
- Molecular chaperones are proteins that prevent protein aggregation.
- Hsp70 recognizes short peptides with sequences that are characteristic of the interior segments of proteins.
- Hsp70 binds and releases protein chains in a cycle that is coupled to ATP binding and hydrolysis.
- The GroEL chaperonin forms a hollow double-ring structure within which protein molecules can fold.
- GroEL works like a two-stroke engine, binding and releasing proteins.

### C. RNA FOLDING

- The electrostatic field around RNA leads to the diffuse localization of metal ions.
- RNA folding can be driven by increasing the concentration of metal ions.
- RNAs form stable secondary structural elements, which increases their tendency to misfold.
- RNA folding is hierarchical, with multiple stable intermediates.
- Collapse is an early event in the folding of RNA.
- RNA folding landscapes are highly rugged.

## Problems

### True/False and Multiple Choice

1. Unlike protein folding, RNA folding is generally hierarchical.

   True/False

2. Which of the following effects favors RNA folding?
   a. Interactions between phosphate groups.
   b. Kinetic traps of alternative structures.
   c. Neutralization of backbone charge by counterions.
   d. The entropy of the native state compared to the unfolded state.

3. Which of the following is least likely to lead to RNA unfolding?
   a. Increasing [$Mg^{2+}$] by 1 mM.
   b. Increasing the temperature to 75°C.
   c. Adding 1 M EDTA.
   d. Adding 6 M urea.

4. Which of the following statements is true at $T_m$ for a monomeric protein or RNA?
   a. $\Delta G_{folding}$ equals 0.
   b. Equal concentrations of folded and unfolded protein are present.
   c. $K_{folding}$ equals 1.
   d. The unfolding curve is at its midpoint.
   e. All of the above.

5. A protein with two-state folding populates two intermediates as folding occurs.

   True/False

6. Which of the following statements about molecular chaperones is *not* true?
   a. They accelerate the folding of client proteins.
   b. They prevent protein aggregation.
   c. They are required for *in vitro*, but not *in vivo*, folding .
   d. They recognize exposed hydrophobic segments.
   e. They can bind and release proteins several times during a folding cycle.

7. Amyloid fibrils are formed by zipper-like hydrogen bonds within and across β sheets.

   True/False

### Fill in the Blank

8. Counterions for nucleic acids carry a __________ charge.

9. In hydroxyl radical footprinting, regions with ________ solvent accessibility are likely to be cleaved.

10. When local secondary structure elements form quickly, protein folding is said to follow a ____________ mechanism. In contrast, when distant contacts are quickly formed without secondary structure formation, folding is said to follow a ____________ mechanism.

11. A mutation that has similar interactions in the folding transition state as in the folded state has a _________ of 1.

12. Folding can be visualized as movement on a multidimensional surface known as a _____________.

### Quantitative/Essay

(Assume $T$ = 300 K and $RT$ = 2.5 kJ•mol$^{-1}$ for all questions.)

13. The concept of hierarchy is important for understanding the differences between RNA and protein folding. How do the two processes differ in structural stability hierarchy and folding kinetics hierarchy?

14. Below are reaction coordinate diagrams for two RNA molecules (A and B). For each molecule of RNA, we present a reaction coordinate for a misfolded conformation (left) and a correctly folded conformation (right). Which of the RNA folding reactions is more likely to be caught in a misfolded kinetic trap? Explain your reasoning.

(A)

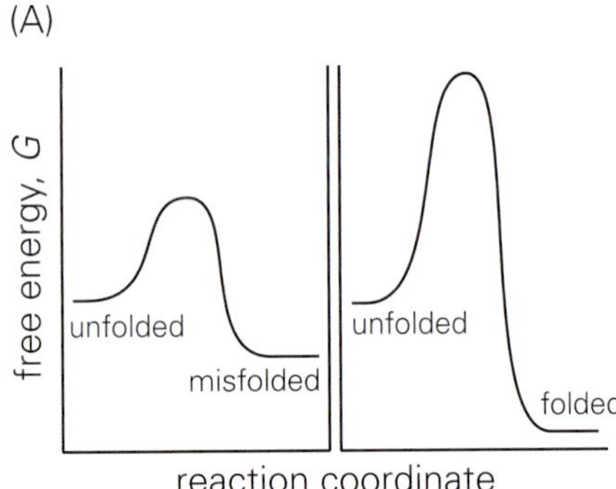


(B)

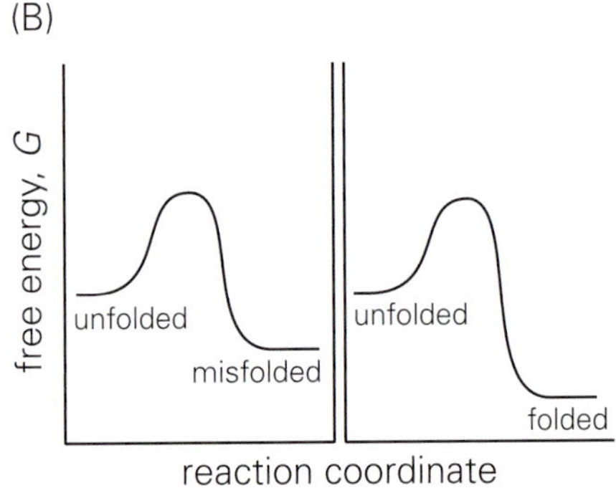

15. Below is a hydroxyl radical footprint, similar to data shown in Figure 18.51 for a 12-nucleotide segment of RNA.

a. Which region becomes protected first?
b. Which region becomes more solvent accessible during the folding reaction?

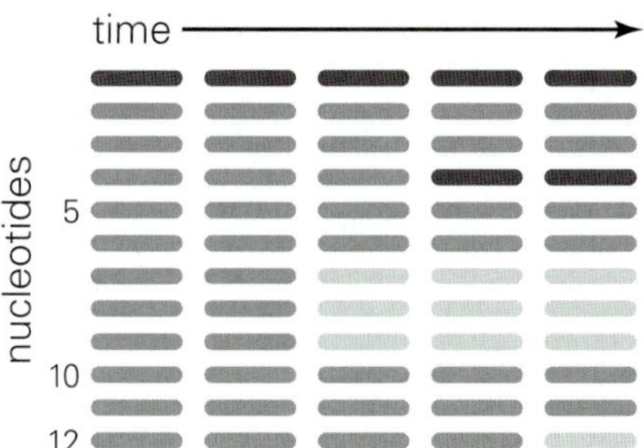


16. How do rugged and smooth energy landscapes differ with regard to the number of local minima, interconversions between minima, and kinetic traps?

17. A mutation of core residues in the *E. coli* protein Rop speeds up the rate of protein folding from 0.013 $sec^{-1}$ to 1.3 $sec^{-1}$.

a. What is the effect of the mutation on the free energy of the transition state relative to the unfolded state?
b. Given that a simple mutation can increase the folding rate, what does this result suggest about the relationship between the protein folding rate and natural selection? (Data based on M. Munson, K.S. Anderson, and L. Regan, *Structure: Folding and Design*, 2: 77–87, 1997.)

18. Shown below are two equally populated alternative conformations of a simplified protein with three β strands. Which conformation is likely to fold and unfold more rapidly?

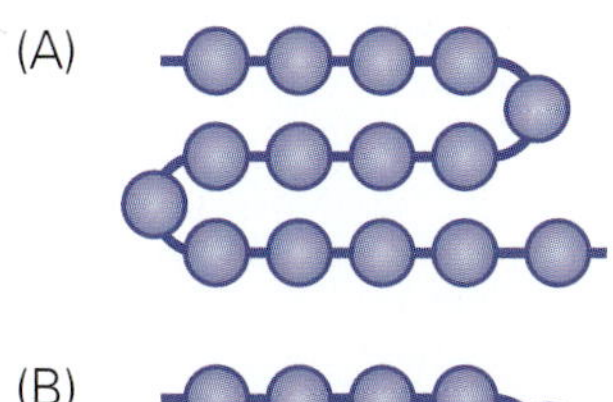


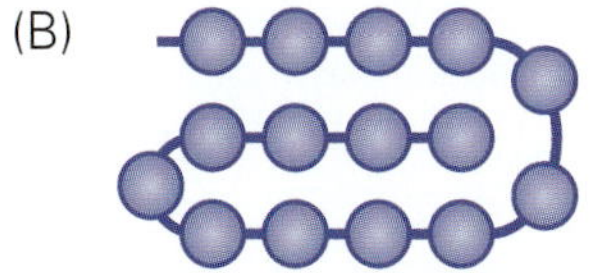


19. The following data were obtained from a mutational analysis on the folding kinetics of the Fyn SH3 domain:

| | $k_f$ ($sec^{-1}$) | $k_u$ ($sec^{-1}$) |
|---|---|---|
| wild type | 30.2 | 0.5 |
| I28A | 1.24 | 1.8 |
| V55A | 29.1 | 18.1 |

Use these data to answer the following questions:
a. What are the $\phi$ values of the two mutants?
b. Which residue forms more native interactions at the transition state? (Data from J.G. Northey, A.A. Di Nardo, and A.R. Davidson, *Nat. Struct. Biol.* 9: 389–402, 2002.)

20. The $\phi$-value analysis of the previous problem is continued for an additional mutant, F4S. This mutation has a $\phi$ value of 0.21. If the folding rate is two times slower for the mutant than for the wild type, what is the unfolding rate of F4S?

21. Why is the rate of protein folding limited to $10^6$ $sec^{-1}$, even in the absence of any specific free-energy barriers?

22. Shown below is a plot of relative fluorescence (reporting on fraction folded) versus time for a protein in the absence (solid line) and presence (dashed line) of a molecular chaperone.

a. How many folding intermediates occur when the protein folds without assistance?
b. What is the effect of the molecular chaperone?

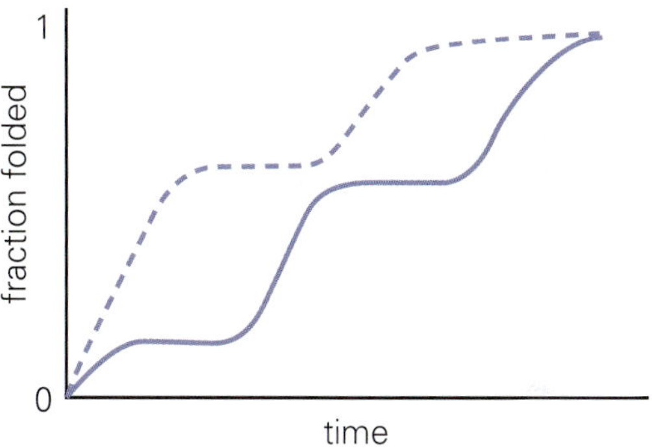


23. How might ATP binding and hydrolysis contribute to the active and passive mechanisms of folding mediated by GroEL-GroES?

24. A solution x-ray scattering experiment for a protein is performed in increasing amounts of urea. The data show a difference in the radius of gyration at 6 M urea and 0.1 M urea.

a. Why is the radius of gyration different between an unfolded and folded protein even though both contain the same number of residues?
b. How will the measured radius of gyration depend on the concentration of urea?

# Further Reading

## General

Fersht A (1999) Structure and Mechanism in Protein Science : A Guide to Enzyme Catalysis and Protein Folding. New York: W.H. Freeman.

Thirumalai D & Woodson SA (1996) Kinetics of folding of proteins and RNA. *Acc. Chem. Res.* 29, 433–439.

## References

### A. How proteins fold

Dill KA (1990) Dominant forces in protein folding. *Biochemistry* 29, 7133–7155.

Dill KA & Chan HS (1997) From Levinthal to pathways to funnels. *Nat. Struct. Biol.* 4, 10–19.

Dobson CM (2003) Protein folding and misfolding. *Nature* 426, 884–890.

Fersht AR (2008) From the first protein structures to our current knowledge of protein folding: delights and scepticisms. *Nat. Rev. Mol. Cell Biol.* 9, 650–654.

Nishimura C, Prytulla S, Dyson HJ & Wright PE (2000) Conservation of folding pathways in evolutionarily distant globin sequences. *Nat. Struct. Biol.* 7, 679–686.

Plaxco KW, Simons KT, Ruczinski I & Baker D (2000) Topology, stability, sequence, and length: Defining the determinants of two-state protein folding kinetics. *Biochemistry* 39, 11177–11183.

Udgaonkar JB (2008) Multiple routes and structural heterogeneities in protein folding. *Annu. Rev. Biophys.* 37, 489–510.

### B. Chaperones for protein folding

Buchner J (1996) Supervising the fold: Functional principles of molecular chaperones. *FASEB J.* 10, 10–19.

Bukau B, Weissman J & Horwich A (2006) Molecular chaperones and protein quality control. *Cell* 125, 443–444.

Chiti F & Dobson CM (2006) Protein misfolding, functional amyloid, and human disease. *Annu. Rev. Biochem.* 75, 333–366.

Flynn GC, Pohl J, Flocco MT & Rothman JE (1991) Peptide-binding specificity of the molecular chaperone BiP. *Nature* 353, 726–730.

Hartl FU (2011) Chaperone assisted protein folding: the path to discovery from a personal perspective. *Nat. Med.* 17, 1206–1210.

Hartl FU (1996) Molecular chaperones in cellular protein folding. *Nature* 381, 571–579.

Horwich AL (2011) Protein folding in the cell: an inside story. *Nat. Med.* 17, 1211–1216.

Prusiner SB (1998) Prions. *Proc. Natl. Acad. Sci. USA* 95, 13363–13368.

Saibil HR (2008) Chaperone machines in action. *Curr. Op. Struct. Biol.* 18, 35–42.

Whitesell L & Lindquist SL (2005) HSP90 and the chaperoning of cancer. *Nat. Rev. Cancer* 5, 761–772.

### C. RNA folding

Baird NJ, Fang XW, Srividya N, Pan T & Sosnick TR (2007) Folding of a universal ribozyme: The ribonuclease P RNA. *Q. Rev. Biophys.* 40(2), 113–161.

Chu VB & Herschlag D (2008) Unwinding RNA's secrets: Advances in the biology, physics, and modeling of complex RNAs. *Curr. Opin. Struct. Biol.* 18, 305–314.

Draper DE (2008) RNA folding: Thermodynamic and molecular descriptions of the roles of ions. *Biophys. J.* 95, 5489–5495.

Li PTX, Vieregg J & Tinoco I (2008) How RNA unfolds and refolds. *Annu. Rev. Biochem.* 77, 77–100.

Pyle AM (2002) Metal ions in the structure and function of RNA. *J. Biol. Inorg. Chem.* 7–8, 679–690.

Thirumalai D & Woodson SA (2000) Maximizing RNA folding rates: A balancing act. *RNA* 6, 790–794.

Shcherbakova I & Brenowitz M (2008) Monitoring structural changes in nucleic acids with single residue spatial and millisecond time resolution by quantitative hydroxyl radical footprinting. *Nat. Protoc.* 3(2), 288–302.

# CHAPTER 19

# Fidelity in DNA and Protein Synthesis

Genomic DNA in cells is present as Watson-Crick base-paired double helices. During replication, this duplex structure must be opened so that polymerase enzymes can access the base sequence in order to synthesize new strands of DNA that are complementary in sequence to the original. DNA replication occurs with extremely high fidelity, with only one base in approximately ten to one hundred million incorrect at the completion of replication. This high accuracy in copying is necessary to prevent mutations of the DNA that can affect the ability of the cell to survive. It is remarkable that nature can maintain very high fidelity while copying DNA at a rate of well over one hundred base pairs per second.

A conceptually related process in cells is the translation of a messenger RNA sequence into a protein by the ribosome. Although errors in this process do not lead to permanent genetic changes (as errors in DNA replication do), incorrect amino acid incorporation can affect protein stability and activity. As discussed in Chapter 1, transfer RNAs act as adaptors that carry the proper amino acid for each codon into the ribosome, allowing it to be added to the growing protein chain. Base-pairing between the three bases of the tRNA anticodon and the codon in the mRNA is a primary element of selecting the correct amino acid for insertion. The ribosomes are very good at selecting the correct amino acids, and we will examine how they accomplish this.

In this chapter, we will first develop a thermodynamic analysis to predict the extent to which the differences in the free energy of pairing for correct versus incorrect bases contribute to the fidelity of DNA replication and codon–anticodon recognition. To do this, we use the ideas of chemical equilibrium to develop a quantitative understanding of the thermodynamic stability of the duplex form of nucleic acids relative to separated single strands. This analysis relates closely to that done in part D of Chapter 10, where we described the stability of the folded state of a protein relative to the unfolded, random coil.

We will see that thermodynamic contributions from base-pairing are far from sufficient to explain the high fidelity of the biological synthesis processes. We must thus also consider the rates of addition of nucleotides to the new copy of DNA, or amino acids to the growing protein chain, including how the enzymes affect the binding of the nucleotide to be added and how they accelerate the chemical step that elongates the growing polymer chain. Additionally, proofreading and error correction by the enzymes are also important. Together these processes give polymerases and ribosomes their remarkable ability to synthesize DNA, RNA, and proteins with minimal introduction of errors.

# A. MEASURING THE STABILITY OF DNA DUPLEXES

## 19.1 The difference in free energy between matched and mismatched base pairs can be determined by measuring the melting temperature of DNA

What is in the intrinsic ability of the DNA template to choose the correctly matched nucleotide at each position, based on the Watson-Crick base-pairing rules? As we discuss in Section 19.9, DNA synthesis involves a sequence of condensation reactions in which nucleotide triphosphates react with the 3′ end of a DNA strand (the primer) that forms a double helix with a longer DNA strand (the template). To restate the question in thermodynamic terms, what is the difference in free energy between correctly matched nucleotide triphosphates forming base pairs with the template, as opposed to incorrectly matched ones?

We could, in principle, measure the free energy of binding of nucleotide triphosphates to DNA, but such measurements are challenging because these interactions are quite weak. An approach that is more accessible to experimentation is to measure, instead, the equilibrium between the single-stranded and double-helical forms of two DNA strands. We can compare the stabilities of duplex DNA molecules in which a terminal, correctly paired nucleotide is replaced by each alternative nucleotide. That is, in one duplex the terminal base pair forms a Watson-Crick pair, whereas in the other three, the terminal base pair is a mismatch. The resulting differences in stability, due to differences in base-pairing and stacking, are a measure of the relative strengths of interaction between the template strand and correctly paired or mismatched bases of the partner strands (**Figure 19.1**). Base-pairing and stacking are also used in selection of the incoming

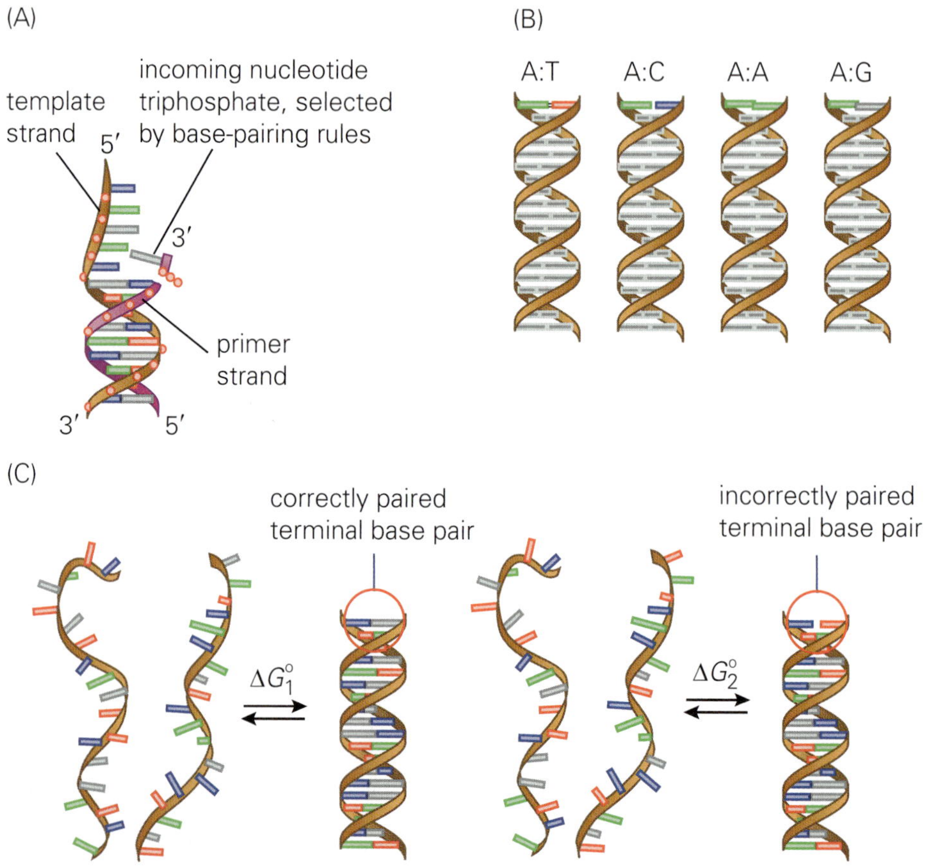

**Figure 19.1 DNA duplexes with mismatches.** (A) DNA with a primer-template junction, corresponding to the substrate in the DNA synthesis reaction. An incoming nucleotide triphosphate is shown, at the step before the chemical reaction to form a new base pair. (B) Duplex DNA with a correctly paired terminal base pair (left) and mismatched terminal base pairs (all others). In practice it is easier to measure the equilibrium between single-stranded and double-stranded DNA, as indicated in (C), than to measure the binding of nucleotide triphosphates to primer–template junctions (shown in A). By comparing the free energies of binding in the two cases shown in C, the free-energy difference between correctly and incorrectly paired nucleotides at primer–template junctions can be estimated.

nucleotides during replication, and so the interactions that determine the stability of the duplexes should be similar to those that determine the binding of nucleotides to be added during DNA replication.

To determine the difference in free energy between correctly matched and wrongly matched sequences, $\Delta G^{o}_{c\rightarrow w}$ , we will compare the two binding equilibria shown in Figure 19.1C. $\Delta G^{o}_{1}$ is the standard free-energy change of double helix formation when the terminal bases of the two DNA stands are correctly paired, and $\Delta G^{o}_{2}$ is the free energy change when the terminal bases are mismatched. The free energy difference between correctly and incorrectly paired bases is then simply given by $\Delta G^{o}_{c\rightarrow w} = \Delta\Delta G^{o} = \Delta G^{o}_{2} - \Delta G^{o}_{1}$.

If we denote one strand in a double-helical piece of DNA as A and the other strand as B, the strands undergo the reaction shown in **Figure 19.2** when they convert between the double-helix and random-coil states. In the single-stranded form, the conformation of each strand is random, whereas the conformation of each strand becomes helical when bound to each other. To denote this explicitly, we write the equilibrium as follows:

$$\underset{\text{double helix}}{H_A \bullet H_B} \xrightleftharpoons{\Delta G^{o}_{H\rightarrow R}} \underset{\text{single strands}}{R_A + R_B} \tag{19.1}$$

where R and H denote the random and helical conformations, respectively (see Figure 19.2). We wish to measure $\Delta G^{o}_{H\rightarrow R}$, which is the standard free-energy change for conversion of the strands from the helical (H) to the random conformation form (R):

$$\Delta G^{o}_{H\rightarrow R} = \Delta H^{o}_{H\rightarrow R} - T\Delta S^{o}_{H\rightarrow R} \tag{19.2}$$

where $\Delta H^{o}_{H\rightarrow R}$ and $\Delta S^{o}_{H\rightarrow R}$ are the changes in the standard enthalpy and entropy upon helix melting, respectively. In our analysis of the free energy of DNA melting, we shall consider not just $\Delta G^{o}_{H\rightarrow R}$ but also the underlying changes in enthalpy and entropy.

Most experimental measurements of helix stability start with double-helical DNA at low temperature and then determine the temperature at which it "melts" into single strands. The **melting temperature of a DNA duplex** ($T_M$) is defined as the temperature at which half of the DNA strands are in the single-stranded form and half are in the double-stranded form (**Figure 19.3**).

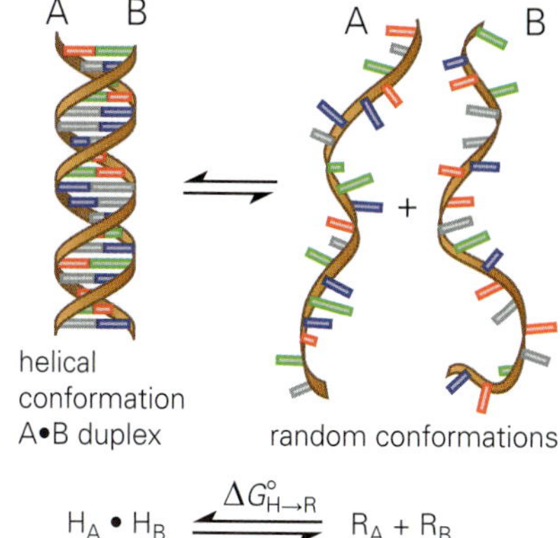


**Figure 19.2 DNA melting equilibrium.** Schematic representation of the equilibrium between single-stranded DNA (denoted "R" because their conformations are random) and double-stranded DNA (denoted "H" for helical). The two strands are labeled A and B.

**Melting temperature of a DNA duplex**

The melting temperature of a DNA duplex is defined as the temperature at which there are equal concentrations of the duplex and the random coil single strands.

## 19.2 DNA melting can be studied by UV absorption spectroscopy

Nucleotides absorb ultraviolet light strongly, but the exact strength of absorption depends on their structural context. The **molar absorptivity** of nucleotides increases when DNA converts from the double-helical form, in which bases are stacked atop one another, to single-stranded forms in which the bases are separated (see Figure 19.3). This phenomenon, arising from **hypochromicity** of the duplex DNA, allows us to determine the extent to which double-helical structure is formed under particular conditions of concentration and temperature. The absorbance of UV radiation by single-stranded DNA is higher than that of double-stranded DNA because the extent of base stacking is much lower in the flexible, random-coil state.

**Molar absorptivity**

Molar absorptivity is a characteristic of a compound that specifies the fraction of light of a particular wavelength that will be absorbed when the light passes through a 1-cm sample of a solution at 1 M concentration.

Figure 19.3 illustrates the change in absorbance at 260 nm (corresponding to the wavelength of maximal light absorbance by nucleotides in the ultraviolet region) as the temperature of a solution of DNA chains (oligonucleotides) is increased. At low temperatures, the absorbance of the sample is low. This corresponds to the situation in which the two strands are zippered up into a double-helical structure, and base stacking reduces the molar absorptivity of the sample. As the temperature is increased, the absorbance of the sample increases gradually, passes through a temperature range of steeper absorbance change and finally reaches a higher plateau value. The increase in the absorbance reflects the loss of helical

**Hypochromicity**

Hypochromicity refers to the absorption of light that is less than expected based on composition. For example, the stacking of bases in helical duplex DNA decreases absorptivity relative to that expected for the same nucleotide strands free in solution, so the duplex is hypochromic relative to single strands.

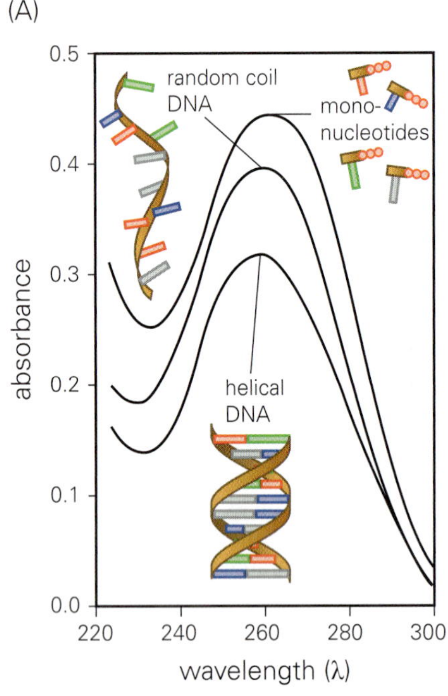


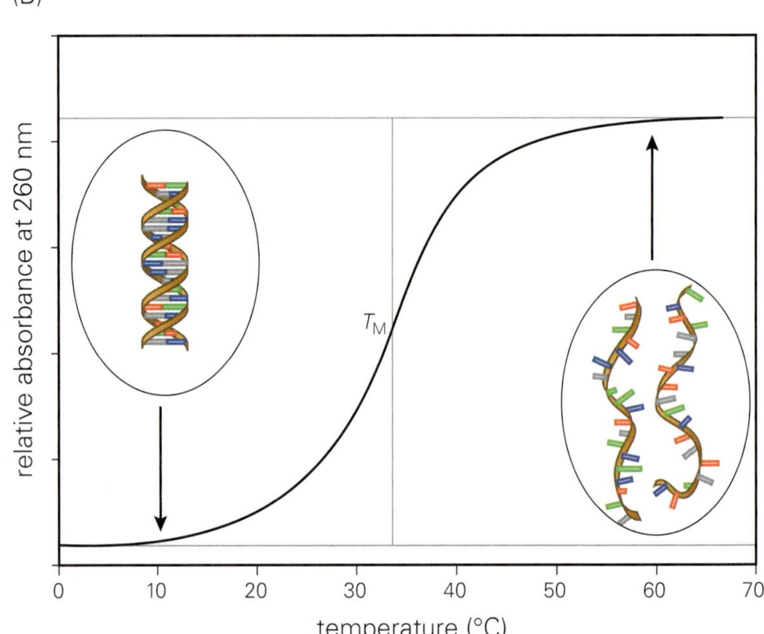


**Figure 19.3 The UV absorbance of DNA.** (A) Absorbance spectra of single-stranded (random-coil) DNA, double-helical DNA, and isolated nucleotides are shown. The maximum absorbance is at ~260 nm in all cases. Helical DNA has a molar absorbance that is lower than that of single-stranded random-coil DNA. (B) The absorbance of a solution of DNA, measured at 260 nm, is plotted as a function of temperature. At low temperatures the DNA is in the double-stranded form, and the molar absorptivity is low. At high temperatures the DNA melts to the single-stranded form, and the molar absorptivity is high. The melting temperature, $T_M$, is the temperature at which half of the DNA strands are in the double-helical conformation and half are in the random-coil conformation. (Adapted from W. Guschlbauer, Nucleic Acid Structure. Heidelberg: Springer-Verlag, 1976. With permission from Springer-Verlag.)

structure, with the sharp transition from low to high absorbance being characteristic of a cooperative structural transition. When one or two base pairs open up, the double helix loses stability, and the rest of the base pairs also open up more easily. This domino-like loss of structural integrity is reminiscent of the transition between the folded and the unfolded states of proteins, as was discussed in part D of Chapter 10.

The melting temperature of duplex DNA ($T_M$) depends on the DNA concentration because one duplex becomes two single strands. According to the principles of chemical equilibrium (see Section 10.7), increasing the total concentration of DNA strands favors formation of the double helix and thereby increases $T_M$. The concentration dependence of the melting temperature is illustrated in **Figure 19.4** for a pair of complementary nine-base oligonucleotides (short pieces of DNA). The melting temperature for the double helix formed by these oligonucleotides increases from ~30°C to ~45°C as the concentration of the oligonucleotides (that is, the total concentration of single strands) is increased from 11 μM to 52 μM (see Figure 19.4).

The determination of melting curves as a function of DNA concentration is very useful because it allows us to derive the free energy of double helix formation ($\Delta G^{o}_{R \to H}$) experimentally, as discussed below. Another common method for determining thermodynamic information about DNA is to use differential scanning calorimetry (discussed in Section 10.25 for proteins).

## 19.3 The changes in enthalpy and entropy associated with DNA melting can be determined from the concentration dependence of melting curves

Let us return to the discussion of the equilibrium between helical (H) and random-coil (R) states for a DNA (Equation 19.1 and Figure 19.2):

$$H_A \bullet H_B \rightleftharpoons R_A + R_B$$

The equilibrium constant, or dissociation constant, for this reaction is denoted $K_D$ and is given by:

$$K_D = \frac{[R_A][R_B]}{[H_A \bullet H_B]} \tag{19.3}$$

where $[H_A \bullet H_B]$ is the concentration of the double-helical form at equilibrium, and $[R_A]$ and $[R_B]$ are the concentrations of the random-coil forms of the A and B strands, respectively. If DNA strands A and B are added in equal amounts to the solution, then $[R_A] = [R_B]$.

There are two important differences between protein unfolding (discussed in part D of Chapter 10) and DNA melting. First, the unfolding of a monomeric protein is a unimolecular event (the reaction is first order, with one folded molecule converting to one unfolded molecule). As a consequence, for the unfolding of a monomeric protein, the value of the dissociation constant at the melting temperature ($T_m$) is 1.0 (that is, the concentration of unfolded protein is equal to that of the folded protein at $T_m$, as illustrated in Figure 10.26). As we shall see, this is not the case for DNA melting, because the reaction is second-order in terms of the concentration of DNA strands.

Second, recall from Section 10.26 that the heat capacities of the folded and unfolded forms of proteins are significantly different, with $\Delta C_P = C_P^u - C_P^f > 0$, where $C_P^u$ and $C_P^f$ are the heat capacities of the unfolded and folded proteins, respectively (see Figure 10.27). This leads to values for $\Delta H^o$ and $\Delta S^o$ that are temperature dependent, giving the protein stability curve shown in Figure 10.30. For DNA duplexes (and RNA as well), it is found experimentally that the value of $\Delta C_P$ is small, and we can assume that $\Delta C_P \approx 0$ for duplex to single strand transitions. This means that the values of $\Delta H^o$ and $\Delta S^o$ are temperature independent for nucleic acids. This makes it convenient to apply a van't Hoff analysis (see Section 10.14) to nucleic acid melting, as discussed below.

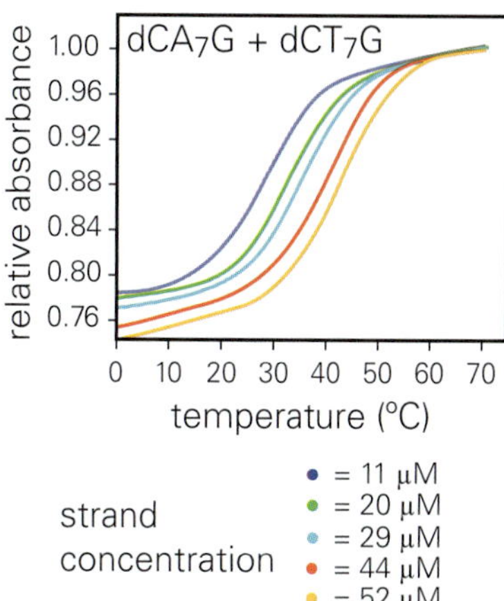


**Figure 19.4 Concentration dependence of DNA melting temperature.** Melting curves for a nine-nucleotide DNA duplex as a function of concentration. As the concentration of the DNA strands increases, the curve shifts to the right because the double-stranded form is favored at higher concentration and therefore melts at a higher temperature. (Adapted from F. Aboul-ela et al. and F.H. Martin, *Nucleic Acids Res.* 13: 4811–4824, 1985. With permission from Oxford University Press.)

Each of the melting curves in Figure 19.4 gives us the fraction of the strands that are in the double-helical form as a function of temperature, at a particular concentration of DNA. We could, in principle, calculate the equilibrium constant at different temperatures from a single melting curve determined for one concentration of DNA, since the equilibrium constant is related to the fraction of the strands that are in the double-helical form, and this can be determined from the absorbance. Such a procedure does not lead to very accurate values, however, because the equilibrium constant only changes substantially in a narrow temperature range near $T_M$. A better approach is to fit each melting curve to obtain the value of the melting temperature, $T_M$ (the temperature of the midpoint, at which half the strands are in the double-helical form). Let us now see how to relate the values of $T_M$ to a van't Hoff analysis.

The key to deriving thermodynamic parameters from melting temperature data is to recognize that, at the melting temperature, the equilibrium constant is related in a very simple manner to the concentration of the reactants. For DNA melting, if $C_{total}$ is the total concentration of DNA strands in the reaction ($C_{total} = [R_A] + [R_B] + 2[H_A \bullet H_B]$), then it turns out that, when the temperature is equal to $T_M$, the equilibrium constant is simply given by:

$$K_D = \frac{C_{total}}{4} \quad \text{(only at } T = T_M) \qquad (19.4)$$

See **Box 19.1** for a justification of Equation 19.4. We can substitute this value for the dissociation constant into the van't Hoff equation (Equation 10.84):

$$\ln K = -\frac{\Delta H^o}{RT} + \frac{\Delta S^o}{R} \qquad \text{(van't Hoff equation)}$$

to get the following equation:

$$\ln K_D = \ln\left(\frac{C_{total}}{4}\right) = \frac{-\Delta H^o_{H\to R}}{RT_M} + \frac{\Delta S^o_{H\to R}}{R} \quad \text{(only at } T = T_M) \qquad (19.5)$$

The $T_M$ value is determined for a series of samples at different, known DNA concentrations. By plotting the relationship between $\frac{1}{T_M}$ and $\ln\left(\frac{C_{total}}{4}\right)$ for such an experiment, we can directly obtain the values of $\Delta H^o_{H\to R}$ and $\Delta S^o_{H\to R}$ for the particular DNA molecules used in the experiment as the slope and intercept of the

## Box 19.1 The relationship between $K_D$ and $C_{total}$ at the melting temperature

In this justification we show that, when the temperature is equal to the melting temperature, $T_M$, the dissociation constant, $K_D$, for the double helix is related to the total concentration of DNA strands, $C_{total}$, as follows:

$$K_D = \frac{C_{total}}{4} \quad (19.1.1)$$

Consider the following equilibrium, where R refers to single-stranded DNA in a random conformation and H refers to strands in the double helical form:

$$R_A + R_B \rightleftharpoons H_A \bullet H_B \quad (19.1.2)$$

This is just a special case of the general binding equilibrium we have discussed earlier, in Chapter 12. For example, when a ligand (L) binds to a protein (P), we write:

$$P + L \rightleftharpoons P \bullet L \quad (19.1.3)$$

and the dissociation constant, $K_D$, for this binding equilibrium is given by:

$$K_D = \frac{[P][L]}{[P \bullet L]} \quad (19.1.4)$$

Recall that an important parameter is the fractional saturation, $f$, of the protein. As shown in Chapter 12 (see Equation 12.11):

$$f = \frac{\frac{[L]}{K_D}}{1 + \frac{[L]}{K_D}} = \frac{\text{concentration of protein bound to ligand}}{\text{total concentration of protein}} \quad (19.1.5)$$

We can recognize the intrinsic similarity of Equation 19.1.2 to Equation 19.1.3 and, for the DNA equilibrium, we can write:

$$f = \frac{\text{concentration of strand A in double-helical form}}{\text{total concentration of strand A}}$$

$$= \frac{[H_A \bullet H_B]}{[R_A] + [H_A \bullet H_B]} \quad (19.1.6)$$

At the melting temperature, the concentration of strand A in the single-stranded form is the same as the concentration of the double-helical form:

$$[R_A] = [H_A \bullet H_B] \text{ (when } T = T_M) \quad (19.1.7)$$

Substituting Equation 19.1.7 in 19.1.6, we see that:

$$f = \frac{[R_A]}{[R_A] + [R_A]} = \frac{1}{2} \text{ (when } T = T_M) \quad (19.1.8)$$

We can rewrite Equation 19.1.5 by using $[R_A]$ for the concentration of the "ligand":

$$f = \frac{\frac{[R_A]}{K_D}}{1 + \frac{[R_A]}{K_D}} = \frac{1}{2}$$

$$\Rightarrow \frac{[R_A]}{K_D} = \frac{1}{2} + \frac{[R_A]}{2K_D} \Rightarrow \frac{1}{2}\frac{[R_A]}{K_D} = \frac{1}{2}$$

$$\Rightarrow [R_A] = K_D \quad \text{(when } T = T_M) \quad (19.1.9)$$

Now we are getting close to the expression we seek. All we need to do is to relate the total concentration of DNA strands ($C_{total}$) to $[R_A]$:

$$C_{total} = [R_A] + [R_B] + 2[H_A \bullet H_B] \quad (19.1.10)$$

Note that $[H_A \bullet H_B]$ is multiplied by 2 because it contains two DNA strands.

Since a common way to set up the experiment is to add equal amounts of the A and B strands, it is usually the case that $[R_A] = [R_B]$. Thus,

$$C_{total} = 2[R_A] + 2[H_A \bullet H_B] \quad (19.1.11)$$

Combining Equations 19.1.7 and 19.1.11, we get:

$$[R_A] = \frac{C_{total}}{4} \quad (19.1.12)$$

From Equation 19.1.9,

$$[R_A] = K_D$$

Thus,

$$K_D = \frac{C_{total}}{4}$$

which is what we sought to justify.

resulting straight line. Examples of melting curves analyzed in this way are given in **Figure 19.5**.

## 19.4 DNA duplexes containing a mismatched base pair at one end are only marginally less stable than duplexes with matched bases

Returning once more to the DNA duplexes that contain mismatches, let us compare the stabilities of duplexes of the form:

```
5′-A C G A A T G G N-3′
   • • • • • • • • •
3′-T G C T T A C C T-5′
```

where **N** is A, T, C, or G. Such a comparison allows us to deduce the extent to which proper base-pairing controls the fidelity of the DNA polymerase in terms of the intrinsic free-energy differences between matched and unmatched base pairs.

When **N** is A (that is, the correct base pair is formed), $\Delta H^{\circ}_{\text{H}\to\text{R}}$ is measured to be +254.0 kJ•mol$^{-1}$. When **N** is G, C, or T (that is, a mismatch is introduced at the end), the values of $\Delta H^{\circ}_{\text{H}\to\text{R}}$ are +248.9, +244.8, and +235.1 kJ•mol$^{-1}$, respectively. Since these changes in enthalpy are in the direction of helix melting, the larger numbers reflect sequences for which the helical form is more stable in terms of energy.

In considering the net stability of the helices, we need the changes in free energy and, hence, must also have the entropy contributions. The values of $\Delta S^{\circ}_{\text{H}\to\text{R}}$, also determined from the van't Hoff analysis, are +711.3, +698.7, +686.2, and +656.9 J•mol$^{-1}$•K$^{-1}$, respectively, for **N** = A, G, C, and T. At 37°C, this means that the values of the change in free energy for converting the double helix to random coil single strands, $\Delta G^{\circ}_{\text{H}\to\text{R}}$ , are 33.5, 32.4, 32.1, and 31.4 kJ•mol$^{-1}$, respectively, for **N** = A, G, C, and T (these values are summarized in **Table 19.1**).

Note that the correctly paired duplex has the largest positive entropy change on melting or, equivalently, the largest loss in entropy upon duplex formation. Thus, in terms of entropy, the correctly paired duplex is the *least favorable* to form. This entropy effect offsets the greater energy ($\Delta H^{\circ}_{\text{H}\to\text{R}}$) required to open the Watson-Crick base pair. As you can see from Table 19.1, the difference in stability for double helices with mismatches in the terminal base pair relative to the correctly paired helices is very small. The maximum value of the difference, 1.7 kJ•mol$^{-1}$, is less than the value of $k_BT$ (~2.5 kJ•mol$^{-1}$ at room temperature). These data make it clear that the intrinsic ability of double-helical DNA to select for correct base pairs cannot underlie the high fidelity of replication, transcription, and translation. We return to this point later, in Section 19.9.

Why is it that the Watson-Crick base-pairing rules lead to such low discrimination between correct and incorrect base pairs in DNA? One reason is that the effective strengths of hydrogen bonds is weakened by competition with water, as we discussed in Chapter 2 (see Figure 2.4). Another reason is that base stacking interactions provide the dominant energetic factor that stabilizes the double helix (see Sections 1.9 and 2.3 and also Section 19.6, below). The terminal base pair is not stabilized by base stacking as much as those in the middle, which contributes to the lack of discrimination at this position.

In the next three sections of this part of the chapter we will analyze the stability of the duplex form of DNA in much the same way that changes in energy and entropy were discussed for protein folding in part D of Chapter 10. If you wish to

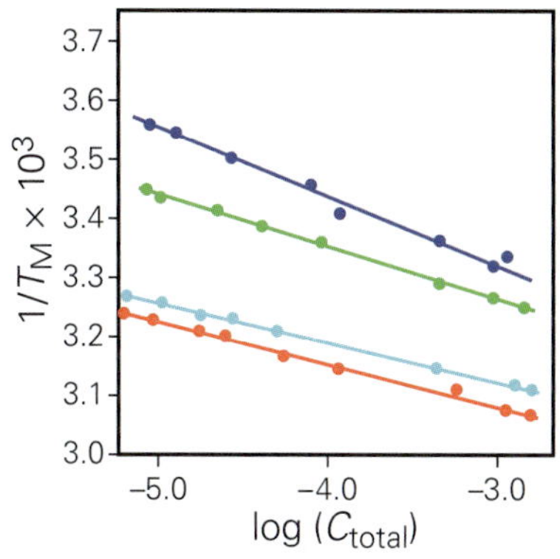


• $dCA_3GA_3G + dCT_3AT_3G$
• $dCA_7G + dCT_3GT_3G$
• $dCA_7G + dCT_7G$
• $dCA_3CA_3G + dCT_3GT_3G$

**Figure 19.5 van't Hoff analysis for four different DNA duplexes.** For each duplex the melting temperature, $T_M$, is determined at different concentrations of DNA (see Figure 19.4). The graph shows the relationship between $1/T_M$ and $\log_{10}(C_{total})$, where $C_{total}$ is the total concentration of DNA strands. (Note that the slope calculated using $\log_{10}$, as in this plot, is reduced by a factor of ln(10) = 2.3 relative to the slope from Equation 19.5). (Adapted from F. Aboul-ela et al. and F.H. Martin, *Nuc. Acids Res*. 13: 4811–4824, 1985. With permission from Oxford University Press.)

**Table 19.1 The changes in enthalpy, entropy, and free energy for the double helix to single strand transition of the DNA duplexes:**

```
5'-A C G A A T G G N-3'
   • • • • • • • •
3'-T G C T T A C C T-5'
```

| Terminal base pair | $\Delta H^{\circ}_{\text{H}\to\text{R}}$ (kJ•mol$^{-1}$) | $\Delta S^{\circ}_{\text{H}\to\text{R}}$ (J•mol$^{-1}$•K$^{-1}$) | $\Delta G^{\circ}_{\text{H}\to\text{R}}$ (kJ•mol$^{-1}$; at 310 K) | $\Delta\Delta G^{\circ}_{\text{H}\to\text{R}}$ (kJ•mol$^{-1}$; at 310 K)[a] |
|---|---|---|---|---|
| A–T | 254.0 | 711.3 | 33.5 | – |
| G–T | 248.9 | 698.7 | 32.4 | 1.0 |
| C–T | 244.8 | 686.2 | 32.1 | 1.4 |
| T–T | 235.1 | 656.9 | 31.4 | 1.7 |

[a]$\Delta\Delta G^{\circ}_{\text{H}\to\text{R}}$ corresponds to the difference in $\Delta G^{\circ}_{\text{H}\to\text{R}}$ for each "wrong" base pair relative to the properly paired A-T. (Based on data in Table 2 in J. Petruska et al. and I. Tinoco Jr., *Proc. Natl. Acad. Sci. USA* 85: 6252–6256, 1988. With permission from the National Academy of Sciences.)

continue instead with the discussion of fidelity, you may skip ahead to part B of this chapter, in which we discuss the mechanism of DNA polymerase.

## 19.5 The entropy of each DNA chain is reduced upon forming a duplex

We begin our analysis of the thermodynamics of DNA melting by estimating the change in the entropy of DNA strands upon helix formation. To do this we consider a very simple model for the melting transition of DNA. The change in entropy due to the DNA alone (ignoring water) is:

$$\Delta S^{o}_{H\rightarrow R} = R\ln\left(\frac{W_R}{W_H}\right) \tag{19.6}$$

where $W_R$ is the number of conformations of the DNA in the random-coil state and $W_H$ is the number of conformations in the helical state. In the double-helical state, there is generally just one stable conformation, so $W_H = 1$, making the value of $\Delta S^o$ equal to $R\ln W_R$. To see how Equation 19.6 arises, you should review the analysis of the entropy of protein chains in Section 10.20.

We can estimate the value of $W_R$ by counting the number of possible conformations about each of the single bonds in the backbone of each nucleotide (there are five such bonds, identified and numbered in **Figure 19.6**), and also two possible sugar conformations (puckers; see Section 2.6) and two possible orientations of the base (anti and syn; see Section 2.5). Structural data suggest that the number of conformers accessed are angle(1): 1; angle(2): 3; angle(3): 3; angle(4): 3; and angle(5): 2. Each of these possibilities is combined with each other and also with the two alternatives each for the sugar pucker and the base orientation, and so the value of $W_R$ for a base pair (consisting of two independent nucleotides) is $[(1)(3)(3)(3)(2)(2)(2)]^2$, giving us an estimate for $\Delta S^{o}_{H\rightarrow R}$ of 89 $J{\bullet}mol^{-1}{\bullet}K^{-1}$ per base pair.

This value for $\Delta S^{o}_{H\rightarrow R}$ probably represents an *upper limit* on the expected entropy change because, in actuality, some combinations of torsion angles would lead to the molecule running into itself and would therefore be disallowed. Nevertheless, the calculation gives us a sense of the maximum value we might expect for the change in conformational entropy of the DNA upon going from the helical to the random-coil state. The fact that each duplex converts to two molecules of single strands that can move independently also increases entropy, by an estimated 190 $J^{-1}{\bullet}mol^{-1}{\bullet}K^{-1}$.

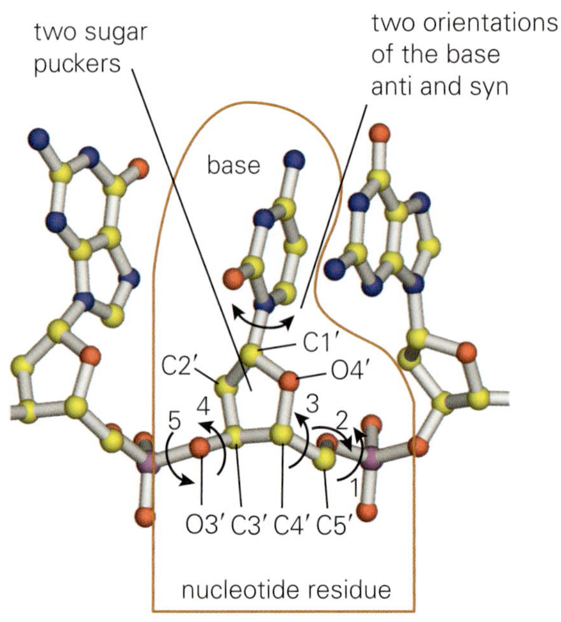


**Figure 19.6 Conformational degrees of freedom in a nucleotide residue in DNA.** The backbone bonds about which rotations can occur are indicated. A simple calculation of entropy assumes that there are three stable values for each of these torsional angles (corresponding to distinct conformations of the DNA backbone). The base can be in one of two orientations, and the sugar can be in one of two stable conformations (puckers).

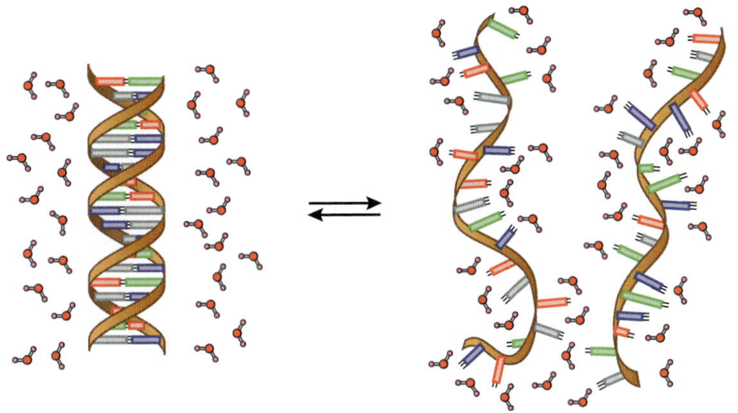

**Figure 19.7 The effect of water on helix formation.** When the DNA double helix melts, the bases interact with water. These interactions order the water in a manner somewhat analogous to the hydrophobic effect, and releasing them favors duplex formation, offsetting the loss in conformational freedom of the DNA in the duplex. Although not as large an effect as in protein folding, the changes in interactions with water play a role in duplex formation.

In addition to the entropy contributions from the DNA itself, there are also contributions from water molecules and ions that bind to the double-helical duplex (an estimated 0.3 $Na^+$ ions per base pair has restricted mobility), but are released upon separation to single strands, entropically favoring the single-strand state. The contribution of these to the entropy change is significant but one that is difficult to estimate from basic principles. Both waters and ions have a favorable enthalpy of interaction with the duplex, which overcomes the unfavorable entropy of localizing them.

When a base pair is broken, the two bases move out of the double helix and into water (**Figure 19.7**). This alters the hydrogen-bonding networks formed by water molecules, but because many of the base atoms have significant partial charge, this effect is not really equivalent to the hydrophobic effect with nonpolar amino acid sidechains. A positive $\Delta C_P$ value is considered to reflect the hydrophobic effect in protein folding and, as noted above, the value of $\Delta C_P$ for nucleic acids is ~0, which means that the solvent entropy change does not arise from the same phenomenon as the hydrophobic effect.

Note that although the entropy change per residue of going from folded helix to random coil is even more unfavorable than for proteins, the favorable enthalpy change per residue is also larger, so that the net free-energy difference that stabilizes the duplex state at normal temperatures is quite small, as is the case for proteins.

## 19.6 The stability of DNA depends on the pattern of base stacks in the duplex

An observation that emerges from analyzing the stability of different DNA oligomers is that the melting temperatures of duplex DNA molecules vary significantly with sequence. DNA duplexes with the same number of A, T, C, and G nucleotides in the duplex, but in different sequences, may have significantly different $T_M$ values. This is an important observation, because it tells us that hydrogen bonding between the bases cannot be a dominant contributor to the stabilization of double helices (the different duplexes all have the same number of hydrogen bonds).

Consider calorimetry measurements on two DNA duplexes, both containing only A-T base pairs (**Figure 19.8**). The first duplex is poly:d(A)-poly:d(T), in which the A residues stack only with A, and T residues with T. The second duplex is poly:d(A•T)-poly:d(A•T), in which A stacks on T and T on A on both strands. Calorimetry measurements provide experimental values of $\overline{\Delta H^o_{H\rightarrow R}}$ and $\overline{\Delta S^o_{H\rightarrow R}}$, where the bar indicates average values of these parameters per base pair in the polymer.

The heats absorbed as a function of temperature for poly:d(A)-poly:d(T) and poly:d(A•T)-poly:d(A•T) are shown in Figure 19.8. The melting temperatures, $T_M$, for these two DNA sequences are 56°C and 48.5°C, respectively, under identical

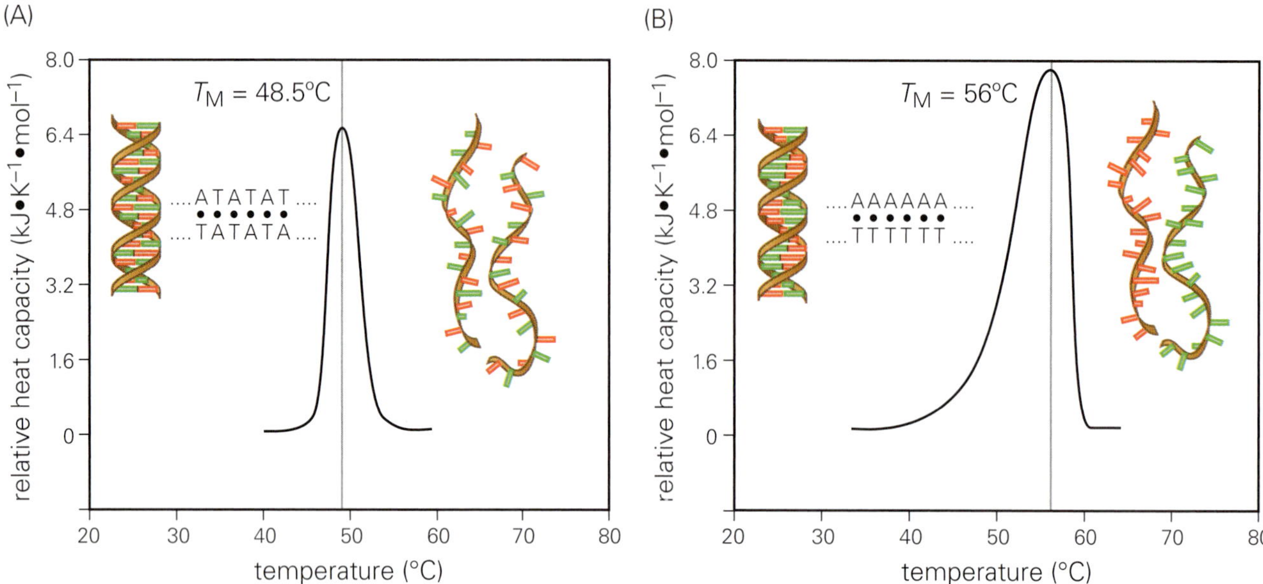


**Figure 19.8 Heat capacity changes upon DNA melting for two different DNA duplexes**. (A) Differential scanning calorimetry for a repeating ATAT sequence. The integrated area under the peaks yields the enthalpy change upon melting. The vertical line indicates the melting temperature. (B) Calorimetric data for polyadenosine (...AAA...) paired with polythymidine (...TTT...). (From L.A. Marky and K.J. Breslauer, *Biopolymers* 21: 2185–2194, 1982. With permission from John Wiley & Sons, Inc.)

buffer conditions. Thus, we can conclude that, despite the identical pattern of hydrogen bonding in the two cases, the stabilities of the two duplexes are indeed different.

Integrating the areas under the heat capacity curves to calculate the total heat for the transitions (see Section 10.25) shows that the value of $\overline{\Delta H^{\circ}_{H\to R}}$ is +36.0 kJ•mol$^{-1}$ for the A•T on A•T stacks and +29.7 kJ•mol$^{-1}$ for the A•T on T•A stacks. The positive sign of the enthalpy changes shows that both DNA duplexes are energetically more favorable than the random coils in terms of enthalpy, but the A on A plus T on T stacks are significantly more stable than the A on T plus T on A.

These data can also be used to estimate the entropy changes for these two sequences. For A•T on A•T and A•T on T•A, the values of $\overline{\Delta S^{\circ}_{H\to R}}$ are +109.2 J•mol$^{-1}$•K$^{-1}$ and +92.5 J•mol$^{-1}$•K$^{-1}$, respectively. In both cases, the positive values for the entropy of dissociation indicate that entropy effects favor dissociation and oppose helix formation. The entropy term is more unfavorable for A•T on A•T stacks than for A•T on T•A stacks.

The effect of entropy is opposite to the enthalpy effect, which favors helix formation by A•T on A•T compared to A•T on T•A. This is a manifestation of a phenomenon that is commonly observed for biological molecules and is referred to as **enthalpy–entropy compensation**. The tighter the interaction between two molecules, the greater is the reduction in entropy when they form a complex, which works against the stability of the interaction. The compensating effects of entropy and enthalpy makes it very difficult to understand and predict the strengths of macromolecular interactions quantitatively.

**Enthalpy–entropy compensation**

Enthalpy–entropy compensation is the tendency of these two thermodynamic parameters to have offsetting effects on stability in a particular system. Tighter binding (higher stability) generally means less entropy in the bound (stable) state and, hence, a less favorable entropy change in getting to that state.

Stacking brings the atoms of successive aromatic bases in the DNA chain close together (**Figure 19.9**). The bases contain nitrogen and oxygen atoms (with different electronegativities than carbon) and so most of the atoms in the bases have a partial charge (**Figure 19.10**). In the right-handed B-DNA helix, the alignment of atoms is such that there are favorable electrostatic interactions between many of the partial charges, but the exact value for the energy of interaction depends on the specific bases involved. Although all base combinations have favorable stacking (that is, have lower energy than separated bases), differences between the energetic contributions in two base stacks can vary by a factor of two. In addition to the electrostatic energy, there is also a favorable contribution from van der Waals interactions, again varying somewhat depending on the bases that are stacked.

Measurements of DNA duplex stability have been made for many different DNA sequences. Because DNA is a linear system (each base stacks only with the

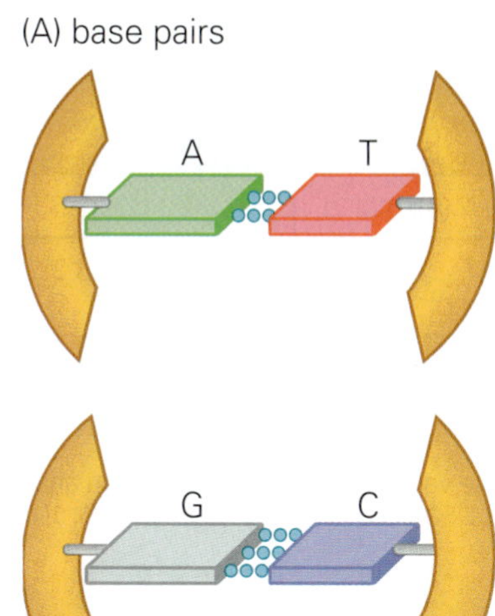


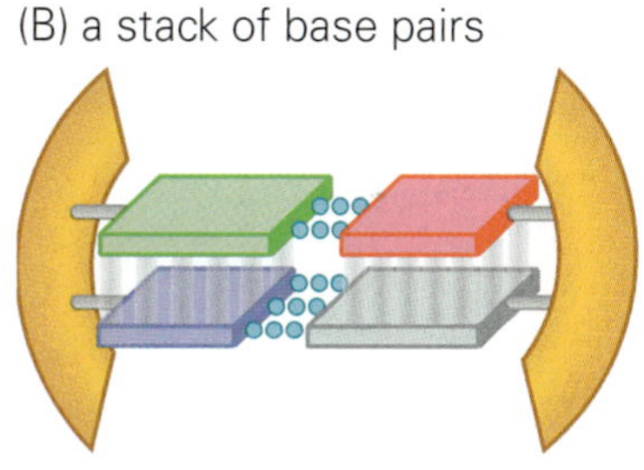


**Figure 19.9 Base pair stacking.** (A) Individual base pairs. (B) A stack of base pairs. Because of the importance of energetically favorable interactions between base pairs that are stacked on top of one another (indicated by the *gray* shaded lines), a base pair stack (rather than just one base pair) is the unit for analyzing the energy of a DNA duplex. There are 10 unique kinds of base pair stacks that can be formed by the Watson-Crick base pairs.

preceding and following bases), the stacking contributions from particular base pairs can be deduced by comparison of related sequences. Different bases have different arrangements of charged atoms, and so the precise value of the stacking energy is sequence dependent (as noted in Chapter 2). It has been found that, to a good approximation, the total stacking energy can be calculated by adding contributions from the different kinds of stacks, which are given in **Table 19.2**. The accuracy of the calculation of DNA stabilization energy can be improved further by considering both immediate-neighbor stacked bases and also the next neighbors on either side.

## 19.7 Base stacking is more important than hydrogen bonding in determining the stability of DNA helices

The complementarity of the hydrogen-bond donors and acceptors in the Watson-Crick partner bases in nucleic acids is a striking feature of the double helix. Hydrogen bonds can be quite strong; for example, calculations show that bringing together a G and a C in a vacuum so that they form hydrogen bonds results in an energetic stabilization of ~46 kJ•mol$^{-1}$. As expected, an A-T base pair is less stable, but still very strong in vacuum, with a stabilization energy of ~25 kJ•mol$^{-1}$. If these values for the stabilization energy were applicable in solution, then hydrogen bond formation would be comparable to base stacking in terms of energetics.

In the single-strand state, the bases are exposed to water, and strong hydrogen bonds are formed between the bases and water molecules (see Figure 2.4). These must be broken when the duplex forms and are replaced by the hydrogen bonds to the partner base, which are also strong. It is not easy to rigorously separate the contributions of hydrogen bonding and stacking to duplex stability, but by considering stacking without hydrogen bonding at the ends of duplexes with a single-base overhang, and by measuring the stability of duplexes with base analogs that cannot hydrogen bond, it has been shown that the inter-base hydrogen bonds

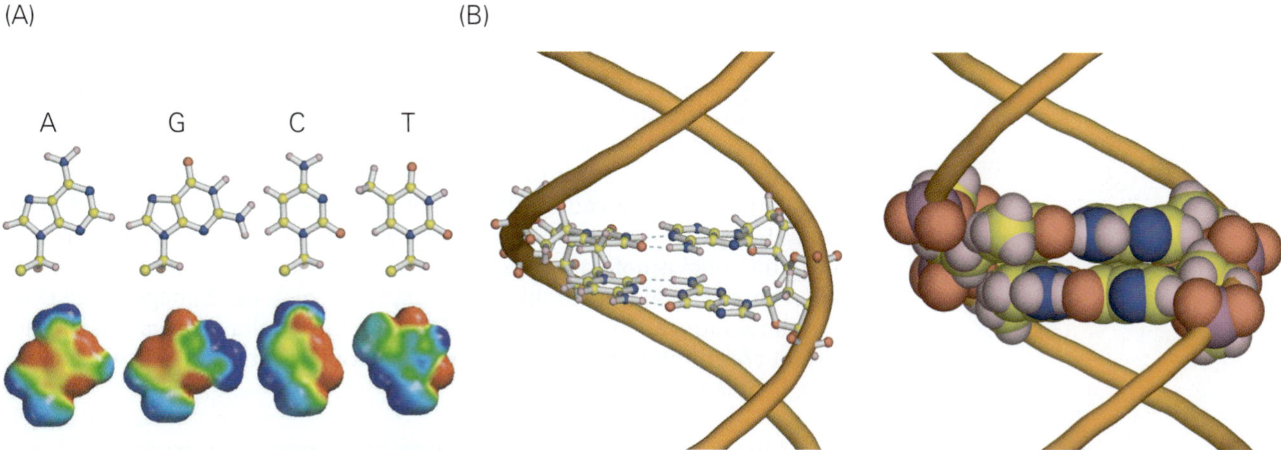


**Figure 19.10 Electrostatics and base stacking**. (A) The molecular surface of each of the bases is shown below the chemical structure of the base. The surface is colored based on the electrostatic potential, with regions of negative charge colored *red* and regions of positive charge colored *blue*. When the bases stack on top of each other, the *red* areas interact favorably with the *blue* areas due to electrostatic complementarity. (B) Stacking of base pairs is shown in the context of a B-form helix. (A, adapted from a figure by Eric Kool.)

**Table 19.2 Changes in enthalpy, entropy, and free energy in 1 M NaCl solution (at 37°C) for stacked DNA base pairs.**

| Base stack energy | 5′GC3′ 3′CG5′ | 5′GG3′ 3′CC5′ | 5′CG3′ 3′GC5′ | 5′GA3′ 3′CT5′ | 5′GT3′ 3′CA5′ | 5′CA3′ 3′GT5′ | 5′CT3′ 3′GA5′ | 5′TA3′ 3′AT5′ | 5′AT3′ 3′TA5′ | 5′AA3′ 3′TT5′ |
|---|---|---|---|---|---|---|---|---|---|---|
| $\Delta H^o$ (kJ•mol$^{-1}$) | –41.0 | –33.5 | –44.3 | –34.3 | –35.1 | –35.6 | –32.6 | –30.1 | –30.1 | –33.1 |
| $\Delta S^o$ (J•mol$^{-1}$•K$^{-1}$) | –102.1 | –83.3 | –114 | –92.9 | –93.7 | –95.0 | –87.9 | –89.1 | –85.4 | –92.9 |
| $\Delta G^o$ (37°C; kJ•mol$^{-1}$) | –9.37 | –7.70 | –9.08 | –5.44 | –6.03 | –6.07 | –5.36 | –2.43 | –3.68 | –4.18 |

These values were determined from analysis of melting data from many DNA oligomers of defined sequence. (Based on Table 1 in H.T. Allawi and J. SantaLucia, Jr. *Biochemistry* 36: 10581–10594, 1997.)

contribute a net energy (the difference relative to hydrogen bonding to water) of 2 to 6 kJ•mol$^{-1}$ per hydrogen bond (three for a G-C pair; two for an A-T pair). The range of values seems to arise from sequence context with stacking and hydrogen bonding competing for their optimum geometry at each site. The energies of hydrogen bonds are considerably smaller than those for base stacking, but do make a substantial contribution to duplex stability and are also very important for the specificity in base-pairing.

## B. FIDELITY IN DNA REPLICATION

The discussion in part A of this chapter has made clear that the intrinsic energy of base-pairing and stacking can only contribute a small amount to the selection of the correct nucleotide at each step of the DNA synthesis process. In this part of the chapter we shall look at how the DNA polymerase enzyme imposes high fidelity on the DNA replication process. We begin by making a connection between the problem of fidelity in replication and the stability of mismatches that we discussed in part A. We then introduce a kinetic model for how DNA polymerase synthesizes DNA and discuss measurements of the rates of different steps. By comparing the rates of key steps, we can understand how very high-fidelity copying of the template is achieved by the polymerase.

DNA replication proceeds with an overall error rate of one incorrect nucleotide incorporated for every $10^7$ to $10^8$ nucleotides added to the new DNA chain. We will see that the intrinsic fidelity of initial DNA synthesis by polymerase accounts for a substantial part but not all of the overall fidelity of the replication process (DNA polymerase makes one error for every $10^4$ to $10^5$ nucleotides). We will then describe a second catalytic activity in polymerases, which depends on an associated exonuclease domain that can remove nucleotides from the 3′ end of the growing chain. This domain does proofreading that decreases the overall error rate by a further hundred-fold. The combination of the activities of the polymerase and the exonuclease still produces an error rate that is higher than observed in the final products of replication. The further suppression of errors occurs through repair systems which recognize and correct mismatched base pairs that have escaped the initial proofreading process. There are enzymes that can repair mismatched bases and also fix problems arising from chemical damage to the DNA, such as those induced by carcinogens. Although these cellular repair systems are important contributors to the overall accuracy of DNA replication, we shall not discuss them in this chapter.

### 19.8 The process of DNA replication is very accurate

The discovery of the double-helical structure of DNA, in which the sequence of nucleotides in one strand is complementary to the sequence in the other, immediately suggested to its discoverers a mechanism for the duplication of genetic information (see Box 1.1). As Watson and Crick proposed, DNA replication does

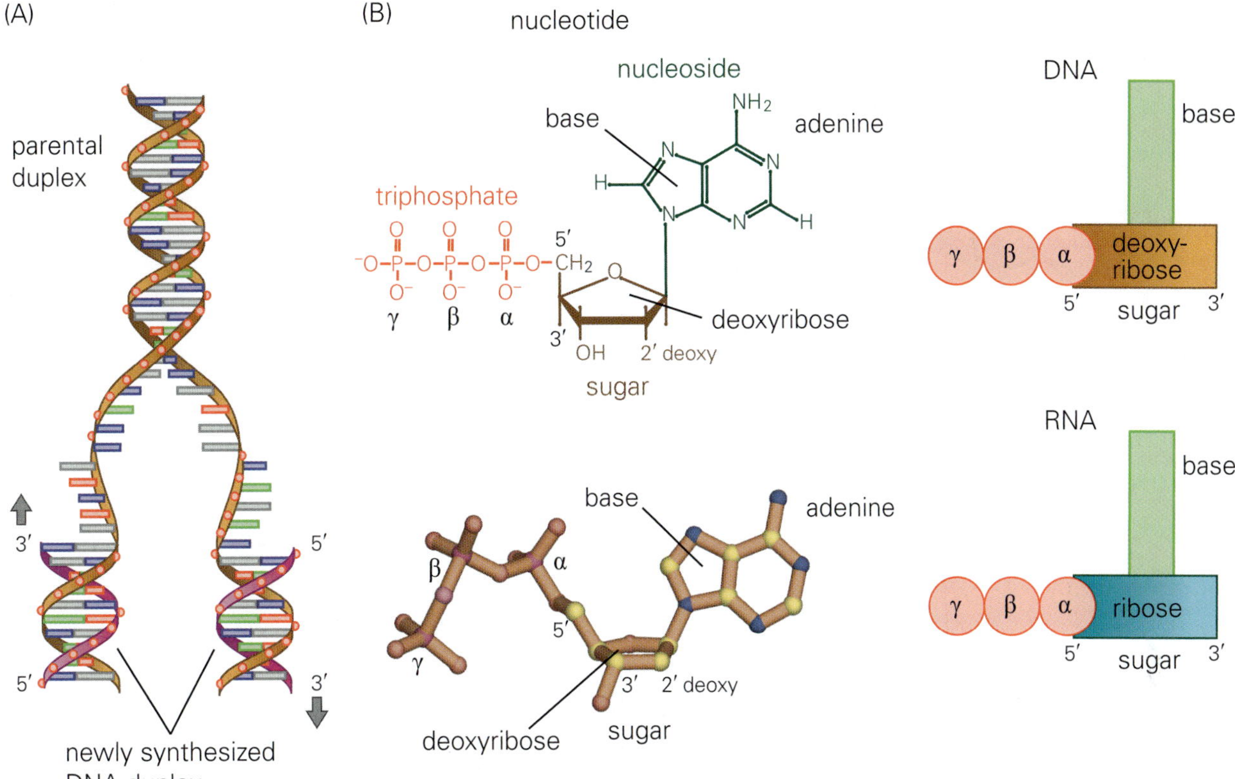


**Figure 19.11 The replication of DNA, as proposed originally by Watson and Crick.** (A) The two strands of DNA from the parent duplex are separated and copied into two daughter strands. The structure of DNA shown here is known as a replication fork, in which two new duplexes are formed by copying templates produced by the unwinding of a single parental duplex. (B) Various representations of the structures of nucleotides used for DNA synthesis.

**Fidelity**

Fidelity in DNA replication is related to the rate at which errors are introduced into the newly synthesized DNA strand, based on the Watson–Crick base-pairing rules. The lower the error rate, the higher the fidelity.

indeed proceed in a way that is conceptually simple: the double-helical structure of DNA opens and each of the two strands serves as a template to generate two new strands that contain nucleotides that are complementary in sequence to those in the original strands (**Figure 19.11**). Through this process, the two duplex DNA products should be identical in sequence.

The process of DNA replication does not occur, however, without help from many proteins. The most critical of these are the **DNA polymerases**, which lower the activation free energy for adding a new nucleotide to the growing DNA chains, greatly accelerating this key step. DNA polymerases are multisubunit protein machines that catalyze the sequential addition of nucleotides to the growing DNA chains in a precise order that is dictated by the Watson-Crick base-pairing rules and the sequence of nucleotides in the parental template (**Figure 19.12**).

DNA polymerases catalyze the synthesis of new DNA strands with great accuracy, with an incorrect nucleotide being inserted into the final product only once for every $10^7$–$10^8$ correct bases. The actual accuracy depends on the particular DNA polymerase being studied and the conditions under which replication occurs. This accuracy in the copying process, or **fidelity**, is of great importance in ensuring the faithful transmission of genetic information from parent cells to their progeny.

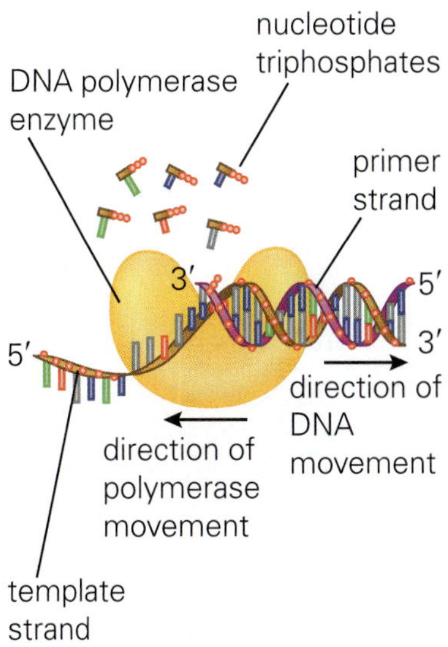


**Figure 19.12 DNA polymerase.** Schematic drawing of a polymerase enzyme (*yellow*) replicating a DNA sequence by using nucleotide triphosphates as substrates that are added to the 3′ end of the growing chain.

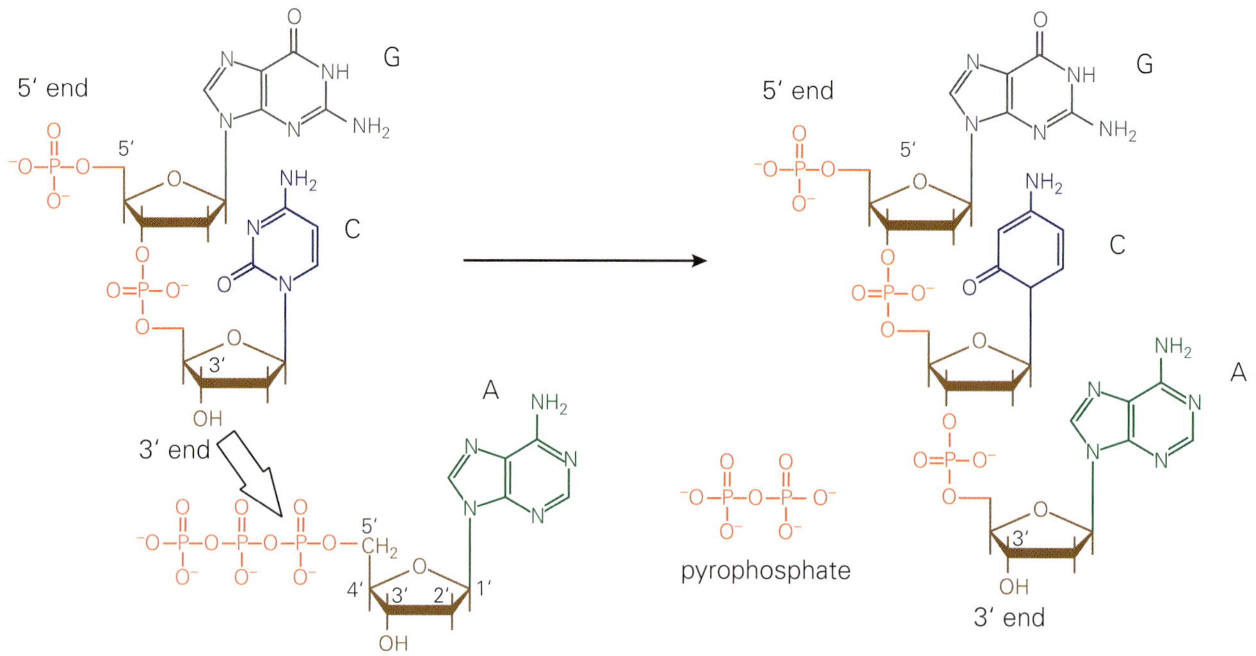

**Figure 19.13 The chemical reaction catalyzed by DNA polymerases.** A chain of nucleotides, running from the 5′ end at the top toward the 3′ end below it, is shown. The free 3′-hydroxyl group of the terminal nucleotide in the chain does a nucleophilic attack on the α phosphate of the incoming nucleoside triphosphate (adenine in this case), resulting in the covalent linkage of the incoming nucleotide to the growing DNA chain.

## 19.9 The energy of DNA base-pairing cannot explain the accuracy of DNA replication

The process of adding nucleotides to a growing DNA chain is shown in **Figure 19.13**. This process follows the Watson-Crick base-pairing rules. Deoxyadenosine (dA) is incorporated opposite deoxythymidine (dT), and vice versa. Likewise, deoxycytidine (dC) is incorporated opposite deoxyguanosine (dG). In this part of the chapter, we shall denote these pairing rules as A-T, T-A, C-G, and G-C, with the implicit assumption that we are dealing with deoxyribonucleotides.

As noted above, it is found experimentally that incorrect (mispaired) nucleotides are incorporated by the DNA polymerase into the growing strand very infrequently. There are three processes that contribute to the overall fidelity of DNA replication: selection of the correct nucleotide to add; faster formation of the covalent bond between the incoming nucleotide and the growing chain (catalyzed by polymerase) for the correct nucleotide; and having the polymerase detect and remove errors.

Can the high level of fidelity in replication originate solely from the intrinsic ability of the DNA template (that is, without assistance from the polymerase) to select the correct incoming nucleotide triphosphate (dNTP)? To address this question in terms of thermodynamics, let us separate the process of adding a nucleotide to the growing chain into two distinct steps. First, a nucleotide triphosphate molecule (dNTP) binds to the growing DNA chain at a primer–template junction (**Figure 19.14**). The dNTP interacts with the base opposite it on the template strand, without forming a covalent bond. Then, in a slow step, a chemical reaction is catalyzed by the enzyme, resulting in the release of pyrophosphate and the formation of a chemical linkage between the new nucleotide and the growing DNA strand (**Figure 19.15**).

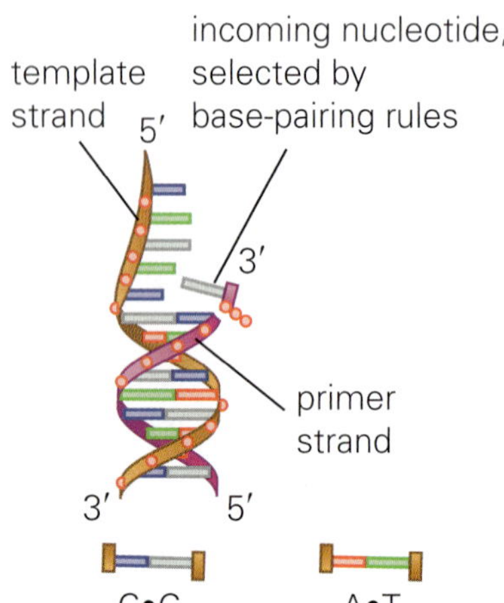


**Figure 19.14 A primer–template junction.** Denoted "P-T" in the text, these are structures in which a single strand of DNA extends beyond a double-helical region. The double-helical segment has a free 3′-OH group at the junction, and the strand bearing the 3′ end is the primer for DNA replication (*purple*). The other strand is the template (*orange*). An incoming nucleotide triphosphate is shown, with its three phosphate groups indicated by *red* circles.

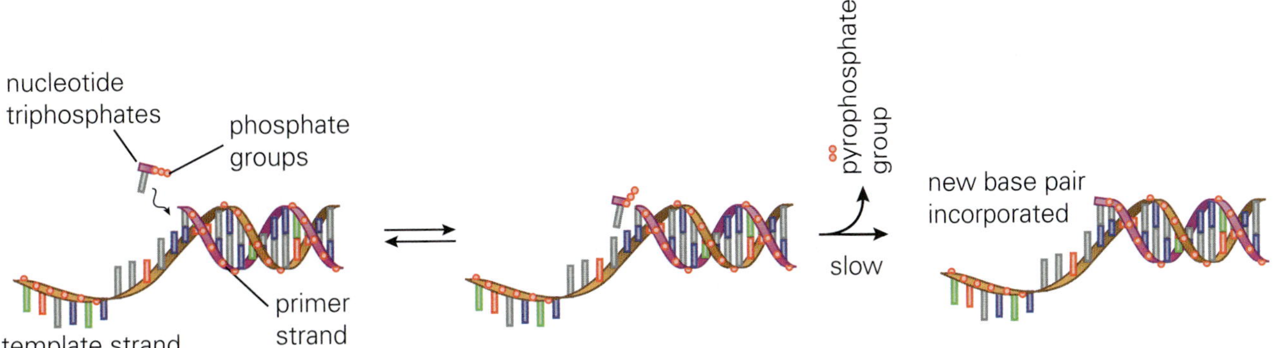


**Figure 19.15 The process of adding a nucleotide to the growing primer strand takes place in two steps.** In a first step, the incoming nucleotide binds to the primer–template junction (P-T). In a second process, a chemical reaction takes place and the nucleotide is incorporated into the growing DNA strand. Because the chemical step is slow, the first step can be considered to come to equilibrium, with the nucleotide binding to and leaving the primer–template junction many times before the chemical reaction takes place.

We will assume that the first step is essentially in equilibrium, with nucleotides binding to and releasing from the template rapidly compared to the chemical step. Either the correct nucleotide or one of the three incorrect ones can bind to the template. With our assumption of equilibrium, the ratio of bases occupying the site will be determined just by the relative binding free energies of correct and incorrect bases.

We can easily estimate the relative free energy of dNTPs binding to the primer–template junction that would be required to explain the observed fidelity. Let us denote the growing DNA chain (the primer–template junction) by P-T, and the incoming deoxyribonucleotide triphosphate by dNTP. The first step, which is the binding equilibrium, is then described by:

$$\text{P-T} + \text{dNTP} \xrightleftharpoons{\Delta G^{\text{o}}} \text{P-T}\bullet\text{dNTP} \tag{19.7}$$

Here $\Delta G^{\text{o}}$ is the standard free-energy change upon binding of the free dNTP to the primer–template junction (P-T). Let us denote the correctly paired nucleotide by $\text{dN}_\text{c}\text{TP}$ and a wrongly paired nucleotide by $\text{dN}_\text{w}\text{TP}$. We can now write two equilibrium reactions, one for the correct (c) nucleotide binding to the primer–template, and one for the wrongly matched one (w):

$$\text{P-T} + \text{dN}_\text{c}\text{TP} \xrightleftharpoons{\Delta G^{\text{o}}_\text{c}} \text{P-T}\bullet\text{dN}_\text{c}\text{TP}$$

and with a corresponding equilibrium constant:

$$K_\text{c} = \frac{[\text{P-T}\bullet\text{dN}_\text{c}\text{TP}]}{[\text{P-T}][\text{dN}_\text{c}\text{TP}]} \tag{19.8}$$

Also,

$$\text{P-T} + \text{dN}_\text{w}\text{TP} \xrightleftharpoons{\Delta G^{\text{o}}_\text{w}} \text{P-T}\bullet\text{dN}_\text{w}\text{TP}$$

and

$$K_\text{w} = \frac{[\text{P-T}\bullet\text{dN}_\text{w}\text{TP}]}{[\text{P-T}][\text{dN}_\text{w}\text{TP}]} \tag{19.9}$$

There are three different nucleotides corresponding to incorrect matches for every correctly paired one, but for simplicity we consider only one of the incorrectly matched ones. We are interested in the process of exchange of the correct nucleotide for an incorrect one:

$$\text{P-T}\bullet\text{dN}_\text{c}\text{TP} + \text{dN}_\text{w}\text{TP} \xrightleftharpoons{\Delta G^{\text{o}}_\text{ex}} \text{P-T}\bullet\text{dN}_\text{w}\text{TP} + \text{dN}_\text{c}\text{TP} \tag{19.10}$$

The equilibrium constant for the exchange of nucleotides is:

$$K_\text{ex} = \frac{[\text{P-T}\bullet\text{dN}_\text{w}\text{TP}][\text{dN}_\text{c}\text{TP}]}{[\text{P-T}\bullet\text{dN}_\text{c}\text{TP}][\text{dN}_\text{w}\text{TP}]} = \frac{K_\text{w}}{K_\text{c}} \tag{19.11}$$

We assume that the free nucleotides are all at the same concentration and that this concentration is higher than the concentration of the primer–template. The

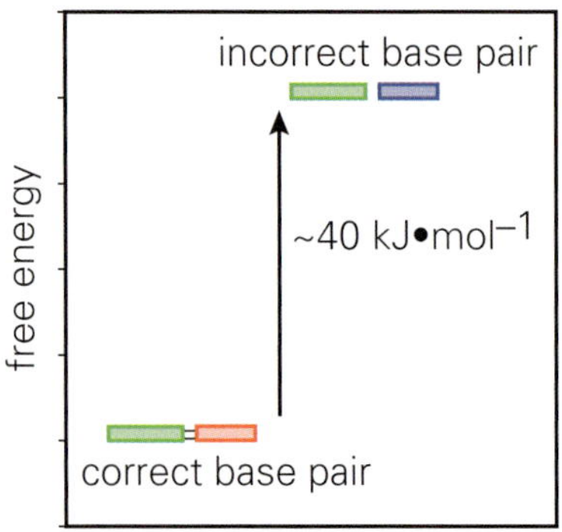


**Figure 19.16 Difference in binding free energy required to explain the fidelity of DNA replication.** If the process of DNA replication were to rely solely on the intrinsic base-pairing ability of the DNA template, then the free-energy difference between correct and incorrect base pairs would have to be at least 40 kJ•mol$^{-1}$ to account for the observed fidelity of DNA replication, at 300 K.

concentrations of the free nucleotides do not change very much during the process, and so $[\text{dN}_\text{C}\text{TP}] \approx [\text{dN}_\text{W}\text{TP}]$. Equation 19.11 then simplifies to:

$$K_{ex} = \frac{[\text{P-T} \bullet \text{dN}_\text{w}\text{TP}]}{[\text{P-T} \bullet \text{dN}_\text{c}\text{TP}]} \tag{19.12}$$

If the fidelity of DNA replication were controlled by the intrinsic ability of the primer–template junction to distinguish between correct and wrong nucleotides, then $K_{ex}$ is the equilibrium constant that governs fidelity. The value of $K_{ex}$ tells us how often a particular wrongly paired nucleotide is bound to the primer–template when compared to the correct one.

DNA replication proceeds with an error rate of approximately 1 in $10^7$, which requires that $K_{ex} \leq 10^{-7}$. The value of this equilibrium constant tells us how much more strongly correct nucleotides would have to bind to the primer–template when compared to wrong nucleotides in order to ensure an error rate of 1 in $10^7$. We can easily convert this value to a difference in free energy of interaction:

$$\Delta G_{ex} = \Delta G_c - \Delta G_w = -RT\ln(K_{ex}) \tag{19.13}$$

If $K_{ex} = 10^{-7}$, then a difference in binding free energy of ~40 kJ•mol$^{-1}$ between correct and incorrect nucleotides would be required to explain the high fidelity of replication at 300 K (**Figure 19.16**). But, as we saw in Section 19.4, experiments indicate that the value of $\Delta G_{ex}$ is really ~1 kJ•mol$^{-1}$, far below the level of discrimination that would be required to explain the fidelity of DNA replication. The actual mechanisms by which DNA polymerases achieve high fidelity must therefore rely on factors other than the differences in the strengths of the base-pairing interactions between correctly and wrongly paired bases. To understand these mechanisms we now describe how DNA polymerases work.

## 19.10 The overall process of DNA synthesis can be described as a series of kinetic steps

The synthesis of DNA by polymerase is an enzyme-catalyzed reaction and, hence, many of the ideas developed in Chapter 16 can be applied to this process. The reaction catalyzed by the polymerase enzyme is illustrated in **Figure 19.17**, which also explains how the progress of the reaction can be monitored experimentally. The full reaction comprises many steps and, to begin the description, we will lay out a basic kinetic scheme for the steps that must occur. The key to analyzing polymerase fidelity is understanding the relative rates of various steps under different circumstances.

We will refer to the primer–template junction at which nucleotides are added to the primer strand as P-T (see Figure 19.17). If the existing primer is *n* nucleotides long, we will designate this $\text{P}_n$-T. The polymerase enzyme will be denoted Pol, and the incoming deoxynucleotide triphosphate will be denoted dNTP.

The processes through which DNA is initially opened, so that polymerase has access to the template, and by which a primer is generated, are very interesting and complex. These are not, however, central to the issue of fidelity in synthesis once the process has started. We will ignore these steps and shall assume that, once the polymerase has bound to the initial primer–template junction ($\text{P}_n$-T•Pol), it remains bound as it synthesizes DNA. Most polymerases are **processive enzymes** (that is, they catalyze the addition of many nucleotides, one after the other, before falling off the template), so this is a valid approach for our analysis.

**Processive enzymes**

Some enzymes, such as DNA polymerase, work on substrates that contain multiple sites where the enzymatically catalyzed reaction can take place. Each nucleotide in the DNA template, for example, serves as a site for the addition of a nucleotide to the growing DNA chain. A processive enzyme catalyzes many rounds of the reaction before dissociating from the template.

As is normal for an enzyme–catalyzed reaction, the process of DNA synthesis is described in the general context of a Michaelis-Menten model for the kinetics, with the nucleotide substrate first binding to the $\text{P}_n$-T•Pol complex. However, in this case there are multiple steps to consider between substrate binding and release of products. These are outlined in **Figure 19.18**.

The first step is the reversible binding of a substrate nucleotide to the primer-template junction. The substrate nucleotide must be a deoxyribonucleotide—the

(A)
radioactive label
primer
template
5′
3′
A
dTTP
Mg2+
T 3′
+ PPi
(B)
primer +1
primer
time of reaction

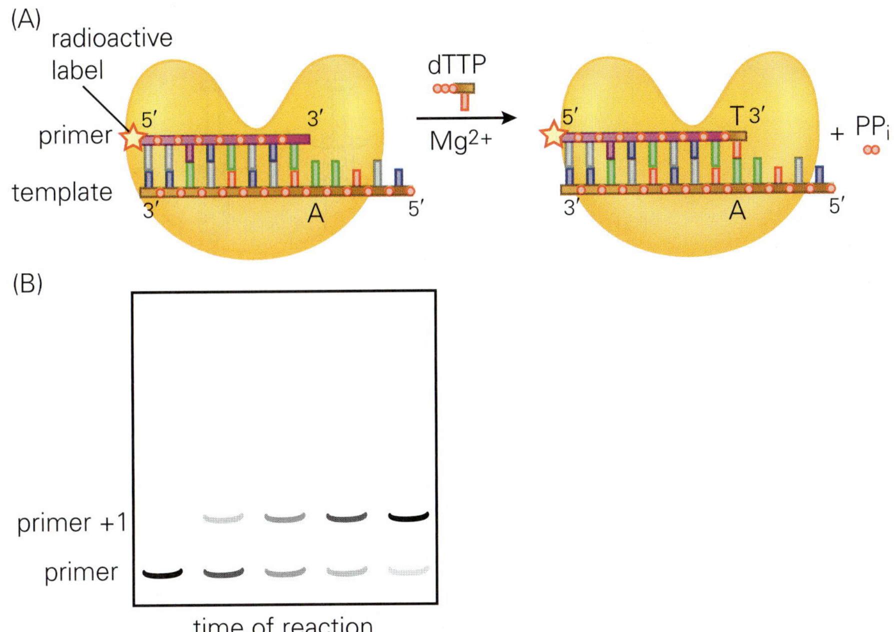

**Figure 19.17 Nucleotide addition by DNA polymerase.** (A) The polymerase enzyme is shown schematically, bound to a primer–template junction (the template is the lower DNA strand and the primer is the upper DNA strand). The first unpaired base on the template is A, so a dTTP will be added and pyrophosphate will be released. (B) This process can be followed by radio-labeling the primer strand, initiating the reaction by addition of dTTP and $Mg^{2+}$ and then, after a period of time, stopping the reaction. The product with T added can be separated from the primer on a gel, as shown here.

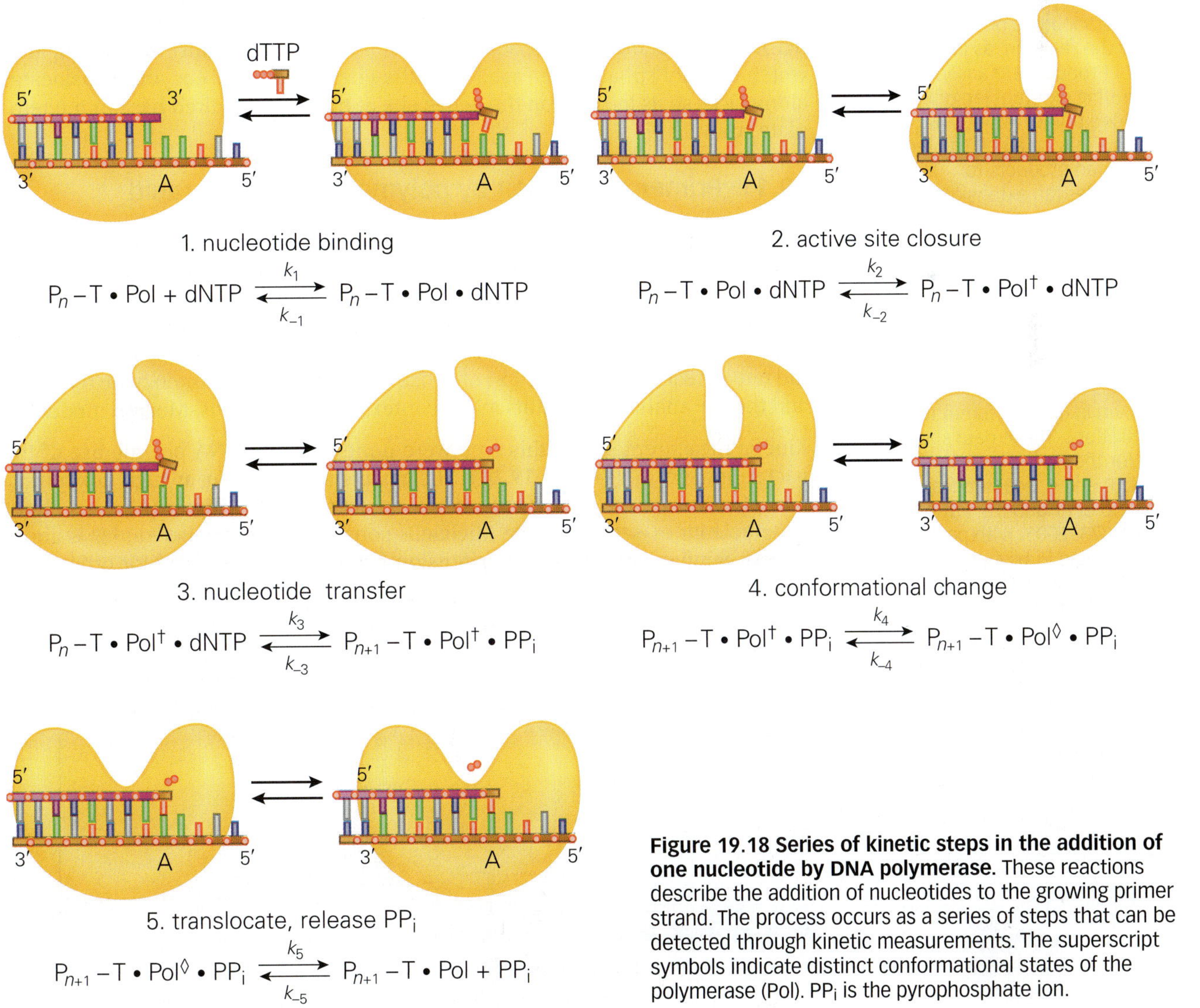

dTTP
5′
3′
A
1. nucleotide binding
$P_n-T\bullet Pol + dNTP \underset{k_{-1}}{\overset{k_1}{\rightleftharpoons}} P_n-T\bullet Pol\bullet dNTP$
2. active site closure
$P_n-T\bullet Pol\bullet dNTP \underset{k_{-2}}{\overset{k_2}{\rightleftharpoons}} P_n-T\bullet Pol^{\dagger}\bullet dNTP$
3. nucleotide transfer
$P_n-T\bullet Pol^{\dagger}\bullet dNTP \underset{k_{-3}}{\overset{k_3}{\rightleftharpoons}} P_{n+1}-T\bullet Pol^{\dagger}\bullet PP_i$
4. conformational change
$P_{n+1}-T\bullet Pol^{\dagger}\bullet PP_i \underset{k_{-4}}{\overset{k_4}{\rightleftharpoons}} P_{n+1}-T\bullet Pol^{\Diamond}\bullet PP_i$
5. translocate, release PPi
$P_{n+1}-T\bullet Pol^{\Diamond}\bullet PP_i \underset{k_{-5}}{\overset{k_5}{\rightleftharpoons}} P_{n+1}-T\bullet Pol + PP_i$

**Figure 19.18 Series of kinetic steps in the addition of one nucleotide by DNA polymerase.** These reactions describe the addition of nucleotides to the growing primer strand. The process occurs as a series of steps that can be detected through kinetic measurements. The superscript symbols indicate distinct conformational states of the polymerase (Pol). $PP_i$ is the pyrophosphate ion.

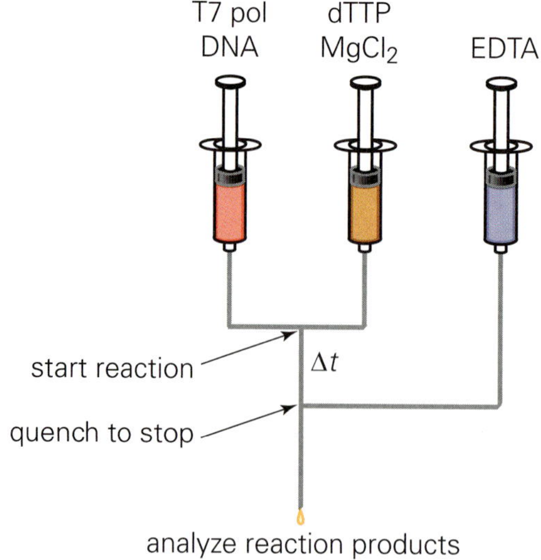


**Figure 19.19 Flow mixer for measuring rapid kinetics.** Solutions of reactants are loaded into pumps that drive the solutions into a mixing region (upper arrow) where the reaction starts. The reaction is then later quenched (for example, by addition of EDTA, which binds all of the $Mg^{2+}$ ions, blocking further reaction), and the product is collected for analysis. The rate of flow of reactants controls how long they stay in the region between mixing and quenching, with slow flow corresponding to longer reaction times times, $\Delta t$. Using a set of different flow rates (corresponding to different reaction times) the rate of reaction can be determined.

polymerase prevents the stable binding of ribonucleotides. The polymerase needs to undergo a substantial conformational change that closes the active site before the chemical step can occur; the evidence for this will be discussed later in the chapter. After the active site is closed, the nucleotide is transferred to the growing primer strand. Forming the bond between nucleotide and primer produces a pyrophosphate ion ($PP_i$). Another conformational change must then occur, after which pyrophosphate is released from the enzyme and the enzyme translocates along DNA so as to position the active site at the next unpaired base in the template strand.

## 19.11 Primer elongation by polymerase is quite rapid

The overall rate at which polymerase elongates a primer can be measured by rapidly mixing a solution of the already formed complex of $P_n$-T•Pol with the nucleotides and $Mg^{2+}$ ions that are required for the reaction. Because the reaction is fairly rapid, this is done with a flow mixing device, such as that shown schematically in **Figure 19.19**. As discussed in Chapter 16, it is generally useful to measure the rate of reaction as a function of substrate concentration for enzyme reactions and then to analyze the results in the context of the Michaelis–Menten model.

Kinetic measurements of this type were done with the DNA polymerase of the bacteriophage T7 (**Figure 19.20**). The results showed that at high concentration of substrate the maximum rate of extension of the primer ($V_{max}$) is about 230 nucleotides•sec$^{-1}$, and the Michaelis constant ($K_M$) for nucleotides is about 11 μM. Similar rates of primer extension have been determined for other DNA

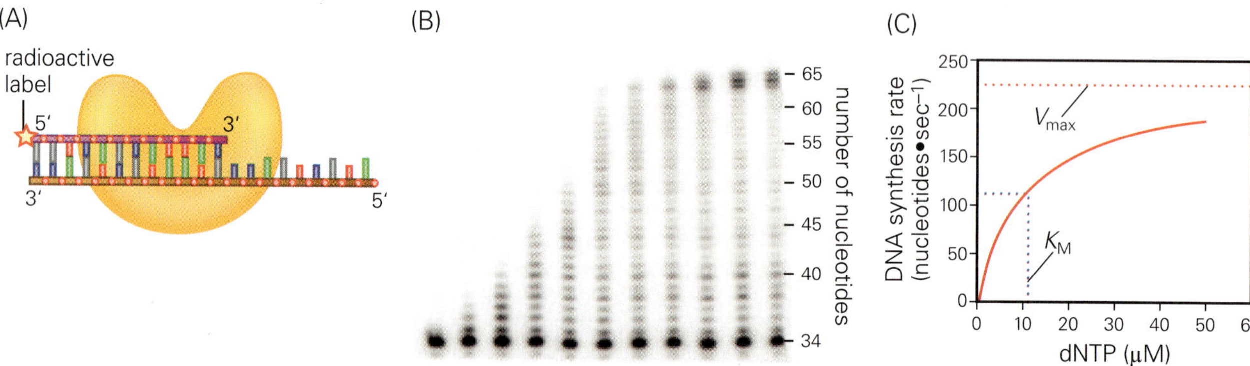


**Figure 19.20 Elongation of primer by T7 DNA polymerase.** (A) A radioactively labeled primer is used to follow the reaction, as in Figure 19.17. (B) Products are analyzed by separating different lengths of extended primer strands by electrophoresis on a gel. The template length limited products to 64 nucleotides (an extension of 30 from the initial primer of 34 nucleotides). All four dNTPs were at 100 μM concentration (well above the value of $K_M$). (C) Michaelis–Menten analysis of the rates determined as a function of substrate dNTP concentration with rate values determined by experiments such as that shown in (B) at different NTP concentrations. These data show that the $K_M$ for the dNTPs in this reaction is ~11 μM. (B and C, adapted from N.M. Stano et al. and S.S. Patel, *Nature* 435: 370–373, 2005. With permission from Macmillan Publishers Ltd.)

polymerase enzymes. These measurements show that the reaction is quite fast. But, because the measurements are made under steady-state conditions with multiple turnovers, they do not help us identify which step in the scheme outlined in Figure 19.18 limits the rate of reaction.

## 19.12 The rate-limiting step in the DNA synthesis reaction is a conformational change in DNA polymerase

To fully understand a coupled set of reactions, such as those set out in Figure 19.18, it is necessary to measure more than just the rate of formation of the final product. This is because the rate of product formation is a complex function of all of the individual rate constants (see Section 16.4, for example). If one step is slower than the others, the overall rate will be determined by that rate-limiting step, but one cannot tell which step this is by simply measuring the overall rate of the reaction. To determine rate constants for specific steps, it is necessary to carry out additional rate measurements that isolate or modify one particular step.

For DNA synthesis by polymerase, the first step in the reaction scheme is the binding of substrate nucleotide (see Figure 19.18 and also **Figure 19.21**, which illustrates the set of reactions using structures of the various polymerase complexes). The reaction was initiated by very rapidly mixing nucleotide with the $P_n$-T•Pol complex and detecting binding by changes in fluorescence of the protein. These measurements showed that the rate constant, $k_1$, is large ($\geq 5 \times 10^7$ $M^{-1}$•$sec^{-1}$), so

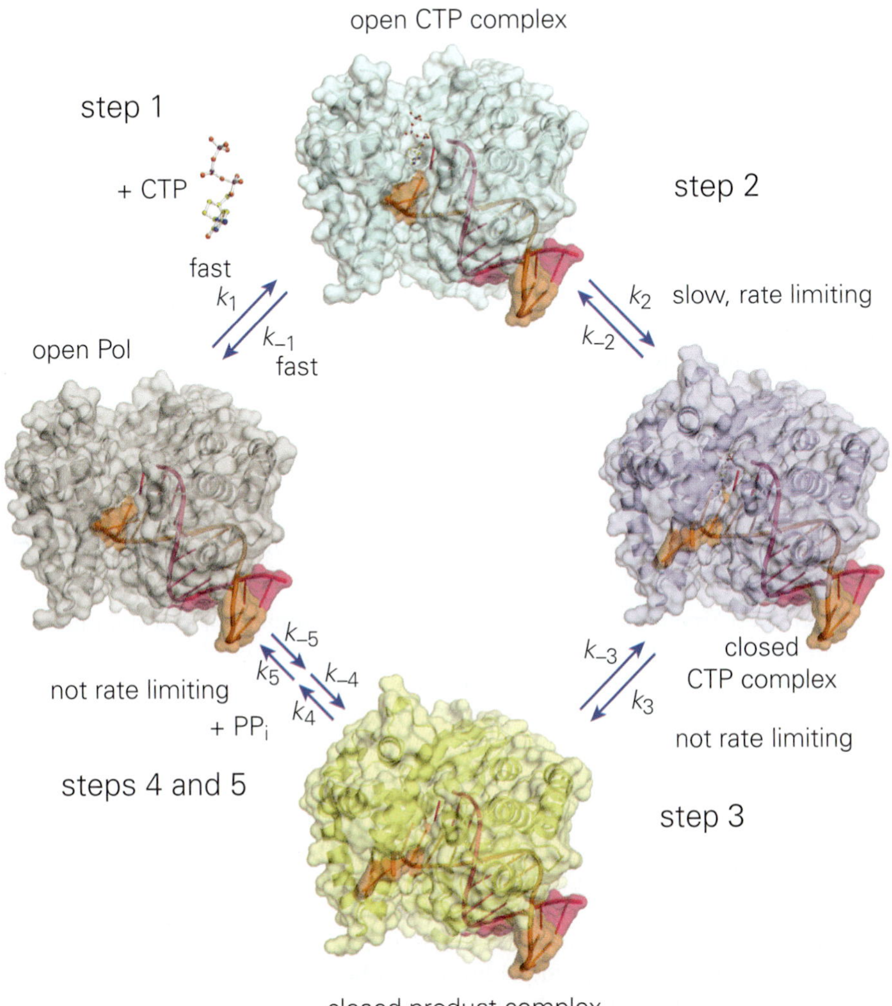


**Figure 19.21 Kinetic steps in the addition of one nucleotide by DNA polymerase.** The reaction cycle shown here is the same as the series of steps illustrated in Figure 19.18. The structures shown are that of a DNA polymerase from a thermophilic bacterium known as *Thermus acquaticus*. The primer and template strands are in *pink* and *orange*, respectively. (PDB codes: 2KTQ and 3KTQ.)

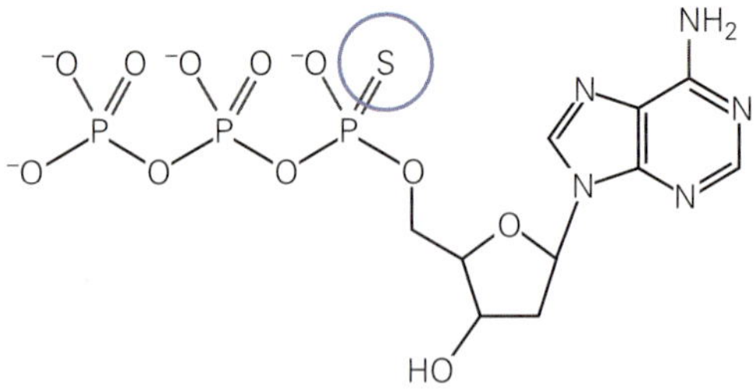

**Figure 19.22 Thio-dATP.** Schematic diagram of the α-thio-dATP nucleotide used to probe the chemical step of the nucleotide addition. The sulfur atom on the phosphate decreases the rate at which displacement at the substituted phosphate occurs by about 60-fold. If displacement is the rate-limiting step, then the overall reaction will slow by this amount.

that at typical nucleotide concentrations in cells, the rate of binding is much faster than the overall rate determined in the experiment shown in Figure 19.20.

The dissociation rate of the nucleotide leaving the polymerase is also quite high ($k_{-1} \geq 10^3$ sec$^{-1}$). Because the overall rate of the reaction (~200 sec$^{-1}$) is much slower than this dissociation rate, this means that nucleotide binding is essentially at equilibrium. As we concluded earlier, the thermodynamic discrimination of correct versus wrong nucleotides cannot explain fidelity, so it is not surprising that this first step does not control the overall rate.

The chemical step of adding the nucleotide to the extending primer strand was probed by comparing the rate of the reaction of a normal nucleotide with one that had a sulfur substituted for an oxygen on the first phosphate of the dNTP being added (for example, thio-dATP, shown in **Figure 19.22**). Measurement of the rate of reaction of such compounds in other kinds of chemical reactions has shown that if the phosphate transfer step is rate limiting, then the sulfur substitution should decrease the rate of the reaction by about 60-fold. But, for the DNA synthesis reactions, experiments showed that the overall rate was reduced only by three-fold when thio-dATP was used as the substrate. Dissection of the individual steps of the scheme shown in Figure 19.18 showed that the rate constant for the chemical step of nucleotide transfer, $k_3$, is ~$9 \times 10^3$ sec$^{-1}$. This relatively large rate constant, and the fact that the thiophosphate substitution does not affect the overall rate very much, shows that the nucleotide transfer step cannot be rate limiting.

After nucleotide transfer to the growing primer strand, there is pyrophosphate ($PP_i$) bound to polymerase, which must be released before the addition cycle can repeat. Release of $PP_i$ requires a conformational change in polymerase to occur before dissociation is possible (step 4 in Figures 19.18 and 19.21). The opening and dissociation steps are reversible, and their combined rates depend on the concentration of pyrophosphate present. A series of rate measurements done at different pyrophosphate concentrations were analyzed to deduce the rates of the conformational change and pyrophosphate release. The finding was that the rate constant for the conformational change is ~$1.2 \times 10^3$ sec$^{-1}$ and that for release of pyrophosphate is $>10^3$ sec$^{-1}$, again both much faster than the overall rate through the pathway, so neither of these steps is rate limiting.

The combined kinetic data show, by a process of elimination, that the rate-limiting step in the overall addition process is a conformational change which must occur after binding of the dNTP that will be transferred to the DNA chain being synthesized. We will return to discuss this important result and understand how it affects the fidelity of DNA synthesis. We will also examine what specific structural changes are associated with this key step in the reaction pathway.

As a final note in the overall scheme, after pyrophosphate is released, it is hydrolyzed by the enzyme pyrophosphatase to give two phosphates in an energetically favorable step (**Figure 19.23**). This removes pyrophosphate and prevents the back reaction in which a base would be displaced from the growing strand by addition of the pyrophosphate. The step of nucleotide transfer to the growing chain is energetically nearly neutral and, without pyrophosphatase, the likelihood of the addition reaction reversing (in the synthesis active site) would be larger.

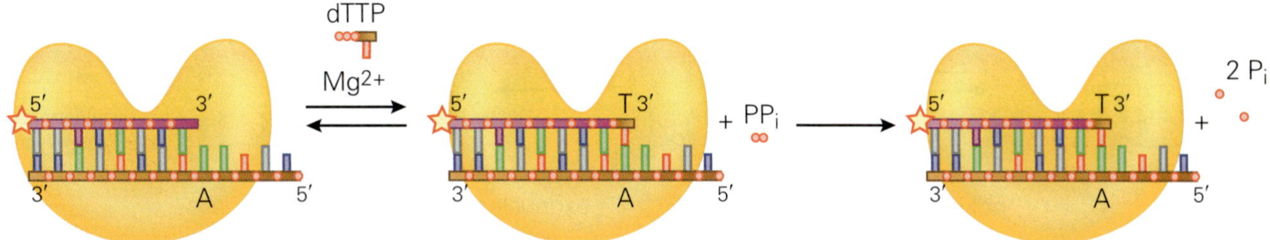


**Figure 19.23 Hydrolysis of pyrophosphate.** The conversion of pyrophosphate to phosphate by the enzyme pyrophosphatase makes the DNA synthesis reaction essentially irreversible. After this happens, cleavage of the newly added nucleotide only occurs through the action of an exonuclease enzyme.

## 19.13 Determination of the values of $V_{max}$ and $K_M$ for the incorporation of correct and incorrect base pairs provides insights into fidelity

The fidelity of DNA replication can be defined as the number of incorrectly incorporated bases (those not following Watson-Crick pairing rules) per correct base in the newly synthesized double helix. As polymerase functions in the cell, all four of the deoxynucleotides are available (at nearly equal concentrations), and the enzyme is operating under steady-state conditions (see the Section 16.6 for a discussion of how the steady-state assumption enters into Michaelis–Menten kinetics). One way of measuring fidelity is to let the reaction run and measure the relative concentration of correct and incorrect bases. But, because the fidelity of DNA polymerase is very high, it can be difficult to measure the very small number of incorrect bases that are incorporated.

To get around this difficulty, we can start by recognizing that fidelity is a measure of the specificity of an enzyme (DNA polymerase in this case) for the correct substrate. As discussed in Section 16.7, the relative rates of reaction, and thus specificity, can be calculated if the values of $k_{cat}$ (or equivalently $V_{max}$ and the enzyme concentration, [Pol]) and $K_M$ are known for both the correct and incorrect nucleotide substrates for different steps in the overall reaction. The usual approach used to characterize the fidelity of DNA polymerase reactions has been to determine these parameters experimentally, which gets around the problem of counting the very small number of mismatches in newly synthesized DNA.

To measure the kinetic parameters for reactions with different nucleotides, the methods described in the previous sections are used. A complex of polymerase with a specific primer–template DNA is generated, but rather than adding all four of the dNTPs that are needed for extension of a template with arbitrary sequence, just one dNTP is added and the rate of addition of one base is measured.

**Table 19.3** lists the kinetic parameters determined when the next base on the template being replicated is a T, so the Watson-Crick partner to be added is an A. From

**Table 19.3 Rates of nucleotide incorporation by DNA polymerase.**

| Base pair being formed | $K_M$ (μM) | Relative $V_{max}$ | Relative rate |
|---|---|---|---|
| A–T | 3.7 ± 0.7 | 1.0 | 1.0 |
| G–T | 4200 ± 900 | 0.24 ± 0.05 | $2 \times 10^{-4}$ |
| C–T | 10,000 ± 4000 | 0.13 ± 0.04 | $5 \times 10^{-5}$ |
| T–T | 10,000 ± 4000 | 0.13 ± 0.04 | $5 \times 10^{-5}$ |

Nucleotide incorporation in:

```
        ↓
5′ .....GG□
3′ .....CCTAG
```

Rates at which nucleotides are inserted into a growing DNA strand (by *Drosophilia* DNA polymerase α, which has no editing or proofreading activity). The rates are for the addition of nucleotides to a primer–template junction with the previously incorporated nucleotides forming correct Watson-Crick pairs. The arrow and box in the schematic on the right indicate the position at which addition occurs. An A at this site to make an A-T base pair is the correct nucleotide, whereas incorporation of a G, C, or T is incorrect. (From J. Petruska et al. and I. Tinoco Jr., *Proc. Natl. Acad. Sci. USA* 85: 6252–6256, 1988. With permission from the National Academy of Sciences.)

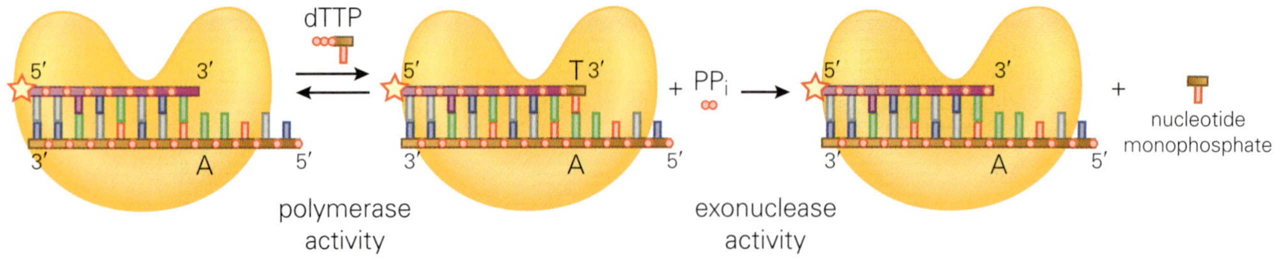


**Figure 19.24 Exonuclease activity of a DNA polymerase.** Newly added nucleotides can be excised from the growing DNA strand by hydrolytic cleavage of the phosphate linkage, catalyzed by an exonuclease domain. The exonuclease domain, which is not shown in this diagram, is either a part of the DNA polymerase or is in a protein that is bound to the polymerase.

the table we see that, when the reaction is done with dATP, the reaction proceeds rapidly and the $K_M$ value for the dATP (3.7 μM) is close to the values described in Section 19.11. For the other three incorrect bases, however, the values of $K_M$ are much higher (4000 to 10,000 μM). The rate of the chemical transfer step (reflected in the $V_{max}$ value) is also somewhat reduced. The effect of the mismatch on $K_M$ thus provides most of the discrimination against incorrect bases, amplified somewhat by the lower rate of addition.

In the last column of Table 19.3, the relative rates of addition for correct and incorrect base pairs for a DNA polymerase operating under typical cellular concentrations of dNTPs are given, reflecting the fidelity of this nucleotide incorporation step. As can be seen, this is quite substantial, giving a much better suppression of misincorporation than provided by the free energy of base-pairing alone. The level of discrimination in this step (1 in ~$10^4$), however, does not reach that achieved for actual DNA replication in the cell.

## 19.14 DNA polymerase has a nuclease activity that can remove bases from the 3′ end of a DNA strand

Since the purpose of DNA polymerase is to synthesize a new DNA strand, it may seem strange that these enzymes have an associated exonuclease activity that can hydrolyze bases from the end of a DNA strand (**Figure 19.24**). If polymerase is added to a primer–template complex without dNTPs, it is found that the 3′ base on the primer strand is slowly removed. This activity is associated with a domain that is separate from the polymerase domains that catalyze DNA synthesis. The end of the DNA must move out of the polymerase active site in order to access the separate exonuclease active site.

It seems that with synthesis proceeding very rapidly (230 nucleotides per second; see Figure 19.20) there would be almost no chance for the nuclease to act on the growing DNA strand. This is, in fact, correct when Watson-Crick base pairs are incorporated. But, when one of the rare incorrect base insertions occurs, the rate of DNA synthesis slows down (see Table 19.3), allowing time for the exonuclease to cleave newly added nucleotides.

The rates of DNA synthesis were measured for primer–template junctions that already have a mismatched base at the last double helical position (that is, an incorrect base has been inserted, and the DNA polymerase must now move on to add another base after the incorrect one). It was found that, for the reaction adding the subsequent base after the error, the $K_M$ value for the correct (Watson-Crick) incoming dNTP was much higher than when addition was occurring after a correct base, as shown in **Table 19.4**.

The rate of the chemical transfer step was also reduced by three- to six-fold when an incorrect base was already incorporated into the DNA. Together these two effects slow the extension after an error by 200- to 2500-fold. This slowing down means that there is a much greater chance that the exonuclease can act on the terminal (incorrect) base and remove it from the 3′ end. The polymerase will then

**Table 19.4 Rates of nucleotide incorporation by DNA polymerase.**

| Base pair at the previous site (*–T) | $K_M$ (μM) | Relative $V_{max}$ | Relative rate[a] |
|---|---|---|---|
| A–T | 2.8 ± 0.7 | 1.0 | 1.0 |
| G–T | 170 ± 30 | 0.30 ± 0.05 | $5 \times 10^{-3}$ |
| C–T | 1150 ± 200 | 0.30 ± 0.04 | $7 \times 10^{-4}$ |
| T–T | 1000 ± 200 | 0.16 ± 0.04 | $4 \times 10^{-4}$ |

Nucleotide incorporation in:

```
          ↓
5′ . . . . . GG*□
3′ . . . . . CCTAG
```

Slowing of the DNA polymerase nucleotide insertion immediately following a mismatched base pair. As for Table 19.3, data are for *Drosophila* DNA polymerase α in reactions where there is no editing or proofreading. The incoming nucleotide is always T opposite A (the correct nucleotide). Rates are compared for either correct or incorrect pairs at the previous position, determined by the base marked by * in the schematic on the right. (From J. Petruska et al. and I. Tinoco Jr., *Proc. Natl. Acad. Sci. USA* 85: 6252–6256, 1988. With permission from the National Academy of Sciences.)

[a]Relative rates at physiological nucleotide concentrations.

just extend the DNA again using the template and, with high probability, will insert the correct base on the second try. The slowing of extension by polymerase after an error, together with the nuclease activity, acts as a proofreading mechanism, removing incorrectly paired nucleotides. The nuclease will occasionally also remove a correctly paired nucleotide, but this makes very little difference in the overall efficiency of the enzyme and is a small price to pay for a large increase in fidelity. Together the intrinsic fidelity of initial incorporation and the proofreading activity give a fidelity of ~1 error in $10^6$–$10^7$ nucleotides.

## 19.15 The structure of DNA polymerase has fingers, palm, and thumb subdomains

Information from many different structures of polymerases with various substrates bound has given us a detailed understanding of how DNA polymerases catalyze DNA synthesis. DNA polymerase I (Pol I) from *E. coli* was the first DNA polymerase to be discovered and its structure and activity have been characterized extensively. We will use this enzyme, as well as its close relatives from the bacterium *Thermus acquaticus* and the bacteriophage T7, to introduce the basic features of polymerase enzymes. DNA polymerase I is one of five polymerases in *E.coli* and is of a type responsible for replicating small segments of DNA as part of DNA repair processes.

First, we will consider a large fragment of DNA polymerase I—called the Klenow fragment. This fragment, shown in **Figure 19.25**A, has the parts of the protein responsible for DNA synthesis and also the exonuclease activity for removing bases from the 3′ end of the elongating primer strand. The full protein has another, different proofreading domain (the first 323 residues of the protein, which need not concern us here), which was removed for both the kinetic and structural studies. Within the Klenow fragment, the N-terminal sequence forms the exonuclease domain, followed by the polymerase domain.

The DNA polymerase domain forms a large U-shaped DNA-binding cleft that resembles a right hand made up of "fingers," "palm," and "thumb" subdomains (Figure 19.25C). The palm subdomain is at the base of the U-shaped structure and consists of a β sheet that contains three acidic residues that are found in all DNA polymerases. These residues bind magnesium and are crucial for enzymatic activity. Both the fingers and thumb subdomains have helical secondary structure, bracketing the palm domain. The 3′ → 5′ proofreading exonuclease domain is located beneath the palm domain in the "standard view" of the polymerase (Figure 19.25B).

## 19.16 DNA polymerase binds DNA using the "palm" and nearly encircles it

When DNA polymerase binds to the primer–template junction, contacts are made by the DNA across the palm, which is positively charged, and also with the fingers and thumb regions. The nature of the interaction with DNA is illustrated in **Figure 19.26**, which shows the structure of a DNA polymerase from a virus known as T7 bacteriophage. The T7 DNA polymerase is closely related to *E. coli* Pol I, and has been crystallized in the active conformation, bound to a primer-template junction and a nucleotide triphosphate. The polymerase forms a tight interface with the DNA, with the "hand" of the polymerase grasping the DNA within the palm. The tip of the thumb subdomain changes conformation upon binding DNA and makes contact with the primer strand of the primer–template.

The structure of the bound DNA is altered somewhat from the conventional B conformation due to these interactions, as will be discussed in more detail below. To accommodate contacts with the protein, the helix axis becomes curved, and there is a significant widening of the minor groove, where many contacts are made. In the complex with polymerase, the 3′ end of the primer DNA strand that is to be elongated is in place in the active site of the enzyme.

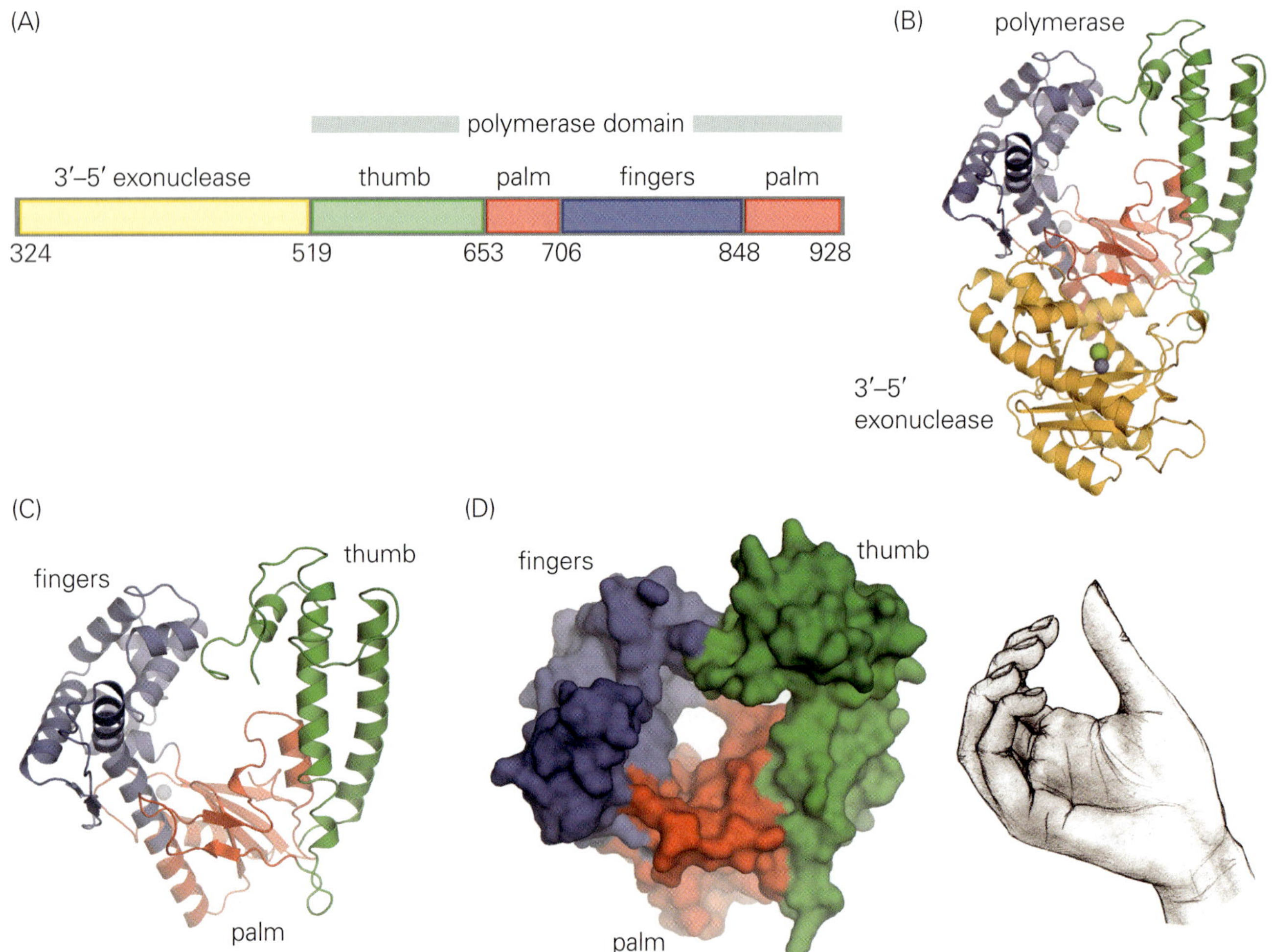


**Figure 19.25 The structure of *E. coli* DNA polymerase I.** (A) Schematic representation of the organization of DNA polymerase I. The segment of the protein shown here is known as the Klenow fragment, and it includes an exonuclease domain and a polymerase domain. (B) Structure of the polymerase. The three subdomains of the polymerase are colored *blue*, *red*, and *green* and the 3′ → 5′ exonuclease domain is *yellow*. (C) Structure of the polymerase domain (3′ → 5′ exonuclease domain removed). (D) Molecular surface of the polymerase domain, the shape of which resembles a cupped right hand. (Adapted from C.A. Brautigam and T.A. Steitz, *J. Mol. Biol.* 277: 363–377, 1998; PDB code: 1KFS.)

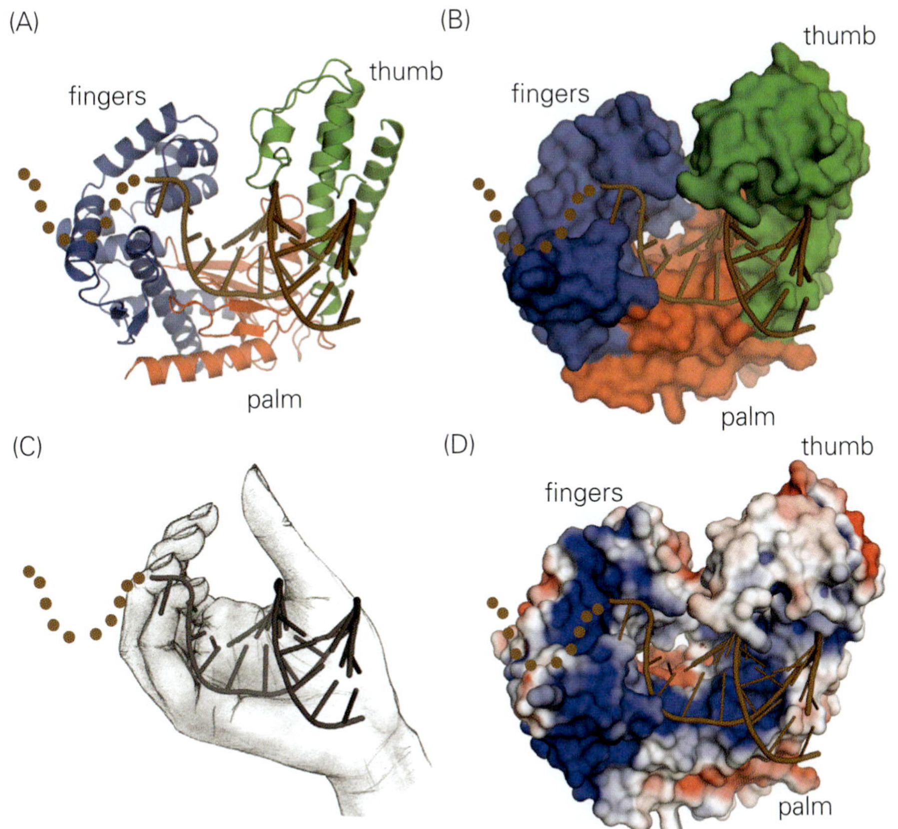


**Figure 19.26 Various representations of the polymerase domain of T7 DNA polymerase bound to DNA.** The structure of the closed complex, ready for catalysis, is shown in several representations. In (D), the molecular surface of the polymerase is colored according to the local electrostatic potential generated by the protein. Note that the path along which DNA binds corresponds to regions of positive electrostatic potential (*blue*). (PDB code: 1T7P.)

## 19.17 The active site of polymerase contains two metals ions that help catalyze nucleotide addition

Enzymes that act on nucleoside triphosphate substrates often require metal ions as cofactors in the active site. These metal ions, by virtue of their positive charge, help bind the phosphate groups of the substrate. In addition, metal ions can play a direct role in catalysis (for example, see Section 16.35 for a discussion of how metal ions are used by RNA enzymes).

DNA polymerases use two divalent metal ions (referred to as metal A and metal B) to stabilize the geometry of the reaction partners and to activate them for participation in the chemical steps of the reaction (**Figure 19.27**). By binding divalent metal ions at the active site of the polymerase, the protein concentrates positive charge where it can optimally facilitate the reaction. The two metal ions, very likely $Mg^{2+}$ under physiological conditions, are bound at the active site of the polymerase by interactions with two conserved acidic residues (a third conserved acid residue is also important, but is not discussed here). These residues are usually aspartate rather than glutamate, probably because the sidechain of Asp is less flexible.

The two metal ions are located close to each other (3–4 Å apart) and are well positioned to stabilize negative charges on the activated hydroxyl group of the terminal nucleotide in the primer strand, the developing negative charge on the α phosphate of the incoming nucleotide, and the negative charges on the β and γ phosphates. An important aspect of the coordination of the metals is that the two conserved aspartate sidechains both bridge the two metal ions, thus providing particularly tight geometrical restraints on the positions of the metals (Figure 19.27B and C).

The reaction mechanism for all known phosphoryl transfer reactions in nucleic acids involves a pentacovalent phosphate transition state and inversion of stereochemical configuration at the phosphorus (Figure 19.27C). In this **two metal ion mechanism**, the metals are positioned with one at each end of the trigonal

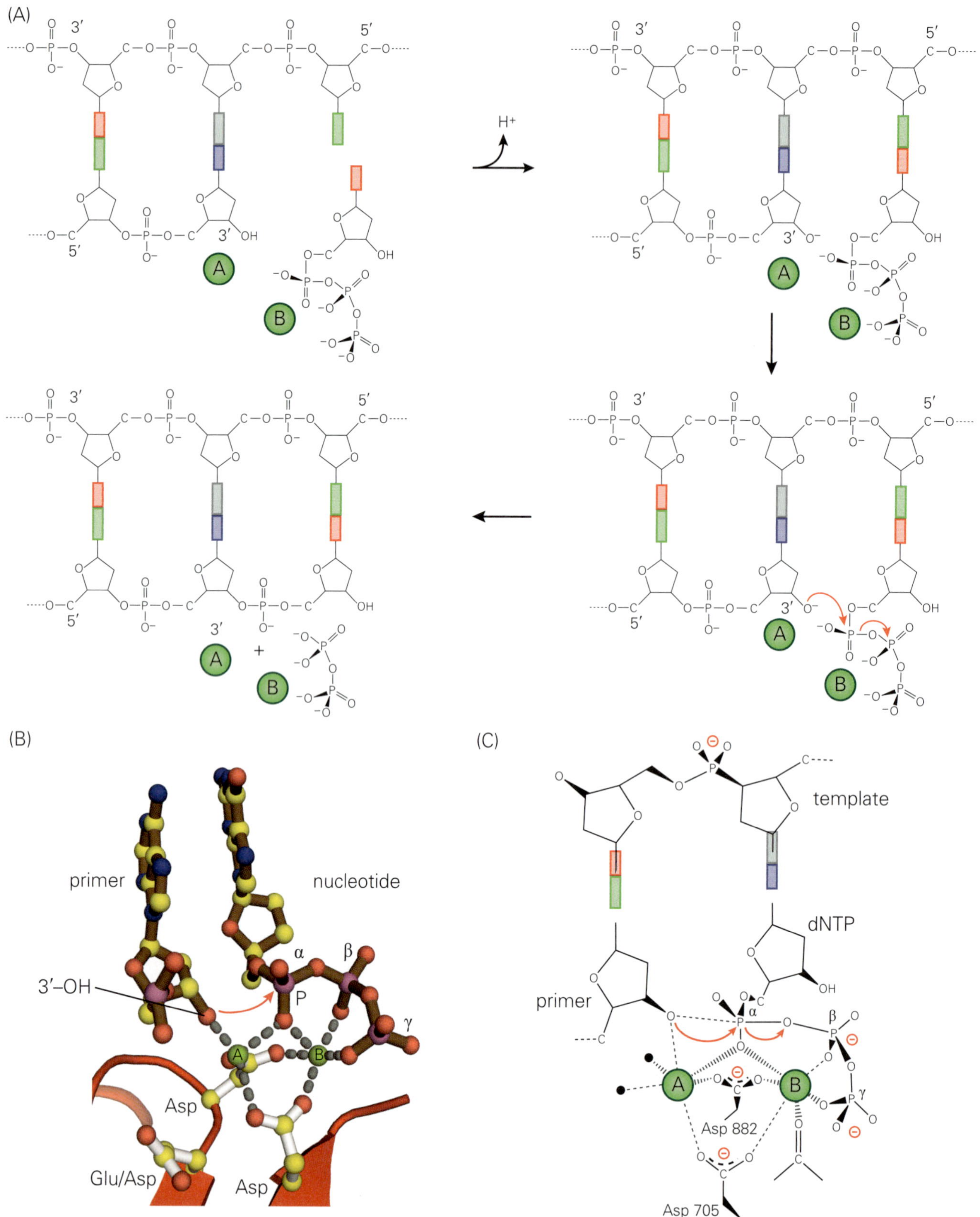


**Figure 19.27 The reaction mechanism for DNA polymerases.** Two metal ions, denoted A and B, bind the incoming nucleotide triphosphate and help activate the attacking 3′-hydroxyl group. (A) Metal ion A is located between the α phosphate of the incoming nucleotide and the 3′ end of the primer strand. The metal ion lowers the affinity of the 3′ OH for the hydrogen, thus facilitating the 3′ O- attack on the α phosphate. Metal B interacts with all three phosphates of the nucleotide. (B) A close-up view of the reaction showing the two metals interacting with two conserved aspartate residues. (C) Schematic diagram corresponding to the structure shown in (B). The template strand is shown here, but not in (B). (Adapted from S. Doublie et al. and T. Ellenberger, *Nature* 391: 251–258, 1998, and T.A. Steitz and J.A. Steitz, *Proc. Natl. Acad. Sci. USA* 90: 6498–6502, 1993; PDB code: 1T7P.)

bipyramidal transition state, interacting with both the attacking and leaving groups found at the two apical positions. Metal ion A facilitates the deprotonation of the sugar hydroxyl group, activating it to become a nucleophile. Metal ion B directly ligates the pyrophosphate leaving group, thus promoting the bond breakage step by neutralizing the developing negative charge. Both metals also ligate one of the nonbridging oxygens of the phosphate being transferred. This stabilizes the developing negative charge in the transition state and helps to position the substrate phosphate group properly in the active site.

## 19.18 A conformational change in DNA polymerase upon binding dNTP contributes to replication fidelity

Structures of a DNA polymerase from the bacterium *Thermus acquaticus* have been determined in several states: empty, bound to primer–template DNA, and simultaneously to DNA and a nucleotide triphosphate (**Figure 19.28**). In order to prevent the nucleotide triphosphate from reacting with the primer strand in the crystal, the dideoxy form of the nucleotide was used (ddNTP). Dideoxy nucleotides lack a hydroxyl group at the 3′ position of the sugar, in addition to not having a hydroxyl group at the 2′ position. Comparison of these structures reveals distinct "open" and "closed" conformations. The polymerase (apo) and polymerase–DNA binary complexes form an "open" conformation (Figure 19.28A and B). In the polymerase ternary complex, the fingers domain undergoes a further

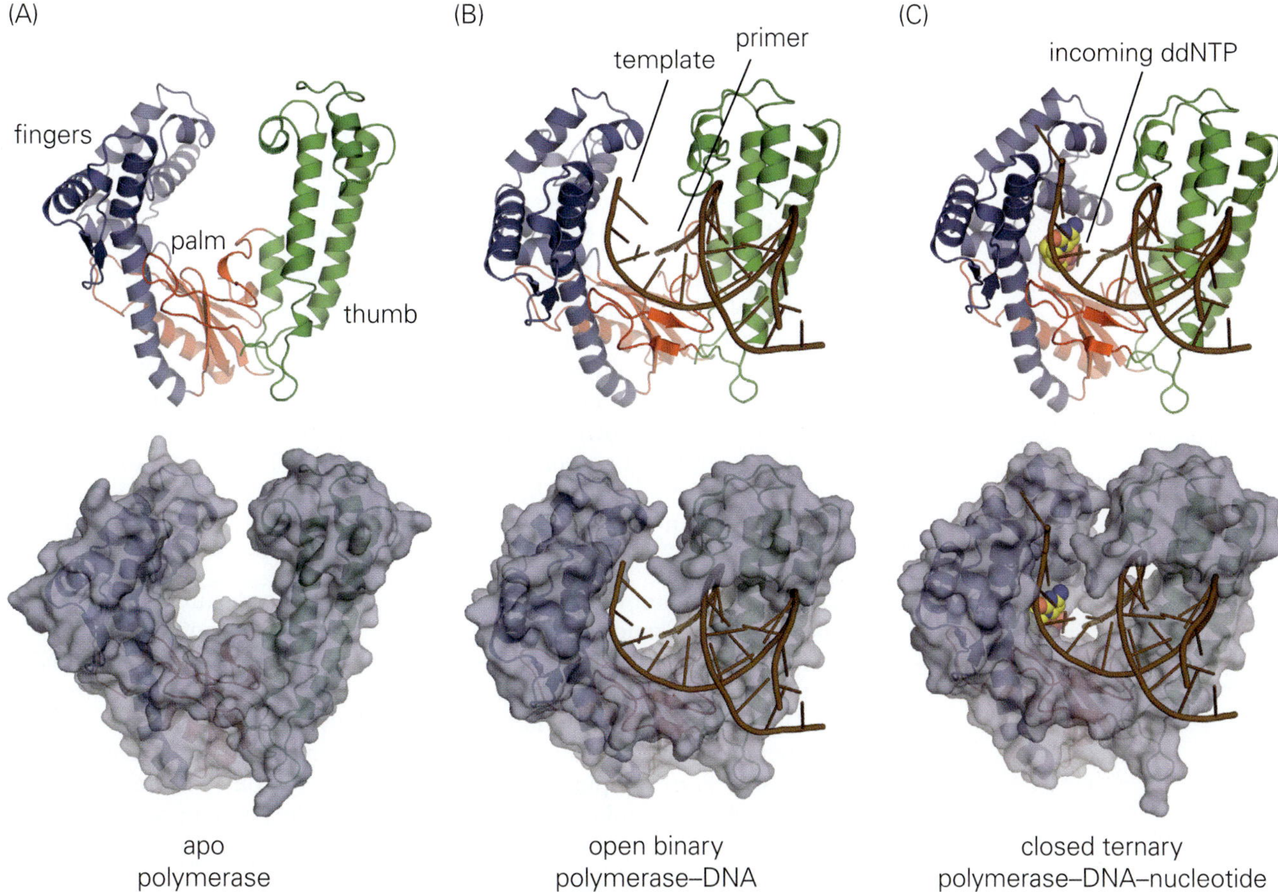


**Figure 19.28 Conformational changes in polymerase upon binding DNA and the correct incoming dNTP.** (A) Structure of the Klenow fragment of the *Thermus acquaticus* DNA polymerase in the apo form—that is, in the absence of nucleotide or primer–template. The protein is in the fully open conformation. (B) Polymerase bound to primer–template, in an open binary complex. (C) Ternary complex, with incoming nucleotide (ddNTP) and the primer–template, in which the catalytic site is fully assembled. The fingers and thumb subdomains close over the DNA and touch each other. In this conformation, the polymerase is fully sensing the shape of the newly formed base pair. (PDB codes: A, 1KTQ; B, 4KTQ; C, 3KTQ.)

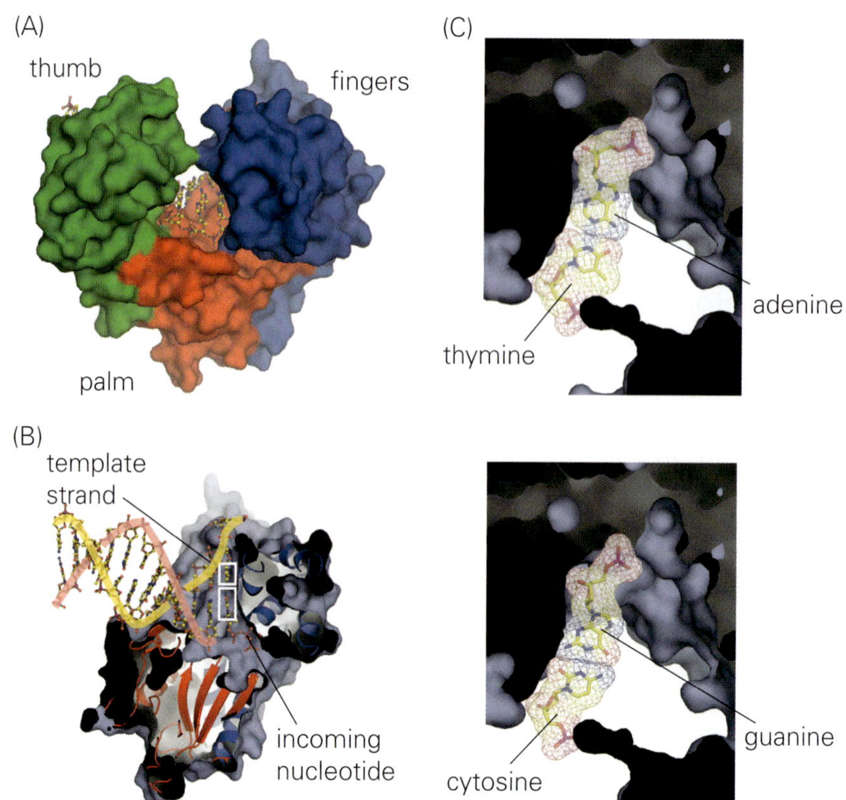


**Figure 19.29 The DNA polymerase grasps the incoming nucleotide in a tight embrace.** (A) Molecular surface of T7 DNA polymerase bound to DNA and incoming nucleotide. This view is from the back of the polymerase with respect to the view in Figure 19.28. In this view, there is a hole in the surface, through which the newly formed base pairs can be seen. (B) Cross-sectional view of the complex, in the same orientation. The surface of the polymerase forms a slot-like binding site into which the new base pair fits snugly. (C) The binding site for the newly formed base pair (looking roughly down the axis of the DNA duplex). Two correctly formed base pairs are shown here: A-T (above) and G-C (below). Both of these fit nicely into the binding site, whereas mismatched base pairs would not. (PDB code: 1T7P.)

closure, a conformational rearrangement that results in the active site becoming completely enclosed by the protein (Figure 19.28C). In this "closed" conformation, the incoming nucleoside triphosphate is held in a pincer-like grip that tightly brackets the newly formed base pair (**Figure 19.29**). This closure of the polymerase around the newly formed base pair is the rate-limiting step in the overall cycle.

The transition from the open DNA complex to the closed conformation in the presence of the dNTP to be added involves the movement of the fingers subdomain as a rigid unit (no internal structural changes—a rigid-body motion) and conformational changes in three hinge regions in the fingers. Together these changes result in a ~40° rotation of the O helix in the fingers domain (**Figure 19.30**). Residues on the O helix have been shown to play important roles in the binding of the incoming nucleotide and its incorporation into the growing DNA strand. In the closed form of the polymerase, the O helix packs against the face of the new base pair, helping to seal the new nucleotide completely into the active site of the polymerase.

A very important point with respect to fidelity is that the closure of the polymerase complex occurs much more easily when the correct nucleotide triphosphate (Watson-Crick paired to the template) is present. The shapes of mispaired dNTPs prevent closure; the details of how polymerase senses shape will be discussed below. Once in the closed conformation, the polymerase active site has all of the components correctly placed to catalyze the transfer of the nucleotide to the primer strand. For example, the location of the metal ions with respect to the incoming nucleotide triphosphate is as described in the previous section.

When incorrect nucleotides are present, usually only partial closing occurs, and the active site is not correctly formed. In this way, the polymerase impedes the incorporation of incorrect nucleotides into the growing DNA strand. The fact that rare additions of incorrect bases to the new strand do occur suggests that there are rare, high-energy fluctuations to a closed state. A partial closure is more likely to be followed by reopening and release of the nucleotide to sample another incoming nucleotide for its ability to pair with the template. As discussed in the first

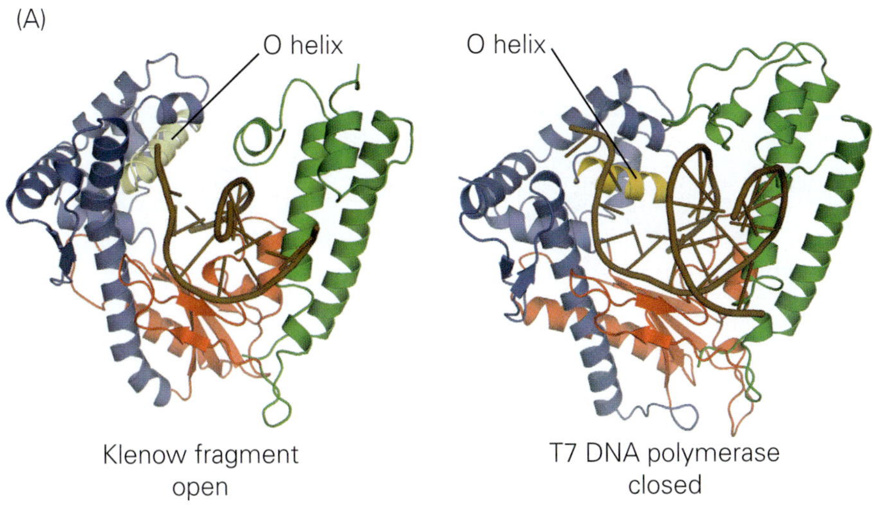


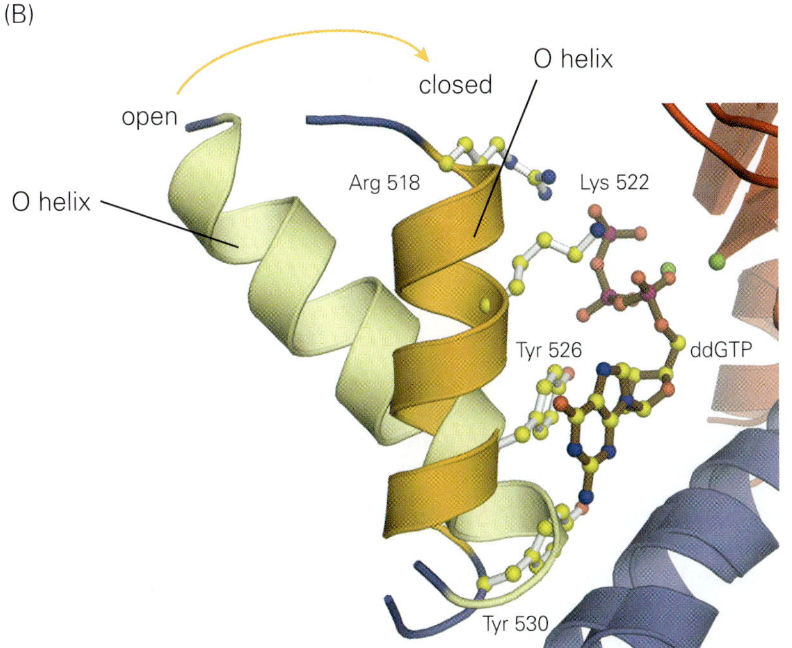


**Figure 19.30 Closure of the polymerase to form the fully assembled catalytic site.** (A) Left: The open binary complex where the polymerase is bound to the primer–template. Right: The ternary closed complex formed upon binding of the primer–template and the incoming dNTP. (B) A close-up view of the conformational change that takes place upon ternary complex formation. The rotation of the fingers domain brings four conserved residues in the O helix into contact with the incoming nucleotide (ddGTP in this case). (PDB codes: 1TAU and 1T7P.)

part of the chapter, the correct pairing in the open state is slightly more favorable energetically than incorrect pairing, so the correct nucleotide is the most likely to bind at the open active site.

## 19.19 DNA polymerases recognize DNA using the backbone and minor groove

To be able to replicate any DNA sequence, the DNA polymerases must interact with their DNA substrates in a sequence-independent manner, but they need to maintain the ability to distinguish correctly paired bases from mismatches. DNA polymerases accomplish sequence-independent recognition of DNA by interacting extensively with the minor groove of the four to five base pairs near the 3′ end that is being extended. The last base pair, at the 3′ end of the primer strand, is located in a pocket that is very closely matched to the shape of Watson-Crick base pairs, as shown in Figure 19.29.

The tight grip that the polymerase exerts on the DNA near the active site results in significant structural distortions in the DNA, which are illustrated in **Figure 19.31**. The contacts between the DNA and the protein that bring about these structural

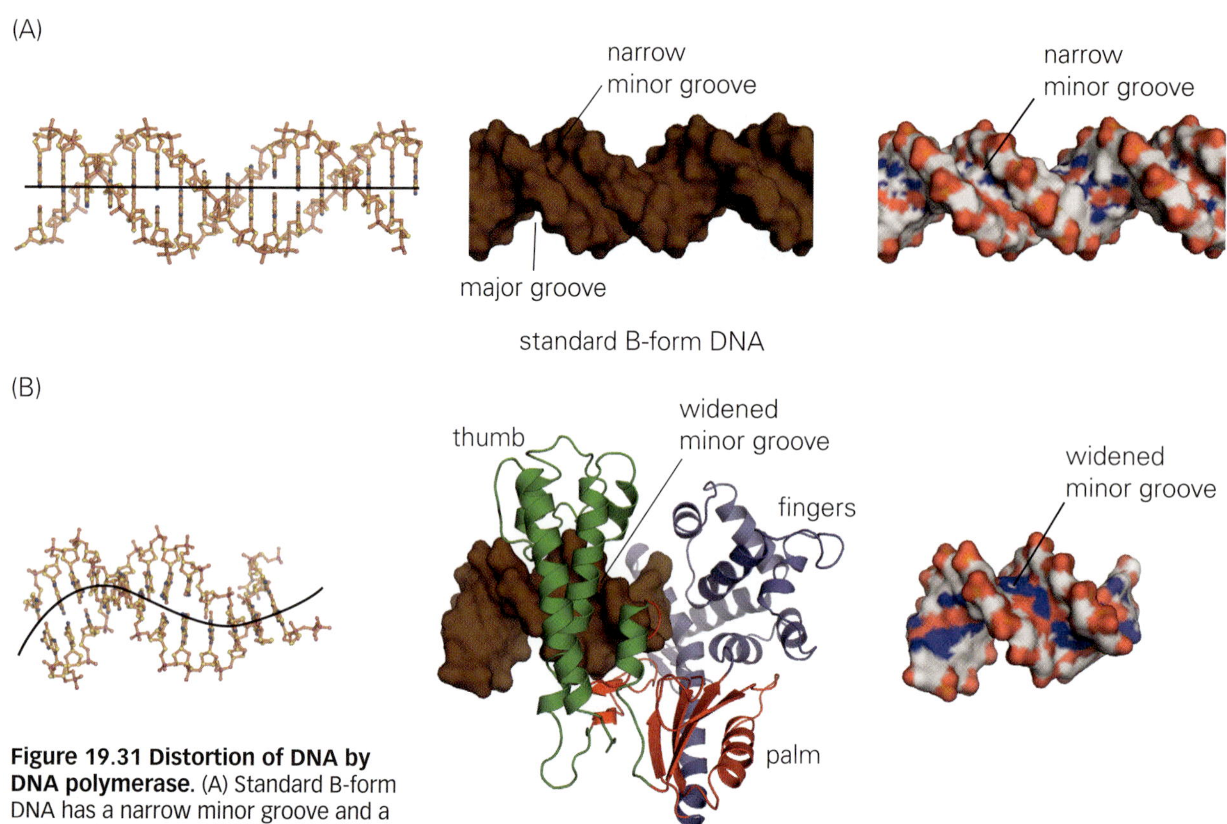


**Figure 19.31 Distortion of DNA by DNA polymerase.** (A) Standard B-form DNA has a narrow minor groove and a wide major groove and the bases are perpendicular to the axis of the double helix. (B) When DNA polymerase binds to DNA, the thumb makes contact with the minor groove and widens it. This bends the axis of the double helix. (PDB code: 4KTQ.)

changes in the DNA are essentially restricted to the sugar-phosphate backbone, while avoiding the major groove (there are a few contacts to the minor groove, noted below, which are important for achieving fidelity).

In contrast to standard B-form DNA, which has a straight helix axis, the axis of DNA bound to DNA polymerase is bent into an S-shape (Figure 19.31B). The distortion in the DNA helix is brought about by tight interactions between the polymerase fingers and thumb subdomains and the DNA. The distortions in the DNA are closely matched by the shape of the protein, so that there is excellent steric complementarity between the protein and the DNA at the site of the incoming nucleotide, as well as several base pairs downstream from it.

The DNA nearest the polymerase active site has a compressed major groove and significantly widened minor groove, somewhat resembling A-form DNA. This allows protein sidechains to more closely approach the edges of the bases at the minor groove. As we discussed in Chapter 2, the pattern of hydrogen-bond donors and acceptors is uniform in the minor groove of DNA, irrespective of the nucleotide sequence (see Figure 2.13). This feature allows the DNA polymerase to hold onto the newly synthesized DNA duplex for any sequence of DNA.

Hydrogen bonding between the protein and the DNA in the minor groove is one check on the formation of properly matched base pairs, because only base pairs with the correct shape will maintain the correct inter-nucleotide spacing to make these contacts. A total of four base pairs at the 3′ end of the primer strand have minor groove contact to T7 DNA polymerase (**Figure 19.32**), and mismatches at any of these positions cause the polymerase to slow down addition of the next nucleotide. The importance of the minor groove contacts between bases and the highly conserved Arg and Gln residues is demonstrated by kinetic studies in which these Arg and Gln residues were mutated. The mutations cause a large decrease in the rate of DNA synthesis and also reduce the affinity of the polymerase for DNA.

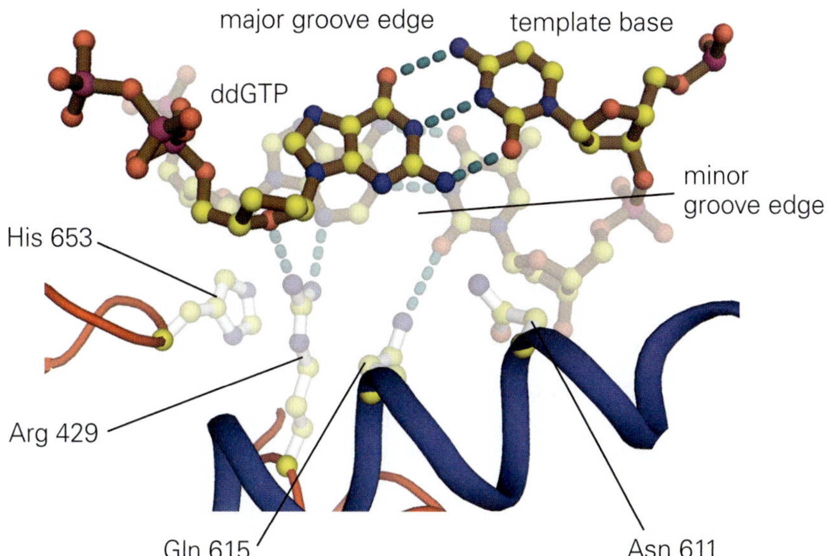


**Figure 19.32 Contacts between the minor groove of the DNA and the nascent base pair in the polymerase active site contribute to fidelity.** The template base and incoming dNTP (dideoxy-GTP, ddGTP, in this case, which acts as a chain terminator) are shown in front. Four conserved residues line the floor of the active site. The Arg and Gln sidechains interact with the N2 and O3 atoms in the minor groove of the base pair at the primer 3′ end, just before the site of nucleotide incorporation. These interactions select for Watson-Crick pairings but are not sequence specific. The structure shown here is that of T7 DNA polymerase. (PDB code: 1T7P.)

## 19.20 DNA polymerases sense the shapes of correctly paired bases

Mismatched nucleotides form hydrogen bonds with each other in duplex DNA and can stack as part of the duplex, but they introduce distortions compared to Watson-Crick pairs (**Figure 19.33**). Using x-ray crystallography, the three-dimensional structures of short pieces of duplex DNA containing various kinds of mismatched base pairs have been determined. The general shape of a base pair can be characterized by three geometrical parameters. First, the width of the base pair

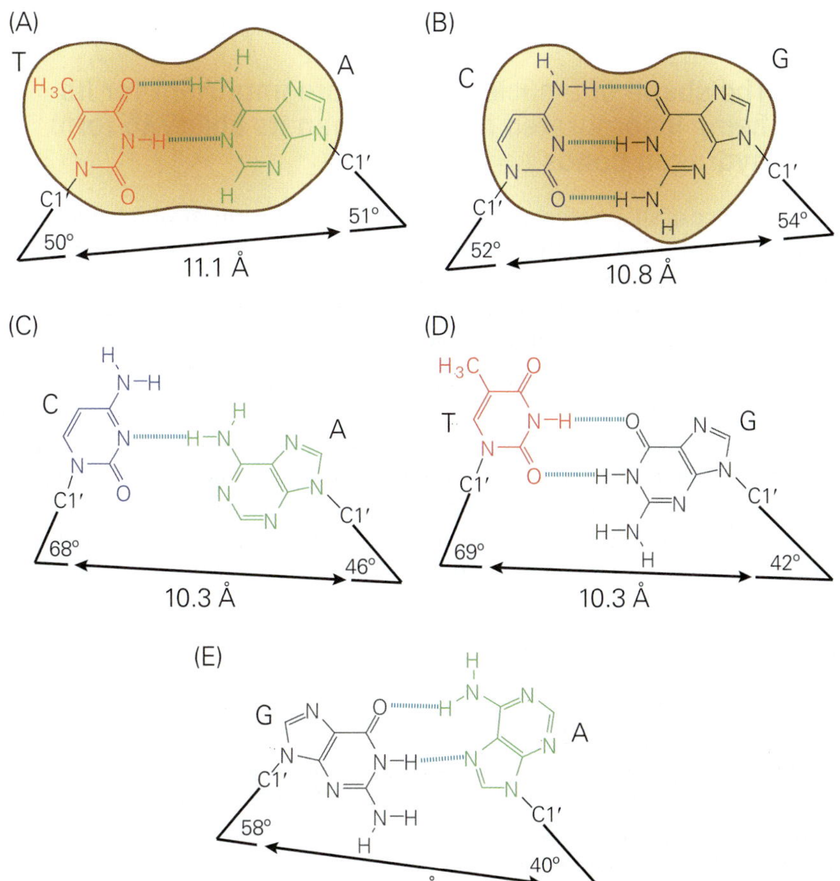


**Figure 19.33 The shapes of Watson-Crick and non-Watson-Crick base pairs.** (A and B) Geometrical parameters describing the shapes of Watson-Crick base pairs. (C, D, and E). Various incorrectly paired nucleotides. (Adapted from W.N. Hunter et al. and O. Kennard, *Nature* 320: 552–555, 1986. With permission from Macmillan Publishers Ltd.)

is indicated by the distance between the two C1′ atoms of the ribose rings of the two nucleotides, to which the bases are attached. The C1′ to C1′ distance is about 10.8 Å to 11.1 Å in correctly formed base pairs. Some mismatched base pairs have similar C1′ to C1′ distances (see Figure 19.33), which is why they can be accommodated uniformly within duplex B-form DNA.

Second, there are two angles between the vector connecting the C1′ atoms of the two ribose rings and the glycosyl bonds of the two nucleotides. The values of these angles are very close to 50° for both nucleotides in Watson-Crick base pairs (see Figure 19.33). This means that correctly formed base pairs are nearly symmetrical—that is, an A-T base pair has a very similar overall shape to a T-A base pair, and both of these are similar in overall shape to G-C and C-G base pairs.

In contrast to the symmetry observed in correctly formed base pairs, the values of this angle range from ~40° to ~70° in mismatched base pairs (see Figure 19.33). The mismatched base pairs are no longer symmetrically disposed with respect to the glycosyl bonds and, given a fixed orientation of DNA, a T-G base pair does not look like a G-T base pair and neither does it look like any of the correctly matched base pairs. The importance of this is that the mismatched pairs do not fit correctly into the active site of the polymerase and prevent it from closing fully, thereby inhibiting the reaction that would result in incorporation of the incorrect nucleotide.

If a mispaired base is incorporated into the newly synthesized strand, the incorporation of the next nucleotide is slowed down significantly (see Section 19.13). This leaves more time for the exonuclease domain to act and remove the incorrect base. Translocation also slows down because of an incorrect fit of mismatched DNA into the extended region of the polymerase that is in contact with the growing DNA. Hence, mismatched bases that distort the DNA to a greater extent prevent translocation for a longer period, and these mispaired bases have a greater chance of being removed before synthesis continues. The relative repair efficiencies of the mispairs shown in Figure 19.33 are 15:5:1 for G-T:A-C:G-A.

## 19.21 The shape of a nucleotide is more important for its being incorporated into DNA than its ability to form hydrogen bonds

The importance of shape recognition in determining the rate at which nucleotides are incorporated into DNA by polymerase has been demonstrated by measuring how well *E. coli* DNA polymerase I utilizes a nucleotide analog that closely resembles thymidine in shape, but which cannot form base-pairing hydrogen bonds. This structural mimic of thymidine has fluorine atoms in place of the oxygen atoms of the thymidine base, and the fluorine atoms are not polarized enough to form hydrogen bonds with adenine or other nucleotides (**Figure 19.34**).

The fluorinated nucleotide analog of dTTP (referred to as dFTP or F; Figure 19.34) can be readily incorporated into oligonucleotides by chemical synthesis. These oligonucleotides can be annealed with complementary strands of DNA to form primer–template junctions to be extended by polymerase. Alternatively, the fluorinated analog, dFTP, can be supplied as a nucleoside triphosphate to normal primer–template junctions.

dFTP

F, $CH_3$, F, $^{-}O-P(=O)(O^{-})-O-P(=O)(O^{-})-O-P(=O)(O^{-})-O-CH_2$, O, OH

dTTP

O, $CH_3$, HN, O, N, $^{-}O-P(=O)(O^{-})-O-P(=O)(O^{-})-O-P(=O)(O^{-})-O-CH_2$, O, OH

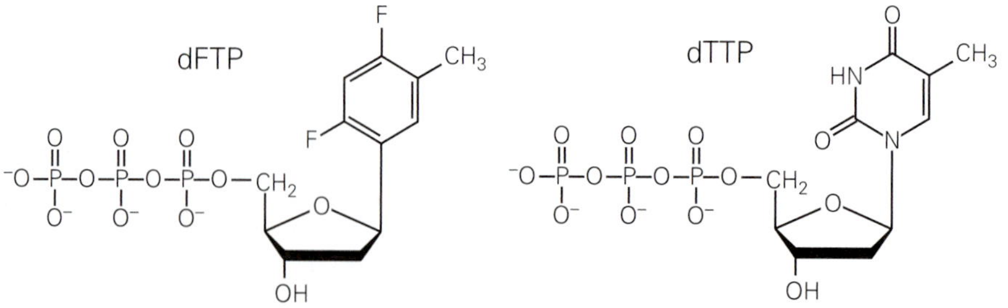

**Figure 19.34 Fluorinated analog of thymidine.** A synthetic nucleotide triphosphate analog, designated dFTP, is shown to the left of the normal deoxythymidine triphosphate, dTTP. The shape of dFTP is essentially identical to dTTP, but all of the hydrogen-bonding capacity of the thymidine base has been removed. Efficient incorporation of this base into DNA by polymerase shows that the shape is more important than the ability to hydrogen bond.

The rates at which DNA polymerase I either incorporated dFTP into the growing strand opposite various normal nucleotides in the template or incorporated normal nucleoside triphosphates into an analog-containing template were measured. The experiments used DNA polymerase I enzyme that lacked exonuclease activity, eliminating its ability to proofread. Remarkably, when the fluorinated analog, F, is placed at the first unpaired template position, DNA polymerase I was still able to very efficiently incorporate adenosine opposite F, despite the lack of hydrogen bonding.

These experiments, and analogous ones with other base mimics, demonstrate convincingly that the ability of the DNA polymerase to work accurately does not rely directly on the formation of Watson-Crick hydrogen bonds between correctly paired bases. Also, because the ability of the fluorinated nucleotide analog to form hydrogen bonds with the DNA polymerase protein is also impaired, interactions between the bases and the protein cannot be crucially important. This leads to the conclusion that the similarity in shape between thymidine and the fluorinated nucleotide must be the central factor in determining the high fidelity with which DNA polymerase I incorporates the correct partners for the nucleotides in the template. Thus, the secret to the high fidelity of DNA polymerase is its ability to work as a Braille reader, choosing the correct nucleotides by shape.

## 19.22 The growing DNA strand can shuttle between the polymerase and exonuclease active sites

The 3′ → 5′ exonuclease domain of polymerase is specific for single-stranded DNA and is able to cleave one nucleotide at a time in the 3′ to 5′ direction. When a mismatched nucleotide is incorporated into the growing DNA strand, as discussed above, the presence of the mismatch slows down the incorporation of the next nucleotide, even when the incoming nucleotide is correctly paired. This stalling is crucial for correcting errors, slowing down the rate of DNA synthesis from ~50–300 nucleotides per second to 0.01–0.1 $sec^{-1}$ at mismatches. This gives the exonuclease domain the chance to remove the mistake. The exonuclease and polymerase domains are far away from each other (30 Å in the case of DNA polymerase I; **Figure 19.35**), so the 3′ end of the primer strand must somehow get from the

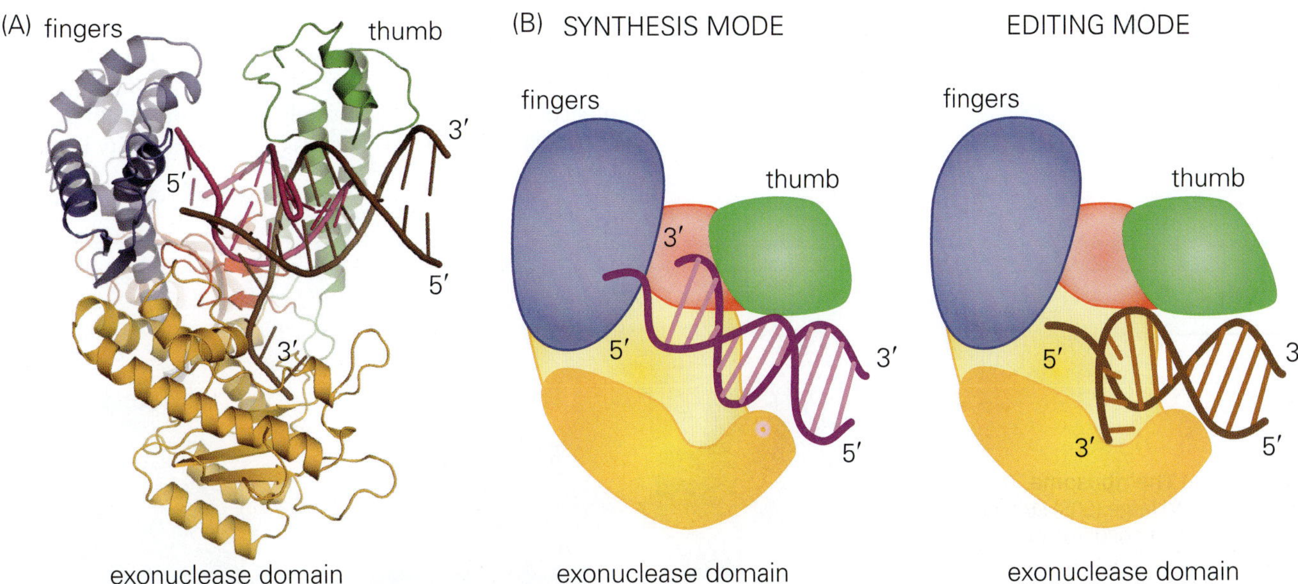


**Figure 19.35 Shuttling of DNA between DNA synthesis and editing modes**. (A). Primer–template DNA bound to the polymerase active site (*purple*) and to the exonuclease active site (*brown*). The structure shown here is a representation of two different crystal structures. (B) The 3′ end of the primer can shuttle into the exonuclease active site without dissociating from the polymerase. Mismatched base pairs that destabilize the duplex shift the equilibrium from the synthesis mode to the editing mode. (PDB codes: 1TAU and 1KLN.)

polymerase site to the exonuclease site. While it is unclear exactly how this happens, the DNA manages to do so without dissociating from the DNA polymerase.

The two active sites—polymerase and 3′ → 5′ exonuclease—compete for the 3′ end of the primer strand, and the overall rate of the reaction is a result of the relative rates of DNA synthesis and editing. DNA partitions between the 3′ exonuclease active site and the polymerase active site with a ratio of about 1:10 for correctly base-paired DNA. Mismatched base pairs destabilize the duplex DNA and thus favor binding of the 3′ single-stranded DNA to the exonuclease active site. In addition, a mismatch causes the polymerase to slow down for as many as five subsequent additions, which also favors the exonuclease reaction.

## C. HOW RIBOSOMES ACHIEVE FIDELITY

We have seen, in part B of this chapter, that DNA polymerases make remarkably few errors during the process of DNA synthesis. Underlying the high fidelity of replication is the ability of these enzymes to recognize the shapes of correctly matched base pairs. Incorrectly paired bases have different shapes, leading to their rejection. The reliance on shape as the principal mechanism for preventing errors turns out to also underlie fidelity in the process of mRNA translation, which we discuss in this part of the chapter.

As described in Part C of Chapter 1, the proteins encoded by the genome are produced in two steps, the first carried out by RNA polymerases, which make a messenger RNA (mRNA) copy of the DNA. The mRNA serves to specify the sequence of a protein to be synthesized by ribosomes. The process of transcription is intrinsically very similar to that of DNA replication, except that ribonucleotides are used in transcription and deoxyribonucleotides in replication. Both processes occur at similar rates, with ~200 nucleotides incorporated per second. Because the mRNA synthesized is only a temporary copy, fidelity is less crucial in transcription than in DNA replication (which generates a permanent, master copy of the genome), and the error rate during transcription, about 1 error in $10^4$ nucleotides added, is roughly 1000 times higher than during DNA synthesis.

The final step of translating the mRNA sequence into a protein by the ribosome is slower, occurring at a rate of ~20 amino acids incorporated per second. Protein synthesis has a fairly low error rate of about 1 in $10^3$. Protein synthesis by ribosomes has been intensively studied biochemically, and detailed structural information on the ribosome has become available recently (**Figure 19.36**). Together, these have allowed the construction of detailed models for the process of protein synthesis. In this section, we will examine this process and how fidelity is maintained.

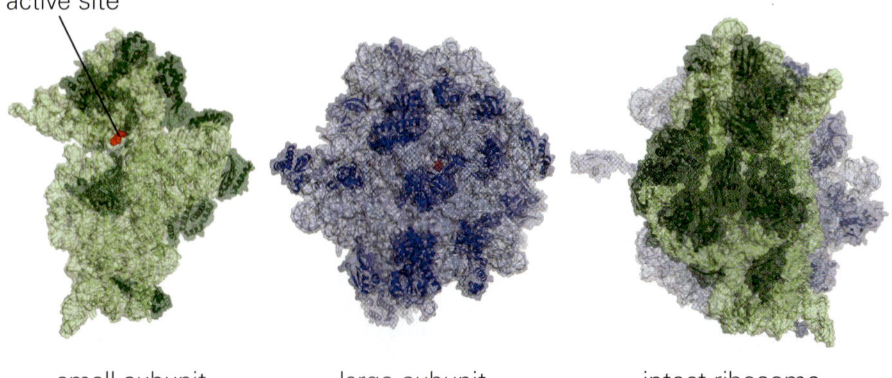


**Figure 19.36 The ribosome.** The small (30S) and large (50S) subunits of the bacterial ribosome combine to form the functional 70S ribosome. See Section 17.29 for a discussion of the Svedberg (S) units used to name the ribosomal subunits. (From the Research Consortium for Structural Biology [Protein Data Bank], David Goodsell.)

## 19.23 The ribosome has two subunits, each of which is a large complex of RNA and proteins

The ribosome consists of two subunits, one large and one small (Figure 19.36). The components of ribosomes are generally identified by their size (specified in Svedberg units as discussed in Section 17.29). We focus on the highly studied bacterial ribosomes, which are good models for the somewhat more complicated eukaryotic ribosomes. Fully assembled bacterial ribosomes are denoted "70S" and are comprised of the large (50S) and small (30S) subunits (see Figure 19.36).

Both subunits of the ribosome are complexes of RNA and proteins in which the nucleic acids and the proteins are associated intimately. In bacterial ribosomes, the large subunit has two RNAs (5S, with 120 nucleotides, and 23S, with 2900 nucleotides) along with 34 proteins. The small subunit has one RNA (16S, with 1540 nucleotides) and 21 proteins.

To begin a cycle of protein synthesis, the two subunits of the ribosome come together with a messenger RNA (mRNA) and a special aminoacyl-tRNA known as initiator tRNA sandwiched between them. This process, corresponding to the initiation of protein synthesis, is regulated to control the levels of proteins present in cells. Here we will focus on the cycle of steps through which translation occurs, once it has been initiated, and on how the ribosome ensures that the right amino acid is being incorporated at each step.

Within the assembled ribosome there are three binding sites for transfer RNAs, which are called the A site (for Aminoacyl), the P site (for Peptidyl), and the E site (for Exit) (**Figure 19.37**). In each of these sites the three bases of a tRNA anticodon interact with the three bases of a codon within the mRNA. tRNAs are shuttled through these sites by a combination of conformational changes in auxiliary proteins (called elongation factors) and the ribosome itself at appropriate points during the cycle.

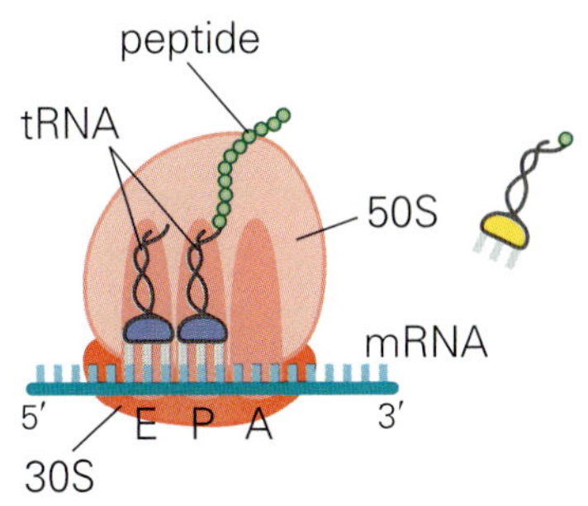


**Figure 19.37 Ribosome during protein synthesis.** This schematic diagram of the ribosome (*pink*) is shown just after the addition of a new amino acid to the growing peptide chain (*green*). There are three binding sites for tRNA on the ribosome, denoted E, P and A. The tRNA in the E site will be ejected when the next tRNA arrives in the A site; the peptide being synthesized is attached to the tRNA in the P site; and the A site is open, awaiting arrival of the next aminoacyl tRNA (*yellow*). The messenger RNA is also shown (*cyan*).

## 19.24 Protein synthesis on the ribosome occurs as a repeated series of steps of tRNA and protein binding, with conformational changes in the ribosome

Each cycle in the protein synthesis process begins with a tRNA that has the covalently attached, growing protein chain occupying the P site, and the tRNA that had brought in the next-to-the-last amino acid in the E site, as shown in Figure 19.37. The A site is open, ready to accept an aminoacyl-tRNA for the next amino acid to be added. Binding of the aminoacyl-tRNA in the A site causes release of the tRNA in the E site.

The tRNAs that are aminoacylated (see Section 1.20) and ready for reaction on the ribosome have a protein known as elongation factor Tu (EF-Tu) associated with them. EF-Tu is a GTPase, that is, a protein that binds to GTP and catalyzes its hydrolysis to GDP. EF-Tu undergoes a conformational change when GTP bound to it is hydrolyzed to GDP, promoting the release of EF-Tu from the ribosome. This property allows EF-Tu to act like a molecular timer that checks that the correct tRNA is bound to the ribosome. We will see, in Section 19.26, that the rate at which EF-Tu hydrolyzes GTP is an important step in controlling fidelity in protein synthesis. This rate is controlled by the recognition by the ribosome of the shape of a correctly docked EF-Tu•GTP•tRNA complex at the A site.

Once the correct tRNA has been bound at the A site and EF-Tu has been released, a conformational change of the ribosome occurs, called **accommodation** (**Figure 19.38**), which brings the amino acid on the A-site tRNA close to the polypeptide on the P-site tRNA. After this step, addition of the amino acid to the growing protein chain occurs. The accommodation step also involves conformational changes in the tRNA, which is described in Section 19.26. These structural changes are a crucial aspect of the mechanism by which the ribosome achieves high fidelity.

Once a new peptide bond has been formed, another elongation factor, EF-G (also a GTPase), binds near the A site. GTP hydrolysis by EF-G triggers a concerted shift in the positions of both the mRNA and the tRNAs. The tRNA that was in the A site (still associated with its codon in the mRNA) moves into the P site, and the previous A-site tRNA (that no longer has the peptide bound) shifts into the E site. The movement of the mRNA that occurs, together with the shifts in tRNA positions, is called **translocation**.

This process brings the next three bases of the mRNA (that is, the next codon) into the A site, and the ribosome is ready to begin the next cycle of amino acid addition (see Figure 19.38). Each step in the cycle, and the structural changes that occur, are discussed in the following sections. We will see that the processes related to GTP hydrolysis by EF-Tu and the subsequent accommodation step are critical for

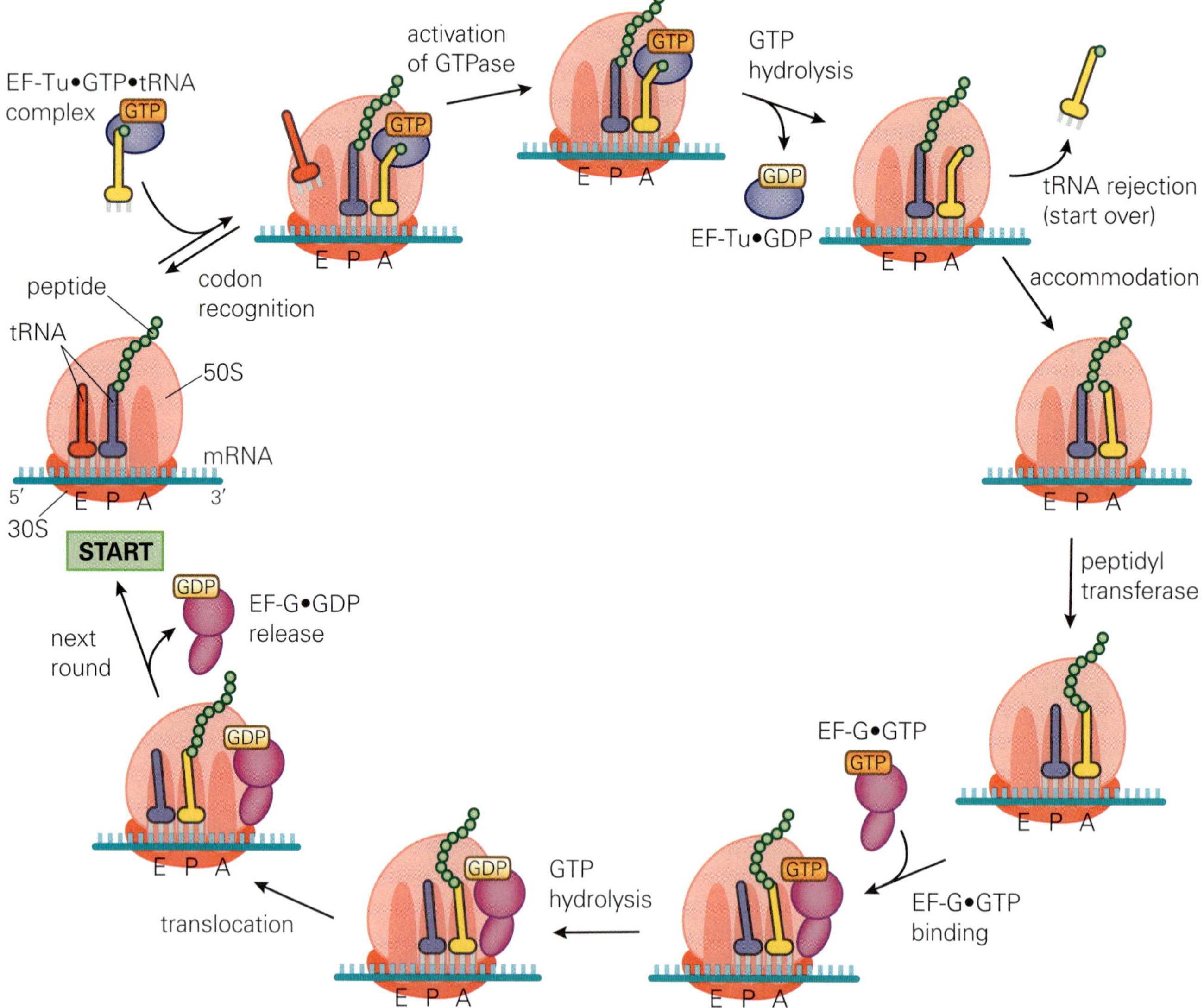


**Figure 19.38 The elongation cycle.** The ribosome (*pink*) is shown with the large subunit above the small subunit and the mRNA (*cyan*) running between them. The three tRNA binding sites: A, P, and E, are shown in a darker shade. tRNA molecules (inverted T shape) bind at these sites. An incoming aminoacyl-tRNA (the small *green* circle represents the amino acid) in complex with EF-Tu•GTP enters the A site. Correct codon–anticodon pairing causes hydrolysis of GTP and release of the aminoacyl end of the tRNA from EF-Tu. The transfer of the peptide chain onto the A-site tRNA occurs at the active site of the enzyme (P), then the mRNA must translocate on the ribosome so that the next mRNA codon is at the A site. Translocation of the tRNAs and mRNA is facilitated by binding of another GTPase, EF-G, which causes the tRNA at the P site to move to the E site and the peptidyl-tRNA at the A site to move to the P site. The ribosome is then ready for the next round of elongation. The tRNA in the E site is released on binding of the next aminoacyl-tRNA to the A site. (Adapted from T.A. Steitz, *Nat. Rev. Mol. Cell Biol.* 9: 242–253, 2008, and T.M. Schmeing and V. Ramakrishnan, *Nature* 461: 1234–1242, 2009.)

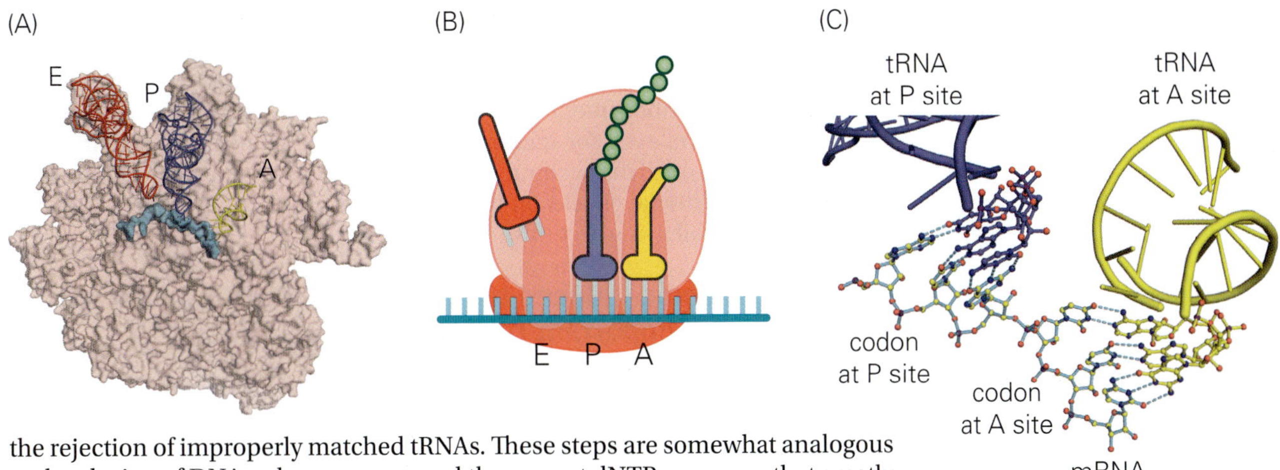


**Figure 19.39 mRNA codons pairing with tRNA anticodons.** (A) Structure of the ribosome, indicating tRNAs at the A, P, and E sites, along with mRNA. Only a portion of the A-site tRNA is shown. (B) Schematic diagram corresponding to the structure shown in (A). (C) Watson-Crick hydrogen bonds between codons at the A and P sites and anticodons in the two tRNAs at these sites. (PDB codes: A, 2J00; C, 2J01.)

the rejection of improperly matched tRNAs. These steps are somewhat analogous to the closing of DNA polymerase around the correct dNTP, a process that greatly enhances the specificity relative to the base-pairing and stacking interactions alone.

## 19.25 Selection of the correct A-site tRNA by base-pairing alone cannot explain ribosome fidelity

When we examined the thermodynamics of base-pairing in DNA, we found that the high fidelity of DNA polymerases cannot be explained by the difference in free energy between correctly and incorrectly paired bases in DNA. You might think that it is easier for the ribosome to select the correct tRNA by base-pairing alone, because three bases are used in the anticodon–codon interaction and the fidelity requirement for protein synthesis is considerably lower than for replication. But this turns out not to be the case.

Structural studies of tRNAs bound to ribosomes show that the mRNA codon and anticodon in the tRNAs do indeed form three base pairs that stack against each other, forming a small segment of duplex RNA (**Figure 19.39**). If you examine the the table of codons shown in Figure 1.48, you will see that there are many codons for different amino acids that differ by only a single base. As a consequence, the ribosome must be able to distinguish between aminoacyl-tRNAs that differ in their anticodon stem by only a single base. The detailed thermodynamics of base interactions are slightly different for interior versus terminal bases, and for RNA instead of DNA, but the energetic difference between a correct Watson-Crick match and single-base mismatch is never sufficient to explain ribosome specificity (error rate of 1 in $10^3$, corresponding to an energy difference of ~18 kJ•mol$^{-1}$). With tRNAs, wobble base pair formation (G paired with U; see Figure 2.32) is also tolerated at the third base of the codon, which complicates the analysis of energetic discrimination.

To anticipate the discussion that follows, when the incoming EF-Tu•GTP•tRNA complex binds to the ribosome, a bend is induced in the tRNA (shown schematically in Figure 19.38). This energetically unfavorable distortion of the tRNA is made possible by the free energy of binding of the mRNA codon with the tRNA anticodon, and also by additional interactions between the tRNA anticodon stem and the ribosome. This bending of the tRNA is necessary for EF-Tu to interact with the ribosome in a way that stimulates GTP hydrolysis and the consequent dissociation of EF-Tu from the tRNA.

After release of EF-Tu, the energetically unfavorable distorted tRNA relaxes into its normal shape, which brings the aminoacyl end of the tRNA into the catalytic center of the ribosome. In other words, the tRNA acts as a molecular spring that requires proper matching of the codon and the anticodon in order to bend and then release EF-Tu before the tRNA falls off the ribosome. If the codon and the anticodon are not properly aligned, then the tRNA cannot bend readily, and the EF-Tu•GTP•tRNA complex is more likely to be rejected before GTP is hydrolyzed.

Thus, one key step in the error-checking mechanism is the ability of the tRNA to bend. Anything that makes the bent state of the tRNA more energetically favorable (for example, mutations in the ribosome, or mutations in the tRNA, or the binding of antibiotics to the ribosome) makes it easier to reach the transition state for GTP hydrolysis by EF-Tu, even for incorrect tRNAs. In such situations, the incorrect tRNA might move on to the accommodation step before being rejected. The ribosome also has additional checks on the correctness of the newly arrived tRNA, as discussed in the following sections.

## 19.26 A ribosome-induced bend in the EF-Tu•tRNA complex plays an important role in generating specificity

A large set of enzymes (the aminoacyl-tRNA synthetases) ensure that each tRNA is "charged" with the appropriate amino acid. This is an aspect of fidelity that we will not address here, but the incorrect charging of tRNAs occurs infrequently and, hence, does not contribute much to errors in translation. As we noted in Section 19.24, there is a critical protein called EF-Tu that is required for protein synthesis on the ribosome. EF-Tu binds GTP (to form an EF-Tu•GTP complex), and this complex interacts with aminoacyl-tRNAs before they bind to the ribosome in the codon recognition step of Figure 19.38. One function of EF-Tu•GTP is in making sure that each tRNA carries the correct amino acid (another part of fidelity that we do not address here).

The individual steps of the elongation cycle (shown in Figure 19.38) have been dissected by kinetic studies, much as we described for the DNA polymerases. The individual kinetic steps of one cycle of elongation are shown in **Figure 19.40**. The

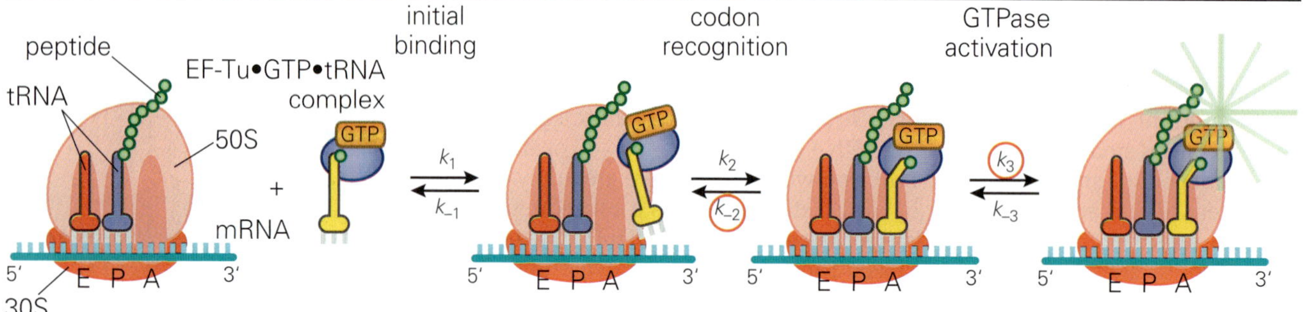


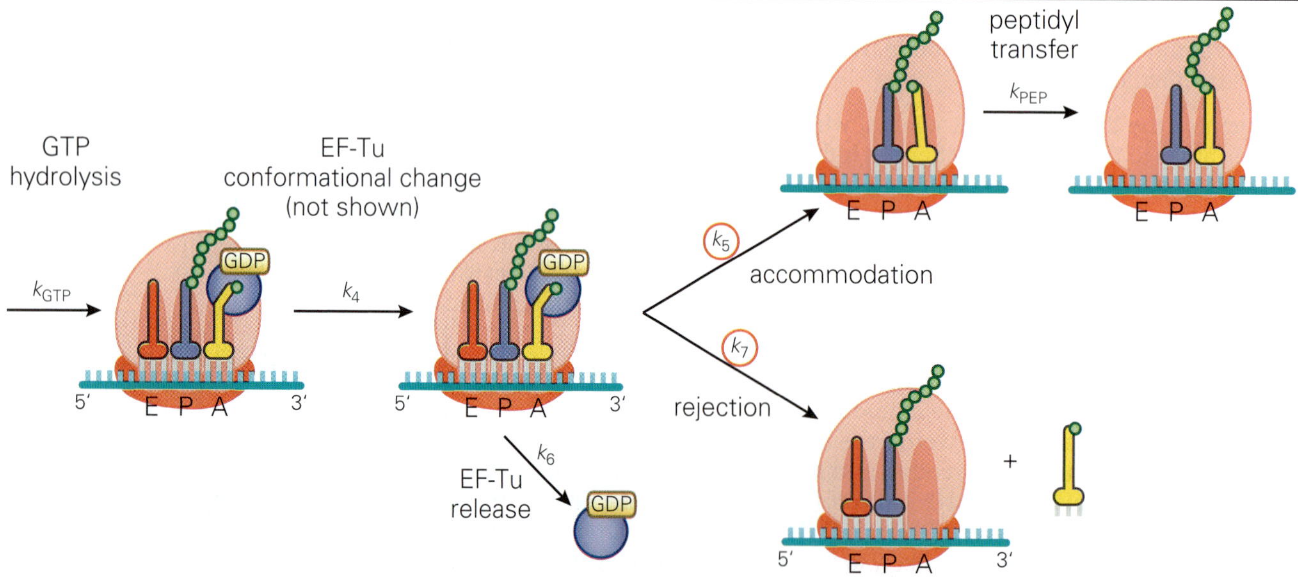


**Figure 19.40 Kinetic steps for the process of tRNA selection by the ribosome.** The rate constants for key steps that discriminate correct vs. incorrect tRNA molecules are circled. These are $k_{-2}$ (dissociation occurs faster for incorrect codons), $k_3$ (GTPase activation occurs preferentially for correct codons), $k_5$ (accommodation is faster for correct codons), and $k_7$ (rejection is faster for incorrect codons). (Adapted from H.S. Zaher and R. Green, *Cell* 136: 746–762, 2009. With permission from Elsevier.)

initial binding of the ternary EF-Tu•GTP•tRNA complex to the A site of the ribosome occurs with relatively low affinity, with a high dissociation rate, and does not involve interaction of the codon and anticodon. In a subsequent kinetic step, the codon and anticodon are brought together. If they are not properly matched, then the complex usually reverts to the loose complex from which full dissociation can occur, and another tRNA can be sampled.

If the codon and anticodon are matched, then a further tightening of the interaction can occur with higher probability (about 50-fold more likely for matched interactions). In this tightly bound state, the tRNA is bent and EF-Tu•GTP interacts with the ribosome in a way that stimulates GTP hydrolysis (explained in Section 19.30), resulting in EF-Tu being bound to GDP.

The altered nucleotide state on EF-Tu drives another conformational change that can have two possible outcomes. When the codon and anticodon are properly matched, then EF-Tu•GDP dissociates and the tRNA shifts its position, moving the aminoacyl end into the peptidyl transfer site (the accommodation step). In this conformation, addition of the amino acid to the protein being synthesized can occur.

If GTP hydrolysis occurs on EF-Tu bound to a tRNA for which the codon-anticodon pair is incorrect, then there is a high probability (~95%) that the tRNA will dissociate, a process called **rejection**. The combination of these two sequential, yet kinetically distinct, steps, each with significant preference for correct codon-anticodon pairings, is called **kinetic proofreading**. The designation *kinetic* is significant in that it is the rate constants for the correct aminoacyl-tRNA interaction that are increased, not the equilibrium constants.

The combination of stimulated GTP hydrolysis and proofreading gives most of the overall ~$10^3$ fidelity for the ribosome incorporation of amino acids. To complete the synthesis cycle after peptidyl transfer, the GTP-bound form of elongation factor G binds near the A site of the ribosome, and hydrolysis of GTP drives a conformational change in which the mRNA and associated tRNAs are moved by three bases to position the next codon in the A site of the ribosome. The binding of EF-G and translocation do not affect fidelity and, hence, will not be discussed further.

## 19.27 The ribosome undergoes conformational changes during the process of tRNA selection

We mentioned earlier, in Section 19.25, that the ribosome induces a bend in the tRNA at the A site. The energy required to enforce this bend comes from many interactions between the ribosome and the tRNA, not just from the codon-anticodon interactions. Structures of the small ribosomal subunit (S30) solved with and without mRNA and tRNAs bound show that there is a significant rearrangement of the ribosomal RNA that accompanies binding (**Figure 19.41**). One such conformational change brings the ribosomal RNA into close contact with the minor groove of the small segment of duplex formed by the codon base-pairing with the anticodon.

The structures of the ribosome with and without mRNA and tRNAs present show that the region of the ribosome called the **decoding center** undergoes a significant change in structure to hold the mRNA and the A-site tRNA close together (see Figure 19.41). In the absence of these substrate RNAs, the decoding center is quite open, with the conserved bases G530 and C1054 on one side and A1492 and A1493 on the other, relatively far apart. In this state, G530 takes on the less common syn conformation.

When tRNA binds to messenger RNA in the decoding center, these bases reorganize, coming closer together to interact with one another and with the small segment of helix formed by the pairing of the codon and anticodon. The importance of these residues in codon recognition is supported by genetic and biochemical studies. The interactions are through the minor groove region of the codon–anticodon helix. Both A1492 and A1493 make contacts through A-minor motifs (see Section 2.26). Contacts are also made to the small ribosome subunit protein S12, which is shown in Figure 19.41. Taken together, these interactions support the bend in the tRNA that is necessary for the activation of GTP hydrolysis by EF-Tu.

## 19.28 Tight interactions in the decoding center can only occur for correct codon–anticodon pairs

The details of the interactions of the ribosome with the base pairs in the codon–anticodon helix are shown in **Figure 19.42**. In addition to van der Waals contacts to the bases, important hydrogen bonds are formed between the ribosomal RNA and the codon–anticodon helix, particularly with the sugars. These contacts

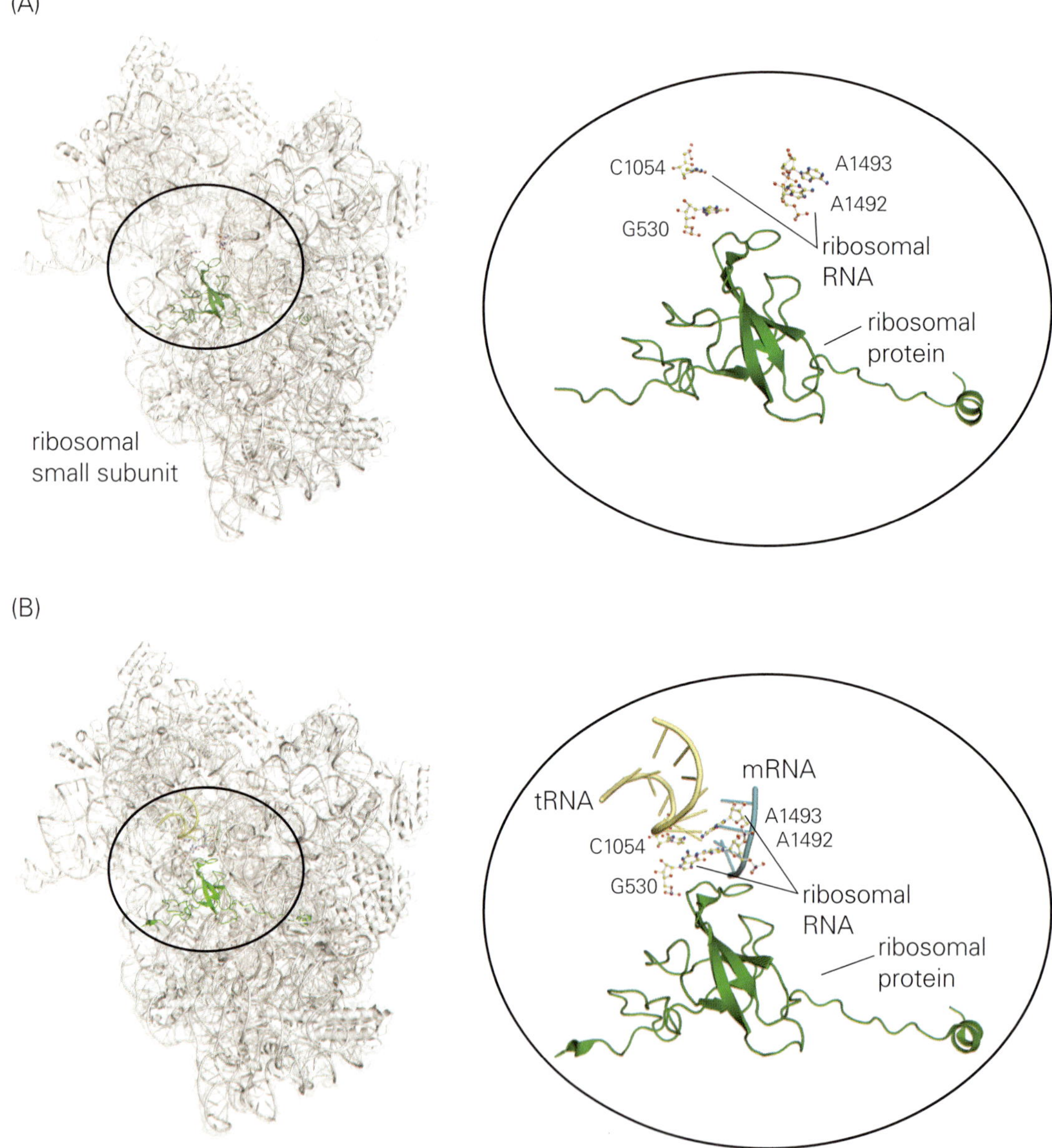


**Figure 19.41 Structural effects of mRNA and tRNA binding.** (A) The small subunit of the ribosome is shown on the left in an empty state. The ribosomal protein S12 (*green*) and a few nucleotides of ribosomal RNA are shown in the expanded view. These elements make up part of the decoding center, where mRNA and cognate tRNAs are bound. (B) The structure of the small subunit bound to mRNA and the anticodon-bearing region of a tRNA is shown on the left. The expanded view is similar to that in (A), except that the mRNA and the tRNA are also shown. The conformation of the ribosomal RNA and the positions of these important bases change dramatically upon interaction with mRNA and tRNA. (Adapted from H.S. Zaher and R. Green, *Cell* 136: 746–762, 2009; PDB codes: 1J5E and 1IBM.)

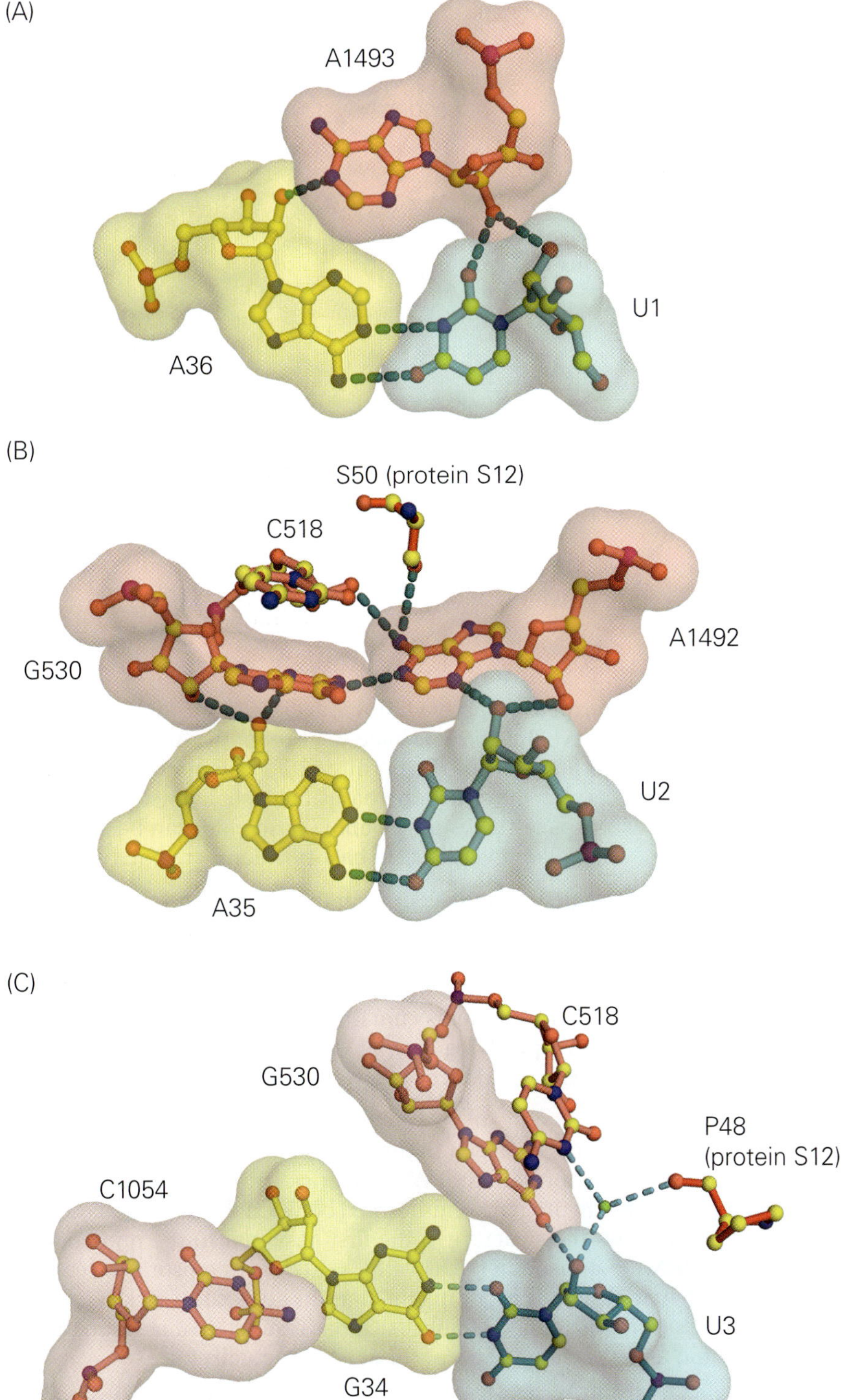


**Figure 19.42 Recognition of the codon–anticodon helix by the ribosome**. (A)–(C) show the first through third base pairs, respectively, of a UUU codon (*blue*) bound to cognate GAA anticodon (*yellow*) from the Phe tRNA stem-loop. Ribosomal elements (*pink*) closely interact with the minor groove at the first two base pairs, (A) and (B), but less so at the third (wobble) base pair (C). (PDB code: 1IBM.)

contribute to recognition of the codon by the anticodon because a precise geometry associated with correct base pairs is required for their formation. The interactions that form are compatible with all four possible Watson-Crick pairs. These interactions stimulate the closure of the ribosome around the codon–anticodon helix, much as correct base-pairing allows closure of the active site of DNA polymerase.

For codon–anticodon recognition on the ribosome, contacts are made at all three base pairs. At the first two positions of the paired region, the contacts are very

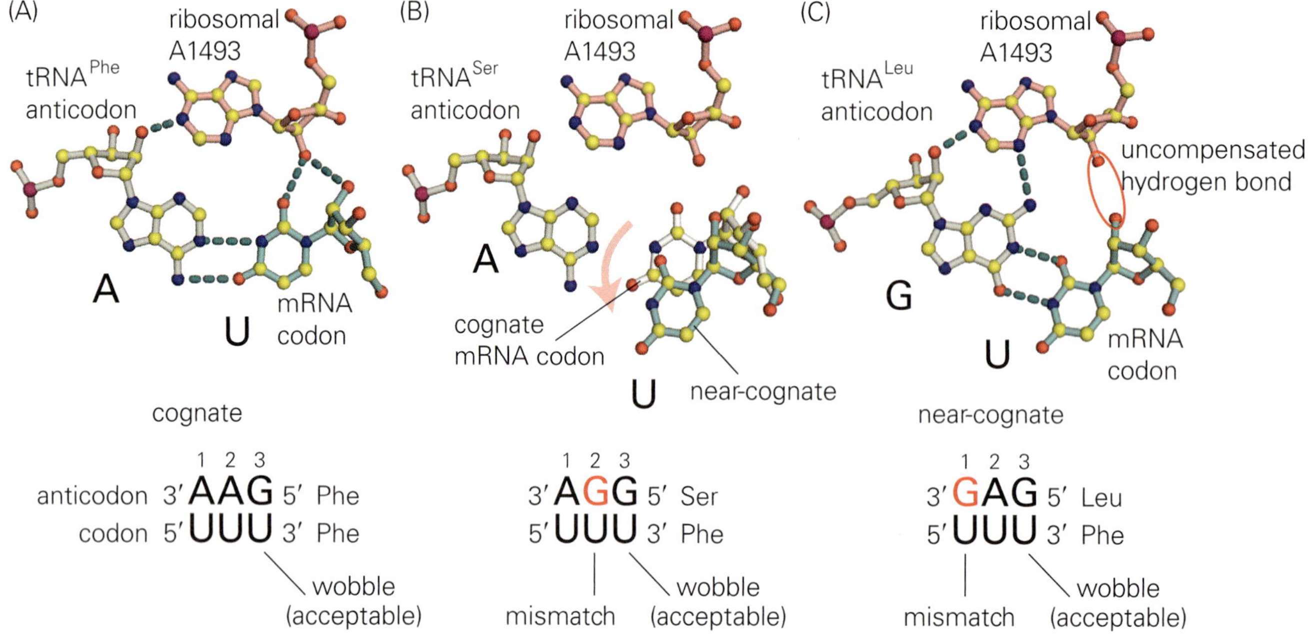


**Figure 19.43 Structural discrimination at the first position of a codon–anticodon pair.** In each of the panels the first position of the codon–anticodon interaction is shown (this corresponds to the nucleotide at the 3′ end of the anticodon). (A) Structure of a cognate interaction between a Phe codon and a $tRNA^{Phe}$ anticodon. (B) Structure of a near-cognate interaction between the Phe codon and a $tRNA^{Ser}$ anticodon. Even though there is no mismatch at the first position, the mismatch at the second position causes a structural distortion that propagates to the first position (shown here). For comparison, the U in the mRNA of the cognate interaction is shown with *white* bonds. (C) Structure of a near-cognate interaction between the Phe codon and an anticodon from $tRNA^{Leu}$. The *red* oval highlights an uncompensated loss of a hydrogen bond, caused by separation of A1493 and the codon. (Adapted from J.M. Ogle et al. and V. Ramakrishnan, *Cell* 111: 721–732, 2002; PDB codes: A, 1IBM; B, 1N32; C, 1N36.)

close and can only be formed properly with a Watson-Crick codon–anticodon pair, because only these have the correct shape to fit properly. At the third position, however, wobble pairing (G with U) is permitted. To accommodate the wobble base pair, the contacts with the anticodon base are looser at the third site (see Figure 19.42C). The extra space shows how wobble base pairs can be accepted, while other mispairs, with wider divergence in base positions, are not.

The effects of mismatches between the codon (in the mRNA) and the anticodon (in the tRNA) have been analyzed by determining structures of ribosomes bound to both cognate and to near-cognate (that is, not perfectly matched) combinations of codons and anticodons. In this study, a UUU codon (specifying phenylalanine) was paired with tRNA segments containing anticodons with sequences GAA (specifying Phe), GGA (specifying Ser), or GAG (specifying Leu). Note that the cognate pairing (UUU with GAA) involves a wobble base pair at the third position.

The structure around the first base pair in the cognate pairing is shown in **Figure 19.43**A. In addition to the formation of Watson-Crick hydrogen bonds between the bases, there are other hydrogen bonds formed between the bases and the sugars. Figure 19.43B shows the structure of the first base pair in the non-cognate paring with the $tRNA^{Ser}$ anticodon (UUU with GGA). It is interesting to see that, even though the first base pair is complementary in a Watson-Crick sense, the mismatch at the second position (U opposite G) has resulted in distortions at the first position, and the expected A-U base pair does not actually form.

In the case of the GAG anticodon ($tRNA^{Leu}$), the mismatch with UUU occurs at the first position, with a G-U pair rather than the correct A-U. With the base mispaired, the position of the pyrimidine base is altered, and a hydrogen bond to the ribosomal RNA is lost (Figure 19.43C). The gap that appears is insufficiently large to insert a water molecule, which leaves the hydroxyl groups poorly solvated, corresponding to a rather high-energy state.

Although the focus of this discussion has been on the codon–anticodon interaction region, it is important to note that correct pairing of the codon and anticodon also induces conformational changes in more distant parts of the ribosome. In particular, a region of the 30S subunit rotates toward the 50S subunit, which closes the inter-subunit interface and brings this region of the 30S subunit into

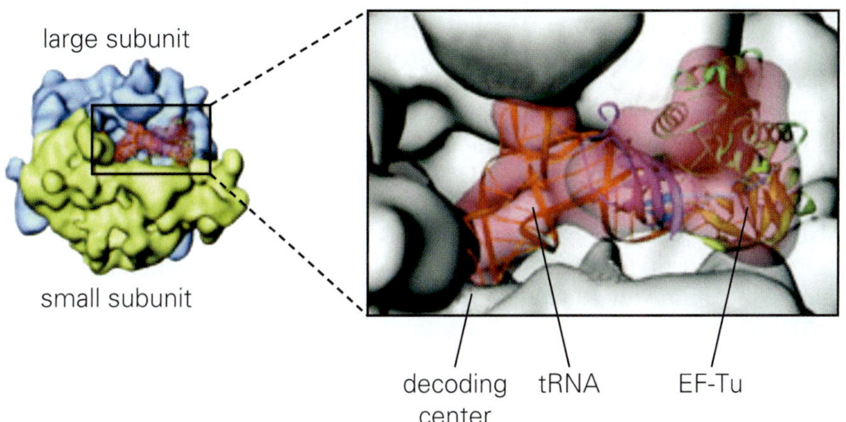


**Figure 19.44 Coupling between the decoding center and EF-Tu.** Electron microscopic reconstruction of the complex of the ribosome with a tRNA bound to EF-Tu. The decoding center, where the codon–anticodon interaction occurs, is located at the opposite end of the tRNA from where EF-Tu is bound. The distance between the decoding center and the catalytic site of EF-Tu, where GTP is hydrolyzed, is ~70 Å. (Adapted from M. Valle et al. and J. Frank, *EMBO J.* 21: 3557–3567, 2002. With permission from the European Molecular Biology Organization.)

contact with EF-Tu, which is important in moving on to the next step in the kinetic scheme. It is known that either mutations in, or the binding of antibiotics to, the ribosome can lead to this domain closure movement, even with an incorrect tRNA bound and hence lead to decreased fidelity in protein synthesis.

## 19.29 Coupling of the decoding center and the GTPase active site of EF-Tu involves multiple conformational rearrangements

We now return to the bending of the tRNA at the A site, which we first discussed in Section 19.25. The structure of the aminoacyl-tRNA is altered considerably upon binding to the ribosome in order to make contacts both at the decoding center (where the anticodon of the tRNA binds to the codon of the mRNA) and at the aminoacyl end (where EF-Tu is bound, **Figure 19.44**). An overlay of the structures of the complex of a tRNA with EF-Tu alone and when bound to the ribosome is shown in **Figure 19.45**. As you can see, binding to the ribosome induces a kink in the stem of the tRNA. A more detailed analysis shows that the distortion includes an untwisting of the anticodon stem, leading to a widening of one of the helices.

The largest changes in the structure of the tRNA are near the anticodon, rather than near EF-Tu, which must be activated. However, these changes in the A-site tRNA structure, induced by closure of the ribosome when codon–anticodon recognition occurs, also cause changes in the conformation of the ribosomal RNA. A specific set of residues in a loop (which is called the sarcin–ricin loop after protein toxins that cleave a base in this loop to inactivate ribosomes) near EF-Tu are altered in position, which alters, in turn, a contact to one critical residue in EF-Tu, His 84. The importance of this histidine sidechain will be made apparent in the discussion that follows.

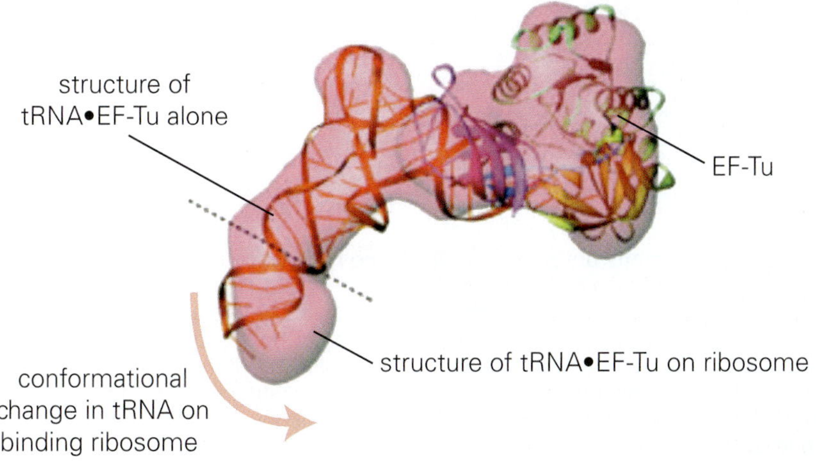


**Figure 19.45 Structures of the EF-Tu$^{GTP}$-tRNA complex on and off the ribosome.** The structure of the EF-Tu complex bound to the ribosome is indicated by the *pink* surface rendering and is based on an electron microscopic reconstruction (see Figure 19.44). The crystal structure of the isolated EF-Tu complex is shown in a ribbon representation. The dashed line indicates a kink in the tRNA bound to the ribosome. (Adapted from M. Valle et al. and J. Frank, *EMBO J.* 21: 3557–3567, 2002. With permission from the European Molecular Biology Organization.)

## 19.30 The active site of EF-Tu needs only a small rearrangement to be activated

Recall from the kinetic scheme shown in Figure 19.40 that once the codon–anticodon interaction is formed at the decoding center, the GTPase activity of EF-Tu is switched on. There are many GTPase proteins in cells for which GTP hydrolysis acts as a conformational switch to alter a particular activity of the protein. In the case of EF-Tu, the GTP-bound state binds to aminoacyl-tRNAs strongly, but the GDP-bound form does not. Thus, GTP hydrolysis acts as a switch to make EF-Tu release the aminoacyl-tRNA, so that it can reorient to position the amino acid it carries for addition to the protein chain.

In the structure of EF-Tu bound to tRNA (but not on the ribosome), most residues that are in conserved sequence elements and also occur in many other GTPases are positioned correctly for EF-Tu to be ready for the hydrolysis of its substrate. There is, however, one key element that is positioned incorrectly—the sidechain of His 84 points away from the terminal phosphate of the GTP. This sidechain acts as a base to activate a water molecule for attack on the GTP $\gamma$ phosphate to release it, triggering the EF-Tu conformational change. When His 84 points away from GTP the hydrolytic activity of EF-Tu is suppressed.

When the aminoacyl-tRNA•EF-Tu•GTP complex is bound to the ribosome, as shown in **Figure 19.46**, with the anticodon fully engaged with the codon in the mRNA, then a propagated structural change in the sarcin–ricin loop repositions

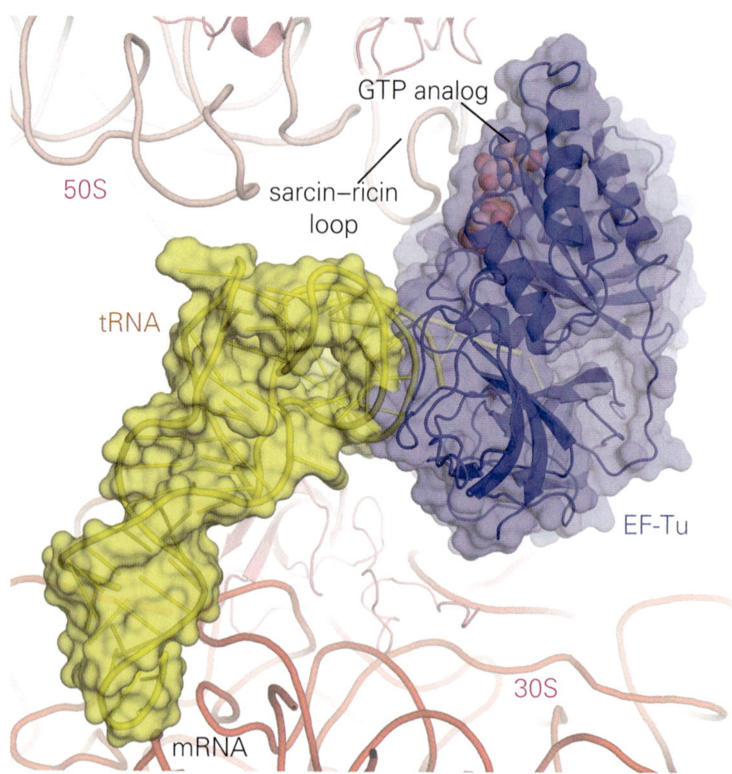


**Figure 19.46 Interactions between the tRNA, the ribosome, and GTP-loaded EF-Tu.** The structure shown here is that of a crystal structure of EF-Tu loaded with a non-hydrolyzable GTP analog and bound to the ribosome. Note the interactions between EF-Tu and a loop on the large subunit known as the sarcin–ricin loop. The tRNA is bent, a deformation that is necessary for simultaneous engagement of the mRNA codon and the sarcin–ricin loop in the ribosome. (Adapted from R.M. Voorhees et al. and V. Ramakrishnan, *Science* 330: 835–838, 2010. With permission from the AAAS; PDB code: 2XQD.)

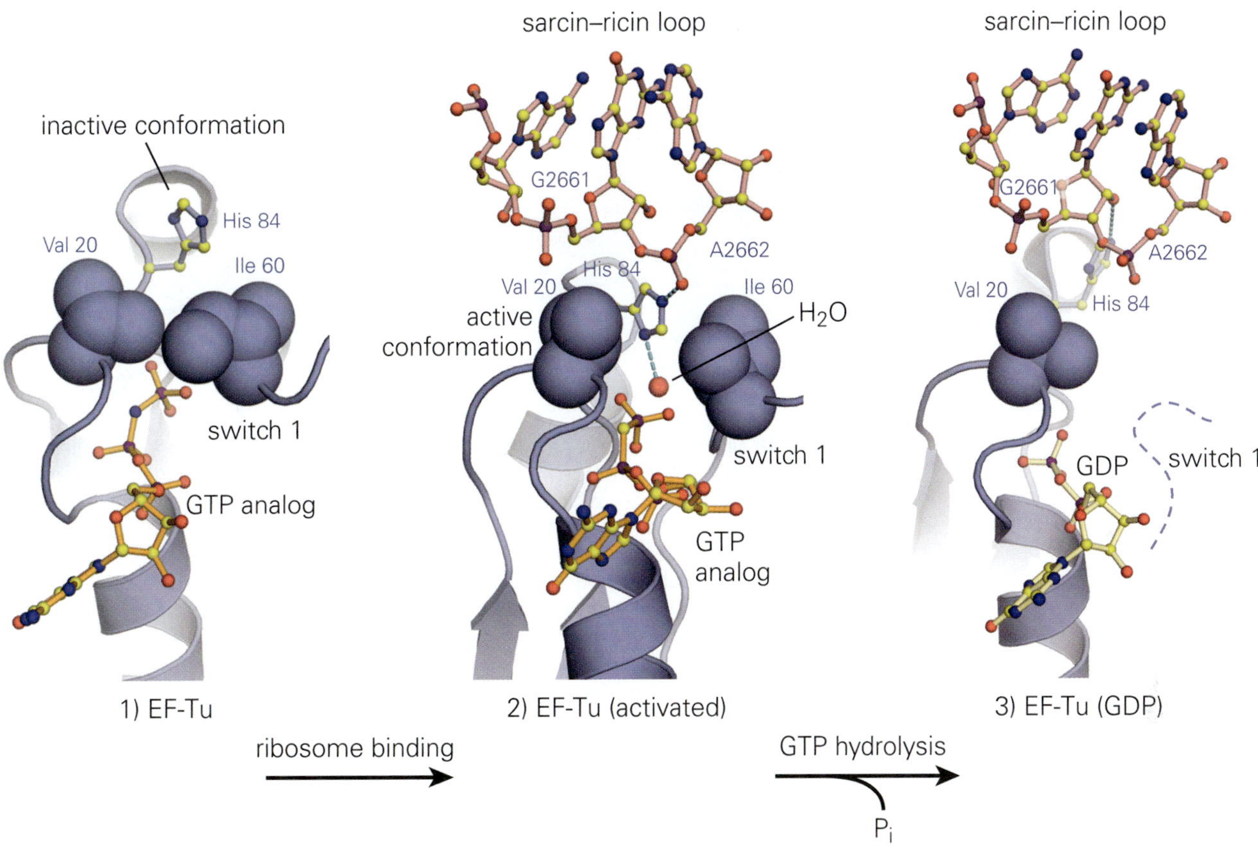


**Figure 19.47 Activation of EF-Tu by ribosome binding**. (1) Prior to ribosome binding, His 84 at the active site of EF-Tu points away from the terminal phosphate group of GTP, which cannot therefore be hydrolyzed (the structure shown here has a non-hydrolyzable GTP analog bound in order to trap this form). (2) When the tRNA-EF-Tu complex is properly docked on the ribosome, the sarcin–ricin loop of the large subunit causes His 84 to swing into position for catalysis. (3) After hydrolysis, GDP is left at the active site of EF-Tu. The structural consequence of this is that an element known as switch 1 becomes disordered, and this reduces the affinity of EF-Tu for the tRNA. (Adapted from R.M. Voorhees et al. and V. Ramakrishnan, *Science* 330: 835–838, 2010. With permission from the AAAS; PDB codes: 1OB2, 2XQD, 3FIC, and 3FIN.)

a phosphate from the ribosomal RNA to interact with His84. This causes the sidechain of His 84 to move so that it is immediately next to a water molecule just beyond the GTP γ phosphate, as shown in **Figure 19.47**. Hydrolysis of GTP occurs rapidly when the histidine sidechain is in this position, and EF-Tu changes its conformation. The change in EF-Tu lowers its affinity for the tRNA, which is released from EF-Tu. In terms of fidelity, the conformational change leading to activation of EF-Tu occurs considerably more rapidly with a correct codon–anticodon pair than when there is even one base of the three mispaired. This kinetic discrimination facilitates the conformational change required for accommodation, the next step in the process.

## 19.31 Release of EF-Tu allows the aminoacyl group of the A-site tRNA to move to the peptidyl transfer center

The hydrolysis of GTP by EF-Tu induces a conformational change that causes release of the aminoacyl-tRNA to which it was bound. After this release step, there are two fates possible for the tRNA. In order to deliver the attached amino acid to the peptidyl transfer center, there is a large conformational change requiring rotation of the tRNA relative to the ribosome, the process referred to earlier as accommodation (see Figures 19.38 and 19.40). This process moves the amino acid attached to the A-site tRNA into the peptidyl transfer center, which also holds the growing polypeptide attached to the P-site tRNA. This step is illustrated in **Figure 19.48.**

The reaction adding the amino acid to the polypeptide is very rapid once accommodation has occurred. Once dissociated from the tRNA, EF-Tu$^{GDP}$ also dissociates from the ribosome in a separate step that is somewhat slower than accommodation. The alternative fate for the released A-site tRNA is dissociation from the ribosome, leaving an empty A site.

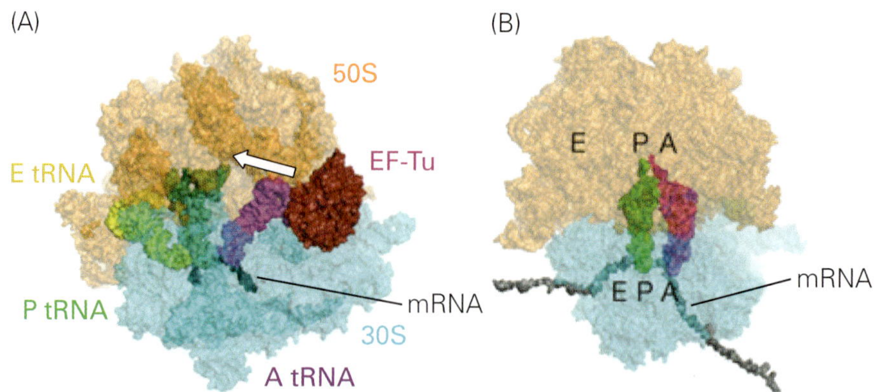


**Figure 19.48 Motion of the A-site tRNA in accommodation.** In the process called accommodation, the tRNA at the A site rotates to position the amino acid it carries in the peptidyl transfer center, which also holds the P-site tRNA with the attached growing polypeptide. (A) Before the movement of the A-site tRNA. (B) After accommodation. The two views of the ribosome are rotated with respect to each other about the vertical axis. (Adapted from T.M. Schmeing and V. Ramakrishnan, *Nature* 461: 1234–1242, 2009. With permission from Macmillan Publishers Ltd.)

The relative probability of accommodation relative to dissociation is determined by the codon–anticodon pairing at the decoding center. If the codon and anticodon are correctly paired (cognate pairing), then accommodation is at least 20 times faster than dissociation (over 95% of the amino acids will be added to the polypeptide being synthesized). However, if there is even one base that is incorrect (a "near-cognate" case; see Figure 19.43), then accommodation becomes very slow and the dissociation rate is high (~50-fold greater than the accommodation rate), suppressing addition of incorrect amino acids.

This "proofreading" step, as it is known, together with the accelerated hydrolysis rate of GTP in EF-Tu for cognate codon–tRNA pairs, explains most of the fidelity in protein synthesis. Indeed, the selectivities of these two sequential steps multiply to give the observed selectivity in amino acid choice. Some recent data indicate that, if an incorrect amino acid is incorporated, then a conformational change in the ribosome increases the likelihood that further errors will be made and that the peptide product will be degraded after release. However, this process and the magnitude of its effect on protein quality control are not yet well understood.

## 19.32 The ribosome catalyzes peptidyl transfer

After accommodation occurs, the A-site tRNA has its aminoacylated terminal base positioned in the peptidyl transfer center together with the P-site tRNA that has attached the peptide chain being synthesized. The synthesis of the new peptide bond occurs through attack of the amino group of the amino acid attached to the A-site tRNA on the ester bond connecting the peptide being synthesized to the P-site tRNA, as shown in **Figure 19.49**.

Comparison of the rate of peptidyl transfer on the ribosome with solution reactions with analogous functional groups indicates that the rate on the ribosome is accelerated by about $3.5 \times 10^6$-fold. This enhancement is independent of pH, so it is unlikely that the ribosome contributes any groups for base catalysis. Measurements of the temperature dependence of the reaction rates show that the ribosome-catalyzed reaction actually has a slightly higher activation energy than the uncatalyzed reaction. This is evident in the graph shown in **Figure 19.50**, where you can see that the slope of log $k$ versus $1/T$ is steeper for the ribosome-catalyzed reaction, corresponding to a higher activation energy ($E_A$ or $\Delta H^{\ddagger}$, refer back to Figure 15.36). This indicates that the ribosome alters the large and unfavorable entropy of activation for the uncatalyzed reaction ($-T\Delta S^{\ddagger} = +55$ kJ•mol$^{-1}$ for the

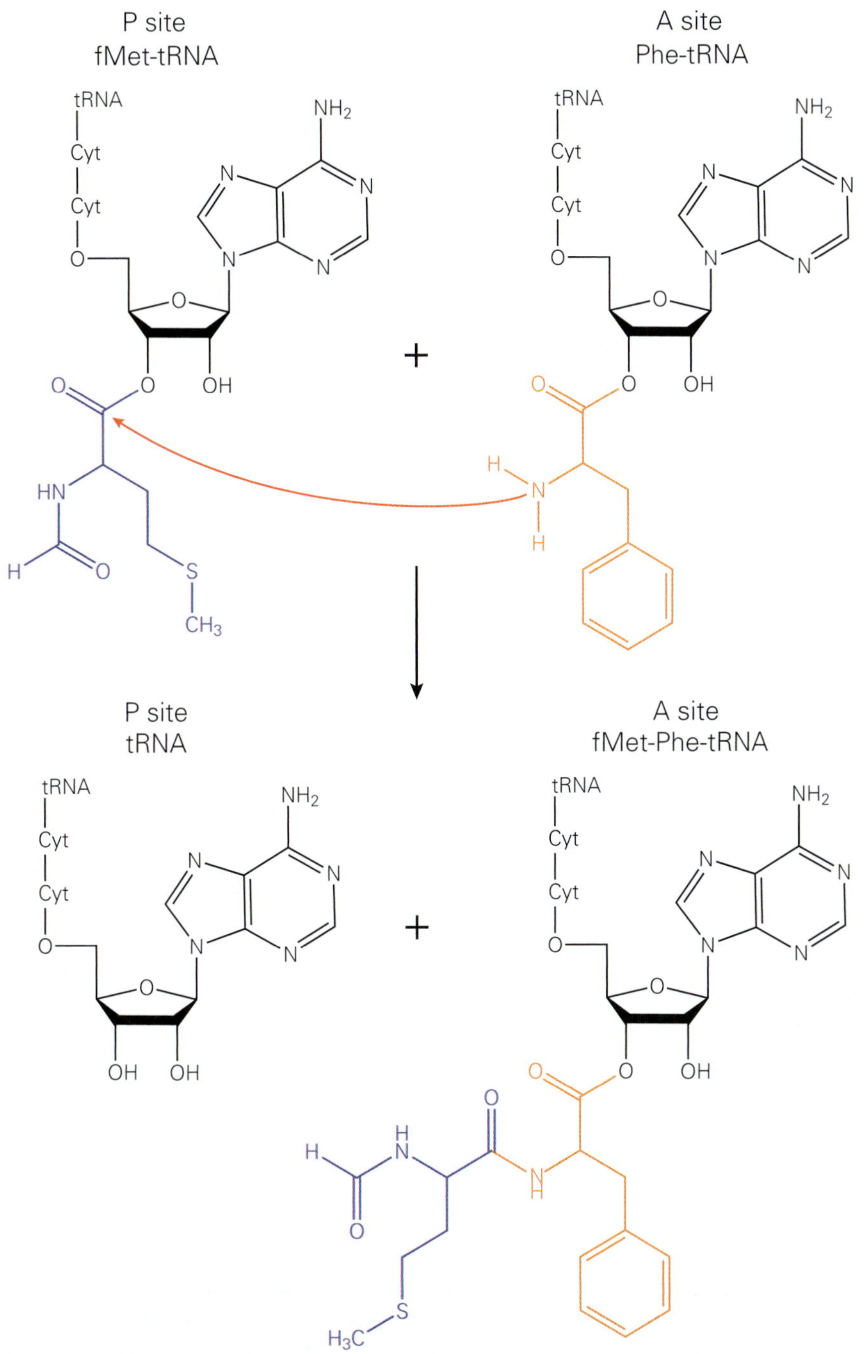

**Figure 19.49 Ribosomal protein synthesis**. Amide bond formation is catalyzed by the peptidyl transferase center of the ribosome. Attack of the α-amino group of the amino-acid-charged A-site tRNA (*orange*) at the ester carbonyl carbon of the peptidyl-tRNA in the P site (*blue*) yields the amide linkage.

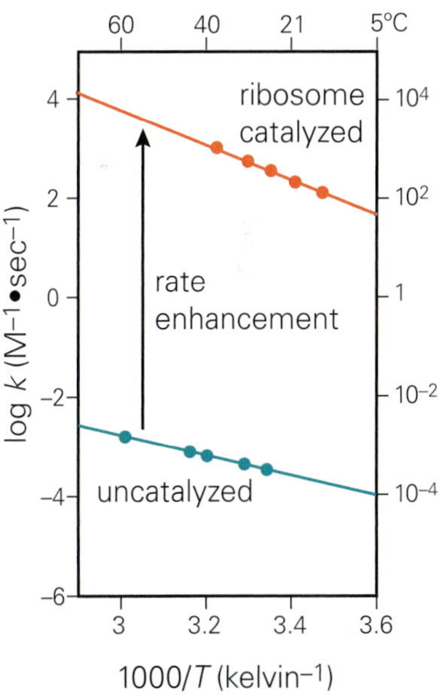


**Figure 19.50 The ribozyme-catalyzed reaction has a higher activation energy**. Temperature dependence of the rate constants for uncatalyzed (*blue* circles) and ribosome-catalyzed (*red* circles) peptide bond formation. Review Figure 15.36 to see how to interpret this graph. (From A. Sievers et al. and R. Wolfenden, *Proc. Natl. Acad. Sci. USA* 101: 7897–7901, 2004. With permission from the National Academy of Sciences.)

uncatalyzed reaction) to be a small, favorable contribution ($-T\Delta S^{\ddagger} = -8$ kJ•mol$^{-1}$ for the reaction on the ribosome). Interactions with the ribosome, and with the other bound substrates, position the incoming amino acid for transfer to the growing peptide chain (**Figure 19.51**). Although this step does not directly affect fidelity, it is important that the ribosome prevents water from attacking the ester bond of the P-site tRNA, because this would release the incomplete protein prematurely, which would be an error in protein synthesis.

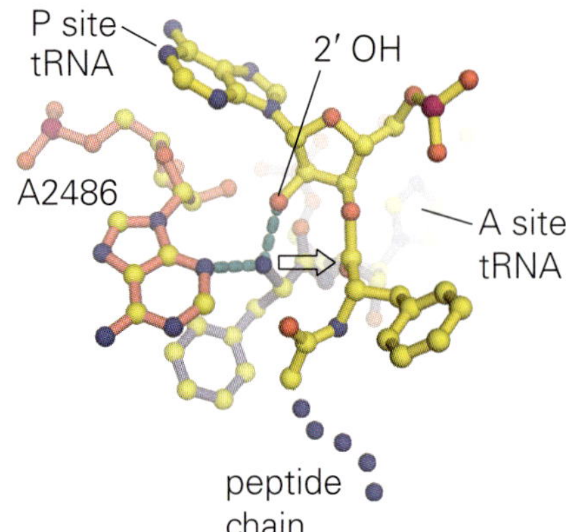


**Figure 19.51 Synthesis of a peptide bond.** This diagram shows the pre-attack configuration of the A-site and P-site tRNAs. For peptidyl transfer, the amino group from the amino acid on the A-site tRNA (*blue* bonds) is well positioned for attack on the ester linkage of the peptide chain to displace the P-site tRNA and elongate the peptide by one residue. (PDB code: 1VQN.)

## Summary

The replication of DNA in living cells proceeds with remarkable accuracy, with the wrong (that is, mismatched) nucleotides being introduced into the newly replicated chain only once per $10^7$–$10^8$ nucleotide additions. This means that the effective free-energy difference between the correct and the incorrect base pair would have to be ~40 kJ•mol$^{-1}$ if we consider base selection to occur through a near-equilibrium process of base-pairing in which nucleotides bind to and dissociate from the growing primer–template junction. The stability of DNA duplexes can be studied experimentally using van't Hoff analysis, from which values for the changes in the free energy, the enthalpy, and the entropy can be derived. This analysis shows that correct matches and incorrect mismatches are actually very close in free energy (the correct base pair is only ~1 kJ•mol$^{-1}$ more stable than the incorrect base pairs). Thus, the intrinsic selectivity of base-pairing cannot explain the fidelity of DNA replication.

The very small difference in free energy between matched and mismatched base pairs is one manifestation of a very general phenomenon exhibited by biological molecules. When the enthalpy of binding becomes more favorable (more negative, as when two nucleotide bases interact more strongly than do two others), binding is opposed by the entropy change of the process, which also becomes more negative. This comes about because of a higher degree of flexibility in most biological molecules, including DNA, when they are not bound to other molecules. This phenomenon of enthalpy–entropy compensation makes it very difficult to predict the strengths of interactions between molecules, but is essential to life because it allows the interactions between molecules to be highly dynamic. If DNA were constructed of rigid building blocks, it would be so strongly held together by the dominant enthalpy of stabilization that the two strands would never "breathe," and the processes of replication and transcription would become impossible.

A variety of DNA polymerases use a common mechanism to catalyze the addition of new nucleotides to the growing chain with high fidelity. A central subdomain of the polymerase, the palm subdomain, contains conserved acidic amino acid residues that bind two divalent metal ions, thereby providing a highly organized binding site for the reactants, in which the negative charges on the phosphates are neutralized and the nucleophilic attack on the incoming nucleoside triphosphate is facilitated. A large conformational change in the structure of the polymerase occurs upon nucleotide binding, resulting in a tightly packed binding site into which mismatched base pairs do not readily fit. By requiring that the enzyme undergo a conformational change that is sensitive to the shape of the newly formed base pair, DNA polymerases overcome the intrinsic difficulty of discriminating between correctly and incorrectly matched nucleotides that are very similar in free energy. In addition, the polymerase stalls after addition of a mismatched nucleotide, allowing the chain being synthesized to exit from the polymerase active site and move to the nearby exonuclease active site to be cleaved, a step referred to as proofreading. The combination of rapid addition only for correct incoming nucleotides and removal of most of the few mistakes that are added to the growing DNA chain together explain the high fidelity of DNA polymerases. This is boosted further by repair systems, which we have not discussed, that can remove most of the synthesis errors that occur.

When the ribosome synthesizes proteins from mRNA, it must also select the correct substrate, the aminoacyl-tRNA matched to the codon in the message. As for

DNA polymerase, the intrinsic free energy of pairing of the codon and anticodon is insufficient to explain the fidelity, in spite of the fact that it is much lower (about 1 in $10^3$ error rate) than for DNA polymerase. The quality of the match, assessed by requiring a snug fit of the short codon–anticodon helix into the ribosome, is transmitted to EF-Tu, a protein bound to the tRNA being evaluated, by the bending of the aminoacyl-tRNA. The bending of the tRNA, and the resulting interactions between EF-Tu and the ribosome, switches on the GTPase activity of EF-Tu. GTP hydrolysis enables a conformational change that releases the aminoacyl end of the tRNA. In a second test for the quality of the codon–anticodon match, correct pairing results in movement of the amino acid into the peptidyl transfer center where it can be added to the growing protein chain. If the codon–anticodon pair is incorrect (in spite of GTP hydrolysis having occurred), then usually the tRNA will dissociate rather than rotate into position for the peptide transfer step. The selectivities at these two successive steps are multiplied to give the overall observed selectivity.

## Key Concepts

### A. MEASURING STABILITY OF DNA DUPLEXES

- The melting of DNA duplexes provides information on base-pair stabilities for matched and mismatched pairs.
- UV absorption spectroscopy can be used to follow DNA duplex melting.
- Melting curves can be used to determine the enthalpy and entropy differences between single strands and duplexes.
- Mismatched base pairs at the terminus of a duplex are only marginally less stable than correctly paired bases.
- The formation of duplex DNA from single strands results in a large loss in entropy.
- The stability of a DNA duplex depends on its sequence.
- Stacking and hydrogen bonding contribute to the stability of the duplex, but stacking is more important.

### B. FIDELITY IN DNA REPLICATION

- DNA polymerase copies DNA sequences very accurately, with only about 1 in $10^7$ bases miscopied.
- The difference in free energy between correct and incorrect bases interacting with the template is far too small to explain the high fidelity of DNA replication.
- The polymerase domain resembles a right hand made up of fingers, palm, and thumb subdomains.
- The polymerase determines the specificity of nucleotide insertion by geometric selection, where the overall shape of the base pair formed is more important than the ability of the new base pair to form hydrogen bonds.
- DNA polymerase undergoes a conformational change to form a closed complex upon binding of the correct nucleotide.
- Two metal ions facilitate the addition of a nucleotide to the 3′ end of the primer.
- Mismatches are removed by the 3′ to 5′ exonuclease activity.
- The overall fidelity of the DNA polymerase includes a contribution from the editing activity of the enzyme.

### C. HOW RIBOSOMES ACHIEVE FIDELITY

- The accuracy of translation depends on the ability of the ribosome to discriminate between cognate tRNAs from near- or non-cognate tRNAs.
- Specific bases in the decoding center of the ribosome monitor the base-pairing geometry or shape of the first two codon–anticodon positions.
- The geometry of base-pairing in the third (wobble) codon position is less stringently monitored.
- When a cognate codon–anticodon interaction is sensed by the ribosome in this way, a conformational change in the 30S subunit takes place from an open to a closed conformation.
- tRNA selection depends on shape recognition and induced fit, much like the selection of the correct nucleotide by DNA polymerase.
- Codon recognition triggers a bending of the aminoacyl-tRNA that is bound to EF-Tu•GTP.
- For GTPase activation to occur in EF-Tu, the tRNA must be bent, and a loop in the ribosome then positions a catalytic histidine sidechain into the catalytic center of EF-Tu, resulting in rapid GTP hydrolysis.
- GDP hydrolysis leads to dissociation of EF-Tu from the ribosome, accommodation of the aminoacyl-tRNA within the active site, and peptidyl transfer.
- The ribosome catalyzes peptidyl transfer by lowering the unfavorable entropy of the transition state.

## Problems

### True/False and Multiple Choice

1. DNA polymerases from both prokaryotes and eukaryotes resemble a right hand made up of fingers, palm, and thumb subdomains.

   True/False

2. Which of the following is *not* a factor that favors DNA synthesis?

   a. The release of pyrophosphate.
   b. The base-stacking interactions.
   c. The change of entropy when the free dNTP is added to the polynucleotide chain.
   d. The formation of hydrogen bonds.
   e. The hydrolysis of pyrophosphate that is released.

3. The 5′ base in the codon is less important for codon–anticodon recognition and is therefore known as the "wobble base."

   True/False

4. The error rate of DNA replication in an *E. coli* strain that lacked the mismatch repair system would increase by what amount?

   a. $\sim 10^9$
   b. $\sim 10^7$–$10^8$
   c. $\sim 10^6$
   d. $\sim 10^2$
   e. $\sim 10^4$–$10^5$

5. Mismatched base pairs are symmetric about the glycosyl bonds.

   True/False

6. Which of the following statements is *not* an effect of codon–anticodon recognition in the ribosome?

   a. The 30S subunit changes from an open to closed conformation.
   b. The aminoacyl-tRNA structure is distorted.
   c. EF-Tu hydrolyzes GTP.
   d. The large subunit dissociates from the small subunit.
   e. EF-Tu dissociates from the ribosome.

7. The large ribsomal subunit is comprised of proteins and RNA, whereas the small subunit is comprised only of RNA.

   True/False

### Fill in the Blank

8. Mismatches can be removed by a 3′ to 5′ ________ enzymatic activity.

9. Conserved ______ residues bind divalent metal ions in the palm domain.

10. DNA polymerase can add a nucleotide to the free ____ hydroxyl group of DNA or RNA.

11. The "________" of the "finger-palm-thumb" subdomain structure for DNA polymerase I inserts into the minor groove of DNA.

12. The movement of mRNA, as the tRNA shifts between sites on the ribosome, is called ______________.

### Quantitative/Essay

(Assume $T = 300$ K and $RT = 2.5$ kJ•mol$^{-1}$ for all questions unless otherwise stated.)

13. A scientist compares a bacterial polymerase (without a proofreading domain) with a yeast polymerase (also without a proofreading domain). The bacterial enzyme has lower fidelity. Surprisingly, the two enzymes have approximately the same value of $K_M$ for correctly matched base pairs. Explain which enzyme, the bacterial or yeast, would likely have a higher $K_M$ for mismatched base pairs?

14. How would inactivating a mutation of pyrophosphatase affect the rate of DNA synthesis in the cell?

15. What significant free-energy barriers affect the rate of DNA synthesis?

16. In the "two-metal" reaction mechanism for nucleic acid phosphoryl transfer, what roles are played by a) only metal ion A, b) only metal ion B, and c) both metal ions?

17. Why does EDTA quench DNA polymerase reactions?

18. a. Why are fluorinated bases readily incorporated during DNA synthesis?

    b. Draw a fluorinated thymidine analog-adenine base pair and label the C1′–C1′ distance and angles between the C1′–C1′ vector and the glycosyl bonds with predicted values.

19. How does the interaction between the O helix and the incoming nucleotide change between the "open" and "closed" forms of DNA polymerase?

20. A scientist probes the thermodynamics of forming an RNA hairpin (H) from the unfolded state (R). She measures $\Delta S^{\circ}_{H\rightarrow R} = 908$ J•K$^{-1}$•mol$^{-1}$ and $\Delta H^{\circ}_{H\rightarrow R} = 295$ kJ•mol$^{-1}$. What are the values of a) the equilibrium constant $K_{H\rightarrow R}$ and b) $\Delta G^{\circ}_{H\rightarrow R}$?

21. The $T_m$ of a DNA duplex is measured to be 55°C at 10 μm. When the concentration is 1000 times higher, the $T_m$ changes to 60°C. What are the $\Delta S$ and $\Delta H$ values for formation of this duplex?

22. Measurements on several variants of a DNA double helix are performed. Although the free energy of double helix formation does not vary significantly, the enthalpy

and entropy measurements show a large range. A scientist observes that mutations that are favorable with respect to binding enthalpy are entropically disfavored. What effect has the scientist observed and why is this phenomenon common in biological molecules?

**23.** A DNA polymerase is isolated and found to have an error rate of 1 in $10^6$.

a. Suppose that the error rate is determined solely by the relative stabilities of incorrect and correct base pairs. What would the difference in free energy between correct and incorrect nucleotides incorporated by the polymerase have to be in order to explain the error rate?

b. Solution studies of isolated oligonucleotides indicate that the energetic difference is actually $-1.2$ kJ•mol$^{-1}$. What is the equilibrium constant of correct–incorrect base pair discrimination based on these solution studies?

c. What other enzymatic activity, in addition to nucleotide insertion, contributes to the increased fidelity of DNA polymerase?

**24.** Both thermodynamics and kinetics play an important role in achieving fidelity. How is kinetic control exercised by a) DNA polymerase and b) the ribosome?

## Further Reading

### General

Bloomfield VA, Crothers DM & Tinoco I Jr (2000) Nucleic Acids. Structures, Properties, and Functions. Sausalito, CA: University Science Books.

Kornberg A & Baker TA (1992) DNA Replication, 2nd ed. New York: W.H. Freeman.

Saenger W (1984) Principles of Nucleic Acid Structure. New York: Springer-Verlag.

### References

#### A. Measuring the stability of DNA duplexes

Petruska J, Goodman MF, Boosalis MS, Sowers LC, Cheong C & Tinoco I Jr (1988) Comparison between DNA melting thermodynamics and DNA polymerase fidelity. *Proc. Natl. Acad. Sci. USA* 85, 6252–6256.

Aboul-ela F, Koh D, Tinoco I Jr & Martin FH (1985) Base-base mismatches. Thermodynamics of double helix formation for $dCA_3XA_3G$ + $dCT_3YT_3G$ (X, Y = A,C,G,T). *Nucleic Acids Res.* 13, 4811–4824.

Lai JS, Qu J & Kool ET (2003) Fluorinated DNA bases as probes of electrostatic effects in DNA base stacking. *Angew. Chem. Int. Ed. Engl.* 42, 5973–5977.

#### B. Fidelity in DNA replication

Doublie S, Tabor S, Long AM, Richardson CC & Ellenberger T (1998) Crystal structure of a bacteriophage T7 DNA replication complex at 2.2 Å resolution. *Nature* 391, 251–258.

Johnson KA (1993) Conformational coupling in DNA polymerase fidelity. *Annu. Rev. Biochem.* 62, 685–713.

Joyce CM & Steitz TA (1994) Function and structure relationships in DNA polymerases. *Annu. Rev. Biochem.* 63, 772–822.

Kunkel TA (2004) DNA replication fidelity. *J. Biol. Chem.* 279, 16895–16898.

Rothwell PJ & Waksman G (2005) Structure and mechanism of DNA polymerases. *Adv. Protein Chem.* 71, 401–440.

#### C. How ribosomes achieve fidelity

Korostelev A & Noller HF (2007) The ribosome in focus: new structures bring new insights. *Trends Biochem. Sci.* 32, 434–441.

Ramakrishnan V (2010) Unraveling the structure of the ribosome (Nobel Lecture). *Angew. Chem. Int. Ed. Engl.* 49, 4355–4380.

Rodnina MV, Beringer M & Wintermeyer W (2006) Mechanism of peptide bond formation on the ribosome. *Quart. Rev. Biophys.* 39, 203–225.

Schmeing TM & Ramakrishnan V (2009) What recent ribosome structures have revealed about the mechanism of translation. *Nature* 461, 1234–1242.

Steitz, TA (2010 From the structure and function of the ribosome to new antibiotics (Nobel Lecture). *Angew. Chem. Int. Ed. Engl.* 49, 4341–4354.

Yonath A (2010) Polar bears, antibiotics and the evolving ribosome (Nobel Lecture). *Angew. Chem. Int. Ed. Engl.* 49, 4381–4398.

Zaher HS & Green R (2009) Fidelity at the molecular level: Lessons from protein synthesis. *Cell* 136, 746–762.

# Glossary

**A-form helix** The standard double-helical conformation of RNA, it is also adopted by DNA. The sugar pucker is C3′ endo.

**A-minor motif** A tertiary interaction that involves the minor groove edge of an adenine residue.

**absolute scale temperature** In the absolute scale for temperature, the freezing point of water is 273.15 K (degrees kelvin).

**accommodation** A conformational change in the ribosome, which brings the amino acid on the A-site tRNA close to the polypeptide on the P-site tRNA (see Figure 19.38).

**acid dissociation constant** The equilibrium constant for the dissociation reaction of an acid, denoted $K_a$.

**acidic** If the proton concentration in a solution is greater than $10^{-7}$ M (that is, the pH is lower than 7.0), the solution is referred to as acidic.

**action potential** A transient change in the voltage difference across the plasma membrane of an axon is called an action potential, also called a *nerve impulse.* Action potentials move along the axon with essentially no attenuation. They are the currency of signal transmission through the axon.

**activation energy** The activation energy, $E_A$, is the minimum increase in potential energy for reactants to be converted to products, and hence the minimum energy that reactants must have to be able to react. The requirement that the reactants have sufficient energy to cross from reactants to products limits the rate of most reactions.

**active site** Within each enzyme there is an active site, which includes a region where the chemical reaction actually occurs (the catalytic site).

**active transport** Active transport is an energy-consuming process through which molecules are moved between different parts of cells. The term "active transport" is also used to mean movement of compounds across membranes against a concentration gradient.

**active transporter** Membrane proteins that use energy to move molecules against a concentration gradient.

**additive property of energy functions** By saying that a molecular property, such as the energy, is additive, we mean that we can calculate the total value of that property by simply adding up various components. For example, the total energy due to pair-wise interactions between atoms is the sum of the individual pair-wise energy terms. This is true only if the interaction between one pair of atoms does not influence the interaction between other pairs of atoms.

**adenine** A substituted purine found in DNA and RNA.

**adenosine platform** Adenosine platforms are formed by two consecutive A residues in a sequence. Rather than stacking in the normal fashion, they form a pseudo-base pair. The two bases are arranged side by side and form hydrogen bonds. The two bases form a platform onto which other RNA elements can stack.

**adiabatic** A process occurring without heat transfer is an adiabatic process.

**affinity** The affinity of a molecular interaction refers to its strength. The greater the decrease in free energy upon binding, the greater the affinity. The specificity of an interaction refers to the relative strength of the interactions made between one protein and alternative ligands. In a highly specific interaction, the free-energy change upon binding to a favored ligand is much greater than that for other ligands. Biologically relevant interactions are usually highly specific, as discussed in more detail in Chapter 13.

**agarose** A gelatinous polysaccharide isolated from seaweed. Agarose is a linear disaccharide repeat polymer of galactose, linked 1⟶4 to 3,6-anhydrogalactose.

**aldehyde** When the oxygen double bond in a sugar is at the end of the chain, the sugar is known as an aldehyde, whereas if the oxygen double bond is somewhere in the middle, the sugar is known as a ketone.

**aldose** A sugar that contains an aldehyde group.

**allosteric effector** Allosteric effectors are molecules that bind to a protein or RNA at a site other than the active site and affect the binding of the main target or substrate of the protein or RNA.

**allosteric enzyme** An allosteric enzyme is one in which the activity of a catalytic center is altered by the binding of molecules at sites other than this catalytic center. Multimeric allosteric enzymes can exhibit cooperativity—that is, the binding of substrate to one active site increases the activity of another active site in the assembly.

**allostery** An allosteric protein or RNA is one in which the activity of the protein or RNA is modulated by interactions that occur at a distance from the active site.

**α/β structure** A common type of protein fold, containing alternating α helices and β strands, which form a central parallel or mixed β sheet surrounded by α helices.

**α helix** Common folding pattern in proteins in which a linear sequence of amino acids folds into a right-handed helix stabilized by internal hydrogen bonding between backbone atoms.

**alternating access mechanism** A mechanism by which active transporter proteins move molecules across the membranes. The transporter cycles between two conformations. In each conformation, the interior binding site for the molecule to be transported is open towards one side of the membrane, but closed towards the other (see Figure 4.85).

**Alzheimer's beta peptide** Fibrillar protein deposits found in the brains of people suffering from Alzheimer's disease are composed primarily of a peptide 39 to 42 residues in length, known as Alzheimer's beta peptide (abbreviated Aβ). This is a fragment from a large, membrane-associated protein, the Alzheimer's precursor protein.

**Alzheimer's disease** A neurodegenerative disease, associated with the formation of fibrillar protein deposits (amyloid plaques) in the brain.

**Alzheimer's precursor protein** See Alzheimer's beta peptide.

**amide plane** The junction between two amino acid residues in a protein is formed by the N–H and C=O groups of the first and second residues, respectively. The four atoms in the peptide group are coplanar, and define the amide plane, which is illustrated in Figure 4.14.

**amino acid residue** Amino acids linked by peptide bonds are called amino acid residues. They are no longer intact amino acids because the amino and carboxyl groups that reacted to form the peptide bonds no longer exist.

**amino acid substitution matrix** A matrix in which each row and column corresponds to one of the 20 amino acids. Each entry in the matrix is related to the probability that one amino acid is replaced by the other in proteins that are related evolutionarily.

**amino-aromatic interaction** An energetically favorable interaction between a positively charged or basic side-chain and the π electrons of an aromatic ring.

**amphipathic** A molecule that has parts that are hydrophilic (water loving) and other parts that are hydrophobic (water hating). Also known as an amphiphilic molecule.

**amphipathic helix** Secondary structural elements that have distinctive faces, one hydrophobic and one hydrophilic, are referred to as amphipathic. The hydrophobic faces of amphipathic sheets and helices can pack against each other to form a hydrophobic core, leaving the hydrophilic faces to interact favorably with water. This is a central principle in the architecture of most globular proteins.

**amphiphiles** Amphiphiles are molecules that have parts that are hydrophilic (loving water) and other parts that are hydrophobic (hating water).

**amyloid** The staining of fibrillar protein aggregates is similar to starch, and hence they became known as amyloid deposits (a misnomer that is now in widespread use). The corresponding medical disorders are called amyloid diseases.

**amylose** A sugar polymer that is highly bent and forms spirals.

**Anfinsen cage** The chambers within the GroEL chaperone are referred to as Anfinsen cages, to emphasize the fact that they allow protein folding to occur according to the dictates of thermodynamics, without interference from other polypeptides.

**anode** Negatively charged electrode.

**anomer** Anomers are forms of sugars that differ only in the equatorial or axial position of the hydroxyl at the position of ring closure. The carbon atom at which the hydroxyl group is in alternative positions is known as the anomeric carbon.

**anomeric carbon** See anomer.

**anti** A nucleotide conformation in which the base is turned away from the sugar. In this conformation, the sugar H1 atom and the base C6 or C8 atoms (for pyrimidines and purines, respectively) are in a *trans* conformation about the glycosidic bond. This is referred to as the anti conformation of the base, and is much more common than the alternate syn conformation.

**anticodon** A transfer RNA for a specific amino acid contains a sequence of three nucleotides at the tip of the stem, known as the anticodon, that is complementary to the mRNA codon for that amino acid.

**apo form** Form of a protein in the absence of a cofactor that is normally bound to it. For example, myoglobin without a heme group is known as apomyoglobin.

**Arrhenius equation** Arrhenius rate behavior gives an exponential dependence of the rate on temperature, $k \propto \exp(-E_A/RT)$.

**aspartate protease** A protease with two aspartic acid sidechains that are critical for catalysis.

**association constant, $K_A$** The equilibrium constant for the reaction describing the binding of two molecules.

**axial** Axial and equatorial are terms that define the directions of substituents of rings such as sugars, relative to the plane of the ring. In a chair conformation, substituents, such as the hydroxyls, can either point out from the ring (termed equatorial) or point up or down (termed axial).

**B-form DNA** B-form DNA is the standard conformation of double-helical DNA. The sugar pucker is C2′ endo, which is inaccessible to RNA. The base is in the anti conformation with respect to the sugar.

**backbone** See peptide backbone.

**bacteriorhodopsin** Bacteriorhodopsin is a proton pump found in certain "light-harvesting" bacteria, and it couples light energy to the generation of a proton gradient across the cell membrane. Bacteriorhodopsin has seven transmembrane helices, and its structure is reminiscent of that of a large family of transmembrane proteins in eukaryotic cells known as G-protein-coupled receptors (GPCRs). GPCRs transduce signals across the membrane. Rhodopsin, for example, is a GPCR that converts the detection of light photons into a neuronal signal.

**base readout** See indirect readout.

**base triple** Base triples are formed when three bases come together in a near-planar arrangement, with hydrogen bonding. Two of the bases usually form a Watson-Crick base pair, and the third base comes in from the major groove side.

**basic** If the proton concentration in a solution is less than $10^{-7}$ M (that is, the pH is greater than 7.0), the solution is referred to as basic.

**β-α-β motif** A common structural motif in proteins. Two adjacent strands of a parallel β sheet are consecutive in the amino acid sequence, and are connected by an α helix that crosses from one edge of the β sheet to the other (see Figure 4.37).

**β hairpin loop or reverse turn** A loop connecting two consecutive strands of an antiparallel β sheet (see Figure 4.25). Such loops are usually quite short, and typically contain only four to six residues.

**bimolecular reaction** A reaction with two reactant molecules in the elementary reaction. See Figure 15.5A.

**binding assay** An experimental procedure to determine the affinity of the interaction between two molecules.

**binding change mechanism** The mechanism by which ATP is synthesized by the enzyme known as ATP synthase. In this mechanism, the conformations of the three catalytic subunits of the enzyme are changed in a cyclic fashion by rotation of a central shaft.

**binding free energy** Free-energy change for the binding of two molecules.

**binding free-energy change** See binding free energy.

**binding isotherm or binding curve** A series of measurements of the extent of binding or the fractional saturation, $f$, as a function of ligand concentration is known as a binding isotherm. Such a measurement can be analyzed to yield the dissociation constant only if all of the measurements are made at the same temperature and, hence, the series of measurements is called an *isotherm*.

**binomial coefficient** The values of the multiplicity $W(M, N)$ are known as binomial coefficients. A simple diagrammatic representation of binomial coefficients for different numbers of trials is given by Pascal's triangle, named after the French mathematician Blaise Pascal (see Figure 7.10).

**binomial distribution** The probability distribution $P(M, N)$ that determines the probability of obtaining $N$ positive outcomes in a trial with $M$ binary events is known as the binomial distribution. When the number of trials, $M$, is large, the binomial distribution is well approximated by a Gaussian distribution.

**binomial statistics** The statistics governing a series of events with binary outcomes (such as coin flips) is called binomial statistics.

**biochemical standard state** The natural standard state for protons in water is that of an aqueous solution at pH 7 ($[H^+] = 10^{-7}$ M). This convention, common in biochemistry, defines the biochemical standard state, and is different from the convention in other branches of chemistry, where the standard states are set to be 1 M for all solutions, including $H^+$ in water.

**biopolymer** Biological macromolecule made of repeating units, such as DNA, RNA, proteins and carbohydrates.

**boat conformation** A nonplanar sugar geometry that has both ends of the ring out of the plane of the central four carbons, but on the same side of the plane rather than opposite (see chair conformation).

**Bohr effect** The reduction in the affinity of hemoglobin for oxygen that occurs at lower pH is known as the Bohr effect, named after the physiologist Christian Bohr. This phenomenon is important because respiration acidifies venous blood, and the Bohr effect facilitates the release of oxygen from hemoglobin.

**Boltzmann constant** Denoted $k_B$, the Boltzmann constant is the gas constant, $R$, divided by Avogadro's number.

**Boltzmann distribution** The Boltzmann distribution is a mathematical expression that tells us the probability of finding molecules in different energy levels at a given temperature and at equilibrium. According to the Boltzmann distribution, the number of molecules, $N$, in a particular energy level decreases exponentially with the energy, $U$, of the energy level:

$$N \propto e^{-\frac{U}{k_B T}}$$

The exponential term is known as the Boltzmann factor, and $k_B$ is the Boltzmann constant.

**Boltzmann factor** See Boltzmann distribution.

**bomb calorimeter** An instrument used to determine the amount of heat released when a defined amount of a compound undergoes complete combustion.

**branching ratio** For a reaction with two products, B and C, the ratio of the concentrations of the two products is known as the branching ratio, [B]/[C].

**Brownian motion** Random diffusive motion of particles.

**bulk water** A term used when describing the interactions between macromolecules, such as proteins, and water. Bulk water refers to water molecules that are too far away from atoms in the macromolecule to be affected by it.

**buried surface area** Protein-protein interfaces are characterized by their buried surface area. This refers to the surface area on the interacting proteins that is accessible to water before the complex is formed, but is inaccessible in the complex.

**C-terminal amino acid** The last amino acid in a protein chain —the one with the free carboxyl group ($-COO^-$)—is known as the C-terminal amino acid.

**cable equation** A differential equation that was developed originally to describe macroscopic electrical cables, such as those used to transmit telephone signals. When applied to axons, this second-order differential equation relates the membrane potential, $E_m$, to the conductances of the sodium and potassium channels ($g_{Na}$ and $g_K$) and to their Nernst potentials ($E_{Na}$ and $E_K$). See Equation 11.74.

**calorimeter** An instrument that measures heat transfer during a process.

**capacitance, *C*** An insulator sandwiched between two conductors is known as a capacitor. A capacitor builds up charge on its surfaces when the conductors are connected to an electrical circuit. The ratio of the maximum charge to the voltage across the conductors is known as the capacitance of the conductor. The capacitance, $C$, is defined by Equation 11.67.

**capacitive current** A transient current that flows when an uncharged capacitor is connected to a battery. The current stops when the capacitor is fully charged. Cell membranes act as electrical capacitors, and transient currents associated with changes in the membrane charge are observed when the membrane potential changes.

**capacitor** An electrical capacitor consists of a pair of conducting plates separated by an insulating material. Cell membranes act like electrical capacitors because the interior of the lipid bilayer is an insulator.

**catalyst** A catalyst is a substance that accelerates the rate of a reaction, but does not appear in the overall balanced reaction. Such substances participate in the reaction mechanism, but are regenerated in their original form in the course of the reaction.

**catalytic efficiency** The catalytic efficiency or specificity constant for an enzyme is the ratio $k_{cat}/K_M$. An enzyme with a higher rate for the chemical step and/or tighter binding of substrate will produce more product per unit time (that is, higher efficiency). For comparing reactions of different substrates with the same enzyme, the relative catalytic efficiencies predict the relative amounts of products formed (that is, the specificity of the enzyme for a substrate).

**catalytic rate constant, $k_{cat}$** The apparent rate constant for an enzyme reaction at high substrate concentration (that is, when the reaction is proceeding with maximal rate).

**catalytic site** See active site.

**cathode** Positively charged electrode.

**cell wall** A cell wall is a polymeric outer layer just outside the cell membrane of some cells (especially in plants, bacteria, fungi, algae, and archaea). The cell wall generally has a high polysaccharide content.

**cellulose** Cellulose, found in plants, is a polymer made from 1→4 linked glucose subunits. The key difference between cellulose and glycogen is that the β anomer of glucose is used to form the linkages in cellulose, rather than the α linkages in glycogen (Figure 3.14). Cellulose is thus polymeric glucose-β(1→4)-glucose, without the branching that glycogen has.

**centipoise** Unit of viscosity, equal to 0.01 poise.

**central dogma of molecular biology** The flow of genetic information is described by the central dogma of molecular biology, which states that information flows from DNA into RNA and then into protein.

**centrifuge** A centrifuge is an instrument that spins a sample very rapidly, making the molecules in it move under the centrifugal force.

**cerebroside** A sphingolipid with a single sugar.

**chair conformation** A nonplanar sugar geometry that has both ends of the ring out of the plane of the central four carbons, but on opposite sides (see boat conformation).

**chaperone** See Molecular chaperone.

**chaperonins** Chaperones that form oligomeric structures with large internal chambers, such as the bacterial protein GroEL.

**chemical potential** The molecular chemical potential, $\mu$, is the rate of change of free energy with the number of molecules at constant temperature, pressure, and number of other molecules:

$$\mu_i = \left(\frac{\partial G}{\partial N_i}\right)_{T,P,N_{i \neq j}}$$

For dilute solutions, $\mu_i$ is the free energy of one $i$-type molecule, or the molar free energy, $G_i$.

**chemical work** Chemical work corresponds to the change in the free energy of the system that results from changes in the number of molecules.

**chemotaxis** A directed movement of bacteria or other cells towards sources of food and away from toxins.

**chiral center** Molecules that form stereoisomers have special atoms known as chiral centers, which are bonded to inequivalent groups of atoms.

**chitin** An extremely abundant polymer (probably second only to cellulose in the biosphere): poly-β(1→4)-*N*-acetylglucosamine.

**chloroplast** Light energy is captured in specialized compartments within plant cells, known as chloroplasts.

**cholesterol** A steroid, containing a conserved core of four fused carbocyclic rings, one with a polar OH group attached and, at the opposite end, one with a short alkyl chain.

**chronic myelogenous leukemia** A cancer involving the proliferation of white blood cells.

***cis*** Atoms or chemical groups that are linked by a covalent bond and face each other across the bond are said to be in a *cis* configuration.

**closed system** In a closed system, matter (molecules) stays inside the system, whereas energy is exchanged with the surroundings (see Figure 6.2).

**coaxial base stacking** When two regular helical elements (A-form for RNA) stack end-to-end, they are said to be coaxially stacked.

**codon** The sequence of messenger RNA is read one codon at a time by the ribosome during the synthesis of proteins. Each codon consists of three consecutive nucleotides that are specific for a given amino acid. There are 64 possible codons, three of which encode stop signals. The mapping of codons to amino acids is called the genetic code.

**coenzyme** A small molecule, such as NADPH and FADH, that provides chemical functionalities that are not available within the protein components of the enzymes that utilize it.

**coiled coil** Two α helices that coil around each other are referred to as a coiled coil. These can either be parallel (when the two protein chains run in the same direction) or antiparallel (when the chains run in opposite directions). Residues at the $i$, $i + 3$, and $i + 7$ positions in a coiled coil face each other at the interface between the helices. These residues are usually hydrophobic.

**cold denaturation** Cold denaturation refers to the unfolding of proteins induced at low temperature. This arises from the $\Delta C_P$ term in the protein stability equation, which correlates with the burial of hydrophobic residues in the folded state of the protein.

**collision rate constant** Rate constant for the collision between molecules due to diffusion.

**combustion reaction** A reaction involving the burning of a compound in oxygen.

**common structural core** When comparing the structures of two proteins, the region of the structure that is similar between the two proteins is called the common structural core (see Figure 5.19). The positions of $C_\alpha$ atoms within the common structural core are closely aligned in the two proteins (that is, the $C_\alpha$ positions deviate by less than ~3 Å).

**competitive inhibition** Competitive inhibition occurs when the inhibiting compound can bind reversibly to the active site of an enzyme and block the binding of substrate, but the compound itself does not undergo a reaction.

**competitive inhibitor** A molecule that blocks the functioning of a protein by displacing a naturally occurring ligand of the protein is known as a competitive inhibitor. A noncompetitive inhibitor is one that binds to some site on the protein other than the binding site of the natural ligand and exerts its influence through an allosteric mechanism.

**conductance** The conductance of a channel, denoted $g$, is the inverse of its electrical resistance, denoted $R$. When describing electrical circuits, it is more common to use resistance because we think of the impedance to current flow due to an element of the circuit. When discussing ion channels, however, it is more intuitive to describe their facilitation of ion flow in terms of the conductance. The units of resistance and conductance are ohms ($\Omega$) and siemens (S), respectively.

**configurational entropy** Entropy arising from positional variation.

**conformational change** A change in structure that arises solely from rotations about covalent bonds is called a conformational change. Conformational rearrangements do not involve breaking and forming covalent bonds, and can often occur readily at room temperature.

**conformational isomer** Two forms of a molecule that are related to each other by rotations about covalent bonds are called conformational isomers.

**conformational selection** Induced-fit binding is thought to involve the binding of a ligand to conformations of the receptor that are populated, although at a low level, even in the absence of ligand (see Figure 12.22). This is referred to as conformational selection by the ligand.

**conjugate base** Dissociation of an acid releases a proton ($H^+$) and a base ($A^-$), which is referred to as the conjugate base of the acid.

**conservative substitution** The replacement of an amino acid residue in one protein by a similar one in another protein is called a conservative substitution. The replacement of serine by threonine, or leucine by isoleucine, are examples of conservative substitutions.

**contact order** The contact order of a protein is a measure of how far apart in primary sequence contacting secondary structure elements are in the native structure, normalized for the length of the protein.

**continuum dielectric model** An electrostatic calculation that approximates the effect of polarizable groups by a choice of suitable dielectric constants for the medium is referred to as a continuum dielectric model for electrostatics.

**cooperative DNA binding** When the binding of one protein to DNA increases the affinity of another protein for DNA, the two proteins are said to bind cooperatively.

**cooperative response** When the binding of a ligand to a protein is ultrasensitive, the binding is said to be cooperative. As more ligand molecules bind to the protein, the saturation of the protein increases more sharply than would be expected for a normal binding event, as if the ligand molecules "cooperate" with each other.

**counterion** Ions with a charge opposite to that of a macromolecule.

**counterion condensation** Counterion condensation is the local concentration of counterions (cations in the case of RNA) in regions of high electrostatic potential around a polymer. These ions are loosely bound but play an important role in stabilizing structures such as helices.

**critical micelle concentration** Amphiphilic compounds are soluble as individual molecules up to a specific concentration, called the critical micelle concentration. When that concentration is exceeded, micelles begin to form.

**cross-β spine** Amyloid fibrils are made up of β strands, packed so that the long axis of the fibrils runs in the plane of the β sheet, with the individual strands running perpendicular to this axis (see Figure 18.29). This kind of structure is called a cross-β spine (the term "cross-β" refers to a characteristic feature of the x-ray diffraction pattern of these fibrils).

**cyclodextrin** A cyclized sugar with six or seven glucose residues.

**cytosine** A substituted pyrimidine base, one of the four bases in DNA and RNA.

**D form** Amino acids, except for glycine, have two stereoisomers, known as the L form and the D form, which are illustrated in Figure 4.13.

**decoding center** Region of the ribosome where the mRNA codons are recognized by anticodons in tRNAs.

**2′-deoxyribonucleotide** The nucleotides that make up DNA (deoxyribonucleic acid) are derived from 2′-deoxyribose, in which the 2′-OH group in ribose is replaced by hydrogen. These are called 2′-deoxyribonucleotides.

**depolarization** The membrane potential is usually negative—that is, the cytoplasmic surface of the cell membrane is usually negatively charged. If there is an influx of positively charged ions, then the inner surface becomes less negative, or may even become positively charged. This is referred to as depolarization. Alternatively, if there is a net outflow of positive ions, then the membrane becomes hyperpolarized.

**desolvation energy** Removal of a charged chemical group, such as an amino acid sidechain or a phosphate group, from water has a large energetic penalty, known as the desolvation energy. The desolvation energy reflects the loss of favorable interactions with water.

**detailed balance** See principle of microscopic reversibility.

**detergent** Amphiphilic molecules that tend to form micelles rather than bilayers.

**dielectric constant** The dielectric constant, denoted $\varepsilon$, is a scale factor that reduces the magnitude of the electrostatic energy as calculated using Coulomb's law. The dielectric constant accounts for the effect of the environment in weakening the interaction between charges.The dielectric constant of bulk water is 80. The interior of a protein, which is slightly polar, has a much smaller dielectric constant (~2.0).

**differential scanning calorimeter** An instrument that is used to measure the difference in heat capacity between the contents of two chambers as the temperature is scanned. The two chambers contain identical solutions, except that one contains a macromolecule of interest in addition to everything else. In this way, the heat capacity of the macromolecule is determined as a function of temperature.

**diffuse ion interaction** A general charge screening of the negative phosphate groups of DNA or RNA by a "sea" of monovalent or divalent cations.

**diffusion** Random motion of molecules, without a driving force.

**diffusion collision** In the process of protein folding by diffusion collision, secondary structural elements form first and undergo collisions with other secondary structure elements to form the native structure.

**diffusion constant** The diffusion constant, $D$, is a parameter that describes the average rate of motion of a molecule in solution, related to its mean-square displacement as a function of time. The units of the diffusion constant are $cm^2\ sec^{-1}$.

**diffusion equation** A mathematical statement of Fick's second law (Equation 17.20). This equation allows us to calculate the net diffusive movement of molecules in response to concentration gradients.

**diffusion-limited reaction** A diffusion-limited reaction is one in which every collision between reactants leads to products. The rate of reaction in this case is limited by the rate of collisions, which is affected just by the rates of diffusion. The rate constant for collisions between molecules in water is about $1 \times 10^{10}\ M^{-1}\ sec^{-1}$. A reaction that occurs with a rate constant close to this value is said to be diffusion limited.

**dihedral angle rotation** See torsion angle.

**direct readout** Recognition of DNA by a protein in which a specific pattern of hydrogen-bond donors, acceptors, and nonpolar groups on the protein is matched to a complementary pattern displayed on the DNA. Also known as base readout.

**dissociation constant, $K_D$** The inverse of the equilibrium constant for the reaction describing the binding of two molecules.

**disulfide bond** A disulfide bond is a covalent bond between the sulfur atoms of two cysteine residues. Disulfide bonds cross-link different parts of a protein chain and can increase the stability of the three-dimensional structure.

**DNA** A polymer of 2′-deoxyribonucleotides.

**DNA polymerase** Enzyme that catalyzes the template-based synthesis of DNA.

**dynamic light scattering** Dynamic light scattering measures fluctuations in the number of scattering molecules in a small volume through their effect on the scattered light intensity. The rate of light intensity fluctuation is directly related to the rate of diffusion of the scattering molecules.

**Eadie–Hofstee plot** A graph of $v_0$ vs. $v_0/[S]$, which for Michaelis–Menten kinetics gives a straight line with slope $-K_M$ and an intercept of $V_{max}$.

**EF hand** A type of helix-loop-helix motif involved in binding calcium ions.

**electric potential difference** Voltage difference.

**electrical work** Work done by charges moving against an electrical potential gradient.

**electrochemical cell** An electrochemical cell is a battery that is powered by connecting two redox reactions into a circuit.

**electrochemical gradient** An electrochemical gradient is a free-energy gradient arising from a combination of the differences in electrical potential and chemical potential across a membrane.

**electrode** A metallic conductor that is in contact with solutions or other non-metallic parts of electrical circuits.

**electronegativity** Electronegativity is a parameter that describes the tendency of an atom to attract electrons towards it when it is participating in a covalent bond. The greater the electronegativity of an atom, the greater its tendency to withdraw electrons.

**electrophoresis** Electrophoresis is the driven movement of charged molecules in an electric field, analogous to movement in a centrifugal field.

**electrophoretic mobility** Movement of molecules in a gel, under the influence of an electric field.

**electrostatic focusing effect** The change in electrostatic potential around a macromolecule that arises because the low dielectric medium inside the macromolecule does not attenuate the electric field generated by charges within it. See Figure 6.38 for an example of an electrostatic focusing effect due to the shape of a protein.

**electrostatic steering** Enhancement of the rate of an enzyme-catalyzed reaction by electrostatic effects that increase the speed at which substrates enter the active site.

**elementary charge** The magnitude of the charge on the electron is $1.602 \times 10^{-19}$ coulombs (C). This is known as the elementary charge. The charges on atoms are described in terms of the elementary charge rather than coulombs. Thus, an atom with a single positive charge is said to have a +1 charge, which is understood to be $1.602 \times 10^{-19}$ C.

**elementary reaction** An elementary reaction represents the most basic step used to describe a reaction process. Many chemical mechanisms result from the combination of several elementary steps.

**empirical potential energy function** These are relatively simple mathematical expressions that allow us to calculate the potential energy of a molecule or a collection of molecules, given the conformation (that is, the internal structure) of each molecule and their relative positions. Empirical potential energy functions are approximations to the true quantum mechanical energy of the system.

**enantiomer** Molecules that are mirror images of each other.

**endo** See sugar pucker.

**energy distribution** The population (number) of molecules in each energy level. The energy distribution does not refer to the identities of specific molecules, just the aggregate number of molecules in each energy level.

**enhanceosome** A complex of several eukaryotic transcription factor proteins bound to a control site on DNA for a gene located nearby, known as an enhancer element. The formation of the enhanceosome is a first step in the initiation of transcription from the gene.

**enhancer element** See enhanceosome.

**enthalpy** The change in enthalpy, $H$, of a system during a process carried out under constant pressure conditions is equal to the heat taken up by the system during the process. Enthalpy is defined as follows: $H = U + PV$. Since biochemical processes usually occur under constant pressure conditions, the change in enthalpy is readily determined if the amount of heat transferred is measured. If no gases are involved, then the change in enthalpy is essentially equal to the change in energy.

**enthalpy of formation** The enthalpy of formation of a compound, denoted $\Delta_f H^o$, is the difference in the enthalpy of one mole of the compound in the standard state and stoichiometric equivalents of the corresponding elements.

**enthalpy–entropy compensation** Enthalpy–entropy compensation is the tendency of these two thermodynamic parameters to have offsetting effects on binding energy or stability in a particular system. Tighter binding (higher stability) generally means less entropy in the bound (stable) state and, hence, a less favorable entropy change in getting to that state.

**entropy** The entropy of a system of molecules is a measure of the disorder in the system. The greater the number of equivalent rearrangements of the molecules, the greater the value of the entropy. Mathematically, the entropy of a system is proportional to the natural logarithm of the multiplicity of the system.

**entropy of formation** The entropy of formation of a compound, denoted $\Delta_f S^o$, is the difference in the entropy of one mole of the compound in the standard state and stoichiometric equivalents of the corresponding elements.

**environment class** Descriptor used to characterize the environment of a position in a protein fold, in the 3D-1D profile method for fold recognition.

**environmental profile** String of environmental classes for each position in a protein fold, in the 3D-1D profile method for fold recognition.

**environmental score** A log-likelihood score that characterizes how well a particular sequence fits with a particular environmental profile, in the 3D-1D profile method.

**enzyme** A protein or RNA catalyst.

**equatorial** See axial.

**equilibrium constant, *K*** See equilibrium constant, $K_{eq}$.

**equilibrium constant, $K_{eq}$** The equilibrium constant, $K_{eq}$, for a reaction is given by:

$$K_{eq} = \frac{[C]_{eq}^{\upsilon_C}\,[D]_{eq}^{\upsilon_D}}{[A]_{eq}^{\upsilon_A}\,[B]_{eq}^{\upsilon_B}}$$

for a simple reaction involving A, B, C, and D. $[A]_{eq}$, $[B]_{eq}$, $[C]_{eq}$, and $[D]_{eq}$ are the molecular concentrations at equilibrium, and $\upsilon_A$, $\upsilon_B$, $\upsilon_C$, and $\upsilon_D$, are the stoichiometric coefficients of the reaction. The equilibrium constant is often simply denoted *K*.

**equilibrium potential of potassium, $E_K$** The membrane potential at which there is no net flow through potassium channels.

**equivalent circuit** An electrical circuit consisting of batteries, resistors, capacitors, and conductors that represents the electrical behavior of an axon.

**exo** See sugar pucker.

**exon** The regions of eukaryotic genes that code for functional RNA or proteins are known as exons. These are interrupted by noncoding regions known as introns. The splicing process removes introns and joins the exons to produce messenger RNA.

**exothermic** A process that releases heat is an exothermic process.

**expansion work** Work done during expansion against an external pressure.

**extensive property** An extensive property of a system depends on the size of the system. Energy is an extensive property, as is the volume. Intensive properties do not depend on the size of the system. The temperature, density, and pressure of the system are all intensive properties.

**factorial** The value of *M*! (pronounced "*M* factorial") is given by:

$$M! = M \times (M-1) \times (M-2) \times \ldots \times 2 \times 1$$

For example, the value of 5! is given by:

$$5! = 5 \times 4 \times 3 \times 2 \times 1$$
$$= 120$$

**farad** Unit of capacitance.

**fatty acid binding protein** A protein that bind to lipids and fatty acids.

**Fick's first law** Fick's laws are differential equations that describe the net movements of molecules in solutions. They can be solved to give the distribution of molecules after a specific time from a given starting distribution. Fick's first law simply states that the flux, $J(x)$, at position $x$ is proportional to the concentration gradient at $x$.

**Fick's second law** A second-order differential equation that describes how the concentration of a molecule changes with position and time, due to a concentration gradient.

**fidelity** Fidelity in DNA replication corresponds to the rate at which bases are incorporated into the new DNA copy that are not the Watson-Crick complement of the base on the template strand. Such errors can introduce mutations into encoded proteins or into sequences that regulate gene expression. The term has a similar meaning when applied to transcription or translation.

**filter binding assay** A binding assay involving a separation step in which one of the components of a binding reaction adhere to filter paper, while the other components do not.

**first law of thermodynamics** Also referred to as the law of conservation of energy, the first law states that energy is neither created nor destroyed in a chemical process.

**first-order kinetics** The kinetics of a reaction for which the rate depends linearly on the concentration of a single reactant.

**Fischer projection** A representation of molecular stereochemistry that is very convenient for comparing the stereochemistry of sugar isomers.

**flippase** Enzymes that recognize and "flip" specific lipids between the inner and outer leaflets of the membrane, a process needed for generating and maintaining membrane asymmetry.

**fluorescence lifetime** The time constant for the disappearance of fluorescence, due to relaxation of the excitation causing the fluorescence.

**fluorescence quencher** A compound, such as acrylamide, to which the excitation energy of a fluorophore is readily transferred. The quencher converts this energy to heat. In this way, fluorescence quenchers reduce the amount of light emitted by fluorescent molecules.

**Fluorescence Recovery After Photobleaching** An experimental method for studying the diffusion of molecules in a lipid bilayer. See Figure 3.40.

**fluorophore** A fluorescent molecule.

**folding funnel** A feature of the free-energy landscape of proteins that direct diverse unfolded structures towards the folded structure.

**fold-recognition algorithm** Fold-recognition or threading algorithms are computational procedures that help predict the three-dimensional structures of proteins of unknown structure whose sequences are not very closely related to proteins that have had their structures determined. These methods take advantage of the fact that any particular protein fold places constraints on the kinds of amino acids that can be accommodated by the fold—that is, they explicitly use knowledge of the three-dimensional structures of proteins.

**forbidden conformation** A molecular conformation in which some atoms are closer than the sum of their van der Waals radii. The van der Waals repulsion energy is so large for such conformations that favorable interactions that may occur cannot overcome the repulsion.

**fractional saturation or fractional occupancy** The fractional saturation, denoted $f$, is the extent to which the binding sites on a protein are filled with ligand. For a protein with a single ligand binding site, the value of $f$ is given by the ratio of the concentration of the protein with ligand bound to the total protein concentration. The fractional saturation is an important parameter because experimentally measurable responses to ligand binding usually depend directly on the fractional saturation.

**free energy** A parameter that incorporates both the energy and the entropy of a system. The free energy of a system always decreases when a process occurs spontaneously, and is at a minimum when the system is at equilibrium. The change in free energy that occurs during a process is equal to the maximum amount of work that can be extracted from the process.

**free-energy change for the reaction or the reaction free energy ($\Delta G$)** $\Delta G$ is the molar free-energy change upon stoichiometric conversion of reactants to products, given their actual nonequilibrium concentrations.

**free-energy landscape** A multidimensional surface, in which the value of the free energy for a particular combination of conformational variables is analogous to elevation in a conventional landscape.

**friction factor** The friction factor is a parameter that relates the size and shape of a molecule to the drag (the resistance to movement) it generates in a fluid.

**fructose** A ketohexose, also often called fruit sugar because of its common occurrence in fruit.

**fucose** Several of the sugars that are present in oligosaccharides arise from relatively small modifications of the simple sugars. For example, replacing the commonly occurring $-CH_2OH$ group with a methyl group ($-CH_3$) gives fucose (which is also called 6-deoxy-L-galactose).

**ganglioside** A sphingolipid with multiple sugars (typically three to eight).

**gating current** A small membrane current that results from a conformational change in voltage-dependent channels when the membrane is depolarized.

**Gaussian distribution or a normal distribution** A probability distribution for the values of a variable, $x$, of the form:

$$P(x) = Ce^{-ax^2}$$

is known as a Gaussian distribution. The parameters $a$ and $C$ are constants that define the width of the distribution and ensure that it is normalized.

**Gaussian function** A Gaussian function of a variable $x$ is an exponential of $-x^2$. Most distributions generated by random processes are Gaussian.

**gel-shift assay** Gel-shift assays are used to determine the affinity of the interaction between proteins and DNA or RNA. Gel-shift assays are done by varying the concentrations of potentially interacting molecules and analyzing the extent to which complexes form by using electrophoresis to separate the species present (Figure 13.38).

**gene** A gene is a segment of DNA that is transcribed into RNA. The RNA could be functional on its own or, if it is a messenger RNA, used for the production of proteins.

**genetic code** The genetic code relates triplets of nucleotides in a gene sequence to each amino acid in a protein sequence.

**Gibbs free energy** The Gibbs free energy, $G$, of a system is given by:

$$G = H - TS$$

For a system at constant pressure and temperature, the value of $G$ always decreases in a spontaneous process.

**globin fold** The all-helical protein fold that is common to globins, such as myoglobin and hemoglobin (see Figure 5.6).

**globular protein** A water-soluble protein that folds into a compact three-dimensional shape.

**glucosamine or galactosamine** Several of the sugars that are present in oligosaccharides arise from relatively small modifications of the simple sugars. For example, replacing one hydroxyl with an amino group gives glucosamine or galactosamine.

**glucose** An aldohexose sugar that is a major energy source in eukaryotic cells.

**glycan** Glycans are sugars or sugar derivatives, and their polymers. The simplest individual subunits of glycans have roughly the formula $(HCOH)_n$ with $n$ between 3 and 9, but most commonly 5, 6, or 7. Many glycans also have sugars with additional functionality relative to these simplest subunits. Glycans are also known as carbohydrates.

**glycerophospholipid** Glycerophospholipids are the most abundant molecules in biological membranes. As the name suggests, they are built from a glycerol unit, $HOCH_2$-CHOH- $CH_2OH$, as shown in Figure 3.28.

**glycogen** Glycogen is a primary energy storage compound in animal cells that can be rapidly broken down to glucose and then biochemically metabolized (that is, "burned") to produce $CO_2$ and water with the release of energy. Glycogen is made up of repeating glucose-$\alpha(1\rightarrow4)$-glucose units, but with occasional glucose-$\alpha(1\rightarrow6)$-glucose branches, yielding a high-molecular-weight polymer (Figure 3.12).

**glycosidic bond or glycosidic linkage** Glycosidic linkages are the attachments between sugars in polysaccharides. They are specified by the atom linked on each sugar and the orientation of the bond to the ring (when needed). These linkages are often called glycosidic bonds.

**GNRA tetraloop** A GNRA tetraloop is a structural motif in RNA that helps induce a sharp turn in the backbone. The name refers to the sequence of the motif, which has a guanine (G) in the first position, any nucleotide (N) in the second position, a purine (R, either A or G) in the third position, and adenine (A) in the fourth position, as illustrated in Figure 2.40.

**graded response** An output function that depends on the input in a hyperbolic fashion, as in a simple binding equilibrium, is known as a graded, or linear, response. In such a response, the output switches from ~10% to ~90% of the maximum response over a 100-fold change in input strength.

**Gram negative** Bacteria were distinguished originally by their color when treated with the "Gram" stain, developed by the bacteriologist Hans Christian Gram. Bacteria with thinner peptidoglycan layers produce light pink staining when treated with the Gram stain, and were classified as Gram negative bacteria. In contrast, bacteria with thicker peptidoglycan layers produce dark purple staining and were classified as Gram positive.

**Gram positive** See Gram negative.

**Greek key motif** A protein structural motif in which four adjacent antiparallel β strands are arranged in a pattern similar to the repeating unit of an ornamental pattern, or fret, used in ancient Greece, called a Greek key (see Figure 4.36).

**GroEL** See chaperonin.

**GroES** A small protein that forms a heptameric assembly that caps the central chamber of GroEL.

**guanine** A substituted pyrimidine base, one of the four bases in DNA and RNA.

**gyrase** A form of topoisomerase enzyme.

**half-life ($t_{1/2}$)** The half-life of a reaction is the time for the concentration of a reactant to drop to half of its initial value. First-order processes are also commonly described by their time constants, which is the time required for the reactant to decay to ~37% of its initial value. The time constant for a first-order reaction is given by $1/k$, where $k$ is the rate constant.

**half-width at half height** The displacement from the origin (mean position) of a probability distribution at which the probability drops to half the value at the origin is called the half-width at half height (see Figure 17.7B).

**hard metal ion** Metal ions, such as $Na^+$, $K^+$, $Mg^{2+}$, and $Cr^{3+}$, that generally favor interacting with oxygen ligands.

**head group** A head group is the hydrophilic part of a lipid, which is attached to the phosphate that bridges to the glycerol and fatty acid chains. The head group remains in contact with water when the lipids are in bilayers or micelles.

**heat capacity** The heat capacity of a system is the amount of heat required to increase the temperature of the system by 1 kelvin. The magnitude of the heat capacity depends on the conditions under which the system is heated. If the heating is carried out under constant-volume conditions, the heat capacity is denoted $C_V$. If, instead, the heating occurs under conditions of constant pressure, the heat capacity is denoted $C_P$.

**heat or work** Energy transfer to and from the system occurs in the form of heat and work. Heat transfer results in increased random motion of molecules. When energy is transferred as work, it results in the ordered movement of some component of the system or the surroundings, such as the movement of a piston.

**heat shock** An increase in ambient temperature that causes proteins to unfold.

**heat-shock protein** A chaperone.

**heat-shock protein 70** A bacterial chaperone that binds to small hydrophobic segments in unfolded proteins. The "70" refers to the molecular weight of the chaperone.

**helix dipole** In an α helix all of the peptide linkages are parallel to each other, as shown in Figure 1.36. As a result, an α helix behaves as a single electrostatic dipole, known as a helix dipole, due to the additive effect of each of the individual dipole moments of the carbonyl and amide groups. The positive and negative ends of the helix dipole are at the N- and C-terminal ends of the helix, respectively.

**helix-loop-helix motif** A common protein structural motif, in which the connection between the two helices is longer than in a helix-turn-helix motif (see Figure 4.33B).

**helix-turn-helix motif** A particularly simple protein structural motif, consisting of two α helices joined by a short turn region (see Figure 4.33A).

**Helmholtz free energy** The Helmholtz free energy, $A$, of a system is given by:

$$A = U - TS$$

For a system at constant volume and temperature, the value of $A$ always decreases in a spontaneous process.

**hemoglobin** The oxygen carrying protein in blood. Human hemoglobin is a tetrameric protein.

**Henderson–Hasselbalch Equation** The Henderson–Hasselbalch equation relates the pH of a solution to the ratio of the concentrations of a weak acid and its conjugate base:

$$\mathrm{pH} = \mathrm{p}K_\mathrm{a} + \log_{10}\left(\frac{[\mathrm{A}^-]}{[\mathrm{HA}]}\right)$$

**heptad repeat** In coiled-coil structures, every seventh residue occupies an identical position along the helix. This leads to a pattern of hydrophobic residues that repeats in groups of seven, known as the heptad repeat. Residues in each repeat are labeled *a* to *g*, and the residues at the *a* and *d* positions are usually hydrophobic and interact with the corresponding residues in the partner helix.

**Hill coefficient** This parameter, named after the physiologist Archibald Hill, measures the steepness of the log-log binding isotherm at the point when the protein is half saturated with ligand. A noncooperative system has a Hill coefficient of unity. Systems exhibiting positive and negative cooperativity have Hill coefficients that are greater than and less than unity, respectively. The Hill coefficient is also referred to as the Hill slope.

**Hill slope** See Hill coefficient.

**holoenzyme** Assembled form of an enzyme with several subunits.

**Hoogsteen base pair** A nonstandard base pair in which the hydrogen-bonding interactions utilize the Watson-Crick base-pairing edge on one base and the edge corresponding to the major groove in the other (see Figure 2.37).

**hydrogen bond** Hydrogen bonds are interactions between polar groups in which a hydrogen atom with a partial positive charge is located close to an atom with a partial negative charge, called the acceptor. The partial positive charge on the hydrogen atom is a consequence of a polarized covalent bond with a more electronegative atom, called the donor. See Figure 1.11.

**hydrogen-bond acceptor** A hydrogen bond is formed when a hydrogen atom bridges two other atoms. The atom that interacts closely with the hydrogen, but is not covalently bonded to it, is called the hydrogen-bond acceptor.

**hydrogen-bond donor** A hydrogen bond is formed when a hydrogen atom bridges two other atoms. The atom that is covalently bonded to the hydrogen is called the hydrogen-bond donor.

**hydrogen exchange** Certain groups in proteins, such as the backbone amide group, can readily exchange protons with the solvent. The rate of exchange of these protons is measured in hydrogen exchange experiments (see Figure 18.12).

**hydrophilic** Hydrophilic groups are polar and they can form hydrogen bonds with water. Molecules that are hydrophilic dissolve readily in water.

**hydrophobic** Hydrophobic groups are nonpolar—that is, they do not form hydrogen bonds. Hydrophobic groups prefer to cluster with other hydrophobic groups rather than interact with water, in part because they cannot form hydrogen bonds with water. Molecules that are hydrophobic do not dissolve readily in water.

**hydrophobic core** The three-dimensional structure of most proteins is characterized by an interior hydrophobic core, consisting mainly of hydrophobic sidechains. The exclusion of these sidechains from water stabilizes the protein.

**hydrophobicity scale** A hydrophobicity scale assigns a numerical value to each amino acid based on its hydrophobicity. There are many different hydrophobicity scales, which use different criteria for assigning a relative value for the hydrophobicity of the amino acids. One commonly used scale relies on the partitioning of amino acids between water and octanol to assign a value for the hydrophobicity.

**hydroxyl radical (•OH) footprinting** Using this method it is possible to distinguish solvent exposed regions of RNA from those that are buried in the folded structure. Strand cleavage of the RNA backbone occurs in the presence of hydroxyl radicals, while residues that are on the interior of a folded RNA are protected from cleavage.

**hyperbolic binding isotherm** The simple non-allosteric binding of a ligand to a protein results in a hyperbolic relationship between the fractional saturation, $f$, and the ligand concentration [L]. Because of this relationship, a simple ligand-binding equilibrium is referred to as hyperbolic binding. Deviation from the hyperbolic shape of the binding curve is evidence for more complicated phenomena, such as allostery or multiple binding sites with different affinities.

**hypochromicity** Hypochromicity refers to the absorption of light that is less than expected based on composition. For example, the stacking of bases in DNA decreases absorptivity relative to that expected for the same nucleotides free in solution, so the duplex is hypochromic relative to single strands.

**$IC_{50}$ value** The concentration of inhibitor that reduces the activity of a protein to half the maximal value (that is, the value seen in the absence of the inhibitor) is known as the $IC_{50}$ value.

**ideal dilute solution** A solution in which the concentration of the solute is so low that individual solute molecules do not influence each other is referred to as an ideal dilute solution.

**ideal monatomic gas** An ideal monatomic gas consists of atoms that do not interact with each other. An ideal gas obeys the ideal gas laws. For example, the pressure, $P$, volume, $V$, and the temperature, $T$, of an ideal gas are related by the following equation: $PV = nRT$ where $n$ is the number of moles of the gas and $R$ is the gas constant. Noble gases, such as argon, are very close to ideal in their behavior.

**indirect readout** When a protein recognizes the ability of a DNA sequence to undergo a particular conformational change, the process is referred to as indirect readout or shape readout. An example is shown in Figure 13.34.

**induced dipole** A dipole is a pair of opposite charges separated by a distance. An induced dipole is the result of a transient separation of charge in an atom that causes the atom to behave as if it were a dipole. Induced dipoles occur when an atom is subjected to an electrical field that causes the redistribution of electrons in an asymmetric way.

**induced fit** In this model for the binding of a ligand to a receptor, the binding site is considered to have some plasticity, allowing it to change somewhat to accommodate binding of the ligand or substrate, like a hand fitting into a stretchy glove. See conformational selection.

**initial velocity, $v_0$** The rate of a reaction, measured during the initial period of the reaction (that is, before substrate is depleted and product builds up).

**inner-sphere coordination** An interaction between a metal ion and a nucleic acid in which the metal ion loses a water ligand and binds directly to the nucleic acid.

**integral membrane protein** Proteins with at least one transmembrane segment are known as integral membrane proteins. The three-dimensional structure of these proteins depends on interactions with the lipids in the membrane.

**intensive property** See extensive property.

**interfacial water** Water molecules at the interfaces between proteins, known as interfacial waters, have quite different properties than water molecules that are not close to the protein (referred to as bulk water).

**intermediate** An intermediate is a chemical species that appears during the reaction but disappears when the reaction is complete.

**internal energy** The potential energy of a molecule.

**intron** Protein-coding regions of genes (known as exons) are interrupted by noncoding regions (known as introns). The splicing process removes introns and joins the exons to produce messenger RNA.

**inversion reaction** Some molecules polarize light. An inversion reaction is one involving such a molecule, during which the polarization of light by the solution switches from counterclockwise to clockwise, or the other way around (that is, the polarization inverts). An example of an inversion reaction is the hydrolysis of sucrose.

**inverted repeat sequence** Many DNA binding proteins, particularly in bacteria, occur as dimers and recognize DNA sequences that have two-fold symmetry, known as inverted repeat sequences, or palindromic sequences. In such cases, each monomer recognizes the same features in each half of the inverted repeat DNA, allowing both higher affinity and sequence specificity to be achieved. An example of this is shown in Figure 13.41.

**ion pair** When two oppositely charged groups are close to each other, the interaction is called an ion pair or a salt bridge.

**ion product of water** The dissociation constant for water, $K_W$, is also called the ion product of water.

$$K_W = [H^+][OH^-] = 1.0 \times 10^{-14}$$

at 25°C. The pH of pure water is:

$$-\log_{10}\sqrt{K_W}.$$

Thus, the pH of pure water is 7.0 at 25°C.

**ionic current** In the analysis of the electrical properties of neurons, this term refers to the current due to the flow of ions through ion channels.

**irreversible inhibitors** A molecule that permanently damages the ability of the enzyme to catalyze reactions is called an irreversible inhibitor. Such inhibitors react covalently with the enzyme.

**isolated system** A system that exchanges neither matter nor energy with the surroundings (see Figure 6.2).

**isothermal process** A process that occurs with no change in temperature.

**isothermal titration calorimetry** A calorimetric method that is used to obtain the enthalpy and entropy changes for a binding reaction.

**ketone** See aldehyde.

**ketose** A sugar that contains a ketone group.

**$K_I$** The inhibitor constant, $K_I$, is the dissociation constant for the interaction between the receptor and the inhibitor.

**kinesin** A molecular motor that moves along protein filaments called microtubules.

**kinetic energy** Energy due to motion.

**kinetic proofreading** A mechanism for increasing the fidelity of processes such as replication or translation. Rejection of the incorrect substrate can occur at two or more consecutive steps in the reaction pathway, each of which proceeds faster if the correct substrate is present. The probability of the incorrect substrate making it through the multiple selection steps becomes very small.

**kinetics** Kinetics is the study of the rates at which chemical and physical processes occur.

**L form** See D form.

**Le Châtelier's principle** This principle that states that if the concentration of one or more molecules in a reaction are changed, the system will adjust itself and reach a new equilibrium so as to balance the change.

**lead compound** The development of drugs for a specific disease begins with the identification of the proteins that are involved in steps critical to disease progression, and then proceeds to the design or discovery of molecules that inhibit the function of these proteins. In the initial steps of this process, the molecules that are first discovered to inhibit the target protein usually do not have all of the properties that are desirable in a drug. Such molecules are called lead compounds.

**lead optimization** The process of changing the chemical structure of a lead compound so as to improve its pharmacological properties, with the aim of converting it into a useful drug.

**leaflet** Lipid bilayers contain two layers of lipids, each of which is called a leaflet.

**lectin** Lectins are proteins that recognize specific sugars, usually binding with moderate affinity and high specificity.

**Lennard-Jones potential** Mathematical expression describing the van der Waals attraction and repulsion between atoms.

**ligand** A molecule that binds to another molecule. When considering the thermodynamics of binding between two kinds of molecules, the ligand is usually taken to be the molecule that is in excess. The other molecule is called the receptor.

**linear response** See graded response.

**Lineweaver-Burk plot** A graph of $1/v_0$ vs. $1/[S]$. If the enzyme obeys Michaelis-Menten kinetics, then this gives a straight line with slope $K_M/V_{max}$ and an intercept of $1/V_{max}$.

**linking number** See writhe.

**lipid** Lipids are amphiphilic molecules (that is, part hydrophobic and part hydrophilic). Lipids are the primary components of biological membranes. Although membranes contain embedded proteins, most of the critical properties of membranes arise from the lipid components.

**lipid bilayer** A lipid bilayer is comprised of lipids packed into two parallel layers with the head groups exposed to water and the alkyl chains packed together away from water. Lipid molecules move relatively freely within the plane, but do not easily flip between the two layers.

**lipid vesicle** Lipid vesicles are closed spherical bilayers, rather like bubbles, but with water inside and outside rather than air (see Figure 3.34).

**lipoprotein** A protein that transport lipids. See Figure 3.48A for an example.

**lock-and-key model** A model for ligand-receptor interactions in which both molecules are assumed to be rigid.

**London dispersion force** The attractive component of van der Waals forces.

**London force** When two neutral atoms approach each other, transient fluctuations in the electron clouds of each atom set up transient dipoles, leading to an attractive force between the atoms. This attractive force is called the London force, or the van der Waals attraction.

**macromolecular crowding** Macromolecular crowding is the effect of macromolecules occupying a significant fraction of the volume of a cell. This increases the effective concentration of proteins, or viewed another way, appears to increase the strength of interactions at a given protein concentration.

**major groove** DNA and RNA double helices have two characteristic grooves, denoted the major and minor grooves. In B-form DNA, the major groove is wide and can accommodate α helices, which is important for the sequence-specific recognition of DNA by proteins. The major and minor grooves can be identified by looking at the connections of the base pairs to the sugars. The major groove is at the convex edge of the base pair, while the minor groove is at the concave edge, as shown in Figure 2.10.

**maltose** A disaccharide, shown in Figure 3.10. Maltose is glucose-alpha(1→4)-glucose.

**mass action ratio** The capacity of a reaction to run in the forward direction is determined by the ratio of $Q$ to $K$, known as the mass action ratio. $Q$ is the reaction quotient and $K$ is the equilibrium constant.

**maximal entropy principle** The maximum entropy priniciple is equivalent to the second law of thermodynamics. Spontaneous change is always in the direction of maximizing entropy, and at equilibrium the entropy has a maximal value.

**maximal velocity ($V_{max}$)** The maximal velocity, $V_{max}$, is the maximum rate of reaction catalyzed by an enzyme at very high substrate concentration.

**mean** The average value of a variable.

**mechanical work** When a system does mechanical work on the surroundings, it causes the ordered movement of some part of the surroundings.

**melting curve** The variation in some property of a macromolecule, such as the heat capacity, molar absorptivity or catalytic activity, as a function of temperature is known as a melting curve.

**melting temperature of a DNA duplex ($T_M$)** The melting temperature of a DNA is defined as the temperature at which there are equal concentrations of the duplex (or other folded state) and the random coil single strands.

**melting temperature of a protein ($T_M$)** At the melting temperature, half the protein molecules in a solution are unfolded. The standard free-energy change for folding, $\Delta G^o$, is zero at $T = T_M$ for a monomeric protein.

**membrane potential** The electrical potential difference across the cell membrane (that is, the difference between the potential inside the cell and the potential outside the cell) is known as the membrane potential. Resting mammalian cells have a membrane potential that is approximately −70 mV, with the interior of the cell at a negative potential with respect to the outside.

**membrane protein** A protein that is associated with lipid bilayers.

**metalloprotease** A protease that requires a metal ion bound to it for catalysis.

**micelle** A micelle is a spherical array of lipids with head groups on the outside and chains on the inside (see Figures 3.36 and 3.37). Micelles are typically formed by single-chain lipids, or by detergents.

**Michaelis constant ($K_M$)** The Michaelis constant, $K_M$, is the concentration of substrate required for the reaction velocity to be ½ $V_{max}$.

**Michaelis–Menten equation** For the simplest enzyme-catalyzed reaction, shown in Equation 16.3, the rate of the reaction under steady-state conditions is given by:

$$v_0 = \frac{V_{max}}{\left(1+\frac{K_M}{[S]}\right)}$$

$V_{max}$ is the maximum rate of the reaction, [S] is the substrate concentration and $K_M$ is the Michaelis constant, which corresponds to the substrate concentration at which the rate is half-maximal.

**Michaelis–Menten kinetics** A simple model for understanding enzyme kinetics, in which the focus is on the formation of a complex between a substrate and an enzyme, followed by a catalytic step.

**microstate** Each particular spatial configuration of atoms, or distribution of atoms in energy levels, that is consistent with the definition of the overall state of the system is referred to as a microstate.

**microtubules** Microtubules are intracellular filaments which are involved in the positioning of organelles in cells, and they are used for transport, by kinesins, of materials within the cell.

**minor groove** See major groove.

**mitochondrion** An organelle that is the principal site of ATP production in eukaryotic cells.

**molar absorptivity** Molar absorptivity is a characteristic of a compound that specifies the fraction of light of a particular wavelength that will be absorbed when the light passes through a 1-cm sample of a solution at 1 M concentration.

**molar free energy** The free energy of one mole of a molecule, under specific conditions.

**molecular chaperone** Molecular chaperones are a group of unrelated protein families whose roles include stabilizing unfolded proteins, assisting in correct folding and assembly of unfolded proteins, unfolding proteins for translocation across membranes, or disassembling complexes or aggregates.

**molecular chemical potential ($\mu_i$)** See chemical potential.

**molecular flux** The flux of molecules in a direction $x$ is the rate at which molecules cross a unit area of an imaginary plane perpendicular to the $x$-axis.

**molecular recognition** The specific binding of two molecules.

**Morse potential** A mathematical expression that describes the energy of a covalent bond.

**multilamellar vesicle** Concentric spheres of bilayers, rather like the layers of an onion.

**multiplicity** The multiplicity of an outcome is the number of different ways in which that outcome can be achieved. For molecular systems, the multiplicity refers to the number of different molecular configurations that are consistent with the macroscopic parameters that define the system (for example, temperature or volume). The multiplicity is denoted by $W$ in this book.

**myelin** A highly insulating layer that forms a sheath around segments of an axon. Myelinated segments of axons do not contain voltage-gated ion channels. These are restricted to the nodes of Ranvier, which interrupt the myelinated segments and are the sites where action potentials are regenerated.

**myoglobin** Oxygen storage protein found in tissues. The fold of myoglobin is similar to that of each of the subunits of hemoglobin.

**N-linked glycosylation** See O-linked glycosylation.

**N-terminal amino acid** The first amino acid in a protein chain—the one with the free amino group ($-NH_3^+$)—is known as the N-terminal amino acid.

**native structure** The native structure of a protein is the conformation that it adopts under normal physiological conditions.

**negative cooperativity** See positive cooperativity.

**Nernst equation** The Nernst equation allows us to calculate the change in voltage of an electrochemical cell as the concentrations of the reacting species change. Another form of the Nernst equation is used in calculating the voltage across biological membranes that arises due to differences in the concentrations of ionic species across the membrane. See Equations 11.50 and 11.66.

**Nernst potential** The membrane potential at which there is no net flux through channels for a particular ion (for example, $K^+$) is known as the Nernst potential for that ion. The Nernst potential depends on the concentrations of the ion inside and outside the cell.

**neutral pH** pH 7, the pH of pure water.

**node of Ranvier** See myelin.

**noncompetitive inhibitor** A noncompetitive inhibitor is a compound whose effects occur through binding to an enzyme somewhere other than the active site. Thus, a noncompetitive inhibitor prevents the chemical reaction from occurring, but does *not* do so by competing with substrate for the active site.

**nonconservative substitution** A substitution of one amino acid for another in a protein chain, where the chemical and structural properties of the original amino acid are not preserved.

**noncovalent complex** An interaction between two molecules that are not linked by covalent bonds.

**noncovalent energy term** A component of an empirical energy function that describes a noncovalent interaction.

**noncovalent interaction** Noncovalent interactions are interactions between atoms that are not covalently bonded to each other. Noncovalent interactions can be attractive or repulsive, and they arise from interactions between transient or stable charges on atoms.

**nonpolar** Molecules or groups of atoms that do not contain strongly polarized covalent bonds are called nonpolar molecules or nonpolar groups.

**nonstandard base pair** A base pair that does not conform to the Watson-Crick base pairing rules.

**normalized** A probability distribution function whose integral over the range of variables is unity.

**nuclear or steroid hormone receptor** A class of transcription factors that are activated by the binding of small molecules, such as steroids.

**nucleation condensation** In the process of nucleation condensation, a general compaction of the protein chain occurs first, leading to the nucleation of the final secondary and tertiary structure.

**nucleoside** The combination of a ribose or deoxyribose sugar and a nucleotide base, without the phosphate groups, is called a nucleoside.

**nucleosome** DNA in eukaryotic cells is packaged into chromatin. The structure of chromatin resembles beads on a string. Each "bead" is called a nucleosome, and it consists of about 146 base pairs of DNA wrapped tightly twice around a protein core.

**nucleotide** A nucleotide consists of a sugar covalently bonded to a phosphate group and to a substituted purine or pyrimidine base.

**nucleotide-binding domain** Protein domain that binds to nucleotides, usually containing a Rossman fold.

**O-linked glycosylation** O-linked and N-linked glycosylation refer to the attachment of polysaccharides through the side chain hydroxyl of serine or threonine (O-linked), or the side chain amide of asparagine (N-linked).

**Ohm's law** Ohm's law states that the voltage difference, $E$, across a resistor is equal to product of the current, $I$, and the resistance, $R$:

$$E = I \times R$$

Ohm's law can also be stated in terms of the conductance, $g$:

$$I = g \times E$$

**Okazaki fragment** At a replication fork, the synthesis of the lagging strand is discontinuous. As a result, discontinuous segments of DNA, called Okazaki fragments, are synthesized one fragment at a time on the lagging strand template.

**oligosaccharide** Oligosaccharides are glycans that are polymers of sugars. Specific small polymers are usually designated by the number of sugar units, such as monosaccharide, disaccharide, trisaccharide, etc. The term polysaccharide is usually used when there are more than about 30 sugar units in the polymer.

**open system** In an open system, molecules and energy can move in and out the system (see Figure 6.2).

**operator sequence** A specific DNA sequence that controls the expression of genes.

**optical tweezers** A method of single-molecule analysis in which a molecule is pulled using sets of lasers.

**optimal bond length** The bond length at which the energy has minimal value.

**ordered binding** An enzyme-catalyzed reaction with multiple substrates in which the order of binding of substrates to the enzyme is important.

**organelle** An organelle is a subcellular structure or chamber, often bounded by a membrane, that is specialized for particular functions. Examples include the nucleus, the endoplasmic reticulum, the mitochondria, the lysosomes, and the Golgi.

**osmosis** Osmosis occurs when a semipermeable membrane separates two parts of a chamber, restricting the movement of one kind of molecule, but allowing free diffusion of another kind of molecule.

**osmotic pressure** The pressure created by movement of molecules into one region due to osmosis.

**outer-sphere coordination** The metal ions that interact with RNA (for example, $Mg^{2+}$) are classified by how directly they interact with the nucleic acid. Diffuse ions have water between the hydrated ion and the nucleic acid. Outer-sphere ions are those where a water that is coordinated to the metal also makes contact to the nucleic acid. Inner-sphere ions make direct contact with the nucleic acid.

**overwound** When extra turns are introduced into a DNA double helix, the helix is said to be overwound.

**oxidative phosphorylation** The process of oxidative phosphorylation couples the oxidation of fuel molecules, such as glucose, to ATP production.

**oxidized** In an electron transfer reaction, the molecule (or ion) that accepts the electron is said to undergo a reduction, whereas the molecule that loses the electron is said to be oxidized. See redox reactions.

**palindromic sequence** See inverted repeat sequence.

**partition function** Denominator in the mathematical expression defining the Boltzmann distribution. The value of the partition function indicates the range of energy levels that are occupied to a significant extent.

**Pascal's triangle** A triangular diagram in which each row corresponds to a series of events with $M$ trials, and the entries in the row are the multiplicities for $N$ positive outcomes ($N = 0, 1, 2, ..., M$). The numbers in Pascal's triangle are known as binomial coefficients.

**passive spread** The propagation of a voltage spike without regeneration is known as passive spread. Passive spread occurs through diffusion, and the amplitude of the action potential decreases during this process.

**passive transport** Passive transport is the movement of molecules through the process of diffusion, while in active transport, movement of molecules in cells is driven by chemical energy—usually the hydrolysis of ATP or other nucleoside triphosphates.

**passive transporter** Transmembrane protein that allows the movement of molecules across the membrane, depending on the concentration gradient. Passive transporters do not consume energy.

**pectin** A polysaccharide, $\alpha(1\rightarrow4)$-galacturonic acid with ~4% branching with neutral sugars, galactose, arabinose, and xylose.

**peptide** When a protein chain contains only a small number of residues (fewer than ~50) it is referred to as a peptide or a polypeptide, with the term "protein" usually being reserved for longer chains.

**peptide backbone** The repetitive sequence of -NH-CH-CO- atoms that runs the length of a protein chain is called the peptide backbone.

**peptide bond** The synthesis of proteins involves a condensation reaction in which the amino group of one amino acid combines with the carboxyl group of another, with the formation of a peptide bond and the elimination of a water molecule.

**peripheral membrane protein** Proteins that are tightly associated with membranes, but do not traverse the membrane, are known as peripheral membrane proteins. Such proteins may be bound to the membrane by noncovalent or covalent interactions.

**persistence length** The persistence length of DNA corresponds to the maximum length of a segment of DNA that behaves as a rather rigid rod. DNA segments that are shorter than the persistence length behave as relatively rigid rods. DNA segments that are longer than the persistence length can be bent.

**$\phi$-value analysis** Different residues in a protein affect the free energy of the transition state for folding to different extents. We can measure the relative contributions of different residues to the transition state by determining a parameter known as the $\phi$-value, which captures the effects of mutations on the folding and unfolding rates (see Equation 18.9).

**phosphodiester linkage** The nucleotides in chains of DNA and RNA are connected by phosphodiester linkages. There are two ester bonds in each linkage. These connect the phosphate group to the 3′-oxygen atom of the first nucleotide and the 5′-oxygen atom of the second nucleotide, respectively.

**phospholipid** The principal lipid components of membranes, in which a phosphorylated polar or charged scaffold, known as the head group, is attached to two hydrocarbon chains.

**"ping-pong" mechanism** A type of enzyme mechanism. A "ping-pong" mechanism is one in which there are two distinct steps required, the first of which produces a modified enzyme.

**p$K_a$ value** The p$K_a$ value for a weak acid is the negative logarithm of the acid dissociation constant ($K_a$):

$$pK_a = -\log_{10} K_a$$

The p$K_a$ is the pH at which the concentration of the acid and its conjugate base are equal.

**poise** Standard unit for viscosity.

**Poisson equation** A differential equation describing the electrostatic potential generated by a set of charges in a region of space where the dielectric constant is not uniform.

**Poisson–Boltzmann equation** A modified form of the Poisson equation that takes account of the redistribution of ions around a macromolecule.

**polar molecules or polar groups** Molecules or groups of atoms containing polarized covalent bonds are called polar molecules or polar groups.

**polarized** A covalent bond is said to be polarized when one of the atoms participating in the bond withdraws electrons towards it.

**polypeptide** When a protein chain contains only a small number of residues (fewer than ~50) it is referred to as a peptide or a polypeptide, with the term "protein" usually being reserved for longer chains.

**polysaccharide** See oligosaccharide.

**positive cooperativity** When two or more ligands bind to a protein in such a way that they mutually reinforce each of the binding affinities, the phenomenon is called positive cooperativity. If the ligands make it more difficult for each other to bind, the phenomenon is called negative cooperativity.

**potential energy** The capacity of a system to do work.

**preequilibrium approximation** A preequilibrium approximation refers to a case in which there are forward and backward reactions that are both much faster than subsequent reaction steps. In this case, the reactants and products for this fast step always remain at relative concentrations determined as if they were fully at equilibrium.

**pre-mRNA** The transcription of eukaryotic genes results in RNA transcripts, known as precursor mRNAs or pre-mRNA, which contain both the exons and introns. The RNAs are then processed in a reaction known as RNA splicing in which the introns are cleaved from the primary transcript and the exon sequences are ligated together.

**primary structure** The amino acid sequence of a protein is referred to as its primary structure. The local conformation of the protein backbone (α helix, β strand, or loop) is the secondary structure. The three-dimensional fold of a protein is the tertiary structure. Finally, the arrangement of subunits in a multi-subunit protein complex is the quaternary structure.

**principle of microscopic reversibility** This principle states that, for a set of cyclic reactions, the rates of the forward and backward reactions must be equal for every individual reaction in the cycle. This is also referred to as the principle of detailed balance.

**prion proteins** An amyloid-forming protein protein that causes transmissible spongiform encephalopathies.

**probability distribution** A mathematical function describing the probability of different outcomes.

**processive enzymes** Some enzymes, such as DNA polymerase, work on substrates that contain multiple sites where the enzymatically catalyzed reaction can take place. Each nucleotide in the DNA template, for example, serves as a site for the addition of a nucleotide to the growing DNA chain. A processive enzyme catalyzes many rounds of the reaction before dissociating from the template.

**protease** An enzyme that cleaves proteins.

**protein** A covalently linked chain of amino acid residues.

**protein fold** The visually recognizable arrangement of α helices and β strands in the three-dimensional structure of a protein is referred to as the protein fold. Different proteins can have the same fold, as is the case for myoglobin and the subunits of hemoglobin (see Figure 4.2).

**protein kinase** These are proteins involved in cellular signaling that phosphorylate serine, threonine, or tyrosine residues in proteins. The human genome contains approximately 500 different protein kinases, all of which are very closely related in their catalytic domain (known as the kinase domain), but which respond to different input signals using additional domains with different functions. Bacterial cells also utilize kinases that phosphorylate histidine and aspartate residues, but these proteins form a distinct family that is unrelated to the protein kinases found principally in eukaryotes.

**pseudo-first-order rate constant** A higher-order reaction that behaves as if it were a first-order reaction. A second-order reaction between two molecules, A and B, becomes pseudo-first-order under a limiting condition (that is, $[A]_0 >> [B]_0$).

**pseudoknot** A pseudoknot is an RNA structure containing a hairpin loop, with a segment at one end of the stem folding back to base-pair with residues in the loop.

**purine** The nucleotide bases in RNA and DNA are substituted forms of two heterocyclic molecules known as pyrimidine and purine. Purine has two fused rings, one with five atoms and one with six.

**pyrimidine** The nucleotide bases in RNA and DNA are substituted forms of two heterocyclic molecules known as pyrimidine and purine. Pyrimidine has a single six-membered aromatic ring, with two nitrogens and four carbons.

**quaternary structure** See primary structure.

**radius of gyration** The radius of gyration, $R_g$, is the r.m.s. displacement of the atoms in a macromolecule from the centroid (see Equation 18.3).

**raft** Certain regions of cell membranes, called patches or rafts, are enriched in particular lipid components, such as glycolipids, sphingolipids, and cholesterol, and specific proteins. These membrane patches have distinct physical properties, including different density and resistance to solubilization by detergents (see Figure 3.43).

**Ramachandran diagram** This diagram shows the energy of an alanine dipeptide as a function of the backbone torsion angles $\phi$ and $\psi$. Allowed combinations of $\phi$ and $\psi$ are those for which the backbone does not run into itself.

**random coil** A random coil polypeptide is one that lacks any well-defined secondary and tertiary structure. In such states polypeptides rapidly fluctuate between different, energetically equivalent conformations.

**random order** An enzyme-catalyzed reaction with multiple substrates in which the order of binding of substrates to the enzyme is not important.

**random walk** A series of discrete steps in which the direction of each step is uncorrelated with the direction of the previous step is known as a random walk.

**rate** The time derivative of concentration, during a reaction.

**rate constant** Proportionality constant, denoted $k$, relating the rate of a reaction to the concentrations of reactants.

**rate law or rate equation** A rate law or *rate equation* is a differential equation that specifies how the reaction rate depends on the concentrations of species present in the reaction mixture including, but not limited to, reactants and products.

**rate-determining step** The rate-determining step in a series of reactions is the slowest step, which therefore ultimately limits the rate of formation of product.

**reaction coordinate** A reaction coordinate is a variable that describes how far a reaction has progressed from separated reactants to final products. The reaction coordinate is usually some combination of distances and angles that change during the reaction.

**reaction order** The order of the reaction is the sum of the concentration exponents in the rate law, which is the number of molecules colliding if it is an elementary reaction. An elementary bimolecular reaction has a reaction order of two.

**reaction progress variable, $\xi$** $\xi$ (Greek letter xi) measures how far a reaction has proceeded. If the stoichiometric coefficient for a reactant $i$ is $\upsilon_i$, then the amount of this reactant that has been consumed is given by $\upsilon_i d\xi$.

**reaction quotient** Denoted $Q$, the reaction quotient is a ratio involving the actual or observed values of the concentrations of the reactants and products, rather than the equilibrium concentrations. For a reaction in which reactants A and B are converted to products C and D, the value of $Q$ is given by:

$$Q = \frac{[\mathrm{C}]_{\mathrm{obs}}^{\upsilon_C}\,[\mathrm{D}]_{\mathrm{obs}}^{\upsilon_D}}{[\mathrm{A}]_{\mathrm{obs}}^{\upsilon_A}\,[\mathrm{B}]_{\mathrm{obs}}^{\upsilon_B}}$$

where the "obs" subscripts indicate that the concentrations are specific to the observed conditions, and the exponents of the concentrations are the stoichiometric coefficients.

**receptor** See ligand.

**redox couple** See redox reactions.

**redox reactions** Chemical reactions that involve the transfer of an electron from one molecule to another are referred to as redox reactions. For example, consider the reaction in which a molecule (M) loses an electron:

$$\mathrm{M} \rightleftharpoons \mathrm{M}^+ + e^-$$

When the reaction proceeds from left to right, the M molecule is said to be oxidized. When the reaction proceeds in the reverse direction the $M^+$ ion is said to be reduced. The oxidized and reduced forms of a molecule, such as M and $M^+$, form a redox couple.

**reduction** See oxidized.

**reduction potential** A parameter that determines the direction of electron flow in coupled oxidation–reduction reactions. The reduction potential of a molecule can be determined by measuring the voltage generated by electrochemical cells containing redox couples formed by the molecule.

**rejection** When applied to the process of mRNA translation on the ribosome, this term refers to the dissociation of a tRNA without the associated amino acid being incorporated into the growing protein chain.

**relaxation method** A relaxation method follows the rate of return to equilibrium after a sudden perturbation that disturbs the equilibrium.

**relaxation time** The time constant for a reaction in which the concentrations of reactants relax from non-equilibrium values to equilibrium.

**replication** Replication is the process by which DNA molecules are duplicated.

**resistor** An element in an electrical circuit that impedes the flow of current.

**resting membrane potential** The membrane potential of a cell at steady state, with no net flux through ion channels, is known as the resting membrane potential, denoted $E_m^*$. The value of $E_m^*$ for a mammalian cell is usually around −70 mV.

**retrovirus** Retroviruses are viruses that store genetic information in RNA, which is then copied by the enzyme reverse transcriptase into DNA when it infects a suitable host cell.

**reversible** A process is referred to as near-equilibrium or reversible if it is carried out so slowly that the driving force is always nearly in balance with the internal forces. When a system moves from one state to another, the amount of heat absorbed and the amount of work done depends on how exactly the transformation occurred. By specifying a near-equilibrium process, we can calculate the heat and work precisely. For such a process, the work done is maximal, given the initial and final states of the system.

**ribonucleoprotein complex** A complex of specific RNAs and proteins.

**ribonucleotide** The nucleotides found in RNA are derived from ribose, a pentose sugar in which hydroxyl groups are attached to the C1′, C2′, C3′, and C5′ carbon atoms. The nucleotide building blocks of RNA are called ribonucleotides, and these have the C1′-OH group of ribose replaced by a base and the C5′-OH group replaced by a phosphate group.

**ribosome** A very large assembly of proteins and RNA responsible for translation, that is, the linkage of amino acids into proteins, based on the sequence of nucleotides in mRNA.

**ribozyme** RNA enzyme.

**ridge into groove** Helical proteins that do not have coiled coils have relatively straight α helices. The sidechains of these α helices form ridges and grooves that are arranged at characteristic angles with respect to the helix axis. α helices pack against each other so that a ridge on one helix inserts into a groove on the other. This results in a limited set of crossing angles between the helices.

**RNA** A polymer of ribonucleotides.

**RNA secondary structural element or motif** The helices in RNA and the structural elements that connect them are called RNA secondary structural elements or motifs, and include helices, single-stranded regions, hairpins, bulges, internal loops, and junctions.

**RNA splicing** In eukaryotic cells, RNA transcripts are processed in a reaction known as RNA splicing, in which the introns are cleaved from the primary transcript and the exon sequences are ligated together.

**root mean square displacement** The square root of the variance of a probability distribution. The root mean square displacement is also a measure of how far particles have moved, on average, due to diffusive motion.

**Rossmann fold** A very commonly occurring protein fold, found in nucleotide-binding domains. The Rossmann fold was the first modular protein domain to be identified (see Figure 4.49B).

**salt bridge** When two oppositely charged groups are close to each other, the interaction is called an ion pair or a salt bridge.

**saturable binding** In a situation where a ligand binds to a single binding site on a protein and no other, all of the protein molecules are bound to ligand at very high ligand concentrations. Increasing the ligand concentration beyond this point does not lead to any increase in protein binding. Such a binding interaction is referred to as being saturable. When a binding isotherm does *not* saturate, even at very high ligand concentrations, then it usually indicates that the ligand is also binding to things other than the protein of interest.

**saturation** When the amount of bound ligand reaches a maximum plateau value and then does not increase further, the binding is said have reached saturation. Saturable binding is a hallmark of specific binding to a single receptor. If there are many non-specific receptors present then the binding may not saturate.

**Scatchard analysis** A simple binding equilibrium between a protein and a ligand results in the hyperbolic binding curve shown in Figure 12.4. Deviations from the hyperbolic curve, however, can be difficult to detect visually. Scatchard analysis involves rearranging the basic binding equation to yield the following form, known as the Scatchard equation:

$$\frac{[\mathrm{L}]_{\mathrm{bound}}}{[\mathrm{L}]} = -\frac{1}{K_{\mathrm{D}}}[\mathrm{L}]_{\mathrm{bound}} + \frac{[\mathrm{P}]_{\mathrm{total}}}{K_{\mathrm{D}}}$$

The Scatchard equation, which is an alternative form of the hyperbolic binding equation, tells us that the concentration of the bound ligand is related linearly to the total protein concentration (Figure 12.10C).

**Scatchard equation** See Scatchard analysis.

**second law of thermodynamics** One statement of the second law is that the combined entropy of the system and the surroundings always increases for a spontaneous process. This is equivalent to saying that the entropy of the system and the surroundings has a maximum value at equilibrium, which is referred to as the maximal entropy principle.

**secondary structure** See primary structure.

**second-order** A reaction with a reaction order of two.

**sedimentation coefficient, *S*** The sedimentation coefficient, *S*, of a macromolecule is inversely related to the friction factor, *f*, of the macromolecule (see Equation 17.65). Because the friction factor depends on the shape, the value of *S* provides information on molecular shape.

**selectivity filter** A structural feature in the central pore of ion channels that determines the specificity of the channel for particular ions.

**self-cleaving ribozyme** A ribozyme that cleaves itself.

**serine protease** Serine proteases are enzymes that cleave peptide bonds. The catalytic center of these enzymes contains a catalytic triad: a serine, a histidine, and an aspartic acid. The hydroxyl group of the serine sidechain attacks the peptide bond that is to be cleaved.

**shape readout** See indirect readout.

**sidechain** The variable R groups of amino acids are called sidechains.

**sigmoid binding curve** When the binding of ligands to a protein is cooperative, the shape of the binding curve (fractional saturation versus ligand concentration) is no longer hyperbolic. Instead, it resembles an "S" shape and is called sigmoid (for sigma, the Greek letter S).

**silent mutation** A mutation in a gene that does not change the sequence of the corresponding protein.

**small-angle x-ray scattering** Small-angle x-ray scattering (SAXS), or solution x-ray scattering, involves measuring the scattering of x-rays by macromolecules in solution. Solution x-ray scattering data can be used to derive the size of the macromolecules (that is, the radius of gyration) as well their shapes.

**$S_N2$-type in-line attack mechanism** $S_N2$-type reactions are nucleophilic substitution reactions that involve simultaneous bond-breaking and bond-making steps. They result in the inversion of stereochemistry at the reaction center.

**soft metal ion** Metal ions, such as $Cu^{2+}$, $Au^+$, $Hg^+$, $Cd^{2+}$, $Pt^{2+}$, and $Mn^{2+}$, that interact more efficiently with sulfur or nitrogen ligands, compared with oxygen.

**space constant** In the description of the electrical properties of axons, the space constant is a parameter that determines the distance over which a voltage perturbation decays.

**specificity** See affinity.

**specificity constant** The ratio of the Michaelis–Menten parameters $k_{cat}$ and $K_M$. For two different substrates, each present at the same low concentration, the relative amounts of product formed will be determined by the corresponding values of $k_{cat}/K_M$ and, hence this ratio is referred to as the specificity constant for the enzyme.

**specificity factor** The ratio of the concentration of ligand bound to the target receptor to the total concentration of ligand bound to all other receptors is known as the specificity factor, denoted $\alpha$.

**sphingolipid** A lipid with one "built-in" hydrophobic chain from the molecule sphingosine that in essence replaces the glycerol unit of glycerophospholipids.

**spongiform neuroencephalopathy** A degenerative neurological disease caused by amyloid plaque formation.

**stabilization energy** For an interaction between two atoms, the stabilization energy is the amount by which the energy at the optimal distance is lower than when the atoms are far apart.

**standard change in enthalpy** The standard change in enthalpy for a reaction is denoted $\Delta H^o$, and this is the change in enthalpy when one mole of reactants is converted to one mole of products under standard conditions.

**standard change in entropy** The standard change in entropy for a reaction is denoted $\Delta S^o$, and this is the change in entropy when one mole of reactants is converted to one mole of products under standard conditions.

**standard deviation** The square root of the variance of a distribution.

**standard electrode potential** An electrode potential that is obtained by combining the half-cell of interest with the standard hydrogen electrode.

**standard free energy of formation** By convention, the molar free energy of a molecule is the standard free-energy change that results from converting stoichiometric amounts of the pure elements that are its constituents into one mole of the molecule of interest under standard conditions. This is referred to as the standard free energy of formation of the molecule ($\Delta_f G^o$).

**standard free-energy change** $\Delta G^o$ is the change in free energy when a molar equivalent of reactants are converted into products under standard conditions of concentration. If the biochemical standard state is used (that is, pH 7.0), the standard free-energy change is denoted $\Delta G^{o\prime}$.

**standard hydrogen electrode** The electrode potentials for various electrodes are referenced to a standard hydrogen electrode. This electrode is illustrated in Figure 11.24, and it reflects the oxidation of hydrogen gas at 1 atm pressure to yield protons and electrons under conditions where the concentration of protons is 1 M at room temperature.

**standard or canonical base pair** The standard base pairs, or Watson-Crick base pairs, are A-T and C-G in DNA and A-U and C-G in RNA. Folded RNA structures often contain alternative base pairs (for example, G-U), which are referred to as nonstandard.

**standard reduction potential** The voltage generated by an electrochemical cell in which the reactants are all at standard concentrations (1 M solution, solid metals) is known as the standard reduction potential. The voltage generated by such a cell depends on the temperature, so standard reduction potentials are usually given for 298 K.

**standard state** The standard state is a condition of defined molecular state, concentration, and pressure that is used as a reference for reporting free energy values. For water soluble molecules, the standard state is usually a 1 M solution of the molecule at 1 atm pressure. The standard state of water is pure water (55 M). The concentration of water is assumed to be constant at 55 M, and the concentration of $H^+$ is $10^{-7}$ M (that is, pH = 7.0). This definition of the standard state is the biochemical standard state. In other branches of chemistry, the standard state proton concentration is 1 M (that is, pH 0).

**standard voltage** The voltage generated by an electrochemical cell in which the concentrations of the reactants in the half-cells are at their standard state concentrations (that is, 1 M solution). This voltage is also referred to as the standard cell potential, and is denoted $\Delta E^o$, or sometimes $E^o$.

**state** A state of a system is characterized by the global properties of the system, such as the temperature, pressure, and number of molecules. A microstate is a specific configuration of molecules that is consistent with the state. Each state corresponds to many different microstates.

**state variable or state function** Properties of the system that are independent of the history of the system are known as state variables or functions. The values of these variables can be calculated from knowledge of the present state of the system, without knowing its history. Energy, volume, temperature, and pressure are state variables. The heat transferred to a system is not a state variable, because its magnitude depends on what has been done to the system in the past.

**statin** A drug that blocks an important step in the synthesis of cholesterol.

**statistical definition of entropy** The Boltzmann constant, $k_B$, relates the multiplicity, $W$, to entropy, $S$, through the statistical definition of entropy: $S = k_B \ln W$.

**statistical thermodynamics** The study of the connection between molecular properties and thermodynamic properties using statistical treatments is called statistical thermodynamics.

**steady state** A set of reactions for which the concentrations of reactions and products do not change with time is said to be in a steady state. The reactants and products do not need to be at equilibrium for a steady state to occur. All that is required for a steady state is that the rates of production of molecules be balanced by the rates at which they are converted to other molecules.

**stereoisomer** Two structures with the same atoms and the same type of chemical bonds, but which cannot be interconverted without breaking and remaking covalent bonds.

**steric effect** Steric effects are the consequences of the van der Waals repulsion between atoms at short distances. These are responsible for many fundamental aspects of the structures of DNA, RNA, and proteins.

**Stern–Volmer equation** An equation that relates the ratio of fluorescence in the presence and absence of a quencher to the rate constants involved in the fluorescence process. See Equation 15.59.

**steroid** A flat, rigid, hydrophobic molecule containing a conserved core of four fused carbocyclic rings (for example, cholesterol).

**Stirling's approximation** Calculations involving the factorials of large numbers are simplified by using Stirling's approximation:

$$\ln n! = n \ln n - n$$

Stirling's approximation can be used when $n$ is larger than about 100, and it becomes more accurate as $n$ becomes larger.

**Stokes' law** Stokes' law states that the friction factor for a molecule moving through a solution is directly proportional to the product of the viscosity and the radius of the molecule (see Equation 17.38).

**Stokes–Einstein equation** The Stokes–Einstein equation states that the diffusion constant for a molecule is directly proportional to the temperature and inversely proportional to the product of the viscosity and the radius of the molecule (see Equation 17.39).

**structural domain** Structural domains are the fundamental units of three-dimensional protein structure. A protein domain is typically 50–200 residues long and contains a well-defined hydrophobic core. Domains are mixed and matched in evolution to produce more complex proteins.

**structural motif** A three-dimensional arrangement of two or more secondary structural elements that is commonly found in many proteins is called a structural motif. Motifs are typically components of larger domains, and more than one motif may be found in a protein domain.

**substitution likelihood ratio** Ratio of the frequency of substitution of one amino acid by another in related proteins, relative to the substitution frequency in unrelated proteins.

**substitution score** The logarithm of the substitution likelihood ratio.

**substrate** A reactant in an enzyme-catalyzed reaction.

**substrate-dependent noncompetitive inhibition** A substrate-dependent noncompetitive inhibitor is a compound whose effects occur through binding to an enzyme–substrate complex, preventing completion of the enzyme reaction and release of the product.

**sugar** See glycan.

**sugar pucker** The sugar ring in polynucleotides is generally nonplanar and it displays a preferred puckering mode, C3′ endo, found in A-form helices, or C2′ endo, found in B-form helices. The terms endo and exo specify the nature of the out-of-plane atom of the sugar ring, with endo indicating displacement toward the side with the C5′ carbon and exo indicating displacement toward the opposite side.

**suicide substrate** A suicide substrate is an enzyme inhibitor that triggers the catalytic cycle just as a substrate does, but then is trapped in a dead-end complex on the enzyme. A suicide substrate is also known as a mechanism-based inhibitor (see Figure 16.21).

**supercoil, or superhelix** When the axis of a helix is coiled rather than straight, the resulting structure is called a supercoil or a superhelix.

**surrounding** Everything outside a system is referred to as the surrounding.

**syn** A nucleotide conformation in which the base is turned toward the sugar. In this conformation, the sugar H1 atom and the base C6 or C8 atoms (for pyrimidines and purines, respectively) are in a *cis* conformation about the glycosidic bond. The syn conformation is less common than the anti conformation.

**system** The region of interest.

**temperature jump** A sudden perturbation in the temperature of a system, used to obtain information of the rates of binding reactions.

**template** A template strand of DNA is one that will be copied in the process of replication or mRNA synthesis. The newly synthesized strand will have a sequence that is complementary to the original (in the Watson-Crick base pairing sense).

**template strand** The order of addition of nucleotides to a growing chain of DNA or RNA is dictated by the sequence of another strand of DNA or RNA, known as the template strand.

**terminal velocity** Under an applied force, such as gravity, an object will accelerate according to Newton's law ($F = ma$) until the friction force equals the applied force, at which point the object reaches a terminal velocity.

**termolecular or third order** A reaction with a reaction order of three.

**tertiary structure** See primary structure.

**thermal energy** The amount of energy that is readily transferred between atoms by random collisions. The value of the thermal energy is given by $k_BT$.

**thermodynamic cycle** Two different pathways that connect the same set of reactants and products, with the same starting and ending conditions, define a thermodynamic cycle. The change in any state variable, such as the free energy, enthalpy, or entropy, must the same following either pathway. Thermodynamic cycles are useful when one of the pathways is experimentally accessible, but the other one is not. Experimental measurements on one pathway provide information about the other pathway. See Figure 9.12 for an example of a thermodynamic cycle.

**thermodynamic definition of entropy** A historical definition of entropy based on heat transfer is known as the thermodynamic definition of entropy:

$$\Delta S = \frac{q_{rev}}{T}$$

where $\Delta S$ is the change in entropy for a reversible (near-equilibrium) process, $q_{rev}$ is the heat transferred, and $T$ is the absolute temperature. This definition of the entropy turns out to be equivalent to the statistical definition.

**thermodynamic hypothesis** Anfinsen's hypothesis that proteins fold spontaneously to a structure with minimal free energy.

**thermodynamic hypothesis in protein folding** The thermodynamic hypothesis refers to the idea that the native structure of a protein is determined solely by the properties of the protein and is not the result of an external template.

**third law of thermodynamics** It is assumed that the entropy of a pure substance is zero when the temperature is absolute zero on the Kelvin scale. This principle is referred to as the third law of thermodynamics.

**3D-1D profile method** A fold-recognition algorithm that converts information about the chemical and structural environments at each position of a known three-dimensional protein structure (the "3D" information) into a one-dimensional list that provides one environmental descriptor per residue (the "1D" list; hence, 3D-1D profile) (see Figure 5.27).

**threading algorithm** See fold-recognition algorithm.

**thymine** A substituted pyrimidine base found in DNA.

**time constant (charging of a capacitor)** During the charging of a capacitor, the time constant is the time taken for the voltage across the capacitor to reach ~63% of its final value. This is also the time taken for the capacitive current to decay to ~37% of its initial value.

**time constant (reaction kinetics)** The time constant for a reaction is the time required for the concentration of a reactant to decay by a factor of $e^{-1}$ (that is, to decay to ~37% of its initial value). For a first-order reaction with a single product, the time constant is also the time taken for the concentration of the product to increase to ~63% of its final value.

**topoisomerase** An enzyme that modifies the supercoiling of DNA.

**torsion angle or dihedral angle** Torsion angles describe the rotation of two covalently linked parts of a molecule with respect to each other, about the axis of the covalent bond. Torsion angles are also called dihedral angles.

***trans*** Atoms or chemical groups that are linked by a covalent bond and are in a staggered configuration across the bond (that is, on opposite sides) are said to be in a *trans* configuration.

**transcription** The sequences of genes are copied during the process of transcription to make RNAs with different kinds of functions.

**transfer RNA (tRNA)** Translation depends upon adapters that match each amino acid to its mRNA codon. The adapters are transfer RNAs or tRNAs, so-called because they transfer amino acids to the growing polypeptide in the ribosome.

**transition state** The transition state is the point at which the maximum potential energy is reached along the reaction coordinate, and it is the point at which the conversion of reactants to product is energetically downhill.

**transition state theory** Transition state theory uses the concept of an equilibrium between the ground state and the transition state for a reaction to predict the temperature dependence of the reaction rate.

**translation** The process by which mRNA directs the assembly of amino acids into protein. This occurs in the ribosome.

**translocation** When applied to a processive enzyme, such as DNA polymerase or the ribosome, translocation refers to the movement of the enzyme as it catalyzes multiple rounds of reaction on a single substrate.

**transmembrane segment** Part of a protein that crosses a lipid bilayer.

**turnover number** The value of the catalytic rate constant, $k_{cat}$, is referred to as the turnover number.

**twist** See writhe.

**two metal ion mechanism** A characteristic aspect of the mechanism of DNA polymerases, in which two metal ions that are bound at the active site are essential for the catalysis of DNA synthesis.

**two-state folding** Two-state folding means that only the unfolded protein and the fully folded protein occur at any significant concentration. Two-state folding kinetics can be fully described with a single rate constant.

**tyrosine kinase** An enzyme that transfers the terminal phosphate group of ATP to tyrosine residues in proteins.

**ultrasensitive** An ultrasensitive system is one in which the response to an input is sharper than expected from a simple binding equilibrium. For example, the response of a protein to a ligand can be defined as the fractional saturation, $f$ (see Chapter 12). When $f$ rises from 0.1 to 0.9 over a less than ~100-fold concentration range of the ligand, the system is said to be ultrasensitive.

**underwound** When turns are removed from a DNA double helix, the helix is said to be underwound.

**unimolecular** A unimolecular or first-order reaction is one in which the rate depends only on the concentration of the reactant to the first power.

**uracil** A substituted pyrimidine, found in RNA.

**van der Waals attraction** When two neutral atoms approach each other, transient fluctuations in the electron clouds of each atom set up transient dipoles, leading to an attractive force between the atoms. This attractive force is called the London force, or the van der Waals attraction.

**van der Waals contact** Two atoms that are separated by the sum of the van der Waals radii are said to be in van der Waals contact.

**van der Waals energy** Energy arising from van der Waals attractions and repulsions.

**van der Waals radius** The van der Waals radius is a measure of the size of an atom. The energy due to the van der Waals attraction between two atoms is optimal when they are separated from each other by the sum of their van der Waals radii. If they move closer, the energy increases sharply.

**van der Waals repulsion** The strong repulsion between atoms when they approach each other so closely that their electronic orbitals overlap.

**van't Hoff equation** The van't Hoff equation gives the temperature dependence of the equilibrium constant, relating the logarithm of $K$ to the values of $\Delta H^{\text{o}}$ and $\Delta S^{\text{o}}$:

$$\ln K = -\frac{\Delta H^{\circ}}{RT} + \frac{\Delta S^{\circ}}{R}.$$

**vectorial transport** The net movement of molecules in one direction is sometimes referred to vectorial transport.

**velocity of the reaction** Reaction rate.

**viscosity** A measure of the drag imposed by a solvent on the movement of a solute. The viscosity of the solvent (given the symbol $\eta$) is the proportionality constant between the momentum flux and the velocity gradient along a direction $z$ orthogonal to the direction of movement, $x$ (see Equation 17.37).

**volt (V)** Standard unit for electrical potential difference.

**voltage clamp** An experimental setup for studying the electrical properties of neurons, in which the membrane potential is under the control of an electrode that runs axially through the interior of the axon. The membrane potential can then be stepped through various values and the currents measured.

**voltage spike** A transient, localized and rapid change in the membrane potential.

**water-soluble protein** A protein with a hydrophilic surface, which dissolves in water when folded properly.

**wobble base pair** A nonstandard base pair, first identified in codon–anticodon interactions. The base pair involves two hydrogen bonds (between G and U, as illustrated in Figure 2.32), but the geometry of the base pair is different from that of a standard base pair.

**writhe** The writhe describes the rotation of the axis of the DNA helix. The linking number counts how many times the DNA strands cross each other. The twist describes how many times the DNA strands cross the helix axis, and it changes depending on how twisted the helix axis is. These three parameters characterize the nature of supercoiling in DNA.

**Z-form DNA (Z-DNA)** A form of DNA in which the double helix is left-handed.

**zero order** A reaction whose rate does not depend on the concentrations of the reactants.

**zinc finger** A zinc finger is a small module, about 25 amino acids plus a short linker, with a fold stabilized by binding of a zinc ion. An example is shown in Figure 13.42.

# Index

Page numbers suffixed by B, F, and T refer to boxes, figures, and tables respectively; vs. indicates a comparison.

## A

## D

## F

## G

## H

## I

## J



## K

## T

## U



## V